Mouse Development

Patterning, Morphogenesis, and Organogenesis

Mouse Development

Patterning, Morphogenesis, and Organogenesis

Edited by

Janet Rossant

Samuel Lunenfeld Research Institute
Mount Sinai Hospital
Toronto, Ontario, Canada

Patrick P. L. Tam

Children's Medical Research Institute
University of Sydney
Westmead, New South Wales, Australia

ACADEMIC PRESS

San Diego   San Francisco   New York   Boston
London   Sydney   Tokyo

Cover photograph: An E13.5 mouse embryo stained for cartilage and nerves. Courtesy of P. Akinwunmi and R. Behringer, MD Anderson Cancer Center, Houston, USA.

This book is printed on acid-free paper.∞

Copyright © 2002 by ACADEMIC PRESS

Academic Press
A division of Harcourt, Inc.
525 B Street, Suite 1900, San Diego, California 92101-4495, USA
http://www.academicpress.com

Academic Press
Harcourt Place, 32 Jamestown Road, London NW1 7BY, UK
http://www.academicpress.com

Library of Congress Catalog Card Number: 2001098457

International Standard Book Number: 0-12-597951-7

PRINTED IN CANADA
02 03 04 05 06 FR 9 8 7 6 5 4 3 2 1

*Dedicated to the memory of
our dear friend and colleague, Rosa Beddington,
for all her great insights into the intricacies of
the early mouse embryo.*

Contents

Contributors   xiii
About the Editors   xvii

I Establishment of Body Patterns

1 Fertilization and Activation of the Embryonic Genome

Davor Solter, Wilhelmine N. de Vries, Alexei V. Evsikov, Anne E. Peaston, Frieda H. Chen, and Barbara B. Knowles

I. Introduction   5
II. Oogenesis   6
III. Meiosis and the Beginning of Oocyte Asymmetry   7
IV. Fertilization   8
V. Transcription andIts Control   8
VI. mRNA Utilization during Oocyte Maturation and Preimplantation Development   10
VII. Gene Expression in the Early Mouse Embryo   11
VIII. Functional Analysis   13
References   15

2 Asymmetry and Prepattern in Mammalian Development

R. L. Gardner

I. Introduction   21
II. Asymmetries in Early Development   23
III. Asymmetry of the Blastocyst   27

IV. Specification of the Polarity of the Anterior-Posterior Axis of the Fetus?   29
V. Conclusions   32
References   33

3 Anterior Posterior Patterning of the Mouse Body Axis at Gastrulation

Stew-Lan Ang and Richard R. Behringer

I. Introduction   37
II. Gastrulation   38
III. The Node: Morphogenesis, Cell Fate, and Cell Movement   38
IV. The Organizer Phenomenon: Conserved Properties of Vertebrate Organizers   40
V. The Vertebrate Organizer is a Dynamic, Nonhomogeneous, and Renewable Cell Population at Gastrulation   40
VI. Insights into the Function of the Mouse Organizer Gained from Genetic and Embryological Studies   41
VII. Genetic Analysis of Organizer Function: Mouse Mutants Showing Defects in Organizer Function   42
VIII. Inhibitory Signals Secreted by the Organizer and Its Derivatives   44
IX. Specification of the Primitive Streak and the Organizer   44
X. Role of the AVE in Anterior Patterning in Mouse   45
XI. Embryological and Genetic Analysis of the Function of the AVE in Anterior Patterning   46

XII. A Model for AVE Function in
     Anterior Patterning   47
XIII. Conclusions and Future Directions 48
     References   49

                    4 Left-Right Asymmetry
                         Hiroshi Hamada
    I. Introduction   55
   II. Morphological Left-Right Asymmetries   56
  III. Genetic/Molecular Pathway Governing Left-
       Right Determination   58
   IV. Molecular Readout of the
       First Asymmetry   61
    V. Role of the Midline   64
   VI. Readout of Left-Right Asymmetry in
       Later Development   65
  VII. Miscellaneous Mutations/
       Gene Factors   67
 VIII. Diversity among Vertebrates   68
   IX. Future Challenges   69
       References   70

      5 Patterning, Regionalization, and Cell
            Differentiation in the Forebrain
          Oscar Marín and John L. R. Rubenstein
    I. Organization of the Forebrain   75
   II. Early Patterning and Regional Specification of
       the Forebrain   78
  III. Morphogenetic Mechanisms in the
       Forebrain   85
   IV. Control of Neurogenesis and Cell-Type
       Specification in the Forebrain   87
       References   97

      6 Establishment of Anterior-Posterior and
   Dorsal-Ventral Pattern in the Early Central
                    Nervous System
                    Alexandra L. Joyner
    I. Overview of Early CNS Development
       and Patterning   107
   II. Anterior-Posterior Patterning
       of the Mesencephalon and
       Metencephalon   110
  III. Hindbrain Anterior-Posterior Patterning
       Involves Segmental Units
       of Development   117

   IV. CNS Dorsal-Ventral Patterning Involves a
       Tug of War between Dorsal and
       Ventral Signaling   120
    V. Summary   122
       References   122

          7 Somitogenesis: Segmentation of the
      Paraxial Mesoderm and the Delineation of
                    Tissue Compartments
               Achim Gossler and Patrick P. L. Tam
    I. Overview of Somite Development   127
   II. Allocation of Progenitor Cells to the
       Paraxial Mesoderm   132
  III. Cells Are in Transit in the Presomitic
       Mesoderm   132
   IV. Regionalized Genetic Activity Points to a
       Prepattern of Prospective Somites   133
    V. Emergence of Anterior-Posterior
       Somite Compartments   134
   VI. Role of Notch Signaling in the Establishment
       of Somite Borders and Anterior-Posterior
       Polarity   134
  VII. A Molecular Clock Operates in the Paraxial
       Mesoderm to Control the Kinetics of
       Somite Formation   138
 VIII. Specification of Lineage Compartments by
       Inductive Interactions   139
   IX. Summary and Open Questions   142
       References   144

      II Lineage Specification and
               Differentiation

          8 Extraembryonic Lineages
            Janet Rossant and James C. Cross
    I. Introduction   155
   II. Early Development of the Trophoblast and
       Primitive Endoderm Lineages   156
  III. Cell Lineage Analysis and the
       Extraembryonic Lineages   156
   IV. Setting Aside the Blastocyst Lineages   158
    V. Molecular Specification of the Blastocyst
       Cell Lineages   159
   VI. Differentiation of the Yolk Sacs   161
  VII. Morphogenetic Events in Development of the
       Chorioallantoic Placenta   161

VIII. Comparative Aspects of Development of
Extraembryonic Membranes   162
IX. Molecular Control of Primitive
Endoderm Development   164
X. Signaling Pathways in Early
Trophoblast Development   166
XI. Control of Spongiotrophoblast and Giant
Cell Fate   168
XII. Trophoblast Giant Cell Development:
Gene Pathways and Control of
Endoreduplication   169
XIII. Initiating Chorioallantroic Fusion   170
XIV. *Gcm1* Regulates the Initiation of
Chorioallantoic Branching   170
XV. Growth Factor Signaling Regulates Branching
Morphogenesis of the Labyrinth   171
XVI. Placental Development and
Pregnancy Complications   173
References   174

9 Germ Cells
Christopher Wylie and Robert Anderson
I. General Concepts   181
II. Early Appearance of Germ Cells in
the Mouse   182
III. Specification of Germ Cells in
the Mouse   183
IV. Migration of Germ Cells   185
V. Motility of Germ Cells   185
VI. Guidance of Germ Cell Migration   186
VII. Adhesive Behavior of Germ Cells
during Migration   187
VIII. Survival and Proliferation of Germ Cells
during Migration   188
References   189

**10 Development of the Vertebrate
Hematopoietic System**
Nancy Speck, Marian Peeters, and Elaine Dzierzak
I. Introduction   191
II. Cellular Aspects of Blood Development in the
Mouse Embryo   192
III. Molecular Genetic Aspects of Blood
Development in the Mouse Embryo   202
IV. Current Cellular and Molecular Conceptual
Frameworks for Hematopoietic
Ontogeny   205

V. Future Directions   206
References   206

11 Vasculogenesis and Angiogenesis
Thomas N. Sato and Siobhan Loughna
I. Introduction   211
II. Overview of Vascular Development   211
III. Generation of Endothelial Cells   212
IV. Vascular Morphogenesis   220
V. Concluding Remarks   228
References   228

12 Stem Cells of the Nervous System
Sean J. Morrison
I. Introduction   235
II. Lineage Determination of Neural
Stem Cells   237
III. Do Stem Cells Retain Broad or Narrow
Neuronal Potentials?   241
IV. Regulation of Neural Stem Cell
Self-Renewal   242
V. Differences between Hematopoietic Stem Cells
and Neural Stem Cells   243
VI. *In Vivo* Function of Neural
Stem Cells   244
VII. Surprising Potential of Neural
Stem Cells   245
VIII. Are Neural Stem Cells Involved
in Disease?   246
IX. Outstanding Issues   248
References   248

**13 Cellular and Molecular Mechanisms
Regulating Skeletal Muscle Development**
Atsushi Asakura and Michael A. Rudnicki
I. Introduction   253
II. Embryonic Origin of Skeletal Muscle   254
III. MyoD Family of Myogenic Regulatory
Factors   256
IV. Muscle-Specific Transcriptional
Regulation   262
V. Inductive Mechanisms
of Myogenesis   262
VI. Specification of Muscle Fiber Types   268
VII. Muscle Regeneration   269
VIII. Conclusion   272
References   272

14 Deconstructing the Molecular Biology of Cartilage and Bone Formation

Benoit de Crombrugghe, Véronique Lefebvre, and Kazuhisa Nakashima

   I. Introduction   279
  II. Sox Transcription Factors: Essential Roles in the Chondrocyte Differentiation Program   281
 III. Parathyroid Hormone-Related Peptide (PTHrP) and Parathyroid Hormone (PTH)/PTHrP Receptor: Gatekeepers of the Zone of Hypertrophic Chondrocytes   284
  IV. FGFs and FGF Receptor 3: Counterintuitive Inhibitors of Chondrocyte Proliferation   286
   V. Ihh: A Central Coordinator of Endochondral Bone Formation   287
  VI. The Two Roles of the Transcription Factor Cbfa1 in Endochondral Bone Formation   288
 VII. Other Transcription Factors Involved in Bone Formation   290
VIII. Gelatinase B and Vascular Endothelial Growth Factor: Additional Coordinators of Endochondral Bone Formation   290
  IX. Conclusion   291
      References   292

III Organogenesis

15 Development of the Endoderm and Its Tissue Derivatives

Brigid L. M. Hogan and Kenneth S. Zaret

   I. Introduction and Overview   301
  II. Endoderm Development prior to Organogenesis   302
 III. Patterning and Differentiation of the Digestive Tract   307
  IV. Development of Tissues That Bud from the Endoderm   310
   V. Perspectives and Remaining Issues on Organogenesis from the Endoderm   322
      References   322

16 Molecular Determinants of Cardiac Development and Congenital Disease

Richard P. Harvey

   I. Introduction   332
  II. Overview of Heart Structure and Development   332
 III. A Conserved Pathway for Cardiac Induction and Morphogenesis   334
  IV. Cardiac Induction: The Role of Endoderm   334
   V. Bone Morphogenetic Proteins as Cardiac Inducing Molecules   336
  VI. Other Factors Involved in Cardiac Induction   336
 VII. A Role for Anterior Visceral Endoderm in Cardiac Induction in the Mouse?   338
VIII. The Heart Morphogenetic Field   339
  IX. The Size and Shape of the Heart Field   339
   X. The Timing and Stability of Cardiac Induction   340
  XI. Migration of Cardiac Precursors   340
 XII. Cellular Proliferation and Death in the Forming Heart   341
XIII. Cardiac Myogenesis   341
XIV. Modulation of Myogenesis in Heart Chambers   343
 XV. Regionality in the Developing Heart   343
XVI. Plasticity of Heart Regionalization   344
XVII. The Segmental Model of Cardiac Morphogenesis   344
XVIII. An Inflow/Outflow Model of Early Heart Tube Patterning   345
XIX. A Role for Retinoic Acid Signaling in Inflow/Outflow Patterning   346
 XX. A Role for the Delta/Notch Pathway in Primary Heart Patterning   347
 XXI. Cardiac Chamber Formation   347
XXII. Ventricular Specification: Knock-Out and Transgenic Phenotypes   348
XXIII. Transcriptional Circuits Acting in Chamber Formation   351
XXIV. The Cardiac Left-Right Axis   351
XXV. Developmental Pathways and Congenital Heart Disease   356
XXVI. Horizons   357
      References   358

17 Sex Determination and Differentiation

Amanda Swain and Robin Lovell-Badge

   I. Introduction   371
  II. Gonad Development   372
 III. Sex Determination   376
  IV. Testis Differentiation   380
   V. Cell Movement and Proliferation in the Early Gonad   382

VI. Ovary Differentiation   384
VII. Sexual Development   384
VIII. Evolution and Sex Determination   386
IX. Conclusion   388
References   389

18 Development of the Excretory System
Gregory R. Dressler

I. Introduction   395
II. Patterning of the Intermediate Mesoderm   396
III. Growth of the Nephric Duct and Ureteric Bud Diverticulum   400
IV. Inductive Interactions   404
V. Mesenchyme-to-Epithelial Conversion   407
VI. Glomerular Development and Vascularization   412
VII. Developmental Basis of Human Renal Disease   414
VIII. Future Perspectives   416
References   416

19 Craniofacial Development
Michael J. Depew, Abigail S. Tucker, and Paul T. Sharpe

I. Introduction   421
II. Primordial Cells of the Head   422
III. Organ Development   433
IV. Conclusion   454
V. Appendix 1: Descriptive Dental Development   454
VI. Appendix 2: Morphological Organization of the Murine Skull   456
VII. Appendix 3: Molecular Regulators of Craniofacial Pattern and Development   465
References   481

20 Pituitary Gland Development
Sally Camper, Hoonkyo Suh, Lori Raetzman, Kristin Douglas, Lisa Cushman, Igor Nasonkin, Heather Burrows, Phil Gage, and Donna Martin

I. Pituitary Gland Anatomy and Function   499
II. Development of the Pituitary Primordia and Cell Specification   500
III. Expansion of Committed Cell Types   510
IV. Conclusion   512
References   513

21 Development of the Eye
Hisato Kondoh

I. Overview of Eye Development   519
II. Development of the Retina   521
III. Lens Development   528
IV. Conservation and Divergence of the Transcriptional Regulatory Systems in the Eye Development   533
References   535

22 Development of the Mouse Inner Ear
Amy E. Kiernan, Karen P. Steel, and Donna M. Fekete

I. Introduction   539
II. Anatomy of the Inner Ear   540
III. Development of the Inner Ear   541
IV. Early Development of the Otic Blacode and Otocyst   542
V. Pattern Formation in the Inner Ear   546
VI. Sensory Differentiation   552
VII. Neurogenesis   558
VIII. The Stria Vascularis   559
IX. Future Directions   560
References   561

23 Integumentary Structures
Carolyn Byrne and Matthew Hardman

I. Introduction   567
II. Mature Skin   569
III. Non-Neural Embryonic Ectoderm   570
IV. Stratification   571
V. Dermal Development   572
VI. Epidermal Appendage Morphogenesis   574
VII. Model for Follicle Formation: The First Dermal Signal   574
VIII. Follicle Spacing   577
IX. Follicle Morphogenesis and Differentiation   578
X. Follicle Morphogenesis and Follicle Cycling   578
XI. Molecular Parallels between Skin Tumorigenesis and Skin Development   579
XII. Early Terminal Differentiation   579
XIII. Regulation of Transit to Late Stages of Terminal Differentiation   580
XIV. Late Terminal Differentiation: Formation of Stratum Corneum and Skin Barrier   581

XV. Periderm Disaggregation   583   Author Index   591
XVI. Conclusions and Future Directions   584   Subject Index   691
    References   584

Contributors

Numbers in parentheses indicate the pages on which the authors' contributions begin.

Robert Anderson (181)
University of Minnesota School of Medicine,
Minneapolis, Minnesota 55455

Siew-Lan Ang (37)
Institut de Génétique et de Biologie Moléculaire et
Cellulaire, CNRS/INSERM, Université Louis Basteur,
67404 Illkirch cedex, C.U. de Strasbourg, France

Atsushi Asakura (253)
Program in Molecular Genetics, Ottawa Hospital
Research Insititute, Ottawa, Ontario, Canada K1H 8L6

Richard R. Behringer (37)
Department of Molecular Genetics, University of Texas
M. D. Anderson Cancer Center, Houston, Texas 77030

Heather Burrows (499)
Graduate Program in Cellular and Molecular Biology,
University of Michigan Medical School, Ann Arbor,
Michigan 48109

Carolyn Byrne (567)
School of Biological Sciences, University of Manchester,
Manchester M13 9PT, United Kingdom

Sally Camper (499)
Department of Human Genetics, University of Michigan
Medical School, Ann Arbor, Michigan 48109

Freida H. Chen (5)
Max Planck Institute of Immunobiology, 79108
Freiburg, Germany

James C. Cross (155)
Department of Biochemistry and Molecular Biology,
University of Calgary Faculty of Medicine, Calgary,
Alberta, Canada T2N 4N1

Lisa Cushman (499)
Department of Human Genetics, University of Michigan
Medical School, Ann Arbor, Michigan 48109

Benoit de Crombrugghe (279)
Department of Molecular Genetics, The University
of Texas M. D. Anderson Cancer Center, Houston,
Texas 77030

Wilhelmine N. De Vries (5)
The Jackson Laboratory, Bar Harbor, Maine 04609

Michael J. Depew (421)
Nina Ireland Laboratory of Developmental
Neurobiology and Department of Oral Biology,
University of California, San Francisco, San Francisco,
California 94143

Kristin Douglas (499)
Department of Human Genetics, University of Michigan
Medical School, Ann Arbor, Michigan 48109

Gregory R. Dressler (395)
Department of Pathology, University of Michigan,
Ann Arbor, Michigan 48109

Elaine Dzierzak (191)
Department of Cell Biology and Genetics, Erasmus
University, Rotterdam 3000, The Netherlands

Alexi V. Evsikov (5)
The Jackson Laboratory, Bar Harbor, Maine 04609

Donna M. Fekete (539)
Department of Biological Sciences, Purdue University,
West Lafayette, Indiana 47907

Phil Gage (499)
Department of Human Genetics, University of Michigan
Medical School, Ann Arbor, Michigan 48109

R. L. Gardner (21)
Department of Zoology, University of Oxford,
Oxford OX1 3PS, United Kingdom

Achim Gossler (127)
Institute fur Molekularbiologie, Medizinische
Hochschule Hannover, D-30625 Hannover, Germany

Hiroshi Hamada (55)
Division of Molecular Biology, Institute for Molecular

and Cellular Biology, Osaka University, Osaka
565-0871, Japan

Matthew Hardman (567)
School of Biological Sciences, University of Manchester,
Manchester M13 9PT, United Kingdom

Richard P. Harvey (331)
The Victor Chang Cardiac Research Institute, St.
Vincent's Hospital, Darlinghurst 2010, Australia; and
Faculties of Medicine and Life Sciences, University of
New South Wales, New South Wales 2052, Australia

Brigid L. M. Hogan (301)
Department of Cell Biology, Howard Hughes Medical
Institute, Vanderbilt University Medical Center,
Nashville, Tennessee 37232

Alexandra L. Joyner (107)
Developmental Genetics Program, Skirball Institute of
Biomolecular Medicine, New York University School
of Medicine, New York, New York 10016

Amy E. Kiernan[1] (539)
MRC Institute of Hearing Research, Nottingham
NG7 2RD, United Kingdom

Barbara B. Knowles (5)
The Jackson Laboratory, Bar Harbor, Maine 04609

Hisato Kondoh (519)
Institute for Molecular and Cellular Biology, Osaka
University, Osaka 565-0871, Japan

Véronique Lefebvre (279)
Department of Biomedical Engineering, Lerner Research
Institute, Cleveland Clinic Foundation, Cleveland, Ohio
44195

Siobhan Loughna (211)
The University of Texas Southwestern Medical Center at
Dallas, Dallas, Texas 75390

Robin Lovell-Badge (371)
Section of Gene Function and Regulation, Chester
Beatty Laboratories, London SW3 6JB, United Kingdom

Oscar Marin (75)
Nina Ireland Laboratory of Developmental
Neurobiology, Department of Psychiatry, Langley Porter
Psychiatric Institute, University of California, San
Francisco, San Francisco, California 94143

Donna Martin (499)
Department of Pediatrics, University of Michigan
Medical School, Ann Arbor, Michigan 48109

Sean J. Morrison (235)
Departments of Internal Medicine and Cell and
Developmental Biology, Howard Hughes Medical
Institute, University of Michigan, Ann Arbor, Michigan
48109

Kazuhisa Nakashima (279)
Department of Molecular Genetics, The University of
Texas M. D. Anderson Cancer Center, Houston, Texas
77030

Igor Nasonkin (499)
Department of Human Genetics, University of Michigan
Medical School, Ann Arbor, Michigan 48109

Anne E. Peaston (5)
The Jackson Laboratory, Bar Harbor, Maine 04609

Marlan Peeters (191)
Department of Cell Biology and Genetics, Erasmus
University, Rotterdam 3000, The Netherlands

Lori Raetzman (499)
Department of Human Genetics, University of Michigan
Medical School, Ann Arbor, Michigan 48109

Janet Rossant (155)
Samuel Lunenfeld Research Institute, Mount Sinai
Hospital, Toronto, Ontario, Canada M5G 1X5

John L. R. Rubenstein (75)
Nina Ireland Laboratory of Developmental
Neurobiology, Department of Psychiatry, Langley Porter
Psychiatric Institute, University of California, San
Francisco, San Francisco, California 94143

Michael A. Rudnicki (253)
Program in Molecular Genetics, Ottawa Hospital
Research Insititute, Ottawa, Ontario, Canada K1H 8L6

Thomas N. Sato (211)
The University of Texas Southwestern Medical Center at
Dallas, Dallas, Texas 75390

Paul T. Sharpe (421)
Department of Craniofacial Development, GKT School
of Dentistry, Guy's Hospital, London SE1 9RT, United
Kingdom

Davor Solter (5)
Max Planck Institute of Immunobiology, 79108
Freiburg, Germany; and The Jackson Laboratory,
Bar Harbor, Maine 04609

Nancy Speck (191)
Department of Biochemistry, Dartmouth Medical
School, Hanover, New Hampshire 03755

Karen P. Steel (539)
MRC Institute of Hearing Research, Nottingham
NG7 2RD, United Kingdom

Hoonkyo Suh (499)
Graduate Program in Neuroscience, University of
Michigan Medical School, Ann Arbor, Michigan 48109

Amanda Swain (371)
Section of Gene Function and Regulation, Chester
Beatty Laboratories, London SW3 6JB, United Kingdom

Patrick P. L. Tam (127)
Embryology Unit, Children's Medical Research Institute,
University of Sydney, Westmead, New South Wales
2145, Australia

Abigail S. Tucker (421)
MRC Centre for Developmental Neurobiology, GKT
School of Biomedical Sciences, Guy's Hospital, London
SE1 1UL, United Kingdom

Chris Wylie (181)
Division of Developmental Biology, Children's Hospital
Research Foundation, Cincinnati, Ohio 45229

Kennneth S. Zaret (301)
Cell and Developmental Biology Program, Fox Chase
Cancer Center, Philadelphia, Pennsylvania 19111

About the Editors

Janet Rossant

Dr. Janet Rossant is Joint Head of the Program in Development and Fetal Health at the Samuel Lunenfeld Research Institute, Mount Sinai Hospital, Toronto and University Professor and Professor in the Department of Molecular and Medical Genetics and the Department of Obstetrics/Gynaecology, University of Toronto. Her research interests center on understanding the genetic control of normal and abnormal development in the early mouse embryo. She uses the powerful techniques available to genetically manipulate the mouse genome to address these problems. Recently, her research has moved in two new directions. First, stem cell research, with her discovery of a novel placental stem cell type, the trophoblast stem cell. Second, genome-wide functional genomics. She directs the Centre for Modelling Human Disease in Toronto, which is undertaking genome-wide mutagenesis in mice to develop new mouse models of human disease.

Dr. Rossant trained at the Universities of Cambridge and Oxford, United Kingdom and has been in Canada since 1977, first at Brock University and then in Toronto. She is a Fellow of both the Royal Societies of London and Canada, an International Scholar of the Howard Huges Medical Institute, and a Distinguished Scientist of the Canadian Institutes of Health Research.

Dr. Rossant is actively involved in the international developmental biology community. She was Editor of the journal *Development* for many years. She has organized a number of international developmental biology meetings, including the International Developmental Biology Congress in 1997. She was President of the Society for Development Biology in 1996/97. She has also been actively involved in public issues related to developmental biology, most recently serving as Chair of the Canadian Institutes of Health Research working group on stem cell research.

Patrick P. L. Tam

Dr. Patrick P. L. Tam is a Senior Principal Research Fellow of the National Health and Medical Research Council (NHMRC) of Australia. He is the Head of the Embryology Research Unit at the Children's Medical Research Institute in Sydney and holds a conjoint appointment of Senior Principal Research Fellow in the Faculty of Medicine, University of Sydney. His research focuses on the elucidation of the cellular and molecular mechanisms of body patterning during mouse development. Dr. Tam pioneers the application of micromanipulation and embryo culture for analyzing tissue potency and lineage specification in normal and mutant embryos. He studies the morphogenetic role of the gastrula organizer in axis formation, and the developmental processes leading to the regionalization of the neural tube, the paraxial mesoderm, and the embryonic gut. His other current research is on the pathogenesis of X-linked diseases using mouse models generated by transgenesis, gene targeting, and chemical mutagenesis.

Dr. Tam received his training in mammalian embryology in Hong Kong, London, and the United States, took a faculty appointment in Anatomy at the Chinese University of Hong Kong, and has been in Australia since 1990 to establish his research laboratory in Sydney. He is an honorary consulting scientist of the Children's Hospital at Westmead, Australia and an Honorary Professor at the University of Hong Kong. He was a recipient of the Symington Memorial Prize in Anatomy of the Anatomical Society of Great Britain and Ireland and the Croucher Foundation Fellowship at Oxford University, and has been holder of the prestigious NHMRC Research Fellowship since 1996.

Dr. Tam was an instructor and lecturer of the teaching course on molecular embryology of the mouse at the Cold Spring Harbor Laboratory, served on numerous grant and fellowship review panels, and was involved with the organization of international conferences on cell and developmental biology. He was an Editor of *Anatomy and Embryology*, and is currently a member of the editorial boards of *Developmental Biology, Mechanisms of Development, International Journal of Developmental Biology, Genesis, Differentiation, The Scientific World,* the *Faculty of 1000,* and the *Highlights Advisory Board of Nature Reviews Neuroscience.*

I

Establishment of Body Patterns

It has been long held that, in contrast to lower vertebrate and invertebrate embryos, the mouse embryo develops from the zygote without any endowment of positional information from the mother or early zygote to specify the orientation and polarity of the prospective embryonic axis. Experimental manipulation of the early preimplantation embryo revealed that early blastomeres are remarkably plastic in their lineage potency and that the assembly of cells into an embryo is highly regulative. These findings lent staunch support to the concept that the preimplantation embryo is a "tabula rasa."

Morphological study of the fertilized oocytes of other mammalian embryos, such as the rat and the goat (De Smedt et al., 2000), however, has shown that localization of cytoplasmic inclusions or cellular organelles occurs, which seems to be consistent with the concept of localized morphogenetic determinants with long lasting impact on the patterning of the embryo.

Some *hint of asymmetry* is present in the fertilized mouse oocyte. The second polar body, which contains the haploid set of maternal genome and associated cytoplasm, is extruded from the animal pole of the fertilized oocyte. The animal-vegetal axis as marked by the position of the polar body is found to be aligned with the plane that is orthogonal to the embryonic-abembryonic axis of the blastocyst (Chapter 2). The other landmark, albeit transient, is the localized protrusion on the oocyte surface known as the fertilization cone where sperm fusion occurs. It has been found that

the meridional plane defined by the circumference passing through the polar body and the sperm entry point is frequently aligned with the plane of the first cleavage of the fertilized oocyte. The membrane of the sperm entry site also stays with the blastomere that preferentially contributes to the inner cell mass and the mural trophectoderm, in other words, cells on the embryonic half of the blastocyst (Piotrowska and Zernicka-Goetz, 2001; Piotrowska et al., 2001). The position of the polar body and the sperm entry point in the fertilized oocyte therefore marks the animal-vegetal axis and the border of the embryonic and abembryonic compartments. It is not known whether the extrusion of polar bodies always happens at a defined site in the oocyte, but the position of the sperm entry site relative to the polar body may vary considerably. This raises the intriguing possibility that there may be no predetermined polarity in the unfertilized oocyte, and the information is acquired on sperm–egg association. The challenge in the future is to identify the subcellar reorganization (e.g., cytoskeletal remodeling and membrane stabilization), trafficking, and localization of morphogenetic determinants either preexisting in the oocyte or encoded by the activated zygotic genome (Chapter 1) that underpin this patterning process.

Preimplantation mouse development is preoccupied with the generation of asymmetry and the delineation of the embryonic and extraembryonic cell lineages (Chapter 8). Recently, results of cell tracing experiments performed on periimplantation mouse embryos (i.e., embryos at blastocyst to pregastrula stages) points strongly to *a relationship between the axes of asymmetry of the blastocyst and the primary body axes* of the early-organogenesis-stage embryo (Chapter 2). The most intriguing finding is the directional coherent movement of the visceral endoderm from the rim of the inner cell mass first to the extraembryonic and then to the posterior region of the gastrula. In the embryonic region, the endoderm at the most distal tip of the embryo seems to be displaced toward the prospective anterior region. This overall anterior shift of the endodermal population is reminiscent of that in the avian embryo, which has been shown to determine the orientation of the body axes (Stern, 2001). In contrast to the increasing knowledge of how cells move and how the embryo changes its shape and tissue architecture, amazingly little is known of the driving forces underlying these developmental phenomena. Some of the factors that might contribute to the morphogenetic forces are the differential rates of cell proliferation in the epiblast and the visceral endoderm, the kinetics of recruitment of cells from the epiblast to the visceral endoderm, the trafficking of cells between the visceral and parietal endoderm, the differential growth of the germ layers, localized epithelial expansion and selective tissue adhesion. In addition, the *impact of the uterine environment*, with respect to the space constraints in the implantation site and the physical and inductive interactions between the implanting embryo and the surrounding uterine tissues, should not be ignored.

Chimera and cell marking experiments have shown extensive mixing of clonal descendants from the inner cell mass to the epiblast, ruling out the possibility that any positional information conferred on inner cell mass cells could have been maintained and utilized for later patterning of the embryonic tissues. The visceral endoderm, where interclonal mixing is more restricted, is a likely *repository of patterning information*. Experimental evidence obtained from embryological experiments (Chapter 2) and analysis of mouse mutant (Chapter 3) has presented a compelling case for a critical role of the visceral endoderm in supporting the differentiation of the epiblast, the initiation of gastrulation, and the induction of anterior cell fates. The anterior region of the visceral endoderm (AVE) is postulated to be a particularly important source of patterning activity (Chapter 3). Whether AVE is absolutely required for the formation of anterior structures has yet to be tested critically, since the findings to date are largely indirect or circumstantial (Chapter 2). Other populations of the visceral endoderm, such as those associated with the primitive streak, might also have roles in patterning (Kalantry *et al.,* 2001). Some indication of the critical role of posterior visceral endoderm is highlighted by the impact of loss of *Wnt3* or *Hnf3β* (*Foxa2*) activity in this tissue (Chapter 3). During gastrulation, the visceral endoderm is displaced by the definitive endoderm recruited from the epiblast. This raises the question of how the patterning information may be relayed to other germ layer tissues after the

departure of the initial signaling tissue. Analysis of the expression of genes like *Hesx1* points to possible crosstalk between the visceral endoderm and the epiblast, such that the epiblast retains and executes the patterning instruction acquired from the endoderm. The role in body patterning of the extraembryonic ectoderm, which is derived from the polar trophectoderm and is closely associated with the proximal region of the embryo until the appearance of the extraembryonic mesoderm, is not known. However, the inductive activity mediated by bone morphogenetic proteins expressed in the extraembryonic ectoderm and the adjacent visceral endoderm is found to be critical for the formation and maintenance of the primordial germ cells (Chapter 9). Recently, activity of Arkadia, which may modulate Nodal signaling in the extraembryonic tissues, has been shown to be critical for the specification of the gastrula organizer in the mouse (Niederlander *et al.,* 2001)

Extraembryonic tissues are not the only source of patterning activity in gastrulation. Indeed, until recently, they were not thought to be a major source of signals at all. The *gastrula organizer*, equivalent to the Spemann's organizer in the frog, was considered paramount. The gastrula organizer is defined by its fate to form the axial mesoderm (prechordal mesoderm and notochord), the expression of a set of genes that are common to those expressed by the organizers of fish, frog, and bird gastrula and, unique to this group of cells, the ability to induce the formation of a partial embryo in the host embryo after transplantation (Camus and Tam, 1999). However, it has been found that induction of anterior characteristics in the new body axis requires more than the organizer and involves the synergistic interaction with the anterior visceral endoderm and the associated anterior epiblast (Chapter 3; De Souza and Niehrs, 2000).

Both the organizer and its axial derivatives are also essential for establishing left-right asymmetry (Chapter 4). Embryos harboring mutations of genes that affect the formation or differentiation of the organizer (Chapter 3) are found to display abnormal specification of left-right body asymmetry (Chapter 4). The node in particular plays a central role as the source of signals initiating the cascade of signaling activity involving fibroblast growth factors (FGFs), Sonic hedgehog (SHH) protein, Nodal, EGF-CFC, and the asymmetrical activation of transcription factors that determine the sidedness of the body and handedness of visceral organs (Chapter 4). A strong correlation is seen between the ciliary activity of cells in the node to the generation of left-right body asymmetry, but it is yet unclear whether this leads to a tidal flow for distributing morphogenetic factors or instead reflects a more global pattern of the cytoskeletal architecture that underpins the lateral asymmetry.

The development of the neural axis begins with the induction of the neural primordium by the combined activity of the mesoderm and endoderm and the organizer (Chapter 3; Camus and Tam, 1999). The morphogenesis of the major regions of the embryonic brain requires the continuous *inductive activity of the axial mesoderm,* especially the prechordal mesoderm and the anterior notochord (Chapter 5). The delineation of the major brain segments is accomplished by regionalized gene activity and the consolidation of the expression domains defining the boundary between the segments (Chapter 6). Finer subdivision of the neural tube results in the delineation of the neuromeres (e.g., prosomeres in the forebrain and rhombomeres in the hindbrain) along the anterior-posterior axis (Chapters 5 and 6). This is accomplished primarily by the patterning activity of the *organizing centers in the neural tube* at the anterior neural ridge, the diencephalon-mesencephalon junction and the midbrain-hindbrain junction (Chapters 5 and 6). The brain organizer acts mainly via FGF signaling, which controls cell proliferation and the activation of segment-specific transcriptional activity determining neuronal fates. How the brain organizers are formed is not known, but surgical ablation of the axial mesoderm results in the loss from the brain of the genetic activity that is characteristic of the organizers (Camus *et al.,* 2000). Whether this mirrors the normal role of the axial mesoderm in the establishment of the brain organizers is not known. In addition to the brain organizers, the paraxial mesoderm of the avian embryo has been shown to impart an instructive effect on the expression of rhombomere-specific *Hox* genes in the hindbrain. This may suggest that the paraxial mesoderm not only influences the development of the peripheral nervous system by determining the segmental migratory paths of the neural crest cells (Chapter 7), but it may also provide the signal that specifies the segmental characteristics of the neural tube. The concept of organizers as key elements in pattern formation is not restricted to the main body axis. The formation and patterning of the limbs also involves localized sources of signals—the apical ectodermal ridge and the zone of polarizing activity—and some of the same signaling pathways including FGF, WNTs, and SHH (Martin, 1999; Dudley and Tabin, 2000; Schaller *et al.,* 2001).

Besides the neuromeric pattern of the cephalic neural tube, the *meristic organization* of the paraxial mesoderm into somites in the trunk and tail and the presence of somitomeres in the cranial region and the presomitic mesoderm highlights the segmental characteristics of the body plan of the mouse embryo. During gastrulation, cells destined for the paraxial mesoderm are allocated to the anterior-posterior axis in the temporal order of their recruitment from the somitic progenitor pool and incorporation to the caudal end of the paraxial mesoderm (Kinder *et al.,* 1999). Oscillatory genetic activity that regulates Notch signaling provides the molecular control of the timing of segmentation of the presomitic mesoderm and the positioning of the intersegmental and half-segment boundary of the somite (Chapter 7).

Patterning of the tissues in the *dorsoventral plane of the embryo* is accomplished by competing inductive signals from the axial mesoderm and the floor plate of the neural tube, and from the roof and dorsal plate of the neural tube (Chapter 6; Jessell, 2000). The same signals are also involved with the specification of the ventromedial (sclerotome) and dorsolateral (dermomyotome) compartments of the somites (Chapter 7). An additional source of regionalization signal emanating from the lateral plate mesoderm further specifies the medial (epaxial) and lateral (hypaxial) myotome of the somites (Chapters 7 and 13). The axial mesoderm also acts to compartmentalize the embryonic gut and its accessory organs into the dorsal (closer to the axial mesoderm) and the ventral (farther from the axial mesoderm) primordia primarily through SHH signaling activity (Chapter 15; Wells and Melton, 1999).

The contents of the seven chapters in this section summarize our current understanding of the cellular and molecular aspects of embryonic patterning from fertilization to early organogenesis. Despite significant gaps in our knowledge about the transition of embryonic architecture and cell lineages from blastocyst to the gastrula and details of the functional interactions of the signaling molecules that leads to the organization of the body plan, several themes of the developmental process can be recognized:

1. The blastocyst architecture and the pattern of allocation of the progeny of blastomeres to the embryonic and extraembryonic lineages may be founded on the asymmetry of the zygote established epigenetically at fertilization.
2. The asymmetry of the blastocyst as defined by the animal-vegetal and embryonic-abembryonic axes may have a consequential relationship to the orientation of three primary embryonic axes.
3. The translation of the blastocyst asymmetry to the polarity and orientation of the embryonic axes is influenced by morphogenetic tissue movements driven by physical constraints and expediency within the embryonic and the uterine confine.
4. There are likely multiple sources of patterning activity from the extraembryonic tissues, intraembryonic sources, and the gastrula organizer. Embryonic patterning is the result of the interplay between synergistic and antagonistic actions of morphogenetic factors from these sources.
5. The patterning activity of the organizer is not restricted to the various forms of gastrula organizer but also their derivatives.
6. Finer patterning of the body plan requires more than the activity of the organizer derivatives and involves other tissues that provide the supplementary patterning instructions.
7. Organizer activities are not restricted to the body, but are also involved in the axes of appendages.

The final outcome of these patterning activities is the establishment of a basic body plan with defined anterior-posterior polarity, the segmentally organized neural tube and paraxial mesoderm, a distinct laterality of organ primordia, and the dorsoventrally compartmentalized neural tube, somites, and endodermal structures.

References

Camus, A., Davidson, B. P., Billiards, S., Khoo, P., Rivera-Perez, J. A., Wakamiya, M., Behringer, R. R., and Tam, P. P. (2000). The morphogenetic role of midline mesendoderm and ectoderm in the development of the forebrain and the midbrain of the mouse embryo. *Development* **127,** 799–813.

Camus, A., and Tam, P. P. L. (1999). The organizer of the gastrulating mouse embryo. *Curr. Topics Dev. Biol.* **145,** 117–153.

De Smedt, V., Szollosi, D., and Kloc, M. (2000). The Balbioni body: Asymmetery in the mammalian oocyte. *Genesis* **26,** 268–312.

De Souza, F. S., and Neihrs, C. (2000). Anterior endoderm and head induction in early vertebrate embryos. *Cell Tissue Res.* **300,** 207–217.

Dudley, A. T., and Tabin, C. J. (2000). Constructive antagonism in limb development. *Curr. Opin. Genet. Dev.* **10,** 387–392.

Jessell, T. M. (2000). Neuronal specification in the spinal cord: Inductive signals and trasnscriptional codes. *Nature Rev. Genet.* **1,** 20–29.

Kalantry, S., Manning, S., Haud, O., Tomihoara-Newberger, C., Lee, H.-G., Fangman, J., Dieteche, C. M., Manova, K., and Lacy, E. (2001). The *amnionless* gene, essential for mouse gastrulation, encodes a visceral-endoderm-specific protein with an extracellular cysteine-rich domain. *Nature Genet.* **27,** 412–416.

Kinder, S. J., Tsang, T. E., Quinlan, G. A., Hadjantonakis, A. K., Nagy, A., and Tam, P. P. L. (1999). The orderly allocation of mesodermal cells to the extraembryonic structures and the anteroposterior axis during gastrulation of the mouse embryo. *Development* **126,** 4691–4701.

Martin, G. R. (1999). The roles of FGFs in the early development of vertebrate limbs. *Genes Dev.* **12,** 1571–1586.

Niederlander, C., Walsh, J. L., Episkopou, V. and Jones, M. J. (2001). Arkadia enhances nodal-related signalling to induce mesendoderm. *Nature* **410,** 830–834.

Piotrowska, K., and Zernicka-Goetz, M. (2001). Role for sperm in spatial patterning of the early mouse embryo. *Nature* **409,** 517–521.

Piotrowska, K., Wianny, F., Pedersen, R. A., and Zernicka-Goetz, M. (2001). Blastomeres arising from the first cleavage division have distinguishable fates in normal mouse development. *Development* **128,** 3739–3748.

Schaller, S. A., Li, S., Ngo-Muller, V., Han, M. J., Omi, M., Anderson, R., and Muneoka, K. (2001). Cell biology of limb patterning. *Int. Rev. Cytol.* **203,** 483–517.

Stern, C. D. (2001). Initial patterning of the central nervous system: How many organizers? *Nature Rev. Neurosci.* **2,** 92–98.

Wells, J. M., and Melton, D. A. (1999). Vertebrate endoderm development. *Annu. Rev. Cell Dev. Biol.* **15,** 393–410.

1

Fertilization and Activation of the Embryonic Genome

Davor Solter,*,† Wilhelmine N. de Vries,† Alexei V. Evsikov,† Anne E. Peaston,†
Frieda H. Chen,* and Barbara B. Knowles†

*Max Planck Institute of Immunobiology, 79108 Freiburg, Germany
†The Jackson Laboratory, Bar Harbor, Maine 04609

I. Introduction

II. Oogenesis

III. Meiosis and the Beginning of Oocyte Asymmetry

IV. Fertilization

V. Transcription and Its Control

VI. mRNA Utilization during Oocyte Maturation and Preimplantation Development

VII. Gene Expression in the Early Mouse Embryo

VIII. Functional Analysis

References

I. Introduction

The full-grown oocyte, arrested in prophase of the first meiotic division, contains all of the molecules that will be utilized to bridge the period of transcriptional silence that begins with the completion of oocyte growth. Under hormonal stimulation the full-grown oocyte begins maturation, completing the first meiosis and the first half of the second meiotic division before arresting in metaphase of the second meiotic division. During this period the extensive stores of maternal messages are selectively utilized, which can result in the synthesis of a new and perhaps different set of proteins. Simultaneously, preexisting maternal proteins can undergo post-translational modification and degradation. These programmed events result in an oocyte that is ready for fertilization. Fertilization initiates a cascade of events, also dependent on protein modifications and on the timely synthesis of new proteins from maternal mRNA stores, that leads to completion of the second meiotic division, remodeling of egg and sperm chromatin, DNA synthesis, entry into the first mitosis, and activation of the embryonic genome. The molecular control of the oocyte-to-embryo transition is underexplored. However, novel strategies are now being employed to identify the genes and their function at the initiation of development.

This chapter covers what is known about the crucial molecular events and processes that lead to the activation of the embryonic genome and the beginning of development. We do not delve into these processes in extensive detail. For details, the reader should consult the following excellent and extensive reviews: Johnson (1981), Kidder (1993), Latham (1999), Nothias *et al.* (1995), Schultz (1999), Telford *et al.* (1990), and Thompson *et al.* (1998).

5

II. Oogenesis

The initial stages of oogenesis take place during fetal development in mammals. After entering the genital ridge, the germ cells begin to divide mitotically and thus create a large supply of oogonia, which then enter into meiosis. A large proportion of these primary oocytes degenerates and only some of them start forming primary follicles. At birth a finite number of primary follicles containing primary oocytes arrested in the prophase of the first meiotic division is available to provide the female with functional gametes for her entire reproductive life. Most of these oocytes also degenerate; only a small proportion ever develops into mature fertilizable ova. The oocytes continue to grow within the follicle, thus gradually creating in their cytoplasm a molecular environment that contains all nutritive and informational elements necessary to start and support the initial stages of development.

Cellular organization and structure are largely dictated by function, thus it is to be expected that cells performing the same function in widely distant species are structurally very similar. However, development in different species can vary greatly and thus it is not surprising that oocytes display a wide variety in size and molecular content. Despite these real and important differences, certain basic principles operate equally in sea urchin, frog, and ostrich oocytes, with diameters ranging from 100 μm to a score of centimeters. One basic but trivial principle is that the content of nutrients in each oocyte is commensurate with the length of time the developing embryo needs to be supported without access to an external food source. The size of oocyte in each species is largely determined by this specific demand.

Another much more interesting basic principle is that each oocyte contains a certain amount of morphogenetic information. Cytoplasmic regionalization and organization are often obvious by inspection alone (Davidson, 1986), but our understanding of the molecular organization of an oocyte and its role in further development has been dramatically expanding in recent years. It is well beyond the scope of this chapter to detail the tremendous amount of genetic and biochemical information as it relates to the molecular biology of the morphogenetic information in the *Drosophila* or *Xenopus* egg, the most intensely studied models. It is clear that correct spatial and temporal control of distribution and utilization of maternal mRNA and protein molecules is the prerequisite for normal embryonic development in these species and in many other nonmammalian species. One example of a mechanism establishing cytoplasmic regionalization involves the regulation of transport and anchorage of specific mRNA molecules to defined positions in the egg. The specific motif in the mRNA is recognized by a chaperone protein with dual or multiple specificities, which recognize both the target mRNA molecule and part of the cellular cytoarchitecture where this particular mRNA molecule is to be localized. Another set of controlling mechanisms most probably involves specific and selective mRNA translation-inhibition in all parts of the egg except where the protein product is spatially and temporally required. For further examples and details, the reader is referred to numerous articles and reviews on the subject (e.g., Danpure, 1995; Gavis, 1997; Holliday, 1989; King *et al.*, 1999; Schnapp, 1999; Solter and Knowles, 1999).

Do these or similar mechanisms also operate during mammalian oogenesis? That is, can we detect cytoplasmic localizations with morphogenetic consequences in the mammalian egg? One should bear in mind that, although the mouse is the most intensely studied of all eutherian species and a large amount of information has been accumulated on mouse development, the question remains: To what extent is early mouse development the universal model for that in all mammals? Nevertheless, to our admittedly limited observational capacities, the fully grown mouse oocyte is radially symmetrical and without obvious morphological differences throughout. After the oocyte matures and progresses through meiosis (see Section III), certain obvious morphological changes take place. However, can there be any elaboration of cytoplasmic diversity during oocyte growth and before maturation in the mouse? If so, does this have any functional significance?

In many invertebrate and vertebrate species the initial expansion phase of gametogenesis (before the initiation of meiosis) involves a series of divisions of a single progenitor cell whose progeny remain connected by intercellular bridges, thus forming a syncytium. Such a group of cells is called a germline cyst, and synchronous divisions and an exchange of material between the cells in the cyst are probably essential for normal gametogenesis. A typical example is *Drosophila* oogenesis, during which 1 of the 16 interconnected cells becomes an oocyte and the remaining 15 are localized on one side of the egg and serve as nurse cells supplying the developing egg with nutrients and localization signals. Pepling and Spradling (1998) observed premeiotic germ cells in clusters in fetal mouse ovaries; these cells were connected by intercellular bridges often containing mitochondria, which would suggest active intercellular transport as observed in the germline cysts of *Drosophila*. They also observed cell cycle synchronization within the cluster. In addition, the number of cells in the cluster frequently corresponded to powers of 2, again suggesting their origin from a single cell. The existence of several mouse mutants involving genes like *Dax1* (Yu *et al.*, 1998) or *rjs/jdf-2* (Lehman *et al.*, 1998), which result in the formation of follicles containing more than one oocyte, suggests the possible presence and important functional role for the germline cyst. Germline cysts are an essential feature of gametogenesis in vertebrate males and in higher insect females where they can easily be observed, since gametogenesis takes place continuously during adult life (Pepling *et al.,* 1999). If germline cysts are formed during oogenesis in mammals, then it is possible that the primary oocyte was one of several sister cells and the remaining cells would be the equivalent of nurse cells. The presence of germline cysts in the ovaries of mammalian females has not yet been established beyond doubt and their possible function is even less clear.

III. Meiosis and the Beginning of Oocyte Asymmetry

Mammalian oocytes are arrested in the diplotene stage of the first meiotic prophase and they continue growing and accumulating RNA, protein, and other molecules until they reach their full size (~80 μm in diameter in the mouse). Fully grown oocytes [also known as germinal vesicle (GV) oocytes because their nuclear membrane is still intact] are able to reinitiate meiosis and undergo maturation, which makes them capable of being fertilized. Follicle cells normally maintain the oocyte in a state of meiotic arrest as evidenced by the observation that oocytes released from follicular cells *in vitro* spontaneously initiate maturation and resumption of meiosis (Edwards, 1965). Maturation *in vivo* is initiated by a surge of luteinizing hormone that binds to receptors on the follicular cells. Follicular cells signal the oocyte through junctional complexes formed between them and oocyte microvilli.

The molecular aspects of maturation and control of meiosis are fairly well understood and, though obvious differences exist between different species, certain basic elements are the same. The central component of the entire process is the M-phase or maturation promoting factor (MPF) (Masui and Markert, 1971), which is composed of protein kinase p34^{cdc2} and the regulatory protein cyclin B. Germinal vesicle breakdown (GVBD) and reinitiation of meiosis are independent of protein synthesis, indicating that the MPF components are present, but inactive, in the GV oocyte. Inactivation is likely mediated by phosphorylation at tyrosine 15 of p34^{cdc2} through the action of WEE1 kinase (Mitra and Schultz, 1996). The p34^{cdc2} is then dephosphorylated, most likely by a mouse homologue of yeast cdc25 kinase, two of which have been described in the mouse oocyte and early embryos (Wickramasinghe *et al.,* 1995). In addition to phosphorylation–dephosphorylation of p34^{cdc2}, its rapid accumulation at the end of oocyte growth probably also contributes to the acquisition of meiotic competence (de Vantéry *et al.,* 1996). The synthesis and accumulation of the regulatory subunit of MPF, cyclin B (de Vantéry *et al.,* 1997; Polanski *et al.,* 1998), represent another controlling factor in oocyte maturation. Because the level of cyclin B mRNA is essentially identical in meiosis-incompetent and meiosis-competent oocytes (de Vantéry *et al.,* 1997), the amount of cyclin B is determined by con-

trolled translation. Translational control of cyclin B synthesis (Tay *et al.,* 2000) involves cytoplasmic polyadenylation of dormant mRNA. This control mechanism is of great significance during the period of transcriptional silence between the cessation of oocyte growth and the activation of the embryonic genome (see Section VI). Several other molecules such as MOS (Gebauer *et al.,* 1994) involved in progression through meiosis are similarly regulated. In mice, MOS is necessary not for GVBD and initiation of meiosis as it is in *Xenopus* (Sagata *et al.,* 1988), but for progression from meiosis I to meiosis II. MOS kinase is an essential part of the cytostatic factor (CSF), which is necessary to stabilize the high level of MPF activity that leads to arrest in metaphase II representing the end of oocyte maturation. In *MOS$^{-/-}$* mice, the oocytes complete meiosis I but instead of going into meiosis II, the nuclei are reformed and the oocytes can undergo parthenogenetic activation (Colledge *et al.,* 1994; Hashimoto *et al.,* 1994).

The cytoplasmic polyadenylation and translation of *c-mos* mRNA in *Xenopus* are regulated through binding of its cytoplasmic polyadenylation element (CPE) by the CPE binding factor (CPEB); phosphorylation of CPEB by Eg2 kinase is essential for it to bind to the CPE of MOS in the *Xenopus* oocyte (Mendez *et al.,* 2000). Two kinases, which share high amino acid homology with Eg2, have recently been described in mouse (AIE1) and human (AIE2) (Tseng *et al.,* 1998). AIE1- related kinases (STK-1 and IAK1/Ayk1) were identified in mouse oocytes (Tseng *et al.,* 1998). In addition, expression sequence tags (ESTs) corresponding to AIE1 and AIE2 were found among the 15,000 ESTs in the Knowles-Solter two-cell stage library (B. B. Knowles and D. Solter, unpublished results). These data suggest that *MOS* may also be regulated during mouse oocyte maturation by a complex sequence of phosphorylation reactions that regulates translational control of its mRNA. Although we have identified some of the molecular events that occur during mouse oocyte maturation and regulate progression through meiosis, many more questions remain.

The radial symmetry of the GV fully grown oocyte is lost during oocyte maturation. A detailed discussion about the possible functional consequences of this loss is provided in Chapter 2. It has recently been shown that even GV oocytes display a nonsymmetrical distribution of leptin and STAT3, both members of the transcriptional activator cascade (Antczak and Van Blerkom, 1997). Both leptin and STAT3 were localized within a subpopulation of follicular cells and a corresponding portion of the oocyte, again suggesting differences within the follicular cell population and the possible existence of nurse cells and germline cysts in mouse oogenesis (see Section II). On completion of oocyte maturation, the location of the first polar body denotes the animal pole and also marks part of the egg surface largely devoid of microvilli (Evans *et al.,* 2000; Van Blerkom and Motta, 1979). It has been demonstrated that cortical endoplasmic reticulum accumulates in every part of the cortex except in the micro-villi-free area around the metaphase spindle, that is, the area from which the second polar body is released (Kline *et al.,* 1999). In mice, sperm entry is restricted to the egg surface containing microvilli, although the functional significance of these domains is open to question since human eggs lack completely a nonmicrovillar region (De Smedt *et al.,* 2000). Another possible example of asymmetry in the eggs of some mammalian species involves the movement of the so-called Balbiani body, which in *Xenopus* localizes to the vegetal pole and likely participates in the formation of germinal plasma (Kloc and Etkin, 1995). It is possible that a similar structure also localizes to the vegetal pole of some mammalian oocytes (De Smedt *et al.,* 2000). Currently there is little evidence for the existence of functionally relevant cytoplasmic localization in the mammalian egg although the question may be clarified as more candidate genes expressed during oogenesis become known and available for investigation.

IV. Fertilization

Only a mature egg arrested in metaphase of the second meiotic division is capable of being fertilized. Fertilization initiates numerous complex changes in all compartments of the resulting conceptus (Snell and White, 1996; Wassarman, 1999) involving all components of the egg as well as the sperm-derived chromatin (Wright, 1999). Although mammalian eggs can be activated by many divergent stimuli, natural fertilization cannot be reproduced in full by artificial activation (Fissore *et al.,* 1999). This certainly contributes to the poor success rate of cloning by nuclear transfer (Solter, 2000). For example, incorporation of the sperm membrane into the oolemma results in an egg membrane block to polyspermy, whereas oocytes activated by intracytoplasmic sperm injection (ICSI) and parthenogenetically activated eggs can be penetrated by additional sperm (Maleszewski *et al.,* 1996).

Following fertilization the arrested meiosis resumes, the second polar body is soon extruded, and the female pronucleus is formed to join the already formed male pronucleus. This process is controlled by the inactivation of MPF, which has been stabilized in oocytes by CSF composed of MOS and other MAP (mitogen-activated protein) kinases. Fertilization results in a rapid decrease in MPF activity followed by a slower decrease in MAP kinase activity (Moos *et al.,* 1995). The role of inactivation of these kinases in completing meiosis and initiating interphase is further demonstrated by the ability of protein kinase inhibitors to initiate the same events (Sun *et al.,* 1998). Changing the metaphase spindle into an anaphase configuration requires the presence of calcium/calmodulin-dependent protein kinase II (CaM kinase II), which is associated with the metaphase spindle; following fertilization calmodulin is very quickly colocalized, presumably leading to the activation of CaM kinase II (Johnson

et al., 1998). Another molecule associated with the meiotic spindle is spindlin (Oh *et al.,* 1997), which is also phosphorylated, depending on the cell cycle stage, and whose phosphorylation is at least in part mediated by the MOS–MAP kinase pathway (Oh *et al.,* 1998). In *MOS*$^{-/-}$ mice, spindlin is hypophosphorylated and, although abundantly present in the cytoplasm, it does not bind to the spindle (Oh *et al.,* 1998). As mentioned before, the eggs of *MOS*$^{-/-}$ females undergo spontaneous activation, and it is possible that the lack of spindlin phosphorylation is one of the elements that destabilizes metaphase II arrest.

Fertilization triggers a series of signaling events accompanied by a series of Ca^{2+} transients (waves of increased Ca^{2+} concentration passing through the egg cytoplasm). Subsequently, the cytoplasm and the nuclear compartments of the egg are remodeled suggesting that Ca^{2+} transients may play a crucial role in the initiation of transcription. Calcium transients following fertilization may be mediated by the soluble sperm protein oscillin (Parrington *et al.,* 1996), though its mechanism of action is not clear. The presence of this protein may explain oocyte activation following ICSI, and it may be possible to use this soluble protein to "normalize" egg activation in nuclear transfer experiments.

One analysis of Ca^{2+} transients demonstrated that they initially originate at the point of sperm penetration, perhaps an immediate effect of oscillin, and their later origin from a site opposite the second polar body (Deguchi *et al.,* 2000). Ca^{2+} waves continue until pronuclear formation and possibly, at a low frequency, until the formation of two-cell embryos. Ca^{2+} transients could have many important and long-range consequences. It was shown that the number of cells in the inner cell mass was higher in embryos parthenogenetically activated by Sr^+ (Ca^{2+} transients present) as compared with those activated by ethanol (Ca^{2+} transients absent) (Bos-Mikich *et al.,* 1997). Ca^{2+} oscillations reduce the effective Ca^{2+} threshold necessary for the activation of transcription factors (Dolmetsch *et al.,* 1998), and it was also suggested that the frequency of oscillation may determine which factors are activated (Dolmetsch *et al.,* 1998). If one extrapolates these results to the fertilized egg, one can envision all kinds of regulatory events that might proceed in waves directionally through the cytoplasm. Temporal changes in Ca^{2+} transients (Tang *et al.,* 2000), and a positional change in their origin could provide both sequential and spatial information.

V. Transcription and Its Control

The period of intensive transcriptional activity during mouse oogenesis ceases when the egg is arrested at the GV-intact, full-grown oocyte stage and only gradually resumes after fertilization. Until the embryonic genome is activated, the processes of egg maturation, completion of meiosis and initial postfertilization events are controlled by maternal molecules (proteins and RNAs) accumulated during oogenesis. These maternal molecules are also responsible for proper embryonic genome activation. This period of development has been extensively studied and certain basic mechanisms are being elucidated (e.g., Latham, 1999; Nothias *et al.,* 1995; G. A. Schultz, 1986; R. M. Schultz, 1999; Telford *et al.,* 1990; Thompson *et al.,* 1998). In perusing the literature on transcription control in early mouse embryos the reader should be aware of a possible source of confusion. Certain authors (Stein and Schultz, 2000; Wiekowski *et al.,* 1997) describe two-cell stage chromatin as being in a transcriptionally repressive state because at that stage transcription is for the first time dependent on the presence of enhancers. Others (Forlani *et al.,* 1998) talk about a transcriptionally repressive state when describing the genome of the early one-cell stage, that is, at a time when there is no observable transcription. This duplication of usage (though both can be justified) is unfortunate and should be resolved, or one could end up saying that DNA synthesis in the zygote leads to the establishment (first definition) or abolishment (second definition) of the transcriptionally repressive state.

It is not entirely clear how transcription is suspended in the full-grown and maturing oocyte and immediately after fertilization, but the absence and/or inhibition of the functional basic transcription machinery may be responsible. Transcription was not observed when transcriptionally competent nuclei were transferred into the cytoplasm of the early zygote, but was detected following transfer into the cytoplasm of the late zygote (Latham *et al.,* 1992). The largest subunit of RNA polymerase II is hyperphosphorylated in the oocyte but becomes dephosphorylated following fertilization (Bellier *et al.,* 1997). The hyperphosphorylated form is thought to be unable to initiate transcription, explaining at least in part the transcriptional silence of the full-grown oocyte. It is unclear whether its dephosphorylation, which takes place several hours after fertilization, is responsible for the short burst of transcription in the late zygote. A gradual reestablishment of the "somatic"-type phosphorylation pattern of RNA polymerase II, and its translocation to the nucleus, takes place in the two-cell stage embryo concomitant with complete embryonic genome activation (Bellier *et al.,* 1997; Worrad *et al.,* 1995).

The mechanism that blocks transcription immediately after fertilization and throughout the first cell cycle is multifactorial and complex. In addition to the changes in RNA polymerase II discussed above, the chromatin status of both the male and female pronucleus is initially incompatible with transcription. The transcription that occurs in the late zygote is more prominent in the male than in the female pronucleus. This disparate rate of transcription has been observed for endogenous genes (Aoki *et al.,* 1997) and also for injected plasmid constructs (Wiekowski *et al.,* 1993). Differential active demethylation of the paternal genome, while the maternal genome remains methylated (Mayer *et al.,* 2000; Oswald *et al.,* 2000), may be one of the factors that affects this differential transcription rate, but this would affect only endogenous genes.

Transcription of both endogenous and injected constructs is probably also affected by the structure of chromatin and the changing rates of histone synthesis. Hyperacetylated histone H4 is preferentially associated with the male pronucleus in the early zygote, and only at the end of the first cell cycle is this form of H4 associated with all chromosomes (Adenot *et al.,* 1997). Histones are obviously translated from maternal mRNA; however, while H3 and H4 are continuously synthesized, synthesis of H1, H2A, and H2B starts only in the late one-cell stage (Wiekowski *et al.,* 1997). Acetylated histone H4 (and also RNA polymerase II) becomes localized to the nuclear periphery in the two-cell stage embryo, and this restricted localization is no longer visible in four-cell or later embryos (Worrad *et al.,* 1995). It is likely that the availability of acetylated histones and their specific localization within the nucleus controls to some extent the nature of genome transcription in the two-cell-stage embryo.

Other structural components of chromatin that may be relevant for the control of transcription are high-mobility-group (HMG) proteins. HMG1 (Spada *et al.,* 1998) and HMG-I/Y (Beaujean *et al.,* 2000; Thompson *et al.,* 1995) have been observed in early mouse embryos. HMG-I/Y translocates to the pronuclei and its accumulation therein is associated with embryonic genome activation. Microinjection of antibody to HMG-I/Y delays the onset of transcription while injection of purified HMG-I/Y protein advances it (Beaujean *et al.,* 2000). In addition, injection of HMG-I/Y also modifies the structure of chromatin as demonstrated by increased DNaseI sensitivity (Beaujean *et al.,* 2000).

In addition to the presence or absence of the basic transcriptional apparatus and chromatin state, other general and specific transcription factors play a role in activation of the embryonic genome. mRNAs for Sp1 and TBP (TATA box binding protein) are present during oocyte maturation and their amount decreases especially in the two-cell stage. A steady increase in abundance follows due to the increase in transcription from the embryonic genome (Worrad and Schultz, 1997). Considering global transcription factors, it is interesting to note that, at least for one gene, the utilization of the TATA-less promoter in preference to the TATA-containing promoter has been observed at the time of genome activation (Davis and Schultz, 2000). This may indicate that the overall transcriptional control at the beginning of genome activation could be subtly different from that usually observed in somatic or later embryonic cells. mTEAD-2 mRNA is present in the oocyte, decreases significantly in abundance at the two-cell stage, and then gradually increases (Kaneko and DePamphilis, 1998; Kaneko *et al.,* 1997). However, the activity of mTEAD was detected only in two-cell-stage embryos at the time of embryonic genome activation (Kaneko *et al.,* 1997). The most likely explanation for this apparent discrepancy is that mTEAD maternal mRNA is stored in a dormant state and is activated for translation by polysomal recruitment at the two-cell stage (Wang and Latham, 2000). Recruitment to polysomes, resulting in translation initiation, is one of several post-transcriptional controlling mechanisms significant at this time in development (see Section VI).

The uncoupling of transcription and translation of endogenous genes at the beginning of embryonic genome activation is an obvious feature of stored maternal mRNA utilization, but it has also been observed with genes transcribed for the first time only after fertilization. It is at present unclear whether endogenous genes, transcribed at the beginning of embryonic genome activation, are immediately translated. This postfertilization uncoupling phenomenon was described only when transgenes (Matsumoto *et al.,* 1994) or injected constructs (Nothias *et al.,* 1996) were analyzed. Analysis of other transcription factors present in the embryo at the time of transition from maternal to zygotic control is just beginning, and as yet only a small number have been described (Bevilacqua *et al.,* 2000; Parrott and Gay, 1998). However, EST analysis of the available cDNA libraries derived from oocyte and early embyos (see Section VII) should proceed rapidly, enabling identification of more transcription factors and the subsequent determination of their targets and functional roles (Bevilacqua *et al.,* 2000).

Another group of transcription factors, which could play a role in controlling gene expression in the perifertilization period, is represented by the gene *Maid* (Hwang *et al.,* 1997). Basic helix–loop–helix (bHLH) factors are present in the oocyte and during early development (Domashenko *et al.,* 1997), and their main function is to activate the genes involved in various differentiation programs. Premature activation of such genes might be detrimental at specific developmental stages. It is possible that *Maid,* which belongs to the Id family of proteins (Hwang *et al.,* 1997), interacts with bHLH factors during oogenesis and in early preimplantation embryos, preventing the premature activation of differentiation-related genes (Norton *et al.,* 1998). As is the case with several other members of the Id gene family (Afouda *et al.,* 1999), *Maid* translation is translationally controlled (Hwang *et al.,* 1997).

VI. mRNA Utilization during Oocyte Maturation and Preimplantation Development

The proper utilization of maternal mRNAs is the key to molecular control of early development. In the development of any multicellular organism, there exists a period of transcriptional silence between the time when the oocyte achieves full growth and the activation of the embryonic genome. This period can either be short, in the range of hours as in the sea urchin, fruit fly, or frog, or several days as is the case in mammals. During this period of transcriptional silence all processes in the egg and embryo are carried out by the molecules stored during oocyte growth. The sequestration of maternal mRNA molecules to within the ooplasm and their activation for translation in precise spatial and temporal sequences suggest their central role in controlling early development (Wickens *et al.,* 1996).

The role of cytoplasmic polyadenylation in controlling the translation of mammalian maternal mRNAs has been extensively documented using tissue plasminogen activator (tPA) as a model (Strickland *et al.,* 1988; Vassalli *et al.,* 1989). In the full-grown oocyte deadenylated tPA mRNA is abundant, stable, and not translated. During oocyte maturation tPA mRNA is polyadenylated, translated, and degraded so that there is no detectable tPA mRNA in the zygote. A short motif in the 3′ untranslated region (3′UTR), the CPE, first identified in the frog (for review, see Richter, 1999) and then in the mouse (Verrotti *et al.,* 1996), was shown to be essential for these changes (Strickland *et al.,* 1988; Vassalli *et al.,* 1989). tPA is only one of many genes whose mRNA is deposited in the growing oocyte and whose translation depends on processes regulating cytoplasmic polyadenylation (Oh *et al.,* 2000). We are just beginning to realize that translational control based on 3′UTR sequences is complex and could be responsible for the precise timing of maternal mRNA translation. CPEB is a protein that binds to CPE in the *Xenopus* oocyte and is essential for polyadenylation. A similar mouse protein, mCPEB, has been identified (Gebauer and Richter, 1996). In *Xenopus,* CPEB binds Maskin, a protein that could interact with the 5′ translation initiation complex, specifically with eIF-4E, thus preventing initiation of translation (Stebbins-Boaz *et al.,* 1999). Thus CPEs, and possibly other motifs in the 3′UTR, could have a dual role: first by ensuring the dormancy of maternal mRNA in the full-grown oocyte and, subsequently, by promoting polyadenylation and translation (Simon and Richter, 1994). The presence of a silencing factor was implicated in the repression followed by translation of cyclin B1 and tPA in *Xenopus* and mouse oocytes, respectively (Barkoff *et al.,* 2000; Stutz *et al.,* 1998). However, derepression and translation of tPA required only the displacement of the putative silencing factor but not the extension of the short polyA tail. We investigated in detail the timing of polyadenylation of mRNAs of spindlin, a gene abundantly expressed in the mouse oocyte and zygote (Oh *et al.,* 1997, 1998). Transcription of *Spin* results in mRNAs of three different sizes, each with the same open reading frame but differing in the length of the 3′UTR. Two of these messages undergo differential cytoplasmic polyadenylation and translation during oocyte maturation and postfertilization (Oh *et al.,* 2000). The longest *Spin* message is polyadenylated, but not translated in the full-grown oocyte. Following initiation of maturation the message is deadenylated and still not translated. However, after fertilization the message is polyadenylated and translated. It has been suggested that ongoing polyadenylation and not a static polyA tail can displace the factors that prevent translation, although the exact mechanism is not fully understood (Stebbins-Boaz *et al.*, 1999). It is thus possible that only deadenylation followed by readenylation can ensure the eventual translation of a long *Spin* message. CPEB-mediated control of translation is crucially important in regulating gene expression during oocyte-to-embryo transition but this mechanism also functions in adult tissues (Wu *et al.*, 1998).

Substantial scope for control is provided by the understanding that translation depends on assembly of multicomponent complexes at the 3′ and 5′ ends of the mRNA molecule, which then leads to RNA unwinding, recruitment to polysomes, and initiation of translation (Coller *et al.*, 1998; Craig *et al.*, 1998; Davis *et al.*, 1996; Deo *et al.*, 1999; Gao *et al.*, 2000; Jacobsen *et al.*, 1999; Laroia *et al.*, 1999; Paynton, 1998). Novel 3′ and 5′ sequence-specific RNA binding proteins can be identified in mammalian oocytes and early embryos. Considering the tremendous importance of these mechanisms in controlling the early development of *Drosophila* and *Xenopus*, it would be surprising if similar molecules and mechanisms were not identified in mammals.

VII. Gene Expression in the Early Mouse Embryo

Analysis of gene expression and gene discovery using early mammalian embryos was, until recently, seriously limited by the scarcity of experimental material. Analysis of protein biosynthesis in oocytes, zygotes, and cleavage-stage mouse embryos by computerized 2-D gel methods suggested that gene expression patterns change substantially during the oocyte-to-embryo transition (Latham *et al.*, 1991). The improvement of molecular biology techniques eventually led to the development of cDNA libraries representing nearly all stages of preimplantation mouse development. Several sets of such libraries are available today (Table I) (Ko *et al.*, 2000; Rothstein *et al.*, 1992, 1993; Sasaki *et al.*, 1998). This material has enabled us to initiate the search for novel genes expressed at this time in development and to attempt to gain some global insight into the control of gene expression.

Three major sets of cDNA libraries representing preimplantation development have been described and a number of ESTs from these libraries have been sequenced (Table I). These sequences are available through several Internet resources (see, for example, http://www.ncbi.nim.nih.gov/dbEST/index.html; http://www.ncbi.nim.nih.gov/UniGene/

Table I   Preimplantation Embryo Libraries with Sequenced ESTs

Library name	Unigene identification number	DbEST identification number	Total number EST	Reference
Knowles-Solter unfertilized egg	89	867	403	*a*
Knowles-Solter mouse 2 cell	88	862	14813	*a*
Knowles-Solter mouse blastocyst B1	85	850	12955	*b*
Knowles-Solter mouse blastocyst B3	94	875	2499	*b*
Mouse unfertilized egg cDNA	151	1389	3096	*c*
Mouse fertilized 1-cell-embryo cDNA	106	1119	3314	*c*
Mouse 2-cell-stage embryo cDNA	149	1382	3687	*c*
Mouse 4-cell-embryo cDNA	175	1524	3011	*c*
Mouse 8-cell-stage embryo cDNA	150	1381	3443	*c*
Mouse 16-cell-embryo cDNA	176	1532	3196	*c*
Mouse 3.5-dpc blastocyst cDNA	102	1021	5692	*c*
Mouse early blastocyst	134	1310	1161	*d*
RIKEN full-length enriched, *in vitro* fertilized eggs	319	2589	7664	*e*
RIKEN full-length enriched, 2-cell eggs	414	Not available	6315	*e*

[a] *Rothstein et al.* (1992)

[b] S.-Y. Hwang, D. Solter and B. B. Knowles (unpublished results). These blastocyst libraries were made from the mRNA from 800 blastocysts using a similar approach as before (Rothstein *et al.*, 1992). Briefly, first-strand cDNA synthesis was primed using SalI-dT (5′-CGGTCGACCGTCGACCGTTTTTTTTTTTTTTTT-3′) primer. dscDNAs were sized; insert >1500 bp and inserts 1500–1000 bp were used to make B1 and B3 libraries, respectively. cDNAs were cloned into NotI/SalI sites of pSPORT vector.

[c] *Ko et al.* (2000)

[d] *Sasaki et al.* (1998)

[e] Relatively little information is provided concerning the construction and analysis of these libraries. For the available information consult http://www.ncbi.nim.nih.gov/UniGene/lbrowse.cgi?ORG=Mm

Mm.Home.html; and http://www.lgsun.grc.nia.nih.gov), thus enabling clustering and library comparison. Only a few attempts have been made to explore this material in depth and even these attempts were limited either to only one library (Sasaki *et al.*, 1998) or to libraries (Ko *et al.*, 2000) from one laboratory.

The confidence level of these analyses is obviously influenced by the reliability and quality of the respective libraries and, because the amount of material used was minimal, all preimplantation embryo libraries probably suffer from a variety of inherent deficiencies. For example, an abundant EST cluster is derived from a bacterial contaminant of the *Escherichia coli* tRNA carrier used in the two-cell-stage library from our laboratory (Rothstein *et al.*, 1992). Two of the three most highly represented EST clusters (3.1 and 1.8% of all ESTs in a 3.5-dpc blastocyst library) correspond to adult hemoglobin chains (Ko *et al.*, 2000), which must be due to blood contamination of the original RNA sample. There is no hemoglobin in another, much larger, blastocyst library (Rothstein *et al.*, 1992). This type of contamination is of serious concern because any other cDNAs from mouse erythrocytes or other somatic cells cannot be distinguished from those of the blastocyst. The oocyte and preimplantation libraries produced in our laboratory were made by standard cloning procedures, the RNA was DNAse treated and the RNA was not PCR amplified (Rothstein *et al.*, 1992). The other published libraries (Ko *et al.*, 2000; Sasaki *et al.*, 1998) and presumably those produced by RIKEN (http://www.ncbi. nim.nih.gov/UniGene/lbrouse.cgi?org=Mm) were made without pronase treatment to remove the zona pellucida (ZP). Removal of the ZP eliminates any possible contamination by adherent somatic cells and also eliminates fragmented oocytes that could otherwise be included in samples of cleavage-stage embryos. Likewise, the libraries described by Ko *et al.* (2000), Sasaki *et al.* (1998), and the RIKEN set were not subjected to DNAse treatment and PCR was used to amplify dsDNA. As a consequence, several clusters containing a large number of ESTs are apparently derived from genomic DNA representing repetitive elements. In addition, the number of ESTs in the clusters is not always a true indication of mRNA abundance, as several abundant clusters contain ESTs, which are all obviously derived from a single amplicon, that is, all sequences have exactly the same 5′ start site.

Considering the difficulties involved in making these libraries, it is not surprising that some artifacts are present. This does not diminish the value of the information we can derive from them but the results from *in silico* analysis must be rigorously checked using freshly isolated embryos. Despite these caveats, analysis of the libraries provides some initial information about gene expression in preimplantation embryos. A large number of abundantly expressed genes, especially in oocyte and two-cell-stage embryos, are unknown (Ko *et al.*, 2000). A significant proportion of highly abundant mRNAs at these stages contain CPEs and the translation

of these CPE-containing mRNAs appears to be regulated (Oh *et al.*, 2000).

The power and possible pitfalls of an *in silico* mining approach for characterizing genes that are important for preimplantation development are exemplified below. We have clustered all the ESTs from our two-cell cDNA library and performed BLAST searches for 500 clusters. Each cluster was tentatively identified to represent a known mouse gene, the mouse homologue of a known gene from another species, or a novel gene. More than half of these were completely unknown or they were previously named genes of unknown function. The most abundant 26 EST clusters from this library were used to construct a "virtual Northern blot" by on-line (http://www.algenes.org/) library analysis (Table II). Four of them, OM2, Pc3B, and two unknowns comprised of 39 and 23 ESTs, were only found in the oocyte and cleavage-stage libraries, suggesting they are products of genes of limited tissue expression. Eight of these gene clusters were not found in the available oocyte and zygote libraries. It may be that these represent the first transcripts from the embryonic genome but more likely it is a reflection of the small library size, the small insert size, and the nonrepresentative nature of the available oocyte and zygote libraries (Table II). Eighteen of these 26 EST clusters (70%) contain CPEs, suggesting that their translation is controlled during the oocyte-to-embryo transition.

To refine the picture of the genes expressed at this time in development additional techniques must also be employed. For example, quantitative amplification and dot blotting have been employed to measure the abundance of transcripts of specific genes throughout early development (Rhambhatla *et al.*, 1995). cDNA arrays may also be employed because it is now possible to produce informative probes from a very limited amount of material (10–100 ng of total RNA) (Luo *et al.*, 1999). Another alternative is to prepare better cDNA libraries. Sequenced ESTs from libraries of growing and full-grown oocytes are needed to represent the pattern of gene expression at the beginning of the period of transcriptional silence.

Existing libraries can be used not only for the global analysis of gene expression and for gene discovery *in silico* (Hwang *et al.*, 1999, 2001), but for more conventional approaches of identifying novel genes. Subtraction hybridizations have been used to identify several previously unknown genes (Temeles *et al.*, 1994), which have not been characterized further, and to identify *Spin* (Oh *et al.*, 1997) and *Maid* (Hwang *et al.*, 1997). Another tool for gene discovery, differential display, has been used on cDNA libraries from embryos or on embryo-isolated RNA material (Zimmermann and Schultz, 1994). The translation initiation factor eIF-4C that is transiently expressed in two-cell-stage embryos and could have a significant role in the activation of the embryonic genome was demonstrated by this technique (Davis *et al.*, 1996) as was *Melk*, a novel member of the Snf1/ AMPK kinase family (Heyer *et al.*, 1997, 1999).

Table II   EST Assemblies That Contain at Least 20 Sequences from Knowles, Solter Two-Cell Embryo cDNA Library[a]

Number of ESTs	Cluster designation	GenBank accession number	Presence of CPE within 120 nt upstream of poly(A) signal[b]	Similar ESTs found in other libraries				
				Oocyte	Zygote	Morula[c]	Blastocyst	Other
113	OM2b	S81935	Yes	+	+	−	−	−
48	Unknown	—	No	+	+	+	+	+
47	Unknown	—	No	−	−	+	−	+
45	Spin	U48972	Yes	+	+	+	+	+
45	Unknown	—	—	+	+	+	−	−
40	M-phase phosphoprotein, MPP6	X98263	Yes	+	+	−	−	+
39	Unknown	—	Yes	+	+	+	−	−
36	Siah-2	Z19581	Yes	+	+	−	+	+
34	Pc3b	AJ005120	Yes	+	+	+	−	−
33	I3 protein	AF106967	Yes	+	−	−	+	+
31	Ubiquitin-conjugating enzyme, ubc4	U62483	No	+	+	+	−	+
28	EI24	U41751	No	−	−	−	+	+
27	Unknown	—	Yes	+	+	+	−	+
26	Nuclear export factor (exportin 1)	Y08614	Yes	−	−	−	+	+
25	Ornithine decarboxylase	M10624	Yes	−	−	+	+	+
25	Unknown	—	Yes	+	+	+	−	+
24	Hmg4	AF022465	Yes	−	−	−	+	+
23	CD63	D16432	Yes	−	−	+	+	+
23	Unknown	AB045323	Yes	+	+	−	+	+
23	Unknown	—	Yes	+	+	−	−	−
23	Cdr2	U88588	Yes	−	−	−	−	+
22	Ing1	AF177757	No	−	+	−	+	+
22	Bmi-1	M64067	No	+	+	−	+	+
22	Unknown	—	Yes	+	+	+	−	+
21	Tcl1b1	AF195488	No	+	+	−	−	+
20	DBF4-related protein	AJ003132	Yes	+	+	−	+	+

[a] Mitochondrial, viral, and contaminant clusters were excluded.

[b] For details, see Oh et al. (2000).

[c] Combined 8- and 16-cell library data.

Analysis of EST libraries demonstrated that a large proportion of abundantly expressed genes, especially in two-cell-stage embryos, is unknown and of unknown function. The characterization and especially the establishment of the functional role of these genes will represent a major effort in the future. In addition to this gene discovery effort, preimplantation embryos are being analyzed for the expression of known genes in order to expand our understanding of their possibly multiple functions. A few examples of such work include the analysis of G proteins (Rambhatla et al., 1995; Williams et al., 1996), the establishment of the role of protein kinase C in preimplantation development (Gallicano et al., 1997; Pauken and Capco, 2000), and the examination of the expression of the genes involved in apoptotic pathways (Jurisicova et al., 1998). An excellent site to find all published information about the genes expressed during mouse preimplantation development is maintained by The Jackson Laboratory (http://www.informatics.jax.org/menus/expression_menus.html).

These few, briefly discussed examples indicate that mammalian preimplantation embryos can be successfully analyzed using all the standard tools of molecular biology and, moreover, that such tools must be applied even when the scarcity of material makes such an application difficult (Oh et al., 1999). It is certainly true that most molecular processes can be analyzed and explained using more accessible models, but the complete understanding of mammalian development cannot be accomplished without the molecular analysis of embryos and the understanding of the function of the genes expressed in embryos.

VIII. Functional Analysis

The loss of the ability of the zygote cytoplasm to reprogram somatic nuclei introduced by nuclear transfer dramatically illustrates its functional distinction from the ooplasm. However, the molecular changes that accompany this functional differentiation are totally unknown. The usual methods used to find mutant genes that function at this time of development, for example, that affect fertility, are impractical. Simple knock-out studies are complicated; namely, if the gene in question is not uniquely expressed in the oocyte and zygote, null mutants may die as embryos, or the females may fail to reach sexual maturity. This is a problem for the mutation, which must be studied in ovulated eggs and embryos. However, there are now methods to engineer the loss or gain of expression of specific genes, and to follow the effect of mutation on timely progression through meiosis/mitosis into preimplantation embryogenesis.

A. *In Vitro* Approaches

These approaches are primarily aimed at inhibiting the translation of mRNAs present in the oocyte, or those that are transcribed following activation of the embryonic genome. Antisense oligonucleotides or antisense RNA directed against coding or 3'UTR sequences have been used to eliminate the expression of a number of genes (Ao *et al.,* 1991; O'Keefe *et al.,* 1989; Strickland *et al.,* 1988; Tay *et al.,* 2000). However, the results obtained from these experiments vary widely in both the level of elimination of gene expression as well as the confidence level of the results. Recently the use of morpholino-oligonucleotides seems to have produced more reliable results (Summerton and Weller, 1997). Morpholino-oligonucleotides seem to act by preventing translation rather than causing RNase H-mediated mRNA cleavage and they must be targeted to a narrow area around the translation start site. Morpholino-oligonucleotides have been used successfully in the analysis of early events during *Xenopus* development (Heasman *et al.,* 2000), but their utility in the analysis of early mammalian development awaits experimental confirmation.

Double-stranded RNA (dsRNA), can produce potent, specific, post-transcriptional gene silencing (Birchler *et al.,* 2000; Bosher and Labouesse, 2000; Fire, 1999) and this method is known as RNA-mediated interference (RNAi). Microinjection of dsRNA has been used to determine specific gene function in disparate animal species, such as nematodes—*Caenorhabditis elegans* (Fire *et al.,* 1998), *Drosophila* (Kennerdell and Carthew, 1998), and zebrafish (Li *et al.,* 2000). The mechanism is incompletely understood, but the intracellular introduction of double-stranded sense/antisense RNA corresponding to a single gene results in degradation of its homologous RNA, thereby silencing expression of the cognate gene (Montgomery *et al.,* 1998; Plasterk and Ketting, 2000; Tuschl *et al.,* 1999; Zamore *et al.,* 2000). The effects of dsRNA appear to be nonstoichiometric in invertebrates, occurring after injection of only a few molecules. The effect seems to be passed from cell to cell and may persist through several mitoses, suggesting replication of some RNAi effector molecules (Birchler *et al.,* 2000; Fire, 1999). Microinjection of dsRNA was recently documented to downregulate specific gene expression in the mouse oocyte and preimplantation embryo, demonstrating in principle that RNAi may be a useful tool to investigate vertebrate oocyte and early zygotic gene product function (Svoboda *et al.,* 2000; Wianny and Zernicka-Goetz, 2000). However, the use of dsRNA injection in studying development has not been tested extensively in vertebrates, and several practical problems remain to be addressed. It is not yet clear whether RNAi is as specific in vertebrates as it appears to be in invertebrates. Some investigators have noted indiscriminate downregulation of gene expression after microinjection of dsRNA in zebrafish embryos (Oates *et al.,* 2000) and mammalian cell lines (Caplen *et al.,* 2000). The RNAi in mouse oocytes appears to require injection of a greater quantity of dsRNA than in invertebrate cells and it may be involved in a stoichiometric process.

Microinjection of proteins is another means of studying specific gene product function, by augmenting the protein content of the cell, by replacing a missing protein, or by inhibiting its activity with inactivating antibody. For example, the function of MPF has been studied by microinjection of purified MPF in mouse oocytes (Nakano and Kubo, 2000); and microinjection of antibodies directed against specific histones has been used to study the relationship between transcriptional control and chromatin structure in mouse oocytes and embryos (Adenot *et al.,* 2000; Beaujean *et al.,* 2000).

These types of approaches are suitable for diverse biochemical investigations, despite the demanding technical nature of microinjection-based methods, and the difficulties associated with collection of sufficient material to enable adequate interpretation of the results. However, methods targeting genes with maternal stores of both mRNA and protein may not effectively eliminate gene function because of persistence of the protein product. The many *in vitro* manipulations (injections of oocytes, *in vitro* fertilization, growth in culture to blastocyst, transfer to the uterus) and the time of exposure of the oocyte and embryo to culture conditions can by themselves affect normal development. This makes attribution of the effect of any perturbation of development, other than blatant lethality, difficult. Furthermore, the experimental manipulations must be repeated each time a new permutation is investigated.

B. *In Vivo* Approaches

The ideal way of initiating a study to find the function of a gene is to eliminate it, or to inhibit its function in a heritable fashion, and then study the phenotype in the intact organism. A tissue-specific gene deletion paradigm is possible in the oocyte, and in the embryo before transcription initiates. Expression of the zona pellucida 3 gene (*Zp3*), which encodes the egg glycoprotein to which the sperm binds (Bleil *et al.,* 1988), begins early in oogenesis and is oocyte restricted (Epifano *et al.,* 1995). *Zp3* mRNA is synthesized as oocytes enter the growth phase about day 8 after birth, reaching a peak at days 10–14 after birth. By the time oocytes are fully grown and transcription has ceased, *Zp3* message is at background levels. Study of transgenic mice carrying the *Zp3* promoter driving expression of antisense RNA directed against maternal transcripts of tPA demonstrated oocyte-specific transgene expression, although the maternal message was not completely eliminated (Richards *et al.*, 1993).

A further elaboration of tissue-specific gene targeting is provided by the recently developed Cre-*lox*P approach. The Cre recombinase of the P1 bacteriophage mediates site-specific recombination between *lox*P sites, the 34-bp sequence at which recombination takes place (Sternberg and Hamilton, 1981). The Cre-*lox*P recognition system was shown to function in eukaryotes *in vitro* (Sauer and Henderson, 1989) and *in vivo* (Medberry *et al.*, 1995), where it is now routinely used for tissue-specific gene targeting. Mice with a homozygous floxed allele (a gene flanked by *lox*P sites), prepared by standard methods in ES cells, can be crossed to a transgenic line expressing the Cre recombinase under the control of any tissue-specific promoter and the gene will be deleted in Cre-expressing but not in Cre-negative cells (Gu *et al.*, 1994). A *Zp3-Cre* transgenic FVB (FVB mouse strain) lineage has been reported (Lewandoski *et al.*, 1997) and we have made several C57BL/6J *Zp3-Cre* transgenic lineages. These *Zp3-Cre* transgenic mice can be used to specifically eliminate a target gene only in the oocyte, thus providing the means to study the role of the maternal mRNA of a specific gene (de Vries *et al.*, 2000). This experimental approach enables us for the first time to explore the oocyte-specific function of otherwise lethal mutations.

An alternative approach is to prevent the normal function of the gene coding for maternal mRNA without an actual mutation. As discussed before, it would be ideal if this could be accomplished by transgenic means so that the experimental material does not depend on repeated injections of interfering molecules with concomitant experimental variations. One possible way to accomplish this is to adopt RNAi methodology to the transgenic situation. RNAi preventing expression of transgenes and their homologous endogenous genes has been observed in transgenic nonmammalian model organisms (Bosher and Labouesse, 2000), and silencing of a transfected gene and its endogenous homologue by a mechanism closely resembling RNAi has been observed in rodent cells (Bahramian and Zarbl, 1999). These observations raise the notion that RNAi might be amenable to deliberate induction by transgenic methods; indeed, heritable, inducible specific downregulation of gene expression has recently been documented in several invertebrates carrying a transgene expressing dsRNA (Bastin *et al.*, 2000; Kennerdell and Carthew, 2000; Lam and Thummel, 2000; Shi *et al.*, 2000; Tavernarakis *et al.*, 2000). Such strategies are conditional in time and can be made tissue-type specific (i.e., oocyte specific by using the *Zp3* promoter) so that the experiment

can be repeated *in vivo* time after time. The mutants can be cryopreserved individually, and if it becomes interesting to study whether a particular gene product functions in a pathway, combinations of mutants can be made.

The mRNAs essential for early development are now being identified and we are just beginning to address their function and role in this process. Recent technological advances hold the promise that we will soon have an enhanced ability to understand their function.

Acknowledgments

D.S. and B.B.K. dedicate this chapter to the memory of G. Christian Overton who worked in concept and in fact on the problem of the initiation of embryogenesis with us. We thank Dr. Christian Stoeckert at the Center for Bioinformatics at the University of Pennsylvania and Dr. Webb Miller, Penn State University, for their helpful advice. Original work presented herein was supported by grants from NIH-NICHD and the Lalor Foundation.

References

Adenot, P. G., Mercier, Y., Renard, J.-P., and Thompson, E. M. (1997). Differential H4 acetylation of paternal and maternal chromatin precedes DNA replication and differential transcriptional activity in pronuclei of 1-cell mouse embryos. *Development (Cambridge, UK)* **124**, 4615–4625.

Adenot, P. G., Campion, E., Legouy, E., Allis, C. D., Dimitrov, S., Renard, J.-P., and Thompson, E. M. (2000). Somatic linker histone H1 is present throughout mouse embryogenesis and is not replaced by variant H1 degrees. *J. Cell Sci.* **113**, 2897–2907.

Afouda, A. B., Reynaud-Deonauth, S., Mohun, T., and Spohr, G. (1999). Localized XId3 mRNA activation in *Xenopus* embryos by cytoplasmic polyadenylation. *Mech. Dev.* **88**, 15–31.

Antczak, M., and Van Blerkom, J. (1997). Oocyte influences on early development: The regulatory proteins leptin and STAT3 are polarized in mouse and human oocytes and differentially distributed within the cells of the preimplantation stage embryo. *Mol. Hum. Reprod.* **3**, 1067–1086.

Ao, A., Erickson, R. P., Bevilacqua, A., and Karolyi, J. (1991). Antisense inhibition of β-glucuronidase expression in preimplantation mouse embryos: A comparison of transgenes and oligodeoxynucleotides. *Antisense Res. Dev.* **1**, 1–10.

Aoki, F., Worrad, D. M., and Schultz, R. M. (1997). Regulation of transcriptional activity during the first and second cell cycles in the preimplantation mouse embryo. *Dev. Biol.* **181**, 296–307.

Bahramian, M. B., and Zarbl, H. (1999). Transcriptional and posttranscriptional silencing of rodent alpha1(I) collagen by a homologous transcriptionally self-silenced transgene. *Mol. Cell. Biol.* **19**, 274–283.

Barkoff, A. F., Dickson, K. S., Gray, N. K., and Wickens, M. (2000). Translational control of cyclin B1 mRNA during meiotic maturation: Coordinated repression and cytoplasmic polyadenylation. *Dev. Biol.* **220**, 97–109.

Bastin, P., Ellis, K., Kohl, L., and Gull, K. (2000). Flagellum ontogeny in trypanosomes studied via an inherited and regulated RNA interference system. *J. Cell Sci.* **113**, 3321–3328.

Beaujean, N., Bouniol-Baly, C., Monod, C., Kissa, K., Jullien, D., Aulner, N., Amirand, C., Debey, P., and Kas, E. (2000). Induction of early transcription in one-cell mouse embryos by microinjection of the nonhistone chromosomal protein HMG-I. *Dev. Biol.* **221**, 337–354.

Bellier, S., Chastant, S., Adenot, P. G., Vincent, M., Renard, J.-P., and Bensaude, O. (1997). Nuclear translocation and carboxyl-terminal domain phosphorylation of RNA polymerase II delineate the two phases of zygotic gene activation in mammalian embryos. *EMBO J.* **16,** 6250–6262.

Bevilacqua, A., Fiorenza, M. T., and Mangia, F. (2000). A developmentally regulated GAGA box-binding factor and Sp1 are required for transcription of the *hsp70.1* gene at the onset of mouse zygotic genome activation. *Development (Cambridge, UK)* **127,** 1541–1551.

Birchler, J. A., Bhadra, M. P., and Bhadra, U. (2000). Making noise about silence: Repression of repeated genes in animals. *Curr. Opin. Genet. Dev.* **10,** 211–216.

Bleil, J. D., Greve, J. M., and Wassarman, P. M. (1988). Identification of a secondary sperm receptor in the mouse egg zona pellucida: Role in maintenance of binding of acrosome-reacted sperm to eggs. *Dev. Biol.* **128,** 376–385.

Bosher, J. M., and Labouesse, M. (2000). RNA interference: Genetic wand and genetic watchdog. *Nat. Cell Biol.* **2,** E31–E36.

Bos-Mikich, A., Whittingham, D. G., and Jones, K. T. (1997). Meiotic and mitotic CA^{2+} oscillations affect cell composition in resulting blastocysts. *Dev. Biol.* **182,** 172–179.

Caplen, N. J., Fleenor, J., Fire, A., and Morgan, R. A. (2000). dsRNA-mediated gene silencing in cultured drosophila cells: A tissue culture model for the analysis of RNA interference. *Gene* **252,** 95–105.

Colledge, W. H., Carlton, M. B. L., Udy, G. B., and Evans, M. J. (1994). Disruption of c-*mos* causes parthenogenetic development of unfertilized mouse eggs. *Nature (London)* **370,** 65–68.

Coller, J. M., Gray, N. K., and Wickens, M. P. (1998). mRNA stabilization by poly(A) binding protein is independent of poly(A) and requires translation. *Genes Dev.* **12,** 3226–3235.

Craig, A. W., Haghighat, A., Yu, A. T., and Sonenberg, N. (1998). Interaction of polyadenylate-binding protein with the eIF4G homologue PAIP enhances translation. *Nature (London)* **392,** 520–523.

Danpure, C. J. (1995). How can the products of a single gene be localized to more than one intracellular compartment? *Trends Cell Biol.* **5,** 230–238.

Davidson, E. H. (1986). "Gene Activity in Early Development." Academic Press, New York.

Davis, W., Jr., and Schultz, R. M. (2000). Developmental change in TATA-box utilization during preimplantation mouse development. *Dev. Biol.* **218,** 275–283.

Davis, W., Jr., De Sousa, P. A., and Schultz, R. M. (1996). Transient expression of translation initiation factor eIF-4C during the 2-cell stage of the preimplantation mouse embryo: Identification by mRNA differential display and the role of DNA replication in zygotic gene activation. *Dev. Biol.* **174,** 190–201.

Deguchi, R., Shirakawa, H., Oda, S., Mohri, T., and Miyazaki, S. (2000). Spatiotemporal analysis of Ca^{2+} waves in relation to the sperm entry site and animal-vegetal axis during Ca^{2+} oscillations in fertilized mouse eggs. *Dev. Biol.* **218,** 299–313.

Deo, R. C., Bonanno, J. B., Sonenberg, N., and Burley, S. K. (1999). Recognition of polyadenylate RNA by the poly(A)-binding protein. *Cell (Cambridge, Mass.)* **98,** 835–845.

De Smedt, V., Szöllösi, D., and Kloc, M. (2000). The Balbiani body: Asymmetry in the mammalian oocyte. *Genesis* **26,** 208–212.

de Vantéry, C., Gavin, A. C., Vassalli, J. D., and Schorderet-Slatkine, S. (1996). An accumulation of p34^{cdc2} at the end of mouse oocyte growth correlates with the acquisition of meiotic competence. *Dev. Biol.* **174,** 335–344.

de Vantéry, C., Stutz, A., Vassalli, J. D., and Schorderet-Slatkine, S. (1997). Acquisition of meiotic competence in growing mouse oocytes is controlled at both translational and posttranslational levels. *Dev. Biol.* **187,** 43–54.

de Vries, W. N., Binns, L. T., Fancher, K. S., Dean, J., Moore, R., Kemler, R., and Knowles, B. B. (2000). Expression of Cre recombinase in mouse oocytes: A means to study maternal effect genes. *Genesis* **26,** 110–112.

Dolmetsch, R. E., Xu, K., and Lewis, R. S. (1998). Calcium oscillations increase the efficiency and specificity of gene expression. *Nature (London)* **392,** 933–936.

Domashenko, A. D., Latham, K. E., and Hatton, K. S. (1997). Expression of myc-family, myc-interacting and myc-target genes during preimplantation mouse development. *Mol. Reprod. Dev.* **47,** 57–65.

Edwards, R. G. (1965). Maturation in vitro of mouse, sheep, cow, pig, rhesus monkey and human ovarian oocytes. *Nature (London)* **208,** 349–351.

Epifano, O., Liang, L.-F., Familari, M., Moos, M. C., Jr., and Dean, J. (1995). Coordinate expression of the three zona pellucida genes during mouse oogenesis. *Development (Cambridge, UK)* **121,** 1947–1956.

Evans, J. P., Foster, J. A., McAvey, B. A., Gerton, G. L., Kopf, G. S., and Schultz, R. M. (2000). Effects of perturbation of cell polarity on molecular markers of sperm-egg binding sites on mouse eggs. *Biol. Reprod.* **62,** 76–84.

Fire, A. (1999). RNA-triggered gene silencing. *Trends Genet.* **15,** 358–363.

Fire, A., Xu, S., Montgomery, M. K., Kostas, S. A., Driver, S. E., and Mello, C. C. (1998). Potent and specific genetic interference by double-stranded RNA in Caenorhabditis elegans. *Nature (London)* **391,** 806–811.

Fissore, R. A., Long, C. R., Duncan, R. P., and Robl, J. M. (1999). Initiation and organization of events during the first cell cycle in mammals: Applications in cloning. *Cloning* **1,** 89–100.

Forlani, S., Bonnerot, C., Capgras, S., and Nicolas, J.-F. (1998). Relief of a repressed gene expression state in the mouse 1-cell embryo requires DNA replication. *Development (Cambridge, UK)* **125,** 3153–3166.

Gallicano, G. I., McGaughey, R., W., and Capco, D. G. (1997). Activation of protein kinase C after fertilization is required for remodeling the mouse egg into the zygote. *Mol. Reprod. Dev.* **46,** 587–601.

Gao, M., Fritz, D. T., Ford, L. P., and Wilusz, J. (2000). Interaction between a poly(A)-specific ribonuclease and the 5′ cap influences mRNA deadenylation rates in vitro. *Mol. Cell* **5,** 479–488.

Gavis, E. R. (1997). Expeditions to the pole: RNA localization in Xenopus and Drosophila. *Trends Cell Biol.* **7,** 485–492.

Gebauer, F., and Richter, J. D. (1996). Mouse cytoplasmic polyadenylylation element binding protein: An evolutionarily conserved protein that interacts with the cytoplasmic polyadenylylation elements of c-*mos* mRNA. *Proc. Natl. Acad. Sci. U.S.A.* **93,** 14602–14607.

Gebauer, F., Xu, W., Cooper, G. M., and Richter, J. D. (1994). Translational control by cytoplasmic polyadenylation of c-*mos* mRNA is necessary for oocyte maturation in the mouse. *EMBO J.* **13,** 5712–5720.

Gu, H., Marth, J. D., Orban, P. C., Mossmann, H., and Rajewsky, K. (1994). Deletion of a DNA polymerase β gene segment in T cells using cell type-specific gene targeting. *Science* **265,** 103–106.

Hashimoto, N., Watanabe, N., Furuta, Y., Tamemoto, H., Sagata, N., Yokoyama, M., Okazaki, K., Nagayoshi, M., Takeda, N., Ikawa, Y., and Aizawa, S. (1994). Parthenogenetic activation of oocytes in c-*mos*-deficient mice. *Nature (London)* **370,** 68–71.

Heasman, J., Kofron, M., and Wylie, C. (2000). β-catenin signaling activity dissected in the early *Xenopus* embryo: A novel antisense approach. *Dev. Biol.* **222,** 124–134.

Heyer, B. S., Warsowe, J., Solter, D., Knowles, B. B., and Ackerman, S. L. (1997). New member of the Snf1/AMPK kinase family, *Melk,* is expressed in the mouse egg and preimplantation embryo. *Mol. Reprod. Dev.* **47,** 148–156.

Heyer, B. S., Kochanowski, H., and Solter, D. (1999). Expression of *Melk,* a new protein kinase, during early mouse development. *Dev. Dyn.* **215,** 344–351.

Holliday, R. (1989). A molecular approach to the problem of positional information in eggs and early embryos. *New Biol.* **1** (3), 337–343.

Hwang, S. Y., Oh, B., Füchtbauer, A., Füchtbauer, E. M., Johnson, K. R., Solter, D., and Knowles, B. B. (1997). Maid: A maternally transcribed novel gene encoding a potential negative regulator of bHLH proteins in the mouse egg and zygote. *Dev. Dyn.* **209,** 217–226.

Hwang, S.-Y., Oh, B., Zhang, Z., Miller, W., Solter, D., and Knowles, B. B. (1999). The mouse *cornichon* gene family. *Dev. Genes Evol.* **209**, 120–125.

Hwang, S. Y., Oh, B., Knowles, B. B., Solter, D., and Lee, J-S. (2001). Expression of genes involved in mammalian meiosis during the transition from egg to embryo. *Mol. Reprod. Dev.* **59**, 144–150.

Jacobsen, S. E., Running, M. P., and Meyerowitz, E. M. (1999). Disruption of an RNA helicase/RNAse III gene in *Arabidopsis* causes unregulated cell division in floral meristems. *Development (Cambridge, UK)* **126**, 5231–5243.

Johnson, J., Bierle, B. M., Gallicano, G. I., and Capco, D. G. (1998). Calcium/Calmodulin-Dependent protein kinase II and calmodulin: Regulators of the meiotic spindle in mouse eggs. *Dev. Biol.* **204**, 464–477.

Johnson, M. H. (1981). The molecular and cellular basis of preimplantation mouse development. *Bio. Rev. Cambridge Philos. Soc.* **56**, 463–498.

Jurisicova, A., Latham, K. E., Casper, R. F., and Varmuza, S. L. (1998). Expression and regulation of genes associated with cell death during murine preimplantation embryo development. *Mol. Reprod. Dev.* **51**, 243–253.

Kaneko, K. J., and DePamphilis, M. L. (1998). Regulation of gene expression at the beginning of mammalian development and the TEAD family of transcription factors. *Dev. Genet.* **22**, 43–55.

Kaneko, K. J., Cullinan, E. B., Latham, K. E., and DePamphilis, M. L. (1997). Transcription factor mTEAD-2 is selectively expressed at the beginning of zygotic gene expression in the mouse. *Development (Cambridge, UK)* **124**, 1963–1973.

Kennerdell, J. R., and Carthew, R. W. (1998). Use of dsRNA-mediated genetic interference to demonstrate that frizzled and frizzled 2 act in the wingless pathway. *Cell (Cambridge, Mass.)* **95**, 1017–1026.

Kennerdell, J. R., and Carthew, R. W. (2000). Heritable gene silencing in drosophila using double-stranded RNA. *Nat. Biotechnol.* **18**, 896–898.

Kidder, G. M. (1993). Genes involved in cleavage, compaction, and blastocyst formation. *In* "Genes in Mammalian Reproduction" (R. B. L Gwatkin, ed.), pp. 45–71. Wiley-Liss, New York.

King, M. L., Zhou, Y., and Bubunenko, M. (1999). Polarizing genetic information in the egg: RNA localization in the frog oocyte. *BioEssays* **21**, 546–557.

Kline, D., Mehlmann, L., Fox, C., and Terasaki, M. (1999). The cortical endoplasmic reticulum (ER) of the mouse egg: Localization of ER clusters in relation to the generation of repetitive calcium waves. *Dev. Biol.* **215**, 431–442.

Kloc, M., and Etkin, L. D. (1995). Two distinct pathways for the localization of RNAs at the vegetal cortex in *Xenopus* oocytes. *Development (Cambridge, UK)* **121**, 287–297.

Ko, M. S. H., Kitchen, J. R., Wang, X., Threat, T. A., Wang, X., Hasegawa, A., Sun, T., Grahovac, M. J., Kargul, G. J., Lim, M. K., Cui, Y., Sano, Y., Tanaka, T., Liang, Y., Mason, S., Paonessa, P. D., Sauls, A. D., DePalma, G. E., Sharara, R., Rowe, L. B., Eppig, J., Morrell, C., and Doi, H. (2000). Large-scale cDNA analysis reveals phased gene expression patterns during preimplantation mouse development. *Development (Cambridge, UK)* **127**, 1737–1749.

Lam, G., and Thummel, C. S. (2000). Inducible expression of double-stranded RNA directs specific genetic interference in *Drosophila. Curr. Biol.* **10**, 957–963.

Laroia, G., Cuesta, R., Brewer, G., and Schneider, R. J. (1999). Control of mRNA decay by heat shock-ubiquitin-proteasome pathway. *Science* **284**, 499–502.

Latham, K. E. (1999). Mechanisms and control of embryonic genome activation in mammalian embryos. *Int. Rev. Cytol.* **193**, 71–124.

Latham, K. E., Garrels, J. I., Chang, C., and Solter, D. (1991). Quantitative analysis of protein synthesis in mouse embryos. I. Extensive reprogramming at the one- and two-cell stages. *Development (Cambridge, UK)* **112**, 921–932.

Latham, K. E., Solter, D., and Schultz, R. M. (1992). Acquisition of a transcriptionally permissive state during the 1-cell stage of mouse embryogenesis. *Dev. Biol.* **149**, 457–462.

Lehman, A. L., Nakatsu, Y., Ching, A., Bronson, R. T., Oakey, R. J., Keiper-Hrynko, N., Finger, J. N., Durham-Pierre, D., Horton, D. B., Newton, J. M., Lyon, M. F., and Brilliant, M. H. (1998). A very large protein with diverse functional motifs is deficient in rjs (runty, jerky, sterile) mice. *Proc. Natl. Acad. Sci. U.S.A.* **95**, 9436–9441.

Lewandoski, M., Wassarman, K. M., and Martin, G. R. (1997). *Zp3-cre,* a transgenic mouse line for the activation or inactivation of loxP-flanked target genes specifically in the female germ line. *Curr. Biol.* **7**, 148–151.

Li, Y.-X., Farrell, M. J., Liu, R., Mohanty, N., and Kirby, M. L. (2000). Double-stranded RNA injection produces null phenotypes in zebrafish. *Dev. Biol.* **217**, 394–405; erratum: *Ibid.* **220**, 432 (2000).

Luo, L., Salunga, R. C., Guo, H., Bittner, A., Joy, K. C., Galindo, J. E., Xiao, H., Rogers, K. E., Wan, J. W., Jackson, M. R., and Erlander, M. G. (1999). Gene expression profiles of laser-captured adjacent neuronal subtypes. *Nat. Med.* **5**, 117–122.

Maleszewski, M., Kimura, Y., and Yanagimacht, R. (1996). Sperm membrane incorporation into oolemma contributes to the oolemma block to sperm penetration: Evidence based on intracytoplasmic sperm injection experiments in the mouse. *Mol. Reprod. Dev.* **44**, 256–259.

Masui, Y., and Markert, C. L. (1971). Cytoplasmic control of nuclear behavior during meiotic maturation of frog oocytes. *J. Exp. Zool.* **177**, 129–145.

Matsumoto, K., Anzai, M., Nakagata, N., Takahashi, A., Takahashi, Y., and Miyata, K. (1994). Onset of paternal gene activation in early mouse embryos fertilized with transgenic mouse sperm. *Mol. Reprod. Dev.* **39**, 136–140.

Mayer, W., Niveleau, A., Walter, J., Fundele, R., and Haaf, T. (2000). Demethylation of the zygotic paternal genome. *Nature (London)* **403**, 501–502.

Medberry, S. L., Dale, E., Qin, M., and Ow, D. W. (1995). Intra-chromosomal rearrangements generated by Cre-lox site-specific recombination. *Nucleic Acids Res.* **23**, 485–490.

Mendez, R., Hake, L. E., Andresson, T., Littlepage, L. E., Ruderman, J. V., and Richter, J. D. (2000). Phosphorylation of CPE binding factor by Eg2 regulates translation of c-*mos* mRNA. *Nature (London)* **404**, 302–307.

Mitra, J., and Schultz, R. M. (1996). Regulation of the acquisition of meiotic competence in the mouse: Changes in the subcellular localization of cdc2, cyclin B1, cdc25C and wee1, and in the concentration of these proteins and their transcripts. *J. Cell Sci.* **109**, 2407–2415.

Montgomery, M. K., Xu, S., and Fire, A. (1998). RNA as a target of double-stranded RNA-mediated genetic interference in *Caenorhabditis elegans. Proc. Natl. Acad. Sci. U.S.A.* **95**, 15502–15507.

Moos, J., Visconti, P. E., Moore, G. D., Schultz, R. M., and Kopf, G. S. (1995). Potential role of mitogen-activated protein kinase in pronuclear envelope assembly and disassembly following fertilization of mouse eggs. *Biol. Reprod.* **53**, 692–699.

Nakano, H., and Kubo, H. (2000). Study of the in vitro maturation of mouse oocytes induced by microinjection of maturation promoting factor (MPF). *J. Assist. Reprod. Genet.* **17**, 67–73.

Norton, J. D., Deed, R. W., Craggs, G., and Sablitzky, F. (1998). Id helix-loop-helix proteins in cell growth and differentiation. *Trends Cell Biol.* **8**, 58–65.

Nothias, J.-Y., Majumder, S., Kaneko, K. J., and DePamphilis, M. L. (1995). Regulation of gene expression at the beginning of mammalian development. *J. Biol. Chem.* **270**, 22077–22080.

Nothias, J.-Y., Miranda, M., and DePamphilis, M. L. (1996). Uncoupling of transcription and translation during zygotic gene activation in the mouse. *EMBO J.* **15**, 5715–5725.

Oates, A. C., Bruce, A. E., and Ho, R. K. (2000). Too much interference: Injection of double-stranded RNA has nonspecific effects in the zebrafish embryo. *Dev. Biol.* **224**, 20–28.

Oh, B., Hwang, S.-Y., Solter, D., and Knowles, B. B. (1997). Spindlin, a major maternal transcript expressed on the mouse during the transition from oocyte to embryo. *Development (Cambridge, UK)* **124,** 493–503.

Oh, B., Hampl, A., Eppig, J. J., Solter, D., and Knowles, B. B. (1998). SPIN, a substrate in the MAP kinase pathway in mouse oocytes. *Mol. Reprod. Dev.* **50,** 240–249.

Oh, B., Hwang, S.-Y., De Vries, W. N., Solter, D., and Knowles, B. (1999). Identification of genes and processes guiding the transition between the mammalian gemete and embryo. *In* "A Comparative Methods Approach to the Study of Oocytes and Embryos" (J. D. Richter, ed.), pp. 101–126. Oxford University Press, New York.

Oh, B., Hwang, S.-Y., McLaughlin, J., Solter, D., and Knowles, B. B. (2000). Timely translation during the mouse oocyte-to-embryo transition. *Development (Cambridge, UK)* **127,** 3795–3803.

O'Keefe, S. J., Wolfes, H., Kiessling, A. A., and Cooper, G. M. (1989). Microinjection of antisense *c-mos* oligonucleotides prevents meiosis II in the maturing mouse egg. *Proc. Natl. Acad. Sci. U.S.A.* **86,** 7038–7042.

Oswald, J., Engemann, S., Lane, N., Mayer, W., Olek, A., Fundele, R., Dean, W., Reik, W., and Walter, J. (2000). Active demethylation of the paternal genome in the mouse zygote. *Curr. Biol.* **10,** 475–478.

Parrington, J., Swann, K., Shevchenko, V. I., Sesay, A. K., and Lai, A. F. (1996). Calcium oscillations in mammalian eggs triggered by a soluble sperm protein. *Nature (London)* **379,** 364–368.

Parrott, J. N., and Gay, N. J. (1998). Expression and subcellular distribution of rel/NFκB transcription factors in the preimplantation mouse embryo: Novel κB binding activities in the blastocyst stage embryo. *Zygote* **6,** 249–260.

Pauken, C. M., and Capco, D. G. (2000). The expression and stage-specific localization of protein kinase C isotypes during mouse preimplantation development. *Dev. Biol.* **223,** 411–421.

Paynton, B. V. (1998). RNA-binding proteins in mouse oocytes and embryos: Expression of genes encoding Y box, DEAD box RNA helicase, and polyA binding proteins. *Dev. Genet.* **23,** 285–298.

Pepling, M. E., and Spradling, A. C. (1998). Female mouse germ cells form synchronously dividing cysts. *Development (Cambridge, UK)* **125,** 3323–3328.

Pepling, M. E., de Cuevas, M., and Spradling, A. C. (1999). Germline cysts: A conserved phase of germ cell development? *Trends Cell Biol.* **9,** 257–262.

Plasterk, R. H. A., and Ketting, R. F. (2000). The silence of the genes. *Curr. Opin. Genet. Dev.* **10,** 562–567.

Polanski, Z., Ledan, E., Brunet, S., Louvet, S., Verlhac, M.-H., Kubiak, J. Z., and Maro, B. (1998). Cyclin synthesis controls the progression of meiotic maturation in mouse oocytes. *Development (Cambridge, UK)* **125,** 4989–4997.

Rambhatla, L., Patel, B., Dhanasekaran, N., and Latham, K. E. (1995). Analysis of G protein α subunit mRNA abundance in preimplantation mouse embryos using a rapid, quantitative RT-PCR approach. *Mol. Reprod. Dev.* **41,** 314–324.

Richards, W. G., Carroll, P. M., Kinloch, R. A., Wassarman, P. M., and Strickland, S. (1993). Creating maternal effect mutations in transgenic mice: Antisense inhibition of an oocyte gene product. *Dev. Biol.* **160,** 543–553.

Richter, J. D. (1999). Cytoplasmic polyadenylation in development and beyond. *Microbiol. Mol. Biol. Rev.* **63,** 446–456.

Rothstein, J. L., Johnson, D., DeLoia, J. A., Skowronski, J., Solter, D., and Knowles, B. B. (1992). Gene expression during preimplantation mouse development. *Genes Dev.* **6,** 1190–1201.

Rothstein, J. L., Johnson, D., Jessee, J., Skowronski, J., DeLoia, J. A., Solter, D., and Knowles, B. B. (1993). Construction of primary and subtracted cDNA libraries from early embryos. *Methods Enzymol.* **225,** 587–610.

Sagata, N., Oskarsson, M., Copeland, T., Brambaugh, J., and Vande Woude, G. F. (1988). Function of c-mos proto-oncogene product in meiotic maturation in Xenopus oocytes. *Nature (London)* **335,** 519–525.

Sasaki, N., Nagaoka, S., Itoh, M., Izawa, M., Konno, H., Carninci, P., Yoshiki, A., Kusakabe, M., Moriuchi, T., Muramatsu, M., Okazaki, Y., and Hayashizaki, Y. (1998). Characterization of gene expression in mouse blastocyst using single- pass sequencing of 3995 clones. *Genomics* **49,** 167–179.

Sauer, B., and Henderson, N. (1989). Cre-stimulated recombination at *loxP*-containing DNA sequences placed into the mammalian genome. *Nucleic Acids Res.* **17,** 147–161.

Schnapp, B. J. (1999). A glimpse of the machinery. *Curr. Biol.* **9,** R725–R727.

Schultz, G. A. (1986). Utilization of genetic information in the preimplantation mouse embryo. *In* "Experimental Approaches to Mammalian Embryonic Development" (J. Rossant and R. A. Pedersen, eds.), pp. 239–265. Cambridge University Press, Cambridge, UK.

Schultz, R. M. (1999). The regulation and reprogramming of gene expression in the preimplantation embryo. *Adv. Dev. Biochem.* **5,** 129–164.

Shi, H., Djikeng, A., Mark, T., Wirtz, E., Tschudi, C., and Ullu, E. (2000). Genetic interference in Trypanosoma brucei by heritable and inducible double-stranded RNA. *RNA* **6,** 1069–1076.

Simon, R. and Richter, J. D. (1994). Further analysis of cytoplasmic polyadenylation in Xenopus embryos and identification of embryonic cytoplasmic polyadenylation element-binding proteins. *Mol. Cell. Biol.* **14,** 7867–7875.

Snell, W. J., and White, J. M. (1996). The molecules of mammalian fertilization. *Cell (Cambridge, Mass.)* **85,** 629–637.

Solter, D. (2000). Mammalian cloning: Advances and limitations. *Nat. Rev. Genet.* **1,** 199–207.

Solter, D., and Knowles, B. B. (1999). Spatial and temporal control of maternal message utilization. *In* "Development: Genetics, Epigenetics and Environmental Regulation" (V. E. A. Russo, D. J. Cove, L. G. Edgar, R. Jaenisch, and F. Salamini, eds.), pp. 389–394. Springer, Berlin.

Spada, F., Brunet, A., Mercier, Y., Renard, J.-P., Bianchi, M. E., and Thompson, E. M. (1998). High mobility group 1 (HMG1) protein in mouse preimplantation embryos. *Mech. Dev.* **76,** 57–66.

Stebbins-Boaz, B., Cao, Q., de Moor, C. H., Mendez, R., and Richter, J. D. (1999). Maskin is a CPEB-associated factor that transiently interacts with elF-4E. *Mol. Cell* **4,** 1017–1027.

Stein, P., and Schultz, R. M. (2000). Initiation of a chromatin-based transcriptionally repressive state in the preimplantation mouse embryo: Lack of a primary role for expression of somatic histone H1. *Mol. Reprod. Dev.* **55,** 241–248.

Sternberg, N., and Hamilton, D. (1981). Bacteriophage P1 site-specific recombination. I. Recombination between loxP sites. *J. Mol. Biol.* **150,** 467–486.

Strickland, S., Huarte, J., Belin, D., Vassalli, A., Rickles, R., and Vassalli, J.-D. (1988). Antisense RNA directed against the 3' noncoding region prevents dormant mRNA activation in mouse oocytes. *Science* **241,** 680–684.

Stutz, A., Conne, B., Huarte, J., Gubler, P., Volkel, V., Flandin, P., and Vassalli, J.-D. (1998). Masking, unmasking, and regulated polyadenylation cooperate in the translational control of a dormant mRNA in mouse oocytes. *Genes Dev.* **12,** 2535–2548.

Summerton, J., and Weller, D. (1997). Morpholino antisense oligomers: Design, preparation, and properties. *Antisense Nucleic Acid Drug Dev.* **7,** 187–195.

Sun, Q.-Y., Luria, A., Rubinstein, S., and Breitbart, H. (1998). Protein kinase inhibitors induce the interphase transition by inactivating mitogen-activated protein kinase in mouse eggs. *Zygote* **6,** 277–284.

Svoboda, P., Stein, P., Hayashi, H., and Schultz, R. M. (2000). Selective reduction of dormant maternal mRNAs in mouse oocytes by RNA interference. *Development (Cambridge, UK)* **127,** 4147–4156.

Tang, T.-S., Dong, J.-B., Huang, X.-Y., and Sun, F.-Z. (2000). Ca²⁺ oscillations induced by a cytosolic sperm protein factor are mediated by a maternal machinery that functions only once in mammalian egg. *Development (Cambridge, UK)* **127,** 1141–1150.

Tavernarakis, N., Wang, S. L., Dorovkov, M., Ryazanov, A., and Driscoll, M. (2000). Heritable and inducible genetic interference by double-stranded RNA encoded by transgenes. *Nat. Genet.* **24**, 180–183.

Tay, J., Hodgman, R., and Richter, J. D. (2000). The control of cyclin B1 mRNA translation during mouse oocyte maturation. *Dev. Biol.* **221**, 1–9.

Telford, N. A., Watson, A. J., and Schultz, G. A. (1990). Transition from maternal to embryonic control in early mammalian development: A comparison of several species. *Mol. Reprod. Dev.* **26**, 90–100.

Temeles, G. L., Ram, P. T., Rothstein, J. L., and Schultz, R. M. (1994). Expression patterns of novel genes during mouse preimplantation embryogenesis. *Mol. Reprod. Dev.* **37**, 121–129.

Thompson, E. M., Legouy, E., Christians, E., and Renard, J.-P. (1995). Progressive maturation of chromatin structure regulates HSP70.1 gene expression in the preimplantation mouse embryo. *Development (Cambridge, UK)* **121**, 3425–3437.

Thompson, E. M., Legouy, E., and Renard, J.-P. (1998). Mouse embryos do not wait for the MBT: Chromatin and RNA polymerase remodeling in genome activation at the onset of development. *Dev. Genet.* **22**, 31–42.

Tseng, T.-C., Chen, S.-H., Hsu, Y.-P. P., and Tang, T. K. (1998). Protein kinase profile of sperm and eggs: Cloning and characterization of two novel testis-specific protein kinases (AIE1, AIE2) related to yeast and fly chromosome segregation regulators. *DNA Cell Biol.* **17**, 823–833.

Tuschl, T., Zamore, P. D., Lehmann, R., Bartel, D. P., and Sharp, P. A. (1999). Targeted mRNA degradation by double-stranded RNA in vitro. *Genes Dev.* **13**, 3191–3197.

Van Blerkom, J., and Motta, P. (1979). "The Cellular Basis of Mammalian Reproduction." Urban & Schwarzenberg, Baltimore and Munich.

Vassalli, J. D., Huarte, J., Belin, D., Gubler, P., Vassalli, A., O'Connell, M. L., Parton, L. A., Rickles, R. J., and Strickland, S. (1989). Regulated polyadenylation controls mRNA translation during meiotic maturation of mouse oocytes. *Genes Dev.* **3**, 2163–2171.

Verrotti, A. C., Thompson, S. R., Wreden, C., Strickland, S., and Wickens, M. (1996). Evolutionary conservation of sequence elements controlling cytoplasmic polyadenylylation. *Proc. Natl. Acad. Sci. U.S.A.* **93**, 9027–9032.

Wang, Q., and Latham, K. E. (2000). Translation of maternal messenger ribonucleic acids encoding transcription factors during genome activation in early mouse embryos. *Biol. Reprod.* **62**, 969–978.

Wassarman, P. M. (1999). Mammalian fertilization: Molecular aspects of gamete adhesion, exocytosis, and fusion. *Cell (Cambridge, Mass.)* **96**, 175–183.

Wianny, F., and Zernicka-Goetz, M. (2000). Specific interference with gene function by double-stranded RNA in early mouse development. *Nat. Cell Biol.* **2**, 70–75.

Wickens, M., Kimble, J., and Strickland, S. (1996). Translational control of developmental decisions. *In* "Translational Control" (M. Matthews, J. Hershey, and N. Sonnenberg, eds.), pp. 411–450. Cold Spring Harbor Lab. Press, Cold Spring Harbor, N. Y.

Wickramasinghe, D., Becker, S., Ernst, M. K., Resnick, J. L., Centanni, J. M., Tessarollo, L., Grabel, L. B., and Donovan, P. J. (1995). Two CDC25 homologues are differentially expressed during mouse development. *Development (Cambridge, UK)* **121**, 2047–2056.

Wiekowski, M., Miranda, M., and DePamphilis, M. L. (1993). Requirements for promoter activity in mouse oocytes and embryos distinguish paternal pronuclei from maternal and zygotic nuclei. *Dev. Biol.* **159**, 366–378.

Wiekowski, M., Miranda, M., Nothias, J.-Y., and DePamphilis, M. L. (1997). Changes in histone synthesis and modification at the beginning of mouse development correlate with the establishment of chromatin mediated repression of transcription. *J. Cell Sci.* **110**, 1147–1158.

Williams, C. J., Schultz, R. M., and Kopf, G. S. (1996). G protein gene expression during mouse oocyte growth and maturation, and preimplantation embryo development. *Mol. Reprod. Dev.* **44**, 315–323.

Worrad, D. M., and Schultz, R. M. (1997). Regulation of gene expression in the preimplantation mouse embryo: Temporal and spatial patterns of expression of the transcription factor Sp1. *Mol. Reprod. Dev.* **46**, 268–277.

Worrad, D. M., Turner, B. M., and Schultz, R. M. (1995). Temporally restricted spatial localization of acetylated isoforms of histone H4 and RNA polymerase II in the 2-cell mouse embryo. *Development (Cambridge, UK)* **121**, 2949–2959.

Wright, S. J. (1999). Sperm nuclear activation during fertilization. *Curr. Top. Dev. Biol.* **46**, 133–178.

Wu, L., Wells, D., Tay, J., Mendis, D., Abbott, M. A., Barnitt, A., Quinlan, E., Heynen, A., Fallon, J. R., and Richter, J. D. (1998). CPEB-mediated cytoplasmic polyadenylation and the regulation of experience-dependent translation of •-CaMKII mRNA at synapses. *Neuron* **5**, 1129-1139.

Yu, R. N., Ito, M., Saunders, T. L., Camper, S. A., and Jameson, J. L. (1998). Role of *Ahch* in gonadal development and gametogenesis. *Nat. Genet.* **20**, 353-357.

Zamore, P. D., Tuschl, T., Sharp, P. A., and Bartel, D. P. (2000). RNAi: Double-stranded RNA directs the ATP-dependent cleavage of mRNA at 21 to 23 nucleotide intervals. *Cell (Cambridge, Mass.)* **101**, 25–33.

Zimmermann, J. W., and Schultz, R. M. (1994). Analysis of gene expression in the preimplantation mouse embryo: Use of mRNA differential display. *Proc. Natl. Acad. Sci. U.S.A.* **91**, 5456–5460.

2

Asymmetry and Prepattern in Mammalian Development

R. L. Gardner

Department of Zoology, University of Oxford, Oxford OX1 3PS, United Kingdom

I. Introduction

II. Asymmetries in Early Development

III. Asymmetry of the Blastocyst

IV. Specification of the Polarity of the Anterior-Posterior Axis of the Fetus?

V. Conclusions

References

I. Introduction

Superovulation, the *in vitro* culture of early conceptuses, and the successful return of those conceptuses to the oviduct or uterus are prominent among the repertoire of basic techniques that made the critical initial stages of mammalian development accessible to experimental investigation. It is easy to forget that before such technical advances had been achieved there was little scope for doing more than looking at conceptuses, either in the living state or after they had been fixed, sectioned, and stained conventionally or histochemically. During this descriptive era, interpretation of what was observed depended largely on extrapolation from findings obtained in experimentally tractable amphibian embryos. Among the principal conclusions to emerge from this ap-

proach was that early development in mammals resembled that in other animals in depending on information that was already present in the egg before it began to cleave. Specifically, the cytoplasm of the egg was held to be divisible into two qualitatively different regions, which, despite variability in the plane of first cleavage, were claimed to segregate to different quartets of blastomeres by the eight-cell stage. So-called "dorsal" cytoplasm was postulated to direct the differentiation of inner cell mass (ICM) cells and "ventral" cytoplasm, trophectoderm cells (e.g., Jones-Seaton, 1950; Dalcq, 1957). Hence, cytoplasmic localization was viewed as the mechanism for producing initial cellular diversification.

However, once they could be grown *in vitro,* preimplantation conceptuses were soon found to be able to develop normally following various perturbations including the loss, gain, or rearrangement of their cells. Such a marked regulative ability was held to be incompatible with prepatterning of the egg, thus placing mammals apart from nearly all other animals (Davidson, 1986). Attention was therefore focused on explaining how asymmetries that are required for cellular diversification might be generated postzygotically. Thus, during much of this experimental era, as throughout the descriptive one preceding it, emphasis was primarily on accounting for how the various cell types of the early conceptus differentiate rather than how they are arranged to produce the characteristic form of the blastocyst and the later fetus.

Hence, the effects of experimental intervention were assessed almost entirely in terms of their consequences for the differentiation of ICM versus trophectoderm cells. Indeed, it is only within the past few years that developmental biologists have directed their attention to studying pattern formation in mammals. The inspiration to do so came largely from advances in understanding of this key aspect of development in other organisms, most notably *Drosophila.*

Apart from the impressive regulative ability of their preimplantation stages, various other reasons for discounting the possibility that early patterning in mammals depends on information in the egg have been presented. One hinges on the undisputed fact that the onset of development of the embryo proper is preceded by a substantial period that is devoted to the differentiation of entirely extraembryonic tissues (Chapter 8). These tissues are vital both for retaining the future fetus in the uterus and for enabling it to exploit the mother to meet its nutritional and other requirements. Thus it has been argued that imposition of such necessary adaptations to viviparity results in so long an interval between fertilization and gastrulation as to make dependence of the latter on information in the egg or zygote most unlikely (Tarkowski and Wroblewska, 1967). This view does, however, disregard the situation in other vertebrates where evidence for sequestration of patterning information in extraembryonic tissues has been found (Kostamarova, 1968; Ho *et al.,* 1999). As discussed later, recent studies provide evidence that extraembryonic tissues are also involved in embryonic patterning in mammals. Parenthetically, problems of terminology are posed by the fact that the initial period of mammalian development is devoted to the formation of extraembryonic rather than embryonic structures. Application of the term *embryo* to stages prior to gastrulation, though widespread, is an inappropriate and confusing use of the term. Even thereafter, it is applicable only to part of the derivatives of the fertilized egg. Adoption of either the term *pro-embryo* or *pre-embryo* has been suggested as a way of avoiding having to distinguish between embryo and embryo proper. However, this seems unnecessary given that the time-honored term *conceptus,* which embraces all the products of conception, already exists, and it will be used in this sense hereafter.

Additional grounds for dismissing the possibility that patterning information is present in the egg in mammals were provided by classical teratological studies. In terms of sensitivity to a variety of physical and chemical teratogens, gastrulation and early organogenesis were recognized as particularly sensitive periods (Saxen and Rapola, 1969; Austin, 1973). The consequences of applying such treatments to preimplantation stages were held to be all-or-none, with development either proceeding entirely normally thereafter or failing altogether around the time of implantation. While this is, of course, precisely how one might expect regulative embryos to respond to damage, not even data available at that

time accorded fully with such a simplistic view. More recently, additional findings have shown that fetal development can indeed be perturbed by subjecting the very early conceptus to altered conditions. These include the demonstration that exposing early mouse zygotes to certain alkylating agents results in a range of malformations among fetuses that survive to a late stage in gestation. Although well-established mutagens, these agents have been shown not to exert their effect through inducing gene mutation or either structural or numerical chromosomal changes. Nevertheless, they evidently act on the zygote directly rather than via altering the maternal environment (Katoh *et al.,* 1989) and, according to the results of zygote reconstitution experiments, their teratogenic action requires exposure of the cytoplasm as well as pronuclei to them (Generoso *et al.,* 1990). Fetal malformation has also been reported after early preimplantation mouse conceptuses have been exposed to other chemicals, including both the lithium ion (Rogers and Varmuza, 1996) and reagents used for cryopreservation (Kola *et al.,* 1988). While epigenetic modification of the genome has been invoked to account for such enduring consequences of early chemical insult, no evidence supporting this notion been offered, nor even a plausible mechanism by which it might occur.

Of particular interest in the present context are the various altered conditions to which the oocyte or very early conceptus is exposed that result in an increase in the incidence of partial or complete duplication of the anterior-posterior (AP) axis of the fetus. Thus, an increase in monozygotic twinning has been recorded repeatedly following *in vitro* fertilization in man (Wenstrom *et al.,* 1993), and also after induction of ovulation with exogenous gonadotrophins, even when the latter was not used in conjunction with the former (Derom *et al.,* 1987). In the mouse, the incidence of blastocysts with two ICMs has been found to be higher following *in vitro* than *in vivo* fertilization (Chida, 1990). This, unlike the increased rate of twinning seen in similar circumstances in man, cannot be attributed to hardening of the zona pellucida and consequent subdivision of the blastocyst through partial herniation of trophectoderm and ICM tissue. Varying degrees of axial duplication or twinning have also been observed when oocytes are deliberately aged before fertilization, both in the rat (Butcher *et al.,* 1969) and the rabbit (Bomsel-Helmreich and Papiernik-Berkhauer, 1976). Indirect evidence also exists for an association between oocyte aging and monozygotic twinning in man (Bomsel-Helmreich and Papiernik-Berkhauer, 1976; Harlap *et al.,* 1985). Whatever their explanation, these and other examples where altered conditions at the very onset of development affect patterning of the fetus must be accommodated within any framework that purports to account for how such patterning is established. Specifically, they have to be reconciled with the considerable regulative properties of the preimplantation conceptus, which have, as discussed earlier, engendered a view of early development in mammals in which the origin of asymmetries, which are crucial for both cellular differentiation and patterning, is attributed to evolving relations between initially naive and equivalent cells.

The aim in this chapter is to look at the various asymmetries that are established before the onset of gastrulation and consider both their possible significance for development of the conceptus or fetus and their origin. Nearly all of the studies that are discussed relate to the mouse simply because of the dearth of relevant work in other species.

II. Asymmetries in Early Development

That the mammalian oocyte is a polarized cell is evident from the slightly eccentric location of its nucleus, the germinal vesicle, before meiosis is resumed. Thereafter, abstriction of the first polar body and, following sperm penetration, the second serves to define a site, which, according to convention, is termed the animal pole. While the animal pole of the mouse oocyte represents the center of a microvillus-free zone that is said to be inimical to sperm attachment (Talansky *et al.,* 1991), this does not appear to be a conserved feature of this region in eutherian mammals (Santella *et al.,* 1992). Again by convention, the part of the surface of the mammalian oocyte or zygote that lies diametrically opposite the animal pole is termed the vegetal pole. However, unlike in many lower vertebrates and invertebrates, the latter is not a center of enrichment for yolk. The only internal indication of polarity of the mouse oocyte relates to the distribution of cortical mitochondria, which extend in a radially symmetrical pattern from the animal pole to about two-thirds of the way to the vegetal pole (Calarco, 1995). This arrangement does not persist through the zygote stage so that enduring regional cytoplasmic differentiation has not been regarded as a consistent feature of the mouse egg except by Dalcq and his colleagues (Dalcq, 1957), whose observations typically related to fixed rather than living material. A localized region of cortical basophilia has recently been described in the rat oocyte, but this becomes less prominent during maturation and seems no longer to be detectable in the zygote (Young *et al.,* 1999). Nothing corresponding to this region has been discerned in living as opposed to fixed and stained material.

Even during cleavage the mouse conceptus does not exhibit an overall asymmetry although, when examined carefully, it is typically found not to be strictly spherical (Gardner, unpublished observations). By the eight-cell stage polarity nonetheless becomes evident at the level of individual blastomeres with respect to both their surface and internal organization (Johnson *et al.,* 1986). Furthermore, intercellular relations become more intimate through the process of compaction, which persists until the blastocyst stage. The polarization of blastomeres at the eight-cell stage, particularly the formation of a heritable microvillus-rich pole at

their outer or apical surface, has been implicated in the genesis of cellular diversification in the morula. Thus, products of fourth and fifth cleavages that retain all or part of this pole remain external while those lacking it are internalized (Johnson and Ziomek, 1981), possibly because its presence renders blastomeres less adhesive. However, given the rather confusing situation regarding polarization of blastomeres in other eutherian mammals, its significance for the differentiation of trophectoderm versus ICM remains uncertain. In the hamster, for example, polarization of blastomeres is also seen at the eight-cell stage, but entails enrichment of microvilli basally and migration of nuclei apically (Suzuki *et al.,* 1999), which is essentially the opposite of what occurs in the mouse (Johnson *et al.,* 1986).

The first stage when the conceptus as a whole is unquestionably asymmetric is the blastocyst. Although details of the process are still obscure, morphogenesis of the blastocyst entails localized accumulation of fluid between outer and inner cells so that attachment between the two populations becomes spatially restricted (Fig. 1). Hence, part of the outer cell layer which is now a differentiated epithelium, the trophectoderm, surrounds the enlarging blastocoelic cavity and the remainder serves as the site of attachment for the compact inner population of ICM cells. Hence a polarized axis, usually referred to as the embryonic–abembryonic (Em.Ab) axis, extends from the center of the polar trophectoderm overlying the ICM to the most distal part of the mural trophectoderm surrounding the blastocoele (Fig. 1). Both the trophectoderm and ICM undergo differentiation with respect to this Em.Ab axis by the late blastocyst stage. In the case of the trophectoderm, only the polar cells continue to proliferate beyond implantation because they retain contact with the ICM, and thus produce all the numerous and diverse trophoblast cells of the postimplantation conceptus. Various pieces of recent evidence argue that this ICM stimulation of polar trophectoderm growth is mediated by FGF4 (Chai *et al.,* 1998; Tanaka *et al.,* 1998; Nichols *et al.,* 1998) (see Chapter 8). Mural cells, in contrast, soon withdraw from cycle to form trophoblastic giant cells via a polytene mode of endoreduplication of the entire genome (Brower, 1987; Varmuza *et al.,* 1988; Keighren and West, 1993; Zybina and Zybina, 1996). Differentiation within the ICM entails delamination of the primitive endoderm, another extraembryonic tissue, on its abembryonic surface. The remaining cells that occupy the region of the tissue next to the polar trophectoderm form the primitive ectoderm or epiblast, a population of cells that appear morphologically undifferentiated and will, among other things, form the entire fetal soma and germline (Fig. 1). Because of this relationship between epiblast and primitive endoderm, the Em.Ab axis of the blastocyst and later conceptus corresponds with the dorsal-ventral (DV) axis of the future embryo.

Until recently, breaking of radial symmetry of the mouse conceptus about its Em.Ab axis was assumed not to occur

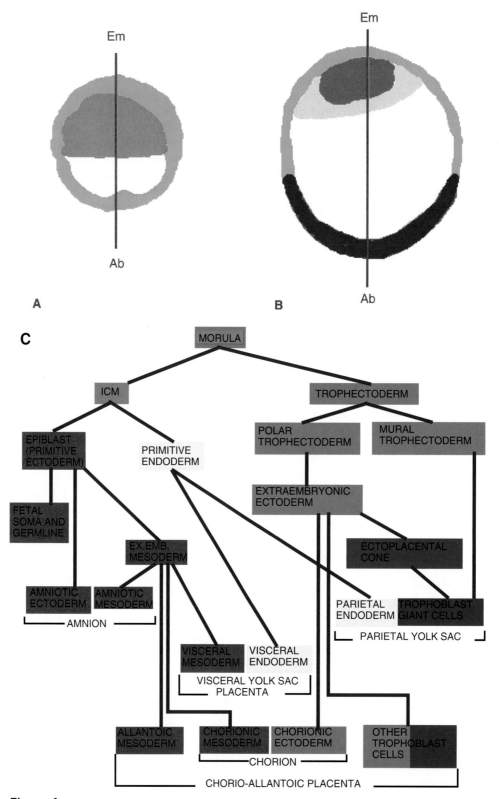

Figure 1 Diagrams to show the arrangement of tissues relative to the embryonic–abembryonic (Em.Ab) axis in the (A) very early versus (B) late blastocyst and (C) their subsequent fate. Proliferative and postmitotic trophectoderm cells and their derivatives are indicated, respectively, in light and dark green, the early ICM in orange, the epiblast and its derivatives in red, and the primitive endoderm and its derivatives in yellow.

until the primitive streak formed at the onset of gastrulation. However, the notion that acquisition of bilateral symmetry is such a late event is no longer tenable. On the basis of histological analysis of blastocysts *in utero,* Smith (1980, 1985) concluded, in agreement with much earlier findings in the rat (Huber, 1915), that the mouse blastocyst is already bilaterally symmetrical orthogonal to its Em.Ab axis before it implants. This conclusion was confirmed in subsequent observations on intact mouse blastocysts, regardless of whether they were freshly recovered from the uterus or had developed *in vitro* from conceptuses explanted early in cleavage (Gardner, 1990, 1997). Hence, the possibility that such bilaterality is simply an artifact of fixation *in situ* can be discounted. In the mouse, even the early blastocyst is typically oval rather than circular in profile when viewed along its Em.Ab axis, and can thus also be assigned an axis of bilateral symmetry that is orthogonal to its Em.Ab axis (Gardner, 1997). At this stage, however, the axis of bilateral symmetry is not perceptibly polarized. It is, furthermore, obscured during blastocyst enlargement but becomes obvious once again by the late blastocyst stage when it also shows obvious polarity (Fig. 2).

Hence, by the time blastulation has occurred, the mouse conceptus exhibits two distinct axes rather than, as formerly supposed, just a single one.

As noted earlier, regional differentiation within both the trophectoderm and ICM occurs along the Em.Ab axis of the blastocyst. What can be said about the acquisition of an axis of bilateral symmetry by the early blastocyst? The question of its significance cannot be addressed definitively until it is known whether this axis is conserved during subsequent development. The polarized bilateral form of the late blastocyst is evident from marked asymmetry of both the trophectoderm and ICM, the most conspicuous feature of which is tilting of the ICM/polar trophectoderm complex relative to the Em.Ab axis (Fig. 2). Following implantation, proximal derivatives of the polar trophectoderm, together with the adjacent ICM-derived visceral endoderm, show similar tilting until beyond the start of gastrulation (Smith, 1980; Gardner *et al.,* 1992). On the basis of her histological analysis, Smith (1980, 1985) concluded not only that this tilt, and hence the bilateral axis of the conceptus, is conserved through to gastrulation but that the AP axis of the nascent embryo bears a

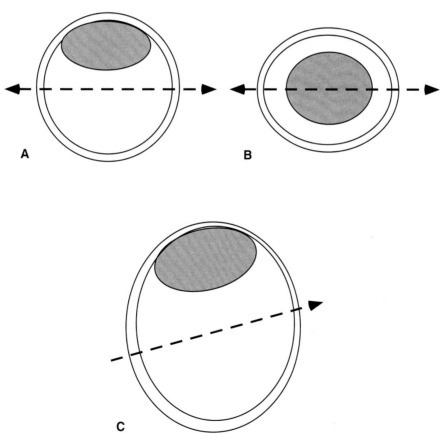

Figure 2 Diagrams illustrating the bilateral axis (dashed line) in the early versus late blastocyst. (A) Side view and (B) embryonic polar view of early blastocyst to show that the axis is not discernibly polarized at this stage. (C) Side view of late blastocyst showing that the bilateral axis has acquired polarity, most obviously through tilting of the ICM/polar trophectoderm complex in relation to the Em.Ab axis.

fixed relationship to it. Specifically, the embryonic AP axis was held to correspond with the proximal tilt of the conceptus both in orientation and polarity. Accordingly, Smith termed the axis of bilateral symmetry of the conceptus its AP axis and defined its anterior surface as the point where the tilted proximal region was most remote from the abembryonic pole (Fig. 2). Given the observed correspondence between the DV axes of the conceptus and embryo, this implied that asymmetries in the late blastocyst anticipated all three axes of the embryo (Smith, 1980, 1985) and, indeed, might be instrumental in specifying them. However, this could not be confirmed in subsequent experiments in which the posterior of conceptuses recovered early in gastrulation was marked before they were fixed and sectioned transversely to the Em.Ab axis. Thus, although the AP axes of the embryo and conceptus were found to share a common orientation, they were as often of the opposite as the same polarity (Gardner et al., 1992). The significance of this finding hinges on whether the orientation of the bilateral axis of the conceptus is conserved through implantation and also, as pointed out recently by Viebahn (1999), whether its polarity, namely, the direction of its tilt, can undergo reversal during the period of morphogenesis prior to gastrulation. Although conservation of both the orientation and polarity of this axis from the late blastocyst stage seemed likely, various problems confront studies aimed at ascertaining whether this is indeed the case. These include the relatively poor spatial resolution that can be achieved through marking individual cells at the blastocyst stage and also dilution of exogenous labels. Nonetheless, recent findings have been reported that are consistent with the notion that the bilateral axis of the conceptus is conserved through implantation from the early blastocyst stage (Weber et al., 1999). How and when this axis acquires polarity are further questions that still need to be addressed and which will be considered in more detail later in this chapter.

What about the origins of the two axes of the blastocyst? Athough the Em.Ab axis has been the subject of a variety of studies, rather little is still known about it. More emphasis has been placed on ascertaining what determines the timing of secretion and accumulation of blastocoelic fluid than on how it is localized. Various factors in timing such as total cells, number of cell cycles, or DNA replications have been discounted (Smith and McLaren, 1977; Alexandre, 1979; Surani et al., 1980; Eglitis and Wiley, 1981; Dean and Rossant, 1984; O'Brien et al., 1984), and the correlation with nuclear/cytoplasmic ratio seems rather marginal (Evsikov et al., 1990).

Blastocoele formation is presaged by the local accumulation of refractile drops, a process that begins very early in cleavage (Calarco and Brown, 1969) and is followed in the late morula by the formation of a similarly refractile cleft or furrow (Wiley and Eglitis, 1980). Although both processes are sensitive to inhibition by agents that disrupt microtubules

(Wiley and Eglitis, 1980), it is still not clear whether siting of the nascent blastocoele is due to the localization of fluid secretion or the retention of attachment between inner and outer cells. Two hypotheses have been advanced to account for the formation of the Em.Ab axis of the blastocyst. Both depend on disparities in cell division cycles, which can be marked by an advanced stage in cleavage (Barlow et al., 1972), to account for siting of the blastocoele and hence the mural trophectoderm. One assigns competence to form the blastocoele to the outer blastomeres that are most advanced in cleavage (Garbutt et al., 1987), while the other ascribes differentiation as mural trophectoderm to those that are least advanced (Surani and Barton, 1984). No compelling evidence exists to support either view, though implicit in both is the notion that specification of the Em.Ab axis of the blastocyst depends on stochastic processes during cleavage. Evidence that the latter is indeed the case would be strengthened if it could be shown that orientation of the Em.Ab axis was essentially random with respect to an enduring landmark that is conserved between conceptuses. Without such a landmark, the possibility that specification of this axis depends on cleavage planes or other basic features of organization of the early conceptus cannot be discounted. Indeed, even assumptions about the equivalence of early blastomeres, and the between-conceptus variability in orientation and timing of cleavage divisions remain insecure (Gardner, 2000b).

The second polar body (Pb), which by definition marks the animal pole of the zygote, seems to persist for an unusually long time in eutherian mammals compared with most other species. In the mouse, for example, almost two-thirds of early blastocysts have an intact Pb (Lewis and Wright, 1935; Gardner, 1997). Furthermore, the location of the Pb is highly nonrandom at this stage, being not only typically at the junction between the polar and mural trophectoderm, but aligned with the blastocyst's bilateral axis (Gardner, 1997). This very circumscribed distribution of surviving Pbs in blastocysts was difficult to reconcile with various earlier claims that these bodies move about freely on the surface of the cleaving conceptus (e.g., Borghese and Cassini, 1963). Not only did cell marking experiments fail to reveal such movement, but investigation of the mode of attachment of the second Pb to the conceptus by morphological, electrophysiological, and mechanical means suggested that it was mediated by a weakly elastic tether, which permits the passage of small molecules and ions (Gardner, 1997). The only plausible candidate for such a tether is the intercellular bridge which continues to unite the products of cell division for some time after cytokinesis is essentially complete. Such a bridge can persist between blastomeres through several cleavages in the mouse (Goodall and Johnson, 1984), and might be expected to endure even longer where one partner is a Pb that is both incapable of further division and relatively inert metabolically. It would seem, therefore, that the second Pb can be used as a marker of the animal pole of the zygote for as long as it survives. Hence, according to the location of the Pb, the axis of bilateral symmetry of the early blastocyst corresponds with the AV axis of the zygote (Gardner, 1997). Furthermore, given that the Em.Ab axis of the blastocyst is orthogonal to its bilateral axis, specification of the two cannot be independent. Such a conserved relationship between axes of the blastocyst and zygote is most unlikely to be fortuitous, and it is therefore difficult to escape the conclusion that, at least in normal undisturbed development, patterning of the blastocyst depends to some extent on spatial information that is already established by, if not before, the onset of cleavage.

At present, neither the nature of the relevant cues, nor the stage at which they operate are known. Indeed, it has yet to be established what the dependence relationship of the axes of the blastocyst is, namely, which axis is specified primarily and which secondarily. This is of relevance to the interpretation of experiments whose purpose has been to ascertain whether subsequent development is perturbed by removing parts of the zygote or by redistributing its contents (Mulnard and Puissant, 1984; Evsikov et al., 1994; Zernicka-Goetz, 1998). As discussed elsewhere (Gardner, 1999), more significance has been attributed to negative findings from these procedures than is warranted, and the studies undertaken so far unquestionably fail to provide compelling grounds for discounting egg organization from playing a patterning role in early development in mammals. The redistribution experiments leave open the question of what cytoplasmic components are affected and whether they recover their original location before cleavage. While further support for the dispensibility of polar regions of the zygote has been presented (Ciemerych et al., 2000), the assumption that patterning information will necessarily be localized to one or the other pole does not accord with experience in certain other organisms (Slack, 1991). It is relevant to note in this context that the plane of bilateral symmetry assigned to the oocyte and zygote by Dalcq and his colleagues (Jones-Seaton, 1950; Dalcq, 1957), was approximately parallel rather than orthogonal to the animal-vegetal (AV) axis. However, the possibility remains that patterning information required to specify the axes of the blastocyst is global rather than localized, depending on the overall polarity of components of the cytoskeleton, for example. Cytoskeletal elements have been described that are not only believed to be peculiar to oocytes and early stages of development in mammals (Capco and McGaughey, 1986; Gallicano et al., 1991, 1992), but which appear to undergo reorganization at the time of major morphogenetic transitions (Schwarz et al., 1995). There is, however, little to be gained from further speculation until a number of issues have been addressed. Not only is there the question, given the fixed relationship between them, as to whether it is the bilateral or Em.Ab axis of the blastocyst that is specified first, but also whether their orientation and polarity are specified simultaneously or sequentially. Thus, one can envisage

a situation where cues provided in the zygote serve to define two opposing sides of the conceptus where the blastocoele could form, the actual choice of which depended on stochastic processes during cleavage. While the bilateral axis appears not to be polarized in the early blastocyst, it clearly is in the late one. Following the demonstration that the flow of cells from polar to mural trophectoderm is polarized during blastocyst growth (Gardner, 2000a), it has been speculated that it is the direction in which this process occurs that accounts for polarity of this axis (see later).

Because the plane of first cleavage is typically parallel to the AV axis of the zygote, it is conceivable that this, rather than axial information in the zygote, could provide the cues necessary for orienting the axes of the blastocyst. One way of addressing this possibility would be to determine whether the Pb is aligned with the bilateral axis in blastocysts resulting from nonmeridional first cleavage. The plane of first cleavage varies naturally, but this tends to be obscured because, except where it is very markedly off axis, the Pbs are usually squeezed into the interblastomeric groove. A common finding in animals in which egg organization clearly has a patterning role is that the fertilizing sperm may trigger the definitive redistribution or segregation of cytoplasmic components. Its site of penetration of the egg may also determine the plane of first cleavage (Slack, 1991). Whether the sperm has any such role in mammals is a question that has yet to be answered definitively. Although claims have been made that there is no restriction on the site of sperm entry in the mouse, except in the immediate environs of the animal pole (Talansky et al., 1991), it has never been shown to be otherwise essentially random. Fertilization of mouse eggs has been accomplished with sperm that have been exposed briefly to tetramethyl rhodamine in order to obtain covalent linkage of this fluorochrome to surface components (Gabel et al., 1979). This results in a persistent focus of cortical fluorescence that can still be discerned as late as the eight-cell stage. If, as seems likely, this focus provides an enduring marker of the site of sperm entry, it could be exploited to investigate more critically whether the sperm might have a patterning role.

III. Asymmetry of the Blastocyst

So far, discussion has centered on asymmetries that might be established at or even before first cleavage and that could therefore specify the relative orientation of the two axes of the blastocyst. Although the Em.Ab axis also has an obvious polarity as soon as blastulation begins, the bilateral axis does not become discernibly polarized until the blastocyst reaches an advanced stage of development. Nevertheless, according to recent work, the blastocyst is polarized in its growth, at least so far as the trophectoderm is concerned. Findings from a variety of studies have shown that cell proliferation in the polar trophectoderm outpaces that in the mural region as the blastocyst expands. In keeping with the traditional notion that the conceptus retains radial symmetry about its Em.Ab until the primitive streak forms, the consequent net polar to mural flow of cells has been assumed to be radially symmetrical. However, when peripheral as opposed to central polar trophectoderm cells were lineage labeled, the resulting clones were not consistently displaced murally. Not only did the majority show no net displacement, but a significant minority clearly moved toward rather than away from the central polar region (Gardner, 1996a). Further experiments were therefore undertaken in which the entire polar trophectoderm was labeled selectively via endocytosis of fluorescent latex microspheres. The results confirmed the impression gained from the earlier clonal analysis that the flow of surplus polar cells into the mural trophectoderm was anisotropic (Gardner, 2000a).

When scored after a period of growth *in vitro* or *in utero* following polar labeling, blastocysts typically displayed a single coherent, crescent-shaped, patch of fluorescence extending into the mural region. Diametrically opposite the point where the patch reached most distally, the mural trophectoderm was usually either entirely unlabeled or had label confined to its most proximal cells (Fig. 3). Hence, egress of cells from the polar trophectoderm was evidently confined to the same limited part of its junction with the mural region throughout the period of post-labeling growth. One possibility was that this region is defined by the absence of junctional cells with a process extending onto the blastocoelic surface of the ICM (Fig. 4), which might otherwise prevent net mural displacement of their clones (Gardner, 1996a). This was investigated by strongly labeling individual junc-

tional cells in relatively early blastocysts with fluorescent lineage labels so that cells with an extension could be distinguished from those without. After overnight culture of blastocysts, the clones formed by junctional cells that had an extension at the time of labeling were as often displaced murally as those that did not (Gardner, 2000a). Hence, such extensions are clearly transient, and other explanations for restriction of the site of egress of polar cells must be sought. The task has been hampered by failure to detect this site in living blastocysts using a variety of optical conditions to detect circumferential variation in the overall shape or direction of elongation of junctional cells (Gardner, 2000a, and unpublished observations).

A clue that the flow might be aligned with the bilateral axis of the early blastocyst came from observations on the exceptional blastocysts in the global polar labeling experiments, which had two patches of fluorescence extending into the mural trophectoderm rather than just one. In all of these blastocysts, the two sites of egress of polar cells were diametrically opposite each other rather than related randomly. Various multiple labeling experiments have been undertaken to investigate the possibility that the direction of polar to mural flow is nonrandom with respect to the bilateral axis of the early blastocyst. One difficulty that is common to these experiments, and to those mentioned earlier concerning conservation of the bilateral axis through to gastrulation, is the limited resolution. This is because the circumference of the early blastocyst is made up of less than 10 trophectoderm cells and labeled single cells seldom form strictly linear clones that are oriented parallel to the Em.Ab axis. Nonetheless, the results show that the direction of polar to mural cell flow is significantly nonrandom and is approximately paral-

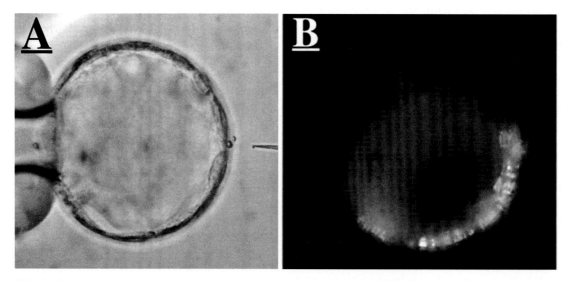

Figure 3 (A) Differential interference contrast and (B) fluorescence images of the axial midplane of a blastocyst cultured overnight following selective labeling of the polar trophectoderm with fluorescent microspheres. Note the obvious circumferential restriction in the spread of labeled cells into the mural trophectoderm.

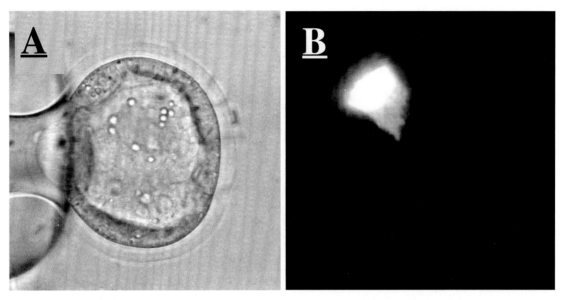

Figure 4 (A) Differential interference contrast and (B) fluorescence images of a day 4 postcoitum blastocyst viewed from the abembryonic pole shortly after labeling of a junctional trophectoderm cell with DiI in soya oil. Note the broad triangular extension of the labeled cell onto the blastocoelic surface of the ICM.

lel to the bilateral axis rather than orthogonal to it (Gardner and Davies, unpublished).

An hypothesis proposed earlier (Gardner, 1998), based on the assumption that the bilateral axis is conserved from the early blastocyst stage, suggests how the direction of polar to mural flow might account for the acquisition of polarity (Fig. 5). However, it is important to note that in cases where the Pb persisted to an advanced stage of blastocyst growth, the flow could be toward or away from this body. Hence, whereas the orientation of this flow may depend on the AV axis of the zygote, its polarity clearly does not. As noted earlier, the polarity of the AP axis of the nascent fetus is likewise not fixed in relation to the bilateral axis of the postimplantation conceptus. Hence, while bilateral symmetry of the blastocyst, which, as noted earlier, is evidently specified before or very early in cleavage, may well play a role in defining the AP axis of the fetus, it can only do so in terms of its orientation rather than its polarity. The AP axis of the nascent fetus is also oriented consistently in relation to the uterine horn, being typically transverse to its long axis. This implies that asymmetry of the blastocyst is matched by corresponding asymmetry of the uterine luminal epithelium at the time of implantation (Gardner, 2000b).

IV. Specification of the Polarity of the Anterior-Posterior Axis of the Fetus?

As discussed in the last section, anterior-posterior differentiation of the AP axis of the fetus clearly cannot be dic-

tated by the polarity of trophectoderm growth, so it is necessary to consider alternative cues on which it might depend. What about other tissues of the blastocyst? Asymmetries in the pattern of expression of a growing number of genes are evident in both the epiblast and primitive endoderm well before gastrulation is apparent morphologically (reviewed in Gardner, 2000b). Although expression of some of these genes is confined to the posterior of the future fetal AP axis, that of others is clearly localized anteriorly. It is difficult to envisage how positional information for specifying the fetal AP axis could be maintained in the epiblast itself given the very extensive cell mixing that occurs in this tissue once it transforms into a pseudostratified epithelium approximately 1 day before gastrulation (Gardner and Cockroft, 1998). The part of the primitive endoderm that continues to invest the epiblast would seem a far more likely repository for such information because it continues to grow coherently (Gardner, 1984; Gardner and Cockroft, 1998). Recently, evidence supporting such a role for this population of cells, for which the term "nascent" visceral endoderm (nVE) would seem most appropriate prior to gastrulation (see Cockroft and Gardner, 1987; Gardner, 2000b), has been provided by a variety of studies employing mainly genetic rather than experimental manipulations.

Although some of the genes expressed in the nVE have been implicated in maintaining the survival of the epiblast, others appear to play more specific roles in securing its regional differentiation. Detailed consideration of the relevant molecular studies will not be undertaken here since this has been attempted elsewhere (Gardner, 2000b) and also features

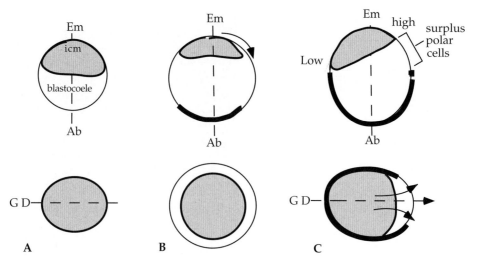

Figure 5   Diagrams of (A) early, (B) mid, and (C) late blastocysts viewed from the side (upper row) and from the embryonic pole (lower row). The early blastocyst (A) is oval in polar view, and its axis of bilateral symmetry (dashed line) is not discernibly polarized. As it expands, the blastocyst temporarily appears circular in polar view (B) before reacquiring an oval shape once it reaches a late stage. In the late as opposed to the early blastocyst, the bilateral axis is obviously polarized through tilting of the ICM/polar trophectoderm complex with respect to the Em.Ab axis (C). Polarization of the blastocyst has been ascribed to two concurrent processes (Gardner, 1998). One is a reduction in the deformability of trophectoderm cells as the time they have resided in the mural trophectoderm increases through deposition of extracellular matrix on their blastocoelic surface. The other is the focal recruitment of cells into the mural from the polar region. As denoted by differences in line thickness, the patch of new mural cells is presumed to be more deformable and thus stretched more readily with the increasing hydrostatic pressure of the expanding blastocoele. Hence, the anterior surface of the trophectoderm and ICM will gradually become tilted away from the abembryonic pole. (Reproduced from Gardner, 1998, with permission from Academic Press.)

in another contribution to this volume (Ang and Behringer, Chapter 3). Rather, the aim is briefly to examine both what is known and what has yet to be established regarding involvement of the nVE in formation of the AP axis of the fetus.

When grafted ectopically in mouse embryos, Hensen's node induces secondary axes lacking head structures (Beddington, 1994). Because grafts of a candidate early gastrula organizer (EGO) behave similarly (Tam *et al.,* 1997), this deficiency does not seem to be attributable simply to use of a late organizer. Rather, these and other findings have prompted the suggestion that the node may function only as a trunk organizer in the mouse and, hence, that a discrete head organizer may reside elsewhere. Attention has been focused on a circumscribed region of the nVE as a likely candidate. This region, which, by virtue of its final location, has been termed the anterior visceral endoderm (AVE), differs from the remainder of the tissue in its pattern of gene expression, both before and after the onset of gastrulation. Not only has it been argued that the AVE is instrumental in anterior patterning, some believe that it may already be engaged in doing so before gastrulation (Beddington and Robertson, 1998, 1999). However, in considering whether the anterior

end of the fetal AP axis may accordingly be specified before the posterior, the fact that expression of several genes is localized to the posterior region of the endoderm or epiblast well before gastrulation should not be ignored (Gardner, 2000b).

One gene in particular is of interest in relation to the developmental history of the AVE, in showing continuity in its localized expression in the nVE from the advanced blastocyst stage through to gastrulation. This is a putative transcription factor, *Hhex,* whose transcripts were first detected in the central part of the primitive endoderm of the implanting blastocyst, and which, following implantation, are confined initially to the derivative of this tissue which occupies the most distal part of the embryonic region of the egg cylinder (Thomas *et al.,* 1998). Beyond 5.5 dpc, expression of *Hhex* within the nVE ceases to be radially symmetrical, and transcripts show a progressive shift to one side of the egg cylinder which, from its relationship to the primitive streak (PS), can be recognized as the anterior once gastrulation begins (Thomas *et al.,* 1998). Evidence that this shift reflects movement of the apical *Hhex*-expressing cells was provided by selectively marking these cells with DiI in early egg cylinders, which were then grown *in vitro.* Although few of the

satisfactorily labeled conceptuses reached the stage of forming a PS, the patch of DiI-positive cells lay opposite this
landmark in each of those that did (Thomas et al., 1998).
Hence, before gastrulation there appears to be an anteriorly
directed movement of nVE cells located initially at the distal
tip of the egg cylinder toward the proximal part of the embryonic region. These cells already differ from those in the
remainder of the tissue, both in gene expression and possibly
also in morphology (Beddington, unpublished observations,
cited in Thomas et al., 1998), before they leave the tip of the
egg cylinder. Early observations suggesting that the future
anterior of the fetal AP axis can be distinguished morphologically before gastrulation in the rabbit (Van Beneden,
1883) have been confirmed recently (Viebahn et al., 1995).

Questions prompted by the foregoing findings include the
following: What role does the AVE play in the differentiation
of the fetal AP axis? How is its translocation to the diametrically opposite side of the egg cylinder to where the PS
forms brought about? Most of the relevant studies relate to
the first question and, at present, rather little can be said
about the second.

Disruption of such genes as *Lim1, Otx2, and Hesx1*
(Shawlot and Behringer, 1995; Acampora et al., 1995; Matsuo et al., 1995; Ang et al., 1996; Dattani et al., 1998), which
are expressed in the AVE once gastrulation has started, produces effects that range from perturbing patterning to causing
various degrees of anterior truncation. That these effects
depend on expression of the genes in the extraembryonic endoderm rather than elsewhere is supported by use of embryonic stress (ES) cells in conjunction with diploid or tetraploid conceptuses to generate chimeras in which mutant cells
are largely confined to wholly extraembryonic tissues versus
derivatives of epiblast. Among the genes that are expressed
specifically in the AVE region of the nVE before gastrulation, the consequences of disruption are known only for *Cer*
and *Hhex*. In the case of *Cer,* neither homozygosity for targeted disruption of the gene (Belo et al., 2000; Shawlot
et al., 2000; Stanley et al., 2000), nor for a radiation-induced
deletion that eliminated it entirely (Simpson et al., 1999) had
any discernible effect on anterior patterning. Abrogating expression of *Hhex* has also been reported recently and impairs
maintenance of differentiation of the forebrain rather than
its initial induction (Martinez Barbera et al., 2000). So far,
therefore, the only genes expressed specifically in the AVE
whose disruption causes anterior truncation are those whose
transcripts have not been detected there before gastrulation.
Furthermore, not all of these are involved in inducing the
anterior central nervous system (CNS). Thus, conceptuses
that are homozygous for disruption of *Hesx1* show aberrant
patterning of the anterior CNS rather than its absence (Dattani et al., 1998), and this is phenocopied by ablating the
AVE early in gastrulation (Thomas and Beddington, 1996).

The case that anterior differentiation of the fetal AP axis
anticipates posterior in the mouse is therefore not compelling, particularly since localized gene expression does not
obviously begin earlier in the former than the latter region
(Liu et al., 1999; Gardner, 2000b). Although evidence from
gene disruption studies supports the contention that the AVE
forms independently of the PS (Ding et al., 1998; Liu et al.,
1999; Sun et al., 1999), whether it also functions independently of derivatives of the latter as a head organizer remains
more contentious. That PS-derived anterior mesendoderm
tissue is involved in anterior neural differentiation was suggested by in vitro tissue recombination studies that were undertaken well before the AVE came to be recognized as a
distinct part of the embryonic visceral endoderm (Ang and
Rossant, 1993; Ang et al., 1994). This view has received further support from the results of two recent studies. In one,
expression of anterior neural markers, or genesis of an entire
secondary fetal axis, was found to depend on grafts containing the EGO in addition to the AVE and anterior ectoderm
(Tam and Steiner, 1999). In the other, expression of *Lhx1* in
both the AVE and PS derivatives was shown to be required
for head development (Shawlot et al., 1999). In the latter
study, it was proposed that the AVE normally induces a labile
state of anterior specification in the adjacent epiblast tissue,
which requires PS-derived tissues to convert it to a stable one
(Shawlot et al., 1999). The notion that induction by the AVE
is labile was invoked to account for the fact that in homozygous *Tdgf1* (formerly *Cripto*) null conceptuses the AVE not
only differentiates without PS formation, albeit distally, but
also induces the epiblast to express anterior neural markers
(Ding et al., 1998). Thus AVE induction is held to become
stable in the absence of *Tdgf1* activity. Widespread induction
of expression of anterior neural structures at the expense of
hindbrain and more posterior ones by the AVE in the apparent absence of PS derivatives has also been reported very
recently in conceptuses that are homozygous for disruption
of the *Fgf8* gene (Sun et al., 1999). This contrasts with the
apparent absence of neural differentiation in mutant homozygotes for a gene implicated in mesoderm induction (Holdener et al., 1994). Hence, without knowing more about how
Tdgf1 acts and, in particular, whether its effect could be mediated by *Fgf8,* the situation regarding the role of the AVE
in head induction remains somewhat confusing.

What is also clear from the phenotype of *Tdgf1* mutant
homozygotes (Ding et al., 1998) is that movement of the
Hhex-positive cells that are initially distalmost in the egg
cylinder is not essential for differentiation of the AVE. Indeed, the phenotypes of both this mutant and that resulting
from disruption of the *Madh2* gene (Waldrip et al., 1998)
have been interpreted to mean that the AP axis of the fetus is
initially aligned with its future DV axis, that is, with the
Em.Ab axis of the conceptus (Fig. 6). Although, how it fi-

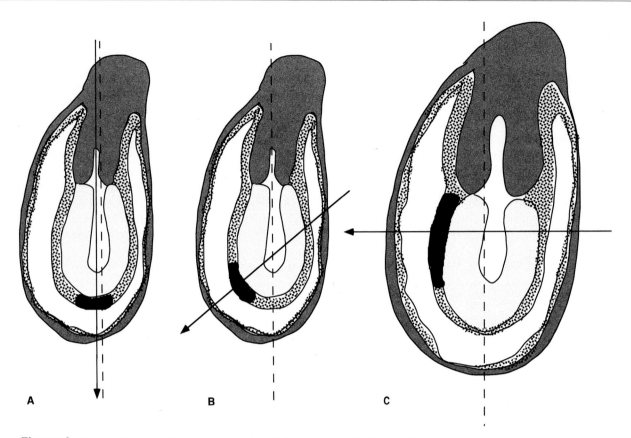

Figure 6 Diagrams illustrating the presumed changing relationship between the Em.Ab of the conceptus (dashed line) and the fetal AP axis (solid line with arrow indicating anterior) in the early postimplantation mouse conceptus. Initially (A) the two axes have the same orientation and the presumptive AVE (black) is distal. The presumptive AVE, and thus the fetal AP axis, then begins to move proximally toward one side of the egg cylinder (B) so that both end up orthogonal to the Em.Ab axis (C). Thereafter, the Em.Ab. axis of the conceptus corresponds with the DV axis of the fetus.

nally becomes orthogonal to the long axis of the egg cylinder is not known, the net movement of cells in the nVE revealed by lineage labeling is clearly implicated. One way in which this polarized movement of cells might be accomplished is through localized growth, which, as discussed earlier, clearly occurs in the trophectoderm. Another is through polarized egress of cells from the nVE during the later stages of parietal endoderm formation (Gardner, 1982; Cockroft and Gardner, 1987). However, there is no evidence to support either possibility at present. Nonetheless, evidence has emerged very recently that demonstrates an intriguing continuity between the polarity of preimplementation and postimplantation conceptuses. This is that the distribution within the nVE of descendants of single peripheral ICM cells lineage labeled in preimplantation blastocysts differed consistently according to whether the labeled cell was located by or opposite the Pb (Weber *et al.,* 1999). Thus, the descendants were found to extend further distally when the labeled cell was located by the surviving Pb than when located opposite it (Weber *et al.,* 1999). If one makes the not unreasonable assumption that this reflects the general movement of

cells within the tissue that is involved in the localization of the AVE, then the AP axis of the embryo may well be related directly to the animal-vegetal axis of the egg.

V. Conclusions

The possibility that developmentally significant prepatterning of the oocyte or zygote occurs in eutherian mammals, as in most other animals, is generally felt to be incompatible with the marked capacity of early conceptuses to regulate their development in response to various perturbations. As discussed in detail elsewhere, there are a number of concerns regarding the interpretation of the relevant experiments undertaken so far that somewhat undermine the force of this view (Gardner, 1996b, 1998, 2000b). There is, furthermore, a growing body of data that is not readily accommodated within conceptual frameworks that envisage emergence of patterning as being wholly dependent on postzygotic events (Gardner, 2000b). Collectively, these data point to regularities which suggest that patterning is indeed rooted in egg organization in eutherian mammals, at least in unperturbed development. That the early mammalian conceptus seems to have evolved efficient ways of normalizing its development following experimental perturbation is likely to complicate this task. Nonetheless, the fact that continuity of asymmetries in the distribution of products of several genes between oocyte and blastocyst has now been demonstrated (Antczuk and Van Blerkom, 1997, 1999) offers the prospect of extending the analysis of prepatterning to the molecular level.

Acknowledgments

I wish to thank Ann Yates and Tim Davies for invaluable assistance in preparing the manuscript, and both the Royal Society and the Wellcome Trust for support.

References

Acampora, D., Mazan, S., Lallemand, Y., Avantaggiato, V., Maury, M., Simeone, A., and Brulet, P. (1995). Forebrain and midbrain regions are deleted Otx 2-/- in mutants due to a defective anterior neurectoderm specification during gastrulation. *Development (Cambridge, UK)* **121**, 3279–3290.

Alexandre, H. (1979). The utilization of an inhibitor of spermidine and spermine synthesis as a tool for the study of the determination of cavitation in the preimplantation mouse embryo. *J. Embryol. Exp. Morphol.* **53**, 145–162.

Ang, S.-L., and Rossant, J. (1993) Anterior mesendoderm induces mouse Engrailed genes in explant cultures. *Development (Cambridge, UK)* **118**, 139–149.

Ang, S.-L., Conlon, R. A., Jin, O., and Rossant, J. (1994). Positive and negative signals from mesoderm regulate the expression of mouse Otx2 in ectoderm explants. *Development (Cambridge, UK)* **120**, 2979–2989.

Ang, S.-L., Jin, O., Rhinn, M., Daigle, N., Stevenson, L., and Rossant, J. (1996). A targeted mouse Otx2 mutation leads to severe defects in gastrulation and formation of axial mesoderm and to deletion of rostral brain. *Development (Cambridge, UK)* **122**, 243–252.

Antczuk, M., and Van Blerkom, J. (1997). Oocyte influences on early development: the regulatory proteins leptin and STAT3 are polarized in mouse and human oocytes and differentially distributed within cells of the preimplantation stage embryo. *Mol. Hum. Reprod.* **12**, 1067–1086.

Antczuk, M., and Van Blerkom, J. (1999). Temporal and spatial aspects of fragmentation in early human embryos- possible effects on developmental competence and association with differential elimination of regulatory proteins from polarized domains. *Hum. Reprod.* **14**, 429–447.

Austin, C. R. (1973). Embryo transfer and sensitivity to teratogenesis. *Nature (London)* **244**, 333–334.

Barlow, P., Owen, D. A. J., and Graham, C. F. (1972). DNA synthesis in the preimplantation embryo. *J. Embryol. Exp. Morphol.* **27**, 431–445.

Beddington, R. S. P. (1994). Induction of a second neural axis by the mouse node. *Development (Cambridge, UK)* **120**, 613–620.

Beddington, R. S. P., and Robertson, E. J. (1998). Anterior patterning in the mouse. *Trends Genet.* **14**, 277–284.

Beddington, R. S. P., and Robertson, E. J. (1999). Axis development and early asymmetry in mammals. *Cell (Cambridge, Mass.)* **96**, 195–209.

Belo, J. A., Bachiller, D., Agius, E., Kemp, C., Borges, A. C., Marques, S., Piccolo, S., and De Robertis, E. M. (2000). Cerberus-like is a secreted BMP and nodal antagonist not essential for mouse development. *Genesis* **26**, 265–270.

Bomsel-Helmreich, O., and Papiernik-Berkhauer, E. (1976). Delayed ovulation and monozygotic twinning. *Acta Genet. Med. Gemellol.* **25**, 73–76.

Borghese, E., and Cassini, A. (1963). Cleavage of the mouse egg. *In* "Cinemicrography in Cell Biology" (G. G. Rose, ed.), pp. 263–277. Academic Press, New York.

Brower, D. (1987). Chromosome organization in polyploid mouse trophoblast nuclei. *Chromosoma* **95**, 76–80.

Butcher, R. L., Blue, J. D., and Fugo, N. W. (1969). Overripeness and the mammalian ova. III. Fetal development at midgestation and at term. *Fertil. Steril.* **20**, 222–231.

Calarco, P. G. (1995). Polarization of mitochondria in the unfertilized mouse oocyte. *Dev. Genet.* **16**, 36–43.

Calarco, P. G., and Brown, E. A. (1969). An ultrastructural and cytological study of preimplantation development in the mouse. *J. Exp. Zool.* **171**, 253–284.

Capco, D. G., and McGaughey, R. W. (1986). Cytoskeletal reorganization during early mammalian development: Analysis using embedment-free sections. *Dev. Biol.* **115**, 446–458.

Chai, N., Patel, Y., Jacobson, K., McMahon, J., McMahon, A., and Rapolee, D. A. (1998). FgF is an essential regulator of the fifth cell division in preimplantation mouse embryos. *Dev. Biol.* **198**, 105–115.

Chida, S. (1990). Monozygous double inner cell masses in mouse blastocysts following fertilization in vitro and in vivo. *J. In Vitro Fertil. Embryo Transf.* **7**, 177–179.

Ciemerych, M. A., Mesnard, D., and Zernicka-Goetz, M. (2000). Animal and vegetal poles of the mouse egg predict the polarity of the embryonic axis, yet are nonessential for development. *Development (Cambridge, UK)* **127**, 3467–3474.

Cockroft, D. L., and Gardner, R. L. (1987). Clonal analysis of the developmental potential of 6th and 7th day visceral endoderm cells in the mouse. *Development (Cambridge, UK)* **101**, 143–155.

Dalcq, A. M. (1957). "Introduction to General Embryology." Oxford University Press, London.

Dattani, M., Martinez-Barbera, J.-P., Thomas, P. Q., Brickman, J. M., Gupta, R., Krauss, S., Wales, J., Hindmarsh, P. C., Beddington, R. S. P., and Robinson, I. C. (1998). Mutations in the homeobox gene Hesx1 associated with septo-optic dysplasia in human and mouse. *Nat. Genet.* **19**, 125–133.

Davidson, E. H. (1986). "Gene Activity in Early Development," 3rd ed. Academic Press, Orlando, FL.

Dean, W. L., and Rossant, J. (1984). Effects of delaying DNA replication on blastocyst formation in the mouse. *Differentiation (Berlin)* **26**, 134–137.

Derom, C., Vlietinck, R., Derom, R., Van Den Berghe, H., and Thiery, M. (1987). Increased monozygotic twinning rate after ovulation induction. *Lancet* **1**, 1236–1238.

Ding, J., Yang, L., Yan, Y.-T., Chen, A., Desai, N., Wynshaw-Boris, A., and Shen, M. M. (1998). Cripto is required for correct orientation of the anterior-posterior axis in the mouse embryo. *Nature (London)* **395**, 702–707.

Eglitis, M. A., and Wiley, L. M. (1981). Tetraploidy and early development: Effects on developmental timing and embryonic metabolism. *J. Embryo. Exp. Morphol.* **66**, 91–108.

Evsikov, S. V., Morozova, L. M., and Solomko, A. P. (1990). The role of the nucleocytoplasmic ratio in development regulation of the early mouse embryo. *Development (Cambridge, UK)* **109**, 323–328.

Evsikov, S. V., Morozova, L. M., and Solomko, A. P. (1994). Role of ooplasmic segregation in mammalian development. *Roux's Arch. Dev. Biol.* **203**, 199–204.

Gabel, C. A., Eddy, E. M., and Shapiro, B. M. (1979). After fertilization, sperm surface components remain as a patch in sea urchin and mouse embryos. *Cell (Cambridge, Mass.)* **18**, 207–215.

Gallicano, G. I., McGaughey, R. W., and Capco, D. G. (1991). Cytoskeleton of the mouse egg and embryo: Reorganization of planar elements. *Cell Motil. Cytoskel.* **18**, 143–154.

Gallicano, G. I., McGaughey, R. W., and Capco, D. G. (1992). Cytoskeletal sheets appear as universal components of mammalian eggs. *J. Exp. Zool.* **263,** 194–203.

Garbutt, C. L., Chisholm, J. C., and Johnson, M. H. (1987). The establishment of the embryonic-abembryonic axis in the mouse embryo. *Development (Cambridge, UK)* **100,** 125–134.

Gardner, R. L. (1982). Investigation of cell lineage and differentiation in the extraembryonic endoderm of the mouse embryo. *J. Embryol. Exp. Morphol.* **68,** 175–198.

Gardner, R. L. (1984). An in situ marker for clonal analysis of development of the extraembryonic endoderm in the mouse *J. Embryol. Exp. Morphol.* **80,** 251–288.

Gardner, R. L. (1990). Location and orientation of implantation. *In* "Establishment of a Successful Human Pregnancy: Sevono Symposium Publications" (R. G. Edwards, ed), Vol. 66, pp. 225–238. Raven Press, New York.

Gardner, R. L. (1996a). Clonal analysis of growth of the polar trophectoderm in the mouse. *Hum. Reprod.* **11,** 1979–1984.

Gardner, R. L. (1996b). Can developmentally significant spatial patterning of the egg be discounted in mammals. *Hum. Reprod. Update* **2,** 3–27.

Gardner, R. L. (1997). The early blastocyst is bilaterally symmetrical and its axis of symmetry is aligned with the animal-vegetal axis of the zygote in the mouse. *Development (Cambridge, UK)* **124,** 289–301.

Gardner, R. L. (1998). Axial relationships between egg and embryo in the mouse. *Curr. Top. Dev. Biol.* **39,** 35–71.

Gardner, R. L. (1999). Scrambled or bisected mouse eggs and the basis of patterning in mammals. *BioEssays* **21,** 271–274.

Gardner, R. L. (2000a). Flow of cells from polar to mural trophectoderm is polarized in the mouse blastocyst. *Hum Reprod.* **15,** 694–701.

Gardner, R. L. (2000b). The initial phase of embryonic patterning in mammals. *Int. Rev. Cytol.* **203,** 233–290.

Gardner, R. L., and Cockroft, D. L. (1998). Complete dissipation of coherent clonal growth occurs before gastrulation in mouse epiblast. *Development (Cambridge, UK)* **125,** 2397–2402.

Gardner, R. L., Meredith, M. M., and Altman, D. G. (1992). Is the anterior-posterior axis of the fetus specified before implantation in the mouse? *J. Exp. Zool.* **264,** 437–443.

Generoso, W. M., Rutledge, J. C., and Aronson, J. (1990). Developmental anomalies: mutational consequences of mouse zygote exposure. *Banbury Rep.* **34,** 311–319.

Goodall, H., and Johnson, M. H. (1984). The nature of intercellular coupling within the preimplantation mouse embryo. *J. Embryol. Exp. Morphol.* **79,** 53–76.

Harlap, S., Shahar, S., and Baras, M. (1985). Overripe ova and twinning. *Am. J. Hum. Genet.* **37,** 1206–1215.

Ho, C. Y., Houart, C., Wilson, S. W., and Steiner, D. Y. (1999). A role for the extraembryonic yolk syncytial layer in patterning the zebrafish embryo suggested by properties of the hex gene. *Curr. Biol.* **9,** 1131–1134.

Holdener, B. C., Faust, C., Rosenthal, N. S., and Magnuson, T. (1994). msd is required for mesoderm induction in mice. *Development (Cambridge, UK)* **120,** 1335–1340.

Huber, G. C. (1915). The development of the albino rat, *Mus norvegicus albinus.* I. From the pronuclear stage to the stage of the mesoderm anlage: End of the first to the end of the 9th day. *J. Morphol.* **26,** 247–358.

Johnson, M. H., and Ziomek, C. A. (1981). The foundation of two distinct lineages within the mouse morula. *Cell (Cambridge, Mass.)* **24,** 71–80.

Johnson, M. H., Chisholm, J. C., Fleming, T. P., and Houliston, E. (1986). A role for cytoplasmic determinants in the development of the mouse early embryo?. *J. Embryol. Exp. Morphol.* **97** (Suppl.), 97–121.

Jones-Seaton, A. (1950). Etude de l'organization cytoplasmic de l'oeuf de rongeurs principalement quant a la basophile ribonucleique. *Arch. Biol.* **61,** 291–444.

Katoh, M., Cacheiro, N. L. A., Cornett, C. V., Cain, K. T., Rutledge, J. C., and Generoso, W. M. (1989). Fetal anomalies produced subsequent to

treatment of zygotes with ethylene oxide or ethyl methanesulfonate are not likely due to the usual genetic causes. *Mutat. Res.* **210,** 337–344.

Keighren, M., and West, J. D. (1993). Analysis of cell ploidy in histological sections of mouse tissued by DNA-DNA in situ hybridization with digoxigenin labelled probes. *Histochem. J.* **25,** 30–44.

Kola, I., Kirby, C., Shaw, J., Davey, A., and Trounson, A. (1988). Vitrification of mouse oocytes results in aneuploid zygotes and malformed fetuses. *Teratology* **38,** 467–474.

Kostamarova, A. A. (1968). The differentiation capacity of isolated loach (*Misgurnis fossilis*) blastoderm. *J. Embryol. Exp. Morphol.* **22,** 407–430.

Lewis, W. H., and Wright, E. S. (1935). On the early development of the mouse egg. *Carnegie Inst. Washington Publ.* **25,** 113–143.

Liu, P., Wakamiya, M., Shea, M. J., Albrecht, U., Behringer, R. R., and Bradley, A. (1999). Requirement for Wnt3 in vertebrate axis formation. *Nat. Genet.* **22,** 361–365.

Martinez Barbera, J. P., Clements, M., Thomas, P., Rodriguez, T., Meloy, D., Kioussis, D., and Beddington, R. (2000). The homeobox gene *Hex* is required in definitive endodermal tissue for normal forebrain, liver and thyroid formation. *Development (Cambridge, UK)* **127,** 2433–2445.

Matsuo, I., Kuratani, S., Kimura, C., Tadaka, N., and Aizawa, S. (1995). Mouse Otx2 functions in the formation and patterning of the rostral head. *Genes Dev.* **9,** 2646-2658.

Mulnard, J. G., and Puissant, F. (1984). Development of mouse embryos after ultracentrifugation at the pronuclei stage. *Arch. Biol.* **97,** 301–315.

Nichols, J., Zevnik, B., Anastassiadis, K., Niwa, H., Klewe-Nebenius, D., Chambers, I., Scholer, H., and Smith, A. (1998). Formation of pluripotent stem cells in the mammalian embryo depends on the POU transcription factor Oct 4. *Cell (Cambridge, Mass.)* **95,** 379–391.

O'Brien, M. J., Critser, E. S., and First, N. L. (1984). Developmental potential of isolated blastomeres from early murine embryos. *Theriogenology* **22,** 601–607.

Rogers, I., and Varmuza, S. (1996). Epigenetic alterations brought about by lithium treatment disrupts mouse embryo development. *Mol. Reprod. Dev.* **45,** 163–170.

Santella, L., Alikani, M., Talansky, B. E., Cohen, J., and Dale B. (1992). Is the human oocyte plasma membrane polarized? *Hum. Reprod.* **7,** 999-1003.

Saxen, L., and Rapola, J. (1969). "Congenital Defects." Holt, Rinehart & Winston, New York.

Schwarz, S. M., Gallicano, G. I., McGaughey, R. W., and Capco, D. G. (1995). A role for intermediate filaments in the establishment of the primitive epithelia during mammalian embryogenesis. *Mech. Dev.* **53,** 305–321.

Shawlot, W., and Behringer, R. R. (1995). Requirement for Lim 1 in head-organizer function. *Nature (London)* **374,** 425–430.

Shawlot, W., Wakamiya, M., Kwan, K. M., Kania, A., Jessell, T. M., and Behringer, R. R. (1999). Lim 1 is required in both primitive streak-derived tissues and visceral endoderm for head formation in the mouse. *Development (Cambridge, UK)* **126,** 4925–4932.

Shawlot, W., Min Deng, J., Wakamiya, M., and Behringer, R. R. (2000). The cerberus-related gene, *Cerr 1,* is not essential for mouse head formation. *Genesis* **26,** 253–258

Simpson, E. H., Johnson, D. K., Hunsicker, P., Suffolk, R., Jordan, S. A., and Jackson, I. J. (1999). The mouse *Cer1* (Cerberus related or homologue) gene is not required for anterior pattern formation. *Dev. Biol.* **213,** 202–206.

Slack, J. M. W. (1991). "From Egg to Embryo: Regional Specification in Early Development," 2nd ed. Cambridge University Press, Cambridge, UK.

Smith, L. J. (1980). Embryonic axis orientation in the mouse and its correlation with blastocysts relationship to uterus: Part I. Relationships between 82 hours and 4 1/2 days. *J. Embryol. Exp. Morphol.* **55,** 257–277.

Smith, L. J. (1985). Embryonic axis orientation in the mouse and its correlation with blastocysts relationship to uterus: Part II. Relationships from 4 1/2 to 9 1/2 days. *J. Embryol. Exp. Morphol.* **89,** 15–35.

Smith, R., and McLaren, A. (1977). Factors affecting the time of formation of the mouse blastocoele. *J. Embryol. Exp. Morphol.* **41,** 79–92.

Stanley, E. G., Biben, C., Allison, J., Hartley, L., Wicks, I. P., Campbell, I. K., MvKinley, M., Barnett, L., Koentgen, F., Robb, L., and Harvey, R. B. (2000). Targeted insertion of a LacZ reporter gene into the mouse *Cer 1* locus reveals complex and dynamic expression during embryogenesis. *Genesis* **26,** 259–264.

Sun, X,. Meyers, E. N., Lewandoski, M., and Martin, G. R. (1999). Targeted disruption of Fgf8 causes failure of cell migration in the gastrulating mouse embryo. *Genes Dev.* **13,** 1834–1846.

Surani, M. A. H., and Barton, S. C. (1984). Spatial distribution of blastomeres is dependent on cell division order and interactions in mouse morulae. *Dev. Biol.* **102,** 335–343.

Surani, M. A. H., Barton, S. C., and Burling, A. (1980). Differentiation of 2-cell and 8-cell mouse embryos arrested by cytoskeletal inhibitors. *Exp. Cell Res.* **125,** 275–286.

Suzuki, H., Azuma, T., Koyama, H., and Yang, X. (1999). Development of cellular polarity of hamster embryos during compaction. *Biol. Reprod.* **61,** 521–526.

Talansky, B. E., Malter, H. E., and Cohen, J. (1991). A preferential site for sperm-egg fusion in mammals. *Mol. Reprod. Dev.* **28,** 183–188.

Tam, P. P. L., and Steiner, K. A. (1999). Anterior patterning by synergistic acitivity of the early gastrula organizer and the anterior germ layer tissues of the mouse embryo. *Development (Cambridge, UK)* **126,** 5171–5179.

Tam, P. P. L., Steiner, K. A., Zhou, S. X., and Quinlan, G. A. (1997). Lineage and functional analyses of the mouse organizer. *Cold Spring Harbor Symp. Quant. Biol.* **62,** 135–144.

Tanaka, S., Kunath, T., Hadjantonakia, A.-K., Nagy, A., and Rossant, J. (1998). Promotion of trophoblast stem cell proliferation by FGF 4. *Science* **282,** 2072–2075.

Tarkowski, A.K., and Wroblewska, J. (1967). Development of blastomeres of mouse eggs isolated at the 4- and 8- cell stage. *J. Embryol. Exp. Morphol.* **18,** 155–180.

Thomas, P. Q., and Beddington, R. S. P. (1996). Anterior primitive endoderm may be responsible for patterning the anterior neural plate in the mouse embryo. *Curr. Biol.* **6,** 1487–1496.

Thomas, P. Q., Brown, A., and Beddington, R. S. P. (1998). Hex: A homeobox gene revealing peri-implantation asymmetry in the mouse embryo and an early transient marker of endothelial cell precursors. *Development (Cambridge, UK)* **125,** 85–94.

Van Beneden, E. (1883). Recherches sur la maturation de l'oeuf et la fecundation. *Arch. Biol.* **4,** 265–640.

Varmuza, S., Prideaux, V., Kothary, R., and Rossant, J. (1988). Polytene chormosomes in mouse trophoblastic giant cells. *Development (Cambridge, UK)* **102,** 127–134.

Viebahn, C. (1999). The anterior margin of the mammalian gastrula: Comparative and phylogenetic aspects of its role in axis formation and head induction. *Curr. Top. Dev. Biol.* **46,** 63–103.

Viebahn, C., Mayer, B., and Hrabe de Angelis, M. (1995). Signs of the principle body axes prior to primitive streak formation in the rabbbit embryo. *Anat. Embryol.* **192,** 159–169.

Waldrip, W. R., Bikoff, E. K., Hoodless, P. A., Wrana, J. L., and Robertson, E. J. (1998). Smad 2 signalling in extraembryonic tissues determines anterior-posterior polarity of the early mouse embryo. *Cell (Cambridge, Mass.)* **92,** 797–808.

Weber, R. J., Pedersen, R. A., Wianny, F., Evans, M. J., and Zernicka-Goetz, M. (1999). Polarity of the mouse embryo is anticipated before implantation. *Development (Cambridge, UK)* **126,** 5591–5598.

Wenstrom, K. D., Syrop, C. H., Hammitt, D. G., and Van Voorhis, B. J. (1993). Increased risk of monchorionic twinning associated with assisted reproduction. *Fertil. Steril.* **60,** 510–514.

Wiley, L. M., and Eglitis, M. A. (1980). Effects of colcemid on cavitation during mouse blastocoele formation. *Exp. Cell Res.* **127,** 89–101.

Young, J. K., Allworth, A. E., and Baker, J. H. (1999). Evidence for polar cytoplasm/nuage in rat oocytes. *Anat. Embryol.* **200,** 43–48.

Zernicka-Goetz, M. (1998). Fertile offspring derived from mammalian eggs lacking either animal or vegetal poles. *Development (Cambridge, UK)* **125,** 4803–4808.

Zybina, E. V., and Zybina, T. G. (1996). Polytene chromosomes in mammalian cells. *Int. Rev. Cytol.* **165,** 53–119.

3

Anterior-Posterior Patterning of the Mouse Body Axis at Gastrulation

Siew-Lan Ang* and Richard R. Behringer†

*Institut de Génétique et de Biologie Moléculaire et Cellulaire, CNRS/INSERM/Université Louis Pasteur,
67404 Illkirch cedex, C.U. de Strasbourg, France
†Department of Molecular Genetics, University of Texas M. D. Anderson Cancer Center, Houston, Texas 77030

I. Introduction

II. Gastrulation

III. The Node: Morphogenesis, Cell Fate, and Cell Movement

IV. The Organizer Phenomenon: Conserved Properties of Vertebrate Organizers

V. The Vertebrate Organizer Is a Dynamic, Nonhomogeneous, and Renewable Cell Population at Gastrulation

VI. Insights into the Function of the Mouse Organizer Gained from Genetic and Embryological Studies

VII. Genetic Analysis of Organizer Function: Mouse Mutants Showing Defects in Organizer Function

VIII. Inhibitory Signals Secreted by the Organizer and Its Derivatives

IX. Specification of the Primitive Streak and the Organizer

X. Role of the AVE in Anterior Patterning in Mouse

XI. Embryological and Genetic Analysis of the Function of the AVE in Anterior Patterning

XII. A Model for AVE Function in Anterior Patterning

XIII. Conclusions and Future Directions

References

I. Introduction

The growth and differentiation of the developing mouse egg cylinder leads to the process of primary body axis formation that sets the stage for tissue differentiation and organogenesis. One of the primary morphogenetic processes that drives development at this stage of embryogenesis is gastrulation. During gastrulation, cell proliferation, cell morphology transitions and differential adhesion, cell migration, and inductive tissue interactions combine to produce the basic body plan of the vertebrate embryo (for recent reviews, see Tam and Behringer, 1997; Beddington and Robertson, 1999). In this chapter, we examine the postimplantation mouse embryo from the initiation of gastrulation to the formation of the initial anterior-posterior (AP) neural axis. Particular attention is focused on the process of gastrulation, the identity of organizing centers, and the genetic pathways that establish the AP body plan.

II. Gastrulation

Gastrulation is a morphogenetic process that results in the formation of mesoderm and the generation of a three germ-layered embryo composed of ectoderm, mesoderm, and endoderm. In the mouse, gastrulation also leads to the formation of a transient specialized structure called the node that generates new tissues with axis patterning activities (see below). In the mouse, gastrulation initiates after implantation at E6.5. By E7.5 a complete intervening mesoderm layer is produced. After E7.5, mesoderm continues to be generated posteriorly by the primitive streak, and finally, by the tail bud.

At the initiation of gastrulation, epiblast cells located at one region around the circumference of the egg cylinder adjacent to the extraembryonic ectoderm undergo an epithelial-to-mesenchymal transition to form a transient embryonic structure called the primitive streak. These epiblast-derived mesenchyme cells move between the visceral endoderm and the epiblast and extraembryonic ectoderm to form embryonic and extraembryonic mesoderm, respectively. The mesoderm cells move laterally in both directions as well as proximally and distally to form the so-called "wings" of mesoderm. As gastrulation proceeds, the primitive streak lengthens toward the distal tip of the egg cylinder and the wings of mesoderm continue to move until they meet at the anterior midline.

Fate mapping studies of the mouse gastrula, using dye labeling or cell transplantation followed by in vitro culture, indicate that as the primitive streak forms and lengthens, mesodermal precursors of different fates emerge in a specific pattern along the AP axis of the primitive streak (Lawson et al., 1991; Parameswaran and Tam, 1995; Tam et al., 1997; Kinder et al., 1999). These studies have demonstrated that the initial cells that ingress through the primitive streak are fated to form extraembryonic mesoderm. By the early streak stage, the posterior region of the primitive streak also contains mesodermal precursors that are fated to become extraembryonic mesoderm. The middle region of the primitive streak produces cells fated to become cardiac and forebrain/cranial mesoderm, whereas the cells in the anteriormost region of the primitive streak generate axial and paraxial mesoderm. At the midstreak and late streak stages, this AP order of mesodermal precursors in the primitive streak is similar to the early streak stage except that lateral mesoderm precursors are located in the streak posterior to the precur-

sors of the heart/cranial and paraxial mesoderm but anterior to the precursors of the extraembryonic mesoderm. Interestingly, this pattern is very similar to that found in chick, suggesting that there is a conservation in the mechanisms that pattern the mesoderm during gastrulation (Kinder et al., 1999). These observations suggest that the time and place where the epiblast cells ingress through the primitive streak during gastrulation specify their mesodermal fate and contribute to the establishment of the body plan. The molecular details of this process remain to be determined.

III. The Node: Morphogenesis, Cell Fate, and Cell Movement

At the anterior end of the primitive streak of the gastrulating mouse embryo (at the late streak stage or E7.5) lies an important structure that has been named the node (Figs. 1B and 1C) (Viebahn, 2001). The node becomes morphologically visible at the headfold stage (E7.75) as an indentation at the distal tip of the embryo. The node has a bilaminar organization with a dorsal cell layer and a ventral cell layer. Scanning electron microscopy reveals that each cell of the ventral layer of the node possesses a single large cilium on its surface (Sulik et al., 1994). Recent studies have shown that the cilia of the node generate a leftward flow of extraembryonic fluid that may be coupled to the establishment of the left-right axis of the embryo (see Chapter 4) (Nonaka et al., 1998).

Fate map studies of the node, performed by dye labeling of node cells in situ (Beddington, 1994; Sulik et al., 1994) and by cell transplantations (Beddington, 1981), have shown that the node gives rise to cell progeny located in the midline of the three germ layers, that is in the ectoderm-derived floor plate, the mesoderm-derived notochord, and the endoderm of the gut. Node derivatives are also found in the cranial mesoderm and somites. Although the node is first clearly recognizable at the late streak stage, lineage and fate map studies have shown that the precursors of the node already reside at the anterior end of the primitive streak in embryos at the early streak stage (E6.5) (Fig. 1A) (Lawson et al., 1991; Tam et al., 1997). These studies also showed that cells located in the anterior region of the early primitive streak have similar fates as node cells, and they contribute to the axial mesendoderm, of the headfold stage (head process or prechordal plate), the heart, the lateral plate mesoderm, and the extraembryonic mesoderm (Tam et al., 1997). Some of the descendants of the early anterior primitive streak also remain within the node at the late streak stage, which in turn contributes to the node of early-somite stage embryos. These results are consistent with the idea that node precursors are already

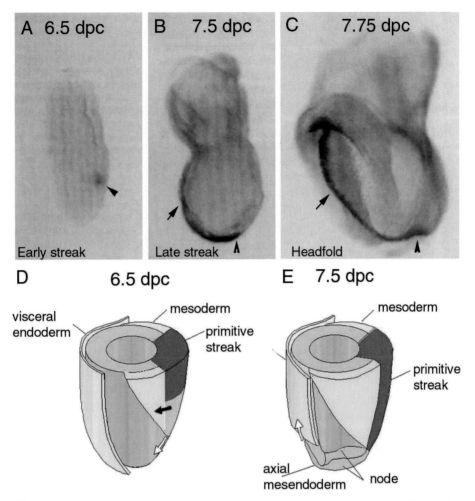

Figure 1 The mouse organizer and its derivatives during gastrulation. A–C *HNF3β* (*Foxa2*) expression (purple staining) marks early gastrula organizer (arrowhead in A), node (arrowhead in B and C), and axial derivatives of the organizer that migrate anteriorly from the node (arrows in B and C). (D and E) Schematic diagrams showing cell movements (arrows) from organizer to generate axial derivatives. (Organizer and derivatives are indicated in pink.)

present in the anterior primitive streak of early streak stage embryos.

It is widely recognized that in mouse, the axial mesoderm originates from the node and migrates anteriorly along the midline of the embryo (Figs. 1D and 1E) (Lawson *et al.*, 1991; Tam *et al.*, 1997). Similarly, in chick, the prechordal plate and the notochord are thought to derive from Hensen's node cells that migrate anteriorly along the midline (Selleck and Stern, 1991; Lemaire *et al.*, 1997). However, part of the axial mesoderm underlying the future forebrain and midbrain of mouse embryos may also derive from the anteriormost region of the lateral mesoderm that migrates as wings

that progressively surround the whole embryo to reach the anterior midline (Tam *et al.*, 1997) (black arrow in Fig. 1D).

In chick, the lengthening of the AP axis of the embryo results in a posterior displacement of the node, that leaves behind mesodermal cells that undergo a morphogenetic movement of convergent extension to form the notochord (Catala *et al.*, 1996). In mouse, some posterior regression of the node must have occurred because the primitive streak becomes progressively shorter during organogenesis (Lawson *et al.*, 1991). However, the extent of posterior displacement of the node is unknown. At later stages, the cells of the tail bud continue to contribute to the notochord and the floor plate in mouse and chick (Catala *et al.*, 1995; Wilson and Beddington, 1996; Knezevic *et al.*, 1998).

IV. The Organizer Phenomenon: Conserved Properties of Vertebrate Organizers

In 1924, Spemann and Mangold reported the existence of an embryonic structure with axis forming and patterning activity. They grafted the dorsal blastoporal lip from a newt gastrula stage embryo into another embryo and found that it induced the formation of a secondary embryo with a normal body pattern (reviewed by Spemann, 1938; Hamburger, 1988). The transplanted tissue not only instructed cells in surrounding tissues to acquire new fates, but also provided global patterning signals regulating the dimension, orientation, and position of the induced tissues along the AP embryonic axis, hence, its name of organizer. Similar grafting paradigms have more recently been used to identify homologous structures with axis-inducing ability in other vertebrate species (Storey *et al.*, 1992, 1995; Beddington, 1994; Shih and Fraser, 1996; Tam *et al.*, 1997; Viebahn, 2001). The axis-inducing organizer is called Hensen's node in chick, the shield in zebrafish, and the node in mouse (reviewed in Lemaire and Kodjabachian, 1996; Camus and Tam, 1999). In the mouse, the activity of anterior primitive streak cells in early streak embryos has also been examined in transplantation experiments. These node precursors have been shown to be able to induce a secondary axis in host embryos (Tam and Steiner, 1999), and have thus has been given the name of early gastrula organizer (EGO). Therefore, the mouse organizer corresponds to the anterior primitive streak at the early streak stage (the EGO) and to the node at the late streak stage.

The organizers from different vertebrate species share additional properties. Cells of the organizer of *Xenopus*, chick, zebrafish, and mouse have similar fates. Their descendants include cells in the midline of the three germ layers and in somites (Lemaire and Kodjabachian, 1996; Camus and Tam, 1999; see above). Organizer cells in different species also express an overlapping set of genes encoding transcription factors and secreted molecules, an indication that the mechanisms of organizer activity are conserved among different vertebrate species. Genes expressed in the EGO and in the node in the mouse are listed in Table I and discussed below. Interestingly, several of these genes are also expressed at the anterior region of the visceral endoderm, the outer cell layer of the early mouse gastrula. This suggested that the anterior visceral endoderm (AVE) also has organizing properties (see below). Gain-of-function experiments in *Xenopus* have shown that ectopic expression of some of these genes is sufficient to induce a secondary axis, suggesting that they

Table I   Genes Expressed in the Mouse Organizer and the AVE during Gastrulation[a]

	Early gastrula		Late gastrula
	Anterior primitive streak	Anterior visceral endoderm	Node
Genes encoding secreted proteins			
Chordin	Yes	No	Yes
Follistatin	Yes	No	No
Noggin	No	No	Yes
Cerberus	Yes	Yes	No
Frzb1	Yes	No	No
Lefty1	No	Yes	No
Nodal	Yes	Yes	Yes
Tdgf1	Yes	No	No
Cfc1	No	Yes	Yes
Shh	No	No	Yes
Dante	No	No	Yes
Genes encoding transcription factors			
Otx2	Yes	Yes	No
Lhx1	Yes	Yes	Yes
Foxa2	Yes	Yes	Yes
Gsc	Yes	Yes	No
Hhex	Yes	Yes	No
T	No	No	Yes

[a]References: *Chordin* (Rhinn *et al.*, 1998; Klingensmith *et al.*, 1999); *Follistatin* (Albano *et al.*, 1994); *Noggin* (Belo *et al.*, 1997; Biben *et al.*, 1998; McMahon *et al.*, 1998; Shawlot *et al.*, 1998); *Frzb1* (Leyns *et al.*, 1997; Wang *et al.*, 1997); *Lefty1* (Oulad-Abdelghani *et al.*, 1998); *Nodal* (Zhou *et al.*, 1993; Varlet *et al.*, 1997); Cripto (*Tdgf1*) (J. Ding *et al.*, 1998); Cryptic (*Cfc1*) (Shen *et al.*, 1997); *Shh* (Echelard *et al.*, 1993); *Dante* (Pearce *et al.*, 1999); *Otx2* (Simeone *et al.*, 1993; Ang and Rossant, 1994); *Lim1* (*Lhx1*) (Barnes *et al.*, 1994; Shawlot and Behringer, 1995); *HNF3β* (*Foxa2*) (Ang *et al.*, 1993; Monaghan *et al.*, 1993; Sasaki and Hogan, 1993); *Gsc* (Blum *et al.*, 1992; Filosa *et al.*, 1997), *T* (Wilkinson *et al.*, 1990); *Hhex* (Thomas *et al.*, 1998).

may be normally involved in the function of the organizer (Harland and Gerhart, 1997; Leyns *et al.*, 1997).

V. The Vertebrate Organizer Is a Dynamic, Nonhomogeneous, and Renewable Cell Population at Gastrulation

In *Xenopus* and chick, the organizer has been shown to have different inducing properties when taken from embryos at different embryonic stages (reviewed in Dias and Schoenwolf, 1990; Lemaire and Kodjabachian, 1996; Harland and Gerhart, 1997). At early stages, the organizer induces head structures, while at later stages it induces more posterior structures, such as trunk and tail. This observation has led to the idea that there are distinct head and trunk organizers in vertebrate embryos. More detailed studies in *Xenopus* have shown that the lower (or anterior) and upper (or posterior) parts of the dorsal blastopore lip induce head and trunk structures, respectively, indicating that there is also compartmentalization of distinct head and body patterning activities within the organizer at a particular stage (Zoltewicz and Gerhart, 1997). In addition, the anterior and posterior domains of the *Xenopus* organizer express different genes. For example, the homeobox genes *Xgsc* and *Xotx2* are predominantly expressed in the anterior domain, whereas *Xbra* and *Xnot* are mainly expressed in the posterior domain (Vodicka and Gerhart, 1995). Whether these different patterns of gene activity are implicated in the distinct inducing activities of the anterior and posterior domains of the organizer remains to be determined. It has been shown, however, that these gene expression domains correlate with the acquisition of distinct cell fates. When transplanted orthotopically into another *Xenopus* embryo, *Xgsc*-expressing and *Xnot*-expressing cells can give rise to prechordal mesoderm and notochord/somites, respectively (reviewed in Harland and Gerhart, 1997).

Whether the organizer of the mouse gastrula also has distinct functional domains is not known. There is nevertheless a clear heterogeneity in gene expression in the mouse organizer at the early gastrula stage. Double labeling studies have shown that the proximal domain of the anterior primitive streak contains cells coexpressing HNF3β (*Foxa2*) and *Lim1* (*Lhx1*), whereas its distal domain contains cells expressing only HNF3β (Perea-Gomez et al., 1999). There are also temporal differences in gene expression between the EGO and the node such that they only have a subset of the genes in common (see Table 1). These changes in gene expression that take place during the development of the organizer may be responsible for the differences in the fate of its cells. For example, cells from the EGO contribute to the prechordal plate, whereas cells from the node lack this ability (Beddington, 1994; Tam et al., 1997; see above).

Dye-labeling experiments of the organizer in chick and mouse embryos have shown that a subset of the labeled cells remains in the node up to 24 hr after labeling (Selleck and Stern, 1991; Beddington, 1994; Tam et al., 1997). These results have been interpreted as evidence for the existence of a resident population of organizer stem cells that, by asymmetric divisions, both self-renew and generate cell progeny contributing to the notochord and the somites. However, a recent study performed in chick has shown that Hensen's node is a dynamic structure in which most cells are constantly being renewed during gastrulation (Joubin and Stern, 1999). As cells leave the node to generate midline tissues, they are replaced by more lateral cells that migrate into the node and acquire node properties after being exposed to inducing signals produced by the middle portion of the primitive streak. These results demonstrate that the chick organizer can be defined as a cell state associated with a particular position and gene expression profile, but that its cellular composition is constantly changing during gastrulation. It is currently not clear whether the mouse node shares these characteristics with the chick organizer.

VI. Insights into the Function of the Mouse Organizer Gained from Genetic and Embryological Studies

The first demonstration that the mouse node could induce a secondary axis was performed by grafting *lacZ*-marked nodes from E7.5 embryos into the lateral regions of the same stage wild-type embryos (Beddington, 1994). The grafted node induced the formation of a secondary axis that was composed of both graft and host tissues. Interestingly, the secondary axes lacked anterior structures, that is the anterior head region. These results indicated that the node of the mouse at E7.5 may only function as a trunk organizer and not as a head organizer. Because studies in *Xenopus* and chick have shown that the organizer initially has head inducing activity, and only later acquires a trunk inducing activity, it was important to address the possibility that the EGO was able to induce a complete axis, including head structures, at the early streak stage. This has recently been performed by transplanting the EGO region into E7.5 mouse embryos, and results similar to those of the node grafting experiments were obtained, with only the induction of hindbrain and spinal cord structures expressing molecular markers for the caudal neural tube (Tam and Steiner, 1999). These studies question whether the mouse EGO or node have head organising activity at any time. However, it is interesting to note that although the EGO alone cannot induce anterior neural tissues, it can induce the expression of the forebrain marker *Otx2* and the midbrain marker *Engrailed* (*En*) when grafted together with both anterior epiblast and AVE from early streak stage embryos (Tam and Steiner, 1999). How these different germ layers synergize with the EGO to induce anterior neural tissues remains to be clarified. One possibility is that the AVE regulates the competence of the anterior epiblast to respond to the neural inducing and patterning signals provided by the EGO. In organizer grafting experiments performed in zebrafish, differences in competence of the ectoderm along the AP axis to respond to organizer signals have been shown to govern the AP character of the induced neural tissues (Koshida et al., 1998). Finally, recent interspecific chimera experiments have provided evidence for anterior neural inductive activity by the E7.5 mouse node when grafted into chick embryos (Knoetgen et al., 2000). It is possible that an E7.5 recipient mouse embryo does not provide a suitable environment for a grafted node to express its full activity.

The ability of the mouse node and node-derived tissues to induce anterior neural tissues has also been examined in experiments in which embryonic tissue fragments are cultured *in vitro* in isolation or in different combinations. The germ layers of gastrulating mouse embryos can be dissociated by enzymatic and mechanical treatment and then recombined and cultured *in vitro* for 1–2 days. When fragments of epiblast from the early streak stage are cultured *in vitro* in isolation they do not express neural fates. Thus, at this early stage the epiblast has not yet been committed toward an anterior neural fate. However, when fragments of the anterior epiblast from mid- to late streak stage embryos are cultured alone, they can express the anterior neural marker *Engrailed* (*En*), suggesting that they have received signals to become committed to anterior neural fates. When fragments of early streak stage anterior epiblast are recombined with mid- to late streak stage anterior mesendoderm (mesoderm and endoderm germ layers remaining associated), expression of the anterior neural markers *Otx2* and *En* is induced in the explants (Ang and Rossant, 1993; Ang *et al.*, 1994). In similar experiments, the node is also able to induce *En* in the anterior epiblast (Klingensmith *et al.*, 1999). These tissue recombination assays therefore suggest that organizer and organizer-derived tissues are indeed able to promote anterior neural differentiation.

VII. Genetic Analysis of Organizer Function: Mouse Mutants Showing Defects in Organizer Function

The winged helix transcription factor *HNF3β* (*Foxa*2) is expressed specifically in the EGO at the early streak stage (Ang and Rossant, 1993; Kaestner *et al.*, 1993; Monaghan,

et al., 1993; Sasaki and Hogan, 1993), and also in the visceral endoderm surrounding the embryo prior to gastrulation (Perea-Gomez, *et al.*, 1999). Later *HNF3β* is also expressed in the node and its derivatives, the notochord and floor plate. Embryos mutant for *HNF3β* do not form these latter structures, but undergo gastrulation to form somites and lateral mesoderm (Fig. 2) (Ang and Rossant, 1994; Weinstein *et al.*, 1994). To determine whether *HNF3β* function is required in the organizer and/or in the visceral endoderm, mouse chimera studies were performed.

When embryonic stem (ES) cells are injected into blastocysts, they preferentially colonize the epiblast and its derivatives (Beddington and Robertson, 1989). This assay creates a chimera in which the extraembryonic tissues are blastocyst-derived and the embryo proper can be predominantly ES cell derived. In contrast, tetraploid embryos, now typically produced by electrofusion of the blastomeres of two-cell stage embryos, preferentially contribute to extraembryonic tissues and not the embryo proper (Nagy *et al.*, 1993; reviewed in Rossant and Spence, 1998). The complementary properties of ES cells and tetraploid cells permit the production of chimeric embryos in which epiblast-derived embryonic tissues are of one genotype and the extraembryonic tissues are of another (Nagy *et al.*, 1990).

Chimeric embryos were produced by aggregating *lacZ*-marked tetraploid wild-type morulas with *HNF3β* mutant ES cells. In such experiments, *HNF3β* was shown to be re-

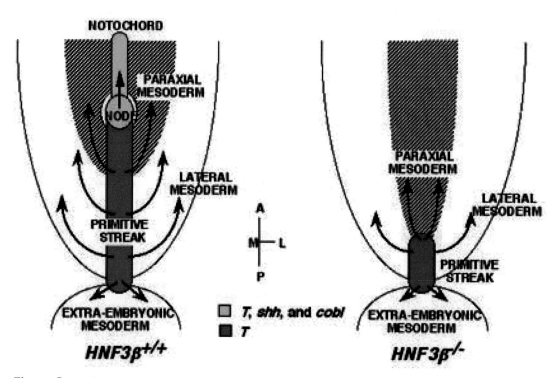

Figure 2 *HNF3β* (*Foxa*2) is required autonomously for the generation of node and its derivatives in the midline. Somitic and lateral mesoderm resulting from cells ingressing through the anterior and middle regions of the streak are formed in *HNF3β* homozygous null mutants. cobl; cordon-bleu (Gasa *et al.*, 1995). (Modified from Yamaguchi *et al.*, 1884.)

quired autonomously in the node and the notochord for the generation of these tissues (Dufort *et al.*, 1998). Despite the absence of a node at the distal tip of the *HNF3β* mutant embryos, a rudimentary axis still formed and a neural tube with an AP pattern was usually observed. However, dorsal-ventral patterning of the neural tube was severely affected. These results indicate that the node is not essential for the induction of neural tissues or initiation of AP patterning of the neural tube, whereas it is required for the generation of axial mesoderm. Lack of dorsal-ventral patterning in the neural tube is the likely consequence of the lack of Sonic hedgehog (SHH), a signaling molecule normally expressed by the axial mesoderm that is essential for the generation and patterning of the ventral neural tube (Ericson *et al.*, 1996; reviewed in Tanabe and Jessell, 1996). However, note that *HNF3β* mutant embryos transiently express several markers of the EGO, such as *Gsc* and *Chrd* (Klingensmith *et al.*, 1999). Thus, it is possible that the EGO may be present and still function in *HNF3β* mutant embryos for neural inducing and AP patterning activity. In summary, the mutation analysis of *HNF3β* has shown that this transcription factor is not required or functions redundantly with other genes in the anterior primitive streak for the early axis inducing activity of the organizer, whereas it is required subsequently for the generation and function of the node. The generation of a neural tube with proper AP pattern in the absence of a node also implies that AP patterning signals are provided early by the mouse organizer, before the formation of the node, or that other tissues can compensate for the loss of the mature organizer.

The analysis of mice mutant in *Wnt3*, which encodes a member of the Wnt family of secreted signaling molecules, has revealed an important role for this protein in the generation of the organizer (Liu *et al.*, 1999). *Wnt3* is expressed in the proximal epiblast directly adjacent to the extraembryonic ectoderm before gastrulation. Its expression is then restricted to the posterior proximal epiblast and its associated posterior visceral endoderm (PVE), and subsequently to the primitive streak and the mesoderm. Later, *Wnt3* is downregulated in the primitive streak and mesoderm. *Wnt3* mutant embryos completely lack an AP axis (defined by region-specific neural markers), and do not express molecular markers of the organizer, suggesting that the organizer is not formed. Interestingly, AVE markers including *Lim1* and *Cerl* are asymmetrically localized, suggesting that AP patterning of the visceral endoderm occurs in the *Wnt3* mutants. The absence of an organizer in *Wnt3* mutants is most likely the cause of the lack of anterior and posterior neural tissue. These results show that *Wnt3* is required for the specification of the organizer and that the organizer is required for the formation of the AP axis of the epiblast. Because pan-neural markers were not examined in these mutants, it remains to be determined whether these mutant embryos make any neural tissue and thus whether the loss of anterior and posterior neural

markers is a consequence of the loss of AP patterning signals, or of the loss of neural inducing signals, or both. The initial expression of *Wnt3* in the proximal epiblast is intriguing because it suggests that the adjacent extraembryonic ectoderm may induce or maintain *Wnt3* expression in this region (Behringer *et al.*, 2000).

The LIM homeodomain transcription factor LIM1 (also called LHX1) is coexpressed with HNF3β in the EGO, the node, and the AVE (Barnes *et al.*, 1994; Shawlot and Behringer, 1995; Perea-Gomez *et al.*, 1999). Mice with a null mutation in *Lim1* lack all head structures anterior to the third rhombomere of the hindbrain (Shawlot and Behringer, 1995). To distinguish between the requirements for *Lim1* in the AVE and in the EGO/node and/or their derivatives, chimeric embryos have been generated with *Lim1* mutant ES cells and wild-type embryos. Chimeric embryos with *Lim1* mutant extraembryonic tissues and predominantly wild-type embryonic tissues lacked anterior head structures, suggesting that the *Lim1*-expressing AVE was required for normal head formation. Interestingly, the complementary chimera approach, generating wild-type extraembryonic tissues and *Lim1* mutant embryonic tissues, also showed a loss of forebrain and midbrain as early as 8.0 dpc (Shawlot *et al.*, 1999). Thus, *Lim1* activity is required both in the AVE and in embryonic tissues.

A number of other mouse mutants also have defects in forebrain development, including holoprosencephaly, cyclopia, and also various craniofacial abnormalities. This is the case of mutants for *Shh* (Chiang *et al.*, 1996), and for SMAD2, a member of the SMADS family of transforming growth factor β (TGF-β)-signaling molecules that is widely expressed in the embryo during gastrulation. *Smad2* chimeric embryos, consisting of *Smad2* mutant and wild-type cells, have holoprosencephaly and also lack expression of *Shh* in the axial mesoderm, which is probably the cause of the brain defects (Heyer *et al.*, 1999). A similar phenotype has been observed in embryos lacking the homeobox gene *Gsc* and carrying only one copy of the *HNF3β* gene (Filosa *et al.*, 1997), in double heterozygous mutants for *HNF3β* and *Nodal* (Collignon *et al.* 1996) and also in double heterozygous mutants for *HNF3β* and *Otx2* (Jin *et al.*, 2000). The homeobox genes *Gsc* and *Otx2* are expressed in the EGO and its derivatives in the anterior mesendoderm, while Nodal is expressed in the EGO and the node (Ang *et al.*, 1994; Filosa *et al.*, 1997; Varlet *et al.*, 1997). The phenotypes of *Smad2* mutants and of *Gsc;HNF3β*, *HNF3β;Nodal* and *HNF3β;Otx2* double mutants therefore also argue that the axial mesoderm is involved in regulating morphogenesis and growth of the forebrain.

The studies summarized above have identified a number of genes that are involved sequentially in the development and the function of the organizer. First, *Wnt3* is required for the establishment of the organizer; then *Lim1* is involved in the head forming activity of the EGO; subsequently, *HNF3β* is required for the formation of the node; then *HNF3β, Shh, Smad2, Nodal, Gsc* and *Otx2* are involved in the generation and/or the function of the node-derived axial mesoderm.

VIII. Inhibitory Signals Secreted by the Organizer and Its Derivatives

A number of secreted proteins, including Noggin, Chordin, Follistatin, Cerberus, Nodal, and Frzb1, are specifically expressed in the node and/or the EGO of the mouse embryo (reviewed by Camus and Tam, 1999; Zhou *et al.,* 1993; Collignon *et al.,* 1996; Leyns *et al.,* 1997; Table 1). In *Xenopus* embryos, *Noggin, Chordin,* and the gene *Xenopus nodal-related 3* (*Xnr3*), also expressed in the organizer (the dorsal blastopore lip), have been shown to induce a secondary axis when ectopically expressed (reviewed in Harland and Gerhart, 1997), suggesting that they play an important role in the activity of the organizer. Biochemical studies have shown that the *Xenopus* factors Noggin, Chordin, Follistatin, and Cerberus all act as antagonists of bone morphogenetic protein (BMP) signaling, by preventing ligands such as BMP2, 4, and/or 7 from interacting with their cognate receptors. *Xnr3* can also block BMP signaling, but its mode of action may be different (Hansen *et al.,* 1997).

Cerberus belongs to the Cerberus/DAN family of secreted molecules (Pearce *et al.,* 1999). The common motif of this new family of molecules resembles the cysteine knot motif found in a number of signaling molecules, including members of the TGF-β superfamily (Biben *et al.,* 1998). Besides attenuating BMP signaling, Cerberus has also been shown to be able to antagonize signaling by Wnt factors and the TGF-β-related protein Nodal (Hsu *et al.,* 1998; Piccolo *et al.,* 1999). In *Xenopus,* blocking BMP function in an animal cap assay results in neural induction (Sasai and De Robertis, 1997; Wilson and Hemmati-Brivanlou, 1997). In chick embryos, grafting Chordin-soaked beads also results in blocking BMP activity, but this is not sufficient by itself to induce neural tissue, although Chordin may be involved in defining the lateral boundaries of the neural plate (reviewed in Streit and Stern, 1999). Considering the potent activities of the above factors for inducing ectopic axes or neural tissue, it was somewhat surprising that in the mouse, null mutations in *Noggin, Follistatin,* and *Cerberus* have not caused defects in neural induction or AP axis formation (Matzuk *et al.,* 1996; McMahon *et al.,* 1998; Belo *et al.,* 2000; Shawlot *et al.,* 2000; Stanley *et al.,* 2000). These mutant mice suggest caution when interpreting overexpression assays or that these factors have overlapping activities whose essential roles may be revealed by the generation of compound mutants.

Recent studies in *Xenopus* have shown that the initiation of head development involves blocking BMP and Wnt signaling (reviewed in Harland and Gerhart 1997; Niehrs, 1999) as well as Nodal signaling (Piccolo *et al.,* 1999). Beside Cerberus, other Wnt antagonists are expressed in the organizer, including Dickkopf1 (Glinka *et al.,* 1998) and Frzb1 (Leyns *et al.,* 1997; Wang *et al.,* 1997). Mouse *Dickkopf1* (*Dkk1*) is expressed in the AVE and the anterior axial mesoderm at the headfold stage (Glinka *et al.,* 1998), whereas *Frzb1* is expressed in the EGO and the anterior mesendoderm underlying the neural plate at the early somite stage (Leyns *et al.,* 1997). The Frzb1 protein, which resembles the extracellular domain of members of the Frizzled-like family of Wnt receptors, can function as a Wnt antagonist when overexpressed by complexing certain Wnt proteins, thereby blocking their interaction with their receptor. Dkk1 belongs to a new family of secreted proteins with cysteine-rich domains (Glinka *et al.,* 1998). The role of Wnt antagonists in head formation in the mouse remains to be established. It is interesting to note, however, that expression of *Dkk1* is absent in the *Otx2* mutant cells found in the anterior axial mesendoderm of chimeric embryos (Perea-Gomez *et al.,* 2000), suggesting that loss of *Dkk1* may contribute to the lack of maintenance of forebrain and midbrain in chimeric *Otx2* mutant embryos.

IX. Specification of the Primitive Streak and the Organizer

In *Xenopus,* fertilization results in a cytoplasmic rearrangement, called cortical rotation, that localizes maternal determinants from the vegetal region of the egg to the prospective dorsal region of the embryo (reviewed in Harland and Gerhart, 1997; Moon and Kimelman, 1998). The dorsal vegetal blastomeres that inherit these determinants constitute the Nieuwkoop center (Gerhart *et al.,* 1989). The Nieuwkoop center is an activity that has been shown to act early, before gastrulation, to induce the Spemann organizer without directly contributing to it. The Nieuwkoop center is thought to involve a synergy between TGF-β and Wnt signaling pathways. Indeed, ectopic expression of Wnt molecules in *Xenopus* embryos leads to induction of dorsal mesoderm and axis duplication (McMahon and Moon, 1989; Smith and Harland, 1991). However, even though maternal *Wnt* mRNAs are present in the frog egg, there is growing evidence that the Wnt pathway may be activated intracellularly, because reagents that should block endogenous Wnt proteins in the egg do not prevent axis formation (reviewed in Harland and Gerhart, 1997; Moon and Kimelman, 1998). Activation of WNT signaling in *Xenopus* embryos results in stabilization and nuclear localization of β-Catenin in dorsal marginal cells. β-catenin is a cytoplasmic protein that is part of the adherens junction; however, in the nucleus it interacts with the LEF1/TCF3 family of transcription factors, resulting in the activation of organizer homeobox genes such as *X-Siamois* and *X-Twin* (reviewed in Moon and Kimelman, 1998). In chick embryos, the grafting of cells expressing the TGF-β-related molecule Vg1, alone or together with cells expressing Wnt1, in lateral or anterior regions of the embryo leads to induction of organizer-specific genes (Seleiro *et al.,* 1996; Shah *et al.,* 1997; Joubin and Stern, 1999; reviewed in Bachvarova, 1999). Presumably, this mimics the activities of the cells coexpressing Wnt8c and Vg1, which are found in two organizer inducing centers in the posterior marginal zone (the Nieuwkoop center of the chick) before gastrulation and in the middle part of the primitive streak during gastrulation. Interestingly, cells in the posterior marginal zone do not contribute to the primitive streak, and therefore these two organizer inducing centers existing before and during gastrulation are not related by lineage (Joubin and Stern, 1999).

Wnt and TGF-β signals have also been implicated in primitive streak and organizer formation in mouse based on the study of Nodal and Wnt3 signaling mutants. The TGF-β superfamily member Nodal is initially widely expressed in the epiblast, and then subsequently becomes restricted to the proximal posterior epiblast in the primitive streak region (Conlon *et al.,* 1994; Collignon *et al.,* 1996). *Nodal* mutants fail to form a primitive streak, although they can express molecular markers of mesoderm such as *Brachyury*. *Cripto* encodes an extracellular and membrane-associated ligand belonging to a family of EGF-CFC proteins, including the zebrafish one-eyed-pinhead (Oep), mouse Cryptic, and frog FRL1, characterized by an EGF-like domain and a cysteine-rich domain called the CFC domain (reviewed in Schier and Shen, 2000). Oep has been shown to be an essential extracellular cofactor of Nodal signaling (Gritsman *et al.,* 1999). Interestingly, mouse mutants for the *Cripto* gene show defects in primitive streak formation, only generating extraembryonic mesoderm (Ding *et al.,* 1998).

Wnt3 mutant embryos lack a primitive steak and an EGO (Liu *et al.,* 1999), providing evidence that Wnt signaling is also involved in the specification of the primitive streak and the organizer in mouse. Additional support for a role of the Wnt pathway in mammalian axis formation is provided by the observation of axis duplications in transgenic mouse embryos with widespread *Cwnt8C* expression (Popperl *et al.,* 1997). In addition, analysis of the *Fused* mutation in mouse has also suggested a role for antagonists of Wnt signaling in regulating organizer formation. The gene *Fused* encodes Axin, an inhibitor of Wnt signaling. Axin displays similarities with the RGS (regulators of G-protein signaling) (Dohlman and Thorners, 1997) and Disheveled proteins (Klingensmith and Nusse, 1994; Sussman *et al.,* 1994), and has been shown to negatively regulate the axis inducing activity of Wnt signaling by down regulating β-catenin activity. Mutation in *Fused* leads to posterior axis duplication in mouse embryos, and forced expression of a mutant mouse Axin

leads to axis duplication in *Xenopus* (Zeng *et al.,* 1997). Taken together, these data suggest that Wnt signaling is required for correct establishment of the organizer.

X. Role of the AVE in Anterior Patterning in Mouse

Because neither the EGO nor the node has been able to induce a complete secondary axis in mice, other tissues with anterior patterning activities may exist in mouse embryos. At the pre- to early streak stage, the visceral endoderm surrounds the epiblast and the extraembryonic ectoderm. Progressive recruitment of epiblast cells to the definitive endoderm eventually excludes the vast majority of visceral endoderm cells from the invaginating foregut and the remainder of the embryo proper and displaces them extraembryonically to generate the yolk sac endoderm (Lawson *et al.,* 1986; Lawson and Pedersen, 1987; Tam and Beddington, 1992). The visceral endoderm plays a major role in uptake and delivery of nutrients to the embryo (reviewed in Bielinska *et al.,* 1999) (see Chapter 8) as well as in the process of cavitation that forms the proamniotic cavity at the center of the embryonic and extraembryonic regions (Coucouvanis and Martin, 1999). Superficially, there do not appear to be any AP distinctions of the visceral endoderm in mouse embryos prior to gastrulation. However, recent studies using *lacZ* and *GFP* transgenes under the control of *Otx2* regulatory sequences suggest that there is an early morphology difference in visceral endoderm cells that will become the AVE (Kimura *et al.,* 2000). In rabbit embryos both the ectoderm and the visceral endoderm are significantly thickened at the anterior margin, forming the anterior margin crescent before primitive streak initiation (Viebahn *et al.,* 1995).

Despite the absence of an obvious morphological landmark, gene expression studies have revealed a distinct molecular heterogeneity along the AP axis of the visceral endoderm in mouse embryos (see Table I). One of the first genes found to be expressed in the AVE was the homeobox gene *Hesx1/Rpx* (Hermesz *et al.,* 1996; Thomas and Beddington, 1996). At later stages, *Hesx1/Rpx* is also expressed in the anterior neural ectoderm (ANE). Another homeobox gene called *Hex* was found to be expressed in the distal visceral endoderm prior to gastrulation but is restricted to the AVE at later stages, suggesting that the distal visceral endoderm cells are the precursors of the AVE (Thomas *et al.,* 1998). Tracing of DiI-injected distal VE cells into later stages reinforced this idea. Later, *Hhex* is expressed in the anterior definitive endoderm. Many genes that encode various secreted molecules and transcription factors have been found to be expressed specifically in the AVE. This molecular heterogeneity between anterior and posterior regions of the visceral endoderm and the morphogenetic movements of the visceral endoderm (Thomas *et al.*, 1998; Weber *et al.*, 1999) suggest that the AVE may be involved in patterning the anterior region of the mouse embryo.

XI. Embryological and Genetic Analysis of the Function of the AVE in Anterior Patterning

To determine if the AVE indeed has a role in anterior neural development, the AVE was physically removed in early gastrulating mouse embryos that were subsequently cultured *in vitro*. Removal of the AVE results in a loss or reduction of the developing forebrain in operated embryos after 1 day of culture, indicating that the AVE is required for anterior neural development (Thomas and Beddington, 1996). To test whether the AVE is also sufficient to induce anterior neuroectoderm, the AVE has been transplanted to lateral regions of the E7.5 embryo (Tam and Steiner, 1999). Interestingly, when the AVE was transplanted alone, anterior neural development was not induced, while, as mentioned earlier, ectopic anterior neural development was observed when the AVE was grafted together with anterior epiblast and the EGO. These results suggest that anterior neural development requires synergistic interactions between the organizer and the anterior germ layers. Similarly, the chick AVE is unable to induce forebrain markers in the chick epiblast (Knoetgen *et al.*, 1999). In contrast, the rabbit AVE grafted in chick embryos can induce forebrain markers in the epiblast (Knoetgen *et al.*, 1999).

Genetic studies have demonstrated a role in anterior neural development for several factors expressed in the AVE. These factors include the transcription factors OTX2 and LIM1, and the TGF-β-related molecule Nodal. Because all of these factors are also expressed in other embryonic cell types (Simeone *et al.*, 1993; Ang *et al.*, 1994; Shawlot and Behringer, 1995; Collignon *et al.*, 1996), chimeric embryos have been used to demonstrate the requirement for these genes in extraembryonic rather than embryonic lineages. Surprisingly, similar chimera studies have shown that both *Hesx1/Rpx* and *Hex* are dispensable in the AVE for anterior patterning (Martinez Barbera *et al.*, 2000a,b).

Otx2 mutant embryos lack the forebrain, midbrain, and anterior hindbrain (Acampora *et al.*, 1995; Ang *et al.*, 1996; Matsuo *et al.*, 1995). Chimeric embryos containing wild-type embryonic tissues and *Otx2* mutant extraembryonic tissues phenocopy the *Otx2* mutants. Conversely, chimeras composed of *Otx2* mutant embryonic tissues and wild-type extraembryonic tissues show an initial rescue of anterior neurectoderm development (Fig. 3), although this tissue is

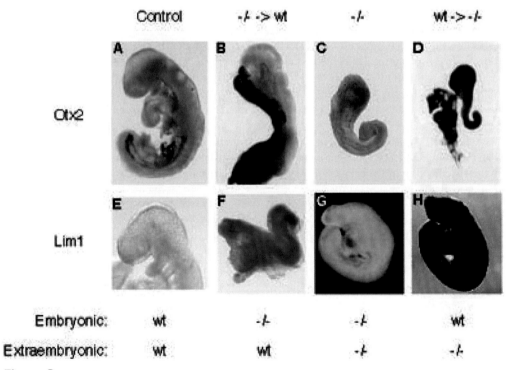

Figure 3 Chimera studies of *Otx2* and *Lim1*. (A) Control chimera at E8.75. (B–D) *Otx2* chimeras at E8.75 dpc. (E) Wild-type E9.5-dpc embryo. (F-H) *Lim1* chimeras at E9.5 dpc. The genotypes of the embryonic and extraembryonic regions of the chimeras are shown below. *Otx2* and *Lim1* are both required in extraembryonic tissues, presumably the visceral endoderm, for the formation of the anterior neural plate (Rhinn *et al.*, 1998; Shawlot *et al.*, 1999). *Lim1* is also required in embryonic tissues for reinforcement or refinement of the anterior neural pattern.

not maintained in embryos at the early somite stage (Rhinn et al., 1998). Thus, initial anterior neural development requires *Otx2* expression in visceral endoderm. The lack of maintenance of anterior neural tissue implies a later role for *Otx2* in the axial mesoderm and/or the neuroectoderm for the specification of the AP identity of the neural plate. Mouse chimeras have also been used to study the role of *Nodal* and *Lim1* in head development. Forebrain development is affected in *Nodal* chimeras containing mutant extraembryonic tissues and a high percentage (80%) of wild-type cells in the embryo, but not in chimeras of the reciprocal type (Varlet et al., 1997). *Lim1* chimeras with mutant visceral endoderm and wild-type embryonic tissues lack forebrain and midbrain (Shawlot et al., 1999). Thus, *Otx2, Nodal,* and *Lim1* all have essential roles in extraembryonic tissues, presumably in the visceral endoderm, for rostral brain development.

As mentioned before, *Hesx1* is expressed in the AVE and also later in the ANE. Mice homozygous for a mutation in the *Hesx1/Rpx* gene have variable defects of the anterior central nervous system and other abnormalities (Dattani et al., 1998). The mutant phenotypes include a reduction in the size of the prosencephalon, anopthalmia or micropthalmia, alterations in nasal development and defects in Rathke's pouch, the precursor of the pituitary gland. Chimera studies demonstrate that *Hesx1/Rpx* is not essential in the AVE for anterior neural development, rather it is required later in the anterior neuroectoderm (Martinez-Barbera et al., 2000b).

Hhex mutant mice also have variable anterior abnormalities that are restricted to the rostral forebrain (Martinez-Barbera et al., 2000b). Analysis of the *Hhex* mutants shows that the future forebrain region is induced properly but it subsequently fails to develop. As with *Hesx1/Rpx, Hex* is also not essential in the AVE for anterior development, but rather is probably required in the definitive endoderm for normal forebrain formation. Thus, the required roles for *Hesx1/Rpx* and *Hex* in anterior neural development reside in embryonic tissues not in the extraembryonic tissues. Perhaps their roles in the AVE are compensated by other genes in their absence.

XII. A Model for AVE Function in Anterior Patterning

Two secreted molecules, Lefty1 and Cerberus, are expressed in the AVE (Meno et al., 1996; Oulad-Abdelghani et al., 1998). Lefty1 belongs to a novel subclass of TGF-β superfamily molecules that includes a closely related mouse factor, Lefty2, and the zebrafish, chick, and *Xenopus* antivin (Thisse and Thisse, 1999; Ishimaru et al., 2000; Cheng et al., 2000) and the human LeftyA/Endometrial bleeding associated factor and LeftyB (Kothapalli et al., 1997; Kosaki et al., 1999). Members of this subclass lack both the large α helix

and the seventh cysteine known to be involved in dimerization of TGF-β molecules, suggesting that these proteins act either as monomers or as noncovalent dimers (Thisse and Thisse, 1999). Overexpression of *antivin* or mouse *Lefty1/Lefty2* in zebrafish embryos leads to depletion of mesoderm and endoderm tissues (Thisse and Thisse, 1999), which can be overcome by coexpression of zebrafish Nodal-related molecules such as Cyclops or Squint (Bisgrove et al., 1999; Meno et al., 1999). *Lefty2* mutant mouse embryos have an expanded primitive streak and excess mesoderm formation (Meno et al., 1999), a phenotype opposite to that of *Nodal* mutants (Conlon et al., 1994). Interestingly, the *Lefty2* mutant phenotype is partially suppressed by heterozygosity of *Nodal*. Altogether, these data strongly support a role for Lefty proteins in antagonizing Nodal signaling during gastrulation. Cerberus, as mentioned previously, also antagonizes Nodal as well as BMP and Wnt signaling pathways. The phenotype of *Nodal* mutant mice is consistent with a role for this gene in the formation of the primitive streak. Thus, Cerberus and Lefty1 expressed in the AVE could, by limiting the activity of Nodal, define the extent of the primitive streak in the epiblast.

Indirect evidence for a role of the AVE in regulating primitive streak formation has recently been obtained in double mutant studies of *HNF3β* and *Lim1* (Perea-Gomez et al., 1999). Double homozygous *HNF3β,Lim1* mutants have an expanded primitive streak and have lost *Cerberus* and *Lefty1* expression in the AVE. An enlarged primitive streak is also observed in *Smad2* mutant embryos (Waldrip et al., 1998). In these mutants, loss of the *Smad2* pathway results in the epiblast adopting an extraembryonic mesoderm fate and absence of Cerberus expression in the visceral endoderm. Moreover, chimera studies demonstrate that *Smad2* is required in extraembryonic tissues, either in the visceral endoderm and/or in the extraembryonic ectoderm, to restrict the site of primitive streak formation. A phenotype similar to that of *Smad2* mutants was also observed in chimeric embryos with a wild-type epiblast and extraembryonic tissues mutant for the activin receptor 1B (Gu et al., 1998), which is a type I transmembrane serine/threonine kinase receptor that mediates activin signaling when complexed with type II activin receptors (reviewed in Massague, 1998). The similarity of the mutant phenotypes of *Smad2* and *Acur1b* chimeras suggests that *Smad2* functions downstream of the 1B receptor. Together, these data suggest that the AVE functions to prevent the anterior epiblast, which normally gives rise to ectoderm and neurectoderm derivatives (Lawson et al., 1991; Quinlan et al., 1995), from acquiring a posterior primitive streak fate. Given the similarities of their mutant phenotypes, *HNF3β, Lim1, Smad2,* activin receptor 1B as well as *Otx2* may act in a common pathway in the AVE to control the expression of *Cerberus* and *Lefty1* (Perea-Gomez et al., 2001).

Another role has been ascribed to the AVE, that of inducing or maintaining *Otx2* expression in the epiblast. In embryos homozygous for a mutation in which the *Otx2* gene is replaced by a *lacZ* reporter, *lacZ* transcripts are not expressed in the epiblast although they are found in distal regions of the visceral endoderm (Acampora *et al.,* 1995). However, epiblast expression directed by *Otx2* regulatory sequences is observed in *Otx2* mutant embryos, when *Otx2* function is replaced only in the visceral endoderm and not in the epiblast by the closely related gene, *Otx1*, in knock-in experiments (Acampora *et al.,* 1998). Thus *Otx2* is required in the visceral endoderm to regulate *Otx2* expression in the epiblast, suggesting that the AVE functions to promote anterior epiblast fate via regulation of *Otx2* expression. The nature of the signaling molecules provided by the AVE to induce or maintain *Otx2* expression in the epiblast remains to be determined.

Taken together, the current data from mouse fit well with a two-step induction or double assurance model for anterior specification (Fig. 4). Spemann (1927) developed the double assurance model to describe a series of inductive events that would ultimately result in the stablization of a particular cell fate. Thomas and Beddington (1996) proposed that the visceral endoderm (the AVE) initiated anterior identity in the adjacent epiblast, an action that was subsequently reinforced by primitive streak-derived mesendoderm as it physically replaces the visceral endoderm because of the morphogenetic movements of gastrulation. Similar conclusions were drawn from the *Lim1* chimera studies described above (Shawlot *et al.,* 1999) and recent studies in the chick (Foley *et al.,*

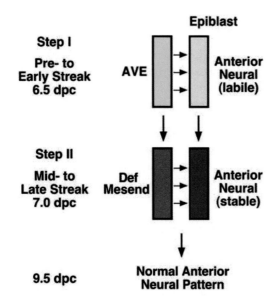

Figure 4 Double assurance model for anterior specification. At the pre- to early streak stage (6.5 dpc), the AVE initiates anterior neural specification of the adjacent epiblast. This anterior identity is labile but can be reinforced and stabilized by the mid- to late streak stage (7.0 dpc) by the definitive mesendoderm that displaces the AVE proximally. (Modified from Shawlot *et al.,* 1999.)

2000). In those studies, it was proposed that LIM1 regulated the expression of anteriorizing signals in both the AVE and streak-derived anterior mesendoderm. The *Hhex* chimera studies also support a reinforcing role for the anterior endoderm for anterior specification (Martinez-Barbera *et al.,* 2000b). The molecular details of this two-step model for anterior identity will surely become more clear in the near future.

XIII. Conclusions and Future Directions

Our knowledge of anterior-posterior patterning in mouse has come a long way since the identification of the mouse organizer by grafting experiments. One of the major findings has been the identification of the visceral endoderm as an essential regulator of anterior patterning. We have some knowledge of the signals produced by the AVE and of the response of the epiblast to these signals based on the results obtained from phenotypic analyses of mutants with loss of anterior neural development. All the data available so far support an essential role of the AVE in anterior development and we have proposed that the AVE functions to promote anterior epiblast fates in part by inducing or maintaining *Otx2* expression and to block posterior development in anterior regions of the embryo. Heterotopic grafting experiments indicate that the mouse AVE is unable to induce anterior neural development. In contrast, the rabbit AVE is sufficient to trigger ectopic anterior neural development. Whether these conflicting results are due to the varied potentials of the AVE in these two mammalian species or due to the different experimental conditions requires further investigation. It is clear, however, from genetic and embryological data that anterior patterning requires synergistic signals from the EGO to refine or maintain anterior differentiation initiated by the AVE. The exact nature of these synergistic interactions and the molecules involved remain to be elucidated.

Another significant contribution to our understanding of AP patterning of vertebrate embryos in recent years has come from the identification of secreted molecules that are capable of inducing ectopic secondary axes in *Xenopus* assays. Interestingly, many of these molecules can function as inhibitory signals regulating the activities of two distinct families of proteins, TGF-β superfamily and Wnt family molecules. However, loss of function studies of these inhibitory molecules in mice have not revealed any defects in axis formation perhaps due to redundant or overlapping functions of these molecules. The recent demonstration that Noggin and Chordin function redundantly in forebrain development in mice (Bachiller *et al.,* 2000) emphasizes the importance of generating other multiple/compound mutant combinations of secreted molecules, to clarify their functions in AP patterning of the embryonic axis. A better understanding of the molecular events involved in anterior patterning would also benefit from the identification of molecules required for neural induction in mouse.

A number of genes have been identified in nonmammalian vertebrates that are either required for normal axis formation or have axis forming activities in overexpression assays, including *Siamois, X-twin, Xnot/floating head/Cnot, dharma/boz,* and *XVg1/CVg1.* Many laboratories have unsuccessfully tried to isolate the mammalian orthologs of these genes. Either these genes are species-specific regulators of axis formation or the mammalian orthologs still remain to be identified. Clearly, the human and mouse genome projects will clarify whether or not these genes exist in mammals.

Another line of study that will likely emerge is the exploitation of these important mouse mutants for differential screens to identify downstream genes (Shimono *et al.,* 1999; Shimono and Behringer, 1999). The strategies used to identify downstream targets are influenced by the amounts of the embryonic tissues required in the differential screens. This is especially a challenge when studying mutants that influence the formation of the organizer and early patterning events. PCR strategies have been successfully employed, utilizing subregions of single wild-type and *Lim1* mutant gastrula stage embryos for subtractions (Shimono and Behringer, 1999). Furthermore, if sufficient amounts of tissue can be obtained, then perhaps expression profiling using DNA microarrays may prove to be an effective means for identifying differentially expressed genes.

The next breakthrough in understanding how the AP axis is established is the determination of the molecular mechanisms involved in the induction of the AVE and the organizer. Nieuwkoop centers, whose activities involve TGF-β and Wnt signaling, are able to induce the organizer and have been identified in frog and more recently in chick. Is there a mouse Nieuwkoop center? Comparative fate mapping and gene expression studies indicate that if a mouse Nieuwkoop center does exist it may reside in the posterior epiblast or the adjacent extraembryonic ectoderm (Tam and Behringer, 1997). Genetic studies have also demonstrated essential roles for the TGF-β superfamily molecule Nodal and Wnt3 in the formation of the primitive streak and organizer. In contrast, nothing is known about how the AVE is induced, except that it is established before and occurs independently of primitive streak formation. Analyses of the regulatory sequences that direct the expression of genes specifically to the AVE may shed some light on this problem. *Cis*-acting DNA elements that direct AVE-specific expression in transgenic mice have recently been identified for *OTX2* (Kimura *et al.,* 2000). Interestingly, OTX2 has been shown to physically interact with LIM1 and HNF3β (Nakano *et al.,* 2000). These findings and other findings should facilitate the definition of the molecular pathways that regulate anterior development.

Acknowledgment

We thank Bill Shawlot for helpful comments.

References

Acampora, D., Mazan, S., Lallemand, Y., Avantaggiato, V., Maury, M., Simeone, A., and Brulet, P. (1995). Forebrain and midbrain regions are deleted in Otx2-/- mutants due to a defective anterior neuroectoderm specification during gastrulation. *Development (Cambridge, UK)* **121,** 3279–3290.

Acampora, D., Avantaggiato, V., Tuorto, F., Briata, P., Corte, G., and Simeone, A. (1998). Visceral endoderm-restricted translation of Otx1 mediates recovery of Otx2 requirements for specification of anterior neural plate and normal gastrulation. *Development (Cambridge, UK)* **125,** 5091–5104.

Albano, R. M., Arkell, R., Beddington, R. S., and Smith, J. C. (1994). Expression of inhibin subunits and follistatin during postimplantation mouse development: Decidual expression of activin and expression of follistatin in primitive streak, somites and hindbrain. *Development (Cambridge, UK)* **120,** 803–813.

Ang, S. L., and Rossant, J. (1993). Anterior mesendoderm induces mouse Engrailed genes in explant cultures. *Development (Cambridge, UK)* **118,** 139–149.

Ang, S. L., and Rossant, J. (1994). HNF-3 beta is essential for node and notochord formation in mouse development. *Cell (Cambridge, Mass.)* **78,** 561–74.

Ang, S. L., Wierda, A., Wong, D., Stevens, K. A., Cascio, S., Rossant, J., and Zaret, K. S. (1993). The formation and maintenance of the definitive endoderm lineage in the mouse: Involvement of HNF3/forkhead proteins. *Development (Cambridge, UK)* **119,** 1301–1315.

Ang, S. L., Conlon, R. A., Jin, O., and Rossant, J. (1994). Positive and negative signals from mesoderm regulate the expression of mouse Otx2 in ectoderm explants. *Development (Cambridge, UK)* **120,** 2979–2989.

Ang, S. L., Jin, O., Rhinn, M., Daigle, N., Stevenson, L., and Rossant, J. (1996). A targeted mouse Otx2 mutation leads to severe defects in gastrulation and formation of axial mesoderm and to deletion of rostral brain. *Development (Cambridge, UK)* **122,** 243–252.

Bachiller, D., Klingensmith, J., Kemp, C., Belo, J. A., Anderson, R. M., May, S. R., McMahon, A. P., Harland, R. M., Rossant, J., and De Robertis, E. M. (2000). The organizer factors Chordin and Noggin are required for mouse forebrain development. *Nature (London)* **403,** 658–661.

Bachvarova, R. F. (1999). Establishment of anterior-posterior polarity in avian embryos. *Curr. Opin. Genet. Dev.* **9,** 411–416.

Barnes, J. D., Crosby, J. L., Jones, C. M., Wright, C. V., and Hogan, B. L. (1994). Embryonic expression of Lim-1, the mouse homolog of Xenopus Xlim-1, suggests a role in lateral mesoderm differentiation and neurogenesis. *Dev. Biol.* **161,** 168–178.

Beddington, S. P. (1981). An autoradiographic analysis of the potency of embryonic ectoderm in the 8th day postimplantation mouse embryo. *J. Embryol. Exp. Morphol.* **64,** 87–104.

Beddington, R. S. (1994). Induction of a second neural axis by the mouse node. *Development (Cambridge, UK)* **120,** 613–620.

Beddington, R. S., and Robertson, E. J. (1989). An assessment of the developmental potential of embryonic stem cells in the midgestation mouse embryo. *Development (Cambridge, UK)* **105,** 733–737.

Beddington, R. S., and Robertson, E. J. (1999). Axis development and early asymmetry in mammals. *Cell (Cambridge, Mass.)* **96,** 195–209.

Behringer, R. R., Wakamiya, M., Tsang, T. E., and Tam, P. P. L. (2000). A flattened mouse embryo: Leveling the playing field. *Genesis* **28,** 23–30.

Belo, J. A., Bouwmeester, T., Leyns, L., Kertesz, N., Gallo, M., Follettie, M., and De Robertis, E. M. (1997). Cerberus-like is a secreted factor with neutralizing activity expressed in the anterior primitive endoderm of the mouse gastrula. *Mech. Dev.* **68,** 45–57.

Belo, J. A., Bachiller, D., Agius, E., Kemp, C., Borges, A. C., Marques, S., Piccolo, S., and De Robertis, E. M. (2000). Cerberus-like is a secreted BMP and nodal antagonist not essential for mouse development. *Genesis* **26,** 265–270.

Biben, C., Stanley, E., Fabri, L., Kotecha, S., Rhinn, M., Drinkwater, C., Lah, M., Wang, C. C., Nash, A., Hilton, D., Ang, S.-L., Mohun, T., and Harvey, R. P. (1998). Murine cerberus homologu mCer-1: A candidate anterior patterning molecule. *Dev. Biol.* **194,** 135–151.

Bielinska, M., Narita, N., and Wilson, D. B. (1999). Distinct roles for visceral endoderm during embryonic mouse development. *Int. J. Dev. Biol.* **43,** 183–205.

Bisgrove, B. W., Essner, J. J., and Yost, H. J. (1999). Regulation of midline development by antagonism of lefty and nodal signaling. *Development (Cambridge, UK)* **126,** 3253–3262.

Blum, M., Gaunt, S. J., Cho, K. W., Steinbeisser, H., Blumberg, B., Bittner, D., and De Robertis, E. M. (1992). Gastrulation in the mouse: The role of the homeobox gene goosecoid. *Cell (Cambridge, Mass.)* **69,** 1097–1106.

Camus, A., and Tam, P. P. (1999). The organizer of the gastrulating mouse embryo. *Curr. Top. Dev. Biol.* **45,** 117–153.

Catala, M., Teillet, M. A., and Le Douarin, N. M. (1995). Organization and development of the tail bud analyzed with the quail- chick chimaera system. *Mech. Dev.* **51,** 51–65.

Catala, M., Teillet, M. A., De Robertis, E. M., and Le Douarin, M. L. (1996). A spinal cord fate map in the avian embryo: While regressing, Hensen's node lays down the notochord and floor plate thus joining the spinal cord lateral walls. *Development (Cambridge, UK)* **122,** 2599–2610.

Cheng, A. M., Thisse, B., Thisse, C., and Wright, C. V. (2000). The lefty-related factor Xatv acts as a feedback inhibitor of Nodal signaling in mesoderm induction and L-R axis development in Xenopus. *Development (Cambridge, UK)* **127,** 1049–1061.

Chiang, C., Litingtung, Y., Lee, E., Young, K. E., Corden, J. L., Westphal, H. and Beachy, P. A. (1996). Cyclopia and defective axial patterning in mice lacking Sonic hedgehog gene function. *Nature (London)* **383,** 407–413.

Collignon, J., Varlet, I., and Robertson, E. J. (1996). Relationship between asymmetric nodal expression and the direction of embryonic turning. *Nature (London)* **381,** 155–158.

Conlon, F. L., Lyons, K. M., Takaesu, N., Barth, K. S., Kispert, A., Herrmann, B., and Robertson, E. J. (1994). A primary requirement for nodal in the formation and maintenance of the primitive streak in the mouse. *Development (Cambridge, UK)* **120,** 1919–28.

Coucouvanis, E., and Martin, G. R. (1999). BMP signaling plays a role in visceral endoderm differentiation and cavitation in the early mouse embryo. *Development (Cambridge, UK)* **126,** 535–546.

Dattani, M. T., Martinez-Barbera, J. P., Thomas, P. Q., Brickman, J. M., Gupta, R., Martensson, I. L., Toresson, H., Fox, M., Wales, J. K., Hindmarsh, P. C., Krauss, S., Beddington, R. S., and Robinson, I. C. (1998). Mutations in the homeobox gene HESX1/Hesx1 associated with septo-optic dysplasia in human and mouse. *Nat. Genet.* **19,** 125–133.

Dias, M. S., and Schoenwolf, G. C. (1990). Formation of ectopic neurepithelium in chick blastoderms: Age-related capacities for induction and self-differentiation following transplantation of quail Hensen's nodes. *Anat. Rec.* **228,** 437–448.

Ding, J., Yang, L., Yan, Y. T., Chen, A., Desai, N., Wynshaw-Boris, A., and Shen, M. M. (1998). Cripto is required for correct orientation of the anterior-posterior axis in the mouse. *Nature (London)* **395,** 702–707.

Dohlman, H. G., and Thorners, J. (1997). RGS proteins and signaling by heterotrimeric G proteins. *J. Biol. Chem.* **272,** 3871–3874.

Dufort, D., Schwartz, L., Harpal, K., and Rossant, J. (1998). The transcription factor HNF3beta is required in visceral endoderm for normal primitive streak morphogenesis. *Development (Cambridge, UK)* **125,** 3015–3025.

Echelard, Y., Epstein, D. J., St.-Jacques, B., Shen, L., Mohler, J., McMahon, J. A., and McMahon, A. P. (1993). Sonic hedgehog, a member of putative signaling molecules is implicated in the regulation of CNS polarity. *Cell* **75,** 1417–1430.

Ericson, J., Morton, S., Kawakami, A., Roelink, H., and Jessell, T. M. (1996). Two critical periods of Sonic Hedgehog signaling required for the specification of motor neuron identity. *Cell (Cambridge, Mass.)* **87,** 661–673.

Filosa, S., Rivera-Perez, J. A., Gomez, A. P., Gansmuller, A., Sasaki, H., Behringer, R. R., and Ang, S. L. (1997). Goosecoid and HNF-3beta genetically interact to regulate neural tube patterning during mouse embryogenesis. *Development (Cambridge, UK)* **124,** 2843–2854.

Foley, A. C., Skromne, I., and Stern, C. D. (2000). Reconciling different models of forebrain induction and patterning: a dual role for the hypoblast. *Development (Cambridge, UK)* **127,** 3839–3854.

Gasca, S., Hill, D. P., Klingensmith, J. and Rossant, J. (1995). Characterization of a gene trap insertion into a novel gene, *cordon-bleu* expressed in axial structures of the gastrulating mouse embryo. *Dev. Genet.* **17,** 141–151.

Gerhart, J., Danilchik, M., Doniach, T., Roberts, S., Rowning, B. and Stewart, R. (1989). Cortical rotation of the Xenopus egg: Consequences for the anteroposterior pattern of embryonic dorsal development. *Development (Cambridge, UK)* **107,** 37–51.

Glinka, A., Wu, W., Delius, H., Monaghan, A. P., Blumenstock, C., and Niehrs, C. (1998). Dickkopf-1 is a member of a new family of secreted proteins and functions in head induction. *Nature (London)* **391,** 357–362.

Gritsman, K., Zhang, J., Cheng, S., Heckscher, E., Talbot, W. S., and Schier, A. F. (1999). The EGF-CFC protein one-eyed pinhead is essential for nodal signaling. *Cell (Cambridge, Mass.)* **97,** 121–132.

Gu, Z., Nomura, M., Simpson, B. B., Lei, H., Feijen, A., van den Eijnden-van Raaij, J., Donahoe, P. K., and Li, E. (1998). The type I activin receptor ActRIB is required for egg cylinder organization and gastrulation in the mouse. *Genes Dev.* **12,** 844–857.

Hamburger, V. (1988). "The Heritage of Experimental Embryology: Hans Spemann and the Organizer." Oxford University Press, New York.

Hansen, C. S., Marion, C. D., Steele, K., George, S., and Smith, W. C. (1997). Direct neural induction and selective inhibition of mesoderm and epidermis inducers by Xnr3. *Development (Cambridge, UK)* **124,** 483–492.

Harland, R., and Gerhart, J. (1997). Formation and function of Spemann's organizer. *Annu. Rev. Cell. Dev. Biol.* **13,** 611–667.

Hermesz, E., Mackem, S., and Mahon, K. A. (1996). Rpx: A novel anterior-restricted homeobox gene progressively activated in the prechordal plate, anterior neural plate and Rathke's pouch of the mouse embryo. *Development (Cambridge, UK)* **122,** 41–52.

Heyer, J., Escalante-Alcalde, D., Lia, M., Boettinger, E., Edelmann, W., Stewart, C. L., and Kucherlapati, R. (1999). Postgastrulation Smad2-deficient embryos show defects in embryo turning and anterior morphogenesis. *Proc. Natl. Acad. Sci. U.S.A.* **96,** 12595–12600.

Hsu, D. R., Economides, A. N., Wang, X., Eimon, P. M., and Harland, R. M. (1998). The Xenopus dorsalizing factor Gremlin identifies a novel family of secreted proteins that antagonize BMP activities. *Mol. Cell.* **1,** 673–683.

Ishimaru, Y., Yoshioka, H., Tao, H., Thisse, B., Thisse, C., C, Wright, V. E., Hamada, H., Ohuchi, H., and Noji, S. (2000). Asymmetric expression of antivin/lefty1 in the early chick embryo. *Mech. Dev.* **90,** 115–118.

Jin, O., Harpal K., Ang, S.-L. and Rossant, J. (2000). *Otx2* and *HNF3β* genetically interact in anterior midline patterning. *Int. J. Dev. Biol.,* (in press).

Joubin, K., and Stern, C. D. (1999). Molecular interactions continuously define the organizer during the cell movements of gastrulation. *Cell (Cambridge, Mass.)* **98,** 559–571.

Kaestner, K. H., Lee, K. H., Schlondorff, J., Hiemisch, H., Monaghan, A. P., and Schutz, G. (1993). Six members of the mouse forkhead gene family are developmentally regulated. *Proc. Natl. Acad. Sci. U.S.A.* **90,** 7628–31.

Kimura, C., Yoshinaga, K., Tian, E., Suzuki, M., Aizawa, S., and Matsuo, I. (2000). Visceral endoderm mediates forebrain development by suppressing posteriorizing signals. *Dev. Biol.* **225,** 304–321.

Kinder, S. J., Tsang, T. E., Quinlan, G. A., Hadjantonakis, A. K., Nagy, A., and Tam, P. P. (1999). The orderly allocation of mesodermal cells to the extraembryonic structures and the anteroposterior axis during gastrulation of the mouse embryo. *Development (Cambridge, UK)* **126,** 4691–4701.

Klingensmith, J., and Nusse, R. (1994). Signaling by wingless in Drosophila. *Dev. Biol.* **166,** 396–414.

Klingensmith, J., Ang, S.-L., Bachiller, D., and Rossant, J. (1999). Neural induction and patterning in the mouse in the absence of the node and its derivatives. *Dev. Biol.* **216,** 535–549.

Klingensmith, J. *et al.* (2001). In press.

Knezevic, V., De Santo, R., and Mackem, S. (1998). Continuing organizer function during chick tail development. *Development (Cambridge, UK)* **125,** 1791–1801.

Knoetgen, H., Viebahn, C., and Kessel, M. (1999). Head induction in the chick by primitive endoderm of mammalian, but not avian origin. *Development (Cambridge, UK)* **126,** 815–825.

Knoetgen, H., Teichmann, U., Wittler, L., Viebahn, C., and Kessel, M. (2000). Anterior neural induction by nodes from rabbits and mice. *Dev. Biol.* **225,** 370–380.

Kosaki, K., Bassi, M. T., Kosaki, R., Lewin, M., Belmont, J., Schauer, G., and Casey, B. (1999). Characterization and mutation analysis of human LEFTY A and LEFTY B, homologus of murine genes implicated in left-right axis development. *Am. J. Hum. Genet.* **64,** 712–721.

Koshida, S., Shinya, M., Mizuno, T., Kuroiwa, A., and Takeda, H. (1998). Initial anteroposterior pattern of the zebrafish central nervous system is determined by differential competence of the epiblast. *Development (Cambridge, UK)* **125,** 1957–1966.

Kothapalli, R., Buyuksal, I., Wu, S. Q., Chegini, N., and Tabibzadeh, S. (1997). Detection of ebaf, a novel human gene of the transforming growth factor beta superfamily association of gene expression with endometrial bleeding. *J. Clin. Invest.* **99,** 2342–2350.

Lawson, K. A., and Pedersen, R. A. (1987). Cell fate, morphogenetic movement and population kinetics of embryonic endoderm at the time of germ layer formation in the mouse. *Development (Cambridge, UK)* **101,** 627–652.

Lawson, K. A., Meneses, J. J., and Pedersen, R. A. (1986). Cell fate and cell lineage in the endoderm of the presomite mouse embryo, studied with an intracellular tracer. *Dev. Biol.* **115,** 325–339.

Lawson, K. A., Meneses, J. J., and Pedersen, R. A. (1991). Clonal analysis of epiblast fate during germ layer formation in the mouse embryo. *Development (Cambridge, UK)* **113,** 891–911.

Lemaire, L., Roeser, T., Izpisua-Belmonte, J. C., and Kessel, M. (1997). Segregating expression domains of two goosecoid genes during the transition from gastrulation to neurulation in chick embryos. *Development (Cambridge, UK)* **124,** 1443–1452.

Lemaire, P., and Kodjabachian, L. (1996). The vertebrate organizer: Structure and molecules. *Trends Genet.* **12,** 525–531.

Leyns, L., Bouwmeester, T., Kim, S. H., Piccolo, S., and De Robertis, E. M. (1997). Frzb-1 is a secreted antagonist of Wnt signaling expressed in the Spemann organizer. *Cell (Cambridge, Mass.)* **88,** 747–756.

Liu, P., Wakamiya, M., Shea, M. J., Albrecht, U., Behringer, R. R., and Bradley, A. (1999). Requirement for Wnt3 in vertebrate axis formation. *Nat. Genet.* **22,** 361–365.

Martinez-Barbera, J. P., Rodriguez, T. A., and Beddington, R. S. (2000a). The homeobox gene hesx1 is required in the anterior neural ectoderm for normal forebrain formation. *Dev. Biol.* **223,** 422–430.

Martinez-Barbera, J. P., Clements, M., Thomas, P., Rodriguez, T., Meloy, D., Kioussis, D., and Beddington, R. S. (2000b). The homeobox gene Hex is required in definitive endodermal tissues for normal forebrain, liver and thyroid formation. *Development (Cambridge, UK)* **127,** 2433–2445.

Massague, J. (1998). TGF-beta signal transduction. *Annu. Rev. Biochem.* **67,** 753–791.

Matsuo, I., Kuratani, S., Kimura, C., Takeda, N., and Aizawa, S. (1995). Mouse Otx2 functions in the formation and patterning of rostral head. *Genes Dev.* **9,** 2646–2658.

Matzuk, M. M., Kumar, T. R., Shou, W., Coerver, K. A., Lau, A. L., Behringer, R. R., and Finegold, M. J. (1996). Transgenic models to study the roles of inhibins and activins in reproduction, oncogenesis, and development. *Recent Prog. Horm. Res.* **51,** 123–154.

McMahon, A. P., and Moon, R. T. (1989). Ectopic expression of the proto-oncogene int-1 in Xenopus embryos leads to duplication of the embryonic axis. *Cell (Cambridge, Mass.)* **58,** 1075–1084.

McMahon, J. A., Takada, S., Zimmerman, L. B., Fan, C. M., Harland, R. M., and McMahon, A. P. (1998). Noggin-mediated antagonism of BMP signaling is required for growth and patterning of the neural tube and somite. *Genes Dev.* **12,** 1438–52.

Meno, C., Saijoh, Y., Fujii, H., Ikeda, M., Yokoyama, T., Yokoyama, M., Toyoda, Y., and Hamada, H. (1996). Left-right asymmetric expression of the TGF beta-family member lefty in mouse embryos. *Nature (London)* **381,** 151–155.

Meno, C., Gritsman, K., Ohishi, S., Ohfuji, Y., Heckscher, E., Mochida, K., Shimono, A., Kondoh, H., Talbot, W. S., Robertson, E. J., Schier, A. F., and Hamada, H. (1999). Mouse Lefty2 and zebrafish antivin are feedback inhibitors of nodal signaling during vertebrate gastrulation. *Mol. Cell.* **4,** 287–298.

Monaghan, A. P., Kaestner, K. H., Grau, E., and Schutz, G. (1993). Postimplantation expression patterns indicate a role for the mouse forkhead/HNF-3 alpha, beta and gamma genes in determination of the definitive endoderm, chordamesoderm and neuroectoderm. *Development (Cambridge, UK)* **119,** 567–578.

Moon, R. T., and Kimelman, D. (1998). From cortical rotation to organizer gene expression: Toward a molecular explanantion of axis specification in Xenopus. *BioEssays* **20,** 536–545.

Nagy, A., Gocza, E., Diaz, E. M., Prideaux, V. R., Ivanyi, E., Markkula, M., and Rossant, J. (1990). Embryonic stem cells alone are able to support fetal development in the mouse. *Development (Cambridge, UK)* **110,** 815–821.

Nagy, A., Rossant, J., Nagy, R., Abramow-Newerly, W., and Roder, J. C. (1993). Derivation of completely cell culture-derived mice from early-passage embryonic stem cells. *Proc. Natl. Acad. Sci. U.S.A.* **90,** 8424–8428.

Nakano, T., Murata, T., Matsuo, I., and Aizawa, S. (2000). OTX2 directly interacts with LIM1 and HNF-3beta. *Biochem. Biophys. Res. Commun.* **267,** 64–70.

Niehrs, C. (1999). Head in the WNT: The molecular nature of Spemann's head organizer. *Trends Genet.* **15,** 314–319.

Nonaka, S., Tanaka, Y., Okada, Y., Takeda, S., Harada, A., Kanai, Y., Kido, M., and Hirokawa, N. (1998). Randomization of left-right asymmetry due to loss of nodal cilia generating leftward flow of extraembryonic fluid in mice lacking KIF3B motor protein. *Cell (Cambridge, Mass.)* **95,** 829–837; erratum: Ibid. **99**(1); 117 (1999).

Oulad-Abdelghani, M., Chazaud, C., Bouillet, P., Mattei, M. G., Dolle, P., and Chambon, P. (1998). Stra3/lefty, a retinoic acid-inducible novel member of the transforming growth factor-beta superfamily. *Int. J. Dev. Biol.* **42,** 23–32.

Parameswaran, M., and Tam, P. P. (1995). Regionalization of cell fate and morphogenetic movement of the mesoderm during mouse gastrulation. *Dev. Genet.* **17,** 16–28.

Pearce, J. J., Penny, G., and Rossant, J. (1999). A mouse cerberus/Dan-related gene family. *Dev. Biol.* **209,** 98–110.

Perea-Gomez, A., Shawlot, W., Sasaki, H., Behringer, R. R., and Ang, S.-L. (1999). HNF3beta and Lim1 interact in the visceral endoderm to regulate primitive streak formation and anterior-posterior polarity in the mouse embryo. *Development (Cambridge, UK)* **126,** 4499–4511.

Perea-Gomez, A., Lawson, K. A., Rhinn M., Zakin, L., Brûlet, P., Mazan, S., and Ang, S.-L. (2001). Otx2 is required for visceral endoderm movement and for the restriction of posterior signals in the epiblast of the mouse embryo. *Development (Cambridge, UK)* **128,** 753–765.

Perea-Gomez, A., Rhinn, M. and Ang, S.-L. (2001).Role of the anterior visceral endoderm in restricting posterior signals in the mouse embryo. *Int. J. Dev. Biol.* **45,** 311–320.

Piccolo, S., Agius, E., Leyns, L., Bhattacharyya, S., Grunz, H., Bouw-meester, T., and De Robertis, E. M. (1999). The head inducer Cerberus is a multifunctional antagonist of Nodal, BMP and Wnt signals. *Nature (London)* **397,** 707–710.

Popperl, H., Schmidt, C., Wilson, V., Hume, C. R., Dodd, J., Krumlauf, R., and Beddington, R. S. (1997). Misexpression of Cwnt8C in the mouse induces an ectopic embryonic axis and causes a truncation of the anterior neuroectoderm. *Development (Cambridge, UK)* **124,** 2997–3005.

Quinlan, G. A., Williams, E. A., Tan, S. S., and Tam, P. P. (1995). Neuroectodermal fate of epiblast cells in the distal region of the mouse egg cylinder: Implication for body plan organization during early embryogenesis. *Development (Cambridge, UK)* **121,** 87–98.

Rhinn, M., Dierich, A., Shawlot, W., Behringer, R. R., Le Meur, M., and Ang, S. L. (1998). Sequential roles for Otx2 in visceral endoderm and neuroectoderm for forebrain and midbrain induction and specification. *Development (Cambridge, UK)* **125,** 845–856.

Rossant, J., and Spence, A. (1998). Chimeras and mosaics in mouse mutant analysis. *Trends Genet.* **14,** 358–363.

Sasai, Y., and De Robertis, E. M. (1997). Ectodermal patterning in vertebrate embryos. *Dev. Biol.* **182,** 5–20.

Sasaki, H., and Hogan, B. L. (1993). Differential expression of multiple fork head related genes during gastrulation and axial pattern formation in the mouse embryo. *Development (Cambridge, UK)* **118,** 47–59.

Schier, A. F., and Shen, M. M. (2000). Nodal signalling in vertebrate development. *Nature (London)* **403,** 385–389.

Seleiro, E. A., Connolly, D. J., and Cooke, J. (1996). Early developmental expression and experimental axis determination by the chicken Vg1 gene. *Curr. Biol.* **6,** 1476–1486.

Selleck, M. A., and Stern, C. D. (1991). Fate mapping and cell lineage analysis of Hensen's node in the chick embryo. *Development (Cambridge, UK)* **112,** 615–626.

Shah, S. B., Skromne, I., Hume, C. R., Kessler, D. S., Lee, K. J., Stern, C. D. and Dodd, J. (1997). Misexpression of chick Vg1 in the marginal zone induces primitive streak formation. *Development (Cambridge, UK)* **124,** 5127–5138.

Shawlot, W., and Behringer, R. R. (1995). Requirement for Lim1 in head-organizer function. *Nature (London)* **374,** 425–430.

Shawlot, W., Deng, J. M., and Behringer, R. R. (1998). Expression of the mouse cerberus-related gene, Cerr1, suggests a role in anterior neural induction and somitogenesis. *Proc. Natl. Acad. Sci. U.S.A.* **95,** 6198–6203.

Shawlot, W., Wakamiya, M., Kwan, K. M., Kania, A., Jessell, T. M., and Behringer, R. R. (1999). Lim1 is required in both primitive streak-derived tissues and visceral endoderm for head formation in the mouse. *Development (Cambridge, UK)* **126,** 4925–4932.

Shawlot, W., Deng, J. M., Wakamiya, M., and Behringer, R. R. (2000). The cerberus-related gene, Cerr1, is not essential for mouse head formation. *Genesis* **26,** 253–258.

Shen, M. M., Wang, H., and Leder, P. (1997). A differential display strategy identifies Cryptic, a novel EGF-related gene expressed in the axial and lateral mesoderm during mouse gastrulation. *Development (Cambridge, UK)* **124,** 429–442.

Shih, J., and Fraser, S. E. (1996). Characterizing the zebrafish organizer: Microsurgical analysis at the early-shield stage. *Development (Cambridge, UK)* **122,** 1313–1322.

Shimono, A., and Behringer, R. R. (1999). Isolation of novel cDNAs by subtractions between the anterior mesendoderm of single mouse gastrula stage embryos. *Dev. Biol.* **209,** 369–380.

Shimono, A., Okuda, T., and Kondoh, H. (1999). N-myc-dependent repression of ndr1, a gene identified by direct subtraction of whole mouse embryo cDNAs between wild type and N-myc mutant. *Mech. Dev.* **83,** 39–52.

Simeone, A., Acampora, D., Mallamaci, A., Stornaiuolo, A., D'Apice, M. R., Nigro, V., and Boncinelli, E. (1993). A vertebrate gene related to orthodenticle contains a homeodomain of the bicoid class and demarcates anterior neuroectoderm in the gastrulating mouse embryo. *Embo J.* **12,** 2735–2747.

Smith, W. C., and Harland, R. M. (1991). Injected Xwnt-8 RNA acts early in Xenopus embryos to promote formation of a vegetal dorsalizing center. *Cell (Cambridge, Mass.)* **67,** 753–765.

Spemann, H. (1927). Neue arbeiten über organisatoren in der tierischen entwicklung. *Naturwissenschaften* **15,** 946–951.

Spemann, H. (1938). "Embryonic Development and Induction." Yale University Press, New Haven, CT.

Spemann, H., and Mangold, H. (1924). Über induktion von embryonalanlagen durch implantation artfremder organisatoren. *Wilhelm Roux' Arch. Entwicklungsmech Org.* **100,** 599–638.

Stanley, E. G., Biben, C., Allison, J., Hartley, L., Wicks, I. P., Campbell, I. K., McKinley, M., Barnett, L., Koentgen, F., Robb, L., and Harvey, R. P. (2000). Targeted insertion of a lacZ reporter gene into the mouse Cer1 locus reveals complex and dynamic expression during embryogenesis. *Genesis* **26,** 259–264.

Storey, K. G., Crossley, J. M., De Robertis, E. M., Norris, W. E., and Stern, C. D. (1992). Neural induction and regionalisation in the chick embryo. *Development (Cambridge, UK)* **114,** 729–741.

Storey, K. G., Selleck, M. A., and Stern, C. D. (1995). Neural induction and regionalization by different subpopulations of cells in Hensen's node. *Development (Cambridge, UK)* **121,** 417–428.

Streit, A., and Stern, C. D. (1999). Neural induction. A bird's eye view. *Trends Genet.* **15,** 20–24.

Sulik, K., Dehart, D. B., Iangaki, T., Carson, J. L., Vrablic, T., Gesteland, K., and Schoenwolf, G. C. (1994). Morphogenesis of the murine node and notochordal plate. *Dev. Dyn.* **201,** 260–278.

Sussman, D. J., Klingensmith, J., Salinas, P., Adams, P. S., Nusse, R., and Perrimon, N. (1994). Isolation and characterization of a mouse homolog of the Drosophila segment polarity gene dishevelled. *Dev. Biol.* **166,** 73–86.

Tam, P. P., and Beddington, R. S. (1992). Establishment and organization of germ layers in the gastrulating mouse embryo. *Ciba Found Symp.* **165,** 27–41; discussion: pp. 42–49.

Tam, P. P., and Behringer, R. R. (1997). Mouse gastrulation: The formation of a mammalian body plan. *Mech. Dev.* **68,** 3–25.

Tam, P. P., Steiner, K. A., Zhou, S. X., and Quinlan, G. A. (1997). Lineage and functional analyses of the mouse organizer. *Cold Spring Harbor Symp. Quant. Biol.* **62,** 135–144.

Tam, P. P., and Steiner, K. A. (1999). Anterior patterning by synergistive activity of the early gastrula organizer and the anterior germ layer tissues of the mouse embryo. *Development (Cambridge, UK)* **126,** 5171–5179.

Tanabe, Y., and Jessell, T. M. (1996). Diversity and pattern in the developing spinal cord. *Science* **274,** 1115–1123; erratum: Ibid. 1997 **276,** 21 (1997).

Thisse, C., and Thisse, B. (1999). Antivin, a novel and divergent member of the TGFbeta superfamily, negatively regulates mesoderm induction. *Development (Cambridge, UK)* **126**, 229–240.

Thomas, P. Q., and Beddington, R. S. (1996). Anterior primitive endoderm may be responsible for patterning the anterior neural plate in the mouse embryo. *Curr. Biol.* **6**, 1487–1496.

Thomas, P. Q., Brown, A., and Beddington, R. S. (1998). Hex: A homeobox gene revealing peri-implantation asymmetry in the mouse embryo and an early transient marker of endothelial cell precursors. *Development (Cambridge, UK)* **125**, 85–94.

Varlet, I., Collignon, J., and Robertson, E. J. (1997). nodal expression in the primitive endoderm is required for specification of the anterior axis during mouse gastrulation. *Development (Cambridge, UK)* **124**, 1033–1044.

Viebahn, C. (2001). Hensen's node. *Genesis* **29**, 96–103.

Viebahn, C., Mayer, B., and Hrabe de Angelis, M. (1995). Signs of the principle body axes prior to primitive streak formation in the rabbit embryo. *Anat. Embryol.* **192**, 159–169.

Vodicka, M. A., and Gerhart, J. C. (1995). Blastomere derivation and domains of gene expression in the Spemann Organizer of *Xenopus laevis*. *Development (Cambridge, UK)* **121**, 3505–3518.

Waldrip, W. R., Bikoff, E. K., Hoodless, P. A., Wrana, J. L., and Robertson, E. J. (1998). Smad2 signaling in extraembryonic tissues determines anterior-posterior polarity of the early mouse embryo. *Cell (Cambridge, Mass.)* **92**, 797–808.

Wang, S., Krinks, M., and Moos, M., Jr. (1997). Frzb-1, an antagonist of Wnt-1 and Wnt-8, does not block signaling by Wnts -3A, -5A, or -11. *Biochem. Biophys. Res. Commun.* **236**, 502–504.

Weber, R. J., Pedersen, R. A., Wianny, F., Evans, M. J., and Zernicka-Goetz, M. (1999). Polarity of the mouse embryo is anticipated before implantation. *Development (Cambridge, UK)* **126**, 5591–5598.

Weinstein, D. C., Ruiz i Altaba, A., Chen, W. S., Hoodless, P., Prezioso, V. R., Jessell, T. M., and Darnell, J. E., Jr. (1994). The winged-helix transcription factor HNF-3 beta is required for notochord development in the mouse embryo. *Cell (Cambridge, Mass.)* **78**, 575–588.

Wilkinson, D. G., Bhatt, S., and Hermann, B. G. (1990). Expression pattern of the mouse *T* gene and its role in mesoderm formation. *Nature (London)* **343**, 657–659.

Wilson, P. A., and Hemmati-Brivanlou, A. (1997). Vertebrate neural induction: Inducers, inhibitors, and a new synthesis. *Neuron* **18**, 699–710.

Wilson, V., and Beddington, R. S. (1996). Cell fate and morphogenetic movement in the late mouse primitive streak. *Mech. Dev.* **55**, 79–89.

Yamaguchi, T. P., Harpal, K., Henkemeyer, M. and Rossant, J. (1994) *fgfr-1* is required for embryonic growth and mesodermal patterning during mouse gastrulation. *Genes Dev.* **8**, 3032–3044.

Zeng, L., Fagotto, F., Zhang, T., Hsu, W., Vasicek, T. J., Perry, W. L., 3rd, Lee, J. J., Tilghman, S. M., Gumbiner, B. M., and Costantini, F. (1997). The mouse Fused locus encodes Axin, an inhibitor of the Wnt signaling pathway that regulates embryonic axis formation. *Cell (Cambridge, Mass.)* **90**, 181–192.

Zhou, X., Sasaki, H., Lowe, L., Hogan, B. L., and Kuehn, M. R. (1993). Nodal is a novel TGF-beta-like gene expressed in the mouse node during gastrulation. *Nature (London)* **361**, 543–547.

Zoltewicz, J. S., and Gerhart, J. C. (1997). The Spemann organizer of Xenopus is patterned along its anteroposterior axis at the earliest gastrula stage. *Dev. Biol.* **192**, 482–491.

4

Left-Right Asymmetry

Hiroshi Hamada

Division of Molecular Biology, Institute for Molecular and Cellular Biology, Osaka University, Osaka 565-0871, Japan

I. Introduction

II. Morphological Left-Right Asymmetries

III. Genetic/Molecular Pathway Governing
Left-Right Determination

IV. Molecular Readout of the First Asymmetry

V. Role of the Midline

VII. Readout of Left-Right Asymmetry in
Later Development

VII. Miscellaneous Mutations/Gene Factors

VIII. Diversity among Vertebrates

IX. Future Challenges

References

I. Introduction

Establishing three body axes—anterior-posterior (AP), dorsal-ventral (DV), and left-right axes—is fundamental to the formation of the body plan. The study of left-right (L-R) asymmetry has been vitalized by the identification of several asymmetrically expressed genes in various vertebrates, which has made molecular/genetic analysis possible for the first time. Expression studies as well as the reverse genetics strategy in mouse and surgical manipulations in chick and frog embryos have defined the specific roles of these genes in the L-R pathway. Analysis of mouse mutants and human genetic disorders with laterality defects has also revealed new genes that are involved in this process.

A number of theoretical models have been proposed to explain how L-R symmetry is broken (Brown and Wolpert, 1990; Brown *et al.,* 1991; Yost, 1991). The most attractive one among those is a so-called "F molecule model" proposed by Brown and Wolpert (1990). The model has three components: (1) *conversion,* in which molecular asymmetry is transmitted to the cellular level; (2) *random generation of asymmetry,* which can be biased by *conversion* to produce asymmetry; and (3) *interpretation,* in which individual organs/cells recognize asymmetric information. *Conversion,* which can be regarded as the initial break in symmetry, involves three hypothetical molecules (Fig. 1) X, Y, and F, a handed molecule. First, cells on both sides of the midline become polarized with respect to the midline (by a substance produced in the midline). As a result, the molecule X present in those cells becomes more concentrated at the side opposite to the midline. Then, the model integrates the F molecule, which has three arms (which look like the letter F), and can be placed in an orientation in cells on both sides such that two of the arms are aligned along the AP and DV axes (thus, the third arm would be automatically aligned along the future L-R axis). One can envisage that the F molecule is a microtubule protein and the Y molecule is cargo carried by a motor along the third arm toward the periphery of the arm. (Indeed, microtubules are known to have minus or plus end

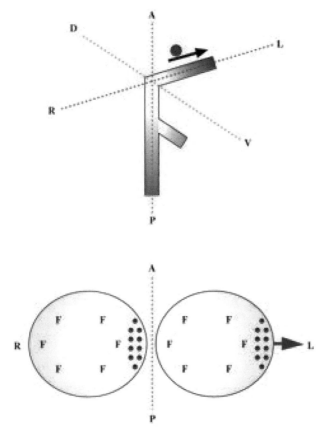

Figure 1 The role of handed molecule (F) in the specification of L-R asymmetry. Three types of hypothetical molecules are found in the cells bordering the midline of the embryo. The polarizing activity emitted by the midline structure repels the yellow molecules to the intracellular sites farthest away from the midline. The localization of the blue molecules, in contrast, is driven by the handed molecule (F) to one side of the cells, which in this model corresponds to the left pole of the cells. As a result, colocalization of the yellow and blue molecules is only found in the cells on the left side of the body axis, leading to the handedness of the body plan. (Adapted from Brown and Wolpert, 1990; Brown *et al.*, 1991.)

polarity, and transport is directional, either anterograde or retrograde.) This would create a distinct difference on both sides; on the left side, molecules X and Y are concentrated at opposite sides of the cells; on the right, X and Y are together on one side. There are numerous examples in which a readout depends on the association of two molecules (for example, X is a precursor of a signaling molecule while Y is its maturation enzyme). The end result is that cells on the right side—but not cells on the left side—can execute a certain activity.

Observations that loss of conversion results not in symmetry but in random asymmetry (as seen in *iv/iv* mutant mice) led researchers to propose a separate mechanism for *random generation of asymmetry.* They speculated that random generation of asymmetry may involve a gradient of a morphogen produced by reaction-diffusion (Turing, 1952; Kaufman *et al.*, 1978). Importantly, this mechanism itself is

random so that a slight initial asymmetry (such as the one generated by conversion) can bias the gradient to one side rather than the other. In *interpretation,* cells on the left or right side of the embryo receive positional information and adopt their fate. This three-step model is still an inspiring model even at the present days as detailed molecular mechanisms are revealed.

In our current understanding, the process by which L-R asymmetry is established can be divided into four steps (Fig. 2): (1) breaking of L-R symmetry in/near the node, (2) transfer of L-R biased signals from the node to the lateral plate, (3) L-R asymmetric expression of signaling molecules such as Nodal and Lefty in the lateral plate, and (4) L-R asymmetric morphogenesis of visceral organs induced by these signaling molecules (for reviews, see King and Brown, 1997; Varlet and Robertson, 1997; Levin and Mercola, 1998a; Harvey, 1998; Beddington and Robertson, 1999; Burdine and Schier, 2000). L-R asymmetry provides a unique opportunity to study the complete cellular and molecular mechanisms of asymmetry generation.

II. Morphological Left-Right Asymmetries

The mammalian body is overtly laterally symmetrical about the midline. However, virtually all organs in viscera are asymmetrical in shape and/or position (Fujinaga, 1997). Each organ must be placed in the correct position in order to utilize the limited space in a body cavity yet keep connections between each other in the most functional way.

What kinds of morphological asymmetries are generated during development? Several different mechanisms come into play depending on the type of organs (Fig. 1, step 4). The first mechanism is achieved by directional looping/coiling/turning. Heart looping from an initially symmetrical cardiac tube is the first overt morphological asymmetry that takes place during mammalian development (see Chapter 19). The digestive tract is generated from the foregut, midgut, and hindgut, through complex looping and rotating processes. The stomach, for example, derives from a part of the foregut. The primitive stomach can be seen as a fusiform dilation of the foregut, which will eventually be displaced to the left probably due to its own and surrounding differential growth changes. Rapid growth of the midgut and dorsal mesentery results in looping, generating a midgut loop (a protrusion called a "physiological hernia"). This midgut loop then rotates in such a direction that the future colon crosses the duodenum on its ventral side. The rotation of midgut loop now determines the overall position of the colon. The spleen is another asymmetric organ. This organ derives from the mesenchyme of the dorsal mesogastrium. The primitive dorsal mesogastrium is initially located at the midline, but later it loops toward the left. This looping (bending) may take place secondary to the dislocation of the stomach. Therefore,

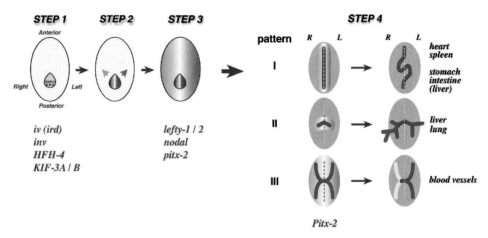

Figure 2 The specification of the L-R axis takes four steps: (1) the break in symmetry, (2) the relay of an asymmetric signal from the node to the lateral plate, (3) activation of asymmetric gene activity (e.g., *nodal* and *lefty2*) in the lateral plate, and (4) situs-specific morphogenesis. The genes that are involved with each of these steps are listed. (Figure composed by Yukio Saijoh.)

the repositioning of the spleen is usually coupled to that of the stomach.

The second mechanism involves differential growth. The best example is the lung. Both sides of the lungs differ in lobation. In mouse, the left lung is monolobed while the right lung has four lobes. Initially, a pair of lung buds of the same size and shape are generated as protrusions of part of the foregut. However, each lung bud will grow and branch differently and finally take on the distinct lobation patterns.

The third mechanism involves differential remodeling. The best example is the vascular system. The vascular system is initially formed symmetrically, but the primitive vascular system undergoes complex remodeling. During this remodeling process, differential regression takes place in some regions so that one side persists while the other side regresses. As a result, some portions of the arteries and veins become asymmetric. For the azygous vein, a vein that collects blood from the chest and returns it to the heart, it is the left side that persists in mouse (in human, the left side remains as the azygous vein while the right side regresses and results in a smaller vein called the hemi-azygous vein). The definitive aortic arch found in adults is derived from the left half of the fourth aortic arch. In the case of the inferior vena cava, on the other hand, it is the right side that persists at the end of development. Therefore, the inferior vena cava in adults is located to the right of the midline. Thus, L-R asymmetry in the vascular system may depend on which side the primordia regresses, a process probably involving programmed cell death.

Another asymmetry, which is apparently coupled with the visceral asymmetry and yet cannot be categorized into any of these mechanisms, is the direction of axial turning (embryonic turning). Slightly after the heart looping, the embryo turns along the AP axis in a clockwise direction (when viewed from the head). As a result, a dorsally flexed position

of the embryo is changed to a ventrally flexed position. Also the clockwise rotation brings the tail, umbilical vein, and allantois of the embryo on the right side and the vitelline vein on the left side. Thus, the direction of the axial turning is frequently used to judge the L-R determination of mouse embryos. Although the mechanistic basis of the axial turning is not clear, it may involve differential mitotic activity on both sides.

Although relative positions and shapes of most of the visceral organs are conserved among vertebrates, there are some apparent differences. The aortic arch is a good example. In mammals, the definitive aortic arch in adults loops to the left because it is derived from the left half of the fourth aortic arch. However, in birds, the right fourth aortic arch is transformed into the definitive aortic arch. In reptiles, the fourth aortic arches of both sides persist and give rise to their characteristic double aortic arch. (Direct comparisons may be inappropriate because development of the heart and aortic arches in these classes are so divergent from the mammalian pattern.) Another example is ovary. In birds, a pair of primordial ovaries and oviducts are formed during development, but after hatching, the left part persists while the right part regresses. There are differences even within the same class of vertebrates. A good example is the lobation part of the lung; a single lobe on the left and four lobes on the right in mouse, two lobes on the left and three lobes on the right in human and horse, three lobes on the left and four lobes on the right in dog and rabbit. The molecular/genetic basis for these differences is currently unknown.

Conceptually, when a gene that functions upstream of a central L-R pathway is impaired, it would result in randomization or complete inversion of the situs. On the other hand, when a gene that functions in a branched pathway is defective, it would result in abnormal positions of some organs but not others (partial inversion). This is exactly what has

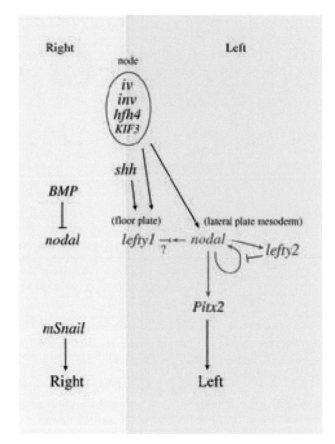

Figure 3 Genetic pathway of L-R determination in the mouse.

been seen in various mouse mutants; the position of visceral organs may be coordinately reversed or independently affected, depending on the type of mutation.

III. Genetic/Molecular Pathway Governing Left-Right Determination

As described above, the whole process of L-R determination can be divided into four steps: (1) break in L-R symmetry in/near the node, (2) transfer of L-R biased signals from the node to the lateral plate, (3) L-R asymmetric expression of signaling molecules in the lateral plate, and (4) L-R asymmetric morphogenesis of visceral organs. Here we describe each step in a developmental sequence, as well as the role of the midline. The genetic pathway generating L-R asymmetry is summarized in Figure 3.

A. Initial Determination: When and Where Is Left-Right Symmetry Broken?

The first overt morphological asymmetry that can be detected during development is the heart looping, which takes place at the early somite stage (6–8 somite stage). Embryonic turning and situs-specific morphogenesis of many other organs follow this. L-R asymmetry in gene expression becomes obvious at the 1–2 somite stage, earlier than the heart looping. Therefore, L-R polarity should be determined sometime before somites appear. Involvement of positional cues from the uterine environment is unlikely because cultured embryos (when dissected at the neural fold stage) can develop the L-R axis normally. It is generally believed that the L-R axis is determined after the AP and DV axes are fixed. Several lines of evidence suggest that the symmetry is first broken, in the mouse, at the neural fold stage (presomite stage) in/near the node. First, Fujinaga and Baden (1991a,b) have employed an embryo culture system, and have determined sensitive stages when rat/mouse embryos develop situs inversu in response to an adrenergic agonist. There is a very narrow time window that is sensitive to the drug, which is the presomite stage. Second, L-R asymmetry in gene expression first becomes apparent in small domains adjacent to the node (*Shh, Fgf8,* and *lefty1* in chick; *nodal, lefty1,* and *Dante* in mouse). However, one cannot rigorously exclude a possibility that L-R symmetry is broken at an earlier stage or may not even involve the asymmetry at the node.

B. Mutations Affecting the Initial Determination (*iv, inv*)

Mutants in which L-R asymmetry is randomized or reversed would be expected to have defects in the initial determination step (conversion in the model by Brown and Wolpert, 1990). In mice, there are several such mutations (Hummel and Chapman, 1959; Yokoyama *et al.,* 1993; Heymer *et al.,* 1997; Melloy *et al.,* 1998), two of the more extensively studied of which are *iv* and *inv.* The *iv* mutation is a classical recessive mutation in which L-R specification is randomized. Thus, in general, about a half of *iv/iv* mice show normal situs, whereas the remaining half show situs inversus. (Although the phenotype of *iv* mutants is often described as this, careful analysis of a large number of *iv* mutant mice has indicated that the phenotype is much more complex; a significant number of *iv* mutants show variable degrees of partial situs inversus instead of total normal situs or total inversed situs: for example, reversed viscera with the inferior vena cava on the normal side; Chapman and Hammel, 1950.) The *legless* mutation is an insertional mutation in which a transgene is integrated into the *iv* gene (*legless* mutants exhibit the truncation of the hindlimb in addition to the randomization of the visceral situs, but the former defect is probably due to disruption of an unrelated gene located nearby). The *inv* mutation is a recessive insertional mutation in which L-R specification is reversed (Yokoyama *et al.*, 1993). This mutant has been identified in a stock of transgenic mouse harboring a tyrosinase transgene (the transgene was inserted within the *inv* gene). Almost all *inv/inv* mice show situs inversus along with kidney defects (the homozygous mice die shortly after birth due to kidney malfunction).

The *iv* encodes Lrd, a protein related to axonemal dynein (Supp *et al.*, 1997). The *inv* gene codes for a protein (called Inversin) that contains ankyrin repeats (Mochizuki *et al.*, 1998; Morgan *et al.*, 1998). Interestingly, *iv* expression is predominantly detected in and near the node at the presomite stage. The original *iv* mutation allele is a point mutation leading to amino acid substitution in the ATP-binding domain of Lrd protein, but a targeted deletion of the *iv* gene results in the identical phenotype (Supp *et al.*, 1999), suggesting that both mutations cause loss of function of Lrd. The *iv* mutation leads to immotile cilia in the node (see Section III.C). Lrd may be a component of a motor that carries

certain molecules essential for the cilial motility. The *inv* is expressed rather ubiquitously at the presomite and early somite stage. However, the subcellular localization and the precise function of the Inversin protein remain unclear.

Other such mutants have been recently generated by gene targeting. Mice deficient in HFH4, KIF3A, or KIF3B all show randomization of the initial L-R axis (Chen *et al.*, 1998; Nonaka *et al.*, 1998; Marszalek *et al.*, 1999; Takeda *et al.*, 1999).

C. Role of the Node Flow

A link between cilia and L-R determination had long been speculated since the discovery of Kartagener syndrome (Kartagener, 1933), a relatively rare genetic disorder in human. This genetic disorder exhibits situs inversion similar to the *iv* mutant mouse accompanied by loss of motility of respiratory cilia and the sperm flagella (Afzelius, 1976). Furthermore, ultrastructural observations confirmed a lack of the dynein arms in the respiratory cilia and in the sperm flagella. Identification of the causative gene of *iv* mutation as a dynein-related gene further strengthened this link. Interestingly, of course, microtubules do show "handedness," reminiscent of the Wolpert and Brown model. However, the connection between ciliary defects and random L-R determination had remained enigmatic.

Recently, an intriguing model was provided by an elegant series of studies (Nonaka *et al.*, 1998; Okada *et al.*, 1999). Cells on the ventral side of the node (node pit cells) are known to possess a monocilium (Sulik *et al.*, 1994; Fig. 4).

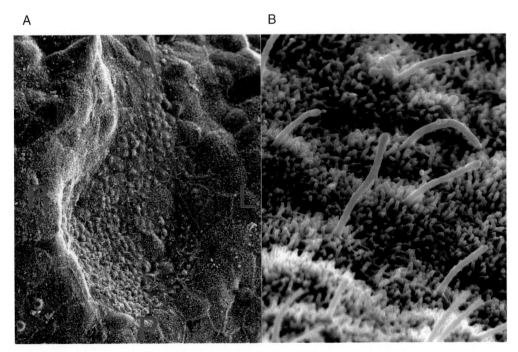

Figure 4 Cells on the ventral side of the node have monocilia. (A) Ventral low-power view of the node. AP and L-R axes are indicated. (B) A high-power view of monocilia of the node. (Photographed by Daisuke Watanabe.)

These cilia, which project into the extraembryonic space, had been previously regarded as nonmotile cilia. (It should be noted that there are two distinct types of cilia; "9+2" cilia containing a ring of 9 peripheral doublets of microtubles plus 2 central microtubles, and "9+0" cilia lacking the central 2 microtubules. The latter are also called *primary* cilia or *solitary* cilia. Cilia on the respiratory epithelium belong to the former, whereas the cilia in the node cells belong to the latter.) However, recent observations have revealed that these cilia are, in fact, motile and rapidly rotate in a clockwise direction (when viewed from the ventral side), which generates a leftward flow of the extraembryonic fluid in the node (Nonaka *et al.*, 1998). This leftward "nodal flow" may transport an unknown factor that acts as a left determinant.

Nodal flow is impaired in several mutant mice that exhibit situs defects, supporting the involvement of the nodal flow in L-R determination. For example, mice deficient in the kinesin superfamily proteins KIF3A or KIF3B show randomized situs. These mice lack the primary cilia in the node and the nodal flow is absent (Nonaka *et al.*, 1998; Takeda *et al.*, 1999; Marszalek *et al.*, 1999). On the other hand, *iv* mutant embryos do possess morphologically normal cilia in the node, but these cilia are immotile (Okada *et al.*, 1999; Supp *et al.*, 1999). As a result, the nodal flow is completely absent in *iv* embryos. Impairment of the nodal flow may also account for the situs defects apparent in other mutant mice that are known to lack cilia in other organs. In *Hfh4* mutants, ciliated cells are absent in the respiratory system similar to that seen in Kartagener syndrome (Chen *et al.*, 1998). In addition, expression of *Lrd* was abolished. Although cilia in the node cells have not been examined, HFH-4 (a winged-helix transcription factor) may regulate the expression of several dynein genes required for 9+0 and 9+2 cilia. An insertional mutation in the mouse, called Tg737, results in L-R randomization and loss of ciliated cells in the node (Murcia *et al.*, 2000). The causative gene encodes a protein with tetratricopeptide repeats (referred to as Polaris). Interestingly, its homolog in *Chlamydomonas* is required for assembly of cilia and flagella (Pazour *et al.*, 2000).

These observations strongly implicate nodal flow in the early step of L-R determination. However, several important questions remain. First, the presence of monocilia in the organizer region (equivalent to the node in mouse) has not yet been demonstrated in any vertebrate other than mouse. Thus, it is unknown whether nodal flow is a common mechanism for generating L-R asymmetry in vertebrates. Also, a soluble factor that is moved by the nodal flow (and acts as a left determinant) remains to be clearly identified. A transforming growth factor β (TGF-β)-related factor GDF1 (Lee, 1990) may be one of such candidates. *Gdf1* is expressed initially throughout the embryo proper and then most prominently in the primitive endoderm of the node. In most *Gdf1-/-* embryos, expression of *nodal*, *lefty2*, and *Pitx2* in the left lateral plate mesoderm (LPM) is absent, suggesting that the left

side signaling pathway fails to take place (Rankin *et al.*, 2000). These observations support an idea with GDF1 being a factor moved by the node flow. Another candidate may be FGF8. *Fgf8* is expressed in cells near the node (cells caudolateral to the node). However, its expression is symmetric. In most *Fgf8-/-* embryos, left-sided expression of *nodal*, *lefty2*, and *Pitx2* is abolished, suggesting that FGF8 acts as a left determinant (Meyer and Martin, 1999).

Although this nodal flow model is a very attractive hypothesis, one can make several arguments against it (Wagner and Yost, 2000). The nodal flow hypothesis cannot simply explain the phenotype of *inv*, an interesting yet puzzling mutation. The *inv* mutation is unique in that almost all of the homozygous animals exhibit situs inversus. One may therefore predict that the node flow in the *inv/inv* embryos would be reversed, but this was not the case (Okada *et al.*, 1999). The mutant mice have cilia that look morphologically normal and are motile (move with the normal gyrating speed of 600 rpm). The nodal flow is affected in a peculiar way; it is leftward but is slow and turbulent (Okada *et al.*, 1999). In the *inv* mutants, the shape of the node is also deformed. The mechanistic basis for this "turbulent" nodal flow is not certain. We also do not know how the turbulent nodal flow would lead to total situs inversus (instead of the randomization of situs) although several complex models have been proposed (Okada *et al.*, 1999).

It is possible that the node flow and node cilia simply serve as flags for other events inside the node cells, and that they do not play any role in L-R determination. Lrd protein is thought to function in cilia, but it may also function in cytoplasm. KIF3B protein has been detected in monocilia of the node cells, but it is also localized to vesicles (Nonaka *et al.*, 1998). These motor proteins may be involved in L-R determination by exerting some function in cytoplasm (such as vesicle transport) rather than in ciliary action.

The node flow hypothesis suggests that the initial break in asymmetry takes place within the node. However, it is possible that the initial break occurs outside the node, earlier than the beginning of the node flow. Several observations in chick and frog suggest this possibility. In *Xenopus*, it has been suggested that the L-R axis may be established as early as the one-cell-stage. Experiments on UV-treated one-cell stage embryos (Yost, 1995) suggest that an L-R prepattern that can be disrupted by loss of microtubules exists in the one-cell-stage embryo. It has been proposed that an active form of Vg1 (a TGF-β family member) may be involved in this prepatterning (Yost, 1991, 1995).

Gap junctional communication (GJC) may function earlier than the role of the node (or the node flow) in frog and chick. In *Xenopus*, inhibition of GJC by pharmacological agents increases the frequency of abnormal heart looping (Levin and Mercola, 1998b). Similarly, treatment of chick embryos with antibodies or antisense oligonucleotides against connexin subunit Cx43 leads to bilateral *nodal* expression (Levin and Mercola, 1999). Gap junctions may allow unidirectional transport of a small molecule, resulting in its accumulation on one side. To achieve this, there must be a barrier that prevents the molecule from traveling back. In chick, the midline may act as this barrier. Mutations in the connexin subunit Cx43 have been identified in human patients showing heterotaxia (Britz-Cunningham *et al.*, 1995). However, subsequent screening of 38 heterotaxy patients failed to detect *Cx43* mutations (Gebbia *et al.*, 1996). Furthermore, L-R defects are not detected in *Cx43* mutant mice (Lo, 1999). Therefore, the role of GJC remains to be established at least in mammals.

IV. Molecular Readout of the First Asymmetry

A. Left-Right Asymmetric Expression of TGF-β Signals, Nodal and Lefty, in the Lateral Plate

Asymmetric signaling molecules in mammals include three TGF-β-related factors—Nodal, Lefty1, and Lefty2—

all of which are expressed on the left half of the lateral plate mesoderm of developing mouse embryos (Fig. 5; Collignon *et al.*, 1996; Lowe *et al.*, 1996; Meno *et al.*, 1996, 1997). Nodal is a typical member of the TGF-β superfamily that seems to transduce signals through its receptor (not established yet but most likely ActRIs and ActRIIs), an EGF-CFC protein as a cofactor (Gritsman *et al.*, 1999), and transcription factors Smad2, Smad4, and FAST2 (for review, see Schier and Shen, 2000). Transcription factors other than FAST may also be involved because Smads are known to interact with a variety of transcription factors (for review, see Whitman, 1998). Lefty1 and Lefty2 comprise a unique subgroup of the TGF-β superfamily (Meno *et al.*, 1996, 1997). They lack a cysteine residue required for dimer formation and also lack the long α helix involved in homodimerization and heterodimerization. These structural considerations suggest that Lefty proteins act differently than other TGF-β family proteins. In fact, as described below, genetic evidence suggests that Lefty proteins act as antagonists against Nodal, being able to bind to receptors but unable to transduce signals.

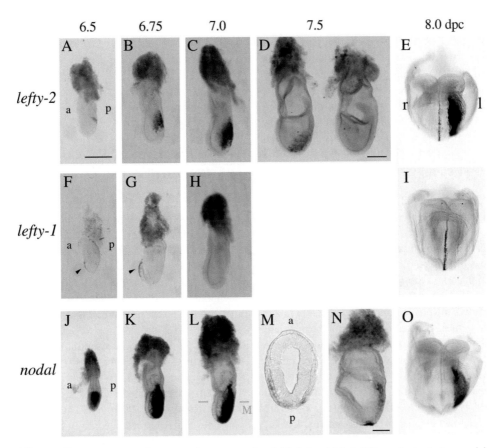

Figure 5 A comparison of the expression patterns of *nodal, lefty1,* and *lefty2.* At the early-streak stage, *nodal* is expressed in the posterior ectoderm and the endoderm (L, M) while *lefty2* is expressed in nascent mesoderm (B, C, D). At E6.5 to E6.75, *lefty1* is expressed in the anterior visceral endoderm (F, G). *nodal* is expressed throughout the epiblast (J), but by E7.75, *nodal* expression is restricted to the posterior region (K). At the early somite stage, *nodal* and *lefty2* are coexpressed in the left LPM (E, O) while *lefty1* is expressed only on the left side of the floor plate (I). (Adapted from Meno *et al.,* 1999.)

nodal is initially expressed throughout the epiblast and primitive endoderm at pregastrulation stages (Fig. 5). At the early somite stage, it is expressed in the LPM and in the node. While its expression in LPM is exclusively on the left side, the expression in the node shows subtle asymmetry (expression on the left side of the node is stronger and wider than that on the right side). The expression in the left LPM is highly transient, and is no longer detectable by the 8 somite stage. The L-R asymmetric expression pattern of *nodal* is conserved among vertebrates (Levin *et al.*, 1995; Collignon *et al.*, 1996; Lowe *et al.*, 1996; Lustig *et al.*, 1996). Left-sided expression of *lefty1*, *lefty2*, and *nodal* is altered in *iv* and *inv* mutant embryos suggesting that Nodal, Lefty1, and Lefty2 all act downstream of the initial asymmetry established by the node.

nodal is essential for mesoderm formation during gastrulation, thus *nodal* null mutants are early embryonic lethal (Zhou *et al.*, 1993; Conlon *et al.*, 1994), preventing direct genetic analysis of the role of *nodal* in L-R patterning in mouse. However, numerous lines of evidence suggest Nodal is a determinant for "leftness." First, misexpression of *nodal* on the right side of chick and *Xenopus* embryos can affect heart looping and gut coiling (Levin *et al.*, 1997a; Sampath *et al.*, 1997). Second, a relationship between aberrant *nodal* expression pattern and L-R defects of the visceral organs is observed in many mutants. For example, mutant mice in which *nodal* is bilaterally expressed (such as *lefty1*−/− mutants) show left isomerism. On the other hand, those mutants lacking *nodal* expression in the LPM (such as *Cryptic* mutant, Yang *et al.*, 1999; Gaio *et al.*, 1999) show right isomerism in the lung. Similarly, mouse embryos lacking an essential component of Nodal signaling such as ActRIIB (likely, a type II receptor for Nodal) exhibit right isomerism (Oh and Li, 1997).

Lefty2, an atypical member of the TGF-β superfamily, is also expressed in the left LPM (Fig. 5). *lefty2* expression begins in newly formed mesoderm at the early primitive streak stage (E6.0). This expression domain is symmetric and disappears by the end of the neural fold stage, but a new expression domain appears in the left LPM at the 2/3-somite stage. The left-sided expression of *lefty2*, like that of *nodal*, is very transient; *lefty2* mRNA is no longer detectable in left LPM by the 6–8 somite stage. *lefty2* null mutant mice are early embryonic lethal due to the lack of its expression at the primitive streak stage, and display an expanded primitive streak and form excess mesoderm. This phenotype is opposite to that of *nodal* mutants that fail to form mesoderm (Conlon *et al.*, 1994), suggesting that the defects of *lefty2* null mutants are due to an increase in Nodal activity. In fact, *nodal* expression is markedly upregulated in the *lefty2* mutant embryo. Furthermore, the *lefty2* mutant phenotype is partially suppressed by heterozygosity for *nodal*, suggesting antagonistic genetic interaction between the two genes. Therefore, Lefty2 antagonizes Nodal signaling and restricts

the range and duration of Nodal activity during gastrulation, perhaps by competing for a common receptor (Meno *et al.*, 1999). However, there has been no direct evidence for physical interaction between Lefty and receptors/coreceptor of Nodal. Thus, it is not certain how Lefty proteins inhibit Nodal signaling.

Direct evidence for the role of Lefty2 in the L-R axis formation is not available because *lefty2* null mutants die before L-R defects can be assessed (Meno *et al.*, 1999). However, it is conceivable that Lefty2 has a similar role in the left LPM: to restrict the duration and site of the action of Nodal, a leftness-inducing signal. To know the exact role of *lefty2* in L-R patterning, it will be necessary to generate another type of *lefty2* mutant in which its asymmetric expression in the left LPM is specifically eliminated. As we will describe later, Lefty1 is not a determinant for "leftness," but is rather a regulator that restricts the expression of *nodal* and *lefty2* to the left side, possibly by generating a midline barrier (Meno *et al.*, 1998).

Nodal homologs exist in all vertebrates analyzed, and they show related left-side expression patterns. Indeed, experimental data in all species analyzed to date places nodal as a critical determinant of left-sidedness. Antivin, which was originally identified as an antagonist of activin in zebrafish (Thisse and Thisse, 1999), is structurally related to Lefty proteins and, like Lefty2, inhibits Nodal signaling (Bisgrove *et al.*, 1999; Meno *et al.*, 1999); the zebrafish *antivin* gene is thus likely a *lefty* ortholog. The *lefty* homologs are also known in *Xenopus* and chick, and they show expression patterns similar to that of *lefty* in mouse.

B. Regulatory Relationship between *nodal* and *lefty2*: Positive and Negative Feedback Loops

Studies on transcriptional regulatory mechanisms of *nodal* and *lefty2* (Adachi *et al.*, 1999; Norris and Robertson, 1999; Saijoh *et al.*, 1999) have demonstrated the presence of a complex yet elegant regulatory loops between the two genes (Figs. 6 and 7). Various genomic regions of *lefty2* have been examined by transgenic assay for an enhancer activity that can drive left-sided expression in the LPM. Such analysis has located an asymmetric enhancer (ASE) in the upstream region, which is essential and sufficient for the left sided expression (Saijoh *et al.*, 1999). *nodal* genomic regions have been similarly searched, and a similar enhancer (ASE) has been found in an intron (Adachi *et al.*, 1999; Norris and Robertson, 1999). *lefty2* ASE and *nodal* ASE share not only the enhancer specificity but also several common sequences, the most remarkable of which is the presence of two copies of AATCCACA in both ASEs. These sequences are essential to the ASE activity because mutations of both of these sequences abolish the ASE activity. By yeast one hybrid cloning, the transcription factor binding to this se-

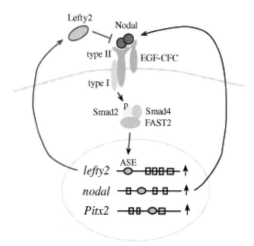

Figure 6 *nodal, lefty2,* and *Pitx2* expression is induced by Nodal signaling. The left-sided expression of *nodal, lefty2,* and *Pitx2* genes is directed by a common ASE enhancer. The ASE enhancer is activated by a Nodal signaling pathway mediated by activin type I and type II receptors, EGF-CFC protein (as a coreceptor), Smad2, Sma4, and FAST2. Lefty2 protein inhibits Nodal signaling probably by interacting with a common receptor. (Based on results of Saijoh *et al.,* 1999, 2000; Adachi *et al.,* 1999; Norris and Robertson, 1999; Shiratori *et al.,* 2001.)

quence has been identified as FAST2, a homolog of FAST-1 (a winged helix transcription factor in frogs that can mediate signaling of activin and TGF-β; Chen *et al.,* 1996). *FAST2* itself is bilaterally expressed in the LPM. Activin and TGF-β can stimulate the ASE activity (in the presence of FAST2). Also, Nodal protein itself can activate ASE in the presence of an EGF-CFC protein such as Cripto or Cryptic (a cofactor of Nodal signaling). Therefore, it is most likely that ASE

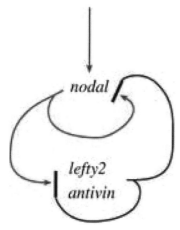

Figure 7 The regulatory relationship between *nodal* and *lefty2* involving activation (red arrow) and inhibition (blue line and bar). *Nodal* expression is activated by a putative signal (green arrow) that has not yet been identified. (Based on results of Saijoh *et al.,* 1999, 2000; Adachi *et al.,* 1999; Meno *et al.,* 1999.)

is in fact a Nodal-responsive element *in vivo.* This implies that amplification of *nodal* and *lefty2* expression throughout LPM is induced by Nodal protein through ASE of each gene.

On the basis of these observations, the following scenario can be proposed for the asymmetric expression of *nodal* and *lefty2* genes in the lateral plate (Fig. 7): The left-sided expression of *nodal* is initially induced by an unknown factor (step 1). The Nodal protein thereby produced autoregulates the *nodal* gene and activates *lefty2* expression in the left LPM (step 2). Components of the Nodal signaling pathway, including an EGF-CFC protein (most likely Cryptic), FAST2, and Smad2–Smad4 would therefore all be required for the maintenance of *nodal* expression and for induction of *lefty2* expression. In fact, although most *Smad2* knockout mice die during early embryonic development due to severe gastrulation defects, some survive and show situs defects (Nomura and Li, 1998; Waldrip *et al.,* 1998). Lefty2 produced as a result of Nodal signaling likely antagonizes Nodal action (step 3), thereby rapidly terminating the left-sided expression of *nodal* and *lefty2.* This feedback mechanism would restrict the range and duration of Nodal signaling in developing embryos.

This model is supported by several additional observations. Domains of *lefty2* expression always overlap with those of *nodal.* Furthermore, in most mouse mutants with situs defects, *lefty2* expression is affected in parallel with *nodal* expression. The model also predicts that the lack of Lefty2 would not affect initial expression of *nodal,* but that it would subsequently result in upregulation of *nodal* expression. This prediction has been borne out by *lefty2* mutant mice (Meno *et al.,* 1999). And in zebrafish, the lack of Oep (an EGF-CFC protein) results in downregulation of *cyc* and *sqt,* two *nodal*-related genes (Meno *et al.,* 1999), consistent with autoregulation of *nodal.* Similarly, the asymmetric expression of *nodal* and *lefty2* in left LPM is abolished in *Cryptic* mutant mice (Yang *et al.,* 1999). However, the relationship between *nodal* and *lefty2* described here may be too simplified. There are some instances where *lefty2* is ectopically expressed on the right side but *nodal* is not. These include Furin-deficient mice (Constam and Robertson, 1999) and tetraploid *HNF3β*⁻ᐟ⁻ aggregation chimeras (Dufort *et al.,* 1998). Conversely, in mutant mice lacking ActRIIB (activin type II receptor B), *lefty2* expression is absent while *nodal* expression is normal (C. Meno, Y. Saijoh, and H. Hamada, unpublished).

The relationship between Nodal and Lefty2 resembles that of two hypothetical molecules that comprise the *reaction-diffusion mechanism* proposed by Turing (1952). This theoretical model requires two (or more) diffusible molecules, one of which is an activator that stimulates both its own synthesis and that of the other (a feedback inhibitor). Also the model requires that an inhibitor diffuse more rapidly than an activator. The *nodal–lefty2* connection may be the first example of the reaction-diffusion mechanisms that exist in developing organisms. In this regard, it is interesting to note that the asymmetric expression of *nodal* and *lefty2* begins as a small domain of the left LPM adjacent to the node and gradually expands in both directions along the AP axis. This rapid expansion of their expression domain may be achieved by the reaction-diffusion mechanism. However, it remains to be seen how far Nodal and Lefty2 proteins can diffuse, and which of these two proteins diffuses more rapidly. It may be that Nodal is a long-range acting molecule but its range of action is limited by Lefty2, a feedback inhibitor that may diffuse more efficiently than Nodal. If the regulatory loops between *nodal* and *lefty* indeed function as the reaction-diffusion mechanism, it may be involved in the generation of random asymmetry mechanism, as proposed by Brown *et al.,* (1991).

C. Signal Transfer from the Node to the Lateral Plate: Link between the Node and the LPM

Thus, L-R biased information generated in/near the node (by nodal flow) must be transferred to the lateral plate, in order to establish the left-sided expression of *nodal* in the left LPM, which is a key step in L-R axis formation. How is the node-derived L-R biased signal (perhaps generated by the nodal flow) relayed to the lateral plate? Also, how is *nodal* expression initiated in the left LPM?

In the chick, signals from the node seem to be relayed at the paraxial mesoderm, which is located between the node and the lateral plate (Pagan-Westphal and Tabin, 1998). The paraxial mesoderm on the left side expresses Caronte (Car), an important relay signal recently identified in chick. Car is a member of the Cerberus/DAN family that can antagonize TGF-β-related ligands (and Wnt ligands as well) by directly interacting with them. Several observations made in chick support that Car plays a role in the signal relay: (1) Car expression on the left side is induced by Shh, (2) misexpression of Car on the right can induce ectopic *nodal* expression, (3) the ability to induce Nodal can be mimicked by Noggin, a specific antagonist against bone morphogenetic protein (BMP), (4) a number of BMPs are bilaterally expressed in LPM, (5) Car protein can physically interact with BMP. In all, these observations suggest the following model: *nodal* expression in the LPM would be bilaterally repressed by BMP signaling. Car, which is induced on the left paraxial mesoderm by Shh, would in turn relieve the repressive action of BMP, resulting in *nodal* activation on the left LPM.

Although this is an attractive model in chick, it is unknown if a similar mechanism operates in mouse (Car orthologs have not been identified in other vertebrates). Another factor possibly involved in the relay mechanism is FGF8. In mouse, *Fgf8* is not expressed asymmetrically. However, mouse transheterozygous mutants harboring mypomorphic and null *Fgf8* alleles show right isomerism in the lungs

(Meyer and Martin, 1999). Most of such embryos seem to have lost expression of the left side-specific genes such as *nodal, lefty2,* and *Pitx2* (although a small population shows bilateral *Pitx2* expression with *nodal* and *lefty2* being expressed normally). Furthermore, implantation of FGF8 beads on the right LPM of mouse embryos can induce ectopic nodal expression. Thus, *Fgf8* appears to act as a *left* determinant in mouse. It is conceivable that mouse *Fgf8,* which is expressed symmetrically in cells within the primitive streak and caudolateral to the node, plays a role as an essential/ubiquitous component of signal relay from the node to LPM. In particular, FGF8 may be a molecule that is unevenly distributed by the leftward nodal flow.

V. Role of the Midline

The midline structures such as the floor plate and notochord are important for maintenance of L-R patterning. They are required to separate the left and right halves of embryos until L-R patterning is completed. This notion has been proposed by embryological experiments in frog and mutant analysis in zebrafish and mouse. For example, surgical removal of the midline structures including the notochord from *Xenopus* embryos randomizes heart looping and gut coiling and leads to bilateral expression of *Xnr-1,* a nodal-related gene in *Xenopus* (Danos and Yost, 1996). Furthermore, mouse (and zebrafish) mutants with defects in the midline structures (*nt, shh, T,* etc.) often exhibit L-R patterning defects with left side-specific genes (such as *nodal, lefty2,* and *Pitx2*) being bilaterally expressed. These observations suggest that the role of the midline is to prevent the right side from acquiring left side identity, perhaps by serving as a barrier to block the spread of signals from the left to the right.

Analysis of *lefty1* mutant mice further supports an important role of the midline (Meno *et al.,* 1998). In contrast to *nodal* and *lefty2, lefty1* is expressed in the prospective floor plate (PFP) on the left side with a narrow time window (between the 2–6 somite stages). The lack of Lefty1 leads to bilateral expression of left-specific genes such as *nodal, lefty2,* and *Pitx2,* and results in pulmonary left isomerism (Meno *et al.,* 1998; Fig. 8). PFP remains morphologically normal in *lefty1* mutants. Thus, it is clear that Lefty1 is not a left side determinant. Rather, this phenotype suggests that the role of Lefty1 is to function as (or induce) the midline barrier that prevents an unknown left side-specific signaling molecule (factor X) from crossing the midline. This notion is supported by observations on other mutants that seem to have midline defects. For example, *shh⁻ᐟ⁻, SIL⁻ᐟ⁻,* and *nt* mutant embryos that show bilateral expression of the left side-specific genes all lack *lefty1* expression in PFP (Izraeli *et al.,* 1999; Meyer and Martin, 1999; Melloy *et al.,* 1998). Thus the phenotype of these mutants can be explained by the lack of Lefty1. L-R defects of *ft* (*fused toe*) mutants (their caus-

wild type   *lefty1* $^{-/-}$

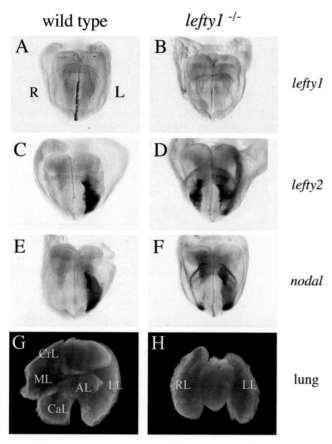

Figure 8 Lefty1 is essential for the midline barrier. *lefty1* is expressed in the prospective floor plate on the left side (A). In mutant embryos lacking *lefty1*, both *nodal* and *lefty2* are bilaterally expressed (D and F). In wild-type embryos, *nodal* and *lefty2* are expressed on the left side (C and E). In the wild-type mouse, the lungs are bilaterally asymmetric, with one lobe on the right and four lobes on the left (G). In the *lefty1* mutant mouse, both left and right lungs have one lobe (H). CaL, caudal lobe; CrL, cranial lobe; ML, medial lobe; LL, left lobe; RL, right lobe. (Adapted from Meno *et al.*, 1998.)

the passage of signals across the midline. In either case, the barrier would have to prevent diffusion of factor X, but the identity of factor X is unknown. Factor X may be Nodal itself. In support of this possibility, recent findings (Meno *et al.*, 2001) suggest that Nodal protein is a long-range acting molecule but its range of action is normally limited by Lefty protein that may diffuse more efficiently than does Nodal. Lefty1 protein expressed in the floor plate may bind to Nodal receptors in the adjacent area and prevent Nodal activity from propagating across the midline.

The midline barrier would be necessary only when the left side signals (such as Nodal and Lefty2) are present. In fact, the timing of *lefty1* expression coincides with that of *nodal* expression. The expression pattern of *lefty1* in PFP suggest that the midline barrier exists only between the 2–6 somite stages, and that it is initially formed in the area near the node and expands anteriorly along the AP axis. Also, indirect evidence points to the fact that the anterior and posterior portions of the midline barrier are formed/maintained by different mechanisms. When Lefty1 is absent, ectopic expression of the left side-specific genes is mainly restricted to the anterior region of the right LPM. Thus, Lefty1 may be a component of the anterior midline barrier, while the midline barrier in the posterior region probably involves a different mechanism (consistently, the posterior border of *lefty1* expression ends at the node). The nature of the posterior midline barrier (posterior to the node) remains unknown. However, excess retinoic acid may specifically perturb the posterior midline function as described below.

VI. Readout of Left-Right Asymmetry in Later Development

A. *Pitx2* on the Left Side, *Sna* on the Right Side

Shortly after *nodal* and *lefty* expression is turned off, situs-specific morphogenesis starts to take place in many primordial organs. L-R asymmetric structures are generated in organs by different mechanism, as described in Section I: rightward looping of the cardiac tube, axial turning of the embryo in a clockwise direction, followed by asymmetric lobation of the lungs, rotation/coiling of the digestive tract, regression of one side of some portions of the vascular system, and so forth. How is this situs-specific morphogenesis executed in response to Nodal signaling? Do different organs respond differently (e.g., do different mechanisms regulate the heart looping and lung lobation)?

Situs-specific morphogenesis in response to L-R asymmetric signals (mainly Nodal) involves Pitx2, a bicoid-type homeobox transcription factor. Asymmetric expression of *Pitx2* begins in the left LPM concomitantly with, but persists for longer than, that of *nodal* and *lefty2*. Thus, *Pitx2* expres-

ative genes are not cloned yet) may be also due to midline barrier defects, because they show bilateral *nodal* and *lefty2* expression and seem to lack *lefty1* expression in PFP (Heymer *et al.*, 1997). In fact, no mutants have been reported that exhibit bilateral expression of *nodal* and *lefty2* yet retain normal *lefty1* expression in PFP.

What does "barrier" mean? How would the midline function as a barrier? We still do not know the answer, but the barrier could be physical, biochemical, or both. In the simplest model, Lefty1 protein itself may function as a barrier molecule that interacts with factor X and prevents its diffusion. Note that the space connecting both sides of the embryo at this stage (the space between PFP and the ventral endoderm) is very narrow. In this regard, PFP would be an ideal place for a barrier molecule to be expressed. Alternatively, Lefty1 may be required to induce or maintain a physical barrier (such as tight junction) that would restrict

sion is still apparent in primordia of most of asymmetric organs at the late somite stage: The sinus venosus, vitelline vein, common atrial chamber, cardinal vein, septum transversus (future liver), and a foregut region corresponding to future lung all bud on the left side (Fig. 9). This L-R asymmetric expression pattern of *Pitx2* is conserved among vertebrates. Observations made in chick and frogs suggest that *Pitx2* expression is induced, directly or indirectly, by Nodal signaling. Thus, ectopic expression of *nodal* in chick embryos is able to induce expression of *Pitx2* on the right side (Logan *et al.*, 1998; Meno *et al.*, 1998; Piedra *et al.*, 1998 ; Ryan *et al.*, 1998; Yoshioka *et al.*, 1998). Ectopic expression of *Pitx2* can be also induced in chick when mouse *lefty2* is misexpressed on the right side at certain stage (Yoshioka *et al.*, 1998), but this induction may not be direct. Transcriptional regulation of the *Pitx2* gene has been recently analyzed (Shiratori *et al.*, 2001). Asymmetric *Pitx2* expression is conferred by an enhancer containing three FAST binding sites and an Nkx2 binding site. The FAST binding sites are essential for initiating the asymmetric expression. The Nkx2 binding site is not essential for the initiation but is indispensable for maintaining the late-stage expression. Therefore, the left-sided expression of *Pitx2* is initiated by Nodal signaling and is maintained by *Nkx2*.

Pitx2 is responsible for generating left side morphology of, at least, some organs. Ectopic expression of *Pitx2* on the right LPM in chick and frog can affect the direction of heart looping and gut coiling. However, the role of Pitx2 has been shown more directly by recent analysis of Pitx2-deficient mice (Gage *et al.* 1999; Kitamura *et al.*, 1999; Lin *et al.*, 1999; Lu *et al.*, 1999). Most obvious L-R defects are found in the lungs, which show typical right isomerism (four lobes on both sides). Therefore, Pitx2 is apparently responsible for generating left side-specific morphology. However, due to additional severe defects (*Pitx2* is also expressed symmetrically in the head mesenchyme at the same stage), L-R defects of *Pitx2* mutants have not been fully analyzed (e.g., other visceral organs such as the stomach, guts, liver, spleen and vascular system have not been analyzed). Thus it is not clear whether Pitx2 is responsible for all the asymmetric organs or only some of them. The role of Pitx2 needs to be examined further by generating/analyzing conditional knock-out mice that simply lack the left side expression of *Pitx2*. Then, the readout of Nodal signaling could be different in different areas. It could involve changes in growth rate, apoptosis, migration or other mechanisms depending on the type of genes induced in a given cell. We do not currently know which genes Pitx2 regulates.

The *cSna* gene, which encodes a zinc finger-type transcription factor related to *Drosophila* Snail, is expressed on the right side of chick embryos (Isaac *et al.*, 1997). *cSna* may be negatively regulated by Nodal because misexpression of Nodal on the right side can abolish *cSna* expression. Also, cSna seems to negatively regulate *Pitx2*, because misexpression of *cSna* on the left side can repress *Pitx2* expression. Therefore, the left-sided *Pitx2* expression may be induced by a double-negative regulatory mechanism; on the left side, Nodal would repress *cSna*, and the absence of cSna in turn would activate *Pitx2* on the left. The mouse homolog of *cSna* (called *mSna*) is also expressed in the LPM on the right side (Sefton *et al.*, 1998). It may participate in right side-specific morphogenesis, although its role has not been directly challenged by genetics.

B. Other Genes Regulated by Nodal, Including *Nkx3.2*

The homeobox gene called *Nkx3.2* (also known as *BapX1*) is also expressed asymmetrically (Rodriguez-Esteban *et al.*, 1999; Schneider *et al.*, 1999). In mouse, it is expressed symmetrically in the paraxial mesoderm but asymmetrically in the LPM; predominantly on the right side. In *inv* mutant embryos, predominant expression in the LPM is found on the left side, indicating that the asymmetric expression of *Nkx3.2* is under the control of current L-R determining pathway. Timing of its asymmetric expression suggests that *Nkx3.2*, like *mSna*, may be negatively regulated by Nodal signaling and may participate in right side-specific morphogenesis. Mutant mice lacking *Nkx3.2* do not display obvious situs defects (Lettice *et al.*, 1999; Tribioli and Lufkin, 1999; Akazawa *et al.*, 2000). However, they exhibit asplenia (the

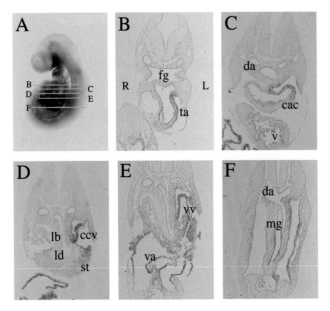

Figure 9 Asymmetric expression of *Pitx2* in various organ primordia. Expression domains of *Pitx2* in E9.5 mouse embryo. Note that Pitx2 is asymmetrically expressed in many organs including the common atrial chamber (cac), common cardinal vein (ccv), lung bud (lb), septum transversum (st), gut mesentery (mg), vitelline vein (vv). da, dorsal aorta; fg, foregut; ld, liver diverticulum; va, vitelline artery. (Adapted from Meno *et al.*, 1998.)

lack of the spleen) and malformation of the gastroduodenal tract, which may be regarded as L-R positional defects. Thus, the role of this gene in L-R asymmetric morphogenesis needs to be examined further.

It is interesting to note that heart looping seems to take place normally in *Pitx2-/-* and *Nkx3.2-/-* mice. Therefore, other genes might be regulated by Nodal signaling, but they have yet to be identified.

VII. Miscellaneous Mutations/ Gene Factors

Several other mutations, genes, and physiological agents are implicated in L-R patterning whose exact roles remain uncertain. Some of them are described next.

A. Retinoic Acid

Retinoic acid (RA), which mainly exists as the all-trans or 9-cis isoform in our body, plays important roles in many aspects of development such as limb formation and AP patterning of the neural tube (Conlon, 1995).

Retinoic acid is also implicated in L-R patterning. Excesses as well as deficiencies in RA have been known to result in laterality defects in vertebrates. In the chick embryo, exogenous RA causes the host heart to loop in an abnormal direction. In mammals, RA exposure during gastrulation results in partial situs inversus of thorax and viscera with a high incidence of total situs inversus (especially in rat). Total situs inversus is also observed in the majority of RA-deficient quail embryos. Effects of RA have been recently analyzed at a molecular level. When mouse embryos are exposed to excess RA during a specific window (at the headfold stage; E7.5), it evokes bilateral expression of *nodal, lefty2,* and *Pitx2* in the posterior LPM; their ectopic expression is restricted to the LPM posterior to the node (Wasiak and Lohnes, 1999). Although excess RA affects expression of midline markers (such as upregulation of HNF3b), the notochord and PFP remain histologically normal and *lefty1* expression in PFP remains normal too. These observations suggest an interesting possibility that excess RA perturbs posterior (posterior to the node) midline function, but not the anterior midline function (the latter requires Lefty1 as described above). On the other hand, the treatment of mouse embryos at the headfold stage with an RA-antagonist inhibits expression of *nodal, lefty1,* and *lefty2* (Chazaud et al., 1999; Tsukui et al., 1999). Mutant mouse embryos lacking *Shh* lose *lefty1* expression in PFP and exhibit bilateral expression of *nodal* and *lefty2,* but application of RA to *Shh* mutant embryos in culture can rescue *lefty1* expression (Tsukui et al., 1999).

These lines of evidence strongly implicate RA in L-R patterning, but the exact role of RA is not clear. Nonetheless, it is most likely that RA acts at multiple steps of the L-R pathway. One of the roles of RA appears to be in the midline barrier function; RA may be required for the anterior midline barrier by positively regulating *lefty1* (interestingly, an RA-responsive element is located in the *lefty1* promoter region although its role in *lefty1* regulation is unknown), whereas RA level may need to be kept low for the posterior midline function. (Coincidentally, *CYP26* encoding an RA-degrading enzyme is highly expressed in the posterior region including the neural plate and hindgut; Fujii et al., 1997). In addition to its role in the midline, loss of *nodal* and *lefty2* expression induced by an RA-antagonist suggests that RA may also be involved in an earlier step of L-R axis formation (for example, in signal transfer from the node to LPM).

B. Zic3

Zic3 is a member of the Zic family encoding a zinc-finger transcription factor with a sequence homology to *Drosophila cubitus interruptus* (Ci) and human GLI. *Zic3* is a causative gene for an X-linked inheritance of situs abnormalities in human (Gebbia et al., 1997). These patients have cardiac malformation associated with complex situs defects, such as asplenia or polysplenia, symmetric liver, intestinal malroration, and abnormal lung lobation. Ci is a member of the Gli family of transcription factors, which mediates the Hedgehog signal and is regulated by Hedgehog in the midline. The finding that Gli-type Zic3 is mutated in human heterotaxy syndrome may suggest the importance of midline signals in L-R patterning in mammals as well. In fact, mouse *Zic3* is expressed in the notochord before *nodal* and *lefty2* are expressed in the left LPM. These considerations suggest that *Zic3* may play a role in maintaining the midline barrier function. However, the phenotype of the patients is not similar to that of mice with midline barrier defects (such as *lefty1* mutants). Alternatively, Zic3 may be required as a transcription factor mediating signals from the node to the LPM. The exact role of Zic3 should be analyzed by fully examining *Zic3* expression profile in mice and and by generating mutant mice lacking this gene.

C. Others

UDP-*N*-acetylglucosamine:α-3-D-mannosideβ-1,2-*N*-acetylglucosaminyl transferase I (GlcNac-TI) is a key enzyme in *N*-linked protein glycosylation. It is encode by a single gene termed **Mgat** in mouse. Mutant mice lacking *Mgat* fail to complete axial turning and exhibit inverted heart looping (Metzler et al., 1994). Thus, impaired glycosylation of unknown signaling molecule involved in L-R patterning may account for the phenotype.

Furin is a member of a family of serine proteases that specifically activates various prohormones and growth factors via cleavage at a cluster of basic amino acid residues.

Furin-deficient embryos fail to undergo axial turning. The heart primordia of *fur* mutants either fail to fuse at the midline or only form a short, unlooped tube (Roebroek *et al.,* 1998). While *nodal* expression in LPM is unperturbed in *fur* mutant embryos, *Pitx2* expression is upregulated in the left LPM, and *Pitx2* and *lefty2* are occasionally expressed bilaterally (Constam and Robertson, 1999). *lefty1* expression in PFP remains normal. Furin may be one of several convertases activating multiple TGF-β-related factors including Nodal, Lefty, and BMPs. In fact, *fur* is coexpressed in the node and LPM with these TGF-β-related genes. Then, the phenotype of *furin* mutants is likely due to impaired maturation of one or more of these factors, but it is not clear which TGF-β factor(s) is responsible for the phenotype; impaired activity of Nodal alone, Lefty2 alone, Lefty1 alone, or any combination of these does not seem to fit with the phenotype. Defective maturation of BMP may exist as the primary cause of the phenotype.

The **NOD** (nonobese diabetic) mouse is an animal model for insulin-dependent diabetes melitus. A high incidence of situs inversus occurs in fetuses when their mothers have already developed diabetes (Morishima *et al.,* 1991). The most frequent phenotype of such fetuses is atrial right isomerism. Thus, maternal diabetes seems to have an influence on L-R asymmetry, but it is not certain which step of the L-R patterning pathway is impaired in NOD embryos.

Hyd is a mutant strain of rat with an X-linked recessive gene (*hyd*). When heterozygous females were mated with wild-type males, about half of male offsprings showed total situs inversus and hydrocephalus while the remaining half of male offsprings were normal (Koto *et al.,* 1987). It appears that L-R determination is reversed in *hyd* mutant males (as in *inv* mice). *Hyd* may play a role similar to that of *inv* in the initial L-R determination.

VIII. Diversity among Vertebrates

As for many other mechanisms regulating developmental processes, the central mechanism regulating L-R asymmetry has been widely conserved in vertebrates. In particular, *nodal, lefty,* and *Pitx2* exhibit similar expression patterns among vertebrates (from zebrafish to mouse) and they appear to have conserved roles in the L-R asymmetry pathway. However, apparent differences exist in the expression patterns (and functions) of some regulatory genes among vertebrates. The full extent to which L-R asymmetry pathways differ among vertebrate species remains to be seen.

The most obvious example is Sonic hedgehog (Shh). In chick, *Shh* is expressed in an apparently asymmetric fashion, on the left side of the Hensen's node beginning at stage 5 (Hamberger-Hamilton stage). Misexpression of Shh on the right side of the node induces ectopic *nodal* expression while administration of an anti-Shh antibody on the left side

downregulates endogenous nodal expression. These results strongly imply that Shh is a left determinant in chick (Levin *et al.,* 1995). On the other hand, in mouse (and any other class of vertebrates), *Shh* is expressed in the midline, but its expression is not L-R asymmetric. Recent analysis of Shh-deficient mice has uncovered situs anomalies including left pulmonary isomerism (Meyer and Martin, 1999). However, these mutant embryos show bilateral expression of *nodal, lefty2,* and *Pitx2,* and lack *lefty1* expression in PFP. Therefore, all the L-R defects seen in *Shh* mutant mice can be explained by the lack of Lefty1 in PFP, and they are most likely due to the failure of the midline barrier. Thus, Shh is a crucial molecule in chick relaying L-R signals from the node to the paraxial mesoderm (or the LPM), whereas in mouse Shh is simply required for maintaining the integrity of PFP (thus, midline barrier).

Another example is *Fgf8*. In chick, *Fgf8* is expressed on the right side of the node (Boettger *et al.,* 1999). Application of FGF8-soaked beads on the left side of the node can abolish *nodal* expression (a left side-specific pathway). Thus, superficially, FGF8 appears to act as a *right* determinant in chick. However, in the mouse it is bilaterally expressed but seems to act as a left determinant (Meyer and Martin, 1999).

Other examples include *Nkx3.2* and *ActRIIa*. *Nkx3.2* shows predominantly *right*-sided expression in the LPM of mouse, as described above. However, in chick, it is asymmetrically expressed predominantly in the *left* half of the anterior LPM and head mesenchyme (Schneider *et al.,* 1999; Rodriguez-Esteban *et al.,* 1999). Misexpression of SHH or Nodal on the right side of chick embryos can upregulate *Nkx3.2* on the right, suggesting that *Nkx3.2* is downstream of the left side-specific pathway (Shh—nodal—) in chick. *Nkx3.2* also shows a left side-dominated expression in the LPM of *Xenopus*. ActRIIa, a receptor for activin, is expressed on the right side of the node in the chick embryo. This initially suggested an involvement of an activin or activin-like molecule in an early step of the L-R pathway in chick. However, ActRIIa or any other receptor for TGF-β-related ligands does not show asymmetric expression in the mouse. Possible involvement of activin in the L-R pathway has been excluded, at least in mouse, by the fact that activin-lacking mice show no signs of L-R defects (Matzuk *et al.,* 1995). Although mice deficient in ActRIIb show pulmonary right isomerism (Oh and Li, 1997), the phenotype of these mice is most likely due to the lack of Nodal signaling (ActRIIB probably functions as a receptor for Nodal).

What would be the significance of these differences among vertebrates? Some of the differences may reflect morphological divergence among vertebrates. This may apply to genes that act downstream of the L-R patterning pathway such as *Nkx3.2*. Other differences (in particular, those found in upstream genes) may simply represent the variation of signal relay mechanisms. Divergent, molecularly different pathways may eventually converge onto the conserved central pathway (*nodal/lefty—Pitx2*), giving rise to the same readout.

Some differences appear to exist even within the same class of vertebrates. Analysis of human *LEFTY* genes (Yashiro *et al.*, 2000) indicates that transcriptional regulation of *LEFTY1* is different from that of mouse *lefty1;* unlike mouse *lefty1,* human *LEFTY1* seems to be expressed in the left LPM and is regulated by an ASE-like enhancer (furthermore, human *LEFTY2* is expressed both in left LPM and PFP). However, the sum level of *lefty/LEFTY* expression in LPM and PFP would be maintained in mouse and human. Thus, if Lefty/LEFTY proteins possess the same intrinsic activity, the outcome would be the same.

IX. Future Challenges

Breaking symmetry represents a fundamental mechanism for many aspects of development. For several reasons, L-R asymmetry provides an ideal model system for studying how symmetry is broken. First, at least in mammals, L-R polarity is determined later than the two other axes. L-R defects are not as life threatening as other patterning defects, so embryos with L-R defects often develop until late stages (the causes of lethality are usually cardiac malformations associated with L-R defects). These situations make experimental approaches easier even in mouse. Secondly, the mechanism of L-R patterning is probably less complex than that of AP or DV patterning. Finally and most importantly, several key genes have already been identified that are known to play a critical role in each step of the L-R pathway. Thus, L-R asymmetry provides a unique opportunity to study a mechanism of asymmetry generation from the beginning to the end. However, this topic is still evolving, and here we envisage several challenging problems.

A. How Is Symmetry Broken in the First Place?

Obviously, the most challenging question concerning axis patterning in general is how symmetry is broken in the first place. The nodal flow seems to be involved in the early step of L-R determination, at least in mammals. However, many questions remain to be answered. One must examine whether the nodal flow is indeed essential for L-R determination by taking more direct approaches such as manipulating the nodal flow. Another immediate goal is to know the identity of a molecule(s) that is presumably transported by the nodal flow. An assay system for such a molecule must be established when candidates are to be tested. There are also "chicken and egg" questions; what determines the direction of the nodal flow? Is the direction a consequence of the shape of the node and/or the direction of cilial rotation? If the latter is the case, then what determines the direction of

cilial gyration? It is not obvious how one can address these questions, but it will be necessary to integrate a variety of approaches including cell biological and biophysical ones.

One other important question is how general the role of node flow is among vertebrates. The organizer region in other vertebrates, which is equivalent to the node in mouse, needs to be examined for the presence of cilia. Perhaps, node flow is not a universal mechanism. If cilia turn out to be absent in the organizer regions of nonmammals, then we must think of some other mechanism for these animals. In frog and chick, gap junction has been proposed to play a role in an early step of L-R determination (Levin and Mercola, 1998b). However, it remains to be seen how communication through gap junctions contributes to L-R specification.

B. How Is a Node-Derived Asymmetric Signal Transferred to the Lateral Plate?

L-R biased information, which is somehow generated in/ near the node, must be transferred to the LPM finally to induce *nodal* expression in the left LPM. There is a long distance from the node to LPM and a question concerns how this signal transfer is achieved over a distance. In a current model proposed in chick, a node-derived signal is relayed via the paraxial mesoderm (Pagan-Westphal and Tabin, 1998), secreting a BMP antagonist, Caronte, which will diffuse to the left LPM and induce *nodal* expression there (Rodriguez Esteban *et al.* 1999; Yokouchi *et al.,* 1999). Although this is an attractive model, it remains to be seen whether a similar pathway exists in other vertebrates.

Obviously, immediate goals would be to search for a Caronte homolog in mouse and to examine its expression and the role. It will be also necessary to examine the transcriptional regulatory mechanism that initiates the left-sided *nodal* expression. This latter approach may reveal the signal relay mechanism and clarify whether it involves BMP and a Caronte-like signal.

C. How Is Situs-Specific Morphogenesis Executed?

What exactly happens to cells in the left LPM in response to Nodal? We know little about the cellular and molecular basis of asymmetric morphogenesis. A current model suggests that left side morphology is coordinately induced by Nodal signaling, whereas the right side morphology is a result of the absence of Nodal. Although *Pitx2* is apparently involved *in situ*-specific morphogenesis, it should be clarified whether *Pitx2* is responsible for generating all the asymmetric organs or only some of them. Because *Pitx2* is a transcription factor, it presumably regulates a group of genes whose products are more directly involved in morphogenesis. In this regard, it is essential to know what genes are regulated by this transcription factor. Mutant mice specifically lacking asymmetric *Pitx2* expression may be useful for searching downstream genes. *Pitx2* may regulate distinct sets of genes in different organs because situs-specific morphogenesis seem to involve different mechanisms (i.e., looping and turning may involve a differential growth rate, whereas regression of one side may involve differential apoptosis).

D. Search for New Genes Involved in a Left-Right Pathway

Although a number of key genes have been identified, many gaps still remain in the whole pathway (especially, those involved in signal transfer from the node to the LPM and those executing situs-specific morphogenesis). Genetic approaches, such as a large-scale screening of mouse mutants with situs defects, identification of causative genes of human laterality disorders, and saturation mutagenesis of zebrafish, may reveal the remaining genes involved in L-R specification. In fact, 21 zebrafish mutants with situs defects have been identified after screening of 279 mutations (Chen *et al.* 1997). Also, novel genes may be identified by systemic screening of L-R asymmetric genes by employing cDNA microarrays.

E. Left-Right Asymmetry in Organs Other than Visceral Organs?

Most of the studies on L-R asymmetry have been limited to visceral organs. Are there L-R asymmetries in organs other than visceral organs? The most obvious candidate is the brain. Subtle morphological asymmetries may exist in human brain (such as in the planum temporale), but most anatomical approaches have proved negative. However, it is generally accepted that both hemispheres of the brains have different functions, a phenomenon known as brain lateralization. Are the functional asymmetries of the brain coupled with visceral organ asymmetry? The information is still limited, but the answer seems to be no. For example, there is no strong correlation between hand preference and visceral situs in human. Also, hearing preference is not reversed in *iv* mice showing situs inversus. The mechanism responsible for brain asymmetries appears to be different from the one for visceral asymmetry, and this remain a fascinating problem for future study. Although no genes in mammals are known to be asymmetrically expressed in developing brain (the anterior border of *nodal/lefty2*-expressing domains in the LPM ends in the heart primordia, whereas the anterior border of the *lefty1* expression domain is located in the hindbrain region), zebrafish *lefty* homologs are indeed expressed asymmetrically in developing habenulae (Concha *et al.,* 2000; Liang *et al.,* 2000).

Subtle asymmetries are also reported in other organs, although their significance is not clear at this time. For ex-

ample, limb buds are generally considered to be mirror image symmetric structures. However, forelimb buds are asymmetric at the beginning of their development; normally, the left forelimb bud is slightly larger than the right counterpart. This may be due to the fact that mesoderm cells in the left LPM that are used to express the left side-specific genes at an earlier stage subsequently contribute to the left limb bud. Curiously, some mutant mice show distinct defects in either side of the limb. In *legless* mutant animals, forelimb malformation asymmetries correlate with visceral situs; forelimb defects are predominantly found on the left side in those with situs inversus, whereas they occurs on the right side in those with normal situs (Schreiner *et al.,* 1993). *Tail short* (*Ts*) mutant mice also show limb defects in an asymmetric fashion. Thus, subtle L-R asymmetry may exist in wider areas than our current understanding leads us to believe.

Acknowledgments

The work performed in H.H.'s lab was supported by a grant from CREST, JST, Japan. I thank members of my lab (especially Yukio Saijoh and Chikara Meno) for their stimulating discussion and for preparing figures.

References

Afzelius, B. A. (1976). A human syndrome caused by immotile cilia. *Science* **193**, 317–319.

Adachi, H., Saijoh, Y., Mochida, K., Ohishi, S., Hashiguchi, H., Hirao, A., and Hamada, H. (1999). Determination of left/right asymmetric expression of *nodal* by a left side-specific enhancer with sequence similarity to a *lefty2* enhancer. *Genes Dev.* **13**, 1589–1600.

Akazawa, H., Komuro, I., Sugitani, Y., Yazaki, Y., Nagai, R., and Noda, T. (2000). Targeted disruption of the homeobox transcription factor Bapx1 results in lethal skeletal dysplasia with asplenia and gastroduodenal malformation. *Gene Cells* **5**, 499–513.

Beddington, R. and Robertson, E. J. (1999). Axis development and early asymmetry in mammals. *Cell (Cambridge, Mass.)* **96**, 195–209.

Bisgrove, B. W., Essner, J. J., and Yost, H. J. (1999). Regulation of midline development by antagonism of *lefty* and *nodal* signaling. *Development (Cambridge, UK)* **126**, 3253–3262.

Boettger, T., Wittler, L., and Kessel, M. (1999). FGF8 functions in the specification of the right body side of the chick. *Curr. Biol.* **9**, 277–290.

Britz-Cunningham, S. H., Shah, M. M., Zuppan, C. W., and Fletcher, W. H. (1995). Mutations of the conexin 43 gap-junction gene in patients with heart malformations and defects of laterality. *N. Engl. J. Med.* **332**, 1323–1329.

Brown, N. A., McCarthy, A., and Wolpert, L. (1991). Development of handed body asymmetry in mammals. *Ciba Found. Symp.* **162**, 182–201.

Brown, N. A., and Wolpert, L. (1990). The development of handedness in left/right asymmetry. *Development (Cambridge, UK)* **109**, 1–9.

Burdine, R. and Schier, A. F. (2000). Conserved and divergent mechanisms in left-right axis formation. *Genes Dev.* **14**, 763–776.

Chang, H., Zwijsen, A., Vogel, H., Huylebroeck, D., and Matzuk, M. M. (2000). Smad5 is essential for left-right asymmetry in mice. *Dev. Biol.* **219**, 71–78.

Chazaud, C., Dolle, P., and Chambon, P. (1999). Retinoic acid is required in the mouse embryo for left-right asymmetry determination and heart morphogenesis. *Development (Cambridge, UK)* **126**, 2589–2596.

Chen, J., Knowles, H. J., Hebert, J. L., and Hackett, B. P. (1998). Mutation of the mouse hepatocyte nuclear factor/forkhead homolog 4 gene results in an absence of cilia and random left-right asymmetry. *J. Clin. Invest.* **102**, 1077–1082.

Chen, J.-N., van Eeden, F. J. M., Warren, K. S., Chin, A., Nusslein-Volhardt, C., Haffter, P., and Fishman, M. C. (1997). Left-right pattern of cardiac BMP4 may drive asymmetry of the heart in zebrafish. *Development (Cambridge, UK)* **124**, 4373–4382.

Chen, X., Rubock, M. J., and Whitman, M. (1996). A transcriptional partner for Mad proteins in TGFβ signaling. *Nature (London)* **383**, 691–696.

Collignon, J., Varlet, I., and Robertson, E.J. (1996). Relationship between asymmetric *nodal* expression and the direction of embryonic turning. *Nature (London)* **381**, 155–158.

Concha, M. L., Burdine, R. D., Russell, C., Schier, A. F., and Wilson, S. W. (2000). A nodal signaling pathway regulates the laterality of neuroanatomical asymmetries in the zebrafish forebrain. *Neuron* **28**, 399–409.

Conlon, R. A. (1995). Retinoic acid and pattern formation in vertebrates. *Trends Genet.* **11**, 314–319.

Conlon, F. L., Lyons, K. M., Takaesu, N., Barth, K. S., Kispert, A. K., Hermann, B., and Robertson, E. J. (1994). A primary requirement of *nodal* in the formation and maintenance of the primitive streak in mouse. *Development (Cambridge, UK)* **120**, 1919–1928.

Constam, D. B. and Robertson, E. J. (1999). Tissue-specific requirement for the proprotein convertase Furin/SPC1 during embryonic turning and heart looping. *Development (Cambridge, UK)* **127**, 245–254.

Danos, M. C., and Yost, H. J. (1996). Role of notochord in specification of cardiac left-right orientation in zebrafish and Xenopus. *Dev. Biol.* **177**, 96–103.

Dufort, D., Schwartz, L., Harpel, K., and Rossant, J. (1998). The transcription factor HNF3β is required in the visceral endoderm for normal primitive streak morphogenesis. *Development (Cambridge, UK)* **125**, 3015–3025.

Fujii, H., Sato, T., Kaneko, S., Gotoh, O., Fujii-Kuriyama, Y., Kato, S., and Hamada, H. (1997). Metabolic inactivation of retinoic acid by a novel P450 differentially expressed in developing mouse embryos. *EMBO J.* **14**, 4163–4173.

Fujinaga, M. (1997). Development of sideness of asymmetric body structures in vertebrates. *Int. J. Dev. Biol.* **41**, 153–186.

Fujinaga, M., and Baden, J. M. (1991a). Evidence for an adrenergic mechanism in the control of body asymmetry. *Dev. Biol.* **143**, 203–205.

Fujinaga, M., and Baden, J. M. (1991b). Critical period of rat development when sideness of body asymmetry is determined. *Teratology* **43**, 95–100.

Gage, P. J., Suh, H., and Camper, S. (1999). Dosage requirement of Pitx2 for development of multiple organs. *Development (Cambridge, UK)* **126**, 4643–4651.

Gaio, U., Schweickert, A., Fischer, A., Garratt, A. N., Muller, T., Ozcelik, C., Lankes, W., Strehle, M., Britsch, S., Blum, M. and Birchmeier, C. (1999). A role of the *cryptic* gene in the correct establishment of the left-right axis. *Curr. Biol.* **9**, 1339–1342.

Gebbia, M., Towbin, J. A., and Casey, B. (1996). Failure to detect connexin 43 mutations in 38 cases of sporadic and familial heterotaxy. *Circulation* **94**, 1909–1912.

Gebbia, M., Ferrero, G. B., Pilia, G., Bassi, M. T., Aylsworth, A. S., Penmann-Splitt, M., Bird, L., Bamforth, J. S., Burn, J., Schlessinger, D., Nelson, D., and Casey, B. (1997). X-linked situs abnormalities result from mutation in ZIC3. *Nat. Genet.* **17**, 305–308.

Gritsman, K., Zhang, J., Cheng, S., Heckscher, E., Talboy, W. S., and Schier, A. F. (1999). The EGF-CFC protein one-eyed pinhead is essential for Nodal signaling. *Cell (Cambridge, Mass.)* **97**, 121–132.

Harvey, R. P. (1998). Links in the left/right axial pathway. *Cell* **94**, 273–276.

Heymer, J., Kuehn, M., and Ruther, U. (1997). The expression pattern of *nodal* and *lefty* in the mouse mutant *Ft* suggests a function in the establishment of handedness. *Mech. Dev.* **66**, 5–11.

Hummel, K. P., and Chapman, D. B. (1959). Visceral inversion and associated anomalies in the mouse. *J. Hered.* **50**, 9–13.

Isaac, A., Sargent, M. G., and Cooke, J. (1997). Control of vertebrate left-right asymmetry by a *Snail*-related zinc finger gene. *Science* **275**, 1301–1304.

Kartagener, M. (1933). Zur Pathogenese der Bronchiektasien bei Situs viscerum inversus. *Beitr. Klin. Tuberk.* **83**, 489–501.

Kaufman, S. A., Shymko, R., and Trabert, K. (1978). Control of sequential compartment formation in Drosophila. *Science* **199**, 259–270.

King, T., and Brown, N. A. (1997). Embryonic asymmetry: Left TGFβ at the right time. Curr. Biol. **7**, 212–215.

Kitamura, K., Miura, H., Miyagawa-Tomita, S., Yanazawa, M., Katoh-Fukui, Y., Suzuki, R., Ohuchi, H., Suehiro, A., Motegi, Y., Nakahara, Y., Kondo, S., and Yokoyama, M. (1999). Mouse Pitx2 deficiency leads to anomalies of the ventral body wall, heart, extra-and periocular mesoderm and right pulmonary isomerism. *Development (Cambridge, UK)* **126**, 5749–5758.

Koto, M., Adachi, J. and Shimizu, A. (1987). A new mutation of primary ciliary diskinesia (PCD) with visceral inversion and hydrocephalus. *Rat News Lett.* **18**, 14–15.

Lee, S.-J. (1990). Identification of a novel member of the transforming growth factor-β superfamily. *Mol. Endocrinol.* **4**, 1034–1040.

Lettice, L. A., Purdie, L. A., Carlson, G. J., Kilanowski, F., Dorin, J., and Hill, R. E. (1999). The mouse bagpipe gene controls development of axial skeleton, skull and spleen. *Proc. Natl. Acad. Sci. U.S.A.* **96**, 9695–9700.

Levin, M., and Mercola, M. (1998a). The compulsion of chirality: Toward an understanding of left-right asymmetry. *Genes Dev.* **12**, 763–769.

Levin, M., and Mercola, M. (1998b). Gap junctions are involved in the early generation of left-right asymmetry. *Dev. Biol.* **203**, 90–105.

Levin, M., and Mercola, M. (1999). Gap junction-mediated transfer of left-right patterning signals in the early chick blastoderm is upstream of Shh asymmetry in the node. *Development (Cambridge, UK)* **126**, 4703–4714.

Levin, M., Johnson, R. L., Stern, C. D., Kuehn, M., and Tabin, C. (1995). Molecular pathway determining left-right asymmetry in chick embryogenesis. *Cell (Cambridge, Mass.)* **82**, 803–814.

Levin, M., Pagan, S., Roberts, D. J., Cooke, J., Kuehn, M. R., and Tabin, C. (1997a). Left/right patterning signals and the independent regulation of different aspects of situs in the chick embryo. *Dev. Biol.* **189**, 57–67.

Liang, J. Q., Etheridge, A., Hantsoo, L., Rubinstein, A. L., Nowak, S. J., Izpisua-Belmonte, J. C., and Halpern, M. E. (2000). Asymmetric nodal signaling in the zebrafish diencephalon positions the pineal organ. *Development (Cambridge, UK)* **127**, 5101–5112.

Lin, C. R., Kioussi, C., O'Connell, S., Briata, P., Szeto, D., Liu, F., Izpisua-Belmonte, J. C., and Rosenfeld, M. G. (1999). Pitx2 regulates lung asymmetry, cardiac positioning and pituitary and tooth morphogenesis. *Nature (London)* **401**, 279–282.

Lo, C.W. (1999). Genes, gene knockouts and mutations in the analysis of gap junctions. *Dev. Genet.* **24**, 1–4.

Logan, M., Pagan-Westphal, S. M., Smith, D. M., Paganessi, L., and Tabin, C. (1998). The transcription factor Pitx2 mediates situs-specific morphogenesis in response to left-right asymmetric signals. *Cell (Cambridge, Mass.)* **94**, 307–317.

Lowe, L., Supp, D. M., Sampath, K., Yokoyama, T., Wright, C. V. E., Potter, S. S., Overbeek, P., and Kuehn, M. R. (1996). Conserved left-right asymmetry of *nodal* expression and alteration in murine situs inversus. *Nature (London)* **381**, 158–161.

Lu, M. F., Pressman, C., Dyer, R., Johnson, R. L., and Martin, J. F. (1999). Function of Rieger syndrome gene in left-right asymmetry and craniofacial development. *Nature (London)* **401**, 276–278.

Lustig, K. D., Kroll, K., Sun, E., Ramos, R., Elmenford, H., and Kirschner, M. W. (1996). A Xenopus nodal-related gene that acts in synergy with noggin to induce complete secondary axis and notochord formation. *Development (Cambridge, UK)* **122**, 3275–3282.

Marszalek, J. R., Ruiz-Lozano, P., Roberts, E., Chien, K. R., and Goldstein, L. S. (1999). Situs inversus and embryonic ciliary morphogenesis defects in mouse mutants lacking the KIF3A subunit of kinesin II. *Proc. Natl. Acad. Sci. U.S.A.* **96,** 5043–5048.

Matzuk, M. M., Kumar, T. R., and Bradley, A. (1995) Different phenotypes for mice deficient in either activins or activin receptor type II. *Nature (London)* **374,** 356–360.

Melloy, P. G., Ewart, J. L., Cohen, M. F., Desmond, M. E., Kuehn, M. R., and Lo, C. W. (1998). No turning, a mouse mutation causing left-right and axial patterning defects. *Dev. Biol.* **193,** 77–89.

Meno, C., Saijoh, Y., Fujii, H., Ikeda, M., Yokoyama, T., Yokoyama, M., Toyoda, Y., and Hamada, H. (1996). Left-right asymmetric expression of the TGFβ-family member *lefty* in mouse embryos. *Nature (London)* **381,** 151–155.

Meno, C., Itoh, Y., Saijoh, Y., Matsuda, Y., Tashiro, K., Kuhara, S., and Hamada, H. (1997). Two closely-related left-right asymmetrically expressed genes, *lefty-1* and *lefty-2:* their distinct expression domains, chromosomal linkage and direct neuralizing activity in *Xenopus* embryos. *Genes Cells* **2,** 513–524.

Meno, C., Shimono, A., Saijoh, Y., Yashiro, K., Oishi, S., Mochida, K., Noji, S., Kondoh, H., and Hamada, H. (1998). *lefty-1* is required for left-right determination as a regulator of *lefty-2* and *nodal. Cell (Cambridge, Mass.)* **394,** 287–297.

Meno, C., Gritman, K., Ohishi, S., Ohfuji, Y., Heckscher, E., Mochida, K., Shimono, A., Kondoh, H., Talbot, W. S., Robertson, E. J., Schier, A. F., and Hamada, H. (1999). Mouse Lefty2 and zebrafish antivin are feedback inhibitors of Nodal signaling during vertebrate gastrulation. *Mol. Cell* **4,** 287–298.

Meno, C., Takeuchi, J., Sakuma, R., Koshiba-Takeuchi, K., Ohishi, S., Saijoh, Y., Miyazaki, J., ten Dejike, P., Ogura, T., and Hamada, H. (2001). Diffusion of Nodal signaling activity in the absence of the feedback inhibitor Lefty2. *Dev. Cell* **1,** 127–138.

Metzler, M., Gerttz, A., Sarkar, M., Schachter, H., Schrader, J. W. and Marth, J. D. (1994). Complex asparagine-linked oligosaccharides are required for morphogenic events during post-implantation development. *EMBO J.* **13,** 2056–2065.

Meyer, E. N., and Martin, G. R. (1999). Differences in left-right axis pathways in mouse and chick: Functions of FGF8 and SHH. *Science* **285,** 403–406.

Mochizuki, T., Saijoh, Y., Tsuchiya, K., Shirayoshi, Y., Takai, S., Taya, C., Yonekawa, H., Yamada, K., Nihei, H., Nakatsuji, N., Overbeek, P., Hamada, H., and Yokoyama, T. (1998). Cloning of *inv,* a gene that controls left/right asymmetry and kidney development. *Nature (London)* **395,** 177–181.

Morgan, D., Turnpenny, L., Goodship, J., Dai, W., Majumder, K., Mattews, L., Gardner, A., Schuster, G., Vien, L., Harrison, W., Elder, F. F. B., Pennman-Splitt, M., Overbeek, P., and Strachan, T. (1998). Inversin, a novel gene in the vertebrate left-right axis pathway, is partially deleted in the inv mouse. *Nat. Genet.* **20,** 149–156.

Morishima, M., Ando, M., and Takao, A. (1991). Visceroatrial heterotaxy syndrome in the NOD mouse with special reference to atrial situs. *Teratology* **44,** 91–100.

Murcia, N. S., Richard, W. G., Yoder, B. K., Mucenski, M. L., Dunlap, J. R., and Woychik, R. P. (2000). The Oak Ridge Polycystic kidney (orpk) disease gene is required for left-right axis determination. *Development (Cambridge, UK)* **127,** 2347–2355.

Nomura, M., and Li, E. (1998). Roles for Smad2 in mesoderm formation, left-right determination and craniofacial development in mice. *Nature (London)* **393,** 786–789.

Nonaka, S., Tanaka, Y., Okada, Y., Takeda, S., Harada, A., Kanai, Y., Kido, M., and Hirokawa, N. (1998). Randomization of left-right asymmetry due to loss of nodal cilia generating leftward flow of extraembryonic fluid in mice lacking KIF3B motor protein. *Cell (Cambridge, Mass.)* **95,** 829–837.

Norris, D. P., and Robertson, E. J. (1999). Node-specific and asymmetric *nodal* expression patterns are controlled by two distinct cis-acting regulatory elements. *Genes Dev.* **13,** 1575–1589.

Oh, S. P., and Li, E. (1997). The signaling pathway mediated by the type IIB activin receptor controls axial patterning and lateral asymmetry in the mouse. *Genes Dev.* **11,** 1812–1826.

Okada, Y., Nonaka, S., Tanaka, Y., Saijoh, Y., Hamada, H., and Hirokawa, N. (1999). Abnormal nodal flow precedes *situs inversus* in *iv* and *inv* mutant mice. *Mol. Cell* **4,** 459–468.

Pagan-Westphal, S. M., and Tabin, C. (1998). The transfer of left-right positional information during chick embryogenesis. *Cell (Cambridge, Mass.)* **93,** 25–35.

Pazour, G. J., Dickert, B. L., Vucica, Y., Seeley, E. S., Rosenbaum, J. L., Witman, G. B., and Cole, D.Æ. (2000). Chlamydomonas IFT88 and its mouse homolog, polycyctic kidney disease gene tg737, are required for assembly of cilia and flagella. *J. Cell Biol.* **151,** 709–718.

Piedra, M. E., Icardo, J. M., Albajar, M., Rodriguez-Rey, J. C., and Ros, M. A. (1998). Pitx2 participates in the late phase of the pathway controlling left-right asymmetry. *Cell (Cambridge, Mass.)* **94,** 319–324.

Rankin, C. T., Bunton, T., Lawler, A. M., and Lee, S.-J. (2000). Regulation of left-right patterning in mice by growth/differentiation factor-1. *Nat. Genet.* **24,** 262–265.

Rodriguez-Esteban, C., Capdevila, J., Economides, A. N., Pascual, J., Ortiz, A., and Izpisua-Belmonte, J.-C. (1999). The novel Cer-like protein Caronte mediates the establishment of embryonic left-right asymmetry. *Nature (London)* **401,** 243–251.

Roebroek, A. J. M., Umans, L., Pauli, I. G. L., Robertson, E. J., van Leiven, F., Van de Ven, W. J. M., and Constam, D. B. (1998). Failure of ventral closure and axial rotation in embryos lacking the protein convertase Furin. *Development (Cambridge, UK)* **125,** 4863–4876.

Ryan, A. K., Blumberg, B., Rodriguez-Esteban, C., Yonei-Tamura, S., Tamura, K., Tsukui, T., de la Pena, J., Sabbagh, W., Greenwald, J., Choe, S., Norris, D., Robertson, E. J. R., Evans, R. M., Rosenfeld, M. G., and Izpisua-Belmonte, J. C. (1998). Pitx2 determines left-right asymmetry of internal organs in vertebrates. *Nature (London)* **394,** 545–551.

Saijoh, Y., Adachi, H., Mochida, K., Ohishi, S., Hirao, A., and Hamada, H. (1999). Distinct transcriptional regulatory mechanisms underlie left-right asymmetric expression of *lefty-1* and *lefty-2. Genes Dev.* **13,** 259–269.

Saijoh, Y., Adachi, H., Sakuma, R., Yeo, C.-Y., Yashiro, K., Watanabe, M., Hashiguchi, H., Mochida, K., Ohishi, S., Kawabata, M., Miyazoni, K., Whitman, M., and Hamada, H. (2000). Left-right asymmetric expression of *lefty2* and *nodal* is induced by a signaling pathway that includes the transcription factor FAST2. *Mol. Cell* **5,** 35–47.

Sampath, K., Cheng, A. M. S., Frisch, A., and Wright, C. V. E. (1997). Functional differences among nodal-related genes in left-right axis determination. *Development (Cambridge, UK)* **124,** 3293–3302.

Schier, A. F., and Shen, M. (2000). Nodal signaling. *Nature (London)* **403,** 385–389.

Schneider, A., Mijalski, T., Schlange, T., Dai, W., Overbeek, P., Arnold, H., and Brand, T. (1999). The homeobox gene *Nkx3.2* is a target of left-right signaling and is expressed on opposite sides in chick and mouse embryos. *Curr. Biol.* **9,** 911–914.

Schreiner, C. M., Scott, W. J., Supp, D. M., and Potter, S. S. (1993). Correlation of forelimb malformation asymmetries with visceral organ situs in the transgenic mouse insertional mutation, *legless. Dev. Biol.* **158,** 560–562.

Sefton, M., Sanchez, S., and Nieto, M. A. (1998). Conserved and divergent roles for members of the *Snail* family of transcription factors in the chick and mouse embryo. *Development (Cambridge, UK)* **125,** 3111–3121.

Shiratori, H., Sakuma, R., Watanabe, M., Hashiguchi, H., Mochida, K., Nishino, J., Sakai, Y., Saijoh, Y., Whitman, M. and Hamada, H. (2001). Two step regulation of asymmetric *Pitx2* expression: Initiation by Nodal signaling and maintenance by Nkx2. *Mol.Cell* **7,** 137–149.

Sulik, K., Dehart, D. B., Inagaki, T., Carson, J. L., Vrablic, T., Gesteland, K., and Schoenwolf, G. C. (1994). Morphogenesis of the murine node and notochordal plate. *Dev. Dyn.* **201,** 260–278.

Supp, D. M., Witte, D. P., Potter, S. S., and Brueckner, M. (1997). Mutation of an axonemal dynein affects left-right asymmetry in *inversus viscerum* mice. *Nature (London)* **389,** 963–966.

Supp, D. M., Brueckner, M., Kuehn, M. R., Witte, D. P., Lowe, L. A. McGrath, J., Corrales, J., and Potter, S. S. (1999). Targeted deletion of the ATP-binding domain of left-right dynein confirms its role in specifying development of left-right asymmetries. *Development (Cambridge, UK)* **126,** 5495–5504.

Takeda, S., Yonekawa, Y., Tanaka, Y., Okada, Y., Nonaka, S., and Hirokawa, N. (1999). Left-right asymmetry and kinesin superfamily protein KIF3A: New insights in determination of laterality and mesoderm induction by *kif3A*-/- mice analysis. *J. Cell Biol.* **145,** 825–836.

Thisse, C., and Thisse, B. (1999). Antivin, a novel and divergent member of the TGF*β* superfamily, negatively regulates mesoderm induction. *Development (Cambridge, UK)* **126,** 229–240.

Tribioli, C. and Lufkin, T. (1999). The murine Bapx1 homeobox gene plays a critical role in embryonic development of the axial skeleton and spleen. *Development (Cambridge, UK)* **126,** 5699–5711.

Tsukui, T., Capdevila, J., Tamura, K., Ruiz-Lozano, P., Rodriguez-Esteban, C., Yonei-Tamura, S., Magallon, J., Chandraratna, R. A., Chien, K., Blumberg, B., Evans, R. M., and Izpisua-Belmonte, J. C. (1999). Multiple left-right asymmetry defects in Shh(-/-) mutant mice unveil a convergence of the shh and retinoic acid pathways in the control of Lefty1. *Proc. Natl. Acad. Sci. USA* **96,** 11376–11381.

Turing, A. M. (1952). The chemical basis of morphogenesis. *Philos. Trans. R. Soc. London, Ser. B* **237,** 37–72.

Varlet, I., and Robertson, E. J. (1997). Left-right asymmetry in vertebrates. *Curr. Opin. Genet. Dev.* **6,** 519–523.

Wagner, M. K., and Yost, H. J. (2000). The role of nodal cilia. *Curr. Biol.* **10,** 149–151.

Waldrip, W., Bikoff, E., Hoodless, P., Wrana, J., and Robertson, E. J. (1998). Smad2 signaling in extraembryonic tissues determines anterior-posterior polarity of the early mouse embryo. *Cell (Cambridge, Mass.)* **92,** 797–808.

Wasiak, S., and Lohnes, D. (1999). Retinoic scid affects left-right patterning. *Dev. Biol.* **215,** 332–342.

Whitman, M. (1998). Smads and early developmental signaling by the TGF*β* superfamily. *Genes Dev.* **12,** 2445–2462.

Yang, Y.-T., Gritman, K., Ding, J., Burdine, R. D., Corrales, J. D., Price, S. M., Talbot, W. S., Schier, A. F., and Shen, M. M. (1999). Conserved requirement for EGF-CFC genes in vertebrate left-right axis formation. *Genes Dev.* **13,** 2527–2537.

Yashiro, K., Saijoh, Y., Sakuma, R., Tada, M., Tomita, N., Amano, K., Matsuda, Y., Monden, M., Okada, S., and Hamada, H. (2000). Distinct transcriptional regulation and phylogenetic divergence of human *LEFTY* genes. *Genes Cells* **5,** 343–357.

Yokouchi, Y., Vogan, K. J., Pearse, R. V., II, and Tabin, C. (1999). Antagonistic signaling by *Caronte*, a novel *Cerberus*-related gene, establishes left-right asymmetric gene expression. *Cell (Cambridge, Mass.)* **98,** 573–583.

Yokoyama, T., Copeland, N. G., Jenkins, N. A., Montgomery, C. A., Elder, F. F. B., and Overbeek, P. (1993). Reversal of left-right asymmetry: A situs inversus mutation. *Science* **260,** 679–682.

Yoshioka, H., Meno, C., Koshiba, K., Sugihara, M., Itoh, H., Ishimaru, Y., Inoue, T., Ohuchi, H., Semina, E. V., Murray, J. C., Hamada, H., and Noji, S. (1998). Pitx2, a bicoid type homeobox gene, is involved in a Lefty-signaling pathway in determination of left-right asymmetry. *Cell (Cambridge, Mass.)* **94,** 299–305.

Yost, H. J. (1991). Development of the left-right axis in amphibians. *Ciba Found. Symp.* **162,** 165–176.

Yost, H. J. (1995). Vertebrate left-right development. *Cell (Cambridge, Mass.)* **82,** 689–692.

Zhou, X., Sasaki, H., Lowe, L., Hogan, B., and Kuehn, M. R. (1993). Nodal is a novel TGF*β*-like gene expressed in the mouse node during gastrulation. *Nature (London)* **361,** 543–547.

5

Patterning, Regionalization, and Cell Differentiation in the Forebrain

Oscar Marín and John L. R. Rubenstein

Nina Ireland Laboratory of Developmental Neurobiology, Department of Psychiatry, Langley Porter Psychiatric Institute, University of California, San Francisco, San Francisco, California 94143

I. Organization of the Forebrain

II. Early Patterning and Regional Specification of the Forebrain

III. Morphogenetic Mechanisms in the Forebrain

IV. Control of Neurogenesis and Cell-Type Specification in the Forebrain

References

I. Organization of the Forebrain

The mammalian forebrain is possibly the most complex of all biological structures. It comprises an intricate aggregation of highly heterogeneous structures that control most aspects of homeostasis, learning, and behavior. This chapter focuses of the developmental strategies that are used to generate anatomically and functionally unique regions within the forebrain.

The forebrain comprises a complex set of structures that derive from the most anterior region of the neural tube, the prosencephalon. Soon after the closure of the neural tube, the primary prosencephalon subdivides into two major components, the caudal diencephalon and the secondary prosencephalon. The core of the secondary prosencephalon is the conventional rostral diencephalon/hypothalamus. The secondary prosencephalon also includes the telencephalic and optic vesicles, which evaginate from the dorsal aspect of the rostral diencephalon. Like more caudal regions of the neural tube, the embryonic forebrain appears to be organized into transverse and longitudinal subdivisions (Fig. 1). Increasing evidence suggests that transversal subdivision of the prosencephalon generates segment-like domains called prosomeres (Bergquist and Källen, 1954; Figdor and Stern, 1993; Puelles and Rubenstein, 1993; Rubenstein *et al.*, 1994, Puelles, 1995). These structures are more evident in the caudal diencephalon, where they constitute prosomeres 1–3. Nevertheless, there is evidence that the rostral diencephalon is similarly organized into prosomeres 4–6. The relationship of these postulated segments to telencephalic subdivisions is still uncertain.

The prosencephalon is also subdivided into longitudinal domains that are related to the longitudinal subdivisions of more caudal regions of the neural tube. From ventral to dorsal these domains are the floor, basal, alar, and roof plates (Fig.1; see Puelles and Rubenstein, 2000, for a detailed dis-

75

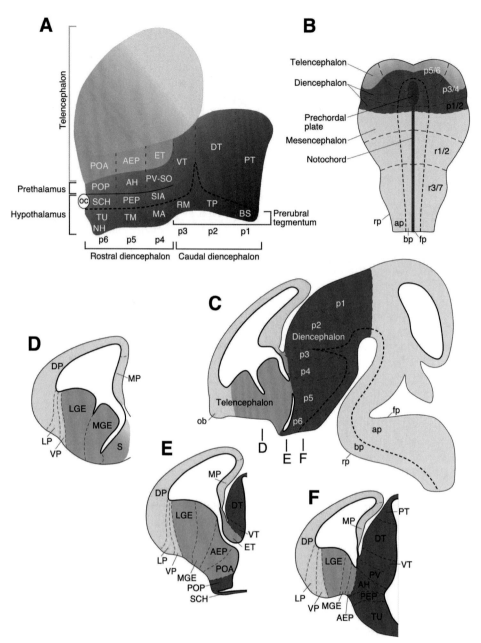

Figure 1 Organization of the prosencephalon. (A) Lateral view of the prosencephalon showing its diencephalic (blue) subdivisions as proposed by the prosomeric model. In panels A–C, transverse subdivisions are separated by red lines and longitudinal subdivisions are separated by black lines (see Puelles and Rubenstein, in press). Note that ventral properties extend dorsally at the zona limitans, a region between p2 and p3. The topologic relationships of the subdivisions in the diencephalon to subdivisions in the telencephalon (green-yellow areas) are uncertain. The rostral diencephalon consists of ventral (hypothalamus) and dorsal (prethalamic) regions. The caudal diencephalon also consists of ventral (prerubral tegmentum) and dorsal (thalamic: VT, DT, and PT) subdivisions. (B) Schema of the neural plate. The axial mesendoderm (prechordal plate and notochord), which lie below the neural plate, are shown in gray. (C) Schema of a saggital section through the brain of an E13.5 mouse showing the prosencephalic subdivisions proposed by the prosomeric model. (D–F) Transversal sections through the telencephalon and diencephalon; the planes of section are indicated in part (C). In the telencephalon, the pallium is depicted in yellow and the subpallium in green; its subdivisions, as defined in Puelles *et al.* (2000), are separated by black dashed lines. Abbreviations: AEP, anterior entopeduncular area; AH, anterior hypothalamus; ap, alar plate; bp, basal plate; BS, basal synencephalon; DP, dorsal pallium; DT, dorsal thalamus; ET, eminentia thalami; LGE, lateral ganglionic eminence; LP, lateral pallium; MA, mammillary region; MGE, medial ganglionic eminence; MP, medial pallium; NH, neurohypophisis; OC, optic chiasm; PEP, posterior entopeduncular area; POA, anterior preoptic area; POP, posterior preoptic area; PT, pretectum; PV-SO, paraventricular nucleus–supraoptic nucleus; p1–p6, prosomeres 1–6; SCH, suprachiasmatic nucleus; RM, retromammillary region; rp, roof plate; S, septum; TM, tuberomammillary region; TP, posterior tuberculum; TU, tuberal region; VP, vental pallium; VT, ventral thalamus.

cussion about this subject). As in caudal regions of the neural tube, the floor and roof plates constitute the ventral and dorsal midline, respectively. The basal plate, on the other hand, forms the prerubral tegmentum in the caudal diencephalon (p1–p3), whereas rostrally, it give rises to the hypothalamus *sensu stricto* (p4–p6), roughly limited dorsally by the optic tract and chiasm. The alar plate of the caudal diencephalon is constituted by the pretectum (p1), the dorsal thalamus/epithalamus (p2), and the ventral thalamus (p3). Anteriorly (p4–p6), the alar plate comprises diverse prethalamic regions (including the eminentia thalami, peduncular, anterior preoptic, and suprachiasmatic areas) of the conventional "hypothalamus" and "subthalamus." These regions separately continue into the optic and telencephalic vesicles.

The topological arrangement of these forebrain subdivisions is a consequence of the location of their primordia in the neural plate. Fate mapping studies have demonstrated that the prosencephalon derives from the anteriormost portion of the neural plate (Couly and Le Douarin, 1987; Eagleson and Harris, 1990; Inoue *et al.,* 2000; Cobos *et al.,* 2001). In this region, the roof, alar and basal plates appear to course in a semicircle around the rostral end of the floor plate (Fig. 1). According to the prosomeric model, transverse boundaries in the forebrain (prosomeres) divide the longitudinal domains orthogonally as radial lines (Puelles and Rubenstein, 1993; Rubenstein *et al.,* 1994; Fig.1). Fate mapping experiments suggest that the telencephalic vesicles derive from the anterolateral neural plate (Fig. 1; reviewed in Rubenstein *et al.,* 1998; Wilson and Rubenstein, 2000). This region includes the lateral part of the anterior neural ridge, the neuroectodermal tissue located at the border of the neural plate (Fig. 1; Couly and Le Douarin, 1988; Eagleson *et al.,* 1995). As described later in this chapter, the anterior neural ridge plays a prominent role in telencephalic development (Graziadei and Monti-Graziadei, 1992; Byrd and Burd, 1993; Shimamura and Rubenstein, 1997; Houart *et al.,* 1998).

A. Organization of the Telencephalon

Regionalization of the telencephalon is a progressive process that becomes apparent very early, during neurulation and the evagination of the telencephalic vesicles (see later section). The two major telencephalic subdivisions are the pallium (roof) and the subpallium (base). The pallium can be conceived as being organized into four main radial subdivisions: the medial, dorsal, lateral, and ventral pallium (Puelles *et al.,* 1999, 2000; Fig. 1). All of these pallial domains form cortical structures (e.g., superficial laminar neuronal zones), but the lateral and ventral pallium parts also develop deep-lying nuclear structures, integrated in the claustro-amygdaloid complex (therefore "pallial" is not synonymous with "cortical"). The medial pallium, or limbic cortex, includes the hippocampal complex and the subicular

regions; the dorsal pallium corresponds to the mesocortex and isocortex (also less appropriately known as neocortex), which develops between the medial and the lateral pallium; the lateral pallium is conceived of as including the primary olfactory cortex, the dorsolateral claustrum, parts of the amygdala (basolateral nucleus, amygdalo-hippocampal area), and the adjacent mesocortical periamygdaloid cortex and entorhinal cortex. The lateral and medial pallia fuse together rostral and caudal to the dorsal pallium, forming an allocortical cortical limbus around the meso- and isocortex. The ventral pallium abuts the subpallium and includes the ventromedial claustrum, parts of the amygdala (lateral, basomedial, and corticolateral nucleus), and parts of the olfactory system (olfactory bulb, anterior olfactory nucleus, nucleus of the lateral olfactory tract) and the dorsalmost septum.

The subpallium is currently conceived as including three primary subdivisions: the striatal, pallidal, and telencephalic stalk domains (Fig. 1), all of which extend medially into the septum. The striatal domain abuts the ventral pallium, and its progenitor zone is known as the lateral ganglionic eminence (LGE). Derivatives from the striatal domain include the caudoputamen nucleus, nucleus accumbens, part of the septum, and central parts of the amygdala. Below the striatal domain lies the pallidal domain, which derives from progenitor cells in the medial ganglionic eminence (MGE), and includes the globus pallidus, ventral pallidum, and part of the bed nucleus of the stria terminalis and septum. Striatal and pallidal cell populations interdigitate in the olfactory tuberculum, at the ventral surface of the telencephalon. The subpallial telencephalic stalk, located close below the pallidal domain, contains the anterior entopeduncular and preoptic areas (AEP and POA) and is the site where the major tracts entering or streaming out of the telencephalon pass through (i.e., the medial and lateral forebrain bundles, the fornix, the amygdalofugal tracts, the anterior commissure). There is also a less defined pallial part of the telencephalic stalk, which connects the pallial amygdala and caudal olfactory and limbic cortex with the eminentia thalami. This area contains the bed nucleus stria terminalis and the bed nucleus stria medullaris. In this view, striatal, pallidal, and pallial domains converge at different levels of the telencephalon to form functionally and embryologically heterogeneous structures, such as the septum, rostrally; the conventional basal ganglia and related pallial nuclei and cortex, at intermediate levels; and the amygdala, caudally (Puelles *et al.,* 1999, 2000; Puelles and Rubenstein, 2000; Fig. 1). Finally, the midline and paramedian areas of the telencephalon form specialized structures. The rostrodorsal midline forms the commissural plate, through which the major commissural pathways decussate (corpus callosum and hippocampal commissure). More caudally, the roof-derived tissues form the choroid plexus, which continues into the choroid plexus of the dorsal diencephalon.

II. Early Patterning and Regional Specification of the Forebrain

A. Patterning Subdivisions of the Prosencephalic Neural Plate

The embryonic forebrain, or prosencephalon, derives from the anterior neural plate. Induction and specification of the anterior neural plate are described in detail in Chapters 3 and 6. In this section we review information regarding how subdivisions of the anterior neural plate are generated and ultimately give rise to subdivisions in the prosencephalon.

There is increasing evidence that patterning of all regions of the neural plate involves two general sets of mechanisms, one that patterns along the anterior-posterior (AP) axis and another that patterns along the medial-lateral (ML) axis. AP patterning generates transverse subdivisions of the neural plate and involves both anteriorizing and posteriorizing signals from multiple tissues (reviewed in Niehrs, 1999; Wilson and Rubenstein, 2000). Mesendodermal tissues underlying the anterior neural plate are implicated in providing vertically transmitted signals (see Chapter 3).

Anteriorizing substances, such as Cerberus and Dickkopf, are secreted proteins that inhibit signaling by transforming growth factor β (TGF-β) [nodal type and bone morphogenetic protein (BMP) type] and/or WNT proteins (reviewed in Niehrs, 1999; Beddington and Robertson, 1999). In mammals, some of these substances are produced in the node (e.g., *noggin* and *chordin*) or in the anterior visceral endoderm (e.g., *cerberus*) (Beddington and Robertson, 1999). Accordingly, mice lacking expression of the BMP antagonists noggin and chordin do not form a forebrain (Bachiller *et al.*, 2000). In addition, recent studies in zebrafish suggest that antivin proteins have anteriorizing properties by virtue of their ability to antagonize activin-related TGF-β molecules (Thisse *et al.*, 2000). The consequences of disrupting the function of WNT antagonists in mice have not yet been reported.

Posterior neural plate fates can be induced by FGFs, WNTs, retinoids, and activin/nodal-related TGF-β molecules. Thus, zebrafish lacking nodal-type TGF-β signals (*cyclops* and *squint*) or lacking a transmembrane protein implicated in TGF-β signaling (*one-eyed pinhead*), have increased forebrain structures (Gritsman *et al.*, 1999). On the other hand, zebrafish carrying the *masterblind* mutation exhibit a severe reduction in anterior forebrain tissues (telencephalon and eyes) and an expansion of posterior forebrain (e.g., pineal gland) tissues (Masai *et al.*, 1997). *Masterblind* appears to promote telencephalic and eye development by inhibiting the expression of the homeodomain containing gene *floating head,* which is required to promote neurogenesis in the posterior diencephalon (Masai *et al.*, 1997).

Once the anterior neural plate is formed, AP patterning within the forebrain appears to be controlled in part by ectodermal tissue at the rostral end of the neural plate (Shimamura and Rubenstein, 1997; Houart *et al.*, 1998; E. Storm, J. L. R. Rubenstein, and G. R. Martin, unpublished observations). Several lines of evidence suggest that the patterning properties of this tissue, known as the anterior neural ridge, are in part mediated by FGF8. For example, reduction in the expression of Fgf8 in the anterior neural ridge leads to rostral truncations and midline defects in the forebrain (Shanmugalingam *et al.*, 2000; E. Storm, J. L. R. Rubenstein, and G. R. Martin, unpublished observations). This phenotype is similar to that described in mice lacking either the winged helix *BF1* (*Foxg1*) or *Vax1* homeodomain transcription factor genes (Xuan *et al.*, 1995; Dou *et al.*, 1999). Both of these genes are expressed in the anterior neural ridge and are activated by FGF8 (Shimamura *et al.*, 1995, Shimamura and Rubenstein, 1997; Hallonet *et al.*, 1998; E. Storm, J. L. R. Rubenstein, and G. R. Martin, unpublished observations).

ML patterning generates longitudinal subdivisions of the neural plate and involves signals from the axial mesendoderm and nonneural ectoderm (reviewed in Rubenstein and Beachy, 1998; Lee and Jessell, 1999; Wilson and Rubenstein, 2000). Medial patterning of the anterior forebrain is primarily regulated by the prechordal mesoderm, whereas medial patterning of more posterior parts of the forebrain may be controlled by the anterior portion of the notochord (Ruiz i Altaba, 1993; Pera and Kessel, 1997; Shimamura and Rubenstein, 1997; Dale *et al.*, 1997; Dodd *et al.*, 1998). Several lines of evidence suggest that the secreted molecule Sonic Hedgehog (SHH) is responsible for the medial patterning activity of the notochord and the prechordal mesoderm. Like the notochord, the prechordal mesoderm (and the dorsal foregut) expresses *Shh* (Echelard *et al.*, 1993; Krauss *et al.*, 1993; Roelink *et al.*, 1994; Marti *et al.*, 1995; Ericson *et al.*, 1995; Shimamura *et al.*, 1995), and SHH can induce expression of medial markers in the anterior neural plate (Ericson *et al.*, 1995; Dale *et al.*, 1997; Pera and Kessel, 1997; Shimamura and Rubenstein, 1997; Kohtz *et al.*, 1998; Qiu *et al.*, 1998). Thus, defects that affect the formation or differentiation of the axial mesendoderm (e.g., *cyclops, nodal,* and *one-eyed pinhead* mutations; Rebagliati *et al.*, 1998; Sampath *et al.*, 1998; Shinya *et al.*, 1999) or that directly disrupt the production or signal transduction of SHH (Barth and Wilson, 1995; Chiang *et al.*, 1996; Goodrich *et al.*, 1997; Cooper *et al.*, 1998; Incardona *et al.*, 1998; Gaffield *et al.*, 1999; Gaiano *et al.*, 1999) affect medial patterning of the forebrain. Severe defects in medial patterning in turn lead to the loss of the prosencephalic basal plate, including a large portion of the hypothalamus, and also affect the development of adjacent tissues. Thus, cyclopia and holoprosencephaly can result from defective patterning of the medial eye-field structures and the basal telencephalon, respectively (Muenke and Beachy, 2000).

Lateral patterning of the anterior neural plate appears to be mediated by members of the TGF-β superfamily, such as BMPs and growth differentiation factors (GDF), largely derived from the neural ridge and nonneural ectoderm flanking the anterior neural plate. For example, nonneural ectoderm or BMPs can induce the expression of dorsal (lateral) markers in anterior neural plate explants (reviewed in Lee and Jessell, 1999). Moreover, recent observations in mutant zebrafish embryos with compromised BMP signaling suggest that BMP activity is required for the specification of all dorsal (lateral) cell fates, including those of the forebrain region (Barth et al., 1999).

The combination of AP and ML patterning mechanisms in the anterior neural plate are hypothesized to generate a gridlike organization of primordial morphogenetic units that contribute to the formation of the major prosencephalic subdivisions. Superimposed on these subdivisions are the primordia of the optic and telencephalic vesicles, olfactory bulbs, and posterior pituitary. The induction and patterning of these evaginating structures is not fully understood, but appears to involve interactions with adjacent tissues, including the lens and olfactory placodes and Rathke's pouch, respectively (Daikoku et al., 1982, 1983; Graziadei and Monti-Graziadei, 1992).

B. Neurulation of the Anterior Neural Plate and Early Morphogenesis of the Prosencephalon

While regional patterning of the prosencephalic neural plate is still taking place, the edges of the neural plate thicken, extend upward and medially, and eventually fuse in the midline to form the neural tube. Through this morphogenetic process, called neurulation, lateral parts of the neural plate become dorsal parts of the neural tube, and medial parts become ventral. Closure of the neural tube does not occur simultaneously throughout the ectoderm, but typically starts at specific locations along the AP axis (Golden and Chernoff, 1993). Anterior closure starts at two different places, one adjacent to the stomodeum (or oral plate; see Chapter 15) and the other at the mesencephalic–diencephalic boundary. Fusion of the dorsal midline (roof plate) progresses caudally and rostrally from these points, respectively, to close the presumptive prosencephalic vesicle. Defects in anterior neurulation lead to an "open" brain or exencephaly, which is found in several mouse mutants including those lacking the transcription factors Pax3 (Epstein et al., 1991), Gli3 (Hui and Joyner, 1993), twist (Chen and Beringer, 1995), AP2 (Schorle et al., 1996; Zhang et al., 1996), Cart1 (Zhao et al., 1996), and Dlx5 (Depew et al., 1999), as well as the PDZ domain protein Shroom (Hildebrand and Soriano, 1999).

As the anterior neural tube closes, the embryonic brain subdivides into three primary vesicles, the prosencephalon, mesencephalon, and rhombencephalon. Subsequently, two sets of vesicles evaginate from the dorsolateral walls of the rostral prosencephalon. The optic vesicles are the first to become readily apparent, but are soon followed by the evagination of the telencephalic vesicles. Several days later, the olfactory bulbs develop as an evagination from the anterior portion of the telencephalic vesicles. The formation of these evaginations is probably mediated by focal increases in cell proliferation (Hentges et al., 1999). Simultaneously, the optic and telencephalic vesicles undergo marked morphogenetic changes. Thus, the rostrodorsal midline of the telencephalon grows relatively little compared to the adjacent neuroepithelium, leading to the formation of a pair of bilateral vesicles separated by the interhemispheric sulcus. In addition, three prominent intraventricular bulges form in the basal or subpallial telencephalon: the septum, LGE, and MGE. These regional thickenings in the wall of the neural tube are probably generated by the combination of rapid localized cell proliferation, the onset of differentiation and the radial migration of postmitotic neurons. In contrast, the telencephalic pallium remains relatively thin for several days, presumably due to its later onset of neurogenesis. In the next sections, we discuss mechanisms that regulate regionalization within the telencephalon.

C. Patterning of the Subpallial Telencephalon

The embryonic subpallium consists of the striatal (LGE), pallidal (MGE), and AEP/POA primordia, as well as parts of the septal and amygdaloid anlages (Fig.1). Each of these primordia expresses a distinct combination of regulatory genes that defines their identity (Figs. 2 and 3). Thus, expression of the Dlx1, Dlx2, Gsh1, Gsh2, Mash1, Nkx2.1, Isl1, Six3, and Vax1 transcription factor genes contribute in defining the identity of progenitor cells in the subpallium (Bulfone et al., 1993; Guillemot and Joyner, 1993; Porteus et al., 1994; Hsieh-Li et al., 1995; Oliver et al., 1995; Valerius et al., 1995; Liu et al., 1997; Hallonet et al., 1998; Eisenstat et al., 1999). Cells that contribute to the subpallium begin to express several of these genes at the neural plate stage (e.g., Otx2, Six3, and Vax1), while the expression of other genes is delayed and follows neurulation. This suggests that patterning of the subpallium is a stepwise process. In addition, some transcription factors have a more restricted pattern of expression within the subpallium and contribute to the development of specific subpallial structures. For instance, the subpallial expression of the Otx2, Nkx2.1 (Ttf1; T/Ebp), and Isl1 homeobox genes in progenitor cells is restricted to the most ventrorostral regions within the subpallium (the septal, pallidal, and AEP/POA primordia, Fig. 2; Simeone et al., 1993; Ericson et al., 1995; Shimamura et al., 1995; O. Marín and J. L. R. Rubenstein, unpublished observations).

Several lines of evidence suggest that SHH has an essential role in basal telencephalic specification (Fig. 4). Shh mu-

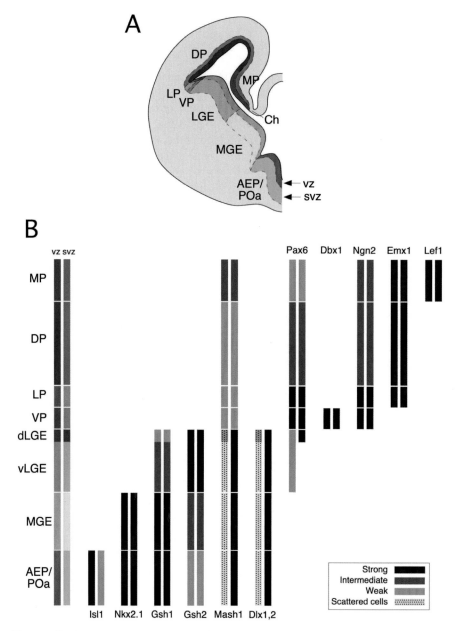

Figure 2 The expression patterns of transcription factors have boundaries that respect pallial and subpallial progenitor domains in the telencephalon. (A) Coronal hemisection of an E14.5 telencephalon showing the progenitor cell domains in different colors. (B) Some genes are expressed only in the subpallium (*Isl1, Nkx2.1, Gsh1/2, Mash1, Dlx1/2*), whereas others are predominantly expressed in the pallium (*Pax6, Dbx1, Ngn2, Emx1, Lef1*). It is suggested that, as in the spinal cord (see Briscoe *et al.*, 2000), the combinatorial expression of these genes may define distinct progenitor domains in the telencephalon. Abbreviations: AEP, anterior entopeduncular area; DP, dorsal pallium; LGE, lateral ganglionic eminence; LP, lateral pallium; MGE, medial ganglionic eminence; MP, medial pallium; POa, anterior preoptic area; svz, subventricular zone; VP, vental pallium; vz, ventricular zone.

tant mice lack a morphologically defined subpallium, do not express basal telencephalic markers (e.g., *Nkx2.1* and *Dlx2*), and appear to ubiquitously express dorsal markers throughout their telencephalon (Chiang *et al.*, 1996; Y. Ohkubo, K. Yun, and J. L. R. Rubenstein, unpublished observations).

Mice lacking the SHH coreceptor patched (which represses SHH signaling), express ventral telencephalic markers (e.g., *Nkx2.1*) in most of the telencephalon (Goodrich *et al.*, 1997). Furthermore, SHH can induce the expression of ventral telencephalic markers such as *Nkx2.1* and *Dlx2* (Ericson *et al.*,

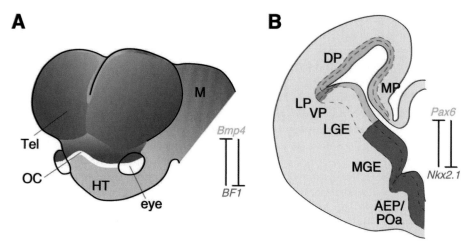

Figure 3 Repressive interactions between regulatory genes contribute to the generation of boundaries in the telencephalon. (A) Reciprocal repression between *Bmp4* and *BF1* is required to establish appropriate dorsoventral patterning in the telencephalon. (B) Interaction between *Nkx2.1* and *Pax6* is required to define independent progenitor cell populations in the LGE and MGE. Abbreviations: AEP, anterior entopeduncular area; DP, dorsal pallium; HT, hypothalamus; LGE, lateral ganglionic eminence; LP, lateral pallium; M, mesencephalon; MGE, medial ganglionic eminence; MP, medial pallium; OC, optic chiasm; POa, anterior preoptic area; Tel, telencephalon; VP, vental pallium.

1995; Dale *et al.*, 1997; Pera and Kessel, 1997; Shimamura and Rubenstein, 1997; Kohtz *et al.*, 1998), and repress the expression of dorsal telencephalic markers, such as *Emx1* and *Tbr1* (Kohtz *et al.*, 1998). Nevertheless, the capacity of SHH to ventralize the telencephalon appears to be limited by a competence period (Kohtz *et al.*, 1998). SHH is also able to induce the expression of the *Vax1* homeobox gene (Hallonet *et al.*, 1999). As noted above, *Vax1* is expressed at neural plate stages in the primordia of the basal telencephalon (Hallonet *et al.*, 1998). *Vax1* mutants exhibit defects in the rostral part of the basal telencephalon (septum and preoptic

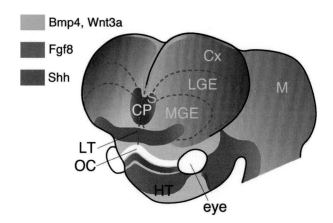

Figure 4 Signaling centers in the prosencephalon. Schematic representation of the expression of patterning molecules in the prosencephalon of the mouse at midgestation. Abbreviations: CP, commissural plate; Cx, cortex; LGE, lateral ganglionic eminence; LT, lamina terminalis; M, mesencephalon; OC, optic chiasm; S, septum.

area), similar to the defects observed in mice lacking telencephalic *Fgf8* (E. Storm, J. L. R. Rubenstein and G. R. Martin, unpublished observations). These experiments suggest a role for both FGF8 and SHH in patterning the rostrobasal telencephalon (Fig. 4).

Various factors control the competence of the telencephalon to be ventralized. For example, *BF1* (*Foxg1*) appears to mediate ventralization of the telencephalon by restricting BMP signaling to the dorsal telencephalon (Fig. 3; Dou *et al.*, 1999). Accordingly, *BF1* (*Foxg1*) mutant mice lack expression of subpallial markers and ectopically express dorsal telencephalic markers including *Bmps* (e.g., *Bmp4*) in most of the telencephalon (Xuan *et al.*, 1995; Dou *et al.*, 1999; Huh *et al.*, 1999).

The development of the LGE, MGE, and AEP/POA primordia is controlled differentially. Shortly after the telencephalic vesicles begin to evaginate, expression of the homeobox gene *Nkx2.1* is induced in a region that includes the MGE, AEP/POA, and part of the septal primordia (Fig. 2; Sussel *et al.*, 1999; Crossley *et al.*, 2001). Soon thereafter, the progenitor zones of the AEP/POA express *Shh* and the homeobox gene *Isl1*, suggesting that these genes may contribute to the unique developmental features of this region (Fig. 2; Ericson *et al.*, 1995; O. Marín and J. L. R. Rubenstein, unpublished observations). *Nkx2.1* has a primary role in ventral specification of the telencephalon, as evidenced by the apparent lack of MGE derivatives (e.g., globus pallidus) and enlargement of the striatum in *Nkx2.1* mutant mice (Kimura *et al.*, 1996; Sussel *et al.*, 1999). Molecular analyses of the ventral telencephalon in *Nkx2.1* mutants suggest that there has been a ventral-to-dorsal respecification of the

MGE, such that it now has properties similar to the LGE (Fig. 2). For instance, expression of genes that are normally restricted to the progenitor zones of the LGE (e.g., *Pax6*) expand into the MGE in *Nkx2.1* mutants with a coincident loss of MGE/POA/AEP specific genes (e.g., *Lhx6, Lhx7, Isl1* and *Shh*) (Sussel *et al.,* 1999; O. Marín and J. L. R. Rubenstein, unpublished observations). The fact that *Shh* expression is lost, while *Dlx* expression is maintained, suggests that Shh expression within the subpallium does not have a major role in telencephalic patterning.

Specification of the LGE appears to involve the *Gsh2* homeobox gene. This gene is expressed in the ventricular and subventricular zones of the LGE and MGE, but with higher levels in the LGE (Fig. 2; Hsieh-Li *et al.,* 1995; K. Yun and J. L. R. Rubenstein, unpublished observations). Mice lacking *Gsh2* develop a small LGE-like structure that lacks *Dlx2* expression and does not form a normal striatum (Szucsic *et al.,* 1997; K. Yun and J. L. R. Rubenstein, unpublished observations). This defect may result from misspecification of the LGE, because *Gsh2* appears to be required for the repression of cortical specification genes in the subpallium (i.e., *Dbx2, neurogenin1;* Fig. 2; Toresson *et al.,* 2000; Yun *et al.,* 2001).

Expression of the *Pax6* paired-homeobox gene also distinguishes the ventricular zone of the LGE from that of the MGE (Sussel *et al.,* 1999; Puelles *et al.,* 2000; Figs. 2 and 3), and may contribute to the specification of the LGE in part through the repression of MGE specific genes. For instance, in *Pax6* mutants, the expression of MGE markers (e.g., *Nkx2.1*) expands dorsally into the LGE (Stoykova *et al.,* 2000). Futhermore, *Pax6* can repress the expression of *Nkx* genes in the hindbrain and spinal cord (Briscoe *et al.,* 1999, 2000).

D. Patterning of the Pallial Telencephalon

The pallial telencephalon consists of several domains that are interposed between the roof and commissural plates and the subpallium (Fig. 2). There is evidence that the same patterning mechanisms that regulate lateral patterning in the neural plate also participate in regional specification of the cortex. Thus, secreted proteins of the TGF-β (BMP, GDF), WNT, and FGF families have been implicated in the dorsal patterning of the telencephalon and its subsequent regionalization. *TGFβ* and *Wnt* genes are predominantly expressed in the dorsomedial region of the telencephalon, while *Fgf8* expression is highest in the rostral midline (Fig. 4).

After neural tube closure, a domain of *Bmp2, 4, 5, 6,* and *7* coexpression defines the dorsomedial region of the telencephalon (Furuta *et al.,* 1997; Fig. 4). As development proceeds, expression of *Bmp4, 5, 6,* and *7* identifies the prospective hippocampus, fimbria, and choroid plexus, whereas *Bmp2* expression becomes restricted to the developing dentate gyrus of the hippocampus (Furuta *et al.,* 1997). The

expression patterns of *Bmps* and *BF1* (*Foxg1*) in the dorsal telencephalon are complementary and are maintained by mutual repression (Fig. 3; Furuta *et al.,* 1997; Dou *et al.,* 1999).

Several lines of evidence suggest that gain of BMP function dorsalizes the telencephalon, whereas the absence of BMP signaling is required for ventral telencephalic development. Ectopic expression of *Bmp4* in *BF1* (*Foxg1*) mutant mice is correlated with a ventral-to-dorsal transformation of the telencephalon (Dou *et al.,* 1999). Implantation of beads soaked in recombinant BMP4 or BMP5 into the neural tube of the chicken forebrain induces dorsal markers (e.g., Wnt4) and represses ventral markers (Golden *et al.,* 1999). The generation of mice bearing mutations in either *Bmp2, 4, 5, 6,* and *7* has proven to be insufficient to understand the role of these genes in the patterning of the forebrain (King *et al.,* 1994; Dudley *et al.,* 1995; Luo *et al.,* 1995; Winnier *et al.,* 1995; Zhang and Bradley, 1996; Solloway *et al.,* 1998), probably due to genetic redundancy as suggested by their overlapping expression patterns. In agreement with this idea, *Bmp5;Bmp7* double mutants exhibit delayed closure of the rostral neural tube and hypoplasia of the telencephalic vesicles (Solloway and Robertson, 1999).

How do cells in the dorsal telencephalon respond to BMP signaling? Several transcription factors have been implicated in the transduction of BMP signals. The expression of BMPs in the dorsomedial telencephalon correlates with high levels of *Msx1* expression, and *in vitro* experiments have shown that BMPs can induce *Msx1* expression (Furuta *et al.,* 1997; Shimamura and Rubenstein, 1997). Expression of *Bmps* also correlates with that of *Hfh4* (*Foxj 1*), a winged helix gene that is involved in choroid plexus differentiation (Lim *et al.,* 1997). However, experiments *in vitro* have suggested that *Hfh4* expression is not induced by BMPs (Furuta *et al.,* 1997), and that other factors may be required for choroid plexus formation.

The WNT family of secreted proteins appears to cooperate with BMPs in the patterning of the pallial telencephalon. There is growing evidence that WNT proteins function in specifying dorsal telencephalic fates and/or in expanding dorsal progenitor cell populations. Like the *Bmp* genes, *Wnt* family members (e.g., *Wnt2a, 3a, 5a, 7a, 7b,* and *8b*) are expressed in nested patterns in the dorsal telencephalon. While some of the *Wnt* genes are broadly expressed in the telencephalon (e.g., *Wnt7b*), others are restricted to specific dorsomedial domains (e.g., *Wnt3a* in the fimbria). The fimbria lies between the hippocampal anlage and more medial tissues (e.g., choroid plexus). Mice lacking *Wnt3a* function do not form a hippocampus (Lee *et al.,* 2000), and there is evidence that this phenotype is in part due to reduced proliferation of the hippocampal primordium.

WNT signals are mediated by the FRIZZLED family of receptors, which act to stabilize β-catenin and propagate the signal intracellularly (reviewed in Wodarz and Nusse, 1998).

Increases in β-catenin concentration lead to the activation of *Lef/Tcf* transcription factors, which mediate many of the responses to WNT signals. *Lef1* is expressed in the dorsocaudal telencephalon, adjacent to the fimbria. *Lef1* null mutant mice lack the hippocampal dentate gyrus (Galceran *et al.*, 2000). In addition, mice carrying a neomorphic allele of *Lef1* (*Lef1-b/b*), which blocks β-catenin activation of LEF/TCF proteins, lack the entire hippocampal formation similar to *Wnt3a* mutant mice (Galceran *et al.*, 2000). Like the *Wnt3a* mutants, *Lef1-b/b* mice have a modest reduction in cell proliferation, which might contribute to the lack of the hippocampal formation.

Inhibition of WNT signaling is mediated by secreted factors of the Frizzled-related protein (SFRP) family, which have an amino-terminal domain highly homologous to the ligand binding domain of Frizzled proteins (reviewed in Capdevilla and Izpisua-Belmonte, 1999). Expression of SFRPs is induced by *Wnt* genes, perhaps to fine-tune the spatial and temporal patterns of Wnt activity. For example, high levels of SFRP2 expression are found at the pallial–subpallial boundary, suggesting that this protein may play a prominent role in the patterning of this region through the modulation of WNT signals from surrounding structures (Kim *et al.*, 2001).

Transduction of patterning signals in pallial progenitor cells is conducted through several known transcription factors (Fig. 2). Mutations in the *neurogenin1, neurogenin2, Pax6,* and *Gli3* genes appear to affect regional specification, because they lead to dorsal-to-ventral transformations in pallial molecular properties. Mice lacking both the *neurogenin1* and *neurogenin2* basic helix–loop–helix genes exhibit a dramatic reduction in the expression of pallial markers (e.g., *Tbr1*) and ectopically express subpallial markers in the pallium (e.g., *Mash1, Dlx1,* and glutamic acid decarboxylase; Fode *et al.*, 2000). The *neurogenin* mutants that also lack *Mash1* do not exhibit this transformation, demonstrating a central role of *Mash1* in subpallial specification. Similarly, ectopic expression of *Mash1* under the control of *neurogenin* promoter leads to the expression of subpallial markers in the cortex (Fode *et al.*, 2000). The cerebral cortex in mice lacking expression of the *Pax6* homeobox gene also develops increased expression of subpallial genes (e.g., *Dlx1;* Stoykova *et al.*, 1996, 1997). It has been suggested that this is due to increased tangential migration of ventrally derived neurons into the cerebral cortex (Chapouton *et al.*, 1999). Nevertheless, more recent studies suggest that the dorsal expression of subpallial markers may also be due to a ventral transformation of progenitor cell identity (Toresson *et al.*, 2000; Stoykova *et al.*, 2000; Yun *et al.*, 2001).

Severe dorsal-to-ventral cortical transformations are observed in mice lacking the *Gli3* zinc finger transcriptional regulator (Franz, 1994; Theil *et al.*, 1999; Tole *et al.*, 2000). *Gli3* mutants lack recognizable dorsomedial structures (fimbria, choroid plexus, and hippocampal complex) and have a

dysmorphic cortex (Franz, 1994; Theil *et al.*, 1999; Tole *et al.*, 2000). In the spinal cord and anterior limb bud, *Gli3* acts as a suppressor of *Shh* expression (Büscher *et al.*, 1997; Masuya *et al.*, 1997; Ruiz i Altaba, 1998). In the forebrain, although we do not know whether *Gli3* interferes directly with *Shh* signaling, the dorsal telencephalon of *Gli3* mutants is partially ventralized, as demonstrated by ectopic expression of subpallial markers (e.g., *Dlx2, Isl1;* Tole *et al.*, 2000). Consistent with these results, the expression of dorsal telencephalic genes, including *Bmp2, Bmp4, Bmp6, Bmp7, Wnt2b, Wnt3a, Wnt5a, Msx1,* and *Emx1,* is lost, and the expression of *Emx2* is delayed and reduced (Grove *et al.*, 1998; Theil *et al.*, 1999; Tole *et al.*, 2000). On the other hand, other genes that are exclusively expressed in the cortex, such as *Wnt8b* and *neurogenin2,* have an almost normal distribution in the cortex of *Gli3* mutants (Tole *et al.*, 2000), suggesting that *Gli3* is not essential in all aspects of pallial patterning.

Gli3 function is required to establish or maintain *Emx* homeobox gene expression (Theil *et al.*, 1999; Tole *et al.*, 2000). Expression of *Emx1* and *Emx2* largely overlaps in the ventricular zone of the pallium, although the domain of *Emx2* expression is broader than that of *Emx1,* extending from the cortical hem into the LGE (Simeone *et al.*, 1992; Gulisano *et al.*, 1996). *Emx1* expression is excluded from the cortical hem, and its ventral boundary does not enter the ventral pallium (Shimamura *et al.*, 1997; Smith-Fernández *et al.*, 1998; Puelles *et al.*, 1999, 2000; Fig. 2). *Emx1* mutants lack the corpus callosum, the main fiber tract connecting the left and right cerebral hemispheres, but have normal histology and molecular properties in the cerebral cortex (Qiu *et al.*, 1996; Yoshida *et al.*, 1997). On the other hand, *Emx2* mutants lack a morphologically identifiable hippocampal dentate gyrus, although all hippocampal fields are present, including a dentate cell population (Pellegrini *et al.*, 1996; Yoshida *et al.*, 1997; Tole *et al.*, 2000).

In contrast, the hippocampal field is missing in mice lacking the LIM homeobox gene *Lhx5* (Zhao *et al.*, 1999). These mice have a dorsal expansion of genes expressed in the hippocampal primordia, resulting in a reduction of *Wnt* and *Bmp* expression in the cortical hem and a disruption of the development of the choroid plexus. As noted above, hippocampal patterning depends on *Wnt* signaling, based on the analysis of *Wnt3a* mutants and in mice homozygous for a neomorphic *Lef1* allele.

E. Anteroposterior Patterning within the Telencephalon

There are clear AP differences in molecular properties, timing of differentiation, and patterns of connectivity between different telencephalic subdivisions and even within individual domains. The mechanisms that generate these differences are just beginning to be understood. It is possible that positional values along the AP dimension arise in part from morphogens produced at the rostral midline (commissural plate), a structure derived from the anterior neural ridge (Fig. 4). As noted earlier, this structure expresses *Fgf8*. It is possible that *Fgf8* expression in the commissural plate has a similar role to *Fgf8* expression in the isthmus, where it is required for AP patterning of the midbrain and anterior hindbrain (Crossley *et al.*, 1996; Lee *et al.*, 1997; Wassarman *et al.*, 1997; Liu *et al.*, 1999; Martínez *et al.*, 1999).

Fgf8 function at the isthmic organizer is partly mediated through its interaction with the homeobox gene *Otx2*, since both genes reciprocally repress each other (Acampora *et al.*, 1997; Liu *et al.*, 1999; Martínez *et al.*, 1999). *Otx1⁻/⁻;Otx2⁺/⁻* mutant mice have severe defects in AP patterning of the midbrain and forebrain. In these mice, the isthmic organizer appears to move rostrally and the pallial telencephalon, which normally expresses *Otx* genes, expresses midbrain markers, such as *En2* and *Wnt1* (Acampora *et al.*, 1997, 1999a; Suda *et al.*, 1997). Furthermore, insertion of beads containing FGF8 in the forebrain of chick embryos alters the pattern of cortical *Otx2* and *Emx2* expression (Crossley *et al.*, 2001). Thus, it appears likely that FGF and OTX proteins have an important role in AP patterning of the telencephalon.

F. Regionalization within the Cerebral Cortex

Considerable controversy has arisen over the mechanisms through which early subdivisions of the cerebral cortex are generated. This has been particularly contentious with regard to the origin of distinct neocortical (isocortical) subdivisions (see Rubenstein *et al.*, 1999, for a review of this subject). One school proposes that mechanisms intrinsic to the cortex play a fundamental role in this process (Rakic, 1988), whereas another proposes that regional identity is primarily controlled by the nature of thalamic axonal inputs that the different neocortical domains receive (O'Leary, 1989). A large body of evidence now suggests that regional subdivisions of the cerebral cortex form before thalamic innervation is received, as evidenced by regionally restricted expression of regulatory genes (Levitt, 1984; Arimatsu *et al.*, 1992; Hatanaka *et al.*, 1994; Mason *et al.*, 1994; Bulfone *et al.*, 1995; Nothias *et al.*, 1998; Paysan and Fritschy, 1998; Donoghue and Rakic, 1999; Grove and Tole, 1999; Mackarehtschian *et al.*, 1999; Nakagawa *et al.*, 1999; Rubenstein *et al.*, 1999). Furthermore, a number of experimental approaches have demonstrated that a substantial degree of cortical regionalization occurs in the absence of thalamic inputs (Wise and Jones, 1978; Cohen-Tannoudji *et al.*, 1994, Levitt *et al.*, 1997; Nothias *et al.*, 1998; Gitton *et al.*, 1999a,b; Miyashita-Lin *et al.*, 1999; Soria and Fairén, 2000). For instance, mice lacking the *Gbx2* homeobox gene have a severe defect in thalamic development and fail to produce thalamic axons that innervate the neocortex. Despite the lack of these

axons, neocortical molecular subdivisions form normally (Miyashita-Lin *et al.*, 1999). Ongoing efforts are beginning to study the intrinsic mechanisms that generate early subdivisions in the cerebral cortex.

G. Induction and Early Morphogenesis of the Olfactory Bulb

The olfactory bulbs are bilateral evaginations of the telencephalic vesicles. Like other forebrain structures, induction of the olfactory bulb requires the interaction of neural and nonneural ectoderm, in this case the olfactory placodes (Stout and Graziadei, 1980; Byrd and Burd, 1993; Graziadei and Monti-Graziadei, 1992, and references therein; Gong and Shipley, 1995). The olfactory placodes derive from the rostrolateral parts of the anterior neural ridge (Jacobson, 1959; Couly and LeDouarin, 1987), whereas the olfactory bulbs derive from a region of the prosencephalic neural plate intercalated between the septal and the cortical anlages (reviewed in Rubenstein *et al.*, 1998; Cobos *et al.*, 2001). Interestingly, *Pax6* mutant mice lack evaginated olfactory bulbs. It is unclear whether this is due to defects in the olfactory placode and/or the telencephalon, since both of these tissues normally express *Pax6* and are abnormal in these mutants (Hogan *et al.*, 1986; Grindley *et al.*, 1995; Anchan *et al.*, 1997). However, an olfactory bulblike structure appears to develop in *Pax6* mutants, implying that this gene is required for olfactory bulb morphogenesis and not its specification (López-Mascaraque *et al.*, 1998). *Gli3* mutant mice also lack olfactory bulbs (Franz, 1994). In addition, there is evidence that retinoids produced by the mesenchyme interposed between the olfactory placode and telencephalon play a role in regulating the process of olfactory bulb morphogenesis (Anchan *et al.*, 1997; Whitesides *et al.*, 1998).

H. Induction and Patterning of the Diencephalon

The diencephalon constitutes the central core of the forebrain, from which the optic and telencephalic vesicles evaginate (see Section I; Fig. 1). It extends from the mesencephalon to the anterior limit of the brain (region of lamina terminalis, the optic chiasm, and the retrochiasmatic hypothalamus). At late neural plate stages, most of the diencephalon is distinguished from the telencephalon by the lack of *BF1* (*Foxg1*) expression. In contrast, the diencephalic territory rostral to the *zona limitans intrathalamica* (ZLI) is characterized by the expression of *BF2*, another winged helix transcription factor (Hatini *et al.*, 1994).

The diencephalon consists of both basal and alar plate domains (Fig. 1). Available evidence suggests that the basal plate tissues are induced and patterned by SHH mediated signals originating from the axial mesendoderm (Ericson *et al.,* 1995). These signals are transduced in part by *Nkx* homeobox genes. At least six genes are known to be expressed in the ventral diencephalon: *Nkx2.1, Nkx2.2, Nkx2.4, Nkx5.1, Nkx5.2,* and *Nkx6.1* (Price *et al.,* 1992; Bober *et al.,* 1994; Rinkwitz-Brandt *et al.,* 1995; Shimamura *et al.,* 1995; Qiu *et al.,* 1998; Marcus *et al.,* 1999; O. Marín and J. L. R. Rubenstein, unpublished observations). To date, hypothalamic defects have only been reported in *Nkx2.1* mutant mice. Morphological analysis of these mutants reveals that most of the ventral hypothalamus is unrecognizable (Kimura *et al.,* 1996), and molecular studies demonstrate early patterning defects (O. Marín and J. L. R. Rubenstein, unpublished observations).

SHH signal transduction in the diencephalon is mediated by *Gli* zinc finger transcription factors. However, while mutation of *Shh* causes the loss of a large portion of the hypothalamus (Chiang *et al.,* 1996), mutation of individual *Gli* genes is less deleterious. Thus, whereas *Gli2* mutant mice have a variable loss of the pituitary, *Gli1;Gli2* double mutants lack a pituitary and exhibit abnormal expression of *Shh* and *Nkx2.1* in the hypothalamus (Park *et al.,* 2000). These studies suggest that *Gli1* and *Gli2* have overlapping functions mediating *Shh* signaling in the diencephalon. In support of a role for *Gli* factors in the development of the diencephalon, a *Gli2* homolog is required for the development of postoptic hypothalamus and the anterior pituitary in zebrafish, based on analysis of *you-too* mutants (Karlstrom *et al.,* 1999). Interestingly, *Fgf8* is also expressed in the anteriormost region of the ventral diencephalon, where it may participate in collaboration with *Shh* in the patterning of the tuberal hypothalamus (Fig. 4; Ye *et al.,* 1998).

The induction and patterning of the ventral hypothalamus is highly related to the development of the pituitary gland (reviewed in Treier and Rosenfeld, 1996; Watkins-Chow and Camper, 1998; Dasen and Rosenfeld, 1999). The pituitary consists of two parts, adenohypophysis and neurohypophysis. The adenohypophysis develops from the middle portion of the anterior neural ridge and encompasses the anterior and intermediate pituitary. The neurohypophysis, in contrast, develops from the adjacent medial neural plate and comprises the posterior pituitary (Couly and LeDouarin, 1985; reviewed in Rubenstein *et al.,* 1998). This subject will be covered in detail in Chapter 20.

Dorsal patterning and regionalization of the diencephalon is mediated by members of the TGF-β, WNT, and FGF families. At least six *Wnt* genes are expressed in partially overlapping domains in the developing diencephalon (Hollyday *et al.,* 1995). *Wnt1, 3a, 4, 5a,* and *8b* are expressed in one or two caudal subdivisions of the developing alar diencephalon, the synencephalon (i.e., pretectum), and posterior parencephalon (i.e., dorsal thalamus), but do not extend rostral to the ZLI. In contrast, *Wnt7b* is expressed dorsally in the anterior parencephalon (i.e., ventral thalamus). *Wnt1;Wnt3a* double mutants have severe hypoplasia of both the midbrain and cau-

dal forebrain (S. M. Lee and A. P. McMahon, personal communication). In addition, *Wnt1* function has been shown to be required for the normal expression of *Sim2* in the diencephalon (Mastick *et al.,* 1996). *Sim2,* a murine homolog of the *Drosophila* single-minded gene, is expressed during early stages of diencephalic regionalization (Fan *et al.,* 1996).

AP patterning of the diencephalon and the formation of the prosomeres is poorly understood. There is evidence for a major transition in AP properties at the ZLI, a transverse boundary region between the primordia of the dorsal and ventral thalamus (Fig. 1; reviewed in Puelles and Rubenstein, 1993; Rubenstein and Beachy, 1998). Caudal to the ZLI, FGF8 can induce the expression of the engrailed homeobox gene and can transform the neural tube to develop into midbrain and cerebellar tissues (Martínez *et al.,* 1999). Anterior to the ZLI, FGF8 does not induce midbrain/cerebellar tissues, but it does induce the expression of the telencephalic marker *BF1 (Foxg1;* Shimamura and Rubenstein, 1997). As the diencephalon matures, genes expressed in the basal plate are expressed within the ZLI (e.g., *Shh, Nkx2.2,* and *Sim1;* Shimamura *et al.,* 1995; Fan *et al.,* 1996), suggesting that the ZLI becomes a patterning center at this stage. The dorsal expansion of *Shh* at the ZLI approaches the diencephalic roof, where *Fgf8* and *Bmp4* are expressed (Crossley *et al.,* 2001). Around the same time, genes such as *Gbx2* and *Dlx2* are expressed in the dorsal thalamus (prosomere 2) and ventral thalamus (prosomere 3), where they are required for differentiation of these primordia (Miyashita-Lin *et al.,* 1999; O. Marín, S. A. Anderson, and J. L. R. Rubenstein, unpublished observations).

Like in other regions of the neuroaxis, *Pax6* expression in the diencephalon appears to control certain aspects of dorsoventral patterning and regionalization. For example, ventral markers of the diencephalon are expressed more dorsally than normal in *Pax6* mutants (Grindley *et al.,* 1997). In addition, defects in the establishment of transverse molecular boundaries in the diencephalon are found in the absence of *Pax6* function (Stoykova *et al.,* 1996, 1997; Grindley *et al.,* 1997; Mastick *et al.,* 1997; Warren and Price, 1997). The ventral thalamus appears to be more severely affected than the dorsal thalamus or the pretectum in *Pax6* mutants (Stoykova *et al.,* 1996), possibly due to the abnormal enlargement of the ZLI in the absence of *Pax6* mutants (Grindley *et al.,* 1997).

III. Morphogenetic Mechanisms in the Forebrain

The morphogenesis of the forebrain is the result of spatial and temporal control of proliferation, cell death (apoptosis), cell migration, axonal growth, and other changes in cell shape. Regional specification mechanisms (described in Section II) act in concert to regulate, directly or indirectly, these cellular aspects of morphogenesis. Here, we review examples of genetic mechanisms that link the programs in regional specification to proliferation and apoptosis.

There is evidence that patterning and precursor cell proliferation in the forebrain are regulated by the same molecules (Fig. 4). For example, WNT signaling through the LEF/TCF transcription factors is implicated in positively regulating proliferation through *c-myc* and *cyclin D1* (He *et al.*, 1998; Tetsu and McCormick, 1999; Shtutman *et al.*, 1999). In addition, as described above, mutations in *Wnt3a* and *Lef1* lead to hypoplasia of the dorsomedial telencephalon, which is correlated with some decrease in the mitotic index, but no measureable change in apoptosis (Lee *et al.*, 2000; Galceran *et al.*, 2000). Since the *Wnt* responsive transcription factors *Tcf3* and *Tcf4* are expressed throughout the primordia of the telencephalon and dorsal thalamus, respectively (Galceran *et al.*, 2000), Wnt signaling may have a general role in regulating proliferation within the forebrain.

Although WNT signaling has been associated with increases in proliferation and perhaps survival, studies in the forebrain suggest that members of the TGF-β superfamily have the opposite effect. For example, local application of beads containing BMPs inhibits progenitor cell proliferation and induces local apoptosis both in telencephalic explants (Furuta *et al.*, 1997) and within the telencephalon *in vivo* (Golden *et al.*, 1999). The BMP-mediated apoptosis is associated with the expression of *Msx1*, a target of BMP signaling (Davidson, 1995; Furuta *et al.*, 1997). This is consistent with the observation that in the dorsal midline of the forebrain, where high levels of BMPs are detected, a significant number of the cells undergo programmed cell death at E10.5 (Furuta *et al.*, 1997). Furthermore, BMP-mediated reduction of the proliferative rate in the dorsalmost region of the telencephalon appears to be associated with the morphogenesis of the choroid plexus (Tao and Lai, 1992; Shimamura *et al.*, 1995; Dou *et al.*, 1999). In contrast to these BMP effects, it is known that *Bmp5* and *Bmp7* are necessary for the proliferation and maintenance of specific cell populations in the forebrain (Solloway and Robertson, 1999).

FGF8 signaling is also implicated in the control of telencephalic growth. Reduction of *Fgf8* expression at the rostrodorsal midline of the telencephalon results in severe hypoplasia of the rostral telencephalon (Shanmugalingam *et al.*, 2000; E. Storm, J. L. R. Rubenstein, and G. R. Martin, unpublished observations). However, it has not been demonstrated that FGF8 directly controls proliferation in the early forebrain. Loss of FGF8 causes hypoplasia through an increase in apoptosis rather than through modulation of proliferation in the mesenchyme of the first branchial arch (Trumpp *et al.*, 1999). Nevertheless, there is evidence that FGF2 (bFGF) can increase the proliferation of neuroepithelial cells. Telencephalic stem cells proliferate *in vitro* in the presence of FGF2 (bFGF) and EGF, both of which signal

through receptor tyrosine kinases (Reynolds *et al.*, 1992; Ghosh and Greenberg, 1995; Kilpatrick and Bartlett, 1995). In addition, mice lacking the *Fgf2* gene have a reduced cortical surface area, possibly due to a progenitor cell proliferation defect (Dono *et al.*, 1998).

Shh mutant mice have a very small forebrain (Chiang *et al.*, 1996). This may be due to the poor health of these embryos, but may also reflect a more direct role for *Shh* in regulating forebrain proliferation. In support of this, activation of the SHH signaling pathway, either through overexpression of SHH or through blocking molecules that inhibit SHH signaling (e.g., mutation of patched), leads to increased proliferation in other regions of the neuroaxis (Jensen and Wallace, 1997; Wechsler-Reya and Scott, 1999; Dahmane and Ruiz i Altaba, 1999; Rowitch *et al.*, 1999; Wallace, 1999). Loss of *Shh* expression in the basal telencephalon may contribute to the severe proliferation defect observed in *BF1* (*Foxg1*) mutants (Dou *et al.*, 1999; Huh *et al.*, 1999).

The aforementioned studies do not establish whether WNT, BMP, FGF, and SHH signaling directly or indirectly affects proliferation and cell survival. Though the mechanisms remain to be elucidated, recent studies suggest that these molecules function together and that the complex interplay between them controls proliferation and apoptosis in the forebrain. In addition to these molecules, other secreted factors have been implicated in regulating proliferation of telencephalic progenitors. For instance, neurotransmitters and peptides, such as γ-aminobutyric acid (GABA), glutamate, and PACAP, can modulate proliferation of progenitors in the telencephalon (LoTurco *et al.*, 1995; Pesce *et al.*, 1996; Sadikot *et al.*, 1998). Recent genetic screens for morphological malformations in the brain of mice and zebrafish have led to the implication of additional factors in proliferation and cell death. For instance, the *flat-top* mutant mouse lacks telencephalic vesicles because it fails to increase the rate of proliferation in the telencephalic primordia (Hentges *et al.*, 1999). However, the gene affected in the *flat-top* mutant remains to be elucidated.

The mechanisms by which secreted molecules regulate cell proliferation and apoptosis in the forebrain are beginning to be elucidated. Several transcription factors have been implicated as modulators of proliferation. In addition to the already described LEF/TCF transcription factors, *BF1* (*Foxg1*) appears to regulate the balance between proliferation and cell death by restricting the expression of *Bmps* primarily to the dorsomedial telencephalon (Dou *et al.*, 1999). *BF1* (*Foxg1*) appears to control cell proliferation in a dose-dependent manner, promoting cell proliferation, at least in part, through the suppression of cyclin-dependent kinase (cdk) inhibitors (Hardcastle and Papalopulu, 2000). *Pax6* mutants exhibit increased cell proliferation and abnormal interkinetic nuclear movements in the developing cortex

(Caric *et al.*, 1997; Götz *et al.*, 1998; Warren *et al.*, 1999). Conversely, loss of *Pax6* expression in the diencephalon leads to a dramatic reduction in proliferation (Warren and Price, 1997). *Lhx2* mutants have hypoplasia of the cortex, which is particularly pronounced in the hippocampus, due to a proliferation defect (Porter *et al.*, 1997). Similarly, *Emx2* mutants appear to have a proliferation defect that strongly affects the hippocampus (Tole *et al.*, 2000).

In addition to transcription factors, protein kinases and proteases of the caspase family have been implicated as downstream effectors of these secreted molecules. Mice lacking both c-Jun-NH(2)-terminal kinases (JNK1 and JNK2) display a dramatic increase in cell death in the forebrain, suggesting that these proteins are involved in transducing survival or anti-apoptotic signals (Kuan *et al.*, 1999; Sabapathy *et al.*, 1999; reviewed in Haydar *et al.*, 1999). On the other hand, disruption of the caspase cascade by target mutagenesis of either *Casp3* or *Casp9* leads to overgrowth of the telencephalon, perhaps due to reduced levels of cell death (Kuida *et al.*, 1996, 1998; reviewed in Haydar *et al.*, 1999). Thus, the correct coordination of different apoptotic signaling pathways during neurogenesis is crucial for the regulation of normal forebrain morphogenesis.

IV. Control of Neurogenesis and Cell-Type Specification in the Forebrain

The processes that control regional specification and differential growth of the prosencephalic primordia act on neuroepithelial progenitor cells that reside in the pseudostratified epithelium lining the ventricular space (Fig. 5). This epithelium is called the ventricular zone (VZ). When progenitor cells become postmitotic, they migrate to the differentiation zone (mantle) that lies under the meningeal (pial) surface. However, some cells that leave the VZ remain mitotically active and enter a secondary proliferative zone (the subventricular zone, SVZ) that lies between the VZ and mantle (Smart, 1976; Halliday and Cepko, 1992; Bhide, 1996; Takahashi *et al.*, 1995a,b). In the striatum, there is evidence that the SVZ is required for the production of later-generated neurons (Sheth and Bhide, 1997; Anderson *et al.*, 1997a).

The generation of neurons or glial cells from neuroepithelial precursors involves successive steps in commitment and differentiation that progressively generate more restricted cell types (reviewed in Anderson and Jan, 1997; Cepko, 1999; Edlund and Jessell 1999). These steps include (1) cell-type

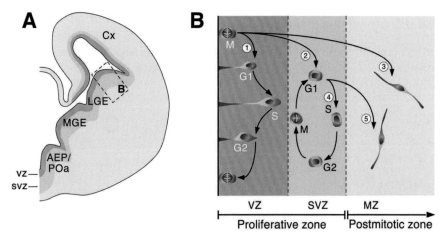

Figure 5 Proliferative regions in the telencephalon. (A) Proliferative regions in the telencephalon are the periventricular zones indicated by their darker shades of gray. The boxed region of the LGE is shown at higher magnification in panel B. (B) Schematic representation of the stratification of the LGE. The LGE is subdivided into three main regions along the ventriculopial axis: ventricular zone (VZ), subventricular zone (SVZ), and marginal or mantle zone (MZ). The lateral ventricle is to the left. The VZ and SVZ are proliferative regions containing mitotically active cells, whereas the MZ contains predominantly postmitotic cells. Cells in the VZ belong to the pseudostratified periventricular epithelium, whereas cells in the SVZ constitute a secondary proliferative population. VZ cells show interkinetic nuclear migration during the various stages of the cell cycle (M, G1, S, G2) indicated by pathway 1. VZ cells may reenter the cell cycle in the SVZ (indicated by pathway 2) or become postmitotic cells in the MZ (indicated by pathway 3). SVZ cells are not organized into pseudostratifed epithelium that undergoes interkinetic nuclear migration (indicated by pathway 4). Postmitotic cells arising from the SVZ migrate to the MZ (indicated by pathway 5). Abbreviations: AEP, anterior entopeduncular area; Cx, cortex; LGE, lateral ganglionic eminence; MGE, medial ganglionic eminence; POa, anterior preoptic area.

specification; (2) withdrawal from the cell cycle; (3) differentiation of its distinct cellular properties; (4) migration to the correct destination; and (5) the production of correct cell–cell contacts (e.g., synapses) through the elaboration of dendritic and axonal processes.

A. Control of Neuronal Subtype Specification

Early steps in cell-type specification may be largely controlled by the process of regional specification. For example, dorsal-ventral (DV) patterning produces longitudinal columns of neuronal progenitors with distinct molecular properties at different DV locations (Shimamura *et al.*, 1995; Tanabe and Jessell, 1996; Lumsden and Krumlauf, 1996). In the spinal cord, distinct cell types form at specific DV positions (Ericson *et al.*, 1997; Briscoe *et al.*, 1999), and appear to derive from distinct groups of progenitors under the control of a specific homeobox gene code (Briscoe *et al.*, 2000; Sander *et al.*, 2001). In a similar manner, certain populations of telencephalic neurons with distinct neurotransmitter phenotypes appear to be derived from progenitors that are located in different regions (Fig. 6). Thus, while glutamatergic cells are generally produced only in pallial areas, most or perhaps all cells expressing GABA are generated in the subpallium (Anderson *et al.*, 1997b, 1999; Casarosa *et al.*, 1999; Lavdas *et al.*, 1999; Sussel *et al.*, 1999) from progenitors expressing genes of the *Dlx* family (Anderson *et al.*, 1997b, 1999; Stühmer *et al.*, 2001). In addition, telencephalic cholinergic neurons appear to be derived exclusively from progenitors expressing *Nkx2.1* (Figs. 2 and 5; Sussel

et al., 1999; Marín *et al.*, 2000). Nevertheless, it is possible that some aspects of forebrain cell-type specification may be regulated independently of regional specification and after the progenitor cell stage. For instance, it is possible that cell identity can be modulated after a postmitotic neuron has migrated from its site of origin to a location that contains specific differentiation signals. Evidence for this type of mechanism exists in other regions of the neuroaxis (Sockanathan and Jessell, 1998; Waid and McLoon, 1998; Cepko, 1999), and transplantation studies suggest that such mechanisms also operate in the forebrain (Fishell, 1995, 1997; Campbell *et al.*, 1995; Brüstle *et al.*, 1997).

Once a region of neuroepithelium is exposed to a differentiation signal, subsets of cells within this zone begin to mature, whereas other cells remain as undifferentiated progenitors. In the forebrain, this process appears to be regulated through mechanisms of lateral inhibition, which are primarily controlled by the Notch signaling pathway. For instance, the *Mash1* bHLH transcription factor is expressed in subsets of progenitor cells in the subpallium, part of the diencephalon, and more caudal regions of the CNS (Fig. 2; Lo *et al.*, 1991; Guillemot and Joyner, 1993; Porteus *et al.*, 1994). Mice lacking *Mash1* fail to express normal levels of Delta1, a Notch ligand (Casarosa *et al.*, 1999), thereby reducing Notch signaling pathway in the proliferative zone. In the absence of Notch signaling, the neuroepithelium prematurely takes on properties of the SVZ. Consequently, *Mash1* mutants fail to produce normal numbers of early born cells, whereas the generation of late born neurons appears normal (Casarosa *et al.*, 1999; Horton *et al.*, 1999; K. Yun and

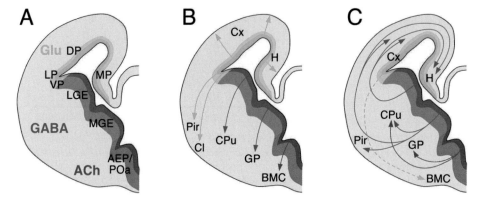

Figure 6 Regionalization of the telencephalon generates neuronal progenitors fated to produce the major neurotransmitters; radial and tangential migrations direct the final positions of these cells. (A) Distinct neurotransmitter phenotypes appear to be specified in different progenitor populations in the telencephalon. Thus, it is suggested that glutamatergic neurons are specified in the pallium, whereas GABAergic, and cholinergic neurons are specified in the subpallium. Migration of glutamatergic, GABAergic, and cholinergic neurons to their final postmitotic destination takes place through radial (B) and tangential (C) routes in the telencephalon. The dotted line indicating a migration of pallial neurons into the BMC in panel C is theoretical. Abbreviations: Ach, acetylcholine; AEP, anterior entopeduncular area; BMC, basal magnocellular complex; Cl, claustrum; Cpu, caudate-putamen nucleus; Cx, cortex; DP, dorsal pallium; GABA, γ-aminobutyric acid; GP, globus pallidus; Glu, glutamate; H, hippocampus; LGE, lateral ganglionic eminence; LP, lateral pallium; MGE, medial ganglionic eminence; MP, medial pallium; Pir, piriform cortex; POa, anterior preoptic area; VP, vental pallium.

J. L. R. Rubenstein, unpublished observations). A complementary phenotype is observed in mice lacking both the *Dlx1* and *Dlx2* homeobox genes (Anderson *et al.*, 1997a), which are coexpressed with *Mash1* in most cells of the SVZ (Porteus *et al.*, 1994). These mutants exhibit increased Notch signaling in the SVZ. As a result, fewer late born neurons fully differentiate, whereas early born neurons appear to be normal (K. Yun and J. L. R. Rubenstein, unpublished observations). These results suggest that bHLH and homeobox genes control the differentiation rate of specific sets of neurons through the modulation of Notch signaling in progenitor cells. Interestingly, Notch1 signaling may also promote the differentiation of radial glia in the forebrain before the onset of neurogenesis (Gaiano *et al.*, 2000).

Members of the POU homeodomain transcription factor genes are additional candidate regulators of neurogenesis in the forebrain. Class III POU proteins are broadly expressed in the neuroepithelium of the forebrain at the time of neurogenesis (He *et al.*, 1989; Alvarez-Bolado *et al.*, 1995) and may be required for neuronal differentiation. For example, *Brn2* is transcribed in the forebrain as early as E8.5 (Schonemann *et al.*, 1995), and inhibition of *Brn2* activity *in vitro* blocks neuronal differentiation at an early stage (Fujii and Hamada, 1993). Moreover, recent data suggest that POU proteins and nuclear hormone receptors cooperate to drive gene expression in the forebrain (Josephson *et al.*, 1998). Thus, *Brn1* and *Brn2* may be required to drive the expression of specific sets of genes that characterize neural stem cells (Josephson *et al.*, 1998). Interestingly, expression of *Brn1* and *Brn2* is downregulated during neuronal differentiation, while other class III POU genes are upregulated, suggesting that different POU proteins may act sequentially to promote neuronal differentiation (Alvarez-Bolado *et al.*, 1995; Josephson *et al.*, 1998; Shimazaki *et al.*, 1999).

B. Control of Differentiation

Once a cell is committed to leave the cell cycle and has been specified to form a particular class of neuron, a progressive process of differentiation guides the cell to its mature phenotype. Genes that control this event may begin to be expressed in progenitor cells, but either they or their targets are expressed in postmitotic neurons to further consolidate the process of cellular maturation. Here we will describe some of the mechanisms that control differentiation in the subpallium, pallium, and diencephalon.

1. Differentiation of Subpallial Neurons

The subpallium generates two main types of neurons: GABAergic and cholinergic (Fig. 6). Available evidence suggests that most, if not all, differentiating GABAergic neurons express the *Dlx* and *Mash1* genes (Porteus *et al.*,

1994; Anderson *et al.*, 1997b; Eisenstat *et al.*, 1999; Fode *et al.*, 2000). Four *Dlx* genes are transcribed in the forebrain in the following temporal sequence: *Dlx2*, *Dlx1*, *Dlx5*, and *Dlx6* (Liu *et al.*, 1997; Eisenstat *et al.*, 1999; Zerucha *et al.*, 2000). They exhibit an overlapping expression pattern that spans the ventriculopial axis, with *Dlx2* being expressed primarily in progenitor cells and *Dlx6* being expressed mostly in postmitotic neurons. There is evidence that the different *Dlx* genes are coexpressed in individual cells (Eisenstat *et al.*, 1999). This coexpression, as well as the sequence similarity of the *Dlx* gene products, may explain why *Dlx1*, *Dlx2*, and *Dlx5* single mutant mice have roughly normal appearing forebrains (Qiu *et al.*, 1995; Anderson *et al.*, 1997a; J. E. Long and J. L. R. Rubenstein, unpublished observations), although *Dlx2* and *Dlx5* mutants have some olfactory bulb defects (Qiu *et al.*, 1995; J. E. Long and J. L. R. Rubenstein, unpublished observations). On the other hand, as noted above, *Dlx1;Dlx2* double mutants have a strong block in differentiation of late born GABAergic neurons (Anderson *et al.*, 1997a), suggesting that redundancy exists among the *Dlx* genes in the subpallium. Although these mutants possess a SVZ, the molecular properties of this proliferative zone are abnormal. For instance, they largely lack expression of *Dlx5* and *Dlx6*, suggesting that *Dlx1* and *Dlx2* are required for their induction. Indeed, evidence suggests that the DLX1 and DLX2 proteins bind an enhancer element in the mouse *Dlx5/6* genes and their zebrafish orthologs (Zerucha *et al.*, 2000). In addition, ectopic expression experiments suggest that DLX2 may be sufficient to induce *Dlx5* expression and specify the GABAergic phenotype (Anderson *et al.*, 1999; T. Stühmer, S. A. Anderson, and J. L. R. Rubenstein, unpublished observations).

The subpallium generates two main types of neurons, projection neurons (those that send their axons outside their local environment) and local circuit neurons (those that make synaptic contacts with nearby neurons, also known as interneurons). Current evidence suggests that projection neurons largely follow radial routes from their progenitor zone into the overlying mantle (Fig. 6). Thus, striatal projection neurons are thought to derive from the LGE (Deacon *et al.*, 1994; Olsson *et al.*, 1995, 1998; Anderson *et al.*, 1997a), whereas most pallidal neurons appear to derive from the MGE (Sussel *et al.*, 1999; O. Marín and J. L. R. Rubenstein, unpublished observations). In addition, the MGE and the adjacent AEP/POA probably give rise to the projection neurons of the cholinergic magnocellular basal forebrain complex (including the basal nucleus of Meynert). Interneurons, on the other hand, appear to follow tangential migration pathways to reach their final destination. Recent studies have demonstrated that most striatal interneurons are derived from the MGE and the adjacent AEP/POA (Sussel *et al.*, 1999; Marín *et al.*, 2000). In addition, it is now well established that cortical interneurons arise through two main tangential migrations, a superficial pathway, which primarily originates in the MGE (Lavdas *et al.*, 1999; Sussel *et al.*, 1999; Wichterle *et al.*, 1999; Anderson *et al.*, 1999, 2001; Pleasure *et al.*, 2000), and a deep pathway, which primarily originates in the LGE (de Carlos *et al.*, 1996; Anderson *et al.*, 1997b, 2001; Tamamaki *et al.*, 1997; Pleasure *et al.*, 2000). Finally, a rostral migratory pathway originating in the region of the LGE and septum is the origin of olfactory bulb interneurons (Lois *et al.*, 1996; Bulfone *et al.*, 1998). Evidence suggests that the secreted molecule SLIT is implicated in these migrations (Wu *et al.*, 1999; Zhu *et al.*, 1999).

The subpallium gives rise to primarily two types of neurons on the basis of their neurotransmitter content, GABAergic neurons and cholinergic neurons (Fig. 6). Both types of neurons are found among projection and local circuit neurons in the basal telencephalon. For example, GABA is expressed in cells derived from the MGE that differentiate either as interneurons in the striatum or as projection neurons in the globus pallidus (Sussel *et al.*, 1999; Marín *et al.*, 2000; O. Marín and J. L. R. Rubenstein, unpublished observations). Mutations in both *Dlx1;Dlx2* or *Mash1* affect the generation of projection neurons and interneurons from both the LGE and MGE. However, while in *Mash1* mutants there is a reduction of early born cells, *Dlx1;Dlx2* mutants exhibit a reduction of late born cells (Anderson *et al.*, 1997a,b; Casarosa *et al.*, 1999; Marín *et al.*, 2000). These mutants have severe reductions in the tangential migrations of GABAergic interneurons into the cerebral cortex, hippocampus, olfactory bulb, and striatum (Fig. 7; Anderson *et al.*, 1997b, 1999; Bulfone *et al.*, 1998; Casarosa *et al.*, 1999; Anderson *et al.*, 2001; Marín *et al.*, 2000; Pleasure *et al.*, 2000). In contrast, the presence of *Nkx2.1* mutants, which selectively affect the MGE but not the LGE, results in a severe reduction in the tangential migration of GABAergic interneurons to the cortex and striatum, but not to the olfactory bulb (Fig. 7; Sussel *et al.*, 1999; Marín *et al.*, 2000). Furthermore, the subpallium of *Nkx2.1* mutants lacks all cholinergic neurons, including both the projection neurons of the basal magnocellular complex and the striatal interneurons (Marín *et al.*, 2000). Thus, there is mounting evidence that, like telencephalic GABAergic interneurons, cholinergic interneurons are specified in one location and migrate tangentially to their final destination (Marín *et al.*, 2000).

As noted earlier, there is evidence that *Dlx1;Dlx2* and *Mash1* mutants affect neurogenesis in the forebrain through disrupting Notch signaling. In addition, these mutants provide insights into other regulatory mechanisms of differen-

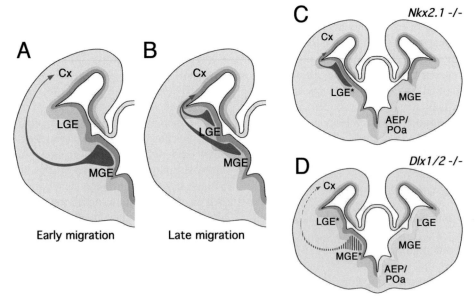

Figure 7 Tangential migration of immature interneurons from the subpallium to the cortex in wild-type and mutant mice. These migrations arise from the MGE and LGE and seed the cortex through at least two spatially and temporally distinct routes (see Anderson *et al.*, submitted). (A) Early during development (E11.5–E14-5) interneurons migrating from the subpallium to the cortex arise primarily from the MGE and follow a superficial route. (B) Later on, migrating cells arise from both the MGE and LGE and follow a deep pathway into the cortical subventricular zone. (C) Migration from the MGE to the cortex is impaired in *Nkx2.1* mutants (the expression of *Nkx2.1* in the proliferative zones of the MGE and AEP/POa is shown in yellow). (D) Migration from both the MGE and LGE to the cortex is impaired in *Dlx1/2* mutants (the expression of *Dlx1/2* in the proliferative zones of the subpallium is shown in yellow). Only a few early born interneurons migrate to the cortex in *Dlx1/2* mutants. Abbreviations: AEP, anterior entopeduncular area; Cx, cortex; LGE, lateral ganglionic eminence; LGE*, mutant lateral ganglionic eminence; MGE*, mutant medial ganglionic eminence; POa, anterior preoptic area.

tiation, because they affect the expression of multiple transcription factors in the basal telencephalon (Anderson *et al.,* 1997a; Marín *et al.,* 2000; K. Yun, O. Marín, S. A. Anderson, and J. L. R. Rubenstein, unpublished observations). For instance, *Lhx2,* which controls proliferation in the VZ (Porter *et al.,* 1997), is ectopically expressed in the SVZ and mantle of *Dlx1;Dlx2* mutants. This suggests that *Dlx1* and/or *Dlx2* may normally be required to repress the expression of factors that regulate the proliferation of progenitor cells, in order that cell cycle exit and neuronal differentiation may proceed.

In *Dlx1;Dlx2* mutants, other transcription factors are not expressed at normal levels in the subpallium, such as *Dlx5, Dlx6, Oct6, Ebf1, ER81, Pbx1, Pbx3, NPAS1,* and *Brn4.* Interestingly, *Brn4* is a POU-class homeodomain gene implicated in striatal cell differentiation (Shimazaki *et al.,* 1999) that can be induced by BDNF and IGF, secreted factors that regulate differentiation of striatal cells (Jones *et al.,* 1994; Beck *et al.,* 1995).

Pbx1 and *Pbx3* homeobox gene expression is also affected in the subpallium of *Dlx1;Dx2* double mutants (O. Marín, Rhee, J. M. Cleary and J. L. R. Rubenstein, unpublished observations). Nuclear transport of PBX-type homeodomain proteins is regulated by binding to MEIS homeodomain proteins (Rieckhof *et al.,* 1997; Pai *et al.,* 1998; Mercader *et al.,* 1999). PBX and MEIS proteins are coexpressed in the LGE but not in the MGE, suggesting that they may play a role in the differentiation of striatal neurons (Toresson *et al.,* 1999a, 2000).

Ebf1 belongs to a small family of transcription factors containing an atypical zinc finger DNA binding domain and an HLH domain (Hagman *et al.,* 1995). *Ebf1* is the only known member of this family to be expressed in the developing striatum (Garel *et al.,* 1997). Loss of function mutants have demonstrated that *Ebf1* is required for the differentiation of striatal neurons in mice. *Ebf1* mutants fail to activate striatal mantle specific genes such as *cad8* or *CRABP1* (cellular retinoid acid binding protein 1), and fail to repress the expression of SVZ markers, like *Oct6* or *Dlx5* (Garel *et al.,* 1999). The effect of the *Ebf1* mutation on *CRABP1* expression is intriguing because retinoid signaling is implicated in regulating the differentiation of striatal cells. Thus, loss of function mutants of retinoic acid receptor genes and *in vitro* experiments provide evidence that retinoid signaling regulates the expression of genes characteristic of mature striatal projection neurons, such as D1 and D2 dopamine receptors, enkephalin, and DARPP32 (Samad *et al.,* 1997; Krezel *et al.,* 1998; Toresson *et al.,* 1999a). Evidence suggests that retinoids are produced by radial glia in the developing striatum (Toresson *et al.,* 1999a).

Additional transcription factors have been found to play a role in the differentiation of striatal neurons. *Sox1* is an Sry-related transcription factor required for the normal development of the olfactory tubercle (Malas *et al.,* 1998), a

part of the ventral striatum. *Ikaros* is a zinc finger protein that is implicated in the regulation of enkephalin expression in the striatum (Ring *et al.,* 1999). In addition, a number of previously characterized genes are expressed in the developing basal ganglia, although their function in this region remains undetermined. These include members of the Eph family of tyrosine kinase receptors and their ligands, the ephrins, which are expressed in partially overlapping patterns in the striatum and may participate in the segregation of distinct functional compartments in the basal ganglia (Janis *et al.,* 1999; Yue *et al.,* 1999). A similar role has been suggested for cadherins in the telencephalon (Stoykova *et al.,* 1997; Korematsu *et al.,* 1998; Hirano *et al.,* 1999).

Normal patterns of gene expression are also disrupted in the subpallium of *Nkx2.1* mutants (Sussel *et al.,* 1999; O. Marín and J. L. R. Rubenstein, unpublished observations). *Nkx2.1* mutants exhibit ectopic expression of *Pax6* and *Oct6* in the VZ and SVZ of the MGE, respectively, suggesting that *Nkx2.1* is required to maintain the identity of MGE progenitor cells. Moreover, these mutants fail to express several LIM homeobox genes (*Isl1, Lhx6,* and *Lhx7*), which are normally expressed in distinct subsets of cells that tangentially migrate from the MGE/AEP/POA into the striatum and cerebral cortex (Marín *et al.,* 2000).

Finally, as in other regions of the neuroaxis, the process of cell differentiation in the basal telencephalon is not dictated only by intrinsic factors, but also by extrinsic cues, such as neurotrophic factors. Thus, developing basal telencephalic neurons express receptors of the Trk family (Vazquez and Ebendal, 1991; Steininger *et al.,* 1993; Barbacid, 1994; Canals *et al.,* 1999; Costantini *et al.,* 1999), and different neurotrophins have been shown to modulate some phenotypic characteristics of striatal and other basal forebrain projection neurons and interneurons (Jones *et al.,* 1994; Li *et al.,* 1995; Arenas *et al.,* 1996; Alderson *et al.,* 1996; Altar *et al.,* 1997; Fagan *et al.,* 1997; Ivkovic *et al.,* 1997).

2. Neuronal Differentiation and Laminar Patterning in the Pallium

Histogenesis of the cerebral cortex, like other dorsal structures such as the tectum of the midbrain, requires the elaboration of a precise laminar architecture. This is an important feature that distinguishes it from the subpallium and deep pallial nuclei (e.g., claustrum), which are organized into nonlaminar modules. Although the number of layers and their morphological characteristics vary in different cortical regions (e.g., three layers in the paleocortex but six in the isocortex), some of the principles underlying the development of this laminar appear to be similar for all cortical regions.

Once the process of regional specification has defined the different types of cortex and their main subdivisions (Figs. 1 and 2; see Section II.D), and neurogenesis of the cerebral cortex has started, different mechanisms coordinate its laminar patterning. The majority of cortical neurons are pyramidal projection neurons that use the excitatory amino acid glutamate as their neurotransmitter, while most of the remaining neurons are GABAergic interneurons. As described earlier, recent evidence suggests that most GABAergic interneurons are generated in the subpallium, whereas the glutamatergic neurons are generated by progenitors in the VZ of the cerebral cortex.

The development of cortical lamination follows a series of stages (Bayer and Altman, 1991). The first neuronal populations that migrate from the proliferative zone constitute the preplate. The preplate is separated from the proliferative zone by the intermediate zone, which contains migratory neurons and axons (Fig. 8). The preplate is composed of Cajal-Retzius and subplate neurons. Subsequent neuronal migrations create the cortical plate, which splits the preplate in two layers, the marginal zone and the subplate. Cajal-Retzius neurons remain near the pial surface in the marginal zone. Thereafter, successive waves of migration position neurons within layers in the cortical plate. Birthdating studies have shown that cortical plate layers are established according to an inside-outside pattern, where the deeper layers contain cells that become postmitotic earlier than the cells in more superficial layers (Fig. 8; Angevine and Sidman, 1961; Rakic, 1974). This suggests that laminar fate is linked to neurogenetic events in the proliferative zone (Caviness, 1982; McConnell, 1989, 1991; McConnell and Kaznowski 1991; Chenn et al., 1997). However, recent high-resolution birthdating studies have shown that neurons born at approximately the same time and from the same region of the proliferative zone may have different laminar fates in the cortex (Takahashi et al., 1999), suggesting that the time of neurogenesis is not sufficient to specify unambiguously laminar position. In contrast, neurons arising within the same cycle of the neurogenetic sequence have common laminar fates independently of their positions in the proliferative zone or their time of neurogenesis, suggesting that the mechanisms that control the cell cycle sequence may regulate the subsequent processes that determine the laminar fate (Takahashi et al., 1999).

The mechanisms that control the differentiation of distinct subtypes of cortical glutamatergic neurons are not understood. There is evidence that extrinsic factors influence progenitor fate during the G1 stage of the cell cycle (reviewed in Chenn et al., 1997). Moreover, it has been hypothesized that progression in the sequence of cell cycles may serve to limit the classes of neurons that can arise within each cycle (Takahashi et al., 1999). Thus, changing environmental cues may modify the developmental programs within progenitors during successive cell cycles. It is possible that the lateral inhibition mechanism, which operates in the basal ganglia, may participate in regulating the timing

and specification of distinct types of neurons in the cortex (see above).

Several transcription factors have been implicated in the development of specific cell types in the cortex (reviewed in Chenn et al., 1997; Rubenstein et al., 1999). Some of these genes are expressed in specific laminar patterns. For example, Tbr1, a T-Box gene, is expressed in early born neocortical and olfactory bulb neurons, where it is required for their development (Bulfone et al., 1995, 1998; Hevner et al., 2001). Tbr1 appears to be essential for the differentiation of Cajal-Retzius neurons. For example, Tbr1 mutants have a reduced expression of reelin, a secreted molecule normally expressed by Cajal-Retzius cells that is required for the radial migration of immature cortical neurons (Hevner et al., 2001), and a similar defect has been recently described in Emx2 mutants (Mallamaci et al., 2000). Remarkably, in vitro experiments suggest that Tbr1 may cooperate to activate transcription of the Reelin gene (Hesch et al., 2000). Tbr1 mutants also have defects in subplate neurons (Hevner et al., 2001), cells that send the earliest cortical axons to the thalamus (McConnell et al., 1989, 1994). Similarly, mice lacking the chicken ovalbumin upstream promoter-transcription factor 1 (COUP-TF1), an orphan nuclear receptor, show abnormal differentiation and premature death of subplate neurons (Zhou et al., 1999). In both Tbr1 and COUP-TF1 mutant mice, loss of subplate neurons is associated with defects in the guidance of thalamocortical projections (Zhou et al., 1999; Hevner et al., 2001), supporting the idea that the subplate has a prominent role in the early development of thalamocortical connectivity (Allendoerfer and Shatz, 1994; Molnár and Blakemore, 1995).

Neurogenesis within subdivisions of the hippocampal complex is under distinct genetic controls. For example, mice homozygous for loss of function mutations in the Lef1, NeuroD, and NEX transcription factors are defective in the differentiation of dentate gyrus granule cells, whereas the CA fields appear to form normally (Galceran et al., 1999; Miyata et al., 1999; Liu et al., 2000; Schwab et al., 2000). On the other hand, mutation of Lhx5 affects migration and differentiation of cells in all fields of the hippocampus (Zhao et al., 1999).

Terminal differentiation of cortical projection neurons requires the formation of lamina-specific dendritic arborizations and appropriate axonal connections. Recently, it has been shown that Notch1 signaling regulates dendritic morphology in cortical neurons (Redmond et al., 2000). Migrating neurons translocate the intracellular domain of Notch1 to the nucleus once they reach the cortical plate. Nuclear Notch1 is implicated in stimulating dendritic branching and reduces dendritic growth in cortical neurons. Similarly, neurotrophins can also regulate dendritic growth in cortical neurons (McAllister et al., 1995, 1996; Yacoubian and Lo, 2000; reviewed in McAllister et al., 1999). Axonal arbori-

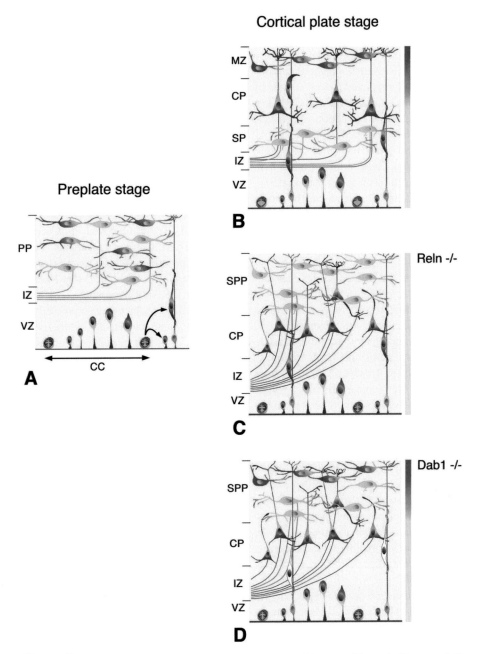

Figure 8 Radial migration generates the laminar organization of the cortex. The cortical layers are laid down successively by the coordinated radial migration of neurons. (A) The preplate is constituted by two main types of early generated neurons; Cajal-Retzius cells, which express Reelin (dark green cells), and subplate neurons (light green cells). The radial migration of cortical neurons (blue cells) requires interaction with radial glia (red cells). (B) Migration of cortical neurons forms the cortical plate, which splits the preplate in two layers, called the marginal zone and the subplate. The marginal zone contains Cajal-Retzius cells that form a gradient of Reelin through the cortical plate (depicted in the green vertical bar on the right of panels B, C, and D). The subplate neurons produce pioneer corticofugal axonal projections that travel through the intermediate zone to the subpallium. (C) In *Reelin* mutants, cortical neurons fail to invade the preplate and the cortical plate develops deep to Cajal-Retzius and preplate cells (superplate). (D) A similar phenotype is found in *Dab1* mutants, in which migrating cortical neurons (white cells) are not able to respond to the gradient of Reelin. Abbreviations: CC, indicates cells at different stages of the cell cycle; CP, cortical plate; IZ, intermediate zone; MZ, marginal zone; PP, preplate; SP, subplate; SPP, superplate; VZ, ventricular zone.

zations of cortical neurons follow a stereotyped, lamina-specific pattern. For example, axons of layer 6 neurons arborize in layer 4, whereas layer 2/3 neurons extend axon collaterals locally and in layer 5. The development of lamina-specific branching appears to be regulated to some extent by the interaction of Eph receptors and ephrins (Castellani *et al.*, 1998). In addition, the axonal projections of cortical neurons are correlated with their laminar fate. For example, pyramidal neurons in layer 5 project to the spinal cord, pons, and superior colliculus, whereas those in layer 6 send their axons to the thalamus. The specification of appropriate axonal projections appears to be largely independent of proper neuronal migration, as demonstrated by the development of normal patterns of axonal connections in mice that have abnormal cortical migration (reviewed in Rakic and Caviness, 1995).

Depending on the region of the cerebral cortex, neurons located in the same layer may project to different sets of targets. For example, layer 5 neurons of the visual cortex project to the superior colliculus, midbrain tegmentum, and pons, whereas layer 5 neurons in the motor cortex project to the midbrain tegmentum, pons, and spinal cord, but not to the superior colliculus. Interestingly, layer 5 neurons from neocortical regions develop their connections as a single population, innervating initially all subcortical regions independently of their region of origin in the cortex. After this step, specific collateral branches are selectively eliminated from the projections of each region to generate the mature set of projections functionally appropriate for the cortical region in which the layer 5 neuron is located (reviewed in Chenn *et al.*, 1997). Although we do not have a complete understanding of the mechanisms that control this process, recent experiments have demonstrated that expression of the *Otx1* homeobox gene in layer 5 is required for the correct pruning of visual cortex axon collaterals (Weimann *et al.*, 1999).

The migration of young neurons derived from cortical progenitors toward the cortical plate is largely dependent on radial glia and Cajal-Retzius cells (Rakic, 1971, 1972; Rakic *et al.*, 1974). Radial glia, whose somata reside in the ventricular zone, extend long processes that reach the pial surface. Mutations in mice and humans are providing insights into the molecular regulation of radial migration (reviewed in Rice and Curran, 1999; Allen and Walsh, 1999; Walsh, 1999). Three types of processes are affected by these mutations: (1) those intrinsic to radial glia, (2) those controlled by the Cajal-Retzius cells, and (3) those intrinsic to the migrating cells.

Pax6, which is expressed in the progenitors of radial glia (Götz *et al.*, 1998), is required for the migration of late born cortical neurons (Schmahl *et al.*, 1993; Caric *et al.*, 1997; Götz *et al.*, 1998; Warren *et al.*, 1999). There is evidence that the migration defect is not due to intrinsic abnormalities in the migrating cells, but rather due to defects in the radial glia (Caric *et al.*, 1997; Götz *et al.*, 1998). Defects in cortical radial glia are also found in mutants of the *ErbB2* and *ErbB4* tyrosine kinase receptor genes, which are associated with abnormal migration in the cerebral cortex and cerebellum, respectively (Anton *et al.*, 1997; Rio *et al.*, 1997).

As the migrating cortical cells approach the Cajal-Retzius cells, they detach from the radial glia. As noted earlier, Cajal-Retzius cells secrete a large protein called Reelin. Mice defective in Reelin function (*reeler* mutants) have cortical lamination defects that roughly correspond to an inversion of the cortical layers (Fig. 8; Caviness, 1982; Hoffarth *et al.*, 1995; Ogawa *et al.*, 1995; Sheppard and Pearlman, 1997; reviewed in Curran and D'Arcangelo, 1998). These results suggest that signaling from the marginal zone directs the laminar organization of the cortex. Recently, it was established that Reelin is a high-affinity ligand for the ApoE receptor 2 (ApoER2) and VLDL receptor (VLDLR), which are expressed by migrating cortical cells (D'Arcangelo *et al.*, 1999; Hiesberger *et al.*, 1999). Mice lacking both of these members of the LDL family of lipoprotein receptors have lamination defects resembling those found in reeler mice (Trommsdorff *et al.*, 1999). Interestingly, the cytoplasmic tails of ApoER2 and VLDLR interact with the Disabled protein (Dab1; Trommsdorff *et al.*, 1998, 1999). Dab1 is a cytoplasmic adapter protein that is also essential for radial migration (Sweet *et al.*, 1996; Howell *et al.*, 1997; Sheldon *et al.*, 1997; Yoneshima *et al.*, 1997; Fig. 8). Remarkably, Reelin signaling through ApoER2 and VLDLR can induce phosphorylation of Dab1 (D'Arcangelo *et al.*, 1999; Hiesberger *et al.*, 1999; Howell *et al.*, 1999), which has been linked to the reorganization of microtubules and microfilaments. Mutations in one of these proteins, Cdk5, or its activator p35, also produce inversion-like migratory defects in the cortex (Ohshima *et al.*, 1996; Chae *et al.*, 1997; Gilmore *et al.*, 1998; Kwon and Tsai, 1998). Similarly, mutations in the *Lis1/PA-FAH1b1* (Reiner *et al.*, 1993) and *doublecortin* (des Portes *et al.*, 1998; Gleeson *et al.*, 1998) genes, originally identified in humans, also disrupt radial migration probably by affecting microtubule polymerization (Sapir *et al.*, 1997; Ware *et al.*, 1999). Finally, several molecules have been implicated in regulating the interaction of migrating cells with the radial glia, such as astrotactin (Zheng *et al.*, 1996) and members of the integrin family. For instance, mice lacking *integrin α3* or *α6* have abnormalities in the laminar organization of the developing cortex (Georges-Labouesse *et al.*, 1998; Anton *et al.*, 1999).

Certain cortical neurons (e.g., GABAergic cells) follow tangential migratory pathways (reviewed in Chenn *et al.*, 1997; Hatten, 1999; Ware *et al.*, 1999; Anderson *et al.*, 1999). As described earlier in this chapter, many tangentially migrating cells in the cortex derive from progenitor cells located in the subpallium (reviewed in Anderson *et al.*, 1999; Figs. 6 and 7; see Section IV.B.1).

3. Cell Differentiation in the Olfactory Bulb

The olfactory bulb is an evaginated cortical structure that contains two main types of neurons, projection neurons (mitral/tufted cells) and interneurons. Recent evidence suggests that like in the neocortex, these neurons originate from different progenitor populations. Thus, olfactory bulb interneurons are born in the SVZ of the LGE, and migrate rostrally through the rostral migratory stream (Lois and Alvarez-Buylla, 1994; Jankovski and Sotelo, 1996; Goldman and Luskin, 1998). Accordingly, *Dlx1;Dlx2* double mutants lack most GABAergic interneurons in the olfactory bulb (Anderson *et al.*, 1997a; Bulfone *et al.*, 1998). In contrast, mice with a targeted mutation in *Tbr1* lack most projection neurons (Bulfone *et al.*, 1998). Cell migration from the SVZ of the LGE to the olfactory bulb may be mediated by the repulsion of neural precursors by the secreted protein *Slit1* (Wu *et al.*, 1999), and utilizes a distinctive migration mechanism, that continues even in mature animals (Doetsch and Alvarez-Buylla, 1996; Lois *et al.*, 1996).

C. Cell Specification and Differentiation in the Diencephalon

The neuroendocrine hypothalamus consists of several distinct neuronal populations: (1) magnocellular neurons of the paraventricular nucleus, which release oxytocin (OT) in the posterior pituitary; (2) magnocellular neurons of the supraoptic nucleus, which release arginine vasopressin (AVP) in the posterior pituitary; (3) parvocellular neurons of the paraventricular nucleus, which release corticotropin-releasing hormone (CRH); (4) parvocellular neurons of the paraventricular nucleus, which release thyrotropin-releasing hormone (TRH); (5) parvocellular neurons of the anterior periventricular and arcuate nuclei, which release somatostatin (SOM); and (6) parvocellular neurons of the anterior periventricular and arcuate nuclei, which release growth hormone-releasing hormone (GHRH). The neurogenesis of the endocrine hypo-

thalamus requires the combined action of several transcription factors. Thus, mice lacking the homeobox containing gene *Otp* fail to complete the differentiation of neuroendocrine cell types present in the paraventricular, supraoptic, anterior periventricular, and arcuate nuclei. Specifically, *Otp* mutant mice lack expression of OT, AVP, CRH, TRH, and SOM, whereas GHRH is normally expressed (Acampora *et al.*, 1999b). Interestingly, loss of the bHLH-PAS transcription factor *Sim1* leads to deficiency of the same cell types missing in the *Otp* mutants (Michaud *et al.*, 1998). Indeed, *Sim1* and *Otp* are largely coexpressed, and both genes are required for *Brn2* expression (Michaud *et al.*, 1998; Acampora *et al.*, 1999b). Lack of *Brn2* function also leads to abnormal differentiation of magnocellular and parvocellular neurons (Nakai *et al.*, 1995; Schonemann *et al.*, 1995). *Brn2*, however, is only required for the differentiation of the OT, AVP, and CRH lineages. Finally, targeted mutation of *Gsh1* results in inappropriate differentiation of GHRH synthesizing neurons (Li *et al.*, 1996), suggesting that *Gsh1* functions in a different specification pathway than *Otp*, *Sim1*, and *Brn2*. Loss of *Lhx4* expression in the hypothalamus also results in deficiency in GHRH expression (Sheng *et al.*, 1997).

Differentiation of the mammillary hypothalamus appears to be under the control of the *Foxb1* winged helix transcription factor (Alvarez-Bolado *et al.*, 2000). Remarkably, normal expression of *Foxb1* and *Pax6* in the diencephalon is required for the development of the mammillo-thalamic tract, but not for the mammillo-tegmental tract (Alvarez-Bolado *et al.*, 2000; Valverde *et al.*, 2000).

Differentiation of the main subdivisions of the thalamus is controlled by distinct sets of transcription factors. The rostral part of the thalamus, in prosomere 3, is generically known as the ventral thalamus. Like the basal telencephalon, this structure produces GABAergic neurons that require both *Dlx1* and *Dlx2*, or *Mash1* gene function for their differentiation (Torii *et al.*, 1999; O. Marín, S. A. Anderson, and J. L. R. Rubenstein, unpublished observations). The caudal part of the thalamus, which is in prosomere 2, is generically known as the dorsal thalamus. Mice lacking the *Gbx2* homeobox gene or that are double mutants for *neurogenin1;neurogenin2* have severe defects in thalamic differentiation (Miyashita-

Glossary

Allocortex
Cortical regions with heterogeneous laminar structure, typically comprising one main layer of projection neurons sandwiched between layers that are rich in axons, dendrites and local circuit neurons.

Anterior neural ridge
Tissue located at the anterior border of the neural plate that contributes to rostrodorsal parts of the prosencephalon.

(continues)

(continued)

Archicortex
Limbic region of the allocortex found in the dorsomedial pallium, which largely consists of the hippocampal complex.

Commissure
Bundle of axons crossing the midline of the brain or spinal cord.

Corpus callosum
Large bundle of commissural axons that interconnects both sides of the cerebral cortex.

Cortex
Laminar neuronal structures that form at the surface of the central nervous system, such as the cerebellum and cerebral cortex.

Diencephalon
Part of the forebrain, or prosencephalon. It consists of caudal and rostral subdivisions. The caudal diencephalon includes alar plate derivatives such as the ventral thalamus, dorsal thalamus, epithalamus, and pretectum, and basal plate derivatives such as the prerubral tegmentum. The rostral diencephalon is comprised of alar plate derivatives including the prethalamic region, and basal derivatives forming the hypothalamus. The optic and telencephalic vesicles are evaginations from the rostral diencephalon.

Floor plate
Specialized cells located in the ventral midline of the neural plate and neural tube that are induced by the underlying notochord.

Forebrain
The prosencephalon, or most anterior region of the brain. It lies rostral to the midbrain, or mesencephalon, and comprises the telencephalon and the diencephalon.

Hypothalamus
Group of nuclei that constitute most of the rostral diencephalon, involved in functions that include the regulation of emotion and body homeostasis through the control of the autonomic nervous system.

Interneuron
Local circuit neuron, sometimes also referred to as Golgi type II neuron. In a general sense, any neuron that lies between an afferent neuron and an effector neuron.

Isocortex
Six-layered part of the dorsal pallium, also known as neocortex.

Mesocortex
Transitional cortex, with stepwise changes in its architectonic lamination pattern ranging from a typical isocortical zone to a characteristic allocortical structure.

Neocortex
See Isocortex.

Neural plate
Thickened ectodermal tissue that forms in the early embryo and constitutes the anlage of the nervous system.

Neural tube
Structure formed by the fusion of the neural folds (edges of the neural plate).

Neuromere
Transverse morphological unit (or segment) of the brain formed during embryonic development.

Notochord
Longitudinal cord of mesendodermal cells that lies below the midline of the neural plate and late under the floor plate; plays an important role in the ventral patterning of the caudal brain and spinal cord.

Olfactory tubercle
Cortical region of the basal telencephalon.

Optic chiasm
Region of the rostral midline where the optic nerve axons cross to the contralateral side of the brain.

Paleocortex
Olfactory region of the allocortex; also known as the piriform cortex.

Pallium
Roof of the telencephalon, it contains both cortical (e.g., hippocampus and neocortex) and deep-lying nuclear structures (e.g., claustrum and parts of the amygdala). Pallium is not synonymous with cortex

(continues)

(continued)

Pallidum

Part of the subpallium and one of the components of the striatopallidal complex. It comprises dorsal (globus pallidus) and ventral (ventral pallidum) parts.

Prechordal plate

Mesendodermal tissues underlying the medial aspect of the anterior neural plate located just anterior to the rostral end of the notochord; it participates in anterior and ventral patterning of the neural plate.

Prosomere

Neuromeric subdivision of the prosencephalon. *See* Neuromere.

Septum

Group of nuclei located in the rostral ventromedial wall of the telencephalon.

Striatum

Part of the subpallium and one of the components of the striatopallidal complex. It comprises deep (caudate nucleus, putamen, nucleus accumbens) and superficial (olfactory tubercle) parts.

Subpallium

Base of the telencephalon, it consists primarily of the basal ganglia (e.g., striatum, globus pallidus, and parts of the septum and amygdala).

Telencephalic stalk

Transition region between the rostral diencephalon and the telencephalon proper. It includes the preoptic area, anterior entopeduncular region, and eminetia thalami.

Telencephalon

Bilateral evaginations form the rostral forebrain. It is comprised of the pallium and the subpallium.

Zonal limitans intrathalamica

Transversal region that separates the dorsal and the ventral thalamus.

References

Acampora, D., Avantaggiato, V., Tuorto, F., and Simeone, A. (1997). Genetic control of brain morphogenesis through Otx gene dosage requirement. *Development (Cambridge, UK)* **124**, 3639–3650.

Acampora, D., Barone, P., and Simeone, A. (1999a). Otx genes in corticogenesis and brain development. *Cereb. Cortex* **9**, 533–542.

Acampora, D., Postiglione, M. P., Avantaggiato, V., Di Bonito, M., Vaccarino, F. M., Michaud, J., and Simeone, A. (1999b). Progressive impairment of developing neuroendocrine cell lineages in the hypothalamus of mice lacking the Orthopedia gene. *Genes Dev.* **13**, 2787–2800.

Alderson, R. F., Wiegand, S. J., Anderson, K. D., Cai, N., Cho, J. Y., Lindsay, R. M., and Altar, C. A. (1996). Neurotrophin-4/5 maintains the cholinergic phenotype of axotomized septal neurons. *Eur. J. Neurosci.* **8**, 282–290.

Allen, K. M., and Walsh, C. A. (1999). Genes that regulate neuronal migration in the cerebral cortex. *Epilepsy Res.* **36**, 143–154.

Allendoerfer, K. L., and Shatz, C. J. (1994). The subplate, a transient neocortical structure: Its role in the development of connections between thalamus and cortex. *Ann. Rev. Neurosci.* **17**, 185–218.

Altar, C. A., Cai, N., Bliven, T., Juhasz, M., Conner, J. M., Acheson, A. L., Lindsay, R. M., and Wiegand, S. J. (1997). Anterograde transport of brain-derived neurotrophic factor and its role in the brain. *Nature (London)* **389**, 856–860.

Alvarez-Bolado, G., Rosenfeld, M. G., and Swanson, L. W. (1995). Model of forebrain regionalization based on spatiotemporal patterns of POU-III homeobox gene expression, birthdates, and morphological features. *J. Comp. Neurol.* **355**, 237–295.

Alvarez-Bolado, G., Zhou, X., Voss, A. K., Thomas, T., and Gruss, P. (2000). Winged helix transcription factor Foxb1 is essential for access of mammillothalamic axons to the thalamus. *Development (Cambridge, UK)* **127**, 1029–1038.

Anchan, R. M., Drake, D. P., Haines, C. F., Gerwe, E. A., and LaMantia, A. S. (1997). Disruption of local retinoid-mediated gene expression accompanies abnormal development in the mammalian olfactory pathway. *J. Comp. Neurol.* **379**, 171–184.

Anderson, D. J., and Jan, Y. N. (1997). The determination of the neural phenotype. *In* "Molecular and Cellular Aproaches to Neural Development" (W. M. Cowan, T. M. Jessell, and S. L. Zipursky, eds.), pp. 26–63. Oxford University Press, New York.

Anderson, S. A., Qiu, M., Bulfone, A., Eisenstat, D. D., Meneses, J., Pedersen, R., and Rubenstein, J. L. R. (1997a). Mutations of the homeobox genes Dlx-1 and Dlx-2 disrupt the striatal subventricular zone and differentiation of late born striatal neurons. *Neuron* **19**, 27–37.

Anderson, S. A., Eisenstat, D. D., Shi, L., and Rubenstein, J. L. R. (1997b). Interneuron migration from basal forebrain to neocortex: Dependence on Dlx genes. *Science* **278**, 474–476.

Anderson, S. A., Mione, M., Yun, K., and Rubenstein, J. L. R. (1999). Differential origins of neocortical projection and local circuit neurons: Role of Dlx genes in neocortical interneuronogenesis. *Cereb. Cortex* **9**, 646–654.

Anderson, S. A., Marín, O., Horn, C., Jennings, K., and Rubenstein, J. L. R. (2001). Distinct cortical migrations from the medial and lateral ganglionic eminences. *Development* **128**, 353–363.

Angevine, J. B., and Sidman, R. L. (1961). Autoradiographic study of cell migration during histogenesis of cerebral cortex in the mouse. *Nature (London)* **192**, 766–768.

Anton, E. S., Marchionni, M. A., Lee, K. F., and Rakic, P. (1997). Role of GGF/neuregulin signaling in interactions between migrating neurons and radial glia in the developing cerebral cortex. *Development (Cambridge, UK)* **124**, 3501–3510.

Anton, E. S., Kreidberg, J. A., and Rakic, P. (1999). Distinct functions of alpha3 and alpha(v) integrin receptors in neuronal migration and laminar organization of the cerebral cortex. *Neuron* **22,** 277–289.

Arenas, E., Akerud, P., Wong, V., Boylan, C., Persson, H., Lindsay, R. M., and Altar, C. A. (1996). Effects of BDNF and NT-4/5 on striatonigral neuropeptides or nigral GABA neurons in vivo. *Eur. J. Neurosci.* **8,** 1707–1717.

Arimatsu, Y., Miyamoto, M., Nihonmatsu, I., Hirata, K., Uratani, Y., Hatanaka, Y., and Takiguchi-Hayashi, K. (1992). Early regional specification for a molecular neuronal phenotype in the rat neocortex. *Proc. Natl. Acad. Sci. U.S.A.* **89,** 8879–8883.

Bachiller, D., Klingensmith, J., Kemp, C., Belo, J. A., Anderson, R. M., May, S. R., McMahon, J. A., McMahon, A. P., Harland, R. M., Rossant, J., and De Robertis, E. M. (2000). The organizer factors Chordin and Noggin are required for mouse forebrain development. *Nature (London)* **403,** 658–661.

Barbacid, M. (1994). The Trk family of neurotrophin receptors. *J. Neurobiol.* **25,** 1386–1403.

Barth, K. A., and Wilson, S. W. (1995). Expression of zebrafish nk2.2 is influenced by sonic hedgehog/vertebrate hedgehog-1 and demarcates a zone of neuronal differentiation in the embryonic forebrain. *Development (Cambridge, UK)* **121,** 1755–1768.

Barth, K. A., Kishimoto, Y., Rohr, K. B., Seydler, C., Schulte-Merker, S., and Wilson, S. W. (1999). Bmp activity establishes a gradient of positional information throughout the entire neural plate. *Development* **126,** 4977–4987.

Bayer, S. A., and Altman, J. (1991). "Neocortical Development." Raven Press, New York.

Beck, K. D., Powell-Braxton, L., Widmer, H. R., Valverde, J., and Hefti, F. (1995). Igf1 gene disruption results in reduced brain size, CNS hypomyelination, and loss of hippocampal granule and striatal parvalbumin-containing neurons. *Neuron* **14,** 717–730.

Beddington, R. S., and Robertson, E. J. (1999). Axis development and early asymmetry in mammals. *Cell (Cambridge, Mass.)* **96,** 195–209.

Bergquist, H., and Källen, B. (1954). Notes on the early histogenesis and morphogenesis of the central nervous system in vertebrates. *J. Comp. Neurol.* **100,** 627–659.

Bhide, P. G. (1996). Cell cycle kinetics in the embryonic mouse corpus striatum. *J. Comp. Neurol.* **374,** 506–522.

Bober, E., Baum, C., Braun, T., and Arnold, H. H. (1994). A novel NK-related mouse homeobox gene: Expression in central and peripheral nervous structures during embryonic development. *Dev. Biol.* **162,** 288–303.

Briscoe, J., Sussel, L., Serup, P., Hartigan-O'Connor, D., Jessell, T. M., Rubenstein, J. L. R., and Ericson, J. (1999). Homeobox gene Nkx2.2 and specification of neuronal identity by graded Sonic hedgehog signalling. *Nature (London)* **398,** 622–627.

Briscoe, J., Pierani, A., Jessell, T. M., and Ericson, J. (2000). A homeodomain protein code specifies progenitor cell identity and neural fate in the ventral neural tube. *Cell (Cambridge, Mass.)* **101,** 435–445.

Brüstle, O., Spiro, A. C., Karram, K., Choudhary, K., Okabe, S., and McKay, R. D. (1997). In vitro-generated neural precursors participate in mammalian brain development. *Proc. Natl. Acad. Sci. U.S.A.* **94,** 14809–14814.

Bulfone, A., Puelles, L., Porteus, M. H., Frohman, M. A., Martin, G. R., and Rubenstein, J. L. R. (1993). Spatially restricted expression of Dlx-1, Dlx-2 (Tes-1), Gbx-2, and Wnt-3 in the embryonic day 12.5 mouse forebrain defines potential transverse and longitudinal segmental boundaries. *J. Neurosci.* **13,** 3155–3172.

Bulfone, A., Smiga, S. M., Shimamura, K., Peterson, A., Puelles, L., and Rubenstein, J. L. R. (1995). T-brain-1: A homolog of Brachyury whose expression defines molecularly distinct domains within the cerebral cortex. *Neuron* **15,** 63–78.

Bulfone, A., Wang, F., Hevner, R., Anderson, S., Cutforth, T., Chen, S.,

Meneses, J., Pedersen, R., Axel, R., and Rubenstein, J. L. R. (1998). An olfactory sensory map develops in the absence of normal projection neurons or GABAergic interneurons. *Neuron* **21,** 1273–1282.

Büscher, D., Bosse, B., Heymer, J., and Rüther, U. (1997). Evidence for genetic control of Sonic hedgehog by Gli3 in mouse limb development. *Mech. Dev.* **62,** 175–182.

Byrd, C. A., and Burd, G. D. (1993). Morphological and quantitative evaluation of olfactory bulb development in Xenopus after olfactory placode transplantation. *J. Comp. Neurol.* **331,** 551–563.

Campbell, K., Olsson, M., and Björklund, A. (1995). Regional incorporation and site-specific differentiation of striatal precursors transplanted to the embryonic forebrain ventricle. *Neuron* **15,** 1259–1273.

Canals, J. M., Checa, N., Marco, S., Michels, A., Pérez-Navarro, E., and Alberch, J. (1999). The neurotrophin receptors trkA, trkB and trkC are differentially regulated after excitotoxic lesion in rat striatum. *Brain Res. Mol. Brain Res.* **69,** 242–248.

Capdevilla, J., and Izpisua-Belmonte, J. C. (1999). Extracellular modulation of the Hedgehog, Wnt and TGF-beta signalling pathways during embryonic development. *Curr. Opin. Genet. Dev.* **9,** 427–433.

Caric, D., Gooday, D., Hill, R. E., McConnell, S. K., and Price, D. J. (1997). Determination of the migratory capacity of embryonic cortical cells lacking the transcription factor Pax-6. *Development (Cambridge, UK)* **124,** 5087–5096.

Casarosa, S., Fode, C., and Guillemot, F. (1999). Mash1 regulates neurogenesis in the ventral telencephalon. *Development (Cambridge, UK)* **126,** 525–534.

Castellani, V., Yue, Y., Gao, P. P., Zhou, R., and Bolz, J. (1998). Dual action of a ligand for Eph receptor tyrosine kinases on specific populations of axons during the development of cortical circuits. *J. Neurosci.* **18,** 4663–4672.

Caviness, V. S., Jr. (1982). Neocortical histogenesis in normal and reeler mice: A developmental study based upon [3H]thymidine autoradiography. *Brain Res* **256,** 293–302.

Cepko, C. L. (1999). The roles of intrinsic and extrinsic cues and bHLH genes in the determination of retinal cell fates. *Curr. Opin. Neurobiol.* **9,** 37–46.

Chae, T., Kwon, Y. T., Bronson, R., Dikkes, P., Li, E., and Tsai, L. H. (1997). Mice lacking p35, a neuronal specific activator of Cdk5, display cortical lamination defects, seizures, and adult lethality. *Neuron* **18,** 29–42.

Chapouton, P., Gärtner, A., and Götz, M. (1999). The role of Pax6 in restricting cell migration between developing cortex and basal ganglia. *Development (Cambridge, UK)* **126,** 5569–5579.

Chen, Z. F., and Behringer, R. R. (1995). Twist is required in head mesenchyme for cranial neural tube morphogenesis. *Genes Dev.* **9,** 686–699.

Chenn, A., Braisted, J. E., McConnell, S. K., and O'Leary, D. D. M. (1997). Development of the cerebral cortex: Mechanisms controlling cell fate, laminar and areal patterning, and axonal connectivity. *In* "Molecular and Cellular Aproaches to Neural Development" (W. M. Cowan, T. M. Jessell, and S. L. Zipursky, eds.), pp. 440–473. Oxford University Press, New York.

Chiang, C., Litingtung, Y., Lee, E., Young, K. E., Corden, J. L., Westphal, H., and Beachy, P. A. (1996). Cyclopia and defective axial patterning in mice lacking Sonic hedgehog gene function. *Nature (London)* **383,** 407–413.

Cobos, I., Shimamura, K., Rubenstein, J. L. R., Martinez, S. and Puelles, L. (2001). Fate map of the avian anterior forebrain at the 4 somite stage, based on the analysis of quail-chick chimeras. *Dev. Biol.* (in press).

Cohen-Tannoudji, M., Babinet, C., and Wassef, M. (1994). Early determination of a mouse somatosensory cortex marker. *Nature (London)* **368,** 460–463.

Cooper, M. K., Porter, J. A., Young, K. E., and Beachy, P. A. (1998). Teratogen-mediated inhibition of target tissue response to Shh signaling. *Science* **280,** 1603–1607.

Costantini, L. C., Feinstein, S. C., Radeke, M. J., and Snyder-Keller, A. (1999). Compartmental expression of trkB receptor protein in the developing striatum. *Neuroscience* **89**, 505–513.

Couly, G. F., and Le Douarin, N. M. (1985). Mapping of the early neural primordium in quail-chick chimeras. I. Developmental relationships between placodes, facial ectoderm, and prosencephalon. *Dev. Biol.* **110**, 422–439.

Couly, G. F., and Le Douarin, N. M. (1988). The fate map of the cephalic neural primordium at the presomitic to the 3-somite stage in the avian embryo. *Development (Cambridge, UK)* **103**, 101–113.

Couly, G. F., and Le Douarin, N. M. (1987). Mapping of the early neural primordium in quail-chick chimeras. II. The prosencephalic neural plate and neural folds: Implications for the genesis of cephalic human congenital abnormalities. *Dev. Biol.* **120**, 198–214.

Crossley, P. H., Martínez, S., and Martin, G. R. (1996). Midbrain development induced by FGF8 in the chick embryo. *Nature (London)* **380**, 66–68.

Crossley, P. H., Martinez, S., Ohkubo, Y., and Rubenstein, J. L. R. (2001). Evidence that coordinate expression of Fgf8, Otx2, Bmp4, and Shh in the rostral prosencephalon define patterning centers for the telencephalic and optic vesicles. *Neuroscience* (in press).

Curran, T., and D'Arcangelo, G. (1998). Role of reelin in the control of brain development. *Brain Res. Brain Res. Rev.* **26**, 285–294.

Dahmane, N., and Ruiz-i-Altaba, A. (1999). Sonic hedgehog regulates the growth and patterning of the cerebellum. *Development (Cambridge, UK)* **126**, 3089–3100.

Daikoku, S., Chikamori, M., Adachi, T., and Maki, Y. (1982). Effect of the basal diencephalon on the development of Rathke's pouch in rats: A study in combined organ cultures. *Dev. Biol.* **90**, 198–202.

Daikoku, S., Chikamori, M., Adachi, T., Okamura, Y., Nishiyama, T., and Tsuruo, Y. (1983). Ontogenesis of hypothalamic immunoreactive ACTH cells in vivo and in vitro: Role of Rathke's pouch. *Dev. Biol.* **97**, 81–88.

Dale, J. K., Vesque, C., Lints, T. J., Sampath, T. K., Furley, A., Dodd, J., and Placzek, M. (1997). Cooperation of BMP7 and SHH in the induction of forebrain ventral midline cells by prechordal mesoderm. *Cell (Cambridge, Mass.)* **90**, 257–269.

D'Arcangelo, G., Homayouni, R., Keshvara, L., Rice, D. S., Sheldon, M., and Curran, T. (1999). Reelin is a ligand for lipoprotein receptors. *Neuron* **24**, 471–479.

Dasen, J. S., and Rosenfeld, M. G. (1999). Combinatorial codes in signaling and synergy: Lessons from pituitary development. *Curr. Opin. Genet. Dev.* **9**, 566–574.

Davidson, D. (1995). The function and evolution of Msx genes: Pointers and paradoxes. *Trends Genet.* **11**, 405–411.

Deacon, T. W., Pakzaban, P., and Isacson, O. (1994). The lateral ganglionic eminence is the origin of cells committed to striatal phenotypes: Neural transplantation and developmental evidence. *Brain Res.* **668**, 211–219.

de Carlos, J. A., López-Mascaraque, L., and Valverde, F. (1996). Dynamics of cell migration from the lateral ganglionic eminence in the rat. *J. Neurosci.* **16**, 6146–6156.

Depew, M. J., Liu, J. K., Long, J. E., Presley, R., Meneses, J. J., Pedersen, R. A., and Rubenstein, J. L. R. (1999). Dlx5 regulates regional development of the branchial arches and sensory capsules. *Development (Cambridge, UK)* **126**, 3831–3846.

des Portes, V., Francis, F., Pinard, J. M., Desguerre, I., Moutard, M. L., Snoeck, I., Meiners, L. C., Capron, F., Cusmai, R., Ricci, S., Motte, J., Echenne, B., Ponsot, G., Dulac, O., Chelly, J., and Beldjord, C. (1998). Doublecortin is the major gene causing X-linked subcortical laminar heterotopia (SCLH). *Hum. Mol. Genet.* **7**, 1063–1070.

Dodd, J., Jessell, T. M., and Placzek, M. (1998). The when and where of floor plate induction. *Science* **282**, 1654–1657.

Doetsch, F., and Alvarez-Buylla, A. (1996). Network of tangential pathways for neuronal migration in adult mammalian brain. *Proc. Natl. Acad. Sci. U.S.A.* **93**, 14895–14900.

Dono, R., Texido, G., Dussel, R., Ehmke, H., and Zeller, R. (1998). Impaired cerebral cortex development and blood pressure regulation in FGF-2-deficient mice. *EMBO J.* **17**, 4213–4225.

Donoghue, M. J., and Rakic, P. (1999). Molecular evidence for the early specification of presumptive functional domains in the embryonic primate cerebral cortex. *J. Neurosci.* **19**, 5967–5979.

Dou, C. L., Li, S., and Lai, E. (1999). Dual role of brain factor-1 in regulating growth and patterning of the cerebral hemispheres. *Cereb. Cortex* **9**, 543–550.

Dudley, A. T., Lyons, K. M., and Robertson, E. J. (1995). A requirement for bone morphogenetic protein-7 during development of the mammalian kidney and eye. *Genes Dev.* **9**, 2795–2807.

Eagleson, G. W., and Harris, W. A. (1990). Mapping of the presumptive brain regions in the neural plate of *Xenopus laevis*. *J. Neurobiol.* **21**, 427–440.

Eagleson, G. W., Ferreiro, B., and Harris, W. A. (1995). Fate of the anterior neural ridge and the morphogenesis of the Xenopus forebrain. *J. Neurobiol.* **28**, 146–158.

Echelard, Y., Epstein, D. J., St.-Jacques, B., Shen, L., Mohler, J., McMahon, J. A., and McMahon, A. P. (1993). Sonic hedgehog, a member of a family of putative signaling molecules, is implicated in the regulation of CNS polarity. *Cell (Cambridge, Mass.)* **75**, 1417–1430.

Edlund, T., and Jessell, T. M. (1999). Progression from extrinsic to intrinsic signaling in cell fate specification: A view from the nervous system. *Cell (Cambridge, Mass.)* **96**, 211–224.

Eisenstat, D. D., Liu, J. K., Mione, M., Zhong, W., Yu, G., Anderson, S., Ghatas, I., Puelles, L., and Rubenstein, J. L. R. (1999). DLX-1, DLX-2, and DLX-5 expression define distinct dtages of basal forebrain differentiation. *J. Comp. Neurol.* **414**, 217–237.

Epstein, D. J., Vekemans, M., and Gros, P. (1991). Splotch (Sp2H), a mutation affecting development of the mouse neural tube, shows a deletion within the paired homeodomain of Pax-3. *Cell (Cambridge, Mass.)* **67**, 767–774.

Ericson, J., Muhr, J., Placzek, M., Lints, T., Jessell, T. M., and Edlund, T. (1995). Sonic hedgehog induces the differentiation of ventral forebrain neurons: A common signal for ventral patterning within the neural tube. *Cell (Cambridge, Mass.)* **81**, 747–756.

Ericson, J., Rashbass, P., Schedl, A., Brenner-Morton, S., Kawakami, A., van Heyningen, V., Jessell, T. M., and Briscoe, J. (1997). Pax6 controls progenitor cell identity and neuronal fate in response to graded Shh signaling. *Cell (Cambridge, Mass.)* **90**, 169–180.

Fagan, A. M., Garber, M., Barbacid, M., Silos-Santiago, I., and Holtzman, D. M. (1997). A role for TrkA during maturation of striatal and basal forebrain cholinergic neurons in vivo. *J. Neurosci.* **17**, 7644–7654.

Fan, C. M., Kuwana, E., Bulfone, A., Fletcher, C. F., Copeland, N. G., Jenkins, N. A., Crews, S., Martínez, S., Puelles, L., Rubenstein, J. L. R., and Tessier-Lavigne, M. (1996). Expression patterns of two murine homologs of Drosophila single-minded suggest possible roles in embryonic patterning and in the pathogenesis of Down syndrome. *Mol. Cell Neurosci.* **7**, 1–16.

Figdor, M. C., and Stern, C. D. (1993). Segmental organization of embryonic diencephalon. *Nature (London)* **363**, 630–634.

Fishell, G. (1995). Striatal precursors adopt cortical identities in response to local cues. *Development (Cambridge, UK)* **121**, 803–812.

Fishell, G. (1997). Regionalization in the mammalian telencephalon. *Curr. Opin. Neurobiol.* **7**, 62–69.

Fode, C., Ma, Q., Casarosa, S., Ang, S. L., Anderson, D. J., and Guillemot, F. (2000). A role for neural determination genes in specifying the dorsoventral identity of telencephalic neurons. *Genes Dev.* **14**, 67–80.

Franz, T. (1994). Extra-toes (Xt) homozygous mutant mice demonstrate a role for the Gli-3 gene in the development of the forebrain. *Acta Anat.* **150**, 38–44.

Fujii, H., and Hamada, H. (1993). A CNS-specific POU transcription factor,

Brn-2, is required for establishing mammalian neural cell lineages. *Neuron* **11**, 1197–1206.

Furuta, Y., Piston, D. W., and Hogan, B. L. (1997). Bone morphogenetic proteins (BMPs) as regulators of dorsal forebrain development. *Development (Cambridge, UK)* **124**, 2203–2212.

Gaffield, W., Incardona, J. P., Kapur, R. P., and Roelink, H. (1999). A looking glass perspective: Thalidomide and cyclopamine. *Cell. Mol. Biol. (Noisy-le-grand)* **45**, 579–588.

Gaiano, N., Kohtz, J. D., Turnbull, D. H., and Fishell, G. (1999). A method for rapid gain-of-function studies in the mouse embryonic nervous system. *Nat. Neurosci.* **2**, 812–819.

Gaiano, N., Nye, J. S., and Fishell, G. (2000). Radial glia identity is promoted by notch1 signaling in the murine forebrain. *Neuron* **26**, 395–404.

Galceran, J., Fariñas, I., Depew, M. J., Clevers, H., and Grosschedl, R. (1999). Wnt3a-/—like phenotype and limb deficiency in Lef1(-/-)Tcf1 (-/-) mice. *Genes Dev.* **13**, 709–717.

Galceran, J., Miyashita-Lin, E. M., Devaney, E., Rubenstein, J. L. R., and Grosschedl, R. (2000). Hippocampus development and generation of dentate gyrus granule cells is regulated by LEF1. *Development (Cambridge, UK)* **127**, 469–482.

Garel, S., Marín, F., Mattéi, M. G., Vesque, C., Vincent, A., and Charnay, P. (1997). Family of Ebf/Olf-1-related genes potentially involved in neuronal differentiation and regional specification in the central nervous system. *Dev. Dyn.* **210**, 191–205.

Garel, S., Marín, F., Grosschedl, R., and Charnay, P. (1999). Ebf1 controls early cell differentiation in the embryonic striatum. *Development (Cambridge, UK)* **126**, 5285–5294.

Georges-Labouesse, E., Mark, M., Messaddeq, N., and Gansmüller, A. (1998). Essential role of alpha 6 integrins in cortical and retinal lamination. *Curr. Biol.* **8**, 983–986.

Ghosh, A., and Greenberg, M. E. (1995). Distinct roles for bFGF and NT-3 in the regulation of cortical neurogenesis. *Neuron* **15**, 89–103.

Gilmore, E. C., Ohshima, T., Goffinet, A. M., Kulkarni, A. B., and Herrup, K. (1998). Cyclin-dependent kinase 5-deficient mice demonstrate novel developmental arrest in cerebral cortex. *J. Neurosci.* **18**, 6370–6377.

Gitton, Y., Cohen-Tannoudji, M., and Wassef, M. (1999a). Specification of somatosensory area identity in cortical explants. *J. Neurosci.* **19**, 4889–4898.

Gitton, Y., Cohen-Tannoudji, M., and Wassef, M. (1999b). Role of thalamic axons in the expression of H-2Z1, a mouse somatosensory cortex specific marker. *Cereb. Cortex* **9**, 611–620.

Gleeson, J. G., Allen, K. M., Fox, J. W., Lamperti, E. D., Berkovic, S., Scheffer, I., Cooper, E. C., Dobyns, W. B., Minnerath, S. R., Ross, M. E., and Walsh, C. A. (1998). Doublecortin, a brain-specific gene mutated in human X-linked lissencephaly and double cortex syndrome, encodes a putative signaling protein. *Cell (Cambridge, Mass.)* **92**, 63–72.

Golden, J. A., and Chernoff, G. F. (1993). Intermittent pattern of neural tube closure in two strains of mice. *Teratology* **47**, 73–80.

Golden, J. A., Bracilovic, A., McFadden, K. A., Beesley, J. S., Rubenstein, J. L. R., and Grinspan, J. B. (1999). Ectopic bone morphogenetic proteins 5 and 4 in the chicken forebrain lead to cyclopia and holoprosencephaly. *Proc. Natl. Acad. Sci. U.S.A.* **96**, 2439–2444.

Goldman, S. A., and Luskin, M. B. (1998). Strategies utilized by migrating neurons of the postnatal vertebrate forebrain. *Trends Neurosci.* **21**, 107–114.

Gong, Q., and Shipley, M. T. (1995). Evidence that pioneer olfactory axons regulate telencephalon cell cycle kinetics to induce the formation of the olfactory bulb. *Neuron* **14**, 91–101.

Goodrich, L. V., Milenkovic, L., Higgins, K. M., and Scott, M. P. (1997). Altered neural cell fates and medulloblastoma in mouse patched mutants. *Science* **277**, 1109–1113.

Götz, M., Stoykova, A., and Gruss, P. (1998). Pax6 controls radial glia differentiation in the cerebral cortex. *Neuron* **21**, 1031–1044.

Graziadei, P. P., and Monti-Graziadei, A. G. (1992). The influence of the olfactory placode on the development of the telencephalon in *Xenopus laevis. Neuroscience* **46**, 617–629.

Grindley, J. C., Davidson, D. R., and Hill, R. E. (1995). The role of Pax-6 in eye and nasal development. *Development (Cambridge, UK)* **121**, 1433–1442.

Grindley, J. C., Hargett, L. K., Hill, R. E., Ross, A., and Hogan, B. L. (1997). Disruption of PAX6 function in mice homozygous for the Pax6Sey-1Neu mutation produces abnormalities in the early development and regionalization of the diencephalon. *Mech. Dev.* **64**, 111–126.

Gritsman, K., Zhang, J., Cheng, S., Heckscher, E., Talbot, W. S., and Schier, A. F. (1999). The EGF-CFC protein one-eyed pinhead is essential for nodal signaling. *Cell (Cambridge, Mass.)* **97**, 121–132.

Grove, E. A., and Tole, S. (1999). Patterning events and specification signals in the developing hippocampus. *Cereb. Cortex* **9**, 551–561.

Grove, E. A., Tole, S., Limon, J., Yip, L., and Ragsdale, C. W. (1998). The hem of the embryonic cerebral cortex is defined by the expression of multiple Wnt genes and is compromised in Gli3-deficient mice. *Development (Cambridge, UK)* **125**, 2315–2325.

Guillemot, F., and Joyner, A. L. (1993). Dynamic expression of the murine Achaete-Scute homolog Mash-1 in the developing nervous system. *Mech. Dev.* **42**, 171–185.

Gulisano, M., Broccoli, V., Pardini, C., and Boncinelli, E. (1996). Emx1 and Emx2 show different patterns of expression during proliferation and differentiation of the developing cerebral cortex in the mouse. *Eur. J. Neurosci.* **8**, 1037–1050.

Hagman, J., Gutch, M. J., Lin, H., and Grosschedl, R. (1995). EBF contains a novel zinc coordination motif and multiple dimerization and transcriptional activation domains. *EMBO J.* **14**, 2907–2916.

Halliday, A. L., and Cepko, C. L. (1992). Generation and migration of cells in the developing striatum. *Neuron* **9**, 15–26.

Hallonet, M., Hollemann, T., Wehr, R., Jenkins, N. A., Copeland, N. G., Pieler, T., and Gruss, P. (1998). Vax1 is a novel homeobox-containing gene expressed in the developing anterior ventral forebrain. *Development (Cambridge, UK)* **125**, 2599–2610.

Hallonet, M., Hollemann, T., Pieler, T., and Gruss, P. (1999). Vax1, a novel homeobox-containing gene, directs development of the basal forebrain and visual system. *Genes Dev.* **13**, 3106–3114.

Hardcastle, Z., and Papalopulu, N. (2000). Distinct effects of XBF-1 in regulating the cell cycle inhibitor p27(XIC1) and imparting a neural fate. *Development (Cambridge, UK)* **127**, 1303–1314.

Hatanaka, Y., Uratani, Y., Takiguchi-Hayashi, K., Omori, A., Sato, K., Miyamoto, M., and Arimatsu, Y. (1994). Intracortical regionality represented by specific transcription for a novel protein, latexin. *Eur. J. Neurosci.* **6**, 973–982.

Hatini, V., Tao, W., and Lai, E. (1994). Expression of winged helix genes, BF-1 and BF-2, define adjacent domains within the developing forebrain and retina. *J. Neurobiol.* **25**, 1293–1309.

Hatten, M. E. (1999). Expansion of CNS precursor pools: A new role for Sonic Hedgehog. *Neuron* **22**, 2–3.

Haydar, T. F., Kuan, C. Y., Flavell, R. A., and Rakic, P. (1999). The role of cell death in regulating the size and shape of the mammalian forebrain. *Cereb. Cortex* **9**, 621–626.

He, T. C., Sparks, A. B., Rago, C., Hermeking, H., Zawel, L., da Costa, L. T., Morin, P. J., Vogelstein, B., and Kinzler, K. W. (1998). Identification of c-MYC as a target of the APC pathway. *Science* **281**, 1509–1512.

He, X., Treacy, M. N., Simmons, D. M., Ingraham, H. A., Swanson, L. W., and Rosenfeld, M. G. (1989). Expression of a large family of POU-domain regulatory genes in mammalian brain development. *Nature (London)* **340**, 35–41.

Hentges, K., Thompson, K., and Peterson, A. (1999). The flat-top gene is required for the expansion and regionalization of the telencephalic primordium. *Development (Cambridge, UK)* **126**, 1601–1609.

Hevner, R. F., Shi, L., Justice, N., Hsueh, Y., Sheng, M., Smiga, S., Bulfone, A., Goffinet, A. M., Campagnoni, A. T., and Rubenstein, J. L. (2001).

Tbr1 regulates differentiation of the preplate and layer 6. *Neuron* **29**, 353–366.

Hiesberger, T., Trommsdorff, M., Howell, B. W., Goffinet, A., Mumby, M. C., Cooper, J. A., and Herz, J. (1999). Direct binding of Reelin to VLDL receptor and ApoE receptor 2 induces tyrosine phosphoryla-tion of disabled-1 and modulates tau phosphory-lation. *Neuron* **24**, 481–489.Hildebrand, J. D., and Soriano, P. (1999). Shroom, a PDZ domain-containing actin-binding protein, is required for neural tube morpho-genesis in mice. *Cell (Cambridge, Mass.)* **99**, 485–497.

Hirano, S., Yan, Q., and Suzuki, S. T. (1999). Expression of a novel proto-cadherin, OL-protocadherin, in a subset of functional systems of the de-veloping mouse brain. *J. Neurosci.* **19**, 995-1005.

Hoffarth, R. M., Johnston, J. G., Krushel, L. A., and van der Kooy, D. (1995). The mouse mutation reeler causes increased adhesion within a subpopulation of early postmitotic cortical neurons. *J. Neurosci.* **15**, 4838–4850.

Hogan, B. L., Horsburgh, G., Cohen, J., Hetherington, C. M., Fisher, G., and Lyon, M. F. (1986). Small eyes (Sey): A homozygous lethal muta-tion on chromosome 2 which affects the differentiation of both lens and nasal placodes in the mouse. *J. Embryol. Exp. Morphol.* **97**, 95–110.

Hollyday, M., McMahon, J. A., and McMahon, A. P. (1995). Wnt expres-sion patterns in chick embryo nervous system. *Mech. Dev.* **52**, 9–25.

Horton, S., Meredith, A., Richarson, J. A., and Johnson, J. E. (1999). Correct coordination of neuronal differentiation events in ventral fore-brain requires the bHLH factor MASH1. *Mol. Cell Neurosci.* **14**, 355–369.

Houart, C., Westerfield, M., and Wilson, S. W. (1998). A small population of anterior cells patterns the forebrain during zebrafish gastrulation. *Na-ture (London)* **391**, 788–792.

Howell, B. W., Hawkes, R., Soriano, P., and Cooper, J. A. (1997). Neuronal position in the developing brain is regulated by mouse disabled-1. *Na-ture (London)* **389**, 733–737.

Howell, B. W., Herrick, T. M., and Cooper, J. A. (1999). Reelin-induced tyrosine phosphorylation of disabled 1 during neuronal positioning. *Genes Dev.* **13**, 643–648.

Hsieh-Li, H. M., Witte, D. P., Szucsik, J. C., Weinstein, M., Li, H., and Potter, S. S. (1995). Gsh-2, a murine homeobox gene expressed in the developing brain. *Mech. Dev.* **50**, 177–186.

Huh, S., Hatini, V., Marcus, R. C., Li, S. C., and Lai, E. (1999). Dorsal-ventral patterning defects in the eye of BF-1-deficient mice associated with a restricted loss of shh expression. *Dev. Biol.* **211**, 53–63.

Hui, C. C., and Joyner, A. L. (1993). A mouse model of greig cephalopoly-syndactyly syndrome: The extra-toesJ mutation contains an intragenic deletion of the Gli3 gene. *Nat. Genet.* **3**, 241–246; erratum: Ibid. **19**(4), 404 (1998).

Incardona, J. P., Gaffield, W., Kapur, R. P., and Roelink, H. (1998). The teratogenic Veratrum alkaloid cyclopamine inhibits Sonic hedgehog sig-nal transduction. *Development (Cambridge, UK)* **125**, 3553–3562.

Inoue, T., Nakamura, S., and Osumi, N. (2000). Fate mapping of the mouse prosencephalic neural plate. *Dev. Biol.* **219**, 373–383.

Ivkovic, S., Polonskaia, O., Fariñas, I., and Ehrlich, M. E. (1997). Brain-derived neurotrophic factor regulates maturation of the DARPP-32 phe-notype in striatal medium spiny neurons: Studies in vivo and in vitro. *Neuroscience* **79**, 509–516.

Jacobson, C.-O. (1959). The localization of the presumptive cerebral re-gions in the neural plate of the axolotl larva. *J. Embryol. Exp. Morph.* **7**, 1–21.

Janis, L. S., Cassidy, R. M., and Kromer, L. F. (1999). Ephrin-A binding and EphA receptor expression delineate the matrix compartment of the striatum. *J. Neurosci.* **19**, 4962–4971.

Jankovski, A., and Sotelo, C. (1996). Subventricular zone-olfactory bulb migratory pathway in the adult mouse: Cellular composition and speci-ficity as determined by heterochronic and heterotopic transplantation. *J. Comp. Neurol.* **371**, 376–396.

Jensen, A. M., and Wallace, V. A. (1997). Expression of Sonic hedgehog

and its putative role as a precursor cell mitogen in the developing mouse retina. *Development (Cambridge, UK)* **124**, 363–371.

Jones, K. R., Fariñas, I., Backus, C., and Reichardt, L. F. (1994). Targeted disruption of the BDNF gene perturbs brain and sensory neuron devel-opment but not motor neuron development. *Cell (Cambridge, Mass.)* **76**, 989–999.

Josephson, R., Müller, T., Pickel, J., Okabe, S., Reynolds, K., Turner, P. A., Zimmer, A., and McKay, R. D. (1998). POU transcription factors control expression of CNS stem cell-specific genes. *Development (Cambridge, UK)* **125**, 3087–3100.

Karlstrom, R. O., Talbot, W. S., and Schier, A. F. (1999). Comparative syn-teny cloning of zebrafish you-too: Mutations in the Hedgehog target gli2 affect ventral forebrain patterning. *Genes Dev.* **13**, 388–393.

Kilpatrick, T. J., and Bartlett, P. F. (1995). Cloned multipotential precursors from the mouse cerebrum require FGF-2, whereas glial restricted pre-cursors are stimulated with either FGF-2 or EGF. *J. Neurosci.* **15**, 3653–3661.

Kim, A. S., Anderson, S. A. Rubenstein, J. L., Lowenstein, D. H., and Plea-sure, S. J. (2001). Pax-6 regulates expression of SFRP-2 and Wnt-7b in the developing CNS. *J. Neurosci.* **21**, RC132.

Kimura, S., Hara, Y., Pineau, T., Fernandez-Salguero, P., Fox, C. H., Ward, J. M., and Gonzalez, F. J. (1996). The T/ebp null mouse: Thyroid-spe-cific enhancer-binding protein is essential for the organogenesis of the thyroid, lung, ventral forebrain, and pituitary. *Genes Dev.* **10**, 60–69.

King, J. A., Marker, P. C., Seung, K. J., and Kingsley, D. M. (1994). BMP5 and the molecular, skeletal, and soft-tissue alterations in short ear mice. *Dev. Biol.* **166**, 112–122.

Kohtz, J. D., Baker, D. P., Corte, G., and Fishell, G. (1998). Regionalization within the mammalian telencephalon is mediated by changes in respon-siveness to Sonic Hedgehog. *Development (Cambridge, UK)* **125**, 5079–5089.

Korematsu, K., Goto, S., Okamura, A., and Ushio, Y. (1998). Heterogeneity of cadherin-8 expression in the neonatal rat striatum: Comparison with striatal compartments. *Exp. Neurol.* **154**, 531–536.

Krauss, S., Concordet, J. P., and Ingham, P. W. (1993). A functionally con-served homolog of the Drosophila segment polarity gene hh is expressed in tissues with polarizing activity in zebrafish embryos. *Cell (Cam-bridge, Mass.)* **75**, 1431–1444.

Krezel, W., Ghyselinck, N., Samad, T. A., Dupé, V., Kastner, P., Borrelli, E., and Chambon, P. (1998). Impaired locomotion and dopamine signaling in retinoid receptor mutant mice. *Science* **279**, 863–867.

Kuan, C. Y., Yang, D. D., Samanta Roy, D. R., Davis, R. J., Rakic, P., and Flavell, R. A. (1999). The Jnk1 and Jnk2 protein kinases are required for regional specific apoptosis during early brain development. *Neuron* **22**, 667–676.

Kuida, K., Zheng, T. S., Na, S., Kuan, C., Yang, D., Karasuyama, H., Rakic, P., and Flavell, R. A. (1996). Decreased apoptosis in the brain and pre-mature lethality in CPP32-deficient mice. *Nature (London)* **384**, 368–372.

Kuida, K., Haydar, T. F., Kuan, C. Y., Gu, Y., Taya, C., Karasuyama, H., Su, M. S., Rakic, P., and Flavell, R. A. (1998). Reduced apoptosis and cytochrome c-mediated caspase activation in mice lacking caspase 9. *Cell (Cambridge, Mass.)* **94**, 325–337.

Kwon, Y. T., and Tsai, L. H. (1998). A novel disruption of cortical devel-opment in p35(-/-) mice distinct from reeler. *J. Comp. Neurol.* **395**, 510–522.

Lavdas, A. A., Grigoriou, M., Pachnis, V., and Parnavelas, J. G. (1999). The medial ganglionic eminence gives rise to a population of early neurons in the developing cerebral cortex. *J. Neurosci.* **19**, 7881–7888.

Lee, K. J., and Jessell, T. M. (1999). The specification of dorsal cell fates in the vertebrate central nervous system. *Ann. Rev. Neurosci.* **22**, 261–294.

Lee, S. M., Danielian, P. S., Fritzsch, B., and McMahon, A. P. (1997). Evi-dence that FGF8 signalling from the midbrain-hindbrain junction regu-lates growth and polarity in the developing midbrain. *Development (Cambridge, UK)* **124**, 959–969.

Lee, S. M., Tole, S., Grove, E., and McMahon, A. P. (2000). A local Wnt-3a signal is required for development of the mammalian hippocampus. *Development (Cambridge, UK)* **127,** 457–467.

Levitt, P. (1984). A monoclonal antibody to limbic system neurons. *Science* **223,** 299–301.

Levitt, P., Barbe, M. F., and Eagleson, K. L. (1997). Patterning and specification of the cerebral cortex. *Ann. Rev. Neurosci.* **20,** 1–24.

Li, H., Zeitler, P. S., Valerius, M. T., Small, K., and Potter, S. S. (1996). Gsh-1, an orphan Hox gene, is required for normal pituitary development. *EMBO J.* **15,** 714–724.

Li, Y., Holtzman, D. M., Kromer, L. F., Kaplan, D. R., Chua-Couzens, J., Clary, D. O., Knüsel, B., and Mobley, W. C. (1995). Regulation of TrkA and ChAT expression in developing rat basal forebrain: Evidence that both exogenous and endogenous NGF regulate differentiation of cholinergic neurons. *J. Neurosci.* **15,** 2888–2905.

Lim, L., Zhou, H., and Costa, R. H. (1997). The winged helix transcription factor HFH-4 is expressed during choroid plexus epithelial development in the mouse embryo. *Proc. Natl. Acad. Sci. U.S.A.* **94,** 3094–3099.

Liu, A., Losos, K., and Joyner, A. L. (1999). FGF8 can activate Gbx2 and transform regions of the rostral mouse brain into a hindbrain fate. *Development (Cambridge, UK)* **126,** 4827–4838.

Liu, J. K., Ghattas, I., Liu, S., Chen, S., and Rubenstein, J. L. R. (1997). Dlx genes encode DNA-binding proteins that are expressed in an overlapping and sequential pattern during basal ganglia differentiation. *Dev. Dyn.* **210,** 498–512.

Liu, M., Pleasure, S. J., Collins, A. E., Noebels, J. L., Naya, F. J., Tsai, M. J., and Lowenstein, D. H. (2000). Loss of BETA2/NeuroD leads to malformation of the dentate gyrus and epilepsy. *Proc. Natl. Acad. Sci. U.S.A.* **97,** 865–870.

Lo, L. C., Johnson, J. E., Wuenschell, C. W., Saito, T., and Anderson, D. J. (1991). Mammalian achaete-scute homolog 1 is transiently expressed by spatially restricted subsets of early neuroepithelial and neural crest cells. *Genes Dev.* **5,** 1524–1537.

Lois, C., and Alvarez-Buylla, A. (1994). Long-distance neuronal migration in the adult mammalian brain. *Science* **264,** 1145–1148.

Lois, C., García-Verdugo, J. M., and Alvarez-Buylla, A. (1996). Chain migration of neuronal precursors. *Science* **271,** 978–981.

López-Mascaraque, L., García, C., Valverde, F., and de Carlos, J. A. (1998). Central olfactory structures in Pax-6 mutant mice. *Ann. N.Y. Acad. Sci.* **855,** 83–94.

LoTurco, J. J., Owens, D. F., Heath, M. J., Davis, M. B., and Kriegstein, A. R. (1995). GABA and glutamate depolarize cortical progenitor cells and inhibit DNA synthesis. *Neuron* **15,** 1287–1298.

Lumsden, A., and Krumlauf, R. (1996). Patterning the vertebrate neuraxis. *Science* **274,** 1109–1115.

Luo, G., Hofmann, C., Bronckers, A. L., Sohocki, M., Bradley, A., and Karsenty, G. (1995). BMP-7 is an inducer of nephrogenesis, and is also required for eye development and skeletal patterning. *Genes Dev.* **9,** 2808–2820.

Mackarehtschian, K., Lau, C. K., Caras, I., and McConnell, S. K. (1999). Regional differences in the developing cerebral cortex revealed by ephrin-A5 expression. *Cereb. Cortex* **9,** 601–610.

Malas, S., Nishiguchi, S., Wood, H., Postlethwaite, M., Constanti, A., and Episkopou, V. (1998). Sox1 deficient mice fail to develop an olfactory tubercle and suffer from spontaneous epileptic seizures. *Mouse Mol. Genet. Meet., Cold Spring Harbor Lab.* p. 187.

Mallamaci, A., Mercurio, S., Muzio, L., Cecchi, C., Pardini, C. L., Gruss, P., and Boncinelli, E. (2000). The lack of Emx2 causes impairment of Reelin signaling and defects of neuronal migration in the developing cerebral cortex. *J. Neurosci.* **20,** 1109–1118.

Marcus, R. C., Shimamura, K., Sretavan, D., Lai, E., Rubenstein, J. L. R., and Mason, C. A. (1999). Domains of regulatory gene expression and the developing optic chiasm: Correspondence with retinal axon paths and candidate signaling cells. *J. Comp. Neurol.* **403,** 346–358.

Marín, O., Anderson, S. A., and Rubenstein, J. L. R. (2000). Origin and molecular specification of striatal interneurons. *J. Neurosci.* **20,** 6063–6076.

Martí, E., Takada, R., Bumcrot, D. A., Sasaki, H., and McMahon, A. P. (1995b). Distribution of Sonic hedgehog peptides in the developing chick and mouse embryo. *Development (Cambridge, UK)* **121,** 2537–2547.

Martínez, S., Crossley, P. H., Cobos, I., Rubenstein, J. L., and Martin, G. R. (1999). FGF8 induces formation of an ectopic isthmic organizer and isthmocerebellar development via a repressive effect on Otx2 expression. *Development (Cambridge, UK)* **126,** 1189–1200.

Masai, I., Heisenberg, C. P., Barth, K. A., Macdonald, R., Adamek, S., and Wilson, S. W. (1997). Floating head and masterblind regulate neuronal patterning in the roof of the forebrain. *Neuron* **18,** 43–57.

Mason, I. J., Fuller-Pace, F., Smith, R., and Dickson, C. (1994). FGF-7 (keratinocyte growth factor) expression during mouse development suggests roles in myogenesis, forebrain regionalisation and epithelial-mesenchymal interactions. *Mech. Dev.* **45,** 15–30.

Mastick, G. S., Fan, C. M., Tessier-Lavigne, M., Serbedzija, G. N., McMahon, A. P., and Easter, S. S. (1996). Early deletion of neuromeres in Wnt-1-/- mutant mice: Evaluation by morphological and molecular markers. *J. Comp. Neurol.* **374,** 246–258.

Mastick, G. S., Davis, N. M., Andrew, G. L., and Easter, S. S., Jr. (1997). Pax-6 functions in boundary formation and axon guidance in the embryonic mouse forebrain. *Development (Cambridge, UK)* **124,** 1985–1997.

Masuya, H., Sagai, T., Moriwaki, K., and Shiroishi, T. (1997). Multigenic control of the localization of the zone of polarizing activity in limb morphogenesis in the mouse. *Dev. Biol.* **182,** 42–51.

McAllister, A. K., Lo, D. C., and Katz, L. C. (1995). Neurotrophins regulate dendritic growth in developing visual cortex. *Neuron* **15,** 791–803.

McAllister, A. K., Katz, L. C., and Lo, D. C. (1996). Neurotrophin regulation of cortical dendritic growth requires activity. *Neuron* **17,** 1057–1064.

McAllister, A. K., Katz, L. C., and Lo, D. C. (1999). Neurotrophins and synaptic plasticity. *Ann. Rev. Neurosci.* **22,** 295–318.

McConnell, S. K. (1989). The determination of neuronal fate in the cerebral cortex. *Trends Neurosci.* **12,** 342–349.

McConnell, S. K. (1991). The generation of neuronal diversity in the central nervous system. *Ann. Rev. Neurosci.* **14,** 269–300.

McConnell, S. K., and Kaznowski, C. E. (1991). Cell cycle dependence of laminar determination in developing neocortex. *Science* **254,** 282–285.

McConnell, S. K., Ghosh, A., and Shatz, C. J. (1989). Subplate neurons pioneer the first axon pathway from the cerebral cortex. *Science* **245,** 978–982.

McConnell, S. K., Ghosh, A., and Shatz, C. J. (1994). Subplate pioneers and the formation of descending connections from cerebral cortex. *J. Neurosci.* **14,** 1892–1907.

Mercader, N., Leonardo, E., Azpiazu, N., Serrano, A., Morata, G., Martínez, C., and Torres, M. (1999). Conserved regulation of proximodistal limb axis development by Meis1/Hth. *Nature (London)* **402,** 425–429.

Michaud, J. L., Rosenquist, T., May, N. R., and Fan, C. M. (1998). Development of neuroendocrine lineages requires the bHLH-PAS transcription factor SIM1. *Genes Dev.* **12,** 3264–3275.

Miyashita-Lin, E. M., Hevner, R., Wassarman, K. M., Martínez, S., and Rubenstein, J. L. R. (1999). Early neocortical regionalization in the absence of thalamic innervation. *Science* **285,** 906–909.

Miyata, T., Maeda, T., and Lee, J. E. (1999). NeuroD is required for differentiation of the granule cells in the cerebellum and hippocampus. *Genes Dev.* **13,** 1647–1652.

Molnár, Z., and Blakemore, C. (1995). How do thalamic axons find their way to the cortex? *Trends Neurosci.* **18,** 389–397.

Muenke, M., and Beachy, P. A. (2000). Genetics of ventral forebrain development and holoprosencephaly. *Curr. Opin. Genet. Dev.* **10,** 262–269.

Nakagawa, Y., Johnson, J. E., and O'Leary, D. D. (1999). Graded and areal expression patterns of regulatory genes and cadherins in embryonic neocortex independent of thalamocortical input. *J. Neurosci.* **19,** 10877–10885.

Nakai, S., Kawano, H., Yudate, T., Nishi, M., Kuno, J., Nagata, A., Jishage, K., Hamada, H., Fujii, H., Kawamura, K. *et al.* (1995). The POU domain transcription factor Brn-2 is required for the determination of specific neuronal lineages in the hypothalamus of the mouse. *Genes Dev.* **9,** 3109–3121.

Niehrs, C. (1999). Head in the WNT: The molecular nature of Spemann's head organizer. *Trends Genet.* **15,** 314–319.

Nothias, F., Fishell, G., and Ruiz i Altaba, A. (1998). Cooperation of intrinsic and extrinsic signals in the elaboration of regional identity in the posterior cerebral cortex. *Curr. Biol.* **8,** 459–462.

Ogawa, M., Miyata, T., Nakajima, K., Yagyu, K., Seike, M., Ikenaka, K., Yamamoto, H., and Mikoshiba, K. (1995). The reeler gene-associated antigen on Cajal-Retzius neurons is a crucial molecule for laminar organization of cortical neurons. *Neuron* **14,** 899–912.

Ohshima, T., Ward, J. M., Huh, C. G., Longenecker, G., Veeranna, Pant, H. C., Brady, R. O., Martin, L. J., and Kulkarni, A. B. (1996). Targeted disruption of the cyclin-dependent kinase 5 gene results in abnormal corticogenesis, neuronal pathology and perinatal death. *Proc. Natl. Acad. Sci. U.S.A.* **93,** 11173–11178.

O'Leary, D. D. (1989). Do cortical areas emerge from a protocortex? *Trends Neurosci.* **12,** 400–406.

Oliver, G., Mailhos, A., Wehr, R., Copeland, N. G., Jenkins, N. A., and Gruss, P. (1995). Six3, a murine homolog of the sine oculis gene, demarcates the most anterior border of the developing neural plate and is expressed during eye development. *Development (Cambridge, UK)* **121,** 4045–4055.

Olsson, M., Campbell, K., Wictorin, K., and Björklund, A. (1995). Projection neurons in fetal striatal transplants are predominantly derived from the lateral ganglionic eminence. *Neuroscience* **69,** 1169–1182.

Olsson, M., Björklund, A., and Campbell, K. (1998). Early specification of striatal projection neurons and interneuronal subtypes in the lateral and medial ganglionic eminence. *Neuroscience* **84,** 867–876.

Pai, C. Y., Kuo, T. S., Jaw, T. J., Kurant, E., Chen, C. T., Bessarab, D. A., Salzberg, A., and Sun, Y. H. (1998). The Homothorax homeoprotein activates the nuclear localization of another homeoprotein, extradenticle, and suppresses eye development in Drosophila. *Genes Dev.* **12,** 435–446.

Park, H. L., Bai, C., Platt, K. A., Matise, M. P., Beeghly, A., Hui, C., Nakashima, M., and Joyner, A. L. (2000). Mouse gli1 mutants are viable but have defects in SHH signaling in combination with a gli2 mutation. *Development (Cambridge, UK)* **127,** 1593–1605.

Paysan, J., and Fritschy, J. M. (1998). GABAA-receptor subtypes in developing brain. Actors or spectators? *Perspect. Dev. Neurobiol.* **5,** 179–192.

Pellegrini, M., Mansouri, A., Simeone, A., Boncinelli, E., and Gruss, P. (1996). Dentate gyrus formation requires Emx2. *Development (Cambridge, UK)* **122,** 3893–3898.

Pera, E. M., and Kessel, M. (1997). Patterning of the chick forebrain anlage by the prechordal plate. *Development (Cambridge, UK)* **124,** 4153–4162.

Pesce, M., Canipari, R., Ferri, G. L., Siracusa, G., and De Felici, M. (1996). Pituitary adenylate cyclase-activating polypeptide (PACAP) stimulates adenylate cyclase and promotes proliferation of mouse primordial germ cells. *Development (Cambridge, UK)* **122,** 215–221.

Pleasure, S. J., Anderson, S., Hevner, R., Bagri, A., Marín, O., Lowenstein, D. H., and Rubenstein, J. L. R. (2000). Cell migration from the ganglionic eminences is required for the development of hippocampal GABAergic interneurons. *Neuron* **28,** 727–740.

Porter, F. D., Drago, J., Xu, Y., Cheema, S. S., Wassif, C., Huang, S. P., Lee, E., Grinberg, A., Massalas, J. S., Bodine, D., Alt, F., and Westphal, H.

(1997). Lhx2, a LIM homeobox gene, is required for eye, forebrain, and definitive erythrocyte development. *Development (Cambridge, UK)* **124,** 2935–2944.

Porteus, M. H., Bulfone, A., Liu, J. K., Puelles, L., Lo, L. C., and Rubenstein, J. L. R. (1994). DLX-2, MASH-1, and MAP-2 expression and bromodeoxyuridine incorporation define molecularly distinct cell populations in the embryonic mouse forebrain. *J. Neurosci.* **14,** 6370–6383.

Price, M., Lazzaro, D., Pohl, T., Mattei, M. G., Rüther, U., Olivo, J. C., Duboule, D., and Di Lauro, R. (1992). Regional expression of the homeobox gene Nkx-2.2 in the developing mammalian forebrain. *Neuron* **8,** 241–255.

Puelles, L. (1995). A segmental morphological paradigm for understanding vertebrate forebrains. *Brain Behav. Evol.* **46,** 319–337.

Puelles, L., and Rubenstein, J. L. R. (1993). Expression patterns of homeobox and other putative regulatory genes in the embryonic mouse forebrain suggest a neuromeric organization. *Trends Neurosci.* **16,** 472–479.

Puelles, L., and Rubenstein, J. L. R. (2000). Forebrain. *In* "Encyclopedia of the Human Brain" (V. A. Ramachandran, Ed.). Academic Press, San Diego (in press).

Puelles, L., Kuwana, E., Puelles, E., and Rubenstein, J. L. R. (1999). Comparison of the mammalian and avian telencephalon from the perspective of gene expression data. *Eur. J. Morphol.* **37,** 139–150.

Puelles, L., Kuwana, E., Puelles, E., Bulfone, A., Shimamura, K., Keleher, J., Smiga, S., and Rubenstein, J. L. R. (2000). Pallial and subpallial derivatives in the embryonic chick and mouse telencephalon, traced by the expression of the genes Dlx-2, Emx-1, Nkx-2.1, Pax-6 and Tbr-1. *J. Comp. Neurol.* **424,** 409–438.

Qiu, M., Bulfone, A., Martínez, S., Meneses, J. J., Shimamura, K., Pedersen, R. A., and Rubenstein, J. L. R. (1995). Null mutation of Dlx-2 results in abnormal morphogenesis of proximal first and second branchial arch derivatives and abnormal differentiation in the forebrain. *Genes Dev.* **9,** 2523–2538.

Qiu, M., Anderson, S., Chen, S., Meneses, J. J., Hevner, R., Kuwana, E., Pedersen, R. A., and Rubenstein, J. L. R. (1996). Mutation of the Emx-1 homeobox gene disrupts the corpus callosum. *Dev. Biol.* **178,** 174–178.

Qiu, M., Shimamura, K., Sussel, L., Chen, S., and Rubenstein, J. L. R. (1998). Control of anteroposterior and dorsoventral domains of Nkx-6.1 gene expression relative to other Nkx genes during vertebrate CNS development. *Mech. Dev.* **72,** 77–88.

Rakic, P. (1971). Guidance of neurons migrating to the fetal monkey neocortex. *Brain Res.* **33,** 471–476.

Rakic, P. (1972). Mode of cell migration to the superficial layers of fetal monkey neocortex. *J. Comp. Neurol.* **145,** 61–83.

Rakic, P. (1974). Neurons in rhesus monkey visual cortex: Systematic relation between time of origin and eventual disposition. *Science* **183,** 425–427.

Rakic, P. (1988). Specification of cerebral cortical areas. *Science* **241,** 170–176.

Rakic, P., and Caviness, V. S., Jr. (1995). Cortical development: View from neurological mutants two decades later. *Neuron* **14,** 1101–1104.

Rakic, P., Stensas, L. J., Sayre, E., and Sidman, R. L. (1974). Computeraided three-dimensional reconstruction and quantitative analysis of cells from serial electron microscopic montages of foetal monkey brain. *Nature (London)* **250,** 31–34.

Rebagliati, M. R., Toyama, R., Haffter, P., and Dawid, I. B. (1998). cyclops encodes a nodal-related factor involved in midline signaling. *Proc. Natl. Acad. Sci. U.S.A.* **95,** 9932–9937.

Redmond, L., Oh, S. R., Hicks, C., Weinmaster, G., and Ghosh, A. (2000). Nuclear Notch1 signaling and the regulation of dendritic development. *Nat. Neurosci.* **3,** 30–40.

Reiner, O., Carrozzo, R., Shen, Y., Wehnert, M., Faustinella, F., Dobyns, W. B., Caskey, C. T., and Ledbetter, D. H. (1993). Isolation of a Miller-

Dieker lissencephaly gene containing G protein beta-subunit-like repeats. *Nature (London)* **364,** 717–721.

Reynolds, B. A., Tetzlaff, W., and Weiss, S. (1992). A multipotent EGF-responsive striatal embryonic progenitor cell produces neurons and astrocytes. *J. Neurosci.* **12,** 4565–4574.

Rice, D. S., and Curran, T. (1999). Mutant mice with scrambled brains: Understanding the signaling pathways that control cell positioning in the CNS. *Genes Dev.* **13,** 2758–2773.

Rieckhof, G. E., Casares, F., Ryoo, H. D., Abu-Shaar, M., and Mann, R. S. (1997). Nuclear translocation of extradenticle requires homothorax, which encodes an extradenticle-related homeodomain protein. *Cell (Cambridge, Mass.)* **91,** 171–183.

Ring, M., Dobi, A., Georgopoulos, K., and Agoston, D. (1999). The role of the limphoid transcription factor ikaros in the developmentalspecification of enkephalinergic neurons in the striatum. *Soc. Neurosci. Abstr.* **25,** 252.

Rinkwitz-Brandt, S., Justus, M., Oldenettel, I., Arnold, H. H., and Bober, E. (1995). Distinct temporal expression of mouse Nkx-5.1 and Nkx-5.2 homeobox genes during brain and ear development. *Mech. Dev.* **52,** 371–381.

Rio, C., Rieff, H. I., Qi, P., Khurana, T. S., and Corfas, G. (1997). Neuregulin and erbB receptors play a critical role in neuronal migration. *Neuron* **19,** 39–50; erratum: Ibid., p. 1349.

Roelink, H., Augsburger, A., Heemskerk, J., Korzh, V., Norlin, S., Ruiz i Altaba, A., Tanabe, Y., Placzek, M., Edlund, T., Jessell, T. M., *et al.* (1994). Floor plate and motor neuron induction by vhh-1, a vertebrate homolog of hedgehog expressed by the notochord. *Cell (Cambridge, Mass.)* **76,** 761–775.

Rowitch, D. H., St.-Jacques, B., Lee, S. M., Flax, J. D., Snyder, E. Y., and McMahon, A. P. (1999). Sonic hedgehog regulates proliferation and inhibits differentiation of CNS precursor cells. *J. Neurosci.* **19,** 8954–8965.

Rubenstein, J. L. R., and Beachy, P. A. (1998). Patterning of the embryonic forebrain. *Curr. Opin. Neurobiol.* **8,** 18–26.

Rubenstein, J. L. R, Martínez, S., Shimamura, K., and Puelles, L. (1994). The embryonic vertebrate forebrain: The prosomeric model. *Science* **266,** 578–580.

Rubenstein, J. L. R., Shimamura, K., Martínez, S., and Puelles, L. (1998). Regionalization of the prosencephalic neural plate. *Ann. Rev. Neurosci.* **21,** 445–477.

Rubenstein, J. L. R, Anderson, S., Shi, L., Miyashita-Lin, E., Bulfone, A., and Hevner, R. (1999). Genetic control of cortical regionalization and connectivity. *Cereb. Cortex* **9,** 524–532.

Ruiz i Altaba, A. (1993). Induction and axial patterning of the neural plate: Planar and vertical signals. *J. Neurobiol.* **24,** 1276–1304.

Ruiz i Altaba, A. (1998). Combinatorial Gli gene function in floor plate and neuronal inductions by Sonic hedgehog. *Development (Cambridge, UK)* **125,** 2203–2212.

Sabapathy, K., Jochum, W., Hochedlinger, K., Chang, L., Karin, M., and Wagner, E. F. (1999). Defective neural tube morphogenesis and altered apoptosis in the absence of both JNK1 and JNK2. *Mech. Dev.* **89,** 115–124.

Sadikot, A. F., Burhan, A. M., Bélanger, M. C., and Sasseville, R. (1998). NMDA receptor antagonists influence early development of GABAergic interneurons in the mammalian striatum. *Brain Res. Dev. Brain Res.* **105,** 35–42.

Samad, T. A., Krezel, W., Chambon, P., and Borrelli, E. (1997). Regulation of dopaminergic pathways by retinoids: Activation of the D2 receptor promoter by members of the retinoic acid receptor-retinoid X receptor family. *Proc. Natl. Acad. Sci. U.S.A.* **94,** 14349–14354.

Sampath, K., Rubinstein, A. L., Cheng, A. M., Liang, J. O., Fekany, K., Solnica-Krezel, L., Korzh, V., Halpern, M. E., and Wright, C. V. (1998). Induction of the zebrafish ventral brain and floorplate requires cyclops/nodal signalling. *Nature (London)* **395,** 185–189.

Sander M., Paydar, s., Ericson, J., Briscoe, J., Berber, E., German, M., Jessell, T. M., and Rubenstein, J. L. R. (2000). Ventral neural patterning by Nkx homeobox genes: Nkx6.1 controls somatic motor neuron and ventral interneuron fates. *Genes Dev.* **14,** 2134–2139.

Sapir, T., Elbaum, M., and Reiner, O. (1997). Reduction of microtubule catastrophe events by LIS1, platelet-activating factor acetylhydrolase subunit. *EMBO J.* **16,** 6977–6984.

Schmahl, W., Knoedlseder, M., Favor, J., and Davidson, D. (1993). Defects of neuronal migration and the pathogenesis of cortical malformations are associated with Small eye (Sey) in the mouse, a point mutation at the Pax-6-locus. *Acta Neuropathol.* **86,** 126–135.

Schonemann, M. D., Ryan, A. K., McEvilly, R. J., O'Connell, S. M., Arias, C. A., Kalla, K. A., Li, P., Sawchenko, P. E., and Rosenfeld, M. G. (1995). Development and survival of the endocrine hypothalamus and posterior pituitary gland requires the neuronal POU domain factor Brn-2. *Genes Dev.* **9,** 3122–3135.

Schorle, H., Meier, P., Buchert, M., Jaenisch, R., and Mitchell, P. J. (1996). Transcription factor AP-2 essential for cranial closure and craniofacial development. *Nature (London)* **381,** 235–238.

Schwab, M. H., Bartholomae, A., Heimrich, B., Feldmeyer, D., Druffel-Augustin, S., Goebbels, S., Naya, F. J., Zhao, S., Frotscher, M., Tsai, M.-J., and Nave, K.-A. (2000). Neuronal basic helix-loop-helix proteins (HEX and BETA2/Neuro D) regulate terminal granule cell differentiation in the hippocampus. *J. Neurosci.* **20,** 3714–3724.

Shanmugalingam, S., Houart, C., Picker, A., Reifers, F., Macdonald, R., Barth, A., Griffin, K., Brand, M., and Wilson, S. W. (2000). Ace/Fgf8 is required for forebrain commisure formation and patterning of the telencephalon. *Development (Cambridge, UK)* **127,** 2549–2561.

Shawlot, W., Wakamiya, M., Kwan, K. M., Kania, A., Jessell, T. M., and Behringer, R. R. (1999). Lim1 is required in both primitive streak-derived tissues and visceral endoderm for head formation in the mouse. *Development (Cambridge, UK)* **126,** 4925–4932.

Sheldon, M., Rice, D. S., D'Arcangelo, G., Yoneshima, H., Nakajima, K., Mikoshiba, K., Howell, B. W., Cooper, J. A., Goldowitz, D., and Curran, T. (1997). Scrambler and yotari disrupt the disabled gene and produce a reeler-like phenotype in mice. *Nature (London)* **389,** 730–733.

Sheng, H. Z., Moriyama, K., Yamashita, T., Li, H., Potter, S. S., Mahon, K. A., and Westphal, H. (1997). Multistep control of pituitary organogenesis. *Science* **278,** 1809–1812.

Sheppard, A. M., and Pearlman, A. L. (1997). Abnormal reorganization of preplate neurons and their associated extracellular matrix: An early manifestation of altered neocortical development in the reeler mutant mouse. *J. Comp. Neurol.* **378,** 173–179.

Sheth, A. N., and Bhide, P. G. (1997). Concurrent cellular output from two proliferative populations in the early embryonic mouse corpus striatum. *J. Comp. Neurol.* **383,** 220–230.

Shimamura, K., and Rubenstein, J. L. R. (1997). Inductive interactions direct early regionalization of the mouse forebrain. *Development (Cambridge, UK)* **124,** 2709–2718.

Shimamura, K., Hartigan, D. J., Martínez, S., Puelles, L., and Rubenstein, J. L. R. (1995). Longitudinal organization of the anterior neural plate and neural tube. *Development (Cambridge, UK)* **121,** 3923–3933.

Shimamura, K., Martinez, S., Puelles, L., and Robenstein, J. L. R. (1997). Patterns of gene expression in the neural plate and neural tube subdivide the embryonic forebrain into transverse and longitudinal domains. *Dev. Neurosci.* **19,** 88–96.

Shimazaki, T., Arsenijevic, Y., Ryan, A. K., Rosenfeld, M. G., and Weiss, S. (1999). A role for the POU-III transcription factor Brn-4 in the regulation of striatal neuron precursor differentiation. *EMBO J.* **18,** 444–456.

Shinya, M., Furutani-Seiki, M., Kuroiwa, A., and Takeda, H. (1999). Mosaic analysis with oep mutant reveals a repressive interaction between floor-plate and non-floor-plate mutant cells in the zebrafish neural tube. *Dev. Growth Differ.* **41,** 135–142.

Shtutman, M., Zhurinsky, J., Simcha, I., Albanese, C., D'Amico, M., Pestell, R., and Ben-Ze'ev, A. (1999). The cyclin D1 gene is a target of the beta-catenin/LEF-1 pathway. *Proc. Natl. Acad. Sci. U.S.A.* **96,** 5522–5527.

Simeone, A., Acampora, D., Gulisano, M., Stornaiuolo, A., and Boncinelli, E. (1992). Nested expression domains of four homeobox genes in developing rostral brain. *Nature (London)* **358,** 687–690.

Simeone, A., Acampora, D., Mallamaci, A., Stornaiuolo, A., D'Apice, M. R., Nigro, V., and Boncinelli, E. (1993). A vertebrate gene related to orthodenticle contains a homeodomain of the bicoid class and demarcates anterior neuroectoderm in the gastrulating mouse embryo. *Embo J.* **12,** 2735–2747.

Smart, I. H. (1976). A pilot study of cell production by the ganglionic eminences of the developing mouse brain. *J. Anat.* **121,** 71–84.

Smith-Fernández, A., Pieau, C., Repérant, J., Boncinelli, E., and Wassef, M. (1998). Expression of the Emx-1 and Dlx-1 homeobox genes define three molecularly distinct domains in the telencephalon of mouse, chick, turtle and frog embryos: Implications for the evolution of telencephalic subdivisions in amniotes. *Development (Cambridge, UK)* **125,** 2099–2111.

Sockanathan, S., and Jessell, T. M. (1998). Motor neuron-derived retinoid signaling specifies the subtype identity of spinal motor neurons. *Cell (Cambridge, Mass.)* **94,** 503–514.

Solloway, M. J., and Robertson, E. J. (1999). Early embryonic lethality in Bmp5 − p7 double mutant mice suggests functional redundancy within the 60A subgroup. *Development (Cambridge, UK)* **126,** 1753–1768.

Solloway, M. J., Dudley, A. T., Bikoff, E. K., Lyons, K. M., Hogan, B. L., and Robertson, E. J. (1998). Mice lacking Bmp6 function. *Dev. Genet.* **22,** 321–339.

Soria, J. M., and Fairén, A. (2000). Cellular mosaics in the rat marginal zone define an early neocortical territorialization. *Cereb. Cortex* **10,** 400–412.

Steininger, T. L., Wainer, B. H., Klein, R., Barbacid, M., and Palfrey, H. C. (1993). High-affinity nerve growth factor receptor (Trk) immunoreactivity is localized in cholinergic neurons of the basal forebrain and striatum in the adult rat brain. *Brain Res.* **612,** 330–335.

Stout, R. P., and Graziadei, P. P. (1980). Influence of the olfactory placode on the development of the brain in *Xenopus laevis* (Daudin). I. Axonal growth and connections of the transplanted olfactory placode. *Neuroscience* **5,** 2175–2186.

Stoykova, A., Fritsch, R., Walther, C., and Gruss, P. (1996). Forebrain patterning defects in Small eye mutant mice. *Development (Cambridge, UK)* **122,** 3453–3465.

Stoykova, A., Götz, M., Gruss, P., and Price, J. (1997). Pax6-dependent regulation of adhesive patterning, R-cadherin expression and boundary formation in developing forebrain. *Development (Cambridge, UK)* **124,** 3765–3777.

Stoykova, A., Treichel, D., Hallonet, M., and Gruss, P. (2000). Pax6 modulates the dorsoventral patterning of the mammalian telencephalon. *J. Neurosci.* **20,** 8042–8050.

Stühmer, T., Puelles, L., Ekker, M., and Rubenstein, J. L. R. (2001). Expression from a Dlx gene enhancer marks adult mouse cortical GABAergic neurons. *Cereb. Cortex* (in press).

Suda, Y., Matsuo, I., and Aizawa, S. (1997). Cooperation between Otx1 and Otx2 genes in developmental patterning of rostral brain. *Mech. Dev.* **69,** 125–141.

Sussel, L., Marin, O., Kimura, S., and Rubenstein, J. L. R. (1999). Loss of Nkx2.1 homeobox gene function results in a ventral to dorsal molecular respecification within the basal telencephalon: Evidence for a transformation of the pallidum into the striatum. *Development (Cambridge, UK)* **126,** 3359–3370.

Sweet, H. O., Bronson, R. T., Johnson, K. R., Cook, S. A., and Davisson, M. T. (1996). Scrambler, a new neurological mutation of the mouse with abnormalities of neuronal migration. *Mamm Genome* **7,** 798–802.

Szucsik, J. C., Witte, D. P., Li, H., Pixley, S. K., Small, K. M., and Potter, S. S. (1997). Altered forebrain and hindbrain development in mice mutant for the Gsh-2 homeobox gene. *Dev. Biol.* **191,** 230–242.

Takahashi, T., Nowakowski, R. S., and Caviness, V. S., Jr. (1995a). The cell cycle of the pseudostratified ventricular epithelium of the embryonic murine cerebral wall. *J. Neurosci.* **15,** 6046–6057.

Takahashi, T., Nowakowski, R. S., and Caviness, V. S., Jr. (1995b). Early ontogeny of the secondary proliferative population of the embryonic murine cerebral wall. *J. Neurosci.* **15,** 6058–6068.

Takahashi, T., Goto, T., Miyama, S., Nowakowski, R. S., and Caviness, V. S., Jr. (1999). Sequence of neuron origin and neocortical laminar fate: Relation to cell cycle of origin in the developing murine cerebral wall. *J. Neurosci.* **19,** 10357–10371.

Tamamaki, N., Fujimori, K. E., and Takauji, R. (1997). Origin and route of tangentially migrating neurons in the developing neocortical intermediate zone. *J. Neurosci.* **17,** 8313–8323.

Tanabe, Y., and Jessell, T. M. (1996). Diversity and pattern in the developing spinal cord. *Science* **274,** 1115–1123.

Tao, W., and Lai, E. (1992). Telencephalon-restricted expression of BF-1, a new member of the HNF-3/fork head gene family, in the developing rat brain. *Neuron* **8,** 957–966.

Tetsu, O., and McCormick, F. (1999). Beta-catenin regulates expression of cyclin D1 in colon carcinoma cells. *Nature (London)* **398,** 422–426.

Theil, T., Alvarez-Bolado, G., Walter, A., and Rüther, U. (1999). Gli3 is required for Emx gene expression during dorsal telencephalon development. *Development (Cambridge, UK)* **126,** 3561–3571.

Thisse, B., Wright, C. V., and Thisse, C. (2000). Activin- and Nodal-related factors control antero-posterior patterning of the zebrafish embryo. *Nature (London)* **403,** 425–428.

Tole, S., Ragsdale, C. W., and Grove, E. A. (2000). Dorsoventral patterning of the telencephalon is disrupted in the mouse mutant extra-toes(J). *Dev. Biol.* **217,** 254–265.

Toresson, H., Mata de Urquiza, A., Fagerström, C., Perlmann, T., and Campbell, K. (1999). Retinoids are produced by glia in the lateral ganglionic eminence and regulate striatal neuron differentiation. *Development (Cambridge, UK)* **126,** 1317–1326.

Toresson, H., Potter, S. S., and Campbell, K. (2000). Genetic control of dorsal-ventral identity in the telencephalon: Opposing roles for Pax6 and Gsh2. *Development* **127,** 4361 d4371.

Toresson, H., Parmar, M., and Campbell, K. (2000). Expression of Meis and Pbx genes and their protein products in the developing telencephalon: Implications for regional differentiation. *Mech. Dev.* **94,** 183–187.

Torii, M., Matsuzaki, F., Osumi, N., Kaibuchi, K., Nakamura, S., Casarosa, S., Guillemot, F., and Nakafuku, M. (1999). Transcription factors Mash-1 and Prox-1 delineate early steps in differentiation of neural stem cells in the developing central nervous system. *Development (Cambridge, UK)* **126,** 443–456.

Treier, M., and Rosenfeld, M. G. (1996). The hypothalamic-pituitary axis: Co-development of two organs. *Curr. Opin. Cell Biol.* **8,** 833–843.

Trommsdorff, M., Borg, J. P., Margolis, B., and Herz, J. (1998). Interaction of cytosolic adaptor proteins with neuronal apolipoprotein E receptors and the amyloid precursor protein. *J. Biol. Chem.* **273,** 33556–33560.

Trommsdorff, M., Gotthardt, M., Hiesberger, T., Shelton, J., Stockinger, W., Nimpf, J., Hammer, R. E., Richardson, J. A., and Herz, J. (1999). Reeler/Disabled-like disruption of neuronal migration in knockout mice lacking the VLDL receptor and ApoE receptor 2. *Cell (Cambridge, Mass.)* **97,** 689–701.

Trumpp, A., Depew, M. J., Rubenstein, J. L. R., Bishop, J. M., and Martin, G. R. (1999). Cre-mediated gene inactivation demonstrates that FGF8 is required for cell survival and patterning of the first branchial arch. *Genes Dev.* **13,** 3136–3148.

Valerius, M. T., Li, H., Stock, J. L., Weinstein, M., Kaur, S., Singh, G., and Potter, S. S. (1995). Gsh-1: A novel murine homeobox gene expressed in the central nervous system. *Dev. Dyn.* **203,** 337–351.

Valverde, F., García, C., López-Mascaraque, L., and De Carlos, J. A. (2000). Development of the mammillothalamic tract in normal and Pax-6 mutant mice. *J. Comp. Neurol.* **419**, 485–504.

Vazquez, M. E., and Ebendal, T. (1991). Messenger RNAs for trk and the low-affinity NGF receptor in rat basal forebrain. *NeuroReport* **2**, 593–596.

Waid, D. K., and McLoon, S. C. (1998). Ganglion cells influence the fate of dividing retinal cells in culture. *Development (Cambridge, UK)* **125**, 1059–1066.

Wallace, V. A. (1999). Purkinje-cell-derived Sonic hedgehog regulates granule neuron precursor cell proliferation in the developing mouse cerebellum. *Curr. Biol.* **9**, 445–448.

Walsh, C. A. (1999). Genetic malformations of the human cerebral cortex. *Neuron* **23**, 19–29.

Ware, M. L., Tavazoie, S. F., Reid, C. B., and Walsh, C. A. (1999). Coexistence of widespread clones and large radial clones in early embryonic ferret cortex. *Cereb. Cortex* **9**, 636–645.

Warren, N., and Price, D. J. (1997). Roles of Pax-6 in murine diencephalic development. *Development (Cambridge, UK)* **124**, 1573–1582.

Warren, N., Caric, D., Pratt, T., Clausen, J. A., Asavaritikrai, P., Mason, J. O., Hill, R. E., and Price, D. J. (1999). The transcription factor, Pax6, is required for cell proliferation and differentiation in the developing cerebral cortex. *Cereb. Cortex* **9**, 627–635.

Wassarman, K. M., Lewandoski, M., Campbell, K., Joyner, A. L., Rubenstein, J. L. R., Martínez, S., and Martin, G. R. (1997). Specification of the anterior hindbrain and establishment of a normal mid/hindbrain organizer is dependent on Gbx2 gene function. *Development (Cambridge, UK)* **124**, 2923–2934.

Watkins-Chow, D. E., and Camper, S. A. (1998). How many homeobox genes does it take to make a pituitary gland? *Trends Genet.* **14**, 284–290.

Wechsler-Reya, R. J., and Scott, M. P. (1999). Control of neuronal precursor proliferation in the cerebellum by Sonic Hedgehog. *Neuron* **22**, 103–114.

Weimann, J. M., Zhang, Y. A., Levin, M. E., Devine, W. P., Brûlet, P., and McConnell, S. K. (1999). Cortical neurons require Otx1 for the refinement of exuberant axonal projections to subcortical targets. *Neuron* **24**, 819–831.

Whitesides, J., Hall, M., Anchan, R., and LaMantia, A. S. (1998). Retinoid signaling distinguishes a subpopulation of olfactory receptor neurons in the developing and adult mouse. *J. Comp. Neurol.* **394**, 445–461.

Wichterle, H., Garcia-Verdugo, J. M., Herrera, D. G., and Alvarez-Buylla, A. (1999). Young neurons from medial ganglionic eminence disperse in adult and embryonic brain. *Nat. Neurosci.* **2**, 461–466.

Wilson, S. W., and Rubenstein, J. L. R. (2000). Induction and dorsoventral patterning of the telencephalon. *Neuron* **28**, 641–651.

Winnier, G., Blessing, M., Labosky, P. A., and Hogan, B. L. (1995). Bone morphogenetic protein-4 is required for mesoderm formation and patterning in the mouse. *Genes Dev.* **9**, 2105–2116.

Wise, S. P., and Jones, E. G. (1978). Developmental studies of thalamocortical and commissural connections in the rat somatic sensory cortex. *J. Comp. Neurol.* **178**, 187–208.

Wodarz, A., and Nusse, R. (1998). Mechanisms of Wnt signaling in development. *Annu. Rev. Cell Dev. Biol.* **14**, 59–88.

Wu, W., Wong, K., Chen, J., Jiang, Z., Dupuis, S., Wu, J. Y., and Rao, Y. (1999). Directional guidance of neuronal migration in the olfactory system by the protein Slit. *Nature (London)* **400**, 331–336.

Xuan, S., Baptista, C. A., Balas, G., Tao, W., Soares, V. C., and Lai, E. (1995). Winged helix transcription factor BF-1 is essential for the development of the cerebral hemispheres. *Neuron* **14**, 1141–1152.

Yacoubian, T. A., and Lo, D. C. (2000). Truncated and full-length TrkB receptors regulate distinct modes of dendritic growth. *Nat. Neurosci.* **3**, 342–349.

Ye, W., Shimamura, K., Rubenstein, J. L. R., Hynes, M. A., and Rosenthal, A. (1998). FGF and Shh signals control dopaminergic and serotonergic cell fate in the anterior neural plate. *Cell (Cambridge, Mass.)* **93**, 755–766.

Yoneshima, H., Nagata, E., Matsumoto, M., Yamada, M., Nakajima, K., Miyata, T., Ogawa, M., and Mikoshiba, K. (1997). A novel neurological mutant mouse, yotari, which exhibits reeler-like phenotype but expresses CR-50 antigen/reelin. *Neurosci. Res.* **29**, 217–223.

Yoshida, M., Suda, Y., Matsuo, I., Miyamoto, N., Takeda, N., Kuratani, S., and Aizawa, S. (1997). Emx1 and Emx2 functions in development of dorsal telencephalon. *Development (Cambridge, UK)* **124**, 101–111.

Yue, Y., Widmer, D. A., Halladay, A. K., Cerretti, D. P., Wagner, G. C., Dreyer, J. L., and Zhou, R. (1999). Specification of distinct dopaminergic neural pathways: Roles of the Eph family receptor EphB1 and ligand ephrin-B2. *J. Neurosci.* **19**, 2090–2101.

Yun, K., Potter, S., and Rubenstein, J. L. (2001). Gsh2 and Pax6 play complementary roles in dorsoventral patterning of the mammalian telencephalon. *Development* **128**, 193–205.

Zerucha, T., Stühmer, T., Hatch, G., Park, B. K., Long, Q., Yu, G., Gambarotta, A., Schultz, J. R., Rubenstein, J. L. R., and Ekker, M. (2000). A highly conserved enhancer in the Dlx5/Dlx6 intergenic region is the site of cross-regulatory interactions between Dlx genes in the embryonic forebrain. *J. Neurosci.* **20**, 709–721.

Zhang, H., and Bradley, A. (1996). Mice deficient for BMP2 are nonviable and have defects in amnion/chorion and cardiac development. *Development (Cambridge, UK)* **122**, 2977–2986.

Zhang, J., Hagopian-Donaldson, S., Serbedzija, G., Elsemore, J., Plehn-Dujowich, D., McMahon, A. P., Flavell, R. A., and Williams, T. (1996). Neural tube, skeletal and body wall defects in mice lacking transcription factor AP-2. *Nature (London)* **381**, 238–241.

Zhao, Q., Behringer, R. R., and de Crombrugghe, B. (1996). Prenatal folic acid treatment suppresses acrania and meroanencephaly in mice mutant for the Cart1 homeobox gene. *Nat. Genet.* **13**, 275–283.

Zhao, Y., Sheng, H. Z., Amini, R., Grinberg, A., Lee, E., Huang, S., Taira, M., and Westphal, H. (1999). Control of hippocampal morphogenesis and neuronal differentiation by the LIM homeobox gene Lhx5. *Science* **284**, 1155–1158.

Zheng, C., Heintz, N., and Hatten, M. E. (1996). CNS gene encoding astrotactin, which supports neuronal migration along glial fibers. *Science* **272**, 417–419.

Zhou, C., Qiu, Y., Pereira, F. A., Crair, M. C., Tsai, S. Y., and Tsai, M. J. (1999). The nuclear orphan receptor COUP-TFI is required for differentiation of subplate neurons and guidance of thalamocortical axons. *Neuron* **24**, 847–859.

Zhu, Y., Li, H., Zhou, L., Wu, J. Y., and Rao, Y. (1999). Cellular and molecular guidance of GABAergic neuronal migration from an extracortical origin to the neocortex. *Neuron* **23**, 473–485.

6

Establishment of Anterior-Posterior and Dorsal-Ventral Pattern in the Early Central Nervous System

Alexandra L. Joyner

Developmental Genetics Program, Skirball Institute of Biomolecular Medicine, New York University School of Medicine, New York, New York 10016

I. **Overview of Early CNS Development and Patterning**

II. **Anterior-Posterior Patterning of the Mesencephalon and Metencephalon**

III. **Hindbrain Anterior-Posterior Patterning Involves Segmental Units of Development**

IV. **CNS Dorsal-Ventral Patterning Involves a Tug of War between Dorsal and Ventral Signaling**

V. **Summary**

References

I. Overview of Early CNS Development and Patterning

The central nervous system (CNS) becomes obvious on the dorsal side of the mouse embryo at approximately 9 days postcoitus (9 dpc) as a tube structure running the length of the body (see *Atlas of Mouse Development,* p. 88, Plate 21a). As with the rest of the embryo, the neural tube forms in an anterior-to-posterior (AP) progression. The neural tube forms from a flat sheet of neural tissue referred to as the neural plate. Induction of these cells begins during gastrulation and involves inhibition of signaling by bone morphogenetic proteins (BMPs) (see Chapter 3). In the next 2 days of development, the neural plate bends ventrally at the midline and the lateral edges rise up and join at the dorsal midline, producing a closed neural tube. Closure of the cephalic neural tube occurs through a specific order of events starting at the 5 somite stage, with the base of the hindbrain closing first and then the midbrain/forebrain junction and the regions between then being "zipped up" (*Atlas of Mouse Development,* Jacobson and Tam, 1981).

At 9.5 dpc when the neural tube has formed along most of the mouse embryo, the CNS has clear pattern along the AP axis in the form of swellings called vesicles (Figs. 1A and 1B). At the rostral end of the neural tube, two vesicles, the telencephalon and diencephalon, make up what is considered the forebrain. The telencephalon becomes divided down the midline by 12.5 dpc (see *Atlas of Mouse Development,* p. 162, Plate 28a), producing two symmetrical anterior telencephalic vesicles that form the cerebral cortex. Posterior to the diencephalon is the mesencephalon, which develops

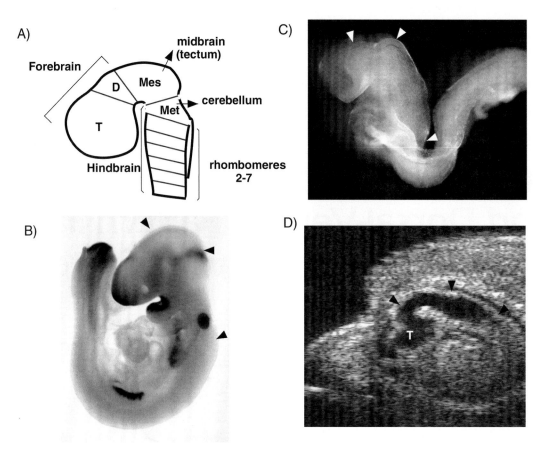

Figure 1 Mouse early neural tube development. (A) Schematic showing the segmental organization of the 9-dpc mouse brain. (B) A 9.5-dpc mouse embryo showing *Fgf8* mRNA expression in black. (C) Early somite stage mouse embryo. Anterior is to the left. (D) UBM midsagittal image of a 9-dpc mouse embryo *in utero*. Arrowheads in B–D indicate, from anterior to posterior, the approximate location of the junctions between the diencephalon/mesencephalon, mesencephalon/metencephalon, and rhombencephalon/spinal cord. T, telencephalon; D, diencephalon; Mes, mesencephalon; Met, metencephalon. (Panels B and C kindly provided by Aimin Liu and Panel D by Daniel Turnbull.)

into the midbrain (tectum dorsally). The rhombencephalon, or hindbrain, is divided into 7 rhombomeres by constrictions in the side of the neural tube. The most anterior rhombomere, the metencephalon, gives rise dorsally to the cerebellum, and posterior to it forms the myelencephalon.

The spinal cord makes up the posterior region of the CNS, extending from the brain to the tip of the tail. Each brain vesicle gives rise to a different adult brain structure, thus from an early stage of AP patterning, the CNS is divided morphologically into regions that will later develop into distinct functional structures. Although the brain vesicles seen at 9.5 dpc are a conspicuous form of AP pattern, in fact, the neural plate at early somite stages (8.5 dpc) already has obvious morphological distinctions and domains of gene expression along the AP axis (see below) that prefigure the vesicles (Fig. 1C). At this stage, the brain makes up close to half the length of the embryo, occupying a much larger proportion of the embryo than it does just a day later. Studies

using frog and chick embryos indicate that all neural plate cells initially are specified to become anterior neural tissue but during a short time period signals from the adjacent mesoderm "caudalize" the cells (for example, Doniach, 1993; Muhr *et al.,* 1997, 1999; Nieuwkoop, 1952). Fibroblast growth factors (FGFs) and retinoic acid (RA) seem to play critical roles in this process, as well as an unknown factor (see Muhr *et al.,* 1999). In addition, anteriorizing factors from the anterior mesendoderm likely induce and/or stabilize anterior cells.

The dorsal-ventral (DV) axis of the neural tube appears symmetrical at early stages, but it too takes on pattern, in general after the AP axis is established. DV pattern can be seen in some regions, such as the forebrain, as alterations in the shape of the ventricular zone(s) (see Chapter 5 for details). In other regions, such as the spinal cord, the main morphological indication of DV pattern is the structure of particular sets of differentiated neurons that have migrated away from the ventricular layer. Despite the morphology of the ventricular layer appearing homogeneous in some regions, the cells are patterned molecularly along the DV axis based on domains of gene expression from at least 9 dpc.

The cells that line the ventricles are called the ventricular zone (VZ) and consist of undifferentiated cells that are the precursors of all differentiated cells in the CNS. VZ cells have the capacity to self renew and to produce differentiated progeny. Secreted factors produced at particular AP or DV positions by both mesenchymal cells adjacent to the neural tube and by cells within it induce VZ cells to differentiate into particular neurons or glial cells. With few exceptions, once neurons begin to differentiate, they migrate away from the VZ and no longer divide (see Chapter 5 for details of CNS differentiation and late development). The cells then take on their final phenotype, which includes the extension of axons and dendrites, thus producing the final pattern of the CNS. This chapter concentrates on the genetic and cellular events that set up the early cues responsible for inducing appropriate differentiation along the AP and DV axes.

In simple terms, the early CNS (8.5 dpc) can be considered as four distinct developmental units: the forebain, mesencephalon and metencephalon (mes/met), myelencephalon, and the spinal cord. Each unit appears to utilize a different set of developmental mechanisms for patterning along the AP axis, whereas the mechanism of DV patterning is similar in the four regions. AP patterning of the myelencephalon involves a segmentation process that divides the region into rhombomeres, with limited communication between them. The mes/met, on the other hand, is patterning along the AP axis by an "organizing center" at the junction between the midbrain and hindbrain. Forebrain patterning appears to have characteristics of both rhombomere and mes/met development, with at least one organizer at the anterior end and division into segmental structures called prosomeres. Less is known about AP patterning of the spinal cord, but signals from the adjacent mesoderm clearly play an important role (see Ensini et al., 1998). The focus of this chapter is on the cellular and genetic mechanisms that control A/P patterning of the mes/met and hindbrain, and DV patterning along the length of the neural tube. Forebrain patterning at early and late stages is covered in Chapter 7.

Approaches to Studying Early Brain Patterning in Mouse

To fully understand patterning of a particular brain region, it is necessary to compile information about the region using a number of key experimental approaches. One is to describe the morphological changes that occur in a tissue across developmental stages. Because mouse embryos can not be observed directly, this usually requires obtaining embryos at different stages of development, by setting up timed pregnancies. Recently, a high-resolution ultrasound imaging approach (ultrasound backscatter microscopy or UBM) has been developed that can be used to follow the dynamic development of the brain of an embryo *in utero* over the period of 8.5 to 12.5 dpc (Fig. 1D) (Turnbull, 1999, 2000; Turnbull *et al.*, 1995). The resolution of UBM images is approximately 60 μm, and the brain ventricles are particularly easy to visualize. If cellular resolution is required, however, then embryos at multiple stages from different mothers must be sacrificed for histology. Use of UBM imaging should be particularly useful for following the development of mutants in which a variable phenotype is observed and it is of interest to relate an early phenotype directly with a particular late one.

For descriptive studies of brain development, marker genes, or genes that mark particular regions or cell types, can be extremely useful in understanding how pattern is imparted on a tissue. An extension of such an analysis is fate mapping studies. In such an approach, a particular group of cells is marked either genetically or by physical means, and the fate of the cells and their descendants determined at a later stage. Site-specific recombination approaches (Dymecki, 2000) using an inducible recombinase such as CRE (for example, Brocard *et al.*, 1997) can now be used to refine fate mapping studies (Kimmel *et al.*, 2000; Zinyk *et al.*, 1998) and determine the fate of groups of cells at different times in development (Fig. 2). The limiting factor to such studies is the promoters that are available to drive CRE expression. By using the gene targeting knock-in approach (Hanks *et al.*, 1995) and an inducible CRE, the fate of cells that express a particular gene at a given time can be determined. Fate mapping studies are extremely important for understanding brain development, because cells in the brain can migrate great distances and thus come under the influence of different cellular environments with time.

Genetic studies that utilize mutants made by gene targeting in embryonic stem cells or by using transgenes are among the most powerful approaches that can be applied to mice. The broad range of gain-of-function mutants that can be made using standard transgenic techniques, combined with loss-of-function mutants made by gene targeting

(continues)

(continued)

(Hasty *et al.,* 2000), allows one to address what processes a specific gene is sufficient to promote, as well as what ones it is required for. With the recent introduction of approaches for making conditional mutants using site-specific recombination (Dymecki, 2000), it is now possible to address such questions not just at one stage in development, but at many stages. UBM-guided injection of CRE-expressing retroviruses into the CNS of mutants that contain a targeted loxP conditional allele allow deletion of the gene at a particular time in development and/or in a particular region of the CNS. However, not all cells in a region of the brain will be infected with the retrovirus, thus the approach is limited to studies where mosaics are valuable. Chimeras between embryos of two different genotypes have proven to be an effective method for studying late functions of genes, when knock-outs of the genes cause early embryonic lethality (Rossant and Spence, 1998). See Chapter 3 for more discussion of the use of chimeras to study development. UBM-guided injections of retroviruses or cells that express a gene of interest can also be used to produce gain-of-function mutants in which the gene is introduced at a particular stage of development and region of the CNS (Gaiano *et al.,* 1999; Liu *et al.,* 1998).

The large number of naturally occurring mouse mutants involved in brain patterning can be used to identify new genes involved in brain development. In addition, studies of some of the interesting mutant phenotypes can shed light on certain developmental processes. Mouse embryos can develop to birth without a brain; however, at birth certain regions of the brain, in particular those that control basic processes such as breathing, heartbeat and feeding, become necessary for survival. Many homozygous mutations that lead to early brain patterning defects therefore cause postnatal lethality and are not likely to be identified as random mutations. Some mutations, however, cause mild behavior defects in the heterozygous state, and thus can be recovered on the basis of a dominant phenotype.

Finally, where possible, it is very useful to use approaches such as cell transplantation studies to study cellular interactions. Heterotopic transplantation experiments can be used to study whether cells are committed to a particular fate or whether they have inducing properties. Although transplantaion studies using mouse embryos are difficult to perform *in utero, in vitro* culturing techniques have been developed for mouse embryos (Sadler and New, 1981) that make it possible to perform short-term transplantation studies. Embryos of 7.5–11.5 dpc can be cultured for 48 hr and cells can be introduced into the brain ventricles of such embryos (Arkell and Beddington, 1997). Furthermore, for long term studies cells can be introduced into the CNS of embryos *in utero* from 8.5 to 13.5 dpc using UBM as a guidance technique (Liu *et al.,* 1998; Olsson *et al.,* 1997). After 11 dpc, the cells can be introduced into specific regions of the brain, whereas earlier the neural tube parenchyma is too thin to incorporate cells and injections must be into the ventricle where the cells integrate randomly into the VZ. These same approaches can be used to inject mutant cells into a normal host to study the fate of mutant cells in a wild-type environment, or vice versa. Explant cultures of brain regions from 8.5- to 12.5-dpc embryos have been used effectively recently to study cellular inter-actions by making tissue recombinations (Ang and Rossant, 1993; Irving and Mason, 1999), or to study the role of a particular secreted factor on patterning and differentiation by applying the factor to the medium or inserting beads soaked in the factor (see Fig. 3D) (Kohtz *et al.,* 1998; Liu *et al.,* 1999; Shimamura and Rubenstein, 1997; Ye *et al.,* 1998). The advantage of doing any of these studies in mouse, rather than in other vertebrate model organisms, is the vast resource of well characterized mouse mutants which can be used as a source of tissue to study the developmental potential of tissues lacking a particular gene, or to begin to study the epistatic relationships of genes.

II. Anterior-Posterior Patterning of the Mesencephalon and Metencephalon

A. Mes/Met AP Patterning Is Regulated by a Centrally Located Organizer

The experimental techniques outlined in the preceding boxed material have been used to varying degrees in mouse to study AP patterning of the mes/met region (reviewed in Liu and Joyner, 2001b). The largest body of information has come from descriptive studies of gene expression and analysis of mouse mutants made by gene targeting. Transplantation and fate mapping studies, done primarily with chick/quail embryos, however, provided the initial intellectual framework for interpreting and extending the genetic studies in mouse. With the advent of more sophisticated genetic approaches for fate mapping and cell transplantation in mouse, a broader range of studies will be done in mouse using both wild-type and mutant tissues.

A number of transplantation studies between chick and quail embryos were done that demonstrated that the mid/

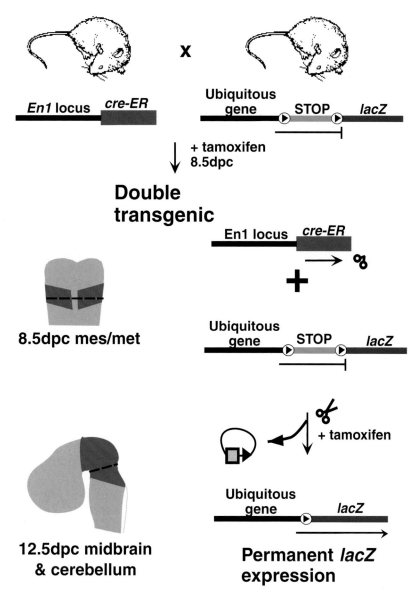

Figure 2 An approach using an inducible CRE protein and a STOP *lacZ* reporter mouse to fate map gene expression regions in the mouse brain.

hindbrain junction, referred to as the isthmus, has the potential to change the fate of host cells if it is inserted in more rostral or caudal positions at the 5–10 somite stage (reviewed in Alvarado-Mallart, 1993; Wassef and Joyner, 1997). When moved to the caudal diencephalon (prosomeres 1 and 2; P1/2) or anterior midbrain, the isthmic tissue induces the surrounding cells to develop into midbrain and isthmus-derived tissues (Fig. 3A). The transplanted tissue appears to maintain its normal fate. Most interesting is the finding that the induced midbrain tissue has an AP polarity that is the reverse of the normal midbrain: Tissue close to the isthmic transplant develops into posterior midbrain cell types, whereas cells at a distance develop into anterior mid-

brain tissues. Thus, isthmic tissue not only can induce formation of midbrain structures, but ones with appropriate AP pattern. The DV polarity also generally is conserved. If isthmic tissue is transplanted into P3 of the diencephalon or into the telencephalon, no midbrain tissue is induced. Furthermore, if the isthmic tissue is transplanted into hindbrain rhombomeres, then cerebellar tissue is induced. These studies demonstrated that isthmic tissue has the potential to induce midbrain or cerebellum development, and has therefore been called the isthmic mes/met organizer. In addition, the studies showed that depending on the AP position of the host cells in the brain, they responded differently to the same isthmic signals.

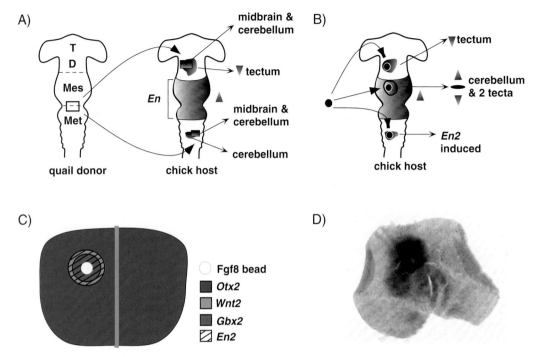

Figure 3 The isthmus and FGF8 can induce midbrain and cerebellum development. (A) Summary of isthmus transplant experiments in chick at early somite stages to different positions indicating the tissues that are induced and the gradient of En expression and AP patterning of the ectopic tectum. (B) Summary of FGF8 bead implant experiments in chick at early somite stages in different positions indicating the tissues that are induced and the gradient of En expression and AP patterning of the ctopic tectum. (C) Summary of induction of gene expression by FGF8-soaked beads placed into 9-dpc anterior midbrain explants. (D) Mouse diencephalic explant treated with FGF8 beads and showing induction of *En2* expression. (Panels C and D kindly provided by Aimin Liu.)

More recent transplant studies have indicated that the situation is more complex (reviewed in Liu and Joyner, 2001b; Joyner *et al.*, 2000). Insertion of mesencephalic or diencephalic tissue into the metencephalon at the ~10 somite stage, or vice versa, results in induction of an ectopic organizer (Hidalgo-Sanchez *et al.*, 1999; Irving and Mason, 1999). These studies indicate that the organizer is formed at the junction between mesencephalic and metencephalic tissue, similar to the way a number of organizers in *Drosophila* embryos are induced at the junction between two territories that express different genes (reviewed by Dahmann and Basler, 1999). Based on these recent studies, in some of the old transplant experiments a new organizer may have been induced, rather than an already existing organizer being transplanted and maintained.

B. Gene Expression Patterns Define Domains within the Mes/Met

Development of AP pattern in the mes/met from presomite stages (7.5 dpc) to the time when differentiated midbrain and cerebellar structures become apparent dorsally (12.5 dpc) can be best visualized using a series of gene expression patterns as markers for patterning of the region (reviewed in Wassef and Joyner, 1997; Joyner *et al.*, 2000). The first known molecular indication of AP patterning in the CNS is a division of the neural plate (and entire epiblast) into two domains: an anterior *Otx2*-positive one and a posterior *Gbx2*-expressing region. These two homeobox-containing genes share a similar gene expression border from 7.5 dpc until at least late gestation stages (Fig. 4A). By 9 dpc, when the isthmus is visible as a constriction, the common border of *Gbx2* and *Otx2* expression is within the isthmus. The finding that *Otx2* and *Gbx2* have gene expression borders in the isthmus and are homeodomain-containing transcription factors raised the question of whether either, or both, genes are involved in inducing the isthmic organizer.

At early somite stages, a number of genes required for mes/met development begin to be expressed. Each gene has a dynamic expression pattern for about a 24-hr period, when the patterns resolve into a more or less stable expression pattern that is maintained until at least 12 dpc (Figs. 4B and 4C). Interestingly, expression of the genes is centered on the *Otx2/Gbx2* border. The paired domain transcription factor PAX2 is the first gene to be expressed, just before the first somite forms, in a broad region spanning what is likely all of the mes/met (Rowitch and McMahon, 1995). The homeo domain transcription factor Engrailed1 (EN1) turns on a few

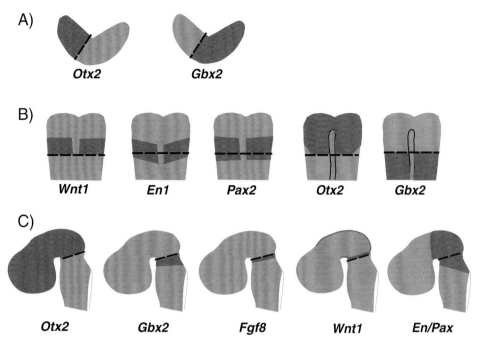

Figure 4 Schematic showing mes/met gene expression patterns during early brain development. Gene expression domains are shown in purple at (A) 7.5 dpc, (B) the 1–2 somite stage, and (C) 9.5 dpc. (Anterior to left in A and C, and to top in B; modified from Wassef and Joyner, 1997.)

hours later in a similar, but smaller, domain. About the same time, the secreted factor WNT1 begins to be expressed in the mesencephalon and by the 3–4 somite stage the secreted factor FGF8 is expressed in the metencephalon (Crossley *et al.,* 1996a). Two genes related to *Pax2* and *En1, Pax5* and *En2,* are then expressed in broad domains across most of the mes/met (Davis and Joyner, 1988; Song *et al.,* 1996).

Between 8.5 and 9.5 dpc the expression domains of *En1/2, Pax2/5, Wnt1,* and *Fgf8* become more restricted. From 9.5 to 12.5 dpc *Wnt1* and *Fgf8* are expressed in adjacent small rings of cells in the isthmus, and *En1/2* and *Pax2/5* are expressed in broader domains in the mes/met, with *En2* and *Pax5* having the broadest domains and *Pax2* the smallest. *Wnt1* has an additional domain of expression along the roof of the midbrain, diencephalon, and neural tube posterior to the metencephalon and one in the ventral midbrain. Between 7.5 and 9.5 dpc, *Gbx2* expression regresses from posterior regions and becomes limited to the metencephalon in a domain broader than *Fgf8,* but smaller than *En2.* Fate mapping studies in which an inducible Cre is expressed in the normal domains of mes/met genes will be important for determining exactly which precursor cells express each gene at any given time in development. For example, does the entire mes/met at any time express *Pax2, En1, En2,* and/or *Pax5* and, if so, at what time during development?

In relation to the finding that isthmic tissue can induce midbrain tissue with normal AP polarity, it is interesting to note that the *En* genes are expressed in decreasing gradients away from the isthmus. Furthermore, this gradient is replicated within 48 hr of an isthmic transplant and the level of *En* expression generally correlates with the AP position of the induced midbrain tissue such that *En* is expressed at higher levels in more posterior midbrain tissues. The same may be true of the metencephalon, but the AP polarity of the cerebellum in transplants has not been determined. Evidence that the *En* genes alone are sufficient to change the AP positional information of a cell comes from experiments in chick where *En1* was overexpressed throughout the midbrain and resulted in posteriorization of the cells (reviewed in Retaux and Harris, 1996).

C. Fate Mapping of the Mes/Met Suggests the Presence of Two Compartments

Chick-quail homotypic transplantation studies have been used to map the fate of the midbrain and cerebellum. One surprise was that at the ~10 somite stage, the cerebellum is derived from cells not only posterior to the isthmus, but also from cells in the posterior metencephalic vesicle (Alvarez Otero *et al.,* 1993; Hallonet and Le Douarin, 1993). In addition, the anterior medial region of the cerebellum was mainly mesencephalon derived, whereas the posterior and lateral cerebellum was derived from the metencephalon. More recent studies that reexamined these findings and noted that the posterior border of *Otx2* expression at these somite stages in chick is not in the isthmus, but anterior to it in the mesencephalic vesicle (Millet *et al.,* 1996). Later, the *Otx2* caudal expression border does lie in the isthmus. By mapping the fate of cells that lie on either side of the *Otx2* posterior border, it was found that the division between adult midbrain and cerebellum tissues is the *Otx2* caudal border (Millet *et al.,* 1996). The studies also indicate that there could be a lineage restriction at the *Otx2* caudal expression border.

In mouse embryos the neural tube closes later than in the chick, and the isthmic construction appears only around 9 dpc. From 9.5 dpc onward the isthmic constriction correlates with the *Otx2/Gbx2* common expression border (S. Millet, personal communication). One study has used a CRE fate mapping approach to study the mid/hindbrain region. Cells expressing CRE from an *En2* enhancer in the mid/hindbrain junction region gave rise to cells in the posterior midbrain and medial cerebellum (Zinyk *et al.,* 1998). The presence of cells in the midbrain and cerebellum could be taken to indicate that the transgene was expressed in *Otx2-* and *Gbx2*-positive cells, or that a lineage restriction is not present in the isthmus of mouse. Studies using an inducible CRE expressed from the *En2* enhancer would allow this question to be addressed.

D. Mouse Mutants Reveal Roles for *Otx2* and *Gbx2* in Positioning the Organizer

Mice lacking *Otx2* have a gastrulation defect and deletion of the brain anterior to rhombomere 3 (r3) (Acampora *et al.,* 1995; Ang *et al.,* 1996; Matsuo *et al.,* 1995), due to a requirement for *Otx2* in the visceral endoderm (see Chapter 5). The role of *Otx2* in mes/met development has been addressed using a number of mouse mutants and by making chimeras with wild-type and *Otx2* null mutant cells (reviewed in Simeone, 1998). In either case, if the visceral endoderm is made up of wild-type cells and the embryo derived from mutant cells, the forebrain and midbrain initially form, but are rapidly lost (Acampora *et al.,* 1998; Rhinn *et al.,* 1998). *Gbx2, Fgf8, Wnt1, En1/2,* and *Pax1/2* are all expressed, but their domains of expression overlap and are shifted to the anterior end of the embryo.

These studies indicate that *Otx2* is not required for induction of mes/met genes, but is required to position the genes, in particular *Wnt1* and *Fgf8,* that normally are expressed at the caudal limit of *Otx2* expression. *Otx2* also is required for later survival of midbrain and forebrain cells and possibly to maintain expression of mes/met genes. Further evidence for these functions has come from studies of $Otx1^{-/+}$ $Otx2^{-/+}$ and $Otx1^{-/-}$ $Otx2^{-/+}$ mutants (Acampora *et al.,* 1997; Suda *et al.,* 1997). Depending on the genetic background, both mutants have an expanded cerebellum and no midbrain or posterior diencephalon (Fig. 5C). At early somite stages the mes/met genes are expressed normally and then quickly their expression domains are shifted anterior into the diencephalon. Finally, misexpression studies showed that *Otx2* is

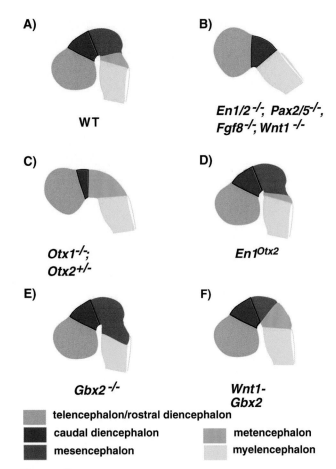

Figure 5 Brain phenotypes of mice carrying mutations in mes/met genes. The $Fgf^{-/-}$ phenotype is based on the phenotype of the hypomorphic $Fgf8^{neo/\Delta}$ compound heterozygous allele (Meyers *et al.,* 1998). It is not known whether the brain defects simply reflect greatly reduced *Fgf8* expression in the isthmus, or also reduced *Fgf8* expression during gastrulation. Two other *Fgf* genes are also expressed in the mes/met that could also function during mes/met development. (Figure provided courtesy of Aimin Liu.)

sufficient to alter the position of the isthmic organizer, as defined by the domains of *Wnt1* and *Fgf8* expression. In mouse, *Otx2* was inserted into the *En1* locus by gene targeting, expanding the *Otx2* expression domain into the dorsal metencephalon (Broccoli *et al.,* 1999). In such mutants, the anterior and medial region of the cerebellum did not develop and the midbrain was expanded posterior (Fig. 5D). At 9.5 dpc, the expression domains of *Fgf8* and *Wnt1* were shifted posterior to correlate with the new posterior border of *Otx2* expression, and *Gbx2* was repressed in the anterior metencephalon. Misexpression of *Otx2* in the hindbrain of chick embryos leads to similar findings (Katahira *et al.,* 2000).

Gbx2 mutants have a phenotype that is somewhat complementary to $Otx1^{-/-}$ $Otx2^{-/+}$ mutants: an expanded midbrain and no cerebellum or structures that develop from r2 and r3 (Wassarman *et al.,* 1997) (Fig. 5E). At early somite stages the posterior border of *Otx2* expression expands caudally to the r3/4 border and the domains of *Fgf8* and *Wnt1* shift posteriorly (Millet *et al.,* 1999). An abnormal situation develops in which *Fgf8* and *Wnt1/Otx2* have overlapping expression domains, although *Wnt1* is still expressed only in *Otx2*-positive cells and the *Fgf8* domain extends posterior to *Otx2*. These studies indicate that *Gbx2*, like *Otx2*, is involved in positioning isthmic genes. Furthermore, misexpression studies showed that *Gbx2* can also alter the position of the organizer. The domain of *Gbx2* expression was expanded into the mesencephalon at 8–9 dpc in transgenic embryos using a *Wnt1* enhancer, and this leads to repression of *Otx2* in the posterior mesencephalon and an anterior shift in the *Wnt1* and *Fgf8* domains to the new *Otx2/Gbx2* border region (Millet *et al.,* 1999). At 9.5 dpc the midbrain is smaller and the hindbrain larger in such transgenics (Fig. 5F). Misexpression of *Gbx2* in the midbrain of chick embryos leads to similar alterations in gene expression (Katahira *et al.,* 2000).

Taken together, these *Otx2* and *Gbx2* mutant studies indicate the presence of an antagonistic feedback loop between the genes (reviewed in Joyner *et al.,* 2000). Furthermore, the border between the expression domains of the two genes seems to position the isthmus and domains of expression of *Wnt1* and *Fgf8* and other mes/met genes. It is interesting to note that *Fgf8* and *Wnt1* are repressed in their normal domains by misexpression of *Otx2* or *Gbx2*, and induced only at the new border between the two genes. It will be interesting to determine whether additional signaling molecules are expressed at the *Otx2/Gbx2* border that stabilize or induce *Wnt1* and *Fgf8* expression in the appropriate cells.

E. Mouse Mutants Identify Genes Required for Mes/Met Development

Gene targeting experiments in embryonic stem (ES) cells have been used to produce loss-of-function mutations in the mes/met genes described above, and a naturally occurring *Wnt1* mutant, *Swaying,* was identified based on its similar phenotype to one *Wnt1* knock-out mouse (reviewed in Joyner, 1996; Wassef and Joyner, 1997; also see Meyers *et al.,* 1998; Schwarz *et al.,* 1997). Mice lacking *En1* and *2, Pax2* and *5,* or *Wnt1* have a similar late mes/met phenotype: an absence of the midbrain and cerebellum (Fig. 5B). As is often the case during mouse development, the pairs of genes *En1/2* and *Pax2/5* that have overlapping expression patterns show redundancy of function. This has been addressed for the *Pax2/5* genes by making double mutants (Schwarz *et al.,* 1997) and for *En1/2* using the gene targeting knock-in approach, as well as by making double mutants.

For the *En1/2* genes, the story seems to be relatively simple. The two genes have similar expression patterns in the early mes/met, except that *En1* is initiated about 6 hr before *En2*. A null mutation in *En1* causes an early loss of a variable amount of the mes/met (Wurst *et al.,* 1994), whereas *En2* null mutants have only a slightly smaller mid-

brain and cerebellum that is not apparent until late gestation (Joyner *et al.,* 1991; Millen *et al.,* 1994). In addition, *En2* mutants have cerebellar patterning defects that could be due to late expression of *En2*, and not *En1*, in the cerebellum (Millen *et al.,* 1995). *En1;En2* double homozygous mutants have no mes/met by E9.5 and although expression of mes/met genes is initiated, by the 11 somite stage *Wnt1* expression is lost and *Fgf8* and *Pax5* are greatly reduced (Liu and Joyner, 2001a). To determine whether EN1 and EN2 proteins are interchangeable, the coding sequences of *En1* were replaced with those of *En2* (Hanks *et al.,* 1995). Such *En1^{En2}* knock-in mice have normal mes/met development, showing that either gene is sufficient for mes/met development, and that one *En* gene must be expressed between the 1 and 5 somite stages for development of most of the midbrain and cerebellum to proceed. Interestingly, *En2* is not able to fully rescue the limb defects seen in *En1* mutants, thus EN1 and EN2 proteins are not fully interchangeable. Furthermore, *Drosophila en* is able to rescue the brain, but not any of the limb defects in *En1* mutants, thus one basic function of the EN family of transcription factors used in brain development has been conserved through evolution (Hanks *et al.,* 1998).

Fgf8 null mutants have gastrulation defects and die before brain AP patterning can be studied, but a hypomorphic mutant in which the bacterial *neo* gene is inserted into an *Fgf8* intron allows enough *Fgf8* to be produced for gastrulation to proceed (Meyers *et al.,* 1998). Embryos that carry this allele over a null allele lack much of the midbrain and cerebellum (Fig. 5B), providing evidence that *Fgf8* is required during mes/met development. The zebrafish mutant *acerebellar (ace)* carries a mutation in *Fgf8* and it has a milder phenotype of loss of only the cerebellum (Reifers *et al.,* 1998). The finding that mutations in *Fgf8* or *Wnt1*, which are expressed in the isthmus, lead to disruption of the midbrain and cerebellum has raised the question of whether these genes could be responsible for the organizing activity of the isthmus.

F. Gain-of-Function Mutants Demonstrate FGF8 Has Organizer Activity

To test whether any of the mes/met genes alone have organizer function, the genes have been misexpressed in the brain, primarily in chick embryos. An early study showed that if beads soaked in FGF8 protein are placed in P1 or P2 of the diencephalon, but not in more anterior positions, then an ectopic midbrain is induced with a reverse polarity to the normal midbrain (Crossley *et al.,* 1996a) (Fig. 3B). *En2* and *Fgf8* itself were the first genes to be induced, and then *Wnt1* expression was altered. In the same studies, FGF8-soaked beads placed in the rhombencephalon had no effect. However, recently, FGF8 was found to induce mes/met genes in the chick or mouse hindbrain (Irving and Mason, 2000; Liu *et al.,* 1999). In addition, FGF8-soaked beads placed in posterior P1 or the anterior midbrain can induce ectopic cerebellum tissue near the bead, in addition to one or two ectopic midbrains at a distance (Fig. 3B) (Martinez *et al.,* 1999). These studies demonstrated that FGF8 can induce both midbrain and cerebellum tissue, depending possibly on the type of host responding tissue or concentration of FGF8.

The phenotypes of transgenic mice expressing *Fgf8* from a *Wnt1* regulatory element indicate that the situation could be more complex. Transgenic embryos expressing FGF8b from a *Wnt1* enhancer showed that FGF8b was a potent inducer of *Gbx2* and *Wnt1,* and repressor of *Otx2* in the mesencephalon and diencephalon (Liu *et al.,* 1999). The late phenotype of such transgenics was a deletion of the midbrain and diencephalon. A similar, but milder, phenotype was seen in some transgenic embryos expressing FGF8a (Liu *et al.,* 1999), whereas the others mainly had an overproliferation of the midbrain and diencephalon (S. M. K. Lee *et al.,* 1997b; Liu *et al.,* 1999). The milder phenotype caused by FGF8a is consistent with *in vitro* assays that have shown FGF8b more effectively binds FGFR2c, R3c, and R4 and induces signaling (MacArthur *et al.,* 1995).

Consistent with the transgenic experiments, when mouse diencephalic brain explants are treated with beads soaked in FGF8b, *Wnt1, Gbx2, Pax5,* and *En1/2* (Fig. 3D) are induced, whereas *Otx2* is repressed in midbrain explants (Liu *et al.,* 1999). *Wnt1* has an interesting response to FGF8b of first being induced in cells around the bead and then only being maintained in cells at a distance from the bead (Liu and Joyner, 2001a). The cumulative expression patterns induced by FGF8b beads is similar to formation of a concentric mes/met border region with *Gbx2* in the center, *Wnt1* in a ring around it, and *Otx2* at a distance, with *En1/2* and *Pax5* expression overlapping with *Gbx2* and *Otx2* (Fig. 3C). Different from the chick, the endogenous *Fgf8* gene is not induced by FGF8b either in mouse embryos or in explant cultures. Perhaps in mouse, one of the other *Fgfs* normally expressed in the isthmus (Xu *et al.,* 2000) is induced by FGF8b.

FGF4 and at least four alternatively spliced forms of FGF8 (a, b, c, and d) can induce mes/met genes and midbrain tissue development in the diencephalon (Crossley *et al.,* 1996b; Shamim *et al.,* 1999). Because two other *Fgfs* are also found to be expressed in the isthmus (Xu *et al.,* 2000), it is not clear which FGF protein(s) are involved in mes/met AP patterning. Furthermore, since *Fgf8* is not expressed in the neural plate until after *Wnt1* and *En1,* it is not clear whether an FGF is responsible for the initial induction of mes/met genes in mouse. In chick, *Fgf4* appears to be expressed in the underlying mesoderm at the early somite stages, thus it could be involved in this induction (Shamim *et al.,* 1999). A comparison of the response of mes/met genes to FGF8 in diencephalic explants taken from *En1;En2* double homozygous mutants argues against an FGF being the molecule that induces *En, Pax, Fgf8,* and *Wnt1* (Liu and

Joyner, 2001a). Induction of *Pax5* by FGF8 is dependent on the *En* genes, whereas *Pax5* is initially expressed in *En1;En2* double mutants. Furthermore, normally in the embryo *En1* is expressed before *En2,* however in explants *En2* is induced before *En1* by FGF8. Finally, to complicate matters, misexpression of *En1/2* or *Pax2/5* in the diencephalon of chick, frog, or fish embryos leads to repression of the diencephalon gene *Pax6* and induction of *Fgf8* and the expected alterations in mes/met gene expression due to *Fgf8* expression (Araki and Nakamura, 1999; Funahashi *et al.,* 1999; Okafuji *et al.,* 1999; Ristoratore *et al.,* 1999; Shamim *et al.,* 1999). Thus, there must be a complicated set of feedback loops between all the mes/met genes. It has therefore been suggested that reciprocal antagonism between *En1/2* and/or *Pax2/5* with *Pax6* defines the midbrain/forebrain junction, perhaps in a way analogous to *Gbx2* and *Otx2* at the midbrain/hindbrain border. However, from E9 onward in mouse *Pax2/5* or *En1/2* and *Pax6* do not share a common border of expression.

G. Model of Mes/Met AP Patterning and Future Directions

A working model for the genetic and cellular events that pattern the mes/met along the AP axis can be developed based on the studies reviewed above. An early event (by 7.5 dpc) is the division of the neural plate into an anterior *Otx2*-positive and posterior *Gbx2*-positive domain (Fig. 4A). The signals that induce and restrict *Otx2* and *Gbx2* expression remain to be identified. Because their domains of expression initially do not exactly abut and appear normal in mutant *Otx2* epiblasts and *Gbx2* null mutants, the mechanism that initially sets their common border region must be independent of any mutual inhibition between the genes. Retinoic acid induces *Gbx2* and represses *Otx2* in early embryos (Ang *et al.,* 1994; Li Y. and A. L. Joyner unpublished observations). Because both retinoic acid and *Fgf8* are expressed in the primitive streak, these molecules may regulate the initial expression of *Otx2* and *Gbx2.* In addition, signals from the anterior visceral endoderm and anterior mesendoderm may be involved in inducing *Otx2* expression in the neuralectoderm.

A next critical stage in mes/met AP patterning is induction of mes/met genes such as *En1/2, Pax2/5, Wnt,* and *Fgf8* in specific domains at early somite stages (Fig. 4B). As discussed above, neither *Gbx2, Otx2* or an FGF seems to be required for this induction, although *Gbx2* and *Otx2* are clearly involved in positioning of the domains. Interactions between all of the genes then seem to refine the domains of expression. FGF8 appears to be high up in the hierarchy of these genetic interactions. Tissue that is *Gbx2* positive then goes on to form cerebellar structures, whereas *Otx2*-positive cells form midbrain structures. The type of differentiation that cells undergo along the AP axis might then be influenced by the concentration of EN protein in each cell type. If a lineage restriction exists at the *Gbx2/Otx2* border, then differential cell adhesion between *Otx2-* and *Gbx2*-positive cells may play a further role in producing a sharp and stable *Otx2/Gbx2* border leading to proper separation of, and patterning within, the midbrain and cerebellum.

Many questions remain to be addressed. For example, what molecule(s) induces the mes/met genes. What molecule(s) then directs AP patterning across the entire midbrain and r1? Is the organizer simply a source of FGF8 or is it all the genes involved in maintaining the complex set of feedback loops that have been identified? FGF8 seems unlikely to be the sole factor, since it does not normally diffuse over the long distance needed to pattern the entire midbrain. Is a signaling cascade that involves a relay system therefore set up? If FGF8 can repress *Otx2* in adjacent cells, why then in the embryo is *Otx2* normally expressed in cells adjacent to *Fgf8* expressing cells? One possibility is that it is a matter of concentration. In the embryo, only the cells expressing *Fgf8* receive a high enough concentration to repress *Otx2*. Alternatively, perhaps FGF8 can only repress *Otx2* in an autocrine manner.

Is WNT1 part of the isthmic organizer activity? Arguing against this are two experiments in which *Wnt1* was misexpressed in the spinal cord or midbrain and WNT1 was found to induce proliferation, but did not appear to alter AP or DV patterning (Adams *et al.,* 2000; Dickinson *et al.,* 1994). The phenotype of *Wnt1* null mutants could be taken as an argument that WNT1 is part of the organizer activity or that WNT1 is required for proliferation of mes/met precursors. Evidence that *Wnt1* might play a regulatory role comes from an experiment in which transgenic mice expressing *En1* from a *Wnt1* enhancer were bred onto a *Wnt1* mutant background (Danielian and McMahon, 1996). In such animals, expression of *En1* was maintained in the *Wnt1* expression domain and this led to a nearly complete rescue of the mes/met defects seen in *Wnt1* mutants. This indicates that *En1* is a target (direct or indirect) of WNT1 signaling and that *En1* is required to maintain expression of mes/met genes. Consistent with this, in *Wnt1-En1* transgenic embryos the anterior midbrain cells that misexpress *En1* ectopically express a marker of posterior midbrain differentiation (S. M. K. Lee *et al.,* 1997). It has been suggested that WNT1 might also play a role in maintaining the border between *Otx2-* and *Gbx2*-positive cells. *Swaying* mutants, which have a point mutation in *Wnt1*, have a hypomorphic phenotype compared to targeted mutants with only a partial deletion of the midbrain and cerebellum (Thomas *et al.,* 1991). Interestingly, in these mutants the *Otx2* caudal border is irregular and *Otx2*-positive and *Otx2*-negative cells are found in the hindbrain and midbrain, respectively (Bally-Cuif *et al.,* 1995). Further elucidation of the molecular basis of this phenotype would be very interesting.

III. Hindbrain Anterior-Posterior Patterning Involves Segmental Units of Development

A series of constrictions marking the rhombomeres of the hindbrain becomes apparent at 9 dpc in the mouse. The rhombomeres form in a characteristic pattern, with rhombomeres 3, 4, and 5 becoming apparent first. The most rostral and caudal rhombomeres then become divided. Cellular and genetic studies have shown that this overt segmentation of the hindbrain reflects an underlying compartmentalization of the hindbrain during early development.

A. Transcription Factor Expression Defines the Rhombomeres

Important clues as to the genetic mechanisms controlling AP patterning of the hindbrain have come from the expression patterns of the *Hox* clusters of homeobox-containing genes. Soon after they were cloned in mouse, it was recognized that they are expressed in specific regions of the posterior neural tube with overlapping expression patterns. Moreover, most of the genes are expressed from the posterior end of the CNS to a specific anterior boundary situated between two rhombomeres in the rhombencephalon (reviewed in Maconochie *et al.,* 1996). Some of the *Hox* genes are expressed at highest levels in particular rhombomeres (Fig. 6). None of the genes are expressed in the metencephalon or more anterior brain regions of early embryos. When the gene expression patterns are compiled, each rhombomere has a specific "Hox code." In addition, a few other transcription factors, such as KROX20 and the *Kreisler* gene product are expressed in specific rhombomeres (Fig. 6). Based on these gene expression patterns and the segmented morphology of the rhombomeres, it was suggested that development of the hindbrain could involve a process of segmentation. Because some of the genes are expressed before the rhombomeres can be distinguished morphologically, these genes could be involved in the segmentation process.

Promoter studies using transgenic mice have been used to identify upstream regulators of the segmental pattern of *Hox* gene expression in early neural tissue. Three groups of key regulators of *Hox* gene expression have been identified. In terms of what sets up the initial broad expression of the early *Hox* genes along the AP axis with a particular anterior border, RA seems to be a major player. RA receptor (RARE) binding sites have been found in the regulatory regions of a number of *Hox* genes and they are required for early expression (reviewed in Maconochie *et al.,* 1996; also see Dupe *et al.,* 1997; Gould *et al.,* 1998; Marshall *et al.,* 1994; Studer *et al.,* 1998). One very interesting finding is that in the context of a *Hoxb4* reporter construct that expresses up to the

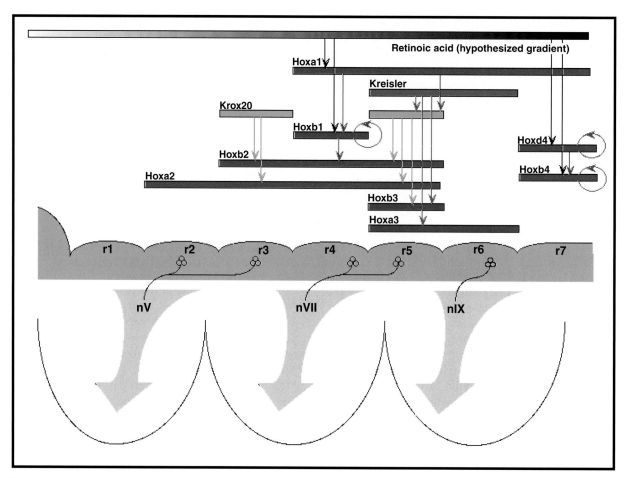

Figure 6 Schematic showing rhombomere 1–7 specific gene expression (colored bars above), motor nerves (nV, nVII and nIX), and migration of neural crest cells (yellow arrows) to the first three branchial arches. Anterior is to the left. An arrow indicates that the gene or retinoic acid likely positively regulates (directly or indirectly) expression of another gene, or itself, based on transgene and mutant analysis. See text for details. (Figure by Cecilia Moens, Fred Hutchison Cancer Institute.)

r6/7 border, exchanging the *Hoxb4* RARE sites with ones from *Hoxb1*, which is expressed to the r3/4 border, extended expression of the *Hoxb4* reporter construct to the r3/4 border. This result indicates that the RARE within a particular *Hox* gene responds to positional cues, possibly in the form of the concentration of RA, and positions the anterior border of *Hox* gene expression. Furthermore, an important tissue source of RA, or a molecule that induces RA synthesis in the neural tube, seems to be the somites (Gould *et al.,* 1998).

Two additional proteins that appear to directly regulate certain *Hox* genes in particular rhombomeres are KROX20 and KREISLER (reviewed in Maconochie *et al.,* 1996; also see Manzanares *et al.,* 1997, 1999a; Nonchev *et al.,* 1996; Prince *et al.,* 1998; Seitanidou *et al.,* 1997). In addition, extensive autoregulation and cross-regulation of *Hox* genes occur in the form of feedback loops (reviewed in Maconochie *et al.,* 1996; also see Gould *et al.,* 1997; Maconochie *et al.,* 1997; Morrison *et al.,* 1997; Studer *et al.,* 1998). For example, due to the existence of multiple regulatory mecha-

nisms for each *Hox* gene, removal of a single regulatory element from an endogenous *Hox* gene locus only has mild and transient effects on expression (Studer *et al.,* 1998). For example, removal of a RARE in one *Hox* gene can be compensated for by cross-regulation of the gene by a paralogous HOX protein, thus resulting in only a delay in initiation of expression of the mutant gene.

B. Rhombomere Junctions Are Regions of Lineage Restriction

If the rhombomeres do develop as suggested through a segmentation process, then the junction between them would be expected to be an area where cells do not mix. This was tested in chick by DiI labeling single cells in a rhombomere and determining whether the resulting progeny cross into neighboring rhombomeres. It was found that after rhombomeres are obvious most clones did not mix between (Birgbauer and Fraser, 1994; Fraser *et al.,* 1990). Cell adhesion has been shown to play an important role in the lineage restriction. For example, it was recently shown in mouse using hindbrain explant cultures that if cells from odd-numbered rhombomeres are transplanted into even-numbered rhombomeres, the cells stay as a clump, whereas if they are transplanted into odd-numbered rhombomeres they mix freely (Manzanares *et al.,* 1999b).

C. Ephrin/Eph Signaling Regulates Rhombomere Border Formation

The Ephrin/Eph families of ligands and receptors have been implicated in controlling the cellular process of hindbrain segmentation based mainly on gain-of-function studies in zebrafish embryos. There are two types of Eph receptors, A and B, based on amino acid homology and specificity of ligand binding (Eph Nomenclature Committee, 1997). The Ephrins are also divided into two groups, the A group being proteins bound to the membrane by a GPI linkage and which bind EphA receptors, and the B group, which is made up of membrane-spanning proteins and binds to EphB receptors. EphA4 is an exception that binds both classes of Ephrins. The B-type Ephrin ligands have the property of bidirectional signaling, since they can both induce signaling in adjacent cells through the receptor they bind and also elicit a signaling cascade within the cell in which they are expressed in response to receptor binding (reviewed in Holder and Klein, 1999; O'Leary and Wilkinson, 1999). EphrinB2 is expressed in r2, r4, and r6, whereas EphA4 is expressed in r3 and r5. It appears that signaling at the rhombomere borders leads to differential cell adhesion and a reduction in cell–cell communication and this leads to a sharp cell border between rhombomeres. The cells at the rhombomere borders have a more elongated shape than other rhombomere cells. Interestingly, based on misexpression studies in zebrafish animal caps, it appears that a restriction to cell mixing can require bidirectional signaling, whereas downregulation of cell communication through gap junctions is unidirectional (Mellitzer *et al.,* 1999). Furthermore, activation of inappropriate Ephrin/Eph signaling within a rhombomere leads to sorting out of the cell misexpressing the Ehp or Ephrin genes to the rhombomere border (Xu *et al.,* 1999). Thus, the normal signaling between adjacent Eph- and Ehprin-expressing cells at the rhombomere borders likely leads to segregation of cells.

D. Mouse Mutants Show Rhombomere Identity Regulates Segmentation and Cell Fate

The phenotypes of a number of *Hox* null mutants shows that these genes control early and late phases of rhombomere

formation and differentiation. Due to the overlap in function and expression of paralogous *Hox* genes, single and double or triple mutants must be studied to uncover the full range of functions of a particular family of *Hox* genes. In general, it appears that combinations of *Hox* genes are initially involved in specifying the appropriate number of cells to make up each rhombomere and then they become involved in maintaining rhombomere integrity, likely through regulation of the *Ephin/Eph* genes. Finally, the genes are involved in differentiation of both the neural crest cells that migrate from specific rhombomeres and neural cells that differentiate within each rhombomere (for example, see Rossel and Capecchi, 1999; Studer *et al.,* 1998). For example, in mice lacking *Hoxa2,* which is expressed up to the r1/2 border, r1 and subsequently the cerebellum are expanded posteriorly, and neurons that normally are derived from r2 are missing (Gavalas *et al.,* 1997). *En2* expression ia also expanded posteriorly into r2. Thus, *Hoxa2* plays roles in inhibiting cerebellum development, possibly through repression of *En2* and *Gbx2* expression, and in promoting appropriate r2 differentiation. A gain-of-function study in chick nicely showed that *Hox* genes can alter the identity of a rhombomere (Bell *et al.,* 1999). When *Hoxb1,* which is expressed at highest levels in r4, was mis-expressed only in r2, the r2 motor axon projections were reorganized to resemble r4 motor axons. Knockout and natural mouse mutations in the *Krox20* and *Kreisler* genes, respectively, have shown that these genes are also involved in development of specific rhombomeres. This is likely at least in part due to their early role in regulating rhombomere-specific expression of particular *Hox* genes (see above) and, in the case of *Krox20, Epha4* as well (Theil *et al.,* 1998).

In summary, AP patterning of the rhombencephalon involves a series of processes, beginning with spatially restricted expression of the *Hox* and other genes in a manner that prefigures where rhombomeres will form. RA seems to be involved in directing the initial gene expression domains, possibly through a mechanism that involves a gradient of RA. Then a complicated set of genetic interactions and feedback loops take over within the CNS to maintain and refine the gene expression patterns (see Fig. 6). In addition, the *Ephrin/Eph* genes are induced in specific alternating rhombomeres and signaling at their borders of expression creates lineage restrictions and stable rhombomere borders. Neural precursors within each rhombomere then differentiate into specific groups of cells and this is likely, at least in part, dictated by the combination of *Hox* genes expressed by a particular cell within a rhombomere. After differentiation begins, some migration of postmitotic cells occurs between rhombomeres. Segmentation of the hindbrain therefore appears to be a transient state, presumably required only to establish an initial framework of AP pattern within the hindbrain.

IV. CNS Dorsal-Ventral Patterning Involves a Tug of War between Dorsal and Ventral Signaling

DV patterning of the entire neural tube is directed primarily by signals from two sources: Dorsal differentiation is initially induced by signals from the adjacent ectoderm and ventral differentiation is likewise induced by signals from the adjacent axial mesoderm (Fig. 7). Interestingly, for both dorsal and ventral CNS patterning, the signaling process is transferred at an early stage from the nonneural tissues to the neural tube and imparted on specialized cells called the roofplate and floorplate, respectively. In addition to inducing dorsal or ventral differentiation, the dorsal and ventral signals appear to repress each other's activity. Numerous studies have shown that the key molecules in dorsal and ventral CNS induction are members of the Transforming Growth Factor (TGF) family and Sonic Hedgehog (SHH), respectively (Lee and Jessell, 1999; Tanabe and Jessell, 1996). There is also at least one parallel pathway regulated by RA that induces interneurons in the middle of the spinal cord (Pierani *et al.,* 1999) and striatal neurons in the forebrain (Toresson *et al.,* 1999).

A. SHH Regulates Ventral CNS Patterning

DV patterning is best understood in the spinal cord. In the ventral part of the spinal cord, motor neurons form dorsolateral to the floorplate and interneurons in intermediate re-

gions. *Shh* is initially expressed in the node and then in the axial mesoderm, which includes the notochord that underlies the spinal cord, midbrain and hindbrain, and the prechordal plate that underlies the forebrain. SHH from the axial mesoderm then induces expression of *Shh* in the floorplate and ventral regions of the forebrain, through a mechanism considered to be homeogenetic induction. A direct transcriptional target of SHH signaling in the floorplate is likely Foxa2 (Sasaki *et al.,* 1997) , which in turn induces *Shh* (Epstein *et al.,* 1999). *In vitro* explant cultures have been used to define some of the ventralizing properties of SHH. Collectively the studies using chick spinal cord explants argue that SHH acts in a concentration-dependent manner to induce specific cell types at different DV positions: High concentrations of SHH induce floorplate marker genes (Foxa2), moderate concentrations induce motor neurons (Isl1), and low concentrations can induce interneurons (Roelink *et al.,* 1995; reviewed in Tanabe and Jessell, 1996). SHH also represses *Pax3* and *7,* which are initially expressed throughout the spinal cord, but later become restricted to dorsal regions.

SHH appears to act in conjunction with other molecules to induce the appropriate cell types at different AP positions. Using lateral neural plate tissue from posterior regions of early somite embryos SHH induces the floorplate marker Foxa2. In contrast, if rostral diencephalic explants are used, high concentrations of SHH induce a ventral diencephalon marker NKX2.1 (Ericson *et al.,* 1995; Muhr *et al.,* 1997; Roelink *et al.,* 1995). When dorsal telencephalon explants are used, striatal markers are induced and which

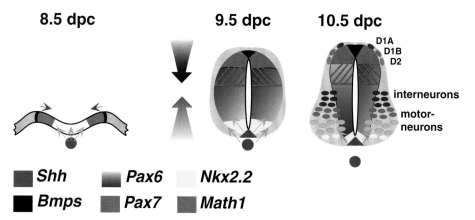

Figure 7 Genetic mechanisms regulating dorsal-ventral patterning of the early spinal cord. Each gene expression domain is shown in the color indicated. The small gray arrows shown at 8.5 dpc indicate that BMPs from the adjacent ectoderm (light green) likely regulate specification of the lateral neural plate cells that will become the dorsal neural tube. The small pink arrows at all stages indicate that SHH is secreted by the notochord (red circles) and floorplate (red triangles). The large black and red graded arrows shown to the left of the 9.5-dpc neural tube indicate that gradients of BMPs and SHH are thought to be involved in regulating differentiation of the different dorsal and ventral cell types, respectively. At 10.5 dpc differentiated neuron types are indicated outside of the ventricular zone. The light gray areas at 9.5 and 10.5 dpc indicate regions containing differentiated cells. (Figure adapted from one provided by Michael Matise.)

markers are induced depends on the age of the explants (Kohtz et al., 1998). Thus, different tissues appear to be competent to differentiate into different cell types in response to SHH, and this may be due to an intrinsic difference in the cells and/or differences in the local cellular environment. Support for the latter has come from explant studies using midbrain and hindbrain tissues where it was found that FGFs can alter the types of genes induced by SHH (Ye et al., 1998).

B. The Gli Genes Act in the SHH Pathway to Pattern the CNS

Many aspects of the mouse genetic cascade downstream of SHH have been found to be similar to the analogous Drosophila Hedgehog pathway (Hh). In fly it has been shown that Hh binds to its receptor Patched (Hh) and through a complicated set of protein–protein interactions results in activation of the transcription factor Ci. Furthermore, Ptc is a negative regulator of the pathway. In the absence of Hh, Ci is cleaved into an N-terminal repressor protein that inhibits transcription of Hh target genes, as well as hh itself. In the presence of Hh, Ci is not cleaved and is modified such that it can induce the transcription of Hh targets including Ptc and the Wnt homolog wingless.

The situation is more complicated in mice. For example, three GLI proteins seem to have different biochemical and genetic functions. In addition, the mouse genes have different but overlapping expression patterns with Gli1 being expressed adjacent to the floorplate, Gli3 in the dorsal CNS, and Gli2 in between (Hui et al., 1994; Platt et al., 1997). Furthermore, unlike ci, Gli1 transcription is activated by SHH and Gli3 is repressed (Hynes et al., 1997; J. Lee et al., 1997). Biochemical and misexpression studies indicate that mouse and human GLI3 acts primarily as a repressor and can be cleaved similar to Ci, whereas GLI1 is primarily an activator and can induce floorplate when expressed in the dorsal CNS (Dai et al., 1999; Hynes et al., 1997; J. Lee et al., 1997a; Ruiz i Altaba, 1998, 1999; Sasaki et al., 1999; Wang et al., 2000). GLI2 appears to have properties of both GLI1 and 3. Studies of the naturally occurring Gli3 null mutant Extra toes (Gli3^{Xt}) and targeted loss-of-function mutations in Gli1 and Gli2 that remove the zinc finger DNA binding domains have shown that the three genes have both unique and overlapping functions. Double mutant analysis indicates that Gli3 primarily has overlapping functions with Gli2, whereas Gli2 has overlapping functions with both Gli3 and Gli1 (Hardcastle et al., 1998; Matise et al., 1998; Mo et al., 1997; Motoyama et al., 1998; Park et al., 2000; reviewed by Matise and Joyner, 1999). The existing Gli1 mutant does not have an obvious phenotype on its own, but has major defects when one allele of Gli2 is also mutant (Park et al., 2000).

Shh mutants lack all ventral cell types up to and including some interneurons in the spinal cord and brain (Chiang et al., 1996; Pierani et al., 1999). Ptch mutants in contrast have a ventralized spinal cord made up primarily of ventral cell types (Goodrich et al., 1997). Gli mutants have aspects of both Shh and Ptc mutant phenotypes. Gli2 mutants and Gli1/2 double mutants lack the floorplate in the spinal cord and ventral-medial cell types in the diencephalon, but retain more lateral and dorsal cells (Ding et al., 1998; Matise et al., 1998; Park et al., 2000). Gli3 mutants in contrast have dorsal CNS defects including ectopic dorsal expression of Shh (Ruiz i Altaba, 1998). It is not clear whether Gli3 is compensating for the loss of Gli1 and 2 in double mutants or whether an alternative pathway can act downstream of SHH in ventral-lateral regions of the CNS, as well as other regions of the embryo. Nevertheless, the Gli1/2 mutant phenotypes indicate that different modes of SHH signaling are used in ventromedial verses ventrolateral cells. Production of additional Gli mutant alleles and knock-ins is needed to reveal the absolute requirement of each gene during development and to determine the degree to which each protein can replace the function of the others.

C. TGF-β Proteins Regulate Dorsal CNS Patterning

Neural crest cells migrate during neural tube closure from the most lateral (which becomes dorsal) regions of the neural plate posterior to the diencephalon (reviewed by Lee and Jessell, 1999). The roof plate then forms in the dorsal midline of the neural tube. Lineage studies in chick and mouse indicate that neural crest and roof plate cells have a common precursor that expresses later markers of both cell types (Echelard et al., 1994; Lee and Jessell, 1999). Dorsal ectoderm and BMPs have been shown to be sufficient to induce neural crest and roofplate cells in chick neural explants as well as upregulating Pax3 and 7 dorsally (Dickinson et al., 1995; Liem et al., 1995, 1997; Selleck and Bronner-Fraser, 1996; Selleck et al., 1998). FGFs and Wnts also have been implicated in generation of dorsal cells, but their roles appear to be indirect with the primary role of FGF being to caudalize neural plate tissue so that it can respond to BMPs, and WNT1 regulating expansion of dorsal precursors (reviewed in Lee and Jessell, 1999).

In the mouse, Bmp2, 4, and 7 are expressed at the right time and place to be the initial dorsalizing signals (Lee and Jessell, 1999). Later, the ectoderm is no longer in contact with the neural tube and a number of transforming growth factor β (TGF-β) family genes, including Bmp7 and Gdf7, are expressed in and around the roofplate. The roofplate has been ablated in mouse embryos using a knock-in approach to express the diphtheria toxin (DT) from the Gdf7 gene that is expressed in the roofplate (Lee *et al.,* 2000). In such embryos the D1A, D1B, and D2 dorsalmost interneurons do not develop, providing further evidence that the roofplate secretes factors necessary for dorsal interneuron induction. *Dreher* homozygous mutant embryos also appear to lack a roofplate but only have a reduction in the dorsalmost interneurons (Millonig *et al.,* 2000). The *Dreher* mutant was recently found to contain a mutation in the *Lmx1a* gene, which encodes a transcription factor. The milder phenotype in *Dreher* mutants compared to mice expressing DT in the roofplate might be due to an additional defect in *Dreher* dorsal mesoderm, such that the dorsal ectoderm stays in contact with the dorsal neural tube longer than normal providing an extra source of BMPS at late stages.

Loss of *Gdf7* in mice leads to loss of a specific set of dorsal interneurons in the spinal cord, the D1A neurons that express *Math1* (Lee *et al.,* 1998), whereas loss of *Bmp7* does not (Lee and Jessel, 1999). These results indicate that the TGF-β factors have distinct as well as redundant roles in generating the different interneurons in the dorsal horns of the spinal cord, rather than that they act in a concentration dependent manner like SHH in the ventral CNS. It appears that the basic mechanism of dorsalization of the CNS is conserved along the AP axis of the neural tube in more anterior regions. TGF-β family members are expressed along the length of the dorsal neural tube and can induce dorsal genes at all levels and loss of the roofplate leads to dorsal defects at all AP levels. Furthermore, BMPs have been implicated in generation of dorsal cerebellar granule cells, via induction of transcription factors such as MATH1and ZIC1, which are required for granule cell development (Alder *et al.,* 1999; Aruga *et al.,* 1998; Ben-Arie *et al.,* 1997).

V. Summary

Early patterning of the mouse neural tube at 7.5–9.5 dpc establishes broad domains along the anterior-posterior axis that later develop into particular functional components of the central nervous system: the forebrain, midbrain, hindbrain, and spinal cord. Patterning of the various regions seems to the regulated by different and independent mechanisms. For example, the midbrain and anterior hindbrain (rhombomere 1) is regulated by an organizer at the midbrain/hindbrain junction, whereas the remaining rhombomeres of the hindbrain are patterned by a segmentation process. Dorsal-ventral patterning begins around embryonic day 8.5 and is regulated by a similar mechanism along the AP axis with TGF molecules inducing dorsal cell types and Sonic Hedgehog inducing ventral cell types. An interaction between dorsalizing and ventralizing molecules and other factors at particular positions along the AP axis likely influences the specific neural cell types that develop in each region of the CNS.

References

Acampora, D., Mazan, S., Lallemand, Y., Avantaggiato, V., Maury, M., Simeone, A., and Brulet, P. (1995). Forebrain and midbrain regions are deleted in Otx2-/- mutants due to a defective anterior neuroectoderm specification during gastrulation. *Development (Cambridge, UK)* **121,** 3279–3290.

Acampora, D., Avantaggiato, V., Tuorto, F., and Simeone, A. (1997). Genetic control of brain morphogenesis through *Otx* gene dosage requirement. *Development (Cambridge, UK)* **124,** 3639–3650.

Acampora, D., Avantaggiato, V., Francesca, T., Briata, P., and Corte, G. (1998). Visceral endoderm-restricted translation of Otx1 mediates recovery of Otx2 requirements for specification of anterior nerual plate and normal gastrulation. *Development (Cambridge, UK)* **125,** 5091–5104.

Adams, K. A., Maida, J. M., Golden, J. A., and Riddle, R. D., (2000). The transcript factor LMx1b maintains Wnt1 expression within the isthmic organizer. *Development (Cambridge, UK)* **127,** 1857–1858.

Alder, J., Lee, K. J., Jessell, T. M., and Hatten, M. E., (1999). Generation of cerebellar granule neurons *in vivo* by transplantation of BMP-treated neural progenitor cells. *Nat. Neurosci.* **2,** 535–540.

Alvarado-Mallart, R. M., (1993). Fate and potentialities of the avian mesencephalic/metencephalic neuroepithelium. *J. Neurobiol.* **24,** 1341–1355.

Alvarez Otero, R., Sotelo, C., and Alvarado-Mallart, R. M., (1993). Chick/quail chimeras with partial cerebellar grafts: An analysis of the origin and migration of cerebellar cells. *J. Comp. Neurol.* **333,** 597–615.

Ang, S.-L., and Rossant, J. (1993). Anterior mesendoderm induces mouse *Engrailed* genes in explant cultures. *Development (Cambridge, UK)* **118,** 139–149.

Ang, S. L., Jin, O., Rhinn, M., Daigle, N., Stevenson, L., and Rossant, J. (1996). A targeted mouse Otx2 mutation leads to severe defects in gastrulation and formation of axial mesoderm and to deletion of rostral brain. *Development (Cambridge, UK)* **122,** 243–252.

Ang, S.-L., Conlon, R. A., Jin, O., and Rossant, J. (1994). Positive and negative signals from mesoderm regulate the expression of mouse Otx2 in ectoderm explants. *Development (Cambridge, UK)* **120,** 2979–2989.

Araki, I., and Nakamura, H. (1999). Engrailed defines the position of dorsal di-mesencephalic boundary by repressing diencephalic fate. *Development (Cambridge, UK)* **126,** 5127–5135.

Arkell, R., and Beddington, R. (1997). BMP-7 influences pattern and growth of the developing hindbrain of mouse embryos. *Development (Cambridge, UK)* **124,** 1–12.

Aruga, J., Minowa, O., Yaginuma, H., Kuno, J., Nagai, T., Noda, T., and Mikoshiba, K. (1998). Mouse Zic1 is involved in cerebellar development. *J. Neurosci.* **18,** 284–293.

Bally-Cuif, L., Cholley, B., and Wassef, M. (1995). Involvement of *Wnt-1* in the formation of the mes/metencephalic boundary. *Mech. Devel.* **53,** 23–34.

Bell, E., Wingate, R. J., and Lumsden, A. (1999). Homeotic transformation of rhombomere identity after localized Hoxb1 misexpression. *Science* **284,** 2168–2171.

Ben-Arie, N., Bellen, H. J., Armstrong, D. L., McCall, A. E., Gordadze, P. R., Guo, Q., Matzuk, M. M., and Zoghbi, H. Y., (1997). *Math1* is essential for genesis of cerebellar granule neurons. *Nature (London)* **390,** 169–172.

Birgbauer, E., and Fraser, S. (1994). Violation of cell lineage restriction compartments in the chick hindbrain. *Development (Cambridge, UK)* **120,** 1347–1356.

Brocard, J., Warot, X., Wendling, O., Messaddeq, N., Vonesch, J.-L., Chambon, P., and Metzger, D. (1997). Spatio-temporally controlled site-specific somatic mutagenesis in the mouse. *Proc. Natl. Acad. Sci. U.S.A.* **94,** 14559–14563K.

Broccoli, V., Boncinelli, E., and Wurst, W. (1999). The caudal limit of Otx2 expression positions the isthmic organizer. *Nature (London)* **401,** 164–168.

Chiang, C., Litingtung, Y., Lee, E., Young, K. E., Corden, J. L., Westphal, H., and Beachy, P. A., (1996). Cyclopia and defective axial patterning in mice lacking Sonic hedgehog gene function. *Nature (London)* **383**, 407–413.

Crossley, P., Martínez, S., and Martin, G. (1996a). Midbrain development induced by FGF8 in the chick embryo. *Nature (London)* **380**, 66–68.

Crossley, P., Minowada, G., MacArthur, C., and Martin, G. (1996b). Roles for FGF8 in the induction, initiation, and maintenance of chick limb development. *Cell (Cambridge, Mass.)* **84**, 127–136.

Dahmann, C., and Basler, K. (1999). Compartment boundaries: At the edge of development. *Trends Genet.* **15**, 320–326.

Dai, P., Akimaru, H., Tanaka, Y., Maekawa, T., Nakafuku, M., and Ishii, S. (1999). Sonic Hedgehog-induced activation of the Gli1 promoter is mediated by GLI3. *J. Biol. Chem.* **274**, 8143–8152.

Danielian, P. S., and McMahon, A. P., (1996). *Engrailed-1* as a target of the Wnt-1 signalling pathway in vertebrate midbrain development. *Nature (London)* **383**, 332–334.

Davis, C. A., and Joyner, A. L., (1988). Expression patterns of the homeo box-containing genes En-1 and En-2 and the proto-oncogene int-1 diverge during mouse development. *Genes Dev.* **2**, 1736–1744.

Dickinson, M. E., Krumlauf, R., and McMahon, A. P. (1994). Evidence for a mitogenic effect of Wnt-1 in the developing mammalian central nervous system. *Development (Cambridge, UK)* **120**, 1453–1471.

Dickinson, M. E., Selleck, M. A., McMahon, A. P., and Bronner-Fraser, M. (1995). Dorsalization of the neural tube by the nonneural ectoderm. *Development (Cambridge, UK)* **121**, 2099–2106.

Ding, Q., Motoyama, J., Gasca, S., Mo, R., Sasaki, H., Rossant, J., and Hui, C. C., (1998). Diminished Sonic hedgehog signaling and lack of floor plate differentiation in Gli2 mutant mice. *Development (Cambridge, UK)* **125**, 2533–2543.

Doniach, T. (1993). Planar and vertical induction of anteroposterior pattern during the development of the amphibian central nervous system. *J. Neurobiol.* **24**, 1256–1275.

Dupe, V., Davenne, M., Brocard, J., Dolle, P., Mark, M., Dierich, A., Chambon, P., and Rijli, F. (1997). In vivo functional analysis of the Hoxa-1 3' retinoic acid response element (3'RARE). *Development (Cambridge, UK)* **124**, 399–410.

Dymecki, S. M., (2000). Site specific recombination in cells and mice. *In* "Gene Targetting: A Practical Approach" (A. Joyner, ed.), pp. 37–96. Oxford University Press, New York.

Echelard, Y., Vassileva, G., and McMahon, A. P., (1994). *Cis*-acting regulatory sequences governing *Wnt-1* expression in the developing mouse CNS. *Development (Cambridge, UK)* **120**, 2213–2224.

Ensini, M., Tsuchida, T. N., Belting, H. G., and Jessell, T. M., (1998). The control of rostrocaudal pattern in the developing spinal cord: Specification of motor neuron subtype identity is initiated by signals from paraxial mesoderm. *Development (Cambridge, UK)* **125**, 969–982.

Eph Nomenclature Committee (1997). Unified nomenclature for Eph family receptors and their ligands, the ephrins. *Cell* **3**, 403–404.

Epstein, D. J., McMahon, A. P., and Joyner, A. L., (1999). Regionalization of Sonic hedgehog transcription along the anteroposterior axis of the mouse central nervous system is regulated by Hnf3-dependent and -independent mechanisms. *Development (Cambridge, UK)* **126**, 281–292.

Ericson, J., Muhr, J., Placzek, M., Lints, T., Jessell, T. M., and Edlund, T. (1995). Sonic hedgehog induces the differentiation of ventral forebrain neurons: A common signal for ventral patterning within the neural tube. **82**(1), 165. *Cell (Cambridge, Mass.)* **81**, 747–756; erratum: *Ibid.*

Fraser, S., Keynes, R., and Lumsden, A. (1990). Segmentation in the chick embryo hindbrain is defined by cell lineage restrictions. *Nature (London)* **344**, 431–435.

Funahashi, J., Okafuji, T., Ohuchi, H., Noji, S., Tanaka, H., and Nakamura, H. (1999). Role of Pax-5 in the regulation of a mid-hindbrain organizer's activity. *Dev. Growth Differ.* **41**, 59–72.

Gaiano, N., Kohtz, J., Turnball, D., and Fishell, G. (1999). A method for rapid gain-of-function studies in the mouse embryonic nervous system. *Nat. Neurosci.* **2**, 812–819.

Gavalas, A., Davenne, M., Lumsden, A., Chambon, P., and Rijli, F. M., (1997). Role of Hoxa-2 in axon pathfinding and rostral hindbrain patterning. *Development (Cambridge, UK)* **124**, 3693–3702.

Goodrich, L. V., Milenkovic, L., Higgins, K. M., and Scott, M. P., (1997). Altered neural cell fates and medulloblastoma in mouse *patched* mutants. *Science* **277**, 1109–1113.

Gould, A., Morrison, M., Sproat, G., White, R. A., and Krumlauf, R. (1997). Positive cross-regulation and enhancer sharing: Two mechnisms of specifying overlapping Hox expression patterns. *Genes Dev.* **11**, 900–913.

Gould, A., Itasaki, N., and Krumlauf, R. (1998). Initiation of rhombomeric Hoxb4 expression requires induction by somites and a retinoid pathway. *Neuron* **21**, 39–51.

Hallonet, M., and Le Douarin, N. (1993). Tracing neuroepithelial cells of the mesencephalic and metencephalic alar plates during cerebellar ontogeny in qual-chick chimaeras. *Eur. J. Neurosci.* **5**, 1145–1155.

Hanks, M. C., Wurst, W., Anson-Cartwright, L., Auerbach, A. B., and Joyner, A. L., (1995). Rescue of the *En-1* mutant phenotype by replacement of En-1 with En-2. *Science* **4**, 679–682.

Hanks, M. C., Loomis, C. A., Harris, E., Tong, C. X., Anson-Cartwright, L., Auerbach, A., and Joyner, A. L. (1998). Drosophila engrailed can substitute for mouse Engrailed1 function in mid-hindbrain, but not limb development. *Development (Cambridge, UK)* **125**, 4521–4530.

Hardcastle, Z., Mo, R., Hui, C. C., and Sharpe, P. T., (1998). The Shh signalling pathway in tooth development: Defects in Gli2 and Gli3 mutants. *Development (Cambridge, UK)* **125**, 2803–2811.

Hasty, P., Abuin, A., and Bradley, A. (2000). Gene targeting, principles, and practice in mammalian cells. *In* "Gene Targeting: A Practical Approach" (A. L. Joyner, ed.), pp. 1–34. Oxford University Press, New York.

Hidalgo-Sanchez, M., Simeone, A., and Alvarado-Mallart, R. (1999). Fg8 and Gbx2 induction concomitant with Otx2 repression is correlated with midbrain-hindbrain fate of caudal prosencephalon. *Development (Cambridge, UK)* **126**, 3191–3203.

Holder, N., and Klein, R. (1999). Eph receptors and ephrins: Effectors of morphogenesis. *Development (Cambridge, UK)* **126**, 2033–2044.

Hui, C. C., Slusarski, D., Platt, K., Homgren, R., and Joyner, A. (1994). Expression of three mouse homologs of the drosophila segment polarity gene *cubitus interrupus*, Gli, Gli-2, and Gli-3, in ectoderm and mesoderm-derived tissues suggests multiple roles during postimplantation development. *Dev. Biol.* **162**, 402–413.

Hynes, M., Stone, D. M., Dowd, M., Pitts-Meek, S., Goddard, A., Gurney, A., and Rosenthal, A. (1997). Control of cell pattern in the neural tube by the zinc finger transcription factor and oncogene *Gli-1. Neuron* **19**, 15–26.

Irving, C., and Mason, I. (1999). Regeneration of isthmic tissue is the result of a specific and direct interaction between rhombomere 1 and midbrain. *Development (Cambridge, UK)* **126**, 3981–3989.

Irving, C., and Mason, I. (2000). Signalling by FGF8 from the isthmus patterns anterior hindbrain and establishes the anterior limit of Hox gene expression. *Development (Cambridge, UK)* **127**, 177–186.

Jacobson, A., and Tam, P. (1982). Cephalic neurulation in the mouse embryo analyzed by SEM and morphometry. *Anatom. Record* **203**, 375–396.

Joyner, A. L., (1996). *Engrailed, Wnt* and *Pax* genes regulate midbrain-hindbrain development. *Trends Genet.* **12**, 15–20.

Joyner, A. L., Herrup, K., Auerbach, A., Davis, C. A., and Rossant, J. (1991). Subtle cerebellar phenotype in mice homozygous for a targeted deletion of the En-2 homeobox. *Science* **251**, 1239–1243.

Joyner, A. L., Liu, A., and Millet, S. (2000). *Otx2, Gbx2,* and *Fgf8* interact to position and maintain a mid-hindbrain organizer. *Curr. Opin. Cell Biol.* **12**, 736–741, Review.

Katahira, T., Sato, T., Sugiyama, S., Okafuji, T., Araki, I., Funahashi, J., and Nakamura, H. (2000). Interaction between Otx2 and Gbx2 defines the organizing center for the optic tectum. *Mech. Dev.* **91,** 43–52.

Kimmel, R. A., Turnbull, D. H., Blanquet, V., Wurst, W., Loomis, C. A., and Joyner, A. L., (2000). Two lineage boundaries coordinate vertebrate apical ectodermal ridge formation. *Genes Dev.* **14,** 1377–1389.

Kohtz, J. D., Baker, D. P., Corte, G., and Fishell, G. (1998). Regionalization within the mammalian telencephalon is mediated by changes in responsiveness to Sonic Hedgehog. *Development (Cambridge, UK)* **125,** 5079–5089.

Lee, J., Platt, K. A., Censullo, P., and Ruiz i Altaba, A. (1997). Gli1 is a target of Sonic hedgehog that induces ventral neural tube development. *Development (Cambridge, UK)* **124,** 2537–2552.

Lee, K. J., and Jessell, T. M., (1999). The specification of dorsal cell fates in the vertebrate central nervous system. *Annu. Rev. Neurosci.* **22,** 261–294.

Lee, K. J., Mendelsohn, M., and Jessell, T. M., (1998). Neuronal patterning by BMPs: A requirement for GDF7 in the generation of a discrete class of commissural interneurons in the mouse spinal cord. *Genes Dev.* **12,** 3394–3407.

Lee, K. J., Dietrich, P., and Jessell, T. M., (2000). Genetic ablation reveals that the roof plate is essential for dorsal interneruon specification. *Nature (London)* **403,** 734–740.

Lee, S. M., K., Danielian, P. S., Fritzsch, B., and McMahon, A. P., (1997). Evidence that FGF8 signalling from the midbrain-hindbrain junction regulates growth and polarity in the developing midbrain. *Development (Cambridge, UK)* **124,** 959–969.

Liem, K. F., Jr., Tremml, G., Roelink, H., and Jessell, T. M., (1995). Dorsal differentiation of neural plate cells induced by BMP-mediated signals from epidermal ectoderm. *Cell (Cambridge, Mass.)* **82,** 969–979.

Liem, K. F., Jr., Tremml, G., and Jessell, T. M., (1997). A role for the roof plate and its resident TGFbeta-related proteins in neuronal patterning in the dorsal spinal cord. *Cell (Cambridge, Mass.)* **91,** 127–138.

Liu, A., and Joyner, A. L., (2001a). EN and GBX2 play essential roles downstream of FGF8 in patterning the mouse mid/hindbrain region. *Development (Cambridge, UK)* **28,** 181–191.

Liu, A., and Joyner, A. L., (2001b). Early anterior/posterio patterning of the midbrain and cerebellum. *Annu. Rev. Neurosci.* **24,** 869–896.

Liu, A., Joyner, A. L., and Turnbull, D. H., (1998). Alteration of limb and brain patterning in early mouse embryos by ultrasound-guided injection of Shh-expressing cells. *Mech. Dev.* **75,** 107–115.

Liu, A., Losos, K., and Joyner, A. L., (1999). FGF8 can activate Gbx2 and transform regions of the rostral mouse brain into a hindbrain fate. *Development (Cambridge, UK)* **126,** 4827–4838.

MacArthur, C. T. P., MacArthur, C. A., Lawshe, A., Xu, J., Santos-Ocampo, S., Heikinheimo, M., Chellaiah, A. T., and Orritz, D. M. (1995). FGF-8 isoforms activate receptor splice forms that are expressed in mesenchymal regions of mouse development. *Development (Cambridge, UK)* **121,** 3603–3613.

Maconochie, M., Nonchev, S., Morrison, A., and Krumlauf, R. (1996). Paralogous Hox genes: Function and regulation. *Annu. Rev. Genet.* **30,** 529–556.

Maconochie, M., Nonchev, S., Studer, M., Chan, S., Popperl, H., Sham, M., Richard, S., and Krumlauf, R. (1997). Cross-regulation of the segment-restricted patterns of HoxB complex: The expression of Hoxb2 in rhombomere 4 is regulated by Hoxb1. *Genes Dev.* **11,** 1885–1895.

Manzanares, M., Cordes, S., Kwan, C. T., Sham, M. H., Barsh, G. S., and Krumlauf, R. (1997). Segmental regulation of Hoxb-3 by kreisler. *Nature (London)* **387,** 191–195.

Manzanares, M., Cordes, S., Ariza-McNaughton, L., Sadl, V., Maruthainar, K., Barsh, G., and Krumlauf, R. (1999a). Conserved and distinct roles of kreisler in regulation of the paralogous Hoxa3 and Hoxb3 genes. *Development (Cambridge, UK)* **126,** 759–769.

Manzanares, M., Trainor, P. A., Nonchev, S., Ariza-McNaughton, L., Brodie, J., Gould, A., Marshall, H., Morrison, A., Kwan, C. T., Sham, M. H.,

Wilkinson, D. G., and Krumlauf, R. (1999b). The role of kreisler in segmentation during hindbrain development. *Dev. Biol.* **211,** 220–237.

Marshall, H., Studer, M., Pöpperl, H., Aparicio, S., Kuroiwa, A., Brenner, S., and Krumlauf, R. (1994). A conserved retinoic acid response element required for early expression of the homeobox gene *Hoxb-1*. *Nature (London)* **370,** 567–571.

Martínez, S., Crossley, P. H., Cobos, I., Rubenstein, J. L., and Martin, G. R., (1999). FGF8 induces formation of an ectopic isthmic organizer and isthmocerebellar development via a repressive effect on Otx2 expression. *Development (Cambridge, UK)* **126,** 1189–1200.

Matise, M. P., and Joyner, A. L., (1999). Gli genes in development and cancer. *Oncogene* **18,** 7852–7859.

Matise, M. P., Epstein, D. J., Park, H. L., Platt, K. A., and Joyner, A. L., (1998). Gli2 is required for induction of floor plate and adjacent cells, but not most ventral neurons in the mouse central nervous system. *Development (Cambridge, UK)* **125,** 2759–2770.

Matsuo, I., Kuratani, S., Kimura, C., and Takeda, N. (1995). Mouse Otx2 functions in the formation and patterning of rostral head. *Genes Dev.* **9,** 2646–2658.

Mellitzer, G., Xu, Q., and Wilkinson, D. G., (1999). Eph receptors and ephrins restrict cell intermingling and communication. *Nature (London)* **400,** 77–81.

Meyers, E. N., Lewandoski, M., and Martin, G. R., (1998). An *Fgf8* mutant allelic series generated by Cre- and Flp-mediated recombination. *Nat. Genet.* **18,** 136–141.

Millen, K. J., Wurst, W., Herrup, K., and Joyner, A. L., (1994). Abnormal embryonic cerebellar development and patterning of postnatal foliation in two mouse *Engrailed-2* mutants. *Development (Cambridge, UK)* **120,** 695–706.

Millen, K. J., Hui, C. C., and Joyner, A. L., (1995). A role for En-2 and other murine homologs of Drosophila segment polarity genes in regulating positional information in the developing cerebellum. *Development (Cambridge, UK)* **121,** 3935–3945.

Millet, S., Bloch-Gallego, E., Simeone, A., and Alvarado-Mallart, R.-M. (1996). The caudal limit of Otx2 gene expression as a marker of the midbrain/hindbrain boundary: A study using in situ hybridisation and chick/quail homotopic grafts. *Development (Cambridge, UK)* **122,** 3785–3797.

Millet, S., Campbell, K., Epstein, D. J., Losos, K., Harris, E., and Joyner, A. L., (1999). A role for Gbx2 in repression of Otx2 and positioning the mid/hindbrain organizer. *Nature (London)* **401,** 161–164.

Millonig, J. H., Millen, K. J., and Hatten, M. E. (2000). The mouse Dreher gene Lmx1a contrlos formation of the roof plate in the vertbrate CNS. *Nature (London)* **403,** 764–769.

Mo, R., Freer, A. M., Zinyk, D. L., Crackower, M. A., Michaud, J., Heng, H. H., Chik, K. W., Shi, X. M., Tsui, L. C., Cheng, S. H., Joyner, A. L., and Hui, C. (1997). Specific and redundant functions of Gli2 and Gli3 zinc finger genes in skeletal patterning and development. *Development (Cambridge, UK)* **124,** 113–123.

Morrison, A., Ariza-McNaughton, L., Gould, A., Featherstone, M., and Krumlauf, R. (1997). *HOXD4* and regulation of the group 4 paralog genes. *Development (Cambridge, UK)* **124,** 3135–3146.

Motoyama, J., Liu, J., Mo, R., Ding, Q., Post, M., and Hui, C. C., (1998). Essential function of Gli2 and Gli3 in the formation of lung, trachea and oesophagus. *Nat. Genet.* **20,** 54–57.

Muhr, J., Jessell, T. M., and Edlund, T. (1997). Assignment of early caudal identity to neural plate cells by a signal from caudal paraxial mesoderm. *Neuron* **19,** 487–502.

Muhr, J., Graziano, E., Wilson, S., Jessell, T. M., and Edlund, T. (1999). Convergent inductive signals specify midbrain, hindbrain, and spinal cord identity in gastrula stage chick embryos. *Neuron* **23,** 689–702.

Nieuwkoop, P. (1952). Activation and organization of the central nervous system in amphibians. *J. Exp. Zool.* **120,** 109.

Nonchev, S., Vesque, C., Maconochie, M., Seitanidou, T., Ariza-McNaughton, L., Frain, M., Marshall, H., Sham, M. H., Krumlauf, R., and

Charnay, P. (1996). Segmental expression of Hoxa-2 in the hindbrain is directly regulated by Krox-20. *Development (Cambridge, UK)* **122**, 543–554.

Okafuji, T., Funahashi, J., and Nakamura, H. (1999). Roles of Pax-2 in initiation of the chick tectal development. *Brain Res. Dev. Brain Res.* **116**, 41–49.

O'Leary, D. D., and Wilkinson, D. G., (1999). Eph receptors and ephrins in neural development. *Curr. Opin. Neurobiol.* **9**, 65–73.

Olsson, M., Campbell, K., and Turnbull, D. H., (1997). Specification of mouse telencephalic and mid-hindbrain progenitors following heterotopic ultrasound-guided embryonic transplantation. *Neuron* **19**, 761–772.

Park, H. L., Bai, C., Platt, K. A., Matise, M. P., Beeghly, A., Hui, C., Nakashima, M., and Joyner, A. L., (2000). Mouse gli1 mutants are viable but have defects in SHH signaling in combination with a gli2 mutation. *Development (Cambridge, UK)* **127**, 1593–1605.

Pierani, A., Brenner-Morton, S., Chiang, C., and Jessell, T. M., (1999). A sonic hedgehog-independent, retinoid-activated pathway of neurogenesis in the ventral spinal cord. *Cell (Cambridge, Mass.)* **97**, 903–915.

Platt, K. A., Michaud, J., and Joyner, A. L., (1997). Expression of the mouse Gli and Ptc genes is adjacent to embryonic sources of hedgehog signals suggesting a conservation of pathways between flies and mice. *Mech. Dev.* **62**, 121–135.

Prince, V. E., Moens, C. B., Kimmel, C. B., and Ho, R. K., (1998). Zebrafish hox genes: Expression in the hindbrain region of wild-type and mutants of the segmentation gene, valentino. *Development (Cambridge, UK)* **125**, 393–406.

Reifers, F., Bohli, H., Walsh, E. C., Crossley, P. H., Stainier, D. Y., and Brand, M. (1998). Fgf8 is mutated in zebrafish acerebellar (ace) mutants and is required for maintenance of midbrain-hindbrain boundary development and somitogenesis. *Development (Cambridge, UK)* **125**, 2381–2395.

Retaux, S., and Harris, W. (1996). Engrailed and retinotectal topography. *Trends Neurosci.* **19**, 542–546.

Rhinn, M., Dierich, A., Shawlot, W., Behringer, R. R., LeMeur, M., and Ang, S.-L. (1998). Sequential roles for Otx2 in visceral endoderm and neuroectoderm for forebrain and midbrain induction and specification. *Development (Cambridge, UK)* **125**, 845–856.

Ristoratore, F., Carl, M., Deschet, K., Richard-Parpaillon, L., Boujard, D., Wittbrodt, J., Chourrout, D., Bourrat, F., and Joly, J. S., (1999). The midbrain-hindbrain boundary genetic cascade is activated ectopically in the diencephalon in response to the widespread expression of one of its components, the medaka gene Ol-eng2. *Development (Cambridge, UK)* **126**, 3769–3779.

Roelink, H., Porter, J., Chiang, C., Tanabe, Y., Chang, D., Beachy, P., and Jessell, T. (1995). Floor plate and motor neuron induction by different concentrations of the amino-terminal cleavage product of Sonic Hedgehog autoproteolysis. *Cell (Cambridge, Mass.)* **81**, 445–455.

Rossant, J., and Spence, A. (1998). Chimeras and mosaics in mouse mutant analysis. *Trends Genet.* **14**, 358–363.

Rossel, M., and Capecchi, M. R., (1999). Mice mutant for both Hoxa1 and Hoxb1 show extensive remodeling of the hindbrain and defects in craniofacial development. *Development (Cambridge, UK)* **126**, 5027–5040.

Rowitch, D., and McMahon, A. (1995). Pax-2 expression in the murine neural plate precedes and encompasses the expression domains of Wnt-1 and En-1. *Mech. Dev.* **52**, 3–8.

Ruiz i Altaba, A. (1998). Combinatorial Gli gene function in floor plate and neuronal induction by Sonic hedgehog. *Development (Cambridge, UK)* **125**, 2203–2212.

Ruiz i Altaba, A. (1999). Gli proteins encode context-dependent positive and negative functions: Implications for development and disease. *Development (Cambridge, UK)* **126**, 3205–3216.

Sadler, T. W., and New, D. (1981). Culture of mouse embryos during neurulation. *J. Embryol. Exp. Morphol.* **66**, 109–116.

Sasaki, H., Hui, C.-C., Nakafuku, M., and Kondoh, H. (1997). A binding site for Gli proteins is essential for *HNF-3b* floor plate enhancer activity in transgenics and can respond to Shh in vitro. *Development (Cambridge, UK)* **124**, 1313–1322.

Sasaki, H., Nishizaki, Y., Hui, C., Nakafuku, M., and Kondoh, H. (1999). Regulation of Gli2 and Gli3 activities by an amino-terminal repression domain: Implication of Gli2 and Gli3 as primary mediators of Shh signaling. *Development (Cambridge, UK)* **126**, 3915–3924.

Schwarz, M., Alvarez-Bolado, G., Urbanek, P., Busslinger, M., and Gruss, P. (1997). Conserved biological function between *Pax-2* and *Pax-5* in midbrain and cerebellum development: Evidence from targeted mutations. *Proc. Natl. Acad. Sci. U.S.A.* **94**, 14518–14523.

Seitanidou, T., Schneider-Maunoury, S., Desmarquet, C., Wilkinson, D. G., and Charnay, P. (1997). Krox-20 is a key regulator of rhombomere-specific gene expressio in the developing hindbrain. *Mech. Dev.* **65**, 31–42.

Selleck, M. A., and Bronner-Fraser, M. (1996). The genesis of avian neural crest cells: A classic embryonic induction. *Proc. Natl. Acad. Sci. U.S.A.* **93**, 9352–9357.

Selleck, M. A., Garcia-Castro, M. I., Artinger, K. B., and Bronner-Fraser, M. (1998). Effects of Shh and Noggin on neural crest formation demonstrate that BMP is required in the neural tube but not ectoderm. *Development (Cambridge, UK)* **125**, 4919–4930.

Shamim, H., Mahmood, R., Logan, C., Doherty, P., Lumsden, A., and Mason, I. (1999). Sequential roles for Fgf4, En1 and Fgf8 in specification and regionalisation of the midbrain. *Development (Cambridge, UK)* **126**, 945–959.

Shimamura, K., and Rubenstein, J. L., (1997). Inductive interactions direct early regionalization of the mouse forebrain. *Development (Cambridge, UK)* **124**, 2709–2718.

Simeone, A. (1998). Otx1 and Otx2 in the development and evolution of the mammalian brain. *EMBO J.* **17** (23), 6790–6798.

Song, D. L., Chalepakis, G., Gruss, P., and Joyner, A. L., (1996). Two Pax-binding sites are required for early embryonic brain expression of an Engrailed-2 transgene. *Development (Cambridge, UK)* **122**, 627–635.

Studer, M., Gavalas, A., Marshall, H., Ariza-McNaughton, L., Rijli, F. M., Chambon, P., and Krumlauf, R. (1998). Genetic interactions between Hoxa1 and Hoxb1 reveal new roles in regulation of early hindbrain patterning. *Development (Cambridge, UK)* **125**, 1025–1036.

Suda, Y., Matsuo, I., and Aizawa, S. (1997). Cooperation between Otx1 and Otx2 genes in developmental patterning of rostral brain. *Mech. Dev.* **69**, 125–141.

Tanabe, Y., and Jessell, T. M., (1996). Diversity and pattern in the developing spinal cord. *Science* **274**, 1115–1123; erratum: *Ibid.* **276**, 21 (1997).

Theil, T., Frain, M., Gilardi-Hebenstreit, P., Flenniken, A., Charnay, P., and Wilkinson, D. G., (1998). Segmental expression of the EphA4 (Sek-1) receptor tyrosine kinase in the hindbrain is under direct transcriptional control of Krox-20. *Development (Cambridge, UK)* **125**, 443–452.

Thomas, K. R., Musci, T. S., Neumann, P. E., and Capecchi, M. R., (1991). Swaying is a mutant allele of the proto-oncogene Wnt-1. *Cell (Cambridge, Mass.)* **67**, 969–976.

Toresson, H., Mata de Urquiza, A., Fagerström, C., Perlmann, T., and Campbell, K. (1999). Retinoids are produced by glia in the lateral ganglionic eminence and regulate striatal neuron differentiation. *Development (Cambridge, UK)* **126**, 1317–1326.

Turnbull, D. H. (1999). In utero ultrasound backscatter microscopy of early stage mouse embryos. *Comput. Med. Imaging Graphics* **23**, 25–31.

Turnbull, D. H., (2000). Ultrasound backscatter microscopy of mouse embryos. *Methods Mol. Biol.* **135**, 235–243.

Turnbull, D. H., Bloomfield, T. S., Baldwin, H. S., Foster, F. S., and Joyner, A. L. (1995). Ultrasound backscatter microscope analysis of early mouse embryonic brain development. *Proc. Natl. Acad. Sci. U.S.A.* **92**, 2239–2243.

Wang, B., Fallon, J. F., and Beachy, P. A., (2000). Hedgehog-regulated processing of Gli3 produces an anterior/posterior repressor gradient in the developing vertebrate limb. *Cell (Cambridge, Mass.)* **100,** 423–434.

Wassarman, K. M., Lewandoski, M., Campbell, K., Joyner, A. L., Rubenstein, J. L., Martínez, S., and Martin, G. R., (1997). Specification of the anterior hindbrain and establishment of a normal mid/hindbrain organizer is dependent on Gbx2 gene function. *Development (Cambridge, UK)* **124,** 2923–2934.

Wassef, M., and Joyner, A. L., (1997). Early mesencephalon/metencephalon patterning and development of the cerebellum. *Perspect. Dev. Neurobiol.* **5,** 3–16.

Wurst, W., Auerbach, A., and Joyner, A. L., (1994). Multiple developmental defects in Engrailed-1 mutant mice: An early mid-hindbrain deletion and patterning defects in forelimbs and sternum. *Development (Cambridge, UK)* **120,** 2065–2075.

Xu, J., Liu, Z., and Ornitz, D. (2000). Temporal and spatial gradients of Fgf8 and Fgf17 regulate proliferation and differentiation of midline cerebellar structures. *Development (Cambridge, UK)* **127,** 1833–1843.

Xu, Q., Mellitzer, G., Robinson, V., and Wilkinson, D. G., (1999). In vivo cell sorting in complementary segmental domains mediated by Eph receptors and ephrins. *Nature (London)* **399,** 267–271.

Ye, W., Shimamura, K., Rubenstein, J. L., R., Hynes, M. A., and Rosenthal, A. (1998). FGF and Shh signals control dopaminergic and serotonergic cell fate in the anterior neural plate. *Cell (Cambridge, Mass.)* **93,** 755–766.

Zinyk, D. L., Mercer, E. H., Harris, E., Anderson, D. J., and Joyner, A. L., (1998). Fate mapping of the mouse midbrain-hindbrain constriction using a site-specific recombination system. *Curr. Biol.* **8,** 665–668.

7

Somitogenesis: Segmentation of the Paraxial Mesoderm and the Delineation of Tissue Compartments

Achim Gossler* and Patrick P. L. Tam†

*Institut für Molekularbiologie, Medizinische Hochschule Hannover, D–30625 Hannover, Germany
†Embryology Unit, Children's Medical Research Institute, University of Sydney, Westmead, New South Wales, Australia

I. Overview of Somite Development

II. Allocation of Progenitor Cells to the Paraxial Mesoderm

III. Cells Are in Transit in the Presomitic Mesoderm

IV. Regionalized Genetic Activity Points to a Prepattern of Prospective Somites

V. Emergence of Anterior-Posterior Somite Compartments

VI. Role of Notch Signaling in the Establishment of Somite Borders and Anterior-Posterior Polarity

VII. A Molecular Clock Operates in the Paraxial Mesoderm to Control the Kinetics of Somite Formation

VIII. Specification of Lineage Compartments by Inductive Interactions

IX. Summary and Open Questions

References

I. Overview of Somite Development

Segmentation is a fundamental developmental process that subdivides the body, or parts thereof, into a series of serially repeated subunits and thereby generates a segmental (meristic) pattern. In the mouse embryo, like other vertebrate embryos, the earliest manifestation of tissue segmentation is the formation of somites during organogenesis. Somites are blocks of mesodermal cells located on either side of the neural tube and the notochord. Together with the mesenchyme that envelops the cephalic neural tube, they constitute the paraxial mesoderm of the embryo. The segmental arrangement of the somites along the anterior-posterior body axis prefigures and underlies the metamerism of the somite-derived vertebral column and epaxial muscles and also determines the segmented arrangement of parts of the peripheral nervous system.

In the mouse embryo, somite formation begins on embryonic day 8 (E8) at the completion of germ layer formation and the commencement of neurulation (see Kaufman and Bard, 1999, for a description of the anatomy of somite formation). Altogether, 63–65 pairs of somites are formed by successive segmentation of the paraxial mesoderm in a craniocaudal progression. About 30 of these are allocated to the body, with the remainder in the tail (Table I). Somites are formed by the epithelialization of groups of mesenchymal cells in the paraxial mesoderm and their separation from the anterior end of the unsegmented presomitic mesoderm (Fig. 1). This is accomplished by changes in cellular behavior and tissue architecture such that clusters of cells in the paraxial mesoderm become separated in a craniocaudal succession along the body axis. Concomitantly, borders between adjacent groups of cells (the space constitutes the intersomitic fissures) are generated in a spatially and temporally precise manner. In the newly segmented somite, the apical cell surface of the epithelium faces the somitocoel (the cavity enclosed by the epithelium), while the basal surface of the epithelium that is covered on all but the ventromedial side by a basement membrane forms the external aspect of the somites. The most rostral somites are formed next to the prospective caudal hindbrain. Successively, one new pair of somites is generated every 1.5–2 hr in the trunk and about 2–3 hr in the tail—until the full complement of somites has formed on about E13.25 to E13.5 (Box 1). During normal development of the mouse embryo, somites at different levels of the body contain different numbers of cells and are of different sizes at the time of segmentation. Both the rate of somite segmentation and the allocation of cells to each somite can be correlated with the rate of tissue recruitment to the paraxial mesoderm, suggesting that somite formation is subject to the same mechanism that globally controls embryonic growth (Flint et al., 1978; Snow and Gregg, 1986; Tam, 1981).

Somite formation and differentiation are continuous processes. While new somites are still being generated posteri-

orly, more anterior (therefore older and more mature) somites begin to differentiate. Therefore, somites located at different positions along the anterior-posterior body axis are at different stages of maturation and differentiation, and within an embryo, every intermediate stage of somite differentiation is represented. The sequence of differentiation events is repeated in every somite of the craniocaudal series (Box 1, "Staging of Somite Development"). Shortly after their formation, epithelial somites differentiate by undergoing localized mesenchymal transformation to form the sclerotome and the myotome (Box 1, "Somite Differentiation").

The initial differentiation of the somite entails the dispersion of the epithelial cells in the ventromedial portion of the somite adjacent to the notochord and the ventral part of the neural tube. This population of cells, which acquires a mesenchymal morphology, together with cells in the somitocoel, forms the sclerotome. The sclerotome of the cranialmost four or five somites (see footnote[a] in Table I) gives rise to the occipital bone, while the sclerotome of the other somites forms the bony and cartilaginous components of the vertebral column (Christ and Ordahl, 1995; Couly et al., 1992; Theiler, 1988). Sclerotome cells that stay lateral to the neural tube form the neural arches, whereas those that migrate ventromedially and surround the notochord give rise to the vertebral bodies and intervertebral disks. In the thoracic region, the ventrolateral sclerotome gives rise to the ribs. Although the basic cellular differentiation pattern of somites at different axial positions is very similar, unique vertebral structures form along the craniocaudal axis, indicating that somites acquire specific identities according to their axial position. Regional specification of somite identity in the

Table I Chronology of Somite Formation in the Mouse

Theiler stage	Age	Total somite number	Occipital	Cervical	Thoracic	Lumbar	Sacral	Caudal
TS12	E8	4	4					
TS13	E8.5	10	4	6				
TS14	E9	17	4[a]	7	6			
TS15	E9.5	25		7	13	1		
TS16	E10	32		7	13	6	2	
TS17	E10.5	38		7	13	6	4	4
TS18	E11	40		Con	13	6	4	8
TS19	E11.5	46		Chn	Con	6	4	12
TS20	E12	50		Car	Chn	Con	Con	16
TS22-23	E13.5	60–65		Crt	Crt	Crt	Crt	26–31 Con/Chn
Average number of cells in recently segmented somite			160	200	520	900	1200	1300– 280[b]

Source: Adapted from Kaufman and Bard, 1999; Tam, 1981.

[a] Based on Kaufman and Bard (1999); five occipital somites have also been reported for the mouse (Theiler, 1988).

[b] Caudal somites are formed with progressively fewer cells as the tail develops. Stages of prevertebrae development: Con, prechondrogenic mass; Chn, chondrocytes of various stages of maturation; Crt, deposition of cartilaginous matrix.

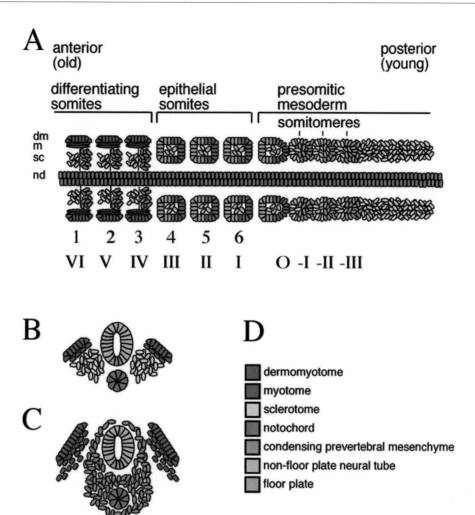

Figure 1 Schematic representation of somitogenesis. (A) During somitogenesis the paraxial mesoderm becomes subdivided into metameric subunits, the somites. Somitogenesis is a continuous process, during which new somites are generated in a strict craniocaudal order while older somites undergo differentiation. Thus, various stages of somitogenesis are simultaneously present in developing embryos, anterior (older) somites being more advanced than posterior (younger) ones. Somite development comprises the (1) formation of somitomeres in the presomitic mesoderm, (2) separation of somites from the presomitic mesoderm, (3) compaction of somites, (4) their epithelialization (formation of the epithelial somite) and compartmentalization, and (5) differentiation into sclerotome, dermatome, and myotome. The numbering system described in Box1 is illustrated here in a six-somite-stage embryo. Arabic numerals (1–6) indicate the numerical order of the somites in the craniocaudal series and Roman numerals (I to VI) refer to the stage of differentiation of the somite according to the staging system of Christ and Ordahl while the somitomeres (or prospective somites) in the presomitic mesoderm are numbered as 0, −I, −II, −III and so on, beginning with the most cranial one which is about to segment (see Box 1). Abbreviations: dm, dermomyotome; m, myotome; sc, sclerotome; nd, notochord. The tissue types in the differentiating somites shown in (B) and (C) are color coded as in the accompanying legend (D).

body axis is likely to be mediated by the activity of the *Hox* gene clusters (e.g., Kessel and Gruss, 1991; reviewed by Burke, 2000) and transforming growth factor β (TGF-β) signaling mechanisms (McPherron *et al.,* 1999; reviewed by Gad and Tam, 1999).

The sclerotome is subdivided into cranial (anterior) and caudal (posterior) parts that differ in gene expression and cellular properties, resulting in functionally distinct tissue compartments. This subdivision was first described by von Ebner (1888), who observed subtle clefts (von Ebner's fissures) separating the cranial and caudal sclerotome in the somites of snake embryos. The sclerotomal subdivision has a major impact on the patterning of the peripheral nervous system by restricting the migration of neural crest cells and axonal outgrowths from the spinal cord to the cranial somite halves. The two sclerotomal compartments also display dif-

Box 1

Staging of Somite Development

To facilitate the comparison of somites at different axial positions in embryos of different developmental ages, a system for staging somites was devised by Christ and Ordahl (1995) for avian embryos and can also be applied to the mouse somites. This staging system takes into account that somites are continuously and progressively differentiating along the craniocaudal axis (the more caudal somites being less mature than the more cranial ones) and the differentiation state of a specific somite is directly correlated with the position in the craniocaudal series relative to the most recently formed somites. Somites are counted in Arabic numerals beginning at the anteriormost somite. In addition, somites are numbered according to their differentiation state in Roman numerals, the most recently formed somite being number I (Fig. 1). For example, the most recently formed somites in a 15-somite- and 30-somite-stage embryo would be somite I/15 and I/30, respectively. Under this staging convention, both the I/15 and I/30 somites are at comparable stage of differentiation. Likewise, the 10th somite in a 15-somite embryo (somite VI/15) is developmentally similar to the 25th somite of the 30-somite embryo (somite VI/30).

In the mouse, six to seven somitomeres, swirls of mesenchymal cells, have been defined in the unsegmented presomitic mesoderm by scanning electron microscopy. The somite staging system can also be extended to the somitomeres. The most cranial and segmenting somitomere (or prospective somites) is referred to as somite SO, followed by S−I, S−II and so on. (Fig. 1) No reference to the somite number of the embryo is required, since the sequence of cellular differentiation and morphogenesis of the somitomeres in the presomitic mesoderm are similar in embryos at different developmental stages (Tam, 1986; Tam and Beddington, 1986).

Somite Number

Variation in the somite number is common in embryos within and between litters of a particular nominal age (E, embryonic day). By regression analysis, the variation of somite numbers (3–8 somites) among embryos in the same litter may represent a difference of 5.6–16.8 hr of development (Tam, 1981; Vickers, 1983). The somite number also overlaps among embryos from different litters that are chonologically apart by 6–10 hr, for example 30-somite embryo is found in both E9.5 and E10.0 litters.

As the embryo develops, differentiation may render the somitic subdivision unrecognizable, and thus scoring the precise number of mature somites becomes impossible. Several morphological landmarks can be used to pinpoint specific somites for scoring purposes. For example, in embryos at E9.5–11, the 1st somite is localized posterior to the otic vesicle. The 8th to 14th somites are next to the forelimb bud, the 25th to 30th somites are next to the hindlimb bud, and the 30th and 31st somites are at the level of the base of the genital tubercle. Using these reference points, the number of somites caudal to these landmarks can then be scored. However, as development progresses, a shift is seen in the position of the limb bud relative to the somites due to differential growth of the paraxial mesoderm and the limbs. This shift might result in some imprecision in somite scoring. Other landmarks, such as the morphology of the skeletal derivatives of the first two cervical somites, and the position of the otic vesicle and the first two cervical spinal ganglia, have been used to precisely identify the cervical and upper thoracic segments in TS15 E11 to TS19 E11.5 embryos (Table I) (Spörle and Schughart, 1997). Using the position of the spinal ganglia as a reference point, this system can be used to score the somites in other parts of the body axis. As a result of the realignment of the dermomyotome and sclerotome (resegmentation: Christ *et al.,* 1998), the intersomitic fissure that separates consecutive sclerotomes is obscured and a new intersegmental boundary is formed. The prevertebrae therefore correspond in number but not in space with the somites (Christ and Ordahl, 1995).

Rate of Segmentation

The first pair of somites is formed at about E7.75. Initially, somites are formed at a fast rate of about one pair in every hour for the first 6–10 somites, every 1.5–2 hr for the trunk somites, and every 2–3 hr for the tail somites (Tam, 1981).

(continues)

ferent developmental fates. Initially, cells from the ventral portion of the sclerotome of each somite first form a morphologically homogeneous mesenchyme around the notochord, but cells from the cranial and caudal compartments remain separate populations. Subsequently, cells from the caudal half of one somite and cells from the cranial half of the adjacent caudal somite form one vertebral body. Cells just anterior to the caudal half of a somite, that is, those in the middle of a somite, give rise to the intervertebral disks that separate the vertebral bodies. Cells in the caudal portion of the sclerotome form the pedicles and the bulk of the lamina of the neural arches (Goldstein and Kalcheim, 1992; Huang *et al.,* 1994, 1996, 2000) and the ribs of the thoracic vertebrae. However, as the vertebral column develops, these caudal-half-segment-derived structures become attached to the cranial aspect of the vertebral body (Christ and Ordahl, 1995; Theiler, 1988; Verbout, 1976). This change in topological relationship, termed *resegmentation* ("Neugliederung"; Remak, 1855) is believed to be caused by the fusion of the caudal sclerotome of one somite with the cranial sclerotome of the next somite in the series (reviewed by Christ *et al.,* 1998). Thus, the boundary between anterior and posterior sclerotomal compartments (intrasomitic border) becomes the intervertebral boundary in the fully developed vertebral column (Fig. 2).

As the sclerotome is formed, the dorsal part of the somite, which is juxtaposed to the surface ectoderm, retains the epithelial organization and becomes the precursor tissue of the dermis and the myotome, known as the dermomyotome. The myotome is formed in several successive waves, which was first demonstrated in the avian embryo and recently also in the mouse. First, the ventromedial margin of epithelial dermomyotome abutting the neural tube bends ventrally and cells emerge from this epithelial lip to form the early myotome that consists of postmitotic pioneer myotomal myocytes that express a *MLC₃F-lacZ* transgene (MLC = myosin light chain; Kahane *et al.,* 1998; Kahane and Kalcheim, 1998; Venters *et al.,* 1999). Subsequently, the rostral, caudal, and ventrolateral lips of the dermomyotome contribute more mitotically quiescent *Myf5* and sMHC (skeletal muscle myosin heavy chain)-expressing myocytes, which intercalate

with pioneer myocytes. They are then followed by a third wave of myoblasts that is mitotically active (Cinnamon *et al.,* 1999; Venters *et al.,* 1999). In the mouse, early myogenic bHLH genes are expressed in the sites that are equivalent to those that form the myotome in avian embryos (Cossu

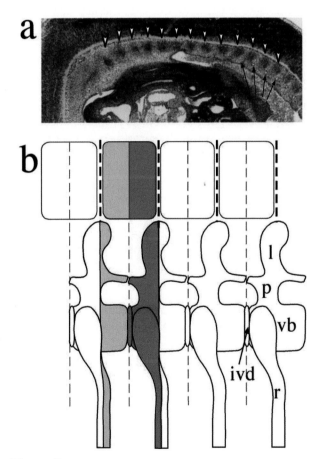

Figure 2 Relation between somitic and vertebral borders. (a) A sagittal section of an E10.5 embryo showing the alternating zones of loosely organized cranial (arrows) and densely packed caudal sclerotome (arrowheads). (b) Schematic drawing showing the vertebral structures derived from the caudal sclerotome (dark red) and cranial sclerotome (pink) and the disparity between intersomitic and intervertebral borders. Abbreviations: l, lamina; p, pedicle; vb, vertebral body; ivd, intervertebral disk; r, rib.

et al., 1996b; Ott *et al.,* 1991; Smith *et al.,* 1994; Spörle *et al.,* 1996; Venters *et al.,* 1999), suggesting that myotome formation is similar in birds and mammals. As more myotome is formed, the central region of the dermomyotome dissociates to form the dermis of the dorsal skin, while the peripheral epithelial regions persist until they are completely converted to the myotome. The dorsomedial myotome gives rise to the epaxial muscles of the back, while the ventrolateral myotome produces the hypaxial muscles of the somatopleure (mesodermal tissues of the body wall) and the muscles of the limb (Cinnamon *et al.,* 1999). The delineation of the sclerotome and the dermomyotome, and the regionalization of myotomal fate in the dermomyotome are the result of a combination of tissue interactions between the somite and the surrounding tissues in the dorsoventral and mediolateral axes of the paraxial mesoderm (see Chapter 13).

In summary, somitogenesis involves a sequence of orchestrated morphogenetic events involving (1) the allocation of somitic precursors to the paraxial mesoderm, (2) the delineation of the segmental units in the presomitic mesoderm, (3) the establishment of anterior and posterior compartments ("segment polarity"), (4) the regionalization of tissue fates in the dorsoventral and mediolateral planes of the differentiating somites, and (5) the rearrangement of the primary segmental pattern during vertebral morphogenesis. Aspects of somitogenesis pertaining to the origin and the potency of progenitor cells, the presegmental pattern in the presomitic mesoderm, and the secondary segmentation are covered by recent reviews (Christ and Ordahl, 1995; Christ *et al.,* 1998; Gossler and Hrabé de Angelis, 1997; Tam *et al.,* 2000; Tam and Trainor, 1994), and details of myogenic and skeletal differentiation are discussed by Asakura and Rudnicki in Chapter 13 and by de Crombrugghe *et al.,* in Chapter 14 of this book. Here we focus on the cellular and molecular aspects of the allocation of progenitor cells to the paraxial mesoderm, the establishment of segment boundaries and tissue compartments, and the regulation of the periodicity of somite formation.

II. Allocation of Progenitor Cells to the Paraxial Mesoderm

During gastrulation, epiblast cells are recruited into the primitive streak and then ingress to form a new germ layer, the mesoderm. The first population of mesoderm cells is primarily fated for the extraembryonic mesoderm of the yolk sac, the chorion, and the amnion. Cells that are allocated to the paraxial mesoderm first ingress at about 10–12 hr after the onset of germ layer formation and constitute the progenitors of the cranial mesenchyme. As gastrulation progresses, more paraxial mesoderm progenitors are recruited and they contribute to successively more posterior regions of the embryo axis (Kinder *et al.,* 1999). By late gastrulation, the majority of somitic precursors resides in the anterior segment

of the primitive streak and in the adjacent epiblast. This progenitor cell population provides the constant source of new paraxial mesoderm for the trunk region during the elongation of the body axis until the primitive streak has completely regressed on closure of the posterior neuropore (Tam, 1984; Tam and Beddington, 1987; Wilson and Beddington, 1996). Subsequently, the progenitor cells for the somites of the tail reside in the tail bud mesenchyme (Tam *et al.,* 2000; Tam and Tan, 1992).

The homology of cells in the primitive streak and the tail bud is highlighted by the expression of a common set of genes such as *brachyury, Wnt3a, Tbx6, Evx1, Fgf3, Fgf4,* and *Fgf8* (Chapman *et al.,* 1996; Crossley and Martin, 1995; Gofflot *et al.,* 1997; Goldman *et al.,* 2000; Herrmann, 1991; Takada *et al.,* 1994). Moreover, progenitor cells in the primitive streak and tail bud mesenchyme display no restriction in the potency to contribute to any specific somites in the body axis. A surprising finding is that the tail bud mesenchyme of the E13.5 embryo that has ceased to form new somites still retains the potency to colonize host somites after transplantation to the primitive streak of younger embryos (Tam and Tan, 1992).

The cessation of somite formation coincides with the regression of a specific region of surface ectoderm ventral to the tail bud known as the ventral ectodermal ridge (VER; Grüneberg, 1956) and the elevation of apoptotic activity in the tail bud mesenchyme (Bellairs, 1986). In the developing tail, genes encoding putative secreted signaling molecules, *Wnt5a, Fgf17,* and *Bmp2,* are expressed in the VER, while the adjacent mesenchyme expresses the bone morphogenetic protein (BMP) antagonist, *Noggin* (Gofflot *et al.,* 1997; Goldman *et al.,* 2000). Regression of the VER therefore may result in the loss of both signaling and antagonist activities, which in turn may lead to the cessation of somitogenesis. Surgical ablation of the VER leads to a downregulation of *Noggin* activity and arrests further somite formation after the preexisting presomitic mesoderm is segmented (Goldman *et al.,* 2000). The cessation of somitogenesis during normal development therefore appears to be brought about by the loss of signaling activity that maintains the tail bud mesenchyme rather than the loss of somitogenic potency of the progenitor cells (Tam and Tan, 1992).

III. Cells Are in Transit in the Presomitic Mesoderm

The presomitic mesoderm can be regarded as a transition zone where cells that are continuously recruited from the progenitor pool in the primitive streak or the tail bud progressively mature and are assembled into somites. Continuous somite formation is maintained by a constant influx of cells to the posterior region of the presomitic mesoderm to replenish the cells that have been deployed for somite formation anteriorly. The removal of the primitive streak or the tail bud results in the termination of somite formation when the existing presomitic mesoderm is exhausted (Tam, 1986; Tam and Beddington, 1986). A continuous supply of cells to the presomitic mesoderm is provided by the proliferation of the pool of progenitors. When these cells are first incorporated into the presomitic mesoderm, they retain some of the molecular properties of cells in the progenitor population. They continue to express a multitude of genes that encode known or putative transcription factors (e.g., *Tbx6, Hox, Foxc2 (Mf1), Cdx1, Cdx4)*, peptide growth factors (e.g., *Fgf3, Fgf4, Fgf5*), and molecules for cell–cell signaling (e.g., *Crabp, Notch1, Fgfr1, Epha4*). The differentiation of cells in the presomitic mesoderm is reflected by the progressive loss of the progenitor-specific gene activity and the acquisition of somite-specific molecular properties by cells in the presomitic mesoderm. Transcripts of several genes that are initially expressed in the newly recruited cells diminish progressively toward the rostral part of the presomitic mesoderm (e.g., *Fgf3*). Conversely, cells in the more rostral part of the presomitic mesoderm increasingly acquire the somite-specific gene activity (e.g., *Fgfr1, Notch1, Epha4, Foxc2, Fgf5, Meox1, Notch2, Rbpsuh, Twist, Scleraxis,* and *Paraxis*). However, other genes that are expressed only in the differentiating somites (such as *Col2a1, Sox9, Pax, Myf5, MyoD, α-cardiac actin,* and *desmin*) are not expressed in the presomitic mesoderm. These patterns of expression are consistent with the gradual maturation and the impending segmentation of the presomitic mesoderm. Irrespective of the roles of these genes in somitogenesis, their overall expression profiles suggest that the presomitic mesoderm constitutes an intermediate tissue type during the transition of the somitic progenitors to the somites.

The phenotypes of mutations in *T, Fgfr1, Wnt3a,* and *Tbx6* have demonstrated an essential requirement for these genes during the formation of the paraxial mesoderm. In the mouse, a loss of *T* gene function leads to failure of axis development and arrested somite formation (Conlon *et al.,* 1995a; Herrmann, 1995; Wilkinson *et al.,* 1990), which is most likely due to impaired migration of mesodermal cells through the primitive streak (Beddington *et al.,* 1992; Wilson *et al.,* 1993). The loss of *Wnt3a* function also appears to affect mesodermal cell migration (Yoshikawa *et al.,* 1997). In *Wnt3a* mutants, presumptive paraxial mesoderm cells accumulate beneath the primitive streak and form ectopic neural tissues (Yoshikawa *et al.,* 1997). Reduced *T* function in *Wnt3a* mutants could have caused the migratory defect in *Wnt3a* mutants since *T* has been shown to be a direct target of *Wnt3a* signals in the paraxial mesoderm (Yamaguchi *et al.,* 1999). The loss of *Tbx6* function leads to the differentiation of presumptive paraxial mesoderm cells into neural tissue resulting in supernumerary neural tubes in the place of somites (Chapman and Papaioannou, 1998). Interestingly, these mutants all fail to form somites in the posterior region

of the body, whereas somites in the occipital and upper cervical regions are apparently unaffected. These findings suggest that the genetic control of somite formation may be different along the craniocaudal body axis. Embryos lacking *Fgfr1* function are grossly abnormal and display more extensive disruption of somitogenesis. Axial mesoderm is formed apparently at the expense of somitic mesoderm, resulting in the complete absence of somites (Yamaguchi *et al.,* 1994). In the chimeras, *Fgfr1* mutant cells accumulate in the primitive streak, failing to colonize the paraxial mesoderm, and differentiate instead to neuroepithelial tissue (Ciruna *et al.,* 1997). Collectively, these results suggest that WNT and FGF signaling regulate the migration of mesodermal cells and the choice between neural and paraxial mesodermal fates.

IV. Regionalized Genetic Activity Points to a Prepattern of Prospective Somites

Cells in the presomitic mesoderm have been shown to display a predictable pattern of development as if they are already endowed with a preset morphogenetic program. In explants of the presomitic mesoderm, somites are formed in the correct craniocaudal order as in the intact embryo. Despite a nearly twofold difference in the size of the presomitic mesoderm of embryos at different developmental stages, explants of this tissue consistently produce six to seven somites in culture (Tam, 1986). By SEM (scanning electron microscopy) analysis, mesenchymal cells in the presomitic mesoderm are found to be arranged in spherical clusters that have been termed somitomeres. The number of somitomeres contained in explants matches the total number of somites formed in culture. By following the course of segmentation of the presomitic mesoderm explants, researchers found that the number of somites that are formed corresponds to the reduction in the number of somitomeres in the tissue (Tam and Trainor, 1994; Tam and Beddington, 1986).

Although the experimental evidence supports the existence of a presegmental pattern of cellular organization that might be manifested morphologically as somitomeres, it is not entirely clear if they represent prospective somites. Cells in the presomitic mesoderm may move over a distance that spans two or more somitomeres, suggesting that the mesodermal cells are not yet definitively allocated to specific somitomeres. However, cell movements are significantly curtailed at the rostral end of the presomitic mesoderm where the next one to two prospective somites are about to segment (Tam, 1988). The trafficking of cells between somitomeres therefore argues against the existence of predetermined cellular segmental units. In the avian species, because of the smaller size of the quail embryo, a fragment of the segmental plate (presomitic mesoderm) is found to contain more somitomeres (or more prospective somites as tested by explant studies) than a fragment of chick segmental plate of similar sizes. However, the replacement of the chick (host) segmental plate with an equivalent size of the quail (donor) segmental plate results in the formation of fewer somites than the number of somitomeres the original quail grafts contain (Packard *et al.,* 1993). This finding suggests that the pattern of segmentation in the quail segmental plate is not irreversibly determined but it can be modified by chick "morphogenetic factors" that control somite sizes.

The consistent generation of a predictable number of somites from presomitic mesoderm explants has been correlated with the expression of genes that are expressed in regionalized patterns in the presomitic mesoderm. Some genes are expressed in regions of the presomitic mesoderm corresponding to the size of one or several segmental units (Table II). In most cases, however, it remains to be shown that these expression domains indeed demarcate prospective segments and that the cells within such a domain will populate the same somite during development. Even if such expression domains indeed delineate future segments, it is unclear whether the gene function would also be confined to one segment, since the domain defined by the transcriptional activity may not temporally or spatially correspond to the tissue domain containing the functional gene products. To date, no gene is known to display an expression pattern that precisely matches the somitomeres in the presomitic mesoderm. The closest to such a pattern of expression is found with the described activities of *her1* and *Notch5* in the zebrafish and *X-Delta2* in *Xenopus.* These genes are expressed in multiple stripes in the presomitic mesoderm. Fate mapping has shown that the *her1*-expressing cells in the presomitic mesoderm colonized every other somite (Müller *et al.,* 1996). This was originally considered to reflect a presegmental pair rule pattern (i.e., expression in every other segment). However, recent observations indicate that *her1* expression is dynamic (J. Campos-Ortega, personal communication). In the mouse, discrete regions of strong expression of a number of genes corresponding in size to either one or several prospective somites are found in the rostral part of the presomitic mesoderm (for examples, see Fig. 3 and Table II). Again, in these cases, the precise correlation between the expression domains and the prospective somites or somitomeres is not fully known.

V. Emergence of Anterior-Posterior Somite Compartments

In the differentiating somite, mesenchymal cells are packed at different densities in the two halves of the sclerotome (Verbout, 1985), which differ in their patterns of gene expression and contribute to different skeletal components of the vertebral column (see Section I). They also show different adhesive, chemotactic, and growth promoting properties that facilitate motor axon outgrowth and neural crest cell

migration in the anterior sclerotome even when the orientation of the two compartments relative to the body axis is reversed experimentally (Keynes and Stern, 1984, 1988; Rickmann *et al.,* 1985; Stern and Keynes, 1982). In the avian embryo, the regionalization of peanut lectin agglutinin affinity (Keynes and Stern, 1988) and cytotactin-mediated cell–cell contact inhibition (Tan *et al.,* 1987) to the anterior and posterior regions of the somite, respectively, is heralded by the localization of these activities to alternating bands of cells in the presomitic mesoderm, suggesting that the anterior-posterior compartments are established prior to somite segmentation.

Consistent with this notion of early specification of anterior-posterior compartments, a number of genes that are involved in intercellular signaling are expressed in a manner consistent with the partitioning of the prospective somites in the presomitic mesoderm. In the absence of morphological landmarks that can be used to define the borders of the presegmental unit, the assignment of regionalized gene expression to specific portions of the prospective somites is extrapolated from the polarized pattern of these genes in the mature somites. Genes such as *Dll1, Jag1, Mesp2, Lnfg* (Fig. 3), and *Hes5* (not shown) are expressed in the rostral part of the presomitic mesoderm in a manner that leaves a gap of no expression adjacent to the intersomitic fissure of the last completely segmented somite. Because some of these genes (e.g., *Dll1* and *Lfng*) continue to be expressed in the posterior region of the somites, it is likely that their earlier expression marks the posterior portion of the prospective or nascent somites (Fig. 3). At present, only *Tbx18* is known to be expressed specifically in the anterior halves of the prospective somites (Kraus *et al.,* 2001). *Epha4* and *Cer1* are expressed in the anterior half of the somite but their expression domain in the presomitic mesoderm does not seem to be confined to a half-segment. It is interesting to note that in mouse embryos *Lnfg* expression is localized to the presumptive posterior half of the forming somite, whereas the chick ortholog is expressed in the anterior half (Forsberg *et al.,* 1998; McGrew *et al.,* 1998). Similarly, the chicken *Hairy2* gene is expressed in anterior somite halves while its mouse ortholog *Hes1* is expressed in the posterior halves (Jouve *et al.,* 2000). This suggests that, although the molecular mechanism of anterior-posterior specification may be conserved, the developmental impact of the genetic activity may have diversified in these two species.

VI. Role of Notch Signaling in the Establishment of Somite Borders and Anterior-Posterior Polarity

Analysis of the patterning defects in somites of mutant mice have revealed a critical role for Notch signaling in the establishment of anterior-posterior (AP) polarity and the

Table II Set of Genes Expressed in the Presomitic Mesoderm, and Nascent and Epithelial Somites[a]

			Gene expression in							
			Presomitic mesoderm			Epithelial somites				
Symbol	Name, synonym(s)	Biochemical function	Whole, uniform	Graded	Regionalized	Whole	Anterior portion	Posterior portion	Notes	References[b]
Bmp5	Bone morphogenetic protein 5	Growth factor					•	•	Confined to the marginal tissues of the somit	1
Cer1	Cerberus 1	Putative Wnt and BMP antagonist			•	•				2
Dll1	delta like 1, delta1	Notch ligand		•				•	Strongest in anterior presomitic mesoderm	3
Dll3	delta like 3, delta3	Notch ligand		•				•		4
Epha4	Sek1, Eph receptor A4	Receptor tyrosine kinase			•					5
Fgf8	fibroblast growth factor 8	Growth factor					•	•	Weak expression in the somite edges	1
Fgfr1	fibroblast growth factor receptor 1	Receptor tyrosine kinase			•		•			6
Foxc1	Mf1, forkhead box C1	Forkhead domain transcription factor	•			•				7,8
Foxc2	Mfh1, forkhead box C2	Forkhead domain transcription factor			•	•				9,10
Hes1	Hairy and enhancer of Split 1	bHLH transcription factor			•				Oscillating expression in the presomitic mesoderm	11
Hes5	Hairy and enhancer of Split 5	bHLH transcription factor			•					12
Jag1	Jagged 1	Notch ligand			•					13
Lfng	Lunatic fringe	Modulator of Notch signaling			•			•	Oscillating expression in the presomitic mesoderm	14
Lmo4	LIM only 4	LIM domain transcription factor								15
Meox1	Mox1, mesenchyme	Homeobox transcription factor homeobox 1								16
Mesp1	Mesoderm posterior 1	bHLH transcription factor			•					17
Mesp2	Mesoderm posterior 2	bHLH transcription factor			•					18
Notch1	Notch gene homolog 1	Receptor			•					19
Notch2	Notch gene homolog 2	Receptor			•					20
Rbpsuh	RBPjk, CBF1, KBF2, recombining binding protein suppressor of hairless	Transcriptional regulator, mediator of Notch signaling	•			•				21
Sna	snail	Zinc finger transcription factor	•			•				21
Tbx6	T-box 6	T-box transcription factor	•							22a
Tbx18	T-box 18	T-box transcription factor			•					22b
Tcf15	paraxis, EC2	bHLH transcription factor			•	•			Expressed later in sclerotome and dermomyotome	23
Terra	terra gene homolog	Zinc finger transcription factor				•			Expressed later in dermomyotome	24
Twist	twist gene homolog	bHLH transcription factor	•			•			The protein is detected in sclerotome and dermomyotome and absent from myotome	25,26
Uncx4.1	Unc4.1 homeobox	Homeobox transcription factor						•	Expressed later in posterior sclerotome	27,28

[a] The expression of most of these genes is not confined to the paraxial mesoderm; expression in other cell types or tissues is not listed.

[b] References cited: Solloway and Robertson (1999); 2. Biben *et al.* (1998); 3. Bettenhausen and Gossler (1995); 4. Dunwoodie *et al.* (1997); 5. Nieto *et al.* (1992); 6. Yamaguchi *et al.* (1992); 7. Sasaki and Hogan (1993); 8. Yoshikawa *et al.* (1997); 9. Kaestner *et al.* (1996); 10. Furumoto *et al.* (1999); 11. Jouve *et al.* (2000); 12. de la Pompa *et al.* (1997); 13. Zhang and Gridley (1998); 14. Forsberg *et al.* (1998); 15. Kenny *et al.* (1998); 16. Candia *et al.* (1992); 17. Saga *et al.* (1996); 18. Saga *et al.* (1997); 19. Réaume *et al.* (1992); 20. Williams *et al.* (1995); 21. Smith *et al.* (1992); 22a. Chapman *et al.* (1996); 22b. Kraus *et al.*, (2001) 23. Burgess *et al.* (1995); 24. Meng *et al.* (1999); 25. Füchtbauer (1995); 26. Gitelman (1997); 27. Neidhardt *et al.* (1997); 28. Mansouri *et al.* (1997).

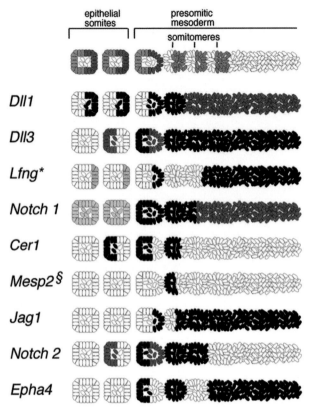

Figure 3 Schematic representation of gene expression patterns in the presomitic mesoderm and somites of mouse embryos. Level of expression are color coded: black, strong expression; gray, moderate levels of expression; light gray to blank, very low to no expression. The precise domains of gene expression and the extent of spatial overlaps between the expression of different genes within the presegmental unit and the somites are not fully delineated in many cases.

maintenance of compartment borders during somitogenesis (for recent reviews, see Pourquié, 2000; Rawls *et al.,* 2000; Table III). The Notch pathway is an evolutionarily conserved cell-to-cell signaling mechanism that regulates cell differentiation in multiple tissues in diverse organisms (reviewed in Artavanis-Tsakonas *et al.,* 1999; Blaumueller and Artavanis-Tsakonas, 1997; Gridley, 1997; Robey, 1997; Simpson, 1997). In *Drosophila,* Notch signaling is involved in the regulation of primary neurogenesis (Campos-Ortega, 1995), cellular differentiation in derivatives of all germ layers (Corbin *et al.,* 1991), epithelial mesenchymal transitions (Hartenstein *et al.,* 1992), and wing margin development (Couso *et al.,* 1995; de Celis and Garcia Bellido, 1994; Doherty *et al.,* 1996). The communication between neighboring cells is brought about by functional interaction of Delta, Serrate, and Notch (Fehon *et al.,* 1990), which are transmembrane proteins with multiple epidermal growth factor (EGF)-like repeats in their extracellular domains (Thomas *et al.,* 1991;

Vässin *et al.,* 1987; Wharton *et al.,* 1985). Notch is proteolytically processed in the Golgi to generate two fragments, one of which contains the transmembrane and cytoplasmic domains and the other the extracellular domain. The two fragments are probably tethered by disulfide bridges and form the functional receptor at the cell surface (Blaumueller *et al.,* 1997; Pan and Rubin, 1997).

In the current model of Notch signaling, Delta and Serrate act as ligands and bind individually to the Notch receptor. In addition, the proteolytic cleavage of Delta in the extracellular domain apparently generates a soluble ligand for Notch (Qi *et al.,* 1999). Signaling of Delta and Serrate can be modulated by Fringe, a membrane-associated extracellular protein (Fleming *et al.,* 1997; Irvine and Wieschaus, 1994; Wu and Rao, 1999). On ligand binding, the intracellular portion of Notch is proteolytically cleaved, translocates to the nucleus, and by complexing with the transcriptional regulator suppressor of hairless [*su (h)*], activates transcription of bHLH genes of the enhancer of split [*e (spl)*] family (Fortini and Artavanis-Tsakonas, 1994; Jarriault *et al.,* 1995, 1998; Kidd *et al.,* 1998; Kopan *et al.,* 1996; Schroeter *et al.,* 1998; Struhl and Adachi, 1998). Their gene products in turn regulate the transcription of other downstream bHLH effector genes. However, there is also evidence that Notch signaling may bypass *su(H)* in the regulation of downstream gene activity (Rusconi and Corbin, 1999; Shawber *et al.,* 1996).

Orthologous components of the *Drosophila* Notch pathway have been isolated in various vertebrates. Generally several copies of these genes are found. In the mouse, there are three reported *Delta* (*Dll1, Dll3,* and *Dll4*), two *Serrate* (*Jagged1* and *2*), four Notch (*Notch1–4*), three fringe (*Lunatic, Radical,* and *Maniac fng*), one *su(h)* (*Rbpsuh*), and several *e(spl)* (*Hes*) genes. In addition, other bHLH genes may also be involved in Notch signaling (see below). Mutations in *Rbpsuh* (Oka *et al.,* 1995), *Notch1* (Conlon *et al.,* 1995b; Swiatek *et al.,* 1994), *Dll1* (Hrabé de Angelis *et al.,* 1997), *Dll3* (Kusumi *et al.,* 1998), and *Lfng* (Evrard *et al.,* 1998; Zhang and Gridley, 1998) lead to specific defects in somitogenesis (Table III). In contrast, mutations in *Notch2* (Hamada *et al.,* 1999), *Jag1* (Xue *et al.,* 1999), *Hes1* (Ishibashi *et al.,* 1995), and *Hes5* (Ohtsuka *et al.,* 1999), which are expressed in the presomitic mesoderm and/or somites, have no effect on somitogenesis. In all mutants with somite defects, somites appear irregular in shape and size. However, significant variations are observed in the nature of somite defects in the different mutants, which might be caused by functional differences among the components of the Notch pathway and the efficacy of the compensatory mechanism that acts to overcome the loss of specific signaling molecules.

In homozygous *Rbpsuh* mutant embryos (Oka *et al.,* 1995), somites lack detectable segment polarity as revealed by loss of *Uncx4*.1 and *Cer1* expression, and the expression of *Lfng* in the presomitic mesoderm is abolished (del Barco

Table III Mutations in Mice Affecting Segmentation and Tissue Patterning in the Somite

| Gene/mutation | Type | | Phenotype/defects in | | |
	Spontaneous	Knock-out	Paraxial mesoderm/somites (A= anterior, P=posterior)	Axial skeleton	References[a]
amputated (am)	•		Reduced somite size, altered adhesive properties of somitic cells	Truncated axis, reduced and fused vertebrae	1,2
crooked tail (Cd)	•		Somite irregularities, fusion, and pycnosis	Vertebral malformations; rib fusion	3
Dll1		•	Disrupted AP patterning (reduction or loss of posterior segment), incomplete epithelialization	Not known (embryonic lethal approximately E12)	4
Dll3[pu] pudgy (pu) likely null allele	•		Disrupted AP patterning (no distinct separation of A and P compartments)	Loss of caudal vertebrae; bifurcated and fused ribs; open neural arches; severe disorganization of vertebrae	5
Foxc1		•	Defective chondrogenic differentiation	Open neural arches; reduced vertebral bodies	6
Foxc2		•	Reduced sclerotome proliferation	Vertebral malformations	7
Lfng		•	Irregular morphology of somites, no distinct separation of A and P compartments	Loss of caudal vertebrae; bifurcated and fused ribs that are not attached to the vertebrae	8,9
Malformed vertebrae (Mv)	•		Somite irregularities and fusion	Severe vertebral malformations	10
Mesp2		•	Disrupted AP somite patterning (reduction or loss of anterior segment)	Axial truncation; severe fusion of vertebral elements	11
Notch1		•	Irregular shape of somites	Not known (embryonic lethal E10–11)	12,13
paraxis		•	No epithelialization of somites	Loss of caudal vertebrae; vertebral fusion (sacral region), bifurcated and fused ribs that are not attached to the vertebrae	14
PS1		•	Reduced Dll1 and Notch1 expression; defective AP patterning	Axial truncation; vertebral fusion	15
rachiterata (rh)	•		Irregular somites and fusion of somites	Fusion of ribs and vertebrae	16
Rbpsuh		•	Disrupted AP patterning (reduction / loss of posterior segment identity)	Not known (embryonic lethal E9–10)	17,18
rib vertebrae (rv)	•		Fused and irregular somites, disrupted AP patterning	Fusion of ribs, neural arches, vertebral bodies	19
rib-fusions (Rf)	•		Indistinct somites with disrupted epithelial morphology	Severe axial truncation and fusion of vertebrae	20
tail kinks (tk)	•		Caudal sclerotome abnormal	Axial truncation; split and fused vertebrae	21
Tbx6		•	Only cervical somites form; posterior somites replaced by neuroectoderm	Not known (embryonic lethal approximately E12)	22
uncx4.1			Reduced cell number in caudal slerotomes	Missing proximal ribs and pedicles (derivatives of posterior sclerotome)	23,24
Wnt3a		•	No somites caudal to ninth somite; homozygous mutant cells do not form somites in chimeras	Not known (embryonic lethal between E10–12)	25,26
Wnt3a[vt], vestigial tail (vt) hypomorphic Wnt3a allele	•		Reduced Wnt3a expression in the tail bud	Loss of caudal vertebrae; deformed sacral vertebrae	27

[a] References cited: 1. Flint and Ede (1978); 2. Flint et al. (1978); 3. Morgan (1954); 4. Hrabé de Angelis et al. (1997); 5. Kusumi et al. (1998); 6. Kume et al. (1998); 7. Winnier et al. (1997); 8. Evrard et al. (1998); 9. Zhang and Gridley (1998); 10. Theiler et al. (1975); 11. Saga et al. (1997); 12. Conlon et al. (1995b); 13. Swiatek et al. (1994); 14. Burgess et al. (1996); 15. Wong et al. (1997); 16. Theiler et al. (1974); 17. Oka et al. (1995); 18. del Barco Barrantes et al. (1999); 19. Theiler and Varnum (1985); 20. Theiler and Stevens (1960); 21. Grüneberg (1955); 22. Chapman and Papaioannou (1998); 23. Leitges et al. (2000); 24. Mansouri et al. (2000); 25. Takada et al. (1994); 26. Yoshikawa et al. (1997); 27. Greco et al. (1996).

Barrantes et al., 1999). In Notch1 mutants, somite segmentation appears to be out of register between the contralateral paraxial mesoderm, and somite borders are misaligned (Conlon et al., 1995b), but segment polarity is essentially normal (del Barco Barrantes et al., 1999). The somites of Dll1 mutant embryos also lack AP polarity, which is already evident before segmentation. Somites are not fully epithelialized and their borders are not maintained (Hrabé de An-

gelis et al., 1997). Dll3 is disrupted in pudgy (pu) mutant mice (Kusumi et al., 1998) and this results in severe malformations of the vertebral column (Grüneberg, 1961). In Dll3[pu] mice, somite polarity is disrupted with Uncx4.1 expression being widespread instead of localized to the posterior compartment of the somite. A similar disorganization of Uncx4.1 expression is found in the somites of Lfng mutants. The expression of Hes5, a downstream target of Notch, is also downregulated in the *Lnfg* mutant somites. This finding suggests that *Lfng* modulates Notch activity in the paraxial mesoderm and thereby leads to the observed patterning defects (Evrard *et al.,* 1998; Zhang and Gridley, 1998). The phenotypic outcome of the *Mesp2* gene mutation strongly suggests that *Mesp2* activity is coupled with Notch signaling. Loss of *Mesp2* function leads to delayed segmentation and the somites acquire only posterior characteristics. *Dll1* expression expands into anterior halves of somites, and *Notch1* expression in the presomitic mesoderm is diminished (Saga *et al.,* 1997). However, a more complex functional relationship might exist between *Mesp2* and Notch signaling because not only is the expression of Notch pathway components affected in *Mesp2* mutants, but *Mesp2* expression is reciprocally affected by the loss of *Dll1* activity in mutant embryos. Despite the defects in the anterior-posterior patterning of the somites in these mutants, appropriate differentiation of the somites into dermomyotome, myotome, and sclerotome seems unhindered. This raises the possibility that the establishment of somitic compartments is not critical for lineage differentiation.

Several important conclusions can be drawn from the analysis of the mutations of the Notch signaling pathway: (1) Notch signaling is critical for the establishment of AP somite polarity and for the establishment of epithelial somites. It is required for positioning or refining somite borders, but not for segmentation per se, because segment borders are always formed. However, note that perturbation of Notch signaling in *Xenopus* or zebrafish embryos in some cases led to the failure of segmentation (Jen *et al.,* 1999; Takke and Campos-Ortega, 1999). (2) The more severe somite phenotype in the *Dll1* mutants compared to *Notch1*-deficient embryos suggests that *Dll1* may signal through several Notch receptors expressed in the paraxial mesoderm and that *Notch2* may compensate for some of the *Notch1* functions. The similar defect in segment polarity in *Rbpsuh* mutant embryos suggests that signals from different Notch receptors converge on *Rbpsuh* in the paraxial mesoderm. (3) Epithelialization is not essential for differentiation of somite derivatives or the formation of segment borders. This concept is further supported by the finding that segmentation of the paraxial mesoderm can occur without any epithelial organization in *Paraxis* mutant embryos (Burgess *et al.,* 1996).

Contrary to the concept that the acquisition of anterior and posterior properties by cells in the somite is essential for establishing intersomitic borders (Meinhardt, 1986), segments can be formed in the absence of any discernible craniocaudal polarity in *Rbpsuh* and *Dll1* mutants. This suggests that a pair-rule mechanism that distinguishes successive (even- and odd-numbered) segments may be operating in the paraxial mesoderm. The apposition of dissimilar (even versus odd) segments (Meinhardt, 1986) may be a potential means of generating tissue borders. However, such two-segment periodicity of gene activity has not been demonstrated in mouse embryos thus far.

The subdivision of somites into anterior and posterior compartments generates alternating AP and posterior-anterior (PA) interfaces. While Notch signaling appears to be instrumental in the generation of borders at the interfaces of the anterior and posterior compartments, it remains unclear how the PA border (the future intersomitic interface) and the AP border (the future intrasomitic interface) can be discriminated in anatomical and molecular terms since the tissue characteristics of the compartments that confront each other at the alternating AP and PA interfaces are apparently the same, but the developmental fates of these borders are different. These differences could be achieved by the site-specific reciprocal signaling of ligands and receptors in the compartments of the segmenting somites, which could result in qualitatively different signals at these interfaces. Alternatively, some yet unidentified cues that are alternately expressed in A and P compartments could determine the identity and the developmental fate of AP (intersegmental) and PA (intersomitic) interfaces.

VII. A Molecular Clock Operates in the Paraxial Mesoderm to Control the Kinetics of Somite Formation

Recently, several dynamic and periodic patterns of transcriptional activity were discovered in the paraxial mesoderm prior to segmentation (reviewed by Pourquié, 2000; Stern and Vasiliauskas, 2000). Cyclical gene expression in the presomitic mesoderm with a periodicity equivalent to the time required to form one somite was first described for cHairy1, a chick homolog of the *Drosophila* pair-rule gene *hairy* (Palmeirim *et al.,* 1997). Each cycle of expression begins in the posterior region of the presomitic mesoderm, sweeps anteriorly through the presomitic mesoderm, and finally consolidates in a band of tissue in the anterior presomitic mesoderm that seems to correspond to the anterior portion of the prospective somite. The final site of cHairy1 transcription therefore appears to mark the position of the next intersomitic boundary. The cyclic expression of c-Hairy1 is independent of protein synthesis, cell movements, propagation of signals, or the physical integrity of the presomitic mesoderm. The mouse *Hes1* gene, which is the ortholog of cHairy2, is found to be expressed cyclically in the presomitic mesoderm like the chick gene (Jouve *et al.,* 2000). In addition to the cHairy and *Hes1* genes, *Lfng* expression oscillates in the presomitic mesoderm of chick and mouse embryos (Forsberg *et al.,* 1998; McGrew *et al.,* 1998). However, like the *Hes1* gene, the final site of *Lfng* expression is restricted to the posterior portion of the most mature prospective somite in the mouse and not to the anterior portion as in the chick. In contrast to the cHairy genes and their mouse orthologs, the maintenance of the cyclic expression of *Lfng* requires protein synthesis. This has led to the suggestion that *Lfng* might be a regulatory target of the bHLH proteins encoded by the cHairy and mouse *Hes1* genes. Because *fringe* acts as a modulator of Notch activity in *Drosophila*, cycling *Lfng* expression directed by the molecular clock thus may result in periodic activation of Notch in the presomitic mesoderm and nascent somites (del Barco Barrantes *et al.*, 1999; Jiang *et al.*, 1998). Interestingly, expression of *Lfng* is not affected in *Hes1* mutant embryos indicating that *Hes1* is not an essential component of the clock (Jouve *et al.*, 2000). In contrast, the expression of *Lfng* and *Hes1* is abolished in *Dll1* mutant embryos (del Barco Barrantes *et al.*, 1999; Jouve *et al.*, 2000)), suggesting that the periodic expression of these genes requires Notch signaling. This raises the possibility that Notch signaling is not just a readout of the clock, but is also an essential component of the molecular oscillator.

The recent finding of cyclical gene activity in the presomitic mesoderm whose periodicity matches the generation of somites (Forsberg *et al.*, 1998; McGrew *et al.*, 1998; McGrew and Pourquié, 1998; Palmeirim *et al.*, 1997) supports models that feature cyclical changes of cell states in the presomitic mesoderm as a means to synchronize groups of cells for somite segmentation (Cooke and Zeeman, 1976; Meinhardt, 1986). One particular model, the "clock and wavefront" model, was originally proposed to explain experimental results obtained with size-manipulated *Xenopus* embryos (Cooke, 1981; Cooke and Zeeman, 1976): When the size of *Xenopus* embryos was reduced at the blastula stage, small embryos developed with normal numbers of somites. The somites in these embryos contained fewer cells than controls, indicating that the cell number per somite was adjusted according to the reduced body size (Cooke, 1975).

The model contains two functional components. The first is a gradient that determines the rates of development of cells in the presomitic mesoderm along the AP axis toward segmentation such that anterior cells acquire the ability to form somites earlier than more posterior cells. This kinematic "wave" of maturation sweeps through the paraxial mesoderm in an AP direction and does not depend on cell movements or the propagation of signals. The second component is a biochemical oscillator. This "cellular clock" operates synchronously throughout the presomitic mesoderm and switches between alternate phases of prohibition and permissiveness for somite formation. Through the activity of other specification processes, a group of cells in the presomitic mesoderm is fated to form a somite. However, they are prevented from segmentation at the inhibitory phase of the oscillator, but are allowed to undergo concerted morphogenetic changes and form a somite at the permissive phase. Thus, the continuous maturation of the paraxial mesoderm along the AP axis is transformed into a discontinuous sequence of segmentation events by the gating mechanism of the oscillator. To explain the size regulation, it was assumed that the kinematic "wave" needs the same time to travel through the body axis in embryos with different body sizes or tissue masses. The regulation of somite formation is thus

achieved by maintaining identical oscillator frequencies and relative speeds of the morphogenetic wave in both normal and manipulated embryos. This ensures that the same number of morphogenetic events will be executed and, hence, the same number of segments will be produced (Cooke, 1977, 1981; Richardson *et al.*, 1998).

Segmentation in *Xenopus* commences at the late neurula stage when all presumptive mesoderm cells have been generated and somite formation (in the trunk) is essentially a process of subdividing the mesoderm. In contrast, somite formation in amniotes commences while new mesoderm is still being generated from the primitive streak and the tail bud and therefore the regulation of somite formation requires some means to predict the amount of precursor tissue that may be available. Segmentation in a continuously growing system therefore may not be entirely accounted for by the "clock and wavefront" model. Nevertheless, size regulation of somites can be achieved in mouse embryos. In homozygous amputated (*am*) E9.5 embryos, the axis is shorter than in wild-type embryos, but mutant embryos contain the same number of somites of reduced size (Flint and Ede, 1978). A similar adjustment of somite size has also been shown in embryos that undergo compensatory tissue growth after the loss of cells induced by DNA synthesis inhibition (Tam, 1981). A positive correlation of somite size (and the number of cells contained in the somite at segmentation), but a negative correlation of the segmentation rate with the rate of axial growth has been observed. This may suggest that somite size and number are regulated by similar mechanisms that controls growth or cell proliferation in the mouse as in several other vertebrates (Richardson *et al.*, 1998). Based on the recent molecular observations it has been suggested that the molecular oscillator also acts as a gating mechanism and that the periodicity of segmentation could be based on counting the number of expression cycles or the titration of the amount of gene products accumulated after each round of gene expression (Cooke, 1998). If this were the case, adjusting the number of cells that oscillate synchronously during allocation of progenitors to the paraxial mesoderm could be a way to achieve size regulation. The problem of size regulation notwithstanding, it is clear now that a molecular oscillator or "segmentation clock" operates in the unsegmented paraxial mesoderm. Recently, the cyclical gene activity associated with Notch signaling has been shown to be coupled with transient activation of *Hox* genes during somite segmentation, suggesting that oscillatory gene activity may regulate the specification of morphological identity of the somites (Zakany *et al.*, 2001).

VIII. Specification of Lineage Compartments by Inductive Interactions

The generation of sclerotome, dermomyotome, and myotome represents tissue patterning along the dorsoventral and mediolateral axes of the epithelial somites. Somites are first partitioned along the dorsoventral axis when the ventromedial portion of the somite loses its epithelial architecture and acquires a mesenchymal morphology to form the sclerotome (Christ and Wilting, 1992; Theiler, 1988). The remaining dorsal part of the somite remains epithelial and becomes the dermomyotome. Further partitioning along the mediolateral axis of the dermomyotome becomes evident following the formation of the myotome for epaxial and hypaxial muscles (Kalcheim *et al.*, 1999; Cinnamon *et al.*, 1999; also see Chapter 13). The formation of the lineage compartments in the somite has been shown to be regulated by inductive interactions with the surrounding tissues (notochord, neural tube, and lateral plate mesoderm) and involves the BMP, WNT, and SHH signaling pathways (for detailed reviews, see Arnold and Braun, 2000; Borycki and Emerson, 2000; Dockter, 2000; Gossler and Hrabe de Angelis, 1997; Tajbakhsh and Buckingham, 2000; Tajbakhsh and Spörle, 1998). Signals derived from the notochord/ventral neural tube promote the formation of sclerotome (Brand-Saberi *et al.*, 1993; Dietrich *et al.*, 1993; Ebensperger *et al.*, 1995; Koseki *et al.*, 1993; Monsoro-Burq *et al.*, 1994; Pourquié

et al., 1993) and myotome (Avery *et al.*, 1956; Buffinger and Stockdale, 1994; Cossu *et al.*, 1996b; Kenny-Mobbs and Thorogood, 1987; Lassar and Münsterberg, 1996; Münsterberg and Lassar, 1995; Stern and Hauschka, 1995; Vivarelli and Cossu, 1986). The ventral signals, however, repress the expression of the dermomyotomal genes *Pax3*, *Pax7*, and *Sim1* (Fan and Tessier Lavigne, 1994). Signals from the dorsal neural tube promote dermomyotome formation and induce the differentiation of epaxial myoblasts (Fan and Tessier Lavigne, 1994), but are not required for the differentiation of lateral (hypaxial and limb) myoblasts (Rong *et al.*, 1992; Spörle *et al.*, 1996). Dermomyotome formation is also induced by signals from the surface ectoderm (Fan and Tessier Lavigne, 1994). Both the surface ectoderm and the lateral plate mesoderm can promote the differentiation of hypaxial (lateral) muscles (Cossu *et al.*, 1996a; Pourquié *et al.*, 1996; Tajbakhsh *et al.*, 1998).

A. Ventralizing Signals (Fig. 4a)

The delineation of the ventral (sclerotome) compartment of the somites is mediated by the activity of Sonic Hedgehog

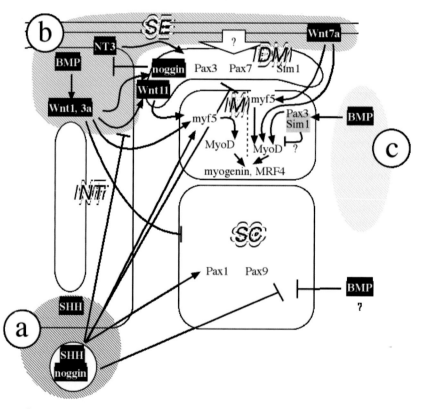

Figure 4 Schematic summary of tissue interactions regulating somite differentiation. The differentiation of somites is regulated by a complex network of agonistic and antagonistic interactions that subdivide the somites along the dorsoventral and mediolateral axes. Three major sources of signals are present: (a) ventralizing signals from the notochord and floor plate (red), (b) dorsalizaing signals from the dorsal neural tube and surface ectoderm (green), and (c) lateralizing signals from the lateral plate mesoderm (yellow). Abbreviations: SE, surface ectoderm; SC, sclerotome; NT, neural tube; M, myotome; DM, dermomyotome.

(*Shh*) (Bumcrot and McMahon, 1995), a vertebrate homolog of the *Drosophila* segment polarity gene *Hedgehog*, which is expressed in the notochord and floor plate (Echelard *et al.*, 1993; Fan *et al.*, 1995; Johnson *et al.*, 1994; Roelink *et al.*, 1994). In explant cultures of paraxial mesoderm, SHH can induce the expression of two sclerotomal genes, *Pax1* and *Nkx3.1* (Fan *et al.*, 1995; Fan and Tessier Lavigne, 1994), and the continuous presence of SHH is required to maintain high levels of *Pax1* expression in somite explants (Münsterberg *et al.*, 1995).

Although SHH appears to be sufficient to promote sclerotome differentiation under experimental conditions, its activity does not seem to be critical for the induction of sclerotome and myotome differentiation in mouse embryos. In the somites of embryos lacking *Shh* activity, *Pax1* expression is initiated but cannot be maintained in the sclerotomes (Chiang *et al.*, 1996). This suggests that other signals possibly derived from the ventral tissues may be critical for the ventralization of the somites, and that SHH is a maintenance signal for sclerotome differentiation. Patterning of the myotome seems to be differentially dependent on SHH activity since *Myf5* and *MyoD* activities are only induced in the hypaxial muscle precursors of the somites of *Shh*-deficient embryos but are absent from the epaxial myotome (Borycki *et al.*, 1999). SHH actvity is therefore essential for the specification of the epaxial myotome but not the hypaxial myogenic progenitor cells. In addition to SHH, the activity of notochord-derived Noggin is required for sclerotome differentiation. Noggin-deficient embryos display defective sclerotome formation varying at different axial levels (McMahon *et al.*, 1998). The action of Noggin is likely to antagonize the BMP activity emanating from the lateral plate mesoderm. BMP4 has been shown to inhibit the differentiation of the sclerotome (Monsoro-Burq *et al.*, 1996), and both BMP2 and BMP4 can inhibit the *Pax1* activity that is induced by SHH (McMahon *et al.*, 1998).

B. Dorsalizing Signals (Fig. 4b)

The patterning of the dorsal compartment of the somite is influenced by the activity of BMP4, Noggin, WNT1, -3a, and -4, and Neurotrophin 3 (NT-3) emanating from the dorsal neural tube. BMP4 in the neural tube induces expression of *Wnt1* and *Wnt3a* (Marcelle *et al.*, 1997), which can substitute for the inductive effect of the neural tube on muscle differentiation in somite explants (Münsterberg *et al.*, 1995). The critical role of *Wnt* signaling is further demonstrated by the action of exogenous FRZB1, an inhibitor of WNT, which disrupts myogenesis (Borello *et al.*, 1999). In addition to their myogenic action, WNT1 and WNT3A can induce WNT11 expression in the dorsomedial lip where WNT11 acts to enhance the involution of myogenic cells (Marcelle *et al.*, 1997). Differentiation of the dermal compartment is also influenced by signals from the dorsal neural tube as revealed by the finding that the ablation of the neural tube

results in the absence of dermis formation. The defects in dermis differentiation can be mimicked by the application of NT-3 blocking antibodies to the embryo. Normal dermis differentiation, however, can be restored in neural tube-ablated embryo by the administration of NT-3 (Brill *et al.*, 1995).

C. Signals Specifying the Lateral Compartment (Fig. 4c).

BMP4, which is expressed in the lateral (plate) mesoderm, has been shown to maintain *Pax3* activity but to suppress the expression of bHLH genes in myogenic precursors in the lateral (hypaxial) myotome compartment (Pourquié *et al.*, 1995; Tajbakhsh and Buckingham, 1994; Williams and Ordahl, 1994). Consistent with the inhibition of myocyte differentiation by BMP activity, the loss of Noggin activity in mutant embryo results in not only the lack of differentiation of hypaxial muscle but a more widespread suppression of myogenic differentiation (McMahon *et al.*, 1998). The effect of BMP4 activity can be mediated by its regulatory action on the expression of *Sim1*, a bHLH gene specifically expressed in the lateral somite derivatives and migrating hypaxial myoblasts (Pourquié *et al.*, 1995). Additional signals that influence hypaxial myotome formation can also be derived from the dorsal ectoderm. The inductive effect of the ectoderm can be mimicked by Wnt7a, which is expressed in the dorsal ectoderm and therefore is likely to be another signaling factor that determines the lateral compartment (Cossu *et al.*, 1996a; Tajbakhsh *et al.*, 1998).

D. Fine Tuning of Signaling through Synergistic and Antagonistic Interactions

More detailed experimental analysis has revealed that the delineation of lineage compartments in the dorsoventral and mediolateral planes of the somite is the outcome of an intricate pattern of synergistic and antagonistic interactions of different signaling processes and pathways (summarized in Fig. 4). The establishment of the ventral (sclerotome) compartment is brought about by the antagonistic interaction of dorsalizing (WNT) and ventralizing (SHH) signals. Wnt signaling may be modulated by secreted Frizzled-related proteins (encoded by *Sfrp* genes) whose expression is upregulated by the N-fragment of SHH (SHH-N) activity (Lee *et al.*, 2000). The activity of SFRP2 reduces the dermomyotome-inducing activity of WNT1 and WNT4 (Marcelle *et al.*, 1997) and leads to enhanced induction of sclerotome differentiation. Another example of counteracting signals is provided by the inhibitory effect of BMP4 on myogenesis. In the lateral somites, myoblast differentiation is suppressed by BMP4 activity from the lateral plate mesoderm (Pourquié *et al.*, 1996; Reshef *et al.*, 1998). To counteract the potential inhibitory effect of neural tube-derived BMP4 signals on epaxial myogenesis, Noggin is produced in the dorsal neural tube (Hirsinger *et al.*, 1997) and in the dorsomedial lip of the dermomyotome (Hirsinger *et al.*, 1997; Tonegawa and Takahashi, 1998). Interestingly, expression of noggin in the dorsomedial lip of the dermomyotome can be induced experimentally by WNT1 (Hirsinger *et al.*, 1997; Reshef *et al.*, 1998) suggesting that BMP4 produced in the neural tube neutralizes its "lateralizing" activity on somitic tissue by indirectly upregulating its own inhibitor via WNT signaling. Similarly, the effects of the lateral plate mesoderm are counteracted by signals from the neural tube, suggesting that along the mediolateral axis (i.e., epaxial versus hypaxial myotome) somites are patterned by antagonistic signals from the neural tube and lateral plate (Hirsinger *et al.*, 1998; Pourquié *et al.*, 1996).

IX. Summary and Open Questions

Somitogenesis is a continuous process that encompasses (1) the specification and recruitment of cells to the paraxial mesoderm, (2) the organization of the presomitic mesoderm into presegmental units and compartmentalization, (3) the generation of morphologically distinct segments, and (4) the generation of distinct differentiated cell types and lineages (Fig. 5). Superimposed on these ongoing and reiterated events is the assigment of positional information that results in the regionalization of the AP axis and leads to the formation of anatomically distinct somite-derived structures in different regions of the body (reviewed in Burke, 2000; Gad and Tam, 1999; Krumlauf, 1994).

Fate mapping of the mouse embryo has provided a detailed elucidation of the successive localization of the somite precursors in the germ layers, the primitive streak, and the tail bud during gastrulation and organogenesis. Lineage analyses suggest that the precursor population behaves as a self-renewing cell population (Fig. 5a) from which new cells are constantly recruited to the presomitic mesoderm to sustain somitogenesis (Fig. 5b). However, neither the size nor the proliferative activity of the precursor population in the primitive streak and the tail bud is known. Both parameters may be critical for controlling the allocation of cells to the somitic mesoderm and the maintenance of the progenitor pool and the determination of somite size. The precursor cells (Fig. 5a) are not fated for specific somites along the AP axis, but are likely to acquire their positional identity after their allocation to the paraxial mesoderm. During development, the precursor cells display no discernible restriction in the somitogenetic potency that may account for the variation in segmental characteristics or cessation of somitogenesis. Recent evidence suggests that the maintenance of the precursor population in the tail bud and the allocation of progenitor cells to the presomitic mesoderm are dependent on TGF-β and FGF signaling. However, the precise molecular and cellular mechanisms that mediate the inductive interac-

tions for the specification of the somite precursors at gastrulation and the maintenance of these precursors during somitogenesis are not known.

After their recruitment to the paraxial mesoderm, cells in the presomitic mesoderm are organized into presegmental units as shown by the meristic pattern of somitomeres (Fig. 5b) and the generation of a constant number of somites from the presomitic mesoderm in explants. This concept is further supported by the regionalized transcriptional activities that appear to delineate domains of prospective somites in the presomitic mesoderm (see Fig. 3). It is an attractive hypothesis that, immediately on their exit from the precursor population, groups of mesodermal cells are allocated *en masse* to prospective somites as the prelude to segmentation. However, there is at present no evidence that cells are permanently assigned to any presegmental unit prior to segmentation. There is also no known gene activity that defines or reveals the organization of the presomitic mesoderm into somitomeres or the corresponding number of prospective somites. The expression patterns of genes that are finally restricted to either one somite or specific halves or regions of the somites indicate that the regionalized gene expression is consolidated in the presomitic mesoderm immediately prior to segmentation (Fig. 5c). Whether such differential genetic activity is strictly correlated with the developmental fate of the cells regarding segmental destination and lineage differentiation is not known. However, it is likely that the enhanced gene activity in the anterior region of the presomitic mesoderm is required for changes in cellular properties such as cell shape, cell adhesion, and the assembly of extracellular matrix. The outcome of these molecular activities may be predominantly associated with epithelialization of the presomitic mesoderm and the generation of borders between tissue compartments. Thus, it will be important to identify the molecules that actually execute the morphogenetic program in the anterior presomitic mesoderm and to define the regulatory networks that control their coordinate expression.

The generation of tissue discontinuity during the segmentation of the mesoderm and the periodicity of segmentation has been attributed to regulation by a rhythmic gating process. The oscillating patterns of gene expression that can be visualized as "waves" of transcriptional activity that spread through the cells in the presomitic mesoderm (Fig. 5d) provide a molecular basis for such a mechanism. It could be envisaged that the products of these oscillating genes or targets of their activity transiently accumulate during each cycle. The concerted activity of a group of cells could then be brought about by the interaction of a wavefront of maturation with these gene products, which thereby synchronizes the activation and execution of the segmentation program in every cell in the population. Alternatively, the products of oscillating genes or targets of their activity could accumulate with each additional cycle until they reach a threshold level that drives the simultaneous execution of the segmentation

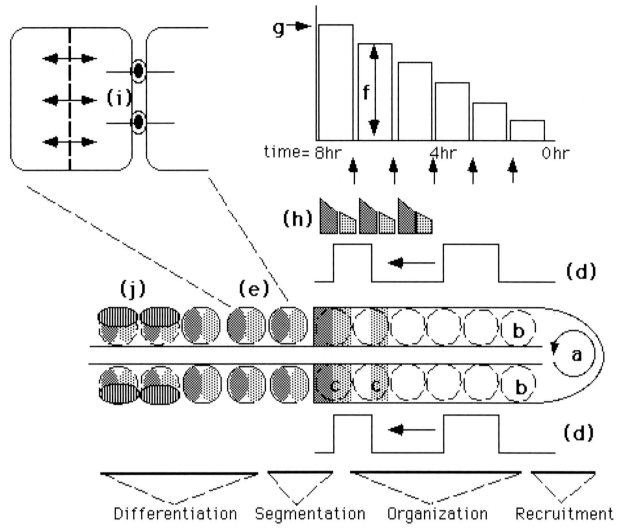

Figure 5 Morphogenetic regulation of somite formation. *Recruitment:* Somitic mesoderm is recruited from the self-renewing pool of somite precursors in the primitive streak and the tail bud (a). The newly recruited cells are allocated to both bands of paraxial mesoderm and are incorporated into (b) prospective somites (or somitomeres). *Organization:* Mesodermal cells progressively mature in the presomitic mesoderm. The position of a particular prospective somite in the presomitic mesoderm shifts anteriorly as the most mature prospective somite segments from the anterior end of this tissue. The expression of some genes becomes localized to bands of mesodermal cells that either correspond to one-somite or half-somite size. This may indicate the consolidation of the physical domain corresponding to the whole or (c) half of the prospective somite. (d) Waves of transcriptional activation are initiated in the somite precursor population or the posterior presomitic mesoderm and appear to spread anteriorly (in the direction of the arrow) as successive groups of cells activate and suppress the expression of the same gene. The progression of waves of transcriptional activity is expected to be bilaterally synchronous [shown in (d) as two identical waves forms] to achieve the bilateral symmetry of the segmental pattern and registering of somite boundaries. The duration of a complete sweep of the wave through the presomitic mesoderm and the interval between consecutive passes of the wavefront (shown for the *Lnfg* gene with a 4-hr cycle and 2-hr interval) may regulate the rate of somite segmentation and the size of the segments. *Segmentation:* Blocks of mesodermal cells (e) are successively separated from the anterior end of the presomitic mesoderm. The timing and coordinate execution of the segmentation program can be achieved by counting the number of gene activation cycles that cells have experienced after their allocation to the presomitic mesoderm or by titrating the quantal amount of gene products that are generated over time (f). Segmentation will be initiated once the gene products in the mesodermal cells have reached the threshold values (g). A periodic gating mechanism (clock) operates to allow the stepwise preparation (multiple arrows along the x axis) and eventual activation of the segmentation process. The anterior and posterior portions of the prospective somites display differential gene activities (h), which herald AP segment polarity and link the molecular clock with Notch signaling, thereby providing a potential molecular basis to couple segmentation and the generation of the AP polarity of the prospective somites. The compartmental interface is maintained by cell–cell interactions that require the Notch signaling. (i) The inter- and intra-somitic interfaces have different morphogenetic roles during the resegmentation of the paraxial mesoderm. Segmentation (or epithelialization) of the paraxial mesoderm, however, is not essential for the differentiation but may have a role in the patterning of the myogenic and skeletogenic derivatives the somite. *Differentiation:* Inductive interactions between somitic compartments and between the somite and the surrounding embryonic structures lead to the establishment of the sclerotome and dermomyotome (j) and subsequent further differentiation.

program in a group of cells (Figs. 5f and 5g). The oscillating patterns of gene expression commence at the caudal end of the presomitic mesoderm, suggesting that cells entrain the oscillator as soon as they are recruited. This raises the question of whether and how recruitment and entrainment are coupled, and whether it is possible to recruit cells to the paraxial mesoderm without subjecting them to the molecular clock, or to effectively eliminate the clock without disrupting the recruitment of cells to the paraxial mesoderm. At present, this cannot be answered unambiguously, although the phenotypes of disrupted Notch signaling may suggest that recruitment and entrainment are independent. This notion is based on the finding that *Dll1* mutants generate paraxial mesoderm despite the loss of *Lfng* and *Hes1* expression. However, it is not known whether these cycling genes are the clockwork or only the readouts of an upstream mechanism that still operates in mutant embryos. The latter notion is supported by the observations that neither the loss of *Lfng* nor of *Hes1*, the two known oscillating genes in mouse, abolishes segmentation or leads to an obvious variation of the rate of segmentation.

The phenotypes caused by mutations of genes encoding components of the Notch signaling pathway have pointed to an essential role of cell-to-cell communication in the establishment and the maintenance of the somite compartments and borders (Fig. 5e). However, the precise mechanism that generates this AP asymmetry is unclear. The oscillating expression of *Lfng* and the disrupted AP somite patterning in Notch pathway mutants suggest that the molecular oscillator may regulate the morphogenetic processes mediated by Notch signaling. Synchronization of the oscillator (Fig. 5d) may be achieved as cells leave the progenitor population and are recruited to the paraxial mesoderm on either side of the body axis (Fig. 5b). Such synchronization may be critical for the generation of bilateral symmetry of the somite pattern.

The processes leading to the formation of epithelial somites and their AP subdivision are largely independent from signals of the surrounding tissues. In contrast, the partitions along the dorsoventral and mediolateral axes and the generation of the differentiated cell lineages (Fig. 5j) are governed by complex networks of agonistic and antagonistic signals that emerge from virtually every tissue surrounding the mature epithelial somites (Fig. 4). Many of the molecular players that act on the differentiation of somites are known but how particular differentiation pathways are initiated is less clear. It is noteworthy that the paraxial mesoderm differentiates in mutants with defective segmental patterning or disrupted compartmentalization, although the spatial arrangement of the differentiated cell types might be abnormal. This suggests that early and late somite patterning processes are not interdependent, and segmentation and compartmentalization are not required to render paraxial mesoderm cells competent to respond to inductive differentiation signals. Rather, segmentation and compartmentalization appear to impose spatial restrictions on the differentiating cell types to set the stage for and ensure the highly ordered outcome of the subsequent morphogenetic program.

Acknowledgments

We wish to thank Bruce Davidson, Jacqueline Gad, Andreas Kispert, Meredith O'Rourke, and Peter Rowe for their critical comments on the manuscript. Our work is supported by the German Research Council (AG), the National Health and Medical Research Council (NHMRC) of Australia, and Mr. James Fairfax (P.P.L.T.). P.P.L.T. is a NHMRC senior principal research fellow.

References

Arnold, H. H., and Braun, T. (2000). Genetics of muscle determination and development. *Curr. Top. Dev. Biol.* **48,** 129–164.

Artavanis-Tsakonas, S., Rand, M. D., and Lake, R. J. (1999). Notch signaling: Cell fate control and signal integration in development. *Science* **284,** 770–776.

Avery, G., Chow, M., and Holtzer, H. (1956). An experimental analysis of the development of the spinal column V. Reactivity of chick somites. *J. Exp. Zool.* **132,** 409–423.

Beddington, R. S., Rashbass, P., and Wilson, V. (1992). Brachyury—a gene affecting mouse gastrulation and early organogenesis. *Development (Cambridge, UK) Suppl.* pp. 157–165.

Bellairs, R. (1986). The tail bud and cessation of segmentation in the chick embryo. *Life Sci.* **118,** 161–188.

Bettenhausen, B., and Gossler, A. (1995). Efficient isolation of novel mouse genes differentially expressed in early postimplantation embryos. *Genomics* **28,** 436–441.

Biben, C. *et al.* (1998). Murine cerberus homologue mCer-1: A candidate anterior patterning molecule. *Dev. Biol.* **194,** 135–151.

Blaumueller, C. M., and Artavanis-Tsakonas, S. (1997). Comparative aspects of Notch signaling in lower and higher eukaryotes. *Perspect. Dev. Neurobiol.* **4,** 325–343.

Blaumueller, C. M., Qi, H., Zagouras, P., and Artavanis-Tsakonas, S. (1997). Intracellular cleavage of Notch leads to a heterodimeric receptor on the plasma membrane. *Cell (Cambridge, Mass.)* **90,** 281–291.

Borello, U., Coletta, M., Tajbakhsh, S., Leyns, L., De Robertis, E. M., Buckingham, M., and Cossu, G. (1999). Transplacental delivery of the Wnt antagonist Frzb1 inhibits development of caudal paraxial mesoderm and skeletal myogenesis in mouse embryos. *Development (Cambridge, UK)* **126,** 4247–4255.

Borycki, A. G., and Emerson, C. P., Jr. (2000). Multiple tissue interactions and signal transduction pathways control somite myogenesis. *Curr. Top. Dev. Biol.* **48,** 165–224.

Borycki, A. G., Brunk, B., Tajbakhsh, S., Buckingham, M., Chiang, C., and Emerson, C. P., Jr. (1999). Sonic hedgehog controls epaxial muscle determination through Myf5 activation. *Development (Cambridge, UK)* **126,** 4053–4063.

Brand-Saberi, B., Ebensperger, C., Wilting, J., Balling, R., and Christ, B. (1993). The ventralizing effect of the notochord on somite differentiation in chick embryos. *Anat. Embryol.* **188,** 239–245.

Brill, G., Kahane, N., Carmeli, C., von Schack, D., Barde, Y. A., and Kalcheim, C. (1995). Epithelial-mesenchymal conversion of dermatome progenitors requires neural tube-derived signals: Characterization of the role of Neurotrophin-3. *Development (Cambridge, UK)* **121,** 2583–2594.

Buffinger, N., and Stockdale, F. E. (1994). Myogenic specification in somites: Induction by axial structures. *Development (Cambridge, UK)* **120,** 1443–1452.

Bumcrot, D. A., and McMahon, A. P. (1995). Somite differentiation. Sonic signals somites. *Curr. Biol.* **5,** 612–614.

Burgess, R., Cserjesi, P., Ligon, K. L., and Olson, E. N. (1995). Paraxis: A basic helix-loop-helix protein expressed in paraxial mesoderm and developing somites. *Dev. Biol.* **168,** 296–306.

Burgess, R., Rawls, A., Brown, D., Bradley, A., and Olson, E. (1996). Requirement of the *paraxis* gene for somite formation and muscoskeletal patterning. *Nature (London)* **384,** 570–573.

Burke, A. C. (2000). Hox genes and the global patterning of the somitic mesoderm. *Curr. Top. Dev. Biol.* **47,** 155–181.

Campos-Ortega, J. A. (1995). Genetic mechanisms of early neurogenesis in *Drosophila melanogaster. Mol. Neurobiol.* **10,** 75–89.

Candia, A. F. *et al.* (1992). *Mox-1* and *Mox-2* define a novel homeobox gene subfamily and are differentially expressed during early mesoderma patterning in mouse embryos. *Development (Cambridge, UK)* **116,** 1123–1136.

Chapman, D. L., and Papaioannou, V. E. (1998). Three neural tubes in mouse embryos with mutations in the T-box gene Tbx6. *Nature (London)* **391,** 695–697.

Chapman, D. L. *et al.* (1996). Expression of the T-box family genes, Tbx1-Tbx5, during early mouse development. *Dev. Dyn.* **206,** 379–390.

Chapman, D. L., Agulnik, I., Hancock, S., Silver, L. M., and Papaioannou, V. E. (1996). Tbx6, a mouse T-Box gene implicated in paraxial mesoderm formation at gastrulation. *Dev. Biol.* **180,** 534–542.

Cheah, K. S., Lau, E. T., Au, P. K., and Tam, P. P. (1991). Expression of the mouse alpha 1(II) collagen gene is not restricted to cartilage during development. *Development* **111,** 945–953.

Chiang, C., Litingtung, Y., Lee, E., Young, K. E., Corden, J. L., Westphal, H., and Beachy, P. A. (1996). Cyclopia and defective axial patterning in mice lacking Sonic hedgehog gene function. *Nature (London)* **383,** 407–413.

Christ, B., and Ordahl, C. P. (1995). Early stages of chick somite development. *Anat. Embryol.* **191,** 381–396.

Christ, B., and Wilting, J. (1992). From somites to vertebral column. *Ann. Anat.* **174,** 23–32.

Christ, B., Schmidt, C., Huang, R., Wilting, J., and Brand-Saberi, B. (1998). Segmentation of the vertebrate body. *Anat. Embryol.* **197,** 1–8.

Cinnamon, Y., Kahane, N., and Kalcheim, C. (1999). Characterization of the early development of specific hypaxial muscles from the ventrolaterial myotome. *Development (Cambridge, UK)* **126,** 4305–4315.

Ciruna, B. G., Schwartz, L., Harpal, K., Yamaguchi, T. P., and Rossant, J. (1997). Chimeric analysis of fibroblast growth factor receptor-1 (Fgfr1). function: A role for FGFR1 in morphogenetic movement through the primitive streak. *Development (Cambridge, UK)* **124,** 2829–2841.

Conlon, F. L., Wright, C. V., and Robertson, E. J. (1995a). Effects of the TWis mutation on notochord formation and mesodermal patterning. *Mech. Dev.* **49,** 201–209.

Conlon, R. A., Réaume, A. G., and Rossant, J. (1995b). *Notch1* is required for the coordinate segmentation of somites. *Development (Cambridge, UK)* **121,** 1533–1545.

Cooke, J. (1975). Control of somite number during morphogenesis of a vertebrate, *Xenopus laevis. Nature (London)* **254,** 196–199.

Cooke, J. (1977). The control of somite number during amphibian development: Models and experiments. *In* "Vertebrate Limb and Somite Morphogenesis," D. A. Ede, *et al.,* eds., pp. 434–448. Cambridge University Press, Cambridge, UK.

Cooke, J. (1981). The problem of periodic patterns in embryos. *Philos. Trans. R. Soc. London, Ser. B.* **295,** 509–524.

Cooke, J. (1998). A gene that resuscitates a theory—somitogenesis and a molecular oscillator. *Trends Genet.* **14,** 85–88.

Cooke, J., and Zeeman, E. C. (1976). A clock and wavefront model for control of the number of repeated structures during animal morphogenesis. *J. Theor. Biol.* **58,** 455–476.

Corbin, V., Michelson, A. M., Abmayr, S. M., Neel, V., Alcamo, E., Maniatis, T., and Young, M. W. (1991). A role for the *Drosophila* neurogenic

genes in mesoderm differentiation. *Cell (Cambridge, Mass.)* **67,** 311–323.

Cossu, G., Kelly, R., Tajbakhsh, S., Di Donna, S., Vivarelli, E., and Buckingham, M. (1996a). Activation of different myogenic pathways: myf-5 is induced by the neural tube and MyoD by the dorsal ectoderm in mouse paraxial mesoderm. *Development (Cambridge, UK)* **122,** 429–437.

Cossu, G., Tajbakhsh, S., and Buckingham, M. (1996b). How is myogenesis initiated in the embryo? *Trends Genet.* **12,** 218–223.

Couly, G. F., Coltey, P. M., and Le Douarin, N. M. (1992). The developmental fate of the cephalic mesoderm in quail-chick chimeras. *Development (Cambridge, UK)* **114,** 1–15.

Couso, J. P., Knust, E., and Martinez-Arias, A. (1995). Serrate and wingless cooperate to induce vestigial gene expression and wing formation in Drosophila. *Curr. Biol.* **5,** 1437–1448.

Crossley, P. H., and Martin, G. R. (1995). The mouse Fgf8 gene encodes a family of polypeptides and is expressed in regions that direct outgrowth and patterning in the developing embryo. *Development (Cambridge, UK)* **121,** 439–451.

de Celis, J. F., and Garcia Bellido, A. (1994). Roles of the Notch gene in Drosophila wing morphogenesis. *Mech. Dev.* **46,** 109–122.

de la Pompa, J. L. *et al.* (1997). Conservation of the Notch signalling pathway in mammalian neurogenesis. *Development (Cambridge, UK)* **124,** 1139–1148.

del Barco Barrantes, I., Elia, A. J., Wünsch, K., Hrabe de Angelis, M., Mak, T. W., Rossant, R., Conlon, R. A., Gossler, A., and de la Pompa, J.-L. (1999). Interaction between L-fringe and Notch signaling in the regulation of boundary formation and posterior identity in the presomitic mesoderm of the mouse. *Curr. Biol.* **9,** 470–480.

Dietrich, S., Schubert, F. R., and Gruss, P. (1993). Altered Pax gene expression in murine notochord mutants: The notochord is required to initiate and maintain ventral identity in the somite. *Mech. Dev.* **44,** 189–207.

Dockter, J. L. (2000). Sclerotome induction and differentiation. *Curr. Top. Dev. Biol.* **48,** 77–127.

Doherty, D., Feger, G., Younger-Shepherd, S., Jan, L. Y., and Jan, Y. N. (1996). Delta is a ventral to dorsal signal complementary to Serrate, another Notch ligand, in *Drosophila* wing formation. *Genes Dev.* **10,** 421–434.

Dunwoodie, S. L., Henrique, D., Harrison, S. M., and Beddington, R. S. (1997). Mouse Dll3: A novel divergent Delta gene which may complement the function of other Delta homologues during early pattern formation in the mouse embryo. *Development (Cambridge, UK)* **124,** 3065–3076.

Ebensperger, C., Wilting, J., Brand Saberi, B., Mizutani, Y., Christ, B., Balling, R., and Koseki, H. (1995). *Pax-1,* a regulator of sclerotome development is induced by notochord and floor plate signals in avian embryos. *Anat. Embryol.* **191,** 297–310.

Echelard, Y., Epstein, D. J., St Jacques, B., Shen, L., Mohler, J., McMahon, J. A., and McMahon, A. P. (1993). Sonic hedgehog, a member of a family of putative signaling molecules, is implicated in the regulation of CNS polarity. *Cell (Cambridge, Mass.)* **75,** 1417–1430.

Evrard, Y. A., Lun, Y., Aulehla, A., Gan, L., and Johnson, R. L. (1998). *Lunatic fringe* is an essential mediator of somite segmentation and patterning. *Nature (London)* **394,** 377–381.

Fan, C. M., and Tessier Lavigne, M. (1994). Patterning of mammalian somites by surface ectoderm and notochord: Evidence for sclerotome induction by a hedgehog homolog. *Cell (Cambridge, Mass.)* **79,** 1175–1186.

Fan, C. M., Porter, J. A., Chiang, C., Chang, D. T., Beachy, P. A., and Tessier-Lavigne, M. (1995). Long-range sclerotome induction by Sonic Hedgehog: Direct role of the amino-terminal cleavage product and modulation by the cyclic AMP signaling pathway. *Cell (Cambridge, Mass.)* **81,** 457–465.

Fehon, R. G., Kooh, P. J., Rebay, I., Regan, C. L., Xu, T., Muskavitch, M. A., and Artavanis-Tsakonas, S. (1990). Molecular interactions between the protein products of the neurogenic loci *Notch* and *Delta,* two

EGF-homologous genes in Drosophila. *Cell (Cambridge, Mass.)* **61**, 523–534.

Fleming, R. J., Gu, Y., and Hukriede, N. A. (1997). Serrate-mediated activation of Notch is specifically blocked by the product of the gene fringe in the dorsal compartment of the Drosophila wing imaginal disc. *Development (Cambridge, UK)* **124**, 2973–2981.

Flint, O. P., and Ede, D. A. (1978). Cell interactions in the developing somite; *in vivo* comparisons between *amputated (am/am)* and normal mouse embryos. *J. Cell Sci.* **31**, 275–291.

Flint, O. P., Ede, D. A., Wilby, O. K., and Proctor, J. (1978). Control of somite number in normal and *amputated* mutant mouse embryos: An experimental and a theoretical analysis. *J. Embryol. Exp. Morphol.* **45**, 189–202.

Forsberg, H., Crozet, F., and Brown, N. A. (1998). Waves of mouse Lunatic fringe expression, in four-hour cycles at two- hour intervals, precede somite boundary formation. *Curr. Biol.* **8**, 1027–1030.

Fortini, M. E., and Artavanis-Tsakonas, S. (1994). The suppressor of hairless protein participates in notch receptor signaling. *Cell (Cambridge, Mass.)* **79**, 273–282.

Füchtbauer, E. M. (1995). Expression of M-twist during postimplantation development of the mouse. *Dev. Dyn.* **204**, 316–322.

Furumoto, T. *et al.* (1999). Notochord dependent expression of MFH-1 and Pax-1 cooperates to maintain the proliferation of sclerotome cells during vertebral column development. *Dev. Biol.* **210**, 15–29.

Gad, J. M., and Tam, P. P. (1999). Axis development: The mouse becomes a dachshund. *Curr. Biol.* **9**, R783–R786.

Gitelman, I. (1997). Twist protein in mouse enbryogenesis. *Dev. Biol.* **189**, 205–214.

Gofflot, F., Hall, M., and Morriss-Kay, G. M. (1997). Genetic patterning of the developing mouse tail at the time of posterior neuropore closure. *Dev. Dyn.* **210**, 431–445.

Goldman, D., Martin, G., and Tam, P. P. L. (2000). Fate and function of ventral ectodermal ridge during mouse tail development. *Development (Cambridge, UK)* **127**, 2113–2123.

Goldstein, R. S., and Kalcheim, C. (1992). Determination of epithelial half-somites in skeletal morphogenesis. *Development (Cambridge, UK)* **116**, 441–445.

Gossler, A., and Hrabe de Angelis, M. (1997). Somitogenesis. *Curr. Top. Dev. Biol.* **38**, 225–287.

Greco, T. L. *et al.* (1996). Analysis of the *vestigial tail* mutation demonstrates that *Wnt-3a* gene dosage regulates mouse axial development. *Genes Dev.* **10**, 313–324.

Gridley, T. (1997). Notch signaling in vertebrate development and disease. *Mol. Cell. Neurosci.* **9**, 103–108.

Grüneberg, H. (1955). Genetical studies on the skeleton of the mouse. XVI. Tail-kinks. *J. Genet.* **53**, 536–550.

Grüneberg, H. (1956). Genetical studies on the skeleton of the mouse. XVIII. Three genes for syndactylism. *J. Genet.* **54**, 113–145.

Grüneberg, H. (1961). Genetical studies on the skeleton of the mouse. XXIX. Pudgy. *Genet. Res.* **2**, 384–393.

Hamada, Y., Kadokawa, Y., Okabe, M., Ikawa, M., Coleman, J. R., and Tsujimoto, Y. (1999). Mutation in ankyrin repeats of the mouse Notch2 gene induces early embryonic lethality. *Development (Cambridge, UK)* **126**, 3415–3424.

Hartenstein, A. Y., Rugendorff, A., Tepass, U., and Hartenstein, V. (1992). The function of the neurogenic genes during epithelial development in the Drosophila embryo. *Development (Cambridge, UK)* **116**, 1203–1220.

Herrmann, B. G. (1991). Expression pattern of the *Brachyury* gene in whole-mount *T^{Wis}/T^{Wis}* mutant embryos. *Development (Cambridge, UK)* **113**, 913–917.

Herrmann, B. G. (1995). The mouse *Brachyury (T)* gene. *Dev. Biol.* **6**, 385–394.

Hirsinger, E., Duprez, D., Jouve, C., Malapert, P., Cooke, J., and Pourquié, O. (1997). Noggin acts downstream of Wnt and Sonic Hedgehog to antago-

nize BMP4 in avian somite patterning. *Development (Cambridge, UK)* **124**, 4605–4614.

Hirsinger, E., Jouve, C., Malapert, P., and Pourquié, O. (1998). Role of growth factors in shaping the developing somite. *Mol. Cell. Endocrinol.* **140**, 83–87.

Hrabé de Angelis, M., McIntyre, J., II, and Gossler, A. (1997). Maintenance of somite borders in mice requires the *Delta* homolog *Dll1. Nature (London)* **386**, 717–721.

Huang, R., Zhi, Q., Wilting, J., and Christ, B. (1994). The fate of somitocoele cells in avian embryos. *Anat. Embryol.* **190**, 243–250.

Huang, R., Zhi, Q., Neubüser, A., Müller, T. S., Brand-Saberi, B., Christ, B., and Wilting, J. (1996). Function of somite and somitocoel cells in the formation of the vertebral motion segment in avian embryos. *Acta Anat.* **155**, 231–241.

Huang, R., Zhi, Q., Schmidt, C., Wilting, J., Brand-Saberi, B., and Christ, B. (2000). Sclerotomal origin of the ribs. *Development (Cambridge, UK)* **127**, 527–532.

Irvine, K. D., and Wieschaus, E. (1994). Fringe, a boundary-specific signaling molecule, mediates interactions between dorsal and ventral cells during Drosophila wing development. *Cell (Cambridge, Mass.)* **79**, 595–606.

Ishibashi, M., Ang, S.-L., Shiota, K., Nakanishi, S., Kageyama, R., and Guillemot, F. (1995). Targeted disruption of mammalian *hairy* and *Enhancer of split* homolog-1 *(HES-1).* leads to up-regulation of neural helix-loop-helix factors, premature neurogenesis, and severe neural tube defects. *Genes Dev.* **9**, 3136–3148.

Jarriault, S., Brou, C., Logeat, F., Schroeter, E. H., Kopan, R., and Israel, A. (1995). Signaling downstream of activated mammalian Notch. *Nature (London)* **377**, 355–358.

Jarriault, S., Le Bail, O., Hirsinger, E., Pourquié, O., Logeat, F., Strong, C. F., Brou, C., Seidah, N. G., and Israel, A. (1998). Delta-1 activation of notch-1 signaling results in HES-1 transactivation. *Mol. Cell. Biol.* **18**, 7423–7431.

Jen, W., Gawantka, V., Pollet, N., Niehrs, C., and Kintner, C. (1999). Periodic repression of notch pathway genes governs the segmentation of xenopus embryos. *Genes Dev.* **13**, 1486–1499.

Jiang, Y.-J., Smithers, L., and Lewis, J. (1998). Vertebrate segmentation: The clock is linked to Notch signaling. *Curr. Biol.* **8**, R868–R871.

Johnson, R. L., Laufer, E., Riddle, R. D., and Tabin, C. (1994). Ectopic expression of *Sonic hedgehog* alters dorsal-ventral patterning of somites. *Cell (Cambridge, Mass.)* **79**, 1165–1173.

Jouve, C., Palmeirim, I., Henrique, D., Beckers, J., Gossler, A., Ish-Horowcz, D., and Pourquié, O. (2000). Notch signaling is required for cyclic expression of the hairy-like gene HES1 in the presomitic mesoderm. *Development (Cambridge, UK)* **127**, 1421–1429.

Kaestner, K. H. *et al.* (1996). Clustered arrangement of winged helix genes fkh-6 and MFH-1: Possible implications for mesoderm development. *Development (Cambridge, UK)* **122**, 1751–1758.

Kahane, N., and Kalcheim, C. (1998). Identification of early postmitotic cells in distinct embryonic sites and their possible roles in morphogenesis. *Cell Tissue Res.* **294**, 297–307.

Kahane, N., Cinnamon, Y., and Kalcheim, C. (1998). The cellular mechanism by which the dermomyotome contributes to the second wave of myotome development. *Development (Cambridge, UK)* **125**, 4259–4271.

Kalcheim, C., Cinnamon, Y., and Kahane, N. (1999). Myotome formation: A multistage process. *Cell Tissue Res.* **296**, 161–173.

Kaufman, M. H., and Bard, J. B. L. (1999). "The Anatomical Basis of Mouse Development." Academic Press, London.

Kenny, D. A., Jurata, L. W., Saga, Y., and Gill, G. N. (1998). Identification and characterization of LMO4, and LMO gene with a novel pattern of expression during embryogenesis. *Preoc. Natl. Acad. Sci. U.S.A.* **95**, 11257–11262.

Kenny-Mobbs, T., and Thorogood, P. (1987). Autonomy of differentiation in avian branchial somites and the influence of adjacent tissues. *Development (Cambridge, UK)* **100**, 449–462.

Kessel, M., and Gruss, P. (1991). Homeotic transformations of murine vertebrae and concomitant alteration of Hox codes induced by retinoic acid. *Cell (Cambridge, Mass.)* **67**, 89–104.

Keynes, R. J., and Stern, C. D. (1984). Segmentation in the vertebrate nervous system. *Nature (London)* **310**, 786–789.

Keynes, R. J., and Stern, C. D. (1988). Mechanisms of vertebrate segmentation. *Development (Cambridge, UK)* **103**, 413–429.

Kidd, S., Lieber, T., and Young, M. W. (1998). Ligand-induced cleavage and regulation of nuclear entry of Notch in Drosophila melanogaster embryos. *Genes Dev.* **12**, 3728–3740.

Kinder, S. J., Tsang, T. E., Quinlan, G. A., Hadjantonakis, A.-K., Nagy, A., and Tam, P. P. L. (1999). The orderly allocation of mesodermal cells to the extraembryonic structures and the anteroposterior axis during gastrulation of the mouse embryo. *Development (Cambridge, UK)* **126**, 4691–4701.

Kopan, R., Schroeter, E. H., Weintraub, H., and Nye, J. S. (1996). Signal transduction by activated mNotch: Importance of proteolytic processing and its regulation by the extracellular domain. *Proc. Natl. Acad. Sci. U.S.A.* **93**, 1683–1688.

Koseki, H., Wallin, J., Wilting, J., Mizutani, Y., Kispert, A., Ebensperger, C., Herrmann, B. G., Christ, B., and Balling, R. (1993). A role for *Pax-1* as a mediator of notochordal signals during the dorsoventral specification of vertebrae. *Development (Cambridge, UK)* **119**, 649–660.

Kraus, F., Haenig, B., and Kispert, A. (2001). Cloning and expression analysis of mouse T-box *Tbx18*. *Mech. Dev.* **100**, 83–86.

Krumlauf, R. (1994). Hox genes in vertebrate development. *Cell (Cambridge, Mass.)* **78**, 191–201.

Kume, T. *et al.* (1998). The forkhead/winged helix gene Mf1 is disrupted in the pleiotropic mouse mutation congenital hydrocephalus. *Cell (Cambridge, Mass.)* **93**, 985–996.

Kusumi, K., Sun, E. S., Kerrebrock, A. W., Bronson, R. T., Chi, D.-C., Bulotsky, M. S., Spencer, J. B., Birren, B. W., Frankel, W. N., and Lander, E. S. (1998). The mouse pudgy mutation disrupts *Delta* homolog *Dll3* and initiation of early somite boundaries. *Nat. Genet.* **19**, 274–278.

Lassar, A. B., and Münsterberg, A. E. (1996). The role of positive and negative signals in somite patterning. *Curr. Opin. Neurobiol.* **6**, 57–63.

Lee, C. S., Buttitta, L. A., May, N. R., Kispert, A., and Fan, C. M. (2000). SHH-N upregulates Sfrp2 to mediate its competitive interaction with WNT1 and WNT4 in the somitic mesoderm. *Development (Cambridge, UK)* **127**, 109–118.

Leitges, M., Neidhardt, L., Haenig, B., Herrmann, B. G., and Kispert, A. (2000). The paired homeobox gene Uncx4.1 specifies pedicles, transverse processes and proximal ribs of the vertebral column. *Development (Cambridge, UK)* **127**, 2259–2267.

Mansouri, A. *et al.* (1997). Paired-related murine homeobox gene expressed in the developing sclerotome, kidney, and nervous system. *Dev. Dyn.* **210**, 53–65.

Mansouri, A., Voss, A. K. Thomas, T., Yokota, Y., and Gruss, P. (2000). The mouse homeobox gene Unvx4.1 acts downstream of Notch and directs the formation of skeletal structures. *Development (Cambridge, UK)* **127**, 2251–2258.

Marcelle, C., Stark, M. R., and Bronner-Fraser, M. (1997). Coordinate actions of BMPs, Wnts, Shh and noggin mediate patterning of the dorsal somite. *Development (Cambridge, UK)* **124**, 3955–3963.

McGrew, M. J., and Pourquié, O. (1998). Somitogenesis: Segmenting a vertebrate. *Curr. Opin. Genet. Dev.* **8**, 487–493.

McGrew, M. J., Dale, J. K., Fraboulet, S., and Pourquié, O. (1998). The lunatic Fringe gene is a target of the molecular clock linked to somite segmentation in avian embryos. *Curr. Biol.* **8**, 979–982.

McMahon, J. A., Takada, S., Zimmerman, L. B., Faq, C.-M., Harland, R. M., and McMahon, A. P. (1998). Noggin-mediated antagonism of BMP signaling is required for growth and patterning of the neural tube and somite. *Genes Dev.* **12**, 1438–1452.

McPherron, A. C., Lawler, A. M., and Lee, S. J. (1999). Regulation of anterior/posterior patterning of the axial skeleton by growth/differentiation factor 11. *Nat. Genet.* **22**, 260–264.

Meinhardt, H. (1986). Models of segmentation. *Life Sci.* **118**, 179–189.

Meng, A., Moore, B., Tang, H., Yuan, B., and Lin, S. (1999). A Drosophila doublesex-related gene, terra, is involved in somitogenesis in vertebrates. *Development (Cambridge, UK)* **126**, 1259–1268.

Monsoro-Burq, A.-H., Bontoux, M., Teillet, M.-A., and Le Douarin, N. M. (1994). Heterogeneity in the development of the vertebra. *Proc. Natl. Acad. Sci. U.S.A.* **91**, 10435–10439.

Monsoro-Burq, A. H., Duprez, D., Watanabe, Y., Bontoux, M., Vincent, C., Brickell, P., and Le Douarin, N. (1996). The role of bone morphogenetic proteins in vertebral development. *Development (Cambridge, UK)* **122**, 3607–3616.

Morgan, W. C. (1954). A new crooked tail mutation involving distinctive pleiotropism. *J. Genet.* **52**, 354–373.

Müller, M., Weizäcker, E., and Campos-Ortega, J. A. (1996). Expression domains of a zebrafish homolog of the *Drosophila* pair-rule gene *hairy* correspond to primordia of alternating somites. *Development (Cambridge, UK)* **122**, 2071–2078.

Münsterberg, A. E., and Lassar, A. B. (1995). Combinatorial signals from the neural tube, floor plate and notochord induce myogenic bHLH gene expression in the somite. *Development (Cambridge, UK)* **121**, 651–660.

Münsterberg, A. E., Kitajewski, J., Bumcrot, D. A., McMahon, A. P., and Lassar, A. B. (1995). Combinatorial signaling by Sonic hedgehog and Wnt family members induces myogenic bHLH gene expression in the somite. *Genes Dev.* **9**, 2911–2922.

Neidhardt, L. M., Kispert, A., and Herrmann, B. G. (1997). A mouse gene of the paired-related homeobox class expressed in the caudal somite compartment and in the developing vertebral column, kidney and nervous system. *Dev. Genes Evol.* **207**, 330–339.

Ng, L. J., Tam, P. P., and Cheah, K. S. (1993). Preferential expression of alternatively spliced mRNAs encoding Type II procollagen with a cysteine-rich amino-propeptide in differentiating cartilage and nonchondrogenic tissues during early mouse development. *Dev. Bio.* **159**, 403–417.

Nieto, M. A., Gilardi Hebenstreit, P., Charnay, P., and Wilkinson, D. G. (1992). A receptor protein tyrosine kinase implicated in the segmental patterning of the hindbrain and mesoderm. *Development (Cambridge, UK)* **116**, 1137–1150.

Ohtsuka, T., Ishibashi, M., Gradwohl, G., Nakanischi, S., Guillemot, F., and Kageyama, R. (1999). *Hes1* and *Hes5* as Notch effectors in mammalian neuronal differentiation. *EMBO J.* **18**, 2196–2207.

Oka, C., Nakano, T., Wakeham, A., de la Pompa, J. L., Mori, C., Sakai, T., Okazaki, S., Kawaichi, M., Shiota, K., Mak, T. W., and Honjo, T. (1995). Disruption of the mouse RBP-J kappa gene results in early embryonic death. *Development (Cambridge, UK)* **121**, 3291–3301.

Ott, M. O., Bober, E., Lyons, G., Arnold, H., and Buckingham, M. (1991). Early expression of the myogenic regulatory gene, *myf-5*, in precursor cells of skeletal muscle in the mouse embryo. *Development (Cambridge, UK)* **111**, 1097–1107.

Packard, D. S., Jr., Zheng, R. Z., and Turner, D. C. (1993). Somite pattern regulation in the avian segmental plate mesoderm. *Development (Cambridge, UK)* **117**, 779–791.

Palmeirim, I., Henrique, D., Ish-Horowicz, D., and Pourquié, O. (1997). Avian hairy gene expression identifies a molecular clock linked to vertebrate segmentation and somitogenesis. *Cell (Cambridge, Mass.)* **91**, 639–648.

Pan, D., and Rubin, G. M. (1997). Kuzbanian controls proteolytic processing of Notch and mediates lateral inhibition during drosophila and vertebrate neurogenesis. *Cell (Cambridge, Mass.)* **90**, 271–280.

Pourquié, O. (2000). Segmentation of the paraxial mesoderm and vertebrate somitogenesis. *Curr. Top. Dev. Biol.* **47**, 81–105.

Pourquié, O., Coltey, M., Teillet, M. A., Ordahl, C., and Le Douarin, N. M. (1993). Control of dorsoventral patterning of somitic derivatives by notochord and floor plate. *Proc. Natl. Acad. Sci. U.S.A.* **90**, 5242–5246.

Pourquié, O., Coltey, M., Bréant, C., and Le Douarin, N. M. (1995). Control of somite patterning by signals from the lateral plate. *Proc. Natl. Acad. Sci. U.S.A.* **92**, 3219–3223.

Pourquié, O., Fan, C. M., Coltey, M., Hirsinger, E., Watanabe, Y., Bréant, C., Francis West, P., Brickell, P., Tessier Lavigne, M., and Le Douarin, N. M. (1996). Lateral and axial signals involved in avian somite patterning: A role for BMP4. *Cell (Cambridge, Mass.)* **84**, 461–471.

Qi, H., Rand, M. D., Wu, X., Sestan, N., Wang, W., Rakic, P., Xu, T., and Artavanis-Tsakonas, S. (1999). Processing of the Notch ligand Delta by the metalloprotease kuzbanian. *Science* **283**, 91–94.

Rawls, A., Wilson-Rawls, J., and Olson, E. N. (2000). Genetic regulation of somite formation. *Curr. Top. Dev. Biol.* **47**, 131–154.

Réaume, A. G., Conlon, R. A., Zirngibl, R., Yamaguchi, T. P., and Rossant, J. (1992). Expression analysis of a *Notch* homologue in the mouse embryo. *Dev. Biol.* **154**, 377–387.

Remak, R. (1855). "Untersuchungen über die Entwicklung der Wirbelthiere." Reimer, Berlin.

Reshef, R., Maroto, M., and Lassar, A. B. (1998). Regulation of dorsal somitic cell fates: BMPs and Noggin control the timing and pattern of myogenic regulator expression. *Genes Dev.* **12**, 290–303.

Richardson, M. K., Allen, S. P., Wright, G. M., Raynaud, A., and Hanken, J. (1998). Somite number and vertebrate evolution. *Development (Cambridge, UK)* **125**, 151–160.

Rickmann, M., Fawcett, J. W., and Keynes, R. J. (1985). The migration of neural crest cells and the growth of motor axons through the rostral half of the chick somite. *J. Embryol. Exp. Morphol.* **90**, 437–455.

Robey, E. (1997). Notch in vertebrates. *Curr. Opin. Genet. Dev.* **7**, 551–557.

Roelink, H., Augsburger, A., Heemskerk, J., Korzh, V., Norlin, S., Ruiz i Altaba, A., Tanabe, Y., Placzek, M., Edlund, T., Jessell, T. M. *et al.* (1994). Floor plate and motor neuron induction by vhh-1, a vertebrate homolog of hedgehog expressed by the notochord. *Cell (Cambridge, Mass.)* **76**, 761–775.

Rong, P. M., Teillet, M. A., Ziller, C., and Le Douarin, N. M. (1992). The neural tube/notochord complex is necessary for vertebral but not limb and body wall striated-muscle differentiation. *Development (Cambridge, UK)* **115**, 657–672.

Rusconi, J. C., and Corbin, V. (1999). A widespread and early requirement for a novel notch function during *Drosophila* embryogenesis. *Dev. Biol.* **215**, 388–398.

Saga, Y. *et al.* (1996). Mesp1: A novel basic helix-loop-helix protein expressed in the nascent mesodermal cells during mouse gastrulation. *Development (Cambridge, UK)* **122**, 2769–2778.

Saga, Y., Hata, N., Koseki, H., and Taketo, M. M. (1997). *Mesp2:* A novel mouse gene expressed in the presegmented mesoderm and essential for segmentation initiation. *Genes Dev.* **11**, 1827–1839.

Sasaki, H., and Hogan, B. L. (1993). Differential expression of multiple fork head related genes during gastrulation and axial pattern formation in the mouse embryo. *Development (Cambridge, UK)* **118**, 47–59.

Schroeter, E. H., Kisslinger, J. A., and Kopan, R. (1998). Notch-1 signaling requires ligand-induced proteolytic release of intracellular domain. *Nature (London)* **393**, 382–386.

Shawber, C., Nofziger, D., Hsieh, J. J., Lindsell, C., Bogler, O., Hayward, D., and Weinmaster, G. (1996). Notch signaling inhibits muscle cell differentiation through a CBF1-independent pathway. *Development (Cambridge, UK)* **122**, 3765–3773.

Simpson, P. (1997). Notch signaling in development. *Perspect. Dev. Neurobiol.* **4**, 297–304.

Smith, D. E., Franco Del Amo, F., and Gridley, T. (1992). Isolation of Sna, a mouse gene homologous to the Drosophila genes snail and escargot: Expression pattern suggests multiple roles during postimplantation development. *Development (Cambridge, UK)* **116**, 1033–1039.

Smith, T. H., Kachinsky, A. M., and Miller, J. B. (1994). Somite subdomains, muscle cell origins, and the four muscle regulatory factor proteins. *J. Cell Biol.* **127**, 95–105.

Snow, M. H. L., and Gregg, B. C. (1986). The programming of vertebral development. *Life Sci.* **118**, 301–311.

Solloway, M. J., and Robertson, E. J. (1999). Early embryonic lethality in Bmp5:Bmp7 double mutant mice suggests functional redundancy within the 60A subgroup. *Development (Cambridge, UK)* **126**, 1753–1768.

Spörle, R., and Schughart, K. (1997). System to identify individual somites and their derivatives in the developing mouse emgryo. *Dev. Dyn.* **210**, 216–226.

Spörle, R., Gunther, T., Struwe, M., and Schughart, K. (1996). Severe defects in the formation of epaxial musculature in open brain (opb) mutant mouse embryos. *Development (Cambridge, UK)* **122**, 79–86.

Stern, C. D., and Keynes, R. J. (1987). Interactions between somite cells: The formation and maintenance of segment boundaries in the chick embryo. *Development (Cambridge, UK)* **99**, 261–272.

Stern, C. D., and Vasiliauskas, D. (2000). Segmentation: A view from the border. *Curr. Top. Dev. Biol.* **47**, 107–129.

Stern, H. M., and Hauschka, S. D. (1995). Neural tube and notochord promote *in vitro* myogenesis in single somite explants. *Dev. Biol.* **167**, 87–103.

Struhl, G., and Adachi, A. (1998). Nuclear access and action of notch in vivo. *Cell (Cambridge, Mass.)* **93**, 649–660.

Swiatek, P. J., Lindsell, C. E., Franco Del Amo, F., Weinmaster, G., and Gridley, T. (1994). *Notch1* is essential for postimplantation development in mice. *Genes Dev.* **8**, 707–719.

Tajbakhsh, S., and Buckingham, M. E. (1994). Mouse limb muscle is determined in the absence of the earliest myogenic factor myf-5. *Proc. Natl. Acad. Sci. U.S.A.* **91**, 747–751.

Tajbakhsh, S., and Buckingham, M. (2000). The birth of muscle progenitor cells in the mouse: spatiotemporal considerations. *Curr. Top. Dev. Biol.* **48**, 225–268.

Tajbakhsh, S., and Spörle, R. (1998). Somite development: Constructing the vertebrate body. *Cell (Cambridge, Mass.)* **92**, 9–16.

Tajbakhsh, S., Borello, U., Vivarelli, E., Kelly, R., Papkoff, J., Duprez, D., Buckingham, M., and Cossu, G. (1998). Differential activation of Myf5 and MyoD by different Wnts in explants of mouse paraxial mesoderm and the later activation of myogenesis in the absence of Myf5. *Development (Cambridge, UK)* **125**, 4155–4162.

Takada, S., Stark, K. L., Shea, M. J., Vassileva, G., McMahon, J. A., and McMahon, A. P. (1994). *Wnt-3a* regulates somite and tailbud formation in the mouse embryo. *Genes Dev.* **8**, 174–189.

Takke, C., and Campos-Ortega, J. A. (1999). Her1, a zebrafish pair-rule like gene, acts downstream of notch signaling to control somite development. *Development (Cambridge, UK)* **126**, 3005–3014.

Tam, P. P. L. (1981). The control of somitogenesis in mouse embryos. *J. Embryol. Exp. Morphol.* **65**(Suppl.), 103–128.

Tam, P. P. L. (1984). The histogenetic capacity of tissues in the caudal end of the embryonic axis of the mouse. *J. Embryol. Exp. Morphol.* **82**, 253–266.

Tam, P. P. L. (1986). A study on the pattern of prospective somites in the presomitic mesoderm of mouse embryos. *J. Embryol. Exp. Morphol.* **92**, 269–285.

Tam, P. P. L. (1988). The allocation of cells in the presomitic mesoderm during somite segmentation in the mouse embryo. *Development (Cambridge, UK)* **103**, 379–390.

Tam, P. P. L., and Beddington, R. S. P. (1986). The metameric organization of the presomitic mesoderm and somite specification in the mouse embryo. *Life Sci.* **118**, 17–36.

Tam, P. P. L., and Beddington, R. S. P. (1987). The formation of mesodermal tissues in the mouse embryo during gastrulation and early organogenesis. *Development (Cambridge, UK)* **99**, 109–126.

Tam, P. P. L., and Tan, S. S. (1992). The somitogenetic potential of cells in the primitive streak and the tail bud of the organogenesis-stage mouse embryo. *Development (Cambridge, UK)* **115**, 703–715.

Tam, P. P. L., and Trainor, P. A. (1994). Specification and segmentation of the paraxial mesoderm. *Anat. Embryol.* **189**, 275–305.

Tam, P. P. L., Goldman, D., Camus, A., and Schoenwolf, G. C. (2000). Early events of somitogenesis in higher vertebrates: Allocation of precursor cells during gastrulation and the organization of a meristic pattern in the paraxial mesoderm. *Curr. Top. Dev. Biol.* **47**, 1–32.

Tan, S. S., Crossin, K. L., Hoffman, S., and Edelman, G. M. (1987). Asymmetric expression in somites of cytotactin and its proteoglycan ligand is correlated with neural crest cell distribution. *Proc. Natl. Acad. Sci. U.S.A.* **84**, 7977–7981.

Theiler, K. (1988). "Vertebral Malformations." Springer-Verlag, Berlin and New York.

Theiler, K., and Stevens, L. C. (1960). The development of rib fusions, a mutation in the house mouse. *Am. J. Anat.* **106**, 171–183.

Theiler, K., and Varnum, D. S. (1985). Development of rib-vertebrae: A new mutation in the house mouse with accessory caudal duplications. *Anat. Embryol.* **173**, 111–116.

Theiler, K., Varnum D., and Stevens, L. C. (1974). Development of rachiterate, a mutation in the house mouse with 6 cervical vertebrae. *Z. Anat. Entwicklungsgesch.* **145**, 75–80.

Theiler, K., Varnum D. S., Southard, J. L., and Stevens, L. C. (1975). Malformed vertebrae: A new mutant with the "wirbel-rippen syndrom" in the mouse. *Anat. Embryol.* **147**, 161–166.

Thomas, U., Speicher, S. A., and Knust, E. (1991). The *Drosophila* gene *Serrate* encodes an EGF-like transmembrane protein with a complex expression pattern in embryos and wing discs. *Development (Cambridge, UK)* **111**, 749–761.

Tonegawa, A., and Takahashi, Y. (1998). Somitogenesis controlled by Noggin. *Dev. Biol.* **202**, 172–182.

Vässin, H., Bremer, K. A., Knust, E., and Campos-Ortega, J. A. (1987). The neurogenic gene Delta of *Drosophila melanogaster* is expressed in neurogenic territories and encodes a putative transmembrane protein with EGF-like repeats. *EMBO J.* **6**, 3431–3440.

Venters, S. J., Thorsteinsdottir, S., and Duxson, M. J. (1999). Early development of the myotome in the mouse. *Dev. Dyn.* **216**, 219–232.

Verbout, A. J. (1976). A critical review of the 'neugliederung' concept in relation to the development of the vertebral column. *Acta Biotheor.* **25**, 219–258.

Verbout, A. J. (1985). The development of the vertebral column. *Adv. Anat. Embryol. Cell Biol.* **90**, 1–122.

Vickers, T. H. (1983). The chronology of somites in rat embryos. *Teratology* **28**, 457–460.

Vivarelli, E., and Cossu, G. (1986). Neural control of early myogenic differentiation in cultures of mouse somites. *Dev. Biol.* **117**, 319–325.

von Ebner, V. (1888). Urwirbel und Neugliederung der Wirbelsäule. *Sitzungsber Akad. Wiss. Wien, Math.-Naturwiss. Kl., Abt. 3* **97**, 194–206.

Wai, A. W. *et al.* (1998). Disrupted expression of matrix genes in the growth plate of the mouse cartilage matrix deficiency (cmd) mutant. *Dev. Gene.* **22**, 349–358.

Wharton, K. A., Johansen, K. M., Xu, T., and Artavanis-Tsakonas, S. (1985). Nucleotide sequence from the neurogenic locus notch implies a gene product that shares homology with proteins containing EGF-like repeats. *Cell (Cambridge, Mass.)* **43**, 567–581.

Wilkinson, D. G., Bhatt, S., and Herrmann, B. G. (1990). Expression pattern of the mouse *T* gene and its role in mesoderm formation. *Nature (London)* **343**, 657–659.

Williams, B. A., and Ordahl, C. P. (1994). Pax-3 expression in segmental mesoderm marks early stages in myogenic cell specification. *Development (Cambridge, UK)* **120**, 785–796.

Williams, R., Lendahl, U., and Lardelli, M. (1995). Complementary and combinatorial patterns of Notch gene family expression during early mouse development. *Mech. Dev.* **53**, 357–368.

Wilson, V., and Beddington, R. S. P. (1996). Cell fate and morphogenetic movement in the late mouse primitve streak. *Mech. Dev.* **55**, 79–89.

Wilson, V., Rashbass, P., and Beddington, R. S. (1993). Chimeric analysis of T (*Brachyury*) gene function. *Development (Cambridge, UK)* **117**, 1321–1331.

Winnier, G. E., Hargett, L., and Hogan, B. L. (1997). The winged helix transcription factor MFH1 is required for proliferation and patterning of paraxial mesoderm in the mouse embryo. *Genes Dev.* **11**, 926–940.

Wong, P. C. *et al.* (1997). Presenilin 1 is required for Notch1 and DII1 expression in the paraxial mesoderm. *Nature (London)* **387**, 288–292.

Wu, J. Y., and Rao, Y. (1999). Fringe: Defining borders by regulating the Notch pathway. *Curr. Opin. Neurobiol.* **9**, 537–543.

Xue, Y., Gao, X., Lindsell, C. E., Norton, C. R., Chang, B., Hicks, C., Gendron-Maguire, M., Rand, E. B., Weinmaster, G., and Gridley, T. (1999). Embryonic lethality and vascular defects in mice lacking the Notch ligand Jagged1. *Hum. Mol. Genet.* **8**, 723–730.

Yamaguchi, T. P., Conlon, R. A., and Rossant, J. (1992). Expression of the fibroblast growth factor receptor FGFR-1/flg during gastrulation and segmentation in the mouse embryo. *Dev. Biol.* **152**, 75–88.

Yamaguchi, T. P., Harpal, K., Henkemeyer, M., and Rossant, J. (1994). *fgfr-1* is required for embryonic growth and mesodermal patterning during mouse gastrulation. *Genes Dev.* **8**, 3032–3044.

Yamaguchi, T. P., Takada, S., Yoshikawa, Y., Wu, N., and McMahon, A. P. (1999). T (*Brachyury*). is a direct target of Wnt3a during paraxial mesoderm specification. *Genes Dev.* **13**, 3185–3190.

Yoshikawa, Y., Fujimori, T., McMahon, A. P., and Takada, S. (1997). Evidence that absence of Wnt-3a signaling promotes neuralization instead of paraxial mesoderm development in the mouse. *Dev. Biol.* **183**, 234–242.

Zakang, J., Kmita, M., Alarcon, P., de la Pompa, J.-L., and Daboule, D. (2001). Localized and transient transcription of *Hox* genes suggests a link between patterning and the segmentation clock. *Cell* **106**, 207–217.

Zhang, N., and Gridley, T. (1998). Defects in somite formation in lunatic fringe-deficient mice. *Nature (London)* **394**, 374–377.

II

Lineage Specification and Differentiation

Part and parcel of the development of a complex organism is the process of setting aside specialized cell lineages that carry out functions needed by the fetus and lay the groundwork for postnatal life. The mechanisms that establish the body plan and lay down the body axis, as described in Part I, occur concurrently with and interact with the mechanisms that set aside cell lineages. In many cases, patterning genes act across cell lineages to impart positional information that allows coordinated development of complex tissues and organs. In Part II, we focus on the events that lead to the specification of different cell lineages and the events that result in the later differentiation of lineage products.

Development is often considered as involving increasing specialization, such that the early cell lineages are typically pluripotent, while later lineages are increasingly more specialized. The blastomeres of the cleavage-stage embryos are known for their plasticity of differentiation. Two of the first cell lineages set aside in mammalian development are the trophoblasts and the primitive endoderm, which are highly specialized cell types, designed to allow the embryo to interact with the maternal environment (Chapter 8). As a result of the derivation of these cell lineages, the remaining cell population in the inner cell mass of the blastocyst and its derivative, the epiblast of the postimplantation embryo, though it remains pluripotent, has become restricted to the formation of cell types other than first the trophoblast and later the primitive endoderm. Although this notion of *progressive restriction in lineage potency* is generally true, this view can be

easily challenged. There is increasing evidence that at least some cells of specialized lineages in the adult animal may also be able to show much broader potential when experimentally challenged. This suggests that lineages of varying potential may be set aside as needed by the specific developmental milieu at the time.

Lineage specification is often associated with *signaling events* leading to expression of lineage-specific transcription factors that may act as "master regulators," determining specialized cell fate. This concept was brought into prominence with the discovery of the so-called myogenic genes, including MyoD and other family members. Transfection of MyoD into fibroblasts in culture was sufficient to transform them into skeletal muscle, suggesting that MyoD alone was capable of activating an entire genetic pathway leading to muscle development. Supporting this model was molecular data showing that MyoD can bind to and transactivate a variety of muscle-specific genes. However, the story is not quite so straightforward *in vivo* (Chapter 13), where the myogenic genes show overlapping functions in the development of the skeletal muscle lineage, and interactions between them and other muscle-specific regulators lead to the final complexity of the skeletal muscle lineage. Indeed, there are few cases in mammals where the initiation of an entire lineage seems to depend on a single regulatory factor. Despite the clear importance of the myogenic genes in development of skeletal muscle, for example, there is no single gene whose expression is either necessary or sufficient for the development of cardiac muscle. Rather, the development of the cardiac lineage seems to depend on interaction between a number of different transcription factors, some of which are cardiac specific and some of which are more general (Chapter 16). Also, although a number of mammalian genes are related to the key neurogenic regulators of invertebrates, there is no single gene that, when mutated, leads to a complete absence of neurons. Instead, genes like neurogenin and neuroD appear to be involved in combinatorial specification of neuronal subtypes (Chapter 12). In other instances, it seems that it is the absence of a factor that is critical for the specification of cell fate. For example, the POU domain transcription factor, *Oct4 (Pou5f1)*, is absolutely required for the development of the pluripotent inner cell mass lineage in the early embryo, the primordial germ cells (Chapter 9), and in the embryonic stem cells derived from these two cell types: Absence of *Oct4* seems to be sufficient to switch cells into the trophoblast lineage instead (Chapter 8). Clearly the mechanisms of lineage specification and differentiation require *complex interactions of positive and negative acting regulatory factors.* However, the general concept that there will be key lineage-specific regulators whose targets are the major structural proteins defining the lineage phenotype continues to form the fundamental intellectual framework for elucidating the molecular pathways of lineage development.

One of the concepts of lineage development is that *lineages may contain stem cells* capable of self-renewal and differentiation into specialized cell types. The classic picture of the stem cell comes from study of the hematopoietic system (Chapter 10), in which a small number of pluripotent cells persist in the bone marrow throughout life, allowing a constant supply of cells for replacement and renewal of the blood system. The therapeutic potential of stem cells is already clear from bone marrow transplantation and has led to considerable interest in determining whether other lineages also contain stem cell populations, which might be exploited for therapeutic purposes. In the past few years, cells with stem-cell-like potential have been isolated from many different organs and tissues, even where there had been little evidence of regenerative potential in the intact organism. Neural stem cells, for example, have been isolated from the adult brain (Chapter 12) despite the limited capacity for repair in the central nervous system.

Although all stem cell populations are characterized by their capacity for self-renewal and differentiation, interesting differences are seen in terms of their extent of self-renewal and their involvement in normal lineage differentiation. Some, like stem cells in the adult hematopoietic tissues (Chapter 10), the vascular endothelium (Chapter 11), and the epidermis (Chapter 22), persist throughout life and play a key role in the normal development of the lineage. Others, like neural crest stem cells (Chapter 12) and the chondrocyte progenitors (Chapter 14), have a life span limited to fetal

and postnatal prepubertal development and then lose self-renewal capacity once the lineage niches have been filled. Interestingly, one of the most transient stem cell populations in the intact embryo is the pluripotent early embryonic cell. Cells of the inner cell mass and early epiblast represent a proliferative and pluripotent population, capable of giving rise to the entire fetus, and yet, by the end of gastrulation, this population rapidly differentiates to establish the definitive germ layer derivatives. However, embryonic stem (ES) cells, derived from the ICM, can be maintained indefinitely in a pluripotent stem cell state *in vitro.* It is curious that the hematopoietic stem cell population has proven difficult to expand *in vitro,* despite its clear ability to persist *in vivo,* while ES cells can be readily maintained *in vitro* although their predecessors are transient *in vivo.* While the biological significance of this is unclear, the practical implications are important. It suggests that it may be possible to get stem cells expansion *in vitro* from cells that would appear to have limited self-renewal capacity, with the right combination of growth factors. In other words, there may be *cells with stem-like properties existing in unexpected places,* including adult tissues not thought to have stem cells.

The whole discussion of lineage specification and stem cell development rests on the notion that lineages, once specified, are closed, with potential restricted to the defined lineage derivatives. Thus, development can be thought of as proceeding through a series of decision points, which progressively restrict cell fate. Indeed, this is the normal pattern observed *in vivo.* However, this unidirectional notion of development does not necessarily require cell-autonomous restriction of cell fate. It merely requires that cells produce the *appropriate response to* the particular *temporal and spatial cues* to which they are exposed. In this model, it would be possible for cells normally restricted to one pathway to change their fate when exposed to different cues. Recent studies have shown that a number of so-called tissue-restricted stem cells have the capacity to contribute to different lineages in various transplant assays. For example, stem cells from muscle have been shown to contribute cells to the brain (Chapter 13); bone-marrow derived stem cells have contributed to vasculature, muscle, and neurons; and neural stem cells have been shown to contribute to multiple lineages in chimeric embryo (Chapter 12). It has also been shown that vascular endothelium may acquire other cell fates *in vivo* and *in vitro* and can differentiate into myocytes (Chapter 11). The therapeutic implications of this are enormous. If one could understand how to control lineage plasticity, one might be able to use stem cells from adult organs as sources of cells for many kinds of tissue repair. At this stage, we do not fully understand how much of adult cell fate restriction is "hardwired" and how much can be altered by relatively simple alterations in the cell environment.

It is certainly likely that stem cells may be more plastic than their differentiated progeny and we can expect considerable focus on the isolation and manipulation of stem cells from many different tissues in the next few years. However, it is not yet clear that an adult stem cell can be totally reprogrammed by its environment and become truly pluripotent, like an embryonic stem cell. It is possible that the potency displayed by these "reprogrammed" cells is restricted by the options of *de novo* cell differentiation allowed by the tissue environment in which they are tested experimentally. The expression of lineage-specific regulatory genes may be difficult to override by extrinsic signals and may need to be manipulated genetically. Identification of the key lineage determination genes again becomes important. Clearly the study of the molecular pathways of lineage development is critical for both understanding normal development and identifying ways in which lineage development can be redirected for therapeutic purposes.

8

Extraembryonic Lineages

Janet Rossant* and James C. Cross†

*Samuel Lunenfeld Research Institute, Mount Sinai Hospital, Toronto, Ontario, Canada M5G 1X5
†Department of Biochemistry and Molecular Biology, University of Calgary Faculty of Medicine, Calgary, Alberta, Canada T2N 4N1

I. Introduction

II. Early Development of the Trophoblast and Primitive Endoderm Lineages

III. Cell Lineage Analysis and the Extraembryonic Lineages

IV. Setting Aside the Blastocyst Lineages

V. Molecular Specification of the Blastocyst Cell Lineages

VI. Differentiation of the Yolk Sacs

VII. Morphogenetic Events in Development of the Chorioallantoic Placenta

VIII. Comparative Aspects of Development of Extraembryonic Membranes

IX. Molecular Control of Primitive Endoderm Development

X. Signaling Pathways in Early Trophoblast Development

XI. Control of Spongiotrophoblast and Giant Cell Fate

XII. Trophoblast Giant Cell Development: Gene Pathways and Control of Endoreduplication

XIII. Initiating Chorioallantoic Fusion

XIV. *Gcm1* Regulates the Initiation of Chorioallantoic Branching

XV. Growth Factor Signaling Regulates Branching Morphogenesis of the Labyrinth

XVI. Placental Development and Pregnancy Complications

References

I. Introduction

The first lineages to differentiate in the mammalian embryo are the trophectoderm and the primitive endoderm, both of which differentiate at the blastocyst stage of development. Experimental studies in the mouse have shown that these early cell lineages give rise to components of the extraembryonic membranes that are used to ensure the survival of the mammalian embryo in the uterine environment, but do not contribute to the fetus itself. There is thus intrinsic interest in understanding the regulation of the development of these tissues because their functions are essential for achieving viviparity in mammals. Disruption in the development and differentiation of these lineages may underlie a variety of problems of human pregnancy, including intrauterine growth retardation and preeclampsia. Evidence is accumulating that in early development these extraembryonic lineages are not just involved in a supporting role for the embryo, but play an active role in providing informational signals to the embryonic lineages to promote correct patterning and differentiation of the embryo itself. Understanding the regulation of the formation, maintenance, and function of these lineages is therefore an important part of understanding the development of the mammalian embryo.

II. Early Development of the Trophoblast and Primitive Endoderm Lineages

The development of the mouse blastocyst involves the generation of the polarized epithelium of the trophectoderm, which pumps fluid to form the internal blastocoel. At one end of the blastocoelic cavity is a nonpolarized clump of cells called the inner cell mass (ICM). Formation of the blastocoel begins around 3 days of development at the 32-cell late morula stage, initially by accumulation of intracellular fluid-filled vesicles that later fuse to form the intercellular blastocoel. The blastocoelic fluid is maintained by active transport of Na^+ ions via a Na^+/K^+ transporter localized on the blastocoelic (basal) side of the trophectoderm layer (MacPhee et al., 2000). Initially all cells of the trophectoderm are apparently similar, but before the embryo implants in the uterus, the trophectoderm cells away from the ICM, the mural trophectoderm, cease cell division but continue to replicate their DNA to begin the process of primary trophoblast giant cell formation (Barlow and Sherman, 1974; Gardner and Davies, 1993). Trophectoderm cells in contact with the ICM continue to proliferate to form the polar trophectoderm. Initially, at the 3.5-day blastocyst stage, the blastocoelic surface of the ICM is covered by finger-like protrusions from the trophectoderm cells at the ICM margin (Fleming et al., 1984). These protrusions recede around the time when the surface cells of the ICM become recognizably distinct from the rest of the ICM to form the primitive endoderm.

The primitive endoderm is a single cell layer on the surface of the ICM distinguishable by a variety of ultrastructural and molecular markers. Between E3.5 and E4.5 the zona pellucida surrounding the blastocyst is lost and the embryo starts to undergo net growth. The primitive endoderm expands laterally and starts to migrate over the surface of the mural trophectoderm to form the parietal endoderm. Thus by the time the embryo implants in the uterus around 4.5-days of gestation, it has formed three distinct cell types and further differentiation has begun in both the trophoblast and primitive endoderm lineages.

Implantation occurs by attachment of the mural trophectoderm cells to the uterine epithelium and subsequent erosion of the epithelial cells and is absolutely dependent on the correct functioning of the trophoblast. Isolated ICMs cannot implant in the uterus and several mouse mutations affecting trophoblast development cause embryonic death by interfering with the implantation process (see later). Uterine receptivity to blastocyst implantation is confined to a short period of the estrus cycle, and is regulated hormonally. Expression of several growth factors has been observed in the uterus at the site of implantation (Carson et al., 2000), and the involvement of heparin-binding epidermal growth factor (HB-EGF) and leukemia inhibitory factor (LIF) in the implantation process has been well studied. It has been proposed that the blastocyst signals the luminal epithelial cells to express HB-EGF at the site of implantation. Secreted HB-EGF can interact with receptors on the blastocyst to stimulate trophoblast proliferation and prepare the blastocyst for implantation (Paria et al., 1999). LIF is also expressed in the uterine stroma around the time of implantation and its expression is independent of the presence of the blastocyst (Bhatt et al., 1991), unlike HB-EGF. Knock-out mice lacking LIF are viable but blastocysts fail to implant, confirming a key role for LIF in preparation of the uterus for implantation (Stewart et al., 1992). HB-EGF fails to be expressed in $Lif^{-/-}$ uteri (Carson et al., 2000) and both pathways probably converge to regulate the production of prostaglandins in the uterus, which promote decidualization and implantation. Mice deficient in COX2, the cyclooxygenase enzyme responsible for prostaglandin production in the uterus, show defective implantation and decidualization (Lim et al., 1997).

Between E4.5 and E6.0, further growth and differentiation of the extraembryonic lineages occurs prior to gastrulation in the embryo itself. The polar trophectoderm of the blastocyst proliferates and pushes the ICM into the blastocoelic cavity to form the so-called egg cylinder stage of development (Copp, 1979). The diploid proliferating trophoblast in apposition to the epiblast of the embryo is called the extraembryonic ectoderm and the trophoblast further away forms the ectoplacental cone that protrudes into the uterine cavity. Trophoblast cells at the edge of the ectoplacental cone cease division but endoreduplicate their DNA to form the secondary giant cells that migrate to surround the entire conceptus and are the layer in direct contact with the uterine tissue. The parietal endoderm lines the giant cell layer to form the parietal yolk sac, while the primitive endoderm overlying the egg cylinder forms the visceral endoderm. These two cell types are quite distinct in morphology; parietal endoderm cells are small, spindle-shaped cells that have been shown to actively migrate over the surface of the parietal yolk sac (Cockroft, 1986), secreting extracellular matrix (ECM) as they go, while visceral endoderm cells form a columnar epithelial layer on the surface of the egg cylinder.

III. Cell Lineage Analysis and the Extraembryonic Lineages

Experimental embryology investigating the prospective fate of the three cell types of the blastocyst and the processes of cellular commitment to the early lineages has been extensive. Some of the earliest blastocyst injection chimera experiments were designed to investigate the fate of these early cell lineages, by injecting genetically marked cells into host blastocysts and following their later distribution (Gardner, 1968). The first experiments used markers, such as isozymal variants of glucose phosphate isometase (GPI) (Chapman *et al.*, 1971), which, while sensitive enough to pick up minor contributions from donor cells, require destruction of tissue integrity for analysis. Later experiments have been performed with a variety of cell autonomous *in situ* markers, such as repetitive nuclear DNA sequences that can be detected by *in situ* hybridization (Lo *et al.*, 1987; Rossant *et al.*, 1983), and more recently, ubiquitously expressed *Escherichia coli* β-galactosidase (Zambrowicz *et al.*, 1997), detectable by *in situ* enzyme assay. All of the experiments revealed the same general findings about the fate of the early cell lineages (Fig. 1), with some finer details added by the later *in situ* experiments (Gardner and Cockroft, 1998).

The exact derivatives of the ICM and the trophectoderm at the E3.5 blastocyst stage are best revealed by blastocyst reconstitution experiments, in which the ICM is excised from one blastocyst, creating a trophectoderm vesicle into which a genetically distinct donor ICM can be injected (Gardner *et al.*, 1973; Papaioannou, 1982). The trophectoderm was found to give rise to the extraembryonic ectoderm, ectoplacental cone, and primary and secondary giant cells after implantation, whereas the ICM contributed to all other extraembryonic lineages and the embryo itself (Fig. 1). The division of the ICM into primitive ectoderm or epiblast and primitive endoderm at E4.5 marks the delineation of another solely extraembryonic lineage. When primitive endoderm cells were injected into host blastocysts, their progeny were detected in the parietal endoderm lining the trophoblast giant cells and the visceral endoderm, which initially covers both embryonic and extraembryonic regions of the pregastrulation embryo (Gardner and Rossant, 1979). However, when fetal tissues were analyzed later in development for contributions from primitive endoderm, none was detected (Gardner, 1984; Gardner and Rossant, 1979). Contributions were only found in the endoderm layer of the visceral yolk sac. This suggested that the definitive endoderm, which forms the gut derivatives of the fetus (See Chapter 15), must derive from the epiblast. This was later confirmed by direct lineage

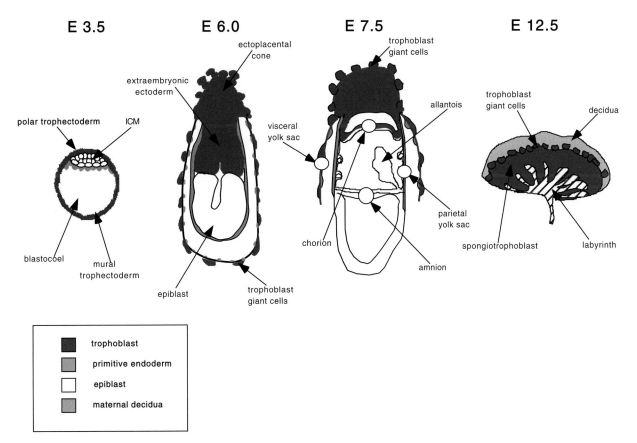

Figure 1 Lineage relationships in the early embryo. Diagrammatic representation of the major derivatives of the trophectoderm and primitive endoderm in early postimplantation stages and the major layers of the later placenta.

tracing experiments in cultured postimplantation embryos, where it was shown that the definitive endoderm arises from the anterior of the primitive streak and displaces the primitive endoderm from the surface of the embryonic region (Lawson and Pedersen, 1987).

Although it is clear that the vast majority of the definitive endoderm is derived from the epiblast, a minor contribution from the primitive endoderm remains a possibility. It is very hard to reconstitute a blastocyst in which the entire primitive ectoderm and endoderm differ in genotype (Gardner et al., 1990) in order to assess exhaustively the possible derivatives of the two layers. Instead, a less direct method has been used in which mouse embryonic stem (ES) cells are combined with tetraploid embryos (Nagy et al., 1990). Mouse ES cells behave generally like epiblast cells in chimeras and rarely contribute to the primitive endoderm or trophectoderm lineages, whereas tetraploid embryos are capable of generating viable primitive endoderm and trophectoderm but not epiblast derivatives. Thus conceptuses in which the epiblast and the two extraembryonic lineages are of different genotypes can be generated (Fig. 2), a technique that has been widely used to assess the embryonic or extraembryonic site of action of particular mutations (Rossant and Spence, 1998). Control experiments with wild-type ES cells and embryos routinely reveal minor contributions of marked cells of the tetraploid component in the hindgut of otherwise ES-derived fetuses at midgestation (Dufort et al., 1998). Although the way these embryos are generated is somewhat artificial, it certainly remains possible that some primitive endoderm cells may persist in the developing gut. Whether they differentiate and function in that environment is unknown.

IV. Setting Aside the Blastocyst Lineages

Many experiments have been performed over the years to address the manner by which cells get set aside to form the three separate lineages at the blastocyst stage, but the story is still unclear. An extensive series of experiments, largely by Martin Johnson's group, has highlighted the process of compaction (Pratt et al., 1982), which occurs at the 8-cell stage, as critical for later blastocyst differentiation (Fig. 3). At compaction, the blastomeres become closely adherent, forming a polarized epithelial ball (Fleming and Johnson, 1988). At the next two cleavage divisions, cells may divide symmetrically or asymmetrically, depending on the direction of the plane of cleavage in relation to the outer polarized region of the blastomere. Asymmetric divisions will generate a polarized outer cell and an apolar inner cell, while symmetric divisions lead to production of two polar cells (Fig. 3) (Johnson and Ziomek, 1981). A combination of the two types of division leads to the development of the blastocyst with the requisite number of apolar ICM cells and the polarized epithelial trophectoderm cells (Fleming, 1987). The use

of asymmetric cell divisions to segregate cell fate is a common theme in developmental biology (Horvitz and Herskowitz, 1992), and the mouse blastocyst represents an interesting example for further study. However, relatively few attempts have been made to explore any possible molecular segregation that may accompany the asymmetric divisions in the morula and be involved in ICM/trophectoderm segregation. Part of the reason for this is that the asymmetric divisions of the early embryo are not entirely invariant. In the intact embryo, the proportion of aymmetric to symmetric divisions at the fourth and fifth cleavage varies from embryo to embryo (Fleming, 1987). In addition, the embryo is highly regulative—an entire embryo can be regenerated from a group of entirely polar or apolar cells (Ziomek et al., 1982), including the early ICM (Handyside, 1978; Rossant and Lis, 1979; Spindle, 1978). However, this simply shows that blastomeres can reestablish polarity if disturbed and does not preclude the importance of molecules segregated during the subsequent asymmetric cell divisions. As pointed out in Chapter 2, the capacity of the early mouse embryo to regulate its development after many kinds of cellular manipulation may distract attention from underlying asymmetries in the intact embryo. The recent reexamination of the possibility that there may be a relationship among the position of the second polar body at the oocyte stage, the axis of the blastocyst, and potentially the anterior-posterior axis of the embryo itself (Gardner et al., 1992; Weber et al., 1999) should open up a new look at all aspects of prepatterning in the early embryo.

Even less is known about the formation of the primitive endoderm at the blastocyst stage. Isolated ICMs from mature E3.5 blastocysts form a complete outer layer of primitive endoderm on their surface, whatever their initial size (Rossant, 1975), suggesting a mechanism of allocation rather similar to the generation of the trophectoderm. When ICMs are isolated at different stages of blastocyst development, early ICMs regenerate trophectoderm on their surface, whereas late ICMs generate primitive endoderm (Rossant and Lis, 1979). Intermediate stages generate a mixed layer (Nichols and Gardner, 1984), suggesting that there may be a specific cell division cycle at which the response of the ICM cell to the same signals shifts from forming trophectoderm to forming primitive endoderm. The exact stage at which the primitive endoderm is allocated *in vivo* is not clear, although marking of single ICM cells at the E3.5 blastocyst stage generates epiblast or primitive endoderm-restricted clones in most cases (Weber et al., 1999). In the E4.5 blastocyst, the primitive endoderm is morphologically distinct and the remaining epiblast appears to be unable to regenerate the outer endoderm layer (Gardner, 1985). Thus, by the time of implantation, most studies suggest that there is little, if any, interchange of cells between the three lineages, and the differentiation of the two extraembryonic lineages has begun.

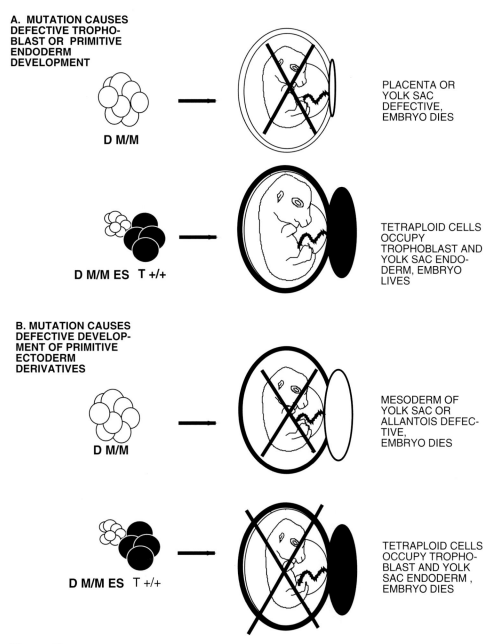

Figure 2 Tetraploid complementation assay. When ES cells are combined with tetraploid mouse embryos, the ES cells contribute solely to primitive ectoderm or epiblast derivatives, whereas the tetraploid cells fill the primitive endoderm and trophectoderm compartments, producing conceptuses in which epiblast can be of one genotype and the extraembryonic lineages are of another. When the ES cells carry a mutation, this allows determination of the primary site of action of a mutation that appears to affect extraembryonic membrane development. Wild-type tetraploid cells will rescue a mutation that causes trophectoderm or primitive endoderm defects but will not rescue a mutation that causes defects in epiblast derivatives such as allantois or yolk sac mesoderm.

V. Molecular Specification of the Blastocyst Cell Lineages

It is clear that, by the late blastocyst stage of development, the three cell lineages—epiblast, primitive endoderm, and trophectoderm—are distinct, with separate future fates in the intact embryo. By analogy with the specification of other cell lineages in development, one might expect to find lineage-specific transcription factors expressed at this stage and required for cell-type specification. A number of transcription factors whose expression becomes lineage restricted at the late blastocyst stage have been defined. How-

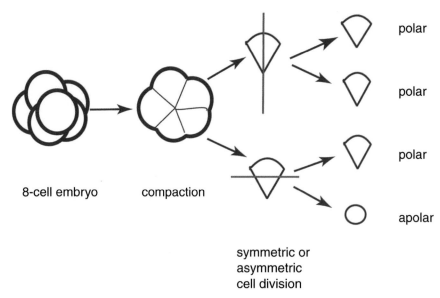

Figure 3 The generation of ICM and trophectoderm by asymmetric cell divisions in the cleavage stage embryo. After compaction, the next cleavage divisions may be symmetrical, dividing the polarized cells equally, or asymmetrical, generating one polar and one apolar progeny. Polar cells will end up as trophectoderm, while apolar cells form ICM. The number of symmetric versus asymmetric divisions at the 8- to 16-cell and the 16- to 32-cell stages varies, but the combined outcomes of the two cleavage divisions result in the correct number of ICM and trophectoderm cells.

ever, genetic analysis of the function of such genes has only to date identified one as being clearly involved and required for cell fate specification at the blastocyst stage. That gene is *Oct4 (Pou5f1)*, a POU domain homeobox transcription factor, whose expression marks the pluripotent cells of early cleavage, the ICM, the epiblast, and then becomes restricted to the germ cell lineage at gastrulation (Palmieri *et al.,* 1994; Scholer *et al.,* 1990). A knock-out of *Oct4* leads to production of blastocysts in which all cells take up the trophectoderm phenotype and no ICM cells are formed (Nichols *et al.,* 1998). *Oct4* thus plays a critical role in specifying ICM cell fate, where it acts in concert with a SOX protein, SOX2, to regulate key ICM products (Botquin *et al.,* 1998), including the growth factor, FGF4 (Yuan *et al.,* 1995).

OCT4 may thus actively promote the ICM fate. To date, there is no knock-out of a trophectoderm or primitive endoderm-specific transcription factor that gives the equivalent phenotype of failure to produce the relevant cell lineage. The earliest genes that have been identified to function in the trophectoderm lineage, such as *Eomes* (Russ *et al.,* 2000) and *Cdx2* (Chawengsaksophak *et al.,* 1997), or in the primitive endoderm lineage, such as *Hnf4, vHnf1,* and *Gata6* (Bielinska *et al.,* 1999) (see below), all allow initial formation of the lineage but are required for later maintenance and/or differentiation. Although these negative data do not preclude the existence of an as yet unidentified positive factor, or the requirement for more than one factor acting in parallel path-

ways, recent evidence has raised the possibility that it is the levels of OCT4 expressed in a given early embryonic cell that directly determine its cell fate at the blastocyst stage. Because *Oct4* is required for ICM development, it is not possible to derive *Oct4* mutant ES cells. However, Niwa and colleagues (2000) devised a strategy to conditionally inactivate *Oct4* in ES cells and showed that this caused the ES cells to differentiate into cells resembling trophoblast giant cells. If the culture conditions were switched to those appropriate for trophoblast stem (TS) cell proliferation (see later), cell lines with all the properties of TS cells could be obtained. This was quite a remarkable result, because ES cells normally do not differentiate into trophoblast cells, although they can make all other cell types of the mouse (Beddington and Robertson, 1989; Rossant *et al.,* 1993). Extrapolation of these results to the embryo itself would suggest that the initiation of the trophoblast pathway at the blastocyst stage may occur once *Oct4* is turned off and the pluripotent embryonic pathway is repressed. Niwa and colleagues (2000) also showed that overexpression of *Oct4* in ES cells tends to drive differentiation of ES cells, suggesting that the relative levels of OCT4 influence the choice between differentiation and stem cell proliferation of ES cells. Further they propose that the same may be true in the embryo itself. Normal levels of *Oct4* determine epiblast fate, whereas the absence of *Oct4* determines trophectoderm fate. Transient elevated levels of *Oct4* reported in the early primitive endoderm may be the critical trigger for endoderm specification. Although this model of graded levels of one transcription factor, OCT4, setting up all three early lineages, is an attractive one, more experimental proof is needed.

VI. Differentiation of the Yolk Sacs

In the postimplantation embryo, the parietal yolk sac is the extraembryonic membrane that contacts uterine tissue directly. It consists of an outer layer of trophoblast giant cells with an inner layer of parietal endoderm cells. The parietal endoderm cells are derived from the primitive endoderm layer of the ICM and migrate away to line the blastocoelic cavity. The parietal endoderm secretes a thick basement membrane layer, Reichert's membrane, that is rich in collagen, laminin, and associated dystroglycan (Hogan *et al.,* 1984). This membrane presumably acts as a protective layer, ensuring the integrity of the maternal–fetal interface and provides an interchange layer for diffusional maternally derived nutrients in the early postimplantation period. Knock-out of the dystroglycan gene leads to disruption of the integrity of Reichert's membrane and early postimplantation death (Williamson *et al.,* 1997). Parietal endoderm appears to be the terminal cell type of the primitive endoderm lineage. If no embryo develops, then all primitive endoderm follows the parietal cell fate. However, contact with embryonic derivatives keeps cells in the visceral pathway. This plasticity in cell fate is further highlighted by the ability of the isolated visceral endoderm to transform to a parietal cell type *in vitro* (Gardner, 1983).

The visceral endoderm (VE) overlying the embryonic region of the embryo is a transient layer that is displaced by the definitive endoderm. However, during its transient existence, it plays critical roles in the anterior patterning of the epiblast and elongation of the primitive streak (see Chapter 3). The VE plays a more prolonged role in the development of the visceral yolk sac, which is formed around E7.5 of development by the association of an outer layer of VE and an inner layer of extraembryonic mesoderm derived from the posterior of the primitive streak. In the mesoderm layer are the yolk sac blood islands that mark the site of the first embryonic hematopoiesis and vasculogenesis. By 8 days of development, the vitelline circulation is established, linking the yolk sac and the embryonic circulation. This allows the visceral yolk sac to be used as an efficient means of exchange of nutrients, oxygen, and waste products between embryo and maternal environment. The VE of the visceral yolk sac is highly specialized for its trophic functions. First it plays an active role in inducing and organizing the development of the underlying vasculature, via its production of the critical growth factor for vascular development, VEGF (see Chapters 10 and 11). Second, it expresses

a large number of proteins involved in active nutrient transport across epithelia, such as apoliprotein and transferrin, and many studies have demonstrated active uptake and transport across the VE layer (Huxham and Beck, 1985; Seibel, 1974). Further, a number of teratogenic insults can be shown to act on nutrient uptake and delivery (Jollie, 1990; Rogers *et al.,* 1985) and lead to abnormalities in development. Knock-out of genes critically involved in lipid transfer, like apolipoprotein B (Farese *et al.,* 1996) and microsomal triglyceride transfer protein (Raabe *et al.,* 1998), can lead to embryo lethality presumably due to failure in nutrient transfer. Many of the proteins expressed by the VE are the same as those produced by the later fetal and adult liver (Thomas *et al.,* 1990), and presumably reflect some shared functions in facilitating nutrient delivery and in detoxifying waste products. The similarities in gene expression between the VE of the yolk sac and the liver extend to the genes involved in their molecular control, as discussed below, raising the possibility that a close evolutionary relationship exists between the primitive and definitive endoderm, despite their temporal and spatial separation in the embryo (see Chapter 15).

VII. Morphogenetic Events in the Development of the Chorioallantoic Placenta

As development proceeds, the extraembryonic ectoderm expands to form the chorionic epithelium, which is lined by a thin layer of mesothelial cells produced as a result of the formation of the amniotic folds. The allantois grows out from the posterior end of the embryo and makes contact with the chorion at E8.5 in mice, an event termed *chorioallantoic fusion* though no actual cell fusion takes place. Within hours of allantoic attachment, folds appear in the chorionic plate in proximity to the fetoplacental blood vessels that grow in from the allantois to generate the fetal components of the placental vascular network. The trophoblast, with its associated fetal blood vessels, undergoes extensive villus branching to create a densely packed structure called the labyrinth. Coincident with the onset of morphogenetic branching, chorionic trophoblast cells begin to differentiate into the three layers of labyrinthine trophoblast cells. Mice have two layers of syncytiotrophoblast cells that are in direct apposition to the endothelial cells of the fetal-derived blood vessels. Syncytiotrophoblast cells are multinucleated cells formed by fusion of postmitotic precursor cells. An additional mononuclear cell type remains outside the syncytiotrophoblast layer, though its origin and functions are unknown.

While the labyrinth is developing, it is supported structurally by the formation of the spongiotrophoblast which form a compact layer of nonsyncytial cells between the labyrinth

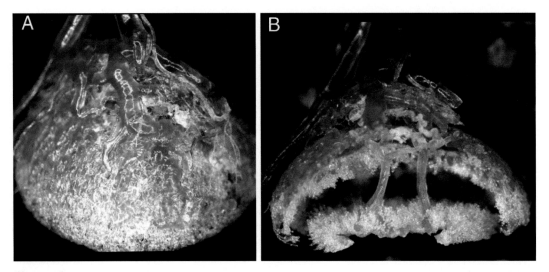

Figure 4 Vascular cast of the mouse placenta showing the maternal arterial and venous flow. (A) Surface view of the placenta from the maternal side, showing the venous outflow cast in blue plastic and the maternal arterial blood supply in pink. (B) Sagittal section showing the maternal arterial sinuses passing through the placenta and the resulting dispersion of maternal blood in the labyrinth (pink). Venous blood is collected at the edge of the labyrinth (blue) and returned to the uterine circulation. The umbilical vessels of the embryo enter at the bottom and interdigitate with the maternal arterial system in a countercurrent distribution, from the edge to the center of the placenta (not shown) (Adamson, Lu, Whiteley, and Cross, with permission).

and the outer giant cells. Although direct lineage analysis has not been performed, it has been assumed, based mainly on histological studies and shared marker gene expression, that the spongiotrophoblast largely derives from the ectoplacental cone cells of earlier stages. The maternal blood supply passes through the spongiotrophoblast by means of large central "arterial" sinuses in which the maternal endothelial cells are eroded away and replaced by trophoblast cells. The maternal blood eventually enters into the tortuous, small spaces of the labyrinth in which it directly bathes the fetal trophoblastic villi, ensuring ease of exchange of materials between the two blood systems. The labyrinthine exchange area is set up on the counter current principle, whereby the small maternal blood channels course in the opposite direction to the capillary blood flow in the fetoplacental circulation (Fig. 4) (Adamson and Cross, in preparation). The placenta undertakes many functions beyond simple exchange between the maternal and fetal environment. For example, the giant and spongiotrophoblast cells produce many gene products including hormones like placental lactogens (Gurtner *et al.,* 1995; Kwee *et al.,* 1995), angiogenic factors like proliferin (Groskopf *et al.,* 1997) and VEGF (Achen *et al.,* 1997; Vuorela *et al.,* 1997), and tissue remodeling factors like matrix metalloproteinases (MMPs) (Teesalu *et al.,* 1999) and urokinase-type plasminogen activator (uPA) (Teesalu *et al.,* 1998). The mouse placenta is thus a highly tuned exchange bed with supporting cells involved in both structural and functional aspects of fetal–maternal exchange.

VIII. Comparative Aspects of Development of Extraembryonic Membranes

Unlike the organization of the basic body plan, which is highly conserved across mammals, the structural aspects and relative roles of the different extraembryonic membranes vary considerably across the mammalian radiation. In rodents the yolk sac comes to surround the entire fetus as the major extraembryonic membrane of the conceptus. It is the sole exchange structure until the maturation of the chorioallantoic placenta and persists as a major exchange layer and protective layer late into pregnancy. In other mammals the role of the yolk sac placenta varies. Marsupials only utilize a yolk sac placenta, while in humans and ruminants, the yolk sac does not surround the fetus, lasts for only a few weeks and is largely vestigial in terms of exchange function. Interestingly, however, it remains as the initial site of embryonic hematopoiesis. The amnion is the major protective membrane of the fetus in primates, enclosing the fetus from an early stage. In rodents, the process of embryonic turning results in the enclosure of both the fetus and the amnion within the visceral yolk sac, such that the amnion is of subsidiary importance. Consistent with this, there are no mouse mutants in which embryonic death has been reported to result from defective amnion function (rather than formation).

The placenta is at least superficially quite distinct among different species. Placentae can be classified by their overall structure and by the number of layers of cells between the maternal and fetal blood systems. Structurally, the rodent and human placentae are both classified as discoid, in which the placenta forms one discrete unit of attachment and association with the uterus. The human placenta consists of several discoid units called cotyledons though they are clustered together. Cotyledonary placentae, as found in ruminants, consist of multiple dispersed sites of attachment of placental cotyledons to the uterine endometrium. Some other species, such as dogs and cats, have zonary placentae in which placental attachment takes place over an equatorial circular band encompassing the entire conceptus. Finally, in species like the pig, horse, and whale, the placenta is diffuse and attachment is distributed over most of the inner surface of the uterus. The second classification system for placentae relates to their cellular structure and the degree to which the trophoblast cells invade and erode maternal tissue (Fig 5). In epitheliochorial placentae, as found in ruminants, pigs and horses, no uterine cell layers are lost and so the uterine epithelium is the maternal layer in contact with the fetal trophoblast. There are thus six layers of cells between the maternal and fetal blood. In endotheliochorial placentae, found in carnivores, the uterine epithelium is broken down, so that fetal and maternal endothelial cells are separated only by trophoblast layers. In hemochorial placentae, such as are found in rodents and humans, there is complete loss of uterine epithelium and breakdown of the endothelial cells of the maternal capillaries, such that the maternal blood directly bathes the surface of the fetal trophoblast. Clearly the placenta is still undergoing evolutionary experimentation and the variations seen are all compatible with normal development. A

phylogeny based solely on similarities of placental structures would suggest some very unusual evolutionary relationships!

However, closer comparison of placental cell types, their functions, and the types of genes and proteins that they express reveals that considerable similarities exist across evolution likely reflecting conserved underlying pathways of regulation. A comparison of human and mouse placentae is particularly important, since it is hoped that the information gleaned on the molecular control of mouse placental development will be applicable to the human (Fig. 6). The outer layer of the rodent placenta is composed of trophoblast giant cells that mediate implantation and invasion into the uterus. The latter behavior is similar, therefore, to the invasive extravillous cytotrophoblast cells in humans. Rodent trophoblast giant cells are less invasive *in vivo* than human extravillous cytotrophoblast cells, which can migrate considerable distances into the uterine wall. However, the molecules that are thought to underlie cell invasion in humans are also expressed by trophoblast giant cells (Cross *et al.*, 1994). One of the striking features of trophoblast giant cells in that they are polyploid (Zybina and Zybina, 1996) due to repeated rounds of DNA replication in the absence of intervening mitoses (endoreduplication) (MacAuley *et al.*, 1998). Human extravillous cytotrophoblast cells (Berezowsky *et al.*, 1995) also become polyploid, albeit to a much lesser extent. The function of the middle layer of the rodent placenta, spongiotrophoblast, is unknown. However, at least some of the spongiotrophoblast cells (and particularly its precursor, the ectoplacental cone) can differentiate into giant cells and in this way are analogous to cells of the cytotrophoblast cell

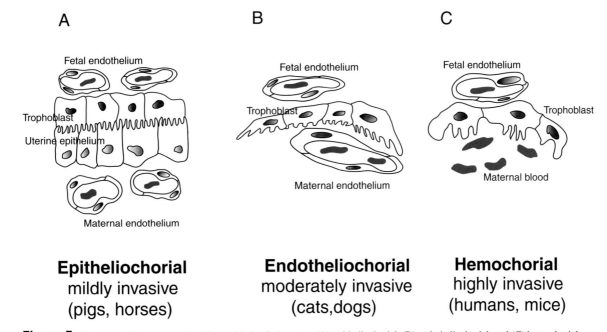

Figure 5 Trophoblast invasiveness in different kinds of placentae: (A) epitheliochorial, (B) endotheliochorial, and (C) hemochorial.

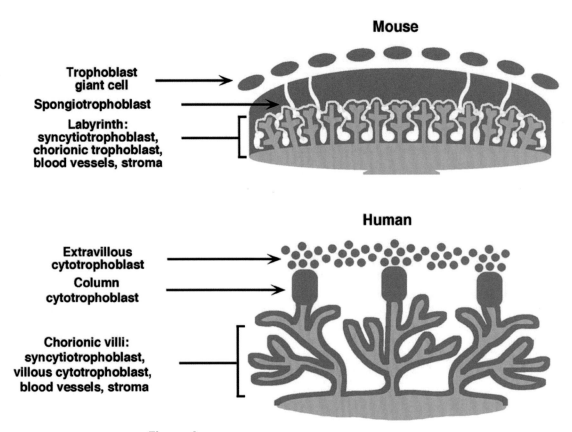

Figure 6 Comparison of mouse and human placental structure.

columns in anchoring villi of the human placenta. The laby-rinth layer of the placenta is completely analogous in func-tion to the floating chorionic villi in the human placenta, in that they provide the large surface areas for nutrient and gas exchange. In both human and rodent placentas, the villi are covered by syncytiotrophoblasts that lie in direct con-tact with maternal blood. As we describe the pathways of molecular control of the primitive endoderm and different mouse trophoblast cell types, one should bear in mind the possible similarities and differences to humans.

IX. Molecular Control of Primitive Endoderm Development

Athough the events initiating the specification of the primitive endoderm lineage are still not clear, recent evi-dence suggests that a hierarchy of transcription factors is involved in establishing functional primitive endoderm in the early postimplantation stages. Interestingly, these factors were largely first identified as potential regulators of defini-tive endoderm differentiation (see Chapter 15). Three factors in particular have been shown to be critical for the early

functional differentiation of the visceral endoderm: HNF4, vHNF1, and GATA6 (Bielinska *et al.,* 1999). HNF4 is a member of the steroid hormone receptor superfamily (Sla-dek *et al.,* 1990) and is a ligand-activated transcriptional regulator. Fatty acyl-CoA thioesters can act as ligands for the receptor (Hertz *et al.,* 1998), and the complex is known to directly regulate a number of liver and visceral endoderm specific genes, such as transferrin, ApoB, and transthyretin (Kuo *et al.,* 1992). *Hnf4* is an early marker of the primitive endoderm lineage, is expressed strongly in the visceral en-doderm of the egg cylinder stage, and is later expressed in the developing gut endoderm and the liver (Duncan *et al.,* 1994). Knock-out of *Hnf4* leads to embryonic death around the time of gastrulation (Chen *et al.,* 1994). Excessive cell death is observed in the developing epiblast soon after im-plantation and, although early mesoderm markers initiate ex-pression, the embryo is highly disorganized and pycnotic. Interestingly, visceral endoderm does form in mutant em-bryos and embryoid bodies and appears to be morphologi-cally normal. Expression of differentiation markers of VE is reduced, however. Tetraploid complementation (Fig. 2) was used to show that the epiblast defects observed were sec-ondary to defective VE. When the VE is wildtype, mutant epiblast can undergo normal gastrulation (Duncan *et al.,*

1997). It seems likely that the main reason for gastrulation failure in *Hnf4* mutants is a failure of the nutrient/metabolite exchange functions of the early visceral endoderm, and the results emphasize the dependence of the development of the embryo itself on the various extraembryonic cell types simply for survival.

Mutations in at least two other transcription factors give phenotypes similar or more severe to the loss of HNF4. GATA6, a member of the zinc finger GATA-binding family of transcription factors, is expressed throughout the early primitive endoderm lineages and also in selected regions of the later definitive endoderm (Morrisey *et al.*, 1996). Mutants die before gastrulation, are severely growth retarded, and have extensive apoptosis in the epiblast (Koutsourakis *et al.*, 1999; Morrisey *et al.*, 1998). The VE forms but is not complete and homozygous ES-derived embryoid bodies apparently fail to produce a clear VE layer (Morrisey *et al.*, 1998). Again, markers of differentiation are dramatically reduced and the defect can be rescued by wild-type VE in chimeras (Koutsourakis *et al.*, 1999). The homeodomain factor, vHNF1, shows similar expression patterns and knockout phenotypes to GATA6 (Barbacci *et al.*, 1999; Coffinier *et al.*, 1999), thus implicating yet another endoderm regulatory molecule in the transcriptional hierarchy of visceral endoderm differentiation. Both vHNF1 and GATA6 have been shown biochemically to directly activate *Hnf4* transcription, and, consistent with this, *Hnf4* is downregulated in both mutants but *Gata6* and *vHnf1* are both still expressed in *Hnf4* mutants. HNF4 itself directly activates the transcription of many of the differentiated products of VE. Thus the beginnings of a regulatory hierarchy of genes involved in primitive endoderm can be developed, with GATA6 and vHNF1 independently regulating *Hnf4* transcription, which then regulates differentiation. This may, however, be an oversimplistic explanation, because other members of the GATA and HNF families and other transcription factor families are also involved in complex cross-regulatory interactions in yolk sac development.

GATA4, another zinc finger transcription factor, is also strongly expressed in the early primitive endoderm lineages as well as precardiac mesoderm (Heikinheimo *et al.*, 1994). Most *Gata4*−/− mice make it through gastrulation but show major defects in ventral closure, by which the visceral yolk sac expands to enclose the foregut within the embryo (Kuo *et al.*, 1997; Molkentin *et al.*, 1997). As a result, the embryos show cardia bifida—two lateral heart tubes—and major defects in foregut development. These defects reside in defective primitive endoderm development, because normal closure and heart development occurs when wild-type VE is provided in chimeras (Narita *et al.*, 1997). Although it has been reported that *Gata4*−/− ES cell-derived embryoid bodies lack VE (Soudais *et al.*, 1995), VE clearly forms in the embryo, suggesting that *Gata4* is required for morphogenetic aspects of yolk sac function rather than its specifica-

tion. Consistent with this, *Gata6* is actually upregulated in GATA4 mutants (Kuo *et al.*, 1997; Molkentin *et al.*, 1997).

The last major family of definitive endoderm/liver-specific transcriptional regulators is the HNF3 subfamily of the winged helix or *Fox* gene family (Kaufmann and Knochel, 1996), and this family also has been shown to play a role in primitive endoderm development. There are three HNF3s in mammals, encoded by *Foxa1, a2,* and *a3*. *Foxa2* or HNF3β, is the family member that seems to play a predominant role in a number of different lineages in the early embryo. It is expressed throughout the developing primitive and definitive endoderm lineages, as well as in the node, the notochord, and floorplate (Ang *et al.*, 1993; Monaghan *et al.*, 1993). Knockouts have major patterning defects due to loss of the node and notochord. They also fail to form foregut and they have abnormal yolk sac morphogenesis and primitive streak extension (Ang and Rossant, 1994; Weinstein *et al.*, 1994). The latter two defects can be rescued by wild-type VE in tetraploid complementation assays. Reverse chimeras, in which the primitive endoderm is mutant and the embryo is wild type, show major defects in morphogenesis, such that the embryo fails entirely to be enclosed within the yolk sac and rapidly degenerates due to failure of nutrition from the yolk sac (Dufort *et al.*, 1998). Embryoid body studies have shown that *Foxa2* mutant embryoid bodies also have defective expression of the differentiation markers known to be regulated by the various endoderm factors, including HNF3. However, the multiple defects seen in the *Foxa2* mutant embryos and in the *Gata4* mutants are unlikely to be simply explained by failure in the hierarchy of regulation of the liver-related specialized products of the VE. It is now known that the VE is not only a major metabolic, nutritive, and waste disposal unit for the early embryo, but it is also a source of signals to the early embryo to pattern the developing body axis (see Chapter 3). The upstream pathways that regulate the local production of signaling molecules in the VE are not known, and it is quite likely that the so-called general VE factors described here may also play a role in the regulation of the signaling properties of this important tissue. Knowledge of other targets regulated by the HNF and GATA gene families will help elucidate how the various properties of the VE are coordinately regulated.

The major gene families involved in both primitive and definitive endoderm development in the mouse show conservation across evolution. The HNF3 winged helix family is represented by the *forkhead* gene in *Drosophila* (Lai *et al.*, 1991) and the *Pha-4* gene in *Caenorhabditis elegans* (Horner *et al.*, 1998; Kalb *et al.*, 1998), both of which play key roles in gut endoderm specification in those species. Similarly, the GATA factor genes, *serpent* in *Drosophila* (Rehorn *et al.*, 1996) and *end-1* (Zhu *et al.*, 1997) and *elt-2* (Fukushige *et al.*, 1998) in *C. elegans* are also involved in endoderm development. It would appear that these genes play a primary, evolutionarily conserved role in definitive gut endoderm development, and that their dual roles in primitive and definitive endoderm in mammals have arisen with the specialization of the visceral endoderm to carry out many of the functions of the later liver and gut. It remains unclear whether further regulatory pathways are specific to the primitive endoderm lineage in the mouse. For example, two genes, *Mixer* (Henry and Melton, 1998) and *Sox17* (Hudson *et al.,* 1997), have been defined in Xenopus as being critical for initiating definitive endoderm development. In mice, a mixer-related gene, *Mml* (Pearce and Evans, 1999), has been isolated and shows expression in visceral endoderm as well as definitive endoderm and some transient mesoderm. Mutation of these genes will be interesting. On the other hand, there are factors identified by their ability to regulate liver-specific genes, such as HNF1 α (Pontoglio *et al.,* 1996) and HNF6 (Jacquemin *et al.,* 2000), whose expression is confined to the gut and whose mutant phenotypes are confined to the later stages of development (Chapter 15). Thus there is clearly more complexity in the regulation of the later definitive endoderm lineage than in the visceral endoderm, but still a remarkable degree of overlapping gene functions is seen.

X. Signaling Pathways in Early Trophoblast Development

Early experiments demonstrated that the trophoblast depended on signals from the ICM and its later derivatives for its proliferation and differentiation. As mentioned earlier, trophectoderm cells away from the ICM at the blastocyst stage fail to proliferate but instead form trophoblast giant cells, which are terminally differentiated cell types. If, however, a new ICM is inserted into a mural trophectodermal vesicle, a new proliferative center is established and normal development can ensue (Gardner *et al.,* 1973). This led to the suggestion that the ICM provides a positive signal to the overlying trophectoderm to promote its proliferation and hence its later differentiation into the major trophoblast cell types. The early postimplantation phases of trophoblast development were also shown to depend on an interaction with embryonic tissues for continued proliferation and differentiation. Extraembryonic ectoderm or ectoplacental cone tissue isolated from early postimplantation stages and transplanted ectopically (Hunt and Avery, 1976) or cultured *in vitro* (Jenkinson and Billington, 1974; Rossant and Ofer, 1977) fails to proliferate and instead forms trophoblast giant cells. Enclosure of extraembryonic ectoderm, but not ectoplacental cone, inside the amniotic cavity allowed some continued proliferation (Ilgren, 1981; Rossant and Tamara-Lis, 1981). These experiments and others, including protein expression profile comparisons, led to the proposal that the polar trophectoderm and the early extraembryonic ectoderm

act as stem cells for the development of the trophoblast lineage (Johnson and Rossant, 1981). Stem cell maintenance would depend on signals from the embryo. As cells moved away from contact with the signaling source, they would progressively differentiate into ectoplacental cone and trophoblast giant cells.

Recent molecular and genetic experiments have strongly suggested that the major signaling pathway involved in this interaction is the fibroblast growth factor (FGF) signaling pathway. The FGF gene family consists of at least 23 members but attention has focused on the role of FGF4 in promoting trophoblast development. FGF4 is expressed throughout early preimplantation development, becomes restricted to the ICM at the blastocyst stage, and is expressed throughout the early postimplantation epiblast (Niswander and Martin, 1992; Rappolee *et al.,* 1994), as predicted for the embryo-derived signal (Fig. 7). Embryos homozygous for a loss of function mutation in *Fgf4* die at implantation and show poor development of all cell lineages (Feldman *et al.,* 1995). However, homozygous *Fgf4* mutant ES cells can be obtained (Wilder *et al.,* 1997), suggesting that the *in vivo* failure of the embryo may be secondary to a failure of trophoblast development. More direct evidence for a role for FGF4 in promoting trophoblast proliferation came from treating wild-type (Chai *et al.,* 1998) or *Oct4* mutant blastocysts (Nichols *et al.,* 1998), which lack ICM, with FGF4 and showing enhanced trophectoderm cell numbers.

These insights led Tanaka *et al.* (1998) to attempt to derive permanent TS cell lines from blastocysts and extraembryonic ectoderm in the presence of FGF4. They were able to derive such lines, provided that the cells were also given conditioned medium from fibroblasts, suggesting that FGF4

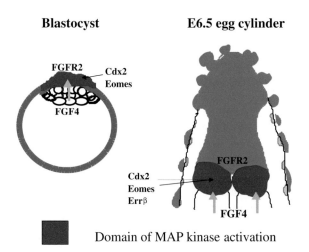

Figure 7   FGF signaling in early trophoblast development. FGF4 from the ICM or epiblast signals to the overlying trophoblast to activate MAP kinase (pink). A number of transcription factors are expressed in the same domain and are likely mediators of FGF signaling in trophoblast stem cells.

may not be the only trophoblast proliferation factor. These so-called TS cell lines are diploid, have an extended proliferative capacity, and express markers of extraembryonic ectoderm and not of the other early cell lineages, but will stop dividing and differentiate into giant cells on removal of FGF4 *in vitro*. They are also capable of colonizing all trophoblast lineages and only trophoblast lineages after introduction into the mouse blastocyst. TS cells are thus a useful new tool for molecular and developmental analysis of the trophoblast lineage.

If FGF4 coming from the embryo is signaling to the overlying trophoblast, then there should be an FGF receptor response mechanism in the trophoblast lineage. All four FGF receptors have been reported to be expressed at the blastocyst stage (Haffner-Krausz *et al.*, 1999; Rappolee *et al.*, 1998) and a dominant-negative *Fgfr* construct, which blocks all receptor signaling, blocks trophoblast proliferation at the blastocyst stage (Chai *et al.*, 1998). In the blastocyst and early postimplantation stages, *Fgfr2* is particularly strongly expressed in the diploid trophoblast (Haffner-Krausz *et al.*, 1999; Ciruna and Rossant, 1999), leading to the suggestion that it could be the major TS cell-specific receptor. This was supported by reports of two mutations in *Fgfr2*, one of which died peri-implantation and showed no trophoblast proliferation (Arman *et al.*, 1998). The other mutation died later but showed placental defects (Xu *et al.*, 1998). We have been able to derive TS cells lacking *Fgfr2* (J. Rossant, unpublished). This suggests that FGF4 signals through more than one receptor in TS cells, and that the first reported mutation in *Fgfr2* (Arman *et al.*, 1998) is likely a dominant negative that may interfere with the function of multiple FGF receptors.

FGF signaling can activate several different downstream signal transduction pathways. However, the downstream signaling response necessary for stem cell maintenance appears to be the canonical MAP kinase cascade. Whole-mouse immunocytochemistry with an antibody to activated MAP kinase is entirely consistent with the model proposed for embryo–trophoblast interactions (Fig. 7). Activation of the MAP kinase cascade occurs in the trophectoderm of the blastocyst and in a layer of extraembryonic ectoderm in proximity to the epiblast in the early postimplantation stages (L. Corson and J. Rossant, unpublished). This domain of MAP kinase activation is thus associated with the proposed FGF-dependent stem cell zone. The stem cell model would also predict that key downstream targets of FGF signaling will be specifically expressed in this domain and will be necessary for trophoblast lineage development.

Interestingly, several genes fit this category. Two of them, *Cdx2* and *Eomes*, are particularly interesting because they belong to transcription factor gene families, members of which have been proposed to be downstream of FGF signaling in other embryonic contexts (Griffin *et al.*, 1998; Isaacs *et al.*, 1998; Smith *et al.*, 1991). *Cdx2* is a member of the caudal-

related gene family and *Eomes* is a member of the T-box gene family. Both are expressed in domains of the developing embryo itself and have been shown to have important functions in embryonic patterning (Beck *et al.*, 1999; Russ *et al.*, 2000). However, both also show strong early expression in the extraembryonic ectoderm, particularly in the regions of MAP kinase activation (Beck *et al.*, 1995; Ciruna and Rossant, 1999). Null mutations in either gene cause peri-implantation mortality, as predicted for a failure of early trophoblast development (Chawengsaksophak *et al.*, 1997; Russ *et al.*, 2000). Mutants form trophectoderm at the blastocyst stage but fail to outgrow or form TS cell lines (C.-A. Mao, D. Strumpf, and J. Rossant, unpublished results), making these genes strong candidates as key regulators of the early trophoblast lineage. Understanding their upstream regulation may provide key insights into early trophoblast development and clues as to how these genes evolved new functions in the placenta.

FGF signaling is not the only signaling pathway involved in early trophoblast stem cell maintenance and differentiation. The estrogen-related receptor, *Errβ*, is an orphan nuclear hormone receptor, expressed in a very similar domain to *Cdx2* and *Eomes* (Pettersson *et al.*, 1996). Mutation of this gene causes embryonic lethality around E9, with complete absence of any of the diploid trophoblast layers of the placenta: the conceptus is surrounded by a thick layer of trophoblast giant cells (Luo *et al.*, 1997). Examination of earlier embryos showed that extraembryonic ectoderm formed normally but failed to be maintained at the stage of chorion formation, with subsequent transformation into giant cells. This suggests that the ERRβ pathway is critical for stem cell maintenance, perhaps downstream or independent of FGF function. It has recently been shown that diethylstilbestrol (DES) can act as a specific antagonist of ERRβ (Tremblay *et al.*, 2001). Treatment of pregnant mice with pharmacological doses of DES can mimic the *Errβ* mutant phenotype. In addition, treatment of TS cells with DES, even in the presence of FGF4, promotes giant cell formation (Tremblay *et al.*, 2001). Whether there is an endogenous agonist or antagonist for ERRβ is currently unknown. However, it is interesting to speculate that there might be an endogenous antagonist, perhaps produced by the maternal environment, that is acting to promote giant cell formation in opposition to the embryonic FGF signal that is promoting stem cell proliferation. The correct balance between these opposing signals would be required for normal development of the early phases of the trophoblast lineage.

Maternal administration of retinoic acid (RA) can also lead to a transformation of diploid trophoblast to the giant cell fate *in vivo* and in TS cells (Yan *et al.*, 2001). It was proposed that endogenous maternal RA from the decidua could be promoting giant cell differentiation. Clearly there needs to be further experimentation to determine the relative roles of maternal and fetal signals in promoting early trophoblast development.

XI. Control of Spongiotrophoblast and Giant Cell Fate

Once the trophoblast lineage begins to differentiate into the structures that will compose the layers of the chorioallantoic placenta, additional genetic pathways come into play, beyond those controlling the stem cell versus giant cell fate. Differentiation of the ectoplacental cone is dependent on the expression of the ETS domain transcription factor, *Ets2*. The *Ets* gene family is a large and diverse one, acting downstream of a variety of signaling pathways, including the FGF pathway. *Ets2* shows very specific expression in the diploid trophoblast in the early postimplantation period and knockout embryos show reduced ectoplacental cone development, leading to embryonic lethality at E8.5 (Yamamoto *et al.*, 1998). Extraembryonic ectoderm development appears normal but ectoplacental cone proliferation and differentiation are affected. In particular, it was shown that the production of matrix metalloproteinases, such as MMP9, is blocked in mutant ectoplacental cone. These molecules are critical for the processes of implantation and invasion into the uterine environment and so this represents a key pathway of trophoblast differentiation. At this stage it is not clear what signaling pathway is upstream of *Ets2* in the trophoblast. The trophoblast defect can be rescued by tetraploid aggregation and resulting mice are viable but have wavy hair, curly whiskers, and abnormal hair follicle development (Yamamoto *et al.*, 1998), reminiscent of mutants in the EGF signaling pathway. EGF signaling is active in the placenta (see below), but it is also possible that FGF signaling is involved in ETS-regulated responses, since FGF failed to induce MMP expression in *Ets2*-deficient fibroblasts (Yamamoto *et al.*, 1998).

The next stage of formation of the spongiotrophoblast or giant cells from the ectoplacental cone is regulated at the transcriptional level by a balance between the action of two basic helix–loop–helix (bHLH) proteins, MASH2 and HAND1, and is affected by EGF signaling and hypoxia. *Mash2* is one of two mammalian homologs of the *Drosophila achaete-scute* genes (Johnson *et al.*, 1990), whose members are involved in neurogenesis in flies. *Mash1* is also involved in various aspects of neural specification in mammals (Guillemot *et al.*, 1993; Lo *et al.*, 1991), but *Mash2* is not detectably expressed in the nervous system. Expression analysis in early development revealed that *Mash2* mRNA is present throughout oogenesis and preimplantation development and becomes restricted to the trophectoderm at the blastocyst stage (Rossant *et al.*, 1998). After implantation, *Mash2* is expressed strongly in all diploid trophoblast cells but not in giant cells (Guillemot *et al.*, 1994). Expression persists until midgestation, with strongest expression in the spongiotrophoblast but starts to be reduced by 12 days and continues to decline thereafter (Nakayama *et al.*, 1997). Knock-out of *Mash2* causes embryonic lethality by E10,

typical of a failure of chorioallantoic development (Guillemot *et al.*, 1994). When heterozygous mutant females were crossed with wild-type males, the heterozygous progeny also showed the same embryonic lethality as homozygotes, which led to the discovery that *Mash2* was an imprinted gene, only expressed from the maternal genome (Guillemot *et al.*, 1995). Examination of the placenta in mutants revealed a complete absence of the spongiotrophoblast layer and apparent expansion of the giant cell layer. The labyrinth was also reduced and presumably not fully functional (Guillemot *et al.*, 1994). Tetraploid aggregates of mutant and wild-type embryos proved that *Mash2* is solely required in the trophoblast lineage, because it was possible to obtain viable, fertile *Mash2*$^{-/-}$ mice (Guillemot *et al.*, 1994). The phenotype of diploid aggregation chimeras, in which mutant and wild-type cells are mixed in both the ICM and trophoblast lineages, revealed that the loss of *Mash2* affects spongiotrophoblast development in a cellautonomous manner. Chimeras could develop with normal placentae but the spongiotrophoblast in such chimeras was always completely wildtype. Giant cells and labyrinthine cells were of mixed mutant and wild-type composition.

These studies demonstrated a distinct role for *Mash2* in formation of the spongiotrophoblast from the ectoplacental cone, but did not explain why no defect was seen earlier in development, when *Mash2* is expressed throughout the developing diploid trophoblast lineage. One possibility was that maternally derived MASH2 protein, generated during oogenesis, bcould play a role in early trophoblast development, even in the absence of zygotic transcription. Availability of viable *Mash2*$^{-/-}$ mice from the tetraploid aggregation experiments allowed this hypothesis to be tested. When *Mash2*$^{-/-}$ females were crossed with *Mash2*$^{+/-}$ males, the homozygous mutant phenotype was not enhanced, showing that maternal MASH2 is dispensable (Rossant *et al.*, 1998). MASH2 function, therefore, seems to be restricted to the development of the spongiotrophoblast, despite its widespread trophoblast expression. It is possible to derive *Mash2*$^{-/-}$ TS cells, consistent with this later role for *Mash2* in trophoblast lineage differentiation, rather than early lineage commitment.

Another bHLH gene identified by its expression in the trophoblast is *Hand1*. It shows some overlap in expression with *Mash2*, but is most strongly expressed in the secondary giant cells (Cross *et al.*, 1995). Unlike *Mash2*, which is restricted in expression to the trophoblast, *Hand1* also shows expression in a number of regions of the embryo including the developing heart and limbs. Knock-out of the gene leads to early embryonic lethality around E8 (Firulli *et al.*, 1998; Riley *et al.*, 1998), and giant cell formation is blocked (Riley *et al.*, 1998). Chimera and tetraploid aggregation studies have shown that this early lethality is a consequence of the defective implantation process initiated by the mutant trophoblast cells (Riley *et al.*, 1998), and the studies have revealed important roles for *Hand1* in cardiac development

(Riley *et al.*, 2000). In a rat trophoblast choriocarcinoma cell line, transfected *Mash2* can inhibit giant cell formation, while *Hand1* enhances giant cell development (Cross *et al.*, 1995). HAND1 can interfere with the ability of MASH2 to transactivate reporter constructs (Scott *et al.*, 2000), suggesting that the balance between spongiotrophoblast formation and giant cell formation can be determined by the balance between MASH2 and HAND1 action. Inhibition of MASH2 function by HAND1 may be critical for giant cell formation. However, inhibiting MASH2 function cannot be the only function of HAND1, because double mutants between *Mash2* and *Hand1* show the same loss of giant cells seen in *Hand1* single mutants. *Hand1* must therefore both positively drive giant cell development and negatively regulate spongiotrophoblast development by inhibiting *Mash2*.

An additional bHLH repressor protein, I-mfa, also appears to be involved in this regulatory interaction. I-mfa is a non-bHLH inhibitor of bHLH function that physically interacts with bHLH proteins and prevents their DNA binding activity (Chen *et al.*, 1996). I-mfa is widely expressed and mutants show a variety of phenotypes, consistent with a role for this gene in regulating multiple bHLH pathways (Kraut *et al.*, 1998). On a C57BL/6 background the mutants show a placental phenotype with reduced numbers of giant cells. *In vitro* studies showed that I-mfa could inhibit the transcriptional activity of MASH2, but did not affect the activity of HAND1 (Kraut *et al.*, 1998). Thus there may be at least two negative regulators of MASH2 function that help determine the balance between proliferation of spongiotrophoblast cells and terminal differentiation of giant cells.

The signaling pathways upstream of the *Mash2/Hand1* interactions in spongiotrophoblast development are not clear. EGF signaling is known to be required for proper spongiotrophoblast development, but it is not clear whether this is mediated by effects on *Mash2/Hand1* expression. Depending on the genetic background, EGF receptor mutants can survive past term or die at different stages of embryonic development. On some backgrounds, embryos develop into late gestation, but show a very specific reduction in the spongiotrophoblast layer (Sibilia and Wagner, 1995; Threadgill *et al.*, 1995). Markers of spongiotrophoblast are reduced, but it is not clear that *Mash2* expression is affected, other than by the loss of some of the cells expressing it. Whether it is maternal or fetal EGF that is required for spongiotrophoblast development is also unclear.

XII. Trophoblast Giant Cell Development: Gene Pathways and Control of Endoreduplication

Trophoblast giant cells are polyploid cells that form as a result of endoreduplication, a process in which rounds of DNA synthesis continue in the absence of intervening mi-

toses. This is a curious phenomenon for two reasons. First, it "violates" one of the central regulatory mechanisms of the mitotic cell cycle—that DNA replication machinery cannot be reset until completion of mitosis. Second, it raises questions as to its function. It is possible that different patterns or levels of gene expression are only possible in polyploid cells compared to their diploid precursors. However, it is notable that the entire genome is replicated during each cycle in trophoblast giant cells, apparently not just the "important" genes. Polyploidy is also observed in other cell types in mammals such as megakaryocytes, cardiomyocytes, chondrocytes, and hepatocytes. There are variations in the extent of DNA amplification and the precise mechanisms occurring in each cell type. Trophoblast giant cells are at the extreme high end in that each cell can have up to $1000\times$ the haploid value of DNA (Zybina and Zybina, 1996; Zybina *et al.*, 1985). The endoreduplicated DNA is in polytene chromosomes in which sister chromatids remain associated after DNA replication (Varmuza *et al.*, 1988; Zybina and Zybina, 1996). By contrast, most hepatocytes are tetraploid or octaploid and show no association of duplicated chromatids. It is clear that endoreduplication is restricted to cells that are no longer dividing. Therefore, it may represent a mechanism by which cells can stop dividing yet continue to grow as required for development of that organ (Edgar and Orr-Weaver, 2001).

The expression and function of several cell cycle regulators changes dramatically during the transition from the mitotic cell cycle to the endocycle in trophoblast cells, as well as factors that govern the transition (MacAuley *et al.*, 1998; Nakayama *et al.*, 1998). The first important feature is that commitment to differentiate into giant cells occurs during the G_2 phase of the cell cycle. Proliferating precursor trophoblast cells enter G_2 and begin to express mitotic cyclins (cyclin B) as they would in a normal cycle. In the G_2 phase, the cells either activate cyclin B/Cdk1 enzyme activity and go through mitosis (to remain as proliferating cells), or arrest and eventually enter a new S phase (to begin the first or "transition" endocycle). Mouse *Sna* gene encodes a zinc finger transcription factor that is related to the *Drosophila Snail* and *Escargot* genes. Loss-of-function mutations in *Escargot* result in ectopic sites of endoreduplication, whereas misexpression of *Escargot* in salivary glands prevents their normal endoreduplication (Fuse *et al.*, 1994). In the placenta, mouse *Sna* expression occurs in the ectoplacental cone/spongiotrophoblast but then is downregulated as these cells differentiate into giant cells (Nakayama *et al.*, 1998). *Sna* limits the commitment of cells to the giant cell fate when overexpressed in proliferating Rcho-1 trophoblast cells. Importantly, *Sna* has no effect on the progression of the endocycle when expressed in already committed giant cells (Nakayama *et al.*, 1998). Unlike *Cdc5*, for example, *Sna/Escargot* probably does not play a role in activation of mitosis. In transfected cells, whereas *Cdc5* shortens the length of G_2 (Bernstein and Coughlin, 1998), overexpression of

Sna has no effect on the timing of the cell cycle phases (Nakayama *et al.,* 1998).

Progression through the mitotic cell cycle and the endocycle is dependent on periodic episodes of cyclin E/Cdk activity (Edgar and Lehner, 1996; Sherr, 1994). How this problem is solved during endoreduplication is intriguing since the endocycle lacks mitosis, an event that is central to the resetting of DNA replication machinery during the mitotic cell cycle. In salivary epithelial cells of *Drosophila,* cyclin-E expression is confined to short pulses during the endocycle, and progress through the endocycle is blocked if cyclin E is continuously expressed (Follette *et al.,* 1998; Weiss *et al.,* 1998). Cyclin E is present throughout much of the endocycle in trophoblast giant cells, however (MacAuley *et al.,* 1998), indicating that modulation of cyclin E/Cdk activity—rather than abundance—may be the major regulatory step. Regulated expression of a Cdk inhibitor during the endocycle could achieve this effect, and the Cdk inhibitor, p57^{Kip2}, is cyclically expressed during the endocycle (Hattori *et al.,* 2000). This creates two distinct gap phases: a G_1-like phase during which p57^{Kip2} levels drop and stay low for several hours in advance of S phase, followed by a G_2-like phase characterized by accumulation of p57^{Kip2} that may terminate cyclin A/ and E/Cdk activities on completion of DNA replication. The regulated degradation and reaccumulation of a Cdk inhibitor to modulate increases in G_1/S Cdk activity, and thus DNA replication, is a novel mechanism for cell cycle regulation.

XIII. Initiating Chorioallantoic Fusion

Many mouse mutants show defects in chorioallantoic fusion. Indeed, this is one of the common causes of midgestation embryonic lethality in the mouse (Copp, 1995). Any mutation that interferes with the growth and development of the allantois will result in failure of chorioallantoic fusion and hence failure of placental development. A number of important embryonic patterning genes, such as the T-box gene, *Brachyury,* and the bone morphogenetic protein genes, *Bmp5* and *7,* show allantoic defects and placental failure when mutated. Some mutant phenotypes have provided more specific insights into the physical process of chorioallantoic attachment. *Vcam1* (Gurtner *et al.,* 1995; Kwee *et al.,* 1995) and *Itga4* loss-of-function mutants (Yang *et al.,* 1995) both show chorioallantoic fusion and/or labyrinth defects. VCAM1 is a cell adhesion molecule that is expressed on the tip of the allantois. It binds to another cell adhesion molecule, α4 integrin, encoded by the *Itga4* gene, which is expressed on the basal surface of the chorion. The similarity of the two phenotypes strongly suggests that binding of the two molecules is a functional component of chorioallantoic fusion. However, the VCAM1/α4 integrin interaction does not appear to be the only cell adhesion mechanism required for

chorioallantoic attachment, since chorioallantoic fusion occurs stochastically in some *Vcam1* mutants. Also, VCAM1/ α4 integrin expression is normal in other mutants showing defective chorioallantoic fusion (Hunter *et al.,* 1999).

XIV. *Gcm1* Regulates the Initiation of Chorioallantoic Branching

Labyrinth development begins early on E9 with the formation of simple villi that consist of folds in the chorion trophoblast surface, which are underlain by stromal cells and blood vessels derived from the allantoic mesoderm. Although this event is often described as "vascular invasion" of the chorion, the trophoblast epithelium is not simply a passive participant but rather dictates where villous branchpoints will occur. As such, labyrinth development is due to active chorioallantoic branching, analogous to the formation of other branched organs like the lung and kidney. Points in the chorionic plate where folding occurs and invagination of the allantoic mesoderm is initiated are preceded by expression of the *Glial cells missing-1* (*Gcm1*) transcription factor gene (Anson-Cartwright *et al.,* 2000) (Fig. 8). *Gcm1* expression appears in focal sites as early as E8.0, well before the allantois makes contact with the chorion. During active branching of the villous tree, *Gcm1*-expressing cells are confined to the tip of elongating branches, where trophoblast cells elongate and fuse to form the syncytiotrophoblast layers. In *Gcm1*-deficient mice, chorioallantoic branching fails to initiate (Anson-Cartwright *et al.,* 2000; Schreiber *et al.,* 2000a) and mutants arrest at the flat chorion stage. In addition, chorionic trophoblast cells fail to differentiate into syncytiotrophoblast (Anson-Cartwright *et al.,* 2000). *Gcm1* continues to be expressed as long as the labyrinth continues to enlarge by villous branching processes (Basyuk *et al.,* 1999).

Gcm1 function defines points in the chorionic plate where chorioallantoic branching and syncytiotrophoblast differentiation occur, prior to and independent of allantoic attachment and invasion. However, several lines of evidence indicate that a reciprocal interaction between allantoic mesoderm and chorionic trophoblast regulates subsequent development. First, early ultrastructural changes in chorionic trophoblast cells appear only after chorioallantoic fusion (Hernandez-Verdun, 1974). Second, chorionic trophoblast cells fail to differentiate and initiate morphogenesis unless/ until the allantois makes contact. For example, the chorion remains flat in several mouse mutants in which chorioallantoic fusion is blocked (Gurtner *et al.,* 1995; Hunter *et al.,* 1999; Kwee *et al.,* 1995; Yang *et al.,* 1995). A molecular explanation for this effect is that, while *Gcm1* expression begins before allantoic contact, maintenance of its expression requires allantoic attachment (Hunter *et al.,* 1999).

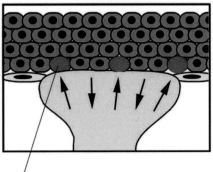

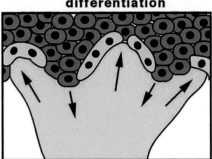

Figure 8 Branchpoint selection by *Gcm1* expression. Domains of *Gcm1* expression mark the sites of villous migration into the labyrinthine trophoblast.

XV. Growth Factor Signaling Regulates Branching Morphogenesis of the Labyrinth

Several signaling systems converge to regulate chorioallantoic development, including many that regulate cell differentiation and tissue morphogenesis in other organs (e.g., FGF, hepatocyte growth factor/HGF, and Wnt) (Hemberger and Cross, 2001) (Fig. 9).

FGF signaling is required for chorioallantoic branching, as shown by severe labyrinth defects in *Fgfr2* mutant mice (Xu *et al.*, 1998). This is consistent with known roles for FGF in regulating epithelial branching in the developing lung (Arman *et al.*, 1999; Nogawa and Ito, 1995) and kidney (Dudley *et al.*, 1999) in mice, and the tracheal system in Drosophila (Skaer, 1997). Aside from *Fgfr2*, mice mutant for other growth factor receptors also show labyrinth defects. These include mice mutant for leukemia inhibitory factor receptor (*Lifr*), epidermal growth factor receptor (*Egfr*), and hepatocyte growth factor (HGF) receptor (*Met*) (Hemberger and Cross, 2001). Like FGFR2, these receptors all have intrinsic or receptor-associated kinases, the activities of which are induced by ligand stimulation. Many mutant mice lacking general signal transduction components that act downstream of these receptors (e.g., *Grb2, Gab1, Sos1, Mek1*) also show specific labyrinth defects (Table I).

Biochemical experiments have suggested that growth factor signaling can induce a range of alternative downstream signaling events, but the importance of each different pathway for mediating the biological effects of the individual growth factors have been elusive. GRB2 is a signaling adaptor protein downstream of several growth factor (including FGF) receptors. *Grb2* null mutants arrest at the blastocyst stage (Cheng *et al.*, 1998), whereas a hypomorphic *Grb2* mutation is associated with chorioallantoic branching defects (Saxton *et al.*, 2001). These data indicate that GRB2 likely mediates both actions of FGF signaling in the trophoblast. In contrast, a null mutation of the alternative signaling adaptor *Gab1* is only associated with chorioallantoic branching defects (Itoh *et al.*, 2000). Downstream of *Grb2*, there are several signaling components the functions of which are not required for trophoblast cell proliferation but are required for branching. These include SOS1 (a guanine nucleotide exchange factor that activates RAS and therefore the MAP kinase pathway) (Qian *et al.*, 2000); MEK1, a kinase discovered for its ability to activate ERK1/2 (Giroux *et al.*, 1999); p38αMAP kinase (Adams *et al.*, 2000); and MEKK3

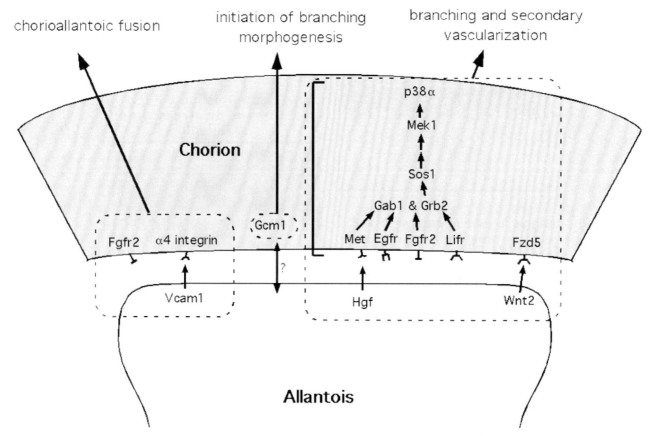

Figure 9   Signaling pathways regulating chorioallantoic morphogenesis.

(Yang *et al.*, 2000), an activator of p38. Together, these data imply that the effects of FGF signaling on early trophoblast stem cell and labyrinth morphogenesis are mediated by distinct downstream signaling pathways.

Wnt signaling through the β-catenin pathway, resulting in activation of the Tcf1/Lef1 transcription factors, has also recently been implicated in both chorioallantoic fusion and the subsequent morphogenesis of the placenta. Although neither *Tcf1* nor *Lef1* mutants have placental phenotypes, *Tcf1;Lef1* compound mutants die at midgestation and show an absence of chorioallantoic fusion (Galceran *et al.*, 1999). The specific Wnt and Wnt receptor (encoded by the *Frizzled* or *Fzd* genes) proteins implicated by this function are unknown. Indeed very few studies have even addressed which Wnt/Fzd pairs are expressed in the extraembryonic tissues around the time of early labyrinth development. After chorioallantoic fusion, *Wnt2* (Monkley *et al.*, 1996) and the Wnt receptor gene *Fzd5* (Ishikawa *et al.*, 2001) are both essential for labyrinth development. Although it has not been shown that Wnt2 can bind to Fzd5, *Wnt2* mRNA is expressed by allantoic mesoderm (Monkley *et al.*, 1996), and *Fzd5* mRNA appears to be expressed in trophoblast cells of the labyrinth (Ishikawa *et al.*, 2001).

Some recent mouse mutants have identified new players in the regulation of trophoblast function and vascular invasion. *Esx1* is a homeobox transcription factor gene that is exclusively expressed in the chorion and later in the trophoblast compartment of the labyrinth (Li *et al.*, 1997). The labyrinth develops in *Esx1* loss-of-function mutants and is actually larger than normal, but vascular density is dramatically reduced (Li and Behringer, 1998). This implies that a signal from the labyrinthine trophoblast, the expression of which is regulated by *Esx1*, in turn regulates vascularization of the labyrinth. Mutations in *Rxra* and β (Wendling *et al.*, 1999) and *Pparγ* (Barak *et al.*, 1999) also implicate nuclear hormone receptor signaling in proper development of the labyrinth, although the targets of these pathways are unclear.

The hypertrophy of the *Esx1* mutant placentas may reflect an attempt to compensate for poor vascularization and the potential associated fetal hypoxia. Mice mutant for the aryl-hydrocarbon receptor nuclear translocator (ARNT) suggest that hypoxia is a normal regulator of placental growth (Adelman *et al.*, 2000; Kozak *et al.*, 1997). These mice die around day 10 of development and show failure of vascularization of the placenta and vascular defects in the embryo and yolk

Table I   Mutations Affecting Development of the Labyrinthine Region of the Placenta

Gene mutation	Gene product	Primary trophoblast defect?	Reference
Arnt	bHLH/PAS transcription factor	yes	Kozak *et al.*, 1997; Adelman *et al.*, 2001
Dlx3	homeobox transcription factor		Morasso *et al.*, 1999
Esx1	homeobox transcription factor	yes	Li and Behringer, 1997
Fgfr2	fibroblast growth factor receptor	yes	Xu *et al.*, 1998
Fra1	AP1transcription factor	yes	Schreiber *et al.*, 2000
Fzd5	Frizzled 5, Wnt receptor		Ishikawa *et al.*, 2001
Gab1	signaling adaptor molecule		Itoh *et al.*, 2000
Gcm1	transcription factor	yes	Anson-Cartwright *et al.*, 2000; Schreiber *et al.*, 2000
Grb2 hypomorph	signaling adaptor molecule		Saxton *et al.*, 2001
Gja7 (Cx45)	connexin, gap junction protein		Kruger *et al.*, 2001
Gjb2 (Cx26)	connexin, gap junction protein		Gabriel *et al.*, 1998
Hgf	hepatocyte growth factor	yes	Uehara *et al.*, 1995
Hsp84-1 (Hsp90b)	heat-shock protein	no	Voss *et al.*, 2000
Itgav (αv integrin)	adhesion molecule		Goh *et al.*, 1997
Junb	AP1transcription factor	yes	Schorpp-Kistner *et al.*, 1999
Lifr	leukemia inhibitory factor receptor		Ware *et al.*, 1995
Map2k1 (Mek1)	MAP kinase cascade	yes	Giroux *et al.*, 1999
Map2k2 (p38alpha)	MAP kinase cascade	yes	Adams *et al.*, 2000
Map3k3 (Mekk3)	MAP kinase cascade		Yang *et al.*, 2000
Met	hepatocyte growth factor receptor	yes	Bladt *et al.*, 1995
Pdgfrα	platelet derived growth factor receptor a		Ogura *et al.*, 1998
Pdgfb	platelet derived growth factor chain B		Ohlsson *et al.*, 1999
Pparγ	nuclear hormone receptor	yes	Barak *et al.*, 1999
Rxrα/β double	nuclear hormone receptor	yes	Wendling *et al.*, 1999
Sos1	GDP/GTP exchange factor		Qian *et al.*, 2000
Tcfeb (Tfeb)	bHLH-Zip transcription factor		Steingrimsson *et al.*, 1998
Vhlh	tumor suppressor		Gnarra *et al.*, 1997
Wnt2	Wingless homolog, secreted glycoprotein		Monkley *et al.*, 1996

sac. Hypoxia has been shown to regulate critical components of vascular development, including both the expression of the VEGF and endothelial-specific response genes, such as *Epas1* and some of the endothelial-specific receptor tyrosine kinases (Maltepe and Simon, 1998). Hypoxia-induced regulation of these genes depends on activation of gene expression by heterodimers of the bHLH-PAS protein, hypoxia-inducible factor 1α (HIF-1α) along with ARNT. The observed failure of vascularization of the placenta in *Arnt* mutants was not therefore too surprising, because so many components of vascular development are hypoxia responsive. However, tetraploid rescue experiments revealed that the placental vascular defect could be rescued by providing wild-type trophoblast (Adelman *et al.*, 2000). Further, hypoxia was shown to directly affect the differentiation of different trophoblast cell types from TS cells. This suggests that the hypoxic environment that occurs during different phases of placental development in mouse and in human may play a direct role in regulating trophoblast differentiation and function, and only secondarily affect vascularization of the villous tree (Adelman *et al.*, 2000). Hypoxia can affect the differentiation of invasive human trophoblast cells *in vitro*

(Caniggia *et al.*, 2000; Genbacev *et al.*, 1997), but the importance of oxygen tension in regulating villous development has not been tested in humans. The studies from mice suggest, however, that there is a complex interplay between the trophoblast and the development of its associated fetal-derived vasculature, in which trophoblast is not just a passive bystander.

Of all the phases of placental development, it is the formation and function of the labyrinth that is the most susceptible to genetic disruption (see Table I for a list of mutations affecting labyrinth development). This reflects both the critical nature of this structure for the survival of the fetus and the complex interplay of maternal, trophoblast, and fetal signals and the responses required for its development.

XVI. Placental Development and Pregnancy Complications

Obviously one of the hopes for genetic studies in mice is that we will gain insights into the molecular basis of diseases that arise from placental defects during intrauterine development. These diseases include "missed abortion" (losses within the first 2 months of gestation), some types of intrauterine growth restriction (IUGR), and preeclampsia. With the comparisons between human and mouse placentae in mind, it is now possible to begin to ask whether the key regulatory molecules involved in the various mouse placental subtypes are expressed and functional in the analogous cell types in humans. Several genes necessary for placental development in mice have been shown to be expressed in an analogous manner in humans including homologs of *Mash2* (Alders *et al.,* 1997; Janatpour *et al.,* 1999), *Hand1* (Knofler *et al.,* 1998), *Gcm1* (Janatpour *et al.,* 1999; Nait-Oumesmar *et al.,* 2000), and *Hgf/Met* (Somerset *et al.,* 1998). Minimal functional data are available as yet, but what is known is largely consistent with mouse studies. For example, hypoxia appears to regulate placental development in both humans and mice (Adelman *et al.,* 2000; Caniggia *et al.,* 2000; Genbacev *et al.,* 1997). Although work is clearly at an early stage, the available data do suggest that key molecular determinants of placental development in mice are likely to be active also in humans.

One of the important principles to emerge from genetic studies in mice is that distinct genetic pathways control different aspects of placental development. Therefore, mutants can specifically be affected in villous/labyrinth morphogenesis or in invasive trophoblast (giant cells) but not usually both. Although missed abortion, some types of IUGR, and preeclampsia conditions are routinely considered as part of a spectrum, the evidence suggests that they are due to distinct pathologies affecting different placental components. Preeclampsia is often associated with defective differentiation/invasion of extravillous cytotrophoblast cells, although the complexity of this disease makes assignment of the primary defect difficult (Cross, 1996). In contrast, with missed abortion (Meegdes *et al.,* 1988; Ornoy *et al.,* 1981; van Lijnschoten *et al.,* 1994) and severe cases of early-onset IUGR (Krebs *et al.,* 1996), the placenta is typically characterized by reduced branching of chorionic villi and the underlying vasculature. Because of the observations in mice, it is likely that these different human placental pathologies have distinct underlying molecular abnormalities. One major step forward will be to start classifying these diseases by different cellular criteria, and using this to narrow the focus on what molecular pathways to test. With respect to villous development specifically, because of the importance of trophoblast cells in directing the development of the underlying vasculature, it will be important to devote attention to this cell type. More studies are urgently needed to elucidate the pathways regulating placental development in mice because this knowledge is likely to impact our understanding of etiological factors leading to early pregnancy loss and pregnancy complications in humans.

Acknowledgments

The authors' work described here is supported by the Canadian Institutes of Health Research (CIHR), J. C. is an investigator of the CIHR and senior scholar of the Alberta Heritage Foundation for Medical Research (AHFMR), and J. R. a distinguished investigator of CIHR and an international scholar of the Howard Hughes Medical Institute.

References

Achen, M. G., Gad, J. M., Stacker, S. A., and Wilks, A. F. (1997). Placenta growth factor and vascular endothelial growth factor are co-expressed during early embryonic development. *Growth Factors* **15,** 69–80.

Adams, R. H., Porras, A., Alonso, G., Jones, M., Vintersten, K., Panelli, S., Valladares, A., Perez, L., Klein, R., and Nebreda, A. R. (2000). Essential role of p38alpha MAP kinase in placental but not embryonic cardiovascular development. *Mol. Cell* **6,** 109–116.

Adelman, D. M., Gertsenstein, M., Nagy, A., Simon, M. C., and Maltepe, E. (2000). Placental cell fates are regulated in vivo by HIF-mediated hypoxia responses. *Genes Dev.* **14,** 3191–3203.

Alders, M., Hodges, M., Hadjantonakis, A. K., Postmus, J., van Wijk, I., Bliek, J., de Meulemeester, M., Westerveld, A., Guillemot, F., Oudejans, C., Little, P., and Mannens, M. (1997). The human Achaete-Scute homologue 2 (ASCL2,HASH2) maps to chromosome 11p15.5, close to IGF2 and is expressed in extravillus trophoblasts. *Hum. Mol. Genet.* **6,** 859–867.

Ang, S.-L., and Rossant, J. (1994). HNF-3beta is essential for node and notochord formation in mouse development. *Cell (Cambridge, Mass.)* **78,** 561–574.

Ang, S.-L., Wierda, A., Wong, D., Stevens, K. A., Cascio, S., Rossant, J., and Zaret, K. S. (1993). The formation and maintenance of the definitive endoderm lineage in the mouse: Involvement of HNF3/forkhead proteins. *Development (Cambridge, UK)* **119,** 1301–1315.

Anson-Cartwright, L., Dawson, K., Holmyard, D., Fisher, S. J., Lazzarini, R. A., and Cross, J. C. (2000). The glial cells missing-1 protein is essential for branching morphogenesis in the chorioallantoic placenta. *Nat. Genet.* **25,** 311–314.

Arman, E., Haffner-Krausz, R., Chen, Y., Heath, J. K., and Lonai, P. (1998). Targeted disruption of fibroblast growth factor (FGF) receptor 2 suggests a role for FGF signaling in pregastrulation mammalian development. *Proc. Natl. Acad. Sci. U.S.A.* **95,** 5082–5087.

Arman, E., Haffner-Krausz, R., Gorivodsky, M., and Lonai, P. (1999). Fgfr2 is required for limb outgrowth and lung-branching morphogenesis. *Proc. Natl. Acad. Sci. U.S.A.* **96,** 11895–11899.

Barak, Y., Nelson, M. C., Ong, E. S., Jones, Y. Z., Ruiz-Lozano, P., Chien, K. R., Koder, A., and Evans, R. M. (1999). PPAR gamma is required for placental, cardiac, and adipose tissue development. *Mol. Cell* **4,** 585–595.

Barbacci, E., Reber, M., Ott, M. O., Breillat, C., Huetz, F., and Cereghini, S. (1999). Variant hepatocyte nuclear factor 1 is required for visceral endoderm specification. *Development (Cambridge, UK)* **126,** 4795–4805.

Barlow, P. W., and Sherman, M. I. (1974). Cytological studies on the organization of DNA in giant trophoblast nuclei of the mouse and the rat. *Chromosoma* **47,** 119–131.

Basyuk, E., Cross, J. C., Corbin, J., Nakayama, H., Hunter, P. J., Nait-Oumesmar, B., and Lazzarini, R. A. (1999). The murine *Gcm1* gene is expressed in a subset of placental trophoblast cells. *Dev. Dynam.* **214,** 303–311.

Beck, F., Erler, T., Russell, A., and James, R. (1995). Expression of Cdx-2 in the mouse embryo and placenta: Possible role in patterning of the extra-embryonic membranes. *Dev. Dynam.* **204,** 219–227.

Beck, F., Chawengsaksophak, K., Waring, P., Playford, R. J., and Furness,

J. B. (1999). Reprogramming of intestinal differentiation and intercalary regeneration in Cdx2 mutant mice. *Proc. Natl. Acad. Sci. U.S.A.* **96,** 7318–7323.

Beddington, R. S. P., and Robertson, E. J. (1989). An assessment of the developmental potential of embryonic stem cells in the midgestation mouse embryo. *Development (Cambridge, UK)* **105,** 733–737.

Berezowsky, J., Zbieranowski, I., Demers, J., and Murray, D. (1995). DNA ploidy of hydatidiform moles and nonmolar conceptuses: A study using flow and tissue section image cytometry. *Mod. Pathol.* **8,** 775–781.

Bernstein, H. S., and Coughlin, S. R. (1998). A mammalian homologue of fission yeast Cdc5 regulates G2 progression and mitotic entry. *J. Biol. Chem.* **273,** 4666–4671.

Bhatt, H., Brunet, L. J., and Stewart, C. L. (1991). Uterine expression of leukemia inhibitory factor coincides with the onset of blastocyst implantation. *Proc. Natl. Acad. Sci. U.S.A.* **88,** 11408–11412.

Bielinska, M., Narita, N., and Wilson, D. B. (1999). Distinct roles for visceral endoderm during embryonic mouse development. *Int. J. Dev. Biol.* **43,** 183–205.

Bladt, F., Riethmacher, D., Isenmann, S., Aguzzi, A., and Birchmeier, C. (1995). Essential role for the c-met receptor in the migration of myogenic precursor cells into the limb bud [see comments]. *Nature (London)* **376,** 768–771.

Botquin, V., Hess, H., Fuhrmann, G., Anastassiadis, C., Gross, M. K., Vriend, G., and Scholer, H. R. (1998). New POU dimer configuration mediates antagonistic control of an osteopontin preimplantation enhancer by Oct-4 and Sox-2. *Genes Dev.* **12,** 2073–2090.

Caniggia, I., Mostachfi, H., Winter, J., Gassmann, M., Lye, S. J., Kuliszewski, M., and Post, M. (2000). Hypoxia-inducible factor-1 mediates the biological effects of oxygen on human trophoblast differentiation through TGFbeta(3). *J. Clin. Invest.* **105,** 577–587.

Carson, D. D., Bagchi, I., Dey, S. K., Enders, A. C., Fazleabas, A. T., Lessey, B. A., and Yoshinaga, K. (2000). Embryo implantation. *Dev. Biol.* **223,** 217–237.

Chai, N., Patel, Y., Jacobson, K., McMahon, J., McMahon, A., and Rappolee, D. A. (1998). FGF is an essential regulator of the fifth cell division in preimplantation mouse embryos. *Dev. Biol.* **198,** 105–115.

Chapman, V. M., Whitten, W. K., and Ruddle, F. H. (1971). Expression of paternal glucose phosphate isomerase-1 (Gpi-1) in preimplantation stages of mouse embryos. *Dev. Biol.* **26,** 153–158.

Chawengsaksophak, K., James, R., Hammond, V. E., Kontgen, F., and Beck, F. (1997). Homeosis and intestinal tumours in Cdx2 mutant mice. *Nature (London)* **386,** 84–87.

Chen, C. M., Kraut, N., Groudine, M., and Weintraub, H. (1996). I-mf, a novel myogenic repressor, interacts with members of the MyoD family. *Cell (Cambridge, Mass.)* **86,** 731–741.

Chen, W. S., Manova, K., Weinstein, D. C., Duncan, S. A., Plump, A. S., Prezioso, V. R., Bachvarova, R. F., and Darnell, J. E., Jr. (1994). Disruption of the HNF-4 gene, expressed in visceral endoderm, leads to cell death in embryonic ectoderm and impaired gastrulation of mouse embryos. *Genes Dev.* **8,** 2466–2477.

Cheng, A. M., Saxton, T. M., Sakai, R., Kulkarni, S., Mbamalu, G., Vogel, W., Totorice, C. G., Cardiff, R. D., Cross, J. C., Muller, W. J., and Pawson, A. J. (1998). Mammalian Grb2 regulates multiple steps in embryonic development, lineage commitment and malignant transformation. *Cell (Cambridge, Mass.)* **95,** 793–803.

Ciruna, B. G., and Rossant, J. (1999). Expression of the T-box gene Eomesodermin during early mouse development. *Mech. Dev.* **81,** 199–203.

Cockroft, D. L. (1986). Regional and temporal differences in the parietal endoderm of the midgestation mouse embryo. *J. Anat.* **145,** 35–47.

Coffinier, C., Thepot, D., Babinet, C., Yaniv, M., and Barra, J. (1999). Essential role for the homeoprotein vHNF1/HNF1beta in visceral endoderm differentiation. *Development (Cambridge, UK)* **126,** 4785–4794.

Copp, A. J. (1979). Interaction between inner cell mass and trophectoderm of the mouse blastocyst. II. The fate of the polar trophectoderm. *J. Embryol. Exp. Morphol.* **51,** 109–120.

Copp, A. J. (1995). Death before birth: Clues from gene knockouts and mutations. *Trends Genet.* **11,** 87–93.

Cross, J. C. (1996). Trophoblast function in normal and preeclamptic pregnancy. *Fet. Mat. Med. Rev.* **8,** 57–66.

Cross, J. C., Werb, Z., and Fisher, S. J. (1994). Implantation and the placenta: Key pieces of the development puzzle. *Science* **266,** 1508–1518.

Cross, J. C., Flannery, M. L., Blanar, M. A., Steingrimsson, E., Jenkins, N. A., Copeland, N. G., Rutter, W. J., and Werb, Z. (1995). Hxt encodes a basic helix–loop–helix transcription factor that regulates trophoblast cell development. *Development (Cambridge, UK)* **121,** 2513–2523.

Dudley, A. T., Godin, R. E., and Robertson, E. J. (1999). Interaction between FGF and BMP signaling pathways regulates development of metanephric mesenchyme. *Genes Dev.* **13,** 1601–1613.

Dufort, D., Schwartz, L., Harpal, K., and Rossant, J. (1998). The transcription factor HNF3beta is required in visceral endoderm for normal primitive streak morphogenesis. *Development (Cambridge, UK)* **125,** 3015–3025.

Duncan, S. A., Manova, K., Chen, W. S., Hoodless, P., Weinstein, D. C., Bachvarova, R. F., and Darnell, J. E., Jr. (1994). Expression of transcription factor HNF-4 in the extraembryonic endoderm, gut, and nephrogenic tissue of the developing mouse embryo: HNF-4 is a marker for primary endoderm in the implanting blastocyst. *Proc. Natl. Acad. Sci. U.S.A.* **91,** 7598–7602.

Duncan, S. A., Nagy, A., and Chan, W. (1997). Murine gastrulation requires HNF-4 regulated gene expression in the visceral endoderm: Tetraploid rescue of Hnf-4(–/–) embryos. *Development (Cambridge, UK)* **124,** 279–287.

Edgar, B. A., and Lehner, C. F. (1996). Developmental control of cell cycle regulators: a fly's perspective. *Science* **274,** 1646–1652.

Edgar, B. A., and Orr-Weaver, T. L. (2001). Endoreplication cell cycles: More for less. *Cell (Cambridge, Mass.)* **105,** 297–306.

Farese, R. V., Jr., Cases, S., Ruland, S. L., Kayden, H. J., Wong, J. S., Young, S. G., and Hamilton, R. L. (1996). A novel function for apolipoprotein B: Lipoprotein synthesis in the yolk sac is critical for maternal–fetal lipid transport in mice. *J. Lipid. Res.* **37,** 347–360.

Feldman, B., Poueymirou, W., Papaioannou, V. E., DeChiara, T. M., and Goldfarb, M. (1995). Requirement of FGF-4 for postimplantation mouse development. *Science* **267,** 246–249.

Firulli, A. B., McFadden, D. G., Lin, Q., Srivastava, D., and Olson, E. N. (1998). Heart and extra-embryonic mesodermal defects in mouse embryos lacking the bHLH transcription factor Hand1. *Nat. Genet.* **18,** 266–270.

Fleming, T. P. (1987). A quantitative analysis of cell allocation to trophectoderm and inner cell mass in the mouse blastocyst. *Dev. Biol.* **119,** 520–531.

Fleming, T. P., and Johnson, M. H. (1988). From egg to epithelium. *Ann. Rev. Cell Biol.* **4,** 459–485.

Fleming, T. P., Warren, P. D., Chisholm, J. C., and Johnson, M. H. (1984). Trophectodermal processes regulate the expression of totipotency within the inner cell mass of the mouse expanding blastocyst. *J. Embryol. Exp. Morphol.* **84,** 63–90.

Follette, P. J., Duronio, R. J., and O'Farrell, P. H. (1998). Fluctuations in cyclin E levels are required for multiple rounds of endocycle S phase in Drosophila. *Curr. Biol.* **8,** 235–238.

Fukushige, T., Hawkins, M. G., and McGhee, J. D. (1998). The GATA-factor elt-2 is essential for formation of the *Caenorhabditis elegans* intestine. *Dev. Biol.* **198,** 286–302.

Fuse, N., Hirose, S., and Hayashi, S. (1994). Diploidy of *Drosophila* imaginal cells is maintained by a transcriptional repressor encoded by escargot. *Genes Dev.* **8,** 2270–2281.

Gabriel, H. D., Jung, D., Butzler, C., Temme, A., Traub, O., Winterhager, E., and Willecke, K. (1998). Transplacental uptake of glucose is decreased in embryonic lethal connexin26-deficient mice. *J. Cell Biol.* **140,** 1453–1461.

Galceran, J., Farinas, I., Depew, M. J., Clevers, H., and Grosschedl, R.

(1999). Wnt3a⁻/⁻-like phenotype and limb deficiency in Lef1(–/–) Tcf1(–/–) mice. *Genes Dev.* **13,** 709–717.

Gardner, R. L. (1968). Mouse chimaeras obtained by the injection of cells into the blastocyst. *Nature (London)* **220,** 596–597.

Gardner, R. L. (1983). Origin and differentiation of extraembryonic tissues in the mouse. *Int. Rev. Exp. Pathol.* **24,** 63–133.

Gardner, R. L. (1984). An *in situ* cell marker for clonal analysis of development of the extraembryonic endoderm in the mouse. *J. Embryol. Exp. Morphol.* **80,** 251–288.

Gardner, R. L. (1985). Regeneration of endoderm from primitive ectoderm in the mouse embryo: Fact or artifact? *J. Embryol. Exp. Morphol.* **88,** 303–326.

Gardner, R. L., and Cockroft, D. L. (1998). Complete dissipation of coherent clonal growth occurs before gastrulation in mouse epiblast. *Development (Cambridge, UK)* **125,** 2397–2402.

Gardner, R. L., and Davies, T. J. (1993). Lack of coupling between onset of giant-transformation and genome endoreduplication in the mural trophectoderm of the mouse blastocyst. *J. Exp. Zool.* **265,** 54–60.

Gardner, R. L., and Rossant, J. (1979). Investigation of the fate of 4–5 day post-coitum mouse inner cell mass cells by blastocyst injection. *J. Embryol. Exp. Morphol.* **52,** 141–52.

Gardner, R. L., Barton, S. C., and Surani, M. A. (1990). Use of triple tissue blastocyst reconstitution to study the development of diploid parthenogenetic primitive ectoderm in combination with fertilization-derived trophectoderm and primitive endoderm. *Genet. Res.* **56,** 209–222.

Gardner, R. L., Papaioannou, V. E., and Barton, S. C. (1973). Origin of the ectoplacental cone and secondary giant cells in mouse blastocysts reconstituted from isolated trophectoderm and inner cell mass. *J. Embryol. Exp. Morphol.* **30,** 561–572.

Gardner, R. L., Meredith, M. R., and Altman, D. G. (1992). Is the anterior-posterior axis of the fetus specified before implantation in the mouse? *J. Exp. Zool.* **264,** 437–443.

Genbacev, O., Zhou, Y., Ludlow, J. W., and Fisher, S. J. (1997). Regulation of human placental development by oxygen tension. *Science* **277,** 1669–1672.

Giroux, S., Tremblay, M., Bernard, D., Cardin-Girard, J. F., Aubry, S., Larouche, L., Rousseau, S., Huot, J., Landry, J., Jeannotte, L., and Charron, J. (1999). Embryonic death of Mek1-deficient mice reveals a role for this kinase in angiogenesis in the labyrinthine region of the placenta. *Curr. Biol.* **9,** 369–372.

Gnarra, J. R., Ward, J. M., Porter, F. D., Wagner, J. R., Devor, D. E., Grinberg, A., Emmert-Buck, M. R., Westphal, H., Klausner, R. D., and Linehan, W. M. (1997). Defective placental vasculogenesis causes embryonic lethality in VHL-deficient mice. *Proc. Natl. Acad. Sci. U.S.A.* **94,** 9102–9107.

Goh, K. L., Yang, J. T., and Hynes, R. O. (1997). Mesodermal defects and cranial neural crest apoptosis in alpha5 integrin-null embryos. *Development (Cambridge, UK)* **124,** 4309–4319.

Griffin, K. J., Amacher, S. L., Kimmel, C. B., and Kimelman, D. (1998). Molecular identification of spadetail: Regulation of zebrafish trunk and tail mesoderm formation by T-box genes. *Development (Cambridge, UK)* **125,** 3379–3388.

Groskopf, J. C., Syu, L. J., Saltiel, A. R., and Linzer, D. I. (1997). Proliferin induces endothelial cell chemotaxis through a G protein-coupled, mitogen-activated protein kinase-dependent pathway. *Endocrinology* **138,** 2835–2840.

Guillemot, F., Lo, L.-C., Johnson, J. E., Auerbach, A., Anderson, D. J., and Joyner, A. L. (1993). Mammalian achaete-scute homologue 1 is required for the early development of olfactory and autonomic neurons. *Cell (Cambridge, Mass.)* **75,** 463–476.

Guillemot, F., Nagy, A., Auerbach, A., Rossant, J., and Joyner, A. L. (1994). Rescue of a lethal mutation in Mash-2 reveals its essential role in extraembryonic development. *Nature (London)* **371,** 333–336.

Guillemot, F., Caspary, T., Tilghman, S. M., Copeland, N. G., Gilbert, D. J., Jenkins, N. A., Anderson, D. J., Joyner, A. L., Rossant, J., and Nagy, A.

(1995). Genomic imprinting of Mash-2, a mouse gene required for trophoblast development. *Nature Genet.* **9,** 235–241.

Gurtner, G. C., Davis, V., Li, H., McCoy, M. J., Sharpe, A., and Cybulsky, M. I. (1995). Targeted disruption of the murine VCAM1 gene: Essential role of VCAM-1 in chorioallantoic fusion and placentation. *Genes Dev.* **9,** 1–14.

Haffner-Krausz, R., Gorivodsky, M., Chen, Y., and Lonai, P. (1999). Expression of Fgfr2 in the early mouse embryo indicates its involvement in preimplantation development. *Mech. Dev.* **85,** 167–172.

Handyside, A. H. (1978). Time of commitment of inside cells isolated from preimplantation mouse embryos. *J. Embryol. Exp. Morphol.* **45,** 37–53.

Hattori, N., Davies, T. C., Anson-Cartwright, L., and Cross, J. C. (2000). Periodic expression of the Cdk inhibitor p57_{Kip2} in trophoblast giant cells defines a G2-like gap phase of the endocycle. *Mol. Biol. Cell* **11,** 1037–1045.

Heikinheimo, M., Scandrett, J. M., and Wilson, D. B. (1994). Localization of transcription factor GATA-4 to regions of the mouse embryo involved in cardiac development. *Dev. Biol.* **164,** 361–373.

Hemberger, M., and Cross, J. C. (2001). Genes governing placental development. *Trends Endocrin. Metab.* **12,** 162–168.

Henry, G. L., and Melton, D. A. (1998). Mixer, a homeobox gene required for endoderm development. *Science* **281,** 91–96.

Hernandez-Verdun, D. (1974). Morphogenesis of the syncytium in the mouse placenta. Ultrastructural study. *Cell Tissue Res.* **148,** 381–396.

Hertz, R., Magenheim, J., Berman, I., and Bar-Tana, J. (1998). Fatty acyl-CoA thioesters are ligands of hepatic nuclear factor-4alpha. *Nature (London)* **392,** 512–516.

Hogan, B. L., Barlow, D. P., and Kurkinen, M. (1984). Reichert's membrane as a model for studying the biosynthesis and assembly of basement membrane components. *Ciba Found. Symp.* **108,** 60–74.

Horner, M. A., Quintin, S., Domeier, M. E., Kimble, J., Labouesse, M., and Mango, S. E. (1998). pha-4, an HNF-3 homologue, specifies pharyngeal organ identity in *Caenorhabditis elegans. Genes Dev.* **12,** 1947–1952.

Horvitz, H. R., and Herskowitz, I. (1992). Mechanisms of asymmetric cell division: Two Bs or not two Bs, that is the question. *Cell (Cambridge, Mass.)* **68,** 237–255.

Hudson, C., Clements, D., Friday, R. V., Stott, D., and Woodland, H. R. (1997). Xsox17alpha and -beta mediate endoderm formation in Xenopus. *Cell (Cambridge, Mass.)* **91,** 397–405.

Hunt, C. V., and Avery, G. B. (1976). The development and proliferation of the trophoblast from ectopic mouse embryo allografts of increasing gestational age. *J. Reprod. Fertil.* **46,** 305–311.

Hunter, P. J., Swanson, B. J., Haendel, M. A., Lyons, G. E., and Cross, J. C. (1999). *Mrj* encodes a DnaJ-related co-chaperone that is essential for murine placental development. *Development (Cambridge, UK)* **126,** 1247–1258.

Huxham, I. M., and Beck, F. (1985). Maternal transferrin uptake by and transfer across the visceral yolk sac of the early postimplantation rat conceptus in vitro. *Dev. Biol.* **110,** 75–83.

Ilgren, E. B. (1981). On the control of the trophoblastic giant-cell transformation in the mouse: Homotypic cellular interactions and polyploidy. *J. Embryol. Exp. Morphol.* **62,** 183–202.

Isaacs, H. V., Pownall, M. E., and Slack, J. M. (1998). Regulation of Hox gene expression and posterior development by the *Xenopus* caudal homologue Xcad3. *EMBO J.* **17,** 3413–3427.

Ishikawa, T., Tamai, Y., Zorn, A. M., Yoshida, H., Seldin, M. F., Nishikawa, S., and Taketo, M. M. (2001). Mouse Wnt receptor gene Fzd5 is essential for yolk sac and placental angiogenesis. *Development (Cambridge, UK)* **128,** 25–33.

Itoh, M., Yoshida, Y., Nishida, K., Narimatsu, M., Hibi, M., and Hirano, T. (2000). Role of Gab1 in heart, placenta, and skin development and growth factor- and cytokine-induced extracellular signal-regulated kinase mitogen-activated protein kinase activation. *Mol. Cell Biol.* **20,** 3695–3704.

Jacquemin, P., Durviaux, S. M., Jensen, J., Godfraind, C., Gradwohl, G.,

Guillemot, F., Madsen, O. D., Carmeliet, P., Dewerchin, M., Collen, D., Rousseau, G. G., and Lemaigre, F. P. (2000). Transcription factor hepatocyte nuclear factor 6 regulates pancreatic endocrine cell differentiation and controls expression of the proendocrine gene ngn3. *Mol. Cell Biol.* **20,** 4445–4454.

Janatpour, M. J., Utset, M. F., Cross, J. C., Rossant, J., Dong, J., Israel, M. A., and Fisher, S. J. (1999). A repertoire of differentially expressed transcription factors that offers insight into mechanisms of human cytotrophoblast differentiation. *Dev. Genet.* **25,** 146–157.

Jenkinson, E. J., and Billington, W. D. (1974). Differential susceptibility of mouse trophoblast and embryonic tissue to immune cell lysis. *Transplantation* **18,** 286–289.

Johnson, J. A., Birren, S. J., and Anderson, D. J. (1990). Two rat homologues of *Drosophila* Achaete-Scute specifically expressed in neuronal precursors. *Nature (London)* **346,** 858–861.

Johnson, M. H., and Rossant, J. (1981). Molecular studies on cells of the trophectoderm lineage of the postimplantation mouse embryo. *J. Embryol. Exp. Morphol.* **61,** 103–116.

Johnson, M. H., and Ziomek, C. A. (1981). The foundation of two distinct cell lineages within the mouse morula. *Cell (Cambridge, Mass.)* **24,** 71–80.

Jollie, W. P. (1990). Effects of sustained dietary ethanol on the ultrastructure of the visceral yolk-sac placenta of the rat. *Teratology* **42,** 541–552.

Kalb, J. M., Lau, K. K., Goszczynski, B., Fukushige, T., Moons, D., Okkema, P. G., and McGhee, J. D. (1998). pha-4 is Ce-fkh-1, a fork head/HNF-3alpha, beta, gamma homologue that functions in organogenesis of the *C. elegans* pharynx. *Development (Cambridge, UK)* **125,** 2171–2180.

Kaufmann, E., and Knochel, W. (1996). Five years on the wings of fork head. *Mech. Dev.* **57,** 3–20.

Knofler, M., Meinhardt, G., Vasicek, R., Husslein, P., and Egarter, C. (1998). Molecular cloning of the human *Hand1* gene/cDNA and its tissue-restricted expression in cytotrophoblastic cells and heart. *Gene* **224,** 77–86.

Koutsourakis, M., Langeveld, A., Patient, R., Beddington, R., and Grosveld, F. (1999). The transcription factor GATA6 is essential for early extraembryonic development. *Development (Cambridge, UK)* **126,** 723–732.

Kozak, K. R., Abbott, B., and Hankinson, O. (1997). ARNT-deficient mice and placental differentiation. *Dev. Biol.* **191,** 297–305.

Kraut, N., Snider, L., Chen, C. M., Tapscott, S. J., and Groudine, M. (1998). Requirement of the mouse I-mfa gene for placental development and skeletal patterning. *EMBO J.* **17,** 6276–6288.

Krebs, C., Macara, L. M., Leiser, R., Bowman, A. W., Greer, I. A., and Kingdom, J. C. P. (1996). Intrauterine growth restriction with absent end-diastolic flow velocity in the umbilical artery is associated with maldevelopment of the placental terminal villous tree. *Am. J. Obstet. Gynecol.* **175,** 1534–1542.

Kruger, O., Plum, A., Kim, J. S., Winterhager, E., Maxeiner, S., Hallas, G., Kirchhoff, S., Traub, O., Lamers, W. H., and Willecke, K. (2000). Defective vascular development in connexin 45-deficient mice. *Development (Cambridge, UK)* **127,** 4179–4193.

Kuo, C. J., Conley, P. B., Chen, L., Sladek, F. M., Darnell, J. E., Jr., and Crabtree, G. R. (1992). A transcriptional hierarchy involved in mammalian cell-type specification. *Nature (London)* **355,** 457–461.

Kuo, C. T., Morrisey, E. E., Anandappa, R., Sigrist, K., Lu, M. M., Parmacek, M. S., Soudais, C., and Leiden, J. M. (1997). GATA4 transcription factor is required for ventral morphogenesis and heart tube formation. *Genes Dev.* **11,** 1048–1060.

Kwee, L., Baldwin, H. S., Shen, H. M., Stewart, C. L., Buch, C., Buck, C. A., and Labow, M. A. (1995). Defective development of the embryonic and extraembryonic circulatory systems in vascular cell adhesion molecule (VCAM-1) deficient mice. *Development (Cambridge, UK)* **121,** 489–503.

Lai, E., Prezioso, V. R., Tao, W. F., Chen, W. S., and Darnell, J. E., Jr.

(1991). Hepatocyte nuclear factor 3 alpha belongs to a gene family in mammals that is homologous to the *Drosophila* homeotic gene fork head. *Genes Dev.* **5,** 416–427.

Lawson, K. A., and Pedersen, R. A. (1987). Cell fate, morphogenetic movement and population kinetics of embryonic endoderm at the time of germ layer formation in the mouse. *Development (Cambridge, UK)* **101,** 627–652.

Li, Y., and Behringer, R. R. (1998). Esx1 is an X-chromosome-imprinted regulator of placental development and fetal growth. *Nat. Genet.* **20,** 309–311.

Li, Y., Lemaire, P., and Behringer, R. R. (1997). Esx1, a novel X chromosome-linked homeobox gene expressed in mouse extraembryonic tissues and male germ cells. *Dev. Biol.* **188,** 85–95.

Lim, H., Paria, B. C., Das, S. K., Dinchuk, J. E., Langenbach, R., Trzaskos, J. M., and Dey, S. K. (1997). Multiple female reproductive failures in cyclooxygenase 2-deficient mice. *Cell (Cambridge, Mass.)* **91,** 197–208.

Lo, C. W., Coulling, M., and Kirby, C. (1987). Tracking of mouse cell lineage using microinjected DNA sequences: Analysis using genomic Southern blotting and tissue-section *in situ* hybridizations. *Differentiation* **35,** 37–44.

Lo, L.-C., Johnson, J. E., Wuenschell, C. W., Saito, T., and Anderson, D. J. (1991). Mammalian achaete-scute homologue 1 is transiently expressed by spatially restricted subsets of early neuroepithelial and neural crest cells. *Genes Dev.* **5,** 1524–1537.

Luo, J., Sladek, R., Bader, J. A., Matthyssen, A., Rossant, J., and Giguere, V. (1997). Placental abnormalities in mouse embryos lacking the orphan nuclear receptor ERR-beta. *Nature (London)* **388,** 778–782.

MacAuley, A., Cross, J. C., and Werb, Z. (1998). Reprogramming the cell cycle for endoreduplication in rodent trophoblast cells. *Mol. Biol. Cell* **9,** 795–807.

MacPhee, D. J., Jones, D. H., Barr, K. J., Betts, D. H., Watson, A. J., and Kidder, G. M. (2000). Differential involvement of Na(+),K(+)-ATPase isozymes in preimplantation development of the mouse. *Dev. Biol.* **222,** 486–498.

Maltepe, E., and Simon, M. C. (1998). Oxygen, genes, and development: An analysis of the role of hypoxic gene regulation during murine vascular development. *J. Mol. Med.* **76,** 391–401.

Meegdes, B. H., Ingenhoes, R., Peeters, L. L., and Exalto, N. (1988). Early pregnancy wastage: Relationship between chorionic vascularization and embryonic development. *Fertil. Steril.* **49,** 216–220.

Molkentin, J. D., Lin, Q., Duncan, S. A., and Olson, E. N. (1997). Requirement of the transcription factor GATA4 for heart tube formation and ventral morphogenesis. *Genes Dev.* **11,** 1061–1072.

Monaghan, A. P., Kaestner, K. H., Grau, E., and Schutz, G. (1993). Postimplantation expression patterns indicate a role for the mouse forkhead/HNF-3 alpha, beta and tau genes in determination of the definitive endoderm, chordamesoderm and neuroectoderm. *Development (Cambridge, UK)* **119,** 567–578.

Monkley, S. J., Delaney, S. J., Pennisi, D. J., Christiansen, J. H., and Wainwright, B. J. (1996). Targeted disruption of the Wnt2 gene results in placentation defects. *Development (Cambridge, UK)* **122,** 3343–3353.

Morasso, M. I., Grinberg, A., Robinson, G., Sargent, T. D., and Mahon, K. A. (1999). Placental failure in mice lacking the homeobox gene Dlx3. *Proc. Natl. Acad. Sci. U.S.A.* **96,** 162–167.

Morrisey, E. E., Ip, H. S., Lu, M. M., and Parmacek, M. S. (1996). GATA-6: A zinc finger transcription factor that is expressed in multiple cell lineages derived from lateral mesoderm. *Dev. Biol.* **177,** 309–322.

Morrisey, E. E., Tang, Z., Sigrist, K., Lu, M. M., Jiang, F., Ip, H. S., and Parmacek, M. S. (1998). GATA6 regulates HNF4 and is required for differentiation of visceral endoderm in the mouse embryo. *Genes Dev.* **12,** 3579–3590.

Nagy, A., Gocza, E., Diaz, E. M., Prideaux, V. R., Ivanyi, E., Markkula, M., and Rossant, J. (1990). Embryonic stem cells alone are able to support fetal development in the mouse. *Development (Cambridge, UK)* **110,** 815–821.

Nait-Oumesmar, B., Copperman, A. B., and Lazzarini, R. A. (2000). Placental expression and chromosomal localization of the human Gcm 1 gene. *J. Histochem. Cytochem.* **48,** 915–922.

Nakayama, H., Liu, Y., Stifani, S., and Cross, J. C. (1997). Developmental restriction of Mash-2 expression in trophoblast correlates with potential activation of the notch-2 pathway. *Dev. Genet.* **21,** 21–30.

Nakayama, H., Scott, I. C., and Cross, J. C. (1998). The transition to endoreduplication in trophoblast giant cells is regulated by the mSNA zinc-finger transcription factor. *Dev. Biol.* **199,** 150–163.

Narita, N., Bielinska, M., and Wilson, D. B. (1997). Wild-type endoderm abrogates the ventral developmental defects associated with GATA-4 deficiency in the mouse. *Dev. Biol.* **189,** 270–274.

Nichols, J., and Gardner, R. L. (1984). Heterogeneous differentiation of external cells in individual isolated early mouse inner cell masses in culture. *J. Embryol. Exp. Morphol.* **80,** 225–240.

Nichols, J., Zevnik, B., Anastassiadis, K., Niwa, H., Klewe-Nebenius, D., Chambers, I., Scholer, H., and Smith, A. (1998). Formation of pluripotent stem cells in the mammalian embryo depends on the POU transcription factor Oct4. *Cell (Cambridge, Mass.)* **95,** 379–391.

Niswander, L., and Martin, G. R. (1992). Fgf-4 expression during gastrulation, myogenesis, limb and tooth development in the mouse. *Development (Cambridge, UK)* **114,** 755–768.

Niwa, H., Miyazaki, J., and Smith, A. G. (2000). Quantitative expression of Oct-3/4 defines differentiation, dedifferentiation or self-renewal of ES cells. *Nat. Genet.* **24,** 372–376.

Nogawa, H., and Ito, T. (1995). Branching morphogenesis of embryonic mouse lung epithelium in mesenchyme-free culture. *Development (Cambridge, UK)* **121,** 1015–1022.

Ogura, Y., Takakura, N., Yoshida, H., and Nishikawa, S. I. (1998). Essential role of platelet-derived growth factor receptor alpha in the development of the intraplacental yolk sac/sinus of Duval in mouse placenta. *Biol. Reprod.* **58,** 65–72.

Ohlsson, R., Falck, P., Hellstrom, M., Lindahl, P., Bostrom, H., Franklin, G., Ahrlund-Richter, L., Pollard, J., Soriano, P., and Betsholtz, C. (1999). PDGFB regulates the development of the labyrinthine layer of the mouse fetal placenta. *Dev. Biol.* **212,** 124–136.

Ornoy, A., Salamon-Arnon, J., Ben-Zur, Z., and Kohn, G. (1981). Placental findings in spontaneous abortions and stillbirths. *Teratology* **24,** 243–252.

Palmieri, S. L., Peter, W., Hess, H., and Scholer, H. R. (1994). Oct-4 transcription factor is differentially expressed in the mouse embryo during establishment of the first two extraembryonic cell lineages involved in implantation. *Dev. Biol.* **166,** 259–267.

Papaioannou, V. E. (1982). Lineage analysis of inner cell mass and trophectoderm using microsurgically reconstituted blastocysts. *J. Embryol. Exp. Morphol.* **68,** 199–209.

Paria, B. C., Elenius, K., Klagsbrun, M., and Dey, S. K. (1999). Heparin-binding EGF-like growth factor interacts with mouse blastocysts independently of ErbB1: A possible role for heparan sulfate proteoglycans and ErbB4 in blastocyst implantation. *Development (Cambridge, UK)* **126,** 1997–2005.

Pearce, J. J., and Evans, M. J. (1999). Mml, a mouse Mix-like gene expressed in the primitive streak. *Mech. Dev.* **87,** 189–192.

Pettersson, K., Svensson, K., Mattsson, R., Carlsson, B., Ohlsson, R., and Berkenstam, A. (1996). Expression of a novel member of estrogen response element-binding nuclear receptors is restricted to the early stages of chorion formation during mouse embryogenesis. *Mech. Dev.* **54,** 211–223.

Pontoglio, M., Barra, J., Hadchouel, M., Doyen, A., Kress, C., Bach, J. P., Babinet, C., and Yaniv, M. (1996). Hepatocyte nuclear factor 1 inactivation results in hepatic dysfunction, phenylketonuria, and renal Fanconi syndrome. *Cell (Cambridge, Mass.)* **84,** 575–585.

Pratt, H. P., Ziomek, C. A., Reeve, W. J., and Johnson, M. H. (1982). Compaction of the mouse embryo: An analysis of its components. *J. Embryol. Exp. Morphol.* **70,** 113–32.

Qian, X., Esteban, L., Vass, W. C., Upadhyaya, C., Papageorge, A. G., Yienger, K., Ward, J. M., Lowy, D. R., and Santos, E. (2000). The Sos1 and Sos2 Ras-specific exchange factors: Differences in placental expression and signaling properties. *EMBO J.* **19,** 642–654.

Raabe, M., Flynn, L. M., Zlot, C. H., Wong, J. S., Veniant, M. M., Hamilton, R. L., and Young, S. G. (1998). Knockout of the abetalipoproteinemia gene in mice: Reduced lipoprotein secretion in heterozygotes and embryonic lethality in homozygotes. *Proc. Natl. Acad. Sci. U.S.A.* **95,** 8686–8691.

Rappolee, D. A., Basilico, C., Patel, Y., and Werb, Z. (1994). Expression and function of FGF-4 in peri-implantation development in mouse embryos. *Development (Cambridge, UK)* **120,** 2259–2269.

Rappolee, D. A., Patel, Y., and Jacobson, K. (1998). Expression of fibroblast growth factor receptors in peri-implantation mouse embryos. *Mol. Reprod. Dev.* **51,** 254–264.

Rehorn, K. P., Thelen, H., Michelson, A. M., and Reuter, R. (1996). A molecular aspect of hematopoiesis and endoderm development common to vertebrates and *Drosophila. Development (Cambridge, UK)* **122,** 4023–4031.

Riley, P., Anson-Cartwright, L., and Cross, J. C. (1998). The Hand1 bHLH transcription factor is essential for placentation and cardiac morphogenesis. *Nat. Genet.* **18,** 271–275.

Riley, P. R., Gertsenstein, M., Dawson, K., and Cross, J. C. (2000). Early exclusion of Hand1-deficient cells from distinct regions of the left ventricular myocardium in chimeric mouse embryos. *Dev. Biol.* **227,** 156–168.

Rogers, J. M., Daston, G. P., Ebron, M. T., Carver, B., Stefanadis, J. G., and Grabowski, C. T. (1985). Studies on the mechanism of trypan blue teratogenicity in the rat developing in vivo and in vitro. *Teratology* **31,** 389–399.

Rossant, J. (1975). Investigation of the determinative state of the mouse inner cell mass. II. The fate of isolated inner cell masses transferred to the oviduct. *J. Embryol. Exp. Morphol.* **33,** 991–1001.

Rossant, J., and Lis, W. T. (1979). Potential of isolated mouse inner cell masses to form trophectoderm derivatives in vivo. *Dev. Biol.* **70,** 255–261.

Rossant, J., and Ofer, L. (1977). Properties of extraembryonic ectoderm isolated from postimplantation mouse embryos. *J. Embryol. Exp. Morphol.* **39,** 183–194.

Rossant, J., and Spence, A. (1998). Chimeras and mosaics in mouse mutant analysis. *Trends Genet.* **14,** 358–363.

Rossant, J., and Tamara-Lis, W. (1981). Effect of culture conditions in diploid to giant-cell transformation in postimplantation mouse trophoblast. *J. Embryol. Exp. Morphol.* **62,** 217–227.

Rossant, J., Vijh, M., Siracusa, L. D., and Chapman, V. M. (1983). Identification of embryonic cell lineages in histological sections of *M. musculus* ↔ *M. caroli* chimeras. *J. Embryol. Exp. Morphol.* **73,** 179–191.

Rossant, J., Merentes-Diaz, E., Gocza, E., Ivanyi, E., and Nagy, A. (1993). Developmental potential of mouse embryonic stem cells. *In* "Serono Symposium on Preimplantation Embryo Development" (B. Bavister, Ed.), pp. 157–164. Springer Verlag, Berlin.

Rossant, J., Guillemot, F., Tanaka, M., Latham, K., Gertenstein, M., and Nagy, A. (1998). Mash2 is expressed in oogenesis and preimplantation development but is not required for blastocyst formation. *Mech. Dev.* **73,** 183–191.

Russ, A. P., Wattler, S., Colledge, W. H., Aparicio, S. A., Carlton, M. B., Pearce, J. J., Barton, S. C., Surani, M. A., Ryan, K., Nehls, M. C., Wilson, V., and Evans, M. J. (2000). Eomesodermin is required for mouse trophoblast development and mesoderm formation. *Nature (London)* **404,** 95–99.

Saxton, T. M., Cheng, A. M., Ong, S.-H., Lu, Y., Sakai, R., Cross, J. C., and Pawson, T. (2001). Gene dosage dependent functions for phosphotyrosine-Grb2 signaling during mammalian tissue morphogenesis. *Curr. Biol.* **11,** 662–670.

Scholer, H. R., Dressler, G. R., Balling, R., Rohdewohld, H., and Gruss, P.

(1990). Oct-4: A germline-specific transcription factor mapping to the mouse t-complex. *EMBO J.* **9**, 2185–2195.

Schorpp-Kistner, M., Wang, Z. Q., Angel, P., and Wagner, E. F. (1999). JunB is essential for mammalian placentation. *EMBO J.* **18**, 934–948.

Schreiber, J., Riethmacher-Sonnenberg, E., Riethmacher, D., Tuerk, E. E., Enderich, J., Bosl, M. R., and Wegner, M. (2000a). Placental failure in mice lacking the mammalian homologue of glial cells missing, GCMa. *Mol. Cell Biol.* **20**, 2466–2474.

Schreiber, M., Wang, Z. Q., Jochum, W., Fetka, I., Elliott, C., and Wagner, E. F. (2000b). Placental vascularisation requires the AP-1 component fra1. *Development (Cambridge, UK)* **127**, 4937–4948.

Scott, I. C., Anson-Cartwright, L., Riley, P., Reda, D., and Cross, J. C. (2000). The Hand1 basic helix-loop-helix transcription factor regulates trophoblast giant cell differentitation via multiple mechanisms. *Mol. Cell Biol.* **20**, 530–541.

Seibel, W. (1974). An ultrastructural comparison of the uptake and transport of horseradish peroxidase by the rat visceral yolk-sac placenta during mid- and late gestation. *Am. J. Anat.* **140**, 213–235.

Semenza, G. L. (2000). HIF-1: Mediator of physiological and pathophysiological responses to hypoxia. *J. Appl. Physiol.* **88**, 1474–1480.

Sherr, C. J. (1994). G1 phase progression: Cycling on cue. *Cell (Cambridge, Mass.)* **79**, 551–555.

Sibilia, M., and Wagner, E. F. (1995). Strain-dependent epithelial defects in mice lacking the EGF receptor. *Science* **269**, 234–238.

Skaer, H. (1997). Morphogenesis: FGF branches out. *Curr. Biol.* **7**, R238–R241.

Sladek, F. M., Zhong, W. M., Lai, E., and Darnell, J. E., Jr. (1990). Liver-enriched transcription factor HNF-4 is a novel member of the steroid hormone receptor superfamily. *Genes Dev.* **4**, 2353–2365.

Smith, J. C., Price, B. M. J., Green, J. B. A., Weigel, D., and Herrmann, B. G. (1991). Expression of a *Xenopus* homologue of Brachyury (T) is an immediate-early response to mesoderm induction. *Cell (Cambridge, Mass.)* **67**, 79–87.

Somerset, D. A., Li, X. F., Afford, S., Strain, A. J., Ahmed, A., Sangha, R. K., Whittle, M. J., and Kilby, M. D. (1998). Ontogeny of hepatocyte growth factor (HGF) and its receptor (c-met) in human placenta: Reduced HGF expression in intrauterine growth restriction. *Am. J. Pathol.* **153**, 1139–1147.

Soudais, C., Bielinska, M., Heikinheimo, M., MacArthur, C. A., Narita, N., Saffitz, J. E., Simon, M. C., Leiden, J. M., and Wilson, D. B. (1995). Targeted mutagenesis of the transcription factor GATA-4 gene in mouse embryonic stem cells disrupts visceral endoderm differentiation in vitro. *Development (Cambridge, UK)* **121**, 3877–3888.

Spindle, A. I. (1978). Trophoblast regeneration by inner cell masses isolated from cultured mouse embryos. *J. Exp. Zool.* **203**, 483–489.

Steingrimsson, E., Tessarollo, L., Reid, S. W., Jenkins, N. A., and Copeland, N. G. (1998). The bHLH-Zip transcription factor Tfeb is essential for placental vascularization. *Development (Cambridge, UK)* **125**, 4607–4616.

Stewart, C. L., Kaspar, P., Brunet, L. J., Bhatt, H., Gadi, I., Kontgen, F., and Abbondanzo, S. J. (1992). Blastocyst implantation depends on maternal expression of leukaemia inhibitory factor. *Nature (London)* **359**, 76–79.

Tanaka, M., Gertsenstein, M., Rossant, J., and Nagy, A. (1997). Mash2 acts cell autonomously in mouse spongiotrophoblast development. *Dev. Biol.* **190**, 55–65.

Tanaka, S., Kunath, T., Hadjantonakis, A. K., Nagy, A., and Rossant, J. (1998). Promotion of trophoblast stem cell proliferation by FGF4. *Science* **282**, 2072–2075.

Teesalu, T., Blasi, F., and Talarico, D. (1998). Expression and function of the urokinase type plasminogen activator during mouse hemochorial placental development. *Dev. Dynam.* **213**, 27–38.

Teesalu, T., Masson, R., Basset, P., Blasi, F., and Talarico, D. (1999). Expression of matrix metalloproteinases during murine chorioallantoic placenta maturation. *Dev. Dynam.* **214**, 248–258.

Thomas, T., Southwell, B. R., Schreiber, G., and Jaworowski, A. (1990).

Plasma protein synthesis and secretion in the visceral yolk sac of the fetal rat: Gene expression, protein synthesis and secretion. *Placenta* **11**, 413–30.

Threadgill, D. W., Dlugosz, A. A., Hansen, L. A., Tennenbaum, T., Lichti, U., Yee, D., LaMantia, C., Mourton, T., Herrup, K., Harris, R. C., Barnard, J. A., Yuspa, S. H., Coffey, R. J., and Magnuson, T. (1995). Targeted disruption of mouse EGF receptor: Effect of genetic background on mutant phenotype. *Science* **269**, 230–234.

Tremblay, G. B., Kunath, T., Bergeron, D., Lapointe, L., Champigny, C., Bader, J. A., Rossant, J., and Giguere, V. (2001). Diethylstilbestrol regulates trophoblast stem cell differentiation as a ligand of orphan nuclear receptor ERRbeta. *Genes Dev.* **15**, 833–838.

Uehara, Y., Minowa, O., Mori, C., Shiota, K., Kuno, J., Noda, T., and Kitamura, N. (1995). Placental defect and embryonic lethality in mice lacking hepatocyte growth factor/scatter factor. *Nature (London)* **373**, 702–705.

van Lijnschoten, G., Arends, J. W., and Geraedts, J. P. (1994). Comparison of histological features in early spontaneous and induced trisomic abortions. *Placenta* **15**, 765–773.

Varmuza, S., Prideaux, V., Kothary, R., and Rossant, J. (1988). Polytene chromosomes in mouse trophoblast giant cells. *Development (Cambridge, UK)* **102**, 127–134.

Voss, A. K., Thomas, T., and Gruss, P. (2000). Mice lacking HSP90beta fail to develop a placental labyrinth. *Development (Cambridge, UK)* **127**, 1–11.

Vuorela, P., Hatva, E., Lymboussaki, A., Kaipainen, A., Joukov, V., Persico, M. G., Alitalo, K., and Halmesmaki, E. (1997). Expression of vascular endothelial growth factor and placenta growth factor in human placenta. *Biol. Reprod.* **56**, 489–494.

Ware, C. B., Horowitz, M. C., Renshaw, B. R., Hunt, J. S., Liggitt, D., Koblar, S. A., Gliniak, B. C., McKenna, H. J., Papayannopoulou, T., Thoma, B., *et al.* (1995). Targeted disruption of the low-affinity leukemia inhibitory factor receptor gene causes placental, skeletal, neural and metabolic defects and results in perinatal death. *Development (Cambridge, UK)* **121**, 1283–1299.

Weber, R. J., Pedersen, R. A., Wianny, F., Evans, M. J., and Zernicka-Goetz, M. (1999). Polarity of the mouse embryo is anticipated before implantation. *Development (Cambridge, UK)* **126**, 5591–5598.

Weinstein, D. C., Ruiz i Altaba, A., Chen, W. S., Hoodless, P., Prezioso, V. R., Jessell, T. M., and Darnell, J. E., Jr. (1994). The winged-helix transcription factor HNF-3 beta is required for notochord development in the mouse embryo. *Cell (Cambridge, Mass.)* **78**, 575–588.

Weiss, A., Herzig, A., Jacobs, H., and Lehner, C. F. (1998). Continuous Cyclin E expression inhibits progression through endoreduplication cycles in *Drosophila*. *Curr. Biol.* **8**, 239–42.

Wendling, O., Chambon, P., and Mark, M. (1999). Retinoid X receptors are essential for early mouse development and placentogenesis. *Proc. Natl. Acad. Sci. U.S.A.* **96**, 547–551.

Wilder, P. J., Kelly, D., Brigman, K., Peterson, C. L., Nowling, T., Gao, Q. S., McComb, R. D., Capecchi, M. R., and Rizzino, A. (1997). Inactivation of the FGF-4 gene in embryonic stem cells alters the growth and/or the survival of their early differentiated progeny. *Dev. Biol.* **192**, 614–629.

Williamson, R. A., Henry, M. D., Daniels, K. J., Hrstka, R. F., Lee, J. C., Sunada, Y., Ibraghimov-Beskrovnaya, O., and Campbell, K. P. (1997). Dystroglycan is essential for early embryonic development: Disruption of Reichert's membrane in Dag1-null mice. *Hum. Mol. Genet.* **6**, 831–841.

Xu, X., Weinstein, M., Li, C., Naski, M., Cohen, R. I., Ornitz, D. M., Leder, P., and Deng, C. (1998). Fibroblast growth factor receptor 2 (FGFR2)-mediated reciprocal regulation loop between FGF8 and FGF10 is essential for limb induction. *Development (Cambridge, UK)* **125**, 753–765.

Yamamoto, H., Flannery, M. L., Kupriyanov, S., Pearce, J., McKercher, S. R., Henkel, G. W., Maki, R. A., Werb, Z., and Oshima, R. G. (1998).

Defective trophoblast function in mice with a targeted mutation of Ets2. *Genes Dev.* **12**, 1315–1326.

Yan, J., Tanaka, S., Oda, M., Makino, T., Ohgane, J., and Shiota, K. (2001). Retinoic acid promotes differentiation of trophoblast stem cells to a giant cell fate. *Dev. Biol.* **235**, 422–432.

Yang, J., Boerm, M., McCarty, M., Bucana, C., Fidler, I. J., Zhuang, Y., and Su, B. (2000). Mekk3 is essential for early embryonic cardiovascular development. *Nat. Genet.* **24**, 309–313.

Yang, J. T., Rayburn, H., and Hynes, R. O. (1995). Cell adhesion events mediated by alpha-4 integrins are essential in placental and cardiac development. *Development (Cambridge, UK)* **121**, 549–560.

Yuan, H., Corbi, N., Basilico, C., and Dailey, L. (1995). Developmental-specific activity of the FGF-4 enhancer requires the synergistic action of Sox2 and Oct-3. *Genes Dev.* **9**, 2635–2645.

Zambrowicz, B. P., Imamoto, A., Fiering, S., Herzenberg, L. A., Kerr, W. G., and Soriano, P. (1997). Disruption of overlapping transcripts in the ROSA beta geo 26 gene trap strain leads to widespread expression of beta-galactosidase in mouse embryos and hematopoietic cells. *Proc. Natl. Acad. Sci. U.S.A.* **94**, 3789–3794.

Zhu, J., Hill, R. J., Heid, P. J., Fukuyama, M., Sugimoto, A., Priess, J. R., and Rothman, J. H. (1997). end-1 encodes an apparent GATA factor that specifies the endoderm precursor in Caenorhabditis elegans embryos. *Genes Dev.* **11**, 2883–2896.

Ziomek, C. A., Johnson, M. H., and Handyside, A. H. (1982). The developmental potential of mouse 16-cell blastomeres. *J. Exp. Zool.* **221**, 345–355.

Zybina, E. V., and Zybina, T. G. (1996). Polytene chromosomes in mammalian cells. *Int. Rev. Cytol.* **165**, 53–119.

Zybina, T. G., Zybina, E. V., and Shtein, G. I. (1985). DNA content of the nuclei of secondary giant cells of the rat trophoblast at different phases of the polytene nucleus cycle. *Tsitologiia* **27**, 957–960.

9

Germ Cells

Chris Wylie* and Robert Anderson†

*Division of Developmental Biology, Children's Hospital Research Foundation, Cincinnati, Ohio 45229
†University of Minnesota School of Medicine, Minneapolis, Minnesota 55455

I. General Concepts

II. Early Appearance of Germ Cells in the Mouse

III. Specification of Germ Cells in the Mouse

IV. Migration of Germ Cells

V. Motility of Germ Cells

VI. Guidance of Germ Cell Migration

VII. Adhesive Behavior of Germ Cells during Migration

VIII. Survival and Proliferation of Germ Cells during Migration

References

I. General Concepts

Germ cells are the embryonic precursors of the gametes. They are set aside from the somatic cell lineages early in the development of most species. In the mouse, the germ cells, once they have formed, migrate through the tissues of the embryo to the gonad primordia (genital ridges), where they coassemble with somatic gonadal cells to form the sex cords. (See Fig. 1 for a scheme of germ cell formation and migration.) The sex cords are the forerunners of the seminiferous tubules of the male gonad or the ovarian follicles of the fe-

male gonad. They become sexually dimorphic as they form in the mouse embryo, between embryonic day 11.5 (E11.5) and E12.5. In addition to their complex differentiation into eggs and sperm, germ cells retain the property of pluripotency, which is required for the gametes to differentiate into new individuals. Germ cells that do not enter the gonad primordia can develop into germ line tumors later in life. Errors in germ cell differentiation can lead to infertility.

Germ cells are the only cells in the body to undergo meiotic cell divisions during their differentiation. This leads to haploidy of the gametes and also generates genetic differences between individuals. Errors of meiotic recombination lead to aneuploidy and to congenital disorders in the offspring. In many species, the differentiation of the female gametes includes the synthesis and storage of molecules (known as cytoplasmic determinants) that control early growth and patterning of the ensuing embryo. The degree to which early development is patterned maternally varies widely between species. In many species, the formation of the germ line itself is controlled by maternal cytoplasmic determinants. So far there is no evidence for this in the mouse.

This review focuses on early events of germ cell formation in the mouse embryo. However, one of the major problems in germ cell biology centers around the apparently different ways in which germ cells form, and so evidence from other species will also be briefly reviewed.

E7.5

E9.5

E10.5

E11.5

♀

♂

E12.5

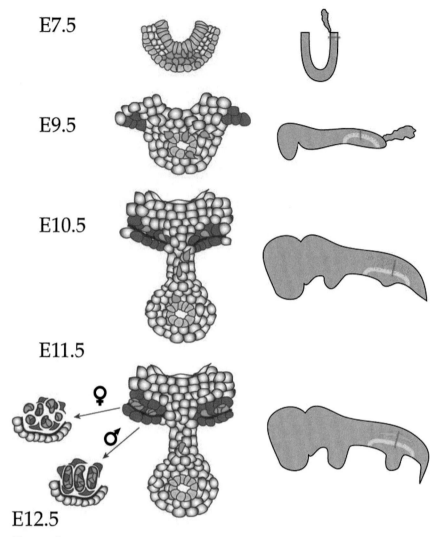

Figure 1 Scheme for primordial germ cell (PGC) migration in the mouse embryo. At each stage, a cross section is shown at the position of the red line in the schematic midsagittal section (artificially straightened out for clarity). PGCs (green) migrate from the primitive streak (blue) into the posterior embryonic endoderm (yellow) around E7.5. They remain in the hindgut until E9.5, when they migrate laterally and dorsally out of the hindgut epithelium, and enter the genital ridges between E10.5 and E11.5, where they join the gonadal somatic cells (brown) and the basement membrane (dull red) to form the sex cords. These cords are sexually dimorphic; in the female they are irregular clumps, whereas in the male they are elongated. The precise boundaries of the gonadal somatic cells are unknown. (Reprinted with the permission of Academic Press from Anderson and Wylie, 2000.)

II. Early Appearance of Germ Cells in the Mouse

In many species (for example, *Drosophila, Caenorhabditis elegans,* and *Xenopus*), germ cells arise in the embryo complete with their own endogenous marker, the germ plasm. This unique cluster of mRNAs and organelles, whose content and organization are only partially understood, has made possible the identification of genes controlling the appearance and early migration of germ cells in these species (for review, see Wylie, 1999; Saffman and Lasko, 1999). In

mammals, however, there is no visible germ plasm, making it impossible to identify any early population of cells as belonging to the germ line before gastrulation. In addition, the germ plasm of species such as *Drosophila* actually specifies the cells that inherit it to enter the germ line. Cells that do not inherit germ plasm do not enter the germ line, giving a somewhat preformationist view of germ cell formation in these species. In mammals, single-cell transplantation analysis (Gardner and Rossant, 1979) and grafting of groups of cells (Tam and Zhou, 1996) have shown that any intraembryonic epiblast cell before gastrulation can enter the germ line. Without markers, or apparent determinants in mammals, an understanding of the origin of the germ line cells, as well as the timing and mechanism of their specification, has had to await techniques such as lineage analysis (Lawson and Hage, 1994) and use of living markers of the early germ cells (Anderson et al., 1999a).

In an elegant series of experiments involving microinjection of individual cells with lineage markers, Lawson and colleagues showed that germ cells arise from a founder population of pluripotential cells in the proximal epiblast, adjacent to the extraembryonic ectoderm (Lawson and Hage, 1994). These pass through the posterior primitive streak and enter several cell lineages including the allantois, the germ line, blood islands, yolk sac mesoderm, and the amnion. At about E7.2, a cluster of cells posterior to the primitive streak, in the root of the developing allantois, was tentatively identified as germ cells based on their expression of the enzyme alkaline phosphatase (Ginsburg et al., 1990). Germ cells in the mouse express tissue nonspecific alkaline phosphatase (TNAP; MacGregor et al., 1995), which has been a useful marker of germ cell migration (Chiquoine, 1954). However, the posterior primitive streak also expresses TNAP, and different studies have suggested different origins of the germ cells, including the posterior primitive streak (Chiquoine, 1954; Copp et al., 1986; Snow, 1981), the yolk sac endoderm (Chiquoine, 1954), and the allanoic mesoderm (Ginsburg et al., 1990).

More recent studies have made use of the expression pattern of the POU domain transcription factor OCT4 (Pou5f1) (also known as Oct3; Okamoto et al., 1990; Rosner et al., 1990; Scholer et al., 1990, 1991). Oct4 is expressed in all pluripotent cells of the early mouse embryo, including all cells of the egg to morula stages, the inner cell mass of the blastula, and the primary ectoderm. Its expression becomes restricted during gastrulation to the primordial germ cells. A truncated Oct4 promoter (Yeom et al., 1996) was used to drive the expression of green fluorescent protein (GFP) in a transgenic mouse line (Anderson et al., 1999a). At early stages of development, GFP expression faithfully reproduced the expression pattern of Oct4 in these mice. During gastrulation, the posterior primitive streak expresses GFP. As cells leave the primitive streak, GFP expression is lost in all but a few cells. These also expressed alkaline phosphatase, and were identified on this basis as the earliest identifiable germ cells.

The trajectories of these cells out of the posterior primitive streak were both surprising and interesting, for they appeared to spread out into all adjacent structures, including the allantois, the definitive endoderm, and the primitive endoderm (Fig. 2). Individual cells were observed directly in living embryos to move from the posterior primitive streak into the definitive endoderm, which will become the hindgut (Anderson et al., 2000). These results help to explain the somewhat diverse origins of germ cells reported previously, and they also suggest that cells are specified to enter the germ line either within the posterior primitive streak or as they leave it.

By E8.5, the hindgut endoderm contains a population of a few hundred germ cells, which are assumed to be the founder cell population of all later germ cells, based on their subsequent movements (see below). However, a population of alkaline phosphatase (AP)/GFP+ cells continues to survive outside the gut, especially in the allantois. The fate of GFP+/AP+ cells outside the hindgut is unknown.

III. Specification of Germ Cells in the Mouse

Two pieces of evidence suggest that in the mouse, cell signaling, rather than the presence of cytoplasmic determinants, specifies cells to enter the germ line. First, single-cell transplantation of inner cell mass cells (Gardner and Rossant, 1979) showed that cells taken randomly from the inner cell mass have the capacity to enter the germ line. Second, grafting of distal epiblast cells (which do not contribute to the germ line) from early gastrulae to more proximal positions caused their descendants to enter the germ line (Tam and Zhou, 1996), showing that it is the position of a cell in the gastrulating epiblast, not its ancestry, that determines germ cell specification.

The nature of one such signal was revealed recently in a careful analysis of the expression pattern and effects of disruption of the bone morphogenetic protein (Bmp4) gene in the mouse (Lawson et al., 1999). In some genetic backgrounds the absence of BMP4 expression causes death during gastrulation. In others, development proceeds further, albeit with abnormal development of many organs. Neither the allantois, nor the germ cells, formed in these homozygous null embryos. In heterozygous $Bmp4^{-/-}$ embryos, the allantois appeared normal, but the number of germ cells was reduced. Regression analysis suggested that this reduction was due to a smaller number of cells entering the germ line. In addition, the expression pattern of BMP4, revealed by knocking in β-galactosidase into the endogenous Bmp4 gene, showed that BMP4 is synthesized by the extraembryonic ectoderm and then mesoderm cells immediately adjacent to the proximal epiblast (shown previously to be the endogenous source of germ cells in the mouse embryo; see above). BMP4 is also synthesized later during development in the allantois, but not in the germ cells themselves.

This study shows that BMP4 signaling at the posterior of the embryo is required for germ cell specification. The interpretation is still complex, however, because it is not yet known if the signaling is direct, nor what BMP4 tells cells to do. There are a number of possibilities: First, BMP signaling could tell cells to be posterior (or ventral, since the germ cells and the allantois are both ventral structures), and germ cell specification could be downstream of this. Second, BMP could prevent proximal epiblast cells from apoptosis, and in its absence the prospective germ cells die. Third, it could instruct proximal epiblast cells directly to enter the germ line.

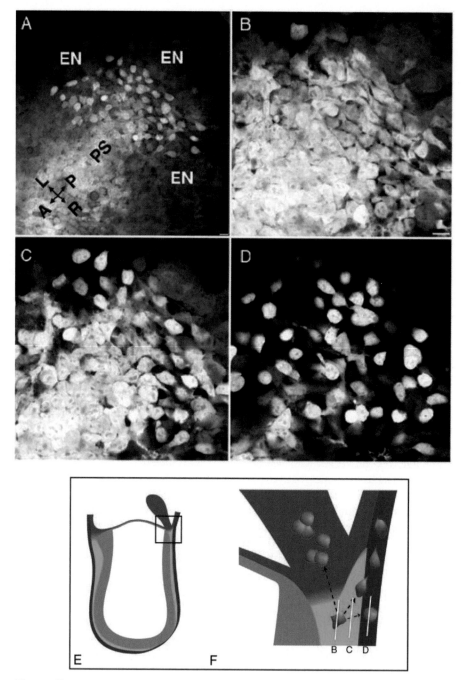

Figure 2 The first emergence of germ cells from the posterior primitive streak. (A) A stack of confocal images through the posterior primitive streak. (B–D) Individual planes of optical section shown in the diagram in panel F (the position of which is shown in panel E). GFP+ cells pass ventrally directly from the posterior primitive streak into the definitive endoderm that will give rise to hindgut. They also pass into other adjacent structures, such as the extraembryonic endoderm and the allantois, shown diagrammatically in panel F. Abbreviations: L, left; R, right; A, anterior; P, posterior. Scale bars = 10 μM. (This figure is modified from Anderson *et al.*, 2000. Parts of this figure are reprinted, with permission, from Anderson *et al.*, 2000.)

This appears at first sight the least likely possibility, since BMP4 signaling is clearly required for the formation of other structures derived from the proximal epiblast (the allantois, for example), so this option requires either a differential response, or differential exposure to the signal, of different proximal epiblast cells to cause them to adopt different fates. Signaling at the posterior end of the gastrula is likely to be complex and represents a major challenge for future studies.

Another major challenge in germ cell biology is to rationalize the seemingly diverse methods of germ cell specification in different species. In many animal groups, including *Drosophila, C. elegans,* and, among the vertebrates, *Xenopus* and possibly zebrafish, germ cells are specified by localized cytoplasmic molecules known as germ plasm (also known as polar granules in *Drosophila* and P granules in *C. elegans*). Germ plasm consists of a large number of proteins and RNAs, some of which are similar in different species. This subject has been reviewed extensively recently (Rongo *et al.,* 1997; Saffman and Lasko, 1999; Seydoux and Strome, 1999; Wylie, 1999), so it is not covered in detail here. The components of germ plasm, the mechanism of its localization and assembly, and the means by which it specifies cells to enter the germ line are only partially understood. However, recent studies have suggested some general principles. First, germ plasm components are transported in the oocyte by polarized cytoskeletal elements (Clark *et al.,* 1994, 1997; Emmons *et al.,* 1995; Erdelyi *et al.,* 1995; Manseau *et al.,* 1996; Pokrywka and Stephenson, 1995; Robb *et al.,* 1996; Therkauf, 1994), based on earlier polarity in the developing oocyte (Heasman *et al.,* 1984). In some species, *Xenopus* and *C. elegans,* for example, further movements of germ plasm occur during the first cell cycle, bringing it to its final location in the cleaving egg. Second, some germ plasm components seem to work by negative regulation. Mutations in the gene *pie-1* in *C. elegans* cause cells that would ordinarily be allocated to the germ line to enter other lineages (Seydoux and Dunn, 1997; Seydoux *et al.,* 1996). The cellular function of *pie-1* includes repression of all RNA polymerase II-based transcription, which is characteristic of the cells in the cleaving egg that will enter the germ line (reviewed in Seydoux and Strome, 1999). When this blockade is lifted, these cells enter other lineages. Negative transcriptional regulation in the germ cell precursor cells has also been reported in *Drosophila* (Seydoux and Dunn, 1997; Zalokar, 1976), and was suggested a number of years ago, based on radio-labeled uridine incorporation, in *Xenopus* (Whitington and Dixon, 1975). It is not clear how negative transcriptional regulation leads to germ cell specification. One possibility is that it preserves a defined early embryonic cell population in a pluripotential state, by making it unable to respond to the cell signaling events that establish other cell lineages (Nieuwkoop and Sutasurya, 1981, p. 186). Third, some germ plasm components positively regulate germ cell specification. In *Drosophila,* the gene *oskar* will cause cells to enter the germ line wherever it is expressed in the early embryo (Ephrussi and Lehmann, 1992).

Do any of these events occur during germ cell specification in the mouse? Although it cannot be formally excluded, current evidence suggests not. First, a number of efforts have been made to identify homologs in the mouse of components of *Drosophila* germ plasm, which have so far proved unsuccessful. Second, the ability of any cell in the mouse inner cell mass, and later the primary ectoderm, to enter the germ line suggests that cytoplasmic determinants inherited from the egg are not involved in this process. Third, there is no

evidence in the mouse for transcriptional blockade in cells that will enter the germ line. Last, there is no evidence of negative specification in the mouse, although it remains a formal possibility.

These apparently completely different mechanisms of germ cell specification are puzzling in light of the fact that the molecules involved in specification of other lineages seem to possess a high degree of evolutionary conservation. A unifying theory may emerge when we know more about exactly how germ plasm components lead to germ cell specification in the species that contain it, and more details of the signaling interactions that lead to germ cell specification in those that do not. In particular, it will be interesting to know if there is any signaling requirement for germ cell formation in germ plasm-bearing species. The only current evidence, from the ectopic expression of the *oskar* gene, suggests that in *Drosophila* the process is cell autonomous.

IV. Migration of Germ Cells

Germ cell specification leads to the expression of a set of genes that control germ cell migration from the posterior primitive streak to the developing gonads. Germ cells are specified 2–3 days before the appearance of the gonad primordia. Some of the intervening period (E7.5–E9.5) is spent lodged in the epithelium of the hindgut. This arises from the posterior definitive endoderm, derived in turn from the primitive streak. From E9.5–E11.5 germ cells leave the hindgut and migrate across the intervening mesenchyme to the gonad primordia, where they coalesce with gonadal somatic cells, derived from the intermediate mesoderm, to form the sex cords (E11.5–E12.5). They become sexually dimorphic as they form. Male sex cords are elongated, whereas female ones form irregular clumps. This process is shown diagrammatically in Fig. 1.

V. Motility of Germ Cells

Evidence now exists that germ cells are actively motile from the time of their specification to the time of gonad colonization. Observation of GFP-expressing germ cells leaving the posterior primitive streak suggests active locomotion (Anderson *et al.,* 2000). In the E7.5–E9.0 period, the hindgut diverticulum forms, and germ cells are carried along with this morphogenetic movement. However, GFP-expressing germ cells in living embryos have the appearance of motile cells, being elongated and extending processes (Anderson *et al.,* 2000), and germ cells explanted from the hindgut at E8.5 are motile in culture (Godin and Wylie, 1991; Godin *et al.,* 1990). However, the morphogenetic movements of the gut have so far made cinematography of germ cells at this stage impossible.

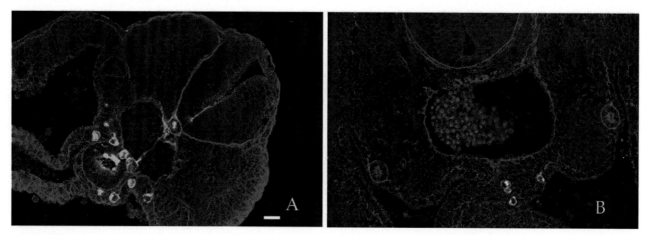

Figure 3 The change in the topographic relationship of hindgut and genital ridges during germ cell migration. (A) Cross section through the hindgut (G) and gonad primordia (GP) at E9.5. Germ cells are stained with anti-SSEA1, and outlines of other cells are stained red with a lectin that binds sialic acid. The first germ cells to leave the gut have only a few cell diameters to travel to the gonad primordia. (B) A similar section at E10.5. The hindgut is out of the picture downward. The germ cells still in the mesentery at E10.5 have many more cell diameters to travel to the gonad primordia (GP). In this panel, the red stain is cross reaction with an anti-laminin antibody. Bar = 20 μm for both panels.

During the E9.0–E9.5 period, germ cells start to leave the hindgut into the adjacent mesenchyme; by E10.5 many of them are in the gonad primordia, and by E11.5 all of them are. This phase of germ cell migration occurs over a period that does not exceed 48 hr, and over a terrain that is rapidly changing due to formation of the hindgut mesentery. The cells leaving the hindgut last have considerably further to go than those leaving earliest (which only have a few cell diameters to travel; see Fig. 3). Not all germ cells reach the gonads, but it is not known whether this is related systematically to the timing of their exit from the gut and therefore the different lengths of the route. Because of this, there is no consensus on the "distance" over which germ cell migration takes place. Germ cell motility is apparently lost as cells reach the gonad, because germ cells removed from the gonad primordia move less actively in the same culture conditions as those removed from the migratory route (Donovan et al., 1986). Not much is known of the mechanism of germ cell motility. They extend extremely long processes (Gomperts et al., 1994), but it is not known if these are functional equivalents of motile filopodia in other cells, because very little study of their motility has been done in vivo. The use of GFP as a living marker of germ cells should permit the analysis of the roles of these long filopodial processes.

VI. Guidance of Germ Cell Migration

The evidence reviewed above suggests that germ cells are motile from the time of their specification. Whether their motility is autonomous or requires signaling from surrounding tissues is unknown. However, it is becoming clear that their guidance to the gonad primordia, and other aspects of

their behavior during migration, does require signals from surrounding cells, which provide both positive and negative signals. Most of the evidence comes from *Drosophila*, where a number of genes required for germ cell migration and guidance have been identified. Some of these were found in a systematic screen for zygotic genes required for germ cell migration (Moore et al., 1998a).

First, several genes are required for germ cell migration itself. Examples include *nanos* (Forbes and Lehmann, 1998; Kobayashi et al., 1996) and polar granule component (*pgc*; Nakamura et al., 1996). Second, genes required for correct patterning of the gut, for example, *serpent* (Reuter, 1994; Warrior, 1994), *huckebein* (Jaglarz and Howard, 1995; Warrior, 1994), and *dorsal* (Warrior, 1994), also affect the ability of germ cells to migrate through its wall. Third, genes expressed in the somatic mesoderm are required for homing of the germ cells to the gonad primordium. These include patterning genes such as *abdominal A, Abdominal B, trithorax, trithoraxgleich,* and *tinman* (for review, see Warrior, 1994; Moore et al., 1998a; Wylie, 1999), which are presumably required upstream of region-specific guidance factors, as well as more specific guidance cue molecules such as *zinc finger homeodomain-1* (Moore et al., 1998a), *columbus,* and *heartless* (Moore et al., 1998b, Van Doren et al., 1998). *columbus,* which encodes the protein HMG CoA reductase, will attract germ cells to other regions of the embryo when ectopically expressed, showing that it is a positive guidance cue for germ cells. Negative guidance cues also exist. The gene *wunen,* which encodes a protein involved in lipid metabolism, is expressed in the gut and is required for movement of the germ cells away from the gut. Germ cells do not colonize gonad primordia that ectopically express *wunen* (Van Doren et al., 1998; Zhang et al., 1997). Not much is known about the receptors expressed by the germ cells that allow them to respond to these cues. It is assumed that they are maternally inherited, and may be revealed by maternal screens for germ cell migration mutants.

It is not yet known if functional homologs of these *Drosophila* genes exist in mice. However, there is some evidence for positive guidance of germ cells to the gonad primordia, since culture medium conditioned by E10.5 gonad primordia acts chemotropically on E8.5 or E10.5 germ cells in culture (Godin *et al.*, 1990). This action is mimicked by purified transforming growth factor β type 1 (TGF-β₁), and antibodies against TGF-β₁ block the chemotropic effects of whole gonad primordia (Godin and Wylie, 1991). These experiments were carried out in crude culture conditions in which somatic cells were present in addition to germ cells. They need to be repeated genetically to identify which member(s) of the TGF-β family may be involved.

VII. Adhesive Behavior of Germ Cells during Migration

It is clear that germ cell adhesiveness must change for germ cells to leave the hindgut, migrate within the mesentery, stop in the gonad primordia, and assemble into sex cords. Germ cells visibly interact with three different types of structures during this process, as discussed next.

1. Germ cells associate with each other. When they arrive at the gonad primordia they cohere with each other to form sex cords. It is likely that this process starts earlier, since germ cells in the mesentery at E10.5 extend long processes that link germ cells into a network of dendritic-looking cells (Gomperts *et al.*, 1994). A similar process takes place in *Drosophila* germ cells (Jaglarz and Howard, 1995). In a search for cell–cell adhesion molecules that may mediate this behavior, mouse germ cells were found to express both E- and P-cadherin during migration (Bendel-Stenzel *et al.*, 2000), as well as the epithelial marker EpCAM (Anderson *et al.*, 1999b). Paradoxically, germ cells do not express E-cadherin when they are in the hindgut, but start to express it after they have emigrated through the hindgut wall (Bendel-Stenzel *et al.*, 2000), suggesting that emigration of the germ cells from the hindgut is not a simple epithelial–mesenchymal transition. Blocking antibodies against E-cadherin (but not P-cadherin) perturb germ cell condensation into sex cords in embryo slice cultures, suggesting that E-cadherin plays a role in this process (Bendel-Stenzel *et al.*, 2000).

2. Germ cells associate with extracellular matrix. As they leave the hindgut at E9.5, laminin, type IV collagen, and fibronectin are expressed interstitially between the cells in the mesentery. As they reach the gonad primordia, laminin is becoming restricted to a prominent basal lamina under the epithelium of the gonad primordia. Arriving germ cells assemble on this line of laminin-positive material. This interaction

with a basal lamina continues. As sex cords assemble from the initial clusters of germ cells, they are continuously surrounded by a laminin-rich basal lamina (Garcia-Castro *et al.*, 1997). Changes in germ cell adhesiveness to extracellular matrix glycoproteins have been shown by quantitative adhesion assays in culture to take place during and after germ cell migration (Garcia-Castro *et al.*, 1997). Attachment to laminin was found to be dependent on both integrin and heparan sulfate proteoglycan (HSPG)-mediated mechanisms during this period (Garcia-Castro *et al.*, 1997).

In an effort to identify the cell surface receptors responsible for these changes in adhesion, germ cells expressing GFP have been isolated by flow cytometry, and either immunochemistry or PCR used to identify integrin subunits or cell surface HSPGs expressed by germ cells during and after migration (Anderson *et al.*, 1999a). Figure 4 shows the integrin expression pattern of germ cells. The principal difference found was that α₆ integrin expression was switched on when germ cells reach the gonad primordia (Anderson *et al.*, 1999a). Germ cells were also found to express the laminin-binding HSPG α-dystroglycan during and after migration (R. Anderson and C. Wylie, unpublished observations).

Once cell surface receptors have been identified on germ cells, their roles can be examined genetically in chimeric

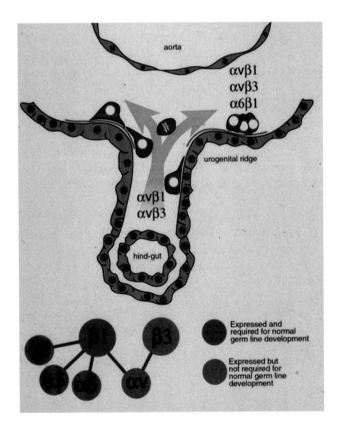

Figure 4 Integrin subunit expression in mouse germ cells during (in the hindgut mesentery) and after (in the urogenital ridge) migration. For details, see Anderson *et al.* (1999a) and text discussion of adhesive behavior.

embryos in which ES cells carrying a homozygous null mutation, and carrying a visible marker expressed in the germ cells (β-galactosidase, or GFP), are injected into wild-type blastulae. Two simple questions can be asked. First do cells enter the germ line in the absence of the target gene and, second, do they migrate correctly to the gonad primordia (since the migration of the mutant cells can be followed by their expression of the marker)? Using this method, α-dystroglycan was shown not to be required for germ cell migration (R. Anderson and C. Wylie, unpublished observations). However, β_1 integrin was required for germ cell migration (Anderson *et al.*, 1999a). The α integrin partner(s) of β_1 integrin have not been identified. Germ cells migrate normally to the gonad primordia in the absence of α_6, α_3, or α_V integrins (Anderson *et al.*, 1999a). This general method should prove useful in the future to identify the roles of individual cell surface receptors on germ cells. Interactions with extracellular membrane glycoproteins are also important in *Drosophila* gonad assembly, because this is disrupted in embryos carrying laminin mutations (Jaglarz and Howard, 1995).

3. *Germ cells associate with somatic cells during sex cord assembly.* The adhesive interactions that allow them to recognize each other, and cohere, are largely unknown. However, it has been suggested that this may be mediated by the interaction between membrane-bound Steel factor on the somatic cells and its receptor c-kit on the germ cells (Pesce *et al.*, 1997). In *Drosophila*, the genes *fear of intimacy* (which encodes an unknown protein) and *clift/eyes absent* (which encodes a transcription factor) are required for condensation of the gonad primordia (Boyle *et al.*, 1997; Moore *et al.*, 1998a,b).

VIII. Survival and Proliferation of Germ Cells during Migration

In the mouse, germ cells divide at a constant cycle time of 16–17 hr from the time they first become recognizable until they reach the gonad (Tam and Snow, 1981). Germ cell numbers and behavior in culture were originally shown to be affected by the addition of serum-free medium conditioned by E10.5 gonad primordia. This observation suggested that proliferation/survival of germ cells during migration is not cell autonomous, but is controlled by signals released by surrounding cells (Godin *et al.*, 1990). Subsequent experiments with purified signaling ligands showed that germ cell migration, proliferation, and survival can all be controlled in culture by known growth factors (summarized in Fig. 5). However, firm genetic evidence for a function *in vivo* exists only for a small number of these, including Steel factor and its receptor c-kit, whose interaction is required for germ cell survival (Dolci *et al.*, 1991; Godin *et al.*, 1991; Matsui *et al.*,

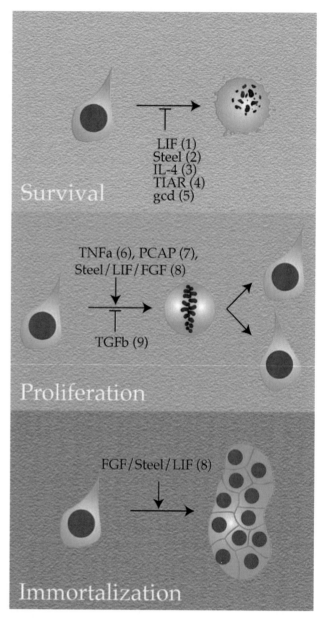

Figure 5 A scheme showing known effects of cytokines on the survival, proliferation, and immortalization of mouse PGCs during migration. Data from (1) Dolci *et al.*, 1993; (2) Godin *et al.*, 1991; Dolci *et al.*, 1991; Matsui *et al.*, 1991; (3) Cooke *et al.*, 1996; (4) Beck *et al.*, 1998; (5) Pellas *et al.*, 1991; (6) Kawase *et al.*, 1994; (7) Pesce *et al.*, 1996; (8) Matsui *et al.*, 1992; Resnick *et al.*, 1992; (9) Godin and Wylie, 1991.

1991), the RNA binding protein TIAR (Beck *et al.*, 1998), and *germ cell deficient* (*gcd*), a gene whose identity is unknown, but in which insertional mutation leads to loss of germ cells during migration (Pellas *et al.*, 1991). Some factors may play more than one role. Members of the TGF-β family negatively regulate proliferation (Godin and Wylie, 1991; Richards *et al.*, 1999), as well as acting chemotropically in culture (Godin and Wylie, 1991), while Steel factor has been proposed to be involved in germ cell survival (for references, see above), adhesion (Pesce *et al.*, 1997), guidance (Kaneko *et al.*, 1991; Marziali *et al.*, 1993; Meininger *et al.*, 1992), and possibly motility, since in severe c-kit alleles, the reduced population of germ cells remains in the hindgut, and does not migrate to the genital ridges (Buehr *et al.*, 1993).

Factors can also act combinatorially; the combination of bFGF, Steel factor, and any member of the IL6/LIF cytokine family, will cause cultured germ cells to form ES cell-like pluripotential, immortal cell lines (Matsui *et al.*, 1992; Resnick *et al.*, 1992). In many cases, it is not known whether exogenously added signaling ligands are acting directly on germ cells in culture, since somatic cells are present also, nor whether germ cells express their receptors. Good genetic evidence for the roles of these factors and their receptors is sorely needed. It is clear, however, from results to date, that most aspects of behavior of germ cells are controlled by locally released signaling molecules.

Acknowledgments

The authors would like to thank the NIH (RO1-HD33440), National Life and Health Insurance Medical Research Fund, and the Harrison Fund for financial support for their work reviewed in this paper.

References

Anderson, R., and Wylie, C. C. (2000). *Int. Rev. Cytol.* **203**, 215–230.

Anderson, R., Faessler, R., Georges-Labouesse, E., Hynes, R., Bader, R., Kriedberg, J., Schaible, K., Heasman, J., and Wylie, C. C. (1999a). Mouse primordial germ cells lacking b1 integrin enter the germ line but fail to migrate normally to the gonads. *Development (Cambridge, UK)* **126**, 1655–1644.

Anderson, R., Schaible, K., Heasman, J., and Wylie, C. C. (1999b). Expression of the homophilic adhesion molecule Ep-CAM in the mammalian germ line. *J. Reprod. Fertil.* **116**, 379–384.

Anderson, R., Copeland, T., Schoeller, H., Heasman, J., and Wylie, C. C. (2000). The onset of germ cell migration in the mouse embryo. *Mech. Dev.* **91**, 61–68.

Beck, A., Miller, I., Anderson, P., and Streuli, M. (1998). RNA-binding protein TIAR is essential for primordial germ cell development. *Proc. Natl. Acad. Sci. U.S.A.* **95**, 2331–2336.

Bendel-Stenzel, M., Gomperts, M., Anderson, R., Heasman, J., and Wylie, C. C. (2000). The role of cadherins during primordial germ cell migration and early gonad formation in the mouse. *Mech. Dev.* 143–152.

Boyle, M., Bonini, N., and DiNardo, S. (1997). Expression and function of *clift* in the development of somatic gonadal precursors within the *Drosophila* mesoderm. *Development (Cambridge, UK)* **124**, 971–982.

Buehr, M., McLaren, A., Bartley, A., and Darling, S. (1993). Proliferation and migration of primordial germ cells in We/We mouse embryos. *Dev. Dyn.* **198**, 182–189.

Chiquoine, A. D. (1954). The identification, origin and migration of the primordial germ cells of the mouse embryo. *Anat. Rec.* **118**, 135–146.

Clark, I., Giniger, E., Ruohola-Baker, H., Jan, L., and Jan, Y. (1994). Transient posterior localization of a kinesin fusion protein reflects anteroposterior polarity of the *Drosophila* oocyte. *Curr. Biol.* **4**, 289–300.

Clark, I., Jan, L., and Jan, Y. (1997). Reciprocal localization of Nod and kinesin fusion proteins indicates microtubule polarity in the *Drosophila*

oocyte, epithelium, neuron and muscle. *Development (Cambridge, UK)* **124**, 461–470.

Cooke, J., Heasman, J., and Wylie, C. C. (1996). The role of interleukin 4 in the regulation of mouse primordial germ cell numbers. *Dev. Biol.* **174**, 14–22.

Copp, A. J., Roberts, H. M., and Polani, P. (1986). Chimaerism of primordial germ cells in the early postimplantation mouse embryo following microsurgical grafting of posterior primitive streak cells in vitro. *J. Embryol. Exp. Morphol.* **95**, 95–115.

Dolci, S., Williams, D. E., Ernst, M. K., Resnick, J. L., Brannan, C. I., Lock, L. F., Lyman, S. D., Boswell, H. S., and Donovan, P. J. (1991). Requirement for mast cell growth factor for primordial germ cell survival in culture. *Nature (London)* **352**, 809–811.

Dolci, S., Pesce, M., and de Felici, M. (1993). Combined action of stem cell factor, leukemia inhibitory factor and cAMP on in vitro proliferation of mouse primordial germ cells. *Mol. Reprod. Dev.* **35**, 134–139.

Donovan, P., Stott, D., Cairns, L., Heasman, J., and Wylie, C. C. (1986). Migratory and nonmigratory mouse primordial germ cells behave differently in culture. *Cell (Cambridge, Mass.)* **44**, 831–838.

Emmons, S., Phan, H., Calley, J., Chen, W., James, B., and Manseau, L. (1995). *Cappucino*, a *Drosophila* maternal effect gene required for polarity of the egg and embryo, is related to the vertebrate *limb deformity* locus. *Genes Dev.* **9**, 2482–2494.

Ephrussi, A., and Lehmann, R. (1992). Induction of germcell formation by oskar. *Nature (London)* **358**, 387–389.

Erdelyi, M., Michon, A., Guichet, A., Glotzer, J., and Ephrussi, A. (1995). Requirement for *Drosophila* cytoplasmic tropomyosin in *oskar* mRNA localization. *Nature (London)* **377**, 524–527.

Forbes, A., and Lehmann, R. (1998). Nanos and pumilio have critical roles in the development and function of *Drosophila* germline stem cells. *Development (Cambridge, UK)* **125**, 679–690.

Garcia-Castro, M., Anderson, R., Heasman, J., and Wylie, C. C. (1997). Interactions between germ cells and extracellular matrix glycoproteins during migration and gonad assembly in the mouse embryo. *J. Cell Biol.* **138**, 471–480.

Gardner, R. L., and Rossant, J. (1979). Investigation of the fate of 4.5 day *post-coitun* mouse inner cell mass cells by blastocyst injection. *J. Embryol. Exp. Morphol.* **52**, 141–152.

Ginsburg, M., Snow, M. H. L., and McLaren, A. (1990). Primordial germ cells in the mouse embryo during gastrulation. *Development (Cambridge, UK)* **110**, 521–528.

Godin, I., and Wylie, C. C. (1991). TGFb inhibits proliferation and has a chemotropic effect on mouse primordial germ cells in culture. *Development (Cambridge, UK)* **113**, 1451–1457.

Godin, I., Wylie, C. C., and Heasman, J. (1990). Genital ridges exert long-range effects on mouse primordial germ cell numbers and direction of migration in culture. *Development (Cambridge, UK)* **108**, 357–363.

Godin, I., Deed, R., Cooke, J., Zsebo, K., Dexter, M., and Wylie, C. C. (1991). Effects of the steel gene product on mouse primordial germ cells in culture. *Nature (London)* **352**, 807–809.

Gomperts, M., Wylie, C. C., and Heasman, J. (1994). Interactions between primordial germ cells play a role in their migration in mouse embryos. *Development (Cambridge, UK)* **120**, 135–141.

Heasman, J., Quarmby, J., and Wylie, C. C. (1984). The mitochondrial cloud of *Xenopus* oocytes: The source of germinal granule material. *Dev. Biol.* **105**, 458–469.

Jaglarz, M., and Howard, K. (1995). The active migration of *Drosophila* primordial germ cells. *Development (Cambridge, UK)* **121**, 3495–3503.

Kaneko, Y., Takenawa, J., Yoshida, O., Fujita, K., Sugimoto, K., Nakayama, H., and Fujita, J. (1991). Adhesion of mouse mast cells to fibroblasts: Adverse effects of Steel (Sl) mutation. *J. Cell. Physiol.* **147**, 224–230.

Kawase, E., Yamamoto, H., Hashimoto, K., and Nakatsuji, N. (1994). Tumor necrosis factor-alpha (TNF-alpha) stimulates proliferation of mouse primordial germ cells in culture. *Dev. Biol.* **161**, 91–95.

Kobayashi, S., Yamada, M., Asaoka, M., and Kitamura, T. (1996). Essential role of the posterior morphogen nanos for germline development in *Drosophila*. *Nature (London)* **380**, 708–711.

Lawson, K., and Hage, M. (1994). Clonal analysis of the origin of primordial germ cells in the mouse. *Ciba Found. Symp.* **182**, 68–84.

Lawson, K., Dunn, N., Roelen, B., Zeinstra, L., Davies, A., Wright, C., Korving, J., and Hogan, B. (1999). BMP4 is required for the generation of primordial germ cells in the mouse embryo. *Genes Dev.* **13**, 424–436.

MacGregor, G., Zambrowicz, B., and Soriano, P. (1995). Tissue nonspecific alkaline phosphatase is expressed in both embryonic and extraembryonic lineagesduring mouse embryogenesis but is not required for migration of primordial germ cells. *Development (Cambridge, UK)* **121**, 1487–1496.

Manseau, L., Calley, J., and Phan, H. (1996). Profilin is required for posterior patterning of the *Drosophila* oocyte. *Development (Cambridge, UK)* **122**, 2109–2116.

Marziali, G., Lazzaro, D., and Sorrentino, V. (1993). Binding of germ cells to mutant sld sertoli cells is defective and is rescued by expression of the transmembrane form of c-kit ligand. *Dev. Biol.* **157**, 182–190.

Matsui, Y., Toksoz, D., Nishikawa, S., Nishikawa, S.-I., Williams, D., Zsebo, K., and Hogan, B. L. M. (1991). Effect of Steel factor and leukaemia inhibitory factor on murine primordial germ cells in culture. *Nature (London)* **353**, 750–752.

Matsui, Y., Zsebo, K., and Hogan, B. L. M. (1992). Derivation of pluripotential stem cells from murine primordial germ cells in culture. *Cell (Cambridge, Mass.)* **70**, 841–847.

Meininger, C. J., Yano, H., Rottapel, R., Bernstein, A., Zsebo, K., and Zetter, B. R. (1992). The c-kit receptor ligand functions as a mast cell chemoattractant. *Blood* **79**, 958–963.

Moore, L., Broihier, H., Van Doren, M., Lunsford, L., and Lehmann, R. (1998a). Identification of genes controlling germ cell migration and embryonic gonad formation in *Drosophila*. *Development (Cambridge, UK)* **125**, 667–678.

Moore, L., Broihier, H., Van Doren, M., and Lehmann, R. (1998b). Gonadal mesoderm and fat body initially follow a common developmental path in *Drosophila*. *Development (Cambridge, UK)* **125**, 837–844.

Nakamura, A., Amikura, R., Mukai, M., Kobayashi, S., and Lasko, P. (1996). Requirement for a noncoding RNA in *Drosophila* polar granules for germ cell establishment. *Science* **274**, 2075.

Nieuwkoop, P., and Sutasurya, L. (1981). "Primordial Germ Cells in the Invertebrates." Cambridge University Press, Cambridge, UK.

Okamoto, K., Okazawa, H., Okuda, A., Sakai, M., Muramatsu, M., and Hamada, H. (1990). A novel octamer binding transcription factor is differentially expressed in mouse embryonic cells. *Cell (Cambridge, Mass.)* **60**, 461–472.

Pellas, T. C., Ramachandran, B., Duncan, M., Pan, S. S., Marone, M., and Chada, K. (1991). Germ-cell deficient (gcd), and insertional mutation manifested as infertility in transgenic mice. *Proc. Natl. Acad. Sci. U.S.A.* **88**, 8787–8791.

Pesce, M., Canipari, R., Ferri, G. L., Siracusa, G., and de Felici, M. (1996). Pituitary adenylate cyclase-activating polypeptide (PACAP) stimulates adenylate cyclase and promotes proliferation of mouse primordial germ cells. *Development (Cambridge, UK)* **122**, 215–221.

Pesce, M., Carlo, A. D., and de Felici, M. (1997). The c-kit receptor is involved in the adhesion of mouse primordial germ cells to somatic cells in culture. *Mech. Dev.* **68**, 37–44.

Pokrywka, N., and Stephenson, E. (1995). Microtubules are a general component of mRNA localization systems in *Drosophila* oocytes. *Dev. Biol.* **167**, 363–370.

Resnick, J. L., Bixler, L. S., Cheng, L., and Donovan, P. J. (1992). Longterm proliferation of mouse primordial germ cells in culture. *Nature (London)* **359**, 550–551.

Reuter, R. (1994). The gene serpent has homeotic properties and specifies endoderm versus ectoderm within the *Drosophila* gut. *Development (Cambridge, UK)* **120**, 1123.

Richards, A., Enders, G., and Resnick, J. (1999). Activin and TGFbeta limit murine primordial germ cell proliferation. *Dev. Biol.* **207**, 470–475.

Robb, D., Heasman, J., Raats, J., and Wylie, C. C. (1996). A kinesin-like protein is required for germ plasm aggregation in *Xenopus*. *Cell (Cambridge, Mass.)* **87**, 823–831.

Rongo, C., Broihier, H., Moore, L., Van Doren, M., Forbes, A., and Lehmann, R. (1997). Germ plasm assembly and germ cell migration in *Drosophila*. *Cold Spring Harbor Symp. Quant. Biol.* **62**, 1–11.

Rosner, M. H., Vigano, M. A., Ozato, K., Timmons, P. M., Poirier, F., Rigby, P. W. J., and Staudt, L. M. (1990). A POU-domain transcriptional factor in early stem cells and germ cells of the mammalian embryo. *Nature (London)* **345**, 686–692.

Saffman, E., and Lasko, P. (1999). Germline development in vertebrates and invertebrates. *Cell Mol. Life Sci.* **55**, 1141–1163.

Scholer, H., Ruppert, S., Suzuki, N., Chowdury, K., and Gruss, P. (1990). New type of POU domain in germ-line specific protein. *Nature (London)* **344**, 435–439.

Scholer, H., Ciesiolka, T., and Gruss, P. (1991). A nexus between Oct-4 and E1A: Implications for gene regulation in embryonic stem cells. *Cell (Cambridge, Mass.)* **66**, 291–304.

Seydoux, G., and Dunn, M. (1997). Transcriptionally repressed germ cells lack a subpopulation of phosphorylated RNA polymerase II in early embryos of *C. elegans* and D. melanogaster. *Development (Cambridge, UK)* **124**, 2191–2201.

Seydoux, G., and Strome, S. (1999). Launching the germline in *Caenorhabditis elegans*: Regulation of gene expression in early germ cells. *Development (Cambridge, UK)* **126**, 3275–3283.

Seydoux, G., Mello, C., Pettit, J., Wood, W., Priess, J., and Fire, A. (1996). Repression of gene expression in the embryonic germ lineage of *C. elegans*. *Nature (London)* **382**, 713–716.

Snow, M. H. L. (1981). Autonomous development of parts isolated from primitive-streak-stage mouse embryos. Is development clonal? *J. Embryol. Exp. Morphol.* **65**, 269–287.

Tam, P. P. L., and Snow, M. H. L. (1981). Proliferation and migration of primordial germ cells during compensatory growth in mouse embryos. *J. Embryol. Exp. Morphol.* **64**, 133–147.

Tam, P. P. L., and Zhou, S. X. (1996). The allocation of epiblast cells to ectodermal and germ line lineages is influenced by the position of the cells in the gastrulating mouse embryo. *Dev. Biol.* **178**, 124–132.

Therkauf, W. (1994). Premature microtubule-dependent cytoplasmic streaming in *cappucino* and *spire* mutant embryos. *Science* **265**, 2093–2096.

Van Doren, M., Broiher, H., Moore, L., and Lehmann, R. (1998). HMG-CoA reductase guides migrating primordial germ cells. *Nature (London)* **396**, 466–469.

Warrior, R. (1994). Primordial germ cell migration and the assembly of the *Drosophila* embryonic gonad. *Dev. Biol.* **166**, 180.

Whitington, P. M., and Dixon, K. E. (1975). Quantitative studies of germ plasm and germ cells during early embryogenesis of *Xenopus laevis*. *J. Embryol. Exp. Morphol.* **33**, 57–74.

Wylie, C. C. (1999). Germ cells. *Cell (Cambridge, Mass.)* **96**, 165–174.

Yeom, Y., Fuhrmann, G., Ovitt, C., Brehm, A., Kazuyuki, O., Gross, M., Hubner, K., and Scholer, H. (1996). Germ line regulatory element of Oct4 specific for the totipotent cycle of embryonic cells. *Development (Cambridge, UK)* **122**, 881–894.

Zalokar, M. (1976). Autoradiographic study of protein and RNA formation during early development of *Drosophila* eggs. *Dev. Biol.* **49**, 97–106.

Zhang, N., Zhang, J., Purcell, K., Cheng, Y., and Howard, K. (1997). The Drosophila protein Wunen repels migrating germ cells. *Nature (London)* **385**, 64–66.

10

Development of the Vertebrate Hematopoietic System

Nancy Speck, * Marian Peeters,† and Elaine Dzierzak†

*Department of Biochemistry, Dartmouth Medical School, Hanover, New Hampshire 03755
†Department of Cell Biology and Genetics, Erasmus University, Rotterdam 3000, The Netherlands

I. Introduction
II. Cellular Aspects of Blood Development in the Mouse Embryo
III. Molecular Genetic Aspects of Blood Development in the Mouse Embryo
IV. Current Cellular and Molecular Conceptual Frameworks for Hematopoietic Ontogeny
V. Future Directions
References

I. Introduction

The adult hematopoietic system is a heterogeneous group of fully differentiated blood cells and their precursors. At its foundation are hematopoietic stem cells (HSCs) from which terminally differentiated blood cells are continuously renewed. The HSCs divide to yield immature progenitors, which in turn progress through many intermediate stages of commitment in a cascade of differentiation events (Fig. 1). The large number of intermediate cell types and the fact that blood cells circulate throughout the embryo present many challenges to studying the embryonic origins and generation of the adult hematopoietic system.

Our current knowledge of embryonic hematopoiesis is drawn from studies in both mammalian and nonmammalian vertebrates. The ability to perform fate mapping experiments, particularly in avian and amphibian embryos, has contributed substantially to our understanding of the mesodermal origins of hematopoietic cells beginning at gastrulation, and has firmly established that multiple independent origins of hematopoiesis occur in the developing embryo. Studies in nonmammalian vertebrates also support the colonization theory of hematopoiesis, which posits that hematopoietic cells emerge from specific locations in the embryo and then seed secondary hematopoietic organs such as the fetal liver and thymus.

Although fate mapping studies are difficult to perform in mammalian embryos because of their in utero development, a significant advantage that mouse affords as a developmental model for hematopoiesis is its rich tradition of in vivo transplantation and in vitro clonal assays for hematopoietic progenitors and stem cells. These assays were particularly useful for investigating the adult hematopoietic system and were logically extended to studies of mammalian embryos. It was established that hematopoiesis in mammalian embryos, like their nonmammalian vertebrate counterparts, originates in multiple anatomically distinct sites in the embryo (Fig. 2), including the yolk sac, the vitelline and umbilical arteries, and in an intraembryonic region containing

191

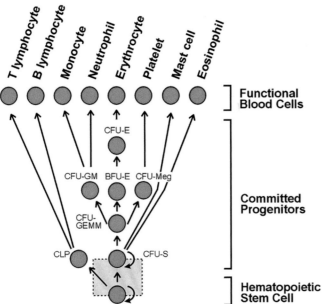

Figure 1 Adult hematopoietic hierarchy. Schematic classification of the differentiation hierarchy of adult hematopoietic cells. The hematopoietic stem cell (HSC) is at the foundation of the hierarchy. This cell can self-renew and give rise to all differentiated lineages of blood cells and cells in the hematopoietic tissues of the adult. Multilineage progenitors and committed progenitors are intermediate cell types in the hierarchy and include CFU-S (colony forming unit–spleen), CLP (common lymphoid progenitor), CFU-GEMM (colony forming unit–granulocyte, erythroid, macrophage, megakaryocyte), CFU-GM (colony forming unit–granulocyte, macrophage), BFU-E (burst forming unit–erythroid), CFU-E (colony forming unit–erythroid), and CFU-MEG (colony forming unit–megakaryocyte). The gray box represents the immature progenitor compartment that has yet to be fully characterized.

the dorsal aorta, gonads, and mesonephroi (AGM). *In vivo* transplantation, the experimental forte of mouse, revealed that the AGM is the first site of HSC initiation and is a potent source of adult hematopoietic cells. New and evolving concepts regarding the hematopoietic potentials of stem cells from other organ systems are now leading to notions of reprogramming, dedifferentiation, and/or maintenance of ontogenic potential (plasticity). This chapter presents the concepts and data past and present that lead to our current understanding of the developmental aspects of hematopoiesis.

II. Cellular Aspects of Blood Development in the Mouse Embryo

A. Adult Hematopoietic Hierarchy

The adult hematopoietic system is composed of many cells positioned in a complex differentiation hierarchy. At least eight different lineages of functional hematopoietic

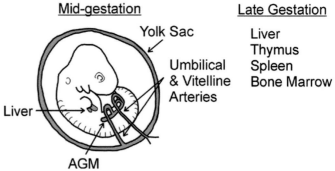

Figure 2 Sites of hematopoiesis in the mouse embryo. Depicted is an E10.5 embryo. At midgestation, (E7–12), several sites of hematopoietic activity are found (shown in orange) and include the yolk sac, AGM, vitelline and umbilical arteries, and the liver. The yolk sac, AGM, and vitelline and umbilical arteries are thought to be the sites that generate the first hematopoietic activity. Subsequently hematopoiesis shifts to secondary hematopoietic tissues including the liver, thymus, spleen, and bone marrow, which are thought to be colonized by cells produced in the primary sites.

cells are formed from a small population of HSCs that reside mainly in the bone marrow of the adult and serve as the renewable source of the adult blood system. The terminally differentiated hematopoietic cells consist primarily of erythrocytes, monocytes, platelets, neutrophils, mast cells, eosinophils, B lymphocytes, and T lymphocytes (Fig. 1). These functional blood cells differentiate from the HSC through a series of committed progenitor cells. Progenitors are classified retrospectively by the functional blood cells they generate, using *in vivo* and *in vitro* clonal assays (Metcalf, 1984). Some of the more frequently used assays that detect the presence of progenitors and/or HSCs are indicated in Table I. These assays yield readouts for progenitors at different stages of commitment, including progenitors that give rise to only one lineage (e.g., CFU-E, CFU-Meg), multilineage progenitors (cells with potential for two or more lineages such as CFU-GEMM and CLP (Kondo *et al.*, 1997) and pluripotent HSCs.

B. Identification of Hematopoietic Progenitors and Stem Cells

Unfortunately, no single descriptive characteristic can identify progenitors or stem cells in the absence of functional assays. Cells that are functionally distinct have overlapping cell surface phenotypes, responses to growth factors, homing abilities, and potential for self-renewal. For example, adult HSCs express Sca-1, c-kit, CD34, CD31 and AA4.1 (Ikuta *et al.*, 1990; Ikuta and Weissman, 1992; Jordan *et al.*, 1990; Spangrude *et al.*, 1988; van der Loo *et al.*, 1995) (see Table II), but these markers are not exclusively expressed on HSCs. Activated T cells and some epithelial cells express Sca-1 (Sinclair and Dzierzak, 1993); c-kit is expressed by hematopoietic progenitors and primordial

Table I Hematopoietic Activities

Cell type	Potency	Assay
Hematopoietic stem cell	Cell at the foundation of the adult differentiation hierarchy	*In vivo* transplantation into hematopoietic-ablated adult recipients leading to long-term, high-level, multilineage repopulation. Such stem cells are capable of self-renewal.
Neonatal repopulating cell	Pre-HSC or cell at the foundation of a neonatal/fetal hierarchy	*In vivo* transplantation into hematopoietic-ablated newborn (within 1 day of birth) recipients leads to long-term, moderate level, multilineage repopulation. These cells are unable to directly repopulate adult recipients.
Multipotential cell	Cell with erythroid-myeloid-lymphoid lineage multipotency	Two-step *in vitro* assay in which cells are clonally expanded in culture. After clonal expansion cells are tested under three sets of conditions leading to B-lymphoid, T-lymphoid, or erythroid-myeloid differentiation. These cells cannot *in vivo* repopulate adult recipients.
CFU-S (colony forming unit–spleen)	Immature progenitor with erythroid and myeloid activity	*In vivo* transplantation into lethally irradiated adult recipients leads to the short-term (within 9–14 days after transplantation) formation of macroscopic colonies.
CFU-C (colony forming unit culture)	More mature progenitor with erythroid and myeloid activity	*In vitro* 5 to 14-day culture in semisolid medium with hematopoietic growth factors leading to the production of colonies containing one or mixed lineages of differentiated erythroid and myeloid cells.
B-lymphoid progenitor	B lymphocytes	Coculture with S17 stromal line and IL-7 leads to differentiation of B cells.
T-lymphoid progenitor	T lymphocytes	Fetal thymic organ culture (FTOC). Fetal thymic rudiments ablated of all progenitors are seeded with test cells in a hanging drop culture. All CD4 and CD8 subsets of T cells are differentiated in such *in vitro* cultures.

germ cells (PGCs) (Keshet *et al.*, 1991; Motro *et al.*, 1991); CD34 is expressed by hematopoietic progenitors, endothelial cells, and muscle cells (Wood *et al.*, 1997; Young *et al.*, 1995); and AA4.1 is expressed by pre-B cells (Cumano *et al.*, 1992). Generally, HSC isolation is performed using a combination of markers to first deplete cell populations of more mature, committed hematopoietic cells (negative selection), then to enrich for the HSCs (positive selection). Adult HSCs do not express a number of cell surface markers characteristic of mature hematopoietic cells, including B220 (B cells), CD3, CD4, and CD8 (T cells), Gr-1 (granulocytes), and Mac1 (monocytes and granulocytes). These markers are often used for negative selection of differentiated blood cells

from the HSC population followed by positive marker selection (Sca-1, c-kit, CD34, and/or AA4.1). Recently, it was shown that the chemical Hoechst 33342 can be used in flow cytometric sorting to enrich HSCs (Goodell *et al.*, 1996) as well as stem cells for other tissues (Gussoni *et al.*, 1999; Jackson *et al.*, 1999). Hoechst 33342 fluorescent marking identifies cells that rapidly efflux small molecules and promises to be an extensively used research tool in stem cell analyses.

Responsiveness to hematopoietic growth factors has also been used to categorise hematopoietic cells (Metcalf, 1984, 1998). Specific growth factors affect distinct classes of progenitor cells *in vitro*. For example, interleukin 7 (IL-7), erythropoietin, granulocyte macrophage colony-stimulating factor (GM-CSF) and macrophage colony-stimulating factor (M-CSF) promote the growth of pre-B lymphocytes, CFU-E, CFU-GM, and CFU-M, respectively. Hematopoietic growth factors thought to affect HSCs and progenitors include IL-3, SCF, IL-6, and thrombopoietin. Cocktails of such factors are often used in *in vitro* cultures to maintain the growth of hematopoietic progenitors and stem cells (Breems *et al.*, 1998). Although certain combinations of growth factors promote the expansion and differentiation of progenitor cells, no special combination has been found that expands the starting stem cell pool, suggesting that there may be separate, yet to be described factors involved in stem cell growth. Targeted disruption of many individual (and in combination) growth factor and growth factor receptor genes has very little effect on the *in vivo* growth and differentiation of hematopoietic

Table II Cell Specificity Markers

Hematopoietic stem cells		Endothelial cells	Hematopoietic/ endothelial cells in arterial clusters
Positive	Negative	Positive	Positive
AA4.1	B220	CD34	AML1/CBFβ
CD34	CD3	Tie-1	CD31
c-kit	CD4	Tie-2	CD34
Sca-1	CD8	VE-cadherin	c-kit
CD31	Gr-1	flk-1	GATA-2
	Mac1	CD31	flk-1
		SCL	SCL
		AA4.1	AA4.1

cells, suggesting functional redundancy (Lieschke *et al.,* 1994; Nilsson *et al.,* 1995). However, some growth factors do provide function in distinct hematopoietic cell subsets *in vivo.* For example, thrombopoietin is used at both ends of the hematopoietic hierarchy—for the maintenance of HSCs as well as for the terminal differentiation of megakaryocytes (Kimura *et al.,* 1998). Thus, growth factor regulation of hematopoiesis is complex and the role of specific growth factors or combinations of growth factors in HSC and hematopoietic progenitor function during adult and embryonic stages, particularly *in vivo,* is unclear.

It has been proposed that the pattern of gene expression could be used to determine a cell's position within the hematopoietic hierarchy. For example, expression of hematopoietic growth factor receptors such as the CSF-1 receptor might characterize a macrophage progenitor while expression of the IL-7 receptor might be used to identify a lymphoid progenitor. A single-cell nested polymerase chain reaction (PCR) approach has recently shown that all progenitor cells may express a whole spectrum of receptors and are thus "primed" for a stochastic event leading to commitment (Hu *et al.,* 1997). As the cell progresses down a particular differentiation pathway, expression of the unnecessary receptors is downregulated. The ability to study global gene expression patterns through the use of gene chip technology could eventually establish a gene expression profile for each class of hematopoietic progenitors and fully differentiated blood cells (Phillips *et al.,* 2000).

In summary, the adult hematopoietic hierarchy is a functional continuum of differentiating cells as defined by *in vitro* and *in vivo* assays with many overlapping and redundant features.

C. Hematopoietic Cells in the Conceptus

1. Differentiated Hematopoietic Cells

How similar is the mammalian hematopoietic system of the embryo to that of the adult? The appearance of functional hematopoietic cells, hematopoietic progenitors and stem cells has been examined both temporally and spatially in mouse (reviewed in Dzierzak *et al.,* 1998) and human embryos (Huyhn *et al.,* 1995; Tavian *et al.,* 1999). Table III summarizes the major landmarks in mouse and human development. Most of the differentiated hematopoietic lineages found in the adult can also be found in the embryo. However, some clear differences exist between embryonic and adult cell types especially in the erythroid lineage. Historically, embryonic erythrocytes that retain their nuclei and express embryonic globin genes were called "primitive" to distinguish them from the enucleated "definitive" erythrocytes that appear somewhat later in development and express fetal and/or adult globin genes. Acknowledgment of these differences has led to a general but confusing terminology

with the term "primitive" referring to the early embryonic hematopoietic system and "definitive" referring to the adult hematopoietic system. Many difficulties in the use of this terminology arise during the intermediate fetal and neonatal stages, and in particular when it is used for other lineages of hematopoietic cells. These difficulties reflect our lack of knowledge concerning the lineage relationships and embryonic origins of many of the hematopoietic cells.

The first and most numerous differentiated hematopoietic cells in the mammalian embryo, primitive erythrocytes, are found in the yolk sac (Fig. 2 and Table III). As the developing intraembryonic vitelline and umbilical vessels link to the yolk sac vessels and the placenta, respectively, primitive erythrocytes begin to circulate throughout the embryo. These cells predominate in the yolk sac until a switch to definitive erythropoiesis occurs. The first cells of the macrophage lineage appear in the yolk sac of the mouse conceptus soon after the primitive erythrocytes. These primitive macrophages are able to colonize other embryonic tissues through the circulation. However, definitive macrophages found during fetal and adult stages do not circulate, and arise from monocytic progenitors that begin to appear slightly later in the fetal liver and yolk sac. Thus, there appear to be two separate waves of erythroid and macrophage production. The production of lymphoid lineage cells occurs only during the later stages of gestation when the fetal liver is the predominant hematopoietic site (Table III). B-lymphoid precursors in the mouse embryo seed the fetal liver after E11, T-lymphoid precursors are found in the thymus beginning at E12–13, and the circulation of mature lymphoid cells begins at E17 (Delassus and Cumano, 1996). The order in which various hematopoietic cells appear in the human embryo and fetus is similar to that in the mouse. The timing of human hematopoietic development is different due to the longer gestation period, and is summarized in Table III.

It is unclear whether all lineages of differentiated hematopoietic cells in the embryo and the adult are functionally equivalent (Bonifer *et al.,* 1998). In fact, at least three functional cell types are unique to the embryo or fetus. The first example is the primitive erythrocyte. In mouse and human embryos, primitive erythrocytes are larger than adult erythrocytes, retain their nuclei, and express embryonic globin genes. In contrast, adult erythrocytes are small, enucleated, and express adult globin genes (Russell, 1979). The functional outcome of these differences is that embryonic hemoglobin has an increased affinity for oxygen as compared to adult hemoglobin, thereby facilitating the transport of oxygen through the placenta to the conceptus. Another example is the TcRγ3 subset of lymphoid cells, that is thought to be important in mucosal immunity, and is produced during fetal but not the adult stages of development (Ikuta *et al.,* 1990). Likewise, the CD5 subset of B lymphocytes is produced predominantly during fetal stages and is also thought to serve a special immune function as compared to adult B cells (Her-

Table III   Landmarks in Mouse (and Human) Hematopoietic Development

Gestation day	Developmental stage	Landmarks (human)	References
E7.0	Early primitive streak	Gastrulation, mesoderm formation	Kaufman (1992)
E7.5 to E8.0	Advanced primitive streak	Yolk sac blood islands with primitive erythrocytes appear (primitive erythrocytes appear in human at E16–20)	Russel and Bernstein (1966); Keleman et al. (1979); Palis et al. (1999); Naito (1993); Naito et al. (1996); Kaufman (1992)
		Cells of primitive circulating macrophage lineage appear in yolk sac (first cells of the human monocytic macrophage lineage appear at 4–5 weeks)	
	1–2 Somite pairs	Yolk sac vasculature and paired dorsal aortae form	
E8.5	8–12 Somite pairs	Onset of circulation, intraembryonic vitelline and umbilical vessels link to yolk sac vessels and placenta, respectively	Garcia-Porrero et al. (1995); Kaufman (1992)
E9.0	15–20 Somite pairs	Pronephric duct and nephric vesicle formation	Kaufman (1992)
E9.5	20–29 Somite pairs	Urogenital ridges and liver form	Kaufman (1992); Johnson and Jones (1973)
		Liver rudiment contains first erythroblast beginning at 28 somite pair stage	
E10 to E12		Switch from primitive to definitive erythropoiesis (in humans this switch is visible at 7–10 weeks in the blood and slightly earlier in the fetal liver)	Keleman et al. (1979); Naito (1993); Naito et al. (1996)
		Monocytic progenitors appear in yolk sac and fetal liver (human primitive macrophages in circulation until 14 weeks and definitive monocytes appear after 11 weeks)	
E10.5	35–39 Somite pairs	HSC generation commences in AGM	Muller et al. (1994); Medvinsky and Dzierzak (1996)
E11	> 40 Somite pairs	HSCs appear in yolk sac and fetal liver	Muller et al. (1994); Medvinsky and Dzierzak (1996)
E12 to E13		B-lymphoid precursors found in fetal liver, T-lymphoid precursors found in thymus (in the human embryo, production of lymphoid cells begins at 7–10 weeks, with small lymphocytes in the circulation beginning at 9–10 weeks)	Delassus and Cumano (1996); Keleman et al. (1979)
E14 to E16		Spleen, omentum, and bone marrow begin to serve as hematopoietic sites	Godin et al. (1999)

zenberg et al., 1986). In addition, during mid- and late gestation some lymphocytes are thought to contribute to the maturation of secondary lymphopoietic organs and microenvironments through inductive signals and "crosstalk" (van Ewijk et al., 2000).

In summary, and as shown in Fig. 2, hematopoietic cells are harbored in several sites in the mammalian embryo and include the yolk sac, fetal liver, and circulation (Delassus and Cumano, 1996). Slightly later in the mouse embryo, beginning at E14–16 and thereafter, the thymus, spleen, omentum, and bone marrow also serve as hematopoietic sites (Godin et al., 1999).

2. Hematopoietic Progenitors and Stem Cells

The locations of precursors for differentiated hematopoietic cells, that is, the hematopoietic progenitors and stem cells, have been mapped temporally and spatially in the mouse conceptus. Recent studies show that the earliest onset of hematopoiesis occurs in the proximal regions of the egg cylinder at midprimitive streak stage (E7.5) (Palis et al., 1999) with the production of primitive erythroid and macrophage precursors. The erythroid and macrophage precursors increase in number in the yolk sac until E8.25 and then sharply decline by E9.0. At no time are these progenitors found in the embryo body. Hence, the yolk sac appears to be the exclusive source of the first wave of primitive erythroid and macrophage progenitors.

Multipotent CFU-C progenitors (granulocyte and macrophage colony-forming cells), CFU-S, and HSCs were first demonstrated to appear within the yolk sac of the mouse conceptus beginning at E7, E8, and E11, respectively, by Moore and Metcalf (1970). The appearance of these hematopoietic activities in the yolk sac was always followed by their appearance 1–2 days later in the circulation as well as in the fetal liver and led to the general notion of the yolk sac origins of adult hematopoiesis (reviewed in Medvinsky *et al.*, 1993). The use of fetal thymic organ cultures (FTOCs) and B-cell stromal cultures in the 1980s enabled other researchers to find T- and B-lymphoid progenitors in the yolk sac beginning at E8.5 (Eren *et al.*, 1987) and E10 (Ogawa *et al.*, 1988), respectively. Although B-lymphoid progenitors were also detected in the embryo body at E9.5 (Ogawa *et al.*, 1988), the preliver embryo body was generally ignored as a hematopoietic site for many years.

In 1993 a potent preliver site of hematopoietic progenitor and stem cell activity was uncovered in the mouse embryo body (Godin *et al.*, 1993; Medvinsky *et al.*, 1993). Dissected caudal regions of embryos, which included the dorsal aorta, gonads, and pro/mesonephroi (AGM; see Fig. 2), contained the more complex and potent hematopoietic progenitor and stem cells characteristic of those found in an adult-type hematopoietic system. A number of different *in vivo* and *in vitro* assays (Fig. 3), together with the careful dissection of the AGM or its earlier ontogenic form, the para-aortic splanchnopleura (PAS) (both are derived from the intraembryonic splanchnopleura; see Kaufman, 1992) at E8.5–9 revealed the presence of CFU-S erythroid-myeloid progenitors (Medvinsky *et al.*, 1993), B-lymphoid progenitors (Godin *et al.*, 1993), multipotential erythroid-myeloid-lymphoid progenitors (Godin *et al.*, 1995), and multipotential cells capable of repopulating neonatal recipients on transplantation (Yoder *et al.*, 1997). These activities generally appeared simultaneously in both the yolk sac and PAS/AGM, and before the onset of fetal liver hematopoiesis. Quantitative comparisons of progenitor numbers in the yolk sac and PAS/AGM revealed that while the number of neonatal repopulating cells was higher in the yolk sac, the number of CFU-S was higher in the AGM (Fig. 4). Furthermore, with the exception of CFU-S, none of these hematopoietic progenitors was capable of giving rise to *in vivo* short-term or long-term hematopoietic repopulation of adult mice.

Efforts to determine when and where long-term adult repopulating HSCs first appear revealed their presence in the AGM region (Muller *et al.*, 1994) and the vitelline and umbilical vasculature (de Bruijn *et al.*, 2000b) at E10.5 (>34 somites) before their appearance in the yolk sac. Furthermore, at later time points the AGM was shown to contain HSCs at a higher frequency than the yolk sac (Muller *et al.*, 1994). While these results demonstrated that the AGM region harbors adult-type HSCs, whether they originate in the AGM region and the vitelline and umbilical arteries remains controversial. This is due in part to the fact that blood freely circulates between the embryo body and the yolk sac starting at E8.5 (Garcia-Porrero *et al.*, 1995; see Fig. 2). In addition, interstitial migration of cells may play a role in hematopoietic cell dispersion thoughout the mouse conceptus as observed in amphibian embryos (Turpen *et al.*, 1981).

D. Origins of Embryonic and Adult Hematopoietic Compartments

1. Hematopoietic Origins in the Avian and Amphibian Species

The yolk sac origin for the adult hematopoietic system was first called into question by findings from embryo culture and grafting experiments performed in nonmammalian vertebrate species. In the mid-1970s, elegant studies on cell fate, morphogenesis, and organogenesis in birds were performed through the use of interspecific grafts (between quail and chick embryos) or intraspecific grafts (between different strains of chicks). Grafting was used to create chimeras in which the embryonic origins of adult blood cells were identified using nucleolar or immunochemical markers to determine from which tissue (endogenous or grafted) the differentiated adult blood cells were derived (Dieterlen-Lievre, 1975; Dieterlen-Lievre and Le Douarin, 1993). For example, a quail embryo body was grafted onto the extraembryonic area of a chick blastodisc, and blood cells in the chimeric fetus or adult animal were examined to determine if they were of quail or chick origin. The combined results of many such experiments (Beaupain *et al.*, 1979; Dieterlen-Lievre, 1975; Dieterlen-Lievre and Martin, 1981; Martin *et al.*, 1978) showed that the first emerging hematopoietic cells are derived from the extraembryonic yolk sac. Slightly later in development, hematopoietic cells derived from both extra-

Figure 3 Explant culture experiments. Schematic of the strategy and results of three experiments examining the embryonic sites where the first hematopoietic progenitors and stem cells are produced. (a) CFU-C production was found in E7 yolk sac explants cultured for 2 days but not in cultured E7 embryo body (EB) explants. The presence of CFU-Cs in the EB of the cultured intact conceptus suggests that CFU-Cs are produced in the yolk sac and subsequently migrate and colonize the EB (Moore and Metcalf, 1970). (b) Multipotential progenitors for the erythroid, myeloid, and lymphoid lineages are first generated in the precirculation (<E8.5) intraembryonic splanchnopleura. After the circulation is established between the yolk sac (YS) and PAS, multipotential progenitors are found in both sites (Cumano *et al.*, 1996). (c) CFU-S and HSC production. Explant cultures of E10 YS and AGM reveal that both tissues contain CFU-S but only the AGM can autonomously produce HSCs. At E11 both cultured tissues contain CFU-S and HSCs, suggesting that AGM-generated HSCs colonize the YS or that the YS can automously produce HSCs at this later time (Medvinsky and Dzierzak, 1996).

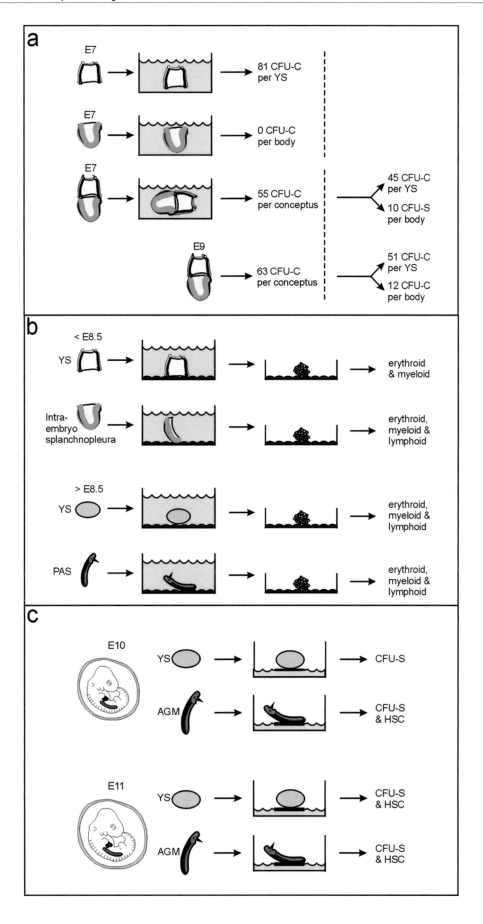

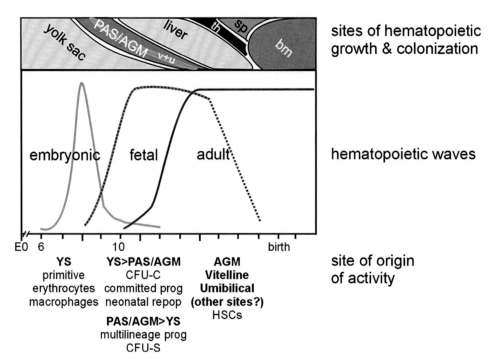

sites of hematopoietic
growth & colonization

hematopoietic waves

site of origin
of activity

Figure 4 Waves of hematopoietic activity in the mouse embryo. At least three waves of hematopoietic activity
are thought to occur in the mouse embryo. The embryonic wave consists of the exclusive production of primitive
erythrocytes and macrophages by the yolk sac. The fetal wave of hematopoiesis includes the production of several
committed and multilineage progenitors and neonatal repopulating cells from the YS and PAS/AGM. Quantitative
levels of progenitors in these two sites of production are compared. The adult wave of hematopoiesis includes the
generation of the first HSCs in the AGM region, and the vitelline and umbilical arteries. Abbreviations: ys, yolk
sac; PAS, para-aortic splanchopleura; AGM, aorta-gonads-mesonephros; th, thymus; sp, spleen; bm, bone marrow.

and intraembryonic sites were found in the fetus. The permanent adult hematopoietic system is derived almost entirely from intraembryonic sites. Thus it appears that cells
derived from the extraembryonic yolk sac contribute only
transiently to embryonic hematopoiesis and become extinct.
Further experimentation suggested that the adult hematopoietic system is derived more specifically from the region
containing the dorsal aorta (Cormier and Dieterlen-Lievre,
1988; Dieterlen-Lievre, 1975). More recent experiments indicate that the allantois is also a potent source of adult blood
cells in the chick (Caprioli *et al.*, 1998).

Amphibian embryo grafting experiments, in which DNA
content was used as the donor graft marker, also demonstrated independent intraembryonic and extraembryonic sites
of hematopoiesis (Kau and Turpen, 1983; Maeno *et al.*,
1985; Turpen *et al.*, 1981). Chimeric frog embryos were
generated by reciprocal grafting of the ventral blood island
region (VBI, yolk sac analog) and the dorsal lateral plate region (DLP, intraembryonic region analog) from diploid and
triploid embryos. These experiments revealed that the VBIs
produce the first hematopoietic cells, and the DLP subsequently generates adult hematopoietic cells. Intrabody hematopoiesis was specifically localized to the developing excretory system near the aortic region, with the most abundant

hematopoiesis found in the pronephros (Turpen and Knudson, 1982). In contrast to the avian species, some VBI-
derived cells are thought to persist into the adult stage in
amphibians and contribute to both red and white blood cell
populations (Kau and Turpen, 1983; Maeno *et al.*, 1985;
Turpen *et al.*, 1997). Recent lineage tracing experiments
have shown that the DLP and the VBIs are derived from distinct blastomeres of the 32-cell-stage embryo, thus clearly
demonstrating two distinct early origins for hematopoietic
cells (Ciau-Uitz *et al.*, 2000).

2. Hematopoietic Origins in Mammals

Explant cultures have been used to study the origins of
hematopoietic progenitors and stem cells in the mouse embryo. Moore and Metcalf (1970) performed the first study
using E7 mouse embryos. After 2 days of culture, abundant granulocyte-macrophage progenitors (CFU-Cs; colony
forming units–culture) were detected in yolk sac explants,
whereas embryo body explants were devoid of this activity
(Fig. 3a). If, however, whole E7 embryos were cultured *in
vitro* for 2 days before separating them into embryo bodies
and yolk sacs, CFU-C activity was also detected in the embryo body. These results suggested that the first CFU-Cs
emerge from the yolk sac and migrate to the embryo proper

between E7 and E9. While these studies examined only one progenitor cell type, the results contributed to the general notion that the yolk sac served as the embryonic source of entire adult hematopoietic system of mammals.

More recently, the source of the first multipotential hematopoietic progenitors was tested by explant cultures of mouse embryonic tissues both pre- and postcirculation (Cumano *et al.*, 1996). An initial organotypic culture step with isolated yolk sac and intraembryonic splanchnopleura explants was followed by a two-step clonal *in vitro* culture system with a readout for erythroid-myeloid, B- and T-lymphoid cells (Fig. 3b). It was found that the intraembryonic splanchnopleura autonomously produces such progenitors beginning at E7.5, one full day earlier than they appear in the yolk sac. These studies not only showed that multipotential progenitors are present first in the embryo body, but also suggested that a more complex hematopoietic progenitor (one with potential for the lymphoid lineage, which is considered to be a more adult-type activity) is present at very early stages of mouse development. This progenitor activity could be demonstrated only in *in vitro* assays. No hematopoietic activity was detected after *in vivo* transplantation, demonstrating that these cells are not fully potent adult-type HSCs.

As shown in Fig. 3c, organotypic cultures were also used to identify the first source of definitive CFU-S and HSCs (Medvinsky and Dzierzak, 1996). Yolk sac, AGM, and liver tissues from E9, E10, and E11 conceptuses were cultured as explants at the air−medium interface for 2–4 days and assayed by *in vivo* transplantation for short-term and long-term hematopoietic repopulation. At early E10 (32–33 somites), CFU-S appear simultaneously in both yolk sac and AGM explants. However, HSCs are detected only in AGM explants from mid-E10 embryos (>34 somites). No HSCs are found in E10.5 yolk sac explants or E9 AGM or yolk sac explants. Thus, the AGM region autonomously generates the first HSCs. Interestingly, while the frequency of HSCs in directly transplanted E10.5 AGM regions is very low [only 3 of about 100 recipients receiving one embryo equivalent of AGM cells is repopulated by HSCs (Muller *et al.*, 1994)], it dramatically increases after explant culture (24 out of 27 recipients are repopulated by HSCs). Therefore, HSCs continue to be induced or expand during the culture period independently of influx from other embryonic tissues. HSCs are found in explants of yolk sac and liver at E11, suggesting that these tissues may be colonized by AGM-generated HSCs. Alternatively, HSCs may emerge autonomously in the yolk sac, but one day later than the AGM.

In summary, the PAS/AGM region in the mouse is the most potent and first autonomous source of multipotential hematopoietic progenitors and adult-type HSCs, strongly suggesting that the mammalian PAS/AGM, like that of nonmammalian vertebrates, is a major source of cells leading to adult hematopoiesis. Hence, strong evolutionary conserva-

tion in the origins of embryonic and adult hematopoietic cells exists across vertebrate species (Dzierzak *et al.*, 1998; Zon, 1995).

3. Embryonic to Definitive Hematopoietic Hierarchy—Evidence for Multiple Hematopoietic Births

In the early/midgestation mammalian embryo, terminally differentiated hematopoietic cells appear temporally before hematopoietic progenitors, and these progenitors in turn appear before HSCs (Fig. 4). Although the adult hematopoietic system is thought to derive solely from the HSCs (Fig. 1), the hematopoietic system in the mammalian embryo must have alternative pathways leading to the rapid and early production of functionally differentiated blood cells. Indeed, primitive erythroid cells are found in the yolk sac of the mouse embryo within 1 day of gastrulation and generation of the first mesoderm (Kaufman, 1992; Russell, 1979), ruling out the possibility that they differentiate through a conventional HSC. Thus, our concept of the adult hematopoietic hierarchy as a continuum from HSCs to functional blood cells does not provide a logical framework for hematopoietic cell development in the early embryo.

Instead, hematopoietic development in the embryo appears to progress via successive births of multiple independent lineages of cells with progressively more complex hematopoietic activities (Fig. 4). Depending on the position and timing of birth, properties such as high proliferative potential, the presence of certain homing receptors, and the ability to self-renew may be sequentially and/or differentially attained. In other words, between E7.5 when the first primitive erythrocytes appear, and E10.5 when the first HSCs appear, the successive waves of hematopoietic progenitors that emerge seem to be less "differentiated" or "less mature," as defined by the adult hematopoietic scheme (Fig. 1). Progenitors committed to a single lineage appear first in development followed by bipotential progenitors, then multilineage progenitors, CFU-S, neonatal repopulating cells, and finally at E10.5 adult HSCs.

There is strong evidence for at least two independent origins of hematopoiesis in nonmammalian and mammalian vertebrates (yolk sac and PAS/AGM). In addition, the vitelline and umbilical arteries in the mouse may be additional sites from which hematopoietic cells are autonomously generated. This is based on the recent demonstration that CFU-S and HSCs are present in these vessels beginning at E10.5 (de Bruijn *et al.*, 2000a,b). In the avian embryo it was recently demonstrated that the allantois is also a hematopoietic site (Caprioli *et al.*, 1998). It is speculated that a close association of mesoderm and endodermal cells is necessary for hematopoietic generation (Pardanaud and Dieterlen-Lievre, 1999). Using a transgenic mouse explant culture system, it was shown that primitive endoderm adjacent to embryonic ectoderm or nascent mesoderm leads to the production of hematopoietic cells and cells with endothelial markers

(Belaoussoff *et al.*, 1998). If indeed a close association of certain mesodermal components in direct or indirect contact with certain endodermal components can lead to the generation of the hematopoietic lineage, it is possible that more sites of hematopoietic birth in the mammalian embryo remain to be uncovered.

In summary, unlike the adult hematopoietic system that is harbored in mainly one site, the bone marrow, the embryonic hematopoietic system appears to emerge through multiple births in anatomically distinct sites.

4. Mesodermal Precursors to the Hematopoietic Lineage

The hemangioblast theory of hematopoietic cell generation is a long-held notion of a common mesodermal source for both the hematopoietic and endothelial lineages in the yolk sac blood islands (Murray, 1932; Sabin, 1920). At present, embryonic stem (ES) cells are the best experimental system for studying hemangioblasts (Choi *et al.*, 1998). Differentiating ES cells contain a unique precursor population, which in response to vascular endothelial growth factor (VEGF) gives rise to blast colonies in semisolid culture medium. Cells from blast colonies can generate both hematopoietic and endothelial cells *in vitro*, suggesting that blast colony-forming cells are hemangioblasts. It is interesting to speculate that hemangioblasts exist not only in the yolk sac but also in PAS/AGM. However unlike the yolk sac in which endothelial cells and primitive erythrocytes appear almost simultaneously, in the PAS/AGM endothelial cells differentiate and form functional arteries before hematopoietic progenitors such as CFU-S and HSCs emerge.

Nevertheless, a close developmental link is seen between endothelial cells and hematopoietic cells in PAS/AGM. Clusters of hematopoietic cells are found on the ventral wall of the dorsal aorta in the AGM region, and in the vitelline and umbilical arteries in close association with the endothelium, at the time when CFU-S and HSCs appear in these sites (Cormier and Dieterlen-Lievre, 1988; Garcia-Porrero *et al.*, 1995; Tavian *et al.*, 1996; Wood *et al.*, 1997; Medvinsky *et al.*, 1993; Muller *et al.*, 1994; de Bruijn *et al.*, 2000a,b). These clusters have been seen in many vertebrate species ranging from sharks to humans. The hematopoietic clusters are arranged in 1–2 rows along the lumenal endothelial lining of these vessels, and some cells at the base of the cluster form tight junctions with the endothelial cells (Garcia-Porrero *et al.*, 1995). Studies in avian and amphibian embryos suggest that the intra-aortic clusters and endothelial cells on the floor of the dorsal aorta share a common precursor in the splanchnopleural mesoderm (Pardanaud *et al.*, 1996) and are derived from the same early stage blastomeres (Ciau-Uitz *et al.*, 2000), respectively. The regulation of endothelial and hematopoietic cell surface markers on endothelial cells and cells in the clusters also suggests a precursor–progeny relationship. For example, before the hematopoietic

clusters appear in the chick, all the endothelial cells lining the dorsal aorta express VEGF-R2, the receptor for the vascular endothelial growth factor. As clusters appear, the cells in the clusters no longer express VEGF-R2, but instead express the pan-leukocyte marker CD45, while the endothelial cells in the dorsal aspect of the aorta continue to express VEGF-R2 (Eichmann *et al.*, 1996; Jaffredo *et al.*, 1998; Pardanaud *et al.*, 1987, 1989). Speculation on these descriptive data have led to the notion that hematopoietic cells bud from the endothelium rather than develop directly from a hemangioblast. In this scenario hemangioblasts could adopt an endothelial fate but retain plasticity for the hematopoietic lineage. An alternative hypothesis is that mesenchymal cells/hemangioblasts located between the endothelial cells or beneath the aorta may be the direct precursors to the emerging hematopoietic clusters.

Two recent experiments support the hypothesis that hematopoietic cells can differentiate directly from endothelial cells. Fate mapping experiments in avian embryos suggest that the hematopoietic clusters on the ventral wall of the dorsal aorta differentiate from hemogenic endothelium (Jaffredo *et al.*, 1998, 2000). Low-density lipoprotein (LDL) linked to the lipophilic fluorescent dye DiI was injected into the heart cavity of chick embryos in order to specifically mark the endothelial cells lining the vascular tree. The LDL-DiI injections were performed at Hamburger and Hamilton (HH) stage 13, 1 day before hematopoietic clusters appear. DiI-positive cells were found in hematopoietic clusters 24–48 hr after the endothelial cells were labeled, suggesting that the hematopoietic clusters are the progeny of the marked endothelial cells. CD45-positive cells that have ingressed into the mesenchyme ventral to the dorsal aorta were also DiI positive, suggesting that endothelial cells are capable of generating cells that migrate from the endothelium both into and away from the aortic lumen. It was demonstrated previously that chick embryos contain diffuse hematopoietic foci in the mesentery underlying the ventral part of the aorta (Dieterlen-Lievre and Martin, 1981). Diffuse foci of cells that express the hematopoietic markers GATA-3, LMO2, and AA4.1 were also found in the mesenchyme ventral and lateral to the aorta in the early stage mouse embryo (Manaia *et al.*, 2000; Petrenko *et al.*, 1999). At present, the function and significance of the para-aortic foci expressing these markers in the mouse are unknown.

The second experiment that supports the notion of a hemogenic endothelium is that VE-cadherin-positive endothelial cells, when sorted from mouse embryos and plated on OP9 stromal cells, yield hematopoietic cells *in vitro* (Nishikawa *et al.*, 1998). VE-cadherin is expressed at the adherens junctions that connect adjacent endothelial cells and is thought not to be expressed on hematopoietic progenitor cells.

Finally, the expression of common markers on endothelial cells and hematopoietic cells supports the notion of a close developmental link between these lineages in both human and mouse embryos (Table II). The cell surface sialo-mucin CD34, adhesion molecule CD31, and the flk-1 receptor tyrosine kinase are expressed by both hematopoietic clusters and endothelial cells but generally not by mesenchymal cells (Wood *et al.*, 1997; Tavian *et al.*, 1996, 1999; Labastie *et al.*, 1998; Cai *et al.*, 2000). The transcription factors SCL and GATA-2 are also expressed by hematopoietic cell clusters and endothelial cells (Tavian *et al.*, 1996, 1999; Labastie *et al.*, 1998). Expression of the extracellular matrix glycoprotein tenascin C in the human embryo is concentrated along the ventral endothelium as well as in hematopoietic clusters (Marshall *et al.*, 1999).

In the mouse, one recently described marker is expressed only by the endothelial cells in the ventral wall of the dorsal aorta where hematopoietic clusters are localized, in some of the mesenchymal cells underlying this region, and in the hematopoietic cell clusters themselves (North *et al.*, 1999). This marker is the AML1 (Runx1 or CBFα2) transcription factor that was shown to be required for definitive adult hematopoiesis (Okuda *et al.*, 1996; Q. Wang *et al.*, 1996a). AML1 is spatially and temporally expressed in sites with demonstrated functional HSCs, suggesting that HSC activity is derived from the hematopoietic clusters, the mesenchymal cells and/or endothelial cells on the ventral part of the dorsal aorta (North *et al.*, 1999). Furthermore, AML1 is required for the formation of the intra-aortic hematopoietic clusters (North *et al.*, 1999). The study of AML1 expression, in combination with that of other more general markers, appears to be promising in deciphering the precursor-progeny relationships of the cells within this complex region.

What types of hematopoietic activities do lumenal clusters contain? In the chick, *in vitro* CFU-Cs were demonstrated to be present in the para-aortic cluster region (Cormier and Dieterlen-Lievre, 1988). While the grafting experiments revealed that the avian adult hematopoietic system originates from the region of the dorsal aorta, the exact location of the the HSCs within the aortic region was not shown. Recent subdissections of the mouse AGM region reveal that the dorsal aorta and its surrounding mesenchyme are indeed the first source of adult-type HSCs (de Bruijn *et al.*, 2000b). As described above, the marker expression patterns attributed to HSCs can be found in the aortic hematopoietic clusters. However, marker expression within the cell clusters is heterogeneous at least for one molecule, glycoprotein IIb-IIIa, suggesting that the clusters contain several types of hematopoietic cells (Ody *et al.*, 1999). Until unique and specific markers for the endothelium, mesenchyme, and cells in the aortic clusters in the AGM region become available, it will not be possible to determine which of these cells types is the source of HSCs.

5. Colonization Theory of Hematopoiesis

How do secondary tissues such as the thymus, spleen, liver, and bone marrow become hematopoietic? Much of our

current thinking is influenced by the observations made in 1967 by Moore and Owen. Using parabiosed chick embryos, these investigators found that hematopoietic tissues were colonized by cells delivered through the blood. Definitive proof that hematopoietic cells in adult tissues are extrinsically derived came from grafting studies where it was found that waves of hematopoietic activity entered and colonized thymus and spleen rudiments of avian embryos during receptive periods (Dieterlen-Lievre, 1975; Martin *et al.*, 1978). Thus thymus and spleen provide the microenvironment for the seeding and differentiation of extrinsic precursors. The sites of *de novo* hematopoietic cell emergence were found to be both the yolk sac and the intraembryonic region containing the dorsal aorta. Interestingly, during embryonic stages a small number of intrabody-derived hematopoietic cells can be found in the yolk sac and a small number of yolk sac-derived macrophage-like (microglial) cells can be found intraembryonically (localized to the eye and brain) (Pardanaud *et al.*, 1987). Thus, migration is not unidirectional and colonization is thought to occur through the circulation of small populations of hematopoietic cells. Similarly, recent experiments in which the prevascularized allantoic bud from quail embryos was grafted into the coelom of chick embryos resulted in hematopoietic and endothelial cells of quail origin in the bone marrow of the chick host (Caprioli *et al.*, 1998). These data suggest that the adult hematopoietic bone marrow is also colonized by precursors that arise *in situ* in the allantois.

In mammals, Cudennec and colleagues (1981) performed one of the experiments that strongly suggested the liver is colonized by cells from the yolk sac. The inclusion of a cell nonpermeable filter between yolk sac and liver explants in culture blocked the influx of cells between the tissues and resulted in no erythropoietic activity in the liver. Colonization was also demonstrated in an *in vivo* mouse model system in which fetal liver rudiments were implanted under the kidney capsule of adult mice. After several days the liver rudiments were examined and found to be colonized by adult hematopoietic cells, thus demonstrating the liver's receptivity to exogenous hematopoietic cells (Johnson and Moore, 1975). These early experiments indicate that hematopoietic colonization of secondary territories most likely plays an important role in mammalian development. A summary of the possible colonization events occurring within the mammalian embryo during ontogeny is shown in Fig. 4.

6. Fate Mapping

Largely absent from all of the studies performed on mammalian embryonic hematopoietic cells is a clear demonstration of precursor–progeny relationships. As emphasized in the previous sections, fate mapping is necessary to determine the true precursor(s) of hematopoietic cells, whether they are hemangioblast, mesenchymal, endothelial, or even other hematopoietic cells. It is possible, for example, that CFU-S at

E11 and E12 are derived from HSCs, whereas CFU-S at E9 and E10 are derived from mesenchymal, hemangioblast, or endothelial cells (de Bruijn *et al.*, 2000a). Without fate mapping we cannot unambiguously determine if the first HSCs observed at E11 in the fetal liver are derived from cells that originated in the PAS/AGM region or from yolk sac hematopoietic progenitors that acquired HSC characteristics in the fetal liver. Injections of cells into embryo or adult recipients reveal the potential of cells, but do not illuminate the developmental process as it occurs in the embryo. For example, it was shown that E8 yolk sac cells injected transplacentally (Toles *et al.*, 1989) or into the amnionic cavity (Weissman *et al.*, 1978) of early stage mouse embryos results in contribution of yolk sac donor cells to the adult hematopoietic system. These experiments demonstrate that E8 yolk sac cells have the potential to contribute to adult hematopoiesis in these experimental settings, but they do not show that E8 yolk sac cells normally contribute to the adult hematopoietic system in an intact organism. While injection of E9 yolk sac cells into the hematopoietically active livers of neonatal mice uncovers their long-term multilineage potential (Yoder *et al.*, 1997), without fate mapping we cannot conclude that the fetal liver plays a role in the maturation of yolk sac progenitors into HSCs. Similarly, direct injection of AGM cells into adult recipients demonstrates that the AGM harbors HSCs. However, it remains possible that the adult hematopoietic system is *de novo* generated by colonization of the bone cavities at later stages of gestation by hemangioblasts that emerge from the AGM or from elsewhere in the embryo. Fate mapping will help resolve issues such as precursor–progeny relationships, homing, colonization, and transient versus long-lived functional hematopoiesis. Cre-LoxP fate mapping may help to resolve this issue.

III. Molecular Genetic Aspects of Blood Development in the Mouse Embryo

A. Developmental Regulation of Hematopoiesis

The molecular genetic basis of hematopoietic development has been of great interest starting from the time when the human β-globin gene was cloned. Since then a wealth of genes important for various aspects of hematopoiesis has been isolated. These include genes encoding growth factors and their receptors, signaling molecules, transcription factors, homing/adhesion molecules, as well as proteins like the globins and immunoglobulins that are involved in specific erythroid and lymphoid cell functions, respectively. Much insight and many general molecular genetic paradigms have emerged from the studies of lineage-specific transcription factors and, in particular, the developmental regulation of the globin gene locus.

The globin gene program exemplifies a tightly controlled system of developmentally expressed genes. The human β-globin locus on chromosome 11 is organized as a linear array of the embryonic ϵ, fetal $^G\gamma$ and $^A\gamma$ and adult β genes ordered along the DNA in the sequence in which they are developmentally expressed (Grosveld *et al.*, 1993). Analyses of promoter regions and enhancers have revealed consensus sequences for putative DNA binding proteins. Many erythroid-specific DNA binding proteins such as GATA-1, EKLF, and Sp-1 have been isolated and shown to be important for the expression of these globin genes and in some cases other erythroid-specific genes (reviewed in Orkin, 1995). Moreover, the strict developmental regulation of embryonic, fetal, and adult globin genes in erythropoiesis is controlled by a powerful regulatory element called the locus control region (LCR). The globin gene expression program may be determined intrisically by the hematopoietic precursor/stem cell or may be influenced by the surrounding microenvironment or both. Today these issues are still controversial and efforts are under way to link gene activity in the nucleus with signals emanating from the inside or outside of the cell.

Genetic studies on the macromolecular level are a focus of attention in the hematopoietic lineages. Chromatin and chromosome structure is known to have important consequences for gene activity. For example, the globin LCR has been implicated in the formation of a holocomplex of transcription factors that maintains the chromatin accessibility of the globin locus in erythroid cells (Wijgerde *et al.*, 1995; Fraser and Grosveld, 1998). It is thought that the chromatin structure is important in the commitment or maintenance of plasticity in specific lineages of cells.

Finally, metaphase spreads of chromosomes from leukemic cells were instrumental in revealing macromolecular changes in genes important in hematopoiesis. The most frequent types of leukemias, acute myelogenous leukemias, were characterized by several common chromosomal translocations/inversions in hematopoietic progenitor cells, for example, t(15:17), t(8:21), inv(16) (Rowley, 1999). The chromosomal breakpoints were mapped and cloned and provided a means by which genes such as the AML1/CBFα transcription factor, involved in both normal and abnormal hematopoiesis, could identified and characterized further (Miyoshi *et al.*, 1991; Liu *et al.*, 1993). Thus, these molecular genetic studies performed at many levels have contributed enormously to our current understanding of developmental hematopoiesis.

B. Genetic Program for Hematopoiesis

The genetic program of hematopoiesis is thought to involve a progressive cascade of gene expression starting from the first signals for hematopoietic lineage specification and proceeding through the final stages of functional blood cell

Table IV   Genetic Requirements in Hematopoietic Development

Target	Mesoderm	YS hematopoiesis and endothelium	Fetal and adult hematopoiesis	Lineage specification	Migration	Proliferation
Time	E7	E9.5–11.5	E10–16			
Gene	FGF	TGF-β_1	SCL	GATA-1 (erythrocytes)	β_1 integrin	LIF
	TGF-β_1	flk-1	GATA-2	Jak2 (erythrocytes)	α_4 intergrin	IL-6
		VEGF	Runx1	EKLF (erythrocytes)		flk-2/flt3
		VE-cadherin	CBFb	Ikaros (T lymphocytes)		
		SCL	c-kit	Pax5 (B lymphocytes)		
		LMO2	SCF			
		c-myb				

production. This cascade has many gene products that serve functions in several stages of hematopoiesis, and also in non-hematopoietic lineages. Examples of genes required for hematopoiesis are provided here (also see Table IV).

1. Genes Involved in Mesoderm Formation and Hematopoietic Specification

A graded expression pattern of many genes acting as positive and negative signals (Stennard and Gurdon, 1997) is thought to define the normal spatial borders of hematopoiesis in different mesodermally derived regions of the conceptus. The transforming growth factor β (TGF-β) and fibroblast growth forming (FGF) families of genes are pivotal in mesoderm and blood formation in *Xenopus* embryos (Dale *et al.,* 1992; Smith and Albano, 1993). BMP4, a TGF-β family member, acts as a ventralizing molecule and induces the expression of Mix.1, a gene that induces hematopoiesis in the Xenopus animal cap assay (Mead *et al.,* 1996; Turpen *et al.,* 1997). Mice deficient for BMP4 suffer embryonic lethality at the time of gastrulation, and have little or no mesoderm (Winnier *et al.,* 1995). The few BMP4-deficient embryos that do survive to postgastrulation stages show profound decreases in mesoderm formation and erythropoiesis in the yolk sac, demonstrating that BMP4 is strictly required for the formation of ventral mesoderm. Indeed, when BMP4 is administered *in vitro* to presumptive anterior headfold tissue from the epiblast, hematopoietic cell production (in CFU-C assays) is induced (Kanatsu and Nishikawa, 1996).

Overlapping with early mesodermal induction events and/ or slightly downstream, other growth factors, receptors, and transcription factors were found to be critical for hematopoietic specification. Both endothelial cells and hematopoietic cells require TGF-β_1. Perinatal lethality occurs in TGF-β_1-deficient embryos between E9.5 and 11.5 (Dickson *et al.,* 1995). Although initial endothelial cell differentiation begins, these cells do not form a vascular network.

Mice null for the receptor tyrosine kinase flk-1 (*Kdr*) also suffer early embryonic lethality between E8.5 and 9.5 (Shalaby *et al.,* 1995). Flk-1-deficient embryos are defective in

yolk sac blood island and vessel formation and hematopoietic progenitor numbers are reduced. Similar defects (although with slightly later embryonic lethality at E11) are observed in mice null for the flk-1 ligand, VEGF (Carmeliet *et al.,* 1996; Ferrara *et al.,* 1996). Knock-in of a *lacZ* marker gene into the flk-1 locus revealed that *lacZ*-expressing cells accumulated in the amnions and never entered into the areas of putative blood island formation in flk-1 null embryos. Flk-1 is also required for formation of the adult hematopoietic system as revealed by chimeras made from flk-1 homozygous null ES cells (Shalaby *et al.,* 1997).

Targeting of the *Scl* gene demonstrated a requirement for the production of all hematopoietic lineages in the embryo (Robb *et al.,* 1995; Shivdasani *et al.,* 1995) and the adult (Porcher *et al.,* 1996; Robb *et al.,* 1996). SCL is also required in the endothelial lineage. It is not required for formation of the small yolk sac vessels. However, it is required for angiogenesis and the formation of vitelline vessels connecting the embryo with the yolk sac (Visvader *et al.,* 1998). Absence of the partner protein of SCL, LMO2, results in an identical phenotype (Warren *et al.,* 1994; Yamada *et al.,* 1998).

2. Gene Affecting Definitive and Fetal Liver Hematopoiesis

After specification of cells to the hematopoietic lineage, genes involved in the generation, maintenance, self-renewal, and/or differentiation of definitive hematopoietic progenitors and stem cells come into play. The GATA-2 transcription factor is important for the proliferation of hematopoietic progenitors (Tsai *et al.,* 1994). GATA-2-deficient embryos generate primitive erythrocytes but in low number, and definitive progenitors (CFU-Cs) in the yolk sac are decreased 100-fold. GATA-2 null embryos exhibit severe fetal liver anemia and embryonic lethality at E10.5. No contribution of GATA-2 null ES cells to the adult hematopoietic system of chimeric mice was observed, suggesting that GATA-2 acts in a cell autonomous fashion in hematopoietic progenitors at all stages of fetal and adult hematopoiesis.

Two genes in the family of core-binding factors (CBFs) are the most frequent targets of chromosomal rearrangements in human leukemias. AML1 (Runx1) and CBFβ form a heterodimeric transcription factor that binds the core enhancer motif present in a number of genes expressed in hematopoietic cells. Both AML1 and CBFβ were shown to be required for definitive hematopoiesis by gene targeting experiments (Okuda et al., 1996; Sasaki et al., 1996; Wang et al., 1996a; Q. Wang et al., 1996b). AML1 and CBFβ null embryos die at E12.5 and suffer from severe fetal liver anemia, and HSCs and definitive hematopoietic progenitors are absent. Furthermore, a hemizygous dose of AML1 affects both the timing and spatial distribution of HSCs and CFU-S within the AGM and yolk sac at E10 and E11 (Cai et al., 2000). AML1 is expressed in hematopoietic clusters, endothelial cells, and mesenchymal cells in the ventral aspect of the dorsal aorta, and in endothelial cells and hematopoietic clusters in the vitelline and umbilical arteries (North et al., 1999). The AML1:CBFβ heterodimer appears to be required for the generation, proliferation, and/or maintenance of the first HSCs as they emerge in the embryo, and perhaps for the differentiation of HSCs to committed definitive hematopoietic progenitors.

Targeted disruption of the c-myb proto-oncogene transcription factor also affects definitive fetal liver hematopoiesis but not yolk sac hematopoiesis (Mucenski et al., 1991). Multipotential granulocyte-macrophage progenitors and adult erythrocytes are decreased (Lin et al., 1996) but other hematopoietic lineages and hematopoiesis in adult chimeric mice must be examined to determine whether a progenitor or stem cell is affected. The available data suggest that c-myb acts downstream of the GATA-2 and AML1: CBFβ transcription factors in definitive hematopoiesis.

Two of the most widely used naturally occurring mouse hematopoietic mutants are the dominant white spotting (W) and Steel (Sl) mouse strains. W mice are affected in CFU-S and mast cell production. The most severe W alleles result in embryonic lethality at around E16. While the Sl mouse strain has virtually the same phenotype, transplantation studies originally showed that these strains complement each other for CFU-S development. The W strain is defective for the c-kit receptor tyrosine kinase, which is expressed on HSCs, and the mutation in the Sl strain disrupts the gene encoding the c-kit ligand [also called steel or stem cell factor (SCF)], which is expressed by stromal cells (Motro et al., 1991; Keshet et al., 1991; and reviewed in Bernstein, 1993; Fleischman, 1993).

3. Genes Affecting Lineage Specification and/or Differentiation

One of the major branch points of the hematopoietic hierarchy is where the lymphoid lineage diverges from the erythroid-myeloid lineage (Fig. 1). Two transcription factors near this branch point are Ikaros and Pax5, which are impli-

cated in the specification of the T- and B-lymphoid lineages, respectively. Ikaros null mice display defects that differentially affect the development of fetal and adult lymphocytes, in that T lymphocytes are absent from the fetus but present in the adult (J. H. Wang et al., 1996). B lymphocytes on the other hand, are absent from both the fetus and the adult. The T lymphocytes in Ikaros null mice that are found postnatally display defects in $\gamma\delta$ T cells and thymic dendritic cells. Pro-B cells from Pax5 null mice are capable of differentiating into many hematopoietic lineages on transplantation into recipient mice, but do not give rise to B-lymphoid cells, suggesting that Pax5 plays an essential role in B-lineage commitment by suppressing alternative lineage choices (Nutt et al., 1999).

Gene targeting has identified many genes essential in the erythroid lineage differentiation program. The most essential erythroid-specific transcription factor is GATA-1 (Fujiwara et al., 1996; Pevny et al., 1991), which is required for the production of both primitive and adult erythrocytes. The GATA-1 null defect has been localized to the proerythroblast stage of differentiation, leading to anemia and embryonic lethality around E10. The Jak2 kinase signal transduction molecule is required for definitive but not primitive erythropoiesis. Targeted disruption of Jak2 results in embryonic lethality at E12.5, with no detectable BFU-E or CFU-E in the fetal liver. Targeted disruptions of the EKLF transcription factor, erythropoietin growth/differentiation factor, and the erythropoietin receptor genes result in embryonic lethality due to the lack of definitive erythropoiesis. These defects manifest themselves slightly later than those in Jak2 mutants in that lethality due to fetal liver anemia occurs at E14–16, at the time of the switch from fetal to adult erythropoiesis.

4. Genes Affecting Hematopoietic Cell Migration

Integrins have been found to play a role in the differentiation and migration of hematopoietic cells, presumably by promoting adhesive interactions between hematopoietic cells and supportive stromal cells. Hematopoietic cells generated in intra- and extraembryonic sites during midgestation are thought to seed the secondary hematopoietic territories: fetal liver, spleen, thymus, and bone marrow. Gene targeting revealed an important role for the β_1 integrin in seeding the fetal liver. Since β_1 integrin null embryos die during preimplantation stages (Fassler and Meyer, 1995; Stephens et al., 1995) embryos chimeric for β_1 integrin null ES cells were generated. These experiments revealed that although a normal number of β_1 integrin null hematopoietic cells were found in the yolk sac and circulation, none were found in the fetal liver, thymus, or bone marrow, suggesting that β_1 integrin is necessary for seeding these secondary hematopoietic sites. Likewise, chimeric mice made with α_4 integrin null ES cells revealed that α_4 integrin is required for the homing of B and T lymphocytes to adult hematopoietic territories, but not to the fetal liver (Arroyo et al., 1996).

5. Genes Thought to Affect the Proliferative Potential of Hematopoietic Progenitor/Stem Cells

The adult hematopoietic system can arise from a single clonally marked transplanted HSC (Lemischka *et al.*, 1986) and hence does not absolutely require the simultaneous function of all HSCs. It appears that a balance between self-renewal and proliferation of HSCs is maintained. Several genes have been implicated through targeted gene disruption to affect the balance between HSC proliferation and self-renewal. Mice lacking the growth factors IL-6 or LIF, or the flk-2/flt-3 receptor tyrosine kinase do not suffer from embryonic lethality or the absence of definitive hematopoietic cells. However, subtle phenotypes are observed after transplantation into normal adult recipients. LIF-deficient mice have decreased numbers of CFU-S and other clonogenic progenitors (Escary *et al.*, 1993), while IL-6 deficient mice have a decrease in CFU-S progenitor numbers and potency of HSCs (Bernad *et al.*, 1994). In contrast, flk-2/flt-3 null mice are defective in myeloid and lymphoid reconstitution but not in CFU-S or pre-CFU-S numbers (Mackarehtschian *et al.*, 1995). Thus, these genes affect maintenance, proliferative potential, and/or self-renewal of hematopoietic progenitors and/or stem cells.

IV. Current Cellular and Molecular Conceptual Frameworks for Hematopoietic Ontogeny

A. Origins

General concepts conserved throughout evolution are evident from the experimental studies of hematopoiesis in vertebrate animal models. The vertebrate hematopoietic system originates within the mesodermal germ layer and its progeny. This property is conserved in invertebrates as well. *Drosophila* embryos have two major types of hematopoietic cells, plasmatocytes and crystal cells, and both cell types differentiate from hemocyte precursors located in the head mesoderm of the *Drosophila* embryo (Tepass *et al.*, 1994). In vertebrates, multiple births of the hematopoietic system occur in anatomically distinct mesodermal sites of the embryo, including the yolk sac, PAS/AGM, and the allantois (the last has been demonstrated in avian but not in murine embryos). The vitelline and umbilical vasculature may also generate hematopoietic cells, and it is speculated that other yet unknown sites may be hemogenic. At this time, most researchers in the field believe that hemangioblasts are a common precursor for at least some hematopoietic and endothelial lineages. But there may be several types of hemangioblasts (or hemogenic precursors) that originate in different anatomical sites during ontogeny. Indeed experiments in avian and mouse embryos have strongly suggested that some endothelial cells are direct precursors of hematopoietic cells. It is also possible that hemogenic potential is harbored in mesodermal/mesenchymal cells. It is important to identify the immediate precursors of hematopoietic cells, and to determine if hematopoietic activities are sequestered in some unexpected sites in the adult.

B. Colonization

Colonization of secondary hematopoietic organs is an important feature of developmental hematopoiesis. Sequential colonization events suggest that the growing embryo needs additional territories to expand hematopoietic cells. Colonization of the fetal liver has been a focus of interest in early stages of hematopoiesis since the appearance of the different hematopoietic activities in this tissue parallels their birth in other sites in the mouse embryo. The first cells colonizing the fetal liver are primitive erythroid cells, followed by committed progenitors, CFU-S, multipotential progenitors, and finally HSCs. In general, these activities appear in the fetal liver 1–2 days after they are detected in the yolk sac and PAS/AGM. These observations strongly suggest that waves of hematopoietic activity seed the liver after migrating from their originating source(s). While it is generally accepted that the secondary hematopoietic tissues are colonized by hematopoietic cells generated elsewhere in the embryo, *in vivo* clonal fate mapping has not been performed. It remains possible that within hematopoietic sites that appear later in development, such as the bone marrow, commitment to the hematopoietic lineage may occur from mesodermal cells or hemangioblasts seeding this tissue rather than from the definitive HSCs generated in the AGM.

C. Genetic Programming

Gene targeting studies in the mouse have revealed a complex genetic cascade that can be generally classified into important developmental milestones. These milestones include formation of the mesoderm, hematopoietic specification, the emergence of primitive and definitive hematopoietic progenitors, and the generation of complex hematopoietic characteristics such as multilineage capability, self-renewing ability, homing to specific hematopoietic territories, and/or high proliferative potential. A major decision point appears to be at the crossroads of primitive embryonic hematopoiesis and definitive adult hematopoiesis as the PAS/AGM becomes maximally active and the fetal liver becomes a secondary hematopoietic territory. Clearly, the initial molecular programs of primitive and definitive hematopoietic cells overlap in their requirements for genes inducing mesoderm formation, some hematopoietic genes, and even some genes for terminal differentiation (particularly in the erythroid lineage). However, the definitive hematopoietic program is much more complex, requiring an abundance of unique genes that promote the establishment of the complete multi-lineage adult hematopoietic hierarchy. Do these genetic programs have a basis in the anatomically separate and independent sites of hematopoietic generation in the conceptus, are they intrinsic to the mesodermal precursors, or are they regulated by the microenvironment? Further studies of the embryonic localization and specific microenvironments required for hematopoietic progenitor/stem cells should allow for the isolation of novel genes and molecules that may be useful for manipulating human hematopoietic cells in a clinical setting. In the future, a database of all expressed sequences in HSCs isolated from bone marrow, fetal liver, and AGM promises to further our understanding and manipulation of the genetic program of HSCs throughout ontogeny (Phillips *et al.,* 2000).

D. Redirecting the Fate of Stem Cells

Only a few years ago it was accepted that stem cells for tissues such as blood, skin, gut, muscle, and neurons in the adult vertebrate were restricted in fate, and able to contribute through differentiation to only one somatic lineage. This paradigm was recently challenged by demonstrations that stem cells isolated from one tissue can form functional cells of another tissue. For example, it has been found that bone marrow-derived cells can contribute to a phagocytic cell type in the brain called microglia (Theele and Streit, 1993) with a small percentage of cells also developing into astrocytes (Eglitis and Mezey, 1997). The reciprocal experiment, testing whether neural stem cells could change fate, found that indeed neural stem cells injected into the circulation of irradiated adult mice contributed to blood cell formation (Bjornson *et al.,* 1999). Thus, the blood cell microenvironment appears capable of redirecting the program of neural stem cells to the blood cell fate. Furthermore, human mesenchymal stem cells, which are thought to be multipotent cells in the adult marrow, replicate as undifferentiated cells in culture but can be induced to differentiate into adipocytic, chondrocytic, or osteocytic lineages (Pittenger *et al.,* 1999). Most recently, bone marrow cells have been shown to take on muscle characteristics (Ferrari *et al.,* 1998; Gussoni *et al.,* 1999) and muscle cells to take on hematopoietic characteristics (Jackson *et al.,* 1999) when transplanted *in vivo* into adult recipient mice. Despite the use of cell populations in these studies, it is interesting to speculate that what were formerly thought to be tissue-restricted stem cells are in fact multipotential cells. Perhaps common features in the genetic programming of the many types of stem cells (hematopoietic stem cells, neural stem cells, mesenchymal stem cells, muscle stem cells) will be found and reveal genes that maintain these cells in an uncommitted state. In such a scenario, plasticity of all these stem cells is maintained and it is the inductive microenvironment that determines the ultimate cell fate. The precise mechanisms of stem cell maintenance and induction remain to be determined.

V. Future Directions

The future of developmental hematopoietic studies in mammals will rely on fate mapping linked to functional hematopoietic readouts. Until now only avian and amphibian embryo grafting methods were efficient for fate mapping. However, whole mouse embryo culture is improving and can be used in DiI marking studies and clonal retroviral marking. Transgenic mouse approaches hold promise for both fate mapping studies and deleting genes in specific cell lineages at defined developmental time points. The cre-lox recombination system together with the induced expression of the mutated ligand binding domain of the estrogen receptor is a two-tiered method of controlling recombination in the targeted population and requires no *in vitro* manipulation. Thus, targeted hematopoietic cells can be followed through all developmental stages *in vivo,* from their sites of origin to sites of colonization.

The genetic programming of hematopoietic cells at all stages of development is only in its infancy. The complex levels of interacting gene programs will become increasingly apparent as more genes are cloned and as more hematopoietic mutant mice are generated. Furthermore, the genes that retain the self-renewing properties and plasticity of stem cells will be important for understanding how stem cells are sequestered throughout life. In summary, the study of hematopoietic development has contributed to many paradigms in stem cell biology and will continue to be on the forefront of stem cell biology through its focused examination of a complex developmental problem. Clinical applications are sure to follow.

References

Arroyo, A. G., Yang, J. T., Rayburn, H., and Hynes, R. O. (1996). Differential requirements for alpha4 integrins during fetal and adult hematopoiesis. *Cell (Cambridge, Mass.)* **85,** 997–1008.

Beaupain, D., Martin, C., and Dieterlen-Lievre, F. (1979). Are developmental hemoglobin changes related to the origin of stem cells and site of erythropoiesis? *Blood* **53,** 212–225.

Belaoussoff, M., Farrington, S. M., and Baron, M. H. (1998). Hematopoietic induction and respecification of A-P identity by visceral endoderm signaling in the mouse embryo. *Development (Cambridge, UK)* **125,** 5009–5018.

Bernad, A., Kopf, M., Kulbacki, R., Weich, N., Koehler, G., and Gutierrez-Ramos, J. C. (1994). Interleukin-6 is required in vivo for the regulation of stem cells and committed progenitors of the hematopoietic system. *Immunity* **1,** 725–731.

Bernstein, A. (1993). Receptor tyrosine kinases and the control of hematopoiesis. *Semin. Dev. Biol.* **4,** 351–358.

Bjornson, C. R., Rietze, R. L., Reynolds, B. A., Magli, M. C., and Vescovi, A. L. (1999). Turning brain into blood: A hematopoietic fate adopted by adult neural stem cells in vivo. *Science* **283,** 534–537.

Bonifer, C., Faust, N., Geiger, H., and Muller, A. M. (1998). Developmental changes in the differentiation capacity of haematopoietic stem cells. *Immunol. Today* **19,** 236–241.

Breems, D. A., Blokland, E. A., Siebel, K. E., Mayen, A. E., Engels, L. J., and Ploemacher, R. E. (1998). Stroma-contact prevents loss of hematopoietic stem cell quality during ex vivo expansion of CD34+ mobilized peripheral blood stem cells. *Blood* **91,** 111–117.

Cai, Z. L., de Bruijn, M., Ma, X., Dortland, B., Luteijn, T., Downing, J. R., and Dzierzak, E. (2000). Haploinsufficiency of AML1/CBFA2 affects the temporal and spatial generation of hematopoietic stem cells in the mouse embryo. *Immunity* **13,** 423–431.

Caprioli, A., Jaffredo, T., Gautier, R., Dubourg, C., and Dieterlen-Lievre, F. (1998). Blood-borne seeding by hematopoietic and endothelial precursors from the allantois. *Proc. Natl. Acad. Sci. U.S.A.* **95,** 1641–1646.

Carmeliet, P., Ferreira, V., Breier, G., Pollefeyt, S., Kieckens, L., Gertsenstein, M., Fahrig, M., Vandenhoeck, A., Harpal, K., Eberhardt, C., Declercq, C., Pawling, J., Moons, L., Collen, D., Risau, W., and Nagy, A. (1996). Abnormal blood vessel development and lethality in embryos lacking a single VEGF allele. *Nature (London)* **380,** 435–439.

Choi, K., Kennedy, M., Kazarov, A., Papadimitriou, J. C., and Keller, G. (1998). A common precursor for hematopoietic and endothelial cells. *Development (Cambridge, UK)* **125,** 725–732.

Ciau-Uitz, A., Walmsley, M., and Patient, R. (2000). Distinct origins of adult and embryonic blood in Xenopus. *Cell (Cambridge, Mass.)* **102,** 787–796.

Cormier, F., and Dieterlen-Lievre, F. (1988). The wall of the chick embryo aorta harbours M-CFC, G-CFC, GM-CFC and BFU-E. *Development (Cambridge, UK)* **102,** 279–285.

Cudennec, C. A., Thiery, J. P., and Le Douarin, N. M. (1981). *In vitro* induction of adult erythropoiesis in early mouse yolk sac. *Proc. Natl. Acad. Sci. U.S.A.* **78,** 2412–2416.

Cumano, A., Paige, C. J., Iscove, N. N., and Brady, G. (1992). Bipotential precursors of B cells and macrophages in murine fetal liver. *Nature (London)* **356,** 612–615.

Cumano, A., Dieterlen-Lievre, F., and Godin, I. (1996). Lymphoid potential, probed before circulation in mouse, is restricted to caudal intraembryonic splanchnopleura. *Cell (Cambridge, Mass.)* **86,** 907–916.

Dale, L., Howes, G., Price, B. M., and Smith, J. C. (1992). Bone morphogenetic protein 4: A ventralizing factor in early Xenopus development. *Development (Cambridge, UK)* **115,** 573–585.

de Bruijn, M. R. T. R., Peeters, M. C. E., Luteijn, T., Visser, P., Speck, N. A., and Dzierzak, E. (2000a). CFU-S$_{11}$ activity does not localize solely with the aorta in the AGM region. *Blood* **96,** 2902–2904.

de Bruijn, M. R. T., R., Speck, N. A., Peeters, M. C. E., and Dzierzak, E. (2000b). Definitive hematopoietic stem cells first emerge from the major arterial regions of the mouse embryo. *EMBO J.* **19,** 2465–2474.

Delassus, S., and Cumano, A. (1996). Circulation of hematopoietic progenitors in the mouse embryo. *Immunity* **4,** 97–106.

Dickson, M. C., Martin, J. S., Cousins, F. M., Kulkarni, A. B., Karlsson, S., and Akhurst, R. J. (1995). Defective haematopoiesis and vasculogenesis in transforming growth factor-beta 1 knock out mice. *Development (Cambridge, UK)* **121,** 1845–1854.

Dieterlen-Lievre, F. (1975). On the origin of haemopoietic stem cells in the avian embryo: An experimental approach. *J. Embryol. Exp. Morphol.* **33,** 607–619.

Dieterlen-Lievre, F., and Le Douarin, N. M. (1993). Developmental rules in the hematopoietic and immune systems of birds: How general are they? *Semin. Dev. Biol.* **4,** 325–332.

Dieterlen-Lievre, F., and Martin, C. (1981). Diffuse intraembryonic hemopoiesis in normal and chimeric avian development. *Dev. Biol.* **88,** 180–191.

Dzierzak, E., Medvinsky, A., and de Bruijn, M. (1998). Qualitative and quantitative aspects of haemopoietic cell development in the mammalian embryo. *Immunol. Today* **19,** 228–236.

Eglitis, M. A., and Mezey, E. (1997). Hematopoietic cells differentiate into both microglia and macroglia in the brains of adult mice. *Proc. Natl. Acad. Sci. U.S.A.* **94,** 4080–4085.

Eichmann, A., Marcelle, C., Breant, C., and Le Douarin, N. M. (1996). Molecular cloning of Quek 1 and 2, two quail vascular endothelial growth factor (VEGF) receptor-like molecules. *Gene* **174,** 3–8.

Eren, R., Zharhary, D., Abel, L., and Globerson, A. (1987). Ontogeny of T cells: Development of pre-T cells from fetal liver and yolk sac in the thymus microenvironment. *Cell Immunol.* **108,** 76–84.

Escary, J. L., Perreau, J., Dumenil, D., Ezine, S., and Brulet, P. (1993). Leukaemia inhibitory factor is necessary for maintenance of haematopoietic stem cells and thymocyte stimulation. *Nature (London)* **363,** 361–364.

Fassler, R., and Meyer, M. (1995). Consequences of lack of beta 1 integrin gene expression in mice. *Genes Dev.* **9,** 1896–1908.

Ferrara, N., Carver-Moore, K., Chen, H., Dowd, M., Lu, L., O'Shea, K. S., Powell-Braxton, L., Hillan, K. J., and Moore, M. W. (1996). Heterozygous embryonic lethality induced by targeted inactivation of the VEGF gene. *Nature (London)* **380,** 439–442.

Ferrari, G., Cusella-De Angelis, G., Coletta, M., Paolucci, E., Stornaiuolo, A., Cossu, G., and Mavilio, F. (1998). Muscle regeneration by bone marrow-derived myogenic progenitors. *Science* **279,** 1528–1530.

Fleischman, R. A. (1993). From white spots to stem cells: The role of the Kit receptor in mammalian development. *Trends Genet.* **9,** 285–290.

Fraser, P. and Grosveld, F. (1998). Locus control regions, chromatin activation and transcription. *Curr. Opin. Cell Biol.* **10,** 361–365.

Fujiwara, Y., Browne, C. P., Cunniff, K., Goff, S. C., and Orkin, S. H. (1996). Arrested development of embryonic red cell precursors in mouse embryos lacking transcription factor GATA-1. *Proc. Natl. Acad. Sci. U.S.A.* **93,** 12355–12358.

Garcia-Porrero, J. A., Godin, I. E., and Dieterlen-Lievre, F. (1995). Potential intraembryonic hemogenic sites at pre-liver stages in the mouse. *Anat. Embryol.* **192,** 425–435.

Godin, I. E., Garcia-Porrero, J. A., Coutinho, A., Dieterlen-Lievre, F., and Marcos, M. A. (1993). Para-aortic splanchnopleura from early mouse embryos contains B1a cell progenitors. *Nature (London)* **364,** 67–70.

Godin, I. E., Dieterlen-Lievre, F., and Cumano, A. (1995). Emergence of multipotent hemopoietic cells in the yolk sac and paraaortic splanchnopleura in mouse embryos, beginning at 8.5 days postcoitus. *Proc. Natl. Acad. Sci. U.S.A.* **92,** 773–777.

Godin, I. E., Garcia-Porrero, J. A., Dieterlen-Lievre, F., and Cumano, A. (1999). Stem cell emergence and hemopoietic activity are incompatible in mouse intraembryonic sites. *J. Exp. Med.* **190,** 43–52.

Goodell, M. A., Brose, K., Paradis, G., Conner, A. S., and Mulligan, R. C. (1996). Isolation and functional properties of murine hematopoietic stem cells that are replicating in vivo. *J. Exp. Med.* **183,** 1797–1806.

Grosveld, F., Dillon, N., and Higgs, D. (1993). The regulation of human globin gene expression. *In* "Baillière's Clinical Haematology: The Haemoglobinopathies" (D. R. Higgs and D. J. Weatherall, eds), Vol. 6, no. 1, pp. 31–66. Balliere Tindall, London.

Gussoni, E., Soneoka, Y., Strickland, C. D., Buzney, E. A., Khan, M. K., Flint, A. F., Kunkel, L. M., and Mulligan, R. C. (1999). Dystrophin expression in the mdx mouse restored by stem cell transplantation. *Nature (London)* **401,** 390–394.

Herzenberg, L. A., Stall, A. M., Lalor, P. A., Sidman, C., Moore, W. A., Parks, D. R., and Herzenberg, L. A. (1986). The Ly-1 B cell lineage. *Immunol. Rev.* **93,** 81–102.

Hu, M., Krause, D., Greaves, M., Sharkis, S., Dexter, M., Heyworth, C., and Enver, T. (1997). Multilineage gene expression precedes commitment in the hemopoietic system. *Genes Dev.* **11,** 774–785.

Huyhn, A., Dommergues, M., Izac, B., Croisille, L., Katz, A., Vainchenker, W., and Coulombel, L. (1995). Characterization of hematopoietic progenitors from human yolk sacs and embryos. *Blood* **86**, 4474–4485.

Ikuta, K., and Weissman, I. L. (1992). Evidence that hematopoietic stem cells express mouse c-kit but do not depend on steel factor for their generation. *Proc. Natl. Acad. Sci. U.S.A.* **89**, 1502–1506.

Ikuta, K., Kina, T., MacNeil, I., Uchida, N., Peault, B., Chien, Y. H., and Weissman, I. L. (1990). A developmental switch in thymic lymphocyte maturation potential occurs at the level of hematopoietic stem cells. *Cell (Cambridge, Mass.)* **62**, 863–874.

Jackson, K. A., Mi, T., and Goodell, M. A. (1999). Hematopoietic potential of stem cells isolated from murine skeletal muscle. *Proc. Natl. Acad. Sci. U.S.A.* **96**, 14482–14486.

Jaffredo, T., Gautier, R., Eichmann, A., and Dieterlen-Lievre, F. (1998). Intraaortic hemopoietic cells are derived from endothelial cells during ontogeny. *Development (Cambridge, UK)* **125**, 4575–4583.

Jaffredo, T., Gautier, R., Brajeul, V., and Dieterlen-Lievre, F. (2000). Tracing the progeny of the aortic hemangioblast in the avian embryo. *Dev. Biol.* **224**, 204–214

Johnson, G.R. and Jones, R.O. (1973). Differentiation of the mammalian hepatic primordium in vitro. I. Morphogenesis and the onset of haematopoiesis. *J. Embryol. Exp. Morphol.* **30**, 83–96.

Johnson, G. R., and Moore, M. A. (1975). Role of stem cell migration in initiation of mouse foetal liver haemopoiesis. *Nature (London)* **258**, 726–728.

Jordan, C. T., McKearn, J. P., and Lemischka, I. R. (1990). Cellular and developmental properties of fetal hematopoietic stem cells. *Cell (Cambridge, Mass.)* **61**, 953–963.

Kanatsu, M., and Nishikawa, S. I. (1996). *In vitro* analysis of epiblast tissue potency for hematopoietic cell differentiation. *Development (Cambridge, UK)* **122**, 823–830.

Kau, C. L., and Turpen, J. B. (1983). Dual contribution of embryonic ventral blood island and dorsal lateral plate mesoderm during ontogeny of hemopoietic cells in *Xenopus laevis. J. Immunol.* **131**, 2262–2266.

Kaufman, M. (1992). "The Atlas of Mouse Development." Academic Press, London.

Kelemen, E., Calvo, W., and Fliedner, T. (1979). "Atlas of Human Hemopoietic Development." Springer-Verlag, Berlin.

Keshet, E., Lyman, S. D., Williams, D. E., Anderson, D. M., Jenkins, N. A., Copeland, N. G., and Parada, L. F. (1991). Embryonic RNA expression patterns of the c-kit receptor and its cognate ligand suggest multiple functional roles in mouse development. *EMBO J.* **10**, 2425–2435.

Kimura, S., Roberts, A. W., Metcalf, D., and Alexander, W. S. (1998). Hematopoietic stem cell deficiencies in mice lacking c-Mpl, the receptor for thrombopoietin. *Proc. Natl. Acad. Sci. U.S.A.* **95**, 1195–1200.

Kondo, M., Weissman, I.L., and Akashi, K. (1997). Identification of clonogenic common lymphoid progenitors in mouse bone marrow. *Cell (Cambridge, Mass.)* **91**, 661–672.

Labastie, M. C., Cortes, F., Romeo, P. H., Dulac, C., and Peault, B. (1998). Molecular identity of hematopoietic precursor cells emerging in the human embryo. *Blood* **92**, 3624–3635.

Lemischka, I. R., Raulet, D. H., and Mulligan, R. C. (1986). Developmental potential and dynamic behavior of hematopoietic stem cells. *Cell (Cambridge, Mass.)* **45**, 917–927.

Lieschke, G. J., Stanley, E., Grail, D., Hodgson, G., Sinickas, V., Gall, J. A., Sinclair, R. A., and Dunn, A. R. (1994). Mice lacking both macrophage- and granulocyte-macrophage colony- stimulating factor have macrophages and coexistent osteopetrosis and severe lung disease. *Blood* **84**, 27–35.

Lin, H. H., Sternfeld, D. C., Shinpock, S. G., Popp, R. A., and Mucenski, M. L. (1996). Functional analysis of the c-myb proto-oncogene. *Curr. Top. Microbiol. Immunol.* **211**, 79–87.

Liu, P., Tarle, S. A., Hajra, A., Claxton, D. F., Marlton, P., Freedman, M., Siciliano, M. J., and Collins, F. S. (1993). Fusion between transcription factor CBF beta/PEBP2 beta and a myosin heavy chain in acute myeloid leukemia. *Science* **261**, 1041–1044.

Mackarehtschian, K., Hardin, J. D., Moore, K. A., Boast, S., Goff, S. P., and Lemischka, I. R. (1995). Targeted disruption of the flk2/flt3 gene leads to deficiencies in primitive hematopoietic progenitors. *Immunity* **3**, 147–161.

Maeno, M., Tochinai, S., and Katagiri, C. (1985). Differential participation of ventral and dorsolateral mesoderms in the hemopoiesis of Xenopus, as revealed in diploid-triploid or interspecific chimeras. *Dev. Biol.* **110**, 503–508.

Manaia, A., Lemarchandel, V., Klaine, M., Max-Audit, I., Romeo, P., Dieterlen-Lievre, F., and Godin, I. (2000). Lmo2 and GATA-3 associated expression in intraembryonic hemogenic sites. *Development (Cambridge, UK)* **127**, 643–653.

Marshall, C. J., Moore, R. L., Thorogood, P., Brickell, P. M., Kinnon, C., and Thrasher, A. J. (1999). Detailed characterization of the human aorta-gonad-mesonephros region reveals morphological polarity resembling a hematopoietic stromal layer. *Dev. Dyn.* **215**, 139–147.

Martin, C., Beaupain, D., and Dieterlen-Lievre, F. (1978). Developmental relationships between vitelline and intra-embryonic haemopoiesis studied in avian 'yolk sac chimaeras'. *Cell Differ.* **7**, 115–130.

Mead, P. E., Brivanlou, I. H., Kelley, C. M., and Zon, L. I. (1996). BMP-4-responsive regulation of dorsal-ventral patterning by the homeobox protein Mix.1. *Nature (London)* **382**, 357–360.

Medvinsky, A. L., and Dzierzak, E. A. (1996). Definitive hematopoiesis is autonomously initiated by the AGM region. *Cell (Cambridge, Mass.)* **86**, 897–906.

Medvinsky, A. L., Samoylina, N. L., Muller, A. M., and Dzierzak, E. A. (1993). An early pre-liver intraembryonic source of CFU-S in the developing mouse. *Nature (London)* **364**, 64–67.

Medvinsky, A. L., Gan, O. I., Semenova, M. L., and Samoylina, N. L. (1996). *Development (Cambridge, UK)* of day-8 colony-forming unit-spleen hematopoietic progenitors during early murine embryogenesis: spatial and temporal mapping. *Blood* **87**, 557–566.

Metcalf, D. (1984). "The Hemopoietic Colony Stimulating Factors." Elsevier, Amsterdam.

Metcalf, D. (1998). The molecular control of hematopoiesis: Progress and problems with gene manipulation. Stem Cells **16**, 314–321.

Miyoshi, H., Shimizu, K., Kozu, T., Maseki, N., Kaneko, Y., and Ohki, M. (1991). t(8;21) breakpoints on chromosome 21 in acute myeloid leukemia are clustered within a limited region of a single gene, AML1. *Proc. Natl. Acad. Sci. U.S.A.* **88**, 10431–10434.

Moore, M. A., and Metcalf, D. (1970). Ontogeny of the haemopoietic system: Yolk sac origin of in vivo and in vitro colony forming cells in the developing mouse embryo. *Br. J. Haematol.* **18**, 279–296.

Moore, M. A., and Owen, J. J. (1967). Experimental studies on the development of the thymus. *J. Exp. Med.* **126**, 715–726.

Motro, B., van der Kooy, D., Rossant, J., Reith, A., and Bernstein, A. (1991). Contiguous patterns of c-kit and steel expression: Analysis of mutations at the W and Sl loci. *Development (Cambridge, UK)* **113**, 1207–1221.

Mucenski, M. L., McLain, K., Kier, A. B., Swerdlow, S. H., Schreiner, C. M., Miller, T. A., Pietryga, D. W., Scott, W. J., Jr., and Potter, S. S. (1991). A functional c-myb gene is required for normal murine fetal hepatic hematopoiesis. *Cell (Cambridge, Mass.)* **65**, 677–689.

Muller, A. M., Medvinsky, A., Strouboulis, J., Grosveld, F., and Dzierzak, E. (1994). Development of hematopoietic stem cell activity in the mouse embryo. *Immunity* **1**, 291–301.

Murray, P. (1932). The development in vitro of the blood of the early chick embryo. *Proc. R. Soc. London* **11**, 497–521.

Naito, M. (1993). Macrophage heterogeneity in development and differentiation. *Arch. Histol. Cytol.* **56**, 331–351.

Naito, M., Umeda, S., Yamamoto, T., Moriyama, H., Umezu, H., Hasegawa, G., Usuda, H., Shultz, L. D., and Takahashi, K. (1996). Development, differentiation, and phenotypic heterogeneity of murine tissue macrophages. *J. Leukocyte Biol.* **59**, 133–138.

Nilsson, S. K., Lieschke, G. J., Garcia-Wijnen, C. C., Williams, B., Tzelepis, D., Hodgson, G., Grail, D., Dunn, A. R., and Bertoncello, I. (1995). Granulocyte-macrophage colony-stimulating factor is not responsible for the correction of hematopoietic deficiencies in the maturing op/op mouse. *Blood* **86,** 66–72.

Nishikawa, S. I., Nishikawa, S., Kawamoto, H., Yoshida, H., Kizumoto, M., Kataoka, H., and Katsura, Y. (1998). *In vitro* generation of lymphohematopoietic cells from endothelial cells purified from murine embryos. *Immunity* **8,** 761–769.

North, T., Gu, T.-L., Stacy, T., Wang, Q., Howard, L., Binder, M., Marin-Padilla, M., and Speck, N. (1999). Cbf__ is required for the formation of intraaortic hematopoietic clusters. *Development (Cambridge, UK)* **126,** 2563–2575.

Nutt, S. L., Heavey, B., Rolink, A. G., and Busslinger, M. (1999). Commitment to the B-lymphoid lineage depends on the transcription factor Pax5. *Nature (London)* **401,** 556–562.

Ody, C., Vaigot, P., Quere, P., Imhof, B. A., and Corbel, C. (1999). Glycoprotein IIb-IIIa is expressed on avian multilineage hematopoietic progenitor cells. *Blood* **93,** 2898–2906.

Ogawa, M., Nishikawa, S., Ikuta, K., Yamamura, F., Naito, M., Takahashi, K., and Nishikawa, S. (1988). B cell ontogeny in murine embryo studied by a culture system with the monolayer of a stromal cell clone, ST2: B cell progenitor develops first in the embryonal body rather than in the yolk sac. *EMBO J.* **7,** 1337–1343.

Okuda, T., van Deursen, J., Hiebert, S. W., Grosveld, G., and Downing, J. R. (1996). AML1, the target of multiple chromosomal translocations in human leukemia, is essential for normal fetal liver hematopoiesis. *Cell (Cambridge, Mass.)* **84,** 321–330.

Orkin, S.H. (1995). Regulation of globin gene expression in erythroid cells. *Eur. J. Biochem.* **15,** 271–281.

Palis, J., Robertson, S., Kennedy, M., Wall, C., and Keller, G. (1999). Development of erythroid and myeloid progenitors in the yolk sac and embryo proper of the mouse. *Development (Cambridge, UK)* **126,** 5073–5084.

Pardanaud, L., and Dieterlen-Lievre, F. (1999). Manipulation of the angiopoietic/hemangiopoietic commitment in the avian embryo. *Development (Cambridge, UK)* **126,** 617–627.

Pardanaud, L., Altmann, C., Kitos, P., Dieterlen-Lievre, F., and Buck, C. (1987). Vasculogenesis in the early quail blastodisc as studied with a monoclonal antibody recognizing endothelial cells. *Development (Cambridge, UK)* **100,** 339–349.

Pardanaud, L., Yassine, F., and Dieterlen-Lievre, F. (1989). Relationship between vasculogenesis, angiogenesis and haemopoiesis during avian ontogeny. *Development (Cambridge, UK)* **105,** 473–485.

Pardanaud, L., Luton, D., Prigent, M., Bourcheix, L. M., Catala, M., and Dieterlen-Lievre, F. (1996). Two distinct endothelial lineages in ontogeny, one of them related to hemopoiesis. *Development (Cambridge, UK)* **122,** 1363–1371.

Petrenko, O., Beavis, A., Klaine, M., Kittappa, R., Godin, I., and Lemischka, I. R. (1999). The molecular characterization of the fetal stem cell marker AA4. *Immunity* **10,** 691–700.

Pevny, L., Simon, M. C., Robertson, E., Klein, W. H., Tsai, S. F., V, D. A., Orkin, S. H., and Costantini, F. (1991). Erythroid differentiation in chimaeric mice blocked by a targeted mutation in the gene for transcription factor GATA-1. *Nature (London)* **349,** 257–260.

Phillips, R. L., Ernst, R. E., Brunk, B., Ivanova, N., Mahan, M. A., Deanehan, J. K., Moore, K. A., Overton, G. C., and Lemischka, I. R. (2000). The genetic program of hematopoietic stem cells. *Science* **288,** 1635–1640.

Pittenger, M. F., Mackay, A. M., Beck, S. C., Jaiswal, R. K., Douglas, R., Mosca, J. D., Moorman, M. A., Simonetti, D. W., Craig, S., and Marshak, D. R. (1999). Multilineage potential of adult human mesenchymal stem cells. *Science* **284,** 143–147.

Porcher, C., Swat, W., Rockwell, K., Fujiwara, Y., Alt, F. W., and Orkin, S. H. (1996). The T cell leukemia oncoprotein SCL/tal-1 is essential for development of all hematopoietic lineages. *Cell (Cambridge, Mass.)* **86,** 47–57.

Robb, L., Lyons, I., Li, R., Hartley, L., Kontgen, F., Harvey, R. P., Metcalf, D., and Begley, C. G. (1995). Absence of yolk sac hematopoiesis from mice with a targeted disruption of the scl gene. *Proc. Natl. Acad. Sci. U.S.A.* **92,** 7075–7079.

Robb, L., Elwood, N. J., Elefanty, A. G., Kontgen, F., Li, R., Barnett, L. D., and Begley, C. G. (1996). The scl gene product is required for the generation of all hematopoietic lineages in the adult mouse. *EMBO J.* **15,** 4123–4129.

Rowley, J. D. (1999). The role of chromosome translocations in leukemogenesis. *Semin. Hematol.* **36**(4, Suppl. 7), 59–72.

Russell, E. S. (1979). Hereditary anemias of the mouse: A review for geneticists. *Adv. Genet.* **20,** 357–459.

Russell, E. S., and Bernstein, S. E. (1966). Blood and blood formation. *In* "Biology of the Laboratory Mouse" (E. L. Green, ed.), pp. 351–372. McGraw-Hill, New York.

Sabin, F. (1920). Studies on the origin of blood vessels and of red blood corpuscles as seen in the living blastoderm of chicks during the second day of incubation. *Carnegie Inst. Washington Publ.* **272** *Contrib. Embryol.* **9,** 214.

Sasaki, K., Yagi, H., Bronson, R. T., Tominaga, K., Matsunashi, T., Deguchi, K., Tani, Y., Kishimoto, T., and Komori, T. (1996). Absence of fetal liver hematopoiesis in mice deficient in transcriptional coactivator core binding factor beta. *Proc. Natl. Acad. Sci. U.S.A.* **93,** 12359–12363.

Shalaby, F., Rossant, J., Yamaguchi, T. P., Gertsenstein, M., Wu, X. F., Breitman, M. L., and Schuh, A. C. (1995). Failure of blood-island formation and vasculogenesis in Flk-1-deficient mice. *Nature (London)* **376,** 62–66.

Shalaby, F., Ho, J., Stanford, W. L., Fischer, K. D., Schuh, A. C., Schwartz, L., Bernstein, A., and Rossant, J. (1997). A requirement for Flk1 in primitive and definitive hematopoiesis and vasculogenesis. *Cell (Cambridge, Mass.)* **89,** 981–990.

Shivdasani, R. A., Mayer, E. L., and Orkin, S. H. (1995). Absence of blood formation in mice lacking the T-cell leukaemia oncoprotein tal-1/SCL. *Nature (London)* **373,** 432–434.

Sinclair, A. M., and Dzierzak, E. A. (1993). Cloning of the complete Ly-6E.1 gene and identification of DNase I hypersensitive sites corresponding to expression in hematopoietic cells. *Blood* **82,** 3052–3062.

Smith, J. C., and Albano, R. M. (1993). Mesoderm induction and erythroid differentiation in early vertebrate development. *Semin. Dev. Biol.* **4,** 315–324.

Spangrude, G. J., Heimfeld, S., and Weissman, I. L. (1988). Purification and characterization of mouse hematopoietic stem cells. *Science* **241,** 58–62.

Stennard F. R. K., and Gurdon, J. B. (1997). Markers of vertebrate mesoderm induction. *Curr. Opin. Genet. Dev.* **7,** 620–627.

Stephens, L. E., Sutherland, A. E., Klimanskaya, I. V., Andrieux, A., Meneses, J., Pedersen, R. A., and Damsky, C. H. (1995). Deletion of beta1 integrins in mice results in inner cell mass failure and peri-implanatation lethality. *Genes Dev.* **9,** 1883–1895.

Tavian, M., Coulombel, L., Luton, D., Clemente, H. S., Dieterlen-Lievre, F., and Peault, B. (1996). Aorta-associated CD34+ hematopoietic cells in the early human embryo. *Blood* **87,** 67–72.

Tavian, M., Hallais, M. F., and Peault, B. (1999). Emergence of intraembryonic hematopoietic precursors in the pre-liver human embryo. *Development (Cambridge, UK)* **126,** 793–803.

Tepass, U., Fessler, L. I., Aziz, A., and Hartenstein, V. (1994) Embryonic origin of hemocytes and their relationship to cell death in *Drosophila. Development (Cambridge, UK)* **120,** 1829–1837.

Theele, D. P., and Streit, W. J. (1993). A chronicle of microglial ontogeny. *Glia* **7,** 5–8.

Toles, J. F., Chui, D. H., Belbeck, L. W., Starr, E., and Barker, J. E. (1989). Hemopoietic stem cells in murine embryonic yolk sac and peripheral blood. *Proc. Natl. Acad. Sci. U.S.A.* **86,** 7456–7459.

Tracey, W. D., Jr., Pepling, M. E., Horb, M. E., Thomsen, G. H., and Gergen, J. P. (1998). A Xenopus homologue of aml-1 reveals unexpected patterning mechanisms leading to the formation of embryonic blood. *Development (Cambridge, UK)* **125,** 1371–1380.

Tsai, F. Y., Keller, G., Kuo, F. C., Weiss, M., Chen, J., Rosenblatt, M., Alt, F. W., and Orkin, S. H. (1994). An early haematopoietic defect in mice lacking the transcription factor GATA-2. *Nature (London)* **371,** 221–226.

Turpen, J. B., and Knudson, C. M. (1982). Ontogeny of hematopoietic cells in Rana pipiens: precursor cell migration during embryogenesis. *Dev. Biol.* **89,** 138–151.

Turpen, J. B., Knudson, C. M., and Hoefen, P. S. (1981). The early ontogeny of hematopoietic cells studied by grafting cytogenetically labeled tissue anlagen: Localization of a prospective stem cell compartment. *Dev. Biol.* **85,** 99–112.

Turpen, J. B., Kelley, C. M., Mead, P. E., and Zon, L. I. (1997). Bipotential primitive-definitive hematopoietic progenitors in the vertebrate embryo. *Immunity* **7,** 325–334.

Van der Loo, J. C. M., Slieker, W. A. T., Kieboom, D., Ploemacher, R. E. (1995) Identification of hematopoietic stem cell subsets on the basis of their primitiveness using antibody ER-MP12. *Blood* **85,** 952–962.

van Ewijk, W., Hollander, G., Terhorst, C., and Wang, B. (2000). Stepwise development of thymic microenvironments in vivo is regulated by thymocyte subsets. *Development (Cambridge, UK)* **127,** 1583–1591.

Visvader, J. E., Fujiwara, Y., and Orkin, S. H. (1998). Unsuspected role for the T-cell leukemia protein SCL/tal-1 in vascular development. *Genes Dev.* **12,** 473–479.

Wang, J. H., Nichogiannopoulou, A., Wu, L., Sun, L., Sharpe, A. H., Bigby, M., and Georgopoulos, K. (1996a). Selective defects in the development of the fetal and adult lymphoid system in mice with an Ikaros null mutation. *Immunity* **5,** 537–549.

Wang, Q., Stacy, T., Binder, M., Marin-Padilla, M., Sharpe, A. H., and Speck, N. A. (1996a). Disruption of the Cbfa2 gene causes necrosis and hemorrhaging in the central nervous system and blocks definitive hematopoiesis. *Proc. Natl. Acad. Sci. U.S.A.* **93,** 3444–3449.

Wang, Q., Stacy, T., Miller, J. D., Lewis, A. F., Gu, T. L., Huang, X., Bush-weller, J. H., Bories, J. C., Alt, F. W., Ryan, G., Liu, P. P., Wynshaw-Boris, A., Binder, M., Marin-Padilla, M., Sharpe, A. H., and Speck, N. A. (1996b). The CBFbeta subunit is essential for CBFalpha2 (AML1) function in vivo. *Cell (Cambridge, Mass.)* **87,** 697–708.

Warren, A. J., Colledge, W. H., Carlton, M. B., Evans, M. J., Smith, A. J., and Rabbitts, T. H. (1994). The oncogenic cysteine-rich LIM domain protein rbtn2 is essential for erythroid development. *Cell (Cambridge, Mass.)* **78,** 45–57.

Weissman, I., Papaioannou, V., and Gardner, R. (1978). Fetal hematopoietic origins of the adult hematolymphoid system. *In* "Differentiation of Normal and Neoplastic Hematopoietic Cells" (B. Clarkson, P. A. Marks, and J. E. Till, eds), Conf. Cell Proliferation, Vol. 5, pp. 33–47. Cold Spring Harbor Lab., Cold Spring Harbor, NY.

Wijgerde, M., Grosveld, F., and Fraser, P. (1995). Transcription complex stability and chromatin dynamics in vivo. *Nature (London)* **21,** 209–213.

Winnier, G., Blessing, M., Labosky, P. A., and Hogan, B. L. (1995). Bone morphogenetic protein-4 is required for mesoderm formation and patterning in the mouse. *Genes Dev.* **9,** 2105–2116.

Wood, H. B., May, G., Healy, L., Enver, T., and Morriss-Kay, G. M. (1997). CD34 expression patterns during early mouse development are related to modes of blood vessel formation and reveal additional sites of hematopoiesis. *Blood* **90,** 2300–2311.

Yamada, Y., Warren, A. J., Dobson, C., Forster, A., Pannell, R., and Rabbitts, T. H. (1998). The T cell leukemia LIM protein Lmo2 is necessary for adult mouse hematopoiesis. *Proc. Natl. Acad. Sci. U.S.A.* **95,** 3890–3895.

Yoder, M. C., Hiatt, K., Dutt, P., Mukherjee, P., Bodine, D. M., and Orlic, D. (1997). Characterization of definitive lymphohematopoietic stem cells in the day 9 murine yolk sac. *Immunity* **7,** 335–344.

Young, P. E., Baumhueter, S., and Lasky, L. A. (1995). The sialomucin CD34 is expressed on hematopoietic cells and blood vessels during murine development. *Blood* **85,** 96–105.

Zon, L. I. (1995). Developmental biology of hematopoiesis. *Blood* **86,** 2876–2891.

11

Vasculogenesis and Angiogenesis

Thomas N. Sato and Siobhan Loughna

The University of Texas Southwestern Medical Center at Dallas, Dallas, Texas 75390

I. Introduction

II. Overview of Vascular Development

III. Generation of Endothelial Cells

IV. Vascular Morphogenesis

V. Concluding Remarks

References

I. Introduction

Formation of the vascular system is recognized as one of the most important events in development and has been a subject of intensive investigations. Although modern analyses of vascular development date back to the beginning of the 20th century, the primary focus was to provide morphological description of the process (Evans, 1909; His, 1900; Sabin, 1917). Two sequential key morphogenic processes underlie vascular development: vasculogenesis (Box 1) and angiogenesis (Box 2). Recently, significant advancement in our understanding of molecular mechanisms for these processes, as well as the emergence of new concepts, has been revolutionizing the field. Application of many mouse genetics tools have played a pivotal role in this recent revolution of the field. In this chapter, we discuss many of the key principles of vascular development with a primary emphasis on the studies accomplished by the use of contemporary mouse genetics and embryological tools.

II. Overview of Vascular Development

Development of virtually all organs is heavily dependent on sufficient nutrient feeding and oxygenation, processes primarily accomplished by the vascular system during embryogenesis. Thus, the normal patterned formation of a functional vascular system is one of the most critical and earliest events to occur during embryogenesis.

The sequences leading to the formation of the vascular system have been classically divided into multiple phases (Fig. 1). The first vascular structure can be identified as early as the gastrulation stage. At this embryonic stage, a subset of resident mesodermal cells differentiates to endothelial cells, which assemble to form an initial vascular network. This vascular network is referred to as the primary capillary plexus (Risau and Flamme, 1995). At this early stage, the entire vascular network is composed of primarily one cell type, the endothelial cell. In both the yolk sac and embryo, subsets of mesodermal cells differentiate to endothelial cells and form cell clusters referred to as *blood islands* (Box 1). These clusters subsequently become connected to each other to form an intricate network of vessels. This initial vessel network (i.e., primary capillary plexus) is characterized by its relatively uniform honeycomb-like capillary channel network (Fig. 1). This process of primary capillary plexus formation from *in situ* differentiating endothelial cells is referred to as *vasculogenesis* (Box 1).

Subsequently, the primary capillary plexus expands by forming additional branches and remodeling its network

Box 1: Vasculogenesis and Blood Islands

Vasculogenesis defines the first morphogenic process during vascular development. Clusters of mesoderm-derived angioblastic cells differentiate to blood and endothelial cells, to form a structure referred to as a *blood island*. In this structure, clusters of blood cells are surrounded by a single layer of endothelial cells. The endothelial cells of these blood islands subsequently coalesce to form a number of initial vascular channels called the *primary capillary plexus*. This process, the formation of the primary capillary plexus, is referred to as *vasculogenesis*. The primary capillary plexus is characterized by a honeycomb-like network of vascular channels of uniform diameter. This vascular network can be clearly identified in the yolk sac and embryo, such as the cephalic region, by 8.0 dpc.

to assume its final form (Risau, 1997). During this later phase, the vessels form a more complex network and cells other than endothelial cells become involved in the process (Fig. 1). This later phase of vascular development is referred to as *angiogenesis* (Box 2). Two distinct mechanisms are proposed for the formation of additional vessel branches

Box 2: Angiogenesis

Angiogenesis was classically defined as a process involving sprouting from preexisting vessels, such as the primary capillary plexus. However, as we learned more details of vessel formation, this process was found to be far more complex than just sprouting of vessels. It has been proposed that the primary capillary plexus expands its network by both sprouting and nonsprouting processes. Furthermore, this newly expanded vascular network is suggested to undergo "pruning," "remodeling," and "maturation" to complete the whole process. At this point, the morphogenic processes following vasculogenesis seem to be complex, and are not critically well defined. Therefore, the authors would like to use the term *angiogenesis* to cover the entire vessel formation process following vasculogenesis including pruning, remodeling, and maturation.

from the primary capillary plexus: sprouting and nonsprouting (Fig. 2). In the sprouting process, an endothelial cell that is already part of a continuous vessel transforms to an elongated shape, invades nearby tissue, and forms an additional vascular channel from the existing one. In the nonsprouting process, surrounding cells of an existing vascular channel invade and intercept the vessel, thus leading to the splitting of one vascular channel into two. Combining both of these processes, the primary capillary plexus transforms to more complex vascular channels.

Finally, these primarily endothelial-based processes are further integrated with a process involving nonendothelial vascular cells during angiogenesis. Vascular channels formed by endothelial cells are now being invested by smooth muscle cells. Smooth muscle cells infiltrate the vascular channels to provide more rigid integrity to the vessels as well as contractility.

In many organs, the vessel phenotypes are further modified. Various organs require distinct morphological and phenotypical characteristics of the vessels in order to support specific developmental processes and physiological functions. This leads to significant variations and specifications among the vessels of different organs and parts of the body.

In the following sections, each process leading to a mature vessel is described in detail. Both cellular and molecular mechanisms are emphasized, with an in-depth discussion of some critical questions.

III. Generation of Endothelial Cells

As briefly outlined in Section II, endothelial cells are the primary cell type involved in the initial phases of vascular development. Generation of endothelial cells is regulated via two processes: differentiation and expansion (Fig. 3). By means of a differentiation program, the endothelial cell lineage is specified. By means of an expansion program, both proliferation and survival of progenitors and endothelial cells are tightly controlled to establish the normal size of the endothelial cell population. In this section, the mechanistic basis of these programs underlying endothelial cell generation is described in detail.

A. Endothelial Cells Are Mesodermal in Origin

All endothelial cells originate from mesoderm during embryogenesis (Figs. 1 and 3). In mice, at around 7.0–7.5 days postcoitus (dpc), the lateral mesoderm produces a population of progenitors called hemangioblasts (Box 3) from which endothelial cells are differentiated. Endothelial cells generated from hemangioblasts at this embryonic stage in mice contribute primarily to the vasculature of extraembryonic yolk sac membrane. Later (7.5–8.0 dpc), as embryonic structures becomes more discrete, scattered clusters

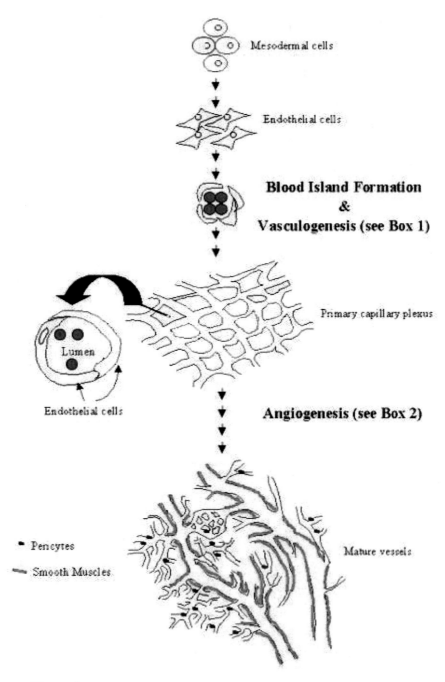

Figure 1   Overview of vascular development. See accompanying text for the description.

of hemangioblasts derived from intraembryonic mesodermal cells become identifiable, and these cells contribute to the intraembryonic vasculature.

B. Regulation of Endothelial Cell Lineage Specification

Molecular mechanisms of neither hemangioblast nor endothelial specification from multipotential mesodermal cells during mouse embryogenesis remain elusive at this time. However, studies with several zebrafish mutants provided some insight into the specific genetic pathways controlling this cell type specification process. A series of genetic studies in zebrafish resulted in an emerging paradigm defining how hemangioblast lineage may be established.

Zebrafish *cloche* is a mutation identified in zebrafish, and no endothelial or hematopoietic cells exist except in the dorsal tip region in this mutant (Stainier *et al.*, 1995). Further-

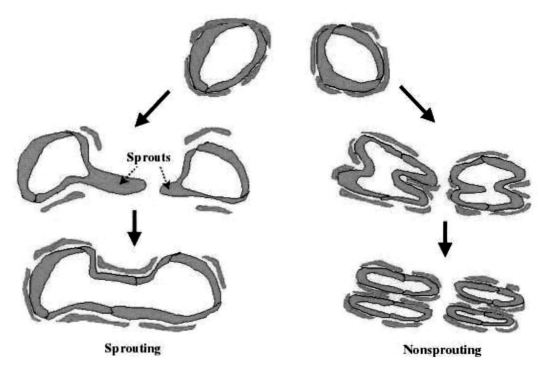

Figure 2   Sprouting and nonsprouting angiogenesis. There are two distinct mechanisms for angiogenesis (i.e., formation of new vessel channels): sprouting and nonsprouting. In sprouting angiogenesis, an endothelial cell (EC) transforms to an elongated shape and invades the immediately proximal space. This involves loosing interactions of surrounding nonendothelial cells such as mesenchymal cells, which normally function as supporting cells for EC. Each EC sprout eventually interacts with another EC sprout and forms a new vascular channel. In nonsprouting angiogenesis, surrounding mesenchymal cells push each EC inward, and the ECs eventually divide one vascular channel into two. These two processes are considered to be the main mechanisms for angiogenesis based on discontinuous and non-real-time histological analyses of vessels undergoing angiogenesis. Definitive description of cellular processes underlying angiogenesis waits for future documentation of angiogenesis in real time.

more, genetic studies in zebrafish showed that the *cloche* function is upstream to the expression of one of the earliest hemangioblast markers, VEGF-R2/flk-1 (*Kdr*) (Box 4) (Liao *et al.,* 1997). Therefore, these studies suggested that the putative *cloche* gene is critical for establishing hemangio-

blast lineage. *Cloche* was also shown to be upstream of two other genes critical for hemangioblast formation: *Scl* (*tal1*) (Box 5) and *Hhex* (Box 6) (E. C. Liao *et al.,* 1998; W. Liao *et al.,* 2000). In a *cloche* mutant, neither *Scl* nor *Hhex* expression is detected. Furthermore, lack of endothelial cells

---

Box 3: Hemangioblasts

A *hemangioblast* is defined as a common precursor cell for both the endothelial and hematopoietic lineages. The existence of this common precursor cell had been speculated based on the evidence from histological and cell lineage studies. In blood islands, endothelial and hematopoietic cells are clustered and colocalized. This observation suggested that these two lineages are derived from a common precursor cell type. In studies of chick embryos, it is also suggested that these two lineages originate from a common precursor cell. In addition, these two cell types share many marker genes, suggesting a close relationship between these two lineages. Fur-

thermore, recent studies have provided additional direct evidence to support the existence of a common precursor for these two lineages. *In vitro,* a precursor cell type derived from embryonic stem cells was isolated and shown to differentiate to both endothelial and hematopoietic cells. This was shown at the single clonal cell level. In addition, it was shown that the expression of VEGF-R1 (*Kdr*) in the mesodermal-derived cells seemed to identify this putative hemangioblast population. Interestingly, some evidence seems to support the notion that VEGF-R1[+] cells are multipotential, instead of bipotential.

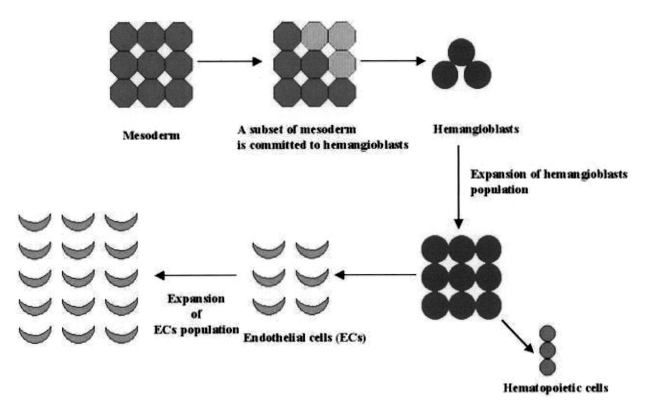

Figure 3 Multilevel regulation for establishing normal size of endothelial population. The final population size of ECs is determined at the multiple level. A subset of mesodermal cells is committed to the hemangioblast lineage, which determines the initial number of hemangioblasts. These hemangioblasts undergo an expansion process that is regulated by subtle balances between the regulatory mechanisms underlying EC survival and proliferation. As a subset of hemangioblasts differentiates to ECs, this process determines the initial number of endothelial cells. The final number of ECs is determined by the subtle balance between EC survival and proliferation signals.

in developing embryos can be rescued by forced expression of either *Scl* or *Hhex* in the *cloche* mutant. Interestingly, forced expression of *Scl* or *Hhex* also results in the expression of the other, suggesting a mutual coregulation of these two transcription factors.

Another potential regulator of endothelial cell lineage specification is bFGF (*Fgf2*). It has been suggested that bFGF plays a role in endothelial and hematopoietic cell differentiation, at least *in vitro* (Flamme and Risau, 1992). Uncommitted mesodermal cells in quail epiblasts were successfully induced to form both hematopoietic and endothelial cells. This was based on the staining of these cells with lineage-specific monoclonal antibodies. However, the precise differentiation stages of mesodermal cells stained by these antibodies were not clearly defined. Therefore, at present, it is difficult to define the mode of bFGF action for endothelial and hematopoietic differentiation. An alternative interpretation of this *in vitro* result is that there are cells already committed to differentiating to the hematopoietic and endothelial cells that do not stain with these lineage-specific antibodies in the epiblast preparation. In any event, further investigation is required to determine whether bFGF itself or other related factors are in fact inducer(s) for endothelial and hematopoietic cell differentiation during normal devel-

opment. It is also important to keep in mind that this is an *in vitro* system and therefore needs confirmation by *in vivo* analyses.

C. Regulation of Normal Endothelial Cell Number

The number of hemangioblasts affects the final size of endothelial cell population and is tightly controlled during vascular development. A reduced number of hemangioblasts (i.e., lack of sufficient number of endothelial cells) leads to undervascularization. The presence of too many hemangioblasts leads to abnormal vascular development.

One of the most extensively studied pathways controlling the endothelial population size is a VEGF receptor family (Box 4). Two receptors among this receptor family seem to be critical in controlling the number of endothelial cells, and they do so in an elegantly orchestrated manner. Hemangioblasts were found to express VEGF-R2 (*Kdr*), one of the receptors for vascular endothelial growth factor (VEGF, *Vegf*). Furthermore, the formation of hemangioblasts was found to be dependent on VEGF in a culture medium. However, VEGF-R2$^{-/-}$ ES cells were able to form hemangioblast

Box 4: VEGF and VEGF-R family

Vascular endothelial growth factor (VEGF) is a family of soluble glycoproteins. The first family member, VEGF-A, was originally isolated as vascular permeability factor (VPF) based on its activity to induce vascular leakiness. Subsequently, it was also shown to induce proliferation of cultured endothelial cells in a cell type specific manner. The gene for VEGF-A encodes three isoforms of VEGF-A, VEGF-164, VEGF-120, and VEGF-188 in mice. All of these isoforms are generated by alternative splicing of the same gene. VEGF-A164 is the most predominant form. Functional differences among these three isoforms remain to be determined. Subsequently, three additional VEGF family members have been identified based on high sequence homology. They are now referred to as VEGF-B, VEGF-C, and VEGF-D, and are all encoded by distinct genes. In addition, placenta growth factor (*Pgf*) is also included in this family as the fifth member. Recently, VEGF related gene was also discovered in Orf virus, a member of the poxvirus family that produces a pustular dermatitis in sheeps, goats, and humans.

VEGFs bind specifically to a family of cell-surface receptors, VEGF receptors (VEGF-Rs). Currently, there are three VEGF-Rs, VEGF-R1 *(Flt1)*, VEGF-R2 (*Kdr* for mouse gene and *KDR* for human gene) and VEGF-R3 *(Flt4)*. VEGF-Rs belong to a family of receptor tyrosine kinases. The extracellular portion of the receptors consists of seven immunoglobulin-like domains. This modular structure is also found in the platelet-derived growth factor (PDGF) receptor family. Overlapping but also distinct receptor-ligand binding specificity exists. VEGF-A binds specifically VEGF-R1 and VEGF-R2. VEGF-B binds to VEGF-R1. VEGF-C (*Vegfc*) binds to VEGF-R2 and VEGF-R3. VEGF-D (*Figf*) binds to VEGF-R2 and VEGF-R3. Recently, another receptor that is not related to VEGF-Rs was discovered and shown to interact with VEGF-A. This receptor, neuropilin-1, was originally implicated in neural growth-cone behavior, particularly growth-cone repulsion. Neuropilin-1 was shown to interact specifically with the VEGF-A164 isoform.

Box 5: *Scl/tal1*

The *Scl/tal1* gene was originally identified through its translocation in acute T-cell lymphoblastic leukemia. It encodes a basic helix–loop–helix transcription factor and is expressed specifically in hematopoietic and endothelial lineages, as well as in the developing brain. Null mutant embryos for the *Scl/tal1* gene exhibit defective erythropoiesis and *Scl/tal1*⁻/⁻ embryonic stem cells do not contribute to any hematopoietic lineages in chimeric mice. Therefore, it is proposed that the *Scl/tal1* function is critical for hematopoietic lineage regulation.

Rescue of the *Scl/tal1* function in the hematopoietic lineage was accomplished by expressing the wild-type *Scl/tal1* gene specifically in this lineage in a null background. Although the hematopoietic defect was completely rescued, these embryos exhibit abnormal yolk sac vascular development. Furthermore, *Scl/tal1*⁻/⁻ endothelial cells failed to contribute to developing yolk sac vessels. However, the observed defects were at the level of endothelial cell organization rather than endothelial cell differentiation.

In addition, forced expression of wild-type *Scl/tal1* in *Cloche* mutant zebrafish (see above) rescued both hematopoietic and endothelial defects in this mutant, suggesting that the *Scl/tal1* function lies downstream to the cloche function and is critical for establishing endothelial cell lineage. This study also implies that target genes for the *Scl/tal1* transcription factor may be critically involved in endothelial lineage establishment. Interestingly in this respect, it has also recently been shown that the function of *Scl/tal1* in regulating hematopoietic and endothelial lineages may be independent from its DNA-binding activity, indicating novel mechanisms of its function.

Box 6: *Hex*

Hex encodes a homeobox containing protein and was initially identified as a hematopoietically expressed homeobox gene. *Hex* was previously identified as being expressed in both hematopoietic and endothelial lineages. It has been shown that *Hex* works as a transcriptional repressor to regulate anterior and dorsoventral patterning. It is also shown that *Hex* functions in early vascular development in both *Xenopus* and zebrafish embryos.

colonies in a VEGF-dependent manner *in vitro* (Schuh *et al.,* 1999).

In this *in vitro* study, a unique system called *blast colony-forming cell (BL-CFC) assay* (Box 7) was used. In this assay, a unique clonal population of transitional precursor cells from the ES cell-derived embryoid bodies was identified. This clonal population of cells was induced to differentiate into both hematopoietic and endothelial cells when cultured in the presence of VEGF. This provided the first proof of the existence of a bipotential hemangioblast population at least *in vitro*.

In vivo, VEGF-R2$^{-/-}$ embryos exhibit complete lack of blood cells and endothelial cells (Shalaby *et al.,* 1995).

Box 7: Blast Colony-Forming Cell (BL-CFC) Assay

This is an *in vitro* assay system allowing the clonal analysis of hematopoietic and endothelial cell lineage specification. Embryonic stem cell-derived embryoid bodies are dissociated into a single-cell suspension and plated, allowing the formation of single-cell-derived colonies (i.e., blast colonies). Among these colonies, some represent a unique characteristic: They exhibit a potential to differentiate to both hematopoietic and endothelial lineages when cultured in the presence of VEGF. Therefore, this *in vitro* system, for the first time, allows for the identification of putative hemangioblasts. Furthermore, this *in vitro* system allows systematic analysis of regulatory mechanisms for endothelial and hematopoietic differentiation from a common precursor by testing various factors in media and deriving blast colonies from ES cells that have mutations of various genes.

VEGF-R2$^{-/-}$ ES cells were used in the study of chimeric embryos, and it was found that these mutant ES cells do not contribute to either endothelial or hematopoietic cells *in vivo* (Shalaby *et al.,* 1997). Based on these results, it has been proposed that VEGF-R2 is probably dispensable for hemangioblast differentiation. However, it may play a critical role in survival and expansion of the hemangioblast population. In addition, VEGF-R2 may be required for the directed migration of hemangioblasts to receive a cue(s) from the appropriate microenvironment, which seems to be essential for hematopoietic differentiation (Hidaka *et al.,* 1999; Shalaby *et al.,* 1997).

In addition to VEGF-R2, another VEGF receptor, VEGF-R1 (*Flt1*), has also been implicated in hemangioblast development. Lack of VEGF-R2 expression results in an increased number of endothelial cells both *in vitro* and *in vivo* (G.-H. Fong *et al.,* 1996; G.-H. Fong *et al.,* 1999). This phenotype was proposed to be a result of increased commitment of mesodermal cells to the hemangioblastic lineage (G.-H. Fong *et al.,* 1999). Interestingly, the lack of a cytoplasmic tyrosine kinase domain for VEGF-R1 does not seem to be critical for its function *in vivo* (Hiratsuka *et al.,* 1998). Thus, it was suggested that VEGF-R1 acts as a "VEGF sink" by binding to extracellular VEGF and making only the optimal concentration of VEGF available to the cells (G.-H. Fong *et al.,* 1999). This particular function of VEGF-R1 is mediated in a cell nonautonomous fashion (G.-H. Fong *et al.,* 1999).

Based on this model, VEGF is considered to be a hemangioblast differentiation inducer. However, *in vitro* differentiation studies have clearly shown that VEGF-R2$^{-/-}$ ES cells can be induced to form hemangioblasts, albeit less efficiently, in a VEGF-dependent manner (Schuh *et al.,* 1999). Therefore, it is possible that VEGF acts through another class of VEGF receptor to regulate hemangioblast formation. Alternatively, it is possible that VEGF acts through VEGF-R2 in normal hemangioblast formation *in vivo* and a putative compensatory mechanism is switched on when VEGF-R2 is absent *in vitro*. It is also possible that VEGF/VEGF-R2 function is to regulate migration and expansion of hemangioblasts, rather than the formation of these cells. Interestingly, aberrant migration of the VEGF-R2$^{-/-}$ ES cells that would normally have expressed this receptor was observed *in vivo* (Shalaby *et al.,* 1997). This suggested that VEGF-R2 controls the directed migration of such cells. This migration enables these cells to receive an appropriate signal from the microenvironment that is necessary for their differentiation. With respect to the possibility that VEGF may regulate the expansion of the hemangioblast population, one needs to also consider the fact that the regulation of hemangioblast formation via VEGF-R1 is independent from cell proliferation and survival.

Clearly, further studies are required to fully understand how endothelial lineage specification and expansion are orchestrated by VEGF and VEGF receptors. Furthermore, it would be essential to understand how the VEGF pathway interacts with other pathways such as Cloche and bFGF in determining the final size of the endothelial population.

D. Diversity of Endothelial Cells

Recently, the existence of many molecularly distinct subclasses of endothelial cells has become evident. Some of them are genetically preprogrammed and the phenotypes of the others are determined by their microenvironment. In this section, representative classes of endothelial cells are described.

1. Arterial versus Venous Endothelial Cells

Although blood vessels have been conventionally classified as arteries or veins based primarily on physiological parameters, it has recently become clear that this distinction between arteries and veins is genetically encoded (Gerety *et al.*, 1999; Wang *et al.*, 1998). Furthermore, this distinction seems to be already established at the very beginning of blood vessel formation during embryogenesis (Gerety *et al.*, 1999; Wang *et al.*, 1998).

It has been found that arterial and venous endothelial cells can be defined by their specific expression of ephrin-B2 (*Efnb2*) and EphB4 (*Ephb4*), respectively (Box 8) (Gerety *et al.*, 1999; Wang *et al.*, 1998). This distinctive expression pattern was detected early, when initial blood vessel formation begins during embryogenesis. This indicates that a molecular distinction between arterial and venous endothelial cells is established during the initial phase of blood vessel formation.

Box 8: Eph and Ephrin Family in Vascular Development

Eph is a class of transmembrane receptors that belongs to a family of receptor tyrosine kinases. The extracellular portion of the receptor consists of an N-terminal globular domain, a cysteine-rich region, and two fibronectin type III domains. The intracellular portion has intrisic tyrosine kinase activity. Ephrins are a family of specific ligands for the Eph receptors. Ephrins are expressed on the cell surface, either by GPI linkage to the membrane or transmembrane domain. This ligand-receptor family has been known to critically regulate developmental processes such as axon guidance and neural cell migration. Recently, some members of this ligand-receptor family were shown to participate in vascular development.

Although this surprising finding is fascinating, many important questions remain unsolved. Assuming that the initial endothelial cells immediately following the differentiation of hemangioblasts do not exhibit a distinctive expression pattern of ephrin-B2 and EphB4, how is this expression pattern subsequently established? Are there any local cues that turn on the expression of ephrin-B2 and/or EphB4? What are the mechanisms underlying this mutually exclusive expression pattern? Alternatively, it is possible that all of the initial endothelial cells generated from hemangioblasts already express either ephrin-B2 or EphB4, with subsets subsequently expressing the other gene. For example, all of the initial endothelial cells may express EphB4 (i.e., they are, in essence, venous type) and, later, subsets of them begin to express ephrin-B2 and assume an arterial phenotype while the remainder continue to express EphB4 to maintain a venous phenotype. It is also possible that there are two distinct sets of hemangioblasts, one expressing ephrin-B2 and the other expressing EphB4. Distinctive expression pattern of ephrin-B2 and EphB4 seems to persist following the completion of blood vessel formation. Therefore, it would be of interest to investigate the mechanisms that underlie the maintenance of this mutually exclusive expression pattern and its relationship to the initial induction of the expression of such markers.

2. Lymphatic Endothelial Cells

Most of the studies of vascular development have focused on blood vessels. However, another circulatory system, the lymphatic vascular system, provides important vessels for cellular waste drainage and circulation of lymphocytes.

VEGF-R3 (*Flt4*) was found to be expressed in venous endothelial cells during early embryonic development, but the expression becomes gradually restricted to lymphatic endothelial cells as the lymphatic vasculature is formed (Kaipainen *et al.*, 1995). This expression study suggested a close lineage relationship between venous and lymphatic endothelial cells.

Furthermore, one of the specific ligands for VEGF-R3, VEGF-C (*Vegfc*), was shown to induce lymphatic angiogenesis when ectopically overexpressed *in vivo* (Jeltsch *et al.*, 1997). This finding indicates that the VEGF-C/VEGF-R3 pathway may be important for lymphatic vessel development. However, VEGF-R3[−/−] embryos die before the first lymphatic vessels form, precluding a possibility to test the potential involvement of this ligand-receptor system in lymphatic vessel formation (Dumont *et al.*, 1998). Future development and analysis of conditional knock-outs for these genes may allow us to test this possibility directly.

More recently, it has been found that one of the homeobox genes, *Prox1*, is also expressed by developing lymphatic endothelial cells in the embryo (Wigle and Oliver, 1999). *Prox1*[−/−] embryos exhibit significant retardation of the lymphatic vessel formation, suggesting that this putative transcription factor may be involved critically in this process (Wigle and Oliver, 1999).

3. Organ-Specific Endothelial Types

As blood vessels form and mature, it is essential that they meet the specific demands of the particular organs being vascularized. It is speculated that this process requires very sophisticated cell communications between vascular cells and surrounding nonvascular organ-specific cells. Therefore, each organ presumably requires a specific input from its vasculature that exhibits a unique structure and physiological function. Although this is an important aspect of vascular development, very little information is available.

a. Brain. Brain vascular endothelial cells form complex tight junctions and also develop specialized transporter systems (Risau *et al.,* 1986a,b; Risau and Wolburg, 1990). These structural and biochemical barriers formed by the brain vascular endothelial cells permit the selective trafficking of chemicals between the circulation and nervous system. Therefore, this unique blood–brain barrier (BBB) protects the nervous system against toxic insults from the circulation. This barrier system has important physiological and therapeutic implications. However, very few molecular mechanisms underlying BBB formation or maintenance are known. *In vitro,* coculturing endothelial cells with astrocytes was shown to induce the tight-junction phenotype of endothelial cells found at the BBB (Neuhaus *et al.,* 1991). Obviously, further studies are necessary to decipher the unique molecular nature of endothelial cells at the BBB.

b. Heart. The heart may require a specialized circulation that is essential for its physiological functions. This requirement may be met by reciprocal cellular and chemical communications between cardiac vessel endothelial cells and the myocardium. While this area has obvious therapeutic importance, it has not been explored extensively at a molecular level. One study has suggested that cardiac vessel endothelial cells are molecularly distinct from other endothelial cell types. A part of the upstream promoter sequences of the von Willebrand factor (*Vwf*) gene was shown to confer cardiac vessel endothelial specific expression when studied in transgenic mice (Aird *et al.,* 1997). This putative unique cardiac vessel endothelial expression was shown to be regulated, in part, by the PDGF-B (*Pdgfb*) pathway (Edelberg *et al.,* 1998).

Sufficient evidence has recently accumulated to further suggest that the distinct nature of cardiac vessel endothelial cell type reflects its unique developmental origin. Chimera and retroviral cell lineage analyses in chick embryos identified epicardial cells as the origin of many of the coronary vessel endothelial cells (Dettman *et al.,* 1998; Mikawa and Fischman, 1992; Mikawa and Gourdie, 1996; Perez-Pomares *et al.,* 1998; Vrancken Peeters *et al.,* 1999). During heart development, a subset of epicardial cells immediately surrounding the outside of the myocardium migrates into the subepicardial space, a space between the epicardium and myocardium. Following the migration, these epicardial-derived cells undergo epithelial-mesenchymal transformation, and subsequently differentiate to hemangioblasts, endothelial cells, and smooth muscle cells, all of which contribute to coronary vessel formation. In mice, this principle seems to hold true as knock-out mice for genes such as *VCAM1* (*Vcam1*), α_4 integrin (*Itga4*), and erythropoietin (*Epo*) exhibit the failure of epicardial development, resulting in the lack of a coronary vasculature (Kwee *et al.,* 1995; Wu *et al.,* 1999; Yang *et al.,* 1995).

Recently, a novel cofactor of GATA transcription factors, FOG-2 (*Zfpm2*), was also shown to be critical for the formation of coronary vessels (Tevosian *et al.,* 2000). In the FOG-2$^{-/-}$ embryo, no coronary vessels (neither endothelial nor smooth muscle cells) were formed. Because epicardium formation in this knock-out embryo is normal, it is suspected that FOG-2 plays a critical role in the epithelial-mesenchymal transformation of the epicardial cells. Interestingly, forced expression of FOG-2 in the myocardium was found to rescue the coronary vessel phenotype in the FOG-2$^{-/-}$. Therefore, it is proposed that FOG-2, together with GATA factors, regulates the expression of a set of genes whose functions are important for paracrine regulation between epicardial cells and myocardial cells to regulate the epithelial-mesenchymal transformation.

c. Eye. Some of the vascular beds form only transiently during development and will eventually regress. Prominent examples are the hyaloid and papillary membrane vessels of the developing eye. These vessels form during embryonic stages but regress postnatally. While the mechanisms underlying regression of these vessels are not clear, it has been suggested that macrophages may participate in this process (Diez-Roux *et al.,* 1999; Lang and Bishop, 1993; Meeson *et al.,* 1996, 1999).

In one experiment, macrophages were specifically ablated by directing diphtheria toxin expression by using macrophage-specific promoter elements in transgenic mice (Lang and Bishop, 1993). In these mice, persistent hyaloid arteries were observed, suggesting that macrophages are required for normal hyaloid artery regression.

4. Hemogenic Endothelial Cells

Hemogenic endothelial cells (see Chapter 10) represent a unique class of progenitors (Smith and Glomski, 1982). It has been suggested that some of the hematopoietic cells in embryos originate from endothelial cells. These hematopoietic cells originate from endothelial cells located at specialized vascular beds, such as the ventral portion of the dorsal aortae, the aorta-genital ridge-mesonephros (AGM) region, and umbilical vessels (Dieterlen-Lievre and Martin, 1981;

Godin *et al.,* 1995; Medvinsky and Dzierzak, 1996; Medvinsky *et al.,* 1993; Tavian *et al.,* 1996). In these vascular beds, subsets of the endothelial cells were shown to "bud out" and become hematopoietic cells (Smith and Glomski, 1982).

One of the most critical genes in this process was recently discovered. *Cbfa2* and *Cbfb* encode two subunits of core-binding factor (CBF), which is required for definitive hematopoiesis during an early embryonic stage (Niki *et al.,* 1997; Okuda *et al.,* 1996; Sasaki *et al.,* 1996; Wang *et al.,* 1996a,b). *Cbfa2* was shown to be expressed by a subset of endothelial cells budding and differentiating to hematopoietic cells (North *et al.,* 1999). Furthermore, *Cbfa2*$^{-/-}$ embryos exhibit a lack of hematopoietic emergence from hemogenic endothelial cells (North *et al.,* 1999). Therefore, this study suggested that Cbfa2 function is to suppress endothelial phenotype and/or induce hematopoietic differentiation. However, a more definitive conclusion may rely on systematic dissection of genetic pathways regulated downstream from *Cbfa2.*

5. Blood-Borne Circulating Endothelial Progenitor Cells in the Adult

Recently, cells that can differentiate to endothelial cells have been isolated from adult blood. These cells were referred to as "blood-borne" endothelial progenitors (Asahara *et al.,* 1997). Subsequently, bone marrow transplantation experiments suggested that these progenitors could be bone marrow derived (Asahara *et al.,* 1999). Furthermore, they were found to participate in blood vessel formation in the adult (Asahara *et al.,* 1999). Although these findings are intriguing and provide a novel concept in the field, so far none of these findings have been confirmed independently. Further investigation is required to confirm the existence of physiologically relevant endothelial progenitors in the adult circulation. One of the first challenges is to define these putative endothelial progenitors at the molecular level. It is also important to reevaluate the source of these putative progenitors. One of the most difficult tasks is to eliminate potential problems resulting from the contamination from tissue-derived endothelial and/or endothelial progenitor cells, and to show that the circulation of such progenitors is physiological.

E. Current View of Endothelial Cell Type Specification

Based on the discoveries outlined in Section III, the current view of endothelial cell type specification is schematically shown in Fig. 4. One new principle in this field learned during the 1990s is that endothelial cells are a highly heterogeneous population of cells. We are just beginning to understand their origins and the regulatory mechanisms that underlie the establishment of this diversity among the endothelial cells. Many of the specific questions for future chal-

lenges are already discussed in this section. However, one of the most urgent goals is to identify further molecular markers for a variety of endothelial cell types that possibly exist and remain to be defined. The ability to separate and characterize one type of endothelial cell from another based on the expression of such marker(s) certainly facilitates the identification of even further heterogeneity of endothelial cells. The regulatory mechanism underlying the establishment and maintenance of heterogeneous endothelial types is virtually unknown and is certainly an area for future challenge.

IV. Vascular Morphogenesis

Establishment of a blood vessel network is a dynamic process. The initial phase is accomplished primarily by endothelial cells. In later phases, other cell types such as smooth muscle cells participate in further shaping up the vessel network. Vascular morphogenesis is operationally divided into two phases: vasculogenesis and angiogenesis. In this section, mechanisms underlying each phase are discussed.

A. Vasculogenesis

The first blood vessel network is formed by assembling primarily endothelial cells into a channel-like structure (Box 1). These vascular channels fuse to each other to form an interconnected network of blood vessels. As described in Section II, this initial vessel network is referred to as the primary capillary plexus (Risau and Flamme, 1995). It is thought that this process involves morphological changes of endothelial cell shape, cell–cell adhesion, cell–matrix interactions, and perhaps some degree of endothelial cell migration.

Numerous studies have led to the implication that formation of this initial network is an intrinsic property of differentiated endothelial cells. Endothelial cells in culture are known to spontaneously form tubule-like structures that resemble the primary capillary plexus observed *in vivo* (Folkman and Haudenschild, 1980). In an explant culture system, bFGF-induced endothelial cell differentiation from mesodermal precursors is always accompanied by the formation of blood islands (Box 1), a precursor structure toward the primary capillary plexus (Flamme and Risau, 1992). The precisely controlled number of existing endothelial cells seems to be essential for producing a normal network of primary capillary plexus (G.-H. Fong *et al.,* 1999). As described in Section III, VEGF-R1 knock-out embryos exhibit a dramatically increased number of endothelial cells. This resulted in an abnormal patterning of the primary capillary plexus in the yolk sac, an extraembryonic membrane where these first vessels form.

A.

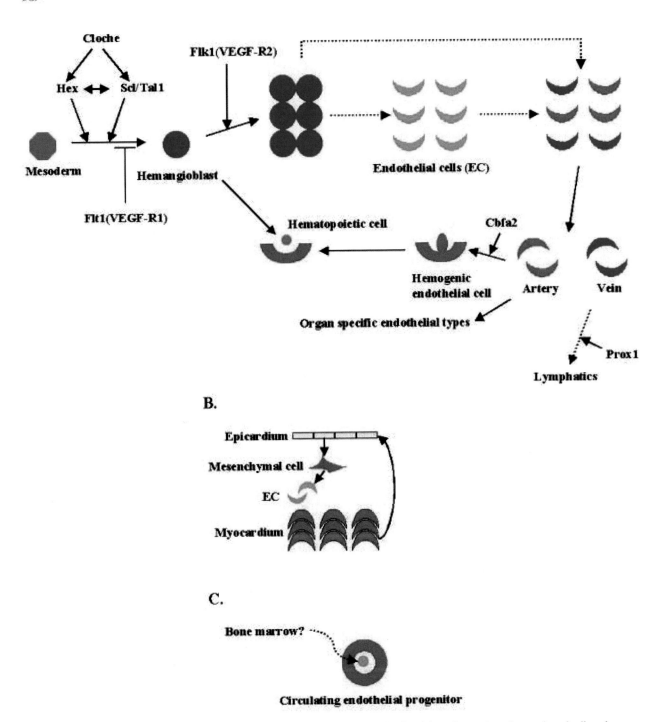

B.

C.

Figure 4 Current model of endothelial cell types specification. (A) Many endothelial cells originate from embryonic mesodermal cells and are generated by sequential cell specification processes. The details are described in the text. (B) In heart, coronary vessel endothelial cells seem to originate from epicardium. A subset of epicardial cells undergoes epithelial-mesenchymal transformation, and these mesenchymal cells contribute to cell types that participate in the formation of coronary vessels including endothelial cells. It is proposed that myocardial cells send a critical paracrine signal(s) to epicardium for its transformation to mesenchymal cells. (C) In the adult, the presence of circulating endothelial cell progenitors, possibly derived from bone marrow, is proposed.

In addition to these cell-biological parameters, it is known that the transforming growth factor β (TGF-β) (*Tgfb1*) signaling pathway may also play an important role in vasculogenesis. ES cells that overexpress the kinase-deficient type II TGF-β receptor (*Tgfbr2*) failed to contribute to a normal primary capillary plexus in the yolk sac of chimeric embryos (Goumans *et al.,* 1999). Similar results were also obtained by studying the differentiation of embryoid bodies from these ES cells *in vitro* (Goumans *et al.,* 1999). This vasculogenic defect was correlated with a significant reduction in the production of extracellular matrix proteins such as fibronectin and laminin (Goumans *et al.,* 1999). These results led to the conclusion that the TGF-β signaling pathway is critical for vasculogenesis, and that this process is either directly or indirectly mediated by controlling the production of extracellular matrix proteins such as fibronectin and laminin. Alternatively, it is also possible that expression of the dominant-negative type II TGF-β receptor leads to buffering of extracellular factors that are not the authentic ligands for this receptor. As a consequence, this may result in the modulation of other signaling pathways that are not related to the authentic TGF-β signaling pathway, but which are critical for vasculogenesis.

At this point, very little is known about the specific factors that control cell-biological processes underlying vasculogenesis. None of the gene knock-out lines analyzed so far exhibit a complete lack of vasculogenesis. This may be due in part to the fact that these morphogenic processes are critically regulated by factors that control fundamental aspects of cell migration, survival, and cell shape control in a wide range of cell types. This problem could be addressed by examining the functions of such factors that control these fundamental cell biological processes during vasculogenesis, via either endothelial-cell-specific gene knock-out or inducible knock-out strategies.

B. Angiogenesis

In contrast to vasculogenesis, the mechanisms of angiogenesis (Box 2) have been more extensively investigated, since it is considered to be a primary process involved in the pathogenesis of many human diseases.

1. Models for Angiogenesis

Subsequent to vasculogenesis, the vascular network expands and remodels. This process is referred to as *angiogenesis* and is thought to involve multiple related but distinguishable phases (Risau, 1997). However, no reliable documentation regarding how the primary capillary plexus expands and remodels its network is currently available. This is in part due to the fact that no studies have recorded ongoing angiogenesis at the real timescale with the single-cell resolution *in vivo*.

In this respect, a useful transgenic model was recently developed. In this transgenic mouse line, green fluorescent protein (GFP) was specifically expressed in virtually all endothelial cells using an endothelial specific promoter/enhancer expression vector (Motoike *et al.,* 2000). This resulted in the labeling of endothelial cells by GFP and the visualization of live fluorescent endothelial cells without any complex pretreatment. This transgenic line of embryo may eventually be useful to visualize and follow the behavior of each endothelial cell with sophisticated imaging methods in combination with an *in vitro* embryo culture system to achieve real-time documentation of angiogenesis.

Although no real-time recording of angiogenesis has been documented so far, substantial information is available regarding how angiogenesis proceeds based on conventional histological examinations. By these studies, multiple distinct processes underlying angiogenesis were identified and each process is discussed in this section.

a. Formation of New Vascular Channels. As discussed in Section II, new vascular channels are formed by both sprouting and nonsprouting mechanisms (Fig. 2). These processes are mediated by dynamic remodeling of the extracellular environment and changes in endothelial motility and shape. Endothelial cells secrete various proteases that digest extracellular matrices. These endothelial extracellular matrices usually keep endothelial cells in place by providing structural integrity and cell survival signals. However, in order to generate new vascular channels, endothelial cells are required to digest and remodel these extracellular matrices to create a microenvironment that allows increased motility. Endothelial extracellular matrix components and their remodeling, which are involved in the formation of new vessels during angiogenesis, have been extensively studied. Many of these components are not unique to endothelial cells, but rather play a general role in morphogenesis for many different cell types and organ systems. Therefore, matrix remodeling is a common biological strategy utilized among many biological systems that require dynamic changes such as angiogenesis. Here, some of the specific components that belong to this category and play roles in new vessel formation are discussed.

Fibronectin (*Fn1*) is one of the major extracellular matrix components (Hynes, 1985). It plays important roles in cell survival, migration, and shape control (Hynes, 1985). Fibronectin null mutants exhibit abnormal formation of blood vessels during mouse embryonic development (George *et al.,* 1997). However, because fibronectin is expressed in many different cell types during early embryonic development, it is difficult to determine which of these defects is directly caused by the absence of fibronectin function. It would be important to address this question by a more sophisticated genetic manipulation method, such as cell type specific and/or inducible knock-out of the fibronectin gene.

Integrins are a family of transmembrane proteins that are specific receptors for extracellular matrix proteins such as fibronectin, laminin, and vitronectin (Hynes, 1992; Hynes *et al.*, 1999). They mediate intracellular signaling pathways that lead to the control of cell survival, migration, and morphogenesis (Hynes, 1992; Hynes *et al.*, 1999). The $\alpha_v(Itga4)\beta_3(Itgb3)$, an integrin complex expressed by endothelial cells, has been suggested to play a key role in angiogenesis (Brooks *et al.*, 1994). The main evidence for this comes from an experiment using the specific peptide inhibitor for this integrin (Brooks *et al.*, 1994; Drake *et al.*, 1995). This peptide inhibitor has been shown to interfere with angiogenesis in the chick embryo, in tumor formation models, and in *in vitro* angiogenic assays (Brooks *et al.*, 1994; Drake *et al.*, 1995).

However, mouse genetic experiments have not supported such a specific role for this integrin in angiogenesis. Both $\alpha_v^{-/-}$ and $\beta_3^{-/-}$ embryos exhibit extensive normal vasculogenesis and angiogenesis, suggesting an alternative explanation for the inhibitory peptide experiment (Bader *et al.*, 1998; Hodivala-Dilke *et al.*, 1999). It is possible that the inhibitory peptide binding to $\alpha_v\beta_3$ integrin complex leads to the modulation of signaling pathways that are not normally under the control of this integrin complex. This ectopic signaling pathway in endothelial cells may be inhibitory for angiogenesis, although the $\alpha_v\beta_3$-integrin complex may not be directly involved in angiogenesis. Further studies will be required to sort out this controversy.

Extracellular proteinases modulate the extracellular microenvironment of vascular cells (Haas and Madri, 1999; Nagase and Woessner, 1999; Werb, 1997; Werb and Chin, 1998). This modulation may provide a permissive environment for angiogenesis. Matrix metalloproteinases represent a family of extracellular proteinases that have been implicated in angiogenesis (Haas and Madri, 1999; Nagase and Woessner, 1999; Werb, 1997; Werb and Chin, 1998). MMP-9/gelatinase B (*Mmp9*) has been shown to play an important role in angiogenesis during bone development (Vu *et al.*, 1998). MMP-9/gelatinase B$^{-/-}$ mice exhibit aberrant angiogenesis, which was shown to be rescued by transplantation with gelatinase-B-expressing wild-type bone marrow cells (Vu *et al.*, 1998). Furthermore, growth plates from MMP-9/gelatinase-B$^{-/-}$ mice were shown to exhibit delayed release of an angiogenic activator (Vu *et al.*, 1998). These results suggest that gelatinase-B produced by bone-marrow-derived cells allows the release of angiogenic factors from these cells that would otherwise be sequestered as an inactive form.

Matrix metalloproteinases have their specific endogenous inhibitors (Haas and Madri, 1999; Nagase and Woessner, 1999; Werb, 1997; Werb and Chin, 1998). Functions of these factors have been studied *in vivo* by generating null mutant mice (Shapiro, 1997). However, it is difficult to assess their specific role in the developing vasculature due to their expression by multiple cell types. Important insights into

the roles of such family of factors in angiogenesis may be gained by generating and analyzing vascular-cell-type-specific knock-out mice in the future.

The matrix remodeling discussed above provides a permissive microenvironment for endothelial cells to form new vascular channels. However, it is believed that subsequent local inductive signals are required to ultimately make endothelial cells form a new vessel. It is proposed that VEGF (Box 4) is one class of soluble factors that may take part in this process. Much evidence supports the notion that VEGF is capable of inducing an angiogenic response by endothelial cells. VEGF was shown to induce endothelial cell proliferation *in vitro* (Leung *et al.*, 1989). Overexpression of VEGF-A in developing skin during postnatal stages results in increased skin vascularization (Thurston *et al.*, 1999). Overexpression of VEGF-C was also shown to result in ectopic angiogenesis of both lymphatic and blood vessels *in vivo* (Jeltsch *et al.*, 1997). Furthermore, VEGF-A gene knockout mice exhibit severely decreased angiogenesis (Carmeliet *et al.*, 1996; Ferrara *et al.*, 1996). Interestingly, reduced angiogenesis was also observed in mice that are heterozygous for the VEGF-A gene mutation, suggesting that a subtle titration of VEGF-A activity is critical for normal angiogenesis (Carmeliet *et al.*, 1996; Ferrara *et al.*, 1996).

Multiple receptors are suspected to mediate the angiogenic function of VEGFs. VEGF-R1 and VEGF-R2 knock-out mice die before active angiogenesis occurs during embryogenesis (Fong *et al.*, 1995; Shalaby *et al.*, 1995), precluding the possibility of using these knock-out mice to study the roles of VEGF in angiogenesis. However, it was shown that overexpressing soluble VEGF-R1 (i.e., blocker of VEGF-R1 binding to VEGF) during postnatal mouse development leads to insufficient blood vessel development and growth retardation of many organs (Gerber *et al.*, 1999). Moreover, it was reported that VEGF-R3 knock-out embryos exhibit retarded embryonic angiogenesis at 9.5 dpc (Dumont *et al.*, 1998). In addition to the classic VEGF receptor family, it was shown that neuropilin-1 (*Nrp*), one of the factors that mediate neuronal cell guidance, also serves as an isoform specific receptor for one alternatively spliced form of VEGF-A, VEGF$_{165}$ (Soker *et al.*, 1998). To investigate the possibility that this unconventional VEGF receptor may also be involved in angiogenesis, neuropilin-1 knock-out mice were generated and characterized (Kawasaki *et al.*, 1999). In this study, neuropilin-1$^{-/-}$ embryos were shown to exhibit defective aortic arch formation (Kawasaki *et al.*, 1999). Although the study failed to pinpoint the exact process that is affected in the formation of the aortic arch vessels, it is possible that the defect may be in the angiogenic response by aortic arch endothelial cells (Kawasaki *et al.*, 1999).

Another class of soluble factors that may be involved in angiogenesis is fibroblast growth factor (FGF). FGF-1 (*Fgf1*) and FGF-2 (*Fgf2*) were found to induce angiogenesis *in vivo* (Jouanneau *et al.*, 1995; Klagsbrun, 1991; Schelling,

1991; Seghezzi et al., 1998). Therefore, this suggested that exogenous FGF can induce angiogenesis by itself or in combination with other endogenous factor(s). However, there are no studies clearly showing that "endogenous" FGFs or FGF receptors are involved in angiogenesis during normal development.

The third class of soluble factors that induces the formation of new vascular channels is the angiopoietins (Box 9) (Davis and Yancopoulos, 1999). Currently, several available data suggest that angiopoietins may be involved in angiogenesis. Ectopic expression of angiopoietin-1/Ang-1 (*Agpt*) *in vivo* is sufficient to induce new blood vessels (Suri *et al.*, 1998). Furthermore, mouse embryos deficient in Ang-1 or its specific receptor, Tie2/Tek (*Tek*), were shown to exhibit retarded angiogenesis (Dumont *et al.*, 1994; Sato *et al.*, 1995; Suri *et al.*, 1996). In addition, ectopic expression of the inhibitory ligand for the Tie2 receptor, angiogpoietin-2/Ang-2 (*Agpt2*), was shown to cause inhibition of embryonic angiogenesis (Maisonpierre *et al.*, 1997). Another member of the Tie receptor family, Tie1 (*Tie1*), was also shown to be important for an angiogenic response of endothelial cells during development (Puri *et al.*, 1995; Sato *et al.*, 1995).

While no ligand(s) have been identified for this receptor, Tie1$^{-/-}$ embryos exhibit severe hemorrhaging and edema (Puri *et al.*, 1995; Sato *et al.*, 1995). Tie1$^{-/-}$ ES cells contributed normally to many vessels except the capillaries of the brain and kidney (Partanen *et al.*, 1996). In these organs, vascularization is thought to be primarily accomplished by sprouting of preexisting vessels (Risau, 1991a,b). These studies suggest that Tie1 may also be involved in angiogenesis in an endothelial cell autonomous manner during embryonic development. In addition, functional interactions have been proposed between Tie1 and Tie2 receptors during blood vessel formation, based on the analyses of double knock-out and compound heterozygous mice for these two related genes (Puri *et al.*, 1999).

Spatial and temporal expression patterns of endogenous angiopoietins suggest an interesting correlation with their potential role in angiogenesis (Holash *et al.*, 1999; Maisonpierre *et al.*, 1997). Ang-1 expression is primarily associated with vessels that have completed their formation and become stable. In contrast, Ang-2 expression is associated with vessels that are either just beginning to form or regressing. When Ang-2 expression is codetected with VEGF, these ves-

Box 9: Angiopoietin and the Tie Receptor Family

Tie receptors were originally discovered as a family of novel receptor tyrosine kinases that is preferentially expressed in endothelial cells. Two related receptors belong to this class of receptor tyrosine kinases: Tie1 (also known as Tie) and Tie2 (also known as Tek). Both receptors exhibit a unique set of domain structures in their extracellular region. It consists of two immunoglobulin-like domains flanking three EGF-like repeats, which are followed by three fibronectin type III repeats. Both of the receptors were found to be expressed mainly in vascular endothelial cells during development and in the adult. Subsequently, these receptors were also shown to be expressed in hematopoietic lineages. This relatively specific expression in endothelial cells suggested important roles in the vascular system. To confirm this possibility, knock-out mice for the genes encoding *Tie1* and *Tie2* were generated and characterized. *Tie1* knock-out mice died between 14.5 dpc and P0 and exhibited severe vascular hemorrhage. This phenotype and the expression of the receptor in endothelial cells suggested that a primary role for *Tie1* is to maintain the integrity of blood vessels. *Tie2* knock-out mice died between 9.5 and 10.5 dpc, and exhibited severe vascular malformation. This phenotype suggested a primary role for Tie2

in the formation of blood vessels during the early stages of vascular development.

As suggested from the unique domain structure of the extracellular part of the receptor, a unique class of soluble ligands for the Tie2 receptor was found to exist. There are four known specific ligands: angiopoietin-1/Ang-1, angiopoietin-2/Ang-2, angiopoietin-3/Ang-3 (identified only in human so far), and angiopoietin-4/Ang-4 (*Agpt4*). They are all secreted glycoproteins that possess a unique domain structure consisting of a coiled-coil domain followed by a fibrinogen-like domain. It has been shown that this coiled-coil domain mediates the multimerization of angiopoietins that seem to mediate bioactivities of the proteins. All of these angiopoietins were shown to bind to the Tie2 receptor *in vitro*, but failed to bind to the Tie1 receptor. In addition to these conventional angiopoietins, several splice variants of the angiopoietins have also been identified. Furthermore, the existence of a few angiopoietin-related proteins has been reported. However, these angiopoietin-related proteins do not exhibit specific binding to the Tie receptors. The functional significance of both splice variants of the angiopoietins and the angiopoietin-related proteins remains unknown.

sels seem to be actively engaged in vessel formation. However, when Ang-2 expression is not associated with VEGF, these vessels tend to be regressing. These studies suggest that Ang-2 may serve as a local angiogenic signal for endothelial cells. Blocking Ang-1 signaling by Ang-2 in an endothelial cell may create a permissive environment for the cell to respond to other angiogenic signals such as VEGF.

Intracellular signaling pathways that may be subjected to this "ying-yang" regulation of the angiogenic phenotypes of endothelial cells by angiopoietins was investigated. It was recently shown that activation of the Tie 2 receptor by Ang-1 leads to activation of the PI3 kinase and Akt/protein kinase B pathway *in vitro* (Kontos *et al.*, 1998; Papapetropoulos *et al.*, 1999, 2000). It is possible that fundamental cell-biological processes directly controlled by this signaling pathway may underlie the determination of the endothelial cell phenotype regarding angiogenic versus nonangiogenic states.

In addition to regulation by soluble factors, certain physiological conditions also play important roles in inducing new vessel formation by endothelial cells. One such condition is hypoxia. One of the most important physiological functions of the circulatory system is to deliver sufficient oxygen to organs during development and in the adult. This leads to a notion that organs that are not fed with sufficient oxygen need more blood vessels. Thus, it is conceivable to imagine that a hypoxic environment induces angiogenesis.

Recently, significant advances have been made in understanding the molecular mechanisms underlying hypoxic regulation of angiogenesis. A family of nuclear receptors, collectively referred to as bHLH/PAS transcription factors, is involved in the hypoxic responses by a variety of cells (Crews, 1998; Crews and Fan, 1999; Semenza, 1999). They have two key domains, a basic helix–loop–helix (bHLH) domain that mediates the interaction with another member of this family and a PAS domain, which seems to be a key element for transcriptional target specificities. This family of transcription factors has been shown to be critically involved in many biological systems including neurogenesis, xenobiotic metabolism, and angiogenesis (Crews, 1998; Crews and Fan, 1999).

Two key bHLH/PAS transcription factors have been clearly shown to be involved in angiogenesis: HIF-1α (*Hif1a*) and ARNT (*Arnt*). HIF-1α$^{-/-}$ mouse embryos die by E11.5 due to several morphogenic defects, including cardiovascular malformations (Iyer *et al.*, 1998; Ryan *et al.*, 1998). Furthermore, HIF-1α$^{-/-}$ ES cells failed to upregulate VEGF in response to hypoxic conditions (Iyer *et al.*, 1998). In addition, HIF-1α$^{-/-}$ tumors failed to support aggressive tumor angiogenesis (Carmeliet *et al.*, 1998; Ryan *et al.*, 1998). It has been suggested that hypoxia-mediated tumor angiogenesis is, in part, regulated by the selective degradation of HIF-1α by p53 (*Trp53*) (Ravi *et al.*, 2000). A more detailed analysis of HIF-1α$^{-/-}$ embryos and ES cells *in vitro* suggested that the lack of HIF-1α leads to the death of mesen-

chymal cells that surround developing embryonic vessels (Kotch *et al.*, 1999). This in turn leads to the failure of normal vessel development in a VEGF-independent manner, perhaps due to the lack of other mesenchyme-derived angiogenic factor(s) (Kotch *et al.*, 1999). These *in vivo* and *in vitro* results suggest that a hypoxic environment is translated to upregulation of angiogenic factors that are required for angiogenesis during development and tumor growth. In addition, HIF-1α forms a functional heterodimer with ARNT. ARNT$^{-/-}$ mouse embryos do not survive past 10.5 dpc, with defective angiogenesis in the yolk sac and placenta (Kozak *et al.*, 1997; Maltepe *et al.*, 1997). These analyses support the notion that both HIF-1α and ARNT play a critical role in embryonic angiogenesis, perhaps as a heterodimer.

b. Segregation of Arteries and Veins. As discussed in Section III.D.1, arterial and venous endothelial identities are established prior to angiogenesis. This fact implies that arterial endothelial cells assemble among themselves to form arteries, and venous endothelial cells do so among themselves. Therefore, two mechanisms are expected to be in place to achieve this segregation process: a mechanism that allows homophilic interactions among each endothelial cell type and a mechanism that prevents interaction of these two endothelial types.

Two molecular pathways are proposed to be critically involved in these processes. Arterial endothelial cells express ephrin-B2 and venous endothelial cells express its receptor, EphB4 (Wang *et al.*, 1998). This transmembrane ligand-receptor pathway has been suggested to operate as a cell–cell repulsion signal (Cook *et al.*, 1998; Mellitzer *et al.*, 1999; Wilkinson, 2000a,b; Xu *et al.*, 1999). Therefore, it is possible that this specific segregation of ephrin-B2 and EphB4 expression pattern ensures the segregation of arterial and venous vessels during angiogenesis. To support this possibility, knock-out mice lacking either ephrin-B2 or EphB4 exhibit abnormal angiogenesis (Gerety *et al.*, 1999; Wang *et al.*, 1998). Most interestingly, it is reported that the borders between arterial and venous channels become less clear in these knock-out mice (Gerety *et al.*, 1999; Wang *et al.*, 1998). Furthermore, it is reported that ectopic expression of ephrin-B2 or dominant negative EphB4 in *Xenopus* embryos results in the ectopic formation of veins into a region where these vessels do not normally invade (Helbling *et al.*, 2000). These experimental results seem to strongly support the idea that this ligand-receptor pathway plays a unique role in segregation of arteries and veins by mediating the repulsion between these two types of vessels during angiogenesis.

In addition to this ligand-receptor pathway, the TGF-β pathway is also shown to be involved in this process. Activin receptor-like kinase-1 (*Acvrl1*) encodes a type I receptor for the TGF-β (*Tgfbr1*) superfamily of growth factors, and the knock-out of the *Acvrl1* gene in mice resulted in the downregulation of ephrin-B2 (arterial-specific marker) and the subsequent failure of shunting between dorsal aortae (artery) and cardinal vein (Urness *et al.*, 2000). This study suggests that the Acvrl1-mediated signal transduction pathway is critical for establishing the identity of arterial endothelial cells and/or segregating arteries and veins.

The third pathway that seems to be involved in this process is the Notch pathway. Notch is a family of transmembrane receptors whose function is primarily regulated by a specific family of ligands (Gridley, 1997). One of the Notch family members, *Notch4*, was identified as an endothelial-specific Notch in mouse development and in adult (Shirayoshi *et al.*, 1997; Uyttendaele *et al.*, 1996). Notch4 is very similar to other mammalian Notch proteins because it contains conserved motifs; however, Notch4 has fewer EGF-like repeats and a shorter intracellular domain than other mouse Notch homologs. These structural differences, along with the endothelial-specific expression of Notch4, suggest that this Notch protein plays a unique role in vascular development. To support this notion, double knock-out embryos for *Notch4* and *Notch1* genes result in vascular malformation (Krebs *et al.*, 2000). Furthermore, a novel Notch ligand, Delta-Like Ligand 4 (*Dll4*) was found to show an expression pattern that is consistent with the vascular phenotype of the *Notch1/Notch4* double knock-out (Krebs *et al.*, 2000; Shutter *et al.*, 2000). Interestingly, the expression of *Dll4* is relatively restricted to arterial endothelial cells (Shutter *et al.*, 2000). Because Notch receptors do not exhibit such specificity, it is possible that arterial expression of Dll4 may elicit specific signals in the regulation of angiogenic pathways in both venous and arterial endothelial cells in a paracrine and autocrine manner, respectively.

c. Establishment of Endothelial Cell–Cell Junctions. As new vascular channels are formed, it becomes essential to establish "leakage-proof" endothelial cell-cell junctions during angiogenesis. Endothelial cells establish these cell–cell junctions, and it is assumed that most of the general cell-junction-forming mechanisms found in other cell types such as epithelial cells are at work. In addition to these general mechanisms, endothelial cells utilize cell type specific mechanisms. Cadherins are a family of transmembrane proteins localized specifically at the cell–cell junction, and mediate Ca^{2+}-dependent homophilic interaction between the cells (Urushihara and Takeichi, 1980). Endothelial cells express a cell type specific cadherin, called VE-cadherin (*Cdh5*) (Breier *et al.*, 1996; Wu and Maniatis, 1999, 2000). A critical role for VE-cadherin in angiogenesis was recently shown by studying VE-cadherin[-/-] mice (Carmeliet *et al.*, 1999). VE-cadherin[-/-] embryos failed to form a normal vascular network at 9.5 dpc. Furthermore, deletion of the cytoplasmic portion of VE-cadherin was sufficient to produce a similar phenotype in vascular network formation (Carmeliet *et al.*, 1999). Therefore, this study suggested an essential role for VE-cadherin in angiogenesis, which may be

regulated by its interaction with cytoplasmic components that lead to various intracellular signaling events in endothelial cells.

d. Recruitment of Smooth Muscle Cells and Pericytes. During angiogenesis, endothelial cells recruit other cell types into the vascular channels. Smooth muscle cells and pericytes become invested around the endothelial cells of medium to large sized vessels and capillaries, respectively. Smooth muscle cells form continuous layers around the vessels, but pericytes are invested around the capillaries only in a discontinuous manner. Smooth muscle cells and pericytes are considered to be related. Both of them provide structural integrity and contractility of the vessels, two hallmarks of the mature vessel. For recruitment of smooth muscle cells/pericytes, reciprocal communications between these cells and endothelial cells are critical. Endothelial cells secrete PDGF-B (*Pdgfb*), which binds and activates the specific receptor PDGFR-β (*Pdgfrb*) expressed on the surface of smooth muscle/pericyte precursors. This paracrine pathway mediates the migration of these precursors into the immediate proximity of the endothelial cells that are already a part of the vessel. This initial recruitment of smooth muscle/pericyte precursors seems to result in the clustering of these cells at the ventral abluminal surface of the vessels (Hungerford *et al.*, 1997; Lee *et al.*, 1997; Sato, 2000). This model is supported by the fact that both PDGF-B[-/-] and PDGFR-β[-/-] embryos lack pericytes along the endothelial vascular channels and vessel integrity was impaired, resulting in microaneurysm (Hellstrom *et al.*, 1999; Lindahl *et al.*, 1997).

Differentiation of these precursors to more mature smooth muscle cells/pericytes is mediated by the TGF-β pathway. It has been shown that smooth muscle progenitor-like cell lines or multipotential mesodermal-derived cell lines can convert to smooth muscle cells on treatment with TGF-β *in vitro* (Hirschi *et al.*, 1998). In contrast to this *in vitro* result, a lack of TGF-β in mice does not seem to interfere with the differentiation of smooth muscle cells (Dickson *et al.*, 1995; Kulkarni *et al.*, 1993). This may be due to the multiplicity of receptor specificity and other factors that belong to the TGF-β family acting in a compensatory manner. In support of this possibility, knock-out mice for endoglin (*Eng*), a gene encoding an extracellular TGF-β binding protein, were shown to exhibit reduced numbers of smooth muscle cells and lacked normal endothelial/smooth muscle cell interaction (Li *et al.*, 1999).

The TGF-β signaling pathway is mediated by a family of signal transduction proteins called SMAD (Derynck *et al.*, 1998; Heldin *et al.*, 1997; Whitman, 1997). Among them, Smad5 has been shown to be critically involved in embryonic vascular development (Chang *et al.*, 1999; Yang *et al.*, 1999); it is thought to transduce signals of the bone morphogenic protein (BMP) pathway (Miyazono, 1999; Raftery and Sutherland, 1999). Smad5 (*Madh5*) null mutant mice exhibit abnormal vascular development, although they seem to have differentiated smooth muscle cells (Chang *et al.,* 1999; Yang *et al.,* 1999). It is possible that Smads have overlapping functions and therefore a single gene knock-out does not cause complete absence of smooth muscle cell differentiation. Such compensatory mechanisms are likely since various TGF-β and related signaling pathways are regulated by numerous and complex interactions involving Smad family members.

Although pericytes are often localized as a single cell on top of the endothelial cells in a discontinuous manner, smooth muscle cells form a complete layer around the vessels. This requires the upward migration of smooth muscle cells initially localized to the ventral abluminal surface along the vessels. This step seems to be mediated by a class of glycolipids-mediated-signaling pathway. Sphingosine phosphate-1 binds and activates a family of G-protein-coupled receptors referred to as the Edg family (Hla *et al.,* 2000). In mice lacking one of this receptor family members, Edg1 (*Edg1*), smooth muscle cells remain clustered at the ventral abluminal surface of the vessels and fail to surround the vessels (Liu *et al.,* 2000). Therefore, it is proposed that this novel ligand-receptor pathway is involved in this later phase of smooth muscle cell investment of the vessel wall.

e. Regression of Vascular Channels. During angiogenesis, some of the vascular channels undergo regression. As discussed in Section III.D.3.c, hyaloid and papillary membrane vascular systems in the eye are the most extensively studied, at least at the morphological level. The molecular basis for the regression of specific vascular channels during vascular development is completely unknown at this time and awaits future investigation.

f. Establishment of Vascular Polarity. The vascular network is a highly polarized structure. Branching points distributed along the vascular channel and the directionality of the new branches form a basis for the polarity. Furthermore, this polarized formation of new branches together with regression of specific vascular branches during embryonic angiogenesis leads to the establishment of a left-right asymmetry of the network. One pathway has been implicated to play a critical role in establishing this left-right asymmetry of the vascular network. Mice doubly mutated for Ang-1 and Tie1 have been shown to exhibit the lack of right-hand-side cardinal veins, but normal left-hand-side cardinal veins (Loughan and Sato, 2001). This suggested that the combinatorial role of Ang-1 and Tie1 is required in the formation of specifically the right-hand-side cardinal veins. Furthermore, this asymmetrical phenotype was shown to correlate with the polarized expression of Ang-1 in the sinus venosus from which the cardinal veins branch. Based on this study, it is anticipated that other pathways may also regulate the establishment of vascular polarity.

2. Clinical Implications

Although the main theme of this chapter is vascular development in mice, many of the findings described above are leading the way to potential therapeutics for human diseases. Tumor angiogenesis is one of these areas. In addition to its developmental roles, the VEGF pathway is implicated in tumor angiogenesis. This pathway has been one of the most popular drug development targets for antiangiogenic therapy in an effort to starve tumor cells to death by cutting the blood vessel supply (T. A. Fong *et al.,* 1999; Millauer *et al.,* 1994, 1996; Witte *et al.,* 1998). In addition, in an experimental animal model for tumor formation, blocking Tie2 receptor function prevented tumor angiogenesis and consequently retarded tumor growth significantly (Lin *et al.,* 1998).

Another area is proangiogenic therapy. The objective of this therapy is to induce blood vessel formation in damaged organs such as after heart failure. Recent studies such as gene transfer, recombinant protein injections, and transgenic analyses have suggested that VEGF and angiopoietins may be potentially useful for this type of therapy (Peters, 1998; Suri *et al.,* 1998; Thurston *et al.,* 1999).

Furthermore, several of the genes described above have been implicated in human genetic diseases related to vascular malformations. Tie2 receptor mutations have been linked to venous malformation disease (Vikkula *et al.,* 1996). Endoglin mutation has been linked to vascular malformation disease associated with hereditary hemorrhagic telangiectasia (HHT1) as discussed above (Guttmacher *et al.,* 1995; McAllister *et al.,* 1994). VHL tumor suppressor gene (*Vhlh*), a critical angiogenesis regulator, is associated with von Hippel-Lindau disease (Gnarra *et al.,* 1996; Linehan *et al.,* 1995). Notch pathways are involved in several human disease conditions (Gridley, 1996, 1997; Joutel and Tournier-Lasserve, 1998).

As we learn more about the mechanisms of vascular development and human genetics in the future, the list of factors and pathways that are critical for both normal development and human disease formation is expected to expand. This may certainly lead to the invention of novel and more effective therapies for many human diseases.

3. Major Questions on Angiogenesis

Based on the most current description of angiogenesis, it is clear that our current knowledge of angiogenesis is composed of only fragmentary information. Furthermore, the detailed biochemical and molecular mechanisms of the biological functions of each factor discussed in this section remain highly speculative. Angiogenesis is clearly a complex and heterogeneous process. Angiogenesis at different developmental stages and in different organs may involve distinct mechanisms. It is likely that even a single process during angiogenesis involves multiple factors and signaling pathways. The specific factors and the physiological environment discussed in this section are only a fraction of the continuously growing list of angiogenic factors. Therefore, it will be our future task to decipher precise mechanisms by which multiple factors and signaling pathways interact to establish new blood vessels. It is also important to understand how angiogenesis utilizes differential mechanisms at each developmental stage and in each organ.

V. Concluding Remarks

As outlined in this chapter, it is clear that the vascular development is regulated by a complex network of gene functions and signaling pathways. The use of modern mouse genetics and embryo manipulation technologies has significantly contributed to our advancement of this field. However, key questions regarding several fundamental aspects of vascular development still remain unanswered. We would like to discuss a couple of them as our concluding remarks.

As discussed in Section III, we are beginning to realize the complexity and heterogeneity of vascular cell types. Our knowledge of the exact origins of various vascular cell types and their eventual fates remains at the preliminary and inconclusive level. Conventionally, this area was studied by using classic chimera and cell transplantation approaches as well as a retrovirus-mediated cell marking system (Mikawa and Fischman, 1992; Mikawa and Gourdie, 1996; Perez-Pomares et al., 1998; Yamashita et al., 2000). These methods will certainly continue to be useful; however, the complex nature of vascular morphogenesis may require more sophisticated methods. Recently, recombinase-based cell fate mapping methods have been applied to study cell lineage regulation in the mammalian system (Chai et al., 2000; Dymecki and Tomasiewicz, 1998; Jiang et al., 2000; Kimmel et al., 2000; Zinyk et al., 1998). This system relies on the cell type specific recombinase (such as Cre and Flp) mediated permanent cell marking. The recombinase can be expressed transgenically by using a promoter that can drive the expression specifically in a progenitor cell population. This recombinase transgenic line is crossed to a Cre-excision reporter transgenic line in which the excision of the sequences flanked by two pairs of loxP (for Cre) or Frt (for Flp) elements leads to the permanent reporter such as lacZ expression in all the descendent cells during development. One advantage of this method is the ability to permanently mark the specific cell lineage of mammalian embryos, which are quite inaccessible to retrovirus marking system. Furthermore, recombinase-based methods allow the analysis of "true" cell fate regulation during normal and physiological development, as opposed to the chimera and cell transplantation methods, which can only measure the "ability" of transplanted cells to contribute to various lineages (Yamashita et al., 2000). Application of such a novel in vivo cell fate mapping system to the

problem in vascular cell lineage regulation is expected to contribute significantly to the advancement of the field.

The second area is that of understanding how multiple pathways regulate vascular development. This is a very complex and challenging subject. However, the recent advent of new mouse genetic manipulation methods is expected to aid this problem. We can now knock out genes in an inducible manner (Rossant and McMahon, 1999). This possibility certainly allows us to address the function of specific genes in developmental stages where conventional gene knock-out strategies could not be applied due to the earlier embryonic lethality. Furthermore, several useful transgenic expression systems now exist that allow the targeting of gene expression in a specific vascular cell type (Kappel et al., 1999; Korhonan et al., 1995; Li et al., 1996; Schlaeger et al., 1995, 1997). The availability of these methods and reagents is expected to facilitate the dissection of complex pathways underlying vascular development by knocking out or overexpressing specific genes in a specific vascular cell type at various developmental stages transiently or permanently.

New genetic and embryo manipulation tools are continuing to be invented in the mouse system. Continuing improvements in our understanding of vascular cell lineage regulation not only contribute to our fundamental understanding of vascular development, but also facilitate the invention of new reagents that may allow us to manipulate the expression of specific genes in a specific vascular cell type. Many intriguing possibilities clearly lie ahead of us in this field and we all look forward to it.

References

Aird, W. C., Edelberg, J. M., Weiler-Guettler, H., Simmons, W. W., Smith, T. W., and Rosenberg, R. D. (1997). Vascular bed-specific expression of an endothelial cell gene is programmed by the tissue microenvironment. J. Cell Biol. 138, 1117–1124.
Asahara, T., Murohara, T., Sullivan, A., Silver, M., van der Zee, R., Li, T., Witzenbichler, B., Schatteman, G., and Isner, J. M. (1997). Isolation of putative progenitor endothelial cells for angiogenesis. Science 275, 964–967.
Asahara, T., Masuda, H., Takahashi, T., Kalka, C., Pastore, C., Silver, M., Kearne, M., Magner, M., and Isner, J. M. (1999). Bone marrow origin of endothelial progenitor cells responsible for postnatal vasculogenesis in physiological and pathological neovascularization. Circ. Res. 85, 221–228.
Bader, B. L., Rayburn, H., Crowley, D., and Hynes, R. O. (1998). Extensive vasculogenesis, angiogenesis, and organogenesis precede lethality in mice lacking all alpha v integrins. Cell (Cambridge, Mass.) 95, 507–519.
Breier, G., Breviario, F., Caveda, L., Berthier, R., Schnurch, H., Gotsch, U., Vestweber, D., Risau, W., and Dejana, E. (1996). Molecular cloning and expression of murine vascular endothelial- cadherin in early stage development of cardiovascular system. Blood 87, 630–641.
Brooks, P. C., Clark, R. A., and Cheresh, D. A. (1994). Requirement of vascular integrin alpha v beta 3 for angiogenesis. Science 264, 569–571.
Carmeliet, P., Ferreira, V., Breier, G., Pollefeyt, S., Kieckens, L., Gertsen-

stein, M., Fahrig, M., Vandenhoeck, A., Harpal, K., Eberhardt, C., Declercq, C., Pawling, J., Moons, L., Collen, D., Risau, W., and Nagy, A.
(1996). Abnormal blood vessel development and lethality in embryos
lacking a single VEGF allele. *Nature (London)* **380,** 435–439.

Carmeliet, P., Dor, Y., Herbert, J. M., Fukumura, D., Brusselmans, K., Dewerchin, M., Neeman, M., Bono, F., Abramovitch, R., Maxwell, P., Koch,
C. J., Ratcliffe, P., Moons, L., Jain, R. K., Collen, D., and Keshet, E.
(1998). Role of HIF-1alpha in hypoxia-mediated apoptosis, cell proliferation and tumour angiogenesis. *Nature (London)* **394,** 485–490.

Carmeliet, P., Lampugnani, M. G., Moons, L., Breviario, F., Compernolle, V., Bono, F., Balconi, G., Spagnuolo, R., Oostuyse, B., Dewerchin, M., Zanetti, A., Angellilo, A., Mattot, V., Nuyens, D., Lutgens, E.,
Clotman, F., de Ruiter, M. C., Gittenberger-de Groot, A., Poelmann, R.,
Lupu, F., Herbert, J. M., Collen, D., and Dejana, E. (1999). Targeted
deficiency or cytosolic truncation of the VE-cadherin gene in mice impairs VEGF-mediated endothelial survival and angiogenesis. *Cell (Cambridge, Mass.)* **98,** 147–157.

Chai, Y., Jiang, X., Ito, Y., Bringas, P., Han, J., Rowitch, D. H., Soriano, P.,
McMahon, A. P., and Sucov, H. M. (2000). Fate of the mammalian cranial neural crest during tooth and mandibular morphogenesis. *Development (Cambridge, UK)* **127,** 1671–1679.

Chang, H., Huylebroeck, D., Verschueren, K., Guo, Q., Matzuk, M. M., and
Zwijsen, A. (1999). Smad5 knockout mice die at mid-gestation due to
multiple embryonic and extraembryonic defects. *Development (Cambridge, UK)* **126,** 1631–1642.

Cook, G., Tannahill, D., and Keynes, R. (1998). Axon guidance to and from
choice points. *Curr. Opin. Neurobiol.* **8,** 64–72.

Crews, S. T. (1998). Control of cell lineage-specific development and transcription by bHLH- PAS proteins. *Genes Dev.* **12,** 607–620.

Crews, S. T., and Fan, C. M. (1999). Remembrance of things PAS: Regulation of development by bHLH-PAS proteins. *Curr. Opin. Genet. Dev.*
9, 580–587.

Davis, S., and Yancopoulos, G. D. (1999). The angiopoietins: Yin and Yang
in angiogenesis. *Curr. Top. Microbiol. Immunol.* **237,** 173–185.

Derynck, R., Zhang, Y., and Feng, X. H. (1998). Smads: Transcriptional
activators of TGF-beta responses. *Cell (Cambridge, Mass.)* **95,** 737–
740.

Dettman, R. W., Denetclaw, W., Jr., Ordahl, C. P., and Bristow, J. (1998).
Common epicardial origin of coronary vascular smooth muscle, perivas-
cular fibroblasts, and intermyocardial fibroblasts in the avian heart. *Dev.
Biol.* **193,** 169–181.

Dickson, M. C., Martin, J. S., Cousins, F. M., Kulkarni, A. B., Karlsson, S.,
and Akhurst, R. J. (1995). Defective haematopoiesis and vasculogenesis
in transforming growth factor-β1 knock out mice. *Development (Cambridge, UK)* **121,** 1845–1854.

Dieterlen-Lievre, F., and Martin, C. (1981). Diffuse intraembryonic hemopoiesis in normal and chimeric avian development. *Dev. Biol.* **88,** 180–
191.

Diez-Roux, G., Argilla, M., Makarenkova, H., Ko, K., and Lang, R. A.
(1999). Macrophages kill capillary cells in G1 phase of the cell cycle
during programmed vascular regression. *Development (Cambridge, UK)*
126, 2141–2147.

Drake, C. J., Cheresh, D. A., and Little, C. D. (1995). An antagonist of
integrin alpha v beta 3 prevents maturation of blood vessels during embryonic neovascularization. *J. Cell Sci.* **108,** 2655–2661.

Dumont, D. J., Anderson, L., Breitman, M. L., and Duncan, A. M. V.
(1994). Assignment of the endothelial-specific protein receptor tyrosine
kinase gene (TEK) to human chromosome 9p21. *Genomics* **23,** 512–
513.

Dumont, D. J., Jussila, L., Taipale, J., Lymboussaki, A., Mustonen, T., Pajusola, K., Breitman, M., and Alitalo, K. (1998). Cardiovascular failure
in mouse embryos deficient in VEGF receptor-3. *Science* **282,** 946–949.

Dymecki, S. M., and Tomasiewicz, H. (1998). Using Flp-recombinase to
characterize expansion of Wnt1-expressing neural progenitors in the
mouse. *Dev. Biol.* **201,** 57–65.

Edelberg, J. M., Aird, W. C., Wu, W., Rayburn, H., Mamuya, W. S., Mercola, M., and Rosenberg, R. D. (1998). PDGF mediates cardiac micro-
vascular communication. *J. Clin. Invest.* **102,** 837–843.

Evans, H. M. (1909). On the development of the aortae, cardinal and umbilical veins, and the other blood vessels of vertebrate embryos from
capillaries. *Anat. Rec.* **3,** 598–519.

Ferrara, N., Carver-Moore, K., Chen, H., Dowd, M., Lu, L., O'Shea, K. S.,
Powell-Braxton, L., Hillan, K. J., and Moore, M. W. (1996). Heterozygous embryonic lethality induced by targeted inactivation of the VEGF
gene. *Nature (London)* **380,** 439–442.

Flamme, I., and Risau, W. (1992). Induction of vasculogenesis and hematopoiesis in vitro. *Development (Cambridge, UK)* **116,** 435–439.

Folkman, J., and Haudenschild, C. (1980). Angiogenesis in vitro. *Nature
(London)* **288,** 551–556.

Fong, G.-H., Rossant, J., and Breitman, M. L. (1995). Role of the Flt-1
receptor tyrosine kinase in regulating the assembly of vascular endothelium. *Nature (London)* **376,** 66–70.

Fong, G.-H., Klingensmith, J., Wood, C. R., Rossant, J., and Breitman,
M. L. (1996). Regulation of *flt-1* expression during mouse embryogene-
sis suggests a role in the establishment of vascular endothelium. *Dev.
Dyn.* **207,** 1–10.

Fong, G.-H., Zhang, L., Bryce, D. M., and Peng, J. (1999). Increased hemangioblast commitment, not vascular disorganization, is the primary
defect in flt-1 knock-out mice. *Development (Cambridge, UK)* **126,**
3015–3025.

Fong, T. A., Shawver, L. K., Sun, L., Tang, C., App, H., Powell, T. J., Kim,
Y. H., Schreck, R., Wang, X., Risau, W., Ullrich, A., Hirth, K. P., and
McMahon, G. (1999). SU5416 is a potent and selective inhibitor of the
vascular endothelial growth factor receptor (Flk-1/KDR) that inhibits
tyrosine kinase catalysis, tumor vascularization, and growth of multiple
tumor types. *Cancer Res.* **59,** 99–106.

George, E. L., Baldwin, H. S., and Hynes, R. O. (1997). Fibronectin are
essential for heart and blood vessel morphogenesis but are dispensable
for initial specification of precursor cells. *Blood* **90,** 3073–3081.

Gerber, H. P., Hillan, K. J., Ryan, A. M., Kowalski, J., Keller, G. A., Rangell, L., Wright, B. D., Radtke, F., Aguet, M., and Ferrara, N. (1999).
VEGF is required for growth and survival in neonatal mice. *Development (Cambridge, UK)* **126,** 1149–1159.

Gerety, S. S., Wang, H. U., Chen, Z. F., and Anderson, D. J. (1999). Symmetrical mutant phenotypes of the receptor EphB4 and its specific transmembrane ligand ephrin-B2 in cardiovascular development. *Mol. Cell* **4,**
403–414.

Gnarra, J. R., Duan, D. R., Weng, Y., Humphrey, J. S., Chen, D. Y., Lee, S.,
Pause, A., Dudley, C. F., Latif, F., Kuzmin, I., Schmidt, L., Duh, F. M.,
Stackhouse, T., Chen, F., Kishida, T., Wei, M. H., Lerman, M. I., Zbar, B.,
Klausner, R. D., and Linehan, W. M. (1996). Molecular cloning of the
von Hippel-Lindau tumor suppressor gene and its role in renal carci-
noma. *Biochim. Biophys. Acta* **1242,** 201–210.

Godin, I., Dieterlen-Lievre, F., and Cumano, A. (1995). Emergence of multipotent hemopoietic cells in the yolk sac and paraaortic splanchnopleura
in mouse embryos, beginning at 8.5 days postcoitus. *Proc. Natl. Acad.
Sci. U.S.A.* **92,** 773–777; erratum: *Ibid.,* p. 10815.

Goumans, M. J., Zwijsen, A., van Rooijen, M. A., Huylebroeck, D., Roelen,
B. A., and Mummery, C. L. (1999). Transforming growth factor-beta
signalling in extraembryonic mesoderm is required for yolk sac vasculogenesis in mice. *Development (Cambridge, UK)* **126,** 3473–3483.

Gridley, T. (1996). Human genetics. Notch, stroke and dementia. *Nature
(London)* **383,** 673.

Gridley, T. (1997). Notch signaling in vertebrate development and disease.
Mol. Cell Neurosci. **9,** 103–108.

Guttmacher, A. E., Marchuk, D. A., and White, R. I., Jr. (1995). Hereditary
hemorrhagic telangiectasia. *N. Engl. J. Med.* **333,** 918–924.

Haas, T. L., and Madri, J. A. (1999). Extracellular matrix-driven matrix
metalloproteinase production in endothelial cells: Implications for angiogenesis. *Trends Cardiovasc. Med.* **9,** 70–77.

Helbling, P. M., Saulnier, D. M., and Brandli, A. W. (2000). The receptor tyrosine kinase EphB4 and ephrin-B ligands restrict angiogenic growth of embryonic veins in *Xenopus laevis. Development (Cambridge, UK)* **127,** 269–278.

Heldin, C. H., Miyazono, K., and ten Dijke, P. (1997). TGF-beta signalling from cell membrane to nucleus through SMAD proteins. *Nature (London)* **390,** 465–471.

Hellstrom, M., Kal, M., Lindahl, P., Abramsson, A., and Betsholtz, C. (1999). Role of PDGF-B and PDGFR-beta in recruitment of vascular smooth muscle cells and pericytes during embryonic blood vessel formation in the mouse. *Development (Cambridge, UK)* **126,** 3047–3055.

Hidaka, M., Stanford, W. L., and Bernstein, A. (1999). Conditional requirement for the Flk-1 receptor in the in vitro generation of early hematopoietic cells. *Proc. Natl. Acad. Sci. U.S.A.* **96,** 7370–7375.

Hiratsuka, S., Minowa, O., Kuno, J., Noda, T., and Shibuya, M. (1998). Flt-1 lacking the tyrosine kinase domain is sufficient for normal development and angiogenesis in mice. *Proc. Natl. Acad. Sci. U.S.A.* **95,** 9349–9354.

Hirschi, K. K., Rohovsky, S. A., and D'Amore, P. A. (1998). PDGF, TGF-beta, and heterotypic cell-cell interactions mediate endothelial cell-induced recruitment of 10T1/2 cells and their differentiation to a smooth muscle fate. *J. Cell Biol.* **141,** 805–814; erratum: *Ibid.,* p. 1287.

His, W. (1900). Lecithoblast und Angioblast der Wirbelthiere. *Abh. K. Sitzungsber. Ges. Wiss., Math.-Phys.* **22,** 171–328.

Hla, T., Lee, M. J., Ancellin, N., Thangada, S., Liu, C. H., Kluk, M., Chae, S. S., and Wu, M. T. (2000). Sphingosine-1-phosphate signaling via the EDG-1 family of G-protein-coupled receptors. *Ann. N.Y. Acad. Sci.* **905,** 16–24.

Hodivala-Dilke, K. M., McHugh, K. P., Tsakiris, D. A., Rayburn, H., Crowley, D., Ullman-Cullere, M., Ross, F. P., Coller, B. S., Teitelbaum, S., and Hynes, R. O. (1999). Beta3-integrin-deficient mice are a model for Glanzmann thrombasthenia showing placental defects and reduced survival. *J. Clin. Invest.* **103,** 229–238.

Holash, J., Maisonpierre, P. C., Compton, D., Boland, P., Alexander, C. R., Zagzag, D., Yancopoulos, G. D., and Wiegand, S. J. (1999). Vessel co-option, regression, and growth in tumors mediated by angiopoietins and VEGF. *Science* **284,** 1994–1998.

Hungerford, J. E., Hoeffler, J. P., Bowers, C. W., Dahm, L. M., Falchetto, R., Shabanowitz, J., Hunt, D. F., and Little, C. D. (1997). Identification of a novel marker for primordial smooth muscle and its differential expression pattern in contractile vs noncontractile cells. *J. Cell Biol.* **137,** 925–937.

Hynes, R. (1985). Molecular biology of fibronectin. *Annu. Rev. Cell Biol.* **1,** 67–90.

Hynes, R. O. (1992). Integrins: Versatility, modulation, and signaling in cell adhesion. *Cell (Cambridge, Mass.)* **69,** 11–25.

Hynes, R. O., Bader, B. L., and Hodivala-Dilke, K. (1999). Integrins in vascular development. *Braz. J. Med. Biol. Res.* **32,** 501–510.

Iyer, N. V., Kotch, L. E., Agani, F., Leung, S. W., Laughner, E., Wenger, R. H., Gassmann, M., Gearhart, J. D., Lawler, A. M., Yu, A. Y., and Semenza, G. L. (1998). Cellular and developmental control of O2 homeostasis by hypoxia- inducible factor 1 alpha. *Genes Dev.* **12,** 149–162.

Jeltsch, M., Kaipainen, A., Joukov, V., Meng, X., Lakso, M., Rauvala, H., Swartz, M., Fukumura, D., Jain, R. J., and Alitalo, K. (1997). Hyperplasia of lymphatic vessels in VEGF-C transgenic mice. *Science* **276,** 1423–1425.

Jiang, X., Rowitch, D. H., Soriano, P., McMahon, A. P., and Sucov, H. M. (2000). Fate of the mammalian cardiac neural crest. *Development (Cambridge, UK)* **127,** 1607–1616.

Jouanneau, J., Moens, G., Montesano, R., and Thiery, J. P. (1995). FGF-1 but not FGF-4 secreted by carcinoma cells promotes in vitro and in vivo angiogenesis and rapid tumor proliferation. *Growth Factors* **12,** 37–47.

Joutel, A., and Tournier-Lasserve, E. (1998). Notch signalling pathway and human diseases. *Semin. Cell Dev. Biol.* **9,** 619–625.

Kaipainen, A., Korhonen, J., Mustonen, T., van Hinsbergh, V. W., M., Fang, G.-H., Dumont, D., Breitman, M., and Alitalo, K. (1995). Expression of the fms-like tyrosine kinase 4 gene becomes restricted to lymphatic eondothelium during development. *Proc. Natl. Acad. Sci. U.S.A.* **92,** 3566–3570.

Kappel, A., Ronicke, V., Damert, A., Flamme, I., Risau, W., and Breier, G. (1999). Identification of vascular endothelial growth factor (VEGF) receptor-2 (Flk-1) promoter/enhancer sequences sufficient for angioblast and endothelial cell-specific transcription in transgenic mice. *Blood* **93,** 4284–4292.

Kawasaki, T., Kitsukawa, T., Bekku, Y., Matsuda, Y., Sanbo, M., Yagi, T., and Fujisawa, H. (1999). A requirement for neuropilin-1 in embryonic vessel formation. *Development (Cambridge, UK)* **126,** 4895–4902.

Kimmel, R. A., Turnbull, D. H., Blanquet, V., Wurst, W., Loomis, C. A., and Joyner, A. L. (2000). Two lineage boundaries coordinate vertebrate apical ectodermal ridge formation. *Genes Dev.* **14,** 1377–1389.

Klagsbrun, M. (1991). Angiogenic factors: Regulators of blood supply-side biology. FGF, endothelial cell growth factors and angiogenesis: A keystone symposium, Keystone, CO, USA, April 1–7, 1991. *New Biol.* **3,** 745–749.

Kontos, C. D., Stauffer, T. P., Yang, W. P., York, J. D., Huang, L., Blanar, M. A., Meyer, T., and Peters, K. G. (1998). Tyrosine 1101 of Tie2 is the major site of association of p85 and is required for activation of phosphatidylinositol 3-kinase and Akt. *Mol. Cell Biol.* **18,** 4131–4140.

Korhonan, J., Lahtinen, I., Halmekyto, M., Alhonen, L., Janne, J., Dumont, D., and Alitalo, K. (1995). Endothelial-specific gene expression directed by the tie gene promoter in vivo. *Blood* **86,** 1828–1835.

Kotch, L. E., Iyer, N. V., Laughner, E., and Semenza, G. L. (1999). Defective vascularization of HIF-1alpha-null embryos is not associated with VEGF deficiency but with mesenchymal cell death. *Dev. Biol.* **209,** 254–267.

Kozak, K. R., Abbott, B., and Hankinson, O. (1997). ARNT-deficient mice and placental differentiation. *Dev. Biol.* **191,** 297–305.

Krebs, L. T., Xue, Y., Norton, C. R., Shutter, J. R., Maguire, M., Sundberg, J. P., Gallahan, D., Closson, V., Kitajewski, J., Callahan, R., Smith, G. H., Stark, K. L., and Gridley, T. (2000). Notch signaling is essential for vascular morphogenesis in mice. *Genes Dev.* **14,** 1343–1352.

Kulkarni, A. B., Huh, C. G., Becker, D., Geiser, A., Lyght, M., Flanders, K. C., Roberts, A. B., Sporn, M. B., Ward, J. M., and Karlsson, S. (1993). Transforming growth factor beta 1 null mutation in mice causes excessive inflammatory response and early death. *Proc. Natl. Acad. Sci. U.S.A.* **90,** 770–774.

Kwee, L., Baldwin, H. S., Shen, H. M., Stewart, C. L., Buck, C., Buck, C. A., and Labow, M. A. (1995). Defective development of the embryonic and extraembryonic circulatory systems in vascular cell adhesion molecule (VCAM-1) deficient mice. *Development (Cambridge, UK)* **121,** 489–503.

Lang, R. A., and Bishop, J. M. (1993). Macrophages are required for cell death and tissue remodeling in the developing mouse eye. *Cell (Cambridge, Mass.)* **74,** 453–462.

Lee, S. H., Hungerford, J. E., Little, C. D., and Iruela-Arispe, M. L. (1997). Proliferation and differentiation of smooth muscle cell precursors occurs simultaneously during the development of the vessel wall. *Dev. Dyn.* **209,** 342–352.

Leung, D. W., Cachianes, G., Kuang, W. J., Goeddel, D. V., and Ferrara, N. (1989). Vascular endothelial growth factor is a secreted angiogenic mitogen. *Science* **246,** 1306–1309.

Li, D. Y., Sorensen, L. K., Brooke, B. S., Urness, L. D., Davis, E. C., Taylor, D. G., Boak, B. B., and Wendel, D. P. (1999). Defective angiogenesis in mice lacking endoglin. *Science* **284,** 1534–1537.

Li, L., Miano, J. M., Mercer, B., and Olson, E. N. (1996). Expression of the SM22α promoter in transgenic mice provides evidence for distinct transcriptional regulatory programs in vascular and visceral smooth muscle cells. *J. Cell Biol.* **132,** 849–859.

Liao, E. C., Paw, B. H., Oates, A. C., Pratt, S. J., Postlethwait, J. H., and

Zon, L. I. (1998). SCL/Tal-1 transcription factor acts downstream of cloche to specify hematopoietic and vascular progenitors in zebrafish. *Genes Dev.* **12,** 621–626.

Liao, W., Bisgrove, B. W., Sawyer, H., Hug, B., Bell, B., Peters, K., Grunwald, D. J., and Stainier, D. Y. R. (1997). The zebrafish gene *cloche* acts upstream of a *flk-1* homologue to regulate endothelial cell differentiation. *Development (Cambridge, UK)* **124,** 381–389.

Liao, W., Ho, C., Yan, Y. L., Postlethwait, J., and Stainier, D. Y. (2000). Hhex and scl function in parallel to regulate early endothelial and blood differentiation in zebrafish. *Development (Cambridge, UK)* **127,** 4303–4313.

Lin, P., Buxton, J. A., Acheson, A., Radziejewski, C., Maisonpierre, P. C., Yancopoulos, G. D., Channon, K. M., Hale, L. P., Dewhirst, M. W., George, S. E., and Peters, K. G. (1998). Antiangiogenic gene therapy targeting the endothelium-specific receptor tyrosine kinase Tie2. *Proc. Natl. Acad. Sci. U.S.A.* **95,** 8829–8834.

Lindahl, P., Johansson, B. R., Leveen, P., and Betsholtz, C. (1999). Pericyte loss and microaneurysm formation in PDGF-B-deficient mice. *Science* **277,** 242–245.

Linehan, W. M., Lerman, M. I., and Zbar, B. (1995). Identification of the von Hippel-Lindau (VHL) gene. Its role in renal cancer. *JAMA, J. Am. Med. Assoc.* **273,** 564–570.

Liu, Y., Wada, R., Yamashita, T., Mi, Y., Deng, C.-X., H0bson, J. P., Rosenfeldt, H. M., Nava, V. E., Chae, S.-S., Lee, M.-J., Liu, C. H., Hla, T., Spiegel, S., and Proia, R. L. (2000). Edg-1, the G-protein-coupled receptor for sphingosine-1-phosphate, is essential for vascular maturation. *J. Clin. Invest.* **106,** 951–961.

Loughan, S., and Sato, T. N. (2001). A combinatorial role of angiopoietin-1 and orphan receptor Tie1 pathways in establishing vascular polarity during angiogenesis. *Mol. Cell* **7,** 233–239.

Maisonpierre, P. C., Suri, C., Jones, P. F., Bartunkova, S., Wiegand, S. J., Radziejewski, C., Compton, D., McClain, J., Aldrich, T. H., Papadapoulos, N., Daly, T. J., Davis, S., Sato, T. N., and Yancopoulos, G. D. (1997). Angiopoietin-2, a natural antagonist for Tie2 that disrupts in vivo angiogenesis. *Science* **277,** 55–60.

Maltepe, E., Schmidt, J. V., Baunoch, D., Bradfield, C. A., and Simon, M. C. (1997). Abnormal angiogenesis and responses to glucose and oxygen deprivation in mice lacking the protein ARNT. *Nature (London)* **386,** 403–407.

McAllister, K. A., Grogg, K. M., Johnson, D. W., Gallione, C. J., Baldwin, M. A., Jackson, C. E., Helmbold, E. A., Markel, D. S., McKinnon, W. C., Murrell, J., McCormick, M. K., Pericak-Vance, M. A., Heutink, P., Oostra, B. A., Haitjema, T., Westerman, C. J., J., Porteous, M. E., Guttmacher, A. E., Letarte, M., and Marchuk, D. A. (1994). Endoglin, a TGF-β binding protein of endothelial cells, is the gene for hereditary haemorrhagic telangiectasia type 1. *Nat. Genet.* **8,** 345–351.

Medvinsky, A., and Dzierzak, E. (1996). Definitive hematopoiesis is autonomously initiated by the AGM region. *Cell (Cambridge, Mass.)* **86,** 897–906.

Medvinsky, A. L., Samoylina, N. L., Muller, A. M., and Dzierzak, E. A. (1993). An early pre-liver intraembryonic source of CFU-S in the developing mouse. *Nature (London)* **364,** 64–67.

Meeson, A. P., Palmer, M., Calfton, M., and Lang, R. A. (1996). A relationship between apoptosis and flow during programmed capillary regression is revealed by vital analysis. *Development (Cambridge, UK)* **122,** 3929–3938.

Meeson, A. P., Argilla, M., Ko, K., Witte, L., and Lang, R. A. (1999). VEGF deprivation-induced apoptosis is a component of programmed capillary regression. *Development (Cambridge, UK)* **126,** 1407–1415.

Mellitzer, G., Xu, Q., and Wilkinson, D. G. (1999). Eph receptors and ephrins restrict cell intermingling and communication. *Nature (London)* **400,** 77–81.

Mikawa, T., and Fischman, D. A. (1992). Retroviral analysis of cardiac morphogenesis: Discontinuous formation of coronary vessels. *Proc. Natl. Acad. Sci. U.S.A.* **89,** 9504–9508.

Mikawa, T., and Gourdie, R. G. (1996). Pericardial mesoderm generates a population of coronary smooth muscle cells migrating into the heart along with ingrowth of the epicardial organ. *Dev. Biol.* **174,** 221–232.

Millauer, B., Shawver, L. K., Plate, K. H., Risau, W., and Ullrich, A. (1994). Glioblastoma growth inhibited in vivo by a dominant-negative Flk-1 mutant. *Nature (London)* **367,** 576–579.

Millauer, B., Longhi, M. P., Plate, K. H., Shawver, L. K., Risau, W., Ullrich, A., and Strawn, L. M. (1996). Dominant-negative inhibition of Flk-1 suppresses the growth of many tumor types in vivo. *Cancer Res.* **56,** 1615–1620.

Miyazono, K. (1999). Signal transduction by bone morphogenetic protein receptors: Functional roles of Smad proteins. *Bone* **25,** 91–93.

Motoike, T., Loughna, S., Perens, E., Roman, B. L., Liao, W., Chau, T. C., Richardson, C. D., Kawate, T., Kuno, J., Weinstein, B. M., Stainier, D. Y., and Sato, T. N. (2000). Universal GFP reporter for the study of vascular development. *Genesis* **28,** 75–81.

Nagase, H., and Woessner, J. F., Jr. (1999). Matrix metalloproteinases. *J. Biol. Chem.* **274,** 21491–21494.

Neuhaus, J., Risau, W., and Wolburg, H. (1991). Induction of blood-brain barrier characteristics in bovine brain endothelial cells by rat astroglial cells in transfilter coculture. *Ann. N.Y. Acad. Sci.* **633,** 578–580.

Niki, M., Okada, H., Takano, H., Kuno, J., Tani, K., Hibino, H., Asano, S., Ito, Y., Satake, M., and Noda, T. (1997). Hematopoiesis in the fetal liver is impaired by targeted mutagenesis of a gene encoding a non-DNA binding subunit of the transcription factor, polyomavirus enhancer binding protein 2/core binding factor. *Proc. Natl. Acad. Sci. U.S.A.* **94,** 5697–5702.

North, T., Gu, T. L., Stacy, T., Wang, Q., Howard, L., Binder, M., Marin-Padilla, M., and Speck, N. A. (1999). Cbfa2 is required for the formation of intra-aortic hematopoietic clusters. *Development (Cambridge, UK)* **126,** 2563–2575.

Okuda, T., van Deursen, J., Hiebert, S. W., Grosveld, G., and Downing, J. R. (1996). AML1, the target of multiple chromosomal translocations in human leukemia, is essential for normal fetal liver hematopoiesis. *Cell (Cambridge, Mass.)* **84,** 321–330.

Papapetropoulos, A., Garcia-Cardena, G., Dengler, T. J., Maisonpierre, P. C., Yancopoulos, G. D., and Sessa, W. C. (1999). Direct actions of angiopoietin-1 on human endothelium: Evidence for network stabilization, cell survival, and interaction with other angiogenic growth factors. *Lab. Invest.* **79,** 213–223.

Papapetropoulos, A., Fulton, D., Mahboubi, K., Kalb, R. G., O'Connor, D. S., Li, F., Altieri, D. C., and Sessa, W. C. (2000). Angiopoietin-1 inhibits endothelial cell apoptosis via the Akt/survivin pathway. *J. Biol. Chem.* **275,** 9102–9105.

Partanen, J., Puri, M. C., Schwartz, L., Fischer, K.-D., Bernstein, A., and Rossant, J. (1996). Cell autonomouse functions of the receptor tyrosine TIE in a late phase of angiogenic capillary growth and endothelial cell survival during murine development. *Development (Cambridge, UK)* **122,** 3013–3021.

Perez-Pomares, J. M., Macias, D., Garcia-Garrido, L., and Munoz-Chapuli, R. (1998). The origin of the subepicardial mesenchyme in the avian embryo: An immunohistochemical and quail-chick chimera study. *Dev. Biol.* **200,** 57–68.

Peters, K. G. (1998). Vascular endothelial growth factor and the angiopoietins: Working together to build a better blood vessel. *Circ. Res.* **83,** 342–343.

Puri, M. C., Rossant, J., Alitalo, K., Bernstein, A., and Partanen, J. (1995). The receptor tyrosine kinase TIE is required for integrity and survival of vascular endothelial cells. *EMBO J.* **14,** 5884–5891.

Puri, M. C., Partanen, J., Rossant, J., and Bernstein, A. (1999). Interaction of the TEK and TIE receptor tyrosine kinases during cardiovascular development. *Development (Cambridge, UK)* **126,** 4569–4580.

Raftery, L. A., and Sutherland, D. J. (1999). TGF-beta family signal transduction in Drosophila development: From Mad to Smads. *Dev. Biol.* **210,** 251–268.

Ravi, R., Mookerjee, B., Bhujwalla, Z. M., Sutter, C. H., Artemov, D., Zeng, Q., Dillehay, L. E., Madan, A., Semenza, G. L., and Bedi, A. (2000). Regulation of tumor angiogenesis by p53-induced degradation of hypoxia- inducible factor 1alpha. *Genes Dev.* **14,** 34–44.

Risau, W. (1991a). Embryonic angiogenesis factors. *Pharmacol. Ther.* **51,** 371–376.

Risau, W. (1991b). Vasculogenesis, angiogenesis and endothelial cell differentiation during embryonic development. *In* "The Development of the Vascular System" (R. N. Feinberg, G. K. Shere, and R. Auerbach, eds.), pp. 58–68. Krager, Basel, Switzerland.

Risau, W. (1997). Mechanisms of angiogenesis. *Nature (London)* **386,** 671–674.

Risau, W., and Flamme, I. (1995). Vasculogenesis. *Annu. Rev. Cell Dev. Biol.* **11,** 73–91.

Risau, W., and Wolburg, H. (1990). Development of the blood-brain barrier. *Trends Neurosci.* **13,** 174–178.

Risau, W., Hallmann, R., and Albrecht, U. (1986a). Differentiation-dependent expression of proteins in brain endothelium during development of the blood-brain barrier. *Dev. Biol.* **117,** 537–545.

Risau, W., Hallmann, R., Albrecht, U., and Henke, F. S. (1986b). Brain induces the expression of an early cell surface marker for blood-brain barrier-specific endothelium. *EMBO J.* **5,** 3179–3183.

Rossant, J., and McMahon, A. (1999). "Cre"-ating mouse mutants—a meeting review on conditional mouse genetics. *Genes Dev.* **13,** 142–145.

Ryan, H. E., Lo, J., and Johnson, R. S. (1998). HIF-1 alpha is required for solid tumor formation and embryonic vascularization. *EMBO J.* **17,** 3005–3015.

Sabin, F. R. (1917). Origin and development of the primitive vessels of the chick and of the pig. *Carnegie Contrib. Embryol.* **6,** 61–124.

Sasaki, K., Yagi, H., Bronson, R. T., Tominaga, K., Matsunashi, T., Deguchi, K., Tani, Y., Kishimoto, T., and Komori, T. (1996). Absence of fetal liver hematopoiesis in mice deficient in transcriptional coactivator core binding factor beta. *Proc. Natl. Acad. Sci. U.S.A.* **93,** 12359–12363.

Sato, T. N. (2000). A new role of lipid receptors in vascular and cardiac morphogenesis. *J. Clin. Invest.* **106,** 939–940.

Sato, T. N., Tozawa, Y., Deutsch, U., Wolburg-Buchholz, K., Fujiwara, Y., Gendron-Maguire, M., Gridley, T., Wolburg, H., Risau, W., and Qin, Y. (1995). Distinct roles of the receptor tyrosine kinases Tie-1 and Tie-2 in blood vessel formation. *Nature (London)* **376,** 70–74.

Schelling, M. E. (1991). FGF mediation of coronary angiogenesis. *Ann. N.Y. Acad. Sci.* **638,** 467–469.

Schlaeger, T. M., Qin, Y., Fujiwara, Y., Magram, J., and Sato, T. N. (1995). Vascular endothelial cell lineage-specific promoter in transgenic mice. *Development (Cambridge, UK)* **121,** 1089–1098.

Schlaeger, T. M., Bartunkova, S., Lawitts, J. A., Teichmann, G., Risau, W., Deutch, U., and Sato, T. N. (1997). Uniform vascular-endothelial-cell-specific gene expression in both embryonic and adult transgenic mice. *Proc. Natl. Acad. Sci. U.S.A.* **94,** 3058–3063.

Schuh, A. C., Faloon, P., Hu, Q. L., Bhimani, M., and Choi, K. (1999). In vitro hematopoietic and endothelial potential of flk-1(-/-) embryonic stem cells and embryos. *Proc. Natl. Acad. Sci. U.S.A.* **96,** 2159–2164.

Seghezzi, G., Patel, S., Ren, C. J., Gualandris, A., Pintucci, G., Robbins, E. S., Shapiro, R. L., Galloway, A. C., Rifkin, D. B., and Mignatti, P. (1998). Fibroblast growth factor-2 (FGF-2) induces vascular endothelial growth factor (VEGF) expression in the endothelial cells of forming capillaries: An autocrine mechanism contributing to angiogenesis. *J. Cell Biol.* **141,** 1659–1673.

Semenza, G. L. (1999). Perspectives on oxygen sensing. *Cell (Cambridge, Mass.)* **98,** 281–284.

Shalaby, F., Rossant, J., Yamaguchi, T. P., Breitman, M., and Schuh, A. C. (1995). Failure of blood island formation, vasculogenesis, and hematopoiesis in flk-1 deficient mice. *Nature (London)* **376,** 62–66.

Shalaby, F., Ho, J., Stanford, W. L., Fischer, K.-D., Schuh, A. C., Schwartz, L., Bernstein, A., and Rossant, J. (1997). A requirement for Flk1 in

primitive and definitive hematopoiesis and vasculogenesis. *Cell (Cambridge, Mass.)* **89,** 981–990.

Shapiro, S. D. (1997). Mighty mice: Transgenic technology "knocks out" questions of matrix metalloproteinase function. *Matrix Biol.* **15,** 527–533.

Shirayoshi, Y., Yuasa, Y., Suzuki, T., Sugaya, K., Kawase, E., Ikemura, T., and Nakatsuji, N. (1997). Proto-oncogene of int-3, a mouse Notch homologue, is expressed in endothelial cells during early embryogenesis. *Genes Cells* **2,** 213–224.

Shutter, J. R., Scully, S., Fan, W., Richards, W. G., Kitajewski, J., Deblandre, G. A., Kintner, C. R., and Stark, K. L. (2000). Dll4, a novel notch ligand expressed in arterial endothelium. *Genes Dev.* **14,** 1313–1318.

Smith, R. A., and Glomski, C. A. (1982). "Hemogenic endothelium" of the embryonic aorta: Does it exist? *Dev. Comp. Immunol.* **6,** 359–368.

Soker, S., Takashima, S., Miao, H. Q., Neufeld, G., and Klagsbrun, M. (1998). Neuropilin-1 is expressed by endothelial and tumor cells as an isoform- specific receptor for vascular endothelial growth factor. *Cell (Cambridge, Mass.)* **92,** 735–745.

Stainier, D. Y., Weinstein, B. M., Detrich, H. W., 3rd, Zon, L. I., and Fishman, M. C. (1995). Cloche, an early acting zebrafish gene, is required by both the endothelial and hematopoietic lineages. *Development (Cambridge, UK)* **121,** 3141–3150.

Suri, C., Jones, P. F., Patan, S., Bartunkova, S., Maisonpierre, P. C., Davis, S., Sato, T. N., and Yancopoulos, G. D. (1996). Requisite role of angiopoietin-1, a ligand for the TIE2 receptor, during embryonic angiogenesis. *Cell (Cambridge, Mass.)* **87,** 1171–1180.

Suri, C., McClain, J., Thurston, G., McDonald, D. M., Zhou, H., Oldmixon, E. H., Sato, T. N., and Yancopoulos, G. D. (1998). Increased vascularization in mice overexpressing angiopoietin-1. *Science* **282,** 468–471.

Tavian, M., Coulombel, L., Luton, D., Clemente, H. S., Dieterlen-Lievre, F., and Peault, B. (1996). Aorta-associated CD34+ hematopoietic cells in the early human embryo. *Blood* **87,** 67–72.

Tevosian, S. G., Deconinck, A. E., Tanaka, M., Schinke, M., Litovsky, S. H., Izumo, S., Fujiwara, Y., and Orkin, S. H. (2000). FOG-2, a cofactor for GATA transcription factors, is essential for heart morphogenesis and development of coronary vessels from epicardium. *Cell (Cambridge, Mass.)* **101,** 729–739.

Thurston, G., Suri, C., Smith, K., McClain, J., Sato, T. N., Yancopoulos, G. D., and McDonald, D. M. (1999). Leakage-resistant blood vessels in mice transgenically overexpressing angiopoietin-1. *Science* **286,** 2511–2514.

Urness, L. D., Sorensen, L. K., and Li, D. Y. (2000). Arteriovenous malformations in mice lacking activin receptor-like kinase-1. *Nat. Genet.* **26,** 328–331.

Urushihara, H., and Takeichi, M. (1980). Cell-cell adhesion molecule: Identification of a glycoprotein relevant to the Ca2+-independent aggregation of Chinese hamster fibroblasts. *Cell (Cambridge, Mass.)* **20,** 363–371.

Uyttendaele, H., Marazzi, G., Wu, G., Yan, Q., Sassoon, D., and Kitajewski, J. (1996). Notch4/int-3, a mammary proto-oncogene, is an endothelial cell-specific mammalian Notch gene. *Development (Cambridge, UK)* **122,** 2251–2259.

Vikkula, M., Boon, L. M., Carraway, K. L., Calvert, J. T., Diamonti, A. J., Goumnerov, B., Pasyk, K. A., Marchuk, D. A., Warman, M. L., Cantley, L. C., Mulliken, J. B., and Olsen, B. R. (1996). Vascular dysmorphogenesis caused by an activating mutation in the receptor tyrosine kinase TIE2. *Cell (Cambridge, Mass.)* **87,** 1181–1190.

Vrancken Peeters, M. P., Gittenberger-de Groot, A. C., Mentink, M. M., and Poelmann, R. E. (1999). Smooth muscle cells and fibroblasts of the coronary arteries derive from epithelial-mesenchymal transformation of the epicardium. *Anat. Embryol.* **199,** 367–378.

Vu, T. H., Shipley, J. M., Bergers, G., Berger, J. E., Helms, J. A., Hanahan, D., Shapiro, S. D., Senior, R. M., and Werb, Z. (1998). MMP-9/gelatinase B is a key regulator of growth plate angiogenesis and apoptosis of hypertrophic chondrocytes. *Cell (Cambridge, Mass.)* **93,** 411–422.

Wang, H. U., Chen, Z. F., and Anderson, D. J. (1998). Molecular distinction and angiogenic interaction between embryonic arteries and veins revealed by ephrin-B2 and its receptor Eph-B4. *Cell (Cambridge, Mass.)* **93**, 741–753.

Wang, Q., Stacy, T., Binder, M., Marin-Padilla, M., Sharpe, A. H., and Speck, N. A. (1996a). Disruption of the Cbfa2 gene causes necrosis and hemorrhaging in the central nervous system and blocks definitive hematopoiesis. *Proc. Natl. Acad. Sci. U.S.A.* **93**, 3444–3449.

Wang, Q., Stacy, T., Miller, J. D., Lewis, A. F., Gu, T. L., Huang, X., Bushweller, J. H., Bories, J. C., Alt, F. W., Ryan, G., Liu, P. P., Wynshaw-Boris, A., Binder, M., Marin-Padilla, M., Sharpe, A. H., and Speck, N. A. (1996b). The CBFbeta subunit is essential for CBFalpha2 (AML1) function in vivo. *Cell (Cambridge, Mass.)* **87**, 697–708.

Werb, Z. (1997). ECM and cell surface proteolysis: Regulating cellular ecology. *Cell (Cambridge, Mass.)* **91**, 439–443.

Werb, Z., and Chin, J. R. (1998). Extracellular matrix remodeling during morphogenesis. *Ann. N.Y. Acad. Sci.* **857**, 110–118.

Whitman, M. (1997). Signal transduction. Feedback from inhibitory SMADs. *Nature (London)* **389**, 549–551.

Wigle, J. T., and Oliver, G. (1999). Prox1 function is required for the development of the murine lymphatic system. *Cell (Cambridge, Mass.)* **98**, 769–778.

Wilkinson, D. G. (2000a). Eph receptors and ephrins: Regulators of guidance and assembly. *Int. Rev. Cytol.* **196**, 177–244.

Wilkinson, D. G. (2000b). Topographic mapping: Organizing by repulsion and competition? *Curr. Biol.* **10**, R447–R451.

Witte, L., Hicklin, D. J., Zhu, Z., Pytowski, B., Kotanides, H., Rockwell, P., and Bohlen, P. (1998). Monoclonal antibodies targeting the VEGF receptor-2 (Flk1/KDR) as an anti-angiogenic therapeutic strategy. *Cancer Metastasis Rev.* **17**, 155–161.

Wu, H., Lee, S. H., Gao, J., Liu, X., and Iruela-Arispe, M. L. (1999). Inactivation of erythropoietin leads to defects in cardiac morphogenesis. *Development (Cambridge, UK)* **126**, 3597–3605.

Wu, Q., and Maniatis, T. (1999). A striking organization of a large family of human neural cadherin-like cell adhesion genes. *Cell (Cambridge, Mass.)* **97**, 779–790.

Wu, Q., and Maniatis, T. (2000). Large exons encoding multiple ectodomains are a characteristic feature of protocadherin genes. *Proc. Natl. Acad. Sci. U.S.A.* **97**, 3124–3129.

Xu, Q., Mellitzer, G., Robinson, V., and Wilkinson, D. G. (1999). In vivo cell sorting in complementary segmental domains mediated by Eph receptors and ephrins. *Nature (London)* **399**, 267–271.

Yamashita, J., Itoh, H., Hirashima, M., Ogawa, M., Nishikawa, S., Yurugi, T., Naito, M., and Nakao, K. (2000). Flk1-positive cells derived from embryonic stem cells serve as vascular progenitors. *Nature (London)* **408**, 92–96.

Yang, J. T., Rayburn, H., and Hynes, R. O. (1995). Cell adhesion events mediated by alpha 4 integrins are essential in placental and cardiac development. *Development (Cambridge, UK)* **121**, 549–560.

Yang, X., Castilla, L. H., Xu, X., Li, C., Gotay, J., Weinstein, M., Liu, P. P., and Deng, C. X. (1999). Angiogenesis defects and mesenchymal apoptosis in mice lacking SMAD5. *Development (Cambridge, UK)* **126**, 1571–1580.

Zinyk, D. L., Mercer, E. H., Harris, E., Anderson, D. J., and Joyner, A. L. (1998). Fate mapping of the mouse midbrain-hindbrain constriction using a site- specific recombination system. *Curr. Biol.* **8**, 665–668.

12

Stem Cells of the Nervous System

Sean J. Morrison

Departments of Internal Medicine and Cell and Developmental Biology, Howard Hughes Medical Institute,
University of Michigan, Ann Arbor, Michigan, 48109

I. Introduction

II. Lineage Determination of Neural Stem Cells

III. Do Stem Cells Retain Broad or Narrow
   Neuronal Potentials?

IV. Regulation of Neural Stem Cell Self-Renewal

V. Differences between Hematopoietic Stem
   Cells and Neural Stem Cells

VI. *In Vivo* Function of Neural Stem Cells

VII. Surprising Potential of Neural Stem Cells

VIII. Are Neural Stem Cells Involved in Disease?

IX. Outstanding Issues

   References

I. Introduction

Like other tissues, the nervous system contains stem cells. Stem cells are self-renewing, multipotent progenitors with the broadest developmental potential in a given tissue at a given time (Morrison *et al.,* 1997a). Neural stem cells give rise to multiple different types of neurons and glia. Many different types of neural stem cells are probably present in different regions of the nervous system that differ in the types of cells they produce. Broadly speaking, neural stem cells can be considered to fall within two general classes:

central nervous system (CNS) stem cells (McKay, 1997; Gage, 1998; Temple and Alvarez-Buylla, 1999) and neural crest stem cells (NCSCs) (Anderson, 1989, 1997). These two classes of stem cells are present in different regions of the nervous system, have different developmental potentials, and give rise to different types of cells *in vivo.* CNS stem cells are born in the ventricular zone of the neural tube and give rise to neurons, astrocytes, and oligodendrocytes in the spinal cord and brain (Fig. 1). Multipotent CNS progenitors self-renew *in vitro* (Davis and Temple, 1994; Gritti *et al.,* 1996), and lineage marking experiments strongly suggest that these cells self-renew *in vivo* as well (Reid *et al.,* 1995; Morshead *et al.,* 1998). NCSCs are also born in the neural tube but migrate throughout the embryo and give rise to the neurons and glia of the peripheral nervous system (PNS) as well as mesectodermal derivatives (such as vascular smooth muscle and bone) in other tissues (Fig. 1). NCSCs self-renew *in vitro* and *in vivo* (Stemple and Anderson, 1992; Morrison *et al.,* 1999). Although CNS stem cells and NCSCs are different types of neural stem cells, important principles in neural stem cell biology can best be illustrated by combining examples from both systems.

A. When and Where Are Neural Stem Cells Present?

CNS stem cells persist throughout life in addition to participating in the formation of the CNS during embryonic

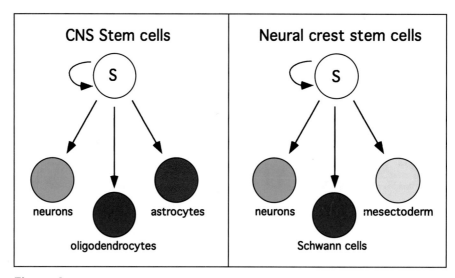

Figure 1 The types of lineages generated by neural stem cells. CNS stem cells and neural crest stem cells are different types of stem cells that can be distinguished based on differences in where they occur *in vivo* and in the types of cells they give rise to. CNS stem cells self-renew and also give rise to neurons, astrocytes, and oligodendrocytes. Astrocytes and oligodendrocytes are the two types of glia from the CNS. Stem cells from different regions of the CNS and from different times during ontogeny give rise to different types of neurons. NCSCs self-renew and also give rise to neurons, Schwann cells, and mesectoderm. Schwann cells are glia from the PNS. Mesectodermal cells include certain bones in the head, dermis in the head, and vascular smooth muscle around the great arteries near the heart. NCSCs give rise to different types of neurons in different regions of the PNS.

development. During fetal development, stem cells are present in the ventricular zone that lines the lumen of the neural tube. These cells are present throughout the neural axis, from regions fated to form the cortex (Davis and Temple, 1994) to regions fated to form the spinal cord (Kalyani *et al.,* 1997; Palmer *et al.,* 1999). In the adult, stem cells continue to be present near the ventricles. The ventricular zone of fetal development is replaced in adults by the ependymal layer (immediately adjacent to the lumen of the ventricle) and the subependymal layer (Barres, 1999). There is evidence for the existence of stem cells in both the ependymal (Johansson *et al.,* 1999) and subependymal layers (Morshead *et al.,* 1994; Doetsch *et al.,* 1999) of the adult CNS. In addition to being present near the ventricles, some stem cells take up residence in the dentate gyrus of the adult hippocampus (Palmer *et al.,* 1997). Indeed, there is evidence that neural stem cells can be generated from many different regions of the adult CNS by culturing cells *in vitro* (Weiss *et al.,* 1996), even from regions of the CNS that show no evidence of cell turnover. One possibility is that rare, latent stem cells persist in a quiescent state throughout the CNS. Another possibility is that certain cells that do not have stem cell properties *in vivo* can sometimes acquire stem cell properties as a result of prolonged culture in high concentrations of mitogens like *basic fibroblast growth factor* (bFGF or FGF-2) (Donovan, 1994; Gage, 1998). To confirm where and when CNS stem cells are present, it is necessary to identify them prospectively (identify markers that predict which cells are stem

cells and which are restricted progenitors) so that they can be studied *in vivo.*

NCSCs are only known to exist during fetal development and there is not yet any convincing evidence for the persistence of NCSCs or the generation of new neurons in the adult PNS. Neural crest cells, including NCSCs, are born in the neural tube but migrate throughout the embryo in early to midstation, depending on the species. Neural crest migration lasts only a short period of time, 24–48 hr in avians and rodents. Multipotent neural crest progenitors have been detected transiently in postmigratory sites such as the skin (Richardson and Sieber-Blum, 1993), the sensory ganglia (Duff *et al.,* 1991; Deville *et al.,* 1992; Hagedorn *et al.,* 1999), the gut (Deville *et al.,* 1994; Lo and Anderson, 1995), and the sympathetic ganglia (Duff *et al.,* 1991); however, multipotent neural crest cells progressively restrict their developmental potential while migrating (Baroffio *et al.,* 1991), and many cells differentiate soon after reaching postmigratory sites (Le Douarin and Dupin, 1992).

Until recently there was no evidence for NCSC self-renewal *in vivo,* and it had been thought that NCSCs differentiate within a few days of migrating; however, NCSCs were recently observed to persist into late gestation by self-renewing within peripheral nerves (Morrison *et al.,* 1999). It remains to be determined whether NCSCs self-renew in other regions of the PNS. Nonetheless, the *in vivo* self-renewal and persistence of NCSCs in nerves suggest that NCSCs may play a more dynamic role in PNS development than previously thought. Even so, there remains no evidence that NCSCs persist or give rise to new neurons in the adult PNS.

B. What Do Stem Cells Do in the Adult CNS?

Stem cells in the adult CNS give rise to neurons that may be involved in learning and memory. In adult mammals, neurons continue to be born in the olfactory bulb and hippocampus (Altman and Das, 1966; Altman, 1969; Kaplan and Hinds, 1977; Eriksson *et al.*, 1998) and perhaps throughout the neocortex (Gould *et al.*, 1999b). Stem cells that reside near the lateral ventricle (Lois and Alvarez-Buylla, 1993; Doetsch *et al.*, 1999; Johansson *et al.*, 1999) undergo differentiation into neurons and glia in the subependymal zone, and then they migrate through what is known as the rostral migratory stream into the olfactory bulb (Lois and Alvarez-Buylla, 1994). Within the olfactory bulb the neurons integrate as inhibitory interneurons called granule cells. The exact function of the newly born granule cells is unknown but increased odorant stimulation promotes neurogenesis (Frazier-Cierpial and Brunjes, 1989; Rosselli-Austin and Williams, 1989; Corotto *et al.*, 1994), and added granule cells may improve odorant discrimination by modulating the signals transmitted from the olfactory bulb to the cortex (Yokoi *et al.*, 1995). In the hippocampus, stem cells reside in the dentate gyrus where they differentiate into granule neurons (Palmer *et al.*, 1997). Again, the function of these new neurons is unknown but there is evidence that hippocampal neurogenesis is associated with environmental enrichment and learning (Kempermann *et al.*, 1997, 1998; Gould *et al.*, 1999a). The discovery that some neurons born in the subventricular zone also incorporate into the neocortex has given further impetus to the idea that adult neurogenesis may be involved in learning and memory (Gould *et al.*, 1999b). The next step will be to determine whether the formation of new neurons in the hippocampus is required for learning or memory.

C. Do All Nervous System Cells Derive from Stem Cells?

It is not yet clear whether all nervous system cells can be derived from stem cells. For example, certain lineages of nervous system cells may derive from progenitors that were born with restricted developmental potential rather than multipotency. Within the CNS, multipotent and restricted progenitors coexist in the ventricular zone (Luskin *et al.*, 1988; Davis and Temple, 1994; Williams and Price, 1995). The simplest interpretation is that the restricted progenitors arise from stem cells, but we cannot refute the possibility that at least certain lineages of CNS cells arise from progenitors that were born committed (Barres, 1999). That is, such cells may have had restricted developmental potentials at the time they acquired neural potential.

Within the PNS there is evidence that some cells arise from committed progenitors that may be geneologically unrelated to NCSCs. The neural crest is composed of a heterogeneous collection of progenitors including multipotent as well as restricted progenitors (Le Douarin, 1986). At least some melanocytes may derive from progenitors that are already committed to this fate at the time they migrate from the neural tube (Erickson and Goins, 1995; Wakamatsu *et al.*, 1998). Similarly, restricted progenitors of sensory neurons are observed among migrating neural crest cells (Greenwood *et al.*, 1999) as well as among developing sensory ganglion cells (Frank and Sanes, 1991). Furthermore, even some premigratory neural crest progenitors (that still reside in the neural tube) are fated to give rise to only sensory ganglion or melanocyte lineage cells (Bronner-Fraser and Fraser, 1989; Serbedzija *et al.*, 1994). These observations suggest the possibility that some committed progenitors may be born in the neural tube as a separate lineage from multipotent NCSCs, even though there is also evidence that multipotent progenitors can retain sensory (Fraser and Bronner-Fraser, 1991) and/or melanocytic (Sieber-Blum and Cohen, 1980; Baroffio *et al.*, 1988; Sieber-Blum, 1989) potential. If certain lineages of cells do not arise from multipotent stem cells, then controls on the overt differentiation of such cells might be quite different than on multipotent progenitors that must undergo lineage determination before differentiating.

II. Lineage Determination of Neural Stem Cells

Environmental factors can promote the generation of a particular cell type from neural stem cells via two very different types of mechanisms: instruction and selection (Fig. 2). Instruction is the process by which a factor promotes differentiation into one lineage at the expense of other lineages by acting on the stem cells and causing them to preferentially undergo commitment. Selection does not act at the level of stem cells but rather affects the survival or proliferation of the committed cells that arise from the stem cells. Thus a factor can selectively promote the generation of a particular lineage by promoting cell survival or proliferation after they differentiate from stem cells (or by killing or impairing the proliferation of other lineages of cells after they arise from stem cells). In many systems, such as in hematopoiesis, it has been difficult to rigorously demonstrate whether particular factors act instructively or selectively because the stem cells cannot be studied *in vitro* with enough precision to determine the mechanism by which factors act. In contrast, outstanding tools are available to study the self-renewal and multilineage differentiation of individual neural stem cells in culture. Thus impressive progress has been made in understanding how lineage determination factors work in the

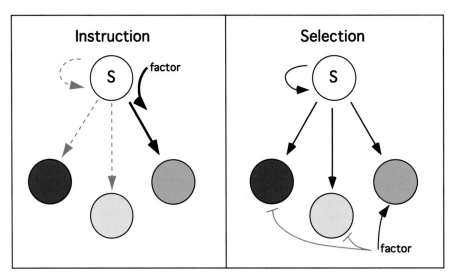

Figure 2 Lineage determination factors can act instructively or selectively. Instructive factors act at the level of stem cells to promote differentiation into one fate at the expense of other fates. For example, BMP2 instructs NCSCs to differentiate into neurons (Shah *et al.*, 1996). Selective factors do not affect stem cell differentiation but increase the numbers of cells from one lineage by promoting their survival or proliferation after they differentiate, or by impairing the survival or proliferation of other lineages of cells after they differentiate. For example, neurotrophins often act selectively to regulate the survival of neurons after they differentiate.

nervous system. As described below, lineage determination factors in the nervous system can act instructively, selectively, or by a combination of those mechanisms. These mechanisms are best illustrated by examples taken from NCSC biology, but other growth factors are thought to have analogous effects on CNS stem cells.

A. Bone Morphogenetic Proteins Instruct Neuronal Differentiation

Bone morphogenetic proteins (BMPs) instruct NCSCs to differentiate into autonomic neurons, thereby regulating the formation of the autonomic nervous system. At approximately E11 of mouse development, when autonomic ganglia are forming, BMP2 and/or BMP4 are expressed around the dorsal aorta, near where sympathetic neurons form, and in the heart and lung, near where parasympathetic neurons form (Bitgood and McMahon, 1995; Lyons *et al.*, 1995; Reissman *et al.*, 1996; Shah *et al.*, 1996). Thus BMPs are expressed at the right time and in the right place to influence autonomic neuron differentiation. To demonstrate that BMPs are sufficient to cause autonomic neuron differentiation *in vivo*, BMP4 was ectopically expressed in the neural crest migration pathway of chick embryos (Reissman *et al.*, 1996). This caused an increase in the number of sympathetic neurons that differentiated in the sympathetic ganglion as well as the formation of additional sympathetic neurons in locations where they do not normally form, such as in the neural crest migration pathway. To demonstrate that BMPs

are necessary for autonomic neuron differentiation, Noggin, a protein that binds and inactivates BMPs (Smith and Harland, 1992; Zimmerman *et al.*, 1996), was overexpressed in chick embryos (Schneider *et al.*, 1999). This greatly impaired the differentiation of sympathetic neurons. Thus BMPs are necessary and sufficient for the generation of sympathetic neurons *in vivo*.

Although *in vivo* studies have shown that BMPs are expressed at the right time and in the right place and are necessary and sufficient for sympathetic neuron differentiation, these studies did not tell us how BMPs regulate sympathetic neuron differentiation. Do BMPs instruct NCSCs to differentiate into neurons, or are they required for the survival of sympathetic neurons? Do BMPs act directly on NCSCs or indirectly by causing other cells in the environment to secrete secondary factors that affect NCSC differentiation? *In vitro* experiments were required to answer these questions. Purified recombinant BMP2 causes NCSCs to differentiate into neurons when added to cultures of purified NCSCs (Shah *et al.*, 1996; Morrison *et al.*, 1999). Thus BMP2 acts directly on NCSCs or their progeny to cause neuronal differentiation. To determine whether BMP2 acts instructively or selectively, purified NCSCs were added to culture at clonal density (~20 stem cells/35-mm dish such that individual stem cells formed spatially separated clones in which the progeny of individual stem cells could be distinguished and observed separately). After 4 hr, when cells had attached to the plate, the live cells were circled by etching on the underside of the culture dish. Then BMP2 was added to some cultures. After only 4 days of incubation the cultures were stained with antibodies against the neuronal marker peripherin. In both control and BMP2-supplemented plates, around 90% of clones survived. In the absence of BMP2 no neurons differentiated (under control conditions neuronal differentiation is not evident for 13 days), but in the presence of BMP2 more than 80% of clones gave rise to neurons. Serial analysis of cultures every 24 hr indicated that only rare dead cells were observed within colonies exposed to BMP2, demonstrating that BMP2 could not be acting selectively in an intraclonal manner by killing cells within clones that did not differentiate into neurons (Shah *et al.*, 1996). Thus BMP2 increases the rate and extent of neurogenesis by stem cells without killing cells and therefore acts instructively (Shah *et al.*, 1996; Morrison *et al.*, 1999).

BMP2 instruction promotes both sympathetic and parasympathetic neuron differentiation in culture, depending on culture conditions (Morrison *et al.*, 2000). Under standard culture conditions BMP2 caused the differentiation of parasympathetic neurons, as judged by their expression of vesicular acetyl choline transferase (VAChT) and their lack of expression of sympathetic markers; however, when BMP2 was added along with forskolin to cultures in a reduced oxygen chamber that more closely approximates physiological oxygen levels, many neurons began expressing sympathetic markers (tyrosine hydroxylase, dopamine-β-hydroxylase). These observations suggest that while BMP2 induces autonomic differentiation, other factors can determine the subtype of neurons that form. We must also remember that the effects of BMPs on cell fate determination are often highly concentration dependent (Mehler *et al.*, 1997; Dale and Wardle, 1999). Therefore, it is possible that further work will show the sympathetic/parasympathetic lineage decision to be determined by different concentrations of BMPs *in vivo*.

BMPs instruct the differentiation of autonomic neurons from NCSCs by inducing the expression of the transcription factor MASH1 (see Fig. 4 in a later section) (Sommer *et al.*, 1995; Shah *et al.*, 1996; Lo *et al.*, 1997). MASH1 is a mammalian basic helix–loop–helix transcription factor with homology to the *Drosophila* proneural *achete-Scute* genes (Johnson *et al.*, 1990, 1992). MASH1 is required for the generation of autonomic neurons in the PNS as well as adrenergic neurons of the CNS (Guillemot *et al.*, 1993; Hirsch *et al.*, 1998). MASH1 initiates autonomic neuron differentiation by promoting the expression of pan-neuronal genes, like peripherin, as well as genes specific to autonomic neurons, like the transcription factor PHOX-2A (Sommer *et al.*, 1995; Lo *et al.*, 1998). In turn, PHOX-2A regulates the expression of additional neuronal genes such as *c-Ret*. Thus these data demonstrate how a cell-extrinsic factor, such as BMP2, can instruct stem cells to differentiate in a lineage-specific manner, by causing a hierarchy of genes to be expressed.

In addition to instructing NCSCs to differentiate into autonomic neurons, BMP2 also causes CNS stem cells to dif-

ferentiate into neurons and astrocytes (Gross *et al.*, 1996; Li *et al.*, 1998; Mabie *et al.*, 1999). Platelet-derived growth factor (PDGF) has also been observed to instruct neuronal differentiation by CNS stem cells (Williams *et al.*, 1997; Johe *et al.*, 1996). The observation of neuronal differentiation in response to BMP2 by both CNS stem cells and NCSCs suggests that different types of neural stem cells can exhibit a similar response to the same lineage determination factor.

B. Neurotrophins Act Selectively to Regulate Neuronal Survival

In contrast to the effects of BMPs, neurotrophins promote the generation of neurons by acting selectively. Neurotrophins are often secreted by the target cells on which neurons synapse and are required for the survival of neurons. There are several neurotrophins including nerve growth factor (NGF), brain-derived neurotrophic factor (BDNF), and neurotrophin 3 (NT-3). All of these factors, alone or in combination, have been implicated in regulating the survival of CNS and PNS neurons. For example, NGF and NT-3 are required for the survival of differentiated sympathetic neurons and their precursors *in vivo* (Francis *et al.*, 1999). The demonstration that mature sympathetic neurons require target-derived NGF and NT-3 for survival indicates that these factors act selectively. NGF and NT-3 do not promote neuronal differentiation by NCSCs (unpublished data) or oligopotent sympathoadrenal lineage progenitors (Anderson and Axel, 1986). This suggests that NGF and NT-3 act selectively but not instructively in promoting the differentiation of sympathetic neurons from neural crest progenitors.

Although neurotrophins act selectively in promoting the generation of sympathetic neurons, they are also capable of acting instructively. NT-3 induces neuronal differentiation by CNS stem cells *in vitro* (Vicario-Abejon *et al.*, 1995). This demonstrates that a cell-extrinsic factor can promote differentiation by acting selectively in one case, while acting instructively in another.

C. Neuregulin Acts Both Instructively and Selectively on Neural Crest Progenitors

Neuregulin (also known as glial growth factor; Marchionni *et al.*, 1993) is expressed on motor axons in the nerve (Meyer and Birchmeier, 1994) and is required for Schwann cell development (Meyer and Birchmeier, 1995; Riethmacher *et al.*, 1997). Neuregulin promotes the survival (Dong *et al.*, 1995) and proliferation (Lemke and Brockes, 1984; Morrissey *et al.*, 1995; Rutkowski *et al.*, 1995) of Schwann cells and their progenitors. These observations suggested that Neuregulin was required to selectively increase the numbers of Schwann cells by promoting their survival and proliferation (Murphy *et al.*, 1996; Topilko *et al.*, 1997b); however,

NCSCs have now been found within the fetal sciatic nerve when Schwann cells are differentiating (Morrison *et al.,* 1999), and Neuregulin is capable of instructing NCSC to differentiate into glia (Shah *et al.,* 1994). These latter observations suggested that Neuregulin might also be required to initiate glial differentiation by acting instructively on NCSCs. In fact, NCSCs in the sciatic nerve probably undergo differentiation within the nerve into both glia and myofibroblasts (Morrison *et al.,* 1999). By studying the effects of Neuregulin on different types of progenitors purified from sciatic nerve we have observed that Neuregulin has different effects on cells at different stages of development: Neuregulin promotes the survival of NCSCs and instructs them to differentiate into glia. Neuregulin promotes the survival and proliferation of glia, but only promotes the proliferation (not survival) of myofibroblasts (Morrison *et al.,* 1999). Thus, Neuregulin can have both selective and instructive effects on the same cell and acts in different ways on cells at different stages of development. The genetic requirement for Neuregulin in nerve development development (Meyer and Birch-meier, 1995; Riethmacher *et al.,* 1997) may represent a complex combination of functions.

D. Neural Stem Cells Differentiate via Progressive Restrictions in Developmental Potential

There are two alternative models by which stem cells can give rise to differentiated cells (Fig. 3). Stem cells can give rise directly to differentiated cells, such that different types of differentiated cells are generated from different mitoses. This mode of division is often observed in invertebrate stem cell lineages in which a stem cell goes through a stereotyped series of asymmetric divisions to generate a differentiated cell and a stem cell in each division [for example, in *Caenorhabditis elegans* germline specification (Mello *et al.,* 1996) or in the *Drosophila* neuroblast lineage (Li *et al.,* 1997)]. Alternatively, stem cells can generate differentiated cells by undergoing an ordered pattern of progressive restrictions in

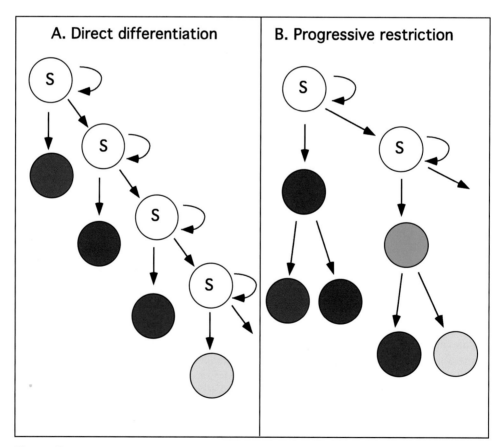

Figure 3 Direct differentiation versus progressive restrictions in developmental potential. Stem cells (uncolored) can either differentiate directly into different types of mature cells (A) or can undergo progressive restrictions in developmental potential (B) by giving rise to oligopotent progenitors (brown and green), which in turn give rise to different types of mature cells. For example, work by Rao and colleagues (1998) has demonstrated that stem cells in the neural tube give rise to a neuronal oligopotent progenitor and a glial oligopotent progenitor, which in turn give rise to different types of neurons and different types of glia respectively (Kalyani *et al.,* 1997).

developmental potential. For example, hematopoietic stem cells give rise to oligopotent lymphoid-committed (Kondo *et al.*, 1997) or myeloid-committed progenitors that in turn become further committed to numerous individual lymphoid (B, T, or dendritic) or myeloid (macrophage, neutrophil, or erythroid) lineages (Morrison *et al.*, 1995b). In this way stem cells give rise to differentiated cells by going through a series of progressive restrictions in developmental potential.

At least some neural stem cells also give rise to differentiated cells by undergoing progressive restrictions in developmental potential. It has long been hypothesized that NCSCs might give rise to multiple lineages of differentiated cells by undergoing progressive restrictions in developmental potential (Anderson, 1989; Le Douarin and Dupin, 1992); however, direct evidence for progressive restrictions by NCSCs *in vivo* is lacking. In contrast, CNS stem cells derived from the neural tube give rise to neurons and glia by first giving rise to oligopotent restricted progenitors. Stem cells from the E10.5 rat neural tube give rise to astrocytes, oligodendrocytes, and neurons with a variety of different neurotransmitter phenotypes (Kalyani *et al.*, 1997). *In vitro* and *in vivo* the stem cells appear to give rise to E-NCAM$^+$ neuronal-committed progenitors as well as A2B5$^+$N-CAM$^-$ glial progenitors (Mayer-Proschel *et al.*, 1997). The E-NCAM$^+$ neuronal-committed progenitors retain the ability to give rise to multiple different classes of neurons (Kalyani *et al.*, 1998). Similarly, the A2B5$^+$N-CAM$^-$ glial committed progenitors in turn can give rise to both oligodendrocytes and astrocytes (Rao *et al.*, 1998). Thus stem cells from the developing spinal cord give rise to specific lineages of neuronal and glial cells by undergoing progressive restrictions in developmental potential marked by the generation of oligopotent progenitors.

III. Do Stem Cells Retain Broad or Narrow Neuronal Potentials?

It is not yet clear whether neural stem cells tend to retain a broad potential to give rise to many different types of neurons, or whether neural stem cells tend to be restricted in the types of neurons they can form. After transplantation, progenitors derived from the telencephalon were able to give rise to neurons throughout the forebrain and midbrain while progenitors from the mesencephalon gave rise to neurons in the midbrain but not in parts of the forebrain (Campbell *et al.*, 1995). Supporting this idea that neural progenitors are at least partially restricted in the types of neurons they can form, telencephalic progenitors differentiated appropriately when transplanted into different dorsoventral positions within the telencephalon, but did not differentiate appropriately when transplanted into different brain regions along the rostrocaudal axis (Fishell, 1995; Na *et al.*, 1998). Even within the adult forebrain, stem cells from one location sometimes cannot give rise to normal neurons in other loca-

tions. Stem cells cultured from the hippocampus gave rise to neurons in both the hippocampus and the olfactory bulb on transplantation (Suhonen *et al.*, 1996). On the other hand, these stem cells gave rise to neurons in the retina but the neurons did not exhibit the normal phenotype of retinal neurons (Takahashi *et al.*, 1998).

NCSCs in the developing PNS may also be heterogeneous with respect to the types of neurons they can form. Postmigratory NCSCs isolated from sciatic nerve gave rise to parasympathetic and sympathetic neurons *in vitro* and *in vivo* but it is not yet clear whether such cells have sensory or enteric potential (Morrison *et al.*, 1999, 2000). Indeed, migrating neural crest progenitors had sensory, parasympathetic, and sympathetic potential upon transplantation, but postmigratory enteric neural crest progenitors had only parasympathetic potential (White and Anderson, 1999). Together the data suggest that while stem cells often have the potential to give rise to multiple different classes of neurons they are at least partially restricted and cannot give rise to all classes of neurons.

The corollary of the observation that different stem cells have different neuronal potentials is that different regions in the CNS and PNS must be formed by different types of stem cells (that are defined by different developmental potentials). Beyond the differences in neuronal potential cited above, NCSCs from different levels of the neural tube along the rostrocaudal axis differ in their potential to give rise to mesectodermal derivatives (Le Douarin, 1982, 1986). Cephalic neural crest gives rise to dermis, cartilage, and bone, even when transplanted to the trunk level; however, trunk neural crest does not contribute to these tissues, even when transplanted to the cephalic level (Le Douarin and Dupin, 1992). These observations raise the possibility that the CNS and PNS may each include multiple classes of stem cells. In all cases these stem cells can give rise to neurons and glia but the types of neurons and glia may differ. The types of glia formed by stem cells from different levels of the neural axis may overlap broadly, with all CNS stem cells giving rise to astrocytes and oligodendrocytes, and all NCSCs giving rise to Schwann cells. The types of neurons formed by stem cells from different rostrocaudal levels may be more restricted. In many cases stem cells may be restricted to forming types of neurons that are specific to particular rostrocaudal regions. That is, although stem cells may be able to engraft in heterotopic regions of the nervous system, the neurons they produce may not exhibit the full phenotypic and functional properties of normal neurons from those regions. Beyond being a way to organize nervous system development, cell-intrinsic differences between neural stem cells in neuronal potential may be a mechanism for generating neuronal diversity: Differences between stem cell populations may interact with environmental differences to cause different types of neurons to differentiate in different places. If neuronal identity were specified by both environmental and stem cell-intrinsic determinants it would greatly simplify the problem of how such a vast diversity of neurons is generated.

IV. Regulation of Neural Stem Cell Self-Renewal

Self-renewal and differentiation are like opposite sides of a coin. They must be regulated in concert because in a given cell at a given time the execution of one excludes the other. Little is known about how self-renewing divisions are regulated in mammalian stem cells. While transcription factors have been identified that are sufficient to cause the lineage-specific differentiation of stem cells, the regulation of self-renewal is much less well defined. Are there single factors that are sufficient to induce all of the machinery necessary for a stem cell to self-renew? A cell extrinsic factor, FGF-2, causes neural stem cells to self-renew in culture (Gritti *et al.*, 1996). In contrast, no cell-intrinsic factors (such as transcription factors) have yet been identified whose expression is sufficient to cause the self-renewal of mammalian neural stem cells. The question of whether a single cell-intrinsic factor is sufficient to cause self-renewing divisions is important because it remains unclear whether the decision to divide in mammalian stem cells is regulated independently of the decision to remain undifferentiated.

Although no genes have yet been shown to cause neural stem cell self-renewal in a cell-intrinsic manner, single gene products can promote the self-renewal of other stem cells. Constitutive activation of the Notch homolog Glp-1 in *C. elegans* germline stem cells is sufficient to cause indefinite self-renewal (Wilson Berry *et al.*, 1997). Overexpression of stabilized β-catenin appears to promote the self-renewal of epidermal stem cells (Zhu and Watt, 1999). Expression of the homeodomain protein HoxB4 in mouse hematopoietic stem cells is sufficient to increase their numbers *in vivo* (Sauvageau *et al.*, 1995); however, this example illustrates the challenges in demonstrating that a factor promotes mammalian stem cell self-renewal because the mechanism by which HoxB4 acts is unknown. Rather than increasing self-renewing divisions by stem cells, it might promote survival or impair differentiation. Nonetheless, if HoxB4, β-catenin, or FGF-2 causes self-renewal by both promoting mitosis and impairing differentiation, then these factors may be a upstream of a genetic program that executes self-renewing divisions by stem cells. Presumably this program would be composed of a hierarchy of genes similar to programs that execute lineage specific differentiation (Fig. 4). If so, one or a few gene products at the top of the hierarchy might be able to induce all of the other genes that are required to execute a self-renewing division. Perhaps some of the downstream gene products in the hierarchy regulate division while others inhibit differentiation. If so, then genes at the top of the hierarchy might promote mitosis as well as maintenance of the stem cell state, whereas downstream genes might regulate one or the other. This would be analogous to the regulation of neuronal differentiation by Mash-1, which couples the induction of pan-neuronal markers with subtype-specific

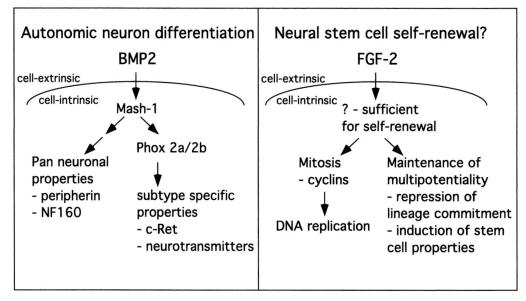

Figure 4 Hierarchies of genes may control stem cell self-renewal as well as differentiation. The first panel depicts a schematic of autonomic neuron differentiation from NCSCs in which BMP2 induces Mash-1 expression. Mash-1 is sufficient to cause autonomic neuron differentiation. The differentiation program involves inducing the expression of many genes, including those that regulate pan-neuronal properties and those that regulate autonomic-specific properties. Because the expression of pan-neuronal genes and subtype-specific genes is regulated differently it appears that Mash-1 must induce the expression of genes in multiple genetically independent pathways. Similarly, we might hypothesize that b-FGF leads to the expression of a transcription factor that is sufficient to cause neural stem cell self-renewal. Multiple genetic pathways may be induced by such a factor, including genes like cyclins that lead to DNA replication and mitosis, and genes that cause the cell to maintain the stem cell state. (The diagram describing neuronal differentiation was adapted from Lo *et al.*, 1998.)

markers by inducing the expression of downstream genes that act through different pathways (Lo *et al.*, 1998).

Maintenance of the stem cell state can be regulated independent of the decision to divide as demonstrated by the function of Pie-1 in the *C. elegans* germline lineage. Pie-1 is a maternally expressed transcription factor that represses the transcription of embryonic genes associated with acquisition of somatic cell fates (Seydoux *et al.*, 1996; Tenenhaus *et al.*, 1998). In *C. elegans* embryos, totipotent germline blastomeres undergo a series of asymmetric divisions to yield one somatic daughter cell and one germline stem cell. Pie-1 is asymmetrically localized to the germline stem cell in each division and appears to maintain totipotency by repressing the transcription of genes that confer a somatic fate (Mello *et al.*, 1996; Batchelder *et al.*, 1999). Loss of Pie-1 function does not prevent mitosis, but both daughter cells assume somatic cell fates. It is not certain whether a mechanism analogous to Pie-1 operates in mammalian stem cells; however, a transcription factor called neuron restrictive silencing factor (NRSF) (Schoenherr and Anderson, 1995; Schoenherr *et al.*, 1996) or REST (Chong *et al.*, 1995) can inhibit neuronal differentiation by repressing the transcription of a range of neuron-specific genes (Chen *et al.*, 1998). Thus maintenance of the stem cell state can be regulated independently of the decision to divide, raising the possibility that stem cell self-renewal may require the coordination of multiple genetic pathways.

Few insights exist into which genes compose the self-renewal program for neural stem cells, though some recently described mutations provide candidates. For example, deletion of the winged helix transcription factor BF-1 reduces the proliferation of undifferentiated neuroepithelial cells in the developing ventral telencephalon and leads to premature neuronal differentiation (Xuan *et al.*, 1995). These observations strongly suggest that BF-1 is required for the normal self-renewal of CNS stem cells. *Flat-top* is a mutation (in an as yet unidentified gene) that was generated by ethyl-nitroso-urea mutagenesis of mouse germ cells in a screen for mutations that affect CNS development (Hentges *et al.*, 1999). The *flat-top* mutation is associated with decreased proliferation of undifferentiated neural progenitors in the embryonic telencephalon, leading to a failure to form normal telencephalic vesicles. As genetic screens for mutations that affect nervous system development progress, and tools for studying stem cell function at the cellular level become increasingly sophisticated, the genetic regulation of neural stem cell self-renewal will be elucidated.

V. Differences between Hematopoietic Stem Cells and Neural Stem Cells

Hematopoietic stem cells were first among stem cells to be purified and extensively characterized (reviewed by Morrison *et al.*, 1995b) (see Chapter 13). As a result, they have been a model on which predictions about the properties of other stem cells have been based. Many hypotheses about neural stem cell function have been inspired by analogy to hematopoietic stem cells. There are similarities between hematopoietic stem cells and neural stem cells, but there are also fundamental differences in developmental strategy between the hematopoietic and nervous systems. If we are to create expectations of neural stem cell function based on the properties of hematopoietic stem cells, then it is critical to understand these parallels and differences.

A. Stem Cell Frequency

The hematopoietic and nervous systems adopted a fundamental difference in strategy with respect to how they use stem cells. Hematopoietic stem cells are rare but give rise to massive numbers of progeny. For example, hematopoietic stem cells account for only 0.04% of mouse fetal liver cells around E12–14 (Morrison *et al.*, 1995a), but in lineage marking experiments individual stem cells from around this stage of development gave rise to very large numbers of progeny by birth (Clapp *et al.*, 1995). In contrast, when stem cells are present in the nervous system, large numbers of stem cells each give rise to only small numbers of progeny. For example, in the E14 rat sciatic nerve, NCSCs account for around 15% of nerve cells (Morrison *et al.*, 1999). But multipotent neural crest or CNS progenitors consistently give rise to fewer than 100 cells in the developing nervous system (Fraser and Bronner-Fraser, 1991; Walsh and Cepko, 1993; Henion and Weston, 1997). Thus the hematopoietic system forms few stem cells but causes them to give rise to enormous numbers of progeny, whereas the nervous system forms many stem cells that each give rise to few progeny. The hematopoietic and nervous systems may be subject to different developmental constraints that select for different strategies.

B. Regional Specialization

A second difference in strategy between the hematopoietic and nervous systems is that while the hematopoietic system goes out of its way to avoid regional specialization, regional specialization is the rule in the nervous system. Hematopoiesis occurs in bone marrow that is distributed among many different bones throughout the body. Yet as far as we know the same cell types are produced in similar proportions by all bone marrow compartments. In addition, the hematopoietic system can be ablated with a lethal dose of radiation and then fully reconstituted by purified hematopoietic stem cells from any bone in the body (Spangrude *et al.*, 1988; Uchida *et al.*, 1994; Osawa *et al.*, 1996; Morrison *et al.*, 1997b).

In contrast, even slight differences in position within the nervous system are associated with functional and phenotypic differences in the cell types that differentiate. This is clearly demonstrated in the case of the neural crest, which gives rise to enteric neurons in the gut, sympathetic neurons in the sympathetic chain, parasympathetic neurons associated with organs, and sensory neurons in the spinal ganglia. Neural stem cells give rise to progeny that are often confined to a small area within the nervous system (Fraser and Bronner-Fraser, 1991; Reid *et al.*, 1995). Because only a restricted subset of neural cells differentiates in any one location within the nervous system, it is likely that neural stem cells have less opportunity than hematopoietic stem cells to undergo multilineage differentiation *in vivo*. However, progenitors from one region of the developing nervous system sometimes fail to give rise to the neurons found in other regions of the nervous system on transplantation (Campbell *et al.*, 1995; Na *et al.*, 1998; Takahashi *et al.*, 1998). This suggests that different lineages of cells within the nervous system derive from different types of stem cells. Indeed, while many if not most migrating neural crest cells are multipotent (Le Douarin and Dupin, 1992; Shah *et al.*, 1994), most neural crest progenitors are fated to give rise to only one type of differentiated cell (Henion and Weston, 1997). Thus regional specialization within the nervous system may prevent multipotent progenitors from exhibiting their full developmental potential.

VI. *In Vivo* Function of Neural Stem Cells

These differences in strategy between the hematopoietic and nervous systems mean that we should be cautious about expecting neural stem cells to have properties analogous to hematopoietic stem cells. A single hematopoietic stem cell can engraft in lethally irradiated mice and give rise to all major lineages of blood cells for the life of the mouse (Morrison and Weissman, 1994; Spangrude *et al.*, 1995; Osawa *et al.*, 1996). This observation has caused some to propose that neural stem cells should also self-renew for long periods of time *in vivo* while giving rise to large numbers of progeny. But these may be unreasonable expectations, in part because most of the nervous system forms during a limited window of fetal and neonatal development, as discussed next.

A. Long-Term *in Vivo* Self-Renewal

A self-renewing division is a mitosis in which the mother cell gives rise to at least one daughter cell that has a developmental potential that is indistinguishable from the mother cell. For example, if a multipotent mother cell gives rise to similarly multipotent daughter cell(s) then the mother cell is a self-renewing stem cell. In contrast, if a multipotent mother cell only gives rise to daughter cells with restricted developmental potentials, then it does not self-renew.

Most neural stem cells probably persist for only short periods of time *in vivo* and therefore the criterion of long-term

self-renewal may not be relevant. The PNS forms between E8 and P14 in mice and rats, in which gestation is 18–21 days (Pham *et al.*, 1991). NCSCs have only been shown to persist until E17 (Morrison *et al.*, 1999). Thus there is no reason to believe that NCSCs should or can self-renew for more than a week or two *in vivo* even though they have the *potential* to self-renew for considerably longer periods of time *in vitro*. Most of the CNS also forms during fetal and neonatal life. Because CNS stem cells persist near the lateral ventricle and in the dentate gyrus of adults, an opportunity may arise for some such cells to exhibit long term self-renewal in those locations; however, like NCSCs, these cells appear to give rise to only small numbers of progeny *in vivo* (Morshead *et al.*, 1998; Doetsch *et al.*, 1999; Johansson *et al.*, 1999) despite their ability to self-renew and give rise to large numbers of progeny *in vitro* (Baroffio *et al.*, 1988; Gritti *et al.*, 1996). Thus while these studies provided evidence that stem cells near the lateral ventricle undergo asymmetric self-renewing divisions *in vivo*, it is not clear whether individual stem cells repeatedly undergo such divisions over long periods of time or whether individual stem cells might only self-renew a few times before differentiating. The available evidence suggests that most neural stem cells self-renew to only a limited extent *in vivo* despite their capacity for extensive self-renewal *in vitro*. Therefore, the criterion of long-term self-renewal *in vivo* may be an unphysiological one for most neural stem cell populations.

Unlike the hematopoietic system, there is no known circumstance in the nervous system in which neural stem cells are required to reconstitute most of the nervous system after injury. *In vivo* assays for hematopoietic stem cells depend on irradiating recipient mice to reduce competition from endogenous stem cells prior to reconstituting with transplanted stem cells. Mammals can live for around 2 weeks after a lethal dose of radiation that destroys most blood cells and the ability to make new blood cells. This provides an opportunity to test whether a small number of transplanted hematopoietic cells have the potential to self-renew, undergo multilineage reconstitution, and give rise to the massive numbers of blood cells that are required to reconstitute the hematopoietic system. While the number of stem cells in the subependymal layer of the lateral ventricle can be reduced by administering cytotoxic compounds prior to transplanting neural progenitors (Doetsch *et al.*, 1999), it is much more difficult in the nervous system to promote the engraftment and proliferation of transplanted progenitors by eliminating endogenous stem cells. Most of the nervous system cannot be experimentally ablated in viable animals or reconstituted by transplanted progenitors. It is possible that in the future an *in vivo* assay will be developed in which a single neural stem cell can be introduced into a tissue that is permissive for neural differentiation, but where there is little competition from endogenous neural progenitors. Only then are we likely to have an opportunity to demonstrate that single neural stem cells consistently undergo multilineage differentiation and give rise to large numbers of progeny *in vivo*.

B. The Relationship between Fetal and Adult Neural Stem Cells

It is sometimes assumed that stem cells in the adult CNS are equivalent to the stem cells that form the CNS during fetal development—that the stem cells in the adult brain are just "left over" from fetal development. In fact, the relationship between stem cells in the adult CNS and stem cells in the fetal CNS is unknown. Stem cells in the adult CNS may be lineally unrelated to fetal CNS stem cells, or they may derive from fetal stem cells. Even if adult CNS stem cells derive from fetal CNS stem cells, the adult cells may be developmentally distinct. For example, fetal CNS stem cells may be able to give rise to certain classes of neurons that adult stem cells cannot. By analogy with the hematopoietic system such a possibility seems likely. Adult hematopoietic stem cells probably derive from fetal liver hematopoietic stem cells (Fleischman *et al.*, 1982; Clapp *et al.*, 1995), but their properties differ in important ways. Certain lineages of lymphocytes are produced by fetal but not adult hematopoietic stem cells (Ikuta *et al.*, 1990; Kantor *et al.*, 1992; Hardy and Hayakawa, 1994). More importantly, fetal hematopoietic stem cells have a greater self-renewal and proliferative potential than adult hematopoietic stem cells (Morrison *et al.*, 1995a; Rebel *et al.*, 1996). Additional work will be required to carefully compare the properties of stem cells from the fetal and adult CNS.

C. Cell Cycle Status

The cell cycle distribution of neural stem cells is often hypothesized to be similar to hematopoietic stem cells. For example, adult CNS stem cells have been predicted to be quiescent *in vivo* by analogy to hematopoietic stem cells. Hematopoietic stem cells are relatively quiescent in normal C57BL adult mice: 75% of hematopoietic stem cells are in G_0 at any one time, and on average only 8% of stem cells divide each day (Cheshier *et al.*, 1999); however, this contrasts with hematopoietic stem cells from fetal liver, which undergo daily self-renewing divisions (Morrison *et al.*, 1995a). Not only does the cell cycle status of hematopoietic stem cells change during ontogeny (Morrison *et al.*, 1996), but it changes in response to injury (Harrison and Lerner, 1991; Morrison *et al.*, 1997c) and differs between mouse strains (deHaan *et al.*, 1997). Thus we should be cautious when it comes to generalizing about the cell cycle regulation of neural stem cells. In fetal rat sciatic nerve, NCSCs are not quiescent but undergo rapid self-renewing divisions (Morrison *et al.*, 1999). In the fetal CNS, analyses of the proliferation of ventricular zone progenitor cells suggest that fetal CNS stem cells are likely proliferative as well (Cai *et al.*,

1997), although this is not certain since stem cells compose a small minority of cells in the ventricular zone (Davis and Temple, 1994). Under normal conditions in the adult CNS there are quiescent as well as actively proliferating progenitors (Morshead *et al.*, 1994, 1998; Johansson *et al.*, 1999). These studies indicate that at least some stem cells in the adult CNS are quiescent but the precise relationship between the proliferating and the quiescent progenitors is uncertain because CNS stem cells have not yet been purified or studied directly. Although we have an incomplete understanding of the cell cycle regulation of neural stem cells it appears the analogy to hematopoietic stem cells is appropriate: fetal neural stem cells tend to be proliferative, while adult CNS stem cells are often quiescent and can be activated in response to injury (Doetsch *et al.*, 1999; Johansson *et al.*, 1999).

VII. Surprising Potential of Neural Stem Cells

A series of recent papers has suggested that cultured CNS stem cells have the potential to give rise to nonneural derivatives in tissues that have been thought to be developmentally unrelated. Other papers suggest that progenitors from tissues outside of the nervous system may retain the potential to make neurons and glia. Are classical lineage relationships incorrect? Or do stem cells retain potentials that they never exhibit during normal development? Or have we identified an intriguing phenomenon in which the developmental potential of cells can be reprogrammed under specific conditions in culture or after transplantation, without reflecting lineage relationships that exist in normal development? The surprising developmental potential of somatic stem cells suggests we may have to redefine our understanding of developmental potential.

Stem cells cultured from the mouse or human CNS have the potential to give rise to nonneural derivatives. CNS stem cells are often isolated by culturing neurospheres, which are spheres of multipotent progenitors that grow out of mixed populations of CNS cells in bFGF-containing media. Such neurospheres are thought of as CNS stem cells but because the neurosphere cells have only been studied in culture, it is not certain whether they have properties that are similar to normal CNS stem cells *in vivo*. Neurosphere cells have been observed to give rise to blood cells on transplantation into irradiated mice (Bjornson *et al.*, 1999), skeletal muscle on coculture with a myogenic cell line or on transplantation into regenerating muscle *in vivo* (Galli *et al.*, 2000), or to colonize tissues of all three germ layers on injection into blastocysts or early chick embryos (Clarke *et al.*, 2000). All of these papers concluded that CNS stem cells retain a much broader developmental potential than previously thought; however, there is concern that CNS progenitors can lose patterning information and acquire a broader developmental potential as a result of being cultured in high concentrations of mitogens such as bFGF (Gage, 1998). Indeed, primordial germ cells have already been demonstrated to be reprogrammed in culture to acquire the properties of embryonic stem cells (Matsui *et al.*, 1992; Donovan, 1994). As a result, it is unknown whether neurosphere cells correspond to normal CNS stem cells, or whether their developmental potential is broadened in culture. When it becomes possible to prospectively identify and purify uncultured CNS stem cells, it will be important to test whether these cells still give rise to blood, muscle, and other somatic lineages on transplantation or whether they must be cultured first in order to exhibit these potentials.

Another example of nervous system progenitors exhibiting unexpected developmental potentials comes from work on oligodendrocyte precursor cells (OPCs). OPCs are progenitors from the optic nerve that have been extensively studied and thought to be glial committed (and therefore could be considered restricted progenitors rather than stem cells). The recent demonstration that OPCs can make neurons after being exposed in culture to a specific series of growth factors, including bFGF, is consistent with the idea that restricted neural progenitors can acquire neuronal potential in culture (Kondo and Raff, 2000). Indeed, the OPCs not only gave rise to neurons, but to neurospheres as well. This bolsters the concern that neurospheres can sometimes be produced in culture from cells that have properties *in vivo* that are different from multipotent CNS stem cells. The demonstration that OPCs can give rise to neurons is important because OPCs can be purified from uncultured postnatal optic nerve by immunopanning. Thus this is the first demonstration that a well-characterized, prospectively identified nervous system progenitor can acquire a broader developmental potential than previously thought in culture. This supports the idea that specific progenitors can exhibit unexpected developmental plasticity in culture but cautions us that such plasticity may not be exhibited during normal development.

Not only can neural stem cells make non-neural cell types, but non-neural stem cells can make neurons and glia. Bone marrow progenitors, which are best known for making blood cells, have recently been observed to contribute to a variety of tissues on transplantation. Cells from mouse bone marrow have been observed to give rise to skeletal muscle (Ferrari *et al.*, 1998; Gussoni *et al.*, 1999), hepatocytes (Peterson *et al.*, 1999; Lagasse *et al.*, 2000), and neurons and glia (Kopen *et al.*, 1999). The bone marrow is known to contain at least two different types of stem cells including hematopoietic stem cells and mesenchymal stem cells. Additionally, it is conceivable that stem cells from other tissues such as liver, nervous system, and muscle may circulate at low levels and be found in the bone marrow. So which stem cells gave rise to the unexpected derivatives? The cells that gave rise to neurons and glia were from cultures of adherent bone marrow stroma, suggesting that they included mesenchymal

stem cells (Kopen *et al.*, 1999). Further work will be required to rigorously establish the identity and the origin of the progenitors with neurogenic potential from bone marrow.

With so many cells exhibiting unexpected developmental potentials, some have begun to question the classical germ layer origin of some tissues. Is it possible that CNS stem cells give rise to blood, muscle, and liver during normal development? This question is testable by CRE-recombinase fate mapping, which can identify all of the progeny produced by a specific lineage of progenitors during normal development, as long as a promoter can be identified that is absolutely specific to that progenitor lineage. Several studies have used this approach to examine the progeny produced by cells that express *Wnt-1*, which include neural stem cells in the midbrain, dorsal neural tube, and neural crest. *Wnt-1* expressing cells were observed to give rise to many expected CNS, PNS, and other neural crest derivatives, but have not yet been observed to give rise to blood, skeletal muscle, liver, kidney, lung, or other unexpected derivatives (Chai *et al.*, 2000; Jiang *et al.*, 2000). This is not a perfect experiment in that not all neural stem cells are marked by *Wnt1* expression, and it is possible that the CNS stem cells with unusually broad developmental potentials derive from lineages that never expressed *Wnt1*. Nonetheless, the failure to yet observe any unusual derivatives from a wide cross section of neural progenitors *in vivo* during normal development contrasts with the spate of papers reporting the broad potential of a variety of cultured CNS stem cells after transplantation. This suggests that the potential exhibited by cultured CNS stem cells may not be exercised during normal development.

Even if most of the surprising neural stem cell potential arises as a result of "de-differentiation" in culture and does not reflect potential that is normally exercised *in vivo,* it is still scientifically and clinically important to understand the phenomenon. On the other hand, if normal somatic stem cells routinely retain much broader developmental potential than they exhibit during normal development, this would pose the question as to whether differentiation is primarily environmentally regulated. That is, are the normal lineage relationships that exist between progenitors in different germ layers governed primarily by controlling the locations and interactions of such progenitors? Maybe the only reason why neural stem cells do not seem to make liver or blood during normal development is that such cells are prevented from accessing hepatic or hematopoietic microenvironments.

VIII. Are Neural Stem Cells Involved in Disease?

A. NCSCs and Childhood Cancer

There is no published evidence that NCSCs can persist postnatally, but if they do, it might have important implications for understanding the origin of certain childhood cancers. A popular hypothesis is that many types of cancer cells derive from the transformation of normal stem cells (Sell and Pierce, 1994). A number of different types of tumors are thought to arise from neural crest derivatives including neuroblastomas, neurofibromas, and Ewing's sarcomas/peripheral neuroectodermal tumors (PNET). Ewing's sarcomas/PNET are particularly primitive in appearance, containing cells that can express neuronal, glial, and mesectodermal markers. This has led to speculation that these tumors are derived from primitive neural crest progenitors (Fujii *et al.,* 1989; Marina *et al.,* 1989). Indeed some speculate that PNS tumors not only derive from the transformation of neural crest progenitors but that the degree of differentiation exhibited by the tumor is determined by the stage of differentiation at which transformation of the normal progenitor occurred (Thiele, 1991). The obvious difficulty with this hypothesis is that, while these tumors occur predominantly in children and adolescents, there is yet no evidence for the persistence of uncommitted neural crest progenitors postnatally. If neural crest progenitors undergo terminal differentiation in midgestation, it is hard to imagine how there is an opportunity for such progenitors to be transformed in a way that leads to tumors that present late in childhood. If, on the other hand, rare NCSCs persist postnatally, then perhaps such cells represent the cellular locus for transforming events that lead to childhood cancers.

B. Adult Neurogenesis and Mental Illness

Given the recent demonstration that neurogenesis continues in regions of the adult human brain that are involved in learning and memory (Eriksson *et al.,* 1998), scientists have started to consider whether changes in neurogenesis might be associated with mental illness (Brown *et al.,* 1999). It has been proposed that stress-induced changes in the hippocampus lead to the development of depression in at-risk individuals (Duman *et al.,* 1997). Indeed, exposure to stress can reduce hippocampal neurogenesis, at least in adult monkeys (Gould *et al.,* 1998). A possible mechanism that might link mental illness with altered neurogenesis comes from corticosteroids, which are elevated in patients with mood disorders (reviewed by Brown *et al.,* 1999) and which can impair hippocampal neurogenesis (Cameron and McKay, 1999). Reduced hippocampal serotonin levels have also been associated with cognitive disorders such as depression and schizophrenia (Cross, 1990) and, when hippocampal serotonin levels are experimentally depleted, neurogenesis is reduced (Brezun and Daszuta, 1999). Much more work will have to be done to determine whether changes in neurogenesis are associated with mental illnesses, let alone to determine whether such changes are causative. Nonetheless these considerations illustrate the range of health issues that may be impacted by the discovery of stem cells and neurogenesis in the adult human brain.

C. Do Neural Stem Cells Respond to Nervous System Injury?

A number of neurodegenerative disorders that afflict older people are caused by the loss of certain types of neurons in the brain. Parkinson's disease is caused by the progressive loss of dopaminergic neurons from the substantia nigra in the striatum (Date, 1996). Huntington's disease is also caused by the loss of striatal neurons (Reddy *et al.,* 1999). Alzheimer's disease is caused by the loss of cholinergic as well as other neurons from the forebrain (Winkler *et al.,* 1998). All of these neurodegenerative diseases have been treated by transplanting neural progenitor cells into the affected areas in an effort to regenerate the lost neurons (Winkler *et al.,* 1998; Bjorklund and Lindvall, 1999; Date and Ohmoto, 1999). Of these disorders, Parkinson's is thought to be the most promising clinical application for transplantation therapies, because the lesion is localized and because it may be sufficient to reintroduce dopaminergic cells without necessarily restoring their normal pattern of projections. Nonetheless, although the transplantation of neural progenitors has proven very effective for ameliorating symptoms in Parkinsonian animal models (Date, 1996), success in patients has not been as consistent (Olanow *et al.,* 1997). Improved techniques and more sophisticated strategies for generating neural progenitors for transplantation may lead to improvement in clinical outcomes.

An outstanding question that has received little attention is whether there is any response of endogenous neural stem cells to neurodegeneration (Lowenstein and Parent, 1999). Do stem cells in the brain attempt to make new neurons to compensate for the neurons that die in Parkinson's, Huntington's, or Alzheimer's disease? At present there is no evidence of this, but dentate gyrus neurogenesis is accelerated in response to seizure and associated neuronal death (Bengzon *et al.,* 1997; Parent *et al.,* 1997, 1998; Scott *et al.,* 1998); furthermore, dentate gyrus neurogenesis is also stimulated in response to mechanical lesions in the granule cell layer (Gould and Tanapat, 1997). If neurogenesis can be stimulated by excitotoxic or mechanical damage, it could also be stimulated by neural degeneration. If endogenous progenitors are activated by neurodegenerative disease then perhaps the clinical course of some diseases is slowed by regeneration. If not, is it because endogenous stem cells do not have the potential to make the types of neurons that are lost in most neurodegenerative disorders? Or are the stem cells in the wrong place and unable to make neurons that can migrate to the sites of the lesions? Could the response of endogenous stem cells to neurodegeneration be pharmacologically enhanced? Do existing pharmacological treatments for mental illness or neurodegeneration impact on the function of stem cells in the brain? These are fundamental questions that will be addressed as more sophisticated tools become available to study the *in vivo* functions of neural stem cells.

IX. Outstanding Issues

Since the pioneering studies of Ramón y Cajál (Ramón y Cajál and May, 1959; cited in Lowenstein and Parent, 1999; and Rakic, 1985), the nervous system was thought to form during embryonic development and then to remain invariant throughout adult life, with no capacity for neurogenesis. Starting in the 1960s with studies by Altman (1962, 1969) and Altman and Das (1966), neurogenesis in certain regions of the adult brains of rodents was recognized. However, it was not until 1992 that stem cells were discovered in the CNS (Reynolds and Weiss, 1992) and in the neural crest (Stemple and Anderson, 1992). Not until 1998 was neurogenesis demonstrated in the adult human brain (Eriksson et al., 1998). Thus the study of neural stem cells is an exciting new field with many unanswered questions and surprises yet to come:

1. *The prospective identification and purification of neural stem cells.* Prospective identification refers to the ability to reliably predict which cells are stem cells and which are other cell types *in vivo* or among freshly dissociated cells, based on marker expression. The importance of prospective identification is illustrated by the experience with hematopoiesis where our understanding of hematopoietic stem cells was greatly accelerated by their prospective identification and flow-cytometric purification (Spangrude et al., 1988). So far, our understanding of neural stem cells has been based mainly on retrospective analyses *in vivo* and on analyses of cultured multipotent progenitors. As a result, fundamental questions about neural stem cells remain unanswered. Do multipotent neural progenitors that are cultured from the adult CNS arise from cells with similar properties *in vivo* or do they arise from more restricted progenitors by dedifferentiation (Gage, 1998)? Do stem cells in the adult brain reside in the ependymal layer (Johansson et al., 1999) or in the subependymal layer (Doetsch et al., 1999), or both? Markers are finally being identified that will allow us to predict which cells are neural stem cells, so that they can be studied as they exist *in vivo*. Uncultured NCSCs were recently purified by flow-cytometry (Morrison et al., 1999). As a result of the ability to prospectively identify these cells it was demonstrated that surprisingly they persist into late gestation by self-renewing in peripheral nerves. These approaches should soon enable the purification of uncultured CNS stem cells as well. As it becomes possible to purify different classes of neural stem cells, to study their properties *in vivo,* and to directly compare the properties of different classes of stem cells our understanding of their biology will quicken.

2. *The potential of neural stem cells.* Is the nervous system formed by many different kinds of stem cells that are independently born in different regions of the nervous system and restricted in the types of neurons they can form? Do cell-intrinsic differences between stem cells from different regions of the nervous system interact with environmental differences to generate neural diversity? We need to determine what types of neurons can be formed by stem cells from different regions of the nervous system, and whether there are cell-intrinsic differences between neural stem cells in terms of self-renewal potential or their response to lineage determination factors.

3. *The genetic regulation of stem cell self-renewal and differentiation.* With the advent of microarray analysis it should be possible to conduct increasingly sophisticated screens for genes whose expression is associated with stem cell self-renewal or differentiation. In combination with analyses of gene function in neural stem cells *in vitro* and in genetically modified mice *in vivo,* such screens may make it possible to decode the genetic programs that regulate stem cell function.

4. *The response of neural stem cells to disease.* It will be important to test whether stem cells in the adult CNS respond to injury and disease, and whether their potential for regeneration can be therapeutically enhanced. The approach of trying to optimize a regenerative response by endogenous neural stem cells would be an important new approach to repairing the nervous system after injury or disease.

References

Altman, J. (1962). Are new neurons formed in the brains of adult mammals? *Science* **135,** 1127–1128.

Altman, J. (1969). Autoradiographic and histological studies of postnatal neurogenesis. IV. Cell proliferation and migration in the anterior forebrain, with special reference to persisting neurogenesis in the olfactory bulb. *J. Comp. Neurol.* **137,** 433–458.

Altman, J., and Das, G. D. (1966). Autoradiographic and histological studies of postnatal neurogenesis. *J. Comp. Neurol.* **126,** 337–390.

Anderson, D. J. (1989). The neural crest cell lineage problem: Neuropoiesis? *Neuron* **3,** 1–12.

Anderson, D. J. (1997). Cellular and molecular biology of neural crest cell lineage determination. *Trends Genet.* **13,** 276–280.

Anderson, D. J., and Axel, R. (1986). A bipotential neuroendocrine precursor whose choice of cell fate is determined by NGF and glucocorticoids. *Cell (Cambridge, Mass.)* **47,** 1079–1090.

Baroffio, A., Dupin, E., and Le Douarin, N. M. (1988). Clone-forming ability and differentiation potential of migratory neural crest cells. *Proc. Natl. Acad. Sci. U.S.A.* **85,** 5325–5329.

Baroffio, A., Dupin, E., and Le Douarin, N. M. (1991). Common precursors for neural and mesectodermal derivatives in the cephalic neural crest. *Development (Cambridge, UK)* **112,** 301–305.

Barres, B. A. (1999). A new role for glia: Generation of neurons! *Cell (Cambridge, Mass.)* **97,** 667–670.

Batchelder, C., Dunn, M. A., Choy, B., Suh, Y., Cassie, C., Shim, E. Y., Shin, T. H., Mello, C., Seydoux, G., and Blackwell, T. K. (1999). Transcriptional repression by the Caenorhabditis elegans germ-line protein PIE-1. *Genes Dev.* **13,** 202–212.

Bengzon, J., Kokaia, Z., Elmer, E., Nanobashvili, A., Kokaia, M., and Lindvall, O. (1997). Apoptosis and proliferation of dentate gyrus neurons after single and intermittent limbic seizures. *Proc. Natl. Acad. Sci. U.S.A.* **94,** 10432–10437.

Bitgood, M. J., and McMahon, A. P. (1995). *Hedgehog* and *Bmp* genes are coexpressed at many diverse sites of cell-cell interaction in the mouse embryo. *Dev. Biol.* **172,** 126–138.

Bjorklund, A., and Lindvall, O. (1999). Transplanted nerve cells survive and are functional for many years. *Laekartidningen* **96,** 3407–3412.

Bjornson, C. R. R., Rietze, R. L., Reynolds, B. A., Magli, M. C., and Vescovi, A. L. (1999). Turning brain into blood: A hematopoietic fated adopted by adult neural stem cells in vivo. *Science* **283**, 534–537.

Brezun, J. M., and Daszuta, A. (1999). Depletion in serotonin decreases neurogenesis in the dentate gyrus and the subventricular zone of adult rats. *Neuroscience* **89**, 999–1002.

Bronner-Fraser, M., and Fraser, S. (1989). Developmental potential of avian trunk neural crest cells in situ. *Neuron* **3**, 755–766.

Brown, E. S., Rush, A. J., and McEwen, B. S. (1999). Hippocampal remodeling and damage by corticosteroids: Implications for mood disorders. *Neuropsychopharmacology* **21**, 474–484.

Cai, L., Hayes, N. L., and Nowakowski, R. S. (1997). Local homogeneity of cell cycle length in developing mouse cortex. *J. Neurosci.* **17**, 2079–2087.

Cameron, H. A., and McKay, R. D. (1999). Restoring production of hippocampal neurons in old age. *Nat. Neurosci.* **2**, 894–897.

Campbell, K., Olsson, M., and Bjorklund, A. (1995). Regional incorporation and site-specific differentiation of striatal precursors transplanted to the embryonic forebrain ventricle. *Neuron* **15**, 1259–1273.

Chai, Y., Jiang, X., Ito, Y., Bringas, P., Han, J., Rowitch, D. H., Soriano, P., McMahon, A. P., and Sucov, H. M. (2000). Fate of the mammalian cranial neural crest during tooth and mandibular morphogenesis. *Development (Cambridge, UK)* **127**, 1671–1679.

Chen, Z.-F., Paquette, A. J., and Anderson, D. J. (1998). NRSF/REST is required in vivo for repression of multiple neuronal target genes during embryogenesis. *Nat. Genet.* **20**, 136–142.

Cheshier, S. H., Morrison, S. J., Liao, X., and Weissman, I. L. (1999). In vivo proliferation and cell cycle kinetics of long-term hematopoietic stem cells. *Proc. Natl. Acad. Sci. U.S.A.* **96**, 3120–3125.

Chong, J. A., Tapia-Ramirez, J., Kim, S., Toledo-Aral, J. J., Zheng, Y., Boutros, M. C., Altshuller, Y. M., Frohman, M. A., Kraner, S. D., and Mandel, G. (1995). REST: A mammalian silencer protein that restricts sodium channel expression to neurons. *Cell (Cambridge, Mass.)* **80**, 949–957.

Clapp, D. W., Freie, B., Lee, W.-H., and Zhang, Y.-Y. (1995). Molecular evidence that in situ-transduced fetal liver hematopoietic stem/progenitor cells give rise to medullary hematopoiesis in adult rats. *Blood* **86**, 2113–2122.

Clarke, D. L., Johansson, C. B., Wilbertz, J., Veress, B., Nilsson, E., Karlstrom, H., Lendahl, U., and Frisen, J. (2000). Generalized potential of adult neural stem cells. *Science* **288**, 1660–1663.

Corotto, F. S., Henegar, J. R., and Maruniak, J. A. (1994). Odor deprivation leads to reduced neurogenesis and reduced neuronal survival in the olfactory bulb of the adult mouse. *Neuroscience* **61**, 739–744.

Cross, A. J. (1990). Serotonin in Alzheimer-type dementia and other dementing illness. *Anna. N.Y. Acad. Sci.* **600**, 405–417.

Dale, L., and Wardle, F. C. (1999). A gradient of BMP activity specifies dorsal-ventral fates in early Xenopus embryos. *Semin. Cell Dev. Biol.* **10**, 319–326.

Date, I. (1996). Parkinson's disease, trophic factors, and adrenal medullary chromaffin cell grafting: Basic and clinical studies. *Brain Res. Bull.* **40**, 1–19.

Date, I., and Ohmoto, T. (1999). Neural transplantation for Parkinson's disease. *Cell. Mol. Neurobiol.* **19**, 67–78.

Davis, A., and Temple, S. (1994). A self-renewing multipotential stem cell in embryonic rat cerebral cortex. *Nature (London)* **372**, 263–266.

deHaan, G., Nijhof, W., and VanZant, G. (1997). Mouse strain-dependent changes in frequency and proliferation of hematopoietic stem cells during aging: Correlation between lifespan and cycling activity. *Blood* **89**, 1543–1550.

Deville, F. S.-S.-C., Ziller, C., and Le Douarin, N. (1992). Developmental potentialities of cells derived from the truncal neural crest in clonal cultures. *Dev. Brain Res.* **66**, 1–10.

Deville, F. S.-S.-C., Ziller, C., and Le Douarin, N. M. (1994). Developmental potentials of enteric neural crest-derived cells in clonal and mass cultures. *Dev. Biol.* **163**, 141–151.

Doetsch, F., Caille, I., Lim, D. A., Garcia-Verdugo, J. M., and Alvarez-Buylla, A. (1999). Subventricular zone astrocytes are neural stem cells in the adult mammalian brain. *Cell (Cambridge, Mass.)* **97**, 703–716.

Dong, J.-M., Smith, P., Hall, C., and Lim, L. (1995). Promoter region of the transcriptional unit for human α1-chimaerin, a neuron-specific GTPase-activating protein for p21rac. *Eur. J. Biochem.* **227**, 636–646.

Donovan, P. J. (1994). Growth factor regulation of mouse primordial germ cell development. *Curr. Top. Dev. Biol.* **29**, 189–225.

Duff, R. S., Langtimm, C. J., Richardson, M. K., and Sieber-Blum, M. (1991). In vitro clonal analysis of progenitor cell patterns in dorsal root and sympathetic ganglia of the quail embryo. *Dev. Biol.* **147**, 451–459.

Duman, R. S., Heninger, G. R., and Nestler, E. J. (1997). A molecular and cellular theory of depression. *Arch. Gen. Psychiatry* **54**, 597–606.

Erickson, C. A., and Goins, T. L. (1995). Avian neural crest cells can migrate in the dorsolateral path only if they are specified as melanocytes. *Development (Cambridge, UK)* **121**, 915–924.

Eriksson, P. S., Perfilieva, E., Bjork-Eriksson, T., Alborn, A.-M., Nordborg, C., Peterson, D. A., and Gage, F. H. (1998). Neurogenesis in the adult human hippocampus. *Nat. Med.* **4**, 1313–1317.

Ferrari, G., Cusella-De Angelis, G., Coletta, M., Paolucci, E., Stornaiuolo, A., Cossu, G., and Mavilio, F. (1998). Muscle regeneration by bone marrow-derived myogenic progenitors. *Science* **279**, 1528–1530.

Fishell, G. (1995). Striatal precursors adopt cortical identities in response to local cues. *Development (Cambridge, UK)* **121**, 803–812.

Fleischman, R. A., Custer, R. P., and Mintz, B. (1982). Totipotent hematopoietic stem cells: Normal self-renewal and differentiation after transplantation between mouse fetuses. *Cell (Cambridge, Mass.)* **30**, 351–359.

Francis, N., Farinas, I., Brennan, C., Rivas-Plata, K., Backus, C., Reichardt, L., and Landis, S. (1999). NT-3, like NGF, is required for survival of sympathetic neurons, but not their precursors. *Dev. Biol.* **210**, 411–427.

Frank, E., and Sanes, J. R. (1991). Lineage of neurons and glia in chick dorsal root ganglia: Analysis in vivo with a recombinant retrovirus. *Development (Cambridge, UK)* **111**, 895–908.

Fraser, S. E., and Bronner-Fraser, M. E. (1991). Migrating neural crest cells in the trunk of the avian embryo are multipotent. *Development (Cambridge, UK)* **112**, 913–920.

Frazier-Cierpial, L., and Brunjes, P. C. (1989). Early postnatal cellular proliferation and survival in the olfactory bulb and rostral migratory stream of normal and unilaterally odor-deprived rats. *J. Comp. Neurol.* **289**, 481–492.

Fuji, Y., Hongo, T., Nakagawa, Y., Nasuda, K., Mizuno, Y., Igarashi, Y., Naito, Y., and Maeda, M. (1989). Cell culture of small round cell tumor originating in the thoracopulmonary region: Evidence for derivation from a primitive pluripotent cell. *Cancer* **64**, 43–51.

Gage, F. H. (1998). Stem cells of the central nervous system. *Curr. Opin. Neurobiol.* **8**, 671–676.

Galli, R., Borello, U., Gritti, A., Minasi, M. G., Bjornson, C., Coletta, M., Mora, M., Cusella De Angelis, M. G., Fiocco, R., Cossu, G., and Vescovi, A. L. (2000). Skeletal myogenic potential of human and mouse neural stem cells. *Nat. Neurosci.* **3**, 986–991.

Gould, E., and Tanapat, P. (1997). Lesion-induced proliferation of neuronal progenitors in the dentate gyrus of the adult rat. *Neuroscience* **80**, 427–436.

Gould, E., Tanapat, P., McEwen, B. S., Flugge, G., and Fuchs, E. (1998). Proliferation of granule cell precursors in the dentate gyrus of adult monkeys is diminished by stress. *Proc. Natl. Acad. Sci. USA* **95**, 3168–3171.

Gould, E., Beylin, A., Tanapat, P., Reeves, A., and Shors, T. J. (1999a). Learning enhances adult neurogenesis in the hippocampal formation. *Nat. Neurosci.* **2**, 260–265.

Gould, E., Reeves, A. J., Graziano, M. S. A., and Gross, C. G. (1999b). Neurogenesis in the neocortex of adult primates. *Science* **286**, 548–552.

Greenwood, A. L., Turner, E. E., and Anderson, D. J. (1999). Identification of dividing, determined sensory neuron precursors in the mammalian neural crest. *Development (Cambridge, UK)* **126**, 3545–3559.

Gritti, A., Parati, E. A., Cova, L., Frolichsthal, P., Galli, R., Wanke, E., Faravelli, L., Morassutti, D. J., Roisen, F., Nickel, D. D., and Vescovi, A. L.
(1996). Multipotential stem cells from the adult mouse brain proliferate
and self-renew in response to basic fibroblast growth factor. *J. Neurosci.*
16, 1091–1100.

Gross, R. E., Mehler, M. F., Mabie, P. C., Zang, Z., Santschi, L., and Kessler, J. A. (1996). Bone Morphogenetic proteins promote astroglial lineage commitment by mammalian subventricular zone progenitor cells.
Neuron **17,** 595–606.

Guillemot, F., Lo, L.-C., Johnson, J. E., Auerbach, A., Anderson, D. J., and
Joyner, A. L. (1993). Mammalian achaete-scute homolog-1 is required
for the early development of olfactory and autonomic neurons. *Cell
(Cambridge, Mass.)* **75,** 463–476.

Gussoni, E., Soneoka, Y., Strickland, C. D., Buzney, E. A., Khan, M. K.,
Flint, A. F., Kunkel, L. M., and Mulligan, R. C. (1999). Dystrophin ex-
pression in the mdx mouse restored by stem cell transplantation. *Nature
(London)* **401,** 390–394.

Hagedorn, L., Suter, U., and Sommer, L. (1999). P0 and PMP22 mark a
multipotent neural crest-derived cell type that displays community effects in response to TGF-β family members. *Development (Cambridge,
UK)* **126,** 3781–3794.

Hardy, R. R., and Hayakawa, K. (1994). CD5 B cells, a fetal B cell lineage.
Adv. Immunol. **55,** 297–339.

Harrison, D. E., and Lerner, C. P. (1991). Most primitive hematopoietic stem
cells are stimulated to cycle rapidly after treatment with 5-fluorouracil.
Blood **78,** 1237–1240.

Henion, P. D., and Weston, J. A. (1997). Timing and pattern of cell fate
restrictions in the neural crest lineage. *Development (Cambridge, UK)*
124, 4351–4359.

Hentges, K., Thompson, K., and Peterson, A. (1999). The flat-top gene is
required for the expansion and regionalization of the telencephalic primordium. *Development (Cambridge, UK)* **126,** 1601–1609.

Hirsch, M. R., Tiveron, M. C., Guillemot, F., Brunet, J. F., and Goridis, C.
(1998). Control of noradrenergic differentiation and Phox2a expression
by MASH1 in the central and peripheral nervous system. *Development
(Cambridge, UK)* **125,** 599–608.

Ikuta, K., Kina, T., MacNeil, I., Uchida, N., Peault, B., Chien, Y. H., and
Weissman, I. L. (1990). A developmental switch in thymic lymphocyte
maturation potential occurs at the level of hematopoietic stem cells. *Cell
(Cambridge, Mass.)* **62,** 863–874.

Jiang, X., Rowitch, D. H., Soriano, P., McMahon, A. P., and Sucov, H. M.
(2000). Fate of the mammalian cardiac neural crest. *Development (Cambridge, UK)* **127,** 1607–1616.

Johansson, C. B., Momma, S., Clarke, D. L., Risling, M., Lendahl, U., and
Frisen, J. (1999). Identification of a neural stem cell in the adult mammalian central nervous system. *Cell (Cambridge, Mass.)* **96,** 25–34.

Johe, K. K., Hazel, T. G., Muller, T., Dugich-Djordjevic, M. M., and Mc-
Kay, R. D. G. (1996). Single factors direct the differentiation of stem
cells from the fetal and adult central nervous system. *Genes Dev.* **10,**
3129–3140.

Johnson, J. E., Birren, S. J., and Anderson, D. J. (1990). Two rat homologues of *Drosophila achaete-scute* specifically expressed in neuronal
precursors. *Nature (London)* **346,** 858–861.

Johnson, J. E., Birren, S. J., Saito, T., and Anderson, D. J. (1992). DNA
binding and transcriptional regulatory activity of mammalian achaete-
scute homologous (MASH) proteins revealed by interaction with a
muscle-specific enhancer. *Proc. Natl. Acad. Sci. U.S.A.* **89,** 3596–3600.

Kalyani, A. J., Hobson, K., and Rao, M. S. (1997). Neuroepithelial stem
cells from the embryonic spinal cord: Isolation, characterization, and
clonal analysis. *Dev. Biol.* **186,** 202–223.

Kalyani, A. J., Piper, D., Mujtaba, T., Lucero, M. T., and Rao, M. S. (1998).
Spinal cord neuronal precursors generate multiple neuronal phenotypes
in culture. *J. Neurosci.* **18,** 7856–7868.

Kantor, A. B., Stall, A. M., Adams, S., Herzenberg, L. A., and Herzenberg,
L. A. (1992). Differential development of progenitor activity for three
B-cell lineages. *Proc. Natl. Acad. Sci. U.S.A.* **89,** 3320–3324.

Kaplan, M. S., and Hinds, J. W. (1977). Neurogenesis in the adult rat: Electron microscopic analysis of light radioautographs. *Science* **197,** 1092–
1094.

Kempermann, G., Kuhn, H. G., and Gage, F. H. (1997). More hippocampal
neurons in adult mice living in an enriched environment. *Nature (London)* **386,** 493–495.

Kempermann, G., Brandon, E. P., and Gage, F. H. (1998). Environmental
stimulation of 129/SvJ mice causes increased cell proliferation and neurogenesis in the adult dentate gyrus. *Curr. Biol.* **8,** 939–942.

Kondo, M., Weissman, I. L., and Akashi, K. (1997). Identification of clonogenic common lymphoid progenitors in mouse bone marrow. *Cell (Cambridge, Mass.)* **91,** 661–672.

Kondo, T., and Raff, M. (2000). Oligodendrocyte precursor cells reprogrammed to become multipotential CNS stem cells. *Science* **289,** 1754–
1757.

Kopen, G. C., Prockop, D. J., and Phinney, D. G. (1999). Marrow stromal
cells migrate throughout forebrain and cerebellum, and they differentiate
into astrocytes after injection into neonatal mouse brains. *Proc. Natl.
Acad. Sci. U.S.A.* **96,** 10711–10716.

Lagasse, E., Connors, H., Al-Dhalimy, M., Reitsma, M., Dohse, M., Osborne, L., Wang, X., Finegold, M., Weissman, I. L., and Grompe, M.
(2000). Purified hematopoietic stem cells can differentiate to hepatocytes in vivo. *Nat. Med.* **6,** 1229–1234.

Le Douarin, N. M. (1982). "The Neural Crest." Cambridge University
Press, Cambridge, UK.

Le Douarin, N. M. (1986). Cell line segregation during peripheral nervous
system ontogeny. *Science* **231,** 1515–1522.

Le Douarin, N. M., and Dupin, E. (1992). Cell lineage analysis in neural
crest ontogeny. *J. Neurobiol.* **24,** 146–161.

Lemke, G. E., and Brockes, J. P. (1984). Identification and purification of
glial growth factor. *J. Neurosci.* **4,** 75–83.

Li, P., Yang, X., Wasser, M., Cai, Y., and Chia, W. (1997). Inscuteable and
Staufen mediate asymmetric localization and segregation of prospero
RNA during Drosophila neuroblast cell divisions. *Cell (Cambridge,
Mass.)* **90,** 437–447.

Li, W., Cogswell, C. A., and LoTurco, J. J. (1998). Neuronal differentiation
of precursors in the neocortical ventricular zone is triggered by BMP.
J. Neurosci. **18,** 8853–8862.

Lo, L., and Anderson, D. (1995). Postmigratory neural crest cells expressing c-RET display restricted developmental and proliferative capacities.
Neuron **15,** 527–539.

Lo, L., Sommer, L., and Anderson, D. J. (1997). MASH1 maintains competence for BMP2-induced neuronal differentiation in post-migratory
neural crest cells. *Curr. Biol.* **7,** 440–450.

Lo, L., Tiveron, M.-C., and Anderson, D. J. (1998). MASH1 activates expression of the paired homeodomain transcription factor Phox2a, and
couples pan-neuronal and subtype-specific components of autonomic
neuronal identity. *Development (Cambridge, UK)* **125,** 609–620.

Lois, C., and Alvarez-Buylla, A. (1993). Proliferating subventricular zone
cells in the adult mammalian forebrain can differentiate into neurons and
glia. *Proc. Natl. Acad. Sci. U.S.A.* **90,** 2074–2077.

Lois, C., and Alvarez-Buylla, A. (1994). Long-distance neuronal migration
in the adult mammalian brain. *Science* **264,** 1145–1148.

Lowenstein, D. H., and Parent, J. M. (1999). Brain, heal thyself. *Science*
283, 1126–1127.

Luskin, M. B., Pearlman, A. L., and Sanes, J. R. (1988). Cell lineage in the
cerebral cortex of the mouse studied in vivo and in vitro with a recombinant retrovirus. *Neuron* **1,** 635–647.

Lyons, K. M., Hogan, B. L. M., and Robertson, E. J. (1995). Colocalization
of BMP 7 and BMP 2 RNAs suggests that these factors cooperatively
mediate tissue interactions during murine development. *Mech. Dev.* **50,**
71–83.

Mabie, P. C., Mehler, M. F., and Kessler, J. A. (1999). Multiple roles of
bone morphogenetic protein signaling in the regulation of cortical cell
number and phenotype. *J. Neurosci.* **19,** 7077–7088.

Marchionni, M. A., Goodearl, A. D. J., Chen, M. S., Bermingham-

McDonogh, O., Kirk, C., Hendricks, M., Danehy, F., Misumi, D., Sud-halter, J., Kobayashi, K., Wroblewski, D., Lynch, C., Baldassare, M., Hiles, I., Davis, J. B., Hsuan, J. J., Totty, N. F., Otsu, M., McBurney, R. N., Waterfield, M. D., Stroobant, P., and Gwynne, D. (1993). Glial growth factors are alternatively spliced erbB2 ligands expressed in the nervous system. *Nature (London)* **362,** 312–318.

Marina, N. M., Etcubanas, E., Parham, D. M., Bowman, L. C., and Green, A. (1989). Peripheral primitive neuroectodermal tumor (peripheral neuroepithelioma) in children. *Cancer* **64,** 1952–1960.

Matsui, Y., Zsebo, K., and Hogan, B. L. M. (1992). Derivation of pluripotential embryonic stem cells from murine primordial germ cells in culture. *Cell (Cambridge, Mass.)* **70,** 841–847.

Mayer-Proschel, M., Kalyani, A. J., Mujtaba, T., and Rao, M. S. (1997). Isolation of lineage-restricted neuronal precursors from multipotent neuroepithelial stem cells. *Neuron* **19,** 773–785.

McKay, R. (1997). Stem cells in the central nervous system. *Science* **276,** 66–71.

Mehler, M. F., Mabie, P. C., Zhang, D., and Kessler, J. A. (1997). Bone morphogenetic proteins in the nervous system. *Trends Neurosci.* **20,** 309–317.

Mello, C. C., Schubert, C., Draper, B., Zhang, W., Lobel, R., and Priess, J. R. (1996). The PIE-1 protein and germline specification in *C. elegans* embryos. *Nature (London)* **382,** 710–712.

Meyer, D., and Birchmeier, C. (1994). Distinct isoforms of neuregulin are expressed in mesenchymal and neuronal cells during mouse development. *Proc. Natl. Acad. Sci. U.S.A.* **91,** 1064–1068.

Meyer, R., and Birchmeier, C. (1995). Multiple essential functions of neuregulin in development. *Nature (London)* **378,** 386–390.

Morrison, S. J., and Weissman, I. L. (1994). The long-term repopulating subset of hematopoietic stem cells is deterministic and isolatable by phenotype. *Immunity* **1,** 661–673.

Morrison, S. J., Hemmati, H. D., Wandycz, A. M., and Weissman, I. L. (1995a). The purification and characterization of fetal liver hematopoietic stem cells. *Proc. Natl. Acad. Sci. U.S.A.* **92,** 10302–10306.

Morrison, S. J., Uchida, N., and Weissman, I. L. (1995b). The biology of hematopoietic stem cells. *Annu. Rev. Cell Dev. Biol.* **11,** 35–71.

Morrison, S. J., Wandycz, A. M., Akashi, K., Globerson, A., and Weissman, I. L. (1996). The aging of hematopoietic stem cells. *Nat. Med.* **2,** 1011–1016.

Morrison, S. J., Shah, N. M., and Anderson, D. J. (1997a). Regulatory mechanisms in stem cell biology. *Cell (Cambridge, Mass.)* **88,** 287–298.

Morrison, S. J., Wandycz, A. M., Hemmati, H. D., Wright, D. E., and Weissman, I. L. (1997b). Identification of a lineage of multipotent hematopoietic progenitors. *Development (Cambridge, UK)* **124,** 1929–1939.

Morrison, S. J., Wright, D., and Weissman, I. L. (1997c). Cyclophosphamide/granulocyte colony-stimulating factor induces hematopoietic stem cells to proliferate prior to mobilization. *Proc. Natl. Acad. Sci. U.S.A.* **94,** 1908–1913.

Morrison, S. J., White, P. M., Zock, C., and Anderson, D. J. (1999). Prospective identification, isolation by flow cytometry, and in vivo self-renewal of multipotent mammalian neural crest stem cells. *Cell (Cambridge, Mass.)* **96,** 737–749.

Morrison, S. J., Csete, M., Groves, A. K., Melega, W., Wold, B., and Anderson, D. J. (2000). Culture in reduced levels of oxygen promotes clonogenic sympathoadrenal differentiation by isolated neural crest stem cells. *J. Neurosci.* **20,** 7370–7376.

Morrissey, T. K., Levi, A. D., Nuijens, A., Sliwkowski, M. X., and Bunge, R. P. (1995). Axon-induced mitogenesis of human Schwann cells involves heregulin and p185erbB2. *Proc. Natl. Acad. Sci. U.S.A.* **92,** 1431–1435.

Morshead, C. M., Reynolds, B. A., Craig, C. G., McBurney, M. W., Staines, W. A., Morassutti, D., Weiss, S., and van der Kooy, D. (1994). Neural stem cells in the adult mammalian forebrain: A relatively quiescent subpopulation of subependymal cells. *Neuron* **13,** 1071–1082.

Morshead, C. M., Craig, C. G., and van der Kooy, D. (1998). In vivo clonal

analyses reveal the properties of endogenous neural stem cell proliferation in the adult mammalian forebrain. *Development (Cambridge, UK)* **125,** 2251–2261.

Murphy, P., Topilko, P., Schneider-Maunoury, S., Seitanidou, T., Baron van Evercooren, A., and Charnay, P. (1996). The regulation of Krox-20 expression reveals important steps in the control of peripheral glial cell development. *Development (Cambridge, UK)* **122,** 2847–2857.

Na, E., McCarthy, M., Neyt, C., Lai, E., and Fishell, G. (1998). Telencephalic progenitors maintain anteroposterior identities cell autonomously. *Curr. Biol.* **8,** 987–990.

Olanow, C. W., Freeman, T. B., and Kordower, J. H. (1997). Neural transplantation as a therapy for Parkinson's disease. *In* "The Basal Ganglia and New Surgical Approaches for Parkinson's Disease" (J. A. Obeso, M. R. DeLong, C. Ohye, and C. D. Marsden, eds.), pp. 249–269. Lippincott-Raven, Philadelphia.

Osawa, M., Hanada, K.-I., Hamada, H., and Nakauchi, H. (1996). Long-term lymphohematopoietic reconstitution by a single CD34-low/negative hematopoietic stem cell. *Science* **273,** 242–245.

Palmer, T. D., Takahashi, J., and Gage, F. H. (1997). The adult rat hippocampus contans primordial neural stem cells. *Mol. Cell. Neurosci.* **8,** 389–404.

Palmer, T. D., Markakis, E. A., Willhoite, A. R., Safar, F., and Gage, F. H. (1999). Fibroblast growth factor-2 activates a latent neurogenic program in neural stem cells from diverse regions of the adult CNS. *J. Neurosci.* **19,** 8487–8497.

Parent, J. M., Yu, T. W., Leibowitz, R. T., Geschwind, D. H., Sloviter, R. S., and Lowenstein, D. H. (1997). Dentate granule cell neurogenesis is increased by seizures and contributes to aberrant network reorganization in the adult rat hippocampus. *J. Neurosci.* **17,** 3727–3738.

Parent, J. M., Janumpalli, S., McNamara, J. O., and Lowenstein, D. H. (1998). Increased dentate granule cell neurogenesis following amygdala kindling in the adult rat. *Neurosci. Lett.* **247,** 9–12.

Peterson, B. E., Bowen, W. C., Patrene, K. D., Mars, W. M., Sullivan, A. K., Murase, N., Boggs, S. S., Greenberger, J. S., and Goff, J. P. (1999). Bone marrow as a potential source of hepatic oval cells. *Science* **284,** 1168–1170.

Pham, T. D., Gershon, M. D., and Rothman, T. P. (1991). Time of origin of neurons in the murine enteric nervous system: Sequence in relation to phenotype. *J. Comp. Neurol.* **314,** 789–798.

Rakic, P. (1985). Limits of neurogenesis in primates. *Science* **227,** 1054–1056.

Ramón y Cajál, S., and May, R. T. (1959). "Degeneration and Regeneration of the Nervous System," Vol. 2. Hafner, New York.

Rao, M. S., Noble, M., and Mayer-Proschel, M. (1998). A tripotential glial precursor cell is present in the developing spinal cord. *Proc. Natl. Acad. Sci. U.S.A.* **95,** 3996–4001.

Rebel, V. I., Miller, C. L., Eaves, C. J., and Lansdorp, P. M. (1996). The repopulation potential of fetal liver hematopoietic stem cells in mice exceeds that of their adult bone marrow counterparts. *Blood* **87,** 3500–3507.

Reddy, P. H., Williams, M., and Tagle, D. A. (1999). Recent advances in understanding the pathogenesis of Huntington's disease. *Trends Neurosci.* **22,** 248–255.

Reid, C. B., Liang, I., and Walsh, C. (1995). Systematic widespread clonal organization in cerebral cortex. *Neuron* **15,** 299–310.

Reissman, E., Ernsberger, U., Francis-West, P. H., Rueger, D., Brickell, P. D., and Rohrer, H. (1996). Involvement of bone morphogenetic protein-4 and bone morphogenetic protein-7 in the differentiation of the adrenergic phenotype in developing sympathetic neurons. *Development (Cambridge, UK)* **122,** 2079–2088.

Reynolds, B. A., and Weiss, S. (1992). Generation of neurons and astrocytes from isolated cells of the adult mammalian central nervous system. *Science* **255,** 1707–1710.

Richardson, M. K., and Sieber-Blum, M. (1993). Pluripotent neural crest cells in the developing skin of the quail embryo. *Dev. Biol.* **157,** 348–358.

Riethmacher, D., Sonnerberg-Riethmacher, E., Brinkmann, V., Yamaai, T., Lewin, G. R., and Birchmeier, C. (1997). Severe neuropathies in mice with targeted mutations in the erbB3 receptor. *Nature (London)* **389,** 725–730.

Rosselli-Austin, L., and Williams, J. (1989). Enriched neonatal odor exposure leads to increased numbers of olfactory bulb mitral and granule cells. *Dev. Brain Res.* **51,** 135–137.

Rutkowski, J. L., Kirk, C. J., Lerner, M. A., and Tennekoon, G. I. (1995). Purification and expansion of human Schwann cells in vitro. *Nature Med.* **1,** 80–83.

Sauvageau, G., Thorsteinsdottir, U., Eaves, C. J., Lawrence, H. J., Largman, C., Lansdorp, P. M., and Humphries, R. K. (1995). Overexpression of HOXB4 in hematopoietic cells causes the selective expansion of more primitive populations in vitro and in vivo. *Genes Dev.* **9,** 1753–1765.

Schneider, C., Wicht, H., Enderich, J., Wegner, M., and Rohrer, H. (1999). Bone morphogenetic proteins are required in vivo for the generation of sympathetic neurons. *Neuron* **24,** 861–870.

Schoenherr, C. J., and Anderson, D. J. (1995). The Neuron-Restrictive Silencer Factor (NRSF): A coordinate repressor of multiple neuron-specific genes. *Science* **267,** 1360–1363.

Schoenherr, C. J., Paquette, A. J., and Anderson, D. J. (1996). Identification of potential target genes for the neuron-restrictive silencer factor. *Proc. Natl. Acad. Sci. U.S.A.* **93,** 9881–9886.

Scott, B. W., Wang, S., Burnham, W. M., Boni, U. D., and Wojtowicz, J. M. (1998). Kindling-induced neurogenesis in the dentate gyrus of the rat. *Neurosci. Lett.* **248,** 73–76.

Sell, S., and Pierce, G. B. (1994). Maturation arrest of stem cell differentiation is a common pathway for the cellular origin of teratocarcinomas and epithelial cancers. *Lab. Investig.* **70,** 6–22.

Serbedzija, G. N., Bronner-Fraser, M., and Fraser, S. E. (1994). Developmental potential of trunk neural crest cells in the mouse. *Development (Cambridge, UK)* **120,** 1709–1718.

Seydoux, G., Mello, C. C., Pettitt, J., Wood, W. B., Priess, J. R., and Fire, A. (1996). Repression of gene expression in the embryonic germ lineage of *C. elegans. Nature (London)* **382,** 713–716.

Shah, N. M., Marchionni, M. A., Isaacs, I., Stroobant, P. W., and Anderson, D. J. (1994). Glial growth factor restricts mammalian neural crest stem cells to a glial fate. *Cell (Cambridge, Mass.)* **77,** 349–360.

Shah, N. M., Groves, A., and Anderson, D. J. (1996). Alternative neural crest cell fates are instructively promoted by TGFβ superfamily members. *Cell (Cambridge, Mass.)* **85,** 331–343.

Sieber-Blum, M. (1989). Commitment of neural crest cells to the sensory neuron lineage. *Science* **243,** 1608–1610.

Sieber-Blum, M., and Cohen, A. (1980). Clonal analysis of quail neural crest cells: They are pluripotent and differentiate in vitro in the absence of non-neural crest cells. *Dev. Biol.* **80,** 96–106.

Smith, W. C., and Harland, R. M. (1992). Expression cloning of noggin, a new dorsalizing factor localized to the Spemann organizer in Xenopus embryos. *Cell (Cambridge, Mass.)* **70,** 829–840.

Sommer, L., Shah, N., Rao, M., and Anderson, D. J. (1995). The cellular function of MASH1 in autonomic neurogenesis. *Neuron* **15,** 1245–1258.

Spangrude, G. J., Heimfeld, S., and Weissman, I. L. (1988). Purification and characterization of mouse hematopoietic stem cells. *Science* **241,** 58–62.

Spangrude, G. J., Brooks, D. M., and Tumas, D. B. (1995). Long-term repopulation of irradiated mice with limiting numbers of purified hematopoietic stem cells: In vivo expansion of stem cell phenotype but not function. *Blood* **85,** 1006–1016.

Stemple, D. L., and Anderson, D. J. (1992). Isolation of a stem cell for neurons and glia from the mammalian neural crest. *Cell (Cambridge, Mass.)* **71,** 973–985.

Suhonen, J. A., Peterson, D. A., Ray, J., and Gage, F. H. (1996). Differentiation of adult hippocampus-derived progenitors into olfactory neurons *in vivo. Nature (London)* **383,** 624–627.

Takahashi, M., Palmer, T. D., Takahashi, J., and Gage, F. H. (1998). Widespread integration and survival of adult-derived neural progenitor cells in the developing optic retina. *Mol. Cell. Neurosci.* **12,** 340–348.

Temple, S., and Alvarez-Buylla, A. (1999). Stem cells in the adult mammalian central nervous system. *Curr. Opin. Neurobiol.* **9,** 135–141.

Tenenhaus, C., Schubert, C., and Seydoux, G. (1998). Genetic requirements for PIE-1 localization and inhibition of gene expression in the embryonic germ lineage of Caenorhabditis elegans. *Dev. Biol.* **200,** 212–224.

Thiele, C. J. (1991). Biology of pediatric peripheral neuroectodermal tumors. *Cancer Metastasis Rev.* **10,** 311–319.

Topilko, P., Levi, G., Merlo, G., Mantero, S., Desmarquet, C., Mancardi, G., and Charnay, P. (1997). Differential regulation of the zinc finger genes Krox-20 and Krox-24 (Egr-1) suggests antagonistic roles in Schwann Cells. *J. Neurosci. Res.* **50,** 702–712.

Uchida, N., Aguila, H. L., Fleming, W. H., Jerabek, L., and Weissman, I. L. (1994). Rapid and sustained hematopoietic recovery in lethally irradiated mice transplanted with purified Thy-1.1loLin-Sca-1+ hematopoietic stem cells. *Blood* **83,** 3758–3779.

Vicario-Abejón, C., Johe, K. K., Hazel, T. G., Collazo, D., and McKay, R. D. G. (1995). Functions of basic fibroblast growth factor and neurotrophins in the differentiation of hippocampal neurons. *Neuron* **15,** 105–114.

Wakamatsu, Y., Mochii, M., Vogel, K. S., and Weston, J. A. (1998). Avian neural crest-derived neurogenic precursors undergo apoptosis on the lateral migration pathway. *Development (Cambridge, UK)* **125,** 4205–4213.

Walsh, C., and Cepko, C. L. (1993). Clonal dispersion in proliferative layers of developing cerebral cortex. *Nature (London)* **362,** 632–635.

Weiss, S., Dunne, C., Hewson, J., Wohl, C., Wheatley, M., Peterson, A. C., and Reynolds, B. A. (1996). Multipotent CNS stem cells are present in the adult mammalian spinal cord and ventricular neuroaxis. *J. Neurosci.* **16,** 7599–7609.

White, P. A., and Anderson, D. J. (1999). In vivo transplantation of mammalian neural crest cells into chick hosts reveals a new autonomic sublineage restriction. *Development (Cambridge, UK)* **126,** 4351–4363.

Williams, B. P., and Price, J. (1995). Evidence for multiple precursor cell types in the embryonic rat cerebral cortex. *Neuron* **14,** 1181–1188.

Williams, B. P., Park, J. K., Alberta, J. A., Muhlebach, S. G., Hwang, G. Y., Roberts, T. M., and Stiles, C. D. (1997). A PDGF-regulated immediate early response initiates neuronal differentiation in ventricular zone progenitor cells. *Neuron* **18,** 553–562.

Wilson Berry, L., Westlund, B., and Schedl, T. (1997). Germ-line tumor formation caused by activation of glp-1, a Caenorhabditis elegans member of the Notch family of receptors. *Development (Cambridge, UK)* **124,** 925–936.

Winkler, J., Thal, L. J., Gage, F. H., and Fisher, L. J. (1998). Cholinergic strategies for Alzheimer's disease. *J. Mol. Med.* **76,** 555–567.

Xuan, S. H., Baptista, C. A., Balas, G., Tao, W. F., Soares, V. C., and Lai, E. (1995). Winged helix transcription factor BF-1 is essential for the development of the cerebral hemispheres. *Neuron* **14,** 1141–1152.

Yokoi, M., Mori, K., and Nakanishi, S. (1995). Refinement of odor molecule tuning by dendrodendritic synaptic inhibition in the olfactory bulb. *Proc. Natl. Acad. Sci. U.S.A.* **92,** 3371–3376.

Zhu, A. J., and Watt, F. M. (1999). β-catenin signalling modulates proliferative potential of human epidermal keratinocytes independently of intercellular adhesion. *Development (Cambridge, UK)* **126,** 2285–2298.

Zimmerman, L. B., Dejesus-Escobar, J. M., and Harland, R. M. (1996). The Spemann organizer signal noggin binds and inactivates bone morphogenetic protein-4. *Cell (Cambridge, Mass.)* **86,** 599–606.

13

Cellular and Molecular Mechanisms Regulating Skeletal Muscle Development

Atsushi Asakura and Michael A. Rudnicki

Program in Molecular Genetics, Ottawa Hospital Research Institute, Ottawa, Ontario, Canada K1H 8L6

I. Introduction

II. Embryonic Origin of Skeletal Muscle

III. MyoD Family of Myogenic Regulatory Factors

IV. Muscle-Specific Transcriptional Regulation

V. Inductive Mechanisms of Myogenesis

VI. Specification of Muscle Fiber Types

VII. Muscle Regeneration

VIII. Conclusion

References

I. Introduction

The study of skeletal muscle differentiation has been intensive since the 1960s because of readily discerned criteria for differentiation (e.g., multinucleated syncytium) and the availability of biochemical markers such as actin and myosin. In addition, the early isolation of cell lines derived from skeletal muscle was essential for cellular and molecular studies that elucidated mechanisms regulating myogenesis. Consequently, the study of vertebrate myogenesis has provided a powerful biological system to investigate the molecular regulation of the developmental program that controls the genesis, growth, migration, and differentiation of an embryonic cell lineage.

Our knowledge of the molecular mechanisms that regulate skeletal myogenesis was dramatically accelerated about a decade ago following the discovery of the MyoD family of transcription factors, termed the myogenic regulatory factors (MRFs). The MRFs comprise a group of skeletal-muscle-specific bHLH (basic helix–loop–helix) transcription factors, consisting of MyoD, Myf5, myogenin, and MRF4. The MRFs activate transcription of muscle-specific genes through binding of DNA motifs called E-boxes. Forced expression of the MRFs in various cell types *in vitro* and *in vivo* induces muscle differentiation. The restricted expression of the MRFs together with their ability to dominantly induce differentiation led to the suggestion that the MyoD family members are master regulators of the muscle developmental program. Subsequently, gene targeting in mice allowed for genetic dissection of these regulatory pathways and clearly defined the roles played by the MyoD family of transcription factors in myogenesis.

During vertebrate development, the mesodermal progenitors of skeletal muscle originate from precursor cells that arise from the somites and prechordal mesoderm. The specification of muscle progenitor cells from the somites is regulated by signals from the neural tube, notochord, and other surrounding tissues that provide both positive and negative cues (see Section V). Indeed, the embryological experiments currently under way to investigate the specification of skeletal muscle arguably represent the state of the art in vertebrate developmental biology.

The purpose of this chapter is to provide an overview of the molecular mechanisms that regulate myogenesis in the mouse. Topics discussed include the embryonic origin of skeletal muscle, the MyoD family of myogenic regulatory factors, muscle-specific transcriptional regulation, inductive mechanisms regulating embryonic myogenesis, fiber-type specification, and the regulation of adult muscle regeneration.

II. Embryonic Origin of Skeletal Muscle

A. The Terminology of Myogenesis

Defined terms are used for particular muscle cell types during myogenesis (Stockdale, 1992; Hauschka, 1996; Bischoff, 1996). During vertebrate development, *muscle progenitor cells,* which still possess developmental plasticity, derive from the mesodermal lineage. Muscle progenitor cells differentiate into myoblasts that express the MRFs, MyoD (*Myod1*) and Myf5. Proliferating *myoblasts* withdraw from the cell cycle to become terminally differentiated *myocytes* that express muscle-specific genes such as *myogenin* (*Myog*), *MRF4/Herculin* (*Myf6*), myosin heavy chain (*MyHC*), and muscle creatine kinase (*MCK, Ckmm*) genes. These mononucleated myocytes fuse with each other to form multinucleated *myotubes* that as mature *muscle fibers* initiate muscle contraction. Skeletal muscle differentiation occurs in several phases. First formed at about 13 days postcoitus (dpc) in the mouse are the *primary fibers,* followed by the *secondary fibers* beginning at about 16 dpc, which develop parallel to and within the same basal lamina of the primary fibers. Elaboration of muscle fiber subtypes begins late in development and extends into the neonatal period. *Fast-twitch (type II)* glycolytic and *slow-twitch (type I)* oxidative muscle fibers differ in their functional properties and in the expression of distinct isoforms of contractile proteins. Last, *satellite cells,* the quiescent stem cell of adult muscle, are activated (initiate proliferation) in response to stress to produce proliferating *myogenic precursor cells (mpc)* that mediate the postnatal growth and regeneration of muscle. *Muscle-derived stem cells* represent the recently described pluripotent stem cells that reside in adult skeletal muscle.

B. Mesodermal Progenitor Cells Are the Origin of Skeletal Muscle Cells

During vertebrate embryogenesis, the musculature of the trunk and limbs is derived from progenitors that originate from the somites, epithelial spheres of paraxial mesoderm that form in a rostrocaudal progression on either side of the neural tube (Christ and Ordahl, 1995). Craniofacial muscles, including muscle masses of the occulo motors and branchial arches, are derived from cephalic paraxial mesoderm (somitomeres), prechordal mesoderm, and occipital somites (a few of the most rostral somites) (Noden, 1991; Trainor *et al.,* 1994). The seven pairs of somitomeres derived from paraxial head mesoderm give rise to the voluntary craniofacial muscles and, together with prechordal mesoderm and the occipital somites, contribute progenitors to the branchial arches to form the muscles of the jaws and neck.

Trunk and limb muscle development is relatively well understood compared to craniofacial myogenesis. During early embryogenesis, trunk mesoderm is separated into four distinct regions: axial, paraxial, intermediate, and lateral mesoderm. The axial mesoderm gives rise to the notochord located beneath the neural plate. The lateral mesoderm becomes the lateral plate that gives rise to cardiac and smooth muscle. The presomitic mesoderm (or unsegmented mesoderm) derived from the paraxial mesoderm undergoes segmentation to form epithelial spheres or somites on either side of the neural tube in a rostrocaudal direction beginning at 7.5 dpc in the mouse (see Chapter 7).

C. Trunk and Limb Muscles Originate from Somite

Cell lineage experiments using chick-quail chimeras and transplantation of rotated somites reveal that cells in newly formed somites retain the potential for all somitic lineages. Therefore, specification of somitic lineages occurs following the exposure of somitic cells to patterning signals derived from the surrounding tissues (Figs. 1A and 1B and see Section V). The ventral part of the newly formed somites deepithelializes to form mesenchymal cells that give rise to the sclerotome. The cells in the sclerotome subsequently migrate around the neural tube as well as laterally to form the cartilage of the vertebral column and ribs (Fig. 1C). Between 8.0 dpc and 8.5 dpc in the mouse, the dorsal part of the most rostral somite (see Plate 15b–c, d in Kaufman, 1992), consisting in part of proliferating cells, maintains an epithelial structure and is called the dermomyotome. Cells in the dorsomedial region of dermomyotome (dorsomedial lip; DML), residing close to neural tube, gradually extend laterally beneath the entire width of the dermomyotome (Figs. 1A–D). By 8.5 dpc in the mouse, DML-derived cells in the most

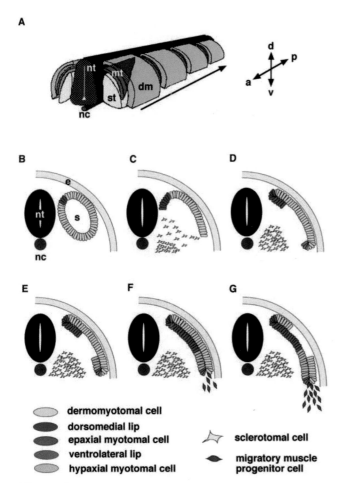

Figure 1 Schematic representation of mouse embryo at thoracic level somites. Somite (s) formation followed by dermomyotome (dm) and myotome (mt) formations proceed in a rostral (a; anterior) to caudal (p; posterior) sequence. By 8.0 dpc, the ventral part of the most rostral somite deepithelializes to form mesenchymal cells, which give rise to the sclerotome (st) (A–C). By 8.5 dpc in the mouse, the dorsal part of the rostral somite maintains a proliferating epithelial structure called the dermomyotome (A–D). Cells in the dorsomedial region of dermomyotome (dorsomedial lip; DML) withdraw from cell proliferation and longitudinally elongate beneath the dorsomedial part of dermomyotome to form the dorsal myotome (epaxial myotome; C–E). At 9.5 dpc, cells in the ventrolateral region of dermomyotome [ventrolateral lip (VLL) or hypaxial somite bud] withdraw from cell proliferation and longitudinally elongate beneath to ventral part of the dermomyotome to form the ventral myotome (hypaxial myotome; D–F). Eventually, both epaxial and hypaxial myotome extend continuously beneath the dermomyotome to form a planar myotomal structure (F). By 11.0 dpc, the center of the dermomyotome has dissociated into dermis while the epithelial DML and VLL still remain and continue to provide myotomal cells (G). At 10.0 dpc, epithelial cells in the VLL at the forelimb bud level delaminate and ventrally migrate out from the dermomyotome into the embryo (migratory muscle progenitor cells; F and G). nt, neural tube; d, dorsal; v, ventral; e, ectodermal layer; nc, notochord.

rostral somite (see Plate 15b–c, d and Plate 18b–e, i in Kaufman, 1992), withdraw from cell proliferation and longitudinally elongate beneath the dorsomedial part of der-

momyotome to form the first terminally differentiated myocytes in a structure called the myotome (Fig. 1D; Williams and Ordahl, 1997; Christ and Ordahl, 1995). The developmental processes of the cells in the ventrolateral region of dermomyotome [ventrolateral lip (VLL) or somite bud] are the mirror image of those of the dorsomedial part (Christ and Ordahl, 1995; Brand-Saberi et al., 1996a). By 9.75 dpc in mouse, VLL-derived cells in the interlimb somite (see Plate 20c–d, e, f, g, h in Kaufman, 1992) begin to withdraw from the cell cycle and longitudinally elongate beneath to the ventrolateral part of the dermomyotome to form the ventral myotome (Figs. 1D and 1E).

The compartment consisting of postmitotic cells beneath the dorsomedial portion of the dermomyotome is called the epaxial myotome and gives rise to the epaxial musculature or muscles of the deep back. The compartment residing beneath the ventrolateral part of the dermomyotome is called the nonmigratory hypaxial myotome and gives rise to lateral skeletal muscles in the trunk such as intercostal muscle and body-wall muscle (Figs. 1E and 1F). Eventually, both epaxial and hypaxial myotomes at the interlimb bud level extend continuously beneath the dermomyotome to form a planar myotomal structure (Figs. 1F and 1G and see Plate 19b–i, Plate 19c–a, and Plate 20c–e, f, g, h in Kaufman, 1992) that expresses muscle structural proteins at the center of myotome. By 12.0 dpc, the central portion of myotomes at trunk level form the first multinucleated myotubes (see Plate 26c and 26d in Kaufman, 1992).

Experimental analysis of avian somitogenesis has revealed unexpected insights into the mechanism of myotome formation (Fig. 2; Denetclaw et al., 1997, 2000; Kahane et al., 1998a,b; Cinnamon et al., 1999; Kalcheim et al., 1999). The first differentiated myocyctes, termed muscle pioneers, are derived from mitotic cells in the DML that withdraw from the cell cycle as they migrate beneath the dermomyotome and translocate to the rostral edge (Figs. 2A and 2B). Subsequently, pioneer cells longitudinally elongate across the rostral to the caudal lip of the dermomyotome to form the primary myotome (Fig. 2C; first wave of myotome development) (Kahane et al., 1998a). At this stage, the nuclei of pioneer myofibers become restricted to the center of the myotome (Fig. 2C). Next, mitotic cells in all four lips (DML, VLL, rostral lip, and caudal lip) of the dermomyotome cease proliferation, migrate beneath the dermomyotome, and translocate into the rostral and caudal edges (Fig. 2D). These migratory cells longitudinally elongate toward the rostral and caudal lips of the dermomyotome between pioneer muscle fibers to form the secondary myotome and constitute the second wave of myotome development (Figs. 2E and 2F; Kahane et al., 1998b). Subsequently, secondary myotome formation continues ventrolaterally to form the hypaxial myotome and medially to increase the thickness of the myotome (Fig. 1F). Following the disap-

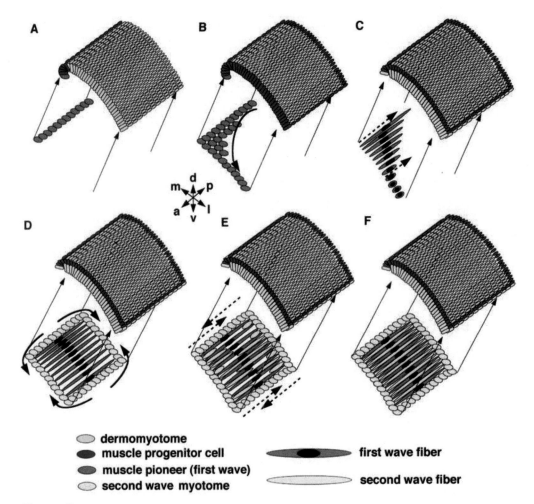

Figure 2 Schematic model representing various stages of avian myotome development. Each view represents the myotomal plane from which the dermomyotome has been lifted apart (indicated by thick arrow in B). The first myotomal cells, termed muscle pioneers, are derived from cells (muscle progenitor cells) in the DML that withdraw from the cell cycle, migrate beneath the dermomyotome (A), and translocate to the rostral edge (indicated by thick arrow in B) of the somite. Subsequently, pioneer cells elongate across the distance spanning from the rostral to the caudal lip of the dermomyotome (indicated by dashed arrows in C) to form the primary myotome (first wave of myotome development). The second wave of myotome is formed by the migration of cells from all four edges of the dermomyotome to the perimeter of the pioneer myotome (D). These cells extend longitudinally in between the pioneer muscle fibers and form the secondary myotome (E, F) (Kahane *et al.,* 1998 a,b: Cinnamon *et al.,* 1999).

pearance of the dermomyotomal lips, a population of mitotically active precursors within the myotome continues to contribute to myotome formation in a process termed third wave of myotome formation.

Much of the skeletal muscle of the body, however, is derived from migratory progenitors that arise as somitically derived multipotential cells. Epithelial cells in the VLL delaminate and ventrally migrate out from the dermomyotome into the embryo (Figs. 1F and 1G and see Fig. 5D in a later section). Such migratory muscle progenitor cells enter the ventral region of trunk and limb buds and continue proliferation. Migratory muscle progenitor cells coalesce as muscle anlagen prior to withdrawal from the cell cycle and terminal

differentiation. Migratory muscle progenitors give rise to the migratory hypaxial muscles: the pectoralis, abdominal, and diaphragm muscles of the trunk, and all limb muscles (Ordahl and Le Douarin, 1992; Ordahl and Williams, 1998).

III. MyoD Family of Myogenic Regulatory Factors

A. The MRFs Are bHLH Transcription Factors

In a series of seminal experiments, Weintraub and his colleagues cloned a bHLH transcription factor called *MyoD* that appeared to function as a master regulator of muscle differentiation (Davis *et al.,* 1987). In vertebrates, four highly related genes, *MyoD, Myf5, myogenin,* and *MRF4,* belong to this class of genes also known as the myogenic regulatory factors (Fig. 3). Forced expression of the MRFs in various cell lines *in vitro* and in different tissue types *in vivo* dominantly induces skeletal muscle differentiation (Weintraub *et al.,* 1991a).

The bHLH region of the MRFs is about 70 amino acid residues and is highly conserved with more than 80% identity in amino acid sequence level. The basic region contains many positively charged amino acid residues and is responsible for DNA binding (Davis *et al.,* 1990). The HLH region is responsible for dimerization with a distinct bHLH class of

E proteins, E12/E47 (*Tcfe2a*), ITF-2 (*Tcf4*), and ME1/HEB (*Tcf12*) (Murre *et al.,* 1989b; Lassar *et al.,* 1991). In general, bHLH protein dimers bind the DNA motif called an E-box with the core consensus sequence CANNTG (Murre *et al.,* 1989a). However, heterodimers of MRFs/E proteins preferentially bind to the sequence CA(G/C)(G/C)TG (Blackwell and Weintraub, 1990). The MRFs efficiently dimerize with E proteins and these heterodimers activate muscle-specific transcription by binding E-boxes present in the regulatory regions of most muscle-specific genes (Hauschka, 1996; Molkentin and Olson, 1996).

Whereas the bHLH region is highly conserved, the conservation of regions outside the bHLH is relatively low among the MRFs, except for some regionally conserved

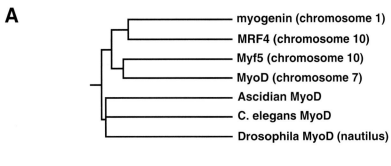

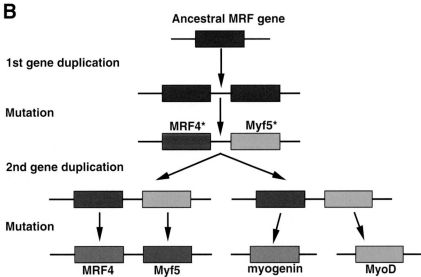

Figure 3 Two phylogenetic groups of the MRFs. (A) Vertebrate MRF genes were derived from a common ancestor by gene duplication. Invertebrates, *Caenorhabditis elegans, Drosophila,* sea urchin, and ascidians have but one MRF gene. Phylogenetic analysis suggests that the duplications occurred before or early during the radiation of the vertebrate. Amino acid sequences of Myf5 and MyoD are more similar to one another (53%) than either is to myogenin (40% and 38%, respectively) or to MRF4 (40% and 43%, respectively). Similarly, myogenin and MRF4 are more related to one another (53%) than to either Myf5 or MyoD. (B) *Myf5* and *MRF4* arose from ancestral MRF gene by gene duplication at the same locus; *myogenin* and *MyoD* arose from *MRF4* and *Myf5,* respectively, after a gene duplication event to a second chromosome, and finally, myogenin and *MyoD* loci were separated. Therefore, sequence homology, phylogeny, and chromosomal location supports the notion that the MRF family falls into two subgroups that arose through evolution by successive gene duplication events (Atchley, 1994).

islands. Experiments using chimeric proteins suggest that the bHLH regions are functionally equivalent between the MRFs (Asakura *et al.,* 1993). Several conserved regions are found outside the bHLH domain including a cystein and histidine-rich (C/H) region just N terminal of the basic domain, as well as box1 and box2 located on the C-terminal side of the HLH domain (Tapscott *et al.,* 1988; Weintraub *et al.,* 1991b). Of these regions only box2 exhibits weak conservation with *C. elegans* and *Drosophila* MyoD, and *Drosophila acheate-scute* complex, neurogenic bHLH genes (Fujisawa-Sehara *et al.,* 1990).

B. Expression of MRF Genes during Development

Analysis of MRF expression in mouse embryos by *in situ* hybridization (Lyons and Buckingham, 1992) indicates that *Myf5* is the first MRF expressed in the dorsomedial region of most rostral somites on embryonic 8.0 dpc (Ott *et al.,* 1991; also see Plate 10–o in Kaufman, 1992). Expression of *Myf5* is maintained in the dorsomedial lip of the dermomyotomes and in the newly formed epaxial myotome (Figs. 4A, 4B, and 4I, and see Plate 15b–b, c, d in Kaufman, 1992). The initial localization of *Myf5* mRNA is significant since the dorsomedial lip is a center for supplying muscle progenitor cells to form the differentiated epaxial myotome (Fig. 4I). Subsequently, expression spreads in a rostral to caudal direction until expression is detected in the whole myotome (Fig. 4J). Thereafter, *Myf5* expression is downregulated in mature myotomes (Fig. 4C) although expression is maintained throughout fetal muscle development.

Myogenin mRNA first appears in the most rostral myotome at 8.5 dpc (Fig. 4E and see Plate 15b–b, and 16b–h, i in Kaufman, 1992) and thereafter spreads caudally (Fig. 4F, Sassoon *et al.,* 1989). *Myogenin* expression is maintained in adult muscle where its expression is localized to motor endplates. *MRF4* mRNA appears transiently in myotome in a rostral to caudal manner from 9.0 dpc to 11.5 dpc (Fig. 4G) and is reexpressed in muscles at 16.0 dpc to become the most abundant MRF expressed after birth (Bober *et al.,* 1991; Hinterberger *et al.,* 1991; Hannon *et al.,* 1992).

MyoD mRNA appears in the dorsomedial edge of myotome at the cervical to thoracic level and the ventrolateral edge of myotome at the interlimb bud level at 9.75 dpc (Figs. 4H and 4I; Sassoon *et al.,* 1989; Faerman *et al.,* 1995; Tajbakhsh *et al.,* 1997; also see Plate 20c–e, f, g, h in Kaufman, 1992). Whereas the onset of *MyoD* expression in the dorsomedial myotomes follows that of *Myf5*, the onset of *MyoD* expression in the ventrolateral myotome is prior to or at the same time as that of *Myf5* (Figs. 4J and 4K). The initial localization of *MyoD* mRNA in the ventrolateral myotome supports the notion that *MyoD* is important in hypaxial myotome formation (Faerman *et al.,* 1995; Cossu *et al.,* 1996; Kablar *et al.,* 1997; Asakura and Tapscott, 1998). By 11.0 dpc, *MyoD* expression spread throughout the myotome and is maintained thereafter throughout development.

Limb and abdominal muscles are derived from muscle progenitor cells migrating from the VLL/somite bud of the dermomyotome (Figs. 2E, 2F, 6; Franz *et al.,* 1993; Bober *et al.,* 1994; Williams and Ordahl, 1994). Cells in the VLL and migrating muscle progenitor cells derived from the VLL do not express any MRF until after they arrive at sites of myogenic differentiation. In the limbs, muscle progenitor cells begin to express *MyoD* and *Myf5* at about the same time (10.5 dpc in forelimb and 11.0 dpc in hindlimb in mouse) (Lyons and Buckingham, 1992). *Myogenin* expression is detected at 11.0 dpc in the forelimb and at 11.5 dpc in the hindlimb.

The complex pattern of MRF expression evident during embryogenesis likely reflects overlapping phases of myogenic determination and differentiation. Consideration of MRF expression in cultured cells supports a division of the MRFs into two groups. Proliferating myoblasts express MyoD and/or Myf5 but do not express myogenin or MRF4 or other markers of differentiation. By contrast, differentiated myocytes express myogenin and MRF4 and downregulate expression of Myf5 and MyoD (Weintraub, 1993). Hence, on this basis the MRF family appears to consist of two groups that likely reflect distinct functional specialization in either proliferating myoblasts (Myf5 and MyoD) or in terminally differentiated myocytes (myogenin and MRF4) (Table I and Fig. 6).

C. Two Functional Groups of MRFs

Gene targeting has allowed a genetic dissection of the roles played by the MRF genes in the determination and differentiation of skeletal muscle during embryonic development. The introduction of null mutations in *Myf5, MyoD, myogenin,* and *MRF4* into the germline of mice has revealed the hierarchical relationships existing among the MRFs, and established that functional redundancy is a feature of the MRF regulatory network (Weintraub, 1993; Megeney and Rudnicki, 1995; Olson *et al.,* 1996).

Newborn mice lacking a functional *MyoD* gene display no overt abnormalities in muscle but express about four-fold higher levels of *Myf5* (Rudnicki *et al.,* 1992). Newborn *Myf5*-deficient animals also exhibit apparently normal muscle, but die after birth due to malformation of the ribs (Braun *et al.,* 1992; Grass *et al.,* 1996). Muscle development in the trunk of embryos lacking *Myf5* is delayed until the onset of *MyoD* expression, which occurs with somewhat delayed kinetics (Braun *et al.,* 1992; Tajbakhsh *et al.,* 1997; Kablar *et al.,* 1997). Importantly, newborn mice deficient in both *Myf5* and *MyoD* are totally devoid of myoblasts and myofibers. Thus, *Myf5* and *MyoD* appear to be required for the determination of muscle progenitor cells and act upstream of *myogenin* and *MRF4* (Rudnicki *et al.,* 1993). The

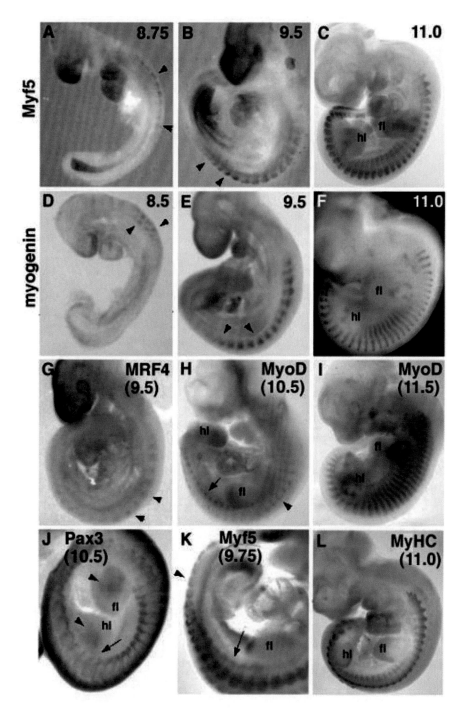

Figure 4 Temporal and spatial expression pattern of mRNAs for *MRFs, Pax3,* and *MyHC. Myf5* mRNA is the first MRF expressed in the dorsomedial region of rostral somites and, subsequently, in the dorsomedial region of dermomyotome and myotome (arrowheads in A, 8.75 dpc; B, 9.5 dpc). At 9.75 dpc, *Myf5* expression appears in the ventral region beneath the dermomyotome (arrow in K). Thereafter (11.0 dpc), *Myf5* expression is downregulated in the central region of mature myotomes (C) where expression of *MyHC* initiates (L). *Myogenin* mRNA first appears in the most rostral myotome at 8.5 dpc (arrowheads in D) and thereafter spreads caudally (E, at 9.5 dpc). Note that both sides of myotomes are displayed in D (arrowheads). By 11.0 dpc, the whole myotomes express *myogenin* (F). *MRF4* mRNA appears transiently in myotome (arrowheads in G, at 9.5 dpc). Last, *MyoD* mRNA appears in the ventro-lateral edge of myotome at the interlimb bud level. By 10.5 dpc (H) *MyoD* is expressed in the dorsomedial edge of myotome at the cervical to thoracic level (arrowhead in H) and the ventrolateral edge of myotome at the interlimb bud level (arrow in H). By 11.5 dpc, the whole myotomes express *MyoD* (I). *Pax3* is detected in the dermomyotomes in trunk level as well as in the dorsal neural tube at 10.5 dpc. At limb bud level, ventral regions of the dermomyotome (VLL) delaminate (arrow in J), and the migratory muscle progenitor cells migrate out from the VLL into the limb buds (arrowheads in J). fl, forelimb bud; hl, hindlimb bud.

Table I Targeted Germline Mutations in Mice Affecting Myogenesis

Targeted mutation	Myogenic phenotype
Myogenic Regulatory Factors	
$MyoD^{-/-}$ (1–5)	Delayed migratory and nonmigratory hypaxial muscle formation; impaired adult muscle regeneration and adult myoblast differentiation; upregulation of Myf5
$Myf5^{-/-}$ (3, 4, 6–8)	Delayed MyoD expression in myotome, epaxial muscle, some nonmigratory hypaxial muscle formation and esophagus transdifferentiation
$myogenin^{-/-}$ (9–11)	Absence of secondary muscle fiber formation; increased myoblast number
$MRF4^{-/-}$ (12)	Upregulation myogenin (mildest allele)
$MyoD^{-/-}:Myf5^{-/-}$ (13,14)	Complete absence of myogenesis
$MyoD^{-/-}:myogenin^{-/-}$ (15)	Phenocopy of $myogenin^{-/-}$
$MyoD^{-/-}:MRF4^{-/-}$ (16)	Phenocopy of $myogenin^{-/-}$
$Myf5^{-/-}:myogenin^{-/-}$ (15)	Phenocopy of $myogenin^{-/-}$
$myogenin^{-/-}:MRF4^{-/-}$ (16)	Phenocopy of $myogenin^{-/-}$
$Myf5^{-/-}:MRF^{-/-}$ (16)	Phenocopy of $Myf5^{-/-}$
$MyoD^{-/-}:myogenin^{-/-}:MRF4^{-/-}$ (17)	Presence of myoblasts; absence of muscle fibers
$Myf5^{-/-}:Splotch (Pax3^{-/-})$ (8)	Absence of trunk and limb muscle
Other Transcription Factors	
$lbx1^{-/-}$ (18–20)	Absence of lateral migratory hypaxial muscle formation
$Mox2$ (21)	Impaired subset of limb muscle formation; decreased Myf5 expression in limb bud
$c\text{-}ski^{-/-}$ (22)	Reduced diameter of muscle fibers; reduction of muscle mass
$paraxis^{-/-}$ (23,24)	Impaired epaxial and nonmigratory hypaxial muscle formation
$paraxis^{-/-}:Myf5^{-/-}$ (24)	Absence of epaxial muscle formation
$Pax7^{-/-}$ (25)	Absence of myogenic satellite cells
Signaling Molecules	
$Shh^{-/-}$ (26,27)	Absence of formation and Myf5 expression in epaxial myotome
$Wnt1^{-/-}:Wnt3a^{-/-}$ (28)	Impaired formation and reduced Myf5 expression in epaxial dermomyotome
$c\text{-}met^{-/-}$ (29,30)	Absence of migratory hypaxial muscle
$SF/HGF^{-/-}$ (30)	Absence of migratory hypaxial muscle
$GDF8^{-/-}$ (31)	Adult muscle hypertrophy
$Noggin^{-/-}$ (32)	Reduced epaxial myotome formation

References cited: 1. Rudnicki *et al.,* 1992, 2. Megeney *et al.,* 1996, 3. Kablar *et al.,* 1997, 4. Kablar *et al.,* 1998, 5. Sabourin *et al.,* 1999, 6. Braun *et al.,* 1992, 7. Tajbakhsh *et al.,* 1997, 8. Kablar *et al.,* 2000, 9. Hasty *et al.,* 1993, 10. Nabeshima *et al.,* 1993, 11. Venuti, 1995, 12. Zhang *et al.,* 1995, 13. Rudnicki *et al.,* 1993, 14. Kablar *et al.,* 1999, 15. Rawls *et al.,* 1995, 16. Rawls *et al.,* 1998, 17. Valdez *et al.,* 2000, 18. Schafer and Braun, 1999, 19. Gross *et al.,* 2000, 20. Brohmann *et al.,* 2000, 21. Mankoo *et al.,* 1999, 22. Berk *et al.,* 1997, 23. Burgess *et al.,* 1996, 24. Wilson-Rawls *et al.,* 1999, 25. Seale *et al.,* 2000, 26. Chiang *et al.,* 1996, 27. Borycki *et al.,* 1999, 28. Ikeya and Takada, 1998, 29. Bladt *et al.,* 1995, 30. Dietrich *et al.,* 1999a, 31. McPherron *et al.,* 1997, 32. McMahon *et al.,* 1998

assertion that Myf5 and MyoD are determination factors is supported by the observation that putative muscle progenitor cells remain multipotential and contribute to nonmuscle tissues in the trunk and limbs of $Myf5^{-/-}:MyoD^{-/-}$ embryos (Kablar *et al.,* 1998).

Mice lacking *myogenin* are immobile and die perinatally due to deficits in myoblast differentiation as evidenced by an almost complete absence of myofibers (Hasty *et al.,* 1993; Nabeshima *et al.,* 1993). However, normal numbers of MyoD-expressing myoblasts are present and these are organized in groups similar to wild-type muscle. *Myogenin*-deficient embryos form primary myofibers normally, but appear unable to form secondary myofibers (Venuti *et al.,* 1995). Moreover, the observation that mice lacking both *myogenin* and *Myf5,* or both *myogenin* and *MyoD,* are identical to mice lacking only *myogenin* has confirmed that *Myf5* and *MyoD* act upstream of *myogenin* (Rawls *et al.,* 1995). Therefore, *myogenin* plays an essential *in vivo* role in the terminal differentiation of myoblasts (Fig. 5).

Mice carrying different targeted *MRF4* mutations display a range of phenotypes consistent with a late role for *MRF4* in the myogenic pathway. Zhang and coworkers observed that mice lacking *MRF4* are viable with seemingly normal muscle and display a fourfold increase in *myogenin* expression (Rawls *et al.,* 1995; Zhang *et al.,* 1995). By contrast, Braun and Arnold (1995) reported a targeted *MRF4* mutation, in which *Myf5* expression (located about 6 kb away) is completely ablated *in cis* (Olson *et al.,* 1996). A third targeted *MRF4* mutation displays an intermediate phenotype due to partial ablation of *Myf5* expression *in cis* (Patapoutian *et al.,* 1995; Yoon *et al.,* 1997; Olson *et al.,* 1996). Interestingly, mice lacking both *MyoD* and *MRF4* display a phenotype similar to the *myogenin*-null phenotype (Rawls *et al.,* 1998). Therefore, *MRF4* function may be substituted by the presence of *myogenin,* but only in the presence of *MyoD* (Fig. 5). Taken together, these data support the hypothesis that the MRFs function as couplets, *Myf5* and *MRF4* to regulate epaxial muscle development,

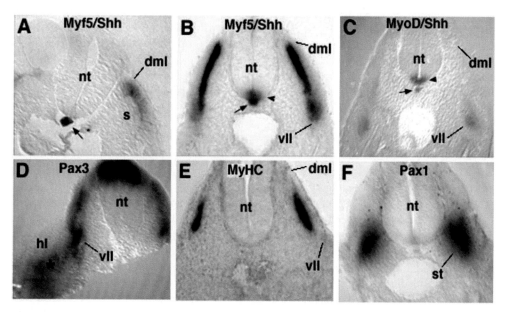

Figure 5 Spatial expression pattern of mRNAs for *Myf5, MyoD, MyHC, Pax3,* and *Pax.* At 10.0 dpc *Myf5* expression appears in the dorsomedial region (dml) of caudal somite (s) at the tail bud region (A). At trunk level, the expression spreads in the whole myotomes (B). *MyoD* mRNA appears in the ventrolateral edge (vll) of myotome but not in the dml at the interlimb bud level (C). *Pax3* mRNA is detected in the dermomyotome and in the dorsal neural tube at forelimb bud level (D). Ventral regions of the dermomyotome (vll) delaminate and the migratory muscle progenitor cells migrate out from the vll into the forelimb bud (asterisk). *MyHC* expression is detected in the mature myotome at the central region but not in the dml or vll region. *Pax1* is expressed in the sclerotome (st) (F). *Shh* (Sonic hedgehog) expression is detected in the midline structures, notochord (arrow), and floor plate (arrow head). nt, neural tube.

and *MyoD* and *myogenin* to regulate hypaxial muscle development.

Triple mutant mice lacking *MyoD, myogenin,* and *MRF4* fail to form differentiated muscle fibers but do contain a normal number of myoblasts (Valdez *et al.,* 2000). This study suggests that *Myf5* alone is insufficient to induce terminal differentiation of muscle, perhaps due to the insufficient levels of MRFs in myoblasts. Alternatively, *Myf5* may possess a limited function inducing terminal differentiation of muscle.

In conclusion, observations from gene-targeting experiments have divided the MRF family into two groups. The primary MRFs, *MyoD* and *Myf5,* are required for myogenic determination. The secondary MRFs, *myogenin* and *MRF4,* act later in the program as differentiation factors (Fig. 6).

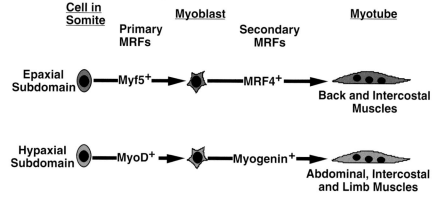

Figure 6 The MyoD family defines discrete myogenic lineages in the myotome and migratory muscle development. Dermomyotomal cells differentiate into myoblasts by the expression of primary MRFs, *Myf5* or *MyoD,* in dorsal subdomain (epaxial myotome) or ventral subdomain (hypaxial myotome), respectively. Thereafter, both *MyoD* and *Myf5* are coexpressed in the same muscle cells. Proliferating myoblasts withdraw from the cell cycle and initiate to express secondary MRFs, *myogenin* and *MRF4* to form multinucleated myotubes. The two subdomains of the myotome give rise to back musculature and intercostal and body musculature, respectively. The migratory hypaxial muscles also give rise to diaphragm and limb musculature (Kablar *et al.,* 1997, 1999; Ordahl and Williams, 1998).

D. *Myf5* and *MyoD* Regulate the Development of Distinct Myogenic Lineages

The temporal-spatial patterns of myogenesis in *Myf5*- and *MyoD*-deficient embryos provides compelling evidence for unique roles for *Myf5* and *MyoD* in the development of epaxial and hypaxial musculature (Kablar *et al.*, 1997). Embryos lacking *MyoD* display normal development of paraspinal and intercostal muscles in the body proper, whereas muscle development in limb buds and branchial arches is delayed by about 2.5 days. By contrast, embryos lacking *Myf5* display normal muscle development in limb buds and branchial arches and a marked delay in development of paraspinal and intercostal muscles. Transdifferentiation of esophageal smooth muscle into skeletal muscle is also dependent on the presence of *Myf5* (Kablar *et al.*, 2000). Although *MyoD* mutant embryos exhibit a delay in the development of limb musculature, the migration of *Pax3* (a paired type homeobox gene) expressing cells into the limb buds and subsequent induction of *Myf5* in muscle progenitor cells occur normally. Therefore, the phenotypes of *Myf5* and *MyoD* mutant mice strongly support the hypothesis that *Myf5* and *MyoD* have unique roles in the regulation of the developmental programs of the myogenic lineages giving rise to epaxial versus hypaxial musculature. Moreover, the observation that mice lacking both *MyoD* and *MRF4* display a phenotype similar to the *myogenin*-null phenotype (Rawls *et al.*, 1998) lends genetic support to the hypothesis that *MyoD* and *myogenin* are together required for the appropriate development of hypaxial musculature (Fig. 6).

IV. Muscle-Specific Transcriptional Regulation

A. Positive Regulators for Muscle-Specific Transcriptions

Analysis of several genes such as *MCK*, *MLC1/3* (*Mylf*), α-actin (*Acta1*), and *myogenin* has revealed the existence of additional *cis*-acting elements involved in muscle-specific transcriptional regulation (Hauschka, 1996; Molkentin and Olson, 1996). The identification of DNA motifs such as the MEF2 (CTA(A/T)4TAG), CArG box (CC(AT)6GG), AT-rich (T(A/T)ATAAT(A/T)A), MCAT (CATTA), and MEF3 (TCAGGT) sites as required elements in muscle-specific regulatory sequences led to the identification of diverse transcription factors that play important roles in regulating transcription in muscle.

Mesodermal cell-specific homeobox protein (mHOX, *Prrx1*) binds the AT-rich site located in the MCK enhancer region and activates MCK gene transcription (Cserjesi *et al.*, 1992). Serum response factor (*Srf*) and factors belonging to the MADS family (MCM1, Agamous, Deficiens, Serum re-

sponse factor) bind the CArG box as well as SRE (serum responsive element) (Mohun *et al.*, 1991). TEF-1 (*Tead1*), a tissue nonspecific transcription factor, binds to the MCAT site (Farrance *et al.*, 1992; Stewart *et al.*, 1994). Six1 (*Six1*), a vertebrate homolog of *Drosophila sine oculis* (*so*) homeobox factor, binds to the MEF3 site and upregulates myogenin and fast muscle specific aldolase (*Aldo1*) promoters (Hidaka *et al.*, 1993; Spitz *et al.*, 1997, 1998).

MEF2 sites are bound by members of the MEF2 family of transcription factors (Yu *et al.*, 1992). In vertebrates, the MEF2 family of genes is composed of MEF2a (*Mef2a*), MEF2b (*Mef2b*), MEF2c (*Mef2c*), and MEF2d (*Mef2d*). MEF2 factors contain a MADS domain, which is essential for dimer formation and DNA binding (Olson *et al.*, 1995; Molkentin and Olson, 1996). MEF2 genes are expressed in heart muscle, internal smooth muscle, and the neural tube as well as in skeletal muscle during early embryogenesis (Edmondson *et al.*, 1994; Subramanian and Nadal-Ginard, 1996). *In vitro* experiments suggest that MEF2 factors function in differentiation rather during skeletal myogenesis as *MEF2* genes are upregulated by expression of MRFs.

V. Inductive Mechanisms of Myogenesis

The molecular mechanisms leading to the *de novo* induction of *Myf5* and *MyoD* transcription in uncommitted progenitor cells during development have remained a central problem in myogenesis. However, several upstream signaling pathways have been implicated as positively or negatively affecting the *de novo* induction of *Myf5* and *MyoD* transcription. For example, Sonic hedgehog (*Shh*) expressed in the floor plate and the notochord and Wnt family members expressed in the dorsal neural tube combinatorially activate myogenesis in the somite (see Fig. 9 in a later section; Johnson *et al.*, 1994; Munsterberg *et al.*, 1995; Stern *et al.*, 1995). Shh also activates *Noggin* expression in the dorsal somite, and Noggin (*Nog*) inhibits the negative activity of lateral-plate-derived BMP4 (bone morphogenetic protein) on myogenesis (Hirsinger *et al.*, 1997; Marcelle *et al.*, 1997; Reshef *et al.*, 1998). Notch signaling pathways also appear to exert negative effects on myogenesis (Kopan *et al.*, 1994; Shawber *et al.*, 1996; Delfini *et al.*, 2000). Although these pathways have been demonstrated to exert effects on the timing and patterning of myogenesis, the molecular mechanisms that activate the *de novo* transcription of *Myf5* and *MyoD* are completely unknown.

A. Neural Tube, Notochord, and Dorsal Ectoderm Induce Myotome Formation

Experiments in which somites were rotated suggest that cells in the newly formed somite are developmentally equivalent and maintain multipotentiality independent of position (Aoyama and Asamoto, 1988; Ordahl and Le Douarin, 1992). However, dermomyotome, myotome, and sclerotome formation obviously proceed in a position-dependent manner presumably mediated by signals from the surrounding tissues. Indeed, *in vitro* experiments reveal that the dorsal neural tube secretes factor(s) capable of inducing myogenic commitment and maintaining myotome differentiation in cocultured somites. The proximity of the dorsal neural tube to the DML of the dermomyotome suggests that signals from the dorsal neural tube induce the formation of the DML and the formation of the myotome from the DML (Bober *et al.,* 1994; Buffinger and Stockdale, 1994; Stern and Hauschka, 1995; Munsterberg and Lassar, 1995; Cossu *et al.,* 1996; Xue and Xue, 1996; Dietrich *et al.,* 1997; Marcelle *et al.,* 1997; Hirsinger *et al.,* 1997).

In organ culture experiments with cocultured neural tube and somites, young neural tube mainly induces *Myf5* expression, whereas mature neural tube mainly induces MyoD expression (Tajbakhsh *et al.,* 1998). The notochord alone induces *Pax1/Pax9,* paired type homeobox genes, and sclerotome (Brand-Saberi *et al.,* 1993, 1996a; Pourquié *et al.,* 1993; Balling *et al.,* 1996). However, inductive (Buffinger and Stockdale, 1994; Bober *et al.,* 1994; Munsterberg and Lassar, 1995; Stern and Hauschka, 1995; Pownall *et al.,* 1996; Xue and Xue, 1996) and suppressive effects (Brand-Saberi *et al.,* 1993; Pourquié *et al.,* 1993; Goulding *et al.,* 1994; Xue and Xue, 1996) of notochord on myogenesis have been observed, and these differences may be generated as a function of the distance between somite and notochord, or they may reflect the developmental stage of notochord. In addition, signals from notochord cooperating with signals from neural tube induce skeletal muscles in somites (Munsterberg and Lassar, 1995; Stern and Hauschka, 1995). In mouse, this cooperation mainly induces *Myf5* gene expression in the somites (Cossu *et al.,* 1996). These results are consistent with the notion that *Myf5* expression is initiated in the DML of the somite and dermomyotome, which are topologically closed to the dorsal part of neural tube. Furthermore, signals from dorsal ectodermal layer above the somites induce dermomyotome and myotome mediated through *MyoD* induction in the mouse somite (Fig. 9; Fan and Tessier-Lavigne, 1994; Cossu *et al.,* 1996, Tajbakhsh *et al.,* 1998).

In mouse, dorsal ectoderm for the most part induces *MyoD* gene expression in the somite (Cossu *et al.,* 1996). These results are consistent with the notion that *MyoD* expression at the interlimb level is initiated in the ventrolateral part of myotome (T. H. Smith *et al.,* 1994; Faerman *et al.,* 1995; Cossu *et al.,* 1996). Lateral plate promotes lateralization of dermomyotome, opposing the dorsal neural tube effect, which promotes medialization of dermomyotome (Pourquié *et al.,* 1995, 1996). Subsequently, muscle differentiation in the dermomyotome is suppressed by lateral plate. This inhibitory effect for muscle differentiation by lat-

eral plate is important since the VLLs of the dermomyotome generate undifferentiated migratory muscle progenitor cells.

B. Cell–Cell Interactions and the Community Effect Regulate Muscle Differentiation

In vitro culture experiments using *Xenopus* and chicken embryos revealed that dissociated blastomeres (*Xenopus*) and epiblasts (chicken) induce *MyoD* expression and subsequently differentiate into skeletal myocytes (Holtzer *et al.,* 1990; Godsave and Slack, 1991; George-Weinstein *et al.,* 1996). Intact blastomeres and epiblasts do not exhibit any *MyoD* expression or muscle differentiation. Similarly, a subset of cells in mouse embryonic brain expressing *Myf5* mRNA but not protein does not differentiate into muscle unless the cells are dissociated (Tajbakhsh *et al.,* 1994; Daubas *et al.,* 2000).

Dissociated presomitic mesodermal cells cocultured with both notochord and neural tube undergo myogenic differentiation. However, a critical mass of mesodermal cells is required in what has been termed the community effect. For example, at least 30–40 dissociated cells derived from mouse presomitic mesoderm (Cossu *et al.,* 1995) and more than 100 dissociated cells from *Xenopus* gastrula are required for the muscle conversion (Gurdon, 1988; Kato and Gurdon, 1993; Gurdon *et al.,* 1993a,b).

Taken together, these observations suggest several common features of the mechanisms that regulate skeletal myogenesis. First, dissociated undetermined cells *in vitro* possess a capacity to readily express *MyoD* and to undergo myogenic commitment. Second, myogenic differentiation requires interactions between a certain critical mass of cells. Therefore, transcriptional regulation of MRF genes is likely regulated by a variety of environmental cues including cell–cell interactions.

The myogenic differentiation of Notch-expressing cells is suppressed following interaction with Delta (*Dll1*)/Jagged (*Jag1*)-expressing cells (lateral inhibition), mediated by direct interaction between Notch and Delta/Jagged that belong to transmembrane proteins (Kopan and Cagan, 1997; Artavanis-Tsakonas *et al.,* 1999). The cytoplasmic domain of Notch is released by the interaction with Delta/Jagged and is translocated into the nucleus. Indeed, expression of the cytoplasmic domain of Notch 1 is sufficient to suppress differentiation of myoblasts *in vitro* (Kopan *et al.,* 1994). Ectopic Delta 1 expression inhibits myogenic differentiation in developing limb muscle (Delfini *et al.,* 2000).

Muscle differentiation is upregulated by cell–cell interactions as mediated by adhesion molecules such as the cadherins in which cell adhesion is achieved by homophilic binding. For example, myoblasts derived from chicken epiblasts express N-cadherin (*Cdh2*), and blocking N-cadherin function with antibodies suppresses muscle differentiation

(George-Weinstein *et al.,* 1997). In addition, expression of a dominant-negative cadherin in *Xenopus* embryos suppresses skeletal muscle differentiation (Holt *et al.,* 1994). Moreover, incubation of antagonistic M-cadherin (*Cdh15*) peptides or antisense RNA inhibits both myoblast fusion and cell cycle withdrawal in conditions that normally promote differentiation (Zeschnigk *et al.,* 1995).

C. Positive (Shh and Wnt Proteins) and Negative (TGF-β family) Factors Regulate Myotome Formation

Recent experiments have revealed the molecular aspects for the positive and negative signals that regulate myotome formation (Currie and Ingham, 1998).

Wnt proteins possess the ability to induce muscle differentiation in somites (Munsterberg *et al.,* 1995; Stern *et al.,* 1995). In addition, notochord provides Shh protein, which can induce both myotomal cells and sclerotomal cells in the somite (Johnson *et al.,* 1994; Fan and Tessier-Lavigne, 1994; Munsterberg *et al.,* 1995). Shh secreted from notochord also induces floor plate in the most ventral region of neural tube and the floor plate also provides Shh (Ingham, 1994). Shh has been suggested to act in diverse roles in somitogenesis, for example as an inductive factor, a proliferation factor, and a cell survival factor (Borycki *et al.,* 1999; Duprez *et al.,* 1998; Teillet *et al.,* 1998). The different roles played by Shh may be due to combinations of different signals and the differences in stage of cell differentiation.

In mouse, cells of the dorsal ectodermal secrete Wnt7a protein, whereas the dorsal neural tube secretes Wnt1. Cultured presegmental mesoderm upregulates *MyoD* expression when exposed to Wnt7a and upregulates *Myf5* when exposed to Wnt1 (Tajbakhsh *et al.,* 1998). These observations are consistent with the relationship between the topologically distinct regions of *Wnt* expression and the induction of *Myf5* in the DML and *MyoD* in VLL (see Fig. 9 in a later section). Moreover, Wnt4, Wnt5a, and Wnt6 all induce expression of both *MyoD* and *Myf5* in the somite. Interestingly, expression of *Myf5* mRNA in the brain is observed close to domains where *Wnt1* and *Shh* are expressed (Daubas *et al.,* 2000).

BMP4, belonging to the transforming growth factor β (TGF-β) family, was identified as a negative regulator of myotome and sclerotome formation. *BMP4* is expressed in the dorsal neural tube, dorsal ectoderm, and lateral plate mesoderm (Pourquié *et al.,* 1996; Marcelle *et al.,* 1997; Maroto *et al.,* 1997; Hirsinger *et al.,* 1997; Reshef *et al.,* 1998; Currie and Ingham, 1998). Therefore, BMP4 appears to maintain and expand the number of precursor cells in the dermomyotome (see Fig. 9 in a later section). However, the DML of the dermomyotome expresses *Wnt11* and BMP4 antagonists *Follistatin* (*Fst*) and *Noggin* (*Nog*). Therefore, BMP4 protein in the DML is inactivated, allowing induction of myotome formation in the dorsomedial region (Amthor *et al.,* 1996; Hirsinger *et al.,* 1997; Marcelle *et al.,* 1997; Reshef *et al.,* 1998). The notochord expresses *Noggin*, which also blocks BMP4 function in the DML (Hirsinger *et al.,* 1997; McMahon *et al.,* 1998).

Myostatin (*Gdf8*), belonging to the TGF-β family, is expressed in the myotome as well as adult skeletal muscle where it acts to negatively control muscle mass. *Myostatin* is mutated in double-muscled cattle (Belgian Blue) in which muscle mass is markedly increased (Grobet *et al.,* 1997; McPherron and Lee, 1997). Similarly, mice lacking the *myostatin* gene display similar muscle hypertrophy (McPherron *et al.,* 1997; Lee and McPherron, 1999).

D. Genetic Analysis of Myogenic Induction

Several gene knock-out experiments have revealed the important roles played by various signaling molecules in myotome formation (Tables I and II). Mice lacking *Wnt1* or *Wnt3a* do not display any defect during somitogenesis. By contrast, mice lacking both *Wnt1* and *Wnt3a* genes display reduced myotome formation (Ikeya and Takada, 1998). In addition, because *Wnt1* and *Wnt3a* are expressed in the dor-

---

Table II   Spontaneous Mutations in Mice and Their Effect on Myogenesis

Spontaneous mutation	Gene	Homozygous myogenic phenotype
Splotch (*Sp*) (1–4)	*Pax3*	Partially impaired myotome formation, epaxial and nonmigratory hypaxial muscle formation; absence of migratory hypaxial formation
Open brain (*opb*) (5)	Unknown (Ch 1)	Absence of Myf5 expression in epaxial myotome; impaired its formation
Danforth's short tail (*Sd*) (6,7)	Unknown (Ch 2)	Absence of Myf5 expression in epaxial myotome; increased apoptosis in epaxial myotome; absence of epaxial muscle
Brachyury curtailed (*T^c*)(7)	*Brachyury T*	Absence of epaxial myotome (heterozygotes)
Pintail (*Pt*) (7)	Unknown (Ch 4)	Absence of epaxial myotome
Truncate (*tc*) (7)	Unknown (Ch 6)	Absence of epaxial myotome

References cited: 1. Franz *et al.,* 1993, 2. Bober *et al.,* 1994, 3. Goulding *et al.,* 1994, 4. Williams and Ordahl, 1994, 5. Sporle *et al.,* 1996, 6. Asakura and Tapscott, 1998, 7. Dietrich *et al.,* 1999b

sal neural tube close to the dorsomedial dermomyotome, loss of both genes induced lateralization of the entire dermomyotome. In the mutant dermomyotome, expression of the lateral dermomyotome marker *Sim1*, a transcription factor homolog of *Drosophila single minded* (*sim*), is medially expanded to the dorsomedial edge, and expression of the medial dermomyotome markers such as *En1*, a mouse homolog of *Drosophila engrailed* (*en*), and dorsomedial markers such as *noggin* and *Wnt11* are not detected.

Mice lacking *Shh* gene display substantial defects in somitogenesis. Initially, a low level of the sclerotomal marker *Pax1* is detected that completely disappears at later times. Therefore, Shh may act as a survival factor or a proliferation factor rather than as an inducer of sclerotome formation (Chiang *et al.*, 1996). During myotome formation, *Myf5* expression is not initiated in the DML of the dermomyotome, but is detected in the hypaxial myotome. *MyoD* expression in the hypaxial myotome is not affected (Chiang *et al.*, 1996; Borycki *et al.*, 1999). These results are consistent with the hypothesis that Shh acts as an inducer of dorsomedial myotome formation, mediated by *Myf5* gene activation by cooperating with Wnt proteins.

Phenotypic characterization of various spontaneous mouse mutants reveals useful insights into the important roles played by the surrounding tissues in myotome formation. In the neural tube mutant mouse, *open brain* (*opb*), *Myf5* gene expression is severely reduced, whereas *MyoD* expression remains normal (Sporle *et al.*, 1996). This result indicates that the dorsal neural tube provides signals (likely Wnts) necessary for the induction of *Myf5* expression in the DML of the dermomyotome, whereas *MyoD* induction in the hypaxial myotome occurs independently of the neural tube.

Danforth's short-tail (*Sd*) mouse is a semidominant mutation that prevents completion of notochord development. In homozygous mutant mice, the notochord completely degenerates by 9.5 dpc, whereas the neural tube and somites continue to form, permitting analysis of somite development in the absence of inductive signals from the midline structure, the notochord, and floor plate. In the somites formed after notochord degeneration, initial *Myf5* expression is significantly reduced (Fig. 7; Asakura and Tapscott, 1998). *Pax1* expression in the sclerotome is not detected. Muscle gene expression including *MyoD* and *Myf5* is normally detected in the hypaxial myotome and there is normal development of both migratory and nonmigratory hypaxial muscles. By contrast, muscle gene expression including *Myf5* is not detected in the epaxial myotome and a high level of apoptosis is detected with significantly decreased formation of epaxial muscles. However, the initial *Pax3* expression in the dermomyotome is relatively normal. Later, the hypaxial myotome is ventrally shifted into the place where the sclerotome normally forms (Asakura and Tapscott, 1998; Dietrich *et al.*, 1999b). These results are similar to the somitic phenotype of *Shh⁻/⁻* mice (Chiang *et al.*, 1996; Borycki *et al.*, 1999) as

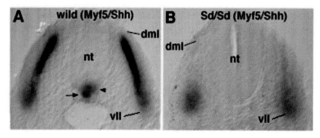

Figure 7 Notochord requires the epaxial myotome formation. Wild-type embryo (A) expresses *Shh* in the notochord (arrow) and floor plate (arrowhead). In *Sd/Sd* embryo (B), *Shh* is not detected in the notochord or floor plate at the caudal level. In wild-type embryo, *Myf5* expression is observed in the entire myotome (A). In contrast, in *Sd/Sd* embryo, the expression is normal in the ventral (hypaxial) myotome but reduced in the dorsal (epaxial) myotome (B) at the caudal level, suggesting an important role of the notochord in the epaxial myotome formation (Asakura and Tapscott, 1998).

well as other distinct mutations affecting notochord development, for example, *Truncate* (*tc*), *Pintail* (*Pt*), and *Brachyury curtailed* (*Tc*) (Dietrich *et al.*, 1999b). Taken together, these results suggest that Shh is the main factor secreted from the notochord with a role in somitogenesis.

The notochord also provides Noggin, which promotes myotome formation, mediated by antagonizing BMP4 in the DML. *Noggin⁻/⁻* mice display a severe reduction in the epaxial myotome (McMahon *et al.*, 1998).

Both *Patched* (*Ptch*) gene, a Shh receptor, and *Gli1* (*Gli*), a vertebrate homolog of *Drosophila Cubitus interruptus* (*ci*), which is a zinc finger type transcription factor, are known to be upregulated by Shh induction. Indeed, these genes are detected in the dermomyotome, sclerotome, and myotome, consistent with a direct role for Shh in sclerotome and myotome induction (Marigo *et al.*, 1996a; Borycki *et al.*, 1997, 2000; Marcelle *et al.*, 1997).

In zebrafish, Shh secreted from the ventral axial structures induces the formation of the slow adaxial muscles. In the notochord mutants, *bozozok* (*boz*) and *no tail* (*ntl*), the adaxial muscle pioneer cells, are absent and slow muscle does not form. However, the formation of fast muscle is unaffected (Devoto *et al.*, 1996; Blagden *et al.*, 1997; Norris *et al.*, 2000; Currie and Ingham, 1998). This observation contrasts with the mouse where, in the absence of the notochord, myotome formation of the DML of the dermomyotome is severely affected. However, slow-type MyHC is detected (Asakura and Tapscott, 1998, in discussion). Therefore, the notochord or Shh appears necessary for a subset of muscle cells, the adaxial muscle pioneer cells in zebrafish, and muscle pioneer cells in the mouse.

E. Migratory Muscle Progenitor Cells

All myotomal cells and migratory muscle progenitor cells in the trunk are derived from the dermomyotome. As

described above, signals such as Shh, Wnts, and BMP4 secreted from the surrounding tissues regulate the development
of the dermomyotome and myotome (Amthor *et al.*, 1998;
Duprez *et al.*, 1998, 1999). Importantly, several transcription
factors that potentially represent the effectors of these signaling molecules have been implicated in regulating the development of the migratory muscle progenitor cells derived
from the VLL of the dermomyotome.

Expression of *Pax3*, a paired type homeobox gene, in the
dermomyotome (Fig. 5D) is activated by BMP4 secreted
from the dorsal ectoderm (Pourquié *et al.*, 1996, Amthor
et al., 1998, 1999). Furthermore, expression of *Pax3* in proliferating muscle progenitor cells arriving in the limb bud is
maintained by *BMP2/4/7* expressed by the surrounding limb
cells. *Pax3* expression suppresses both MRF expression and
muscle differentiation. Therefore, BMP proteins positively
stimulate *Pax3* expression in muscle progenitor cells in both
the dermomyotome and the limb bud.

In the spontaneous mouse mutant *Splotch* (*Pax3*^{Sp}) carrying a defective *Pax3* gene, homozygous mutants die during
embryogenesis. Homozygous *Splotch* mice display malformation of neural tissues and neural crest-derived tissues, decreased elongation of dermomyotome and myotome, and absence of delamination at the DML of the dermomyotome
(Table II; Franz *et al.*, 1993; Goulding *et al.*, 1994; Bober
et al., 1994; Williams and Ordahl, 1994). Subsequently,
epaxial muscles of the deep back and nonmigratory hypaxial
muscles, such as intercostal and abdominal muscles, are reduced and migratory hypaxial muscles at trunk level such as
limb, pectoralis muscles, and diaphragm are completely absent (Tremblay *et al.*, 1998). Because *Pax3* is expressed in
both dermomyotome and migrating muscle progenitor cells,
Pax3 may regulate both short-range migration, such as that
occurring during myotome formation, as well as long-range
migration of muscle progenitor cells into the limbs (Tremblay *et al.*, 1998).

Pax7, a *Pax3* related gene, is also expressed in the similar
region of *Pax3* expression, suggesting roles for *Pax7* in development of migratory cells during somitogenesis. Interestingly, the ascidian has a single *Pax3/7* related gene, which
is expressed in the neural tube, like vertebrates, but not in
the somite (Wada *et al.*, 1997). Therefore, because migratory muscle lineages as well as migratory neural crest cells
emerged during early vertebrate evolution, an additional expression domain of *Pax3/7* in the vertebrate somite may
have resulted in subsequent emergence of novel migratory
muscle lineages, which form craniofacial and limb bud/
paired fin muscles (Gee, 1994; Neyt *et al.*, 2000).

Many cell migration processes have been shown to require specific receptor-ligand interactions. Indeed, the migration of muscle progenitor cells requires a functional c-met-
SF/HGF cascade (Brand-Saberi *et al.*, 1996b; Heymann
et al., 1996). Mice lacking *c-met*, a receptor tyrosine kinase (*Met*), or lacking its ligand, scatter factor/hepatocyte

growth factor (*SF/Hgf*), display similar muscle phenotypes
to *Splotch* mice; that is, an absence of migratory hypaxial
muscle formation as well as defective delamination at the
VLL of their dermomyotome. Therefore, c-met or SF/Hgf
plays an essential role in the delamination and subsequent
migration of muscle progenitor cells (Table I; Bladt *et al.*,
1995; Maina *et al.*, 1996; Dietrich *et al.*, 1999a). Because
c-met is expressed in the dermomyotome and migratory
muscle progenitor cells and *SF/Hgf* is expressed in the limb
buds, loss of Pax3 function may disturb the c-met-SF/Hgf
cascade. However, in *Splotch* mice, reduced *c-met* expression is still detected in the dermomyotome and a subset of
migratory cells is detected in limb bud, suggesting that Pax3
does not entirely regulate *c-met* expressions in the dermomyotome (Yang *et al.*, 1996; Scaal *et al.*, 1999). In addition,
in *Splotch* mice, the presence of migratory cells that might
be derived from the dermomyotome and which do not have
myogenic potential suggests the existence of a novel Pax3-
independent migratory population (Mennerich *et al.*, 1998).
Interestingly, a population migrating from somite has been
reported to differentiate into the angioblast lineage, which
forms the endothelial cells of the blood vessels (Wilting
et al., 1995). However, cell lineage analysis of the novel
Pax3$^-$:c-met$^+$ population remains to be conducted.

Recently, a new marker for the migratory muscle progenitor cells has been identified. *Lbx1* (*Lbx1h*), a vertebrate
homolog of *Drosophila* homeobox protein *ladybird* (*lb*), is
expressed in the ventrolateral region of the dermomyotome
as well as in migratory muscle progenitor cells. In *Splotch*
mice, *Lbx1* gene expression is not detected in the dermomyotome at trunk level. However, expression of *Lbx1* is
detected in the occipital somites but migration of tongue
muscle progenitors is delayed (Mennerich *et al.*, 1998).
These results suggest a possible role for Pax3 as an upstream
regulator of *Lbx1* gene in the dermomyotome at trunk level.
Mice lacking *Lbx1* gene display a complete absence of lateral limb (extensor) muscle formation (Table I; Schafer and
Braun, 1999; Gross *et al.*, 2000; Brohmann *et al.*, 2000).
However, in *Lbx1*-deficient mice formation of ventral limb
(flexor) muscle, tongue muscle, and diaphragm is normal,
suggesting that *Lbx1* is required for the migration of a subset
of muscle progenitor cells.

F. Transcription Factors Regulating Myotome Formation

As described above, signaling molecules such as Shh,
Wnts, and BMPs act to regulate the induction of *Myf5* and
MyoD, leading to subsequent myogenic determination. However, the specific transcription factors that directly activate
MyoD or *Myf5* transcription remain unidentified. Nevertheless, the transcriptional regulatory elements of MRF genes
have been analyzed using *in vivo* transgenic mice to address
this question.

For example, analysis of transgenic mice carrying *MyoD* gene regulatory regions driving bacterial β-galactosidase (*lacZ*) genes, demonstrated that two different regions, a 258-bp core enhancer and a 700-bp distal regulatory region (DRR), located at −20 and −5 kb upstream of the transcriptional initiation site, respectively, are involved in mouse *MyoD* gene activation (Fig. 8). Transgenic mice carrying a *lacZ* gene driven by 6-kb upstream *MyoD* gene containing the DRR (*MyoD6.0-lacZ*) revealed that 6 kb upstream of mouse *MyoD* gene contains elements required for *MyoD* gene expression in a subset of myotomal cells, which appear to represent differentiated myocytes. (Figs. 8A and 8B; Asakura *et al.*, 1995). By contrast, transgenic mice carrying the core enhancer and −2.5-kb promoter region of *MyoD* gene driving *lacZ* gene (*MyoDcore/2.5-lacZ*) showed that the core enhancer is sufficient for initiation of *MyoD* gene transcription in transgenic mice during early myotome formation (10–11 dpc) (Goldhamer *et al.*, 1992, 1995). In addition, *lacZ*-expressing cells in trunk and limb were detected in *MyoD^−/−:Myf5^−/−* mutant embryos, suggesting that the core enhancer contains elements that are required for the initial activation of *MyoD* transcription in muscle progenitor cells (Kablar *et al.*, 1998). From 11 to 12 dpc, the expressions of *lacZ* in myotomes of *MyoD6.0-lacZ* (Fig. 8B) and *MyoDcore/2.5-lacZ* (Fig. 8C) appear mutually exclusive, suggesting that both the core and DRR enhancers together cooperatively regulate mouse *MyoD* gene expression during myotome formation (Kablar *et al.*, 1999). However, transcription factors involved in *MyoD* gene expression mediated through those enhancers remain to be elucidated. Nevertheless, several transcription factors have been suggested to exert positive or negative effects on MRF transcription.

Avian and mouse *Msx1* is expressed in the ventrolateral region of dermomyotome and limb bud and appears to negatively regulate *MyoD* transcription and muscle differentia-

tion (Bendall *et al.*, 1999; Houzelstein *et al.*, 1999). Suppression of *MyoD* transcription is mediated by directly binding to the core enhancer of the *MyoD* gene (Woloshin *et al.*, 1995). Although Msx1 is a negative regulator of *MyoD* transcription, it represents the only factor known that directly regulates *MyoD* transcription during embryogenesis.

The mouse homeobox gene *Mox2* (*Meox2*) is expressed in the dermomyotome as well as in migratory muscle progenitor cells. Mice lacking *Mox2* showed downregulated *Myf5* expression in limb buds and a subsequent loss of a subset of limb muscles (Table I; Mankoo *et al.*, 1999). However, *MyoD* expression is relatively normal. Because mice lacking *Myf5* display normal limb muscle development (Kablar *et al.*, 1997), Mox2 does not appear to act as a simple upstream regulatory factor regulating *Myf5* transcription.

The M-twist (*Twist*) related bHLH gene *Paraxis* (*Tcf15*) is expressed in the dermomyotome during mouse embryogenesis (Burgess *et al.*, 1995). Mice lacking *Paraxis* display abnormal somite formation. However, muscle differentiation occurs within the disorganized myotome (Burgess *et al.*, 1996). The migration of epaxial myotomal cells is disorganized and nonmigratory hypaxial myotome formation is delayed by several days, due to a delay in *MyoD* expression (Wilson-Rawls *et al.*, 1999). The formation of migratory muscles is also delayed. Compound mutant mice lacking both *Paraxis* and *Myf5* display delayed *MyoD* expression in the myotome and an absence of epaxial muscle formation, such as back muscle, and reduced formation of a subset of hypaxial muscles, such as the proximal region of intercostal muscles and appendicular muscles (Wilson-Rawls *et al.*, 1999). These results suggest that Paraxis is an important regulator of a subset of the muscle progenitor cells derived from the dermomyotome (Table I).

Mice lacking both *Myf5* and *Pax3* (*Splotch*) display a complete absence of *MyoD* expression and an absence of

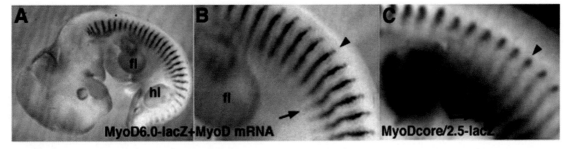

Figure 8 Transcriptional analyses of *MyoD* gene using transgenic mice carrying *lacZ* reporter genes. Two elements, a core enhancer (258 bp) located at −20 kb and a DRR (700 bp) located at −5 kb of the initiation site of *MyoD* gene, are required for mouse *MyoD* gene expression. (A and B) Double staining with X-gal for *lacZ* expression (blue) and *in situ* hybridization for *MyoD* expression (dark brawn) in embryos (11.5 dpc) from transgenic mice carrying 6 kb upstream containing the DRR driving *lacZ* gene (*MyoD6.0-lacZ*). (C) Whole-mount X-gal staining for *lacZ* expression (blue) in embryo (11.5 dpc) from transgenic mice carrying 258-bp core enhancer driving *lacZ* gene (*MyoDcore/2.5-lacZ*). *MyoD* mRNA is expressed in the entire myotomes and limb buds (Panels A and B and see Fig. 4I). Expression of *MyoD6.0-lacZ* is detected in the medial region of myotome but not in the limb bud (A, B). In contrast, *MyoDcore/2.5-lacZ* is expressed in the dorsal and ventral regions of myotome, and limb bud (C) (Asakura *et al.*, 1995; Kablar *et al.*, 1999).

myogenesis in the trunk. This result led to the suggestion that both Myf5 and Pax3 are upstream regulators of MyoD (Table I; Tajbakhsh *et al.*, 1997). Indeed, ectopic expression of Pax3 induces skeletal myogenesis in nonmuscle tissues in avian embryos (Maroto *et al.*, 1997). However, ectopic expression of Pax3 in C2C12 myoblasts efficiently inhibits muscle differentiation (Epstein *et al.*, 1995). In addition, coexpression of *MyoD* and *Pax3* is not observed during mouse myotome formation (Williams and Ordahl, 1994). Mutations in human *Pax3* have been identified and are responsible for Waardenburg syndrome (WS) in which tissues derived from neural crest cells are markedly affected. Importantly, fusions between Pax3 or Pax7 and FKHR (a forkhead related transcription factor) are frequently identified in human alveola rhabdomyosarcoma (RMS). Pax3-FKHR or Pax7-FKHR acts as a potent activated form of Pax3 or Pax7. Therefore, activated Pax3/7 appear to promote RMS cell proliferation or suppress terminal differentiation (Merlino and Helman, 1999). Therefore, Pax3 may act as a indirect upstream factor that induces migration or other cellular changes to facilitate a subsequent induction of *MyoD* transcription.

Recent experiments in chicken embryos have demonstrated that novel transcriptional regulators, Six1, Eya, and Dach, are involved in myotome formation and may act as upstream regulators of the MRFs. Six1 is the vertebrate homolog of *Drosophila* homeobox protein *sine oculis* (*so*). Eya is the vertebrate homolog of *Drosophila* *eyes absent* (*eya*). Dach is the vertebrate homolog of *Drosophila* *dachshund* (*dac*), which exhibits partial homology to the ski/sno protein, which is known to regulate adult muscle mass (Table I; Colmenares and Stavnezer, 1989; Berk *et al.*, 1997). Six1 protein is a transcription factor that binds the MEF3 site originally identified as a regulatory element in the fast muscle-specific *aldolase* (Hidaka *et al.*, 1993; Salminen *et al.*, 1996; Spitz *et al.*, 1997) and *myogenin* genes (Spitz *et al.*, 1998). Recent experiments, using the avian system demonstrated that Dach2 and Eya act as transcriptional cofactors that physically interact with either Six1 or Dach proteins (Heanue *et al.*, 1999; Relaix and Buckingham, 1999). Dach2, Eya2, and Six1 are expressed in the developing somite, the DML of the dermomyotome, the myotome, and the migratory muscle progenitor cells (Relaix and Buckingham, 1999). Ectopic expression experiments have demonstrated that Dach2 and Pax3 can induce each other's transcription in the somite, suggesting the presence of a positive feedback loop among the genes (Fig. 9). Dach2/Eya2 or Six1/Eya2 complexes can induce ectopic myogenesis, including *MRF* and *MyHC* expression within the somite (Heanue *et al.*, 1999). Because ectopic expression of Pax3 can induce ectopic myogenesis in nonmuscle tissue, the downstream targets of Pax3 may include *Dach2/Eya2/Six1* complexes, which then initiate the muscle differentiation program (Fig. 9). However, it remains to be shown whether the *de novo* induction of *MyoD* or *Myf5* gene is directly

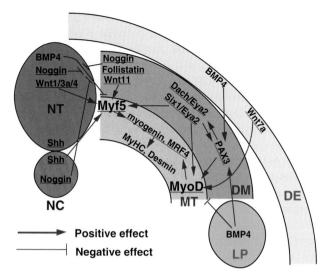

Figure 9 Secreted factors and transcription factors regulate the myotome formations. Secreted factors from surrounding tissues, the neural tube (NT), notochord (NC), dorsal ectoderm (DE), and lateral plate (LP) act to pattern the dermomyotome (DM) and myotome (MT). Positive factors for myotome formation are underlined.

activated by these complexes and whether *MyoD* or *Myf5* regulatory regions contain MEF3 sites that are bound by Six1 protein. Interestingly, ski (*Ski*) and sno (*Skir*) proteins, components of a histone deacetylase complex, suppress the TGF-β/BMP signaling cascade mediated by blocking Smad2 (*Madh2*), Smad3 (*Madh3*), or Smad4 (*Madh4*) proteins (Luo *et al.*, 1999). In addition, ectopic expression of both ski and sno induces myogenesis or muscle hypertrophy (Colmenares and Stavnezer, 1989; Berk *et al.*, 1997). Therefore, Dach2/Eya2 complex may initiate myogenesis by blocking the BMP signaling cascade that normally suppresses myotome formation. In the future, targeted mutations of *Dach*, *Eya*, and *Six* genes and *in vitro* transcriptional activation assays will reveal the genetic hierarchy that regulates myotome formation and MRF gene regulation.

VI. Specification of Muscle Fiber Types

In adult skeletal muscle, each muscle contains different proportions of distinct muscle fiber types. In the murine hindlimb for example, the lateral gastrocnemius mainly consists of fast-twitch fibers and the soleus mainly consists of slow-twitch fibers. The fiber type is judged by the physiological aspect and a distinct MyHC isoform profile. While the slow-twitch fibers consist of oxidative fibers (type I fiber) that express slow MyHC, MyHC-β/MyHC type I (*MHCb*), the fast-twitch fibers consist of glycolytic fibers (type II fiber) that express fast MyHC including type IIA (*Myh2*), type IIB (*Myh4*), and type IIX (*Myh1*) MyHC (Stockdale, 1992; Hughes and Salinas, 1999; Talmadge, 2000).

During myotome formation, embryonic and slow isoforms of MyHC, *MyHC-emb* (*Myh3*), and *MyHC-β*, respectively, are detected in the early differentiating myotome by 9.5 dpc in mouse, and subsequently the perinatal isoform of MyHC, *MyHC-pn* (*Myh8*) is detected at 10.5 dpc (Lyons *et al.*, 1990). However, the distribution of these isoforms is uniform in the myotome, suggesting fiber type is not determined in the myotome. By contrast, during zebrafish myotome formation, adaxial muscle that is proximal to the notochord initiates slow-type MyHC expression and these cells migrate laterally within the myotome (Devoto *et al.*, 1996; Blagden *et al.*, 1997). Shh secreted from the notochord and floor plate induces slow muscle fiber type. Subsequently, myotomal cells expressing fast-type MyHC reside in the medial part of myotome. Therefore, in zebrafish development of distinct muscle fiber types is initiated at an early stage of myotome formation (Currie and Ingham, 1998).

During fetal myogenesis, *MyHC-emb* and *MyHC-pn* genes are expressed at high levels. By contrast, fast type MyHC, *MyHC-IIB*, *MyHC-IIA*, and *MyHC-IIX* mRNA is only detectable at quite low levels from 14.5 dpc to birth. *MyHC-β* is expressed throughout fetal myogenesis at medium level. After birth the relative proportions of MyHC isoforms are changed dramatically. Both expressions of *MyHC-emb* and *MyHC-pn* are downregulated in leg muscles, although a high level of *MyHC-pn* expression is maintained during neonatal stage. Instead, expression of slow-type (*MyHC-β*) or fast-type (*MyHC-IIA, IIB,* or *IIX*) isoform is upregulated in each muscle fiber during postnatal development to constitute slow-, fast-, or mixed-fiber-type muscle (Lu *et al.*, 1999).

During avian or rodent limb muscle development, primary muscle fibers are uniformly formed, and then rapidly acquire slow or fast features depending on their location within the limb. Subsequently, secondary fibers form in association within the primary fiber scaffold (Stockdale, 1992; Hughes and Salinas, 1999). The determination of the fiber types is dependent on both cell lineage and local signals. For example, embryonic myoblasts differentiate into small myotubes that express slow-type MyHC and form primary fibers. By contrast, fetal myoblasts mainly express fast-type MyHC and form secondary fibers. In addition, clonal culture experiments reveal the presence of cell populations that form distinct fiber types (Miller and Stockdale, 1986a,b; Hughes and Blau, 1992; DiMario *et al.*, 1993). However, implantation experiments using distinct cell populations suggest that both cell lineage and local signals affect muscle fiber types (Robson and Hughes, 1996, 1999). Interestingly, two distinct muscle populations, first slow lineage and second fast lineage, may enter the limb buds migrating from dermomyotome (Seed and Hauschka, 1984; Van Swearingen and Lance-Jones, 1995). During fetal myogenesis, motorneurons innervate muscle fibers and the resulting electric activity also affects fiber type specification. The signal from nerve is strong enough to override the cell lineage or local cues,

which determines fiber types, to induce slow/fast MyHC changes (Gundersen, 1998; Hughes and Salinas, 1999; Talmadge, 2000). Recent work demonstrated that a signaling pathway of calcineurin (*Ppp3ca*), a cyclosporin-sensitive, calcium-regulated serine/threonine phosphatase may regulate manifestation of the slow muscle fiber type (Chin *et al.*, 1998; Wu *et al.*, 2000; Olson and Williams, 2000; Talmadge, 2000). The activated calcineurin dephosphorylates the phosphorylated NFATc (*Nfatc1*), allowing NFATc to enter the nucleus and cooperate with MEF2 to activate slow-fiber specific genes.

Whereas MyoD is mainly detected in fast-type fibers, myogenin is mainly detected in slow-type fibers in mouse adult muscle (Hughes *et al.*, 1993). *MyoD*$^{-/-}$ mice display subtle shifts in fiber type of fast muscles toward a slower character and a shift of slow muscles toward a faster phenotype (Hughes *et al.*, 1997). Ectopic expression of myogenin in adult skeletal muscles induces an increase in slow fiber composition. These results suggest that MyoD and myogenin are somehow involved in development or maintenance of muscle fiber types but likely via an indirect mechanism (Hughes *et al.*, 1999).

VII. Muscle Regeneration

A. Satellite Cells in Adult Muscle

Muscle satellite cells are a distinct lineage of myogenic cells responsible for mediating the postnatal growth and repair of adult skeletal muscle. Satellite cells reside beneath the basal lamina of adult skeletal muscle, closely juxtaposed against skeletal muscle fibers. Satellite cells are normally mitotically quiescent, but are activated (initiate proliferation) in response to stress induced by weight bearing or other trauma such as injury, to mediate the postnatal growth and regeneration of muscle. The progeny of activated satellite cells, termed myogenic precursor cells (MPCs), undergo multiple rounds of cell division prior to their terminal differentiation to form new muscle fibers (Fig. 10B). Satellite cells account for 2–5% of sublaminal nuclei in adult muscle (2 months in mice). The number of quiescent satellite cells in adult muscle remains relatively constant over multiple cycles of degeneration and regeneration, indicating an inherent capacity for self-renewal (Grounds *et al.*, 1992; Bischoff, 1996; Seale and Rudnicki, 2000).

B. MRFs in Satellite Cell Activation and Differentiation

The MRF expression program during satellite cell activation, proliferation, and differentiation is analogous to the program manifested during the embryonic development of skeletal muscle. Quiescent satellite cells express no detect-

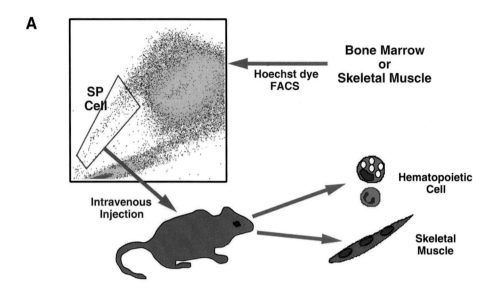

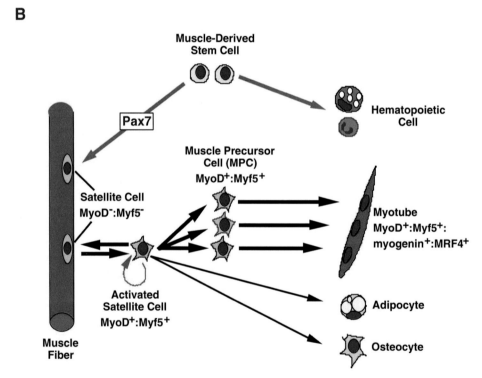

Figure 10 Satellite cells and pluripotential muscle-derived stem cells. (A) Pluripotent muscle-derived stem cells, also called side population (SP) cells, were identified in adult skeletal muscle tissue by fluorescence activated cell sorting (FACS) on the basis of Hoechst dye exclusion. Following intravenous injection, both muscle-derived and marrow-derived stem SP cells give rise to skeletal muscle cells and hematopoietic cells (Gussoni *et al.*, 1999; Jackson *et al.*, 1999). (B) Mitotically quiescent satellite cells are activated (initiated the cell division) and initiate expressions of MyoD and/or Myf5 during muscle regeneration. The self-renewal activated satellite cells give rise to the muscle precursor cells (MPCs) or go back to the quiescent satellite cells as a stem cell compartment. The MPCs cease the cell proliferation to initiate expression of myogenin, MRF4, and several muscle structural proteins then fuse each other to form the multinucleated myotubes. Satellite cells also possess a potential to give rise to adipocytes and osteocytes. Muscle-derived stem cells possess a potential to differentiate into both skeletal muscle cells and hematopoietic cells. *Pax7* gene knock-out experiments suggest that myogenic specification of pluripotent muscle-derived stem cells requires *Pax7* gene (Seale *et al.*, 2000).

able MRFs but do express c-met and M-cadherin (Irintchev *et al.*, 1994; Cornelison and Wold, 1997). During *in vitro* single muscle fiber culture and *in vivo* muscle regeneration, in which satellite cells initiate their activation and entrance into the cell cycle, either *MyoD* or *Myf5* is rapidly upregulated concomitant with entrance of the cell cycle, followed soon after by coexpression of *MyoD* and *Myf5*. Following proliferation, *myogenin* and *MRF4* are expressed in cells beginning their differentiation program (C. K. Smith *et al.*, 1994; Yablonka-Reuveni and Rivera, 1994; Cornelison and Wold, 1997; Cornelison *et al.*, 2000). Therefore, MyoD and Myf5 appear to play an early role in the satellite cell developmental program (Fig. 10B).

Mice lacking *MyoD* gene display marked deficits in satellite cell function (Table I). *MyoD* mutant mice interbred with *mdx* mice, which are a model for Duchenne's and Becker's muscular dystrophies, exhibit increased penetrance of the *mdx* phenotype characterized by muscle atrophy and increased myopathy leading to premature death (Megeney *et al.*, 1996). In spite of the presence of morphologically normal satellite cells in *MyoD*$^{-/-}$ muscles, muscle from *MyoD*$^{-/-}$ mice displays a strikingly reduced capacity for regeneration following injury. Whereas growing *MyoD*$^{-/-}$ myoblasts express normal levels of c-met, expression of M-cadherin and desmin is notably reduced. By contrast, Myf5 expression is markedly upregulated in these cells. Under conditions that normally induce differentiation of wild-type myoblasts, *MyoD*$^{-/-}$ myoblasts continue to proliferate and exhibit delayed differentiation. Supporting this, *MyoD*$^{-/-}$ myoblasts show delayed induction of *myogenin* and *MRF4*. Therefore, *MyoD*$^{-/-}$ myoblasts display characters that are more primitive than wild-type myoblasts and may represent an intermediate stage between a satellite cell and a myogenic precursor cell (Sabourin *et al.*, 1999; Cornelison *et al.*, 2000; Seale and Rudnicki, 2000).

C. Pluripotent Muscle-Derived Stem Cells

Tissue-specific stem cells have been found in various adult organs where they participate in replacement of differentiated cells for physiological turnover or injury. Recent experiments have demonstrated that tissue-specific stem cells appear to have a wider differentiation capability than previously thought. Therefore, many or all tissues may contain a population of pluripotent stem cells that differentiates in an appropriate manner in response to growth factors and signals provided by host tissues (Fuchs and Segré, 2000; Seale and Rudnicki, 2000; Cossu and Mavilio, 2000; Orkin, 2000).

The pluripotent nature of adult stem cells isolated from diverse tissues raises the possibility of stem cell therapy for a variety of degenerative diseases including muscular dystrophy. Ferrari and colleagues were the first to demonstrate that marrow-derived cells could participate in muscle regeneration (Ferrari *et al.*, 1998). Subsequently, Gussoni *et al.*

(1999) demonstrated that highly purified hematopoietic stem cells recruited to skeletal muscle of *mdx* mice resulted in a partial restoration of dystrophin expression (Gussoni *et al.*, 1999). Gussoni *et al.* (1999) and Jackson *et al.* (1999) further identified the presence of side population (SP) cells, enriched in pluripotent muscle-derived stem cells, isolated in adult muscle tissue by fluorescence activated cell sorting (FACS) on the basis of Hoechst dye exclusion (Fig. 10A; Gussoni *et al.*, 1999; Jackson *et al.*, 1999). Following intravenous injection, muscle-derived SP cells efficiently participate in regeneration of skeletal muscle, appear to give rise to myogenic satellite cells, and reconstitute the complete repertoire of the hematopoietic systems (Gussoni *et al.*, 1999; Seale and Rudnicki, 2000).

Cell culture experiments have demonstrated that continuous myoblast lines derived from satellite cells can transdifferentiate into adipoblasts and osteoblasts following treatment with adipogenetic inducers such as thiazolidinedione (Teboul *et al.*, 1995) or by blocking the Wnt signaling pathway (Ross *et al.*, 2000), and BMP2 (Katagiri *et al.*, 1994), respectively. Several *in vivo* observations have postulated the existence of pluripotential stem cells within skeletal muscle. For example, expansion of adipose tissue within skeletal muscles occurs in response to denervation (Dulor *et al.*, 1998) and in some muscle diseases, including muscular dystrophy (Lin *et al.*, 1969) and mitochondrial myopathy (DiMauro *et al.*, 1980). Moreover, transplantation of BMP into adult skeletal muscles can induce ectopic osteogenesis within muscle (Urist and Strates, 1971). Similarly, *BMP2/4* have been implicated in ectopic ossification of muscles in a human disease, fibrodysplasia ossificans progressiva (FOP) (Shafritz *et al.*, 1996). However, it remains unknown whether satellite cells are the origin of adipocytes and osteocytes *in vivo* (Fig. 10).

Satellite cells first appear in the limbs of 17.5 dpc mouse embryos and appear to constitute a myogenic cell lineage distinct from the embryonic muscle lineages derived from somites and prechordal mesoderm. Resent work has revealed that clonal skeletal myogenic cells that closely resemble satellite cell-derived myogenic precursors are readily isolated from the embryonic dorsal aorta of mouse embryos *in vitro* (De Angelis *et al.*, 1999; Ordahl, 1999; Cossu and Mavilio, 2000). Aorta-derived myogenic cells express MyoD and/or Myf5 and several endothelial markers that are also expressed in activated satellite cells. Myoid cells in thymus, which express MyoD and myogenin are also a candidate source of myogenic stem cells (Grounds *et al.*, 1992; Wong *et al.*, 1999; Wakkach *et al.*, 1999). However, the presence of myogenic cells in dorsal aorta and thymus does not preclude the possibility of a somitic origin for satellite cell progenitors.

Recently, we found that *Pax7* is expressed in both quiescent and proliferating satellite cells. Strikingly, mutant mice lacking *Pax7* display a complete absence of satellite cells. Nevertheless, muscle-derived SP cells are present in

normal proportions (Table I; Seale *et al.,* 2000). These results strongly suggest that muscle-derived stem cells are a
distinct stem cell population from satellite cells which yet
possess mesenchymal differentiation plasticity. Additional
experiments have revealed that isolated satellite cells readily
give rise to colonies composed of myocytes, adipocytes, or
osteocytes, whereas muscle-derived SP cells efficiently
formed colonies of hematopoietic cells *in vitro* (A. Asakura
and M. A. Rudnicki unpublished data). Taken together, these
data support the hypothesis that pluripotent muscle-derived
stem cells are the progenitors of myogenic satellite cells and
suggest that the mechanism that induces *Pax7* expression in
pluripotent stem cells is under strict control (Fig. 10B).

VIII. Conclusion

The MRFs play key regulatory roles in the development
of skeletal muscle during embryogenesis and in postnatal
growth and repair of muscle. Recent experiments have revealed several key regulatory molecules including both transcription factors such as Pax3 and Six and signaling molecules such as Shh, Wnts, and BMPs, which positively and
negatively regulate muscle development. In addition, experiments using the avian system have exposed revealing insights into the dynamic mechanisms regulating myotome
formation. In the near future, the molecular mechanisms
regulating myotome formation will certainly be elucidated
by combining information from different animal systems, including the continued genetic analysis afforded by gene targeting in mice.

Recent studies have demonstrated that a novel population
of muscle-derived stem cells gives rise to hematopoietic lineages and also participates in muscle regeneration following
intravenous transplantation. The molecular mechanisms that
control the specification of adult pluripotent stem cells into
differentiated cells types must involve induction of developmental control genes such as *Pax7* in the satellite cell
myogenic lineage. However, how these mechanisms function to effect this induction remains entirely unclear. Nevertheless, the rapid progress in understanding the mechanisms
regulating myogenic stem cell specification and function in
adult muscle will clearly lead to new insights in the analogous mechanisms regulating embryonic myogenesis.

Acknowledgments

We thank Boris Kablar and Robert Perry for critical comments on the
manuscript.

This work was supported by grants to M.A.R. from the Muscular Dystrophy Association, the National Institutes of Health, and the Canadian Institutes of Health Research. A.A. is supported by a development grant from
the Muscular Dystrophy Association. M.A.R. is a research scientist of the
Canadian Institutes of Health Research and is a member of the Canadian
Genetic Disease Network.

References

Amthor, H., Connolly, D., Patel, K., Brand-Saberi, B., Wilkinson, D. G.,
Cooke, J., and Christ, B. (1996). The expression and regulation of follistatin and a follistatin-like gene during avian somite compartmentalization and myogenesis. *Dev. Biol.* **178,** 343–362.

Amthor, H., Christ, B., Weil, M., and Patel, K. (1998). The importance of
timing differentiation during limb muscle development. *Curr. Biol.* **8,**
642–652.

Amthor, H., Christ, B., and Patel, K. (1999). A molecular mechanism enabling continuous embryonic muscle growth—a balance between proliferation and differentiation. *Development (Cambridge, UK)* **126,** 1041–
1053.

Aoyama, H., and Asamoto, K. (1988). Determination of somite cells: In-
dependence of cell differentiation and morphogenesis. *Development
(Cambridge, UK)* **104,** 15–28.

Artavanis-Tsakonas, S., Rand, M. D., and Lake, R. J. (1999). Notch signaling: Cell fate control and signal integration in development. *Science* **284,**
770–776.

Asakura, A., and Tapscott, S. J. (1998). Apoptosis of epaxial myotome in
Danforth's short-tail (Sd) mice in somites that form following notochord
degeneration. *Dev. Biol.* **203,** 276–289.

Asakura, A., Fujisawa-Sehara, A., Komiya, T., and Nabeshima, Y. (1993).
MyoD and myogenin act on the chicken myosin light-chain 1 gene as
distinct transcriptional factors. *Mol. Cell. Biol.* **13,** 7153–7162.

Asakura, A., Lyons, G. E., and Tapscott, S. J. (1995). The regulation of
MyoD gene expression: Conserved elements mediate expression in embryonic axial muscle. *Dev. Biol.* **171,** 386–398.

Atchley, W. R., Fitch, W. M., and Bronner-Fraser, M. (1994). Molecular
evolution of the MyoD family of transcription factors. *Proc. Natl. Acad.
Sci. U.S.A.* **91,** 11522–11526.

Balling, R., Helwig, U., Nadeau, J., Neubuser, A., Schmahl, W., and Imai, K.
(1996). Pax genes and skeletal development. *Ann. N.Y. Acad. Sci.* **785,**
27–33.

Bendall, A. J., Ding, J., Hu, G., Shen, M. M., and Abate-Shen, C. (1999).
Msx1 antagonizes the myogenic activity of Pax3 in migrating limb
muscle precursors. *Development (Cambridge, UK)* **126,** 4965–4976.

Berk, M., Desai, S. Y., Heyman, H. C., and Colmenares, C. (1997). Mice
lacking the ski proto-oncogene have defects in neurulation, craniofacial,
patterning, and skeletal muscle development. *Genes Dev.* **11,** 2029–
2039.

Bischoff, R. (1996). The satellite cell and muscle regeneration. *In* "Myology" (A. G. Engel and A. Franzini-Armstrong, eds.), pp. 97–118. Mc-
Graw-Hill.

Blackwell, T. K., and Weintraub, H. (1990). Differences and similarities
in DNA-binding preferences of MyoD and E2A protein complexes revealed by binding site selection. *Science* **250,** 1104–1110.

Bladt, F., Riethmacher, D., Isenmann, S., Aguzzi, A., and Birchmeier, C.
(1995). Essential role for the c-met receptor in the migration of myogenic precursor cells into the limb bud. *Nature (London)* **376,** 768–
771.

Blagden, C. S., Currie, P. D., Ingham, P. W., and Hughes, S. M. (1997).
Notochord induction of zebrafish slow muscle mediated by Sonic hedgehog. *Genes Dev.* **11,** 2163–2175.

Bober, E., Lyons, G. E., Braun, T., Cossu, G., Buckingham, M., and Arnold,
H. H. (1991). The muscle regulatory gene, Myf-6, has a biphasic pattern
of expression during early mouse development. *J. Cell Biol.* **113,** 1255–
1265.

Bober, E., Franz, T., Arnold, H. H., Gruss, P., and Tremblay, P. (1994).
Pax-3 is required for the development of limb muscles: A possible role
for the migration of dermomyotomal muscle progenitor cells. *Development (Cambridge, UK)* **120,** 603–612.

Borycki, A. G., Strunk, K. E., Savary, R., and Emerson, C. P., Jr. (1997).
Distinct signal/response mechanisms regulate pax1 and QmyoD activa-

tion in sclerotomal and myotomal lineages of quail somites. *Dev. Biol.* **185,** 185–200.

Borycki, A. G., Brunk, B., Tajbakhsh, S., Buckingham, M., Chiang, C., and Emerson, C. P., Jr. (1999). Sonic hedgehog controls epaxial muscle determination through Myf5 activation. *Development (Cambridge, UK)* **126,** 4053–4063.

Borycki, A., Brown, A. M., and Emerson, C. P., Jr. (2000). Shh and Wnt signaling pathways converge to control Gli gene activation in avian somites. *Development (Cambridge, UK)* **127,** 2075–2087.

Brand-Saberi, B., Ebensperger, C., Wilting, J., Balling, R., and Christ, B. (1993). The ventralizing effect of the notochord on somite differentiation in chick embryos. *Anat. Embryol.* **188,** 239–245.

Brand-Saberi, B., Wilting, J., Ebensperger, C., and Christ, B. (1996a). The formation of somite compartments in the avian embryo. *Int. J. Dev. Biol.* **40,** 411–420.

Brand-Saberi, B., Muller, T. S., Wilting, J., Christ, B., and Birchmeier, C. (1996b). Scatter factor/hepatocyte growth factor (SF/HGF) induces emigration of myogenic cells at interlimb level in vivo. *Dev. Biol.* **179,** 303–308.

Braun, T., and Arnold, H. H. (1995). Inactivation of Myf-6 and Myf-5 genes in mice leads to alterations in skeletal muscle development. *EMBO J.* **14,** 1176–1186.

Braun, T., Rudnicki, M. A., Arnold, H. H., and Jaenisch, R. (1992). Targeted inactivation of the muscle regulatory gene Myf-5 results in abnormal rib development and perinatal death. *Cell (Cambridge, Mass.)* **71,** 369–382.

Brohmann, H., Jagla, K., and Birchmeier, C. (2000). The role of Lbx1 in migration of muscle precursor cells. *Development (Cambridge, UK)* **127,** 437–445.

Buffinger, N., and Stockdale, F. E. (1994). Myogenic specification in somites: Induction by axial structures. *Development (Cambridge, UK)* **120,** 1443–1452.

Burgess, R., Cserjesi, P., Ligon, K. L., and Olson, E. N. (1995). Paraxis: A basic helix-loop-helix protein expressed in paraxial mesoderm and developing somites. *Dev. Biol.* **168,** 296–306.

Burgess, R., Rawls, A., Brown, D., Bradley, A., and Olson, E. N. (1996). Requirement of the paraxis gene for somite formation and musculoskeletal patterning. *Nature (London)* **384,** 570–573.

Chiang, C., Litingtung, Y., Lee, E., Young, K. E., Corden, J. L., Westphal, H., and Beachy, P. A. (1996). Cyclopia and defective axial patterning in mice lacking Sonic hedgehog gene function. *Nature (London)* **383,** 407–413.

Chin, E. R., Olson, E. N., Richardson, J. A., Yang, Q., Humphries, C., Shelton, J. M., Wu, H., Zhu, W., Bassel-Duby, R., and Williams, R. S. (1998). A calcineurin-dependent transcriptional pathway controls skeletal muscle fiber type. *Genes Dev.* **12,** 2499–2509.

Christ, B., and Ordahl, C. P. (1995). Early stages of chick somite development. *Anat. Embryol.* **191,** 381–396.

Cinnamon, Y., Kahane, N., and Kalcheim, C. (1999). Characterization of the early development of specific hypaxial muscles from the ventrolateral myotome. *Development (Cambridge, UK)* **126,** 4305–4315.

Colmenares, C., and Stavnezer, E. (1989). The ski oncogene induces muscle differentiation in quail embryo cells. *Cell (Cambridge, Mass.)* **59,** 293–303.

Cornelison, D. D., and Wold, B. J. (1997). Single-cell analysis of regulatory gene expression in quiescent and activated mouse skeletal muscle satellite cells. *Dev. Biol.* **191,** 270–283.

Cornelison, D. D., Olwin, B. B., Rudnicki, M. A., and Wold, B. J. (2000). MyoD(-/-) satellite cells in single-fiber culture are differentiation defective and MRF4 deficient. *Dev. Biol.* **224,** 122–137.

Cossu, G., and Mavilio, F. (2000). Myogenic stem cells for the therapy of primary myopathies: Wishful thinking or therapeutic perspective? *J. Clin. Invest.* **105,** 1669–1674.

Cossu, G., Kelly, R., Di Donna, S., Vivarelli, E., and Buckingham, M. (1995). Myoblast differentiation during mammalian somitogenesis is dependent upon a community effect. *Proc. Natl. Acad. Sci. U.S.A.* **92,** 2254–2258.

Cossu, G., Kelly, R., Tajbakhsh, S., Di Donna, S., Vivarelli, E., and Buckingham, M. (1996). Activation of different myogenic pathways: Myf-5 is induced by the neural tube and MyoD by the dorsal ectoderm in mouse paraxial mesoderm. *Development (Cambridge, UK)* **122,** 429–437.

Cserjesi, P., Lilly, B., Bryson, L., Wang, Y., Sassoon, D. A., and Olson, E. N. (1992). MHox: A mesodermally restricted homeodomain protein that binds an essential site in the muscle creatine kinase enhancer. *Development (Cambridge, UK)* **115,** 1087–1101.

Currie, P. D., and Ingham, P. W. (1998). The generation and interpretation of positional information within the vertebrate myotome. *Mech. Dev.* **73,** 3–21.

Daubas, P., Tajbakhsh, S., Hadchouel, J., Primig, M., and Buckingham, M. (2000). Myf5 is a novel early axonal marker in the mouse brain and is subjected to post-transcriptional regulation in neurons. *Development (Cambridge, UK)* **127,** 319–331.

Davis, R. L., Weintraub, H., and Lassar, A. B. (1987). Expression of a single transfected cDNA converts fibroblasts to myoblasts. *Cell (Cambridge, Mass.)* **51,** 987–1000.

Davis, R. L., Cheng, P. F., Lassar, A. B., and Weintraub, H. (1990). The MyoD DNA binding domain contains a recognition code for muscle-specific gene activation. *Cell (Cambridge, Mass.)* **60,** 733–746.

De Angelis, L., Berghella, L., Coletta, M., Lattanzi, L., Zanchi, M., Cusella-De Angelis, M. G., Ponzetto, C., and Cossu, G. (1999). Skeletal myogenic progenitors originating from embryonic dorsal aorta coexpress endothelial and myogenic markers and contribute to postnatal muscle growth and regeneration. *J. Cell Biol.* **147,** 869–878.

Delfini, M., Hirsinger, E., Pourquié, O., and Duprez, D. (2000). Delta1-activated Notch inhibits muscle differentiation without affecting Myf5 and Pax3 expression in chick limb myogenesis. *Development (Cambridge, UK)* **127,** 5213–5224.

Denetclaw, W. F., and Ordahl, C. P. (2000). The growth of the dermomyotome and formation of early myotome lineages in thoracolumbar somites of chicken embryos. *Development (Cambridge, UK)* **127,** 893–905.

Denetclaw, W. F., Jr., Christ, B., and Ordahl, C. P. (1997). Location and growth of epaxial myotome precursor cells. *Development (Cambridge, UK)* **124,** 1601–1610.

Devoto, S. H., Melancon, E., Eisen, J. S., and Westerfield, M. (1996). Identification of separate slow and fast muscle precursor cells in vivo, prior to somite formation. *Development (Cambridge, UK)* **122,** 3371–3380.

Dietrich, S., Schubert, F. R., and Lumsden, A. (1997). Control of dorsoventral pattern in the chick paraxial mesoderm. *Development (Cambridge, UK)* **124,** 3895–3908.

Dietrich, S., Abou-Rebyeh, F., Brohmann, H., Bladt, F., Sonnenberg-Riethmacher, E., Yamaai, T., Lumsden, A., Brand-Saberi, B., and Birchmeier, C. (1999a). The role of SF/HGF and c-Met in the development of skeletal muscle. *Development (Cambridge, UK)* **126,** 1621–1629.

Dietrich, S., Schubert, F. R., Gruss, P., and Lumsden, A. (1999b). The role of the notochord for epaxial myotome formation in the mouse. *Cell. Mol. Biol. (Noisy-le-grand)* **45,** 601–616.

DiMario, J. X., Fernyak, S. E., and Stockdale, F. E. (1993). Myoblasts transferred to the limbs of embryos are committed to specific fibre fates. *Nature (London)* **362,** 165–167.

DiMauro, S., Mendell, J. R., Sahenk, Z., Bachman, D., Scarpa, A., Scofield, R. M., and Reiner, C. (1980). Fatal infantile mitochondrial myopathy and renal dysfunction due to cytochrome-c-oxidase deficiency. *Neurology* **30,** 795–804.

Dulor, J. P., Cambon, B., Vigneron, P., Reyne, Y., Nougues, J., Casteilla, L., and Bacou, F. (1998). Expression of specific white adipose tissue genes in denervation- induced skeletal muscle fatty degeneration. *FEBS Lett.* **439,** 89–92.

Duprez, D., Fournier-Thibault, C., and Le Douarin, N. (1998). Sonic Hedgehog induces proliferation of committed skeletal muscle cells in the chick limb. *Development (Cambridge, UK)* **125,** 495–505.

Duprez, D., Lapointe, F., Edom-Vovard, F., Kostakopoulou, K., and Robson, L. (1999). Sonic hedgehog (SHH) specifies muscle pattern at tissue and cellular chick level, in the chick limb bud. *Mech. Dev.* **82,** 151–163.

Edmondson, D. G., Lyons, G. E., Martin, J. F., and Olson, E. N. (1994). Mef2 gene expression marks the cardiac and skeletal muscle lineages during mouse embryogenesis. *Development (Cambridge, UK)* **120,** 1251–1263.

Epstein, J. A., Lam, P., Jepeal, L., Maas, R. L., and Shapiro, D. N. (1995). Pax3 inhibits myogenic differentiation of cultured myoblast cells. *J. Biol. Chem.* **270,** 11719–11722.

Faerman, A., Goldhamer, D. J., Puzis, R., Emerson, C. P., Jr., and Shani, M. (1995). The distal human myoD enhancer sequences direct unique muscle-specific patterns of lacZ expression during mouse development. *Dev. Biol.* **171,** 27–38.

Fan, C. M., and Tessier-Lavigne, M. (1994). Patterning of mammalian somites by surface ectoderm and notochord: Evidence for sclerotome induction by a hedgehog homolog. *Cell (Cambridge, Mass.)* **79,** 1175–1186.

Farrance, I. K., Mar, J. H., and Ordahl, C. P. (1992). M-CAT binding factor is related to the SV40 enhancer binding factor, TEF-1. *J. Biol. Chem.* **267,** 17234–17240.

Ferrari, G., Cusella-De Angelis, G., Coletta, M., Paolucci, E., Stornaiuolo, A., Cossu, G., and Mavilio, F. (1998). Muscle regeneration by bone marrow-derived myogenic progenitors. *Science* **279,** 1528–1530.

Franz, T., Kothary, R., Surani, M. A., Halata, Z., and Grim, M. (1993). The Splotch mutation interferes with muscle development in the limbs. *Anat. Embryol.* **187,** 153–160.

Fuchs, E., and Segré, J. A. (2000). Stem cells: A new lease on life. *Cell (Cambridge, Mass.)* **100,** 143–155.

Fujisawa-Sehara, A., Nabeshima, Y., Hosoda, Y., and Obinata, T. (1990). Myogenin contains two domains conserved among myogenic factors. *J. Biol. Chem.* **265,** 15219–15223.

Gee, H. (1994). Vertebrate morphology. Return of the amphioxus. *Nature (London)* **370,** 504–505.

George-Weinstein, M., Gerhart, J., Reed, R., Flynn, J., Callihan, B., Mattiacci, M., Miehle, C., Foti, G., Lash, J. W., and Weintraub, H. (1996). Skeletal myogenesis: The preferred pathway of chick embryo epiblast cells in vitro. *Dev. Biol.* **173,** 279–291.

George-Weinstein, M., Gerhart, J., Blitz, J., Simak, E., and Knudsen, K. A. (1997). N-cadherin promotes the commitment and differentiation of skeletal muscle precursor cells. *Dev. Biol.* **185,** 14–24.

Godsave, S. F., and Slack, J. M. (1991). Single cell analysis of mesoderm formation in the Xenopus embryo. *Development (Cambridge, UK)* **111,** 523–530.

Goldhamer, D. J., Faerman, A., Shani, M., and Emerson, C. P., Jr. (1992). Regulatory elements that control the lineage-specific expression of myoD. *Science* **256,** 538–542.

Goldhamer, D. J., Brunk, B. P., Faerman, A., King, A., Shani, M., and Emerson, C. P., Jr. (1995). Embryonic activation of the myoD gene is regulated by a highly conserved distal control element. *Development (Cambridge, UK)* **121,** 637–649.

Goulding, M., Lumsden, A., and Paquette, A. J. (1994). Regulation of Pax-3 expression in the dermomyotome and its role in muscle development. *Development (Cambridge, UK)* **120,** 957–971.

Grass, S., Arnold, H. H., and Braun, T. (1996). Alterations in somite patterning of Myf-5-deficient mice: A possible role for FGF-4 and FGF-6. *Development (Cambridge, UK)* **122,** 141–150.

Grobet, L., Martin, L. J., Poncelet, D., Pirottin, D., Brouwers, B., Riquet, J., Schoeberlein, A., Dunner, S., Menissier, F., Massabanda, J., Fries, R., Hanset, R., and Georges, M. (1997). A deletion in the bovine myostatin gene causes the double-muscled phenotype in cattle. *Nat. Genet.* **17,** 71–74.

Gross, M. K., Moran-Rivard, L., Velasquez, T., Nakatsu, M. N., Jagla, K., and Goulding, M. (2000). Lbx1 is required for muscle precursor migra-

tion along a lateral pathway into the limb. *Development (Cambridge, UK)* **127,** 413–424.

Grounds, M. D., Garrett, K. L., and Beilharz, M. W. (1992). The transcription of MyoD1 and myogenin genes in thymic cells in vivo. *Exp. Cell Res.* **198,** 357–361.

Gundersen, K. (1998). Determination of muscle contractile properties: The importance of the nerve. *Acta Physiol. Scand.* **162,** 333–341.

Gurdon, J. B. (1988). A community effect in animal development. *Nature (London)* **336,** 772–774.

Gurdon, J. B., Lemaire, P., and Kato, K. (1993a). Community effects and related phenomena in development. *Cell (Cambridge, Mass.)* **75,** 831–834.

Gurdon, J. B., Kato, K., and Lemaire, P. (1993b). The community effect, dorsalization and mesoderm induction. *Curr. Opin. Genet. Dev.* **3,** 662–667.

Gussoni, E., Soneoka, Y., Strickland, C. D., Buzney, E. A., Khan, M. K., Flint, A. F., Kunkel, L. M., and Mulligan, R. C. (1999). Dystrophin expression in the mdx mouse restored by stem cell transplantation. *Nature (London)* **401,** 390–394.

Hannon, K., Smith, C. K. D., Bales, K. R., and Santerre, R. F. (1992). Temporal and quantitative analysis of myogenic regulatory and growth factor gene expression in the developing mouse embryo. *Dev. Biol.* **151,** 137–144.

Hasty, P., Bradley, A., Morris, J. H., Edmondson, D. G., Venuti, J. M., Olson, E. N., and Klein, W. H. (1993). Muscle deficiency and neonatal death in mice with a targeted mutation in the myogenin gene. *Nature (London)* **364,** 501–506.

Hauschka, S. D. (1996). The embryonic origin of muscle. *In* "Myology" (A. G. Engel and A. Franzini-Armstrong, eds.), pp. 3–73. McGraw-Hill.

Heanue, T. A., Reshef, R., Davis, R. J., Mardon, G., Oliver, G., Tomarev, S., Lassar, A. B., and Tabin, C. J. (1999). Synergistic regulation of vertebrate muscle development by Dach2, Eya2, and Six1, homologs of genes required for Drosophila eye formation. *Genes Dev.* **13,** 3231–3243.

Heymann, S., Koudrova, M., Arnold, H., Koster, M., and Braun, T. (1996). Regulation and function of SF/HGF during migration of limb muscle precursor cells in chicken. *Dev. Biol.* **180,** 566–578.

Hidaka, K., Yamamoto, I., Arai, Y., and Mukai, T. (1993). The MEF-3 motif is required for MEF-2-mediated skeletal muscle-specific induction of the rat aldolase A gene. *Mol. Cell. Biol.* **13,** 6469–6478.

Hinterberger, T. J., Sassoon, D. A., Rhodes, S. J., and Konieczny, S. F. (1991). Expression of the muscle regulatory factor MRF4 during somite and skeletal myofiber development. *Dev. Biol.* **147,** 144–156.

Hirsinger, E., Duprez, D., Jouve, C., Malapert, P., Cooke, J., and Pourquié, O. (1997). Noggin acts downstream of Wnt and Sonic Hedgehog to antagonize BMP4 in avian somite patterning. *Development (Cambridge, UK)* **124,** 4605–4614.

Holt, C. E., Lemaire, P., and Gurdon, J. B. (1994). Cadherin-mediated cell interactions are necessary for the activation of MyoD in Xenopus mesoderm. *Proc. Natl. Acad. Sci. U.S.A.* **91,** 10844–10848.

Holtzer, H., Schultheiss, T., Dilullo, C., Choi, J., Costa, M., Lu, M., and Holtzer, S. (1990). Autonomous expression of the differentiation programs of cells in the cardiac and skeletal myogenic lineages. *Ann. N.Y. Acad. Sci.* **599,** 158–169.

Houzelstein, D., Auda-Boucher, G., Cheraud, Y., Rouaud, T., Blanc, I., Tajbakhsh, S., Buckingham, M. E., Fontaine-Perus, J., and Robert, B. (1999). The homeobox gene Msx1 is expressed in a subset of somites, and in muscle progenitor cells migrating into the forelimb. *Development (Cambridge, UK)* **126,** 2689–2701.

Hughes, S. M., and Blau, H. M. (1992). Muscle fiber pattern is independent of cell lineage in postnatal rodent development. *Cell (Cambridge, Mass.)* **68,** 659–671.

Hughes, S. M., and Salinas, P. C. (1999a). Control of muscle fibre and motoneuron diversification. *Curr. Opin. Neurobiol.* **9,** 54–64.

Hughes, S. M., Taylor, J. M., Tapscott, S. J., Gurley, C. M., Carter, W. J., and Peterson, C. A. (1993). Selective accumulation of MyoD and myogenin mRNAs in fast and slow adult skeletal muscle is controlled by innervation and hormones. *Development (Cambridge, UK)* **118,** 1137–1147.

Hughes, S. M., Koishi, K., Rudnicki, M., and Maggs, A. M. (1997). MyoD protein is differentially accumulated in fast and slow skeletal muscle fibres and required for normal fibre type balance in rodents. *Mech. Dev.* **61,** 151–163.

Hughes, S. M., Chi, M. M., Lowry, O. H., and Gundersen, K. (1999b). Myogenin induces a shift of enzyme activity from glycolytic to oxidative metabolism in muscles of transgenic mice. *J. Cell Biol.* **145,** 633–642.

Ikeya, M., and Takada, S. (1998). Wnt signaling from the dorsal neural tube is required for the formation of the medial dermomyotome. *Development (Cambridge, UK)* **125,** 4969–4976.

Ingham, P. W. (1994). Pattern formation. Hedgehog points the way. *Curr. Biol.* **4,** 347–350.

Irintchev, A., Zeschnigk, M., Starzinski-Powitz, A., and Wernig, A. (1994). Expression pattern of M-cadherin in normal, denervated, and regenerating mouse muscles. *Dev. Dyn.* **199,** 326–337.

Jackson, K. A., Mi, T., and Goodell, M. A. (1999). Hematopoietic potential of stem cells isolated from murine skeletal muscle. *Proc. Natl. Acad. Sci. U.S.A.* **96,** 14482–14486.

Johnson, R. L., Laufer, E., Riddle, R. D., and Tabin, C. (1994). Ectopic expression of Sonic hedgehog alters dorsal-ventral patterning of somites. *Cell (Cambridge, Mass.)* **79,** 1165–1173.

Kablar, B., Krastel, K., Ying, C., Asakura, A., Tapscott, S. J., and Rudnicki, M. A. (1997). MyoD and Myf-5 differentially regulate the development of limb versus trunk skeletal muscle. *Development (Cambridge, UK)* **124,** 4729–4738.

Kablar, B., Asakura, A., Krastel, K., Ying, C., May, L. L., Goldhamer, D. J., and Rudnicki, M. A. (1998). MyoD and Myf-5 define the specification of musculature of distinct embryonic origin. *Biochem. Cell. Biol.* **76,** 1079–1091.

Kablar, B., Krastel, K., Ying, C., Tapscott, S. J., Goldhamer, D. J., and Rudnicki, M. A. (1999). Myogenic determination occurs independently in somites and limb buds. *Dev. Biol.* **206,** 219–231.

Kablar, B., Tajbakhsh, S., and Rudnicki, M. A. (2000). Transdifferentiation of esophageal smooth to skeletal muscle is myogenic bHLH factor-dependent. *Development (Cambridge, UK)* **127,** 1627–1639.

Kahane, N., Cinnamon, Y., and Kalcheim, C. (1998a). The origin and fate of pioneer myotomal cells in the avian embryo. *Mech. Dev.* **74,** 59–73.

Kahane, N., Cinnamon, Y., and Kalcheim, C. (1998b). The cellular mechanism by which the dermomyotome contributes to the second wave of myotome development. *Development (Cambridge, UK)* **125,** 4259–4271.

Kalcheim, C., Cinnamon, Y., and Kahane, N. (1999). Myotome formation: A multistage process. *Cell Tissue Res.* **296,** 161–173.

Katagiri, T., Yamaguchi, A., Komaki, M., Abe, E., Takahashi, N., Ikeda, T., Rosen, V., Wozney, J. M., Fujisawa-Sehara, A., and Suda, T. (1994). Bone morphogenetic protein-2 converts the differentiation pathway of C2C12 myoblasts into the osteoblast lineage. *J. Cell Biol.* **127,** 1755–1766; erratum: *Ibid.*

Kato, K., and Gurdon, J. B. (1993). Single-cell transplantation determines the time when Xenopus muscle precursor cells acquire a capacity for autonomous differentiation. *Proc. Natl. Acad. Sci. U.S.A.* **90,** 1310–1314.

Kaufman, M. H. (1992). "The Atlas of Mouse Development." Academic Press, San Diego.

Kopan, R., and Cagan, R. (1997). Notch on the cutting edge. *Trends Genet.* **13,** 465–467.

Kopan, R., Nye, J. S., and Weintraub, H. (1994). The intracellular domain of mouse Notch: A constitutively activated repressor of myogenesis directed at the basic helix-loop-helix region of MyoD. *Development (Cambridge, UK)* **120,** 2385–2396.

Lassar, A. B., Davis, R. L., Wright, W. E., Kadesch, T., Murre, C., Voron-

ova, A., Baltimore, D., and Weintraub, H. (1991). Functional activity of myogenic HLH proteins requires hetero- oligomerization with E12/E47-like proteins in vivo. *Cell (Cambridge, Mass.)* **66,** 305–315.

Lee, S. J., and McPherron, A. C. (1999). Myostatin and the control of skeletal muscle mass. *Curr. Opin. Genet. Dev.* **9,** 604–607.

Lin, C. H., Hudson, A. J., and Strickland, K. P. (1969). Fatty acid metabolism in dystrophic muscle *in vitro. Life Sci.* **8,** 21–26.

Lu, B. D., Allen, D. L., Leinwand, L. A., and Lyons, G. E. (1999). Spatial and temporal changes in myosin heavy chain gene expression in skeletal muscle development. *Dev. Biol.* **216,** 312–326.

Luo, K., Stroschein, S. L., Wang, W., Chen, D., Martens, E., Zhou, S., and Zhou, Q. (1999). The Ski oncoprotein interacts with the Smad proteins to repress TGFbeta signaling. *Genes Dev.* **13,** 2196–2206.

Lyons, G. E., and Buckingham, M. E. (1992). Developmental regulation of myogenesis in mouse. *Semin. Dev. Biol.* **3,** 243–253.

Lyons, G. E., Ontell, M., Cox, R., Sassoon, D., and Buckingham, M. (1990). The expression of myosin genes in developing skeletal muscle in the mouse embryo. *J. Cell Biol.* **111,** 1465–1476.

Maina, F., Casagranda, F., Audero, E., Simeone, A., Comoglio, P. M., Klein, R., and Ponzetto, C. (1996). Uncoupling of Grb2 from the Met receptor in vivo reveals complex roles in muscle development. *Cell (Cambridge, Mass.)* **87,** 531–542.

Mankoo, B. S., Collins, N. S., Ashby, P., Grigorieva, E., Pevny, L. H., Candia, A., Wright, C. V., Rigby, P. W., and Pachnis, V. (1999). Mox2 is a component of the genetic hierarchy controlling limb muscle development. *Nature (London)* **400,** 69–73.

Marcelle, C., Stark, M. R., and Bronner-Fraser, M. (1997). Coordinate actions of BMPs, Wnts, Shh and noggin mediate patterning of the dorsal somite. *Development (Cambridge, UK)* **124,** 3955–3963.

Marigo, V., Johnson, R. L., Vortkamp, A., and Tabin, C. J. (1996a). Sonic hedgehog differentially regulates expression of GLI and GLI3 during limb development. *Dev. Biol.* **180,** 273–283.

Maroto, M., Reshef, R., Munsterberg, A. E., Koester, S., Goulding, M., and Lassar, A. B. (1997). Ectopic Pax-3 activates MyoD and Myf-5 expression in embryonic mesoderm and neural tissue. *Cell (Cambridge, Mass.)* **89,** 139–148.

McMahon, J. A., Takada, S., Zimmerman, L. B., Fan, C. M., Harland, R. M., and McMahon, A. P. (1998). Noggin-mediated antagonism of BMP signaling is required for growth and patterning of the neural tube and somite. *Genes Dev.* **12,** 1438–1452.

McPherron, A. C., and Lee, S. J. (1997). Double muscling in cattle due to mutations in the myostatin gene. *Proc. Natl. Acad. Sci. U.S.A.* **94,** 12457–12461.

McPherron, A. C., Lawler, A. M., and Lee, S. J. (1997). Regulation of skeletal muscle mass in mice by a new TGF-beta superfamily member. *Nature (London)* **387,** 83–90.

Megeney, L. A., and Rudnicki, M. A. (1995). Determination versus differentiation and the MyoD family of transcription factors. *Biochem. Cell. Biol.* **73,** 723–732.

Megeney, L. A., Kablar, B., Garrett, K., Anderson, J. E., and Rudnicki, M. A. (1996). MyoD is required for myogenic stem cell function in adult skeletal muscle. *Genes Dev.* **10,** 1173–1183.

Mennerich, D., Schafer, K., and Braun, T. (1998). Pax-3 is necessary but not sufficient for lbx1 expression in myogenic precursor cells of the limb. *Mech. Dev.* **73,** 147–158.

Merlino, G., and Helman, L. J. (1999). Rhabdomyosarcoma—working out the pathways. *Oncogene* **18,** 5340–5348.

Miller, J. B., and Stockdale, F. E. (1986a). Developmental regulation of the multiple myogenic cell lineages of the avian embryo. *J. Cell Biol.* **103,** 2197–2208.

Miller, J. B., and Stockdale, F. E. (1986b). Developmental origins of skeletal muscle fibers: Clonal analysis of myogenic cell lineages based on expression of fast and slow myosin heavy chains. *Proc. Natl. Acad. Sci. U.S.A.* **83,** 3860–3864.

Mohun, T. J., Chambers, A. E., Towers, N., and Taylor, M. V. (1991). Expression of genes encoding the transcription factor SRF during early development of *Xenopus laevis:* Identification of a CArG box-binding activity as SRF. *EMBO J.* **10,** 933–940.

Molkentin, J. D., and Olson, E. N. (1996). Defining the regulatory networks for muscle development. *Curr. Opin. Genet. Dev.* **6,** 445–453.

Munsterberg, A. E., and Lassar, A. B. (1995). Combinatorial signals from the neural tube, floor plate and notochord induce myogenic bHLH gene expression in the somite. *Development (Cambridge, UK)* **121,** 651–660.

Munsterberg, A. E., Kitajewski, J., Bumcrot, D. A., McMahon, A. P., and Lassar, A. B. (1995). Combinatorial signaling by Sonic hedgehog and Wnt family members induces myogenic bHLH gene expression in the somite. *Genes Dev.* **9,** 2911–2922.

Murre, C., McCaw, P. S., and Baltimore, D. (1989a). A new DNA binding and dimerization motif in immunoglobulin enhancer binding, daughterless, MyoD, and myc proteins. *Cell (Cambridge, Mass.)* **56,** 777–783.

Murre, C., McCaw, P. S., Vaessin, H., Caudy, M., Jan, L. Y., Jan, Y. N., Cabrera, C. V., Buskin, J. N., Hauschka, S. D., Lassar, A. B. *et al.* (1989b). Interactions between heterologous helix-loop-helix proteins generate complexes that bind specifically to a common DNA sequence. *Cell (Cambridge, Mass.)* **58,** 537–544.

Nabeshima, Y., Hanaoka, K., Hayasaka, M., Esumi, E., Li, S., and Nonaka, I. (1993). Myogenin gene disruption results in perinatal lethality because of severe muscle defect. *Nature (London)* **364,** 532–535.

Neyt, C., Jagla, K., Thisse, C., Thisse, B., Haines, L., and Currie, P. D. (2000). Evolutionary origins of vertebrate appendicular muscle. *Nature (London)* **408,** 82–86.

Noden, D. M. (1991). Vertebrate craniofacial development: The relation between ontogenetic process and morphological outcome. *Brain, Behav. Evol.* **38,** 190–225.

Norris, W., Neyt, C., Ingham, P. W., and Currie, P. D. (2000). Slow muscle induction by Hedgehog signalling in vitro. *J. Cell Sci.* **113,** 2695–2703.

Olson, E. N., and Williams, R. S. (2000). Remodeling muscles with calcineurin. *BioEssays* **22,** 510–519.

Olson, E. N., Perry, M., and Schulz, R. A. (1995). Regulation of muscle differentiation by the MEF2 family of MADS box transcription factors. *Dev. Biol.* **172,** 2–14.

Olson, E. N., Arnold, H. H., Rigby, P. W., and Wold, B. J. (1996). Know your neighbors: Three phenotypes in null mutants of the myogenic bHLH gene MRF4. *Cell (Cambridge, Mass.)* **85,** 1–4.

Ordahl, C. P. (1999). Myogenic shape-shifters. *J. Cell Biol.* **147,** 695–698.

Ordahl, C. P., and Le Douarin, N. M. (1992). Two myogenic lineages within the developing somite. *Development (Cambridge, UK)* **114,** 339–353.

Ordahl, C. P., and Williams, B. A. (1998). Knowing chops from chuck: Roasting myoD redundancy. *BioEssays* **20,** 357–362.

Orkin, S. H. (2000). Stem cell alchemy. *Nat. Med.* **6,** 1212–1213.

Ott, M. O., Bober, E., Lyons, G., Arnold, H., and Buckingham, M. (1991). Early expression of the myogenic regulatory gene, myf-5, in precursor cells of skeletal muscle in the mouse embryo. *Development (Cambridge, UK)* **111,** 1097–1107.

Patapoutian, A., Yoon, J. K., Miner, J. H., Wang, S., Stark, K., and Wold, B. (1995). Disruption of the mouse MRF4 gene identifies multiple waves of myogenesis in the myotome. *Development (Cambridge, UK)* **121,** 3347–3358.

Pourquié, O., Coltey, M., Teillet, M. A., Ordahl, C., and Le Douarin, N. M. (1993). Control of dorsoventral patterning of somitic derivatives by notochord and floor plate. *Proc. Natl. Acad. Sci. U.S.A.* **90,** 5242–5246.

Pourquié, O., Coltey, M., Bréant, C., and Le Douarin, N. M. (1995). Control of somite patterning by signals from the lateral plate. *Proc. Natl. Acad. Sci. U.S.A.* **92,** 3219–3223.

Pourquié, O., Fan, C. M., Coltey, M., Hirsinger, E., Watanabe, Y., Bréant, C., Francis-West, P., Brickell, P., Tessier-Lavigne, M., and Le Douarin, N. M. (1996). Lateral and axial signals involved in avian somite patterning: A role for BMP4. *Cell (Cambridge, Mass.)* **84,** 461–471.

Pownall, M. E., Strunk, K. E., and Emerson, C. P., Jr. (1996). Notochord

signals control the transcriptional cascade of myogenic bHLH genes in somites of quail embryos. *Development (Cambridge, UK)* **122,** 1475–1488.

Rawls, A., Morris, J. H., Rudnicki, M., Braun, T., Arnold, H. H., Klein, W. H., and Olson, E. N. (1995). Myogenin's functions do not overlap with those of MyoD or Myf-5 during mouse embryogenesis. *Dev. Biol.* **172,** 37–50.

Rawls, A., Valdez, M. R., Zhang, W., Richardson, J., Klein, W. H., and Olson, E. N. (1998). Overlapping functions of the myogenic bHLH genes MRF4 and MyoD revealed in double mutant mice. *Development (Cambridge, UK)* **125,** 2349–2358.

Relaix, F., and Buckingham, M. (1999). From insect eye to vertebrate muscle: Redeployment of a regulatory network. *Genes Dev.* **13,** 3171–3178.

Reshef, R., Maroto, M., and Lassar, A. B. (1998). Regulation of dorsal somitic cell fates: BMPs and Noggin control the timing and pattern of myogenic regulator expression. *Genes Dev.* **12,** 290–303.

Robson, L. G., and Hughes, S. M. (1996). The distal limb environment regulates MyoD accumulation and muscle differentiation in mouse-chick chimaeric limbs. *Development (Cambridge, UK)* **122,** 3899–3910.

Robson, L. G., and Hughes, S. M. (1999). Local signals in the chick limb bud can override myoblast lineage commitment: Induction of slow myosin heavy chain in fast myoblasts. *Mech. Dev.* **85,** 59–71.

Ross, S. E., Hemati, N., Longo, K. A., Bennett, C. N., Lucas, P. C., Erickson, R. L., and MacDougald, O. A. (2000). Inhibition of adipogenesis by Wnt signaling. *Science* **289,** 950–953.

Rudnicki, M. A., Braun, T., Hinuma, S., and Jaenisch, R. (1992). Inactivation of MyoD in mice leads to up-regulation of the myogenic HLH gene Myf-5 and results in apparently normal muscle development. *Cell (Cambridge, Mass.)* **71,** 383–390.

Rudnicki, M. A., Schnegelsberg, P. N., Stead, R. H., Braun, T., Arnold, H. H., and Jaenisch, R. (1993). MyoD or Myf-5 is required for the formation of skeletal muscle. *Cell (Cambridge, Mass.)* **75,** 1351–1359.

Sabourin, L. A., Girgis-Gabardo, A., Seale, P., Asakura, A., and Rudnicki, M. A. (1999). Reduced differentiation potential of primary MyoD–/– myogenic cells derived from adult skeletal muscle. *J. Cell Biol.* **144,** 631–643.

Salminen, M., Lopez, S., Maire, P., Kahn, A., and Daegelen, D. (1996). Fast-muscle-specific DNA-protein interactions occurring in vivo at the human aldolase A M promoter are necessary for correct promoter activity in transgenic mice. *Mol. Cell. Biol.* **16,** 76–85.

Sassoon, D., Lyons, G., Wright, W. E., Lin, V., Lassar, A., Weintraub, H., and Buckingham, M. (1989). Expression of two myogenic regulatory factors myogenin and MyoD1 during mouse embryogenesis. *Nature (London)* **341,** 303–307.

Scaal, M., Bonafede, A., Dathe, V., Sachs, M., Cann, G., Christ, B., and Brand-Saberi, B. (1999). SF/HGF is a mediator between limb patterning and muscle development. *Development (Cambridge, UK)* **126,** 4885–4893.

Schafer, K., and Braun, T. (1999). Early specification of limb muscle precursor cells by the homeobox gene Lbx1h. *Nat. Genet.* **23,** 213–216.

Seale, P., and Rudnicki, M. A. (2000a). A new look at the origin, function, and "Stem-Cell" status of muscle satellite cells. *Dev. Biol.* **218,** 115–124.

Seale, P., Sabourin, L. A., Girgis-Gabardo, A., Mansouri, A., Gruss, P., and Rudnicki, M. A. (2000). Pax7 is required for the specification of myogenic satellite cells. *Cell (Cambridge, Mass.)* **102,** 777–786.

Seed, J., and Hauschka, S. D. (1984). Temporal separation of the migration of distinct myogenic precursor populations into the developing chick wing bud. *Dev. Biol.* **106,** 389–393.

Shafritz, A. B., Shore, E. M., Gannon, F. H., Zasloff, M. A., Taub, R., Muenke, M., and Kaplan, F. S. (1996). Overexpression of an osteogenic morphogen in fibrodysplasia ossificans progressiva. *N. Engl. J. Med.* **335,** 555–561.

Shawber, C., Nofziger, D., Hsieh, J. J., Lindsell, C., Bogler, O., Hayward, D., and Weinmaster, G. (1996). Notch signaling inhibits muscle cell differentiation through a CBF1-independent pathway. *Development (Cambridge, UK)* **122**, 3765–3773.

Smith, C. K., 2nd, Janney, M. J., and Allen, R. E. (1994). Temporal expression of myogenic regulatory genes during activation, proliferation, and differentiation of rat skeletal muscle satellite cells. *J. Cell. Physiol.* **159**, 379–385.

Smith, T. H., Kachinsky, A. M., and Miller, J. B. (1994). Somite subdomains, muscle cell origins, and the four muscle regulatory factor proteins. *J. Cell Biol.* **127**, 95–105.

Spitz, F., Salminen, M., Demignon, J., Kahn, A., Daegelen, D., and Maire, P. (1997). A combination of MEF3 and NFI proteins activates transcription in a subset of fast-twitch muscles. *Mol. Cell. Biol.* **17**, 656–666.

Spitz, F., Demignon, J., Porteu, A., Kahn, A., Concordet, J. P., Daegelen, D., and Maire, P. (1998). Expression of myogenin during embryogenesis is controlled by Six/sine oculis homeoproteins through a conserved MEF3 binding site. *Proc. Natl. Acad. Sci. U.S.A.* **95**, 14220–14225.

Sporle, R., Gunther, T., Struwe, M., and Schughart, K. (1996). Severe defects in the formation of epaxial musculature in open brain (opb) mutant mouse embryos. *Development (Cambridge, UK)* **122**, 79–86.

Stern, H. M., and Hauschka, S. D. (1995). Neural tube and notochord promote in vitro myogenesis in single somite explants. *Dev. Biol.* **167**, 87–103.

Stern, H. M., Brown, A. M., and Hauschka, S. D. (1995). Myogenesis in paraxial mesoderm: Preferential induction by dorsal neural tube and by cells expressing Wnt-1. *Development (Cambridge, UK)* **121**, 3675–3686.

Stewart, A. F., Larkin, S. B., Farrance, I. K., Mar, J. H., Hall, D. E., and Ordahl, C. P. (1994). Muscle-enriched TEF-1 isoforms bind M-CAT elements from muscle-specific promoters and differentially activate transcription. *J. Biol. Chem.* **269**, 3147–3150.

Stockdale, F. E. (1992). Myogenic cell lineages. *Dev. Biol.* **154**, 284–298.

Subramanian, S. V., and Nadal-Ginard, B. (1996). Early expression of the different isoforms of the myocyte enhancer factor-2 (MEF2) protein in myogenic as well as nonmyogenic cell lineages during mouse embryogenesis. *Mech. Dev.* **57**, 103–112.

Tajbakhsh, S., Vivarelli, E., Cusella-De Angelis, G., Rocancourt, D., Buckingham, M., and Cossu, G. (1994). A population of myogenic cells derived from the mouse neural tube. *Neuron* **13**, 813–821.

Tajbakhsh, S., Rocancourt, D., Cossu, G., and Buckingham, M. (1997). Redefining the genetic hierarchies controlling skeletal myogenesis: Pax-3 and Myf-5 act upstream of MyoD. *Cell (Cambridge, Mass.)* **89**, 127–138.

Tajbakhsh, S., Borello, U., Vivarelli, E., Kelly, R., Papkoff, J., Duprez, D., Buckingham, M., and Cossu, G. (1998). Differential activation of Myf5 and MyoD by different Wnts in explants of mouse paraxial mesoderm and the later activation of myogenesis in the absence of Myf5. *Development (Cambridge, UK)* **125**, 4155–4162.

Talmadge, R. J. (2000). Myosin heavy chain isoform expression following reduced neuromuscular activity: Potential regulatory mechanisms. *Muscle Nerve* **23**, 661–679.

Tapscott, S. J., Davis, R. L., Thayer, M. J., Cheng, P. F., Weintraub, H., and Lassar, A. B. (1988). MyoD1: A nuclear phosphoprotein requiring a Myc homology region to convert fibroblasts to myoblasts. *Science* **242**, 405–411.

Teboul, L., Gaillard, D., Staccini, L., Inadera, H., Amri, E. Z., and Grimaldi, P. A. (1995). Thiazolidinediones and fatty acids convert myogenic cells into adipose-like cells. *J. Biol. Chem.* **270**, 28183–28187.

Teillet, M., Watanabe, Y., Jeffs, P., Duprez, D., Lapointe, F., and Le Douarin, N. M. (1998). Sonic hedgehog is required for survival of both myogenic and chondrogenic somitic lineages. *Development (Cambridge, UK)* **125**, 2019–2030.

Trainor, P. A., Tan, S. S., and Tam, P. P. (1994). Cranial paraxial mesoderm: Regionalisation of cell fate and impact on craniofacial development in mouse embryos. *Development (Cambridge, UK)* **120**, 2397–2408.

Tremblay, P., Dietrich, S., Mericskay, M., Schubert, F. R., Li, Z., and Paulin, D. (1998). A crucial role for Pax3 in the development of the hypaxial musculature and the long-range migration of muscle precursors. *Dev. Biol.* **203**, 49–61.

Urist, M. R., and Strates, B. S. (1971). Bone morphogenetic protein. *J. Dent. Res.* **50**, 1392–1406.

Valdez, M. R., Richardson, J. A., Klein, W. H., and Olson, E. N. (2000). Failure of Myf5 to support myogenic differentiation without myogenin, MyoD, and MRF4. *Dev. Biol.* **219**, 287–298.

Van Swearingen, J., and Lance-Jones, C. (1995). Slow and fast muscle fibers are preferentially derived from myoblasts migrating into the chick limb bud at different developmental times. *Dev. Biol.* **170**, 321–337.

Venuti, J. M., Morris, J. H., Vivian, J. L., Olson, E. N., and Klein, W. H. (1995). Myogenin is required for late but not early aspects of myogenesis during mouse development. *J. Cell Biol.* **128**, 563–576.

Wada, H., Holland, P. W., Sato, S., Yamamoto, H., and Satoh, N. (1997). Neural tube is partially dorsalized by overexpression of HrPax-37: The ascidian homologue of Pax-3 and Pax-7. *Dev. Biol.* **187**, 240–252.

Wakkach, A., Poea, S., Chastre, E., Gespach, C., Lecerf, F., De La Porte, S., Tzartos, S., Coulombe, A., and Berrih-Aknin, S. (1999). Establishment of a human thymic myoid cell line. Phenotypic and functional characteristics. *Am. J. Pathol.* **155**, 1229–1240.

Weintraub, H. (1993). The MyoD family and myogenesis: Redundancy, networks, and thresholds. *Cell (Cambridge, Mass.)* **75**, 1241–1244.

Weintraub, H., Davis, R., Tapscott, S., Thayer, M., Krause, M., Benezra, R., Blackwell, T. K., Turner, D., Rupp, R., Hollenberg, S. *et al.* (1991a). The myoD gene family: Nodal point during specification of the muscle cell lineage. *Science* **251**, 761–766.

Weintraub, H., Dwarki, V. J., Verma, I., Davis, R., Hollenberg, S., Snider, L., Lassar, A., and Tapscott, S. J. (1991b). Muscle-specific transcriptional activation by MyoD. *Genes Dev.* **5**, 1377–1386.

Williams, B. A., and Ordahl, C. P. (1994). Pax-3 expression in segmental mesoderm marks early stages in myogenic cell specification. *Development (Cambridge, UK)* **120**, 785–796.

Williams, B. A., and Ordahl, C. P. (1997). Emergence of determined myotome precursor cells in the somite. *Development (Cambridge, UK)* **124**, 4983–4997.

Wilson-Rawls, J., Hurt, C. R., Parsons, S. M., and Rawls, A. (1999). Differential regulation of epaxial and hypaxial muscle development by paraxis. *Development (Cambridge, UK)* **126**, 5217–5229.

Wilting, J., Brand-Saberi, B., Huang, R., Zhi, Q., Kontges, G., Ordahl, C. P., and Christ, B. (1995). Angiogenic potential of the avian somite. *Dev. Dyn.* **202**, 165–171.

Woloshin, P., Song, K., Degnin, C., Killary, A. M., Goldhamer, D. J., Sassoon, D., and Thayer, M. J. (1995). MSX1 inhibits myoD expression in fibroblast x 10T1/2 cell hybrids. *Cell (Cambridge, Mass.)* **82**, 611–620.

Wong, A., Garrett, K. L., and Anderson, J. E. (1999). Myoid cell density in the thymus is reduced during mdx dystrophy and after muscle crush. *Biochem. Cell. Biol.* **77**, 33–40.

Wu, H., Naya, F. J., McKinsey, T. A., Mercer, B., Shelton, J. M., Chin, E. R., Simard, A. R., Michel, R. N., Bassel-Duby, R., Olson, E. N., and Williams, R. S. (2000). MEF2 responds to multiple calcium-regulated signals in the control of skeletal muscle fiber type. *EMBO J.* **19**, 1963–1973.

Xue, X. J., and Xue, Z. G. (1996). Spatial and temporal effects of axial structures on myogenesis of developing somites. *Mech. Dev.* **60**, 73–82.

Yablonka-Reuveni, Z., and Rivera, A. J. (1994). Temporal expression of regulatory and structural muscle proteins during myogenesis of satellite cells on isolated adult rat fibers. *Dev. Biol.* **164**, 588–603.

Yang, X. M., Vogan, K., Gros, P., and Park, M. (1996). Expression of the met receptor tyrosine kinase in muscle progenitor cells in somites and limbs is absent in Splotch mice. *Development (Cambridge, UK)* **122**, 2163–2171.

Yoon, J. K., Olson, E. N., Arnold, H. H., and Wold, B. J. (1997). Different MRF4 knockout alleles differentially disrupt Myf-5 expression: Cis-regulatory interactions at the MRF4/Myf-5 locus. *Dev. Biol.* **188,** 349–362.

Yu, Y. T., Breitbart, R. E., Smoot, L. B., Lee, Y., Mahdavi, V., and Nadal-Ginard, B. (1992). Human myocyte-specific enhancer factor 2 comprises a group of tissue-restricted MADS box transcription factors. *Genes Dev.* **6,** 1783–1798.

Zeschnigk, M., Kozian, D., Kuch, C., Schmoll, M., and Starzinski-Powitz, A. (1995). Involvement of M-cadherin in terminal differentiation of skeletal muscle cells. *J. Cell Sci.* **108,** 2973–2981.

Zhang, W., Behringer, R. R., and Olson, E. N. (1995). Inactivation of the myogenic bHLH gene MRF4 results in up-regulation of myogenin and rib anomalies. *Genes Dev.* **9,** 1388–1399.

14

Deconstructing the Molecular Biology of Cartilage and Bone Formation

Benoit de Crombrugghe, Véronique Lefebvre,[1] and Kazuhisa Nakashima

Department of Molecular Genetics, The University of Texas M.D. Anderson Cancer Center, Houston, Texas 77030

I. Introduction

II. Sox Transcription Factors: Essential Roles in the Chrondrocyte Differentiation Program

III. Parathyroid Hormone-Related Peptide (PTHrP) and Parathyroid Hormone (PTH)/ PTHrP Receptor: Gatekeepers of the Zone of Hypertrophic Chondrocytes

IV. FGFs and FGF Receptor 3: Counterintuitive Inhibitors of Chondrocyte Proliferation

V. Ihh: A Central Coordinator of Endochondral Bone Formation

VI. The Two Roles of the Transcription Factor Cbfa1 in Endochondral Bone Formation

VII. Other Transcription Factors Involved in Bone Formation

VIII. Gelatinase B and Vascular Endothelial Growth Factor: Additional Coordinators of Endochondral Bone Formation

IX. Conclusion

References

I. Introduction

More than 95% of the skeleton in most vertebrates is formed by endochondral ossification, that is, through a cartilage intermediate. The other bones, mainly some craniofacial bones, are formed by intramembranous ossification, a process in which bones form directly from condensations of mesenchymal cells without a cartilage intermediate. Whereas chondrocytes are the cells in cartilage, bones have both bone-forming osteoblasts, which arise from mesenchymal cells, and bone remodeling osteoclasts, which arise from hematopoietic cells. These two cell types coexist and reach a functional steady state characterized by an equilibrium between bone synthesis and bone destruction. It has been proposed that chondrocytes and osteoblasts arise from a common progenitor that has the potential to differentiate into either cell type (Fang and Hall, 1997). Some cartilage is not replaced by bone and persists as permanent cartilage into adulthood such as in the nose, ears, and throat and also at the

---

[1] Present address: Department of Biomedical Engineering, Lerner Research Institute, Cleveland Clinic Foundation, Cleveland, Ohio 44195

articular surfaces, which line the extremities of many bones. Why chondrocytes in these cartilages do not develop into bones is not understood.

Both endochondral and intramembranous ossification are initiated by signals that provide patterning information to the mesenchyme to generate the primordia of individual skeletal elements. These molecular signals outline the three-dimensional coordinates that determine the shape of mesenchymal condensations (Johnson and Tabin, 1997; Tickle, 1995). In mesenchymal condensations of membranous skeletal elements, cells differentiate into osteoblasts, whereas in endochondral skeletal elements cells in condensations differentiate into chondrocytes to form the cartilages that will later be replaced by bones. Among the molecules that pattern skeletal elements are secreted polypeptides of the bone morphogenetic protein (BMP) family, transforming growth factor β (TGF-β) superfamily, Wnt family, and Hedgehog family, and fibroblast growth factors (FGFs) (DeLise *et al.,* 2000; Erlebacher *et al.,* 1995), as well as transcription factors of the Pax, Hox, homeodomain-containing, basic helix–

loop–helix, and forkhead families (Karsenty, 1999; Olsen *et al.,* 2000).

In the multistep process of endochondral bone development (Fig. 1) after cells in mesenchymal condensations differentiate into chondrocytes and synthesize abundant amounts of cartilage-specific extracellular matrix (ECM) proteins, these cells sustain a series of additional changes as they form the growth plate of endochondral bones (Erlebacher *et al.,* 1995). Chondrocytes first flatten and undergo a unidirectional proliferation, resulting in parallel stacks of dividing cells. This process is largely responsible for the longitudinal growth of bones. Cells at the bottom of these stacks then exit the cell cycle and change their genetic program in successive steps to become prehypertrophic and then hypertrophic. The ECM around the most differentiated hypertrophic chondrocytes becomes mineralized before they undergo apoptosis and are replaced by bone cells. It is this characteristic cellular organization that gives the growth plate its unique aspects (Fig. 1). Several secreted polypeptides have key roles in these processes. In addition, components of the

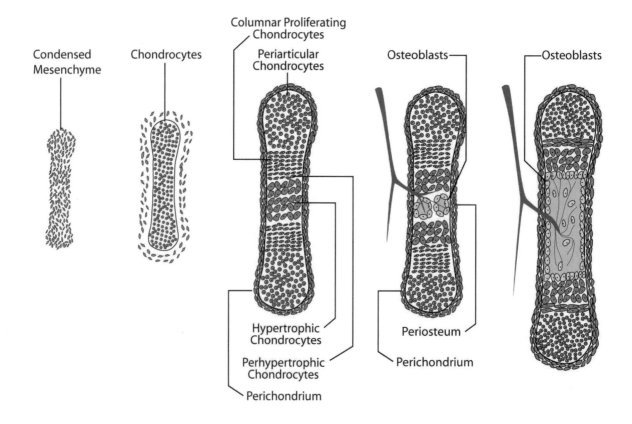

Figure 1 The steps involved in endochondral bone formation.

cartilage ECM play crucial roles in either modulating or maintaining the correct spatial and temporal differentiation of chondrocytes in the growth plate (Mundlos and Olsen, 1997). When the growth plate is being established, the perichondrium, a specialized structure consisting of thin layers of mesenchymal cells surrounding cartilage, forms. In the centers of skeletal elements, where the perichondrium flanks the zone of hypertrophic chondrocytes, perichondrial cells differentiate into osteoblasts. These osteoblasts invade the mineralized cartilage matrix of the hypertrophic zone together with blood vessels and osteoclasts or a subpopulation of these cells called chondroclasts (Vu et al., 1998). These cells, which derive from monocytes/macrophages, degrade the mineralized cartilage matrix, and the remnants of this matrix provide a scaffold for the deposition of a bone-specific matrix by osteoblasts. At the same time osteoblasts in the periosteum produce a similar matrix to form bone collars. Therefore, the overall strategy of endochondral bone formation is that chondrocytes generate a dynamic cartilage skeleton containing vertical columns of proliferating cells, which after exiting the cell cycle, undergo hypertrophy and ultimately die while the cartilage matrix is being degraded. Using the remnants of the degraded cartilage matrix as scaffold, invading osteoblasts deposit a new, bone-specific matrix. Thus, in endochondral bone formation two distinct differentiation pathways, chondrocyte and osteoblast differentiation, interconnect and must be coordinated.

Although intramembranous ossification appears to be very different from endochondral ossification, osteoblasts in the membranous skeleton are indistinguishable from those in skeletal elements formed by endochondral ossification, suggesting that common factors control the differentiation of osteoblasts in the two modes of bone formation.

This chapter focuses on transcription factors and secreted molecules that control the differentiation of chondrocytes and osteoblasts and that link and coordinate these two pathways in endochondral bone formation. Our discussion is largely based on the results of genetic studies in humans and mice. Due to space limitations many other essential aspects of bone formation, including the differentiation of osteoclasts (Suda et al., 2001), the role of the multiple patterning factors that control the shape, size, and number of skeletal elements (Johnson and Fabin, 1997), the formation of joints, and the roles that ECM components have in endochondral bone formation (Olsen et al., 2000), are not covered here.

II. Sox Transcription Factors: Essential Roles in the Chrondrocyte Differentiation Program

Both the skeletal anomalies in the human genetic disease campomelic dysplasia (CD), which is caused by heterozy-gous mutations in the transcription factor SOX9, and the pattern of expression of Sox9 during chondrogenesis in mouse embryos suggest that Sox9 has a major role in chondrocyte differentiation. CD is characterized by hypoplasia of most endochondral skeletal elements and is due to haploinsufficiency of SOX9 (Foster et al., 1994; Meyer et al., 1997; Wagner et al., 1994). The Sox9 gene is expressed in all chondroprogenitor cells in mouse embryos and in chondrocytes not in hypertrophic chondrocytes. Consistent with the non-skeletal symptoms associated with a proportion of CDs, such as XY sex reversal and heart, kidney, and other anomalies, Sox9 is also expressed in the primitive gonads of both males and females and later only in male gonads, and in heart, kidney neuronal tissues, and pancreas (Ng et al., 1997; Wright et al., 1995; Zhao et al., 1997). Thus, Sox9 expression is not restricted to cells of the chondrocyte lineage.

Members of the Sox family of transcription factors have a high-mobility group (HMG) box DNA binding domain whose sequence displays at least 50% identity with that of the sex-determining factor SRY. The Sox family is divided into several subgroups based on sequence identities both within and outside the HMG box domain (Wegner, 1999). Like other Sox proteins, Sox9 binds to a specific sequence in the minor groove of DNA. Sox9 also contains a potent transcription activation domain located at its carboxyl end (Fig. 2).

The creation of a Sox9-null allele in embryonic stem (ES) cells led to the generation of heterozygous Sox9 mice with the skeletal anomalies of CD, including perinatal death, hypoplasia of endochondral bones, bending and angulation of long bones, and cleft palate (Bi et al., 2001). Analysis of the abnormal phenotypes of these mutants during embryonic development strongly suggested that two steps in chondrocyte differentiation are sensitive to Sox9 dosage. One is mesenchymal condensation: Condensations are smaller and delayed, producing the hypoplasia of cartilages and endochondral bones characteristic of CD. In addition, because the zones of hypertrophic chondrocytes are larger in mutant embryos than in wild-type embryos and because premature mineralization occurs in a number of bones in mutants, Sox9 may also have a distinct role in regulating the rate at which chondrocytes differentiate into hypertrophic chondrocytes.

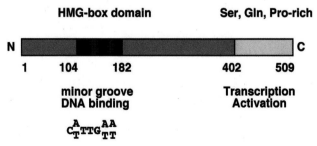

Figure 2 Structure of Sox9. The DNA-binding domain consists of an HMG box that binds to a specific sequence in the minor groove of DNA.

Expression of *Sox9* is turned off in all hypertrophic chondrocytes and SOX9 protein is no longer detectable in these cells; hence, the premature mineralization must be a consequence of the haploinsufficiency of Sox9 in cells that are the progenitors of hypertrophic chondrocytes.

Because heterozygous *Sox9* mutant mice die at birth, homozygous *Sox9* null mutants cannot be generated by the conventional mating of heterozygous mutants. To examine the effects caused by the complete absence of *Sox9* on chondrogenesis *in vivo*, mouse chimeras were generated by injecting *Sox9* homozygous mutant ES cells into wild-type blastocysts. The mutant cells in these chimeras are easily identified because the *lacZ* gene, which had been knocked in in the *Sox9* locus, is expressed in a *Sox9*-specific pattern. Analysis of these chimeras revealed that the *Sox9* mutant cells are excluded first from all wild-type chondrogenic mesenchymal condensations and later from all cartilages (Bi *et al.*, 1999). In addition, the mutant cells do not express chondrocyte marker genes for type II (*Col2a1*), IX (*Col9a2*), and XI (*Col11a2*) collagens and aggrecan that are normally expressed in mesenchymal condensations. However, before chondrogenic mesenchymal condensation occurs, the *Sox9* null mutant cells are intermingled with wild-type cells in these chimeras. Therefore, the segregation of mutant cells from wild-type cells occurs at mesenchymal condensation. Furthermore, in teratomas derived from homozygous *Sox9* mutant ES cells, but not in teratomas derived from heterozygous *Sox9* mutant ES cells, no cartilage forms although these tumors contain a variety of other tissues normally found in teratomas. These experiments indicated that SOX9 is required for chondrocyte differentiation and cartilage formation and that SOX9 is needed for formation of chondrogenic mesenchymal condensations (Fig. 3). Why SOX9 is necessary for mesenchymal condensation is not yet understood, but one can speculate that SOX9 may control genes encoding cell-surface proteins needed for mesenchymal condensation. However, genes controlled by *Sox9* in this process have not

been identified. In addition, *Sox9* controls the genes *Col2a1, Col9a2, Col11a2,* and aggrecan, at least to produce the low levels of expression in chondrocyte precursors before chondrogenic mesenchymal condensation occurs.

Sox9 binding sites have been identified in the chondrocyte-specific enhancers of several genes, including *Col2a1* (Lefebvre *et al.*, 1997), *Col11a2* (Bridgewater *et al.*, 1998), and CDRAP (Xie *et al.*, 1999). These enhancers, which direct expression of reporter genes in cartilages of transgenic mice, are activated by Sox9 both *in vitro* and *in vivo*. In addition, because mutations in these enhancers that abolish *Sox9* DNA binding also prevent these enhancers from directing chondrocyte-specific reporter gene expression *in vivo*, one can assume that these enhancers are direct downstream targets of SOX9. Thus SOX9 appears to be a genuine differentiation factor needed for differentiation of all cartilages and probably controls a wide array of target genes.

Several other important issues concerning the role of SOX9 in chondrogenesis remain unresolved. In mouse chimeras derived from *Sox9*$^{-/-}$ ES cells, the β-gal-actosidase-expressing *Sox9*-null cells migrate to their correct chondrogenic location, indicating that the *Sox9* homozygous mutant cells are specified and respond to upstream signals. However, the factors that specify these chondroprogenitor cells, which are genetically upstream of *Sox9*, still need to be identified. BMPs have been shown to increase *Sox9* expression (Murtaugh *et al.*, 1999; Semba *et al.*, 2000; Zehentner *et al.*, 1999), and a similar role can be attributed to Sonic hedgehog (Murtaugh *et al.*, 1999), but it is unclear whether signaling by these molecules acts directly on the *Sox9* gene. Specific transcription factors also probably have important roles in the control of *Sox9* expression. It has not yet been unambiguously established whether SOX9 has a function in the differentiation of condensed mesenchymal cells into chondrocytes. *In vitro* results suggest that SOX9 has a critical role in this differentiation step but this hypothesis has not yet been tested *in vivo*. In addition, the expression and roles of SOX9 in

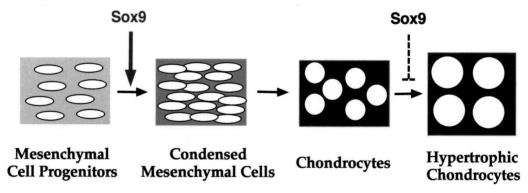

Figure 3 Experiments with mouse chimeras have indicated that SOX9 is required for formation of chondrogenic mesenchymal condensations. Analysis of *Sox9* heterozygous mice suggested that SOX9 inhibits the maturation of chondrocytes into hypertrophic chondrocytes. SOX9 is believed to have a role in the overt differentiation of cells in mesenchymal condensations into chondrocytes, but this has not been proven *in vivo*.

several nonchondrogenic tissues suggest that other transcription factors are needed to specifically activate genes that are characteristic of the chondrocytic lineage in mesenchymal condensations and eventually in later steps of the chondrogenic pathway. Other SOX proteins have been shown to interact with transcription factors of certain families, which could increase their specificity in activating target genes (Ambrosetti *et al.,* 1997; Kamachi *et al.,* 1999, 2000). Analogous partner proteins may interact with SOX9 during chondrogenesis. Overall, expression of SOX9 is obviously less cell specific than expression of the basic helix–loop–helix transcription factors that control myoblast differentiation.

The genes for two other *Sox* family members, L-*Sox5* and *Sox6,* are expressed with *Sox9* during chondrogenesis (Lefebvre *et al.,* 1998). L-SOX5 is a larger isoform of SOX5, a polypeptide found exclusively in testis. Like *Sox9,* expression of the *Sox5* and *Sox6* genes is completely turned off in hypertrophic chondrocytes. Both genes are also expressed in several other nonchondrogenic tissues. No human or murine genetic diseases caused by mutations in *Sox5* or *Sox6* have been identified. L-SOX5 and SOX6, which have a high degree of sequence identity, belong to a different subgroup of SOX proteins than SOX9 and have no sequence homology with SOX9 except in the HMG box. The HMG boxes of L-SOX5 and SOX6, which are almost identical, show only 50% identity with the HMG box of SOX9. In addition, L-SOX5 and SOX6 contain a very highly conserved coiled-coil domain (Fig. 4) through which they form homodimers and heterodimers with each other. These dimers bind much more efficiently to pairs of HMG box-binding sites than to single sites. In contrast, SOX9 binds as efficiently to single HMG box sites in DNA than to double sites. Unlike SOX9, L-

SOX5 and SOX6 do not contain a transcriptional activation domain. However, in DNA transfection experiments, L-SOX5 and SOX6 cooperate with SOX9 to activate endogenous *Col2a1* and aggrecan genes (Lefebvre *et al.,* 1998). This cooperation must be produced by a mechanism that does not require a transcription activation domain in L-SOX5 and SOX6.

Sox5-null mutant mice and *Sox6*-null mutant mice are born with relatively mild skeletal anomalies (Smits *et al.,* 2001). *Sox5*-null mice die in the perinatal period with cleft palate, hypoplasia of the rib cage, and a shorter chondrocranium, whereas *Sox6*-null mice display sternum defects and die within 3 weeks after birth. In contrast, double *Sox5-Sox6*-null fetuses die *in utero* with very severe defects in cartilage formation (Smits *et al.,* 2001). In these double mutants, although normal chondrogenic mesenchymal condensation occurs, overt chondrocyte differentiation, characterized by the abundant deposition of ECM components, does not take place. Expression of genes for cartilage ECM components such as *Col2a1, Col9a2, Col11a2,* aggrecan, and link protein is severely reduced and does not show the robust increase typically observed in wild-type embryos when cells in mesenchymal condensations differentiate into chondrocytes. In addition, specific markers for proliferating chondrocytes such as *matrilin, comp,* and *epiphycan* genes are virtually not expressed. Thus, L-SOX5 and SOX6 are required for the overt differentiation of chondrocytes and for the production of the characteristic cartilage ECM. In the double mutants, the levels of *Sox9* mRNA are, however, comparable to those observed in wild-type cartilage, indicating that expression of *Sox9* does not require L-SOX5 and SOX6. However, the converse, that Sox9 might control the

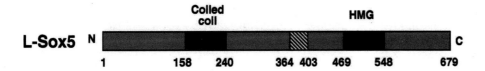

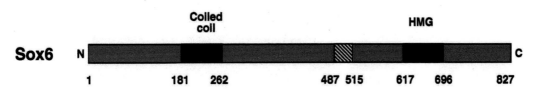

Figure 4 Comparison of the structures of L-SOX5 and SOX6. The coiled-coil domain is a dimerization domain that is highly conserved (91%) between the two polypeptides. No dimers form in mutants lacking the coiled-coil domain. L-SOX5 and SOX6 bind much more efficiently to pairs of HMG sites in DNA than to single sites. A second coiled-coil domain, much less conserved and indicated by blue stripes, is present in both L-SOX5 and SOX6; it is not sufficient for dimerization. The HMG-box domains of L-SOX5 and SOX6 have 93% identity but only 50% identity to the Sox9 HMG box. Overall, L-SOX5 and SOX6 have a sequence identity of 67% with each other.

expression of *Sox5* and *Sox6* during chondrogenesis, is possible.

The cells in the mesenchymal condensations of the double *Sox5-Sox6*-null mutants do not proliferate until they have accumulated a thin layer of ECM. L-SOX5 and SOX6 may either directly control cell-cycle genes or control cellular proliferation indirectly through effects on the synthesis of cartilage ECM proteins. Despite their inability to undergo chondrocyte differentiation, *Sox5*$^{-/-}$, *Sox6*$^{-/-}$ cells activate typical markers of prehypertrophic and hypertrophic chondrocytes although morphological hypertrophy is very limited. This maturation of mutant cells starts only after some extracellular matrix accumulates but then rapidly spreads to the entire cartilages. This late maturation also results in induction of bone collar formation, but there are no primary ossification centers. Similar growth plate and bone formation defects were described in mice lacking one of the main cartilage matrix components (collagen type 2, aggrecan, link protein) (Li *et al.*, 1995; Watanabe *et al.*, 1994; Watanabe and Yamada, 1999). It is, therefore, likely that the severe matrix deficiency of double *Sox5-Sox6*-null cartilages contributed a large part of these late defects. Thus, because of the severe defects in chondrocyte differentiation, the deficiency in cell proliferation and the late but precipitous maturation of double *Sox5-Sox6*-null mutants, the characteristic cellular organization of growth plates does not occur (Smits *et al.*, 2001).

Because the phenotype of the double *Sox5-Sox6*-null mutants is much more severe than the sum of the skeletal defects in single *Sox5*- and *Sox6*-null mice, L-Sox5 and Sox 6 must have largely redundant functions in chondrogenesis. This hypothesis is supported by the high degree of sequence identity of L-SOX5 and SOX6 and their coexpression in chondrogenic tissues. DNA binding sites for L-SOX5 and SOX6 containing pairs of HMG-like sites have been identified in the chondrocyte-specific enhancers of the *Col2a1* and *Col11a2* genes (Bridgewater *et al.*, 1998; Lefebvre *et al.*,

1998). These genes, and presumably many other genes preferentially or specifically expressed in chondrocytes, are, therefore, likely direct targets for L-Sox5 and Sox6.

In summary, SOX9, and L-SOX5 and SOX6 have different essential functions in cartilage formation as they control distinct steps in the chondrocyte differentiation pathway (Fig. 5). Sox9 is needed for chondrogenic mesenchymal condensation, whereas L-SOX5 and SOX6 are not. L-SOX5 and SOX6 are required after mesenchymal condensation, when high levels of expression of chondrocyte marker genes such as *Col2a1, aggrecan,* and others are needed. The biochemical properties of L-SOX5 and SOX6 are consistent with a model whereby these transcription factors act as architectural proteins that organize DNA and chromatin to facilitate the recruitment of other transcription factors to target genes and eventually, in cooperation with SOX9, help induce high levels of expression of these chondrocyte target genes (Fig. 6). That L-SOX5 and SOX6 have no transcription activation domain, unlike SOX9, which has a strong transcription activation domain, supports the notion of architectural function.

Other Sox proteins have also been implicated in determination of cell fate of various lineages, so the transcription factors of the SOX family probably evolved into true differentiation factors. The functions of SOX9, L-SOX5, and SOX6 in mice clearly demonstrate their essential role in determining the fate of chondrocytes.

III. Parathyroid Hormone-Related Peptide (PTHrP) and Parathyroid Hormone (PTH)/ PTHrP Receptor: Gatekeepers of the Zone of Hypertrophic Chondrocytes

The major function of PTHrP in the growth plate is to regulate the rate at which columnar proliferating chondro-

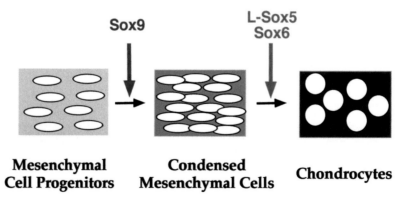

Figure 5 The characterization of double *Sox5-Sox6*-mutants has indicated that L-SOX5 and SOX6 are required for the overt differentiation of chondrocytes. The two proteins have largely redundant roles during chondrogenesis. SOX9 is needed for formation of mesenchymal condensations.

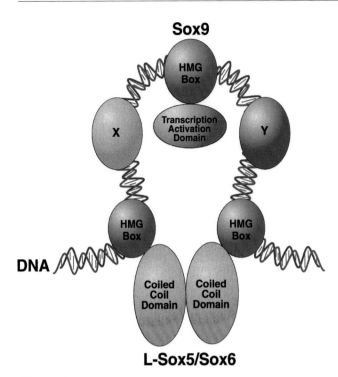

Figure 6 Hypothetical model of the role of L-SOX5 and SOX6 as architectural proteins. The representations of L-SOX5 and SOX6 have been simplified to their DNA binding and dimerization domains. According to this model, L-SOX5 and SOX6, which lack transcriptional activation domains, would organize the DNA to facilitate recruitment of other transcription factors such as X, Y, and SOX9 to form a transcriptionally active enhanceosome. Clusters of HMG box binding sites in chondrocyte-specific enhancers of *Col2a1* and *Col11a2* suggest that this model may account for the activation of these and other genes in cartilages.

cytes exit the cell cycle and are converted into postproliferative hypertrophic chondrocytes (Kronenberg and Chung, 2001). This major checkpoint in endochondral bone formation has a key role in regulating longitudinal bone growth. Indeed, once chondrocytes hypertrophy, they are destined for rapid cell death and replacement by bone cells. As gatekeepers, PTHrP and its receptor control the size of the pool of proliferating chondrocytes and the rate at which chondrocytes hypertrophy. PTHrP is a polypeptide of 84 amino acids; the sequence of the N-terminal 34 amino acids is very similar to that of PTH (Kronenberg and Chung, 2001). In the developing skeleton, *PTHrP* is expressed mainly in periarticular chondrocytes, whereas transcripts for its receptor, the PTH/PTHrP receptor (PPR), are found at low levels in periarticular chondrocytes and columnar proliferating chondrocytes and at much higher levels in prehypertrophic chondrocytes.

PPR is also expressed in osteoblasts. Outside the skeleton, *PTHrP* is widely expressed, whereas *PPR* is expressed in skin and kidney (Lanske and Kronenberg, 1998). *PTHrP*-null mice are dwarfs; their chondrocytes stop proliferating prematurely, and the columns of proliferating chondrocytes

in the growth plate are short and irregular (Karaplis *et al.*, 1994). *PPR*-null mice have a similar phenotype (Lanske *et al.*, 1996). Moreover, humans with a null mutation in PPR have similar anomalies (Jobert *et al.*, 1998; Karperien *et al.*, 1999). In contrast, overexpression of *PTHrP* in chondrocytes lengthens the columns of proliferating chondrocytes, and delays the conversion of these cells into hypertrophic chondrocytes and the formation of bone (Schipani *et al.*, 1997; Weir *et al.*, 1996). Expression of *PTHrP* in chondrocytes is dependent on Indian hedgehog (Ihh), a secreted polypeptide of the Hedgehog family, and these two polypeptides establish a feedback loop. *Ihh* is expressed at the prehypertrophic–hypertrophic boundary so that cells that escape the inhibitory action of PTHrP in the growth plate express *Ihh,* which in turn stimulates *PTHrP* expression by periarticular chondrocytes (Figure 5) (Vortkamp *et al.*, 1996). We discuss other features of Ihh later.

PPR is a seven-transmembrane receptor that is coupled to two different G proteins, one linked to the pathway that activates protein kinase A (PKA), and the other linked to the activation of phospholipase C (Lanske and Kronenberg, 1998). PKA appears to mediate the inhibition by PTHrP of the maturation of chondrocytes into hypertrophic chondrocytes. Upon binding of PTHrP, PPR triggers activation of adenylyl cyclase and, through an increase in the intracellular concentrations of cyclic AMP, activates the cyclic AMP-dependent PKA. This activation presumably results in the phosphorylation of a number of target proteins, including transcription factors. Sox9 has two consensus PKA phosphorylation sites and can be phosphorylated by PKA both in cell-free systems and in intact cells in transfection experiments (Huang *et al.*, 2000). This phosphorylation increases the DNA binding and transcriptional activity of *Sox9*. Furthermore, in DNA transfection experiments, PTHrP treatment increases the phosphorylation of SOX9 (as detected by a phosphorylated SOX9-specific antibody directed against a phosphorylated SOX9 peptide encompassing one of the two consensus PKA sites of SOX9) (Huang *et al.*, 2001). PTHrP also increases the transcriptional activity of *Sox9,* and this increase is mediated by the phosphorylation of the PKA consensus sites of SOX9. This increase in transcription activity does not occur with a SOX9 mutant in which the two consensus PKA phosphorylation sites are mutated. This PTHrP-dependent increase in transcriptional activity is mediated by PKA, as it is inhibited by a PKA-specific inhibitor.

In immunohistochemical experiments using the SOX9 phospho-specific antibody, phosphorylated SOX9 is found mainly in the prehypertrophic zone of the cartilage growth plate, an area in which *PPR* is strongly expressed (Huang *et al.*, 2000). Because no phosphorylation of SOX9 was detected in the growth plates of *PPR*-null mutant mice, although SOX9, as expected, is present in all chondrocytes in the growth plate of the mutants, except hypertrophic chondrocytes, these experiments strongly suggested that SOX9 is a target of PTHrP signaling. Therefore, a model was proposed whereby SOX9 mediates at least some of the effects of PTHrP in the growth plate. The PTHrP-mediated increase in transcriptional activity of *Sox9*, probably together with that of other transcription factors, would help maintain the chondrocyte phenotype of cells in the prehypertrophic zone and inhibit their maturation to hypertrophic chondrocytes (Fig. 7).

These experiments suggest a link between PTHrP, a hormone known to control an important checkpoint in the pathway of chondrogenesis, and *Sox9*, a key transcription factor required for chondrocyte differentiation. That expression of *Sox9* and also of *Sox5* and *Sox6* is shut off in hypertrophic chondrocytes is consistent with the hypothesis that these genes have a role in controlling the transition from prehypertrophic to hypertrophic chondrocytes. This hypothesis is also supported by the finding that in heterozygous *Sox9* mutant mice the zone of hypertrophic chondrocytes is enlarged. L-SOX5 and SOX6 also contain one potential PKA phosphorylation site, but whether their phosphorylation has a role in transcription has not been examined. Additional experiments are needed to determine the role of PTHrP in the activation of Sox9 *in vivo*.

The activity of the ubiquitously expressed transcription factor CREB (cyclic AMP responsive element-binding protein) is increased as a result of phosphorylation by PKA at Ser133 (Gonzalez *et al.*, 1989). A similar mechanism increases the activity of the two related transcription factors, cyclic AMP responsive element modulator (CREM) and activating transcription factor 1 (ATF1). Therefore, one might have expected that the phosphorylation of these transcrip-

tion factors in chondrocytes would be dependent on PTHrP signaling. Surprisingly, in the chondrocytes of *PTHrP*-null mutant mice, CREB is phosphorylated at Ser133, and the percentage of chondrocytes in which CREB Ser133 is phosphorylated is similar to that of wild-type embryos (Long *et al.*, 2001). Similarly, in *PTHrP*-null embryos, phosphorylation at a conserved PKA consensus site in CREM and ATF1 is unaffected. This suggested that cellular signaling pathways other than those triggered by PTHrP control the phosphorylation and activity of CREB family members in the cartilage growth plate. Furthermore, in an experimental model of mouse dwarfism caused by the expression in chondrocytes of a dominant negative CREB that decreased the activity of endogenous CREB, CREM, and ATF1, chondrocyte proliferation and Ihh signaling are inhibited. Therefore, in chondrocytes, the physiological function of the phosphorylation of CREB at Ser133 might be to increase expression of *Ihh*, but this phosphorylation would be independent of PTHrP signaling.

IV. FGFs and FGF Receptor 3: Counterintuitive Inhibitors of Chondrocyte Proliferation

FGFs and FGF receptor 3 (FGFR3) have major roles in endochondral bone formation. Indeed, constitutively active mutations in FGFR3 are the cause of achondroplasia, hypochondroplasia, and thanatophoric dysplasia, three closely related genetic diseases that cause severe dwarfism in humans (Bellus *et al.*, 1995; Rousseau *et al.*, 1994, 1995; Shiang *et al.*, 1994; Tavormina *et al.*, 1995). Mice expressing activated FGFR3 mutations have a dwarfism phenotype similar to that of people with these diseases, much slower growth of the columnar proliferating chondrocytes, and a smaller zone of hypertrophic chondrocytes (Chen *et al.*, 2001; Li *et al.*, 1999; Naski *et al.*, 1998). Interestingly, in these mice there is also decreased expression of *Ihh* and *Patched* (*Ptch*), a downstream target of Ihh. In contrast, mice null for FGFR3 display skeletal overgrowth (Colvin *et al.*, 1996; Deng *et al.*, 1996). Therefore, a normal function of FGF signaling in chondrocytes is to inhibit chondrocyte proliferation. FGF signaling activates several signaling pathways, including the signal transducer and activator of transcription 1 (STAT1) pathway (Sahni *et al.*, 1999; Su *et al.*, 1997). Treatment of chondrocytes in culture with FGF induces the phosphorylation of STAT1, its translocation to the nucleus, and increased expression of the cell-cycle inhibitor *p21*. Similar results were obtained in cells transfected with a thanatophoric dysplasia mutant of FGFR3. Moreover, in organ culture, metatarsals from *Stat1*-null mutant mice are much less sensitive to the growth-inhibitory effects of FGF than are wild-type metatarsals (Sahni *et al.*, 1999). Although *Stat1*-null mutant mice do not have skeletal abnormalities (Durbin *et al.*,

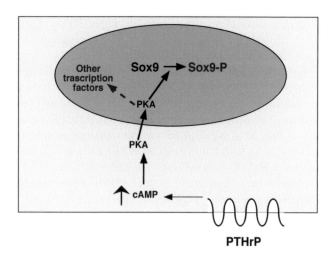

Figure 7 The PKA pathway, which is activated by binding of PTHrP to the PPR. Activation of PKA results in phosphorylation of Sox9. Biochemical experiments have shown that phosphorylated SOX9 binds more efficiently to DNA and is more active in transcription, suggesting that PKA-phosphorylated SOX9 mediates some of the effects of PTHrP, by helping maintain cells as chondrocytes and preventing them from becoming hypertrophic chondrocytes.

1996; Meraz *et al.,* 1996), these experiments suggested that Stat1 activation may be involved in the dwarf phenotype of achondroplasia, hypochondroplasia, and the thanatophoric dysplasias.

Recently, expression of *Sox9* was shown to be increased by FGFs in both chondrocytes and C3H10T1/2 mesenchymal cells in culture. Upregulation of *Sox9* by FGFs in these cells is mediated by the mitogen-activated protein (MAP) kinase pathway (Murakami *et al.,* 2000). FGF treatment of chondrocytes increases phosphorylation of the MAP kinases, ERK1 and ERK2 (extracellular regulated kinase 1 and 2), and the increase in SOX9 levels induced by FGF is inhibited by a specific inhibitor of MAPK-K (kinase kinase) (MEK1). Moreover, by using a functional assay for the transcriptional activity of SOX9, it was shown that FGF markedly increases this activity. The transcriptional activity of *Sox9* is also increased by activating mutations of FGFR3 that cause the human dwarfism syndromes. Furthermore, a constitutively active MEK1 also strongly enhances this activity, whereas a MAPK phosphatase (CL100/MKP-1) inhibits the FGF stimulation of the transcriptional activity of Sox9. Thus, increases in the activity of the MAP kinase pathway and the expression of *Sox9* may have a role, perhaps in collaboration with STAT1, in the dwarfism and inhibition of chondrocyte proliferation in the growth plate caused by activating mutations in FGFR3. The inhibition of chondrocyte proliferation could be due to a combination of increased Stat1 activity, increased MAP Kinase pathway activity, increased *Sox9* expression, and decreased *Ihh* expression. In myoblasts, expression of another key differentiation factor, *MyoD,* inhibits cell cycle progression (Parker *et al.,* 1995).

V. Ihh: A Central Coordinator of Endochondral Bone Formation

Ihh is a secreted polypeptide of the Hedgehog family that has multiple roles during skeletogenesis in both the chondrocyte and osteoblast differentiation pathways. *Ihh* is first expressed in skeletal elements in cells that overtly differentiate into chondrocytes right after mesenchymal condensation. *Ihh* expression is then downregulated in the central cells of these cartilages and later is restricted to the prehypertrophic–hypertrophic boundary of growth plates (St. Jacques *et al.,* 1999). Ihh, like the other members of the Hedgehog family, signals through a complex cellular signaling pathway that includes *Ptc* and Smoothened receptors and downstream Gli transcription factors (McMahon, 2000). *Ptch* expression is strongly stimulated by Hedgehog family members, so the extent of *Ptch* expression in skeletal primordia can be considered the area of active signaling by Ihh. In chick embryos, overexpression of *Ihh* using a retroviral vector increases the expression of *PTHrP* in periarticular chondrocytes, markedly slowing hypertrophic chondrocyte maturation and bone

formation (Vortkamp *et al.,* 1996). It is not clear, however, whether in the growth plate the activation of *PTHrP* expression by Ihh is a direct effect or is mediated by an intermediate signal.

Loss of Ihh function in *Ihh*-null mice illustrates the multiple roles of Ihh (St. Jacques *et al.,* 1999). These mice have severe dwarfism of all axial and appendicular skeletal elements, but the absence of Ihh has no effect on intramembranous ossification of craniofacial skeletal elements. In *Ihh*-null mutants, chondrocyte proliferation is markedly reduced, and there are no typical stacks of proliferating chondrocytes, indicating that Ihh contributes to normal chondrocyte proliferation. The ability of Ihh to stimulate cell proliferation is also a property of another member of the Hedgehog family, Sonic hedgehog, which (in addition to its other actions) increases cell proliferation in many tissues during mouse development (Bellusci *et al.,* 1997; Fan *et al.,* 1995; Wechsler-Reya and Scott, 1999). Moreover, mutations that constitutively activate the Hedgehog signaling pathway are associated with several human cancers including basal cell carcinoma of the skin and glioblastoma (Gailani *et al.,* 1996; Hahn *et al.,* 1996; Johnson *et al.,* 1996; Raffel *et al.,* 1997), another indication of Hedgehog's role in cell proliferation. The severely reduced proliferation rate of chondrocytes in *Ihh*-null mutants is independent of PTHrP signaling. Indeed, overexpression of a constitutively active *PPR* in chondrocytes of *Ihh*-null mutants does not abolish the short-limbed dwarfism and decreased proliferation rate of chondrocytes of these mice (Karp *et al.,* 2000).

The organization of the cartilage growth plate of *Ihh*-null embryos is also severely disrupted by the presence of hypertrophic chondrocytes near the articular surface, where no hypertrophic chondrocytes are seen in wild-type embryos (St. Jacques *et al.,* 1999). This abnormal and ectopic maturation of hypertrophic chondrocytes in the growth plate is due to the lack of PTHrP in the growth plate of *Ihh*-null embryos and can be corrected by restoring PTHrP signaling in chondrocytes (Karp *et al.,* 2000). In addition, *Ihh*-null mice have no cortical or trabecular bone in the endochondral skeleton, indicating that Ihh is required for osteoblast differentiation in endochondral bones. Ihh signaling may be essential for the maturation of the layer of osteogenic cells in the periosteum, which are the precursors of osteoblasts in the endochondral skeleton. In agreement with this hypothesis, *Cbfa1,* a transcription factor required for osteoblast differentiation, is not expressed in the perichondrium of *Ihh*-null embryos. Furthermore, experiments in mouse chimeras supported the hypothesis that Ihh signals the site in the perichondrium where mesenchymal cells differentiate into osteoblasts to form cortical bone and invade the mineralized matrix of hypertrophic chondrocytes (Chung *et al.,* 2001).

IHH is thus a key secreted polypeptide that controls at least three different critical steps in the endochondral growth plate: (1) proliferation of chondrocytes and the establish-

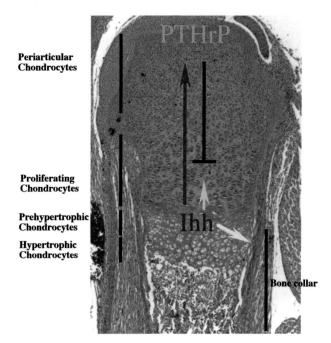

Periarticular Chondrocytes

Proliferating Chondrocytes

Prehypertrophic Chondrocytes

Hypertrophic Chondrocytes

Figure 8 Functions of IHH and PTHrP in the growth plate of endochondral bones. Ihh is needed for chondrocyte proliferation (yellow arrow), for expression of PTHrP (black arrow), and for osteoblast differentiation (white arrow). The function of PTHrP is to inhibit the transition of prehypertrophic chondrocytes into hypertrophic chondrocytes.

ment of characteristic columns of proliferating chondrocytes in the cartilage growth plate, a critical process for the growth of endochondral bones; (2) induction of *PTHrP* expression by periarticular chondrocytes, which inhibits the conversion of proliferating and prehypertrophic chondrocytes into hypertrophic chondrocytes and maintains an appropriate pool of proliferating chondrocytes; and (3) activation of the differentiation of osteoblast precursors in the perichondrium flanking the zone of hypertrophic chondrocytes. Therefore, signaling by Ihh coordinates several essential steps that are part of both the chondrocyte differentiation pathway and the osteoblast pathway in endochondral bone formation (Fig. 8).

Although Ihh has an essential role in osteoblast differentiation of endochondral bones, membranous ossification is apparently not affected in *Ihh*-null embryos. It is possible that other members of the Hedgehog family may have a role that would parallel that of IHH in endochondral ossification, but there is no evidence to support this hypothesis.

VI. The Two Roles of the Transcription Factor Cbfa1 in Endochondral Bone Formation

Cbfa1, or AML3/Runx2, is a member of the Runt-domain family of transcription factors (Fig. 9). The Runt do-

main is a 128- to 130-residue-long DNA binding domain (Ito, 1999). In mammals the Runt-domain family contains three members: Cbfa1; AML1/Runx1, a critical regulator of hematopoiesis; and AML2/Runx3. The *Drosophila* Runt gene is involved in neurogenesis and sex determination, whereas a related Runt-domain-containing transcription factor in flies controls eye, antenna, and tarsal claw development. The Runx polypeptides form heterodimers with a single ubiquitous polypeptide called polyoma enhancer binding protein 2β, which does not bind to DNA by itself but increases the efficiency of DNA binding of the heterodimers. Cbfa1 contains active transcription activation domains, one of which is rich in glutamine and alanine residues.

In mouse embryos, expression of *Cbfa1* begins as early as embryonic day 9.5 in the notochord, and later occurs in prechondrogenic mesenchymal condensations (Otto *et al.,* 1997). Around embryonic day13.5, when both the perichondrium and the organization of the different cellular layers of the cartilage growth plate are beginning to be established, *Cbfa1* is expressed in the perichondrium/periosteum and soon thereafter in prehypertrophic and hypertrophic chondrocytes. Later, *Cbfa1* is expressed in osteoblasts in all bones, whether formed by endochondral or by intramembranous ossification (Ducy *et al.,* 1997).

Heterozygous mutations in the *Cbfa1* gene are the cause of the human genetic disease cleidocranial dysplasia, which is characterized by hypoplastic clavicles, large open spaces between the frontal and parietal bones of the skull, and other skeletal dysplasias (Mundlos *et al.,* 1997). The disease is due to haploinsufficiency of Cbfa1. In *Cbfa1*-null mice, osteoblast differentiation is blocked; neither endochondral nor intramembranous bone form (Komori *et al.,* 1997; Otto *et al.,* 1997). Hence, Cbfa1 is required for osteoblast differentiation.

In these embryos, cells of the periosteum/perichondrium do not invade the matrix of hypertrophic chondrocytes, and no osteoclasts/chondroclasts are found in this matrix, so there is no degradation of the cartilage matrix. In addition, in *Cbfa1*-null mice, there is much less mineralization of the distal hypertrophic zone of the endochondral skeleton (Inada *et al.,* 1999; Komori *et al.,* 1997). Furthermore, in a subset of bone primordia, such as those of the humerus and femur, there is no expression of markers of prehypertrophic and hypertrophic chondrocytes, such as *PPR, Ihh,* and *Col10a1,* indicating that the maturation of chondrocytes into hypertrophic chondrocytes is inhibited (Inada *et al.,* 1999; Kim *et al.,* 1999). These observations strongly suggested that in addition to its essential function in osteoblast differentiation, Cbfa1 also has a role in the differentiation of hypertrophic chondrocytes. In skeletal elements of *Cbfa1*-null mice in which hypertrophic chondrocytes are present, the absence of *Cbfa1* might be compensated for by other transcription factors.

To distinguish between the roles of Cbfa1 in osteoblast differentiation and in hypertrophic chondrocyte maturation,

Cbfa1/Pebp2aA/Runx2

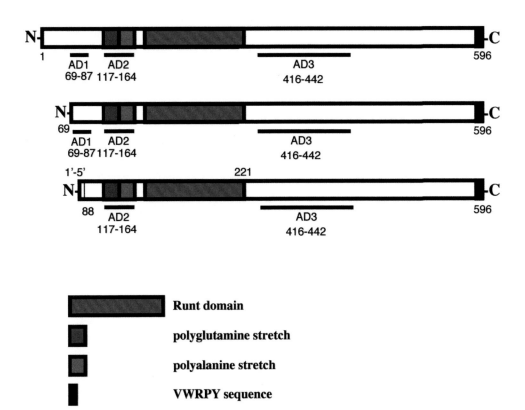

Figure 9 Cbfa1 isoforms. Three isoforms with different N-terminal 1 have been described. The first two differ in the length of the N-terminal part of Cbfa1. In the third isoform, the N-terminal 5 amino acids (1′–5′) are different than in other two isoforms. AD, transcription activation domains. The VWRPY sequence is used for interactions with other polypeptides.

transgenic mice overexpressing Cbfa1 in chondrocytes were generated. These mice have accelerated endochondral ossification (Takeda *et al.,* 2001; Ueta *et al.,* 2001). Overexpression of *Cbfa1* in the chondrocytes of *Cbfa1*-null embryos partially rescued the phenotype of *Cbfa1*-deficient embryos (Takeda *et al.,* 2001). Characteristics of the partial rescue were hypertrophic chondrocyte differentiation in endochondral skeletal elements in which such differentiation was defective in the parent *Cbfa1*-null embryos, vascular endothelial growth factor expression by hypertrophic chondrocytes, vascular invasion of the matrix of hypertrophic chondrocytes, and the presence of multinucleated osteoclasts/chondroclasts in the hypertrophic cartilage matrix, all of which were defective in *Cbfa1*-null embryos. However, no bone formation or osteoblast differentiation occurred in the rescued embryos. Hence, in the rescued animals, the defects in hypertrophic chondrocyte differentiation were corrected, but not the absence of osteoblast differentiation. Thus, Cbfa1 has two distinct functions in bone formation. First, it has an essential role in the differentiation of mesenchymal progenitors into osteoblasts both in endochondral and intra-

membranous skeletons. The second function of Cbfa1 is to stimulate hypertrophic chondrocyte differentiation, which prepares the cartilage skeleton for its subsequent invasion by osteoblasts and its replacement by a bone-specific matrix (Fig. 10).

Cbfa1 binding sites have been identified in the promoters/enhancers of several genes (including *Col1a1, Col1a2, osteocalcin,* and *matrix metalloproteinase 13*) preferentially or specifically expressed in osteoblasts. Mutations in Cbfa1 binding sites that inhibit Cbfa1 binding also decrease the activity of these promoters/enhancers in DNA transfection experiments, suggesting that these genes may be direct targets for Cbfa1 (Ducy *et al.,* 1997; Jimenez *et al.,* 1999).

The establishment of other cell lineages derived from the mesoderm, such as myoblasts (Black and Olson, 1998; Olson and Klein, 1998), adipocytes (Rosen and Spiegelman, 2000), and chondrocytes (Bi *et al.,* 1999; Smits *et al.,* 2001), involves multistep pathways in which several transcription factors control successive steps of differentiation programs. One might, therefore, predict that there are additional steps in the osteoblast differentiation pathway and that other tran-

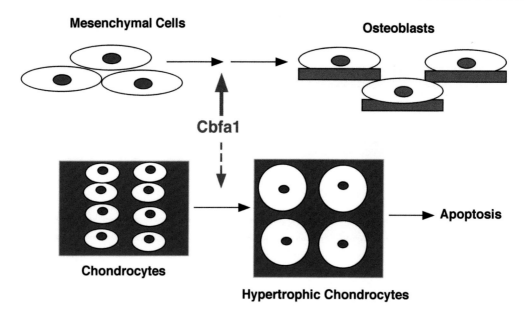

Figure 10 Cbfa1 is required for osteoblast differentiation and stimulates hypertrophic chondrocyte differentiation.

scription factors, in addition to Cbfa1, might also be required for osteoblast differentiation.

VII. Other Transcription Factors Involved in Bone Formation

Many other transcription factors also have important roles in bone formation. We discuss two examples here. MSX1 and MSX2 are homeodomain-containing transcription factors with critical roles in skeletal development (Satokata *et al.,* 2000; Satokata and Maas, 1994). *Msx2*-null mice have defects in the bones of the skull as a result of defective proliferation of osteoprogenitor cells during morphogenesis of calvaria (Satokata *et al.,* 2000). These mice also have defective endochondral ossification and a postnatal decrease in *Cbfa1* expression in the long bones. Interestingly, double-null *Msx2-Msx1* mutants have considerably worse defects than do *Msx2*-null mutants, including an absence of ossification of the membranous bones in the head. However, it is difficult to establish whether MSX1 and MSX2 are true differentiation factors; the phenotypes of *Msx1* and *Msx2* mouse mutants suggest that these proteins act primarily as patterning molecules in the development of the skeleton.

DLX5 and DLX6 are members of the Distal-less homeodomain-containing family of transcription factors and are coexpressed in developing bones. Dlx5 and Dlx6 also have a complex expression pattern in neuronal tissues (Chen *et al.,* 1996; Simeone *et al.,* 1994). *Dlx5*-null mutant mice have craniofacial anomalies affecting derivatives of the branchial arches (Acampora *et al.,* 1999; Depew *et al.,* 1999). In addition, the periosteum is thinner in the long bones of these mutants. Expression of *Cbfa1* is, however, not affected in

Dlx5-null mutants. DLX5 and DLX6 probably have redundant functions, so a full understanding of their role in mouse embryos will require inactivation of both genes.

Overall, the major function of MSX1, MSX2, and DLX5, like those of many other homeodomain-containing proteins, appears to be patterning of the embryo. It is, of course, possible that these transcription factors also directly or indirectly modulate expression and/or activity of genuine differentiation molecules. One of the challenges of this field is to determine how polypeptides that have a role in patterning skeletal elements influence the expression or activity of transcription factors that control the differentiation and cell fate of the skeletal cells.

VIII. Gelatinase B and Vascular Endothelial Growth Factor: Additional Coordinators of Endochondral Bone Formation

Many additional molecules are probably needed to correctly coordinate hypertrophic chondrocyte maturation and bone formation. One of these is gelatinase B, also called matrix metalloproteinase 9 (MMP9), a secreted metalloproteinase expressed in osteoclast/chondroclasts and also in other nonskeletal tissues (Vu and Werb, 2000). The metalloproteinases are a family of secreted proteins that control the turnover of ECM components. In mice lacking MMP9, the zone of hypertrophic chondrocytes is much larger than in wild-type mice during bone development, at least temporarily, probably because of apoptosis hypertrophic chondrocytes, vascularization of the degraded cartilage matrix and

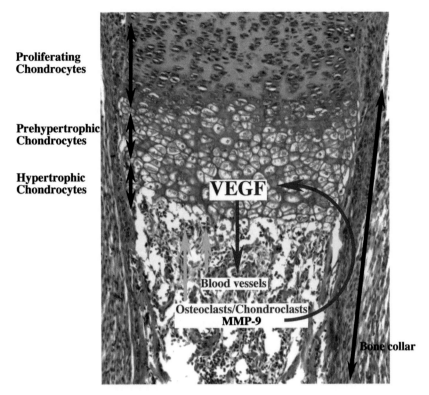

Proliferating
Chondrocytes

Prehypertrophic
Chondrocytes

Hypertrophic
Chondrocytes

VEGF

Blood vessels

Osteoclasts/Chondroclasts
MMP-9

Bone collar

Figure 11 VEGF, expressed in hypertrophic chondrocytes, is an angiogenic factor. Blood vessels in the periosteum together with osteoclasts/chondroclasts and osteoblasts invade the cartilage matrix. MMP9 may help release VEGF from sequestration in the ECM of hypertrophic chondrocytes. Alternatively, MMP9 could be an apoptotic signal for hypertrophic chondrocytes or participate in the degradation of the ECM of hypertrophic chondrocytes.

ossification are delayed (Vu *et al.,* 1998). When placed in culture, the growth plates of these mutants show a delayed release of angiogenic activity. One possible explanation for the abnormal phenotype of these mutants is that MMP9 has some role in the degradation of cartilage matrix. A delay in matrix degradation could retard apoptosis of hypertrophic chondrocytes and inhibit invasion of the cartilage matrix by osteoblasts, osteoclasts, and blood vessels. Alternatively, MMP9 could be an apoptosis signal for hypertrophic chondrocytes at the chondrocyte–bone boundary. A delay in apoptosis of these cells would also slow invasion and bone deposition. Finally, MMP9 could release angiogenic molecules sequestered in the matrix (Engsig *et al.,* 2000). A delay in blood vessel formation would also delay invasion of the matrix and bone development.

Another crucial molecule needed for invasion of the hypertrophic cartilage matrix and subsequent formation of bone trabeculae is vascular endothelial growth factor (VEGF). The levels of VEGF transcripts are much higher in terminal hypertrophic chondrocytes than in proliferating and periarticular chondrocytes (Gerber *et al.,* 1999). Injecting a soluble chimeric VEGF receptor with the ability to sequester VEGF into 24-day-old mice considerably increases the zone of hypertrophic chondrocytes. In the treated mice, blood vessels

fail to invade the zone of hypertrophic chondrocytes, and there are many fewer osteclasts/chondroclasts expressing *MMP9* at the chondrocyte–bone boundary. In contrast, the differentiation and proliferation of chondrocytes, and the maturation of hypertrophic chondrocytes is normal. Overall, the phenotype of the mice treated with the soluble VEGF receptor is very similar to those of *MMP9*-null mice. Therefore, VEGF produced by hypertrophic chondrocytes plays an essential role in recruitment of endothelial cells, induction of blood vessel formation, and invasion of the matrix of hypertrophic zones by these blood vessels. Without blood vessels, osteoclasts and osteoblasts cannot invade this matrix. Thus, the roles of MMP9 and VEGF in endochondral bone formation are illustrative examples in support of the notion that a variety of molecules are likely to be critical to ensure correct coordination of the successive steps of endochondral bone formation (Fig. 11).

IX. Conclusion

Recent human and mouse genetics studies have continued to reveal a wide range of new molecules that participate in the control of skeletogenesis. These molecules include signaling molecules, proteins involved in transport of small molecules across membranes, and transcription factors (DeChiara *et al.*, 2000; Reichenberger *et al.*, 2001; Ueki *et al.*, 2001) [References will be added]. In addition, the systematic mutagenesis studies being performed in mice (Hrabe de Angelis and Balling, 1998; Justice, 2000) [examples will be added] should produce a wealth of new mutants that will lead to the identification of polypeptides that affect skeletal development and to a better understanding of the nature of the molecules that control the physiological functions of the constituent cells of the skeleton. The deconstruction of cartilage and bone is therefore just beginning.

Acknowledgments

Work in the authors' laboratory was supported by NIH grant AR42919 to Benoit de Crombrugghe and NIH grant AR 46249 to Véronique Lefebvre. We are grateful to Janie Finch for editorial assistance.

References

Acampora, D., Merlo, G. R., Paleari, L., Zerega, B., Postiglione, M. P., Mantero, S., Bober, E., Barbieri, O., Simeone, A., and Levi, G. (1999). Craniofacial, vestibular and bone defects in mice lacking the Distal-less-related gene Dlx5. *Development (Cambridge, UK)* **126,** 3795–3809.

Ambrosetti, D. C., Basilico, C., and Dailey, L. (1997). Synergistic activation of the fibroblast growth factor 4 enhancer by Sox2 and Oct-3 depends on protein–protein interactions facilitated by a specific spatial arrangement of factor binding sites. *Mol. Cell Biol.* **17,** 6321–6329.

Bellus, G. A., McIntosh, I., Smith, E. A., Aylsworth, A. S., Kaitila, I., Horton, W. A., Greenhaw, G. A., Hecht, J. T., and Francomano, C. A. (1995). A recurrent mutation in the tyrosine kinase domain of fibroblast growth factor receptor 3 causes hypochondroplasia. *Nat. Genet.* **10,** 357–359.

Bellusci, S., Furuta, Y., Rush, M. G., Henderson, R., Winnier, G., and Hogan, B. L. (1997). Involvement of Sonic hedgehog (Shh) in mouse embryonic lung growth and morphogenesis. *Development (Cambridge, UK)* **124,** 53–63.

Bi, W., Huang, W., Whitworth, D. J., Deng, J. M., Zhang, Z., Behringer, R. R., and de Crombrugghe, B. (2001). Haploinsufficiency of Sox9 results in defective cartilage primordia and premature skeletal mineralization. *Proc. Natl. Acad. Sci. U.S.A.* **98,** 6698–6703.

Bi, W., Deng, J. M., Zhang, Z., Behringer, R. R., and de Crombrugghe, B. (1999). Sox9 is required for cartilage formation. *Nat. Genet.* **22,** 85–89.

Black, B. L., and Olson, E. N. (1998). Transcriptional control of muscle development by myocyte enhancer factor-2 (MEF2) proteins. *Annu. Rev. Cell Dev. Biol.* **14,** 167–196.

Bridgewater, L. C., Lefebvre, V., and de Crombrugghe, B. (1998). Chondrocyte-specific enhancer elements in the Col11a2 gene resemble the Col2a1 tissue-specific enhancer. *J. Biol. Chem.* **273,** 14998–15006.

Chen, L., Li, C., Qiao, W., Xu, X., and Deng, C. (2001). A Ser(365)→Cys mutation of fibroblast growth factor receptor 3 in mouse downregulates Ihh/PTHrP signals and causes severe achondroplasia. *Hum. Mol. Genet.* **10,** 457–465.

Chen, X., Li, X., Wang, W., and Lufkin, T. (1996). Dlx5 and Dlx6: An evolutionary conserved pair of murine homeobox genes expressed in the embryonic skeleton. *Ann. N.Y. Acad. Sci. U.S.A.* **785,** 38–47.

Chung, U. I., Schipani, E., McMahon, A. P., and Kronenberg, H. M. (2001). Indian hedgehog couples chondrogenesis to osteogenesis in endochondral bone development. *J. Clin. Invest.* **107,** 295–304.

Colvin, J. S., Bohne, B. A., Harding, G. W., McEwen, D. G., and Ornitz, D. M. (1996). Skeletal overgrowth and deafness in mice lacking fibroblast growth factor receptor 3. *Nat. Genet.* **12,** 390–397.

DeChiara, T. M., Kimble, R. B., Poueymirou, W. T., Rojas, J., Masiakowski, P., Valenzuela, D. M., and Yancopoulos, G. D. (2000). Ror2, encoding a receptor-like tyrosine kinase, is required for cartilage and growth plate development. *Nat. Genet.* **24,** 271–274.

DeLise, A. M., Fischer, L., and Tuan, R. S. (2000). Cellular interactions and signaling in cartilage development. *Osteoarthrit. Cartilage* **8,** 309–334.

Deng, C., Wynshaw-Boris, A., Zhou, F., Kuo, A., and Leder, P. (1996). Fibroblast growth factor receptor 3 is a negative regulator of bone growth. *Cell (Cambridge, Mass.)* **84,** 911–921.

Depew, M. J., Liu, J. K., Long, J. E., Presley, R., Meneses, J. J., Pedersen, R. A., and Rubenstein, J. L. (1999). Dlx5 regulates regional development of the branchial arches and sensory capsules. *Development (Cambridge, UK)* **126,** 3831–3846.

Ducy, P., Zhang, R., Geoffroy, V., Ridall, A. L., and Karsenty, G. (1997). Osf2/Cbfa1: A transcriptional activator of osteoblast differentiation. *Cell (Cambridge, Mass.)* **89,** 747–754.

Durbin, J. E., Hackenmiller, R., Simon, M. C., and Levy, D. E. (1996). Targeted disruption of the mouse Stat1 gene results in compromised innate immunity to viral disease. *Cell (Cambridge, Mass.)* **84,** 443–450.

Engsig, M. T., Chen, Q. J., Vu, T. H., Pedersen, A. C., Therkidsen, B., Lund, L. R., Henriksen, K., Lenhard, T., Foged, N. T., Werb, Z., and Delaisse, J. M. (2000). Matrix metalloproteinase 9 and vascular endothelial growth factor are essential for osteoclast recruitment into developing long bones. *J. Cell Biol.* **151,** 879–890.

Erlebacher, A., Filvaroff, E. H., Gitelman, S. E., and Derynck, R. (1995). Toward a molecular understanding of skeletal development. *Cell (Cambridge, Mass.)* **80,** 371–378.

Fan, C. M., Porter, J. A., Chiang, C., Chang, D. T., Beachy, P. A., and Tessier-Lavigne, M. (1995). Long-range sclerotome induction by sonic hedgehog: Direct role of the amino-terminal cleavage product and modulation by the cyclic AMP signaling pathway. *Cell (Cambridge, Mass.)* **81,** 457–465.

Fang, J., and Hall, B. K. (1997). Chondrogenic cell differentiation from membrane bone periostea. *Anat. Embryol. (Berl.)* **196,** 349–362.

Foster, J. W., Dominguez-Steglich, M. A., Guioli, S., Kowk, G., Weller, P. A., Stevanovic, M., Weissenbach, J., Mansour, S., Young, I. D., Goodfellow, P. N., *et al.* (1994). Campomelic dysplasia and autosomal sex reversal caused by mutations in an SRY-related gene. *Nature (London)* **372,** 525–530.

Gailani, M. R., Stahle-Backdahl, M., Leffell, D. J., Glynn, M., Zaphiropoulos, P. G., Pressman, C., Unden, A. B., Dean, M., Brash, D. E., Bale, A. E., and Toftgard, R. (1996). The role of the human homolog of *Drosophila* patched in sporadic basal cell carcinomas. *Nat. Genet.* **14,** 78–81.

Gerber, H. P., Vu, T. H., Ryan, A. M., Kowalski, J., Werb, Z., and Ferrara, N. (1999). VEGF couples hypertrophic cartilage remodeling, ossification and angiogenesis during endochondral bone formation. *Nat. Med.* **5,** 623–628.

Gonzalez, G. A., Yamamoto, K. K., Fischer, W. H., Karr, D., Menzel, P., Biggs, W., 3rd, Vale, W. W., and Montminy, M. R. (1989). A cluster of phosphorylation sites on the cyclic AMP-regulated nuclear factor CREB predicted by its sequence. *Nature (London)* **337,** 749–752.

Hahn, H., Wicking, C., Zaphiropoulous, P. G., Gailani, M. R., Shanley, S., Chidambaram, A., Vorechovsky, I., Holmberg, E., Unden, A. B., Gillies, S., Negus, K., Smyth, I., Pressman, C., Leffell, D. J., Gerrard, B., Goldstein, A. M., Dean, M., Toftgard, R., Chenevix-Trench, G., Wainwright, B., and Bale, A. E. (1996). Mutations of the human homolog of *Drosophila* patched in the nevoid basal cell carcinoma syndrome. *Cell (Cambridge, Mass.)* **85,** 841–851.

Hrabe de Angelis, M., and Balling, R. (1998). Large scale ENU screens in the mouse: Genetics meets genomics. *Mutat. Res.* **400,** 25–32.

Huang, W., Chung, U. I., Kronenberg, H. M., and de Crombrugghe, B. (2001). The chondrogenic transcription factor Sox9 is a target of signaling by the parathyroid hormone-related peptide in the growth plate of endochondral bones. *Proc. Natl. Acad. Sci. U.S.A.* **98**, 160–165.

Huang, W., Zhou, X., Lefebvre, V., and de Crombrugghe, B. (2000). Phosphorylation of SOX9 by cyclic AMP-dependent protein kinase A enhances SOX9's ability to transactivate a Col2a1 chondrocyte-specific enhancer. *Mol. Cell Biol.* **20**, 4149–4158.

Inada, M., Yasui, T., Nomura, S., Miyake, S., Deguchi, K., Himeno, M., Sato, M., Yamagiwa, H., Kimura, T., Yasui, N., Ochi, T., Endo, N., Kitamura, Y., Kishimoto, T., and Komori, T. (1999). Maturational disturbance of chondrocytes in Cbfa1-deficient mice. *Dev. Dyn.* **214**, 279–290.

Ito, Y. (1999). Molecular basis of tissue-specific gene expression mediated by the runt domain transcription factor PEBP2/CBF. *Genes Cells* **4**, 685–696.

Jimenez, M. J., Balbin, M., Lopez, J. M., Alvarez, J., Komori, T., and Lopez-Otin, C. (1999). Collagenase 3 is a target of Cbfa1, a transcription factor of the runt gene family involved in bone formation. *Mol. Cell Biol.* **19**, 4431–4442.

Jobert, A. S., Zhang, P., Couvineau, A., Bonaventure, J., Roume, J., Le Merrer, M., and Silve, C. (1998). Absence of functional receptors for parathyroid hormone and parathyroid hormone-related peptide in Blomstrand chondrodysplasia. *J. Clin. Invest.* **102**, 34–40.

Johnson, R. L., and Tabin, C. J. (1997). Molecular models for vertebrate limb development. *Cell (Cambridge, Mass.)* **90**, 979–990.

Johnson, R. L., Rothman, A. L., Xie, J., Goodrich, L. V., Bare, J. W., Bonifas, J. M., Quinn, A. G., Myers, R. M., Cox, D. R., Epstein, E. H., Jr., and Scott, M. P. (1996). Human homolog of patched, a candidate gene for the basal cell nevus syndrome. *Science* **272**, 1668–1671.

Justice, M. J. (2000). Capitalizing on large-scale mouse mutagenesis screens. *Nat. Rev. Genet.* **1**, 109–115.

Kamachi, Y., Uchikawa, M., and Kondoh, H. (2000). Pairing SOX off: With partners in the regulation of embryonic development. *Trends Genet.* **16**, 182–187.

Kamachi, Y., Cheah, K. S., and Kondoh, H. (1999). Mechanism of regulatory target selection by the SOX high-mobility-group domain proteins as revealed by comparison of SOX1/2/3 and SOX9. *Mol. Cell Biol.* **19**, 107–120.

Karaplis, A. C., Luz, A., Glowacki, J., Bronson, R. T., Tybulewicz, V. L., Kronenberg, H. M., and Mulligan, R. C. (1994). Lethal skeletal dysplasia from targeted disruption of the parathyroid hormone-related peptide gene. *Genes Dev.* **8**, 277–289.

Karp, S. J., Schipani, E., St-Jacques, B., Hunzelman, J., Kronenberg, H., and McMahon, A. P. (2000). Indian hedgehog coordinates endochondral bone growth and morphogenesis via parathyroid hormone related-protein-dependent and -independent pathways. *Development (Cambridge, UK)* **127**, 543–548.

Karperien, M., van der Harten, H. J., van Schooten, R., Farih-Sips, H., den Hollander, N. S., Kneppers, S. L., Nijweide, P., Papapoulos, S. E., and Lowik, C. W. (1999). A frame-shift mutation in the type I parathyroid hormone (PTH)/PTH-related peptide receptor causing Blomstrand lethal osteochondrodysplasia. *J. Clin. Endocrinol. Metab.* **84**, 3713–3720.

Karsenty, G. (1999). The genetic transformation of bone biology. *Genes Dev.* **13**, 3037–3051.

Kim, I. S., Otto, F., Zabel, B., and Mundlos, S. (1999). Regulation of chondrocyte differentiation by Cbfa1. *Mech. Dev.* **80**, 159–170.

Kingsley, D. M. (2001). Genetic control of bone and joint formation. *Novartis Found. Symp.* **232**, 213–222.

Komori, T., Yagi, H., Nomura, S., Yamaguchi, A., Sasaki, K., Deguchi, K., Shimizu, Y., Bronson, R. T., Gao, Y. H., Inada, M., Sato, M., Okamoto, R., Kitamura, Y., Yoshiki, S., and Kishimoto, T. (1997). Targeted disruption of Cbfa1 results in a complete lack of bone formation owing to maturational arrest of osteoblasts. *Cell (Cambridge, Mass.)* **89**, 755–764.

Kronenberg, H. M., and Chung, U. (2001). The parathyroid hormone-related protein and Indian hedgehog feedback loop in the growth plate. *Novartis Found. Symp.* **232**, 144–152; discussion, 152–157.

Lanske, B., and Kronenberg, H. M. (1998). Parathyroid hormone-related peptide (PTHrP) and parathyroid hormone (PTH)/PTHrP receptor. *Crit. Rev. Eukaryot. Gene Exp.* **8**, 297–320.

Lanske, B., Karaplis, A. C., Lee, K., Luz, A., Vortkamp, A., Pirro, A., Karperien, M., Defize, L. H., Ho, C., Mulligan, R. C., Abou-Samra, A. B., Juppner, H., Segre, G. V., and Kronenberg, H. M. (1996). PTH/PTHrP receptor in early development and Indian hedgehog-regulated bone growth. *Science* **273**, 663–666.

Lefebvre, V., Li, P., and de Crombrugghe, B. (1998). A new long form of Sox5 (L-Sox5), Sox6 and Sox9 are coexpressed in chondrogenesis and cooperatively activate the type II collagen gene. *EMBO J.* **17**, 5718–5733.

Lefebvre, V., Huang, W., Harley, V. R., Goodfellow, P. N., and de Crombrugghe, B. (1997). SOX9 is a potent activator of the chondrocyte-specific enhancer of the pro alpha1(II) collagen gene. *Mol. Cell Biol.* **17**, 2336–2346.

Li, C., Chen, L., Iwata, T., Kitagawa, M., Fu, X. Y., and Deng, C. X. (1999). A Lys644Glu substitution in fibroblast growth factor receptor 3 (FGFR3) causes dwarfism in mice by activation of STATs and ink4 cell cycle inhibitors. *Hum. Mol. Genet.* **8**, 35–44.

Li, S. W., Prockop, D. J., Helminen, H., Fassler, R., Lapvetelainen, T., Kiraly, K., Peltarri, A., Arokoski, J., Lui, H., Arita, M., *et al.* (1995). Transgenic mice with targeted inactivation of the Col2 alpha 1 gene for collagen II develop a skeleton with membranous and periosteal bone but no endochondral bone. *Genes Dev.* **9**, 2821–2830.

Long, F., Schipani, E., Asahara, H., Kronenberg, H., and Montminy, M. (2001). The CREB family of activators is required for endochondral bone development. *Development (Cambridge, UK)* **128**, 541–550.

McMahon, A. P. (2000). More surprises in the Hedgehog signaling pathway. *Cell (Cambridge, Mass.)* **100**, 185–188.

Meraz, M. A., White, J. M., Sheehan, K. C., Bach, E. A., Rodig, S. J., Dighe, A. S., Kaplan, D. H., Riley, J. K., Greenlund, A. C., Campbell, D., Carver-Moore, K., DuBois, R. N., Clark, R., Aguet, M., and Schreiber, R. D. (1996). Targeted disruption of the Stat1 gene in mice reveals unexpected physiologic specificity in the JAK-STAT signaling pathway. *Cell (Cambridge, Mass.)* **84**, 431–442.

Meyer, J., Sudbeck, P., Held, M., Wagner, T., Schmitz, M. L., Bricarelli, F. D., Eggermont, E., Friedrich, U., Haas, O. A., Kobelt, A., Leroy, J. G., Van Maldergem, L., Michel, E., Mitulla, B., Pfeiffer, R. A., Schinzel, A., Schmidt, H., and Scherer, G. (1997). Mutational analysis of the SOX9 gene in campomelic dysplasia and autosomal sex reversal: Lack of genotype/phenotype correlations. *Hum. Mol. Genet.* **6**, 91–98.

Mundlos, S., and Olsen, B. R. (1997). Heritable diseases of the skeleton. Part II: Molecular insights into skeletal development–matrix components and their homeostasis. *FASEB J.* **11**, 227–233.

Mundlos, S., Otto, F., Mundlos, C., Mulliken, J. B., Aylsworth, A. S., Albright, S., Lindhout, D., Cole, W. G., Henn, W., Knoll, J. H., Owen, M. J., Mertelsmann, R., Zabel, B. U., and Olsen, B. R. (1997). Mutations involving the transcription factor CBFA1 cause cleidocranial dysplasia. *Cell (Cambridge, Mass.)* **89**, 773–779.

Murakami, S., Kan, M., McKeehan, W. L., and de Crombrugghe, B. (2000). Up-regulation of the chondrogenic Sox9 gene by fibroblast growth factors is mediated by the mitogen-activated protein kinase pathway. *Proc. Natl. Acad. Sci. U.S.A.* **97**, 1113–1118.

Murtaugh, L. C., Chyung, J. H., and Lassar, A. B. (1999). Sonic hedgehog promotes somitic chondrogenesis by altering the cellular response to BMP signaling. *Genes Dev.* **13**, 225–237.

Naski, M. C., Colvin, J. S., Coffin, J. D., and Ornitz, D. M. (1998). Repression of hedgehog signaling and BMP4 expression in growth plate cartilage by fibroblast growth factor receptor 3. *Development (Cambridge, UK)* **125**, 4977–4988.

Ng, L. J., Wheatley, S., Muscat, G. E., Conway-Campbell, J., Bowles, J., Wright, E., Bell, D. M., Tam, P. P., Cheah, K. S., and Koopman, P. (1997). SOX9 binds DNA, activates transcription, and coexpresses with type II collagen during chondrogenesis in the mouse. *Dev. Biol.* **183**, 108–121.

Olsen, B. R., Reginato, A. M., and Wang, W. (2000). Bone development. *Annu. Rev. Cell Dev. Biol.* **16**, 191–220.

Olson, E. N., and Klein, W. H. (1998). Muscle minus myoD. *Dev. Biol.* **202**, 153–156.

Otto, F., Thornell, A. P., Crompton, T., Denzel, A., Gilmour, K. C., Rosewell, I. R., Stamp, G. W., Beddington, R. S., Mundlos, S., Olsen, B. R., Selby, P. B., and Owen, M. J. (1997). Cbfa1, a candidate gene for cleidocranial dysplasia syndrome, is essential for osteoblast differentiation and bone development. *Cell (Cambridge, Mass.)* **89**, 765–771.

Parker, S. B., Eichele, G., Zhang, P., Rawls, A., Sands, A. T., Bradley, A., Olson, E. N., Harper, J. W., and Elledge, S. J. (1995). p53-independent expression of p21Cip1 in muscle and other terminally differentiating cells. *Science* **267**, 1024–1027.

Raffel, C., Jenkins, R. B., Frederick, L., Hebrink, D., Alderete, B., Fults, D. W., and James, C. D. (1997). Sporadic medulloblastomas contain PTCH mutations. *Cancer Res.* **57**, 842–845.

Reichenberger, E., Tiziani, V., Watanabe, S., Park, L., Ueki, Y., Santanna, C., Baur, S. T., Shiang, R., Grange, D. K., Beighton, P., Gardner, J., Hamersma, H., Sellars, S., Ramesar, R., Lidral, A. C., Sommer, A., Raposo do Amaral, C. M., Gorlin, R. J., Mulliken, J. B., and Olsen, B. R. (2001). Autosomal dominant craniometaphyseal dysplasia is caused by mutations in the transmembrane protein ANK. *Am. J. Hum. Genet.* **68**, 1321–1326.

Rosen, E. D., and Spiegelman, B. M. (2000). Molecular regulation of adipogenesis. *Annu. Rev. Cell Dev. Biol.* **16**, 145–171.

Rousseau, F., Saugier, P., Le Merrer, M., Munnich, A., Delezoide, A. L., Maroteaux, P., Bonaventure, J., Narcy, F., and Sanak, M. (1995). Stop codon FGFR3 mutations in thanatophoric dwarfism type 1. *Nat. Genet.* **10**, 11–12.

Rousseau, F., Bonaventure, J., Legeai-Mallet, L., Pelet, A., Rozet, J. M., Maroteaux, P., Le Merrer, M., and Munnich, A. (1994). Mutations in the gene encoding fibroblast growth factor receptor-3 in achondroplasia. *Nature (London)* **371**, 252–254.

Sahni, M., Ambrosetti, D. C., Mansukhani, A., Gertner, R., Levy, D., and Basilico, C. (1999). FGF signaling inhibits chondrocyte proliferation and regulates bone growth through the STAT-1 pathway. *Genes Dev.* **13**, 1361–1366.

Satokata, I., and Maas, R. (1994). Msx1 deficient mice exhibit cleft palate and abnormalities of craniofacial and tooth development. *Nat. Genet.* **6**, 348–356.

Satokata, I., Ma, L., Ohshima, H., Bei, M., Woo, I., Nishizawa, K., Maeda, T., Takano, Y., Uchiyama, M., Heaney, S., Peters, H., Tang, Z., Maxson, R., and Maas, R. (2000). Msx2 deficiency in mice causes pleiotropic defects in bone growth and ectodermal organ formation. *Nat. Genet.* **24**, 391–395.

Schipani, E., Lanske, B., Hunzelman, J., Luz, A., Kovacs, C. S., Lee, K., Pirro, A., Kronenberg, H. M., and Juppner, H. (1997). Targeted expression of constitutively active receptors for parathyroid hormone and parathyroid hormone-related peptide delays endochondral bone formation and rescues mice that lack parathyroid hormone-related peptide. *Proc. Natl. Acad. Sci. U.S.A.* **94**, 13689–13694.

Semba, I., Nonaka, K., Takahashi, I., Takahashi, K., Dashner, R., Shum, L., Nuckolls, G. H., and Slavkin, H. C. (2000). Positionally-dependent chondrogenesis induced by BMP4 is co-regulated by Sox9 and Msx2. *Dev. Dyn.* **217**, 401–414.

Shiang, R., Thompson, L. M., Zhu, Y. Z., Church, D. M., Fielder, T. J., Bocian, M., Winokur, S. T., and Wasmuth, J. J. (1994). Mutations in the

transmembrane domain of FGFR3 cause the most common genetic form of dwarfism, achondroplasia. *Cell (Cambridge, Mass.)* **78**, 335–342.

Simeone, A., Acampora, D., Pannese, M., D'Esposito, M., Stornaiuolo, A., Gulisano, M., Mallamaci, A., Kastury, K., Druck, T., Huebner, K., *et al.* (1994). Cloning and characterization of two members of the vertebrate Dlx gene family. *Proc. Natl. Acad. Sci. U.S.A.* **91**, 2250–2254.

Smits, P., Li, P., Mandel, J., Zhang, Z., Den, J. M., Behringer, R. R., de Crombrugghe, B., and Lefebvre, V. (2001). The transcription factor L-Sox5 and Sox6 are essential for cartilage formation. *Dev. Cell* (in press).

St. Jacques, B., Hammerschmidt, M., and McMahon, A. P. (1999). Indian hedgehog signaling regulates proliferation and differentiation of chondrocytes and is essential for bone formation. *Genes Dev.* **13**, 2072–2086.

Su, W. C., Kitagawa, M., Xue, N., Xie, B., Garofalo, S., Cho, J., Deng, C., Horton, W. A., and Fu, X. Y. (1997). Activation of Stat1 by mutant fibroblast growth-factor receptor in thanatophoric dysplasia type II dwarfism. *Nature (London)* **386**, 288–292.

Suda T., Kobayashi K., Jimi E., Udagawa N., and Takahashi N. (2001). The molecular basis of osteoclast differentiation and activation. *Novartis Found. Symp.* **232**, 235–247.

Takeda, S., Bonnamy, J. P., Owen, M. J., Ducy, P., and Karsenty, G. (2001). Continuous expression of Cbfa1 in nonhypertrophic chondrocytes uncovers its ability to induce hypertrophic chondrocyte differentiation and partially rescues Cbfa1-deficient mice. *Genes Dev.* **15**, 467–481.

Tavormina, P. L., Shiang, R., Thompson, L. M., Zhu, Y. Z., Wilkin, D. J., Lachman, R. S., Wilcox, W. R., Rimoin, D. L., Cohn, D. H., and Wasmuth, J. J. (1995). Thanatophoric dysplasia (types I and II) caused by distinct mutations in fibroblast growth factor receptor 3. *Nat. Genet.* **9**, 321–328.

Tickle, C. (1995). Vertebrate limb development. *Curr. Opin. Genet. Dev.* **5**, 478–484.

Ueki, Y., Tiziani, V., Santanna, C., Fukai, N., Maulik, C., Garfinkle, J., Ninomiya, C., doAmaral, C., Peters, H., Habal, M., Rhee-Morris, L., Doss, J. B., Kreiborg, S., Olsen, B. R., and Reichenberger, E. (2001). Mutations in the gene encoding c-Abl-binding protein SH3BP2 cause cherubism. *Nat. Genet.* **28**, 125–126.

Ueta, C., Iwamoto, M., Kanatani, N., Yoshida, C., Liu, Y., Enomoto-Iwamoto, M., Ohmori, T., Enomoto, H., Nakata, K., Takada, K., Kurisu, K., and Komori, T. (2001). Skeletal malformations caused by overexpression of Cbfa1 or its dominant negative form in chondrocytes. *J. Cell Biol.* **153**, 87–100.

Vortkamp, A., Lee, K., Lanske, B., Segre, G. V., Kronenberg, H. M., and Tabin, C. J. (1996). Regulation of rate of cartilage differentiation by Indian hedgehog and PTH-related protein. *Science* **273**, 613–622.

Vu, T. H., and Werb, Z. (2000). Matrix metalloproteinases: Effectors of development and normal physiology. *Genes Dev.* **14**, 2123–2133.

Vu, T. H., Shipley, J. M., Bergers, G., Berger, J. E., Helms, J. A., Hanahan, D., Shapiro, S. D., Senior, R. M., and Werb, Z. (1998). MMP-9/gelatinase B is a key regulator of growth plate angiogenesis and apoptosis of hypertrophic chondrocytes. *Cell (Cambridge, Mass.)* **93**, 411–422.

Wagner, T., Wirth, J., Meyer, J., Zabel, B., Held, M., Zimmer, J., Pasantes, J., Bricarelli, F. D., Keutel, J., Hustert, E., *et al.* (1994). Autosomal sex reversal and campomelic dysplasia are caused by mutations in and around the SRY-related gene SOX9. *Cell (Cambridge, Mass.)* **79**, 1111–1120.

Watanabe, H., and Yamada, Y. (1999). Mice lacking link protein develop dwarfism and craniofacial abnormalities. *Nat. Genet.* **21**, 225–229.

Watanabe, H., Kimata, K., Line, S., Strong, D., Gao, L. Y., Kozak, C. A., and Yamada, Y. (1994). Mouse cartilage matrix deficiency (cmd) caused by a 7 bp deletion in the aggrecan gene. *Nat. Genet.* **7**, 154–157.

Wechsler-Reya, R. J., and Scott, M. P. (1999). Control of neuronal precursor proliferation in the cerebellum by Sonic hedgehog. *Neuron* **22**, 103–114.

Wegner, M. (1999). From head to toes: The multiple facets of Sox proteins. *Nucleic Acids Res.* **27**, 1409–1420.

Weir, E. C., Philbrick, W. M., Amling, M., Neff, L. A., Baron, R., and Broadus, A. E. (1996). Targeted overexpression of parathyroid hormone-related peptide in chondrocytes causes chondrodysplasia and delayed endochondral bone formation. *Proc. Natl. Acad. Sci. U.S.A.* **93,** 10240–10245.

Wright, E., Hargrave, M. R., Christiansen, J., Cooper, L., Kun, J., Evans, T., Gangadharan, U., Greenfield, A., and Koopman, P. (1995). The Sry-related gene Sox9 is expressed during chondrogenesis in mouse embryos. *Nat. Genet.* **9,** 15–20.

Xie, W. F., Zhang, X., Sakano, S., Lefebvre, V., and Sandell, L. J. (1999). Trans-activation of the mouse cartilage-derived retinoic acid-sensitive protein gene by Sox9. *J. Bone Miner. Res.* **14,** 757–763.

Zehentner, B. K., Dony, C., and Burtscher, H. (1999). The transcription factor Sox9 is involved in BMP-2 signaling. *J. Bone Miner. Res.* **14,** 1734–1741.

Zhao, Q., Eberspaecher, H., Lefebvre, V., and De Crombrugghe, B. (1997). Parallel expression of Sox9 and Col2a1 in cells undergoing chondrogenesis. *Dev. Dyn.* **209,** 377–386.

III

Organogenesis

Early postimplantation development of the mouse is pre-occupied with the formation of the extraembryonic structures including the placenta, the yolk sac, the allantois, and the amnion (Chapter 8). Collectively, these tissues function to sustain the viability of the embryo by acting primarily as the fetal-maternal interface through which the nourishment and growth promoting factors reach the embryo and the metabolic waste is exported to the maternal compartment. To fulfill this critical role, the extraembryonic tissues form the blood vessels of the placental, umbilical, and vitelline vasculature and act as the first source of hematopoietic cells. Concomitant with the formation of the extraembryonic circulatory system, a parallel intraembryonic vascular network is established by the morphogenesis of the heart (Chapter 16) and vascular formation (Chapter 11). In addition, intraembryonic hematopoietic progenitors progressively take over as the major source of blood cells (Chapter 10). Studies on the impact of loss of function of various genes that are active in early postimplantation development reveal that a normal circulatory system is crucial for the viability of the embryo, whereas defective formation of other organs can be tolerated during prenatal life. In contrast, viability of the mice after birth is determined by the ability of the animal to obtain nourishment and remove metabolic waste, to perceive and respond to environmental factors, and to maintain homeostasis and normal body function. These are accomplished by the effective and concerted functioning of the multitude of specialized organs in the body. It is absolutely crucial that the vital organs be fully installed in the fetus in preparation for an independent existence after birth.

Formation of the Organ Primordium

Despite the significant variation in the architecture and tissue composition of the organs, the formation of the different organs follows remarkably similar strategy. Organogenesis begins with the *assembly of progenitor tissues into organ primordia,* which usually comprise mesenchymal and/or epithelial tissues derived from different germ layers. For example, the skin is composed of the ectodermal (epidermis) and mesodermal (dermis) tissues; the lung and liver contain the epithelium of the embryonic gut endoderm and the mesenchyme from the lateral plate or the intermediate mesoderm; and the pituitary gland is composed of ectodermal progenitors from the oral (surface) ectoderm and the neuroectoderm of the diencephalon. Other organs consist of a more complex collection of germ layer derivatives. The branchial arches that give rise to face and neck structures contain mesenchyme derived from the paraxial mesoderm and neural crest, and epithelium from the surface and oral ectoderm, and foregut endoderm. The auditory organ is made up of the ectodermal, mesodermal, and endodermal tissues of the branchial arch, branchial groove, and pharyngeal pouch. The otic vesicles derive from the ectodermal placode and neural crest cells. Furthermore, the primordia of some organs are formed by the allocation of the specialized germ layer derivatives within each body regions, such as the juxtaposition of the metanephric bud and the nephrogenic mesenchyme, the endodermal hepatic diverticulum with the mesenchyme of the septum transversum, and the optic evagination with the lens placode.

At the formative phase of the organ primordium, active *cell proliferation* occurs *to build up a critical tissue mass* for morphogenesis and for the specification of the diverse cell types that are unique to the organ. The multiplication of cells may be achieved in waves during the formative phase of organogenesis (e.g., anterior pituitary, cartilage, dentition, lens fiber cells, retina of the eye, heart, limb, lens, lung, and kidney) or throughout the life of the organ (e.g., liver, skin, and gut). Crucial to the initiation and maintenance of active tissue growth is the *vascularization* of the organ primordium (e.g., brain, liver, bone, gonads, and pituitary), which provides the conduit for the transport of nutrients and the growth regulating factors to the developing organ.

The acquisition of the lineage characteristics by cells in the organ primordium is influenced by *inductive interactions* that are mediated by cell–cell signaling, cell matrix interaction, or tissue induction. Examples are found in the endodermal induction of the heart mesoderm, the local induction of the otic placode in the surface ectoderm by the hindbrain, the induction of middle ear ossicle condensation by the branchial arch tissues, the induction of the pituitary by positive and negative signals from the hypothalamus, the induction of tooth bud by the odontogenic mesenchyme, the interactions between pigmented and neural retina, and the reciprocal induction of the tubular epithelium and the investing mesenchyme in the primordium of the lung, the mammary gland, and the kidney.

Common Morphogenetic and Molecular Paradigms in Organ Formation

The nine chapters in this section of the book provide an overview of the formation of a wide variety of organs: the gut and the associated organs (lung, liver, and pancreas, Chapter 15), the heart (Chapter 16), the gonads (Chapter 17), the kidney and the urinary structures (Chapter 18), the craniofacial structures (Chapter 19), the pituitary gland (Chapter 20), the sensory organs (eye, Chapter 21; ear, Chapter 22), and the skin and integumentary derivatives (Chapter 23). Despite the evident diversity of the anatomical and functional characteristics of these organs, some basic morphogenetic and genetic repertoires are universally employed for the formation of these organs.

Remodeling of Progenitor Tissues

During the assembling of the organ primordia, the progenitor tissues often undergo extensive architectural changes such as *epithelial-mesenchymal transformation* and the *condensation* of cells in the organ primordium. The neural crest cells that contribute to the craniofacial structures lose their epithelial morphology as they emerge from the lateral mar-

gin of the neural plate and acquire a mesenchymal morphology. When they have migrated to the appropriate sites, neural crest cells congregate to form the primordia of the cranial nerve ganglia underneath the ectodermal placode and the osteogenic condensation of the facial skeleton in the subectodermal region of the branchial arch. The formation of the skeleton in the limb and the vertebral column involves the condensation of the chondrogenic mesenchyme within which chondrogenesis takes place. The formation of the muscles of the limb begins with the transformation of the epithelial dermomyotome into the mesenchymal myotomal cells. After migration to the limb bud, the mesenchymal myoblasts aggregate and then fuse to form myotubes. During the formation of the metanephros, the intermediate mesoderm in the proximity of the ureteric bud condenses near the tip of the bud under the influence of cadherins and is restructured into epithelial tubules. In the cardiovascular system, the mesenchymal endothelial precursors are organized into epithelial rudiments that coalesce to form the endothelial tubes of the vessels and the endocardium of the heart.

Budding and Branching Morphogenesis

In many organs, the first phase of morphogenesis is the *formation of a bud rudiment* that projects into the surrounding mesenchymal tissues. The lung is formed by the budding of the tracheal portion of the embryonic foregut into the prospective lung mesenchyme. The thyroid, the liver and the pancreas all begin their development by the formation of an endodermal diverticulum that grows into the splanchnic mesoderm of the embryonic foregut and midgut. The kidney is formed by the extension of the endodermal ureteric bud from the urogenital sinus into the nephrogenic intermediate mesoderm. Inward growth of the ectoderm from the oral ectoderm and the epidermis into the branchial arch mesenchyme and the dermis heralds the formation of the tooth and the submandibular gland, and the hair, respectively. For nontubular organs, the invaginating bud typically proliferates to form a column of solid tissues and differentiates *in situ* at the site of budding (e.g., tooth, liver, and hair). In some cases, the bud bifurcates to generate two rudiments that develop into a paired organ (e.g., the thyroid gland). For tubular organs, a patent epithelial tube extends deeply into the mesenchyme and *branching* of the end of the ingrowing rudiment results in an increasing number of growing points for further elongation and branching. In the tubular organs, branching may be dichotomous or ramifying in configuration and is regulated by *regionalized cellular proliferation and apoptosis* and *polarized growth* of the tubules until a network of large and small ducts is built, eventually resulting in a mature gland. The formation of the inner ear involves the formation of a vesicle from the invaginating otic placode and the branching and polarized growth and cell death of the otic epithelium to form the canals and diverticula and the organ of Corti.

Establishment of Functional Compartments

The proper functioning of a complex organ depends on the integration of the activity of its components. A prerequisite for an effective integration is the organization of cellular components into *functionally and anatomically defined compartments.*

Several examples of *compartmentalization of tissues along the major axes* of the organs are found. In the neural tube, specific groups of afferent and efferent neurones are regionalized in the dorsal-ventral plane and in major segments along the anterior-posterior axis of the brain and the spinal cord (Chapters 5 and 6). Specialization of the gut along its oral (anterior)–anal (posterior) axis is evident from segment-specific histological features of the mucosa and submucosal musculature, and the formation of different glandular structures and the associated endodermal organs (Chapter 15). In the jaws, the tooth buds are spaced along the proximal-distal axis of the jaw and each of them forms teeth of different morphology but appropriate to their position in the dentition (Chapter 19). In the heart tube, precursors of the adult heart chambers are found initially along its long axis. The serial subdivision of the functional components of the heart is transformed into a parallel left and right partition through region-specific growth and morphogenesis and the stereotypic looping of the heart tube (Chapter 16).

Functional specialization of the organ may be coupled to the *zonation or laminar organization* of the tissue compartments. The zonation of tissues in the adrenal cortex reflects the superficial to deep order of the formation of the tissue layers during organogenesis and the progressive displacement of tissue toward the medulla and concomitant expression of different endocrine phenotypes of the cortical cells in the adult gland. Similarly, the stratification of the epidermis also reveals the sequence of differentiation of the epidermal cells both during development and in the adult mice (Chapter 23). In the central nervous system, zonation in cerebral cortex, the cerebellum, and the spinal cord is the result of the sequence of differentiation of the neuroepithelium, and the temporal order of generation and the pattern of radial and tangential migration of the neuroblasts. A similar pattern of radial migration and differentiation of progenitor cells also underpins the organization of layers of neurons and supporting cells in the neural retina (Chapter 21).

In other organ primordia, a distinct *partitioning of cells with common functional properties* occurs, such as the segregation of cells that produce different trophic hormones in the pituitary, the separation of somatic and germ cells in the gonads, and the spatial arrangement of the germinal cells and mature fiber cells in the developing lens.

Themes and Variations of the Molecular Control of Organogenesis

The adoption of similar morphogenetic strategies in the formation of organs is paralleled by the activation of similar molecular functions. These common molecular activities often involve the activation of related genes and pathways but result in morphological and functional attributes that are unique to specific organs.

Much intercellular signaling has been shown to be involved with *inductive interactions* at the formative stage of organ development. Specific responses to these signals leads to the *specification* of the different precursor cells (bone morphogenetic proteins for myocardium, lens, pituitary, and inner ear; platelet-derived growth factors for mesangial cells of the nephron; Sonic hedgehog factor in the induction of Vax-expressing ventral cells in the eye) and the *initiation of cell differentiation* [fibroblast growth factors on lens cells, otic placode, and pituitary cells; WNT (wingless) factors on cells of the nephric tubule, pituitary, tooth germ, and otic placode]. Signaling pathways are also involved in mediating epithelial-mesenchymal interaction, developmental competence, cell proliferation, and the apoptosis required for the completion of organogenesis.

Many lineage-specific genes, most of them encoding transcription factors, are expressed in response to inductive signals as cells differentiate and acquire organ-specific phenotypes. For example, *Pod1* is expressed in podocytes of the nephrons, *Sox9* in the Sertoli cells, and *Pdx, Pax4, Pax6, Isl,* and *Nkx* in the pancreatic islet. The expression of *Pit1, Ptx1, Ptx2, Rpx, Lhx3,* and *Lhx4* transcription factors in the pituitary not only marks the various types of endocrine cells in the pituitary, but their synergistic activity when coexpressed is critical for the specification of these cells. Similar synergistic interaction of *Sox5, Sox6,* and *Sox9* is also important for the specification of chondrocyte and bone cells.

Activation of transcription factors and downstream components is also associated with the *establishment of the functional compartments* in different organs. The segmental specific characteristic of the hindbrain, the spinal cord, and the vertebrae can be correlated with the domain of the *Hox* gene in the anterior-posterior axis of the neural tube and the paraxial mesoderm. The activity of the homeobox-containing genes in the branchial arch influences the patterning of the dentition. Compartmentalization of the fetal gut is regulated by the ParaHox genes, and that of the limb bud by the *Hoxd* genes. Finally, in the eye, the differential expression of *Pax6* and *Rx* is associated with the dorsal-ventral (the *Tbx5* and *Vax2*-expressing region) and nasal-temporal (the *BF1* and *BF2* domain) patterning of the neural retina.

Of special note is the *activation of similar molecular pathways* involving different members of the same gene family in unrelated organs. For example, the combination of *Pax-*

Eya-Six-Dach activity is associated with eye development, muscle differentiation, and kidney morphogenesis, whereas *Eya-Prx-Pax* activity is involved in pituitary and inner ear development. *Sox9* is essential for the development of the gonad and the skeleton, while *Sox1* and *Sox2* are key players in neural tube and eye development. This may suggest that the development of different organs involves the deployment of a limited number of common molecular pathways, which activate different organ-specific downstream responses. However, unique molecular mechanisms also control the development of specific organs and the details are outlined in Chapters 15 through 23 in this part of the book.

15

Development of the Endoderm and Its Tissue Derivatives

Brigid L. M. Hogan* and Kenneth S. Zaret†

*Howard Hughes Medical Institute, Department of Cell Biology, Vanderbilt University Medical Center, Nashville, Tennessee 37232
†Cell and Developmental Biology Program, Fox Chase Cancer Center, Philadelphia, Pennsylvania 19111

I. Introduction and Overview

II. Endoderm Development prior to Organogenesis

III. Patterning and Differentiation of the Digestive Tract

IV. Development of Tissues That Bud from the Endoderm

V. Perspectives and Remaining Issues on Organogenesis from the Endoderm

References

I. Introduction and Overview

The definitive endoderm is a population of multipotent stem cells allocated as one of the primary germ layers during gastrulation. Initially an epithelial cup, it is rapidly organized into a tube that runs along the anterior-posterior (AP) axis of the embryo. The endoderm gives rise to the major cell types of many internal organs, including the thyroid, thymus, lung, stomach, liver, pancreas, intestine, and bladder. Even the prostate and urethra are of endodermal origin (Kurzrock *et al.*, 1999). Most of these tissues have secretory and/or absorptive functions and play important roles in controlling body metabolism. Their development poses many interesting biological problems, over and above the basic questions of how endoderm cell fate, proliferation, and differentiation are controlled. For example, the formation of the different organ systems involves remarkably diverse morphogenetic programs, which are only just beginning to be understood at the molecular level. The lung, pancreas, and prostate all originate as simple buds that subsequently develop into branched organs with very different three-dimensional structures. The development of the intestine, on the other hand, involves the regional elaboration of the lumenal surface into crypts and villi.

During the past decade, significant progress has been made in identifying the hierarchy of genes that regulates the differentiation of endodermal stem cells into the many different cell types present in each organ. In addition, modern molecular and cell biology techniques have added new dimensions to classical tissue recombination, grafting, and *in vitro* culture experiments carried out with chick and mouse embryos in the 1960s. The blending of these approaches has led to numerous insights into the mechanisms of organogenesis. Interest in this area has intensified because processes that govern the early development of endoderm-derived tissues may be recapitulated during tissue regeneration in diseased adult organs. Resident stem cell activation and cell transplant therapies are also viewed as important areas of investigation. This chapter reviews progress in endoderm development and highlights areas of possible relevance to medical science as well as questions that remain to be answered in endoderm biology.

II. Endoderm Development prior to Organogenesis

A. Relation to the Yolk Sac Visceral Endoderm

The first endoderm population—the primitive endoderm—develops on the surface of the inner cell mass of the blastocyst. It gives rise to the extraembryonic parietal endoderm and the visceral endoderm (VE) of the yolk sac (see Chapter 8). There appears to be no simple lineage relationship between the VE and the so-called definitive endoderm, or DE, that gives rise to the endoderm-derived tissues of the adult mouse. As described below, cell marking studies have shown that virtually all of the definitive gut endoderm is derived from the epiblast, a cup-shaped cell layer that is surrounded by the VE (Fig. 1). Although the VE is only transiently associated with the epiblast, it nevertheless plays a crucial role in the AP patterning of the embryo (see Chapters 3 and 6). Also, the VE expresses a large number of secreted serum proteins and transcription factors in common with the fetal gut and liver (e.g., Meehan *et al.*, 1984) and is consequently thought to nurture the embryo, both before and after the placental connections are established (Duncan *et al.*, 1997). In fact, it has been

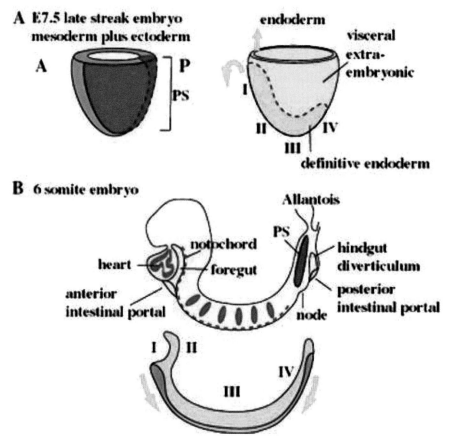

Figure 1. Formation of definitive endoderm. (A) Mouse embryo at the late streak stage, separated into an inner layer of ectoderm plus mesoderm (left) and an outer single-layered epithelial cup of endoderm (right). The endoderm consists of visceral extraembryonic endoderm that is being displaced proximally (arrow) and is contiguous distally with the definitive endoderm. The junction is shown schematically (dotted line). The endoderm will subsequently fold ventrally (arrow) in the anterior and posterior regions to form the foregut pocket and (hindgut diverticulum), respectively. (B) Six-somite-stage embryo separated into ectoderm and mesoderm (upper) and endoderm (lower). The embryo is folding in the direction of the arrows both anteriorly and posteriorly (arrows) and the open tube will close ventrally. I, II, III, and IV indicate the endodermal precursor tissues of the floor of the foregut (I), dorsal part of the foregut (II), midgut (III), and hindgut (IV) (Lawson *et al.*, 1986).

challenging to find gene products that distinguish the VE from anterior DE and fetal hepatocytes. The targeted inactivation of genes encoding various "gut-" or "liver-enriched" transcription factors, such as HNF4, GATA6, and HNF1β, causes defects in the VE leading to early embryonic lethality (Chen *et al.*, 1994; Morrisey *et al.*, 1998; Barbacci *et al.*, 1999). These findings undoubtedly reflect the need of both the VE and the liver to express genes that control secreted serum proteins and metabolite absorption and processing. The recent ability to derive endoderm-like tissue from pluripotential human embryonic stem cells in culture (Thomson *et al.*, 1998) underscores the importance of understanding the full developmental capabilities of the VE for potential cell therapeutic approaches. Studies with the chick embryo have shown that extraembryonic yolk sac endoderm and allantois endoderm can differentiate morphologically into various gut phenotypes when grafted into the developing embryo (Masui, 1981; Yasugi *et al.*, 1985). However, this morphological differentiation is not apparently accompanied by appropriate changes in gene expression.

B. Formation of the Definitive Endoderm during Gastrulation

Fate mapping experiments in the mouse have shown that the definitive endoderm is derived from cells that occupy a localized region of the prestreak epiblast (Fig. 1A). During gastrulation these cells transit the primitive steak and intercalate with the VE. They then spread anteriorly and laterally over the ventral surface of the developing embryo, so that by the head fold stage the DE consists of about 500 cells organized into a single-layered, cuplike sheet (Fig. 1B). The DE is thought to displace the VE proximally, so that normally only a few VE descendants are found in the anterior foregut.

Evidence that the DE arises in the epiblast came from an elegant series of experiments in which Lawson *et al.* (1991) injected an intracellular label to follow the fate of epiblast cells of the prestreak, early streak, and late streak stages (Fig. 1A). Injected embryos were cultured *in vitro* for 24 hr and the positions of the injected cells were scored histologically. Similar strategies have been conducted with chick embryos (Rosenquist, 1971; Stern and Canning, 1990; Schoenwolf *et al.*, 1992). For prestreak and early streak injections of mouse embryos, cells labeled at the anterior of the primitive streak gave rise to DE. Interestingly, this domain corresponds to the domain of the epiblast in which the gene encoding the winged helix transcription factor, HNF3β, is expressed (Sasaki and Hogan, 1993; Monaghan *et al.*, 1993; Ang *et al.*, 1993). *Hnf3β* (Foxa2) expression is subsequently found in the nascent DE and the mesendoderm of the head process, notochord, and ventral node that arise during gastrulation. Studies of chimeric mouse embryos made from wild-type tetraploid embryos and *Hnf3β* $^{-/-}$ mutant embry-

onic stem (ES) cells have shown that the gene is required for the development of the foregut and midgut endoderm. In these chimeras, the wild-type, tetraploid VE is not displaced proximally by mutant DE cells and so populates the much reduced foregut pocket at E9.0 (Dufort *et al.*, 1998). An unresolved question is whether diploid VE would exhibit the developmental capacities of the definitive endoderm if not displaced from the foregut.

Recent studies have revealed a host of signaling molecules and transcription factors that control the formation of the DE. Expression of fibroblast growth factor receptor 1 (FGFR1) is essential for endoderm and mesoderm development, indicating that responsiveness to FGF is critical during gastrulation (Ciruna *et al.*, 1997). Studies with chimeras have provided evidence for a role for *Smad2* (*Madh2*), a component of the transforming growth factor β (TGF-β) family signaling pathway. This is shown by the fact that *Smad2* homozygous null mutant cells fail to colonize the endoderm during gastrulation, whereas they do colonize the mesoderm (Tremblay *et al.*, 2000). Since *Smad3* (*Madh3*) does not compensate for the deficiency, TGF-β family signaling via SMAD2 must be a key parameter for epiblast cells to give rise to endoderm versus mesoderm. TGF-β signaling has also been found to be critical during early endoderm formation in *Xenopus* (Henry *et al.*, 1996; Zorn *et al.*, 1999). For the mouse, much more must be learned about where allocation of the endoderm lineage takes place and whether it is before or during the transit of epiblast-derived cells through the primitive streak.

In *Xenopus*, injection of mRNAs for the HMG domain protein, XSOX17α/β, or the homeodomain protein, MIXER, is sufficient to direct an endodermal cell fate in animal cap cells (Hudson *et al.*, 1997; Henry and Melton, 1998), with MIXER being able to induce the *Xsox17* genes and not vice versa. However, *Xsox17* genes are expressed earlier in Xenopus development than *Mixer*, suggesting that the genes are induced independently and then Mixer maintains *Xsox17* expression (Yasuo and Lemaire, 1999). MIXER is part of a growing "Mix" subfamily of paired-type homeodomain proteins that are being discovered in *Xenopus* endoderm development (Rosa, 1989; Ecochard *et al.*, 1998; Tada *et al.*, 1998). In zebrafish, Nodal signaling induces the Mix-like gene *bonnie and clyde*, which is critical for the generation of endoderm (Alexander and Stanier, 1999; Kikuchi *et al.*, 2000). In the mouse, the expression of one Mix-related gene, *Mml*, appears restricted to the mesoderm (Pearce and Evans, 1999), which is consistent with the dual function of the *Xenopus* Mix proteins in both mesoderm and endoderm development (Ecochard *et al.*, 1998; Tada *et al.*, 1998). Additional *Xenopus* genes for endoderm development include the maternal VegT transcription factor (Zhang *et al.*, 1998) and its homeobox target genes *Bix1* and *Bix4* (Tada *et al.*, 1998, Casey *et al.*, 1999). Clearly, it will be important to evaluate the roles of homologs of these genes in mouse endoderm development.

At present the origin of the endoderm cells in the ventral hindgut is not known. The hindgut develops from a tube or diverticulum that appears in the five somite embryo (Fig. 1B). It is possible that by about E8.5, the hindgut represents a separate cell lineage and that no further endoderm is generated from the primitive streak (Wilson and Beddington, 1996). It is also possible that some of the ventral hindgut is derived from the posterior VE (K. Lawson, personal communication). Consistent with these theories is the finding described above that while *Hnf3β* is required to make foregut and midgut endoderm, it is not required to make hindgut endoderm, indicating different mechanisms of genetic control (Dufort *et al.*, 1998). Given that the hindgut is the progenitor of important structures such as the bladder, urethra, and rectum, it would be useful to know more about the development of the hindgut diverticulum. In addition, little is known about the morphogenetic movements by which the allantois, which is at the posterior of the primitive streak at the six somite stage (Fig. 1B), becomes located ventral and anterior to the future cloaca, after embryo turning. The allantois contains an extension of the hindgut known as the urachus.

1. Early AP Patterning and Regionalization of the Endoderm

Lawson *et al.* (1986) also carried out lineage labeling with late streak mouse embryos. By 24 hr in culture, these embryos had generated about 6–10 somites and undergone extensive morphogenetic movements, including the formation of the foregut and hindgut pockets (anterior and posterior intestinal portals; Fig. 1B). Cells injected in the most anterior endoderm at the midline (zone I, Fig. 1B) ended up in the ventral foregut, the presumptive progenitor population of the liver, ventral pancreas, stomach, and lungs. This location results from the anterior endoderm folding and looping under the developing heart, with which it is transiently in contact before the development of the septum transversum. Cells labeled in zone II became located in the dorsal foregut, which presumably gives rise to the esophagus, stomach, dorsal pancreas, and duodenum. Zone IV cells contributed to epithelial cells in the midgut and dorsal hindgut, which form the intestine.

Clues regarding the AP patterning of the endoderm in the gastrulation stage embryo are beginning to emerge. Both the anterior VE (AVE) and the node are signaling sources that help define the most anterior structures of the embryo, while signals from the primitive streak may pattern the early posterior endoderm (Thomas *et al.*, 1998; Bachiller *et al.*, 2000; Wells and Melton, 2000). Lineage studies have shown that the cells of AVE originate over the distal tip of the prestreak epiblast and migrate anteriorly. During their migration the AVE cells express the homeodomain gene *Hhex* and they maintain this expression at the very anterior of the embryo,

before finally being displaced proximally into the extraembryonic region at the late streak stage (Thomas *et al.*, 1998).

During gastrulation, the AVE expresses a variety of genes encoding transcription factors and secreted signaling factors and antagonists (see Chapter 3; Bielinska *et al.*, 1999). Interestingly, many of these genes are also expressed in the anterior DE, further underscoring the developmental similarities of these tissues, despite their apparent lineage differences. By contrast, the posterior DE at the late streak stage expresses *intestinal fatty acid binding protein* gene (*Ifabp*) and the caudal homeobox gene *Cdx2* (Wells and Melton, 2000). *In vitro* culture experiments have shown that the initial AP patterning of the DE is already established at the late streak stage. However, it can be altered by culturing isolated endoderm with mesoderm from different regions. For example, endoderm anterior of the node can be posteriorized by coculture with primitive streak mesoderm or FGF4, a factor produced by the streak (Wells and Melton, 2000). This initial patterning is later refined by genes expressed within the endoderm and mesoderm (see below).

C. Gastrulation to Midgestation: General Aspects of Endoderm Development

1. Morphogenesis and Spatial Relation to the Mesoderm

During gastrulation, the future anterior and trunk endoderm is contiguous dorsally with the future notochord (Fig. 1B), and it is therefore often referred to as "mesendoderm." At the head fold and early somite stages (E7.5–8.0), the foregut and hindgut pockets form, with their ventral surfaces expanding toward the midline, generating wide anterior and posterior intestinal portals. These morphogenetic movements and the subsequent deepening of the gut pockets create spatial distinctions between ventral and dorsal domains of endoderm, so that the ventral and dorsal foregut endoderm are now exposed to different mesodermal influences. For example, only the ventral endoderm is apposed to the cardiogenic mesoderm, whereas only the dorsal endoderm is apposed to the notochord. By the eight-somite stage (E8.5), the foregut pocket is deep and the specification of the liver, pancreas, and thyroid has just occurred, as described in detail below. At this stage, the anterior of the foregut is sealed at the buccopharyngeal membrane.

More laterally in the trunk region, the endoderm is associated with the somites and intermediate and lateral plate mesoderm. The notochord detaches from the endoderm in an AP sequence and the lateral plate mesoderm splits into the splanchnopleure and somatopleure layers, and the paired dorsal aortae form near the dorsal side of the endoderm tube, which gradually moves away from the notochord. The splanchnic mesoderm gives rise to the mesoderm of the gut posterior to the branchial arches, including the muscle layers around the intestines, and of organs such as the lung and ventral pancreas. The origin of the surface mesothelial layer that faces the peritoneal cavity is not known.

The turning of the embryo, from the 9- to 20-somite stages (E8.5–9.0) allows the lateral walls of the endoderm sheet to move together and fuse, generating a tube. The ventral closing of the gut tube is likely to involve extensive growth and interactions among all three layers (ectoderm, endoderm, and mesoderm). During midgestation, the increase in capacity of the body cavity does not keep up with the rapid growth of the intestines. Consequently, there is a natural and transient herniation of the gut outside the ventral body wall until around E18, when the body wall normally closes up. The endoderm of the midgut is also transiently connected with that of the yolk sac through the vitelline duct, and the mesoderm of the unclosed ventral body wall is continuous with the mesoderm of the amnion (Fig. 2). Several mutants have been described with failure in ventral body wall closure. For example, embryos lacking the genes encoding *Bmp1* have herniation of the gut at birth, probably due to a defect in the collagen fibers in the mesoderm splanchnic mesoderm and the mesoderm around the umbilicus (Suzuki *et al.*, 1996).

As shown in Fig. 2, there are differences in the relative growth of different regions of the embryo between the 12- and 55-somite stages so that the position of the endodermal organs relative to the forming vertebrae changes as development proceeds. During this period, the buccopharyngeal membrane breaks down to create continuity between the pharynx and the amniotic cavity. By the 30- to 34-somite stages (E10), the endoderm at the caudal region of the hindgut contacts the surface ectoderm, forming the cloacal membrane. Later, the cloacal membrane undergoes remodeling to give rise to separate openings of the urethra and rectum.

2. AP Regionalization of the Gut Tube

As we saw in Section II.B.1, the initial patterning of the definitive endoderm into broad presumptive territories is established at the late streak stage. This pattern is later refined between about E8.5 and E12.5 into regions corresponding to specific organs of the gut tube. This is achieved by the combined activity of signaling molecules and transcription factors, the latter including members of the Hox, ParaHox, Pax, and Nkx families. These are expressed in spatially restricted patterns, either in the endoderm or surrounding splanchic or neural crest mesenchyme, or in both. Reciprocal interactions between the endoderm and mesoderm and the dynamic nature of the transcription patterns sometimes make it difficult to assign specific roles to endodermal versus mesodermal gene expression in gut specification. In addition, the results of experimental tissue recombination experiments to study interactions sometimes vary, depending on the species in which they are done and the precise stage of development at which the samples are obtained.

The role of *Hox* genes in gut specification in the mouse has been reviewed (Sekimoto *et al.*, 1998; Pitera *et al.*, 1999; Beck *et al.*, 2000). The expression patterns at E12.5 are summarized in Fig. 3. Note that the Hox genes are expressed solely in the mesoderm or in both the mesoderm and the endoderm. Also note that the AP expression of some genes, such as *Hoxa3, Hoxa7, Hoxb6,* and *Hoxb8* is punctuated by a domain where the gene is not expressed. At an earlier stage, from about E8.5, *Hoxa1, Hoxa2, Hoxa3,* and *Hoxb1* are expressed in the anterior endoderm of the pharynx and foregut, as well as in the surrounding neural crest mesenchyme and splanchnic mesoderm. Embryos lacking a functional *Hoxa3* gene lack a thymus (Manley and Capecchi, 1995), but it is not known whether the primary defect resides in the endoderm, the mesenchyme, or both (see Section IV.A.2 below). Recent experiments show that inhibition of retinoic acid signaling, by treatment with a pan-RAR inhibitor, changes the expression pattern of a number of genes including anterior *Hox* and *Pax1* and *Pax9* genes, and genes encoding FGF3 and FGF8 in the endoderm of the embryonic pharynx and foregut (Wendling *et al.*, 2000). Treated embryos have major defects in the morphogenesis of tissues derived from the branchial region and anterior foregut. The striking alteration in the expression of genes in the endoderm of treated embryos reveals how strongly the endoderm influences the behavior and fate of the surrounding mesenchyme (Wendling *et al.*, 2000).

In the hindgut, deletion of *Hoxd13,* which is normally expressed in the smooth muscle layers of the rectum, leads to defects in anal sphincter morphogenesis (Kondo *et al.*, 1996). Further deletion of *Hoxd4, -8, -9, -10,* and *-11,* in addition to *Hoxd13,* causes the absence of the ileocecal sphincter between the large and small intestine (Zákány and Duboule, 1999). Thus, Hox gene expression appears to regulate the formation of functional boundaries, that is, sphincters, between different regions of the gut tube (Roberts *et al.*, 1995).

ParaHox genes, including *Pdx1, Cdx1,* and *Cdx2,* are so named because they were found to lie in an evolutionarily conserved position outside of the traditional Hox cluster (Brooke *et al.*, 1998). They are expressed in the endoderm and, like the Hox cluster, are arranged in the AP orientation in which they are expressed in the embryo (Fig. 3). ParaHox homologs in mice play clear developmental roles in endoderm patterning. For example, mice heterozygous for null mutation in *Cdx2,* which is normally expressed in the endoderm of the posterior gut, develop polyps in the proximal colon composed of endoderm with more anterior phenotypes characteristic of stomach and small intestine (F. Beck *et al.*, 1999). It has been argued that this represents a homeotic transformation of the *Cdx2* nonexpressing colon into stomach, with intercalary regeneration between (F. Beck *et al.*, 1999). *Pdx1* is expressed in the endoderm of the pancreatic buds and duodenum and is critical for pancreas development (see Section IV.D.2 below).

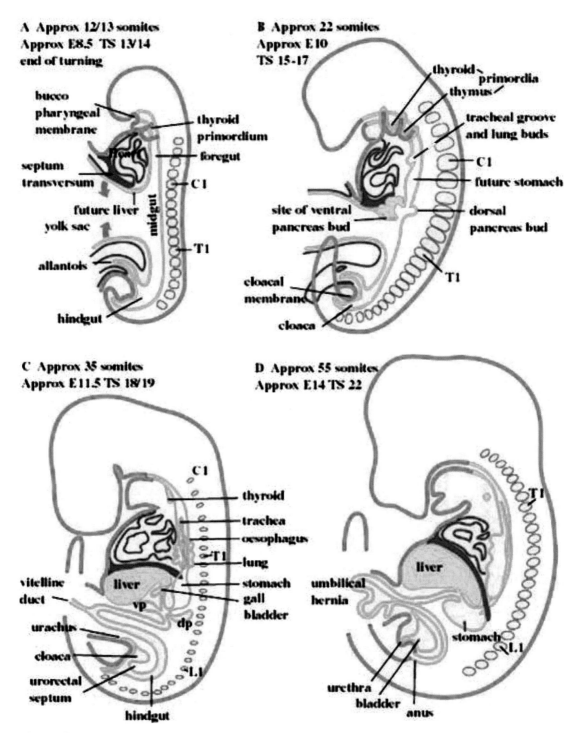

Figure 2. Development of mouse endodermal organs from E8.5 to E14.0. Schematic parasagittal sections showing derivatives of the definitive endoderm in yellow, ectoderm in green, and mesoderm in red. Note the changing position of the endodermal organs relative to the first cervical (C1), thoracic (T1), and lumbar (L1) vertebrae as development proceeds. (A) Approximately E8.5. The buccopharyngeal membrane is still intact. The ventral body wall and open midgut tube are closing (orange arrows). (B) Approximately E10. Buccopharyngeal membrane has broken down but cloacal membrane is still intact. (C) Approximately E11.5. The diaphragm is beginning to form and to separate the thoracic and peritoneal cavities. vp, ventral pancreas bud; dp, dorsal pancreas bud. The urorectal septum is beginning to separate the future bladder from the rectum. (D) Approximately E14.0 The rectal membrane has broken down.

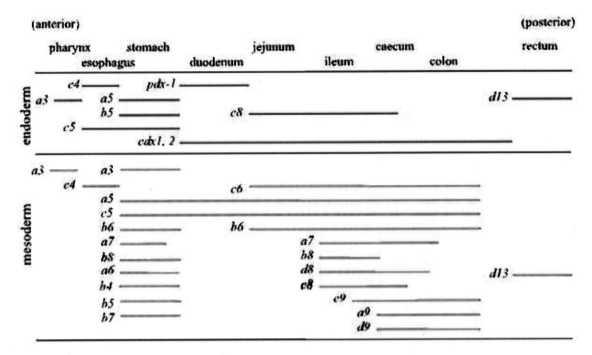

Figure 3. Hox gene expression in gut development. Horizontal lines depict the extent of expression of Hox and ParaHox genes in the AP orientation of the E12.5 gut. Expression in the endoderm is depicted in red; mesoderm, green. (Data adapted from Sekimoto *et al.*, 1998, and Beck *et al.*, 2000.)

III. Patterning and Differentiation of the Digestive Tract

The digestive tract develops from a simple tube with the typical inner endodermal epithelial layer surrounded by splanchnic mesoderm. Its function is to digest food and to absorb nutrients and water. The four major regions are the esophagus, stomach, small intestine (including the duodenum, jejunum and ileum), and large intestine (including the cecum, colon, and rectum). They are separated from each other by muscular sphincters that control the passage of material from one section to the next, while the pharyngoesophageal and anal sphincters control the passage of food into and out of the system (Fig. 4).

The intestines are suspended in the body cavity by extensions of the peritoneal mesoderm known as the mesentery, in which the major blood vessels are found. Beginning about E10.5, the midgut (the region of the intestinal tract between the distal duodenum and the distal third of the colon) does not fit into the peritoneal cavity and protrudes outside the ventral body wall, forming the so-called physiological hernia. During their reentry into the cavity, a process that is complete by E16.5, the intestines are rotated and folded into a specific packing pattern. The vitelline duct (Fig. 2) normally closes around the 25-somite stage, but if it fails to do so properly, a defect known as Meckel's diverticulum may develop. Interestingly, the Meckel's defect is due to intestinal cells at the position of the former vitelline duct differentiat-

ing into other gut-derived cell types, including gastric, pancreatic, and hepatic cells (Garretson and Frederich, 1990; St. Vil *et al.*, 1991). Recent studies indicate that these other cells types develop ectopically due to a local absence of restrictive signals from the mesenchyme that normally limit gut differentiation to intestinal cell types (Bossard and Zaret, 2000).

During development, the undifferentiated splanchnic mesoderm becomes organized into distinct radial layers (Fig. 5). Immediately underneath the gut epithelium is the lamina propria, which is rich in connective tissue, blood vessels, and lymphatics. It is delimited by a layer of smooth muscle known as the muscularis mucosae. The next layer is the highly cellular submucosa, and it is surrounded by an inner circular muscle layer and an outer longitudinal muscle layer. Finally, there is a mesothelial layer facing the peritoneal cavity. The digestive tract is innervated by the enteric nervous system, derived from the neural crest. There are two major enteric plexi; one lying between the circular and longitudinal muscle layers (the myentric plexus) and the other in the submucosa (Fig. 5). The axons form multiple connections and coordinate the contraction and relaxation of the muscles during peristalsis.

Both the endoderm and mesoderm in different regions of the gut become highly specialized for different functions. For example, in the small intestine the epithelium is thrown into circular or spiral folds with numerous long villi that maximize the surface area for nutrient absorption, while in the sphincters the circular muscle layer is particularly thick. The development of these regional specializations, which

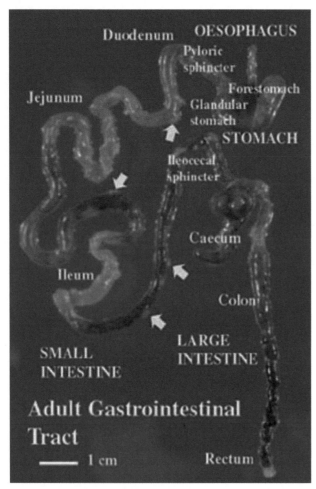

Figure 4. Adult mouse gastrointestinal tract. Photograph of the gastrointestinal tract dissected free of the mesentry from an adult female mouse. Sphincters are indicated by red arrows and Peyer's patches by yellow arrows. Peyer's patches are specialized aggregates of lymphoid cells in the lamina propria. They are overlaid by epithelium containing specialized M cells, which are responsible for transporting microorganisms across the epithelium.

become manifest predominantly after E14.5, in preparation for birth, depends on reciprocal interactions between the endoderm and mesoderm. Although most studies in this area have been done with chick embryos (see, for example, Ishii *et al.,* 1998; Roberts *et al.,* 1998, and references therein), progress is being made in understanding these interactions at a molecular and cellular level in the mouse. It is important to note that significant changes occur in the morphology and gene expression profiles of the mammalian small and large intestines between birth and weaning. These temporal changes are associated with the switch from milk to solid food and are regulated in part by changes in glucocorticoid levels, as mediated through the mesenchyme. The changes are unique to mammalian development, and it is important to bear this in mind when comparing experiments carried out with chick or mouse embryos designed to elucidate interactions be-

tween epithelium and mesenchyme (Kedinger *et al.,* 1986, 1987a,b,c; Haffen *et al.,* 1987; Duluc *et al.,* 1994).

A. Development of the Stomach

The mouse stomach is divided into a proximal forestomach (fundus) and a distal glandular stomach (antrum). In the neonatal and adult mouse the lining of the forestomach, like the esophagus, consists of numerous folds of stratified, keratinized, squamous epithelium (Fukamachi *et al.,* 1979; Lyons *et al.,* 1989). This contrasts vividly with the glandular stomach, which has a simple epithelial lining and numerous gastric pits and glands. The differentiation of the mouse stomach epithelium is first noted around E12.5. Keratinization of the forestomach is first seen at E16.5, while primitive gastric glands can be detected 1 day earlier (Fukamachi *et al.,* 1979). *In vitro* culture experiments indicate that stomach endoderm cannot proliferate or differentiate in the absence of mesenchyme. Moreover, the actual developmental fate of stomach epithelium from E11.5 or later cannot be altered significantly by heterotypic recombination with mesenchyme from either forestomach or glandular stomach (Fukamachi *et al.,* 1979).

Studies in the mouse have shown that the three muscle layers—the circular layer, the longitudinal layer, and the muscularis mucosae—form at different times in the fore and glandular stomach. The circular layer forms in both regions first, at around E11–13, at a specific distance from the epithelium. The outer longitudinal layer appears in the forestomach at E15 but not in the glandular stomach until around birth. The muscularis mucosae immediately adjacent to the epithelium also forms neonatally, at about the same time in both regions (Takahashi *et al.,* 1998).

If the E11 stomach is cultured intact *in vitro,* the muscle layers differentiate in the normal spatial pattern, but if the endoderm is removed, no smooth muscle formation is seen at all. Recombining stomach epithelium with the mesenchyme restores muscle development in the correct spatial pattern (Takahashi *et al.,* 1998). Tracheal endoderm from E11 embryonic lung is also able to induce smooth muscle differentiation in stomach mesenchyme, but in a different pattern. These results suggest that factors produced by the stomach epithelium regulate the specific timing and localization of smooth muscle differentiation. Candidate inducing molecules are Sonic hedgehog (SHH) and Indian hedgehog (IHH), the genes for which are expressed in overlapping domains in the gut epithelium, including the stomach epithelium (Ramalho-Santos *et al.,* 2000). Homozygous null *Shh* mutant embryos show overproliferation of the epithelium in the glandular region, but this epithelium has the phenotype of small intestine. There is also a reduction in the amount of muscle in the mutant mesenchyme, but, interestingly, no change in the expression of *Bmp4,* a potential downstream target of hedgehog signaling (A. McMahon, personal communication).

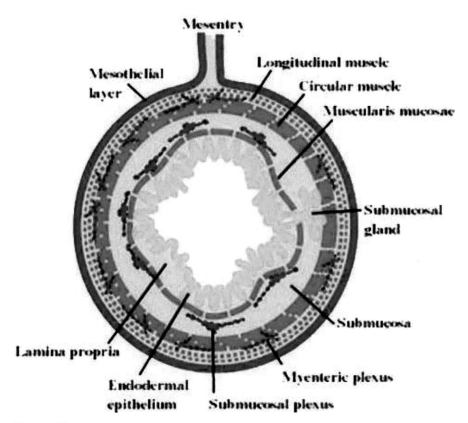

Figure 5. Schematic cross section through adult intestine showing radial organization of the mesodermal layers. The inner endodermal epithelium (yellow) is underlaid by several layers of mesodermal tissues, including smooth muscle (muscularis mucosae, circular muscle, and longitudinal muscle). The whole gut is covered in a thin mesothelium and is suspended in the peritoneal cavity by the mesentery. Two neural plexi are present, one underneath the muscularis mucosae and one between the circular and longitudinal muscle layers.

Studies with chick embryos have suggested that *Bmp4* transcription in the duodenal mesenchyme regulates the formation of the pyloric sphincter and the characteristic expression in this region of *Nkx2-5* (Smith and Tabin, 1999a).

B. Small Intestine: Duodenum, Jejunum, and Ileum

At E9.0, expression of the *Pdx1* homeobox gene marks the duodenal region of the closing gut tube (Offield *et al.,* 1996), in addition to the hepatic duct, which opens into the gut at the level of the duodenum. In embryos lacking a functional P*dx1* gene, the duodenal columnar epithelium is transformed into a more bile ductlike cuboidal epithelium (Offield *et al.,* 1996). These mutants also lack Brunner's glands normally located at the rostral aspect of the duodenum.

By E16, posterior to the duodenum, the gut tube lumen widens and the epithelial lining develops circular or spiral folds with numerous long villi that greatly increase the absorptive surface area of the small intestine. Each villus has a core of mesenchyme containing blood vessels that carry ab-

sorbed nutrients to the liver for processing, and a single lymphatic vessel. Between the villi are the epithelial crypts that penetrate deep into submucosa. Each crypt is surrounded by a sheath of mesenchymal cells.

The sphincter between the ileum and the cecum has a thick circular muscle layer. It fails to form in embryos homozygous for a deletion encompassing *Hoxd4, -8, -9, -10,* and *-11,* in addition to *Hoxd13* (Zákány and Duboule, 1999).

Two of the most interesting aspects of the endodermally derived epithelium of the small intestine are the stem cell system and the crypt-villus axis (Gordon and Hermiston, 1994). Each crypt is monoclonal, so that all cells in it are descended from a single precursor, but each villus receives cells from several crypts. The crypt contains four to five long-lived multipotent stem cells near (but not at) the base that, like all stem cells, both self-renew and give rise to differentiated descendants. They continuously give rise to the four distinct epithelial lineages of the small intestine: the absorptive enterocytes and the goblet, enteroendocrine, and Paneth cells. Each mature cell type expresses characteristic marker genes. For example, the enterocytes express microvillar hydrolases such as sucrase, lactase, and maltase. The first three cell types differentiate as their precursors move in a column up the crypt and surrounding villi. Once they reach the top the mature cells undergo apoptosis and are shed. By contrast, the precursors of the mature Paneth cells migrate into the very base of the crypt where they reside for some time before being turned over. The control of stem cell self-renewal, proliferation, and commitment to different cell types, and the orderly migration of cells from crypt to villus tip, are all carefully regulated to provide a constant flow of cells for intestinal function. These processes have been studied both in transgenic mice (e.g., Hermiston and Gordon, 1995; Nomura *et al.*, 1998; Stappenbeck and Gordon, 2000) and in homozygous null mutants. The latter studies have identified genes regulating intestinal development that act either in the endoderm (e.g., *Ihh, Shh, Pdx1, Beta2/NeuroD,* and *ngn3*), in the mesoderm (e.g., *Nkx2.3* and *Foxl1/Fkh6*), or in both (e.g., *Hes1*).

Ihh is expressed in the intervillus region in E18.5 embryos, and homozygous null mutants have smaller villi, reduced proliferation, and reduced number of enteroendocrine cells (Ramalho-Santos *et al.*, 2000). By contrast *Shh*$^{-/-}$ embryos show overgrowth of the villi in the duodenum. In addition, duodenal enteroendocrine cells (making serotonin, secretin, and cholocystokinin, CCK) are greatly reduced in number. Similarly, mutants for the *Beta2/NeuroD* gene completely lack intestinal secretin and CCK cells in addition to having defects in islet endocrine cell development (Naya *et. al.*, 1997). Most recently, *Ngn3* null mutant embryos were shown to lack expression of *Beta2/NeuroD* throughout the islets and intestine (Gradwohl *et al.*, 2000) whereas *Hes1* mutants showed accelerated differentiation of endocrine cell types in the intestine as well as pancreas (Jensen *et al.*, 2000). These findings demonstrate the use of similar mechanisms for differentiation of gut and pancreas endocrine cells.

The *Nkx2-3* gene is expressed in gut mesenchyme. Homozygous inactivation causes retarded morphogenesis of villus formation in the intestinal epithelium and, consequently, lethality in most animals by weaning (Pabst *et al.*, 1999). However, the survivors through weaning exhibit crypt cell derivatives that hyperproliferate and, by apparent compensation, migrate more rapidly to the elongated villus tip, thereby balancing the excess flow of cells from the crypt compartment into the epithelium. Another critical mesodermal transcription factor gene is *Foxl1* (formerly *Fkh6*), which helps control the proliferation of crypt cells and whose inactivation, like that of *Nkx2-3*, results in intestinal cell hyperproliferation and lengthening of the villi (Kaestner *et al.*, 1997).

C. Large Intestine: Caecum and Colon

One of the major functions of the large intestine is water absorption. In the adult, the lumen is wide and there are few folds or villi, and the crypts are shallow. The muscle layers

are also thicker than in the small intestine. Transiently after birth the epithelium of the colon is more like the small intestine in morphology and expression of genes related to milk utilization (Duluc *et al.*, 1994).

Ihh is expressed throughout the epithelium of the colon and homozygous null mutants have a dilated colon with reduced muscle layers and enteric ganglia (Ramalho-Santos *et al.*, 2000). Hox gene control of the ileocecal and anal sphincters was discussed earlier in Section II.C.2.

D. Rectum

The epithelium of the rectum contains a large number of mucin-secreting goblet cells. The distal end is surrounded by two important sphincters. The outer one is composed of voluntary muscle, whereas the inner one is generated by the enlargement of the circular and longitudinal smooth muscle layers. Abnormalities in the thickness and organization of the smooth muscle layers is seen in *Hox* null mutants (*Hoxd12* and *Hoxd13*)(Kondo *et al.*, 1996) and transgenic mice misexpressing Hox genes (Warot *et al.*, 1997).

Analysis of *Shh* and *Gli* mutants has revealed an important role for Shh and its signaling pathway in the development of the rectum and anus. In *Shh*$^{-/-}$ null mutants the colon ends in a blind sac and there is no anus at all. *Gli3*$^{-/-}$ mutants have anal stenosis and ectopic anus, *Gli2*$^{-/-}$ mutants have imperforate anus and rectourethral fistula, and *Gli2*$^{-/-}$ *Gli3*$^{+/-}$ embryos have abnormalities in the development of the cloaca (Kimmel *et al.*, 2000).

IV. Development of Tissues That Bud from the Endoderm

In this section, the development of endoderm organs budding from the primitive gut is discussed in detail, roughly in the AP order in which they are found along the body axis. Our emphasis will differ somewhat for each tissue type, to illustrate different aspects of development such as morphogenesis and tissue organization (lung, intestines), mechanisms of transcriptional control (liver), and cell signaling for tissue growth (thyroid, liver) and cell type diversification and lineage (pancreas).

A. The Thyroid Gland

The thyroid anlage begins to develop at E8.5 as a thickening of the posterior region of the pharyngeal endoderm. The thyroid bud then dissociates from the endodermal epithelium and during the next 5 days migrates caudally to the anterior wall of the trachea, where the thyroid progenitors begin to differentiate into follicular cells. During this period the thyroid primordia remain as a single bud connected to the pharyngeal floor via the thyroglossal duct. By E13.5, the thyroglossal duct bifurcates into the two lobes of the thyroid, and by E14.5, cells from the ultimobranchial body, a mesenchymal neural crest derivative, mix with the thyroid follicular cells. The latter begin to express genes encoding proteins such as thyroglobulin, thyroperoxidase, and the thyroid stimulating hormone receptor. At E17.5, a marked expansion of the differentiated follicular cells occurs so that the resulting thyroid gland can produce sufficient thyroid hormone to modulate body metabolism. Thyroid stimulating hormone released from the hypophysis into the bloodstream causes the thyroid gland to synthesize and secrete the thyroid hormone precursor thyroxine (T4) and its active metabolite triiodothyronine (T3). The thyroid follicular cells, which secrete thyroid hormones, form a single layer with apical microvilli that line the thyroid follicles. The follicular cells are thus specialized to express proteins that metabolize tyrosine to the thyroid hormones and that carry the hormones in the bloodstream. Secreted T4 and T3 generally affect the differentiation and metabolic rate of various cells of the body.

1. Control of Early Thyroid Development

The first evidence of thyroid commitment in the mouse is the expression of the genes for the homeobox factor *Titf1* (Ttf1, Nkx-2.1, T/ebp) (Lazzaro *et al.,* 1991; Kimura *et al.,* 1996), the forkhead/Fox factor *Ttf2* (now *Foxe1*) (Zannini *et al.,* 1997), and the paired-homeobox factor *Pax8* (Mansouri *et al.,* 1998) in a thickening of the pharyngeal endoderm floor. Although all of these transcription factor genes are expressed at about the time of thyroid specification, none of them individually is critical for the initial budding of the thyroid anlage. Homozygous inactivation of the *Foxe1* gene causes a failure of the thyroid bud cells to migrate to the trachea, perhaps due to a defect in detachment from the pharyngeal endoderm (De Felice *et al.,* 1998). *Foxe1* may be specifically important for cell migration because the gene is normally expressed from E8.5 to E13.5, the period during which the thyroid cells migrate (Zannini *et al.,* 1997). Homozygous null *Foxe1* mouse mutants also exhibit cleft palate, which is strikingly similar to the thyroid agenesis and cleft palate seen in humans with a missense mutation in *FOXE1* (Clifton-Bligh *et al.,* 1998).

Titf1 gene inactivation blocks thyroid development after the cells move to the trachea, whereupon the cells apparently atrophy and there is an absence of a thyroid gland and neonatal lethality (Kimura *et al.,* 1996). *Titf1*$^{-/-}$ embryos also have severe defects of the lung, ventral forebrain, and pituitary, which complicate the cause for lethality. In *Pax8* mutant homozygotes, like that for *Titf1* mutants, the thyroid bud forms and migrates, but there is an absence of mature thyroid follicular cells (Mansouri *et al.,* 1998; Macchia *et al.,* 1998). However, treating the *Pax8*$^{-/-}$ mice with thyroxine rescues the postnatal lethality, demonstrating greater specificity of PAX8 function for the thyroid lineage. For both PAX8 and TTF1, the function of the proteins appears to be in promot-

ing the survival and/or expansion of the thyroid follicular cells.

The HEX transcription factor appears to function upstream of *Titf1* and *Foxe1*, because neither of these genes is activated in the embryonic thyroid region in homozygous *Hhex* null mutants (Martinez-Barbera *et al.,* 2000). However, a small thyroid primordium does form in the *Hhex*$^{-/-}$ embryos at E9.5, showing that HEX is critical for early differentiation but not specification of the thyroid gland.

2. Control of Early Thyroid Growth

The *Hox3a* gene is expressed in the neural crest and paraxial mesoderm-derived cells of the pharyngeal arches, and in the pharyngeal endoderm; and *Hox3a* null mutant embryos exhibit thyroid hypoplasia and are athymic (Manley and Capecchi, 1995). The thyroid defect is in both the endoderm and ultimobranchial derivatives, both of which normally express HOX3A. Cell labeling studies show that migration of the neural crest cells is not markedly altered in the homozygous null embryos, indicating that the mutation affects the ability of the neural crest cells to differentiate or their ability to induce other cells to differentiate. The thyroid defect is variably penetrant in the *Hox3a*$^{-/-}$ embryos, and the phenotype can be exacerbated in double homozygotes between *Hox3a* and *Hox3b* or *Hox3d* (Manley and Capecchi, 1998). Neither these compound mutants, nor embryos with the genotype *Hox3*$^{+/-}$ *Hox3b*$^{-/-}$ *Hox3d*$^{-/-}$, have thyroid agenesis, and the remnant thyroids do express some differentiated thyroid products. Because only the *Hox3a* paralog is expressed in the pharyngeal endoderm, the double mutant data indicate that *Hox3b* and *Hox3d* function in the mesenchyme to help control thyroid development.

Mutations in the genes encoding thyroid stimulating hormone receptor (TSHR) also lead to hypoplasia of the differentiated thyroid gland (Beamer *et al.,* 1981; Sunthornthepvarakul *et al.,* 1995). TSHR is activated about E15. Inactivation of either TSHR or PAX8 leads to defects in early thyroid growth; thus perhaps PAX8 mediates signaling via TSHR in this context (see Macchia *et al.,* 1999).

B. Development of the Lung

The development of the mammalian lung culminates in the formation of huge numbers of terminal respiratory alveoli specialized for highly efficient gas exchange. The molecular basis of lung development has been the topic of a number of reviews (Hogan *et al.,* 1997; Hogan and Yingling, 1998; Hogan, 1999; Perl and Whitsett, 1999; Warburton *et al.,* 2000; Cardoso, 2000; see also Metzger and Krasnow, 1999, for comparison with *Drosophila* tracheal development). Several of these reviews describe the classical tissue recombination experiments that first established the importance of dynamic interactions between the embryonic mesenchyme and endoderm for lung morphogenesis. This background information is therefore not repeated here. A database of lung development can be found at http://www.ana.ed.ac.uk/anatomy/database/lungbase/lunghome.html.

1. Early Development of the Trachea and Primary Lung Buds

The mouse lung arises from the ventral foregut of the approximately E9.5 embryo (21- to 29-somite stage). The primordia of the trachea and the respiratory tree both consist of an inner endodermal layer surrounded by splanchnic mesoderm. However, they have quite distinct origins. The trachea is separated from the esophagus by means of a longitudinal septation of the foregut (Sutcliff and Hutchins, 1994), while the rest of the lung develops from two ventral buds that form at the posterior end of the trachea (Fig. 2). These buds undergo repetitive outgrowth and lateral and dichotomous branching, driven by reciprocal interactions between the distal mesenchyme and endoderm. They develop asymmetrically, so that in the mouse the right bud gives rise to four lobes, whereas the left bud gives rise to only one. The branching morphogenesis of each bud is stereotypic and invariant, at least until the formation of the terminal sacs and alveoli, which occurs between E17.5 and P5 in the mouse (Ten Have-Opbroek, 1981, 1991).

The mechanisms responsible for the initial specification of the lung primordia are not yet known. One of the earliest indications of the site of primary bud formation is the expression of *Bmp4* in two patches of the ventral foregut mesoderm (Weaver *et al.*, 2000). Genetic analysis has provided evidence for a role of several genes in the very early development of the trachea and primary lung buds after specification. These are *Shh, Gli2* and *Gli3, Titf1* (also known as *Ttf1, Nkx2-1, T/ebp), Fgf10,* and *Fgfr2,* as discussed below. Evidence also points to an important role for retinoic acid (RA) in foregut development in general and lung development in particular.

The signaling factor, Sonic hedgehog (Shh) is expressed in the ventral foregut endoderm and throughout the endoderm of the primary buds and early respiratory tree, with highest levels in the most distal tips. By contrast, the gene encoding the patched receptor (Ptch) for SHH is expressed in the adjacent mesoderm (Bellusci *et al.*, 1997a). In *Shh*$^{-/-}$ embryos, the trachea and primary buds fail to separate from the esophagus, resulting in a condition similar to the human birth defect tracheoesophageal fistula (Skandalakis *et al.*, 1994; Litingtung *et al.*, 1998; Pepicelli *et al.*, 1998). The primary buds also remain as small sacs and do not grow out or branch. Abnormal tracheal and bud morphogenesis is seen in *Gli2*$^{-/-}$ *Gli3*$^{-/-}$ and *Gli2*$^{-/-}$ *Gli3*$^{+/-}$ compound mutants, probably reflecting a function for Gli proteins in the Shh signaling pathway in the foregut mesoderm (Motoyama *et al.*, 1998). Embryos mutant or defective in signaling for either *Fgf10* or *Fgfr2* (specifically, the IIIb splice variant) survive to birth and develop a trachea and primary bronchi, but no

respiratory tree. The similarity of these phenotypes reflects the very localized expression of *Fgf10* in the distal mesoderm around the early lung buds and *Fgfr2* in the endoderm (Peters *et al.*, 1994; Bellusci *et al.*, 1997b; Min *et al.*, 1998; Sekine *et al.*, 1999; Arman *et al.*, 1999; De Moerlooze *et al.*, 2000) (see below). As described in Section IV.A.1, the gene encoding the homeodomain protein, TITF1 (also known as NKX2.1 and T/EBP), is expressed in the foregut endoderm and is required for the development of the thyroid, thymus, and lung (Kimura *et al.*, 1996). In *Titf1*$^{-/-}$ embryos the trachea and esophagus do not separate and there are only very small or grossly cystic lung buds that do not express *Bmp4* in the endoderm (Kimura *et al.*, 1996; Minoo *et al.*, 1999). Finally, embryos doubly homozygous for null mutations in the genes encoding RARa and RARß2 have serious abnormalities in early lung morphogenesis (Mendelsohn *et al.*, 1994). In a recent comprehensive study, evidence was presented for RA function at several different stages of lung development (Malpel *et al.*, 2000). It was found that *Raldh2,* which encodes a critical enzyme in the RA synthetic pathway, is expressed extensively in the foregut around the time of lung bud formation and subsequently in the lung mesenchyme during the pseudoglandular stage (see below). Moreover, analysis of embryos transgenic for a RARE-lacZ reporter gene showed that signaling through RA receptors was active in the lung from E10 up to at least E14, with the maximal activity shifting from the mesenchyme and endoderm to the pleura. Exogeneous RA added to lung organ cultures results in proximalization of the endoderm (Cardoso *et al.*, 1996; Packer *et al.*, 2000). Although the precise mechanisms underlying these effects of RA are unknown, there is some evidence that RA affects *Hox* gene expression in the lung (Packer *et al.*, 2000). Some information about the temporal and spatial pattern of Hox gene expression in the developing lung is available, but a systematic and precise study has not yet been made. Embryos homozygous for null mutations in *Hoxa5* have abnormal lung morphogenesis (Kappen, 1996; Bogue *et al.*, 1996; Cardoso *et al.*, 1996; Aubin *et al.*, 1997; Packer *et al.*, 2000).

2. Overview of Subsequent Development of the Respiratory System

Following the establishment of the trachea and primary buds, development of the respiratory tree transits through four stages (Ten Have-Opbroek, 1981, 1991). (1) pseudoglandular (E9.5–16.5), characterized by the arborization of the primary buds and a relatively undifferentiated distal endoderm; (2) canalicular (E16.5–17.5), when the distal endoderm begins to form terminal sacs and their vascularization begins; (3) terminal saccular (E17.5–P5), characterized by a thinning of the mesenchyme and an increase in the number of terminal sacs, their extensive vascularization, and the differentiation of the endoderm into type I and type II pneumocytes; and (4) alveolar (P5–30), when the terminal sacs develop into mature alveolar ducts and alveoli. Note that alveolarization of the human lung, unlike that of the mouse, begins before birth.

3. Proximal-Distal Differentiation of the Lung Endoderm

The epithelial cells of the primary lung buds are columnar and relatively undifferentiated, with a cytoplasm rich in glycogen granules. Between E10.5 and E14.5 the proximal-distal differentiation of the lung endoderm becomes evident, both at the morphological and molecular levels. The proximal endoderm (the bronchi and bronchioles) contain four distinct cell types. The most abundant are ciliated cells that express the forkhead gene, *Hfh4*, a regulator of cilia formation (Hackett *et al.,* 1996; Chen *et al.,* 1998). Interspersed among these are smaller numbers of mucus-secreting cells and Clara cells, which secrete a protein called Clara-cell 10-kDa protein or CC10. By E14.5 small clusters of neuroendocrine cells (neuroendocrine bodies or NEBs) can be detected in the primary and secondary bronchi. NEBs secrete peptides such as bombesin and calcitonin gene-related peptide, but their precise function is unknown. They also express high levels of *Delta1* (Dll1), encoding a surface ligand for Notch that may function in determining cell fate (Post *et al.,* 2000). In support of this idea, embryos homozygous for a null mutation in *Mash1,* which encodes a basic helix–loop–helix (bHLH) transcription factor commonly downstream of Delta-Notch signaling, lack lung neurendocrine cells altogether (Borges *et al.,* 1997).

During the pseudoglandular stage, the distal epithelial cells express the gene for surfactant protein C (Sp-C), initially at low levels and then at higher levels after E15. In later stages of lung development, however, expression of *Sp-C* becomes restricted to the type II alveolar pneumocytes (Wert *et al.,* 1993). The epithelial cells also express uniformly *Fgfr2* and *Hnf3α* (*Foxa1*) and *β* (*Foxa2*), with the latter probably playing a role in regulating cell differentiation (Zhou *et al.,* 1996, 1997). However, as discussed in Section IV.B.6 below, the epithelial and mesenchymal cells at the very tips of the extending branches and buds have a specific pattern of gene transcription, related to intercellular signaling pathways. Genes expressed at high levels in the epithelium at the tips include *Shh, Bmp4, Wnt7b, Fgfr4,* and *c-fos.* By contrast, the mesenchyme around the tips expresses high levels of *Ptch, Fgf10* and *Pod1* (Molinar-Rode *et al.,* 1993; Bellusci *et al.,* 1996, 1997b; Hogan *et al.,* 1997; Park *et al.,* 1998; Quaggin *et al.,* 1999; Weaver *et al.,* 1999, 2000, and reviews cited in Section IV.B). The distal tips are thus thought to be organizing centers that play key roles in regulating both the epithelial-mesenchymal interactions that drive branching morphogenesis, and the proximal-distal patterning of the lung epithelium.

At the canalicular stage the distal epithelial cells become flattened, and by the saccular stage they have differentiated into two distinct cell types. The type II pneumocytes, as described above, are cuboidal and express very high levels of Sp-C and have large cytoplasmic secretory vacuoles (lamellar bodies) containing whorls of surfactant protein that can also be seen in the lumen of the tubes. By contrast the type I pneumocytes are very flattened and do not express Sp-C. There is some evidence that type II cells can give rise to type I cells *in vivo* and *in vitro.* This has been inferred, for example, from pulse-chase tritiated thymidine labeling studies following NO$_2$-induced damage in the adult rat (Evans *et al.,* 1973) or oxygen-induced damage in the mouse (Adamson and Bowden, 1974). For *in vitro* studies, see Danto *et al.* (1995). In the adult lung, the type I cells lie in close apposition to the capillaries of the alveoli, whereas the type II cells tend to be clustered at the junction between the alveolar ducts and the alveoli.

The final alveolar stage involves the formation and maturation of the alveoli. This is a complex process and one that is essential for postnatal survival. It involves, among other things, the formation of primary and then secondary septa that subdivide the saccules into multiple alveoli. Initially, the interstitial mesenchyme layer of each septum is quite substantial and contains a capillary network that is not in very tight contact with the endodermal cells. As postnatal development proceeds, the number of interstitial cells declines dramatically and the highly attenuated type I pneumocytes become very closely apposed to capillary endothelial cells, with only fused basal laminae between them. In the mature lung, the tips of the septa contain smooth muscle cells and bundles of elastin fibers. In *Pdgfa$^{-/-}$* mice the smooth muscle cells and elastin fibres are absent and septae fail to form, leading to postnatal death (Bostrom *et al.,* 1996; Lindahl *et al.,* 1997). During the pseudoglandular stage the growth factor, PDGF-A, is expressed by the distal epithelial cells and the receptor, PDGFR-A, by the surrounding mesenchymal cells. These mesenchyme cells normally differentiate into smooth muscle cells and migrate around the forming alveoli later in development (Souza *et al.,* 1995).

The periphery of the lung is occupied by a layer of connective tissue covered in a surface mesothelium. These layers are called the pleura and they are rich in lymphatic vessels. The mesothelium expresses *Fgf9* (Colvin *et al.,* 1999).

Although lung growth and branching morphogenesis have been well studied during the pseudoglandular stage (see below), relatively little is known about what triggers the switch from pseudoglandular to saccular stages, and postnatal alveolar maturation. The latter process probably involves FGF signaling since the lungs of mice homozygous null for genes encoding both FGF receptor 3 and FGF receptor 4 are normal at birth but are blocked in alveolarization (Weinstein *et al.,* 1998). Genetic and expression analysis also points to a role for glucocorticoids, TGF-β_3, EGF, amphiregulin and $\alpha_3 \beta_1$ integrin in mouse lung development and maturation

(Warburton *et al.*, 1992; Cole *et al.*, 1995; Kaartinen *et al.*, 1995; Kreidbderg *et al.*, 1996; Schuger *et al.*, 1996; Wu and Santoro, 1996; Miettinen *et al.*, 1997).

4. Differentiation of the Lung Mesoderm

Initially, the splanchnic mesoderm around the primary lung buds appears relatively undifferentiated although it may be a site of vasculogenesis (see below) and later of smooth muscle differentiation (see above). Up to about E11.5, high levels of expression of *Bmp4* are seen in the ventral mesoderm but the significance of this localization is not known (Weaver *et al.*, 1999). By birth many different cell types are present in the mesoderm. For example, the mesenchyme gives rise to cartilage rings around the trachea and primary and secondary bronchi, and to smooth muscle, collagen fibrils, and other matrix components outside the continuous basal lamina of the proximal endoderm. By contrast, there is no organized basal lamina around the distal tips of extending buds. A role has been proposed for Tgfβ in matrix synthesis and deposition in the embryonic lung (Heine *et al.*, 1990).

5. Lung Vascular Development

Given its function, it is no surprise that the lung is a highly vascularized organ (see also Chapter 11). The endothelial cells of the capillary networks of the mature alveoli are tightly apposed to the attenuated type I epithelial cells for efficient gas exchange.

Two processes are involved in generating the blood vessels of the lung: vasculogenesis, in which the splanchnic mesoderm gives rise to endothelial cells organized into sinusoids or plexi, and angiogenesis, in which new vessels sprout from the pulmonary arteries and veins. These two systems, which are generated contemporaneously with endodermal branching, begin to connect around E13–14 (deMello *et al.*, 1997). Remodeling continues as the blood flow increases. VEGF mRNA is expressed in the endodermal cells and over-expression gives increased vascularization and abnormal morphogenesis (Zeng *et al.*, 1998).

Vessel remodeling also occurs postnatally, particularly in the septa of the alveoli. Here, capillaries fuse to give rise to a vascular network in which there is a minimal amount of extracellular material or cells interposed between the endothelial and pulmonary epithelial cells.

6. Experimental Approaches toward an Integrated Model for the Role of Signaling Pathways in Lung Branching Morphogenesis and the Proximal-Distal Differentiation of the Endoderm

The distal tips of lung buds and branches at the pseudoglandular stage are characterized by spatially restricted and often high levels of expression of specific genes encoding components of intercellular signaling pathways and likely downstream targets. Rates of cell proliferation are also higher

in the relatively unspecialized columnar epithelial cells in distal tips than in the more differentiated proximal cells (e.g., Mollard and Dziadek, 1998). A variety of experimental approaches has been used to analyze the significance of distal tip cells in lung development. Among these approaches is the expression of transgenes in embryos using a regulatory region of the human *SP-C* gene (Wert *et al.*, 1993). This region drives gene expression in the distal endoderm, with high levels being reached after E15.5. Several genes encoding components of intercellular signaling pathways have been expressed with this cassette, including *Shh, Bmp4,* Δn, *Bmpr1a,* and *noggin,* which encodes a protein that binds various bone morphogenetic proteins (BMPs) and antagonizes their activities (Bellusci *et al.*, 1996, 1997a; Weaver *et al.*, 1999). A second approach has been to culture intact lungs dissected from E11.5 embryos on nucleopore filters and to add purified factors, inhibitors, antisense RNA, or adenovirus vectors to the growth medium (for example, Zhao *et al.*, 1998; Tefft *et al.*, 1999). Under these conditions all cells will be exposed to the additives, including the endoderm and mesoderm, as well as the outer mesothelial cells, that normally express *Fgf9* (Arman *et al.*, 1999; Colvin *et al.*, 1999). Other experimental conditions have been devised to culture endoderm alone, in the absence of any mesoderm. For example, isolated endoderm tips can be embedded in a mound of Matrigel™, an extracellular matrix largely composed of laminin, type IV collagen, fibronectin, and proteoglycan. Endoderm cultured in this way requires added FGF for cell proliferation (Nogawa and Ito, 1995; Bellusci *et al.*, 1997b; Weaver *et al.*, 2000). If the source of FGF10 is spatially localized on a heparin bead, then the endoderm bud will both proliferate and move toward the bead, eventually surrounding it. During the chemotaxis the rate of cell proliferation is not uniform within the bud, but highest in the distal cells closest to the bead (Weaver *et al.*, 2000). The authors combined the technique with the use of a *Bmp4^{lacZ}* reporter allele to follow the expression of *Bmp4* in the endoderm during outgrowth in reponse to an FGF10 soaked bead. *Bmp4^{lacZ}* is expressed at high levels in endoderm closest to the source of FGF10 and is upregulated in these cells in response to the growth factor. However, while FGF10 protein promotes outgrowth of an endoderm bud and its chemotaxis toward the bead, purified BMP2 or BMP4 protein has a very different effect, causing the epithelial tube to flatten and inhibiting its ability to reach and surround the FGF bead. Thus, FGF10 and BMP4 have opposing roles during lung bud morphogenesis (Weaver *et al.*, 2000). Taken together, all of these different experimental approaches have led to the conclusion that the distal lung tips act as organizing centers controlling the stereotypic pattern of bud formation and outgrowth characteristic of the pseudoglandular stage, as well as the proximal-distal differentiation of the endoderm. A simple model summarizing these results is shown in Fig. 6.

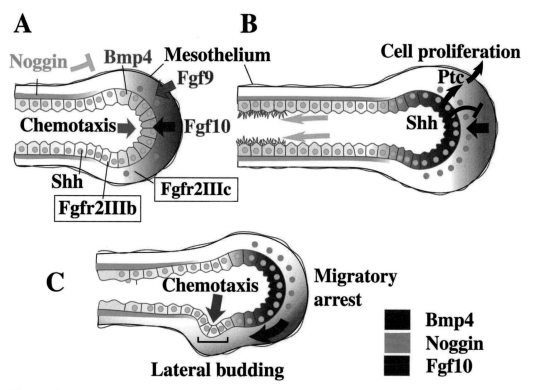

Figure 6. Model for the dynamic interaction of growth factors in lateral budding and proximal-distal patterning of the lung. (A) Bud shortly after initiation of outgrowth. *Fgf10* is transcribed at high levels in distal mesenchyme (red) and acts on the endoderm (red arrow), probably through Fgfr2IIIb receptor isoform to upregulate *Bmp4* (blue). Evidence suggests that FGF10 also promotes both the proliferation of the endoderm and its distal chemotaxis (blue arrow) (Park *et al.*, 1998; Weaver *et al.*, 2000). The mesothelial layer expresses *Fgf9* (pink) (Arman *et al.*, 1999; Colvin *et al.*, 1999) and most likely acts on the underlying mesenchyme, which expresses the receptor isoform, Fgfr2IIIc (pink arrow). Distal mesenchyme expresses *Sprouty4* (*Spry4*) while expression of *Sprouty2* (*Spry2*) is observed in distal epithelium (data not shown; de Maximy *et al.*, 1999; Tefft *et al.*, 1999; S. Bellusci, personal communication). The gene encoding noggin is expressed in the distal mesenchyme (green dots) (Weaver *et al.*, 1999) and at higher levels near the proximal endoderm (M. Weaver, unpublished results, from analysis of lungs of Nog*lacz* embryos). Noggin protein most likely inhibits *Bmp4* made by the endoderm, although the mesenchyme also expresses *Bmp5* (King *et al.*, 1994). Noggin may titrate the available level of *Bmp4* and so control the balance between the antagonistic activitities of FGF10 and BMP4 on the epithelium (Weaver *et al.*, 2000). (B) As outgrowth proceeds, the expression of *Bmp4* increases in the distal endoderm, possibly slowing outgrowth. Epithelial cells at the distal tip, which are exposed to both FGF10 and BMP4, remain undifferentiated, but as cells move proximally (yellow arrows) they differentiate into proximal cell types (ciliated yellow cells). Studies with transgenic embryos support the idea that one function of SHH made by the endoderm is to promote proliferation of the mesenchyme through PTCH and possibly *Gli-dependent* pathways (black arrow). The gene encoding Hedgehog Interacting Protein, Hip, is also expressed in the mesenchyme around the distal endoderm (Chuang and McMahon, 1999). *Shh* downregulates *Fgf10* expression (Bellusci *et al.*, 1997a; Lebeche *et al.*, 1999). As a result, *Fgf10* expression gradually decreases at the tip but is upregulated laterally, in this case asymmetrically, by unknown mechanisms. (C) In response to the lateral *Fgf10* expression, a new lateral bud is initiated (gray arrow), but only where the level of *Bmp4* expression falls below a threshold in the endoderm (bracket). The possible mechanisms underlying dichotomous branching are not shown here but are discussed in Weaver *et al.* (2000) and Lebeche *et al.* (1999).

7. Are There Lung Stem Cells?

A stem cell is defined as one that undergoes both self-renewal and gives rise to differentiated cell types (see Section II overview). Studies indicate that type II cells of the adult lung can give rise to type I cells either *in vitro* or *in vivo* following tissue damage (Section IV.B.3). However, there is no unambiguous evidence that continual replacement occurs of differentiated type I alveolar cells by descendants of an undifferentiated stem cell population in the normal adult lung. It has been speculated that such cells, if they exist, would reside in the alveolar ducts at the junctional zone between the cuboidal cells of the terminal bronchioles and the flattened type I cells of the alveoli (Magdaleno *et al.*, 1998; Emura, 1997).

C. Development of the Liver

1. Overview of Liver Development and Function

The liver develops from the ventral foregut endoderm and it rapidly becomes one of the largest fetal organs. In the mouse, a liver tissue bud develops from the ventral foregut at E9.0, and by E9.5, hepatoblasts from the bud migrate in a cordlike fashion into the surrounding loose mesenchyme of the septum transversum. The migrating hepatoblasts intermingle with the mesenchymal cells, coalescing around spaces in the mesenchyme that ultimately become the sinusoids through which the blood traverses in the mature liver. The liver forms a distinct organ in the mouse by E10.5, when it begins its early function as a site for hematopoiesis. Blood cells migrate to the liver initially from the yolk sac and later from the aorta-mesonephros-gonad region. At the same time, the hepatoblasts begin to differentiate into hepatocytes and biliary cells. The critical role of the liver in fetal hematopoiesis means that genes affecting liver development, when mutated, often lead to prenatal anemia or death and are thus easily identified. This simple phenotypic assay has led to the discovery of diverse signaling pathways that coordinately promote liver organogenesis, as described below. In the perinatal period, many new genes are activated in hepatocytes so that the cells can perform their postnatal roles of secreting serum transport proteins, enzymes, and bile, storing glucose as glycogen, and controlling metabolite and potential toxin levels in the bloodstream.

The adult liver consists of four lobes surrounded by Glisson's capsule, a thin mesothelial connective tissue. Hepatocytes are arranged in broad plates of cells with an apical (secretory) surface facing the small bile canaliculi between the cells and basolateral surfaces on either side that each face a fenestrated layer of endothelial cells. The latter line the sinusoidal spaces, and thus much of the hepatocyte cell surface is exposed to the blood. The hepatic portal vein brings nutrient-rich blood from the intestinal tract and the hepatic artery brings well-oxygenated blood from the heart; in both cases the vessels arborize into the hepatic sinusoids. Bile components secreted from hepatocytes enter the bile canaliculi and flow into successively larger ducts, eventually draining into the intestine via the hepatic duct. Although hepatocytes constitute about 80% of the liver's cells, endothelial cells, Kupffer cells (resident macrophages), and lipocytes (also known as stellate or Ito cells) with large stores of retinoids are also critical for organ function.

2. Establishing Endodermal Competence for Liver Development

Tissue transplantation studies in the chick and mouse have shown that only the prospective anterior-ventral domain of the gut endoderm has the capacity for extensive hepatic differentiation, that is, to the stage where the hepato-cytes can store glycogen (Le Douarin, 1975; Fukuda-Taira, 1981). While this would suggest that a prepattern exists for the liver, more recent studies indicate that although the ventral foregut endoderm uniquely possesses full hepatic competence, the liver is apparently not a default fate of the cells (Deutsch et al., 2000; Section IV.D.1 below). Endoderm along virtually the dorsal gut axis has the potential to activate some early liver genes, though not to undergo full hepatic differentiation (Gualdi et al., 1996; Bossard and Zaret, 1998). The competence of the dorsal-posterior endoderm to initiate liver gene expression, when this domain would normally become the intestine, may be an evolutionary remnant of the primordial gut. Primitive metazoans with a simple gut tube (e.g., *Amphioxus*) have diverse digestive functions that, in higher organisms, are distributed into different organ systems. Nonetheless, the ability of solely the ventral foregut endoderm to elicit terminal hepatic differentiation indicates some level of prior patterning in this endodermal domain.

How is a domain of hepatic competence established in the ventral foregut? Inactivation of the mouse genes for transcription factors HNF3β or GATA-4 leads to severe defects in the formation of the ventral foregut endoderm and mortality prior to hepatic specification (Ang and Rossant, 1994; Weinstein et al., 1994; Dufort et al., 1998; Kuo et al., 1997; Molkentin et al., 1997). Both of these factors are also active in adult liver and are required for many liver-specific genes (Costa et al., 1989; Laverriere et al., 1994), suggesting that their continuous expression from precursor tissue to mature organ may reflect a function at the prepattern stage. This possibility is attractive because HNF3 and GATA factor DNA binding sites are occupied on the liver-specific serum albumin gene in the endoderm, prior to transcriptional activation of the gene or hepatic commitment (Gualdi et al., 1996; Bossard and Zaret, 1998). The function of these factors in silent chromatin may be to cause genes and, thereby, the endoderm cells, to be competent for tissue specification (Zaret, 1999; Cirillo and Zaret, 1999). The generality of this model remains to be tested, but it is interesting to note that other genes expressed in the endoderm or its tissue derivatives utilize adjacent HNF3 and GATA sites (Denson et al., 2000; S. Beck et al., 1999; Al-azzeh et al., 2000). Furthermore, both HNF3- and GATA-related genes are necessary for gut development in flies and worms (Weigel et al., 1989; Mango et al., 1994; Kalb et al., 1998; Horner et al., 1998; Rehorn et al., 1996; Zhu et al., 1997), underscoring the general importance of the factors in endodermal competence.

HEX is a divergent homeobox transcription factor that, like HNF3β and GATA-4, is expressed continuously from the ventral foregut endoderm to the adult liver stages (Hromas et al., 1993; Thomas et al., 1998). Homozygous inactivation of *Hex* leads to the earliest known specific defect in liver organogenesis, though it should be noted that *Hex* mutants are also defective in thyroid development (Martinez-Barbera et al., 2000); again, suggesting a role in endodermal competence. In *Hex* mutant embryos, some proliferation of the hepatic endoderm is seen but little expansion of the cell population and, most strikingly, an absence of hepatic gene expression. Thus, *Hex* is required for proper hepatic differentiation. Interestingly, the small amount of hepatic endoderm proliferation that occurs in *Hex* mutant homozygotes still allows a liver capsule to form and subsequent hematopoietic cell invasion of the liver. Significant liver morphogenesis also occurs when several other liver regulatory factor genes are inactivated (see below). In all such cases a liver structure forms, but it is very loosely populated by hepatocytes and thus fails to provide an appropriate environment for hematopoiesis. Early studies with chick embryos also showed that when hepatic mesenchyme is transplanted without hepatocytes, it can develop into a liver-like structure (Le Douarin, 1975). Taken together, these studies show that gross liver morphogenesis is not dependent on filling of the liver capsule by hepatocytes.

3. Mesodermal Signals That Specify Hepatogenesis of the Endoderm

Chick embryo tissue transplant experiments showed that close interactions with cardiac mesoderm are necessary for the ventral foregut endoderm to initiate liver development (Le Douarin, 1975; Fukuda-Taira, 1981). Similar results have been obtained with the mouse (Houssaint, 1980), where it was found that while the ventral foregut endoderm at the two- to six-somite stages is incapable of hepatic differentiation, in isolation, such endoderm from the seven- to eight-somite stages was capable of doing so (Gualdi *et al.,* 1996). At the seven- to eight-somite stages, there is a burst of expression of FGF in the cardiac mesoderm (Crossley and Martin, 1995; Zhu *et al.,* 1996; Jung *et al.,* 1999) and FGF receptors 1 and 4 are expressed in the ventral foregut endoderm (Stark *et al.,* 1991; Ciruna *et al.,* 1997). Using a tissue explant culture system, FGF signaling from the cardiac mesoderm was found to be necessary and sufficient to induce hepatic gene expression in the ventral foregut endoderm (Jung *et al.,* 1999). These latter studies also found that the induction of hepatic gene expression by certain FGFs could be induced independently of the induction of hepatic endoderm outgrowth, the latter being elicited by other FGFs and growth factors expressed by the cardiac mesoderm. Thus, in terms of forming the initial tissue bud, distinct signals control the two processes of new tissue-specific gene expression and morphogenesis.

4. Cell Interactions That Promote Organogenesis from the Liver Bud

Due to the relatively undifferentiated state of the liver bud cells at E9.5, they are referred to as hepatoblasts. At E10.5–13.5, the hepatoblasts give rise to both hepatocytes and biliary cells (Shiojiri, 1984; Germain *et al.,* 1988). The cells immediately caudal to the liver bud, or the caudal portion of

the liver bud, contribute to the common bile duct and gall bladder.

As the hepatoblasts from the liver bud migrate into the surrounding loose mesenchyme of the septum transversum, at E9.5, other cell types begin contributing to the liver mass. These include mesenchyme cells and endothelial cells that surround the spaces within the septum transversum, thus forming the hepatic sinusoids (Medlock and Haar, 1983). By E10.5, a morphologically definable liver begins to form and hematopoietic cell invasion occurs. Blood cells arrive first from the yolk sac (Johnson and Moore, 1975) and subsequently from the aorta-gonad-mesonephros region (Medvinsky and Dzierzak, 1996). By E12.5 the liver is a major site of erythropoiesis with more than half of its cells derived from the hematopoietic lineage (Paul *et al.,* 1969).

From E10.5 to E11.5, the nuclear receptor family member HNF4 induces many apolipoproteins, metabolic enzymes, and serum factors, but not the earliest liver-specific genes such as serum albumin and α-fetoprotein (Li *et al.,* 2000), and thus HNF4 helps promote the maturation of hepatoblasts into hepatocytes. Further maturation signals are provided by the transcription factor MTF-1 at E13–15.5; it activates genes involved in metal homeostasis and oxidation-reduction (Günes *et al.,* 1998).

Gene inactivation studies have shown that a host of other signaling molecules and transcription factors coordinately promotes the morphogenesis of the liver bud into the organ proper. Prox1 is one of the initial activities required to expand the hepatoblast population from the liver bud. In *Prox1* mutant embryos, the hepatoblasts initially differentiate and proliferate, but they do not migrate into the surrounding hepatic mesenchyme (Sosa-Pineda *et al.,* 2000). While a small liver forms in Prox$^{-/-}$ embryos, most of it is devoid of hepatoblasts; hepatoblast migration is impaired because there is an excess of E-cadherin on the mutant cells. In addition, β_1 integrin, an extracellular matrix (ECM) receptor subunit, is required for growth of the hepatoblasts (Fässler and Meyer, 1995). These findings emphasize the importance of cell–cell interactions and ECM remodeling during early organ growth.

Many other mouse mutations lead to defects in hepatoblast and liver growth and reflect the complex signaling between the hepatoblasts and hepatic mesenchyme cells. The mutants can be grouped by the different stages in fetal liver development in which a growth phenotype is manifest and by different signaling pathways in which they are known to function. For example, the transcription factors HLX (Hentsch *et al.,* 1996) and JUMONJI (JMJ) (Motoyama *et al.,* 1997) are required in the mesenchyme at the E10.5 and E13.5 stages, respectively, for liver growth. Hepatocyte growth factor (HGF) expression in the hepatic mesenchyme and HGF receptor (C-MET) expression in the hepatoblasts are also required for liver growth (Schmidt *et al.,* 1995; Bladt *et al.,* 1995), as is N-MYC transcription factor in the Glisson's capsule cells

(Sawai *et al.,* 1993; Giroux and Charron, 1998). Mutations in *Hgf, c-Met,* and *jmj* lead to hepatocyte apoptosis, as does mutation of the signaling kinase *Sek1* (*Map2k4*) gene (Nishina *et al.,* 1999), indicating the essential nature of proper mesenchymal signals. The SEK1 requirement is earlier than that for the other factors, suggesting that there are sequential growth signals from the mesenchyme. In sum, diverse, sequential signaling pathways coordinately promote growth of the nascent hepatocyte cell population.

Considering the large percentage of fetal liver cells that are hematopoietic, it is not unexpected that the latter are also critical signaling sources. Hepatocytes that are deficient in the gp130 receptor for oncostatin M, a cytokine secreted by hematopoietic cells, fail to multiply properly (Kamiya *et al.,* 1999). Similarly, defects in hematopoietic expression of *myb* or mutations in *Rb* also lead to improper liver growth (Mucenski *et al.,* 1991; Jacks *et al.,* 1992; Lee *et al.,* 1992).

The transcription factors JUN (Hilberg *et al.,* 1993), Rel A (Beg *et al.,* 1995) and XBP-1 (Reimold *et al.,* 2000) are required in hepatocytes to promote proliferation, and therefore they presumably respond to autocrine or paracrine signals. Mutation of genes for these factors also leads to apoptosis of hepatocytes and these phenotypes occur at slightly different stages within E12–16, suggesting the involvement of different pathways.

In summary, much is known about cell signals and intracellular transduction pathways for liver organogenesis. It seems likely that so many signals are required at so many stages for liver growth and differentiation because of the unusually diverse sets of biochemical pathways involved in liver function. Future studies will be directed toward understanding how the diverse pathways are orchestrated to create a functioning liver. This information will bear on our ability to manipulate and reconstitute hepatic cells during various physiological and pathological states.

5. Regenerative Capacity of the Liver

The liver possesses the remarkable capacity to regenerate after partial hepatectomy or chronic tissue damage. This aspect of liver biology has been studied intensely for insights into autocrine and paracrine signaling, the balance between cell growth, differentiation, and carcinogenesis, and the mechanisms of stem cell activation. Transgenic mouse and gene inactivation studies have shown that shortly after partial hepatectomy, tumor necrosis factor α (TNF-α) and interleukin 6 (IL-6) signaling leads to the activation of immediate early genes in hepatocytes and the stimulation of hepatic cell division (Cressman *et al.,* 1996; Yamada *et al.,* 1998). Liver cell transplant models have shown that differentiated hepatocytes, which are normally quiescent, are sufficient to replenish cells in the organ without needing to invoke a role for liver stem cells (Rhim *et al.,* 1994; Overturf *et al.,* 1997). However, the situation appears different for chronic liver damage models, which employ treatments with chemical

agents such as carbon tetrachloride, mutations or drugs that cause the buildup of toxic metabolic intermediates, or the transgenic expression of enzymes that cause liver damage. In such cases, small, undifferentiated cells associated with the bile ducts, referred to as "oval," hepatic epithelial, or hepatic stem cells, can become activated and give rise to well-differentiated hepatocellular carcinoma. The mechanisms of oval or hepatic stem cell activation and liver regeneration have been investigated in detail and the reader is referred to recent reviews for further information (e.g., Michalopolous and DeFrances, 1997; Fausto, 2000). In summary, the liver contains two sources of cells for self-renewal, hepatocytes and oval (or epithelial) cells, which are activated by different stimuli.

The pancreas is another potential source of hepatocytes. Chronic dietary damage can lead to the appearance of hepatocytes within the pancreas (Rao *et al.,* 1989; Dabeva *et al.,* 1997). Transgenic expression of an FGF in the pancreas can also lead to the appearance of hepatocytes therein (Krakowski *et al.,* 1999). Although the cell source of the ectopic hepatocytes within the pancreas is not clear, these findings illustrate the plasticity of endoderm-derived tissues.

6. Hepatogenic Capacity of Nonendodermal Tissues

Recent studies have led to the surprising discovery that stem cells from other germ layer derivatives are also capable of generating hepatocytes. Purified populations of hematopoietic stem cells can reconstitute hepatocytes in the liver after bone marrow transplantation (Peterson *et al.,* 1999; Lagasse *et al.,* 2000). Furthermore, individual clones of neural stem cells, when grown into neurospheres in culture, mixed with mouse blastocyst cells, and implanted into foster mothers, can colonize liver and other endodermal tissues in the resulting embryos (Clarke *et al.,* 2000). While the efficiency is quite low, it is nonetheless striking that nonendodermal stem cells can differentiate into hepatocytes. Clearly much more needs to be done to understand what normally restricts cell fates to particular germ layer lineages and how these restrictions can be circumvented for possible therapeutic benefit.

D. Development of the Pancreas

The pancreas develops separately from the dorsal and ventral domains of the foregut endoderm, with both sites budding at E9.5 at the level of the prospective duodenum. The distinct origin yet similar AP position of appearance of the pancreatic buds is interesting, given that the ventral and dorsal foregut arise from different domains of endoderm during gastrulation (see Section II.B.1 above). While the dorsal pancreatic bud emerges as a separate entity, the ventral bud emerges adjacent to the liver and part of the ventral pancreatic bud contributes to the main pancreatic duct. The latter, in turn, connects to the common bile duct and, hence, to the duodenum. Rotation of the duodenal and stomach region by E11 brings the two pancreatic domains in contact so that they form a single gland. Although in humans the adult pancreas is a well-defined organ, in the mouse it is relatively disperse and can be difficult to distinguish from connective tissue adhering to the duodenal region of the gut.

The pancreas serves a dual glandular function. An exocrine component is organized into acini that secrete digestive enzymes into the intestine, via the pancreatic duct, and an endocrine component is organized into islets of Langerhans, within the exocrine tissue, that secrete hormones into the blood. The exocrine acini develop at the termini of ducts. Enzyme secretion is regulated by hormones from the small intestine that are typically released after a meal. The four types of pancreatic endocrine cells are defined by their unique expression of certain hormones. Beta cells produce insulin and comprise the majority of cells in the center of the islet in the mouse. Alpha cells make glucagon and help define the islet periphery, while delta cells, making somatostatin, and pancreatic polypeptide cells are scattered around the periphery of the islet. There is intense interest in pancreas development because of the large number of humans with diabetes and the high degree of mortality associated with pancreatic cancer. The pancreas therefore represents an experimental system where, like other endodermal organs, it is hoped that developmental principles will apply to organ regeneration.

1. Emergence of the Dorsal and Ventral Pancreatic Buds

The study of early pancreas development *in vitro* was one of the original experimental models for organogenesis from the endoderm (Golosow and Grobstein, 1962; Wessells and Cohen, 1967; Spooner *et al.*, 1970). Also, these studies were among the first to carefully document the relevant phases *in vivo* (with Pictet *et al.*, 1972; Rugh, 1968). Morphologically, the dorsal pancreatic bud appears as an evagination of the duodenal epithelium, at about 25 somites, and the ventral pancreatic bud appears slightly later, at 28–30 somites. Endodermal cells at this level of the duodenum are columnar and, during the 12- to 20-somite period, mesodermal cells begin to increase in density along the side of the gut endoderm. During this period, the notochord is displaced locally from the endoderm by the dorsal aorta. Clear mesodermal cell concentration occurs lateral and dorsal to the prospective pancreatic endoderm at about the 25-somite stage, and a few hours later the pancreatic bulges begin to appear (E9.5). By E10.5, the dorsal pancreatic rudiment starts to constrict at its base and grows to the left; the ventral pancreas follows shortly thereafter. From this stage onward, the pancreatic epithelium branches extensively with the apical (secretory) cell surface facing into the aforementioned pancreatic ducts. By E14, much histological differentiation and vascularization have occurred, and by E16, the gland is finely branched, with a distinct lumen. Endocrine cells can be detected as

early as E10.5, while at E14.5 they increase greatly in number. By E16–18.5, islets begin to appear.

Tissue explant studies showed that both the dorsal and ventral endoderm is committed to become the pancreas by about the 8-somite stage (Wessels and Cohen, 1967; Spooner *et al.*, 1970), which is far earlier than the morphological events described above. Indeed, mRNAs for insulin, glucagon, and somatostatin become detectable within the prospective pancreatic endoderm at the 10- to 20-somite stages, which is earlier than evident cytodifferentiation (Gittes and Rutter, 1992), and the homeobox factor genes *Pdx1* and *Hlxb9* are activated in the pancreatic domains at the 8-somite stage (Offield *et al.*, 1996; Alhgren *et al.*, 1996; Li *et al.*, 1999). Thus these latter genes almost certainly respond to the primary signals for pancreatic specification. Endodermal competence for pancreatic development is probably established shortly after gastrulation, because presomitic endoderm from E7.5, cultured *in vitro*, can express *Pdx1* (Wells and Melton, 2000). However, *Pdx1* expression was only detected in these studies when endoderm was cocultured in association with mesoderm and ectoderm, demonstrating that early patterning by soluble factors from other germ layers is critical for pancreatic competence.

Despite the common timing and AP position of dorsal and ventral pancreatic specification, differences do exist in the underlying mechanisms of control. Inhibitor studies with chick embryo explants have shown that signals from the notochord can promote dorsal pancreas budding by repressing endodermal expression of *Shh* (Kim *et al.*, 1997; Hebrok *et al.*, 1998; Kim and Melton, 1998). By contrast, notochord does not influence ventral pancreatic development (Kim *et al.*, 1997; Hebrok *et al.*, 1998). Although *Shh* is essential for the morphogenesis of the trachea, lung, and esophagus, the initial development of the pancreas and liver are unaffected by a *Shh* null allele (Chiang *et al.*, 1996; Litingtung *et al.*, 1998). However, ectopic *Shh* expression in the prospective pancreatic regions perturbs the differentiation of pancreatic mesenchyme (Apelqvist *et al.*, 1997).

Genetic studies in mouse show that the dorsal pancreatic bud fails to emerge in homozygous mutants for the homeodomain genes *Isl1* and *Hlxb9*, but the ventral bud still develops (Ahlgren *et al.*, 1997; Li *et al.*, 1999; Harrison *et al.*, 1999). Nonetheless, *Hlxb9* is expressed transiently in both the dorsal and ventral buds, from E8.5 to E9.5, and then later in β cells. *Isl1*, a member of the LIM homeodomain subfamily, is first expressed in the dorsal pancreatic mesenchyme and then later in the pancreatic endoderm. Tissue recombination experiments show that its mesodermal expression is necessary for the development of the dorsal exocrine pancreas, but apparently not via changes in *Shh* expression (Ahlgren *et al.*, 1997). *Hlxb9* is required for the expression of *Isl1*, *Pdx1*, and *Nkx2.2* in the dorsal, but not ventral, endoderm (Li *et al.*, 1999), placing *Hlxb9* upstream of these genes within dorsal pancreas development. *Hlxb9* mutant embryos did not exhibit ectopic *Shh* expression or changes in the expression of *Isl1* in the pancreatic mesenchyme. Together, these studies help define a regulatory network for dorsal pancreatic specification (Fig. 7).

Recent explant studies indicate that the ventral foregut endoderm, when cultured *in vitro*, initiates pancreas development by default (Deutsch *et al.,* 2000). The isolated ventral endoderm also fails to express *Shh*. This is in contrast to the isolated dorsal endoderm, in which *shh* expression must be repressed to permit pancreas development (Hebrok *et al.,* 1998; Kim and Melton, 1998). Cardiac/FGF signaling diverts the ventral endoderm cells from the default pancreatic fate to a hepatic fate (Deutsch *et al.,* 2000; see Section IV.C.2 above) and induces *Shh* expression in the process. In summary, although endogenous *Shh* expression is inhibitory for both dorsal and ventral pancreas specification, there are clear distinctions with regard to how *shh* is regulated and the dorsal and ventral pancreatic tissue domains are specified.

2. Morphogenetic Growth of the Pancreatic Buds

Other genes are required for pancreatic morphogenesis after the initial budding of dorsal and ventral domains. The most prominent of these is *Pdx1,* which, as noted above (Section II.C.2), is part of the evolutionarily conserved ParaHox cluster. In *Pdx1* homozygotes, the rostral duodenum fails to differentiate, the ventral pancreatic bud either dies off, shortly after formation, or the cells become incorporated into the bile duct, and the dorsal pancreatic bud is stunted, which together result in an apancreatic phenotype (Jonsson *et al.,* 1994; Offield *et al.,* 1996). Still, *Pdx1* homozygotes initiate glucagon and insulin expression in their pancreatic buds and the pancreatic mesenchyme develops normally. This situation is reminiscent of that described above for mutations that block early liver development, where defects in hepatic endoderm are uncoupled from the ability of the hepatic mesenchyme to differentiate (see end of Section IV.C.1). The extant data clearly indicate that *Pdx1* is required for the pancreatic bud to respond to further morphogenetic signals from the mesenchyme (Ahlgren *et al.,* 1996; Offield *et al.,* 1996). But considering that the initial *Pdx1* expression domain includes the pancreas, pancreatic ducts, and duodenum, that *Pdx1* expression precedes that of other pancreatic transcription factors except *Hlxb9,* and that later factors such as *Isl1* and *Ngn3* promote *Pdx1* expression (Ahlgren *et al.,* 1997; Apelqvist *et al.,* 1999), it seems likely that *pdx1* participates in a cross-regulatory network of genes that together help define the initial morphogenesis of the pancreatic bud (Fig. 7).

3. Formation of the Pancreatic Cell Lineages

Although similarities in endocrine and neural cell metabolism originally led to the hypothesis that endocrine cells are derived from the neural crest (Pearse, 1966), in grafting experiments, neural crest cells do not contribute to the endocrine pancreas (Le Douarin and Teillet, 1973). Direct cell marking studies have definitively shown that both the pancreatic exocrine and endocrine cell lineages arise from the endoderm (Percival and Slack, 1999). Also, the four endocrine hormones are coexpressed in individual cells prior to formation of cell types expressing enyzmes exclusively (Alpert *et al.,* 1988). Finally, chimeric mouse embryo experiments showed that islets of endocrine cells are polyclonal in origin (Deltour *et al.,* 1991). Taken together, these findings indicate that a population of multipotent progenitor cells from the endoderm gives rise to the different pancreatic cell lineages.

Currently, there appear to be two distinct means by which the exocrine and endocrine lineages are generated from the initial PDX1- and HlxB9-positive progenitor cell. The first, which was discovered in embryo tissue recombination stud-

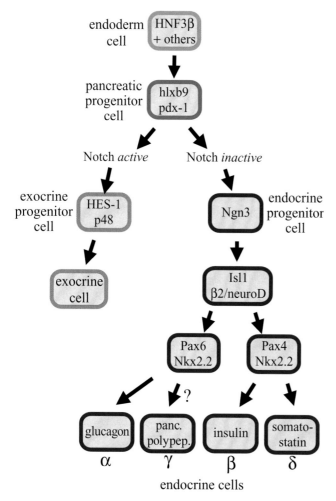

Figure 7. Generation of the cell lineages of the pancreas. Transcription factor genes required for the development of each of the pancreatic lineages are shown within the oval depictions of cells. Note that the early events in the endoderm lineage apply primarily to dorsal pancreas development. See text for details.

ies years ago, is a requirement for mesenchymal signals to promote exocrine cell development. The second, which has been unveiled only recently, is the governance by Notch signaling of "lateral" specification to the exocrine cell type. In both cases, the formation of exocrine cells must be promoted by cell signaling pathways or the endocrine lineage predominates. Also, transcription factors have been identified that control the differentiation of the various cell types.

Rutter and colleagues (1978; Gittes *et al.*, 1996) showed that early pancreatic mesenchymal cells secrete a factor that promotes exocrine acinar development and the growth of the immature gland. Pancreatic mesenchyme expresses follistatin, an inhibitor of TGF-β family members that, in isolation, mimics the exocrine-promoting and endocrine-repressing activities of the mesenchyme (Miralles *et al.*, 1998). Because activin and BMP7 are expressed by the pancreatic epithelium (Furukawa *et al.*, 1995; Lyons *et al.*, 1995) and purified TGF-β promotes endocrine development *in vitro* (Sanvito *et al.*, 1994; Mashima *et al.*, 1996), the balance of activators and inhibitors of TGF-β signaling appears critical to the relative sizes of the exocrine and endocrine compartments. In addition, FGF signaling promotes the exocrine lineage in midgestation embryos (Miralles *et al.*, 1999). The expression of these signaling factors might be regulated by *Isl1* in the mesenchyme, which, as noted in Section III.C.1 above, is necessary for the dorsal exocrine pancreas (Ahlgren *et al.*, 1997).

Another mechanism for generating cell type heterogeneity within the pancreatic progenitor population is via the Notch pathway. Cells that stochastically begin to differentiate may express high levels of Notch ligand (e.g., Delta or Serrate), which activates the Notch receptor in neighboring cells and thereby suppresses them from obtaining the same fate. The *Delta-like gene 1 (Dll1)* is expressed in the prospective dorsal pancreatic endoderm and homozygous *Dll1−/−* embryos exhibit premature differentiation to the endocrine fate (Apelqvist *et al.*, 1999; Lammert *et al.*, 2000). Similar results were obtained with embryos homozygous mutant for the transcriptional mediator of Notch signaling, *RBP-J$_K$*, which normally stimulates the expression of the bHLH repressor gene, *Hes1*. In the absence of *Hes1*, premature differentiation to the endocrine fate depletes the pancreatic progenitor cell population (Jensen *et al.*, 2000). *Hes1* normally represses *neurogenin-3* or *Ngn3* (*Atoh5*), a bHLH gene that initially is expressed in scattered cells of the dorsal and ventral pancreatic epithelium at E9.5–10.5, but that declines in expression by late gestation (Apelqvist *et al.*, 1999; Gradwohl *et al.*, 2000; Schwitzgebel *et al.*, 2000). In turn, *ngn3* is required for all four of the endocrine cell lineages (Gradwohl *et al.*, 2000). *Ngn3* is activated by the *HNF-6 (Onecut1)* gene, a cuthomoeodomain transcription factor (Rausa *et al.*, 1997; Landry *et al.*, 1997), and *HNF-6−/−* mice lack islets of Langerhans and are diabetic (Jacquemin *et al.*, 2000). However, it

is currently not known how Notch signaling might control HNF-6. In summary, Notch pathway activation activates *RBP-J$_K$*, which activates HES1, which represses *Ngn3*, which excludes an endocrine cell fate.

Notch signaling also directly promotes the exocrine cell fate, at least in part, by increasing the expression (Apelqvist *et al.*, 1999) of the p48 bHLH subunit of the transcription factor PTF1 (Krapp *et al.*, 1996). PTF1 activates various genes in the acinar pancreas (Cockell *et al.*, 1989) and homozygous inactivation of the *p48* gene completely deletes the exocrine lineage (Krapp *et al.*, 1998). Interestingly, in *p48−/−* embryos the acinar defect is sufficiently detrimental to pancreas morphogenesis that by E16 the endocrine cells begin to migrate to the spleen, ultimately resulting in postnatal lethality (Krapp *et al.*, 1998). On the other hand, in *Ngn3* mutants no endocrine cells or islets of Langerhans are detectable, yet the pancreas still develops as an organ (Gradwohl *et al.*, 2000). These studies show that the acinar component of the gland is most critical for its overall morphogenesis, and perhaps explain why most of the early signaling in pancreatic development is proexocrine.

Certain proendocrine transcription factor genes, including *Pdx1, Isl1*, and the homeodomain genes *Nkx2-2* (Sussel *et al.*, 1998) and *Nkx6.1* (Inoue *et al.*, 1997; Oster *et al.*, 1998) are first expressed throughout the pancreatic endoderm and then later are resolved into certain endocrine cell lineages. As expected, these genes are repressed by HES1 during Notch activation in exocrine precursors (Jensen *et al.*, 2000). *Isl1* is required within the pancreatic endoderm for the generation of endocrine cells (Ahlgren *et al.*, 1997). Despite the very early expression of *Nkx2-2*, the endocrine lineage is specified in homozygous null mutants and islets develop, but the islet cells do not produce any of the endocrine hormones (Sussel *et al.*, 1998). Thus, *Nkx2-2* promotes terminal endocrine cell differentiation. Induction of *Ngn3*, during generation of the endocrine lineage, results in activation of genes for the paired box-homeodomain factors *Pax4* and *Pax6* and the bHLH factor *β2/neuroD* in an endocrine stem cell population (Sosa-Pineda *et al.*, 1997; St. Onge *et al.*, 1997; Naya *et al.*, 1997; Gradwohl *et al.*, 2000). Each of these genes helps define particular endocrine lineages and/or islet morphology (Fig. 7). *Pax4* is required for the formation of β and δ cells at the expense of α cells (Sosa-Pineda *et al.*, 1997), suggesting that α cells may be a default. Still, *Pax6* is required for the formation of α cells and for the proper development of an islet-like structure (St. Onge *et al.*, 1997; Sander *et al.*, 1997). *Pax4/Pax6* double homozygotes contain some *Pdx1* positive endocrine progenitors in their pancreas but lack any production of endocrine hormones. In *β2/neuroD* homozygotes, all four pancreatic hormones are produced by their respective cell types, but β cells are greatly reduced in number and the remaining endocrine cells fail to form islets; instead, the endocrine cells form small clusters that spread through the acini and the mice become diabetic (Naya *et al.,* 1997). In addition, N-cam$^{-/-}$ embryos exhibit a loss of localization of α cells within the islet (Esni *et al.,* 1999). Clearly it will be of interest to identify additional cell adhesion molecules and other proteins that help coordinate endocrine cell organization with hormone production.

4. Pancreatic Ducts as a Possible Source of a Stem Cell Compartment

Mature pancreatic islet cells in the adult have little replicative capacity. Recently, a mouse model has been developed whereby pancreatic duct cells replenish islets that are damaged by the inflammatory expression of an interferon γ transgene (IFNg) in β cells (Sarvetnick *et al.,* 1988; Gu and Sarvetnick, 1993). In these mice, the duct cells undergo proliferation and subsequent differentiation to endocrine cells. Interestingly, the pancreatic ducts of IFNg-transgenic mice, but not control mice, express the *Pdx1* gene, consistent with the activation of an early endocrine stem cell (Kritzik *et al.,* 1999). In a pancreatectomy model, *Pdx1* expression increases after a burst of replication of pancreatic duct cells (Sharma *et al.,* 1999). An interesting issue relevant to these studies is whether the cells that become *Pdx1* positive are transdifferentiating duct cells or derivatives of a stem cell normally resident in the ducts. In either case, it seems clear that understanding the developmental potential of different pancreatic cells will be useful in controlling islet regeneration in clinical situations.

V. Perspectives and Remaining Issues on Organogenesis from the Endoderm

Although much is known about the way that different organs develop from the definitive endoderm, many more genes need to be identified before most of the relevant control pathways are known. Consider that of the 16 aforementioned genes that are critical for early liver development, only 2, *Hgf* and *c-Met,* are clearly within the same pathway and yield virtually the same mutant phenotype when mutated; thus we are quite distant from having defined all the relevant genes. Clearly, diverse developmental mechanisms are at play in the different organs. Whereas the lung and pancreas undergo extensive branching morphogenesis, the liver and thyroid only exhibit simple budding and the intestine elongates along the AP axis and folds internally. On the other hand, we can discern some common themes for the genesis of the different organs.

Each endodermal-derived organ begins to bud from the gut tube during the relatively narrow interval between E8.5 and E9.5 in the mouse. Although intestinal differentiation begins considerably later, the gut tube closes and begins to elongate greatly during the period between E8.5 and E10.5, so that it can be considered to initiate its development simul-

taneously with the other endoderm-derived organs. At present we do not understand the mechanisms underlying the temporal coordination of the morphogenesis of the different regions.

Detailed studies of liver and thyroid anlagen indicate that a portion of the endoderm begins to express tissue-specific transcription factors and genes, and coincidentally or shortly thereafter, these cells proliferate more rapidly and move into the surrounding mesenchyme, generating a tissue bud. For both tissues, the cells invade the mesenchyme in a cordlike fashion. For the pancreas, lung, and intestine, the early expression of relevant genes in the endodermal domain is followed by lateral growth of the cell population within the epithelium. The general mechanisms that elicit changes in cell proliferation are probably similar for either budding or lateral expansion of endoderm, but are as yet unknown. On the other hand, the growth of a cell population laterally within the epithelium (pancreas, lung, intestine) must require different controls over cell–cell interactions and extracellular matrix remodeling, compared to that for the cordlike migration of cells into the mesenchyme (thyroid, liver). Also, lateral growth of the epithelium can result in either outfolding and subsequent branching (pancreas, lung) or longitudinal growth (intestine), further illustrating the diversity of morphogenetic control.

Currently there is strong focus on how the diverse morphogenetic events in endoderm development are initiated. Although each tissue primordium expresses a unique combination of early genes, some primordia express genes in common. For example, *Hex* expression is critical for both the nascent thyroid and liver. Less is known about genes that cause the initial budding of the pharynx and lung, or gut tube elongation. Clearly the particular combination of genes expressed by the different endodermal domains will dictate different developmental outcomes. However, cell-extrinsic or local environmental factors play critical roles in influencing the formation of cell lineages within a tissue compartment as well as the initial patterning of the early endoderm. Each aspect of endoderm development must, therefore, be considered in the context of inducing signals, responding genes and their functions, and the cellular/tissue environment within which morphogenesis will occur.

References

Acampora, D., Mazan, S., Lallemand, Y., Avantaggiato, V., Maury, M., Simeone, A., and Brulet, P. (1995). Forebrain and midbrain regions are deleted in *Otx2–/–* mutants due to a defective anterior neuroectoderm specification during gastrulation. *Development (Cambridge, UK)* **121,** 3279–3290.

Adamson, I. Y., and Bowden, D. H. (1974). The type 2 cell as progenitor of alveolar epithelial regeneration. A cytodynamic study in mice after exposure to oxygen. *Lab. Invest.* **30,** 35–42.

Ahlgren, U., Jonsson, J., and Edlund, H. (1996). The morphogenesis of the pancreatic mesenchyme is uncoupled from that of the pancreatic epithelium in IPF1/PDX1-deficient mice. *Development (Cambridge, UK)* **122,** 1409–1416.

Ahlgren, U., Pfaff, S. L., Jessell, T. M., Edlund, T., and Edlund, H. (1997). Independent requirement for ISL1 in formation of pancreatic mesenchyme and islet cells. *Nature (London)* **385,** 257–260.

Al-azzeh, E., Fegert, P., Blin, N., and Gött, P. (2000). Transcription factor GATA-6 activates expression of gastroprotective trefoil genes *TFF1* and *TFF2*. *Biochim. Biophys. Acta* **1490,** 324–332.

Alexander, J., and Stainier, D. Y. (1999). A molecular pathway leading to endoderm formation in zebrafish. *Curr Biol* **9,** 1147–1157.

Alpert, S., Hanahan, D., and Teitelman, G. (1988). Hybrid insulin genes reveal a developmental lineage for pancreatic endocrine cells and imply a relationship with neurons. *Cell (Cambridge, Mass.)* **53,** 295–308.

Ang, S.-L., and Rossant, J. (1994). *HNF-3β* is essential for node and notochord formation in mouse development. *Cell (Cambridge, Mass.)* **78,** 561–574.

Ang, S.-L., Wierda, A., Wong, D., Stevens, K. A., Cascio, S., Rossant, J., and Zaret, K. S. (1993). The formation and maintenance of the definitive endoderm lineage in the mouse: Involvement of HNF3/*forkhead* proteins. *Development (Cambridge, UK)* **119,** 1301–1315.

Ang, S.-L., Conlon, R. A., Jin, O., and Rossant, J. (1994). Positive and negative signals from mesoderm regulate the expression of mouse Otx2 in ectoderm explants. *Development (Cambridge, UK)* **120,** 2979–2989.

Apelqvist, A., Ahlgren, U., and Edlund, H. (1997). Sonic hedgehog directs specialized mesoderm differentiation in the intestine and pancreas. *Curr. Biol.* **7,** 801–804.

Apelqvist, A., Li, H., Sommer, L., Beatus, P., Anderson, D. J., Honjo, T., Hrabé de Angelis, M., Lendahl, U., and Edlund, H. (1999). Notch signalling controls pancreatic cell differentiation. *Nature (London)* **400,** 877–881.

Arman, E., Haffner-Kraus, R., Gorivodsky, M., and Lonai, P. (1999). Fgfr2 is required for limb outgrowth and lung branching morphogenesis. *Proc. Natl. Acad. Sci. U.S.A.* **96,** 11895–11899.

Aubin, J., Lemieux, M., Tremblay, M., Berard, J., and Jeannotte, L. (1997). Early postnatal lethality in Hoxa-5 mutant mice is attributable to respiratory tract defects. *Dev. Bio.* **192,** 432–445.

Bachiller, D., Klingensmith, J., Kemp, C., Belo, J. A., Anderson, R. M., May, S. R., McMahon, J. A., McMahon, A. P., Harland, R. M., Rossant, J. *et al.* (2000). The organizer factors Chordin and Noggin are required for mouse forebrain development. *Nature (London)* **403,** 658–661.

Barbacci, E., Reber, M., Ott, M. O., Breillat, C., Huetz, F., and Cereghini, S. (1999). Variant hepatocyte nuclear factor 1 is required for visceral endoderm specification. *Development (Cambridge, UK)* **126,** 4795–4805.

Beamer, W. J., Eicher, E. M., Maltais, L. J., and Southard, J. L. (1981). Inherited primary hypothyroidism in mice. *Science* **212,** 61–63.

Beck, F., Chawengsaksophak, K., Waring, P., Playford, R. J., and Furness, J. B. (1999). Reprogramming of intestinal differentiation and intercalary regeneration in Cdx2 mutant mice. *Proc. Natl. Acad. Sci. U.S.A.* **96,** 7318–7323.

Beck, F., Tata, F., and Chawengsaksophpak, K. (2000). Homeobox genes and gut development. *BioEssays* **22,** 431–441.

Beck, L., and Markovich, D. (2000). The mouse Na(+)-sulfate cotransporter gene Nas1. Cloning, tissue distribution, gene structure, chromosomal assignment, and transcriptional regulation by vitamin D. *J. Biol. Chem.* **275,** 11880–11890.

Beck, S., Sommer, P., dos Santos Silva, E., Blin, N., and Gott, P. (1999). Hepatocyte nuclear factor 3 (winged helix domain). activates trefoil factor gene *TFF1* through a binding motif adjacent to the TATAA box. *DNA Cell Biol.* **18,** 157–164.

Beg, A. A., Sha, W. C., Bronson, R. T., Ghosh, S., and Baltimore, D. (1995). Embryonic lethality and liver degeneration in mice lacking the RelA component of NF-KB. *Nature (London)* **376,** 167–170.

Bellusci, S., Henderson, R., Winnier, G., Oikawa, T., and Hogan, B. L. M. (1996). Evidence from normal expression and targeted misexpression that Bone Morphogenetic protein-4 (Bmp-4) plays a role in mouse

embryonic lung morphogenesis. *Development (Cambridge, UK)* **122,** 1693–1702.

Bellusci, S., Furuta, Y., Rush, M. G., Henderson, R., Winnier, G., and Hogan, B. L. M. (1997a). Involvement of Sonic hedgehog (Shh) in mouse embryonic lung growth and morphogenesis. *Development (Cambridge, UK)* **124,** 53–63.

Bellusci, S., Grindley, J., Emoto, H., Itoh, N., and Hogan, B. L. M. (1997b). Fibroblast growth factor 10 (FGF10) and branching morphogenesis in the embryonic mouse lung. *Development (Cambridge, UK)* **124,** 4867–4878.

Belo, J. A., Bouwmeester, T., Leyns, L., Kertesz, N., Gallo, M., Follettie, M., and De Robertis, E. M. (1997). Cerberus-like is a secreted factor with neutralizing activity expressed in the anterior primitive endoderm of the mouse gastrula. *Mech. Dev.* **68,** 45–57.

Biben, C., Stanley, E., Fabri, L., Kotecha, S., Rhinn, M., Drinkwater, C., Lah, M., Wang, C.-C., Nash, A., Hilton, D. *et al.* (1998). Murine cerberus homologue mCer-1: A candidate anterior patterning molecule. *Dev. Biol.* **194,** 135–151.

Bielinska, M., Narita, N., and Wilson, D. B. (1999). Distinct roles for visceral endoderm during embryonic mouse development. *Int. J. Dev. Biol.* **43,** 183–205.

Bladt, F., Riethmacher, D., Isenmann, S., Aguzzi, A., and Birchmeier, C. (1995). Essential role for the c-met receptor in the migration of myogenic precursor cells into the limb bud. *Nature (London)* **376,** 768–772.

Bogue, C. W., Lou, L. J., Vasavada, H., Wilson, C. M., and Jacobs, H. C. (1996). Expression of Hoxb genes in the developing mouse foregut and lung. *Am. J. Respir. Cell Mol.* Biol. **15,** 163–171.

Borges, M., Linnoila, R., van de Velde, H., Chen, H., Nelkin, B., Mabry, M., Baylin, S., and Ball, D. (1997). An *achaete-scute* homologue essential for neuroendocrine differentiation in the lung. *Nature (London)* **386,** 852–855.

Bossard, P., and Zaret, K. (1998). GATA transcription factors as potentiators of gut endoderm differentiation. *Development (Cambridge, UK)* **125,** 4909–4917.

Bossard, P., and Zaret, K. S. (2000). Repressive and restrictive mesodermal interactions with gut endoderm: A possible explanation for nonintestinal cell types in Meckel's Diverticulum. *Development (Cambridge, UK)* **127,** 4915–4923.

Bostrom, H., Willetts, K., Pekny, M., Leveen, P., Lindahl, P., Hedstrand, H., Pekna, M., Hellstrom, M., Gebre-Medhin, S., Schalling, M. *et al.* (1996). PDGF-a signaling is a critical event in lung alveolar myofibroblast development and alveogenesis. *Cell (Cambridge, Mass.)* **85,** 863–873.

Brooke, N. M., Garcia-Fernández, J., and Holland, P. W. J. (1998). The ParaHox gene cluster is an evolutionary sister of the Hox gene cluster. *Nature (London)* **392,** 920–922.

Cardoso, W. V., Mitsialis, S. A., Brody, J. S., and Williams, M. C. (1996). Retinoic acid alters the expression of pattern-related genes in the developing rat lung. *Dev. Dyn.* **207,** 47–59.

Casey, E. S., Tada, M., Fairclough, L., Wylie, C. C., Heasman, J., and Smith, J. C. (1999). Bix4 is activated directly by VegT and mediates endoderm formation in Xenopus development. *Development (Cambridge, UK)* **126,** 4193–4200.

Chen, J., Knowles, H. J., Hebert, J. L., and Hackett, B. P. (1998). Mutation of the mouse hepatocyte nuclear factor/forkhead homologue 4 gene results in an absence of cilia and random left-right asymmetry. *J. Clin. Invest.* **102,** 1077–1082.

Chen, W. S., Manova, K., Weinstein, D. C., Duncan, S. A., Plump, A. S., Prezioso, V. R., Bachvarova, R. F., and Darnell, J. E., Jr. (1994). Disruption of the HNF-4 gene, expressed in visceral endoderm, leads to cell death in embryonic ectoderm and impaired gastrulation of mouse embryos. *Genes Dev.* **8,** 2466–2477.

Chiang, C., Litingtung, Y., Lee, E., Young, K. E., Corden, J. L., Westphal, H., and Beachy, P. A. (1996). Cyclopia and defective axial patterning in mice lacking *Sonic hedgehog* gene function. *Nature (London)* **383,** 407–412.

Cho, K. W., Blumberg, B., Steinbeisser, H., and De Robertis, E. M. (1991).

Molecular nature of Spemann's organizer: The role of the Xenopus homeobox gene goosecoid. *Cell (Cambridge, Mass.)* **67**, 1111–1120.

Chuang, P. T., and McMahon, A. P. (1999). Vertebrate Hedgehog signalling modulated by induction of a Hedgehog- binding protein. *Nature (London)* **397**, 617–621.

Cirillo, L. A., and Zaret, K. S. (1999). An early developmental transcription factor complex that is more stable on nucleosome core particles than on free DNA. *Mol. Cell* **4**, 961–969.

Ciruna, B. G., Schwartz, L., Harpal, K., Yamaguchi, T. P., and Rossant, J. (1997). Chimeric analysis of *fibroblasts growth factor receptor-1 (Fgfr-1)* function: A role for FGFR1 in morphogenetic movement through the primitive streak. *Development (Cambridge, UK)* **124**, 2829–2841.

Clarke, D. L., Johansson, C. B., Wilbertz, J., Veress, B., Nilsson, E., Karlstrom, H., Lendahl, U., and Frisen, J. (2000). Generalized potential of adult neural stem cells. *Science* **288**, 1660–1663.

Clifton-Bligh, R. J., Wentworth, J. M., Heinz, P., Crisp, M. S., John, R., Lazarus, J. H., Ludgate, M., and Chatterjee, V. K. (1998). Mutation of the gene encoding human TTF-2 associated with thyroid agenesis, cleft palate and choanal atresia. *Nat. Genet.* **19**, 399–401.

Cockell, M., Stevenson, B. J., Strubin, M., Hagenbuchle, O., and Wellauer, P. K. (1989). Identification of a cell-specific DNA-binding activity that interacts with a transcriptional activator of genes expressed in the acinar pancreas. *Mol. Cell. Biol.* **9**, 2464–2476.

Cole, T. J., Blendy, J. A., Monaghan, A. P., Krieglstein, K., Schmid, W., Aguzzi, A., Fantuzzi, G., Hummler, E., Unsicker, K., and Schutz, G. (1995). Targeted disruption of the glucocorticoid receptor gene blocks adrenergic chromaffin cell development and severely retards lung maturation. *Genes Dev.* **9**, 1608–1621.

Colvin, J. S., Feldman, B., Nadeau, J. H., Goldfarb, M., and Ornitz, D. M. (1999). Genomic organization and embryonic expression of the mouse fibroblast growth factor 9 gene. *Dev. Dyn.* **216**, 72–88.

Costa, R. H., Grayson, D. R., and Darnell, J. E., Jr. (1989). Multiple hepatocyte-enriched nuclear factors function in the regulation of transthyretin and alpha 1-antitrypsin genes. *Mol. Cell. Biol.* **9**, 1415–1425.

Cressman, D. E., Greenbaum, L. E., DeAngelis, R. A., Ciliberto, G., Furth, E. E., Poli, V., and Taub, R. (1996). Liver failure and defective hepatocyte regeneration in interleukin-6-deficient mice. *Science* **274**, 1379–1383.

Crossley, P. H., and Martin, G. R. (1995). The mouse *Fgf8* gene encodes a family of polypeptides and is expressed in regions that direct outgrowth and patterning in the developing embryo. *Development (Cambridge, UK)* **121**, 439–451.

Dabeva, M. S., Hwang, S. G., Vasa, S. R., Hurston, E., Novikoff, P. M., Hixson, D. C., Gupta, S., and Shafritz, D. A. (1997). Differentiation of pancreatic epithelial progenitor cells into hepatocytes following transplantation into rat liver. *Proc. Natl. Acad. Sci. U.S.A.* **94**, 7356–7361.

Danto, S. I., Shannon, J. M., Borok, Z., Zabski, S. M., and Crandall, E. D. (1995). Reversible transdifferentiation of alveolar epithelial cells. *Am. J. Respir. Cell Mol. Biol.* **12**, 497–502.

De Felice, M., Ovitt, C., Biffali, E., Rodriguez-Mallon, A., Arra, C., Anastassiadis, K., Macchia, P. E., Mattei, M. G., Mariano, A., Scholer, H. *et al.* (1998). A mouse model for hereditary thyroid dysgenesis and cleft palate. *Nat. Genet.* **19**, 395–398.

Deltour, L., Leduque, P., Paldi, A., Ripoche, M. A., Dubois, P., and Jami, J. (1991). Polyclonal origin of pancreatic islets in aggregation mouse chimaeras. *Development (Cambridge, UK)* **112**, 1115–1121.

de Maximy, A. A., Nakatake, Y., Moncada, S., Itoh, N., Thiery, J. P., and Bellusci, S. (1999). Cloning and expression pattern of a mouse homologue of Drosopila sprouty in the mouse embryo. *Mech. Dev.* **81**, 213–216.

deMello, D. E., Sawyer, D., Galvin, N., and Reid, L. M. (1997). Early fetal development of lung vasculature. *Am. J. Respir. Cell Mol. Biol.* **16**, 568–581.

De Moerlooze, L., Spencer-Dene, B., Revest, J.-M., Hajihosseini, M., Rosewell, I., and Dickson, C. (2000). An important role for the IIIb isoform of fibroblast growth factor receptor 2 (FGFR2) in mesenchymal-epithelial signalling during mouse organogenesis. *Development (Cambridge, UK)* **127**, 484–492.

Denson, L. A., McClure, M. H., Bogue, C. W., Karpen, S. J., and Jacobs, H. C. (2000). HNF3beta and GATA-4 transactivate the liver-enriched homeobox gene, Hex. *Gene* **246**, 311–320.

Deutsch, G., Jung, J., Zheng, M., Lóra, J., and Zaret, K. S. (2001). A bipotential precursor population for pancreas and liver within the embryonic endoderm. *Development (Cambridge, UK)* **128**, 871–888.

Dufort, D., Schwartz, L., Harpal, K., and Rossant, J. (1998). The transcription factor HNF3β is required in visceral endoderm for normal primitive streak morphogenesis. *Development (Cambridge, UK)* **125**, 3015–3025.

Duluc, I., Freund, J. N., Leberquier, C., and Kedinger, M. (1994). Fetal endoderm primarily holds the temporal and positional information required for mammalian intestinal development. *J. Cell Biol.* **126**, 211–221.

Duncan, S. A., Nagy, A., and Chan, W. (1997). Murine gastrulation requires HNF-4 regulated gene expression in the visceral endoderm: Tetraploid rescue of Hnf-4(–/–) embryos. *Development (Cambridge, UK)* **124**, 279–287.

Dunwoodie, S. L., Rodriguez, T. A., and Beddington, R. S. (1998). Msg1 and Mrg1, founding members of a gene family, show distinct patterns of gene expression during mouse embryogenesis. *Mech. Dev.* **72**, 27–40.

Ecochard, V., Cayrol, C., Rey, S., Foulquier, F., Caillol, D., Lemaire, P., and Duprat, A. M. (1998). A novel Xenopus mix-like gene milk involved in the control of the endomesodermal fates. *Development (Cambridge, UK)* **125**, 2577–2585.

Emura, M. (1997). Stem cells of the respiratory epithelium and their *in vitro* cultivation. *In Vitro Cell. Dev. Biol.* **33**.

Esni, F., Taljedal, I. B., Perl, A. K., Cremer, H., Christofori, G., and Semb, H. (1999). Neural cell adhesion molecule (N-CAM) is required for cell type segregation and normal ultrastructure in pancreatic islets. *J. Cell Biol.* **144**, 325–337.

Evans, M. J., Cabral, L. J., Stephens, R. J., and Freeman, G. (1973). Renewal of alveolar epithelium in the rat following exposure to NO2. *Am. J. Pathol.* **70**, 175–198.

Fässler, R., and Meyer, M. (1995). Consequences of lack of β1 integrin gene expression in mice. *Genes Dev.* **9**, 1896–1908.

Fausto, N. (2000). Liver regeneration. *J. Hepatol.* **32**, 19–31.

Fukamachi, H., Mizuno, T., and Takayama, S. (1979). Epithelial-mesenchymal interactions in differentiation of stomach epithelium in fetal mice. *Anat. Embryol.* **157**, 151–160.

Fukuda-Taira, S. (1981). Hepatic induction in the avian embryo: Specificity of reactive endoderm and inductive mesoderm. *J. Embryol. Exp. Morphol.* **63**, 111–125.

Furukawa, M., Eto, Y., and Kojima, I. (1995). Expression of immunoreactive activin A in fetal rat pancreas. *Endocrinol. J.* **42**, 63–68.

Gannon, M., and Bader, D. (1995). Initiation of cardiac differentiation occurs in the absence of anterior endoderm. *Development (Cambridge, UK)* **121**, 2439–2450.

Garretson, D. C., and Frederich, M. E. (1990). Meckel's diverticulum. *Am. Fam. Physician* **42**, 115–119.

Germain, L., Blouin, M. J., and Marceau, N. (1988). Biliary epithelial and hepatocytic cell lineage relationships in embryonic rat liver as determined by the differential expression of cytokeratins, α-fetoprotein, albumin, and cell surface-exposed components. *Cancer Res.* **48**, 4909–4918.

Giroux, S., and Charron, J. (1998). Defective development of the embryonic liver in N-*myc*-deficient mice. *Dev. Biol.* **195**, 16–28.

Gittes, G. K., and Rutter, W. J. (1992). Onset of cell-specific gene expression in the developing mouse pancreas. *Proc. Natl. Acad. Sci. U.S.A.* **89**, 1128–1132.

Gittes, G. K., Galante, P. E., Hanahan, D., Rutter, W. J., and Debase, H. T. (1996). Lineage-specific morphogenesis in the developing pancreas: role of mesenchymal factors. *Development (Cambridge, UK)* **122**, 439–447.

Glinka, A., Wu, W., Delius, H., Monaghan, A. P., Blumenstock, C., and Niehrs, C. (1998). Dickkopf-1 is a member of a new family of secreted proteins and functions in head induction. *Nature (London)* **391,** 357–362.

Golosow, N., and Grobstein, C. (1962). Epitheliomesenchymal interaction in pancreatic morphogenesis. *Dev. Biol.* **4,** 242–255.

Gordon, J. I., and Hermiston, M. L. (1994). Differentiation and self-renewal in the mouse gastrointestinal epithelium. *Curr. Opin. Cell Biol.* **6,** 795–803.

Gradwohl, G., Dierich, A., LeMeur, M., and Guillemot, F. (2000). neurogenin3 is required for the development of the four endocrine cell lineages of the pancreas. *Proc. Natl. Acad. Sci. U.S.A.* **97,** 1607–1611.

Grindley, J. C., Bellusci, S., Perkins, D., and Hogan, B. L. M. (1997). Evidence for the involvement of the Gli gene family in embryonic mouse lung development. *Dev. Biol.* **188,** 337–348.

Gu, D., and Sarvetnick, N. (1993). Epithelial cell proliferation and islet neogenesis in IFN-g transgenic mice. *Development (Cambridge, UK)* **118,** 33–46.

Gualdi, R., Bossard, P., Zheng, M., Hamada, Y., Coleman, J. R., and Zaret, K. S. (1996). Hepatic specification of the gut endoderm *in vitro:* cell signalling and transcriptional control. *Genes Dev.* **10,** 1670–1682.

Günes, C., Heuchel, R., Georgiev, O., Müller, K.-H., Lichtlen, P., Blüthmann, H., Marino, S., Aguzzi, A., and Schaffner, W. (1998). Embryonic lethality and liver degeneration in mice lacking the metal-responsive transcriptional activator MTF-1. *EMBO J.* **17,** 2846–2854.

Hackett, B. P., Bingle, C. D., and Gitlin, J. D. (1996). Mechanisms of gene expression and cell fate determination in the developing pulmonary epithelium. *Annu. Rev. Physiol.* **58,** 51–71.

Haffen, K., Kedinger, M., and Simon-Assmann, P. (1987). Mesenchyme-dependent differentiation of epithelial progenitor cells in the gut. *J. Pediatr. Gastroenterol. Nutr.* **6,** 14–23.

Harrison, K. A., Thaler, J., Pfaff, S. L., Gu, H., and Kehrl, J. H. (1999). Pancreas dorsal lobe agenesis and abnormal islets of Langerhans in H1xb9-deficient mice. *Nat. Genet.* **23,** 71–75.

Hebrok, M., Kim, S. K., and Melton, D. A. (1998). Notochord repression of endodermal sonic hedgehog permits pancreas development. *Genes Dev.* **12,** 1705–1713.

Heine, U. I., Munoz, E. F., Flanders, K. C., Roberts, A. B., and Sporn, M. B. (1990). Colocalization of TGF-beta 1 and collagen I and III, fibronectin and glycosaminoglycans during lung branching morphogenesis. *Development (Cambridge, UK)* **109,** 29–36.

Henry, G. L., and Melton, D. A. (1998). *Mixer,* a homeobox gene required for endoderm development. *Science* **281,** 91–96.

Henry, G. L., Brivanlou, I. H., Kessler, D. S., Hemmati-Brivanlou, A., and Melton, D. A. (1996). TGF-beta signals and a pattern in *Xenopus laevis* endodermal development. *Development (Cambridge, UK)* **122,** 1007–1015.

Hentsch, B., Lyons, I., Ruili, L., Hartley, L., Lints, T. J., Adams, J. M., and Harvey, R. P. (1996). *Hlx* homeo box gene is essential for an inductive tissue interaction that drives expansion of embryonic liver and gut. *Genes Dev.* **10,** 70–79.

Hermiston, M. L., and Gordon, J. I. (1995). In vivo analysis of cadherin function in the mouse intestinal epithelium: Essential roles in adhesion, maintenance of differentiation, and regulation of programmed cell death. *J. Cell Biol.* **129,** 489–506.

Hilberg, F., Aguzzi, A., Howells, N., and Wagner, E. F. (1993). c-Jun is essential for normal mouse development and hepatogenesis. *Nature (London)* **365,** 179–181.

Hogan, B. L. (1999). Morphogenesis. *Cell (Cambridge, Mass.)* **96,** 225–233.

Hogan, B. L. M., and Yingling, J. M. (1998). Epithelial/mesenchymal interactions and branching morphogenesis of the lung. *Curr. Opin. Genet. Dev.* **8,** 481–486.

Hogan, B. L. M., Grindley, J., Bellusci, S., Dunn, N. R., Emoto, H., and Itoh, N. (1997). Branching morphogenesis of the lung: New models for

a classical problem. *Cold Spring Harbor Symp. Quant. Biol.* **62,** 249–256.

Horner, M. A., Quintin, S., Domeier, M. E., Kimble, J., Labouesse, M., and Mango, S. E. (1998). *pha-4,* an *HNF-3* homolog, specifies pharyngeal organ identity in *Caenorhabditis elegans. Genes Dev.* **12,** 1947–1952.

Houssaint, E. (1980). Differentiation of the mouse hepatic primordium. I. An analysis of tissue interactions in hepatocyte differentiation. *Cell Differ.* **9,** 269–279.

Hromas, R., Radich, J., and Collins, S. (1993). PCR cloning of an orphan homeobox gene (PRH) preferentially expressed in myeloid and liver cells. *Biochem. Biophys. Res. Commun.* **195,** 976–983.

Hudson, C., Clements, D., Friday, R. V., Stott, D., and Woodland, H. R. (1997). Xsox17α and -β mediate endoderm formation in Xenopus. *Cell (Cambridge, Mass.)* **91,** 406.

Inoue, H., Rudnick, A., German, M. S., Veile, R., Donis-Keller, H., and Permutt, M. A. (1997). Isolation, characterization, and chromosomal mapping of the human Nkx6.1 gene (NKX6A), a new pancreatic islet homeobox gene. *Genomics* **40,** 367–370.

Ishii, Y., Rex, M., Scotting, P. J., and Yasugi, S. (1998). Region-specific expression of chicken Sox2 in the developing gut and lung epithelium: Regulation by epithelial-mesenchymal interactions. *Dev. Dyn.* **213,** 464–475.

Jacks, T., Fazeli, A., Schmitt, E. M., Bronson, R. T., Goodell, M. A., and Weinberg, R. A. (1992). Effects of an *Rb* mutation in the mouse. *Nature (London)* **359,** 295–300.

Jacquemin, P., Durviaux, S. M., Jensen, J., Godfraind, C., Gradwohl, G., Guillemot, F., Madsen, O. D., Carmeliet, P., Dewerchin, M., Collen, D. *et al.* (2000). Transcription factor hepatocyte nuclear factor 6 regulates pancreatic endocrine cell differentiation and controls expression of the proendocrine gene ngn3. *Mol. Cell. Biol.* **20,** 4445–4454.

Jensen, J., Pedersen, E. E., Galante, P., Hald, J., Heller, R. S., Ishibashi, M., Kageyama, R., Guillemot, F., Serup, P., and Madsen, O. D. (2000). Control of endodermal endocrine development by Hes-1. *Nat. Genet.* **24,** 36–44.

Johnson, G. R., and Moore, M. A. (1975). Role of stem cell migration in initiation of mouse foetal liver haemopoiesis. *Nature (London)* **258,** 726–728.

Jonsson, J., Carlsson, L., Edlund, T., and Edlund, H. (1994). Insulin-promoter-factor 1 is required for pancreas development in mice. *Nature (London)* **371,** 606–609.

Jung, J., Zheng, M., Goldfarb, M., and Zaret, K. S. (1999). Initiation of mammalian liver development from endoderm by fibroblasts growth factors. *Science* **284,** 1998–2003.

Kaartinen, V., Voncken, J. W., Shuler, C., Warburton, D., Bu, D., Heisterkamp, N., and Groffen, J. (1995). Abnormal lung development and cleft palate in mice lacking TGF-B3 indicates defects of epithelial-mesenchymal interaction. *Nat. Genet.* **11,** 415–421.

Kaestner, K., Silberg, D., Traber, P., and Schutz, G. (1997). The mesenchymal winged helix transcription factor *Fkh6* is required for the control of gastrointestinal proliferation and differentiation. *Genes Dev.* **11,** 1583–1595.

Kalb, J. M., Lau, K. K., Goszczynski, B., Fukushige, T., Moons, D., Okkema, P. G., and McGhee, J. D. (1998). *pha-4* is *Ce-fkh-1,* a *fork head*/HNF-1α, β, γ homolog that functions in organogenesis of the *C. elegans* pharynx. *Development (Cambridge, UK)* **125,** 2171–2180.

Kamiya, A., Kinoshita, T., Ito, Y., Matsui, T., Morikawa, Y., Senba, E., Nakashima, K., Taga, T., Yoshida, K., Kishimoto, T. *et al.* (1999). Fetal liver development requires a paracrine action of oncostatin M through the gp130 signal transducer. *EMBO J.* **18,** 2127–2136.

Kappen, C. (1996). Hox genes in the lung. *Am. J. Respir. Cell Mol. Biol.* **15,** 156–162.

Kedinger, M., Simon-Assmann, P. M., Lacroix, B., Marxer, A., Hauri, H. P., and Haffen, K. (1986). Fetal gut mesenchyme induces differentiation of cultured intestinal endodermal and crypt cells. *Dev. Biol.* **113,** 474–483.

Kedinger, M., Haffen, K., and Simon-Assmann, P. (1987a). Intestinal tissue and cell cultures. *Differentiation (Berlin)* **36,** 71–85.

Kedinger, M., Simon-Assmann, P., Alexandre, E., and Haffen, K. (1987b). Importance of a fibroblastic support for *in vitro* differentiation of intestinal endodermal cells and for their response to glucocorticoids. *Cell Differ.* **20,** 171–182.

Kedinger, M., Simon-Assmann, P., and Haffen, K. (1987c). Growth and differentiation of intestinal endodermal cells in a coculture system. *Gut* **28,** 237–241.

Kikuchi, Y., Trinh, L. A., Reiter, J. F., Alexander, J., Yelon, D., and Stainier, D. Y. (2000). The zebrafish bonnie and clyde gene encodes a Mix family homeodomain protein that regulates the generation of endodermal precursors. *Genes Dev.* **14,** 1279–1289.

Kim, S. K., and Melton, D. A. (1998). Pancreas development is promoted by cyclopamine, a hedgehog signaling inhibitor. *Proc. Natl. Acad. Sci. U.S.A.* **95,** 13036–13041.

Kim, S. K., Hebrok, M., and Melton, D. A. (1997). Notochord to endoderm signaling is required for pancreas development. *Development (Cambridge, UK)* **124,** 4243–4252.

Kimmel, S. G., Mo, R., Hui, C. C., and Kim, P. C. (2000). New mouse models of congenital anorectal malformations. *J. Pediatr. Surg.* **35,** 227–230; discussion: pp. 230–221.

Kimura, S., Hara, Y., Pineau, T., Fernandez-Salguero, P., Fox, C. H., Ward, J. M., and Gonzalez, F. J. (1996). The T/ebp null mouse: Thyroid-specific enhancer-binding protein is essential for the organogenesis of the thyroid, lung, ventral forebrain, and pituitary. *Genes Dev.* **10,** 60–69.

King, J., Marker, P. C., Seung, K. S., and Kingsley, D. M. (1994). BMP5 and the molecular, skeletal, and soft-tissue alterations in *short ear* mice. *Dev. Biol.* 112–122.

Kondo, T., Dollé, P., Zákány, J., and Duboule, D. (1996). Function of posterior *HoxD* genes in the morphogenesis of the anal sphincter. *Development (Cambridge, UK)* **122,** 2651–2659.

Krakowski, M. L., Kritzik, M. R., Jones, E. M., Krahl, T., Lee, J., Arnush, M., Gu, D., and Sarvetnick, N. (1999). Pancreatic expression of keratinocyte growth factor leads to differentiation of islet hepatocytes and proliferation of duct cells. *Am. J. Pathol.* **154,** 683–691.

Krapp, A., Knofler, M., Frutiger, S., Hughes, G. J., Hagenbuchle, O., and Wellauer, P. K. (1996). The p48 DNA-binding subunit of transcription factor PTF1 is a new exocrine pancreas-specific basic helix-loop-helix protein. *EMBO J.* **15,** 4317–4329.

Krapp, A., Knofler, M., Ledermann, B., Burki, K., Berney, C., Zoerkler, N., Hagenbuchle, O., and Wellauer, P. K. (1998). The bHLH protein PTF1-p48 is essential for the formation of the exocrine and the correct spatial organization of the endocrine pancreas. *Genes Dev.* **12,** 3752–3763.

Kreidberg, J., Donovan, M., Goldstein, S., Rennke, H., Shepherd, K., Jones, R., and Jaenisch, R. (1996). Alpha 3 beta 1 integrin has a crucial role in kidney and lung organogenesis. *Development (Cambridge, UK)* **122,** 3537–3547.

Kritzik, M. R., Jones, E., Chen, Z., Krakowski, M., Krahl, T., Good, A., Wright, C., Fox, H., and Sarvetnick, N. (1999). PDX-1 and Msx-2 expression in the regenerating and developing pancreas. *J. Endocrinol.* **163,** 523–530.

Kuo, C. T., Morrisey, E. E., Anandappa, R., Sigrist, K., Lu, M. M., Parmacek, M. S., Soudais, C., and Leiden, J. M. (1997). GATA4 transcription factor is required for ventral morphogenesis and heart tube formation. *Genes Dev.* **11,** 1048–1060.

Kurzrock, E. A., Baskin, L. S., and Cunha, G. R. (1999). Ontogeny of the male urethra: Theory of endodermal differentiation. *Differentiation (Berlin)* **64,** 115–122.

Lagasse, E., Connors, H., Al-Dhalimy, M., Reitsma, M., Dohse, M., Osborne, L., Wang, X., Finegold, M., Weissman, IlL., and Grompe, M. (2000). Purified hematopoietic stem cells can differentiate into hepatocytes *in vivo. Nat. Med.* **6,** 1229–1235.

Lammert, E., Brown, J., and Melton, D. A. (2000). Notch gene expression during pancreatic organogenesis. *Mech. Dev.* **94,** 199–203.

Landry, C., Clotman, F., Hioki, T., Oda, H., Picard, J. J., Lemaigre, F. P., and Rousseau, G. G. (1997). HNF-6 is expressed in endoderm derivatives and nervous system of the mouse embryo and participates to the cross-regulatory network of liver-enriched transcription factors. *Dev. Biol.* **192,** 247–257.

Laverriere, A. C., MacNeill, C., Mueller, C., Poelmann, R. E., Burch, J. B. E., and Evans, T. (1994). GATA-4/5/6, a subfamily of three transcription factors transcribed in developing heart and gut. *J. Biol. Chem.* **269,** 23177–23184.

Lawson, K. A., Meneses, J. J., and Pedersen, R. A. (1986). *Cell (Cambridge, Mass.)* fate and cell lineage in the endoderm of the presomite mouse embryo, studied with an intracellular tracer. *Dev. Biol.* **115,** 325–339.

Lawson, K. A., Meneses, J. J., and Pedersen, R. A. (1991). Clonal analysis of epiblast fate during germ layer formation in the mouse embryo. *Development (Cambridge, UK)* **113,** 891–911.

Lazzaro, D., Price, M., De Felice, M., and Di Lauro, R. (1991). The transcription factor TTF-1 is expressed at the onset of thyroid and lung morphogenesis and in restricted regions of the foetal brain. *Development (Cambridge, UK)* **113,** 1093–1104.

Lebeche, D., Malpel, S., and Cardoso, W. V. (1999). Fibroblast growth factor interactions in the developing lung. *Mech. Dev.* **86,** 125–136.

Le Douarin, N. M. (1975). An experimental analysis of liver development. *Med. Biol.* **53,** 427–455.

Le Douarin, N. M., and Teillet, M. A. (1973). The migration of neural crest cells to the wall of the digestive tract in avian embryo. *J. Embryol. Exp. Morphol.* **30,** 31–48.

Lee, E. Y., Chang, C. Y., Hu, N., Wang, Y. C., Lai, C. C., Herrup, K., Lee, W. H., and Bradley, A. (1992). Mice deficient for Rb are nonviable and show defects in neurogenesis and haematopoiesis. *Nature (London)* **359,** 288–294.

Li, H., Arber, S., Jessell, T. M., and Edlund, H. (1999). Selective agenesis of the dorsal pancreas in mice lacking homeobox gene Hlxb9. *Nat. Genet.* **23,** 67–70.

Li, J., Ning, G., and Duncan, S. A. (2000). Mammalian hepatocyte differentiation requires the transcription factor HNF-4α. *Genes Dev.* **14,** 464–474.

Lindahl, P., Karlsson, L., Hellstrom, M., Gebre-Medhin, S., Willetts, K., Heath, J., and Betsholtz, C. (1997). Alveogenesis failure in PDGF-A-deficient mice is coupled to lack of distal spreading of alveolar smooth muscle cell progenitors during lung development. *Development (Cambridge, UK)* **124,** 3943–3953.

Litingtung, Y., Lei, L., Westphal, H., and Chiang, C. (1998). Sonic hedgehog is essential to foregut development. *Nat. Genet.* **20,** 58–61.

Lyons, K. M., Pelton, R. W., and Hogan, B. L. M. (1989). Patterns of expression of murine Vgr-1 and BMP-2a suggest that TGFβ-like genes coordinately regulate aspects of embryonic development. *Genes Dev.* **3,** 1657–1668.

Lyons, K. M., Hogan, B. L., and Robertson, E. J. (1995). Colocalization of BMP 7 and BMP 2 RNAs suggests that these factors cooperatively mediate tissue interactions during murine development. *Mech. Dev.* **50,** 71–83.

Macchia, P. E., Lapi, P., Krude, H., Pirro, M. T., Missero, C., Chiovato, L., Souabni, A., Baserga, M., Tassi, V., Pinchera, A. *et al.* (1998). PAX8 mutations associated with congenital hypothyroidism caused by thyroid dysgenesis. *Nat. Genet.* **19,** 83–86.

Macchia, P. E., Felice, M. D., and Lauro, R. D. (1999). Molecular genetics of congenital hypothyroidism. *Curr. Opin. Genet. Dev.* **9,** 289–294.

Magdaleno, S. M., Barrish, J., Finegold, M. J., and DeMayo, F. J. (1998). Investigating stem cells in the lung. *Adv. Pediatr.* **45,** 363–396.

Malpel, S., Mendelsohn, C., and Cardoso, W. V. (2000). Regulation of retinoic acid signaling during lung morphogenesis. *Development (Cambridge, UK)* **127,** 3057–3067.

Mango, S. E., Lambie, E. J., and Kimble, J. (1994). The alpha-4 gene is required to generate the pharyngeal primordium of Caenorhabditis elegans. *Development (Cambridge, UK)* **120,** 3019–3031.

Manley, N. R., and Capecchi, M. R. (1995). The role of Hoxa-3 in mouse thymus and thyroid development. *Development (Cambridge, UK)* **121**, 1989–2003.

Manley, N. R., and Capecchi, M. R. (1998). Hox group 3 paralogs regulate the development and migration of the thymus, thyroid, and parathyroid glands. *Dev. Biol.* **195**, 1–15.

Mansouri, A., Chowdhury, K., and Gruss, P. (1998). Follicular cells of the thyroid gland require Pax8 gene function. *Nat. Genet.* **19**, 87–90.

Martinez-Barbera, J. P., Clements, M., Thomas, P., Rodriguez, T., Meloy, D., Kioussis, D., and Beddington, R. S. (2000). The homeobox gene hex is required in definitive endodermal tissues for normal forebrain, liver and thyroid formation. *Development (Cambridge, UK)* **127**, 2433–2445.

Mashima, H., Ohnishi, H., Wakabayashi, K., Mine, T., Miyagawa, J., Hanafusa, T., Seno, M., Yamada, H., and Kojima, I. (1996). Betacellulin and activin A coordinately convert amylase-secreting pancreatic AR42J cells into insulin-secreting cells. *J. Clin. Invest.* **97**, 1647–1654.

Masui, T. (1981). Differentiation of the yolk-sac endoderm under the influence of the digestive-tract mesenchyme. *J. Embryol. Exp. Morphol.* **62**, 277–289.

Medlock, E. S., and Haar, J. L. (1983). The liver hemopoietic environment: I. Developing hepatocytes and their role in fetal hemopoiesis. *Anat. Rec.* **207**, 31–41.

Medvinsky, A., and Dzierzak, E. (1996). Definitive hematopoiesis is autonomously initiated by the AGM region. *Cell (Cambridge, Mass.)* **86**, 897–906.

Meehan, R. R., Barlow, D. P., Hill, R. E., Hogan, B. L., and Hastie, N. D. (1984). Pattern of serum protein gene expression in mouse visceral yolk sac and foetal liver. *EMBO J.* **3**, 1881–1885.

Mendelsohn, C., Lohnes, D., Decimo, D., Lufkin, T., LeMeur, M., Chambon, P., and Mark, M. (1994). Function of the retinoic acid receptors (RARs) during development (II). Multiple abnormalities at various stages of organogenesis in RAR double mutants. *Development (Cambridge, UK)* **120**, 2749–2771.

Metzger, R. J., and Krasnow, M. A. (1999). Genetic control of branching morphogenesis. *Science* **284**, 1635–1639.

Michalopoulos, G. K., and DeFrances, M. C. (1997). Liver regeneration. Science **276**, 60–66.

Miettinen, P., Warburton, D., Bu, D., Zhao, J.-S., Berger, J., Minoo, P., Koivisto, T., Allen, L., Dobbs, L., Werb, Z. *et al.* (1997). Impaired lung branching morphogenesis in the absence of functional EGF receptor. *Dev. Biol.* **186**, 224–236.

Min, H., Danilenko, D. M., Scully, S. A., Bolon, B., Ring, B. D., Tarpley, J. E., DeRose, M., and Simonet, W. S. (1998). Fgf-10 is required for both limb and lung development and exhibits striking functional similarity to drosophila branchless. *Genes Dev.* **12**, 3156–3161.

Minoo, P., Su, G., Drum, H., Bringas, P., and Kimura, S. (1999). Defects in tracheoesophageal and lung morphogenesis in Nkx2.1 (–/–) mouse embryos. *Dev. Biol.* **209**, 60–71.

Miralles, F., Czernichow, P., and Scharfmann, R. (1998). Follistatin regulates the relative proportions of endocrine versus exocrine tissue during pancreatic development. *Development (Cambridge, UK)* **125**, 1017–1024.

Miralles, F., Czernichow, P., Ozaki, K., Itoh, N., and Scharfmann, R. (1999). Signaling through fibroblast growth factor receptor 2b plays a key role in the development of the exocrine pancreas. *Proc. Natl. Acad. Sci. U.S.A.* **96**, 6267–6272.

Molinar-Rode, R., Smeyne, R. J., Curran, T., and Morgan, J. I. (1993). Regulation of proto-oncogene expression in adult and developing lungs. *Mol. Cell Biol.* **13**, 3213–3220.

Molkentin, J. D., Lin, Q., Duncan, S. A., and Olson, E. N. (1997). Requirement of the transcription factor GATA4 for heart tube formation and ventral morphogenesis. *Genes Dev.* **11**, 1061–1072.

Mollard, R., and Dziadek, M. (1998). A correlation between epithelial proliferation rates, basement membrane component localization patterns, and morphogenetic potential in the embryonic mouse lung. *Am. J. Respir. Cell Mol. Biol.* **19**, 71–82.

Monaghan, A. P., Kaestner, K. H., Grau, E., and Schütz, G. (1993). Postimplantation expression patterns indicate a role for the mouse forkhead/HNF-3α, β, and γ genes in determination of the definitive endoderm, chordamesoderm and neuroectoderm. *Development (Cambridge, UK)* **119**, 567–578.

Monaghan, A. P., Kioschis, P., Wu, W., Zuniga, A., Bock, D., Poustka, A., Delius, H., and Niehrs, C. (1999). Dickkopf genes are co-ordinately expressed in mesodermal lineages. *Mech. Dev.* **87**, 45–56.

Morrisey, E. E., Tang, Z., Sigrist, K., Lu, M. M., Jiang, F., Ip, H. S., and Parmacek, M. S. (1998). GATA6 regulates HNF4 and is required for differentiation of visceral endoderm in the mouse embryo. *Genes Dev.* **12**, 3579–3590.

Motoyama, J., Kitajima, K., Kojima, M., Kondo, S., and Takeuchi, T. (1997). Organogenesis of the liver, thymus and spleen is affected in *jumonji* mutant mice. *Mech. Dev.* **66**, 27–37.

Motoyama, J., Liu, J., Mo, R., Ding, Q., Post, M., and Hui, C. C. (1998). Essential function of Gli2 and Gli3 in the formation of lung, trachea and oesophagus. *Nat. Genet.* **20**, 54–57.

Mucenski, M. L., McLain, K., Kier, A. B., Swerdlow, S. H., Schreiner, C. M., Miller, T. A., Pietryga, D. W., Scott, W. J., Jr., and Potter, S. S. (1991). A functional c-myb gene is required for normal murine fetal hepatic hematopoiesis. *Cell (Cambridge, Mass.)* **65**, 677–689.

Naya, F. J., Huang, H. P., Qiu, Y., Mutoh, H., DeMayo, F. J., Leiter, A. B., and Tsai, M. J. (1997). Diabetes, defective pancreatic morphogenesis, and abnormal enteroendocrine differentiation in BETA2/neuroD-deficient mice. *Genes Dev.* **11**, 2323–2334.

Nishina, H., Vaz, C., Billia, P., Nghiem, M., Sasaki, T., Pompa, J. L. D. l., Furlonger, K., Paige, C., Hui, C., Fischer, K. D. *et al.* (1999). Defective liver formation and liver cell apoptosis in mice lacking the stress signaling kinase SEK1/MKK4. *Development (Cambridge, UK)* **126**, 505–516.

Nogawa, H., and Ito, T. (1995). Branching morphogenesis of embryonic mouse lung epithelium in mesenchyme-free culture. *Development (Cambridge, UK)* **1221**, 1051–1022.

Nomura, S., Esumi, H., Job, C., and Tan, S.-S. (1998). Lineage and clonal development of gastric glands. *Dev. Biol.* **204**, 124–135.

Offield, M. F., Jetton, T. L., Labosky, P. A., Ray, M., Stein, R. W., Magnuson, M. A., Hogan, B. L., and Wright, C. V. (1996). PDX-1 is required for pancreatic outgrowth and differentiation of the rostral duodenum. *Development (Cambridge, UK)* **122**, 983–995.

Oster, A., Jensen, J., Serup, P., Galante, P., Madsen, O. D., and Larsson, L. I. (1998). Rat endocrine pancreatic development in relation to two homeobox gene products (Pdx-1 and Nkx 6.1). *J. Histochem. Cytochem.* **46**, 707–715.

Overturf, K., al-Dhalimy, M., Ou, C. N., Finegold, M., and Grompe, M. (1997). Serial transplantation reveals the stem-cell like regenerative potential of adult mouse hepatocytes. *Am. J. Pathol.* **151**, 1273–1280.

Pabst, O., Zweigerdt, R., and Arnold, H. H. (1999). Targeted disruption of the homeobox transcription factor Nkx2-3 in mice results in postnatal lethality and abnormal development of small intestine and spleen. *Development (Cambridge, UK)* **126**, 2215–2225.

Packer, A. I., Mailutha, K. G., Ambrozewicz, L. A., and Wolgemuth, D. J. (2000). Regulation of the Hoxa4 and Hoxa5 genes in the embryonic mouse lung by retinoic acid and TGFβ1: Implications for lung development and patterning. *Dev. Dyn.* **217**, 62–74.

Park, W. Y., Miranda, B., Lebeche, D., Hashimoto, G., and Cardoso, W. V. (1998). FGF-10 is a chemotactic factor for distal epithelial buds during lung development. *Dev. Biol.* **201**, 125–134.

Paul, J., Conkie, D., and Freshney, R. I. (1969). Erythropoietic cell population changes during the hepatic phase of erthropoiesis in the fetal mouse. *Cell Tissue Kinet.* **2**, 283–294.

Pearce, J. J., and Evans, M. J. (1999). Mml, a mouse Mix-like gene expressed in the primitive streak. *Mech. Dev.* **87**, 189–192.

Pearse, A. G. (1966). 5-hydroxytryptophan uptake by dog thyroid 'C' cells, and its possible significance in polypeptide hormone production. *Nature (London)* **211,** 598–600.

Pepicelli, C. V., Lewis, P. M., and McMahon, A. P. (1998). Sonic hedgehog regulates branching morphogenesis in the mammalian lung. *Curr. Biol.* **8,** 1083–1086.

Percival, A. C., and Slack, J. M. (1999). Analysis of pancreatic development using a cell lineage label. *Exp. Cell Res.* **247,** 123–132.

Perl, A.-K., and Whitsett, J. A. (1999). Molecular mechanisms controlling lung morphogenesis. *Clin. Genet.* **56,** 14–27.

Peters, K., Werner, S., Liao, X., Wert, S., Whitsett, J., and Williams, L. (1994). Targeted expression of a dominant negative FGF receptor blocks branching morphogenesis and epithelial differeniation of the mouse lung. *EMBO J.* **13,** 3296–3301.

Petersen, B. E., Bowen, W. C., Patrene, K. D., Mars, W. M., Sullivan, A. K., Murase, N., Boggs, S. S., Greenberger, J. S., and Goff, J. P. (1999). Bone marrow as a potential source of hepatic oval cells. *Science* **284,** 1168–1170.

Pictet, R. L., Clark, W. R., Williams, R. H., and Rutter, W. J. (1972). An ultrastructural analysis of the developing embryonic pancreas. *Dev. Biol.* **29,** 436–467.

Pitera, J. E., Smith, V. V., Thorogood, P., and Milla, P. J. (1999). Coordinated expression of 3′ Hox genes during murine embryonal gut developmentz: An enteric Hox code. *Gastroenterology* **117,** 1339–1351.

Post, L. C., Ternet, M., and Hogan, B. L. M. (2000). Notch/Delta expression in the developing mouse lung. *Mech. Dev.* **98,** 95–98.

Quaggin, S. E., Schwartz, L., Cui, S., Igarashi, P., Deimling, J., Post, M., and Rossant, J. (1999). The basic-helix-loop-helix protein pod1 is critically important for kidney and lung organogenesis. *Development (Cambridge, UK)* **126,** 5771–5783.

Ramalho-Santos, M., Melton, D. A., and McMahon, A. P. (2000). Hedgehog signals regulate multiple aspects of gastrointestinal development. *Development (Cambridge, UK)* **127,** 2763–2772.

Rao, M. S., Dwivedi, R. S., Yelandi, A., Subbarao, V., Tan, X., Usman, M. I., Thangada, S., Nemali, M. R., Kumar, S., and Scarpelli, D. G. (1989). Role of periductal and ductalar epithelial cells of the adult rat pancreas in pancreatic hepatocyte lineage: A change in the differentiation commitment. *Am. J. Pathol.* **134,** 1069–1086.

Rausa, F., Samadani, U., Ye, H., Lim, L., Fletcher, C. F., Jenkins, N. A., Copeland, N. G., and Costa, R. H. (1997). The cut-homeodomain transcriptional activator HNF-6 is coexpressed with its target gene HNF-3 beta in the developing murine liver and pancreas. *Dev. Biol.* **192,** 228–246.

Rehorn, K.-P., Thelen, H., Michelson, A. M., and Reuter, R. (1996). A molecular aspect of hematopoiesis and endoderm development common to vertebrates and *Drosophila. Development (Cambridge, UK)* **122,** 4023–4031.

Reimold, A. M., Etkin, A., Clauss, I., Perkins, A., Friend, D. S., Zhang, J., Horton, H. F., Scott, A., Orkin, S. H., Byrne, M. C. *et al.* (2000). An essential role in liver development for transcription factor XBP-1. *Genes Dev.* **14,** 152–157.

Rhim, J. A., Sandgren, E. P., Degen, J. L., Palmiter, R. D., and Brinster, R. L. (1994). Replacement of diseased mouse liver by hepatic cell transplantation. *Science* **263,** 1149–1152.

Roberts, D. J., Johnson, R. L., Burke, A. C., Nelson, C. E., Morgan, B. A., and Tabin, C. (1995). Sonic hedgehog is an endodermal signal inducing *Bmp-4* and *Hox* genes during induction and regionalization of the chick hindgut. *Development (Cambridge, UK)* **121,** 3163–3174.

Roberts, D. J., Smith, D. M., Goff, D. J., and Tabin, C. J. (1998). Epithelial-mesenchymal signaling during the regionalization of the chick gut. *Development (Cambridge, UK)* **125,** 2791–2801.

Rosa, F. M. (1989). Mix.1, a homeobox mRNA inducible by mesoderm inducers, is expressed mostly in the presumptive endodermal cells of Xenopus embryos. *Cell (Cambridge, Mass.)* **57,** 965–974.

Rosenquist, G. C. (1971). The location of the pregut endoderm in the chick embryo at the primitive streak stage as determined by radioautographic mapping. *Dev. Biol.* **26,** 323–335.

Rugh, R. (1968). "The Mouse: Its Reproduction and Development." Burgess, Minneapolis, MN.

Rutter, W. J., Pictet, R. L., Harding, J. D., Chirgwin, J. M., MacDonald, R. J., and Prybyla, A. E. (1978). An analysis of pancreatic development: Role of mesenchymal factor and other extracellular factors. *In* "Molecular Control of Proliferation and Differentiation" (J. Papaconstantinou and W. Rutter, eds.), pp. 205–227. Academic Press, New York.

Sander, M., Neubuser, A., Kalamaras, J., Ee, H. C., Martin, G. R., and German, M. S. (1997). Genetic analysis reveals that PAX6 is required for normal transcription of pancreatic hormone genes and islet development. *Genes Dev.* **11,** 1662–1673.

Sanvito, F., Herrera, P. L., Huarte, J., Nichols, A., Montesano, R., Orci, L., and Vassalli, J. D. (1994). TGF-beta 1 influences the relative development of the exocrine and endocrine pancreas *in vitro. Development (Cambridge, UK)* **120,** 3451–3462.

Sarvetnick, N., Liggitt, D., Pitts, S. L., Hansen, S. E., and Stewart, T. A. (1988). Insulin-dependent diabetes mellitus induced in transgenic mice by ectopic expression of class II MHC and interferon-gamma. *Cell (Cambridge, Mass.)* **52,** 773–782.

Sasaki, H., and Hogan, B. L. M. (1993). Differential expression of multiple fork head related genes during gastrulation and pattern formation in the mouse embryo. *Development (Cambridge, UK)* **118,** 47–59.

Sawai, S., Shimono, A., Wakamatsu, Y., Palmes, C., Hanaoka, K., and Kondoh, H. (1993). Defects of embryonic organogenesis resulting from targeted disruption of the N-*myc* gene in the mouse. *Development (Cambridge, UK)* **117,** 1445–1455.

Schmidt, C., Bladt, F., Goedecke, S., Brinkmann, V., Zschiesche, W., Sharpe, M., Gherardi, E., and Birchmeier, C. (1995). Scatter factor/hepatocyte growth factor is essential for liver development. *Nature (London)* **373,** 699–702.

Schoenwolf, G. C., Garcia-Martinez, V., and Dias, M. S. (1992). Mesoderm movement and fate during avian gastrulation and neurulation. *Dev. Dyn.* **193,** 235–248.

Schuger, L., Johnson, G. R., Gilbride, K., Plowman, G. D., and Madel, R. (1996). Amphiregulin in lung branching morphogenesis: Interaction with heparan sulfate proteoglycan modulates cell proliferation. *Development (Cambridge, UK)* **122,** 1759–1767.

Schultheiss, T. M., Xydas, S., and Lassar, A. B. (1995). Induction of avian cardiac myogenesis by anterior endoderm. *Development (Cambridge, UK)* **121,** 4203–4214.

Schwitzgebel, V. M., Scheel, D. W., Conners, J. R., Kalamaras, J., Lee, J. E., Anderson, D. J., Sussel, L., Johnson, J. D., and German, M. S. (2000). Expression of neurogenin3 reveals an islet cell precursor population in the pancreas. *Development (Cambridge, UK)* **127,** 3533–3542.

Sekimoto, T., Yoshinobu, K., Yoshida, M., Kuratani, S., Fujimoto, S., Araki, M., Tajima, N., Araki, K., and Yamamura, K.-I. (1998). Region-specific expression of murine *Hox* genes implies the *Hox* code-mediated patterning of the digestive tract. *Genes Cells* **3,** 51–64.

Sekine, K., Ohuchi, H., Fujiwara, M., Yamasaki, M., Yoshizawa, T., Sato, T., Tagishita, N., Matsui, D., Koga, Y., Itoh, N. *et al.* (1999). FGF10 is essential for the limb and lung development. *Nat. Genet.* **21,** 138–141.

Sharma, A., Zangen, D. H., Reitz, P., Taneja, M., Lissauer, M. E., Miller, C. P., Weir, G. C., Habener, J. F., and Bonner-Weir, S. (1999). The homeodomain protein IDX-1 increases after an early burst of proliferation during pancreatic regeneration. *Diabetes* **48,** 507–513.

Shawlot, W., Deng, J. M., and Behringer, R. R. (1998). Expression of the mouse cerberus-related gene, Cerr1, suggests a role in anterior neural induction and somitogenesis. *Proc. Natl. Acad. Sci. U.S.A.* **95,** 6198–6203.

Shawlot, W., Min Deng, J., Wakamiya, M., and Behringer, R. R. (2000). The cerberus-related gene, cerr1, is not essential for mouse head formation. *Genesis* **26,** 253–258.

Shiojiri, N. (1984). Analysis of differentiation of hepatocytes and bile duct cells in developing mouse liver by albumin immunofluorescence. *Dev. Growth Differ.* **26,** 555–561.

Skandalakis, J. E., Gray, S. W., and Symbas, P. (1994). The trachea and the lungs. *In* "Embryology for Surgeons" (J. E. Skandalakis and S. W. Gray, eds.), pp. 414–450. Williams & Wilkins, Baltimore, MD.

Smith, D. M., and Tabin, C. J. (1999a). BMP signalling specifies the pyloric sphincter. *Nature (London)* **402**, 748–749.

Smith, D. M., and Tabin, C. J. (1999b). Chick Barx2b, a marker for myogenic cells also expressed in branchial arches and neural structures. *Mech. Dev.* **80**, 203–206.

Sosa-Pineda, B., Chowdhury, K., Torres, M., Oliver, G., and Gruss, P. (1997). The Pax4 gene is essential for differentiation of insulin-producing beta cells in the mammalian pancreas. *Nature (London)* **386**, 399–402.

Sosa-Pineda, B., Wigle, J. T., and Oliver, G. (2000). Hepatocyte migration during liver development requires prox1. *Nat. Genet.* **25**, 254–255.

Souza, P., Kuliszewski, M., Wang, J., Tseu, I., Tanswell, A. K., and Post, M. (1995). PDGF-AA and its receptor influence early lung branching via an epithelial-mesenchymal interaction. *Development (Cambridge, UK)* **121**, 2559–2567.

Spooner, B. S., Walther, B. T., and Rutter, W. J. (1970). The development of the dorsal and ventral mammalian pancreas in vivo and in vitro. *J. Cell Biol.* **47**, 235–246.

Stappenbeck, T. S., and Gordon, J. I. (2000). Rac1 mutations produce aberrant epithelial differentiation in the developing and adult mouse small intestine. *Development (Cambridge, UK)* **127**, 2629–2642.

Stark, K. L., McMahon, J. A., and McMahon, A. P. (1991). FGFR-4, a new member of the fibroblast growth factor receptor family, expressed in the definitive endoderm and skeletal muscle lineages of the mouse. *Development (Cambridge, UK)* **113**, 641–651.

Stern, C. D., and Canning, D. R. (1990). Origin of cells giving rise to mesoderm and endoderm in chick embryo. *Nature (London)* **343**, 273–275.

St. Onge, L., Sosa-Pineda, B., Chowdhury, K., Mansouri, A., and Gruss, P. (1997). Pax6 is required for differentiation of glucagon-producing alpha-cells in mouse pancreas. *Nature (London)* **387**, 406–409.

St. Vil, D., Brandt, M. L., Panic, S., Bensoussan, A. L., and Blanchard, H. (1991). Meckel's diverticulum in children: A 20-year review. *J. Pediatr. Surg.* **26**, 1289–1292.

Sunthornthepvarakul, T., Gottschalk, M., Hayashi, Y., and Refetoff, S. (1995). Resistance to thyrotropin caused by mutations in the thyrotropin-receptor gene. *N. Engl. J. Med.* **332**, 155–160.

Sussel, L., Kalamaras, J., Hartigan-O'Connor, D. J., Meneses, J. J., Pedersen, R. A., Rubenstein, J. L., and German, M. S. (1998). Mice lacking the homeodomain transcription factor Nkx2.2 have diabetes due to arrested differentiation of pancreatic beta cells. *Development (Cambridge, UK)* **125**, 2213–2221.

Sutcliff, K. S., and Hutchins, G. M. (1994). Septation of the respiratory and digestive tracts in human embryos: Crucial role of the tracheoesophageal sulcus. *Anat. Rec.* **238**, 237–247.

Suzuki, N., Labosky, P., Furuta, Y., Hargett, L., Dunn, R., Fogo, A., Takahara, K., Peters, D., Greenspan, D., and Hogan, B. (1996). Failure of ventral body wall closure in mouse embryos lacking a procollagen C-proteinase encoded by *Bmp1*, a mammalian gene related to *Drosophila tolloid*. *Development (Cambridge, UK)* **122**, 3587–3595.

Tada, M., Casey, E. S., Fairclough, L., and Smith, J. C. (1998). Bix1, a direct target of Xenopus T-box genes, causes formation of ventral mesoderm and endoderm. *Development (Cambridge, UK)* **125**, 3997–4006.

Takahashi, Y., Imanaka, T., and Takano, T. (1998). Spatial pattern of smooth muscle differentiation is specified by the epithelium in the stomach of mouse embryo. *Dev. Dyn.* **212**, 448–460.

Tefft, J. D., Lee, M., Smith, S., Leinwand, M., Zhao, J., Bringas, P., Jr., Crowe, D. L., and Warburton, D. (1999). Conserved function of mSpry-2, a murine homolog of Drosophila sprouty, which negatively modulates respiratory organogenesis. *Curr. Biol.* **9**, 219–222.

Ten Have-Opbroek, A. A. W. (1981). The development of the lung in mammals: An analysis of concepts and findings. *Am. J. Anat.* **162**, 201–219.

Ten Have-Opbroek, A. A. W. (1991). Lung development in the mouse embryo. *Exp. Lung Res.* **17**, 111–130.

Thomas, P. Q., and Beddington, R. S. P. (1996). Anterior primitive endoderm may be responsible for patterning the anterior neural plate in the mouse embryo. *Curr. Biol.* **6**, 1487–1496.

Thomas, P. Q., Brown, A., and Beddington, R. S. P. (1998). *Hex*: A homeobox gene revealing peri-implantation asymmetry in the mouse embryo and an early transient marker of endothelial cell precursors. *Development (Cambridge, UK)* **125**, r85–94.

Thomson, J. A., Itskovitz-Eldor, J., Shapiro, S. S., Waknitz, M. A., Swiergiel, J. J., Marshall, V. S., and Jones, J. M. (1998). Embryonic stem cell lines derived from human blastocysts. *Science* **282**, 1145–1147.

Tremblay, K. D., Hoodless, P. A., Bikoff, E. K., and Robertson, E. J. (2000). Formation of the definitive endoderm in mouse is a Smad2 dependent process. *Development (Cambridge, UK)* **127**, 3079–3090.

Warburton, D., Seth, R., Shum, L., Horcher, P. G., Hall, F. L., Werb, Z., and Slavkin, H. C. (1992). Epigenetic role of epidermal growth factor expression and signalling in embryonic mouse lung morphogenesis. *Dev. Biol.* **149**, 123–133.

Warburton, D., Schwarz, M., Tefft, D., Flores-Delgado, G., Anderson, K. D., and Cardoso, W. V. (2000). The molecular basis of lung morphogenesis. *Mech. Dev.* **92**, 55–81.

Warot, X., Fromental-Ramain, C., Fraulob, V., Chambon, P., and Dolle, P. (1997). Gene dosage-dependent effects of the Hoxa-13 and Hoxd-13 mutations on morphogenesis of the terminal parts of the digestive and urogenital tracts. *Development (Cambridge, UK)* **124**, 4781–4791.

Weaver, M., Yingling, J. M., Dunn, N. R., Bellusci, S., and Hogan, B. L. (1999). Bmp signaling regulates proximal-distal differentiation of endoderm in mouse lung development. *Development (Cambridge, UK)* **126**, 4005–4015.

Weaver, M., Dunn, N. R., and Hogan, B. L. M. (2000). Bmp4 and Fgf10 play opposing roles during lung bud morphogenesis. *Development (Cambridge, UK)* **127**, 2695–2704.

Weigel, D., Jürgens, G., Küttner, F., Seifert, E., and Jäckle, H. (1989). The homeotic gene *fork head* encodes a nuclear protein and is expressed in the terminal regions of the Drosophila embryo. *Cell (Cambridge, Mass.)* **57**, 645–658.

Weinstein, D. C., Ruiz i Altaba, A., Chen, W. S., Hoodless, P., Prezioso, V. R., Jessell, T. M., and Darnell, J. E., Jr. (1994). The winged-helix transcription factor *HNF-3β* is required for notochord development in the mouse embryo. *Cell (Cambridge, Mass.)* **78**, 575–588.

Weinstein, M., Xu, X., Ohyama, K., and Deng, C.-X. (1998). FGFR-3 and FGFR-4 function cooperatively to direct alveogenesis in the murine lung. *Development (Cambridge, UK)* **125**, 3615–3623.

Wells, J. M., and Melton, D. A. (2000). Early mouse endoderm is patterned by soluble factors from adjacent germ layers. *Development (Cambridge, UK)* **127**, 1563–1572.

Wendling, O., Dennefeld, C., Chambon, P., and Mark, M. (2000). Retinoid signaling is essential for patterning the endoderm of the third and fourth pharyngeal arches. *Development (Cambridge, UK)* **127**, 1553–1562.

Wert, S. E., Glasser, S. W., Korfhagen, T. R., and Whitsett, J. A. (1993). Transcriptional elements from the human SP-C gene direct expression in the primordial respiratory epithelium of transgenic mice. *Dev. Biol.* **156**, 426–443.

Wessels, N. K., and Cohen, J. H. (1967). Early pancreas organogenesis: Morphogenesis, tissue interactions, and mass effects. *Dev. Biol.* **15**, 237–270.

Wilson, V., and Beddington, R. S. P. (1996). Cell fate and morphogenetic movement in the late mouse primitive streak. *Mech. Dev.* **55**, 79–89.

Wu, J. E., and Santoro, S. A. (1996). Differentiatl expression of integrin α subunits supports distinct roles during lung branching morphogenesis. *Dev. Dyn.* **206**, 169–181.

Yamada, Y., Webber, E. M., Kirillova, I., Peschon, J. J., and Fausto, N. (1998). Analysis of liver regeneration in mice lacking type 1 or type 2 tumor necrosis factor receptor: requirement for type 1 but not type 2 receptor. *Hepatology* **28**, 959–970.

Yasugi, S., Matsushita, S., and Mizuno, T. (1985). Gland formation induced in the allantoic and small-intestinal endoderm by the proventricular mesenchye is not coupled with pepsinogen expression. *Differentiation (Berlin)* **30,** 47–52.

Yasuo, H., and Lemaire, P. (1999). A two-step model for the fate determination of presumptive endodermal blastomeres in Xenopus embryos. *Curr. Biol.* **9,** 869–879.

Zákány, J., and Duboule, D. (1999). Hox genes and the making of sphincters. *Nature (London)* **401,** 761–762.

Zannini, M., Avantaggiato, V., Biffali, E., Arnone, M. I., Sato, K., Pischetola, M., Taylor, B. A., Phillips, S. J., Simeone, A., and Di Lauro, R. (1997). TTF-2, a new forkhead protein, shows a temporal expression in the developing thyroid which is consistent with a role in controlling the onset of differentiation. *EMBO J.* **16,** 3185–3197.

Zaret, K. (1999). Developmental competence of the gut endoderm: Genetic potentiation by GATA and HNF3/fork head proteins. *Dev. Biol.* **209,** 1–10.

Zeng, X., Wert, S. E., Federici, R., Peters, K. G., and Whitsett, J. A. (1998). VEGF enhances pulmonary vasculogenesis and disrupts lung morphogenesis in vivo. *Dev. Dyn.* **211,** 215–227.

Zhang, J., Houston, D. W., King, M. L., Payne, C., Wylie, C., and Heasman, J. (1998). The role of maternal VegT in establishing the primary germ layers in Xenopus embryos. *Cell (Cambridge, Mass.)* **94,** 515–524.

Zhao, J., Lee, M., Smith, S., and Warburton, D. (1998). Abrogation of Smad3 and Smad2 and of Smad4 gene expression positively regulates murine embryonic lung branching morphogenesis in culture. *Dev. Biol.* **194,** 182–195.

Zhou, L., Dey, C. R., Wert, S. E., and Whitsett, J. A. (1996). Arrested lung morphogenesis in transgenic mice bearing an SP-C-TGF-B1 chimeric gene. *Dev. Biol.* **175,** 227–238.

Zhou, L., Dey, C. R., Wert, S. E., Yan, C., Costa, R., and Whitsett, J. A. (1997). Hepatocyte nuclear factor-3β limits cellular diversity in the developing respiratory epithelium and alters lung morphogenesis in vivo. *Dev. Dyn.* **210,** 305–314.

Zhu, J., Hill, R. J., Heid, P. J., Fukuyama, M., Sugimoto, A., Priess, J. R., and Rothman, J. H. (1997). *end-1* encodes an apparent GATA factor that specifies the endoderm precursor in *Caenorhabditis elegans* embryos. *Genes Dev.* **11,** 2883–2896.

Zhu, X., Sasse, J., McAllister, D., and Lough, J. (1996). Evidence that fibroblast growth factors 1 and 4 participate in regulation of cardiogenesis. *Dev. Dyn.* **207,** 429–438.

Zorn, A. M., Butler, K., and Gurdon, J. B. (1999). Anterior endomesoderm specification in *Xenopus* by Wnt/β signalling pathways. *Dev. Biol.* **209,** 282–297.

16

Molecular Determinants of Cardiac Development and Congenital Disease

The Victor Chang Cardiac Research Institute, St. Vincent's Hospital, Darlinghurst 2010 Australia; and Faculties of Medicine and Life Sciences, University of New South Wales, New South Wales 2052, Australia

I. Introduction

II. Overview of Heart Structure and Development

III. A Conserved Pathway for Cardiac Induction and Morphogenesis

IV. Cardiac Induction: The Role of Endoderm

V. Bone Morphogenetic Proteins as Cardiac Inducing Molecules

VI. Other Factors Involved in Cardiac Induction

VII. A Role for Anterior Visceral Endoderm in Cardiac Induction in the Mouse?

VIII. The Heart Morphogenetic Field

IX. The Size and Shape of the Heart Field

X. The Timing and Stability of Cardiac Induction

XI. Migration of Cardiac Precursors

XII. Cellular Proliferation and Death in the Forming Heart

XIII. Cardiac Myogenesis

XIV. Modulation of Myogenesis in Heart Chambers

XV. Regionality in the Developing Heart

XVI. Plasticity of Heart Regionalization

XVII. The Segmental Model of Cardiac Morphogenesis

XVIII. An Inflow/Outflow Model of Early Heart Tube Patterning

XIX. A Role for Retinoic Acid Signaling in Inflow/Outflow Patterning

XX. A Role for the Delta/Notch Pathway in Primary Heart Patterning

XXI. Cardiac Chamber Formation

XXII. Ventricular Specification: Knock-Out and Transgenic Phenotypes

XXIII. Transcriptional Circuits Acting in Chamber Formation

Copyright © 2002 Academic Press
All rights of reproduction in any form reserved.

XXIV. The Cardiac Left-Right Axis

XXV. Developmental Pathways and Congenital Heart Disease

XXVI. Horizons

References

I. Introduction

The development of the mammalian heart is an exquisite example of how form is established during organ ontogeny, and how relationships between form and function are realized. Although heart development has been studied at the morphological level since antiquity, its patterning principles are still far from clear. Nevertheless, details of the genes that guide heart myogenesis and morphogenesis are emerging at a rapid rate, contributing clues to patterning events and substantially informing our understanding of congenital, acquired, and maladaptive cardiac diseases. Remarkably, many of the regulatory genes that act in mammalian heart development also function during formation of the primitive heartlike organ of the fruitfly, *Drosophila*, indicating that the complex form of the mammalian heart has evolved through refinement and embellishment of an ancient and conserved

regulatory network. The combination of the power of *Drosophila* genetics within a multispecies approach, with genomics information, rapidly advancing technologies for imaging and analysis of cardiovascular function in embryonic and adult mice, and high-throughput genetic screening, set the stage for an attack on congenital heart disease of an unprecedented scale. This chapter outlines recent progress on dissecting the molecular programs that guide heart formation and morphogenesis, with emphasis on cardiac induction and the emergence of pattern through cellular and molecular interactions within the heart field and early heart tube. The information provides an essential framework that will allow genetic programs to be meaningfully linked to the emergence of heart form and function and to the interpretation and amelioration of congenital heart disease.

II. Overview of Heart Structure and Development

The mammalian heart is a fluid pump composed of a highly modified muscular vessel. Integral to its form are separate but anatomically fused units serving the systemic and pulmonary circulations of the body (Fig.1). Its correct structure and function rely on the coordinated development of atrial and ventricular myocardium, endocardium, inlet

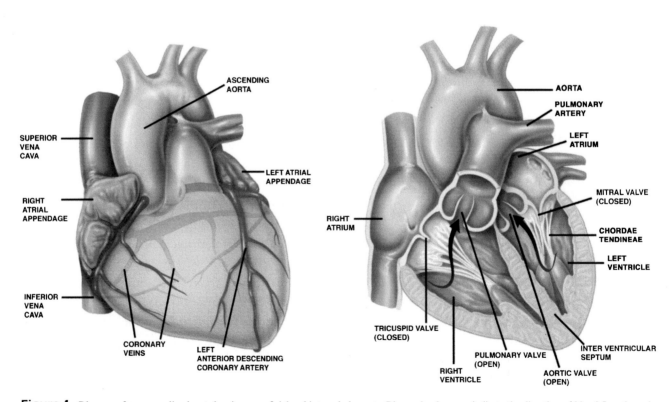

Figure 1 Diagram of a mammalian heart showing superficial and internal elements. Blue and red arrows indicate the direction of blood flow through right and left ventricles toward the valves of the pulmonary artery and aorta, respectively. (Reproduced with permission from Edwards Life Sciences LLC.)

and outlet vessels, valves, septa, coronary circulation, fibrous skeleton, and conduction system. Perhaps more than any other organ, the heart has strict constraints on structural or functional deviation from normal, and its sensitivity to genetic perturbation is reflected in the high prevalence of congenital cardiac abnormalities in humans (Hoffman, 1995a,b).

Cells that contribute to the mammalian heart can be detected by fate mapping techniques as two roughly cohesive and laterally arranged populations in the epiblast layer of the postimplantation egg cylinder (Lawson *et al.*, 1991; Parameswaran and Tam, 1995; Tam and Schoenwolf, 1999). Precursors of myocardium and endocardium can be identified in the same region (Tam and Schoenwolf, 1999). These populations are located close to the region that will become the most anterior portion of the primitive streak. About 12 hr after the onset of gastrulation, cardiac mesoderm, along with head mesoderm and foregut endoderm, ingresses through the primitive streak and migrates to the anteriormost reaches of the forming embryo, condensing around E7.0–7.5 into a crescent-shaped epithelium (the cardiac crescent) (DeRuiter *et al.*, 1992; Kaufman and Navaratnam, 1981).

Soon after arriving at this position, cardiogenic mesoderm begins expressing the first molecular markers of cardiac specification, such as the transcription factor genes *Nkx2-5, Gata4,* and *Mef2b/c* (Edmondson *et al.*, 1994; Heikinheimo *et al.*, 1994; Lints *et al.*, 1993; Molkentin *et al.*, 1996). Endocardial cells appear around this time between the future myocardial layer and the subjacent endoderm (Linask and Lash, 1993). From E8.0, the bilateral cardiac progenitors coalesce at the ventral midline and fuse to form the linear heart tube (Figs. 2A and 2B). This transient structure is composed of an endocardial tube shrouded by a myocardial epithelium, the latter remaining open dorsally and attached to the ventral foregut via the (nonmyogenic) dorsal mesocardium.

At E8–8.5, the linear heart tube then begins the process of looping morphogenesis (Harvey, 1998a), which sets the stage for formation, positioning, and integration of cardiac chambers (Figs. 2C–F). During looping, the heart tube elongates and adopts a spiral form with the arterial end sweeping rightward before returning to the midline (Fig. 2F). Furthermore, the venous input and atrial regions are forced dorsally and cranially relative to the ventricles, so that by E11.5, left

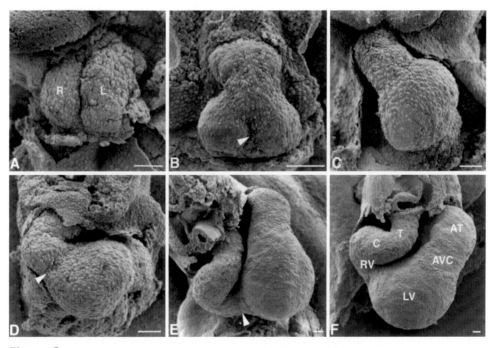

Figure 2 Heart development in the mouse. Scanning electron micrographs of embryonic hearts from Swiss albino mice. (A) Fusing cardiac primordia at ~E8.0. (B) Linear tubular heart (~E8.25) formed after fusion of left and right cardiac primordia. Arrowhead indicates fusion still in progress. (C) The onset of cardiac looping (~E8.25–8.5) during which the caudal region of the heart is displaced leftward through asymmetric morphogenesis, while the outflow region loops toward the right. The bulbous left ventricular primordium is clearly evident. (D) More advanced stage of cardiac looping (~E8.5). Arrowhead indicates the interventricular sulcus. (E) With further bending and rotation, the heart at ~E9.0 is thrown into a more prominent loop. Arrowhead indicates the interventricular sulcus corresponding to the position of the developing interventricular septum. (F) At ~E9.5–10.0, the heart approaches the adult form. Bar = 50 μm. Abbreviations: AT, atrium; AVC, atrioventricular canal; C, conus; L, left cardiac primordium; LV, left ventricle; R, right cardiac primordium; RV, right ventricle; T, truncus. (Reproduced and adapted from Icardo, 1997, with permission from Cambridge University Press.)

and right atria are positioned above and in register with left and right ventricles, respectively. At the same time, the conotruncal region becomes wedged ventrally between left and right ventricles. The looped heart has thus achieved a form that very nearly approximates that of the adult heart (cf. Figs. 1 and 2F). Substantial and complex remodeling of the heart tube leads to elaboration of its internal elements, including valves (Lamers *et al.*, 1995; Mjaatvedt *et al.*, 1999; Wenink and Anderson, 1987), septa (Webb *et al.*, 1998), and conduction system (Moorman *et al.*, 1998; Rentschler *et al.*, 2001; Takebayashi-Suzuki *et al.*, 2000). The epicardium, a derivative of the septum transversum, envelops the heart and gives rise to the coronary circulation and interstitial fibroblasts (Dettman *et al.*, 1998; Mikawa, 1999). Neural crest cells infiltrate the conotruncus and contribute to aorticopulmonary septation, conotruncal cushions, and cardiac ganglia (Jiang *et al.*, 2000; Kirby, 1998).

III. A Conserved Pathway for Cardiac Induction and Morphogenesis

A significant advance in recent years is the finding that genetic pathways that control cardiac development have been conserved in evolution across wide species barriers (Bodmer, 1993; Bodmer and Frasch, 1999; Harvey, 1996; Lints *et al.*, 1993). Although the hearts of vertebrates and invertebrates have never been considered homologous in anatomical terms, they are nonetheless pulsatory, chambered, vessel-like structures that share certain developmental, anatomical, and ultrastructural characteristics (Bodmer and Frasch, 1999; Harvey, 1996). Their common link in evolution may lie, along with visceral muscles, as far back as the ancient myoepithelial derivatives of coelomic epithelium in sedentary organisms (Harvey, 1996; Hirakow, 1985; Ranganayakulu *et al.*, 1998). The *Drosophila* heart is a linear muscular vessel that draws hemolymph in through openings called ostia, then pumps it unidirectionally around an open body cavity by peristalsis (Bodmer and Frasch, 1999; Rugendorff *et al.*, 1994). A larger collecting chamber ("ventricle") is separated from the narrower "aorta" by a valve. The first insights into genetic control of heart formation came with the discovery of the *Drosophila* gene *tinman*, which encodes a transcription factor belonging to the NK2 class of the homeodomain superfamily (Bodmer, 1993; Bodmer *et al.*, 1990; Harvey, 1996; Kim and Nirenberg, 1989). During fly development, *tinman* is expressed in early mesoderm, then in dorsal mesoderm that gives rise to the cardiac, visceral, and dorsal body wall muscles. The gene is essential for formation of those muscle types, as well as certain ventral body wall muscles and glial cells (Azpiazu and Frasch, 1993; Gorczyka *et al.*, 1994). Thus, *tinman* functions initially as a primary patterning gene within mesoderm, setting up the potential for specification of restricted lineages in-

cluding those of the heart. It also apparently acts later as a cardiac differentiation factor (Kremser *et al.*, 1999).

Numerous homologs of *tinman* have been described in mice (Biben *et al.*, 1998a; Komuro and Izumo, 1993; Lints *et al.*, 1993) and other vertebrates (Evans, 1999; Harvey, 1996). Of these, the *Nkx2-5* gene is expressed in the paired cardiac progenitor cells of all models studied (Harvey, 1996). *Nkx2-5* is essential for heart formation in frog embryos and for early heart tube morphogenesis in the mouse (Fu *et al.*, 1998; Grow and Krieg, 1998; Lyons *et al.*, 1995). The finding that mammalian heart development utilizes *tinman*-like relatives is unlikely to be trivial. In fact, since this gene was cloned, considerable evidence has accumulated that truly homologous genetic pathways underlie cardiogenesis in flies and vertebrates (Fig. 3). For example, *tinman* expression is maintained in dorsal mesoderm by the secreted factors decapentaplegic (dpp) and Screw, which are expressed in dorsal ectoderm (Frasch, 1995; Yin and Frasch, 1998). As described below, cognates of dpp and Screw, the bone morphogenetic proteins (BMPs) 2 and 4, induce and/or maintain *Nkx2-5* expression in vertebrate cardiac progenitors. Furthermore, transcription factors of the homeodomain, LIM homeodomain, zinc finger, MADS box, basic helix–loop–helix (bHLH), T-box, and steroid receptor classes are expressed in common in the hearts of flies and vertebrates (Black and Olson, 1999; Fossett *et al.*, 2000; Gajewski *et al.*, 1999, 2000; Griffin *et al.*, 2000; Harvey, 1996; Meins *et al.*, 2000; Moore *et al.*, 2000; Pereira *et al.*, 1999). A mutually beneficial cross-fertilization between these systems is occurring, and there is little doubt that analysis of genes and pathways acting in *Drosophila* heart formation will profoundly inform our understanding of cell lineage determination and genetic regulation in the mammalian heart. The fascinating hint that aspects of inflow/outflow patterning in the *Drosophila* heart have been conserved in mammals (Lo and Frasch, 2001; Pereira *et al.*, 1999) opens up an entirely new perspective on the issue of cross-species conservation.

IV. Cardiac Induction: The Role of Endoderm

Cardiac induction describes the process of stable establishment of the cardiac lineages and is an early event in the life of the embryo. Induction occurs within an anterior embryonic niche that is established as a result of regionalization of the embryonic axes (Marvin *et al.*, 2001; Tzahor and Lassar, 2001). In the frog and chick models, induction occurs across the gastrulation period (Ladd *et al.*, 1998; Nascone and Mercola, 1995; Yuan and Schoenwolf, 1999). However, cell grafting experiments in chick and mouse embryos have shown that gastrulation serves principally to disseminate mesendodermal cell populations to their proper locations, and is not required per se for commitment or differentiation

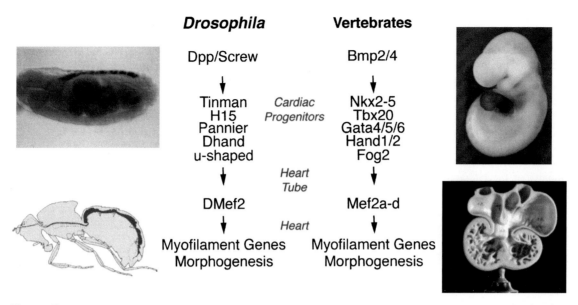

Figure 3 Evolutionary conservation of genetic pathways for heart development. The figure shows cognate genes in *Drosophila* and vertebrates organized into simplistic parallel pathways based on their approximate temporal sequence of action. Accompanying images depict developing and adult hearts in flies and mammals. In the *Drosophila* embryo (top left), the heart (dorsal vessel) is highlighted using *in situ* hybridization with a *tinman* gene probe (purple). (Reproduced with permission from R. Bodmer.) In the adult fly (bottom left), the heart is shaded red. In the E9.0 mouse embryo (top right), the ventricular chamber is highlighted by immunostaining for myosin light chain 2V (brown), while the sinoatrial region is revealed by blue staining reaction for β-galactosidase (*LacZ*) activity expressed from a transgene. (Reproduced from Xavier-Neto *et al.*, 1999, with permission from Academic Press.) A Bleck Schmidt model of the adult human heart (bottom right) is also shown. (Image supplied by Antoon Moorman.)

of descendant lineages (Tam and Schoenwolf, 1999). Thus, epiblast cells will differentiate into cardiac tissue if grafted to the cardiac region of the gastrula and, remarkably, cells from the cardiac region of the gastrula, if grafted into epiblast, will regastrulate and contribute to mesodermal tissues including heart.

Specification of heart mesoderm appears to require at least two (although probably more) independent inductive events. In the frog model, specification requires signaling from both the organizer and anterior endoderm (Jacobson and Sater, 1988; Kao and Elinson, 1988; Nascone and Mercola, 1995; Sater and Jacobson, 1990b; Yuan and Schoenwolf, 1999), and candidate molecules for both of these signals have now been identified and are discussed in detail below. From a historical perspective, the role of the endoderm in heart induction has received much attention. From the beginning of gastrulation, heart mesoderm lies in close apposition to anterior pharyngeal endoderm (Mohun and Leong, 1999; Nascone and Mercola, 1996; Tam and Schoenwolf, 1999), and studies in various models have highlighted multiple roles for this tissue in heart induction (Arai *et al.*, 1997; Auda-Boucher *et al.*, 2000; Jacobson and Sater, 1988; Lough and Sugi, 2000; Marvin *et al.*, 2001; Nascone and Mercola, 1995; Sater and Jacobson, 1990b; Schultheiss *et al.*, 1995; Sugi and Lough, 1994; Tzahor and Lassar, 2001; Yatskievych *et al.*, 1997). In the mouse, the presence of endo-

derm in explants from mid- to late streak stages stimulates adoption of a myocardial fate (Auda-Boucher *et al.*, 2000), although a separate study found that the presence of endoderm in later E7.5 explants facilitated only muscle beating and was not essential for induction of cardiac gene expression (Arai *et al.*, 1997).

As highlighted by these examples, there has been considerable variation in findings in the classical and even more recent literature on whether endoderm is inductive or facilitative for cardiogenesis (Lough and Sugi, 2000; Mohun and Leong, 1999; Nascone and Mercola, 1996). This can be explained, in part, by the fact that endoderm promotes myofibrillar organization and contractility in tissue already possessing cardiogenic potential (Jacobson's "formative influences"), and many early studies scored only beating as the sole experimental outcome (Jacobson and Sater, 1988; Lough and Sugi, 2000; Mohun and Leong, 1999; Nascone and Mercola, 1996). The interpretation of experiments has no doubt been influenced by the differences in methods and timing of assays performed in different laboratories, difficulties associated with explantation techniques in some models, and from the fact that endoderm may be both inductive and supportive for cardiogenesis. In addition, endoderm may not be the only source of cardiogenic factors (see below). Nevertheless, recent studies showing that anterior endoderm can induce cardiogenesis in posterior noncardiogenic mesoderm strongly suggests an instructive role for this tissue (Marvin et al., 2001; Nascone and Mercola, 1995; Schneider and Mercola, 2001; Schultheiss et al., 1995, 1997). Manipulation experiments in chick and frog embryos collectively suggest that the organizer anteriorizes endoderm, thus establishing its capacity for cardiac induction (Inagaki et al., 1993; Schneider and Mercola, 2001; Yuan and Schoenwolf, 1999). In the chick, endoderm is also necessary for formation of endocardial cells (Linask and Lash, 1998; Lough and Sugi, 2000; Sugi and Markwald, 1996), highlighting the diversity and complexity of the interactions between endoderm and mesoderm during heart formation.

V. Bone Morphogenetic Proteins as Cardiac Inducing Molecules

There is considerable evidence that BMP signaling is an essential component of induction and/or maintenance of cardiogenesis in vertebrates. As noted above, dpp and Screw, as well as various components of their downstream signaling pathways, are required for cardiogenesis in Drosophila (Frasch, 1995; Xu et al., 1998b; Yin and Frasch, 1998). In the mouse and chick, BMPs 2 and 4, cognates of Drosophila dpp/Screw, and their potential heterodimeric partners, BMPs 5 and 7, are variously expressed from gastrulation in the anterior region (Andree et al., 1998; Lawson et al., 1999; Schultheiss et al., 1997; Solloway and Robertson, 1999; Zhang and Bradley, 1996). BMPs are prominently expressed in the anterior endoderm implicated in cardiac induction, but also in other anterior tissues, such as the anterior extraembryonic mesoderm and endoderm, anterior embryonic ectoderm, and in cardiac mesoderm itself. BMPs 2 and 4 can induce heart muscle formation in noncardiogenic anterior-medial mesoderm of chick HH stage 6 embryos, as judged by expression of the cardiac markers Nkx2-5, Gata4, and myosin heavy chains, and by elaboration of a beating phenotype (Schultheiss et al., 1997). No such induction was observed in posterior mesoderm treated in the same way (Schultheiss et al., 1997), implying that an additional anteriorly localized factor, or removal of a posterior inhibitor, is necessary for cardiogenesis (Marvin et al., 2001).

Importantly, the BMP inhibitor, noggin, completely blocks cardiogenesis in anterior explants of chick embryos (Ladd et al., 1998; Schultheiss et al., 1997), and in mouse P19 teratocarcinoma cells undergoing dimethyl sulfoxide (DMSO)-induced cardiogenesis (Monzen et al., 1999). Although the outcome of mutation of BMP factors in mice is complicated by early lethality and redundancy (Solloway and Robertson, 1999), some embryos carrying null Bmp2 alleles lack a heart completely (Zhang and Bradley, 1996). Likewise, zebrafish swirl embryos, mutant for Bmp2 (Kishimoto et al., 1997), and Xenopus embryos overexpressing dominant-negative type I and II BMP receptors (Shi et al.,

2000), lack cardiac tissue. The majority of mouse Bmp2 mutants do form some cardiac tissue, albeit localized ectopically (Mishina et al., 1995), and a similar phenotype was seen in Smad5 mutants, implying a role for this BMP Smad in mediating the inductive signal (Chang et al., 1999). Smad6, encoding an inhibitory Smad, is also induced during BMP-mediated cardiogenesis, presumably to modulate the BMP response (Yamada et al., 1999). The role of dpp and BMPs in cardiogenesis is likely to be direct, since cis-regulation of tinman is mediated by binding of Medea, a Drosophila Smad protein (Xu et al., 1998b), and Smad-like binding sites have been found in mouse and frog Nkx2-5 cis-regulatory sequences (Schwartz and Olson, 1999; Sparrow et al., 2000).

VI. Other Factors Involved in Cardiac Induction

Several studies have implicated factors in addition to BMPs in cardiac induction, acting either alone or in combination. A key recent finding is that while overexpression of dominant-negative BMP receptors in Xenopus blocks heart formation and the late phase of Nkx2-5 expression, the early phase of Nkx2-5 expression, lasting from induction until just prior to differentiation, was unaffected (Shi et al., 2000). This suggests that in vivo, BMPs maintain but do not induce Nkx2-5 expression, analogous to the maintenance of tinman expression in dorsal mesoderm by dpp and Screw (Frasch, 1995; Yin and Frasch, 1998). Other factors that might be involved in cardiac induction are discussed briefly below.

A. Wnts and Anti-Wnts

Formation of the Drosophila heart requires not only Dpp and tinman, but also the membrane-tethered signaling molecule Wingless (Wu et al., 1995). Wingless is normally expressed in transverse stripes in the ectoderm that exactly overlie regions that form cardioblasts in mesoderm. Here, wingless activates the winged helix factor gene sloppy paired, required for specification of heart cells (Lee and Frasch, 2000; Reichmann et al., 1997; Yin and Frasch, 1998). The segmented expression and essential requirement for Wingless in heart formation reflects the segmental origin of certain mesodermally derived organ primordia in Drosophila, a situation quite distinct from that in mammals. Nevertheless, murine Wingless relatives Wnt2 and Wnt11 are expressed strongly in cardiac crescent mesoderm (Kispert et al., 1996; Monkley et al., 1996), and Wnt11 appears to be essential for formation of cardiocytes after differentiation of a quail precardiac mesodermal cell line (Eisenberg et al., 1997). Furthermore, in the chick, Wnt11 can induce cardiogenesis in posterior (noncardiac) mesendodermal explants (Eisenberg and Eisenberg, 1999).

In distinct contrast, however, *Wnt1* and *-3a,* expressed in dorsal-anterior neurectoderm in the chick, inhibit cardiogenesis in otherwise competent paraxial mesoderm (Marvin *et al.,* 2001; Tzahor and Lassar, 2001), in line with other studies that have shown the presence of inhibitors of cardiogenesis in neural tissue (Climent *et al.,* 1995; Jacobson and Sater, 1988; Raffin *et al.,* 2000). Furthermore, antagonists of WNT1 and -3A are expressed in the anterior endoderm associated with the cardiac mesoderm in chick embryos, and in the organizer of frog embryos, suggesting that inhibition of these Wnt factors is required for cardiac induction. Indeed, Wnt antagonists can induce cardiogenesis in posterior noncardiogenic mesoderm in frog and chick embryos (Marvin *et al.,* 2001; Schneider and Mercola, 2001). The seemingly opposite effects of Wnt proteins on cardiac induction can be reconciled by observations that different Wnt classes have different and antagonistic signaling properties (Kuhl *et al.,* 2000). One class, including WNT1, -3A, and -8, acts via the canonical Wnt pathway involving the receptor frizzled, stabilization and nuclear localization of β-catenin, and activation of transcription factors of the LEF/TCF family (the Wnt/β-catenin pathway).

Another class, including Wnt5 and 11, uses distinct frizzled family receptors and acts in a pathway coupling heterotrimeric G proteins, phospholipase C and Ca^{2+} release, to activation of protein kinases PKC and CamKII (the Wnt/Ca^{2+} pathway) (Kuhl *et al.,* 2000). The Wnt/Ca^{2+} pathway has been shown in other systems to regulate cytoskeletal dynamics, cell movement, planar cell polarity, and mitotic spindle orientation (Kuhl *et al.,* 2000; Tada and Smith, 2000). Whether the induction of cardiogenesis in chick posterior mesodermal explants by Wnt11 is mediated directly by the Wnt/Ca^{2+} pathway, or indirectly due to the mutual antagonism between the two Wnt signaling pathways is not known (Kuhl *et al.,* 2000). Nevertheless, removal of the potential for Wnt/β-catenin signaling by expression of Wnt antagonists in the cardiogenic region appears to be a key requirement for cardiac induction. In *Drosophila* heart formation, Wingless acts through the Wnt/β-catenin pathway (Park *et al.,* 1996), making it likely that the roles of wingless signaling in heart formation in flies and vertebrates are not strictly homologous. Since targeted mutagenesis of mouse *Wnt11* alone does not show a cardiac phenotype (Monkley *et al.,* 1996), genetic proof of a role for the Wnt/Ca^{2+} pathway in cardiac induction in mammals awaits further study.

B. Activin

Chick epiblast can form foci of beating cardiac muscle at low frequency when explanted and cultured *in vitro* (Montgomery *et al.,* 1994; Yatskievych *et al.,* 1997), but this frequency is dramatically increased by the inclusion of posterior hypoblast (Ladd *et al.,* 1998; Yatskievych *et al.,* 1997). Mouse epiblast plus hypoblast (primitive endoderm) ex-plants also form cardiac muscle at moderate frequency, although this tissue is not fully determined because it fails to differentiate into cardiocytes autonomously when grafted into a chick host (Auda-Boucher *et al.,* 2000). The factor(s) responsible for the hypoblast-mediated specification may include activin, acting via induction of BMPs, although the effects in chick embryos can be mimicked by both transforming growth factor β (TGF-β) and fibroblast growth factor 4 (FGF4) (Ladd *et al.,* 1998; Yatskievych *et al.,* 1997). One interpretation of these results is that in explant culture, activin promotes formation of mesoderm and endoderm of dorsal-anterior character in epiblast (Logan and Mohun, 1993), creating a local environment permissive for BMP-mediated cardiogenesis (Ladd *et al.,* 1998; Nascone and Mercola, 1995; Sater and Jacobson, 1990b; Yatskievych *et al.,* 1997). Indeed, inhibitor studies indicate that activin and BMP act sequentially (Ladd *et al.,* 1998), and BMP actually inhibits the pro-cardiogenic activity of activin at earlier times (Ladd *et al.,* 1998).

C. Fibroblast Growth Factors

FGFs are expressed in anterior-lateral endoderm and cardiac mesoderm in the mouse and chick (Crossley and Martin, 1995; Lough and Sugi, 2000). Furthermore, FGF4, in conjunction with BMP2, can induce heart muscle in explants of chick posterior noncardiac mesoderm from late gastrulation stages (Ladd *et al.,* 1998; Lough *et al.,* 1996). An anecdotal report claims that heart formation in chick embryos is inhibited in the presence of blocking antibodies against FGF2 or its receptor (Gordon-Thomson and Fabian, 1994), although a separate study reported only reduced proliferation and delayed contractility (Zhu *et al.,* 1999). The *Fgf8* gene is involved in several primary inductive processes during embryogenesis and is also expressed in cardiogenic mesoderm in several vertebrates including mouse (Crossley and Martin, 1995; Reifers *et al.,* 2000). Mutation of the *Fgf8* gene in zebrafish leads to dysregulation of the cardiac transcription factor genes *Nkx2.5* and *Gata4,* and formation of abnormal hearts with a diminished ventricle (Reifers *et al.,* 2000), suggesting a direct role for *Fgf8* in cardiogenesis. However, it is difficult at present to judge whether these effects are secondary to reduced migration of precardiac mesoderm from the primitive streak, as clearly seen in *Fgf8* knockout mice (Sun *et al.,* 1999). An array of knockouts in other FGF genes or their receptors has thus far failed to show *or* refute a role for FGF signaling in cardiac induction in the mouse (Lough and Sugi, 2000).

D. Opioids

A recent finding has suggested a role for natural endorphins in cardiogenesis (Ventura and Maioli, 2000). The dynorphin B peptide is expressed in cardiomyocytes and can act in an autocrine and possibly intracrine manner via nuclear receptors to modulate contractility and gene expression (Ventura *et al.*, 1998). At the membrane, dynorphin B acts via κ-class opioid receptors. In the mouse P19 system, expression of the preprodynorphin gene (*Pdyn*) increases on induction of heart muscle formation with DMSO, and antagonist studies show that κ-opioid signaling is essential for this process. Furthermore, the dynorphin B peptide can induce cardiogenesis in the absence of DMSO.

E. CFC Factors

Intense interest surrounds the function of the so-called CFC (*Cripto-F*RL-1-Cryptic) family of secreted factors, which includes mammalian Cripto and Cryptic (Dono *et al.*, 1993; Shen *et al.*, 1997), zebrafish one-eyed pinhead (Gritsman *et al.*, 1999), and frog FRL-1 (Kinoshita *et al.*, 1995). These factors, which carry a CFC homology domain and a modified EGF domain, have been variously implicated in cell migration, mitogenesis, branching morphogenesis, endoderm formation, and left-right asymmetry (Ding *et al.*, 1998; Payrieras *et al.*, 1998; Salomon *et al.*, 1999; Xu *et al.*, 1999; Yan *et al.*, 1999). Genes encoding Cripto (*Tdfg1*) and Cryptic (*Cfc1*) are strongly expressed in cardiac crescent mesoderm of mouse embryos after earlier expression in epiblast and axial mesectoderm (Ding *et al.*, 1998; Dono *et al.*, 1993; Shen *et al.*, 1997). *Tdfg1* mutant embryos show severely disturbed anterior-posterior patterning due to a failure of cell migration within the primitive endoderm and epiblast of the egg cylinder (Ding *et al.*, 1998; Xu *et al.*, 1999). However, these mutant embryos, as well as *in vitro* adherent cultures struck from them, can form some mesoderm, although apparently not cardiac mesoderm (Xu *et al.*, 1999). Furthermore, *Tdfg1* mutant embryonic stem (ES) cell-derived embryoid bodies can express the early mesodermal marker *brachyury,* and the heart marker *Nkx2-5,* but fail to express differentiation markers of heart muscle (Xu *et al.*, 1998a).

Compelling biochemical and genetic evidence suggests an essential facilitative role for CFC factors in signaling by the TGF-β family member, nodal (Gritsman *et al.*, 1999; Kumar *et al.*, 2000; Shen and Schier, 2000; Shiratori *et al.*, 2000), itself essential for mesendoderm formation (Conlon *et al.*, 1994; Zhou *et al.*, 1993) and a key component of the embryonic left-right asymmetry pathway (Capdevila *et al.*, 2000; Lowe *et al.*, 2001). However, other studies suggest inhibitory influences on nodal signaling (Kiecker *et al.*, 2000), autonomous functions (Kiecker *et al.*, 2000; Shen and Schier, 2000), as well as positive and/or negative influences on signaling through tyrosine kinase receptors specific for heregulin (Salomon *et al.*, 1999), FGF (Kiecker *et al.*, 2000; Kinoshita *et al.*, 1995), and BMP (Kiecker *et al.*, 2000). Given the broad roles for these signaling pathways in embryogenesis, CFC proteins may confer specificity in cer-

tain local settings, potentially including BMP and/or FGF-mediated cardiac induction.

F. Cerberus

The *Xenopus* protein cerberus is a member of the DAN/cerberus family of cysteine knot proteins and is a secreted antagonist of BMP, nodal, as well as Wnt signaling (Hsu *et al.*, 1998; Piccolo *et al.*, 1999; Takahashi *et al.*, 2000). Ectopic expression of cerberus represses mesoderm formation and induces additional headlike structures containing organized brain and foregut (Bouwmeester *et al.*, 1996). The murine cognate of cerberus is unable to induce ectopic heads (Belo *et al.*, 2000; Biben *et al.*, 1998b), perhaps because it lacks anti-WNT activity (Belo *et al.*, 2000), which is necessary for head formation (Glinka *et al.*, 1997). Frog cerberus is expressed in the deep endoderm that has heart-inducing activity (Schneider and Mercola, 1999). In *Xenopus* animal cap assays, both frog and mouse cerberus strongly induce neural tissue, consistent with BMP inhibition, but paradoxically also induce the cardiac marker *Nkx2-5* but not markers of heart differentiation (Biben *et al.*, 1998b; Bouwmeester *et al.*, 1996). This result is thus difficult to explain given the known inhibitory activities of cerberus-like proteins. It is possible that cerberus has unique activities favoring early steps in cardiac induction, including formation of anterior endoderm, in which *Nkx2-5* is also expressed.

VII. A Role for Anterior Visceral Endoderm in Cardiac Induction in the Mouse?

In the frog, the endodermal cells required for cardiac induction lie deep within the gastrula, and are marked by expression of the gene *cerberus* (Schneider and Mercola, 1999), which, as noted above, encodes an antagonist of BMP, WNT, and nodal signaling (Piccolo *et al.*, 1999). This deep endodermal population was previously proposed to be analogous to the anterior visceral endoderm (AVE) of the mouse gastrula, which has head-inducing and patterning activities (see Chapter 3), on the basis of their common expression of *cerberus*-like genes (Beddington and Robertson, 1999; Jones *et al.*, 1999). Although it was subsequently shown that head organizer activity in the frog resides in mesendodermal layers more superficial to the deep heart-inducing endoderm (Schneider and Mercola, 1999), the possibility that the mouse AVE has cardiac-inducing activity analogous to the frog deep endoderm remains a possibility worthy of exploration.

The AVE is initially juxtaposed to epiblast before it is displaced into the extraembryonic region by advancing definitive endoderm (Lawson and Pedersen, 1987), although there is a period during which it may contact the anterior cardiogenic mesoderm. The gene *Cited2,* which encodes a transcriptional regulator, is expressed at the most anterior extremity of the AVE, then later in the presumptive cardiac mesoderm and septum transversum that come to lie beneath it (Dunwoodie *et al.,* 1998). BMP2, involved in induction or maintenance of the cardiogenic program (see above), is expressed in proximal AVE and cardiac mesoderm in a pattern remarkably similar to that of *Cited2* (Arkell, 1996; Zhang and Bradley, 1996). Furthermore, the CITED2 relative, CITED1 (Dunwoodie *et al.,* 1998), has been shown to bind SMAD4 and augment its transcriptional activity *in vitro* (Shioda *et al.,* 1998). A credible, although still speculative, hypothesis is that *Cited2* expression marks the site of BMP2 activity in AVE and anterior mesoderm, and plays some role in regulating that activity. Thus, cardiac induction through BMP signaling may be primed by signals from the AVE, as well as definitive endoderm and other embryonic and extraembryonic tissues, a notion compatible with the general concept that anterior patterning instigated by the AVE is supported and extended by derivatives of the streak and node (Beddington and Robertson, 1998).

VIII. The Heart Morphogenetic Field

Many of the issues of induction and cell fate determination described above can be framed in terms of the morphogenetic field concept of heart formation (Fishman and Chien, 1997; Jacobson and Sater, 1988). A morphogenetic field describes a "dynamic region of developmental potency" for formation of an organ or structure, and can be larger than the area fated to that particular tissue (Jacobson and Sater, 1988). Furthermore, fields are usually regulative, in the sense that cells within the field can be mobilized to compensate if definitive progenitors are damaged or removed. The boundaries of a particular field most likely represent boundaries between different fields, the position and size of which may be governed by the broader patterning principles that organize the embryonic axes (Fishman and Chien, 1997). For example, with respect to the heart, concordance between congenital defects of the heart and forelimbs in humans and other models, and the common expression of certain regulatory genes in those same tissues (Bruneau *et al.,* 1999; Wilson, 1998; Yelon *et al.,* 2000), suggests the existence of adjacent, overlapping, and/or analogous heart and forelimb morphogenetic fields.

The heart morphogenetic field concept has recently been reevaluated in fish, frog and chick embryos (Ehrman and Yutzey, 1999; Raffin *et al.,* 2000; Serbedzija *et al.,* 1998). In *Xenopus,* the heart field occupies a region of anterior-ventral and anterior-lateral cells that closely corresponds to the expression domain of *Nkx2-5* (Raffin *et al.,* 2000; Tonissen *et al.,* 1994). Moreover, the *Nkx2-5* territory defines a region

that transiently possesses heart-forming potency and regulative ability. With time, however, potency becomes narrowed to the definitive heart-forming cells, without loss of the broader *Nkx2-5* expression domain (Raffin *et al.,* 2000; Sater and Jacobson, 1990a). The restriction in the field therefore occurs downstream of *Nkx2-5,* most likely as a result of inhibitory influences from the organizer, notochord, neurectoderm, and/or myocardium (Raffin *et al.,* 2000; Schneider and Mercola, 2001; Schultheiss *et al.,* 1997).

The tissues that participate in regulation if definitive precursors are damaged or lost normally form the dorsal mesocardium and dorsal pericardial mesoderm, both nonmyogenic tissues (Raffin *et al.,* 2000). The gene *Serrate1* is expressed in these tissues and signaling through the Serrate1/Notch1 pathway appears to be involved in their progressive loss of cardiogenic potency (Rones *et al.,* 2000). In the chick and mouse, a regulatory heart field has not been identified. In the chick, head mesoderm medial to the cardiac crescent, which falls under the inhibitory influences of Wnt/β-catenin signaling from the neural plate, possesses cardiogenic potential if separated from neural plate (Tzahor and Lassar, 2001), yet does not have regulatory ability *in situ* (Ehrman and Yutzey, 1999). However, the *Nkx2-5* domain in mouse and chick embryos encompasses both dorsal mesocardium and dorsal pericardial mesoderm (Schultheiss *et al.,* 1995; R. Harvey, unpublished data), tissues that transiently possess heart-forming potency and regulative ability in frogs. Thus, the regulatory features of the heart field defined in this species may also apply in birds and mammals.

IX. The Size and Shape of the Heart Field

Numerous antagonistic influences on cardiac induction have now been identified in various systems. A key component of cardiac induction is the removal of the potential for Wnt/β-catenin signaling by expression of Wnt antagonists in adjacent tissues (Marvin *et al.,* 2001; Schneider and Mercola, 2001). Wnt factors inhibit cardiogenesis upstream of *Nkx2-5.* In *Xenopus,* the caudal boundary of *Nkx2-5* expression may also be influenced by the antagonistic effects of high levels of GATA factors induced by retinoic acid (Jiang *et al.,* 1999). Signals from the notochord inhibit cardiogenesis downstream of *Nkx2-5* in both zebrafish and chick embryos (Goldstein and Fishman, 1998; Schultheiss *et al.,* 1997) and the Serrate1/Notch1 system, implicated in restriction of the cardiac field, also inhibits downstream of *Nkx2-5* (Rones *et al.,* 2000; Sparrow *et al.,* 2000). The net influence of these antagonistic signaling pathways may establish the size, shape, and regulatory dynamics of the heart field (Fishman and Chien, 1997; Rosenthal and Xavier-Neto, 2000). They may also affect regionalization with the forming heart tube itself (see below).

X. The Timing and Stability of Cardiac Induction

As discussed above, cardiac induction in frogs requires signals from the organizer and the adjacent deep endodermal layer (Nascone and Mercola, 1995). Timed removal of tissues from explants has shown that the role of the organizer is largely completed by the beginning of gastrulation (stage 10), while that of the endoderm persists for some time, until after stage 10.5 (Nascone and Mercola, 1995). The stability of commitment to a cardiac fate has been assessed in chick embryos by culturing cells from the cardiogenic region at clonal density, and using this assay the myocardial lineage was judged to be determined by midgastrulation (Montgomery et al., 1994). Mouse–chick grafting experiments show determination at the mid- to late streak stages (Auda-Boucher et al., 2000). However, cells can be diverted from a cardiogenic fate by drugs such as TPA and BrdU (Gonzalez-Sanchez and Bader, 1990; Montgomery et al., 1994), as well as the anti-BMP factor noggin (Schlange et al., 2000), until as late as the onset of myogenic differentiation. This suggests considerable plasticity in the inductive process.

XI. Migration of Cardiac Precursors

Two waves of mesodermal cell migration can be distinguished during heart development: the first mobilizes cells from the primitive streak toward the anterior and anterior-lateral regions of the gastrula where cardiac induction occurs; the second directs convergence of cardiac progenitors to the ventral midline during heart tube formation. Involvement of the FGF signaling system in cell migration within mesoderm is a recurrent theme in diverse systems (Blelloch et al., 1999; Gisselbrecht et al., 1996; Zelzer and Shilo, 2000). Genetic experiments in flies show a role for the FGF receptor gene, heartless, acting through Ras, in dorsal migration of nascent mesoderm (Beiman et al., 1996; Gisselbrecht et al., 1996). Mutant mesoderm fails to contact the dorsal ectodermal zone of dpp expression, an essential requirement for cardiac induction (Frasch, 1995). A number of mouse mutations, including knockouts in Fgf8 and the bHLH Mesp genes, form mesoderm that is blocked or delayed in its migration away from the streak (Kitajima et al., 2000; Sun et al., 1999).

In the milder case of the Mesp1 mutation, delay causes bizarre cardiac defects consistent with varying degrees of cardia bifida (Saga et al., 1999). Importantly, Fgf4 was found to be down-regulated in both Fgf8 and Mesp knockouts, suggesting a role for this FGF in migration of cells from the streak. It is possible that the mutations also affect migration of cardiac progenitors from the anterior primitive streak into the heart tube. Mouse Fgf8, for example, is ex-

pressed in the primitive streak, then later in cardiac crescent mesoderm (Crossley and Martin, 1995), and compound heterozygotes carrying a hypomorphic Fgf8 allele over a null allele have poorly developed heart tubes (Meyers et al., 1998). Furthermore, using aggregation chimeras, a cell autonomous requirement for Mesp1 and Mesp2 in migration of cardiac precursor cells into the heart has been demonstrated (Kitajima et al., 2000). However, whether these phenotypes relate to the late wave of cardiac cell migration in addition to the earlier wave from the streak remains to be rigorously determined.

During migration of the cardiogenic plate to the midline, it becomes split into splanchnic and somatic layers by imposition of the intraembryonic coelum. This process potentially involves polar expression of a Na$^+$K$^+$-ATPase (Linask and Lash, 1998). The splanchnic layer forms the cuboidal epithelium that gives rise to myocardial and endocardial cells, while the somatic layer gives rise to pericardial mesoderm and ceases to express cardiac markers. The timing and direction of movement of cardiac progenitors toward the midline (DeHaan, 1963) depends on the graded distribution of fibronectin in extracellular matrix, deposited at the mesodermal/endodermal interface. Inhibition of the interaction between fibronectin and its integrin receptor in the chick using blocking antibodies or RGD peptide (Linask and Lash, 1998), or in mice using gene targeting (George et al., 1997), leads to varying degrees of cardia bifida. In a subset of mice mutant for the fibronectin gene (Fn1) on a 129Sv genetic background, cardiac cells reach the anterior and express myosins normally, but never move from the crescent (George et al., 1997). In chick embryos, cavitation, epithelialization, and even differentiation of cardiac progenitors also depends on the calcium-dependent adhesion molecule N-cadherin (Linask et al., 1997; Linask and Lash, 1998). In mice mutant for the N-cadherin gene (Cdh2), cavitation and cardiac differentiation proceed normally, but the adhesive integrity of the myocardial epithelium is not maintained and the heart tube essentially disintegrates (Radice et al., 1997).

Remarkably, enforced expression of Wnt antagonists in posterior marginal zone mesendoderm of frog embryos induces cardiogenesis and formation of a myocardial tube lined with endothelial cells (Schneider and Mercola, 2001). This finding suggests strong self-organizing principles for heart tube formation. The signals that establish the graded distribution of fibronectin are unknown, but may be extrinsic to the mesendoderm, since rotation of these layers at HH stage 5 in the chick leads to respecification of the fibronectin gradient and normal heart tube formation (Linask and Lash, 1998). Nonetheless, the endoderm would appear to play a dominant role in migration of heart precursors, since at least five of the eight zebrafish mutations causing cardia bifida also show severe endodermal abnormalities (see Kupperman et al., 2000). Mouse and chick embryos lacking the zinc finger transcription factor gene, Gata4, also display cardia bifida (Kuo *et al.,* 1997; Molkentin *et al.,* 1997), a phenotype that may stem from *Gata4* expression in endoderm (Narita *et al.,* 1997). Within the mesodermal layer, loss of expression of the bHLH factor Hand2 causes cardia bifida in zebrafish (Yelon *et al.,* 2000), although not in mice (Srivastava *et al.,* 1997). An interesting recent finding is that the cardia bifida evident in the *miles apart* strain of zebrafish is due to mutation of the gene encoding the sphingosine-1-phosphate sphingolipid receptor (Kupperman *et al.,* 2000), implicating a new pathway in cardiac morphogenesis.

XII. Cellular Proliferation and Death in the Forming Heart

In skeletal muscle, proliferation and myogenesis are mutually exclusive events and the bHLH myogenic regulatory factor myogenin orchestrates the transition between the two states (Halevy *et al.,* 1995). In developing cardiac muscle, however, proliferation and differentiation initially occur in tandem, although as myogenic maturation increases, proliferative capacity decreases (Rumyantsev, 1977). An intrinsic developmental clock has been proposed to regulate the progressive withdrawal of cardiac cells from the cell cycle during fetal life, culminating in a limited period of endoreduplication (Burton *et al.,* 1999; Rumyantsev, 1977). There are many levels on which spatial control of cell division is pertinent to heart development (MacLellan and Schneider, 1999). For example, heart size in relation to body size is controlled by an IGF-1-dependent pathway (Lembo *et al.,* 1996; MacLellan and Schneider, 1999); proliferative rate is dramatically modulated during early stages of heart development and is greater at the outer curvature during cardiac looping morphogenesis (Sissman, 1966; Stalsberg, 1969b; Thompson *et al.,* 1990); the *Nmyc1* gene is required for proliferative expansion of the compact layer of the ventricles (Moens *et al.,* 1993).

Several other genes have been implicated genetically in regulating growth of the ventricles, including *Rxra* (Sucov *et al.,* 1994), *Il6st* (Yoshida *et al.,* 1996), *Adrbk1* (Jaber *et al.,* 1996), *Nf1* (Brannan *et al.,* 1994), *Tead1* (Chen *et al.,* 1994), *Wt1* (Kreidberg *et al.,* 1993), *Zfpm2* (Svensson *et al.,* 2000; Tevosian *et al.,* 2000), and *Aldh1a7* (Niederreither *et al.,* 2001). However, their varied influences on ventricular growth are highlighted by the fact that the phenotypes of some of the above mutations are due to indirect effects (Chen *et al.,* 1998; Hirota *et al.,* 1999; Moore *et al.,* 1999; Subbarayan *et al.,* 2000; Tran and Sucov, 1998) that lead to premature differentiation (Kastner *et al.,* 1997) and diminishment of energy metabolism (Ruiz-Lozano *et al.,* 1998). Indications that placental insufficiency underlies these effects (Barak *et al.,* 1999; Wendling *et al.,* 1999) has profound implications for how we view the etiology of certain congenital ventricular insufficiencies in humans.

Apoptosis is also a prominent feature of heart development, with some 31 zones of cell death highlighted in the developing chick heart (Pexieder, 1975). Roles in shortening of the conotruncus, formation of the cardiac valves, and controlling the influx of cardiac neural crest have been proposed (Fisher *et al.,* 2000; Pexieder, 1975; Poelmann and Gittenberger-de Groot, 1999; Watanabe *et al.,* 1998). Curiously, targeted mutation of genes encoding FADD (*Fadd*) and caspase 8 (*Casp8*), which act in a common death pathway, show thin ventricular walls and poor trabeculation (Varfolomeev *et al.,* 1998; Yeh *et al.,* 2000), although, as for other thin-walled phenotypes, the effects may not be intrinsic.

XIII. Cardiac Myogenesis

Prior to fusion of heart progenitors at the midline, various myofilament protein isoform genes are activated, marking the onset of myogenic differentiation (Christoffels *et al.,* 2000a; de Jong *et al.,* 1997; Franco *et al.,* 1998; Lyons, 1994). Myogenesis occurs in a craniocaudal wave along the early heart tube (Gonzalez-Sanchez and Bader, 1990; Han *et al.,* 1992; Litvin *et al.,* 1992; Montgomery *et al.,* 1994). During looping, some of the myofilament isoform genes are up- or downregulated in chamber primordia as they become evident or sometime after (Franco *et al.,* 1998; Kelly *et al.,* 1999), highlighting the different functional requirements of those chambers and also the multilayered nature of the cardiomyogenic program. Due to the lack of amenable cardiogenic cell lines, we know little about the mechanisms guiding cardiac myogenesis in vertebrates, and it has thus far been impossible to dissociate myogenesis from morphogenesis in mice using genetic approaches. However, promoter studies have implicated numerous transcription factors, including members of the MEF2, SRF, and GATA families, in myogenic differentiation. These families are discussed briefly next.

A. MEF2

The MEF2 proteins are members of the MADS-Box superfamily of transcription factors, and are distinguished from other MADS-Box members by the presence of the highly conserved MEF2 domain (Black and Olson, 1998, 1999). The MADS-Box and MEF2 domains together mediate DNA binding, dimerization, and interaction with cofactors. Studies in *Drosphila* have revealed an essential role for its single *D-Mef2* gene in differentiation of all muscle lineages (Bour *et al.,* 1995; Lilly *et al.,* 1995; Ranganayakulu *et al.,* 1995), although, interestingly, no myogenic role was found for the single *C. elegans MEF2* gene (Dichoso *et al.,* 2000). Mammals have four *MEF2* genes (A–D), which give rise to numerous splice variants with considerable potential for functional diversity through heterodimerization (Black and Olson, 1999). The *Mef2b* and *Mef2c* genes are expressed in the cardiogenic region during the first wave of regulatory gene expression that distinguishes heart progenitors from surrounding mesoderm (Edmondson *et al.,* 1994; Molkentin *et al.,* 1996). In *Mef2c* knockout mice, cardiac morphogenesis is severely disrupted and numerous myogenic genes, including those encoding myosin heavy chain alpha (*Myhca*), myosin light chain 1A (*Myla*), and α-cardiac actin (*Actc1*), are downregulated (Lin *et al.,* 1997). MEF2 was actually discovered as a DNA-binding activity, and was later found to be important for transactivation of many if not most cardiac and skeletal muscle genes (Black and Olson, 1999; Cserjesi and Olson, 1991; Gossett *et al.,* 1989).

The activity of MEF2 proteins is highly dependent on signaling input, providing a mechanism for linking muscle differentiation to sequences of inductive events. In muscle, MEF2 is regulated by calcium/calmodulin-dependent protein kinases (CaMKs), the mitogen and stress-activated protein kinase p38, and the phosphatase calcineurin, through changes in the phosphorylation state of the DNA-binding and transactivation domains (Lu *et al.,* 2000; Wu *et al.,* 2000). CaMK-mediated signaling activates MEF2 by promoting export of class II histone deacetylases from the nucleus, which bind to MEF2C proteins and inhibit their transcriptional activity on DNA (McKinsey *et al.,* 2000). Raf signaling may regulate nuclear import of MEF2 itself (Winter and Arnold, 2000). MEF2 proteins bind a variety of general as well as cardiac-restricted coactivators and repressors (Chen *et al.,* 2000; Molkentin *et al.,* 1995; Morin *et al.,* 2000; Sparrow *et al.,* 1999; Zhang *et al.,* 2000). In cardiac muscle, MEF2 binds directly to the zinc finger factor GATA4 (see below) and via this association can be recruited to GATA-dependent promoters that lack MEF2 binding sites (Morin *et al.,* 2000). MEF2, as well as GATA4 and Nkx2-5, can independently promote cardiac myogenesis in the P19 system in the absence of DMSO, suggesting that a strong positive cross-regulatory network that includes MEF2 initiates the cardiomyogenic program (Grepin *et al.,* 1997; Skerjanc *et al.,* 1998).

B. SRF

Serum response factor (SRF) is a MADS-Box family transcription factor that *in vitro* directs growth factor-mediated induction of immediate-early genes by binding to defined CArG-box sites within their promoters (for references, see Arsenian *et al.,* 1998). SRF is broadly expressed and in the embryo is essential for mesoderm formation (Arsenian *et al.,* 1998). The protein is, however, highly enriched in developing cardiac and skeletal muscle and is essential for skeletal muscle differentiation *in vitro* (Croissant *et al.,* 1996; Soulez *et al.,* 1996; Wei *et al.,* 1998). Many muscle genes may be regulated by SRF, although those encoding both cardiac and skeletal α-actins have been most characterized (Chen *et al.,* 1996; MacLellan *et al.,* 1996; Wei *et al.,*

2000). SRF serves as a template for oligomerization of a host of transcription factors (Arsenian *et al.,* 1998; Hill *et al.,* 1995; Treisman *et al.,* 1998), which includes, in cardiac muscle, the homeodomain factor Nkx2-5 and zinc finger factor GATA4 (Narasimhaswamy *et al.,* 2000; Sepulveda *et al.,* 1998). Like MEF2, SRF activity is signal dependent, being stimulated by members of the mitogen-activated protein kinase family (Arsenian *et al.,* 1998) and small GTPases of the RhoA/Rac/CDC42 family (Hill *et al.,* 1995; Wei *et al.,* 1998, 2000). Rho/Rac/CDC42-dependent regulation of SRF activity occurs at the level of actin treadmilling, with a buildup in the cellular pool of G-actin serving as a negative stimulus (Hill *et al.,* 1995). This mode of control is likely to occur in muscle cells (Hill *et al.,* 1995), linking the progression of differentiation to the state of the cytoskeleton and myofilament.

C. GATA Factors

The cardiac GATA factors (GATA-4, -5 and -6) are members of the type IV zinc finger-containing family of DNA binding transcription factors (Charron and Nemer, 1999; Molkentin, 2000; Parmacek and Leiden, 1999). They play central roles in cardiac myogenesis and morphogenesis, as well as development of other mesodermal and endodermal lineages (Molkentin, 2000). All are expressed in the developing heart from early times (Heikinheimo *et al.,* 1994; Morrisey *et al.,* 1997) and participate in positive and/or negative cross-regulatory myogenic loops involving the GATA genes themselves (Kuo *et al.,* 1997; Reiter *et al.,* 1999), the homeobox gene *Nkx2-5* (Davis *et al.,* 2000; Grepin *et al.,* 1997; Jiang *et al.,* 1999; Lien *et al.,* 1999; Molkentin *et al.,* 2000; Searcy *et al.,* 1998; Skerjanc *et al.,* 1998; Sparrow *et al.,* 2000), the bHLH factor gene *Hand2* (McFadden *et al.,* 2000), and the gene encoding the nuclear ankyrin repeat protein CARP (Kuo *et al.,* 1999).

GATA factors have been shown to be important for the regulation of a host of cardiac myogenic genes, including those that encode myofilament proteins, membrane receptors, ion channels, trancription factors, and endocrine hormones (Molkentin, 2000). Although the cardiac GATA factors may have overlapping functions (Charron *et al.,* 1999; Durocher *et al.,* 1997; Jiang *et al.,* 1998; Morrisey *et al.,* 1996), individual factors appear to act preferentially on certain promoter/enhancers due to differences in DNA binding site affinity (Charron *et al.,* 1999) and/or interaction with cofactors (Durocher *et al.,* 1997). An intriguing dose-dependent connection is seen between GATA factors and regulation of the cell cycle: While GATA-6 can positively activate transcription of target genes as effectively as GATA-4 *in vitro* (Durocher *et al.,* 1997; Morrisey *et al.,* 1996), levels of GATA-6 in *Xenopus* embryo hearts actually decline at the onset of differentiation and overexpression inhibits heart muscle differentiation in favor of prolonged proliferation of its precursors (Gove *et al.*, 1997). A dose-dependent cell cycle effect of the hematopoietic GATA protein, GATA-1, has also been documented (Whyatt *et al.*, 1997). Depletion of GATA-4 from P19 cells using antisense RNA expression causes extensive apoptosis in DMSO-induced cardioblasts (Grepin *et al.*, 1997), and this is also seen if GATA factors are depleted from foregut endoderm, a tissue with multiple roles in heart development (Ghatpande *et al.*, 2000).

GATA factors can interact and synergize with a range of transcription factors *in vitro*. GATA-4 and GATA-6 can interact with each other (Charron *et al.*, 1999), while GATA-4, but not GATA-6, can associate with the homeodomain factor Nkx2-5 (Durocher *et al.*, 1997; Lee *et al.*, 1998; Sepulveda *et al.*, 1998; Shiojima *et al.*, 1999). GATA-4 also interacts with the cardiac transcription factors NFATc and MEF2 (Molkentin *et al.*, 1998; Morin *et al.*, 2000), as well as the zinc finger factor, FOG-2 (friend of GATA-2) (J.-R. Lu *et al.*, 1999; Svensson *et al.*, 1999; Tevosian *et al.*, 1999). FOG-2 is structurally and functionally related to FOG, a cofactor of the hemopoietic GATA factor, GATA-1 (Tevosian *et al.*, 1999; Tsang *et al.*, 1998). The *Drosophila* FOG protein, termed u-shaped, physically and genetically interacts with the cardiac GATA factor, pannier, and is essential for determination of sensory bristle pattern (Cubadda *et al.*, 1999; Haenlin *et al.*, 1997) and the proper number of cardiac cells in the heart (Fossett *et al.*, 2000). How FOG proteins work is currently enigmatic, since they can act as either activators or repressors of GATA factor activity (Cubadda *et al.*, 1999; Fossett *et al.*, 2000; Haenlin *et al.*, 1997; Huggins *et al.*, 2000; J.-R. Lu *et al.*, 1999; Svensson *et al.*, 1999; Tevosian *et al.*, 1999, 2000). Mouse embryos lacking the FOG2 gene (*Zfpm2*) show a spectrum of cardiac abnormalities similar to tetralogy of Fallot or tricuspid atresia syndrome, and also a block in development of the coronary circulation at the level of its epicardial precursors (Svensson *et al.*, 2000; Tevosian *et al.*, 2000). This phenotype is quite distinct from that of the *Gata4* mutation, in which cardia bifida results from defects in endodermal development (Kuo *et al.*, 1997; Molkentin *et al.*, 1997; Narita *et al.*, 1997). Thus, it is evident that transcriptional complexes containing GATA and FOG factors form and function in a highly context-specific manner. Indeed, when an amino acid in GATA-4 essential for its interaction with FOG2 is mutated by gene targeting, endodermal development is normal and hearts have a phenotype similar to those lacking FOG2, although with additional conotruncal defects, suggesting the involvement of multiple FOG factors (Crispino *et al.*, 2001).

XIV. Modulation of Myogenesis in Heart Chambers

Most myogenic isoform genes are expressed across the whole heart tube before being up- or downregulated in dif-

ferent chambers according to functional need (Franco *et al.*, 1998; Lyons, 1994). These genes appear to be responding to the chamber-specific regulatory circuits established during early heart tube patterning (discussed in more detail below). Transcriptional repression and mutually antagonistic pathways appear to play key roles in setting up these patterns, and promoter studies hint at the types of genes involved. For example, downregulation of the quail atrial slow myosin heavy chain 3 gene (*slow MyHC3*) in the forming ventricles requires a repressor element that resembles a vitamin D/retinoic acid receptor binding motif (Wang *et al.*, 1998; Xavier-Neto *et al.*, 1999). The homeodomain gene, *Irx4*, may be involved in this repressive pathway, since deletion of *Irx4* in mice or dominant-negative inhibition in chick leads to inappropriate activation of atrial genes in the ventricle (Bao *et al.*, 1999; Bruneau *et al.*, 2001). Conversely, enforced expression of *Irx4* in the atria induces ventricular gene expression (Bao *et al.*, 1999). Other studies suggest repressive pathways involving vitamin D and thyroid hormone (Edwards *et al.*, 1994; Li and Gardner, 1994; Molkentin *et al.*, 1994), and the orphan nuclear receptor COUP-TFI (Guo *et al.*, 2000).

XV. Regionality in the Developing Heart

Region-specific anatomical or molecular characteristics in a forming organ are potential clues to patterning principles. For the heart, these principles are still far from clear.

Regionality in the developing heart can be visualized in a number of ways. Functionally, it is evident as a cardiocyte beat rate gradient along the anterior-posterior (AP) axis of the forming heart (Satin *et al.*, 1988). This gradient is accompanied by opposing gradients of expression of the genes *Atp2a2* and *Pln*, encoding proteins involved in excitation/contraction coupling (Moorman *et al.*, 1995, 2000). The beat rate gradient in chick myocardium is set up well before heart tube formation (Satin *et al.*, 1988) and is evident electrophysiologically even before excitation/contraction coupling is achieved (Kamino, 1991; Van Mierop, 1967).

Regionality can also be seen by the restricted or graded expression patterns of myofilament and regulatory genes (Biben and Harvey, 1997; Bruneau *et al.*, 1999, 2000a; Chin *et al.*, 2000; Nakagawa *et al.*, 1999; O'Brien *et al.*, 1993; Pereira *et al.*, 1999), as well as numerous transgenes (Firulli and Olson, 1997; Kelly *et al.*, 1999). These patterns help us consider how the building blocks of the heart are laid down (see below), although analysis of transgenic patterns has revealed an intriguing complexity in transcriptional control in the developing heart (Firulli and Olson, 1997; Kelly *et al.*, 1999). Patterns suggest that control of cardiac gene expression is "modular" in the sense that expression of individual genes is regulated in different heart regions by distinct *cis*-elements (Firulli and Olson, 1997; Kelly *et al.*, 1999;

McFadden *et al.,* 2000; Schwartz and Olson, 1999). Thus, the complete expression pattern of at least some cardiac genes appears to be a composite of subpatterns controlled by different regulatory modules (Fig. 4A). This situation may reflect the regional diversification of regulatory mechanisms that likely occurred during addition of new "anatomical modules" to the heart in the course of vertebrate evolution (Fishman and Olson, 1997). Transcription factors in particular may have evolved dedicated functions requiring more intricate regulation in different heart regions. The chick *GATA-6* gene, for example, is expressed across the whole heart tube, yet carries a regulatory module that, in transgenic mice, is expressed only in myocardium of the atrioventricular canal overlying the endocardial cushions (He and Burch, 1997). This myocardium is uniquely specialized for induction of cushion formation (Mjaatvedt *et al.,* 1999). That several mutant strains of zebrafish show cardiac phenotypes restricted to specific chambers or other structures in the heart has also been interpreted as evidence for modular gene regulation (Fishman and Olson, 1997).

The pattern of expression of one skeletal muscle myosin gene and cognate transgene in the forming heart appears to nicely demonstrate the concept of evolutionary modules. The *MLC3F-nlacZ-2E* transgenic line carries the fast skeletal myosin light chain-3F promoter and 3' enhancer driving *LacZ,* and shows expression in left ventricle and right atrium linked through the atrioventricular canal (Franco *et al.,* 1997) (Fig. 4B). This pattern has been proposed to be vestigial, reflecting the most ancient anatomical regions of the heart tube; that is, those that existed in ancestral vertebrates prior to duplication of atria and ventricles (Kelly *et al.,* 1999). This notion is perhaps supported by the fact that the endogenous gene is expressed in an identical fash-

ion, yet has no function since no protein is produced (Kelly *et al.,* 1998).

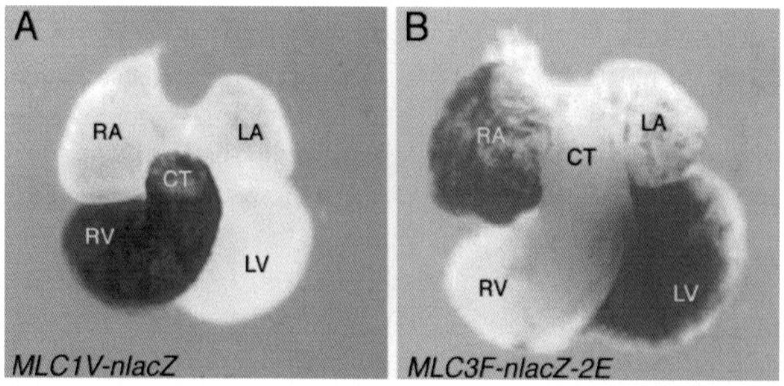

XVI. Plasticity of Heart Regionalization

The stability of regionalization in the heart provides clues to the timing of patterning processes. In the chick and fish, fate mapping experiments demonstrate that progenitors of the rostral and caudal regions of the heart are spatially distinct from gastrulation onward and perhaps earlier (Garcia-Martinez and Schoenwolf, 1993; Rosenquist and DeHaan, 1966; Stainier *et al.,* 1993). However, the rostrocaudal identity of myocytes is not yet fixed at the early to mid primitive streak stages (HH stages 3–4) (Inagaki *et al.,* 1993). Plasticity is also evident later at the early cardiogenic plate stage (HH stages 5–7), since the graded electrical properties of cardiomyocytes can be reset if myocytes are repositioned within the heart field (Satin *et al.,* 1988). In the mouse, transplant experiments show that atrial identity is fixed at least as early as E10.5 (Gruber *et al.,* 1998), although probably much earlier since AP perturbations to the heart tube conferred by exogenous retinoic acid occur only if precursors are treated at the cardiogenic plate stage (E7.5), but not later (Xavier-Neto *et al.,* 1999). An important finding in the chick is that myocytes positioned caudally within the cardiogenic plate (HH stages 4–8) already have some notion of their positional identity, in that they remain programmed to express an atrial-specific myosin heavy chain gene after explant culture (Yutzey *et al.,* 1994). It is therefore during this period, and potentially earlier (Rosenthal and Xavier-Neto, 2000), that patterning elements begin to show their influence.

XVII. The Segmental Model of Cardiac Morphogenesis

The heart has long been considered a segmented structure, stemming from observations of a series of cavities matched by external constrictions in the lumen of the early looping chick and human heart (Davis, 1927; de la Cruz *et al.,* 1989; Markwald *et al.,* 1998; Stalsberg, 1969a). These landmarks demarcate what are commonly held to be the basic building blocks of the embryonic heart: sinus venosus, common atrium, atrioventricular canal, embryonic left ventricle, embryonic right ventricle, and conotruncus (Davis, 1927; de la Cruz, 1998; de la Cruz and Sanchez-Gomez, 1998). The term *segment* has been used in these studies to mean primarily a unit of structure rather than a metamere (Harvey, 1998b; Markwald *et al.,* 1998). However, because heart regions form progressively along the AP axis, with different birth dates and possibly conception dates, the process has been likened to somitogenesis (Markwald *et al.,* 1998).

Figure 4 Modular regionalized expression of transgenes in the mouse heart. Dissected mouse hearts at E10.5 showing staining for β-galactosidase expressed from the *MLC1V-nLacZ* (A) and *MLC3F-nLacZ-2E* (B) transgenes. Expression in panel A is driven by one regulatory module of the *MLC1V* gene. Expression in panel B mimics the endogenous pattern of expression of the MLC3F gene and is proposed to highlight the most ancient anatomical regions of the mammalian heart (Kelly *et al.,* 1999). Abbreviations: CT, conotruncus; LA, left atrium; LV, left ventricles; RA, right atrium; RV, right ventricle. (Images supplied by Robert Kelly and Margaret Buckingham.)

XVIII. An Inflow/Outflow Model of Early Heart Tube Patterning

Although viewing the heart as an array of segments is developmentally appealing, the constrictions that mark segment boundaries are only apparent at the outer curvature of the looping heart. Furthermore, patterns of gene expression in general do not support progressive formation of anatomical segments, as in somitogenesis (Christoffels *et al.*, 2000a; Harvey, 1998b). Rather, marker studies hint at an initial binary segmental process leading to demarcation of inflow and outflow regions of the forming heart. Inflow/outflow patterning may be a prelude to further patterning events. Unlike most myofilament genes, which are initially expressed across the whole myocardium, the *Mylpc* gene, encoding myosin light chain 2V, is expressed in a region-specific manner from the outset (Christoffels *et al.*, 2000a; Franco *et al.*, 2000; Lyons *et al.*, 1995; Ross *et al.*, 1996). Expression begins in the late cardiac crescent in two small bilateral domains, then expands to encompass the cranial half of the linear heart tube including the atrioventricular canal, right and left ventricles, and conotruncus ("outflow" region).

The homeobox transcription factor gene, *Irx4*, is expressed in a pattern very similar to that of *Mylpc* (Bruneau *et al.*, 2000a; Christoffels *et al.*, 2000a,b). Conversely,

genes encoding the orphan nuclear receptor COUP-TFII (*Nr2f2*) and retinaldehyde dehydrogenase-2 (*Aldh1a7*), and transgenes *SMyHC3-HAP* and *RARβ-RARE-LacZ*, are expressed only in the sinoatrial ("inflow") region in the early heart tube (Moss *et al.*, 1998; Pereira *et al.*, 1999; Xavier-Neto *et al.*, 1999). Certain myofilament proteins are also expressed specifically in inflow (sinoatrial) or outflow (ventricles plus conotruncus) regions of the hearts of fish, frog, and chick embryos (Bisaha and Bader, 1991; Stainier and Fishman, 1992; Yutzey *et al.*, 1994; T. Mohun, personal communication), suggesting that an anatomical and functional inflow/outflow demarcation is a conserved feature of vertebrate heart evolution. Even though precursors of the conotruncal endocardium (Noden, 1991), and possibly myocardium (de la Cruz *et al.*, 1997; Markwald *et al.*, 1998; Viragh and Challice, 1973), have their origins outside of the cardiac crescent, these tissues are included in expression domains that mark the outflow region. The inflow/outflow concept may be highly simplistic, since other regionalities along the AP axis become evident at later times. For example, the so-called conduction ring surrounding the interventricular foramen (Fig. 5A) may be established by a primary AP patterning event (Wessels *et al.*, 1992). The apparent boundaries and gradients marked by numerous transgenes (see Section XV) may betray elements of this system.

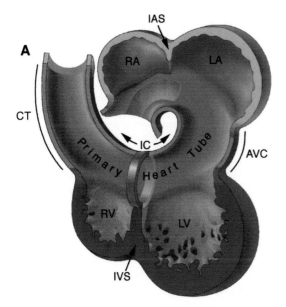
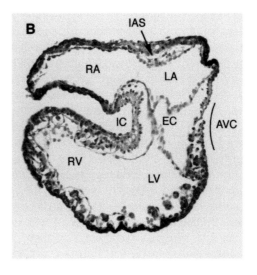

Figure 5 Ballooning model of cardiac chamber formation. This model predicts that myocardium of the heart chambers (working myocardium) is specialized in phenotype and arises at distinct locations at the outer curvature of the early looping heart (Christoffels *et al.*, 2000a). (A) Graphic depiction of the ballooning model, showing working myocardium of the atria (blue) and ventricles (red) arising at the outer curvature of the primary heart tube (purple). The inner curvature (IC) is smooth walled and unsegmented and does not develop trabeculae. (Adapted from Moorman *et al.*, 2000, and reproduced with permission from Academic Press.) (B) Expression of the gene *Csl* marking presumptive working myocardium of the atrial and ventricular chambers of an E10.0 mouse heart (Palmer *et al.*, 2000). Note expression is absent from the nontrabeculated portion of the atrioventricular canal, inner curvature, and forming interatrial septum. Abbreviations: AVC, atrioventricular canal; EC, endocardial cushions; IAS, interatrial septum; IC, inner curvature; IVS, interventricular septum; LA, left atrium; LV, left ventricle; RA, right atrium; RV, right ventricle.

XIX. A Role for Retinoic Acid Signaling in Inflow/Outflow Patterning

Vitamin A is an essential and pleiotropic micronutrient in development, growth, and fertility (Zile, 1998), acting principally through its oxidation form, retinoic acid (RA) (Ross *et al.*, 2000). It has long been appreciated that vitamin A deficiency during pregnancy results in fetal death and severe congenital abnormalities, with retinoid excess also highly teratogenic (Ross *et al.*, 2000). The cardiovascular system is extremely sensitive to perturbations of retinoid homeostasis (Kubalak and Sucov, 1999). Rat and avian models of vitamin A deficiency, as well as single and multiple knock-outs of members of the nuclear retinoic acid receptor (RAR) and retinoid X receptor (RXR) gene families, have revealed roles for retinoids in sinoatrial development, left-right asymmetry, ventricular chamber growth, interventricular septation, and development of the conotruncus and aortic arches (Kastner *et al.*, 1997; Kostetskii *et al.*, 1999; Lee *et al.*, 1997; Wilson and Warkany, 1949; Zile, 1998). The knock-out studies seem to confirm both the complexity and extent of retinoid involvement in virtually all aspects of cardiac morphogenesis through pathways that are quantitatively unique (Kastner *et al.*, 1997; Lee *et al.*, 1997).

Several studies point to a key role for RA in specification of the inflow (sinoatrial) region of the heart. The distribution of retinaldehyde dehydrogenase type 2 (RALDH2), the enzyme responsible for virtually all RA synthesis in the embryo (Ross *et al.*, 2000), and the expression pattern of a *LacZ* transgene responsive to RA signaling, show that RA synthesis and response are initially restricted to the sinoatrial region of the forming heart (Moss *et al.*, 1998; Xavier-Neto *et al.*, 1999). Indeed, it has been suggested that sinoatrial heart progenitors migrate through a field of RA synthesis located lateral and caudal to the node during gastrulation (Niederreither *et al.*, 1997; Rosenthal and Xavier-Neto, 2000; Xavier-Neto *et al.*, 2000). Hearts from vitamin A-deficient quail embryos lack a sinus venosus and are closed caudally (Kostetskii *et al.*, 1999). A similar loss of sinoatrial tissue occurs in mouse embryos cultured in disulfiram, a RA synthesis-inhibitor (Xavier-Neto *et al.*, 1999), or in BMS493, a pan-RAR antagonist (Chazaud *et al.*, 1999). Furthermore, embryos mutant for the RALDH2 gene (*Aldh1a7*) have unlooped hearts with an indistinct hypoplastic sinoatrial region (Niederreither *et al.*, 1999, 2001) (Figs. 6D and 6E.). In contrast to these studies in which RA synthesis or response is inhibited, excess retinoid delivered at the cardiogenic plate stage leads to the opposite phenotype, an expansion of the sinoatrial region and "atrialization" of ventricular tissue (Xavier-Neto *et al.*, 1999). Collectively, the data suggest a role for RA in the specification, proliferation, and/or survival of sinoatrial tissue within the cardiac progenitor field.

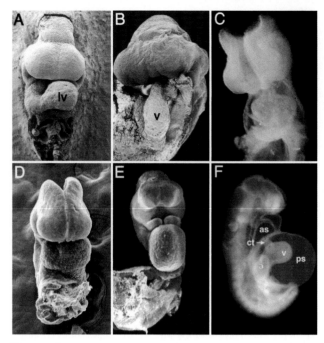

Figure 6   Cardiac defects in mutant mouse strains. (A, B) Scanning electron micrographs (SEMs) of a wild-type embryo at E9.0 and an embryo mutant for the *Mef2c* gene, respectively. The mutant heart fails to undergo rightward looping and there is no evidence of a right ventricular chamber. (Reproduced from Black and Olson, 1999, with permission from Academic Press.) (C) Embryo carrying a null mutation in the *Nkx2-5* gene at E8.5. The mutant heart fails to undergo looping and ventricular chambers do not differentiate. (D, E) SEMs of a wild-type embryo at E8.5 and an embryo mutant for the RALDH2 gene after removal of the pericardial layer. The mutant heart is dilated, does not loop, and has an indistinct sinoatrial region. (Reproduced from Niederreither *et al.*, 2001, with permission from the Company of Biologists Limited.) (F) Embryo mutant for the *Hand2* gene at E9.0. The mutant heart tube lacks a discernible right ventricle and ends at the outflow region as a blind-ended sack, unconnected to branchial arch arteries. Abbreviations: a, atrium; as, aortic sac; ct, conotruncus; lv, left ventricle; ps, pericardial sac; v, ventricle. (Reproduced from Srivastava *et al.*, 1997, with permission from Nature America.)

Clues have emerged as to the identity of genes controlled by RA during sinoatrial development. In chick embryos, a number of *Hox* genes have been shown to be expressed in the heart-forming region and upregulated by RA (Searcy and Yutzey, 1998; Sundin and Eichele, 1992). The gene for the orphan nuclear receptor COUP-TFII (*Nr2f2*) is expressed broadly in mesoderm, but only in the sinoatrial region of the forming heart (Pereira *et al.*, 1999). *Nr2f2* expression is retinoid responsive and required, at least in tumor tissue, for RA-mediated upregulation of the retinoic acid receptor β gene (*Rarb*) (Lin *et al.*, 2000). Importantly, targeted deletion of *Nr2f2* leads to hearts with a severely hypoplastic sinoatrial phenotype (Pereira *et al.*, 1999). Another potentially important gene in sinoatrial development encodes the T-box transcription factor, TBX5, mutated in Holt-Oram syndrome in humans (Basson *et al.*, 1999). *Tbx5* is expressed in a graded pattern across the AP axis of the heart, high in the sinoatrial region (Bruneau *et al.*, 1999). The elevated caudal expression is downregulated in *Aldh1a7* knock-out mice (Niederreither *et al.*, 2001), suggesting a role for RA in setting up the graded pattern. Indeed, *Tbx5* has been shown to be RA-responsive, and transgenic overexpression of *Tbx5* in the ventricles during development induces abnormal ventricle morphology and downregulation of ventricle-specific gene expression (Liberatore *et al.*, 2000). Finally, the GATA zinc finger transcription factor genes are also RA responsive (Ghatpande *et al.*, 2000; Jiang *et al.*, 1999; Kostetskii *et al.*, 1999). *Gata4* is expressed across the whole heart tube and, as for *Tbx5*, is higher in the sinoatrial region (Molkentin *et al.*, 1997). This pattern may also be set by caudal RA synthesis, because *Gata4* is downregulated in the caudal heart region in vitamin A-deficient quail embryos (Kostetskii *et al.*, 1999).

One role for RA in sinoatrial development may be to inhibit myogenic differentiation for a defined period, perhaps promoting proliferation and allowing other patterning events to occur. Exogenous RA inhibits heart differentiation in frog embryos upstream of *Nkx2-5* (Drysdale *et al.*, 1997), potentially mediated by induction of higher levels of expression of GATA factor genes (Jiang *et al.*, 1999). As noted above, GATA factors have dose-dependent effects on both proliferation and differentiation. Importantly, a major consequence of loss of *Aldh1a7* or of RAR/RXR genes in mice is premature differentiation of ventricular tissue (Kastner *et al.*, 1997; Niederreither *et al.*, 2001). In contrast, however, overexpression of the RA-responsive gene *Tbx5* in chick embryonic hearts and cultured cells represses cell proliferation in a cell nonautonomous manner (Hatcher *et al.*, 2001). *In vitro* studies suggest other molecular mechanisms for how RA might affect myogenesis in the caudal heart: The sinoatrial-specific nuclear receptor COUP-TFII can antagonize GATA-4-mediated transcription through a direct association with FOG2 (Huggins *et al.*, 2000), while the related COUP-TFI can antagonize expression of the calreticulin gene (*Calr*) through an Nkx2-5-like binding element (Guo *et al.*, 2000).

A curious feature of retinoid function during embryogenesis is that virtually all of the cardiac morphological and molecular defects elicited by retinoid deficiency, whether generated by dietary or genetic means, can be rescued by the addition of exogenous RA (Niederreither *et al.*, 1999; Zile, 1998). Because the whole embryo appears to be responsive to RA signaling (Moss *et al.*, 1998; although see Chazaud *et al.*, 1999), these findings suggest that RA acts as an essential although spatially nonspecific factor for sinoatrial development. Another interesting finding is that expression of a *LacZ* insertional transgene normally expressed in the right ventricle and conus is expanded caudally in the presence of RA given to mothers from E8.5 (Yamamura *et al.*, 1997). While the time of exposure in this experiment is outside the window of sensitivity for the "atrialization" effect described above (Xavier-Neto *et al.*, 1999), the finding may reflect later functions of RA in the outflow region.

XX. A Role for the Delta/Notch Pathway in Primary Heart Patterning

The Notch/Delta pathway functions in control of differentiation, cell type specification, and the generation of tissue boundaries and is central to the clock mechanism that guides segment formation during somitogenesis (Pourquié, 2000). As such, it would be an interesting candidate pathway for involvement in primary inflow/outflow patterning in the vertebrate heart tube. The mammalian *Hey* genes are homologs of the *Drosophila* Hairy/Enhancer of Split-related genes, which encode transcriptional repressors acting downstream in the Delta/Notch signaling pathway. Two of the three known *Hey* genes are expressed in the early heart in a chamber-specific manner, *Hey1* in the atria and *Hey2* in the ventricles (Chin *et al.*, 2000; Leimeister *et al.*, 1999; Nakagawa *et al.*, 1999). All three *Hey* genes have been shown to be responsive to Notch signaling and, in line with *Drosophila* counterparts, Hey proteins act as transcriptional repressors (Chin *et al.*, 2000; Nakagawa *et al.*, 2000).

A role for Notch/Delta signaling and the *Hey* genes in inflow/outflow patterning has yet to be proven; however, it is noteworthy that these genes are among few that show regional expression in the cardiac crescent (Nakagawa *et al.*, 1999) and Hey proteins associate with the bHLH Hand factors (see below), implicated in heart chamber formation (Firulli *et al.*, 2000). Interestingly, heterozygous mutations in the gene encoding Jagged1 (*Jag1*), a mammalian homolog of Delta, underlie cases of Alagille syndrome and autosomal dominant tetralogy of Fallot in humans (Eldadah *et al.*, 2001; Li *et al.*, 1997; Oda *et al.*, 1997). Furthermore, mice lacking the *Psen1* and *Psen2* genes, encoding presenilin proteins involved in processing and activation of membrane-bound Notch, have small unlooped heart tubes (Donoviel *et al.*, 1999).

XXI. Cardiac Chamber Formation

How are definitive cardiac chambers formed in response to inflow/outflow patterning? As noted above, the constrictions in the heart tube that have been interpreted as segmental boundaries form only on its outer (or greater) curvature (Christoffels *et al.*, 2000a). Other manifestations of cardiac chamber identity, such as formation of the spongiform myocardial trabeculae that line the cavities of the ventricular chambers, also occur only at the outer curvature (Fig. 5A). The inner (or lesser) curvature remains unsegmented and smooth walled. These anatomical and other functional considerations have led to the proposal that heart chambers arise as a specialized "working myocardium" at distinct locations on the outer curvature of the looping heart tube (Christoffels *et al.*, 2000a; de Jong *et al.*, 1992, 1997; Harvey, 1998b; Moorman *et al.*, 2000). Cell marking experiments show that the outer curvature of the looping heart is derived from the original ventral surface of the linear heart tube (de la Cruz *et al.*, 1989, 1997), the inner curvature deriving from the dorsal surface.

Because chamber myocardium can be perceived as expanding outward (or "ballooning") from the primary heart tube during the process of looping, the model has been dubbed the "ballooning model" of cardiac chamber formation (Christoffels *et al.*, 2000a) (Fig. 5A). The inner curvature can be regarded in this model as a hinge point for the pronounced outward growth of chamber myocardium (Markwald *et al.*, 1998). The model highlights the importance and integration of dorsal-ventral (DV) and AP patterning information for chamber formation. Genes encoding the bHLH transcription factor HAND1, transcription factor CITED1, natriuretic hormone ANF (*Nppa*), gap junction protein connexin 43 (*Gja1*), and small muscle-specific protein Chisel (*Csl*; Figure 5B) all show expression in AP and outer curvature-restricted domains that match, to a first approximation, the position of anatomical chamber primordia (Biben and Harvey, 1997; Christoffels *et al.*, 2000a; Delorme *et al.*, 1997; Dunwoodie *et al.*, 1998; Palmer *et al.*, 2001). Conversely, expression of a *LacZ* transgenic marker driven by *cis* elements of the versican gene (*Cspg2*) is concentrated at the inner curvature of the right ventricle and conus (Yamamura *et al.*, 1997).

These findings demonstrate that distinct genetic programs are activated in precursors of chamber working myocardium and that these precursors are not arranged segmentally along the AP axis of the heart tube (Christoffels *et al.*, 2000a). The programs active within working myocardium presumably guide chamber morphogenesis, as well as the myogenic and electrical specialization of chamber muscles. The observations suggest a view of heart patterning and chamber formation that departs from the segmental concept. Myocardial tissue associated with nonchamber components of the heart tube (inflow vessels, atrioventricular canal, conotruncus, and inner curvature; Figs. 5A and 5B) can be regarded as persisting regions of the primary heart tube, as opposed to distinct anatomical segments (Christoffels *et al.*, 2000a; de Jong *et al.*, 1992; Harvey, 1998b; Moorman *et al.*, 2000). That is not to say that these zones are merely structural. Indeed, they participate profoundly in subsequent heart morphogenesis, directing alignment of chamber communications (Kim *et al.*, 2001), integration of neural crest and other mesenchymal populations into the body of the heart (Jiang *et al.*, 2000; Kim *et al.*, 2001; Kirby, 1998; Wessels *et al.*, 2000), induction of endocardial cushions and valves (Markwald *et al.*,

1998), and formation of elements of the conduction system (Moorman and Lamers, 1999; Rentschler *et al.*, 2001).

The zones of chamber working myocardium arise from within inflow and outflow regions previously specified. The spatial information that leads to chamber demarcation is not known, although we can reasonably expect that this event is a convergence point for AP, DV, and perhaps left-right (LR) patterning information in the forming heart (see below). A two-step model of heart patterning involving inflow/outflow specification, then definition of chamber working myocardium, provides a framework for interpretation of mutant phenotypes that affect chamber formation (see below).

XXII. Ventricular Specification: Knock-Out and Transgenic Phenotypes

The phenotypes of several targeted mutations and a transgenic insertion have highlighted genes involved in specification of ventricular chambers and have hinted at some of the patterning principles involved. These are outlined next.

A. Nkx2-5

Nkx2-5 is a mammalian homolog of the *Drosophila* homeobox gene *tinman*, essential for specification of cardiac and visceral muscles in the fly (see Section III). *Nkx2-5* itself is essential although not sufficient for heart formation in frogs, as judged by dominant-negative inhibition and overexpression experiments (Cleaver *et al.*, 1996; Fu and Izumo, 1995; Grow and Krieg, 1998). In the mouse, this gene plays a critical role in ventricular chamber specification and appears less important for formation of heart lineages and the heart tube. In mice carrying homozygous *Nkx2-5* mutations, an apparently normal linear heart tube forms, but looping is arrested at an early stage and virtually all downstream morphogenetic events, including trabeculation and formation of endocardial cushions, are blocked (Biben *et al.*, 2000; Lyons *et al.*, 1995; Tanaka *et al.*, 1999) (Fig. 6C). While most myofilament genes are robustly expressed in the mutant hearts (Lyons *et al.*, 1995), other classes of genes are dysregulated. This include genes encoding the bHLH transcription factor Hand1 (Biben and Harvey, 1997), homeodomain transcription factor Irx4 (Bruneau *et al.*, 2000a), CITED family transcription factor CITED1 (Dunwoodie *et al.*, 1998; R. P. Harvey and C. Biben, unpublished observations), natriuretic hormone ANF (Biben *et al.*, 1997), actin cross-linking molecule SM22α (Biben *et al.*, 1997), and small muscle-specific protein Chisel (Csl), implicated in cytoskeletal dynamics (Palmer *et al.*, 2001). Transcription factor genes *Nmyc* and *Msx2* are also downregulated, potentially the result of arrested morphogenesis (Tanaka *et al.*, 1999). Although the patterns of gene dysregulation are complex, it is notable that markers normally restricted to the outer curvature are all but abolished in the ventricular region. Furthermore, SM22α, which is normally expressed across the whole heart, is downregulated only at the outer curvature of the ventricles (R. Harvey and C. Biben, unpublished data). The expression boundaries of *Irx4* and *Mylpc,* which define the outflow region of the heart, are established normally, but overall levels of expression are much lower than in wild-type hearts (Biben *et al.,* 2000; Bruneau *et al.,* 2000a; Lyons *et al.,* 1995; Tanaka *et al.,* 1999). These patterns suggest that inflow/outflow patterning can be established but its constituent genes not upregulated, and that ventricular chamber specialization is totally blocked (Harvey *et al.,* 1999). *Nkx2-5* appears to be fundamental to ventricular chamber specification, and further dissection of its role will be necessary to understand genetic control of heart patterning.

B. Hand Genes

The bHLH genes *Hand1* and *Hand2* are expressed in developing mammalian hearts from cardiac crescent stages, as well as in a variety of other tissues (Cross *et al.,* 1995; Cserjesi *et al.,* 1995; Hollenberg *et al.,* 1995). Hand genes can homodimerize and heterodimerize with other bHLH proteins of both class A and B, and may act, in part, as transcriptional repressors (Bounpheng *et al.,* 2000; Firulli *et al.,* 2000; Scott *et al.,* 2000). The genomes of all vertebrates studied carry both genes, with the exception of zebrafish, in which *Hand1* has been not been found (Yelon *et al.,* 2000). A single Hand gene is also found in *Drosophila* (Moore *et al.,* 2000). Both genes exist in *Xenopus,* although *Hand2* does not appear to be expressed in the heart at appreciable levels (Yelon *et al.,* 2000; see also Angelo *et al.,* 2000). Hand genes have been implicated in trophoblast giant cell differentiation (Cross *et al.,* 1995), limb and fin bud patterning (Charite *et al.,* 2000; Fernandez-Teran *et al.,* 2000; Yelon *et al.,* 2000), sympathetic and enteric neuronal differentiation (Howard *et al.,* 1999), and cardiovascular development (Firulli *et al.,* 1998; Riley *et al.,* 1998, 2000; Srivastava *et al.,* 1995, 1997; Thomas *et al.,* 1998a; Yamagishi *et al.,* 2000; Yelon *et al.,* 2000). The first suggestion of a role for Hand genes in heart development came from anti-sense inhibition studies in chick embryos, in which blockade of both *Hand1* and *Hand2,* each expressed throughout the heart tube, inhibited cardiac looping and subsequent heart tube development (Srivastava *et al.,* 1995). In zebrafish, mutations in *Hand2* lead to hypoplastic and poorly patterned hearts in which expression of the T-box factor gene *Tbx5* cannot be maintained (Yelon *et al.,* 2000).

In the mouse, *Hand1* and *Hand2* are expressed in the heart in dynamic and strikingly different patterns (Biben and Harvey, 1997; Thomas *et al.,* 1998b). *Hand1* is expressed on the ventral surface of the caudal region of the linear heart tube encompassing precursors of the left (systemic) ventricle

and, transiently, in precursors of the atrioventricular canal, atrium and sinus venosus. As noted above, this pattern was the first molecular evidence for a DV patterning in the early heart tube (Biben and Harvey, 1997). Expression also occurs in the forming aortic sac. During looping, expression becomes restricted to the outer curvature of the left ventricle in myocardial and pericardial layers, and later, to the outer curvature of the right ventricle and conotruncus. *Hand2,* in contrast, is expressed throughout the myocardial, endocardial, and pericardial layers of the linear heart tube, with expression in myocardium evident as a craniocaudal gradient (Biben and Harvey, 1997). This graded pattern is maintained and enhanced during looping such that the most prominent expression occurs in the right (pulmonary) ventricle and conotruncus.

General support for a role in chamber formation has come from knock-out mice. Mice homozygous for a targeted deletion of *Hand1* show abnormal heart tube patterning and arrested looping compatible with failure to form a left ventricle (Firulli *et al.,* 1998; Riley *et al.,* 1998). Chimera studies show that *Hand1* mutant cells are excluded from the outer curvature of the left ventricular myocardium when in competition with wild-type cells (Riley *et al.,* 2000). Conversely, *Hand2* mutant hearts show defects suggestive of the loss of the right ventricle (Fig. 6F). Abnormal branchial artery development leads additionally to formation of a blind-ended heart tube with severely dilated aortic sac (Srivastava *et al.,* 1997). In such hearts, the zinc finger transcription factor gene *Gata4* is globally downregulated, suggesting a fundamental role for HAND2 in the core myogenic program (Section XII). Trabeculation in the remaining left ventricle is also diminished, a possible consequence of abnormal development of the endocardium (Meyer and Birchmeier, 1995), in which *Hand2* is highly expressed. Genetic crosses between *Hand2* and *Nkx2-5* knock-out mice (the latter showing downregulation of *Hand1* in the forming heart), produce embryos with hearts that appear to completely lack a ventricular myocardium (Yamagishi *et al.,* 2001).

The above findings clearly have implications for chamber specification and heart patterning, and potentially the etiology of human congenital conditions involving hypoplastic ventricles (Tchervenkov *et al.,* 2000). The recent finding that Hand proteins can associate with the cardiac bHLH Hey factors, themselves showing AP restriction in expression and responsiveness to Notch signaling (Firulli *et al.,* 2000), suggests roles for both classes of protein in heart regionalization and/or boundary formation. However, there are subtleties and discrepancies in this genetic system that should not be overlooked. First, *Hand1* and *Hand2* actually have very different expression patterns in mice that are only superficially complementary. Furthermore, the varying gene number and the widely differing modes of expression of Hand genes in different vertebrate models suggest a complex evolution. In the original antisense experiments in chick embryos, both

Hand genes needed to be inhibited before heart defects were revealed (Srivastava *et al.,* 1997). In mammals, however, each gene has an essential function and loss of the right ventricle in *Hand2* mutants is accompanied by cell death (Srivastava, 1999), whereas this is not the case in *Hand1* knock-out mice in which the left ventricle is lost (Riley *et al.,* 1998).

As for many biological systems, these discrepancies will probably be understood only in the context of evolution. The hearts of extant lower vertebrates have a single ventricle, as do most reptiles, although crocodilian hearts have divided ventricles like those of birds and mammals (Farrell, 1997; Van Mierop and Kutsche, 1984). Although the morphological details of this evolutionary progression are not clear, the appearance of separate systemic and pulmonary circuits in terrestrial vertebrates may have depended on genetic pathways involving duplicated *Hand* genes. These genes were most likely functionally redundant initially, but may have evolved specialized functions in mammals. It is even possible that different species achieved certain aspects of chamber formation and looping using distinct spatial or molecular mechanisms. Nonetheless, further dissection of the function and regulation of the Hand genes will give us new and important insights into heart patterning in mice and humans and to its evolutionary progression.

C. MEF2

As described in Section XII, the Mef2 genes occupy a key position in the core myogenic program of the heart and other muscles. *Mef2c* and *Mef2b* are expressed from early cardiac crescent stages, along with *Nkx2-5* and *Gata* genes. *Mef2c* knock-out mice have hearts that are severely dysmorphogenic, showing arrested looping, a single hypoplastic left ventricle with poor trabeculation, no sinus venosus, and disorganized myocardium and endocardium (Lin *et al.,* 1997) (Figs. 6A and 6B). One feature of these hearts is the downregulation of a number of myogenic genes and a sevenfold upregulation of *Mef2b.* Importantly, however, the *Hand2* gene is expressed normally in the linear heart, but is downregulated during cardiac looping. Because *Hand1* knockout hearts lack a right ventricle, this may be one reason for the lack of this same ventricle in the *Mef2c* mutants. The *Hand1* gene, normally left ventricle specific, was expressed robustly throughout the mutant heart tube, confirming that the remaining ventricle was left (systemic) in character. Thus, *Mef2c* and *Hand2* have essential roles in specification of the right ventricular chamber.

D. Versican

The space between the myocardium and endocardium of the heart is filled with an extracellular matrix termed *cardiac jelly* (Mjaatvedt *et al.,* 1999). This matrix thickens and becomes more complex in regions of endocardial cush-

ion formation, where endocardial cells undergo epithelial-mesenchymal transition and migrate into the cushion matrix to form progenitors of the cardiac valves (Mjaatvedt *et al.,* 1999). One of the major components of this matrix is a family of large chondroitin sulfate proteoglycans (CSPGs) that aggregate with hyaluronic acid (HA). One member of this family is versican/PG-M, and its gene (*Cspg2*) was disrupted by an insertional transgene carrying *Hoxa1* cis-regulatory sequences linked to a *LacZ* reporter (Mjaatvedt *et al.,* 1998). The *LacZ* transgene appears to faithfully replicate the graded expression of the versican gene across the myocardium, with highest levels in the conus and right ventricle (Henderson and Copp, 1998; Yamamura *et al.,* 1997; Zanin *et al.,* 1999). Prominent expression also surrounds the endocardial cushions of the atrioventricular canal. In the developing hearts of homozygous transgenic mice, the right ventricle and conus are severely underdeveloped or missing, and endocardial cushions of the conus and atrioventricular canal are completely absent (Yamamura *et al.,* 1997). The loss of the right ventricle resembles that of the *Hand2* and *Mef2c* mutant hearts, suggesting that all three genes might lie in a common pathway. Indeed, the AP graded expression pattern of versican mRNA and protein may reflect the similarly graded pattern of *Hand2* (Biben and Harvey, 1997).

CSPGs are complex proteins with HA and glycosaminoglycan-binding domains, as well as Ig-like, EGF-like, lectin-like and complement regulatory protein-like domains. The presence of these proteins in the cell matrix can have profound effects on cell growth and differentiation. Versican has been shown to affect cell proliferation via EGF-like domains (Zhang *et al.,* 1998), and also to bind adhesion molecules, chemokines, growth factors, and HA receptors (Hirose *et al.,* 2001; Kawashima *et al.,* 2000; Zou *et al.,* 2000). Thus, this CSPG may signal directly to myocardial or endocardial cells, or localize and/or stabilize other factors required for their growth and survival. Versican is a morphogenic molecule that most likely sits downstream of the Hand2 and/or Mef2c transcriptional regulatory pathways guiding heart patterning.

E. Irx4

The mammalian Irx genes are related to the Iroquois family of homeobox genes in *Drosophila,* which encode transcription factors involved in tissue specification and patterning (Christoffels *et al.,* 2000b). The *Irx4* gene has been studied in most detail (Bao *et al.,* 1999; Bruneau *et al.,* 2000a,b, 2001; Christoffels *et al.,* 2000a,b). Expression begins in the outflow region of the cardiac crescent and is subsequently expressed in the outflow region of the linear and looping heart tube (conotruncus, ventricles, and atrioventricular canal) with highest levels in the ventricles (Bruneau *et al.,* 2000a; Christoffels *et al.,* 2000a,b).

As noted above, this pattern is downregulated in *Nkx2-5* and *Hand2* knock-out mice (Bruneau *et al.,* 2000a). In *Irx4-*

deficient mice, ventricular chamber formation appears normal, although the bHLH factor *Hand1* gene is significantly downregulated, and an atrial chamber-specific transgene *SMyHC3-HAP* is inappropriately expressed in the ventricles (Bruneau *et al.,* 2001). The related *Irx2* gene, expressed to the side of the interventricular groove, is upregulated. These and perhaps other changes in gene expression lead to the development of a ventricular cardiomyopathy and compensatory hypertrophy in the postnatal period. Interestingly, the inappropriate expression of *SMyHC3-HAP* in the ventricles resembles the normal pattern of *Hand1,* which is downregulated, suggesting that *Irx4* may function through *Hand1* in repressing atrial gene expression in the ventricles.

XXIII. Transcriptional Circuits Acting in Chamber Formation

We are still far from being able to draw transcriptional circuit diagrams for lineage specification and chamber formation in the mammalian heart. The ultimate diagram will identify genes involved in the core myogenic program, specialized myogenesis in heart chambers, heart patterning, boundary formation, and morphogenesis of chambers and other regions. It will also describe the specification and development of the nonmyogenic tissues of the heart and inductive interactions between these regions and the myocardium. The modular development of the heart will be implicit, as will the modifications that come to bear on developmental programs by changing hemodynamics and the LR axis (see below and Chapter 4). The task seems daunting. At present, we can merely indicate our first impressions gleaned from the analysis of knock-out mice and *cis*-regulatory regions of key genes (see Table I).

Figure 7 indicates some of the interactions within the forming heart discussed in the text above. The impression is of a highly interactive and mutually supportive network of transcription factor genes acting in both myogenic and patterning pathways (Fig. 7A). We can anticipate that this first impression will be fleshed out at a rapid rate, although eventually the problem will become too complex to visualize with our intuitive tools, and might demand the application of the science of bioinformatics.

XXIV. The Cardiac Left-Right Axis

One of the most fascinating aspects of cardiac biology is the key role played by the left-right axis in heart morphogenesis (Harvey, 1998a). During its formation the heart adopts a spiral form through a process called *cardiac looping,* or *looping morphogenesis,* and thus departs from its original bilateral morphological symmetry. This occurs in concert with the appearance of other LR body asymmetries, such as a directional handedness in embryonic turning and asymmetries in the formation of vessels, visceral organs and the brain (Liang *et al.,* 2000; Winer-Muram, 1995). The normal direction of heart looping is considered to be "rightward," in that the coiling of the ventricles and conotruncus is to the right when viewing the embryo from the ventral aspect.

During development, looping sets the stage for alignment, integration, and remodeling of the cardiac chambers (Mjaatvedt *et al.,* 1999) and confers efficiencies in patterns of flow (Kilner *et al.,* 2000). An important point to appreciate is that the progenitors of the left (systemic) and right (pulmonary) ventricles are established initially as AP neighbors and are repositioned into a LR arrangement by the process of looping. Thus, each ventricle is made up of contributions (not necessarily equal) from the left and right cardiac progenitor pools. LR allocations in the sinoatrial region are more complex. While the two horns of the sinus venosus derive from left and right cardiac progenitor pools, respectively, both end up draining into the right atrium. Nevertheless, the right atrial appendage (the working component of the right atrium) appears to derive solely from the right progenitor pool, and likewise the left atrium from the left progenitor pool (Campione *et al.,* 2001). Laterality confers each atrial appendage with stereotypical right or left morphology (Brown and Anderson, 1999). The interatrial septum primum appears to be left in origin (Campione *et al.,* 2001; Wessels *et al.,* 2000). Although the precise cellular movements involved in setting up the LR map have not been studied in detail, it is clear that LR asymmetries in the heart are evident before overt looping. In the chick, cardiac progenitors from the left caudal region beat fastest (Satin *et al.,* 1988) and endothelial precursors are more abundant on the right (Markwald, 1995), suggesting maturational gradients (Corballis and Morgan, 1978).

During heart formation in the mouse and zebrafish, asymmetric morphogenesis leads to displacement of the caudal region of the heart tube to the left, a process referred to as the leftward shift or jog (Biben and Harvey, 1997; Chen *et al.,* 1997). This has the effect of orienting the outflow region of the heart toward the right, and may be a key determinant of the direction of looping (Biben and Harvey, 1997; Chen *et al.,* 1997). The asymmetric morphogenesis in the caudal region is at least in part intrinsic: A swelling is evident on the left side of the caudal region of the forming mouse heart prior to overt looping, and cell marking shows that this structure contributes to the atrioventricular canal (Brown and Anderson, 1999). On its opposite side, a large ingression forms (the atrioventricular sulcus), placing the body of the atrioventricular canal substantially leftward.

The consequences of disturbed laterality in humans and mice have been described in detail (Bowers *et al.,* 1996;

Table I Summary of Key Mutant Phenotypes Affecting Early Heart Development and Patterning Discussed in the Text

Process	Gene	Relevant mutant phenotype	Reference
Cardiac induction	Bmp2	No hearts in a proportion of embryos; ectopically located hearts	Mishina et al. (1995)
	Smad5	Ectopically located hearts	Chang et al. (1999)
	Tdgf1	Some mesoderm formed but no cardiac markers	Xu et al. (1999)
Cell migration	Fgf8	Compound heterozygotes of a hypomorphic allele over a null allele show poorly developed heart tubes	Meyers et al. (1998)
	Mesp1	Cardiobifida	Saga et al. (1999)
	Fn1	In a subset of embryos on 129Sv background, cardiac cells differentiate but do not move from the cardiac crescent	George et al. (1997)
	Gata4	Cardiobifida due to defects in endoderm	Kuo et al. (1997); Molkentin et al. (1997); Narita et al. (1997)
Myocardial growth	Rxra	Thin-walled myocardium and a constellation of septal and conotruncal abnormalities; precocious differentiation of myocardium; diminishment of energy metabolism; defects potentially extrinsic to the heart	Chen et al. (1998); Gruber et al. (1996); Kastner et al. (1997); Ruiz-Lozano et al. (1998); Subbarayan et al. (2000); Sucov et al. (1994)
	Il6st	Hypoplastic ventricle; septal and trabecular defects	Yoshida et al. (1996)
	Adrbk1	Thin-walled myocardium	Jaber et al. (1996)
	Nf1	Thin-walled myocardium	Brannan et al. (1994)
	Nmyc1	Lack of expansion of the ventricular compact layer	Moens et al. (1993)
	Tead1	Thin-walled myocardium; reduced trabeculae	Chen et al. (1994)
	Wt1	Hypoplastic left ventricle; thin-walled myocardium; potentially secondary to defects in epicardium	Kreidberg et al. (1993); Moore et al. (1999)
	Zfpm2	Cardiac defects resembling tetralogy of Fallot or triscupid atresia syndrome with epicardial abnormalities	Svensson et al. (2000); Tevosian et al. (2000)
	Aldh1a7	Unlooped hearts with prematurely differentiated myocytes	Niederreither et al. (2001)
	Gata4	Mutation in FOG2 binding site leads to phenotype similar to that of Zfpm2	Crispino et al. (2001)
Cardiac myogenesis	Mef2c	Disrupted heart tube morphogenesis; myogenenic genes Myhca, Myla, and Actc downregulated	Kin et al. (1997)
	Irx4	Inappropriate activation of atrial genes in the ventricle; downregulation of the Hand1 gene	Bruneau et al. (2001)
RA and patterning of the inflow region	Aldh1a7	Unlooped heart tube; indistinct and hypoplastic sinoatrial region	Niederreither et al. (2001)
	Nr2f2	Unlooped heart tubes with hypoplastic sinoatrial region	Pereira et al. (1999)
Notch signaling in the heart	Psen1/2	Small unlooped heart tubes	Donoviel et al. (1999)
Cardiac chamber formation	Nkx2-5	Unlooped heart tubes; Irx4 and Mylpc expression downregulated; no markers of ventricular differentiation	Biben et al. (2000); Bruneau et al. (2000a); Lyons et al. (1995); Tanaka et al. (1999)
	Hand1	Unlooped heart tubes lacking left ventricle	Firulli et al. (1998); Riley et al. (1998)
	Hand2	Unlooped heart tubes lacking right ventricle; downregulation of Gata4	Srivastava et al. (1997)
	Nkx2-5/Hand2	Heart tubes lacking ventricular myocardium	Yamagishi et al. (2001)
	Mef2c	Heart tubes lack right ventricle	Lin et al. (1997)
	Cspg2	Heart tubes with underdeveloped right ventricle and conotruncus; endocardial cushions absent	Yamamura et al. (1997)
Cardiac LR axis	Invs	inv/inv strain; situs inversus	Mochizuki et al. (1998); Morgan et al. (1998)
	Dnahc11	iv/iv strain; heterotaxia; high frequency of cardiac abnormalities	Icardo and Sanchez de Vega (1991); Layton et al. (1980); Seo et al. (1992); Supp et al. (1997)
	Nodal	Compound heterozygotes between hypomorphic and null alleles show randomized and abnormal looping and structural abnormalities such as dextrocardia and mesocardia, transposition of great vessels and ventricular septal defects	Lowe et al. (2001)

(continues)

Table I *(continued)*

Process	Gene	Relevant mutant phenotype	Reference
	Pitx2	Abnormal heart loops; right ventricular hyperplasia, malpositioning of the ventricles, common atrioventricular junction, ventricular and atrial septal defects, and double outlet right ventricle	Gage *et al.* (1999); Kitamura *et al.* (1999); Lin *et al.* (1999); M. F. Lu *et al.* (1999)
	Acvr2b	Abnormal heart loops; dextrocardia and mesocardia, right atrial isomerism, transposition of the great arteries, and atrial and ventricular septal defects	Oh and Li *et al.* (1997)
	Cfc1	Randomization of heart looping; abnormal heart loops; dextrocardia and mesocardia, transposition of great arteries, right atrial isomerism and atrial septal defects	Gaio *et al.* (1999); Yan *et al.* (1999)
	Gdf1	Dextrocardia, transposition of great vessels, atrial and ventricular septal defects and persistant vena cava	Rankin *et al.* (2000)
	T	Midline looping hearts	King *et al.* (1998)
Congenital heart disease models	*Nkx2-5*	As in humans, heterozygotes show atrial septal defects and conduction abnormality	Biben *et al.* (2000)
	Tbx5	Heterozygotes show Holt-Oram syndrome-like defects	Bruneau *et al.* (2001b)
	Tbx1	Heterozygotes and homozygotes show velocardiofacial/DiGeorge syndrome-like defects	Jerome and Papaioannou (2001); Lindsay *et al.* (2001); Merscher *et al.* (2001)

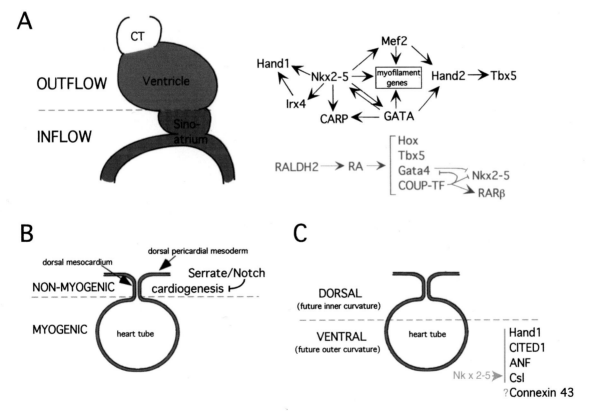

Figure 7 Genes and gene pathways active in vertebrate heart development. The figure depicts genes and/or genetic interactions relative to the anterior-posterior and dorsal-ventral axes of the developing heart, as suggested by a variety of studies in different vertebrate models (see text). (A) The AP axis of the heart indicating inflow/outflow regions (see Sections XVIII and XIX). For simplicity, this figure depicts interactions between transcription factor genes and/or their encoded proteins, although other key elements (RALDH2 and RA) are included. The RALDH2 pathway is in blue to indicate its restriction to the sinoatrial region. (B, C) Transverse sections of the heart tube indicating myogenic/nonmyogenic, and DV domains, respectively, and genes displaying regionally restricted expression and/or function in these domains. Note that most or all ventral markers in panel C require the homeodomain factor Nkx2-5 (in green to indicate it is not ventrally restricted) for proper expression. Abbreviation: CT, conotruncus.

Brown and Anderson, 1999; Campbell and Deuchar, 1966; Icardo and Sanchez de Vega, 1991; Layton *et al.*, 1980; Merklin and Verano, 1963; Min *et al.*, 2000; Seo *et al.*, 1992; Winer-Muram, 1995) (see Chapter by Hamada). Defects fall broadly into three categories: (1) *Situs inversus (totalis),* in which the normal arrangement of body organs (situs solitus) is totally reversed in a mirror-image fashion. In the case of the heart, looping is therefore leftward and the systemic and pulmonary ventricles have a reversed LR arrangement. This pattern is seen in inv/inv mice, which carry a mutation in the inversin gene (*Invs*) (Mochizuki *et al.,* 1998; Morgan *et al.,* 1998). (2) *Heterotaxia,* in which each organ apparently develops randomly with respect to situs and independently of other organs, a condition seen in several mouse strains, including the well-characterized iv/iv strain, mutant for the LR dynein gene (*Dnahc11*) (Supp *et al.,* 1997). The striking range of cardiac defects seen in this strain is evidence that in many cases, organ situs is incomplete or confused (Icardo and Sanchez de Vega, 1991; Layton *et al.,* 1980; Seo *et al.,* 1992). (3) *Isomerism,* in which organs that are normally LR asymmetric in their development show bilateral symmetry, either left or right in nature. In the heart, isomerism is manifested primarily in the atria, which can show bilateral left or right appendage morphology, as well as a tendency toward a predominantly bilateral left or right venous pattern (Min *et al.,* 2000). Heterotaxia and isomerism are associated with a high incidence of mortality and morbidity due to discordant cardiovascular development.

The heart is one of the recipients of LR information established earlier in development and is one site in which asymmetric signaling is converted into asymmetric morphogenesis. In recent years, dramatic advances have been seen in our understanding of the embryonic laterality pathway at the molecular level (Chapter 4) (Fig. 8A). Distinct pathways develop on both the left and right sides of the embryo: These are mutually antagonistic and culminate in expression of transcription factors in organ primordia (Burdine and Schier, 2000; Capdevila *et al.,* 2000). Chick and mouse homologs of the *Drosophila* gene Snail (*SnR* and *Sna,* respectively), which encode Zn^{2+} finger transcription factors, are expressed in heart cells and become enriched on the right side in a brief temporal window (Patel *et al.,* 1999; Sefton *et al.,* 1998). Conversely, genes for the TGF-β family signaling molecule, nodal, and the Rieger syndrome homeodomain transcription factor Pitx2, are expressed on the left side of the heart tube (Campione *et al.,* 1999; Collignon *et al.,* 1996; Gage *et al.,* 1999; Kitamura *et al.,* 1999; Levin *et al.,* 1995; Lin *et al.,* 1999; Logan *et al.,* 1998; Lowe *et al.,* 1996, 2001; M. F. Lu *et al.,* 1999; Piedra *et al.,* 1998; Ryan *et al.,* 1998; Yoshioka *et al.,* 1998). Bilateral expression of either nodal or Pitx2 in the heart region induces randomized looping, and in many cases bilaterally symmetrical hearts (Levin *et al.,* 1997; Logan *et al.,* 1998; Ryan *et al.,* 1998), indicating a key role in determination of heart situs. High-dose antisense oligonu-

cleotide inhibition of chick *SnR* leads to randomization of heart situs and isomerisms, as well as bilateral expression of *Pitx2,* indicating that one function of *SnR* is to inhibit *Pitx2* expression on the right (Patel *et al.,* 1999), although this may be indirect (Shiratori *et al.,* 2000). Nodal represses *SnR* on the left (Patel *et al.,* 1999), while simultaneously activating *Pitx2* (Shiratori *et al.,* 2000) (Fig. 8A).

What are the mechanisms of heart looping and what genes lie downstream of nodal, Pitx2, and Snail? Three separate inputs can now be considered to contribute to heart looping. First, looping is, to some extent, a mechanical imperative, due to elongation of the heart tube within a confined space and influences from embryonic torsion and flexure (Patten, 1922). Second, looping is driven in part by increased proliferation and perhaps changes in cell shape at the outer curvature (Manasek, 1976; Sissman, 1966), a product of specification of ventricular ballooning regions (see Section

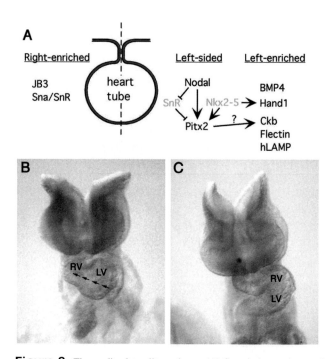

Figure 8 The cardiac laterality pathway. (A) Genetic interactions and gene expression on the left and right side of the developing heart. SnR and Nkx2-5 are in green to indicate that they are not left sided. (B, C) Wild-type embryo and an embryo mutant for the cryptic gene (*Cfc1*), respectively, at E8.5. Arrows in panel B represent the approximate point of fusion between right and left cardiac progenitors and highlight the in-line relationship between presumptive left and right ventricles at this stage. While the heart loop shown in panel B may appear superficially to be reversed, it is clearly not a mirror-image reversal of the normal pattern, as expected in situs inversus. The loop is C-shaped and the presumptive left ventricle is displaced leftward, as in normal hearts. Note that presumptive left and right ventricles remain in an AP arrangement. This heart phenotype appears typical in mutant models in which the flow of laterality information to the heart has been interrupted and correlates with a defined spectrum of defects in more mature hearts (see Section XXIV). Abbreviations: LV, left ventricle; RV, right ventricle. (Reproduced and adapted from Yan *et al.,* 1999, with permission from Cold Spring Harbor Laboratory Press.)

XXI above). Third, input from the LR pathway refines the shape and sets the direction of the loop (see below).

Numerous cellular mechanisms have been proposed to influence or direct the LR aspect of the looping process, including extrinsic factors and hemodynamics, as well as asymmetric proliferation, changes in cell shape, adhesion, matrix rigidity, contractility, and stress-strain relationships (de la Cruz, 1998; Harvey, 1998a; Itasaki *et al.*, 1991; Manasek, 1983; Manasek *et al.*, 1984; Stalsberg, 1970). While some of these mechanisms have lost favor (de la Cruz and Sanchez-Gomez, 1998), we still know very little about the real events underpinning looping. The issue of asymmetric proliferation, while attractive, remains unresolved. *Pitx2* knock-out embryos, which fail to turn, show an increase in *Pitx2-LacZ*-positive cells in the left caudal lateral plate mesoderm compared to heterozygotes (Kitamura *et al.*, 1999), suggesting that the *Pitx2* represses cell proliferation on the left. However, this does not fit with findings that proliferation is greatest in the left body wall normally, a situation proposed to drive the turning process (Miller and White, 1998). Furthermore, *Pitx2* is expressed on the outer curvature of the stomach, where proliferation would be expected to be greatest, but at the inner curvature of the heart, where it is least (Campione *et al.*, 1999; Sissman, 1966). No differences in the rate of proliferation were found between left and right heart primordia of chick embryos, despite asymmetric LR deployment of mesoderm to the heart and the fact that left and right bifid hearts differ in size (Lepori, 1967; Stalsberg, 1969a). Asymmetric proliferation has been found, however, in the caudal region of mouse hearts, highest on the left (N. Brown and D. Belomo, personal communication; see Harvey, 1998a).

Low-dose antisense inhibition of chick SnR does not alter left-sided *Pitx2* expression, yet still leads to heart inversions, demonstrating that the function of the right-sided pathway goes beyond inhibition of *Pitx2* and/or other left-sided genes (Isaac *et al.*, 1997; Patel *et al.*, 1999). Snail-related proteins are potent repressors, and mouse Sna has been shown to repress, in addition to *Pitx2*, the E-cadherin gene (*Cdh1*), and in doing so promotes epithelial-mesenchymal transition (Cano *et al.*, 2000). Thus, Snail proteins may transiently modulate cell adhesion on the right during heart looping. Disruption of actin filaments on the right side of the chick heart inhibits looping (Itasaki *et al.*, 1991).

Knock-out mice for *Nodal* and *Pitx2* have been reported. Complete deletion of *Nodal* is embryonic lethal around gastrulation due to defects in mesendoderm formation (Collignon *et al.*, 1996; Conlon *et al.*, 1994; Zhou *et al.*, 1993). However, a proportion of embryos that are compound heterozygous for a floxed *Nodal* allele, which shows lower than normal expression, and a null allele, gastrulate normally (Lowe *et al.*, 2001). These embryos completely lack expression of *Nodal*, *Pitx2*, and another left-sided marker, *leftb* (encoding lefty-2), in left lateral plate mesoderm and heart, and

are therefore deficient in signaling to the left side. In these embryos, the direction of looping is randomized, and more mature hearts have severe structural abnormalities (Lowe *et al.*, 2001). *Pitx2* mutants also show severe and complex cardiovascular abnormalities, similar to those conferred by *Nodal* mutation (Gage *et al.*, 1999; Kitamura *et al.*, 1999; Lin *et al.*, 1999; M. F. Lu *et al.*, 1999). These include right ventricular hypoplasia, malpositioning of the ventricles, common atrioventricular junction and valve, ventricular septal defect, common atrium, and double outlet right ventricle. However, loss of *Pitx2* does not affect the direction of heart looping, suggesting that the functions of *Nodal* are not all channeled through *Pitx2*. This is supported by uncoupling of the handedness of *Pitx2* expression and the direction of looping in the iv/iv and N-cadherin mutant strains (Campione *et al.*, 2001; Garcia-Castro *et al.*, 2000).

Several genes are transcribed at higher levels on the left or right sides of the heart and may act downstream of Snail, nodal, and/or Pitx2. In both zebrafish and frog hearts, BMP4 is transiently expressed or upregulated along the left side (Breckenridge *et al.*, 2001; Chen *et al.*, 1997). Left-sided BMP expression is downstream of nodal and sonic hedgehog, conserved molecules in the left-sided pathway (Burdine and Schier, 2000). Bilateral expression of BMP4 randomizes looping (Breckenridge *et al.*, 2001), while inhibition of BMP signaling by enforced expression of the BMP antagonist noggin or a dominant-negative BMP4 receptor completely blocks looping (Breckenridge *et al.*, 2001; Chen *et al.*, 1997). Thus, BMP appears to be a key cardiac-intrinsic mediator of looping in these species. In the mouse, the bHLH transcription factor gene *Hand1* is expressed bilaterally in the ventral and caudal part of the heart tube (see Sections XXI–XXIII), but has also been observed to be upregulated on the left side during early heart looping (Biben and Harvey, 1997; Chazaud *et al.*, 1999) and in the anterior portion of the left ventricular wall in well-looped hearts (Riley *et al.*, 2000), a region corresponding to the original left side (Campione *et al.*, 2001).

Expression of the frog *Hand1* gene has also been observed to be bilateral as well as left side enriched in a complex pattern (Mohun *et al.*, 2000; Sparrow *et al.*, 1998). Hand1 is also responsive to BMP signaling (Sparrow *et al.*, 1998). Although left-enhanced expression of *Hand1* has not been uniformly observed (Angelo *et al.*, 2000; Thomas *et al.*, 1998b), chimera studies in mice in which ES cells null for *Hand1* were injected into normal blastocysts show exclusion of mutant cardiocytes from the left side of the linear heart tube and the anterior (originally left-sided) part of the left ventricle in the looped heart (Riley *et al.*, 1999), supporting the idea that *Hand1* is necessary for interpretation of LR information. Since the sorting out of wild-type and mutant cells in chimeras implies changes in cell adhesion, transient and spatially restricted disruption of adhesion may underpin the looping process. Cell matrix molecules have also been observed to be transiently and asymmetrically expressed in chick heart cells, JB3 on the right, and flectin and hLAMP on the left (Smith *et al.*, 1997; Tsuda *et al.*, 1996). In human fetal hearts, the B isoform of creatine kinase is expressed at higher levels in descendants of the left cardiac progenitors, including the left atrial appendage and septum primum (Wessels *et al.*, 2000). This pattern is very similar to that of *Pitx2* (Campione *et al.*, 2001; Franco *et al.*, 2000), suggesting that the creatine kinase B gene (*Ckb*) is a direct Pitx2 target gene.

The connections between the above genes and expression of *SnR/Sna, Nodal,* and *Pitx2* remain to be determined. It is noteworthy, however, that correct expression of some laterality genes in the heart requires the homeodomain factor Nkx2-5 (Fig. 8A), which is not expressed in an asymmetric fashion (Lints *et al.*, 1993). *Nkx2-5* knock-out mice have hearts that show an initial caudal asymmetry but do not loop (Harvey *et al.*, 1999; Lyons *et al.*, 1995), and *Hand1* expression is severely downregulated (Biben and Harvey, 1997). Furthermore, while induction of *Pitx2* in left lateral plate mesoderm occurs in response to nodal signaling via the Smad2/FAST2 transcription factor pathway (Nomura and Li, 1998; Saijoh *et al.*, 2000), maintenance of *Pitx2* after *Nodal* expression has declined requires a high-affinity Nkx2 homeodomain binding site in its *cis*-regulatory region (Shiratori *et al.*, 2000).

One final issue relates to the observation of randomized morphological asymmetry and organ situs in humans and animal models defective in laterality. Randomization has played a strong part in the formulation of theories designed to account for the formation and properties of the LR axis (Brown and Wolpert, 1990; Capdevila *et al.*, 2000; Corballis and Morgan, 1978; Levin and Mercola, 1998). Central to these theories is the notion that molecular and cellular asymmetries can be established intrinsically although randomly within embryonic structures, and that laterality provides a biasing mechanism, resulting in consistent handedness. It is now clear that randomized asymmetry is often the result of a randomized output of the LR axial system itself. In many experimental and genetic models of perturbed laterality, the downstream genes in the laterality pathway (*SnR/Sna, Nodal, lefty2,* and/or *Pitx2*) are expressed in a confused fashion, either on the left, right, both sides or not at, in varying combinations (Burdine and Schier, 2000; Garcia-Castro *et al.*, 2000; Isaac *et al.*, 1997). This also occurs spontaneously at a low frequency in normal colonies of *Xenopus laevis* (Lohr *et al.*, 1997). The randomization of organs situs that results is not, therefore, due to the fact that organ primordia are "agnostic" with regard to laterality information. However, it may now be possible to assess what happens if organs *are* agnostic.

In mouse embryos mutant for *T, Gdf1, Cfc1, Pitx2,* and *Acvr2b* (Gaio *et al.*, 1999; King *et al.*, 1998; Kitamura *et al.*, 1999; Oh and Li, 1997; Rankin *et al.*, 2000; Yan *et al.*, 1999),

as well as in embryos carrying compound *Nodal* alleles (Lowe *et al.*, 2001), the flow of laterality information, at least to the left, is blocked rather than randomized. In some cases the direction of looping has been observed to be random (King *et al.*, 1998; Lowe *et al.*, 2001; Yan *et al.*, 1999), while in others only to the right (Gaio *et al.*, 1999; Oh and Li, 1997). However, a consistent feature in these mutants is the absence of true situs inversus, and the presence of C-shaped heart loops that are highly abnormal and retain left and right ventricles in an AP arrangement (Fig. 8B). Some hearts in *T* mutants form midline or "ventral" loops, suggesting that the direction of looping in such hearts has not been determined. Left and right looping in this situation may be due solely to mechanical imperatives and growth at the outer curvature. The overlapping constellation of abnormalities reported in more mature hearts in these mutants is similar: rightward or midline heart apex, atrial and ventricular septal defects, double outlet right ventricle, transposition of the great arteries, common atrioventricular canal, right atrial isomerism, and persistence of the left inferior vena cava. While many questions remain, particularly relating to the extent and impact of right-sided signaling within these models, it is reasonable to conclude that the LR axial pathway is not just a biasing mechanism, but is indeed a dedicated architect of the cardiac loop. In its absence, looping is crude and defective, leading to severe heart defects incompatible with life.

XXV. Developmental Pathways and Congenital Heart Disease

Congenital heart defects (CHDs) in humans occur in nearly 1% of live births (Hoffman, 1995a). Since the first surgical interventions for patent ductus arteriosis and aortic coarctation, tremendous advances have been made in the diagnosis and treatment of congenital heart malformations, and even highly complex abnormalities, such as tetralogy of Fallot, can be repaired with excellent medium-term results (Hirsch *et al.*, 2000). However, progress in understanding the genetic basis of CHD has been slow (Chien, 2000; Payne *et al.*, 1995). This may be due to many factors, not the least of which is the difficulty in linking complex cardiac phenotypes, often occurring in constellations, with pathogenetic and developmental mechanisms (Marino and Digilio, 2000).

Understanding the complex genetic basis of CHD is one of the greatest challenges in this field. To have any predictive power, we must not only fully understand conditions caused by single gene mutations that have a large phenotypic effect, but also genetic modifiers of those conditions, defects caused by multiple genes, and defects in which both genes and environment play a significant role. Furthermore, methodologies to predict epigenetic influences on congenital disease linked to the probabilistic nature of gene expression (Cook *et al.*, 1998; Fiering *et al.*, 2000) and/or influences of mosaically expressed retrotransposons (Whitelaw and Martin, 2001) remain to be established. As described in Section XII, a certain proportion of CHDs affecting ventricular growth may be secondary to placental insufficiency (Barak *et al.*, 1999; Wendling *et al.*, 1999), and establishing key genetic or clinical indicators of this class may lead to effective therapies based on growth factor or stem cell support (Tanaka *et al.*, 1998).

Although this review has focused on the genetic determinants of lineage specification and patterning in the myocardium, many CHDs will be due to abnormalities in heart tube remodeling and its inherent paracrine signaling pathways, such as those underpinning endocardial-myocardial interactions important for trabeculation and myocardial growth (Myer and Birchmeier, 1995), myocardial-endocardial interactions that induce cushion and valve formation (Mjaatvedt *et al.*, 1999), and neural crest-myocardial interactions involved in conotruncal and aortic arch development (Kirby, 1998).

Significant progress has been made in identifying single genes causative in CHD, and in establishing analogous mouse models. Autosomal dominant mutations in the human cognate of the mouse homeobox gene *Nkx2-5* underlie familial cases of atrial septal defect accompanied by progressive atrioventricular conduction block at high penetrance, and a variety of other abnormalities at lower penetrance including tetralogy of Fallot, Ebstein's anomaly, double outlet right ventricle, left ventricular hypertrophy, subvalvular aortic stenosis, and tricuspid valve abnormality (Benson *et al.*, 1998, 1999; Schott *et al.*, 1998). Mice heterozygous for *Nkx2-5* null mutations show ASD and conduction defects, although with much lower penetrance and expressivity than is seen in humans (Biben *et al.*, 2000). Strain-specific variance in the atrial septal phenotype holds promise for identification of genetic modifiers of CHD (Biben *et al.*, 2000). Mutations in the human *TBX-5* gene underlie the autosomal dominant Holt-Oram syndrome, which shows, in addition to forelimb abnormalities, atrial and ventricular septal defects and tetralogy of Fallot most commonly, but also conduction disease, hypoplastic left heart, mitral valve prolapse, and atrial isomerism (Basson *et al.*, 1997; Q. Y. Li *et al.*, 1997). *Tbx-5* knock-out mice show a similar phenotype in heterozygous form, and are lethal as homozygotes (Bruneau *et al.*, 2001b).

Human velocardiofacial/DiGeorge syndrome (VCFS/DGS) is characterized by a hemizygous 1.5- to 3.0-Mb deletion on chromosome 22q11, and a complex constellation of abnormalities with variable penetrance, including cardiac conotruncal and aortic arch defects, attributable to defective neural crest development (Scambler, 2000). A monumental effort involving gene knock-outs and chromosome engineering in mice has now implicated the T-box transcription factor gene *Tbx1* as a strong VCFS/DGS candidate (Jerome and Papaioannou, 2001; Lindsay *et al.*, 2001; Merscher *et al.*,

2001). *Tbx1* is expressed in pharyngeal arch mesenchyme before neural crest infiltration. Mutation of a second gene from the 22q11 deletion region, *Crkol,* implicated in growth factor and focal adhesion signaling and expressed in neural crest, also confers VCFS/DGS features, although not in a dose-sensitive manner (Guris *et al.*, 2001). The *Crkol* mutant phenotype, together with findings that *Tbx1* mutations were undetected in more than 100 patients with VCFS/DGS features but not carrying the large 22q11 deletion (Lindsay *et al.*, 2001), supports the idea that VCFS/DGS can be generated independently by mutation of distinct genes and/or by genetic interactions between multiple genes (Lindsay *et al.*, 2001; Scambler, 2000). Trisomy 16 mice have abnormalities that overlap with Down's syndrome (trisomy 22) and VCFS/DGS in humans (Waller *et al.*, 2000). Finally, specific gene mutations have been identified in humans with disturbed LR body and heart asymmetry. The *ZIC3* gene, encoding a zinc finger transcription factor and implicated in the LR pathway in *Xenopus* (Kitaguchi *et al.*, 2000), is mutated in humans with situs ambiguous (heterotaxia) and situs inversus (Gebbia *et al.*, 1997). A mutation in *NODAL,* implicated as a key determinant of the left side in mice and other vertebrates (Capdevila *et al.*, 2000; Collignon *et al.*, 1996; Lowe *et al.*, 2001), was also found in an affected individual in one *Zic3* family. Mutations have also been found in the EGF-CFC gene *CFC1* (encoding Cryptic), *LEFTY A,* and *ACVR2B* (encoding activin receptor IIB) in patients with disturbed laterality (Bramford *et al.*, 2000; K. Kosaki *et al.*, 1999; R. Kosaki *et al.*, 1999). These genes have been positioned in the laterality pathway at or upstream of the heart in mice (Meno *et al.*, 1998; Oh and Li, 1997; Yan *et al.*, 1999).

XXVI. Horizons

We currently have a rough draft of the genetic pathways underpinning heart development and CHD. However, the immediate future will see progress in this field on an unprecedented scale. The human and mouse genomes will be fully sequenced and annotated, and the cardiac "transcriptome" at embryonic and adult stages will be defined and available for expression profiling. Single nucleotide polymorphisms and genomic information will allow genes underpinning CHD to be defined with greater ease. In mice, saturation mutagenesis and high-throughput phenotypic screening will produce many new genes for analysis. Furthermore, rapid developments in noninvasive imaging technologies such as NMR and echocardiography, and invasive and noninvasive methods for analysis of cardiac function in embryos and adults that approach those available for humans, will revolutionise our capacity to examine the etiology, progression, and ultimate nature of CHD.

Combined with advancing genetic technologies for conditional expression and deletion of genes in mice, cardiac patterning and morphogenesis will be dissected in detail and links to CHD will become clearer. The use of *in vitro* stem cell-based systems for analysis of developmental pathways in the heart will see greater use and be linked to stem cell-based therapies for the failing heart. These tools will be most effectively deployed in dedicated cardiovascular research centers in which the void that still exists between clinical and basic scientists can be effectively bridged.

Acknowledgments

This work is dedicated to Benjamin Jack Harvey. I thank Nicola Groves, Mark Solloway, David Elliott and Christine Biben for help with graphics, and Rolf Bodmer, Pascal Dolle, Edwards Life Sciences, Jose Icardo, Robert Kelly, Antoon Moorman, Eric Olson, Nadia Rosenthal, and Michael Shen for supplying images.

References

Andree, B., Duprez, D., Vorbusch, B., Arnold, H.-H., and Brand, T. (1998). BMP-2 induces ectopic expression of cardiac lineage markers and interferes with somite formation in chicken embryos. *Mech. Dev.* **70**, 119–131.

Angelo, S., Lohr, J., Lee, K. H., Ticho, B. S., Breitbart, R. E., Hill, S., Yost, H. J., and Srivastava, D. (2000). Conservation of sequence and expression of Xenopus and zebrafish dHand during cardiac, branchial arch and lateral mesoderm development. *Mech. Dev.* **95**, 231–237.

Arai, A., Yamamoto, K., and Toyama, J. (1997). Murine cardiac progenitor cells require visceral embryonic endoderm and primitive streak for terminal differentiation. *Dev. Dyn.* **210**, 344–353.

Arkell, R. M. (1996). Functional analysis of TGF-beta related molecules during early mouse development. Ph.D. thesis, National Institute of Medical Research, UK.

Arsenian, S., Weinhold, B., Oelgeschlager, M., Ruther, U., and Nordheim, A. (1998). Serum response factor is essential for mesoderm formation during mouse embryogenesis. *EMBO J.* **17**, 6289–6299.

Auda-Boucher, G., Bernard, B., Fontaine-Perus, J., Rouaud, T., Mericksay, M., and Gardahaut, M.-F. (2000). Staging of the commitment of murine cardiac cell progenitors. *Dev. Biol.* **225**, 214–225.

Azpiazu, N., and Frasch, M. (1993). *Tinman* and *bagpipe:* Two homeo box genes that determine cell fates in the dorsal mesoderm of *Drosophila. Genes Dev.* **7**, 1325–1340.

Bao, Z.-Z., Bruneau, B. G., Seidman, J. G., Seidman, C. E., and Cepko, C. L. (1999). Regulation of chamber-specific gene expression in the developing heart by Irx4. *Science* **283**, 1161–1164.

Barak, Y., Nelson, M. C., Ong, E. S., Jones, Y. Z., Ruiz-Lozano, P., Chien, K. R., Koder, A., and Evans, R. M. (1999). PPARγ is required for placental, cardiac, and adipose tissue development. *Mol. Cell* **4**, 585–595.

Basson, C. T., Bachinsky, D. R., Lin, R. C., Levi, T., Elkins, J. A., Soults, J., Grayzel, D., Kroumpouzou, E., Traill, T. A., Leblanc-Straceski, J., Renault, B., Kucherlapati, R., Seidman, J. G., and Seidman, C. E. (1997). Mutations in human Tbx5 cause limb and cardiac malformation in Holt-Oram syndrome. *Nat. Genet.* **15**, 30–35.

Basson, C. T., Huang, T., Lin, R. C., Bachinsky, D. R., Weremowicz, S., Vaglio, A., Bruzzone, R., Quadrelli, R., Lerone, M., Romeo, G., Silengo, M., Pereira, A., Krieger, J., Mesquita, S. F., Kamisago, M., Morton, C. C., Pierpont, M. E. M., Muller, C. W., Seidman, J. G., and Seidman, C. E. (1999). Different TBX5 interactions in heart and limb defined by Holt-Oram syndrome mutations. *Proc. Natl. Acad. Sci. U.S.A.* **96**, 2919–2924.

Beddington, R. S. P., and Robertson, E. J. (1998). Anterior patterning in the mouse. *Trends Genet.* **14**, 177–284.

Beddington, R. S. P., and Robertson, E. J. (1999). Axis development and early asymmetry in mammals. *Cell (Cambridge, Mass.)* **96**, 195–209.

Beiman, M., Shilo, B.-Z., and Volk, T. (1996). Heartless, a *Drosophila* FGF receptor homologue, is essential for cell migration and establishment of several mesodermal lineages. *Genes Dev.* **10**, 2993–3002.

Belo, J. A., Bachiller, D., Agius, E., Kemp, C., Borges, A. C., Marques, S., Piccolo, S., and De Robertis, E. M. (2000). *Cerberus-like* is a secreted BMP and nodal antagonist not essential for mouse development. *Genesis* **26**, 265–270.

Benson, D. W., Sharkey, A., Fatkin, D., Lang, P., Basson, C. T., McDonough, B., Strauss, A. W., Seidman, J. G., and Seidman, C. E. (1998). Reduced penetrance, variable expressivity, and genetic heterogeneity of familial atrial septal defects. *Circulation* **97**, 2043–2048.

Benson, D. W., Silberbach, G. M., Kavanaugh-McHugh, A., Cottrill, C., Zhang, Y., Riggs, S., Smalls, O., Johnson, M. C., Watson, M. S., Seidman, J. G., Seidman, C. E., Plowden, J., and Kugler, J. D. (1999). Mutations in the cardiac transcription factor Nkx2.5 affect diverse cardiac developmental pathways. *J. Clin. Invest.* **104**, 1567–1573.

Biben, C., and Harvey, R. P. (1997). Homeodomain factor Nkx2-5 controls left-right asymmetric expression of bHLH *eHand* during murine heart development. *Genes Dev.* **11**, 1357–1369.

Biben, C., Palmer, D. A., Elliot, D. A., and Harvey, R. P. (1997). Homeobox genes and heart development. *Cold Spring Harbor Symp. Quant. Biol.* **63**, 395–403.

Biben, C., Hatzistavrou, T., and Harvey, R. P. (1998a). Expression of *NK-2* class homeobox gene *Nkx2-6* in foregut endoderm and heart. *Mech. Dev.* **73**, 125–127.

Biben, C., Stanley, E., Fabri, L., Kotecha, S., Rhinn, M., Drinkwater, C., Lah, M., Wang, C.-C., Nash, A., Hilton, D., Ang, S.-L., Mohun, T., and Harvey, R. P. (1998b). Murine cerberus homologue mCer-1: A candidate anterior patterning molecule. *Dev. Biol.* **194**, 135–151.

Biben, C., Weber, R., Kestevan, S., Stanley, E., McDonald, L., Elliott, D. A., Barnett, L., Koentgen, F., Robb, L., Feneley, M., and Harvey, R. P. (2000). Cardiac septal and valvular dysmorphogenesis in mice heterozygous for mutations in the homeobox gene Nkx2-5. *Circ. Res.* **87**, 888–895.

Bisaha, J. G., and Bader, D. (1991). Identification and characterisation of a ventricle-specific avian myosin heavy chain, VMHC1: Expression in differentiating cardiac and skeletal muscle. *Dev. Biol.* **148**, 355-3364.

Black, B. L., and Olson, E. N. (1998). Trancriptional control of muscle development by myocyte enhancer factor-2 (MEF2) Proteins. *Annu. Rev. Cell Dev. Biol.* **14**, 167–196.

Black, B. L., and Olson, E. N. (1999). Control of cardiac development by the MEF2 family of transcription factors. *In* "Heart Development" (R. P. Harvey and N. Rosenthal, eds.), pp. 131–142. Academic Press, San Diego, CA.

Blelloch, R., Newman, C., and Kimble, J. (1999). Control of cell migration during *Caenorhabditis elegans* development. *Curr. Opin. Cell Biol.* **11**, 608–613.

Bodmer, R. (1993). The gene *tinman* is required for specification of the heart and visceral muscles in *Drosophila. Development (Cambridge, UK)* **118**, 719–729.

Bodmer, R., and Frasch, M. (1999). Genetic determination of *Drosophila* heart development. *In* "Heart Development" (R. P. Harvey and N. Rosenthal, eds.), pp. 65–90. Academic Press, San Diego, CA.

Bodmer, R., Jan, L. Y., and Jan, Y. N. (1990). A new homeobox-containing gene, *msh-2,* is transiently expressed early during mesoderm formation in *Drosophila. Development (Cambridge, UK)* **110**, 661–669.

Bounpheng, M. A., Morrish, T. A., Dodds, S. G., and Christy, B. A. (2000). Negative regulation of selected bHLH proteins by eHand. *Exp. Cell Res.* **257**, 320–331.

Bour, B. A., O'Brien, M. A., Lockwood, W. L., Goldstein, E. S., Bodmer, R., Tagher, P. H., Abmayr, S. M., and Nguyen, H. T. (1995). Drosophila

MEF2, a transcription factor that is essential for myogenesis. *Genes Dev.* **9,** 730–741.

Bouwmeester, T., Kim, S.-H., Sasai, Y., Lu, B., and De Robertis, E. M. (1996). Cerberus is a head-inducing secreted factor expressed in the anterior endoderm of Spemann's organizer. *Nature (London)* **382,** 596–601.

Bowers, P. N., Brueckner, M., and Yost, H. J. (1996). Laterality disturbances. *Prog. Pediatr. Cardiol.* **6,** 53–62.

Bramford, R. N., Roessler, E., Burdine, R. D., Saplakoglu, U., de la Cruz, J., Splitt, M., Towbin, J., Bowers, P., Marino, B., Schier, A. F., Shen, M. M., Muenke, M., and Casey, B. (2000). Loss-of-function mutations in the EGF-CFC gene *CFC1* are associated with human left-right laterality defects. *Nat. Genet.* **26,** 365–369.

Brannan, C. I., Perkins, A. S., Vogel, K. S., Ratner, N., Nordlund, M. L., Reid, S. W., Buchberg, A. M., Jenkins, N. A., Parada, L. F., and Copeland, N. G. (1994). Targeted disruption of the neurofibromatosis type-1 gene leads to developmental abnormalities in heart and various neural crest-derived tissues. *Genes Dev.* **8,** 1019–1029.

Breckenridge, R. A., Mohun, T. J., and Amaya, E. (2001). A role for BMP signalling in heart looping morphogenesis in *Xenopus. Dev. Biol.* **232,** 191–203.

Brown, N., and Anderson, R. H. (1999). Symmetry and laterality in the human heart: Developmental implications. *In* "Heart Development" (R. P. Harvey and N. Rosenthal, eds.), pp. 447–461. Academic Press, San Diego, CA.

Brown, N. A., and Wolpert, L. (1990). The development of handedness in left/right asymmetry. *Development (Cambridge, UK)* **109,** 1–9.

Bruneau, B. G., Logan, M., Davis, N., Levi, T., Tabin, C. J., Seidman, J. G., and Seidman, C. E. (1999). Chamber-specific cardiac expression of *Tbx5* and heart defects in Holt-Oram syndrome. *Dev. Biol.* **211,** 100–108.

Bruneau, B. G., Bao, Z.-Z., Tanaka, M., Schott, J.-J., Izumo, S., Cepko, C. L., Seidman, J. G., and Seidman, C. E. (2000). Cardiac expression of the ventricle-specific homeobox gene *Irx4* is modulated by Nkx2-5 and dHand. *Dev. Biol.* **217,** 266–277.

Bruneau, B. G., Bao, Z.-Z., Fatkin, D., Xavier-Neto, J., Georgakopaoulis, D., Maguire, C. T., Berul, C. I., Kass, D. A., de Bold, M. L. K., de Bold, A. J., Conner, D. A., Rosenthal, N., Cepko, C. L., Seidman, C. E., and Seidman, J. G. (2001). Cardiomyopathy in *Irx4*-deficient mice is preceded by abnormal ventricular gene expression. *Mol. Cell. Biol.* **21,** 1730–1736.

Bruneau, B. G., Nemer, G., Schmitt, J. P., Charron, F., Robitaille, L., Caron, S., Conner, D. A., Gessler, M., Nemer, M., Seidman, C. E., and Seidman, J. G. (2001b). A Murine model of Holt-Oram syndrome defines roles of the T-Box transcription factor Tbx5 in cardiogenesis and disease. *Cell* **106,** 709–721.

Burdine, R. D., and Schier, A. F. (2000). Conserved and divergent mechanisms in left-right axis formation. *Genes Dev.* **14,** 763–776.

Burton, P. B. J., Raff, M. C., Kerr, P., Yacoub, M. H., and Barton, P. J. R. (1999). An intrinsic timer that controls cell-cycle withdrawal in cultured cardiac myocytes. *Dev. Biol.* **216,** 659–670.

Campbell, M., and Deuchar, D. C. (1966). Dextrocardia and isolated laevocardia II. Situs inversus and isolated dextrocardia. *Br. Heart J.* **28,** 472–478.

Campione, M., Steinbeisser, H., Schweickert, A., Deissler, K., van Bebber, F., Lowe, L. A., Nowotschin, S., Viebahn, C., Haffter, P., Kuehn, M. R., and Blum, M. (1999). The homeobox gene *Pitx2*: Mediator of asymmetric left-right signalling in vertebrate heart and gut looping. *Development (Cambridge, UK)* **126,** 1225–1234.

Campione, M., Ros, M. A., Icardo, J. M., Piedra, E., Christoffels, V. M., Schweickert, A., Blum, M., Franco, D., and Moorman, A. F. M. (2001). *Pitx2* expression defines a left cardiac lineage of cell: Evidence for atrial and ventricular molecular isomerism. *Dev. Biol.* **231,** 252–264.

Cano, A., Perez-Moreno, M. A., Rodrigo, I., Locascio, A., Blanco, M. J., del Barrio, M. G., Portillo, F., and Nieto, M. A. (2000). The transcription factor Snail controls epithelial-mesenchymal transitions by repressing E-cadherin expression. *Nat. Cell Biol.* **2,** 76–83.

Capdevila, J., Vogan, K. J., Tabin, C. J., and Izpisua Belmonte, J.-C. (2000). Mechansims of left-right determination in vertebrates. *Cell (Cambridge, Mass.)* **101,** 9–21.

Chang, H., Huylebroeck, D., Verschueren, K., Guo, Q., Matzuk, M. M., and Zwijsen, A. (1999). Smad5 knockout mice die at mid-gestation due to multiple embryonic and extraembryonic defects. *Development (Cambridge, UK)* **126,** 1631–1642.

Charite, J., McFadden, D. G., and Olson, E. N. (2000). The bHLH transcription factor dHand controls *Sonic hedgehog* expression and establishment of the zone of polarizing activity during limb development. *Development (Cambridge, UK)* **127,** 2461–2470.

Charron, F., and Nemer, M. (1999). GATA transcription factors and cardiac development. *Semin. Cell Dev. Biol.* **10,** 85–91.

Charron, F., Paradis, P., Bronchain, O., Nemer, G., and Nemer, M. (1999). Cooperative interaction between GATA-4 and GATA-6 regulates myocardial gene expression. *Mol. Cell. Biol.* **19,** 4355–4365.

Chazaud, C., Chambon, P., and Dollé, P. (1999). Retinoic acid is required in the mouse embryo for left-right asymmetry determination and heart morphogenesis. *Development (Cambridge, UK)* **126,** 2589–2596.

Chen, C. Y., Croissant, J., Majesky, M., Topouzis, S., McQuinn, T., Frankovsky, M. J., and Schwartz, R. J. (1996). Activation of the cardiac alpha-actin promoter depends upon serum response factor, Tinman homologue, Nkx-2.5, and intact serum response elements. *Dev. Genet.* **19,** 119–130.

Chen, J., Kubalak, S. W., and Chien, K. R. (1998). Ventricular muscle-restricted targeting of the RXRα gene reveals a noncell autonomous requirement in cardiac chamber morphogenesis. *Development (Cambridge, UK)* **125,** 1943–1949.

Chen, J.-N., van Eeden, J. M., Warren, K. S., Chin, A., Nusslein-Volhard, C., Haffter, P., and Fishman, M. C. (1997). Left-right pattern of cardiac *BMP4* may drive asymmetry of the heart in zebrafish. *Development (Cambridge, UK)* **124,** 4373–4382.

Chen, S. L., Dowhan, D. H., Hosking, B. M., and Muscat, G. E. O. (2000). The steriod receptor coactivator, GRIP-1, is necessary for MEF-2C dependent gene expression and skeletal muscle differentiation. *Genes Dev.* **14,** 1209–1228.

Chen, Z., Freidrich, G. A., and Soriano, P. (1994). Transcriptional enhancer factor 1 disruption by a retroviral gene trap leads to heart defects and embryonic lethality in mice. *Genes Dev.* **8,** 2293–2301.

Chien, K. R. (2000). Genomic circuits and the integrative biology of cardiac diseases. *Nature (London)* **407,** 227–232.

Chin, M. T., Maemura, K., Fukumoto, S., Jain, M. K., Layne, M. D., Watanabe, M., Hseih, C.-M., and Lee, M.-E. (2000). Cardiovascular basic helix loop helix factor 1, a novel trancriptional repressor expressed preferentially in the developing and adult cadiovascular system. *J. Biol. Chem.* **275,** 6381–6387.

Christoffels, V. M., Habets, P. E. M., H., Franco, D., Campione, M., de Jong, F., Lamers, W. H., Bao, Z.-Z., Palmer, S., Biben, C., Harvey, R. P., and Moorman, A. F. M. (2000a). Chamber formation and morphogenesis in the developing mammalian heart. *Dev. Biol.* **223,** 266–278.

Christoffels, V. M., Keijser, A. G. M., Houweling, A. C., Clout, D. E. W., and Moorman, A. F. M. (2000b). Patterning the embryonic heart: Identification of five mouse *Iroquois* homeobox genes in the developing heart. *Dev. Biol.* **224,** 263–274.

Cleaver, O. B., Patterson, K. D., and Krieg, P. A. (1996). Overexpression of the *tinman*-related genes *XNkx-2.5* and *XNkx-2.3* in *Xenopus* embryos results in myocardial hyperplasia. *Development (Cambridge, UK)* **122,** 3549–3556.

Climent, S., Sarasa, M., Villar, J. M., and Murillo-Ferrol, N. L. (1995). Neurogenic cells inhibit the differentiation of cardiogenic cells. *Dev. Biol.* **171,** 130–148.

Collignon, J., Varlet, I., and Robertson, E. J. (1996). Relationship between asymmetric *nodal* expression and the direction of embryonic turning. *Nature (London)* **381,** 155–158.

Conlon, F. L., Lyons, K. M., Takaesu, N., Barth, K. S., and Kispert, A.

(1994). A primary requirement for *nodal* in the formation and maintenance of the primitive streak in the mouse. *Development (Cambridge, UK)* **120**, 1919–1928.

Cook, D. L., Gerber, A. N., and Tapscott, S. J. (1998). Modeling stochastic gene expression: Implications for haploinsufficiency. *Proc. Natl. Acad. Sci. U.S.A.* **95**, 15641–15646.

Corballis, M. C., and Morgan, M. J. (1978). On the biological basis of human laterality. Evidence of a maturational left to right gradient. *Behav. Brain Sci.* **2**, 261–269.

Crispino, J. D., Lodish, M. B., Thurberg, B. L., Litovsky, S. H., Collins, T., Molkentin, J. D., and Orkin, S. H. (2001). Proper coronary vascular development and heart morphogenesis depend on interaction of GATA-4 with FOG cofacts. *Genes Dev.* **15**, 839–844.

Croissant, J. D., Kim, J.-H., Eichele, G., Goering, L., Lough, J., Prywes, R., and Schwartz, R. J. (1996). Avian serum response factor expression restricted primarily to musclel cell lineages is required for α-*Actin* gene transcription. *Dev. Biol.* **177**, 250–264.

Cross, J. C., Flannery, M. L., Blanar, M. A., Steingrimsson, E., Jenkins, N. A., Copeland, N. G., Rutter, W. J., and Werb, Z. (1995). *Htx* encodes a basic helix-loop-helix transcription factor that regulates trophoblast cell development. *Development (Cambridge, UK)* **121**, 2513–2523.

Crossley, P. H., and Martin, G. R. (1995). The mouse *Fgf8* gene encodes a family of polypeptides and is expressed in regions that direct outgrowth and patterning in the mouse embryo. *Development (Cambridge, UK)* **121**, 439–451.

Cserjesi, P., and Olson, E. N. (1991). Myogenin induces the myocyte-specific enhancer binding factor MEF-2 independently of other muscle-specific gene products. *Mol. Cell. Biol.* **11**, 4854–4862.

Cserjesi, P., Brown, B., Lyons, G. E., and Olson, E. N. (1995). Expression of the novel basic helix-loop-helix gene *eHAND* in neural crest derivatives and extraembryonic membranes during mouse development. *Dev. Biol.* **170**, 664–678.

Cubadda, Y., Heitzler, P., Ray, R. P., Bourouis, M., Ramain, P., Gelbart, W., Simpson, P., and Haenlin, M. (1999). *u-shaped* encodes zinc finger protein that regulates the proneural genes achaete and scute during the formation of bristles in *Drosophila*. *Genes Dev.* **11**, 3083–3095.

Davis, C. L. (1927). Development of the human heart from its first appearance to the stage found in embryos of twenty paired somites. *Contrib. Embryol.* **19**, 245–284.

Davis, D. L., Wessels, A., and Burch, J. B. E. (2000). An Nkx-dependent enhancer regulates cGATA-6 gene expression during early stages of heart development. *Dev. Biol.* **217**, 310–322.

DeHaan, R. L. (1963). Organization of the cardiogenic plate in the early chick embryo. *Acta Embryol. Morphol. Exp.* **6**, 26–38.

de Jong, F., Opthof, T., Wilde, A. A. M., Janse, M. J., Charles, R., Lamers, W. H., and Moorman, A. F. M. (1992). Persisting zones of slow impulse conduction in developing chicken hearts. *Circ. Res.* **71**, 240–250.

de Jong, F., Viragh, S., and Moorman, A. F. M. (1997). Cardiac development: A morphologically integrated molecular approach. *Cardiol. Young* **7**, 131–146.

de la Cruz, M. V. (1998). Torsion and looping of the cardiac tube and primitive cardiac segments. Anatomical manifestations. *In* "Living Morphogenesis of the Heart" (M. V. de la Cruz and R. R. Markwald, eds.), pp. 99–119. Birkhaeuser, Berlin.

de la Cruz, M. V., and Sanchez-Gomez, C. (1998). Straight tube heart. Primitive cardiac cavities vs. primitive cardiac segments. *In* "Living Morphogenesis of the Heart" (M. V. de la Cruz and R. R. Markwald, eds.), pp. 85–98. Birkhaeuser, Berlin.

de la Cruz, M. V., Sanchez-Gomez, C., and Palomino, M. A. (1989). The primitive cardiac regions in the straight tube heart (Stage 9⁻) and their anatomical expression in the mature heart: An experimental study in the chick embryo. *J. Anat.* **165**, 121–131.

de la Cruz, M. V., Castillo, M. M., Villavicencio, L. G., Valencia, A., and Moreno-Rodriguez, R. A. (1997). Primitive interventricular septum, its

primordium, and its contribution in the definitive interventricular septum: In vitro labelling study in the chick embryo heart. *Anat. Rec.* **247**, 512–520.

Delorme, B., Dahl, E., Jarry-Guichard, T., Briand, J.-P., Willecke, K., Gros, D., and Theveniau-Ruissy, M. (1997). Expression pattern of connexin gene products at the early developmental stages of the mouse cardiovascular system. *Circ. Res.* **81**, 423–437.

DeRuiter, M. C., Poelmann, R. E., VanderPlas-de Vries, I., Mentink, M. M. T., and Gittenberger-de Groot, A. C. (1992). The development of the myocardium and endocardium in mouse embryos. *Anat. Embryol.* **185**, 461–473.

Dettman, R. W., Denetclaw, W., Jr., Ordahl, C. P., and Bristow, J. (1998). Common epicardial origin of coronary vascular smooth muscle, perivascular fibroblasts, and intermyocardial fibroblasts in the avian heart. *Dev. Biol.* **193**, 169–181.

Dichoso, D., Brodigan, T., Chwoe, K. Y., Lee, J. S., Llacer, R., Park, M., Corsi, A. K., Kostas, S. A., Fire, A., Ahnn, J., and Krause, M. (2000). The MADS-box factor CeMEF2 is not essential for *Caenorhabditis elegans* myogenesis and development. *Dev. Biol.* **223**, 431–440.

Ding, J., Yang, L., Yan, Y.-T., Chen, A., Desai, N., Wynshaw-Boris, A., and Shen, M. M. (1998). *Cripto* is required to correct orientation of the anterior-posterior axis in the mouse embryo. *Nature (London)* **395**, 702–707.

Dono, R., Scalera, L., Pacifico, F., Acampora, D., Persico, M. G., and Simeone, A. (1993). The murine cripto gene: Expression during mesoderm induction and early heart morphogenesis. *Development (Cambridge, UK)* **118**, 1157–1168.

Donoviel, D. B., Hadjantonakis, A.-K., Ikeda, M., Zheng, H., Hyslop, P. S. G., and Bernstein, A. (1999). Mice lacking both presenilin genes exhibit early embryonic patterning defects. *Genes Dev.* **13**, 2801–2810.

Drysdale, T. A., Patterson, K. D., Saha, M., and Krieg, P. A. (1997). Retinoic acid can block differentiation of the myocardium after heart specification. *Dev. Biol.* **188**, 205–215.

Dunwoodie, S. L., Rodriguez, T. A., and Beddington, R. S. P. (1998). Msg1 and mrg1, founding members of a gene family, show distinct patterns of gene expression during mouse embryogenesis. *Mech. Dev.* **72**, 27–40.

Durocher, D., Charron, F., Warren, R., Schwartz, R. J., and Nemer, M. (1997). The cardiac transcription factors Nkx2-5 and GATA-4 are mutual cofactors. *EMBO J.* **16**, 5687–5696.

Edmondson, D. G., Lyons, G. E., Martin, J. F., and Olson, E. N. (1994). Mef2 gene expression marks the cardiac and skeletal muscle lineages during mouse embryogenesis. *Development (Cambridge, UK)* **120**, 1251–1263.

Edwards, J. G., Bahl, J. J., Flink, I. L., Cheng, S. Y., and Morkin, E. (1994). Thyroid hormone influences beta myosin heavy chain (beta MHC) expression. *Biochem. Biophys. Res. Commun.* **199**, 1482–1488.

Ehrman, L. A., and Yutzey, K. (1999). Lack of regulation in the heart forming region of avian embryos. *Dev. Biol.* **207**, 163–175.

Eisenberg, C. A., and Eisenberg, L. M. (1999). WNT11 promotes cardiac tissue formation of early mesoderm. *Dev. Dynamics* **216**, 45–58.

Eisenberg, C. A., Gourdie, R. G., and Eisenberg, L. M. (1997). Wnt-11 is expressed in early avian mesoderm and required for the differentiation of the quail mesoderm cell line QCE-6. *Development (Cambridge, UK)* **124**, 525–536.

Eldadah, Z. A., Hamosh, A., Biery, N. J., Montgomery, R. A., Duke, M., Elkins, R., and Dietz, H. C. (2001). Familial tetralogy of fallot caused by mutation in the jagged1 gene. *Hum. Mol. Genet.* **10**, 163–169.

Evans, S. (1999). Vertebrate tinman homologues and cardiac differentiation. *Semin. Cell Dev. Biol.* **10**, 73–83.

Farrell, A. P. (1997). Evolution of cardiovascular systems: Insights into ontogeny. *In* "Development of Cardiovascular Systems: Molecules to Organisms" (W. W. Burggren and B. B. Keller, eds.), pp. 101–113. Cambridge University Press, Cambridge, UK.

Fernandez-Teran, M., Piedra, M. E., Kathiriya, I. S., Srivastava, D., Rodriguez-Rey, J. C., and Ros, M. A. (2000). Role of dHand in the anterior-posterior polarization of the limb bud: Implications for the Sonic hedgehog pathway. *Development (Cambridge, UK)* **127**, 2133–2142.

Fiering, S., Whitelaw, E., and Martin, D. I. K. (2000). To be or not to be active: The stochastic nature of enhancer action. *Bioessays* **22**, 381–387.

Firulli, A. B., and Olson, E. N. (1997). Modular regulation of muscle gene transcription: A mechanism for muscle cell diversity. *Trends Genet.* **13**, 364–369.

Firulli, A. B., McFadden, D. G., Lin, Q., Srivastava, D., and Olson, E. N. (1998). Heart and extra-embryonic mesoderm defects in mouse embryos lacking the bHLH transcription factor Hand1. *Nat. Genet.* **18**, 266–270.

Firulli, B. A., Hadzic, D. B., McDaid, J. R., and Firulli, A. B. (2000). The basic helix-loop-helix transcription factors *dHand* and *eHand* exhibit dimerisation characteristics that suggest complex regulation of function. *J. Biol. Chem.* **275**, 33567–33573.

Fisher, S. A., Langille, B. L., and Srivastava, D. (2000). Apoptosis during cardiovascular development. *Circ. Res.* **87**, 856–864.

Fishman, M. C., and Chien, K. R. (1997). Fashioning the vertebrate heart: Earliest embryonic decisions. *Development (Cambridge, UK)* **124**, 2099–2117.

Fishman, M. C., and Olson, E. N. (1997). Parsing the heart: Genetic modules for organ assembly. *Cell (Cambridge, Mass.)* **91**, 153–156.

Fossett, N., Zhang, Q., Gajewski, K., Choi, C. Y., Kim, Y., and Schultz, R. A. (2000). The multiple zinc-finger protein U-shaped functions in heart cell specification in the *Drosophila* embryo. *Proc. Natl. Acad. Sci. U.S.A.* **97**, 7348–7353.

Franco, D., Kelly, R., Lamers, W. H., Buckingham, M., and Moorman, A. F. M. (1997). Regionalized transcriptional domains of myosin light chain 3f transgenes in the embryonic mouse heart: Morphogenetic implications. *Dev. Biol.* **188**, 17–33.

Franco, D., Lamers, W. H. L., and Moorman, A. F. M. (1998). Patterns of gene expression in the developing heart: Towards a morphologically integrated transcriptional model. *Cardiovasc. Res.* **38**, 25–53.

Franco, D., Campione, M., Kelly, R., Zammit, P. S., Buckingham, M., Lamers, W. H., and Moorman, A. F. M. (2000). Multiple transcriptional domains, with distinct left and right components, in the atrial chambers of the developing heart. *Circ. Res.* **87**, 984–991.

Frasch, M. (1995). Induction of visceral and cardiac mesoderm by ectopic Dpp in the early Drosophila embryo. *Nature (London)* **374**, 464–467.

Fu, Y., and Izumo, S. (1995). Cardiac myogenesis: Overexpression of *XCsx2* or *XMEF2A* in whole *Xenopus* embryos induces the precocious expression of XMHCα gene. *Roux's Arch. Dev. Biol.* **205**, 198–202.

Fu, Y., Yan, W., Mohun, T. J., and Evans, S. M. (1998). Vertebrate *tinman* homologues *XNkx2-3* and *XNkx2-5* are required for heart formation in a functionally redundant manner. *Development (Cambridge, UK)* **125**, 4439–4449.

Gage, P. J., Suh, H., and Camper, S. (1999). Dosage requirements of *Pitx2* for development of multiple organs. *Development (Cambridge, UK)* **126**, 4643–4651.

Gaio, U., Schweickert, A., Fischer, A., Garratt, A. N., Muller, T., Ozcelik, C., Lankes, W., Strehle, M., Britsch, S., Blum, M., and Birchmeier, C. (1999). A role of the cryptic gene in the correct establishment of the left-right axis. *Curr. Biol.* **9**, 1339–1342.

Gajewski, K., Fossett, N., Molkentin, J. D., and Schulz, R. A. (1999). The zinc finger proteins pannier and GATA4 function as cardiogenic factors in drosophila. *Development (Cambridge, UK)* **126**, 5679–5688.

Gajewski, K., Choi, C. Y., Kim, Y., and Schulz, R. A. (2000). Genetically distinct cardial cells within the *Drosophila* heart. *Genesis* **28**, 36–43.

Garcia-Castro, M., Vielmetter, E., and Bronner-Fraser, M. (2000). N-cadherin, a cell adhesion molecule involved in establishment of embryonic left-right asymmetry. *Science* **288**, 1047–1051.

Garcia-Martinez, V., and Schoenwolf, G. C. (1993). Primitive-streak origin of the cardiovascular system in avian embryos. *Dev. Biol.* **159**, 706–719.

Gebbia, M., Ferrero, G. B., Pilia, G., Bassi, M. T., Aylsworth, A. S., Penman-Splitt, M., Bird, L. M., Bamforth, J. S., Burn, J., Schlessinger, D., Nelson, D. L., and Casey, B. (1997). X-linked *situs* abnormalities result from mutations in *ZIC3*. *Nat. Genet.* **17**, 305–308.

George, E. L., Baldwin, H. S., and Hynes, R. O. (1997). Fibronectins are essential for heart and blood vessel morphogenesis but are dispensable for initial specification of precursor cells. *Blood* **90**, 3073–3081.

Ghatpande, S., Ghatpande, A., Zile, M., and Evans, T. (2000). Anterior endoderm is sufficient to rescue foregut apoptosis and heart tube morphogenesis in an embryo lacking retinoic acid. *Dev. Biol.* **219**, 59–70.

Gisselbrecht, S., Skeath, J. B., Doe, C. Q., and Michelson, A. M. (1996). *Heartless* encodes a fibroblastic growth factor receptor (DFR1/DFGF-R2) involved in the directional migration of early mesodermal cells in the *Drosophila* embryo. *Genes Dev.* **10**, 3003–3017.

Glinka, A., Wu, W., Onichtchouk, D., Blumenstock, C., and Niehrs, C. (1997). Head induction by simultaneous repression of Bmp and Wnt signalling in *Xenopus*. *Nature (London)* **389**, 517–519.

Goldstein, A. M., and Fishman, M. C. (1998). Notochord regulates cardiac lineages in zebrafish development. *Dev. Biol.* **201**, 247–252.

Gonzalez-Sanchez, A., and Bader, D. (1990). *In vitro* analysis of cardiac progenitor cell differentiation. *Dev. Biol.* **139**, 197–209.

Gorczyka, M. G., Phyllis, R. W., and Budnik, V. (1994). The role of *tinman*, a mesodermal cell fate gene, in axon pathfinding during the development of the transverse nerve in *Drosophila*. *Development (Cambridge, UK)* **120**, 2143–2152.

Gordon-Thomson, C., and Fabian, B. C. (1994). Hypoblastic tissue and fibroblast growth factor induce blood tissue (haemoglobin) in the early chick embryo. *Development (Cambridge, UK)* **120**, 3571–3579.

Gossett, L. A., Kelvin, D. J., Sternberg, E. A., and Olson, E. N. (1989). A new myocyte-specific enhancer-binding factor that recognizes a conserved element associated with multiple muscle-specific genes. *Mol. Cell. Biol.* **9**, 5022–5033.

Gove, C., Walmsley, M., Nijjar, S., Bertwistle, D., Guille, M., Partington, G., Bomford, A., and Patient, R. (1997). Over-expression of GATA-6 in Xenopus embryos blocks differentiation of heart precursors. *EMBO J.* **16**, 355–368.

Grepin, C., Nemer, G., and Nemer, M. (1997). Enhanced cardiogenesis in embryonic stem cells overexpressing the GATA-4 transcription factor. *Development (Cambridge, UK)* **124**, 2387–2395.

Griffin, K. J. P., Stoller, J., Gibson, M., Chen, S., Yelon, D., Stainier, D. Y. R., and Kimelman, D. (2000). A conserved role for *H15*-related T-box transcription factors in zebrafish and *Drosophila* heart formation. *Dev. Biol.* **218**, 235–247.

Gritsman, K., Zhang, J., Cheng, C., Heckscher, E., Talbot, W. S., and Schier, A. F. (1999). The EGF-CFC protein one-eyed-pinhead is essential for nodal signaling. *Cell (Cambridge, Mass.)* **97**, 121–132.

Grow, M. W., and Krieg, P. A. (1998). Tinman function is essential for vertebrate heart development: Elimination of cardiac differentiation by dominant inhibitory mutants of the *tinman*-related genes, *XNkx2-3* and *XNkx2-5*. *Dev. Biol.* **204**, 187–196.

Gruber, P. J., Kubalak, S. W., and Chien, K. R. (1998). Downregulation of atrial markers during cardiac chamber morphogenesis is irreversible in murine embryos. *Development (Cambridge, UK)* **125**, 4427–4438.

Guo, L., Lynch, J., Nakamura, K., Fleigel, L., Kasahara, H., Izumo, S., Komuro, I., Agellon, L. B., and Michalak, M. (2001). COUP-TF1 antagonizes Nkx2.5-mediated activation of the calreticulin gene during cardiac development. *J. Biol. Chem.* **276**, 2797–2801.

Guris, D. L., Fantes, J., Tara, D., Druker, B. J., and Imamoto, A. (2001). Mice lacking the homologue of the human 22q11.2 gene *CRKL* phenocopy neurocristopathies of DiGeorge syndrome. *Nat. Genet.* **27**, 293–298.

Haenlin, M., Cubadda, Y., Blondeau, F., Heitzler, P., Lutz, Y., Simpson, P., and Ramain, P. (1997). Trancriptional activity of Pannier is regulated negatively by heterodimerisation of the GATA DNA-binding domain with a cofactor encoded by the *u-shaped* gene of Drosophila. *Genes Dev.* **11**, 3096–3108.

Halevy, O., Novitch, B. G., Spicer, D. B., Skapek, S. X., Rhee, J., Hannon, G. J., Beach, D., and Lassar, A. B. (1995). Correlation of terminal cell cycle arrest of skeletal muscle with induction of p21 by MyoD. *Science* **267**, 1018–1021.

Han, Y., Dennis, J. E., Cohen-Gould, L., Bader, D. M., and Fishman, D. A. (1992). Expression of sarcomeric myosin in the presumptive myocardium of chicken embryos occurs within six hours of myocyte commitment. *Dev. Dyn.* **193**, 257–265.

Harvey, R. P. (1996). *NK-2* homeobox genes and heart development. *Dev. Biol.* **178**, 203–216.

Harvey, R. P. (1998a). Cardiac looping: An uneasy deal with laterality. *Semin. Cell Dev. Biol.* **9**, 101–108.

Harvey, R. P. (1999). Seeking a regulatory roadmap for heart morphogenesis. *Semin. Cell Dev. Biol.* **10**, 99–107.

Harvey, R. P., Biben, C., and Elliot, D. A. (1999). Transcriptional control and pattern formation in the developing vertebrate heart: Studies on NK-2 class homeodomain factors. *In* "Heart Development" (R. P. Harvey and N. Rosenthal, eds.), pp. 111–129. Academic Press, San Diego, CA.

Hatcher, C. J., Kim, M.-S., Hah, C. S., Goldstein, M. M., Wong, B., Mikawa, T., and Basson, C. T. (2001). TBX5 transcription factor regulates cell proliferation during cardiogenesis. *Dev. Biol.* **230**, 177–188.

He, C.-Z., and Burch, J. B. E. (1997). The chicken GATA-6 locus contains multiple control regions that confer distinct patterns of heart region-specific expression in transgenic mouse embryos. *J. Biol. Chem.* **272**, 28550–28556.

Heikinheimo, M., Scandrett, J. M., and Wilson, D. B. (1994). Localization of transcription factor GATA-4 to regions of the mouse embryo involved in cardiac development. *Dev. Biol.* **164**, 361–373.

Henderson, D. J., and Copp, A. J. (1998). Versican expression is associated with chamber specification, septation, and valvulogenesis in the developing mouse heart. *Circ. Res.* **83**, 523–532.

Hill, C. S., Wynne, J., and Treisman, R. (1995). The Rho family GTPases RhoA, Rac1 and CDC42Hs regulate transcriptional activation by SRF. *Cell (Cambridge, Mass.)* **81**, 1159–1170.

Hirakow, R. (1985). The vertebrate heart in phylogenetic relation to the prochordates. *In* "Vertebrate Morphology" (H.-R. Duncher and G. Fleischer, eds.), Vol. 30. Fischer Verlag, Stuttgart.

Hirose, J., Kawashima, H., Yoshie, O., Tashiro, K., and Miyasaka, M. (2001). Versican interacts with chemokines and modulates cellular responses. *J. Biol. Chem.* **276**, 5228–5234.

Hirota, H., Chen, J., Betz, U. A., Rajewsky, K., Gu, Y., Ross, J. J., Muller, W., and Chien, K. R. (1999). Loss of a gp130 cardiac muscle cell survival pathway is a critical event in the onset of heart failure during biomechanical stress. *Cell (Cambridge, Mass.)* **97**, 189–198.

Hirsch, J. C., Mosca, R. S., and Bove, E. L. (2000). Complete repair of tetralogy of Fallot in the neonate: Results in the modern era. *Ann. Surg.* **232**, 508–514.

Hoffman, J. I. (1995a). Incidence of congenital heart disease: I. Postnatal incidence. *Pediatr. Cardiol.* **16**, 103–113.

Hoffman, J. I. (1995b). Incidence of congenital heart: II. Prenatal Incidence. *Pediatr. Cardiol.* **16**, 155–165.

Hollenberg, S. M., Sternglanz, R., Cheng, P. F., and Weintraub, H. (1995). Identification of a new family of tissue-specific basic helix-loop-helix proteins with a two hybrid system. *Mol. Cell. Biol.* **15**, 3813–3822.

Howard, M., Foster, D. N., and Cserjesi, P. (1999). Expression of *HAND* gene products may be sufficient for the differentiation of avian neural crest-derived cells into catecholaminergic neurons in culture. *Dev. Biol.* **215**, 62–77.

Hsu, D. R., Economides, A. N., Wang, X., Eimon, P. M., and Harland, R. M. (1998). The *Xenopus* dorsalizing factor Gremlin identifies a novel family of secreted proteins that antagonize BMP activities. *Mol. Cell* **1**, 673–696.

Huggins, G. S., Bacani, C. J., and Leiden, J. M. (2000). Friend of GATA 2 (Fog-2) interacts with coup-TF2 to repress transcription. *Circulation, Suppl. II* **102**, II-219.

Icardo, J. M. (1997). Morphogenesis of vertebrate hearts. *In* "Development of Cardiovascular Systems: Molecules to Organisms" (W. W. Burggren and B. B. Keller, eds.), pp. 114–126. Cambridge University Press, Cambridge, UK.

Icardo, J. M., and Sanchez de Vega, M. J. (1991). Spectrum of heart malformations in mice with *Situs solitus, Situs inversus,* and associated visceral heterotaxia. *Circulation* **84**, 2547–2558.

Inagaki, T., Garcia-Martinez, V., and Schoenwolf, G. C. (1993). Regulative ability of the prospective cardiogenic and vasculogenic areas of the primitive streak during avian gastrulation. *Dev. Dyn.* **197**, 57–68.

Isaac, A., Sargent, M. G., and Cooke, J. (1997). Control of vertebrate left-right asymmetry by a *Snail*-related zinc finger gene. *Science* **275**, 1301–1304.

Itasaki, N., Nakamura, H., Sumida, H., and Yasuda, M. (1991). Actin bundles on the right side in the caudal part of the heart tube play a role in dextro-looping in the embryonic chick heart. *Anat. Embryol.* **183**, 29–39.

Jaber, M., Koch, W. J., Rockman, H., Smith, B., Bond, R. A., Sulik, K. K., Ross, J., Lefkowitz, R. J., Caron, M. G., and Giros, R. (1996). Essential role of beta-adrenergic receptor kinase 1 in cardiac development and function. *Proc. Natl. Acad. Sci. U.S.A.* **93**, 12974–12979.

Jacobson, A. G., and Sater, A. K. (1988). Features of embryonic induction. *Development (Cambridge, UK)* **104**, 341–359.

Jerome, L. A., and Papaioannou, V. E. (2001). DiGeorge syndrome phenotype in mice mutant for the T-box gene, *Tbx1. Nat. Genet.* **27**, 286–291.

Jiang, Y., Trazami, S., Burch, J. B. E., and Evans, T. (1998). Common role for each of the cGATA-4/5/6 genes in the regulation of cardiac morphogenesis. *Dev. Genet.* **22**, 263–277.

Jiang, Y., Drysdale, T. A., and Evans, T. (1999). A role for GATA-4/5/6 in the regulation of Nkx2.5 expression with implications for patterning of the precardiac field. *Dev. Biol.* **216**, 57–71.

Jiang, X., Rowitch, D. H., Soriano, P., McMahon, A. P., and Sucov, H. M. (2000). Fate of the mammalian cardiac neural crest. *Development (Cambridge, UK)* **127**, 1607–1616.

Jones, M. C., Broadbent, J., Thomas, P. Q., Smith, J. C., and Beddington, R. S. P. (1999). An anterior signalling centre in *Xenopus* revealed by the homeobox gene *XHex. Curr. Biol.* **9**, 946–954.

Kamino, K. (1991). Optical approaches to ontogeny of electrical activity and related functional organization during early heart development. *Physiol. Rev.* **71**, 53–91.

Kao, K. R., and Elinson, R. P. (1988). The entire mesodermal mantle behaves as Spemann's organizer in dorsoanterior enhanced *Xenopus laevis* embryos. *Dev. Biol.* **127**, 64–77.

Kastner, P., Messaddeq, N., Mark, M., Wendling, O., Grondona, J. M., Ward, S., Ghyselinck, N., and Chambon, P. (1997). Vitamin A deficiency and mutations of RXRα, RXRβ and RARα lead to early differentiation of embryonic ventricular cardiomyocytes. *Development (Cambridge, UK)* **124**, 4749–4758.

Kaufman, M. H., and Navaratnam, V. (1981). Early differentiation of the heart in mouse embryos. *J. Anat.* **133**, 235–246.

Kawashima, H., Hirose, M., Hirose, J., Nagakubo, D., Plaas, A. H., and Miyasaka, M. (2000). Binding of a large chondroitin sulphate/dermatan sulphate proteoglycan, versican, to L-selectin, P-selectin, and CD44. *J. Biol. Chem.* **275**, 35448–35456.

Kelly, R. G., Zammit, P. S., Mouly, V., Butler-Browne, G., and Buckingham, M. E. (1998). Dynamic left/right regionalization of endogenous myosin light chain 3F transcripts in the developing mouse heart. *J. Mol. Cell. Cardiol.* **30**, 1067–1081.

Kelly, R. G., Franco, D., Moorman, A. F. M., and Buckingham, M. (1999). Regionalization of transcriptional potential in the myocardium. *In* "Heart Development" (R. P. Harvey and N. Rosenthal, eds.), pp. 333–355. Academic Press, San Diego, CA.

Kiecker, C., Muller, F., Wu, W., Glinka, A., Strahle, U., and Niehrs, C. (2000). Phenotypic effects in *Xenopus* and zebrafish suggest that *one-eyed pinhead* functions as antagonist of BMP signalling. *Mech. Dev.* **94,** 37–46.

Kilner, P. J., Yang, G.-Z., Wilkes, A. J., Mohiaddin, R. H., Firmin, D. N., and Yacoub, M. H. (2000). Asymmetric redirection of flow through the heart. *Nature (London)* **404,** 759–761.

Kim, J.-S., Viragh, S., Moorman, A. F. M., Anderson, R. H., and Lamers, W. H. (2001). Development of the myocardium of the atrioventricular canal and the vestibular spine in the human heart. *Circ. Res.* **88,** 395–402.

Kim, Y., and Nirenberg, M. (1989). *Drosophila* NK-homeobox genes. *Proc. Natl. Acad. Sci. U.S.A.* **86,** 7716–7720.

King, T., Beddington, R. S. P., and Brown, N. A. (1998). The role of the brachyury gene in heart development and left-right specification in the mouse. *Mech. Dev.* **79,** 29–37.

Kinoshita, N., Minshull, J., and Kirschner, M. W. (1995). The identification of two novel ligands of the FGF receptor by a yeast screening method and their activities in Xenopus development. *Cell (Cambridge, Mass.)* **83,** 621–630.

Kirby, M. L. (1998). Contribution of neural crest to heart and vessel morphology. *In* "Heart Development" (R. P. Harvey and N. Rosenthal, eds.), pp. 179–177. Academic Press, San Diego, CA.

Kishimoto, Y., Lee, K.-H., Zon, L., Hammerschmidt, M., and Schulte-Merker, S. (1997). The molecular nature of zebrafish *swirl*: BMP2 function is essential during early dorsoventral patterning. *Development (Cambridge, UK)* **124,** 4457–4466.

Kispert, A., Vainio, S., Shen, L., Rowitch, D. H., and McMahon, A. P. (1996). Proteoglycans are required for maintenance of Wnt-11 expression in the ureter tips. *Development (Cambridge, UK)* **122,** 3627–3637.

Kitaguchi, T., Nagai, T., Nakata, K., Aruga, J., and Mikoshiba, K. (2000). Zic3 is involved in the left-right specification of the Xenopus embryo. *Development (Cambridge, UK)* **127,** 4787–4795.

Kitajima, S., Takagi, A., Inoue, T., and Saga, Y. (2000). MesP1 and MesP2 are essential for the development of cardiac mesoderm. *Development (Cambridge, UK)* **127,** 3215–3226.

Kitamura, K., Miura, H., Miyagawa-Tomita, S., Yanazawa, M., Katoh-Fukui, Y., Suzuki, R., Ohuchi, H., Suehiro, A., Motegi, Y., Nakahara, Y., Kondo, S., and Yokayama, M. (1999). Mouse Pitx2 deficiency leads to anomalies of the ventral body wall, heart, extra- and periocular mesoderm and right pulmonary isomerism. *Development (Cambridge, UK)* **126,** 5749–5758.

Komuro, I., and Izumo, S. (1993). *Csx:* A murine homeobox-containing gene specifically expressed in the developing heart. *Proc. Natl. Acad. Sci. U.S.A.* **90,** 8145–8149.

Kosaki, K., Bassi, M. T., Kosaki, R., Lewin, M., Belmont, J., Schauer, G., and Casey, B. (1999). Characterisation and mutation analysis of human *LEFTY A* and *LEFTY B*, homologues of murine genes implicated in left-right axis development. *Am. J. Hum. Genet.* **64,** 712–721.

Kosaki, R., Gebbia, M., Kosaki, K., Lewin, M., Bowers, P., Towbin, J. A., and Casey, B. (1999). Left-right axis malformations associated with mutations in *ACVR2B*, the gene for human activin receptor type IIB. *Am. J. Med. Genet.* **82,** 70–76.

Kostetskii, I., Jiang, Y., Kostetskaia, E., Yuan, S., Evans, T., and Zile, M. (1999). Retinoid signalling required for normal heart development regulates GATA-4 in a pathway distinct from cardiomyocyte differentiation. *Dev. Biol.* **206,** 206–218.

Kreidberg, J. A., Sariola, H., Loring, J. M., Maeda, M., Pelletier, J., Housman, D., and Jaenisch, R. (1993). WT-1 is required for early kidney development. *Cell (Cambridge, Mass.)* **74,** 679–691.

Kremser, T., Gajewski, K., Schulz, R. A., and Renkawitz-Pohl, R. (1999). Tinman regulates the transcription of the *β3 tubulin* gene (*βTub60D*) in the dorsal vessel of *Drosophila*. *Dev. Biol.* **216,** 327–339.

Kubalak, S. W., and Sucov, H. M. (1999). Retinoids in heart development. *In* "Heart Development" (R. P. Harvey and N. Rosenthal, eds.), pp. 209–219. Academic Press, San Diego, CA.

Kuhl, M., Sheldahl, L. C., Park, M., Miller, J. R., and Moon, R. T. (2000). The Wnt/Ca^{2+} pathway. *Trends Genet.* **16,** 279–283.

Kumar, A., Novoselov, V., Celeste, A. J., Wolfman, N. M., ten Dijke, P., and Kuehn, M. R. (2001). Nodal signaling utilizes activin/TGF-β receptor regulated Smads. *J. Biol. Chem.* **276,** 656–661.

Kuo, C. T., Morrisey, E. E., Anandappa, R., Sigrist, K., Lu, M. M., Parmacek, M. S., Soudais, C., and Leiden, J. M. (1997). GATA4 transcription factor is required for ventral morphogenesis and heart tube formation. *Genes Dev.* **11,** 1048–1060.

Kuo, H.-C., Chen, J., Ruiz-Lozano, P., Zou, Y., Nemer, M., and Chien, K. R. (1999). Control of segmental expression of the cardiac-restricted ankyrin repeat protein gene by distinct regulatory pathways in murine cardiogenesis. *Development (Cambridge, UK)* **126,** 4223–4234.

Kupperman, E., An, S., Osborne, N., Waldron, S., and Stainier, D. Y. (2000). A sphingosine-1-phosphate receptor regulates cell migration during vertebrate development. *Nature (London)* **406,** 192–195.

Ladd, A. N., Yatskievych, T. A., and Antin, P. B. (1998). Regulation of avian cardiac myogenesis by activin/TGFβ and bone morphogenetic proteins. *Dev. Biol.* **204,** 407–419.

Lamers, W. H., Viragh, S., Wessels, A., Moorman, A. F. M., and Anderson, R. H. (1995). Formation of the tricuspid valve in the human heart. *Circulation* **91,** 111–121.

Lawson, K. A., and Pedersen, R. A. (1987). Cell fate, morphogenetic movement and population kinetics of embryonic endoderm at the time of germ layer formation in the mouse. *Development (Cambridge, UK)* **101,** 627–652.

Lawson, K. A., Meneses, J. J., and Pederson, R. A. (1991). Clonal analysis of epiblast during germ layer formation in the mouse embryo. *Development (Cambridge, UK)* **113,** 891–911.

Lawson, K. A., Dunn, N. R., Roelen, B. A. J., Zeinstra, L. M., Davis, A. M., Wright, C. V. E., Korving, J. P. W. F. M., and Hogan, B. L. M. (1999). Bmp4 is required for the generation of primordial germ cells in the mouse embryo. *Genes Dev.* **13,** 424–436.

Layton, W. M., Manasek, M. D., and Manasek, D. M. D. (1980). Cardiac looping in early *iv/iv* mouse embryos. *In* "Etiology and Morphogenesis of Congenital Heart Disease" (R. Van Pragh, ed.), pp. 109–126. Futura, Mount Kisco, NY.

Lee, H.-H., and Frasch, M. (2000). Wingless effects mesodermal patterning and ectoderm segmentation events via induction of its downstream target *sloppy paired*. *Development (Cambridge, UK)* **127,** 5497–5508.

Lee, R. Y., Jiangming, L., Evans, R. M., Giguère, V., and Sucov, H. M. (1997). Compartment-selective sensitivity of cardiovascular morphogenesis to combinations of retinoic acid receptor gene mutations. *Circ. Res.* **80,** 757–764.

Lee, Y., Shioi, T., Kasahara, H., Jobe, S. M., Wiese, R. J., Markham, B. E., and Izumo, S. (1998). The cardiac tissue-restricted homeobox protein Csx/Nkx2.5 physically associates with the zinc finger protein GATA4 and cooperatively activates atrial natriuretic factor gene expression. *Mol. Cell. Biol.* **18,** 3120–3129.

Leimeister, C., Externbrink, A., Klamt, B., and Gessler, M. (1999). Hey genes: A novel subfamily of *hairy-* and *Enhancer of split* related genes specifically expressed during mouse embryogenesis. *Mech. Dev.* **85,** 173–177.

Lembo, G., Rockman, H. A., J.J., H., Steinmetz, H., Koch, W. J., Ma, L., Prinz, M. P., Ross, J., Chien, K. R., and Powell-Braxton, L. (1996). Elevated blood pressure and enhanced myocardial contratility in mice with severe IGF-1 deficiency. *J. Clin. Invest.* **98,** 2648–2655.

Lepori, N. G. (1967). Research on heart development in chick embryo under normal and experimental conditions. *Monit. Zool. Ital.* **1,** 159–183.

Levin, M., and Mercola, M. (1998). The compulsion of chirality: Towards an understanding of left-right asymmetry. *Genes Dev.* **12,** 763–769.

Levin, M., Johnson, R. L., Stern, C. D., Kuehn, M., and Tabin, C. (1995). A molecular pathway determining left-right asymmetry in chick embryogenesis. *Cell (Cambridge, Mass.)* **82,** 803–814.

Levin, M., Pagan, S., Roberts, D. J., Cooke, J., Kuehn, M. R., and Tabin, C. J. (1997). Left/right patterning signals and the independent regulation of different aspects of *situs* in the chick embryo. *Dev. Biol.* **189,** 57–67.

Li, L., Krantz, I. D., Deng, Y., Genin, A., Banta, A. B., Collins, C. C., Qi, M., Trask, B. J., Kuo, W. L., Cochran, J., Costa, T., Pierpont, M. E. M., Rand, E. B., Piccoli, D. A., Hood, L., and Spinner, N. B. (1997). Alagille syndrome is caused by mutations in *Jagged1,* which encodes a ligand for Notch1. *Nat. Genet.* **16,** 243–250.

Li, Q., and Gardner, D. G. (1994). Negative regulation of the human atrial natriuretic peptide gene by 1,25-dihydroxyvitamin D3. *J. Biol. Chem.* **269,** 4934–4939.

Li, Q. Y., Newbury-Ecob, R. A., Terrett, J. A., Wilson, D. I., Curtis, A. R. J., Yi, C. H., Gebuhr, T., Bullen, P. J., Robson, S. C., Strachan, T., Bonnet, D., Lyonnet, S., Young, I. D., Raeburn, A., Buckler, A. J., Law, D. J., and Brook, J. D. (1997). Holt-Oram syndrome is caused by mutations in *TBX5,* a member of the Brachyury (T) gene family. *Nat. Genet.* **15,** 21–29.

Liang, J. O., Etheridge, A., Hantsoo, L., Rubinstein, A. L., Nowak, S. J., Izpisua Belmonte, J. C., and Halpern, M. E. (2000). Asymmetric nodal signalling in the zebrafish diencephalon positions the pineal organ. *Development (Cambridge, UK)* **127,** 5101–5112.

Liberatore, C. M., Searcy-Schrick, R. D., and Yutzey, K. E. (2000). Ventricular expression of *tbx5* inhibits normal heart chamber development. *Dev. Biol.* **223,** 169–180.

Lien, C.-L., Wu, C., Mercer, B., Webb, R., Richardson, J. A., and Olson, E. N. (1999). Control of early cardiac-specific transcription of *Nkx2-5* by a GATA-dependent enhancer. *Development (Cambridge, UK)* **126,** 75–84.

Lilly, B., Zhao, B., Ranganayakulu, R., Paterson, B. M., Schulz, R. A., and Olson, E. N. (1995). Requirement of MADS domain transcription factor D-MEF2 for muscle formation in Drosophila. *Science* **267,** 688–693.

Lin, B., Chen, G.-Q., Xiao, D., Kolluri, S. K., Cao, X., Su, H., and Zhang, X.-K. (2000). Orphan receptor COUP-TF is required for induction of retinoic acid receptor *β,* growth inhibition, and apoptosis by retinoic acid in cancer cells. *Mol. Cell. Biol.* **20,** 957–970.

Lin, C. R., Kioussi, C., O'Connel, S., Briata, P., Szeto, D., Liu, F., Izpisua-Belmonte, J. C., and Rosenfelt, M. G. (1999). *Pitx2* regulates lung asymmetry, cardiac positioning and pituitary and tooth morphogenesis. *Nature (London)* **401,** 279–282.

Lin, Q., Schwarz, J., Bucana, C., and Olson, E. (1997). Control of mouse cardiac morphogenesis and myogenesis by transcription factor MEF2C. *Science* **276,** 1404–1407.

Linask, K. K., and Lash, J. W. (1993). Early heart development: Dynamics of endocardial cell sorting suggests a common origin with cardiomyocytes. *Dev. Dyn.* **195,** 62–66.

Linask, K. K., and Lash, J. W. (1998). Morphoregulatory mechanisms underlying early heart development: Precardial stages to the looping, tubular heart. *In* "Living Morphogenesis of the Heart" (M. V. de la Cruz and R. R. Markwald, eds.), pp. 1–41. Birkhaeuser, Boston.

Linask, K. K., Knudsen, K. A., and Gui, Y.-H. (1997). N-cadherin-catenin interaction: Necessary component of cardiac cell compartmentalization during early vertebrate heart development. *Dev. Biol.* **185,** 148–164.

Lindsay, E. A., Vitelli, F., Su, H., Morishima, M., Huynh, T., Pramparo, T., Jurecic, V., Ogunrinu, G., Sutherland, H. F., Scambler, P. J., Bradley, A., and Baldini, A. (2001). *Tbx1* haploinsufficiency in the DiGeorge syndrome region causes aortic arch defects in mice. *Nature (London)* **410,** 97–101.

Lints, T. J., Parsons, L. M., Hartley, L., Lyons, I., and Harvey, R. P. (1993). *Nkx-2.5:* A novel murine homeobox gene expressed in early heart progenitor cells and their myogenic descendants. *Development (Cambridge, UK)* **119,** 419–431.

Litvin, J., Montgomery, M., Gonzalez-Sanchez, A., Bisaha, J. G., and Bader, D. (1992). Commitment and differentiation of cardiac myocytes. *Trends Cardiovasc. Med.* **2,** 27–32.

Lo, P. C., and Frasch, M. (2001). A role for the COUP-TF-related gene *seven-up* in the diversification of cardioblast identities in the dorsal vessel of *Drosophila. Mech. Dev.* **104,** 49–60.

Logan, M., and Mohun, T. (1993). Induction of cardiac muscle differentiation in isolated animal pole explants of *Xenopus laevis* embryos. *Development (Cambridge, UK)* **118,** 865–875.

Logan, M., Pagan-Westphal, S. M., Smith, D. M., Paganessi, L., and Tabin, C. J. (1998). The transcription factor *Ptx2* mediates situs-specific morphogenesis in response to left-right asymmetric signals. *Cell (Cambridge, Mass.)* (this issue).

Lohr, J. L., Danos, M. C., and Yost, H. J. (1997). Left-right asymmetry of a nodal-related gene is regulated by dorsoanterior midline structures during *Xenopus* development. *Development (Cambridge, UK)* **124,** 1465–1472.

Lough, J., and Sugi, Y. (2000). Endoderm and heart development. *Dev. Dyn.* **217,** 327–342.

Lough, J., Barron, M., Brogley, M., Sugi, Y., Bolender, D. L., and Zhu, X. (1996). Combined BMP-2 and FGF-4, but neither factor alone, induces cardiogenesis in non-precardiac embryonic mesoderm. *Dev. Biol.* **178,** 198–202.

Lowe, L. A., Supp, D. M., Sampath, K., Yokoyama, T., Wright, C. V. E., Potter, S. S., Overbeek, P., and Kuehn, M. R. (1996). Conserved left-right asymmetry of *nodal* expression and alterations in murine *situs inversus. Nature (London)* **381,** 158–161.

Lowe, L. A., Yamada, S., and Kuehn, M. R. (2001). Genetic dissection of nodal function in patterning the mouse embryo. *Development (Cambridge, UK)* **128,** 1831–1843.

Lu, J.-R., McKinsey, T. A., Xu, H., Wang, D.-Z., Richardson, J. A., and Olson, E. N. (1999). FOG-2, a heart- and brain-enriched cofactor for GATA transription factors. *Mol. Cell. Biol.* **19,** 4495–4502.

Lu, J.-R., McKinsey, T. A., Nicol, R. L., and Olson, E. N. (2000). Signal-dependent activation of the MEF2 transcription factor by dissociation from histone deacetylases. *Proc. Natl. Acad. Sci. U.S.A.* **97,** 4070–4075.

Lu, M. F., Pressman, C., Dyer, R., Johnson, R. L., and Martin, J. F. (1999). Function of Rieger syndrome gene in left-right asymmetry and craniofacial development. *Nature (London)* **401,** 276–278.

Lyons, G. E. (1994). In situ analysis of the cardiac muscle gene program during embryogenesis. *Trends Cardiovasc. Med.* **4,** 70–77.

Lyons, I., Parsons, L. M., Hartley, L., Li, R., Andrews, J. E., Robb, L., and Harvey, R. P. (1995). Myogenic and morphogenetic defects in the heart tubes of murine embryos lacking the homeobox gene *Nkx2-5. Genes Dev.* **9,** 1654–1666.

MacLellan, P. P., Belaguli, N. S., Schwartz, R. J., and Schneider, M. D. (1996). Serum response factor mediates AP-1-dependent induction of the skeletal alpha-actin promoter in ventricular myocytes. *J. Biol. Chem.* **271,** 10827–10833.

MacLellan, W. R., and Schneider, M. D. (1999). The cardiac cell cycle. *In* "Heart Development" (R. P. Harvey and N. Rosenthal, eds.), pp. 405–427. Academic Press, San Diego, CA.

Manasek, F. J. (1976). Heart development: Interactions involved in cardiac morphogenesis. *In* "The Cell Surface in Animal Embryogenesis and Development" (G. Poste and G. C. Nicholson, eds.), pp. 545–598. North-Holland Publ., Amsterdam.

Manasek, F. J. (1983). Control of early embryonic heart morphogenesis: A hypothesis. *Ciba Found. Symp.* **100,** 4–19.

Manasek, F. J., Kulikowski, R. R., Nakamura, A., Nguyenphuc, Q., and Lacktis, J. W. (1984). Early heart development: A new model of cardiac morphogenesis. *In* "Growth of the Heart in Health and Disease" (R. Zak, ed.), pp. 105–131. Raven Press, New York.

Marino, B., and Digilio, M. C. (2000). Congenital heart disease and genetic syndromes: Specific correlation between cardiac phenotype and genotype. *Cardiovasc. Pathol.* **9,** 303–315.

Markwald, R. R. (1995). Overview: Formation and early morphogenesis of the primary heart tube. *In* "Developmental Mechanisms of Heart Disease" (E. B. Clark, R. R. Markwald, and A. Takao, eds.), pp. 3–27. Futura Publ. Co., Armonk, NY.

Markwald, R. R., Truck, T., and Moreno-Rodriguez, R. (1998). Formation and septation of the tubular heart: Integrating the dynamics of morphology with emerging concepts. *In* "Living Morphogenesis of the Heart" (M. de la Cruz and R. R. Markwald, eds.), pp. 43–84. Birkhaeuser, Berlin.

Marvin, M. J., Di Rocco, G., Gardiner, A., Bush, S. M., and Lassar, A. B. (2001). Inhibition of Wnt activity induces heart formation from posterior mesoderm. *Genes Dev.* **15,** 316–327.

McFadden, D. G., Charite, J., Richardson, J. A., Srivastava, D., Firulli, A. B., and Olson, E. N. (2000). A GATA-dependent right ventricular enhancer controls *dHAND* trancription in the developing heart. *Development (Cambridge, UK)* **127,** 5331–5341.

McKinsey, T. A., Zhang, C.-L., Lu, J., and Olson, E. N. (2000). Signal-dependent nuclear export of a histone deacetlyase regulates muscle differentiation. *Nature (London)* **408,** 106–111.

Meins, M., Henderson, D. J., Bhattacharya, S. S., and Sowden, J. C. (2000). Characterization of the human *TBX20* gene, a new member of the T-box gene family closely related to the *Drosophila H15* gene. *Genomics* **67,** 317–332.

Meno, C., Shimono, A., Saijoh, Y., Yashiro, K., Mochida, K., Oishi, S., Noji, S., Kondoh, H., and Hamada, H. (1998). *lefty-1* is required for left-right determination as a regulator of *lefty-2* and nodal. *Cell (Cambridge, Mass.)* **94,** 287–297.

Merklin, R. J., and Verano, N. R. (1963). Situs inversus and cardiac defects. A study of III cases of reversed asymmetry. *J. Thorac. Cardiovasc. Surg.* **45,** 334–342.

Merscher, S., Funke, B., Epstein, J. A., Heyer, J., Puech, A., Lu, M. M., Xavier, R. J., Damay, M. B., Russell, R. G., Factor, S., Tokooya, K., St. Jore, B., Lopez, M., Pandita, R. J., Lia, M., Carrion, D., Xu, H., Schorle, H., Kobler, J. B., Scambler, P., Wynshaw-Boris, A., Skoultchi, A. I., Morrow, B. E., and Kucherlapati, R. (2001). *TBX1* is responsible for cardiovascular defects in Velo-Cario-Facial/DiGeorge Syndrome. *Cell (Cambridge, Mass.)* **104,** 619–629.

Meyer, D., and Birchmeier, C. (1995). Multiple essential functions of neuregulin in development. *Nature (London)* **378,** 386–390.

Meyers, E. N., Lewandoski, M., and Martin, G. R. (1998). An *Fgf8* mutant allelic series generated by Cre- and Flp-mediated recombination. *Nat. Genet.* **18,** 136–141.

Mikawa, T. (1999). Cardiac lineages. *In* "Heart Development" (R. P. Harvey and N. Rosenthal, eds.), pp. 19–33. Academic Press, San Diego, CA.

Miller, S. A., and White, R. D. (1998). Right-left asymmetry of cell proliferation predominates in mouse embryos undergoing clockwise axial rotation. *Anat. Rec.* **250,** 103–108.

Min, J.-Y., Kim, C.-Y., Oh, M. H., Chun, Y. K., Suh, Y.-L., Kang, I.-S., Lee, H.-J., and Seo, J.-W. (2000). Arrangement of the systemic and pulmonary venous components of the atrial chambers in hearts with isomeric atrial appendages. *Cardiol. Young* **10,** 396–404.

Mishina, Y., Suzuki, A., Ueno, N., and Behringer, R. R. (1995). *Bmpr* encodes a type I bone morphogenetic protein receptor that is essential for gastrulation during mouse embryogenesis. *Genes Dev.* **9,** 3027–3037.

Mjaatvedt, C. H., Yamamura, H., Capehart, A. A., Turner, D., and Markwald, R. R. (1998). The *Cspg2* gene, disrupted in the *hdf* mutant, is required for right cardiac chamber and endocardial cushion formation. *Dev. Biol.* **202,** 56–66.

Mjaatvedt, C. H., Yamamura, H., Wessels, A., Ramsdell, A., Turner, D., and Markwald, R. R. (1999). Mechanisms of segmentation, septation, and remodelling of the tubular heart: Endocardial cushion fate and cardiac

looping. *In* "Heart Development" (R. P. Harvey and N. Rosenthal, eds.), pp. 159–177. Academic Press, San Diego, CA.

Mochizuki, T., Saijoh, Y., Tsuchiya, K., Shirayoshi, Y., Takai, S., Taya, C., Yonekawa, H., Yamada, K., Nihei, H., Nakatsuji, N., Overbeek, P. A., Hamada, H., and Yokayama, T. (1998). Cloning of *inv,* a gene that controls left/right asymmetry and kidney development. *Nature (London)* **395,** 177–181.

Moens, C. B., Stanton, B. R., Parada, L. F., and Rossant, J. (1993). Defects in heart and lung development in compound heterozygotes for two different targeted mutations at the *N-myc* locus. *Development (Cambridge, UK)* **119,** 485–499.

Mohun, T. J., and Leong, L. M. (1999). Heart formation and the heart field in amphibian embryos. *In* "Heart Development" (R. P. Harvey and N. Rosenthal, eds.), pp. 37–49. Academic Press, San Diego, CA.

Mohun, T. J., Leong, L. M., Weninger, W. J., and Sparrow, D. B. (2000). The morphology of heart development in *Xenopus laevis. Dev. Biol.* **218,** 74–88.

Molkentin, J. D. (2000). The zinc finger-containing transcription factors GATA-4, -5, and -6: Ubiquitously expressed regulators of tissue-specific gene expression. *J. Biol. Chem.* **275,** 38949–38952.

Molkentin, J. D., Kalvakolanu, D. V., and Markham, B. E. (1994). Transcription factor GATA-4 regulates cardiac muscle-specific expression of the α-myosin heavy-chain gene. *Mol. Cell. Biol.* **14,** 4947–4957.

Molkentin, J. D., Black, B. L., Martin, J. F., and Olson, E. N. (1995). Cooperative activation of muscle gene expression by Mef2 and myogenic bHLH proteins. *Cell (Cambridge, Mass.)* **83,** 1125–1136.

Molkentin, J. D., Firulli, A. B., Black, B. L., Martin, J. F., Hustad, C. M., Copeland, N., Jenkins, N., Lyons, G., and Olson, E. (1996). MEF2B is a potent transactivator expressed in early myogenic lineages. *Mol. Cell. Biol.* **16,** 3814–3824.

Molkentin, J. D., Lin, Q., Duncan, S. A., and Olson, E. N. (1997). Requirement of the transcription factor GATA4 for heart tube formation and ventral morphogenesis. *Gene Dev.* **11,** 1061–1072.

Molkentin, J. D., Lu, J.-R., Antos, C. L., Markham, B., Richardson, J., Robbins, J., Grant, S. R., and Olson, E. N. (1998). A calcineurin-dependent transcriptional pathway for cardiac hypertrophy. *Cell (Cambridge, Mass.)* **93,** 215–228.

Molkentin, J. D., Antos, C., Mercer, B., Taigen, T., Miano, J. M., and Olson, E. N. (2000). Direct activation of a *GATA6* cardiac enhancer by Nkx2.5: Evidence for a reinforcing regulatory network of Nkx2.5 and GATA transcription factors in the developing heart. *Dev. Biol.* **217,** 301–309.

Monkley, S. J., Delaney, S. J., Pennisi, D. J., Christiansen, J. H., and Wainwright, B. J. (1996). Targeted disruption of the *Wnt2* gene results in placentation defects. *Development (Cambridge, UK)* **122,** 3343–3353.

Montgomery, M. O., Litvin, J., Gonzalez-Sanchez, A., and Bader, D. (1994). Staging of commitment and differentiation of avian cardiac myocytes. *Dev. Biol.* **164,** 63–71.

Monzen, K., Shiojima, I., Hiroi, Y., Kudoh, S., Oka, T., Takimoto, E., Hayashi, D., Hosoda, T., Habara-Ohkubo, A., Nakaoka, T., Fujita, T., Yazaki, Y., and Komuro, I. (1999). Bone morphogenetic proteins induce cardiomyocyte differentiation through the mitogen-activated protein kinase kinase kinase TAK1 and cardiac transcription factors Csx/Nkx-2.5 and GATA-4. *Mol. Cell. Biol.* **19,** 7096–7105.

Moore, A. W., McInnes, L., Kreidberg, J., Hastie, N. D., and Schedl, A. (1999). YAC complementation shows a requirement for Wt1 in the development of epicardium, adrenal gland and throughout nephrogenesis. *Development (Cambridge, UK)* **126,** 1845–1857.

Moore, A. W., Barbel, S., Jan, L. Y., and Jan, Y. N. (2000). A genomewide survey of basic helix-loop-helix factors in Drosophila. *Proc. Natl. Acad. Sci. U.S.A.* **97,** 10436–10441.

Moorman, A. F. M., and Lamers, W. H. (1999). Development of the conduction system of the vertebrate heart. *In* "Heart Development" (R. P. Harvey and N. Rosenthal, eds.), pp. 195–207. Academic Press, San Diego, CA.

Moorman, A. F. M., Vermeulen, J. L. M., Koban, M. U., Schwartz, K., Lamers, W. H., and Boheler, K. R. (1995). Patterns of expression of sarcoplasmic reticulum Ca^{2+} ATPase and phospholamban mRNAs during rat heart development. *Circ. Res.* **76**, 616–625.

Moorman, A. F. M., de Jong, F., Denyn, M. M. F., J., and Lamers, W. H. (1998). Development of the cardiac conduction system. *Circ. Res.* **82**, 629–644.

Moorman, A. F. M., Schumacher, C. A., de Boer, P. A. J., Hagoort, J., Bezstarosti, K., van den Hoff, M. J. B., Wagenaar, G. T. M., Lamers, J. M. J., Wuytack, F., Christoffels, V. M., and Fiolet, J. W. T. (2000). Presence of functional sarcoplasmic reticulum in the developing heart and its confinement to chamber myocardium. *Dev. Biol.* **223**, 279–290.

Morgan, D., Turnpenny, L., Goodship, J., Dai, W., Majumder, K., Matthews, L., Gardner, A., Schuster, G., Vien, L., Harrison, W., Elder, F. F. B., Penman-Splitt, M., Overbeek, P., and Strachen, T. (1998). Inversin, a novel gene in the vertebrate left-right axis pathway, is partially deleted in the *inv* mouse. *Nat. Genet.* **20**, 149–156.

Morin, S., Charron, F., Robitaille, L., and Nemer, M. (2000). GATA-dependent recruitment of MEF2 proteins to target promoters. *EMBO J.* **19**, 2046–2055.

Morrisey, E. E., Ip, H. S., Lu, M. M., and Parmacek, M. S. (1996). GATA-6—a zinc finger transcription factor that is expressed in multiple cell lineages derived from lateral mesoderm. *Dev. Biol.* **177**, 309–322.

Morrisey, E. E., Ip, H. S., Tang, Z., Lu, M. M., and Parmacek, M. S. (1997). GATA-5: A transcriptional activator expressed in a novel and temporally and spatially-restricted pattern during embryonic development. *Dev. Biol.* **183**, 21–36.

Moss, J. B., Xavier-Neto, J., Shapiro, M. D., Nayeem, S. M., McCaffery, P., Drager, U. C., and Rosenthal, N. (1998). Dynamic patterns of retinoic acid synthesis and response in the developing mammalian heart. *Dev. Biol.* **199**, 55–71.

Myer, D., and Birchmeier, C. (1995). Multiple essential functions of neuregulin in development. *Nature (London)* **378**, 386–390.

Nakagawa, O., Nakagawa, M., Richardson, J. A., Olson, E. N., and Srivastava, D. (1999). HRT1, HRT2, and HRT3: A new subclass of bHLH transcription factors marking specific cardiac, somitic, and pharyngeal arch segments. *Dev. Biol.* **216**, 72–84.

Nakagawa, O., McFadden, D. G., Nakagawa, M., Yanagisawa, H., Hu, T., Srivastava, D., and Olson, E. N. (2000). Members of the HRT family of basic helix-loop-helix proteins act as transcriptional repressors downstream of Notch signalling. *Proc. Natl. Acad. Sci. U.S.A.* **97**, 13655–13660.

Narasimhaswamy, S., Belaguli, J. L., Sepulveda, V. N., Charron, R., Nemer, M., and Schwartz, R. J. (2000). Cardiac tissue enriched factors serum response factor and GATA-4 are mutual cofactors. *Mol. Cell. Biol.* **20**, 7550–7558.

Narita, N., Bielinska, M., and Wilson, D. B. (1997). Wild-type endoderm abrogates the ventral developmental defects associated with GATA-4 deficiency in the mouse. *Dev. Biol.* **189**, 270–274.

Nascone, N., and Mercola, M. (1995). An inductive role for the endoderm in *Xenopus* cardiogenesis. *Development (Cambridge, UK)* **121**, 515–523.

Nascone, N., and Mercola, M. (1996). Endoderm and cardiogenesis: New insights. *Trends Cardiovasc. Med.* **6**, 211–216.

Niederreither, K., McCaffery, P., Drager, U. C., Chambon, P., and Dollé, P. (1997). Restricted expression and retinoic acid-induced downregulation of the retinaldehye dehydrogenase type 2 (RALDH-2) gene during mouse development. *Mech. Dev.* **62**, 67–78.

Niederreither, K., Subbarayan, V., Dollé, P., and Chambon, P. (1999). Embryonic retinoic acid synthesis is essential for early mouse post-implantation development. *Nat. Genet.* **21**, 444–448.

Niederreither, K., Vermot, J., Messaddeq, N., Schuhbaur, B., Chambon, P., and Dollé, P. (2001). Embryonic retinoic acid synthesis is essential for

heart morphogenesis in the mouse. *Development (Cambridge, UK)* **128**, 1019–1031.

Noden, D. M. (1991). Origins and patterning of avian outflow tract endocardium. *Development (Cambridge, UK)* **111**, 867–876.

Nomura, M., and Li, N. (1998). Smad2 role in mesoderm formation, left-right patterning and craniofacial development. *Nature (London)* **393**, 786–790.

O'Brien, T. X., Lee, K. J., and Chien, K. R. (1993). Positional specification of ventricular myosin light chain-2 expression in the primitive murine heart tube. *Proc. Natl. Acad. Sci. U.S.A.* **90**, 5157–5161.

Oda, T., Elkahloun, A. G., Pike, B. L., Okajima, K., Krantz, I. D., Genin, A., Piccoli, D. A., Meltzer, P. S., Spinner, N. B., and Collins, F. S. (1997). Mutations in the human *Jagged1* gene are responsible for Alagille syndrome. *Nat. Genet.* **16**, 235–242.

Oh, S. P., and Li, E. (1997). The signaling pathway mediated by the type IIB activin receptor controls axial patterning and lateral asymmetry in the mouse. *Genes Dev.* **11**, 1812–1826.

Palmer, S., Groves, N., Schindeler, A., Yeoh, T., Biben, C., Wang, C.-C., Sparrow, D. B., Barnett, L., Jenkins, N., Copeland, N., Koentgen, F., Mohun, T., and Harvey, R. P. (2001). The small muscle-specific protein Csl modifies cell shape and promotes myocyte fusion in an IGF-1 dependent manner. *J. Cell Biol.* **153**, 985–997.

Parameswaran, M., and Tam, P. P. (1995). Regionalization of cell fate and morphogenetic movements of the mesoderm during mouse gastrulation. *Dev. Genet.* **17**, 16–28.

Park, M., Wu, X., Golden, K., Axelrod, J. D., and Bodmer, R. (1996). The wingless pathway is directly involved in *Drosophila* heart development. *Dev. Biol.* **177**, 104–116.

Parmacek, M. S., and Leiden, J. M. (1999). GATA transcription factors and cardiac development. In "Heart Development" (R. P. Harvey and N. Rosenthal, eds.), pp. 291–306. Academic Press, San Diego, CA.

Patel, K., Isaac, A., and Cooke, J. (1999). Nodal signalling and the roles of the transcription factors SnR and Pitx2 in vertebrate left-right asymmetry. *Curr. Biol.* **9**, 609–612.

Patten, B. M. (1922). The formation of the cardiac loop in the chick. *Am. J. Anat.* **30**, 373–379.

Payne, R. M., Johnson, M. C., Grant, J. W., and Strauss, A. W. (1995). Towards a molecular understanding of congenital heart disease. *Circulation* **91**, 494–504.

Payrieras, N., Strahle, U., and Rosa, F. (1998). Conversion of zebrafish blastomeres to an endodermal fate by TGF-β-related signalling. *Curr. Biol.* **8**, 783–786.

Pereira, F. A., Qiu, Y., Zhou, G., Tsai, M. J., and Tsai, S. Y. (1999). The orphan nuclear receptor COUP-TFII is required for angiogenesis and heart development. *Genes Dev.* **13**, 1037–1049.

Pexieder, T. (1975). Cell death in morphogenesis and teratogenesis of the heart. *Adv. Anat. Embryol. Cell Biol.* **51**, 3–100.

Piccolo, S., Agius, E., Leyns, L., Bhattacharyya, S., Grunz, H., Bouwmeester, T., and De Robertis, E. M. (1999). The head inducer Cerberus is a multifunctional antagonist of Nodal, BMP and Wnt signals. *Nature (London)* **397**, 707–710.

Piedra, M. E., Icardo, J. M., Albajar, M., Rodriguez-Rey, J. C., and Ros, M. A. (1998). *Ptx2* participates in the late phase of the pathway controlling left-right asymmetry. *Cell (Cambridge, Mass.)* **94**, 319–324.

Poelmann, R. E., and Gittenberger-de Groot, A. C. (1999). A subpopulation of apoptosis-prone cardiac neural crest cells targets to the venous pole: Multiple functions in heart development? *Dev. Biol.* **207**, 271–286.

Pourquié, O. (2000). Vertebrate segmentation: Is cycling the rule? *Curr. Opin. Cell Biol.* **12**, 747–751.

Radice, G. L., Rayburn, H., Matsunami, H., Knudsen, K. A., Takeichi, M., and Hynes, R. O. (1997). Developmental defects in mouse embryos lacking N-cadherin. *Dev. Biol.* **181**, 64–78.

Raffin, M., Leong, L. M., Rones, M. S., Sparrow, D., Mohun, T., and Mercola, M. (2000). Subdivision of the cardiac Nkx2.5 expression domain into myogenic and nonmyogenic compartments. *Dev. Biol.* **218**, 326–340.

Ranganayakulu, G., Zhao, B., Dakodis, A., Molkentin, J. D., Olson, E. N., and Schulz, R. A. (1995). A series of mutations in D-MEF2 transcription factor reveals multiple functions in larval and adult myogensis in *Drosophila. Dev. Biol.* **171**, 169–181.

Ranganayakulu, G., Elliot, D. A., Harvey, R. P., and Olson, E. N. (1998). Divergent roles for *NK-2* class homeobox genes in cardiogenesis in flies and mice. *Development (Cambridge, UK)* **125**, 3037–3048.

Rankin, C. T., Bunton, T., Lawler, A. M., and Lee, S. J. (2000). Regulation of left-right patterning in mice by growth/differentiation factor-1. *Nat. Genet.* **24**, 262–265.

Reichmann, V., Irion, U., Wilson, R., Grosskortenhaus, A., and Leptin, M. (1997). Control of cell fates and segmentation in the Drosophila mesoderm. *Development (Cambridge, UK)* **124**, 2915–2922.

Reifers, F., Walsh, E. C., Leger, S., Stainier, D. Y. R., and Brand, M. (2000). Induction and differentiation of the zebrafish heart requires fibroblast growth factor 8 (*fgf8/acerebellar*). *Development (Cambridge, UK)* **127**, 225–235.

Reiter, J. F., Alexander, J., Rodaway, A., Yelon, D., Patient, R., Holder, N., and Stainier, D. Y. R. (1999). Gata5 is required for development of the heart and endoderm in zebrafish. *Genes Dev.* **13**, 2983–2995.

Rentschler, S., Vaidya, D. M., Tamaddon, H., Degenhardt, K., Sassoon, D., Morley, G. E., Jalife, J., and Fishman, G. I. (2001). Visualisation and functional characterisataion of the developing murine cardiac conduction system. *Development (Cambridge, UK)* **128**, 1785–1792.

Riley, P., Anson-Cartwright, L., and Cross, J. C. (1998). The Hand1 bHLH transcription factor is essential for placentation and cardiac morphogenesis. *Nat. Genet.* **18**, 271–275.

Riley, P. R., Gertsenstein, M., Dawson, K., and Cross, J. C. (2000). Early exclusion of hand1-deficient cells from distinct regions of the left ventricular myocardium in chimeric mouse embryos. *Dev. Biol.* **227**, 156–168.

Riley, P. R., Gertsenstein, M., Dawson, K., and Cross, J. C. (2000). Early exclusion of Hand1-deficient cells from distinct regions of the left ventricular myocardium in chimeric mouse embryos. *Dev. Biol.* **227**, 156–168.

Rones, M. S., McLaughlin, K. A., Raffin, M., and Mercola, M. (2000). Serrate and Notch specify cell fates in the heart field by suppressing cardiomyogenesis. *Development (Cambridge, UK)* **127**, 3865–3876.

Rosenquist, G., and DeHaan, R. (1966). Migration of precardiac cells in the chick embryo: A radioautographic study. *Carnegie Contrib. Embryol.* **263**, 111–123.

Rosenthal, N., and Xavier-Neto, J. (2000). From the bottom of the heart: Anteroposterior decisions in cardiac muscle differentiation. *Curr. Opin. Cell Biol.* **12**, 742–746.

Ross, R. S., Navankasattusas, S., Harvey, R. P., and Chien, K. R. (1996). An HF-1a/HF-1b/MEF-2 combinatorial element confers cardiac ventricular specificity and establishes an anterior-posterior gradient of expression via an *Nkx2-5* independent pathway. *Development (Cambridge, UK)* **122**, 1799–1809.

Ross, S. A., McCaffery, R. J., Drager, U., and De Luca, L. M. (2000). Retinoids in embryonal development. *Physiol. Rev.* **80**, 1021–1054.

Rugendorff, A., Younossi-Hartenstein, A., and Hartenstein, V. (1994). Embryonic origin and differentiation of the *Drosophila* heart. *Roux's Arch. Dev. Biol.* **203**, 266–280.

Ruiz-Lozano, P., Smith, S. M., Perkins, G., Kubalak, S. W., Boss, G. W., Sucov, H. M., Evans, R. M., and Chien, K. R. (1998). Energy deprivation and a deficiency in downstream metabolic target genes during the onset of embryonic heart failure in RXRalpha–/– embryos. *Development (Cambridge, UK)* **125**, 533–544.

Rumyantsev, P. P. (1977). Interrelations of the proliferation and differentiation processes during cardiac myogenesis and regeneration. *Int. Rev. Cytol.* **51**, 187–273.

Ryan, A. K., Blumberg, B., Rodrigez-Esteban, C., Yonei-Tamura, S., Tamura, K., Tsukui, T., de la Pena, J., Sabbagh, W., Greewald, J., Choe, S., Norris, D. P., Robertson, E. J., Evans, R. M., Rosenfeld, M. G., and Izpisua Belmonte, J. C. (1998). Pitx2 determines left-right asymmetry of internal organs in vertebrates. *Nature (London)* **394**, 545–551.

Saga, Y., Miyagawa-Tomita, S., Takagi, A., Kitajima, S., Miyazaki, J., and Inoue, T. (1999). MesP1 is expressed in the heart precursor cells and required for the formation of a single heart tube. *Development (Cambridge, UK)* **126**, 3437–3447.

Saijoh, Y., Adachi, H., Sakuma, R., Yeo, C.-Y., Yashiro, K., Watanabe, M., Hashiguchi, H., Mochida, K., Ohishi, S., Kawabata, M., Miyazono, K., Whitman, M., and Hamada, H. (2000). Left-right asymmetric expression of *lefty2* and *nodal* is induced by a signaling pathway that includes the transcription factor FAST2. *Mol. Cell* **5**, 35–47.

Salomon, D. S., Bianco, C., and De Santis, M. (1999). Cripto: A novel epidermal growth factor (EGF)-related peptide in mammary gland development and neoplasia. *BioEssays* **21**, 61–70.

Sater, A. K., and Jacobson, A. G. (1990a). The restriction of the heart morphogenetic field in *Xenopus laevis. Dev. Biol.* **140**, 328–336.

Sater, A. K., and Jacobson, A. G. (1990b). The role of the dorsal lip in induction of heart mesoderm in *Xenopus laevis. Development (Cambridge, UK)* **108**, 461–470.

Satin, J., Fujii, S., and DeHaan, R. L. (1988). Development of cardiac beat rate in early chick embryos is regulated by regional cues. *Dev. Biol.* **129**, 103–113.

Scambler, P. J. (2000). The 22q11 deletion syndromes. *Hum. Mol. Genet.* **9**, 2421–2426.

Schlange, T., Andree, B., Arnold, H.-H., and Brand, T. (2000). BMP2 is required for early heart development during a distinct time period. *Mech. Dev.* **91**, 259–270.

Schneider, V. A., and Mercola, M. (1999). Spatially distinct head and heart inducers within the *Xenopus* organiser region. *Curr. Biol.* **9**, 800–809.

Schneider, V. A., and Mercola, M. (2001). Wnt antagonism initiates cardiogenesis in *Xenopus laevis. Genes Dev.* **15**, 304–325.

Schott, J.-J., Benson, D. W., Basson, C. T., Pease, W., Silberach, G. M., Moak, J. P., Maron, B. J., Seidman, C. E., and Seidman, J. G. (1998). Congential heart disease caused by mutations in the transcription factor *NKX2-5. Science* **281**, 108–111.

Schultheiss, T. M., Xydas, S., and Lassar, A. B. (1995). Induction of avian cardiac myogenesis by anterior endoderm. *Development (Cambridge, UK)* **121**, 4203–4214.

Schultheiss, T. M., Burch, J. B. E., and Lassar, A. B. (1997). A role for bone morphogenetic proteins in the induction of cardiac myogenesis. *Genes Dev.* **11**, 451–462.

Schwartz, R. J., and Olson, E. N. (1999). Building the heart piece by piece: Modularity of cis-elements regulating *Nkx2-5* transcription. *Development (Cambridge, UK)* **126**, 4187–4192.

Scott, I. C., Anson-Cartwright, L., Riley, R., Reda, D., and Cross, J. C. (2000). The HAND1 basic helix-loop-helix transcription factor regulates trophoblast differentiation via multiple mechanisms. *Mol. Cell. Biol.* **20**, 530–541.

Searcy, R. D., and Yutzey, K. E. (1998). Analysis of *Hox* gene expression during early avian heart development. *Dev. Biol.* **21**, 82–91.

Searcy, R. D., Vincent, E. B., Liberatore, C. M., and Yutzey, K. E. (1998). A GATA-dependent *nkx-2.5* regulatory element activates early cardiac gene expression in transgenic mice. *Development (Cambridge, UK)* **125**, 4461–4470.

Sefton, M., Sanchez, S., and Nieto, M. A. (1998). Conserved and divergent roles for members of the *Snail* family of transcription factors in the chick and mouse embryos. *Development (Cambridge, UK)* **125**, 3111–3121.

Seo, J. W., Brown, N. A., Ho, S. Y., and Anderson, R. H. (1992). Abnormal laterality and congenital cardiac abnormalities. Relations of visceral and cardiac morphologies in the iv/iv mouse. *Circulation* **86**, 642–650.

Sepulveda, J. L., Belaguli, N., Nigam, V., Chen, C.-Y., Nemer, M., and Schwartz, R. J. (1998). GATA-4 and Nkx2-5 coactivate Nkx-2 DNA binding targets: Role for regulating early cardiac gene expression. *Mol. Cell. Biol.* **18,** 3405–3415.

Serbedzija, G. N., Chen, J.-N., and Fishman, M. C. (1998). Regulation in the heart field of zebrafish. *Development (Cambridge, UK)* **125,** 1095–1101.

Shen, M. M., and Schier, A. F. (2000). The EGF-CFC gene family in vertebrate development. *Trends Genet.* **16,** 303–309.

Shen, M. M., Wang, H., and Leder, P. (1997). A differential display strategy identifies *Cryptic,* a novel EGF-related gene expressed in the axial and lateral mesoderm during mouse gastrulation. *Development (Cambridge, UK)* **124,** 429–442.

Shi, Y., Katsev, S., Cai, C., and Evans, S. (2000). BMP signaling is required for heart formation in vertebrates. *Dev. Biol.* **224,** 226–237.

Shioda, T., Lechleider, R. J., Dunwoodie, S. L., Li, H., Yahata, T., De Caestecker, M. P., Fenner, M. H., Roberts, A. B., and Isselbacher, K. J. (1998). Transcriptional activating activity of Smad4: Roles of SMAD hetero-oligomerization and enhancement by an associating transactivator. *Proc. Natl. Acad. Sci. U.S.A.* **95,** 9785–9790.

Shiojima, I., Komuro, I., Oka, T., Hiroi, Y., Mizuno, T., Takimoto, E., Monzen, K., Aikawa, H., Yamazaki, T., Kudoh, S., and Yazaki, Y. (1999). Context-dependent transcriptional cooperation mediated by cardiac transcription factors Csx/Nkx-2.5 and GATA-4. *J. Biol. Chem.* **274,** 8231–8239.

Shiratori, H., Sakuma, R., Watanabe, M., Hashiguchi, H., Mochida, K., Sakai, Y., Nishino, Y., Whitman, M., and Hamada, H. (2000). Two-step regulation of left-right asymmetric expression of *Pitx2:* Initiation by Nodal signalling and maintenance by Nkx2. *Mol. Cell* **7,** 139–149.

Sissman, N. J. (1966). Cell multiplication rates during development of the primitive cardiac tube in the chick embryo. *Nature (London)* **210,** 504–507.

Skerjanc, I. C., Peteropoulos, H., Ridgeway, A. G., and Wilton, S. (1998). Myoctye enhancer factor 2C and Nkx2-5 up-regulate each other's expression and initiate cardiomyogenesis in P19 cells. *J. Biol. Chem.* **273,** 34904–34910.

Smith, S. M., Dickman, E. D., Thompson, R. P., Sinning, A. R., Wunsch, A. M., and Markwald, R. R. (1997). Retinoic acid directs cardiac laterality and the expression of early markers of precardiac asymmetry. *Dev. Biol.* **182,** 162–171.

Solloway, M. J., and Robertson, E. J. (1999). Early embryonic lethality in *Bmp5;Bmp7* double mutant mice suggests functional redundancy with the 60A subgroup. *Development (Cambridge, UK)* **126,** 1753–1768.

Soulez, M., Tuil, D., Kahn, A., and Gilgenkrantz, H. (1996). The serum response factor (SRF) is needed for muscle-specific activation of CArG boxes. *Biochem. Biophys. Res. Commun.* **219,** 418–422.

Sparrow, D. B., Kotecha, S., Towers, N., and Mohun, T. J. (1998). *Xenopus eHAND:* A marker for the developing cardiovascular system of the embryo that is regulated by bone morphogenetic proteins. *Mech. Dev.* **71,** 151–163.

Sparrow, D. B., Miska, E. A., Langley, E., Reynaud-Deonauth, S., Kotecha, S., Towers, N., Spohr, G., Kouzarides, T., and Mohun, T. J. (1999). MEF-2 function is modified by a novel co-repressor, MITR. *EMBO J.* **18,** 5085–5098.

Sparrow, D. B., Kotecha, S., Chenleng, C., Latinkic, B., Cooper, B., Towers, N., Evans, S. M., and Mohun, T. J. (2000). Regulation of *tinman* homologues in *Xenopus* embryos. *Dev. Biol.* **227,** 65–79.

Srivastava, D. (1999). HAND Proteins: Molecular mediators of cardiac development and congenital heart disease. *Trends Cardiovasc. Med.* **9,** 11–18.

Srivastava, D., Cserjesi, P., and Olson, E. N. (1995). A subclass of bHLH proteins required for cardiac morphogenesis. *Science* **270,** 1995–1999.

Srivastava, D., Thomas, T., Lin, Q., Brown, D., and Olson, E. N. (1997). Regulation of cardiac mesodermal and neural crest development by the bHLH transcription factor, dHAND. *Nat. Genet.* **16,** 154–160.

Stainier, D. Y. R., and Fishman, M. C. (1992). Patterning the zebrafish heart tube: Acquisition of anteroposterior polarity. *Dev. Biol.* **153,** 91–101.

Stainier, D. Y. R., Lee, R. K., and Fishman, M. C. (1993). Cardiovascular development in the zebrafish I. Myocardial fate map and heart tube formation. *Development (Cambridge, UK)* **119,** 31–40.

Stalsberg, H. (1969a). The origin of heart asymmetry: Right and left contributions to the early chick embryo heart. *Dev. Biol.* **19,** 109–129.

Stalsberg, H. (1969b). Regional mitotic activity in the precardiac mesoderm and differentiating heart tube in the chick embryo. *Dev. Biol.* **20,** 18–45.

Stalsberg, H. (1970). Mechanism of dextral looping of the embryonic heart. *Am. J. Cardiol.* **25,** 265–271.

Subbarayan, V., Mark, M., Messdeq, N., Rustin, P., Chambon, P., and Kastner, P. (2000). RXRalpha overexpression in cardiomyocytes causes dilated cardiomyopathy but fails to rescue myocardial hypoplasia in RXRalpha-null fetuses. *J. Clin. Invest.* **105,** 387–394.

Sucov, H. M., Dyson, E., Gumeringer, C. L., Price, J., Chien, K. R., and Evans, R. M. (1994). RXRa mutant mice establish a genetic basis for vitamin A signaling in heart morphogenesis. *Genes Dev.* **8,** 1007–1018.

Sugi, Y., and Lough, J. (1994). Anterior endoderm is a specific effector of terminal cardiac myocyte differentiation in cells from the embryonic heart forming region. *Dev. Dyn.* **200,** 155–162.

Sugi, Y., and Markwald, R. R. (1996). Formation and early morphogenesis of endocardial endothelial precursor cells and the role of endoderm. *Dev. Biol.* **175,** 66–83.

Sun, X., Meyers, E. N., Lewandoski, M., and Martin, G. R. (1999). Targeted disruption of *Fgf8* causes failure of cell migration in the gastrulating mouse embryo. *Genes Dev.* **13,** 1834–1846.

Sundin, O., and Eichele, G. (1992). An early marker of axial pattern in the chick embryo and its respecification by retinoic acid. *Development (Cambridge, UK)* **114,** 841–852.

Supp, D. M., Witte, D. P., Potter, S. S., and Brueckner, M. (1997). Mutation of an axonemal dynein affects left-right asymmetry in *inversus viscerum* mice. *Nature (London)* **389,** 963–966.

Svensson, E. C., Tufts, R. L., Polk, C. E., and Leiden, J. M. (1999). Molecular cloning of *FOG-2:* A modulator of transcription factor *GATA-4* in cardiomyocytes. *Proc. Natl. Acad. Sci. U.S.A.* **96,** 956–961.

Svensson, E. C., Huggins, G. S., Lin, H., Clendenin, C., Jiang, F., Tufts, R., Dardik, F. B., and Leiden, J. M. (2000). A syndrome of tricuspid atresia in mice with a targeted mutation of the gene encoding Fog-2. *Nat. Genet.* **25,** 353–356.

Tada, M., and Smith, J. C. (2000). *Xwnt11* is a target of *Xenopus* Brachyury: Regulation of gastrulation movements via Dishevelled, but not through the canonical Wnt pathway. *Development (Cambridge, UK)* **127,** 2227–2238.

Takahashi, S., Yokata, C., Takano, K., Tanegashima, K., Onuma, Y., Goto, J.-I., and Asashima, M. (2000). Two novel *nodal*-related genes initiate early inductive events in Xenopus Neiuwkoop center. *Development (Cambridge, UK)* **127,** 5319–5329.

Takebayashi-Suzuki, K., Yanagisawa, M., Gourdie, R. G., Kanzawa, N., and Mikawa, T. (2000). In vivo induction of cardiac Purkinje fiber differentiation by co-expression of preproendothelin-1 and endothelin converting enzyme-1. *Development (Cambridge, UK)* **127,** 3523–3532.

Tam, P. L., and Schoenwolf, G. C. (1999). Cardiac fate maps: lineage allocation, morphogenetic movement, and cell commitment. *In* "Heart Development" (R. P. Harvey and N. Rosenthal, eds.), pp. 3–18. Academic Press, San Diego, CA.

Tanaka, M., Chen, Z., Bartunkova, S., Yamasaki, N., and Izumo, S. (1999). The cardiac homeobox gene *Csx/Nkx2.5* lies genetically upstream of multiple genes for heart development. *Development (Cambridge, UK)* **126,** 1269–1280.

Tanaka, S., Kunath, T., Jadjantonakis, A.-K., Nagy, A., and Rossant, J. (1998). Promotion of trophoblast stem cell proliferation by FGF4. *Science* **282,** 2072–2075.

Tchervenkov, C. I., Jacobs, M. L., and Tahta, S. A. (2000). Congenital heart surgery nomenclature and database project: Hypoplastic left heart syndrome. *Ann. Thorac. Surg.* **69,** S170–S179.

Tevosian, S. G., Deconinck, A. E., Cantor, A. B., Rieff, H. I., Fujiara, Y., Corfas, G., and Orkin, S. H. (1999). *FOG*-2: A novel *GATA*-family cofactor related to multiple zinc finger proteins Friend of *GATA*-1 and U-shaped. *Proc. Natl. Acad. Sci. U.S.A.* **96,** 950–955.

Tevosian, S. G., Deconinck, A. E., Tanaka, M., Schinke, M., Litovsky, S. H., Izumo, S., Fujiwara, Y., and Orkin, S. H. (2000). *FOG*-2, a cofactor for *GATA* transcription factors, is essential for heart morphogenesis and development of the coronary vessels from epicardium. *Cell (Cambridge, Mass.)* **101,** 729–739.

Thomas, T., Kurihara, H., Yamagishi, H., Kurihara, Y., Yazaki, Y., Olson, E. N., and Srivastava, D. (1998a). A signalling cascade involving endothelin-1, dHand and Msx1 regulates development of neural-crest-derived branchial arch mesenchyme. *Development (Cambridge, UK)* **125,** 3005–3014.

Thomas, T., Yamagishi, H., Overbeek, P. A., Olson, E. N., and Srivastava, D. (1998b). The bHLH factors, dHAND and eHAND, specify pulmonary and systemic cardiac ventricles independent of left-right sidedness. *Dev. Biol.* **196,** 228–236.

Thompson, R. P., Lindroth, J. R., and Wong, Y.-M. M. (1990). Regional differences in DNA-synthetic activity in the preseptation myocadium of the chick. *In* "Developmental Cardiology: Morphogenesis and Function" (E. B. Clark and A. Takao, eds.), pp. 219–234. Futura Publ. Co., Mount Kisco, NY.

Tonissen, K. F., Drysdale, T. A., Lints, T. J., Harvey, R. P., and Krieg, P. A. (1994). *XNkx2.5*, a *Xenopus* gene related to *Nkx-2.5* and *tinman*: Evidence for a conserved role in cardiac development. *Dev. Biol.* **162,** 325–328.

Tran, C. M., and Sucov, H. M. (1998). The RXRalpha gene functions in a non-cell-autonomous manner during mouse embryogenesis. *Development (Cambridge, UK)* **125,** 1951–1956.

Treisman, R., Alberts, A. S., and Sahai, E. (1998). Regulation of SRF activity by Rho family GTPases. *Cold Spring Harbor Symp. Quant. Biol.* **63,** 643–651.

Tsang, A. P., Fujiwara, Y., Hom, D. B., and Orkin, S. H. (1998). Failure of megakaryopoiesis and arrested erythropoiesis in mice lacking the *GATA*-1 transcriptional cofactor *FOG*. *Genes Dev.* **12,** 1176–1188.

Tsuda, T., Philp, N., Zile, M. H., and Linask, K. K. (1996). Left-right asymmetric localization of flectin in the extracellular matrix during heart looping. *Dev. Biol.* **173,** 39–50.

Tzahor, E., and Lassar, A. B. (2001). Wnt signals from the neural tube block ectopic cardiogenesis. *Genes Dev.* **15,** 255–260.

Van Mierop, L. H. S. (1967). Location of pacemaker in the chick embryo heart at the time of initiation of heartbeat. *Am. J. Physiol.* **211,** 407–415.

Van Mierop, L. H. S., and Kutsche, L. M. (1984). Comparative anatomy and embryology of the ventricles and arterial pole of the vertebrate heart. *In* "Congenital Heart Disease: Causes and Processes" (J. J. Nora and A. Takao, eds.), pp. 459–479. Futura Publ. Co., Mt. Kisco, NY.

Varfolomeev, E. E., Schuchmann, M., Luria, V., Chiannikulchai, N., Beckmann, J. S., Mett, I. L., Rebrikov, D., Brodianski, V. M., Kemper, O. C., Kollet, O., Lapidot, T., Soffer, D., Sobe, T., Avraham, K. B., Goncharov, T., Holtmann, H., Lonai, P., and Wallach, D. (1998). Targeted disruption of the mouse caspase 8 gene ablates cell death induction by the TNF receptors, Fas/Apo1, and DR3 and is lethal perinatally. *Immunity* **9,** 267–276.

Ventura, C., and Maioli, M. (2000). Opioid peptide gene expression primes cardiogenesis in embryonal pluripotent stem cells. *Circ. Res.* **87,** 189–194.

Ventura, C., Maioli, M., Pintus, G., Posadino, A. M., and Tadolini, B. (1998). Nuclear opioid receptors activate opioid peptide gene trancription in isolated myocardial nuclei. *J. Biol. Chem.* **273,** 13383–13386.

Viragh, S., and Challice, C. E. (1973). Origin and differentiation of cardiac muscle cells in the mouse. *J. Ultrastruct. Res.* **42,** 1–24.

Waller, R. R., McQuinn, T., Phelps, A. L., Markwald, R. R., Lo, C. W., Thompson, R. P., and Wessels, A. (2000). Conotruncal anomalies in the trisomy 16 mouse: An immunohistochemical analysis with emphasis on the involvement of the neural crest. *Anat. Rec.* **260,** 279–293.

Wang, G. F., Nikovits, J., W., Schleinitz, M., and Stockdale, F. E. (1998). A positive GATA element and a negative vitamin D receptor-like element control atrial chamber-specific expression of a slow myosin heavy-chain gene during cardiac morphogenesis. *Mol. Cell. Biol.* **18,** 6023–6034.

Watanabe, M., Choudhry, A., Berlan, M., Singal, A., Siwik, E., Mohr, S., and Fisher, S. A. (1998). Developmental remodelling and shortening of the cardiac outflow tract involves myocyte programmed cell death. *Development (Cambridge, UK)* **125,** 3809–3820.

Webb, S., Brown, N. A., and Anderson, R. H. (1998). Formation of the atrioventricular septal structures in the normal mouse. *Circ. Res.* **82,** 645–656.

Wei, L., Zhou, W., Croissant, J. D., Johansen, F.-E., Prywes, R., Balasubramanyam, A., and Schwartz, R. J. (1998). RhoA signaling via serum response factor plays an obligatory role in myogenic differentiation. *J. Biol. Chem.* **273,** 30287–30294.

Wei, L., Zhou, W., Wang, L., and Schwartz, R. J. (2000). β_1-Integrin and PI 3-kinase regulate RhoA-dependent activation of skeletal α-actin promoter in myoblasts. *Am. J. Physiol. (Heart Circ. Physiol.)* **278,** H1736–H1743.

Wendling, O., Chambon, P., and Mark, M. (1999). Retinoid X receptors are essential for early mouse development and placentogenesis. *Proc. Natl. Acad. Sci. U.S.A.* **96,** 547–551.

Wenink, A. C., and Anderson, R. H. (1987). Embryology of the heart. *In* "Paediatric Cardiology" (R. H. Anderson, E. A. Shinebourne, F. J. Macartney, and M. Tynan, eds.), Vol. 1, pp. 83–105. Churchill Livingstone, Edinburgh and London.

Wessels, A., Vermeulen, J. L. M., Verbeek, F. J., Viragh, S., Kalman, F., Lamers, W. H., and Moorman, A. F. M. (1992). Spatial distribution of "tissue-specific" antigens in the developing human heart and skeletal muscle. III: An immunohistochemical analysis of the distribution of the neural tissue antigen GIN2 in the embryonic heart; implications for the development of the atrioventricular conduction system. *Anat. Rec.* **232,** 97–111.

Wessels, A., Anderson, R. H., Markwald, R. R., Webb, S., Brown, N. A., Viragh, S. Z., Moorman, A. F. M., and Lamers, W. H. (2000). Atrial development in the human heart: An immunohistochemical study with emphasis on the role of mesenchymal tissues. *Anat. Rec.* **259,** 288–300.

Whitelaw, E., and Martin, D. I. K. (2001). Retrotransposons as epigenetic mediators of phenotypic variation in mammals. *Nat. Genet.* **27,** 361–365.

Whyatt, D. J., Karis, A., Harkes, I. C., Verkerk, A., Gillemans, N., Elefanty, A. G., Vario, G., Ploemacher, R., Grosveld, F., and Philipsen, S. (1997). The level of the tissue-specific factor GATA-1 affects the cell-cycle machinery. *Genes Funct.* **1,** 11–24.

Wilson, G. N. (1998). Correlated heart/limb anomalies in mendelian syndromes provide evidence for a cardiomelic developmental field. *Am. J. Med. Genet.* **76,** 297–305.

Wilson, J. G., and Warkany, J. (1949). Aortic-arch and cardiac abnormalities in the offspring of vitamin A-deficient rats. *Am. J. Anat.* **85,** 113–155.

Winer-Muram, H. T. (1995). Adult presentation of heterotaxic syndromes and related complexes. *J. Thorac. Imaging* **10,** 43–57.

Winter, B., and Arnold, H. H. (2000). Activated Raf kinase inhibits muscle differentiation through a MEF2-dependent mechanism. *J. Cell Sci.* **113,** 4211–4220.

Wu, H., Naya, F. J., McKinsey, T. A., Mercer, B., Shelton, J. M., Chin, E. R., Simard, A. R., Michel, R. N., Bassel-Duby, R., Olson, E. N., and Williams, R. S. (2000). MEF2 responds to multiple calcium-regulated signals in the control of skeletal muscle fibre type. *EMBO J.* **19,** 1963–1973.

Wu, X., Golden, K., and Bodmer, R. (1995). Heart development in *Drosophila* requires the segment polarity gene *wingless*. *Dev. Biol.* **169**, 619–628.

Xavier-Neto, J., Neville, C. M., Shapiro, M. D., Houghton, L., Wang, G. F., Nikovits, W., Jr., Stockdale, F. E., and Rosenthal, N. (1999). A retinoic acid-inducible transgenic marker of sino-atrial development in the mouse heart. *Development (Cambridge, UK)* **126**, 2677–2687.

Xavier-Neto, J., Shapiro, M. D., Houghton, L., and Rosenthal, N. (2000). Sequential programs of retinoic acid synthesis in the myocardial and epicardial layers of the developing avian heart. *Dev. Biol.* **219**, 129–141.

Xu, C., Liguori, G., Adamson, E. D., and Persico, M. G. (1998a). Specific arrest of cardiogenesis in cultured embryonic stem cells lacking Cripto-1. *Dev. Biol.* **196**, 237–247.

Xu, X., Yin, Z., Hudson, J. B., Ferguson, E. L., and Frasch, M. (1998b). Smad proteins act in combination with synergistic and antagonistic regulators to target Dpp responses to the *Drosophila* mesoderm. *Genes Dev.* **12**, 2354–2370.

Xu, C., Liguori, G., Persico, M. G., and Adamson, E. D. (1999). Abrogation of the *Cripto* gene in mouse leads to failure of postgastrulation morphogenesis and lack of differentiation of cardiocytes. *Development (Cambridge, UK)* **126**, 483–494.

Yamada, M., Szendo, P. I., Prokscha, A., Schwartz, R. J., and Eichele, G. (1999). Evidence for a role of Smad6 in chick cardiac development. *Dev. Biol.* **215**, 48–61.

Yamagishi, H., Olson, E. N., and Srivastava, D. (2000). The basic helix-loop-helix transcription factor, dHand, is required for vascular development. *J. Clin. Invest.* **105**, 261–270.

Yamagishi, H., Yamagishi, C., Wakagawa, O., Harvey, R. P., Olson, E. N., and Srivastava, D. (2001). The combined activities of NKX2-5 and d-Hand are essential for cardiac ventricle formation. *Dev. Biol.* in press.

Yamamura, H., Zhang, M., Markwald, R. R., and Mjaatvedt, C. H. (1997). A heart segmental defect in the anterior-posterior axis of a transgenic mutant mouse. *Dev. Biol.* **186**, 58–72.

Yan, Y.-T., Gritsman, K., Ding, J., Burdine, R. D., Corrales, J. D., Price, S. M., Talbot, W. S., Schier, A. F., and Shen, M. M. (1999). Conserved requirement for *EGF-CFC* genes in vertebrate left-right axis formation. *Genes Dev.* **13**, 2527–2537.

Yatskievych, T. A., Ladd, A. N., and Antin, P. B. (1997). Induction of cardiac myogenesis in avian pregastrula epiblast: The role of the hypoblast and activin. *Development (Cambridge, UK)* **124**, 2561–2570.

Yeh, W.-C., de la Pompa, L., McCurrach, M. E., Shu, H.-B., Elia, A. J., Shahinian, A., Ng, M., Wakeham, A., Khoo, W., Mitchell, K., El-Deiry, W. S., Lowe, S. W., Goeddel, D. V., and Mak, T. W. (2000). FADD: Essential for embryo development and signalling from some, but not all, inducers of apoptosis. *Science* **279**, 1954–1958.

Yelon, D., Ticho, B., Halpern, M. E., Ruvinsky, I., Ho, R. K., Silver, L. M., and Stainier, D. Y. R. (2000). The bHLH transcription factor Hand2 plays parallel roles in zebrafish heart and pectral fin development. *Development (Cambridge, UK)* **127**, 2573–2582.

Yin, Z., and Frasch, M. (1998). Regulation and function of tinman during dorsal mesoderm induction and heart specification in Drosophila. *Dev. Genet.* **22**, 187–200.

Yoshida, K., Taga, T., Saito, M., Suematsu, S., Kumanogoh, A., Tanaka, T., Fujiwara, H., Hirata, M., Yamagami, T., Nakahata, T., Hirabayashi, T., Yoneda, Y., Wang, W. Z., Mori, C., Shiota, K., Yoshida, N., and Kishimoto, T. (1996). Targeted disruption of gp130, a common signal transducer for the interleukin 6 family of cytokines, leads to myocardial and hematological disorders. *Proc. Natl. Acad. Sci. U.S.A.* **93**, 407–411.

Yoshioka, H., Meno, C., Koshiba, K., Sugihara, M., Itoh, H., Ishimaru, Y., Inoue, T., Ohuchi, H., Semina, E. V., Murray, J. C., Hamada, H., and Noji, S. (1998). *Ptx2*, a bicoid type homeobox gene, is involved in a lefty-signalling pathway in determination of left-right asymmetry. *Cell (Cambridge, Mass.)* **94**, 299–305.

Yuan, S., and Schoenwolf, G. C. (1999). Reconstitution of the organizer is both sufficient and required to re-establish a fully patterned body plan in avian embryos. *Development (Cambridge, UK)* **126**, 2461–2473.

Yutzey, K. E., Rhee, J. T., and Bader, D. (1994). Expression of the atrial-specific myosin heavy chain AMHC-1 and the establishment of anteroposterior polarity in the developing chicken heart. *Development (Cambridge, UK)* **120**, 871–883.

Zanin, M. K. B., Bundy, J., Ernst, H., Wessels, A., Conway, S. J., and Hoffman, S. (1999). Distinct spatial and temporal distributions of aggrecan and versican in the embryonic chick heart. *Anat. Rec.* **256**, 366–380.

Zelzer, E., and Shilo, B. Z. (2000). Cell fate choices in Drosophila tracheal morphogenesis. *BioEssays* **22**, 219–226.

Zhang, C. L., McKinsey, T. A., Lu, J., and Olson, E. N. (2001). Association of COOH-terminal-binding protein (CEBP) and MEF2-interacting transcription repressor (MITR) contributes to repression of the MEF2 transcription factor. *J. Biol. Chem.* **276**, 35–39.

Zhang, H., and Bradley, A. (1996). Mice deficient for BMP2 are nonviable and have defects in amnion/chorion and cardiac development. *Development (Cambridge, UK)* **122**, 2977–2986.

Zhang, Y., Cao, L., Yang, B. L., and Yang, B. B. (1998). The G3 domain of versican enhances cell proliferation via epidermal growth factor-like motifs. *J. Biol. Chem.* **273**, 21342–21351.

Zhou, X., Sasaki, H., Lowe, L., Hogan, B. L. M., and Kuehn, M. R. (1993). Nodal is a novel TGF-β-like gene expressed in the mouse node during gastrulation. *Nature (London)* **361**, 543–547.

Zhu, X., Sasse, J., and Lough, J. (1999). Evidence that FGF receptor signalling is necessary for endoderm-regulated development of precardiac mesoderm. *Mech. Ageing Dev.* **108**, 77–85.

Zile, M. H. (1998). Vitamin A and embryonic development: An overview. *J. Nutr.* **128**, 455S–458S.

Zou, K., Muramatsu, H., Ikematsu, S., Sakuma, S., Salama, R. H., Shinomura, T., Kimata, K., and Muramatsu, T. (2000). A heparin-binding growth factor, midkine, binds to a chondroitin sulphate proteoglycan, PG-M/versican. *Eur. J. Biochem.* **267**, 4046–4053.

17

Sex Determination and Differentiation

Amanda Swain and Robin Lovell-Badge

Section of Gene Function and Regulation, Chester Beatty Laboratories, London SW3 6JB, United Kingdom

I. Introduction

II. Gonad Development

III. Sex Determination

IV. Testis Differentiation

V. Cell Movement and Proliferation in the Early Gonad

VI. Ovary Differentiation

VII. Sexual Development

VIII. Evolution and Sex Determination

IX. Conclusion

References

I. Introduction

The existence of two sexes is a common rule in the animal kingdom. The mechanism by which sexual dimorphism is achieved, however, is very different between species. The primary determinant can be genetic, as occurs in most animals, or environmental as in the case of certain reptile species where the temperature of the incubation of the egg determines the sex of the embryo. Even after the sex of the individual has been established, it can either be fixed, as in most cases, or change during its lifetime, which occurs in certain fish and snail species. Understanding the nature and consequence of this diversity has been a major focus of scientific study.

In mammals, the primary sex determinant is genetic and dependent on the Y chromosome. If the Y chromosome is present, the animal will develop as a male, if absent, as a female. Furthermore, the only gene on the Y chromosome that is needed to determine male development is *Sry*, which is present in most mammals. Therefore in normal conditions, females have two X chromosomes, each of which is inherited from father or mother, and males have one X chromosome, inherited from their mother and one Y chromosome inherited from their father. The X contains many genes not present on the Y, so to achieve equal doses of X-linked gene products in males and females, mammals have evolved a mechanism that silences transcription from genes on one X chromosome, a process called X-inactivation. In contrast to flies and worms, mammalian sex determination is not linked to dosage compensation and these processes occur independently of each other. Not all genes are subject to X-inactivation, however, so it is possible that X-linked gene dosage may contribute to differences between sexes. This is evident in marsupials where pouch development in the female depends on having two X chromosomes and scrotum development in the male on having only one, even though gonadal sex is determined by the presence or absence of the

Y (Butler *et al.*, 1999; Renfree and Short, 1988; Oreal *et al.*, 1988).

The genetic sex of the mammalian embryo is therefore established at conception. However, it is not until the period when organogenesis is well under way, around 12 days of gestation in the mouse, that the phenotypic differences between males and females begin to be apparent. These are first observed specifically within the developing gonads. The gonads develop into either testes or ovaries depending on their genotype, XY or XX, respectively. Further sexual development is then established by diffusible products of the gonads that will transmit sex-specific signals to the rest of the body. This means that the primary sex determinant in mammals, *Sry*, drives the switch toward testis rather than ovary development during gonad differentiation (Jost, 1953).

II. Gonad Development

The gonad originates from the urogenital system identically in males and females (called the sexually indifferent phase) and is then modulated by the action of *Sry* in the male so that it develops into a testis rather than an ovary. The urogenital system derives from the intermediate mesoderm, two columns of tissue lateral to the paraxial mesoderm on either side of the midline, which gives rise to the nephrogenic cords. These cords differentiate into three systems along the rostrocaudal axis: pronephros, mesonephros, and metanephros (Fig. 1). These systems are initially composed of mesenchyme and the nephric duct, the drainage component, which runs the length of the embryo starting in the trunk region approximately at the level of the forelimbs and ending in the cloaca. The pronephros is vestigial in mammals and regresses around 10 dpc in the mouse. The mesonephros can serve as an embryonic kidney in some mammalian species, such as human, but in the mouse it is not functional. This system is made up of a series of tubules with some of them connected to the mesonephric or Wolffian duct, which is derived from the nephric duct. These mesonephric tubules regress during embryogenesis with complete degradation by E15 of mouse embryogenesis. The kidneys arise from the metanephros driven by the interaction between the metanephric mesenchyme and the ureteric buds, which are derived from the nephric ducts posterior to the mesonephros (see Chapter 18).

The gonads develop from each mesonephros on the ventromedial side of the nephrogenic cord and are first visible at approximately 10 days of mouse development (Fig. 1). The

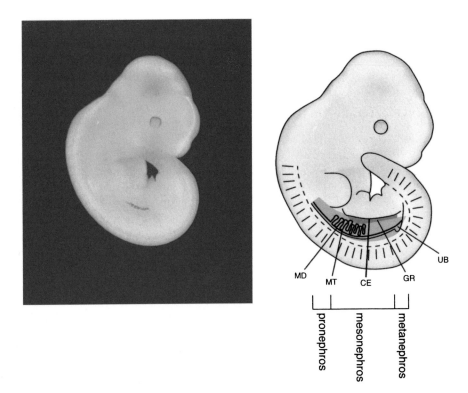

Figure 1 The developing urogenital system in an E11.5 mouse embryo. On the left, a transgenic embryo showing genital ridge expression of β-galactosidase driven by *Dax1* regulatory sequences (see Swain *et al.*, 1998). MD, mesonephric duct; MT, mesonephric tubules; CE, coelomic epithelium; GR, genital ridge; UB, ureteric bud.

early gonads are referred to as genital ridges and are composed of somatic cells derived from the coelomic epithelia and the mesonephros, and primordial germs cells that have immigrated from extragonadal sites (see Chapter 9). The signals that initiate gonad development are not known nor is the origin of all somatic cells, which have been proposed to be derived from the coelomic epithelium, mesonephric mesenchyme, or the mesonephric duct and tubules (Byskov and Hoyer, 1994). Mice lacking PAX2, a member of the paired-box family of transcription factors, show degeneration of the mesonephric duct and tubules soon after they are formed (Torres *et al.*, 1995). Gonad development in these mutant mice is unaffected, suggesting that the ducts and tubules are not essential for this process. *In vitro* organ cultures, where isolated immature mouse gonads are allowed to develop on agar blocks, have been used to study the origin of gonad cells. These studies showed that cells from the coelomic epithelium, which covers the gonadal primordium, contribute to the somatic cells of the gonad but do not prove that they arise exclusively from this source (Karl and Capel, 1998).

Two major cell lineages are found in the somatic cells of the indifferent gonad, the supporting cell lineage and the steroidogenic lineage. It is not clear whether these lineages arise from the same original early gonadal cell type or whether they are separate lineages that are either present from the initial phase or arrive at the gonad at different stages of development. Each of these lineages is thought to have a bipotential fate depending on the sex of the organ in which they are found (Fig. 2). The supporting cells, which surround and nurture the germ cells, will differentiate into Sertoli cells in the testis and follicle cells in the ovary. The steroidogenic cells, which produce the hormones necessary for sexual development, will differentiate into Leydig cells in the testis and theca cells in the ovary. The somatic cells of the gonad also include stromal cells and endothelial cells that contribute to the structure and function of the organ and the coelomic epithelia that eventually give rise to the tunica. These cell types are present in both the testis and ovary although their numbers and precise arrangement differ. There is, however, an additional connective tissue lineage, that of the peritubular myoid cells, that is unique to the testis. This is not present at indifferent stages, however, because the cells only enter the gonad from the mesonephros after Sertoli cells have begun to differentiate (Martineau *et al.*, 1997; Tilmann and Capel, 1999).

The mesonephros contains two sets of ducts: the mesonephric or Wolffian ducts derived from the nephric ducts and the paramesonephric or Müllerian ducts, which originate by invagination of the coelomic epithelium. Differentiation of the Wolffian ducts is dependent on androgens secreted by embryonic Leydig cells (George and Wilson, 1994). In the female, where sexual hormones are not produced during development, the duct degenerates starting around E15. In the male, the Wolffian duct differentiates into the epididymus, vas deferens, and seminal vesicles, which all form part of the genital tract that transports mature sperm from the testis to the urethra (Fig. 3). The Müllerian duct will differentiate into the oviduct, uterus, and upper vagina during female de-

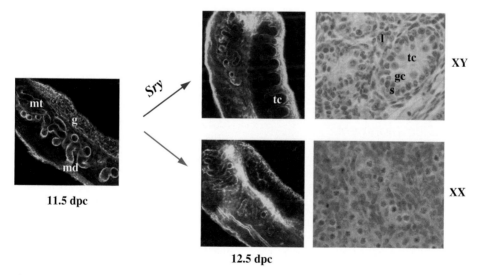

Figure 2 Gonadal morphogenesis during sex determination. Confocal images of laminin staining reveal complex mesonephric tubules, but relatively little structure to the indifferent gonad at E11.5. By E12.5, however, the XY gonad has become very organized, with the testis cord structure being clearly visible. There is no laminin deposited between germ cells and Sertoli cells, hence the clear spaces within the cords. (Adapted from Karl and Capel, 1995, Figs. 2 and 3, pp. 238 and 239.) On the right, eosin and hematoxylin-stained sections reveal the cellular organization within the gonad. Mt, mesonephric tubules; md, mesonephric duct; g, genital ridge; tc, testis cord; gc, germ cells; s, Sertoli cells; l, Leydig cells.

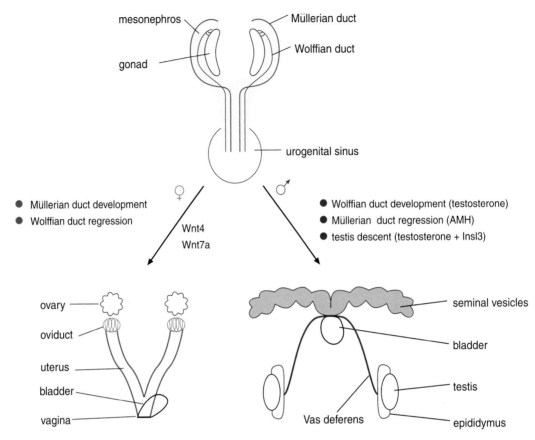

Figure 3 Development of the reproductive system. (Top) Indifferent stage at about E11.5. (Bottom) Adult female and male.

velopment (Fig. 3). To prevent this from happening in the male, the Sertoli cells in the testis produces a diffusible transforming growth factor β (TGF-β)-related growth factor, anti-Müllerian hormone (AMH) or Müllerian inhibiting substance (MIS), that will induce degeneration of this duct (Behringer, 1995; Josso and Picard, 1986; Jost, 1953).

A. Genes Implicated in Early Gonad Development

The generation of mouse mutants has been invaluable in identifying genes involved in the early steps of gonad development. The severe gonadal phenotype generally associated with this group of mutations only allows an assessment of the role of the corresponding genes in early gonad differentiation and not in downstream events such as ovary or testis development. For this, conditional mutations need to be created such that genes can be inactivated in the tissue and developmental stage of choice.

1. Genes Involved in Urogenital Differentiation

Mutations in genes implicated in the differentiation of the intermediate mesoderm and urogenital system as a whole tend to be associated with defects in both early gonad and

kidney development. These genes include *Lim1* (*Lhx1*), *Emx2,* and *WT1. Lim1* and *Emx2,* which encode homeobox containing proteins, are expressed in the derivatives of the nephric duct, whereas *WT1* is expressed in the mesenchyme and coelomic epithelium of the urogenital system (Fujii *et al.,* 1994; Miyamoto *et al.,* 1997; Armstrong *et al.,* 1993; Pelletier *et al.,* 1991b). Most mice homozygous for deletions in *Lim1* die around 10 days of gestation but the few pups that are born have no kidneys or gonads (Shawlot and Behringer, 1995). A similar phenotype is seen in all mice lacking *Emx2,* (Miyamoto *et al.,* 1997). Alkaline phosphatase staining, which is used to identify primordial germ cells, revealed that germ cell migration was normal in *Emx2* mutant embryos. This suggests that the initial specification of the gonad, which presumably is needed for the final stages of germ cell migration, is not affected by the lack of *Emx2.* Interestingly, the *Pax-2* gene has a similar expression pattern in the urogenital system to *Lim1* and *Emx2* but the phenotype of the *Pax2*-deficient mice is different in that they develop gonads but no kidneys (Torres *et al.,* 1995). Further studies are needed to understand the role of *Lim1* and *Emx2* in mesonephros and gonad development.

The Wilms' tumor-associated gene (*WT1*) has been associated with defects in urogenital development in both mouse

Table I Breakdown of Gene Involvement at Specific Stages of Gonadal Development

Genes involved in early gonad development	
Associated with gonad and kidney development	Associated with gonad development
WT1	SF1
Emx2	M33
Lim1	Lhx9

Genes involved in sex determination	
Associated with testis development	Associated with ovary development
Sry	Dax1
SF1	

Genes involved in testis development	
Associated with Sertoli cell differentiation	Associated with Leydig cell differentiation
Sox9	
WT1	SF1
GATA4	Wnt4
SF1	

Genes involved in duct development	
Associated with Müllerian duct development	Associated with Wolffian duct development
Wnt4	
Wnt7a	
Amh	AR
Amh receptor	

and human. Mice with a disrupted *WT1* gene lack gonads and development of the kidney is arrested (Kreidberg *et al.,* 1993). The mesonephroi of these deficient mice develop normally although the number of tubules is reduced because those unattached to the mesonephric duct fail to form (Sainio *et al.,* 1997). During embryogenesis, *WT1* is expressed throughout the intermediate mesoderm at 9 days of development and later in the mesonephros and gonad and the differentiating mesenchyme of the metanephros (Armstrong *et al.,* 1993; Pelletier *et al.,* 1991a). The continued association of the expression of *WT1* with both ovary and testis development suggests that this gene is important throughout gonad differentiation.

Mutations in *WT1* are implicated in three different but related syndromes in humans (Little *et al.,* 1999; Reddy and Licht, 1996). Heterozygous deletions that include *WT1,* called the WAGR (Wilms' tumor, aniridia, genitourinary anomaly, mental retardation) syndrome, are associated with mild genitourinary malformations and predisposition to Wilms' tumors, a childhood kidney tumor. Denys-Drash

syndrome (DDS) patients display early-onset renal failure, Wilms' tumor, and genital anomalies, with XY individuals being most affected and sometimes showing ambiguous or female genitalia. Point mutations in one allele of *WT1* leading to amino acid substitutions or premature termination of the protein are associated with the DDS phenotype (Pelletier *et al.,* 1991b). These mutant proteins cannot bind DNA and it has been proposed that they interfere with the function of the wild-type protein and, therefore, via a dominant-negative effect, induce a more severe phenotype than is seen in WAGR patients (M. Little *et al.,* 1995; M. H. Little *et al.,* 1993). XY patients with Frasier syndrome develop as females with defects in early gonad development, and they also display renal failure, but later in life than the DDS patients. These individuals carry splice site mutations that prevent the production of the isoform containing the amino acids KTS in one allele of *WT1* (Barbaux *et al.,* 1997). This suggests that the phenotype in these patients is due to changes in isoform ratio, rather than to the production of a mutant protein and, therefore, several different functions of the protein may be involved in the disease.

2. Genes Involved in Gonad Differentiation

Steroidogenic factor 1 (*SF1*) and *Lhx9* are the only genes known so far to be expressed specifically in the early gonad at the indifferent stage in both sexes and to be required for development of the genital ridge. Mice with a disrupted *SF1* gene show normal early gonad development, including colonization by germ cells, but differentiation is arrested around E11–11.5 and the cells degenerate via apoptosis (Luo *et al.,* 1994). This suggests that *SF1* is not involved in the initial stages of specifying gonad identity; rather it is required for differentiation and/or maintenance and growth of the somatic cells already present in the early indifferent gonad. Expression studies in the rat, using an antibody to SF1, have shown that the protein is present at the initial phase of gonad development and even before the gonad can be seen morphologically, where it may be marking the gonad primordium (Hatano *et al.,* 1996; Morohashi, 1997). It was observed that the early SF1 positive cells first comprised a coherent group that in later stages divided into two, one group forming part of the gonad and the other part of the adrenal. This supports the proposal that the somatic cells of the gonad and adrenal cortical cells are derived from the same primordia (Hatano *et al.,* 1996, and references therein). Consistent with this expression pattern, *SF1*-deficient mice also lack adrenals.

SF1 is a member of a subfamily of nuclear receptors, the orphan receptors, for which no activating ligand has been found. This transcription factor has a highly conserved DNA-binding domain composed of two zinc fingers and it was first identified as an activator of genes involved in steroid biosynthesis in different steroid hormone producing cells *in vitro* (Ikeda *et al.,* 1993; Morohashi *et al.,* 1993). Steroidogenic cells of the gonad and cortical cells of the adrenal produce steroids during embryogenesis, therefore it is thought that *SF1* is directly involved in regulation of steroid production in these organs during development. Expression studies have shown that *SF1* is not only expressed in the gonads and adrenals but also in the pituitary and hypothalamus (Hatano *et al.*, 1994; Ikeda *et al.*, 1994; Ingraham *et al.*, 1994; Morohashi *et al.*, 1994). Consistent with this, the *SF1*-deficient mice have impaired pituitary function and ventromedial hypothalamic structure (Ikeda *et al.*, 1995; Ingraham *et al.*, 1994; Shinoda *et al.*, 1995). These studies show that *SF1* has a role in the reproductive axis at multiple levels. A heterozygous mutation in SF1 that abolishes DNA binding of the protein was found in one XY human individual showing sex reversal. This indicates that the role of *SF1* is conserved in humans but that it is perhaps more sensitive to gene dosage than in mice, where animals with a heterozygous disruption in this gene appear normal (Achermann *et al.*, 1999).

Lhx9 is a member of the LIM class of homeobox proteins, and mice with a disruption in this gene show defects in early gonad development (Birk *et al.*, 2000). The phenotype is evident around E12 when the gonads fail to proliferate, as measured by BrdU staining, compared to wild-type littermates, but in contrast to *SF1*-deficient mice, there seems to be no evidence of apoptosis. Analysis of gene expression in *Lhx9* mutant embryos showed a decrease in *SF1* expression at E11.5, while levels of *WT1* were not altered. This decrease was not observed in other regions of *SF1* expression such as the adrenals where *Lhx9* is not expressed. Also, *Lhx9* expression is not affected in gonads from *SF1*-deficient mice at the same stage. These data suggest that *Lhx9* acts in the same cells and pathway as *SF1* in the indifferent gonad but is independent of *WT1*. In contrast to all the genes that have been described in this section, gonad development is the only process that is affected by the lack of *Lhx9*.

3. Genes Affecting Gonad Development

Analysis of mice and humans deficient in the activity of certain factors implicated in global transcriptional regulation and chromatin domain organization has uncovered a role for these factors in early gonad development and/or sex determination. *M33*, a gene related to the *Drosophila* polycomb group genes (PcG), is thought to have a role in gonad development. Mice deficient for *M33* show retarded gonad development as early as E11.5 (Katoh-Fukui *et al.*, 1998). The phenotype of the gonads at birth in these mice is variable. The ovaries of the XX animals range from small to absent. The XY animals display different degrees of sex reversal, with none of the mice having bilateral testes and most showing some female-specific development. This phenotype suggests that *M33* acts early in gonad development upstream of the trigger of testis development by *Sry*. The PcG proteins in *Drosophila* are part of a system that controls the expression of homeotic genes, which specify positional information in the embryo. The role of PcG proteins is to repress gene ex-

pression by forming a multimeric complex that compacts chromatin, making it inaccessible to transcription factors.

The trithorax proteins are also part of this system and they act antagonistically to the PcG genes by maintaining active homeotic gene transcription. It is thought that these proteins might be operating in the same way in the mouse. Mice deficient in PcG-related genes such as *M33* and *Bmi-1* show homeotic transformations of the axial skeleton, which are intensified when both genes are absent (Bel *et al.*, 1998; Core *et al.*, 1997; Katoh-Fukui *et al.*, 1998; van der Lugt *et al.*, 1994). Absence of the trithorax group-related gene *MLL* in mice also causes homeotic transformations, but the shift in homeotic expression is generally in the opposite direction to the PcG proteins (Yu *et al.*, 1995). However, when mice are deficient for both *Bmi-1* and *MLL* the transformations are absent, suggesting that the genes act antagonistically to determine segmental identity (Hanson *et al.*, 1999). Further studies are needed to determine whether the position of the gonad along the anterior-posterior axis of the urogenital system is established by combinatorial action of homeotic genes that are regulated by genes such as *M33*.

Another transcriptional regulator, *ATRX*, has been implicated in sex determination in humans. XY patients with mutations in this X-linked gene show sex reversal, which is associated with other phenotypes such as mental retardation (Ion *et al.*, 1996). The ATRX protein contains a highly conserved PHD (plant homeodomain-like) domain, present in many chromatin-associated proteins, and a carboxy-terminal domain similar to that of the SNF2 family of helicase/ATPases. Studies on this protein have suggested that it acts as a global transcriptional regulator through an effect on chromatin remodeling (Gibbons *et al.*, 2000). The mouse ATRX homolog was found to have a widespread pattern of expression during the early stages of embryogenesis (Stayton *et al.*, 1994). Interestingly, a homolog to the ATRX gene was found on the Y chromosome of marsupials. This ATRY gene expression is specific to the testis, suggesting it is important in marsupial testis development (Pask *et al.*, 2000a). The human and mouse studies suggest that ATRX has a role in many different processes during embryogenesis and therefore conditional mutations in this gene in the mouse will be necessary to determine its role in early gonad development.

III. Sex Determination

A. *SRY*

The study of human individuals with defects in sexual development has been important in the identification of genes involved in sex determination. The *SRY* gene (*Sex determining Region on the Y*) was isolated by studying XX male patients with a small part of the Y chromosome transferred onto an X chromosome by abnormal X–Y interchange during male meiosis. *SRY* was found within the minimum region of Y-specific DNA required for male development (Sinclair *et al.,* 1990). This positional cloning strategy was supported by looking for sequences conserved on the Y chromosome of other mammals and by looking at cases of sex reversal in mice (Gubbay *et al.,* 1990). Analysis of human patients with mutations in *SRY,* who have a female phenotype but an XY genotype, has shed light on the functional regions in the protein (Goodfellow and Lovell-Badge, 1993; Pontiggia *et al.,* 1994). Apart from helping to identify the gene, the mouse has provided an important working model to study the process of mammalian sex determination (Gubbay *et al.,* 1990). A transgenic experiment determined that *Sry* is the only gene on the Y chromosome that is necessary and sufficient to establish male development. XX animals transgenic for a 14-kb fragment of Y chromosome that contains only the mouse *Sry* gene develop as males (Koopman *et al.,* 1991) (Fig. 4). These transgenic males are infertile, however, because spermatogenesis is blocked due to the presence of two X chromosomes and the lack of other Y-linked genes (Burgoyne *et al.,* 1992).

The control of sex determination by *SRY* is likely to occur at the level of transcription of critical target genes. The SRY protein contains an HMG box type of DNA-binding domain similar to that found in a number of other transcription factors. This domain is thought to be the main functional part of the protein because analysis of human XY female patients has shown that, with one exception, all mutations in *SRY* cluster within this domain (Harley *et al.,* 1992). In addition, the HMG domain is relatively conserved, unlike the rest of the protein, which is very dissimilar between all but closely related species (Tucker and Lundrigan, 1993; Whitfield *et al.,* 1993). HMG box domains are known to bind DNA in the minor groove and to bend the DNA to acute angles (Pevny and Lovell-Badge, 1997). This has led to the idea that SRY might act by changing the chromatin configuration and in this way influencing neighboring interactions between DNA and proteins. Consistent with this, mutations in sex-reversed patients can affect DNA bending and binding independently (Pontiggia *et al.,* 1994; Werner *et al.,* 1995).

While the HMG box is clearly essential, it may be more sensitive to point mutations than other parts of the SRY protein. However, the rate of change of the nonbox regions in evolution has led to suggestions that they are subject to selection and may have a species-specific role (Tucker and Lundrigan, 1993; Whitfield *et al.,* 1993). In addition, further transgenic experiments in mice have suggested that regions of the protein outside the HMG box domain are important for the sex-determining function of SRY. The mouse SRY protein contains a glutamine-rich region with weak transactivating properties, which is not conserved in other mammals; indeed it is of variable length in different mouse species. XX transgenic mice that should produce a truncated protein lacking this region, created by engineering a transla-

A

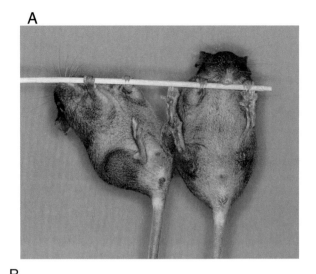

B

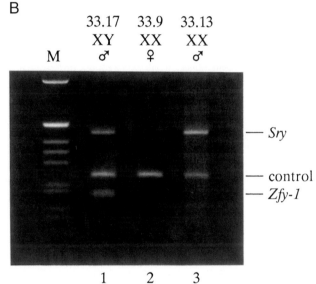

Figure 4 XX male sex reversal with an *Sry* transgene. (A) On the left is a control XY male (m33.17) and on the right, his littermate (m33.13), an XX male transgenic for a 14-kb genomic fragment containing only the mouse *Sry* gene. The external genitalia are indistinguishable as are all internal reproductive organs. The only exception is the testes, which, completely lacking germ cells, are much smaller in the sex-reversed male (not shown). (B) PCR analysis of genomic DNA from these two mice as well as from a control XX sibling. Bands for *Sry* and for the DNA loading control (myogenin) are present in XX male m33.13, but the Y chromosome marker gene *Zfy1,* is missing. M, size markers. (Adapted from Koopman *et al.,* 1991, from which further details may be obtained.)

tional stop codon within the SRY open reading frame, do not exhibit sex reversal in contrast to the nonmutated transgene (Bowles *et al.,* 1999).

Studies in the mouse have shown that *Sry* is expressed exclusively in the male genital ridge starting at about 10.5 days of embryogenesis. Transcript levels reach a peak at E11.5 and then decline to undetectable levels by E12.5 (Hacker *et al.,* 1995; Jeske *et al.,* 1995). When looked at in detail by whole mount *in situ* hybridization, there appears to be a wave of expression starting at the anterior end of the gonad and ending at the posterior end, so that each cell may express *Sry* for only a few hours (Swain *et al.,* 1998). This pattern suggests that *Sry* acts to trigger testis differentiation and is not involved in the subsequent development of the organ or maintenance of any testis-specific patterns of gene expression. Chimeric mice derived from aggregating early XX and XY embryos have been used to determine the cell lineages that are involved in sex determination. These studies showed that almost all Sertoli cells in the chimeric testis had an XY genotype, whereas other lineages showed no chromosomal bias (Palmer and Burgoyne, 1991a). The action of *Sry* is therefore thought to trigger Sertoli cell differentiation from the supporting cell lineage, and these cells in turn direct the differentiation of the rest of the cell lineages along the male pathway.

Different strains of mice that exhibit sex reversal under certain conditions have been used to study the mechanism of *Sry* action. The *Sry* gene from some strains of *M. musculus domesticus* are considered weak alleles because they cause delays in testis formation and sometimes complete sex reversal. Comparison of sequences between strong and weak alleles of *domesticus Sry* has revealed differences in regions of the protein within and outside the HMG box domain. However, there is no strong correlation between a particular change and the severity of sex reversal (Albrecht and Eicher, 1997; Carlisle *et al.,* 1996; Coward *et al.,* 1994). On the other hand, expression studies have found a correlation between levels of *Sry* mRNA and sex reversal, the weaker alleles showing lower levels of *Sry* (Nagamine *et al.,* 1999). Deletions of regions of the Y chromosome at some distance from the *Sry* locus, thought to allow the spread of centromeric heterochromatin, can also reduce levels of *Sry* expression and lead to sex reversal (Capel *et al.,* 1993b). These studies support the proposal that there is a window of time in the development of the gonad where SRY can act to ensure testis differentiation (Burgoyne and Palmer, 1991). Therefore if *Sry* does not achieve a certain level of expression at a particular time in gonad development, testis differentiation will be delayed and ovary development will occur.

B. Other Candidate Genes Involved in Sex Determination

1. *Tda* Loci

A well-studied sex reversal system in mice is that involving the weak *domesticus Sry* allele from a subspecies originating in the Val Poschiavino in Switzerland (Eicher *et al.,* 1982). This poschiavinus *Sry* allele (Sry^{POS}) was found to cause a 14-hr delay in testis development when present on most genetic backgrounds (Palmer and Burgoyne, 1991b). On a C57BL/6 background, however, Sry^{POS} fails to trigger proper testis development in XY animals and the gonads de-

velop either as ovotestes, with both ovarian and testicular components, or as ovaries. Ovotestes frequently fail to maintain both components. This is a problem that can confound studies of isolated cases of sex reversal, particularly those affecting humans. However, in the study of Sry^{POS} it has been useful because if the ovarian component is lost, fertile males may result. When these are outbred to a different strain, all XY^{POS} animals develop as normal males. However, by backcrossing these to C57BL/6 females it was possible to carry out a genome scan, correlating sex reversal with the presence of C57BL/6 chromosome segments, to map the genes involved in modulating SRY action (Eicher *et al.,* 1995). These studies have identified three autosomal loci, which map to chromosome 4 (*tda1*), 2 (*tda2*), and 5 (*tda3*). The genes associated with these loci have not yet been identified. It has been proposed that these loci code for factors that promote ovary development, those derived from C57BL/6 being stronger acting (Burgoyne and Palmer, 1991). However, expression studies on SRY^{POS}, which show that the levels of *Sry* mRNA are lower in the C57BL/6 background when compared to other genetic backgrounds, suggest that these loci might be involved in regulating *Sry* transcription (Nagamine *et al.,* 1999).

2. *Dax1*

It is very likely that some genes act to promote female development. The best described candidate for such a gene is *Dax1*, an X-linked member of the nuclear hormone receptor superfamily. Expression studies in the mouse have shown that *Dax1* is present in the genital ridge at the same time as *Sry,* and within the same cell type (the supporting cell precursors) but in both sexes (Swain *et al.,* 1996). In humans, *DAX1* maps to the minimum region of the X chromosome found to be duplicated in a group of XY patients that display female development (Bardoni *et al.,* 1994; Muscatelli *et al.,* 1994; Zanaria *et al.,* 1994). It was therefore proposed as a candidate gene responsible for the dosage-sensitive sex reversal (DSS) syndrome and considered to be an ovarian-determining gene that could somehow interfere with SRY action when present in a double dose. When extra copies of *Dax1* were introduced as transgenes into XY mice, it was found that they could delay testis cord formation in animals with strong *Sry* alleles and induce sex reversal when tested against weak alleles of *Sry* (Swain *et al.,* 1998) (Fig. 5). This shows that *Dax1* acts antagonistically to *Sry* and can have an antitestis function. However, although these experiments provide proof that DAX1 is responsible for DSS, they did not address whether it has a normal role in ovary formation. Disruption of the *Dax1* gene in mice in fact revealed no effect on either ovary or testis development, which suggests that the gene has no critical function as an ovarian determinant, although it could conceivably be part of a redundant system (Yu *et al.,* 1998).

Dax1 is thought to act in the same pathway as *SF1*. Their expression patterns during embryogenesis correlate quite

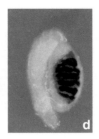

Figure 5 Effect of *Dax1* overexpression on gonadal development in XY[POS] embryos on a mixed background. *Amh* has been used as a marker of Sertoli cell differentiation. (A) Nontransgenic XX control ovary. (B–E) Gonads from transgenic XY[POS] embryos. The latter have a range of phenotypes from almost complete ovary, with a small amount of testicular tissue in panel B, through clear ovostestes in panels C and D and an almost normal testis in panel E. The testicular portion would probably have been lost from the gonad in panel A, if this embryo had been allowed to develop further, giving an XY female. Likewise, that in panel D could have lost the ovarian portion. The gonads were isolated from stage-matched E14.5 embryos derived by crossing XX females of the transgenic line 1812, which express *Dax1* at levels five times higher than the endogenous gene, with XY[POS] males. The *poschiavinus Sry* gene on the Y[POS] chromosome acts as a weak allele, especially on some genetic backgrounds, but in this experiment, control, nontransgenic XY[POS] gonads never showed any ovarian tissue. (Adapted from Swain *et al.,* 1998, from which further details may be obtained.)

closely in the gonad, adrenal, pituitary and hypothalamus (Ikeda *et al.,* 1996). Comparison of the *Dax1*- and *SF1*-deficient mice shows that these genes are involved in the development and function of the same tissues. However, the phenotypes in these mice are different, indicating a complex interaction between the two genes. This difference is seen in the phenotype associated with the gonad and adrenal. Disruption of *SF1* leads to absence of adrenals and gonads, whereas disruption of *Dax1* does not affect development of either. After birth, however, *Dax1*-deficient mice do display abnormalities in these tissues. Males are infertile due to a block in spermatogenesis and normal regression of the fetal zone of the adrenal cortex does not occur in both sexes, even though this does not seem detrimental to the function of the organ (Yu *et al.,* 1998).

DAX1 is an unusual member of the nuclear hormone receptor family in that it does not possess a classic zinc finger DNA-binding domain, but contains instead a novel domain with a repeated motif, which is linked to a ligand-binding domain. As in the case of SF1, it is not known whether a ligand is involved in the action of this protein. *In vitro* studies have shown that DAX1 can inhibit transcriptional activation mediated by SF1 (Yu *et al.,* 1998). The mechanism of inhibition has not been established but might involve protein–protein interactions where regions of the ligand-binding domain of DAX1 act as transcriptional silencers or recruit other repressors to the complex. The novel domain of DAX1 has been shown to bind hairpin-loop structures in the DNA *in vitro* suggesting that this protein might be sensitive to chromatin configuration (Zazopoulos *et al.,* 1997). This is not a conserved part of the protein, however, because it is largely missing in the avian gene (Smith *et al.,* 2000). Phosphorylation of SF-1 by the mitogen-activated protein kinase pathway has been observed to affect SF1 action *in vitro,*

suggesting a complex interaction between these two nuclear hormone receptors and the membrane signaling pathway (Hammer *et al.,* 1999). These interactions, however, have not yet been confirmed in an *in vivo* situation.

Analysis of the promoter for *Dax1* in adrenal- or gonad-specific tissue culture cells has indicated that SF1 activates *Dax1* expression (Burris *et al.,* 1995; Ikeda *et al.,* 1996; Kawabe *et al.,* 1999; Yu *et al.,* 1998). Putative SF1 binding sites have been found upstream of the *Dax1* transcriptional start site, and mutations in these sites can lead to a lowering of *Dax1* expression, as measured by the activity of reporter genes linked to *Dax1* promoter elements, in cells. However, SF1 is not essential for *Dax1* expression because SF1-deficient mice show *Dax1* expressing cells in the degenerating gonad albeit at lower levels (Ikeda *et al.,* 1996; Kawabe *et al.,* 1999). These studies suggest that SF1 is a master regulator of the development of organs such as the gonad and adrenal and that fine-tuning of the molecular pathway is achieved by increasing the levels of repressors such as DAX1 that will modulate SF1 action when needed.

Another factor implicated in the regulation of *Dax1* is WT1. Putative WT1 binding sites are present upstream of the *Dax1* transcriptional start site, and *in vitro* studies have shown that the form of WT1 that lacks KTS will bind to these sites (Kim *et al.,* 1999). This isoform will activate transcription of the *Dax1* gene in tissue culture cells, including the endogenous gene. WT1 is present in the developing gonad in both sexes at the time of *Dax1* expression. However, WT1 is also expressed in the mesonephros and metanephros, which are not sites of *Dax1* expression. This suggests that WT1 interacts with other transcription factors, including SF1, to achieve correct expression of *Dax1*. *In vivo* studies are needed to confirm the role of WT1 in the regulation of *Dax1* and to determine how the latter is involved in sex determination.

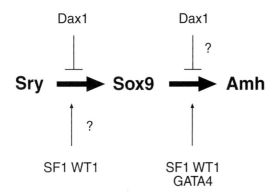

Figure 6 Model for molecular interactions within the supporting cell lineage during sex determination. Expression of *Sry* leads to the activation of *Sox9*, which in turn leads to the activation of Sertoli cell-specific gene expression, as illustrated by *Amh*. Various factors help to promote this pathway, such as SF1, WT1, and GATA4, whereas others, such as DAX1, try to repress it. In the absence of *Sry*, the "repressors" are sufficiently robust to prevent *Sox9* expression and Sertoli cell differentiation. There is reasonable evidence for some of the interactions in this model, whereas others, marked with question marks, have not been demonstrated *in vivo*.

It remains to be established whether the antagonistic action of DAX1 in sex determination involves SF1. As the genital ridge develops, SF1 expression is upregulated in the testis, specifically in Sertoli and Leydig cells, when compared to the ovary (Hatano *et al.,* 1994; Shen *et al.,* 1994). The fact that SF1 is present at the same time as SRY and DAX1 in the early genital ridge and is later expressed in testicular Sertoli cells suggests that SF1 plays a role in testis differentiation. Conversely, *Dax1* expression is downregulated in the developing testis while it continues to be expressed in the ovary (Swain *et al.,* 1996). This is consistent with DAX1 having an antitestis function by inhibiting SF1 action in the early genital ridge, at least when the former is overexpressed. In normal XY male development, the role of SRY might then be to inhibit DAX1 repression and allow SF1 to activate testis differentiation, although any molecular mechanism as to how this might happen awaits finding critical target genes for SRY (Fig. 6). Also, as suggested above, DAX1 might not be the only repressor of SF1 action. Tissue- and stage-specific mutations in *SF1* are needed to address whether this gene is directly involved in sex determination, because they will allow the assessment of its function at later stages of gonadal development, for example, at the time when *Sry* acts.

IV. Testis Differentiation

A. Sertoli Cell Differentiation

1. *Sox9*

The transient nature of *Sry* expression in the mouse implies that one of the first downstream events that must occur after the switch to testis development is the activation of

genes involved in defining and maintaining Sertoli cell identity. A candidate for this type of downstream gene is *Sox9*. This is a gene with critical functions in several systems, notably in developing cartilage, where it is involved in specifying the chondrocyte lineage and in regulating genes such as type II collagen (Bell *et al.,* 1997; Bi *et al.,* 1999; Foster *et al.,* 1994; Wagner *et al.,* 1994). However, it also seems to play a vital role in testis development. Expression studies in the mouse have shown that *Sox9* is present at low levels in both sexes in the early genital ridge. At E11.5, when the level of *Sry* expression is at its peak, *Sox9* expression is upregulated in the male gonad and turned off in the female gonad (Kent *et al.,* 1996; Morais da Silva *et al.,* 1996). *Sox9* expression is then associated with Sertoli cells throughout their life and is never seen in the ovary. This pattern is consistent with *Sox9* being directly regulated by the action of SRY and suggests that it might be important in determining Sertoli cell fate. Importantly, mutations in one allele of the gene in humans is associated with male to female sex reversal in XY individuals suffering from the severe dwarfism syndrome, campomelic dysplasia (CD) (Foster *et al.,* 1994; Wagner *et al.,* 1994).

Analysis of the *Sox9* promoter will determine the factors that drive expression of this gene in the male genital ridge. One possible model, which relates to the discussion above, is that SF1 is involved in the activation of this gene, whereas DAX1 and other factors are repressors. SRY might then antagonize this repressor action, possibly through bending DNA and changing the chromatin structure (Fig. 6). Testing this model, however, depends on identifying the critical gonad-specific regulatory region of *Sox9*. *In vitro* studies suggested that there was such an element very close to the transcription start site (Kanai and Koopman, 1999). However, this has not been borne out by *in vivo* studies. There is evidence from human CD patients that the regulation of *SOX9* may depend on elements spread out over a long distance, because translocation breakpoints have been identified more than 1 Mb upstream (Kwok *et al.,* 1995; Pfeifer *et al.,* 1999). Studies using human *SOX9* YAC constructs to express a reporter gene in transgenic mice also support this view, as one with 450 kb 5′ failed to give testis expression (Wunderle *et al.,* 1998). Recent findings with a chance integration event in a transgenic experiment have also been interpreted to suggest a very distant regulatory region (Bishop *et al.,* 2000). A tyrosinase transgene integration in a region 1.3 Mb upstream of the *Sox9* gene was found to be associated with ectopic expression of Sox9 in the XX genital ridge. Interestingly, the transgenic XX embryos developed as males. However, as yet there is nothing to rule out a general long-range effect on expression of *Sox9* in many if not all tissues, which would be consistent with a case of human XX male sex reversal involving a duplication of the entire SOX9 locus (Huang *et al.,* 1999). Nevertheless, both of these latter two studies suggest that when *Sox9* is expressed at higher than normal levels, it may be sufficient to ensure proper testis development. It may therefore be the only critical gene downstream of SRY required for Sertoli cell and testis differentiation.

SOX9 protein is a transcription factor that is related to SRY is that it contains a similar HMG box DNA-binding domain. It differs from SRY in that it also contains a strong transcriptional activation domain in the carboxy terminus (Kamachi *et al.*, 1999; Ng *et al.*, 1997; Sudbeck *et al.*, 1996). In addition, the entire protein is highly conserved throughout vertebrate evolution, unlike SRY, which, apart from being highly divergent, has been found only in eutherian and metatherian mammals. Different regions of the SOX9 protein were found to be targets for mutations in patients showing male to female sex reversal, implicating several domains, including the HMG domain and the activation domain, in its function in sexual development (Kwok *et al.*, 1995; Meyer *et al.*, 1997). The nature of the mutations implies that the phenotype in these patients is due to haploinsufficiency for *SOX9*. This suggests that levels of SOX9 are critical for its function in testis development and the protein must achieve a critical threshold in the gonad to be effective.

2. *Amh* Regulation

One of the functions of Sertoli cells is to produce AMH, a member of the TGF-β family of proteins that induces regression of the Mullerian ducts in the male embryo. The *Amh* gene is first expressed at about E12, specifically in Sertoli cells, and expression continues until birth (Hacker *et al.*, 1995; Munsterberg and Lovell-Badge, 1991). *Amh* expression is absent in the developing female gonad. The regulation of this gene has been intensely studied, because it is an ideal system for identifying transcription factors that are important in Sertoli cell function (Fig. 6). The *Amh* locus is located close to a gene encoding a spliceosome-associated protein, *SAP62,* the *Amh* transcriptional start site being about 300 bp downstream of the *SAP62* polyadenylation site (Dresser *et al.*, 1995). It is thought that the promoter elements that drive *Amh* expression to the developing Sertoli cells are located close to the transcriptional start site. Consistent with this, transgenic mice with 180 bp upstream of the transcriptional start site of *Amh* linked to the human growth hormone gene showed expression of this marker in the Sertoli cells of the developing gonad (Giuili *et al.*, 1997).

Analysis of the sequence of the promoter region of mouse *Amh* revealed a consensus-binding site for proteins containing an HMG box domain, which is conserved in other mammals. *In vitro* studies have shown that SRY can bind to this sequence but only weakly (Haqq *et al.*, 1993; Mertin *et al.*, 1999; Morais da Silva, 1998). The expression patterns of both genes suggests that *Amh* is not a direct downstream target of SRY because the onset of *Amh* expression is at least 24 hr later than *Sry* and continues to be expressed in Sertoli cells when *Sry* transcripts are not present. A better candidate

for binding to the HMG domain consensus site is SOX9, which is present when *Amh* is activated and remains in Sertoli cells throughout their development. *In vitro* studies have shown that SOX9 will bind efficiently to this site and activate transcription of the *Amh* gene in tissue culture cells (De Santa Barbara *et al.*, 1998). An elegant study in mouse introduced a mutation that abolished binding of SOX9 to this site, by targeted mutagenesis within the context of the endogenous *Amh* promoter (Arango *et al.*, 1999). XY animals homozygous for this mutation did not express *Amh* and therefore showed no regression of the Mullerian ducts, developing instead both female and male reproductive tracts. This indicates that SOX9 is likely to be an essential regulator of *Amh* expression and that it must bind its target DNA to function, rather than merely relying on protein–protein interactions. Perhaps the latter is because it also needs to bend DNA to orchestrate local chromatin architecture and the interaction of other transcription factors.

A conserved binding site for SF1 is also found in the *Amh* promoter (Shen *et al.*, 1994). SF1 is present in early Sertoli cells but is also found in other cell lineages of the developing testis such as Leydig cells, which do not express *Amh*. In tissue culture cells, SF1 has been shown to cooperate with SOX9 in the activation of the *Amh* promoter (De Santa Barbara *et al.*, 1998). When the SF1 binding site is mutated in both alleles by targeted mutagenesis, such that SF1 can no longer bind, *Amh* is still expressed, although at significantly reduced levels (Arango *et al.*, 1999). These mice still produce enough AMH to induce Mullerian duct regression. However, when one such *Amh* allele with the SF1 binding site mutation is combined with an *Amh* null allele, defects in Mullerian duct regression are observed. These studies suggest that SF1 is involved in the regulation of *Amh* expression but in a different way to SOX9. The role of SF1 might be as a quantitative regulator of *Amh* transcription, to increase levels above those produced by the action of SOX9, while not being essential itself for expression. Alternatively, SF1 action might not be entirely dependent on binding to its target DNA, and interactions with neighboring proteins such as SOX9 may suffice. Recently, an additional more distal SF1 binding site has been recognized (Watanabe *et al.*, 2000). This site may be responsible for the low level of *Amh* expression seen when the proximal SF1 site is mutated. It may also be responsible for *Amh* expression at later stages of testis development and/or in the postnatal ovary.

Two other factors, WT1 and GATA4, have also been implicated in the regulation of *Amh* expression. These studies, however, have only been carried out with *in vitro* systems and *in vivo* validation is necessary to establish the role of these two proteins in early testis development. For WT1, it has been shown that the isoform that lacks KTS can potentiate the activation of *Amh* transcription by SF1 in tissue culture cells (Nachtigal *et al.*, 1998). This synergistic interaction is inhibited by DAX1. In this context, WT1 seems to act as a coactivator, because it is unable to directly activate or bind to *Amh* reporter constructs on its own. In contrast, conserved binding sites for the GATA family of transcription factors is found in the *Amh* promoter (Viger *et al.*, 1998; Watanabe *et al.*, 2000; Beau *et al.*, 2000). Expression studies showed that GATA4 is the only member of this family to be present in the developing gonads (Viger *et al.*, 1998). GATA4 protein binds to the consensus-binding site and activates expression of a reporter construct. Both WT1 and GATA4 are expressed in the developing Sertoli cells of the testis but also in other tissues, WT1 in the mesonephros and GATA4 in interstitial cells of the testis and in the ovary. This suggests that they do not activate *Amh* on their own but rather act in concert with other tissue-specific factors such as SOX9 and SF1 to achieve a unique Sertoli-specific pattern of expression (Fig. 6).

3. *Dmrt1*, a New Candidate Gene

A recent addition to candidates involved in testis development is the *Dmrt1* gene. This gene encodes a protein with a DM domain, which is a novel DNA-binding domain found in other proteins known to be involved in sexual development of flies and worms (Raymond *et al.*, 1998). In humans, *DMRT1* maps to chromosome 9p21, a region associated with XY female sex reversal. Patients with heterozygous translocations or deletions, which either remove *DMRT1* or could affect its expression, are affected. However, point mutations in the gene have not yet been found in other affected patients, which might be expected if simple haploinsufficiency is responsible for the sex reversal (Ottolenghi *et al.*, 2000; Raymond *et al.*, 1998). Expression of *Dmrt1* in the mouse shows that it is present in the early genital ridge in both sexes, but as development proceeds its expression is higher in the testis than the ovary (De Grandi *et al.*, 2000; Raymond *et al.*, 1999; Smith *et al.*, 1999). In the testis, *Dmrt1* expression is associated with Sertoli and germ cells. Targeted mutation of this gene in mice, however, shows that it has a role in postnatal rather than embryonic testis differentiation (Raymond *et al.*, 2000). After 10 days of postnatal development, the testis from mutant mice show a reduced number of germ cells and the Sertoli cells fail to mature. In contrast to the situation in humans, this phenotype is only seen when the animal is homozygous for the mutation. This suggests that the sex reversal seen in humans is due to more than the loss of *DMRT1*. Further studies are needed to establish what these other events are and how they fit in with the role of *DMRT1*.

V. Cell Movement and Proliferation in the Early Gonad

The most striking consequences of *Sry* action are the rapid changes that occur in cell proliferation, movement, and organization, which give the testis a very different mor-

phology compared to the ovary. The latter changes relatively little in comparison to the indifferent gonad until much later in development. Using BrdU as a marker of dividing cells, it was shown that one of the earliest events to occur after *Sry* action is the male-specific increase in proliferation of the somatic cells of the gonad. The XY gonad doubles in size with respect to the XX gonad within about 24 hr. The dividing cells were initially concentrated in the coelomic epithelium but later were seen throughout the testis (Schmahl *et al.*, 2000). Pulse/chase BrdU labeling showed that Sertoli cells were derived from the first wave of highly proliferating cells. This suggests that *Sry* is acting, either directly or indirectly, on cells of the coelomic epithelia to increase their proliferation.

Another clear morphological difference between the male and female gonad is the development of structures termed testis cords, which first appear at approximately 12.5 days of development and which eventually mature into the seminiferous tubules of the adult testis (Fig. 2). These cords comprise the Sertoli cells, which form an epithelial layer enclosing the germ cells, while the Leydig cells and various connective tissue cells are excluded to the interstitium. Surrounding the Sertoli cells lie the peritubular myoid (PTM) cells, which are unique to the testis and contribute to the structure of the cords by laying down basal lamina together with the Sertoli cells (Fig. 2). The testis also shows increased vasculature with respect to the ovary, most notably a prominent blood vessel running around its periphery. This may facilitate the efficient export to the rest of the embryo of the various effectors of male differentiation, such as testosterone.

Organ culture experiments have shown that the mesonephros is essential for testis cord formation (Buehr *et al.*, 1993; Merchant-Larios *et al.*, 1993). Studies using mesonephroi derived from mice carrying an ubiquitously expressed *LacZ* gene have shown that mesonephric cells migrate into the developing testis, at about E11.5, and give rise to the myoid cells and to some endothelial and associated cells (Martineau *et al.*, 1997) (Fig. 7). This type of migration only occurs when the cultured gonad is derived from an XY embryo. Experiments done with gonads derived from animals with weak alleles of *Sry* have shown that mesonephric cell migration is only seen in the portion of the gonad that gives rise to testicular cords (Albrecht *et al.*, 2000). This suggests that mesonephric cell migration is dependent on the action of SRY and that the Sertoli cells of the early testis signal the mesonephric cells to migrate into the gonad. The molecular identity of this signal is not known, but it seems able to act over considerable distances.

Further organ culture studies have indicated that cell interactions are important in establishing or maintaining Sertoli cell fate. Using a "sandwich" technique, where an XX gonad is placed between a mesonephros and an XY gonad before culturing, it was shown that cells from the mesonephros are able to migrate into the XX gonad and induce cord-

XX
gonad

XY
gonad

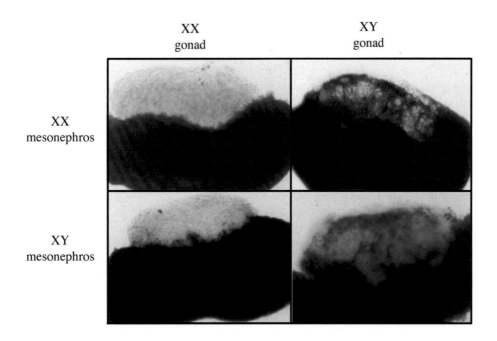

XX
mesonephros

XY
mesonephros

Figure 7 Cell migration from the mesonephros into the testis, but not the ovary. XX and XY genital ridges from E11.5 embryos were combined in an organ culture system with mesonephroi from either XX or XY embryos carrying the ROSA26 ubiquitous β-galactosidase marker gene. The migrating cells differentiate into several cell types, notably the peritubular myoid cells, which are unique to the testis and which are required for testis cord formation. There is also a significant contribution to the developing vascular system. (From Martineau *et al.,* 1997, Fig. 2, p. 960.)

like structures. Moreover, *Sox9* expression was activated and *Dax1* expression was repressed in the XX gonad. This implies that the supporting cells of the XX gonad were being induced to differentiate as Sertoli cells by the migrating cells (Tilmann and Capel, 1999). This did not occur if cell migration was blocked with a filter, showing that cell–cell interactions are involved in the induction. The ability of the XY gonad to trigger migration of the appropriate cells and of the XX gonad to respond in this way was dependent on their developmental stage, suggesting that competence to form cords is restricted to a relatively short interval.

XX Sertoli cells have been observed in mouse chimeras made with XX and XY cells; however, more than 90% of Sertoli cells had an XY genotype (Palmer and Burgoyne, 1991a). Perhaps in this case the XX cells from the supporting cell lineage are at a disadvantage because they have not received the initial instruction from *Sry* to differentiate along the male pathway. They will not (at least initially) produce the factor that attracts the migrating cells and therefore will be less likely to encounter them. In the sandwich cultures, the migrating cells, presumably myoid cell precursors, first encounter only XX cells. These then readily form Sertoli-like cells, which may then produce yet more of the migration factor. These studies suggest a model, whereby in normal testis development, SRY initiates Sertoli cell differentiation, perhaps by allowing high levels of *Sox9* expression. SOX9 then directly or indirectly activates expression of the factor

that signals the migration of the cells from the mesonephros to the gonad. These migrating cells in turn interact with the Sertoli cells to maintain *Sox9* expression, and differentiate as myoid cells. The two cell types then together form testis cords. The noncell autonomous regulatory loop suggested in this model provides a way to maintain Sertoli cell differentiation in the absence of SRY (which is expressed for only a brief period), and to generally ensure that the whole gonad follows the same coordinated pattern of differentiation (Fig. 8).

A. Germ Cells

Once Sertoli cell differentiation and cord formation have begun, the remaining bipotential cell types within the testis are triggered to follow the male pathway. This is true of germ cells, which in the testis arrest in mitosis (at about E13.5), resuming divisions and progressing through spermatogenesis only after birth. In the ovary the germ cells enter (and then arrest in) meiosis shortly after about E13.5. This behavior is independent of the chromosomal sex of the germ cell; it is the gonadal environment that matters. Several observations suggest that it is being enclosed by testicular cells, probably Sertoli cells, that induces the germ cells to follow the male pathway. *In vivo*, when XY germ cells fail to enter the genital ridge, for example, if they are within the adrenal gland, they also enter meiosis early and attempt to make oocytes (McLaren, 1995). *In vitro* culture assays, where germ cells are combined with somatic cells from the testis, ovary, or lung, again show it is only the embryonic testis that can direct the differentiation of germ cells to a spermatogenic fate. The early entry into meiosis is therefore a program that is intrinsic to the germ cells and which needs to be overridden in the testis. Moreover, these experiments were able to demonstrate that the signal inducing germ cells to enter mitotic arrest was limited to a brief period just after E11.5 (McLaren, 1995; McLaren and Southee, 1997). The molecular nature of this influence is not known (see Chapter 9), but it is clearly important to initiate spermatogenesis. On the other hand, germ cells are not required for testis cord formation, because these develop normally in mouse mutants, such as in the tyrosine kinase receptor, cKit, where germ cells fail to colonize the genital ridge (McLaren, 1995).

The essential role of Sertoli cells is to sustain germ cells throughout their development and later during spermatogenesis. A factor thought to be important in this process is Desert hedgehog (Dhh), a member of the hedgehog family of molecules that signal at close range. Sertoli cells of the early testis express *Dhh,* but it is not found in the ovary (Bitgood *et al.,* 1996). *Dhh* transcripts are seen at E11.5, suggesting that the gene is activated as a direct response to the factors that initiate Sertoli cell differentiation. Consistent with the role of *Dhh* in germ cell function, mice deficient for the gene are infertile due to the failure of spermatogonial cell division after birth (Bitgood *et al.,* 1996). The phenotype is very late compared to the early onset of *Dhh* expression. The action of *Dhh* on germ cells is thought to be indirect, however, as the expression of *Ptch1,* a likely receptor and target of hedgehog signaling, is confined to the interstitial cells of the testis, which include Leydig cells.

B. Leydig Cells

The initiation of production of testosterone by Leydig cells occurs around E13, 12–24 hr after cord formation is established. It therefore seemed likely that Sertoli cells direct their differentiation. The origin of Leydig cells is unclear; there is some evidence that these cells can migrate into the gonad from the mesonephros with the PTM cells (Merchant-Larios and Moreno-Mendoza, 1998). Alternatively, the precursors of these cells could be present in the gonad when it first develops, but testosterone production is not activated until the testis differentiates. In support of this last model, mice deficient for Wnt4, a member of the Wnt family of secreted signaling molecules, show activation of enzymes involved in testosterone biosynthesis in the XX gonad and masculinization of the Wolffian duct (Vainio *et al.,* 1999). Expression studies have shown that *Wnt4* is present in the mesonephros and gonad during the initial stages of urogenital development. As gonad development proceeds *Wnt4* expression is downregulated in the testis, whereas it stays on in the ovary. This suggests that WNT4 represses testosterone

production, perhaps by preventing steroidogenic cell differentiation (Fig. 6). In normal males, *Sry* relieves this repression by inhibiting *Wnt4* expression either directly or indirectly.

Not much is known about the factors that are involved in early Leydig cell differentiation and function. SF1 is thought to be an activator of testosterone production in these cells. *In vitro* studies have shown that this transcription factor can regulate the expression of genes encoding the enzymes involved in testosterone biosynthesis (Morohashi *et al.,* 1993). *SF1* is expressed in the embryonic Leydig cells as they develop (Shen *et al.,* 1994). However, it is also expressed in the Sertoli cells of the embryonic testis suggesting that other factors are needed to convey tissue-specific expression of the steroidogenic genes. There is also some evidence that *Dax1* is expressed in early Leydig cells although at much lower levels (Ikeda *et al.,* 1996). The role of DAX1 in this context is not known as studies have suggested that DAX1 acts to inhibit SF1 action. Conditional mutations of *SF1* and *Dax1* specifically in early Leydig cells are needed to address their role.

VI. Ovary Differentiation

No genes have been identified as having a specific role in ovary development. Analysis of mice lacking germ cells, such as those homozygous for mutations in the White-spotting (cKit) or Steel (Kit-ligand) genes, shows that these cells are essential for proper ovary differentiation (McLaren, 1991, 1995). The ovaries in these mice develop to about the stage when germ cells would enter meiosis in wild-type embryos, but then differentiation is arrested. Subsequent degeneration will give rise to streak gonads, which contain only stromal tissue. A different phenotype is observed if germ cells are lost after they have arrived in the genital ridge and have entered into meiosis. In these conditions the ovaries acquire cordlike structures, and activation of testis-specific genes and production of AMH and testosterone is observed. This phenomenon has been termed *transdifferentiation* and occurs naturally in freemartin cattle or in experimental situations where the ovary experiences a high level of AMH, such as in transgenic mice misexpressing the gene (Behringer *et al.,* 1990). Alternatively, it may be seen when fetal ovaries are grafted under the kidney capsule of male or female mouse hosts (Taketo and Merchant-Larios, 1986). Although the latter situation is not well understood, *Sox9* expression correlates with the appearance of testis cordlike structures (Morais da Silva *et al.,* 1996).

VII. Sexual Development

The mammalian embryo will develop as a female if it possesses ovaries or in the total absence of gonads. It is the

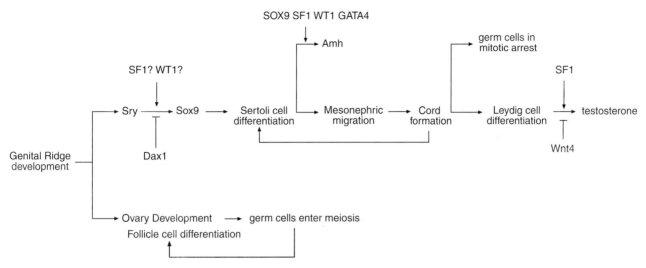

Figure 8 Summary of critical events in sex determination and gonadal differentiation. The embarrassing lack of detail in the female pathway reflects our ignorance.

role of the embryonic testis to impose male sexual development on this basic female body plan. It does this by producing at least three diffusible factors that establish the differentiation of male-specific structures in the developing embryo. These are AMH, testosterone, and INSL3. AMH is produced by the Sertoli cells and induces regression of the Müllerian ducts. Testosterone is produced by the Leydig cells and is responsible for the development of the male reproductive tract, male genitalia, and male-specific regions of the brain. INSL3 is produced by the Leydig cells and together with testosterone is involved in the process of testis descent. In this way, the testis regulates male sexual development throughout embryogenesis.

A. Reproductive Tract Development

Pax2 is the only known gene to affect both Müllerian and Wolffian duct development. Embryos deficient for *Pax-2* show initial development of both these ducts but they degenerate soon after they are formed (Torres *et al.*, 1995). *Pax2* is expressed in the epithelium of the ducts during embryogenesis, but also in the mesenchyme of the early stages of the urogenital system. Therefore, it is not clear whether the loss of this gene is affecting the ducts directly or through a defect in neighboring cells.

1. Müllerian Duct Differentiation

The female reproductive tract is derived from the Müllerian duct, which develops into the oviduct, uterus, and upper vagina in a rostrocaudal order. Two members of the Wnt family of secreted signaling molecules are thought to be important in Müllerian duct development, WNT4 and WNT7A (Fig. 3). The genes associated with both of these factors are expressed in the Müllerian duct epithelium when the ducts first appear and remain associated with these ducts through-

out their development (Parr and McMahon, 1998; Vainio *et al.*, 1999). Mice deficient for *Wnt4* fail to develop Müllerian ducts, showing that this gene is essential for their formation (Vainio *et al.*, 1999). In contrast, *Wnt7a*-deficient mice do develop Müllerian ducts and the females have reproductive tracts, but their structure is abnormal (Parr and McMahon, 1998). Further analysis has shown that there is a defect in the identity of the different regions of the tract; the oviducts have acquired uterine characteristics and the uterus has acquired some vaginal characteristics (Miller and Sassoon, 1998). *Hox* genes, which confer positional information in the embryo, were expressed at early stages of duct development, but not at later stages in the *Wnt7a* mutant mice. This suggests that *Wnt7a* is involved in maintaining rather than establishing the identity of different regions of the female reproductive tract along the anterior-posterior axis.

In the male, the Müllerian ducts regress around 14 days of development. It is clear that AMH induces this regression, because males deficient in this factor develop two sets of reproductive ducts, male and female (Behringer *et al.*, 1994). Although these mice can produce functional sperm, most are infertile because of the physical constraints imposed by the development of both sets of tracts. In support of the central role of AMH in Müllerian duct regression, transgenic female mice that ectopically express high levels of human AMH lack reproductive tracts (Behringer *et al.*, 1990). TGF-β family members signal through two membrane-bound serine/threonine kinases, type I and II. The type II receptor for AMH has been identified, and mice deficient for this receptor show an identical phenotype to those lacking AMH itself (Mishina *et al.*, 1996). Expression studies in rat and mouse have shown that the AMH type II receptor is present in the mesenchyme that surrounds the Müllerian duct epithelium, indicating that regression of the duct is triggered indirectly (Baarends *et al.*, 1994; Teixeira and Donahoe, 1996). Persistent Müllerian ducts are also found in *Wnt7a*-deficient mice implicating this factor in Müllerian duct regression (Parr and McMahon, 1998). Analysis of these mice during embryogenesis showed that the duct epithelium develops normally. However, *Wnt7a*-deficient mice show no AMH type II receptor expression during development, indicating that WNT7A might act as a signal from the epithelium to the surrounding mesenchyme to direct the regression of the Müllerian duct.

2. Wolffian Duct Differentiation

Testosterone is the major steroid produced by the embryonic testis. It acts on tissues to induce differentiation by binding to the androgen receptor, an X-linked member of the nuclear hormone receptor family, which regulates gene expression when bound to its ligand. In this way, the Wolffian duct is induced to differentiate into the epidydimus, vas deferens, and seminal vesicles (Fig. 3). A spontaneous point mutation in the androgen receptor gene that leads to a truncation of the protein has been found in a strain of mice. The mutation, named *Tfm* for testicular feminization, leads to a phenotype similar to that found in humans deficient for the same gene, where XY individuals have testes and no female reproductive tract, but fail to develop either Wolffian duct derivatives or male external genitalia (Gaspar *et al.*, 1991; Lyon and Hawkes, 1970).

3. Testis Descent

Although the male and female gonads initially develop at identical positions within the body, in the adult, the ovary is located close to the kidney while the testis is found in the scrotum (Fig. 3). The differentiation of the cranial suspensory ligament (CSL) and the gubernaculum establish this sexually dimorphic position of the ovary and testis. These structures develop within the context of the genital mesentery, which connects the gonads and genital ducts to the abdominal wall. Testis descent occurs in two phases, one between E15.5 and E17.5 and the other between 2 and 3 weeks after birth. During the first phase, the development of the gubernaculum and regression of the CSL results in the transabdominal movement of the testis to the inguinal region. In the second phase the gubernaculum regresses and the testis descends to the scrotum. Mice deficient for *Insl3* showed bilateral cryptorchid testes located high in the abdomen, indicating that this gene is essential for testis descent (Nef and Parada, 1999; Zimmermann *et al.*, 1999). Further analysis of these mice showed a defect in gubernaculum differentiation, however CSL regression still took place. Defects in testis descent are also found in *Tfm* mice where the differentiation of both CSL and gubernaculum results in the testes being located at an intermediate position between ovaries and testes of wild-type animals. When the *Tfm* mice were bred with the *Insl3*-deficient mice, the double mutant animals showed completely undescended testes located in a position comparable to ovaries in wild-type mice (Zimmer-

mann *et al.*, 1999). These results show that the initial phase of testis descent is dependent on *Insl3*, to ensure gubernaculum differentiation, and the androgen receptor to ensure CSL regression.

VIII. Evolution and Sex Determination

An amazing diversity of sex-determining mechanisms is present within the animal kingdom and it seems paradoxical that such a vital mechanism should evolve so rapidly. The mechanism is, however, critical to reproductive success. Therefore, if a situation arises that favors change, for example, a system involving a biased rather than equal sex ratio, then it will be subject to selection. The sex-determining genes themselves can also show rapid divergence in sequence, in their regulation, and in their relative importance in the process. The most important constraint is that fertility is not affected. This is a problem in mammals, where XX male mice are usually sterile and XY females are sterile or have reduced fertility; the former because they lack other Y genes required for fertility and because two X chromosomes are incompatible with spermatogenesis and the latter most likely because of a problem of chromosome pairing (Burgoyne *et al.*, 1992; Lovell-Badge and Robertson, 1990).

Several views have been proposed regarding how the X and the Y chromosome evolved, but it is thought to be intimately linked with the appearance of *Sry* as a sex-determining gene within the past 200 million years, specifically within the ancestors of mammals. *Sry* is likely to have evolved from *Sox3*, a gene found throughout vertebrate taxa (Collignon *et al.*, 1996; Foster *et al.*, 1994; Graves, 1998). The two genes have very similar HMG box domains and *Sox3* is X-linked in mammals. *Sox3* has been found on the autosome in the marsupials and does not play a role in sex determination (Pask *et al.*, 2000). It is plausible, then, that the X and Y evolved from a pair of autosomes, when *Sry* diverged from *Sox3*, having acquired its sex-determining function. The ancestor to mammals could then have had an environmental-based sex-determining mechanism or one based on different heteromorphic chromosomes. Some authors have proposed that the ancestral mammal had a sex-determining mechanism that was dependent on X chromosome dosage, which was then overridden by the appearance of *Sry* (Tiersch *et al.*, 1991). This idea came in part from the finding that in marsupials (metatheria), development of structures such as the scrotum and pouch are dependent on the number of X chromosomes rather than on presence or absence of *Sry*. However, the way these structures develop seems quite unique to marsupials and the underlying dosage mechanism is probably also a marsupial adaptation (Butler *et al.*, 1999). An *Sry* gene has so far not been found in monotremes (prototheria), which diverged from eutherian mammals 170 million year ago (Pask and Graves, 1999). The Y chromosome may have evolved more than once or the role of *Sry* as a sex-determining gene may have been taken over on the mono-treme Y by another gene. It would be interesting to see if the ATRY gene is on the monotreme Y chromosome, if this has a role in sex determination. See Delbridge and Graves (1999) and Lahn and Page (1999) for discussions on X and Y chromosome evolution.

How did *Sry* acquire its sex-determining role? There is so far no evidence that *Sox3* acts in sex determination in any species. Studies in mouse, chick, and marsupials suggest instead that *Sox3* is likely to have a conserved function within the developing central nervous system, where it is expressed at high levels (Collignon *et al.*, 1996; Zhu *et al.*, 1999, Pask *et al.*, 2000b). Perhaps the ancestral *Sry* gene acquired its sex-determining function simply through a change in regulation, such that it became expressed within the genital ridge. It would then have lost its role in the central nervous system as *Sox3* became subject to X-inactivation (Lahn and Page, 1999). The *Sry* sequence is then probably free to evolve as long as sex determination is not compromised. This may be why it has been found to be so rapidly evolving, yet is present in most eutherian mammals (the exception being some species of mole-vole as discussed below).

When *Sry* genes from evolutionarily distant mammals are compared, essentially the only conserved region is the HMG box. For example, with the mouse and humans genes, the rest of the open reading frame, untranslated regions of the transcript, and flanking genomic DNA all differ so extensively that they would be considered different genes if there was not functional data to equate the two (Tucker and Lundrigan, 1993; Whitfield *et al.*, 1993; Hacker *et al.*, 1995). When comparing the sequence of SRY in different closely related species a high rate of divergence was also found in the regions of the protein outside the HMG box domain. This was reflected in a very high rate of nonsynonymous substitutions, suggesting that either these regions have no functional significance or that directional selection had occurred (Tucker and Lundrigan, 1993; Whitfield *et al.*, 1993). Perhaps sequence divergence leads to a newly acquired male-specific function, which provides a selective advantage. *Sry* is expressed during spermatogenesis in the mouse and in several other species. A role in this process could increase reproductive fitness and therefore be subject to selection, explaining the high rate of nonsynonomous substitutions. This is very speculative, however, and for the mouse, there is evidence that *Sry* has no critical role in spermatogenesis (Lovell-Badge and Roberston, 1990; Capel *et al.*, 1993a).

Dax1 is also poorly conserved among mammalian species, although not to the extent of *SRY* (Swain *et al.*, 1996). Comparisons of cases of sex reversal in humans and mice involving DAX1 and SRY suggest that the proteins have the same function in the two species, but that the dynamics of their activity and interaction are different (Bardoni *et al.*, 1994; Swain *et al.*, 1998). Simple differences in the time at

which one begins to be expressed relative to the other may explain why a twofold increase in DAX1 expression appears to be sufficient to cause sex reversal in humans, but not in mice, unless it is tested against a variant *Sry* allele. Differences in gene expression patterns of DAX1 and SRY in the developing gonad between human (Salas-Cortes *et al.*, 1999; Hanley *et al.*, 2000) and mouse have been observed. It is therefore very likely that there are more or less subtle differences in the way the genes are regulated among different species. However, care has to be taken in interpreting phenotypic data from cases of sex reversal in humans, because it is only possible to examine the individual a long time after the critical events have occurred.

DAX1 is found in the chicken, where it shows a similar expression pattern to that in mouse, but the sequence is quite divergent and lacks much of the putative DNA binding domain (Smith *et al.*, 2000). It may have a similar function to the mammalian protein, with respect to its interaction with SF1, but how it relates to the sex determination pathway is not known as birds do not possess an SRY gene. Moreover, *Dax1* is autosomal in birds and marsupials (Pask *et al.*, 1997), suggesting that the sex reversal seen in humans and mice, when *Dax1* is expressed at higher than normal levels, does not reflect any such ancestral mechanism.

In contrast to *Sry* and *Dax1*, *SOX9* is highly conserved among different vertebrates in both its coding sequence and its sexually dimorphic pattern of expression within the gonad, where it always marks Sertoli cell differentiation. This is seen in animals such as chickens, with a ZZ/ZW sex-determining system, and in turtles and alligators, which have environmental sex-determining mechanisms (Kent *et al.*, 1996; Morais da Silva *et al.*, 1996; Moreno-Mendoza *et al.*, 1999; Spotila *et al.*, 1998; Western *et al.*, 1999b). In alligators, however, SOX9 comes on after the temperature-sensitive commitment to male development (Western *et al.*, 1999a); and in chickens, AMH expression precedes the upregulation of *Sox9* in the testis (Oreal *et al.*, 1998). These data are in agreement with a general model that genes at the top of a cascade are the last to have evolved and therefore are the least conserved among species (Wilkins, 1995). It also suggests that different genes active within the early gonad may take on more or less important roles in sex determination, depending on which have been selected as rate limiting in the process.

Another striking example of conservation is that of the *Dmrt1* gene. As discussed above, its location on human chromosome 9p21, to a region associated with sex reversal, and its pattern of expression in the mouse both strongly argue for an involvement in sex determination and/or differentiation in mammals. The gene sequence and its gonadal-specific expression pattern are also highly conserved among other vertebrates. These include chickens with ZZ/ZW and reptiles, with environmental sex-determining systems, respectively (De Grandi *et al.*, 2000; Raymond *et al.*, 1998,

1999; Smith *et al.,* 1999; Guioli *et al.,* 2000; Kettlewell *et al.,* 2000). Interestingly, *Dmrt1* maps to the Z chromosome in birds, making it a candidate for a dosage-dependent sex-determining gene, especially as higher transcript levels are associated with early male (ZZ) genital ridge development (De Grandi *et al.,* 2000; Nanda *et al.,* 1999; Raymond *et al.,* 1999; Smith *et al.,* 1999). In alligators and turtles a higher level of expression also correlates with male-specific temperatures and this appears to precede the high levels of *Sox9* associated with Sertoli cell differentiation (Kettlewell *et al.,* 2000).

Even more intriguingly, *Dmrt1* was found because of its homology with *mab-3* and *dsx,* genes found in the sex-determining cascades of *Caenorhabditis elegans* and *Drosophila,* respectively. All of the genes possess DM types of DNA-binding domain, first defined by comparisons of the *mab-3* and *dsx* open reading frames (Raymond *et al.,* 1998). On the other hand, while this seems to break the rule that there is no conservation of genes involved in sex determination between invertebrates and vertebrates, the exact relationship between them is not clear. The *mab-3* and *dsx* genes are both placed downstream in their respective sex-determining cascades and have similar biochemical and genetic properties, including regulation of yolk protein gene expression and sex-specific peripheral nervous system development (Raymond *et al.,* 1998; Yi and Zarkower, 1999). The studies in vertebrates, however, suggest that *Dmrt1* is located nearer the top of the cascade and is unlikely to have either of the functions mentioned above. It is possible, therefore, that the apparent conservation is merely a coincidence, where by chance (or because it has special as yet unknown properties) a gene containing a DM type of DNA-binding domain has simply been coopted for a role in sex determination. This is possible, because it is now clear that several genes in both vertebrates and invertebrates possess DM domains, but seem to have no role in sex determination. This would parallel the finding that SRY and the *Schizosaccharomyces pombe* mating type gene Mc both have HMG box DNA-binding domains (Sinclair *et al.,* 1990). More work needs to be done on the exciting family of vertebrate *Dmrt* genes to establish their role in sexual development in the different species.

The existence of mammals with exceptional sex determination systems indicates how precariously balanced the mechanism is and how rapidly evolution can take place. For example, the majority of fertile female moles have ovotestes, despite having a normal XX karyotype and no *Sry* gene (Jimenez *et al.,* 1996; Whitworth *et al.,* 1999). Perhaps the testosterone produced by the testicular component gives the females a degree of masculine characteristics appropriate for their peculiar lifestyle. Wood lemmings, on the other hand, have a system that favors female development. Apart from normal X and Y chromosomes, a variant X chromosome

(termed X*), segregates at high frequency within the population and overrides the action of *Sry,* to give X*Y females (Fredga, 1983). These are fertile due to a special nondisjunction mechanism. An excess of females may allow rapid increase in population numbers after periodic crashes due to lack of resources. Different sex-determining systems have even been observed within a family, where some mole-vole (*Ellobius*) species have a normal XY sex chromosome-based system, whereas closely related species have lost not just *Sry,* but the complete Y chromosome (Just *et al.,* 1995). Perhaps a gene such as *Sox9,* or a regulator of *Sox9* expression, has now become the sex-determining switch. It is difficult to imagine how this evolved, because other XX male mammals are invariably sterile. However, understanding the genetic basis of these systems will provide invaluable insight into the process of sex determination.

IX. Conclusion

Analysis of how *Sry* acts in the process of sex determination has shed light on how cell fate decisions can be made in the embryo. This factor seems to act as a trigger that initiates the process of cell fate determination leading to Sertoli cell differentiation. But this initial decision is only a tentative one, which needs to be reinforced by a series of events that fix cell fate, making a stable system within which the development of the rest of the organ can take place. We have also gained a clearer idea of how transcription factors can work to regulate differentiation. Specificity is achieved by having the right combination of factors all present within the same cell. These factors, which may include WT1, SF1, and GATA4, restrict *Sry* expression to a precise place and time in development, but they also play subsequent roles to ensure that Sertoli cell-specific gene expression is maintained. To avoid development of gonads with mixed structures (ovotestes), the switch triggered by SRY is of an all or nothing type; either there is enough SRY present to ensure testis development or there is not and ovary development ensues. This may be reinforced by the recruitment of cells into the testis, which have a strong effect in ensuring that SOX9 is expressed and therefore that Sertoli cell fate is maintained. Factors that affect the transcriptional control achieved by SRY, either by controlling *Sry* expression or the action of the protein, are part of a delicate regulatory balance that can be disrupted by small changes, such as a twofold decrease in levels of protein.

Studies on sex determination also provide insight into the process of evolution. In contrast to many systems in embryogenesis, sex-determining mechanisms are poorly conserved throughout the animal kingdom and are frequently where the most variety is displayed. The comparison to other animals has allowed us to identify the components of the system that are conserved and therefore might share a common role and those that are not conserved that must somehow deliver functions that are unique to each species.

The mouse has proven to be an invaluable tool in the study of sex determination and sexual differentiation. The advent of genetic techniques in the mouse such as targeted mutagenesis and the use of transgenics has allowed the verification of models that have arisen out of comparative studies with other organisms and has uncovered novel genes involved in the process. This has moved the study of sex determination out of the realm of description into more functional studies. It has also meant that many new and exciting players in the process have been identified and their roles have been determined. Future progress will also capitalize on advances in genomics, in microarray-based differential expression screens, and in mutagenesis screens to identify other critical genes (Bowles *et al.,* 2000; Grimmond *et al.,* 2000; Wertz and Herrmann, 2000). However, the study of any new genes cannot be done without greater understanding of the developmental system in which they act. Many questions remain regarding the underlying embryology and cell biology of the gonads. Therefore, more detailed studies are needed, starting with the mutant mice that already exist, to acquire full understanding of this basic process of reproduction.

Acknowledgments

We thank Blanche Capel for critical reading of the manuscript. We thank past and present members of the Lovell-Badge and Swain laboratories for their help and support. We thank Joe Brock for creating the figures. Financial support from the MRC and the Louis Jeantet Foundation are gratefully acknowledged.

References

Achermann, J. C., Ito, M., Hindmarsh, P. C., and Jameson, J. L. (1999). A mutation in the gene encoding steroidogenic factor-1 causes XY sex reversal and adrenal failure in humans. *Nat. Genet.* **22,** 125–126.

Albrecht, K. H., and Eicher, E. M. (1997). DNA sequence analysis of Sry alleles (subgenus Mus) implicates misregulation as the cause of C57BL/6J-Y(POS) sex reversal and defines the SRY functional unit. *Genetics* **147,** 1267–1277.

Albrecht, K. H., Capel, B., Washburn, L. L., and Eicher, E. M. (2000). Defective mesonephric cell migration is associated with abnormal testis cord development in C57BL/6J XY(*Mus domesticus*) mice. *Dev. Biol.* **225,** 26–36.

Arango, N. A., Lovell-Badge, R., and Behringer, R. R. (1999). Targeted mutagenesis of the endogenous mouse Mis gene promoter: In vivo definition of genetic pathways of vertebrate sexual development. *Cell (Cambridge, Mass.)* **99,** 409–419.

Armstrong, J. F., Pritchard-Jones, K., Bickmore, W. A., Hastie, N. D., and Bard, J. B. (1993). The expression of the Wilms' tumour gene, WT1, in the developing mammalian embryo. *Mech. Dev.* **40,** 85–97.

Baarends, W. M., van Helmond, M. J., Post, M., van der Schoot, P. J., Hoogerbrugge, J. W., de Winter, J. P., Uilenbroek, J. T., Karels, B., Wilming, L. G., Meijers, J. H. *et al.* (1994). A novel member of the transmembrane serine/threonine kinase receptor family is specifically expressed in the gonads and in mesenchymal cells adjacent to the mullerian duct. *Development (Cambridge, UK)* **120,** 189–197.

Barbaux, S., Niaudet, P., Gubler, M. C., Grunfeld, J. P., Jaubert, F., Kuttenn, F., Fekete, C. N., Souleyreau-Therville, N., Thibaud, E., Fellous, M., and McElreavey, K. (1997). Donor splice-site mutations in WT1 are responsible for Frasier syndrome. *Nat. Genet.* **17,** 467–470.

Bardoni, B., Zanaria, E., Guioli, S., Floridia, G., Worley, K. C., Tonini, G., Ferrante, E., Chiumello, G., McCabe, E. R., Fraccaro, M. *et al.* (1994). A dosage sensitive locus at chromosome Xp21 is involved in male to female sex reversal. *Nat. Genet.* **7,** 497–501.

Beau, C., Rauch, M., Joulin, V., Jegou, B., and Guerrier, D. (2000). GATA-1 is a potential repressor of anti-Mullerian hormone expression during the establishment of puberty in the mouse. *Mol. Reprod. Dev.* **56,** 124–138

Behringer, R. R. (1995). The mullerian inhibitor and mammalian sexual development. *Philos. Trans. R. Soc. London, Ser. B* **350,** 285–288; discussion: p. 289.

Behringer, R. R., Cate, R. L., Froelick, G. J., Palmiter, R. D., and Brinster, R. L. (1990). Abnormal sexual development in transgenic mice chronically expressing mullerian inhibiting substance. *Nature (London)* **345,** 167–170.

Behringer, R. R., Finegold, M. J., and Cate, R. L. (1994). Mullerian-inhibiting substance function during mammalian sexual development. *Cell (Cambridge, Mass.)* **79,** 415–425.

Bel, S., Core, N., Djabali, M., Kieboom, K., Van der Lugt, N., Alkema, M. J., and Van Lohuizen, M. (1998). Genetic interactions and dosage effects of Polycomb group genes in mice. *Development (Cambridge, UK)* **125,** 3543–3551.

Bell, D. M., Leung, K. K., Wheatley, S. C., Ng, L. J., Zhou, S., Ling, K. W., Sham, M. H., Koopman, P., Tam, P. P., and Cheah, K. S. (1997). SOX9 directly regulates the type-II collagen gene. *Nat. Genet.* **16,** 174–178.

Bi, W., Deng, J. M., Zhang, Z., Behringer, R. R., and de Crombrugghe, B. (1999). Sox9 is required for cartilage formation. *Nat. Genet.* **22,** 85–89.

Birk, O. S., Casiano, D. E., Wassif, C. A., Cogliati, T., Zhao, L., Zhao, Y., Grinberg, A., Huang, S., Kreidberg, J. A., Parker, K. L., Porter, F. D., and Westphal, H. (2000). The LIM homeobox gene Lhx9 is essential for mouse gonad formation. *Nature (London)* **403,** 909–913.

Bishop, C. E., Whitworth,, D. J., Qin, Y., Agoulnik, I. U. Harrison, W. R., Behringer, R. R., and Overbeek, P. A. (2000). A transgenic insertion upstream of Sox9 is associated with dominant XX sex reversal in the mouse. *Nat. Genet.* **26,**490–494.

Bitgood, M. J., Shen, L., and McMahon, A. P. (1996). Sertoli cell signaling by Desert hedgehog regulates the male germline. *Curr. Biol.* **6,** 298–304.

Bowles, J., Cooper, L., Berkman, J., and Koopman, P. (1999). Sry requires a CAG repeat domain for male sex determination in *Mus musculus. Nat. Genet.* **22,** 405–408.

Bowles, J., Bullejos, M., and Koopman, P. (2000) A subtractive gene expression screen suggests a role for vanin-1 in testis development in mice. *Genesis* **27,** 124–135.

Buehr, M., Gu, S., and McLaren, A. (1993). Mesonephric contribution to testis differentiation in the fetal mouse. *Development (Cambridge, UK)* **117,** 273–281; erratum: *Ibid.* **118**(4), 1384 (1993).

Burgoyne, P. S., and Palmer, S. J. (1991). The genetics of XY sex reversal in the mouse and other mammals. *Dev. Biol.* **2,** 277–284.

Burgoyne, P. S., Mahadevaiah, S. K., Sutcliffe, M. J., and Palmer, S. J. (1992). Fertility in mice requires X-Y pairing and a Y-chromosomal gene mapping to the long arm. *Cell (Cambridge, Mass.)* **71,** 391–398.

Burris, T. P., Guo, W., Le, T., and McCabe, E. R. (1995). Identification of a putative steroidogenic factor-1 response element in the DAX-1 promoter. *Biochem. Biophys. Res. Commun.* **214,** 576–581.

Butler, C. M., Shaw, G., and Renfree, M. B. (1999). Development of the penis and clitoris in the tammar wallaby, *Macropus eugenii. Anat. Embryol.* **199,** 451–457.

Byskov, A. G., and Hoyer, P. E. (1994). Embryology of mammalian gonads and ducts. *In* "The Physiology of Reproduction" (E. Knobil and J. D. Neill, eds.), p. 487. Raven Press, New York.

Capel, B., Rasberry, C., Dyson, J., Bishop, C. E., Simpson, E., Vivian, N., Lovell-Badge, R., Rastan, S., and Cattanach, B. M. (1993a). Deletion of Y chromosome sequences located outside the testis determining region can cause XY female sex reversal. *Nat. Genet.* **5,** 301–307.

Capel, B., Swain, A., Nicolis, S., Hacker, A., Walter, M., Koopman, P., Goodfellow, P., and Lovell-Badge, R. (1993b). Circular transcripts of the testis-determining gene Sry in adult mouse testis. *Cell (Cambridge, Mass.)* **73,** 1019–1030.

Carlisle, C., Winking, H., Weichenhan, D., and Nagamine, C. M. (1996). Absence of correlation between Sry polymorphisms and XY sex reversal caused by the M. m. domesticus Y chromosome. *Genomics* **33,** 32–45.

Collignon, J., Sockanathan, S., Hacker, A., Cohen-Tannoudji, M., Norris, D., Rastan, S., Stevanovic, M., Goodfellow, P. N., and Lovell-Badge, R. (1996). A comparison of the properties of Sox-3 with Sry and two related genes, Sox-1 and Sox-2. *Development (Cambridge, UK)* **122,** 509–520.

Core, N., Bel, S., Gaunt, S. J., Aurrand-Lions, M., Pearce, J., Fisher, A., and Djabali, M. (1997). Altered cellular proliferation and mesoderm patterning in Polycomb-M33-deficient mice. *Development (Cambridge, UK)* **124,** 721–729.

Coward, P., Nagai, K., Chen, D., Thomas, H. D., Nagamine, C. M., and Lau, Y. F. (1994). Polymorphism of a CAG trinucleotide repeat within Sry correlates with B6.YDom sex reversal. *Nat. Genet.* **6,** 245–250.

De Grandi, A., Calvari, V., Bertini, V., Bulfone, A., Peverali, G., Camerino, G., Borsani, G., and Guioli, S. (2000). The expression pattern of a mouse doublesex-related gene is consistent with a role in gonadal differentiation. *Mech. Dev.* **90,** 323–326.

Delbridge, M. L., and Graves, J. A. (1999). Mammalian Y chromosome evolution and the male-specific functions of Y chromosome-borne genes. *Rev. Reprod.* **4,** 101–109.

De Santa Barbara, P., Bonneaud, N., Boizet, B., Desclozeaux, M., Moniot, B., Sudbeck, P., Scherer, G., Poulat, F., and Berta, P. (1998). Direct interaction of SRY-related protein SOX9 and steroidogenic factor 1 regulates transcription of the human anti-Mullerian hormone gene. *Mol. Cell. Biol.* **18,** 6653–6665.

Dresser, D. W., Hacker, A., Lovell-Badge, R., and Guerrier, D. (1995). The genes for a spliceosome protein (SAP62) and the anti-Mullerian hormone (AMH) are contiguous. *Hum. Mol. Genet.* **4,** 1613–1618.

Eicher, E. M., Washburn, L. L., Whitney, J. B. D., and Morrow, K. E. (1982). *Mus poschiavinus* Y chromosome in the C57BL/6J murine genome causes sex reversal. *Science* **217,** 535–537.

Eicher, E. M., Shown, E. P., and Washburn, L. L. (1995). Sex reversal in C57BL/6J-YPOS mice corrected by a Sry transgene. *Philos. Trans. R. Soc. London, Ser. B* **350,** 263–268; discussion: pp. 268–269.

Foster, J. W., and Graves, J. A. (1994). An SRY-related sequence on the marsupial X chromosome: Implications for the evolution of the mammalian testis-determining gene. *Proc. Natl. Acad. Sci. U.S.A.* **91,** 1927–1931.

Foster, J. W., Dominguez-Steglich, M. A., Guioli, S., Kowk, G., Weller, P. A., Stevanovic, M., Weissenbach, J., Mansour, S., Young, I. D., Goodfellow, P. N. *et al.* (1994). Campomelic dysplasia and autosomal sex reversal caused by mutations in an SRY-related gene. *Nature (London)* **372,** 525–530.

Fredga, K. (1983). Aberrant sex chromosome mechanisms in mammals. Evolutionary aspects. *Differentiation (Berlin)* **23,** S23–S30.

Fujii, T., Pichel, J. G., Taira, M., Toyama, R., Dawid, I. B., and Westphal, H. (1994). Expression patterns of the murine LIM class homeobox gene lim1 in the developing brain and excretory system. *Dev. Dyn.* **199,** 73–83.

Gaspar, M. L., Meo, T., Bourgarel, P., Guenet, J. L., and Tosi, M. (1991). A single base deletion in the Tfm androgen receptor gene creates a short-lived messenger RNA that directs internal translation initiation. *Proc. Natl. Acad. Sci. U.S.A.* **88,** 8606–8610.

George, F. W., and Wilson, J. D. (1994). Sex determination and differentiation. *In* "The Physiology of Reproduction" (E. Knobil and J. Neill, eds.), 1, pp. 3–28. Raven Press, New York.

Gibbons, R. J., McDowell, T. L., Raman, S., O'Rourke, D. M., Garrick, D., Ayyub, H., and Higgs, D. R. (2000). Mutations in ATRX, encoding a SWI/SNF-like protein, cause diverse changes in the pattern of DNA methylation. *Nat. Genet.* **24,** 368–371.

Giuili, G., Shen, W. H., and Ingraham, H. A. (1997). The nuclear receptor SF-1 mediates sexually dimorphic expression of Mullerian Inhibiting Substance, in vivo. *Development (Cambridge, UK)* **124,** 1799–1807.

Goodfellow, P. N., and Lovell-Badge, R. (1993). SRY and sex determination in mammals. *Annu. Rev. Genet.* **27,** 71–92.

Graves, J. A. (1998). Interactions between SRY and SOX genes in mammalian sex determination. *BioEssays* **20,** 264–269.

Grimmond, S., Van Hateren, N., Siggers, P., Arkell, R., Larder, R., Soares, M. B., de Fatima Bonaldo, M., Smith, L., Tymowska-Lalanne, Z., Wells, C., and Greenfield, A. (2000). Sexually dimorphic expression of protease nexin-1 and vanin-1 in the developing mouse gonad prior to overt differentiation suggests a role in mammalian sexual development. *Hum. Mol. Genet.* **9,** 1553–1560.

Gubbay, J., Collignon, J., Koopman, P., Capel, B., Economou, A., Munsterberg, A., Vivian, N., Goodfellow, P., and Lovell-Badge, R. (1990). A gene mapping to the sex-determining region of the mouse Y chromosome is a member of a novel family of embryonically expressed genes. *Nature (London)* **346,** 245–250.

Hacker, A., Capel, B., Goodfellow, P., and Lovell-Badge, R. (1995). Expression of Sry, the mouse sex determining gene. *Development (Cambridge, UK)* **121,** 1603–1614.

Hammer, G. D., Krylova, I., Zhang, Y., Darimont, B. D., Simpson, K., Weigel, N. L., and Ingraham, H. A. (1999). Phosphorylation of the nuclear receptor SF1 modulates cofactor recruitment: Integration of hormone signalling in reproduction and stress. *Mol. Cell* **3,** 521–526.

Hanley, N. A., Hagan, D. M., Clement-Jones, M., Ball, S. G., Strachan, T., Salas-Cortes, L., McElreavey, K., Lindsay, S., Robson, S., Bullen, P., Ostrer, H., and Wilson, D. I. (2000). SRY, SOX9, and DAX1 expression patterns during human sex determination and gonadal development. *Mech. Dev.* **91,** 403–407.

Hanson, R. D., Hess, J. L., Yu, B. D., Ernst, P., van Lohuizen, M., Berns, A., van der Lugt, N. M., Shashikant, C. S., Ruddle, F. H., Seto, M., and Korsmeyer, S. J. (1999). Mammalian Trithorax and polycomb-group homologs are antagonistic regulators of homeotic development. *Proc. Natl. Acad. Sci. U.S.A.* **96,** 14372–14377.

Haqq, C. M., King, C. Y., Donahoe, P. K., and Weiss, M. A. (1993). SRY recognizes conserved DNA sites in sex-specific promoters. *Proc. Natl. Acad. Sci. U.S.A.* **90,** 1097–1101.

Harley, V. R., Jackson, D. I., Hextall, P. J., Hawkins, J. R., Berkovitz, G. D., Sockanathan, S., Lovell-Badge, R., and Goodfellow, P. N. (1992). DNA binding activity of recombinant SRY from normal males and XY females. *Science* **255,** 453–456.

Hatano, O., Takayama, K., Imai, T., Waterman, M. R., Takakusu, A., Omura, T., and Morohashi, K. (1994). Sex-dependent expression of a transcription factor, Ad4BP, regulating steroidogenic P-450 genes in the gonads during prenatal and postnatal rat development. *Development (Cambridge, UK)* **120,** 2787–2797.

Hatano, O., Takakusu, A., Nomura, M., and Morohashi, K. (1996). Identical origin of adrenal cortex and gonad revealed by expression profiles of Ad4BP/SF-1. *Genes Cells* **1,** 663–671.

Huang, B., Wang, S., Ning, Y., Lamb, A. N., and Bartley, J. (1999). Autosomal XX sex reversal caused by duplication of SOX9. *Am. J. Med. Genet.* **87,** 349–353.

Ikeda, Y., Lala, D. S., Luo, X., Kim, E., Moisan, M. P., and Parker, K. L. (1993). Characterization of the mouse FTZ-F1 gene, which encodes a key regulator of steroid hydroxylase gene expression. *Mol. Endocrinol.* **7,** 852–860.

Ikeda, Y., Shen, W. H., Ingraham, H. A., and Parker, K. L. (1994). Developmental expression of mouse steroidogenic factor-1, an essential regulator of the steroid hydroxylases. *Mol. Endocrinol.* **8,** 654–662.

Ikeda, Y., Luo, X., Abbud, R., Nilson, J. H., and Parker, K. L. (1995). The nuclear receptor steroidogenic factor 1 is essential for the formation of the ventromedial hypothalamic nucleus. *Mol. Endocrinol.* **9,** 478–486.

Ikeda, Y., Swain, A., Weber, T. J., Hentges, K. E., Zanaria, E., Lalli, E., Tamai, K. T., Sassone-Corsi, P., Lovell-Badge, R., Camerino, G., and Parker, K. L. (1996). Steroidogenic factor 1 and Dax-1 colocalize in multiple cell lineages: Potential links in endocrine development. *Mol. Endocrinol.* **10,** 1261–1272.

Ingraham, H. A., Lala, D. S., Ikeda, Y., Luo, X., Shen, W. H., Nachtigal, M. W., Abbud, R., Nilson, J. H., and Parker, K. L. (1994). The nuclear receptor steroidogenic factor 1 acts at multiple levels of the reproductive axis. *Genes Dev.* **8,** 2302–2312.

Ion, A., Telvi, L., Chaussain, J.L., Galacteros, F., Valayer, J., Fellous, M., and McElreavey, K. (1996). A novel mutation in the putative DNA helicase XH2 is responsible for male-to-female sex reversal associated with an atypical form of the ATR-X syndrome. *Am. J. Hum. Genet.* **58,** 1185–1191.

Jeske, Y. W., Bowles, J., Greenfield, A., and Koopman, P. (1995). Expression of a linear Sry transcript in the mouse genital ridge. *Nat. Genet.* **10,** 480–482.

Jimenez, R., Alarcon, F. J., Sanchez, A., Burgos, M., and De La Guardia, R. D. (1996). Ovotestis variability in young and adult females of the mole *Talpa occidentalis* (insectivora, mammalia). *J. Exp. Zool.* **274,** 130–137.

Josso, N., and Picard, J. Y. (1986). Anti-Mullerian hormone. *Physiol. Rev.* **66,** 1038–1090.

Jost, A. (1953). Problems in fetal endocrinology; the gonadal and hypophyseal hormones. *Recent Prog. Horm. Res.* **8,** 379–418.

Just, W., Rau, W., Vogel, W., Akhverdian, M., Fredga, K., Graves, J. A., and Lyapunova, E. (1995). Absence of Sry in species of the vole Ellobius. *Nat. Genet.* **11,** 117–118.

Kamachi, Y., Cheah, K. S., and Kondoh, H. (1999). Mechanism of regulatory target selection by the SOX high-mobility-group domain proteins as revealed by comparison of SOX1/2/3 and SOX9. *Mol. Cell. Biol.* **19,** 107–120.

Kanai, Y., and Koopman, P. (1999). Structural and functional characterization of the mouse Sox9 promoter: Implications for campomelic dysplasia. *Hum. Mol. Genet.* **8,** 691–696.

Karl, J., and Capel, B. (1995). Three-dimensional structure of the developing mouse genital ridge. *Philos. Trans. R. Soc. London* **350,** 235–242.

Karl, J., and Capel, B. (1998). Sertoli cells of the mouse testis originate from the coelomic epithelium. *Dev. Biol.* **203,** 323–333.

Katoh-Fukui, Y., Tsuchiya, R., Shiroishi, T., Nakahara, Y., Hashimoto, N., Noguchi, K., and Higashinakagawa, T. (1998). Male-to-female sex reversal in M33 mutant mice. *Nature (London)* **393,** 688–692.

Kawabe, K., Shikayama, T., Tsuboi, H., Oka, S., Oba, K., Yanase, T., Nawata, H., and Morohashi, K. (1999). Dax-1 as one of the target genes of Ad4BP/SF-1. *Mol. Endocrinol.* **13,** 1267–1284.

Kent, J., Wheatley, S. C., Andrews, J. E., Sinclair, A. H., and Koopman, P. (1996). A male-specific role for SOX9 in vertebrate sex determination. *Development (Cambridge, UK)* **122,** 2813–2822.

Kettlewell, J. R., Raymond, C. S., and Zarkower, D. (2000). Temperature-dependent expression of turtle Dmrt1 prior to sexual differentiation. *Genesis* **26,** 174–178.

Kim, J., Prawitt, D., Bardeesy, N., Torban, E., Vicaner, C., Goodyer, P., Zabel, B., and Pelletier, J. (1999). The Wilms' tumor suppressor gene (wt1) product regulates Dax-1 gene expression during gonadal differentiation. *Mol. Cell. Biol.* **19,** 2289–2299.

Koopman, P., Gubbay, J., Vivian, N., Goodfellow, P., and Lovell-Badge, R. (1991). Male development of chromosomally female mice transgenic for Sry. *Nature (London)* **351,** 117–121.

Kreidberg, J. A., Sariola, H., Loring, J. M., Maeda, M., Pelletier, J., Housman, D., and Jaenisch, R. (1993). WT-1 is required for early kidney development. *Cell (Cambridge, Mass.)* **74,** 679–691.

Kwok, C., Weller, P. A., Guioli, S., Foster, J. W., Mansour, S., Zuffardi, O., Punnett, H. H., Dominguez-Steglich, M. A., Brook, J. D., Young, I. D. et al. (1995). Mutations in SOX9, the gene responsible for Campomelic dysplasia and autosomal sex reversal. *Am. J. Hum. Genet.* **57,** 1028–1036.

Lahn, B. T., and Page, D. C. (1999). Four evolutionary strata on the human X chromosome. *Science* **286,** 964–967.

Larsson, S. H., Charlieu, J. P., Miyagawa, K., Engelkamp, D., Rassoulzadegan, M., Ross, A., Cuzin, F., van Heyningen, V., and Hastie, N. D. (1995). Subnuclear localization of WT1 in splicing or transcription factor domains is regulated by alternative splicing. *Cell (Cambridge, Mass.)* **81,** 391–401.

Little, M., Holmes, G., Bickmore, W., van Heyningen, V., Hastie, N., and Wainwright, B. (1995). DNA binding capacity of the WT1 protein is abolished by Denys-Drash syndrome WT1 point mutations. *Hum. Mol. Genet.* **4,** 351–358.

Little, M., Holmes, G., and Walsh, P. (1999). WT1: What has the last decade told us? *BioEssays* **21,** 191–202.

Little, M. H., Williamson, K. A., Mannens, M., Kelsey, A., Gosden, C., Hastie, N. D., and van Heyningen, V. (1993). Evidence that WT1 mutations in Denys-Drash syndrome patients may act in a dominant-negative fashion. *Hum. Mol. Genet.* **2,** 259–264.

Luo, X., Ikeda, Y., and Parker, K. L. (1994). A cell-specific nuclear receptor is essential for adrenal and gonadal development and sexual differentiation. *Cell (Cambridge, Mass.)* **77,** 481–490.

Lyon, M. F., and Hawkes, S. G. (1970). X-linked gene for testicular feminization in the mouse. *Nature (London)* **227,** 1217–1219.

Martineau, J., Nordqvist, K., Tilmann, C., Lovell-Badge, R., and Capel, B. (1997). Male-specific cell migration into the developing gonad. *Curr. Biol.* **7,** 958–968.

McLaren, A. (1991). Development of the mammalian gonad: The fate of the supporting cell lineage. *BioEssays* **13,** 151–156.

McLaren, A. (1995). Germ cells and germ cell sex. *Philos. Trans. R. Soc. London, Ser. B* **350,** 229–233.

McLaren, A., and Southee, D. (1997). Entry of mouse embryonic germ cells into meiosis. *Dev. Biol.* **187,** 107–113.

Merchant-Larios, H., and Moreno-Mendoza, N. (1998). Mesonephric stromal cells differentiate into Leydig cells in the mouse fetal testis. *Exp. Cell Res.* **244,** 230–238.

Merchant-Larios, H., Moreno-Mendoza, N., and Buehr, M. (1993). The role of the mesonephros in cell differentiation and morphogenesis of the mouse fetal testis. *Int. J. Dev. Biol.* **37,** 407–415.

Mertin, S., McDowall, S. G., and Harley, V. R. (1999). The DNA-binding specificity of SOX9 and other SOX proteins. *Nucleic Acids Res.* **27,** 1359–1364.

Meyer, J., Sudbeck, P., Held, M., Wagner, T., Schmitz, M. L., Bricarelli, F. D., Eggermont, E., Friedrich, U., Haas, O. A., Kobelt, A., Leroy, J. G., Van Maldergem, L., Michel, E., Mitulla, B., Pfeiffer, R. A., Schinzel, A., Schmidt, H., and Scherer, G. (1997). Mutational analysis of the SOX9 gene in campomelic dysplasia and autosomal sex reversal: Lack of genotype/phenotype correlations. *Hum. Mol. Genet.* **6,** 91–98.

Miller, C., and Sassoon, D. A. (1998). Wnt-7a maintains appropriate uterine patterning during the development of the mouse female reproductive tract. *Development (Cambridge, UK)* **125,** 3201–3211.

Mishina, Y., Rey, R., Finegold, M. J., Matzuk, M. M., Josso, N., Cate, R. L., and Behringer, R. R. (1996). Genetic analysis of the Mullerian-inhibiting substance signal transduction pathway in mammalian sexual differentiation. *Genes Dev.* **10,** 2577–2587.

Miyamoto, N., Yoshida, M., Kuratani, S., Matsuo, I., and Aizawa, S. (1997). Defects of urogenital development in mice lacking Emx2. *Development (Cambridge, UK)* **124,** 1653–1664.

Morais da Silva, S., Hacker, A., Harley, V., Goodfellow, P., Swain, A., and Lovell-Badge, R. (1996). Sox9 expression during gonadal development implies a conserved role for the gene in testis differentiation in mammals and birds. *Nat. Genet.* **14,** 62–68.

Moreno-Mendoza, N., Harley, V. R., and Merchant-Larios, H. (1999). Differential expression of SOX9 in gonads of the sea turtle *Lepidochelys olivacea* at male- or female-promoting temperatures. *J. Exp. Zool.* **284,** 705–710.

Morohashi, K. (1997). The ontogenesis of the steroidogenic tissues. *Genes Cells* **2,** 95–106.

Morohashi, K., Zanger, U. M., Honda, S., Hara, M., Waterman, M. R., and Omura, T. (1993). Activation of CYP11A and CYP11B gene promoters by the steroidogenic cell-specific transcription factor, Ad4BP. *Mol. Endocrinol.* **7,** 1196–1204.

Morohashi, K., Iida, H., Nomura, M., Hatano, O., Honda, S., Tsukiyama, T., Niwa, O., Hara, T., Takakusu, A., Shibata, Y. *et al.* (1994). Functional difference between Ad4BP and ELP, and their distributions in steroidogenic tissues. *Mol. Endocrinol.* **8,** 643–653.

Munsterberg, A., and Lovell-Badge, R. (1991). Expression of the mouse anti-mullerian hormone gene suggests a role in both male and female sexual differentiation. *Development (Cambridge, UK)* **113,** 613–624.

Muscatelli, F., Strom, T. M., Walker, A. P., Zanaria, E., Recan, D., Meindl, A., Bardoni, B., Guioli, S., Zehetner, G., Rabl, W. *et al.* (1994). Mutations in the DAX-1 gene give rise to both X-linked adrenal hypoplasia congenita and hypogonadotropic hypogonadism. *Nature (London)* **372,** 672–676.

Nachtigal, M. W., Hirokawa, Y., Enyeart-VanHouten, D. L., Flanagan, J. N., Hammer, G. D., and Ingraham, H. A. (1998). Wilms' tumor 1 and Dax-1 modulate the orphan nuclear receptor SF-1 in sex-specific gene expression. *Cell (Cambridge, Mass.)* **93,** 445–454.

Nagamine, C. M., Morohashi, K., Carlisle, C., and Chang, D. K. (1999). Sex reversal caused by *Mus musculus domesticus* Y chromosomes linked to variant expression of the testis-determining gene Sry. *Dev. Biol.* **216,** 182–194.

Nanda, I., Shan, Z., Schartl, M., Burt, D. W., Koehler, M., Nothwang, H., Grutzner, F., Paton, I. R., Windsor, D., Dunn, I., Engel, W., Staeheli, P., Mizuno, S., Haaf, T., and Schmid, M. (1999). 300 million years of conserved synteny between chicken Z and human chromosome 9. *Nat. Genet.* **21,** 258–259.

Nef, S., and Parada, L. F. (1999). Cryptorchidism in mice mutant for Insl3. *Nat. Genet.* **22,** 295–299.

Ng, L. J., Wheatley, S., Muscat, G. E., Conway-Campbell, J., Bowles, J., Wright, E., Bell, D. M., Tam, P. P., Cheah, K. S., and Koopman, P. (1997). SOX9 binds DNA, activates transcription, and coexpresses with type II collagen during chondrogenesis in the mouse. *Dev. Biol.* **183,** 108–121.

Oreal, E., Pieau, C., Mattei, M. G., Josso, N., Picard, J. Y., Carre-Eusebe, D., and Magre, S. (1998). Early expression of AMH in chicken embryonic gonads precedes testicular SOX9 expression. *Dev. Dyn.* **212,** 522–532.

Ottolenghi, C., Veitia, R., Quintana-Murci, L., Torchard, D., Scapoli, L., Souleyreau-Therville, N., Beckmann, J., Fellous, M., and McElreavey, K. (2000). The region on 9p associated with 46,XY sex reversal contains several transcripts expressed in the urogenital system and a novel doublesex-related domain. *Genomics* **64,** 170–178.

Palmer, S. J., and Burgoyne, P. S. (1991a). In situ analysis of fetal, prepuberal and adult XX----XY chimaeric mouse testes: Sertoli cells are predominantly, but not exclusively, XY. *Development (Cambridge, UK)* **112,** 265–268.

Palmer, S. J., and Burgoyne, P. S. (1991b). The *Mus musculus domesticus* Tdy allele acts later than the *Mus musculus musculus* Tdy allele: A basis for XY sex-reversal in C57BL/6-YPOS mice. *Development (Cambridge, UK)* **113,** 709–714.

Parr, B. A., and McMahon, A. P. (1998). Sexually dimorphic development of the mammalian reproductive tract requires Wnt-7a. *Nature (London)* **395,** 707–710.

Pask, A., and Graves, J. A. (1999). Sex chromosomes and sex-determining genes: Insights from marsupials and monotremes. *Cell Mol. Life. Sci.* **55,** 864–875.

Pask, A., Renfree, M. B., and Marshall Graves, J. A. (2000a). The human sex-reversing ATRX gene has a homolog on the marsupial Y chromosome, ATRY: Implications for the evolution of mammalian sex determination. *Proc. Natl. Acad. Sci. U.S.A.* **97,** 13198–13202.

Pask, A. J., Harry, J. L., Renfree, M. B., and Marshall Graves, J. A. (2000b). Absence of SOX3 in the developing marsupial gonad is not consistent with a conserved role in mammalian sex determination. *Genesis* **27,** 145–152.

Pelletier, J., Bruening, W., Li, F. P., Haber, D. A., Glaser, T., and Housman, D. E. (1991a). WT1 mutations contribute to abnormal genital system development and hereditary Wilms' tumour. *Nature (London)* **353,** 431–434.

Pelletier, J., Schalling, M., Buckler, A. J., Rogers, A., Haber, D. A., and Housman, D. (1991b). Expression of the Wilms' tumor gene WT1 in the murine urogenital system. *Genes Dev.* **5,** 1345–1356.

Pevny, L. H., and Lovell-Badge, R. (1997). Sox genes find their feet. *Curr. Opin. Genet. Dev.* **7,** 338–344.

Pfeifer, D., Kist, R., Dewar, K., Devon, K., Lander, E. S., Birren, B., Korniszewski, L., Back, E., and Scherer, G. (1999). Campomelic dysplasia translocation breakpoints are scattered over 1 Mb proximal to SOX9: Evidence for an extended control region. *Am. J. Hum. Genet.* **65,** 111–124.

Pontiggia, A., Rimini, R., Harley, V. R., Goodfellow, P. N., Lovell-Badge, R., and Bianchi, M. E. (1994). Sex-reversing mutations affect the architecture of SRY-DNA complexes. *EMBO J.* **13,** 6115–6124.

Raymond, C. S., Shamu, C. E., Shen, M. M., Seifert, K. J., Hirsch, B., Hodgkin, J., and Zarkower, D. (1998). Evidence for evolutionary conservation of sex-determining genes. *Nature (London)* **391,** 691–695.

Raymond, C. S., Kettlewell, J. R., Hirsch, B., Bardwell, V. J., and Zarkower, D. (1999). Expression of Dmrt1 in the genital ridge of mouse and chicken embryos suggests a role in vertebrate sexual development. *Dev. Biol.* **215,** 208–220.

Raymond, C. S., Murphy, M. W., O'Sullivan, M. G., Bardwell, V. J., and Zarkower, D. (2000). Dmrt1, a gene related to worm and fly sexual regulators, is required for mammalian testis differentiation. *Genes Dev.* **14,** 2587–2595.

Reddy, J. C., and Licht, J. D. (1996). The WT1 Wilms' tumor suppressor gene: How much do we really know? *Biochim. Biophys. Acta* **1287,** 1–28.

Renfree, M. B., and Short, R. V. (1988). Sex determination in marsupials: Evidence for a marsupial-eutherian dichotomy. *Philos. Trans. R. Soc. London, Ser. B* **322,** 41–53.

Sainio, K., Hellstedt, P., Kreidberg, J. A., Saxen, L., and Sariola, H. (1997). Differential regulation of two sets of mesonephric tubules by WT-1. *Development (Cambridge, UK)* **124,** 1293–1299.

Salas-Cortes, L., Jaubert, F., Barbaux, S., Nessmann, C., Bono, M. R., Fellous, M., McElreavey, K., and Rosemblatt, M. (1999). The human SRY protein is present in fetal and adult Sertoli cells and germ cells. *Int. J. Dev. Biol.* **43,** 135–140.

Schmahl, J., Eicher, E. M., Washburn, L. L., and Capel, B. (2000). Sry induces cell proliferation in the mouse gonad. *Development (Cambridge, UK)* **127,** 65–73.

Shawlot, W., and Behringer, R. R. (1995). Requirement for Lim1 in head-organizer function. *Nature (London)* **374,** 425–430.

Shen, W. H., Moore, C. C., Ikeda, Y., Parker, K. L., and Ingraham, H. A.

(1994). Nuclear receptor steroidogenic factor 1 regulates the mullerian inhibiting substance gene: A link to the sex determination cascade. *Cell (Cambridge, Mass.)* **77**, 651–661.

Shinoda, K., Lei, H., Yoshii, H., Nomura, M., Nagano, M., Shiba, H., Sasaki, H., Osawa, Y., Ninomiya, Y., Niwa, O. *et al.* (1995). Developmental defects of the ventromedial hypothalamic nucleus and pituitary gonadotroph in the Ftz-F1 disrupted mice. *Dev. Dyn.* **204**, 22–29.

Sinclair, A. H., Berta, P., Palmer, M. S., Hawkins, J. R., Griffiths, B. L., Smith, M. J., Foster, J. W., Frischauf, A. M., Lovell-Badge, R., and Goodfellow, P. N. (1990). A gene from the human sex-determining region encodes a protein with homology to a conserved DNA-binding motif. *Nature (London)* **346**, 240–244.

Smith, C. A., McClive, P. J., Western, P. S., Reed, K. J., and Sinclair, A. H. (1999). Conservation of a sex-determining gene. *Nature (London)* **402**, 601–602.

Smith, C. A., Clifford, V., Western, P. S., Wilcox, S. A., Bell, K. S., and Sinclair, A. H. (2000). Cloning and expression of a DAX1 homolog in the chicken embryo. *J. Mol. Endocrinol.* **24**, 23–32.

Spotila, L. D., Spotila, J. R., and Hall, S. E. (1998). Sequence and expression analysis of WT1 and Sox9 in the red-eared slider turtle, Trachemys scripta. *J. Exp. Zool.* **281**, 417–427.

Stayton, C. L., Dabovic, B., Gulisano, M., Gecz, J., Broccoli, V., Giovanazzi, S., Bossolasco, M., Monaco, L., Rastan, S., Boncinelli, E. *et al.* (1994). Cloning and characterization of a new human Xq13 gene, encoding a putative helicase. *Hum. Mol. Genet.* **3**, 1957–1964.

Sudbeck, P., Schmitz, M. L., Baeuerle, P. A., and Scherer, G. (1996). Sex reversal by loss of the C-terminal transactivation domain of human SOX9. *Nat. Genet.* **13**, 230–232.

Swain, A., Zanaria, E., Hacker, A., Lovell-Badge, R., and Camerino, G. (1996). Mouse Dax1 expression is consistent with a role in sex determination as well as in adrenal and hypothalamus function. *Nat. Genet.* **12**, 404–409.

Swain, A., Narvaez, V., Burgoyne, P., Camerino, G., and Lovell-Badge, R. (1998). Dax1 antagonizes Sry action in mammalian sex determination. *Nature (London)* **391**, 761–767.

Taketo, T., and Merchant-Larios, H. (1986). Gonadal sex reversal of fetal mouse ovaries following transplantation into adult mice. *Prog. Clin. Biol. Res.* **217A**, 171–174.

Teixeira, J., and Donahoe, P. K. (1996). Molecular biology of MIS and its receptors. *J. Androl.* **17**, 336–341.

Tilmann, C., and Capel, B. (1999). Mesonephric cell migration induces testis cord formation and Sertoli cell differentiation in the mammalian gonad. *Development (Cambridge, UK)* **126**, 2883–2890.

Torres, M., Gomez-Pardo, E., Dressler, G. R., and Gruss, P. (1995). Pax-2 controls multiple steps of urogenital development. *Development (Cambridge, UK)* **121**, 4057–4065.

Tucker, P. K., and Lundrigan, B. L. (1993). Rapid evolution of the sex determining locus in Old World mice and rats. *Nature (London)* **364**, 715–717.

Vainio, S., Heikkila, M., Kispert, A., Chin, N., and McMahon, A. P. (1999). Female development in mammals is regulated by Wnt-4 signalling. *Nature (London)* **397**, 405–409.

van der Lugt, N. M., Domen, J., Linders, K., van Roon, M., Robanus-Maandag, E., te Riele, H., van der Valk, M., Deschamps, J., Sofroniew, M., van Lohuizen, M. *et al.* (1994). Posterior transformation, neurological abnormalities, and severe hematopoietic defects in mice with a targeted deletion of the bmi-1 proto-oncogene. *Genes Dev.* **8**, 757–769.

Viger, R. S., Mertineit, C., Trasler, J. M., and Nemer, M. (1998). Transcription factor GATA-4 is expressed in a sexually dimorphic pattern during mouse gonadal development and is a potent activator of the Mullerian inhibiting substance promoter. *Development (Cambridge, UK)* **125**, 2665–2675.

Wagner, T., Wirth, J., Meyer, J., Zabel, B., Held, M., Zimmer, J., Pasantes, J., Bricarelli, F. D., Keutel, J., Hustert, E. *et al.* (1994). Autosomal sex reversal and campomelic dysplasia are caused by mutations in and around the SRY-related gene SOX9. *Cell (Cambridge, Mass.)* **79**, 1111–1120.

Wai-Sum, O., Short, R. V., Renfree, M. B., and Shaw, G. (1988). Primary genetic control of somatic sexual differentiation in a mammal. *Nature (London)* **331**, 716–717.

Watanabe, K., Clarke, T. R., Lane, A. H., Wang, X., and Donahoe, P. K. (2000). Endogenous expression of Mullerian inhibiting substance in early postnatal rat sertoli cells requires multiple steroidogenic factor-1 and GATA-4-binding sites. *Proc. Natl. Acad. Sci. U.S.A.* **97**, 1624–1629.

Werner, M. H., Huth, J. R., Gronenborn, A. M., and Clore, G. M. (1995). Molecular basis of human 46X,Y sex reversal revealed from the three-dimensional solution structure of the human SRY-DNA complex. *Cell (Cambridge, Mass.)* **81**, 705–714.

Wertz, K., and Herrmann, B. G (2000). Large-scale screen for genes involved in gonad development. *Mech. Dev.* **98**, 51–70.

Western, P. S., Harry, J. L., Graves, J. A., and Sinclair, A. H. (1999a). Temperature-dependent sex determination: Upregulation of SOX9 expression after commitment to male development. *Dev. Dyn.* **214**, 171–177.

Western, P. S., Harry, J. L., Graves, J. A., and Sinclair, A. H. (1999b). Temperature-dependent sex determination in the American alligator: AMH precedes SOX9 expression. *Dev. Dyn.* **216**, 411–419.

Whitfield, L. S., Lovell-Badge, R., and Goodfellow, P. N. (1993). Rapid sequence evolution of the mammalian sex-determining gene SRY. *Nature (London)* **364**, 713–715.

Whitworth, D. J., Licht, P., Racey, P. A., and Glickman, S. E. (1999). Testis-like steroidogenesis in the ovotestis of the European mole, *Talpa europaea. Biol. Reprod.* **60**, 413–418.

Wilkins, A. S. (1995). Moving up the hierarchy: A hypothesis on the evolution of a genetic sex determination pathway. *BioEssays* **17**, 71–77.

Wunderle, V. M., Critcher, R., Hastie, N., Goodfellow, P. N., and Schedl, A. (1998). Deletion of long-range regulatory elements upstream of SOX9 causes campomelic dysplasia. *Proc. Natl. Acad. Sci. U.S.A.* **95**, 10649–10654.

Yi, W., and Zarkower, D. (1999). Similarity of DNA binding and transcriptional regulation by *Caenorhabditis elegans* MAB-3 and *Drosophila melanogaster* DSX suggests conservation of sex determining mechanisms. *Development (Cambridge, UK)* **126**, 873–881.

Yu, B. D., Hess, J. L., Horning, S. E., Brown, G. A., and Korsmeyer, S. J. (1995). Altered Hox expression and segmental identity in Mll-mutant mice. *Nature (London)* **378**, 505–508.

Yu, R. N., Ito, M., Saunders, T. L., Camper, S. A., and Jameson, J. L. (1998). Role of Ahch in gonadal development and gametogenesis. *Nat. Genet.* **20**, 353–357.

Zanaria, E., Muscatelli, F., Bardoni, B., Strom, T. M., Guioli, S., Guo, W., Lalli, E., Moser, C., Walker, A. P., McCabe, E. R. *et al.* (1994). An unusual member of the nuclear hormone receptor superfamily responsible for X-linked adrenal hypoplasia congenita. *Nature (London)* **372**, 635–641.

Zazopoulos, E., Lalli, E., Stocco, D. M., and Sassone-Corsi, P. (1997). DNA binding and transcriptional repression by DAX-1 blocks steroidogenesis. *Nature (London)* **390**, 311–315.

Zhu, L., Belo, J. A., De Robertis, E. M., and Stern, C. D. (1999). Goosecoid regulates the neural inducing strength of the mouse node. *Dev. Biol.* **216**, 276–281.

Zimmermann, S., Steding, G., Emmen, J. M., Brinkmann, A. O., Nayernia, K., Holstein, A. F., Engel, W., and Adham, I. M. (1999). Targeted disruption of the Insl3 gene causes bilateral cryptorchidism. *Mol. Endocrinol.* **13**, 681–691.

18

Development of the Excretory System

Gregory R. Dressler

Department of Pathology, University of Michigan, Ann Arbor, Michigan 48109

I. Introduction

II. Patterning of the Intermediate Mesoderm

III. Growth of the Nephric Duct and Ureteric
Bud Diverticulum

IV. Inductive Interactions

V. Mesenchyme-to-Epithelial Conversion

VI. Glomerular Development
and Vascularization

VII. Developmental Basis of Human
Renal Disease

VIII. Future Perspectives

References

I. Introduction

Perhaps more so than any other tissue, the development of the urogenital system in higher vertebrates illustrates quite clearly how "ontogeny follows phylogeny." In mammals, the embryonic kidney follows a developmental progression that mimics its evolutionary origin, proceeding through a series of transition structures before generating the adult kidney. Although detailed embryology and anatomy can be gleaned from a variety of excellent sources (Carlson, 1988; Kaufman, 1992), this chapter provides a brief introduction to the basic design of the mouse urogenital system and the relationships between the parts.

In all mammals, the excretory system begins as a single epithelial duct formed from the intermediate mesoderm, a region between the paraxial, or somitic, mesoderm and the lateral plate mesoderm (Fig. 1). As this nephric duct extends caudally, it induces a linear array of epithelial tubules, which extend medioventrally and are thought to derive from periductal mesenchyme. In mammals, a graded evolution of renal tubule development occurs with the most anterior, or pronephric tubules, being very rudimentary, and the more posterior, mesonephric tubules becoming well developed. In contrast, the pronephros of the zebrafish larvae is a fully developed, functional filtration unit with a single midline glomerulus (Drummond *et al.,* 1998). Amphibian embryos such as *Xenopus laevis* have bilateral pronephric glomeruli and tubules that are functional until replaced by a mesonephric kidney in the tadpole (Vize *et al.,* 1997). In fact, it is not altogether obvious in mammals where to draw the distinction between pronephric tubules and mesonephric tubules. Although mature mesonephric tubules are characterized by a vascularized glomerulus at the proximal end of the tubule that empties into the nephric duct, the most anterior and

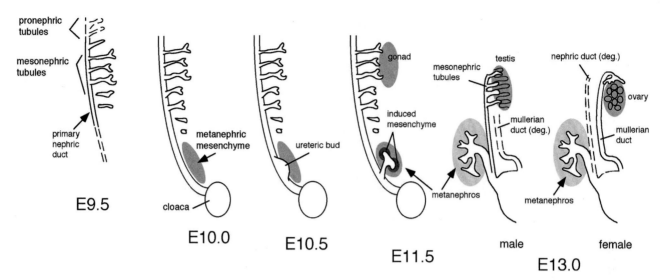

Figure 1   The three stages of kidney development in the mouse. Between E9 and E13, the three embryonic kidneys are formed in a sequential manner along the anterior-posterior axis. As the primary nephric duct extends caudally, pro- and mesonephric tubules are induced along the periductal mesenchyme. By E10, the nephric duct has reached the cloaca, the mesonephros is present as a linear array of tubules, and the metanephric mesenchyme remains uninduced at the posterior end. By E10.5, ureteric bud growth becomes evident and invasion into the mesenchyme is complete by E11.5. In response to inductive signals, the metanephric mesenchyme aggregates around the ureteric bud tips. In turn, the ureteric bud undergoes repeated dichotomous branching morphogenesis. By E13, the Müllerian duct is observed running parallel to the posterior nephric duct. The male mesonephric tubules, adjacent to the testis, will form the epididymis and vas deferens, as the Müllerian duct degenerates. The female Müllerian duct will form the oviduct and uterus, while the nephric duct and mesonephric tubules have completely degenerated.

posterior mesonephric tubules are more rudimentary, with the posterior tubules not connected to the duct at all.

The adult kidney, or metanephros, is formed at the caudal end of the nephric duct when an outgrowth, called the ureteric bud or metanephric diverticulum, extends into the surrounding metanephric mesenchyme. This process begins at E10.5 and requires signals emanating from the mesenchyme for bud initiation. Inductive signals from the bud begin the process of converting the metanephric mesenchyme to epithelium. Mesenchymal cells aggregate around the tips of the bud, form a primitive polarized epithelial vesicle, and undergo a well-defined morphological series of events leading to the mature nephron (Fig. 2). Reciprocal inductive signals, derived from the mesenchyme, promote growth and branching of the ureteric bud. Branching is dichotomous and results in new aggregates forming at the tips of the branches and further induction of nephrons. This repeated branching and induction results in the formation of nephrons along the radial axis of the kidney, with the oldest nephrons being more medullary and the younger nephrons located toward the periphery.

It was generally believed that most of the epithelium of the nephron was derived from the metanephric mesenchyme, whereas the ureteric bud epithelium generates the collecting ducts and the most distal tubules. As will be discussed in following sections, more precise cell lineage studies have challenged this hypothesis. However, not all the mesenchyme becomes induced and converted to epithelium, with some cells contributing to the interstitial mesenchyme, or

stroma, and others generating endothelial cells of the renal vasculature. Mesenchyme-derived epithelial cells become highly specialized and express markers specific for glomerular podocytes, proximal tubules, cells of the ascending and descending limbs of Henle's loop, and distal tubules. How cells are instructed to differentiate along individual pathways remains obscure. The analysis of gene expression patterns and the generation of many mouse mutants, with distinct renal phenotypes, have provided a wealth of data with respect to early inductive events, mesenchyme-to-epithelium conversion, and branching morphogenesis. Much of the available data are accessible on-line through the Kidney Development Database (http://www.ana.ed.ac.uk/anatomy/database/kidbase/kidhome.html) (Davies, 1999), a systematic compilation that is searchable and frequently updated.

II. Patterning of the Intermediate Mesoderm

The complex pattern of gene expression during kidney development begins with the specification of the intermediate mesoderm, which already expresses a number of unique markers prior to the initiation of urogenital development. Molecular genetic analysis has confirmed the essential functions of a variety of early patterning genes in the intermediate mesoderm. Yet, the genetic network that positions the urogenital tract, restricts the cell lineages of the developing

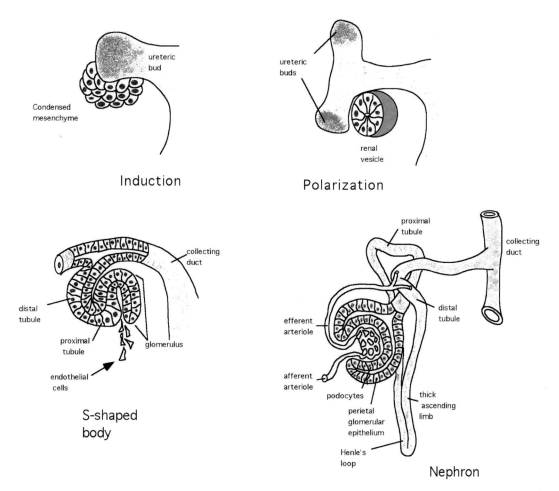

Figure 2 The conversion of metanephric mesenchyme to epithelium. The sequential stages in the formation of the nephron are outlined. Induction by the ureteric bud results in mesenchymal aggregates forming at the bud tips. The aggregates become polarized and form a primitive epithelial renal vesicle. That part of the renal vesicle adjacent to the ureteric bud tip fuses to the collecting duct primordium to form a continuous lumen at the S-shaped body stage. The ureteric bud has grown outward to induce the next generation of nephrons along the radial axis of the kidney. Furthest from the collecting duct, endothelial precursor cells invade the glomerular cleft. The developing nephron is fully elaborated as glomerular vascularization and proximal tubule extension begins.

kidney, and remodels the epithelium to generate the precise three-dimensional architecture of the nephron still remains obscure.

A. Early Regionalization

Among the two earliest genes expressed as makers of nephric progenitors are *lim1* and *Pax2*. The *lim1* gene encodes a homeobox containing protein that is expressed in the developing head and the intermediates mesoderm, extending along the anterior-posterior axis. At E8.5, *lim1* expression is found in both the somatopleure and the splanchnopleure of the lateral plate mesoderm but becomes restricted to the nephrogenic cord by E9.5 (Tsang *et al.*, 2000). Based on studies using a *LacZ*-tagged lim1 null allele (Tsang *et al.*, 2000), the gene appears essential for the aggregation of these early mesodermal cells into the nephric cord. Furthermore,

lim1 seems to restrict the fate of the mesodermal precursors to preferentially generate intermediate and lateral plate mesoderm. In the absence of lim1 activity, formation of the nephric duct is very rudimentary and expression of the *Pax2* gene is suppressed.

The *Pax2* gene is also essential for early urogenital development. *Pax2* expression is first detected in the intermediate mesoderm, prior to formation of the nephric duct, beginning at approximately E8.5 (Fig. 3). In the chick, transplantation experiments suggest that *Pax2* expression and specification of the nephric cord is dependent on signals emanating from the trunk paraxial mesoderm (Mauch *et al.*, 2000). Despite its early expression pattern, *Pax2* is not required for initiation and extension of the nephric duct. Mouse mutants lacking *Pax2* show an epithelial duct extending caudally toward the cloaca, from E9.5 to E11.5. However, by E11.5 the nephric duct starts to degenerate. *Pax2* mutants show no evidence

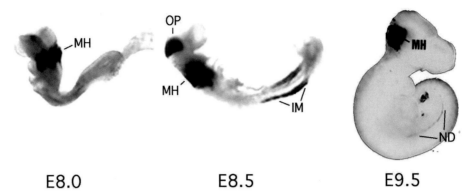

Figure 3 Early expression pattern of *Pax2* as marked by *Pax2/lacZ* transgene. Approximately 4 kb of upstream regulatory sequences from the mouse *Pax2* gene were fused to the *Escherichia coli* β-galactosidase (*lacZ*) coding sequences. The expression of *lacZ* mimics the early pattern of *Pax2* expression, as shown by whole-mount staining for *lacZ* activity. At E8, approximately 0–4 somites, *Pax2* expression is prominent in the midbrain-hindbrain (MH) junction. By E8.5 (4–8 somites), *Pax2* expression is detectable in the intermediate mesoderm (IM) prior to nephric duct formation, in the midbrain-hindbrain junction and in and around the optic placode (OP). By E9.5, *Pax2* expression in the nephric duct (ND) is evident.

of mesonephric tubule formation, completely lack the metanephros, and are deficient in the sex-specific epithelial components derived from the intermediate mesoderm (Torres *et al.,* 1995). Thus, the periductal mesenchyme is unable to generate mesonephric tubules or cannot respond to signals derived from the nephric duct. Metanephric mesenchyme isolated from *Pax2* mutants is also unable to respond to inductive signals from heterologous tissues in the organ culture assay (G. R. Dressler, unpublished observation). Thus it appears that *Pax2* is necessary for specifying the region of intermediate mesoderm destined to undergo mesenchyme-to-epithelium conversion. Furthermore, *Pax2* mutant nephric duct does not respond to GDNF and shows no evidence of ureteric bud outgrowth despite expression of at least some markers of normal ductal epithelium. In the nephric duct epithelium, *Pax2* is also required for maintenance, or responsiveness perhaps through regulation of the glial-derived neurotrophic factor (GDNF) receptor, RET.

In the developing metanephros, *Pax2* expressing cells are closely associated with the ureteric bud tips. This observation led to the proposition that *Pax2* expression was activated by inductive signals emanating from the bud (Dressler *et al.,* 1990; Dressler and Douglass, 1992). Further support for the dependence of *Pax2* expression on inductive signals came from studies with the *Danforth's short tail* (*Sd*) mouse, in which *Pax2* expression was detected in the nephric duct and ureteric bud but not in the mesenchyme, presumably due to lack of ureteric bud invasion (Phelps and Dressler, 1993). Yet, the primary defect in the *Sd* mouse is posterior degeneration of the notocord. Thus, gene expression in the intermediate mesoderm could be affected by loss of notocord-derived patterning signals. Until recently, it has proved difficult to separate *Pax2* expression in the mesenchyme from ureteric bud invasion, since it is not easy to define the mesenchyme morphologically prior to E11. However, re-

cently we observed *Pax2* expression in the metanephric mesenchyme of E11.5 RET homozygous mutants, which have no ureteric bud but essentially wild-type mesenchyme (Fig. 4). Thus, *Pax2* expression in the metanephric mesenchyme predates ureteric bud invasion and marks the posterior intermediate mesoderm as the metanephric anlagen.

The Pax gene family encodes transcription factors essential for the development of the eye, the vertebral column, certain derivatives of the neural crest, B lymphocytes, the thyroid, and various neural structures (Dahl *et al.,* 1997; Noll, 1993; Stuart *et al.,* 1993). In mice and humans, Pax genes are haploinsufficient, indicating a strict requirement for Pax gene dosage during normal development. The phenotype of heterozygous *Pax* mutant individuals is variable and less severe than that found among homozygotes. Human mutations in the *PAX2* gene have been described. These heterozygous individuals have a variety of renal defects, including small, poorly differentiated kidneys and reflux of urine from the bladder (Sanyanusin *et al.,* 1995; Schimmenti *et al.,* 1997).

Within the mesonephric and metanephric mesenchymal cells along much of the nephric duct, expression of the mouse *Eya1* gene is prevalent. *Eya1* is a mammalian homolog of the *Drosophila eyes absent* gene and is a member of a small gene family. In humans, mutations in the *Eya1* gene are associated with branchio-oto-renal syndrome, a complex multifaceted phenotype (Abdelhak *et al.,* 1997). In mice homozygous for an *Eya1* mutation, kidney development is arrested at E11 because ureteric bud growth is inhibited and the mesenchyme remains uninduced, though *Pax2* and *WT1* expression appears normal (Xu *et al.,* 1999). However, two other markers of the metanephric mesenchyme, *Six2* and *GDNF* expression, are lost in the *Eya1* mutants. The loss of *GDNF* expression most probably underlies the failure of ureteric bud growth (see Section III). However, it is not clear if

anti-Pax2   anti-Cytokeratin

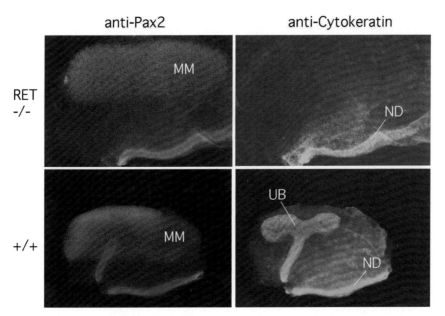

Figure 4 *Pax2* marks the metanephric mesenchyme even in the absence of inductive signals. The figure shows whole-mount antibody staining for *Pax2* (red-orange) and cytokeratins (green) in E11.5 metanephroi. The region in and around the metanephric mesenchyme was dissected from wild-type and RET homozygous null mutants. Note the growth and branching of the ureteric bud, which is absent in the RET mutants. Pax2 protein is evident even in the absence of the ureteric bud, as indicated by nuclear staining of the metanephric mesenchyme of RET mutants.

the mesenchyme is competent to respond to inductive signals if a wild-type inducer were to be used *in vitro*.

The Eyes Absent gene family is part of a conserved network that underlies cell specification in several other developing tissues. Eya proteins share a conserved domain but lack DNA-binding activity. The Eya proteins interact directly with the Six family of DNA-binding proteins. Mammalian Six genes are homologs of the *Drosophila sina oculis* homeobox gene. This cooperative interaction between Six and Eya proteins is necessary for nuclear translocation and transcriptional activation of Six target genes (Ohto *et al.*, 1999) (see Fig. 5). Eya proteins in turn can also bind to the dachshund family of proteins. In *Drosophila* eye development (Pignoni *et al.*, 1997) and during muscle cell specification in the chick embryo (Heanue *et al.*, 1999), the regulatory network involving Pax, Eya, Six, and dachshund genes is well conserved. Based on conservation of expression patterns and mutant phenotypes, a similar regulatory network has been proposed in the developing mouse kidney (Xu *et al.*, 1999), with *Pax2* being upstream of *Six2* and *Eya1*, although a dachshund family homolog has not yet been described in the kidney.

B. Posterior Specification of the Metanephric Mesenchyme

As the mesonephric tubules develop along the anterior-posterior axis of the nephric cord and the nephric duct ex-

tends to the cloaca, the most posterior part of the duct is not associated with mesonephric tubules at all; rather it runs adjacent to an aggregate of cells called the metanephric mesenchyme (Fig. 1). This metanephric mesenchyme is morphologically identifiable by E10.5–11. Even in *Pax2* mutants, the metanephric mesenchyme appears as a distinct aggregate of cells apart from the surrounding mesoderm. Several genes are now known to mark this mesenchyme and to play critical roles in the survival and response of this mesenchyme to inductive signals.

The Wilms' tumor suppressor gene, *WT1*, is one of the early markers of the metanephric mesenchyme and is essential for its survival. Wilms' tumor is an embryonic kidney neoplasia that consists of undifferentiated mesenchymal cells, poorly organized epithelium, and surrounding stromal cells. Because the neoplastic cells of the tumor are able to differentiate into a wide variety of cell types, the genes responsible for Wilms' tumor were thought to be important regulators of early kidney development (van Heyningen and Hastie, 1992). Expression of *WT1* is regulated spatially and temporally in a variety of tissues and is further complicated by the presence of at least four isoforms, generated by alternative splicing. In the developing kidney, WT1 can be found in the uninduced metanephric mesenchyme and in differentiating epithelium after induction (Armstrong *et al.*, 1992; Pritchard-Jones *et al.*, 1990). Early expression of WT1 may be mediated by PAX2 (Dehbi *et al.*, 1996). Initial expression levels are low in the metanephric mesenchyme, but become upregulated at the S-shaped body stage in the precursor cells

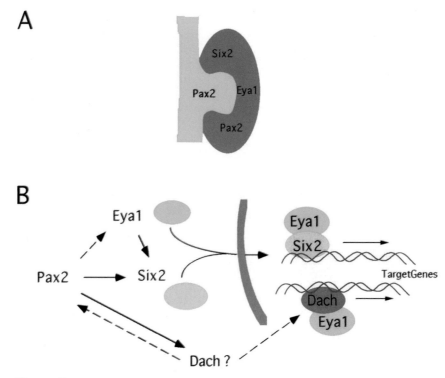

Figure 5 Interactions among the Pax, Eya, Six, and dachshund gene families. (A) In the E11 metanephros, Pax2 expression is present in the ureteric bud and the mesenchyme, whereas Six2 and Eya1 are present only in the mesenchyme. (B) Current model of genetic interactions based on work in the fly, chick, and mouse embryos. Pax genes are required for the activation of Six and dachsund genes. Eya, which may also be regulated by Pax, functions to translocate Six to the nucleus to cooperatively bind a target sequence and activate transcription. Eya can also interact with dachshund to activate other sets of target genes.

of the glomerular epithelium, the podocytes (Fig. 6). High WT1 levels persists in the podocytes of the glomerulus into adulthood.

The impact of *WT1* on kidney development is still under intense investigation, though some conclusions can be drawn from the phenotypes of both mouse and human mutations. The homozygous null *WT1* mouse has complete renal agenesis (Kreidberg *et al.,* 1993), because the ureteric bud fails to grow out of the nephric duct and the metanephric mesenchyme undergoes apoptosis. The arrest of ureteric bud growth appears to be noncell autonomous and is most probably due to lack of signaling by the *WT1* mutant mesenchyme. Yet, the mesenchyme is unable to respond to inductive signals even if a wild-type inducer is utilized in the organ culture assay. Taken together, it appears that *WT1* is required early in the mesenchyme to promote cell survival, such that cells can respond to inductive signals and express ureteric bud growth promoting factors. At later stages of nephron development, increased WT1 expression levels in the podocyte precursor cells correlate with repression of the *Pax2* gene (Ryan *et al.,* 1995). WT1 binds to the promoters of a variety of genes, including *Pax2, IGF2,* and the IGF2 receptor, by direct contact of the zinc finger DNA binding domain to a conserved GC rich nanomer. Much of the data

indicate that the amino terminal region of WT1 acts as a transcription repression domain (Madden *et al.,* 1991), consistent with the increased levels of growth factor expression observed in Wilms' tumors. Yet in other cases, WT1 is able to function as a transcription activator (Lee *et al.,* 1999). Furthermore, specific isoforms of WT1 proteins have been colocalized to the mRNA splicing machinery suggesting a post-transcriptional role (Davies *et al.,* 1998). Thus, many questions still remain to be addressed. The expression levels of *WT1* are precisely regulated during kidney development and may indicate different functions depending on the threshold level of protein present and the ratio of individual isoforms.

III. Growth of the Nephric Duct and Ureteric Bud Diverticulum

A. Initiation of the Nephric Duct

The first epithelial structure formed within the intermediate mesoderm is the pronephric duct. Because this duct extends caudally it is also known as the mesonephric duct or the Wolffian duct. All three of these terms describe essen-

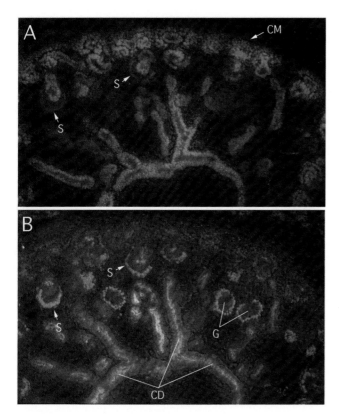

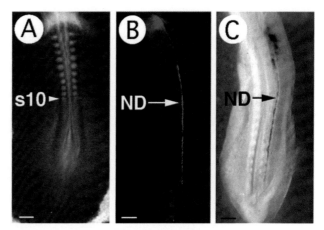

Figure 7 Origin of the nephric duct in the chick embryo. Cell lineage analysis has proven notoriously difficult in the mouse. However, the rapid development and accessibility of the chick embryo makes it much more amenable to lineage tracing. (A) In the experiments of Obara-Ishihara *et al.,* (1999) dI was injected adjacent to somite 10 of the stage 10 chick embryo. After 24 hours further incubation, the nephric duct extends caudally and is almost completely labeled with diI(B). Lineage tracing using a *lacZ*-expressing retrovirus shows similar results (C). These experiments demonstrate that most, if not all, of the nephric duct is derived from the intermediate mesoderm around somite 10. Thus, the nephric duct precursor cells are specified in a restricted position and grow caudally by extension and proliferation, not by recruiting posterior mesodermal cells. (Photomicrograph courtesy of D. Herzlinger, Cornell University.)

Figure 6 Expression of WT1 and Pax2 in nephron development. Serial sections are shown from a kidney at E14.5. Antibody staining (A) for Pax2 (orange) and E-cadherin (green) and (B) for WT1 (orange) and E-cadherin (green). All stages of nephron development are shown, from undifferentiated mesenchyme along the periphery to maturing glomeruli, tubules, and collecting ducts located more centrally. The ureteric bud has branched repeatedly at this point and the major collecting ducts (CD) are now evident. WT1 expression is low in the mesenchyme, but increases at the S-shaped body stage (S) in the cells of the glomerular cleft. These are the podocyte precursors, which surround the developing glomerulus (G) before migrating into the glomerular tuft. Pax2 expression is high in the condensing mesenchyme (CM) and its early derivatives but becomes more restricted as the nephron matures. In the developing podocytes, Pax2 levels decrease as WT1 goes up. Also note the lack of Pax2 in the interstitium, or stroma.

tially the same structure as it extends along the anterior-posterior (AP) axis. I prefer to use *nephric duct* as a generic term for this epithelial tube, because this is less confusing to the reader. I believe the terminology should reflect the origin and function, not necessarily the position along the axis, although this may be upsetting to some more classically schooled readers. Nevertheless, the position of origin of the nephric duct has not been precisely determined in the mouse. Epithelium can be seen in the intermediate mesoderm at E9.0 in the developing trunk and extending caudally. By E9.5 this epithelium has canalized and extends toward the cloaca. Whether the nephric duct epithelium extends by proliferation of the initiated duct epithelium or by recruitment of mesenchyme cells as it extends caudally has been the subject of some debate.

In the chick embryo, more detailed lineage tracing data are now available (Obara-Ishihara *et al.,* 1999) (Fig. 7). By

injecting DiI lineage markers into the intermediate mesoderm, Obara-Ishihara *et al.* mapped the origin of the duct cells to the region adjacent to somite 10 of the chick embryo. At later developmental stages, most, if not all, of the nephric duct epithelium was derived from the *Pax2* and *Sim1* positive intermediate mesoderm cells located adjacent to somites 9–11. Thus, nephric duct extension is promoted by proliferation of existing epithelium formed at a specific position along the AP axis and does not require additional mesenchymal recruitment. Furthermore, initiation of nephric duct epithelium requires signals derived from the overlying surface ectoderm, because removal of surface ectoderm in the chick inhibits duct formation. This block in nephric duct formation can be rescued by bone morphogenic protein 4 (BMP4), suggesting that this is the signal emanating from the ectoderm necessary for epithelial duct initiation (Obara-Ishihara *et al.,* 1999).

B. The RET/GDNF Pathway Determines Ureteric Bud Outgrowth

Morphological observations indicated that the development of the metanephros begins on invasion of the ureteric bud into the adjacent metanephric mesenchyme. This occurs at approximately E11 in the mouse and E13 in the rat. The ureteric bud begins as a slight bulge in the posterior nephric duct, protruding toward the adjacent mesenchyme (see Fig. 8A). As the bulge narrows to a ductal epithelial structure, it invades the mesenchyme and forms an ampulla

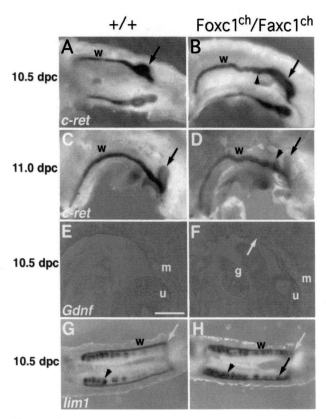

+/+   Foxc1^ch/Faxc1^ch

10.5 dpc
c-ret

11.0 dpc
c-ret

10.5 dpc
Gdnf

10.5 dpc
lim1

Figure 8 Fox genes regulate ureteric bud positioning and GDNF expression. The *Foxc1* gene encodes a forkhead/winged helix transcription factor. Loss-of-function results in ureteric bud abnormalities as indicated in panels B, D, F, H. Panels A–D are stained for c-ret expression to mark the nephric duct and ureteric bud. Note that in the *Foxc1* mutant (B), the early ureteric bud (E10.5) extends more anterior (arrowhead) resulting in a broadening of the nephric duct, compared to wild type (A). The position of the normal bud is marked with an arrow. By E11, this broadened ureteric bud results in a supernumerary bud (arrowhead, D). Expression of GDNF is shifted more anterior in *Foxc1* mutants (arrow, F), compared to controls (E). The mesonephroi are marked by expression of lim1 in panels G and H. Note the more posterior mesonephric tubules in the *Foxc1* mutant (black arrow, H), whereas the wild-type tubules do not extend as far posterior (G). The position of the ureteric bud is marked by a yellow arrow. (Original figure in Kume *et al.*, 2000; reproduced by courtesy of B. Hogan, Vanderbilt University.)

within the center of the mesenchymal cylinder. The ampulla grows larger until the first branch point is evident within the distal midline of the bud. The bud then forms a T-shaped structure, such that the top of the T is parallel to the AP axis of the embryo and the mesenchymal cylinder. At the tips of the T, new ampullae appear prior to specification of the second branch points.

Although the nephric duct extends along much of the AP axis, from the midthoracic region to the cloaca, the ureteric bud always branches out in the same position adjacent to the mesenchyme. How is this achieved? The genetic and biochemical pathways that determine ureteric bud outgrowth have recently been revealed, primarily through the use of gene targeting in the mouse. The c-ret gene encodes a recep-

tor type tyrosine kinase (RET), expressed along the entire nephric duct at E10–11 and is critical for proper outgrowth of the ureteric bud (Schuchardt *et al.*, 1994). Loss of RET function in the mouse results in severe renal agenesis due in large part to a complete failure of the ureteric bud to grow. In the small percentage of embryos that do exhibit some bud growth, the mesenchyme appears induced but subsequent branching morphogenesis is arrested. The *RET*^−/− mutant mesenchyme retains the ability to respond to wild-type inductive signals and can differentiate into epithelial cell types (Schuchardt *et al.*, 1996). The phenotype is cell autonomous in that loss of RET inhibits growth and branching of the ureteric bud epithelium, exclusively where the receptor is expressed. Thus, it appears that RET is necessary to receive and transduce a signal required for ureteric bud outgrowth and branching.

Classical experiments by Grobstein implied that the mesenchymal signal required for branching, the presumptive RET ligand, was produced by the metanephric mesenchyme. This was confirmed by multiple genetic and biochemical studies. Although for many years an orphan receptor, the ligand for RET was deduced in part by genetic experiments that addressed the role of the neurotrophin GDNF in mouse development. The generation of a GDNF null mouse by three independent laboratories produced a phenotype almost identical to the RET homozygous null mouse (Moore *et al.*, 1996; Pichel *et al.*, 1996; Sanchez *et al.*, 1996). Taken together with the embryonic expression pattern, which showed GDNF expression in the metanephric mesenchyme adjacent to the tips of the ureteric bud (Hellmich *et al.*, 1996), the genetics clearly pointed to a receptor ligand relationship between GDNF and RET. Furthermore, GDNF could physically associate with RET and activate the intracellular kinase domain of the RET protein (Jing *et al.*, 1996; Trupp *et al.*, 1996; Vega *et al.*, 1996). *In vitro* organ culture experiments demonstrated that GDNF could stimulate branching morphogenesis of the ureteric bud and its derivatives (Vega *et al.*, 1996). At the biochemical level, the receptor–ligand relationship was not so clear. Rather than binding to RET directly, GDNF binds a high-affinity receptor, called GDNFRα or GFRα1, a Gpi-linked protein expressed at the surface of both neural and renal precursor cells (Jing *et al.*, 1996; Treanor *et al.*, 1996). Activation of RET requires a complex of GDNF, GFRα1, and RET, the stoichiometry of which remains to be determined. Yet, several issues remained. Was GDNF just a mitogen that could stimulate proliferation of the ureteric bud epithelium? Or was the GDNF/RET pathway part of a directional guidance mechanism that controlled bud growth into the proper position within the mesenchyme?

Once the GDNF/RET connection had been clarified, the nature of GDNF could be addressed more directly in biological model systems. Within the context of the nephric duct and surrounding mesoderm, a localized source of GDNF was sufficient to induce ureter-like outgrowths from ectopic sites

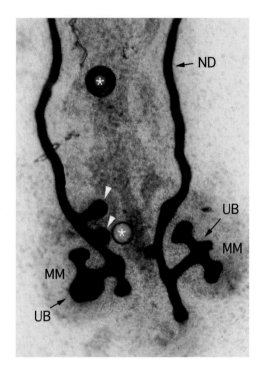

Figure 9 GDNF promotes ureteric bud outgrowth *in vitro*. In this experiment, the developing excretory system was dissected out at E10 and placed on a transwell filter. The tissue is from the *Pax2/lacZ* transgenic mouse that marks the nephric duct. Two beads (asteriks) soaked in recombinant GDNF were placed along the midline, between the left and right nephric ducts. After 24 hr in culture, the normal ureteric bud (UB) grows laterally into the metanephric mesenchyme. However, the posterior GDNF bead stimulates ectopic ureteric buds growing toward the midline (arrowheads). The bead placed more anterior does not stimulate bud growth, suggesting that there may be local inhibitory factors that limit the activity of GDNF or that this anterior duct has lost competence to respond to GDNF (G. R. Dressler, unpublished.)

along the nephric duct (Fig. 9) (Sainio *et al.*, 1997). Furthermore, in a renal epithelial cell culture model, activation of RET resulted in increased cell motility and chemoattraction to a localized source of GDNF (Tang *et al.*, 1998). In this model, GDNF plus GFRα1 could activate RET and cause cells to lose their adhesive junctions, reorganize the actin cytoskeleton, and extend long lamellipodia and filopodia. These activated RET cells became very motile and would migrate toward a source of GDNF embedded within a collagen gel. Yet, activation of RET proved not to be mitogenic in this system. Taken together, these data support the idea that GDNF is a target-derived guidance cue that determines when and where the ureteric bud grows out of the nephric duct.

C. Inhibitors and Enhancers of Ureteric Bud Growth and Branching

Although genetic analysis reveals the GDNF/GFRα1/ RET signaling complex as necessary for bud growth, additional factors may regulate growth and branching independently or by modifying the response to RET activation. One

such potential inhibitor is BMP4. In *Bmp4* heterozygous mice, ectopic budding of the ureter is often observed, resulting in a high frequency of renal abnormalities with similarities to human congenital anomalies of the kidney and urinary tract (CAKUT) syndrome (Miyazaki *et al.*, 2000). In the developing metanephros, BMP4 can inhibit the expression of Wnt-11 at the tips of the branching ureteric buds. Because activation of RET, through GDNF, enhances Wnt-11 expression in this same region, BMP4 appears to negate the effects of GDNF without lowering expression of GDNF. Thus, BMP4 expression along the nephric duct at E11 and in the stromal cells surrounding the mature stalks of the ureteric epithelium at later times can limit branching morphogenesis more precisely to only the tips of the bud.

Retinoids and their receptors also affect ureteric bud branching by modifying the RET signaling pathway. It is well documented that vitamin A deficiency results in severe renal defects (Lelievre-Pegorier *et al.*, 1998). In organ culture, retinoic acid stimulates expression of RET and can dramatically increase the number of ureteric bud branch points by threefold (Vilar *et al.*, 1996). The effects of retinoids appear to be mediated by the surrounding stromal cell population, which expresses the retinoic acid receptors α and β2. Both *RARα* and *RARβ2* homozygous mutant mice have no significant renal defects. However, double homozygotes mutant for both *RARα* and *RARβ2* exhibit severe growth retardation in the kidney that can be explained by downregulation of RET and suppression of ureteric bud branching morphogenesis (Mendelsohn *et al.*, 1999). The *in vivo* source of retinoic acid remains to be determined and it is not clear how this affects the ability of the stroma to support branching of the ureteric bud epithelium. It may be that RAR signaling is required for stromal cell expansion as the kidney grows.

An intrinsic factor expressed in the ureteric bud and necessary for branching is the homeobox gene *Emx2*. *Emx2* is expressed in the nephric duct and in the ureteric bud as it invades the mesenchyme at E11. However, *Emx2* homozygous mutants fail to develop beyond the stage of ureteric bud invasion at E11.5, resulting in complete bilateral agenesis (Miyamoto *et al.*, 1997). Strikingly, the mutant ureteric bud grows out of the duct and invades the mesenchyme but fails to form a characteristic ampulla, a step just prior to the formation of the first branch point. At E11.5, *Emx2* mutants exhibit a reduction in the expression of RET and PAX2 in the bud epithelium and spotty expression of PAX2 in the mesenchyme. The *Emx2* mutant mesenchyme retains expression of most markers at E10–11 and is able to respond to wild-type inductive signals, indicating that the primary defect is in the ureteric bud epithelium. Thus, *Emx2* is required for maintaining the proliferation of the ureteric bud after invasion of the mesenchyme, perhaps by controlling the responses to mesenchymal-derived proliferation and branching signals.

If GDNF is a target-derived guidance cue for ureteric bud growth, what controls the spatial and temporal pattern of GDNF expression? The developmental defects observed in the mouse mutants of the forkhead/winged helix genes *Foxc1* and *Foxc2* shed light on this question (Kume *et al.,* 2000). *Foxc1* and *Foxc2* have similar expression domains in the presomitic and intermediate mesoderm, as early as E8.5. As nephric duct extension progresses, Foxc1 is expressed in a dorsoventral gradient with the highest levels near the neural tube and lower levels in the BMP4 positive ventrolateral regions. In *Foxc1* homozygous null mutants, GDNF expression in the posterior intermediate mesoderm, which presumably marks the metanephric mesenchyme, extends more anteriorly (Fig. 8). This results in a broader ureteric bud forming along the AP axis and eventual duplication of ureters. Similar defects are observed in compound heterozygotes of *Foxc1* and *Foxc2,* indicating some redundancy and gene dosage effects. Thus, *Foxc1* and *Foxc2* act to restrict GDNF expression to the most posterior intermediate mesoderm and thus limit ureteric bud outgrowth to a specific region along the nephric duct. Strikingly, the bifurcated ureters, hydroureters, and lobular or duplex kidneys observed in *Foxc1/c2* mutant mice are in many ways similar to the human congenital renal abnormalities frequently found in the pediatric clinic.

In addition to the aforementioned effectors, various other growth factors have been implicated in the control of branching morphogenesis. Prominent among these is hepatocyte growth factor (HGF) and its receptor, the c-met tyrosine kinase. Using the *in vitro* organ culture model, both gain-of-function experiments with exogenous HGF and loss-of-function experiments with neutralizing antibodies point to a multifaceted role for HGF in epithelial differentiation and branching morphogenesis (Woolf *et al.,* 1995). However, genetic analysis in mice does not yet support a role for this signaling pathway in early renal development, because neither *Hgf* (Schmidt *et al.,* 1995; Uehara *et al.,* 1995) or *c-met* mutants (Bladt *et al.,* 1995) have an observable renal phenotype. One caveat though is that *Hgf* and *c-met* homozygous mutants are embryonic lethal at approximately E14.5. Although renal patterning is well established at this stage, more subtle defects might not be detected until later developmental stages.

IV. Inductive Interactions

A. Classical Models

The principles of epithelial and mesenchymal cell interactions, which are necessary for the survival and differentiation of both cell types, are ideally illustrated in the developing kidney, or metanephros. The conceptual framework for kidney induction dates back to the work of Grobstein (1956) who utilized primarily *in vitro* organ culture methods to formulate his conclusions (Fig. 10). By microdisection of the

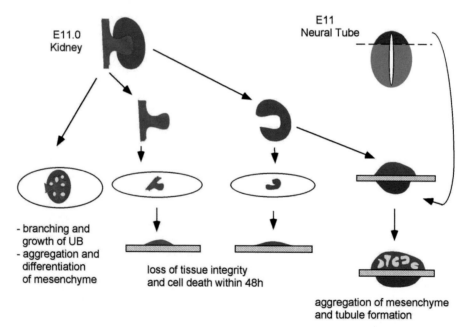

Figure 10 Development of the *in vitro* organ culture model for induction. To test the interdependence of metanephric mesenchyme and ureteric bud, Grobstein (1956) separated the two tissues from an E11 kidney rudiment. The tissues were cultured on nucleopore filters suspended above a defined medium. Under these conditions, the whole kidney rudiment would undergo branching and mesenchyme-to-epithelial conversion. However, isolated ureteric bud or mesenchyme alone would not survive. The transfilter induction method utilizes dorsal neural tube as a heterologous inducer of the mesenchyme. The inducing tissue is placed beneath the filter and is able to provide inductive signals through the filter pores. In response to permissive inductive signals, the metanephric mesenchyme develops tubular epithelial structures (yellow).

ureteric bud from the metanephric mesenchyme, at a time when the mesenchyme had not shown any measurable response to the bud (E11.0), and culturing the two tissues separately and in combination, Grobstein concluded that neither bud epithelia nor metanephric mesenchyme were competent to proliferate or differentiate *in vitro* in the absence of inductive signals emanating from the other. Thus the mesenchyme provided essential signals to promote growth and branching of the ureteric bud, whereas the bud epithelium provided signals to promote aggregation and differentiation of mesenchyme into the epithelium of the developing nephron. Strikingly, the metanephric mesenchyme could be induced to form epithelium *in vitro* by tissues other than the ureteric bud. The dorsal embryonic neural tube was a powerful inducer of kidney mesenchyme that was technically easier to work with in the coculture assay.

Despite the heterologous nature of the inducing tissues examined, the metanephric mesenchyme would always form characteristic renal epithelium, suggesting that the inductive signals derived from the ureteric bud were primarily permissive rather than instructive. This distinction, first proposed by Saxen (1987), means that metanephric mesenchyme is already fated to become renal epithelium, needing only a signal to initiate and maintain the process of differentiation, and can not be reprogrammed to generate other epithelial cell types by heterologous inducers. That the metanephric mesenchyme is patterned with respect to surrounding intermediate mesoderm is evident morphologically, because the mesenchyme is already a distinct aggregate of cells adjacent to the nephric duct, and by the expression of specific genes such as *WT1, Gdnf,* and *Pax2*.

The use of the transfilter organ culture model proved instrumental in defining the properties of the inductive signals emanating from the ureteric bud and spinal cord (Saxen, 1987). By decreasing the pore size of the filter, it could be determined that the inducers were not small, freely diffusible molecules. Rather, cellular processes from the inducing tissue permeate the filters and are thought to be essential for mediating induction. Thus, the idea of cell contact, being a necessary event for induction, was established. Early attempts to use conditioned media from spinal cord or ureteric bud also failed. However, more recent experiments using conditioned media from rat pituitary (Perantoni *et al.,* 1991) or from ureteric bud cell lines (Barasch *et al.,* 1999; Karavanova *et al.,* 1996), in conjunction with survival factors such as fibroblast growth factor (FGF) and transforming growth factor alpha (TGF-α), indicate that inducing activity could be found in a soluble form.

If cell contact were indeed necessary, as much of the data indicate, how was the inductive signal able to penetrate the mesenchyme such that cells not immediately around the bud could also be induced? Using the transfilter model, a series of experiments by Saxen and colleagues put two hypotheses to the test. The signal may penetrate several cell layers deep into the mesenchyme by attachment to cellular processes or extracellular matrix that extends into the mesenchymal aggregate. Alternatively, mesenchyme cells immediately abutting the ureteric bud, once induced, could pass on the inductive signals to neighboring uninduced cells. The later model was disproved by recombining induced mesenchyme, which had been exposed for 24 hr to spinal cord, with uninduced mesenchyme. Only the mesenchyme originally exposed to the inducer was able to form tubules, whereas the uninduced mesenchyme remained as such (Saxen and Saksela, 1971).

B. Wnt Signaling and Mesenchymal Induction

Wnt genes encode a family of secreted peptides that are known to function in the development of many tissues (Miller *et al.,* 1999; Wodarz and Nusse, 1998). The prototype Wnt gene, *Drosophila wingless,* is required for segment polarity, for patterning of the imaginal disks, and for development of the heart and nervous system. In the mouse 18 Wnt genes have been identified to date (for the latest update, see the WNT gene home page at http://www.stanford.edu/~rnusse/wntwindow.html). At least 3 Wnt genes are expressed in a dynamic pattern during urogenital development. The WNT proteins are secreted but are generally not freely diffusible, rather they seem to be associated with extracellular matrix through interactions with heparin sulfated glycoproteins. In the best characterized Wnt signaling pathway, binding of WNT proteins to the Frizzled family of transmembrane receptors activates the cytoplasmic protein dishevelled and inhibits the kinase, GSK3 (Fig. 11). GSK3 normally phosphorylates β-catenin such that it can be targeted to the ubiquitin degradation pathway via the β-catenin binding proteins axin and APC. If GSK3 activity is inhibited, concentrations of β-catenin rise such that it can now be found in the nucleus complexed with members of the TCF/Lef1 family of transcription factors. In the absence of β-catenin, the TCF/Lef proteins can bind DNA and repress transcription, whereas the β-catenin/TCF complex is a potent activator. Alternative Wnt signaling pathways can work through the c-Jun N-terminal kinase (JNK), perhaps through interactions with protein kinase C. However, the details of these pathways are just now beginning to emerge.

That Wnt signaling may be involved in early kidney inductive interactions was first suggested by Herzlinger *et al.* (1994). In the classical transfilter assay, metanephric mesenchyme was induced and able to form tubules when exposed to cultured cells expressing the *Wnt1* gene. Other members of the WNT family are also able to promote mesenchyme induction (Kispert *et al.,* 1998) (Fig. 12) suggesting that WNT proteins are the active inducers found in the spinal cord, the tissue of choice as a heterologous inducer in the transfilter system. The role of WNT proteins is further supported by the ability of lithium chloride (Davies and Garrod, 1995), a specific inhibitor of the GSK3 kinase (Klein and

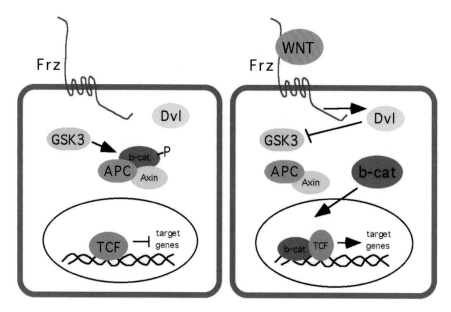

Figure 11 Schematic outline of the Wnt signaling pathway. In the simplest model of Wnt signaling, GSK3 phosphorylates β-catenin in the absence of extracellular Wnt protein. Phosphorylated β-catenin binds to axin and APC and is shuttled to the ubiquitin degradation pathway. In the absence of β-catenin, the DNA-binding proteins of the TCF/LEF family block transcription of target genes. Binding of WNT to the Frizzled family of transmembrane proteins activates the dishevelled (Dvl) protein, which inhibits GSK3 activity. The unphosphorylated β-catenin accumulates and is translocated to the nucleus and interacts with the TCF/LEF family of DNA-binding proteins. The β-catenin/TCF complex is a potent activator of target gene transcription. (Adapted from the Wnt gene homepage, http://www.stanford.edu/~rnusse/wntwindow.html.)

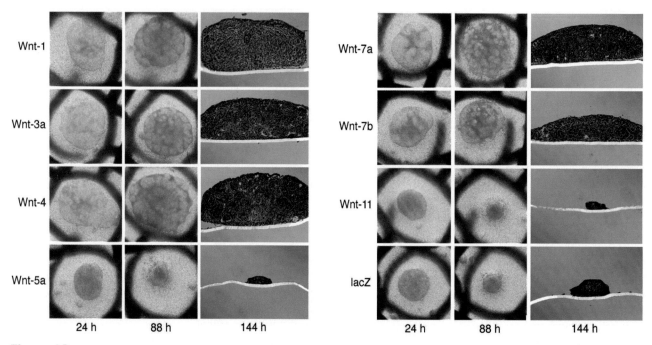

Figure 12 Mesenchyme-to-epithelial conversion as mediated by WNT signals. Metanephric mesenchyme were isolated from E11 kidney rudiments and placed in the transfilter culture assay (Kispert *et al.,* 1998). On the opposing side of the filters were NIH 3T3 cells expressing various Wnt proteins or *lacZ* as a control. The mesenchymal rudiments were photographed over time and sectioned after 6 days in culture. Many Wnt proteins are potent inducers of the mesenchyme, these include Wnt-1, -3a, -4, -7a, and -7b. However, Wnt-11 and Wnt-5a show no inductive capacity. This may reflect a fundamental divergence of the WNT signaling pathways, as described in other systems. (Figure courtesy of A. Kispert, MPI for Immunobiology, Freiburg.)

Melton, 1996), to mimic an inductive signal and promote aggregation of the mesenchyme.

The WNT4 protein is a potent inducer and is expressed in mesenchymal aggregates *in vivo* (Fig. 12). Mice homozygous for a *Wnt4* mutation exhibit renal agenesis due to growth arrest shortly after branching of the ureteric bud (Stark *et al.*, 1994). Although some mesenchymal aggregation has occurred, there is no evidence of cell differentiation into a polarized epithelial vesicle. Expression of *Pax2* is maintained but reduced, whereas the related gene *Pax8* is not activated. It would appear that *Wnt4* is a secondary inductive signal in the mesenchyme that propagates, or maintains the primary induction response. This question then remains: Which Wnt protein, if any, is the inducer found at the ureteric bud tips? The *Wnt11* gene is expressed in a restricted manner at the tips of the growing ureteric bud and as such would be prime candidate (Kispert *et al.*, 1996). Unfortunately, *Wnt11* is one of the few family members unable to induce metanephric mesenchyme *in vitro* (Kispert *et al.*, 1998). The expression of *Wnt11* mRNA is upregulated by GDNF (Pepicelli *et al.*, 1997) and requires sulfated proteoglycans for high-level maintenance at the ureteric bud tips (Kispert *et al.*, 1996). Inhibition of the sulfotransferases by the addition of chlorate reduces *Wnt11* mRNA levels and inhibits ureteric bud branching, yet has little effect on epithelial cell differentiation from the mesenchyme. A mutation of the gene encoding heparan sulfate 2-sulfotransferase also results in renal agenesis in homozygous mice (Bullock *et al.*, 1998), primarily because the ureteric bud does not undergo branching morphogenesis once the mesenchyme has been invaded. In these mutant mice, GDNF, RET, and WNT11 are all downregulated but *Pax2* levels are normal. Taken together, it would appear that WNT11 and its potential interactions with heparan sulfated glycoproteins may be more important for ureteric bud branching than induction of the mesenchyme. However, the final judgment on WNT11 is still out and will require genetic analysis.

C. Induction of Tubules by Soluble Factors

The search for the inductive signal that promotes mesenchyme to epithelial conversion has been a long and convoluted journey. Although early transfilter experiments ruled out the possibility of soluble, diffusible factors, several reports have demonstrated mesenchyme induction using a cocktail of growth factors and extracts. Initial studies focused on rat pituitary extracts in combination with epidermal growth factor and extracellular matrix to show epithelial conversion of rat metanephric mesenchyme (Perantoni *et al.*, 1991). More refined studies utilized basic FGF (bFGF) and TGF-α together with conditioned media from a ureteric bud cell line. The most recent demonstration of epithelial conversion of rat mesenchyme used bFGF, TGF-α, and leukocyte inhibitory factor (LIF), which was purified from the ac-

tive, inducing fraction of conditioned media from a ureteric bud cell line (Barasch *et al.*, 1999).

Although these data are thought provoking and are cause to reexamine some of the early transfilter experiments, a number of caveats make the interpretation difficult. All of these reports were done with rat tissue and, at least in one case, the corresponding mouse mesenchyme showed no response (Perantoni *et al.*, 1991). Genetic analyses indicate no real renal phenotype in bFGF mutant mice and only a slight reduction in kidney size, with no real effect on induction or epithelial differentiation, in mutants of the LIF receptor gp130. Because all of these experiments are done with mesenchyme that has been microdissected away from the ureteric bud, the most difficult conceptual problem to address is the potential for growth and survival factors to enhance or maintain a signal that has already affected a small population of mesenchymal cells. Classical transfilter induction experiments indicated a 24-hr time of contact was required before a full response was seen in the mesenchyme. Perhaps the appropriate mixture of growth and survival factors can significantly reduce this time such that at the time of dissection and explant, sufficient inductive signaling has already occurred. These issues need to be clarified before definitive conclusions can be draw regarding the role of LIF and FGFs during primary induction. Nevertheless, FGF and the interleukin 6 (IL-6) family of cytokines, of which LIF is a member, may provide important survival or secondary signals for the expansion of the mesenchyme and its derivatives.

V. Mesenchyme-to-Epithelial Conversion

The response of the metanephric mesenchyme to inductive signals has been the subject of many years of investigation (Ekblom, 1989). The mesenchyme must undergo extensive remodeling, as the induced cells proliferate, form adhesive junctions, become polarized, and assemble the epithelial basement membrane. Changes in the expression of epithelial specific cellular adhesion molecules and extracellular matrix must be accompanied by activation or suppression of transcription factors in response to inductive signals. How the inductive signals are integrated into the genetic program of the epithelial cell remains poorly understood. Yet the expression of new genes in response to induction and the role of cell–cell and cell–matrix adhesion molecules are becoming more evident.

A. Epithelium versus Stroma

Although the process of mesenchyme-to-epithelial conversion has received the most scrutiny over the years, it must be appreciated that not all of the metanephric mesenchyme is fated to become epithelium. In fact, the full potential of the metanephric mesenchyme and how that potential is partitioned is just beginning to be realized. As for most other questions in the kidney development field, the issues are being addressed with a combination of genetic and tissue culture models.

The transcription factor BF-2 is expressed in uninduced mesenchyme and becomes restricted to those cells not undergoing epithelial conversion after induction (Hatini *et al.,* 1996). *BF-2* (*foxd1*) expression is found along the periphery of the kidney and in the interstitial mesenchyme, or stroma. There is little overlap between *BF-2* and the *Pax2* expression domain, prominent in the condensing pretubular aggregates. Although clear lineage analysis is still lacking, the expression patterns are consistent with the interpretation that mesenchyme cells may already be partitioned into a *BF-2* positive stromal precursor and a *Pax2* positive epithelial precursor prior to, or shortly after, induction. Mouse mutants in *BF-2* exhibit severe developmental defects in the kidney that point to an essential role for *BF-2* in maintaining growth and structure (Hatini *et al.,* 1996). Early ureteric bud growth and branching are unaffected, as is the formation of the first mesenchymal aggregates. However, at later stages (E13–14) these mesenchymal aggregates fail to differentiate into comma and S-shaped bodies at a rate similar to wild-type. Branching of the ureteric bud is greatly reduced at this stage, resulting in the formation of fewer new mesenchymal aggregates. The fate of the initial aggregates is not fixed, because some are able to form epithelium and almost all express the appropriate early markers, such as *Pax2, Wnt4,* and *WT1.* Nevertheless, it appears that the *BF-2* expressing stromal lineage is necessary to maintain growth of both ureteric bud epithelium and mesenchymal aggregates. Perhaps factors secreted from the stroma provide survival or proliferation cues for the epithelial precursors, in the absence of which the non-self-renewing population of mesenchyme is exhausted.

Both FGFs and the family of BMPs are secreted signals that affect cell fate and development in the early metanephros. BMP7 is expressed in the ureteric bud epithelium and in the induced mesenchyme and has been implicated as an inductive signal for mesenchyme-to-epithelium conversion (Vukicevic *et al.,* 1996). However, *Bmp7* mutants are developmentally arrested after the induction phase, exhibit ureteric bud branching morphogenesis, and express early markers of induced mesenchyme (Dudley *et al.,* 1995; Luo *et al.,* 1995). Thus it seems unlikely that secreted BMP7 from the ureteric bud is the primary inductive signal. More recent experiments demonstrate that BMP7 promotes survival of uninduced metanephric mesenchyme *in vitro* (Dudley *et al.,* 1999). This survival function is enhanced by the addition of FGF2. In fact, FGF2 is necessary to maintain the ability of the mesenchyme to respond to inductive signals. BMP7 alone inhibits apoptosis but is not sufficient to enable mesenchyme to undergo tubulogenesis at some later time. Within induced mesenchyme, FGF2 and BMP7 appear to reduce the proportion of mesenchyme that undergoes tubulo-

genesis while increasing the population of BF-2 positive stromal cells (Dudley *et al.,* 1999). All indications thus far point to a delicate balance between a self-renewing population of stromal and epithelial progenitor cells, the proportion of which must be well regulated by both autocrine and paracrine factors.

B. Aggregation and Cell Adhesion

The cadherin gene family encodes calcium-dependent cellular adhesion molecules that promote homophilic adhesion between cells. Cadherins are transmembrane proteins, whose extracellular domains contain multiple copies of the so-called cadherin repeat (Gumbiner, 1996; Koch *et al.,* 1999). Cadherins are classified into type I and type II subfamilies based on structural features of the extracellular domain and functional properties of the intracellular domain. The extracellular domain of the classical (type I) cadherins, which include E-, N-, P-, and R-cadherin, contains five cadherin repeats, the first of which mediates specific homophilic interactions. These sequences promote interactions between cadherins of the same type on opposing cells (in trans) and possibly in the same cell (in *cis*) (Shapiro *et al.,* 1995). In addition, the type I cadherins contain intracellular domains that mediate binding to the actin-based cytoskeleton via a group of linker proteins, the catenins (Kemler and Ozawa, 1989; Ozawa *et al.,* 1989), and promote *cis* interactions for clustering cadherins at the cell surface (Katz *et al.,* 1998; Yap *et al.,* 1997, 1998). Cooperativity achieved by interactions of cadherins on opposing cells in *trans,* along with the interactions in *cis,* is thought to generate a strong adhesive force that can drive selective cell–cell aggregation. In addition, the binding and clustering of cadherins into adhesive plaques appears to be the underlying basis of the adherents junction, a specialized site of cell–cell contact that binds cells into solid tissue (Gumbiner, 1996). In epithelia, disruption of this junction has a profound effect on the maintenance of other junctional contacts, and ultimately to a loss of epithelial polarity and integrity. Many of these features also apply to the type II subfamily of cadherins. Although less well characterized, cadherin-6, cadherin-11, and F-cadherin are closely related to the classical cadherins in both structure and function (Cho *et al.,* 1998; Hoffmann and Balling, 1995; Kimura *et al.,* 1995). The type II cadherins also mediate cell adhesion via a homophilic binding interaction, and bind to the actin-based cytoskeleton via the catenins (Inoue *et al.,* 1997; Nakagawa and Takeichi, 1995).

In the developing kidney, cadherin expression is dynamic and complex (Fig. 13). The uninduced metanephric mesenchyme expresses cadherin-11, whereas the ureteric bud expresses E-cadherin (Cho *et al.,* 1998). When the metanephric mesenchyme aggregates in response to inducing signals from the ureteric bud, the expression of cadherin-11 is downregulated and the expression of R-cadherin and cadherin-6

A

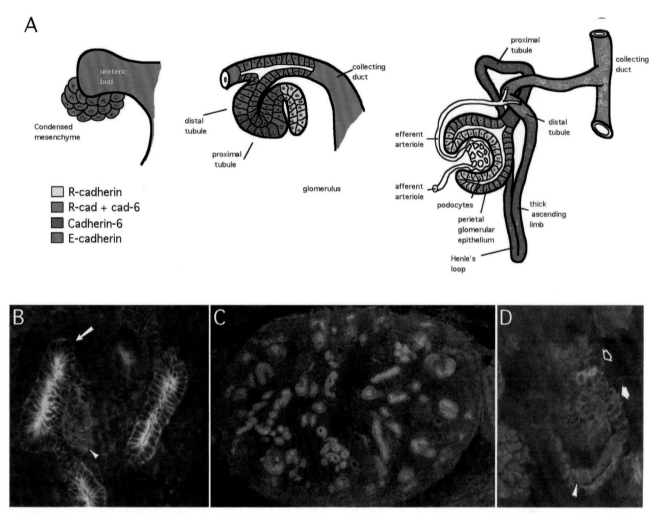

R-cadherin
R-cad + cad-6
Cadherin-6
E-cadherin

Figure 13 Cadherin gene expression in the developing nephron. The dynamic expression of cell adhesion molecules during nephron development may provide some of the driving forces behind cell movement and tissue remodeling. (A) A summary of cadherin gene expression during early development indicates the expression of E-cadherin, R-cadherin, and cadherin-6. Not shown are cadherin-11, a mesenchymal specific cadherin expressed prior to induction and in the interstitial mesenchyme, and P-cadherin, expressed in the visceral glomerular epithelium at later stages. (B) Expression of E-cadherin (green) and cadherin-6 (red) in the ureteric bud (arrow) and in the mesenchymal aggregates (arrowhead) from E12 kidney. Note that the aggregate expresses E-cadherin adjacent to the bud and cadherin-6 furthest from the bud tip. (C) A section through an E14 kidney stained for E-cadherin (green), cadherin-6 (red), or coexpression (yellow). More mature proximal tubules in the very cortex are switching from cadehrin-6 to E-cadherin (yellow), whereas ureteric bud-derived epithelium continues to express E-cadherin (green) and mesenchymal-derived epithelium along the periphery expresses cadherin-6 (red). (D) An S-shaped body at E14 shows a clear demarcation between E-cadherin (green) and cadherin-6 (red, cell surface). The developing glomerular tuft (arrowhead) is marked antibodies against WT1 (red, nuclear). The solid arrow marks proximal tubule precursor cells, whereas the open arrow marks the distal tubules leading to the collecting duct.

is upregulated (Cho *et al.*, 1998; Rosenberg *et al.*, 1997). This onset of R-cadherin and cadherin-6 expression correlates with formation of the polarized epithelium, the renal vesicle, and the deposition of a basement membrane. The renal vesicle differentiates further by forming the comma-shaped and then the S-shaped body. At the S-shaped body stage, R-cadherin is primarily restricted to the epithelial structures oriented furthest from the ureteric bud, E-cadherin is expressed in the bud and most distal tubules, and cadherin-6 is found between the E-cadherin and R-cadherin domains in cells destined to become proximal tubules (Cho *et al.*,

1998; Rosenberg *et al.*, 1997). As tubulogenesis proceeds, cadherin-6 becomes restricted to the developing proximal tubules, including the ascending and descending limbs of Henle's loop and R-cadherin becomes restricted to the developing cells of the future Bowman's capsule and the adjoining proximal tubule cells. As cadherin-6 expression is suppressed, E-cadherin becomes more prominent in the mature proximal tubules and P-cadherin is evident in the glomerular cleft and subsequent visceral glomerular epithelium. Thus, during kidney development, dynamic changes in the expression of the type II and type I cadherins correlate with tissue remodeling, suggesting strongly that cadherins mediate at least some aspects of the differential cell adhesion required for proper nephrogenesis.

How do specific cadherins regulate tissue remodeling? This question has been addressed *in vitro* and *in vivo* for a limited set of cadherin proteins. Cadherin-mediated cell adhesion can be neutralized *in vitro* using the Fab fragments of antibodies against the extracellular domains. In kidney organ cultures, anti-cadherin-6 Fabs inhibit mesenchymal cell adhesion and the subsequent formation of comma-shaped and S-shaped structures, without affecting ureteric bud branching (Cho *et al.*, 1998). However, the effect of cadherin-6 loss-of-function in the mouse is not nearly as striking (Mah *et al.*, 2000). In cadherin-6 mutant kidneys, mesenchymal aggregates that express R-cadherin and the zona occludins protein ZO-1, but have not yet formed a laminin containing basement membrane, are overrepresented, indicating that a delay occurs in the conversion of the mesenchyme to a fully polarized epithelium. This delay compromises the ability of the mesenchymal-derived epithelium to fuse with the ureteric bud-derived epithelium, resulting in a high frequency of dead-ending nephrons. Although there are some similarities in the cadherin-6 loss-of-function phenotypes observed *in vitro* and *in vivo*, it is clear that the organ culture assay is much more sensitive to perturbations. *In vitro*, development proceeds at a slower rate, as cell proliferation is minimized and certain cell types, such as the vasculature, are lacking altogether. With respect to E-cadherin, neutralizing antibodies have no apparent effect on the conversion of mesenchyme to epithelium (Vestweber *et al.*, 1985). Unfortunately, mice with mutations in E-cadherin are early embryonic lethal due to a lack of compaction at the morula stage (Larue *et al.*, 1994), making analysis of kidney development dependent on the generation of conditional mutants.

C. Cell Matrix Interactions

Essential functions for several subunits of the integrin proteins during kidney development have been revealed by genetic analysis in the mouse and by *in vitro* organ culture experiments. The integrins are heterodimeric transmembrane receptors that interact with the extracellular matrix (ECM) and the cytoskeleton (Giancotti and Ruoslahti, 1999). The integrin dimer consists of an α and β subunit, whereby a single β subunit is able to heterodimerize with multiple α subunits. Binding specificity to the ECM is determined by the combination of the α and the β subunits, whereas the β subunit interacts with cytoplasmic signal transduction molecules, such as the integrin-linked kinases and the focal adhesion kinases. At least four different α subunits, together with the β_1 subunit, are found in different cell types of the developing kidney. At early times, $\alpha_8\beta_1$ integrin is expressed in the mesenchyme of the uregenital tract and at particularly high levels in induced mesenchyme surrounding the ureteric

bud tips. Mice homozygous for an α_8 mutant allele exhibit severe renal agenesis, although with a varying degree of penetrance (Muller *et al.*, 1997). The majority of newborn α_8 mutants had bilateral agenesis and no urethras present. However, some mutant animals had a single kidney rudiment and about 25% had a single, apparently normal kidney. The primary defect was apparent at E11 with 100% of the α_8 mutants exhibiting a ureter that had not invaded the mesenchyme and undergone branching morphogenesis. Thus, $\alpha_8\beta_1$ expression in the mesenchyme provides a permissive environment for optimal ureteric bud growth and invasion. It is not clear whether this permissive environment is due to interactions of $\alpha_8\beta_1$ integrin with the ECM of the ureteric bud directly or if $\alpha_8\beta_1$ affects the expression of bud guidance or survival factors, such as GDNF, secreted by the mesenchyme.

A more subtle defect in branching morphogenesis and collecting duct patterning is observed in mice carrying a mutation for the α_3 integrin subunit (Kreidberg *et al.*, 1996). The expression pattern of $\alpha_3\beta_1$ is more complex and encompasses the more developed epithelium of Bowman's capsule, the glomerular podocytes, and distal tubules. The overall number of medullary collecting ducts is reduced in the α_3 mutant animals, although the nephron number appeared unaffected. This finding was somewhat counterintuitive given that the number of nephrons generally reflects the degree of branching morphogenesis, as each new branch induces new nephrons along the radial axis of the kidney. In the developing lungs of α_3 mutants, branching was also reduced, indicating some common mode of integrin function in ureteric bud-derived collecting ducts and bronchial epithelium. The $\alpha_3\beta_1$-integrin complex is necessary for maintaining the integrity of the glomerular basement membrane, which appears discontinuous in the mutants. The question still remains of how these integrins function during development. Are they driving forces behind tissue remodeling, promoting cell migration and differentiation through intracellular signaling pathways in response to cell–matrix interactions? Or, do integrins maintain an established pattern by the nucleation of matrix factors to organize and maintain the extracellular environment?

D. Role of Epithelial Basement Membranes

The polarized epithelial cells of the renal tubules have apical and basolateral sides, with the apical sides facing the lumen of the tubules and the basal side bound to a basement membrane that completely surrounds the outside of individual tubules. This basement membrane consists of collagens, laminins, entactin, and sulfated proteoglycans. During development, the composition of the tubular and glomerular basement membranes shifts both spatially and temporally such that specific isoforms of both collagen and laminin are found in different parts of the developing nephron. Alterations in the composition of the basement membranes, by genetic mutations, can have dramatic effects on both early and late development and have also been linked to human disease processes. Shortly after induction of the mesenchyme, the assembly of the epithelial basement membrane can be detected as early as the renal vesicle stage, surrounding the newly polarized epithelium. The composition of this basement membranes undergoes changes with respect to the types of collagens and laminins expressed (Miner, 1999).

Laminins are heterotrimers consisting of a single α, β, and γ chain. In the kidney, laminin-1 $(\alpha_1\beta_1\gamma_1)$, is a major component of basement membranes at most stages of development. *In vitro*, antibodies against the α_1 chain of laminin-1 inhibit the conversion of mesenchyme to epithelium, presumably by disturbing the formation of the epithelial basement membrane (Klein *et al.*, 1988). Expression of the laminin β_1 and γ_1 chains can already be detected in the uninduced mesenchyme in a filamentous pattern. Thus, the synthesis of the laminin α_1 chain appears to be the rate-limiting step in basement membrane assembly at the earliest stages of epithelial cell polarization.

At the S-shaped body stage, the glomerular epithelial precursors begin to express laminin-10 $(\alpha_5\beta_1\gamma_1)$, which is replaced by laminin-11 $(\alpha_5\beta_2\gamma_1)$ in the mature glomerular basement membrane (Miner *et al.*, 1997). This shift in laminin isoform expression is necessary for maintaining the filtration barrier in the glomerulus. Mice homozygous for mutations in the laminin β_2 chain develop severe proteinuria, which is indicative of a glomerular filtration defect (Noakes *et al.*, 1995). The podocyte cells of the glomerulus are abnormal and fused, although the ultrastructure of the basement membrane seems normal. The laminin β_1 chain is found in the mature glomerular basement membrane of the β_2 mutants, suggesting that downregulation or turnover of β_1 is linked to β_2 expression. However, the β_1 chain does not functionally compensate for the lack of β_2. The mutants suggest a role for the β_2 chain in regulating the differentiation of the podocyte cell. Thus, the developing epithelia respond to changes in the extracellular matrix, perhaps through the binding of integrins to specific laminin chains and activation of the intracellular integrin linked signal transduction pathways.

The laminin α_5 chain has an even more pronounced role in glomerular development, as revealed by gene targeting in the mouse (Fig. 14). Bilateral or unilateral renal agenesis is observed in about 20% of mice homozygous for a *Lama5* mutant allele (Miner and Li, 2000). In all of the kidneys that develop further, the glomerulus is grossly affected. The α_5 chain is expressed at the capillary loop stage when endothelial cells are first observed invading the glomerular cleft. This most likely reflects a shift from laminin-1 expression to laminin-10 and -11. At this time, the *WT1* gene is highly expressed in the podocyte precursor cells, which initially form a single, crescent-shaped layer along the cleft and then

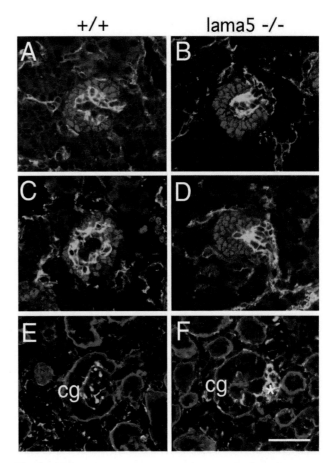

Figure 14 Role of laminin α_5 in glomerular development. Sections through the developing glomeruli of wild-type and *lama5* homozygous null mutants are stained for WT1 (red) and PECAM (green) in panels A–D. The WT1 positive podocytes form a crescent-shaped structure that becomes infiltrated with endothelial cells, expressing PECAM (A, B). As the glomerular tuft begins to form (C), endothelial cells in the *lama5* mutant are extruded from the glomerulus (D). Panels E and F are stained for laminin-1 (red) and desmin (green) to mark the glomerular basement membrane and the mesangial cells, respectively. The mesangial cells are also found dispersed in the developing glomerular tuft (E) of control kidneys, but are excluded from the tuft in *lama5* mutant kidneys. Thus, switching of laminin isoforms within the glomerular basement membrane is necessary for correct spatial organization of the individual cell types within the glomerular tuft. (Micrograph courtesy of J. Miner, Washington University, St. Louis; Miner and Li, 2000.)

completely outline the developing Bowman's capsule. Strikingly, *lama5* mutants exhibit multiple layers of podocyte precursors that ultimately extrude the invading endothelial cells from the glomerular cleft. Mesangial cells are a smooth-muscle-like cell that provide structural support for the capillary loops and are also displaced in *Lama5* mutants, resulting in an avascular glomerulus.

In vitro analysis reveals intrinsic defects in branching morphogenesis of ureteric bud epithelium from *Lama5* mutant mice, much more so than is observed in *in vivo*. The increased sensitivity to loss of laminin α_5 *in vitro* points to a potential limitation of the organ culture system. Outside of the whole embryo environment, the developing kidney rudiment grows slower and appears less able to compensate for this specific laminin mutation. This case also underscores the increasing number of discrepancies found between *in vivo* and *in vitro* analyses using the organ culture system. Just as in the experiments with cadherin-6, it would seem that growth in culture of whole kidneys is much more sensitive to protein or gene disruption by both genetic and physical means, such as antibody inhibition or antisense mRNA strategies.

The second major structural component of the basement membrane is collagen IV. The six distinct chains of collagen IV, α_1–α_6, form a triple helical rodlike structure that is organized into a meshwork through trimer–trimer interactions. In the rodent kidney, different collagen IV isoforms are expressed depending on the type of basement membrane and the developmental stage (Miner and Sanes, 1994). The α_1 and α_2 chains are first detected in the basement membrane of the renal vesicle, the comma-shaped bodies, and the S-shaped bodies. In the developing glomerulus, α_1 and α_5 are detected at the capillary loop stage, whereas mature glomerular basement membranes contain mostly α_3, α_4, and α_5. This developmental switch in collagen IV chain expression is necessary to preserve glomerular basement membrane function and integrity in adult kidneys. Mouse mutants in the α_3 chain of collagen IV (Miner and Sanes, 1996) exhibit a phenotype strikingly similar to human Alport syndrome (Barker *et al.*, 1990; Hudson *et al.*, 1993), which is characterized by decreased glomerular filtration, increased mesangial matrix deposition, glomerular sclerosis, and tubule atrophy. In humans, mutations in any of the α_3, α_4, or α_5 chains can cause Alport syndrome, because they affect the assembly of the mature heterotrimer. Instead, collagen IV α_1 and α_2 are found in the Alport syndrome kidneys, which seems to increase the susceptibility of the glomerular basement membrane to damage.

VI. Glomerular Development and Vascularization

The mammalian glomerulus is the filtration unit of the kidney. The unique structure of the glomerulus is intricately linked to its ability to retain large macromolecules within the circulating bloodstream while allowing for rapid diffusion of ions and small molecules into the urinary space. The glomerulus consists of four major cell types, the endothelial cells of the microvasculature, the mesangial cells, the podocyte cells of the visceral epithelium, and the parietal epithelium (Fig. 15). The glomerular basement membrane separates the endothelial cells of the capillary tufts from the urinary space. The basement membrane is covered with podocyte cells and their interdigitated foot-processes, such that

a slit pore is formed between the foot-processes. The slit pore is a unique intercellular junction between podocyte cells that lacks the typical protein composition of an epithelial cell zona occludins. Rather, it is a porous junction whose protein composition had remained in large part unknown. The recent discovery of the gene responsible for Finnish-type nephrotic syndrome (Kestila *et al.*, 1998) and the localization of the encoded protein, Nephrin, to the slit pore (Holzman *et al.*, 1999; Ruotsalainen *et al.*, 1999) have provided the first real insight into the molecular structure of the filtration barrier.

The development of the glomerular architecture and the origin of the individual cell types are just beginning to be understood. The epithelial cells of the glomerulus are largely derived from condensing metanephric mesenchyme. However, the origin of the endothelium and the mesangium is less clear. At the S-shaped body stage, the glomerular cleft becomes evident at the most proximal part of the S-shaped body, furthest from the ureteric bud epithelium. The expression of high levels of WT-1 protein in the presumptive podocyte precursor cells marks this region clearly. Endothelial precursor cells migrate into this glomerular cleft and begin the process of vascularization. The origin of the endothelial cells has been the subject of much debate. Under normal growth conditions, kidneys cultured *in vitro* do not exhibit signs of vascularization. However, hypoxygenation or treatment with vascular endothelial growth factor (VEGF) promotes survival or differentiation of endothelial precursors in whole-organ culture (Tufro *et al.*, 1999; Tufro-McReddie *et al.*, 1997). *In vivo* transplantation experiments using *lacZ* expressing donors or hosts also indicate that the E11.5 kidney rudiment has the potential to generate endothelial cells, although recruitment of endothelium is also observed from exogenous tissue depending on the environment (Hyink *et al.*, 1996; Robert *et al.*, 1996, 1998) (Fig. 16). When E11.5 metanephroi are grafted into the anterior eye chamber or under the capsules of adult kidneys, endothelial cells were derived from the graft exclusively. However, when E11.5 metanephroi are grafted under the capsule of a newborn kidney, invasion of endothelial precursor cells from the host is observed. Thus, angioblasts from the developing host kidney can populate the E11.5 kidney rudiment. Yet, cells within the E11.5 kidney have the ability to differentiate along the endothelial lineage. Thus it appears that by E11.5, the metanephros is a heterogeneous mixture of endothelial, epithelial, and stromal precursor cells. The latter hypothesis is supported by staining of E12 kidneys for the expression of Flk-1 (*Kdr*), a receptor for VEGF. Using a *lacZ* knock-in allele for Flk-1 (*Kdr*), Robert *et al.* (1998) showed that presumptive angioblasts are dispersed along the periphery of the E12 kidney mesenchyme, with some positive cells invading the mesenchyme along the aspect of the growing ureteric bud. At later stages, Flk-1 positive angioblasts were localized to the nephrogenic zone, the developing glomerular cleft of

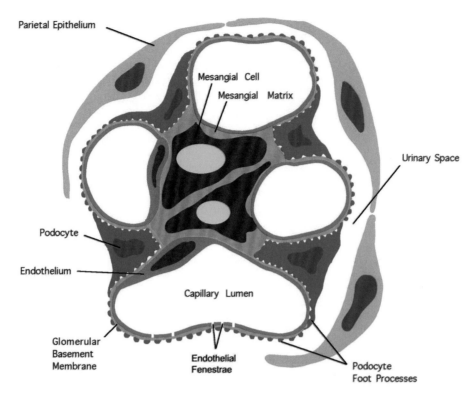

Figure 15 Cross section through the mature glomerulus. The filtration unit of the kidney is the glomerular tuft, a complex three-dimensional structure consisting of four major cell types. A fenestrated capillary endothelium and the matrix-producing mesangial cells are surrounded by the glomerular basement membrane. On the contrary side of the glomerular basement membrane are the podocytes, specialized epithelial cells with an elaborate network of interdigitated cellular processes, called foot-processes. The interdigitated foot-processes form a fine sieve, which acts as the filtration barrier between the plasma and the urinary space. The pores of the filtration barrier are located between individual foot-processes and are called the slit diaphragms. Surrounding the urinary space is the parietal epithelium, which traps the filtrate and connects to the proximal tubules where resorption of electrolytes, water, and macromolecules can begin.

the S-shaped bodies, and the more mature capillary loops, whereas VEGF localizes to the parietal and visceral glomerular epithelium. Injection of neutralizing VEGF antibodies into newborn mice, at a time when nephrogenesis is still on-going, disturbs vascular growth and glomerular architecture (Kitamoto *et al.,* 1997). Taken together, the data point to a potential role for VEGF in recruiting angioblasts and/or stimulating growth and differentiation of these endothelial precursors.

Genes that regulate the glomerular podocyte and mesangial cell lineages have also been characterized in the mouse. The basic helix–loop–helix protein POD1 is expressed in epithelial precursor cells and in more mature interstitial mesenchyme. At later developmental stages, POD1 is restricted to the podocytes. In mice homozygous for a *Pod1* null allele, podocyte development appears arrested (Quaggin *et al.,* 1999). Normal podocytes flatten and wrap their foot-processes around the glomerular basement membrane. *Pod1* mutant podocytes remain more columnar and fail to fully develop foot-processes. Because Pod1 is expressed in epithelial

precursors and in the interstitium, it is unclear whether these podocyte effects are due to a general developmental arrest because of the stromal environment or a cell autonomous defect within the *Pod1* mutant podocyte precursor cells.

The mesangial cells are located between the capillary loops of the glomerular tuft and have been referred to as specialized pericytes. The pericytes are found within the capillary basement membranes and have contractile abilities, much like a smooth muscle cell. Whether the mesangial cell is derived from the endothelial or epithelial lineage remains unclear. However, genetic and chimeric analyses in the mouse have revealed a clear role for the platelet-derived growth factor receptor (PDGFr) and its ligand (PDGF). In mice deficient for either PDGF or PDGFr (Leveen *et al.,* 1994; Soriano, 1994), a complete absence of mesangial cells results in glomerular defects, including the lack of microvasculature in the tuft. PDGF is expressed in the developing endothelial cells of the glomerular tuft, whereas the receptor is found in the presumptive mesangial cell precursors. Using embryonic stem cell chimeras of *Pdgfr* −/− and +/+ geno-

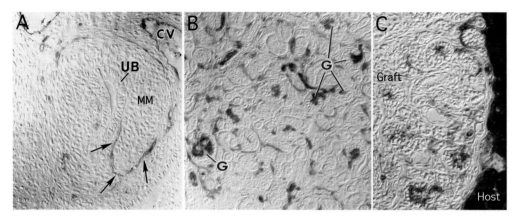

Figure 16 Flk-1 expression and endogenous origin of the renal vasculature. (A) The expression of the VEGF receptor Flk-1 was determined by staining for *lacZ* activity in a heterozygous Flk-1^*lacZ*/+ embryo at E11.5. The *lacZ* gene was inserted into the Flk-1 locus as part of a gene targeting strategy. In a transverse section, Flk-1 expressing cells surround the metanephros and are seen entering the developing kidney along the aspect of the ureteric bud (UB). The metanephric mesenchyme (MM) and the cardinal vein (CV) are indicated. (B) An E12 metanephros from a heterozygous Flk-1^*lacZ*/+ embryo was cultured *in vitro* for 6 days then grafted into the anterior chamber of the eye of a wild-type host. Staining for *lacZ* shows many positive endothelial cells in the vasculature of the glomeruli and surrounding tubules, indicating the cells were derived from the graft. Thus, even after 6 days *in vitro*, the metanephros retains the ability to vascularize once the appropriate environment is achieved. (C) A wild-type E12 metanephros was grafted under the renal capsule of a newborn ROSA26 mouse and cultured for 7 days. *LacZ* positive angioblasts, derived from the host, have invaded the graft and contributed to the vasculature. Endothelial precursor cells from the newborn host kidney have the ability to populate the graft, whereas the adult kidney does not retain this potential (not shown). (Micrographs courtesy of B. Robert and D. Abrahamson, University of Kansas Medical Center; Robert *et al.,* 1996, 1998.)

types, only the wild-type cells could contribute to the mesangial lineage (Lindahl *et al.,* 1998). This cell autonomous effect indicates that signaling from the developing vasculature promotes proliferation and/or migration of the mesangial precursor cells. Also, expression of PDGF and smooth muscle actin supports a model where mesangial cells are derived from smooth muscle of the afferent and efferent arterioles during glomerular maturation (Lindahl *et al.,* 1998).

VII. Developmental Basis of Human Renal Disease

In addition to deciphering the basic mechanisms of patterning and differentiation in a complex organ system, the study of organ development can also illuminate the pathogenic processes underlying human disease. It is becoming increasingly clear that developmental control genes play decisive roles in disease processes, through either aberrant expression or loss of function. The previous sections have already discussed two examples of renal diseases due to mutations in *WT1,* Wilms' tumor, and collagen IV, Alport's syndrome. Given the prevalence of both acute and chronic renal disease in the human population, the molecular mechanisms underlying pathogenesis and the potential interactions with developmental regulatory networks are rapidly expanding areas of interest.

A. Polycystic Kidney Disease

Polycystic kidney disease (PKD) is among the most common genetic disorders in the adult population. PKD is characterized by the accumulation of large fluid-filled cysts within the renal tubules. Cyst formation is thought to occur through a combination of mechanisms that include proliferation of epithelium, blockage of tubules, reversal or loss of epithelial polarity, and fluid transport into the cystic cavity. It has been proposed that the renal cystic epithelial cell has dedifferentiated to a more embryonic phenotype, leading to mislocalization of apical and basolateral cell surface markers and increased expression of specific basement membrane components (Calvet, 1993). In humans, autosomal dominant PKD (ADPKD) has been associated with mutations in at least two genes, *ADPKD1* (Consortium, 1994, 1995) and *ADPKD2* (Mochizuki *et al.,* 1996), that encode large transmembrane proteins, polycystin-1 and polycystin-2, which are known to associate (Qian *et al.,* 1997). Mutations in *PKD1* or *PKD2* account for about 95% of the clinical PKD cases. In individuals carrying a single mutant allele of the *PKD1* gene, clonally derived cysts accumulate from single epithelial cells that have lost the remaining normal allele (Qian *et al.,* 1996). This two-hit model of ADPKD suggests a tumor-suppressor-like function. Yet, the frequency of second hits must be very high given the numbers of cysts found in ADPKD patients. One possible factor affecting the mutational frequency of *PKD1* in humans is the presence of a closely linked pseudogene, with large regions of sequence identity. Unequal crossing over or gene conversion between *PKD1* and its associated pseudogene could account for this high mutational rate (Watnick *et al.*, 1998).

In the mouse, the homologous genes encoding polycystin-1 and -2 have been knocked out and in both cases cystic phenotypes are observed (Lu *et al.*, 1997; Wu *et al.*, 1998). *PKD1* homozygous mouse mutants already develop cysts at E15.5, within the developing proximal tubules, underscoring the essential developmental function of polycystin-1 (Fig. 17). The size of polycystin-1 (around 400 kDa) and the presence of multiple repeat domains has made raising specific antibodies difficult. Thus, there has been some confusion regarding the developmental expression and cellular localization of the protein. The most convincing data indicate that polycystin-1 is found in the developing collecting ducts, but not in the tips of the ureteric bud epithelium, and in the epithelium of the S-shaped body that is fated to become parietal glomerular epithelium and proximal tubules (Geng *et al.*, 1997; Ward *et al.*, 1996). How loss of polycystin-1 affects epithelial cell maturation is still unclear. The large extracellular domain may interact with a wide variety of matrix and cell surface proteins, whereas the intracellular domain has been linked, at least indirectly, to β-catenin and the Wnt signaling pathway (Kim *et al.*, 1999). Yet the cell biology of polycystin-1 and -2 requires much more extensive investigation. How the polycystins affect apical and basolateral epithelial cell polarity during development and whether this is linked to proliferation and active fluid secretion into the cysts remain to be determined. Nevertheless, the cystic mice will be important models for the development of new therapies that may slow the progression of this debilitating disease.

At least five other mouse loci have been identified that can generate a PKD phenotype on disruption (Schieren *et al.*, 1996). Among the best characterized mouse PKD mutants is the *congenital polycystic kidney* (*cpk*) mouse (Gattone *et al.*, 1996). The *cpk* gene is recessive and maps to mouse chromosome 12, unlinked to either polycystin-1 or -2. Homozygous *cpk* animals develop cysts from birth and die at approximately 3 weeks from acute renal failure. At the cellular level, the *cpk* mutant mouse exhibits many of the same traits found in human PKD, including mislocalization of the EGF receptor (Orellana *et al.*, 1995) and altered deposition of basement membrane proteins (Ojeda, 1999; Rocco *et al.*, 1992; Taub *et al.*, 1990). In *cpk* mice and in human juvenile cystic disease, expression of *Pax2* correlates with proliferation of the cystic epithelium and underscores the embryonic nature of the cysts. Strikingly, *cpk* mice carrying only one *Pax2* allele show a less severely progressing form of the disease (Ostrom *et al.*, 2000). Thus, it appears that a gene necessary for proliferation and specification of the epithelial lineage is also required for aberrant growth of the cystic epithelium.

B. Wilms' Tumor

Human genetic analysis revealed a region around chromosome 11p13–11p15 that was consistently linked to familial Wilms' tumor, an embryonal kidney carcinoma. The first candidate human gene, *WT1*, was identified by positional cloning from the 11p13 locus (Call *et al.*, 1990; Gessler *et al.*, 1990). Deletions or mutations in *WT1* are associated with approximately 10–15% of human Wilms' tumors (Haber and Housman, 1992).

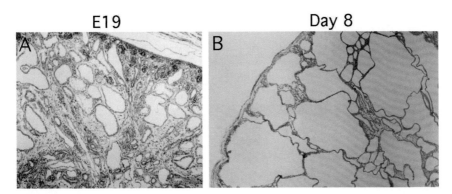

Figure 17 Rapid development of renal cysts in PKD1 mutant mice. The mouse *PKD1* gene encodes polycystin-1, the protein most frequently associated with human autosomal dominant polycystic kidney disease type 1 (ADPKD1). Homozygous mice carrying a loss-of-function *PKD1* allele already develop polycystic kidney disease *in utero*, as shown in panel A at E19. The developing cysts are both medullary and cortical. (B) In the newborns, the kidneys are almost completely cystic by 8 days. Thus, polycystin-1 has an essential role in cyst prevention during the later stages of tubulogenesis. (Micrograph courtesy of J. Zhou, Harvard Medical School.)

As discussed in the previous section, *WT1* is an essential regulator of early kidney development. In humans, loss of heterozygosity in Wilms' tumor, by somatic mutations in the induced and proliferating renal epithelium, demonstrates a tumor suppressor function for *WT1* that generates a phenotype quite different from the homozygous null mouse. In humans carrying a single *WT1* mutant allele, a second mutation can occur in the remaining normal allele to initiate tumorigenesis from a single renal epithelial cell during kidney development. This has not been observed in mice heterozygous for *WT1*, carrying either the engineered mutation or a large deletion of the *WT1* locus (Glaser *et al.,* 1990). Two possible explanations could account for this difference. The increased size of the human kidney, with as much as 1000 times the number of cells, provides much more opportunity for a single second hit. Alternatively, the absolute mutation frequency may be lower among mice than in humans due to other variables, such as DNA mismatch repair efficiency or environmental factors. In any event, the *WT1* mutant cell is in an environment that supports clonal expansion of this neoplastic cell. Furthermore, the second hit may occur in an induced, proliferating mesenchymal cell that no longer requires WT1 to rescue the mesenchyme from apoptosis.

A second type of *WT1* mutation is associated with Denys-Drash syndrome, a dominant disorder that encompasses urogenital abnormalities, nephrotic syndrome, and Wilms' tumor. Specific mutations or truncations in the zinc finger DNA-binding domain are thought to be dominant because they may change the binding specificity of the WT1 protein and activate new target genes (Patek *et al.,* 1999; Pelletier *et al.,* 1991). Alternatively, given the ability of WT1 to repress and activate depending on the promoter context, the possibility of dominant mutations relieving repression or enhancing activation must be considered.

VIII. Future Perspectives

The molecular genetic basis of organ development presents a variety of challenges for the future. The preceding decade has seen the power of genetics and modern molecular biology used to reexamine old problems. In many respects, our knowledge base is vastly greater, due to increasing numbers of genes and proteins known to function during many stages of kidney development. Yet some of the old problems remain. Induction of the mesenchyme is still not well defined. At best, it appears to be a multistep process requiring some type of contact-dependent signal and growth or survival factors derived from the ureter and/or the mesenchyme. How the multiple cell lineages are defined and how they interact is still unclear. Clearly, the relationship among stromal, epithelial, and endothelial precursor populations is critical. All of these lineages must be derived from either a single, or multiple, self-renewing stem cell population. If

any cells are exhausted before kidney development is complete, the effects are observed in the entire structure and not just in the missing lineage.

The kidney also has the capacity to regenerate proximal tubule cells once exposed to a toxic or ischemic injury. These regenerating proximal tubule cells express at least some embryonic markers and can respond to secreted factors, such as BMP7 (Vukicevic *et al.,* 1998). Why are these proximal tubule cells able to regenerate, whereas cells of the glomerulus are not? Even proximal tubule cells reach a point of no return, when damage is so extensive that the ability to regenerate is lost. At what point do cells lose this capacity? Is it because a population of preexisting stem cells in the adult organ becomes depleted, or do terminally differentiated cells have the ability to regress to a more embryonic state only within the appropriate environment for regeneration?

As kidney development proceeds, little information is available regarding the specification of epithelial subtypes. The expression of physiologically important transporters varies greatly along the nephron and implies differential regulation of gene expression. What are the mechanisms underlying these aspects of terminal epithelial cell differentiation? Are these mechanisms dependent on the position of cells along the proximal-distal axis of the tubule? With respect to the pathogenesis of human renal disease, it is these terminally differentiated cells that have genetic or biochemical lesions, which result in proliferation, dedifferentiation, and misexpression of physiologically relevant molecules. Thus in addition to a more complete understanding of how specialized cell types differentiate within a complex organ system, unraveling such key developmental control pathways will enable investigators to design novel strategies for tissue engineering, regeneration, or tumor suppression that can have an enormous impact in clinical medicine.

Acknowledgments

I thank the members of my lab who have contributed comments and suggestions to the chapter, especially Eun Ah Cho, Sanj Patel, and Pat Brophy. I also thank Dale Abrahamson, Brigid Hogan, Doris Herzlinger, Andreas Kispert, Jeff Miner, Barry Robert, and Jing Zhou for sharing data and figures. G.R.D. is supported by grants DK51043, DK54723, and DK54740 from the National Institutes of Health.

References

Abdelhak, S., Kalatzis, V., Heilig, R., Compain, S., Samson, D., Vincent, C., Weil, D., Cruaud, C., Sahly, I., Leibovici, M., Bitner-Glindzicz, M., Francis, M., Lacombe, D., Vigneron, J., Charachon, R., Boven, K., Bedbeder, P., Van Regemorter, N., Weissenbach, J., and Petit, C. (1997). A human homologue of the *Drosophila* eyes absent gene underlies branchio- oto-renal (BOR) syndrome and identifies a novel gene family. *Nat. Genet.* **15,** 157–164.

Armstrong, J. F., Pritchard-Jones, K., Bickmore, W. A., Hastie, N. D., and Bard, J. B. L. (1992). The expression of the Wilms' tumor gene, *WT1,* in the developing mammalian embryo. *Mech. Dev.* **40,** 85–97.

Barasch, J., Yang, J., Ware, C. B., Taga, T., Yoshida, K., Erdjument-Bromage, H., Tempst, P., Parravicini, E., Malach, S., Aranoff, T., and Oliver, J. A. (1999). Mesenchymal to epithelial conversion in rat metanephros is induced by LIF. *Cell (Cambridge, Mass.)* **99,** 377–386.

Barker, D. F., Hostikka, S. L., Zhou, J., Chow, L. T., Oliphant, A. R., Gerken, S. C., Gregory, M. C., Skolnick, M. H., Atkin, C. L., and Tryggvason, K. (1990). Identification of mutations in the COL4A5 collagen gene in Alport syndrome. *Science* **248,** 1224–1227.

Bladt, F., Riethmacher, D., Isenmann, S., Aguzzi, A., and Birchmeier, C. (1995). Essential role for the c-met receptor in the migration of myogenic precursor cells in the limb bud. *Nature (London)* **376,** 768–771.

Bullock, S. L., Fletcher, J. M., Beddington, R. S., and Wilson, V. A. (1998). Renal agenesis in mice homozygous for a gene trap mutation in the gene encoding heparan sulfate 2-sulfotransferase. *Genes Dev.* **12,** 1894–1906.

Call, K. M., Glaser, T., Ito, C. Y., Buckler, A. J., Pelletier, J., Haber, D. A., Rose, E. A., Kral, A., Yeger, H., Lewis, W. H., Jones, C., and Housman, D. E. (1990). Isolation and characterization of a zinc finger polypeptide gene at the human chromosome 11 Wilms tumor locus. *Cell (Cambridge, Mass.)* **60,** 509–520.

Calvet, J. P. (1993). Polycystic kidney disease: Primary extracellular matrix abnormality or defective cellular differentiation? *Kidney Int.* **43,** 101–108.

Carlson, B. M. (1988). "Patten's Foundations of Embryology." McGraw-Hill, New York.

Cho, E. A., Patterson, L. T., Brookhiser, W. T., Mah, S., Kintner, C., and Dressler, G. R. (1998). Differential expression and function of cadherin-6 during renal epithelium development. *Development (Cambridge, UK)* **125,** 4806–4815.

Consortium, T. E. P. K. D. (1994). The polycystic kidney disease 1 gene encodes a 14 kb transcript and lies within a duplicated region on chromosome 16. The European Polycystic Kidney Disease Consortium. *Cell (Cambridge, Mass.)* **77,** 881–894; errata: *Ibid.* **78**(4), 724 (1994); **81**(7), 1170 (1995).

Consortium, T. I. P. K. D. (1995). Polycystic kidney disease: The complete structure of the PKD1 gene and its protein. The International Polycystic Kidney Disease Consortium. *Cell (Cambridge, Mass.)* **81,** 289–298.

Dahl, E., Koseki, H., and Balling, R. (1997). Pax genes and organogenesis. *BioEssays* **19,** 755–765.

Davies, J. A. (1999). The Kidney Development Database. *Dev. Genet.* **24,** 194–198.

Davies, J. A., and Garrod, D. R. (1995). Induction of early stages of kidney tubule differentiation by lithium ions. *Dev. Biol.* **167,** 50–60.

Davies, R. C., Calvio, C., Bratt, E., Larsson, S. H., Lamond, A. I., and Hastie, N. D. (1998). WT1 interacts with the splicing factor U2AF65 in an isoform-dependent manner and can be incorporated into spliceosomes. *Genes Dev.* **12,** 3217–3225.

Dehbi, M., Ghahremani, M., Lechner, M., Dressler, G., and Pelletier, J. (1996). The paired-box transcription factor, PAX2, positively modulates expression of the Wilms' tumor suppressor gene (WT1). *Oncogene* **13,** 447–453.

Dressler, G. R., and Douglass, E. C. (1992). *Pax-2* is a DNA-binding protein expressed in embryonic kidney and Wilms tumor. *Proc. Natl. Acad. Sci. U.S.A.* **89,** 1179–1183.

Dressler, G. R., Deutsch, U., Chowdhury, K., Nornes, H. O., and Gruss, P. (1990). *Pax2,* a new murine paired-box containing gene and its expression in the developing excretory system. *Development (Cambridge, UK)* **109,** 787–795.

Drummond, I. A., Majumdar, A., Hentschel, H., Elger, M., Solnica-Krezel, L., Schier, A. F., Neuhauss, S. C., Stemple, D. L., Zwartkruis, F., Rangini, Z., Driever, W., and Fishman, M. C. (1998). Early development of the zebrafish pronephros and analysis of mutations affecting pronephric function. *Development (Cambridge, UK)* **125,** 4655–4667.

Dudley, A. T., Lyons, K. M., and Robertson, E. J. (1995). A requirement for bone morphogenetic protein-7 during development of the mammalian kidney and eye. *Genes Dev.* **9,** 2795–2807.

Dudley, A. T., Godin, R. E., and Robertson, E. J. (1999). Interaction between FGF and BMP signaling pathways regulates development of metanephric mesenchyme. *Genes Dev.* **13,** 1601–1613.

Ekblom, P. (1989). Developmentally regulated conversion of mesenchyme to epithelium. *FASEB J.* **3,** 2141–2150.

Gattone, V. H., 2nd, MacNaughton, K. A., and Kraybill, A. L. (1996). Murine autosomal recessive polycystic kidney disease with multiorgan involvement induced by the cpk gene. *Anat. Rec.* **245,** 488–499.

Geng, L., Segal, Y., Pavlova, A., Barros, E. J., Lohning, C., Lu, W., Nigam, S. K., Frischauf, A. M., Reeders, S. T., and Zhou, J. (1997). Distribution and developmentally regulated expression of murine polycystin. *Am. J. Physiol.* **272,** F451–F459.

Gessler, M., Poustka, A., Cavenee, W., Neve, R. L., Orkin, S., and Bruns, G. A. P. (1990). Homozygous deletion in Wilms' tumors of a zinc-finger gene identified by chromosome jumping. *Nature (London)* **343,** 774–778.

Giancotti, F. G., and Ruoslahti, E. (1999). Integrin signaling. *Science* **285,** 1028–1032.

Glaser, T., Lane, J., and Housman, D. E. (1990). A mouse model for the aniridia-Wilms' tumor deletion syndrome. *Science* **250,** 823–827.

Grobstein, C. (1956). Trans-filter induction of tubules in mouse metanephric mesenchyme. *Exp. Cell Res.* **10,** 424–440.

Gumbiner, B. M. (1996). Cell adhesion: The molecular basis of tissue architecture and morphogenesis. *Cell (Cambridge, Mass.)* **84,** 345–357.

Haber, D. A., and Housman, D. E. (1992). Role of WT1 gene in Wilms' tumor. *Cancer Surv.* **12,** 105–117.

Hatini, V., Huh, S. O., Herzlinger, D., Soares, V. C., and Lai, E. (1996). Essential role of stromal mesenchyme in kidney morphogenesis revealed by targeted disruption of Winged Helix transcription factor BF-2. *Genes Dev.* **10,** 1467–1478.

Heanue, T. A., Reshef, R., Davis, R. J., Mardon, G., Oliver, G., Tomarev, S., Lassar, A. B., and Tabin, C. J. (1999). Synergistic regulation of vertebrate muscle development by Dach2, Eya2, and Six1, homologs of genes required for Drosophila eye formation. *Genes Dev.* **13,** 3231–3243.

Hellmich, H. L., Kos, L., Cho, E. S., Mahon, K. A., and Zimmer, A. (1996). Embryonic expression of glial cell-line derived neurotrophic factor (GDNF) suggests multiple developmental roles in neural differentiation and epithelial-mesenchymal interactions. *Mech. Dev.* **54,** 95–106.

Herzlinger, D., Qiao, J., Cohen, D., Ramakrishna, N., and Brown, A. M. (1994). Induction of kidney epithelial morphogenesis by cells expressing Wnt-1. *Dev. Biol.* **166,** 815–818.

Hoffmann, I., and Balling, R. (1995). Cloning and expression analysis of a novel mesodermally expressed cadherin. *Dev. Biol.* **169,** 337–346.

Holzman, L. B., St. John, P. L., Kovari, I. A., Verma, R., Holthofer, H., and Abrahamson, D. R. (1999). Nephrin localizes to the slit pore of the glomerular epithelial cell. *Kidney Int.* **56,** 1481–1491.

Hudson, B. G., Reeders, S. T., and Tryggvason, K. (1993). Type IV collagen: Structure, gene organization, and role in human diseases. Molecular basis of Goodpasture and Alport syndromes and diffuse leiomyomatosis. *J. Biol. Chem.* **268,** 26033–26036.

Hyink, D. P., Tucker, D. C., St. John, P. L., Leardkamolkarn, V., Accavitti, M. A., Abrass, C. K., and Abrahamson, D. R. (1996). Endogenous origin of glomerular endothelial and mesangial cells in grafts of embryonic kidneys. *Am. J. Physiol.* **270,** F886–F899.

Inoue, T., Chisaka, O., Matsunami, H., and Takeichi, M. (1997). Cadherin-6 expression transiently delineates specific rhombomeres, other neural tube subdivisions, and neural crest subpopulations in mouse embryos. *Dev. Biol.* **183,** 183–194.

Jing, S., Wen, D., Yu, Y., Holst, P. L., Mei, Y. L., Fang, M., Tamir, R., Antonio, L., Hu, Z., Cupples, R., Louis, J.-C., Hu, S., Altrock, B. W., and Fox, G. M. (1996). GDNF-induced activation of the Ret protein tyrosine kinase is mediated by GDNFR-α, a novel receptor for GDNF. *Cell (Cambridge, Mass.)* **85,** 1113–1124.

Karavanova, I. D., Dove, L. F., Resau, J. H., and Perantoni, A. O. (1996). Conditioned medium from a rat ureteric bud cell line in combination with bFGF induces complete differentiation of isolated metanephric mesenchyme. *Development (Cambridge, UK)* **122,** 4159–4167.

Katz, B. Z., Levenberg, S., Yamada, K. M., and Geiger, B. (1998). Modulation of cell-cell adherens junctions by surface clustering of the N-cadherin cytoplasmic tail. *Exp. Cell Res.* **243,** 415–424.

Kaufman, M. H. (1992). "The Atlas of Mouse Development." Academic Press, San Diego, CA.

Kemler, R., and Ozawa, M. (1989). Uvomorulin-catenin complex: Cytoplasmic anchorage of a Ca2+-dependent cell adhesion molecule. *BioEssays* **11,** 88–91.

Kestila, M., Lenkkeri, U., Mannikko, M., Lamerdin, J., McCready, P., Putaala, H., Ruotsalainen, V., Morita, T., Nissinen, M., Herva, R., Kashtan, C. E., Peltonen, L., Holmberg, C., Olsen, A., and Tryggvason, K. (1998). Positionally cloned gene for a novel glomerular protein—nephrin—is mutated in congenital nephrotic syndrome. *Mol. Cell* **1,** 575–582.

Kim, E., Arnould, T., Sellin, L. K., Benzing, T., Fan, M. J., Gruning, W., Sokol, S. Y., Drummond, I., and Walz, G. (1999). The polycystic kidney disease 1 gene product modulates Wnt signaling. *J. Biol. Chem.* **274,** 4947–4953.

Kimura, Y., Matsunami, H., Inoue, T., Shimamura, K., Uchida, N., Ueno, T., Miyazaki, T., and Takeichi, M. (1995). Cadherin-11 expressed in association with mesenchymal morphogenesis in the head, somite, and limb bud of early mouse embryos. *Dev. Biol.* **169,** 347–358.

Kispert, A., Vainio, S., Shen, L., Rowitch, D. H., and McMahon, A. P. (1996). Proteoglycans are required for maintenance of Wnt-11 expression in the ureter tips. *Development (Cambridge, UK)* **122,** 3627–3637.

Kispert, A., Vainio, S., and McMahon, A. P. (1998). Wnt-4 is a mesenchymal signal for epithelial transformation of metanephric mesenchyme in the developing kidney. *Development (Cambridge, UK)* **125,** 4225–4234.

Kitamoto, Y., Tokunaga, H., and Tomita, K. (1997). Vascular endothelial growth factor is an essential molecule for mouse kidney development: Glomerulogenesis and nephrogenesis. *J. Clin. Invest.* **99,** 2351–2357.

Klein, G., Langegger, M., Timpl, R., and Ekblom, P. (1988). Role of laminin A chain in the development of epithelial cell polarity. *Cell (Cambridge, Mass.)* **55,** 331–341.

Klein, P. S., and Melton, D. A. (1996). A molecular mechanism for the effect of lithium on development. *Proc. Natl. Acad. Sci. U.S.A.* **93,** 8455–8459.

Koch, A. W., Bozic, D., Pertz, O., and Engel, J. (1999). Homophilic adhesion by cadherins. *Curr. Opin. Struct. Biol.* **9,** 275–281.

Kreidberg, J. A., Sariola, H., Loring, J. M., Maeda, M., Pelletier, J., Housman, D., and Jaenisch, R. (1993). WT1 is required for early kidney development. *Cell (Cambridge, Mass.)* **74,** 679–691.

Kreidberg, J. A., Donovan, M. J., Goldstein, S. L., Rennke, H., Shepherd, K., Jones, R. C., and Jaenisch, R. (1996). Alpha 3 beta 1 integrin has a crucial role in kidney and lung organogenesis. *Development (Cambridge, UK)* **122,** 3537–3547.

Kume, T., Deng, K., and Hogan, B. L. (2000). Murine forkhead/winged helix genes Foxc1 (Mf1) and Foxc2 (Mfh1) are required for the early organogenesis of the kidney and urinary tract. *Development (Cambridge, UK)* **127,** 1387–1395.

Larue, L., Ohsugi, M., Hirchenhain, J., and Kemler, R. (1994). E-cadherin null mutant embryos fail to form a trophectoderm epithelium. *Proc. Natl. Acad. Sci. U.S.A.* **91,** 8263–8267.

Lee, S. B., Huang, K., Palmer, R., Truong, V. B., Herzlinger, D., Kolquist, K. A., Wong, J., Paulding, C., Yoon, S. K., Gerald, W., Oliner, J. D., and Haber, D. A. (1999). The Wilms tumor suppressor WT1 encodes a transcriptional activator of amphiregulin. *Cell (Cambridge, Mass.)* **98,** 663–673.

Lelievre-Pegorier, M., Vilar, J., Ferrier, M. L., Moreau, E., Freund, N., Gilbert, T., and Merlet-Benichou, C. (1998). Mild vitamin A deficiency leads to inborn nephron deficit in the rat. *Kidney Int.* **54,** 1455–1462.

Leveen, P., Pekny, M., Gebre-Medhin, S., Swolin, B., Larsson, E., and Betsholtz, C. (1994). Mice deficient for PDGF B show renal, cardiovascular, and hematological abnormalities. *Genes Dev.* **8,** 1875–1887.

Lindahl, P., Hellström, M., Kalen, M., Karlsson, L., Pekny, M., Pekna, M., Soriano, P., and Betsholtz, C. (1998). Paracrine PDGF-B/PDGF-Rbeta signaling controls mesangial cell development in kidney glomeruli. *Development (Cambridge, UK)* **125,** 3313–3322.

Lu, W., Peissel, B., Babakhanlou, H., Pavlova, A., Geng, L., Fan, X., Larson, C., Brent, G., and Zhou, J. (1997). Perinatal lethality with kidney and pancreas defects in mice with a targetted Pkd1 mutation. *Nat. Genet.* **17,** 179–181.

Luo, G., Hofmann, C., Bronckers, A. L. J., J., Sohocki, M., Bradley, A., and Karsenty, G. (1995). BMP-7 is an inducer of nephrogenesis, and is also required for eye development and skeletal patterning. *Genes Dev.* **9,** 2808–2820.

Madden, S. L., Cook, D. M., Morris, J. F., Gashier, A., Sukhatme, V. P., and Rauscher, F. J., III (1991). Transcriptional repression mediated by the WT1 Wilms' tumor gene product. *Science* **253,** 1550–1553.

Mah, S. P., Saueressig, H., Goulding, M., Kintner, C., and Dressler, G. R. (2000). Kidney development in cadherin-6 mutants: Delayed mesenchyme-to-epithelial conversion and loss of nephrons. *Dev. Biol.* **223,** 38–53.

Mauch, T. J., Yang, G., Wright, M., Smith, D., and Schoenwolf, G. C. (2000). Signals from trunk paraxial mesoderm induce pronephros formation in chick intermediate mesoderm. *Dev. Biol.* **220,** 62–75.

Mendelsohn, C., Batourina, E., Fung, S., Gilbert, T., and Dodd, J. (1999). Stromal cells mediate retinoid-dependent functions essential for renal development. *Development (Cambridge, UK)* **126,** 1139–1148.

Miller, J. R., Hocking, A. M., Brown, J. D., and Moon, R. T. (1999). Mechanism and function of signal transduction by the Wnt/beta-catenin and Wnt/Ca2+ pathways. *Oncogene* **18,** 7860–7872.

Miner, J. H. (1999). Renal basement membrane components. *Kidney Int.* **56,** 2016–2024.

Miner, J. H., and Li, C. (2000). Defective glomerulogenesis in the absence of laminin alpha5 demonstrates a developmental role for the kidney glomerular basement membrane. *Dev. Biol.* **217,** 278–289.

Miner, J. H., and Sanes, J. R. (1994). Collagen IV alpha 3, alpha 4, and alpha 5 chains in rodent basal laminae: Sequence, distribution, association with laminins, and developmental switches. *J. Cell Biol.* **127,** 879–891.

Miner, J. H., and Sanes, J. R. (1996). Molecular and functional defects in kidneys of mice lacking collagen alpha 3(IV): Implications for Alport syndrome. *J. Cell Biol.* **135,** 1403–1413.

Miner, J. H., Patton, B. L., Lentz, S. I., Gilbert, D. J., Snider, W. D., Jenkins, N. A., Copeland, N. G., and Sanes, J. R. (1997). The laminin alpha chains: Expression, developmental transitions, and chromosomal locations of alpha1-5, identification of heterotrimeric laminins 8-11, and cloning of a novel alpha3 isoform. *J. Cell Biol.* **137,** 685–701.

Miyamoto, N., Yoshida, M., Kuratani, S., Matsuo, I., and Aizawa, S. (1997). Defects of urogenital development in mice lacking Emx2. *Development (Cambridge, UK)* **124,** 1653–1664.

Miyazaki, Y., Oshima, K., Fogo, A., Hogan, B. L. M., and Ichikawa, I. (2000). Bone morphogenetic protein 4 regulates the budding site and elongation of the mouse ureter. *J. Clin. Invest.* **105,** 863–873.

Mochizuki, T., Wu, G., Hayashi, T., Xenophontos, S. L., Veldhuisen, B., Saris, J. J., Reynolds, D. M., Cai, Y., Gabow, P. A., Pierides, A., Kimberling, W. J., Breuning, M. H., Deltas, C. C., Peters, D. J., and Somlo, S. (1996). PKD2, a gene for polycystic kidney disease that encodes an integral membrane protein. *Science* **272,** 1339–1342.

Moore, M. W., Klein, R. D., Farinas, I., Sauer, H., Armanini, M., Phillips, H., Reichardt, L. F., Ryans, A. M., Carver-Moore, K., and Rosenthal, A. (1996). Renal and neuronal abnormalities in mice lacking GDNF. *Nature (London)* **382,** 76–79.

Muller, U., Wang, D., Denda, S., Meneses, J. J., Pedersen, R. A., and Reichardt, L. F. (1997). Integrin alpha8beta1 is critically important for epithelial-mesenchymal interactions during kidney morphogenesis. *Cell (Cambridge, Mass.)* **88,** 603–613.

Nakagawa, S., and Takeichi, M. (1995). Neural crest cell-cell adhesion controlled by sequential and subpopulation-specific expression of novel cadherins. *Development (Cambridge, UK)* **121**, 1321–1332.

Noakes, P. G., Miner, J. H., Gautam, M., Cunningham, J. M., Sanes, J. R., and Merlie, J. P. (1995). The renal glomerulus of mice lacking s-laminin/laminin beta 2: Nephrosis despite molecular compensation by laminin beta 1. *Nat. Genet.* **10**, 400–406.

Noll, M. (1993). Evolution and role of Pax genes. *Cur. Opin. Genet. Dev.* **3**, 595–605.

Obara-Ishihara, T., Kuhlman, J., Niswander, L., and Herzlinger, D. (1999). The surface ectoderm is essential for nephric duct formation in intermediate mesoderm. *Development (Cambridge, UK)* **126**, 1103–1108.

Ohto, H., Kamada, S., Tago, K., Tominaga, S. I., Ozaki, H., Sato, S., and Kawakami, K. (1999). Cooperation of six and eya in activation of their target genes through nuclear translocation of Eya. *Mol. Cell Biol.* **19**, 6815–6824.

Ojeda, J. L. (1999). Abnormal tenascin expression in murine autosomal recessive polycystic kidneys. *Nephron* **82**, 261–269.

Orellana, S. A., Sweeney, W. E., Neff, C. D., and Avner, E. D. (1995). Epidermal growth factor receptor expression is abnormal in murine polycystic kidney. *Kidney Int.* **47**, 490–499.

Ostrum, L., Tang, M. J., Gruss, P., and Dressler, G. R. (2000). Reduced *Pax2* gene dosage increases apoptosis and slows the progression of renal cystic disease. *Dev. Biol.* **219**, 250–258.

Ozawa, M., Baribault, H., and Kemler, R. (1989). The cytoplasmic domain of the cell adhesion molecule uvomorulin associates with three independent proteins structurally related in different species. *EMBO J.* **8**, 1711–1717.

Patek, C. E., Little, M. H., Fleming, S., Miles, C., Charlieu, J. P., Clarke, A. R., Miyagawa, K., Christie, S., Doig, J., Harrison, D. J., Porteous, D. J., Brookes, A. J., Hooper, M. L., and Hastie, N. D. (1999). A zinc finger truncation of murine WT1 results in the characteristic urogenital abnormalities of Denys-Drash syndrome. *Proc. Natl. Acad. Sci. U.S.A.* **96**, 2931–2936.

Pelletier, J., Bruening, W., Kashtan, C. E., Mauer, S. M., Manivel, J. C., Striegel, J. E., Houghton, D. C., Junien, C., and Habib, R. (1991). Germline mutations in the Wilms' tumor suppressor gene are associated with abnormal urogenital development in Denys-Drash syndrome. *Cell (Cambridge, Mass.)* **67**, 437–447.

Pepicelli, C. V., Kispert, A., Rowitch, D. H., and McMahon, A. P. (1997). GDNF induces branching and increased cell proliferation in the ureter of the mouse. *Dev. Biol.* **192**, 193–198.

Perantoni, A. O., Dove, L. F., and Williams, C. L. (1991). Induction of tubules in rat metanephric mesenchyme in the absence of an inductive tissue. *Differentiation (Berlin)* **48**, 25–31.

Phelps, D. E., and Dressler, G. R. (1993). Aberrant expression of Pax-2 in Danforth's short tail (Sd) mice. *Dev. Biol.* **157**, 251–258.

Pichel, J. G., Shen, L., Sheng, H. Z., Granholm, A.-C., Drago, J., Grinberg, A., Lee, E. J., Huang, S. P., Saarma, M., Hoffer, B. J., Sariola, H., Westphal, H. *et al.* (1996). Defects in enteric innervation and kidney development in mice lacking GDNF. *Nature (London)* **382**, 73–76.

Pignoni, F., Hu, B., Zavitz, K. H., Xiao, J., Garrity, P. A., and Zipursky, S. L. (1997). The eye-specification proteins So and Eya form a complex and regulate multiple steps in Drosophila eye development. *Cell (Cambridge, Mass.)* **91**, 881–891; erratum: *Ibid.* **92**(4), 585 (1998).

Pritchard-Jones, K., Fleming, S., Davidson, D., Bickmore, W., Porteous, D., Gosden, C., Bard, J., Buckler, A., Pelletier, J., Housman, D., van Heyningen, V., and Hastie, N. (1990). The candidate Wilms' tumor gene is involved in genitourinary development. *Nature (London)* **346**, 194–197.

Qian, F., Watnick, T. J., Onuchic, L. F., and Germino, G. G. (1996). The molecular basis of focal cyst formation in human autosomal dominant polycystic kidney disease type I. *Cell (Cambridge, Mass.)* **87**, 979–987.

Qian, F., Germino, F. J., Cai, Y., Zhang, X., Somlo, S., and Germino, G. G. (1997). PKD1 interacts with PKD2 through a probable coiled-coil domain. *Nat. Genet.* **16**, 179–183.

Quaggin, S. E., Schwartz, L., Cui, S., Igarashi, P., Deimling, J., Post, M., and Rossant, J. (1999). The basic-helix-loop-helix protein pod1 is critically important for kidney and lung organogenesis. *Development (Cambridge, UK)* **126**, 5771–5783.

Robert, B., St. John, P. L., Hyink, D. P., and Abrahamson, D. R. (1996). Evidence that embryonic kidney cells expressing flk-1 are intrinsic, vasculogenic angioblasts. *Am. J. Physiol.* **271**, F744–F753.

Robert, B., St. John, P. L., and Abrahamson, D. R. (1998). Direct visualization of renal vascular morphogenesis in Flk1 heterozygous mutant mice. *Am. J. Physiol.* **275**, F164–F172.

Rocco, M. V., Neilson, E. G., Hoyer, J. R., and Ziyadeh, F. N. (1992). Attenuated expression of epithelial cell adhesion molecules in murine polycystic kidney disease. *Am. J. Physiol.* **262**, F679–F686.

Rosenberg, P., Esni, F., Sjodin, A., Larue, L., Carlsson, L., Gullberg, D., Takeichi, M., Kemler, R., and Semb, H. (1997). A potential role of R-cadherin in striated muscle formation. *Dev. Biol.* **187**, 55–70.

Ruotsalainen, V., Ljungberg, P., Wartiovaara, J., Lenkkeri, U., Kestila, M., Jalanko, H., Holmberg, C., and Tryggvason, K. (1999). Nephrin is specifically located at the slit diaphragm of glomerular podocytes. *Proc. Natl. Acad. Sci. U.S.A.* **96**, 7962–7967.

Ryan, G., Steele-Perkins, V., Morris, J., Rauscher, F. J., III, and Dressler, G. R. (1995). Repression of Pax-2 by WT1 during normal kidney development. *Development (Cambridge, UK)* **121**, 867–875.

Sainio, K., Suvanto, P., Davies, J., Wartiovaara, J., Wartiovaara, K., Saarma, M., Arume, U., Meng, X., Lindahl, M., Pachnis, V., and Sariola, H. (1997). Glial-cell-line-derived neurotrophic factor is required for bud initiation from ureteric epithelium. *Development (Cambridge, UK)* **124**, 4077–4087.

Sanchez, M. P., Silos-Santiago, I., Frisen, J., He, B., Lira, S. A., and Barbacid, M. (1996). Renal agenesis and the absence of enteric neurons in mice lacking GDNF. *Nature (London)* **382**, 70–73.

Sanyanusin, P., McNoe, L. A., Sullivan, M. J., Weaver, R. G., and Eccles, M. R. (1995). Mutation of PAX2 in two siblings with renal-coloboma syndrome. *Hum. Mol. Genet.* **4**, 2183–2184.

Saxen, L. (1987). Organogenesis of the kidney. *In* "Developmental and Cell Biology Series 19" (P. W. Barlow, P. B. Green, and C. C. White, eds.). Cambridge University Press, Cambridge, UK.

Saxen, L., and Saksela, E. (1971). Transmision and spread of embryonic induction II. Exclusion of an assimilatory transmission mechanism in kidney tubule induction. *Exp. Cell Res.* **66**, 369–377.

Schieren, G., Pey, R., Bach, J., Hafner, M., and Gretz, N. (1996). Murine models of polycystic kidney disease. *Nephrol. Dial. Transplant.* **11**, 38–45.

Schimmenti, L. A., Cunliffe, H. E., McNoe, L. A., Ward, T. A., French, M. C., Shim, H. H., Zhang, Y.-H., Proesmans, W., Leys, A., Byerly, K.A., Braddock, S. R., Masuno, M., Imaizumi, K., Devriend, T. K., and Eccles, M. R. (1997). Further delineation of renal-colombo syndrome in patients with extreme variability of phenotype and identical PAX2 mutations. *Am. J. Hum. Genet.* **60**, 968–978.

Schmidt, C., Bladt, F., Goedecke, S., Brinkmann, V., Zschlesche, W., Sharpe, M., Gherardi, E., and Birchmeier, C. (1995). Scatter factor/hepatocyte growth factor is essential for liver development. *Nature (London)* **373**, 699–702.

Schuchardt, A., D'Agati, V., Larsson-Blomberg, L., Costantini, F., and Pachnis, V. (1994). Defects in the kidney and enteric nervous system of mice lacking the tyrosine kinase receptor Ret. *Nature (London)* **367**, 380–383.

Schuchardt, A., D'Agati, V., Pachnis, V., and Costantini, F. (1996). Renal agenesis and hypodysplasia in ret-k⁻ mutant mice result from defects in ureteric bud development. *Development (Cambridge, UK)* **122**, 1919–1929.

Shapiro, L., Fannon, A. M., Kwong, P. D., Thompson, A., Lehmann, M. S., Grubel, G., Legrand, J. F., Als-Nielsen, J., Colman, D. R., and Hendrickson, W. A. (1995). Structural basis of cell-cell adhesion by cadherins. *Nature (London)* **374**, 327–337.

Soriano, P. (1994). Abnormal kidney development and hematological disorders in PDGF beta- receptor mutant mice. *Genes Dev.* **8,** 1888–1896.

Stark, K., Vainio, S., Vassileva, G., and McMahon, A. P. (1994). Epithelial transformation of metanephric mesenchyme in the developing kidney regulated by Wnt-4. *Nature (London)* **372,** 679–683.

Stuart, E. T., Kioussi, C., and Gruss, P. (1993). Mammalian pax genes. *Annu. Rev. Genet.* **27,** 219–236.

Tang, M. J., Worley, D., Sanicola, M., and Dressler, G. R. (1998). The RET-glial cell-derived neurotrophic factor (GDNF) pathway stimulates migration and chemoattraction of epithelial cells. *J. Cell Biol.* **142,** 1337–1345.

Taub, M., Laurie, G. W., Martin, G. R., and Kleinman, H. K. (1990). Altered basement membrane protein biosynthesis by primary cultures of cpk/cpk mouse kidney. *Kidney Int.* **37,** 1090–1097.

Torres, M., Gomez-Pardo, E., Dressler, G. R., and Gruss, P. (1995). Pax-2 controls multiple steps of urogenital development. *Development (Cambridge, UK)* **121,** 4057–4065.

Treanor, J. J. S., Goodman, L., de Sauvage, F., Stone, D. M., Poulsen, K. T., Beck, C. D., Gray, C., Armanini, M. P., Pollock, R. A., Hefti, F., Phillips, H. S., Goddard, A., Moore, M. W., Buj-Bello, A., Davies, A. M., Asai, N., Takahashi, M., Vandlen, R., Henderson, C. E., and Rosenthal, A. (1996). Characterization of a multicomponent receptor for GDNF. *Nature (London)* **382,** 80–83.

Trupp, M., Arenas, E., Fainzilber, M., Nilsson, A.-S., Sieber, B.-A., Grigoriou, M., Kilkenny, C., Salazar-Grueso, E., Pachnis, V., Arumae, U., Sariola, H., Saarma, M., and Ibañez, C. F. (1996). Functional receptor for GDNF encoded by the c-ret proto-oncogene. *Nature (London)* **381,** 785–789.

Tsang, T. E., Shawlot, W., Kinder, S. J., Kobayashi, A., Kwan, K. M., Schughart, K., Kania, A., Jessell, T. M., Behringer, R. R., and Tam, P. P. (2000). Lim1 activity is required for intermediate mesoderm differentiation in the mouse embryo. *Dev. Biol.* **223,** 77–90.

Tufro, A., Norwood, V. F., Carey, R. M., and Gomez, R. A. (1999). Vascular endothelial growth factor induces nephrogenesis and vasculogenesis. *J. Am. Soc. Nephrol.* **10,** 2125–2134.

Tufro-McReddie, A., Norwood, V. F., Aylor, K. W., Botkin, S. J., Carey, R. M., and Gomez, R. A. (1997). Oxygen regulates vascular endothelial growth factor-mediated vasculogenesis and tubulogenesis. *Dev. Biol.* **183,** 139–149.

Uehara, Y., Minowa, O., Mori, C., Shiota, K., Kuno, J., Noda, T., and Kitamura, N. (1995). Placental defect and embryonic lethality in mice lacking hepatocyte growth factor/scatter factor. *Nature (London)* **373,** 702–705.

van Heyningen, V., and Hastie, N. (1992). Wilms' tumour: Reconciling genetics and biology. *Trends Genet.* **8,** 16–21.

Vega, Q. C., Worby, C. A., Lechner, M. S., Dixon, J. E., and Dressler, G. R. (1996). Glial cell line-derived meurotrophic factor activates RET and promotes kidney morphogenesis. *Proc. Natl. Acad. Sci. U.S.A.* **93,** 10657–10661.

Vestweber, D., Kemler, R., and Ekblom, P. (1985). Cell-adhesion molecule uvomorulin during kidney development. *Dev. Biol.* **112,** 213–221.

Vilar, J., Gilbert, T., Moreau, E., and Merlet-Benichou, C. (1996). Metanephros organogenesis is highly stimulated by vitamin A derivatives in organ culture. *Kidney Int.* **49,** 1478–1487.

Vize, P. D., Seufert, D. W., Carroll, T. J., and Wallingford, J. B. (1997). Model systems for the study of kidney development: Use of the pronephros in the analysis of organ induction and patterning. *Dev. Biol.* **188,** 189–204.

Vukicevic, S., Kopp, J. B., Luyten, F. P., and Sampath, T. K. (1996). Induction of nephrogenic mesenchyme by osteogenic protein 1 (bone morphogenetic protein 7). *Proc. Natl. Acad. Sci. U.S.A.* **93,** 9021–9026.

Vukicevic, S., Basic, V., Rogic, D., Basic, N., Shih, M. S., Shepard, A., Jin, D., Dattatreyamurty, B., Jones, W., Dorai, H., Ryan, S., Griffiths, D., Maliakal, J., Jelic, M., Pastorcic, M., Stavljenic, A., and Sampath, T. K. (1998). Osteogenic protein-1 (bone morphogenetic protein-7) reduces severity of injury after ischemic acute renal failure in rat. *J. Clin. Invest.* **102,** 202–214.

Ward, C. J., Turley, H., Ong, A. C., Comley, M., Biddolph, S., Chetty, R., Ratcliffe, P. J., Gattner, K., and Harris, P. C. (1996). Polycystin, the polycystic kidney disease 1 protein, is expressed by epithelial cells in fetal, adult, and polycystic kidney. *Proc. Natl. Acad. Sci. U.S.A.* **93,** 1524–1528.

Watnick, T. J., Gandolph, M. A., Weber, H., Neumann, H. P., and Germino, G. G. (1998). Gene conversion is a likely cause of mutation in PKD1. *Hum. Mol. Genet.* **7,** 1239–1243.

Wodarz, A., and Nusse, R. (1998). Mechanisms of Wnt signaling in development. *Annu. Rev. Cell Dev. Biol.* **14,** 59–88.

Woolf, A. S., Kolatsi-Joannou, M., Hardman, P., Andermarcher, E., Moorby, C., Fine, L. G., Jat, P. S., Noble, M. D., and Gherardi, E. (1995). Roles of hepatocyte growth factor/scatter and the met receptor in the early development of the metanephros. *J. Cell Biol.* **128,** 171–184.

Wu, G., D'Agati, V., Cai, Y., Markowitz, G., Park, J. H., Reynolds, D. M., Maeda, Y., Le, T. C., Hou, H., Jr., Kucherlapati, R., Edelmann, W., and Somlo, S. (1998). Somatic inactivation of Pkd2 results in polycystic kidney disease. *Cell (Cambridge, Mass.)* **93,** 177–188.

Xu, P. X., Adams, J., Peters, H., Brown, M. C., Heaney, S., and Maas, R. (1999). Eya1-deficient mice lack ears and kidneys and show abnormal apoptosis of organ primordia. *Nat. Genet.* **23,** 113–117.

Yap, A. S., Brieher, W. M., and Gumbiner, B. M. (1997). Molecular and functional analysis of cadherin-based adherens junctions. *Annu. Rev. Cell Dev. Biol.* **13,** 119–146.

Yap, A. S., Niessen, C. M., and Gumbiner, B. M. (1998). The juxtamembrane region of the cadherin cytoplasmic tail supports lateral clustering, adhesive strengthening, and interaction with p120ctn. *J. Cell Biol.* **141,** 779–789.

19

Craniofacial Development

Michael J. Depew,* Abigail S. Tucker,† and Paul T. Sharpe**

*Nina Ireland Laboratory of Developmental Neurobiology and Department of Oral Biology, University of California, San Francisco, San Francisco, California 94143; †MRC Centre for Developmental Neurobiology, GKT School of Biomedical Sciences, Guy's Hospital, London SE1 1UL, United Kingdom; **Department of Craniofacial Development, GKT School of Dentistry, Guy's Hospital, London SE1 9RT, United Kingdom

I. Introduction

II. Primordial Cells of the Head

III. Organ Development

IV. Conclusion

V. Appendix 1: Descriptive Dental Development

VI. Appendix 2: Morphological Organization of the Murine Skull.

VII. Appendix 3: Molecular Regulators of Craniofacial Pattern and Development

References

I. Introduction

Historically, research on craniofacial development and on the evolution of cranial diversity has been concerned with (1) the nature of the fundamental organization and structure of the head and (2) the nature, origin, and differentiation of the cephalic cells forming cranial structures (see Gregory, 1904; Kingsbury, 1904; Goodrich, 1958; Jollie, 1977; Patterson, 1977; Romer, 1977; Moore, 1981; Gans and Northcutt, 1983; de Beer, 1985; Kuhn and Zeller, 1987; Alberch and Kaller, 1988; Thorogood, 1988; Northcutt, 1990; Noden, 1991; Beresford, 1993; Gans, 1993; Hanken and Hall, 1993;

Kuratani *et al.*, 1997, 1999). The classical view of the former concern suggested that the vertebrate head is metamerically segmented in a manner similar to the trunk (as exemplified in the "vertebral theory of the skull" put forth by Göethe, Okën and Owen). Evidence for organized segmentation, or the repetition of structures, was found, for example, in the presence of a number of paraxial, mesodermal somites rostral to the otic vesicle (somitomerism) and in the mesenchyme of the branchial arches (branchiomerism; Goodrich, 1958; Moore, 1981; de Beer, 1985; Northcutt, 1990). Individual nerves, muscle groups, and cartilaginous bodies in the skull were viewed with respect to their placement within a generally segmented bauplan. In the trunk, however, segmental control lies within the meristic somites; in contrast, in the head primary segmentation appears to occur in the neural tissues (neuromeres) and branchial arches and pouches. Of the latter concern, moreover, it was believed that the tissues of the head adhered to the germ theory of tissue origins (i.e., cranial bones and cartilage were strictly mesodermal derivatives).

Although further investigation into cranial cell origins revealed that much of the cephalic hard tissues is generated not by mesodermal cells but rather by vertebrate-synapomorphic cranial neural crest (CNC) cells, more recent research on vertebrate craniogenesis has centered around modern corollaries of these classic concerns. For example, as colonizers of the viscerocranium, neurocranium, and dermatocranium, the

421

degree to which cell fate and regional pattern is generated before, during, and/or after the migration of CNC cells has been actively sought (Noden, 1983b, 1991; Hall, 1987, 1999; Lumsden, 1988; Couly et al., 1993; Graham et al., 1996; Langille and Hall, 1993; Hanken and Thorogood, 1993; Thorogood, 1993; Osumi-Yamashita et al., 1994, 1997a,b; Köntges and Lumsden, 1996; Schilling, 1997; Hunt et al., 1998a; Le Douarin and Kalcheim, 1999). So too has been molecular evidence for the generation of pattern (organization) within the developing cranium—particularly within the developing neural tube (e.g., Lumsden and Keynes, 1989; Rubenstein et al., 1994; Lumsden and Krumlauf, 1996; LaBonne and Bronner-Fraser, 1998; 1999), CNC (e.g., Hunt et al., 1991a,b, 1998b; Sharpe, 1995; Qiu et al., 1995, 1997), cranial ectoderm (Couly and Le Douarin, 1990; Thesleff and Sharpe, 1997; Trumpp et al., 1999), and chordomesoderm (Beddington and Robertson, 1999; Camus et al., 2000; Tam et al., 2000).

This research has established that these tissues exhibit unique spatiotemporal patterns of gene expression correlating with cell and tissue fate and identity, as well as the coordination of tissue integration (e.g., the precise correspondence of bone with nerve and muscle). These genes encompass a number of putative regulatory genes, including a number encoding for secreted growth factors (and their receptors) or transcription factors, expressed in the developing neurectoderm (e.g., members of the *Hox, Otx, Emx, Pax, Nkx, Wnt, Bmp, Fgf, BF,* and *Dlx* gene families), CNC (e.g., *Dlx* (*1, 2, 3, 5, 6,* and *7*), *Msx* (*1, 2,* and *3*), *Prx* (*MHox, S8*), *Hox, Pbx, Gsc, Endothelin-1, Otx2,* and others), and the ectodermal epithelium (e.g., *Bmp, Fgf, Shh, Pitx, Tgf-β, Egf, Pdgf,* and *Endothelin-1*). The products of such genes are thought to regulate and control the regional specification of craniofacial structures through epitheliomesenchymal crosstalk: Some instruct identity, behavior, proliferative state, and/or death, while others instruct competence to receive and act on instruction (Hanken and Thorogood, 1993; Langille and Hall, 1993; Thorogood, 1993; Webb and Noden, 1993; Thesleff and Sharpe, 1997; Francis-West et al., 1998). Because differential gene expression is thought to be a determinant of cell fate and identity, the pattern of expression in the developing head is therefore indicative of the organization and subsequent development of the structures of the skull.

II. Primordial Cells of the Head

The origins and development of cells that form the tissues of the vertebrate head have attracted the increasing attention of developmental biologists during the last 20 years. The craniogenic processes involved must accomplish the generation of cell and tissue types, in the appropriate location and

number, and then integrate them. The correct bone, tooth, muscle, nerve, and vessel must be generated with fidelity. Craniofacial structures are derived from ectodermal (neural, placodal, oral, epidermal and CNC), mesodermal (paraxial and somitic), and endodermal (pharyngeal) sources. Each of these tissue types will be discussed with respect to roles they play in cranial development. These tissues types act to generate a number of recognizable early embryonic structures, including the branchial arches (BA) and the primary sensory and cranial neurogenic placodes.

The BA are iterative, metameric structures composed of epithelially covered mesenchymal bars of paraxial mesoderm and CNC that arch around the ventrolateral aspects of the primordial oral cavity (stomodeum) and pharyngeal foregut. Each BA is delimited from another externally by ectodermally lined pharyngeal (visceral) grooves, or clefts, and internally by endodermally lined pharyngeal pouches. These are named in relation to their position relative to the BA they separate; hence, the first pharyngeal grooves and pouches separate BA1 and BA2, the second separate BA2 from BA3, and so on. Thus, the demarcation of the BA occurs at sites of epithelial–epithelial contact.

The BA temporally develop in a craniocaudal sequence. The first, and most prominent, to appear (~E8.0) is BA1. This initially appears as a straight bar of tissue caudolateral to the stomodeum. This BA is unique in that its proximal end forms a growth center, the maxillary process (prominence, or branch), which begins to grow toward the optic and olfactory regions. This results in a bent, C-shaped arch. The distal end of this arch, the mandibular process, grows toward the ventral midline. In all, there are six pairs of BA, though BA5 is only rudimentary. It is difficult to externally discriminate individually BA4–6 in most mammals, including mice.

Experimental evidence, principally in chick and amphibian embryos, has suggested that an early ectodermal field is established from which the primary sensory placodes—olfactory, optic and otic—will later develop (Jacobson, 1963; Couly and Le Douarin, 1985, 1987; Torres and Giráldez, 1998). An initial stage of placodal competence exists in which ectoderm, if placed in the appropriate environment, can support any of these three sense organs. This stage is followed by a process of field specification in which the identity of each sense organ is established and committed. Regional and cellfate specification then follow suit. This ectodermal field eventually also yields the ganglionic placodes. Around each of the primary sensory placodes, cartilaginous capsules will form. Each capsule makes a significant contribution to the development of the skull. For mice this is particularly true of the nasal capsules, which develop within the frontonasal processes that form around the olfactory placode. The development of the eye and ear is described in Chapters 21 and 22, respectively while olfactory development is discussed briefly below.

A. Cranial Neural Crest

CNC cells originate at the ectodermal-neurectodermal junction of the developing brain from the diencephalon of the forebrain (FB) to the caudal boundary of the hindbrain (HB) and spinal neural tube (reviewed by Baker and Bronner-Fraser, 1997a,b; Le Douarin and Kalcheim, 1999; Hall, 1999). The experimental confrontation of different tissues in *in vivo, in vitro,* and microculture assays indicates that neural crest (NC) induction involves the apposition surface ectoderm (SE) and dorsal neural plate (NP) in a time-dependent manner (Rollhäuser-ter-Horst, 1979; Moury and Jacobson, 1989; Dickinson *et al.,* 1995; Liem *et al.,* 1995; Selleck and Bronner-Fraser, 1995; Mancilla and Mayor, 1996; La-Bonne and Bronner-Fraser, 1998; Ruffins and Bronner-Fraser, 2000). Whether NC induction is due to a specific *inductor/evocator,* or rather is a reflection of differential responsiveness to neural induction, is an issue not fully resolved (Raven and Kloos, 1945; Baker and Bronner-Fraser, 1997a; LaBonne and Bronner-Fraser, 1998; Hall, 1999). Moreover, it seems that both the NP and SE are competent to produce NC (Moury and Jacobson, 1990; Selleck and Bronner-Fraser, 1995), indicating a reciprocity of signal transduction between the SE and NP (Le Douarin and Kalcheim, 1999).

Lineage and cell fate analyses have demonstrated a great array of cellular fates arising from CNC cells, including neurons, neuroglia, smooth muscle cells, calcitonin-producing C cells, melanocytes, adipocytes, tendon connective tissue cells, general mesenchymal cells, fibroblasts, cementoblasts, odontoblasts, chondroblasts, chondrocytes, secondary chondroblasts (in secondary cartilage), osteoblasts, and osteocytes (Thorogood, 1993; Hall, 1999; Le Douarin and Kalcheim, 1999). It is the unique ability of CNC cells to develop into hard skeletal tissue that differentiates the CNC from the truncal neural crest (TNC) (although dental tissues have been experimentally induced in TNC; see Lumsden, 1988); elsewhere in the body it is the mesoderm that yields cartilage and bone. That the NC can form mesenchymal derivatives was first noted by Kastschenko (1888) in selacians, and then by Goronowitsch (1892,1893) in teleosts and birds. Platt (1893) found that these mesenchymal cells formed the cartilages of the BA and created the term *mesectoderm* for the mesenchyme of ectodermal origin. Today, such NC-derived mesenchymal cells are known as *ectomesenchyme.*

Two main controversies have arisen surrounding the role of CNC in head development: the first regards the relative contributions of ectomesenchymal and mesodermal mesenchyme to the skull, in particular to the calvarial bones; the second regards the extent to which the ectomesenchyme forming cranial structures is prepatterned. These issues are currently being resolved with the aid of molecular biology and improved lineage tracing techniques.

1. CNC Migration

After induction, NC cells undergo extensive epithelial-to-mesenchymal transitions characterized by delamination and migration of cells. Members of the *Slug/Snail* zinc finger transcription factor gene family are some of the earliest genes expressed in NC cells and have been implicated in NC emigration: Antisense oligonucleotides to *Slug* disrupt the emigration of NC cells in chick and frog embryos (Nieto *et al.,* 1994; Nagai *et al.,* 1997; Carl *et al.,* 1999). The exact nature of *Slug*'s role in murine NC migration is unclear, however, because loss of functional alleles in mice has no effect on NC cells (Nieto *et al.,* 1994; Jiang *et al.,* 1998; Carl *et al.,* 1999). The early patterns of expression of the related gene, *Snail,* are swapped with *Slug* in mice relative to chicks, such that *Snail* is expressed in both premigratory and migratory NC, while *Slug* only in migratory NC (Sefton *et al.,* 1998). Thus *Snail* might be compensating for the loss of *Slug* expression in *Slug*$^{-/-}$ embryos. *Snail* may act through the repression of *E-cadherin* (Cano *et al.,* 2000), and upstream of *Slug/Snail* may lie *Neolin-1* and *-2,* encoding for secreted proteins whose overexpression induces and prolongs extensive NC emigration (Barembaum *et al.,* 2000). Overexpression of *Neolin-1* upregulates *Slug* expression without modulating the expression of dorsal neural tube (NT) markers (e.g., *Pax3, Wnt1,* and *Bmp4*).

A significant postinductive event is the elevation and eventual closure of the neural tube (NT) (reviewed by Harris and Juriloff, 1999; Juriloff and Harris, 2000). The murine NT consists of four craniocaudal zones—roughly corresponding to FB, midbrain (MB), HB, and spinal cord—within which elevation and closure proceeds longitudinally from discrete initiation sites. These elevation zones appear to be genetically independent though similar molecular and cellular mechanisms may be involved.

Mouse and rat CNC cells start to migrate, at headfold stages long before NT closure is complete. This is in contrast to the chick or rabbit where the neural folds close prior to NC emigration. The NC cells follow distinct, idiosyncratic pathways to their target destinations according to their axial origins. The most rostral CNC cells of the MB and FB migrate around the prosencephalon to form the frontonasal and associated regions; the more caudal MB crest and those from the HB migrate ventrolaterally to populate the BA (Johnston, 1966; Couly *et al.,* 1993; Selleck *et al.,* 1993; Köntges and Lumsden, 1996), filling BA1 by E9 (Lumsden, 1984). As a mobile cellular population, control of the NC migratory route and endpoint is critical, and both the migratory environment and NC perception of it are potential regulatory points in craniofacial development (Erickson and Perris, 1993; Langille and Hall, 1993; Robinson *et al.,* 1997). This is exemplified by *rSey* mutant embryos, which possess mutant alleles of *Pax6* (Matsuo *et al.,* 1993). While *Pax6* is not expressed in the migratory CNC, it is expressed in the optic and olfactory placodes. DiI labeling has shown that the CNC emigrates from the MB aggregates adjacent to the optic placode and does not enter the frontonasal processes. Wild-type MB crest cells injected into mutants have the same abnormal migration, indicating that it is the subectodermal migratory pathway rather than the crest itself that is responsible for the phenotype (Halder *et al.*, 1995). Defects in pathfinding are also seen in CNC cells lacking the neuregulin receptor *Erbb4* (Golding *et al.*, 2000). This defect is noncellautonomous, and wild-type and mutant crest cells both migrate in a pattern consistent with the host environment. ErbB4 signaling in the HB, therefore, provides patterning information that is essential for the proper migration of NC cells.

TNC migration is highly influenced by the paraxial somitic mesoderm (Bronner-Fraser, 1986). In the head, FB, MB and rostral HB crest migration is confronted not by somites but by a loose mesodermal mesenchyme and the route is essentially subectodermal (Meier and Tam, 1982; Nichols, 1986; Langille and Hall, 1993; Le Douarin and Kalcheim, 1999). Prior to migration, the space eventually occupied by the CNC is filled with an extracellular matrix (ECM), which is rich in hyaluronic acid secreted by the epithelium (Solursh *et al.*, 1979; Nichols, 1986). The vagal, postotic CNC migratory streams are distinct as they are confronted by the first four somites. Here, three spatiotemporal streams of vagal crest are evident: a lateral stream into BA4/5, a lateral stream along the aortic arches eventually yielding mesenchyme of the heart outflow tracts, and a ventromedial stream into the pharynx and foregut yielding pharyngeal mesenchyme and the enteric nervous system (Kuratani, 1997; Le Douarin and Kalcheim, 1999).

ECM mixture and subconcentrations appear to be important factors in migration, and a basic concern is what cell–ECM and cell–cell interactions occur during this migration and their relative impacts (Erickson and Perris, 1993; Hall, 1999; Le Douarin and Kalcheim, 1999). Fibronectin, hyaluronic acid, collagens I to IX, XI, XII, and XIV, laminin-1, vitronectin, tenascin, cytotactin, biglycan, aggrecan, versican, perlecan, decorin, fibromodulin, fibulin, fibrillin, nidogen, and link protein have all been implicated in regulation of NC migration pathways (reviewed by Perris, 1997; Le Douarin and Kalcheim, 1999). These ECM components may create both permissive and restrictive, passive and instructive, attractive and repulsive, short- and long-range pathways (Robinson *et al.*, 1997). Cooperative activities of integrin receptors may be the key mediators of migration on the various ECM substrata (Hynes, 1996; Testaz *et al.*, 1999). NC cells contribute, moreover, to their own microenvironment as they migrate, as, for example, by their expression of MMP-2 and ADAM (adhesion/de-adhesion) ECM modulating proteins (Alfandari *et al.*, 1997; Hall, 1999). Moreover, polarity (i.e.,

unidirectionality) in migration exists because cells generally do not reverse course for long (but see Kulesa and Fraser, 2000). Overall, greater understanding of the precise nature and significance of ECM composition, cell–ECM and cell–cell interactions in CNC migratory control is essential.

A further important issue is that of the cessation of CNC migration (Hall, 1999). As important as the inducement to emigrate may be the need to stop doing so at the appropriate location. Do cells simply run into each other and cease migrating based on cellular concentration? Do they stop because of a physical barrier? Is cessation of migration an active process informed by the ECM or other cells? Investigating the factors controlling the local accumulation of subpopulations of CNC, Thorogood (1988, 1993) showed that CNC that will form the chondrocranium (the embryonic cartilaginous skeleton) accumulate in areas rich in collagen II. This led to his "flypaper" model, which suggested that the neurepithelium deposits the collagen, which acts essentially as a flypaper trap for the migratory CNC. However, loss of collagen II in mice does not appear to affect the initial patterning of the chondrocranium (Brown *et al.*, 1981; Rintala *et al.*, 1997; Pace *et al.*, 1997). Conceptually, such a model may still hold if the inductor of the localized collagen II produces other, localized factors capable of such a trap. Moreover, as suggested by Hall (1999), a two-step program may be in effect: One step traps cells and exposes them to a mitogen capable of generating a cell population of appropriate size and another step follows whereby a differentiative signal is expressed.

2. CNC Fate Map

The contributions of CNC cells to the skeleton and teeth of the frontonasal, trabecular/prechordal, optic/orbital, otic, BA, and calvarial regions have been determined through cell marking, xenoplastic transplantation and extirpation studies done on avian, amphibian, and rodent embryos (reviewed by Noden, 1978; Hall, 1999; Le Douarin and Kalcheim, 1999). Essentially corroborating avian and amphibian studies, studies on mouse and rat embryos in culture have demonstrated that the CNC, which contributes to frontonasal and trabecular structures, comes from mixed FB (diencephalic) and MB crest (Bartelmez, 1962; Noden, 1978, Nichols, 1981, 1986; Tan and Morriss-Kay, 1986; Serbedzija *et al.*, 1992; Matsuo *et al.*, 1993; Hall, 1983; Osumi-Yamashita *et al.*, 1994, 1997a,b; Trainor and Tam, 1995; Köntges and Lumsden, 1996). The diencephalic crest migrates ventrally as a sheet extended from the dorsal NT to the level of the eye and fills in both periocular and medial and lateral frontonasal processes (FNP) between the surface ectoderm and the telencephalon (Osumi-Yamashita *et al.*, 1994, 1997a,b). MB crest migrates ventrolaterally to the FNP, periocular, maxillary, and mandibular branches of BA1 and trigeminal ganglion (Fig. 1). That these cell populations from distinct axial levels

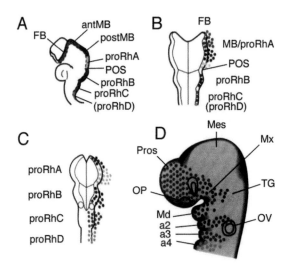

Figure 1 Migration patterns of cranial crest cells in mouse and rat embryos. (A, B) Lateral and dorsal views at 5–6 ss. (C) Dorsal view at 8 ss. (D) Lateral view at the pharyngula stage. (A) At the time of mammalian cranial crest cell emigration, four morphological units are present in the rostral neural plate, from anterior to posterior: forebrain (FB); midbrain (anterior MB and posterior MB) + presumptive prorhombomere A, prorhombomere B, and presumptive prorhombomere D. The preotic sulcus (POS) is an obvious landmark in the hindbrain. (B) Mammalian cranial crest cells emigrate from the neuroepiethelium before its closure. (C) At 8 ss crest free zones divide the exiting crest streams up into three streams in the hindbrain. (D) Streams of crest from the hindbrain and midbrain reach the branchial arches. Mx, maxilla; Man, mandible; a2,3,4, branchial arches 2, 3, and 4; OP, olfactory placode; OV, otic vesicle; TG, trigeminal; Pros, prosencephalon; Mes, mesencephalon. (Reproduced with permission from Osumi-Yamashita *et al.*, 1997b.)

are mixed rostrally may be a significant reflection of their organization and patterning—distinct from the HB crest which appears not to mix (see below).

The HB is clearly segmented into 7–8 units, or rhombomeres (Vaage and Weiss, 1969; Lumsden and Keynes, 1989), and migration of rhombencephalic crest in mice also appears to be subectodermal (Trainor *et al.*, 1994; Trainor and Tam, 1995). In a manner coincident with their craniocaudal origins, three basic streams of cells are seen: one entering BA1, one BA2, and a third entering BA3 (Hunt *et al.*, 1991b; Lumsden *et al.*, 1991; Serbedzija *et al.*, 1992; Trainor and Tam, 1995). Cells originating from rhombomere 1 (R1) and R2 contribute (with MB crest) to BA1; BA2 is colonized by R4; and R6/7/8 contribute the crest to the third and more caudal branchial arches. BA1 crest leaves an open NT at the 4- to 5-somite stage, while the BA2 crest starts emigrating between the 5+- and 6-somite stage and after the neural folds have begun to rise (Verwoerd and van Oostrum, 1979; Nichols, 1981). Contrary to BA1, an overt subectodermal space is not seen in BA2. Although R1 and R2 cells migrate into BA1 and to the trigeminal ganglion, they do not colonize the frontonasal region (Osumi-Yamashita *et al.*, 1994).

In all, BA1 is filled in a rostrocaudal, distoproximal (ventrodorsal) manner; that is, mesencephalic crest enters first followed by rhombencephalic CNC.

In *Xenopus*, CNC streams are contiguous and become separated only at later migratory stages (Sadaghion and Thiebaud, 1987). Although small numbers of R3 and R5 crest appear to join up with the even-numbered rhombomeric streams (Sechrist *et al.*, 1993) and make a limited contribution to the arches (Köntges and Lumsden, 1996), in mice three discrete streams are patent and separated from each other by crest-free regions adjacent to R3 and R5. A number of mechanisms have been put forth to explain both the relative lack of cells migrating next to R3 and R5 as well as the general rostrocaudal segregation of rhombencephalic crest. Graham *et al.* (1993, 1996) presented evidence that the even-numbered rhombomeres in chicks regulate apoptotic cell death in R3 and R5 by BMP-induced expression of *Msx2*; segregation of CNC would therefore be an inherited character of the mechanisms controlling axial specification of the neural plate.

Chick-quail fate map studies have shown that the rhombencephalic NC segragation is preserved throughout craniofacial development (Köntges and Lumsden, 1996). Cryptic intraskeletal boundaries that do not correspond to any anatomical landmark are found between the R1/2, R4 and R6/7 crest. Selective death prior to emigration does not explain, however, such a lack of subsequent mixing of NC during migration, leading to the suggestion that CNC from different axial levels has intrinsic sorting capacities. Currently, relatively few molecules have been identified as mitigating CNC mixing during migration (Golding *et al.*, 2000; Wilkinson, 2000). Recent work has identified members of the large *Eph* membrane bound receptor tyrosine kinase gene family and their Ephrin ligands as regulators of NC migration and segregation (reviewed by Wilkinson, 2000). Moreover, *Collapsin-1/Semaphorin-III* and its receptor gene, *Neuopillin-1* (encoding for additional proteins implicated in cellular guidance), are expressed in NC from R1, R3, and R5 and under the ectoderm in R1/2, R4, and R6, respectively (Eickholt *et al.*, 1999). Notably, these proteins do not appear to be in place during mesencephalic crest migration.

In the BA, subsets of CNC separate from the main streams and infiltrate the mesodermal cores (Noden, 1983b; Trainor *et al.*, 1994; Trainor and Tam, 1995). Regulation of these subsets of cells is uncharacterized. The functional nature of these cells may be clearer: in the chick, a constrained pattern of cranial skeletomuscular connectivity has been shown such that CNC cells not only form the cartilage and bone but also the attachment sites and connective tissue fasciae of the mesodermal-derived muscle. Thus, muscle connective tissue is exclusively anchored to skeletal tissue derived from the same initial axial (rostrocaudal) crest population (Köntges and Lumsden, 1996). These axial constraints would allow changes in the shape of the visceral skeleton without rendering them nonfunctional because skeletal elements would be linked to the neurocranium via muscle connective tissue of the same axial origin. This might explain how the BA2-derived retroarticular process and attached *M. depressor mandibulae* were lost, and how the *M. sylohyoideus* moved its attachment site from the lower jaw to the styloid process during mammalian evolution (Allin, 1975).

Not all of the skeletal elements of the head are derived from the NC; a proportion originates from the mesodermal lineage. Grafting experiments in chicks demonstrate that the skull receives contributions from three sources of cells. Though the frontal and parietal bones were originally considered to be of mesodermal origin (Le Lievre, 1978), the dermis and rostral membrane bones of the skull vault (including the frontal and parietal bones and their sutures), as well as the BA-derived viscerocranium, were shown to be CNC derived (Couly *et al.,* 1993). The chondrocranium contains both ectomesenchymal and mesodermal cells. The paraxial mesoderm of the metencephalic level forms the supraoccipital and portions of the otic capsule, while the first 5 somites contribute to the exoccipital, basioccipital, and other portions of the otic capsule. The boundary between the mesodermal and ectomesenchymal parts of the chondrocranial base has been found to correspond to the cranialmost level reached by the notochord at the hypophysis (which is housed in the sella turcica of man or chicks, though mice, with flattened cranial bases, do not possess true sellae; Augier, 1931; Kuratani, 1989). Thus, in chicks, the nasal capsules are CNC derived, the orbital cartilages are of mixed cephalic paraxial mesoderm and CNC origin, and the otic capsule of mixed CNC, cephalic, and somitic mesoderm origin. Care should be taken, however, not to overextrapolate the relative contributions of these cell types in chicks to those in mice, in particular with regard to the orbital cartilages and posterior chondrocranium.

Although the viscerocranium (the BA-derived chondrocranium) is thought to be CNC derived, all of its chondrocytes may not be. In the mouse this has been suggested by the use of a dual *Cre-loxP*/ROSA26 (R26R) conditional reporter transgene system generating transgenic mice in which CNC cells are labeled with *lacZ* (Chai *et al.,* 2000). Transgenic mice with a *Wnt1*-promoter driving *Cre-recombinase* were bred with R26R reporter mice to demonstrate the contribution of the CNC to BA1 structures, including Meckel's cartilage and the dentition. Notably, not all of Meckel's cartilage was made of labeled cells. Chai *et al.* hypothesized that the *lacZ*-negative cells were derived from the ventral NT. Such ventral NT cells (VENT) have been observed in the chick to migrate late in development, long after the cessation of CNC emigration, and to take part in cartilage development (Sohal *et al.,* 1999). The role these cells play in skeletal development, however, is so far unknown.

3. Pattern Formation and CNC

Three positions have generally been taken on when and where CNC-derived skeletal form is specified: (1) CNC cells are prepatterned at their source according to their axial origins; (2) CNC cells acquire their developmental instructions during migration; or (3) CNC cells remain a naive population until they reach their final destinations whereupon they receive local instructions from other cells or the ECM. In the former, prepattern position specification is intrinsic to the structure-forming cells, while in the latter two epigenetic situations arise via cell–cell and/or cell–ECM interactions after the onset of crest emigration. Hence, epigenetic specification is extrinsic, and imposed on the structure-forming cells. Experimental evidence for all three of these possibilities has been published, and they are not mutually exclusive. Thus, development of cranial hard tissues may involve any one or a combination of these three mechanisms.

Several pieces of experimental evidence have been proposed to support the idea of prepatterning of CNC cells. One consists of the now classic chick-quail grafting experiments of Drew Noden. In these experiments, when premigratory anterior NT was grafted to more posterior positions, CNC cells produced skeletal elements associated with anterior fates; that is, heterotopic chick-quail grafts of BA1 crest to BA2 or BA3 locations yielded donor site (BA1) specific structures (Noden, 1983b). The crest appeared not only to determine the shape of the elements it formed, but also to control the patterning of the mesodermal cells that migrated with the NC into the arch as the muscle pattern was of a more anterior character as well (Noden, 1986). This work corroborated earlier heterospecific grafts between urodele and anuran amphibians, which resulted in donor-specific morphologies (reviewed by Hall, 1999). Other results from Noden's experiments, however, suggest a degree of plasticity in the crest. Grafts of rostral mesencephalon gave the same phenotype as grafts of caudal mesencephalon; that is, they formed BA1 structures associated with R1/2 crest, rather than producing duplicated nasal capsular and maxillary structures as might have been predicted given the fate map. Different manipulations (in particular those addressing the regulative capacity of CNC) involving rotation of the NT, rerouting of crest by adjacent deletions, or heterotopic transplantation have resulted in variable behavior in the relocated crest (Couly *et al.,* 1998, Saldivar *et al.,* 1996, 1997; Hunt *et al.,* 1998b). Thus, the degree of prepattern depends on the experiment performed, with the size of the cell community being challenged playing an important part (Trainor and Krumlauf, 2000). These experimental manipulations of CNC fate, moreover, are further consistent with the notion that the CNC rostral to R3 acts as a distinct unit, under the influence of different head organizers, from the caudal, *Hox*-positive CNC (see below).

A second source of experimental evidence for prepatterning involves an interpretation of *Hox* expression patterns in the HB, wherein members of the *Hox* gene family are expressed following the boundaries of distinct rhombomeres. Alteration of *Hox* expression patterns within the HB results in a change of the neuronal identity characteristic of each rhombomere, suggesting that regional identity and pattern are regulated by *Hox* genes. This control has been extrapolated to regulation of the associated CNC. The fact that CNC cells migrating from different rhombomeres generally retain their expression of *Hox* genes, and that these CNC cells then follow different migration routes (subsequently giving each BA an unique combination, or code, of *Hox* gene expression), has been suggested to provide a mechanism for prepatterning (Hunt *et al.,* 1991a,b,c). The ability of NC cells to retain their host *Hox* code of origin when grafted to new anteroposterior positions supports this idea (Guthrie *et al.,* 1992; Prince and Lumsden, 1994; Couly *et al.,* 1996, 1998), although a few studies have shown plasticity of the NC *Hox* code upon grafting (Salvidar *et al.,* 1996; Trainor and Krumlauf, 2000). Further suggestive evidence for a prepattern mechanism has come from knock-out studies, such as for *Hoxa2*. *Hoxa2* is expressed in the HB with its anterior limit at the R1-R2 boundary. The CNC emigrating from R2, however, turns off *Hoxa2,* leading to a first arch completely devoid of *Hox* gene expression (Prince and Lumsden, 1994), while the crest migrating into the second and more caudal arches strongly expresses *Hoxa2*. *Hoxa2* mutants have a mirror image homeotic transformation of BA2 skeletal elements into duplications of proximal BA1 elements (Rijli *et al.,* 1993; Gendron-Maguire *et al.,* 1993). *Hoxa2* would therefore appear to be a selector gene for second arch fate. Knockouts of other *Hox* genes, such as for *Hoxa1;Hoxb1* and *Hoxa3* have produced a loss of cartilage elements in the second and third arch, respectively, but not a conversion to a more anterior fate (Chisaka and Capecchi, 1991; Gavalas *et al.,* 1998). Thus, while *Hoxa2* may act as a selector gene, specifying arch fate, other *Hox* genes may act to generate or maintain a specific population of cells. *Hox* genes are not expressed in CNC cells rostral to R3, exerting their influence on craniofacial regions only caudal to BA1. Hence, most of the mouse skull develops outside of the influence of *Hox*-positive CNC, and patterning mechanisms of these rostral tissues are likely to be distinct from those influencing the caudal HB. A number of other homeobox-containing genes that regulate craniofacial development (e.g., *Dlx, Otx,* and *Gsc*) are, however, expressed in these more rostral regions.

Arguments in support for epigenetic patterning of the CNC have also been published (reviewed by Hall, 1999). One source of support is the finding that CNC have a self-regulative ability: Ablation of the neural fold leads to a filling in by CNC from other axial levels and yet structures form normally; and in grafts containing excess numbers of cells,

the cells will decrease their rate of proliferation to make normal structures. That epigenetic factors affect craniofacial development is clearly exemplified by the necessity of biomechanical forces generated by cranial muscular contractions for the development of the secondary cartilages—which are necessary for proper mandibular growth—of the condyle and angle of the dentary (reviewed by Herring, 1993; Fang and Hall, 1997).

4. Sources of Patterning Information

Arguments for prepattern or epigenesis in CNC elaboration of form both necessitate proximate sources of information. This might come, for example, in concert with, or secondary to, specification of the neurectoderm, surface ectoderm, endoderm, or mesoderm. Moreover, CNC populations need not be specified by the same mechanisms. Although regulation of the *Hox* genes themselves may be an open question, after intensive investigation they have been implicated in the molecular control of HB organization and CNC caudal to R2 (Rijli *et al.,* 1998; Gavalas and Krumlauf, 2000). Less intensive attention perhaps has been paid to the regulation of the CNC, which populates BA1, the frontonasal/prechordal regions, and the calvarial dermatocranium.

It has been recognized for some time that the vertebrate chondrocranium can be segregated into chordal and prechordal regions: Chordal regions derive from paraxial mesoderm and somites apparently influenced and patterned by signals from the notochord and extend to the notochordal tip at the hypophysis (e.g., the perioccipital, parachordal regions), while prechordal regions derive from CNC that has migrated to lie beneath the developing forebrain. The integration of these regions at the cephalic flexure normally proceeds fluidly. Moreover, it has been proposed that vertebrates possess a "new head" in large part due to the presence of this CNC and the additional presence of sensory placodes in the vertebrate head added onto the segmented trunk (Gans and Northcutt, 1983). Experimental embryological and genetic studies suggest that separate head and trunk patterning organizers likely exist (or at the least a system to protect the head from the trunk), and it appears that signals from the anterior visceral endoderm (AVE), anterior neural ridge (ANR), and the mesendoderm (prechordal plate, PcP) are needed for the complete refinement of head structures (Spemann, 1938; Shimamura and Rubenstein, 1997; Beddington and Robertson, 1999; Knoetgen *et al.,* 1999a,b; Bachiller *et al.,* 2000; Camus *et al.,* 2000; Foley *et al.,* 2000; Tam *et al.,* 2000). Thus, from the outset, anterior head organization in the mouse is distinct from that of the somitic posterior head and trunk. The salient question surrounding craniofacial development is how does this separate head organizer inform craniofacial organization? How does the AVE or the mesendoderm inform craniofacial regional specification? Is it primary or secondary to the development of the neuraxis?

What is the spatiotemporal flow of information? It is clear, however, that factors which do regulate neuraxial formation also regulate prechordal (and BA) skeletal development. A number of genes encoding for secreted molecules or transcription factors are expressed in these early organizer tissues, including *Shh, Nodal, Noggin, Chordin, Smad2, Otx2, Lim1 (Lhx1), Hnf3β (foxa2), Brachyury, Hhex, Hesx1,* and *Gsc* (Camus and Tam, 1999; Bachiller *et al.,* 2000; Schier and Shen, 2000). Mutations in these genes generally result in two (likely related) types of defects: (1) failure of overall head development and (2) failure of midline coordination rostral to the pituitary. Of note, these organizer defects generally involve structures rostral to R3.

The isthmus (midbrain-hindbrain junction) contains organizing capacity (Martinez *et al.,* 1995; Joyner, 1996). Positioning and maintenance of this isthmic organizer (IsO) involves a number of genes (most of which have been knocked out) including *En1* and *En2, Fgf8, Gbx2, Otx1,* and *Otx2, Pax 2* and *Pax5,* and *Wnt1* (McMahon and Bradley, 1990; Thomas and Capecchi, 1990; McMahon *et al.,* 1992; Millen *et al.,* 1994; Acampora *et al.,* 1995, 2000; J. Lee *et al.,* 1997; Urbánek *et al.,* 1997; A. Liu *et al.,* 1999; Martinez *et al.,* 1999; Irving and Mason, 2000). WNT1, essential for the elaboration of MB and rostralmost HB development, is secreted in the rostral isthmus, where it may function to maintain expression of *En1* and *En2, Otx1* and *Otx2;* FGF8 is secreted in the caudal isthmus, where it maintains and may induce metencephalic *Gbx2* and represses *Otx2* (perhaps through *Gbx2*) and the rostral *Hox* genes (McMahon and Bradley, 1990; Thomas and Capecchi, 1990; Ikeya *et al.,* 1997; Y. H. Liu *et al.,* 1999; Martinez *et al.,* 1999; Millet *et al.,* 1999; Irving and Mason, 2000). Disruption of these genes either shifts the position of the IsO or results in its loss (reviewed by Y. H. Liu *et al.,* 1999; Acampora *et al.,* 2000). Although a skeletal analysis of these disruptions has generally not been presented, skeletal defects are known to occur. For example, *Wnt1* is necessary for proper development of the parietal bone (Ikeya *et al.,* 1997); compound *Wnt1$^{-/-}$; 3a$^{-/-}$* mutants appear to regulate the development of dorsolateral NT derivatives, including the CNC (Ikeya *et al.,* 1997); and *Gbx2$^{-/-}$* mice lack interparietal bones and possess diminished supraoccipitals and expanded metopic sutures (Wassarman *et al.,* 1997; M. J. Depew and J. L. R. Rubenstein, unpublished observations). The exact origin and nature of these skeletal defects needs clarification: Are they directly due to a role in the skeletogenesis of these bones or indirectly due to respecification of neuraxial or CNC identity with subsequent respecification of the overlying dermatocranium (the intramembranously derived skeleton). Whether similar defects are found in the *En1$^{-/-}$;2$^{-/-}$* and *Pax2$^{-/-}$;5$^{-/-}$* mice is unclear. In all, the relationship of these two head organizing centers to cranial skeletogenesis needs to be further addressed.

B. Mesoderm

Paraxial mesodermal and CNC cells of the same axial, segmental position share common destinations (Trainor and Tam, 1995). A two-segment periodicity (similar to the origins of the branchial motor neurons and the distribution of the rhombencephalic CNC) has been observed in the cranial mesoderm by fluorescent cell labeling (Trainor *et al.,* 1994). Unlike within the BA (where the CNC essentially surrounds a mesodermal core), facial, periotic, periocular, and cervical NC and mesodermal mesenchyme mix extensively.

It has been proposed that the paraxial mesoderm is organized into seven loosely packed cell clusters called somitomeres (Meier, 1979, 1982; Meier and Tam, 1982). Thus, segregation and segmentation—characteristics of the truncal paraxial mesoderm—also exist in the head. Although it is well established that the HB is segmented into rhombomeres and to a lesser extent the FB into prosomeres (Vaage and Weiss, 1969; Lumsden and Keynes, 1989; Rubenstein *et al.,* 1994), the nature of the segmentation of the cephalic paraxial mesoderm into somitomeres is less clear. Unlike within the rhombomeres, lineage restriction and molecular heterogeneity of cells have not been demonstrated in the cephalic somitomeres. Heterotopic grafting of paraxial mesodermal cells to different regions of the cranial mesoderm in the mouse showed no restriction in cell potency, revealing considerable plasticity in the fate of the cranial mesoderm (Trainor *et al.,* 1994). Thus, the mesoderm appears not to retain its fate of origin on grafting (i.e., it is not prepatterned) as was seen by the rostral CNC, further arguing that it is indeed the NC that sets the pattern within the mesoderm (Noden, 1983a, 1986). Recent evidence, however, has revealed a potential patterning role for the cranial mesoderm. CNC cells from mouse rhombencephalic levels were shown to display a surprising degree of variability with respect to their retention of expression of *Hox* code of axial origin when transposed in small groups (Trainor and Krumlauf, 2000). However, when mesoderm cells were cotransposed, the crest cells maintained their *Hox* code of origin. Thus, the paraxial mesoderm may be able to impart as well as receive patterning information.

Cranial paraxial mesoderm forms craniofacial muscles (see Table I), but not the NC-derived connective tissue fasciae (Köntges and Lumsden, 1996). Furthermore, although a comigration of NC and paraxial mesoderm into the BA occurs, some of the muscles formed in the face are actually derived from somitic mesodermal cells having migrated into the cranial region (Mayo *et al.,* 1992). Chick-quail chimeras have revealed the complex and highly specific migration pathway taken by the precursor myoblasts (Mackenzie *et al.,* 1998). The cells migrate out of the somite, around the pharyngeal endoderm, then course rostrad in a tight band up to the developing tongue skeleton. No developmental

Table I   Generalized Mammalian Craniofacial Muscles and Their Theoretical Cellular Origins, Embryonic Migration, Function, Muscular Origin, Insertion, and Innervation

Muscle	Cell origin	Migration	Function	Muscle origin	Muscle insertion	Innervation
Auricularis (anterior, posterior, superior)	Somitomere (~4)	Into BA2	Ear movement; facial expression	Otic capsule (mastoid process), galea aponeurotica	Auricle	CN VII
Buccinator	Somitomere (~4)	Into BA2	Cheek movement; facial expression	Maxilla, mandible and pteregomandibular raphe	Fascia of the Obicularis oris	CN VII
Buccopharyngeus	Occipital somites 2–4	Into BA4	Pharyngeal constrictor (superior)	Pterygomandibular raphe, medial pterygoid plate, pterygoid hamulus	Median raphe of pharynx and pharyngeal tubercle	CN X/XI
Capitis (obliquus, rectus, semispinalis, splenius) and trapezius	Occipital/cervical somites	—	Cranial movement	Cervical vertibrae	Occipital	C1–4
Cervicoauricularis	Somitomere (~4)	—	Elevates, rotates ears; facial expression:	Median raphe of neck	Base, anterior of auricle/pinnae	CN VII
Ciliaris	Prechordal meso.	—	Relaxes suspensory ligament of the lens	Scleral spur	Ciliary process	CN III
Clavotrapezius	Cervical somites	—	Cranial movement	Occipital	Clavicle	CN XI, C3–4
Cleidomastoideus	Cervical somites	—	Cranial movement	Clavicle	Mastoid process, occipital and temp. ridge	CN XI, C2–3
Constrictor pharyngis inferior (thyro- and cricopharyngeus)	Occipital somites	—	Pharyngeal constriction (inferior)	Thyroid and cricoid cartilage	Median raphe of pharynx	CN X
Constrictor pharyngis medius (chondro- and ceratopharyngeus)	Occipital somites	—	Pharyngeal constriction (medial)	Lesser and greater horns of hyoid, stylohyoid ligament	Median raphe of pharynx	CN X/XI
Corrugator	Somitomere (~4)	Into BA2	Movement of eyebrow; facial expression	Frontal (medial supercillary arch)	Skin of eyebrow, Orbicularis oculi	CN VII
Depressor labii inferioris (quadratus labii inferioris)	Somitomere (~4)	Into BA2	Depression of lower lip; facial expression	Anterior surface of mandible	Fascia of the Obicularis oris, skin of lower lip	CN VII
Depressor nasi	Somitomere (~4)	Into BA2	Movement of nose; facial expression	Premaxillae	Ala and septum of nose	CN VII
Dilatores nasi	Somitomere (~4)	Into BA2	Movement of nose; facial expression	Premaxillae, nasal mesenchyme	Nasal cartilage	CN VII
Digastricus, anterior belly	Somitomere (~3)	Into BA1	Depression of mandible, hyoid elevation; mastication	Dentary below the mandibular symphasis	Intermediate digastric tendon and hyoid, mylohyoideus	CN V
Digastricus, posterior belly	Somitomere (~4)	Into BA2	Mastication; jaw opening	Mastoid notch of occipital	Intermediate digastric tendon and hyoid, stylohyoideus connective tissue	CN VII
Frontalis (epicranius)	Somitomere (~4)	Into BA2	Movement of scalp; facial expression	Galea aponeurotica	Fascia of orbicularis oculi and skin of scalp	CN VII
Frontoscutularis	Somitomere (~4)	—	Ear movement; facial expression	Frontal	Base of auricle/pinnae	CN VII
Genioglossus	Occipital somites (3–4)	—	Protrusion and depression of the tongue	Mental spine on mandibular symphasis	Inferior tongue	CN XII
Geniohyoideus	Occipital somites	—	Hyoid elevation and mandibular depression	Mental spines of mandible	Body of hyoid	C 1
Glossopharyngeus	Occipital somites	—	Pharyngeal constrictor (superior)	Muccous membrane of mouth	Posterior raphe of pharynx	CN X/XI
Hyoglossus	Occipital somites (3–5)	—	Depression and retraction of the tongue	Body and greater horn of hyoid	Lateral tongue	CN XII

(continues)

Table I *(continued)*

Muscle	Cell origin	Migration	Function	Muscle origin	Muscle insertion	Innervation
Interscutularis	Somitomere (~4)	—	Ear movement; facial expression	Parietal	Base of auricle/pinnae	CN VII
Laryngeal, intrinsic (human crico-, thyro arytenoid, *et al.*)	Occipital somites (2–5)	Into BA6	Movement of laryngeal apparatus	Laryngeal cartilages	Laryngeal cartilages	CN X
Levator auris longus	Somitomere (~4)	—	Ear movement; facial expression	Spines and bases of C1–4	Base of auricle/pinnae	CN VII
Levator labii superioris	Somitomere (~4)	Into BA2	Elevation of upper lip; facial expression	Inferior margin of the orbit	Skin of upper lip	CN VII
Levator nasolabialis (quadratus labii superioris)	Somitomere (~4)	—	Dilation of nostril and drawing back of labial commissure; facial expression	Base of the zygomatic arch at the infra-orbital foramen of maxillae	Lateral nostril and upper lip, obicularis oris	CN VII
Levator palpebrae superioris	Prechordal meso.	—	Elevation of upper eyelid	Apex of orbit	Orbicularis oculi, conjunctiva, superior tarsal plate	CN III
Levator veli palatini	Occipital somites 2–4	Into BA4	Elevation of soft palate	Petrous process of temporal, auditory tube cartilage	Palatine aponeurosis and contralateral muscle	CN X
Linguae (intrinsic muscles of the tongue: longitudinalis superior and inferior, transversus and verticalis linguae)	Occipital somites (1–3)	Tongue	Movement of the tongue	Regional connective tissues within tongue (poorly defined)	Regional connective tissues, within tongue (poorly defined)	CN XII
Masseter, ant-deep	Somitomere (~3)	Into BA1	Mastication; jaw closure	Maxillae, orbit	Anterior, massateric ridge of dentary (after passing through the infra-orbital foramen)	CN V
Masseter, ant-sup.	Somitomere (~3)	Into BA1	Mastication; jaw closure	Lateral maxilla	Angular process of the dentary	CN V
Masseter, pos-deep	Somitomere (~3)	Into BA1	Mastication; jaw closure	Zygomatic arch (jugal)	Lateral ramus of the dentary	CN V
Masseter, pos-sup.	Somitomere (~3)	Into BA1	Mastication; jaw closure	Zygomatic arch (entire length)	Lateral, massateric ridge of dentary	CN V
Mylohyoideus	Somitomere 3	Into BA1	Hyoid elevation, tongue support, depression of mandible	Mylohyoid line of mandible at the level of the third molar	Median raphe of contralateral muscle, body and greater horn of hyoid	CN V
Mylopharyngeus	Occipital somites	—	Pharyngeal constriction (superior)	Mylo-hyoid line of dentary (internal)	Posterior raphe of pharynx	CN X
Nasalis (compressor naris)	Somitomere (~4)	Into BA2	Nasal movement; facial expression	Premaxilla	Nasal connective tissue	CN VII
Orbicularis oculi	Somitomere (~4)	Into BA2	Closure of eye	Medial wall of orbit, medial palpebral ligament	Skin of lateral cheek, and lateral palpebral raphe (poorly defined)	CN VII
Orbicularis oris	Somitomere (~4)	Into BA2	Oral movement; facial expression	Fascia of orbicularis oculi, other regional fascia (poorly defined)	Nasal septum, premaxilla, symphasis of dentary	CN VII
Obliquus inferior	Prechordal meso	—	Movement of the eye	Floor of orbit (lateral to nasolacrimal canal)	Lateral surface of the sclera	CN III
Obliquus superior	Somitomere (~2)	—	Movement of the eye	Body of the sphenoid	Superior surface of the sclera	CN IV
Occipitalis (epicranius)	Somitomere (~4)	Into BA2	Scalp movement; facial expression	Superior nuchal line of occipital	Galea aponeurotica	CN VII
Omohyoideus	Occipital somites	—	Depression/stabilization of hyoid	Superior border of scapula and suprascapular ligament	Body hyoid	C1–3 (ansa hypoglossi)
Palatoglossus	Occipital somites	—	Movement of the tongue	Palatal Aponeurosis	Lateral part of tongue	CN X
Pharyngeopalatinus (palatopharyngeus)	Somitomere 7	Into BA3	Pharyngeal and laryngeal movement	Hard palatal shelf, auditory tube	Thyroid cartilage, aponeurosis of pharynx	CN X
Platysma (cranial, superficial, cervical)	Somitomere (~4)	Into BA2	Mandibular depression; facial expression	Fascia over deltoid and pectoral muscles	Inferior border of mandible, fascia and skin of lower face	CN VII
Pterygoideus, medial	Somitomere (~3)	Into BA1	Mastication: Elevation and protraction of mandible	Lateral pterygoid plate, palatine bone and maxilla	Ramus and angle of mandible	CN V

Pterygoideus, lateral	Somitomere (~3)	Into BA1	Lateral pterygoid plate, greater wing of sphenoid	Mastication: jaw opening	Medial condylar process of the mandible	CN V
Pterygopharyngeus	Occipital somites	—	Posterior raphe of pharynx	Pharyngeal constriction (superior)	Pterygoid	CN X
Pyramidalis nasi	Somitomere (~4)	—	Nasal connective tissue	Nasal movement; facial expression	Epicranius connective tissue	CN VII
Rectus, lateral	Somitomeres 3	—	Tendonous ring around optic foramen	Movement of eyeball	Lateral surface of sclera	CN VI
Rectus: superior, inferior, medial	Prechordal meso.		Tendonous ring around optic foramen	Movement of eyeball	Sclera of eyeball	CN III
Risorius	Somitomere (~4)	Into BA2	Masseteric fascia	Facial expression	Fascia of the Obicularis oris	CN VII
Stapedius	Somitomere (~4)	Into BA2	Canal of otic capsule	Stapeal movement	Neck of stapes	CN VII
Sternofacialis	Somitomere (~4)	Into BA2	Auricular cartilage	Facial expression	Sternal manubrium	CN VII
Sternohyoideus	Somitomere 5/6?	Into BA3/4?	Sternal manubrium/medial clavicle	Depression of hyoid, inspiration	Body and greater horn of hyoid	C1–3 (ansa hypoglossi)
Sternomastoideus	Occipital somites	—	Sternal manubrium	Cranial movement	Mastoid and paroccipital processes	CN XI, C2–3
Sternothyroideus	Somitomere 5/6?	—	Sternal manubrium and costal cartilage	Depression of thyroid cartilage	Thyroid cartilage	C1–3 (ansa hypoglossi)
Styloideus	Somitomere (~4)	Into BA3/4?	Postero-lateral styloid process, paroccipital process			CN VII
Stylohyoideus	Somitomere (~4)	Into BA2		Hyoid retraction and elevation, tongue support	Greater horn and body of the hyoid	CN VII
Styloglossus	Occipital somites	Tongue	Anterior side of the styloid process	Movement of the tongue	Lateral tongue	CN XII
Stylopharyngeus	Somitomere (~4/5?)	Into BA3	Medial side of the styloid process	Laryngeal elevation	Thyroid cartilage	CN IX
Temporalis	Somitomere (~3)	Into BA1	Temporal fossa and temporal fascia	Mastication	Medial and lateral ramus of the dentary, coronoid process	CN V
Tensor tympani	Somitomere (~3)	Into BA1	Cartilage of auditory tube, greater wing of the sphenoid	Movement of malleus and tympanic membrane	Manubrium of malleus	CN V
Tensor veli palatini	Somitomere (~3)	Into BA1	Medial pterygoid plate and spine of sphenoid and lateral wall of auditory tube	Tension of soft palate, opening of auditory tube	Hard palatal shelf and aponeurosis of soft palate	CN V
Thyrohyoideus	Occipital somites	—	Thyroid cartilage	Depression/stabilization of hyoid, elevation of larynx	Inferior border of hyoid body	C1
Transversus mandibularis	Somitomere (~2)	Into BA1	Median raphe of the contralateral muscle	Mastication; mandibular support, tongue elevation and support	Dentary, posterior to the mandibular symphasis of the other	CN V
Trapezius	Cervical somites	—	Occipital	Cranial movement	Cervical vertibrae	CN XI, C1–4
Triangularis	Somitomere (~4)	Into BA2	Fascia of the Obicularis oris, skin	Oral movement; facial expression	Dentary	CN VII
Uvulae	Occipital somites	—	Posterior nasal septum	Shortening of the uvalae	Mucosa of the uvula	CN XI
Zygomaticus	Somitomere (~4)	Into BA2	Zygomatic arch of Maxillae	Facial expression	Fascia of the Obicularis oris and overlying skin	CN VII

changes are noted when occipital somites are replaced with more caudal trunk somites, indicating the extrinsic nature of the migration cues (Mackenzie *et al.,* 1998). This migration route is also taken by the hypoglossal nerve and circumpharyngeal NC, though ablations of these populations have no effect on muscle cell migration.

C. Ectoderm

Cranial ectoderm is a critical component of the craniogenesis machinery, contributing both directly and indirectly to the development of craniofacial structures. It has been suggested that the ectodermal field can be segregated into *ectomeres,* though the functional significance to these divisions remains to be determined.

The cranial surface ectoderm has been shown to play a vital role in the differentiation of CNC into skeletal tissue. In amphibian and chick studies, isolated CNC does not appear to self-differentiate *in vitro* to form primary cartilage, dentine or intramembranous bone (Thorogood, 1993; Hall, 1999), and there is compelling extirpation and tissue recombination evidence that focal (localized) epitheliomesenchymal tissue interactions have a fundamental role in leading ectomesenchyme to differentiate into these hard tissues (reviewed by Hall, 1987, 1999; Lumsden, 1988; Thorogood, 1993). For example, extirpation of surface or neural ectoderm leads to an absence of associated dermal bones (Schowing, 1968a,b). Tissue recombination studies conducted by Hall and colleagues suggest that the basal lamina of a mitotically active ectoderm is necessary for CNC osteogenesis; the ectoderm need not be cephalic because limb bud, dorsal trunk, and periscleral epithelium can induce osteogenesis in mandibular ectomesenchyme (Tyler and Hall, 1977; Bradamante and Hall, 1980; Hall *et al.,* 1983; Tyler, 1983). Thorogood and colleagues have extensively studied the chondrogenic promoting capacities of the collagen II-rich ECM secreted by the neurepithelium at the condensation sites of the chondrocranium (reviewed by Thorogood, 1993). Though a capacity as a direct inductor has been ruled out, this collagen II-rich ECM may play a role as part of a three-dimensional repository for ECM-sequestered chondrogenic factors (Bissell and Barcellos-Hoff, 1987; Thorogood, 1993; Hall, 1999). The potential to form cartilage, but apparently not intramembranous bone, may be informed while the CNC is still associated with the NT or soon thereafter (Hall and Tremaine, 1979; Bee and Thorogood, 1980; Hall, 1980a,b, 1999). It has further been demonstrated that the skeletogenic factors associated with the craniofacial ectoderm can be either matrix mediated or diffusible (Bee and Thorogood, 1980; Thorogood and Smith, 1984).

These studies further address an acknowledged but perhaps underappreciated aspect of craniofacial development: the temporal delay between the advent of the chondrocranium (the initial, cartilaginous embryonic cranial skeletal

structures) and the dermatocranium (the perinatal skull that arises with the advent of the nascent dentition and the intramembranously ossified elements of the skull that develop around the chondrocranium). This is likely to be under the control of the ectoderm (Hall, 1987). It may be that the one is necessary for the development of the other, as is suggested by the work of Corsin (1966, 1975, 1977) showing the need of amphibian BA dermatocranium for the presence of the viscerocranium. The embryonic skull has an unique set of functional demands for which the cartilaginous skull is best fit to fulfill, and likewise for the perinatal dermatocranium built around the chondrocranium (Hanken and Thorogood, 1993; Presley, 1993). Clearly it is advantageous to temporally regulate the onset of each (Presley, 1993). How this is achieved is not yet clear. Thus, the skeletogenic promoting factors of the embryonic skull are not uniform with regard to mechanism, and the diversity in cellular fate is due to the influences of other cells and the extracellular environment.

Significantly, regulation of these events in BA1—where cartilage, membrane bone, and teeth all form—must be highly regulated spatially and temporally. For example, teeth develop on the rostral (oral) surface of the developing mandibular primordium, whereas bone and cartilage develop more caudally (aborally). The different positional fates of these CNC cells must be determined early in the formation of the primordium. Expression of the closely related LIM domain homeobox genes *Lhx6* and *Lhx7* is restricted to oral ectomesenchyme of the mandibular and maxillary processes and complements that of *Gsc,* which is expressed in aboral ectomesenchyme (Tucker *et al.,* 1999). The ectoderm appears to be involved in inducing both oral and aboral mesenchymal gene expression. The ectoderm expresses a wide range of signaling molecules, including FGFs, BMPs, WNTs, and HHs, and it is the restriction of *Fgf8* expression to the oral (and pericleftal) ectoderm that appears to set up the anterior-posterior (AP) axis of BA1 (Grigoriou *et al.,* 1998; Trumpp *et al.,* 1999). The restriction of *Gsc* expression to aboral mesenchyme involves repression by *Lhx6/7* expressing cells, although the mechanism that restricts *Lhx6/7* expression to oral mesenchyme is independent of *Gsc* and is more probably related to the distance from the source of FGF8. Targeted mutations in *Lhx6* or *Lhx7,* however, do not result in dental defects; defects may only be revealed when these mutations are combined (Zhao *et al.,* 1999; V. Pachnis, personal communication). Mutations in *Gsc* do, however, lead to mandibular bone defects (see below) but the teeth develop normally (Rivera-Pérez *et al.,* 1995; Yamada *et al.,* 1995). *Endothelin-1* expression in the entire mandibular epithelium appears to act as a maintenance factor for *Gsc* expression (Tucker *et al.,* 1999) and it is regulated in part by FGF8 (Trumpp *et al.,* 1999); as with *Gsc,* both *Endothelin-1* and *Endothelin receptor A* knock-outs have mandibular defects where bone is affected but dental development essentially is not (Kurihara *et al.,* 1994; Clouthier *et al.,* 1998).

More specific roles for the ectoderm in dental development are discussed later in this chapter.

The expression of signaling molecules in the BA ectoderm has been shown to be independent of the NC, such that when the NC is ablated the expression of signaling molecules still comes on in a defined spatially restricted pattern (Veitch *et al.,* 1999). Thus, it would appear that although the skeletal identity of an arch might be determined by the NC cells, the AP polarity within an arch is determined by the ectoderm. This perhaps explains why, when R1 and R2 (which both produce crest destined for the BA1) are rotated, no change in first arch AP pattern is seen (Noden, 1983b; unpublished observations). This may also explain why, when mesencephalic CNC is grafted caudally (Noden, 1983b), or in the *Hoxa2* knock-out (Rijli *et al.,* 1993; Gendron Maguire *et al.,* 1993), the duplicated first arch elements have a mirror image symmetry. In addition to expressing signaling molecules involved in local epithelial-mesenchymal interactions, BA ectoderm also expresses a range of transcription factors, including *Hox* genes. *Hoxa2,* for example is expressed in the ectoderm of BA2 as well as its CNC. Such ectodermal expression comes on after the NC has migrated and was initially thought to be induced by the crest (Hunt *et al.,* 1991a). However, the ectoderm's *Hox* code has now been shown to be independent of the NC: When non-*Hox*-expressing crest replaces *Hox*-expressing crest, the ectoderm still turns on its normal *Hox* code (Couly *et al.,* 1998). Arch identity may therefore come from a combination of patterning information from the CNC and patterning information in the tissues into which it migrates, possibly explaining why *Hox*-expressing NC appears unable to form cartilage elements when placed in a normally *Hox*-devoid setting (Couly *et al.,* 1998; Grammatopoulos *et al.,* 2000). Hence, the ectoderm may impart positional information and the CNC its interpretation, necessitating rigorous investigation of the proximate regulators of ectodermal gene expression.

D. Endoderm

Pharyngeal endoderm lines the internal surface of the BA, and the pouches between the BA yield a variety of structures. In the first pharyngeal pouch (between BA1 and BA2), the endoderm forms the pharyngeotympanic tube and tympanic cavity, which meets up with the ectodermal, first pharyngeal cleft-derived external auditory meatus. Where these two epithelial structures join, the ear drum forms. Between the second and third arch, the endoderm invaginates to form the palatine tonsil. Between the third and fourth arch the endoderm invagination forms the inferior parathyroid gland and thymus, while at the base of BA4 the superior parathyroid gland and ultimobranchial body form. The thyroid descends in front of the pharynx until it reaches its final position in front of the trachea. It remains connected to the tongue by a narrow canal, the thyroglossal duct, which even-

tually regresses. The origin of the migration route is marked by a persistent blind opening at the back of the tongue, the foramen caecum. The endoderm of the floor of the pharynx proliferates to help generate—with contributions from each of the arches—the tongue.

Other than these derivatives, the endoderm has a limited contribution itself to craniofacial tissues but has an important role as an inducer, for example, through its induction of the epibranchial placodes (see below). The epibranchial placodes contain the gustatory sensory neurons that link up with the taste buds in the tongue (Northcutt and Barlow, 1998). Therefore, in this case, by inducing the sensory neurons, the endoderm appears to be inducing its own afferent innervation. Endodermal markers, such as *Bmp7*, are also retained after CNC cell ablation, indicating that, like the ectoderm surrounding the crest, its patterning information is independent (Veitch *et al.,* 1999).

III. Organ Development

A. Mouth Development

Mouth development begins with the formation of the stomodeum (mouth pit), an ectodermal depression around which the facial primordia grow and extend to create the oral cavity. An early regional landmark is an area of cellular apposition (here an epithelial–epithelial contact) that develops just caudal to the forming cardiac tissue between the prechordal/foregut mesendoderm and the stomodeal ectoderm at the cranial end of what will be the foregut (Schwind, 1928; Waterman, 1977). The division between ectoderm and endoderm is not readily distinguishable until after this cellular apposition—the buccopharyngeal membrane—forms (~E8.25). The buccopharyngeal membrane breaks down by E8.75 through a process of cell rearrangement and death, at which point once again there is no morphological or histological distinction between ectoderm and endoderm. With the breakdown of the buccopharyngeal membrane an obvious mouth is distinguishable, consisting exclusively of the oropharynx. Rupture of the membrane is critical to organismal survival and may be a developmentally important early barrier to mesenchymal migration across the ventral midline (Waterman and Schoenwolf, 1980).

At E8.75, the development of the first structure from the stomodeal ectoderm, the pituitary, begins when a diverticulum (Rathke's pouch) grows cranially from the oropharynx toward the infundibulum growing toward it from the floor of the third ventricle of the brain. The first characterized molecular marker of the stomodeal ectoderm is the homeobox gene *Pitx2*, which is initially expressed throughout the ectoderm from E8.5 and gradually becomes refined to presumptive dental ectodermal development (Gage *et al.,* 1999; Lin *et al.,* 1999; Lu *et al.,* 1999a). Mutations in the human *PITX2* gene result in Reiger syndrome which affects several cranio-facial tissues, including teeth. Mouse knock-outs of *Pitx2* surprisingly show tooth development arrested at the early bud stage, indicating that dental ectoderm does form in the absence of *Pitx2*.

B. Secondary Palate Development

The primary palate develops from the medial frontonasal processes and is a transient structure from which the upper lip and the premaxillary palate (anterior to the secondary palate) form. The secondary (or definitive) palate, which forms the roof of the mouth, develops from the maxillary processes of BA1. Bilateral secondary palatal shelves initially arise from the maxillary processes at E12. They begin as vertical (dorsoventral) projections down the sides of the developing tongue whose growth is driven by mechanisms that probably involve controlled cellular proliferation. At a particular time in development (E13.5), the shelves rapidly elevate to a horizontal position above the tongue. The medial edges approach each other toward the midline where they contact and fuse. The epithelial edges of the apposing shelves fuse to form a seam, rapidly degenerating by processes that may involve apoptosis, epithelial-to-mesenchymal transitions, retractions of epithelial cells to the dorsal and ventral aspects of the shelves, or possibly any combination of these. Epithelial seam degeneration allows the underlying mesenchymal cells to contact along the horizontal plane thus forming a continuous structure. As the seam epithelial cells degenerate, the epithelia on the nasal aspect of the palate differentiate into pseudostratified ciliated columnar cells, while those on the oral aspect of the palate, become stratified squamous cells. Osteogenesis begins at sites in the anterior mesenchyme of the palate, forming the maxillary and palatine palatal shelves of the hard palate. Myogenesis occurs in the posterior third of the palate giving rise to the soft palate.

Because cleft palate—with or without cleft lip—is the most common human birth defect, the biology of palatal development has attracted much attention (Ferguson, 1988). Cleft palate can result from disturbances at any phase of development up to epithelial seam degradation. In humans cleft palate has long been recognized as a multifactorial disorder and despite major efforts little progress has been made via the human genetics route toward understanding palatal development. The multifactorial nature of cleft palate has been confirmed by the many mouse gene knock-outs that give rise to cleft palates; even so, in only a few cases has a direct primary association been established. Because palatal development is a very dynamic process, any disturbance in the development of surrounding tissues can affect shelf elevation and/or fusion resulting in a cleft. Indeed, even minor disturbances affecting the timing of elevation, for example, can ultimately result in a cleft. Thus, the outcome of many disturbances in facial development can lead to the same phenotype, namely, a cleft, but via what can be very different mechanisms.

One molecule identified as having an essential primary role in palatal shelf development is transforming growth factor β3 (*Tgfb3*). *Tgfb3* is expressed in medial edge palatal shelf epithelium at the time of fusion and degradation, and *Tgfb3* knock-out mice develop cleft palate and die within 24 hr of birth (Kaartinen *et al.,* 1995; Proetzel *et al.,* 1995). Unlike knock-outs in many other genes that produce cleft palate, *Tgfb3*$^{-/-}$ mice have no other obvious craniofacial defects. Palatal shelves isolated from *Tgfb3*$^{-/-}$ mutant embryos are unable to fuse *in vitro* when cultured but are able to fuse when placed next to wild-type or heterozygous shelves (Taya *et al.,* 1999). Addition of exogenous TGF-β3 or other TGF-βs to *Tgfb3*$^{-/-}$ palatal shelves in culture enables the shelves to fuse. These *in vitro* experiments confirm a primary role for TGF-β3 in palatal shelf fusion although the biochemical details of this action remained to be determined.

As described above, palatal shelves can be cultured as isolated explants using methods similar to those used for teeth. An advantage of this culture system is that shelves from mouse mutants with cleft palates can be placed alongside those from wild-type embryos to see if they are capable of fusing. In addition, rescue of a fusion defect by the specific addition of factors into these cultures provides a powerful method of extending genetic analysis and developing possible nonsurgical correction approaches. These types of *in vitro* experiments are important for determining whether any cleft palate phenotype results from failure of elevation or fusion and if the phenotype is primary or secondary.

C. Neurogenic Placodes

Another unique feature of craniofacial development relates to the development of sensory neurons. In the trunk all sensory neurons develop from NC, while in the head the earliest differentiating neurons in the sensory ganglia are derived from specialized ectodermal fields, the dorsolateral and epibranchial placodes (collectively known as the neurogenic or ganglionic placodes) (Hörstadius, 1950; D'Amico-Martel and Noden, 1983; Begbie *et al.,* 1999). The dorsolateral placodes (the vestibular and trigeminal) develop alongside the central nervous system, while the epibranchial placodes (the geniculate, pertrosal, and nodose) develop at the base of each BA at the top of the branchial clefts. The significance of this dual origin of sensory ganglia is unknown, but they have been found to respond differently to neurotropic factors such as NFG (Lindsay *et al.,* 1985). The placodal neurons differentiate early and establish both peripheral and central projections before the NC-derived neurons initiate axonogenesis (Webb and Noden, 1993). They play an important role in establishing the peripheral projections of the NC-derived ganglia, because in their absence these are abnormal

(Moody and Heaton, 1983). The epibranchial placodes are induced to form by BMP7 signaling from the pharyngeal endoderm (Begbie *et al.*, 1999). Trunk ectoderm, unlike cranial ectoderm, is unable to respond to this inductive signal *in vitro*, although *in vivo* studies have shown that trunk ectoderm is able to contribute neurons to the distal Xth ganglion when grafted to the position of the nodose placode *in vivo* (Vogel and Davies, 1993). The dorsolateral placodes, however, appear to have a different system of induction. The trigeminal, for example, does not form in proximity to a source of endoderm. Instead, induction by the NC appears more likely (Stark *et al.*, 1997). The fact that the two neurogenic placode types use distinct developmental pathways is emphasized by the fact that the dorsolateral placodes are lost in *neurogenin 1* knock-out mice, while the epibranchial placodes are lost in *neurogenin 2* knock-outs (Fode *et al.*, 1998; Ma *et al.*, 1998).

D. Olfactory development

The olfactory placodes (OfP) develop bilaterally in the rostroventral surface ectoderm over the prosencephalon. A focal thickening of the ectoderm is the initial histologic indication of the onset of placodogenesis (Jacobson, 1963; Verwoerd and van Oostrom, 1979). Fate mapping using chick-quail chimeras has demonstrated that the placodal epithelium descends from the ectoderm of the ANR of the rostral neural fold (Couly and Le Douarin, 1985, 1987, 1990). However, that the OfP is formed from the fringes of the neural plate, and not from the adjacent non-neural ectoderm, has recently been shown in zebrafish (Whitlock and Westerfield, 2000).

Transplantation and recombination experiments suggest that placodogenic inductive processes involve the presumptive nasal epithelium, chordomesoderm, and endoderm during gastrulation, and the nasal epithelium and the telencephalic primordia later (Webb and Noden, 1993). After induction, the placode invaginates to form a pit. The ectoderm of this invagination will yield the respiratory, olfactory, and vomeronasal (Jacobson's) organ (VNO) epithelia. The neuroblasts of the olfactory nerves (CN I) do not delaminate, but develop *in situ* within this pit sending axonal projections to the FB. The nervus terminalis (CN 0) develops out of the medial placodal epithelium (Webb and Noden, 1993). Cells immunocytochemically identified as producing gonadotropin releasing hormone/leutenizing hormone releasing hormone (GnRH/LHRH) delaminate from the epithelium and migrate into the FB (Schwanzel-Fukuda and Pfaff, 1989; Wray *et al.*, 1989). The path of the nervus terminalis is marked by these LHRH-positive cells. Couly and Le Douarin (1985) provide evidence that CN I myelin-producing cells also arise from the placode.

Olfactory pit invagination is a morphogenic process coupled with the formation of a ring of tissue surrounding the pit, delineating a medial frontonasal prominence (MFP) and a lateral frontonasal prominence (LFP). The basal lamina of the OfP, and later the olfactory and respiratory epithelium, contains collagen II, foreshadowing the sites of chondrogenesis of the nasal capsules and basal plate formation (Croucher and Tickle, 1989; Thorogood, 1993). Extirpation experiments provide evidence that the epithelium of the invaginated olfactory pit, moreover, induces chondrogenesis in the CNC mesenchyme to form the nasal capsule (Corsin, 1971). The MNP will support the induction of CNC (posterior prosencephalic and anterior mesencephalic) to yield medial capsular structures (e.g., rostral nasal septum, paraseptal cartilages etc.; see below); the LFP supports the induction of the lateral capsular structures (e.g., paries nasi, turbinals, etc.) (Osumi-Yamashita *et al.*, 1997a,b). The trabecular basal plate to which the nasal capsules are ultimately connected (see below) develops as a rostrad extension from initially paired struts (trabecula cranii) developing adjacent to the prechordal plate anterior to the notochord. A mechanism is in place that directs the meeting and fusion at the midline of these paired trabecula cranii. This mechanism is likely to involve signals from the prechordal plate and developing FB. The premaxillae (forming the primary palate) and upper incisors develop out of the more superficial CNC in this region. A transient constriction of epithelium, the nasolacrimal groove, separates the LFP from the expanding maxillary branch of the first branchial arch.

Analysis of olfactory development, perhaps lagging behind that of optic and otic development, has generally addressed the developmental relationships of the olfactory bulbs, olfactory neurons, VNO, and LHRH cells, their neurogenesis and targeting, as well as (to a lesser extent) the role played by the intervening ectomesenchyme (Keverne, 1999; Lin and Ngai, 1999; Mombaerts, 1999; Mori *et al.*, 1999). Mutations of a number of genes affect the development of olfactory structures, including of *BF-1 (forg1)*, *Dlx5*, *Gsc*, *Hesx1*, *Otx2*, *RAR*, and *Pax6* (see below). Some of these genes are ectodermally expressed, such as *BF-1*, *Dlx5*, and *Pax6*, while others are primarily mesenchymally expressed, such as *Gsc* and *RAR*. As with the BA, epithelial–mesenchymal interactions appear vital to proper nasal development. For example, the nasal capsular defects in $Gsc^{-/-}$ mice are, in effect, a subset of the defects seen in the $Dlx5^{-/-}$ mice; not surprisingly, ectodermal loss of *Dlx5* expression leads to loss of mesenchymal *Gsc* expression, further suggesting that ectodermal regulation of the mesenchyme is necessary for proper capsulogenesis (Depew *et al.*, 1999; see below). Similarly, a loss of placodal *Pax6* leads to a loss of *Msx1* expression and retinoic acid signaling in the underlying frontonasal mesenchyme (Grindley *et al.*, 1995; Anchan *et al.*, 1997; Enwright and Grainger, 2000). In general, however, analysis of early nasal development has been obscured by the tight developmental relationship of the rostral neuraxis and the OfP.

E. Submandibular Gland Development

A number of important glands develop within the craniofacial tissues, the best studied of which is perhaps the submandibular gland (SMG). Among other things, study of these glands has provided a wealth of information of the nature of branching morphogenesis (Hieda and Nakanishi, 1997; Jaskoll and Melnick, 1999). SMG development begins around E11.5 as an in-growth of the oral ectoderm into the mandibular mesenchyme in an initial "bud stage." The invaginating epithelium proliferates to form an elongate, solid epithelial stalk terminating in a bulb. With continued proliferation and end-bud branching, along with selective growth inhibition and cell death, a network of large and small ducts is built, eventually resulting in a mature gland.

F. Tooth Development

Teeth are remarkable structures. They contain the hardest substance in the human body, enamel; their preservation in the fossil record provides the main source material for palaeontologists and anthropologists; and their unique function in feeding means tooth shape and organization are powerful driving forces in evolution. The origins of teeth have been traced back to the earliest vertebrates where skin denticles of extinct fishes are thought to have "moved" into the oral cavity as jaws evolved (reviewed by Butler, 1995). Early jawed fish (e.g., acanthodians) possessed denticles in a variety of shapes attached to the endoskeleton of the jaw. The similarity of these denticles to modern teeth suggests that they could have been precursors of modern teeth. It has been suggested that teeth have evolved from a specific group of denticles similar to those in the oropharyngeal region of agnathan (jawless fish) vertebrates (Smith and Coates, 1998). One classic theory suggests that the evolution of dermal denticles into teeth coincided with the evolution of the mandibular arch into a jaw; contrary to this it has been suggested that oropharyngeal denticles were specialized feeding structures prior to the evolution of jaws in early fishes and thus teeth may have evolved before jaws (Smith and Coates, 2000).

Historically, dental research has focused on the teeth themselves and how to repair and preserve them in adults. More recently, the embryonic development of teeth has attracted increasing interest not only for an understanding of how these important organs develop but also because tooth development offers a powerful experimental system to address general questions in organogenesis. Mice have thecodont dentition (teeth occupying bony sockets) which exhibits heterodonty; that is, they have teeth of different shapes: in this case, two incisors (which continuously grow throughout life) and six molars in each jaw. Mice do not develop canine or premolar teeth; instead, the incisors and molars are separated by a region, the diastema, devoid of teeth. Mice only develop one set of teeth (monophyodont dentition), unlike human dentition, which is diphyodont. In considering the differences between the human (and many other vertebrates) and murine dentition, and the fact that mice lack canines and premolars, the extrapolation of data from mice to humans and other mammals should be approached with caution. However, the early development of teeth in most species, at least histologically, is fundamentally the same.

Teeth develop on the oral surfaces of the facial processes (mandibular, maxillary, and frontonasal) from interactions between the oral epithelium and underlying CNC. The early epithelial–mesenchymal interactions led to the formation of epithelial tooth buds surrounded by condensed mesenchyme cells. The formation of epithelial buds is a common feature of the development of several organs such as hair, lung, kidney, and sweat glands and not surprisingly many of the same signaling pathways are involved. In mammals, the position at which an epithelial tooth bud invaginates into the ectomesenchyme specifies the type of tooth that develops. Tooth buds forming in distal regions of the facial processes will develop as monocuspids (incisors and canines), whereas tooth buds in more proximal regions will develop with multiple cusps (premolars and molars). Tooth type is thus intrinsically linked to tooth position to produce the characteristic pattern of tooth types, the dentition.

1. Origins of Tooth Cells

Teeth have several unique cell types that are found nowhere else in the body. These include ameloblasts, odontoblasts, and cementoblasts. The differentiated cell types that are responsible for secreting and organizing the specific hard tissues of teeth develop early in embryogenesis. Ameloblasts secrete enamel matrix and are derived from oral ectodermal cells. Odontoblasts produce dentine and develop from CNC cells, as do all other supporting dental cells. Ameloblasts are the only cells remaining in teeth at birth that are derived from the ectoderm. The stages of tooth development are outlined in Appendix 1.

Tooth development is initiated by local interactions between oral (stomodeal) ectoderm cells and the underlying CNC. These first interactions are specific and unique to these two tissues. Recombination experiments using explant cultures and anterior eye transfers have identified a requirement for these two tissues in tooth development. Thus, recombinations of stomodeal ectoderm with any CNC cell population can support tooth development. Recombinations between stomodeal ectoderm and non-NC mesenchyme cannot support tooth development. Recombinations between non-stomodeal ectoderm and CNC cells cannot support tooth development. Similar experiments (see below) also show that premigratory CNC cells can support tooth development. Thus, teeth can only form when stomodeal ectoderm is in contact with ectomesenchyme and the only site of such contact in mouse embryos is the developing oral cavity.

Lineage tracing using DiI in rodent embryos has identified the axial origins of the CNC cells that contribute to tooth formation (Nichols, 1981; Tan and Morriss-Kay, 1986; Fukiishi and Morriss-Kay, 1992; Osumi-Yamashita *et al.,* 1994, 1997a; Imai *et al.,* 1996). These studies have defined the caudal MB and rostral HB as the source of odontogenic NC cells for mandibular incisors and molars and maxillary molars. Premaxillary incisors develop on the frontonasal process, and, though the origins of these odontogenic cells have not been directly mapped, studies in avian embryos implicate the rostral MB and FB as the source of these cells.

2. Patterning of Dentition

Patterning of the dentition leads to the development of different shapes (types) of teeth in their correct positions in the jaws. In mice this pattern consists of incisors distally and molars proximally. A theory for dental patterning based on a field model was proposed by Butler (reviewed by ten Cate, 1995). This model predicts that diffusible morphogens determine areas within the jaw in which incisors, canines, and molars will develop. Alternatively, Osborn (1978, cited by ten Cate, 1995) proposed a clonal model in which isolated clones of ectomesenchymal cells specify each different type of tooth. The first model to be proposed based on experimental evidence was the homeobox code model, which proposes that the restricted domains of homeobox gene products are responsible for specifying the pattern of the dentition.

The homeobox code model was based on observations of the spatially restricted expression of several homeobox genes in ectomesenchymal cells prior to E11 (Sharpe, 1995; Thomas and Sharpe, 1998). The early expression of *Msx1* and *Msx2* prior to the initiation of tooth germs is restricted to distal, midline ectomesenchyme in regions where incisor but not molar teeth will develop, while *Dlx1* and *Dlx2* are expressed in ectomesenchyme cells where molars but not incisors will develop (MacKenzie *et al.,* 1991; Qiu *et al.,* 1995, 1997). These expression domains are broad and do not exactly correspond to presumptive molar and incisor odontogenic cells. Rather, they are considered to define broad territories. Expression of *Barx1* overlaps with *Dlx1/2* and corresponds closely to ectomesenchymal cells that will develop into molars. The homeobox code model proposes that the overlapping domains of these (and other genes) provide the positional information for tooth type morphogenesis.

The first support for this model came from the dental phenotype of *Dlx1/2*$^{-/-}$ compound mutant mice where development of maxillary molar teeth is arrested at the epithelial thickening stage (Qiu *et al.,* 1997; Thomas *et al.,* 1997). As predicted by the code model incisor development was normal in these mice; normal development of mandibular molars (not predicted by the code) was assumed to result from the functional redundancy with other *Dlx* genes such as *Dlx5* and *Dlx6* that are expressed in ectomesenchyme in the mandibular primordium. Maxillary molar teeth are re-

placed by ectopic cartilage in *Dlx1/2*$^{-/-}$ mice, suggesting that, in the absence of these genes, ectomesenchymal cells become reprogrammed from an odontogenic to a chondrogenic phenotype.

The phenotype of *Msx1*$^{-/-}$*;2*$^{-/-}$ compound mutant embryos shows development of all teeth arrested at the epithelial thickening stage, suggesting a functionally redundant role for these genes in tooth bud formation (Satokata *et al.,* 2000). This phenotype has incorrectly been interpreted as indicating that *Msx* and *Dlx* functions are mechanistically the same in tooth development. This conclusion ignores the fact that the *Dlx1/2*$^{-/-}$ mutant phenotype only affects maxillary molars, whereas in *Msx1*$^{-/-}$*;2*$^{-/-}$ mutants all teeth are affected. Significantly, the lack of expression of *Dlx* genes in presumptive incisor mesenchyme prior to initiation is consistent with *Dlx* genes having a specific role in molar tooth development. The reported presence of ectopic bone in molar regions of newborn *Msx1*$^{-/-}$ mice (Satokata and Maas, 1994) has also been suggested to indicate a common mechanism with *Dlx* genes since in *Dlx1/2*$^{-/-}$ mutants, maxillary molar teeth are replaced by ectopic cartilage. However, whereas it is very clear that chondrogenesis replaces odontogenesis in *Dlx1/2*$^{-/-}$ mutant maxillary molar mesenchyme, the cartilage does not mineralize and the occasional formation of ectopic bone in *Msx1*$^{-/-}$ mutants is most likely to be a result of effects on alveolar bone development. Thus taking into account all the available evidence it is likely that *Msx* and *Dlx* genes function through different mechanisms to regulate tooth formation. *Dlx* genes have a clearly established role in patterning molar tooth development, *Msx* genes are clearly required for the bud-to-cap transition and may also have a role in bud formation and possibly incisor patterning based on the odontogenic homeobox code. This early role for *Msx1* specifically in incisor development is supported by the observation that whereas ectopic BMP4 can rescue molar tooth development in *Msx1* mutants, it does not rescue incisor development (Zhao *et al.,* 2000). Thus the requirement for *Msx1* in incisor development is different from that in molar development and is not mediated by BMPs. This is consistent with the early, incisor region expression pattern of *Msx1*, which is induced by BMP4 but which does not correlate with downstream activation of *Bmp4* in the mesenchyme.

Strong functional support for the code model came from misexpression of *Barx1* in distal ectomesenchyme cells, which resulted in incisor tooth germs developing as molars (Tucker *et al.,* 1998a). *Barx1* expression was found to be localized to proximal ectomesenchyme (molar) by a combination of positive and negative signals from the oral ectoderm. FGF8 localized in proximal ectoderm induces *Barx1* expression while BMP4 in the distal ectoderm represses *Barx1* expression. *Barx1* expression was experimentally induced in distal (presumptive incisor) ectomesenchyme by inhibition of BMP signaling following implantation of Noggin (a BMP antagonist) beads. This also had the effect of re-

pressing *Msx* gene expression, which is induced in distal ectomesenchyme by BMP4 and thus it is not yet established
whether the transformation of incisors into molars requires
misexpression of *Barx1* alone or whether it also needs accompanied loss of *Msx* gene expression. However, the fact
that tooth type could be transformed in this way provides
powerful support for the model.

3. Instructive Signals for Patterning

Recombinations between oral ectoderm and ectomesenchyme have been carried out to determine the origin of
the instructive information for dental patterning. Dryburgh
(1967) published an abstract outlining preliminary work in
which he recombined incisor and molar epithelium and mesenchyme from E10–14 embryos. It was found that when
molar epithelium was recombined with incisor mesenchyme,
a molar tooth formed; and when incisor epithelium was recombined with molar mesenchyme, an incisor formed. This
led to the conclusion that the epithelium was responsible for
determining the type/shape of a tooth. Further experiments
by Miller (1969), in which recombinations were cultured on
the chick chorioallantois, supported Dryburgh's work. Similar recombinations revealed that at E11–12, the ectoderm determined the tooth type.

The results of a series of classic experiments by Kollar
and Baird (1969, 1970a,b) disagreed with the previously reported findings that epithelium was responsible for patterning. The work of Miller (1969) and Dryburgh (1967) was
repeated at E13–16, and it was shown that molar epithelium
recombined with incisor mesenchyme resulted in incisor
teeth and incisor epithelium recombined with molar mesenchyme resulted in molar teeth (Kollar and Baird, 1969). Kollar and Baird performed recombinations between dental tissues and nondental tissues and after an initial *in vitro* culture
period, transferred the tissue *in oculo*. In the first set of experiments, lip furrow epithelium (E15 and E16) was recombined with either molar or incisor mesenchyme of the same
stage (Kollar and Baird, 1970a). The shape of the teeth was
determined by the origin of the mesenchyme. In addition,
when E16 cervical loop epithelium from an incisor was recombined with incisor mesenchyme, a fully formed incisor
tooth formed; and when recombined with molar mesenchyme, a fully formed molar tooth formed. Further experiments used tissue from the hairless (plantar) surface of the
foot in combination with dental tissues (Kollar and Baird,
1970b). At either E14, or E15, dental epithelium, when recombined with foot mesenchyme, showed no tooth development; however, when plantar epithelium was combined
with dental mesenchyme, tooth development occurred (Kollar and Baird, 1970b).

The apparent conflict of whether the ectoderm or ectomesenchyme provides the instructive information for patterning
has now been resolved by studying the temporal regulation
of homeobox gene expression in ectomesenchyme by ecto-

dermal signals. Removal of the ectoderm from E10 mandibular arch explants resulted in loss of expression of ectomesenchymal homeobox gene expression within 6 hr,
indicating that expression requires signals produced by the
ectoderm. Expression could be restored by implantation of
beads soaked in FGF8, a factor expressed in oral ectoderm
at this time (see below). Expression of *Dlx1/2, Msx,* and
Barx1 was seen around the implanted beads regardless of
their position in the explant, indicating that all ectomesenchymal cells at this time are competent to respond to FGF8
and implying that the CNC cells are not entirely prepatterned
(Ferguson *et al.,* 2000). When this experiment was repeated
at E10.5 ectomesenchymal gene expression was again lost
following removal of ectoderm but this time implantation
of FGF8 beads only restored expression in the original domains. Thus at E10.5 ectomesenchymal cell competence to
express homeobox genes in response to FGF8 has become
restricted to those cells that expressed the gene at E10. By
E11 removal of ectoderm had no effect of ectomesenchymal
gene expression showing that by this stage expression is independent of ectodermal signals. These results provide a
molecular understanding of the control of dental patterning and an explanation for the conflicting recombination results. The distoproximal (incisor-molar) spatial domains of
homeobox gene expression (homeobox code) are produced
in response to spatially restricted ectodermal signals acting
on pluricompetent ectomesenchymal cells. Recombinations
carried out before E10.5 will therefore show the instructive
influence of ectoderm on tooth shape, whereas those carried
out after E10.5 will show an instructive influence of ectomesenchyme since by this stage expression is independent
of ectodermal signals.

4. Initiation

Well in advance of the first morphological signs of tooth
development, processes are occurring that will determine
where a tooth will develop and which cells will contribute to
the developing tooth germs. It has previously been suggested
that nerves may play a role in determining the sites of tooth
development (reviewed by Kollar and Lumsden, 1979). This
suggestion was based mainly on descriptive data which
showed that the trigeminal nerve arose at E9 and fibers
branched from this nerve by E9.5 into presumptive tooth
forming regions (Lumsden, 1982). Neuronal induction of
odontogenesis was disproved when dennervated E9 and E10
mandibular processes developed teeth in the absence of innervation (Lumsden and Buchanan, 1986).

More recently it has been suggested that the foregut endoderm may be involved in establishing the sites of tooth
development. Imai and coworkers (1996) performed lineage
labeling experiments in which they infected rat foregut endoderm with adenovirus-inserted *lacZ*. After 9 days of culture (3 days of whole embryo rolling culture and 6 days
of mandibular organ culture), foregut endoderm cells were

shown to be immediately adjacent to developing tooth germs. The conclusion was that dental epithelium originates from the oral ectoderm in a region adjacent to foregut endoderm and that this endoderm does not directly contribute to the developing tooth. However, no direct experimental evidence was presented to suggest that dental epithelium could not arise in oral ectoderm in areas not adjacent to foregut endoderm.

Recombination experiments have been the main tool with which the capacity to initiate tooth development has been studied. Wagner (1955, cited in ten Cate, 1995) worked on anuran (whose "teeth" are merely keratinous appendages and are not true teeth) and urodele larvae (which develop true teeth) (ten Cate, 1995). He transplanted anuran NC cells under the newt oral epithelium and found that teeth developed where the dental papilla (NC-derived) was of frog origin and the dental organ (ectodermal) was from the urodele (Wagner, 1955). When the reciprocal experiment was performed (i.e., newt NC cells transplanted under anuran oral epithelium), teeth did not develop. This suggested that although anuran ectomesenchyme was capable of forming teeth, anurans lacked a signal from the oral epithelium that would normally initiate tooth development.

Lumsden (1988) performed informative experiments in which he either (1) recombined oral epithelium with CNC cells, TNC cells, or limb mesenchyme, or (2) recombined odontogenic mesenchyme with limb epithelium. These recombinations were performed with E9 or E10 tissue, then transferred *in vivo* into the anterior chamber of the mouse eye. Lumsden did not observe any signs of tooth development in the grafts of odontogenic mesenchyme with limb epithelium, limb epithelium with cranial neural crest cells, or, perhaps surprisingly, oral epithelium with limb mesenchyme. Lumsden did observe tooth development when oral epithelium was recombined with CNC cells or TNC cells. This work suggests that oral ectoderm can signal to mesenchyme and initiate tooth development as long as the mesenchyme is of neural crest origin; however, mandibular ectomesenchyme can only participate in tooth development if it is in combination with oral epithelium.

Experiments carried out by Mina and Kollar (1987) supported the view that the oral epithelium was the instructive component with regard to initiation. Mina and Kollar performed recombination experiments with mouse tissue between the ages of E9.0 and E13.0. They recombined oral epithelium with second branchial arch (nonodontogenic) neural crest-derived mesenchyme and found that when the recombinations were grafted *in oculo*, normal tooth development occurred with grafts between E9 and E11; the incidence of tooth development decreased in the E12 recombinations and no teeth were formed from the E13.0 experiments. Conversely, mandibular odontogenic mesenchyme recombined with second arch epithelium did not give rise to tooth rudiments at any age of the recombinations.

BMP4 was the first secreted signaling molecule to be identified that had a potential role in tooth initiation (Vainio *et al.,* 1993). *Bmp4* shows a dynamic expression pattern during early tooth development. It is first detectable at E10–10.5 in distal midline ectoderm where it is involved in controlling the spatial domains of ectomesenchyme expression of homeobox genes such as *Msx1* and *Barx1* (see above, below). At E11 this expression resolves into patches of ectodermal expression that correspond to the sites of tooth formation, and by E11.5 expression in the ectoderm is downregulated and transferred to the underlying ectomesenchyme. Throughout these changes in expression, *Bmp4* is intimately linked with expression of *Msx1* (Tucker *et al.,* 1998b). Thus, at E10.5 BMP4 induces *Msx1* expression in the underlying mesenchyme and at E11 is responsible for maintaining *Msx1* expression specifically in the ectomesenchyme at the sites of tooth formation. The transfer of *Bmp4* expression to the ectomesenchyme at E11.5 is dependent on *Msx1*. *Bmp4* expression is thus linked with tooth initiation but it is not known whether it is required for this process. *Bmp2* is coexpressed with *Bmp4* at E11 in dental epithelium, but unlike Bmp4, expression does not shift to dental mesenchyme but remains in the epithelium at the bud and cap stages (Vanio *et al.,* 1993; Vaahtokari *et al.,* 1996a; Dassule and McMahon, 1998). Unfortunately, both $Bmp2^{-/-}$ and $Bmp4^{-/-}$ mutant mice are early embryonic lethal and loss-of-function analysis is therefore limited (Winnier *et al.,* 1995; Zhang and Bradley, 1996).

Expression of *Shh* is localized to the presumptive dental ectoderm at E11, and is thus another good signaling candidate for tooth initiation. $Shh^{-/-}$ mice have little development of the facial processes and thus any role in tooth initiation cannot be identified from these (see Appendix 3; Chiang *et al.,* 1996). Mutations in *Gli* genes do suggest a role in early tooth development since $Gli2^{-/-};3^{-/-}$ double mutant embryos do not produce any recognizable tooth buds (Hardcastle *et al.,* 1998). Addition of SHH-soaked beads to oral ectoderm can induce local epithelial cell proliferation to produce invaginations that are reminiscent of tooth buds but which do not develop into teeth (Hardcastle *et al.,* 1998). *Shh* thus appears to have a role in stimulating epithelial cell proliferation and its local expression at the sites of tooth development implicate SHH signaling in tooth initiation. It is unclear to what extent the BMP and SHH pathways interact during initiation since all are coexpressed in epithelial cells and it is not until E12.5 that *Bmp4* expression shifts to the mesenchyme. BMPs and SHH are not capable of reciprocal induction of expression (Dassule and McMahon, 1998). However there are clear interactions between BMP and SHH signaling pathways at the bud stage (see below).

Lef1 is a member of the HMG family of nuclear proteins that includes the TCF proteins, known to be nuclear mediators of Wnt signaling. *Lef1* is first expressed in dental epithelial thickenings and on bud formation shifts to being expressed in the condensing mesenchyme. In *Lef1* knock-out mice, all dental development is arrested at the bud stage; recombination assays, however, have identified the requirement for *Lef1* in the dental epithelium as occurring earlier, prior to bud initiation (van Genderen *et al.*, 1994; Kratochwil *et al.*, 1996). Ectopic expression of *Lef1* in oral epithelium has also been shown to result in ectopic tooth formation, albeit in a single transgenic mouse (Zhou *et al.*, 1995). *Lef1* expression does not appear to be linked to BMP signaling, and tooth arrest at the bud stage may be rescued by a source of fibroblast growth factor (FGF) signaling (K. Kratochwil, unpublished, referenced in Bei *et al.*, 2000). *Lef1* is coexpressd with *Wnt10b* in tooth epithelial thickenings and thus may lie downstrem of signaling by this ligand. Wnt10b has been shown to be capable of inducing *Lef1* expression in tooth mesenchyme (Dassule and McMahon, 1998).

Expression of several genes in ectomesenchyme mark the sites of tooth germ initiation. These include *Pax9* and *activin-A* both of which are expressed beginning around E11 in small localized groups of cells corresponding to where tooth epithelium will invaginate to form buds. In the case of *Pax9* it has been shown that antagonistic interactions between FGF8 and BMP4 from oral ectoderm may act to localize *Pax9* to these sites and thus this was proposed to be a mechanism specifying the sites of tooth formation (Neubüser *et al.*, 1997). Significantly, however tooth buds form in *Pax9* mutant embryos. *Activin-A* expression is not regulated by the same mechanism suggesting that such FGF8/BMP4 interactions may not have a direct role in tooth initiation (Ferguson *et al.*, 1998).

5. Regionalization of Oral and Dental Ectoderm

Because regionally restricted expression of signaling protein genes in oral ectoderm controls dental initiation and patterning, it follows that the mechanisms that control the regional restriction of ectodermal signals need to be understood. At present there is no experimental evidence to explain how the ectodermal domains of *Fgf8* and *Bmp4* are generated. Some limited progress has been made toward understanding how *Shh* expression is restricted to dental ectoderm. During *Drosophila* segmentation, interactions between Hh and wingless signaling are involved in ectodermal cell boundary specification. Several *Wnt* genes are expressed during tooth development and one, *Wnt7b*, has a reciprocal expression pattern to *Shh* in oral ectoderm (Sarkar and Sharpe, 1999; Sarkar *et al.*, 2000). *Wnt7b* is expressed throughout the oral ectoderm except for presumptive dental ectoderm where *Shh* is expressed. The possible interactions between these two pathways was investigated by misexpressing *Wnt7b* in presumptive dental ectoderm using a murine retrovirus. Expression of *Wnt7b* in the *Shh* domain of ectoderm resulted in loss of *Shh* and *Ptc* expression and failure of tooth bud formation. This repression of Shh expression by Wnt7b appeared to be specific since expression of

genes regulated by other ectodermal signals was not affected. More significantly, tooth development could be rescued by addition of exogenous SHH to *Wnt7b*-infected explants. This suggests that *Wnt7b* acts to repress *Shh* expression in oral ectoderm and thus the boundaries between oral and dental ectoderm are maintained by an interaction between Wnt and Shh signaling similar to ectodermal boundary maintenance in segmentation in *Drosophila*.

6. Bud-to-cap Transition

The transition from bud to cap marks the onset of morphological differences between tooth germs that will give rise to different types of teeth. Because several gene knockouts have resulted in tooth development being arrested at the bud stage, much attention has focused on this transition. *Msx1* is expressed with *Bmp4* in the mesenchymal cells that condense around tooth buds. *Msx1*[−/−] embryos have tooth development arrested at the bud stage, and *Bmp4* expression is lost from the mesenchyme suggesting that *Msx1* is required for *Bmp4* expression (Y. Chen *et al.*, 1996). BMP4 is able to maintain *Msx1* expression in wild-type tooth bud mesenchyme indicating a positive feedback loop between the two where BMP4 induces its own expression via Msx1. Tooth development can be rescued in *Msx1*[−/−] embryos by addition of exogenous BMP4 (Y. Chen *et al.*, 1996, Bei *et al.*, 2000). BMP4 is also required at the bud stage to induce expression of an early EK marker gene, *p21*, suggesting that BMP4 is a key mesenchymal-to-epithelial signal that is required for buds to progress into caps.

BMP4 is also capable of inducing *Dlx2* and *Lef1* expression in bud mesenchyme but these actions are independent of *Msx1* (Bei and Maas, 1998). FGF3 also probably plays a role at the bud stage because *Fgf3* expression in mesenchyme is reduced in *Msx1*[−/−] embryos and FGF3 is able to maintain *Msx1* expression in wild-type embryos (Bei and Maas, 1998). The lack of a dental phenotypes in FGF3[−/−] mutant embryos does not support a role in tooth development, though functional redundancy with other FGFs, such as FGF7, may be important. *Bmp4* expressed in the bud mesenchyme is required to maintain *Bmp2* and *Shh* expression in the epithelium. Loss of *Bmp4* expression in *Msx1* mutants is accompanied by loss of *Shh* expression at E12.5, which can be restored by exogenous BMP4 (Zhang *et al.*, 2000; Zhao *et al.*, 2000). Blocking *Shh* function with neutralizing antibodies also results in loss of *Bmp2* expression, suggesting that *Shh* and *Bmp2* may be in the same pathway and that downregulation of *Bmp2* in *Msx1* mutants may be downstream of the loss of *Shh* (Zhang *et al.*, 2000). Paradoxically, overexpression of *Bmp4* in wild-type tooth buds leads to a repression of *Shh* expression that has been suggetsed to imply that the levels of *Shh* expression are tuned by BMP4 concentration (Zhang *et al.*, 2000).

The requirement of *Shh* for normal tooth development has been contentious largely because facial development is too disrupted in *Shh* mutant embryos to draw any conclusions about tooth development. The loss of Shh signaling at different stages of tooth development, either by addition of neutralizing agents *in vitro* or by Cre-mediated excision *in vivo*, has identifed distinct time-dependent requirements for Shh. Blocking Shh signaling using neutralizing antobodies or forskolin shows that at E11–12 Shh is required for dental epithelium proliferation to form tooth buds whereas blocking at E13 affects tooth bud morphology but these buds can still form teeth (Cobourne *et al.*, 2001). Genetic disruption of Shh signaling from E12.5 by Cre-mediated excision of targeted *Shh* null alleles results in a disruption of molar tooth morphology but cytodifferentiation appears normal, suggesting that Shh has a major role at the cap stage of development (Dassule *et al.*, 2000).

Another homeobox gene with a role in the bud-to-cap transition is *Pax9*. *Pax9* is expressed in bud stage mesenchyme and also earlier in similar domains to *activin-βA* and *Msx1* in patches of mesenchyme that mark the sites of tooth formation. *Pax9*$^{-/-}$ mutant embryos have all teeth arrested at the bud stage. Despite being coexpressed, early *activin-βA* expression is not affected in *Pax9*$^{-/-}$ embryos and *Pax9* expression is not affected in *activin-βA*$^{-/-}$ embryos (Matzuk *et al.*, 1995b). These two genes that are essential for tooth development to progress beyond the bud stage thus appear to function independently. There are, however, changes in expression of other genes such as *Bmp4*, *Msx1* and *Lef1* in *Pax9*$^{-/-}$ tooth bud mesenchyme (Peters *et al.*, 1998).

7. The Enamel Knot

The enamel knot (EK) is made up of transient clusters of epithelial cells visible in sections of molar cap stage tooth germs (see Fig. 5 later in this chapter). The EK was believed to have some ill-defined physical role in cusp formation but it was not until the expression of *Msx2* was shown to be localized to the EK at the cap stage that interest in this structure was renewed (MacKenzie *et al.*, 1992). The significance of the EK has now been established due largely to the work of Irma Thesleff in identifying it as a signaling center with many similar characteristics to other signaling centers such as the ZPA, notochord, and floor plate (Vaahtokari *et al.*, 1996b). EK precursor cells can first be detected at the tip of the tooth buds by expression of *p21*, followed shortly after by *Shh* (Jernvall *et al.*, 1998). By the cap stage, when it is visible histologically, the EK expresses genes for many signaling molecules including *Bmp2*, *Bmp4*, *Bmp7*, *Fgf9*, *Wnt10b*, and *Shh* (Thesleff and Sharpe, 1997) Three-dimensional reconstructions of the expression of these genes have revealed highly dynamic spatial and temporal nested patterns in the rodlike structure of the EK. On the whole, receptors for the EK signals are localized in the epithelial cells surrounding the EK, consistent with a highly active signaling center. Moreover analyses of cell proliferation and apoptosis reveal a low rate of proliferation of EK cells that rapidly disappear

toward the end of the cap stage via apoptosis (Vaahtokari *et al.*, 1994).

The exact physical role of the EK is not yet established but changes in its morphology in tooth germs of spontaneous mouse mutants with abnormal molar cusp formation such as *Tabby* (*Eda*) and *downless* (*dl*) have started to reveal some detail of this remarkable structure. *Tabby* mice have abnormal molar tooth cusps (in addition to other abnormalities most particularly in hair and sweat gland formation), and the mutated gene has been identified as a protein that has homology with the *Tnf* ligand family and is a transmembrane protein that probably trimerizes through an internal collagen-like domain. The *Tabby* gene (*Ta*) is analogous to the human gene *ED1*, which is responsible for X-linked hypohidrotic (anhydrotic) ectodermal dysplasia (XLHED) (Christ-Siemens-Touraine syndrome) (Kere *et al.*, 1996). The *Tabby* phenotype is indistinguishable from that seen for mutations in *downless* (Philips, 1960; Grüneberg, 1965). The *downless* gene has recently been identified as encoding a novel tumor necrosis factor (TNF) receptor (Headon and Overbeek, 1999). Mutations in the human homolog of mouse *dl* cause ARHED (autosomal recessive HED) and ADHED (autosomal dominant HED), which are clinically indistinguishable from the more common XLHED (Monreal *et al.*, 1999). *Ta* is expressed in the outer enamel epithelium but not the inner enamel epithelium, whereas *dl* is expressed in the EK. This suggests that if these proteins are going to interact directly at least one of them must be secreted. In *Tabby* mice the EKs are smaller than in wild type but have a normal shape (Pispa *et al.*, 1999). In *downless* mice, however, the EKs are of normal size but the structure of the EK is different with the cells being arranged into an elongated "ropelike" structure (Tucker *et al.*, 2000). Thus despite the apparently identical tooth phenotypes and the interactive structure of the proteins, the molecular biology indicates a more complex mechanism of action. The fact that the abnormal EKs are evident in both *Tabby* and *downless* provides evidence that this structure is involved in cusp formation.

Incisor tooth germs also have EKs distinguishable by expression of the signaling genes but not evident morphologically. Significantly the difference between monocuspid incisors and multicuspid molars is the formation of secondary EKs at the tips of the forming cusps in bell stage molars that are not present in incisors.

G. Skeletal Development

The vertebrate skull has been a focal point for the investigation of vertebrate development and evolution for more than a century. Note that most vertebrate adaptive transitions involve a role for the skull (Hanken and Thorogood, 1993), a fact that, along with its relative preservation in the fossil record and the ability to reconstruct developing skulls (either with wax models made from sections or differential staining of bone and cartilage), continues to make it a vital structure for investigation of development and evolution. Studies of cranial development provide perhaps the best means of a continuum of investigation and thought from the early masters of evolution and development to the modern world of molecular genetics and cellular biology. The use of Alizarin red and Alcian blue for differential staining of bone and cartilage (a system that allows for the clear three-dimensional identification of both normal and abnormal cranial skeletal development) is a nearly ubiquitous feature of any knockout paper. Unfortunately, there is often little appreciation or understanding of the intricacies and nuances of vertebrate cranial development (or of the classic comparative, evolutionary, and genetic studies) in many discussions of knockouts. In large part this is due to the lack of a clear text describing the developing murine skull and the scattered and abstruse nature of the classic descriptions of cranial development. In Appendix 2 we attempt to address this situation with a brief description of the embryonic and perinatal cranial skeletons (further details of which can be found in Flower, 1885; Jenkinson, 1911; Fawcett, 1917, 1922; Kesteven, 1926; Thomas, 1926; Broom, 1930; Johnson, 1933; Parrington and Westoll, 1940; Greene, 1955; Goodrich, 1958; Grüneberg, 1963; Youssef, 1966, 1969; Barghusen and Hopson, 1979; Sher, 1971; Presley and Steel, 1976, 1978; Moore, 1981; de Beer, 1985; Frick, 1986; Rugh, 1968; Kuhn and Zeller, 1987; Zeller, 1987; Allin and Hopson, 1992; Novacek, 1993; Kuratani, 1989; Kuratani et al., 1997, and references therein).

1. Regulation of Skeletogenesis

Well over 150 heritable defects in murine skeleto-odontogenesis have been recorded (Kalter, 1980; Li and Olsen, 1997) and craniofacial malformations occur in one third of all human congenital maladies (Gorlin et al., 1990; Thorogood, 1997)—attesting to the complexities of craniofacial skeletal and dental formation, growth, and homeostasis (Mundlos and Olsen, 1997a,b). Such deficiencies may be either osteochondrodysplasias, affecting elements (and teeth) in a generalized fashion, or dysosteochondroses, only affecting particular skeletal elements (or teeth) while leaving all others unchanged (reviewed by Mundlos and Olsen, 1997a,b). Moreover, skeletal ontogenesis is lifelong, with shifts in balance of growth and resorption, modeling and remodeling (Frost, 1969; Glimcher, 1989).

Each step in skeletogenesis is (in large part) characterized by differential cellular behaviors, whether, for example, that means proliferation, death, transcription, translation and secretion, division, and/or migration (Hall and Miyake, 1992, 2000). As agents of change, both morphogenesis (detected as change in an organism over a lifetime) and evolution (detected as change in a population over generations) share regulation of these behaviors in part by controlling the timing (Thompson, 1942; Gould, 1977; Richtsmeier, 1992;

Hall, 1999). Local alterations in the timing (heterochrony) of this control, say, either in induction or response, can result in new morphology and perhaps organisms. Thus, understanding craniofacial morphogenesis or evolution is an exercise in understanding the local regulation of cellular behaviors and the resulting development of structures.

Although mechanisms of skeletogenesis are described in detail in Chapter 14, it should be axiomatic that misregulation at any step in these processes can have consequences for craniofacial development (Grüneberg, 1963). This is well exemplified by the range of cranial defects present when misregulation of endochondral ossification occurs leading to malformations. Such malformations can occur, for example, through (1) deficiencies in the formation and maturation of the ECM, as seen with *collagen XI a1* in *cho* mice (Li et al., 1995), various *collagen II a1* mutations, including the *Dmm* mice (Brown et al., 1981; Li et al., 1995; Pace et al., 1997; Rintala et al., 1997), and *aggrecan* in *cmd* mice (Rittenhouse et al., 1978; Watanabe et al., 1994); or (2) deficiencies in cellular maturation in endochondral ossification, as seen with $PTHrP^{-/-}$ and $PTH/PTHrP\text{-}Receptor^{-/-}$ mutant mice, which exhibit diminished chondrocyte proliferation and accelerated hypertrophy (Karaplis et al., 1994; Lanske et al., 1996), and $Ihh^{-/-}$ mutant mice (St. Jacques et al., 1999; Karp et al., 2000). Skeletal defects in the skull are obvious in mice with either a block in osteoblast maturation [*Cbfa1*, or *Cliedocranio dysostosis* (*Ccd*) mice; Komori et al., 1997; Otto et al., 1997] or in bone-resorbing osteoclasts (e.g., *c-fos*, Johnson et al., 1992, Wang et al., 1992; *c-src*, Soriano et al., 1991; *M-CSF/op*, Marks and Lane, 1976; Yoshida et al., 1990; *NF-κB*, Franzoso et al., 1997). Table II lists selected genes implicated as regulators of skeletal ontogeny whose mutations affect craniofacial development.

2. Molecular Evidence of Regulation of Craniofacial Development

An exponentially increasing number of genes is being implicated in the regulation of pattern and development of craniofacial structures. Above we have related the essential requirement for epithelial–mesenchymal cross-talk for the realization of craniofacial development. How this requirement is elaborated into discrete, morphologically stereotypic structures is a central focus in craniofacial research. The embryonic complexity of the cranial region, as evidenced by the number of cellular origins, potentially instructive and responsive tissues, and craniofacial primordia involved, has complicated the search for spatiotemporal manifestations of coordinating tissues. Because the skull is a composite structure, this elaboration of signal and response into discrete, morphologically stereotypic elements may be considered with respect to a number of interrelated tasks (Fig. 2a), including the establishment of (1) inter-BA identity such that each BA within the series is unique; (2) intra-BA identity such that each element within a given BA has an unique

Table II Selected Regulators of Skeletogenesis Revealed through Genetic Analysis

Gene	Principal defects	Citation
Aggrecan	*Cmd* mice, with depleted ECM and disorganized arrangement of chondrocytes	Rittenhouse *et al.* (1978); Watanabe *et al.* (1994)
α-globin	Hydrop fatalis, due to loss of haemaglobin	Pászty *et al.* (1995)
bm	Sulfate transport and metabolism affecting proteoglycans in the ECM in diastrophic dysplasia and *bm* mice	Mundlos and Olsen (1997a,b); Penny-packer *et al.* (1981)
Bpa	Bpa (bare patches) mice, chondrodysplasia	Happle *et al.* (1983)
Cbfa1	*Ccd* mice, wormian bones, expanded fontanelles, delayed ossification of calvarial elements; diminished clavicles	Sillence *et al.* (1987); Komori *et al.* (1997); Otto *et al.* (1997)
c-fos	Osteopetrosis due to a failure of resorption and remodeling	Grigoriadis *et al.* (1994); Wang *et al.* (1992)
Collagen I	Osteogenesis imperfecta	Mundlos and Olsen (1997a,b); Stacey *et al.* (1988); Bonadio *et al.* (1990)
Collagen II a1	Dmm/Dmm mice, with a paucity of collagen fibrils in the ECM and retention of procollagen II in the endoplasmic reticulum	Brown *et al.* (1981); Pace *et al.* (1997)
Collagen II a1	Older *Del1* mice	Rintala *et al.* (1997)
Collagen IX a1		Nakata *et al.* (1993); Fässler *et al.* (1994)
Collagen X a1		Mundlos and Olsen (1997a,b); Jacenko *et al.* (1993)
Collagen XI a1	*cho/cho* mice, with misassemblage of collagen II fibrils, disorganized chondrocytes in growth centers and deceased numbers of HTC	Seegmiller and Fraser (1977); Li *et al.* (1995)
c-src	Osteopetrosis due to a failure of resorption and remodeling	Soriano *et al.* (1991)
GHRH-Receptor	Growth hormone has no receptor in the pituitary and dwarfism ensues	Godfrey *et al.* (1993); Wajnrajch *et al.* (1996)
Hyp	Osteomalacia and altered magnesium metabolism in X-linked hypophosphatemic mice	Meyer *et al.* (1979)
Ihh	Inappropriate timing of skeletal maturation	St. Jacques *et al.* (1999)
Csf1/op	Osteopetrosis due to a failure of resorption and remodeling	Kaku *et al.* (1999)
PTH	Inappropriate timing of skeletal maturation	Karaplis *et al.* (1994)
PTH/PTHrP-R	Inappropriate timing of skeletal maturation	Lanske *et al.* (1996)
Mo	Copper transport	Camakaris *et al.* (1979) (cit: Sillence *et al.* (1987)

identity; (3) the neurocranium–parachordal, prechordal/trabecular, and sensory capsular; and (4) the calvarium.

a. Establishing Inter-BA Identity. It has generally been thought that the BA are segmental, metameric (branchiomeric) structures (Barghusen and Hopson, 1979; Langille and Hall, 1989). Following the paradigm established for the vertebrae (and, in effect, the limb as well), the generation of BA identity has been thought to be regulated by *Hox* gene expression such that the meristic, overlapping pattern of gene expression of this gene family correlates with the identity of the branchiomeres (Fig. 2b). In effect, distinct populations of uniquely *Hox*-expressing CNC (perhaps with the complementation of *Hox*-positive ectoderm) are hypothesized to endow inter-BA identity. The most compelling arguments for such a "*Hox* code" have come from the knock-out studies of *Hoxa2* (see Appendix 3 and above; Rijli *et al.,* 1993; Gendron-Maquire *et al.,* 1993). Although *Hoxa2* is expressed in R2 (transiently) and more caudal rhombomeres, BA1 CNC is *Hoxa2* negative. BA2, in contrast, expresses *Hoxa2* in both the CNC and ectoderm. *Hoxa2*−/− mutants

exhibit mirror image homeotic-like transformations of the BA2-derived structures into proximal BA1-like structures (Fig. 2i). These BA1-like transformations are limited, however, to proximal, pericleftal structures and not the entire arch. Targeted disruption of other rostral *Hox* genes, *Hoxa1*, for example, disrupts BA2-derived structures (though defects are generally dysmorphologies and not AP arch transformations), as well as the supraoccipital arch and associated structures (Chisaka and Capecchi, 1991; Chisaka *et al.,* 1992; Lufkin *et al.,* 1991, 1992; Mark *et al.,* 1993). Compound *Hoxa1*−/−;2−/− mutants have exacerbated defects in relevant structures (Barrow and Capecchi, 1999; Fig. 11j). Further, indirect evidence for *Hox* regulation of inter-BA identity comes from analysis of *Pbx1*−/− mutants (see Appendix 3; Fig. 11r; Selleri *et al.*, 2001). *Pbx1* is the vertebrate homolog of the *Drosophila* gene *extradenticle* (*exd*), which functions genetically in parallel with *Hox* genes to alter the morphological consequences of Hox activities in flies (Peifer and Weischaus, 1990; Rauskolb and Weischaus, 1994). Nuclear PBX binds DNA with Hox partners thus forming complexes on appropriate DNA sites (Berthelsen *et al.,*

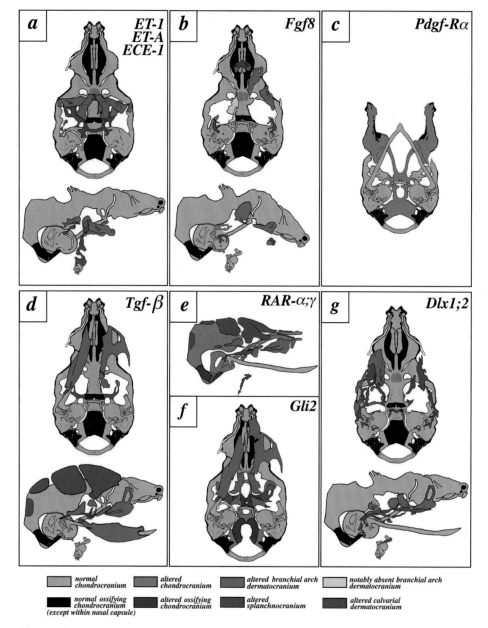

Figure 2 Schema of selected mutants with craniofacial defects revealed through genetic analysis of murine development. The schema represent the authors' interpretations of published descriptions and figures. Further description and details can be found within the text. Follow the color key for interpretation of defects. As a rule, where representation of elements would obscure that of other elements, unilateral representations of skeletal alterations are depicted.

Abbreviations of morphological structures in Figs. 2, 4, 5, 6, 7

acc,   alicochlear commissure

agp,   angular process

(continues)

(continued)

ahy,   ala hypochiasmata

alat,   anterolateral process of ala temporalis

alf,   alisphenoid foramen

(continues)

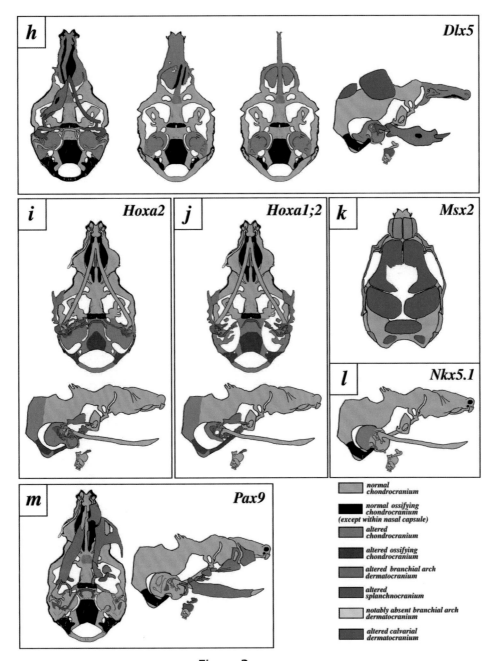

Figure 2 *(continued)*

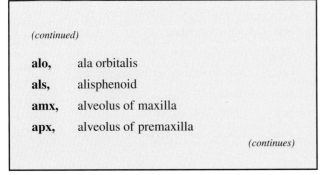

(continued)

alo, ala orbitalis

als, alisphenoid

amx, alveolus of maxilla

apx, alveolus of premaxilla

(continues)

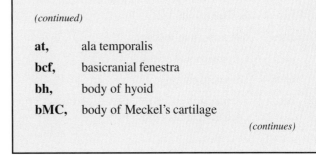

(continued)

at, ala temporalis

bcf, basicranial fenestra

bh, body of hyoid

bMC, body of Meckel's cartilage

(continues)

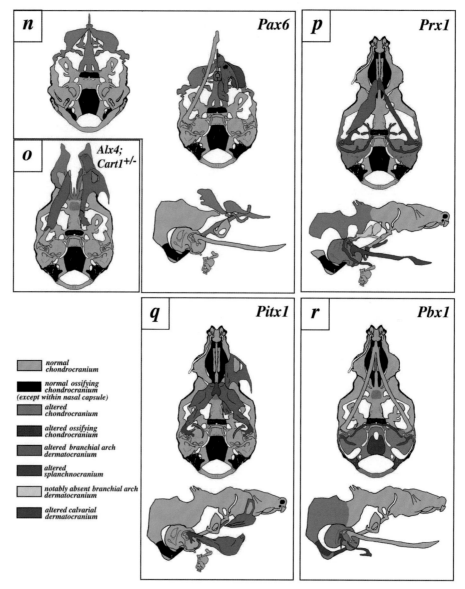

Figure 2 *(continued)*

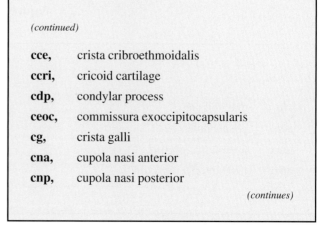

(continued)

bo,	basioccipital
bs,	basisphenoid
btp,	basitrabecular process
cac,	commissura alicochlearis
cart,	aretynoid cartilage
ccc,	commissura chordocochlearis, cordo-cochlear commissure

(continues)

(continued)

cce,	crista cribroethmoidalis
ccri,	cricoid cartilage
cdp,	condylar process
ceoc,	commissura exoccipitocapsularis
cg,	crista galli
cna,	cupola nasi anterior
cnp,	cupola nasi posterior

(continues)

(continued)

cop,	commissura orbitoparietalis, orbitoparietal commissure
cpr,	crista parotica, parotic crest
cptg,	catilago pterygoidea
crp,	coronoid process
Cs,	coronal suture
csc,	commissura sphenocochlearis, sphenocochlear commissure
csf,	commissura suprafacialis, suprafacial commissure
cthy,	cartilago thyroidea
ctra,	tracheal cartilage
cuco,	cupola cochlearis
dnp,	ductus nasopalatinis, nasopalatine duct
dnt,	dentary
eo,	exoccipital
etb,	ethmoturbinale
etm,	ectotympanic
fbc,	fissura basicapsularis, basicapsular fissure
fbs,	fissura basalis nasi, basal nasal fissure
fct,	foramen caroticum, carotid foramen
fds,	fenestra dorsalis
fed,	foramen endolymphaticum, endolymphatic foramen
fep,	foramen epiphinale, epiphinale foramen
fhg,	foramen hypoglossum, hypoglossal foramen
fhyp,	fossa hypophyseosa, hypophyseal fossa
fioc,	fissura inferior occipitocapsularis, recessus supralaris
fin,	fossa incudis
fj; mf,	foramen jugulare, jugular foramen
fla,	foramen lucerum anterior
fmg,	foramen magnum
fmt,	fissura metotica, metotic fissure
fmx,	frontal process of maxilla
fn,	fenestra nasi
fns,	fenestra nasi superior

(continues)

(continued)

fon,	fissura orbitonasalis, orbitonasal fissure
fop,	foramen opticum, optic foramen
fov,	foramen ovale
fovl,	fenestra ovalis
fpl,	foramen perilymphaticum, perilymphatic foramen
fpx,	frontal process of premaxilla
fr,	frontal
fro,	fenestra rotundra
fsa,	fossa subarcuata, subarcuate fossa
fsf,	foramen suprafacialis, suprafacial foramen
fsoc,	fissura superior occipitocapsularis
fsp,	fenestra sphenoparietalis
ftb,	frontoturbinale
fve,	fenestra vestibulii
gdnl,	groove ductus nasolacrymale
ghh,	greater horn of the hyoid, magnum cornu hyoidea??
gn,	goniale
gtn,	groove tectum nasi
hrt,	heart
iin,	incisura incudis
in,	incus
ina,	incissive alveolus of dentary
inf,	incissive foramen
iof,	infraorbital foramen
ip,	interparietal
jg,	jugal
la,	lacrimal
laat,	lamina ascendens ala temporalis
lap,	lamina parietalis
lcb,	lamina cribrosa, cribriform plate
lfp,	lateral frontonasal process
lhh,	lesser horn of hyoid
ll,	lower incisor
lo,	lamina obturans

(continues)

(continued)

lon,	lamina orbitonasalis
lta,	lamina transversus anterior
Ls,	lambdoidal suture
ma,	malleus
mai,	meatus auditoria interni
mes,	mesethmoid
mfp,	medial frontonasal process (prominance)
mm,	manubrium mallii, manubrium of the malleus
moa,	molar alveolus of dentary
Ms,	metopic suture
mtf,	mental foramen
mx,	maxilla
na,	nasal
ocd,	occipital condyle
ow,	oval window
pai,	processus alaris inferior
pao,	planum antorbitale
pas,	processus alris superior
paso,	processus anterior supraoccipitalis
pbvm,	processus brevis mallii
pc,	paraseptal cartilages
pca,	pars canalicularis
pchp,	parachordal plate
pco,	pars cochlearis
pdl,	prominentia ductus lymphaticum
pfi,	processus folii
pfsa,	prominentia fossa subarcuata
pl,	palatine
pma,	processus mastoideus
poc,	pila occipitalis
pop,	processus operculum
pmx,	premaxilla
pn,	paries nasi
pp,	parietal plate, lamina parietalis
ppa,	prominentia pars anterior
ppcd,	processus paracondylis

(continues)

(continued)

ppi,	prominentia pars intermedia
ppmx,	palatal process of maxilla
pppl,	palatal process of palatine
pppr,	palatal process of premaxilla
ppro,	pila preoptica
ppso,	pila postoptica
ppt,	processus pterygoidis ala temporalis
pr,	parietal
prat,	processus alaris trabecularis
prco,	prominentia cochlea
pre,	processus recessus
ps,	presphenoid
pscca,	prominentia semicircularis anterior
psccp,	prominentia semicircularis posterior
psccl,	prominentia semicircularis lateralis
ptaa,	processus transversus anterior??
ptg,	pterygoid
ptp,	pterygoid process of ala temporalis
puas,	prominentia utriculoampularis superior
rpMC,	rostral process of Meckel's cartilage
rw,	round window
rtp,	retrotympanic process
sal,	sulcus anterior lateralis
sec,	sphenethmoid commissura
soa,	supraoccipital arch
sof,	sphenorbital fenestra, sphenoparietasl fenestra
sn,	septum nasi
so,	supraoccipital
sp,	processus styloidius
spl,	sulcus posterior lateralis
sq,	squamosal
sql,	squamosal lamina
Ss,	sagittal suture
st,	stapes
tbp,	trabecular basal plate
tb-ns,	trabecular plate-nasal septum

(continues)

(continued)

tgt,	tegman tympani
tm,	taenia marginalis
tn,	tectum nasi
tso,	tectum synoticum, tectum posterius
uI,	upper incissor
vm,	vomer
Vg,	ganglia trigeminalis
za,	zona anularis
zpmx,	zygomatic process of maxilla
zps,	zygomatic process of squamosal
2agp,	secondary cartilage of angular parocess
2cdp,	secondary cartilage of condylar process
2cSs,	secondary cartilage of sagittal suture

1998a,b; Jacobs *et al.*, 1999; Ferretti *et al.*, 2000). As with the *Hoxa2*$^{-/-}$ mutants, splanchnocranial structures derived from BA2 and the first pharyngeal groove display striking morphological alterations in *Pbx1*$^{-/-}$ mutants (which die ~E16.0; Selleri *et al.*, 2001). Morphologically, the lesser horns of the hyoid are transformed into elongated cartilages whose structures are reminiscent of Meckel's cartilage or of the suspensorial BA2 derivatives of certain nonmammalian vertebrates. Although both *Hoxa2*$^{-/-}$ and *Pbx1*$^{-/-}$ mutants display transformations of BA2 structures into BA1-like structures, they are distinct in nature: *Hoxa2*$^{-/-}$ mutants have transformations that are more proximal-BA1 in nature, and include a distinct mirror image duplication around the first cleft, while *Pbx1*$^{-/-}$ mutants have transformations of a more distal-BA1 nature. Homeotic transformations of other BA, for example, BA3 to BA2, have yet, however, to be demonstrated for any knock-out (*Hox* or otherwise), and while a Hox code may regulate inter-BA identity other, non-Hox-associated factors are likely involved. Moreover, chick-quail grafting experiments (see above), complemented by work suggesting that *Fgf8* expression at the isthmus restricts the rostral extent of *Hox* gene expression in the rhombencephalon and possibly the CNC, suggest that BA1 is under a distinct regulation from the other BA, and is thus unique within the BA series to begin with (Fig. 2b; Noden, 1983b; Hunt *et al.*, 1998b; Irving and Mason, 2000). Finally, fuller appreciation of what sets inter-BA identity may only come with an understanding of the nature of the final arbiter setting the number of BAs which develop.

b. Establishing Intra-BA Identity. The generation of intra-BA identity can be viewed as the culmination of processes forming polarities [PD, AP (oral-aboral), and ML] within each BA, processes perhaps most complex and well studied within BA1. Paleontological and comparative embryological studies have suggested that the prototypical BA may have been comprised of five PD-oriented chondrocranial and associated dermal elements that would have been repeated in each BA (Goodrich, 1958; Barghusen and Hopson, 1979; Moore, 1981; de Beer, 1985; Hildebrand, 1988; Langille and Hall, 1989). This fundamental bauplan has tended toward modification in all major vertebrate groups, where in mammals, for example, the BA1 chondrocranium is represented by only two components: the maxillary and mandibular derivatives of the palatoquadrate and Meckel's cartilage, respectively. Moreover, within BA1 at least, an AP axis must be established in order that the teeth appear on the appropriate, functional surface; and, because, for example, the palatal shelves need to meet at the midline, the ML axis must be regulated as well. While it is not entirely clear how the establishment of each axis is determined, or if/how each developing axis informs the establishment of the other axes, it seems clear that intra-BA identity is informed by multiple epithelial signals (BMP, FGF, SHH, ET-1, and WNT driven) that either induce and/or maintain discrete populations of BA ectomesenchyme (as assessed by unique profiles of gene expression).

The epithelium covering the BA, including both the ectodermal and endodermal linings, is characterized by distinct spatiotemporal fields of gene expression (see above). These genes generally encode secreted factors (e.g., *Bmps, ET-1, Fgfs, Shh*) but not exclusively so (e.g., *Pitx*). Each BA receives signals from at least two sources centered around the pharyngeal clefts: one along the anterior (rostral) and another along the posterior (caudal) surface (Fig. 2c). For BA1, the anterior source emanates from the oral ectoderm. Patterning information in the BA comes from a balance of these sources. The targeted loss of *Hoxa2*, leading to a mirror image duplication of proximal mandibular structures in BA2, provides evidence that the cleft is at least a point source of patterning information read by the mesenchyme; *Hoxa2*$^{-/-}$ mutants, however, do not show a full PD transformation of BA2 (see Appendix 3 and Fig. 11i; Rijli *et al.*, 1993; Gendron-Maquire *et al.*, 1993). A conditional loss of function of *Fgf8* in BA1 ectoderm (*Fgf8*$^{Nes-Cre-/-}$ mice; see Appendix 3; Fig. 11b; Trumpp *et al.*, 1999) further supports this idea of multiple signals. *Fgf8* is normally expressed in the epithelium of the oral ectoderm as well as the peri-cleftal ectoderm and endoderm (Crossley and Martin, 1995; Trumpp *et al.*, 1999). In *Fgf8*$^{Nes-Cre-/-}$ mutants (achieved by driving expression of *Cre* under a *Nestin* promoter to flox *Fgf8*), *Fgf8* is lost within the BA1 ectoderm with the exception of a transient posterior (aboral) pericleftal patch of expression. This results in a near complete loss of all BA1 skeletal structures. The exceptions are informative: one, a malleus (with a body, manubrium, and processus brevis) develops in the mesenchyme underlying this posterior patch; and two, the distal mandibular midline develops with a rostral process of Meckel's cartilage, incisors, and associated dermal bone as well as maxillary bone associated with the nasal capsules. Mutation of *ET-1*, another epithelial factor, also leads to drastic loss of maxillary and proximal mandibular structures while the distal mandibular structures form (though they may be aberrantly fused; see Appendix 3 and Fig. 11a; Kurihara *et al.*, 1994). *ET-1* is also expressed within the mesodermal core of the BA, though discrimination of the respective roles of its expression within ectodermal and mesodermal cell populations remains to be achieved. *ET-1* expression is lost in BA1 ectoderm of the *Fgf8* mutants—except within the transient posterior patch of *Fgf8* expression (Trumpp *et al.*, 1999). Thus, it appears that multiple (e.g., both anterior and posterior) sources of FGF8 inform intra-BA identity, though not of the entire PD axis because the distal midline is clearly under a distinct regulatory regime. Principal candidates for this role, at least within BA1, come from studies of *Shh*, *Bmps*, and associated factors. For example, although loss of *Bmp4*, which marks the midline ectoderm, results in embryonic lethality, BMP-soaked beads can induce the expression of *Msx* within mandibular mesenchymal explants (see above). *Msx1* and *Msx2* are expressed in this midline tissue, and their potential induction by BMP signals may be significant because mutation of *Msx1* leads to midline defects including incisors (though proximal teeth are also affected; see Appendix 3; Satokata and Maas, 1994). Such defects appear to be more severe in compound *Msx1*$^{-/-}$*2*$^{-/-}$ mutants (Satokata *et al.*, 2000). Moreover, when the BMP antagonists *Noggin* and *Chordin* are both mutated a general, catastrophic loss of midline structures results (Bachiller *et al.*, 2000). Within the oral ectoderm, furthermore, experimental evidence suggests that these epithelial factors regulate the dental fields that establish molar (proximal) versus incisor (distal) fields: Antagonism of BMP signaling along the midline results in ectopic expression of *Barx1*, loss of *Msx1*, and the transformation of tooth type (see above; Tucker *et al.*, 1998a). Unfortunately, as with *Fgf8*, the proximate regulation of *Bmp* epithelial expression has yet to be clarified.

The distinction of PD domains of epithelial gene expression is reflected by similar, discrete populations within the underlying ectomesenchyme. One discrete ectomesenchymal population is defined by the expression of the *Dlx* genes (Fig. 10d; see Appendix 3). This gene family has been implicated in the elaboration of element morphology along the PD axis (Qiu *et al.*, 1995, 1997; Depew *et al.*, 1999). In mice, six *Dlx* genes have been described: *Dlx1*, *Dlx2*, *Dlx3*, *Dlx5*, *Dlx6*, and *Dlx7* (Dollé *et al.*, 1992; Robinson and Mahon, 1994; Simeone *et al.*, 1994; Qiu *et al.*, 1995, 1997; Stock *et al.*, 1996). Studies of genomic organization indicate that

the *Dlx* genes are arranged as tightly linked, convergently transcribed (tail-to-tail) pairs (*Dlx1* and *Dlx2*, *Dlx5* and *Dlx6*, and *Dlx3* and *Dlx7*) located adjacent to *Hox* gene clusters. Analysis of DNA sequence similarity and chromosomal location indicates that the *Dlx* genes can be placed in two paralogous groups: *Dlx1*, *Dlx6*, and *Dlx7* in one, and *Dlx2*, *Dlx5*, and *Dlx3* in the other. Moreover, tightly linked *Dlx* pairs appear to share regulatory regions and are expressed in similar patterns within the BA, placodes, and embryo as a whole (Dollé *et al.*, 1992; Akimenko *et al.*, 1994; Simeone *et al.*, 1994; Robinson and Mahon, 1994; Qiu *et al.*, 1995, 1997; Anderson *et al.*, 1997a,b; Yang *et al.*, 1998; Depew *et al.*, 1999; Eisenstat *et al.*, 1999; Zerucha *et al.*, 2000). Importantly for BA development, paralogous *Dlx* genes share nested expression patterns within the developing BA: *Dlx1/2* are expressed throughout most of the BA ectomesenchyme (i.e., both proximally and distally within the BA), while *Dlx5/6* and *Dlx3/7* share progressively restricted domains distally. These genes are not mesenchymally expressed, however, along the distal midline of the arches. The correlation of the nested *Dlx* gene family expression with a PD skeletal series led to the hypothesis that a combinatorial *Dlx* code regulates the establishment of the distinct skeletal elements within a particular BA unit. This idea has been genetically examined in *Dlx1*, *Dlx2*, *Dlx1/2*, and *Dlx5* mutant mice (see Appendix 3 and Figs. 10g,h; Qiu *et al.*, 1995, 1997; Acampora *et al.*, 1999; Depew *et al.*, 1999) where *Dlx1* and *Dlx2* have been shown to regulate proximal (e.g., maxillary) BA development and *Dlx5* to regulate (among other things) proximal mandibular (distal BA) structures. That *Dlx1*$^{-/-}$, *Dlx2*$^{-/-}$, and *Dlx1/2*$^{-/-}$ mutants evince no change of morphology where other *Dlx* genes are expressed, suggest that these other *Dlx* genes, in particular *Dlx5* and *Dlx6*, compensate for the loss. In line with this, a genetic interaction between *Dlx2* and *Dlx5* is seen in compound *Dlx2*$^{-/-}$;*5*$^{-/-}$ mutant mice in which the mandibular (i.e., distal BA) structures—except the distal-most midline, as represented by the incisor fields—are drastically altered (M. J. Depew and J. L. R. Rubenstein, unpublished observations). A further prediction of this combinatorial model is that loss of *Dlx5* and *Dlx6*, resulting in duplicated regions whereupon only *Dlx1* and *Dlx2* are expressed, should proximalize a segment of the skeletal series. This prediction is currently being tested in *Dlx5/6*$^{-/-}$ double mutants (M. J. Depew, T. Lufkin, and J. L. R. Rubenstein, unpublished data). Regulation of *Dlx* expression remains an open question. While FGF8-soaked beads are capable of inducing *Dlx2* and *Dlx5* in mandibular mesenchyme, mutants in which *Fgf8* has been floxed in BA1 epithelium maintain early (E9.5) *Dlx* expression patterns. BMP-soaked beads are also capable of inducing *Dlx* expression, but *Bmp* expression domains in the BA are generally not in register with mesenchymal *Dlx* expression patterns and assessment of *Dlx* expression in *Bmp* mutants has not been generally available (Fig. 2d).

PD polarity within the arches patently extends beyond the expression domain of the *Dlx* genes to the distal midline (as well as to the olfactory placode in BA1). A number of other genes are mesenchymally expressed along the distal midline, often members of a family of genes likewise distally expressed [e.g., *Msx1* and *Msx2*; *Prx1* and (to a much lesser extent) *Prx2*; *Alx3*, *Alx4*, and *Cart1*; *dHAND* and *eHAND*]. Distal midline skeletal defects often accompany mutations in such genes, including in the *Prx1⁻/⁻*, *Prx1⁻/⁻;2⁻/⁻*, *Msx1⁻/⁻*, *Msx1⁻/⁻;2⁻/⁻*, *Alx4⁻/⁻*, *Cart1⁻/⁻*, *Otx2⁺/⁻*, and *Oto* mutants (see Appendix 3 and Figs. 10o,p; Satokata and Maas, 1994; Acampora *et al.*, 1995; Martin *et al.*, 1995; Matsuo *et al.*, 1995; Ang *et al.*, 1996; Qu *et al.*, 1998, 1999; ten Berge *et al.*, 1998b; Lu *et al.*, 1999b; Zoltewicz *et al.*, 1999; Satokata *et al.*, 2000). Defects are usually associated with hypoplasia of the midline though (perhaps unexpectedly) proximal portions of the BA may be nearly normal despite the absence of the midline. Mutations of the distal midline factor *dHAND*, however, die early (E11) but have hypoplastic BA (Srivastava *et al.*, 1995, 1997; Thomas *et al.*, 1998). While most patterns of gene expression assayed in these mutants appear normal, *Msx1* expression is decreased (Thomas *et al.*, 1998). Additional genes are expressed in regions within *Dlx*-positive domains but may not extend throughout the proximal most or distalmost mesenchyme, including *Lhx6* and *Lhx7*, *Barx1*, *Pax9*, and *Gsc*. Expression of some of these are affected in *Dlx* mutants (e.g., *Gsc* in *Dlx5⁻/⁻* and *Barx1* in *Dlx1/2⁻/⁻* mutants; Depew *et al.*, 1999; Thomas *et al.*, 1997) while others less so. At the other end of the PD axis in BA1 lies the olfactory apparatus. Numerous genes are distinctly expressed either within the epithelium of the placode and pit (e.g., *Dlx*, *Fgf8*, *Pax*, *Pitx*, *Shh*, and *Bmp*) or the underlying mesenchyme (e.g., *Msx*, *Gsc*, *Alx*, *Pax*, *RAR*, and *Prx*), where the general correlation of gene expression within the distal midline and the olfactory-maxillary conjunction is striking and may reflect a mechanism to maintain dental occlusion/alignment of the maxillae (BA1 derived), premaxillae (frontonasal prominence derived), and the mandibulae (see above and Appendix 3 for further details). While many genes have been identified as required for some aspect of intra-BA elaboration (Fig. 2e), their developmental interrelationships are only rudimentarily understood.

c. Forming the Neurocranium. Proper generation of the neurocranium involves forming the parachordal and trabecular basal plates, the paired sensory capsules, and their eventual integration to form a structural unit (see Appendix 2). Signals from the notochord (e.g., SHH; see Appendix 3) and a competent adjacent mesoderm (e.g., expressing the appropriate *Gli*, *Pax*, and *Nkx* genes) appear essential for parachordal development (see Appendix 3; Chiang *et al.*, 1996; Mo *et al.*, 1997; Lettice *et al.*, 1999; Peters *et al.*, 1999; Tribioli and Lufkin, 1999). The development of the neurocranium anterior to the notochord is rather complex as competent

chordomesodermal, AVE, facial ectodermal, olfactory placodal, neurepithelial, and CNC tissues are each essential for trabecular basal plate and/or nasal capsular formation (optic and otic capsular development are touched on elsewhere). This region (essentially the rostral head) subsequently is sensitive to perturbation. Many types of disruption occur in the neurocranium anterior to the hypophysis in mutant mice, including loss of the nasal capsules without full loss of associated dermal bone or midline (trabecular) structures (e.g., *Pax6⁻/⁻*; Roberts, 1967; Hogan *et al.*, 1986; Hill *et al.*, 1991; Matsuo *et al.*, 1993; Grindley *et al.*, 1995; Kaufman *et al.*, 1995; Osumi *et al.*, 1997; Osumi-Yamashita *et al.*, 1997b; Quinn *et al.*, 1997; Sander *et al.*, 1997; Sosa-Pineda *et al.*, 1997); asymmetric loss of the nasal capsular structures (*Hesx1⁻/⁻*; Dattani *et al.*, 1998) with associated trabecular defects (e.g., *Dlx5⁻/⁻*; Depew *et al.*, 1999); loss of medial nasal capsular structures without particular trabecular defects (e.g., *Pax7⁻/⁻*, *Pax9⁻/⁻*; Mansouri *et al.*, 1996; Peters *et al.*, 1998); loss of, or collapse about, the midline trabecular structures resulting in holoprosencephaly and cyclopsia (e.g., *Oto*, *Shh⁻/⁻*, *Noggin⁻/⁻*; *Chordin⁻/⁻*; Chiang *et al.*, 1996; Zoltewicz *et al.*, 1999; Bachiller *et al.*, 2000); separation but not loss of trabeculae without loss of capsules (e.g., *PDGF-Rα⁻/⁻*, *Alx4⁻/⁻*; *Cart1⁻/⁻*, *Gli2⁻/⁻*); separation with loss of both trabecular and capsular structures (e.g., *Otx2⁺/⁻*, *RARα⁻/⁻;γ⁻/⁻*; Lohnes *et al.*, 1994; Acampora *et al.*, 1995; Mark *et al.*, 1995; Matsuo *et al.*, 1995; Ang *et al.*, 1996); disruption of both trabecular and capsular structures without specific midline separation (e.g., *Wnt5a⁻/⁻*, *Gli3⁻/⁻*; Mo *et al.*, 1997; Yamaguchi *et al.*, 1999); and total loss of both trabecular and capsular structures (e.g., *Lim⁻/⁻*; *Otx2⁺/⁻*; Acampora *et al.*, 1995; Matsuo *et al.*, 1995; Shawlot and Behringer, 1995; Ang *et al.*, 1996). Moreover, the integration of the CNC-derived trabecular plate and taenia marginalis (see Appendix 2) with the mesoderm-derived parachordal plate may variously be affected (e.g., *Bmp7⁻/⁻*; *Dlx1/2⁻/⁻*; *Msx1⁻/⁻*; *Otx1⁻/⁻*; Satokata and Maas, 1994; Dudley *et al.*, 1995; Luo *et al.*, 1995; Suda *et al.*, 1996; Qiu *et al.*, 1997). Overall, anterior neurocranial development follows a delicate balance of factors tending to expand the midline and those tending to collapse, or constrict, it (i.e., a balance of anterior DV and ML patterning mechanisms; Fig. 2e). Untangling the relative patterning contributions of the chordomesodermal, AVE, facial ectodermal, olfactory placodal, foregut endodermal, neural plate/neurepithelial, and CNC tissues (and genes expressed therein) to the patterning and development of the skull anterior to the hypophysis remains a principal challenge in craniofacial research.

d. Forming the Calvarium: Cranial Articulations and Suturogenesis. The growth of the dermatocranial calvarial bones (see Appendix 2) begins as mesenchymal precursors condense at specific sites on the dorsal surface of the head and produce an osseous matrix. These condensations extend outward, eventually meeting at the presumptive sutures. Growth then becomes localized to the periphery of the extending bone, at the lateral margins of the sutures known as the osteogenic front (Johansen and Hall, 1982). The location of cells within the osteogenic front appears to be correlated with their state of differentiation, with the lateralmost cells expressing terminal differentiation markers such as bone sialoprotein.

Although an increasing number of genes are being implicated in calvarial development (see Table III), to date our understanding is limited of the cellular and molecular regulation of the individual calvarial elements, why they form where they do, and what establishes their boundaries, growth, and sutural articulations. Sutural closure involves a transition from synchondroses (bones connected via cartilage) or syndesmoses (bones connected through dense fibrous connective tissue) to synostoses (Zimmerman *et al.*, 1998). Early in development the calvarial elements are principally syndesmotic, being "open" to allow for growth. The normal development of the calvarium requires mechanisms that ensure that the rate of growth and morphology of the elements is coordinated with that of the brain and jaws (Iseki *et al.*, 1999), and involves the tissues between them, including the meninges. The dura mater is integral to the maintenance of a suture because without it the suture fails to remain open and will synostose (Opperman *et al.*, 1993, 1995, 1998; Kim *et al.*, 1998). Light and electron microscopic examinations of sutural maturation reveal a transition from a bone forming front to a bone–tendon junction accompanied by the replacement of osteoblasts (in part through apoptosis) and mineralizing osteoid to fibroblasts and nonmineralizing, straight collagen bundles (Zimmerman *et al.*, 1998). The development, form and time of sutural closure have traditionally been thought to result from an interplay been hereditary determination and functional adaptation to brain growth (Young, 1959; Moss *et al.*, 1972; Oudhof, 1982; Herring, 1993; Jabs *et al.*, 1993; Bellus *et al.*, 1996a; Zimmerman *et al.*, 1998). For example, calvarial elemental morphology and sutural sequence and maturation were influenced by cranial contents in studies of experimentally induced macrocephaly (hydrocephaly) and microcephaly (Young, 1959).

Each suture delineates a growth center (Iseki *et al.*, 1999). Different sutures share some signaling mechanisms though developmental specificity is reflected by the uniqueness of the affected suture in craniosynostotic syndromes (conditions of premature sutural closure; see Table III). Clues to the molecular and cellular events involved in sutural growth and closure have come from genetic and experimental analysis.

Msx2 is expressed in some of the cells of the osteogenic front and dura mater. Boston type craniosynostosis has been shown to be due to a single amino acid substitution in *MSX2*, which produces a dominant positive (Jabs *et al.*, 1993). Consistent with this, overexpression of *Msx2* causes bone cells to invade the sutures (Dodig *et al.*, 1999). This appears to be due to a transient retardation of osteogenic cell differentiation within the suture, leading to an increased pool of proliferative osteogenic cells and ultimately to an increase in bone growth (Y. H. Liu *et al.*, 1999). *Msx2* has been shown to be regulated by BMP signaling during suture development (Kim *et al.*, 1998). *Bmp2* and *Bmp4* are expressed in the osteogenic front and sutural mesenchyme, and it has been proposed that they regulate the balance between the undifferentiated and differentiated states of osteogenic cells via regulation of *Msx*. BMP4 beads cause an increase in tissue volume when placed on the osteogenic front or midsuture area, but do not affect sutural closure (Kim *et al.*, 1998).

Activated mutations in the *FGF Receptors 1–3* lead to craniosynostosis, including Apert, Jackson-Weiss, and Crouzon and Pfeifer syndromes (reviewed by Webster and Donoghue, 1997; Wilkie, 1997; Burke *et al.*, 1998). FGF4 beads are able to induce *Msx1*, but not *Msx2,* in sutural tissue, and lead to accelerated sutural closure when placed on the osteogenic fronts (Kim *et al.*, 1998). FGF2 beads have been shown to upregulate *Fgfr1*, downregulate *Fgfr2*, and inhibit cell proliferation (Iseki *et al.*, 1999). *Fgf2* is abundant in skeletal membranes and is upregulated within sutures as they fuse, while *Fgf9* is expressed throughout the calvarial mesenchyme (Iseki *et al.*, 1999; Mehrara *et al.*, 1998; Kim *et al.*, 1998). *Fgfr2* is intensively expressed in the osteogenic fronts in proliferating osteoprogenitor cells. The onset of proliferation is preceded by *Fgfr2* downregulation with a corresponding upregulation of *Fgfr1*. As the differentiation marker *osteopontin* turns on, *Fgfr1* is downregulated. *Fgfr3* is expressed in cranial cartilages, including the plate of cartilage underlying the coronal suture, as well as in osteogenic cells, suggesting a dual role (Iseki *et al.*, 1999). The phenotype of the *FGFR3* mutation is similar to that of Saethre-Chotzen syndrome. This syndrome is caused by a loss-of-function mutation in *TWIST,* a gene encoding a basic helix–loop–helix (bHLH) protein. The *Drosophila* ortholog of TWIST regulates transcription of *heartless,* a homolog of vertebrate *Fgfrs;* Saethre-Chotzen syndrome is thus likely to have a defect in FGF signaling (El Ghouzzi *et al.*, 1997). Overall, experimental assessment of coronal and sagittal sutural development has suggested a mechanism involving the balanced coupling of the recruitment of cells into osteogenic differentiation (involving *Fgfr1* and ligands) and the maintenance of osteoprogenitor cells in a proliferative state associated with the outer osteogenic front (involving *Fgfr2*, its ligands, and possibly *BMP, Msx, Twist,* and *Id1;* Kim *et al.*, 1998; Iseki *et al.*, 1999; Rice *et al.*, 2000).

Other signaling molecules have also been shown to play a role in suture development. *Shh* and its receptor gene, *patched,* are expressed in a mosaic pattern along osteogenic fronts at the end of embryonic development, by which time the expression of *Msx2, Bmp2,* and *Bmp4* are reducing (Kim *et al.*, 1998). *Tgf-β1–3* also exhibit distinct patterns of expression during sutural formation (Opperman *et al.*, 1997).

Table III Selected Regulators of Skeletogenesis Revealed through Genetic Analysis

Gene	Principal defects	Citation
Alx4	Undermineralized, diminished parietals	Qu *et al.* (1997)[a]
Alx4/Cart1 compound	Undermineralized diminished frontals, parietals, supraoccipital, and squamosals	Qu *et al.* (1999)[a]
Bapx1	Agenic supraoccipital	Lettice *et al.* (1999)[a]; Tribioli and Lufkin (1999)[a]
BMP-1	Deficiency of frontal, parietal, interparietal, and expanded associated fontanelles	Suzuki *et al.* (1996)[a]
Cbfa-1 (Ccd)	Arrested development of osteogenic tissues; cleidocranio dysostosis	Sillence *et al.* (1987); Komori *et al.* (1997)[a]; Otto *et al.* (1997)[a]
Crtl1	Domed skull; misshapen parietal and frontal	Watanabe and Yamada (1999)[a]
Csf1/osteopetrosis	Misregulated remodeling with stenosis and synostosis of the sagittal suture in maturing mice	Kaku *et al.* (1999)[a]
Dlx2	Split and duplicated squamosal, loss of lamina obturans	Qu *et al.* (1995)[a]
Dlx1/2	Split and duplicated squamosal, loss of lamina obturans	Qu *et al.* (1997)[a]
Dlx5	Undermineralized parietal and interparietals	Acampora *et al.* (1999)[a]; Depew *et al.* (1999)[a]
Fbn1	Craniosynostoses: Marfanoid, Shprintzen, Goldberg	Sood *et al.* (1996)[b]
FGF2	Enlarged occipital, chondrodysplasia	Coffin *et al.* (1995)[c]
FGF3/4 (Bulgy-eye)	Coronal craniosynostoses: Crouzon	Carlton *et al.* (1998)
Fgf8	Loss of lamina obturans, dysmorphic squamosal	Trumpp *et al.* (1999)[b]
Fgfr1	Craniosynostoses: Pfeiffer	Muenke *et al.* (1994)[b]; Bellus *et al.* (1996a,b)[b]
Fgfr2	Coronal Craniosynostoses: Crouzon, Pfeiffer, Jackson-Weiss, Apert and Beare-Stevenson	Jabs *et al.* (1994)[b]; Reardon *et al.* (1994)[b]; Lajeunie *et al.* (1995)[b]; Neilson and Fricsel (1995)[b]; Rutland *et al.* (1995)[b]; Wilkie *et al.* (1995)[b]; Bellus *et al.* (1996a,b)[b]; Galvin *et al.* (1996)[b]
Fgfr3	Muenke craniosynostosis, achondroplasia and macrocephaly	Rousseau *et al.* (1994)[b]; Shiang *et al.* (1994)[b]; Naski *et al.* (1996)[b]; Deng *et al.* (1996)[b]
Fidget	Fusion of parietosquamosal suture and presphenoid and basisphenoid	Truslove (1956)
Gbx2	Agenic interparietal, diminished supraoccipital	Wasserman *et al.* (1997); M. J. Depew (personal observation))
Gli2	Deficiency and delayed ossification of frontal, parietal, interparietal	Mo *et al.* (1997)[a]
Gli3/Xt	Failure of skull vault formation	Johnson (1967)[a]; Mo *et al.* (1997)[a]
IKK1/IKKα (Chuk)	Domed skull with ectopic infraparietal, dysmorphic zygomatic	McLaughlin *et al.* (1997)[a]; Hu *et al.* (1999)[a]; Li *et al.* (1999)[a]; Takeda *et al.* (1999)[a]
Lmx1a/dreher	Dysmorphic interparietal with fenestra, dysmorphic lambdoidal and parieto-interparietal sutures	Bierwolf (1958)[a]; Manzaneres *et al.*, 2000)[a]; Millonig *et al.*, 2000)[a]
M33/bmi1 (Pc)	Split supraoccipital	Bel *et al.* (1998)[a]
Mf1/Foxc1 cong. hydro-cephalus	Hydrocephalus, with no bony calvarium covering the bulging cerebral hemispheres and ectopic midline frontal, disorganized neurocranium	Grüneberg (1953); Grüneberg and McKramaratne (1974); Kume *et al.* (1998)
Msx1/Msx2 compound		
Msx2	Craniosynostosis: Boston type Enlarged parietal foramen, enlarged parietosagittal fontanelle	Jabs *et al.* (1993, 1994)[b,d]; Y. H. Liu *et al.* (1999); Wilkie *et al.* (2000)[b,e]; Satokata *et al.* (2000)[a]
P107/p130	Diminished supraoccipital/interparietal	Cobrinik and Knechel (1996)[a]
Pax6 (Small eye)	Dysmorphic frontal, parietal, and squamosal	Kaufman *et al.* (1996); Osumi-Yamashita *et al.* (1997b)
Prx1(Mhox)	Agenic supraoccipital (including tectum)	
Pugnose	Domed skull; misshapen, shortened parietal and frontal	Kidwell *et al.* (1961)[a]
Querkopf	Dysmorphic frontal, parietal, occipital, and associated sutures (not craniosynostotic)	Thomas *et al.* (2000)[a]
RARα/γ		Lohnes *et al.* (1994)
Ryk	Minor dysmorphogenesis of vault elements	Halford *et al.* (2000)
Tgf-β2	Reduced and dysmorphic (undermineralized?) frontal, parietal, interparietal, squamosal and occipital	Sanford *et al.* (1997)[a]
Twist	Coronal craniosynostosis: Saethre-Chotzen syndrome	El Ghouzzi *et al.* (1997)[a,b]; Howard *et al.* (1997)[a,b]; Bourgeois *et al.* (1998)[a]
Wnt5a		Yamaguchi *et al.* (1999)[a]

[a] Indicates knock-out, conditional knock-out, mutagenesis, or classical mutant stock.

[b] Indicates human studies.

[c] Indicates overexpression.

[d] Indicates a dominant negative allele.

[e] Indicates haploinsufficiency.

IV. Conclusion

In this chapter, we have focused on the organization, development, and pattern of craniofacial tissues. We have seen that in order to generate a functional skull, an embryo must specify its body axes, along with its CNC, mesoderm, ectoderm, and endoderm. CNC must "migrate" correctly in concert with the ectoderm; CNC must stop at appropriate locations; cells must differentiate and condense; cells must be competent to recognize change of status at each junction and be aware of their position; cells must know how to respond appropriately to apoptotic signals; and cells must behave differentially in order to generate morphology. How this occurs is regulated by both genetically intrinsic factors (e.g., gene expression eventually regulating the capacity to signal, competence to receive signals, etc.) and extrinsic factors (e.g., when and where cells get signals, how loudly the signals come/go, etc.). Though we can now recognize the need to fulfill each of the above functions, the particulars of each step are not always clear, and answers to many of the basic questions that have been asked for a century regarding craniofacial development remain incomplete: What dictates where each element will form and how many blastamata it will incorporate? What sets its boundaries and eventual shape? What is the nature of the fundamental organization of the skull? What is the nature and origin of the cells forming cranial structures? What is the precise nature of the interplay of genetic and epigenetic regulators of craniofacial patterning? While strides have been made in answering the question "What is the molecular evidence for the generation of pattern (organization) within the developing cranium, especially within the ectoderm, neural tube, and CNC?," the ever proximate establishment (induction) of regional gene expression in these tissues remains a goal. For example, what is it that induces the expression of *Fgf8* in the facial ectoderm and by what mechanism(s) is the nested pattern of *Dlx* gene expression in the BA established? The realization of the vast contribution of the CNC to the skull revitalized studies of craniogenesis; yet, we still seek to understand the mechanisms that are involved in the generation of the CNC from the ectoderm, and how they possibly impart positional information, cell fate, and regional pattern. Studies of the CNC still address basic questions: What are the relative contributions of CNC and ectoderm to patterning? Do separate patterning mechanisms exist for CNC rostral to R3 relative to that caudal to R2? What is the significance of the uniqueness of BA1 relative to the other BA? What causes the CNC in the maxillary BA1 to proliferate toward the eye and nose? What cell–cell and cell–ECM interactions are critical for the elaboration of craniofacial development, and do the epithelial–mesenchymal interactions necessary for dentoskeletal development unmask pattern, generate pattern, or both? What establishes the precise, idiosyncratic and stereotypical patterns of cell migration? What regulates the subset of cells that enters the mesodermal core of the BA? Investigation of the neuraxial patterning has further revealed pertinent questions: What roles do the isthmic organizer, anterior visceral endoderm, anterior neural ridge, and chordomesoderm play in patterning craniofacial hard tissues? Clearly, the combination of an array of critical questions and the accessibility of cranial hard tissues to examination will make the next century of investigation on craniogenesis as productive as the last.

Acknowledgments

P. T. S. would like to acknowledge the support of the MRC, Wellcome Trust, and BBSRC. M. J. D. would like to thank John L. R. Rubenstein and Darrell Evans for their time and input, and the support of the March of Dimes, NIDR grant T32 DE07204 and ARCS. A. S. T would like to acknowledge the Wellcome Trust for its support

V. Appendix 1: Descriptive Dental Development

A. Epithelial Thickening Stage

CNC cells migrating to BA1 complete their journey by E9 (Lumsden and Buchanan, 1986). The first morphological signs of tooth development can be seen at approximately E11.0 (E11.5 for the molars) as localized thickenings of the oral epithelium; hence, this stage of tooth development is known as the *epithelial thickening* stage. Hay (1961) reported that the mouse mandibular tooth germs arose from three separate in-growths of the oral epithelium into the mesenchyme. She observed that the molars developed from the two proximal in-growths on each side of the jaw and that both incisors formed from one anterior, distal-midline thickening. This thickened epithelium is comprised of approximately five layers of cuboidal cells at its thickest part; by comparison, the rest of the oral epithelium is composed of one or two layers of cuboidal cells (Cohn, 1957). The nuclei of the cuboidal cells in the thickening are oriented perpendicularly to the basement membrane (Fig. 3) (Peterkova et al., 1993). The epithelial thickening yields the dental lamina on the lingual aspect and the vestibular lamina on the buccal aspect. The vestibular lamina forms a sulcus between the cheek and the teeth. The dental lamina gives rise to the teeth. As the dental lamina thickens, the deep cells of the lamina remain cuboidal in shape with their nuclei oriented perpendicular to the basement membrane. However, the epithelial cells on the oral surface of the dental lamina form a layer of flattened squamous cells whose long axes of their nuclei are parallel to the oral surface (Peterkova et al., 1993). Three-dimensional reconstructions of the premaxillary incisor region reveal multiple dental laminae, which are perhaps remnants of incisors lost by mice during their evolution (Peterkova et al., 1993). These dental laminae (six in total) are

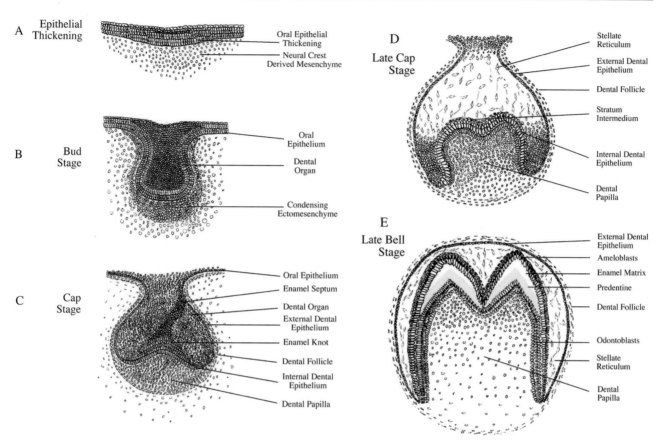

Figure 3 Tooth development. The diagrams represent frontal sections of developing tooth germs. (A) The epithelial thickening stage of tooth development. The oral epithelium has thickened in presumptive tooth-forming areas. (B) The bud stage of tooth development. The dental lamina has invaginated into the underlying ectomesenchyme and the mesenchymal cells have begun to group together to form the dental papilla. (C) The early cap stage of tooth development. Morphogenesis of the tooth germs has begun, the ectomesenchyme has condensed to form the dental papilla. Internal dental epithelial cells can be distinguished from external dental epithelial cells; see text. The transient, putative signaling center, the enamel knot, is visible at this stage and is associated with another transient structure, the enamel septum. (D) The late cap stage of tooth development. Epithelial cells of the dental organ have formed the stellate reticulum. Internal dental epithelial cells become more columnar in shape, whereas the external dental epithelial cells are cuboidal. (E) The late bell stage of tooth development. Internal dental epithelial cells have differentiated into ameloblasts that secrete enamel matrix. Dental papilla cells have differentiated into odontoblasts that give rise to predentine.

all found to contribute to the further development of the premaxillary incisor. A further seven transitory tooth primordia were found between the premaxillary incisors and maxillary molars in the diastemal regions (Peterkova *et al.,* 1995, 1996; Lesot *et al.,* 1998). However, these all regressed (probably by apoptosis). It has been suggested that these transitory primordia might be anlagen of ancestral canines and premolars. Three-dimensional reconstruction of mandibular diastemal and molar regions showed that, unlike in the upper jaw diastema, only one shallow lamina forms (Lesot *et al.,* 1998).

B. Bud Stage

After the dental lamina has formed, further invagination of the epithelium into the underlying ectomesenchyme results in an epithelial structure that resembles a bud shape; hence, this is the "bud" stage (reached for incisors at E13

and molars at E13.5–14.0). The epithelial invagination is now termed the dental organ. The peripheral cells of the organ remain cuboidal, whereas the cells on the oral surface are squamous and those internal to the peripheral epithelial cells are densely packed and round in shape (Fig. 3B). The formation of the epithelial bud is associated with a localized condensation of the underlying ectomesenchyme. Most of the mesenchyme condenses at the base of the dental organ forming the dental papilla.

C. Cap Stage

The next stage of tooth development is the "cap" stage; incisors reach an early cap stage by E14.0 and molars by E14.5. The dental organ changes in shape, looking less bud-like and more conical with an indentation on its deep surface (Fig. 3C). Peripheral epithelial cells are now different in appearance. External dental epithelial cells are cuboidal in shape, having a very large nucleus and little cytoplasm, whereas internal dental epithelial cells are slightly more columnar in shape with a central nucleus. The region where the internal epithelial cells meet the external epithelial cells is called the cervical loop. The dental organ sits on the dental papilla, now a "ball" of densely packed ectomesenchyme. Both the papilla and the dental organ are encased in the dental follicle, which consists of condensed mesenchyme cells, oriented in a radial pattern. At this stage of tooth development, the enamel knot (EK) first becomes evident as a transient localized thickening of internal dental epithelial cells that has been proposed to act as a signaling center with a putative role in cusp formation (Butler, 1956; Vaahtokari *et al.*, 1996a). The enamel septum is another transient structure of the dental organ present at the early cap stage of molar tooth development. The septum is a strand of epithelial cells that runs from the EK to the external dental epithelial cells, dividing the dental organ into two parts. The function of the enamel septum is as yet unknown. At this early cap stage the incisor tooth germs begin to differ in shape slightly from the molars. In the incisors the labial projection of the cap stage tooth grows more posteriorly (more anteriorly in the maxillary incisors) than the lingual projection of the dental organ.

By E15, the central cells of the dental organ become recognizable as the stellate reticulum (Fig. 3D). The stellate reticulum cells are star shaped with large intercellular spaces potentially playing a role in supplying enamel-secreting ameloblasts with nutrients. By E16.0 in the incisors and E17.0 in the molars, the stratum intermedium begins to be recognizable from the internal dental epithelial cells as a layer of flattened epithelial cells between the stellate reticulum and the internal dental epithelium. The internal dental epithelial cells increase in height and become preameloblasts.

D. Bell Stage

Incisors reach the bell stage by E17.0 and molars by E17.5–18.0. The dental papilla cells differentiate into odontoblasts, beginning in the most anterior mesenchymal cells (Fig. 3E). The external dental epithelial cells decrease in thickness and become a one or two cuboidal cell layer. The preameloblasts almost double in height and differentiate into ameloblasts, their nuclei peripherally placed; this differentiation occurs in the most anterior regions first. There are areas of internal dental epithelial cells in both the incisors and molars that do not differentiate into enamel-secreting ameloblasts. The lingual side of the incisors does not become coated with enamel because internal dental epithelial cells do not differentiate into ameloblasts on this side of the incisor tooth. In addition, in the molars the tips of the cusps do not become coated with enamel; again, this is as a result of internal dental epithelial cells failing to differentiate into ame-

loblasts. At E17.0 these nondifferentiating internal dental epithelial cells are at their greatest height. In subsequent stages of development, they diminish and become cuboidal in shape, then flattened, and eventually merge with adjacent connective tissue cells.

By E18 in the incisors and E19 in molars, the odontoblasts begin to secrete predentine. The dentine laid down on the labial side of the incisors forms a much thicker layer than that on the lingual side. After a further 24 hr of development, predentine becomes calcified and enamel matrix is secreted by the ameloblasts. Enamel matrix is only laid down in areas where ameloblast differentiation has occurred, therefore, in the incisors, no enamel is seen on the most anterior labial side, the incisal edge, or the lingual side of the tooth. Calcification of the enamel matrix occurs postnatally and the incisors and first molars erupt by 20 days after birth (P20). When mineralization of dentine and enamel occurs the shape of the tooth becomes fixed.

E. Second and Third Molar Development

The second and third molars develop when the jaw of the mouse has elongated enough to accommodate their large size. The first signs of the second molar, the dental lamina, can be seen at E15.5 forming as an outgrowth of the first molar tooth germ epithelium. By E18.5 the second molar is at the cap stage of tooth development and these teeth erupt approximately 25 days after birth (P25). The lamina of the third set of molars is not apparent until the fourth day of postnatal development; they reach the cap stage by P7–9 and the bell stage by P10; the third set of molars have erupted by the 35th day postnatal. A percentage of wild-type mice may fail to develop third molars (Grüneberg, 1963).

More than 200 genes have been identified as having been expressed during tooth development, and their expression patterns are available on a database (http://www.honeybee .helsinki.fi/toothexp). Molecular markers exist for all cell types and stages of tooth development.

VI. Appendix 2: Morphological Organization of the Murine Skull

The murine skull is a complex, composite, modular assemblage of skeletal elements with diverse developmental origins that encases the brain, its associated primary sensory organs, and the oral and respiratory openings. Ontogenetically it can be viewed as three separate skulls: embryonic, perinatal, and adult (Moore, 1981; Zeller, 1987; Presley, 1993). The initial, embryonic cranial skeletal structures to appear are the cartilaginous (chondrocranial) elements, whose development reflects the bauplan of the vertebrate skull (Fawcett, 1927; Goodrich, 1958; Barghusen and Hopson, 1979; Moore, 1981; de Beer, 1985; Kuhn and Zeller, 1987; Zeller, 1987; Northcutt, 1990; Gans, 1993; Hanken and Thorogood, 1993; Novacek, 1993; Kuratani *et al.,* 1997). The perinatal skull arises with the advent of the nascent dentition and the intramembranously ossified elements of the skull (the dermatocranium) which develop around the chondrocranium. The adult skull, or syncranium, develops with the refined modeling and remodeling of the cranial elements to fit the functional demands of the adult.

A. General Description: Units and Divisions

Historically a number of terms have been used to describe divisions and units within the skull based on embryonic origin, function, or mode of ossification. The chondrocranium (endocranium) is composed of those structures, which develop as cartilaginous units. These units have numerous possible fates: endochondral ossification, direct investment by dermal bone, degeneration, or transdifferentiation. Two chondrocranial components are distinguished: the neurocranium and the splanchnocranium (viscerocranium). The neurocranium is based on a functional division and includes those structures that support the central nervous system (CNS), and the primary sensory organs. The neurocranium can be further subdivided into parachordal, trabecular (prechordal or achordal), and sensory capsular divisions. The viscerocranium is composed of those elements derived from the branchial (visceral or pharyngeal) arches, which give rise to the masticatory, pharyngeal, and laryngeal apparatuses. In mice, as in eutherians in general, splanchnocranial elements also form portions of the cranial side wall (see below). The dermatocranium (exocranium) is composed of those intramembranous elements that surround and develop in coordination with the chondrocranium and dorsal CNS.

B. Overview of the Embryonic Chondrocranium

The early (~E13) embryonic neurocranium develops out of the coalescence of a number of chondrogenic centers (Fig. 4). The floor is subdivided rostrocaudally at the hypophysis into the parachordal basal plate (pars parachordalis) and the trabecular basal plate (pars trabecularis, trabecula cranii or prechordal plate). This division is fundamental in nature, dividing the NC-derived from the mesoderm-derived structures. As the name suggests, the parachordal cartilage develops adjacent to, and appears to be induced by, the notochord (Pourquié *et al.,* 1993; Martaugh *et al.,* 1999). The rostralmost end forms from the paired parachordal cartilages proper, while the caudal end forms from the somite-derived occipital cartilages. Initially, the chondrogenic centers of the paired parachordals are separate, but they coalesce medially; the rostral gap between them, the basicranial fenestra, may

perinatally persist as a small hole. The occipital cartilages develop in greater proximity to the notochord and no gap is seen. Caudally, lateral alae will extend to meet at the dorsal midline. The caudal basisphenoid and basioccipital will form within the parachordal basal plate. The trabecular basal plate has a complex evolutionary genesis and is formed by a number of chondrogenic centers. The caudalmost end is formed by the acrochordal and polar cartilages which eventually form as a plate (hypophyseal lamina) across the midline just anterior to the rostralmost end of the parachordal plate (clearly seen ~E14.0). Rostral to the acrochordals are the associated, paired trabecular cartilages. In the mouse, the trabecular cartilages are closely apposed at the midline so as to usually appear as a single midline cartilaginous rod. The trabecular and acrochordal cartilages quickly come into continuity, yielding the bodies of the basisphenoid (caudad) and presphenoid (rostrad). These structures will further conjoin with the parachordal cartilage culminating in a continuous basal plate (central stem). At the rostralmost tip of the notochord, within the basisphenoid where the pituitary forms from Rathke's pouch, a hypophyseal fenestra may be seen early and a hypophyseal fossa later. Hence, the mature basisphenoid, on which lies the hypophysis, is both a mesodermal and an ectomesenchymal element. The basisphenoid extends bilateral basitrabecular processes (processus alaris), which conjoin with the alicochlear commissures and alae temporalis (see below). The trabecular rod extends anteriorly to contribute to the midline nasal septum (septum nasi) to which the nasal cartilages join, forming the mesethmoid under the olfactory bulbs. The trabecular plate thus provides the support for CNS structures from the hypophysis to the olfactory bulbs. To either side of the basal plate are the paired nasal, orbital, and otic sensory capsules. The nasal and most of the orbital cartilages appear to be NC derived while the otic is principally mesodermal in origin. Ventral and lateral to the neurocranium the elements of the splanchnocranium develop, the most prominent of these being Meckel's cartilage (MC)—the cartilaginous core of the mandibular branch of the first branchial arch (BA1). [Terminology for the branchial/visceral/pharyngeal arches, and for the skull in general, grew out of a number of distinct anatomic and palaeontologic traditions and suffers from a lack of cohesion. For example, for some the first "branchial arch" develops caudal to the "mandibular" (BA1) and "hyoid" (BA2) arches and is equivalent to the third branchial arch of the terminology employed here.]

C. Regional Morphology

1. Nasoethmoidal (Frontonasal) Region (Regio Ethmoidalis): Chondrocranial

The nasoethmoidal (frontonasal) region is comprised of those structures which support the olfactory apparatus, in-

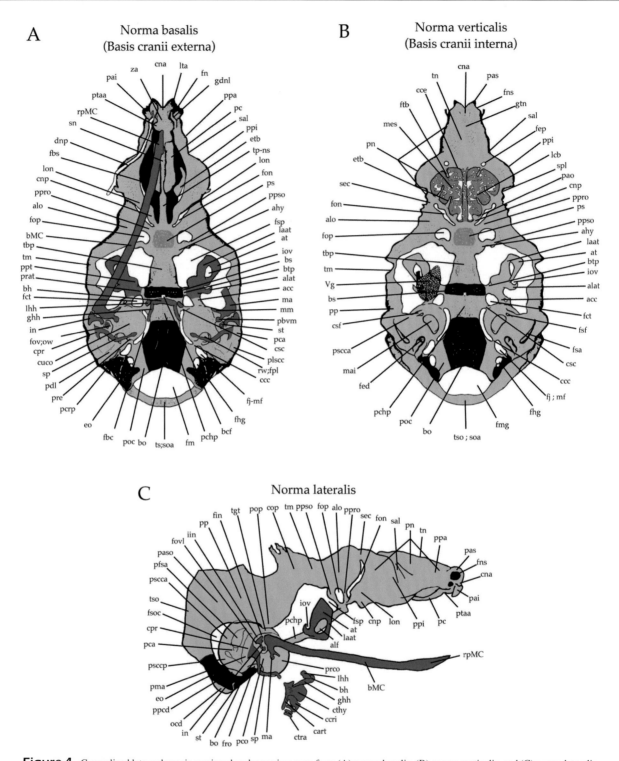

Figure 4 Generalized late embryonic murine chondrocranium seen from (A) norma basalis, (B) norma verticalis, and (C) norma lateralis. Dark blue represents splanchnocranial elements; turquoise represents the cribriform plate. See list of abbreviations for structural identification.

cluding both respiratory and olfactory epithelium, the VNO, the olfactory bulbs, and rostral palate. The regio ethmoidalis of the chondrocranium is represented therefore by the nasal capsules and portions of the trabecula cranii (Fig. 4). The nasal capsules themselves are paired cartilaginous sacs

formed around the invaginating olfactory pits. They fuse to the trabecular nasal septum at the midline, thereafter forming a continuous dorsal roof (tectum nasi), a fenestrated ventral floor (solum nasi), and side walls (paries nasi). The antero-rostral ends of the capsules yield the cupula nasi anterior, and the anterodorsal tectum is lined with a median sulcus dorsalis nasi which forms a shallow rostrocaudal groove. Paired projections at the rostralmost end, the processus alaris superior, form the posterosuperior margins of the narial openings (fenestra narina). On each capsule, a processus alaris inferior is separated from the superior by the fenestra superior nasi. The caudal end of the tectum is perforated at the foramina epiphaniale, which transmit the lateral branches of the ethmoidal (nasal) nerve of the ophthalmic (profundal) branch of the trigeminal (V1). These run from within to without the nasal cavity under cover of the nasal bones. The caudal border of the tectum, the crista cribroethmoidalis, meets the obliquely oriented paired laminae cribrosae (cribriform plates), which underlie the olfactory bulbs. These are mammalian neomorphs. Numerous olfactory foramina (foramina olfactoria) perforate this cartilaginous plate. At the dorsal margin of the plate, just ventral to the crista cribroethmoidale, are the cribroethmoidal foramina, through which pass into the nasal cavity a few fila olfactoria (CN I) and the parent trunks of the ethmoidal nerves (Fig. 5). Where the posterior end of the paries nasi meets the caudomedial end of the lamina cribrosa (the lamina infracribrosa) the planum antorbitale (cartilago antorbitalis, or lamina orbitonasalis) forms. This meets the trabecular plate (here, the interorbital septum) medially, the lamina infracribrosa superiorly and the lamina transversalis posterior inferiorly (see below) thus forming the cupula posterior. The paries nasi are divided into an anterior wall (pars anterior nasi, the interior of which is the pars maxillonasoturbinalis), an intermediate wall (pars intermedia, the interior of which is the pars lateralis) and a

posterior wall (pars posterior nasi, or pars ethmoidale, the interior of which is the pars ethmoturbinalis). The pars anterior is perforated by the fenestra dorsalis. The pars intermediale contains a laterally expanded prominentia lateralis, in correspondence to the internal recessus frontoturbinalis (wherein lies the crista semicirularis) and is separated from the pars anterior by the sulcus anteriolateralis. The external surface of the caudolateral prominentia extends a small spur. The anterior nasal floor is formed by the paired lamina transversalis anterior, which meet at the midline to form the zona anularis. Projecting from the lateral margins of the lamina transversalis anterior are the processus alaris inferior (see above). Caudal to this, on either side of the nasal septum (septum nasi), are the curled paraseptal cartilages, which house the VNO and represent the junction of the medial nasal capsule and the septum. Further caudad the floor takes contribution from the paired lamina transversalis posterior, extensions from the planum antorbitale representing the posteromedial nasal capsules. Between the medial septum nasi and paraseptal cartilages and the lateral paries nasi is an anteroposterior vacuity—the fenestra basalis. Within the interior of the nasal capsules a number of turbinal (turbinate) cartilages form. These are thin, convoluted shelves that greatly increase the surface area covered by the nasal epithelium. The anteriormost turbinals, the atrioturbinale (where the paries nasi meets the lamina transversalis anterior) and marginoturbinale (caudad to the atrioturbinale), develop from the nasal floor. These are followed caudad by the (inferior) maxilloturbinale and (superior) nasoturbinale which project from the pars intermediale. A cartilaginous crest, the crista semicircularis, projects inward from the prominentia lateralis. From the paries nasi and lamina cribrosa arise the frontoturbinale (superiorly) and several ethmoturbinale. The turbinals are separated by various recesses. The mesethmoid is formed by the elements of the cribriform plate, nasal septum, and associated turbinals.

2. Nasoethmoidal (Frontonasal) Region (Regio Ethmoidalis): Dermatocranial

A number of dermatocranial elements are associated with the nasoethmoidal region (Figs. 6 and 7). These include the nasal, frontal, lacrimal, premaxillary, vomer, and maxillary bones. The nasal bones develop as paired elements over most of the tectum nasi. Their rostral borders are free and just caudal to the processus alaris superior, while their caudal borders suture with the frontals. Medially they contact the maxillae and premaxillae. The frontals are expansive, paired bones overlying the rostrodorsal telencephalon as well as the hindmost nasal capsule, mesethmoid (and the olfactory bulbs) medially, and the ala orbitalis (see below) laterally. They articulate rostrad with the nasals, caudad with the parietals (the coronal suture), and mediad with each other (metopic fontanelle, thereafter suture). Interfrontal bones can variably be seen within the rostral metopic suture in certain backgrounds. Ventrolaterally the frontals form the interior

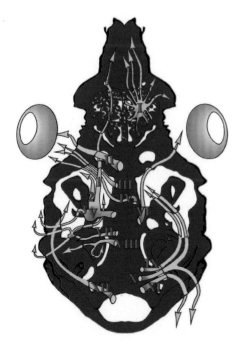

Figure 5 Exit points of the cranial nerves (green) relative to the late embryonic murine chondrocranium (purple).

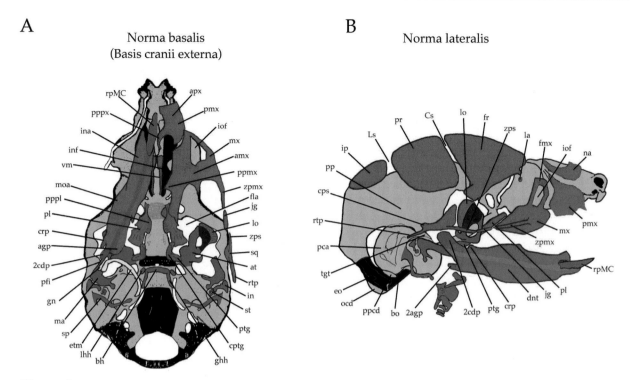

Figure 6 Generalized late embryonic murine dermatocranium relative to the underlying chondrocranium seen from (A) norma basalis and (B) norma lateralis. Red depicts the ossifying elements of the dermatocranium, blue depicts the chondrocranium, and yellow depicts the future position of late ossifying elements. See list of abbreviations for structural identification.

orbital wall dorsal and anterior to the optic foramen of the orbitosphenoids (see below). Within the orbit, they articulate with the lacrimals, orbito- and alisphenoids, maxillae, palatines, and squamosals. Ventrolateral to the nasals, the premaxillae develop along the pars anterior of the capsular side wall. Each supports three main divisions: the body and alveolus, the palatine (palatal) process, and the nasofrontal process. Each body and alveolus houses an upper incisor and extends caudad to a maxilla. The palatine process represents the primary palate and is extended mediad toward its contralateral partner, with a rostrocaudal expansion at the midline, and caudad to the vomer. Perinatally, the palatal process underlies the paraseptal cartilages and the VNO. The lacrimal, a small laminar bone, forms along the rostrodorsal orbit adjacent to the posterior of the prominentia lateralis of the pars intermedia. It houses the lacrimal glands and represents the caudodorsal end of the nasolacrimal duct. It articulates with the frontal processes of the maxillae and the frontals. The maxillae develop out of the maxillary branch of BA1 but their morphogenesis is intimately tied to that of the frontonasal prominences, which give rise to the nasal capsule. It is the largest bone of the upper face and takes part in forming the oral, nasal, and orbital cavities. It consists of a body and a number of processes: alveolar, frontal (ascending), palatine, spheno-orbital, and zygomatic. The body will form the lateral osseous wall of the maxillary recess (undeveloped early on). The alveolar process extends caudad and houses the upper molar dentition. The zygomatic process lies lateral

to the alveolar and is a crest of bone that articulates with the jugal to contribute to the inferorostral lateral skeletal orbit. From the root of the zygomatic process and the body arises the dorsal frontal process, which eventually articulates with the frontal. This process develops around the prominentia lateralis. An expansive fenestration, the infraorbital foramen, dominates the inferior aspect of the frontal process; the maxillary division of the trigeminal (CN V2) and the nasolacrimal duct pass through. The palatine process runs toward the midline from the caudal end of the maxilla. At the midline the process runs rostrad toward the palatine process of the premaxillae. The fenestrations at either side of this meeting yield the incisive foramen. The spheno-orbital process is the flat sagittaly oriented process caudal to the frontal process which contacts the sphenoid and inferoposterior frontal. The vomers are paired, midline dermal bones that (usually) will eventually fuse across the midline. In shape they resemble a cup under the lower margins of the septum nasi. Rostrad, they extend to the posterior extremities of the palatine processes of the premaxillae between the paraseptal cartilages. Caudad, the vomer stretch laterally to the lamina transversalis posterior and the nasopharyngeal openings between the rostral vertical lamellae of the palatine.

3. Orbitotemporal Region (Regio Orbitotemporalis): Chondrocranial

The orbitotemporal region is represented by those structures which develop in support of the optic apparatuses and

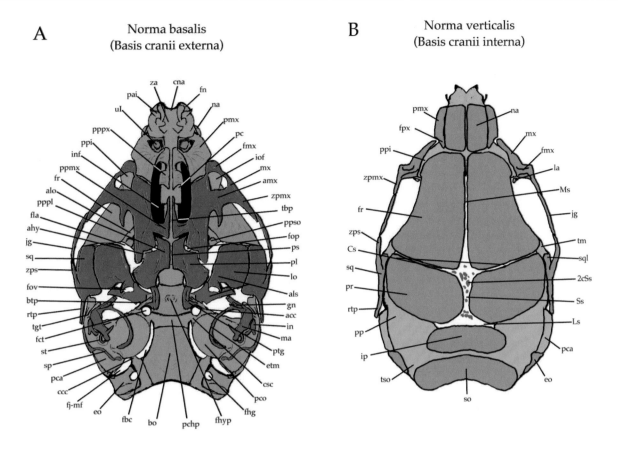

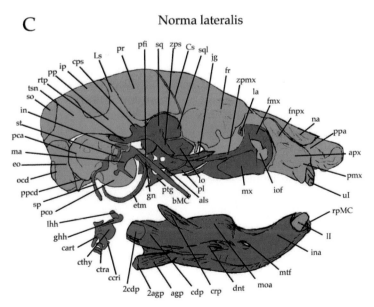

Figure 7 Neonatal murine cranial hard tissue seen from (A) norma basalis, (B) norma verticalis, and (C) norma lateralis. Dark red represents the branchial arch dermatocranial elements; light red represents the remaining ossifying centers of the dermatocranium and chondrocranium. Orange depicts the dental tissues. See list of abbreviations for structural identification.

structures within the cavum epiptericum, effectively forming the neurocranial side walls and bridging the nasoethmoidal and otic regions (Figs. 4, 6 and 7). A principal cartilaginous structure is a dorsal, rostrocaudally oriented strut running from the nasal capsule to the occipital arch and connected to the basal plate by mediolateral struts. The dorsal strut is divided into a rostral ala orbitalis and a caudal taenia marginalis (processus marginalis, or orbitoparietal commissure). From the ala orbitalis to the trabecular basal plate at the presphenoid run two pillars (radix): an anterior pila preoptica (preoptic root, or pillar) and a posterior pila postoptica (postoptic root/pillar, or pila metoptica). The dorsorostral border of the ala orbitalis connects with the nasal capsule via the sphenethmoidal commissure (commissura orbitonasalis). The ala orbitalis and its pillars comprise the orbitosphenoid. The orbitosphenoid and the presphenoid (within the trabecular basal plate) thus delimit the optic foramen for the second cranial nerves (CN II; Fig. 5). The orbitonasal fissure (fissura orbitonasalis), through which the ethmoidal branch runs dorsad (reentering the cavum cranii extradurally) along the cribriform plate to eventually exit through the cribroethmoidal foramen, separates the nasal capsule from the orbitosphenoid. In mice the mature pila postoptica is formed by an extension from the ala orbitalis (the posterior radix) which meets a separate cartilaginous body, the ala hypochiasmata, which extends to the basal plate. The posterior of the taenia marginalis forms an expansive parietal plate (lamina parietalis) dorsal to the otic capsule. Adding to the chondrocranial side wall of eutherians are the ala temporalis (see below), which is derived from the maxillary branch of BA1, and the mammalian neomorphic tegmen tympani of the otic capsule (see below). The cellular origins of the tegmen tympani, unique to therian mammals and intimately connected to the middle ear bones and geniculate ganglion, are uncharacterized and may well prove to be CNC derived. Expansion of the mammalian brain was accommodated in part by modification of the cavum epiptericum and cavum supracochleare (which is to the geniculate ganglion as the ala temporalis is to the trigeminal). This was achieved through the loss of portions of the original, primary side wall (including the pila antotica, which normally would run from the posterior taenia marginalis of the basitrabecular process at the basisphenoid) and the incorporation of the splanchnocranial ala temporalis and the tegmen tympani.

4. Orbitotemporal Region (Regio Orbitotemporalis): Dermatocranial

The intramembranous bones associated with the orbital region include those described in association with the nasal capsule (lacrimal, maxillae, and frontal) and the BA-associated jugal, squamosal, palatine, and alisphenoid (Figs. 6 and 7). The jugal (zygomatic or malar) are small rostrocaudally elongate dermal bones connecting the maxillary zygomatic process (overlying it) and the squamosal zygomatic process

(underlying it). The three together form the zygomatic arch, the lateral orbital skeleton. The squamosals are key integrating elements contributing to the orbit, the temporal side wall and the primary jaw articulation (but secondary craniomandibular joint). They possess a central body from which a number of processes and lamina extend. The zygomatic process runs rostrolaterally to articulate with the jugal. The squamosal lamina extends anteriorly to meet the lamina obturans (LO, or anterior lamina) of the alisphenoid and the frontal. A retrotympanic process runs caudad to overlie the tegmen tymapani and the middle ear elements. The sphenotic lamina extends ventrad toward the cupula cochlea and the sphenoid. The caudal process extends dorsocaudad between the squamosal lamina and the retrotympanic processes. The body forms at the confluence of these, forming the ventrally oriented glenoid cavity for the articulation with the condylar process of the dentary (mandible). The palatines make minor contributions to the caudoventral orbit. They are composed of horizontal and vertical lamellae. The horizontal lamellae form the posterior hard palate, the floor of the nasopharyngeal duct, and the anterior margin of the internal naries. The vertical lamellae extend up around the nasopharyngeal duct, where they eventually become closely related to the lamina transversalis posterior and the vomers. The sphenoid is a composite structure formed from trabecular, parachordal, orbital, and splanchnocranial cartilage, as well as the mammalian neomorphic LO. The LO, an appositional, investing, dermal bone, named after the homologous structure in monotreme mammals, spreads through the spheno-obturatory membrane and into the ala temporalis. The ala temporalis and the lamina obturans together form the alisphenoid. Its anterior border is initially separated from the pila postoptica by the large sphenoparietal foramen, which later persists as the sphenoidal fissure (sphenorotundal, superior orbital fissure, or foramen lacerum anterior). Cranial nerves III, IV, V1, V2, and VI all pass through (Fig. 5).

5. Otic Region (Regio Otica)

The otic region is extremely complex morphologically and is principally represented by capsular structures that fuse to support the auditory and vestibular apparatuses. The otic capsule (capsula auditiva) itself develops around the otic vesicle and has two main components: the pars canalicularis, which supports the vestibular semicircular canals and utriculus, and the pars cochlearis, which supports the cochlear duct and sacculus. The pars canalicularis is dorsal, lateral, and caudal to the medioventrally oriented pars cochlearis. The intracranial surface of the otic capsule is dominated by the rostromedial interior acoustic meatus (for the exit of CN VIII; Fig. 5), the caudolateral prominence of the endolymphatic duct (which passes through the foramen endolymphaticum), and the caudolateral fossa subarcuata (which houses the dura around the cerebellar paraflocculus). The internal acoustic meatus is split by the crista falciformis into a foramen acousticum superior (for the vestibular division of CN VIII) and the foramen acousticum inferior (for the cochlear division). A cartilaginous commissure, the commissura suprafacialis (praefacialis), forms a cartilaginous bridge on the dorsorostral border of the capsule beneath which passes the facial nerve (CN VII; Fig. 5). At the copula cochlea (apex cochlea), the rostromedial pars cochlearis connects to the ala temporalis via the alicochlear commissure. It is further connected to the parachordal basal plate via the anterior sphenocapsular commissure and the posterior chordocapsular commissure (connecting the medial edge of the processus recessus and the lateral edge of the parachordal plate). Between these commissures runs the basicapsular fissure. The alicochlear and sphenocapsular commissures delimit the carotid foramen. The extracranial surface is dominated by two fenestrations. The first is the fenestra ovalis (vestibuli, or orale), into which the stapes is lodged, on the lateral surface at the junction of the pars canalicularis and pars cochlearis. The second, on the ventral, lateral and caudal surface of the capsule, is the fenestra rotunda (cochlea, or tympani) for the secondary tympanic membrane and housing the scala tympani and aquaductus cochleae (for the communication of the perilymph and the cerebrospinal fluid in the subarachnoid space). It is covered in part by a shelf, the processus recessus, and less directly by a portion of the splanchnocranially derived styloid process. The aquaductus cochlea and processus recessus as such are considered mammalian neomorphs. The fenestra and aquaductus are contiguous with the jugular foramen (posterior lacerate foramen), together forming the metotic fissure. The jugular vein and glossopharyngeal, vagal, and spino-accessory cranial nerves (CN IX, X and XI) pass through here (Fig. 5). The styloid process is fused to the otic capsule at the crista parotica, and the eustachian tube passes forward in the interval between the styloid and the pars cochlearis. The crista parotica is related to the prominence of the lateral semicircular canal at the junction of the pars canalicularis and pars cochlearis, which overlies the elements of the middle ear. The rostral extension of this prominence forms the tegmen tympani, a mammalian neomorph, which overhangs the incus and malleus. The tegmen is the lateral wall of the cavum supracochleare, which lodges the geniculate ganglion of the facial nerve. The caudal end of the prominence forms the mastoid process, which is weakly developed in mice. The incus and malleus are housed in the fossa incudis and epitympanic recess, cavitations created by the overhanging prominences. The remainder of the pars canalicularis is dominated by the prominences of the anterior and posterior semicircular canals and the subarcuate prominence. The otic capsule fuses eventually near the paracondylar process of the exoccipital cartilage (see below) at the exoccipitocapsular commissure and to the tectum posterior of the supraoccipital arch via the supraoccipitocapsular commissure. These commissures help delimit two eutherian neomorphic fissurae, a dorsal superior occipitocapsular fissure

and a ventral inferior exoccipitocapsular fissure. Each capsule generally begins to ossify at three or four periotic centers to form the uniform petrosal: (1) from the prootic, anterior to the fenestra ovalis, over the suprafacial commissure and covering the ampulla of the superior semicircular canal; (2) from the opisthotic, enveloping the ventral cochlea and around the fenestra rotundra; (3) from the epiotic, in the posterior capsule around the mastoid process and posterior semicircular canal; and (4) from the pterotic, when forming as a separate center, in the lateral surface of the capsule juxtaposed to the tympanic.

6. Occipital Region (Regio Occipitalis)

The occipital region forms the caudal end of the chondrocranium. Within the caudal parachordal basal plate forms the basioccipital. Posterolaterally oriented occipital pillars (pilae occipitalis) lead to the dorsolaterally located exoccipital bones. Between the exoccipital and basioccipital the hypoglossal foramen can be found (for CN XII). Cartilaginous protrusions on the rostral and caudal borders of the exoccipital form the paracondylar (parocciptial process) and occipital condyle cartilages, respectively. Continuous dorsally, the occipital cartilages form a complete ring around the foramen magnum. The dorsal aspect of this ring is the tectum synoticum (tectum posterior) within which the supraoccipital will form. An anterior extension of tectum, the anterior supraoccipital process, runs to form a continuous cartilaginous sheet with the parietal plate.

7. Splanchnocranium

The splanchnocranial elements arise from the branchial arches and principally give rise to the masticatory, pharyngeal, and laryngeal apparatuses (Fig. 4). These elements have been under extensive evolutionary selective pressures and have subsequently undergone complex morphologic modification. The first branchial arch of mice is the largest and has two major components: the maxillary and mandibular branch derivatives of the palatoquadrate cartilage (PQ) and Meckel's cartilage (MC), respectively. In mice, the PQ derivatives are the ala temporalis and the incus (but see Zeller, 1987, regarding other possible remnants), which are united at the blastemal stage but condense into separate cartilaginous elements. As a cranial structure, the ala temporalis has a distinguished, complex evolutionary history. It is attached to the basitrabecular process of the cranial base at the basisphenoid. The ala temporalis is comprised of a number of structures: a horizontal lamina, a pterygoid process, an ascending process (lamina ascendens or anteromedial process) and an anterolateral process. At the rostromedial end of its fusion to the basitrabecular process forms the pterygoid process. The horizontal lamina underlies the body of the trigeminal ganglion. The ascending process extends dorsolaterally and is pierced by the alisphenoid foramen (for the stapedial, infraorbital artery). Between the ascending and the anterolateral processes is the incissura ovale, a cleft through which passes the mandibular branch of the trigeminal (CN V3). With the eventual investment of the lamina obturans, this will become an ossified foramen ovale. The ophthalmic and maxillary branches (CN V1 and V2) run rostromedial to the ascending process and exit the skull through the sphenoidal fissure (see above; Fig. 4). The incus (quadrate homolog) forms two primary synovial joints: The body extends to the malleus and the crus longus to the stapes. A crus brevis is lodged into the fossa incudis. The mandibular branch of BA1 is represented by the large rostrocaudally oriented MC. Though continuous, MC has three principal parts: the anterodistal, midline rostral process (Meckelian symphasis), the extended intermediate body, and the caudoproximal malleus (articular homolog). Principal characters of the malleus include a neck, head, manubrium, and processus brevis. The head is in continuity with the body of MC and articulates with the incus (the primary craniomandibular joint). The neck (or synapsid retroarticular process) is angled ventrad relative to MC and accepts the inserts of the tensor tympani muscles. The end of the neck yields the bulbous processus brevis (muscular process) and the manubrium (a neomorph), which runs down and forward following the line of the spiral sulcus of the cochlea. The body of MC degenerates and the surrounding, sheathing tissue forms the sphenomandibular ligament and malleal anterior ligament.

The second branchial arch (BA2) splanchnocranium is represented by four structures: the stapes, the styloid process, the lesser horns of the hyoid, and the dorsoproximal hyoid body. The proximal end of BA2 at blastemal stages has two connected processes, a medial and a lateral: a medial one pierced by the stapedial artery and yielding the stapes (hyomandibular homolog), and a lateral one, Reichert's cartilage (RC, hyoid cornu), fused to the otic capsule at the crista parotica. The head of the stapes articulates with the incus, the neck receives the stapedial muscle insertions, the two crurae encircle either side of the stapedial foramen, and the base, or foot plate, is ligamentously attached to the margins of the fenestra ovalis. The proximal end of RC is formed by the tympanohyal (laterohyal, or suaropodial extrastapes), which is fused to the otic capsule at the crista parotica, and the proximodistally elongated stylohyal. Together, these form the styloid process. The styloid process is united to the lesser horns (cornu) of the hyoid (ceratohyal homolog) via the stylohyoid ligament. The lesser horns meet the lateral ends of hyoid body. The body is a composite of BA2 (dorsoproximal body; certato and hypohyal) and BA3 (ventrodistal body; basihyal or basibranchial) element. The greater horns (cornu) of the hyoid, which extend caudad from the lateral margins of the body, are also BA3 derived. The fourth and more caudal BA gives rise to the laryngeal cartilages. The thyroid cartilage (BA4) is a concave element composed of two lamellae (alae) that meet at the midline. From the posterior upper margin project (toward the greater horns of the

hyoid) the superior cornu, while the inferior cornu project from the posterior lower margin. Within the concavity of the thyroid, in the back of the larynx, are the paired arytenoid cartilages (BA4–6). Articulating with the inferior arytenoid borders is the cricoid cartilage (BA4–6).

8. Branchial Arch Dermatocranium

The first branchial arch yields a number of dermal ossifications, many of which have been presented in relation to the nasoethmoidal (maxillae) and orbitotemporal regions (jugal, squamosal, palatine and alisphenoid). In addition to these, BA1 generates the pterygoid, dentary, gonial, and ectotympanic bones (Figs. 2, 6 and 7). The pterygoids develop in close association with the lateral basisphenoid and ala temporalis on the ventral neurocranium. They form L-shaped structures caudal to the palatines and ventral to the basitrabecular processes, occupying the lateral margins of the nasopharyngeal openings. In the body of murine pterygoids, as in many mammals, a pterygoid cartilage, composed of secondary cartilage, transiently forms around E16. The dentary is the largest dermal bone and forms the lower jaw (mandible), developing around most of the length of MC. As the name implies it bears the teeth of the lower jaw: an elongate incisor, surrounded by an incisive alveolus, and three molars housed in a molar alveolus. An extended diastema separates the two, and they converge about a central body where the mental foramen exits buccally. Lingually, there is a mandibular foramen. The proximal end is dominated by three processes: the anteriormost coronoid, the condylar, and the posteriormost angular. The coronoid forms a caudodorsal strut toward the zygomatic arch. The condyle runs back toward the squamosal (the secondary craniomandibular joint), and contains at its proximal heart a large secondary cartilage mass. The angle, likewise maintaining a secondary cartilage proximally, runs caudally. The gonial (prearticular) is an investing bone that lies on the dorsodistal surface of the malleus. It will invade the malleus and become the processus folii (anterior, or gracilis) of the mature malleus at the point where the malleus separates from MC; hence (like the maxillary alisphenoid) the mature mandibular malleus is a composite endochondral (articular) and dermal (prearticular) bone. Underlying the same position on the malleus is the anterior process of the ectotympanic (angular). The ventral posterior (reflected lamina) process curves back caudad as a ring toward the styloid process, thus delimiting the tympanic membrane.

9. Calvarial Dermatocranium

By osteological tradition, the calvarium is the cranium minus the face (splanchnocranium) while the calotte is the calvarium minus the neurocranial base. Thus, the calvarial dermatocranium includes the frontals, parietals, interparietal, lamina obturans of the alisphenoids, and squamosals. The parietals are large, rectangular bones caudal to the frontals and dorsal to the squamosals. They come together me-diad at the sagittal suture, and form the coronal suture with the frontals and the lambdoidal suture with the interparietal. The interparietal lies dorsal to the otic capsule and just rostral to the endochondral supraoccipital.

The calvarium develops from the mesenchyme encompassing the brain. This mesenchyme has two layers: an inner endomenix and an outer ectomenix. The endomenix includes the deep pia mater, which covers the brain, and the arachnoid, through which the cerebrospinal fluid and blood supply interface. The ectomenix likewise has two closely apposed layers: a deeper dura mater and a more superficial mesenchyme in which the skeletal elements form. The dura acts as the endosteum, or endocranial periosteum, of the calvarium. The ectomenix has both chondrogenic and osteogenic capacity. The dura mater has three principal septa subdividing the brain: the falx cerebri, the falx cerebelli, and the tentorium cerebelli. These structures have organized fiber bundles related and attached to the developing sutures. The eventual adult cranial form is shaped in part by directed forces due to growth of the brain and constraints of these fibers. Secondary cartilage, which is derived from the application of biomechanical forces to particular ectomesenchyme in the head, can often be seen perinatally in the sagittal and lambdoidal sutures.

VII. Appendix 3: Molecular Regulators of Craniofacial Pattern and Development

A. Genes Encoding Secreted Signaling and Related Factors

1. Endothelin-1 (ET-1), ET-A, ECE-1

ET-1, initially reported as a potent vasoconstrictor produced by endothelial cells, is one of several closely related peptide ligands that bind to the G-protein-coupled receptor ET-A (Kurihara et al., 1994; Clouthier et al., 1998). Biosynthesis of active ET-1 occurs in a two-step process: First, a furin-like protease cleaves preproendothelin-1; second, a cleavage by a metalloprotease, endothelin converting enzyme 1 (ECE-1), generates active ET-1 (Yanagisawa et al., 1998). ET-1 mRNA is detected in the ectoderm and endoderm of BA1–3 (Kurihara et al., 1995; Clouthier et al., 1998). It is also detected in the paraxial mesodermal core of BA1 and BA2 and the aortic arch artery and cardiac tissues. Conversely, the ET-1 receptor gene, ET-A, is expressed in migrating and postmigratory rhombencephalic CNC of the first three arches. ECE-1 is found expressed in both paraxial and CNC BA mesenchyme, neurepithelium, and both ectodermal and endodermal BA epithelia (Yanagisawa et al., 1998). Targeted disruption of ET-1, ET-A and ECE-1 each yields mice virtually identical in craniofacial defects (Kurihara et al., 1994; Clouthier et al., 1998; Yanagisawa et al.,

1998) (Fig. 2a). Neonates are micrognathic, with variable midline fusion of the lower jaw, and the pinnae and ventral neck are hypoplastic. The proximal dentary is agenic, and the body of MC, malleus, gonial, ectotympanic, incus, stapes (usually), and (presumably) the styloid process are absent. The external auditory meatus is missing. The rostral process of MC, distal dentary, and teeth (which are abnormally embedded in loose mesenchyme) are, however, present. The dentary does not extend to the squamosal at the otic capsule, instead articulating with a hypoplastic jugal. The dermal bone of the medial dentary fuses with the basal plate at the basisphenoid. The hyoid is severely malformed, forming a ringlike structure fused to the thyroid and the basisphenoid near the ala temporalis and pterygoids. The alisphenoid, palatine, pterygoid, and squamosal bones are underdeveloped. Muscles of the tongue are severely diminished, and submandibular and sublingual glands are lacking. In ET-1$^{-/-}$ embryos, distal midline expression of Hand2 and Hand1 is decreased but present (Thomas et al., 1998). The sucker (suc) gene encodes the zebrafish homolog of ET-1; suc mutants fail to induce or maintain expression of dlx2, dlx3, msxE, gsc, EphA3, or dHAND in ventral (distal) BA1 and BA2 tissues (Miller et al., 2000). The micrognathia of ET-A$^{-/-}$ embryos is not due to a failure of CNC to migrate to the BA based on AP2, Dlx2 and Crabp1 expression, though Dlx2 (and Barx1) expression is downregulated in the BA2 (only) by E10.5 (Clouthier et al., 2000). The BA are, however, smaller in these mutant embryos, and by E11.5 fewer BrdU positive cells and more apoptotic cells are detected in the BA. Neither Gsc nor Dlx3 (both ectodermal and ectomesenchymal) expression is detected in the BA of ET-A$^{-/-}$ embryos at E10.5 (Clouthier et al., 1998, 2000). A delay and downregulation of dHAND and eHAND expression, but not the associated gene Ufd1, are also seen. Dlx1 and Prx1 mRNA patterns, however, appear similar to wild-type embryos.

2. Epidermal Growth Factor (Egf) and Transforming Growth Factor α (TGF-α)

EGF, one of the first growth factors identified, was isolated from murine submaxillary glands (Cohen, 1962), while TGF-α was isolated from murine-sarcoma-virus transformed fibroblast cultures (De Larco and Todaro, 1978); both appear to bind the same receptor (Egfr). In vitro, EGF acts as an inhibitor of chondrogenesis of mandibular arch mesenchyme, including BMP-stimulated chondrogenesis (Coffin-Collins and Hall, 1989; Nonaka et al., 1999). Both Tgf-α$^{-/-}$ and Egfr$^{-/-}$ mice exhibit craniofacial defects, including cleft palate and defects in MC (Miettinen et al., 1999).

3. Fibroblast Growth Factors (Fgf), Fgfr1–3 and Sprouty

There are at least 14 mammalian fibroblast growth factor (Fgf) genes based on the presence of a core sequence encoding receptor-binding and heparin-binding domains (Lewandowski *et al.,* 1997). Most gene products, but not all (e.g., FGF1/aFGF and FGF2/bFGF) are secreted and avidly bind molecules at the cell surface, in particular heparan sulfate proteoglycans. FGF signaling is known to regulate craniofacial development: for example, *Fgf3*$^{-/-}$ mutants have dysmorphic otic capsules (Represa *et al.,* 1991; Mansour *et al.,* 1993; McKay *et al.,* 1996) and disruptions of the three known receptor tyrosine kinase (RTK) *FGF Receptors* lead to craniosynostosis or achondroplasia (reviewed by Webster and Donoghue, 1997; Wilkie, 1997; Burke *et al.,* 1998; see Table III).

Functional analysis of *Fgf8* provides perhaps the clearest demonstration of FGF involvement in craniofacial development. FGF8 was originally identified as a secreted androgen-induced growth factor (Tanaka *et al.,* 1992). It has a dynamic spatiotemporal pattern of expression in a number of known signaling centers which regulate pattern and morphogenesis, including the primitive streak, the apical ectodermal ridge of the limb bud, the midbrain-hindbrain isthmus, and the intraembryonic coelom (to form the heart), olfactory placodes, pharyngeal clefts (both endoderm and ectoderm) and oral ectoderm of BA1 (Heikinheimo *et al.,* 1994; Crossley and Martin, 1995; Trumpp *et al.,* 1999). To circumvent the embryonic death at gastrulation seen in *Fgf8*$^{-/-}$ mutants (Sun *et al.,* 1999), Trumpp *et al.,* (1999) employed a *Cre/loxP* strategy to inactivate *Fgf8* within the ectoderm of BA1 by E9.0. Significantly, gene inactivation in the caudal, pericleftal BA1, however, occurred later than in the rest of BA1 (Trumpp *et al.,* 1999). This strategy resulted in an agnathic mouse (*Fgf8*$^{Nes\text{-}Cre-/-}$) with a dramatic loss of most of the BA1-derived skeletal elements (Fig. 2b). The entire maxillary (incus and ala temporalis) and most of the mandibular splanchnocranium failed to form. Of Meckel's cartilage, only the rostral process (with associated dermal alveolar bone and diminutive incisors) and most of the malleus (including a well-formed manubrium and processus brevis) developed. The elongate body of MC did not. Most of the BA1 dermatocranial elements were agenic (e.g., palatine, pterygoid, lamina obturans, jugal) or severely hypoplastic (e.g., maxilla, dentary, ectotympanic-goniale). By E9.5, BA1 is clearly hypoplastic in these mutants. Based on *Cad6, CRABPI,* and *AP2.2* expression, CNC migrate to BA1 and, in an assay of proliferation, the ratio of BrdU-labeled cells to unlabeled cells was similar to wild-type embryos. However, extensive BA1 cell death was detected in *Fgf8*$^{Nes\text{-}Cre-/-}$ mice, suggesting that FGF8 acts as a survival factor in BA1.

Evidence for disruption of patterning in both the ectoderm and mesenchyme in *Fgf8*$^{Nes\text{-}Cre-/-}$ mice revealed that BA1 ectoderm maintained oral expression of *Pitx1* but lost *ET-1* (Trumpp *et al.,* 1999). Moreover, FGF8 dependent and independent regions of gene expression were found: The distal midline, which will eventually yield the rostral process,

incisors, and associated dermal bone, maintained expression of both *Pax9* and *Bmp4,* while the ectoderm of the molar field associated with the body of MC and much of the dentary lost *Pax9* expression. Expression of *Msx1,* a BMP4-inducible gene, was maintained in the underlying midline mesenchyme as was *Msx2, eHAND* and *dHAND.* Loss of *Msx1* leads to a loss of those mandibular midline structures that persist in the *Fgf8*$^{Nes\text{-}Cre-/-}$ mice—namely, the rostral process and incisors (Satokata and Maas, 1994). At E9.5, the FGF8-inducible genes *Dlx2* and *Dlx5* were appropriately expressed in the proximodistal axis of BA1, further suggesting that BA1 retained some appropriate spatial information. Whether FGF8 is necessary for later maintenance of the *Dlx* genes in the BA is uncertain. Mesenchymal expression of *Lhx7,* another FGF8-inducible gene, was maintained though that of the related gene, *Lhx6,* was not. Moreover, with a pertinent exception, mesenchymal expression of *Gsc* and *Barx1* was lost. This exception is in mesenchyme beneath the above-mentioned patch of caudal, peri-cleftal BA1 *Fgf8* expression. The ectoderm over this patch also maintained *ET-1* expression.

Those BA1 elements that do form in *Fgf8*$^{Nes\text{-}Cre-/-}$ mice are noteworthy. The portions of the maxillae that form in these mutants normally develop in close association with the nasal capsules and frontonasal processes—an *Fgf8*-positive region—which may have influenced their growth. The malleus and rudimentary ossification in the region of the juxtaposition of the ectotympanic, gonial, and malleal processus folii developed out of a region of BA1 in which the gene inactivation in the caudal, pericleftal BA1 occurred later than the rest of BA1. Expression of *ET-1, Barx1,* and *Gsc* in this spot remained: hence, the cells contributing to these structures may have been exposed to an elaborated FGF8 signal at the critical point of their ontogeny. The persistence of *Bmp* and *Msx* expression in the mandibular midline suggests the presence of an *Fgf8*-independent, *Bmp/Msx*-dependent field at the midline (Trumpp *et al.,* 1999). Last, data from the *Fgf8*$^{Nes\text{-}Cre-/-}$ mice support the idea that, as outgrowths, the BA receive patterning signals from *both* the rostral and caudal associated epithelium, as well as possibly from the olfactory field.

Members of the *Sprouty* (*Spry*) gene family encode RTK inhibitors, including *Fgfr* RTKs (Hacohen *et al.,* 1998; Casci *et al.,* 1999; Kramer *et al.,* 1999). Moreover, *Spry2* has been identified as a regulatory target of FGF8 signaling in BA1 (Minowada *et al.,* 1999; Trumpp *et al.,* 1999). Although overexpression of *Spry2* appears to lead to chondrodysplasia, craniofacial defects due a loss of function have yet to be presented (Minowada *et al.,* 1999).

4. *Pdgf, Pdgf-Rα,* and *Ph*

Platelet-derived growth factors (PDGFs) are disulfide-bonded homo- or heterodimers of "A" and "B" polypeptide chains. There are two RTKs for these ligands, PDGF-

Rα and -Rβ, which likewise can homo- or heterodimerize upon ligand binding (Seifert *et al.*, 1989). Homozygous mice carrying targeted deletions of *Pdgf-Rα* die during embryonic development with neural tube defects and subepidermal blebs and bleeding, particularly within the head (Soriano, 1997). Both dorsal and ventral skull structures are unfused along the rostral midline (Fig. 2c). Nasal capsules, with relatively intact paries nasi, develop and fuse to trabecula cranii—which, however, fail to fuse and are separated by a large gap. The face is thus cleft, as are the frontals and parietals. Other ventral neurocranial defects may also occur. Though largely undetailed, it seems likely that BA and otic structures are also extensively affected. Increased cell death in the cephalic region and branchial arches was detected in regions of migrating CNC. This supports a model of PDGF action as a survival factor, although loss of *Pdgfa* or *Pdgfb* does not appear to lead to skeletal defects of the nature seen in *Pdgfra$^{-/-}$* mice (Levéen *et al.*, 1994; Boström *et al.*, 1996; Soriano, 1997). Neuronal CNC derivatives in *Pdgfra$^{-/-}$* mice appear to migrate properly based on neurofilament assays. The *Patch* (*Ph*) mouse mutant, which presents a cleft face in particular backgrounds, is thought to have a deletion encompassing the *Pdgfra* gene (Orr-Urtreger *et al.*, 1992).

5. *Transforming Growth Factor β (Tgf-β)/Bone Morphogenetic Protein (Bmp)* **Superfamily, Inhibitors, and Effectors of Signaling**

The *Tgf-β* superfamily is an extensive gene family encoding secreted proteins that regulate cellular behavior, including migration, proliferation, differentiation, ECM production, apoptosis and regional patterning (reviewed by Kingsley, 1994; Hogan, 1996). This superfamily includes *Tgf-βs, Bmps, Gdfs, activins, inhibins, nodal,* and *Mullerian-inhibiting substance.*

TGF- proteins were initially named for their ability to promote anchorage-independent growth of fibroblasts in culture, and they are expressed in the craniofacial mesenchyme (Roberts *et al.*, 1981; Millan *et al.*, 1991; Pelton *et al.*, 1991). Targeted disruption of *Tgfb1* results in embryonic lethality by E10.5 (Dickson *et al.*, 1995). *Tgfb1* mRNA is detected in the preimplantation blastocyst and persists in multiple tissues through adulthood (Sanford *et al.*, 1997). Among the sites of expression are chondrocytes and perichondria, osteocytes of the head and optic and otic sensory epithelia. *Tgfb2* null neonates die with multiple craniofacial defects (Sanford *et al.*, 1997). The frontals, squamosals, parietals, and interparietal are dysgenic with reduced ossification and expanded fontanelles (Fig. 2d). The supraoccipital has no ossification and the exoccipital and basioccipital are connected. A malformed ala temporalis is present, extending two small ossicles laterad; the dermal lamina obturans is absent. A percentage have a cleft secondary palate. The mandible is reduced, in particular proximally where the angle is

absent and the condyle and coronoid much diminished. Defects are observed in the pars cochlearis. With the loss of functional *Tgfb3*, the medial edge epithelia of the palatal shelves fail to adhere and eliminate as a seam, resulting in a clefting of the palate (Kaartinen *et al.*, 1995; Proetzel *et al.*, 1995).

Bone morphogenetic proteins (BMPs) were identified as components of bone extracts with the capacity to induce ectopic bone and cartilage when implanted in rodents (Sampath and Reddi, 1981; Wozney *et al.*, 1988). *Bmp1*, the homolog of the *Drosophila* gene *tolloid* (*tld*), encodes a secreted procollagen C-proteinase with this inductive capacity (Wozney *et al.*, 1988; Suzuki *et al.*, 1996). *Bmp1$^{-/-}$* mice have malformed frontals, parietals, and interparietals along their sagittal sutures. *Bmp2* is expressed in the ectomesenchyme and *Bmp4* in discrete regions of the overlying ectoderm (in particular along the oral midline) of the early craniofacial primordia, as well as dorsal midline of the neural tube (Francis-West *et al.*, 1994; Bennett *et al.*, 1995). The effect of genetic inactivation of either *Bmp2* or *Bmp4*, however, is embryonic death (Winnier *et al.*, 1995; Zhang and Bradley, 1996). Ectopic application of BMP2 or BMP4 in mandibular explants has been shown to induce the expression of *Msx1* and *Msx2* (see above; Barlow and Francis-West, 1997; Tucker *et al.*, 1998b). *Bmp5* is the gene mutated in the *short ear* mouse (Kingsley *et al.*, 1992). As the name implies, loss of functional *Bmp5* leads to shortened BA-derived auditory pinnae. Functional loss of *Bmp6* appears to only affect the sternum (Solloway *et al.*, 1998). *Bmp7* is expressed in many cranial tissues, including in the stomodeal and surface ectoderm covering BA1, over and between the frontonasal processes, in the otic vesicle, the commissural plate between the telencephalic vesicles, and the NT at the cephalic flexure (Lyons *et al.*, 1995). *Bmp7* is required for the development of structures around the interface of the mesodermal and CNC skeleton. *Bmp7$^{-/-}$* mutant mice have an extensive fenestra in the caudal basisphenoid at the hypophysis (Dudley *et al.*, 1995; Luo *et al.*, 1995). The ala temporalis is dysmorphic, apparently lacking an anterolateral process and supporting an aberrant alicochlear commissure with an ectopic strut, which runs to the middle ear and fuses to the otic capsule. The lamina obturans is malformed, as is the pterygoid. Any defects in the otic or nasal capsule have not been reported. *Bmp5$^{-/-}$;7$^{-/-}$* compound mutants die at E10.5 with striking defects in tissues where both are expressed, including the rostral NT, the BA, and CNC (Solloway and Robertson, 1999). *Gdf5*, along with *Bmp5*, has been implicated in joint development (Storm and Kingley, 1996). *Activin βA* is expressed in the mesenchyme of the facial primordia (Feijen *et al.*, 1994; Roberts and Barth, 1994) and its knock-out yields mice with cleft palates and without mandibular incisors (Matzuk *et al.*, 1995a,c).

Nodal mRNA is detected early, marking the cells of the primitive streak. Nodal signaling is required for cephalic development, being necessary for the formation of the primitive streak and most mesoderm and anterior neural tissues (Conlon *et al.,* 1994; Varlet *et al.,* 1997; reviewed by Schier and Shen, 2000).

Noggin, chordin, and *follistatin* encode for proteins that appear to act as specific antagonists of TGF-β family signaling. *Noggin* is expressed early in the anterior primitive streak, node and mesendoderm, and later in condensing and developing cartilaginous tissues (Brunet *et al.,* 1998; Bachiller *et al.,* 2000). Targeted disruption, theoretically leading to enhanced BMP signaling, leads to excessive chondrogenesis and disrupted skeletal morphogenesis (Zimmerman *et al.,* 1996; Brunet *et al.,* 1998). The skull and cervical vertebrae of mutants are nearly normal at birth (Brunet *et al.,* 1998), though defects in MC and other chondrocranial elements may be evident with further characterization. The remainder of the axial and the appendicular skeleton is malformed with multiple failures of joint formation. *Chordin* is likewise expressed early in the primitive streak, node, and mesendoderm (Bachiller *et al.,* 2000). *Chordin*$^{-/-}$ mutant mice display inner and outer ear defects as well as abnormalities of pharyngeal and cardiovascular organization (Bachiller *et al.,* 2000). Compound *Noggin*$^{-/-}$;*Chordin*$^{-/-}$ mutants (similar to *Shh*$^{-/-}$ mutants, see below) have cyclopsia with agnathia and a single nasal pit (Bachiller *et al.,* 2000). Although neither *Noggin* nor *Chordin* are expressed in the AVE, expression of *Hesx1* in the AVE fails to be maintained in the compound mutants. Using a knock-in approach, Kanzler *et al.* (2000) utilized an enhancer of the *Hoxa2* gene to drive expression of *Xnoggin* (*Xenopus noggin*) in the BA to attempt to address the potential roles of BMP2 and BMP4 in regional development. This particular enhancer is active in the premigratory and migratory CNC of the BA2 and caudal arches (Nonchev *et al.,* 1996). Affected embryos have hypoplastic BA2, BA3, and BA4, apparently not due to increased apoptosis but to a loss of migrating cells. The associated skeletal elements are malformed. *Follistatin*$^{-/-}$ mutants have, among other defects, cleft palates and lack lower incisors (Matzuk *et al.,* 1995b).

Smad proteins are cytoplasmic mediators of TGF-β superfamily signaling through receptor serine/threonine kinases (reviewed by Heldin *et al.,* 1997). *Smad4* associates with *Smad1* and *Smad5* in response to BMP signaling, while Smad4–Smad2 and Smad3 associations occur in response to TGF- and activin signaling; these complexes translocate to the nucleus where complexes may act as either transcriptional activators or repressors (Heldin *et al.,* 1997; Gripp *et al.,* 2000). *Smad4*$^{-/-}$ mutants have undifferentiated visceral endoderm, lack mesoderm, and have anterior patterning problems (Sirard *et al.,* 1998). Inactivation of *Smad2* results in embryonic lethality (by E10.5) with defects in the visceral endoderm, epiblast, extraembryonic ectoderm, and an absence of embryonic mesoderm (Nomura and Li, 1998; Waldrip *et al.,* 1998), suggestive of a role in the organization

of the primitive germ layers prior to gastrulation. A percentage of *Smad2* heterozygotes are micrognathic (or agnathic) and may be eyeless. These phenotypes (severe cyclopsia and truncation of the rostral head) are compounded with transheterozygote *Smad2*$^{+/-}$;*Nodal*$^{+/-}$ mutants, indicating a genetic interaction if not a signaling relationship (Nomura and Li, 1998). *Activin Receptor IIA*$^{-/-}$;*IIB*$^{-/-}$ mutants, moreover, appear like *Smad2*$^{+/-}$;*Nodal*$^{+/-}$ compound mutants (Matzuk *et al.,* 1995c; Nomura and Li, 1998). Nuclear Smads interact with a number of cofactors, including TGIF and ski, thereby acting as transcriptional repressors (Gripp *et al.,* 2000). Mutations in the human *TGIF* gene, which is expressed in craniofacial and forebrain tissues, result in hypotelorism and holoprosencephaly (Gripp *et al.,* 2000). *Ski* knock-out mutants exhibit variable upper facial clefting, neural tube closure, neurocranial base, and dentary defects (Berk *et al.,* 1997).

6. *Retinoic Acid, RAR, RXR*

Vitamin A (retinol) is crucial for normal pre- and postnatal growth and survival, acting as a regulator of gene expression and morphogenesis (reviewed by Morriss-Kay, 1993; Sporn *et al.,* 1994; Brickell and Thorogood, 1997; Morriss-Kay and Ward, 1999). Retinoic acid (RA) is generated as an active derivative of vitamin A. RA acts as a lipidsoluble ligand for two families of nuclear receptors, RARs and RXRs, which act as transcriptional regulators and are expressed in many embryonic tissues including the CNC (Mangelsdorf and Evans, 1995). RA balance in the developing embryo is critical, because either a deficiency or an excess leads to developmental defects. The effects of vitamin A deficiency (VAD) syndrome include microphthalmia and cleft lip, palate, and/or face—defects that can be prevented with RA administration (Mark *et al.,* 1995; Morriss-Kay and Skolova, 1996). Ectopic exposure to RA leads to distinct, complex craniofacial malformations, including facial clefting and branchial arch alterations, in a manner that depends on dosage, location, and gestational time of exposure (Webster *et al.,* 1986; Wedden *et al.,* 1988; Morriss-Kay, 1993; Grant *et al.,* 1997; Mallo, 1997). Excess RA administered early (presomite stages) leads to rostral shifts in the position of the otic vesicle and BA, partial fusion of trigeminal and facial ganglia and the first and second BA, and a diminished preotic hindbrain. Such treatments appear to alter HB segmentation, as assayed by rhombencephalic markers (Morriss-Kay *et al.,* 1991; Holder and Hill, 1991; Conlon and Rossant, 1992; Marshall *et al.,* 1992). Of particular interest is the range (from shortened to lost) of MC morphology seen with the concomitant appearance of an ectopic MC-like structure in the maxillary BA1 (Morriss-Kay, 1993). The molecular nature of this shifting morphology is unclear but may be accompanied by changes in *Dlx* gene family expression (see below).

Both of the RA receptors, RAR and RXR, exist in three forms—α, β, and γ—with isoforms of each. For RAR, this includes $\alpha 1$ and $\alpha 2$, $\beta 1$–$\beta 4$, and $\gamma 1$ and $\gamma 2$. *RARα* transcripts are apparently nearly ubiquitous developmentally, *RARβ* more restricted, and *RARγ* in specific mesenchymal populations where there are precartilaginous condensations, including the frontonasal and branchial arch ectomesenchyme and the mesodermal limb buds (but apparently not in the FB, MB, or HB regions of presumptive CNC origin) (Dollé *et al.,* 1989, 1990; Ruberte *et al.,* 1990, 1991; Leroy *et al.,* 1991; Mendelsohn *et al.,* 1991, 1994a). *In vivo* functional studies of these genes have been addressed via targeted mutation, where the range of defects seen in the various mutations recapitulates nearly all of the defects associated with VAD syndrome (Ghyselinck *et al.,* 1997). Mice deficient for individual receptor isoforms, such as *RAR$\alpha 1$* (Li *et al.,* 1993; Lufkin *et al.,* 1993), *RAR$\beta 2$* (Mendelsohn *et al.,* 1994b), or *RAR$\gamma 2$* (Lohnes *et al.,* 1993), appear normal; however, when all isoforms of *RARα*, *RARβ* or *RARγ* are targeted, postpartum lethality and growth disturbances ensue (Lohnes *et al.,* 1993, 1994). The greatest malformations occur with compound mutants, especially with *RAR$\alpha^{-/-}$;$\gamma^{-/-}$* double mutant mice, suggesting some redundancy in RAR transduction of RA signaling. *RAR$\alpha^{-/-}$;$\gamma^{-/-}$* double mutant mice are clearly discernible from their littermates by their external features: diminished eyes, shortened snout and median facial cleft, occasional exencephaly, and agenic auditory pinnae (Lohnes *et al.,* 1994; Mark *et al.,* 1995) (Fig. 2e). Massive cell death is seen in the frontonasal CNC at E10.5, and although an olfactory pit forms the frontonasal processes are fused to the ipsilateral maxillary process and never at the midline. Thus, RA signaling is required for proper coalescence of the facial primordia. Consequently, nasoethmoidal development is severely deficient: The trabecular basal plate is widely split and the nasal capsules and mesethmoid are represented by rods of cartilage (laterocaudal rudiments) without any midline structures. The associated dermatocranial elements (premaxillae, nasals, vomers, lacrimals, frontals) are partially or completely agenic. The orbitotemporal region and those BA derived elements that develop in close association with the nasal capsules (including the maxillae, palatines, and sphenoid) are deficient and malformed. The incus is fused to the ala temporalis of the alisphenoid. The mandibular BA1 elements are relatively spared though not perfectly normal. The elements of the calvarium are diminished and undermineralized. Hypoplastic otic vesicles seen at E10.5 eventually yield small and incomplete otic capsules. Notably, the meninges chondrify. BA2- and BA3-derived skeletal elements are either unidentifiable (e.g., the stapes) or severely malformed (e.g., the hyoid; Mendelsohn *et al.,* 1994a; Mark *et al.,* 1995). *RAR$\alpha 1^{-/-}$;$\alpha 2^{+/-}$;$\gamma^{-/-}$* mutants, which are less severely affected than the complete double mutants, exhibit an ectopic pillar running parallel to the trabecular basal plate from the dysmorphic orbital cartilages

(i.e., the pila postoptica) to the basisphenoid. Thus, many CNC-derived skeletal elements are altered in these mice. Less dramatic malformations are seen in the *RAR$\beta^{-/-}$* (all isoforms) mutants, which are exacerbated in the compound *RAR$\alpha^{-/-}$;$\beta^{-/-}$* mutants This includes ocular defects and agenesis of the postoptic pillar and zygomatic process of the squamosal, hypoplasia of the ethmoturbinals and aberrant gonials (Ghyselinck *et al.,* 1997). Moreover, the hyoid and laryngeal cartilages of the various compound mutants are malformed (Mendelsohn *et al.,* 1994a; Ghyselinck *et al.,* 1997).

Various additional proteins are known to bind RA or retinol, including the cytoplasmic cellular retinol-binding proteins (CRBPI and II) and cytoplasmic retinoic acid binding proteins (CRABPI and II; Sporn *et al.,* 1994, Morriss-Kay and Ward, 1999). CRBPI and II likely act to store and regulate biologically active retinoids. The CRABPs do not appear to be essential for craniofacial development, as *CRABPI$^{-/-}$;II$^{-/-}$* mutants are normal in this respect (Gorry *et al.,* 1994; Lampron *et al.,* 1995).

Retinaldehyde dehydrogenase 2 (*Raldh2*) has a high substrate affinity for retinaldehyde, an intermediate product of retinol oxidation to RA (Wang *et al.,* 1996; D. Zhao *et al.,* 1996). *Raldh2* expression has been correlated spatiotemporally with sites of RA synthesis (Niederreither *et al.,* 1997, 1999; Berggren *et al.,* 1999). *Raldh2$^{-/-}$* mice die at ~E10.5, exhibiting, among other traits, disruptions of otic vesicular and BA development (Niederreither *et al.,* 1999). Although the isthmic organizer at the MB and R1/2 is apparently normal in the absence of RA synthesis (e.g., *Ffg8, Pax2, Gbx2, Engrailed2,* and *Meis2* appear to be expressed normally), the caudal HB is altered throughout (as assayed by *EphA2, EphA4, Fgf3,* regional *Hox* genes, *Krox-20, kreisler,* and *Wnt8a* expression) and associated with increased neuroepithelial and ectomesenchymal cell death (Niederreither *et al.,* 2000).

The spatiotemporal elaboration of RA signaling and cellular response continues to be clarified, in particular as it relates to the development of the olfactory axis (placode, nasal ectomesenchyme, and FB), optic and otic tissues, the rhombencephalon and CNC (e.g., Colbert *et al.,* 1993; Mallo, 1997; Anchan *et al.,* 1997; Choo *et al.,* 1998; Gavalas *et al.,* 1998; Whitesides *et al.,* 1998; Dupé *et al.,* 1999; Niederreither *et al.,* 1997; Zetterström *et al.,* 1999; Enwright and Grainger, 2000). For example, RA signaling is impaired in the olfactory and optic systems in the *Pax6$^{Sey/Sey}$* (Anchan *et al.,* 1997; Enwright and Grainger, 2000). The identification of RA-responsive elements (RAREs) in the promoters of HB *Hox* genes has highlighted their roles in regulating HB patterning and development (reviewed by Morrison, 1998; Morriss-Kay and Ward, 1999; Gavalas and Krumlauf, 2000). When administered to the facial primordia of chicks, excess RA has been shown to inhibit expression of some potential signaling molecules (*Shh* and its receptor, *Ptc*) but not others (*Ffg8*) (Helms *et al.,* 1997).

7. Hedgehogs (Shh, Ihh, and Dhh), Receptors (Ptc) and Effectors of Signaling (Gli)

There are at least three members of the mammalian *Hedgehog* gene family of secreted signaling molecules, *Desert (Dhh), Indian (Ihh),* and *Sonic (Shh) hedgehog,* of which *Shh* has been most extensively studied. *Dhh* is expressed in the gut and Sertoli cells of the testis (Bitgood and McMahon, 1995). *Dhh*$^{-/-}$ mutant mice have a blockage of spermatocyte differentiation and do not appear to have craniofacial defects (Bitgood *et al.,* 1996). IHH is involved in the processing of endochondral ossification (Bitgood and McMahon, 1995; Vortkamp *et al.,* 1996; St. Jacques *et al.,* 1999). Analysis of the *Ihh*$^{-/-}$ mice suggests that IHH acts with PTHrP and the PTH/PTHrP receptor to regulate the transition of proliferating chondrocytes into hypertrophic chondrocytes and development of osteoblasts (St. Jacques *et al.,* 1999). As such, transitions are essential to normal cranial development, in particular in the neurocranium, the skulls of *Ihh*$^{-/-}$ mice are affected although the exact defects remain to be characterized (St. Jacques *et al.,* 1999).

SHH is a secreted protein with autoproteolytic activity that, in conjunction with covalent addition of cholesterol, results is both long- and short-range acting peptides (Lee *et al.,* 1994; Porter *et al.,* 1996; Hammerschmidt *et al.,* 1997). (Of note, it appears that cholesterol may be needed in both the SHH-secreting cell and the target cell; Cooper *et al.,* 1998). *Shh* is expressed in many ventral, midline tissues including the node, head process, notochord, and floor plate and ventral forebrain (Echelard *et al.,* 1993; Martí *et al.,* 1995). In the developing craniofacial primordia, *Shh* is generally restricted to epithelial layers—often in sites of epithelial–epithelial contacts (Echelard *et al.,* 1993; Wall and Hogan, 1995; Helms *et al.,* 1997). These include the stomodeal ectoderm around the buccopharyngeal membrane, the pharyngeal endoderm, the ectodermal–endodermal boundary of the second pharyngeal cleft/pouch, the ectoderm of the merging frontonasal processes, and the forming dentition. Both gain- and loss-of-function studies suggest that SHH has organizing potential, in particular in the midline mesoderm (i.e., the notochord and prechordal plate) and the ZPA of limb buds (Riddle *et al.,* 1993; Johnson *et al.,* 1994; Chiang *et al.,* 1996). Targeted disruption of *Shh* has devastating effects on cranial development: severe cyclopic holoprosencephaly (i.e., loss of midline structures) and, despite early (~E9.5) BA development, a near complete loss of the BA-derived structures (Chiang *et al.,* 1996). Thus, the entire chondrocranium (parachordal, trabecular, and splanchnocranial) is compromised. In all, these defects appear to result from a loss of SHH in midline and unelaborated dorsoventral/mediolateral cephalic development. A proboscislike "nasal" protrusion forms, however, without discernible skeletal elements. The early loss of midline SHH obscures whatever role that ectodermal SHH signals may play in the refinement of later craniofacial development. Experimental perturbation of SHH signaling in chicks suggests that it may modulate craniofacial morphogenesis through cellular proliferation (Hu and Helms, 1999; Ahlgren and Bronner-Fraser, 1999). Tissue-specific loss of *Shh* in craniofacial tissues will help define it roles.

The SHH receptor, *patch (Ptch),* is a multipass transmembrane molecule that acts as a negative modulator of SHH signaling (reviewed by Hammerschmidt *et al.,* 1997; Ming *et al.,* 1998). PTCH further binds to smoothened (*Smo*), another transmembrane protein with homology to G-protein-coupled receptors (Alcedo *et al.,* 1996; van den Heuvel and Ingham, 1996). Upon binding SHH, Ptc releases Smo, which is then freed to transmit the signal. *Ptch* is, moreover, a SHH-inducible gene and is expressed in target tissues of SHH signaling (Hahn *et al.,* 1996a; Goodrich *et al.,* 1996; Platt *et al.,* 1997). Haploinsufficiency of *PTCH* in humans or mice leads to Gorlin syndrome, characterized by basal cell carcinomas, overgrowth of body size, jaw cysts, and other skeletal anomalies (Hahn *et al.,* 1996b, 1998; Unden *et al.,* 1996; reviewed by Ming *et al.,* 1998). Homozygous null mutants are exencephalic embryonic lethals (E9.5; Hahn *et al.,* 1998).

The *Gli* zinc finger transcription factor gene family has been implicated in the transduction of SHH signaling (Domingeuz *et al.,* 1996; Marigo *et al.,* 1996; Aza-Blanc *et al.,* 1997; Büscher *et al.,* 1997; Grindley *et al.,* 1997; J. Lee *et al.,* 1997; Mo *et al.,* 1997). At gastrulation *Gli1, Gli2,* and *Gli3* are all widely expressed throughout the ectoderm and mesoderm (Hui *et al.,* 1994). During later stages of development, *Gli1* is found in condensing mesenchyme and perichondria of skeletal elements (Walterhouse *et al.,* 1993; Hui *et al.,* 1994). *Gli2* and *Gli3* are localized to the undifferentiated mesenchyme of skeletal elements and then are downregulated. All three are seen in the craniofacial mesenchyme (Hui *et al.,* 1994). *Gli2*$^{-/-}$ mutants are microcephalic, with disruptions in frontonasal and trabecular development (Mo *et al.,* 1997). The nasal capsules are smaller and the alveoli of the premaxillae are hypoplastic and fail to form incisors, suggesting a direct role in incisor development (Fig. 2f). The trabecular basal plate is variably cleft and proper midline fusion fails to occur. This clefting greatly disrupts the development of the basisphenoid, presphenoid and associated, adjacent structures. The palate follows suit and is also cleft. The parachordal plate is also variably cleft. Lateral commissures to the otic capsule, though aberrant, appear to be maintained however. In all, this suggests that *Gli2* mediates the coordination of midline development of regions influenced by both the notochord and prechordal plate, consistent with a role in the modulation of SHH signaling. *Gli2*$^{-/-}$ mice also present deficiencies and delayed ossifications of the frontal, parietal, and interparietal bones and often lack ectotympanic bones (Mo *et al.,* 1997). The dentary, moreover, is slightly smaller. Disruption of *Gli3,* the gene mutant in the *extra toes (Xt)* mutant and two human syndromes (Greig cephalopoly-syndactyly and Pallister-Hall), yields deficiencies of neuro-cranium, including all sensory capsules (Johnson, 1967; Hui and Joyner, 1993; Kang *et al.,* 1997; Mo *et al.,* 1997). *Gli3*$^{Xt/Xt}$ mice have parietal brain hernias, with a diminished FB and a bulging MB and associated failure of calvarial devel-opment. The olfactory bulbs are agenic and nasoethmoidal defects are evident: Enlarged maxillae and premaxillae de-velop around hypoplastic rostral nasal capsules and hyper-plastic posterior capsule, ala orbitalis, and mesethmoid. De-formed paraseptal cartilages form adjacent to short and broad septum nasi. The olfactory nerves are enormous, pass-ing through enlarged but deformed cribriform plates—only to stack up against the FB. The pars canalicularis is hypo-plastic: The posterior semicircular canal is present but with a small lumen; the superior only has a lumen posteriorly; and no lateral canal develops (Johnson, 1967). The sensory capsular defects are presaged by defective placodogenesis. Moreover, this is a failure of the trabecular and parachordal plates (which are broadened) to unite. Compound *Gli2*$^{-/-}$*; Gli3*$^{Xt/Xt}$ mutants have been generated (Mo *et al.,* 1997) but die at ~E10.5. *Gli2*$^{-/-}$*;Gli3*$^{Xt/+}$ and *Gli2*$^{+/-}$*;Gli3*$^{Xt/Xt}$ mu-tants exhibit exacerbated skeletal defects, in particular hy-poplasia of the dentary processes (angular, condylar, and co-ronoidal). Hence, there is evidence of the involvement of the HH signaling system in craniofacial development.

8. *Wnt* and *Lef1*

Wnt genes, for which there are at least 16 in mice, encode cysteine-rich secreted signaling proteins. WNT signals have been implicated in the development of the paraxial meso-derm, NT, and craniofacial primordia. *Wnt1* is essential for the elaboration of MB and rostralmost HB development, and there are defects in the parietals (McMahon and Brad-ley, 1990; Thomas and Capecchi, 1990; Ikeya *et al.,* 1997). *Wnt1* and *Wnt3a* are both expressed in the dorsal NT: com-pound *Wnt1*$^{-/-}$*;3a*$^{-/-}$ mutants appear to regulate the devel-opment of dorsolateral neural tube derivatives, including the CNC (Ikeya *et al.,* 1997). The heads of these mutants are small and dysmorphic, compounding the losses generally associated with the *Wnt1*$^{-/-}$ mutants. Expression of *AP-2, CRABPI,* and *TRP-2,* various markers of NC, suggests that fewer NC cells form and NC progenitors may be retained at the dorsal NT in these mutants. Based on *Bmp7* expression, dorsal BMP signaling is not lost. Neurogenic, gliogenic, me-lanogentic, and skeletogenic CNC deficiencies are seen. The stapes, body, and greater horns of the hyoid are absent and the thyroid cartilage is dysmorphic. Elements of the neuro-cranial basal plate, the otic capsule, the squamosal, and alis-phenoid are all dysmorphic in these compound mutants. *Wnt5a* is also required for craniofacial development (Yama-guchi *et al.,* 1999). It is expressed in the MB and in outgrow-ing regions of the frontonasal processes, adjacent maxillary and distal mandibular BA1, as well as the limbs and genital tubercle. Functional loss of *Wnt5a* results in a truncation of

outgrowing structures, including those in the skull. Although the cranial skeletal defects are not clearly characterized, the nasal capsules, jaws and calotte are truncated, in a manner perhaps reminiscent of the *amputated* (*amp*) mouse mutant (Flint and Ede, 1978). WNT signaling generally involves β-catenin/Tcf/Lef1 mediators. *Lef1*$^{-/-}$ mutants have defects in their dental, and associated skeletal, development (Kratoch-wil *et al.,* 1996).

B. Genes Related to Regulative Transcription Factors

1. *Dlx*

The *Dlx* homeodomain transcription factor gene family, vertebrate homologs of the *Drosophila* gene *Distal-less,* has been implicated in bringing about this latter task (Qiu *et al.,* 1995, 1997; Depew *et al.,* 1999). In *Drosophila* and a num-ber of other invertebrates, *Distal-less* is thought to control the proximodistal development of appendages (Cohen *et al.,* 1989; Panganiban *et al.,* 1997). In mice, six *Dlx* genes have been described: *Dlx1, Dlx2, Dlx3, Dlx5, Dlx6,* and *Dlx7* (Dollé *et al.,* 1992; Robinson and Mahon, 1994; Simeone *et al.,* 1994; Qiu *et al.,* 1995, 1997; Stock *et al.,* 1996). Stud-ies of genomic organization indicate that the *Dlx* genes are arranged as tightly linked, convergently transcribed (tail-to-tail) pairs associated with *Hox* clusters: *Dlx1* and *Dlx2* (*HoxD*), *Dlx3* and *Dlx7* (*HoxB*), and *Dlx5* and *Dlx6* (*HoxA*) (reviewed by Stock *et al.,* 1996). Based on DNA sequence similarity and chromosomal location, the *Dlx* genes can be placed in two paralogous groups: *Dlx1, Dlx6,* and *Dlx7* in one, and *Dlx2, Dlx5,* and *Dlx3* in the other. Moreover, tightly linked *Dlx* pairs appear to share regulatory regions and are expressed in similar patterns within the BA, placodes and embryo as a whole (Dollé *et al.,* 1992; Akimenko *et al.,* 1994; Simeone *et al.,* 1994; Robinson and Mahon, 1994; Qiu *et al.,* 1995, 1997; Anderson *et al.,* 1997a,b; Yang *et al.,* 1998; Depew *et al.,* 1999; Eisenstat *et al.,* 1999; Zerucha *et al.,* 2000). Importantly for BA development, paralogous *Dlx* genes share nested expression patterns within the devel-oping BA: *Dlx1/2* are expressed throughout most of the BA ectomesenchyme (i.e., both proximally and distally within the BA), whereas *Dlx5/6* and *Dlx3/7* share progressively re-stricted domains distally.

Evidence principally from paleontological series and com-parative embryology has suggested that the prototypical BA may have been comprised of five proximodistally oriented chondrocranial elements (a pharyngeobranchial, an epibran-chial, a ceratobranchial, a hypobranchial, and a basibran-chial) and associated dermal bones, which would have been repeated in each BA (Goodrich, 1958; Barghusen and Hop-son, 1979; Moore, 1981; de Beer, 1985; Hildebrand, 1988; Langille and Hall, 1989). This fundamental bauplan has tended toward modification in all major vertebrate groups, where in mammals, for example, BA1 is represented by only two components: the maxillary and mandibular derivatives of the palatoquadrate (epibranchial homolog) and Meckel's cartilage (ceratobranchial homolog) respectively. The correlation of the nested *Dlx* gene family expression with a proximodistal skeletal series suggests the possibility that a combinatorial *Dlx* code may regulate the establishment of the distinct skeletal elements within a particular BA unit. Hence, in such a code model, the particular combination of *Dlx* genes expressed along the proximodistal axis of a BA determines the morphological outcome of each skeletal element.

This idea has been genetically examined in *Dlx1, Dlx2, Dlx1/2,* and *Dlx5* mutant mice (Qiu *et al.,* 1995, 1997; Acampora *et al.,* 1999; Depew *et al.,* 1999). In *Dlx1* $^{-/-}$ mice, chondrocranial and dermatocranial structures derived from the proximal BA are affected (Qiu *et al.,* 1997). The majority of the ala temporalis, except the lateralmost tip of the ascending lamina (which fails to undergo ossification), is absent. The lamina obturans develops, but suffers for the loss of the associated ala temporalis; the pterygoids are pushed rostrad. The basisphenoid forms, but also suffers from the loss of the conjoining ala temporalis. The stapes is a nodule without a stapedial foramen, and occasionally the styloid process is diminished with a gap midway down its length. In the *Dlx2* $^{-/-}$ mice, these trends are exacerbated (Qiu *et al.,* 1995). The alisphenoid consists of persistent, aberrant cartilages and small dermal ossifications in the large field usually occupied by the dermal lamina obturans. An ala temporalis, as such, never forms, nor does the alicochlear commissure; the cartilage of the basitrabecular process of the basisphenoid remains and contributes to an expanded osseous strut running from the basisphenoid toward the middle ear. The incus remains juxtaposed to the malleus but is dysmorphic and fused to an ectopic cartilage(s), which extends rostrad ventral to the taenia marginalis. These splanchnocranial modifications are reminiscent in position and articulation of the presumed palatoquadrate structures of early synapsids. The palatine is pushed rostrad and the pterygoid is dysmorphic, connecting to the aberrant basitrabecular projection. The secondary palate is cleft, but the maxillae bear teeth. The zygomatic processes of the maxillae are shortened. In place of the dermal squamosal and jugal are generally four ossifications: two dorsal laminar plates and two ventral plates with rostrad dermal "zygomatic process"-like projections. The anterior of these projections usually completes the zygomatic arch and may represent the cells normally allocated to the formation of the jugal. The stapes also lacks a stapedial foramen and the styloid is truncated and severed near the crista parotica. The skeletal alterations of the double *Dlx1/2* $^{-/-}$ mutants are greater still (Qiu *et al.,* 1997) (Fig. 2g). The ectopic, palatoquadrate-like cartilage seen in *Dlx2* $^{-/-}$ mutants is vastly expanded at the expense of the tooth-bearing regions of the maxillae. Moreover, the

taenia marginalis at its caudal end near the tegmen tympani extends a cartilaginous strut toward the neurocranial base (M. J. Depew and J. L. R. Rubenstein, unpublished observations).

Thus, clear modification of individual elements within the BA occurs in *Dlx1* $^{-/-}$, *Dlx2* $^{-/-}$, and *Dlx1/2* $^{-/-}$ mutant mice, as predicted by a combinatorial model: change of proximal *Dlx* gene code leads to change in proximal structures. However, although *Dlx1* and *Dlx2* are expressed throughout most of the proximodistal axis of the BAs, mutations of these genes fail to disrupt distal BA structures (e.g., malleus, dentary, gonial, tympanic, lesser hyoid horns, etc.). The absence of phenotypic change distally has been seen as either a matter of not exerting biological, regulatory functions in these distal domains or genetic compensation by other, distally (mandibularly) restricted *Dlx* genes (Qiu *et al.,* 1995, 1997).

Further consistent with the model, targeted disruption of *Dlx5* leads to defects in the development of distal BA structures, principally in BA1 (Acampora *et al.,* 1999; Depew *et al.,* 1999). *Dlx5* is expressed both in the olfactory and otic placodes, their derived epithelia, as well as in the CNC of the mandibular arch (Simeone *et al.,* 1994; Qiu *et al.,* 1997; Yang *et al.,* 1998). *Dlx5* $^{-/-}$ mutants die shortly after birth, approximately a quarter being exencephalic. Nonexencephalic mutant mice have hypomineralized parietals and interparietals, and all mutants have regional defects in their nasal and otic capsules and proximal mandibular arch structures (Fig. 2h). Meckel's cartilage is shortened and its path back toward the middle ear is disrupted. At a point near the proximocaudal end of the dentary, it sharply deviates laterad only to abruptly reorient caudomedially again for a short distance whereupon it splits. A medial branch forms a strut toward the pterygoid, basisphenoid, and ala temporalis, while a lateral one runs (at the level of the processus folii) to the malleus. By P0, this deviated cartilage is invested by ectopic intramembranous bone which may also invest, or form a synovial joint with, the pterygoids. This ectopic bone also forms a synovial joint with the misshapen gonial, and sutures with the anterior crus of the tympanic. The malleus has a smaller than normal head and is caudally extended and thickened at the level of the manubrium. The tympanic is slightly smaller and thicker. A short and dysmorphic (mostly at the proximal end) dentary develops around the abnormal Meckel's cartilage. The proximal lamina of the coronoid is absent, and the condylar and angular processes are shortened, misshapen, and juxtaposed. Hence, disruption of the distally expressed *Dlx5* is consistent with a *Dlx* code. Clearly, neither *Dlx1* nor *Dlx2* is capable of compensation for a loss of *Dlx5* in distal BA development. A role for *Dlx2* in the distal BA has been revealed, however, through the generation of compound *Dlx2* $^{-/-}$;*5* $^{-/-}$ mice (M. J. Depew and J. L. R. Rubenstein, 2000). For example, these mice are severely micrognathic and possess cleft mandibles; with the exception of the distalmost, incisor bearing portions of the dentary, mandibular arch derivatives are agenic or hypoplastic. This, then is consistent with a *Dlx* code model.

The model further suggests that a set, intra-BA morphology results from the region where *Dlx1* and *Dlx2* alone form the code. Expansion of this region, at the expense of other *Dlx* genes, namely, *Dlx5* and *Dlx6*, would duplicate this solely *Dlx1/2*-positive region. This, in turn, should result in the transformation of morphology of the duplicated region into something like that seen where the code contains just *Dlx1* and *Dlx2*. In essence, the hypothesized code predicts that loss of *Dlx5/6* should result in a proximalization of the distal BA. The generation of *Dlx5/6*$^{-/-}$ mutant mice now allows for the testing of this prediction (M. J. Depew, J. L. R. Rubenstein and T. Lufkin, unpublished data).

Dlx5 mRNA is also detected in distinct ectodermal tissues, including the ANR, neural folds along the AP axis, and olfactory and otic placodes (Akimenko *et al.*, 1994; Simeone *et al.*, 1994; Zhao *et al.*, 1994b; Liu *et al.*, 1997; Yang *et al.*, 1998; Depew *et al.*, 1999). Hence, it is unsurprising that *Dlx5* is also essential for the development of the nasal and auditory capsules (Acampora *et al.*, 1999; Depew *et al.*, 1999). In *Dlx5* mutant newborns, chondrocranial and dermatocranial defects are observed in the nasal capsule and mesethmoid. In severe cases, there is symmetric, near aplasia of the capsule and mesethmoid. A midline cartilaginous rodlike extension of the rostral trabecular cartilage forms but fails to develop a dorsoventrally expanded nasal septum. Small, cartilaginous spicules are the only remnants of the paries nasi. The posterior capsule and mesethmoid are reduced to thin, planar orbitonasal laminae. In such cases, there is no evidence for the development of a nasal epithelium, nor of a true tectum or solum, including the paraseptal cartilages and associated VNO. The majority of *Dlx5*$^{-/-}$ mutants (85%) are, however, less severely affected. Notably, in these cases a pronounced asymmetry in capsular development, 90% with the right side being more hypoplastic, occurs. The trabecular plate-nasal septum, which is compressed dorsoventrally, deviates to the right to occupy the position of the right cribriform plate and posterior nasal capsular wall. Olfactory foramen are not seen on the right and very few on the left. The tectum, solum, and paries nasi and associated turbinates (e.g., frontoturbinates and ethmoturbinates) are hypoplastic. In these less severe cases, rudimentary branches of the nasal epithelium are seen, as are hypoplastic rostral turbinal cartilages. Cartilages in the floor of the nasal capsule are hypoplastic and sometimes associated with a rudimentary VNO. The dermal bones that encase the nasal capsule (e.g., nasals, premaxillae, maxillae, vomers, and lacrimals) are subsequently dysmorphic and small. *Gsc* mutants also have dysmorphic turbinates within their nasal capsules, as well as altered mandibular branchial arch, ear and mesethmoid morphologies (Rivera-Pérez *et al.*, 1995; Yamada

et al., 1995, 1997; Belo *et al.*, 1998), and at E10.5 *Dlx5*$^{-/-}$ embryos show a severe decrease, or loss, of *Gsc* expression within the mesenchyme of the branchial arches and the frontonasal prominences.

Clearly, the potential for asymmetric development of the frontonasal prominences exists: for example, midline frontonasal prominence-derived structures (e.g., the blow hole) are normally asymmetrically located in certain whales (Raven and Gregory, 1933; Klima, 1987) and the frontonasal prominence-derived upper incisors of the narwhal, *Monodon monoceros* L, are asymmetrically developed (Eales, 1950). Though there has been recent success in identifying the molecular bases for left-right asymmetry in several organ systems (Levin and Mercola, 1998; Ramsdell and Yost, 1998), it is not yet clear how these findings explain the asymmetry in the *Dlx5*$^{-/-}$ mutants.

Moreover, the pars canalicularii of the otic capsules are malformed and hypoplastic in the *Dlx5*$^{-/-}$ mutants as the anterior and posterior semicircular canals (SCC) do not form. The lateral SCC has a complete, but shortened, cartilaginous canal. The maculae and papillae are present though a proper endolymphatic duct is not. The perilymphatic duct, however, is roughly the size of the utricle. The pars cochlearis is smaller; the cochlea completes roughly only one coil (compared to the usual one and a half; Sher, 1971), and the fenestra cochlea is anomalous in size and orientation.

2. *dHAND/eHAND, Id2,* and *Twist: bHLH-Related Proteins*

dHAND and *eHAND* encode bHLH transcription factors generally coexpressed in the distal midline mesenchyme of BA1 and BA2, aortic arch arteries, and cardiac mesoderm (Cserjesi *et al.*, 1995; Srivastava *et al.*, 1995). Targeted deletion of *dHAND* leads to embryonic death at E11.0 principally due to cardiac failure, though the BA are hypoplastic by E9.5 due to increased apoptosis (Srivastava *et al.*, 1995, 1997; Thomas *et al.*, 1998). *Msx1*, but not *Msx2, Dlx2, Prx1, eHAND*, or *ET-1*, is downregulated in the BA distal midline of *dHAND*$^{-/-}$ embryos (Thomas *et al.*, 1998). *Twist* mRNA is detected in the cephalic mesenchyme and CNC of the BA, and in primary calvarial osteoblasts (Chen and Behringer, 1995). Despite not being expressed in the neural plate or tube, *twist* null mice exhibit MB-FB neural tube closure defects (Chen and Behringer, 1995). BA1 morphogenesis is also affected. In chimerical studies, *twist*$^{-/-}$ cranial cells failed to elaborate a fully mesenchymal phenotype, suggesting a role for cranial mesenchyme in the realization of NT closure. *TWIST* heterozygocity has been implicated in Saethre-Chotzen craniosynostosis in humans, and dysmorphic squamosal and interparietal bones in mice (El Ghouzzi *et al.*, 1997). Id proteins are negative regulators of bHLH transcription factors, and *Id2* mRNA has been identified as an early marker of emigrating CNC (Martinsen and Bronner-Fraser,

1998). *Id2* knock-out mice, however, have few if any cranial skeletal defects (Lasorella *et al.,* 2000).

3. *Hox*

The vertebrate *Hox* genes are related to the homeotic selector (*HOM*) genes in *Drosophila* (reviewed by Hunt and Krumlauf, 1992; Krumlauf, 1993). The *Hox* genes are arranged in four complexes on four separate chromosomes. Studies of sequence homology suggest that there are 13 paralogous groups of Hox genes. Of importance, like the *HOM* genes in *Drosophila,* there is a colinearity between the order in which the genes are expressed at their anterior boundaries and the order within the complex in which they appear—a feature suggestive of a primary role in the control of metameric units. To this end, a great deal of attention has been placed on *Hox* gene regulation of the meristic rhombomeric divisions and BA caudal to BA1 (Krumlauf, 1993; Rijli *et al.,* 1998). With regards to craniofacial development, the most compelling arguments for a *Hox* code regulating inter-BA identity have come from the knock-out studies of *Hoxa2* (see also above; Rijli *et al.,* 1993; Gendron-Maquire *et al.,* 1993). Although *Hoxa2* is expressed in R2 and more caudal rhombomeres, BA1 CNC is *Hoxa2* negative. BA2, in contrast, expresses *Hoxa2* in both the CNC and ectoderm. *Hoxa2*$^{-/-}$ mutants exhibit mirror image homeotic-like transformations of the BA2-derived structures into proximal BA1-like structures (Fig. 2i). The stapes and styloid process fail to form; instead, duplicated of proximal MC and associated structures (i.e., tympanic, gonial, malleus and processes folii) appear. The transformed structures are fused to the normal, BA1-derived structures which they replicate. This is likewise the case for the incus and squamosal, while an ectopic cartilaginous strut runs toward the basisphenoid. That these duplications arise has been suggested as evidence for a *Hox* regulation of BA2 identity. These BA1-like transformations are limited, however, to proximal, pericleftal structures and not the entire arch. That there is a mirror image duplication suggests the likelihood that the CNC receive positional information around a point source in the pericleftal ectoderm. Other cranial structures, such as the otic capsules and the supraoccipital arch are also affected by the loss of *Hoxa2*. Targeted disruption of other rostral *Hox* genes, *Hoxa1* for example, disrupts BA2-derived structures (though defects are generally dysmorphologies and not AP arch transformations), as well as the supraoccipital arch and associated structures (Chisaka and Capecchi, 1991; Chisaka *et al.,* 1992; Lufkin *et al.,* 1991, 1992; Mark *et al.,* 1993). Compound *Hoxa1*$^{-/-}$*;2*$^{-/-}$ mutants have exacerbated defects in relevant structures (Barrow and Capecchi, 1999) (Fig. 2j).

4. *Msx*

Members of the vertebrate *Msx* gene family were originally cloned by homology to the *Drosophila* gene *muscle segment homeobox* (*msh*) (reviewed by Davidson, 1995).

Mesenchymal *Msx1* or *Msx2* can be induced by various epithelia, or ectopic BMP or FGF application (Bei and Maas, 1998). Typically, *Msx1* and *Msx2* are either coexpressed or reciprocally expressed in ectoderm and mesenchyme of the developing embryo, as well as in the developing neural tube. Targeted inactivation of *Msx1* yields neonates with craniofacial defects (Satokata and Maas, 1994). The secondary palate is cleft, and both upper and lower teeth and associated alveoli fail to form. The maxillae are fused to the premaxillae, and the basisphenoid (at the basitrabecular process) supports ectopic cartilaginous struts running toward the otic capsules. Rostral processes of MC are present but aberrant, and the malleus lacks a processus brevis. The nasals are shortened, the parietals overlap, and the metopic suture is expanded. An interfrontal bone is seen. Disruption of *Msx2* function has been shown to affect the sutures and elements of the developing calvarium (see above; Jabs *et al.,* 1993; Liu *et al.,* 1995; Y. H. Liu *et al.,* 1999; Wilkie *et al.,* 2000; Satokata *et al.,* 2000). Although the defects remain to be fully characterized, compound *Msx1*$^{-/-}$*;2*$^{-/-}$ mutants have severe malformations of both the chondrocranium and dermatocranium (Satokata *et al.,* 2000).

5. *Nkx genes: Nkx2.1, 5.1, 6.1,* and *Bapx1*

The vertebrate *Nkx* genes are the homologs of the invertebrate *NK* genes implicated in ventral neuronal development (Kim and Nirenberg, 1989). Although the *Nkx* genes in vertebrates regulate ventral neural development, they also play a part in craniofacial development: *Nkx2.1*$^{-/-}$ (a.k.a. *T/ebp*$^{-/-}$) mutant mice have disrupted thyroid cartilage development (Kimura *et al.,* 1996); *Nkx5.1*$^{-/-}$ (a.k.a. *Hmx3*$^{-/-}$) mutants exhibit defects of the pars canalicularis, failing to develop semicircular canals (Hadrys *et al.,* 1998; Wang *et al.,* 1998) (Fig. 111), and *Nkx6.1*$^{-/-}$ mutants lack hypoglossal canals due to a disrupted nervous system (Sander *et al.,* 2000; M. J. Depew, M. Sander, and, J. L. R. Rubenstein, unpublished observations). Moreover, *Bapx1*, the homolog of the *Drosophila* gene *bagpipe* (*bap*), is expressed in the developing paraxial mesoderm (sclerotome), splanchnic mesoderm around the gut endoderm, BA1, and eventually in most cartilaginous condensations (Lettice *et al.,* 1999; Tribioli and Lufkin, 1999). *Bapx1* appears to be essential for the transition from prechondrocyte to chondrocyte in perinotochordal mesenchyme (including within the parachordal basal plate) and notochord ontogeny (Lettice *et al.,* 1999; Tribioli and Lufkin, 1999). The basioccipital and basisphenoid are diminished and dysmorphic in *Bapx1*$^{-/-}$ mutant mice. The occipital arch lacks the supraoccipital and the exoccipitals are dysmorphic. The hyoid body also fails to ossify. Expression patterns of *Pax1, Mfh1, Gli2,* and *Shh,* all genes involved in the development of perinotochordal ossifications, are unaffected; however, *Ihh, Fgfr3,* and *Sox9* expressions are reduced and *Bmp4* (normally perichondrally restricted) regionally expanded.

6. *Paired-Box* (*Pax*) Genes

According to Galliot *et al.* (1999), the large paired-class gene families consist of three basic subgroups: (1) the paired-box *Pax* genes, (2) the *Aristaless*-related Q50 homeodomain genes (*Alx, Cart1, Prx,* and *Hesx1*), and (3) the K50 homeodomain genes (*Otx, Gsc,* and *Pitx*). The *Pax* genes encode transcription factors that possess paired boxes (highly conserved DNA-binding domains) and may have a S50-type homeodomain (reviewed by Dahl *et al.,* 1997; Peters *et al.,* 1998; Galliot *et al.,* 1999). Based on genomic organization and structural motifs, the nine mammalian *Pax* genes fall into four subgroups: (1) *Pax1* and *Pax9;* (2) *Pax2, Pax5,* and *Pax8;* (3) *Pax3* and *Pax7;* and (4) *Pax4* and *Pax6.* Mutations in many of these genes affect craniofacial development. As a family, they are broadly expressed in the developing CNS, sensory placodes, and migrating CNC.

Pax1 and *Pax9* are ontogenetically expressed in similar but not identical patterns (Deutsch *et al.,* 1988; Timmons *et al.,* 1994; Neubüser *et al.,* 1995). Both genes are expressed in the developing sclerotome and the pharyngeal endoderm, where it has been proposed that a *Pax1/9* progenitor gene was instrumental in the formation and patterning of the pharyngeal pouches (Holland and Holland, 1996; Ogasawara *et al.,* 2000). However, only *Pax9* is expressed in the CNC of BA1 and the medial olfactory pit. *Pax1,* the gene mutated in the *undulated* mouse, is involved in the development of the paraxial mesoderm that forms the vertebrae (Wright, 1947; Grüneberg, 1954; Balling *et al.,* 1988; Wilm *et al.,* 1998); *Pax1$^{-/-}$;9$^{-/-}$* compound mutants have exacerbated vertebral malformations (Peters *et al.,* 1999). Targeted disruption of *Pax9* leads to distinct cranial skeletal and glandular malformations (Peters *et al.,* 1998). The secondary palatal shelves elevate but are cleft. Dental development is arrested at the bud stage, where mesenchymal expressions of *Bmp4, Msx1,* and *Lef1* are substantially downregulated by E13.5. A rostral process of MC may develop though the alveolar dental bone of either does not. The dentary also lacks a coronoid process. The pterygoid is malformed and, except for the ascending process, the ala temporalis is absent. The ectotympanic is truncated, while the styloid process was either extended to the lesser horns of the hyoid (and ossified midway along) or lacking entirely. In either case, both the lesser and greater horns of the hyoid are small and dysmorphic, as are the laryngeal cartilages (despite a lack of *Pax9* expression from E12.5 to E18.5; Peters *et al.,* 1998). The glands derived from the third and fourth pharyngeal pouches (the thymus, parathyroid, and ultimobranchial bodies) fail to develop in *Pax9$^{-/-}$* mice. Moreover, the nasal capsule is affected because it appears that the midline paraseptal cartilages do not form (Fig. 3, Peters *et al.,* 1998). The BA1 expression of *Pax9* is regulated, in part, by FGF8 and perhaps medially by BMP (Neubüser *et al.,* 1997; Trumpp *et al.,* 1999).

Pax2 is expressed, from the FB to the rostral HB during neuralation, but becomes reduced to a thin stripe at the MB-HB boundary by E9.5 (Püschel *et al.,* 1992; Torres *et al.,* 1995). It is also expressed in the developing optic and otic tissues. *Pax2$^{-/-}$* mice have midbrain exencephaly and disrupted optic and otic neural development, although the effect on the optic and otic capsules is uncharacterized (Epstein *et al.,* 1991; Torres *et al.,* 1995). The related *Pax5* gene is also required for the elaboration of MB patterning (Urbanek *et al.,* 1997). *Pax8* (Mansouri *et al.,* 1998) is required in thyroid follicular cells. It remains to be seen whether *Pax2/5/8* expression at the MB-HB boundary regulates calvarial or other skeletal development.

In the head, *Pax7* is expressed in the developing neural tube, migrating CNC streams over the FB, MB, and HB, and the epithelium of the medial olfactory pit (Jostes *et al.,* 1990; Mansouri *et al.,* 1996). In *Pax7$^{-/-}$* mice, the ventral, midline nasal capsular structures are hypoplastic, as are the maxillae and lacrimals (Mansouri *et al.,* 1996). The paralogous gene, *Pax3,* is mutated in the *splotch* (*Sp*) mutant mice (Auerbach, 1954; reviewed by Bober *et al.,* 1994). *Pax3$^{-/-}$* mice typically suffer NT defects, deficiencies in NC derivatives (such as Schwann cells and melanocytes), and cardiac NC migration problems (Bober *et al.,* 1994; Conway *et al.,* 1997). *Versican,* a chondroitin sulfate proteoglycan thought to be inhibitory to NC migration, is upregulated in *splotch* mutants, suggesting that *Pax3* may regulate *versican* to promote NC migration (Henderson *et al.,* 1997). *PAX3* haploinsufficiency in humans and mice results in Waardenberg syndrome, which is characterized by slight craniofacial defects (reviewed by Asher *et al.,* 1996). A role for *Pax3* in the induction of the trigeminal ganglion in chicks has also been proposed (Stark *et al.,* 1997; Baker *et al.,* 1999).

Although both *Pax4* and *Pax6* are involved in pancreatic development, *Pax6*—the mouse and rat *Small eyes* (*Sey*) gene—has a well-characterized, central role in craniofacial development (Roberts, 1967; Hogan *et al.,* 1986; Hill *et al.,* 1991; Matsuo *et al.,* 1993; Grindley *et al.,* 1995; Kaufman *et al.,* 1995; Osumi *et al.,* 1997; Osumi-Yamashita *et al.,* 1997a; Quinn *et al.,* 1997; Sander *et al.,* 1997; Sosa-Pineda *et al.,* 1997). In *Pax6$^{Sey/Sey}$* mice, catastrophic malformations of nasoethmoidal development occur due to olfactory and optic placodal deficiencies (Fig. 2n). The trabecular basal plate forms but is truncated. Although the rostral extension is fused at the midline, the caudal end is abnormally bifurcated at E15.5, attaching to the acrochordal cartilage at its lateral extremities. Thus, a caudal gap is initially seen. This gap is filled in part by a rostral cartilaginous extension from the acrochordal cartilage, giving the region an early triton-like appearance. The trabecular plate extends as a rod beyond the cavum cranii but is severely truncated. Olfactory bulbs do not form and the rostral FB rests not on mesethmoidal structures but on the ala orbitalii, which are enlarged and extend to meet at the rostral midline. This meeting is represented by a dorsal spur. The nasal capsules are nearly agenic, being represented by short, thin cartilaginous rods extending in parallel to the midline trabecular plate. These rods may fuse caudally to the trabecular cartilage and the dysmorphic preoptic pillars, which variably reach the ala orbitalis. The ala hypochiasmatica and postoptic pillar develop, forming a rudimentary optic foramen, despite the lack of an eye or optic nerve. An orbitonasal fissure exists by default. The regional dermatocranium is further compromised. Nasal bones do not form: The dorsal nasal region is represented by premaxillae that meet each other at the midline and the frontals caudally. Ventrally, the premaxillae meet without generating a true primary palate. Supernumerary incisors are frequently found in the premaxillae. The lacrimals of *Pax6*$^{Sey/Sey}$ mice are smaller. The maxillae form just ventral to the short, thin lateral nasal rods. Vomers develop, but are compressed and in close association with the maxillae and premaxillae. The secondary palate is shortened, and, as a nasopharynx fails to appear, represents an extended oropharynx. The mandibular BA1 is relatively normal and subsequently extends well beyond the upper jaw. Moreover, there are no hypoglossal canals in the exoccipitals because there are no hypoglossal nerves (Osumi *et al.,* 1997; Osumi-Yamashita *et al.,* 1997a). Other cranial nerves in the *Pax6*$^{Sey/Sey}$ mice are affected as well; for example, the abducens nerves (CN VI) are lacking and the ophthalmic nerve (CN V$_1$) misrouted. Assays of CNC migration have suggested that while FB CNC migration occurs, MB CNC fail to migrate to the frontonasal region (Matsuo *et al.,* 1993; Osumi-Yamashita *et al.,* 1997a). Instead these cells remained near the eye. *Pax6,* which is not expressed in the migrating CNC, is expressed in both the optic and olfactory placodal ectoderm, suggesting that *Pax6* regulates CNC behavior in a noncell autonomous manner. To this extent, *Msx1* expression and RA signaling appear to be lost in the underlying frontonasal ectomesenchyme, and cell death is increased (Grindley *et al.,* 1995; Anchan *et al.,* 1997; Enwright and Grainger, 2000). It is significant that despite the failure of nasal capsulogenesis, premaxillae and vomers develop. Hence, despite a deficient olfactory placode, patterning mechanisms are in place for the dermatocranium and associated dentition. Likewise, the orbital cartilages are able to develop despite the absence of an eye.

7. *Paired-Like, Aristaless-Related Q50 Subclass (Alx, Cart1, Hesx1, and Prx)*

Aristaless 3 (Alx3), Alx4, and *Cartilage homeoprotein 1 (Cart1)* represent a structurally related paired-like subgroup of genes (Meijlink *et al.,* 1999). All three are expressed in postmigratory CNC mesenchyme over the MB and/or FB, and within the frontonasal processes and BA beginning at ~E9.0 (G. A. Zhao *et al.,* 1994a; Q. Zhao *et al.,* 1996; Qu *et al.,* 1997; ten Berge *et al.,* 1998a). *Cart1* is also expressed

around the pharyngeal clefts and, as its appellation implies, within cartilaginous elements. Within the BA1 and BA2, expression is restricted to the distal midline, in a pattern generally more distal to the *Dlx* genes. Functional studies of *Alx* have been concluded. *Alx4* mutations, of which *Strong's luxoid (lst)* is one, result in various skeletal defects, the most significant in the head perhaps being the decreased size of the parietal and increased sagittal fontanelle (Qu *et al.,* 1998). *Cart1*$^{-/-}$ mutant mice are exencephalic, perhaps due to the increased cell death observed in the mesenchyme over the FB-MB boundary (a malady apparently reversible by the addition of folic acid; Q. Zhao *et al.,* 1996). Other reported cranial defects may be secondary to the exencephaly (Qu *et al.,* 1998). Genetic interactions between the structurally related *Alx4* and *Cart1* are revealed in the compound mutants in which two significant defects were evinced (Qu *et al.,* 1999). The first is the clefting of the upper face in the *Alx4*$^{-/-}$;*Cart1*$^{-/-}$, *Alx4*$^{+/-}$;*Cart1*$^{-/-}$, and *Alx4*$^{-/-}$;*Cart1*$^{+/-}$ mice (Fig. 11o). This clefting occurred in the trabecular plate contributing to the septum nasi and, unlike many similar midline defects, was distinctly rostral to the presphenoid; hence, the defect appears capsular in nature. The second significant defect is the loss of most of the midline of the mandible. No exacerbation of the exencephaly from the *Cart1* mutation is detected. Molecular analysis demonstrated that neither *Shh* nor *Ptch* expression was disrupted in these compound mutants. The close developmental relationship seen between *aristaless* and *Distal-less* in Drosophila leg development has yet to be tested in mice.

Hesx1 may act as a transcriptional repressor that is expressed in the anterior visceral endoderm during gastrulation and subsequently in the prechordal plate and anterior neurectoderm (Dattani *et al.,* 1998). Expression is gradually restricted to the oral ectoderm of Rathke's pouch. *Hesx1*$^{-/-}$ mutants have variable severity in defects, but generally have smaller heads (dorsal midline telencephalic hypoplasia), short noses, and microophthalmia (Dattani *et al.,* 1998). Although uncharacterized, skeletal defects are apparent. Asymmetries in defects have been reported but any sidedness proclivities have not been. Chimerical studies suggest that expression in the anterior visceral endoderm is dispensable (Martinez-Barbera *et al.,* 2000).

Three mammalian *Paired-related, Prx1–3,* are known, each linked to craniofacial development (Meijlink *et al.,* 1999). *Prx1* (formerly *Mhox, Phox*) expression can be seen at E9.5 within the ectomesenchyme of the BA1, where it is restricted to the outermost domains of the arch: the distal mandibular midline and maxillary branch near the developing olfactory placodes (Cserjesi *et al.,* 1992; Kuratani *et al.,* 1994; ten Berge *et al.,* 1998a; Lu *et al.,* 1999a). The large initial gap in expression subsequently collapses. *Prx1* is weakly expressed in a thin stripe within BA2 and the ventral mesenchyme of the frontonasal processes (ten Berge *et al.,*

1998a). *Prx1* and *Prx2* (formerly *S8*) are generally co-expressed in these regions, though *Prx2* has been reported to be expressed in the ectoderm and endoderm of the second and third pharyngeal clefts and pouches (ten Berge *et al.,* 1998a). *Prx1*⁻/⁻ mice exhibit both chondrocranial and dermatocranial defects (Martin *et al.,* 1995; ten Berge *et al.,* 1998a) (Fig. 2p). The supraoccipital arch is absent, and the cranial side wall is greatly modified. The parietal plate is malformed, and the posterior taenia marginalis extends a cartilaginous strut ventrad to conjoin with the dysmorphic incus. Two rudiments are all that remain of the squamosal and lamina obturans. Likewise, the jugal is lacking. The dentary is truncated and the condylar, coronoidal, and angular processes are hypoplastic; hence, no jaw articulation forms. The teeth are spared. The ala temporalis is represented by hypoplastic cartilage bodies. *Prx1*⁻/⁻ mice also have cleft palates with hypoplastic pterygoids and palatines. MC is affected: The body is truncated and not strongly continuous with the rostral process. At the processus folii, an ectopic cartilaginous projection forms, perhaps consisting or the cells normally allocated to the gonial and ectotympanic, neither of which develop. The malleus is malformed—the body thickened and the processus brevis and manubrium diminished. There is no incudostapedial articulation, the stapes being but a process fused to the styloid process—itself chondrified completely through to the lesser horns of the hyoid. Although defects in the tegmen tympani are not detailed, the otic capsule as a whole is hypoplastic. Although no defects are seen in *Prx2*⁻/⁻ mice, compound mutant mice demonstrate a genetic interaction (ten Berge *et al.,* 1998a; Lu *et al.,* 1999a). *Prx1*⁺/⁻;*2*⁻/⁻ mice have a cleft palate and styloid processes chondrified to the hyoid. Compound homozygous mutants evince exacerbated *Prx1*⁻/⁻ phenotypes. The elements of the palate are further hypoplastic, and the styloid chondrification encompasses the lesser horns. The zygomatic arch is completely ablated. While the malleus and ectopic process form as in the *Prx1*⁻/⁻ mutants, the entire body of MC is absent in *Prx1*⁻/⁻;*2*⁻/⁻ mice. A rostral process develops, though the incisors are either absent or fused into a single tooth, representing a collapse of the mandibular midline. This is accompanied by a transient midline patch of *Fgf8* expression overlying a single *Pax9*-positive field and a loss of medial *Bmp4* (ten Berge *et al.,* 1998a). Despite a loss of the midline dentary, the molar fields survive; further hypoplasticity of the proximal dentary is seen. In 8% of the mutants a cleft mandible is seen. The otic capsule lacks the lateral semicircular canals and the auricular pinnae are hypoplastic. Murine *Prx3* is expressed in frontonasal and branchial arch tissues, but to date no functional studies have been conducted (Meijlink *et al.,* 1999). Mutation of human *PRX3* may be responsible for Cornelia de Lange syndrome, which is characterized by mental retardation, and craniofacial, eye, and limb defects (Meijlink *et al.,* 1999).

8. *Paired-Like* (or *Bicoid-Like*) *K50 Subclass* (*Gsc, Otx,* and *Pitx*)

Goosecoid (*Gsc*) was originally isolated in a screen of a *Xenopus* cDNA library generated from the dorsal lip of the blastopore (Blumberg *et al.,* 1991). Its activation by activin and its ability to induce a second body axis suggested a possible role as a vertebrate gastrula organizer (Cho *et al.,* 1991). In mice, *Gsc* is expressed transiently at the anterior end of the primitive streak, in line with the head process, during gastrulation (Blum *et al.,* 1992; Gaunt *et al.,* 1993; Belo *et al.,* 1998). *Gsc* expression persists in the prechordal plate—expression suggestive of a role in axial patterning. A late phase, beginning after E10.0 is seen in the mesenchyme of the frontonasal prominences and around the first pharyngeal cleft (Gaunt *et al.,* 1993). Despite initial expectations, *Gsc*⁻/⁻ mice pass through gastrulation with only minor defects associated with the prechordal region: The base of the preoptic pillar is fused to the planum antorbitale at the posterior nasal capsules, as perhaps is an ectopic extension from the ala orbitalis (Figure 1, Belo *et al.,* 1998). Both the vomers and the nasal capsules, which lack turbinals, are hypoplastic (Yamada *et al.,* 1995; Rivera-Perez *et al.,* 1995, 1999). The septum nasi fails to meet the palate. Nasal-associated dermatocranial elements (i.e., maxillae and frontals) are subsequently affected. There is evidence that *Gsc* acts cell autonomously in the mesenchyme (Rivera-Perez *et al.,* 1999) though the interplay between the later frontonasal and earlier prechordal plate *Gsc*-positive cells needs further clarification with regard to nasoethmoidal development. A genetic interaction between *Gsc* and another organizer expressing gene, the winged helix protein encoding gene *Hnf-β3,* was however revealed in an analysis of *Gsc*⁻/⁻;*Hnf-3β*⁺/⁻ mice. These mice evince an early reduction and ventralization of the FB accompanied by a downregulation of *Shh* and *Fgf8* expression (Filosa *et al.,* 1997). BA derivatives, moreover, are affected by the loss of functional *Gsc.* Palatal development occurs though the maxillae, palatines, and pterygoids are malformed. The lamina obturans is dysmorphic, as is the ala temporalis (which is split). In each ear a tubotympanic recess forms but the external acoustic meatus does not extend far internally; thus there is no tympanic membrane or ectotympanic. The malleus lacks a processus brevis and the manubrium is reduced. The dentary is truncated and the angular and coronoidal processes hypoplastic. The dentary juxtaposed to MC, which supports aberrant muscular attachments, forms a groove. A general underdevelopment of the cranial muscles occurs. Gsc expression in the frontonasal mesenchyme appears to be under the regulation of *Dlx5* in the overlying olfactory epithelium (Depew *et al.,* 1999). Epithelial FGF8 and ET-1 (through mesenchymal ET-A) and mesenchymal *Dlx5* have been implicated in the regulation of *Gsc* in the BA, though how these proteins interrelate is undetermined (Tucker *et al.,* 1998b; Depew *et al.,* 1999; Clouthier *et al.,* 1998).

Otx1 and *Otx2* are the mammalian homologs of the *Drosophila* transcription factor gene *orthodenticle,* which regulates anterior head development. It has been proposed that elaborated roles for *Otx* genes were also instrumental in the development of the vertebrate head (Kuratani *et al.,* 1997; Williams and Holland, 1998; Tomsa and Langeland, 1999). *Otx2* is expressed early in the anterior visceral endoderm and entire epiblast prior to gastrulation and the axial mesendoderm and rostral neural plate afterward (Simeone *et al.,* 1993). Thus a progressive anteriorization of expression occurs, eventually leading to restricted expression in FB and MB (to its junction with the HB), foregut pharyngeal endoderm, BA1 ectoderm, and mesencephalic CNC, the thyroid rudiment, the anterior cephalic ectoderm and the anterior notochord by E9.5 (Ang *et al.,* 1994; Acampora *et al.,* 1995; Matsuo *et al.,* 1995; Kimura *et al.,* 1997). Both *Otx2* and *Otx1* are expressed on the developing olfactory, optic and otic sensory organs. *Otx2*$^{-/-}$ mutants display an aberrant mesodermal organization and an absence of the rostral neurectoderm, entirely lacking a head rostral to rhombomere 3 — that is, the FB, MB, and that part of the HB apparently under the influence of the IsO (Acampora *et al.,* 1995; Matsuo *et al.,* 1995; Ang *et al.,* 1996; reviewed by Acampora *et al.,* 2000). Chimeral studies have revealed the necessity of both an *Otx2*-posititve AVE and rostral neural plate (Rhinn *et al.,* 1998, 1999). *Otx2*$^{+/-}$ mice have phenotypes ranging from the lack of a head to externally appearing normal (see Matsuo *et al.,* 1995). The less devastating defects variably include microophthalmia, anophthalmia, micrognathia, agnathia, shortened noses, holoprosencephaly, and acephaly. Minimally, the defects include discontinuities, or gaps, within the trabecular basal plate at the basisphenoid. Maximally, mutant skulls are characterized by an absence of elements rostral to the alisphenoid (Matsuo *et al.,* 1995). Whether any particular skeletal defect in *Otx2*$^{+/-}$ mutants arises from defects in neural plate organization, defects in CNC or both is unclear. Two aspects of the *Otx2*$^{+/-}$ phenotypes, both involving the midline, are of particular note: (1) the variable nature of the trabecular basal plate fusion rostrocaudally and (2) the range of loss of mandibular structures at the distal midline and maintenance of proximal structures. Unlike *Otx2*$^{-/-}$ mutants, *Otx1*$^{-/-}$ mutants are born, pass through a transient prepubescent dwarfism, reach adulthood, and exhibit epilepsy (Suda *et al.,* 1996; Acampora *et al.,* 1996; 1998). These mice have a number of other defects, including a loss of lacrimal and Hardinan glands and lateral semicircular canals (Morsli *et al.,* 1999). The orbital cartilages are dysmorphic and an ectopic strut from the alicochlear commissure runs toward the middle ear (Suda *et al.,* 1996).

Pitx1 (*Ptx1*) and *Pitx2* (*Ptx2, Otlx2, Brx1,* or *Reig*) are characterized by their expression in the stomodeal ectoderm, being named for their expression in the pituitary (re-

viewed by Gage *et al.,* 1999; Meijlink *et al.,* 1999). In addition to the stomodeal ectoderm and its derivatives, including Rathke's pouch and the oral epithelium, *Pitx1* is expressed in the rostroventral foregut in the region of the buccopharyngeal membrane (Lanctôt *et al.,* 1997). Within BA1, mRNA is detected in a central stripe of the mandibular mesenchyme, eventually being detected in associated portions of Meckel's cartilage and the tongue musculature. It is also expressed in the epithelium of the olfactory placode, pit, and capsule. *Pitx1*$^{-/-}$ mutants are normal until E12.5 but eventually become severely micrognathic, exhibiting a cleft palate, a bifurcated tongue, and appendicular skeletal defects (Lanctôt *et al.,* 1999; Szeto *et al.,* 1999). Distal midline BA1 structures and all teeth are relatively spared though structures of the proximal mandible are hypoplastic and dysmorphic (Fig. 2q). Extensive ossification occurs in and around MC through to the processes folii; the malleus and other middle ear elements appear, however, unaffected. The gonial is absent, the ectotympanic is smaller (not reaching MC), and submandibular glands fail to form. Any relationship of the MC ossification to the micrognathia and/or the endochondral ossification defects in the limbs is unclear. Expression patterns of *Msx1, Msx2, Bmp2, Bmp4, Gsc, Shh,* and *Pitx2* are unaltered in BA1, though other markers of mandibular development, in particular *Dlx5* and *Prx1,* are uncharacterized. Morphogenesis of Rathke's pouch-derived adenohypophysis is unaffected. Any defects in the olfactory derivatives remain unreported. *Pitx2* (the homolog of the gene responsible for Rieger syndrome in humans) regulates left-right body asymmetry and craniofacial development (Semina *et al.,* 1996; Gage *et al.,* 1999; Lin *et al.,* 1999; Lu *et al.,* 1999a). Like *Pitx1,* it is expressed in the stomodeal ectoderm, its derivatives, and associated craniofacial mesenchyme. *Pitx2*$^{-/-}$ mutant mice die between E9.25 to E15, exhibiting a hypoplastic BA1 and arrested development of the pituitary and the teeth at the bud stage. *Pitx2*$^{+/-}$ mice have maloccluded teeth (Gage *et al.,* 1999). The BA1 defects are accompanied at E10.5 by a decreased ectodermal *Fgf8* expression, and a lateral expansion of ectodermal *Bmp4* and mesenchymal *Msx1* and *Msx2*. *Pitx2* also appears to be necessary in the periocular ectomesenchyme.

9. *Pbx*

The three members of the *Pbx* gene family, *Pbx1–3,* are TALE (*t*hree *a*mino acid *l*oop *e*xtension) class homeodomain proteins implicated as *Hox* gene cofactors (Knoepfler and Kamps, 1995; Lu *et al.,* 1995). *Pbx1* is the vertebrate homolog of the *Drosophila* gene *extradenticle* (*exd*), which functions genetically in parallel with *Hox* genes to alter the morphological consequences of Hox activities in flies (Peifer and Weischaus, 1990; Rauskolb and Weischaus, 1994). Pbx/exd proteins bind DNA cooperatively with those Hox proteins that contain characteristic tryptophan-bearing dimerization motif proteins (Mann and Chan, 1996). Pbx functionality, including nuclear entry and/or retention, further appears to be dependent on interactions with other TALE class proteins such as *Meis* and *Pknox1/Prep1*, TALE proteins evolutionarily related to, but distinct from, *Pbx/exd* (Bürglin, 1997; Chang *et al.*, 1997; Chen *et al.*, 1997; Knoepfler *et al.*, 1997; Berthelsen *et al.*, 1998a,b). Nuclear Pbx binds DNA with Hox partners thus forming trimeric complexes on appropriate DNA sites (Berthelsen *et al.*, 1998a,b; Jacobs *et al.*, 1999; Ferretti *et al.*, 2000). Nuclear Pbx1 is detectable in the ectomesenchyme and ectoderm of BA2 and around the first pharyngeal groove at E11.5, while the ectomesenchyme of BA1 shows much lower nuclear immunoreactivity (Selleri *et al.*, 2001). Splanchnocranial structures derived from BA2 and the first pharyngeal groove display striking morphological changes in *Pbx1*$^{-/-}$ mutants (which die ~E16.0; Selleri *et al.*, 2001). Morphologically, the lesser horns of the hyoid are transformed into elongated cartilages whose structures are reminiscent of Meckel's cartilage or of the suspensorial BA2 derivatives of certain nonmammalian vertebrates. These ectopic cartilages extend from the hyoid body, to which they are fused, to ectopic flanges on the styloid processes. The styloid processes themselves are thickened and truncated; in some ways each structurally reminiscent of a malleus. *Pbx1*$^{-/-}$ mutants lack stapes and oval windows in their otic capsules. Small ectopic cartilaginous structures form adjacent to each malleus, themselves slightly dysmorphic at their articulations with the incui and along their caudal borders. The pinnae of the ears (which develop around the first pharyngeal groove) are hypoplastic and malformed. The otic capsule is smaller as is the thinned supraoccipital arch. Although both *Hoxa2*$^{-/-}$ and *Pbx1*$^{-/-}$ mutants display transformations of BA2 structures into BA1-like structures, they are distinct in nature: *Hoxa2*$^{-/-}$ mutants have transformations that are more proximal-BA1 in nature, and include a distinct mirror image duplication around the first cleft, whereas *Pbx1*$^{-/-}$ mutants have transformations of a more distal-BA1 nature. Anterior-like transformations of BA2-derived structures in *Pbx1*$^{-/-}$ embryos provide support to the idea that the splanchnocrania of the first and second arches share common developmental programs (Rijli *et al.*, 1993).

10. Winged Helix/Forkhead Genes [BF-1 (Foxg1), Mfh1 (Foxg2), and MF1 (FoxC1)]

Winged helix/forkhead (WH) proteins are transcription factors that share an evolutionarily conserved DNA-binding domain (reviewed by Kaufmann and Knochel, 1996). *Brain factor-1 (BF-1)* is expressed in both the forebrain neuroepithelium and the neurogenic placodal ectoderm (Hatini *et al.*, 1999). The epithelium of the olfactory placodes proliferates less in *BF-1*$^{-/-}$ mice, and the nasal and orbital capsules (and associated structures) are hypoplastic at birth (Xuan *et al.*, 1995; Hatini *et al.*, 1999). *Mesenchyme forkhead-1 (Mfh1)* is also required for cranial development, although reports of the defects due to targeted disruption differ (Iida *et al.*, 1997;

Winnier *et al.*, 1997). *Mfh1* is expressed early in non-notochordal mesoderm surrounding the node and notochord, and later in presomitic and cephalic mesoderm. By E11.5 it is expressed in the mesenchyme of the BA, frontonasal processes and periocularly. During organogenesis, *Mfh1* mRNA is detected in developing cartilaginous tissues, kidneys, and the vascular system (Miura *et al.*, 1993; Kaestner *et al.*, 1996). *Mfh1*$^{-/-}$ mice generally die during gestation, with few surviving to P0. These mice have cleft secondary palates. The basisphenoid and pterygoids are hypoplastic, as are the exoccipitals basioccipital, interparietal, and squamosals. The presphenoid is unmineralized. The supraoccipital appears to be agenic. The dentary is malformed and the malleus misshapen and reduced. The gonial is malformed and the incus and stapes may be missing. The caudal alisphenoid (including the ala temporalis and lamina obturans) and otic capsules are hypoplastic. Thus, both mesodermal and ectomesenchymal elements are affected. Importantly, *Mfh1*$^{-/-}$ mice have defects in the developing cardiovascular system, but any possible link to the skeletal defects is unclarified. *Mf1* is the gene disrupted in the *congenital hydocephalus* (*ch*) mouse mutant (Grüneberg, 1953; Grüneberg and Wickramaratne, 1974; Green, 1970; Kume *et al.*, 1998) and is significant in a number of respects. This mouse mutant takes its name from the hydrocephalus manifested by enlarged and hemorrhagic cerebral hemispheres and lack of a skull vault. *Mf1* expression highly overlaps that of *Mfh1*, being detected in presomitic mesoderm, somites, cephalic mesoderm, and BA and periocular ectomesenchyme (Sasaki and Hogan, 1993; Kume *et al.*, 1998). *Mf1* is expressed in condensing prechondrogenic mesenchyme of the appendicular and axial skeleton, subsequently becoming restricted to the perichondrium. The chondrocranium (along with hyoid and laryngeal cartilage) is hypoplastic. For example, the orbital capsular and associated sphenoidal cartilage are malformed and the hypophyseal foramen remains as an enlarged gap. There are also dermatocranial defects; for example, the zygomatic processes of the maxillae are hyperplastic. Significantly, *Mf1* is expressed in both the arachnoid and dura mater of the meninges. Both are deficient: The arachnoid is thinner and has decreased expression of *prostaglandin D synthase* (an important component of the cerebrospinal fluid) and the cells of the dura are disorganized. Both attributes likely contribute to the hydrocephalus and the loss of dermal bones of the calvarium. *Mf1* is thus one of the few characterized genes that regulates meningeal development. WH proteins have been associated with downstream events in TGF-β signaling (X. Chen *et al.*, 1996), and *Mf1*$^{-/-}$ chondrogenic tissues are less responsive to BMP2 or TGF-β1 in micromass cultures (Kume *et al.*, 1998). Last, other WH genes are integral to cranial development, as seen for example with the functional loss of *Fkh10* resulting in a disrupted otic capsule (Hulander *et al.*, 1998).

A schema of the genetic interactions regulating branchial arch development is represented in Fig. 8.

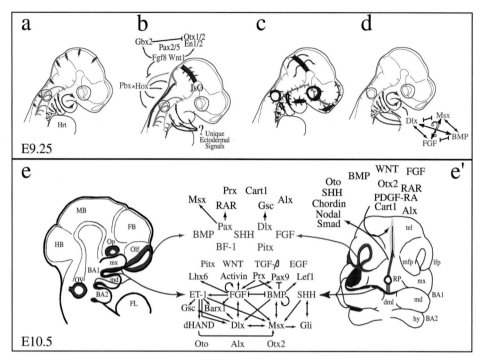

Figure 8 Schema of patterning tasks and positional information within the developing craniofacial primordia. (a) E.9.25 embryo depicting three tasks important to the realization of BA development, including the establishment of a cellular population (red arrows, representing CNC as a source), inter-BA identity (green arrows), and intra-BA identity (blue arrows). (b) Establishment of inter-BA identity (green arrows) is achieved in part through a balance of *Hox* regulation of the HB and associated CNC (blue), *Hox* cofactors such as *Pbx,* and the establishment of the isthmic organizer (IsO) at the MB-HB boundary. The positioning and maintenance of the IsO involves a number of genes, including *Otx1* and *Otx2, En1* and *En2, Pax2* and *Pax5, Gbx2, Wnt1* (red), and *Fgf8* (purple). This focal source of FGF8, then, appears to repress the anterior expression of *Hox* genes thereby contributing to inter-BA identity by helping generate a *Hox*-negative BA1. Inter-BA identity is likely to be further informed by unique ectoderm/endodermal signals. (c) There are multiple sources of positional information within the developing head: signals emanate from the notochord (red), tissues anterior to the notochord (represented by green) including the prechordal plate, and the IsO, sensory placodes, endodermal/ectodermal pharyngeal cleft tissues, and the oral ectoderm (all in purple). Ultimately, development and morphogenesis of craniofacial skeletal elements is directed by the regional balances of signaling information from each of these source tissues. BA1 CNC, for example, receives signaling information at least from the pharyngeal clefts, the oral ectoderm and the optic/olfactory placodes. (d) Intra-BA identity is driven in large part by the proximodistally nested expression of the *Dlx* gene family, possibly regulated by FGF8 (in blue), and distal midline signals, such as BMP, likely regulating *Msx* (in red). (e, e′) Selected genes involved in craniofacial development. (e) Genes expressed in the epithelium of the BA (in purple) and sensory placodes (here, principally the olfactory placode and derivatives, in blue) encode numerous secreted signaling factors (e.g., BMP, ET-1, FGF, Shh, and Wnt) and transcription factors (e.g., *Dlx, Pax,* and *Pitx*) implicated in craniofacial development. Within the BA, evidence of potential regulatory relationships (positive represented by arrows; negative by T-bars), as suggested by both loss- and gain-of-function assays, is beginning to accrue. (See text and Appendix 3 for details.) Numerous secreted factors (in purple) are thought, or hypothesized, to modify the expression of other factors believed to be effectors of craniofacial development (in black). The interrelationships of all of these genes, including those genes highlighted in red with those in purple and black, need further elucidation. (e′) Within the anterorostral head, forces tending to expand the midline (outward arrow, apparently *Oto, Shh, Chordin, Noggin, Nodal,* and *Smad* mediated) are balanced by those tending to collapse the midline (inward arrow, likely *Otx2, RAR, Pdgf-Rα, Cart1,* and *Alx* mediated). Red represents sources of signals associated with the notochord, while green and yellow represent tissues with signaling information anterodorsal to the notochord, including anterior neural ridge derived tissues. Rathke's pouch is depicted in blue-gray. BA1, first branchial arch; BA2, second branchial arch; dml, distal midline of BA; FB, forebrain; FL, forelimb; HB, hindbrain; Hrt, heart; hy, hyoid primordium; IsO, isthmic organizer; lfp, lateral frontonasal prominence/process; MB, midbrain; md, mandibular arch primordium of BA1; mfp, medial frontonasal prominence/process; mx, maxillary arch primordium of BA1; olf, olfactory/nasal primordium; Op, optic primordium; OV, otic vesicular; RP, Rathke's pouch; tel, telencephalic bulge.

References

Acampora, D., Mazan, S., Avantaggiato, V., Barone, P., Turoto, F., Lallemand, Y., Brulet, P., and Simeone, A. (1996). Epilepsy and brain abnormalities in mice lacking the Otx1 gene. *Nat. Genet.* **14,** 218–222.

Acampora, D., Mazan, S., Lallemand, Y., Avantaggiato, V., Maury, M., Simeone, A., and Brûlet, P. (1995). Forebrain and midbrain regions are deleted in Otx2−/− mutants due to a defective anterior neuroectoderm specification during gastrulation. *Development (Cambridge, UK)* **121,** 3279–3290.

Acampora, D., Mazan, S., Tuorto, F., Avantaggiato, V., Tremblay, J. J., Lazzaro, D., Di Carlo, A., Mariano, A., Macchia, P. E., Corte, G., Macchia, V., Drouin, J., Brulet, P., and Simeone, A. (1998). Transient dwarfism and hypogonadism in mice lacking Otx1 reveal prepubescent stage-specific control of pituitary levels of GH, FSH and LH. *Development (Cambridge, UK)* **125,** 1229–1239.

Acampora, D., Merlo, G. R., Paleari, L., Zerega, B., Postiglione, M. P., Mantero, S., Bober, E., Barbieri, O., Simeone, A., and Levi, G. (1999). Craniofacial, vestibular and bone defects in mice lacking the Distal-less-related gene Dlx5. *Development (Cambridge, UK)* **126,** 3795–3809.

Acampora, D., Gulisano, M., and Simeone, A. (2000). Genetic and molecular roles of Otx homeodomain proteins in head development. *Gene* **246,** 23–35.

Ahlgren, S. C., and Bronner-Fraser, M. (1999). Inhibition of sonic hedgehog signaling in vivo results in craniofacial neural crest cell death. *Curr. Biol.* **9,** 1304–1314.

Akimenko, M. A., Ekker, M., Wegner, J., Lin, W., and Westerfield, M. (1994). Combinatorial expression of three zebrafish genes related to distal-less: Part of a homeobox gene code for the head. *J. Neurosci.* **14,** 3475–3486.

Alberch, P., and Kollar, E. (1988). Strategies of head development: Workshop report. *Development (Cambridge, UK)* **103**(Suppl.), 25–30.

Alcedo, J., Ayzenzon, M., Von Ohlen, T., Noll, M., and Hooper, J. E. (1996). The Drosophila smoothened gene encodes a seven-pass membrane protein, a putative receptor for the hedgehog signal. *Cell (Cambridge, Mass.)* **86,** 221–232.

Alfandari, D., Wolfsberg, T. G., White, J. M., and DeSimone, D. W. (1997). ADAM 13, a novel ADAM expressed in somitic mesoderm and neural crest cells during *Xenopus laevis* development. *Dev. Biol.* **182,** 314–330.

Allin, E. F. (1975). Evolution of the mammalian middle ear. *J. Morphol.* **147,** 403–37.

Allin, E. F., and Hopson, A. (1992). Evolution of the auditory system in synapsida ("Mammal-like reptiles" and primitive mammals) as seen in the fossil record. *In* "The Evolutionary Biology of Hearing" (D. B. Webster, R. R. Fay, and A. N. Popper, eds.), pp. 587–614. Springer-Verlag, New York.

Anchan, R. M., Drake, D. P., Haines, C. F., Gerwe, E. A., and LaMantia, A. S. (1997). Disruption of local retinoid-mediated gene expression accompanies abnormal development in the mammalian olfactory pathway. *J. Comp. Neurol.* **379,** 171–184.

Anderson, S. A., Eisenstat, D. D., Shi, L., and Rubenstein, J. L. (1997a). Interneuron migration from basal forebrain to neocortex: Dependence on Dlx genes. *Science* **278,** 474–476.

Anderson, S. A., Qiu, M., Bulfone, A., Eisenstat, D. D., Meneses, J., Pedersen, R., and Rubenstein, J. L. (1997b). Mutations of the homeobox genes Dlx-1 and Dlx-2 disrupt the striatal subventricular zone and differentiation of late born striatal neurons. *Neuron* **19,** 27–37.

Ang, S. L., Conlon, R. A., Jin, O., and Rossant, J. (1994). Positive and negative signals from mesoderm regulate the expression of mouse Otx2 in ectoderm explants. *Development (Cambridge, UK)* **120,** 2979–2989.

Ang, S. L., Jin, O., Rhinn, M., Daigle, N., Stevenson, L., and Rossant, J. (1996). A targeted mouse Otx2 mutation leads to severe defects in gastrulation and formation of axial mesoderm and to deletion of rostral brain. *Development (Cambridge, UK)* **122,** 243–252.

Asher, J. H., Jr., Harrison, R. W., Morell, R., Carey, M. L., and Friedman, T. B. (1996). Effects of Pax3 modifier genes on craniofacial morphology, pigmentation, and viability: A murine model of Waardenburg syndrome variation. *Genomics* **34,** 285–298.

Auerbach, R. (1954). Analysis of the developmental effects of a lethal mutation in the house mouse. *J. Exp. Zool.* **127,** 305–329.

Aza-Blanc, P., Ramirez-Weber, F.-A., Laget, M.-P., Schwartz, C., and Kornberg, T. B. (1997). Proteolysis that is inhibited by Hedgehog targets Cubitus interruptus protein to the nucleus and converts it to a repressor. *Cell (Cambridge, Mass.)* **89,** 1043–1053.

Bachiller, D., Klingensmith, J., Kemp, C., Belo, J. A., Anderson, R. M., May, S. R., McMahon, J. A., McMahon, A. P., Harland, R. M., Rossant, J., and De Robertis, E. M. (2000). The organizer factors Chordin and Noggin are required for mouse forebrain development. *Nature (London)* **403,** 658–661.

Baker, C. V., and Bronner-Fraser, M. (1997a). The origins of the neural crest. Part I: Embryonic induction. *Mech. Dev.* **69,** 3–11.

Baker, C. V., and Bronner-Fraser, M. (1997b). The origins of the neural crest. Part II: An evolutionary perspective. *Mech. Dev.* **69,** 13–29.

Baker, C. V., Bronner-Fraser, M., Le Douarin, N. M., and Teillet, M. A. (1997). Early- and late-migrating cranial neural crest cell populations have equivalent developmental potential in vivo. *Development (Cambridge, UK)* **124,** 3077–3087.

Baker, C. V., Stark, M. R., Marcelle, C., and Bronner-Fraser, M. (1999). Competence, specification and induction of Pax-3 in the trigeminal placode. *Development (Cambridge, UK)* **126,** 147–156.

Balling, R., Deutsch, U., and Gruss, P. (1988). Undulated, a mutation affecting the development of the mouse skeleton, has a point mutation in the paired box of Pax 1. *Cell (Cambridge, Mass.)* **55,** 531–535.

Barembaum, M., Moreno, T. A., LaBonne, C., Sechrist, J., and Bronner-Fraser, M. (2000). Noelin-1 is a secreted glycoprotein involved in generation of the neural crest. *Nat. Cell Biol.* **2,** 219–225.

Barghusen, H. R., and Hopson, A. (1979). The endoskeleton: The comparative anatomy of the skull and visceral skeleton. *In* "Hyman's Comparative Anatomy" (M. Wake, ed.). University of Chicago Press, Chicago.

Barlow, A. J., and Francis-West, P. H. (1997). Ectopic application of recombinant BMP-2 and BMP-4 can change patterning of developing chick facial primordia. *Development (Cambridge, UK)* **124,** 391–398.

Bartelmez, G. W. (1962). The proliferation of neural crest from forebrain levels in the rat. *Contribu. Embryol.* **37,** 3–12.

Beddington, R. S., and Robertson, E. J. (1999). Axis development and early asymmetry in mammals. *Cell (Cambridge, Mass.)* **96,** 195–209.

Bee, J., and Thorogood, P. (1980). The role of tissue interactions in the skeletogenic differentiation of avian neural crest cells. *Dev. Biol.* **78,** 47–62.

Begbie, J., Brunet, J. F., Rubenstein, J. L., and Graham, A. (1999). Induction of the epibranchial placodes. *Development (Cambridge, UK)* **126,** 895–902.

Bei, M., and Maas, R. (1998). FGFs and BMP4 induce both Msx1-independent and Msx1-dependent signaling pathways in early tooth development. *Development (Cambridge, UK)* **125,** 4325–4333.

Bei, M. Kratochwil, K. and Maas, R. (2000). BMP4 rescues a noncell-autonomous function of Msx1 in tooth development. *Development (Cambridge, UK)* **127,** 4711–4718.

Bel, S., Coré, N., Djabali, M., Kieboom, K., Van der Lugt, N., Alkema, M. J., and Van Lohuizen, M. (1998). Genetic interactions and dosage effects of Polycomb group genes in mice. *Development (Cambridge, UK)* **125,** 3543–3451.

Bellus, G. A., Gaudenz, K., Zackai, E. H., Clarke, L. A., Szabo, J., Francomano, C. A., and Muenke, M. (1996a). Identical mutations in three different fibroblast growth factor receptor genes in autosomal dominant craniosynostosis syndromes. *Nat. Genet.* **14,** 174–176.

Bellus, G. A., McIntosh, I., Szabo, J., Aylsworth, A., Kaitila, I., and Francomano, C. A. (1996b). Hypochondroplasia: Molecular analysis of the fibroblast growth factor receptor 3 gene. *Ann. N.Y. Acad. Sci.* **785,** 182–187.

Belo, J. A., Leyns, L., Yamada, G., and De Robertis, E. M. (1998). The prechordal midline of the chondrocranium is defective in Goosecoid-1 mouse mutants. *Mech. Dev.* **72,** 15–25.

Bennett, J. H., Hunt, P., and Thorogood, P. (1995). Bone morphogenetic protein-2 and -4 expression during murine orofacial development. *Arch. Oral Biol.* **40,** 847–854.

Beresford, W. A. (1993). Cranial skeletal tissues: Diversity and evolutionary trends. *In* "The Skull" (J. Hanken and B. K. Hall, eds.), pp. 69–130. University of Chicago Press, Chicago.

Berggren, K., McCaffery, P., Dräger, U., and Forehand, C. J. (1999). Differential distribution of retinoic acid synthesis in the chicken embryo as determined by immunolocalization of the retinoic acid synthetic enzyme, RALDH-2. *Dev. Biol.* **210,** 288–304.

Berk, M., Desai, S. Y., Heyman, H. C., and Colmenares, C. (1997). Mice lacking the ski proto-oncogene have defects in neurulation, craniofacial, patterning, and skeletal muscle development. *Genes Dev.* **11,** 2029–2039.

Berthelsen, J., Zappavigna, V., Mavilio, F., and Blasi, F. (1998a). Prep1, a novel functional partner of Pbx proteins. *EMBO J.* **17,** 1423–1433.

Berthelsen, J., Zappavigna, V., Ferretti, E., Mavilio, F., and Blasi, F. (1998b). The novel homeoprotein Prep1 modulates Pbx-Hox protein cooperativity. *EMBO J.* **17,** 1434–1445.

Bierwolf, D. (1958). Die embryogenese des hydrocephalus und der kleinhirnmissbildungen beim dreherstamm der hausmaus. *Morphol. Jahr.* **99,** 542–612.

Bissell, M. J., and Barcellos-Hoff, M. H. (1987). The influence of extracellular matrix on gene expression: Is structure the message? *J. Cell Sci., Suppl.* **8,** 327–343.

Bitgood, M. J., and McMahon, A. P. (1995). Hedgehog and Bmp genes are coexpressed at many diverse sites of cell-cell interaction in the mouse embryo. *Dev. Biol.* **172,** 126–138.

Blum, M., Gaunt, S. J., Cho, K. W., Steinbeisser, H., Blumberg, B., Bittner, D., and De Robertis, E. M. (1992). Gastrulation in the mouse: The role of the homeobox gene goosecoid. *Cell (Cambridge, Mass.)* **69,** 1097–1106.

Blumberg, B., Wright, C. V., De Robertis, E. M., and Cho, K. W. (1991). Organizer-specific homeobox genes in *Xenopus laevis* embryos. *Science* **253,** 194–196.

Bober, E., Franz, T., Arnold, H. H., Gruss, P., and Tremblay, P. (1994). Pax-3 is required for the development of limb muscles: A possible role for the migration of dermomyotomal muscle progenitor cells. *Development (Cambridge, UK)* **120,** 603–612.

Bonadio, J., Saunders, T. L., Tsai, E., Goldstein, S. A., Morris-Wiman, J., Brinkley, L., Dolan, D. F., Altschuler, R. A., Hawkins, J. E., Jr., Bateman, J. F. *et al.* (1990). Transgenic mouse model of the mild dominant form of osteogenesis imperfecta. *Proc. Natl. Acad. Sci. U.S.A.* **87,** 7145–7149.

Boström, H., Willetts, K., Pekny, M., Levéen, P., Lindahl, P., Hedstrand, H., Pekna, M., Hellström, M., Gebre-Medhin, S., Schalling, M., Nilsson, M., Kurland, S., Törnell, J., Heath, J. K., and Betsholtz, C. (1996). PDGF-A signaling is a critical event in lung alveolar myofibroblast development and alveogenesis. *Cell (Cambridge, Mass.)* **85,** 863–873.

Bourgeois, P., Bolcato-Bellemin, A. L., Danse, J. M., Bloch-Zupan, A., Yoshiba, K., Stoetzel, C., and Perrin-Schmitt, F. (1998). The variable expressivity and incomplete penetrance of the twist-null heterozygous mouse phenotype resemble those of human Saethre-Chotzen syndrome. *Hum. Mol. Genet.* **7,** 945–957.

Bradamante, Z., and Hall, B. K. (1980). The role of epithelial collagen and proteoglycan in the initiation of osteogenesis by avian neural crest cells. *Anat. Rec.* **197,** 305–315.

Brickell, P., and Thorogood, P. (1997). Retinoic acid and retinoic acid receptors in craniofacial development. *Semin. Cell Dev. Biol.* **8,** 437–443.

Bronner-Fraser, M. (1986). Analysis of the early stages of trunk neural crest migration in avian embryos using monoclonal antibody HNK-1. *Dev. Biol.* **115,** 44–55.

Broom, R. (1930). "The origin of the Human Skeleton: An introduction to Human Osteology." Witherby, London.

Brown, K. S., Cranley, R. E., Greene, R., Kleinman, H. K., and Pennypacker, J. P. (1981). Disproportionate micromelia (Dmm): An incomplete dominant mouse dwarfism with abnormal cartilage matrix. *J. Embryol. Exp. Morphol.* **62,** 165–182.

Brunet, L. J., McMahon, J. A., McMahon, A. P., and Harland, R. M. (1998). Noggin, cartilage morphogenesis, and joint formation in the mammalian skeleton. *Science* **280,** 1455–1457.

Bürglin, T. R. (1997). Analysis of TALE superclass homeobox genes (MEIS, PBC, KNOX, Iroquois, TGIF) reveals a novel domain conserved between plants and animals. *Nucleic Acids Res.* **25,** 4173–4180.

Burke, D., Wilkes, D., Blundell, T. L., and Malcolm, S. (1998). Fibroblast growth factor receptors: Lessons from the genes. *Trends Biochem. Sci.* **23,** 59–62.

Büscher, D., Bosse, B., Heymer, J., and Rüther, U. (1997). Evidence for genetic control of Sonic hedgehog by Gli3 in mouse limb development. *Mech. Dev.* **62,** 175–182.

Butler, P. M. (1995). Ontogenetic aspects of dental evolution. *Int. J. Dev. Biol.* **39,** 25–34.

Camakaris, J., Mann, J. R., and Danks, D. M. (1979). Copper metabolism in mottled mouse mutants: Copper concentrations in tissues during development. *Biochem. J.* **180,** 597–604.

Camus, A., Davidson, B. P., Billiards, S., Khoo, P., Rivera-Pérez, J. A., Wakamiya, M., Behringer, R. R., and Tam, P. P. (2000). The morphogenetic role of midline mesendoderm and ectoderm in the development of the forebrain and the midbrain of the mouse embryo. *Development (Cambridge, UK)* **127,** 1799–1813.

Cano, A., Pérez-Moreno, M. A., Rodrigo, I., Locascio, A., Blanco, M. J., del Barrio, M. G., Portillo, F., and Nieto, M. A. (2000). The transcription factor snail controls epithelial-mesenchymal transitions by repressing E-cadherin expression. *Nat. Cell Biol.* **2,** 76–83.

Carl, T. F., Dufton, C., Hanken, J., and Klymkowsky, M. W. (1999). Inhibition of neural crest migration in Xenopus using antisense slug RNA. *Dev. Biol.* **213,** 101–115.

Carlton, M. B., Colledge, W. H., and Evans, M. J. (1998). Crouzon-like craniofacial dysmorphology in the mouse is caused by an insertional mutation at the Fgf3/Fgf4 locus. *Dev. Dyn.* **212,** 242–249.

Casci, T., Vinós, J., and Freeman, M. (1999). Sprouty, an intracellular inhibitor of Ras signaling. *Cell (Cambridge, Mass.)* **96,** 655–665.

Chai, Y., Jiang, X., Ito, Y., Bringas, P., Jr., Han, J., Rowitch, D. H., Soriano, P., McMahon, A. P., and Sucov, H. M. (2000). Fate of the mammalian cranial neural crest during tooth and mandibular morphogenesis. *Development (Cambridge, UK)* **127,** 1671–1679.

Chang, C. P., Jacobs, Y., Nakamura, T., Jenkins, N. A., Copeland, N. G., and Cleary, M. L. (1997). Meis proteins are major in vivo DNA binding partners for wild-type but not chimeric Pbx proteins. *Mol. Cell. Biol.* **17,** 5679–5687.

Chen, H., Rossier, C., Nakamura, Y., Lynn, A., Chakravarti, A., and Antonarakis, S. E. (1997). Cloning of a novel homeobox-containing gene, PKNOX1, and mapping to human chromosome 21q22. 3. *Genomics* **41,** 193–200.

Chen, X., Rubock, M. J., and Whitman, M. (1996). A transcriptional partner for MAD proteins in TGF-beta signalling. *Nature (London)* **383,** 691–696; erratum: *Ibid.* **384,** 648 (1996).

Chen, Y., Bei, M., Woo, I., Satokata, I., and Maas, R. (1996). Msx1 controls inductive signalling in mammalian tooth morphogenesis. *Development* **122,** 3035–3044.

Chen, Z. F., and Behringer, R. R. (1995). Twist is required in head mesenchyme for cranial neural tube morphogenesis. *Genes Dev.* **9,** 686–699.

Chiang, C., Litingtung, Y., Lee, E., Young, K. E., Corden, J. L., Westphal, H., and Beachy, P. A. (1996). Cyclopia and defective axial patterning in mice lacking Sonic hedgehog gene function. *Nature (London)* **383,** 407–413.

Chisaka, O., and Capecchi, M. R. (1991). Regionally restricted developmental defects resulting from targeted disruption of the mouse homeobox gene hox-1.5. *Nature (London)* **350,** 473–479.

Chisaka, O., Musci, T. S., and Capecchi, M. R. (1992). Developmental defects of the ear, cranial nerves and hindbrain resulting from targeted disruption of the mouse homeobox gene Hox-1. 6. *Nature (London)* **355,** 516–520.

Cho, K. W., Blumberg, B., Steinbeisser, H., and De Robertis, E. M. (1991). Molecular nature of Spemann's organizer: The role of the Xenopus homeobox gene goosecoid. *Cell (Cambridge, Mass.)* **67,** 1111–1120.

Choo, D., Sanne, J. L., and Wu, D. K. (1998). The differential sensitivities of inner ear structures to retinoic acid during development. *Dev. Biol.* **204,** 136–150.

Clouthier, D. E., Hosoda, K., Richardson, J. A., Williams, S. C., Yanagisawa, H., Kuwaki, T., Kumada, M., Hammer, R. E., and Yanagisawa, M. (1998). Cranial and cardiac neural crest defects in endothelin-A receptor-deficient mice. *Development (Cambridge, UK)* **125,** 813–824.

Clouthier, D. E., Williams, S. C., Yanagisawa, H., Wieduwilt, M., Richardson, J. A., and Yanagisawa, M. (2000). Signaling pathways crucial for craniofacial development revealed by endothelin-A receptor-deficient mice. *Dev. Biol.* **217,** 10–24.

Cobrinik, D., Lee, M. H., Hannon, G., Mulligan, G., Bronson, R. T., Dyson, N., Harlow, E., Beach, D., Weinberg, R. A., and Jacks, T. (1996). Shared role of the pRB-related p130 and p107 proteins in limb development. *Genes Dev.* **10,** 1633–1644.

Coffin, J. D., Florkiewicz, R. Z., Neumann, J., Mort-Hopkins, T., Dorn, G. W., 2nd, Lightfoot, P., German, R., Howles, P. N., Kier, A., O'Toole, B. A. *et al.* (1995). Abnormal bone growth and selective translational regulation in basic fibroblast growth factor (FGF-2) transgenic mice. *Mol. Biol. Cell* **6,** 1861–1873.

Coffin-Collins, P. A., and Hall, B. K. (1989). Chondrogenesis of mandibular mesenchyme from the embryonic chick is inhibited by mandibular epithelium and by epidermal growth factor. *Int. J. Dev. Biol.* **33,** 297–311.

Cohen, S. (1962). Isolation of a mouse submaxillary gland protein accelerating incisor eruption and eyelid opening in the newborn anima. *J. Biol. Chem.* **237,** 1555–1562.

Cohen, S. M., Brönner, G., Küttner, F., Jürgens, G., and Jäckle, H. (1989). Distal-less encodes a homoeodomain protein required for limb development in Drosophila. *Nature (London)* **338,** 432–434.

Cohn, S. A. (1957). Development of the molar teeth in the albino mouse. *Am. J. Anat.* **101,** 295–320.

Colbert, M. C., Linney, E., and LaMantia, A. S. (1993). Local sources of retinoic acid coincide with retinoid-mediated transgene activity during embryonic development. *Proc. Natl. Acad. Sci. U.S.A.* **90,** 6572–6576.

Conlon, F. L., Lyons, K. M., Takaesu, N., Barth, K. S., Kispert, A., Herrmann, B., and Robertson, E. J. (1994). A primary requirement for nodal in the formation and maintenance of the primitive streak in the mouse. *Development (Cambridge, UK)* **120,** 1919–1928.

Conlon, R. A., and Rossant, J. (1992). Exogenous retinoic acid rapidly induces anterior ectopic expression of murine Hox-2 genes in vivo. *Development (Cambridge, UK)* **116,** 357–368.

Conway, S. J., Henderson, D. J., and Copp, A. J. (1997). Pax3 is required for cardiac neural crest migration in the mouse: Evidence from the splotch (Sp2H) mutant. *Development (Cambridge, UK)* **124,** 505–514.

Cooper, M. K., Porter, J. A., Young, K. E., and Beachy, P. A. (1998). Teratogen-mediated inhibition of target tissue response to Shh signaling. *Science* **280,** 1603–1607.

Corsin, J. (1966). The development of the osteocranium of *Pleurodeles waltlii* Michahelles. *J. Morphol.* **119,** 209–216.

Corsin, J. (1971). Infuence des placodes olfactives et des ébauches optiques sur la morphagéneśe du squelette cranïen chez *Pleurodeles waltii* michah. *Ann. Embryol. Morphog.* **1,** 41–48.

Corsin, J. (1975). Différenciation in vitro de cartilage à patir des crètes neu-

rales céphaliques chez *Pleurodeles waltii* Michah. *J. Embryol. Exp. Morphol.* **33,** 335–342.

Corsin, J. (1977). Le matériel extracellulaire au cours du développement du chondrocrane des amphibiens. Mise en place et constitution. *J. Embryol. Exp. Morphol.* **38,** 139–149.

Couly, G. F., and Le Douarin, N. M. (1985). Mapping of the early neural primordium in quail-chick chimeras. I. Developmental relationships between placodes, facial ectoderm, and prosencephalon. *Dev. Biol.* **110,** 422–439.

Couly, G. F., and Le Douarin, N. M. (1987). Mapping of the early neural primordium in quail-chick chimeras. II. The prosencephalic neural plate and neural folds: Implications for the genesis of cephalic human congenital abnormalities. *Dev. Biol.* **120,** 198–214.

Couly, G. F., and Le Douarin, N. M. (1990). Head morphogenesis in embryonic avian chimeras: Evidence for a segmental pattern in the ectoderm corresponding to the neuromeres. *Development (Cambridge, UK)* **108,** 543–558.

Couly, G. F., Coltey, P. M., and Le Douarin, N. M. (1993). The triple origin of skull in higher vertebrates: A study in quail-chick chimeras. *Development (Cambridge, UK)* **117,** 409–429.

Couly, G. F., Grapin-Botton, A., Coltey, P., and Le Douarin, N. M. (1996). The regeneration of the cephalic neural crest, a problem revisited: The regenerating cells originate from the contralateral or from the anterior and posterior neural fold. *Development (Cambridge, UK)* **122,** 3393–3407.

Couly, G. F., Grapin-Botton, A., Coltey, P., Ruhin, B., and Le Douarin, N. M. (1998). Determination of the identity of the derivatives of the cephalic neural crest: Incompatibility between Hox gene expression and lower jaw development. *Development (Cambridge, UK)* **125,** 3445–3459.

Crossley, P. H., and Martin, G. R. (1995). The mouse Fgf8 gene encodes a family of polypeptides and is expressed in regions that direct outgrowth and patterning in the developing embryo. *Development (Cambridge, UK)* **121,** 439–451.

Croucher, S. J., and Tickle, C. (1989). Characterization of epithelial domains in the nasal passages of chick embryos: Spatial and temporal mapping of a range of extracellular matrix and cell surface molecules during development of the nasal placode. *Development (Cambridge, UK)* **106,** 493–509.

Cserjesi, P., Lilly, B., Bryson, L., Wang, Y., Sassoon, D. A., and Olson, E. N. (1992). MHox: A mesodermally restricted homeodomain protein that binds an essential site in the muscle creatine kinase enhancer. *Development (Cambridge, UK)* **115,** 1087–1101.

Cserjesi, P., Brown, D., Lyons, G. E., and Olson, E. N. (1995). Expression of the novel basic helix-loop-helix gene eHAND in neural crest derivatives and extraembryonic membranes during mouse development. *Dev. Biol.* **170,** 664–678.

Dahl, E., Koseki, H., and Balling, R. (1997). Pax genes and organogenesis. *BioEssays* **19,** 755–765.

D'Amico-Martel, A., and Noden, D. M. (1983). Contributions of placodal and neural crest cells to avian cranial peripheral ganglia. *Am. J. Anat.* **166,** 445–468.

Dassule, H. R., and McMahon, A. P. (1998). Analysis of epithelial-mesenchymal interactions in the initial morphogenesis of the mammalian tooth. *Dev. Biol.* **202,** 215–227.

Dassule, H. R., Lewis, P., Bei, M., Maas, R. and McMahon, A. P. (2000). Sonic hedgehog regulates growth and morphogenesis of the tooth. *Development (Cambridge, UK)* **127,** 4775–4785.

Dattani, M. T., Martinez-Barbera, J. P., Thomas, P. Q., Brickman, J. M., Gupta, R., Mårtensson, I. L., Toresson, H., Fox, M., Wales, J. K., Hindmarsh, P. C., Krauss, S., Beddington, R. S., and Robinson, I. C. (1998). Mutations in the homeobox gene HESX1/Hesx1 associated with septo-optic dysplasia in human and mouse. *Nat. Genet.* **19,** 125–133.

Davidson, D. (1995). The function and evolution of Msx genes: Pointers and paradoxes. *Trends Genet.* **11,** 405–411.

de Beer, G. (1985). "The Development of the Vertebrate Skull. University of Chicago Press, Chicago.

De Larco, J. E., and Todaro, G. J. (1978). Growth factors from murine sarcoma virus-transformed cells. *Proc. Natl. Acad. Sci. U.S.A.* **75,** 4001–4005.

Deng, C., Wynshaw-Boris, A., Zhou, F., Kuo, A., and Leder, P. (1996). Fibroblast growth factor receptor 3 is a negative regulator of bone growth. *Cell (Cambridge, Mass.)* **84,** 911–921.

Depew, M. J., and Rubenstein, J. L. R. (2001). In preparation.

Depew, M. J., Liu, J. K., Long, J. E., Presley, R., Meneses, J. J., Pedersen, R. A., and Rubenstein, J. L. (1999). Dlx5 regulates regional development of the branchial arches and sensory capsules. *Development (Cambridge, UK)* **126,** 3831–3846.

Deutsch, U., Dressler, G. R., and Gruss, P. (1988). Pax 1, a member of a paired box homologous murine gene family, is expressed in segmented structures during development. *Cell (Cambridge, Mass.)* **53,** 617–625.

Dickinson, M. E., Selleck, M. A., McMahon, A. P., and Bronner-Fraser, M. (1995). Dorsalization of the neural tube by the non-neural ectoderm. *Development (Cambridge, UK)* **121,** 2099–2106.

Dickson, M. C., Martin, J. S., Cousins, F. M., Kulkarni, A. B., Karlsson, S., and Akhurst, R. J. (1995). Defective haematopoiesis and vasculogenesis in transforming growth factor-beta 1 knock out mice. *Development (Cambridge, UK)* **121,** 1845–1854.

Dodig, M., Tadic, T., Kronenberg, M. S., Dacic, S., Liu, Y. H., Maxson, R., Rowe, D. W., and Lichtler, A. C. (1999). Ectopic Msx2 overexpression inhibits and Msx2 antisense stimulates calvarial osteoblast differentiation. *Dev. Biol.* **209,** 298–307.

Dollé, P., Ruberte, E., Kastner, P., Petkovich, M., Stoner, C. M., Gudas, L. J., and Chambon, P. (1989). Differential expression of genes encoding alpha, beta and gamma retinoic acid receptors and CRABP in the developing limbs of the mouse. *Nature (London)* **342,** 702–705.

Dollé, P., Ruberte, E., Leroy, P., Morriss-Kay, G., and Chambon, P. (1990). Retinoic acid receptors and cellular retinoid binding proteins. I. A systematic study of their differential pattern of transcription during mouse organogenesis. *Development (Cambridge, UK)* **110,** 1133–1151.

Dollé, P., Price, M., and Duboule, D. (1992). Expression of the murine Dlx-1 homeobox gene during facial, ocular and limb development. *Differentiation (Berlin)* **49,** 93–99.

Dominguez, M., Brunner, M., Hafen, E., and Basler, K. (1996). Sending and receiving the hedgehog signal: Control by the Drosophila Gli protein cubitus interruptus. *Science* **272,** 1621–1625.

Dryburgh, L. C. (1967). Epigenetics of early tooth development in the mouse. *J. Dent. Res.* **46,** 1264 (abstr.).

Dudley, A. T., Lyons, K. M., and Robertson, E. J. (1995). A requirement for bone morphogenetic protein-7 during development of the mammalian kidney and eye. *Genes Dev.* **9,** 2795–2807.

Dunlop, L. L., and Hall, B. K. (1995). Relationships between cellular condensation, preosteoblast formation and epithelial-mesenchymal interactions in initiation of osteogenesis. *Int. J. Dev. Biol.* **39,** 357–371.

Dupé, V., Ghyselinck, N. B., Thomazy, V., Nagy, L., Davies, P. J. A., Chambon, P., and Mark, M. (1999). Essential roles of retinoic acid signaling in interdigital apoptosis and control of BMP-7 expression in mouse autopods. *Dev. Biol.* **208,** 30–43.

Eales, N. B. (1950). The skull of the foetal narwhal, *Monodon monoceros* L. *Philos. Trans. Roy. Soc. London, Ser. B* **235,** 1–33.

Echelard, Y., Epstein, D. J., St-Jacques, B., Shen, L., Mohler, J., McMahon, J. A., and McMahon, A. P. (1993). Sonic hedgehog, a member of a family of putative signaling molecules, is implicated in the regulation of CNS polarity. *Cell (Cambridge, Mass.)* **75,** 1417–1430.

Eickholt, B. J., Mackenzie, S. L., Graham, A., Walsh, F. S., and Doherty, P. (1999). Evidence for collapsin-1 functioning in the control of neural crest migration in both trunk and hindbrain regions. *Development (Cambridge, UK)* **126,** 2181–2189.

Eisenstat, D. D., Liu, J. K., Mione, M., Zhong, W., Yu, G., Anderson, S. A., Ghattas, I., Puelles, L., and Rubenstein, J. L. (1999). DLX-1, DLX-2,

and DLX-5 expression define distinct stages of basal forebrain differentiation. *J. Comp. Neurol.* **414,** 217–237.

El Ghouzzi, V., Le Merrer, M., Perrin-Schmitt, F., Lajeunie, E., Benit, P., Renier, D., Bourgeois, P., Bolcato-Bellemin, A. L., Munnich, A., and Bonaventure, J. (1997). Mutations of the TWIST gene in the Saethre-Chotzen syndrome. *Nat. Genet.* **15,** 42–46.

Enwright, J. F., 3rd, and Grainger, R. M. (2000). Altered retinoid signaling in the heads of small eye mouse embryos. *Dev. Biol.* **221,** 10–22.

Epstein, D. J., Vekemans, M., and Gros, P. (1991). Splotch (Sp2H), a mutation affecting development of the mouse neural tube, shows a deletion within the paired homeodomain of Pax-3. *Cell (Cambridge, Mass.)* **67,** 767–774.

Erickson, C. A., and Perris, R. (1993). The role of cell-cell and cell-matrix interactions in the morphogenesis of the neural crest. *Dev. Biol.* **159,** 60–74.

Fang, J., and Hall, B. K. (1997). Chondrogenic cell differentiation from membrane bone periostea. *Anat. Embryol.* **196,** 349–362.

Fässler, R., Schnegelsberg, P. N., Dausman, J., Shinya, T., Muragaki, Y., McCarthy, M. T., Olsen, B. R., and Jaenisch, R. (1994). Mice lacking alpha 1 (IX) collagen develop noninflammatory degenerative joint disease. *Proc. Natl. Acad. Sci. U.S.A.* **91,** 5070–5074.

Fawcett, E. (1917). The primordial cranium of microtus amphibius (waterrat), as determined by sections and a model of the 25mm stage. With comparative remarks. *J. Anatomy* **51,** 309–359.

Fawcett, E. (1922). The primordial cranium of Xerus (spiny squirrel) at the 17 and 19 millimeters stages. *J. Anatomy* **57,** 221–237.

Feijen, A., Goumans, M. J., and van den Eijnden-van Raaij, A. J. (1994). Expression of activin subunits, activin receptors and follistatin in postimplantation mouse embryos suggests specific developmental functions for different activins. *Development (Cambridge, UK)* **120,** 3621–3637.

Ferguson, C., Tucker, A. S., Christensen, L., Lau, A., Matsuk, M. M., and Sharpe, P. T. (1998). Activin is an essential early mesenchymal signal in tooth development that is required for patterning of the murine dentition. *Genes Dev.* **12,** 2636–2649.

Ferguson, C., Tucker, A. S., and Sharpe, P. T. (2000). Temporo-spatial cell interactions regulating mandibular and maxillary arch patterning. *Development (Cambridge, UK)* **127,** 403–412.

Ferretti, E., Marshall, H., Pöpperl, H., Maconochie, M., Krumlauf, R., and Blasi, F. (2000). Segmental expression of Hoxb2 in r4 requires two separate sites that integrate cooperative interactions between Prep1, Pbx and Hox proteins. *Development (Cambridge, UK)* **127,** 155–166.

Filosa, S., Rivera-Pérez, J. A., Gómez, A. P., Gansmuller, A., Sasaki, H., Behringer, R. R., and Ang, S. L. (1997). Goosecoid and HNF-3beta genetically interact to regulate neural tube patterning during mouse embryogenesis. *Development (Cambridge, UK)* **124,** 2843–2854.

Flint, O. P., and Ede, D. A. (1978). Facial development in the mouse; a comparison between normal and mutant (amputated) mouse embryos. *J. Embryol. Exp. Morphol.* **48,** 249–267.

Flower, W. H. (1885). "An Introduction to the Osteology of the Mammalia." Macmillan, London.

Fode, C., Gradwohl, G., Morin, X., Dierich, A., LeMeur, M., Goridis, C., and Guillemot, F. (1998). The bHLH protein NEUROGENIN 2 is a determination factor for epibranchial placode-derived sensory neurons. *Neuron* **20,** 483–494.

Foley, A. C., Skromne, I., and Stern, C. D. (2000). Reconciling different models of forebrain induction and patterning: A dual role for the hypoblast. *Development (Cambridge, UK)* **127,** 3839–3854.

Francis-West, P. H., Tatla, T., and Brickell, P. M. (1994). Expression patterns of the bone morphogenetic protein genes Bmp-4 and Bmp-2 in the developing chick face suggest a role in outgrowth of the primordia. *Dev. Dyn.* **201,** 168–178.

Francis-West, P., Ladher, R., Barlow, A., and Graveson, A. (1998). Signalling interactions during facial development. *Mech. Dev.* **75,** 3–28.

Franzoso, G., Carlson, L., Xing, L., Poljak, L., Shores, E. W., Brown, K. D., Leonardi, A., Tran, T., Boyce, B. F., and Siebenlist, U. (1997). Requirement for NF-kappaB in osteoclast and B-cell development. *Genes Dev.* **11**, 3482–3496.

Frick, H. (1986). Zur Entwicklung des Knorpelschadels der Albinomaus. *In* "Craniogenesis and Craniofacial Growth" (G.-H. Schumacher, ed.), Vol. 262, pp. 305–317. Deutsche Akademie der Naturforscher Leopoldina, Halle.

Frost, H. M. (1969). Tetracycline-based histological analysis of bone remodeling. *Calcif. Tissue Res.* **3**, 211–237.

Fukiishi, Y., and Morriss-Kay, G. M. (1992). Migration of cranial neural crest cells to the pharyngeal arches and heart in rat embryos. *Cell Tissue Res.* **268**, 1–8.

Gage, P. J., Suh, H., and Camper, S. A. (1999). The bicoid-related Pitx gene family in development. *Mamm. Genome* **10**, 197–200.

Galliot, B., de Vargas, C., and Miller, D. (1999). Evolution of homeobox genes: Q50 Paired-like genes founded the Paired class. *Dev. Genes Evol.* **209**, 186–197.

Galvin, B. D., Hart, K. C., Meyer, A. N., Webster, M. K., and Donoghue, D. J. (1996). Constitutive receptor activation by Crouzon syndrome mutations in fibroblast growth factor receptor (FGFR)2 and FGFR2/Neu chimeras. *Proc. Natl. Acad. Sci. U.S.A.* **93**, 7894–7899.

Gans, C. (1993). Evolutionary origin of the vertebrate skull. *In* "The Skull" (J. Hanken and B. K. Hall, eds.), Vol. 2, pp. 1–35. University of Chicago Press, Chicago.

Gans, C., and Northcutt, R. G. (1983). Neural crest and the origin of vertebrates: A new head. *Science* **220**, 268–274.

Gaunt, S. J., Blum, M., and De Robertis, E. M. (1993). Expression of the mouse goosecoid gene during mid-embryogenesis may mark mesenchymal cell lineages in the developing head, limbs and body wall. *Development (Cambridge, UK)* **117**, 769–778.

Gavalas, A., and Krumlauf, R. (2000). Retinoid signalling and hindbrain patterning. *Curr. Opin. Genet. Dev.* **10**, 380–386.

Gavalas, A., Studer, M., Lumsden, A., Rijli, F. M., Krumlauf, R., and Chambon, P. (1998). Hoxa1 and Hoxb1 synergize in patterning the hindbrain, cranial nerves and second pharyngeal arch. *Development (Cambridge, UK)* **125**, 1123–1136.

Gendron-Maguire, M., Mallo, M., Zhang, M., and Gridley, T. (1993). Hoxa-2 mutant mice exhibit homeotic transformation of skeletal elements derived from cranial neural crest. *Cell (Cambridge, Mass.)* **75**, 1317–1331.

Ghyselinck, N. B., Dupé, V., Dierich, A., Messaddeq, N., Garnier, J. M., Rochette-Egly, C., Chambon, P., and Mark, M. (1997). Role of the retinoic acid receptor beta (RARbeta) during mouse development. *Int. J. Dev. Biol.* **41**, 425–447.

Glimcher, M. J. (1989). Mechanism of calcification: Role of collagen fibrils and collagen-phosphoprotein complexes in vitro and in vivo. *Anat. Rec.* **224**, 139–153.

Godfrey, P., Rahal, J. O., Beamer, W. G., Copeland, N. G., Jenkins, N. A., and Mayo, K. E. (1993). GHRH receptor of little mice contains a missense mutation in the extracellular domain that disrupts receptor function. *Nat. Genet.* **4**, 227–232.

Golding, J. P., Trainor, P., Krumlauf, R., and Gassmann, M. (2000). Defects in pathfinding by cranial neural crest cells in mice lacking the neuregulin receptor ErbB4. *Nat. Cell Biol.* **2**, 103–109.

Goodrich, E. S. (1958). "Studies on the Structure and Development of Vertebrates." Dover, New York.

Goodrich, L. V., Johnson, R. L., Milenkovic, L., McMahon, J. A., and Scott, M. P. (1996). Conservation of the hedgehog/patched signaling pathway from flies to mice: Induction of a mouse patched gene by Hedgehog. *Genes Dev.* **10**, 301–312.

Gorlin, R. J., Cohen, M. M., and Levin, L. S. (1990). "Syndromes of the Head and Neck." Oxford University Press, New York.

Goronowitsch, N. (1892). Die axiale und die laterale Kopfmetamerie der Vögelembryonen. Die Rolle der sog. 'Ganglienleisten' im Aufbaue der Nervenstämme. *Anat. Anz.* **7**, 454–464.

Goronowitsch, N. (1893). Weiters über die ektodermal Entstehung von Skeletanlagen im Kopfe der Wirbeltiere. *Morphol. Jahrb.* **20**, 425–428.

Gorry, P., Lufkin, T., Dierich, A., Rochette-Egly, C., Décimo, D., Dollé, P., Mark, M., Durand, B., and Chambon, P. (1994). The cellular retinoic acid binding protein I is dispensable. *Proc. Natl. Acad. Sci. U.S.A.* **91**, 9032–9036.

Gould, S. J. (1977). "Ontogeny and Phylogeny." Belknap Press of Harvard University Press, Cambridge, MA.

Graham, A., Heyman, I., and Lumsden, A. (1993). Even-numbered rhombomeres control the apoptotic elimination of neural crest cells from odd-numbered rhombomeres in the chick hindbrain. *Development (Cambridge, UK)* **119**, 233–245.

Graham, A., Koentges, G., and Lumsden, A. (1996). Neural crest apoptosis and the establishment of craniofacial pattern: An honorable death. *Mol. Cell. Neurosci.* **8**, 76–83.

Grammatopoulos, G. A., Bell, E., Toole, L., Lumsden, A. and Tucker, A. S. (2000). Homeotic transfromation of branchial arch identity after Hoxa2 overexpression. *Development (Cambridge, UK)* **127**, 5355–5365.

Grant, J. H., Maggio-Price, L., Reutebuch, J., and Cunningham, M. L. (1997). Retinoic acid exposure of the mouse on embryonic day 9 selectively spares derivatives of the frontonasal neural crest. *J. Craniofacial Genet. Dev. Biol.* **17**, 1–8.

Green, M. C. (1970). The developmental effects of congenital hydrocephalus (ch) in the mouse. *Dev. Biol.* **23**, 585–608.

Greene, E. C. (1955). "Anatomy of the Rat." Hafner, New York.

Gregory, W. K. (1904). The relations of the anterior visceral arches to the chondrocranium. *Biol. Bul. (Woods Hole, Mass)* **7**, 55–69.

Grigoriadis, A. E., Wang, Z. Q., Cecchini, M. G., Hofstetter, W., Felix, R., Fleisch, H. A., and Wagner, E. F. (1994). c-Fos: A key regulator of osteoclast-macrophage lineage determination and bone remodeling. *Science* **266**, 443–448.

Grigoriou, M., Tucker, A. S., Sharpe, P. T., and Pachnis, V. (1998). Expression and regulation of Lhx6 and Lhx7, a novel subfamily of LIM homeodomain encoding genes, suggests a role in mammalian head development. *Development (Cambridge, UK)* **125**, 2063–2074.

Grindley, J. C., Davidson, D. R., and Hill, R. E. (1995). The role of Pax-6 in eye and nasal development. *Development (Cambridge, UK)* **121**, 1433–1442.

Grindley, J. C., Bellusci, S., Perkins, D., and Hogan, B. L. (1997). Evidence for the involvement of the Gli gene family in embryonic mouse lung development. *Dev. Biol.* **188**, 337–348.

Gripp, K. W., Wotton, D., Edwards, M. C., Roessler, E., Ades, L., Meinecke, P., Richieri-Costa, A., Zackai, E. H., Massagué, J., Muenke, M., and Elledge, S. J. (2000). Mutations in TGIF cause holoprosencephaly and link NODAL signalling to human neural axis determination. *Nat. Genet.* **25**, 205–208.

Grüneberg, H. (1953). Genetical studies on the skeleton of the mouse. VII. Congenital hydrocephalus. *J. Genet.* **51**, 327–358.

Grüneberg, H. (1954). Genetical studies on the skeleton of the mouse. XII, The development of undulated. *J. Genet.* **52**, 441–455.

Grüneberg, H. (1963). "The pathology of Development; A study of Inherited Skeletal Disorders in Animals." Blackwell Scientific Publications, Oxford.

Grüneberg, H., and Wickramaratne, G. A. (1974). A re-examination of two skeletal mutants of the mouse, vestigial-tail (vt) and congenital hydrocephalus (ch). *J. Embryol. Exp. Morphol.* **31**, 207–222.

Guthrie, S., Muchamore, I., Kuroiwa, A., Marshall, H., Krumlauf, R., and Lumsden, A. (1992). Neuroectodermal autonomy of Hox-2.9 expression revealed by rhombomere transpositions. *Nature (London)* **356**, 157–159.

Hacohen, N., Kramer, S., Sutherland, D., Hiromi, Y., and Krasnow, M. A. (1998). Sprouty encodes a novel antagonist of FGF signaling that patterns apical branching of the Drosophila airways. *Cell (Cambridge, Mass.)* **92**, 253–263.

Hahn, H., Christiansen, J., Wicking, C., Zaphiropoulos, P. G., Chidambaram, A., Gerrard, B., Vorechovsky, I., Bale, A. E., Toftgard, R., Dean, M., and Wainwright, B. (1996a). A mammalian patched homolog is expressed in target tissues of sonic hedgehog and maps to a region associated with developmental abnormalities. *J. Biol. Chem.* **271,** 12125–12128.

Hahn, H., Wicking, C., Zaphiropoulous, P. G., Gailani, M. R., Shanley, S., Chidambaram, A., Vorechovsky, I., Holmberg, E., Unden, A. B., Gillies, S., Negus, K., Smyth, I., Pressman, C., Leffell, D. J., Gerrard, B., Goldstein, A. M., Dean, M., Toftgard, R., Chenevix-Trench, G., Wainwright, B., and Bale, A. E. (1996b). Mutations of the human homolog of Drosophila patched in the nevoid basal cell carcinoma syndrome. *Cell (Cambridge, Mass.)* **85,** 841–851.

Hahn, H., Wojnowski, L., Zimmer, A. M., Hall, J., Miller, G., and Zimmer, A. (1998). Rhabdomyosarcomas and radiation hypersensitivity in a mouse model of Gorlin syndrome. *Nat. Med.* **4,** 619–622.

Halford, M. M., Armes, J., Buchert, M., Meskenaite, V., Grail, D., Hibbs, M. L., Wilks, A. F., Farlie, P. G., Newgreen, D. F., Hovens, C. M., and Stacker, S. A. (2000). Ryk-deficient mice exhibit craniofacial defects associated with perturbed Eph receptor crosstalk. *Nat. Genet.* **25,** 414–418.

Hall, B. K. (1980a). Viability and proliferation of epithelia and the initiation of osteogenesis within mandibular ectomesenchyme in the embryonic chick. *J. Embryol. Exp. Morphol.* **56,** 71–89.

Hall, B. K. (1980b). Tissue interactions and the initiation of osteogenesis and chondrogenesis in the neural crest-derived mandibular skeleton of the embryonic mouse as seen in isolated murine tissues and in recombinations of murine and avian tissues. *J. Embryol. Exp. Morphol.* **58,** 251–264.

Hall, B. K. (1983). Cell-tissue interactions: A rationale and resume. *J. Craniofacial Genet. Dev. Biol.* **3,** 75–82.

Hall, B. K. (1987). Tissue interactions in the development and evolution of the vertebrate head. *In* "Developmental and Evolutionary Aspects of the Neural Crest" (P. F. A. Maderson, ed.), pp. 215–259. Wiley, New York.

Hall, B. K. (1999). "The Neural Crest in Development and Evolution." Springer, New York.

Hall, B. K., and Hanken, J. (1985). Preface: The developmnet of the vertebrate skull. *In* "The Development of the Vertebrate Skull" (G. de Beer, ed.). University of Chicago Press, Chicago.

Hall, B. K., and Miyake, T. (1992). The membranous skeleton: The role of cell condensations in vertebrate skeletogenesis. *Anat. Embryol.* **186,** 107–124.

Hall, B. K., and Miyake, T. (2000). All for one and one for all: Condensations and the initiation of skeletal development. *BioEssays* **22,** 138–147.

Hall, B. K., and Tremaine, R. (1979). Ability of neural crest cells from the embryonic chick to differentiate into cartilage before their migration away from the neural tube. *Anat. Rec.* **194,** 469–475.

Hall, B. K., Van Exan, R. J., and Brunt, S. L. (1983). Retention of epithelial basal lamina allows isolated mandibular mesenchyme to form bone. *J. Craniofacial Genet. Dev. Biol.* **3,** 253–267.

Hammerschmidt, M., Brook, A., and McMahon, A. P. (1997). The world according to hedgehog. *Trends Genet.* **13,** 14–21.

Hanken, J., and Hall, B. K. (1993). Mechanisms of skull diversity and evolution. *In* "The Skull" (J. Hanken and B. K. Hall, eds.), Vol. 1, pp. 1–36. University of Chicago Press, Chicago.

Hanken, J., and Thorogood, P. (1993). Evolution and development of the vertebrate skull: The role of pattern formation. *Trends Ecol. Evol.* **8,** 9–15.

Happle, R., Phillips, R. J., Roessner, A., and Jünemann, G. (1983). Homologous genes for X-linked chondrodysplasia punctata in man and mouse. *Hum. Genet.* **63,** 24–27.

Hardcastle, Z., Mo, R., Hui, C.-C., and Sharpe, P. T. (1998). The Shh signalling pathway in tooth development: Defects in Gli2 and Gli3 mutants. *Development (Cambridge, UK)* **125,** 2803–2811.

Harris, M. J., and Juriloff, D. M. (1999). Mini-review: Toward understanding mechanisms of genetic neural tube defects in mice. *Teratology* **60,** 292–305.

Hatini, V., Ye, X., Balas, G., and Lai, E. (1999). Dynamics of placodal lineage development revealed by targeted transgene expression. *Dev. Dyn.* **215,** 332–343.

Hay, M. F. (1961). The development *in vivo* and *in vitro* of the lower incisor and molars of the mouse. *Arch. Oral Biol.* **3,** 86–109.

Headon, D. J., and Overbeek, P. A. (1999). Involvement of a novel TNF receptor homolog in hair follicle development. *Nat. Genet.* **22,** 370–374.

Heikinheimo, M., Lawshé, A., Shackleford, G. M., Wilson, D. B., and MacArthur, C. A. (1994). Fgf-8 expression in the post-gastrulation mouse suggests roles in the development of the face, limbs and central nervous system. *Mech. Dev.* **48,** 129–138.

Heldin, C. H., Miyazono, K., and ten Dijke, P. (1997). TGF-beta signalling from cell membrane to nucleus through SMAD proteins. *Nature (London)* **390,** 465–471.

Helms, J. A., Kim, C. H., Hu, D., Minkoff, R., Thaller, C., and Eichele, G. (1997). Sonic hedgehog participates in craniofacial morphogenesis and is down-regulated by teratogenic doses of retinoic acid. *Dev. Biol.* **187,** 25–35.

Henderson, D. J., Ybot-Gonzalez, P., and Copp, A. J. (1997). Overexpression of the chondroitin sulphate proteoglycan versican is associated with defective neural crest migration in the Pax3 mutant mouse (splotch). *Mech. Dev.* **69,** 39–51.

Herring, S. (1993). Epigenetic and functional influences on skull growth. *In* "The Skull" (J. Hanken and B. K. Hall, eds.), Vol. 1, pp. 153–206. University of Chicago Press, Chicago.

Hieda, Y., and Nakanishi, Y. (1997). Epithelial morphogenesis in mouse embryonic submandibular gland: Its relationships to the tissue organization of epithelium and mesenchyme. *Dev. Growth Differ.* **39,** 1–8.

Hildebrand, M. (1988). "Analysis of Vertebrate Structure." Wiley, New York.

Hill, R. E., Favor, J., Hogan, B. L., Ton, C. C., Saunders, G. F., Hanson, I. M., Prosser, J., Jordan, T., Hastie, N. D., and van Heyningen, V. (1991). Mouse small eye results from mutations in a paired-like homeobox-containing gene. *Nature (London)* **354,** 522–525; erratum: *Ibid.* **355,** 750 (1992).

Hogan, B. L. (1996). Bone morphogenetic proteins in development. *Curr. Opin. Genet. Dev.* **6,** 432–438.

Hogan, B. L., Horsburgh, G., Cohen, J., Hetherington, C. M., Fisher, G., and Lyon, M. F. (1986). Small eyes (Sey): A homozygous lethal mutation on chromosome 2 which affects the differentiation of both lens and nasal placodes in the mouse. *J. Embryol. Exp. Morphol.* **97,** 95–110.

Holder, N., and Hill, J. (1991). Retinoic acid modifies development of the midbrain-hindbrain border and affects cranial ganglion formation in zebrafish embryos. *Development (Cambridge, UK)* **113,** 1159–1170.

Holland, L. Z., and Holland, N. D. (1996). Expression of AmphiHox-1 and AmphiPax-1 in amphioxus embryos treated with retinoic acid: Insights into evolution and patterning of the chordate nerve cord and pharynx. *Development (Cambridge, UK)* **122,** 1829–1838.

Hörstadius, S. (1950). "The Neural Crest." Oxford University Press, London.

Howard, T. D., Paznekas, W. A., Green, E. D., Chiang, L. C., Ma, N., Ortiz de Luna, R. I., Garcia Delgado, C., Gonzalez-Ramos, M., Kline, A. D., and Jabs, E. W. (1997). Mutations in TWIST, a basic helix-loop-helix transcription factor, in Saethre-Chotzen syndrome. *Nat. Genet.* **15,** 36–41.

Hu, D., and Helms, J. A. (1999). The role of sonic hedgehog in normal and abnormal craniofacial morphogenesis. *Development (Cambridge, UK)* **126,** 4873–4884.

Hu, Y., Baud, V., Delhase, M., Zhang, P., Deerinck, T., Ellisman, M., Johnson, R., and Karin, M. (1999). Abnormal morphogenesis but intact IKK activation in mice lacking the IKKalpha subunit of IkappaB kinase. *Science* **284,** 316–320.

Hui, C. C., and Joyner, A. L. (1993). A mouse model of greig cephalopoly-syndactyly syndrome: The extra-toesJ mutation contains an intragenic deletion of the Gli3 gene. *Nat. Genet.* **3**, 241–246; erratum: *Ibid.* **19**(4), 404 (1998).

Hui, C. C., Slusarski, D., Platt, K. A., Holmgren, R., and Joyner, A. L. (1994). Expression of three mouse homologs of the Drosophila segment polarity gene cubitus interruptus, Gli, Gli-2, and Gli-3, in ectoderm- and mesoderm-derived tissues suggests multiple roles during postimplantation development. *Dev. Biol.* **162**, 402–413.

Hulander, M., Wurst, W., Carlsson, P., and Enerbäck, S. (1998). The winged helix transcription factor Fkh10 is required for normal development of the inner ear. *Nat. Genet.* **20**, 374–376.

Hunt, P., and Krumlauf, R. (1991). Deciphering the Hox code: Clues to patterning branchial regions of the head. *Cell (Cambridge, Mass.)* **66**, 1075–1078.

Hunt, P., and Krumlauf, R. (1992). Hox codes and positional specification in vertebrate embryonic axes. *Ann. Rev. Cell Biol.* **8**, 227–256.

Hunt, P., Gulisano, M., Cook, M., Sham, M. H., Faiella, A., Wilkinson, D., Boncinelli, E., and Krumlauf, R. (1991a). A distinct Hox code for the branchial region of the vertebrate head. *Nature (London)* **353**, 861–864.

Hunt, P., Whiting, J., Muchamore, I., Marshall, H., and Krumlauf, R. (1991b). Homeobox genes and models for patterning the hindbrain and branchial arches. *Development (Cambridge, UK), Suppl.* **1**, 187–196.

Hunt, P., Wilkinson, D., and Krumlauf, R. (1991c). Patterning the vertebrate head: Murine Hox 2 genes mark distinct subpopulations of premigratory and migrating cranial neural crest. *Development (Cambridge, UK)* **112**, 43–50.

Hunt, P., Clarke, J. D., Buxton, P., Ferretti, P., and Thorogood, P. (1998a). Segmentation, crest prespecification and the control of facial form. *Eur. J. Oral Sci.* **106**(Suppl. 1), 12–18.

Hunt, P., Clarke, J. D., Buxton, P., Ferretti, P., and Thorogood, P. (1998b). Stability and plasticity of neural crest patterning and branchial arch Hox code after extensive cephalic crest rotation. *Dev. Biol.* **198**, 82–104.

Hynes, R. O. (1996). Targeted mutations in cell adhesion genes: What have we learned from them? *Dev. Biol.* **180**, 402–412.

Iida, K., Koseki, H., Kakinuma, H., Kato, N., Mizutani-Koseki, Y., Ohuchi, H., Yoshioka, H., Noji, S., Kawamura, K., Kataoka, Y., Ueno, F., Taniguchi, M., Yoshida, N., Sugiyama, T., and Miura, N. (1997). Essential roles of the winged helix transcription factor MFH-1 in aortic arch patterning and skeletogenesis. *Development (Cambridge, UK)* **124**, 4627–4638.

Ikeya, M., Lee, S. M., Johnson, J. E., McMahon, A. P., and Takada, S. (1997). Wnt signalling required for expansion of neural crest and CNS progenitors. *Nature (London)* **389**, 966–970.

Imai, H., Osumi-Yamashita, N., Ninomiya, Y., and Eto, K. (1996). Contribution of early-emigrating midbrain crest cells to the dental mesenchyme of mandibular molar teeth in rat embryos. *Dev. Biol.* **176**, 151–165.

Irving, C., and Mason, I. (2000). Signalling by FGF8 from the isthmus patterns anterior hindbrain and establishes the anterior limit of Hox gene expression. *Development (Cambridge, UK)* **127**, 177–186.

Iseki, S., Wilkie, A. O., and Morriss-Kay, G. M. (1999). Fgfr1 and Fgfr2 have distinct differentiation- and proliferation-related roles in the developing mouse skull vault. *Development (Cambridge, UK)* **126**, 5611–5620.

Jabs, E. W., Müller, U., Li, X., Ma, L., Luo, W., Haworth, I. S., Klisak, I., Sparkes, R., Warman, M. L., Mulliken, J. B. *et al.* (1993). A mutation in the homeodomain of the human MSX2 gene in a family affected with autosomal dominant craniosynostosis. *Cell (Cambridge, Mass.)* **75**, 443–450.

Jabs, E. W., Li, X., Scott, A. F., Meyers, G., Chen, W., Eccles, M., Mao, J. I., Charnas, L. R., Jackson, C. E., and Jaye, M. (1994). Jackson-Weiss and Crouzon syndromes are allelic with mutations in fibroblast growth factor receptor 2. *Nat. Genet.* **8**, 275–279.

Jacenko, O., LuValle, P., Solum, K., and Olsen, B. R. (1993). A dominant negative mutation in the alpha 1 (X) collagen gene produces spondylo-metaphyseal defects in mice. *Prog. Clin. Biol. Res.* **383B**, 427–436.

Jacobs, Y., Schnabel, C. A., and Cleary, M. L. (1999). Trimeric association of Hox and TALE homeodomain proteins mediates Hoxb2 hindbrain enhancer activity. *Mol. Cell. Biol.* **19**, 5134–5142.

Jacobson, A. G. (1963). The determination and positioning of the nose, lens and ear. I. Interactions within the ectoderm, and between the ectoderm and the underlying tissues. *J. Exp. Zool.* **154**, 273–284.

Jaskoll, T., and Melnick, M. (1999). Submandibular gland morphogenesis: Stage-specific expression of TGF-alpha/EGF, IGF, TGF-beta, TNF, and IL-6 signal transduction in normal embryonic mice and the phenotypic effects of TGF-beta2, TGF-beta3, and EGF-r null mutations. *Anat. Rec.* **256**, 252–268.

Jenkinson, J. W. (1911). The development of the ear bones in the mouse. *J. Anat. Physiol.* **45**.

Jernvall, J., Åberg, T., Kettunen, P., Keränen, S. and Thesleff, I. (1998). The life history of an embryonic signalling center: BMP4 induces p21 and is associated with apoptosis in the mouse tooth enamel knot. *Development (Cambridge, UK)* **125**, 161–169.

Jiang, R., Lan, Y., Norton, C. R., Sundberg, J. P., and Gridley, T. (1998). The Slug gene is not essential for mesoderm or neural crest development in mice. *Dev. Biol.* **198**, 277–285.

Johansen, V. A., and Hall, S. H. (1982). Morphogenesis of the mouse coronal suture. *Acta Anat.* **114**, 58–67.

Johnson, D. R. (1967). Extra-toes: A new mutant gene causing multiple abnormalities in the mouse. *J. Embryol. Exp. Morphol.* **17**, 543–581.

Johnson, M. L. (1933). The time and order of appearance of ossification centers in the albino mouse. *Am. J. Anat.* **52**, 241–271.

Johnson, R. L., Laufer, E., Riddle, R. D., and Tabin, C. (1994). Ectopic expression of Sonic hedgehog alters dorsal-ventral patterning of somites. *Cell (Cambridge, Mass.)* **79**, 1165–1173.

Johnson, R. S., Spiegelman, B. M., and Papaioannou, V. (1992). Pleiotropic effects of a null mutation in the c-fos proto-oncogene. *Cell (Cambridge, Mass.)* **71**, 577–586.

Johnston, M. C. (1966). A radioautographic study of the migration and fate of cranial neural crest cells in the chick embryo. *Anat. Rec.* **156**, 143–155.

Jollie, M. (1977). Segmentation of the vertebrate head. *Am. Zool.* **17**, 323–333.

Jostes, B., Walther, C., and Gruss, P. (1990). The murine paired box gene, Pax7, is expressed specifically during the development of the nervous and muscular system. *Mech. Dev.* **33**, 27–37.

Joyner, A. L. (1996). Engrailed, Wnt and Pax genes regulate midbrain—hindbrain development. *Trends Genet.* **12**, 15–20.

Juriloff, D. M., and Harris, M. J. (2000). Mouse models for neural tube closure defects. *Hum. Mol. Genet.* **9**, 993–1000.

Kaartinen, V., Voncken, J. W., Shuler, C., Warburton, D., Bu, D., Heisterkamp, N., and Groffen, J. (1995). Abnormal lung development and cleft palate in mice lacking TGF-beta 3 indicates defects of epithelial-mesenchymal interaction. *Nat. Genet.* **11**, 415–421.

Kaestner, K. H., Bleckmann, S. C., Monaghan, A. P., Schlöndorff, J., Mincheva, A., Lichter, P., and Schütz, G. (1996). Clustered arrangement of winged helix genes fkh-6 and MFH-1: Possible implications for mesoderm development. *Development (Cambridge, UK)* **122**, 1751–1758.

Kaku, M., Kawata, T., Kawasoko, S., Fujita, T., Tokimasa, C., and Tanne, K. (1999). Remodeling of the sagittal suture in osteopetrotic (op/op) mice associated with cranial flat bone growth. *J. Craniofacial Genet. Dev. Biol.* **19**, 109–112.

Kalter, H. (1980). A compendium of the genetically induced congenital malformations of the house mouse. *Teratology* **21**, 397–429.

Kang, S., Graham, J. M., Olney, A. H., and Biesecker, L. G. (1997). GLI3 frameshift mutations cause autosomal dominant Pallister-Hall syndrome. *Nat. Genet.* **15**, 266–268.

Kanzler, B., Foreman, R. K., Labosky, P. A., and Mallo, M. (2000). BMP signaling is essential for development of skeletogenic and neurogenic cranial neural crest. *Development (Cambridge, UK)* **127**, 1095–1104.

Karaplis, A. C., Luz, A., Glowacki, J., Bronson, R. T., Tybulewicz, V. L., Kronenberg, H. M., and Mulligan, R. C. (1994). Lethal skeletal dysplasia from targeted disruption of the parathyroid hormone-related peptide gene. *Genes Dev.* **8**, 277–289.

Karp, S. J., Schipani, E., St.-Jacques, B., Hunzelman, J., Kronenberg, H., and McMahon, A. P. (2000). Indian hedgehog coordinates endochondral bone growth and morphogenesis via parathyroid hormone related-protein-dependent and -independent pathways. *Development (Cambridge, UK)* **127**, 543–548.

Kastschenko, N. (1888). Zur Entwicklungsgeschichte der Selachierembryos. *Anat. Anz.* **3**, 445–467.

Kaufman, M. H., Chang, H. H., and Shaw, J. P. (1995). Craniofacial abnormalities in homozygous Small eye (Sey/Sey) embryos and newborn mice. *J. Anat.* **186**, 607–617.

Kaufmann, E., and Knöchel, W. (1996). Five years on the wings of fork head. *Mech. Dev.* **57**, 3–20.

Kere, J., Srivasta, A. K., Montonen, O., Zonana, J., Thomas, N., Ferguson, B., Munoz, F., Morgan, D., Clarke, A., Baybayan, P., Chen, E. Y., Ezer, S., Saarialho-Kere, U., de la Chapelle, A. and Schlessinger, D. (1996). X-linked anhidrotic (hypohidrotic) ectodermal dysplasia is caused by a mutation in a novel transmembrane protein. *Nat. Genet.* **13**, 409–416.

Kesteven, H. L. (1926). The homology of the ala temporalis and the alisphenoid bone. *J. Anat.* **61**, 112–166.

Keverne, E. B. (1999). The vomeronasal organ. *Science* **286**, 716–720.

Kidwell, J. F., Gowen, J. W., and Stradler, J. (1961). Pugnose—a recessive mutation in linkage group 3 of mice. *J. Hered.* **52**, 145–148.

Kim, H. J., Rice, D. P., Kettunen, P. J., and Thesleff, I. (1998). FGF-, BMP- and Shh-mediated signalling pathways in the regulation of cranial suture morphogenesis and calvarial bone development. *Development (Cambridge, UK)* **125**, 1241–1251.

Kim, Y., and Nirenberg, M. (1989). Drosophila NK-homeobox genes. *Proc. Natl. Acad. Sci. U.S.A.* **86**, 7716–7720.

Kimura, C., Takeda, N., Suzuki, M., Oshimura, M., Aizawa, S., and Matsuo, I. (1997). Cis-acting elements conserved between mouse and pufferfish Otx2 genes govern the expression in mesencephalic neural crest cells. *Development (Cambridge, UK)* **124**, 3929–3941.

Kimura, S., Hara, Y., Pineau, T., Fernandez-Salguero, P., Fox, C. H., Ward, J. M., and Gonzalez, F. J. (1996). The T/ebp null mouse: Thyroid-specific enhancer-binding protein is essential for the organogenesis of the thyroid, lung, ventral forebrain, and pituitary. *Genes Dev.* **10**, 60–69.

Kingbury, B. F. (1904). Branchiomerism and the theory of head segmentation. *J. Morphol.* **42**, 83–109.

Kingsley, D. M. (1994). What do BMPs do in mammals? Clues from the mouse short-ear mutation. *Trends Genet.* **10**, 16–21.

Kingsley, D. M., Bland, A. E., Grubber, J. M., Marker, P. C., Russell, L. B., Copeland, N. G., and Jenkins, N. A. (1992). The mouse short ear skeletal morphogenesis locus is associated with defects in a bone morphogenetic member of the TGF beta superfamily. *Cell (Cambridge, Mass.)* **71**, 399–410.

Klima, M. (1987). Morphogenesis of the nasal structures of the skulls in toothed whales (Odontoceti). *In* "Morphogenesis of the Mammalian Skull" (U. Zeller and H. Kuhn, eds.), pp. 105–122. Parey, Hamburg and Berlin.

Knoepfler, P. S., and Kamps, M. P. (1995). The pentapeptide motif of Hox proteins is required for cooperative DNA binding with Pbx1, physically contacts Pbx1, and enhances DNA binding by Pbx1. *Mol. Cell. Biol.* **15**, 5811–5819.

Knoepfler, P. S., Calvo, K. R., Chen, H., Antonarakis, S. E., and Kamps, M. P. (1997). Meis1 and pKnox1 bind DNA cooperatively with Pbx1

utilizing an interaction surface disrupted in oncoprotein E2a-Pbx1. *Proc. Natl. Acad. Sci. U.S.A.* **94**, 14553–14558.

Knoetgen, H., Teichmann, U., and Kessel, M. (1999a). Head-organizing activities of endodermal tissues in vertebrates. *Cell. Mol. Biol.* **45**, 481–492.

Knoetgen, H., Viebahn, C., and Kessel, M. (1999b). Head induction in the chick by primitive endoderm of mammalian, but not avian origin. *Development (Cambridge, UK)* **126**, 815–825.

Kollar, E. J., and Baird, G. R. (1969). The influence of the dental papilla on the development of tooth shape in embryonic mouse tooth germs. *J. Embryol. Exp. Morphol.* **21**, 131–148.

Kollar, E. J., and Baird, G. R. (1970a). Tissue interactions in embryonic mouse tooth germs. I. Reorganization of the dental epithelium during tooth-germ reconstruction. *J. Embryol. Exp. Morphol.* **24**, 159–171.

Kollar, E. J., and Baird, G. R. (1970b). Tissue interactions in embryonic mouse tooth germs. II. The inductive role of the dental papilla. *J. Embryol. Exp. Morphol.* **24**, 173–186.

Komori, T., Yagi, H., Nomura, S., Yamaguchi, A., Sasaki, K., Deguchi, K., Shimizu, Y., Bronson, R. T., Gao, Y. H., Inada, M., Sato, M., Okamoto, R., Kitamura, Y., Yoshiki, S., and Kishimoto, T. (1997). Targeted disruption of Cbfa1 results in a complete lack of bone formation owing to maturational arrest of osteoblasts. *Cell (Cambridge, Mass.)* **89**, 755–764.

Köntges, G., and Lumsden, A. (1996). Rhombencephalic neural crest segmentation is preserved throughout craniofacial ontogeny. *Development (Cambridge, UK)* **122**, 3229–3242.

Kramer, S., Okabe, M., Hacohen, N., Krasnow, M. A., and Hiromi, Y. (1999). Sprouty: A common antagonist of FGF and EGF signaling pathways in Drosophila. *Development (Cambridge, UK)* **126**, 2515–2525.

Kratochwil, K., Dull, M., Farinas, I., Galceran, J., and Grosschedl, R. (1996). Lef1 expression is activated by BMP-4 and regulates inductive tissue interactions in tooth and hair development. *Genes Dev.* **10**, 1382–1394.

Krumlauf, R. (1993). Hox genes and pattern formation in the branchial region of the vertebrate head. *Trends Genet.* **9**, 106–112.

Kuhn, H., and Zeller, U. (1987). The cavum epiptericum in monotremes and therian mammals. *In* "Morphogenesis of the Mammalian Skull" (U. Zeller and H. Kuhn, eds.), pp. 51–70. Parey, Hamburg and Berlin.

Kulesa, P. M., and Fraser, S. E. (2000). In ovo time-lapse analysis of chick hindbrain neural crest cell migration shows cell interactions during migration to the branchial arches. *Development (Cambridge, UK)* **127**, 1161–1172.

Kume, T., Deng, K. Y., Winfrey, V., Gould, D. B., Walter, M. A., and Hogan, B. L. (1998). The forkhead/winged helix gene Mf1 is disrupted in the pleiotropic mouse mutation congenital hydrocephalus. *Cell (Cambridge, Mass.)* **93**, 985–996.

Kuratani, S. (1989). Development of the orbital region in the chondrocranium of *Caretta caretta*. Reconsideration of the vertebrate neurocranium configuration. *Anat. Anz.* **169**, 335–349.

Kuratani, S. (1997). Spatial distribution of postotic crest cells defines the head/trunk interface of the vertebrate body: Embryological interpretation of peripheral nerve morphology and evolution of the vertebrate head. *Anat. Embryol.* **195**, 1–13.

Kuratani, S., Martin, J. F., Wawersik, S., Lilly, B., Eichele, G., and Olson, E. N. (1994). The expression pattern of the chick homeobox gene gMHox suggests a role in patterning of the limbs and face and in compartmentalization of somites. *Dev. Biol.* **161**, 357–369.

Kuratani, S., Matsuo, I., and Aizawa, S. (1997). Developmental patterning and evolution of the mammalian viscerocranium: Genetic insights into comparative morphology. *Dev. Dyn.* **209**, 139–155.

Kuratani, S., Horigome, N., and Hirano, S. (1999). Developmental morphology of the head mesoderm and reevaluation of segmental theories of the vertebrate head: Evidence from embryos of an agnathan vertebrate, *Lampetra japonica*. *Dev. Biol.* **210**, 381–400.

Kurihara, Y., Kurihara, H., Suzuki, H., Kodama, T., Maemura, K., Nagai, R., Oda, H., Kuwaki, T., Cao, W. H., Kamada, N. *et al.* (1994). Elevated blood pressure and craniofacial abnormalities in mice deficient in endothelin-1. *Nature (London)* **368**, 703–710.

Kurihara, Y., Kurihara, H., Oda, H., Maemura, K., Nagai, R., Ishikawa, T., and Yazaki, Y. (1995). Aortic arch malformations and ventricular septal defect in mice deficient in endothelin-1. *J. Clin. Invest.* **96**, 293–300.

LaBonne, C., and Bronner-Fraser, M. (1998). Neural crest induction in Xenopus: Evidence for a two-signal model. *Development (Cambridge, UK)* **125**, 2403–2414.

LaBonne, C., and Bronner-Fraser, M. (1999). Molecular mechanisms of neural crest formation. *Ann. Rev. Cell Dev. Biol.* **15**, 81–112.

Lajeunie, E., Le Merrer, M., Bonaïti-Pellie, C., Marchac, D., and Renier, D. (1995). Genetic study of nonsyndromic coronal craniosynostosis. *Am. J. Med. Genet.* **55**, 500–504.

Lampron, C., Rochette-Egly, C., Gorry, P., Dollé, P., Mark, M., Lufkin, T., LeMeur, M., and Chambon, P. (1995). Mice deficient in cellular retinoic acid binding protein II (CRABPII) or in both CRABPI and CRABPII are essentially normal. *Development (Cambridge, UK)* **121**, 539–548.

Lanctôt, C., Lamolet, B., and Drouin, J. (1997). The bicoid-related homeoprotein Ptx1 defines the most anterior domain of the embryo and differentiates posterior from anterior lateral mesoderm. *Development (Cambridge, UK)* **124**, 2807–2817.

Lanctôt, C., Moreau, A., Chamberland, M., Tremblay, M. L., and Drouin, J. (1999). Hindlimb patterning and mandible development require the Ptx1 gene. *Development (Cambridge, UK)* **126**, 1805–1810.

Langille, R. M., and Hall, B. K. (1989). Developmental processes, developmental sequences and early vertebrate phylogeny. *Biol. Rev. Cambridge Philos. Soc.* **64**, 73–91.

Langille, R. M., and Hall, B. K. (1993). Pattern formation and the neural crest. *In* "The Skull" (J. Hanken and B. K. Hall, eds.), Vol. 1, pp. 77–111. University of Chicago Press, Chicago.

Lanske, B., Karaplis, A. C., Lee, K., Luz, A., Vortkamp, A., Pirro, A., Karperien, M., Defize, L. H. K., Ho, C., Mulligan, R. C., Abou-Samra, A. B., Jüppner, H., Segré, G. V., and Kronenberg, H. M. (1996). PTH/PTHrP receptor in early development and Indian hedgehog-regulated bone growth. *Science* **273**, 663–666.

Lasorella, A., Noseda, M., Beyna, M., and Iavarone, A. (2000). Id2 is a retinoblastoma protein target and mediates signalling by Myc oncoproteins. *Nature (London)* **407**, 592–598.

Le Douarin, N., and Kalcheim, C. (1999). "The Neural Crest." Cambridge University Press, Cambridge, UK and New York.

Lee, J., Platt, K. A., Censullo, P., and Ruiz i Altaba, A. (1997). Gli1 is a target of Sonic hedgehog that induces ventral neural tube development. *Development (Cambridge, UK)* **124**, 2537–2552.

Lee, J. J., Ekker, S. C., von Kessler, D. P., Porter, J. A., Sun, B. I., and Beachy, P. A. (1994). Autoproteolysis in hedgehog protein biogenesis. *Science* **266**, 1528–1537.

Lee, S. M., Danielian, P. S., Fritzsch, B., and McMahon, A. P. (1997). Evidence that FGF8 signalling from the midbrain-hindbrain junction regulates growth and polarity in the developing midbrain. *Development (Cambridge, UK)* **124**, 959–969.

Le Lievre, C. S. (1978). Participation of neural crest-derived cells in the genesis of the skull in birds. *J. Embryol. Exp. Morphol.* **47**, 17–37.

Lindsay, R. M., Barde, Y. A., Davies, A. M., and Rohrer, H. (1985). Differences and similarities in the neurotrophic growth factor requirements of sensory neurons derived from neural crest and neural placode. *J. Cell Sci., Suppl.* **3**, 115–129.

Leroy, P., Krust, A., Zelent, A., Mendelsohn, C., Garnier, J. M., Kastner, P., Dierich, A., and Chambon, P. (1991). Multiple isoforms of the mouse retinoic acid receptor alpha are generated by alternative splicing and differential induction by retinoic acid. *EMBO J.* **10**, 59–69.

Lesot, H., Peterková, R., Viriot, L., Vonesch, J. L., Turecková, J., Pe-

terka, M., and Ruch, J. V. (1998). Early stages of tooth morphogenesis in mouse analyzed by 3D reconstructions. *Eur. J. Oral Sci.* **106**(Suppl. 1), 64–70.

Lettice, L. A., Purdie, L. A., Carlson, G. J., Kilanowski, F., Dorin, J., and Hill, R. E. (1999). The mouse bagpipe gene controls development of axial skeleton, skull, and spleen. *Proc. Natl. Acad. Sci. U.S.A.* **96**, 9695–9700.

Levéen, P., Pekny, M., Gebre-Medhin, S., Swolin, B., Larsson, E., and Betsholtz, C. (1994). Mice deficient for PDGF B show renal, cardiovascular, and hematological abnormalities. *Genes Dev.* **8**, 1875–1887.

Levin, M., and Mercola, M. (1998). The compulsion of chirality: Toward an understanding of left-right asymmetry. *Genes Dev.* **12**, 763–769.

Lewandoski, M., Meyers, E. N., and Martin, G. R. (1997). Analysis of Fgf8 gene function in vertebrate development. *Cold Spring Harbor Symp. Quant. Biol.* **62**, 159–168.

Li, E., Sucov, H. M., Lee, K. F., Evans, R. M., and Jaenisch, R. (1993). Normal development and growth of mice carrying a targeted disruption of the alpha 1 retinoic acid receptor gene. *Proc. Natl. Acad. Sci. U.S.A.* **90**, 1590–1594.

Li, Y., and Olsen, B. R. (1997). Murine models of human genetic skeletal disorders. *Matrix Biol.* **16**, 49–52.

Li, Y., Lacerda, D. A., Warman, M. L., Beier, D. R., Yoshioka, H., Ninomiya, Y., Oxford, J. T., Morris, N. P., Andrikopoulos, K., Ramirez, F. *et al.* (1995). A fibrillar collagen gene, Col11a1, is essential for skeletal morphogenesis. *Cell (Cambridge, Mass.)* **80**, 423–430.

Liem, K. F., Jr., Tremml, G., Roelink, H., and Jessell, T. M. (1995). Dorsal differentiation of neural plate cells induced by BMP-mediated signals from epidermal ectoderm. *Cell (Cambridge, Mass.)* **82**, 969–979.

Lin, C. R., Kioussi, C., O'Connell, S., Briata, P., Szeto, D., Liu, F., Izpisúa-Belmonte, J. C., and Rosenfeld, M. G. (1999). Pitx2 regulates lung asymmetry, cardiac positioning and pituitary and tooth morphogenesis. *Nature (London)* **401**, 279–282.

Lin, D. M., and Ngai, J. (1999). Development of the vertebrate main olfactory system. *Curr. Opin. Neurobiol.* **9**, 74–78.

Liu, A., Losos, K., and Joyner, A. L. (1999). FGF8 can activate Gbx2 and transform regions of the rostral mouse brain into a hindbrain fate. *Development (Cambridge, UK)* **126**, 4827–4838.

Liu, J. K., Ghattas, I., Liu, S., Chen, S., and Rubenstein, J. L. (1997). Dlx genes encode DNA-binding proteins that are expressed in an overlapping and sequential pattern during basal ganglia differentiation. *Dev. Dyn.* **210**, 498–512.

Liu, Y. H., Tang, Z., Kundu, R. K., Wu, L., Luo, W., Zhu, D., Sangiorgi, F., Snead, M. L., and Maxson, R. E. (1999). Msx2 gene dosage influences the number of proliferative osteogenic cells in growth centers of the developing murine skull: A possible mechanism for MSX2-mediated craniosynostosis in humans. *Dev. Biol.* **205**, 260–274.

Lohnes, D., Kastner, P., Dierich, A., Mark, M., LeMeur, M., and Chambon, P. (1993). Function of retinoic acid receptor gamma in the mouse. *Cell (Cambridge, Mass.)* **73**, 643–658.

Lohnes, D., Mark, M., Mendelsohn, C., Dollé, P., Dierich, A., Gorry, P., Gansmuller, A., and Chambon, P. (1994). Function of the retinoic acid receptors (RARs) during development (I). Craniofacial and skeletal abnormalities in RAR double mutants. *Development (Cambridge, UK)* **120**, 2723–2748.

Lu, M. F., Cheng, H. T., Kern, M. J., Potter, S. S., Tran, B., Diekwisch, T. G., and Martin, J. F. (1999a). prx-1 functions cooperatively with another paired-related homeobox gene, prx-2, to maintain cell fates within the craniofacial mesenchyme. *Development (Cambridge, UK)* **126**, 495–504.

Lu, M. F., Pressman, C., Dyer, R., Johnson, R. L., and Martin, J. F. (1999b). Function of Rieger syndrome gene in left-right asymmetry and craniofacial development. *Nature (London)* **401**, 276–278.

Lu, Q., Knoepfler, P. S., Scheele, J., Wright, D. D., and Kamps, M. P. (1995). Both Pbx1 and E2A-Pbx1 bind the DNA motif ATCAATCAA cooperatively with the products of multiple murine Hox genes, some of which are themselves oncogenes. *Mol. Cell. Biol.* **15**, 3786–3795.

Lufkin, T., Dierich, A., LeMeur, M., Mark, M., and Chambon, P. (1991). Disruption of the Hox-1.6 homeobox gene results in defects in a region corresponding to its rostral domain of expression. *Cell (Cambridge, Mass.)* **66**, 1105–1119.

Lufkin, T., Mark, M., Hart, C. P., Dollé, P., LeMeur, M., and Chambon, P. (1992). Homeotic transformation of the occipital bones of the skull by ectopic expression of a homeobox gene. *Nature (London)* **359**, 835–841.

Lufkin, T., Lohnes, D., Mark, M., Dierich, A., Gorry, P., Gaub, M. P., LeMeur, M., and Chambon, P. (1993). High postnatal lethality and testis degeneration in retinoic acid receptor alpha mutant mice. *Proc. Natl. Acad. Sci. U.S.A.* **90**, 7225–7229.

Lumsden, A. (1984). Tooth morphogenesis: Contributions of the cranial neural crest in mammals. *In* "Tooth Morphogenesis and Differentiation" (A. Belcourt and J. V. Ruch, eds.) Colloq. INSERM 125. INSERM, Paris.

Lumsden, A., and Keynes, R. (1989). Segmental patterns of neuronal development in the chick hindbrain. *Nature (London)* **337**, 424–428.

Lumsden, A., and Krumlauf, R. (1996). Patterning the vertebrate neuraxis. *Science* **274**, 1109–1115.

Lumsden, A., Sprawson, N., and Graham, A. (1991). Segmental origin and migration of neural crest cells in the hindbrain region of the chick embryo. *Development (Cambridge, UK)* **113**, 1281–1291.

Lumsden, A. G. (1988). Spatial organization of the epithelium and the role of neural crest cells in the initiation of the mammalian tooth germ. *Development (Cambridge, UK)* **103**(Suppl.), 155–169.

Lumsden, A. G. and Buchanan, J. A. (1986). An experimental study of timing and topography of early tooth development in the mouse embryo with an analysis of the role of innervation. *Arch. Oral Biol.* **31**, 301–311.

Lumsden, A. G. S., (1982). The developing innervation of the lower jaw and its relation to the formation of tooth germs in mouse embryos. *In* "Teeth, Function and Evolution" (Kurten, ed.), pp. 33–43. Columbia University Press.

Luo, G., Hofmann, C., Bronckers, A. L., Sohocki, M., Bradley, A., and Karsenty, G. (1995). BMP-7 is an inducer of nephrogenesis, and is also required for eye development and skeletal patterning. *Genes Dev.* **9**, 2808–2820.

Lyons, K. M., Hogan, B. L., and Robertson, E. J. (1995). Colocalization of BMP 7 and BMP 2 RNAs suggests that these factors cooperatively mediate tissue interactions during murine development. *Mech. Dev.* **50**, 71–83.

Ma, Q., Chen, Z., del Barco Barrantes, I., de la Pompa, J. L. and Anderson, D. J. (1998). Neurogenin-1 is essential for the determination of neuronal precursors for proximal cranial sensory ganglia. *Neuron* **20**, 469–482.

Ma, Q., Fode, C., Guillemot, F., and Anderson, D. J. (1999). Neurogenin1 and neurogenin2 control two distinct waves of neurogenesis in developing dorsal root ganglia. *Genes Dev.* **13**, 1717–1728.

MacKenzie, A. L., Leeming, G. L., Jowett, A. K., Ferguson, M. W. J., and Sharpe, P. T. (1991). The homeobox gene Hox 7.1 has specific regional and temporal expression patterns during early murine craniofacial embryogenesis, especially tooth development in vivo and in vitro. *Development (Cambridge, UK)* **111**, 269–285.

MacKenzie, A. L., Ferguson, M. W. J., and Sharpe, P. T. (1992). Expression patterns of the homeobox gene Hox-8 in the mouse embryo suggest a role in specifying tooth initiation and shape. *Development (Cambridge, UK)* **115**, 403–420.

Mackenzie, S., Walsh, F. S., and Graham, A. (1998). Migration of hypoglossal myoblast precursors. *Dev. Dyn.* **213**, 349–358.

Mallo, M. (1997). Retinoic acid disturbs mouse middle ear development in a stage-dependent fashion. *Dev. Biol.* **184**, 175–186.

Mancilla, A., and Mayor, R. (1996). Neural crest formation in *Xenopus laevis:* Mechanisms of Xslug induction. *Dev. Biol.* **177**, 580–589.

Mangelsdorf, D. J., and Evans, R. M. (1995). The RXR heterodimers and orphan receptors. *Cell (Cambridge, Mass.)* **83**, 841–850.

Mann, R. S., and Chan, S. K. (1996). Extra specificity from extradenticle: The partnership between HOX and PBX/EXD homeodomain proteins. *Trends Genet.* **12**, 258–262.

Mansour, S. L., Goddard, J. M., and Capecchi, M. R. (1993). Mice homozygous for a targeted disruption of the proto-oncogene int-2 have developmental defects in the tail and inner ear. *Development (Cambridge, UK)* **117**, 13–28.

Mansouri, A., Stoykova, A., Torres, M., and Gruss, P. (1996). Dysgenesis of cephalic neural crest derivatives in Pax7−/− mutant mice. *Development (Cambridge, UK)* **122**, 831–838.

Mansouri, A., Chowdhury, K., and Gruss, P. (1998). Follicular cells of the thyroid gland require Pax8 gene function. *Nat. Genet.* **19**, 87–90.

Manzanares, M., Trainor, P., Ariza-McNaughton, L., Nonchev, S., and Krumlauf, R. (2000). Dorsal patterning defects in the hindbrain, roof plate and skeleton in the dreher (dr-j) mouse mutant. *Mech. Dev.* **94**, 147–156.

Marigo, V., Johnson, R. L., Vortkamp, A., and Tabin, C. J. (1996). Sonic hedgehog differentially regulates expression of GLI and GLI3 during limb development. *Dev. Biol.* **180**, 273–283.

Mark, M., Lufkin, T., Vonesch, J. L., Ruberte, E., Olivo, J. C., Dollé, P., Gorry, P., Lumsden, A., and Chambon, P. (1993). Two rhombomeres are altered in Hoxa-1 mutant mice. *Development (Cambridge, UK)* **119**, 319–338.

Mark, M., Lohnes, D., Mendelsohn, C., Dupé, V., Vonesch, J. L., Kastner, P., Rijli, F., Bloch-Zupan, A., and Chambon, P. (1995). Roles of retinoic acid receptors and of Hox genes in the patterning of the teeth and of the jaw skeleton. *Int. J. Dev. Biol.* **39**, 111–121.

Marks, S. C., Jr., and Lane, P. W. (1976). Osteopetrosis, a new recessive skeletal mutation on chromosome 12 of the mouse. *J. Here.* **67**, 11–18.

Marshall, H., Nonchev, S., Sham, M. H., Muchamore, I., Lumsden, A., and Krumlauf, R. (1992). Retinoic acid alters hindbrain Hox code and induces transformation of rhombomeres 2/3 into a 4/5 identity. *Nature (London)* **360**, 737–741.

Martí, E., Takada, R., Bumcrot, D. A., Sasaki, H., and McMahon, A. P. (1995). Distribution of Sonic hedgehog peptides in the developing chick and mouse embryo. *Development (Cambridge, UK)* **121**, 2537–2547.

Martin, J. F., Bradley, A., and Olson, E. N. (1995). The paired-like homeo box gene MHox is required for early events of skeletogenesis in multiple lineages. *Genes Dev.* **9**, 1237–1249.

Martinez, S., Marín, F., Nieto, M. A., and Puelles, L. (1995). Induction of ectopic engrailed expression and fate change in avian rhombomeres: Intersegmental boundaries as barriers. *Mech. Dev.* **51**, 289–303.

Martinez, S., Crossley, P. H., Cobos, I., Rubenstein, J. L., and Martin, G. R. (1999). FGF8 induces formation of an ectopic isthmic organizer and isthmocerebellar development via a repressive effect on Otx2 expression. *Development (Cambridge, UK)* **126**, 1189–2000.

Martinez-Barbera, J. P., Rodriguez, T. A., and Beddington, R. S. (2000). The homeobox gene Hesx1 is required in the anterior neural ectoderm for normal forebrain formation. *Dev. Biol.* **223**, 422–430.

Martinsen, B. J., and Bronner-Fraser, M. (1998). Neural crest specification regulated by the helix-loop-helix repressor Id2. *Science* **281**, 988–991.

Matsuo, I., Kuratani, S., Kimura, C., Takeda, N., and Aizawa, S. (1995). Mouse Otx2 functions in the formation and patterning of rostral head. *Genes Dev.* **9**, 2646–2658.

Matsuo, T., Osumi-Yamashita, N., Noji, S., Ohuchi, H., Koyama, E., Myokai, F., Matsuo, N., Taniguchi, S., Doi, H., Iseki, S. *et al.* (1993). A mutation in the Pax-6 gene in rat small eye is associated with impaired migration of midbrain crest cells. *Nat. Genet.* **3**, 299–304.

Matzuk, M. M., Kumar, T. R., Vassalli, A., Bickenbach, J. R., Roop, D. R., Jaenisch, R., and Bradley, A. (1995a). Functional analysis of activins during mammalian development. *Nature (London)* **374,** 354–356.

Matzuk, M. M., Kumar, T. R., and Bradley, A. (1995b). Different phenotypes for mice deficient in either activins or activin receptor type II. *Nature (London)* **374,** 356–360.

Matzuk, M. M., Lu, N., Vogel, H., Sellheyer, K., Roop, D. R., and Bradley, A. (1995c). Multiple defects and perinatal death in mice deficient in follistatin. *Nature (London)* **374,** 360–363.

Mayo, M. L., Bringas, P. Jr, Santos, V., Shum, L. and Slavkin, H. C. (1992). Desmin expression during early mouse tongue morphogenesis. *Int. J. Dev. Biol.* **36,** 255–263.

McClearn, D., and Noden, D. M. (1988). Ontogeny of architectural complexity in embryonic quail visceral arch muscles. *Am. J. Anat.* **183,** 277–293.

McKay, I. J., Lewis, J., and Lumsden, A. (1996). The role of FGF-3 in early inner ear development: An analysis in normal and kreisler mutant mice. *Dev. Biol.* **174,** 370–378.

McMahon, A. P., and Bradley, A. (1990). The Wnt-1 (int-1) proto-oncogene is required for development of a large region of the mouse brain. *Cell (Cambridge, Mass.)* **62,** 1073–1085.

McMahon, A. P., Joyner, A. L., Bradley, A., and McMahon, J. A. (1992). The midbrain-hindbrain phenotype of Wnt-1-/Wnt-1- mice results from stepwise deletion of engrailed-expressing cells by 9.5 days postcoitum. *Cell (Cambridge, Mass.)* **69,** 581–595.

Mehrara, B. J., Mackool, R. J., McCarthy, J. G., Gittes, G. K., and Longaker, M. T. (1998). Immunolocalization of basic fibroblast growth factor and fibroblast growth factor receptor-1 and receptor-2 in rat cranial sutures. *Plast. Reconstr. Surg.* **102,** 1805–1817; discussion: p. 1818.

Meier, S. (1979). Development of the chick embryo mesoblast. Formation of the embryonic axis and establishment of the metameric pattern. *Dev. Biol.* **73,** 24–45.

Meier, S. (1980). Development of the chick embryo mesoblast: Pronephros, lateral plate, and early vasculature. *J. Embryol. Exp. Morphol.* **55,** 291–306.

Meier, S. (1982). The distribution of cranial neural crest cells during ocular morphogenesis. *In* "Clinical, Structural and Biochemical Advances in Hereditary Eye Disorders" (D. L. Daentl, ed.), Liss, New York.

Meier, S., and Tam, P. P. (1982). Metameric pattern development in the embryonic axis of the mouse. I. Differentiation of the cranial segments. *Differentiation (Berlin)* **21,** 95–108.

Meijlink, F., Beverdam, A., Brouwer, A., Oosterveen, T. C., and Berge, D. T. (1999). Vertebrate aristaless-related genes. *Int. J. Dev. Biol.* **43,** 651–663.

Mendelsohn, C., Ruberte, E., LeMeur, M., Morriss-Kay, G., and Chambon, P. (1991). Developmental analysis of the retinoic acid-inducible RAR-beta 2 promoter in transgenic animals. *Development (Cambridge, UK)* **113,** 723–734.

Mendelsohn, C., Larkin, S., Mark, M., LeMeur, M., Clifford, J., Zelent, A., and Chambon, P. (1994a). RAR beta isoforms: Distinct transcriptional control by retinoic acid and specific spatial patterns of promoter activity during mouse embryonic development. *Mech. Dev.* **45,** 227–241.

Mendelsohn, C., Mark, M., Dollé, P., Dierich, A., Gaub, M. P., Krust, A., Lampron, C., and Chambon, P. (1994b). Retinoic acid receptor beta 2 (RAR beta 2) null mutant mice appear normal. *Dev. Biol.* **166,** 246–258.

Miettinen, P. J., Chin, J. R., Shum, L., Slavkin, H. C., Shuler, C. F., Derynck, R., and Werb, Z. (1999). Epidermal growth factor receptor function is necessary for normal craniofacial development and palate closure. *Nat. Genet.* **22,** 69–73.

Millan, F. A., Denhez, F., Kondaiah, P., and Akhurst, R. J. (1991). Embryonic gene expression patterns of TGF beta 1, beta 2 and beta 3 suggest different developmental functions in vivo. *Development (Cambridge, UK)* **111,** 131–143.

Millen, K. J., Wurst, W., Herrup, K., and Joyner, A. L. (1994). Abnormal embryonic cerebellar development and patterning of postnatal foliation in two mouse Engrailed-2 mutants. *Development (Cambridge, UK)* **120,** 695–706.

Miller, C. T., Schilling, T. F., Lee, K., Parker, J., and Kimmel, C. B. (2000). sucker encodes a zebrafish Endothelin-1 required for ventral pharyngeal arch development. *Development (Cambridge, UK)* **127,** 3815–3828.

Miller, W. A. (1969). Inductive changes in early tooth development. I. A study of mouse tooth development on the chick chorioallantois. *J. Dent. Res.* **48,** 719–725.

Millet, S., Campbell, K., Epstein, D. J., Losos, K., Harris, E., and Joyner, A. L. (1999). A role for Gbx2 in repression of Otx2 and positioning the mid/hindbrain organizer. *Nature (London)* **401,** 161–164.

Millonig, J. H., Millen, K. J., and Hatten, M. E. (2000). The mouse Dreher gene Lmx1a controls formation of the roof plate in the vertebrate CNS. *Nature (London)* **403,** 764–769.

Mina, M. and Kollar, E. J. (1987). The induction of odontogenesis in nondental mesenchyme combined with early murine mandibular arch epithelium. *Arch. Oral Biol.* **32,** 123–127.

Ming, J. E., Roessler, E., and Muenke, M. (1998). Human developmental disorders and the Sonic hedgehog pathway. *Mol. Med. Today* **4,** 343–349.

Minowada, G., Jarvis, L. A., Chi, C. L., Neubüser, A., Sun, X., Hacohen, N., Krasnow, M. A., and Martin, G. R. (1999). Vertebrate Sprouty genes are induced by FGF signaling and can cause chondrodysplasia when overexpressed. *Development (Cambridge, UK)* **126,** 4465–4475.

Miura, N., Wanaka, A., Tohyama, M., and Tanaka, K. (1993). MFH-1, a new member of the fork head domain family, is expressed in developing mesenchyme. *FEBS Lett.* **326,** 171–176.

Mo, R., Freer, A. M., Zinyk, D. L., Crackower, M. A., Michaud, J., Heng, H. H., Chik, K. W., Shi, X. M., Tsui, L. C., Cheng, S. H., Joyner, A. L., and Hui, C. (1997). Specific and redundant functions of Gli2 and Gli3 zinc finger genes in skeletal patterning and development. *Development (Cambridge, UK)* **124,** 113–123.

Mombaerts, P. (1999). Seven-transmembrane proteins as odorant and chemosensory receptors. *Science* **286,** 707–711.

Monreal, A. W., Ferguson, B. M., Headon, D. J., Street, S. L., Overbeek, P. A. and Zonana, J. (1999). Mutations in the human homolog of the mouse dl cause autosomal recessive and dominant hypohidrotic ectodermal dysplasia. *Nat. Gen.* **22,** 366–369.

Moody, S. A. and Heaton, M. B. (1983). Developmental relationships between trigeminal ganglia motorneurones in chick embryos. I. Ganglion development is necessary for motorneurone migration. *J. Comp. Neurol.* **213,** 327–343.

Moore, W. J. (1981). "The Mammalian Skull." Cambridge University Press, Cambridge, UK and New York.

Mori, K., Nagao, H., and Yoshihara, Y. (1999). The olfactory bulb: Coding and processing of odor molecule information. *Science* **286,** 711–715.

Morrison, A. D. (1998). 1 + 1 = r4 and much much more. *BioEssays* **20,** 794–797.

Morriss-Kay, G. (1993). Retinoic acid and craniofacial development: Molecules and morphogenesis. *BioEssays* **15,** 9–15.

Morriss-Kay, G. M., and Sokolova, N. (1996). Embryonic development and pattern formation. *FASEB J.* **10,** 961–968.

Morriss-Kay, G. M., and Ward, S. J. (1999). Retinoids and mammalian development. *Int. Rev. Cytol.* **188,** 73–131.

Morriss-Kay, G. M., Murphy, P., Hill, R. E., and Davidson, D. R. (1991). Effects of retinoic acid excess on expression of Hox-2.9 and Krox-20 and on morphological segmentation in the hindbrain of mouse embryos. *EMBO J.* **10,** 2985–2995.

Morsli, H., Tuorto, F., Choo, D., Postiglione, M. P., Simeone, A., and Wu, D. K. (1999). Otx1 and Otx2 activities are required for the normal development of the mouse inner ear. *Development (Cambridge, UK)* **126,** 2335–2343.

Moss, M. L., Meehan, M. A., and Salentijn, L. (1972). Transformative and translative growth processes in neurocranial development of the rat. *Acta Anat.* **81,** 161–182.

Moury, J. D., and Jacobson, A. G. (1989). Neural fold formation at newly created boundaries between neural plate and epidermis in the axolotl. *Dev. Biol.* **133,** 44–57.

Moury, J. D., and Jacobson, A. G. (1990). The origins of neural crest cells in the axolotl. *Dev. Biol.* **141,** 243–253.

Muenke, M., Schell, U., Hehr, A., Robin, N. H., Losken, H. W., Schinzel, A., Pulleyn, L. J., Rutland, P., Reardon, W., Malcolm, S. *et al.* (1994). A common mutation in the fibroblast growth factor receptor 1 gene in Pfeiffer syndrome. *Nat. Genet.* **8,** 269–274.

Mundlos, S., and Olsen, B. R. (1997a). Heritable diseases of the skeleton. Part I, Molecular insights into skeletal development-transcription factors and signaling pathways. *FASEB J.* **11,** 125–132.

Mundlos, S., and Olsen, B. R. (1997b). Heritable diseases of the skeleton. Part II, Molecular insights into skeletal development-matrix components and their homeostasis. *FASEB J.* **11,** 227–233.

Nagai, T., Aruga, J., Takada, S., Günther, T., Spörle, R., Schughart, K., and Mikoshiba, K. (1997). The expression of the mouse Zic1, Zic2, and Zic3 gene suggests an essential role for Zic genes in body pattern formation. *Dev. Biol.* **182,** 299–313.

Nakata, K., Ono, K., Miyazaki, J., Olsen, B. R., Muragaki, Y., Adachi, E., Yamamura, K., and Kimura, T. (1993). Osteoarthritis associated with mild chondrodysplasia in transgenic mice expressing alpha 1(IX) collagen chains with a central deletion. *Proc. Natl. Acad. Sci. U.S.A.* **90,** 2870–2874.

Naski, M. C., Wang, Q., Xu, J., and Ornitz, D. M. (1996). Graded activation of fibroblast growth factor receptor 3 by mutations causing achondroplasia and thanatophoric dysplasia. *Nat. Genet.* **13,** 233–237.

Neilson, K. M., and Friesel, R. E. (1995). Constitutive activation of fibroblast growth factor receptor-2 by a point mutation associated with Crouzon syndrome. *J. Biol. Chem.* **270,** 26037–26040.

Neubüser, A., Koseki, H., and Balling, R. (1995). Characterization and developmental expression of Pax9, a paired-box-containing gene related to Pax1. *Dev. Biol.* **170,** 701–716.

Neubüser, A., Peters, H., Balling, R., and Martin, G. R. (1997). Antagonistic interactions between FGF and BMP signaling pathways: A mechanism for positioning the sites of tooth formation. *Cell (Cambridge, Mass.)* **90,** 247–255.

Nichols, D. H. (1981). Neural crest formation in the head of the mouse embryo as observed using a new histological technique. *J. Embryol. Exp. Morphol.* **64,** 105–120.

Nichols, D. H. (1986). Mesenchyme formation from the trigeminal placodes of the mouse embryo. *Am. J. Anat.* **176,** 19–31.

Niederreither, K., McCaffery, P., Dräger, U. C., Chambon, P., and Dollé, P. (1997). Restricted expression and retinoic acid-induced downregulation of the retinaldehyde dehydrogenase type 2 (RALDH-2) gene during mouse development. *Mech. Dev.* **62,** 67–78.

Niederreither, K., Subbarayan, V., Dollé, P., and Chambon, P. (1999). Embryonic retinoic acid synthesis is essential for early mouse postimplantation development. *Nat. Genet.* **21,** 444–448.

Niederreither, K., Vermot, J., Schuhbaur, B., Chambon, P., and Dollé, P. (2000). Retinoic acid synthesis and hindbrain patterning in the mouse embryo. *Development (Cambridge, UK)* **127,** 75–85.

Nieto, M. A., Sargent, M. G., Wilkinson, D. G., and Cooke, J. (1994). Control of cell behavior during vertebrate development by Slug, a zinc finger gene. *Science* **264,** 835–839.

Noden, D. M. (1978). The control of avian cephalic neural crest cytodifferentiation. I. Skeletal and connective tissues. *Dev. Biol.* **67,** 296–312.

Noden, D. M. (1983a). The embryonic origins of avian cephalic and cervi-cal muscles and associated connective tissues. *Am. J. Anat.* **168,** 257–276.

Noden, D. M. (1983b). The role of the neural crest in patterning of avian cranial skeletal, connective, and muscle tissues. *Dev. Biol.* **96,** 144–165.

Noden, D. M. (1986). Patterning of avian craniofacial muscles. *Dev. Biol.* **116,** 347–356.

Noden, D. M. (1991). Vertebrate craniofacial development: The relation between ontogenetic process and morphological outcome. *Brain, Behav. Evol.* **38,** 190–225.

Nomura, M., and Li, E. (1998). Smad2 role in mesoderm formation, left-right patterning and craniofacial development. *Nature (London)* **393,** 786–790.

Nonaka, K., Shum, L., Takahashi, I., Takahashi, K., Ikura, T., Dashner, R., Nuckolls, G. H., and Slavkin, H. C. (1999). Convergence of the BMP and EGF signaling pathways on Smad1 in the regulation of chondrogenesis. *Int. J. Dev. Biol.* **43,** 795–807.

Nonchev, S., Vesque, C., Maconochie, M., Seitanidou, T., Ariza-McNaughton, L., Frain, M., Marshall, H., Sham, M. H., Krumlauf, R., and Charnay, P. (1996). Segmental expression of Hoxa-2 in the hindbrain is directly regulated by Krox-20. *Development (Cambridge, UK)* **122,** 543–554.

Northcutt, R. G. (1990). Ontogeny and phylogeny: A re-evaluation of conceptual relationships and some applications. *Brain, Behav. Evol.* **36,** 116–140.

Northcutt, R. G., and Barlow, L. A. (1998). Amphibians provide new insights into taste-bud development. *Trends Neurosci.* **21,** 38–43.

Novacek, M. J. (1993). Patterns of diversity in the mammalian skull. *In* "The Skull" (J. Hanken and B. K. Hall, eds.), Vol. 2, pp. 438–545. University of Chicago Press, Chicago.

Ogasawara, M., Shigetani, Y., Hirano, S., Satoh, N., and Kuratani, S. (2000). Pax1/Pax9-Related genes in an agnathan vertebrate, *Lampetra japonica:* Expression pattern of LjPax9 implies sequential evolutionary events toward the gnathostome body plan. *Dev. Biol.* **223,** 399–410.

Opperman, L. A., Sweeney, T. M., Redmon, J., Persing, J. A., and Ogle, R. C. (1993). Tissue interactions with underlying dura mater inhibit osseous obliteration of developing cranial sutures. *Dev. Dyn.* **198,** 312–322.

Opperman, L. A., Passarelli, R. W., Morgan, E. P., Reintjes, M., and Ogle, R. C. (1995). Cranial sutures require tissue interactions with dura mater to resist osseous obliteration in vitro. *J. Bone and Min. Res.* **10,** 1978–1987.

Opperman, L. A., Nolen, A. A., and Ogle, R. C. (1997). TGF-beta 1, TGF-beta 2, and TGF-beta 3 exhibit distinct patterns of expression during cranial suture formation and obliteration in vivo and in vitro. *J. Bone and Mineral Res.* **12,** 301–310.

Opperman, L. A., Chhabra, A., Nolen, A. A., Bao, Y., and Ogle, R. C. (1998). Dura mater maintains rat cranial sutures in vitro by regulating suture cell proliferation and collagen production. *J. Craniofacial Genet. Dev. Biol.* **18,** 150–158.

Orr-Urtreger, A., Bedford, M. T., Do, M. S., Eisenbach, L., and Lonai, P. (1992). Developmental expression of the alpha receptor for platelet-derived growth factor, which is deleted in the embryonic lethal Patch mutation. *Development (Cambridge, UK)* **115,** 289–303.

Osborn, J. W. (1978). Morphogenetic gradients: Fields versus clones. *In* "Development, Function and Evolution of Teeth" (P. M. Butler and K. A. Joysey, eds.), pp. 171–201. Academic Press, New York.

Osumi, N., Hirota, A., Ohuchi, H., Nakafuku, M., Iimura, T., Kuratani, S., Fujiwara, M., Noji, S., and Eto, K. (1997). Pax-6 is involved in the specification of hindbrain motor neuron subtype. *Development (Cambridge, UK)* **124,** 2961–2972.

Osumi-Yamashita, N., Kuratani, S., Ninomiya, Y., Aoki, K., Iseki, S., Chareonvit, S., Doi, H., Fujiwara, M., Watanabe, T., and Eto, K. (1997a).

Cranial anomaly of homozygous rSey rat is associated with a defect in the migration pathway of midbrain crest cells. *Dev. Growth Differ.* **39**, 53–67.

Osumi-Yamashita, N., Ninomiya, Y., Doi, H., and Eto, K. (1994). The contribution of both forebrain and midbrain crest cells to the mesenchyme in the frontonasal mass of mouse embryos. *Dev. Biol.* **164**, 409–419.

Osumi-Yamashita, N., Ninomiya, Y., and Eto, K. (1997b). Mammalian craniofacial embryology in vitro. *Int. J. Dev. Biol.* **41**, 187–194.

Otto, F., Thornell, A. P., Crompton, T., Denzel, A., Gilmour, K. C., Rosewell, I. R., Stamp, G. W., Beddington, R. S., Mundlos, S., Olsen, B. R., Selby, P. B., and Owen, M. J. (1997). Cbfa1, a candidate gene for cleidocranial dysplasia syndrome, is essential for osteoblast differentiation and bone. *Cell (Cambridge, Mass.)* **89**, 765–771.

Oudhof, H. A. (1982). Sutural growth. *Acta Anat.* **112**, 58–68.

Pace, J. M., Li, Y., Seegmiller, R. E., Teuscher, C., Taylor, B. A., and Olsen, B. R. (1997). Disproportionate micromelia (Dmm) in mice caused by a mutation in the C-propeptide coding region of Col2a1. *Dev. Dyn.* **208**, 25–33.

Panganiban, G., Irvine, S. M., Lowe, C., Roehl, H., Corley, L. S., Sherbon, B., Grenier, J. K., Fallon, J. F., Kimble, J., Walker, M., Wray, G. A., Swalla, B. J., Martindale, M. Q., and Carroll, S. B. (1997). The origin and evolution of animal appendages. *Proc. Natl. Acad. Sci. U.S.A.* **94**, 5162–5166.

Parrington, F. R., and Westoll, T. S. (1940). On the evolution of the mammalian palate. *Philos. Trans. R. Soc. London, Ser B* **230**, 305–355.

Pászty, C., Mohandas, N., Stevens, M. E., Loring, J. F., Liebhaber, S. A., Brion, C. M., and Rubin, E. M. (1995). Lethal alpha-thalassaemia created by gene targeting in mice and its genetic rescue. *Nat. Genet.* **11**, 33–39.

Patterson, C. (1977). Cartilage bones, dermal bones, and membrane bones, or the exoskeleton versus the endoskeleton. *In* "Problems in Vertebrate Evolution" (S. M. Andrews, R. S. Miles, and A. D. Walker, eds.), pp. 77–121. Academic Press, New York.

Peifer, M., and Wieschaus, E. (1990). Mutations in the Drosophila gene extradenticle affect the way specific homeo domain proteins regulate segmental identity. *Genes Dev.* **4**, 1209–1223.

Pelton, R. W., Johnson, M. D., Perkett, E. A., Gold, L. I., and Moses, H. L. (1991). Expression of transforming growth factor -beta 1, -beta 2, and -beta 3 mRNA and protein in the murine lung. *Am. J. Respir. Cell Mol. Biol.* **5**, 522–530.

Pennypacker, J. P., Kimata, K., and Brown, K. S. (1981). Brachymorphic mice (bm/bm): A generalized biochemical defect expressed primarily in cartilage. *Dev. Biol.* **81**, 280–287.

Perris, R. (1997). The extracellular matrix in neural crest-cell migration. *Trends Neurosci.* **20**, 23–31.

Peterkova, R., Peterka, M., Vonesch, J. L. and Ruch, J. V. (1993). Multiple developmental origin of the upper incisor in mouse: Histological and computer assisted 3-D-reconstruction studies. *Int. J. Dev. Biol.* **37**, 581–588.

Peterkova, R., Peterka, M., Vonesch, J. L. and Ruch, J. V. (1995). Contribution of 3-D computer-assisted reconstructions to the study of the initial steps of mouse odontogenesis. *Int. J. Dev. Biol.* **39**, 239–247.

Peterkova, R., Lesot, H., Vonesch, J. L., Peterka, M. and Ruch, J. V. (1996). Mouse molar morphogenesis revisited by three dimensional reconstruction. I. Analysis of initial stages of the first upper molar development revealed two transient buds. *Int. J. Dev. Biol.* **40**, 1009–1016.

Peters, H., Neubüser, A., Kratochwil, K., and Balling, R. (1998). Pax9-deficient mice lack pharyngeal pouch derivatives and teeth and exhibit craniofacial and limb abnormalities. *Genes Dev.* **12**, 2735–2747.

Peters, H., Wilm, B., Sakai, N., Imai, K., Maas, R., and Balling, R. (1999). Pax1 and Pax9 synergistically regulate vertebral column development. *Development (Cambridge, UK)* **126**, 5399–5408.

Philips, R. J. S. (1960). *Mouse News Lett.* **23**, 29.

Pispa, J., Jung, H.-S., Jernvall, J., Kettunen, P., Mustonen, T., Tabata, M. J., Kere, J., and Thesleff, I. (1999). Cusp patterning defects in Tabby mouse teeth and its partial rescue by FGF. *Dev. Biol.* **216**, 521–534.

Platt, J. B. (1893). Ectodermal origin of the cartilages of the head. *Anat. Anz.* **8**, 506–509.

Platt, K. A., Michaud, J., and Joyner, A. L. (1997). Expression of the mouse Gli and Ptc genes is adjacent to embryonic sources of hedgehog signals suggesting a conservation of pathways between flies and mice. *Mech. Dev.* **62**, 121–135.

Porter, J. A., Young, K. E., and Beachy, P. A. (1996). Cholesterol modification of hedgehog signaling proteins in animal development. *Science* **274**, 255–259.

Presley, R. (1993). Preconception of adult structural pattern in the analysis of the developing skull. *In* "The Skull" (J. Hanken and B. K. Hall, eds.), Vol. 1, pp. 347–377. University of Chicago Press, Chicago.

Presley, R., and Steel, F. L. D., (1976). On the homology of the alisphenoid. *J. Anat.* **121**, 441–459.

Presley, R., and Steel, F. L. D., (1978). The pterygoid and ectopterygoid in mammals. *Anat. Embryol.* **154**, 95–110.

Prince, V., and Lumsden, A. (1994). Hoxa-2 expression in normal and transposed rhombomeres: Independent regulation in the neural tube and neural crest. *Development (Cambridge, UK)* **120**, 911–923.

Proetzel, G., Pawlowski, S. A., Wiles, M. V., Yin, M., Boivin, G. P., Howles, P. N., Ding, J., Ferguson, M. W., and Doetschman, T. (1995). Transforming growth factor-beta 3 is required for secondary palate fusion. *Nat. Genet.* **11**, 409–414.

Püschel, A. W., Westerfield, M., and Dressler, G. R. (1992). Comparative analysis of Pax-2 protein distributions during neurulation in mice and zebrafish. *Mech. Dev.* **38**, 197–208.

Qiu, M., Bulfone, A., Martinez, S., Meneses, J. J., Shimamura, K., Pedersen, R. A., and Rubenstein, J. L. (1995). Null mutation of Dlx-2 results in abnormal morphogenesis of proximal first and second branchial arch derivatives and abnormal differentiation in the forebrain. *Genes Dev.* **9**, 2523–2538.

Qiu, M., Bulfone, A., Ghattas, I., Meneses, J. J., Christensen, L., Sharpe, P. T., Presley, R., Pedersen, R. A., and Rubenstein, J. L. (1997). Role of the Dlx homeobox genes in proximodistal patterning of the branchial arches: Mutations of Dlx-1, Dlx-2, and Dlx-1 and -2 alter morphogenesis of proximal skeletal and soft tissue structures derived from the first and second arches. *Dev. Biol.* **185**, 165–184.

Qu, S., Niswender, K. D., Ji, Q., van der Meer, R., Keeney, D., Magnuson, M. A., and Wisdom, R. (1997). Polydactyly and ectopic ZPA formation in Alx-4 mutant mice. *Development (Cambridge, UK)* **124**, 3999–4008.

Qu, S., Tucker, S. C., Ehrlich, J. S., Levorse, J. M., Flaherty, L. A., Wisdom, R., and Vogt, T. F. (1998). Mutations in mouse Aristaless-like4 cause Strong's luxoid polydactyly. *Development (Cambridge, UK)* **125**, 2711–2721.

Qu, S., Tucker, S. C., Zhao, Q., deCrombrugghe, B., and Wisdom, R. (1999). Physical and genetic interactions between Alx4 and Cart1. *Development (Cambridge, UK)* **126**, 359–369.

Quinn, J. C., West, J. D., and Kaufman, M. H. (1997). Genetic background effects on dental and other craniofacial abnormalities in homozygous small eye (Pax6Sey/Pax6Sey) mice. *Anat. Embryol.* **196**, 311–321.

Ramsdell, A. F., and Yost, H. J. (1998). Molecular mechanisms of vertebrate left-right development. *Trends Genet.* **14**, 459–465.

Rauskolb, C., and Wieschaus, E. (1994). Coordinate regulation of downstream genes by extradenticle and the homeotic selector proteins. *EMBO J.* **13**, 3561–3569.

Raven, C. P., and Kloos, J. (1945). Induction by medial and lateral pieces of the archenteron roof with special reference to the determination of the neural crest. *Acta Neerl. Morphol.* **5**, 348–362.

Raven, H. C., and Gregory, W. K. (1933). The spermaceti organ and nasal passages of the sperm whale (*Physeter catodon*) and other odontocetes. *Am. Mus. Novit.* **677,** 1–18.

Reardon, W., Winter, R. M., Rutland, P., Pulleyn, L. J., Jones, B. M., and Malcolm, S. (1994). Mutations in the fibroblast growth factor receptor 2 gene cause Crouzon syndrome. *Nat. Genet.* **8,** 98–103.

Represa, J., León, Y., Miner, C., and Giraldez, F. (1991). The int-2 proto-oncogene is responsible for induction of the inner ear. *Nature (London)* **353,** 561–563.

Rhinn, M., Dierich, A., Shawlot, W., Behringer, R. R., Le Meur, M., and Ang, S. L. (1998). Sequential roles for Otx2 in visceral endoderm and neuroectoderm for forebrain and midbrain induction and specification. *Development (Cambridge, UK)* **125,** 845–856.

Rhinn, M., Dierich, A., Le Meur, M., and Ang, S. (1999). Cell autonomous and noncell autonomous functions of Otx2 in patterning the rostral brain. *Development (Cambridge, UK)* **126,** 4295–4304.

Rice, D. P., Aberg, T., Chan, Y., Tang, Z., Kettunen, P. J., Pakarinen, L., Maxson, R. E., and Thesleff, I. (2000). Integration of FGF and TWIST in calvarial bone and suture development. *Development (Cambridge, UK)* **127,** 1845–1855.

Richtsmeier, J. T. (1992). Evolutionary biology and craniofacial development and dysmorphology. *In* "Complex Craniofacial Problems: A Guide to Analysis and Treatment" (C. R. Dufresne, B. S. Carson, and S. J. Zinreich, eds.), pp. 13–26. Churchill-Livingstone, New York.

Riddle, R. D., Johnson, R. L., Laufer, E., and Tabin, C. (1993). Sonic hedgehog mediates the polarizing activity of the ZPA. *Cell (Cambridge, Mass.)* **75,** 1401–1416.

Rijli, F. M., Mark, M., Lakkaraju, S., Dierich, A., Dollé, P., and Chambon, P. (1993). A homeotic transformation is generated in the rostral branchial region of the head by disruption of Hoxa-2, which acts as a selector gene. *Cell (Cambridge, Mass.)* **75,** 1333–1349.

Rijli, F. M., Gavalas, A., and Chambon, P. (1998). Segmentation and specification in the branchial region of the head: The role of the Hox selector genes. *Int. J. Dev. Biol.* **42,** 393–401.

Rintala, M., Metsäranta, M., Säämänen, A. M., Vuorio, E., and Rönning, O. (1997). Abnormal craniofacial growth and early mandibular osteoarthritis in mice harbouring a mutant type II collagen transgene. *J. Anat.* **190,** 201–208.

Rittenhouse, E., Dunn, L. C., Cookingham, J., Calo, C., Spiegelman, M., Dooher, G. B., and Bennett, D. (1978). Cartilage matrix deficiency (cmd): A new autosomal recessive lethal mutation in the mouse. *J. Embryol. Exp. Morphol.* **43,** 71–84.

Rivera-Pérez, J. A., Mallo, M., Gendron-Maguire, M., Gridley, T., and Behringer, R. R. (1995). Goosecoid is not an essential component of the mouse gastrula organizer but is required for craniofacial and rib development. *Development (Cambridge, UK)* **121,** 3005–3012.

Rivera-Pérez, J. A., Wakamiya, M., and Behringer, R. R. (1999). Goosecoid acts cell autonomously in mesenchyme-derived tissues during craniofacial development. *Development (Cambridge, UK)* **126,** 3811–3821.

Roberts, A. B., Anzano, M. A., Lamb, L. C., Smith, J. M., and Sporn, M. B. (1981). New class of transforming growth factors potentiated by epidermal growth factor: Isolation from nonneoplastic tissues. *Proc. Natl. Acad. Sci. U.S.A.* **78,** 5339–5343.

Roberts, R. C. (1967). Small-eyes, a new dominant mutant in the mouse. *Genet. Res.* **9,** 121–122.

Roberts, V. J., and Barth, S. L. (1994). Expression of messenger ribonucleic acids encoding the inhibin/activin system during mid- and late-gestation rat embryogenesis. *Endocrinology (Baltimore)* **134,** 914–923.

Robinson, G. W., and Mahon, K. A. (1994). Differential and overlapping expression domains of Dlx-2 and Dlx-3 suggest distinct roles for Distalless homeobox genes in craniofacial development. *Mech. Dev.* **48,** 199–215.

Robinson, V., Smith, A., Flenniken, A. M., and Wilkinson, D. G. (1997). Roles of Eph receptors and ephrins in neural crest pathfinding. *Cell Tissue Res.* **290,** 265–274.

Rollhäuser-ter Horst, J. (1979). Artificial neural crest formation in amphibia. *Anat. Embryol.* **157,** 113–120.

Romer, A. S. (1977). The vertebrate as dual animal: Somatic and visceral. *Evol. Biol.* **6,** 121–156.

Rousseau, F., Bonaventure, J., Legeai-Mallet, L., Pelet, A., Rozet, J. M., Maroteaux, P., Le Merrer, M., and Munnich, A. (1994). Mutations in the gene encoding fibroblast growth factor receptor-3 in achondroplasia. *Nature (London)* **371,** 252–254.

Rubenstein, J. L., Martinez, S., Shimamura, K., and Puelles, L. (1994). The embryonic vertebrate forebrain: The prosomeric model. *Science* **266,** 578–580.

Rubenstein, J. L., Shimamura, K., Martinez, S., and Puelles, L. (1998). Regionalization of the prosencephalic neural plate. *Ann. Rev. Neurosci.* **21,** 445–477.

Ruberte, E., Dollé, P., Krust, A., Zelent, A., Morriss-Kay, G., and Chambon, P. (1990). Specific spatial and temporal distribution of retinoic acid receptor gamma transcripts during mouse embryogenesis. *Development (Cambridge, UK)* **108,** 213–222.

Ruberte, E., Dollé, P., Chambon, P., and Morriss-Kay, G. (1991). Retinoic acid receptors and cellular retinoid binding proteins. II. Their differential pattern of transcription during early morphogenesis in mouse embryos. *Development (Cambridge, UK)* **111,** 45–60.

Ruffins, S., and Bronner-Fraser, M. (2000). A critical period for conversion of ectodermal cells to a neural crest fate. *Dev. Biol.* **218,** 13–20.

Rugh, R. (1968). "The Mouse: Its Reproduction and Development." Burgess, Minneapolis, MN.

Rutland, P., Pulleyn, L. J., Reardon, W., Baraitser, M., Hayward, R., Jones, B., Malcolm, S., Winter, R. M., Oldridge, M., Slaney, S. F. *et al.* (1995). Identical mutations in the FGFR2 gene cause both Pfeiffer and Crouzon syndrome phenotypes. *Nat. Genet.* **9,** 173–176.

Sadaghion, B. and Thiébaud, C. H. (1987). Neural crest development in the *Xenopus laevis* embryo, studied by interspecific transplantation and scanning electron microscopy. *Dev. Biol.* **124,** 91–110.

Saldivar, J. R., Krull, C. E., Krumlauf, R., Ariza-McNaughton, L., and Bronner-Fraser, M. (1996). Rhombomere of origin determines autonomous versus environmentally regulated expression of Hoxa-3 in the avian embryo. *Development (Cambridge, UK)* **122,** 895–904.

Saldivar, J. R., Sechrist, J. W., Krull, C. E., Ruffins, S., and Bronner-Fraser, M. (1997). Dorsal hindbrain ablation results in rerouting of neural crest migration and changes in gene expression, but normal hyoid development. *Development (Cambridge, UK)* **124,** 2729–2739.

Sampath, T. K., and Reddi, A. H. (1981). Dissociative extraction and reconstitution of extracellular matrix components involved in local bone differentiation. *Proc. Natl. Acad. Sci. U.S.A.* **78,** 7599–7603.

Sander, M., Neubüser, A., Kalamaras, J., Ee, H. C., Martin, G. R., and German, M. S. (1997). Genetic analysis reveals that PAX6 is required for normal transcription of pancreatic hormone genes and islet development. *Genes Dev.* **11,** 1662–1673.

Sander, M., Paydar, S., Ericson, J., Briscoe, J., Berber, E., German, M., Jessell, T. M., and Rubenstein, J. L. (2000). Ventral neural patterning by Nkx homeobox genes: Nkx6.1 controls somatic motor neuron and ventral interneuron fates. *Genes Dev.* **14,** 2134–2139.

Sanford, L. P., Ormsby, I., Gittenberger-de Groot, A. C., Sariola, H., Friedman, R., Boivin, G. P., Cardell, E. L., and Doetschman, T. (1997). TGFbeta2 knockout mice have multiple developmental defects that are nonoverlapping with other TGFbeta knockout phenotypes. *Development (Cambridge, UK)* **124,** 2659–2670.

Sarkar, L., and Sharpe, P. T. (1999). Expression of Wnt signalling pathway genes during tooth development. *Mech. Dev.* **85,** 197–200.

Sarkar, L., Cobourne, M., Naylor, S., Smalley, M., Dale, T., and Sharpe, P. T. (2000). Conservation of wingless/hedgehog interactions in ectodermal boundary formation during mammalian tooth development. *Proc. Natl. Acad Sci. U.S.A.* **97,** 4520–4524.

Sasaki, H., and Hogan, B. L. (1993). Differential expression of multiple fork head related genes during gastrulation and axial pattern formation in the mouse embryo. *Development (Cambridge, UK)* **118**, 47–59.

Satokata, I., and Maas, R. (1994). Msx1 deficient mice exhibit cleft palate and abnormalities of craniofacial and tooth development. *Nat. Genet.* **6**, 348–356.

Satokata, I., Ma, L., Ohshima, H., Bei, M., Woo, I., Nishizawa, K., Maeda, T., Takano, Y., Uchiyama, M., Heaney, S., Peters, H., Tang, Z., Maxson, R., and Maas, R. (2000). Msx2 deficiency in mice causes pleiotropic defects in bone growth and ectodermal organ formation. *Nat. Genet.* **24**, 391–395.

Schier, A. F., and Shen, M. M. (2000). Nodal signalling in vertebrate development. *Nature (London)* **403**, 385–389.

Schilling, T. F. (1997). Genetic analysis of craniofacial development in the vertebrate embryo. *BioEssays* **19**, 459–468.

Schowing, J. (1968a). Influence inductrice de l'encéphale embryonnaire sur le développement du crane chez le poulet. I. Influence de l'excision des terrotoires nerveux antérieurs sur le développement cranïen. *J. Embryol. Exp. Morphol.* **19**, 9–22.

Schowing, J. (1968b). Mise en évidence du role inducteur de l'encéphale dans l'ostéogénesè du crane embryonnaire de poulet. *J. Embryol. Exp. Morphol.* **19**, 83–94.

Schwanzel-Fukuda, M., and Pfaff, D. W. (1989). Origin of luteinizing hormone-releasing hormone neurons. *Nature (London)* **338**, 161–164.

Schwind, J. L. (1928). The development of the hypophysis cerebri of the albino rat. *Am. J. Anat.* **41**, 295–315.

Sechrist, J., Serbedzija, G. N., Scherson, T., Fraser, S. E., and Bronner-Fraser, M. (1993). Segmental migration of the hindbrain neural crest does not arise from its segmental generation. *Development (Cambridge, UK)* **118**, 691–703.

Seegmiller, R. E., and Fraser, F. C. (1977). Mandibular growth retardation as a cause of cleft palate in mice homozygous for the chondrodysplasia gene. *J. Embryol. Exp. Morphol.* **38**, 227–238.

Sefton, M., Sánchez, S., and Nieto, M. A. (1998). Conserved and divergent roles for members of the Snail family of transcription factors in the chick and mouse embryo. *Development (Cambridge, UK)* **125**, 3111–3121.

Seifert, R. A., Hart, C. E., Phillips, P. E., Forstrom, J. W., Ross, R., Murray, M. J., and Bowen-Pope, D. F. (1989). Two different subunits associate to create isoform-specific platelet-derived growth factor receptors. *J. Biol. Chem.* **264**, 8771–8778.

Selleck, M. A., and Bronner-Fraser, M. (1995). Origins of the avian neural crest: The role of neural plate-epidermal interactions. *Development (Cambridge, UK)* **121**, 525–538.

Selleck, M. A., Scherson, T. Y., and Bronner-Fraser, M. (1993). Origins of neural crest cell diversity. *Dev. Biol.* **159**, 1–11.

Selleri, L., Depew, M. J., Jacobs, Y., Chandra, S., Rubenstein, J. L. R., O'Gorman, S., and Cleary, M. (2001). Requirement for Pbx1 in patterning of the skeleton and programming of chondrocyte maturation. Submitted for publication.

Semina, E. V., Reiter, R., Leysens, N. J., Alward, W. L., Small, K. W., Datson, N. A., Siegel-Bartelt, J., Bierke-Nelson, D., Bitoun, P., Zabel, B. U., Carey, J. C., and Murray, J. C. (1996). Cloning and characterization of a novel bicoid-related homeobox transcription factor gene, RIEG, involved in Rieger syndrome. *Nat. Genet.* **14**, 392–399.

Serbedzija, G. N., Bronner-Fraser, M., and Fraser, S. E. (1992). Vital dye analysis of cranial neural crest cell migration in the mouse embryo. *Development (Cambridge, UK)* **116**, 297–307.

Sharpe, P. T. (1995). Homeobox genes and orofacial development. *Connect. Tissue Res.* **32**, 17–25.

Sher, A. E. (1971). The embryonic and postnatal development of the inner ear of the mouse. *Acta Oto-Laryngol., Suppl.* **285**, 1–77.

Shiang, R., Thompson, L. M., Zhu, Y. Z., Church, D. M., Fielder, T. J., Bocian, M., Winokur, S. T., and Wasmuth, J. J. (1994). Mutations in the transmembrane domain of FGFR3 cause the most common genetic form of dwarfism, achondroplasia. *Cell (Cambridge, Mass.)* **78**, 335–342.

Shimamura, K., and Rubenstein, J. L. (1997). Inductive interactions direct early regionalization of the mouse forebrain. *Development (Cambridge, UK)* **124**, 2709–2718.

Sillence, D. O., Ritchie, H. E., and Selby, P. B. (1987). Animal model: Skeletal anomalies in mice with cleidocranial dysplasia. *Am. J. Med. Genet.* **27**, 75–85.

Simeone, A., Acampora, D., Mallamaci, A., Stornaiuolo, A., D'Apice, M. R., Nigro, V., and Boncinelli, E. (1993). A vertebrate gene related to orthodenticle contains a homeodomain of the bicoid class and demarcates anterior neuroectoderm in the gastrulating mouse embryo. *EMBO J.* **12**, 2735–2747.

Simeone, A., Acampora, D., Pannese, M., D'Esposito, M., Stornaiuolo, A., Gulisano, M., Mallamaci, A., Kastury, K., Druck, T., Huebner, K. *et al.* (1994). Cloning and characterization of two members of the vertebrate Dlx gene family. *Proc. Natl. Acad. Sci. U.S.A.* **91**, 2250–2254.

Sirard, C., de la Pompa, J. L., Elia, A., Itie, A., Mirtsos, C., Cheung, A., Hahn, S., Wakeham, A., Schwartz, L., Kern, S. E., Rossant, J., and Mak, T. W. (1998). The tumor suppressor gene Smad4/Dpc4 is required for gastrulation and later for anterior development of the mouse embryo. *Genes Dev.* **12**, 107–119.

Smith, M. M. and Coates, M. I. (1998). Evolutionary origins of the vertebrate dentition: Phylogenetic patterns and developmental evolution. *Eur. J. Oral Sci.* **106**,(Suppl. 1), 482–500.

Smith, M. M., and Coates, M. I. (2000). Evolutionary origins of teeth and jaws: Developmental models and phylogenetic patterns. *In* "Development, Function and Evolution of Teeth" (M. F. Teaford, M. M. Smith, and M. W. J. Ferguson, eds.), pp. 133–151. Cambridge University Press, Cambridge, UK.

Sohal, G. S., Ali, M. M., Ali, A. A., and Dai, D. (1999). Ventrally emigrating neural tube cells contribute to the formation of Meckel's and quadrate cartilage. *Dev. Dyn.* **216**, 37–44.

Solloway, M. J., and Robertson, E. J. (1999). Early embryonic lethality in Bmp5;Bmp7 double mutant mice suggests functional redundancy within the 60A subgroup. *Development (Cambridge, UK)* **126**, 1753–1768.

Solloway, M. J., Dudley, A. T., Bikoff, E. K., Lyons, K. M., Hogan, B. L., and Robertson, E. J. (1998). Mice lacking Bmp6 function. *Dev. Genet.* **22**, 321–339.

Solursh, M., Fisher, M., and Singley, C. T. (1979). The synthesis of hyaluronic acid by ectoderm during early organogenesis in the chick embryo. *Differentiation (Berlin)* **14**, 77–85.

Sood, S., Eldadah, Z. A., Krause, W. L., McIntosh, I., and Dietz, H. C. (1996). Mutation in fibrillin-1 and the Marfanoid-craniosynostosis (Shprintzen-Goldberg) syndrome. *Nat. Genet.* **12**, 209–211.

Soriano, P. (1997). The PDGF alpha receptor is required for neural crest cell development and for normal patterning of the somites. *Development (Cambridge, UK)* **124**, 2691–2700.

Soriano, P., Montgomery, C., Geske, R., and Bradley, A. (1991). Targeted disruption of the c-src proto-oncogene leads to osteopetrosis in mice. *Cell (Cambridge, Mass.)* **64**, 693–702.

Sosa-Pineda, B., Chowdhury, K., Torres, M., Oliver, G., and Gruss, P. (1997). The Pax4 gene is essential for differentiation of insulin-producing beta cells in the mammalian pancreas. *Nature (London)* **386**, 399–402.

Spemann, H. (1938). "Embryonic Development and Induction." Yale University Press, New Haven, CT.

Sporn, M. B., Roberts, A. B., and Goodman, D. S. (1994). "The Retinoids: Biology, Chemistry, and Medicine." Raven Press, New York.

Srivastava, D., Cserjesi, P., and Olson, E. N. (1995). A subclass of bHLH proteins required for cardiac morphogenesis. *Science* **270**, 1995–1999.

Srivastava, D., Thomas, T., Lin, Q., Kirby, M. L., Brown, D., and Olson, E. N. (1997). Regulation of cardiac mesodermal and neural crest development by the bHLH transcription factor, dHAND. *Nat. Genet.* **16**, 154–160.

Stacey, A., Bateman, J., Choi, T., Mascara, T., Cole, W., and Jaenisch, R. (1988). Perinatal lethal osteogenesis imperfecta in transgenic mice bearing an engineered mutant pro-alpha 1(I) collagen gene. *Nature (London)* **332**, 131–136.

Stark, M. R., Sechrist, J., Bronner-Fraser, M., and Marcelle, C. (1997). Neural tube-ectoderm interactions are required for trigeminal placode formation. *Development (Cambridge, UK)* **124**, 4287–4295.

St.-Jacques, B., Hammerschmidt, M., and McMahon, A. P. (1999). Indian hedgehog signaling regulates proliferation and differentiation of chondrocytes and is essential for bone formation. *Genes Dev.* **13**, 2072–2086.

Stock, D. W., Ellies, D. L., Zhao, Z., Ekker, M., Ruddle, F. H., and Weiss, K. M. (1996). The evolution of the vertebrate Dlx gene family. *Proc. Natl. Acad. Sci. U.S.A.* **93**, 10858–10863.

Storm, E. E., and Kingsley, D. M. (1996). Joint patterning defects caused by single and double mutations in members of the bone morphogenetic protein (BMP) family. *Development (Cambridge, UK)* **122**, 3969–3979.

Suda, Y., Matsuo, I., Kuratani, S., and Aizawa, S. (1996). Otx1 function overlaps with Otx2 in development of mouse forebrain and midbrain. *Genes Cells* **1**, 1031–1044.

Sun, X., Meyers, E. N., Lewandoski, M., and Martin, G. R. (1999). Targeted disruption of Fgf8 causes failure of cell migration in the gastrulating mouse embryo. *Genes Dev.* **13**, 1834–1846.

Suzuki, N., Labosky, P. A., Furuta, Y., Hargett, L., Dunn, R., Fogo, A. B., Takahara, K., Peters, D. M., Greenspan, D. S., and Hogan, B. L. (1996). Failure of ventral body wall closure in mouse embryos lacking a procollagen C-proteinase encoded by Bmp1, a mammalian gene related to Drosophila tolloid. *Development (Cambridge, UK)* **122**, 3587–3595.

Szeto, D. P., Rodriguez-Esteban, C., Ryan, A. K., O'Connell, S. M., Liu, F., Kioussi, C., Gleiberman, A. S., Izpisúa-Belmonte, J. C., and Rosenfeld, M. G. (1999). Role of the Bicoid-related homeodomain factor Pitx1 in specifying hindlimb morphogenesis and pituitary development. *Genes Dev.* **13**, 484–494.

Takeda, K., Takeuchi, O., Tsujimura, T., Itami, S., Adachi, O., Kawai, T., Sanjo, H., Yoshikawa, K., Terada, N., and Akira, S. (1999). Limb and skin abnormalities in mice lacking IKKalpha. *Science* **284**, 313–316.

Tam, P. P., Goldman, D., Camus, A., and Schoenwolf, G. C. (2000). Early events of somitogenesis in higher vertebrates: Allocation of precursor cells during gastrulation and the organization of a meristic pattern in the paraxial mesoderm. *Curr. Top. Dev. Biol.* **47**, 1–32.

Tan, S. S., and Morriss-Kay, G. M. (1986). Analysis of cranial neural crest cell migration and early fates in postimplantation rat chimaeras. *J. Embryol. Exp. Morphol.* **98**, 21–58.

Tanaka, A., Miyamoto, K., Minamino, N., Takeda, M., Sato, B., Matsuo, H., and Matsumoto, K. (1992). Cloning and characterization of an androgen-induced growth factor essential for the androgen-dependent growth of mouse mammary carcinoma cells. *Proc. Natl. Acad. Sci. U.S.A.* **89**, 8928–8932.

Taya, Y., O'Kane, S., and Ferguson, M. W. J. (1999). Pathogenesis of cleft palate in TBF-b3 knockout mice. *Development (Cambridge, UK)* **126**, 3869–3879.

ten Berge, D., Brouwer, A., el Bahi, S., Guénet, J. L., Robert, B., and Meijlink, F. (1998a). Mouse Alx3: An aristaless-like homeobox gene expressed during embryogenesis in ectomesenchyme and lateral plate mesoderm. *Dev. Biol.* **199**, 11–25.

ten Berge, D., Brouwer, A., Korving, J., Martin, J. F., and Meijlink, F. (1998b). Prx1 and Prx2 in skeletogenesis: Roles in the craniofacial region, inner ear and limbs. *Development (Cambridge, UK)* **125**, 3831–3842.

ten Cate, A. R. (1995). The experimental investigation of odontogenesis. *Int. J. Dev. Biol.* **39**, 5–11.

Testaz, S., Delannet, M., and Duband, J. (1999). Adhesion and migration of avian neural crest cells on fibronectin require the cooperating activities of multiple integrins of the (beta)1 and (beta)3 families. *J. Cell Sci.* **112**, 4715–4728.

Thesleff, I., and Sharpe, P. (1997). Signalling networks regulating dental development. *Mech. Dev.* **67**, 111–123.

Thesleff, I., Vaahtokari, A., and Partanen, A. M. (1995). Regulation of organogenesis. Common molecular mechanisms regulating the development of teeth and other organs. *Int. J. Dev. Biol.* **39**, 35–50.

Thomas, B. L., and Sharpe, P. T. (1998). Patterning of the murine dentition by homeobox genes. *Eur. J. Oral Sci.* **106**(Suppl. 1), 48–54.

Thomas, B. L., Tucker, A. S., Qiu, M., Ferguson, C. A., Hardcastle, Z., Rubenstein, J. L., and Sharpe, P. T. (1997). Role of Dlx-1 and Dlx-2 genes in patterning of the murine dentition. *Development (Cambridge, UK)* **124**, 4811–4818.

Thomas, K. R., and Capecchi, M. R. (1990). Targeted disruption of the murine int-1 proto-oncogene resulting in severe abnormalities in midbrain and cerebellar development. *Nature (London)* **346**, 847–850.

Thomas, K. R., Musci, T. S., Neumann, P. E., and Capecchi, M. R. (1991). Swaying is a mutant allele of the proto-oncogene Wnt-1. *Cell (Cambridge, Mass.)* **67**, 969–976.

Thomas, L. J. (1926). ossification centers of the petrosal bone of a mouse. *Anat. Rec.* **30**, 59–68.

Thomas, T., Kurihara, H., Yamagishi, H., Kurihara, Y., Yazaki, Y., Olson, E. N., and Srivastava, D. (1998). A signaling cascade involving endothelin-1, dHAND and msx1 regulates development of neural-crest-derived branchial arch mesenchyme. *Development (Cambridge, UK)* **125**, 3005–3014.

Thomas, T., Voss, A. K., Chowdhury, K., and Gruss, P. (2000). Querkopf, a MYST family histone acetyltransferase, is required for normal cerebral cortex development. *Development (Cambridge, UK)* **127**, 2537–2548.

Thompson, D. A. W., (1942). "On Growth and Form." The Universtiy Press, Cambridge, England.

Thorogood, P. (1988). The developmental specification of the vertebrate skull. *Development (Cambridge, UK)* **103**(Suppl.), 141–153.

Thorogood, P. (1993). Differentiation and morphogenesis of cranial skeletal tissues. *In* "The Skull" (J. Hanken and B. K. Hall, eds.), Vol. 1, pp. 112–152. University of Chicago Press, Chicago.

Thorogood, P. (1997). "Embryos, Genes, and Birth Defects." Wiley, Chichester and New York.

Thorogood, P., and Smith, L. (1984). Neural crest cells: The role of extracellular matrix in their differentiation and migration. *Prog. Clin. Biol. Res.* **151**, 171–185.

Timmons, P. M., Wallin, J., Rigby, P. W., and Balling, R. (1994). Expression and function of Pax 1 during development of the pectoral girdle. *Development (Cambridge, UK)* **120**, 2773–2785.

Tomsa, J. M., and Langeland, J. A. (1999). Otx expression during lamprey embryogenesis provides insights into the evolution of the vertebrate head and jaw. *Dev. Biol.* **207**, 26–37.

Torres, M., and Giráldez, F. (1998). The development of the vertebrate inner ear. *Mech. Dev.* **71**, 5–21.

Torres, M., Gómez-Pardo, E., Dressler, G. R., and Gruss, P. (1995). Pax-2 controls multiple steps of urogenital development. *Development (Cambridge, UK)* **121**, 4057–4065.

Trainor, P., and Krumlauf, R. (2000). Plasticity in mouse neural crest cells reveals a new patterning role for cranial mesoderm. *Nat. Cell Biol.* **2**, 96–102.

Trainor, P. A., and Tam, P. P. (1995). Cranial paraxial mesoderm and neural crest cells of the mouse embryo: Co-distribution in the craniofacial mesenchyme but distinct segregation in branchial arches. *Development (Cambridge, UK)* **121**, 2569–2582.

Trainor, P. A., Tan, S. S., and Tam, P. P. (1994). Cranial paraxial mesoderm: Regionalization of cell fate and impact on craniofacial development in mouse embryos. *Development (Cambridge, UK)* **120**, 2397–2408.

Tribioli, C., and Lufkin, T. (1999). The murine Bapx1 homeobox gene plays a critical role in embryonic development of the axial skeleton and spleen. *Development (Cambridge, UK)* **126**, 5699–5711.

Trumpp, A., Depew, M. J., Rubenstein, J. L., Bishop, J. M., and Martin, G. R. (1999). Cre-mediated gene inactivation demonstrates that FGF8 is required for cell survival and patterning of the first branchial arch. *Genes Dev.* **13**, 3136–3148.

Truslove, G. M. (1956). The anatomy and development of the fidget mouse. *J. Genet.* **54.**

Tucker, A. S., Matthews, K. L., and Sharpe, P. T. (1998a). Transformation of tooth type induced by inhibition of BMP signaling. *Science* **282**, 1136–1138.

Tucker, A. S., Al Khamis, A., and Sharpe, P. T. (1998). Interactions between Bmp4 and Msx1 act to restrict gene expression to odontogenic mesenchyme. *Dev. Dyn.* **212**, 533–539.

Tucker, A. S., Yamada, G., Grigoriuo, M., Pachnis, V., and Sharpe, P. T. (1999). Fgf-8 determines rostral-caudal polarity in the first branchial arch. *Development (Cambridge, UK)* **126**, 51–61.

Tucker, A. S., Headon, D. J., Schneider, P., Ferguson, B., Overbeek, P., Tschopp, J., and Sharpe, P. T. (2000). Edar/Eda interactions regulate enamel knot formation in tooth morphogenesis. *Development (Cambridge, UK)* **127**, 4691–4700.

Tyler, M. S. (1983). Development of the frontal bone and cranial meninges in the embryonic chick: An experimental study of tissue interactions. *Anat. Rec.* **206**, 61–70.

Tyler, M. S., and Hall, B. K. (1977). Epithelial influences on skeletogenesis in the mandible of the embryonic chick. *Anat. Rec.* **188**, 229–239.

Unden, A. B., Holmberg, E., Lundh-Rozell, B., Stähle-Bäckdahl, M., Zaphiropoulos, P. G., Toftgård, R., and Vorechovsky, I. (1996). Mutations in the human homolog of Drosophila patched (PTCH) in basal cell carcinomas and the Gorlin syndrome: Different in vivo mechanisms of PTCH inactivation. *Cancer Res.* **56**, 4562–4565.

Urbánek, P., Fetka, I., Meisler, M. H., and Busslinger, M. (1997). Cooperation of Pax2 and Pax5 in midbrain and cerebellum development. *Proc. Natl. Acad. Sci. U.S.A.* **94**, 5703–5708.

Vaage, J., and Weiss, D. W. (1969). Immunization against spontaneous and autografted mouse mammary carcinomas in the autochthonous C3H-Crgl mouse. *Cancer Res.* **29**, 1920–1926.

Vaahtokari, A., Åberg., T., and Thesleff. I. (1996a). Apoptosis in the developing tooth: Association with an embryonic signalling centre and suppression by EGF and FGF-4. *Development (Cambridge, UK)* **122**, 121–129.

Vaahtokari, A., Åberg., T., Jernvall, J., Keränen, S., and Thesleff. I. (1996b). The enamel knot as a signalling centre in the developing mouse tooth. *Mech. Dev.* **54**, 39–43.

Vainio, S., Karavanova, I., Jowett, A., and Thesleff, I. (1993). Identification of Bmp4 as a signal mediating secondary induction between epithelial and mesenchymal tissues during early tooth development. *Cell (Cambridge, Mass.)* **75**, 45–58.

van den Heuvel, M., and Ingham, P. W. (1996). smoothened encodes a receptor-like serpentine protein required for hedgehog signalling. *Nature (London)* **382**, 547–551.

van Genderen, C., Okamura, R. M., Farinas, I., Quo, R., Parslow, T. G., Bruhn, L., and Grosschedl, R. (1994). Development of several organs that require inductive epithelial-mesenchymal interactions is impaired in *LEF-1* deficient mice. *Genes Dev.* **8**, 2691–2703.

Varlet, I., Collignon, J., and Robertson, E. J. (1997). Nodal expression in the primitive endoderm is required for specification of the anterior axis during mouse gastrulation. *Development (Cambridge, UK)* **124**, 1033–1044.

Veitch, E., Begbie, J., Schilling, T. F., Smith, M. M., and Graham, A. (1999). Pharyngeal arch patterning in the absence of neural crest. *Curr. Biol.* **9**, 1481–1484.

Verwoerd, C. D., and van Oostrom, C. G. (1979). Cephalic neural crest and placodes. *Adv. Anat., Embryol. Cell Biol.* **58**, 1–75.

Vogel, K. S., and Davies, A. M. (1993). Heterotopic transplantation of presumptive placodal ectoderm changes the fate of sensory neuron precursors. *Development (Cambridge, UK)* **119**, 263–276.

Vortkamp, A., Lee, K., Lanske, B., Segré, G. V., Kronenberg, H. M., and Tabin, C. J. (1996). Regulation of rate of cartilage differentiation by Indian hedgehog and PTH-related protein. *Science* **273**, 613–622.

Wagner, G. (1955). Chimaerishe Zahnanlagen aus *Triton*-Schmelzorgan und *Bombinator*-Papille. Mit Beobachtungen über die Entwicklung von Kiemenzahnchen und Mundsinnesknospen in den *Triton*-Larven. *J. Embryol. Exp. Morphol.* **3**, 160–188.

Wajnrajch, M. P., Gertner, J. M., Harbison, M. D., Chua, S. C., Jr., and Leibel, R. L. (1996). Nonsense mutation in the human growth hormone-releasing hormone receptor causes growth failure analogous to the little (lit) mouse. *Nat. Genet.* **12**, 88–90.

Waldrip, W. R., Bikoff, E. K., Hoodless, P. A., Wrana, J. L., and Robertson, E. J. (1998). Smad2 signaling in extraembryonic tissues determines anterior-posterior polarity of the early mouse embryo. *Cell (Cambridge, Mass.)* **92**, 797–808.

Wall, N. A., and Hogan, B. L. (1995). Expression of bone morphogenetic protein-4 (BMP-4), bone morphogenetic protein-7 (BMP-7), fibroblast growth factor-8 (FGF-8) and sonic hedgehog (SHH) during branchial arch development in the chick. *Mech. Dev.* **53**, 383–392.

Walterhouse, D., Ahmed, M., Slusarski, D., Kalamaras, J., Boucher, D., Holmgren, R., and Iannaccone, P. (1993). Gli, a zinc finger transcription factor and oncogene, is expressed during normal mouse development. *Dev. Dyn.* **196**, 91–102.

Wang, W., Van De Water, T., and Lufkin, T. (1998). Inner ear and maternal reproductive defects in mice lacking the Hmx3 homeobox gene. *Development (Cambridge, UK)* **125**, 621–634.

Wang, X., Penzes, P., and Napoli, J. L. (1996). Cloning of a cDNA encoding an aldehyde dehydrogenase and its expression in *Escherichia coli*. Recognition of retinal as substrate. *J. Biol. Chem.* **271**, 16288–16293.

Wang, Z. Q., Ovitt, C., Grigoriadis, A. E., Möhle-Steinlein, U., Rüther, U., and Wagner, E. F. (1992). Bone and haematopoietic defects in mice lacking c-fos. *Nature (London)* **360**, 741–745.

Wassarman, K. M., Lewandoski, M., Campbell, K., Joyner, A. L., Rubenstein, J. L., Martinez, S., and Martin, G. R. (1997). Specification of the anterior hindbrain and establishment of a normal mid/hindbrain organizer is dependent on Gbx2 gene function. *Development (Cambridge, UK)* **124**, 2923–2934.

Watanabe, H., and Yamada, Y. (1999). Mice lacking link protein develop dwarfism and craniofacial abnormalities. *Nat. Genet.* **21**, 225–229.

Watanabe, H., Kimata, K., Line, S., Strong, D., Gao, L. Y., Kozak, C. A., and Yamada, Y. (1994). Mouse cartilage matrix deficiency (cmd) caused by a 7 bp deletion in the aggrecan gene. *Nat. Genet.* **7**, 154–157.

Waterman, R. E. (1977). Ultrastructure of oral (buccopharyngeal) membrane formation and rupture in the hamster embryo. *Dev. Biol.* **58**, 219–229.

Waterman, R. E., and Schoenwolf, G. C. (1980). The ultrastructure of oral (buccopharyngeal) membrane formation and rupture in the chick embryo. *Anat. Rec.* **197**, 441–470.

Webb, J. F., and Noden, D. M. (1993). Ectodermal placodes: Contributions to the development of the vertebrate head. *Am. Zool.* **33**, 434–447.

Webster, M. K., and Donoghue, D. J. (1997). FGFR activation in skeletal disorders: Too much of a good thing. *Trends Genet.* **13**, 178–182.

Webster, W. S., Johnston, M. C., Lammer, E. J., and Sulik, K. K. (1986). Isotretinoin embryopathy and the cranial neural crest: An in vivo and in vitro study. *J. Craniofacial Genet. Dev. Biol.* **6**, 211–222.

Wedden, S. E., Ralphs, J. R., and Tickle, C. (1988). Pattern formation in the facial primordia. *Development (Cambridge, UK)* **103**(Suppl.), 31–40.

Whitesides, J., Hall, M., Anchan, R., and LaMantia, A. S. (1998). Retinoid signaling distinguishes a subpopulation of olfactory receptor neurons in the developing and adult mouse. *J. Comp. Neurol.* **394**, 445–461.

Whitlock, K. E., and Westerfield, M. (2000). The olfactory placodes of the zebrafish form by convergence of cellular fields at the edge of the neural plate. *Development (Cambridge, UK)* **127**, 3645–3653.

Wilkie, A. O. (1997). Craniosynostosis: Genes and mechanisms. *Hum. Mol. Genet.* **6**, 1647–1656.

Wilkie, A. O., Slaney, S. F., Oldridge, M., Poole, M. D., Ashworth, G. J., Hockley, A. D., Hayward, R. D., David, D. J., Pulleyn, L. J., Rutland, P. *et al.* (1995). Apert syndrome results from localized mutations of FGFR2 and is allelic with Crouzon syndrome. *Nat. Genet.* **9**, 165–172.

Wilkinson, D. G. (2000). Eph receptors and ephrins: Regulators of guidance and assembly. *Int. Rev. Cytol.* **196**, 177–244.

Williams, N. A., and Holland, P. W. (1998). Gene and domain duplication in the chordate Otx gene family: Insights from amphioxus Otx. *Mol. Biol. Evol.* **15**, 600–607.

Wilm, B., Dahl, E., Peters, H., Balling, R., and Imai, K. (1998). Targeted disruption of Pax1 defines its null phenotype and proves haploinsufficiency. *Proc. Natl. Acad. Sci. U.S.A.* **95**, 8692–8697.

Winnier, G., Blessing, M., Labosky, P. A., and Hogan, B. L. (1995). Bone morphogenetic protein-4 is required for mesoderm formation and patterning in the mouse. *Genes Dev.* **9**, 2105–2116.

Winnier, G. E., Hargett, L., and Hogan, B. L. (1997). The winged helix transcription factor MFH1 is required for proliferation and patterning of paraxial mesoderm in the mouse embryo. *Genes Dev.* **11**, 926–940.

Wizenmann, A., and Lumsden, A. (1997). Segregation of rhombomeres by differential chemoaffinity. *Mol. Cell. Neurosci.* **9**, 448–459.

Wozney, J. M., Rosen, V., Celeste, A. J., Mitsock, L. M., Whitters, M. J., Kriz, R. W., Hewick, R. M., and Wang, E. A. (1988). Novel regulators of bone formation: Molecular clones and activities. *Science* **242**, 1528–1534.

Wray, S., Grant, P., and Gainer, H. (1989). Evidence that cells expressing luteinizing hormone-releasing hormone mRNA in the mouse are derived from progenitor cells in the olfactory placode. *Proc. Natl. Acad. Sci. U.S.A.* **86**, 8132–8136.

Wright, M. E. (1947). Undulated: A new genetic factor in *Mus musculus* affecting the spine and tail. *Heredity* **10**, 137–141.

Xuan, S., Baptista, C. A., Balas, G., Tao, W., Soares, V. C., and Lai, E. (1995). Winged helix transcription factor BF-1 is essential for the development of the cerebral hemispheres. *Neuron* **14**, 1141–1152.

Yamada, G., Mansouri, A., Torres, M., Stuart, E. T., Blum, M., Schultz, M., De Robertis, E. M., and Gruss, P. (1995). Targeted mutation of the murine goosecoid gene results in craniofacial defects and neonatal death. *Development (Cambridge, UK)* **121**, 2917–2922.

Yamada, G., Ueno, K., Nakamura, S., Hanamure, Y., Yasui, K., Uemura, M., Eizuru, Y., Mansouri, A., Blum, M., and Sugimura, K. (1997). Nasal and pharyngeal abnormalities caused by the mouse goosecoid gene mutation. *Biochem. Biophys. Res. Commun.* **233**, 161–165.

Yamaguchi, T. P., Bradley, A., McMahon, A. P., and Jones, S. (1999). A Wnt5a pathway underlies outgrowth of multiple structures in the vertebrate embryo. *Development (Cambridge, UK)* **126**, 1211–1223.

Yanagisawa, H., Yanagisawa, M., Kapur, R. P., Richardson, J. A., Williams, S. C., Clouthier, D. E., de Wit, D., Emoto, N., and Hammer, R. E. (1998). Dual genetic pathways of endothelin-mediated intercellular signaling revealed by targeted disruption of endothelin converting enzyme-1 gene. *Development (Cambridge, UK)* **125**, 825–836.

Yang, L., Zhang, H., Hu, G., Wang, H., Abate-Shen, C., and Shen, M. M. (1998). An early phase of embryonic Dlx5 expression defines the rostral boundary of the neural plate. *J. Neurosci.* **18**, 8322–8330.

Yoshida, H., Hayashi, S., Kunisada, T., Ogawa, M., Nishikawa, S., Okamura, H., Sudo, T., and Shultz, L. D. (1990). The murine mutation osteopetrosis is in the coding region of the macrophage colony stimulating factor gene. *Nature (London)* **345**, 442–444.

Young, R. W. (1959). The inlfuence of cranial contents on postnatal growth of the skull in the rat. *Am. J. Anat.* **105**, 383–415.

Youssef, E. H. (1966). The chondrocranium of the albino rat. *Acta Anat.* **64**, 586–617.

Youssef, E. H. (1969). Development of the membrane bones and ossification of the chondrocranium in the albino rat. *Acta Anat.* **72**, 603–623.

Zeller, U. (1987). Morphogenesis of the mammalian skull with special reference to *Tupaia*. *In* "Morphogenesis of the Mammalian Skull" (U. Zeller and H. Kuhn, eds.), pp. 17–50. Parey, Hamburg and Berlin.

Zerucha, T., Stühmer, T., Hatch, G., Park, B. K., Long, Q., Yu, G., Gambarotta, A., Schultz, J. R., Rubenstein, J. L., and Ekker, M. (2000). A highly conserved enhancer in the Dlx5/Dlx6 intergenic region is the site of cross-regulatory interactions between Dlx genes in the embryonic forebrain. *J. Neurosci.* **20**, 709–721.

Zetterström, R. H., Lindqvist, E., de Urquiza, A. M., Tomac, A., Eriksson, U., Perlmann, T., and Olson, L. (1999). Role of retinoids in the CNS: differential expression of retinoid binding proteins and receptors and evidence for presence of retinoic acid. *Eur. J. Neurosci.* **11**, 407–416.

Zhang, H., and Bradley, A. (1996). Mice deficient for BMP2 are nonviable and have defects in amnion/chorion and cardiac development. *Development (Cambridge, UK)* **122**, 2977–2986.

Zhang, Y., Zhang, Z., Zhao, X., Yu, X., Hu, Y., Geronimo, B., Fromm, S. H., and Chen, Y. (2000). A new function of BMP4: Dual role for BMP4 in regulation of Sonic hedgehog expression in the mouse tooth germ. *Development (Cambridge, UK)* **127**, 1431–1443.

Zhao, D., McCaffery, P., Ivins, K. J., Neve, R. L., Hogan, P., Chin, W. W., and Dräger, U. C. (1996). Molecular identification of a major retinoic-acid-synthesizing enzyme, a retinaldehyde-specific dehydrogenase. *Eur. J. Biochem.* **240**, 15–22.

Zhao, G. Q., Eberspaecher, H., Seldin, M. F., and de Crombrugghe, B. (1994a). The gene for the homeodomain-containing protein Cart-1 is expressed in cells that have a chondrogenic potential during embryonic development. *Mech. Dev.* **48**, 245–254.

Zhao, G. Q., Zhao, S., Zhou, X., Eberspaecher, H., Solursh, M., and de Crombrugghe, B. (1994b). rDlx, a novel distal-less-like homeoprotein is expressed in developing cartilages and discrete neuronal tissues. *Dev. Biol.* **164**, 37–51.

Zhao, Q., Behringer, R. R., and de Crombrugghe, B. (1996). Prenatal folic acid treatment suppresses acrania and meroanencephaly in mice mutant for the Cart1 homeobox gene. *Nat. Genet.* **13**, 275–283.

Zhao, X., Zhang, Z., Song, Y., Zhang, X., Zhang, Y., Hu, Y., Fromm, S. H., and Chen, Y. (2000). Transgenically ectopic expression of Bmp4 to the Msx1 mutant dental mesenchyme restores downstream gene expression but represses Shh and Bmp2 in the enm]amel knot of wild type tooth germ. *Mech. Dev.* **99**, 29–38.

Zhao, Y., Guo, Y.-J., Tomac, A. C., Taylor, N. R., Grinberg, A., Lee, E. J., Huang, S., and Westphal, H. (1999). Isolated cleft palate in mice with a targeted mutation of the LIM homeobox gene Lhx8. *Proc. Natl. Acad. Sci U.S.A.* **96**, 15002–15006.

Zhou, P., Byrne, C., Jacobs, J., and Fuchs, E. (1995) Lymphoid enhancer factor 1 directs hair follicle patterning and epithelial cell fate. *Genes Dev.* **9**, 700–713.

Zimmerman, B., Moegelin, A., de Souza, P., and Bier, J. (1998). Morphology of the development of the sagittal suture of mice. *Anat. Embryol.* **197**, 155–165.

Zimmerman, L. B., De Jesús-Escobar, J. M., and Harland, R. M. (1996). The Spemann organizer signal noggin binds and inactivates bone morphogenetic protein 4. *Cell (Cambridge, Mass.)* **86**, 599–606.

20

Pituitary Gland Development

Sally Camper,* Hoonkyo Suh,** Lori Raetzman,* Kristin Douglas,* Lisa Cushman,*
Igor Nasonkin,* Heather Burrows,‡ Phil Gage,* and Donna Martin†

*Department of Human Genetics, †Department of Pediatrics, **Graduate Program in Neuroscience, and ‡Graduate Program in
Cellular and Molecular Biology, University of Michigan Medical School, Ann Arbor, Michigan 48109

I. **Pituitary Gland Anatomy and Function**

II. **Development of the Pituitary Primordia and Cell Specification**

III. **Expansion of Committed Cell Types**

IV. **Conclusion**

References

I. Pituitary Gland Anatomy and Function

A. Anatomy

The pituitary gland is a small endocrine gland located at the base of the brain, under the optic chiasm and the arachnoid membrane, in a bony indentation called the sella turcica (Fig. 1). Its name comes from the Greek *hypophysis,* which means undergrowth. The role of the pituitary gland is the regulated synthesis and secretion of polypeptide hormones that are essential for the development and function of many other organs in the body (Table I). Releasing hormones and inhibiting factors reach the anterior pituitary via hypothalamic neurons that terminate in the capillary beds of the median eminence, just dorsal to the pituitary gland. These capillary beds are connected to the hypophyseal portal vessels that nourish the anterior pituitary. In response to these stimu-

latory factors, pituitary hormones are released into hypophyseal portal blood vessels and carried through the blood stream to their target organs. Hormones produced by the target organs feed back on the pituitary and the brain to maintain homeostasis. Hypo- or hyperpituitarism phenotypes result from depressed or enhanced hormone production, respectively.

The rodent pituitary is composed of three lobes. The posterior pituitary (*neurohypophysis* or *pars nervosa*) is derived from neural ectoderm and contains the nerve terminals that secrete oxytocin and vasopressin. These hormones are synthesized in magnocellular neurons whose cell bodies reside in the paraventricular nucleus of the hypothalamus. Both vasopressin and oxytocin are synthesized as prepropeptides and are transported in neural processes to the posterior pituitary, where their secretion and action are analogous to conventional neurotransmitters. The intermediate lobe (*pars intermedia*) and anterior lobe are both derived from oral ectoderm. The bulk of the anterior lobe (*pars distalis)* is at the same level as the intermediate lobe, but a portion of it known as the *pars tuberalis* extends dorsally along the pituitary stalk. Both the anterior and intermediate lobes produce pro-opiomelanocortin (POMC), which is subject to differential cleavage in each lobe to produce many different hormones including adrenocorticotropin (ACTH), melanocyte-stimulating hormone (MSH), and endorphins. In some species the anatomical distinction between the intermediate and

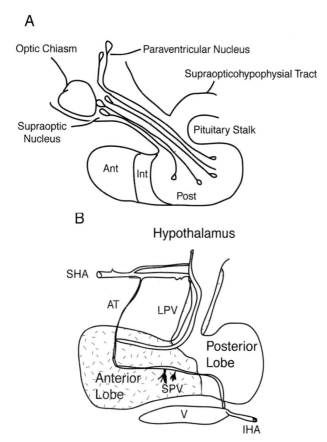

Figure 1 Pituitary gland anatomy. (A) The mature pituitary gland includes the anterior (ant), intermediate (int), and posterior (post) lobes. It is situated at the base of the brain, just beneath the optic chiasm, and is connected to the brain via the pituitary stalk. The cell bodies of the neurons that produce oxytocin and vasopressin are found in the supraoptic nucleus and the paraventricular nucleus, respectively, and these neurons terminate in the posterior lobe. (B) The pituitary gland is a highly vascularized tissue. The superior hypophyseal artery (SHA) transports blood to the hypothalamus. It bifurcates into the trabecular artery (AT) and the long portal vessels (LPV), which bring hypothalamic releasing factors to the anterior pituitary. Pituitary hormones are secreted into the short portal vessels (SPV) and inferior hypophyseal artery (IHA) for transport to the target organs.

anterior lobes of the pituitary is blurred. For example, the bovine intermediate lobe contains a patch of anterior pituitary cells called the cone of Wulzen, and in humans the intermediate and anterior lobes are not as distinct as in rodents. Despite these anatomical differences in the mature gland, the development of the organ is essentially the same in all organisms examined.

B. Hormone Functions

Oxytocin and vasopressin, produced by the posterior pituitary, are considered neuropeptides and have widespread roles in animal and human development (Acher and Chauvet, 1995; McCarthy and Altemus, 1997). Vasopressin regu-

lates body volume status by stimulating small blood vessel contraction and reabsorption of water through kidney tubule cells, whereas oxytocin promotes milk release from the mammary gland and stimulates cervical dilatation.

The anterior and intermediate lobes coordinate many body functions in their role as the primary site of regulation for the peripheral endocrine organs (Table I). POMC is the precursor protein for ACTH and β endorphins, produced by anterior pituitary corticotropes, and for MSH and γ and α endorphin, produced by melanotropes (de Wied, 1999). Most of the anterior pituitary hormones influence both the development and function of their target organs. ACTH stimulation is important for development of the adrenal gland and release of glucocorticoids from the adrenal gland (Muglia *et al.,* 1995). The α-glycoprotein subunit (αGSU or *Cga*) is the common subunit of the heterodimeric glycoprotein hormones thyroid-stimulating hormone (TSH), luteinizing hormone (LH) and follicle-stimulating hormone (FSH). αGSU is required for secretion of these hormone heterodimers and for their biological activity, but hormone specificity is dictated by their β subunits. TSH plays a critical role in stimulation of thyroid hormone production, which is necessary for normal brain development and body growth. By E17.5 in the mouse TSH is needed for organization of thyroid follicles and growth of the thyroid gland (Beamer and Cresswell, 1982). Growth hormone (GH) regulates body growth as well, through stimulation of insulin-like growth factor production in the liver. Prolactin (PRL) stimulates the production of milk, and controls osmoregulation and behavior in both males and females. GH and PRL are evolutionarily related, as are the gonadotropins, LH and FSH. Development of the gonads after birth is dependent on LH and FSH, as is the production of steroid hormones in the testis and ovary (Kumar *et al.,* 1997; Weiss *et al.,* 1992). All vertebrate species make very similar polypeptide hormones, although the function can vary from species to species. For example, bovine GH has lactogenic properties and prolactin regulates osmolarity in fish and song behavior in birds (Bern and Nicoll, 1968).

II. Development of the Pituitary Primordia and Cell Specification

A. Overview

The morphologic changes that take place as the pituitary gland develops are well documented in classical papers and books (Kaufman, 1992; Rugh, 1990; Schwind, 1928). However, our understanding of pituitary development in molecular terms exploded during the 1990s (Dattani and Robinson, 2000; Kioussi, 1999; Parks *et al.,* 1997; Sheng and Westphal, 1999). This chapter is an expansion and update of our earlier

Table I Pituitary Cell Types, Biological Function, and Disease Phenotypes

Cell types	Hormone product	Target		Regulator		Phenotype	
		Tissue	Response	Positive	Negative	Hypopituitarism	Hyperpituitarism
Posterior (*pars nervosa*)							
SON, PVN, magnocellular neuron terminals	Oxytocin	Mammary, uterus	Milk let-down contractions	Suckling activation of stretch receptors	Progesterone and endogenous opioids	Inability to lactate	Galactorrhea
SON/PVN magnocellular neuron terminals	Vasopressin	Kidney	Reabsorption of water	Hypovolemia	Hyperosmolarity	Diabetes insipidus	SIADH
Intermediate lobe (*pars intermedia*)							
Melanotropes	Melanocyte-stimulating hormone (MSH) and γ- and α-endorphins	[a]					Hyperpigmentation, adrenal insufficiency
Anterior lobe (*pars distalis, pars tuberalis*)							
Corticotrope	Adrenocorticotropin (ACTH)	Adrenal cortex	Glucocorticoid production	Corticotropin-releasing hormone (CRH)	Corticosteroids	Adrenal hypoplasia	Cushing disease
Thyrotrope	Thyroid-stimulating hormone (TSH)	Thyroid	Thyroid hormone (T_3, T_4) production	Thyrotropin-releasing hormone (TRH)	T_3, T_4	Thyroid hypoplasia, dwarfism, cretinism, hypothyroidism	Thyroid hyperplasia, hyperthyroidism
Somatotrope	Growth hormone (GH)	Liver and other tissues	Growth	Growth hormone-releasing hormone (GHRH)	Somatostatin, insulin-like growth factor	Dwarfism	Gigantism, acromegaly
Lactotrope	Prolactin (PRL)	Mammary gland	Milk production	Estrogen, TRH	Dopamine	Failure to lactate	Galactorrhea, infertility in females, reduced libido in males
Gonadotrope	Luteinizing hormone (LH), follicle-stimulating hormone (FSH)	Gonads	Spermatogenesis and ovulation	Gonadotropin-releasing hormone (GnRH)	Gonadal steroids, inhibins	Sexual immaturity infertility	Precocious puberty

[a] Melanotropes have pleiotropic functions (for review, see Strand, 1999).

reviews of anterior pituitary development (Burrows *et al.,* 1999; Watkins-Chow and Camper, 1998). Spatial cues established in determining the anterior-posterior and dorsal-ventral axes of the embryo provide the initial foundation for pituitary organogenesis. Fate mapping studies have shown that the anterior neural ridge gives rise to the anterior pituitary, the nasal cavity, and the olfactory placode (Couly and Le Douarin, 1985, 1987, 1988). The more posterior neural plate develops into the hypothalamus, posterior pituitary, optic vesicles, and ventral forebrain. These studies illustrate that initial pituitary patterning occurs very early.

Extrinsic factors secreted by surrounding tissues and inductive interactions between the neural and oral ectoderm produce opposing dorsal-to-ventral and ventral-to-dorsal gradients that act to induce transcription factor expression in spatially distinct areas of the developing pituitary (Treier and Rosenfeld, 1996). Combinations of transcription factors act synergistically to stimulate transcription of individual hormone genes in specialized, spatially restricted patches of cells (Watkins-Chow and Camper, 1998). As the pituitary continues to develop, spatial patterns of expression are lost. Stimulatory and inhibitory factors produced by the hypothalamus and a variety of end organs regulate the proliferation of each cell type in response to the needs of the organism.

B. Morphology during Ontogeny

Development of the posterior lobe occurs concurrently with the anterior and intermediate lobes. The posterior pituitary lobe and ventral hypothalamus are both derived from neural ectoderm. The ventral diencephalon evaginates to produce the infundibulum, which gives rise to the posterior pituitary.

Development of the anterior and intermediate lobes can be arbitrarily divided into steps: the thickening of the oral ectoderm and initial invagination to produce a rudimentary pouch, the formation of a definitive pouch as it extends and maintains contact with the evaginating neural ectoderm, the separation of the pouch from the mouth, and the expansion of the anterior aspect of the pouch through cell proliferation (Fig. 2) (Sheng *et al.,* 1997). Thickening of the oral ectoderm occurs at embryonic day 8 (E8) in the mouse. Rathke's pouch separates from the oral ectoderm at E12.5 and the cells on the ventral side of the pouch undergo proliferation to generate the anterior lobe, whereas cells on the opposite side form the intermediate lobe. The developing anterior pituitary is in close juxtaposition with the ventral diencephalon (developing hypothalamus) dorsally as well as the oral ectoderm ventrally. By E16.5 each of the five hormone-producing cell types of the anterior lobe—corticotropes, thyrotropes, somatotropes, lactotropes, and gonadotropes—have differentiated and produce their characteristic hormone products.

C. Appearance of Hormones

The transcripts for the hormone products of specialized pituitary cells are considered to be the major differentiation markers. The receptors for some of the hypothalamic-releasing hormones, that is, *Gnrhr* and *Ghrhr,* are expressed in a cell-specific manner, but their transcripts appear at approximately the same time as the hormone transcripts. Identification of additional cell-specific markers would be an advance for understanding pituitary development.

During mouse hypothalamic development, vasopressin gene expression is observed first at day 13.5 in the supraoptic nucleus (SON) and at day 14.5 in the paraventricular nucleus (PVN) (Jing *et al.,* 1998). Oxytocin is expressed in the PVN at day 15.5, and in the SON at day 18.5. By day 15.5 the pattern of vasopressin and oxytocin expression within the PVN and SON matches the general adult pattern of cell-specific expression within the PVN and SON. Vasopressin is also expressed in the dorsomedial suprachiasmatic nucleus of the adult rodent (Young and Zoeller, 1987).

Transcription of each of the hormones of the anterior and intermediate lobes occurs in a temporal and spatial pattern (Japon *et al.,* 1994; Simmons *et al.,* 1990). *Cga* transcripts are detected throughout the oral ectoderm that is fated to become Rathke's pouch at E9.5–11.5. As these transcripts become restricted to the rostral tip of the developing organ, POMC transcription is initiated dorsally to CGA in the ventromedial zone at E12.5. β-endorphin expression is also discernable in the anterior pituitary at E12.5. Two spatially distinct populations of thyrotropes develop within the anterior pituitary (Drolet *et al.,* 1991). The first to appear is in the rostral tip of *Cga*-positive cells (presumptive *pars tuberalis*) at E14.5, but the majority of thyrotropes appear in the caudomedial area of the gland at E16.5 (presumptive *pars distalis*) (Fig. 2). The relationship between these two TSHβ positive cell populations is not clear, but the thyrotropes in the caudomedial region are thought to be the functional population. GH and PRL appear in the caudomedial region dorsal to TSHβ at E15.5, but PRL levels remain low until after birth (Hooghe-Peters *et al.,* 1988). LHβ and FSHβ transcripts appear in quick succession (E16.5 and E17.5, respectively) in the most ventral aspect of the gland. In the intermediate lobe POMC-expressing cells appear at E14.5 and dynorphin, met-enkephalin, and β endorphin appear at E12.5 (Elkabes *et al.,* 1989).

D. Timing of Cell Proliferation and Vascularization

There appear to be two initial waves of cell proliferation in the developing pituitary. The first occurs at E12–14.5 and the second at E16.5–18.5 (Ikeda and Yoshimoto, 1991). Colabeling studies with bromodeoxyuridine (BrdU) and immunohistochemical staining have demonstrated that

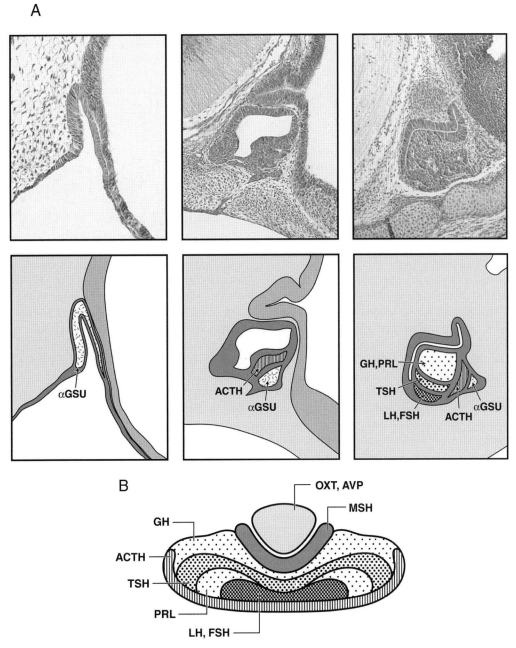

Figure 2 Morphology of pituitary development. (A) Sagittal view of pituitary primordia at E10.5, E12.5, and E16.5 in brightfield (top) and in diagrammatic form (bottom). At E10.5 the oral ectoderm (dark gray) has invaginated to produce Rathke's pouch and expresses *Cga* (αGSU) throughout the pouch. By E12.5 the neuralectoderm at the base of the diencephalon has evaginated (light gray) and is close to the side of the pouch that will become the intermediate lobe. The prospective anterior lobe is detached from the roof of the mouth and separated from it by the proliferation of cartilage. The size of the anterior lobe has increased through cell proliferation, and transcripts for *Pomc*, the gene encoding the precursor for ACTH, appear dorsal to the region expressing *Cga*. By E16.5 additional cell proliferation has occurred and each of the hormone transcripts is detectable in a spatially restricted pattern along the dorsal-ventral axis. Within the caudomedial region GH and PRL appear most dorsally, TSH is medial and the gonadotropins, LH and FSH, are located most ventrally. (B) Coronal view of the developing pituitary at E16.5 reveals the spatially restricted pattern of hormone transcripts. At this stage the posterior pituitary (light gray) is located dorsal to the intermediate (dark gray) and anterior (patterned) lobes. The posterior pituitary contains the nerve terminals that secrete oxytocin (OXT) and vasopressin (AVP), the intermediate lobe produces MSH, and the anterior lobe produces GH, PRL, TSH, LH and FSH, and ACTH. The spatial distinctions in anterior pituitary cell populations are lost as the animal matures.

the majority of the dividing cells have not yet accumulated significant amounts of hormones. This aspect of normal differentiation may explain why some immortalized cell lines make low levels of hormone or lack some of the terminal differentiation markers (Akerblom *et al.*, 1990; Alarid *et al.*, 1996; Lew *et al.*, 1993; Windle *et al.*, 1990).

Vascularization of the anterior pituitary gland and development of its portal blood supply from the hypothalamus occur near the onset of pituitary differentiation marker expression (Fig. 1). Portal vessels extend from the oral plexus (Kaufman, 1992). The long portal vessels originate from the median eminence and upper portion of pituitary stalk and supply the rostral regions of the anterior pituitary. The short portal vessels connect caudal areas of the anterior lobe with the posterior lobe.

The superior and anterior regions of the adenohypophysis are vascularized at E17 in the mouse, and the adult pattern is achieved by postnatal day 5 (Dearden and Holmes, 1976). More complete studies in the rat (Szabo, 1987; Szabo and Csanyi, 1982), indicate that the vascular connections are established between the median eminence and the anterior pituitary gland at E12. The primary portal veins appear at E13 and the vascularization of the *pars distalis* is visible on E15. The hypothalamic-adenohypophyseal portal circulation are functional by E13. In general, angiogenesis results in stimulation of endothelial cells to proliferate, migrate, and differentiate into blood vessels. These steps can be induced by vascular endothelial growth factor (VEGF), which mediates fibroblast growth factor 2 (FGF2)-induced angiogenesis (Gerwins *et al.*, 2000; Seghezzi *et al.*, 1998). FGFs are also produced in regions surrounding the developing pituitary and have important developmental roles. This suggests that there may be interactions between the process of vascularization and cell specialization. It is intriguing that lactotropes predominate in highly vascular areas of the anterior pituitary, whereas somatotropes are more distant from vascular areas (Sasaki and Iwana, 1988). It is also noteworthy that VEGF and FGF are both produced by the folliculo-stellate cells of the anterior pituitary (Hentges *et al.*, 2000; Lloyd, 1993). These cells have distinct processes that extend between the hormone-producing cells. They do not contain any of the major pituitary hormones and have no secretory granules. They have been attributed supporting and paracrine roles for production of anterior pituitary hormones because of the growth factors and cytokines that they produce. The role of these cells in normal pituitary development is not yet clear.

Estrogen may have an important role in pituitary angiogenesis. Estrogen receptor (ER) α and β are expressed in the anterior pituitary of the rat beginning at E17 and E12, respectively, suggesting that ERβ might be involved in cytodifferentiation (Nishihara *et al.*, 2000). *Pit1* expression is regulated by estrogen, and *Pit1*^{dw} mutant mice exhibit retarded development of endothelial cells between E15 and

E18 (Wilson and Christensen, 1983). Estrogen also induces prolactinomas and neovascularization in Fisher 344 rats. The increased vascularization is accompanied by increases in VEGF and the VEGF receptor (Banerjee *et al.*, 1997; Heaney *et al.*, 1999). Estrogen also induces the pituitary tumor transformining gene (*Pttg*), coincident with increases in FGF and VEGF (Heaney *et al.*, 1999). These observations suggest that VEGF-related proteins, FGFs, and estrogen are important in the development of normal pituitary vascularization, which is a critical aspect of organ function. In addition these normal developmental processes may be recapitulated in the development of adenomas and tumors.

E. Inducing Factors from the Hypothalamus and Surrounding Mesenchyme

The development of the pituitary gland and the hypothalamus are codependent (Parks *et al.*, 1997; Treier and Rosenfeld, 1996). Removal of the midline of the anterior neural ridge, the tissue that will give rise to the pituitary, at the 2- to 4-somite stage of the chick embryo results in failure of the entire pituitary to form (ElAmraoui and Dubois, 1993). This is mimicked in mice mutated for *Titf1*, a gene that is expressed in the developing brain, but not in the pituitary (Takuma *et al.*, 1998). The failure of brain development in *Titf1* mutants results in pituitary aplasia, confirming the requirement for inductive interactions between the oral and neural ectoderm and identifying at least one of the transcription factors that is important for this process.

There is a critical developmental period during which the hypothalamus and pituitary are competent to send and receive signals. These time windows have been explored in amphibians and birds, and are beginning to be examined in the mouse (Gleiberman *et al.*, 1999; Kawamura and Kikuyama, 1995). The existence of species differences in these timing events will become more clear as the timing of molecular marker expression is worked out in species other than mice. Each of the anterior pituitary cell types is able to differentiate *in vitro* from rat pituitary anlage isolated at the time the pouch is budding off the oral ectoderm (E12–13, 27-somite stage) (Dubois and Hemming, 1991; Nemeskery *et al.*, 1976). These studies indicate that no additional inductive hypothalamic signals or cell migration is required beyond this stage for cell specification in the anterior lobe of rodents.

Signaling molecules secreted by the ventral diencephalon (dorsal signals) and mesenchymal cells ventral to the pituitary are involved in the early patterning of Rathke's pouch (Table II) (Ericson *et al.*, 1998; Gleiberman *et al.*, 1999; Takuma *et al.*, 1998; Treier *et al.*, 1998). The dorsal signals include bone morphogenetic protein 4 (BMP4), fibroblast growth factor 8 (FGF8), and WNT5A (Ericson *et al.*, 1998; Takuma *et al.*, 1998; Treier *et al.*, 1998). The ventral signals include BMP2 and sonic hedgehog (SHH). These opposing

Table II   Signaling Molecules in Pituitary Development

Signaling molecule	Directionality of signal	Timing of expression	Localization of expression
BMP4	Dorsal	E8.5–13.5	Ventral diencephalon, infundibulum
FGF8	Dorsal	E9.25–14.5	Infundibulum
WNT5A	Dorsal	E9.5–14.5	Widely in diencephalon
WNT4	Dorsal	E9.5–14.5	Oral ectoderm, Rathke's pouch
BMP2	Ventral	E10.5–12.0	Mesenchymal cells ventral to pituitary, Rathke's pouch
SHH	Ventral	E9.5–10.5	Oral ectoderm adjacent to Rathke's pouch

Source: Data compiled from Ericson *et al.* (1998) and Treier *et al.* (1998).

signals are integrated to restrict the pattern of early, critical transcription factors, including *Isl1, Lhx3,* and *Gata2.*

The oral ectoderm expresses *Shh,* except in the region destined to become Rathke's pouch (Treier *et al.,* 1998). This striking expression pattern suggests that *Shh* is important for creating a boundary that defines the pouch structure. Ventral signals from the notochord may also contribute to the invagination of the oral ectoderm to form the rudimentary pouch (Gleiberman *et al.,* 1999). Genes of the *Gli* gene family are responsive to *Shh* signaling and have been shown to be important for normal pituitary development in zebrafish and mice, lending support to the idea that *Shh* signaling is an important part of pituitary patterning (Karlstrom *et al.,* 1999; Park *et al.,* 2000).

Both BMP2 and BMP4 are important for pituitary patterning. BMP2 is expressed by the mesenchymal cells ventral to the pituitary. Pituitary explants exposed to BMP2 express *Isl1, Cga,* and *Tshb* (Ericson *et al.,* 1998). BMP2 is an effective inducer of differentiation only for pouches younger than E11.5, indicating that BMP2 responsiveness is lost during development. BMP4-deficient mice exhibit failure of the oral ectoderm to thicken and form Rathke's pouch (Takuma *et al.,* 1998). Ectopic expression of the BMP antagonist, Noggin, in the oral ectoderm and pouch of transgenic mice leads to an arrest in pituitary development shortly after pouch formation, confirming the importance of BMP signaling (Treier *et al.,* 1998).

Many FGFs and FGF receptors are expressed in the pituitary gland and ventral aspect of diencephalon. The involvement of the *Fgf* gene family in pituitary development is evident from analysis of FGF receptor mutants. Among *Fgf2* receptor isoforms, *Fgfr2* (IIIb) is mainly expressed in epithelia, and has four known ligands (FGF1, 3, 7, and 10). Mice deficient in *Fgfr2* (IIIb) lack the anterior pituitary gland, have a poorly formed infundibulum, and exhibit massive apoptosis, indicating that FGFs act on pituitary gland in early embryogenesis (De Moerlooze *et al.,* 2000).

Analysis of FGF expression patterns and phenotypes of genetically engineered mice suggests that several FGF genes may have roles in Rathke's pouch development but functional redundancy probably minimizes the impact of indi-

vidual FGF gene mutations. FGF10 is a known ligand for FGFR2 (IIIb), and it is strongly expressed in the infundibulum (De Moerlooze *et al.,* 2000). Targeted mutations in *Fgf10* disrupted formation of lungs and limbs (Sekine *et al.,* 1999) and pituitary abnormalities have been observed also (Ohuchi *et al.,* 2000). *Fgf2* and *Fgf4* are found in various pituitary adenomas, especially in prolactinomas, and stimulate transcription of rat prolactin, indicating that they might be involved in the direct regulation of hormone production *in vivo* (Atkin *et al.,* 1993; Schweppe *et al.,* 1997). *Fgf2*-deficient mice, however, have reduced neuronal density in the motor cortex, but no obvious pituitary deficits (Ortega *et al.,* 1998). These observations suggest that other *Fgfs* expressed in the infundibulum may compensate for the loss of *Fgf2* function.

Fgf8 is strongly expressed in the infundibulum from E9.25 to E14.5, and it is important for both the specification and differentiation of dorsal pituitary cell types and the inhibition of ventral cell type determination (De Moerlooze *et al.,* 2000; Takuma *et al.,* 1998; Treier *et al.,* 1998). However, FGF8 is not a ligand for *Fgfr2* (IIIb), suggesting the presence and importance of other FGF receptors in the pouch. The importance of FGF8 in pituitary development is evident from analysis of thyroid-specific enhancer binding protein (*Titf1*) mutant mice. *Titf1* is expressed in the diencephalon, not the pituitary. In *Titf1* mutant mice a rudimentary pouch initially forms, but development is arrested and the pouch undergoes apoptosis (Takuma *et al.,* 1998). The *Titf1* −/− mice fail to express FGF8, suggesting that *Titf1* regulates FGF8 expression in the infundibulum either directly or indirectly, and that FGF8 is important for pouch survival and expansion. FGF8 is sufficient for activation of the early pituitary transcription factor, *Lhx3,* in the pouch and inhibiting expression of *Isl1* in the dorsal part of the pouch (Ericson *et al.,* 1998). Thus, FGF8 is a dorsalizing signal that begins to pattern the pituitary by initiating domains of *Isl1* and *Lhx3* expression. In addition, FGF8 signaling appears to maintain cells in a proliferative state, providing undifferentiated cells to populate the pituitary.

Wnt signaling molecules are also emerging as contributors to pituitary development. Loss of WNT4 leads to a hypocellular pituitary affecting all cell types except cortico-tropes (Treier *et al.,* 1998). WNT5A is expressed broadly throughout the ventral diencephalon from E9.5 to E14.5, but its role in pituitary development is unknown (Yamaguchi *et al.,* 1999).

F. Integration of the Signaling Molecules and Transcription Factor Activation

Dorsal and ventral signals are integrated in the develop-ing pituitary to produce discrete domains of transcription factor expression. Early in pituitary development (E9.0) the transcription factors *Hesx1, Pitx1,* and *Pitx2* are expressed broadly in the oral ectoderm, while the expression of another set of genes, *Lhx3, Isl1, Pax6,* and *Six3,* is restricted to the developing pouch (Dasen and Rosenfeld, 1999). Later in de-velopment (E11.0) transcription factor expression within the pituitary becomes restricted into domains reflecting the dor-sal and ventral positions. The transcription factors *P-Frk1, Isl1, Gata2,* and *Brn4* are expressed more ventrally than *Prop1, Pax6, Six3,* and *Nkx3.1.*

The mechanistic links that integrate the activation of tran-scription factors with patterns of signaling molecules are currently being elucidated. BMP4 is required for initial *Isl1* expression in oral ectoderm cells that will form Rathke's pouch (Ericson *et al.,* 1998). FGF8 is necessary for the acti-vation of *Lhx3* expression, which is essential for further pi-tuitary differentiation. Dorsal FGF8 signaling is integrated with ventral BMP2 signaling to restrict the expression pat-terns of both *Isl1* and *Lhx3* to discrete patches of cells within the pituitary. The ventrally located *Isl1*-expressing cells later express *Cga* and *Tshb,* and their location suggests they be-come the rostral tip thyrotrope population. Later, cortico-trope differentiation is restricted by *Lhx3* to an intermediate region of the pouch near the rostral tip thyrotropes due to opposing gradients of secreted factors from the infundibu-lum and ventral mesenchymal cells. BMP2 plays a role in initiating a gradient of the transcription factor *Gata2,* which is critical for differentiation of both gonadotropes and thy-rotropes (Dasen *et al.,* 1999). Hence, signaling from adjacent tissues is temporally dynamic and crucial for multiple steps in early pituitary gland organogenesis.

G. Genetic Alterations That Demonstrate the Roles of Specific Genes

Human genetic diseases and mutations in mice have dem-onstrated the importance of many genes in pituitary devel-opment. Diseases affecting the intermediate and posterior pituitary are fewer in number than those affecting the ante-rior pituitary. Hyperplasia of the intermediate or posterior pituitary has not been described, although it is a common finding for the anterior lobe. Defects in vasopressin synthe-sis or function present as central diabetes insipidus or the

syndrome of inappropriate antidiuretic hormone (SIADH). There are reports of autosomal dominant familial *diabetes insipidus,* with mutations in the vasopressin gene (Calvo *et al.,* 1998; Gagliardi *et al.,* 1997; Rutishauser *et al.,* 1996). Diabetes insipidus can also be present in individuals with Langerhans-cell histiocytosis (Dunger *et al.,* 1989). One case report described a family with combined hypopituita-rism and central diabetes insipidus (Yagi *et al.,* 1994). Hy-pothalamic hamartoblastoma and hypopituitarism are part of the Pallister-Hall syndrome, also characterized by multiple congenital anomalies including defects of the limb and geni-tourinary tract (Pallister *et al.,* 1989).

Several genes critical for hypothalamic development have recently been discovered (Burbach *et al.,* 1998). The tran-scription factor genes *Otp, Sim1,* and *Brn2* are crucial in di-recting terminal differentiation of the PVN and SON, and for maintaining the neuronal cell lineages that secrete oxytocin and vasopressin (Acampora *et al.,* 1999; Michaud *et al.,* 1998; Nakai *et al.,* 1995; Schonemann *et al.,* 1995). Brain 2 (*Brn2*), a POU-homeodomain transcription factor, is impor-tant for the terminal differentiation and survival of the PVN and SON, and it is critical for *CRH* production. Netrin-1 and *Deleted in Colon Cancer* (DCC) are also essential for normal hypothalamic development. Mice with mutant netrin-1 and DCC exhibit ectopic, ventromedially located VP and OT neurons in the hypothalamus, consistent with abnormalities in axonal pathfinding and neuronal position during develop-ment (Deiner and Sretavan, 1999). Finally, the mouse *Mf3* (*Foxb1*) gene, also known as *Fkh5* and *HFH-e5.1,* encodes a winged helix/forkhead transcription factor. *Mf3⁻ᐟ⁻* females cannot eject their milk supply to feed their pups, a nursing defect that is correctable by interperitoneal injections of oxytocin, suggesting that *Mf3* is required for normal hypo-thalamus development, postnatal growth, and lactation (La-bosky *et al.,* 1997).

Brain 4 (*Brn4*) is a member of the POU-homeodomain gene family that is widely expressed in the developing cen-tral nervous system, but is subsequently restricted to distinct regions, including hypothalamic nuclei, the otic vesicle, and Rathke's pouch (Mathis *et al.,* 1992; Ryan and Rosenfeld, 1997). *Brn4* expression is reduced in mice with hypopituita-rism due to mutations in *Prop1* (Sornson *et al.,* 1996). Inac-tivating mutations in *Brn4* cause auditory deficits in humans and mice, but pituitary defects have not been reported (Mi-nowa *et al.,* 1999; Phippard *et al.,* 1999; Piussan *et al.,* 1995). The overlapping expression of *Brn4* and *Brn2* sug-gests that these genes may compensate for one another in hypothalamic or pituitary development.

Significant overlap exists in diseases affecting both the pi-tuitary and the hypothalamus (Dattani and Robinson, 2000). This overlap is not unexpected given the reciprocal induc-tive forces that act to stimulate development of both tissues (Parks *et al.,* 1997; Treier and Rosenfeld, 1996). Ectopic placement of posterior pituitary tissue, detected by head magnetic resonance imaging, is the most common finding in humans with clinical signs and symptoms of anterior pituitary dysfunction (Chen *et al.*, 1999; Denton *et al.*, 1996). *Titf1*, *Gsh1*, and *Brn2* are involved in hypothalamic development and have a secondary effect on pituitary development, but undoubtedly more genes in this category have yet to be identified.

The thyroid transcription factor 1 gene, *Titf1*, is normally expressed in regions including the brain and posterior pituitary. Targeted disruption of *Titf1* leads to pituitary aplasia, demonstrating the importance of this homeodomain factor in the formation of Rathke's pouch (Takuma *et al.*, 1998). Mice with a targeted deletion of the homeobox gene *Gsh1* lack GHRH, which leads to a dramatic reduction in both somatotropes and lactotropes and reduction in LH (Li *et al.*, 1996).

Mammalian pituitary development is not essential for viability to term (Jost, 1954), but pituitary defects that produce adrenal insufficiency are lethal after birth. Congenital absence of the anterior pituitary gland is rare in humans and results in neonatal lethality if untreated. Because pituitary aplasia is lethal, spontaneous mutants in mice have not contributed to identification of the genes involved in early pituitary development. However, many genes required for early pituitary patterning are being identified by other methods. Many of the genes affecting pituitary development very early also affect the development of other body structures or organs. These genes include *Isl1*, *Hesx1*, *Pitx1* and *Pitx2*, *Lhx3* and *Lhx4*, and *Pax6*. Pleiotropic effects of genes regulating early pituitary patterning are not unexpected because of the common origin of the anterior pituitary, nasal cavity, olfactory placode, the hypothalamus, posterior pituitary, optic vesicles and ventral forebrain from the anterior neural ridge (Couly and Le Douarin, 1985, 1987, 1988). Indeed, many of these genes affect more than one of these anterior structures.

Pitx1 and *Pitx2* encode proteins that are members of the bicoid-related subclass of homeodomain proteins (Gage *et al.*, 1999b). Both genes are expressed in the branchial arches, eye, teeth, and hindlimb mesenchyme, although each gene has some unique sites of expression. Homozygotes for *Pitx1* null alleles exhibit a multitude of defects including lethal cleft palate, hindlimb defects, and mild pituitary alterations (Lanctot *et al.*, 1997; Szeto *et al.*, 1999). Each of the anterior pituitary cell types becomes specified in *Pitx1* null mice, but expansion of the gonadotrope population is reduced significantly and thyrotropes are slightly reduced in number (Szeto *et al.*, 1999). There is little or no effect on somatotropes, and corticotropes may be increased slightly. This suggests that *Pitx1* is required for the regulated expansion of committed cell types or for partitioning of the precursor cells among the various lineages.

Haploinsufficiency for *PITX2* causes Rieger syndrome in humans (Semina *et al.*, 1996). This disorder is typified by varied severity of eye, tooth, and craniofacial defects, as well as umbilical hernia. Rieger syndrome is genetically heterogeneous. Some families have been shown to have short stature due to failure of the pituitary to release growth hormone in response to stimulation (Feingold *et al.*, 1969; Jorgenson *et al.*, 1978; Sadeghi-Nedjad and Senior, 1974; Semina *et al.*, 1997), but none of these families has been shown to have *PITX2* mutations to date. The critical role of *Pitx2* in development of the pituitary and other organs has been demonstrated by gene targeting (Gage *et al.*, 1999a). *Pitx2* null mutant mice develop a small, definitive pouch but no pouch expansion. Expression analysis of the mutants suggests that *Pitx2* is critical for initiation or maintenance of *Hesx1* transcription but has little influence on *Lhx3* expression (Fig. 3). Recently, we have demonstrated that pituitary cell specification is particularly sensitive to lowered levels of PITX2 (H. Suh and S. A. Camper, unpublished). Homozygotes for a hypomorphic allele of *Pitx2* fail to activate *Gata2* expression in the caudomedial area, and *Egr1*, *Lhb*, and *Fshb* transcripts are not detectable, suggesting that gonadotropes are absent. *Pit1* is also reduced in *Pitx2* hypomorphic mice, as are *Gh* and *Tshb*. Taken together, these observations suggest that anterior pituitary cytodifferentiation is critically dependent on the level of PITX2 and that lesions in *PITX2* could result in severe short stature and infertility.

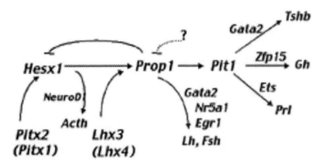

Figure 3 Genetic hierarchy. *Pit1* is required for *Tshb*, *Gh*, and *Prl* expression and for commitment and expansion of the cells that produce these three hormones. *Pit1* acts synergistically with other transcription factors to generate cell-specific expression within the *Pit1* lineage. The genes encoding *Pit1* interacting factors are *Gata2*, *Zfp15*, and *Ets* for thyrotropes, somatotropes, and lactotropes, respectively. *Prop1* is required for activation of *Pit1* but *Pit1* is not required for repression of *Prop1* (Gage, 1996a) (Cushman *et al.*, 2001). *Prop1* deficiency results in reduction of gonadotropin gene expression (*Lh* and *Fsh*). Gata2, Nr5a1 (SF1), and Egr1 are important for final differentiation of gonadotrope cells. Corticotropes apparently constitute a separate cell lineage because *Pomc* (ACTH) is expressed in many mutants that fail to support differentiation into other pituitary cell types. For example, *Lhx3* is not required for expression of ACTH. Homeobox genes from the same family have overlapping functions early in pituitary differentiation. These pairs include *Lhx3* and *Lhx4* and also *Pitx1* and *Pitx2*. The gene shown in parentheses has the more modest effect on pituitary development of the pair. *Hesx1* has variable effects on pituitary development, but formation of Rathke's pouch is affected, suggesting that it acts early in the pathway. *Hesx1* is normally extinguished at the time *Prop1* expression peaks, and it fails to be extinguished in *Prop1*-deficient mice.

Epistasis analysis of *Pitx1* and *Pitx2* reveals overlapping functions essential for viability (P. J. Gage, J. Drouin, and S. A. Camper, unpublished). Double heterozygotes for null alleles of each gene are not viable, but double heterozygotes for the null allele of *Pitx1* and hypomorphic allele of *Pitx2* are viable and normal. Intercross of these mice demonstrated a dosage-sensitive effect of the two genes on both formation and expansion of the pouch. The early activation and broader expression of *Pitx1* and *Pitx2* relative to *Lhx3* and *Lhx4* suggests that the *Pitx* genes may be upstream regulators of *Lhx3* and/or *Lhx4*. Dependency of *Lhx3* expression on *Pitx1* was suggested in cell culture experiments but not substantiated in mouse knock-outs (Szeto *et al.*, 1999; Tremblay *et al.*, 1998). Perhaps *Pitx1* and *Pitx2* have functional overlap in maintaining *Lhx3* expression, and compensation in cell culture is hampered by inadequate levels of PITX2. This issue will be resolved in analysis of *Pitx1* and *Pitx2* double mutants.

Otx1 is a homeodomain transcription factor with very similar DNA-binding specificity to the *Pitx* family (Acampora *et al.*, 1998). It is expressed in the pituitary gland from birth through adulthood and is able to transctivate *Gh*, *Cga*, *Lhb*, and *Fshb*. A role for *Otx1* in somatotrope and gonadotrope function was demonstrated by analysis of knock-out mice that exhibit transient dwarfism and hypogonadism at the prepubescent stage. GH, FSH, and LH are reduced transiently and selectively in *Otx1*-deficient mice with the expected consequences of reduced growth and hypogonadism. By 3 months of age the *Otx1*-deficient mice exhibited catch-up growth and normal gonadal function, suggesting that the role of *Otx1* is most critical during youth and young adulthood. Perhaps the *Pitx* gene family compensates for lack of *Otx1* at other times.

Hesx1, or *Rpx*, is important for the development of all five anterior pituitary cell types, in addition to other structures. *Hesx1* is expressed in anterior structures during gastrulation, and later its expression becomes restricted to Rathke's pouch. Downregulation of *Hesx1* expression at E13.5 is thought to be important for cellular differentiation (Hermesz *et al.*, 1996). This idea is supported by the observation that *Hesx1* expression is persistent in *Prop1*$^{df/df}$ mice, which have pituitary hypoplasia and multiple pituitary hormone deficiencies (Gage *et al.*, 1996a). HESX1 has amino acid similarity with the engrailed repressor domain and can antagonize *Prop1* function (Sornson *et al.*, 1996). Disruption of *Hesx1* results in midline defects that include aberrant pouch formation and septo-optic dysplasia in humans and mice (Dattani *et al.*, 1998). The phenotype of *Hesx1* homozygous mutant mice varies from embryonic lethal to adult viable, and a fraction of heterozygotes is affected, suggesting that genetic and environmental factors and/or stochastic events (chance) have a strong influence on the phenotype. The effect of *Hesx1* deficiency on transcription of other pituitary developmental control genes has not yet been reported, but

this information will be useful in interpreting the position of *Hesx1* in the genetic hierarchy of pituitary development (Fig. 3).

The LIM-homeodomain genes, *Isl1*, *Lhx3*, and *Lhx4*, have been demonstrated to have important roles in pituitary development. *Isl1* is initially expressed throughout Rathke's pouch, but becomes restricted to the ventral domain where *Cga* is expressed. *Isl1*-deficient mice form a rudimentary pouch that fails to differentiate into any of the five specialized cell types of the anterior pituitary, demonstrating the critical importance of *Isl1* (Takuma *et al.*, 1998). *Lhx2* can transcriptionally activate pituitary-specific genes (Roberson *et al.*, 1994). *Lhx2* knock-out mice have defects in eye and forebrain development, but pituitary development has not been described, leaving the role of this gene in pituitary development unclear (Porter *et al.*, 1997).

Lhx3 is expressed in the developing Rathke's pouch. Homozygotes with a targeted deletion of *Lhx3* are stillborn or die within 24 hr of birth (Sheng *et al.*, 1996). Most tissues that express *Lhx3* appear grossly normal, but lack of *Lhx3* expression causes development of both the anterior and intermediate lobes to fail. Rathke's pouch forms, but it does not undergo a characteristic thickening by E12.5, apparently due to a lack of cell proliferation. *Pomc* transcripts are detectable in *Lhx3* mutants during gestation, but the domain of expression is reduced, suggesting that the corticotrope lineage is specified normally but fails to proliferate or survive. *In situ* hybridization studies reveal that *Lhx3* is necessary for maintaining expression of *Hesx1* (Fig. 3). Two *LHX3* mutations have been shown to result in hypopituitarism in humans (Netchine *et al.*, 2000). In these patients the function of all anterior pituitary cell types is lost except corticotropes. The patients have normal adrenal function and exhibit growth insufficiency, hypogonadotropic hypogonadism, and absence of prolactin. They also exhibit limited neck rotation that may be related to the role of *LHX3* in motor neuron differentiation (Sharma *et al.*, 1998).

Targeted disruption of a related gene, *Lhx4*, results in mice with mild hypopituitarism and neonatal lethality (Sheng *et al.*, 1997). Rathke's pouch forms in *Lhx4*$^{-/-}$ mice, indicating that *Lhx4* is not required for induction of the pouch. *Lhx4* is necessary for expansion of specialized cells. Mutants have reduced levels of GH, PRL, and LH, and a slight reduction in other anterior pituitary hormones. The somatotrope deficiency in *Lhx4*$^{-/-}$ mice is partly attributable to the hypothalamic deficiency in growth hormone-releasing hormone (GHRH). This hypothesis is supported by the reduction in somatotropes observed in mice lacking functional growth hormone-releasing hormone receptor, *Ghrhr*$^{lit/lit}$ (Godfrey *et al.*, 1993; Lin *et al.*, 1993).

Lhx3 and *Lhx4* exhibit functional overlap in formation and expansion of Rathke's pouch as described for *Pitx1* and *Pitx2*. *Lhx3*$^{-/-}$, *Lhx4*$^{-/-}$ mice display an early arrest of pituitary development that is more severe than either single mutant (Sheng *et al.*, 1997). The formation of a pouch in the single mutants suggests that either gene can function to induce a definitive, although somewhat abnormal, pouch, but this process fails earlier due to reduction in the combined levels of *Lhx3* and *Lhx4*. The role of *Lhx3* is more significant than that of *Lhx4*, in that at least one functional copy of *Lhx3* is required for pouch formation (Fig. 3).

Pax6 is a paired homeodomain transcription factor that is important for development of the eye placode and certain structures of the forebrain and diencephalon including the neurons that produce GnRH. In the hindbrain and spinal cord it controls motor neuron identity by mediating a gradient of *Shh* signaling. The function of *Pax6* in the pituitary gland is somewhat controversial (Bentley *et al.*, 1999; Kioussi, 1999). *Pax6* transcripts are expressed in the dorsal aspect of Rathke's pouch (Kioussi, 1999). In *Pax6* mutant mice, the thyrotrope lineage is expanded, while the more dorsal somatotrope and lactotrope lineages are reduced. The most ventral lineage, the gonadotropes, appears to be expanded based on the marker *Nr5a1*, although differentiation of these cells is incomplete. *Pax6* expression is influenced by levels of SHH in some organs. This suggests that *Pax6* is repressed in the ventral pouch where SHH levels would be expected to be highest, and that *Pax6* represses the spread of ventralizing factors like BMP2, BRN4, ISL1, and GATA2.

Two transcription factor genes, *Prop1* and *Pit1,* appear to have roles only in the pituitary gland. The *paired*-like homeobox gene *Prop1* is expressed in the developing pituitary gland only from E10 to E15 (Sornson *et al.*, 1996). Ames dwarf mice have a partial loss of function mutation in *Prop1* that causes a failure of differentiation of somatotropes, lactotropes, and thyrotropes and a reduction in gonadotropin production (Tang *et al.*, 1993). Human PROP1 mutations confer a great deal of variability in the age of onset, degree of hypopituitarism, and number of hormones affected (Dattani and Robinson, 2000). The majority of affected individuals lack gonadotropins, which are present but reduced in *Prop1$^{df/df}$* mice. Some individuals develop ACTH deficiency as well, suggesting that PROP1 may be important for survival or differentiation of all anterior pituitary cell types (Pernasetti *et al.*, 2000). *Prop1$^{df/df}$* mice have an abnormal convoluted pouch morphology by E12.5 suggesting that *Prop1* may be important in formation of Rathke's pouch (Gage *et al.*, 1996a). *Prop1$^{df/df}$* mice exhibit reduced activation of the imprinted gene *neuronatin*, enhanced expression of *Brn4*, and no change in *Msx1, Lhx3,* or *Pitx1* expression. They fail to activate *Pit1* and extinguish *Hesx1*, placing *Prop1* in the genetic hierarchy downstream of *Hesx1* and upstream of *Pit1* (Fig. 3). There are rare clusters of *Pit1* expressing cells in Ames dwarf mutants, consistent with the idea that the *Prop1^{df}* allele has residual transcriptional activity. All of the possible pairwise combinations of somatotropes, lactotropes, and thyrotropes have been observed in these cell clusters, consistent with the hypothesis that these

three cell types arise from a common progenitor that expresses *Pit1*.

Two loss-of-function alleles of the POU-homeodomain transcription factor gene, *Pit1* (*Pit1^{dw}* and *Pit1^{dw-J}*), result in anterior pituitary hypocellularity due to failure of somatotrope, lactotrope, and thyrotrope differentiation (Camper *et al.*, 1990; Li *et al.*, 1990). These three cell types are therefore referred to as the *Pit1* lineage. In addition to its role in the proliferation and expansion of the cell types of this lineage, *Pit1* is also a direct transcriptional activator of *Gh, Prl,* and *Tshb* (Bodner *et al.*, 1988; Castrillo *et al.*, 1991; Crenshaw *et al.*, 1989; Gordon *et al.*, 1993, 1997; Ingraham *et al.*, 1988). In the absence of PIT1, differentiation to gonadotropes is enhanced, presumably because PIT1 normally acts synergistically with GATA2 to promote differentiation to the thyrotrope fate (Dasen *et al.*, 1999). Several families with mutations in *POU1F1*, the human homolog of *Pit1*, lack growth hormone, prolactin, and TSH (Parks *et al.*, 1997; Pfäffle *et al.*, 1996). Both dominant and recessive mutations have been characterized, as well as some partial loss-of-function alleles that have less severe hypothyroidism or hypopituitarism (Cohen *et al.*, 1995; Radovick *et al.*, 1992). Differing phenotypes of individuals carrying the same mutation suggest that the effect of *POU1F1* mutations is dependent on genetic background (Cohen *et al.*, 1995).

H. Synergistic Interactions between Transcription Factors during Cell Specification

Synergistic interactions between transcription factors are important in the specialization of each of the anterior pituitary cell types. Interactions have been described between factors that specify corticotrope fate, gonadotrope fate, and differentiation of multipotent *Pit1*-expressing progenitor cells to thyrotrope, somatotrope, and lactotrope fates (Borrelli, 1994; Dasen and Rosenfeld, 1999; Rhodes *et al.*, 1994).

Several transcription factors and secreted factors have been implicated in corticotrope differentiation by transcription analysis in cell culture. NEUROD1 is a helix–loop–helix transcription factor expressed by E11.5 and it is expressed in pituitary corticotropes (de la Pompa *et al.*, 1997; Poulin *et al.*, 1997). NEUROD1 can form heterodimers with other helix–loop–helix transcription factors, and its most likely partner in the pituitary gland is PAN1 (German *et al.*, 1992; Poulin *et al.*, 1997). This complex interacts with PITX proteins and NURR77 (*Nr4a1*) (Zetterstrom *et al.*, 1996). The roles of NEUROD1, PAN1, and NURR77 in corticotrope specification have not yet been delineated, but the hypothesis that they have a role in corticotrope fate is supported by the critical role of these genes in the development of other organs.

Gonadotrope differentiation involves *Pitx1, Pitx2, Otx1, Nr5a1, Egr1,* and *Prop1*. Steroidogenic factor 1 (SF1 or

Nr5a1) is expressed in the developing gonads, adrenals, pituitary, and ventromedial hypothalamus (Lala *et al.*, 1992). SF1 is required for development of the adrenals, gonads, and hypothalamus (Luo *et al.*, 1994). Neonatal lethality results unless newborns are supplemented with glucocorticoids. Pituitary LH and FSH are not detectable, but pulsatile administration of GnRH is sufficient to activate transcription of these genes, suggesting that SF1 is not absolutely essential for gonadotrope specification (Ikeda *et al.*, 1995).

The spatial expression of the zinc finger transcription factor *Egr1* during development is intriguing and unique among the known pituitary transcription factors. It is expressed in the posterior pituitary at E14.5, in each of the lobes by E18.5, and is highly expressed in the anterior lobe with lower expression in the intermediate lobe of adults (Topilko *et al.*, 1998). Inactivation of *Egr1* results in failure to produce LH, although FSH is unaffected (Lee *et al.*, 1996; Topilko *et al.*, 1998). Thus, *Egr1* is essential for one of the terminal differentiation steps of gonadotropes. FSH-specific transcription factors have not been identified, but signaling through the activin receptor II is required for FSH production (Matzuk *et al.*, 1995).

EGR1 acts synergistically with PITX1 and SF1 to enhance *Lhb* transcription, suggesting that all three transcription factors may be required for maintenance of gonadotrope function (Tremblay *et al.*, 1998, 1999). *Egr1* is a downstream effector of GnRH (Tremblay *et al.*, 1998). The GnRH receptor signals via a G-protein-linked phospholipase C/inositol triphosphate intracellular signaling pathway. Protein kinase C has been shown to enhance *Egr1*-mediated transcription of *LHb*, suggesting the possibility that *Egr1* activity may be modulated by phosphorylation. Phosphorylation alters the DNA-binding specificity of *Pit1*, suggesting that phosphorylation may be important for several different transcription factors that regulate pituitary differentiation (Kapiloff *et al.*, 1991). Consistent with this hypothesis, mice lacking protein tyrosine phosphatase σ (PTPσ) exhibit neuroendocrine dysplasia (Elchebly *et al.*, 1999). Lactotropes, thyrotropes and gonadotropes are decreased in number, while corticotropes are increased. The PTPσ is proposed to stimulate cell–cell interactions between epithelial cells of the pouch and the surrounding neuroepithelium and to promote detachment of highly proliferative cells.

Gata2 is important in the cell fate decisions that generate thyrotropes and gonadotropes. It is transcribed in the rostral tip of the developing pituitary at E10.5 (Dasen *et al.*, 1999). This domain of expression spreads throughout the most ventral region of the developing anterior pituitary beginning at E14.5. In cell culture experiments GATA2 has been shown to act synergistically with PIT1 to enhance transcription of the *Tshb* gene (Gordon *et al.*, 1997). The critical role of *Gata2* in development of the hematopoietic system makes it difficult to test its role in the pituitary directly (Tsai *et al.*, 1994). Dominant negative and ectopic expression studies in transgenic mice suggest that *Gata2* is an early participant in the cell fate choice between thyrotropes and gonadotropes (Dasen *et al.*, 1999). Coexpression of *Gata2* and *Pit1* appears sufficient to specify a thyrotrope fate, while expression of *Gata2* without *Pit1* is consistent with gonadotrope specification. Recent studies suggest that *Pitx2* is essential for activation of *Gata2*, *Egr1*, and the gonadotropin subunit genes *Lhb* and *Fshb*, placing *Pitx2* upstream of these genes in the genetic hierarchy (H. Suh and S. A. Camper, unpublished).

An immortalized cell line was derived from transgenic mice that express SV40 T-antigen under the control of the *Pit1* promoter (Lew *et al.*, 1993). These cells express *Pit1* but not the *Pit1* target genes, *Gh*, *Prl*, or *Tshb*, suggesting that PIT1 alone is insufficient for differentiation of the *Pit1* lineage. FGF2 has recently been shown to stimulate differentiation of these cells to lactotropes (Lopez-Fernandez *et al.*, 2000). FGF2 appears to stimulate the phosphorylation of endogenous *Ets* factors via the *ras* signaling pathway. The activated *Ets* factors work together with *Pit1* to promote lactotrope differentiation. FGF2 stimulation of *Ets* factors appears to be required for differentiation of lactotropes and subsequent responsiveness to nerve growth factor (NGF). FGF4 has also been shown to stimulate prolactin production (Shimon *et al.*, 1996). These studies suggest that FGF2 and/or FGF4 may be involved in lactotrope differentiation in normal mouse development, as has been shown for FGF8 (Takuma *et al.*, 1998).

Production of GH in somatotropes involves *Pit1*, *Zfp15*, and the nuclear receptors for retinoic acid (RA) and thyroid hormone (T3). *Zfp15* is necessary for GH expression in transgenic mice, suggesting that *Zfp15* may participate in the specialization of the somatotropes (Lipkin *et al.*, 1993). *Pit1*, a key transcription factor for commitment and expansion of somatotropes, acts cooperatively with the T3 and RA receptors to enhance transcription of the GH gene (Palomino *et al.*, 1998). A complete absence of thyroid hormone receptors results in a significant reduction in somatotrope cell number, confirming the importance of cooperation between PIT1 and the thyroid hormone receptors (Gothe *et al.*, 1999).

III. Expansion of Committed Cell Types

Transgene ablation studies have suggested that the majority of the anterior pituitary cell types do not depend on any other differentiated cell type within the same lobe for commitment, survival, or function. For example, expression of a *herpes simplex virus 1 thymidine kinase* transgene (*HSV1-tk*) under the control of the *POMC* promoter only results in the loss of corticotropes, indicating that these cells are neither an obligate precursor of any of the other cell types nor a source of critical inducing or maintenance factors for the other anterior pituitary cell types (Allen *et al.*, 1995). Similarly, diphtheria toxin (DT-A) ablation of gonadotropes alone, or gonadotropes and thyrotropes together, does not appear to disrupt the development of other cell types, although fewer lactotropes are present in gonadotrope ablation experiments (Burrows *et al.*, 1996; Kendall *et al.*, 1991; Markkula *et al.*, 1994; Seuntjens *et al.*, 1999). DT-A expression under the control of the *Tshβ* promoter leads to ablation of thyrotropes, somatotropes, and lactotropes (Maki *et al.*, 1994), suggesting that both somatotropes and lactotropes could be thyrotrope dependent. The cell-specific expression and developmental timing of the *Tshb* promoter, however, has not been established. Later studies have demonstrated that thyrotropes are apparently not an obligate precursor of somatotropes because thyrotrope ablation using a well-characterized *Cga* promoter driving DT-A expression leaves somatotropes intact (Burrows *et al.*, 1996).

There are two exceptions to the general observation that expansion of individual cell types is independent. The stimulatory effect of thyroid hormone on expansion of committed lactotropes suggests an indirect dependence of lactotropes on thyrotropes (Kendall *et al.*, 1995; Stahl *et al.*, 1999). Several lines of evidence support the hypothesis that lactotrope development requires somatotropes. Elegant transgene ablation studies using both DT-A and *HSV1-tk* have demonstrated that lactotropes are largely dependent on, or derived from, somatotropes. Toxin expression under the control of the *Gh* promoter results in severe reduction of both cell types (Behringer *et al.*, 1988; Borrelli *et al.*, 1989). Using the *HSV1-tk* inducible transgene system, removal of the anti-herpes drug revealed that somatotropes and lactotropes can recover and repopulate the pituitary. In these experiments lactotropes have a much longer time course of recovery than somatotropes, consistent with lactotrope differentiation via a somatotrope intermediary (Borrelli *et al.*, 1989). These studies also indicate that a stem cell population remains in the adult organ.

Hormonal control of pituitary cell proliferation is initiated during embryogenesis and continues through adulthood. The ability of anterior pituitary cells to proliferate in response to demand for a specific hormone is remarkable (Gage *et al.*, 1996b; Kendall *et al.*, 1995; Stahl *et al.*, 1999). For example, pregnancy and lactation induce a profound expansion of the lactotrope population, and hypertrophy and hyperplasia of the thyrotropes and gonadotropes occur in response to thyroid or gonadal failure (Stahl *et al.*, 1999). The capability for expansion apparently changes with age. For example, hypothyroidism in young mice causes global expansion of thyrotropes, while adult mice undergo focal hyperplasia (Kendall *et al.*, 1995). The expansion of individual cell types in adults may involve some of the same genes that regulate differentiation and proliferation during development.

Experiments in adult transgenic mice have suggested that the majority of lactotropes are postmitotic (Borrelli *et al.*, 1989; Carbajo-Perez *et al.*, 1989; Ikeda and Yoshimoto,

1991). Expansion of the lactotrope population occurs through proliferation of a somatomammotrope intermediate, a cell that expresses both GH and PRL, during neonatal life and pregnancy (Goluboff and Ezrin, 1969). This cell makes up 10–15% of the cells in the neonatal mouse pituitary (Frawley and Bloockfor, 1989) and 30–50% of the lactotropes in the adult rat (Hoeffler *et al.*, 1985). A rat anterior pituitary bipotential tumor cell line, GH_3, probably represents a somatomammotrope intermediate cell type. Treatment of GH_3 cells with GHRH causes differentiation into somatotropes and GH secretion. In contrast, treatment with nerve growth factor (NGF) promotes lactotrope differentiation and PRL production (Missale *et al.*, 1994). Lactotrope proliferation is also stimulated by transforming growth factor α (TGF-α), TGF-β, and estrogen (Lloyd, 1983; McAndrew *et al.*, 1995; Phelps and Hymer, 1983; Hentges *et al.*, 2000). The stimulatory effect of estrogen may be mediated through TGF-α. TGF-β acts by stimulating production of FGFs from folliculo-stellate cells. The FGFs probably act upstream of NGF.

Normal expansion of the lactotrope population requires thyroid hormone (Stahl *et al.*, 1999), although the mechanism of action is unknown. A variety of studies have indicated that lactotrope differentiation is also promoted by epidermal growth factor, vasoactive intestinal polypeptide, and insulin (Felix *et al.*, 1995; Goda *et al.*, 1998; Kakeya *et al.*, 2000; Woods and Porter, 1998). Many of these agents are likely to influence early cellular differentiation.

While the majority of pituitary hormones are secreted in response to positive regulation by the hypothalamus, lactotrope proliferation and prolactin secretion are under negative control by dopamine. The inhibitory effect of dopamine on lactotrope proliferation was confirmed in mice with inactivating mutations in the dopamine transporter (*Dat1*) and dopamine receptor (*Drd2*) (Bossé *et al.*, 1997; Saiardi *et al.*, 1997). *Dat1*$^{-/-}$ animals have elevated dopamine levels that cause lactotrope hypoplasia, while *Drd2*$^{-/-}$ animals have lactotrope hyperplasia.

Normal expansion of the somatotrope population requires thyroid hormone and GHRHR signaling (Mayo, 1996; Stahl *et al.*, 1999). Delivery of GHRH by the portal circulatory system from the hypothalamus to the pituitary normally results in GH secretion, enhanced *Gh* transcription, and increased proliferation of somatotrope cells. Somatotrope deficiency and growth insufficiency occur in mice and humans as a result of mutations in the growth hormone-releasing hormone receptor gene (*Ghrhr*) (Baumann and Maheshwari, 1997; Wajnrajch *et al.*, 1996). The receptor is G protein coupled such that GHRH binding causes a stimulation of adenylate cyclase and increase in intracellular cAMP levels. Conversely, constitutive activation of the GHRHR signaling pathway in humans and mice leads to somatotrope hyperplasia, GH-producing adenomas, gigantism, and acromegaly (Burton *et al.*, 1991; Mayo *et al.*, 1988; Stefaneanu and Kovacs, 1993). Close to 40% of GH-secreting tumors from acromegalic patients are thought to have somatic mutations in α_s, the GTP-binding subunit of the GHRHR associated G protein (Landis *et al.*, 1990). The critical role of GHRHR signaling in the regulation of somatotrope proliferation serves as a general model for receptor-mediated regulation of cellular proliferation. Further studies are necessary to determine whether a common mechanism exists for any of the pituitary cell types.

The majority of thyrotropes arise from *Pit1* expressing progenitor cells. When high demand for thyrotropes occurs in neonates, thyrotrope proliferation appears to occur at the expense of the somatotropes and lactotropes (Kendall *et al.*, 1995). However, in adult mice somatotropes may be recruited to produce thyrotropin, just as lactotropes apparently can arise from somatomammotropes. Thyrosomatotropes appear in response to hypothyroidism but disappear quickly when thyroid hormone is restored (Horvath *et al.*, 1990). Production of thyrotropin is stimulated by the hypothalamic peptide thyrotropin-releasing hormone (TRH). Recent studies have indicated that TRH is not necessary for thyrotrope differentiation or for TSH production (Yamada *et al.*, 1997). TRH-deficient mice have transient hypothyroidism, but an unknown compensatory mechanism results in elevated TSH, which supports catch-up growth. However, humans with TRH receptor mutations have profound hypothyroidism, suggesting that TRH receptor signaling is important for thyrotrope function. Targeted disruption of the TRH receptor gene may be necessary to assess the importance of the receptor on thyrotrope and differentiation (Collu *et al.*, 1997).

Gonadotropes develop in hypothyroid mice, but thyroid hormone is required for expansion of the cell population (Stahl *et al.*, 1999). Gonadotrope proliferation and function is also controlled by the hypothalamic peptide, gonadotropin-releasing hormone (GnRH). In mice lacking GnRH, gonadotrope cells are present but they produce very little LH and FSH (Mason *et al.*, 1986a,b). Gonadotropes are also sensitive to the level of gonadal steroids and undergo hypertrophy and hyperplasia in gonadectomized animals (Ibrahim *et al.*, 1986). These results suggest that gonadotropes require both thyroid hormone and GnRH to respond to low levels of gonadal steroids. FGF2 has also been implicated as a trophic factor for gonadotropes (Schechter *et al.*, 1995). *Nhlh2* is expressed in embryonic and adult pituitary gland and in the developing nervous system (Good *et al.*, 1997). Mutations in *Nhlh2* affect gonadal development in males more severely than mutant females. The loss of function is attributed to gonadotrope dependence on *Nhlh2;* however, the relative importance of *Nhlh2* for GnRH and gonadotropin production is not clear.

Corticotrope expansion is sensitive to several mitogenic factors although more experiments are necessary to clarify the role of each factor. Leukemia inhibitory factor (LIF) is produced in the pituitary gland and has been shown to en-

hance corticotrope differentiation (Stefana *et al.*, 1996). However, transgenic mice expressing LIF develop Rathke's cysts and exhibit hypopituitarism that affects all hormone-producing cell types (Akita *et al.*, 1997). Rathke's cysts are a common type of human pituitary adenoma, with cellular features suggesting alteration of cell fate. Cells within the cysts express some neural markers and are morphologically similar to ciliated epithelium of the airway. Analysis of LIF expression in development would be helpful for assessing the role of LIF in normal pituitary development and in corticotrope differentiation. Some cytokines stimulate ACTH secretion and POMC transcription, but their precise role in corticotrope differentiation is not clear. Interleukins 1, 2, 6, and 11 may regulate the development and function of differentiated pituitary cells in a paracrine mode (Auernhammer and Melmed, 1999). CRH is mitogenic for pituitary corticotropes, but CRH-deficient mice exhibit normal corticotrope differentiation suggesting CRH is not a critical factor (Gertz *et al.*, 1987; Muglia *et al.*, 1995).

A. Multihormonal Cells

Multihormonal cell types exist in cells in normal tissue, in established cell lines, and in pituitary adenomas. The bipotential somatomammotrope and thyrosomatotrope have been discussed. In addition to these apparent normal precursor cell intermediates, some perplexing multihormonal combinations have been observed. For example, some corticotropes coexpress ACTH in conjunction with GH, PRL, or TSH (Childs *et al.*, 1989; Ishikawa *et al.*, 1977). These multihormonal cells may represent minor alternative differentiation pathways or rare abnormal cells. Alternatively, the existence of these cells could indicate that the more specialized single hormone-producing cells are derived from a common multihormonal progenitor. However, developmental studies suggest that the specialized cells arise in distinct, spatially restricted regions. It is possible that renewal of specialized cell types occurs differently than initial specialization and/or that the specialized cells are very related, requiring only minor changes in gene expression to support expression of multiple hormones. Clearly, the nature of multihormonal cells requires further study.

IV. Conclusion

The complete set of genes involved in inducing and specifying the hypophyseal placode, which forms Rathke's pouch, remains to be identified. These early steps in pituitary patterning will be difficult to dissect with mouse genetics because many of the genes involved in early patterning have broad effects on structures arising from the anterior neural ridge. Cell-specific knock-outs and explant studies may be the approaches that are best suited to explore the earliest steps directly, although paradigms developed in other organs or other animals may be useful.

An intriguing feature of pituitary development is the initial organization of the cell types into discrete patches and the loss of this spatial organization as the organ expands. These features suggest that cell–cell contacts or extracellular matrix may provide an important role early in the differentiation process. Very little research has been done on the role of extracellular matrix in pituitary development.

Many transcription factors critical for pituitary differentiation have been identified and placed within a genetic hierarchy. A challenge for the future is to move from our simple understanding of the genetic hierarchy to demonstration of molecular interactions in transcriptional control. A challenge in establishing molecular links arises due to the paucity of cell lines that can be induced to differentiate into more than one pituitary cell type. Moreover, none of the available cell lines expresses early markers like *Prop1* or *Hesx1*. Thus, although genetics suggests a hierarchical relationship between *Hesx1*, *Prop1*, and *Pit1*, molecular links have not been firmly established. The existence of additional genes regulating intermediate steps and additional coactivators is likely. Future development of pituitary-stage-specific cDNA libraries and use of gene array technology may be useful in identifying additional target genes of the known transcription factors. Nevertheless, the pituitary stands out as an organ for which the differentiation markers are linked to known transcription factors, and the influence of specific signaling molecules on patterns of transcription factor expression has been established.

It is encouraging that mouse models for hypopituitarism and hyperpituitarism correspond well to homologous conditions in humans. In many cases discovery of genes in mice has led directly to molecular understanding of human birth defects. Some of the molecules demonstrated to have critical roles in development have been implicated in promotion of pituitary tumors and adenomas. Hopefully the future will bring additional mouse models to help understand the basis for human disease, and the mechanisms controlling the specialization and proliferation of pituitary cell types.

Acknowledgments

We have used official gene names wherever possible and refer readers to the mouse genome database for alternative names and additional references: http:/www.informatics.jax. org (Mouse Genome Database). We have also used the standard nomenclature of italics for gene names with human genes being all uppercase and mouse genes with the first letter uppercase and the remainder lowercase. Proteins appear as all uppercase without italics.

References

Acampora, D., Mazan, S., Tuorto, F., Avantaggiato, V., Tremblay, J. J., Lazzaro, D., Carlo, A. D., Mariano, A., Macchia, P. E., Corte, G., Macchia, V., Drouin, J., Brûlet, P., and Simeone, A. (1998). Transient dwarfism and hypogonadism in mice lacking *Otx1* reveal prepubescent stage-specific control of pituitary levels of GH, FSH and LH. *Development (Cambridge, UK)* **125**, 1229–1239.

Acampora, D., Postiglione, M. P., Avantaggiato, V., Di Bonito, M., Vaccarino, F. M., Michaud, J., and Simeone, A. (1999). Progressive impairment of developing neuroendocrine cell lineages in the hypothalamus of mice lacking the Orthopedia gene. *Genes Dev.* **13**, 2787–2800.

Acher, R., and Chauvet, J. (1995). The neurohypophysial endocrine regulatory cascade: Precursors, mediators, receptors, and effectors. *Front. Neuroendocrinol.* **16**, 237–289.

Akerblom, I. E., Ridgway, E. C., and Mellon, P. L. (1990). An α-subunit-secreting cell line derived from a mouse thyrotrope tumor. *Mol. Endocrinol.* **4**, 589–596.

Akita, S., Readhead, C., Stefaneanu, L., Fine, J., Tampanaru-Sarmesiu, A., Kovacs, K., and Melmed, S. (1997). Pituitary-directed leukemia inhibiotry factor transgene forms Rathke's cleft cysts and impairs adult pituitary function. *J. Clin. Invest.* **99**, 2462–2469.

Alarid, E. T., Windle, J. J., Whyte, D. B., and Mellon, P. L. (1996). Immortalization of pituitary cells at discrete stages of development by directed oncogenesis in transgenic mice. *Development (Cambridge, UK)* **122**, 3319–3329.

Allen, R. G., Carey, C., Parker, J. D., Mortrud, M. T., Mellon, S. H., and Low, M. J. (1995). Targeted ablation of pituitary pre-proopiomelanocortin cells by herpes simplex virus-1 thymidine kinase differentially regulates mRNAs encoding the adrenocorticotropin receptor and aldosterone synthase in the mouse adrenal gland. *Mol. Endocrinol.* **9**, 1005–1016.

Atkin, S. L., Landolt, A. M., Jeffreys, R. V., Diver, M., Radcliffe, J., and White, M. C. (1993). Basic fibroblastic growth factor stimulates prolactin secretion from human anterior pituitary adenomas without affecting adenoma cell proliferation. *J. Clin. Endocrinol. Metab.* **77**, 831–837.

Auernhammer, C., and Melmed, S. (1999). Interleukin-11 stimulates pro-opiomelanocortin gene expression and adrenocorticotropin secretion in corticotroph cells: Evidence for a redundant cytokine network in the hypothalamo-pituitary-adrenal axis. *Endocrinology (Baltimore)* **140**, 1559–1566.

Banerjee, S. K., Sarkar, D. K., Weston, A. P., De, A., and Campbell, D. R. (1997). Over expression of vascular endothelial growth factor and its receptor during the development of estrogen-induced rat pituitary tumors may mediate estrogen-initiated tumor angiogenesis. *Carcinogenesis (London)* **18**, 1155–1161.

Baumann, G., and Maheshwari, H. (1997). The Dwarfs of Sindh: Severe growth hormone (GH) deficiency caused by a mutation in the GH-releasing hormone receptor gene. *Acta Paediat.* **423**, 33–38.

Beamer, W. G., and Cresswell, L. A. (1982). Defective thyroid ontogenesis in fetal hypothyroid (*hyt/hyt*) mice. *Anat. Rec.* **202**, 387–393.

Behringer, R. R., Mathews, L. S., Palmiter, R. D., and Brinster, R. L. (1988). Dwarf mice produced by genetic ablation of growth hormone-expressing cells. *Genes Dev.* **2**, 453–461.

Bentley, C. A., Zidehsarai, M. P., Grindley, J. C., Parlow, A. F., Barth-Hall, S., and Roberts, V. J. (1999). *Pax6* is implicated in murine pituitary endocrine function. *Endocrine* **10**, 171–177.

Bern, H. A., and Nicoll, C. S. (1968). The comparative endocrinology of prolactin. *Recent Prog. Horm. Res.* **24**, 681–720.

Bodner, M., Castrillo, J.-L., Theill, L. E., Deerink, T., Ellisman, M., and Karin, M. (1988). The pituitary-specific transcription factor GHF-1 is a homeobox-containing protein. *Cell (Cambridge, Mass.)* **55**, 505–518.

Borrelli, E. (1994). Pitfalls during development: Controlling differentiation of the pituitary gland. *Trends Genet.* **10**, 222–224.

Borrelli, E., Heyman, R. A., Arias, C., Sawchenko, P. E., and Evans, R. M.

(1989). Transgenic mice with inducible dwarfism. *Nature (London)* **339**, 538–541.

Bossé, R., Fumagalli, F., Jaber, M., Giros, B., Gainetdinov, R. R., Wetsel, W. C., Missale, C., and Caron, M. G. (1997). Anterior pituitary hypoplasia and dwarfism in mice lacking the dopamine transporter. *Neuron* **19**, 127–138.

Burbach, J. P., van Schaick, H., Lopes da Silva, S., Asbreuk, C. H., and Smidt, M. P. (1998). Hypothalamic transcription factors and the regulation of the hypothalamo-neurohypophysial system. *Adv. Exp. Med. Biol.* **449**, 29–37.

Burrows, H. L., Birkmeier, T. S., Seasholtz, A. F., and Camper, S. A. (1996). Targeted ablation of cells in the pituitary primordia of transgenic mice. *Mol. Endocrinol.* **10**, 1467–1477.

Burrows, H. L., Douglas, K. R., Seasholtz, A. F., and Camper, S. A. (1999). Genealogy of the anterior pituitary gland: Tracing a family tree. *Trends Endocrinol. Metab.* **10**, 343–352.

Burton, F. H., Hasel, K. W., Bloom, F. E., and Sutcliffe, J. G. (1991). Pituitary hyperplasia and gigantism in mice caused by a cholera toxin transgene. *Nature (London)* **350**, 74–77.

Calvo, B., Bilbao, J. R., Urrutia, I., Eizaguirre, J., Gaztambide, S., and Castano, L. (1998). Identification of a novel nonsense mutation and a missense substitution in the vasopressin-neurophysin II gene in two Spanish kindreds with familial neurohypophyseal diabetes insipidus. *J. Clin. Endocrinol. Metab.* **83**, 995–997.

Camper, S. A., Saunders, T. L., Katz, R. W., and Reeves, R. H. (1990). The *Pit-1* transcription factor gene is a candidate for the Snell dwarf mutation. *Genomics* **8**, 586–590.

Carbajo-Perez, E., Motegi, M., and Watanabe, Y. G. (1989). Cell proliferation in the anterior pituitary of mice during growth. *Biomed. Res.* **10**, 275–281.

Castrillo, J.-L., Theill, L., and Karin, M. (1991). Function of the homeodomain protein GHF1 in pituitary cell proliferation. *Science* **253**, 197–199.

Chen, S., Leger, J., Garel, C., Hassan, M., and Czernichow, P. (1999). Growth hormone deficiency with ectopic neurohypophysis: Anatomical variations and relationship between the visibility of the pituitary stalk asserted by magnetic resonance imaging and anterior pituitary function. *J. Clin. Endocrinol. Metab.* **84**, 2408–2413.

Childs, G. V., Vestlund, K. N., and Unabia, G. (1989). Characterization of anterior pituitary target cells for arginine vasopressin: Including cells that store adrenocorticotrophin, thyrotrophin, and both hormones. *Endocrinology (Baltimore)* **125**, 554–559.

Cohen, L. E., Wondisford, F. E., Salvatoni, A., Maghnie, M., Brucker-Davis, F., Weintraub, B. D., and Radovick, S. (1995). A "hot spot" in the Pit-1 gene responsible for combined pituitary hormone deficiency: Clinical and molecular correlates. *J. Clin. Endocrinol. Metab.* **80**, 679–684.

Collu, R., Tang, J., Castagne, J., Lagace, G., Masson, N., Huot, C., Deal, C., Delvin, E., Faccenda, E., Eidne, K. A., and Van Vliet, G. (1997). A novel mechansim for isolated central hypothyroidism: Inactivating mutations in the thyrotropin-releasing hormone receptor gene. *J. Clin. Endocrinol. Metab.* **82**, 1361–1365.

Couly, G. F., and Le Douarin, N. M. (1985). Mapping of the early neural primordium in quail-chick chimeras. I. Developmental relationships between placodes, facial ectoderm, and proencephalon. *Dev. Biol.* **110**, 422–439.

Couly, G. F., and Le Douarin, N. M. (1987). Mapping of the early neural primordium in quail-chick chimeras. II. The proencephalic neural plate and neural folds: Implications for the genesis of cephalic human congenital abnormalities. *Dev. Biol.* **120**, 198–214.

Couly, G. F., and Le Douarin, N. M. (1988). The fate map of the cephalic neural primordium at the presomititc to the 3-somite stage in the avian embryo. *Development (Cambridge, UK)* **103**, Suppl., 101–113.

Crenshaw, E. B. I., Kalla, K., Simmons, D. M., Swanson, L. W., and Rosen-

feld, M. G. (1989). Cell-specific expression of the prolactin gene in transgenic mice is controlled by synergistic interactions between promoter and enhancer elements. *Genes Dev.* **3**, 959–972.

Cushman, L. J., Watkins-Chow, D. E., Brinkmeier, M. L., Raetzman, L. T., Radak, A. L., Lloyd, R. V., and Camper, S. A. (2001). Persistent *Prop1* expression delays gonadotrope differentiation and enhances pituitary tumor susceptibility. *Hum. Mole. Gene.* **10**, 1141–1153.

Dasen, J. S., and Rosenfeld, M. G. (1999). Combinatorial codes in signaling and synergy: Lessons from pituitary development. *Curr. Opin. Genet. Dev.* **9**, 566–574.

Dasen, J. S., O'Connell, S. M., Flynn, S. E., Treier, M., Gleiberman, A. S., Szeto, D. P., Hooshmand, F., Aggrawal, A. K., and Rosenfeld, M. G. (1999). Reciprocal interactions of Pit1 and GATA2 mediate signaling gradient-induced determination of pituitary cell types. *Cell (Cambridge, Mass.)* **97**, 587–598.

Dattani, M. T., and Robinson, I. C. (2000). The molecular basis for developmental disorders of the pituitary gland in man. *Clin. Genet.* **57**, 337–346.

Dattani, M. T., Martinez-Barbera, J.-P., Thomas, P. Q., Brickman, J. M., Gupta, R., Mårtensson, I.-L., Toresson, H., Fox, M., Wales, J. K. H., Hindmarsh, P. C., Krauss, S., Beddington, R. S. P., and Robinson, I. C. A. F. (1998). Mutations in the homeobox gene *HESX1/Hesx1* associated with septo-optic dysplasia in human and mouse. *Nat. Genet.* **19**, 125–133.

Dearden, N. M., and Holmes, R. L. (1976). Cyto-differentiation and portal vascular development in the mouse adenohypophysis. *J. Anat.* **121**, 551–569.

Deiner, M. S., and Sretavan, D. W. (1999). Altered midline axon pathways and ectopic neurons in the developing hypothalamus of netrin-1- and DCC-deficient mice. *J. Neurosci.* **19**, 9900–9912.

de la Pompa, J. L., Wakeham, A., Correia, K. M., Samper, E., Brown, S., Aguilera, R. J., Nakano, T., Honjo, T., Mak, T. W., Rossant, J., and Conlon, R. A. (1997). Conservation of the Notch signalling pathway in mammalian neurogenesis. *Development (Cambridge, UK)* **124**, 1139–1148.

De Moerlooze, L., Spencer-Dene, B., Revest, J., Hajihosseini, M., Rosewell, I., and Dickson, C. (2000). An important role for the IIIb isoform of fibroblast growth factor receptor 2 (FGFR2) in mesenchymal-epithelial signalling during mouse organogenesis. *Development (Cambridge, UK)* **127**, 483–492.

Denton, E. R., Powrie, J. K., Ayers, A. B., and Sonksen, P. H. (1996). Posterior pituitary ectopia and hypopituitarism—magnetic resonance appearances of four cases and a review of the literature. *Br. J. Radiol.* **69**, 402–406.

de Wied, D. (1999). Behavioral pharmacology of neuropeptides related to melanocortins and the neurohypophyseal hormones. *Eur. J. Pharmacol.* **375**, 1–11.

Drolet, D. W., Scully, K. M., Simmons, D. M., Wegner, M., Chu, K., Swanson, L. W., and Rosenfeld, M. G. (1991). TEF, a transcription factor expressed specifically in the anterior pituitary during embryogenesis, defines a new class of leucine zipper proteins. *Genes Dev.* **5**, 1739–1753.

Dubois, P., and Hemming, F. (1991). Fetal development and regulation of pituitary cell types. *J. Electron Microsc. Techn.* **19**, 2–20.

Dunger, D. B., Broadbent, V., Yeoman, E., Seckl, J. R., Lightman, S. L., Grant, D. B., and Pritchard, J. (1989). The frequency and natural history of diabetes insipidus in children with Langerhans-cell histiocytosis. *N. Engl. J. Med.* **321**, 1157–1162.

ElAmraoui, A., and Dubois, P. M. (1993). Experimental evidence for the early commitment of the presumptive adenohypophysis. *Neuroendocrinology* **58**, 609–615.

Elchebly, M., Wagner, J., Kennedy, T. E., Lanctôt, C., Michaliszyn, E., Itie, A., Drouin, J., and Tremblay, M. L. (1999). Neuroendocrine dysplasia in mice lacking protein tyrosine phosphatase σ. *Nat. Genet.* **21**, 330–338.

Elkabes, S., Loh, Y. P., Nieburgs, A., and Wray, S. (1989). Prenatal ontogenesis of pro-opiomelanocortin in the mouse central nervous system and pituitary gland: An in situ hybridization and immunocytochemical study. *Brain Res. Dev. Brain Res.* **46**, 85–95.

Ericson, J., Norlin, S., Jessell, T. M., and Edlund, T. (1998). Integrated FGF and BMP signaling controls the progression of progenitor cell differentiation and the emergence of pattern in the embryonic anterior pituitary. *Development (Cambridge, UK)* **125**, 1005–1015.

Feingold, M., Shiere, F., Fogels, H. R., and Donaldson, D. (1969). Rieger's syndrome. *Pediatrics* **44**, 564.

Felix, R., Meza, U., and Cota, G. (1995). Induction of classical lactotropes by epidermal growth factor in rat pituitary cell cultures. *Endocrinology (Baltimore)* **136**, 939–946.

Frawley, L. S., and Bloockfor, F. R. (1989). Mammosomatotropes: Presence and functions in normal and neoplastic pituitary tissue. *Endocr. Rev.* **12**, 337–355.

Gage, P. J., Brinkmeier, M. L., Scarlett, L. M., Knapp, L. T., Camper, S. A., and Mahon, K. A. (1996a). The Ames dwarf gene, *df*, is required early in pituitary ontogeny for the extinction of *Rpx* transcription and initiation of lineage specific cell proliferation. *Mol. Endocrinol.* **10**, 1570–1581.

Gage, P. J., Roller, M. L., Saunders, T. L., Scarlett, L. M., and Camper, S. A. (1996b). Anterior pituitary cells defective in the cell autonomous factor, *df*, undergo cell lineage specification but not expansion. *Development (Cambridge, UK)* **122**, 151–160.

Gage, P. J., Suh, H., and Camper, S. A. (1999a). Dosage requirement of Pitx2 for development of multiple organs. *Development (Cambridge, UK)* **126**, 4643–4651.

Gage, P. J., Suh, H., and Camper, S. A. (1999b). Review: The *bicoid*-related Pitx gene family in development. *Mamm. Genome,* **10**, pp. 197–200.

Gagliardi, P. C., Bernasconi, S., and Repaske, D. R. (1997). Autosomal dominant neurohypophyseal diabetes insipidus associated with a missense mutation encoding Gly23—>Val in neurophysin II. *J. Clin. Endocrinol. Metab.* **82**, 3643–3646.

German, M. S., Wang, J., Chadwick, R. B., and Rutter, W. J. (1992). Synergistic activation of the insulin gene by a LIM-homeodomain protein and a bsic helix-loop-helix protein: Building a functional insulin mini-enhancer complex. *Genes Dev.* **6**, 2165–2176.

Gertz, B. J., Contreras, L. N., McComb, D. J., Kovacs, K., Tyrrell, J. B., and Dallman, M. F. (1987). Chronic administration of corticotropin-releasing factor increases pituitary corticotroph number. *Endocrinology (Baltimore)* **120**, 381–388.

Gerwins, P., Skoldenberg, E., and Claesson-Welsh, L. (2000). Function of fibroblast growth factors and vascular endothelial growth factors and their receptors in angiogenesis. *Crit. Rev. Oncol. Hematol.* **34**, 185–194.

Gleiberman, A. S., Fedtsova, N. G., and Rosenfeld, M. G. (1999). Tissue interactions in the induction of anterior pituitary: Role of the ventral diencephalon, mesenchyme, and notochord. *Dev. Biol.* **213**, 340–353.

Goda, H., Sakai, T., Kurosumi, M., and Inoue, K. (1998). Prolactin-producing cells differentiate from G0/G1 arrested somatotrophs in vitro: An analysis of cell cycle phases and mammotroph differentiation. *Endocr. J.* **45**, 725–735.

Godfrey, P., Rahal, J. O., Beamer, W. G., Copeland, N. G., Jenkins, N. A., and Mayo, K. E. (1993). GHRH receptor of *little* mice contains a missense mutation in the extracellular domain that disrupts receptor function. *Nat. Genet.* **4**, 227–231.

Goluboff, L., and Ezrin, C. (1969). Effect of pregnancy on the somatotroph and the prolactin cell of the human adenohypophysis. *J. Clin. Endocrinol. Metab.* **29**, 1533–1538.

Good, D. J., Porter, F. D., Mahon, K. A., Parlow, A. F., Westphal, H., and Kirsch, I. R. (1997). Hypogonadism and obesity in mice with a targeted deletion of the *Nhlh2* gene. *Nat. Genet.* **15**, 397–401.

Gordon, D. F., Haugen, B. R., Sarapura, V. D., Nelson, A. R., Wood, W. M., and Ridgway, E. C. (1993). Analysis of Pit-1 in regulating mouse TSHβ promoter activity in thyrotropes. *Mol. Cell. Endocrinol.* **96**, 75–84.

Gordon, D. F., Lewis, S., Haugen, B. R., James, R., McDermott, M., Wood, W. M., and Ridgeway, E. C. (1997). Pit-1 and GATA-2 interact and functionally cooperate to activate the thyrotropin β-subunit promoter. *J. Biol. Chem.* **272**, 24339–24347.

Gothe, S., Wang, Z., Ng, L., Kindblom, J. M., Barros, A. C., Ohlsson, C., Vennstrom, B., and Forrest, D. (1999). Mice devoid of all known thyroid hormone receptors are viable but exhibit disorders of the pituitary-thyroid axis, growth, and bone maturation. *Genes Dev.* **13**, 1329–1341.

Heaney, A. P., Horwitz, G. A., Wang, Z., Singson, R., and Melmed, S. (1999). Early involvement of estrogen-induced pituitary tumor transforming gene and fibroblast growth factor expression in prolactinoma pathogenesis. *Nat. Med.* **5**, 1317–1321.

Hentges, S., Boyadjieva, N., and Sarkar, D. K. (2000). Transforming growth factor-beta 3 stimulates lactotrope cell growth by increasing basic vibroblast growth factor from folliculo-stellate cells. *Endocrinology (Baltimore)* **141**, 859–867.

Hermesz, E., Mackem, S., and Mahon, K. A. (1996). *Rpx:* A novel anterior-restricted homeobox gene progressively activated in the prechordal plate, anterior neural plate, and Rathke's pouch of the mouse embryo. *Development (Cambridge, UK)* **122**, 41–52.

Hoeffler, J. P., Bloockfor, F. R., and Frawley, L. S. (1985). Ontogeny of prolactin cells in neonatal rats: Initial prolactin secretors also release growth hormone. *Endocrinology (Baltimore)* **117**, 187–195.

Hooghe-Peters, E., Belayew, A., Herregodts, P., Velkeniers, B., Smets, G., Marital, J., and Vanhaelst, L. (1988). Discrepancy between prolactin (PRL) messenger ribonucleic acid and PRL content in rat fetal pituitary cells: Possible role of dopamine. *Mol. Endocrinol.* **2**, 1163–1168.

Horvath, E., Lloyd, R. V., and Kovacs, K. (1990). Propylthiouracyl-induced hypothyroidism results in reversible transdifferentiation of somatotrophs into thyroidectomy cells. *Lab. Invest.* **63**, 511–520.

Ibrahim, S. N., Moussa, S. M., and Childs, G. V. (1986). Morphometric studies of rat anterior pituitary cells after gonadectomy: Correlation of changes of gonadotropes with the serum levels of gonadotropins. *Endocrinology (Baltimore)* **119**, 629–637.

Ikeda, H., and Yoshimoto, T. (1991). Developmental changes in proliferative activity of cells of the murine Rathke's pouch. *Cell Tissue Res.* **263**, 41–47.

Ikeda, Y., Luo, X., Abbud, R., Nilson, J. H., and Parker, K. L. (1995). The nuclear receptor steroidogenic factor 1 is essential for the formation of the ventromedial hypothalamic nucleus. *Mol. Endocrinol.* **9**, 478–486.

Ingraham, H. A., Chen, R., Mangalam, H. J., Elsholtz, H. P., Flynn, S. E., Lin, C. R., Simmons, D. M., Swanson, L., and Rosenfeld, M. G. (1988). A tissue-specific transcription factor containing a homeodomain specifies a pituitary phenotype. *Cell (Cambridge, Mass.)* **55**, 519–529.

Ishikawa, H., Shiino, M., Arimura, A., and Rennels, E. G. (1977). Functional clones of pituitary cells derived from Rathke's pouch epithelium of fetal rats. *Endocrinology (Baltimore)* **100**, 1227–1230.

Japon, M. A., Rubinstein, M., and Low, M. J. (1994). In situ hybridization analysis of anterior pituitary hormone gene expression during fetal mouse development. *J. Histochem. Cytochem.* **42**, 1117–1125.

Jing, X., Ratty, A. K., and Murphy, D. (1998). Ontogeny of the vasopressin and oxytocin RNAs in the mouse hypothalamus. *Neurosci. Res.* **30**, 343–349.

Jorgenson, R., Levin, L., Cross, H., Yoder, F., and Kelly, T. (1978). The Rieger Syndrome. *Am. J. Med. Genet.* **2**, 307–318.

Jost, A. (1954). Hormonal factors in the development of the fetus. *Cold Spring Harbor Symp. Quant. Biol.* **19**, 167–181.

Kakeya, T., Takeuchi, S., and Takahashi, S. (2000). Epidermal growth factor, insulin and estrogen stimulated development of prolactin-secreting cells in cultures of GH3 cells. *Cell Tissue Res.* **299**, 237–243.

Kapiloff, M., Farkash, Y., Wegner, M., and Rosenfeld, M. (1991). Variable effects of phosphorylation of Pit-1 dictated by the DNA response elements. *Science* **253,** 786–789.

Karlstrom, R. O., Talbot, W. S., and Schier, A. F. (1999). Comparative synteny cloning of zebrafish *you-too:* Mutations in the Hedgehog target *Gli2* affect ventral forebrain patterning. *Genes Dev.* **13,** 388–393.

Kaufman, M. H. (1992). "The Atlas of Mouse Development." Academic Press, San Diego, CA.

Kawamura, K., and Kikuyama, S. (1995). Induction from posterior hypothalamus is essential for the development of the pituitary proopiomelacortin (POMC) cells of the toad (*Bufo japonicus*). *Cell Tissue Res.* **279,** 233–239.

Kendall, S. K., Saunders, T. L., Jin, L., Lloyd, R. V., Glode, L. M., Nett, T. M., Keri, R. A., Nilson, J. H., and Camper, S. A. (1991). Targeted ablation of pituitary gonadotropes in transgenic mice. *Mol. Endocrinol.* **5,** 2025–2036.

Kendall, S. K., Samuelson, L. C., Saunders, T. L., Wood, R. I., and Camper, S. A. (1995). Targeted disruption of the pituitary glycoprotein hormone alpha-subunit produces hypogonadal and hypothyroid mice. *Genes Dev.* **9,** 2007–2019.

Kioussi, C. (1999). *Pax6* is essential for establishing ventral-dorsal cell boundaries in pituitary gland development. *Proc. Natl. Acad. Sci. U.S.A.* **96,** 14378–14382.

Kumar, T., Wang, Y., Lu, N., and Matzuk, M. (1997). Follicle stimulating hormone is required for ovarian follicle maturation but not male fertility. *Nat. Genet.* **15,** 201–204.

Labosky, P. A., Winnier, G. E., Jetton, T. L., Hargett, L., Ryan, A. K., Rosenfeld, M. G., Parlow, A. F., and Hogan, B. L. (1997). The winged helix gene, Mf3, is required for normal development of the diencephalon and midbrain, postnatal growth and the milk-ejection reflex. *Development (Cambridge, UK)* **124,** 1263–1274.

Lala, D. S., Rice, D. S., and Parker, K. L. (1992). Steroidogenic factor 1, a key regulator of steroidogenic enzyme expression, is the mouse homolog of *fushi tarazu*-factor 1. *Mol. Endocrinol.* **6,** 1249–1258.

Lanctot, C., Lamolet, B., and Drouin, J. (1997). The bicoid-related homeoprotein Ptx1 defines the most anterior domain of the embryo and differentiates posterior from anterior lateral mesoderm. *Development (Cambridge, UK)* **124,** 2807–2817.

Landis, C. A., Harsh, G., Lyons, J., Davis, R. L., McCormick, F., and Bourne, H. R. (1990). Clinical characteristics of acromegalic patients whose pituitary tumors contain mutant Gs protein. *J. Clin. Endocrinol. Metab.* **71,** 1416–1420.

Lee, S. L., Sadovsky, Y., Swirnoff, A. H., Polish, J. A., Goda, P., Gavrilina, G., and Milbrandt, J. (1996). Luteinizing hormone deficiency and female infertility in mice lacking the transcription factor NGF1-A (Egr-1). *Science* **273,** 1219–1221.

Lew, D., Brady, H., Klausing, K., Yaginuma, K., Theill, L. E., Stauber, C., Karin, M., and Mellon, P. L. (1993). GHF-1 promoter-targeted immortalization of a somatotropic progenitor cell results in dwarfism in transgenic mice. *Genes Dev.* **7,** 683–693.

Li, H., Zeitler, P. S., Valerius, M. T., Small, K., and Potter, S. S. (1996). Gsh-1, an orphan Hox gene, is required for normal pituitary development. *EMBO J.* **15,** 714–724.

Li, S., Crenshaw, E. B., 3rd, Rawson, E. J., Simmons, D. M., Swanson, L. W., and Rosenfeld, M. G. (1990). Dwarf locus mutants lacking three pituitary cell types result from mutations in the POU-domain gene *Pit-1. Nature (London)* **347,** 528–533.

Lin, S.-C., Lin, C. R., Gukovsky, I., Lusis, A. J., Sawchenko, P. E., and Rosenfeld, M. G. (1993). Molecular basis of the *little* mouse phenotype and implications for cell type-specific growth. *Nature (London)* **364,** 208–213.

Lipkin, S. M., Näär, A. M., Kalla, K. A., Sack, R. A., and Rosenfeld, M. G. (1993). Identification of a novel zinc finger protein binding a conserved element critical for Pit-1-dependent growth hormone gene expression. *Genes Dev.* **7,** 1674–1687.

Lloyd, R. V. (1983). Estrogen-induced hyperplasia and neoplasia in the rat anterior pituitary gland: An immunohistochemical study. *Am. J. Pathol.* **113,** 198–206.

Lloyd, R. V. (1993). Surgical pathology of the pituitary gland. *In* "Major Problems in Pathology" (V. A. Livolsi, ed.), p. 257. Saunders, Philadelphia.

Lopez-Fernandez, J., Palacios, D., Castillo, A. I., Tolon, R. M., Aranda, A., and Karin, M. (2000). Differentiation of lactotrope precursor GHFT cells in response to fibroblast growth factor-2. *J. Biol. Chem.* (in press).

Luo, X., Ikeda, Y., and Parker, K. L. (1994). A cell-specific nuclear receptor is essential for adrenal and gonadal development and sexual differentiation. *Cell (Cambridge, Mass.)* **77,** 481–490.

Maki, K., Miyoshi, I., Kon, Y., Yamashita, T., Sasaki, N., Aoyama, S., Takahashi, E., Namioka, S., Hayashizaki, Y., and Kasai, N. (1994). Targeted pituitary tumorigenesis using the human thyrotropin beta-subunit chain promoter in transgenic mice. *Mol. Cell. Endocrinol.* **105,** 147–154.

Markkula, M., Kananen, K., Paukku, T., Mannisto, A., Loune, E., Frojdman, K., Pelliniemi, L. J., and Huhtaniemi, I. (1994). Induced ablation of gonadotropins in transgenic mice expressing Herpes simplex virus thymidine kinase under the FSH beta-subunit promoter. *Mol. Cell. Endocrinol.* **108,** 1–9.

Mason, A. J., Hayflick, J. S., Zoeller, R. H., Young, W. S., III, Phillips, H. S., Nikolics, K., and Seeburg, P. H. (1986a). A deletion truncating the gonadotropin-releasing hormone gene is responsible for hypogonadism in the *hpg* mouse. *Science* **234,** 1366–1371.

Mason, A. J., Pitts, S. L., Nikolics, K., Szonyi, E., Wilcox, J. N., Seeburg, P. H., and Stewart, T. A. (1986b). The hypogonadal mouse: Reproductive functions restored by gene therapy. *Science* **234,** 1372–1378.

Mathis, J. M., Simmons, D. M., He, X., Swanson, L. W., and Rosenfeld, M. G. (1992). Brain 4: A novel mammalian POU domain transcription factor exhibiting restricted brain-specific expression. *EMBO J.* **11,** 2551–2561.

Matzuk, M., Kumar, T., and Bradley, A. (1995). Different phenotypes for mice deficient in either activins or activin receptor type II. *Nature (London)* **374,** 356–360.

Mayo, K. E. (1996). A little lesson in growth regulation. *Nat. Genet.* **12,** 8–9.

Mayo, K. E., Hammer, R. E., Swanson, L. W., Brinster, R. L., Rosenfeld, M. G., and Evans, R. M. (1988). Dramatic pituitary hyperplasia in transgenic mice expressing a human growth hormone-releasing factor gene. *Mol. Endocrinol.* **2,** 606–612.

McAndrew, J., Paterson, A. J., Asa, S. L., McCarthy, K. J., and Kudlow, J. E. (1995). Targeting of transforming growth factor-alpha expression to pituitary lactotrophs in transgenic mice results in selective lactotroph proliferation and adenomas. *Endocrinology (Baltimore)* **136,** 4479–4488.

McCarthy, M. M., and Altemus, M. (1997). Central nervous system actions of oxytocin and modulation of behavior in humans. *Mol. Med. Today* **3,** 269–275.

Michaud, J. L., Rosenquist, T., May, N. R., and Fan, C. M. (1998). Development of neuroendocrine lineages requires the bHLH-PAS transcription factor SIM1. *Genes Dev.* **12,** 3264–3275.

Minowa, O., Ikeda, K., Sugitani, Y., Oshima, T., Nakai, S., Katori, Y., Suzuki, M., Furukawa, M., Kawase, T., Zheng, Y., Ogura, M., Asada, Y., Watanabe, K., Yamanaka, H., Gotoh, S., Nishi-Takeshima, M., Sugimoto, T., Kikuchi, T., Takasaka, T., and Noda, T. (1999). Altered cochlear fibrocytes in a mouse model of DFN3 nonsyndromic deafness. *Science* **285,** 1408–1411.

Missale, C., Boroni, F., Sigala, S., Zanellato, A., Toso, R. D., Balsari, A., and Spano, P. (1994). Nerve growth factor directs differentiation of the bipotential cell line GH-3 into the mammotroph phenotype. *Endocrinology (Baltimore)* **135,** 290–298.

Mouse Genome Database (1996). "Mouse Genome Informatics," M. 3.1. The Jackson Laboratory, Bar Harbor, ME. World Wide Web: URL: http://www.informatics.jax.org/

Muglia, L. J., Jacobson, L., Dikkes, P., and Majzoub, J. A. (1995). Corticotropin-releasing hormone deficiency reveals fetal but not adult glucocorticoid need. *Nature (London)* **373,** 427–432.

Nakai, S., Kawano, H., Yudate, T., Nishi, M., Kuno, J., Nagata, A., Jishage, K., Hamada, H., Fujii, H., Kawamura, K., *et al.* (1995). The POU domain transcription factor Brn-2 is required for the determination of specific neuronal lineages in the hypothalamus of the mouse. *Genes Dev.* **9,** 3109–3121.

Nemeskery, A., Nemeth, A., Setalo, G., Vigh, S., and Halasz, B. (1976). Cell differentiation of the fetal anterior pituitary in vitro. *Cell Tissue Res.* **170,** 263–273.

Netchine, I., Sobrier, M. L., Krude, H., Schnabel, D., Maghnie, M., Marcos, E., Duriez, B., Cacheux, V., Moers, A., Goossens, M., Gruters, A., and Amselem, S. (2000). Mutations in LHX3 result in a new syndrome revealed by combined pituitary hormone deficiency. *Nat. Genet.* **25,** 182–186.

Nishihara, E., Nagayama, Y., Inoue, S., Hiroi, H., Muramatsu, M., Yamashita, S., and Koji, T. (2000). Ontogenetic changes in the expression of estrogen receptor alpha and beta in rat pituitary gland detected by immunohistochemistry. *Endocrinology (Baltimore)* **141,** 615–620.

Ohuchi, H., Hori, Y., Yamasaki, M., Harada, H., Sekine, K., Kato, S., and Itoh, N. (2000). FGF10 acts as a major ligand for FGF Receptor 2 IIIb in mouse multiorgan development. *Biochem. Biophys. Res. Commun.* **277,** 643–649.

Ortega, S., Ittmann, M., Tsang, S. H., Ehrlich, M., and Basilico, C. (1998). Neuronal defects and delayed wound healing in mice lacking fibroblast growth factor 2. *Proc. Natl. Acad. Sci. U.S.A.* **95,** 5672–5677.

Pallister, P. D., Hecht, F., and Herrman, J. (1989). Three additional cases of the congenital hypothalamic "hamartoblastoma" (Pallister-Hall) syndrome. *Am. J. Med. Genet.* **33,** 500–501.

Palomino, T., Sanchez-Pacheco, A., Pena, P., and Aranda, A. (1998). A direct protein-protein interaction is involved in the cooperation between thyroid hormone and retinoic acid receptors and the transcription factor GHF-1. *FASEB J.* **12,** 1201–1209.

Park, H. L., Bai, C., Platt, K. A., Matise, M. P., Beeghly, A., Hui, C. C., and Nakashima, M. (2000). Mouse *Gli1* mutants are viable but have defects in *SHH* signaling in combination with *Gli2* mutation. *Development (Cambridge, UK)* **127,** 1593–1605.

Parks, J., Adess, M., and Brown, M. (1997). Genes regulating hypothamic and pituitary development. *Acta Paediat. Suppl.* **423,** 28–32.

Pernasetti, F., Toledo, S. P., Vasilyev, V. V., Hayashida, C. Y., Cogan, J. D., Ferrari, C., Lourenco, D. M., and Mellon, P. L. (2000). Impaired adrenocorticotropin-adrenal axis in combined pituitary hormone deficiency caused by a two-base pair deletion (301–302delAG) in the prophet of Pit-1 gene. *J. Clin. Endocrinol. Metab.* **85,** 390–397.

Pfäffle, R., Kim, C., Otten, B., Wit, J., Eiholzer, U., Heimann, G., and Parks, J. (1996). Pit-1: Clinical aspects. *Horm. Res.* **45**(Suppl. 1), 25–28.

Phelps, C., and Hymer, W. C. (1983). Characterization of estrogen-induced adenohypophyseal tumors in the Fischer 344 rat. *Neuroendocrinology* **37,** 23–31.

Phippard, D., Lu, L., Lee, D., Saunders, J. C., and Crenshaw, E. B., 3rd (1999). Targeted mutagenesis of the POU-domain gene Brn4/Pou3f4 causes developmental defects in the inner ear. *J. Neurosci.* **19,** 5980–5989.

Piussan, C., Hanauer, A., Dahl, N., Mathieu, M., Kolski, C., Biancalana, V., Heyberger, S., and Strunski, V. (1995). X-linked progressive mixed deafness: A new microdeletion that involves a more proximal region in Xq21. *Am. J. Hum. Genet.* **56,** 224–230.

Porter, F., Drago, J., Xu, Y., Cheema, S., Wassif, C., Huang, S.-P., Lee, E., Grinberg, A., Massalas, J., Bodine, D., Alt, F., and Westphal, H. (1997). *Lhx2,* a LIM homeobox gene, is required for eye, forebrain, and definitive erythrocyte development. *Development (Cambridge, UK)* **124,** 2935–2944.

Poulin, G., Turgeon, B., and Drouin, J. (1997). NeuroD1/β2 contributes to cell-specific transcription of the proopiomelanocortin gene. *Mol. Cell. Biol.* **17,** 6673–6682.

Radovick, S., Nations, M., Du, Y., Berg, L. A., Weintraub, B. D., and Wondisford, F. E. (1992). A mutation in the POU-Homeodomain of Pit-1 responsible for combined pituitary hormone deficiency. *Science* **257,** 1115–1117.

Rhodes, S. J., DiMattia, G. E., and Rosenfeld, M. G. (1994). Transcriptional mechanisms in anterior pituitary cell differentiation. *Curr. Opin. Genet. Dev.* **4,** 709–717.

Roberson, M. S., Schoderbek, W. E., Tremml, G., and Maurer, R. A. (1994). Activation of the glycoprotein hormone α-subunit promoter by a LIM-homeodomain transcription factor. *Mol. Cell. Biol.* **14,** 2985–2993.

Rugh, R. (1990). "The Mouse: Its Reproduction and Development." Oxford University Press, Oxford.

Rutishauser, J., Boni-Schnetzler, M., Boni, J., Wichmann, W., Huisman, T., Vallotton, M. B., and Froesch, E. R. (1996). A novel point mutation in the translation initiation codon of the pre-pro-vasopressin-neurophysin II gene: Cosegregation with morphological abnormalities and clinical symptoms in autosomal dominant neurohypophyseal diabetes insipidus. *J. Clin. Endocrinol. Metab.* **81,** 192–198.

Ryan, A. K., and Rosenfeld, M. G. (1997). POU domain family values: Flexibility, partnerships, and developmental codes. *Genes Dev.* **11,** 1207–1225.

Sadeghi-Nejad, A., and Senior, B. (1974). Autosomal dominant transmission of isolated growth hormone deficiency in iris-dental dysplasia (Rieger's Syndrome). *J. Pediatr.* **85,** 644–648.

Saiardi, A., Bozzi, Y., Baik, J.-H., and Borrelli, E. (1997). Antiproliferative role of dopamine: Loss of D2 receptors causes hormonal dysfunction and pituitary hyperplasia. *Neuron* **19,** 115–126.

Sasaki, F., and Iwana, Y. (1988). Correlation of spatial differences in concentrations of prolactin and growth hormone cells with vascular pattern in the female mouse adenohypiphysis. *Endocrinology (Baltimore)* **122,** 1622–1630.

Schechter, J., Stauber, C., Windle, J. J., and Mellon, P. (1995). Basic fibroblast growth factor: The neurotrophic factor influencing the ingrowth of neural tissue into the anterior pituitary of α-T7 transgenic mice? *Neuroendocrinology* **61,** 622–627.

Schonemann, M. D., Ryan, A. K., McEvilly, R. J., O'Connell, S. M., Arias, C. A., Kalla, K. A., Li, P., Sawchenko, P. E., and Rosenfeld, M. G. (1995). Development and survival of the endocrine hypothalamus and posterior pituitary gland requires the neuronal POU domain factor Brn-2. *Genes Dev.* **9,** 3122–3135.

Schweppe, R. E., Frazer-Abel, A. A., Gutierrez-Hartmann, A., and Bradford, A. P. (1997). Functional components of fibroblast growth factor (FGF) signal transduction in pituitary cells. Identification of FGF response elements in the prolactin gene. *J. Biol. Chem.* **272,** 30852–30859.

Schwind, J. L. (1928). The development of the hypophysis cerebri of the albino rat. *Am. J. Anat.* **41,** 295–315.

Seghezzi, G., Patel, S., Ren, C. J., Gualandris, A., Pintucci, G., Robbins, E. S., Shapiro, R. L., Galloway, A. C., Rifkin, D. B., and Mignatti, P. (1998). Fibroblast growth factor-2 (FGF-2) induces vascular endothelial growth factor (VEGF) expression in the endothelial cells of forming capillaries: An autocrine mechanism contributing to angiogenesis. *J. Cell Biol.* **141,** 1659–1673.

Sekine, K., Ohuchi, H., Fujiwara, M., Yamasaki, M., Yoshizawa, T., Sato, T., Yagishita, N., Matsui, D., Koga, Y., Itoh, N., and Kato, S. (1999). Fgf10 is essential for limb and lung formation. *Nat. Genet.* **21,** 138–141.

Semina, E. V., Reiter, R., Leysens, N. J., Alward, W. L., Small, K. W., Datson, N. A., Siegel-Bartelt, J., Bierke-Nelson, D., Bitoun, P., Zabel, B. U., Carey, J. C., and Murray, J. C. (1996). Cloning and characterization of a novel bicoid-related homeobox transcription factor gene, RIEG, involved in Rieger syndrome. *Nat. Genet.* **14,** 392–399.

Semina, E. V., Reiter, R. S., and Murray, J. C. (1997). Isolation of a new homeobox gene belonging to the Pitx/Rieg family: Expression during lens development and mapping to the aphakia region on mouse chromosome 19. *Hum. Mol. Genet.* **6,** 2109–2116.

Seuntjens, E., Vankelecom, H., Quaegebeur, A., Vande Vijver, V., and Denef, C. (1999). Targeted ablation of gonadotrophs in transgenic mice affects embryonic development of lactotrophs. *Mol. Cell. Endocrinol.* **150,** 129–139.

Sharma, K., Sheng, H. Z., Lettieri, K., Li, H., Karavanov, A., Potter, S., Westphal, H., and Pfaff, S. L. (1998). LIM homeodomain factors Lhx3 and Lhx4 assign subtype identities for motor neurons. *Cell (Cambridge, Mass.)* **95,** 817–828.

Sheng, H., Moriyama, K., Yamashita, T., Li, H., Potter, S., Mahon, K., and Westphal, H. (1997). Multistep control of pituitary organogenesis. *Science* **278,** 1809–1812.

Sheng, H. Z., and Westphal, H. (1999). Early steps in pituitary organogenesis. *Trends Genet.* **15,** 236–240.

Sheng, H. Z., Zhadanov, A. B., Mosinger, B. J., Fujii, T., Bertuzzi, S., Grinberg, A., Lee, E. J., Huang, S. P., Mahon, K. A., and Westphal, H. (1996). Specification of pituitary cell lineages by the LIM homeobox gene Lhx3. *Science* **272,** 1004–1007.

Shimon, I., Huttner, A., Said, J., Spirina, O. M., and Melmed, S. (1996). Heparin-binding secretory transforming gene (hst) facilitates rat lactotrope cell tumorigenesis and induces prolactin gene transcription. *J. Clin. Inves.* **97,** 187–195.

Simmons, D. M., Voss, J. W., Ingraham, H. A., Holloway, J. M., Broide, R. S., Rosenfeld, M. G., and Swanson, L. W. (1990). Pituitary cell phenotypes involve cell-specific *Pit-1* mRNA translation and synergistic interactions with other classes of transcription factors. *Genes Dev.* **4,** 695–711.

Sornson, M. W., Wu, W., Dasen, J. S., Flynn, S. E., Norman, D. J., O'Connell, S. M., Gukovsky, I., Carrière, C., Ryan, A. K., Miller, A. P., Zuo, L., Gleiberman, A. S., Andersen, B., Beamer, W. G., and Rosenfeld, M. G. (1996). Pituitary lineage determination by the *Prophet* of *Pit-1* homeodomain factor defective in Ames dwarfism. *Nature (London)* **384,** 327–333.

Stahl, J., Kendall, S., Brinkmeier, M., Greco, T., Watkins-Chow, D., Canpos-Barros, A., Lloyd, R., and Camper, S. (1999). Thyroid hormone is essential for pituitary somatotropes and lactotropes. *Endocrinology (Baltimore)* **140,** 1884–1892.

Stefana, B., Ray, D. W., and Melmed, S. (1996). Leukemia inhibitory factor induces differentiation of pituitary corticotroph function: An immunoneuroendocrine phenotypic switch. *Proc. Natl. Acad. Sci. U.S.A.* **93,** 12502–12506.

Stefaneanu, L., and Kovacs, K. (1993). Light microscopic special stains and immunochemistry in the diagnosis of pituitary adenomas. *In* "Surgical Pathology of the Pituitary Gland" (R. V. Lloyd, ed.), pp. 34–51. Saunders, Philadelphia.

Strand, F. L. (1999). New vistas for melanocortins. Finally, an explanation for their pleiotropic functions. *Ann. N.Y. Acad. Sci.* **897,** 1–16.

Szabo, K. (1987). Origin of the adenohypopheseal vessels in the rat. *J. Anat.* **154,** 229–235.

Szabo, K., and Csanyi, K. (1982). The vascular architecture of the developing pituitary-median eminence in the rat. *Cell Tissue Res.* **224,** 563–577.

Szeto, D. P., Rodriquez-Estaban, C., Ryan, A. K., O'Connell, S. M., Liu, F., C., K., Gleiberman, A. S., Izpisua-Belmonte, J. C., and Rosenfeld, M. G. (1999). Role of Bicoid-related homeodomain factor Pitx1 in specifying hindlimb morphogenesis and pituitary development. *Genes Dev.* **13,** 484–494.

Takuma, N., Sheng, H. Z., Furuta, Y., Ward, J. M., Sharma, K., Hogan, B. L. M., Pfaff, S. L., Westphal, H., Kimura, S., and Mahon, K. A. (1998). Formation of Rathke's pouch requires dual induction from the diencephalon. *Development (Cambridge, UK)* **125,** 4835–4840.

Tang, K., Bartke, A., Gardiner, C. S., Wagner, T. E., and Yun, J. S. (1993). Gopnadotropin secretion, synthesis, and gene expression in human growth hormone transgenic mice and in Ames dwarf mice. *Endocrinology (Baltimore)* **132,** 2518–2524.

Topilko, P., Schneider-Maunoury, S., Levi, G., Trembleau, A., Gourdji, D., Driancourt, M.-A., Rao, C. V., and Charnay, P. (1998). Multiple pituitary and ovarian defects in *Krox-24 (NGF1-A, Egr-1)*-targeted mice. *Mol. Endocrinol.* **12,** 107–122.

Treier, M., and Rosenfeld, M. G. (1996). The hypothalamic-pituitary axis: Co-development of two organs. *Curr. Opin. Cell Biol.* **8,** 833–843.

Treier, M., Gleiberman, A. S., O'Connell, S. M., Szeto, D. P., McMahon, J. A., McMahon, A. P., and Rosenfeld, M. G. (1998). Multistep signaling requirements for pituitary organogenesis in vivo. *Genes Dev.* **12,** 1691–1704.

Tremblay, J. J., Lanctôt, C., and Drouin, J. (1998). The pan-pituitary activator of transcription, Ptx1 (pituitary homeobox 1), acts in synergy with SF-1 and Pit1 and is an upstream regulator of the lim-homeodomain gene Lim3/Lhx3. *Mol. Endocrinol.* **12,** 428–441.

Tremblay, J. J., Marcil, A., Gauthier, Y., and Drouin, J. (1999). Ptx1 regulates Sf-1 activity by an interaction that mimics the role of the ligand-binding domain. *EMBO J.* **18,** 3431–3441.

Tsai, F.-Y., Keller, G., Kuo, F. C., Weiss, M., Chen, J., Rosenblatt, M., Ait, F. W., and Orkin, S. H. (1994). An early haematopoietic defect in mice lacking the transcription factor GATA-2. *Nature (London)* **371,** 221–226.

Wajnrajch, M. P., Gertner, J. M., Harbison, M. D., Chua, S. C. J., and Leibel, R. L. (1996). Nonsense mutation in the human growth hormone-releasing hormone receptor causes growth failure analogous to the little (*lit*) mouse. *Nat. Genet.* **12,** 88–90.

Watkins-Chow, D. E., and Camper, S. A. (1998). How many homeobox genes does it take to make a pituitary gland? *Trends Genet.* **14,** 284–290.

Weiss, J., Axelrod, L., Whitcomb, R. W., Harris, P. E., Crowley, W. F., and Jameson, J. L. (1992). Hypogonadism caused by a single amino acid substitution in the β-subunit of luteinizing hormone. *N. Eng. J. Med.* **326,** 179–183.

Wilson, D. B., and Christensen, E. (1983). Fine structure of pituitary blood vessels in embryos of the dwarf (dw) mutant mouse. *Virchows Arch. B* **43,** 317–325.

Windle, J. J., Weiner, R. I., and Mellon, P. L. (1990). Cell lines of the pituitary gonadotrope lineage derived by targeted oncogenesis in transgenic mice. *Mol. Endocrinol.* **4,** 597–603.

Woods, K. L., and Porter, T. E. (1998). Ontogeny of prolcatin secreting cells during chick embryonic development: Effect of vasoactive intestinal peptide. *Gen. Comp. Endocrinol.* **112,** 240–246.

Yagi, H., Nagashima, K., Miyake, H., Tamai, S., Onigata, K., Yutani, S., and Kuroume, T. (1994). Familial congenital hypopituitarism with central diabetes insipidus. *J. Clin. Endocrinol. Metab.* **78,** 884–889.

Yamada, M., Saga, Y., Shibusawa, N., Hirato, J., Murakami, M., Iwasaki, T., Hashimoto, K., Satoh, T., Wakabayashi, K., Taketo, M. M., and Mori, M. (1997). Tertiary hypothyroidism and hyperglycemia in mice with targeted disruption of the thyrotropin-releasing hormone gene. *Proc. Natl. Acad. Sci. U.S.A.* **94,** 10862–10867.

Yamaguchi, T. P., Bradley, A., McMahon, A. P., and Jones, S. (1999). A Wnt5a pathway underlies outgrowth of multiple structures in the vertebrate embryo. *Development (Cambridge, UK)* **126,** 1211–1223.

Young, W. S. D., and Zoeller, R. T. (1987). Neuroendocrine gene expression in the hypothalamus: In situ hybridization histochemical studies. *Cell. Mol. Neurobiol.* **7,** 353–366.

Zetterstrom, R. H., Solomin, L., Mitsiadis, T., Olson, L., and Perlmann, T. (1996). Retinoid X receptor heterodimerization and developmental expression distinguish the orphan nuclear receptors NGFI-B, Nurr1, and Nor1. *Mol. Endocrinol.* **10,** 1656–1666.

21

Development of the Eye

Hisato Kondoh

Institute for Molecular and Cellular Biology, Osaka University, Osaka 565-0871, Japan

I. **Overview of Eye Development**

II. **Development of the Retina**

III. **Lens Development**

IV. **Conservation and Divergence of the Transcriptional Regulatory Systems in Eye Development**

References

I. Overview of Eye Development

The basic structure of the eye develops from two major components, the retina and the lens, which interact during development. Morphological development of the eye tissues in embryos is summarized in Fig. 1. Readers are advised to refer to Pei and Rhodin (1970) and Kaufman (1992) for morphological details.

The retina initially develops as a lateral protrusion of the ventral telencephalon called the optic vesicle. The distal part of the vesicle makes contact with the lateral head ectoderm for a short period, and this contact induces lens differentiation in the head ectoderm: The optic vesicle-contacted area thickens to form the lens placode, which then invaginates, is pinched off from the ectodermal layer, and generates a lens vesicle (Fig. 1A).

Among the remaining areas of the head ectoderm, that facing the lens vesicle develops as the cornea epithelium rather than the epidermis. The inner side of the cornea epithelium is populated by neural crest-derived cells that form the cornea stroma (Fig. 1B, E11.5 and thereafter).

The lens vesicle develops into a lens consisting of anteriorly placed lens epithelium and posterior lens fiber cells (Fig. 1B, E11.5 and thereafter). The cells of the former compartment remain mitotic while those of the latter are postmitotic. Lens differentiation is accompanied by initiation of lens-specific crystallin gene expression.

As the lens placode invaginates, so does the distal part of the optic vesicle, and the optic cup is formed (Fig. 1B, E10.5). Invagination of the optic vesicle is dependent on the lens tissue. Absence of the lens from the early stage of eye development, either as the result of surgical removal or of genetic manipulation, results in later development of retina layers with extensive undulation (Ashery-Padan *et al.*, 2000; Coulombre and Coulombre, 1964). Three components are derived from the optic vesicle: the internal layer of the optic cup, which develops into the neural retina; the external layer, which becomes the pigmented epithelium; and the optic stalk, which later provides the route for the optic nerve to the brain (Fig. 1B, E11.5 and thereafter). The anterior margin of the optic cup later develops into the iris and the ciliary body (Fig. 1B, P1.5).

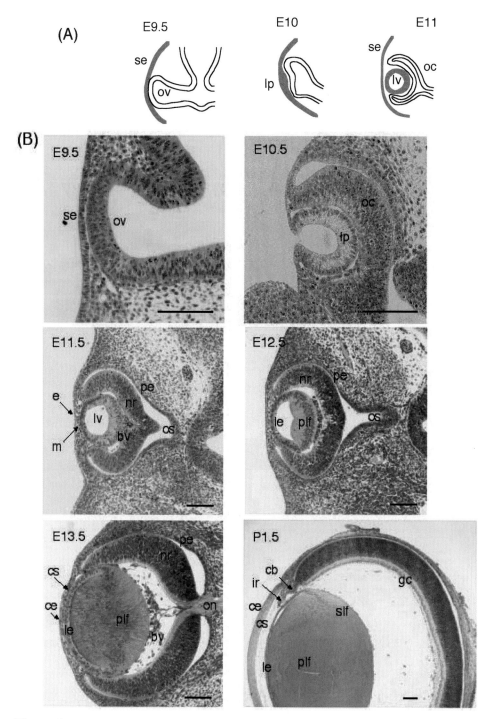

Figure 1 Early stages of eye development. (A) Three major steps of lens induction, indicated by the schemes of E9.5 (9.5 dpc), E10, and E11. Development of optic vesicle (ov) results in its contact with surface ectoderm (se) and lens development initiates (E9.5). The area of surface ectoderm contacted by optic vesicle thickens and forms lens placode (lp) (E10). Invagination of the lens placode forms lens vesicle (lv) (E11). (B) Histological sections of developing eyes at E9.5, E10.5, E11.5, E12.5, E13.5, and P1.5 (postnatal day 1.5), stained with hematoxylin-eosin. Bar = 100 μm. At E9.5, optic vesicle (ov) makes contact with surface ectoderm (se). By E10.5, a series of dramatic morphological changes takes place: The surface ectoderm thickens to form lens placode (lp), the placode invaginates, and the invagination is transformed into lens vesicle by closure of the external opening. In parallel, optic vesicle also invaginates and forms the optic cup (oc). At E11.5, the deeper side of the lens vesicle (lv) becomes thickened and lens fiber differentiation begins. The optic vesicle now develops into three components of the eye: neural retina (nr), pigmented epithelium (pe), and optic stalk (os). Blood vessels (bv) come into the space between the lens vesicle and neural retina. The space between the lens vesicle and ectoderm (e) is populated

II. Development of the Retina

Development of the retina proceeds in multiple steps: determination of the retina potential in the primitive central nervous system (CNS), development of the optic vesicle, differentiation of the optic cup into three components (i.e., optic stalk, pigmented epithelium, and neural retina), and development of neural retina layers. These steps are basically regulated by sets of transcription factors, whose roles are assessed by their spatiotemporal expression patterns, by the phenotype of the relevant mutant mice, by the effect of overexpression in *Xenopus* or fish (zebrafish and *Medaka*) embryos, and by the cellular phenotype of retrovirus-infected cell clones.

A. Genetic Determination of Retina Potential in the Anterior Neuroepithelium

Otx2 is a bicoid-type homeodomain protein and is a homolog of the *Drosophila* orthodenticle gene product. *Otx2* is broadly expressed in the anterior CNS in the forebrain-midbrain region (Acampora *et al.*, 1995; Ang *et al.*, 1996; Matsuo *et al.*, 1995), and its expression continues in the optic vesicle. Its involvement in eye retina development is suggested by its expression domain in the CNS and by the phenotype of the *Otx2* knockout mice. *Otx2* homozygotes lack the rostral portion of the brain (Acampora *et al.*, 1995; Ang *et al.*, 1996; Matsuo *et al.*, 1995), but heterozygotes display microphthalmia or anophthalmia, indicating that *Otx2* is involved in retina development.

Otx2 presumably defines the region of the early CNS that initially has the potency of the retina development as discussed later in conjunction with the activity of *Six6*. Then, the genes coding for homeodomain-containing transcription factors, *Rx* (also called Rax, with Paired-type homeodomain), *Six3* (with Six- and homeodomains), *Lhx2* (with Lim- and homeodomains), and *Pax6* (with Paired- and homeodomains) become expressed together with extensive overlap of the regions of expression in the early forebrain (Furukawa *et al.*, 1997a; Grindley *et al.*, 1995; Mathers *et al.*, 1997; Oliver *et al.*, 1995; Porter *et al.*, 1997). These appear to be involved in providing the forebrain with the potential of retina development in the following stages.

In the mouse, expression of these transcription factors covers a wide region of the forebrain not necessarily restricted to the future retina, but their involvement in retina development is demonstrated by the knock-out mouse phenotype described below. In other vertebrates, amphibians in particular, transcription factors like *Pax6* and *Rx* are expressed in overlapping domains in the medial portion of the forebrain, the *retina field*. The retina field is split by the influence of the prechordal plate in the ventral midline of the brain. Studies using amphibian and chick embryos indicated that surgical ablation of the prechordal plate resulted in fusion of the retina fields and cyclopia (Li *et al.*, 1997). In the mouse, targeted mutation of Shh also caused a single median retina and cyclopia (Chiang *et al.*, 1996), and the failure of separation of the *Rx* or *Pax6* domain (Li *et al.*, 1997). Shh overexpression represses expression of *Rx* and *Pax6* and leads to the loss of the eyes in newt embryos (Mizuno *et al.*, 1997). The importance of Shh signaling from the ventral median tissues, however, is not limited to this early phase of eye development but continues into later phases in which the optic stalk portion is derived from the optic vesicle (Macdonald *et al.*, 1995), as discussed below.

Among the transcription factors expressed in the forebrain in the mouse and in retina fields in amphibia, *Rx* and *Six3* appear to have determinative roles in defining the retina primordia. Support of this notion is provided by different lines of observations. For instance, *Rx* knock-out mouse embryos totally lack the optic vesicle (Mathers *et al.*, 1997). Overexpression of *Six3* in *Medaka* and *Xenopus* embryos (Bernier *et al.*, 2000; Loosli *et al.*, 1999) resulted in enlargement of the retina and production of the ectopic retina. Knock-out/mutant mice of *Pax6* (Grindley *et al.*, 1995) and *Lhx2* (Porter *et al.*, 1997) somehow produce the optic vesicle, but these mutant vesicles are defective in subsequent development, indicating that these genes have essential roles in a later phase of retina development. Expression of these two genes is not interdependent, because *Pax6* is expressed in the optic vesicle of the homozygous *Lhx2* mutant embryo and vice versa (Porter *et al.*, 1997). *Lhx2* homozygous embryos have optic vesicles that are small protrusions of the ventral diencephalon which never reach the head surface ectoderm (Porter *et al.*, 1997), whereas *Pax6* homozygous embryos have initially enlarged optic vesicles that initially make contact with the head ectoderm but then

---

by mesenchymal cells (m) of neural crest origin. At E12.5, lens fiber differentiation progresses, and lens epithelium (le) and primary lens fibers (plf) become distinct. At E13.5, the primary lens fibers (plf) fully develop and make contact with lens epithelium (le). Cornea is now formed consisting of cornea epithelium (ce) and cornea stroma (cs). In neural retina (nr) ganglion cells begin to differentiate along the inner surface and extend the axons along the optic stalk as optic nerves (on). Blood vessels (bv) develop close to the lens. After this stage, secondary lens fibers (slf) begin to develop surrounding the primary lens fibers (plf) as seen in the P1.5 specimen; ciliary body (cb) and iris (ir) develop at the edge of the neural retina; and cornea stroma (cs) thickens. The layer of ganglion cells (gc) becomes clearly separated from other cells at this stage.

regress (Grindley *et al.,* 1995). No lens is induced in this situation.

Six6/Optx2 (Jean *et al.,* 1999; Lopez-Rios *et al.,* 1999; Toy and Sundin, 1999) is closely related to *Six3* in the homeodomain and the SIX domain, but is divergent in other domains. *Six6/Optx2* is expressed much later in the neural retina than *Six3.* Overexpression of *Six6/Optx2* in the *Xenopus* embryo causes enlargement of the retinas, analogous to Six3 overexpression and development of ectopic retina tissues through activation of *Rx, Six3,* and *Pax6* in the ectopic sites of the brain (Bernier *et al.,* 2000; Zuber *et al.,* 1999). The result suggests that *Six6/Optx2* is a maintenance regulator for the neural retina trait, but with a feedback loop to the earlier transcriptional regulation system involving *Rx, Six3,* and *Pax6.* It is intriguing that this ectopic activation of *Rx/Six3/Pax6* and later ectopic retina production occurred only in the *Otx2*-expressing domain of the neuroectoderm (Bernier *et al.,* 2000), implying that the initial retina potency of the neuroectoderm requires the activity of *Otx2.*

An interesting result is reported in which *Six6/Optx2* fused with the Engrailed repression domain behaved similar to *Six6/Optx2* itself, while fusion with the VP16 activation domain counteracted the *Six3* function when overexpressed in *Xenopus* embryos (Zuber *et al.,* 1999). This result indicates that the primary action of *Six6/Optx2* is transcriptional repression.

B. Development of the Three Major Derivatives of the Optic Vesicle: Neural Retina, Pigmented Epithelium, and Optic Stalk

As the optic vesicle develops (Fig. 2A), the most distally located part at the beginning is displaced dorsally in the following phase, as the result of more active growth of the more ventral parts. Invagination of the vesicle then begins from the ventral aspect until the internal layer (prospective neural retina) and the external layer (prospective pigmented epithelium) appose. The ventralmost region of the vesicle proliferates vigorously, expands the two layers of cells, and generates the ventral part of the optic cup. Thus, the ventral optic cup is made from the ventral extension of the nasal and temporal halves of the optic vesicle, which meet and form a fissure (choroid fissure) along the ventral midline of the optic vesicle (Fig. 2A). The same ventral extension of the two-layered cell sheet also occurs in the optic stalk. Therefore, the choroid fissure is continuous from the optic stalk to the optic vesicle. This fissure is later closed, but if it fails in association with the ventral defect of the optic cup/retina, the fissure remains and coloboma results.

In this way, the three structural components of the optic cup are established: the inner layer, which forms the neural retina; the outer layer, which forms the pigmented epithelium; and the optic stalk, which later differentiates into the glial cells of the optic nerve. Identity of these three components appears to be determined by sets of transcription factors that have either synergistic or counteracting effects (Fig. 2B).

The optic stalk expresses *Vax1* and *Pax2,* which determine the stalk and inhibit the retina characteristics. These transcription factors are also involved in closure of the retinal fissure at the ventral side. The transcription factors in the retina part, PAX6 and RX, counteract the stalk transcription factors, and the boundary between the retina and the stalk, called the *optic disk,* seems to be determined by the counterbalance of these stalk-promoting and retina-promoting factors (Fig. 2B). Overexpression of *Pax2* in zebrafish embryos and of *Vax1* in the *Xenopus* embryos resulted in reduction of *Pax6* and *Rx* expression and inferior development of the retina portion (Hallonet *et al.,* 1999; Macdonald *et al.,* 1995). *Vax1* knock-out mouse embryos have retinal neural tissue, which replaces the optic stalk (Hallonet *et al.,* 1999), whereas *Pax2* mutant mice develop coloboma and have pigmented epithelial tissue in place of the optic stalk (Favor *et al.,* 1996). Therefore, *Vax1* and *Pax2* together inhibit retinal development: *Vax1* inhibits the neural retina development, while *Pax2* inhibits the pigment retina development in the optic stalk. Expression domains of these transcription factors appear to be either directly or indirectly influenced by the secreted signaling proteins, for example, *Shh* emanating from the ventral diencephalon.

Expression of *Vax1* and *Pax2* in the optic stalk is dependent on the *Shh* expression along the ventral midline of the forebrain. Overexpression of *Shh* in zebrafish and *Xenopus* embryos increased the domain of optic stalk expressing *Vax1* and *Pax2* and inhibited Pax6 and Rx, resulting in hypomorphism of the retinal components (Hallonet *et al.,* 1999; Macdonald *et al.,* 1995). Knock-out mouse of the winged helix transcription factor BF-1 resembles the *Vax1* knock-out mouse in eye development (Huh *et al.,* 1999), which is accounted for by the loss of the Shh expression in the forebrain domain in the mutant. It is interesting to note that *Gli3* (*Xt, extratoes*) mutant embryos also display a similar phenotype to Pax2 mutant in the optic stalk (Franz and Besecke, 1991) (Fig. 3). Because GLI3 activates genes in response to the Shh signal (Sasaki *et al.,* 1999), it is speculated that the *Gli3* (*Xt*) mutation affects the optic stalk development through regulation of Vax1 and Pax2.

Distinction between neural retina and pigmented epithelium seems to depend on interaction with the surrounding tissue and on a balance of differentially expressed transcription factors. Rx is initially essential for all retina components as indicated by the knockout mouse phenotype, but seems to have a role in the pigmented epithelium development in the later stages. Overexpression of *Rx* in the developing retina in *Xenopus* caused ectopic pigmented epithelium development in the neural tissue (Mathers *et al.,* 1997). On the contrary, overexpression of *Six3* and *Six6* caused hyperplasia of the

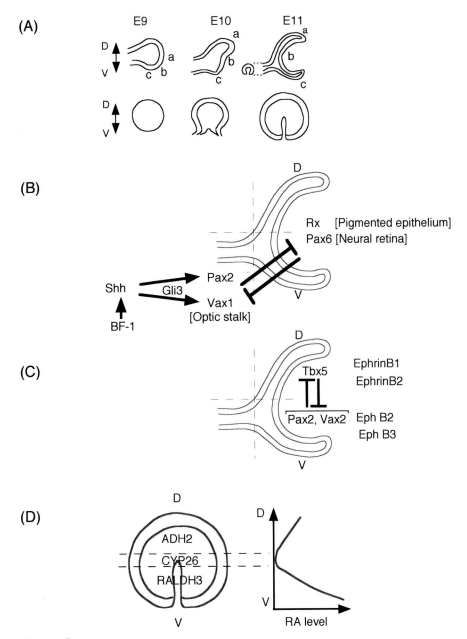

Figure 2 Morphogenesis and polarity of retina development. (A) The scheme of early morphogenesis. Development from optic vesicle to optic cup is schematically illustrated as cross sections (upper) and a lateral view (lower) representing the E9, E10, and E11 stages. The most distal part of the vesicle (a) develops into the most dorsal part of the cup, and more proximal parts of the ventral vesicle (b and c) develop into the medial and ventral parts of the cup. (B) Three major derivatives of optic vesicle—neural retina, pigmented epithelium, and optic stalk—are determined by the balance of synenergizing and counteracting transcription factors. Optic stalk is determined by Pax2 and Vax1, while neural retina and pigmented epithelium are determined by Pax6 and Rx. The boundary of the retina and the stalk (optic disc) is determined by the counteraction of Pax2/Vax1 and Pax6/Rx. In the retina, domination of Rx promotes development of pigmented epithelium, while that of Pax6 promotes neural retina development. Expression of Pax2 and Vax1 in the optic stalk is dependent on Shh expression along the ventral midline of the diencephalon, which is dependent on the activity of BF-1. (C) Dorsoventral patterning by counteraction between Tbx5 and Pax2/Vax2. (D) Hypothetical distribution of retinoic acid (RA) along dorsoventral axis of the retina (right), created by two synthetic enzyme activity (ADH2 and RALDH3) and a degrading enzyme CYP26 (left panel).

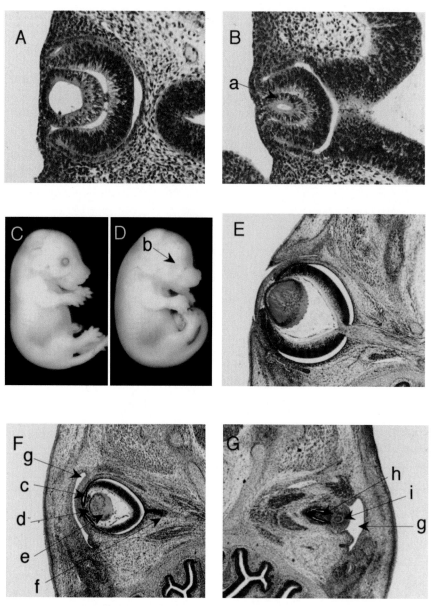

Figure 3 *Gli3* (*Xt*) mutant eyes, as example of genetic defects in eye development: microphthalmia, defective lens development, and pigmentation of the optic stalk and neural retina compartment. (A) Wild-type and (B) *Gli3* homozygous mutant eyes at E10.5$^+$ when the lens vesicle (lv) has just been formed. The mutant lens vesicle (a in B) is significantly smaller than the normal wild-type lens. (C) Wild-type and (D) *Gli3* homozygous mutant embryos at E15.5. The eyes of *Gli3* homozygotes are hardly visible by external inspection (b in D). (E–G) Cross sections through the eyes of E15.5 embryos of (E) wild-type and (F, G) *Gli3* homozygotes. (F) and (G) are left and right eyes, respectively, of the same embryo showing two extremities in the range of *Gli3* homozygous eye phenotype. In (F), lens and neural retina have developed although the structures are smaller than normal. Nevertheless, there are several notable defects in development of eye structures. Pigmented epithelium extends into the neural retina territory through the distal edge of the retina, resulting in absence of the ciliary body (c). Histogenesis of the cornea stroma is abnormal and the stroma cells invade the eye chamber (d). Lens is small and arrangement of the lens fiber nuclei is irregular (e). The optic stalk consists of pigmented cells and is not associated with the optic nerve (f). Cornea with an anterior cavity is embedded deep under the skin without sign of eyelid opening (g). (G) The most severe eye phenotype showing the entire retina tissue occupied by pigmented epithelium (h) and, possibly due to the lack of neural retina, the lens has not developed beyond lens vesicle stage (i). (Data are from the unpublished work of R. Sekido and H. Kondoh.)

neural retina and suppression of the pigmented epithelium (Bernier *et al.*, 2000; Loosli *et al.*, 1999; Zuber *et al.*, 1999). In human, some cases of bilateral anophthalmia have been attributed to haploinsufficiency of *Six6* (Gallardo *et al.*, 1999).

It is also known that the bHLHZip transcription factor MITF is expressed in and required for differentiation of pigmented epithelium (Hodgkinson *et al.*, 1993). Mutations in the mouse (mi) cause microphthalmia and poor pigmentation (Hodgkinson *et al.*, 1993; Steingrimsson *et al.*, 1996), and those in human cause Waardenburg syndrome type 2 with aberrant pigmentation (Tassabehji *et al.*, 1994). It is shown that a hypomorphic mutation in MITF in Japanese quail, called *silver*, causes transformation of the pigmented epithelium into neural retina tissue (Mochii *et al.*, 1998).

C. Dorsoventral and Nasotemporal Patterning of the Retina

As the retina is formed, dorsoventral and nasotemporal polarities develop in the neural retina that are later projected to the visual cortexes for spatial recognition of objects. A part of the system to ensure the spatially organized projection of the optic nerves is attributed to graded distribution of Ephrin/Eph (Ephrin receptor) systems among the neural retina cell populations (Drescher *et al.*, 1997; O'Leary and Wilkinson, 1999). For instance, expression of Ephrin B1 and B2 occurs in the dorsal half of the neural retina with the gradient highest in the dorsal margin, while EphB2 and B3 are expressed in the inverse orientation in the ventral half of the neural retina.

Several transcriptional regulators are expressed with dorsoventral asymmetry; TBX5 and VAX2/VAX are examples of these (Fig. 2C). TBX5 is a T-box protein expressed in the dorsal half of the neural retina. Expression of *Vax2* (or *Vax* in the chicken), structurally related to *Vax1*, is expressed in the ventral half of the neural retina (Barbieri *et al.*, 1999; Ohsaki *et al.*, 1999; Schulte *et al.*, 1999).

When *Tbx5* was overexpressed in the developing retina of the chicken embryo by electroporation, the ventral retina also assumed the morphology of the dorsal retina, resulting in dorsoventrally symmetrical development of the retina (Koshiba-Takeuchi *et al.*, 2000). Expression of the ventral retina-associated genes *Vax* and *Pax2* was suppressed, while expression of the dorsal genes Ephrin B1 and Ephrin B2 was extended to the ventral side. In addition, analysis of retinotectal projection of such embryos demonstrated that ventrally originated axons also project to the region of the optic tectum where dorsal axons normally project. Therefore, Tbx5 is one of the determinants of the dorsal characters of the neural retina.

Overexpression of *Vax2/Vax* in the chicken embryonic retina or *Xenopus* embryos, on the other hand, caused ventralization of the neural retina, suppression of *Tbx5*, Ephrin B1, Ephrin B2, and dorsal extension of the expression of

Pax2 (Barbieri *et al.*, 1999; Schulte *et al.*, 1999). Tectal projection from the dorsal retinal axons also became abnormal (Schulte *et al.*, 1999). Therefore, Tbx5 and Vax2/Vax repress each other and determine the overall dorsoventral polarity of the neural retina (Fig. 2C).

The mechanism determining the nasotemporal polarity of the retina is less well characterized, but two winged helix transcription factors BF-1 and BF-2 seem to account for at least a part of the mechanism (Yuasa *et al.*, 1996). In the retina, BF-1 is expressed in the nasal half of the retina, while BF-2 is in the temporal half (Hatini *et al.*, 1994; Yuasa *et al.*, 1996). Overexpression of these transcription factors in the developing chick retina resulted in misprojection of the retinal axons along the nasotemporal axis (Yuasa *et al.*, 1996).

D. Role of Retinoic Acid in Retina Development

The neural retina is noted for its high content of retinoic acid (McCaffery *et al.*, 1992). Retinoic acid is synthesized from retinal by the catalytic action of aldehyde dehydrogenases, at least two being present in higher vertebrates (McCaffery *et al.*, 1992; Yoshida *et al.*, 1998), and it is degraded by CYP26 (P450RA) (Fujii *et al.*, 1997; McCaffery *et al.*, 1999). The activities of these two classes of enzymes in a cell are believed to be determinative in executing retinoic acid-dependent transcriptional regulation. Nuclear receptors (RARs and RXRs) are more widely expressed than the enzyme genes (Kastner *et al.*, 1997; McCaffery *et al.*, 1999). Two enzyme genes for retinoic acid synthesis, ADH2 and RALDH3 (AldDH6), are expressed in the dorsal and ventral halves of the developing neural retina with two peaks of activity at the dorsal and ventral margins, respectively (McCaffery *et al.*, 1992; Grun *et al.*, 2000; Mic *et al.*, 2000; Suzuki *et al.*, 2000) (Fig. 2D). The degrading enzyme CYP26 is expressed in the middle (Fujii *et al.*, 1997) where there is a gap of retinoic acid synthetic activity, suggesting that there are two dorsoventral gradients of retinoic acid formed in the neural retina, one from the dorsal margin to the midpoint, and the other from the ventral margin to the midpoint (McCaffery *et al.*, 1999) (Fig. 2D).

Mutant mice of nuclear retinoid receptors have a greater developmental defect in the ventral side of the neural retina (Kastner *et al.*, 1997). Exogenous administration of retinoic acid to zebrafish embryos resulted in ventralization of the dorsal retina, while a retinoic acid synthesis inhibitor caused a defect in ventral retina development (Marsh-Armstrong *et al.*, 1994). These observations emphasize the importance of the retinoic acid in ventral retina development. However, considering the fact that development of this part of the retina depends more on tissue growth than does the dorsal part as discussed above (Fig. 2A), and that two retinoic acid gradients are probably formed in the neural retina (Fig. 2D), the observations appear to indicate that a higher concentration of retinoic acid leads to development of the ventral retina. The dorsal and medial parts of the neural retina thus may be determined by intermediate and low concentrations of retinoic acid.

E. Cell Type Specification in the Neural Retina

Histogenesis of the neural layer in the retina follows the same general principle as the neural tube in that mitotic progenitor cells are located in the ventricular zone, which in the retina corresponds to the zone facing the pigmented epithelium. However, the uniqueness of the neural retina is that the earlier differentiated cells are deposited in the layer more distal from the ventricular zone, the opposite of the order of layer deposition in other parts of the CNS. The neuronal cell types are also unique to the retina: Photoreceptors (rods and cones), interneurons (horizontal cells, bipolar cells, and amacrine cells), and the ganglion cells which send axons through the optic nerve tract to the visual cortex of the brain (Fig. 4). The Mueller cells, the only intrinsic glial cells in the neural retina span the neural layers, and probably play the role of radial glia in other parts of the CNS. Many transcription factors are associated with these cell types, and their significance in the given cell type is indicated by respective mutant mice (Table I) and by overexpression experiments. Early neuroge-

nesis in retina seems to depend on a set of basic helix–loop–helix (bHLH) transcription factors, MATH5 (Brown *et al.,* 1998), neurogenin 2 (Sommer *et al.,* 1996), NeuroD (Gradwohl *et al.,* 1996), and HES1 (Brown *et al.,* 1998).

Before differentiation of the neurons, MATH5, a bHLH transcription factor related to *Drosophila atonal,* is widely expressed in the neural retina starting at E11 of the mouse (Brown *et al.,* 1998). Interestingly, *Math5* expression does not take place in all the cells, but *Math5* positive cells are interspersed with negative cells. Importantly, expression of *Math5* depends on the activity of *Pax6*. In *Pax6* (*Sey*) heterozygous animals, Math5 expression in the neural retina is reduced to half, and is totally lost in *Pax6* homozygotes. In contrast, expression of Hes1, another proneural bHLH transcription factor, is increased in the *Pax6* (*Sey*) mutants reciprocally with the *Pax6* gene dosage.

Math5 expression is followed by that of *Ngn2* and *NeuroD* (Brown *et al.,* 1998). Hes1 expression starts the latest around E14 (Tomita *et al.,* 1996). Expression of Math5 and Ngn2 is diminished by birth, but that of *NeuroD* and *Hes1* continues to the postnatal period (Brown *et al.,* 1998). It is speculated that Math5 and Ngn2 primarily regulate differentiation of early differentiating cells, that is, ganglion cells, horizontal cells, and cones, whereas NeuroD and HES1 may regulate

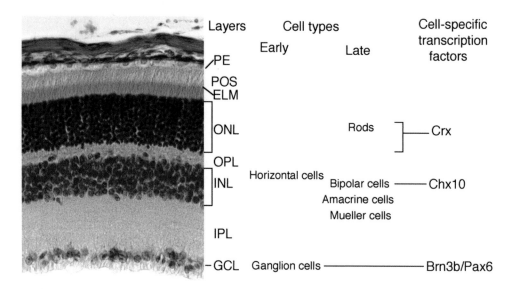

Figure 4 Cell types of the retina. Mature neural retina has regularly arranged layers of neurons. Three nucleus-containing layers are found from the inner side: the ganglion cell layer (GCL); the inner nuclear layer (INL), containing cell bodies of amacrine cells, bipolar cells, and horizontal cells; and the outer nuclear layer (ONL) containing cell bodies of the photoreceptor cells, rods, and cones. The inner plexiform layer (IPL) and outer plexiform layer (OPL) separating these nuclear layers consist mainly of neurites of the neurons. Photoreceptor outer segments (POS) are arranged on the external side of the outer nuclear layer and are faced by the monolayer of pigmented epithelial (PE) cells. Mueller (glial) cells span all layers of neurons to the external limiting membrane (ELM) and place their nuclei in the inner nuclear layer. Neural retina neurons are classified into two groups by the timing of differentiation: those differentiating relatively early (i.e., ganglion cells, horizontal cells, and cones), and those differentiating relatively late (i.e., amacrine cells, bipolar cells, rods, and Mueller cells). These two groups appear to be related to expression of bHLH transcription factors (see text). Cell types exist in which certain homeodomain transcription factors are specifically expressed. Mutant mice of these transcription factors confirm cell-specific action of these transcription factors in the retina (Table I).

Table I Representative Knock-Out/Mutant Mice of Retina and Lens Development

Mutated gene	Mutant phenotype in the retina	References
Otx2	Microphthalmia, anophthalmia	Acampora *et al.* (1995); Ang *et al.* (1996); Matsuo *et al.* (1995)
Rx	No optic vesicle	Mathers *et al.* (1997)
Lhx2	Abortive optic vesicle	Porter *et al.* (1997)
Pax6	Abnormal optic vesicle	Grindley *et al.* (1995); Hill *et al.* (1991); Hogan *et al.* (1986)
Vax1	Stalk with retinal tissue/coloboma	Hallonet *et al.* (1999)
Pax2	Stalk with PE tissue/coloboma	Favor *et al.* (1996)
Gli3	Stalk with PE tissue/coloboma	Franz and Besecke (1991)
BF-1	Stalk with retinal tissue/coloboma	Huh *et al.* (1999)
Brn-3b	Loss of ganglion cells	Gan *et al.* (1996, 1999)
Chx10	No bipolar cells	Burmeister *et al.* (1996)
Crx	No photoreceptor outer segments	Furukawa *et al.* (1999)
Hes1	Premature differentiation of late cells	Tomita *et al.* (1996)
	Mutant phenotype in the lens	
Pax6	No lens placode	Fujiwara *et al.* (1994); Grindley *et al.* (1995); Hill *et al.* (1991); Hogan *et al.* (1986)
Eya1	No phenotype	Xu *et al.* (1999)
c-Maf	Low αA, β, and γ crystallins	Kawauchi *et al.* (1999); Kim *et al.* (1999); Ring *et al.* (2000)
Sox1	Residual γ crystallins	Nishiguchi *et al.* (1998)
Prox1	No lens fibers	Wigle *et al.* (1999)
Pitx3	Arrest at the lens vesicle stage	Semina *et al.* (2000)
Foxe3	Loss of lens epithelium	Blixt *et al.* (2000)

differentiation of late differentiating cells, that is, bipolar cells, amacrine cells, rods, and Mueller cells (Brown *et al.,* 1998). Premature differentiation of rods was observed in *Hes1* knock-out mice (Tomita *et al.,* 1996).

The first differentiating cell type, ganglion cells (Fig. 1, P1.5), depends on *Brn3b (Brn3.2/Pou4f2)* for survival. Brn3b knock-out mice lose 70% of their ganglion cells (Gan *et al.,* 1996), but detailed analysis indicated that this is due to the apoptotic loss of the mutant ganglion cells after deposition in the normal cell layer (Gan *et al.,* 1999).

Expression of Chx10, a Paired-type homeodomain protein, initially occurs widely in the neural retina, and later becomes restricted to the bipolar cells. A nonsense mutation causing protein truncation in the homeodomain was found in the *or (ocular retardation)* mutant mouse, which totally lacks bipolar cells in addition to showing hypocellularity of the retina (Burmeister *et al.,* 1996). Chx10 therefore is essential for differentiation of the bipolar cells.

The most externally located photoreceptors express the OTX-related homeodomain protein CRX, which binds regulatory sequences of photoreceptor-specific genes (e.g., opsins). A dominant-negative form of CRX expressed in the retina by retroviral transfection resulted in dysgenesis of the outer segments of the photoreceptors (Furukawa *et al.,* 1997b), and *Crx* mutant mice totally lack the outer segments of the photoreceptor cells (Furukawa *et al.,* 1999).

Marquart *et al.,* (2001) ablated *Pax6* gene in the anterior half of the neural retina of mouse embryo after E10.5, taking advantage of the Cre-LoxP recombination system. Thickness of the neural retina was normal but only amacrine cells were produced and filled the *Pax6*-ablated anterior half. The loss of *Pax6* resulted in downregulation of expression of the transcription factors, Math5, Brn3b, Ngn2, as well as bHLH factor Mash1, whereas the expression of NeuroD persisted. This confirmed that many transcription factors defining retinal cell types depend on the activity of Pax6, and that one function of NeuroD in the retina is to promote amacrine cell differentiation.

F. Retinal Stem Cells

Cell lineage and clonal analyses of development of the retinal cells have been successfully carried out by labeling progenitors of various stages of development using replication-incompetent retroviral vectors (Turner *et al.,* 1990). During the earliest phase of retina development following optic cup formation, extensive cell mixing occurs among the retinal progenitors, but once cell differentiation initiates, clones derived from a given progenitor make a coherent and tangentially arranged cell cluster (Fekete *et al.,* 1994). An important feature of retina development revealed by the cell-labeling technique is that cell type determination is totally independent of the cell lineage (Turner *et al.,* 1990). The progenitor cells remain multipotential and both daughter cells continue the cell division mode. Even at a late stage of retina development, a single progenitor can still produce multiple cell types of neurons and glia, indicating it possesses stem cell characteristics (Turner and Cepko, 1987). Thus, although various cell types of the retina depend on particular transcription factors, these factors probably do not form regulatory circuits by themselves but respond to the external signals organized in the spatio-temporal order of retina development.

In the mouse and other mammals, development of the eye is complete before adulthood and no substantial growth of tissues occurs thereafter. In amphibia and fish, however, the eyes continue to grow throughout the lifespan, and the ciliary margin of the retina expressing Pax6, Six3, and Rx1 is the source of the retinal progenitors (stem cells) (Perron *et al.,* 1998). It was recently shown that the mammalian ciliary margin also contains cell populations with the potential of retinal stem cells, but the potential is inhibited under ordinary conditions (Tropepe *et al.,* 2000; Ahmad *et al.,* 2000). The pigmented part of the ciliary margin of adult mouse eyes can give rise to virtually an entire repertoire of neural retina and pigmented epithelial cells under clonal culture conditions.

G. Multiple Participations of a Single Transcription Factor Gene in Eye Development and Gene Dosage Effects: Pax6 as an Example

Transcription factors *Pax6, Rx,* and *Six3* are involved in retina development from the earliest period of definition of the retina potential in the CNS, through the step of derivation of the three components from the optic vesicle, to the stage of development of neural cell identity. Pax6 and Six3 are also employed in lens development. Pax6 is also required for lens placode development, and presumably in the maintenance of the cell state of lens epithelium.

Probably because of this multiple participation, the effect of *Pax6* is highly dosage sensitive. Mutations of *Pax6* are discovered in classic *Small eye* (*Sey*) mutant mice, which develop microphthalmia in the heterozygous animals and anophthalmia in the homozygous embryos (Hill *et al.,* 1991; Hogan *et al.,* 1986). The human counterpart of *Sey* is the *aniridia* mutation, which causes aniridia and/or related symptoms in heterozygous patients (Ton *et al.,* 1991). Thus, haploid dosage of the *Pax6* gene is not sufficient for full execution of its functions. Different haploid phenotypes among animals may be explained by involvement of *Pax6* in multiple steps and tissues that differ in their level of Pax6 requirement.

There is an interesting observation that transgenic mice harboring multiple copies of *Pax6* on yeast artificial chromosome display microphthalmia analogous to *Sey* (Schedl *et al.,* 1996). Therefore, the proper level of *Pax6* gene activity is, in fact, critical for normal eye development.

An additional level of complexity exists in the activity of the transcriptional regulator genes. PAX6 proteins exist in two forms, which are derived from two alternatively spliced mRNA isoforms, orthodox and 5a, carrying an additional exon sequence inserted in the C-terminal half of the paired domain, which is the major DNA-binding domain of Pax6 (Epstein *et al.,* 1994). This insertion dramatically alters the DNA-binding specificity of the encoded proteins, implying that these Pax6 proteins have two distinct repertoires of downstream regulatory target genes. Although the actual function of the 5a form of *Pax6* has not been clarified, conservation of this isoform among vertebrate species indicates its significance.

Thus, participation of a single transcription factor gene in multiple steps and possible multifunctionality of a regulatory gene must be seriously considered when interpreting the data of knock-out mice or those of gene overexpression.

III. Lens Development

A. Morphological and Biochemical Events

Lens development may be separated into two phases, first from lens placode to lens vesicle, and second from lens vesicle to mature lens consisting of the lens epithelium positioned anteriorly and lens fibers facing the neural retina through the acellular vitreous body (Figs. 1 and 5). Invagination of the lens placode to form the lens vesicle is analogous to the neural tube closure and separation from the dorsal ectoderm and involves new expression of N-cadherin (Hatta *et al.,* 1987; Takeichi, 1991). Once the vesicle is formed, the posterior half of the vesicle initiates a series of dramatic changes, which are presumably in response to signaling molecules emanating from the optic cup/neural retina: withdrawal from the cell cycle, expression of fiber-specific crystallins, vertical elongation of the cells on the posterior side to form lens fibers, expression of the lens-specific intermediate filament proteins filensin/CP49 (Sandilands *et al.,* 1995), and expression of the major intrinsic protein specific to lens fiber membrane (Pisano and Chepelinsky, 1991).

This first population of lens fiber cells is called *primary lens fibers.* The anterior part of the lens vesicle becomes the lens epithelium, which has the potency to divide, particularly in the zone close to the equator that forms a boundary between the epithelium and the fibers. In this zone, the cells are pushed into the posterior compartment where they cease mitosis, elongate, and constitute secondary lens fibers surrounding the primary lens fibers. This withdrawal from the cell cycle in the lens fibers is dependent on Rb (Morgenbesser *et al.,* 1994) and Cdk inhibitors, Cdkn1b (p27^{Kip1}) and Cdkn1c (p57^{Kip2}) (Zhang *et al.,* 1998).

Crystallins are major soluble proteins of the lens, and provide the lenses with transparency and a high refractive power. They are present in the lens fibers at a high concentration, and even a slight precipitation of the proteins, which may be initiated by an imbalance of crystallins, can cause a cataract. An αA crystallin-lacking mutant mouse line was produced (Brady *et al.,* 1997), where lenses of more or less normal morphology were formed but cataracts developed. Another important characteristic of crystallins is their stability. There is

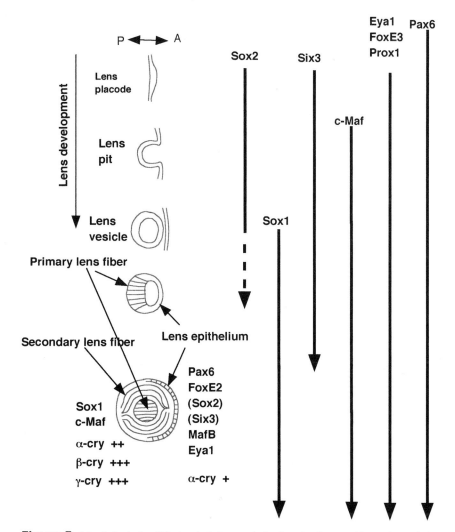

Figure 5 Morphological and biochemical changes during lens development. Expression periods of various transcription factor genes, indicated on the right, are correlated with morphological development of the lens shown on the left. Epithelium-specific and fiber-specific expression of these transcription factors are indicated in the scheme at the lower left corner. Those in parentheses indicate a low expression level. Relative expression levels of crystallin genes (cry) are also indicated by plus marks.

little turnover of crystallin proteins, and crystallins synthesized early in life are carried over throughout the life span.

During lens development, expression of crystallin genes occurs sequentially. In the mouse there are two α-crystallin genes, six γ-crystallin genes (Goring *et al.*, 1992), and a similar number of β-crystallin genes. αA and αB crystallins are expressed from the stage of invaginating lens vesicle and continue to be expressed in both lens epithelium and lens fibers; by contrast, β crystallins and γ crystallins are expressed only during lens fiber development (Kondoh, 1999). In the chicken, γ crystallins are replaced by δ crystallin (δ1 crystallin) coded by a single gene (Kondoh *et al.*, 1991). In contrast to other crystallins in the mouse, δ-crystallin expression in the chicken initiates at the lens placode stage and is highly augmented in the lens fibers. The crystallin genes

are potential targets of the transcription factors that play a role in lens development.

B. Lens Induction

Ectodermal origin of lens and possible interaction with primitive retina had already been noted by the 19th century, but Spemann's experiment reported in 1901 using frog *Rana fusca* demonstrated for the first time the inductive influence of the retina on lens development. He demonstrated that when retina primordia are removed early enough before the contact to the head ectoderm occurs, no lenses develop. Lewis (1904) slightly later observed in frog and salamander embryos that tight contact between the optic cup and the head ectoderm is required for lens induction to occur. A modern version of the Spemann experiment occurred genetically, in Ap2 knock-out mice, where optic vesicles were misoriented without contact with the head ectoderm, resulting in the absence of lens induction (Schorle *et al.,* 1996; Zhang *et al.,* 1996).

Controversy surrounded whether the contact of the lens vesicle to the head ectoderm is required for initiation of lens development, with the variation in the conclusions stemming from the method of retina primordia ablation and interpretation of the data. It has been noted that retinal primordia of the forebrain extensively regenerates after tissue ablation, which may have caused ambiguous situations. In any event, contact of the optic vesicle with the head ectoderm is definitely required for initiation of lens development (Mizuno *et al.,* 1998; Zygar *et al.,* 1998). Under a condition in which lens development somehow initiates, it is noted that at least one of the GroupB *Sox* genes is activated in the region contacted by the optic vesicle, *Sox2* in the case of the mouse, *Sox2* and *Sox3* in the case of the chicken (Kamachi *et al.,* 1998), and *Sox3* in the case of *Xenopus* (Zygar *et al.,* 1998). When the optic vesicle makes contact with the head ectoderm, the surgical separation of these tissues is not possible without use of enzymes degrading extracellular matrices. Thus, if a secretory molecule is involved in lens induction, it must act within a short distance.

For some decades, the optic vesicle was believed to induce a lens when apposed to any place of embryonic ectoderm. This belief, however, no longer exists because Grainger's group, employing a reliable cell-labeling technique, demonstrated that the old data were likely due to incomplete separation of the tissues before grafting of the tissues (Henry and Grainger, 1987).

The competence of the ectoderm to develop into lens tissue in response to the optic vesicle is confined to the head ectoderm. The entire mechanism that defines the competence has not yet been fully resolved, but evidence indicates that expression of Pax6 in the head ectoderm is highly involved.

From before apposition by the optic vesicle, *Pax6* is broadly expressed in the head ectoderm including the prospective lens area (Grindley *et al.,* 1995; Li *et al.,* 1994; Walther and Gruss, 1991), and this expression is not affected by the presence or absence of the optic vesicle (Kamachi *et al.,* 1998; Li *et al.,* 1994).

In homozygous *Sey* mutant mouse embryos lacking *Pax6* gene activity, optic vesicles are formed and, although slightly more expanded than normal vesicles, make contact with the head ectoderm but do not induce thickening of lens placode or any other lens traits (Grindley *et al.,* 1995; Hill *et al.,* 1991; Hogan *et al.,* 1986). Because both the optic vesicle and the responding head ectoderm express Pax6, it is not clear from the mutant phenotype which of the eye components is responsible for the absence of lens development. Using rat *Sey* mutant materials, an organ culture experiment was performed by recombining *Sey* homozygote head ectoderm

and wild-type optic vesicle, and vice versa (Fujiwara *et al.,* 1994). Lenses developed from the ectoderm if the ectoderm was derived from the wild type, regardless of the genotype of the optic vesicle; on the other hand, no lens was induced from the mutant ectoderm even by combination with wild-type optic vesicle. Therefore, *Pax6* expression is an absolute requirement for the head ectoderm to be induced for lens development, but is not required for the inducing activity of the optic vesicle. Thus, *Pax6* provides responsiveness of the head ectoderm to the inductive influence of the optic vesicle.

C. From Lens Placode to Lens Vesicle

Once *Pax6*-positive head ectoderm is contacted by the optic vesicle, the contacted area initiates expression of a few transcription factors: *Sox2* (in the case of mouse), *Six3, Eya1* (Xu *et al.,* 1997), *Prox1* (Wigle *et al.,* 1999), and *Foxe3* (Blixt *et al.,* 2000). Because Sox2 is commonly expressed among sensory placodes (Uchikawa *et al.,* 1999), it is speculated that it is involved in the placodal nature of the tissue.

The problem of how expression and function of these induced transcription factors is dependent on the activity of Pax6 in the head ectoderm is obscured because of involvement of Pax6 in multiple tissues and stages in eye development. Observations in *Sey* (*Pax6*) homozygotes could reflect either the cell autonomous effect of the absence of PAX6 in the ectoderm or an indirect effect caused by the defect of the CNS compartments. To circumvent the problem, *Pax6* was inactivated only in the head ectoderm using a combination of a Cre recombinase transgene and LoxP-flanked *Pax6* locus (Ashery-Padan *et al.,* 2000). Cre recombinase was driven by the lens enhancer of the *Pax6* gene itself, which is active in the head ectoderm before the stage of contact by the optic vesicle. This resulted in initiation of *Pax6* expression in the ectoderm at E9, and in loss of Pax6 activity at E9.5 when the optic vesicle contacts with the ectoderm. Under the condition of ectoderm-specific *Pax6* ablation, *Sox2* expression was induced in the ectoderm, while expression of *Six3* and *Prox1* was lost. Therefore, at least at stages later than E9.5, expression of *Sox2* is independent of *Pax6,* while *Six3* and *Prox1* depend on Pax6 for their expression.

Under the condition that *Sox2* is induced in the head ectoderm in the absence of *Pax6,* no lens placode develops (Ashery-Padan *et al.,* 2000). An interesting comparison is the BMP4 mutant where *Pax6* is expressed in the head ectoderm but *Sox2* is not induced by contact with the optic vesicle, and lens placode fails to develop (Furuta and Hogan, 1998). These observations argue that both Pax6 and Sox2 are required for development of the lens placode.

Kamachi *et al.,* (2001) demonstrated that Pax6 and Sox2 form a co-DNA-binding molecular complex, and this complex activates lens-specific genes and initiates lens placode development, thus providing the basis for the requirement of Pax6 and Sox2.

The optic vesicle-contacted area of the head ectoderm area expressing *Pax6* and *Sox2* together begins to thicken and forms the lens placode. The placode invaginates and develops into the lens. When this invagination occurs, expression of c-Maf and α crystallin begins (Kawauchi *et al.*, 1999; Kim *et al.*, 1999; Ring *et al.*, 2000), and when the lens vesicle is formed, *Sox2* expression largely ceases and is replaced by *Sox1* expression (Kamachi *et al.*, 1998; Nishiguchi *et al.*, 1998). As the lens fibers differentiate, expression of β and γ crystallins initiates in the fiber compartment.

The lens phenotypes of the *Prox1* knock-out mouse embryo and *Foxe3* mutant mouse indicate that major functions of Prox1 and FoxE3 correspond to later phases of lens development, in spite of the early onset of gene expression (Blixt *et al.*, 2000; Wigle *et al.*, 1999).

It is notable that *Six3* and *Eya1* are expressed from the early placodal stage of the lens in a *Pax6*-dependent fashion (Ashery-Padan *et al.*, 2000; Xu *et al.*, 1997), but the significance of their expression in lens development has not yet been determined, in spite of the implied analogy with the case of *toy*-dependent expression of *so, eya,* and *dac* in *Drosophila* (Chen *et al.*, 1997; Pignoni *et al.*, 1997). *Eya1* knockout mouse does not show major eye defects except for an open eyelid (Xu *et al.*, 1999); expression of *Dach* (mouse dac homolog) is neither *Pax6* dependent nor occurs at the placode stage of the lens (Hammond *et al.*, 1998), while the effect of *Six3* overexpression on lens development is not significant (Loosli *et al.*, 1999).

D. Sox, Maf, and Crystallin Gene Regulation

SOX and MAF have been identified as the transcriptional regulators of the crystallin genes. Through the analysis of lens-specific regulation of the δ-crystallin enhancer of the chicken (Hayashi *et al.*, 1987), an essential element was identified that is bound and activated by Sox proteins (Kamachi and Kondoh, 1993; Kamachi *et al.*, 1995). This element is activated by GroupB SOX proteins, SOX1, SOX2, and SOX3, but not by other SOX proteins (Kamachi *et al.*, 1998, 1999). The same SOX proteins also regulate γ-crystallin genes, the mammalian counterpart of δ crystallin. A conserved and essential DNA motif in the promoters of the γ-crystallin genes is the binding site of the SOX proteins that activate the promoters (Kamachi *et al.*, 1995). SOX1 strongly, and SOX2 and SOX3 modestly, activate these crystallin elements. Therefore, the transition from Sox2 to Sox1 in the mouse lens development (Kamachi *et al.*, 1998) will augment γ-crystallin expression.

Sox2$^{-/-}$ embryos die soon after implantation without providing information on eye development (Pevny *et al.*, 1998), but the phenotype of *Sox1* knock-out mice confirmed the essential function of SOX genes in crystallin gene regulation (Nishiguchi *et al.*, 1998). The major phenotype of *Sox1* null mutant mice is the arrest of lens development at the stage of primary lens fiber elongation. Analysis of crystallin gene transcripts indicates that all γ-crystallin genes A to F are severely downregulated, while expression of α and β crystallins is not significantly affected. A low level γ-crystallin gene expression is observed at the initial stage, but becomes totally attenuated by birth. The arrest of morphological development may be accounted for by the absence of continued γ crystallin synthesis, but loss of other structural components is not ruled out. In any case, expression profiles of γ-crystallin genes in *Sox1* null mutant mice must be understood considering the fact that *Sox2*, sharing almost the same characteristics, is expressed in the immediately prior phase of lens development (Kamachi *et al.*, 1998). It is likely that some γ-crystallin gene expression is initiated by the effect of normal *Sox2*, but after attenuation of *Sox2* expression during lens development, *Sox1/2*-dependent transcription of γ-crystallin gene is lost in Sox1 knock-out mice.

In the mouse (and rat) lenses, (c-Maf) is expressed in the lens fibers and Mafb in the lens epithelium (Kawauchi *et al.*, 1999; Yoshida *et al.*, 1997). A third Maf family protein Nrl is also expressed in late phases (possibly in the postnatal stage) (Kawauchi *et al.*, 1999; Liu *et al.*, 1996). Involvement of Maf family bZip transcription factors was indicated by the fact that Maf family proteins bind an element of chicken α-crystallin promoter, which shows lens specificity (Ogino and Yasuda, 1998). c-Maf also binds the Maf-consensus-related sequence in the γ-crystallin promoters present between TATA box and the Sox binding site, which was previously recognized as the γF-1 binding site (Liu *et al.*, 1991). Retrospectively from the results of (*Maf*) knock-out mice, it was also shown that MAF binds the promoter sequences of some β-crystallin promoters (Ring *et al.*, 2000).

Three groups reported targeted disruption of the *c-Maf* gene (Kawauchi *et al.*, 1999; Kim *et al.*, 1999; Ring *et al.*, 2000). In these homozygous mice, lens vesicles are normally formed but elongation of primary lens fibers is largely inhibited. Lens defect is the major phenotype of the homozygous mice, although most of the animals are lost by weaning. There are some discrepancies concerning the effect of *c-Maf* null condition on expression of crystallin genes, but the consensus would be that αA-crystallin expression is reduced, αB-crystallin expression less affected, and expression of the majority of β- and γ-crystallin genes significantly downregulated. It is also noted that *c-Maf* deficiency does not affect expression of other transcription factor genes, such as *Sox1, Sox2, Pax6, Eya1, Eya2,* or *Prox1* (Ring *et al.*, 2000). These results indicate that expression of β and γ crystallins is largely dependent on the activity of *c-Maf*. Considering the result of *Sox1* knock-out mouse, γ-crystallin genes are dependent on both *Sox* and *c-Maf* regulations.

Maf proteins heterodimerize with some other bZip-type transcription factors. ATF-4/CREB2 is one such example. The ATF-4 knock-out mouse develop microphthalmia due to apoptosis of the secondary lens fibers after E15 (Tanaka *et al.,* 1998). This phenomenon can be explained in the context of crystallin regulation by Maf-related transcription factors, implying that Maf/ATF-4 heterodimers play a role late in lens development.

In the chicken (Ogino and Yasuda, 1998), Maf family protein L-Maf has been expressed solely in early lens development, in contrast to multiple expression sites of other Maf protein genes (Ogino and Yasuda, 1998). In addition, ectopic expression of L-Maf is reported to cause ectopic development of the lens tissue in the head ectoderm and cultured neural retina. The protein nature of L-Maf is similar to c-Maf, and its expression during lens development is almost identical to that of c-Maf. Considering the phenotype of c-Maf knock-out mouse, L-Maf may have a redundant function with c-Maf in the chicken lens.

An analysis of ectopic lens development induced by L-Maf indicated that it is restricted to certain predisposed sites expressing *Sox2* (Ogino and Yasuda, 2000). In the head ectoderm, the ectopic lenses arise in a wide stripe on the ventral side of the normal eyes where *Sox2* is strongly expressed (Kondoh, 1999). The neural retina expressing Sox2 and transdifferentiating into lens by itself (Okada *et al.,* 1979) is the one that responds to exogenous L-Maf. These observations and those described above argue for the model that the combination of a *Maf, Sox2,* and *Pax6* is a minimal requirement for lens development.

E. Maintenance of the Lens Epithelial Cell State

Once a lens is made, *Pax6* is expressed more stongly in the epithelial cell layer than in lens fibers (Kondoh, 1999). Although the certain crystallin genes in lens epithelial cells can be activated by Pax6 (Sharon-Friling *et al.,* 1998), fiber-specific β-crystallin genes are repressed by Pax6. This may account for the onset of β-crystallin expression in lens fibers as following the derepression by Pax6 in the epithelial cells (Duncan *et al.,* 1998). Thus, Pax6 may function as a maintenance factor of the lens epithelium.

A forkhead transcription factor gene *Foxe3* is expressed in the lens epithelium and is mutated in *dysgenetic lens* (*dyl*) mouse mutant (Blixt *et al.,* 2000). The mutations in the DNA-binding domain result in failure of the lens vesicle closure and in premature differentiation/apoptosis of the lens epithelial cells. FoxE3 thus represents another case of transcription factors maintaining the state of lens epithelium. In *Xenopus,* a forkhead gene Lens1 related to FoxE1 seems to play an analogous role (Kenyon *et al.,* 1999).

F. Genetic Requirements for Lens Fiber Development

Prox1 expression starts from the lens placode stage, and continues into later development with a preferential expression site at the transition zone from the lens epithelium to the

secondary fibers, called the bow region (Oliver *et al.,* 1993; Tomarev *et al.,* 1996). Lens development in the *Prox1* null mouse (Wigle *et al.,* 1999) is arrested at the vesicle stage more severely than the cases of *Sox1* or *c-Maf* null lenses. In spite of this, expression of crystallins is observed in the late vesicle stage, indicating that the defect of *Prox1* mutant lies not in the crystallin regulation but in the transition from the epithelium to the fiber state. In support of this, the posterior side of the mutant vesicle continues DNA synthesis without expression of Cdk inhibitors Cdkn1b (p27^{Kip1}) and Cdkn1c (p57^{Kip2}), which normally occurs to arrest DNA synthesis in the fiber compartment (Zhang *et al.,* 1998); the expression of E-cadherin, normally restricted to the anterior epithelial compartment, persists to the posterior compartment (Wigle *et al.,* 1999). Considering that *Prox1* expression is prominent in the transition zone, it is conceivable that Prox1 regulates the expression of a reception system to signals derived from the optic cup such as fibroblast growth factors (FGFs).

Arrest of lens development at the lens vesicle stage has been found in aphakia mutant mice (Varnum and Stevens, 1968). *Gli3* (*Xt*) mutants display analogous arrest of lens development at the vesicle stage (Franz and Besecke, 1991). In chimera mice derived from normal and aphakia embryos, aphakia cells are excluded from the lens, indicating cell-autonomous defect of the aphakia homozygous cells (Liegeois *et al.,* 1996). Genetic mapping of the mutation revealed that it has a deletion in the 5′ regulatory region of the *Pitx3* gene causing a serious reduction of *Pitx3* expression in the lens (Semina *et al.,* 2000). Thus, homedomain protein PITX3 is now registered as an essential transcription factor for lens development. In the case of *Gli3,* homozygotes display not only the lens defect but the abnormal retina development as shown in Fig. 3. Therefore, it is possible that the arrest of lens development in *Gli3* homozygotes is secondary to the defect of the retina.

G. Extracellular Signals in Lens Induction and Lens Development

Little is known about the inductive signals for lens differentiation. BMP family proteins, though they themselves are not direct inducers of the lens, are shown to be involved in the lens induction process. BMP4 null embryos die when optic vesicle makes contact with the head ectoderm, but organ culture allows analysis of the following steps of eye development. It was found that Pax6 is expressed in the ectoderm, but *Sox2* is not induced after the vesicle-ectoderm contact, resulting in the failure of lens induction. This defect is rescued by exogenous BMP4 (Furuta and Hogan, 1998). BMP7 null mutant embryos display an eye defect similar to *Pax6* homozygotes: an initially expanded optic vesicle and absence of lens placode development, although not in full penetrance. A key to understanding this phenotype is the finding that *Pax6* expression is impaired in the head ectoderm of the BMP7 mutant (Wawersik *et al.,* 1999). These

BMP mutants suggest a genetic interaction between *Pax6* and *Sox2*, which are essential elements of the initiation of lens development.

FGFs are likely key players for induction of lens fiber differentiation. Strong support for this is the observation that a low level of FGF1/2 stimulated cell growth but a high level of the same growth factors induced fiber differentiation in lens epithelium explant placed in culture (McAvoy and Chamberlain, 1989). Using transgenic mouse, various FGF proteins were overexpressed in the lens and found to be effective in inducing lens fiber differentiation in the epithelium compartment (Lovicu and Overbeek, 1998). However, the expression pattern is compatible with regulation of lens development only for FGF1 and FGF2. These or uncharacterized new members of FGFs are believed to regulate fiber differentiation possibly through FGF receptor 2 (Lovicu and Overbeek, 1998).

Other growth factors such as insulin-like growth factors (IGFs) (Beebe *et al.*, 1987) and platelet-derived growth factors (PDGFs) (Brewitt and Clark, 1988) are also implicated in the lens development, but their precise roles remain to be defined.

Experimental disturbance of these growth factors during eye development causes cataracts in a broad sense. Cataract is an opacity developed in the lens. The transparency of the normal lens cells is maintained by the monophasic state of the highly concentrated crystallin and other lens proteins in the cytoplasm, which may be disturbed by aging process, imbalance of crystallins as exemplified by α-crystallin knockout mice (Brady *et al.*, 1997), or failure to regulate the ion composition. Association of the congenital cataracts with the latter two causes are often noted (Francis *et al.*, 1999). Note that experimental cataract based on growth factor disturbance includes other cases, such as cell death in the lens and delamination of the epithelium.

H. Multiple Pathways of Lens Development

Induction of the head ectoderm by the optic vesicle is not the only possible pathway of lens differentiation. There are three major instances where lenses arise from tissues distinct from the embryonic ectoderm. First, in certain amphibian and fish species, lens regenerates from dorsal iris or outer cornea after lens injury (Bosco, 1986; Eguchi, 1988; Yamada and McDevitt, 1984). In recent molecular analyses, it was demonstrated that the transcription factor genes employed in normal lens development are activated in these regeneration processes (Mizuno *et al.*, 1999; Schaefer *et al.*, 1999). Second, neural retina of avian embryos readily "transdifferentiates" into lens cells under appropriate culture conditions (Okada *et al.*, 1979). It is speculated that sharing of a number of transcription factors between the lens and neural retina (e.g., PAX6, SOX2, SIX3) is the basis for this transition of the cell state (Kamachi *et al.*, 1998). The third and most intriguing instance is the recent discovery that lens cells differ-

entiate from the primordium of the adenohypophysis, the epithelium of Rathke's pouch, in mutant animals where the Shh signal is impeded (Kondoh *et al.*, 2000). Examples are *yot* zebrafish mutants where dominant-negative GLI2 apparently interferes with all Gli activities (Karlstrom *et al.*, 1999) and *talpid³* chicken mutant (Ede and Kelly, 1964) where the cells show no response to Shh (Lewis *et al.*, 1999). It is known that adenohypophysis primordium expresses a number of transcription factors found in the lens, and it is speculated that Shh signaling through GLI proteins inhibits the potential of lens differentiation possessed by the adenohypophysis primordium (Kondoh *et al.*, 2000).

Potentially, multiple developmental pathways give rise to lens differentiation. In all of these cases, expression of a set of transcription factors necessary for lens development, either by *de novo* activation or by carryover from a previous cell state, seems to be the basis of lens differentiation through normal or "transdifferentiation" pathways (Kondoh *et al.*, 2000). The mechanism of normal lens induction dependent on the activity of retina primordium is thus the mechanism used to achieve precise positioning of the lens tissue in front of the retina.

IV. Conservation and Divergence of the Transcriptional Regulatory Systems in Eye Development

The compound eyes of insects and the single eyes of vertebrates have long been viewed as a typical example of convergent evolution where tissues of completely different origins exhibit the same function. However, the demonstration of *Pax6* function in eye development across the animal kingdom indicates that compound eyes and simple eyes share some aspect of evolutionary origin (Gehring and Ikeo, 1999).

A. Pax6 and Eya/Six/Dach Pathway

Sey (*small eye*) mutants of the mouse have a defect in Pax6 function. In *Drosophila*, the *ey* (*eyeless*) gene also codes for a transcription factor with paired box and homeobox similar to Pax6, and its genetic defects result in apoptotic loss of the eye imaginal disks and in the absence of the eye tissues in adults: compound eyes as well as single eyes. Not only did the wild-type *ey* transgene suppress the mutation, but its ectopic expression initiated ectopic development of the compound eyes in a few sites (Halder *et al.*, 1995), for example, antennae and the medial portion of the wings, which are correlated with the site of dpp expression (Chen *et al.*, 1999). Furthermore, expression of mouse *Pax6* in *Drosophila* using the same expression system as *ey* transgene similarly rescued the *ey* mutant phenotype or induced ectopic eye development, demonstrating the conservation of a Pax6-dependent mechanism of eye development (Halder

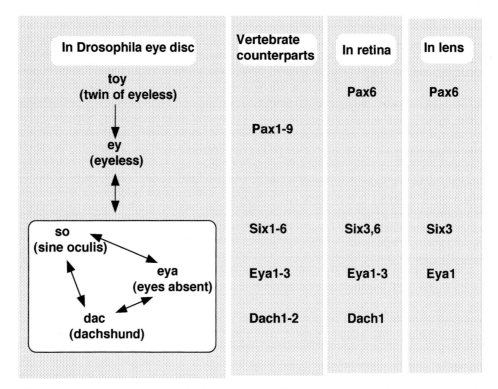

Figure 6 Regulatory genes for eye development in *Drosophila,* and their homologs in vertebrates. Genetic interactions in eye development in *Drosophila* are indicated in the left panel, and expression of corresponding genes in mouse eyes are indicated in the right panels.

et al., 1995). Later analysis revealed that *Drosophila* has another Pax6-related transcription factor gene *toy* (*twin of eyeless*) that is much closer to *Pax6* and acts upstream of *ey* (Czerny *et al.,* 1999). Evidence indicates that mouse *Pax6* expressed in *Drosophila* mimics *toy* rather than *ey* (Czerny *et al.,* 1999), but the initial observation that *Pax6* functions in the *Drosophila* eye system is significant.

Production of ectopic eyes in *Drosophila* is not, however, the unique property of *Pax6/ey* gene activity. It has also been demonstrated that *so* (*sin oculis*) and *eya* (*eyes absent*) together (Pignoni *et al.,* 1997), or *dac* (dachshund) and *eya* together (Chen *et al.,* 1997) promote ectopic eye development after forced expression of these genes, in tissues more widely distributed than in the case of *ey* expression. So, Eya and Dac are thought to constitute a ternary complex in which the homeodomain protein So provides the DNA-binding component (Chen *et al.,* 1997; Pignoni *et al.,* 1997). This complex has a feedback loop with *ey,* where *ey* activates *so* and *eya,* and conversely *eya* and *so* activate *ey.* *Toy* is upstream of this feedback loop (Fig. 6). Therefore, any genetic manipulations that activate this regulatory loop would result in ectopic eye development in the tissue areas where other conditions such as dpp expression (Chen *et al.,* 1999) are met.

In the mouse there are homologs of *eya* (*Eya1* to *Eya3*), *so* (*Six1* to *Six6*), and *dac* (*Dach1* and *Dach2*) identified. Lens placode expresses *Eya1* and *Six3,* while optic vesicle

expresses *Eya1, Eya2, Eya3, Six3,* and *Dach1* (Hammond *et al.,* 1998; Oliver *et al.,* 1995; Xu *et al.,* 1997). In the lens placode, *Six3* expression is dependent on *Pax6* (Ashery-Padan *et al.,* 2000). However, in the retina, which is considered to be analogous to *Drosophila* eye, expression of *Eya, Six,* and *Dach* does not appear to be coordinated: *Eya1* expression is dependent on *Pax6* (Xu *et al.,* 1997), but that of *Six3* or *Dach1* is not (Hammond *et al.,* 1998; Oliver *et al.,* 1995). *Eya1,* the major *Eya* in the eye, is dispensable for eye development in the mutant mouse (Xu *et al.,* 1999).

B. Ectopic Eye Development

Inspired by the ectopic eye development induced by *ey* or *Pax6* in *Drosophila* (Halder *et al.,* 1995), several analogous attempts have been made in vertebrates. Overexpression of mouse *Six3* in *Medaka* embryo resulted in infrequent lens development in otic vesicles (Oliver *et al.,* 1996), which is partly ascribed to abundant expression of *Sox2* in the otic vesicle (Kondoh, 1999). In *Xenopus,* it is reported that injection of RNA coding for FLAG-tagged PAX6 into cleavage stage embryos can induce expression of βB1-crystallin in cell aggregates associated with a rostral ectoderm (Altmann *et al.,* 1997). The same group also reported that overexpression of *Pax6* RNA could induce ectopic retina development with a sharp dependence on the amount of injected RNA (Chow *et al.,* 1999), and the ectopic retinas are occasionally associated with ectopic lenses. As discussed before, overexpression of *Six3* in *Medaka* (Loosli *et al.,* 1999) and *Xenopus* (Bernier *et al.,* 2000) resulted in ectopic retina development particularly in the midbrain tissue. *Six6/Optx2(Six9)* exhibits an analogous but more dramatic effect of ectopic retina induction (Bernier *et al.,* 2000; Zuber *et al.,* 1999). In this last case, ectopic retina is often associated with the lenses. All of these observations suggest that using *Xenopus* (and *Medaka*) embryos, overexpression of *Pax6/Six3/Six6* can activate the endogenous program of retina development, presumably because embryo tissues of these animals are highly sensitive to this sort of genetic perturbation. It is remarkable that even in *Xenopus* embryos, ectopic retina/lens tissues develop only in the head region. It is the combined action of multiple transcription factors that regulates development of complex organs such as eye.

References

Acampora, D., Mazan, S., Lallemand, Y., Avantaggiato, V., Maury, M., Simeone, A., and Brulet, P. (1995). Forebrain and midbrain regions are deleted in Otx2−/− mutants due to a defective anterior neuroectoderm specification during gastrulation. *Development (Cambridge, UK)* **121,** 3279–3290.

Ahmad, I., Tang, L., and Pham, H. (2000). Identification of neural progenitors in the adult mammalian eye. *Biochem. Biophys. Res. Commun.* **270,** 517–521.

Altmann, C. R., Chow, R. L., Lang, R. A., and Hemmati-Brivanlou, A. (1997). Lens induction by Pax-6 in *Xenopus laevis. Dev. Biol.* **185,** 119–123.

Ang, S. L., Jin, O., Rhinn, M., Daigle, N., Stevenson, L., and Rossant, J. (1996). A targeted mouse Otx2 mutation leads to severe defects in gastrulation and formation of axial mesoderm and to deletion of rostral brain. *Development (Cambridge, UK)* **122,** 243–252.

Ashery-Padan, R., Marquardt, T., Zhou, X., and Gruss, P. (2000). Pax6 activity in the lens primordium is required for lens formation and for correct placement of a single retina in the eye. *Genes Dev.* **14,** 2701–2711.

Barbieri, A. M., Lupo, G., Bulfone, A., Andreazzoli, M., Mariani, M., Fougerousse, F., Consalez, G. G., Borsani, G., Beckmann, J. S., Barsacchi, G., Ballabio, A., and Banfi, S. (1999). A homeobox gene, vax2, controls the patterning of the eye dorsoventral axis. *Proc. Natl. Acad. Sci. U.S.A.* **96,** 10729–10734.

Beebe, D. C., Silver, M. H., Blecker, K. S., van Wyk, J. J., Svoboda, M. E., and Zelenka, P. S. (1987). Lentropin, a protein that controls lens fiber formation, is related functionally and immunologically to the insulin-like growth factors. *Proc. Natl. Acad. Sci. U.S.A.* **84,** 2327–2330.

Bernier, G., Panitz, F., Zhou, X., Hollemann, T., Gruss, P., and Pieler, T. (2000). Expanded retina territory by midbrain transformation upon overexpression of Six6 (Optx2) in Xenopus embryos. *Mech. Dev.* **93,** 59–69.

Blixt, A., Mahlapuu, M., Aitola, M., Pelto-Huikko, M., Enerback, S., and Carlsson, P. (2000). A forkhead gene, FoxE3, is essential for lens epithelial proliferation and closure of the lens vesicle. *Genes Dev.* **14,** 245–254.

Bosco, L. (1986). Transdifferentiation of ocular tissues in larval *Xenopus laevis. Differentiation (Berlin)* **39,** 4–15.

Brady, J. P., Garland, D., Duglas-Tabor, Y., Robison, W. G., Jr., Groome, A., and Wawrousek, E. F. (1997). Targeted disruption of the mouse alpha A-crystallin gene induces cataract and cytoplasmic inclusion bodies containing the small heat shock protein alpha B-crystallin. *Proc. Natl. Acad. Sci. U.S.A.* **94,** 884–889.

Brewitt, B., and Clark, J. I. (1988). Growth and transparency in the lens, an epithelial tissue, stimulated by pulses of PDGF. *Science* **242,** 777–779.

Brown, N. L., Kanekar, S., Vetter, M. L., Tucker, P. K., Gemza, D. L., and Glaser, T. (1998). Math5 encodes a murine basic helix-loop-helix transcription factor expressed during early stages of retinal neurogenesis. *Development (Cambridge, UK)* **125,** 4821–4833.

Burmeister, M., Novak, J., Liang, M. Y., Basu, S., Ploder, L., Hawes, N. L., Vidgen, D., Hoover, F., Goldman, D., Kalnins, V. I., Roderick, T. H., Taylor, B. A., Hankin, M. H., and McInnes, R. R. (1996). Ocular retardation mouse caused by Chx10 homeobox null allele: Impaired retinal progenitor proliferation and bipolar cell differentiation. *Nat. Genet.* **12,** 376–384.

Chen, R., Amoui, M., Zhang, Z., and Mardon, G. (1997). Dachshund and eyes absent proteins form a complex and function synergistically to induce ectopic eye development in Drosophila. *Cell (Cambridge, Mass.)* **91,** 893–903.

Chen, R., Halder, G., Zhang, Z., and Mardon, G. (1999). Signaling by the TGF-beta homolog decapentaplegic functions reiteratively within the network of genes controlling retinal cell fate determination in Drosophila. *Development (Cambridge, UK)* **126,** 935–943.

Chiang, C., Litingtung, Y., Lee, E., Young, K. E., Corden, J. L., Westphal, H., and Beachy, P. A. (1996). Cyclopia and defective axial patterning in mice lacking Sonic hedgehog gene function. *Nature (London)* **383,** 407–413.

Chow, R. L., Altmann, C. R., Lang, R. A., and Hemmati-Brivanlou, A. (1999). Pax6 induces ectopic eyes in a vertebrate. *Development (Cambridge, UK)* **126,** 4213–4222.

Coulombre, A. J., and Coulombre, J. L. (1964). Lens development. I. Role of the lens in eye growth. *J. Exp. Zool.* **156,** 39–47.

Czerny, T., Halder, G., Kloter, U., Souabni, A., Gehring, W. J., and Busslinger, M. (1999). twin of eyeless, a second Pax-6 gene of Drosophila, acts upstream of eyeless in the control of eye development. *Mol. Cell* **3,** 297–307.

Drescher, U., Bonhoeffer, F., and Muller, B. K. (1997). The Eph family in retinal axon guidance. *Curr. Opin. Neurobiol.* **7,** 75–80.

Duncan, M. K., Haynes, J. I., 2nd, Cvekl, A., and Piatigorsky, J. (1998). Dual roles for Pax-6: A transcriptional repressor of lens fiber cell-specific beta-crystallin genes. *Mol. Cell. Biol.* **18,** 5579–5586.

Ede, D. A., and Kelly, W. A. (1964). Developmental abnormalities in the head region of the *talpid³* mutant of the fowl. *J. Embryol. Exp. Morphol.* **12,** 161–182.

Eguchi, G. (1988). Cellular and molecular background of Wolffian lens regeneration. *Cell Differ. Dev.* **25**(Suppl.), 147–158.

Epstein, J. A., Glaser, T., Cai, J., Jepeal, L., Walton, D. S., and Maas, R. L. (1994). Two independent and interactive DNA-binding subdomains of the Pax6 paired domain are regulated by alternative splicing. *Genes Dev.* **8,** 2022–2034.

Favor, J., Sandulache, R., Neuhauser-Klaus, A., Pretsch, W., Chatterjee, B., Senft, E., Wurst, W., Blanquet, V., Grimes, P., Sporle, R., and Schughart, K. (1996). The mouse Pax2(1Neu) mutation is identical to a human PAX2 mutation in a family with renal-coloboma syndrome and results in developmental defects of the brain, ear, eye, and kidney. *Proc. Natl. Acad. Sci. U.S.A.* **93,** 13870–13875.

Fekete, D. M., Perez-Miguelsanz, J., Ryder, E. F., and Cepko, C. L. (1994). Clonal analysis in the chicken retina reveals tangential dispersion of clonally related cells. *Dev. Biol.* **166,** 666–682.

Francis, P. J., Berry, V., Moore, A. T., and Bhattacharya, S. (1999). Lens biology: Development and human cataractogenesis. *Trends Genet.* **15,** 191–196.

Franz, T., and Besecke, A. (1991). The development of the eye in homozygotes of the mouse mutant Extra-toes. *Anat. Embryol.* **184,** 355–361.

Fujii, H., Sato, T., Kaneko, S., Gotoh, O., Fujii-Kuriyama, Y., Osawa, K., Kato, S., and Hamada, H. (1997). Metabolic inactivation of retinoic acid by a novel P450 differentially expressed in developing mouse embryos. *EMBO J.* **16,** 4163–4173.

Fujiwara, M., Uchida, T., Osumi-Yamashita, N., and Eto, K. (1994). Uchida rat (rSey): A new mutant rat with craniofacial abnormalities resembling those of the mouse Sey mutant. *Differentiation (Berlin)* **57**, 31–38.

Furukawa, T., Kozak, C. A., and Cepko, C. L. (1997a). Rax, a novel paired-type homeobox gene, shows expression in the anterior neural fold and developing retina. *Proc. Natl. Acad. Sci. U.S.A.* **94**, 3088–3093.

Furukawa, T., Morrow, E. M., and Cepko, C. L. (1997b). Crx, a novel otx-like homeobox gene, shows photoreceptor-specific expression and regulates photoreceptor differentiation. *Cell (Cambridge, Mass.)* **91**, 531–541.

Furukawa, T., Morrow, E. M., Li, T., Davis, F. C., and Cepko, C. L. (1999). Retinopathy and attenuated circadian entrainment in Crx-deficient mice. *Nat. Genet.* **23**, 466–470.

Furuta, Y., and Hogan, B. (1998). BMP4 is essential for lens induction in the mouse embryo. *Genes Dev.* **12**, 3764–3775.

Gallardo, M. E., Lopez-Rios, J., Fernaud-Espinosa, I., Granadino, B., Sanz, R., Ramos, C., Ayuso, C., Seller, M. J., Brunner, H. G., Bovolenta, P., and Rodriguez de Cordoba, S. (1999). Genomic cloning and characterization of the human homeobox gene SIX6 reveals a cluster of SIX genes in chromosome 14 and associates SIX6 hemizygosity with bilateral anophthalmia and pituitary anomalies. *Genomics* **61**, 82–91.

Gan, L., Xiang, M., Zhou, L., Wagner, D. S., Klein, W. H., and Nathans, J. (1996). POU domain factor Brn-3b is required for the development of a large set of retinal ganglion cells. *Proc. Natl. Acad. Sci. U.S.A.* **93**, 3920–3925.

Gan, L., Wang, S. W., Huang, Z., and Klein, W. H. (1999). POU domain factor Brn-3b is essential for retinal ganglion cell differentiation and survival but not for initial cell fate specification. *Dev. Biol.* **210**, 469–480.

Gehring, W. J., and Ikeo, K. (1999). Pax 6: Mastering eye morphogenesis and eye evolution. *Trends Genet.* **15**, 371–377.

Goring, D. R., Breitman, M. L., and Tsui, L. C. (1992). Temporal regulation of six crystallin transcripts during mouse lens development. *Exp. Eye Res.* **54**, 785–795.

Gradwohl, G., Fode, C., and Guillemot, F. (1996). Restricted expression of a novel murine atonal-related bHLH protein in undifferentiated neural precursors. *Dev. Biol.* **180**, 227–241.

Grindley, J. C., Davidson, D. R., and Hill, R. E. (1995). The role of Pax-6 in eye and nasal development. *Development (Cambridge, UK)* **121**, 1433–1442.

Grun, F., Hirose, Y., Kawauchi, S., Ogura, T., and Umesono, K. (2000). ALDH6, a cytosolic retinaldehyde dehydrogenase prominently expressed in sensory neuroepithelia during development. *J. Biol. Chem.* **275**, 41210–41218.

Halder, G., Callaerts, P., and Gehring, W. J. (1995). Induction of ectopic eyes by targeted expression of the eyeless gene in Drosophila. *Science* **267**, 1788–1792.

Hallonet, M., Hollemann, T., Pieler, T., and Gruss, P. (1999). Vax1, a novel homeobox-containing gene, directs development of the basal forebrain and visual system. *Genes Dev.* **13**, 3106–3114.

Hammond, K. L., Hanson, I. M., Brown, A. G., Lettice, L. A., and Hill, R. E. (1998). Mammalian and Drosophila dachshund genes are related to the Ski proto-oncogene and are expressed in eye and limb. *Mech. Dev.* **74**, 121–131.

Hatini, V., Tao, W., and Lai, E. (1994). Expression of winged helix genes, BF-1 and BF-2, define adjacent domains within the developing forebrain and retina. *J. Neurobiol.* **25**, 1293–1309.

Hatta, K., Takagi, S., Fujisawa, H., and Takeichi, M. (1987). Spatial and temporal expression pattern of N-cadherin cell adhesion molecules correlated with morphogenetic processes of chicken embryos. *Dev. Biol.* **120**, 215–227.

Hayashi, S., Goto, K., Okada, T. S., and Kondoh, H. (1987). Lens-specific enhancer in the third intron regulates expression of the chicken delta 1-crystallin gene. *Genes Dev.* **1**, 818–828.

Henry, J. J., and Grainger, R. M. (1987). Inductive interactions in the spatial and temporal restriction of lens-forming potential in embryonic ectoderm of *Xenopus laevis. Dev. Biol.* **124**, 200–214.

Hill, R. E., Favor, J., Hogan, B. L., Ton, C. C., Saunders, G. F., Hanson, I. M., Prosser, J., Jordan, T., Hastie, N. D., and van Heyningen, V. (1991). Mouse small eye results from mutations in a paired-like homeobox-containing gene. *Nature (London)* **354**, 522–525; erratum: **355**, 750 (1992).

Hodgkinson, C. A., Moore, K. J., Nakayama, A., Steingrimsson, E., Copeland, N. G., Jenkins, N. A., and Arnheiter, H. (1993). Mutations at the mouse microphthalmia locus are associated with defects in a gene encoding a novel basic-helix-loop-helix-zipper protein. *Cell (Cambridge, Mass.)* **74**, 395–404.

Hogan, B. L., Horsburgh, G., Cohen, J., Hetherington, C. M., Fisher, G., and Lyon, M. F. (1986). Small eyes (Sey): A homozygous lethal mutation on chromosome 2 which affects the differentiation of both lens and nasal placodes in the mouse. *J. Embryol. Exp. Morphol.* **97**, 95–110.

Huh, S., Hatini, V., Marcus, R. C., Li, S. C., and Lai, E. (1999). Dorsal-ventral patterning defects in the eye of BF-1-deficient mice associated with a restricted loss of shh expression. *Dev. Biol.* **211**, 53–63.

Jean, D., Bernier, G., and Gruss, P. (1999). Six6 (Optx2) is a novel murine Six3-related homeobox gene that demarcates the presumptive pituitary/hypothalamic axis and the ventral optic stalk. *Mech. Dev.* **84**, 31–40.

Kamachi, Y., and Kondoh, H. (1993). Overlapping positive and negative regulatory elements determine lens-specific activity of the delta 1-crystallin enhancer. *Mol. Cell Biol.* **13**, 5206–5215.

Kamachi, Y., Sockanathan, S., Liu, Q., Breitman, M., Lovell-Badge, R., and Kondoh, H. (1995). Involvement of SOX proteins in lens-specific activation of crystallin genes. *EMBO J.* **14**, 3510–3519.

Kamachi, Y., Uchikawa, M., Collignon, J., Lovell-Badge, R., and Kondoh, H. (1998). Involvement of Sox1, 2 and 3 in the early and subsequent molecular events of lens induction. *Development (Cambridge, UK)* **125**, 2521–2532.

Kamachi, Y., Cheah, K. S., and Kondoh, H. (1999). Mechanism of regulatory target selection by the SOX high-mobility-group domain proteins as revealed by comparison of SOX1/2/3 and SOX9. *Mol. Cell Biol.* **19**, 107–120.

Kamachi, Y., Uchikawa, M., Tanouchi, A., Sekido, R., and Kondoh, H. (2001). Pax6 and SOX2 form a co-DNA-binding partner complex that regulates initiation of lens development. *Genes Dev.* **15**, 1272–1286.

Karlstrom, R. O., Talbot, W. S., and Schier, A. F. (1999). Comparative synteny cloning of zebrafish you-too: Mutations in the Hedgehog target gli2 affect ventral forebrain patterning. *Genes Dev.* **13**, 388–393.

Kastner, P., Mark, M., Ghyselinck, N., Krezel, W., Dupe, V., Grondona, J. M., and Chambon, P. (1997). Genetic evidence that the retinoid signal is transduced by heterodimeric RXR/RAR functional units during mouse development. *Development (Cambridge, UK)* **124**, 313–326.

Kaufman, M. H. (1992). "The Atlas of Mouse Development." Academic Press, London.

Kawauchi, S., Takahashi, S., Nakajima, O., Ogino, H., Morita, M., Nishizawa, M., Yasuda, K., and Yamamoto, M. (1999). Regulation of lens fiber cell differentiation by transcription factor c-Maf. *J. Biol. Chem.* **274**, 19254–19260.

Kenyon, K. L., Moody, S. A., and Jamrich, M. (1999). A novel *forkhead* gene mediates early steps during *Xenopus* lens formation. *Development (Cambridge, UK)* **126**, 5107–5116.

Kim, J. I., Li, T., Ho, I. C., Grusby, M. J., and Glimcher, L. H. (1999). Requirement for the c-Maf transcription factor in crystallin gene regulation and lens development. *Proc. Natl. Acad. Sci. U.S.A.* **96**, 3781–3785.

Kondoh, H. (1999). Transcription factors for lens development assessed in vivo. *Curr. Opin. Genet. Dev.* **9**, 301–308.

Kondoh, H., Araki, I., Yasuda, K., Matsubasa, T., and Mori, M. (1991). Expression of the chicken 'delta 2-crystallin' gene in mouse cells: Evidence for encoding of argininosuccinate lyase. *Gene* **99**, 267–271.

Kondoh, H., Uchikawa, M., Yoda, H., Takeda, H., Furutani-Seiki, M., and Karlstrom, R. O. (2000). Mutations in Gli mediated hedgehog signaling lead to lens transdifferentiation from the adenohypophysis anlage. *Mech. Dev.* **96**, 165–174.

Koshiba-Takeuchi, K., Takeuchi, J. K., Matsumoto, K., Momose, T., Uno, K., Hoepker, V., Ogura, K., Takahashi, N., Nakamura, H., Yasuda, K., and Ogura, T. (2000). Tbx5 and the retinotectum projection. *Science* **287,** 134–137.

Lewis, K. E., Drossopoulou, G., Paton, I. R., Morrice, D. R., Robertson, K. E., Burt, D. W., Ingham, P. W., and Tickle, C. (1999). Expression of ptc and gli genes in talpid3 suggests bifurcation in Shh pathway. *Development (Cambridge, UK)* **126,** 2397–2407.

Lewis, W. H. (1904). Experimental studies on the development of the eye in amphibia. I. On the origin of the lens. *Rana palustris. Am. J. Anat.* **3,** 503–536.

Li, H., Tierney, C., Wen, L., Wu, J. Y., and Rao, Y. (1997). A single morphogenetic field gives rise to two retina primordia under the influence of the prechordal plate. *Development (Cambridge, UK)* **124,** 603–615.

Li, H. S., Yang, J. M., Jacobson, R. D., Pasko, D., and Sundin, O. (1994). Pax-6 is first expressed in a region of ectoderm anterior to the early neural plate: Implications for stepwise determination of the lens. *Dev. Biol.* **162,** 181–194.

Liegeois, N. J., Horner, J. W., and DePinho, R. A. (1996). Lens complementation system for the genetic analysis of growth, differentiation, and apoptosis in vivo. *Proc. Natl. Acad. Sci. U.S.A.* **93,** 1303–1307.

Liu, Q. R., Tini, M., Tsui, L. C., and Breitman, M. L. (1991). Interaction of a lens cell transcription factor with the proximal domain of the mouse gamma F-crystallin promoter. *Mol. Cell Biol.* **11,** 1531–1537.

Liu, Q. R., Ji, X., Breitman, M. L., Hitchcock, P. F., and Swaroop, A. (1996). Expression of the bZIP transcription factor gene Nrl in the developing nervous system. *Oncogene* **12,** 207–211.

Loosli, F., Winkler, S., and Wittbrodt, J. (1999). Six3 overexpression initiates the formation of ectopic retina. *Genes Dev.* **13,** 649–654.

Lopez-Rios, J., Gallardo, M. E., Rodriguez de Cordoba, S., and Bovolenta, P. (1999). Six9 (Optx2), a new member of the six gene family of transcription factors, is expressed at early stages of vertebrate ocular and pituitary development. *Mech. Dev.* **83,** 155–159.

Lovicu, F. J., and Overbeek, P. A. (1998). Overlapping effects of different members of the FGF family on lens fiber differentiation in transgenic mice. *Development (Cambridge, UK)* **125,** 3365–3377.

Macdonald, R., Barth, K. A., Xu, Q., Holder, N., Mikkola, I., and Wilson, S. W. (1995). Midline signalling is required for Pax gene regulation and patterning of the eyes. *Development (Cambridge, UK)* **121,** 3267–3278.

Marquardt, T., Ashery-Padan, R., Andrejewski, N., Scardigli, R., Guillemot, F., and Gruss, P. (2001). Pax6 is required for the multipotent state of retinal progenitor cells. *Cell* **105,** 43–55.

Marsh-Armstrong, N., McCaffery, P., Gilbert, W., Dowling, J. E., and Drager, U. C. (1994). Retinoic acid is necessary for development of the ventral retina in zebrafish. *Proc. Natl. Acad. Sci. U.S.A.* **91,** 7286–7290.

Mathers, P. H., Grinberg, A., Mahon, K. A., and Jamrich, M. (1997). The Rx homeobox gene is essential for vertebrate eye development. *Nature (London)* **387,** 603–607.

Matsuo, I., Kuratani, S., Kimura, C., Takeda, N., and Aizawa, S. (1995). Mouse Otx2 functions in the formation and patterning of rostral head. *Genes Dev.* **9,** 2646–2658.

McAvoy, J. W., and Chamberlain, C. G. (1989). Fibroblast growth factor (FGF) induces different responses in lens epithelial cells depending on its concentration. *Development (Cambridge, UK)* **107,** 221–228.

McCaffery, P., Lee, M. O., Wagner, M. A., Sladek, N. E., and Drager, U. C. (1992). Asymmetrical retinoic acid synthesis in the dorsoventral axis of the retina. *Development (Cambridge, UK)* **115,** 371–382.

McCaffery, P., Wagner, E., O'Neil, J., Petkovich, M., and Drager, U. C. (1999). Dorsal and ventral rentinoic territories defined by retinoic acid synthesis, break-down and nuclear receptor expression. *Mech. Dev.* **85,** 203–214.

Mic, F. A., Molotkov, A., Fan, X., Cuenca, A. E., and Duester, G. (2000). RALDH3, a retinaldehyde dehydrogenase that generates retinoic acid, is expressed in the ventral retina, otic vesicle and olfactory pit during mouse development. *Mech. Dev.* **97,** 227–230.

Mizuno, M., Takabatake, T., Takahashi, T. C., and Takeshima, K. (1997). Pax-6 gene expression in newt eye development. *Dev. Genes Evol.* **207,** 167–176.

Mizuno, N., Mochii, M., Takagi, C., Takahashi, T. C., Eguchi, G., and Okada, T. S. (1998). A critical role for the optic vesicle in lens development; a reinvestigation of free lens formation in *Cynops pyrrhogaster*. *Differentiation (Berlin)* **63,** 247–252.

Mizuno, N., Mochii, M., Yamamoto, T. S., Takahashi, T. C., Eguchi, G., and Okada, T. S. (1999). Pax-6 and Prox 1 expression during lens regeneration from Cynops iris and Xenopus cornea: Evidence for a genetic program common to embryonic development. *Differentiation (Berlin)* **65,** 141–149.

Mochii, M., Ono, T., Matsubara, Y., and Eguchi, G. (1998). Spontaneous transdifferentiation of quail pigmented epithelial cell is accompanied by a mutation in the Mitf gene. *Dev. Biol.* **196,** 145–159.

Morgenbesser, S. D., Williams, B. O., Jacks, T., and DePinho, R. A. (1994). p53-dependent apoptosis produced by Rb-deficiency in the developing mouse lens. *Nature (London)* **371,** 72–74.

Morrow, E. M., Furukawa, T., Lee, J. E., and Cepla, C. L. (1999). NeuroD regulates multiple functions in the developing neural retina in rodent. *Development* **126,** 23–36.

Nishiguchi, S., Wood, H., Kondoh, H., Lovell-Badge, R., and Episkopou, V. (1998). Sox1 directly regulates the gamma-crystallin genes and is essential for lens development in mice. *Genes Dev.* **12,** 776–781.

Ogino, H., and Yasuda, K. (1998). Induction of lens differentiation by activation of a bZIP transcription factor, L-Maf. *Science* **280,** 115–118.

Ogino, H., and Yasuda, K. (2000). Sequential activation of transcription factors in lens induction. *Dev., Growth Differ.* **42,** 437–448.

Ohsaki, K., Morimitsu, T., Ishida, Y., Kominami, R., and Takahashi, N. (1999). Expression of the Vax family homeobox genes suggests multiple roles in eye development. *Genes Cells* **4,** 267–276.

Okada, T. S., Yasuda, K., Araki, M., and Eguchi, G. (1979). Possible demonstration of multipotential nature of embryonic neural retina by clonal cell culture. *Dev. Biol.* **68,** 600–617.

O'Leary, D. D., and Wilkinson, D. G. (1999). Eph receptors and ephrins in neural development. *Curr. Opin. Neurobiol.* **9,** 65–73.

Oliver, G., Sosa-Pineda, B., Geisendorf, S., Spana, E. P., Doe, C. Q., and Gruss, P. (1993). Prox 1, a prospero-related homeobox gene expressed during mouse development. *Mech. Dev.* **44,** 3–16.

Oliver, G., Mailhos, A., Wehr, R., Copeland, N. G., Jenkins, N. A., and Gruss, P. (1995). Six3, a murine homolog of sine oculis gene demarcates the most anterior border of the developing neural plate and is expressed during eye development. *Development (Cambridge, UK)* **121,** 4045–4055.

Oliver, G., Loosli, F., Koster, R., Wittbrodt, J., and Gruss, P. (1996). Ectopic lens induction in fish in response to the murine homeobox gene Six3. *Mech. Dev.* **60,** 233–239.

Pei, Y. F., and Rhodin, J. A. (1970). The prenatal development of the mouse eye. *Anat. Rec.* **168,** 105–125.

Perron, M., Kanekar, S, Vetter, M. L., and Harris, W. A. (1998). The genetic sequence of retinal development in the ciliary margin of the *Xenopus* eye. *Dev. Biol.* **199,** 185–200.

Pevny, L. H., Sockanathan, S., Placzek, M., and Lovell-Badge, R. (1998). A role for SOX1 in neural determination. *Development (Cambridge, UK)* **125,** 1967–1978.

Pignoni, F., Hu, B., Zavitz, K. H., Xiao, J., Garrity, P. A., and Zipursky, S. L. (1997). The eye-specification proteins So and Eya form a complex and regulate multiple steps in Drosophila eye development. *Cell (Cambridge, Mass.)* **91,** 881–891; erratum: **92**(4), 585 (1998).

Pisano, M. M., and Chepelinsky, A. B. (1991). Genomic cloning, complete nucleotide sequence, and structure of the human gene encoding the major intrinsic protein (MIP) of the lens. *Genomics* **11,** 981–990.

Porter, F. D., Drago, J., Xu, Y., Cheema, S. S., Wassif, C., Huang, S. P., Lee, E., Grinberg, A., Massalas, J. S., Bodine, D., Alt, F., and Westphal, H. (1997). Lhx2, a LIM homeobox gene, is required for eye, forebrain, and definitive erythrocyte development. *Development (Cambridge, UK)* **124,** 2935–2944.

Ring, B. Z., Cordes, S. P., Overbeek, P. A., and Barsh, G. S. (2000). Regulation of mouse lens fiber cell development and differentiation by the Maf gene. *Development (Cambridge, UK)* **127,** 307–317.

Sandilands, A., Prescott, A. R., Carter, J. M., Hutcheson, A. M., Quinlan, R. A., Richards, J., and FitzGerald, P. G. (1995). Vimentin and CP49/filensin form distinct networks in the lens which are independently modulated during lens fibre cell differentiation. *J. Cell Sci.* **108,** 1397–1406.

Sasaki, H., Nishizaki, Y., Hui, C., Nakafuku, M., and Kondoh, H. (1999). Regulation of Gli2 and Gli3 activities by an amino-terminal repression domain: Implication of Gli2 and Gli3 as primary mediators of Shh signaling. *Development (Cambridge, UK)* **126,** 3915–3924.

Schaefer, J. J., Oliver, G., and Henry, J. J. (1999). Conservation of gene expression during embryonic lens formation and cornea-lens transdifferentiation in *Xenopus laevis. Dev. Dyn.* **215,** 308–318.

Schedl, A., Ross, A., Lee, M., Engelkamp, D., Rashbass, P., van Heyningen, V., and Hastie, N. D. (1996). Influence of PAX6 gene dosage on development: Overexpression causes severe eye abnormalities. *Cell (Cambridge, Mass.)* **86,** 71–82.

Schorle, H., Meier, P., Buchert, M., Jaenisch, R., and Mitchell, P. J. (1996). Transcription factor AP-2 essential for cranial closure and craniofacial development. *Nature (London)* **381,** 235–238.

Schulte, D., Furukawa, T., Peters, M. A., Kozak, C. A., and Cepko, C. L. (1999). Misexpression of the Emx-related homeobox genes cVax and mVax2 ventralizes the retina and perturbs the retinotectal map. *Neuron* **24,** 541–553.

Semina, E. V., Murray, J. C., Reiter, R., Hrstka, R. F., and Graw, J. (2000). Deletion in the promoter region and altered expression of Pitx3 homeobox gene in aphakia mice. *Hum. Mol. Genet.* **9,** 1575–1585.

Sharon-Friling, R., Richardson, J., Sperbeck, S., Lee, D., Rauchman, M., Maas, R., Swaroop, A., and Wistow, G. (1998). Lens-specific gene recruitment of zeta-crystallin through Pax6, Nrl-Maf, and brain suppressor sites. *Mol. Cell Biol.* **18,** 2067–2076.

Sommer, L., Ma, Q., and Anderson, D. J. (1996). neurogenins, a novel family of atonal-related bHLH transcription factors, are putative mammalian neuronal determination genes that reveal progenitor cell heterogeneity in the developing CNS and PNS. *Mol. Cell. Neurosci.* **8,** 221–241.

Spemann, H. (1901). Ueber Correlationen in der Entwickelung des Auges. *Verh. Anat. Ges.* **15,** 61–79.

Steingrimsson, E., Nii, A., Fisher, D. E., Ferre-D'Amare, A. R., McCormick, R. J., Russell, L. B., Burley, S. K., Ward, J. M., Jenkins, N. A., and Copeland, N. G. (1996). The semidominant Mi(b) mutation identifies a role for the HLH domain in DNA binding in addition to its role in protein dimerization. *EMBO J.* **15,** 6280–6289.

Suzuki, R., Shintani, T., Sakuta, H., Kato, A., Ohkawara, T., Osumi, N., and Noda, M. (2000). Identification of RALDH-3, a novel retinaldehyde dehydrogenase, expressed in the ventral region of the retina. *Mech. Dev.* **98,** 37–50.

Takeichi, M. (1991). Cadherin cell adhesion receptors as a morphogenetic regulator. *Science* **251,** 1451–1455.

Tanaka, T., Tsujimura, T., Takeda, K., Sugihara, A., Maekawa, A., Terada, N., Yoshida, N., and Akira, S. (1998). Targeted disruption of ATF4 discloses its essential role in the formation of eye lens fibres. *Genes Cells* **3,** 801–810.

Tassabehji, M., Newton, V. E., and Read, A. P. (1994). Waardenburg syndrome type 2 caused by mutations in the human microphthalmia (MITF) gene. *Nat. Genet.* **8,** 251–255.

Tomarev, S. I., Sundin, O., Banerjee-Basu, S., Duncan, M. K., Yang, J. M., and Piatigorsky, J. (1996). Chicken homeobox gene Prox 1 related to Drosophila prospero is expressed in the developing lens and retina. *Dev. Dyn.* **206,** 354–367; erratum: *Ibid.* **207**(1), 120 (1996).

Tomita, K., Ishibashi, M., Nakahara, K., Ang, S. L., Nakanishi, S., Guillemot, F., and Kageyama, R. (1996). Mammalian hairy and Enhancer of

split homolog 1 regulates differentiation of retinal neurons and is essential for eye morphogenesis. *Neuron* **16,** 723–734.

Ton, C. C. T., Hirvonen, H., Miwa, H., Weil, M. M., Monaghan, P., Jordan, T., van Heyningen, V., Hastie, N. D., Meijers-Heijboer, H., Drechsler, M., Royer-Pokora, B., Collins, F., Swaroop, A., Strong, L. C., and Saunders, G. F. (1991). Positional cloning and characterization of a paired box- and homeobox-containing gene from aniridia region. *Cell (Cambridge, Mass.)* **67,** 1059–1072.

Toy, J., and Sundin, O. H. (1999). Expression of the optx2 homeobox gene during mouse development. *Mech. Dev.* **83,** 183–186.

Tropepe, V., Coles, B. L., Chiasson, B. J., Horsford, D. J., Elia, A. J., McInnes, R. R., and van der Kooy, D. (2000). Retina stem cells in the adult mammalian eye. *Science* **287,** 2032–2036.

Turner, D. L., and Cepko, C. L. (1987). A common progenitor for neurons and glia persists in rat retina late in development. *Nature (London)* **328,** 131–136.

Turner, D. L., Snyder, E. Y., and Cepko, C. L. (1990). Lineage-independent determination of cell type in the embryonic mouse retina. *Neuron* **4,** 833–845.

Uchikawa, M., Kamachi, Y., and Kondoh, H. (1999). Two distinct subgroups of Group B Sox genes for transcriptional activators and repressors: Their expression during embryonic organogenesis of the chicken. *Mech. Dev.* **84,** 103–120.

Varnum, D. S., and Stevens, L. C. (1968). Aphakia, a new mutation in the mouse. *J. Hered.* **59,** 147–150.

Walther, C., and Gruss, P. (1991). Pax-6, a murine paired box gene, is expressed in the developing CNS. *Development (Cambridge, UK)* **113,** 1435–1449.

Wawersik, S., Purcell, P., Rauchman, M., Dudley, A. T., Robertson, E. J., and Maas, R. (1999). BMP7 acts in murine lens placode development. *Dev. Biol.* **207,** 176–188.

Wigle, J. T., Chowdhury, K., Gruss, P., and Oliver, G. (1999). Prox1 function is crucial for mouse lens-fibre elongation. *Nat. Genet.* **21,** 318–322.

Xu, P. X., Woo, I., Her, H., Beier, D. R., and Maas, R. L. (1997). Mouse Eya homologs of the Drosophila eyes absent gene require Pax6 for expression in lens and nasal placode. *Development (Cambridge, UK)* **124,** 219–231.

Xu, P. X., Adams, J., Peters, H., Brown, M. C., Heaney, S., and Maas, R. L. (1999). Eya1-deficient mice lack ears and kidneys and show abnormal apoptosis of organ primordia. *Nat. Genet.* **23,** 113–117.

Yamada, T., and McDevitt, D. S. (1984). Conversion of iris epithelial cells as model of differentiation control. *Differentiation (Berlin)* **27,** 1–12.

Yoshida, A., Rzhetsky, A., Hsu, L. C., and Chang, C. (1998). Human aldehyde dehydrogenase gene family. *Eur. J. Biochem.* **251,** 549–557.

Yoshida, K., Imaki, J., Koyama, Y., Harada, T., Shinmei, Y., Oishi, C., Matsushima-Hibiya, Y., Matsuda, A., Nishi, S., Matsuda, H., and Sakai, M. (1997). Differential expression of maf-1 and maf-2 genes in the developing rat lens. *Invest. Ophthalmol. Visual Sci.* **38,** 2679–2683.

Yuasa, J., Hirano, S., Yamagata, M., and Noda, M. (1996). Visual projection map specified by topographic expression of transcription factors in the retina. *Nature (London)* **382,** 632–635.

Zhang, J., Hagopian-Donaldson, S., Serbedzija, G., Elsemore, J., Plehn-Dujowich, D., McMahon, A. P., Flavell, R. A., and Williams, T. (1996). Neural tube, skeletal and body wall defects in mice lacking transcription factor AP-2. *Nature (London)* **381,** 238–241.

Zhang, P., Wong, C., DePinho, R. A., Harper, J. W., and Elledge, S. J. (1998). Cooperation between the Cdk inhibitors p27(KIP1) and p57(KIP2) in the control of tissue growth and development. *Genes Dev.* **12,** 3162–3167.

Zuber, M. E., Perron, M., Philpott, A., Bang, A., and Harris, W. A. (1999). Giant eyes in Xenopus laevis by overexpression of XOptx2. *Cell (Cambridge, Mass.)* **98,** 341–352.

Zygar, C. A., Cook, T. L., and Grainger, R. M., Jr. (1998). Gene activation during early stages of lens induction in Xenopus. *Development (Cambridge, UK)* **125,** 3509–3519.

22

Development of the Mouse Inner Ear

Amy E. Kiernan,*,1 **Karen P. Steel,*** and **Donna M. Fekete†**

*MRC Institute of Hearing Research, Nottingham NG7 2RD, United Kingdom,
†Department of Biological Sciences, Purdue University, West Lafayette, Indiana 47907

I. Introduction

II. Anatomy of the Inner Ear

III. Development of the Inner Ear

IV. Early Development of the Otic Placode and Otocyst

V. Pattern Formation in the Inner Ear

VI. Sensory Differentiation

VII. Neurogenesis

VIII. The Stria Vascularis

IX. Future Directions

References

I. Introduction

The inner ear is one of the most complex and intricate structures that forms in mammals. Understandably, many developmental biologists in the past have avoided tackling this structure in exchange for simpler models, thereby leaving much of mammalian inner ear development a mystery. For-

tunately, we have learned a certain amount about the early development of the ear from studies in other vertebrates, where manipulation is much easier than in the mouse. Some of this work will be reviewed in the first part of this chapter. However, with the advent of molecular techniques in the mouse and other vertebrates, we are now beginning to scratch the surface of how a complex three-dimensional structure such as the inner ear is formed beyond the otic vesicle stage. By using principles learned from simpler structures such as the vertebrate limb and applying them to the ear, we are starting to develop models of how the broad regions of the ear might be patterned. In addition, by drawing parallels from well-studied genetic models such as *Drosophila* we are also gaining an understanding of how the fine-grained patterning of sensory cells may occur. One lucky feature for unraveling mammalian inner ear development is that mice are particularly sensitive to vestibular defects and will exhibit strange behaviors, such as circling or head-shaking, if a vestibular defect exists. Thus, even genes that are not expected to be involved in ear development can be easily discovered, and several examples can already be found in the literature. Unfortunately, genes involved in cochlear development cannot be uncovered this way unless they are also involved in vestibular development (which many are), although simple tests for severe hearing deficits can be used.

---

1 Present address: The Jackson Laboratory, Bar Harbor, Maine 04609.

II. Anatomy of the Inner Ear

The mammalian inner ear is composed of two diverse functional parts, the cochlea, which is the auditory portion of the ear, and the vestibule, which functions in detecting gravity and linear and rotational motion required for balance (Fig. 1). The sensory receptor cell that performs these many diverse functions is the hair cell. Hair cells, along with their associated supporting cells, make up a sensory patch, and six different sensory patches are located throughout the ear. Within the cochlea lies a single sensory patch, the organ of Corti, which is responsible for transducing sound waves into neuronal impulses. Two different types of hair cells are found in this organ, inner hair cells (IHCs) and outer hair cells (OHCs), which differ in both morphology and function (Fig. 1D). By sending information back to the brain, IHCs act as the traditional receptor cells in the organ of Corti. In contrast, the OHCs are motile cells that primarily receive in-

put from the brain and are thought to function as a cochlear amplifier (Dallos, 1992; Davis, 1983).

Within the vestibule there are five sensory organs that can be of two types: cristae or maculae (Figs. 1B and 1C). Cristae are humplike organs that lie at the base of each of the three semicircular canals. Maculae, of which there are two—saccular and utricular—are flat organs located in the central region of the vestibule. As in the auditory system, two types of hair cells are found in the vestibular organs, type I and type II (shown in Fig. 1C), although the functional significance of having these two cell types is not yet clear.

In addition to being located within the skull, the inner ear is itself encased in a bony shell known as the otic capsule. Within the enclosed epithelial compartments of the ear is the endolymph, a specialized fluid with the unusual ionic composition of high [K^+] and low [Na^+], an essential medium for normal hair cell transduction. Between the otic capsule and the central endolymphatic compartments lies a fluid

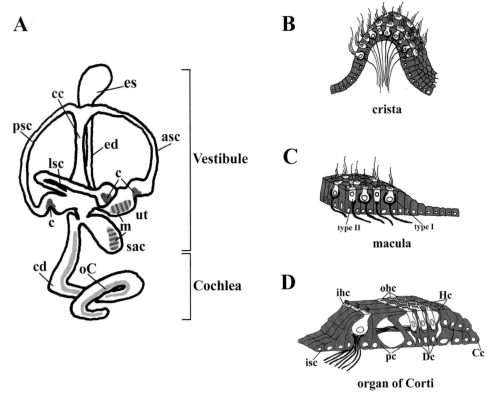

Figure 1 Structure of the mouse inner ear. (A) Structure of the mouse inner ear showing the endolymphatic compartments from a lateral view. The cristae are shown in dark gray, the maculae in stripes, and the organ of Corti in light gray. (B–D) Schematic drawings of the three types of sensory epithelium found in the inner ear (their locations in the ear are indicated in panel A). Support cells are shown in dark gray, nerve fibers in black, and the hair cells in white. The different types of hair cells and supporting cells are indicated. asc, anterior semicircular canal; c, crista; Cc, Claudius' cell; cc, common crus; cd, cochlear duct; Dc, Deiter's cells; ed, endolymphatic duct; es, endolymphatic sac; Hc, Hensen cell; Ihc, inner hair cell; Isc, inner sulcus cell; lsc, lateral semicircular canal; m, maculae; oC, organ of Corti; Ohc, outer hair cells; psc, posterior semicircular canal, pc, pillar cells; sac, sacculus; ut, utricle.

called perilymph, which has low [K$^+$] and high [Na$^+$], similar to most extracellular fluids.

III. Development of the Inner Ear

The inner ear is first identifiable as a thickening in the ectoderm adjacent to the neural plate in the prospective hindbrain region. In mouse this thickening or placode can be observed between embryonic day E8 and E8.5, in an embryo with 4–11 pairs of somites (Fig. 2A; Kaufman, 1992, pp. 50–51; Theiler, 1989, stage 13). It is generally believed that the otic placode is entirely of ectodermal origin. How-

ever, a recent morphological study has suggested that some of the otic placode may be derived from the neural folds in the chick (Mayordomo *et al.,* 1998), an assertion that will require detailed fate mapping to confirm. Once established, the placode begins to invaginate, forming a structure called the otic pit or cup (Fig. 2B; E9; 13–20 somites; Kaufman, 1992, pp. 60–61; Theiler, 1989, stage 14). This pit continues to deepen until a complete vesicle or otocyst is formed (Fig. 2D; E9.5; 21–29 somites; Kaufman, 1992, pp. 78–79; Theiler, 1989, stage 15). During the period of late otic pit and early otocyst formation, cells in the anteroventral portion of the vesicle delaminate from the epithelium and migrate medially. These cells will coalesce to form the eighth cranial

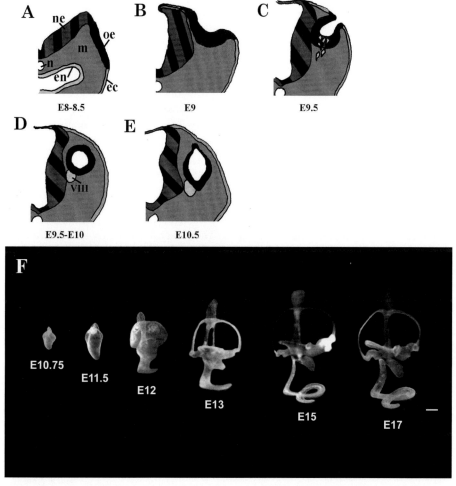

Figure 2 Development of the mouse inner ear. (A–E) Schematic diagrams of the inner ear in cross section as it develops from otic placode (A), to otic pit (B, C), to otocyst (D, E) stages. The surrounding tissues that may play a role in the induction and subsequent development of the ear are indicated. Dorsal is up and lateral is to the right. (F) Lateral views of paint-filled inner ears from otocyst stages and continuing at regular intervals until the adult form of the ear is attained at E17. ec, ectoderm (light gray); en, endoderm (light gray); m, mesenchyme (dark gray); n, notochord (white); ne, neuroectoderm (dark gray stripes); oe, otic ectoderm (black); VIII, eighth cranial ganglia, shown migrating from the otic epithelium in (C). Bar in panel F = 200 μm. (Panel F was taken with permission from Morsli *et al.,* 1998.)

ganglion and their peripheral dendrites will later innervate the sensory epithelium of the ear (Figs. 2C–E, Carney and Silver, 1983; Li *et al.*, 1978).

After the otic vesicle stage of development, ear morphology becomes quite complex and is better viewed in whole-mount situations (Fig. 2F; Morsli *et al.*, 1998). Morphogenetic changes in the otic vesicle can first be observed by E10.5 and these include the elongation of the ventral portion of the vesicle so that it resembles more of an oval shape, as well as a small projection dorsally that will later develop into the endolymphatic duct and sac (Fig. 2E; Kaufman, 1992, pp. 116–117; Theiler, 1989, stage 17). In addition the epithelium does not show a uniform thickness at this time, because the ventral regions are much thicker than the dorsal regions. The semicircular canals evaginate dorsally and are formed by E13 (Fig. 2F), although they continue to grow and thicken until birth. The cochlea extends first ventrally (E11.5) but then begins coiling at E12 and continues until E17.5, by which time it has reached its full 1.5 turns (Fig. 2F; Morsli *et al.*, 1998; Sher, 1971).

IV. Early Development of the Otic Placode and Otocyst

A. Specification of the Otic Placodes

The question of how and when the otic placodes are specified has been the subject of much study throughout the twentieth century. Because the ear vesicle does not form when prospective ear ectoderm is grown in isolation or in foreign environments (see, for example, Gallagher *et al.*, 1996; Jacobson, 1963; Waddington, 1937; Yntema, 1939), these studies have focused on which of the surrounding tis-

sue or tissues might be inducing its formation and at what point in development. The tissues that may potentially play a role in ear induction include the neural plate, mesoderm, notochord, and endoderm (Figs. 2A and 3). Prior to the advent of molecular techniques, these questions were primarily addressed through transplantation, ablation, and explantation experiments in amphibia or chick. These studies revealed that otic determination is not the result of a single inductive event but rather takes place over a period of time and is likely to require a number of different inductive signals (Jacobson, 1966; reviewed in Torres and Giraldez, 1998).

Questions of when instructive signals emanate from the surrounding tissues, and when the ectoderm becomes competent to respond to these instructions, were addressed by transplanting foreign or prospective ear ectoderm at different ages into the otic region of the amphibian embryo. These studies revealed that inductive signals probably occur between mid- and late gastrulation, because ear ectoderm during this time period showed a greater propensity to form otic vesicles than foreign ectoderm or ear ectoderm obtained from earlier stages (Gallagher *et al.*, 1996; Yntema, 1933). Furthermore, ectoderm derived from regions around the prospective ear was more likely to form ear vesicles than ectoderm from farther away, indicating that the first step in otic determination is the specification of an otic field (Fig. 3C). The experiments in amphibia by Jacobson (1963, 1966) have suggested that prior to the establishment of an otic field, there is first a placodal field, which includes a band of ectoderm that lies in a semicircle around the neural plate encompassing the prospective ear, nose, and lens placode (Fig. 3B). These experiments demonstrated that when head ectoderm is rotated 180 degrees, the sensory organs will be respecified at their correct positions in relation to the embryo, indicating

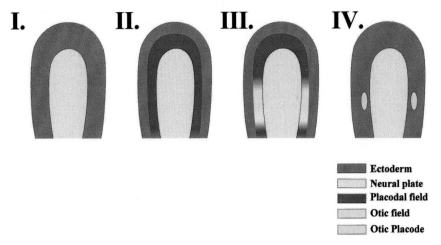

Figure 3 Model for the stages of induction of the inner ear. (I–IV) Schematic diagrams of dorsal views of the anterior portion of an early to late stage neurula amphibian embryo. Anterior is up. The first stage of the induction of the inner ear may be the specification of a placodal competent area (II; red). Secondly, within the placodal field, an otic competent area is specified (III; yellow), and finally an otic placode is formed within the otic field (IV; gray).

that the placodal ectoderm is multipotent during early development. Torres and Giraldez (1998) have taken this idea further and suggested that general placodal competence is maintained by a set of genes (discussed below in the section titled Genes Expressed in the Otic Placode) that are then localized in unique combinations to either the nasal, lens, or otic areas. This issue of multiplacodal competence has been addressed recently in the chicken through misexpression of the otic placode-inducing molecule Fgf3 (Vendrell *et al.,* 2000). These experiments demonstrated that ectopic otic placodes could be induced in a large area of ectoderm that extends both anterior and posterior to the normal position of the otic placode. However, otic placodes did not form in the most anterior ectoderm, including the region occupied by the lens and nasal placodes, suggesting that some regions of the head ectoderm are not otic competent. However, Groves and Bronner-Fraser (2000) showed in the chick that anterior epiblast can form otic placodes when grafted in the correct place, suggesting that either the Fgf3 experiments were not performed early enough before otic competence is lost or other inducing factors are required.

Recent evidence indicates that perhaps *all* ectoderm is competent to respond to *specific* placodal inducers in early embryos and that neither a general placodal inducer nor a multiplacodal state may exist. Rather, it may be the *loss* of competence in all but a few patches that determines the final location of the neurogenic placodes. These patches are presumed to survive based on their proximity to inductive sources that are *specific* to the different placodes (Graham and Begbie, 2000). As such, the idea that all neurogenic placodes have a common developmental and evolutionary origin has been challenged precisely because the inductive mechanisms controlling their specification into different types of placodes are so variable.

Classical transplantation studies have examined which of the surrounding tissues were responsible for the induction of the otic placode. These experiments were performed by explanting the prospective ear ectoderm in conjunction with one or a number of its potential inducers, or by ablating one or a number of its potential inducers. Results of these experiments established that, surprisingly, several of the surrounding tissues, including mesoderm, notochord, and neural tube, could induce an otocyst (see, for example, Jacobson, 1966; Kohan, 1944; Waddington, 1937; Yntema, 1950). It was shown, however, that the most normal differentiation of the otocyst occurred when several potential inducing tissues, such as mesoderm, endoderm, and neural tube, were included in the explant (Jacobson, 1966). In addition, it was revealed that these tissues were not equipotent in their inducing ability; for example, neural tube by itself produced better differentiation of the otic vesicle than mesoderm combined with endoderm. More recently, these questions have been addressed in zebrafish by transplants of germring tissue (early mesoderm and endoderm) or hindbrain tissues. These ex-

periments confirmed the earlier transplantation studies by showing that both types of tissues could induce ectopic otic vesicles (Woo and Fraser, 1997, 1998). Interestingly though, the hindbrain could not induce vesicles during its very early stages of development (80% epiboly), whereas germring tissue demonstrated this ability, suggesting that the nature and/or amount of the inducing signals differs between these two tissues during development.

The question of inductive influences has also been addressed through genetic studies in the zebrafish (Mendonsa and Riley, 1999). Using a number of zebrafish mutants in which one or a number of the potential inducing tissues was either missing or abnormal, Mendonsa and Riley (1999) have examined whether otic placode development was affected. Results of these experiments demonstrated that mutants, such as *cyclops* or *one-eyed pinhead,* that have a partial or complete absence of prechordal mesendoderm experienced a delay in otic placode development, and the morphology of the otic vesicle was abnormal. Moreover, the amount of the delay was dependent on the amount of prechordal mesendoderm that was missing. Mutations that affect the development of the notochord, such as *no tail* and *floating head,* did not affect the timing or morphology of the inner ear. The role of the hindbrain was examined in the mutant *valentino,* a mutation that affects rhombomeres 5 and 6 and is a homolog of the mouse mutant *kreisler.* Interestingly, no delay was found in the appearance of the otic placode, although subsequent patterning of the vesicle was abnormal.

These results have revealed an important role for the mesoderm in the correct timing of otic placode formation and support the manipulation studies that indicated an early role for the mesoderm, although the identity of the mesoderm-inducing factor(s) remains unknown. However, the fact that the placode does eventually form in these mesoderm mutants suggests that other tissues (such as hindbrain) can induce otic placodes, albeit at a later time point. Indeed, a recent report performed in the chicken revealed that otic induction requires at least *two* signals: FGF19, expressed in the mesoderm, and Wnt8c, expressed by the neural tube (Ladher *et al.,* 2000). This study demonstrated that only by combining these two molecules could previously uncommitted explanted ectoderm express a full complement of otic markers. However, despite the expression of several early otic markers, the induced ears did not produce hair cells. Presumably, the specification of hair cells from naive ectoderm requires additional factors, or the amount and timing of FGF19 and/or Wnt8c expression in the cultures is inappropriate. Interestingly, FGF19 was found to be expressed in the neural tube at slightly later time points, providing a possible explanation for the observation that the hindbrain alone can eventually induce an otic placode. It will be important to examine the expression of these two molecules in mice, to see whether they are likely to be playing a similar synergistic role in mouse otic induction.

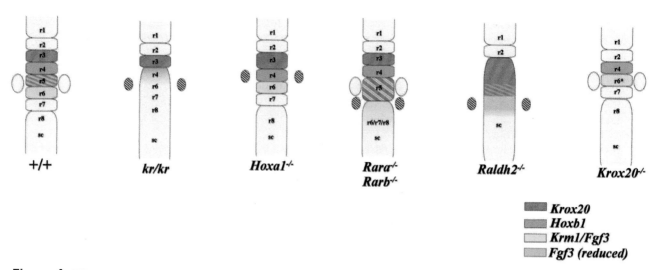

Figure 4 Effects on the hindbrain and inner ear of mutations that disrupt the patterning of the hindbrain. Schematic diagrams of dorsal views of hindbrains and otocysts (ovoids) from wild-type and various mutants. Rhombomeres (r) are numbered from r1 to the spinal cord (sc). Dysmorphic otocysts are indicated by stripes. Colored rhombomeres indicate the presence of the expression of the designated gene. *Note:* The yellow r6 rhombomere in the Krox20 mutant represents only the expression of *Fgf3*, since the expression of *Krm1* was not investigated.

B. Role of the Hindbrain in Otic Patterning: Mouse Mutants

Perhaps the best evidence of the influence of the hindbrain on otic patterning comes from preexisting mouse mutants or, more recently, from mouse knock-outs that show defects in the development of both the hindbrain and otocyst (Fig. 4). This influence is proven most clearly in mutants such as *kreisler* and *Hoxa1$^{-/-}$* (Figs. 4B and 4C), in which the respective genes are never normally expressed in the ear, thereby demonstrating that the otic defects could only be derived from abnormal inductive signals. Other mutants, such as those deficient in certain retinoid signaling proteins (Figs. 4D and 4E), also show hindbrain and ear defects. Although it is possible that retinoic acid is required directly for early otocyst development, the location as well as type of defects reported in these mutants is consistent with the ear abnormalities resulting secondarily from hindbrain disturbances.

As described in Chapter 6, during early development the hindbrain transiently displays repeating metameric units called rhombomeres that appear to function as compartments in the early hindbrain (Lumsden and Krumlauf, 1996). The otocyst develops adjacent to rhombomeres (R) 5 and 6 in normal animals, and so these are the rhombomeres most likely to play a role in ear development. Many genes expressed in the hindbrain appear to respect rhombomeric boundaries, thereby conferring each rhombomere with a unique genetic identity. It is therefore relatively simple to ascertain whether a rhombomere is missing or duplicated simply by examining gene expression patterns in the hindbrain. Interestingly, it is generally those mutants in which a rhombomere is missing or expanded that tend to affect the

development of the ear. Below are summaries of some of the hindbrain mutants that have been described to date and the effects of these mutations on the development of the inner ear (see also Fig. 4).

1. Kreisler

The *kreisler* (*kr*) mutation was identified in an X-ray mutagenesis experiment because the mouse showed hyperactivity and circling behavior (Hertwig, 1942), indicative of a balance defect of the inner ear. Analysis of the *kr/kr* mutants showed that the hindbrain did not exhibit its usual rhombomeric bulges caudal to R3, and the otic vesicle was displaced laterally and subsequently developed into a large cystic structure (Deol, 1964; Hertwig, 1942). Deol suggested, based on the transplantation experiments performed in birds and amphibia, that the malformed ear may be a secondary effect that is caused by defects present in the hindbrain. His theory was proven correct when Cordes and Barsh (1994) cloned the gene at the kr locus, *Krm1,* and showed that it was expressed in the hindbrain adjacent to the developing ear, but not in the otic vesicle itself. *Krm1* is a transcription factor of the bZIP family and its expression domain includes R5 and R6. The most recent analysis of the *kr/kr* hindbrain, using transgenic mouse lines carrying Hox/lacZ reporters and by examining genes expressed in discrete rhombomeres, revealed that R5 failed to form and R6, although present, was abnormal (Fig. 4B; Manzanares *et al.,* 1999). According to Deol (1964) sensory areas do form in *kr/kr* animals, usually maculae and some cristae, although their size and position is abnormal. Additionally the otic capsule was rarely complete and often displayed gaps on the medial side of the cochlear duct.

2. *Hoxa1*

Two different knockouts of the *Hoxa1* gene were produced (Chisaka *et al.,* 1992; Lufkin *et al.,* 1991) and both showed defects in the hindbrain and inner ear, as well as in some of the derivatives of the neural crest. Analysis of the hindbrain in these mutants revealed that R5 failed to form, R3 was expanded, and R4 was reduced (Fig. 4C; Carpenter *et al.,* 1993; Mark *et al.,* 1993; Rossel and Capecchi, 1999). The otic vesicles of the mutants were displaced laterally and rostrally and developed adjacent to r4, rather than in their normal position next to R5 and R6. During subsequent development of the mutant ears, the endolymphatic ducts failed to form and, similar to *kr* mutants, the ears developed into large cystic structures, although much variation was observed. Some sensory development was reported to occur, although it included primarily vestibular epithelium such as cristae and maculae similar to the *kr/kr* mutant (Lufkin *et al.,* 1991). Double knock-outs of both *Hoxa1* and *Hoxb1*, which led to a more sizable expansion of R3 and an absence or severe reduction of R4 in addition to R5, were not reported to have more severe otic defects than *Hoxa1* mutants alone, although the ear abnormalities in the double mutants were more penetrant (Gavalas *et al.,* 1998; Rossel and Capecchi, 1999).

3. Mutants Deficient in Retinoic Acid Signaling Components

Retinoic acid (RA) is a well-known signaling molecule that appears to be involved in the patterning of numerous regions of the developing embryo including the hindbrain (Morriss-Kay and Ward, 1999). Several recent knockouts that affect RA signaling have produced patterning defects in the hindbrain and inner ear. The analysis of these rhombomeric defects (location and type) as well as the similarities of the inner ear defects to other hindbrain mutants (*Kr* and *Hoxa1*) suggests that the otic abnormalities are likely to be due to the hindbrain defects.

In double knock-outs of the retinoic acid receptors, Rara and Rarb, R5 appeared enlarged and R5/6/7 boundaries were abnormal (Fig. 4D; Dupe *et al.,* 1999). Interestingly, ectopic otic vesicles were often observed. These otic vesicles formed from an enlarged otic pit, which may be induced due to an enlargement of R5. However, further development of these ectopic vesicles was not supported and the ear subsequently developed normally in these mutants.

In addition to these receptor knock-outs, RA production has also been affected through disruption of the retinaldehyde dehydrogenase 2 (*Raldh2*) gene, an enzyme required for much of the RA synthesis in the embryo. In addition to other defects (Niederreither *et al.,* 1999), this knock-out resulted in a severe disruption of the caudal hindbrain (Niederreither *et al.,* 2000). Specifically, the rhombomeres posterior to R2 did not appear to form normally and gene expression studies suggested there was an "anteriorization"

of the caudal hindbrain. As observed in other hindbrain mutants, the otic vesicle was small and displaced laterally (Fig. 4E). Unfortunately, an analysis of the mature labyrinth was precluded because the embryos died at E10.5. However, gene expression studies of the otic vesicle using the regionalized homeobox-containing genes, *Pax2* and *Hmx3*, showed there were likely to be severe problems in otic patterning, because *Pax2* expression was absent and *Hmx3* transcripts failed to localize normally in mutant vesicles.

4. Fgf3: An Ear Inducing Molecule?

One of the best candidates for at least one of the signals emanating from the hindbrain is a member of the fibroblast growth factor family, FGF3. FGF3 was originally postulated to be involved in ear induction because it showed strong expression in the otic region of the hindbrain during the period of early ear development (Wilkinson *et al.,* 1988). *In vitro* experiments in chick demonstrated that otic vesicle formation could be prevented using *Fgf3*-targeted antisense oligonucleotides and antibodies (Represa *et al.,* 1991). More recently it was shown that misexpression of *Fgf3* in the chick caused ectopic otic placodes to form (Vendrell *et al.,* 2000). These experiments provide strong evidence for a role for FGF3 as an otic inducer. However, a targeted disruption of the *Fgf3* gene in the mouse displayed no defects in otic vesicle formation, although its subsequent patterning was severely disrupted (Mansour, 1994). These conflicting results can be explained in two ways: (1) The mouse and chicken differ in the identity of their otic inducer(s) or (2) another molecule can compensate for FGF3's normal inducer role in the mouse knock-out. Although the latter explanation cannot be ruled out, expression studies suggest that the former may be true because in the mouse *Fgf3* expression in R5 and R6 is not heightened until the otic placodes are already morphologically identifiable (McKay *et al.,* 1996). In contrast, *Fgf3* expression in chickens is observed in the hindbrain at a time more compatible with otic induction (Mahmood *et al.,* 1995).

The otic phenotype in *Fgf3*$^{-/-}$ mouse mutants was similar in some respects to the other hindbrain mutants in that the endolymphatic duct (ED) failed to form. However, the phenotype did not appear to be as severe as those described for *kr, Hoxa1,* and RA mutants. Early, the morphology of the otic vesicle appeared normal in that it was not displaced or small. Later, most of the ear structures were present but were described as distended and swollen, suggesting endolymphatic hydrops. These observations led Mansour (1994) and colleagues to speculate that the role of FGF3 produced by the hindbrain is to induce the formation of the endolymphatic appendage, and on failing this in *Fgf3*$^{-/-}$ mutants, endolymph production is not properly regulated and the ear becomes swollen. A complication of this interpretation is that *Fgf3* is also expressed in the otic vesicle, indicating that a patterning mechanism autonomous to the otic vesicle may be responsible for the otic defects present in *Fgf3*$^{-/-}$ mutants. However, McKay *et al.* (1996) argue that in fact it is the expression in the hindbrain that is critical for ED formation, because *kr/kr* mutants, which also fail to form an ED, showed decreased expression of *Fgf3* in the hindbrain but not in the otic vesicle.

Further support for a role for FGF3 in the development of the ED has been demonstrated by overexpression of the gene in chicken, which resulted in an elongated ED (Vendrell *et al.,* 2000). A recent knock-out of the IIIb isoform of the *Fgfr2* gene revealed that the effects of Fgf3 in the ear are likely to be mediated via this receptor, since some of the otic defects (including absence of an ED) observed in these mutant mice are very similar to those described for *Fgf3 −/−* mutants (De Moerlooze *et al.,* 2000).

When taken together, although these hindbrain mutants reveal a role for the hindbrain in otic development, its role in the induction of the otic placode or vesicle remains controversial, since vesicles, albeit abnormal, form in all the hindbrain mutants. It is possible, however, that a mutation that interferes with the otic-inducing signal has not yet been discovered. Furthermore, in *Rara* and *Rarb* double mutants, an expanded otic pit forms next to an enlarged R5, suggesting the two phenomenon are causally related. One possibility is that the hindbrain, although it does not initiate otic placode formation, can influence the size of the placode or otic pit. This interpretation would be consistent with the fact that when R5 (or R5 and R6) is missing, the otic vesicle is small. These data suggest a particularly key role for R5 involvement in the signals emanating from the hindbrain, although this interpretation may be influenced by the fact that R5 is the rhombomere that is most often missing. Furthermore a disruption of the *Krox20* gene results in a loss of R3 and R5, but the otic vesicle develops normally (Fig. 4F). This apparent discrepancy can be explained in two ways: (1) R6, which appears unaffected in Krox20 mutants, is sending the required signals; or (2) the required signals could be sent from R5 prior to its loss, since R5 is initially established but then later lost in *Krox20 −/−* animals (Schneider-Maunoury *et al.,* 1993, 1997; Swiatek and Gridley, 1993).

Further analysis of the role of the hindbrain in ear induction would be facilitated by better analysis of the mutant ear phenotypes on a similar genetic background, because it is not clear exactly which mutations produce the most severe otic phenotypes. It is clear, for example, that FGF3 is one of the signals coming from the hindbrain since its absence disrupts otic patterning. It is likely, however, that FGF3 is not the only signal emanating from the hindbrain since all the other mutants, such as *kr/kr, Hoxa1 −/−, Ralh2 −/−, Rara −/−/ Rarb −/−,* exhibit more severe otic defects despite the fact that *Fgf3* expression is reduced or missing in their respective hindbrains. Gene expression studies on the mutant otic vesicles, such as those performed in the *Ralh2 −/−* mutants, would further elucidate the effects of hindbrain signaling on otic patterning.

C. Genes Expressed in the Otic Placode

As yet, despite the number of mouse (Steel, 1995; Steel *et al.,* 2001) and zebrafish (Malicki *et al.,* 1996; Whitfield *et al.,* 1996) mutants that have been examined for ear defects, no gene has been found to be essential for otic placode formation. This suggests that there is either some functional redundancy regarding placodal specification or that the gene(s) involved is/are also essential for early embryogenesis, such that the mutations cause very early lethality. A number of genes are expressed at placodal stages (see Bussoli *et al.,* 2000; Torres and Giraldez, 1998), although only a handful are expressed early enough and in a sufficiently nonregionalized manner (i.e., throughout the otic placode) to represent good candidates for placode specification.

The majority of genes found to mark the early otic placode in its entirety are transcription factors, mostly homeobox-containing, including *Dlx3* (chicken; Pera and Kessel, 1999), *Pax2* (chicken; Groves and Bronner-Fraser, 2000; zebrafish; Krauss *et al.,* 1991), *Pax8* (Xenopus; Heller and Brandli, 1999), *Lmx1* (chicken; Giraldez, 1998; Torres and Giraldez, 1998), *Sox9* (mouse; Heller and Brandli, 1999), *Sox2/3/21* (chicken; Groves and Bronner-Fraser, 2000; Uchikawa *et al.,* 1999), *Six4* (chicken; Esteve and Bovolenta, 1999), *Gbx2* (chicken; Shamim and Mason, 1998), and *Gata3* (chicken; Sheng and Stern, 1999). In addition to these transcription factors, two secreted factors, *Fgf3* (mouse; Mahmood *et al.,* 1996; McKay *et al.,* 1996) and *Bmp7* (chicken: Groves and Bronner-Fraser, 2000; Oh *et al.,* 1996), and a receptor, *Epha4* (mouse; Nieto *et al.,* 1992), are also expressed throughout the placode at early time points. Interestingly, several of these genes, including *Dlx3* (Akimenko *et al.,* 1994), *cSix4* (Esteve and Bovolenta, 1999), and *cGata3* (Sheng and Stern, 1999) have been reported to mark the entire placodal area (Fig. 3B) before being restricted to the otic field. These expression data lend support to Torres and Giraldez's model (1998) in which each placode (ear, lens, or nose) is specified by a unique combination of genes that are first expressed throughout the entire placodal region, but then are subsequently restricted to one or two of the placodes.

V. Pattern Formation in the Inner Ear

One remaining challenge is to determine the *rules* by which gene expression patterns get interpreted and converted into complex pattern formation in the ear. It seems obvious that the *process* of patterning will be manifested by differential spatial and temporal control of cell behavior (such as cell shape changes or migration), as well as cell number (involving both proliferation and programmed cell death). Descriptive details about the control of cell number continue to be refined and are discussed first. However, the *mechanisms* by which these processes are controlled, the involvement of cell–cell signaling, and the site(s) of action and range of diffusible signals remain murky. The degree to which patterning information arises directly within the ear epithelium versus being specified and/or maintained from the surrounding tissues, especially the mesenchyme, remains an open question.

A. Ear Morphogenesis and the Control of Cell Proliferation and Death

Morphogenesis and growth of the inner ear is undoubtedly influenced by cell number, which will be manifest through regulation of cell proliferation and cell death. As the otic placode deepens to form a cup and ultimately pinches off to form a vesicle, cell numbers continue to rise. The topology of dissociating the otic vesicle from the overlying ectoderm requires epithelial cells at the junction to change their cell–cell associations, and in fact this process is accompanied by a focus of cell death at the junction (Alvarez and Navascues, 1990; Lang *et al.*, 2000; Marovitz *et al.*, 1977; Represa *et al.*, 1990).

As the ear progresses through the critical stages of ear morphogenesis, cell proliferation and cell death are spatially regulated. These two processes have been compared systematically in the chick ear (Lang *et al.*, 2000). One surprising finding was that differential outgrowth of the endolymphatic duct, canal plates, and cochlear duct was not necessarily correlated with increased cell proliferation in the growing regions at early stages. Rather, the ventral half of the vesicle contains a disproportionate number of the dividing cells, whereas outgrowth of the dorsal half appears to involve thinning of the epithelial surface with only a modest increase in cell proliferation. The ventral proliferation associated with cochlear duct outgrowth in mouse is discussed below.

As morphogenesis proceeds, the spatial patterning of proliferation gets progressively more complex, as might be expected. Some of the regions of reduced cell proliferation can be explained by the early withdrawal of sensory organs from the cell cycle. Others can be correlated with areas that will undergo programmed cell death (such as the ventromedial wall of otocyst and, in the chick, the fusion plates of the semicircular canals). The function of programmed cell death, particularly within the ventromedial wall where it is especially robust in birds and mammals, remains a mystery. This focus of cell death is located near the proximal end of the cochlear duct and, intriguingly, is not apparent in lower vertebrates that lack a cochlea (Bever and Fekete, 1999).

B. Semicircular Canal Morphogenesis

Morphogenesis of the semicircular canals is a stunning example of complex tissue remodeling to generate rather precise three-dimensional structures. In general, a canal arises initially as an outpocketing from otocyst, called a *canal pouch*. The two vertical canals develop from a single dorsally directed pouch, whereas the horizontal (lateral) canal develops from a separate laterally directed pouch (Fig. 2F, E12). To form a canal, two apposing surfaces approach each other over a relatively broad area near their centers and then fuse into a single epithelial layer. Experiments in the frog suggest that the driving force for the approach phase is dependent on hyaluronan secretion from the canal plates into the underlying mesenchyme (Haddon and Lewis, 1991).

In the mouse this fusion process begins first in the anterior canal, followed closely by the posterior and then lateral canal (Martin and Swanson, 1993). Once fusion has occurred, the fusion plate cells disappear to leave only the rim around the outside of each canal (Fig. 2F, E13). In the chick, this rim (the duct proper) has an increased level of cell proliferation as the canal grows significantly in size. In the canal fusion plates of the chick ear, programmed cell death appears to play an important role in removal of fusion plate cells (Fekete *et al.*, 1997). However, in the mouse this remodeling may be accomplished by a different mechanism, because extensive programmed cell death appears to be lacking in this locale (Martin and Swanson, 1993; Nishikori *et al.*, 1999). Instead, cells of the fusion plate may be resorbed into the expanding rim of the canals.

It has been shown recently that the laminin-like molecule netrin1 (*Ntn1*) plays a critical role in the canal fusion process (Salminen *et al.*, 2000). This result came as something of a surprise because prior to this report, netrins were known primarily for their role in axon guidance. Mice generated from a gene-trapping experiment were identified as having a disruption in the *Ntn1* gene and did not develop lateral or posterior canals, and the anterior canal was small. During development the fusion process did not proceed normally in any of the canals in the mutant. Expression studies showed that *Ntn1* is expressed in the central portions of the canal, consistent with a proposed role in the canal fusion process. However, the fact that the anterior canal does form in the *Ntn1* mutants, albeit abnormally, suggests that *Ntn1* cannot be the sole mediator of the canal fusion event.

C. Gene Expression Domains and Morphogenesis

There is an expanding list of genes that are expressed in or around the otic epithelium in spatially restricted patterns (Fig. 5) prior to overt morphogenesis (e.g., canal formation or cochlear duct elongation) and long before overt cellular differentiation (e.g., recognizable hair cells). A majority of these genes contain homeoboxes, and thus are presumed to be transcription factors. Genes that exhibit regionalized expression at early stages should be considered candidates for

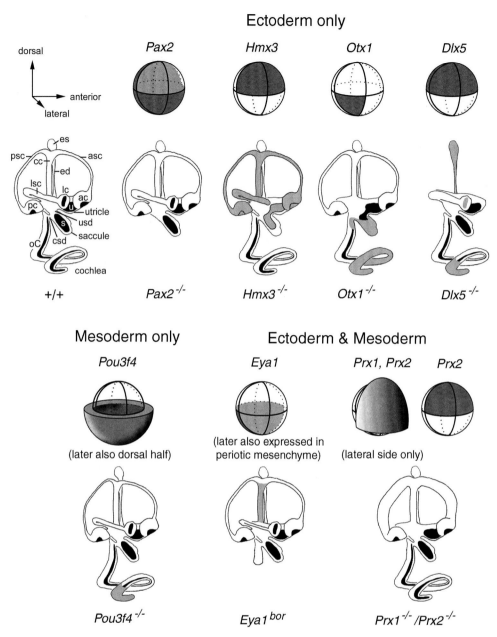

Figure 5 Potential compartments in the early otocyst. The mouse otic vesicle is represented as a hollow sphere with proposed compartment boundaries segregating it into dorsal-ventral, anterior-posterior, and medial-lateral halves. (Top) The expression domains of several genes known to be involved in ear morphogenesis are shown from a lateral perspective. The precision of the compartment boundaries with respect to the gene expression domains is hypothetical, as in most cases the degree of overlap between the different genes has either not been definitely established or is not as precise as that shown (see text). (Bottom) A schematic of the inner ear phenotypes that result when the gene is disrupted, shown from a lateral perspective. Structural defects that are highly penetrant are indicated either as altered or missing structures. When the penetrance of a structural defect or deficiency is variable, the affected region is indicated with gray shading, but appears normal in morphology. See text for references and further details. ac, anterior crista; asc, anterior semicircular canal; cc, common crus; csd, cochlear saccular duct; ed, endolymphatic duct; es, endolymphatic sac; lc, lateral crista; lsc, lateral semicircular canal; oC, organ of Corti; pc, posterior crista; psc, posterior semicircular canal; S, saccular macula; U, utriclar macula; usd, utricular saccular duct.

the regulation and specification of patterning information. Experimental support for such a role is currently limited to a handful of cases in which gene knock-out gives an ear phenotype; these are discussed below.

Genes that are normally expressed in the otocyst itself may be easier to interpret with respect to these deletion phenotypes, since the requirement for the gene product is more likely to be cell autonomous, particularly in the case of transcription factors. Genes that are expressed exclusively in the surrounding mesenchyme, but still can alter patterning in the otic epithelium, lend support to the idea that mesenchymal-epithelial interactions also play an important role in the growth and patterning of the membranous labyrinth.

1. Genes Expressed in Ectoderm That Affect Ear Morphogenesis

Thus far, most of the genes found to affect early ear morphogenesis are expressed within the otic epithelium. These genes are expressed in asymmetrical domains in the very early otocyst (and several are expressed prior to this). The fact that their expression domains at otocyst stages roughly correspond to the location of the defect when the gene is removed (Fig. 5) suggests that these early expression domains may play an important role in the gross patterning of the inner ear. For example, *Pax2* is expressed in the medial and ventral portions of the otocyst, and two mouse mutants that affect the *Pax2* gene show a complete agenesis of the cochlea, a ventral ear structure (Favor *et al.,* 1996; Torres *et al.,* 1996). Similarly, mice deleted for *Dlx5* (expressed dorsally) do not develop the anterior and posterior semicircular canals, structures that are located in the dorsal part of the adult inner ear (Acampora *et al.,* 1999; Depew *et al.,* 1999).

Loss of *Hmx3,* which is normally expressed dorsolaterally, can affect all three semicircular canals, although the lateral and posterior canals are most often abnormal (Hadrys *et al.,* 1998). A milder allele showed absence of only the lateral crista and ampullae as well as a continuity between the saccular and utricular maculae, perhaps reflecting a failure of the utriculosaccular duct to undergo constriction (Wang *et al.,* 1998). *Otx1* is expressed in a discrete wedge in the posterior-ventral-lateral part of the otocyst. Deletion of *Otx1* leads to loss of the lateral canal and crista, as well as the utriculosaccular and cochleosaccular ducts. When the *Otx1* knock-out is combined with the loss of one *Otx2* allele (*Otx2*$^{+/-}$), most of the defects are more penetrant and more severe, particularly the cochlear dysmorphogenesis. An effort was made to see whether the Otx2 protein could substitute for *Otx1* by cloning the human *OTX2* gene into the disrupted *Otx1* locus of the knock-out mouse (Morsli *et al.,* 1999). *OTX2* was able to partially rescue all but the lateral canal phenotype, suggesting a unique role for the OTX1 protein in lateral canal morphogenesis.

2. Genes Expressed in Mesoderm That Affect Ear Morphogenesis

It is generally accepted that the mesoderm plays an important role in sculpting the inner ear epithelium, although exactly how this is done and which genes play a role remain largely unknown. A study in the chicken demonstrated the importance of the mesenchyme by showing that when otic vesicles, devoid of mesenchyme, were transplanted to foreign mesenchyme, they failed to undergo any of the major morphogenetic events, such as cochlear outgrowth or semicircular canal formation (Swanson *et al.,* 1990). Interestingly, although most of the overt morphogenesis of the inner ear failed to occur in the transplants, most cellular differentiation, including sensory, proceeded normally.

Unfortunately, despite the probable importance of the mesenchyme, there is only one mouse inner ear mutant, the *Pou3f4* mouse knock-out, in which the affected gene's exclusively mesenchymal expression pattern may suggest a role in mesenchymal-epithelial interactions. *Pou3f4* is expressed in the ventral mesenchyme at early otocyst stages, although later its expression spreads throughout the entire periotic mesenchyme. *Pou3f4* mutants show hypomorphic bone formation, manifested as a thinning of the otic capsule and a widening of certain bony foramena (Phippard *et al.,* 1999). A possible disruption of mesenchymal-epithelial interactions is suggested by the hypomorphic development of the cochlear duct, which does not express *Pou3f4* yet fails to complete its normal 1.75 turns in most *Pou3f4* knock-out mice. It is possible, however, that *Pou3f4* is not involved directly in mesenchymal-epithelial interactions, but that its effects are derived secondarily due to failure of proper mesenchymal differentiation. For example, the report of abnormal spiral ligament cells, which are derived from the mesenchyme, associated with reduced endocochlear potential in the cochlear duct, suggests an alternative mechanism for the cochlear defects (Minowa *et al.,* 1999).

3. Genes Expressed in Ectoderm and Mesoderm That Affect Ear Morphogenesis

At least two examples are known thus far of mutants whose genes are expressed in both the ectoderm and mesoderm, complicating the interpretation of whether the inner ear defects are due to failed mesenchymal inductive interactions or whether they are due to failed otocyst-autonomous patterning events. However, in the case of *Prx1* and *Prx2* double knock-outs, which lead to loss of the lateral canal, thickening of the anterior and posterior canals, and a reduction in the size of the otic capsule, it is likely that the defects arose due to failed mesenchymal-epithelial interactions, because both genes only colocalize in the lateral periotic mesenchyme (ten Berge *et al.,* 1998).

In the case of the *Eya1* gene, which is expressed in both the ear ectoderm at early otocyst stages and later throughout the mesenchyme, it is more difficult to resolve whether its main effect on inner ear morphogenesis is mediated through its ectodermal expression, inductively through its mesenchymal expression, or both. Total loss of expression of *Eya1* results in one of the most severe inner ear phenotypes described to date, as the otocyst fails to undergo all structural specializations, including formation of the otic ganglion (Xu *et al.,* 1999). The defect is greater than what might be predicted by the gene's early expression domain, which is restricted to the ventral half of the vesicle and the cells of the developing eighth cranial ganglion at E10.5 (Kalatzis *et al.,* 1998). Nonetheless, the temporal expression of the gene in ectoderm versus mesenchyme suggests that the profound morphogenetic block in the knockout is due to a direct effect in the ectoderm rather than an indirect effect from the mesenchyme. A hypomorphic allele (*Eya1^{bor}*) has also been identified that results in a severe truncation of the cochlear duct, while leaving the vestibular portion of the ear relatively intact (Johnson *et al.,* 1999). *Eya1* is of particular interest because it underlies an autosomal dominant human syndrome, branchio-oto-renal (BOR) syndrome, which causes early developmental defects in all three divisions of the ear (outer, middle, and inner) as well as kidney defects (Abdelhak *et al.,* 1997).

In the chick it has been shown that Bmp4 may play an important role in signaling from the mesenchyme, as mesenchymally implanted beads soaked in noggin, a Bmp4 antagonist, impaired inner ear morphogenesis (Chang *et al.,* 1999; Gerlach *et al.,* 2000). The noggin beads seemed to exert their strongest effect on the semicircular canal development, consistent with Bmp4 expression in the mesenchyme surrounding the future canal portion of the otocyst. However, Bmp4 is also expressed within the epithelium of the ear, making it unclear exactly which Bmp4-expressing areas were impaired by the noggin beads.

D. Regionalized Expression of Genes and Possible Compartments

It is worthwhile to consider some aspects of the regionalized gene expression domains in the context of pattern formation. Once the neural progenitors have emigrated from the inner ear, the remaining otocyst segregates into approximately seven gross structural elements: the endolymphatic duct/sac, the three semicircular canals with their ampullary enlargements, the utricle, the saccule, and the cochlear duct. One sensory organ will arise within each of the compartments, with the exception of the endolymphatic duct/sac.

One simplistic mechanism to specify the major parts of the ear would be to put each under independent genetic control. In such a model, the early otic vesicle would consist of a mosaic of seven distinct compartments, each giving rise to one of the seven structures recognized in the adult. There would be cochlea-specifying genes, saccule-specifying

genes, and so on. In the parlance of the field, such genes would be called *selector genes,* because they select a specific identity for the cells that express them that is ultimately interpreted as patterning (Lawrence and Struhl, 1996). In the extreme, each piece of the mosaic would be expected to represent a lineage-restricted compartment, defined by differential gene expression and reinforced by precise lineage-restriction boundaries across which cells would not mix. The question to be asked is whether our current knowledge of lineage compartments, gene expression domains, and the effects of gene knock-outs supports such a model for the ear.

With respect to the existence of lineage compartments, there are no data of sufficient precision to draw any conclusions. In the mouse, Li *et al.* (1978) cultured sectors of the otic vesicle, and followed their subsequent differentiation to generate a specification map for the mouse otocyst. Their data are consistent with a gross regionalization of the otocyst: dorsal structures (such as the endolymphatic duct and semicircular canals) arise from the dorsal otocyst, while ventral structures (the cochlea and saccule) arise from the ventral otocyst. Similar generalities can be made for anterior versus posterior and medial versus lateral tissues, but there is significant overlap in the fates of the eight dissected sectors. Interpretation of these results may be confounded by the phenomenon of embryonic regulation, whereby when one part of an embryonic field is removed, and another can alter its prospective fate to partially or completely fill in the missing part. Another concern is that it might be necessary for different parts of the otocyst to signal to one other in order for the normal pattern to manifest itself. Growing portions of the otocyst in isolation would necessarily disrupt such inductive signaling. Direct fate mapping data, whereby small groups of cells are labeled and allowed to develop *in situ,* are needed but are only just beginning to be compiled for other species, including the chicken (Brigande *et al.,* 2000a).

With respect to gene expression, a precise description of potential compartments is also lacking. Side-by-side comparisons of gene expression domains in single organisms are needed to address this issue with certainty. Relatively few genes have been explicitly compared in the mouse with the goal of determining whether there is overlap or whether the genes meet at precise boundaries. For example, three different members of the *Iroquois* family of transcription factors appear to have different, but partially overlapping, expression domains in the E9.5 ventral otocyst (Bosse *et al.,* 1997). One day later, the three genes are expressed throughout the otocyst. Likewise, *Otx1* and *Otx2* display nested expression domains in the ventrolateral otocyst at E9.5 (Morsli *et al.,* 1999). Along the medial-lateral axis, *Hmx3* (a lateral gene), *Epha4* (*Sek*), and *Pax2* (medial genes) have nearly mirror-image expression patterns, although they apparently overlap slightly at the limits of their expression in the dorsal and ventral otocyst at E10.5 (Rinkwitz-Brandt *et al.,* 1996).

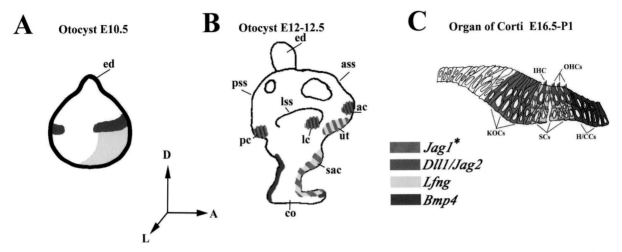

Figure 6 Dynamic patterns of expression of markers of the sensory organs during development. Expression data were accumulated from several different sources (see text for details), so the exact stage cannot be specified. A and B are lateral views of the otocyst. *Note: Jag1 is probably expressed at E10.5 based on other species, but the pattern in mouse has not yet been demonstrated. A, anterior; ac, anterior crista; ass, anterior semicircular canal; co, cochlea; D, dorsal; ed, endolymphatic duct; KOCs, Kölliker's organ cells; H/CCs, Hensen's and/or Claudius' cells; IHCs, inner hair cells; L, lateral; lc, lateral crista; lss, lateral semicircular canal; OHCs, outer hair cells; pc, posterior crista; pss, posterior semicircular canal; sac, saccular macula; SCs, supporting cells; ut, utricular macula. (The outline of the organ of Corti in panel C was reprinted from Lim and Rueda, 1992. Copyright © 1992, with permission from Elsevier Science.)

Although the cochlea arises from the ventral otocyst, it is difficult to identify a cochlea-specific compartment based on the current set of known gene expression domains. More likely, cochlear development requires a more complex arrangement and is defined by several genes. It is notable that the epithelial-derived components of the cochlear duct are divided longitudinally into a number of distinct tissue types (e.g., modiolus, inner sulcus, inner hair cells, tunnel of Corti, outer hair cells, outer sulcus, stria vascularis, and Reissner's membrane). Some of the nested and/or partially overlapping gene expression domains in the ventral otocyst appear to presage what will become the long axis of the cochlea. This raises the possibility that specific combinations of genes may generate unique domains that are required to pattern the location and/or specification of the distinct anatomical components observed later (as seen in Fig. 6C).

In contrast to the overlapping gene expression domains that seem to predominate in the ventral otocyst, the dorsal otocyst may be more orderly. Kiernan *et al.* (1998) showed that the endolymphatic duct of the chicken arises just medial to a lateral domain defined by *SOHo1,* as shown schematically in Fig. 7. Brigande *et al.* (2000b) further showed that the *SOHo1* domain meets the PAX2 domain at the dorsal pole of the ear, with no overlap in the expression of the two genes. That is, they appear to define a mediolateral (ML) boundary at the dorsal pole. Recent fate mapping of the chicken otic cup suggests that there is a ML lineage boundary at this same location, and it intersects an AP boundary that bisects the endolymphatic duct (Brigande *et al.*, 2000b). These lineage boundaries are positioned such that cell–cell signaling across them might be involved in specifying the

location of endolymphatic duct outgrowth. At the present time, there is no direct evidence to support or refute this idea.

In conclusion, the possibility that the otocyst is segmented into compartments, and that these compartments acquire their identity through selector genes, is still only a

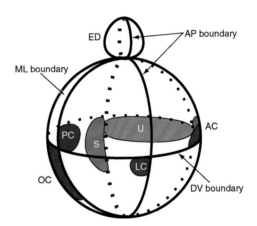

Figure 7 Model for sensory organ specification. One possible scenario for how the different sensory organs of the mouse inner ear may be induced by their proximity to the compartment boundaries shown in Fig. 5. This figure is primarily derived by analogy to studies in the chick, where both the endolymphatic duct and several sensory organs were shown to arise at the mediolateral (ML) boundary (Brigande *et al.*, 2000b; Kiernan *et al.*, 1997). In the mouse, the lateral crista has been shown to form at the boundary of the Otx1 expression domain in anteroventrolateral otocyst (Morsli *et al.*, 1999). An anterior-posterior (AP) boundary that bisects the endolymphatic duct was demonstrated by fate mapping the chick otic cup (Brigande *et al.*, 2000b). AC, anterior crista; AP, anterior-posterior; DV, dorsal-ventral; ED, endolymphatic duct; LC, lateral crista; ML, medial-lateral; OC, organ of Corti; PC, posterior crista; S, saccule; U, utricle.

working model. The fact that the loss-of-function phenotypes often manifest missing structural elements is intriguing. Ultimately, though, the best evidence in support of the model would be if genetic manipulations could be designed that gave rise to homeotic transformations, that is, phenotypes in which one part of the ear is substituted for another. Barring that, it will be important to establish with certainty that there are precise lineage boundaries that correspond in some meaningful way to the gene expression boundaries, and that the two are mutually interdependent for appropriate pattern formation.

E. Compartment Boundaries and the Specification of Sensory Organs

There is a stronger case to be made for a role for gene expression boundaries and the specification of sensory organ location. During early ear patterning in the mouse, several genes appear to label some or all of the sensory organ anlage prior to overt differentiation (Fig. 6B), including the secreted factor, *Bmp4* (marking cristae; Morsli *et al.*, 1998), the Notch-pathway modulator *Lunatic fringe* (marking maculae and the organ of Corti; Morsli *et al.*, 1998), and the Notch ligand, *Jagged1* (marking all the sensory regions; Morrison *et al.*, 1999). Together these markers have pushed back the time at which sensory anlage can now be recognized, and they suggest that the specification of sensory fate is one of the earliest events in ear morphogenesis. Shortly after otic vesicle closure, at about the time that ganglion cells begin to emigrate ventrally and the endolymphatic duct begins to evaginate dorsally, two distinct foci of *Bmp4* gene expression can be found at the anterior and posterior poles of the otocyst (Fig. 6A).

Interestingly, *Bmp4*-positive foci originate near the edge (or boundary) of several broader gene expression domains, including *Lunatic fringe* in the anteroventral otocyst (Fig. 6A) and *SOHo1* (an *Hmx* gene family member) in the lateral domain in the chick (Kiernan *et al.*, 1997). The anlagen of the lateral crista is located at the boundary of the *Otx1* domain (Morsli *et al.*, 1999). These data provide circumstantial evidence in support of the idea that the specification of sensory organ location may be mediated by the intersection of boundaries of genes with broader domains of expression. This was formally proposed as a theoretical possibility (Fekete, 1996). Here we update our model of the placement of the organs with respect to known or presumed gene expression boundaries (compare Figs. 5 and 7); we emphasize that these relationships have only been definitely demonstrated with respect to PAX2, *SOHo-1*, or *Otx1* (Fekete and Gao, 2000; Kiernan *et al.*, 1997; Morsli *et al.*, 1999). Note that the boundaries may be established and maintained by multiple genes, such that elimination of any one gene may not necessarily result in the absence or mislocation of a particular sense organ that arises at that boundary.

The model is based on the idea that short-range signals may diffuse across the boundaries that dictate a sensory organ fate to the cells receiving the signal. Diffusion across a single boundary would be expected to specify an elongated sensory patch (such as shown for the cochlea and utricle). In contrast, the intersection of two orthogonally arrayed boundaries might generate a focal region where signals arising from the two compartments are colocalized, and the responding tissue might then be specified to generate a spatially punctate organ (such as the cristae). This model could explain both the positioning and the overall shape of the initial sensory field that is generated. Finally, the responding tissue might derive further information about the type of sensory organ to generate by the combination of genes expressed in the larger compartment of which it is a member.

VI. Sensory Differentiation

Prior to differentiation, hair cells and supporting cells undergo their final mitosis in the cochlear and vestibular sensory regions between E11.5 and birth, although auditory and vestibular sensory regions exhibit different spatiotemporal patterns of terminal division and differentiation. In the cochlea there is a gradient of hair and supporting cell terminal mitosis starting at the apex at approximately E11.5, and continuing around the coil until, finally, the hair cells at the base undergo their final division between E14.5 and E15.5 (Ruben, 1967). These data led Ruben (1967) to hypothesize the existence of a zone of mitosis at the base of the cochlea, near the saccule, in which the first cells to emerge from this zone would be the cells at the apex and the production of increasing numbers of cells would account for the outgrowth of the cochlea. Marovitz and Shugar (1976) showed that this zone exists by demonstrating that cells in mitosis were only observed in this region, independent of the time point in development that was examined. Interestingly, differentiation proceeds in the opposite direction; hair cells in the base exhibited the first signs of differentiation immediately after their final mitosis, followed by the hair cells in the middle and apical regions (Lim and Anniko, 1985; Sher, 1971).

In addition to this longitudinal gradient of differentiation, there is also a horizontal gradient of differentiation from the modiolar to the lateral side of the cochlea, in which the IHCs begin differentiation before the OHCs. In the vestibular sensory regions there is a center to periphery gradient of terminal mitosis (Sans and Chat, 1982), with the peak of terminal mitosis occurring around E13.5–15.5 (Ruben, 1967). Unlike the organ of Corti, however, differentiation in the vestibular epithelia appears to follow the same spatiotemporal pattern as that observed for the final mitosis (Mbiene and Sans, 1986).

At present we do not know the signals that initiate and regulate hair cell differentiation in these prescribed patterns, and as yet gradients of signaling molecules in these different sensory patches have not been found. One obvious candidate for this differentiation signal was the invading afferent nerve fibers, as their arrival was coincident with the first signs of hair cell differentiation (Sher, 1971). However, depriving the sensory epithium of innervation does not affect hair cell differentiation, demonstrating that the afferent nerve fibers are not the cue for hair cell differentiation (Corwin and Cotanche, 1989; Swanson *et al.,* 1990; Van De Water, 1976).

A. The Sensory Mosaic: A Model of Lateral Inhibition?

The sensory epithelium of the inner ear is composed of two basic cell types, the hair cell and the supporting cell. Exactly how these two cell fates are specified is unknown; although it has been shown that that the decision appears to be a binary one made at or after the final division, as lineage analysis has demonstrated that two-cell clones contain both a hair cell and a supporting cell (Fekete *et al.,* 1998). These two cell types are arranged in a highly organized and reproducible pattern, such that one hair cell is always surrounded by supporting cells. This alternating pattern of cell types led several researchers to hypothesize that the hair/supporting cell decision may be made through lateral inhibition, a process in which a cell that adopts a certain cell fate inhibits its neighbors from adopting the same cell fate (Corwin *et al.,* 1991; Lewis, 1991). Early indications that this mechanism may be operating in the sensory epithelium came from studies in the embryonic cochlea in which nearby supporting cells were found to differentiate into hair cells after laser ablation of individual hair cells (Kelley *et al.,* 1995). More recently an immortalized supporting cell line has been shown to give rise to hair cells (Lawlor *et al.,* 1999). These data suggest that supporting cells have the ability to differentiate into hair cells (at least early in development), but normally do not do so due to lateral inhibition supplied by neighboring hair cells.

In *Drosophila* and *Caenorhabditis elegans,* lateral inhibition has been shown to be mediated by cell-cell interactions using the Notch-signaling pathway (Artavanis-Tsakonas *et al.,* 1995). The vertebrate homologs of the many components of the Notch-signaling pathway have been cloned and show expression in many vertebrate systems including the ear (Chan and Jan, 1999). By applying the principles learned from *Drosophila* about lateral inhibition and Notch signaling, a simple model of lateral inhibition can be devised to explain the hair and supporting cell patterns within the sensory epithelium of the ear (Figs. 8Ai–iii, Adam *et al.,* 1998; Lewis, 1991). In this model, cells in the sensory epithelium represent an equivalence group and express both Notch and Delta (or any Notch ligand). For unknown or stochastic reasons, certain cells in the epithelium begin to express higher levels of the ligand and activate Notch in the surrounding cells. This initial imbalance is amplified due to a inhibitory feedback loop in which activation of Notch results in a down-regulation of Delta. High levels of Notch activation eventually cause a downregulation of prosensory genes within these cells, causing them to differentiate as supporting cells, while cells expressing Delta differentiate as hair cells.

Expression studies demonstrating that Notch-signaling components were expressed within the expected cells in the sensory epithelium provided the first strong evidence for lateral inhibition within the inner ear. For example, the Notch ligands *Delta-like homolog 1* (*Dll1*) and *Jagged2* (*Jag2*) are expressed in the nascent hair cells within the organ of Corti (Fig. 6C; Lanford *et al.,* 1999; Morrison *et al.,* 1999), and *Notch1* is expressed within the supporting cells (Lanford *et al.,* 1999). In addition, the spatiotemporal expression patterns of *Notch1, Dll1,* and *Jag2* during development are consistent with a role in cell fate decisions. *Notch1* is expressed initially throughout the organ of Corti at E14.5, but is downregulated in hair cells as they differentiate. In addition, both *Dll1* and *Jag2* are expressed first within the inner hair cells between E14.5 and E15.5, and then later in the outer hair cells by E16.5 (Fig. 6C), reflecting the normal horizontal gradient of differentiation that is known to occur. *Dll1* has also been shown to be expressed in the nascent hair cells of the vestibular epithelium as early as E12.5 in the cristae.

One aspect of the expression patterns that is not consistent with the original model is that, of the Notch ligands examined thus far that localize to hair cells (*Dll1* and *Jag2*), neither is initially expressed throughout all cells of the epithelium. Therefore, it remains unclear exactly how these Notch ligands are expressed specifically in hair cells. One possibility is that there is an initial bias—possibly through lineage-based determinants such as Numb—as to which cells will express *Dll1* or *Jag2* and become hair cells (Campos-Ortega, 1997; Jan and Jan, 1998).

If the simple model for lateral inhibition within the inner ear is correct, disruption of Notch signaling should lead to an overproduction of hair cells at the expense of supporting cells. Unfortunately the proposed role for *Dll1* within the ear has not yet been tested, because mice deleted for *Dll1* die during early embryonic life (Hrabé de Angelis *et al.,* 1997; Swiatek *et al.,* 1994). However, *Jag2* knock-out mice survive long enough to study sensory differentiation within the ear, and these mice showed extra rows of both inner and outer hair cells (Fig. 8C; Lanford *et al.,* 1999). Interestingly, the normal spacing between hair cells was preserved, indicating that most of the supporting cells were present. These data suggest that either *Jag2* is not the primary Notch ligand regulating the decision between hair cells and supporting cells in the organ of Corti, or perhaps more likely, there is some redundancy in the system (such as *Dll1*). Similar to *Dll1* knock-outs, the *Notch1* knock-outs also die very early in development (Swiatek *et al.,* 1994). Recently, however, the cochleas of *Notch1* heterozygotes were examined, and this

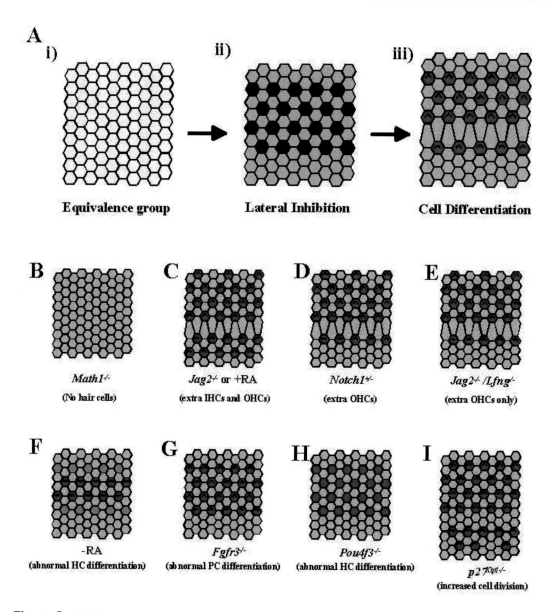

Figure 8 Model for creating the sensory mosaic and phenotypes of mutants that disrupt the mosaic and/or sensory differentiation. (A–G) Schematic diagrams of surface views of the organ of Corti. Inner hair cells are identifiable by the crescent shape of their stereocilia and outer hair cells are depicted by their V-shaped bundles. Pillar cells are depicted as elongated cells that lie between the first row of outer hair cells and the row of inner hair cells. (A–I) All the cells of the epithelium are initially equipotent (white cells) and express Notch and Delta (or any Notch ligand). (ii) For unknown or stochastic reasons, certain cells begin to express the membrane-bound ligand Delta (black) at higher levels than in surrounding cells and activate Notch (gray) in the these cells. Notch and Delta negatively feed back on each other, so that cells that express higher levels of Delta, downregulate their levels of Notch and vice versa, thereby allowing initially small differences between cells to become amplified. (iii) Notch activation results in a signaling cascade that eventually leads to a downregulation of proneural or prosensory genes within these cells, causing them to differentiate as supporting cells (tan), while cells expressing more Delta differentiate as hair cells (red). (B–I) Mutants that show disruptions in the cellular pattern of the organ of Corti due either to defects in cell fate decisions (B–E), defects in differentiation of cell types (F–H), or defects in the control of proliferation (I). Pink cells in F are cells that express some hair cells markers but not others. The straight black lines on some of the hair cells represent the abnormal hair bundle morphology. +RA or −RA refers to experiments in which retinoic acid has been added (+) or depleted (−) in the organ of Corti. See text for references. PC, pillar cells; HC, hair cells.

analysis revealed that extra rows of outer hair cell were formed (Zhang *et al.*, 2000), demonstrating that the gene dosage of the *Notch1* product is important to produce the correct pattern in the organ of Corti.

A more dramatic result has been demonstrated using a *Notch*-directed antisense oligonucleotide in cultures of the developing rat organ of Corti, in which several extra rows were obtained for both inner and outer hair cells (Zine *et al.*, 2000). A similar, although slightly less pronounced effect was obtained using a *Jag1*-directed oligonucleotide (Zine *et al.*, 2000). This was an interesting result considering that, unlike the other Notch ligands, *Jag1* is expressed in the supporting cells of the organ of Corti (i.e., the same cells as Notch1), making its role somewhat unclear. These data indicate that *Jag1* may play a role in promoting Notch activation rather than suppressing it (lateral induction).

A knock-out of *Hes1*, a homolog of the *Drosophila hairy* and *enhancer-of-split* genes and a probable downstream target of Notch, also results in a slight excess of hair cells in the utricle and organ of Corti (Zheng *et al.*, 2000). Interestingly, although the *Lfng* knock-outs show no cochlear defects, the double knock-out of both *Jag2* and *Lfng* suppresses the effect of the *Jag2* knock-out on the inner hair cells (Zhang *et al.*, 2000). These results indicate that *LFng* probably acts to suppress *Jag2*-independent signaling, such as signaling via *Dll1* or *Jag1*. Extra hair cell rows appear to be a common effect of disruption of the Notch-signaling components in the cochlea. Interestingly, extra rows are known to occur sometimes even in control cochleas, particularly at the apex. This suggests that the patterning of the hair cells into only four rows is a tenuous feat, and only achieved when all the signaling components are working correctly.

As previously mentioned, lateral inhibition is thought to result in the downregulation of proneural genes, which in *Drosophila* involves members of the basic helix–loop–helix (bHLH) transcription factor family and includes the genes of the *achaete-scute* complex and *atonal* (Campuzano and Modolell, 1992). Within the inner ear, sensory regions express a homolog of the *Drosophila* gene *atonal*, *Math1*, and *Math1*$^{-/-}$ mutants do not produce hair cells (Fig. 8B; Bermingham *et al.*, 1999). The sensory regions of these mutants showed that supporting cells were present, although it was not clear whether they were present in greater numbers. However, the absence of excess cell death indicated that there may have been a cell fate switch from hair cells to supporting cells in *Math1*$^{-/-}$ inner ear epithelia. In addition, overexpression of *Math1* in the cochlea in which perinatal cells from the greater epithelial ridge were converted to hair cells indicates that *Math1* alone can decide the hair cell fate (Zheng and Gao, 2000). Interestingly, it has been shown more recently that extra hair cells cannot be produced in cochleas that are overexpressing *Hes1*, suggesting there is a balance between *Math1* and factors that inhibit hair cell differentiation such as *Hes1* (Zheng *et al.*, 2000).

Simple lateral inhibition, however, cannot explain all aspects of sensory patterning, particularly in the mammalian organ of Corti where the epithelium is much more organized and specialized than in the vestibular organs or the sensory epithelia of other vertebrates. Many questions remain unanswered including these: What limits the sensory domains? How are inner hair cells generated before outer hair cells? How are inner and outer hair cells generated in such perfect rows? Examination of the hair cells and supporting cells in the chicken cochlea revealed that the sensory mosaic is perfected by cell rearrangement (Goodyear and Richardson, 1997). However, the mammalian organ of Corti shows near-perfect alignment of hair cell rows from very early in their development (E16.5; A. Kiernan, unpublished observations), indicating that much of the patterning in this organ may occur prior to hair cell differentiation. Extra rows of either or both inner and outer hair cells occur in *Jag2*$^{-/-}$ cochleas (Lanford *et al.*, 1999), *Notch1*$^{+/-}$ (Zhang *et al.*, 2000) cochleas, and in cochleas that had been exposed to retinoic acid (Fig. 8C; Kelley *et al.*, 1993), indicating that many of the cells outside the normal sensory domain have the ability to adopt the hair cell fate, but are normally prevented from doing so by as yet unknown mechanisms.

Interestingly, the expression domains of a number of other genes in the Notch pathway as well as other signaling molecules during hair cell genesis in the cochlea suggest a more complex control over when and where hair cell differentiation takes place than classical lateral inhibition. For example Jagged1 (*Jag1*), another Notch ligand, and Lunatic fringe (*Lfng*), a molecule that modulates the Notch pathway (Irvine, 1999), are initially expressed throughout the prospective sensory epithelium of the organ of Corti, and later become restricted to the supporting cells (Fig. 6C; Morrison *et al.*, 1999; Morsli *et al.*, 1998). These genes may be involved in defining the borders of the sensory regions, because it is becoming increasingly clear that Notch signaling is involved in establishing boundaries between fields of cells (lateral induction) as well as playing its more well-known role of selecting between alternative cell fates (Artavanis-Tsakonas *et al.*, 1999; Bray, 1998). Interestingly, *Bmp4* is expressed along the lateral wall of the cochlea early in development adjacent to *Lfng* and *Jag1* expression, marking the boundary between the outer hair cell region and the future Hensen's and Claudius' cell region (Fig. 6C; Morsli *et al.*, 1998; Takemura *et al.*, 1996), raising the possibility that this molecule too may be involved in defining sensory/nonsensory borders.

B. Hair Cell Differentiation

Once the decision to become a hair cell is made, possibly through lateral inhibitory mechanisms, a number of steps still remain to be taken before a mature hair cell is formed. An ever-increasing number of early markers of hair cell differentiation have been identified. These include the transcription factor Pou4f3 (Erkman *et al.,* 1996; Xiang *et al.,* 1997), structural markers such as myosin VIIa and myosin VI (Chen and Segil, 1999; Xiang *et al.,* 1998), and calcium-binding proteins such as calretinin, calmodulin, and parvalbumin (Dechesne *et al.,* 1994; Xiang *et al.,* 1998; Zheng and Gao, 1997). Probably one of the earlier genes required for hair cell differentiation is the gene encoding the POU domain transcription factor *Pou4f3* (*Brn3.1/Brn-3c*). A targeted disruption of this gene resulted in what appeared to be a total failure of hair cell differentiation, because none of the cells displayed stereocilia or nuclei in the correct hair cell position in the lumenal layer of the epithelium (Erkman *et al.,* 1996; Xiang *et al.,* 1997). These results suggested that perhaps Pou4f3 played a role in the initial fate determination of hair cells. However, further analysis earlier in development revealed that hair cells were initially formed in the *Pou4f3⁻/⁻* sensory organs, since there were some cells that expressed early markers of hair cell differentiation, including myosins VI and VIIa as well as calretinin and parvalbumin (Fig. 8F; Xiang *et al.,* 1998). Furthermore, approximately 25% of the cells in the epithelium at E16.5 did show some morphological characteristics of hair cells, including the correct position of the nuclei. Numbers of these hair cell-like cells were reduced over time, suggesting that these cells were gradually dying, and these results were consistent with the increased levels of cell death that were observed. Expression studies have shown that *Pou4f3* is expressed immedi-

ately after the hair cells become postmitotic, and some expression was observed in cells present in the supporting cell layer. These data suggest that *Pou4f3* may be expressed even before prospective hair cells migrate from the lower level of the epithelium to the lumen.

As previously mentioned, myosin VIIa and myosin VI, two unconventional myosins, also mark the nascent hair cells very early in their development. The onset of their expression has been shown to occur 1 day after *Pou4f3*, and expression of these myosins was never observed in the supporting cell layer, indicating these myosins are expressed only after lumenal migration (Xiang *et al.,* 1998). Mutations in *Myo7a* and *Myo6* have been identified in the mouse mutants shaker1 and Snell's waltzer, respectively, and both mutants show stereocilia defects in the cochlea (Self *et al.,* 1998, 1999). Interestingly, these mouse mutants displayed different phenotypes, indicating they are likely to be playing different roles in stereocilia development. For example, shaker1 mutants showed early disorganization of the stereocilia (Fig. 9D), whereas Snell's waltzer mutant hair cells demonstrate early stereocilia fusion (Fig. 9C). Thus, myosin VIIa appears to be involved in hair bundle organization, whereas myosin VI seems to be involved in maintaining stereocilia integrity, possibly through anchoring of the apical membrane.

A third unconventional myosin, myosin XV, may also be involved in stereocilia development (Probst *et al.,* 1998). Shaker2 mutants, which are deaf and circling and display

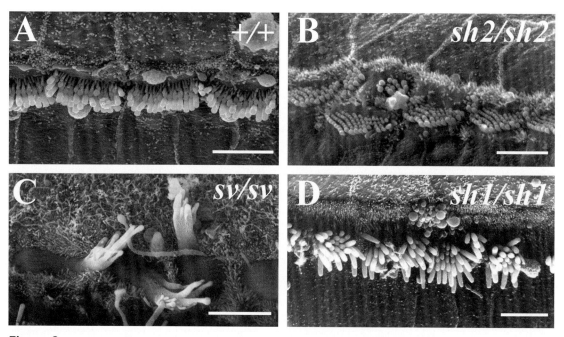

Figure 9 Scanning electron micrographs of the inner hair cells of several myosin mutants. (A) Control at postnatal day (P) 20 demonstrating the normal appearance of the inner hair cell bundle morphology. (B) Shaker2 homozygous mutant at P30. Note the short, stubby stereocilia, multiple rows, and V shape of the bundle, which is not normally observed in mature inner hair cells. (C) Snell's waltzer homozygote inner hair cells at P7. Note the fused stereocilia. (D) Shaker1 (*Myo7a⁸¹⁶ˢᴮ*) homozygote inner hair cells at P12. Note the disorganization of the stereocilia. Bar = 5 μm.

short, stubby stereocilia (Fig. 9B), were shown to have a defective *Myo15* gene. In addition to the short stereocilia, shaker2 homozygotes also displayed long abnormal actin-containing structures that protruded from the base of the inner hair cells. These results indicate that myosin XV may play a more general role in actin organization in the hair cell, although it is not clear as yet whether myosin XV is required during development or for later maintenance of the actin cytoskeleton or both. However, *Myo15* transcripts have been detected as early as E15.5 in the mouse cochlea (Anderson *et al.,* 2000), and mice and humans who have this defective gene are deaf at an early stage, indicating it is likely that myosin XV is required during the development of the hair cell.

Stereocilia production is clearly a complex process and undoubtedly involves numerous other genes in addition to those myosins already mentioned. A potentially important involvement of the extracellular matrix in hair bundle development has been implicated by the finding that stereocilia are not produced normally in the utricles of mice deficient in $\alpha_8\beta_1$ integrin (Littlewood Evans and Muller, 2000). In addition, several mouse mutations, including Jackson shaker (*js*), bustling (v^{bus}), and tailchaser (*Tlc*), have been described as specifically affecting hair bundle formation, and all have been localized to distinct regions of the genome, indicating many different genes are involved (Kiernan *et al.,* 1999; Kitamura *et al.,* 1992; Moriyama *et al.,* 1997; Yonezawa *et al.,* 1996, 1999). In addition to these mouse mutants, several zebrafish mutants have also been found to display abnormal hair bundle morphology (Nicolson *et al.,* 1998). Unfortunately, despite the number of mutations that affect stereociliagenesis, hair bundle development has not been well studied in the mouse. Stereocilia development has been more thoroughly investigated in the chicken cochlea where it was revealed that bundle growth could be divided into three stages consisting of two phases of elongation separated by a phase of increasing stereocilia width (Tilney *et al.,* 1992). In contrast, hair bundle studies in the hamster cochlea have shown that growth in both dimensions happens concurrently (Kaltenbach *et al.,* 1994), suggesting that stereocilia growth in the chicken and mammals happens somewhat differently. Nonetheless, certain features of stereocilia development appear similar among all types of hair cells and between mammals and birds. These features include an initial overproduction of stereocilia, increased growth of the stereocilia nearest the kinocilium so that graded heights are achieved, and finally, reabsorption of the excess stereocilia.

In addition to Notch signaling, retinoic acid signaling also appears to play an important role in the development of the organ or Corti, although it is not yet clear whether it plays a role in cell fate decisions or in early differentiation events (Raz and Kelley, 1999). Several components of the RA pathway are present in the mouse cochlea including retinoic acid (Dolle *et al.,* 1994; Kelley *et al.,* 1993; Raz and Kelley, 1999; Romand *et al.,* 1998), its receptors (Dolle *et al.,* 1990), cel-

lular retinol binding protein (CRBP; Dolle *et al.,* 1990), and cellular retinoic acid binding protein (CRABP; Dolle *et al.,* 1990). Exogenous RA causes supernumerary hair cells to develop in the organ of Corti in a time- and dose-dependent manner (Fig. 8C; Kelley *et al.,* 1993). These new hair cells did not arise from cell division, indicating that RA may play a role in the cell fate decisions within the cochlear epithelium. Recently, retinoic acid receptor expression has been examined in more detail within the ventral cochlear epithelium and all show a dynamic pattern of expression during the time period of hair and supporting cell differentiation (E13–17; Raz and Kelley, 1999). RARα, RXRα, and RXRγ are initially expressed widely throughout the ventral cochlear epithelium at E14, but by E17 were upregulated in the region of the organ of Corti. By P3, all receptors were expressed at higher levels in hair cells than in supporting cells. Specific blocking of the RARα receptor or in the production of RA results in a significant decrease in the number of cells that develop as hair cells (Fig. 8D; Raz and Kelley, 1999). Furthermore, those small number of cells that did differentiate as hair cells did not exhibit normal cellular or bundle morphologies. Examination of early markers of hair cell differentiation showed that both myosin VIIa and Pou4f3 expression were disrupted.

Interestingly, myosin VI expression was not abnormal. In contrast to the experiments with exogenous RA, these data suggested that RA may not be involved in the initial fate determination of hair cells, since hair cells begin to form normally in cochleas that have been blocked for RA signaling, as assessed by myosin VI expression. However, further work needs to be done to confirm this result, because RA signaling may not be blocked completely in these cultures, and myosin VI may be activated normally even with very low levels of RA.

C. Supporting Cell Differentiation

Unfortunately, supporting cell differentiation has not been studied in as much detail as hair cell differentiation. Because of this, many of the roles for the supporting cells throughout the inner ear are not known, although one important role for them seems to be in secreting the extracellular structures, such as the tectorial membrane, otolithic membranes, or cupulae, that lie above the sensory epithelium in the cochlea, maculae, and cristae, respectively (Lim and Rueda, 1992; Riley and Grunwald, 1996). Unlike the vestibular regions, the supporting cells within the cochlea are quite specialized and bear different names depending on their morphology and position (Slepecky, 1996). For example, the pillar cells are the supporting cells that develop between the inner and outer hair cells and form the tunnel of Corti. Their differentiation, which occurs primarily postnatally, leads to a wide separation between the inner and outer hair cells. Deiter's cells, another supporting cell type, surround the base of the outer hair cells, and extend processes to the lumenal surface. Immediately next to these cells toward the lateral wall are Hensen's cells, followed by Claudius' cells. Immediately surrounding the inner hair cells are the inner border and phalangeal cells and medial to these cells are the cells of the inner sulcus.

The differentiation of these supporting cells appears to depend on the correct differentiation of the hair cells, since in mutants where hair cell differentiation is disrupted, such as in the *Math1*$^{-/-}$ and the *Pou4f3*$^{-/-}$ cochleas, many of the supporting cells do not exhibit their normal morphology (Bermingham *et al.*, 1999; Erkman *et al.*, 1996; Figs. 8B and E; Xiang *et al.*, 1997). As previously discussed, several early markers of the supporting cells have recently been found (Fig. 6C) including *Lfng* (Deiter's and pillars; Morsli *et al.*, 1998), *Jag1* (Deiter's, pillars, and Kölliker's organ cells; Morrison *et al.*, 1999), and *Bmp4* (Hensen's and/or Claudius' cells; Morsli *et al.*, 1998). In addition, the p75 low-affinity neurotrophin receptor has been shown to specifically mark the pillar cells in developing rats (Wheeler *et al.*, 1994).

The roles of these genes in the development of supporting cells is not yet known, but their identification as early markers of these different cells types will be helpful in the analysis of present and future mouse mutations that show disruptions in the differentiation of the organ of Corti. In addition to these developmental genes, a number of cytoskeletal proteins have been found to be specific to certain types of mammalian supporting cells (Slepecky, 1996), including cytokeratin and modified tubulin, although the specificity and onset of expression of most of these proteins have not been thoroughly investigated in the mouse.

Pillar cells have been shown to require FGF signaling for their differentiation, since mice with a disruption in the fibroblast growth factor receptor 3 (*Fgfr3*) gene were deaf and did not have normal pillar cells (Fig. 8E; Colvin *et al.*, 1996). The pillar cells in *Fgfr3*$^{-/-}$ cochleas appeared immature and displayed a Deiter-like appearance. Several other cochlear regions, including the spiral vessel and the mesothelial cells beneath the organ of Corti, appeared immature in *Fgfr3*$^{-/-}$ cochleas, and outer hair cell innervation was reduced, indicating that FGF signaling is required for several aspects of cochlear maturation. *Fgfr3* has been shown to be expressed in the pillar and Deiter's cells (Pirvola *et al.*, 1995), demonstrating that the pillar cell abnormality is likely to be a primary defect of the *Fgfr3* disruption. The ligand that normally activates the FGFR3 receptor is unknown, as are the cells that produce it.

D. Control of Cell Proliferation in the Organ of Corti

Unlike the sensory epithelia of other vertebrates and to some degree the mammalian vestibular organs, the mature mammalian organ of Corti shows no capacity for regeneration after damage due to environmental or genetic factors (Stone *et al.*, 1998). The permanent loss of hair cells is likely

to be a major factor in the pathogenesis of genetic and environmental deafness, and much interest has been focused on the possible reasons for this lack of regenerative ability. Recently, two studies of the organ of Corti in a mouse devoid of the cyclin-dependent kinase inhibitor, p27^{Kip1}, have revealed at least one of the factors contributing to the organ's lack of proliferative ability (Chen and Segil, 1999; Lowenheim *et al.*, 1999). The onset of p27^{Kip1} expression was shown to occur sometime between E12.5 and E14.5 in the nascent organ of Corti (Chen and Segil, 1999), consistent with the time of withdrawal from the cell cycle (Ruben, 1967). Initially, p27^{Kip1} expression was observed throughout the sensory region, but by E16.5 expression was concentrated in the supporting cell region, indicating that once hair cells start to differentiate they downregulate expression of p27^{Kip1} (Chen and Segil, 1999). p27^{Kip1} continued to be expressed in supporting cells throughout adult stages (Lowenheim *et al.*, 1999). Both studies reported an increase in the mitotic activity of the organ of Corti prenatally and to a lesser degree postnatally. This mitotic activity resulted in an increased number of supporting cells in the *p27*$^{Kip1-/-}$ cochleas, although only one study reported excess hair cells (Fig. 8G; Chen and Segil, 1999). This difference was probably due to the fact that many of the hair cells were dying and Lowenheim *et al.* (1999) only examined the organs at adult stages. These results suggest that p27^{Kip1} is an important factor in the regulation of cell proliferation in the mammalian cochlea. Tight control of cell proliferation appears to be necessary for the correct cytodifferentiation and functioning of the organ of Corti, since the structure of the *p27*$^{Kip1-/-}$ cochleas was abnormal and the mice were severely hearing impaired (Lowenheim *et al.*, 1999).

However, inactivation of the *p27*^{Kip1} gene did not result in completely uncontrolled growth in the organ of Corti, particularly at postnatal stages, indicating other factors are involved. Indeed, recent evidence suggests that p19^{Ink4d} (another cell cycle regulator) may partially compensate for the loss of p27^{Kip1}, because double-null mice showed increased mitotic activity in the hair cell region after birth when compared to *p27*^{Kip1} mutants alone (Chen *et al.*, 2000).

VII. Neurogenesis

Sensory neurons can arise from two different sources: neural crest or placode. Although fate mapping of the otic ganglia has not yet been performed in the mouse, studies using quail-chick chimeras have shown that in birds the afferent neurons that innervate the ear (the eighth cranial ganglia) are largely otic placode derived, although the neural crest contributes to the glial population (D'Amico-Martel and Noden, 1983). Unfortunately little is known about how the prospective neurons are specified within the otic vesicle, although recent molecular evidence implicates an involvement of the Notch-signaling pathway. Expression data have shown that *Dll1* is expressed in the anteroventral portion of the otocyst (Ma *et al.,* 1998; Morrison *et al.,* 1999), in agreement with the site of neuroblast emigration (Carney and Silver, 1983). This expression has been studied in more detail in the chicken, where *Delta1* expression has been shown to coincide with early neuronal markers, although the two are never coexpressed within the same cell, suggesting that *Delta1* is downregulated prior to neuronal differentiation (Adam *et al.,* 1998).

Based on expression, *Dll1* appears to be used in both hair cell and neuronal production. However, it is unlikely that the same proneural gene is used in both cases, because studies of *Math1* mutants have not reported a reduction in the neuronal population (Bermingham *et al.,* 1999). However, studies have shown that another *atonal*-related gene, neurogenic differentiation 3 (*Neurod3* or *Ngn1*), is required for the generation of neuronal precursors within the inner ear (Ma *et al.,* 1998). *Neurod3* was shown to be expressed in the ventral portion of the otocyst, in the same regions as *Delta1,* and mice devoid of Neurod3 did not develop any sensory neurons (Ma *et al.,* 1998). Interestingly, *Delta1* expression was not present in the *Neurod3⁻/⁻* otocysts, indicating that *Neurod3* positively regulates *Delta1,* as would be expected of a proneural gene. A more recent analysis of the *Neurod3⁻/⁻* inner ears demonstrated that the sensory patches were smaller, particularly the saccular maculae (Ma *et al.,* 2000). This result raises the intriguing possibility that some hair cells and sensory neurons share a common progenitor, although lineage analysis has not yet addressed this question. Another bHLH gene, *Neurod1,* appears to be important for neuronal development because *Neurod1* mutant inner ears show a severe reduction in their innervation (Liu *et al.,* 2000). Developmental studies revealed that this reduction was due to failed delamination of the neuroblasts rather than specification, indicating that *Neurod1* is downstream of *Neurod3.*

One area of neuronal development in the ear that remains largely unknown is the method by which outgrowing neurites find their way back to the ear to innervate their correct targets. As yet no mutants have been identified in which the neurons are generated correctly, but subsequently cannot find their hair cell targets. *In vitro* studies have shown that the otocyst secretes a factor that promotes neurite outgrowth, although the identity of this factor is not yet known (Bianchi and Cohan, 1993). One interesting observation made by Carney and Silver (1983) is that axon outgrowth seemed to occur along the same path that the neuroblasts use to migrate. Thus, they hypothesized that the neuroblasts express or secrete a factor or matrix as they are migrating, thereby guiding the neurites back to their correct target. This hypothesis would require that hair cells and neurons be generated from the same region of the otocyst, which may be the case at least for some sensory areas, since *Delta1* (marking the neurogenic area) and *Lfng* (marking at least some of the sensory areas) are both expressed in the anteroventral portion of the early otocyst (Morrison *et al.,* 1999; Morsli *et al.,* 1998).

Once the axons arrive at the otocyst, their survival depends on one of two neurotrophic substances, neurotrophin 3 (NTF3/NT3) and brain-derived neurotrophic factor (BDNF), which are secreted by the sensory epithelium. Mice that lack both NTF3 and BDNF, or their receptors, NTRK3 (TrkC) and NTRK2 (TrkB), respectively, demonstrate no afferent innervation at birth (Ernfors *et al.,* 1995; reviewed in Fritzsch *et al.,* 1995, 1997), although one study reported only a 40% loss of afferent neurons (Minichiello *et al.,* 1995). However, single knock-outs of either *Bdnf/Ntrk2* or *Ntf3/Ntrk3* had different effects on the ear. For example, *Bdnf* or *Ntrk2* null mutants showed a large reduction in the number of vestibular ganglion cells, which resulted in a complete absence of afferent innervation to the semicircular canals, and reduced innervation to the maculae (Bianchi *et al.,* 1996; Conover *et al.,* 1995; Ernfors *et al.,* 1994, 1995; Fritzsch *et al.,* 1995; Minichiello *et al.,* 1995; Schimmang *et al.,* 1995). The effects on the cochlea are somewhat controversial, because one group reported an absence of innervation to the outer hair cells (Schimmang *et al.,* 1995), whereas another group only found a reduction in the apical portion of the cochlea (Fritzsch *et al.,* 1997). In contrast, *Ntf3* or *Ntrk3* null mutants showed a severe loss of spiral ganglion cells but only a mild reduction in the number of vestibular neurons (Ernfors *et al.,* 1995; Farinas *et al.,* 1994; Minichiello *et al.,* 1995; Schimmang *et al.,* 1995).

These differences can be explained in part by the differential expression patterns displayed by *Bdnf* and *Ntf3* in the sensory regions of the ear, since *Bdnf* alone is expressed in the cristae of the semicircular canals (Pirvola *et al.,* 1992, 1994). Most of the other sensory areas express both *Bdnf* and *Ntf3* (Schecterson and Bothwell, 1994; Wheeler *et al.,* 1994; Ylikoski *et al.,* 1993), suggesting that some neurons may require a specific factor (Fritzsch *et al.,* 1998). However, recent analysis of the expression patterns of *Bdnf* and *Ntf3* during development has suggested this is not the case, since both neurotrophic factors initially show dynamic, nonoverlapping expression patterns in the developing cochlea (Fritzsch *et al.,* 2000). These data indicate that the specific losses that are observed in the mutants are due to lack of expression of either *Bdnf* or *Ntf3* factor at a specific *time,* rather than a requirement for a specific factor. Preliminary results from a transgenic mouse in which *Bdnf* replaces *Ntf3* showed full rescue of the *Ntf3* phenotype, supporting the hypothesis that inner ear neurons do not specifically require one neurotrophin over the other (Coppola *et al.,* 2000).

VIII. The Stria Vascularis

The stria vascularis lies on the lateral wall of the cochlear duct, and has a vital role in generating the high resting potential, the endocochlear potential, in the endolymph filling the scala media. It also contributes to the unusual high-potassium, low-sodium content of the endolymph. As its name suggests, the stria vascularis contains an extensive capillary bed, presumably to supply the high-energy demands of generating the endocochlear potential. The remainder of the stria is composed of three cell layers of divergent origins. The marginal cells line the lumenal surface and are derived from otic epithelium. The basal cells lie several cells thick on the lateral (outer) side of the stria, next to the fibrocytes of the spiral ligament, and these are derived from the mesenchymal cells that surround the otic vesicle at early stages of development. Scattered between the marginal and basal cell layers lie the intermediate cells. These are derived from migratory neural crest and are a specialized form of melanocyte (Hilding and Ginzberg, 1977; Steel et al., 1992).

As with other cochlear structures, the stria in the basal turn develops ahead of the apical turn. The stria first appears as a patch of electron-dense cuboidal epithelial cells sitting on a basal lamina on the cochlear wall opposite the organ of Corti. Around the time of birth, these future marginal cells start to extend fine processes from their basolateral membranes, and these ultimately begin to push through to interdigitate with processes from the basal cells, while the basal lamina between these two layers breaks down. The basal cells at the same time are condensing from the previously loosely packed mesenchymal cells, distinguishing them from the fibrocytes of the spiral ligament, which remain loosely packed. The fine membranous processes of the marginal cells next fill with mitochondria, and extend further down into the basal cell layer, and gap junctions can be seen to develop between adjacent basal cells. Blood vessels are incorporated into the stria during this period (Hilding, 1969; Kikuchi and Hilding, 1966; Steel and Barkway, 1989; Thorn and Schinko, 1985).

The melanocytes migrate as melanoblasts from the neural crest, and can be detected in the mesenchyme in the vicinity of the otic vesicle from around E10 using a specific melanocyte marker (Cable et al., 1995; Steel et al., 1992). During the next 2 or 3 days, they accumulate in clusters in the mesenchyme immediately next to the developing stria, and by E16.5, the labeled melanoblasts become very closely associated with the stria. In young mice, the melanocytes look very similar to normal melanocytes elsewhere in the body, growing extensive dendrites that become filled with melanosomes. However, as the mice get older, their strial melanocytes appear to become more specialized, apparently producing fewer melanosomes, and in some strains of mouse, some of these cells accumulate large amounts of electron-dense material (Cable and Steel, 1991).

The function of the melanocytes is not known, but in mutants which have no melanocytes in their strias, such as several alleles of dominant spotting (Kit) and steel (Mgf), no endocochlear potential is generated (Cable et al., 1994; Steel

et al., 1987). These data suggest that melanocytes are essential for the generation of this resting endocochlear potential, and explain the link between coat spotting and deafness seen in many different mammals including humans. In the absence of melanocytes, the marginal cells extend their basolateral processes as usual, but these fail to interdigitate extensively with the underlying cells, and after some weeks these processes retract (Steel and Barkway, 1989). The abnormal development and function of the stria in these spotting mutants is not caused by lack of pigment, because the stria develops and functions normally in albino animals. Melanocyte deficiencies are likely to be the cause of deafness in humans with Waardenburg syndrome, because all of the genes that have been shown to have a role in this disease, including PAX3, MITF, EDN3, and EDNRB, are known to be involved in neural crest development (Read and Newton, 1997). Apart from the genes involved in melanocyte development, little is known about the molecular basis of strial development.

IX. Future Directions

Clearly our understanding of the molecular processes that underlie inner ear development in the mouse remains in its infancy. However, due in large part to the contribution of studies of mouse knock-outs and mutants (Fekete, 1999), we are starting to gain a rudimentary understanding of how a complex structure such as the inner ear might be patterned. Large gaps in our knowledge remain including the answers to the following questions: What is/are the molecule(s) that induce ear formation? How are the parts of the ear specified and are boundaries important? How large a role does lateral inhibition play in creating the cellular pattern in the sensory regions of the inner ear? Because the inner ear is a fairly late-developing structure in the mouse, some of these questions may be unanswerable by straightforward loss-of-function studies due to early lethality, highlighting the importance of creating conditional mutants.

Additionally, the issue of genetic redundancy can be addressed through creating combinations of mouse mutants, such as those created for the neurotrophins and their receptors. Misexpression experiments are also needed, especially regarding the question of boundaries. For example, moving these boundaries or creating new ones may be more informative than removing them altogether. Finally, because of the ease with which vestibular mutations and severe hearing deficits can be identified in the mouse, phenotypic approaches should be useful for identifying genes involved in ear development, such as large-scale mutagenesis schemes (see, for example, Brown and Nolan, 1998; Hrabé de Angelis and Balling, 1998). A mutagenesis approach will not only increase our catalog of genes involved in ear development, but also enrich for the types of mutations that can be produced for each gene, thereby creating both hypomorphic and hypermorphic alleles that may give new insights into gene function.

Acknowledgments

We would like to thank Doris Wu for donating the picture of the paint-filled mouse inner ears (Fig. 2F) and Yehoash Raphael for donating the shaker2 scanning electron micrograph (Fig. 9B). We would also like to thank Bernd Fritzsch and Rob Grainger for helpful discussions regarding various parts of the chapter and Michele Miller Bever for valuable comments on the entire manuscript. This work was supported by the European Commission (contract CT97-2715), the MRC, and Defeating Deafness. D.M.F. is supported in part by research grant 1-FY99-0483 from the March of Dimes Birth Defects Foundation and by the NIH (DC02756).

References

Abdelhak, S., Kalatzis, V., Heilig, R., Compain, S., Samson, D., Vincent, C., Weil, D., Cruaud, C., Sahly, I., Leibovici, M., Bitner-Glindzicz, M., Francis, M., Lacombe, D., Vigneron, J., Charachon, R., Boven, K., Bedbeder, P., Van Regemorter, N., Weissenbach, J., and Petit, C. (1997). A human homologue of the Drosophila eyes absent gene underlies branchio-oto-renal (BOR) syndrome and identifies a novel gene family. *Nat. Genet.* **15**, 157–164.

Acampora, D., Merlo, G. R., Paleari, L., Zerega, B., Postiglione, M. P., Mantero, S., Bober, E., Barbieri, O., Simeone, A., and Levi, G. (1999). Craniofacial, vestibular and bone defects in mice lacking the distal-less-related gene dlx5. *Development (Cambridge, UK)* **126**, 3795–3809.

Adam, J., Myat, A., Roux, I. L., Eddison, M., Henrique, D., Ish-Horowicz, D., and Lewis, J. (1998). Cell fate choices and the expression of Notch, Delta and Serrate homologues in the chick inner ear: Parallels with Drosophila sense-organ development. *Development (Cambridge, UK)* **125**, 4645–4654.

Akimenko, M. A., Ekker, M., Wegner, J., Lin, W., and Westerfield, M. (1994). Combinatorial expression of three zebrafish genes related to distal-less: Part of a homeobox gene code for the head. *J. Neurosci.* **14**, 3475–3486.

Alvarez, I. S., and Navascues, J. (1990). Shaping, invagination, and closure of the chick embryo otic vesicle: Scanning electron microscopic and quantitative study. *Anat. Rec.* **228**, 315–326.

Anderson, D. W., Probst, F. J., Belyantseva, I. A., Fridell, R. A., Beyer, L., Martin, D. M., Wu, D., Kachar, B., Friedman, T. B., Raphael, Y., and Camper, S. A. (2000). The motor and tail regions of myosin XV are critical for normal structure and function of auditory and vestibular hair cells. *Hum. Mol. Genet.* **9**, 1729–1738.

Artavanis-Tsakonas, S., Matsuno, K., and Fortini, M. E. (1995). Notch signaling. *Science* **268**, 225–232.

Artavanis-Tsakonas, S., Rand, M. D., and Lake, R. J. (1999). Notch signaling: Cell fate control and signal integration in development. *Science* **284**, 770–776.

Bermingham, N. A., Hassan, B. A., Price, S. D., Vollrath, M. A., Ben-Arie, N., Eatock, R. A., Bellen, H. J., Lysakowski, A., and Zoghbi, H. Y. (1999). Math1: An essential gene for the generation of inner ear hair cells. *Science* **284**, 1837–1841.

Bever, M. M., and Fekete, D. M. (1999). Ventromedial focus of cell death is absent during development of Xenopus and zebrafish inner ears. *J. Neurocytol.* **28**, 781–793.

Bianchi, L. M., and Cohan, C. S. (1993). Effects of the neurotrophins and CNTF on developing statoacoustic neurons: Comparison with an otocyst-derived factor. *Dev. Biol.* **159**, 353–365.

Bianchi, L. M., Conover, J. C., Fritzsch, B., DeChiara, T., Lindsay, R. M., and Yancopoulos, G. D. (1996). Degeneration of vestibular neurons in

late embryogenesis of both heterozygous and homozygous BDNF null mutant mice. *Development (Cambridge, UK)* **122**, 1965–1973.

Bosse, A., Zulch, A., Becker, M. B., Torres, M., Gomez-Skarmeta, J. L., Modolell, J., and Gruss, P. (1997). Identification of the vertebrate Iroquois homeobox gene family with overlapping expression during early development of the nervous system. *Mech. Dev.* **69**, 169–181.

Bray, S. (1998). Notch signalling in Drosophila: Three ways to use a pathway. *Semin. Cell Dev. Biol.* **9**, 591–597.

Brigande, J. V., Iten, L. E., and Fekete, D. M. (2000a). A fate map of chick otic cup closure reveals lineage boundaries in the dorsal otocyst. *Dev. Biol.* **227**, 256–270.

Brigande, J. V., Kiernan, A. E., Gao, X., Iten, L. E., and Fekete, D. M. (2000b). Molecular genetics of pattern formation in the ear: Do compartment boundaries play a role? *Proc. Natl. Acad. Sci. U.S.A.* **97**, 11700–11706.

Brown, S. D. M., and Nolan, P. M. (1998). Mouse mutagenesis-systematic studies of mammalian gene function. *Hum. Mol. Genet.* **7**, 1627–1633.

Cable, J., and Steel, K. P. (1991). Identification of two types of melanocyte within the stria vascularis of the mouse inner ear. *Pigm. Cell Res.* **4**, 87–101.

Cable, J., Huszar, D., Jaenisch, R., and Steel, K. P. (1994). Effects of mutations at the W locus (c-kit) on inner ear pigmentation and function in the mouse. *Pigm. Cell Res.* **7**, 17–32.

Cable, J., Jackson, I. J., and Steel, K. P. (1995). Mutations at the W locus affect survival of neural crest-derived melanocytes in the mouse. *Mech. Dev.* **50**, 139–150.

Campos-Ortega, J. A. (1997). Asymmetric division: Dynastic intricacies of neuroblast division. *Curr. Biol.* **7**, R726–R728.

Campuzano, S., and Modolell, J. (1992). Patterning of the Drosophila nervous system: The achaete-scute gene complex. *Trends Genet.* **8**, 202–208.

Carney, P. R., and Silver, J. (1983). Studies on cell migration and axon guidance in the developing distal auditory system of the mouse. *J. Comp. Neurol.* **215**, 359–369.

Carpenter, E. M., Goddard, J. M., Chisaka, O., Manley, N. R., and Capecchi, M. R. (1993). Loss of Hox-A1 (Hox-1.6) function results in the reorganization of the murine hindbrain. *Development (Cambridge, UK)* **118**, 1063–1075.

Chan, Y. M., and Jan, Y. N. (1999). Conservation of neurogenic genes and mechanisms. *Curr. Opin. Neurobiol.* **9**, 582–588.

Chang, W., Nunes, F. D., De Jesus-Escobar, J. M., Harland, R., and Wu, D. K. (1999). Ectopic noggin blocks sensory and nonsensory organ morphogenesis in the chicken inner ear. *Dev. Biol.* **216**, 369–381.

Chen, P., and Segil, N. (1999). p27(Kip1) links cell proliferation to morphogenesis in the developing organ of Corti. *Development (Cambridge, UK)* **126**, 1581–1590.

Chen, P., Zindy, F., Roussel, M., and Segil, N. (2000). Synergistic action of Ink and Cip/Kip CDK inhibitors in the developing organ of Corti. *Assoc. Res. Otolaryngol.* **23** (Abstr. No. 510); also available at: http://www.aro.org.

Chisaka, O., Musci, T. S., and Capecchi, M. R. (1992). Developmental defects of the ear, cranial nerves and hindbrain resulting from targeted disruption of the mouse homeobox gene Hox-1.6. *Nature (London)* **355**, 516–520.

Colvin, J. S., Bohne, B. A., Harding, G. W., McEwen, D. G., and Ornitz, D. M. (1996). Skeletal overgrowth and deafness in mice lacking fibroblast growth factor receptor 3. *Nat. Genet.* **12**, 390–397.

Conover, J. C., Erickson, J. T., Katz, D. M., Bianchi, L. M., Poueymirou, W. T., McClain, J., Pan, L., Helgren, M., Ip, N. Y., Boland, P. *et al.*, (1995). Neuronal deficits, not involving motor neurons, in mice lacking BDNF and/or NT4. *Nature (London)* **375**, 235–238.

Coppola, V., Kucera, J., Palko, M. E., Martinez-De Velasco, J., Lyons, W. E., Fritzsch, B., and Tessarollo, L. (2000). Dissection of NT-3 functions in vivo by gene replacement strategy. *Soc. Neurosci. Abstr.* Available at: http://sfn.scholarone.com.

Cordes, S. P., and Barsh, G. S. (1994). The mouse segmentation gene kr encodes a novel basic domain-leucine zipper transcription factor. *Cell (Cambridge, Mass.)* **79,** 1025–1034.

Corwin, J. T., and Cotanche, D. A. (1989). Development of location-specific hair cell stereocilia in denervated embryonic ears. *J. Comp. Neurol.* **288,** 529–537.

Corwin, J. T., Jones, J. E., Katayama, A., Kelley, M. W., and Warchol, M. E. (1991). Hair cell regeneration: The identities of progenitor cells, potential triggers and instructive cues. *In* "Regeneration of Vertebrate Sensory Cells" (G. R. Bock and J. Whelan, eds.), pp. 103–130. Wiley, Chichester.

Dallos, P. (1992). The active cochlea. *J. Neurosci.* **12,** 4575–4585.

D'Amico-Martel, A., and Noden, D. M. (1983). Contributions of placodal and neural crest cells to avian cranial peripheral ganglia. *Am. J. Anat.* **166,** 445–468.

Davis, H. (1983). An active process in cochlear mechanics. *Hear. Res.* **9,** 79–90.

Dechesne, C. J., Rabejac, D., and Desmadryl, G. (1994). Development of calretinin immunoreactivity in the mouse inner ear. *J. Comp. Neurol.* **346,** 517–529.

De Moerlooze, L., Spencer-Dene, B., Revest, J., Hajihosseini, M., Rosewell, I., and Dickson, C. (2000). An important role for the IIIb isoform of fibroblast growth factor receptor 2 (FGFR2) in mesenchymal-epithelial signalling during mouse organogenesis. *Development (Cambridge, UK)* **127,** 483–492.

Deol, M. S. (1964). The abnormalities of the inner ear in *kr* mice. *J. Embryol. Exp. Morphol.* **12,** 475–490.

Depew, M. J., Liu, J. K., Long, J. E., Presley, R., Meneses, J. J., Pedersen, R. A., and Rubenstein, J. L. (1999). Dlx5 regulates regional development of the branchial arches and sensory capsules. *Development (Cambridge, UK)* **126,** 3831–3846.

Dolle, P., Ruberte, E., Leroy, P., Morriss-Kay, G., and Chambon, P. (1990). Retinoic acid receptors and cellular retinoid binding proteins. I. A systematic study of their differential pattern of transcription during mouse organogenesis. *Development (Cambridge, UK)* **110,** 1133–1151.

Dolle, P., Fraulob, V., Kastner, P., and Chambon, P. (1994). Developmental expression of murine retinoid X receptor (RXR) genes. *Mech. Dev.* **45,** 91–104.

Dupe, V., Ghyselinck, N. B., Wendling, O., Chambon, P., and Mark, M. (1999). Key roles of retinoic acid receptors alpha and beta in the patterning of the caudal hindbrain, pharyngeal arches and otocyst in the mouse. *Development (Cambridge, UK)* **126,** 5051–5059.

Erkman, L., McEvilly, R. J., Luo, L., Ryan, A. K., Hooshmand, F., O'Connell, S. M., Keithley, E. M., Rapaport, D. H., Ryan, A. F., and Rosenfeld, M. G. (1996). Role of transcription factors Brn-3.1 and Brn-3.2 in auditory and visual system development. *Nature (London)* **381,** 603–606.

Ernfors, P., Lee, K. F., and Jaenisch, R. (1994). Mice lacking brain-derived neurotrophic factor develop with sensory deficits. *Nature (London)* **368,** 147–150.

Ernfors, P., Van De Water, T., Loring, J., and Jaenisch, R. (1995). Complementary roles of BDNF and NT-3 in vestibular and auditory development. *Neuron* **14,** 1153–1164.

Esteve, P., and Bovolenta, P. (1999). cSix4, a member of the six gene family of transcription factors, is expressed during placode and somite development. *Mech. Dev.* **85,** 161–165.

Farinas, I., Jones, K. R., Backus, C., Wang, X. Y., and Reichardt, L. F. (1994). Severe sensory and sympathetic deficits in mice lacking neurotrophin-3. *Nature (London)* **369,** 658–661.

Favor, J., Sandulache, R., Neuhauser-Klaus, A., Pretsch, W., Chatterjee, B., Senft, E., Wurst, W., Blanquet, V., Grimes, P., Sporle, R., and Schughart, K. (1996). The mouse Pax2(1Neu) mutation is identical to a human PAX2 mutation in a family with renal-coloboma syndrome and results in developmental defects of the brain, ear, eye, and kidney. *Proc. Natl. Acad. Sci. U.S.A.* **93,** 13870–13875.

Fekete, D. M. (1996). Cell fate specification in the inner ear. *Curr. Opin. Neurobiol.* **6,** 533–541.

Fekete, D. M. (1999). Development of the vertebrate ear: Insights from knockouts and mutants. *Trends Neurosci.* (in press).

Fekete, D. M., and Gao, X. (2000). A model for compartments and boundaries in the developing inner ear. *In* "Molecular and Cell Biology of the Ear" (D. Lim, ed.). Plenum, New York (in press).

Fekete, D. M., Homburger, S. A., Waring, M. T., Riedl, A. E., and Garcia, L. F. (1997). Involvement of programmed cell death in morphogenesis of the vertebrate inner ear. *Development (Cambridge, UK)* **124,** 2451–2461.

Fekete, D. M., Muthukumar, S., and Karagogeos, D. (1998). Hair cells and supporting cells share a common progenitor in the avian inner ear. *J. Neurosci.* **18,** 7811–7821.

Fritzsch, B., Silos-Santiago, I., Smeyne, R., Fagan, A. M., and Barbacid, M. (1995). Reduction and loss of inner ear innervation in trkB and trkC receptor knockout mice: A whole-mount DiI and scanning electron microscopic analysis. *Audiol. Neurosci.* **1,** 401–417.

Fritzsch, B., Silos-Santiago, I., Bianchi, L. M., and Farinas, I. (1997). The role of neurotrophic factors in regulating the development of inner ear innervation. *Trends Neurosci.* **20,** 159–164.

Fritzsch, B., Barbacid, M., and Silos-Santiago, I. (1998). The combined effects of trkB and trkC mutations on the innervation of the inner ear. *Int. J. Dev. Neurosci.* **16,** 493–505.

Fritzsch, B., Silos-Santiago, I., Farinas, I., and Jones, K. (2000). Neurotrophins and neurotrophin receptors involved in supporting afferent inner ear innervation. *In* "The Neurotrophins" (I. Machete, ed.), pp. 197–211. Salzburger and Graham.

Gallagher, B. C., Henry, J. J., and Grainger, R. M. (1996). Inductive processes leading to inner ear formation during Xenopus development. *Dev. Biol.* **175,** 95–107.

Gavalas, A., Studer, M., Lumsden, A., Rijli, F. M., Krumlauf, R., and Chambon, P. (1998). Hoxa1 and Hoxb1 synergize in patterning the hindbrain, cranial nerves and second pharyngeal arch. *Development (Cambridge, UK)* **125,** 1123–1136.

Gerlach, L. M., Hutson, M. R., Germiller, J. A., Nguyen-Luu, D., Victor, J. C., and Barald, K. F. (2000). Addition of the BMP4 antagonist, noggin, disrupts avian inner ear development. *Development (Cambridge, UK)* **127,** 45–54.

Giraldez, F. (1998). Regionalized organizing activity of the neural tube revealed by the regulation of lmx1 in the otic vesicle. *Dev. Biol.* **203,** 189–200.

Goodyear, R., and Richardson, G. (1997). Pattern formation in the basilar papilla: Evidence for cell rearrangement. *J. Neurosci.* **17,** 6289–6301.

Graham, A., and Begbie, J. (2000). Neurogenic placodes: A common front. *Trends Neurosci.* **23,** 313–316.

Groves, A. K., and Bronner-Fraser, M. (2000). Competence, specification and commitment in otic placode induction. *Development (Cambridge, UK)* **127,** 3489–3499.

Haddon, C. M., and Lewis, J. H. (1991). Hyaluronan as a propellant for epithelial movement: The development of semicircular canals in the inner ear of Xenopus. *Development (Cambridge, UK)* **112,** 541–550.

Hadrys, T., Braun, T., Rinkwitz-Brandt, S., Arnold, H. H., and Bober, E. (1998). Nkx5-1 controls semicircular canal formation in the mouse inner ear. *Development (Cambridge, UK)* **125,** 33–39.

Heller, N., and Brandli, A. W. (1999). Xenopus Pax-2/5/8 orthologues: Novel insights into Pax gene evolution and identification of Pax-8 as the earliest marker for otic and pronephric cell lineages. *Dev. Genet.* **24,** 208–219.

Hertwig, P. (1942). Neue Mutationen und Koppellungsgruppen bei der Hausmaus. *Z. indukt. Abstamm. Vererbungsl.* **80,** 327–354.

Hilding, D. A. (1969). Electron microscopy of the developing hearing organ. *Laryngoscope* **79,** 1691–1704.

Hilding, D. A., and Ginzberg, R. D. (1977). Pigmentation of the stria vascularis. The contribution of neural crest melanocytes. *Acta Oto-Laryngol.* **84,** 24–37.

Holme, R. H., Bussoli, T. J., and Steel, K. P. (2001). Table of gene expression in the developing ear. Available at: http://www.ihr.mrc.ac.uk/hereditary/genetable/search.html.

Hrabé de Angelis, M., and Balling, R. (1998). Large scale ENU screens in the mouse: Genetics meets genomics. *Mutat. Res.* **400**, 25–32.

Hrabé de Angelis, M., McIntyre, J., 2nd, and Gossler, A. (1997). Maintenance of somite borders in mice requires the Delta homologue DII1. *Nature (London)* **386**, 717–721.

Irvine, K. D. (1999). Fringe, Notch, and making developmental boundaries. *Curr. Opin. Genet. Dev.* **9**, 434–441.

Jacobson, A. G. (1963). The determination and positioning of the nose, lens, and ear. I. Interactions with the ectoderm, and between the ectoderm and underlying tissues. *J. Exp. Zool.* **154**, 273–283.

Jacobson, A. G. (1966). Inductive processes in embryonic development. *Science* **152**, 25–34.

Jan, Y. N., and Jan, L. Y. (1998). Asymmetric cell division. *Nature (London)* **392**, 775–778.

Johnson, K. R., Cook, S. A., Erway, L. C., Matthews, A. N., Sanford, L. P., Paradies, N. E., and Friedman, R. A. (1999). Inner ear and kidney anomalies caused by IAP insertion in an intron of the Eya1 gene in a mouse model of BOR syndrome. *Hum. Mol. Genet.* **8**, 645–653.

Kalatzis, V., Sahly, I., El-Amraoui, A., and Petit, C. (1998). Eya1 expression in the developing ear and kidney: Towards the understanding of the pathogenesis of Branchio-Oto-Renal (BOR) syndrome. *Dev. Dyn.* **213**, 486–499.

Kaltenbach, J. A., Falzarano, P. R., and Simpson, T. H. (1994). Postnatal development of the hamster cochlea. II. Growth and differentiation of stereocilia bundles. *J. Comp. Neurol.* **350**, 187–198.

Kaufman, M. H. (1992). "The Atlas of Mouse Development." Academic Press, San Diego, CA.

Kelley, M. W., Xu, X. M., Wagner, M. A., Warchol, M. E., and Corwin, J. T. (1993). The developing organ of Corti contains retinoic acid and forms supernumerary hair cells in response to exogenous retinoic acid in culture. *Development (Cambridge, UK)* **119**, 1041–1053.

Kelley, M. W., Talreja, D. R., and Corwin, J. T. (1995). Replacement of hair cells after laser microbeam irradiation in cultured organs of corti from embryonic and neonatal mice. *J. Neurosci.* **15**, 3013–3026.

Kiernan, A. E., Nunes, F., Wu, D. K., and Fekete, D. M. (1997). The expression domain of two related homeobox genes defines a compartment in the chicken inner ear that may be involved in semicircular canal formation. *Dev. Biol.* **191**, 215–229.

Kiernan, A. E., Zalzman, M., Fuchs, H., Hrabé de Angelis, M., Balling, R., Steel, K. P., and Avraham, K. B. (1999). Tailchaser (Tlc): A new mouse mutation affecting hair bundle differentiation and hair cell survival. *J. Neurocytol.* **28**, 969–985.

Kikuchi, K., and Hilding, D. A. (1966). The development of the stria vascularis in the mouse. *Acta Oto-Laryngol.* **62**, 277–291.

Kitamura, K., Kakoi, H., Yoshikawa, Y., and Ochikubo, F. (1992). Ultrastructural findings in the inner ear of Jackson shaker mice. *Acta Oto-Laryngol.* **112**, 622–627.

Kohan, R. (1944). The chordomesoderm as an inducer of the ear vesicle. *C. R. Acad. Sci. URSS* **45**, 39–41.

Krauss, S., Johansen, T., Korzh, V., and Fjose, A. (1991). Expression of the zebrafish paired box gene pax[zf-b] during early neurogenesis. *Development (Cambridge, UK)* **113**, 1193–1206.

Ladher, R. K., Anakwe, K. U., Gurney, A. L., Schoenwolf, G. C., and Francis-West, P. H. (2000). Identification of synergistic signals initiating inner ear development. *Science* **290**, 1965–1967.

Lanford, P. J., Lan, Y., Jiang, R., Lindsell, C., Weinmaster, G., Gridley, T., and Kelley, M. W. (1999). Notch signalling pathway mediates hair cell development in mammalian cochlea. *Nat. Genet.* **21**, 289–292.

Lang, H., Bever, M. M., and Fekete, D. M. (2000). Cell proliferation and cell death in the developing chick inner ear: Spatial and temporal patterns. *J. Comp. Neurol.* **417**, 205–220.

Lawlor, P., Marcotti, W., Rivolta, M. N., Kros, C. J., and Holley, M. C. (1999). Differentiation of mammalian vestibular hair cells from con-

ditionally immortal, postnatal supporting cells. *J. Neurosci.* **19**, 9445–9458.

Lawrence, P. A., and Struhl, G. (1996). Morphogens, compartments, and pattern: Lessons from drosophila? *Cell (Cambridge, Mass.)* **85**, 951–961.

Lewis, J. (1991). Rules for the production of sensory cells. *In* "Regeneration of Vertebrate Sensory Cells" (G. R. Bock and J. Whelan, eds.), pp. 25–53. Wiley, Chichester.

Li, C. W., Van De Water, T. R., and Ruben, R. J. (1978). The fate mapping of the eleventh and twelfth day mouse otocyst: An in vitro study of the sites of origin of the embryonic inner ear sensory structures. *J. Morphol.* **157**, 249–267.

Lim, D. J., and Anniko, M. (1985). Developmental morphology of the mouse inner ear. A scanning electron microscopic observation. *Acta Oto-Laryngol., Suppl.* **422**, 1–69.

Lim, D. J., and Rueda, J. (1992). Structural development of the cochlea. *In* "Development of Auditory and Vestibular Systems 2" (R. Romand, ed.), pp. 33–58. Elsevier, Amsterdam.

Littlewood Evans, A., and Muller, U. (2000). Stereocilia defects in the sensory hair cells of the inner ear in mice deficient in integrin alpha8beta1. *Nat. Genet.* **24**, 424–428.

Liu, M., Pereira, F. A., Price, S. D., Chu, M., Shope, C., Himes, D., Eatock, R. A., Brownell, W. E., Lysakowski, A., and Tsai, M. J. (2000). Essential role of BETA2/NeuroD1 in development of the vestibular and auditory systems. *Genes Dev.* **14**, 2839–2854.

Lowenheim, H., Furness, D. N., Kil, J., Zinn, C., Gultig, K., Fero, M. L., Frost, D., Gummer, A. W., Roberts, J. M., Rubel, E. W., Hackney, C. M., and Zenner, H. P. (1999). Gene disruption of p27(Kip1) allows cell proliferation in the postnatal and adult organ of corti. *Proc. Natl. Acad. Sci. U.S.A.* **96**, 4084–4088.

Lufkin, T., Dierich, A., LeMeur, M., Mark, M., and Chambon, P. (1991). Disruption of the Hox-1.6 homeobox gene results in defects in a region corresponding to its rostral domain of expression. *Cell (Cambridge, Mass.)* **66**, 1105–1119.

Lumsden, A., and Krumlauf, R. (1996). Patterning the vertebrate neuraxis. *Science* **274**, 1109–1115.

Ma, Q., Chen, Z., del Barco Barrantes, I., de la Pompa, J. L., and Anderson, D. J. (1998). Neurogenin1 is essential for the determination of neuronal precursors for proximal cranial sensory ganglia. *Neuron* **20**, 469–482.

Ma, Q., Anderson, D. J., and Fritzsch, B. (2000). Neurogenin 1 null mutant ears develop fewer, morphologically normal hair cells in smaller sensory epithelia devoid of innervation. *J. Assoc. Res. Otolaryngol.* **2**, 129–143.

Mahmood, R., Kiefer, P., Guthrie, S., Dickson, C., and Mason, I. (1995). Multiple roles for FGF-3 during cranial neural development in the chicken. *Development (Cambridge, UK)* **121**, 1399–1410.

Mahmood, R., Mason, I. J., and Morriss-Kay, G. M. (1996). Expression of Fgf-3 in relation to hindbrain segmentation, otic pit position and pharyngeal arch morphology in normal and retinoic acid-exposed mouse embryos. *Anat. Embryol.* **194**, 13–22.

Malicki, J., Schier, A. F., Solnica-Krezel, L., Stemple, D. L., Neuhauss, S. C., Stainier, D. Y., Abdelilah, S., Rangini, Z., Zwartkruis, F., and Driever, W. (1996). Mutations affecting development of the zebrafish ear. *Development (Cambridge, UK)* **123**, 275–283.

Mansour, S. L. (1994). Targeted disruption of int-2 (fgf-3) causes developmental defects in the tail and inner ear. *Mol. Reprod. Dev.* **39**, 62–67; discussion: pp. 67–68.

Manzanares, M., Trainor, P. A., Nonchev, S., Ariza-McNaughton, L., Brodie, J., Gould, A., Marshall, H., Morrison, A., Kwan, C. T., Sham, M. H., Wilkinson, D. G., and Krumlauf, R. (1999). The role of kreisler in segmentation during hindbrain development. *Dev. Biol.* **211**, 220–237.

Mark, M., Lufkin, T., Vonesch, J. L., Ruberte, E., Olivo, J. C., Dolle, P., Gorry, P., Lumsden, A., and Chambon, P. (1993). Two rhombomeres are altered in Hoxa-1 mutant mice. *Development (Cambridge, UK)* **119**, 319–338.

Marovitz, W. F., and Shugar, J. M. (1976). Single mitotic center for rodent cochlear duct. *Ann. Otol., Rhinol., Laryngol.* **85,** 225–233.

Marovitz, W. F., Khan, K. M., and Schulte, T. (1977). Ultrastructural development of the early rat otocyst. *Ann. Otol., Rhinol., Laryngol.* **86,** (Suppl.), 9–28.

Martin, P., and Swanson, G. J. (1993). Descriptive and experimental analysis of the epithelial remodellings that control semicircular canal formation in the developing mouse inner ear. *Dev. Biol.* **159,** 549–558.

Mayordomo, R., Rodriguez-Gallardo, L., and Alvarez, I. S. (1998). Morphological and quantitative studies in the otic region of the neural tube in chick embryos suggest a neuroectodermal origin for the otic placode. *J. Anat.* **193,** 35–48.

Mbiene, J. P., and Sans, A. (1986). Differentiation and maturation of the sensory hair bundles in the fetal and postnatal vestibular receptors of the mouse: A scanning electron microscopy study. *J. Comp. Neurol.* **254,** 271–278.

McKay, I. J., Lewis, J., and Lumsden, A. (1996). The role of FGF-3 in early inner ear development: An analysis in normal and kreisler mutant mice. *Dev. Biol.* **174,** 370–378.

Mendonsa, E. S., and Riley, B. B. (1999). Genetic analysis of tissue interactions required for otic placode induction in the zebrafish. *Dev. Biol.* **206,** 100–112.

Minichiello, L., Piehl, F., Vazquez, E., Schimmang, T., Hokfelt, T., Represa, J., and Klein, R. (1995). Differential effects of combined trk receptor mutations on dorsal root ganglion and inner ear sensory neurons. *Development (Cambridge, UK)* **121,** 4067–4075.

Minowa, O., Ikeda, K., Sugitani, Y., Oshima, T., Nakai, S., Katori, Y., Suzuki, M., Furukawa, M., Kawase, T., Zheng, Y., Ogura, M., Asada, Y., Watanabe, K., Yamanaka, H., Gotoh, S., Nishi-Takeshima, M., Sugimoto, T., Kikuchi, T., Takasaka, T., and Noda, T. (1999). Altered cochlear fibrocytes in a mouse model of DFN3 nonsyndromic deafness. *Science* **285,** 1408–1411.

Moriyama, K., Hashimoto, R., Hanai, A., Yoshizaki, N., Yonezawa, S., and Otani, H. (1997). Degenerative hairlets on the vestibular sensory cells in mutant bustling (BUS/Idr) mice. *Acta Oto-Laryngol.* **117,** 20–24.

Morrison, A., Hodgetts, C., Gossler, A., Hrabé de Angelis, M., and Lewis, J. (1999). Expression of Delta1 and Serrate1 (Jagged1) in the mouse inner ear. *Mech. Dev.* **84,** 169–172.

Morriss-Kay, G. M., and Ward, S. J. (1999). Retinoids and mammalian development. *Int. Rev. Cytol.* **188,** 73–131.

Morsli, H., Choo, D., Ryan, A., Johnson, R., and Wu, D. K. (1998). Development of the mouse inner ear and origin of its sensory organs. *J. Neurosci.* **18,** 3327–3335.

Morsli, H., Tuorto, F., Choo, D., Postiglione, M. P., Simeone, A., and Wu, D. K. (1999). Otx1 and Otx2 activities are required for the normal development of the mouse inner ear. *Development (Cambridge, UK)* **126,** 2335–2343.

Nicolson, T., Rusch, A., Friedrich, R. W., Granato, M., Ruppersberg, J. P., and Nusslein-Volhard, C. (1998). Genetic analysis of vertebrate sensory hair cell mechanosensation: The zebrafish circler mutants. *Neuron* **20,** 271–283.

Niederreither, K., Subbarayan, V., Dolle, P., and Chambon, P. (1999). Embryonic retinoic acid synthesis is essential for early mouse postimplantation development. *Nat. Genet.* **21,** 444–448.

Niederreither, K., Vermot, J., Schuhbaur, B., Chambon, P., and Dolle, P. (2000). Retinoic acid synthesis and hindbrain patterning in the mouse embryo. *Development (Cambridge, UK)* **127,** 75–85.

Nieto, M. A., Gilardi-Hebenstreit, P., Charnay, P., and Wilkinson, D. G. (1992). A receptor protein tyrosine kinase implicated in the segmental patterning of the hindbrain and mesoderm. *Development (Cambridge, UK)* **116,** 1137–1150.

Nishikori, T., Hatta, T., Kawauchi, H., and Otani, H. (1999). Apoptosis during inner ear development in human and mouse embryos: An analysis by computer-assisted three-dimensional reconstruction. *Anat. Embryol.* **200,** 19–26.

Oh, S. H., Johnson, R., and Wu, D. K. (1996). Differential expression of bone morphogenetic proteins in the developing vestibular and auditory sensory organs. *J. Neurosci.* **16,** 6463–6475.

Pera, E., and Kessel, M. (1999). Expression of DLX3 in chick embryos. *Mech. Dev.* **89,** 189–193.

Phippard, D., Lu, L., Lee, D., Saunders, J. C., and Crenshaw, E. B., 3rd (1999). Targeted mutagenesis of the POU-domain gene Brn4/Pou3f4 causes developmental defects in the inner ear. *J. Neurosci.* **19,** 5980–5989.

Pirvola, U., Ylikoski, J., Palgi, J., Lehtonen, E., Arumae, U., and Saarma, M. (1992). Brain-derived neurotrophic factor and neurotrophin 3 mRNAs in the peripheral target fields of developing inner ear ganglia. *Proc. Natl. Acad. Sci. U.S.A.* **89,** 9915–9919.

Pirvola, U., Arumae, U., Moshnyakov, M., Palgi, J., Saarma, M., and Ylikoski, J. (1994). Coordinated expression and function of neurotrophins and their receptors in the rat inner ear during target innervation. *Hear. Res.* **75,** 131–144.

Pirvola, U., Cao, Y., Oellig, C., Suoqiang, Z., Pettersson, R. F., and Ylikoski, J. (1995). The site of action of neuronal acidic fibroblast growth factor is the organ of Corti of the rat cochlea. *Proc. Natl. Acad. Sci. U.S.A.* **92,** 9269–9273.

Probst, F. J., Fridell, R. A., Raphael, Y., Saunders, T. L., Wang, A., Liang, Y., Morell, R. J., Touchman, J. W., Lyons, R. H., Noben-Trauth, K., Friedman, T. B., and Camper, S. A. (1998). Correction of deafness in shaker-2 by an unconventional myosin in a BAC transgene. *Science* **280,** 1444–1447.

Raz, Y., and Kelley, M. W. (1999). Retinoic acid signaling is necessary for the development of the organ of Corti. *Dev. Biol.* **213,** 180–193.

Read, A. P., and Newton, V. E. (1997). Waardenburg syndrome. *J. Med. Genet.* **34,** 656–665.

Represa, J., Leon, Y., Miner, C., and Giraldez, F. (1991). The int-2 protooncogene is responsible for induction of the inner ear. *Nature (London)* **353,** 561–563.

Represa, J. J., Moro, J. A., Gato, A., Pastor, F., and Barbosa, E. (1990). Patterns of epithelial cell death during early development of the human inner ear. *Ann. Otol., Rhinol., Laryngol.* **99,** 482–488.

Riley, B. B., and Grunwald, D. J. (1996). A mutation in zebrafish affecting a localized cellular function required for normal ear development. *Dev. Biol.* **179,** 427–435.

Rinkwitz-Brandt, S., Arnold, H. H., and Bober, E. (1996). Regionalized expression of Nkx5-1, Nkx5-2, Pax2 and sek genes during mouse inner ear development. *Hear. Res.* **99,** 129–138.

Romand, R., Sapin, V., and Dolle, P. (1998). Spatial distributions of retinoic acid receptor gene transcripts in the prenatal mouse inner ear. *J. Comp. Neurol.* **393,** 298–308.

Rossel, M., and Capecchi, M. R. (1999). Mice mutant for both Hoxa1 and Hoxb1 show extensive remodeling of the hindbrain and defects in craniofacial development. *Development (Cambridge, UK)* **126,** 5027–5040.

Ruben, R. J. (1967). Development of the inner ear of the mouse: A radioautographic study of terminal mitoses. *Acta Oto-Laryngol., Suppl.* **220,** 1–44.

Salminen, M., Meyer, B. I., Bober, E., and Gruss, P. (2000). Netrin 1 is required for semicircular canal formation in the mouse inner ear. *Development (Cambridge, UK)* **127,** 13–22.

Sans, A., and Chat, M. (1982). Analysis of temporal and spatial patterns of rat vestibular hair cell differentiation by tritiated thymidine radioautography. *J. Comp. Neurol.* **206,** 1–8.

Schecterson, L. C., and Bothwell, M. (1994). Neurotrophin and neurotrophin receptor mRNA expression in developing inner ear. *Hear. Res.* **73,** 92–100.

Schimmang, T., Minichiello, L., Vazquez, E., San Jose, I., Giraldez, F., Klein, R., and Represa, J. (1995). Developing inner ear sensory neurons require TrkB and TrkC receptors for innervation of their peripheral targets. *Development (Cambridge, UK)* **121,** 3381–3391.

Schneider-Maunoury, S., Topilko, P., Seitandou, T., Levi, G., Cohen-

Tannoudji, M., Pournin, S., Babinet, C., and Charnay, P. (1993). Disruption of Krox-20 results in alteration of rhombomeres 3 and 5 in the developing hindbrain. *Cell (Cambridge, Mass.)* **75,** 1199–1214.

Schneider-Maunoury, S., Seitanidou, T., Charnay, P., and Lumsden, A. (1997). Segmental and neuronal architecture of the hindbrain of Krox-20 mouse mutants. *Development (Cambridge, UK)* **124,** 1215–1226.

Self, T., Mahony, M., Fleming, J., Walsh, J., Brown, S. D., and Steel, K. P. (1998). Shaker-1 mutations reveal roles for myosin VIIA in both development and function of cochlear hair cells. *Development (Cambridge, UK)* **125,** 557–566.

Self, T., Sobe, T., Copeland, N. G., Jenkins, N. A., Avraham, K. B., and Steel, K. P. (1999). Role of myosin VI in the differentiation of cochlear hair cells. *Dev. Biol.* **214,** 331–341.

Shamim, H., and Mason, I. (1998). Expression of Gbx-2 during early development of the chick embryo. *Mech. Dev.* **76,** 157–159.

Sheng, G., and Stern, C. D. (1999). Gata2 and gata3: Novel markers for early embryonic polarity and for non-neural ectoderm in the chick embryo. *Mech. Dev.* **87,** 213–216.

Sher, A. E. (1971). The embryonic and postnatal development of the inner ear of the mouse. *Acta Oto-Laryngol., Suppl.* **285,** 1–77.

Slepecky, N. B. (1996). Structure of the mammalian cochlea. *In* "The Cochlea" (P. Dallos, A. N. Popper, and R. R. Fay, eds.), Vol. 8, pp. 44–129. Springer-Verlag, New York.

Steel, K. P. (1995). Inherited hearing defects in mice. *Annu. Rev. Genet.* **29,** 675–701.

Steel, K. P., and Barkway, C. (1989). Another role for melanocytes: Their importance for normal stria vascularis development in the mammalian inner ear. *Development (Cambridge, UK)* **107,** 453–463.

Steel, K. P., Barkway, C., and Bock, G. R. (1987). Strial dysfunction in mice with cochleo-saccular abnormalities. *Hear. Res.* **27,** 11–26.

Steel, K. P., Davidson, D. R., and Jackson, I. J. (1992). TRP-2/DT, a new early melanoblast marker, shows that steel growth factor (c-kit ligand) is a survival factor. *Development (Cambridge, UK)* **115,** 1111–1119.

Steel, K. P., Erven, A., and Kiernan, A. E. (2001). Mice as models for human hereditary deafness. *In* "Springer Handbook of Auditory Research" (B. J. Keats, A. N. Popper, and R. R. Fay, eds.). Springer, New York (in press).

Stone, J. S., Oesterle, E. C., and Rubel, E. W. (1998). Recent insights into regeneration of auditory and vestibular hair cells. *Curr. Opin. Neurol.* **11,** 17–24.

Swanson, G. J., Howard, M., and Lewis, J. (1990). Epithelial autonomy in the development of the inner ear of a bird embryo. *Dev. Biol.* **137,** 243–257.

Swiatek, P. J., and Gridley, T. (1993). Perinatal lethality and defects in hindbrain development in mice homozygous for a targeted mutation of the zinc finger gene Krox20. *Genes Dev.* **7,** 2071–2084.

Swiatek, P. J., Lindsell, C. E., del Amo, F. F., Weinmaster, G., and Gridley, T. (1994). Notch1 is essential for postimplantation development in mice. *Genes Dev.* **8,** 707–719.

Takemura, T., Sakagami, M., Takebayashi, K., Umemoto, M., Nakase, T., Takaoka, K., Kubo, T., Kitamura, Y., and Nomura, S. (1996). Localization of bone morphogenetic protein-4 messenger RNA in developing mouse cochlea. *Hear. Res.* **95,** 26–32.

ten Berge, D., Brouwer, A., Korving, J., Martin, J. F., and Meijlink, F. (1998). Prx1 and Prx2 in skeletogenesis: Roles in the craniofacial region, inner ear and limbs. *Development (Cambridge, UK)* **125,** 3831–3842.

Theiler, K. (1989). "The House Mouse." Springer-Verlag, New York.

Thorn, L., and Schinko, I. (1985). Light- and electronmicroscopic investigations of the development of the stria vascularis in the ductus cochlearis of the guinea pig fetus. *Acta Anat.* **124,** 159–166.

Tilney, L. G., Tilney, M. S., and DeRosier, D. J. (1992). Actin filaments, stereocilia, and hair cells: How cells count and measure. *Annu. Rev. Cell Biol.* **8,** 257–274.

Torres, M., and Giraldez, F. (1998). The development of the vertebrate inner ear. *Mech. Dev.* **71,** 5–21.

Torres, M., Gomez-Pardo, E., and Gruss, P. (1996). Pax2 contributes to inner ear patterning and optic nerve trajectory. *Development (Cambridge, UK)* **122,** 3381–3391.

Uchikawa, M., Kamachi, Y., and Kondoh, H. (1999). Two distinct subgroups of Group B Sox genes for transcriptional activators and repressors: Their expression during embryonic organogenesis of the chicken. *Mech. Dev.* **84,** 103–120.

Van De Water, T. R. (1976). Effects of removal of the statoacoustic ganglion complex upon the growing otocyst. *Ann. Otol., Rhinol., Laryngol.* **85,** 2–31.

Vendrell, V., Carnicero, E., Giraldez, F., Alonso, M. T., and Schimmang, T. (2000). Induction of inner ear fate by FGF3. *Development (Cambridge, UK)* **127,** 2011–2019.

Waddington, C. H. (1937). The determination of the auditory placode in the chick. *J. Exp. Biol.* **14,** 232–239.

Wang, W., Van De Water, T., and Lufkin, T. (1998). Inner ear and maternal reproductive defects in mice lacking the Hmx3 homeobox gene. *Development (Cambridge, UK)* **125,** 621–634.

Wheeler, E. F., Bothwell, M., Schecterson, L. C., and von Bartheld, C. S. (1994). Expression of BDNF and NT-3 mRNA in hair cells of the organ of Corti: Quantitative analysis in developing rats. *Hear. Res.* **73,** 46–56.

Whitfield, T. T., Granato, M., van Eeden, F. J., Schach, U., Brand, M., Furutani-Seiki, M., Haffter, P., Hammerschmidt, M., Heisenberg, C. P., Jiang, Y. J., Kane, D. A., Kelsh, R. N., Mullins, M. C., Odenthal, J., and Nusslein-Volhard, C. (1996). Mutations affecting development of the zebrafish inner ear and lateral line. *Development (Cambridge, UK)* **123,** 241–254.

Wilkinson, D. G., Peters, G., Dickson, C., and McMahon, A. P. (1988). Expression of the FGF-related proto-oncogene int-2 during gastrulation and neurulation in the mouse. *EMBO J.* **7,** 691–695.

Woo, K., and Fraser, S. E. (1997). Specification of the zebrafish nervous system by nonaxial signals. *Science* **277,** 254–257.

Woo, K., and Fraser, S. E. (1998). Specification of the hindbrain fate in the zebrafish. *Dev. Biol.* **197,** 283–296.

Xiang, M., Gan, L., Li, D., Chen, Z. Y., Zhou, L., O'Malley, B. W., Jr., Klein, W., and Nathans, J. (1997). Essential role of POU-domain factor Brn-3c in auditory and vestibular hair cell development. *Proc. Natl. Acad. Sci. U.S.A.* **94,** 9445–9450.

Xiang, M., Gao, W. Q., Hasson, T., and Shin, J. J. (1998). Requirement for Brn-3c in maturation and survival, but not in fate determination of inner ear hair cells. *Development (Cambridge, UK)* **125,** 3935–3946.

Xu, P. X., Adams, J., Peters, H., Brown, M. C., Heaney, S., and Maas, R. (1999). Eya1-deficient mice lack ears and kidneys and show abnormal apoptosis of organ primordia. *Nat. Genet.* **23,** 113–117.

Ylikoski, J., Pirvola, U., Moshnyakov, M., Palgi, J., Arumae, U., and Saarma, M. (1993). Expression patterns of neurotrophin and their receptor mRNAs in the rat inner ear. *Hear. Res.* **65,** 69–78.

Yntema, C. L. (1933). Experiments on the determination of the ear ectoderm in the embryo of the *Amblystoma punctatum. J. Exp. Zool.* **65,** 317–357.

Yntema, C. L. (1939). Self-differentiation of the heterotopic ear ectoderm in the embryo of the *Amblystoma punctatum. J. Exp. Zool.* **80,** 1–15.

Yntema, C. L. (1950). An analysis of induction of the ear from foreign ectoderm in the embryo of *Amblystoma punctatum. J. Exp. Zool.* **113,** 211–244.

Yonezawa, S., Nodasaka, Y., Kamada, T., Fujita, S. C., Kato, K., Yamada, Y., Ogasawara, N., and Shoji, R. (1996). Cochlear histopathology of the mutant bustling mouse, BUS/Idr. *Acta Oto-Laryngol.* **116,** 409–416.

Yonezawa, S., Yoshiki, A., Hanai, A., Matsuzaki, T., Matsushima, J., Kamada, T., and Kusakabe, M. (1999). Chromosomal localization of a gene responsible for vestibulocochlear defects of BUS/Idr mice: Identification as an allele of waltzer. *Hear. Res.* **134,** 116–122.

Zhang, N., Martin, G. V., Kelley, M. W., and Gridley, T. (2000). A mutation in the lunatic fringe gene suppresses the effects of a jagged2 mutation on inner hair cell development in the cochlea. *Curr. Biol.* **10,** 659–662.

Zheng, J. L., and Gao, W. Q. (1997). Analysis of rat vestibular hair cell development and regeneration using calretinin as an early marker. *J. Neurosci.* **17,** 8270–8282.

Zheng, J. L., and Gao, W. Q. (2000). Overexpression of math1 induces robust production of extra hair cells in postnatal rat inner ears. *Nat. Neurosci.* **3,** 580–586.

Zheng, J. L., Shou, J., Guillemot, F., Kageyama, R., and Gao, W. (2000). Hes1 is a negative regulator of inner ear hair cell differentiation. *Development (Cambridge, UK)* **127,** 4551–4560.

Zine, A., Van De Water, T. R., and de Ribaupierre, F. (2000). Notch signaling regulates the pattern of auditory hair cell differentiation in mammals. *Development (Cambridge, UK)* **127,** 3373–3383.

23

Integumentary Structures

Carolyn Byrne and Matthew Hardman

School of Biological Sciences, University of Manchester, Manchester M13 9PT, United Kingdom

I. Introduction

II. Mature Skin

III. Non-Neural Embryonic Ectoderm

IV. Stratification

V. Dermal Development

VI. Epidermal Appendage Morphogenesis

VII. Model for Follicle Formation: The First Dermal Signal

VIII. Follicle Spacing

IX. Follicle Morphogenesis and Differentiation

X. Follicle Morphogenesis and Follicle Cycling

XI. Molecular Parallels between Skin Tumorigenesis and Skin Development

XII. Early Terminal Differentiation

XIII. Regulation of Transit to Late Stages of Terminal Differentiation

XIV. Late Terminal Differentiation: Formation of Stratum Corneum and Skin Barrier

XV. Periderm Disaggregation

XVI. Conclusions and Future Directions

References

I. Introduction

The integument, derived from the Latin *integumentum* meaning a "covering," includes the skin and its appendages—hair, nails, and glands. The integument provides the principal barrier between internal body structures and the environment. Part of the intense drive to understand integument biology arises from the need for treatments for human dermatological conditions and the feasibility of gene therapy in this most accessible of organs, expanding interest in transdermal delivery of DNA and drugs, and the need to appreciate and treat problems of immature skin in premature infants. A range of cosmetic applications will also arise from understanding the development and biology of integument and it appendages.

Skin appendage morphogenesis and epidermal terminal differentiation, resulting in formation of the protective stratum corneum and barrier, are currently the subject of intensive research. This chapter places these events in the context of skin development from early organogenesis to birth.

Studies of mammalian integument development benefit from both the tractability of mouse to genetic manipulation and a long history of research exploiting experimental accessibility of chick skin. In addition, murine developmental studies and transgenic/gene knock-out models are providing

Table I Comparison of Key Stages during Skin Development in Mouse, Human, and Chicken

Stage of epidermal development	Mouse[a]			Human[b]			Chicken[c]		
	Theiler's stage[d]	Age (days)	Size (mm)	Carnegie stage[e]	Age (days)	Size (mm)/ crown-rump length	Hamburger and Hamilton stage[f]	Age (days) (postincubation)	Size (mm)
Single-layered embryonic ectoderm	13–16	8.5–10	1.5–2.6	10–22	21–55	2–28	3–6	0.5–1	2.5–4
Gut forms	12–14	8–9	1.6–2.3	8–10	14–21	1–2	10–14	1.5–2	5–7
Neural tube closure	13–15	8.5–9.5	1.7–2.5	10–12	21–28	2–5	12–17	2–2.5	6–7
Ectoderm + periderm	14–20	9–12	2.6–7	22–23	55–70	23–61	6–26	1–5	2.6–11.3
Intermediate layer forms	20–23	12–15	7–9.1	No staging system for fetal period	70–90	61–110	29–34	6–8	15–21
Enters terminal differentiation	23	15	9.1		100	120	40–41	14–15	60–65
Initial barrier forms	24–25	16–17	12.2–14.2		147–168	200–230	43	17–17.5	70–75
Periderm dissociates	25–26	17–18	14.2–17.8		175–182	240–250	44	17.5–18.5	75–80
Competent skin barrier	26	18	17.8		210	280	—	—	—

[a] Compiled from Hardman et al. (1998); Kaufman (1992); Weiss and Zelickson (1975a,b,c).
[b] Compiled from Hardman et al. (1999); Moore (1988); Holbrook and Odland (1980).
[c] Compiled from Sengel (1976); Prakkal and Maltosy (1968).
[d] From Theiler (1989).
[e] From Bullen and Wilson (1997).
[f] From Hamburger and Hamilton (1951).

significant insight into human disease. Hence, developmental stages and descriptions are cross-referenced among species (Table I).

Last, an extensive literature documents regulation of epidermal proliferation and terminal differentiation in adult skin, including provision of structural proteins or "markers" for specific epidermal differentiative stages. Cultured keratinocyte models closely replicate adult keratinocyte terminal differentiation and are yielding insights into regulation of differentiation. Sequential change during epidermal development broadly mirrors spatial change from basement membrane to the stratum corneum in mature skin (Fig. 1). Hence, the developing and differentiating systems are compared.

II. Mature Skin

The primary purpose of mature skin is to provide a protective barrier to the environment. In addition, it forms appendages (hairs, nails, glands) that provide sensory, thermoregulatory, and sociosexual functions. During development, ectoderm- and mesenchyme-derived (or cranial neural crest-derived; Le Douarin *et al.,* 1993) dermis are juxtaposed to produce skin. Skin comprises an outer epidermal layer in contact with basement membrane, and an inner dermis (Fig. 1). It is the epidermal component that provides barrier function via the outer, impermeable stratum corneum. The dermis provides support and nourishment.

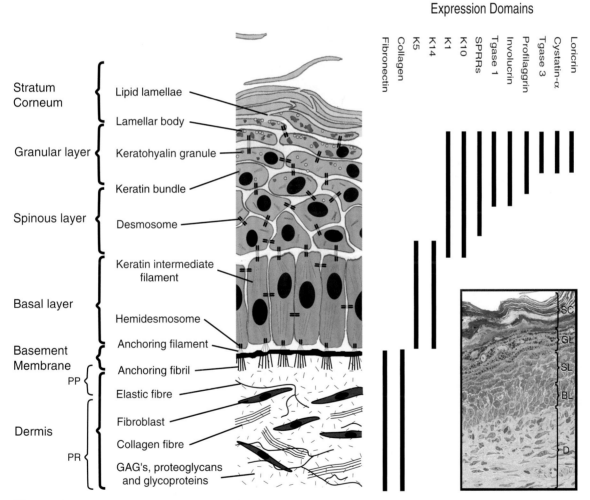

Figure 1 Diagrammatic representation of adult skin. Structurally the skin can be envisaged as three components: dermis, basement membrane, and epidermis. The epidermis can be further subdivided into four layers, each representing a specific terminal differentiation stage. The domains of expression for structural proteins specific to particular layers or strata are shown. An adult histological section is included, clearly demonstrating the specific skin layers. GAG, glycosaminoglycan; K5, K14, K1, K10, keratins 5, 14, 1, and 10; PP, pars papillaris; PR, pars reticularis; Tgase, transglutaminase.

Skin serves additional functions, including an essential immunoregulatory role (Robert and Kupper, 1999; Kupper, 1996), and respiratory, secretory, excretory, and biosynthetic roles (e.g., biosynthesis of vitamin D). Keratinocytes, the major epidermal cell, derive from stem cells located in a basal proliferative epidermal layer (Watt and Hogan, 2000; Fuchs and Segré, 2000). Keratinocytes derived from stem cell division, termed transit amplifying cells, undergo limited proliferation in the basal layer then withdraw from the cell cycle. In response to unknown cues, the G_0/G_1 basal layer keratinocytes downregulate basement membrane integrin receptors, lose contact with basement membrane, and are displaced outward. Outward-bound keratinocytes alter their program of gene expression to produce protein and lipid components of the outer, protective stratum corneum. Keratin, the major protein product of epidermal keratinocytes, protects epidermis from mechanical stress (Fuchs, 1996) and participates in formation of stratum corneum (Candi et al., 1998).

Stratum corneum cells (squames) are anucleate and flattened, filled internally with aggregated keratin matrix, and surrounded by a tough, impermeable cornified envelope that is cross-linked to external lipid. Outer squames are shed into the environment to be replaced by keratinocytes migrating from the basal layer. This dynamic cycle of keratinocyte production, differentiation, and shedding, is called *terminal differentiation* or *keratinization.* One of the key challenges in the epidermal field is to understand how the balance between proliferation and differentiation is achieved.

Epidermal cells adhere to each other mainly via adherens junctions and desmosomes (intercellular adhesion junctions) and to basement membrane via integrin receptors. Basement membrane is anchored to underlying dermis via anchoring fibrils (Fig. 1). The principal dermal cells, fibroblasts, sit in a matrix of collagen, elastic fibers, and fibronectin. Epidermis and dermis interact during development and the adult hair cycle to produce hair, gland, and nail skin appendages.

The epidermal-dermal interface can comprise a series of ridges and hollows, particulary in nonhairy skin. Dermal protrusions into the epidermis are known as dermal papillae and epidermal protrusions into dermis are called rete ridges. Gross architecture of skin is important because it has been proposed that dermal papillae/rete ridges provide microenvironments for epidermal stem cells and transit amplifying cells (Cotsarelis et al., 1990, and references within; Jensen et al., 1999; Lavker and Sun, 2000). Evidence for stem cell microenvironments in skin derives from selective labeling of slow-cycling stem cells in epidermis, and localization of cells with surface properties (e.g., integrin levels) that have been defined as characteristic of stem cells via *in vitro* studies (Jones et al., 1995; Tani et al., 2000).

Recently a follicular origin for epidermal stem cells has been proposed (Taylor et al., 2000; Lavker and Sun, 2000) in hair-bearing skin. *In vivo* stem cells are slow cycling and hence retain DNA label, permitting detection. Follicular stem cells reside in the "bulge" area of the hair follicle (Cotsarelis et al., 1990; Morris and Potten, 1999) and are bipotent, or able to differentiate along both epidermal and follicular pathways. Follicular bulge-derived stem cells can migrate horizontally into interfollicular epidermal locations where they provide a source for epidermal renewal.

III. Non-Neural Embryonic Ectoderm

Epidermis derives from the outer, surface cell layer (ectoderm) of the postgastrulation embryo. Ectoderm consists of a single layer of histologically undifferentiated epithelial cells. Single-layered nonneural ectoderm stratifies regionally to produce an outer surface layer of flattened cells, the periderm (Fig. 2; Theiler, 1989, stages 16–20; Kaufman, 1992, Plate 46, a–d).

Periderm is a transitory embryonic ectodermal cell layer that provides an interface between embryo and amniotic fluid during most of epidermal development *in utero.* Periderm will develop and differentiate in tandem with underlying ectoderm and will be shed into the amniotic fluid later in gestation (Boneko and Merker, 1988; Hoyes, 1968). A marker protein for periderm is keratin 6 (Mazzalupo and Coulombe, 2000, and references within). The function of periderm is still unknown. It has been proposed to play a protective role, sheltering the fragile, underlying ectoderm. In humans exfoliated periderm mixes with secreted sebum to form vernix caseosa, a slippery white epidermal covering thought to act as a lubricant during birth (Agorastos et al., 1997). Periderm may play an analogous role in mouse.

Occasionally in humans periderm persists until birth, forming a fetal "cocoon." A child born with this is termed a *collodion baby* (Akiyama et al., 1997; OMIM®, MIM numbers 245130 and 604780). Invariably, in these cases, the underlying epidermis is abnormal. Periderm appears to have an interactive role with amniotic fluid during development. The outer surface of the cells are blebbed, and covered with microvilli to increase cell surface area, and transport vesicles are evident intracelluarly. Periderm could absorb nutrients from the amniotic fluid to supply epidermal demands prior to formation of the skin capillary network (Hoyes, 1968).

Early ectoderm resembles simple epithelial cells both morphologically and biochemically, because it produces simple epithelial-specific keratins K8 and K18 (Jackson et al., 1981; Moll et al., 1982; Thorey et al., 1993), though the functional significance, if any, is obscure because keratin 8/18 knockout mice do not display obvious skin developmental defects (Baribault et al., 1994; Magin et al., 1998).

The regional nature of postgastrulation skin becomes apparent at this stage. Ectoderm morphology varies with body site, an extreme example being the thin, stretched ectoderm overlying the neural tube. This results in developmental de-

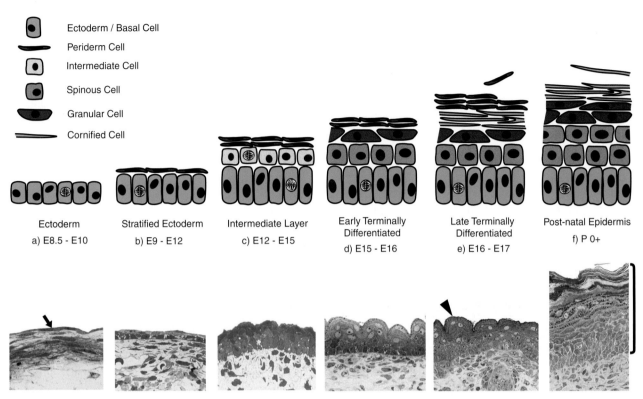

Figure 2 Diagrammatic representation of epidermal development from single-layered ectoderm (E8.5–10) to stratified postnatal skin (P0+). (a) Epidermis derives from single-layered non-neural embryonic ectoderm. Initial mitoses are parallel to the skin resulting in ectodermal expansion. (b) At around E9 regional stratification gives rise to an outer layer, the periderm. (c) Around E12 the orientation of basal layer mitoses changes and many are now at right angles to the epidermis, which results in stratification and formation of the intermediate layer. (d) By E15 cells from the intermediate layer enter terminal differentiation and markers such as K1/K10 are induced. (e) By E16 terminal differentiation is essentially complete, marked by the appearance of a specialized outer epidermal layer, the stratum corneum (arrowhead). Toludine blue-stained histological sections accompanying each stage demonstrate dramatic increase in epidermal thickness from single-layered ectoderm (arrow) to adult (bracket) and characteristic structural changes.

lay manifested as a middorsal stripe of lagging skin persisting to very late (E16) in gestation. Other changes (periderm formation and induction of K5 and K14 keratins characteristic of mature epidermis; Fig. 1) occur regionally (Byrne *et al.*, 1994).

Regional ectodermal development could be due to (1) ectodermal heterogeneity, possibly mediated by *Hox* genes that are expressed regionally in ectoderm (Chuong *et al.*, 1990; Couly and Le Douarin, 1990; Hunt *et al.*, 1991; Kanzler *et al.*, 1994). Murine branchial ectoderm expresses the *Hox* code corresponding to underlying neural crest-derived tissue (Hunt *et al.*, 1991) and it has been proposed that surface ectoderm covering craniofacial primordia forms part of the *ectomeres* or developmental units comprising the central nervous system, neural crest and surface ectoderm (Couly and Le Douarin, 1990). (2) It could also be a result of programming by heterogenous underlying mesenchyme. As described below for follicle morphogenesis, development of skin and its appendages is regulated by underlying dermis. Dermis derives from diverse mesodermal sources (somites, lateral plate, neural crest).

Ectodermal cells also display regional morphogenetic activity. Ectodermal signaling to underlying tissue, such as limb mesoderm (Johnson and Tabin, 1997), nephric duct (Obara-Ishihara *et al.*, 1999), or lateral plate mesoderm (Funayama *et al.*, 1999), is necessary for organ specification, determination, or morphogenesis. At this stage of development ectoderm and mesenchyme act as a regionally heterogeneous signaling organ rather than an inert surface covering.

IV. Stratification

Further stratification (Theiler, 1989, stages 20–23; Kaufman, 1992, Plate 46, c–f) produces an intermediate proliferative ectodermal layer, sometimes called the *stratum intermedium* (Hansen, 1947; Sengel, 1976; Fig. 2). However, the resultant multilayered structure still bears little resemblance to adult epidermis. Stratification also occurs regionally. Transcriptional control of stratification could be mediated by p63, a homolog of the tumor suppressor p53, because epidermis of homozygous *p63* gene-targeted mice remains single layered (Mills *et al.*, 1999; Yang *et al.*, 1999), probably arresting at E9.5 (Theiler, 1989, stages 16–17).

After stratification, mitoses are present throughout the basal layer, intermediate layer and periderm. It is probable that early embryonic stratification is achieved by mitotic activity, rather than via cell migration as in mature epidermis (Smart, 1970; Sengel, 1976). In postnatal epidermis and single-layered embryonic ectoderm, mitoses are mainly oriented with their spindle at right angles to the basal lamina or basement membrane, so that mitotic activity contributes to expansion of the basal layer (Fig. 2). Hence, during earliest-ectodermal development, mitotic activity contributes to ectodermal expansion that can accommodate the rapidly expanding embryo. However, concomitant with embryonic stratification the orientation of a majority of mitotic spindles becomes parallel to the basal lamina, so that mitotic activity leads to stratification (Smart, 1970; Fig. 2). Embryonic proliferative stratification differs radically from stratification associated with terminal differentiation, where withdrawal from the cell cycle precedes stratification. Mitotic spindle orientation will revert to early embryonic pattern later in development, as ectoderm transits to terminally differentiating epidermis.

During this phase of development, the proliferative activity of basal and intermediate layers is paralleled by basal and intermediate layer expression of keratins K14 and K5 (Byrne *et al.*, 1994). These keratins are usually regarded as "markers" of the proliferative basal layer in mature epidermis (Fig. 1; Fuchs, 1996), so K5/K14 expression in the suprabasal layer (upper epidermal layers, not in contact with basement membrane) emphasizes the unique nature of this transitory embryonic skin stage. Adult suprabasal-associated keratins (K1 and K10, Fig. 1) are not yet present. During embryogenesis expression patterns of keratins reflect differentiative changes but probably have no functional significance until birth when epidermal keratin null mice show stratum-specific blistering (Fuchs, 1996; Magin, 1998).

In addition, the transit to stratification is marked by a change in integrin gene expression and protein localization. Integrins are heterodimeric (α and β subunit) transmembrane glycoproteins proposed to function in adhesion, cell survival, migration, and epidermal terminal differentiation (Watt and Hertle, 1994; Fuchs *et al.*, 1997). The $\alpha_6\beta_4$ subunits form part of the specialized hemidesmosome receptor for extracellular laminin from the basement membrane (Fig. 1) and bind the intermediate filament network intracelluarly, while $\alpha_3\beta_1$ subunits also provide receptors for extracellular laminin but bind the actin cytoskeleton intracelluarly.

The $\alpha_6\beta_4$ integrins adopt a pericellular organization in basal cells prior to stratification, then relocate to the basement membrane zone during stratification, concomitant with hemidesmosome formation (Hertle *et al.*, 1991). Relocation probably correlates with adoption by ectodermal cells of laminin-based adhesion (Di Persio *et al.*, 2000)

V. Dermal Development

Dermal influences control appendage (hair, nail, gland, teeth) development and epidermal differentiation (Sengel, 1976; see below), so development of dermis is key to understanding development of the integument. Dermis has diverse origins. Murine dorsal dermis derives from the dermamyotome compartment of somites (somitic mesenchyme); ventral and limb dermis derive from lateral plate mesenchyme (somatic mesenchyme); and much cranial mesenchyme derives from neural crest (Couly *et al.*, 1992; Osumi-Yamashita *et al.*, 1994).

Skin appendages, such as hair, glands, and scales, are characteristic of each body region. Dermis controls regional-specific appendage development and transplantation experiments show that this ability is acquired by mesenchymal tissue extremely early in chick development (Hamburger and Hamilton, 1951, stages 14–23 or 2–4 days of chick development; Mauger, 1972; Dhouaily, 1998). Since early stages of appendage formation are conserved in mouse and chick (see below), it is possible that analogous early regional mesenchymal specification occurs in mouse. In addition, chick neural tube programs this regional appendage-forming activity in somitic mesoderm (Mauger, 1972), although the mechanism is unknown.

The cellular component of dermis consists mostly of fibroblasts, which are responsible for secretion of dermal constituents. The extracellular space between dermal fibroblast cells, with its network of macromolecules, constitutes the extracellular matrix (ECM). Mature dermis is composed primarily of fibroblast cells, collagen, elastin fibers, and an interfibrillar glycosaminoglycans (GAG)/proteoglycan gel that imparts massive water holding capacity (Fig. 1). Collagen (predominantly type 1 in adult) imparts tensile strength. Interconnecting the collagen bundles are networks of elastin fibers, composed of elastin and fibrillin, which restore the normal collagen structure following deformation. Epidermal nutrient requirements are supplied entirely by the dermis, which also acts to cushion against mechanical injury.

The aqueous portion of the ECM permits rapid diffusion of nutrients, hormones, and metabolites and creates turgor pressure to withstand considerable compressive force. GAGs, linked to protein as structurally diverse proteoglycans, assemble into chains and form gels of varying pore size. These act like sieves to regulate molecular traffic. They also play a major role in chemical signaling, binding various secreted signaling molecules, enzymes, and inhibitors (Keene *et al.*, 1997). Specific proteoglycans play important roles in regulating dermal architecture. For example, *decorin* knock-out mice exhibit fragile skin due to abnormal fibril cross-linking (Danielson *et al.*, 1997).

Collagen fibrils of multiple subtypes aggregate to form fibers or bundles (Vuorio and de Crombrugghe, 1990).

Dermal fibers mainly comprise collagens type I, III, with minor components, such as type V (Mauger *et al.*, 1987). Type 1 collagen is predominant in adult skin. Type III collagen is reported to predominate in human fetal skin (e.g., Epstein, 1974), however, both collagen I and III transcripts are expressed coordinately in dermal precursors from very early in mouse development (E8.5; Theiler, 1989, stage 13; Niederreither *et al.*, 1994). The multitude of dermatological disorders resulting from collagen defects (Uitto, 1999; e.g., OMIM, MIM numbers 120180 and 120150) combined with specific transgenic mouse models highlight the complex nature of fibrillogenesis. Type III collagen knock-out mice develop serious skin lesions (Liu *et al.*, 1997) and distribution and size of all types of collagen fibrils is abnormal, confirming a role for type III collagen in mediating fibrillogenesis during development. Knock-out mice for the minor type V collagen exhibit a similar phenotype of weakened skin and severe scarring (Andrikopoulos *et al.*, 1995); however, these mice specifically display reduced adult collagen I fibril formation, implicating type V collagen in the regulation of type I collagen fibril formation.

Murine dermal development has been systematically described (Van Exan and Hardy, 1984) and categorized into four developmental stages. Like epidermal stratification, dermal maturation is regional in mouse and human, where cranial dermis is first to differentiate (Aktan *et al.*, 1999). However, in contrast to epidermis, developing dermis displays an outward-to-inward developmental gradient, that is, fibroblasts and matrix adjacent to the basement membrane are more mature than those in the deeper dermis.

The boxed material provides a systematic description of the stages of murine dermal development (Van Exan and Hardy, 1984) derived from the whisker pad, a region that develops precociously. Dermal development over the remainder of the body is regional and may be retarded in comparison. The dermis greatly increases in thickness as it matures (around a sevenfold increase from stage 1 to stage 4, depending on body region) and stratifies, giving rise to the presumptive hypodermis (Smith and Holbrook, 1982; Van Exan and Hardy, 1984). Dermal expansion leads to decreased fibroblast density. This produces gaps that are filled by newly synthesized fibers and fibrils resulting in a remarkably constant dermal density throughout development.

At the epidermal-dermal junction, extracellular matrix organizes into a thin sheet, the basement membrane (Burgeson and Christiano, 1997; Bruckner-Tuderman, 1999). The basement membrane prevents contact between dermal fibroblasts and the overlying keratinocytes, selectively inhibiting molecular signaling. It contains laminin, nidogen, heparan sulfate proteoglycan, the specialized network-forming type IV and anchoring fibril forming type VII collagens, and originates from components secreted by both keratinocytes and fibroblasts.

Murine Dermal Development

Stage 1
Prior to 13 days gestation (Theiler, 1989, stage 21; Figs. 2a and 2b) the dermis remains essentially undifferentiated mesenchyme.

Stage 2
Between 14 and 15 days (Theiler, 1989, stages 22–23; Fig. 2d) mesenchymal cells adopt a "fibroblast-like" appearance with prominent Golgi apparatus and endoplasmic reticulum (ER), indicating preparations for the synthesis of matrix constituents. Within the matrix small (20-nm) fibrils begin to appear.

Stage 3
By 16 days (Theiler, 1989, stages 24–25; Fig. 2e) fibroblasts have matured and Golgi, ER, and vesicles have become more numerous, indicating increased proteoglycan synthesis and export. The extracellular matrix undergoes rapid change marked by the appearance of 40-nm fibrils. The majority of matrix fibrils are now arranged to form large fibrous bundles.

Stage 4
Matrix fibrils (Theiler, 1989, stage 25+, Fig. 2f) steadily increase in number and thickness (40–60 nm) forming much longer, denser fibres than at stage 3. Matrix differentiation and maturation continues up to, and following, birth.

Integrin α_1, α_3 and α_5 subunits dimerize with a β_1 subunit in keratinocytes to form receptors for collagen, laminin-5, and fibronectin extracellularly and attach the actin network intracellularly. The $\alpha_6\beta_4$ subunits form part of the hemidesmosomes, binding laminin extracellularly and intermediate filaments inside the keratinocyte.

In mature skin integrins localize to the basal surface of basal layer keratinocytes ($\alpha_6\beta_4$) or are expressed pericellularly in basal cells ($\alpha\beta_1$ heterodimers). Keratinocytes have been reported to display heterogeneous β_1 subunit levels, with stem cells located within β_1-rich regions of epidermis and transit amplifying cells (proliferative stem cell progeny) displaying lower levels of β_1 integrin, which is inactive in keratinocytes committed to terminal differentiation (Watt and Hogan, 2000).

Integrin receptors play a role in development of basement membrane. Mice lacking α_3 integrin, a component of the major keratinocyte laminin receptor $\alpha_3\beta_1$, present defects in basement membrane formation (DiPersio *et al.*, 1997) starting at the point when skin has stratified and is about to enter terminal differentiation. Mice lacking α_5 integrin, part of the $\alpha_5\beta_1$ fibronectin receptor, are also embryonic lethal with mesodermal defects (Yang *et al.*, 1993) and are probably defective in fibronectin matrix assembly during development (Wennerberg *et al.*, 1996). Mice lacking β_1 integrin present defects in processing of laminin-5 (Brakebush *et al.*, 2000; Raghavan *et al.*, 2000).

VI. Epidermal Appendage Morphogenesis

During the stratification period in mice, epithelial appendages (hairs, nails, glands, teeth; Theiler, 1989, stages 21–25; Kaufman, 1992, Plate 46, e–l) form from ectoderm and mesenchyme by a well-characterized program of cell movements and proliferative change (Fig. 3). The initial stages of all appendage development (e.g., stages 0–3; Davidson and Hardy, 1952; Fig. 3) are similar, whereas later stages diverge to give rise to different appendage types.

Appendage development involves a binary decision by ectodermal cells regarding epidermal or appendage fate, followed by morphogenesis and differentiation. *In situ* hybridization and immunohistochemical studies show that major signaling molecules associated with follicle morphogenesis are conserved between mouse and chick (see below), and mouse–chick tissue recombinations can successfully support early stages of appendage development (Dhouailly *et al.*, 1998). Hence, the following description of follicle formation will draw from both systems.

Traditional approaches (Sengel, 1990) involved use of epithelial-mesenchymal separation followed by recombination of tissue from different developmental stages, body sites, or species to decipher contribution of each tissue to appendage development. Morphological analysis of the appendage formed by the epithelial-mesenchymal recombinants can demonstrate the controlling tissue at each developmental stage because follicle morphology and density vary over the body surface in a species and developmental-stage-specific manner (Fig. 3). These studies showed that follicle morphogenesis involves a scheme of reciprocal interaction or signaling between mesenchyme-dermis and epithelium (Sengel, 1976; Hardy, 1992; Dhouailly *et al.*, 1998; Chuong and Widelitz, 1998; Paus and Cotsarelis, 1999; Paus *et al.*, 1999a).

Mesenchyme signals first to ectoderm with the message "make an appendage," and it seems that any ectodermal tissue (and some nonectodermal epithelial tissues) is competent to respond. This primary message is probably transient and can operate across species. Ectoderm responds to the primary message by forming an ectodermal placode or cluster of elongated cells (Fig. 3, stage 0). The ectodermal placode becomes partly autonomous and can signal back to

unspecified dermis to render it morphogenetically active. Mesenchymal cells aggregate under the placode and form a signaling center that persists in the mature follicle as the dermal papilla (Wessells, 1965; Sengel, 1976; Fig. 3). Subsequent dermal messages to ectoderm can include information (and apparently detailed instructions) about making region-specific and species-specific appendage types. Subsequent morphological change accompanying follicle formation is described in Fig. 3 using the staging of Hardy (1969), as modified by Paus *et al.* (1999a). Recently many of the molecular players in this story have been identified (Oro and Scott, 1998).

VII. Model for Follicle Formation: The First Dermal Signal

Initiation of follicle formation is controlled by dermis. The succession of events in follicle formation has been dissected in chick epidermis because here appendages form sequentially within a two-dimensional array of rows and files called feather "tracts" (Fig. 4), whereas the pattern and sequence of follicle formation is less ordered and defined in mice (except for specialized areas such as the vibrissae pad; Fig. 5). Placodes form first in an initial row. A wave of appendage-specifying ability or "morphogenetic activity" has been proposed to propagate from row to row across the tract, with new rows induced by the preceding row (Sengel, 1976; Patel *et al.*, 1999), and experimental evidence for waves is provided using spatially separated incisions to disrupt waves in embryonic skin organ culture (Patel *et al.*, 1999). Alternatively, a temporal differentiation gradient is proposed across the array (Davidson, 1983) or a moving "morphogenetic stripe" (Jiang *et al.*, 1999). The morphogenetic stripe confers competence for follicle formation as underlying mesenchymal cell density increases. This latter theory receives experimental support from tissue reconstitution experiments where disaggregated and reconstituted mesenchymal cells from a tract support simultaneous follicle formation as soon as a permissive concentration of mesenchymal cells is provided (Jiang *et al.*, 1999). Propagation of the "morphogenetic stripe" or establishment of the temporal differentiation gradient still requires explanation.

The mode of follicle formation via waves or differentiation gradients, as illustrated by feather morphogenesis, is probably conserved in mammals. Murine vibrissae follicles form as an array (Fig. 5; Kaufman, 1992, plate 49). Embryonic hair follicle formation occurs first at day 13 (Theiler, 1989, stage 21) on the shoulder in mouse and successive waves of follicle formation move caudally, anteriorly, ventrally, and dorsally (Mann, 1962; Kaufman, 1992). Waves of guard hairs (tylotrichs or monotrichs; Fig. 5) appear first, followed by awls at day 16 (Theiler, 1989, stage 24) and auchennes/zigzags at day 18 (Theiler, 1989, stage 26). A cra-

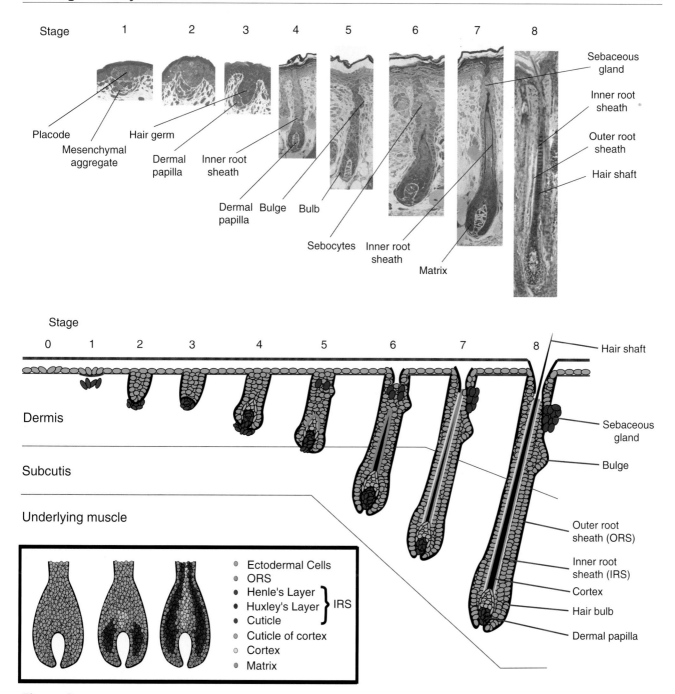

Figure 3 The eight stages of hair follicle development described by Davidson and Hardy (1952). Stage 0: Ectodermal placode formation. Stage 1: Dermal fibroblasts aggregate below the epidermal placode. Stage 2: Prominent hair germ. Stage 3: Hair peg cavity formed by invasion of aggregated dermal fibroblasts. Stage 4: Inner root sheath cone appears. Stage 5: Bulbous peg stage, inner root sheath elongates and the bulge appears. Stage 6: Early sebaceous gland, hair shaft appears and dermal papilla becomes enclosed. Stage 7: Further elongation and hair shaft development. Stage 8: Hair shaft emerges. Note the significant increase in length following stage 5. Inset, post–stage 3 upward movement and differentiation of files of cells from the germinative matrix population (pink) gives rise to all the structures of the mature hair follicle (see key).

nial-to-caudal wave of follicle formation also occurs in human infants (Holbrook *et al.,* 1988). The presence of waves of developmental change is also important when considering follicle cycling and late terminal differentiation of epidermis (see below).

The identity of the first dermal signal is unknown, however, the signal could be a Wnt molecule or another activator of β-catenin signaling (Noramly *et al.,* 1999). Wnts are a family of evolutionarily conserved secreted factors homologous to the *Drosophila* wingless protein (Peifer and Polarkis,

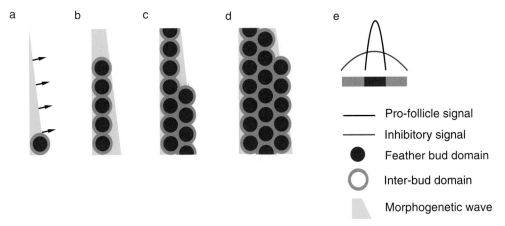

Figure 4 Model of feather tract patterning proposed by Jung *et al.* (1998) and Noramly *et al.* (1999). This model incorporates both lateral inhibition and reaction–diffusion mechanisms. (a) Within each tract a long-range dermal inductive signal crosses the skin as a moving morphogenetic front (arrows). (a, b) The dermal inductive signal induces ectodermal feather placodes, initially within a single row in a rostral-to-caudal progression. Induced ectodermal feather placodes express a local profollicular signal that determines placodal size and a diffusable inhibitory signal that determines placodal spacing by lateral inhibition (see e). (c, d) New placodes only form when the long-range inductive signal passes the individual zones of placodal inhibition.

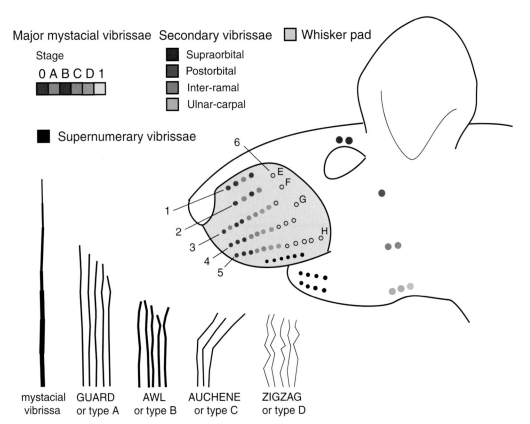

Figure 5 Pelage and vibrissae are the best characterized and most frequently studies of the eight hair types present in the mouse. The mouse coat, composed of pelage hairs and covering the majority of the body, presents several hair types. Fine hairs (auchene, zigzag) are thinner, with less medulla space and one or more bends. The coarser guard and awl hairs are thicker and significantly longer with more medulla cells. Note the smaller size compared to mystacial vibrissae hair. Mouse vibrissae follicles can be subdivided into major mystacial (colored), secondary (blue shades), and supernumerary (black) groups. In an E12 mouse embryo the individual vibrissae display a caudorostral development gradient (Van Exan and Hardy, 1984). The most mature follicles (Davidson and Hardy, 1952, stage 1) are at the caudal edge of the whisker pad in horizontal rows 3 to 5 and vertical row number 6. Conversely, the least mature follicles (stage 1) are those at the rostralmost edge of the whisker pad.

2000; Nusse, 1999). Signaling by Wnt stabilizes β-catenin, which then translocates to the nucleus where it can interact with Lef1/TCF (lymphoid enhancer-binding factor 1/T-cell-specific transcription factor) to regulate expression of downstream genes.

Experimental activation of epithelial β-catenin signaling (by expression of N-terminal truncated, constitutively stabilized forms) induces ectopic follicles in both mouse and chick skin (Gat et al., 1998; Noramly et al., 1999). Conversely, downregulation of β-catenin signaling (through Lef1 knock-out; van Genderen et al., 1994) results in loss of vibrissae and some pelage follicles in mice (Fig. 5), whereas expression of Lef1 in epidermis of transgenic mice can cause ectopic follicle formation (Zhou et al., 1995).

Nuclear-localized β-catenin, indicative of β-catenin signaling, appears diffusely across the skin of the feather tract as one of the earliest indicators of feather formation (Noramly et al., 1999). As the placodes form, β-catenin signaling concentrates at the nascent placodes. Experimental activation of ectodermal β-catenin signaling in chick skin not only results in new placode formation in skin areas that could not normally support follicle formation but induces a battery of downstream genes normally associated with follicle formation (Noramly et al., 1999). It is proposed that the initial dermal signal that induces ectodermal β-catenin signaling has been experimentally short-circuited, resulting in ectoderm that has become an autonomous signaling center able to support follicle formation, a stage predicted by early skin biologists. The identity of the initial dermal signal awaits experimental verification.

Nuclear-localized β-catenin expression is not the only molecular change that occurs ahead of the wave of placode formation. Bmp2, Bmp7, Lnfg, Wnt7a, and follistatin (Jung et al., 1998; Crowe et al., 1998; Noramly and Morgan, 1998; Noramly et al., 1999; Widelitz et al., 1999; Patel et al., 1999) also appear diffusely in the ectoderm prior to placode formation. In fact, ectopic β-catenin signaling is reported to induce Bmp2, Bmp7, and Lnfg (Noramly et al., 1999) and ectopic follistatin expression to induce Bmp2, Bmp4, Bmp7, and Shh (Patel et al., 1999). These molecules will later localize to the ectodermal placode. In addition, the Notch family members and their ligands (Notch1, Notch2, Delta1, Serrate1, and Serrate2) are expressed diffusely first, then later localize to the placode (Chen et al., 1997; Crowe et al., 1998).

The appearance of these molecules in a diffuse wave in the ectoderm has led to the proposal that the initial dermal inducer acts homogeneously and uniformly on tract ectoderm (Fig. 4) rather than inducing single follicles in a localized manner. If so, then spacing of the follicles would be mediated by a separate localized response of ectoderm (Jung et al., 1998; Noramly and Morgan, 1998; Noramly et al., 1999; see below).

VIII. Follicle Spacing

Murine pelage follicles are regularly spaced during development and are initially patterned hexagonally, as in chick skin (see Zhou et al., 1995). Spacing of follicles and placodes within arrays is thought to be self-organizing and operates by lateral inhibition from previously formed placodes (Claxton, 1966; Ede, 1972; Davidson, 1983). Molecules that localize to the nascent follicles include Fgf4, Bmp2, 4 and 7; follistatin, noggin; Wnt7a, Wnt 11; Wnt10b; Notch1 and 2/Delta and Shh and patched (Ptch) (Bitgood and McMahon, 1995; Chuong et al., 1996; Ting-Berreth and Chuong, 1996; Crowe et al., 1998; Jung et al., 1998; Morgan et al., 1998; Noramly and Morgan, 1998; St. Jacques et al., 1998; Patel et al., 1999). Ectopic expression studies show that Fgf4, follistatin, noggin, and Shh are activators of follicle development, while BMPS are follicle inhibitors. The profollicular FGFS are thought to antagonize BMPS. Follistatin and noggin bind to BMP molecules and inhibit BMP "proepidermal" activity, in a manner reminiscent of early neural induction (Normaly and Morgan, 1998; Jung et al., 1998; Patel et al., 1999; Jiang et al., 1999; Botchkarev et al., 1999).

Current models for follicle spacing place both activators and inhibitors within a single source and propose differences in diffusion capabilities between activators and inhibitors (Fig. 4). Activators act locally, while inhibitors diffuse over a greater distance. This concept involves application of the reaction–diffusion model for generation of a repetitive pattern to skin (Nagorcka and Mooney, 1985). Activators determine the size of the follicle domain, while inhibitors define follicle spacing.

Locally acting Bmp inhibitors (follistatin, noggin) neutralize Bmp activity within the placode, leaving follicle activators (e.g., FGF4) to predominate (Fig. 4). Bmps, the follicle inhibitors, are proposed to have greater diffusion capabilities and are dominant and active in the interfollicular space where they inhibit additional follicle formation within their zone of influence. The ability of Bmps to diffuse over distances as high as 20 cell diameters (Nellen et al., 1996) lends some support to this theory. In contrast, follistatin bound to the GAG heparin sulphate in the extracellular matrix would be locally sequestered (Hashimoto et al., 1997). Activity of factors within the follicle could be reinforcing. For example, follistatin upregulates profollicular Fgf4 expression (Patel et al., 1999). The model, largely derived in chick, has received confirmation in mouse by demonstration of retarded follicle formation in noggin knock-out mice and opposing and interacting effects of exogenous Noggin and BMP4 on follicle size in mouse embryonic skin organ culture (Botchkarev et al., 1999).

A number of simplified models for follicle induction have been proposed (e.g., Fig. 4; Jung et al., 1998; Noramly and Morgan, 1998; Oro and Scott, 1998; Jung and Chuong,

1998; Noramly *et al.,* 1999) that are compatible with early models derived from epithelial-mesenchymal recombination experiments. The first dermal message is homogeneous (consistent with diffuse streaks of signaling molecules preceding the placode induction wave). However, both positive and negative signals within the ectoderm restrict response to the primary dermal signal. Locally acting positive signals and diffusing negative signals derive from the placode. The negative signal is proposed to be BMP2, which patterns the ectoderm via lateral inhibition from placodal sources (Jung *et al.,* 1998; Noramly and Morgan, 1998). Positive signals counteract the negative signals locally and provide new sources of positive signaling molecules that reinforce the proplacodal message and induce early morphogenesis.

IX. Follicle Morphogenesis and Differentiation

Follicle morphogenesis and differentiation are being studied in mouse (Philpott and Paus, 1998) using skin culture models and the theories verified in transgenic and knock-out mice. The same molecules involved in follicle induction and spacing are reused in later stages of follicle morphogenesis. A full molecular description is beyond the scope of this report. However, analysis of the role of Shh, one of the best characterized molecules, in follicular differentiation illustrates how information from multiple sources gives an integrated view of the functional role of SHH.

After establishment of the placode, Shh signaling within the epithelial component of the follicle is proposed to control proliferation and epithelial growth in hair/feathers (Ting-Berreth and Chuong, 1996; Morgan *et al.,* 1998; St. Jacques *et al.,* 1998; Chuang and McMahon, 1999; Jung *et al.,* 1998) and additional epithelial appendages (e.g., Dassule and McMahon, 1998; Jung *et al.,* 1999).

Shh$^{-/-}$ murine embryos show histologically normal follicular development during early stages of hair follicle morphogenesis (Chuang and McMahon, 1999; St. Jacques *et al.,* 1998). Epithelial placodes and adjacent mesenchymal aggregates form, but subsequent morphogenesis is arrested, despite expression of some advanced-stage follicle differentiation markers. Reduction in proliferation is seen within the invaginating ectoderm. SHH promotes keratinocyte proliferation as well as opposing p21-mediated withdrawal of differentiating keratinocytes from the cell cycle in culture models (Fan and Khavari, 1999). This is consistent with the reduction in proliferation in knock-out mice.

Expression studies bear out this proliferation-inducing role for SHH in late follicle development. Shh is expressed in epithelial cells, first within the central region of the plaque, and later in the invaginating epithelial cells that maintain contact with the mesenchymal aggregate as it matures and forms the dermal papilla; that is, it is expressed within cells that will eventually form the epithelial matrix (Fig. 3). Prolonged expression within the developing matrix cells could indicate repeated or sustained signaling.

As the matrix matures, SHH changes expression patterns and, in hair, localizes to specific matrix lineages that give rise to an inner root sheath compartment (Bitgood and McMahon, 1995; Fig. 3). Hence, during late stages of hair development, Shh probably adopts a new role in elaboration of the inner root sheath lineages. A specific role for Shh in late stages of feather or hair maturation is also indicated by expression analysis (Ting-Berreth and Chuong, 1996; Morgan *et al.,* 1998). However, functional confirmation is frustrated by growth arrest or distortion at early stages of appendage development in *Shh* misexpression studies in chick and embryonic lethality of *Shh*$^{-/-}$ mice.

X. Follicle Morphogenesis and Follicle Cycling

In adult mice follicles undergo cycles of regression and re-formation called the hair cycle. The adult hair cycle consists of anagen (growth phase), catagen (regression phase), and telogen (resting phase; Paus and Cotsarelis, 1999). During catagen the lower portion of the follicle regresses via apoptosis of keratinocytes and movement of the dermal component (dermal papilla) up beneath the bulge (Fig. 3; Paus and Cotsarelis, 1999). Molecular interaction between the dermal cells and the bulge will result in formation of the new follicle.

It is the length of anagen that determines hair length and this varies over the surface of the mouse and between strains. For example, *Fgf5*$^{-/-}$ mice have an extended anagen reproducing the long-haired phenotype of the angora mouse mutant (Hebert *et al.,* 1994), while other transgenic/knockout mice can have shorten hairs resulting from truncated anagen (e.g., Charpentier *et al.,* 2000, and references within). Human conditions (alopecia, hirsutism; e.g., OMIM, MIM numbers 300042 and 142625) can arise from irregularities in the cycle. Regulation of follicle cycling is the subject of intense speculation and a "hair cycle clock" is proposed (Paus *et al.,* 1999b).

During postnatal development in the mouse, the first few hair cycles are synchronous then later become asynchronous, as in human. The progression of follicles through the stages of the hair cycle occurs regionally in adults, resulting in waves of follicle cycling. Postnatal moving fronts or waves associated with follicle cycling (cranial to caudal, and ventral to dorsal) are well documented in adult mouse (Powell and Rogers, 1990, and references within).

Many of the molecular players associated with follicle development are also involved in regulation of follicle cycling. Onset of anagen may reproduce the changes associated with follicle development (signaling between dermal and epithelia cells). For example, ectopic *Shh* expression in adult mice forces resting stage (telogen) follicles back into the growth phase (anagen; Sato *et al.*, 1999).

XI. Molecular Parallels between Skin Tumorigenesis and Skin Development

Constitutive activation of two of the major skin development signaling pathways (Shh and Wnt pathways) leads to skin tumors in adult. Adult epidermis and its appendages continually cycle, providing ample opportunity for pathway activation to reprogram cell fate.

Mutations in the SHH receptor *Ptch* underlie basal cell Nevus (Gorlin) syndrome (OMIM, MIM number 109400), a disease causing developmental defects and increased frequency of tumors, such as basal cell carcinoma (Hahn *et al.*, 1996, 1999; Johnson *et al.*, 1996). In addition, *ptc* mutation or allele loss occurs at high frequency in sporadic basal cell carcinomas (BCC; Johnson *et al.*, 1996; Gailani *et al.*, 1996). *ptc* is a negative regulator of Shh signaling and can be considered a tumor suppressor.

Conversely, Shh pathway activators should predispose to BCC. In fulfillment of this prediction it was found that overexpression of *Shh* in transgenic mouse skin results in "tumorlike" epidermal growths emanating from follicles with histological and behavioral similarities to BCC (Oro *et al.*, 1997). Similar results are obtained in human (Fan *et al.*, 1997) and chick skin (Ting-Berreth and Chuong, 1996; Morgan *et al.*, 1998). Hence, pathway activation seems to predispose to tumorigenesis via increased proliferation and expansion of developing or cycling hair follicles. This is consistent with the role of Shh in follicle development.

GLI1 transcription factor activation is a downstream outcome of Shh signaling and, interestingly, overexpression of GLI1 in amphibian skin also induces epithelial tumors, with similarity to mammalian BCC (Dahmane *et al.*, 1997). This strongly suggests that pathway activation, by any means, underlies BCC. Therefore, mutations that activate the Shh pathway should cause tumors. A conserved Shh mutation has been detected in BCC and breast carcinoma (Oro *et al.*, 1997; OMIM, MIM number 602968), and activating mutations in smoothened (*smo*, a membrane receptor normally repressed by ptc binding, but activated after Shh binding to ptc) occur in BCC (Xie *et al.*, 1998). Hence, *Gli1* transcriptional activation of (as yet unknown) downstream genes may underlie BCC. To date, known targets activated by Gli1 include *Ptch*, *Gli1* itself, and *Hnf3β (Foxa2)* in certain tissues.

The Wnt signaling pathway also appears upregulated in epithelial tumors. Stabilization of *β*-catenin, through truncation of the *β*-catenin phosphorylation site, causes ectopic follicle formation as described previously, but also causes tumors in transgenic mice (Gat *et al.*, 1998). A variety of tumors result from constitutive signaling through stabilized *β*-catenin, most notably colon cancers where mutation in APC (adenomatous polyposis coli—a pathway component involved in degradation of *β*-catenin; OMIM, MIM number 175100) can promote stabilization and constitutive signaling (Kinzler and Vogelstein, 1996). Constitutive activation of the pathway through *β*-catenin stabilization can also result in tumors in a variety of cell types (Gat *et al.*, 1998, and references within) and skin pilomatricoma tumors show a high frequency of mutation in the N-terminal phosphorylation site of *β*-catenin that targets *β*-catenin for degradation (Chan *et al.*, 1999; OMIM, MIM number 132600).

The downstream targets of *β*-catenin signaling involved in tumorigenesis are uncertain. Identification of *β*-catenin/LEF/TCF downstream targets *Xnr-3* (McKendry *et al.*, 1997), *Siamois* (Brannon *et al.*, 1997; Fan *et al.*, 1998), *Twin* (Laurent *et al.*, 1997), and possibly *Noggin* (Tao *et al.*, 1999) reflects *β*-catenin-mediated specification of dorsal body tissue, because these molecules locate to and induce Speman organizer activity, though *Noggin* has been shown to have a role in skin development.

Examples of additional direct downstream genes include the proto-oncogene *cMyc* (He *et al.*, 1998) and *cyclin D1*, both regulators of cell cycle progression (Tetsu and McCormick, 1999; Shtutman *et al.*, 1999). This is consistent with the potential for nonmembrane *β-catenin* to act as an oncogene (Eastman and Grosschedl, 1999; Nusse, 1999).

XII. Early Terminal Differentiation

Following stratification in mouse (E14–16, Theiler, 1989, stages 22–24; Kaufman, 1992, Plate 46, e–j) basal ectodermal cells enter a program of terminal differentiation, marking the point where ectoderm begins to resemble postnatal epidermis. Terminal differentiation is the process whereby mitotically active cells in the basal layer cease proliferation, downregulate integrin receptors, lose contact with basement membrane, and move to the surface (Fig. 1). Surface-bound cells alter their program of gene expression to produce protein and lipid required for assembly of surface stratum corneum. Stratum corneum is the tough, water-impermeable structure that provides the major environmental barrier in postnatal epidermis and its production is the endpoint—and apparent purpose—of epidermal terminal differentiation. Adoption of a terminal differentiation program is associated with imposition of the mature pattern of marker distribution. Hence, K5/K14 keratin markers become basally associated and K1/K10 keratins are induced in suprabasal layers (Fig. 1).

Regulation of embryonic transit to terminally differentiating epidermis is not understood. However, by analogy with control of terminal differentiation in adult skin it seems probable that the process is mediated by a change in the intracellular calcium level resulting in cell cycle withdrawal

(Dotto, 1999). Addition of calcium to cultured murine keratinocytes induces withdrawal from the cell cycle followed by induction of differentiation markers (Hennings *et al.*, 1980). There is an extracellular calcium gradient in differentiated layers of adult skin, peaking at the outer granular layer (Menon *et al.*, 1992; Forslind *et al.*, 1997; Mauro *et al.*, 1998) that probably mediates terminal differentiation. However, establishing when the calcium gradient forms during development is hampered by insensitivity of detection methods. An epidermal calcium gradient has been demonstrated in fetal rats (Elias *et al.*, 1998a), although the gradient is first detected late in terminal differentiation.

A calcium-sensing membrane receptor has been detected in murine epidermis (Oda *et al.*, 2000, and references within). Knock-out of a form of this receptor, which is most abundantly expressed in basal keratinocytes, results in defects in differentiation and downregulation of epidermal loricrin, a late differentiation marker (Oda *et al.*, 2000; Fig. 1). However, the role of this receptor in transduction of the calcium signal cannot be fully appreciated yet as the gene has an additional splice variant unaffected in the knock-out.

Withdrawal of keratinocytes from the cell cycle is also regulated by *c-Myc* (Gandarillas and Watt, 1997), which could thereby stimulate onset of terminal differentiation. Myc-Max heterodimers predominate in the epidermal basal layer and Mad-Max heterodimers predominate suprabasally (Gandarillas and Watt, 1995; Hurlin *et al.*, 1995; Vastrik *et al.*, 1995).

XIII. Regulation of Transit to Late Stages of Terminal Differentiation

Terminal differentiation in culture models appears to have two phases; an early phase, marked by cell cycle arrest and induction of the keratin stratification markers K1 and K10, and a late phase characterized by upregulation of late terminal differentiation markers (e.g., loricrin, filaggrin; Fig. 1). Evidence for distinct stages includes, first, ability to pharmacologically dissect skin differentiation in cultured keratinocytes. Induction of early stage markers (K1 and K10) is phorbol ester independent, whereas late stage markers can be induced by the phorbol ester TPA (12-O-tetradecanoylphorbol-13-acetate via activation of protein kinase C; Dlugosz and Yuspa, 1993). Secondly, a rise in p21 accompanies keratinoctye cell cycle withdrawal soon after calcium induction of differentiation. However, p21 protein levels later fall. This fall in p21 protein level is necessary because p21 inhibits late stage terminal differentiation (Di Cunto *et al.*, 1998).

If induction of differentiation in keratinocyte culture represents a useful model for fetal skin development, then adoption of the late phase of terminal differentiation probably corresponds to transit to late terminal differentiation at E15–16 (Theiler stages 24–25) during mouse embryogenesis.

This stage is marked by morphological change and appearance of a granular layer in fetal skin. Late marker induction (e.g., loricin) occurs in fetal skin at this time (Byrne *et al.*, 1994: Bickenbach *et al.*, 1995).

The calcium gradient peaks at the granular layer in adult skin and calcium signaling, probably through protein kinase C induction, causes downregulation of early terminal differentiation markers, transit to late stages of terminal differentiation, and induction of late markers (Dlugosz and Yuspa, 1993). Cholesterol sulfate, a metabolite of epidermal cholesterol, is also most abundant in the granular cells of adult epidermis and, in cultured keratinoctyes, cholesterol sulfate induces expression of PKCη and δ isoforms and terminal differentiation (Ohba *et al.*, 1998). PKCη is predominantly expressed in differentiating epithelial tissues, including the upper granular layer of epidermis (Kuroki *et al.*, 2000). Cholesterol sulfate itself activates terminal differentiation in cultured cells and, via PKCη or δ isoforms, induces *TGase1* gene expression. As described below, activation of TGase1 enzyme is key to the formation of the epidermal barrier, the endpoint of epidermal terminal differentiation.

IKKα (*Chuk*) null mice appear developmentally arrested between terminal differentiation stages, that is, after epidermal stratification and expression of early stage keratin differentiation markers, but before onset of late terminal differentiation and barrier formation (Hu *et al.*, 1999; Takeda *et al.*, 1999; Li *et al.*, 1999).

IKKα is a component of the IκB kinase that phosphorylates cytoplasmic IκB, leading to its degradation. Degradation of IκB results in nuclear translocation and activation of NF-κB transcription factors. In mature epidermis NF-κB is located mainly cytoplasmically (i.e., inactively) in basal epidermal cells and in the nucleus in suprabasal layers (Seitz *et al.*, 1998). In addition IKKα is expressed suprabasally in adult (Takeda *et al.*, 1999). The localization and knock-out data suggest that upregulation of *NF-κB* transcriptional activity regulates induction of late stage terminal differentiation during development. However, this interpretation is complicated by the finding that down- or upregulation of NF-κB transcription factors via dominant-negative *IκB* and *p50* overexpression in transgenic mice affects keratinocyte proliferation without affecting differentiation (Seitz *et al.*, 1998). Hence, IKKα (IKK1) could induce late stages of terminal differentiation through regulation of NF-κB or via a novel epidermal function (Li *et al.*, 1999).

Other transcription factors regulating late stages of terminal differentiation and barrier formation include homeobox factor DLX3, which is expressed in the epidermal granular layer. Although the *Dlx3* null mouse is uninformative due to early embryonic lethality, overexpression of *Dlx3* in basal epidermal cells induces premature terminal differentiation and induction of late stage terminal differentiation markers loricrin and filaggrin (Morasso *et al.*, 1996). A member of the Krüppel-like gene family of transcription factors, *Klf4* is also expressed in upper spinous/granular layers of epidermis at E15–16 (at the point fetal epidermis transits to late stage terminal differentiation; Segré *et al.*, 1999; Ohnishi *et al.*, 2000) and *Klf4* null mice appear to misregulate cornified envelope precursor genes and have a defect in very late stages of terminal differentiation (Segré *et al.*, 1999). In addition, a range of steroid hormone transcription factors can stimulate or retard terminal differentiation and barrier formation in a pleiotropic manner. Hence, targeting of dominant-negative retinoic acid receptors (Saitou *et al.*, 1995; Imakado *et al.*, 1995) to skin under control of keratin promoters causes defects in epidermal differentiation and barrier formation, implying that signaling through retinoic acid receptors stimulates barrier formation. Signaling through the glucocorticoid, thyroid, and PPARα receptors also appears to accelerate terminal differentiation (Azsterbaum *et al.*, 1993; Hanley *et al.*, 1997, 1998). Sex steroid hormones estrogen and testosterone stimulate and inhibit late stages of terminal differentiation, respectively, in a global, pleiotropic manner producing a sex-specific difference in the rate of terminal differentiation and barrier formation *in vivo* (Hanley *et al.*, 1996).

XIV. Late Terminal Differentiation: Formation of Stratum Corneum and Skin Barrier

The primary function of the integument is protective and the end product of terminal differentiation is the stratum corneum, a highly specialized cell layer adapted to this purpose (Theiler, 1989, stages 24–25; Kaufman, 1992, Plate 46, i–l). Structurally the stratum corneum has been compared to "bricks and mortar" (Elias, 1983), the "bricks" being flattened, anucleate cornified cells composed internally of aggregated keratin with an external 15-nm-thick cornified cell envelope (Nemes and Steinert, 1999), in turn surrounded by a ceramide capsule (lipid-bound envelope; Swartzendruber *et al.*, 1987). The "mortar" consists of a heterogeneous mixture of predominantly nonpolar lipid, arranged to form a complex lamellar bilayer structure (Downing, 1992). Because of these intercellular lipid structures and the high degree of interdigitation of cornified cells (Fig. 1) water molecules are forced to follow an extremely convoluted and time-consuming path to the epidermal surface. Hence, stratum corneum provides a highly competent barrier to water loss. Transgenic or gene-targeted mice lacking barrier function die soon after birth (Imakado *et al.*, 1995; Saitou *et al.*, 1995; Matsuki *et al.*, 1998; Tarutani *et al.*, 1997; Mills *et al.*, 1999; Yang *et al.*, 1999; Segré *et al.*, 1999; Hu *et al.*, 1999; Takeda *et al.*, 1999; Li *et al.*, 1999), and damage to the barrier causes skin disease in humans (Roop, 1995; e.g., OMIM, MIM numbers 308100, 146700, 242300, and 604117).

Extrusion of extracellular lipid and formation of the lipid-bound cornified envelope are key steps in the acquisition of barrier function and these steps occur in granular layer keratinocytes. Late stage terminal differentiation markers loricrin and small proline-rich region proteins (SPRRs) are abundantly expressed in cornified envelope precursors. Filaggrin, another protein used as a late stage marker, is involved in aggregating keratin filaments in the terminally differentiated cell and is also cross-linked to the envelope.

Key cornified envelope precursor proteins are initially sequestered in keratohyalin granules (Fig. 1; Manabe and O'Guin, 1994, and references within), which increase in size and move to the cell periphery prior to envelope formation. Cornified envelope construction initiates adjacent to the plasma membrane surrounding desmosomal junctions and involves sequential cross-linking of protein and lipid components by transglutaminase enzymes (Steinert and Marekov, 1999) in response to a rise in intracellular Ca^{2+} levels.

The first sign of cornified envelope assembly is localization of envoplakin and periplakin adjacent to the interdesmosomal plasma membrane (plasma membrane uninvolved in desmosome adhesive complexes). The N terminus of periplakin associates with the cortical actin cytoskeleton at membrane microvilli (DiColandrea *et al.*, 2000). Envoplakin colocalizes with periplakin via heterodimerization and is involved in the very first transglutaminase-mediated crosslinks. Involucrin localizes at the plasma membrane by interaction with membrane-bound transglutaminase (TGase) 1 enzyme (Nemes *et al.*, 1999a). Envoplakin, periplakin, and involucrin initially form islands of protein employing an increasing range of cross-links (Steinert and Marekov, 1999). Addition of cystatin-α (Takahashi *et al.*, 1992) and crosslinking to desmosomal components, such as desmoplakin (Robinson *et al.*, 1997), form the initial envelope scaffold, a continuous cross-linked layer encompassing the entire cell membrane.

In tandem with initial scaffold formation, SPRR proteins in the cytoplasm become cross-linked by the cytoplasmically localized TGase3 enzyme to form small oligomers. SPRRs act to cross-bridge other envelope constituents, fortifying the envelope structure (Steinert *et al.*, 1998). The SPRR oligomers bind loricrin molecules and are subsequently incorporated into the cytoplasmic face of the cornified envelope. Loricrin binding is a key step in envelope formation, because loricrin will constitute the majority of the final cornified envelope. Relatively late in envelope formation filaggrin molecules are proteolytically processed, bind keratin molecules internally, and attach to the outer envelope surface (Simon *et al.*, 1996).

Surprisingly, testing the above model *in vivo* has been hampered by lack of significant barrier defects when major cornified envelope structural proteins are deleted. Mice deficient in involucrin, the currently accepted major early envelope scaffold protein, appear normal, with no detrimental effects (Djian *et al.*, 2000). Mice deficient in loricrin, which accounts for around 70% of the final adult cornified envelope mass, display mild erythroderma and slightly delayed barrier formation (Koch *et al.*, 2000). Compensatory mechanism and redundancy have been invoked to explain the normal, or near normal, phenotype of these mice (Steinert, 2000).

Lipid-containing lamellar bodies first appear in the suprabasal layers (Fig. 1). In the granular layer, lamellar bodies increase in number and move to the cell periphery. Immediately prior to cornified envelope formation keratinocytes extrude lamellar body contents into the extracellular space (Elias *et al.*, 1998b). While the cornified envelope is being deposited, the extruded lamellar body contents are gradually rearranged by enzymatic processes and lipid interactions from an amorphous mixture to the intercellular nonpolar lipid lamellar sheets characteristic of mature stratum corneum (Elias *et al.*, 1998b; McIntosh *et al.*, 1996). A lipid capsule forms around the extracellular surface of the mature cornified envelope (Behne *et al.*, 2000). Again TGase1 cross-links the lipid capsule to the external face of the cornified envelope (Nemes *et al.*, 1999b), forming a watertight seal or barrier.

Finally, the barrier is probably self-assembling once the building blocks are brought together in the environment of the granular cell. The process is possibly triggered via high concentrations of extracellular calcium because transglutaminase enzymes are calcium dependent (Nemes and Stei-

nert, 1999). The key role of transglutanimases is confirmed by the transglutaminase 1 knock-out mouse, which shows neonatal mortality due to massive barrier defects (Matsuki *et al.*, 1998).

The barrier forms regionally at E16 (Theiler, 1989, stage 24) in mice (Fig. 6; Hardman *et al.*, 1998) first at conserved initiation sites. Then apparent waves of barrier formation move round the animal's body, converging on ventral and dorsal midlines. This mode of barrier formation is conserved in mammals, including the human preterm infant, although patterns of barrier formation differ (Hardman *et al.*, 1999).

The barrier formation pattern is associated with late-stage epidermal terminal differentiation, because this pattern is also revealed when assaying additional, earlier changes associated with late stages of terminal differentiation, including induction of late stage markers such as SPRR genes (Marshall *et al.*, 2000). There is a gradient of differentiation states in fetal skin associated with the dorsoventral barrier front (Fig. 6), so that epidermis is least differentiated ventrally and more mature dorsally. Hence, it follows that late stages of terminal differentiation must initiate (or be relieved from repression) regionally, at initiation sites, and that the waves of change crossing skin represent conversion of fetal epidermis to late stage terminal differentiation at E14–16 (Theiler, 1989, stages 22–24).

The moving fronts associated with late stage terminal differentiation resemble the moving morphogenetic wave as-

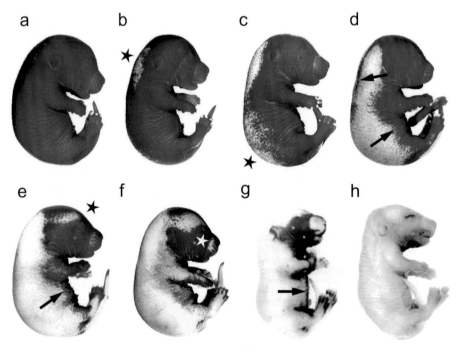

Figure 6 The skin barrier forms in a patterned manner at 16 days gestation. (a–h) Increasing gestational age; *a* <E16; *h* >E17. Specific reproducible dorsal initiation sites initially acquire barrier characteristics (asterisks). Barrier then spreads around the body as moving fronts, which meet at both dorsal and ventral midlines (arrows).

Table II Tbase Links to Transgenic and Knock-out Mice Included in Text

Proteins	Gene knock-out	References	Tbase accession number	Results
BMP4	—	Winnier et al. (1995)	2826	Mesodermal abnormalities, early embryonic lethal
FGF5	—	Hebert et al. (1994)	2071	Abnormally long hair; follicle regression delayed
K8	—	Baribault et al. (1994)	802	Embryonic lethal, no obvious skin defects
K18	—	Magin et al. (1998)	—	Viable, no obvious skin defects
p63	—	Yang et al. (1999); Mills et al. (1999)	5325, 5298–5301	Postnatal lethal, absent epidermis and appendages including all hair follicles
p50	Overexp	Seitz et al. (1998)	—	Epidermal hypoplasia and growth inhibition
Integrin α_6	—	Georges-Labouesse et al. (1996)	3874	Loosely attached skin, blistering, β_4 integrin missing
Integrin β_4	—	van der Neut et al. (1996); Dowling et al. (1996)	3866	No hemidesmosomes, severe phenotype similar to integrin α_6 mice
Fibrillin	—	Pereira et al. (1997)	4557	Postnatal lethal, absence of fibrillin-1 skin fibrils results in no skin phenotype
Fibronectin	—	George et al. (1998)	2583	Defects in mesoderm, neural tube, and vascular development in embryos
Collagen III a1	—	Liu et al. (1997)	4204	Open wound skin lesions, number of collagen fibrils substantially reduced
Collagen V a2	Structurally mutated	Andrikopoulos et al. (1995)	1775	Severely fragile skin; scarring and bleeding
Decorin	—	Danielson et al. (1997)	—	Fragile skin, reduced tensile strength, abnormal collagen morphology
Integrin β_1	—	Fassler and Mayer (1995)	3359	Postimplantation lethal
Integrin β_1	Conditional	Brakebusch et al. (2000); Raghavan et al. (2000)	—	Disrupted basement membrane assembly, abnormal hemidesmosomes, hair follicle defects and decreased proliferation
Integrin α_3	—	Kreidberg et al. (1996)	5004	Microblisters of the basement membrane, actin distruption
Integrin α_5	—	Yang et al. (1993)	—	Embryonic lethal, mesodermal defects
LEF-1	—	van Genderer et al. (1994)	1999	Abnormal hair follicles, whisker placodes absent
Noggin	—	Botchkarev et al. (1999)	5137	Retarded hair follicle induction
SHH	—	St. Jacques et al. (1998); Chuang and McMahon (1999); Chiang et al. (1996)	4017	Arrested embryonic hair follicle development; downregulation of K14; loss of hair follicle polarity; hairless pigmented, thickened epidermis
SHH	K14 overexp	Oro et al. (1997)	4376	Induced basal cell carcinoma and skin erosions
KLF4	—	Segré et al. (1999)	5465	Die shortly after birth due to loss of skin barrier function; aberrant cornified envelopes
DLX3	K5 overexp	Morasso et al. (1996)	—	Highly abnormal skin with disturbed barrier function
Tgase-1	—	Matsuki et al. (1998)	4829	Erythrodermic and taut skin with coarse wrinkles, death occurs within several hours of birth
Involucrin	—	Dijan et al. (2000)	—	Normal development
Loricrin	—	Koch et al. (2000)	—	Shiny, translucent skin and delayed barrier formation
IKKα	—	Hu et al. (1999); Takeda et al. (1999); Li et al. (1999)	—	Taught hyperproliferative hyperplastic skin, severe limb abnormalities
Calcium sensing receptor	—	Oda et al. (2000)	—	Hyperkeratosis, flattened nucleated layers, reduced loricrin expression
Pig-a	K5:Cre	Tarutani et al. (1997)	5506	Die with severely abnormal stratum corneum, lack of lipid extrusion

sociated with follicle formation earlier in skin development, and it is possible that the processes are mechanistically similar. It is still unclear why developmental change in embryonic and fetal skin precedes via waves or moving fronts.

XV. Periderm Disaggregation

The final major integument change before birth is disaggregation of periderm (Theiler, 1989, stages 25–26). Differentiation of periderm cells has been occurring in tandem with epidermal differentiation. The regression of periderm cells is characterized by withering of surface blebs, organelle breakdown, and reduction in microvilli number (Hol-brook and Odland, 1975). It has been shown recently that peridermal cells form cornified envelopes that are similar to those in epidermis (Akiyama et al., 1999).

If periderm does have an interactive role with amniotic fluid, then following barrier initiation at embryonic day 16 it would become redundant. Immediately following barrier formation, periderm disaggregates from fetal mouse skin (Hardman et al., 1999). Interestingly, the pattern of periderm disaggregation follows the late terminal differentiation pattern, showing that the developmental gradient in murine skin, highlighted at the time of barrier formation, persists at this later stage of development. Periderm disaggregation and barrier formation occur at a time when major changes in amniotic fluid composition occur in human; in particular, levels of urea rise sharply due to output from the newly functioning kidneys. It is possible that the timing is more than coincidental and barrier forms in time to assume a more robust role in protection of fetal skin.

XVI. Conclusions and Future Directions

A molecular analysis of development of integument is strongest for skin appendage formation and emerging for areas such as adoption of late terminal differentiation. Present challenges in the skin field include extending the analysis to neglected areas such as formation of periderm, control of stratification, control of dermal differentiation, and initiation of terminal differentiation. Much of the historic work on follicle development, emphasizing heterogeneity and regional morphogenetic activity of embryonic skin, is being confirmed and elaborated by modern approaches. The regional nature of developing skin and its appendages may arise via nonhomogeneous modes of developmental change, as illustrated by waves of developmental change associated with follicle formation, follicle cycling, and barrier formation. A major challenge is to understand why ectodermal cells do not undergo homogeneous change during development and why, instead, fronts or waves of change appear to program ectoderm/fetal epidermis at different developmental stages.

References

Agorastos, T., Hollweg, G., Grussendorf, E. I., and Papaloucas, A. (1988). Features of vernix caseosa cells. *Am. J. Perinatol.* **5**, 253–259.

Akiyama, M., Shimizu, H., Yoneda, K., and Nishikawa, T. (1997). Collo-

dion baby: Ultrastructure and distribution of cornified cell envelope proteins and keratins. *Dermatology.* **195**, 164–168.

Akiyama, M., Smith, L. T., Yoneda, K., Holbrook, K. A., Hohl, D., and Shimizu, H. (1999). Periderm cells form cornified cell envelope in their regression process during human epidermal development. *J. Invest. Dermatol.* **112**, 903–909.

Aktan, M., Buyukmumcu, M., Seker, M., Mermer, S., and Duman, S. (1999). Morphometric analyses of the development of dermis in human fetuses. *Kaibogaku Zasshi* **74**, 639–642.

Andrikopoulos, K., Liu, X., Keene, D. R., Jaenisch, R., and Ramirez, F. (1995). Targeted mutation in the Col5a2 gene reveals a regulatory role for type V collagen during matrix assembly. *Nat. Genet.* **9**, 31–36.

Azsterbaum, M., Feingold, K. R., Menon, G. K., and Williams, M. L. (1993). Glucocorticoids accelerate fetal maturation of the epidermal permeability barrier in the rat. *J. Clin. Invest.* **91**, 2703–2708.

Baribault, H., Penner, J., Iozzo, R. V., and Wilson-Heiner, M. (1994). Colorectal hyperplasia and inflammation in keratin 8-deficient FVB/N mice. *Genes Dev.* **15**, 2964–2973.

Behne, M., Uchida, Y., Seki, T., De Montellano, P. O., Elias, P. M., and Holleran, W. M. (2000). Omega-hydroxylceramides are required for corneocyte lipid envelope (CLE) formation and normal epidermal permeability barrier function. *J. Invest. Dermatol.* **114**, 185–192.

Bickenbach, J. R., Greer, J. M., Bundman, D. S., Rothnagel, J. A., and Roop, D. R. (1995) Loricrin expression is coordinated with other epidermal proteins and the appearance of lipid lamellar granules during development. *J. Invest. Dermatol.* **104**, 405–410.

Bitgood, M. J., and McMahon, A. P. (1995). Hedgehog and Bmp genes are coexpressed at many diverse sites of cell-cell interaction in the mouse embryo. *Dev. Biol.* **172**, 126–138.

Boneko, V. M., and Merker, H. J. (1988). Development and morphology of the periderm of mouse embryos. (days 9–12 of gestation). *Acta Anat.* **133**, 325–336.

Botchkarev, V. A., Botchkarev, N. V., Roth, W., Nakamura, M., Chen, L.-H., Herzog, W., Linder, G., McMahon, J. A., Peters, C., Lauster, R. *et al.* (1999). Noggin is a mesenchymally derived stimulator of hair-follicle induction. *Nat. Cell Biol.* **1**, 158–164.

Brakebusch, C., Grose, R., Quondamatteo, F., Ramirez, A., Jorcano, J. L., Pirro, A., Svensson, M., Herken, R., Sasaki, T., Timpl, R., Werner, S., and Fassler, R. (2000) Skin and hair follicle integrity is crucially dependent on beta 1 integrin expression on keratinocytes. *EMBO J.* **19**, 3990–4003.

Brannon, M., Gomperts, M., Sumoy, L., Moon, R. T., and Kimelman, D. (1997). A beta-catenin/XTcf-3 complex binds to the siamois promoter to regulate dorsal axis specification in Xenopus. *Genes Dev.* **11**, 2359–2370.

Bruckner-Tuderman, L. (1999). Biology and pathology of the skin basement membrane zone. *Matrix Biol.* **18**, 3–4.

Bullen, P., and Wilson, D. (1997). The carnegie staging of human embryos: A practical guide. *In* "Molecular Genetics of Early Human Development" (T. Strachan, S. Lindsay, and D. Wilson, eds.). BIOS Scientific Publishers, Oxford.

Burgeson, R. E. (1993). Type VII collagen, anchoring fibrils, and epidermolysis bullosa. *J. Invest. Dermatol.* **101**, 252–255.

Burgeson, R. E., and Christiano, A. M. (1997). The dermal-epidermal junction. *Curr. Opin. Cell Biol.* **9**, 651–658.

Byrne, C., Tainsky, M., and Fuchs, E. (1994). Programming gene expression in developing epidermis. *Development (Cambridge, UK)* **120**, 2369–2383.

Candi, E., Tarcsa, E., di Giovanna, J. J., Complton, J. G., Elias, P. M., Marekov, L. N., and Steinert, P. M. (1998). A highly conserved lysine residue on the head domain of type II keratins is essential for attachment of keratin intermediate filaments to the cornified cell envelope through isopeptide cross-linking by transglutaminases. *Proc. Natl. Acad. Sci. U.S.A.* **95**, 2067–2072.

Chan, E. F., Gat, U., McNiff, J. M., and Fuchs, E. (1999). A common human skin tumour is caused by activating mutations in beta-catenin. *Nat. Genet.* **21,** 410–413.

Chang, C., and Hemmati-Brivanlou, A. (1998). Cell fate determination in embryonic ectoderm. *J. Neurobiol.* **36,** 128–151.

Charpentier, E., Lavker, R. M., Acquister, E., and Cowin, P. (2000). Plakoglobin supresses epithelial proliferation and hair growth in vivo. *J. Cell Biol.* **149,** 503–520.

Chen, C. W., Jung, H. S., Jiang, T. X., and Chuong, C. M. (1997). Asymmetric expression of Notch/Delta/Serrate is associated with the anterior-posterior axis of feather buds. *Dev. Biol.* **188,** 181–187.

Chuang, P. T., and McMahon, A. P. (1999). Vertebrate Hedgehog signalling modulated by induction of a Hedgehog-binding protein. *Nature (London)* **397,** 617–621.

Chuong, C.-M., and Widelitz, R. B. (1998). Feather morphogenesis: A model of the formation of epithelial appendages. In "Molecular Basis of Epithelial Appendage Morphogensis" (C.-M. Chuong, ed.), pp. 57–74. R. G. Landes, Austin, TX.

Chuong, C.-M., Oliver, G., and Ting, S. A. (1990). Gradients of homeoproteins in developing feather buds. *Development (Cambridge, UK)* **110,** 1021–1030.

Chuong, C.-M., Widelitz, R. B., Ting-Berreth, S., and Jiang, T. X. (1996). Early events during avian skin appendage regeneration: Dependence on epithelial-mesenchymal interaction and order of molecular reappearance. *J. Invest. Dermatol.* **107,** 639–646.

Claxton, J. H. (1966). The hair follicle group in mice. *Anat. Rec.* **154,** 195–208.

Cotsarelis, G., Sun, T. T., and Lavker, R. M. (1990). Label-retaining cells reside in the bulge area of pilosebaceous unit: Implications for follicular stem cells, hair cycle, and skin carcinogenesis. *Cell (Cambridge, Mass.)* **61,** 1329–1337.

Couly, G. F., and Le Douarin, N. M. (1990). Head morphogenesis in embryonic avian chimeras: Evidence for a segmental pattern in the ectoderm corresponding to the neuromeres. *Development (Cambridge, UK)* **108,** 543–558.

Couly, G. F., Coltey, P. M., and Le Douarin, N. M. (1992). The developmental fate of the cephalic mesoderm in quail-chick chimeras. *Development (Cambridge, UK)* **114,** 1–15.

Crowe, R., Henrique, D., Ish-Horowicz, D., and Niswander, L. (1998). A new role for Notch and Delta in cell fate decisions: Patterning the feather array. *Development (Cambridge, UK)* **125,** 767–775.

Dahmane, N., Lee, J., Robins, P., Heller, P., Ruiz, I., and Altaba, A. (1997). Activation of the transcription factor Gli1 and the Sonic hedgehog signalling pathway in skin tumours. *Nature (London)* **389,** 876–881.

Danielson, K. G., Baribault, H., Holmes, D. F., Graham, H., Kadler, K. E., and Iozzo, R. V. (1997). Targeted disruption of decorin leads to abnormal collagen fibril morphology and skin fragility. *J. Cell Biol.* **136,** 729–743.

Dassule, H. R., and McMahon, A. P. (1998). Analysis of epithelial-mesenchymal interactions in the initial morphogenesis of the mammalian tooth. *Dev. Biol.* **202,** 215–227.

Davidson, D. (1983). The mechanism of feather pattern development in the chick. II. Control of the sequence of pattern formation. *J. Exp. Embryol. Morphol.* **74,** 261–273.

Davidson, P., and Hardy, M. H. (1952). The development of mouse vibrissae in vivo and in vitro. *J. Anat.* **86,** 342–360.

Denning, M. F., Kazanietz, M. G., Blumberg, P. M., and Yuspa, S. H. (1995). Cholesterol sulfate activates multiple protein kinase C isoforms and induces cell differentiation in cultured murine keratinocytes. *Cell, Growth Differ.* **6,** 1619–1626.

Dhouailly, D., Prin, F., Kanzler, B., and Viallet, J. P. (1998). Variations in cutaneous appendages: Regional specification and cross-species signals. In "Molecular Basis of Epithelial Appendage Morphogenesis" (C.-M. Chuong, ed.), pp. 45–56. R. G. Landes, Austin, TX.

DiColandrea, T., Karashima, T., Maatta, A., and Watt, F. M. (2000). Subcellular distribution of envoplakin and periplakin: Insights into their role as precursors of the epidermal cornified envelope. *J. Cell Biol.* **151,** 573–585.

Di Cunto, F., Topley, G., Calautti, E., Hsiao, J., Ong, L., Seth, P. K., and Dotto, G. P. (1998). Inhibitory function of p21Cip/WAF1 in differentiation of primary mouse keratinocytes independent of cell cycle control. *Science* **280,** 1069–1072.

DiPersio, C. M., Hodivala-Dilke, K. M., Jaenisch, R., Kreidberg, J. A., and Hynes, R. O. (1997). Alpha3beta1 Integrin is required for normal development of the epidermal basement membrane. *J. Cell Biol.* **137,** 729–742.

DiPersio, C. M., van der Neut, R., Georges-Labouesse, E., Kreidberg, J. A., Sonnenberg, A., and Hynes, R. O. (2000). Alpha3beta1 and alpha6beta4 integrin receptors for laminin-5 are not essential for epidermal morphogenesis and homeostasis during skin development. *J. Cell Sci.* **113,** 3051–3062.

Djian, P., Easley, K., and Green, H. (2000). Targeted ablation of the murine involucrin gene. *J. Cell Biol.* **151,** 381–388.

Dlugosz, A. A., and Yuspa, S. H. (1993). Coordinate change in gene expression which mark the spinous to granular cell transition in epidermis are regulated by protein kinase C. *Cell (Cambridge, Mass.)* **120,** 217–225.

Dotto, G. P. (1999). Signal transduction pathways controlling the switch between keratinocyte growth and differentiation. *Crit. Rev. Oral Biol. Med.* **10,** 442–457.

Dowling, J., Yu, Q. C., and Fuchs, E. (1996). Beta4 integrin is required for hemidesmosome formation, cell adhesion and cell survival. *J. Cell Biol.* **134,** 559–572.

Downing, D. T. (1992). Lipid and protein structures in the permeability barrier of mammalian epidermis. *J. Lipid Res.* **33,** 301–313.

Duplan Perrat, F., Damour, O., Montrocher, C., Peyrol, S., Grenier, G., Jacob, M. P., and Braye, F. (2000). Keratinocytes influence the maturation and organization of the elastin network in a skin equivalent. *J. Invest. Dermatol.* **114,** 365–370.

Eastman, Q., and Grosschedl, R. (1999). Regulation of LEF-1/TCF transcription factors by wnt and other signals. *Curr. Biol.* **11,** 233–240.

Ede, D. A. (1972). Cell behaviour and embryonic development. *Int. J. Neurosci.* **3,** 165–174.

Elias, P. M. (1983). Epidermal lipids, barrier function and desquamation. *J. Invest. Dermatol.* **80,** 44–49.

Elias, P. M., Nau, P., Hanley, K., Cullander, C., Crumrine, D., Bench, G., Sideras-Haddad, E., Mauro, T., Williams, M., and Feingold, K. R. (1998a). Formation of the epidermal calcium gradient coincides with key milestones of barrier ontogenesis in the rodent. *J. Invest. Dermatol.* **110,** 399–404.

Elias, P. M., Cullander, C., Mauro, T., Rassner, U., Kömüves, L., Brown, B. E., and Menon, G. K. (1998b). The secretory granular cell: The outermost granular cell as a specialized secretory cell. *J. Invest. Dermatol., Symp. Proc.* **3,** 87–100.

Epstein, E. H. (1974). Alpha(3)3 human skin collagen. Release by peptin digestion and preponderance in fetal life. *J. Biol. Chem.* **249,** 3225–3231.

Fan, H., and Khavari, P. A. (1999). Sonic hedgehog opposes epithelial cell cycle arrest. *J. Cell Biol.* **147,** 71–76.

Fan, H., Oro, A. E., Scott, M. P., and Khavari, P. A. (1997). Induction of basal cell carcinoma features in transgenic human skin expressing Sonic Hedgehog. *Nat. Med.* **3,** 788–792.

Fan, M. J., Gruning, W., Walz, G., and Sokol, S.Y. (1998). Wnt signalling and transcriptional control of Siamois in Xenopus embryos. *Proc. Natl. Acad. Sci. U.S.A.* **95,** 5626–5631.

Fassler, R., and Mayer, M. (1995). Consequences of lack of β_1 integrin gene expression in mice. *Genes Dev.* **9,** 1896–1908.

Favier, B., Fliniaux, I., Thelu, J., Viallet, J. P., Demarchez, M., Jahoda, C. A., and Dhouailly, D. (2000). Localization of members of the notch system and the differentiation of vibrissa hair follicles: Receptors, ligands, and fringe modulators. *Dev. Dyn.* **218,** 426–437.

Forslind, B., Lindberg, M., Roomans, G. M., Pallon, J., and Werner-Linde, Y. (1997). Aspects of the physiology of human skin: Studies using particle probe analysis. *Microsc. Res. Tech.* **38,** 378–386.

Fuchs, E. (1996). The cytoskeleton and disease: Genetic disorders of intermediate filaments. *Annu. Rev. Genet.* **30,** 197–231.

Fuchs, E., and Segré, J. A. (2000). Stem cells: A new lease on life. *Cell (Cambridge, Mass.)* **100,** 143–155.

Fuchs, E., Dowling, J., Segré, J., Lo, S. H., and Yu, Q.-C. (1997). Integrators of epidermal growth and differentiation: Distinct functions for β1 and β4 integrins. *Curr. Opin. Genet. Dev.* **7,** 672–682.

Funayama, N., Sato, Y., Matsumoto, K., Ogura, T., and Takahashi, Y. (1999). Coelom formation: Binary decision of the lateral plate mesoderm is controlled by the ectoderm. *Development (Cambridge, UK)* **126,** 4129–4138.

Gailani, M. R., Ståhle-Bäckdahl, M., Leffell, D. J., Glynn, M., Zaphiropoulos, P. G., Pressman, C., Undén, A. B., Dean, M., Brash, D. E., Bale, A. E. *et al.* (1996). The role of the human homologue of Drosophila patched in sporadic basal cell carcinomas. *Nat. Genet.* **14,** 78–81.

Gandarillas, A., and Watt, F. M. (1995). Changes in expression of members of the fos and jun families and myc network during terminal differentiation of human keratinocytes. *Oncogene* **11,** 1403–1407.

Gandarillas, A., and Watt, F. M. (1997). c-Myc promotes differentiation of human epidermal stem cells. *Genes Dev.* **11,** 2869–2882.

Gat, U., DasGupta, R., Degenstein, L., and Fuchs, E. (1998). De Novo hair follicle morphogenesis and hair tumors in mice expressing a truncated beta-catenin in skin. *Cell (Cambridge, Mass.)* **95,** 605–614.

Georges-Labouesse, E., Messaddeq, N., Yehia, G., Cadalbert, L., Dierich, A., and Le Meur, M. (1996). Absence of integrin alpha 6 leads to epidermolysis bullosa and neonatal death in mice. *Nat. Genet.* **13,** 370–373.

Hahn, H., Wicking, C., Zaphiropoulous, P. G., Gailani, M. R., Shanley, S., Chidambaram, A., Vorechovsky, I., Holmberg, E., Unden, A. B., Gillies, S. *et al.* (1996). Mutations of the human homologue of Drosophila patched in the nevoid basal cell carcinoma syndrome. *Cell (Cambridge, Mass.)* **85,** 841–851.

Hahn, H., Wojnowski, L., Miller, G., and Zimmer, A. (1999). The patched signalling pathway in tumorigenesis and development: Lessons from animal models. *J. Mol. Med.* **77,** 459–468.

Hamburger, V., and Hamilton, H. (1951). A series of normal stages in the development of the chick embryo. *J. Exp. Morphol.* **88,** 49–92.

Hanley, K., Rassner, U., Jiang, Y., Vansomphone, D., Crumrine, D., Komuves, L., Elias, P. M., Feingold, K. R., and Williams, M. L. (1996). Hormonal basis for the gender difference in epidermal barrier formation in the fetal rat: Acceleration by estrogen and delay by testosterone. *J. Clin. Invest.* **97,** 2576–2584.

Hanley, K., Jiang, Y., Crumrine, D., Bass, N. M., Appel, R., Elias, P. M., Williams, M. L., and Feingold, K. R. (1997). Activators of the nuclear hormone receptors PPARalpha and FXR accelerate the development of the fetal epidermal permeability barrier. *J. Clin. Invest.* **100,** 705–712.

Hanley, K., Feingold, K. R., Kömüves, L. G., Elias, P. M., Muglia, L. J., Majzoub, J. A., and Williams, M. L. (1998). Glucocorticoid deficiency delays stratum corneum maturation in the fetal mouse. *J. Invest. Dermatol.* **111,** 440–444.

Hansen, J. (1947). The histogenesis of epidermis in the mouse. *J. Anat.* **81,** 174–197.

Hardman, M. J., Sisi, P., Banbury, D. N., and Byrne, C. (1998). Patterned acquisition of barrier function during development. *Development (Cambridge, UK)* **125,** 1541–1552.

Hardman, M. J., Moore, L., Ferguson, M. W. J., and Byrne, C. (1999). Barrier formation in the human fetus is patterned. *J. Invest. Dermatol.* **113,** 1106–1114.

Hardy, M. H. (1969). The differentiation of hair follicles and hairs in organ culture. *Adv. Biol. Skin.* **9,** 35–60.

Hardy, M. H. (1992). The secret life of the hair follicle. *Trends Genet.* **8,** 159–166.

Hashimoto, O., Nakamura, T., Shoji, H., Shimasaki, S., Hayashi, Y., and Sugino, H. (1997). A novel role for follistatin, an activin-binding protein, in the inhibition of activin action in rat pituitary cells: Endocytic degradation of activin and its acceleration by follistatin associated with cell-surface heparin sulphate. *J. Biol. Chem.* **272,** 13835–13842.

He, T., Sparks, A., Rago, C., Hermeking, H., Zawel, L., da Costa, L., Morin, P., Vogelstein, B., and Kinzler, K. (1998). Identification of c-MYC as a target of the APC pathway. *Science* **281,** 1509–1512.

Hebert, J. M., Rosenquist, T., Gortz, J., and Martin, G. R. (1994). FGF5 as a regulator of the hair growth cycle: Evidence from targeted and spontaneous mutations. *Cell (Cambridge, Mass.)* **78,** 1017–1025.

Hennings, H., Michael, D., Cheng, C., Steinert, P., Holbrook, K., and Yuspa, S. H. (1980). Calcium regulation of growth and differentiation of mouse epidermal cells in culture. *Cell (Cambridge, Mass.)* **19,** 245–254.

Hertle, M. D., Adams, J. C., and Watt, F. M. (1991). Integrin expression during human epidermal development in vivo and in vitro. *Development (Cambridge, UK)* **112,** 193–206.

Holbrook, K. A., and Odland, G. F. (1975). The fine structure of developing human epidermis: Light, scanning and transmission electron microscopy of the periderm. *J. Invest. Dermatol.* **65,** 16–38.

Holbrook, K. A., and Odland, G. F. (1980). Regional development of the human epidermis in the first trimester embryo and the second trimester fetus (ages related to the time of amniocentesis and fetal biopsy). *J. Invest. Dermatol.* **74,** 161–168.

Holbrook, K. A., Fisher, C., Dale, B. A., and Hartley, R. (1988). Morphogenesis of the hair follicle during ontogeny of human skin. *In* "The Biology of Wool and Hair" (G. E. Rogers, ed.), pp. 15–35. Chapman & Hall, London and New York.

Hoyes, A. D. (1968). Electron microscopy of the surface layer (periderm) of human foetal skin. *J. Anat.* **103,** 321–336.

Hu, Y., Baud, V., Delhase, M., Zhang, P., Deerinck, T., Ellisman, M., Johnson, R., and Karin, M. (1999). Abnormal morphogenesis but intact IKK activation in mice lacking the IKKalpha subunit of IkappaB kinase. *Science* **284,** 316–320.

Hunt, P., Wilkinson, D., and Krumlauf, R. (1991). Patterning the vertebrate head: Murine Hox 2 genes mark distinct subpopulations of premigratory and migrating cranial neural crest. *Development (Cambridge, UK)* **112,** 43–50.

Hurlin, P. J., Foley, K. P., Ayer, D. E., Eisenman, R. N., Hanahan, D., and Arbeit, J. M. (1995). Regulation of Myc and Mad during epidermal differentiation and HPV-associated tumorigenesis. *Oncogene* **11,** 2487–2501.

Imakado, S., Bickenbach, J., Bundman, D. S., Rothnagel, J. A., Attar, P. S., Wang, X.-J., Wlaczak, V. R., Wisniewski, S., Pote, J., Gordon, J. S. *et al.* (1995). Targeted expression of a dominant-negative retinoic acid receptor mutant in the epidermis of transgenic mice results in loss of barrier function. *Genes Dev.* **9,** 317–329.

Jackson, B. W., Grund, C., Winter, S., Franke, W. W., and Illmensee, K. (1981). Formation cytoskeletal elements during murine embryogenesis. II. Epithelial differentiation and intermediate filaments in early postimplantation embryos. *Differentiation (Berlin)* **20,** 203–216.

Jensen, U. B., Lowell, S., and Watt, F. M. (1999). The spatial relationship between stem cells and their progeny in the basal layer of human epidermis: A new view based on whole-mount labeling and lineage analysis. *Development (Cambridge, UK)* **126,** 2409–2418.

Jiang, T. X., Jung, H. S., Widelitz, R. B., and Chunong, C. M. (1999). Self-organization of periodic patterns by dissociated feather mesenchymal cells and the regulation of size, number and spacing primordia. *Development* **126,** 4997–5009.

Johnson, R. L., and Tabin, C. (1997). Molecular models for vertebrate limb development. *Cell (Cambridge, Mass.)* **19,** 979–990.

Johnson, R. L., Rothman, A. L., Xie, J., Goodrich, L. V., Bare, J. W., Bonifas, J. M., Quinn, A. G., Myers, R. M., Cox, D. R., Epstein, E. H. J. *et al.*

(1996). Human homolog of patched, a candidate gene for the basal cell nevus syndrome. *Science* **272**, 1668–1671.

Jones, P. H., Harper, S., and Watt, F. M. (1995). Stem cell patterning and fate in epidermis. *Cell (Cambridge, Mass.)* **80**, 83–93.

Jung, H.-S., and Chuong, C. M. (1998). Periodic pattern formation of feathers. *In* "Molecular Basis of Epithelial Appendage Morphogenesis" (C. M. Chuong, ed.), pp. 359–366. R. G. Landes, Austin, TX.

Jung, H.-S., Francis-West, P. H., Widelitz, R. B., Jiang, T. X., Ting-Berreth, S., Tickle, C., Wolpert, L., and Chuong, C. M. (1998). Local inhibitory action of BMPs and their relationships with activators in feather formation: Implications for periodic patterning. *Dev. Biol.* **196**, 11–23.

Jung, H.-S., Oropeza, V., and Thesleff, I. (1999). Shh, Bmp-2, Bmp-4 and Fgf-8 are associated with initiation and patterning of mouse tongue papillae. *Mech. Dev.* **81**, 179–182.

Kanzler, B., Viallet, J. P., LeMouellic, H., and Dhouailly, D. (1994). Differential expression of two different homeobox gene families during integument morphogenesis. *Int. J. Dev. Biol.* **38**, 633–640.

Kaufman, M. (1992). "The Atlas of Mouse Development." Academic Press, London.

Keene, D. R., Marinkovich, M. P., and Sakai, L.Y. (1997). Immunodissection of the connective tissue matrix in human skin. *Microsc. Res. Tech.* **38**, 394–406.

Kinzler, K. W., and Vogelstein, B. (1996). Lessons from hereditary colorectal cancer. *Cell (Cambridge, Mass.)* **87**, 159–170.

Koch, P. J., de Viragh, P. A., Scharer, E., Bundman, D., Longley, M. A., Bickenbach, J., Kawachi, Y., Suga, Y., Zhou, Z., Huber, M., Hohl, D., Kartasova, T., Jarnik, M., Steven, A. C., and Roop, D. R. (2000). Lessons from loricrin-deficient mice. Compensatory mechanisms maintaining skin barrier function in the absence of a major cornified envelope protein. *J. Cell Biol.* **151**, 389–400.

Kupper, T. S. (1996). The utility of transgenic mouse models in the study of cutaneous immunology and inflammation. *J. Dermatol.* **23**, 741–745.

Kuroki, T., Ikuta, T., Kashiwagi, M., Kawabe, S., Ohba, M., Huh, N., Mizuno, K., Ohno, S., Yamada, E., and Chida, K. (2000). Cholesterol sulfate, an activator of protein kinase C mediating squamous cell differentiation: A review. *Mutat. Res.* **462**, 189–195.

Laurent, M., Blitz, I., Hashimoto, C., Rothbacher, U., and Cho, K.-Y. (1997). The Xenopus gene Twin mediates Wnt induction of Goosecoid in establishment of Speman's organizer. *Development (Cambridge, UK)* **124**, 4905–4916.

Lavker, R. M., and Sun, T.-T. (2000). Epidermal stem cells: Properties, markers, and location. *Proc. Natl. Acad. Sci. U.S.A.* **97**, 13473–13475.

Le Douarin, N. M., Ziller, C., and Couly, G. F. (1993). Patterning of neural crest derivatives in the avian embryo: In vivo and in vitro studies. *Dev. Biol.* **159**, 24–49.

Li, L., Tucker, R. W., Hennings, H., and Yuspa, S. (1995). Inhibitors of the intracellular Ca^{2+} ATPase in cultured mouse keratinocytes reveal components of terminal differentiation that are regulated by distinct intracellular Ca^{2+} compartments. *Cell, Growth Differ.* **6**, 1171–1184.

Li, Q., Lu, Q., Hwang, J. Y., Buscher, D., Lee, K.-F., Izpisua-Belmonte, J. C., and Verma, I. M. (1999). IKK1-deficient mice exhibit abnormal development of skin and skeleton. *Genes Dev.* **13**, 1322–1328.

Liu, X., Wu, H., Byrne, M., Krane, S., and Jaenisch, R. (1997). Type III collagen is crucial for collagen fibrillogenesis and for normal cardiovascular development. *Proc. Natl. Acad. Sci. U.S.A.* **94**, 1852–1856.

Magin, T. M. (1998). Lesson from keratin transgenic and knockout mice. *In* "Subcellular Biochemistry" (Herrmann and Harris, eds.), Vol. 31, pp. 141–172. Plenum, New York.

Magin, T. M., Schroder, R., Leitgeb, S., Wanninger, F., Zatloukal, K., Grund, C., and Melton, D. W. (1998). Lesson from keratin 18 knockout mice: Formation of novel keratin filaments, secondary loss of keratin 7 and accumulation of liver-specific keratin-8 positive aggregates. *J. Cell Biol.* **140**, 1441–1451.

Manabe, M., and O'Guin, W. M. (1994). Existence of trichohyalin-kerato-

hyalin hybrid granules: Co-localization of two major intermediate filament associated proteins in non follicular epithelia. *Differentiation (Berlin)* **58**, 65–76.

Mann, S. J. (1962). Prenatal formation of hair follicle types. *Anat. Rec.* **144**, 135–141.

Marshall, D. M., Hardman, M. J., and Byrne, C. R. (2000). SPRR1 gene induction and barrier formation occur as co-ordinated moving fronts in terminally differentiating epithelia. *J. Invest. Dermatol.* **114**, 967–975.

Matsuki, M., Yamashta, F., Ishida-Yamamoto, A. Yamada, K., Kinoshita, C., Fushiki, S., Ueda, E., Morishima, Y., Tabata, K., Yasuno, H. *et al.* (1998). Defective stratum corneum and early neonatal death in mice lacking the gene for transglutaminase 1. *Proc. Natl. Acad. Sci. U.S.A.* **95**, 1044–1049.

Mauger, A. (1972). Role du mesoderme somitique dans le développement du plumage dorsal chez l'embryon de poulet. II. Régionalisation du mesoderme plumigène. *J. Embryol. Exp. Morphol.* **28**, 343–366.

Mauger, A., Emonard, H., Hartmann, D. J., Foidart, J. M., and Sengel, P. (1987). Immunofluorescent localization of collagen types I, III and IV, fibronectin, laminin, and basement membrane proteoglycan in developing mouse skin. *Roux's Arch. Dev. Biol.* **196**, 295–302.

Mauro, T., Bench, G., Sidderas-Haddad, E., Feingold, K., Elias, P., and Cullander, C. (1998). Acute barrier perturbation abolishes the Ca^{2+} and K^+ gradients in murine epidermis: Quantitative measurement using PIXE. *J. Invest. Dermatol.* **111**, 1198–1201.

Mazzalupo, S., and Coulombe, P. A. (2000). A reporter transgene based on a human keratin 6 promoter is specifically expressed in the periderm of mouse embryos. *Mech. Dev.* **100**, 65–69.

McIntosh, T. J., Stewart, M. E., and Downing, D. T. (1996). X-ray diffraction analysis of isolated skin lipids: Reconstitution of intercellular lipid domains. *Biochemistry* **35**, 3649–3653.

McKendry, R., Hsu, S. C., Harland, R. M., and Grosschedl, R. (1997). LEF-1/TCF proteins mediate wnt-inducible transcription from the Xenopus nodal-related 3 promoter. *Dev. Biol.* **192**, 420–421.

Menon, G., Elias, P. M., Lee, S. H., and Feingold, K. R. (1992). Localization of calcium in murine epidermis following disruption and repair of the permeability barrier. *Cell Tissue Res.* **270**, 503–512.

Mills, A. A., Zheng, B., Wang, X.-J., Vogel, H., Roop, D. R., and Bradley, A. (1999). p63 ia a p53 homologue required for limb and epidermal morphogenesis. *Nature (London)* **398**, 708–713.

Moll, R., Moll, I., and Wiest, W. (1982). Changes in the pattern of cytokeratin polypeptides in epidermis and hair follicles during skin development in human fetuses. *Differentiation (Berlin)* **23**, 170–178.

Moore, K. L. (1988). "The Developing Human: Clinically Orientated Embryology." Saunders, Philadelphia.

Morasso, M. I., Markova, N. G., and Sargent, T. D. (1996). Regulation of epidermal differentiation by a Distal-less homeodomain gene. *J. Cell Biol.* **135**, 1879–1887.

Morgan, B. A., Orkin, R. W., Noramly, S., and Perez, A. (1998). Stage-specific effects of sonic hedgehog expression in the epidermis. *Dev. Biol.* **201**, 1–12.

Morris, R. J., and Potten, C. S. (1999). Highly persistent label-retaining cells in the hair follicles of mice and their fate following induction of anagen. *J. Invest. Dermatol.* **112**, 470–475.

Nagorcka, B. N., and Mooney, J. R. (1985). The role of a reaction-diffusion system in the initiation of primary hair follicles. *Theor. Biol.* **114**, 243–272.

Nellen, D., Burke, R., Struhl, G., and Basler, K. (1996). Direct and long-range action of a DPP morphogen gradient. *Cell (Cambridge, Mass.)* **85**, 357–368.

Nemes, Z., and Steinert, P. M. (1999). Bricks and mortar of the epidermal barrier. *Exp. Mol. Med.* **31**, 5–19.

Nemes, Z., Marekov, L. N., and Steinert, P. M. (1999a). Involucrin cross-linking by transglutaminase 1. Binding to membranes directs residue specificity. *J. Biol. Chem.* **274**, 11013–11021.

Nemes, Z., Marekov, L. N., Fésüs, L., and Steinert, P. M. (1999b). A novel function for transglutaminase 1: Attachment of long-chain omega-hydroxyceramides to involucrin by ester bond formation. *Proc. Natl. Acad. Sci. U.S.A.* **96,** 8402–8407.

Niederreither, K., D'Souza, R., Metsaranta, M., Eberspaecher, H., Toman, P. D., Vuorio, E., and de Crombrugge, B. (1994). Coordinate patterns of expression of type I and type III collagens during mouse development. *Matrix Biol.* **14,** 705–713.

Noramly, S., and Morgan, B. A. (1998). BMPs mediate lateral inhibition at successive stages in feather tract development. *Development (Cambridge, UK)* **125,** 3775–3787.

Noramly, S., Freeman, A., and Morgan, B. A. (1999). β-catenin signalling can initiate feather bud development. *Development (Cambridge, UK)* **126,** 3509–3521.

Nusse, R. (1999). WNT targets. Repression and activation. *Trends Genet.* **15,** 1–3.

Obara-Ishihara, T., Kuhlman, J., Niswander, L., and Herzlinger, D. (1999). The surface ectoderm is essential for nephric duct formation in intermediate mesoderm. *Development (Cambridge, UK)* **126,** 1103–1108.

Oda, Y., Tu, C. L., Chang, W., Crumine, D., Komuves, L., Mauro, T., Elias, P. M., and Bikle, D. D. (2000). The calcium sensing receptor and its alternatively spliced form in murine epidermal differentiation. *J. Biol. Chem.* **275,** 1183–1190.

Ohba, M., Ishino, K., Kashiwagi, M., Kawabe, S., Chida, K., Huh, N.-H., and Kuroki, T. (1998). Induction of differentiation in normal human keratinocytes by adenovirus-mediated introduction of the η and δ isoforms of protein kinase C. *Mol. Cell. Biol.* **18,** 5199–5207.

Ohnishi, S., Laub, F., Matsumoto, N., Asaka, M., Ramirez, F., Yoshida, T., and Terada, M. (2000). Developmental expression of the mouse gene coding for the Krupple-like transcription factor KLF5. *Dev. Dyn.* **217,** 421–429.

Oro, A. E., and Scott, M. P. (1998). Splitting hairs: Dissecting roles of signalling systems in epidermal development. *Cell (Cambridge, Mass.)* **95,** 575–578.

Oro, A. E., Higgins, K. M., Hu, Z., Bonifas, J. M., Epstein, E. H. J., and Scott, M. P. (1997). Basal cell carcinomas in mice overexpressing sonic hedgehog. *Science* **276,** 817–821.

Osumi-Yamashita, N., Ninomiya, Y., Doi, H., and Eto, K. (1994). The contribution of both forebrain and midbrain crest cells to the mesenchyme in the frontonasal mass of mouse embryos. *Dev. Biol.* **164,** 409–419.

Pasquali-Ronchetti, I., and Baccarani-Contri, M. (1997). Elastic fiber during development and ageing. *Microsc. Res. Tech.* **38,** 428–435.

Patel, K., Makarenkova, H., and Jung, H. S. (1999). The role of long range, local and direct signalling molecules during chick feather bud development involving the BMPs, Follistatin and the Eph receptor tyrosine kinase Eph-A4. *Mech. Dev.* **86,** 51–62.

Paus, R., and Cotsarelis, G. (1999). The biology of hair follicles. *N. Engl. J. Med.* **341,** 491–497.

Paus, R., Foitzik, K., Welker, P., Bulfone-Paus, S., and Eichmuller, S. (1997). Transforming growth factor-beta receptor type I and type II expression during murine hair follicle development and cycling. *J. Invest. Dermatol.* **109,** 518–526.

Paus, R., Muller-Rover, S., van der Veen, C., Maurer, M., Eichmuller, S., Ling, G., Hofmann, U., Foitzik, K., Mecklenburg, L., and Hnadkiski, B. (1999a). A comprehensive guide for the recognition and classification of distinct stages of hair follicle morphogenesis. *J. Invest. Dermatol.* **113,** 523–532.

Paus, R., Muller-rover, S., and Botchkarev, V. A. (1999b). Chronobiology of the hair follicle: Hunting the "hair cycle clock." *J. Invest. Dermatol., Symp. Proc.* **4,** 338–345.

Peifer, M., and Polarkis, P. (2000). Wnt signalling in oncogenesis and embryogenesis—a look outside the nucleus. *Science* **287,** 1606–1609.

Philpott, M., and Paus, R. (1998). Principles of hair follicle morphogenesis.

In "Molecular Basis of Epithelial Appendage Morphogenesis" (C. M. Chuong, ed.), pp. 75–110. R. G. Landes, Austin, TX.

Powell, B. C., and Rogers, G. (1990). Cyclic hair loss and regrowth in transgenic mice overexpressing an intermediate filament gene. *EMBO J.* **9,** 1485–1493.

Prakkal, P. F., and Maltosy, A. G. (1968). An electron microscopic study of developing chick skin. *J. Ultrastruct. Res.* **23,** 403–416.

Raghavan, S., Bauer, C., Mundschau, G., Li, Q., and Fuchs, E. (2000). Conditional ablation of beta1 integrin in skin. Severe defects in epidermal proliferation, basement membrane formation, and hair follicle invagination. *J. Cell Biol.* **150,** 1149–1160.

Robert, C., and Kupper, T. S. (1999). Inflammatory skin diseases, T cells and immune surveillance. *N. Engl. J. Med.* **9,** 1817–1828.

Robinson. N. A., Lapic, S., Welter, J. F., and Eckert, R. L. (1997). S100A11, S100A10, annexin I, desmosomal proteins, small proline-rich proteins, plasminogen activator inhibitor-2, and involucrin are components of the cornified envelope of cultured human epidermal keratinocytes. *J. Biol. Chem.* **272,** 12035–12046.

Roop, D. (1995). Defects in the barrier. *Science* **267,** 474.

Saitou, M., Sugai, S., Tanaka, T., Shimouchi, K., Fuchs, E., Narumiya, S., and Kakizuka, A. (1995). Inhibition of skin development by targeted expression of a dominant-negative retinoic acid receptor. *Nature (London)* **374,** 159–162.

Sato, N., Leopold, P. L., and Crystal, R. G. (1999). Induction of the hair growth phase in postnatal mice by localized transient expression of Sonic hedgehog. *J. Clin. Invest.* **104,** 855–864.

Sayama, K., Hanakawa, Y., Shirakata, Y., Yamasaki, K., Sawada, Y., Sun, L., Yamanishi, K., Hidenori, I., and Hashimoto, K. (2000). Apoptosis signal regulating kinase 1 (ASK1) is an intracellular inducer of keratinocyte differentiation. *J. Biol. Chem.* (in press).

Segré, J. A., Bauer, C., and Fuchs, E. (1999). Klf4 is a transcription factor required for establishing the barrier function of the skin. *Nat. Genet.* **22,** 356–360.

Seitz, C. S., Lin, Q., Deng, H., and Khavari, P. A. (1998). Alterations in NF-kappaB function in transgenic epithelial tissue demonstrate a growth inhibitory role for NF-kappaB. *Proc. Natl. Acad. Sci. U.S.A.* **95,** 2307–2312.

Sengel, P. (1976). "Morphogenesis of Skin." Cambridge University Press, Cambridge, London, and New York.

Sengel, P. (1990). Pattern formation in skin development. *Int. J. Dev. Biol.* **32,** 33–50.

Shtutman, M., Zhurinsky, J., Simcha, I., Albanese, C., D'Amico, M., Pestell, R., and Ben-Ze'ev, A. (1999). The cyclin D1 gene is a target of the beta-catenin/LEF-1 pathway. *Proc. Natl. Acad. Sci. U.S.A.* **96,** 5522–5527.

Simon, M., Haftek, M., Sebbag, M., Montezin, M., Girbal-Neuhauser, E., Shmitt, D., and Serre, G. (1996). Evidence that filaggrin is a component of the cornified cell envelope in human planter epidermis. *Biochem. J.* **317,** 173–177.

Smart, I. H. (1970). Variation in plane of cell cleavage during the process of stratification in mouse epidermis. *Br. J. Dermatol.* **82,** 276–282.

Smith, L. T., and Holbrook, K. A. (1982). Development of dermal connective tissue in human embryonic and fetal skin. *Scanning Electron Microsc.,* Pt. 4, pp. 1745–1751.

Song, H., Wang, Y., and Goetinck, P. (1996). Fibroblast growth factor 2 can replace ectodermal signalling for feather development. *Proc. Natl. Acad. Sci. U.S.A.* **93,** 10246–10249.

Steinert, P. M. (2000). The complexity and redundancy of epithelial barrier function. *J. Cell Biol.* **151,** F5–F7.

Steinert, P. M., and Marekov, L. N. (1999). Initiation of assembly of the cell envelope barrier structure of stratified squamous epithelia. *Mol. Cell. Biol.* **10,** 4247–4261.

Steinert, P. M., Candi, E., Kartasova, T., and Marekov, L. (1998). Small proline-rich proteins are cross-bridging proteins in the cornified cell envelopes of stratified squamous epithelia. *J. Struct. Biol.* **122,** 76–85.

St. Jacques, B., Dassule, H. R., Karavanova, I., Botchkarev, V. A., Li, J., Danielian, P. S., McMahon, J. A., Lewis, P. M., Paus, R., and McMahon, A. P. (1998). Sonic hedgehog signalling is essential for hair development. *Curr. Biol.* **8,** 1058–1068.

Swartzendruber, D. C., Wertz, P. W., Madison, K. C., and Downing, D. T. (1987). Evidence that the corneocyte has a chemically bound lipid envelope. *J. Invest. Dermatol.* **88,** 709–713.

Takahashi, M., Tezuka, T., and Kutunumu, N. (1992). Phosphorylated cystatin alpha is a natural substrate of epidermal transglutaminase for formation of skin cornified envelope. *FEBS Lett.* **308,** 79–82.

Takeda, K., Takeuchi, O., Tsujimura, T., Itami, S., Adachi, O., Kawai, T., Sanjo, H., Yoshikawa, K., Terada, N., and Akira, S. (1999). Limb and skin abnormalities in mice lacking IKKalpha. *Science* **284,** 313–316.

Tani, H., Morris, R. J., and Kaur, P. (2000). Enrichment for murine keratinocyte stem cells based on cell surface phenotype. *Proc. Natl. Acad. Sci. U.S.A.* **97,** 10960–10965.

Tao, Q. H., Yang, J., Mei, W. Y., Geng, X., and Ding, X. Y. (1999). Cloning and analysing of 5′ flanking region of Xenopus organizer gene noggin. *Cell Res.* **9,** 209–216.

Tarutani, M., Itami, S., Okabe, M., Ikawa, M., Tezuka, T., Yoshikawa, K., Kinoshita, T., and Takeda, J. (1997). Tissue-specific knockout of the mouse Pig—a gene reveals important roles for GPI-anchored proteins in skin development. *Proc. Natl. Acad. Sci. U.S.A.* **94,** 7400–7405.

Taylor, G., Lehrer, M. S., Jensen, P. J., Sun T.-T., and Lavker, R. M. (2000). Involvement of follicular stem cells in forming not only the follicle but also the epidermis. *Cell (Cambridge, Mass.)* **102,** 451–461.

Tennenbaum, T., Li, L., Bélanger, A. J., De Luca, L. N., and Yuspa, S. H. (1996). Selective changes in laminin adhesion and $\alpha6\beta4$ integrin regulation are associated with the initial steps in keratinocyte maturation. *Cell, Growth Differ.* **7,** 615–628.

Tetsu, O., and McCormick, F. (1999). Beta-catenin regulates expression of cyclin D1 in colon carcinoma cells. *Nature (London)* **398,** 422–426.

Theiler, J. (1989). "The House Mouse: Atlas of Embryonic Development." Spinger-Verlag, New York.

Thorey, I. S., Meneses, J. J., Nezanov, N., Kulesh, D. A., Pedersen, R. A., and Oshima, R. G. (1993). Embryonic expression of human keratin 18 and K18-beta-galactosidase fusion gene in transgenic mice. *Dev. Biol.* **160,** 519–532.

Ting-Berreth, S. A., and Chuong, C. M. (1996). Sonic Hedgehog in feather morphogenesis: Induction of mesenchymal condensation and association with cell death. *Dev. Dyn.* **207,** 157–170.

Uitto, J. (1999). Heritable connective tissue disorders. *Adv. Exp. Med. Biol.* **455,** 15–21.

van der Neut, R., Krimpenfort, P., Calafat, J., Niessen, C. M., and Sonnenberg, A. (1996). Epithelial detachment due to absence of hemidesmosomes in integrin beta 4 null mice. *Nat. Genet.* **13,** 366–369.

Van Exan, R. J., and Hardy, M. H. (1984). The differentiation of the dermis in the laboratory mouse. *Am. J. Anat.* **169,** 149–164.

van Genderen, C., Okamura, R. M., Farinas, I., Quo, R. G., Parslow, T. G.,

Bruhn, L., and Grosschedl, R. (1994). Development of several organs that require inductive epithelial-mesenchymal interactions is impaired in LEF-1-deficient mice. *Genes Dev.* **8,** 2691–2703.

Vastrik, I., Kaipainen, A., Penttila, T. L., Lymboussakis, A., Alitalo, R., Parvinen, M., and Alitalo, K. (1995). Expression of the mad gene during cell differentiation in vivo and its inhibition of cell growth in vitro. *J. Cell Biol.* **128,** 1197–1208.

Vuorio, E., and de Croombrugghe, B. (1990). The family of collagen genes. *Annu. Rev. Biochem.* **59,** 837–872.

Watt, F. M., and Hertle, M. D. (1994). Keratinocyte integrins. *In* "The Keratinocyte Handbook" (I. M. Leigh, E. B. Lane, and F. M. Watt, eds.), pp. 153–164. Cambridge University Press, Cambridge, UK.

Watt, F. M., and Hogan, B. L. M. (2000). Out of eden: Stem cells and their niches. *Science* **287,** 1427–1433.

Weiss, L. W., and Zelickson, A. S. (1975a). Embryology of the epidermis: Ultrastructural aspects. I. Formation and early development in the mouse with mammalian comparisons. *Acta Derm.-Venereol.* **55,** 161–168.

Weiss, L. W., and Zelickson, A. S. (1975b). Embryology of the epidermis: Ultrastructural aspects. II. Period of differentiation in the mouse with mammalian comparisons. *Acta Derm.-Venereol.* **55,** 321–329.

Weiss, L. W., and Zelickson, A. S. (1975c). Embryology of the epidermis: Ultrastructural aspects. III. Maturation and primary appearance of dendritic cells in the mouse with mammalian comparisons. *Acta Derm.-Venereol.* **55,** 431–442.

Wennerberg, K., Lohikangas, L., Gullberg, D., Pfaff, M., Johansson, S., and Fassler R. (1996). Beta 1 integrin-dependent and -independent polymerization of fibronectin. *J. Cell Biol.* **132,** 227–238.

Wessells, N. (1965). Morphology and proliferation during early feather development. *Dev. Biol.* **12,** 131–153.

Widelitz, R. B., Jiang, T. X., Chen, C. W., Stott, N. S., and Chuong, C. M. (1999). Wnt-7a in feather morphogenesis: Involvement of anterior-posterior asymmetry and proximal-distal elongation demonstrated with an in vitro reconstitution model. *Development (Cambridge, UK)* **126,** 2577–2587.

Xie, J., Murone, M., Luoh, S. M., Ryan, A., Gu, Q., Zhang, C., Bonifas, J. M., Lam, C. W., Hynes, M., Goddard, A. *et al.* (1998). Activating smoothened mutations in sporadic basal-cell carcinoma. *Nature (London)* **391,** 90–92.

Yang, A., Schweitzer, R., Sun, D., Kaghad, M., Walker, N., Bronson, R. T., Tabin, C., Sharpe, A., Caput, D., Crum, C. *et al.* (1999). p63 is essential for regenerative proliferation in limb, carniofacial and epithelial development. *Nature (London)* **398,** 714–718.

Yang, J. T., Rayburn, H., and Hynes, R. O. (1993). Embryonic mesodermal defects in alpha 5 integrin-deficient mice. *Development (Cambridge, UK)* **119,** 1093–1105.

Zhou, P., Byrne, C., Jacobs, J., and Fuchs, E. (1995). Lymphoid enhancer factor 1 directs hair follicle patterning and epithelial cell fate. *Genes Dev.* **9,** 700–713.

Zhu, A. J., Haase, I., and Watt, F. M. (1999). Signalling via $\beta1$ integrins and mitogen-activated protein kinase determines human epidermal stem cell fate in vitro. *Proc. Natl. Acad. Sci. U.S.A.* **96,** 6728–6733.

Author Index

Italic numerals indicate authors cited in references.

A

Abate-Shen, C., 267, *272,* 450, 471, 472, 473, *498*
Abbondanzo, S. J., 156, *179*
Abbott, B., 172, 173, *177,* 225, *230*
Abbott, M. A., 11, *19*
Abbud, R., 376, *391,* 510, *515*
Abdelhak, S., 398, *416,* 550, *561*
Abdelilah, S., 546, *563*
Abe, E., 271, *275*
Abel, L., 196, *207*
Åberg, T., 439, 441, 452, 456, *487, 494, 497*
Abmayr, S. M., 136, *145*
Abou-Rebyeh, F., 260, 266, *273*
Abou-Samra, A. B., 285, *293,* 442, 443, *489*
Abrahamson, D. R., 412, 414, *417, 419*
Abramovitch, R., 225, *229*
Abramow-Newerly, W., 42, *51*
Abrass, C. K., 412, *417*
Abuin, A., 110, *123*
Abu-Shaar, M., 91, *103*
Acampora, D., 31, *33,* 40, 46, 48, *49, 52,* 79, 83, 84, 95, *97, 104,* 114, *122,* 290, *292, 294, 322,* 338, *360,* 428, 450, 451, 453, 471, 472, 473, 478, *481, 495,* 506, 508, *513,* 521, 527, *535,* 549, *561*
Accavitti, M. A., 412, *417*
Achen, M. G., 162, *174*
Acher, R., 500, *513*
Achermann, J. C., 376, *389*
Acheson, A. L., 91, *97,* 227, *231*
Ackerman, S. L., 12, *16*
Acquister, E., 578, *585*
Adachi, A., 136, *148*
Adachi, E., 443, *492*

Adachi, H., 356, *367*
Adachi, O., 452, *496,* 580, 581, 583, *589*
Adachi, T., 79, *99*
Adam, J., 553, 559, *561*
Adamek, S., 78, *102*
Adams, J., 550, *565*
Adams, J. C., 572, *586*
Adams, J. M., 317, *325,* 398, 399, *420,* 527, 530, 533, *538*
Adams, K. A., 117, *122*
Adams, P. S., 45, *52*
Adams, R. H., 171, 173, *174*
Adams, S., 245, *250*
Adamson, E. D., 338, 352, *370*
Adamson, I. Y., 313, *322*
Adelman, D. M., 172, 173, 174, *174*
Adenot, P. G., 9, 14, *15*
Ades, L., 468, *485*
Adess, M., 500, 504, 506, 509, *517*
Adham, I. M., 386, *393*
Afford, S., 174, *179*
Afouda, A. B., 10, *15*
Agani, F., 225, *230*
Agellon, L. B., 343, 347, *361*
Aggrawal, A. K., 506, 509, *514*
Agius, E., 31, *33,* 44, *50, 52,* 338, *358, 366*
Agorastos, T., 570, *584*
Agoston, D., 91, *103*
Agoulnik, I. U., 380, *389*
Aguet, M., 223, *229,* 287, *293*
Aguila, H. L., 244, *252*
Aguilera, R. J., 509, *514*
Agulnik, I., 132, *145*
Aguzzi, A., 173, *175,* 260, 266, *272,* 314, 317, 318, *323, 324, 325,* 404, *417*
Ahlgren, S. C., 470, *481*

Ahlgren, U., 319, 320, 321, *322, 323*
Ahmed, A., 174, *179*
Ahmed, M., 470, *497*
Ahnn, J., 341, *360*
Ahrlund-Richter, L., 173, *178*
Aikawa, H., 343, *368*
Aird, W. C., 219, *228, 229*
Ait, F. W., 510, *518*
Aitola, M., 527, 530, 532, *535*
Aizawa, S., 7, *16*, 31, *34*, 46, 49, *51*, 83, 84, *105, 106*, 114, *125*, 374, *392*, 403, *418*, 421, 451, 457, 478, *488, 490, 496*, 521, 527, *537*
Akashi, K., 194, *208*, 241, 245, *250, 251*
Akerblom, I. E., 504, *513*
Akerud, P., 91, *98*
Akhurst, R. J., 203, *207*, 226, *229*, 467, *484, 491*
Akhverdian, M., 388, *391*
Akimaru, H., 121, *123*
Akimenko, M. A., 450, 470, 473, *481*, 546, *561*
Akira, S., 452, *496*, 531, *538*, 580, 581, 583, *589*
Akita, S., 512, *513*
Akiyama, M., 570, 583, *584*
Aktan, M., 573, *584*
Alarcon, F. J., 388, *391*
Alarid, E. T., 504, *513*
Al-azzeh, E., 316, *323*
Albajar, M., 354, *366*
Albanese, C., 86, *104*, 579, *588*
Albano, R. M., 40, *49*, 203, *209*
Alberch, J., 91, *98*
Alberch, P., 421, *481*
Alberts, A. S., 342, *369*
Alborn, A.-M., 237, 247, 248, *249*
Albrecht, K. H., 378, 382, *389*
Albrecht, U., 31, *34*, 43, 45, *51*, 219, *232*
Albright, S., 288, *293*
Alcamo, E., 136, *145*
Alcedo, J., 470, *481*
Alder, J., 122, *122*
Alderete, B., 287, *294*
Alders, M., 174, *174*
Alderson, R. F., 91, *97*
Al-Dhalimy, M., 246, *250*, 318, *326, 327*
Aldrich, T. H., 224, *231*
Alexander, C. R., 224, *230*
Alexander, J., 303, *323, 326*, 342, *367*
Alexander, W. S., 194, *208*
Alexandre, E., 308, *326*
Alexandre, H., 26, *33*
Alfandari, D., 424, *481*
Alhonen, L., 228, *230*
Ali, A. A., 426, *495*
Ali, M. M., 426, *495*

Alikani, M., 23, *34*
Alitalo, K., 162, *179*, 218, 223, 224, 228, *229, 230, 231*, 580, *589*
Alitalo, R., 580, *589*
Alkema, M. J., 376, *389*, 453, *481*
Al Khamis, A., 439, 467, 478, *497*
Allen, D. L., 269, *275*
Allen, K. M., 94, *97, 100*
Allen, L., 314, *327*
Allen, R. E., 271, *276*
Allen, R. G., 510, *513*
Allen, S. P., 139, *148*
Allendoerfer, K. L., 92, *97*
Allin, E. F., 426, 442, *481*
Allis, C. D., 14, *15*
Allison, J., 31, *34*
Allworth, A. E., 23, *35*
Alonso, M. T., 543, 545, 546, *565*
Alpert, S., 320, *323*
Als-Nielsen, J., 408, *419*
Alsonso, G., 171, 173, *174*
Alt, F. W., 87, 91, *103, 106*, 203, 204, *209, 210*, 220, *233*, 508, *517*, 521, 527, *537*
Altaba, A., 579, *585*
Altar, C. A., 91, *97, 98*
Altemus, M., 500, *516*
Altieri, D. C., 225, *231*
Altman, D. G., 25, *34*, 158, *176*
Altman, J., 92, *98*, 237, 248, *248*
Altmann, C., 200, 201, *209*
Altmann, C. R., 533, *535*
Altrock, B. W., 402, *417*
Altschuler, R. A., 443, *482*
Altshuller, Y. M., 243, *249*
Alvarado-Mallart, R. M., 111, 112, 113, 114, *122, 123, 124*
Alvarez, I. S., 541, 547, *561, 564*
Alvarez, J., 289, *293*
Alvarez-Bolado, G., 83, 89, 95, *97, 105*, 115, *125*
Alvarez-Buylla, A., 90, 95, *99, 102, 106*, 235, 236, 237, 244, 245, 248, *249, 251, 252*
Alvarez Otero, R., 113, *122*
Alward, W. L., 478, *495*, 507, *517*
Amacher, S. L., 167, *176*
Amaya, E., 355, *359*
Ambrosetti, D. C., 283, 286, 292, *294*
Ambrozewicz, L. A., 312, *327*
Amijura, R., 186, *190*
Amini, R., 83, *106*
Amirand, C., 9, 14, *15*
Amling, M., 285, *294*
Amoui, M., 530, 533, *535*
Amri, E. Z., 271, *277*
Amselem, S., 508, *516*

Amthor, H., 264, 266, *272*
An, S., 340, 341, *363*
Anandappa, R., 165, *177,* 316, *326,* 342, 343, 352, *363*
Anastassiadis, C., 160, *175*
Anastassiadis, K., 23, *34,* 160, 166, *178,* 311, *324*
Ancellin, N., 227, *230*
Anchan, R., 469, *497*
Anchan, R. M., 84, *97, 106,* 435, 469, 476, *481*
Andermarcher, E., 404, *420*
Andersen, B., 506, 508, 509, *518*
Anderson, D. J., 83, 87, 88, 89, 95, *97, 99, 102,* 109, 114,
    126, 168, *176, 177,* 218, 225, 228, *229, 232, 233,* 235,
    236, 237, 238, 239, 240, 241, 242, 243, 244, 245, 248,
    248, 249, 250, 251, 252, 320, 321, *323, 328,* 435, *490,*
    526, *538,* 559, *563*
Anderson, D. M., 193, 204, *208*
Anderson, D. W., 557, *561*
Anderson, J. E., 260, 271, *275, 277*
Anderson, K. D., 91, *97,* 311, *329*
Anderson, L., 224, *229*
Anderson, P., 188, *189*
Anderson, R., 3, *4,* 182, 183, 184, 185, 187, 188, *189*
Anderson, R. H., 334, 348, 351, 352, 354, *359, 363, 367,*
    369
Anderson, R. M., 48, *49,* 78, *98,* 304, *323*
Anderson, S. A., 79, 83, 84, 85, 87, 88, 89, 90, 91, 92, 94,
    95, *97, 98, 99, 102, 103, 104,* 450, 470, 471, *481, 484*
Andreazzoli, M., 525, *535*
Andree, B., 336, 340, *358, 367*
Andresson, T., 7, *17*
Andrew, G. L., 85, *102*
Andrews, J. E., 334, 343, 345, 348, 349, 352, 356, *364,*
    380, 387, *391*
Andrikopoulos, K., 442, 443, *489,* 573, 583, *584*
Ang, S.-L., 31, *33,* 40, 41, 42, 43, 44, 46, 47, *49, 50, 51, 52,*
    83, 89, 95, *99,* 110, 114, 116, *122, 125,* 136, *146,* 165,
    174, 303, 316, *323,* 338, *358,* 451, 477, 478, *481, 484,*
    494, 521, 526, 527, *535, 538*
Angel, P., 173, *179*
Angellilo, A., 226, *229*
Angelo, S., 349, 355, *358*
Angevine, J. B., 92, *97*
Anniko, M., 552, *563*
Anson-Cartwright, L., 109, 115, *123,* 168, 169, 170, 173,
    174, 176, 178, 179, 349, 350, 352, *367*
Antczak, M., 7, *15*
Antczuk, M., 33, *33*
Antin, P. B., 334, 335, 336, 337, *363, 370*
Anton, E. S., 94, *97, 98*
Antonarakis, S. E., 479, *482, 488*
Antonio, L., 402, *417*
Antos, C., 342, 343, *365*
Anzai, M., 10, *17*

Anzano, M. A., 467, *494*
Ao, A., 14, *15*
Aoki, F., 9, *15*
Aoki, K., 422, 424, 435, 437, 475, 476, *492*
Aoyama, H., 263, *272*
Aoyama, S., 511, *516*
Aparicio, S. A., 117, *124,* 160, 167, *178*
Apelqvist, A., 319, 320, 321, *323*
App, H., 217, 220, *229*
Appel, R., 581, *586*
Arai, A., 335, *358*
Arai, Y., 262, 268, *274*
Araki, I., 112, 114, 115, 116, *122, 123,* 528, *536*
Araki, K., 305, 307, *328*
Araki, M., 305, 307, *328,* 531, 534, *537*
Aranda, A., 510, *516, 517*
Arango, N. A., 381, *389*
Aranoff, T., 405, 407, *416, 417*
Arbeit, J. M., 580, *586*
Arber, S., 319, *326*
Arenas, E., 91, *98,* 402, *420*
Arends, J. W., 173, *179*
Argilla, M., 219, *229, 231*
Arias, C., 511, *513*
Arias, C. A., 89, 95, *104,* 506, *517*
Arimatsu, Y., 84, *98, 100*
Arimura, A., 512, *515*
Arita, M., 284, *293*
Ariza-McNaughton, L., 117, 118, 119, *124, 125,* 426, 427,
    453, 468, *490, 492, 494,* 544, *563*
Arkell, R., 40, *49,* 110, *122,* 390
Arkell, R. M., 339, *358*
Arman, E., 167, 171, *174,* 312, 314, 315, *323*
Armanini, M., 402, *418, 420*
Armes, J., 453, *486*
Armstrong, D. L., 122, *122*
Armstrong, J. F., 374, 375, *389,* 399, *416*
Arnheiter, H., 525, *536*
Arnold, H. H., 85, *98, 103,* 132, 140, *144, 147,* 258, 260,
    263, 264, 266, *272, 273, 274, 276, 277,* 310, *327,* 336,
    340, 342, *358, 367, 369,* 475, *482,* 549, 550, *562, 564*
Arnone, M. I., 311, *330*
Arnould, T., 415, *418*
Arnush, M., 318, *326*
Arokoski, J., 284, *293*
Aronson, J., 22, *34*
Arra, C., 311, *324*
Arroyo, A. G., 204, *206*
Arsenian, S., 342, *358*
Arsenijevic, Y., 89, 91, *104*
Artavanis-Tsakonas, S., 136, *144, 145, 146, 148, 149,* 263,
    272, 553, 555, *561*
Artemov, D., 225, *231*

Artinger, K. B., 121, *125*
Aruga, J., 122, *122*, 357, *363*, 423, *492*
Arumae, U., 402, *420*, 559, *564, 565*
Arume, U., 403, *419*
Asa, S. L., 511, *516*
Asada, Y., 506, *516*, 549, *564*
Asahara, H., 286, *293*
Asahara, T., 220, *228*
Asai, N., 402, *420*
Asaka, M., 581, *588*
Asakura, A., 258, 260, 261, 262, 264, 265, 267, 271, *272,*
    275, 276
Asamoto, K., 263, *272*
Asano, S., 220, *231*
Asaoka, M., 186, *190*
Asashima, M., 338, *368*
Asavaritikrai, P., 87, 94, *105*
Asbreuk, C. H., 506, *513*
Ashby, P., 260, 267, *275*
Asher, J. H., Jr., 475, *481*
Ashery-Padan, R., 519, 530, 533, *535*
Ashworth, G. J., 453, 466, 474, *498*
Atkin, S. L., 505, *513*
Attar, P. S., 581, *586*
Aubin, J., 312, *323*
Aubry, S., 171, 173, *176*
Auda-Boucher, G., 267, *274*, 335, 337, 340, *358*
Audero, E., 266, *275*
Auerbach, A. B., 109, 112, 115, *123, 126*, 168, *176*, 239,
    250
Auerbach, R., 475, *481*
Auernhammer, C., 512, *513*
Augsburger, A., 78, *104*, 141, *148*
Aulehla, A., 136, 137, 138, *145*
Aulner, N., 9, 14, *15*
Aurrand-Lions, M., 376, *390*
Austin, C. R., 22, *33*
Austin, R. J., *251*
Avantaggiato, V., 31, *33*, 46, 48, *49*, 84, 95, *97*, 114, *122,*
    311, *322, 330*, 428, 451, 478, *481*, 506, 508, *513*, 521,
    527, *535*
Avery, G. B., 140, *144*, 166, *176*
Avner, E. D., 415, *419*
Avraham, K. B., 341, *369*, 556, *565*
Axel, R., 90, 92, 95, *98*, 239, *249*
Axelrod, J. D., 337, *366*
Axelrod, L., 500, *518*
Ayer, D. E., 580, *586*
Ayers, A. B., 507, *514*
Aylor, K. W., 412, *420*
Aylsworth, A. S., 286, 288, *292, 293*, 357, *361*, 453, *481*
Ayuso, C., 525, *535*
Ayyub, H., 376, *390*
Ayzenzon, M., 470, *481*

Aza-Blanc, P., 470, *481*
Aziz, A., 205, *209*
Azpiazu, N., 91, *102*, 334, *358*
Azsterbaum, M., 581, *584*
Azuma, T., 23, *35*

B

Baarends, W. M., 385, *389*
Babakhanlou, H., 415, *418*
Babinet, C., 84, *98*, 165, 166, *175, 178*, 546, *565*
Bacani, C. J., 343, 347, *362*
Baccarani-Contri, M., *588*
Bach, E. A., 287, *293*
Bach, J. P., 166, *178*, 415, *419*
Bachiller, D., 31, *33*, 43, 44, 48, *49, 50, 51*, 78, *98*, 304,
    323, 338, *358*, 427, 428, 450, 451, 468, *481*
Bachinsky, D. R., 346, 357, *358*
Bachman, D., 271, *273*
Bachvarova, R. F., 45, *49*, 164, *175*, 303, *323*
Back, E., 380, *392*
Backus, C., 91, *101*, 239, *249*, 549, *562*
Bacou, F., 271, *273*
Bader, B. L., 223, *228, 230*
Bader, D., *324*, 337, 340, 341, 344, 345, *358, 361, 364,*
    365, 370
Bader, D. M., 341, *362*
Bader, J. A., 167, *177, 179*
Bader, R., 183, 187, 188, *189*
Baeuerle, P. A., 381, *393*
Bagchi, I., 156, *175*
Bahl, J. J., 343, *360*
Bahramian, M. B., 15, *15*
Bai, C., 85, *103*, 121, *125*, 505, *517*
Baik, J.-H., 511, *517*
Baird, G. R., 438, *488*
Baker, C. V., 423, 475, *481*
Baker, D. P., 78, 81, *101*, 110, 121, *124*
Baker, J. H., 23, *35*
Balas, G., 78, 81, *106*, 243, *252*, 479, *486, 498*
Balasubramanyam, A., 342, *369*
Balbin, M., 289, *293*
Balconi, G., 226, *229*
Baldassare, M., *106*
Baldini, A., 353, 357, *364*
Baldwin, H. S., 109, *125*, 162, 170, *177*, 219, 222, *229,*
    230, 340, 352, *361*
Baldwin, M. A., 227, *231*
Bale, A. E., 287, *292*, 470, *485, 486*, 579, *586*
Bales, K. R., 258, *274*
Ball, D., 313, *323*
Ball, S. G., 387, *390*
Ballabio, A., 525, *535*
Balling, R., 140, *144, 145, 147*, 160, *178*, 263, *272, 273,*

292, *292*, 398, 408, *417*, 440, 441, 451, 475, *481*, *483*, *492*, *493*, *496*, *498*, 551, 552, 557, 560, *563*
Bally-Cuif, L., 117, *122*
Balsari, A., 511, *516*
Baltimore, D., 257, *275*, 318, *323*
Bamforth, J. S., 357, *361*
Banbury, D. N., 568, 582, *586*
Banerjee, S. K., 504, *513*
Banerjee-Basu, S., 532, *538*
Banfi, S., 525, *535*
Bang, A., 522, 525, 533, *538*
Banta, A. B., 347, *364*
Bao, Y., 452, *492*
Bao, Z.-Z., 341, 343, 345, 347, 348, 349, 350, 351, 352, *358, 359*
Baptista, C. A., 78, 81, *106*, 243, *252*, 479, *498*
Baraitser, M., 453, *494*
Barak, Y., 172, 173, *174*, 341, 357, *358*
Barald, K. F., 550, *562*
Baras, M., 22, *34*
Barasch, J., 405, 407, *416*
Barbacci, E., 165, *174*, 303, *323*
Barbacid, M., 91, *98, 99, 105*, 402, *419*, 559, *562*
Barbaux, S., 375, 387, *389, 392*
Barbe, M. F., 84, *101*
Barbel, S., 334, 349, *365*
Barbieri, A. M., 525, *535*
Barbieri, O., 290, *292*, 450, 453, 472, 473, *481*, 549, *561*
Barbosi, E., 547, *564*
Barcellos-Hoff, M. H., 432, *482*
Bard, J. B., 374, 375, *389, 399, 419*
Bard, J. B. L., 127, 128, 131, *146*, 399, *416*
Barde, Y. A., 141, *144*, 434, *489*
Bardeesy, N., 379, *391*
Bardoni, B., 378, 387, *389, 392, 393*
Bardwell, V. J., 382, 387, 388, *392*
Bare, J. W., 287, *293*, 579, *586*
Barembaum, M., 423, *481*
Barghusen, H. R., 442, 443, 449, 456, 470, *481*
Baribault, H., 408, *419*, 570, 572, 583, *584, 585*
Barker, D. F., 412, *417*
Barker, J. E., 202, *209*
Barkoff, A. F., 10, *15*
Barkway, C., 560, *565*
Barlett, P. F., 86, *101*
Barlow, A. J., 422, 467, *481, 484*
Barlow, D. P., 161, *176*, 302, *327*
Barlow, L. A., 433, *492*
Barlow, P. W., 26, *33*, 156, *174*
Barltey, J., 380, *390*
Barnard, J. A., 169, *179*
Barnes, J. D., 40, 43, *49*
Barnett, L. D., 31, *34*, 203, *209*, 345, 348, 349, 352, 353, 357, *358, 366*

Barnitt, A., 11, *19*
Baroffio, A., 236, 237, 244, *249*
Baron, M. H., 200, *206*
Baron, R., 285, *294*
Barone, P., 84, *97*, 478, *481*
Barr, K. J., 156, *177*
Barra, J., 165, 166, *175, 178*
Barres, B. A., 236, 237, *249*
Barrish, J., 315, *326*
Barron, M., 337, *364*
Barros, A. C., 510, *515*
Barros, E. J., 415, *417*
Barsachi, G., 525, *535*
Barsh, G. S., 118, *124*, 527, 530, 531, *537*, 544, *562*
Bar-Tana, J., 164, *176*
Bartel, D. P., 14, *19*
Bartelmez, G. W., 424, *481*
Barth, K. A., 78, 79, *98, 102*, 521, 522, *537*
Barth, K. S., 45, 47, *50*, 338, 355, *359*, 468, *483*
Barth, S. L., 467, *494*
Barth-Hall, S., 509, *513*
Bartholomae, A., 92, *104*
Bartke, A., 509, *518*
Bartley, A., 189, *189*
Barton, P. J. R., 341, *359*
Barton, S. C., 26, *35*, 157, 158, 160, 166, 167, *176, 178*
Bartunkova, S., 224, 228, *231, 232*, 348, 349, 352, *368*
Baserga, M., 311, *326*
Basic, N., 416, *420*
Basic, V., 416, *420*
Basilico, C., 160, 166, *178, 180*, 283, 286, *292, 294*, 505, *517*
Baskin, L. S., 301, *326*
Basler, K., 112, *123*, 470, *484*, 577, *587*
Bass, N. M., 581, *586*
Bassel-Duby, R., 269, *273, 277*, 342, *369*
Basset, P., 162, *179*
Bassi, M. T., 47, *51*, 357, *361, 363*
Basson, C. T., 346, 347, 357, 358, *362, 367*
Bastin, P., 15, *15*
Basu, S., 527, *535*
Basyuk, E., 170, *174*
Batchelder, C., 243, *249*
Bateman, J. F., 443, *482, 495*
Batourina, E., 403, *418*
Baud, V., 453, *486*, 580, 581, 583, *586*
Bauer, C., 574, 581, 583, *588*
Baum, C., 85, *98*
Baumann, G., 511, *513*
Baunoch, D., 225, *231*
Baur, S. T., 292, *294*
Baybayan, P., 441, *488*
Bayer, S. A., 92, *98*
Baylin, S., 313, *323*

Beach, D., 341, *361, 483*
Beachy, P. A., 43, *50,* 78, 80, 84, 85, 86, *98, 101, 102, 104,* 120, 121, *123, 125,* 141, *145,* 287, 292, 319, *323,* 439, 451, 470, *482, 483, 489, 493,* 521, *535*
Beamer, W. G., 443, *485,* 500, 506, 508, 509, *513, 515, 518*
Beamer, W. J., 311, *323*
Beatus, P., 320, 321, *323*
Beau, C., 382, *389*
Beaujean, N., 9, 14, *15*
Beaupain, D., 196, 201, *206, 208*
Beavis, A., 200, *209*
Beck, A., 188, *189*
Beck, C. D., 402, *420*
Beck, F., 160, 161, 167, *174, 175, 176,* 305, 307, *323*
Beck, K. D., 91, *98*
Beck, L., *323*
Beck, S., 316, *323*
Beck, S. C., 206, *209*
Becker, D., 226, *230*
Becker, M. D., 550, *561*
Becker, S., 7, *19*
Beckers, J., 134, 135, 138, 139, *146*
Beckmann, J. S., 341, *369,* 382, *392,* 525, *535*
Bedbeder, P., 398, *416,* 550, *561*
Beddington, R. S., 31, *34,* 40, 45, 46, 47, 48, *49, 50, 51, 52,* 78, *98,* 110, *122,* 133, 135, *144, 145,* 165, *177,* 288, *294,* 311, 316, *324, 327,* 353, 356, *363,* 407, *417,* 422, 427, 442, 443, 451, 453, 476, *481, 483, 490, 493*
Beddington, R. S. P., 30, 31, *33, 35,* 130, 132, 133, *148, 149,* 160, *175,* 304, 316, *329,* 338, 339, 348, *358, 360, 362,* 508, *514*
Beddington, S. P., 37, 38, 40, 41, 42, *49*
Bedford, M. T., 467, *492*
Bedi, A., 225, *231*
Bee, J., 432, *481*
Beebe, D. C., 532, *535*
Beeghly, A., 85, *103,* 121, *125,* 505, *517*
Beesley, J. S., 82, 86, *100*
Beg, A. A., 318, *323*
Begbie, J., 433, 434, 435, *481, 497,* 543, *562*
Begley, C. G., 203, *209*
Behne, M., 582, *584*
Behringer, R. R., 3, *4,* 31, *34,* 37, 38, 39, 40, 41, 42, 43, 44, 45, 46, 47, 48, 49, *49, 50, 51, 52,* 79, *98, 104, 106,* 114, *125,* 172, 173, *177,* 260, *277,* 281, 282, 283, 284, 289, *292, 294, 328,* 336, 352, *365,* 374, 380, 381, 385, *389, 391, 392,* 397, *420,* 422, 427, 432, 473, 476, 477, 478, *482, 484, 494, 498,* 511, *513*
Bei, M., 290, *294,* 437, 440, 441, 450, 451, 466, 474, *481, 482, 483, 495*
Beier, D. R., 442, 443, *489,* 530, 533, *538*
Beighton, P., 292, *294*
Beilharz, M. W., 269, 271, *274*
Beiman, M., 340, *358*

Bekku, Y., 223, *230*
Bel, S., 376, *389, 390,* 453, *481*
Belaguli, J. L., 342, *366*
Belaguli, N., 342, 343, *364, 367*
Bélanger, A. J., *589*
Bélanger, M. C., 86, *104*
Belaoussoff, M., 200, *206*
Belayew, A., 502, *515*
Belbeck, L. W., 202, *209*
Beldjord, C., 94, *99*
Belin, D., 10, 14, *18, 19*
Bell, B., 214, *230*
Bell, D. M., 281, *293,* 380, 381, *389, 392*
Bell, E., 119, *122,* 433, *485*
Bell, K. S., 379, 387, *393*
Bellairs, R., 132, *144*
Bellen, H. J., 122, *122,* 555, 558, 559, *561*
Bellier, S., 9, *15*
Bellus, G. A., 286, *292,* 452, 453, *481*
Bellusci, S., 287, *292,* 311, 312, 313, 314, 315, *323, 324, 325, 329,* 470, *485*
Belmont, J., 47, *51,* 83, *98,* 357, *363*
Belo, J. A., 31, *33,* 40, 44, 48, *49, 50,* 78, *98,* 304, *323,* 338, *358,* 387, *393,* 473, 477, *482*
Belting, H. G., 109, *123*
Belyantseva, I. A., 557, *561*
Ben-Arie, N., 122, *122,* 555, 558, 559, *561*
Bench, G., 580, *585, 587*
Bendall, A. J., 267, *272*
Bendel-Stenzel, M., 187, *189*
Benezra, R., 257, *277*
Bengzon, J., 247, *249*
Benit, P., 452, 453, *484*
Bennett, C. N., 271, *276*
Bennett, D., 442, 443, *494*
Bennett, J. H., 467, *482*
Bensaude, O., 9, *15*
Benson, D. W., 357, *358, 367*
Bensoussan, A. L., 307, *329*
Bentley, C. A., 509, *513*
Ben-Ze'ev, A., 86, *104,* 579, *588*
Benzing, T., 415, *418*
Ben-Zur, Z., 174, *178*
Berard, J., 312, *323*
Berber, E., 474, *494*
Beresford, W. A., 421, *482*
Berezowsky, J., 163, *175*
Berg, L. A., 509, *517*
Berge, D. T., 476, 477, 478, *491*
Berger, J., 314, *327*
Berger, J. E., 223, *232,* 281, 291, *294*
Bergeron, D., 167, *179*
Bergers, G., 223, *232,* 281, 291, *294*
Berggren, K., 469, *482*

Berghella, L., 271, *273*

Bergquist, H., 75, *98*

Berk, M., 260, 268, *272, 468, 482*

Berkenstam, A., 167, *178*

Berkman, J., 377, *389*

Berkovic, S., 94, *100*

Berkovitz, G. D., 377, *390*

Berlan, M., 341, *369*

Berman, I., 164, *176*

Bermingham, N. A., 555, 558, 559, *561*

Bern, H. A., 500, *513*

Bernad, A., 205, *206*

Bernard, B., 335, 337, 340, *358*

Bernard, D., 171, 173, *176*

Bernasconi, S., 506, *515*

Berney, C., 321, *326*

Bernier, G., 521, 522, 525, 533, *535, 536*

Berns, A., 376, *390*

Bernstein, A., 189, *190,* 193, 203, 204, *206, 208, 209,* 217, 224, *230, 231, 232,* 347, 352, *360*

Bernstein, H. S., 169, *175*

Bernstein, S. E., 195, *209*

Berrih-Aknin, S., 271, *277*

Berry, V., 533, *535*

Berta, P., 377, 381, 388, *390, 393*

Berthelsen, J., 443, 449, 478, *482*

Berthier, R., 226, *228*

Bertini, V., 382, 387, 388, *390*

Bertonocello, I., 194, *208*

Bertuzzi, S., 508, *517*

Bertwistle, D., 343, *361*

Berul, C. I., 343, 350, 351, 352, *359*

Besecke, A., 522, 527, 532, *535*

Bessarab, D. A., 91, *103*

Betsholtz, C., 173, *178,* 313, *326,* 413, 414, *418,* 467, *482, 489*

Bettenhausen, B., 135, *144*

Betts, D. H., 156, *177*

Betz, U. A., 341, *362*

Bever, M. M., 547, *561, 563*

Beverdam, A., 476, 477, 478, *491*

Bevilacqua, A., 10, 14, 15, *16*

Bevolenta, P., 522, *537*

Beyer, L., 557, *561*

Beylin, A., 237, *250*

Beyna, M., 474, *489*

Bezstarosti, K., 343, 345, 348, *366*

Bhadra, M. P., 14, *16*

Bhadra, U., 14, *16*

Bhatt, H., 156, *175, 179*

Bhatt, S., 133, *149*

Bhattacharya, S. S., 334, *365,* 533, *535*

Bhattacharyya, S., 44, *52,* 338, *366*

Bhide, P. G., 87, *98, 104*

Bhimani, M., 217, *232*

Bhujwalla, Z. M., 225, *231*

Bi, W., 281, 282, 289, *292,* 380, *389*

Biancalana, V., 506, *517*

Bianchi, L. M., 549, 559, *561, 562*

Bianchi, M. E., 9, *18*

Bianco, C., 338, *367*

Biben, C., 31, *34,* 35, 40, 44, *50,* 135, *144, 323,* 334, 338, 341, 343, 345, 347, 348, 349, 350, 351, 352, 353, 355, 356, 357, *358, 359, 362, 366*

Bickenbach, J. R., 467, *490,* 580, 581, 582, 583, *584, 586, 587*

Bickmore, W. A., 374, 375, *389, 391,* 399, *416, 419*

Biddolph, S., 415, *420*

Bielinska, M., 45, *50,* 160, 164, 165, *175, 178, 179, 323,* 341, 343, 352, *366*

Bier, J., 452, *498*

Bierke-Nelson, D., 478, *495,* 507, *517*

Bierle, B. M., 8, *17*

Bierwolf, D., 453, *482*

Biery, N. J., 347, *360*

Biesecker, L. G., 470, *487*

Biffali, E., 311, *324, 330*

Bigby, M., 204, *210*

Biggs, W. III, 286, *292*

Bikle, D. D., 580, 583, *588*

Bikoff, E. K., 31, *35,* 303, *329,* 467, 468, *495, 497*

Bilbao, J. R., 506, *514*

Billia, P., 318, *327*

Billiards, S., 3, *4,* 422, 427, *482*

Billington, W. D., 166, *177*

Binder, M., 201, 204, *209, 210,* 220, *231, 232, 233*

Bingle, C. D., 313, *325*

Binns, L. T., 15, *16*

Birchler, J. A., 14, *16*

Birchmeier, C., 173, *175,* 260, 266, *272, 273,* 317, *323, 328,* 349, 353, 356, 357, *361, 365, 366,* 404, *417, 419*

Bird, L. M., 357, *361*

Birgbauer, E., 118, *122*

Birk, O. S., 376, *389*

Birkmeier, T. S., 511, *513*

Birren, B., 380, *392*

Birren, B. W., 136, 137, *147*

Birren, S. J., 168, *177,* 239, *250*

Bisaha, J. G., 341, 345, *358, 364*

Bisgrove, B. W., 47, *50,* 214, *230*

Bishop, C. E., 378, 380, 387, *389, 390*

Bishop, J. M., 86, *105,* 219, *230,* 422, 432, 449, 450, 453, 466, 475, *497*

Bissell, M. J., 432, *482*

Bitgood, M. J., 238, *249,* 384, *389,* 470, *482,* 577, 578, *584*

Bitner-Glindzicz, M., 398, *416,* 550, *561*

Bitoun, P., 478, *495,* 507, *517*

Bittner, A., 12, *17*

Bittner, D., 40, *50,* 477, *482*
Bixler, L. S., 188, 189, *190*
Bjork-Eriksson, T., 237, 247, 248, *249*
Björklund, A., 88, 89, *98, 103,* 241, 244, 247, *249*
Bjornson, C. R., 206, *206,* 246, *249, 250*
Black, B. L., 289, *292,* 334, 341, 342, 346, *358, 365*
Blackwell, T. K., 243, *249,* 257, *272, 277*
Bladt, F., 173, *175,* 260, 266, *272, 273,* 317, *323, 328,* 404, *417, 419*
Blagden, C. S., 265, 269, *272*
Blakemore, C., 92, *102*
Blanar, M. A., 168, 169, *175,* 225, *230,* 349, *360*
Blanc, L., 267, *274*
Blanchard, H., 307, *329*
Blanco, M. J., 355, *359,* 423, *482*
Bland, A. E., 467, *488*
Blanquet, V., 109, *123,* 228, *230,* 522, 527, *535,* 549, *562*
Blasi, F., 162, *179,* 443, 449, 478, 479, *482, 484*
Blau, H. M., 269, *274*
Blaumueller, C. M., 136, *144*
Blecker, K. S., 532, *535*
Bleckmann, S. C., 479, *487*
Bleil, J. D., 14, *16*
Blelloch, R., 340, *358*
Blendy, J. A., 314, *324*
Blessing, M., 82, *106,* 203, *210,* 439, 467, *498*
Bliek, J., 174, *174*
Blin, N., 316, *323*
Blitz, I., 579, *587*
Blitz, J., 264, *274*
Bliven, T., 91, *97*
Blixt, A., 527, 530, 532, *535*
Bloch-Gallego, E., 114, *124*
Bloch-Zupan, A., 451, 453, 468, 469, *482, 490*
Blokland, E. A., 193, *207*
Blondeau, F., 343, *361*
Bloockfor, F. R., 511, *514, 515*
Bloom, F. E., 511, *514*
Bloomfield, T. S., 109, *125*
Blouin, M. J., 317, *324*
Blue, J. D., 22, *33*
Blum, M., 40, *50,* 351, 353, 354, 355, 356, *359, 361,* 432, 473, 477, *482, 485, 498*
Blumberg, B., 40, *50, 323,* 354, *367,* 477, *482, 483*
Blumberg, P. M., *585*
Blumenstock, C., 44, *50, 325,* 338, *361*
Blundell, T. L., 452, 466, *482*
Blüthmann, H., 317, *325*
Boak, B. B., 226, *230*
Boast, S., 205, *208*
Bober, E., 85, *98, 103,* 132, *147,* 258, 263, 264, 266, *272, 276,* 290, *292,* 450, 453, 472, 473, 475, *481, 482,* 547, 549, 550, *561, 562, 564*
Bocian, M., 286, *294*

Bock, D., *327*
Bodine, D. M., 87, 91, *103,* 196, 202, *210,* 508, *517,* 521, 527, *537*
Bodmer, R., 334, 335, 336, 337, 341, *358, 366, 370*
Bodner, M., 509, *513*
Boerm, M., 172, 173, *180*
Boettinger, E., 43, *50*
Boggs, S. S., 246, *251*
Bogler, O., 136, *148,* 262, *276*
Bogue, C. W., 312, 316, *323, 324*
Boheler, K. R., 343, *366*
Bohlen, P., 227, *233*
Bohli, H., 115, *125*
Bohne, B. A., 286, *292,* 558, *561*
Boivin, G. P., 434, 453, 467, *493, 494*
Boizet, B., 381, *390*
Boland, P., 224, *230,* 559, *561*
Bolcato-Bellemin, A. L., 452, 453, *482, 484*
Bolender, D. L., 337, *364*
Bolon, B., 312, *327*
Bomford, A., 343, *361*
Bomsel-Helmreich, O., 22, *33*
Bonadio, J., 443, *482*
Bonafede, A., 266, *276*
Bonaïti-Pellie, C., 453, *489*
Bonaventure, J., 285, 286, *293, 294,* 452, 453, *484, 494*
Boncinelli, E., 40, 46, *52, 83,* 92, *100, 102, 103, 104,* 114, *122,* 376, *393,* 478, *495*
Bond, R. A., 341, 352, *362*
Boneko, V. M., 570, *584*
Bonhoeffer, F., 525, *535*
Boni, J., 506, *517*
Boni, U. D., 247, *252*
Bonifas, J. M., 287, *293,* 579, 583, *586, 588, 589*
Bonifer, C., 194, *207*
Bonini, N., 188, *189*
Boni-Schnetzler, M., 506, *517*
Bonnamy, J. P., 289, *294*
Bonnano, J. B., 11, *16*
Bonneaud, N., 381, *390*
Bonnerj-Weir, S., 322, *328*
Bonnerot, C., 9, *16*
Bonnet, D., 357, *364*
Bono, F., 225, 226, *229*
Bono, M. R., 387, *392*
Bontoux, M., 140, 141, *147*
Boon, L. M., 227, *232*
Borello, U., 140, 141, *144, 148,* 246, *250,* 263, 264, *277*
Borg, J. P., 94, *105*
Borges, A. C., 31, *33,* 44, *50,* 338, *358*
Borges, M., 313, *323*
Borghese, E., 26, *33*
Bories, J. C., 204, *210,* 220, *233*
Borok, Z., 313, *324*

Boroni, F., 511, *516*
Borrelli, E., 91, *101, 104,* 509, 511, *513, 517*
Borsani, G., 382, 387, 388, *390,* 525, *535*
Borycki, A. G., 140, 141, *144,* 260, 264, 265, *272*
Bosco, L., 534, *535*
Bosher, J. M., 14, 15, *16*
Bosl, M. R., 170, 173, *179*
Bos-Mikich, A., 8, *16*
Boss, G. W., 341, 352, *367*
Bossard, P., 307, 316, 317, *323, 325*
Bosse, A., 550, *561*
Bosse, B., 83, *98,* 470, *482*
Bossé, R., 511, *513*
Bossolasco, M., 376, *393*
Bostrom, H., 173, *178,* 313, *323,* 467, *482*
Boswell, H. S., 188, *189*
Botchkarev, N. V., 577, 583, *584*
Botchkarev, V. A., 577, 578, 583, *584, 588, 589*
Bothwell, M., 558, 559, *564, 565*
Botkin, S. J., 412, *420*
Botquin, V., 160, *175*
Boucher, D., 470, *497*
Bouillet, P., 40, 47, *51*
Boujard, D., 116, *125*
Bouniol-Baly, C., 9, 14, *15*
Bounpheng, R., 339, *358*
Bour, B. A., 341, *358*
Bourcheix, L. M., 200, *209*
Bourgarel, P., 386, *390*
Bourgeois, P., 452, 453, *482, 484*
Bourne, H. R., 512, *516*
Bourouis, M., 343, *360*
Bourrat, F., 116, *125*
Boutros, M. C., 243, *249*
Bouwmeester, T., 40, 44, *49, 51, 52, 323,* 338, *359, 366*
Bove, E. L., 356, *362*
Boven, K., 398, *416,* 550, *561*
Bovolenta, P., 525, *535,* 546, *562*
Bowden, D. H., 313, *322*
Bowen, W. C., 246, *251*
Bowen-Pope, D. F., 467, *495*
Bowers, C. W., 226, *230*
Bowers, P., 357, *363*
Bowers, P. N., 351, 357, *359*
Bowles, J., 281, *293,* 377, 381, *389, 391, 392*
Bowman, A. W., 174, *177*
Boyadjieva, N., 504, 511, *515*
Boyce, B. F., 442, *484*
Boylan, C., 91, *98*
Boyle, M., 188, *189*
Boynton, S., *251*
Bozic, D., 408, *418*
Bozzi, Y., 511, *517*
Bracilovic, A., 82, 86, *100*

Bradamante, Z., 432, *482*
Braddock, S. R., 398, *419*
Bradfield, C. A., 225, *231*
Bradford, A. P., 505, *517*
Bradley, A., 31, *34,* 43, 45, *51,* 82, *106,* 110, *123,* 137, 138, *145,* 250, 260, 267, *273, 274,* 287, *294,* 318, *326,* 336, 339, 353, 357, *364, 370,* 408, *418,* 428, 439, 441, 442, 443, 451, 453, 467, 468, 470, 471, 477, *490, 491, 495, 498,* 506, 510, *516, 518,* 571, 581, 583, *587*
Brady, G., 193, *207*
Brady, H., 504, 510, *516*
Brady, J. P., 528, 533, *535*
Brady, R. O., 94, *103*
Braisted, J. E., 92, 94, *98*
Brajeul, V., 200, *208*
Brakebusch, C., 574, 583, *584*
Brambaugh, J., 7, *18*
Bramford, R. N., 357, *359*
Brand, M., 115, *125,* 337, *367,* 546, *565*
Brand, T., 336, 340, *358, 367*
Brandli, A. W., 225, *229,* 546, *562*
Brandon, E. P., 237, *250*
Brand-Saberi, B., 130, 131, 132, 140, *144, 145, 146,* 255, 260, 263, 264, 266, *272, 273, 276, 277*
Brandt, M. L., 307, *329*
Brannan, C. I., 188, *189,* 341, 352, *359*
Brannon, M., 579, *584*
Brash, D. E., 287, *292,* 579, *586*
Bratt, E., 400, *417*
Braun, T., 85, *98,* 140, *144,* 258, 260, 266, *272, 273, 274, 275, 276,* 549, *562*
Bray, S., 555, *561*
Braye, F., *585*
Brdoigan, T., 341, *360*
Bréant, C., 140, 141, 142, *147,* 200, *207,* 263, 264, 266, *276*
Breckenridge, R. A., 355, *359*
Breems, D. A., 193, *207*
Brehm, A., 183, *190*
Breier, G., 203, *207,* 223, 226, 228, *228, 230*
Breillat, C., 165, *174,* 303, *323*
Breitbart, H., 8, *18*
Breitbart, R. E., 349, 355, *358*
Breitman, M. L., 203, *209,* 217, 218, 223, 224, *229, 230, 232,* 528, 531, *536*
Bremer, K. A., 136, *149*
Brennan, C., 239, *249*
Brenner, S., 117, *124*
Brenner-Morton, S., 88, *99,* 120, 121, *125*
Brent, G., 415, *418*
Breuning, M. H., 414, *418*
Breviario, F., 226, 228, *229*
Brewer, G., 11, *17*
Brewitt, B., 532, *535*

Brezun, J. M., 247, *249*

Briand, J.-P., 348, *360*

Briata, P., 48, *49,* 114, *122,* 353, 354, 355, *364,* 433, 478, *489*

Bricarelli, F. D., 281, *293, 294,* 380, 381, *391, 393*

Brickell, P., 140, 141, 142, *147,* 263, 264, 266, *276,* 468, *482*

Brickell, P. D., 238, *251*

Brickell, P. M., 201, *208,* 467, *484*

Brickman, J. M., 31, *33,* 47, *50,* 451, 476, *483,* 508, *514*

Bridgewater, L. C., 282, 284, *292*

Brieher, W. M., 408, *420*

Brigande, J. V., 548, 550, 551, *561*

Brigman, K., 166, *179*

Brill, G., 141, *144*

Brilliant, M. H., 6, *17*

Bringas, P., 228, 229, 246, *249,* 312, *327*

Bringas, P., Jr., 314, 315, *329,* 426, 428, *482, 491*

Brinkley, L., 443, *482*

Brinkman, A. O., 386, *393*

Brinkmann, V., 317, *328,* 404, *419*

Brinkmeier, M. L., 507, 508, 511, 512, *514, 518*

Brinster, R. L., 318, *328,* 374, 385, *389,* 511, *513, 516*

Brion, C. M., 443, *493*

Briscoe, J., 80, 82, 88, *98, 99,* 474, *494*

Bristow, J., 219, *229,* 334, *360*

Britsch, S., 353, 356, *361*

Brivanlou, I. H., 203, *208,* 303, *325*

Broadbent, J., 338, *362*

Broadbent, V., 506, *514*

Broadus, A. E., 285, *294*

Brocard, J., 109, 117, *122, 123*

Broccoli, V., 83, *100,* 114, *122,* 376, *393*

Brodianski, V. M., 341, *369*

Brodie, J., 119, *124,* 544, *563*

Brody, J. S., 311, 312, *323*

Brogley, M., 337, *364*

Brohmann, H., 260, 266, *273*

Broide, R. S., 502, *518*

Broihier, H., 185, 186, 188, *190*

Bronchain, O., 342, 343, *359*

Broncinelli, E., 422, 427, 433, *487*

Bronckers, A. L., 408, *418,* 451, 467, *490*

Brönner, G., 469, *483*

Bronner-Fraser, M., 121, *123, 125,* 141, *147,* 237, 243, 244, *249, 252,* 262, 263, 264, 265, *275,* 355, 356, *361,* 422, 423, 424, 425, 426, 427, 435, 470, 475, *481, 482, 484, 489, 490, 494, 495, 496,* 543, 546, *562*

Bronson, R. T., 6, *17,* 94, *98, 105,* 136, 137, *147,* 204, *209,* 220, *232,* 285, 288, *293,* 318, *323, 325,* 442, 443, 453, *483, 488*

Brook, A., 470, *486*

Brooke, B. S., 226, *230*

Brooke, J. D., 357, *364*

Brooke, N. M., 305, *323*

Brookes, A. J., 416, *419*

Brookhiser, W. T., 408, 409, 410, *417*

Brooks, D. M., 244, *252*

Brooks, P. C., 223, *228*

Broom, R., 442, *482*

Brose, K., 193, *207*

Brou, C., 136, *146*

Brouwer, A., 451, 476, 477, 478, *491, 496,* 549, *565*

Brouwers, B., 264, *274*

Brower, D., 23, *33*

Brown, A. G., 530, 533, *536*

Brown, A. M., 30, 31, *35,* 262, 264, 265, *272, 277,* 304, 316, *329,* 405, *417*

Brown, B., 349, *360,* 582, *585*

Brown, D., 137, 138, *145,* 250, 267, *273,* 341, 346, 349, 350, 352, *368,* 451, 473, *483, 495*

Brown, E. A., 26, *33*

Brown, E. S., 247, *249*

Brown, G. A., 376, *393*

Brown, J., 321, *326*

Brown, J. D., 405, *418*

Brown, K. D., 442, *484*

Brown, K. S., 424, 442, 443, *482, 493*

Brown, M., 500, 504, 506, 509, *517*

Brown, M. C., 398, 399, *420,* 527, 530, 533, *538,* 550, *565*

Brown, N. A., 134, 135, 138, 139, *146,* 334, 348, 351, 352, 353, 354, 356, *359, 363, 367, 369*

Brown, N. L., 526, 527, *535*

Brown, S., 509, *514*

Brown, S. D., 556, *565*

Brown, S. D. M., 560, *561*

Browne, C. P., 204, *207*

Brownell, W. E., 559, *563*

Bruce, A. E., 14, *17*

Brucker-Davis, F., 509, *514*

Bruckner-Tuderman, L., 573, *584*

Brueckner, M., 351, 352, 354, *359, 368*

Bruening, W., 375, *392,* 416, *419*

Bruhn, L., 440, *497,* 583, *589*

Brulet, P., 31, *33,* 44, 46, 48, *49, 52,* 94, *106,* 114, *122,* 205, *207,* 322, 428, 451, 478, *481,* 508, *513,* 521, 527, *535*

Bruneau, B. G., 339, 343, 345, 347, 348, 349, 350, 351, 352, 353, 357, *358, 359*

Brunet, A., 9, *18*

Brunet, J. F., 239, *250,* 434, 435, *481*

Brunet, L. J., 156, *175, 179,* 468, *482*

Brunet, S., 7, *18*

Brunjes, P. C., 237, *249*

Brunk, B., 141, *144,* 194, 206, *209,* 267, *274*

Brunner, H. G., 525, *535*

Brunner, M., 470, *484*

Bruns, G. A. P., 415, *417*

Brunt, S. L., 432, *486*

Brusselmans, K., 225, *229*
Brüstle, O., 88, *98*
Bruzzone, R., 346, *358*
Bryce, D. M., 217, 227, *229*
Bryson, L., 262, *273,* 476, *483*
Bu, D., 314, *325, 327,* 434, 467, *487*
Bubunenko, M., 6, *17*
Bucana, C., 172, 173, *180,* 342, 350, 352, *364*
Buch, C., 162, 170, *177*
Buchanan, J. A., 438, 454, *490*
Buchberg, A. M., 341, 352, *359*
Buchert, M., 79, *104,* 453, *486*
Buck, C. A., 162, 170, *177,* 200, 201, *209,* 219, *230*
Buckert, M., 529, *537*
Buckingham, M., 131, 132, 140, 141, *144, 145, 147, 148,*
      258, 260, 263, 264, 268, *272, 273, 276, 277,* 341, 343,
      344, 345, 356, *361, 362*
Buckingham, M. E., 140, 141, *148,* 258, 267, 269, *274,*
      275, 344, *362*
Buckler, A. J., 357, *364,* 374, 375, *392,* 399, 415, *417, 419*
Budnik, V., 334, *361*
Buehr, M., 189, *189,* 382, *389, 391*
Buffinger, N., 140, *144,* 263, *273*
Buj-Bello, A., 402, *420*
Bulfone, A., 76, 77, 79, 82, 83, 84, 85, 87, 88, 89, 90, 91,
      92, 95, *97, 98, 99, 103, 104,* 382, 387, 388, *390,* 422,
      437, 450, 451, 470, 471, 472, *481, 493,* 525, *535*
Bulfone-Paus, S., *588*
Bullejos, M., *389*
Bullen, P. J., 357, *364,* 387, *390,* 568, *584*
Bullock, S. L., 407, *417*
Bulotsky, M. S., 136, 137, *147*
Bumcrot, D. A., *105,* 140, 141, *145, 147,* 262, 264, *275,*
      470, *490*
Bundman, D. S., 580, 581, 582, 583, *584, 586, 587*
Bundy, J., 350, *370*
Bunton, T., 353, 356, *367*
Burbach, J. P., 506, *513*
Burch, J. B. E., 316, *326,* 336, 339, 342, 344, *360, 362, 367*
Burd, G. D., 77, 84, *98*
Burdine, R. D., 338, 353, 354, 355, 356, 357, *359, 370*
Burgeson, R. E., 573, *584*
Burgess, R., 135.138.137, *145,* 250, 267, *273*
Bürglin, T. R., 479, *482*
Burgos, M., 388, *391*
Burgoyne, P. S., 372, 377, 378, 379, 383, 386, 387, *389,*
      392, 393
Burhan, A. M., 86, *104*
Burke, A. C., 129, 142, *145,* 305, *328*
Burke, D., 452, 466, *482*
Burke, R., 577, *587*
Burki, K., 321, *326*
Burley, S. K., 11, *16,* 525, *538*
Burling, A., 26, *35*

Burmeister, M., 527, *535*
Burn, J., 357, *361*
Burnham, W. M., 247, *252*
Burris, T. P., 379, *389*
Burrows, H. L., 502, 511, *513*
Burt, D. W., 388, *392,* 534, *536*
Burton, F. J., 511, *514*
Burton, P. B. J., 341, *359*
Burtscher, H., 282, *295*
Buscher, D., 83, *98,* 470, *482,* 580, 581, 583, *587*
Bush, S. M., 334, 335, 336, 337, 339, *365*
Bushweller, J. H., 204, *210,* 220, *233*
Busslinger, M., 115, *125,* 204, *209,* 428, 475, *497,* 533, *535*
Bussoli, T. J., 546, *561*
Butcher, R. L., 22, *33*
Butler, C. M., 372, 386, *389*
Butler, K., 303, *330*
Butler, P. M., 436, 456, *482*
Butler-Brown, G., 344, *362*
Buttitta, L. A., 141, *147*
Butzler, C., 173, *175*
Buxton, J. A., 227, *231*
Buxton, P., 422, 426, 449, *487*
Buyukmumcu, M., 573, *584*
Buyuksal, I., 47, *51*
Buzney, E. A., 193, 206, *207,* 246, *250,* 270, 271, *274*
Byerly, K. A., 398, *419*
Byrd, C. A., 77, 84, *98*
Byrne, C., 440, *498,* 568, 571, 572, 577, 580, 582, 583,
      584, 586, 587, 589
Byrne, M. C., 318, *328,* 583, *587*
Byskov, A. G., 373, *390*

C

Cable, J., 560, *561*
Cabral, L. J., 313, *324*
Cabrera, C. V., 257, *276*
Cacheiro, N. L. A., 22, *34*
Cacheux, V., 508, *516*
Cachianes, G., 223, *230*
Cadalbert, L., 583, *586*
Cagan, R., 263, *275*
Cai, C., 336, *368*
Cai, J., 528, *535*
Cai, L., 245, *249*
Cai, N., 91, *97*
Cai, Y., 240, *250,* 414, 415, *418, 419, 420*
Cai, Z. L., 201, 204, *207*
Caille, I., 236, 237, 244, 245, 248, *249*
Caillol, D., 303, *324*
Cain, K. T., 22, *34*
Cairns, L., 185, *189*

Calafat, J., 583, *589*

Calarco, P. G., 23, 26, *33*

Calautti, E., 580, *585*

Calfton, M., 219, *231*

Call, K. M., 415, *417*

Callaerts, P., 533, *536*

Callahan, R., 226, *230*

Calley, J., 185, *189, 190*

Callihan, B., 263, *274*

Calo, C., 442, 443, *494*

Calvari, C., 382, 387, 388, *390*

Calvert, J. T., 227, *232*

Calvet, J. P., 414, *417*

Calvio, C., 400, *417*

Calvo, B., 506, *514*

Calvo, K. R., 479, *488*

Calvo, W., 195, *208*

Camakaris, J., 443, *482*

Cambon, B., 271, *273*

Camerino, G., 372, 378, 379, 380, 382, 384, 387, 388, *390, 391, 393*

Cameron, H. A., 247, *249*

Campbell, D., 287, *293*, 504, *513*

Campbell, I. K., 31, *34*, 115, *124*

Campbell, K., 82, 83, 84, 88, 89, 91, *98, 103, 105, 106,* 110, 114, 120, *125,* 161, *179,* 241, 244, *249,* 428, 453, *491, 497*

Campbell, M., 354, *359*

Camper, S., 6, *19,* 85, *106,* 353, 354, 355, *361,* 378, 379, *393,* 433, 478, *485,* 502, 507, 508, 509, 511, 512, *513, 514, 515, 518,* 556, 557, *561, 564*

Campion, E., 14, *15*

Campione, M., 341, 345, 347, 348, 350, 351, 354, 355, 356, *359, 361*

Campos-Barros, A., 511, 512, *518*

Campos-Ortega, J. A., 134, 136, 138, *145, 147, 148, 149,* 553, *561*

Campuzano, S., 555, *561*

Camus, A., 2, 3, *4,* 40, 44, *50,* 132, *148,* 422, 427, *482, 496*

Canals, J. M., 91, *98*

Candi, E., 570, 581, *584, 588*

Candia, A. F., 135, *145,* 260, 267, *275*

Caniggia, I., 173, 174, *175*

Canipari, R., 86, *103,* 188, *190*

Cann, G., 266, *276*

Canning, D. R., 303, *329*

Cano, A., 355, *359,* 423, *482*

Cantley, L. C., 227, *232*

Cantor, A. B., 343, *369*

Cao, L., 350, *370*

Cao, Q., 10, 11, *18*

Cao, W. H., 432, 450, 465, *489*

Cao, X., 346, *364*

Cao, Y., 558, *564*

Capco, D. G., 8, 13, *16, 17, 18,* 27, *33, 34*

Capdevila, J., 338, 354, 356, 357, *359*

Capdevilla, J., 83, *98*

Capecchi, M. R., 117, 119, *125,* 166, *179,* 305, 311, *327,* 427, 428, 443, 466, 471, 474, *483, 490, 496,* 544, 545, *561, 564*

Capehart, A. A., 350, *365*

Capel, B., 373, 377, 378, 381, 382, 383, 387, *389,* 390, 391, *392, 393*

Capgras, S., 9, *16*

Caplen, N. J., 14, *16*

Caprioli, A., 199, 201, *207,* 1987

Capron, F., 94, *99*

Caras, I., 84, *102*

Carbajo-Perez, E., 511, *514*

Cardell, E. L., 453, 467, *494*

Cardiff, R. D., 171, *175*

Cardin-Girard, J. F., 171, 173, *176*

Cardoso, W. V., 311, 312, 313, 315, *323, 326, 327, 329*

Carey, C., 510, *513*

Carey, J. C., 478, *495,* 507, *517*

Carey, M. L., 475, *481*

Carey, R. M., 412, *420*

Caric, D., 87, 94, *98, 105*

Carl, M., 116, *125*

Carl, T. F., 423, *482*

Carlisle, C., 378, *390, 392*

Carlo, A. D., 188, 189, *190,* 508, *513*

Carlson, B. M., 395, *417*

Carlson, G. J., 451, 453, 474, *489*

Carlson, L., 442, *484*

Carlsson, B., 167, *178*

Carlsson, L., 320, *325,* 409, *419*

Carlsson, P., 479, *487,* 527, 530, 532, *535*

Carlton, M. B., 7, *16,* 160, 167, *178,* 203, *210,* 453, *482*

Carmeli, C., 141, *144*

Carmeliet, P., 166, *176,* 203, *207,* 223, 225, 226, *228, 229,* 321, *325*

Carney, P. R., 542, 559, *561*

Carnicero, E., 543, 545, 546, *565*

Carninci, P., 11, 12, *18*

Caron, M. G., 341, 352, *362,* 511, *513*

Carpenter, E. M., 545, *561*

Carraway, K. L., 227, *232*

Carre-Eusebe, D., 372, 387, *392*

Carrière, C., 506, 508, 509, *518*

Carrion, D., 353, 357, *365*

Carroll, P. M., 15, *18*

Carroll, S. B., 471, *493*

Carroll, T. J., 395, *420*

Carrozzo, R., 94, *103*

Carson, D. D., 156, *175*

Carson, J. L., 38, *52*

Carter, J. M., 527, 530, 531, *537*

Carter, W. J., 268, 269, *274*
Carthew, R. W., 14, 15, *17*
Carver, B., 161, *178*
Carver-Moore, K., 203, *207*, 223, *229*, 287, *293*, 402, *418*
Casagranda, F., 266, *275*
Casares, F., 91, *103*
Casarosa, S., 83, 88, 89, 90, 95, *98, 99, 105*
Casci, T., 466, *482*
Cascio, S., 165, *174*, 303, *323*
Cases, S., 161, *175*
Casey, B., 47, *51*, 357, *359, 361, 363*
Casey, E. S., 303, *323, 329*
Casiano, D. E., 376, *389*
Caskey, C. T., 94, *103*
Caspary, T., 168, *176*
Casper, R. F., 13, *17*
Cassidy, R. M., 91, *101*
Cassie, C., 243, *249*
Cassini, A., 26, *33*
Castagne, J., 512, *514*
Castano, L., 506, *514*
Casteilla, L., 271, *273*
Castilla, L. H., 226, 227, *233*
Castillo, A. I., 510, *516*
Castillo, M. M., 345, 348, *360*
Castrillo, J.-L., 509, *513, 514*
Catala, M., 39, 40, *50*, 200, *209*
Cate, R. L., 374, 385, *389*
Cattanach, B. M., 378, 387, *390*
Caudy, M., 257, *276*
Caveda, L., 226, *228*
Cavenee, W., 415, *417*
Caviness, V. S., Jr., 87, 92, 94, *98, 103, 105*
Cayrol, C., 303, *324*
Cecchi, C., 92, *102*
Cecchini, M. G., 443, *485*
Celeste, A. J., 338, *363*, 467, *498*
Censullo, P., 121, *124*
Centanni, J. M., 7, *19*
Cepko, C. L., 87, 88, *98, 100*, 243, *252*, 343, 345, 348, 349, 350, 351, 352, *358, 359*, 521, 525, 527, *535, 537, 538*
Cereghini, S., 165, *174*, 303, *323*
Cerretti, D. P., 91, *106*
Chada, K., 188, *190*
Chadwick, R. B., 509, *515*
Chae, S.-S., 227, 230, *231*
Chae, T., 94, *98*
Chai, N., 23, *33*, 166, 167, *175*
Chai, Y., 228, 229, 246, *249*, 426, *482*
Chakravarti, A., 479, *482*
Chalepakis, G., 113, *125*
Challice, C. E., 345, *369*
Chamberlain, C. G., 532, *537*
Chamberland, M., 478, *489*

Chambers, A. E., 262, *275*
Chambers, I., 23, *34*, 160, 166, *178*
Chambon, P., 40, 47, *51*, 91, *101, 104*, 109, 117, 118, 119, *122, 123, 125*, 172, 173, *179*, 305, 310, 312, *327, 329*, 341, 346, 347, 352, 355, 357, *359, 362, 366, 368, 369*, 427, 433, 443, 449, 451, 453, 468, 469, 470, 474, 479, *484, 485, 489, 490, 491, 492, 494*, 525, *536*, 544, 545, 557, *562, 563, 564*
Champigny, C., 167, *179*
Chan, E. F., 579, *584*
Chan, S., 117, 118, *124*
Chan, S. K., 478, *490*
Chan, W., 164, *175*, 302, *324*
Chan, Y., 452, *494*
Chan, Y. M., 553, *561*
Chandra, S., 443, 449, 479, *495*
Chang, B., 136, *149*
Chang, C., 11, *17, 101*, 525, *538, 585*
Chang, C. P., 476, *482*
Chang, C. Y., 318, *326*
Chang, D., *101*, 120, *125*
Chang, D. K., 378, *392*
Chang, D. T., 141, *145*, 287, *292*
Chang, H., 226, 227, *229*, 336, 352, *359*
Chang, H. H., 451, 475, *488*
Chang, L., 87, *104*
Chang, W., 550, *561*, 580, 583, *588*
Channon, K. M., 227, *231*
Chapman, D. L., 132, 133, 135, 137, *145*
Chapman, V. M., 157, *175, 178*
Chapouton, P., 83, *98*
Charachon, R., 398, *416*, 550, *561*
Chareonvit, S., 422, 424, 435, 437, 475, 476, *492*
Charite, J., 342, 344, 349, *359, 365*
Charles, R., 348, *360*
Charlieu, J. P., *391*, 416, *419*
Charnas, L. R., 453, *487*
Charnay, P., 91, *100*, 118, 119, *124, 125*, 135, *147*, 468, *492*, 510, *518*, 546, *564, 565*
Charpentier, E., 578, *585*
Charron, F., 342, 343, *359, 360, 366*
Charron, J., 171, 173, *176*, 318, *324*
Charron, R., 342, *366*
Chastant, S., 9, *15*
Chastre, E., 271, *277*
Chat, M., 552, *564*
Chatterjee, B., 522, 527, *535*, 549, *562*
Chatterjee, V. K., 311, *324*
Chau, T. C., 222, *231*
Chaussain, J. L., 376, *391*
Chauvet, J., 500, *513*
Chawengsaksophak, K., 160, 167, *174, 175*, 305, 307, *323*
Chazaud, C., 40, 47, *51*, 346, 347, 355, *359*

Cheah, K. S., 281, 283, *293,* 380, 381, *389, 391, 392,* 531, *536*

Checa, N., 91, *98*

Cheema, S. S., 87, 91, *103,* 508, *517,* 521, 527, *537*

Chegini, N., 47, *51*

Chelly, J., 94, *99*

Chem L., 164, *177*

Chen, A., 31, *33,* 40, 45, *50,* 338, *360*

Chen, C. M., 169, *175, 177*

Chen, C. T., 91, *103*

Chen, C. W., 577, *585, 589*

Chen, C.-Y., 342, 343, *359, 367*

Chen, D., 268, *275,* 378, *390*

Chen, D. Y., 227, *229*

Chen, E. Y., 441, *488*

Chen, F., 227, *229*

Chen, G.-Q., 346, *364*

Chen, H., 203, *207,* 223, *229,* 313, *323,* 479, *482, 488*

Chen, J., 90, 95, *106,* 203, *209,* 313, *323,* 341, 342, 343, 352, *359, 362, 363,* 510, *518*

Chen, J.-N., 339, 351, 355, *359, 368*

Chen, L., 286, *292, 293*

Chen, L.-H., 577, 583, *584*

Chen, P., 556, 558, *561*

Chen, Q. J., 291, *292*

Chen, R., 509, *515,* 530, 533, *535*

Chen, S.-H., 7, *19,* 79, 83, 89, 90, 92, 95, *98, 102, 103*

Chen, S. L., 334, 342, 343, *359, 361,* 473, 474, *489,* 507, *514*

Chen, W. S., 164, 165, *175, 177, 179,* 185, *189,* 303, 316, *323, 329,* 453, *487*

Chen, X., 290, *292,* 479, *482*

Chen, Y., 167, *174, 176,* 437, 440, *482, 498*

Chen, Z., 322, *326,* 341, 348, 349, 352, *359, 368,* 435, *490,* 559, *563*

Chen, Z. F., 473, *482*

Chen, Z.-F., 79, *98,* 218, 225, *229,* 232, 243, *249*

Chen, Z. Y., 556, 558, *565*

Chenevix-Trench, G., 287, *292,* 470, *486*

Cheng, A. M., 47, *50,* 78, *104,* 171, 173, *175, 178*

Cheng, C., 338, *361,* 580, *586*

Cheng, H. T., 433, 476, 477, 478, *489*

Cheng, L., 188, 189, *190*

Cheng, P. F., 257, 258, *273, 277,* 349, *362*

Cheng, S., 45, *50,* 78, *100*

Cheng, S. H., 121, *124,* 451, 453, 470, 471, *491*

Cheng, S. Y., 343, *360*

Cheng, Y., 186, *190*

Chenleng, C., 336, 339, 342, *368*

Chenn, A., 92, 94, *98*

Chepelinsky, A. B., 528, *537*

Cheraud, Y., 267, *274*

Cheresh, D. A., 223, *228, 229*

Chernoff, G. F., 79, *100*

Cheshier, S. H., 245, *249*

Chetty, R., 415, *420*

Cheung, A., 468, *495*

Chhabra, A., 452, *492*

Chi, C. L., 466, *491*

Chi, D.-C., 136, 137, *147*

Chi, M. M., 269, *274*

Chia, W., 240, *250*

Chiang, C., 43, *50,* 78, 80, 85, 86, *98, 101,* 120, 121, *123, 125,* 141, *144, 145,* 260, 265, *273,* 287, 292, 312, 319, *323, 326,* 439, 451, 470, *482,* 521, *535*

Chiang, L. C., 453, *486*

Chiannikulchai, N., 341, *369*

Chida, K., 580, *587, 588*

Chida, S., 22, *33*

Chidambaram, A., 287, *292,* 470, *485, 486*

Chien, K. R., 172, 173, *174,* 339, 341, 342, 343, 344, 345, 352, 356, 357, *358, 359, 361, 362, 363, 366, 367, 368*

Chien, Y. H., 192, 194, *207,* 245, *250*

Chik, K. W., 121, *124,* 451, 453, 470, 471, *491*

Chikamori, M., 79, *99*

Childs, G. V., 512, *514, 515*

Chin, A., 351, 355, *359*

Chin, E. R., 269, *273, 277,* 342, *369*

Chin, J. R., 223, *233,* 465, *491*

Chin, M. T., 347, *359*

Chin, N., 384, 385, *393*

Chin, W. W., 469, *498*

Ching, A., 6, *17*

Chiovato, L., 311, *326*

Chiquoine, A. D., 183, *189*

Chirgwin, J. M., 321, *328*

Chisaka, O., 408, *417,* 427, 443, 474, *483,* 544, 545, *561*

Chisholm, J. C., 23, 26, *34,* 156, *175*

Chiumello, G., 378, 387, *389*

Cho, E. A., 408, 409, 410, *417*

Cho, E. S., 402, *417*

Cho, J., 91, *97,* 286, *294*

Cho, K. W., 40, *50,* 323, 477, *482, 483*

Cho, K.-Y., 579, *587*

Choe, S., 354, *367*

Choi, C. Y., 334, 343, *361*

Choi, J., 263, *274*

Choi, K., 200, *207,* 217, *232*

Choi, T., 443, *495*

Cholly, B., 117, *122*

Chong, J. A., 243, *249*

Choo, D., 469, 478, *483, 491,* 541, 542, 550, 551, 552, 555, 558, 559, *564, 579*

Choudhary, K., 88, *98*

Choudhry, A., 341, *369*

Chourrout, D., 116, *125*

Chow, L. T., 412, *417*

Chow, M., 140, *144*

Chow, R. L., 533, *535*

Chowdhury, K., 311, 321, *327, 329,* 398, *417,* 451, 453, 475, *490, 495, 496,* 527, 530, 532, *538*

Chowdury, K., 183, *190*

Choy, B., 243, *249*

Christ, B., 128, 130, 131, 132, 140, *144, 145, 146, 147,* 254, 255, 263, 264, 266, *272, 273, 276, 277*

Christensen, E., 504, *518*

Christensen, L., 422, 437, 440, 450, 451, 471, 472, *484, 493*

Christiano, A. M., 573, *584*

Christians, E., 9, *19*

Christiansen, J., 281, *294,* 470, *485*

Christiansen, J. H., 172, 173, *177,* 336, 337, *365*

Christie, S., 416, *419*

Christoffels, V. M., 341, 343, 345, 347, 348, 350, 351, 355, 356, *359, 366*

Christofori, G., 322, *324*

Christy, B. A., 349, *358*

Chu, K., 502, *514*

Chu, M., 559, *563*

Chua, S. C., Jr., 443, *497*

Chua-Couzens, J., 91, *101*

Chuang, P. T., 578, 583, *585*

Chui, D. H., 202, *209*

Chun, Y. K., 354, *365*

Chung, P. T., 315, *324*

Chung, U. I., 285, 287, *292, 293*

Chuong, C. M., 571, 574, 577, 578, 579, *585, 586, 587, 589*

Church, D. M., 286, *294*

Chwoe, K. Y., 341, *360*

Chyung, J. H., 282, *293*

Ciau-Uitz, A., 198, 200, *207*

Ciemerych, M. A., 27, *33*

Ciesiolka, T., 183, *190*

Ciliberto, G., 318, *324*

Cinnamon, Y., 131, 132, 140, *145, 146,* 255, 256, *273, 275*

Cirillo, L. A., 316, *324*

Ciruna, B. G., 133, *145,* 167, *175,* 303, 317, *324*

Claesson-Welsh, L., 504, *515*

Clapp, D. W., 243, 244, *249*

Clark, I., 185, *189*

Clark, J. I., 532, *535*

Clark, R. A., 223, *228,* 287, *293*

Clark, W. R., 319, *328*

Clarke, A. R., 416, *419,* 441, *488*

Clarke, D. L., 236, 237, 244, 245, 246, 248, *249, 250,* 318, *324*

Clarke, J. D., 422, 426, 449, *487*

Clarke, L. A., 452, 453, *481*

Clarke, T. R., 381, 382, *393*

Clary, D. O., 91, *101*

Clausen, J. A., 87, 94, *105*

Clauss, I., 318, *328*

Claxton, D. F., 202, *208*

Claxton, J. H., 577, *585*

Cleary, M. L., 443, 449, 476, 479, 482, *487, 495*

Cleaver, O. B., 348, *359*

Clemente, H. S., 200, 201, *209,* 220, *232*

Clement-Jones, M., 387, *390*

Clements, D., 166, *176,* 303, *325*

Clements, M., 31, *34,* 46, 47, 48, *51,* 311, 316, *327*

Clendenin, C., 341, 343, 352, *368*

Clevers, H., 92, *100,* 172, *175*

Clifford, J., 469, *491*

Clifford, V., 379, 387, *393*

Clifton-Bligh, R. J., 311, *324*

Climent, S., 337, *359*

Clore, G. M., 377, *393*

Closson, V., 226, *230*

Clotman, F., 226, 229, 321, *326*

Clout, D. E. W., 341, 345, 347, 348, 350, *359*

Clouthier, D. E., 432, 465, 478, *483, 498*

Coates, M. I., 436, *495*

Cobos, I., 84, 85, *102,* 116, *124,* 428, *490*

Cobourne, M., 440, *494*

Cobrinik, D., *483*

Cochran, J., 347, *364*

Cockell, M., 321, *324*

Cockroft, D. L., 29, 32, *33, 34,* 156, 157, *175, 176*

Coerver, K. A., 44, *51*

Coffey, R. J., 169, *179*

Coffin, J. D., 286, *293,* 453, *483*

Coffin-Collins, P. A., 465, *483*

Coffinier, C., 165, *175*

Cogan, J. D., 509, *517*

Cogliati, T., 376, *389*

Cogswell, C. A., 239, *250*

Cohan, C. S., 549, *561*

Cohen, A., 237, *252*

Cohen, D., 405, *417*

Cohen, J., 23, 27, *34, 35,* 84, *101,* 451, 475, *486,* 527, 528, 530, *536*

Cohen, J. H., 319, *329*

Cohen, L. E., 509, *514*

Cohen, M. M., 442, *485*

Cohen, R. I., 167, 171, 173, *179*

Cohen, S., 465, *483*

Cohen, S. M., 471, *483*

Cohen-Gould, L., 341, *362*

Cohen-Tannoudji, M., 84, *98, 100,* 386, 387, *390,* 546, *564, 565*

Cohn, D. H., 286, *294*

Cohn, S. A., 454, *483*

Colbert, M. C., 469, *483*

Cole, T. J., 314, *324*

Cole, W., 443, *495*

Cole, W. G., 288, *293*

Coleman, J. R., 136, *146*, 316, 317, *325*
Coletta, M., 141, *144*, 206, *207*, 246, *249, 250*, 271, *273, 274*
Colledge, W. H., 7, *16*, 160, 167, *178*, 203, *210*, 453, *482*
Collen, D., 166, *176*, 203, *207*, 223, 225, 226, *228, 229*, 321, *325*
Coller, B. S., 223, *230*
Coller, J. M., 11, *16*
Colligan, J., 377, *390*
Collignon, J., 43, 44, 45, 46, *50*, 354, 355, 357, *359*, 386, 387, *390*, 468, *497*, 529, 530, 531, 534, *536*
Collins, A. E., 92, *102*
Collins, C. C., 347, *364*
Collins, F. S., 202, *208*, 347, *366*, 528, *538*
Collins, N. S., 260, 267, *275*
Collins, S., 316, *325*
Collins, T., 343, 352, *360*
Collu, R., 512, *514*
Colman, D. R., 408, *419*
Colmenares, C., 260, 268, *272, 273*, 468, *482*
Coltey, M., 140, 141, 142, *147*, 263, 264, 266, *276*
Coltey, P. M., 128, *145*, 422, 423, 426, 427, 433, *483*, 572, *585*
Colvin, J. S., 286, *292, 293*, 313, 314, 315, *324*, 558, *561*
Comley, M., 415, *420*
Comoglio, P. M., 266, *275*
Compain, S., 398, *416*, 550, *561*
Compernolle, V., 226, *229*
Complton, J. G., 570, *584*
Compton, D., 224, *230, 231*
Concordet, J. P., 78, *101*, 262, 268, *276*
Conkie, D., 317, *327*
Conley, P. B., 164, *177*
Conlon, F. L., 45, 47, *50*, 133, 137, *145*, 338, 355, *359*, 468, *483*
Conlon, R. A., 31, *33*, 42, 46, *49*, 116, *122*, 135, 136, 137, 139, *145, 148, 149, 323*, 468, 478, *481, 483*, 509, *514*
Conne, B., 10, *18*
Conner, A. S., 193, *207*
Conner, D. A., 343, 350, 351, 352, 353, 357, *359*
Conner, J. M., 91, *97*
Conners, J. R., 321, *328*
Connolly, D., 264, *272*
Connolly, D. J., 45, *52*
Connors, H., 246, *250*, 318, *326*
Conover, J. C., 549, 559, *561*
Consalez, G. G., 525, *535*
Consortium, T. E. P. K. D., 414, *417*
Constam, D. B., *50*
Constanti, A., 91, *102*
Contreras, L. N., 512, *515*
Conway, S. J., 350, *370*, 475, *483*
Conway-Campbell, J., 281, *293*, 381, *392*
Cook, D. L., 356, *359*

Cook, D. M., 400, *418*
Cook, G., 225, *229*
Cook, M., 422, 427, 433, *487*
Cook, S. A., 94, *105*, 550, *563*
Cook, T. L., 529, *538*
Cooke, J., 45, *52*, 139, 142, *145, 146*, 185, 188, *189*, 262, 263, 264, *272, 274*, 354, 355, 356, *362, 364, 366*, 423, *492*
Cookingham, J., 442, 443, *494*
Cooper, B., 336, 339, 342, *368*
Cooper, E. C., 94, *100*
Cooper, G. M., 7, 14, *16, 18*
Cooper, J. A., 94, *100, 101, 104*
Cooper, L., 281, *294*, 377, *389*
Cooper, M. K., 78, *98*, 470, *483*
Copeland, N. G., 78, 79, 81, 85, *99, 100, 103*, 168, 169, 173, *175, 176, 179*, 193, 204, *208*, 321, *328*, 341, 342, 345, 348, 349, 352, *359, 360, 365, 366*, 411, *418*, 443, 467, 476, *482, 485, 488*, 508, *515*, 521, 525, 533, *536, 537, 538*, 556, *565*
Copeland, T., 7, *18*, 183, 184, 185, *189*
Copp, A. J., 156, 170, *175*, 183, *189*, 350, *362*, 475, *483, 486*
Copperman, A. B., 174, *178*
Coppola, V., 559, *562*
Corballis, M. C., 351, 356, *360*
Corbel, C., 201, *209*
Corbi, N., 160, *180*
Corbin, J., 170, *174*
Corbin, V., 136, *145, 148*
Corden, J. L., 43, *50*, 78, 80, 85, 86, *98*, 121, *123*, 141, *145*, 260, 265, *273*, 319, *323*, 439, 451, 470, *482*, 521, *535*
Cordes, S. P., 118, *124*, 527, 530, 531, *537*, 544, *562*
Coré, N., 376, *389, 390*, 453, *481*
Corfas, G., 94, *104*, 343, *369*
Corley, L. S., 471, *493*
Cormier, F., 198, 200, 201, *207*
Cornelison, D. D., 271, *273*
Cornett, C. V., 22, *34*
Corotto, F. S., 237, *249*
Corrales, J. D., 338, 353, 354, 356, 357, *370*
Correia, K. M., 509, *514*
Corsi, A. K., 341, *360*
Corsin, J., 432, 435, *483*
Corte, G., 48, *49*, 78, 81, *101*, 110, 114, 121, *122, 124*, 478, *481*, 508, *513*
Cortes, F., 201, *208*
Corwin, J. T., 553, 555, 557, *562, 563*
Cossu, G., 131, 140, 141, *144, 145, 148, 149*, 206, *207*, 246, *249, 250*, 258, 260, 263, 264, 268, 271, *272, 273, 274, 277*
Costa, M., 263, *274*
Costa, R. H., 82, *102*, 313, 316, 321, *324, 328, 330*
Costa, T., 347, *364*

Costantini, F., 204, *209,* 402, *419*
Costantini, L. C., 91, *98*
Cota, G., 511, *514*
Cotanche, D. A., 553, *562*
Cotsarelis, G., 570, 574, 578, *585, 588*
Cottrill, C., 357, *358*
Coucouvanis, E., 45, *50*
Coughlin, S. R., 169, *175*
Coulling, M., 157, *177*
Coulombe, A., 271, *277*
Coulombe, P. A., 570, *587*
Coulombel, L., 194, 200, 201, *207, 209,* 220, *232*
Coulombre, A. J., 519, *535*
Coulombre, J. L., 519, *535*
Couly, G. F., 77, 84, 85, *98,* 128, *145,* 422, 423, 426, 427, 433, 435, *483,* 502, 507, *514,* 569, 571, 572, *585, 587*
Cousins, F. M., 203, *207,* 226, *229,* 467, *484*
Couso, J. P., 136, *145*
Coutinho, A., 196, *207*
Couvineau, A., 285, *293*
Cova, L., 235, 242, 244, *250*
Coward, P., 378, *390*
Cowin, P., 578, *585*
Cox, D. R., 287, *293,* 579, *586*
Cox, R., 269, *275*
Crabtree, G. R., 164, *177*
Crackower, M. A., 121, *124,* 451, 453, 470, 471, *491*
Craggs, G., 10, *17*
Craig, A. W., 11, *16*
Craig, C. G., 235, 236, 244, 245, *251*
Craig, S., 206, *209*
Crair, M. C., 92, *106*
Crandall, E. D., 313, *324*
Cranley, R. E., 424, 442, 443, *482*
Cremer, H., 322, *324*
Crenshaw, E. B. I., 509, *514*
Crenshaw, E. B. III, 506, 509, *516, 517,* 549, *564*
Cressman, D. E., 318, *324*
Cresswell, L. A., 500, *513*
Crews, S., 85, *99,* 225, *229*
Crisp, M. S., 311, *324*
Crispino, J. D., 343, 352, *360*
Critcher, R., 380, *393*
Critser, E. S., 26, *34*
Croisille, L., 194, *207*
Croissant, J. D., 342, *359, 360, 369*
Crompton, T., 288, *294,* 442, 443, 453, *493*
Crosby, J. L., 40, 43, *49*
Cross, A. J., 247, *249*
Cross, H., 507, *515*
Cross, J. C., 163, 168, 169, 170, 171, 173, 174, *174, 175, 176, 177, 178, 179,* 349, 350, 352, 355, *360, 367*
Crossin, K. L., 134, *148*
Crossley, J. M., 40, *52,* 115, 116, *123, 125*

Crossley, P. H., 84, 85, *99, 102,* 113, 116, *123, 124,* 132, *145,* 317, *324,* 337, 340, *360,* 428, 449, 466, *483, 490*
Croucher, S. J., 435, *483*
Crowe, D. L., 314, 315, *329*
Crowe, R., 577, *585*
Crowley, D., 223, *228, 230*
Crowley, W. F., 500, *518*
Crozet, F., 134, 135, 138, 139, *146*
Cruaud, C., 398, *416,* 550, *561*
Crumine, D., 580, 583, *588*
Crumrine, D., 580, 581, *585, 586*
Crystal, R G., 579, *588*
Cserjesi, P., *102,* 135, *145,* 262, 267, *273,* 342, 349, 350, *360, 362, 368,* 451, 473, 476, *483, 495*
Csete, M., 239, 241, *251*
Cubadda, Y., 343, *360, 361*
Cudennec, C. A., 201, *207*
Cuenca, A. E., 525, *537*
Cuesta, R., 11, *17*
Cui, S., 313, *328,* 413, *419*
Cui, Y., 11, 12, *17*
Cullander, C., 580, 582, *585, 587*
Cullinan, E. B., 9, 10, *17*
Cumano, A., 193, 194, 195, 196, 199, *207,* 220, *229*
Cunha, G. R., 301, *326*
Cunliffe, H. E., 398, *419*
Cunniff, K., 204, *207*
Cunningham, J. M., 411, *419*
Cunningham, M. L., 468, *485*
Cupples, R., 402, *417*
Curran, T., 94, *99, 103, 104,* 313, *327*
Currie, P. D., 264, 265, 266, 269, *272, 273, 276*
Curtis, A. R. J., 357, *364*
Cusella-De Angelis, G., 206, *207,* 246, *249,* 263, 271, *274, 277*
Cusella De Angelis, M. G., 246, *250,* 271, *273*
Cusmai, R., 94, *99*
Custer, R. P., 245, *249*
Cutforth, T., 90, 92, 95, *98*
Cuzin, F., *391*
Cvekl, A., 532, *535*
Cybulsky, M. I., 162, 170, *176*
Czernichow, P., 321, *327,* 507, *514*
Czerny, T., 533, *535*

D

Dabeva, M. S., 318, *324*
Dabovic, B., 376, *393*
Dacic, S., 452, *484*
da Costa, L. T., 86, *100,* 579, *586*
Daegelen, D., 262, 268, *276*
D'Agati, V., 402, *419*
Dahl, E., 348, *360,* 398, *417,* 475, *483, 498*

Dahl, N., 506, *517*
Dahm, L. M., 226, *230*
Dahmane, N., 86, *99, 579, 585*
Dahmann, C., 112, *123*
Dai, D., 426, *495*
Dai, P., 121, *123*
Dai, W., 352, 354, *366*
Daigle, N., 31, *33, 46, 49, 451, 478, 481,* 521, 527, *535*
Daikoku, S., 79, *99*
Dailey, L., 160, *180,* 283, *292*
Dakodis, A., 341, *367*
Dalcq, A. M., 21, 23, 27, *33*
Dale, B., 23, *34*
Dale, B. A., 575, *586*
Dale, E., 15, *17*
Dale, J. K., 78, 81, *99,* 134, 138, 139, *147*
Dale, L., 203, *207,* 239, *249*
Dale, T., 440, *494*
Dallman, M. F., 512, *515*
Dallos, P., 540, *562*
Daly, T. J., 224, *231*
Damay, M. D., 353, 357, *365*
Damert, A., 228, *230*
D'Amico, M., 86, *104,* 579, *588*
D'Amico-Martel, A., 434, *483,* 558, *562*
D'Amore, P. A., 226, *230*
Damour, O., *585*
Danielian, P. S., 84, *101,* 116, 117, *123, 124, 489,* 577, 578, 583, *589*
Daniels, K. J., 161, *179*
Danielson, K. G., 572, 583, *585*
Danilchik, M., 44, *50*
Danilenko, D. M., 312, *327*
Danks, D. M., 443, *482*
Danos, M. C., 356, *364*
Danpure, C. J., 6, *16*
Danse, J. M., 453, *482*
Danto, S. I., 313, *324*
D'Apice, M. R., 40, 46, *52,* 79, *104,* 478, *495*
D'Arcangelo, G., 94, *99, 104*
Dardik, F. B., 341, 343, 352, *368*
D'Argati, C., 415, *420*
Darimont, B. D., 379, *390*
Darling, S., 189, *189*
Darnell, J. E., Jr., 164, 165, *175, 177, 179,* 303, 316, *323, 324, 329*
Das, G. D., 237, 248, *248*
Das, S. K., 156, *177*
Dasen, J. S., 85, *99,* 506, 508, 509, 510, *514, 518*
DasGupta, R., 577, 579, *586*
Dashner, R., 282, *294,* 465, *492*
Dassule, H. R., 439, 440, 441, *483,* 577, 578, 583, *585, 589*
Daston, G. P., 161, *178*
Daszuta, A., 247, *249*

Date, I., 247, *249*
Dathe, V., 266, *276*
Datson, N. A., 478, *495,* 507, *517*
Dattani, M. T., 31, *33,* 47, *50,* 451, 476, *483,* 500, 506, 508, 509, *514*
Dattatreyamurty, B., 416, *420*
Daubas, P., 263, 264, *273*
Dausman, J., 427, *484*
Davenne, M., 117, 119, *123*
Davey, A., 22, *34*
David, D. J., 453, 466, 474, *498*
Davidson, B. P., 3, *4, 50,* 422, 427, *482*
Davidson, D., 84, 86, 94, *99, 100, 104,* 399, *419,* 474, *483,* 574, 575, 576, 577, *585*
Davidson, D. R., 435, 451, 468, 475, 476, *485, 491,* 521, 522, 527, 530, *536,* 560, *565*
Davidson, E. H., 6, *16,* 21, *33*
Davies, A. M., 183, *190,* 402, *420,* 434, 435, *489, 497*
Davies, J. A., 396, 403, 405, *417, 419*
Davies, P. J. A., 469, *484*
Davies, R. C., 400, *417*
Davies, T. C., 170, *176*
Davies, T. J., 156, *176*
Davis, A. M., 235, 236, 237, 245, *249,* 336, *363*
Davis, C. A., 112, 113, 115, *123*
Davis, C. L., 344, *360*
Davis, D. L., 342, *360*
Davis, E. C., 226, *230*
Davis, F. C., 527, *535*
Davis, H., 540, *562*
Davis, M. B., 86, *102*
Davis, N. M., 85, *102,* 339, 343, 347, *359*
Davis, R. J., 87, *101,* 257, 258, 268, *274, 277,* 399, *417*
Davis, R. L., 257, 258, *273, 275, 277,* 512, *516*
Davis, S., 224, *229, 231, 232*
Davis, V., 162, 170, *176*
Davis, W., Jr., 9, 11, 12, *16*
Davisson, M. T., 94, *105*
Dawid, I. B., 78, *103,* 374, *390*
Dawson, K., 169, 170, 173, *174, 178,* 349, 355, *367*
De, A., 504, *513*
Deacon, T. W., 89, *99*
Deal, C., 512, *514*
de Amo, F. F., 553, *565*
Dean, J., 14, 15, *16*
Dean, M., 287, *292,* 470, *485, 486,* 579, *586*
Dean, W. L., 9, *18,* 26, *33*
Deanehan, J. K., 194, 206, *209*
De Angelis, L., 271, *273*
DeAngelis, R. A., 318, *324*
Dearden, N. M., 504, *514*
Debase, H. T., 321, *324*
de Beer, G., 421, 442, 449, 457, 471, *484*
Debey, P., 9, 14, *15*

Deblandre, G. A., 226, *232*

de Boer, P. A. J., 343, 345, 348, *366*

de Bold, A. J., 343, 350, 351, 352, *359*

de Bold, M. L. K., 343, 350, 351, 352, *359*

de Bruijn, M., 194, 199, 201, 204, *207*

de Bruijn, M. R. T. R., 196, 199, 200, 201, 202, *207*

De Caestecker, M. P., 339, *368*

de Carlos, J. A., 84, 90, 95, *99, 102, 105*

de Celis, J. F., 136, *145*

Dechesne, C. J., 556, *562*

DeChiara, T. M., 166, *175*, 292, *292*, 549, *561*

Decimo, D., 312, *327*

Décimo, D., 469, *485*

Declercq, C., 203, *207*, 223, *228*

Deconinck, A. E., 219, *232*, 341, 343, 352, *369*

de Crombrugghe, B., 79, *106*, 281, 282, 283, 284, 285, 287, 289, *292, 293, 294,* 295, 380, *389,* 476, *498,* 572, *589*

deCrombrugghe, B., 451, 453, 473, 476, *493, 498*

de Crombrugghe, B., 573, *588*

de Cuevas, M., 7, *18*

Deed, R. W., 10, *17,* 185, 188, *189*

Deerinck, T., 453, *486,* 580, 581, 583, *586*

Deerink, T., 509, *513*

de Fatima Bonaldo, M., *390*

De Felice, M., 311, *324, 326*

de Felici, M., 86, *103,* 188, 189, *189, 190*

Defize, L. H., 285, *293,* 442, 443, *489*

DeFrances, M. C., 318, *327*

Degen, J. L., 318, *328*

Degenhardt, K., 334, 348, *367*

Degenstein, L., 577, 579, *586*

Degnin, C., 267, *277*

De Grandi, A., 382, 387, 388, *390*

Deguchi, K., 204, *209,* 220, *232,* 288, *293,* 442, 443, 453, *488*

Deguchi, R., 8, *16*

deHaan, G., 245, *249*

DeHaan, R. L., 340, 343, 344, 351, *360, 367*

Dehart, D. B., 38, *52*

Dehbi, M., 399, *417*

Deimling, J., 313, *328,* 413, *419*

Deiner, M. S., 506, *514*

Deissler, K., 354, 355, *359*

Dejana, E., 226, *228,* 229

De Jesus-Escobar, J. M., 238, *252,* 468, *498,* 550, *561*

de Jong, F., 334, 341, 345, 347, 348, 350, *359, 360, 366*

de la Chapelle, A., 441, *488*

de la Cruz, J., 357, *359*

de la Cruz, M. V., 344, 345, 348, 355, *360*

De La Guardia, R. D., 388, *391*

Delaisse, J. M., 291, *292*

Delaney, S. J., 172, 173, *177,* 336, 337, *365*

Delannet, M., 424, *496*

de la Pena, J., 354, *367*

de la Pompa, J. L., 135, 136, 137, 139, *145, 147,* 341, *370,* 435, 468, *490, 495,* 509, *514,* 559, *563*

de La Porte, S., 271, *277*

De Larco, J. E., 465, *484*

Delassus, S., 194, 195, *207*

del Barco, J. L., 135, *145*

del Barco Barrantes, I., 136, 137, 139, *145,* 435, *490,* 559, *563*

del Barrio, M. G., 355, *359,* 423, *482*

Delbridge, M. L., 387, *390*

Delezoide, A. L., 286, *294*

Delfini, M., 262, 263, *273*

Delhase, M., 453, *486,* 580, 581, 583, *586*

DeLise, A. M., 280, *292*

Delius, H., 44, *50,* 325, *327*

DeLoia, J. A., 11, 12, *18*

Delorme, B., 348, *360*

Deltas, C. C., 414, *418*

Deltour, L., 320, *324*

De Luca, L. M., 346, *367, 589*

Delvin, E., 512, *514*

Demarchez, M., *585*

de Maximy, A. A., 315, *324*

DeMayo, F. J., 310, 315, 321, 322, *326, 327*

DeMello, D. E., 314, *324*

Demers, J., 163, *175*

de Meulemeester, M., 174, *174*

Demignon, J., 262, 268, *276*

De Moerlooze, L., 312, *324,* 505, *514,* 546, *562*

De Montellano, P. O., 582, *584*

de Moor, C. H., 10, 11, *18*

Den, J. M., 283, 284, 289, *294*

Denda, S., 410, *418*

Denef, C., 511, *517*

Denetclaw, W., Jr., 219, *229,* 255, *273,* 334, *360*

Deng, C., 167, 171, 173, *179,* 286, *292, 294,* 453, *484*

Deng, C.-X., 226, 227, *231, 233,* 286, *293,* 313, *329*

Deng, H., 580, 583, *588*

Deng, J. M., 40, 44, *52,* 281, 282, 289, *292, 328,* 380, *389*

Deng, K. Y., 402, 404, *418,* 453, 479, *488*

Deng, Y., 347, *364*

Dengler, T. J., 225, *231*

Denhez, F., 467, *491*

den Hollander, N. S., 285, *293*

Dennefeld, C., 305, *329*

Denning, M. F., *585*

Dennis, J. E., 341, *362*

Denson, L. A., 316, *324*

Denton, E. R., 507, *514*

Denyn, M. M. F. J., 334, *366*

Denzel, A., 288, *294,* 442, 443, 453, *493*

Deo, R. C., 11, *16*

Deol, M. S., 544, *562*

DePalma, G. E., 11, 12, *17*

DePamphilis, M. L., 5, 9, 10, *17, 19*
Depew, M. J., 79, 86, 92, *99, 100, 105,* 172, *175,* 290, *292,*
    422, 432, 435, 443, 449, 450, 451, 453, 466, 471, 472,
    473, 475, 477, 478, 479, *484, 495, 497,* 549, *562*
DePinho, R. A., 528, 532, *536, 537, 538*
de Ribaupierre, F., 555, *566*
De Robertis, E. M., 31, *33,* 39, 40, 44, 48, *49, 50, 51, 52,*
    78, *98,* 141, *144, 323,* 338, *358, 359, 366,* 387, *393,*
    432, 473, 477, *482, 483, 485, 498*
Derom, C., 22, *33*
Derom, R., 22, *33*
DeRose, M., 312, *327*
DeRosier, D. J., 557, *565*
DeRuiter, M. C., 333, *360*
de Ruiter, M. C., 226, *229*
Derynck, R., 226, *229,* 280, *292,* 465, *491*
Desai, N., 31, *33,* 40, 45, *50,* 338, *360*
Desai, S. Y., 260, 268, *272,* 468, *482*
de Santa Barbara, P., 381, *390*
De Santis, M., 338, *367*
De Santo, R., 40, *51*
de Sauvage, F., 402, *420*
Deschamps, J., 376, *393*
Deschet, K., 116, *125*
Desclozeaux, M., 381, *390*
Desguerre, I., 94, *99*
DeSimone, D. W., 424, *481*
Desmadryl, G., 556, *562*
Desmarquet, C., 118, *125*
De Smedt, V., 1, *4,* 8, *16*
De Sousa, P. A., 11, 12, *16*
De Souza, F. S., 2, *4*
de Souza, P., 452, *498*
des Portes, V., 94, *99*
D'Esposito, M., 290, *294,* 450, 471, 472, 473, *495*
Detrich, H. W. III, 213, *232*
Dettman, R. W., 219, *229,* 334, *360*
Deuchar, D. C., 354, *359*
Deutsch, G., 316, 320, *324*
Deutsch, U., 224, 228, *232,* 398, *417,* 475, *481, 484*
Devaney, E., 83, 86, *100*
de Vantéry, C., 7, *16*
de Vargas, C., 475, *485*
Deville, F.S.-S., 236, *249*
Devine, W. P., 94, *106*
de Viragh, P. A., 582, 583, *587*
Devon, K., 380, *392*
Devor, D. E., 173, *176*
Devoto, S. H., 265, 269, *273*
Devriend, T. K., 398, *419*
de Vries, W. N., 13, 15, *16, 18*
Dewar, K., 380, *392*
Dewerchin, M., 166, *176,* 225, 226, *229,* 321, *325*
Dewhirst, M. W., 227, *231*

de Wied, D., 500, *514*
de Winter, J. P., 385, *389*
de Wit, D., 465, *498*
Dexter, M., 185, 188, *189,* 194, *207*
Dey, C. R., 313, *330*
Dey, S. K., 156, *175, 177, 178*
Dhanasekaran, N., 12, 13, *18*
Dhouailly, D., 571, 572, 574, *585, 587*
Diamonti, A. J., 227, *232*
Dias, M. S., 40, *50,* 317, *328*
Diaz, E. M., 42, *51,* 158, *177*
Di Bonito, M., 95, *97,* 506, *513*
Di Carlo, A., 478, *481*
Dichoso, D., 341, *360*
Dickinson, M. E., 117, 121, *123,* 423, *484*
Dickman, E. D., 342, *368*
Dickson, C., 84, *102,* 312, *324,* 505, *514,* 545, 546, *562,*
    563, 565
Dickson, K. S., 10, *15*
Dickson, M. C., 203, *207,* 226, *229,* 467, *484*
DiColandrea, T., 581, *585*
Di Cunto, F., 580, *585*
Di Donna, S., 140, *145,* 258, 263, *273*
Diekwisch, T. G., 433, 476, 477, 478, *489*
Dierich, A., 40, 46, 47, *52,* 114, 117, *123, 125,* 310, 321,
    325, 427, 433, 435, 443, 449, 451, 453, 469, 474, 478,
    479, *484, 485, 489, 490, 491, 494,* 544, 545, *563,* 583,
    586
Dieteche, C. M., 2, *4*
Dieterlen-Lievre, F., 195, 196, 198, 199, 200, 201, *206,*
    207, 208, 209, 219, 220, *229, 232*
Dietrich, P., 122, *124*
Dietrich, S., 140, *145,* 260, 263, 264, 265, 266, *273, 277*
Dietz, H. C., 347, *360,* 453, *495*
Diez-Roux, G., 219, *229*
Dighe, A. S., 287, *293*
Digilio, M. C., 356, *365*
di Giovanna, J. J., 570, *584*
Dikkes, P., 94, *98,* 510, 512, *516*
Dikoff, E. K., 82, *105*
Di Lauro, R., 85, *103,* 311, *326, 330*
Dillehay, L. E., 225, *231*
Dillon, N., 202, *207*
Dilullo, C., 263, *274*
DiMario, J. X., 269, *273*
DiMario, S., 271, *273*
DiMattia, G. E., 509, *517*
Dimitrov, S., 14, *15*
Dimitrov, S., 14, *15*
DiNardo, S., 188, *189*
Dinchuk, J. E., 156, *177*
Ding, J., 31, *33,* 40, 45, *50,* 267, *272,* 338, 353, 354, 356,
    357, *360, 370,* 434, 467, *493*
Ding, Q., 121, *123, 124,* 312, *327*
Ding, X. Y., 579, *589*

DiPersio, C. M., 572, 574, *585*
Di Rocco, G., 334, 335, 336, 337, 339, *365*
Diver, M., 505, *513*
Dixon, J. E., 402, *420*
Dixon, K. E., 185, *190*
Djabali, M., 376, *389, 390,* 453, *481*
Djian, P., 582, 583, *585*
Djikeng, A., 15, *18*
Dlugosz, A. A., 169, *179,* 580, *585*
Do, M. S., 467, *492*
doAmaral, C., 292, *294*
Dobbs, L., 314, *327*
Dobi, A., 91, *103*
Dobson, C., 203, *210*
Dobyns, W. B., 94, *100, 103*
Dockter, J. L., 140, *145*
Dodd, J., 45, *52,* 78, 81, *99,* 403, *418*
Dodds, S. G., 349, *358*
Dodig, M., 452, *484*
Doe, C. Q., 340, *361,* 532, *537*
Doetsch, F., 95, *99,* 236, 237, 244, 245, 248, *249*
Doetschman, T., 434, 453, 467, *493, 494*
Doherty, D., 136, *145*
Doherty, P., 116, *125,* 425, *484*
Dohlman, H. G., 45, *50*
Dohse, M., 246, *250,* 318, *326*
Doi, H., 11, 12, *17,* 422, 423, 424, 425, 435, 437, 451, 475, 476, 478, *490, 492, 493,* 572, *588*
Doig, J., 416, *419*
Dolci, S., 188, *189*
Dollé, P., 40, 47, *51,* 117, *123,* 305, 310, *326, 329,* 341, 346, 347, 352, 355, *359, 366,* 427, 433, 443, 449, 450, 451, 453, 469, 470, 474, 479, *484, 485, 489, 490, 491, 492, 494,* 545, 557, *562, 563, 564*
Dolmetsch, R. E., 8, *16*
Domashenko, A. D., 10, *16*
Domeier, M. E., 165, *176,* 316, *325*
Domen, J., 376, *393*
Dominguez, M., 470, *484*
Dominguez-Steglich, M. A., 281, *292,* 380, *390*
Dommergues, M., 194, *207*
Donahoe, P. K., 47, *50,* 381, 382, 385, *390, 393*
Donaldson, D., 507, *514*
Dong, J., 174, *177*
Doniach, T., 44, *50,* 108, *123*
Donis-Keller, H., 321, *325*
Donlan, D. F., 443, *482*
Dono, R., 86, *99,* 338, *360*
Donoghue, D. J., 452, 453, 466, *485, 497*
Donoghue, M. J., 84, *99*
Donovan, M. J., 314, *326,* 410, *418*
Donovan, P. J., 7, *19,* 185, 188, 189, *189, 190,* 236, 246, *249*
Donoviel, D. B., 347, 352, *360*

Dony, C., 282, *295*
Dooher, G. B., 442, 443, *494*
Dor, Y., 225, *229*
Dorai, H., 416, *420*
Dorin, J., 451, 453, 474, *489*
Dorn, G. W. II, 453, *483*
Dorovkov, M., 15, *18*
Dortland, B., 201, 204, *207*
Doss, J. B., 292, *294*
dos Santos Silva, E., 316, *323*
Dotto, G. P., 580, *585*
Dou, C. L., 78, 81, 82, 86, *99*
Douglas, K. R., 502, *513*
Douglas, R., 206, *209*
Douglass, E. C., 398, *417*
Dove, L. F., 405, 407, *417, 419*
Dowd, M., 121, *123,* 203, *207,* 223, *229*
Dowhan, D. H., 342, 343, *359*
Dowling, J., 572, 583, *585, 586*
Dowling, J. E., 525, *537*
Downing, D. T., 581, 582, *585, 587, 589*
Downing, J. R., 201, 204, *207, 209,* 220, *231*
Doyen, A., 166, *178*
Drager, U. C., 345, 346, 347, *366, 367,* 469, *482, 492, 498,* 525, *537*
Drago, J., 87, 91, *103,* 402, *419,* 508, *517,* 521, 527, *537*
Drake, C. J., 223, *229*
Drake, D. P., 84, *97,* 435, 469, 476, *481*
Draper, B., 240, 243, *251*
Drescher, U., 525, *535*
Dreschler, M., 528, *538*
Dresser, D. W., 381, *390*
Dressler, G. R., 160, *178,* 373, 374, 385, *393,* 398, 399, 400, 402, 403, 408, 409, 410, *417, 418, 419, 420,* 475, *484, 493, 496*
Dreyer, J. L., 91, *106*
Driancourt, M.-A., 510, *518*
Driever, W., 395, *417,* 546, *563*
Drinkwater, C., 40, 44, *50, 323,* 338, *358*
Driscoll, M., 15, *18*
Driver, S. E., 14, *16*
Drolet, D. W., 502, *514*
Drossopoulou, G., 534, *536*
Drouin, J., 478, *481, 489,* 507, 508, 509, 510, *513, 514, 516, 517, 518*
Druck, T., 290, *294,* 450, 471, 472, 473, *495*
Druffel-Augustin, S., 92, *104*
Druker, B. J., *361*
Drum, H., 312, *327*
Drummond, I. A., 395, 415, *417, 418*
Dryburgh, L. C., 438, *484*
Drysdale, T. A., 339, 342, 347, *360, 362, 369*
D'Souza, R., 573, *588*
Du, D., 469, 470, *486*

Du, Y., 509, *517*
Duan, D. R., 227, *229*
Duband, J., 424, *496*
Dubois, P., 320, *324, 504, 514*
DuBois, R. N., 287, *293*
Duboule, D., 85, *103,* 305, 309, 310, *326, 330,* 450, *484*
Dubourg, C., 198, 199, 201, *207*
Ducy, P., 288, 289, *292, 294*
Dudley, A. T., 3, *4,* 82, *99, 105,* 171, *175,* 408, *417,* 451, 467, *484, 495,* 532, *538*
Dudley, C. F., 227, *229*
Duester, G., 525, *537*
Duff, R. S., 236, *249*
Dufort, D., 43, *50,* 158, 165, *175,* 303, 304, 316, *324*
Dufton, C., 423, *482*
Duglas-Tabor, Y., 528, 533, *535*
Duh, F. M., 227, *229*
Duke, M., 347, *360*
Dulac, C., 201, *208*
Dulac, O., 94, *99*
Dull, M., 440, 471, *488*
Dulor, J. P., 271, *273*
Duluc, I., 308, 310, *324*
Duman, R. S., 247, *249*
Duman, S., 573, *584*
Dumenil, D., 205, *207*
Dumont, D. J., 218, 223, 224, 228, *229, 230*
Duncan, A. M. V., 224, *229*
Duncan, M. K., 188, *190,* 532, *535, 538*
Duncan, R. P., 8, *16*
Duncan, S. A., 164, 165, *175, 177,* 302, 303, 316, 317, *323, 324, 326, 327,* 340, 343, 347, 352, *365*
Dunger, D. B., 506, *514*
Dunlop, L. L., *484*
Dunn, A. R., 194, *208*
Dunn, I., 388, *392*
Dunn, L. C., 442, 443, *494*
Dunn, M. A., 185, *190,* 243, *249*
Dunn, N. R., 183, *190,* 311, 312, 313, 314, 315, *325, 329,* 336, *363*
Dunn, R., 305, *329,* 453, 467, *496*
Dunne, C., 236, *252*
Dunner, S., 264, *274*
Dunwoodie, S. L., 135, *145, 324,* 339, 348, *360, 368*
Dupé, V., 91, *101,* 117, *123,* 451, 468, 469, *484, 485, 490,* 525, *536, 545, 562*
Dupin, E., 236, 237, 241, 244, *249, 250*
Duplan Perrat, F., *585*
Duprat, A. M., 303, *324*
Duprez, D., 140, 141, 142, *146, 147, 148,* 262, 263, 264, 266, *273, 274, 277,* 336, *358*
Dupuis, S., 90, 95, *106*
Durand, B., 469, *485*
Durbin, J. E., 286, *292*

Durham-Pierre, D., 6, *17*
Duriez, B., 508, *516*
Durocher, D., 342, 343, *360*
Duronio, R. J., 170, *175*
Durviaux, S. M., 166, *176,* 321, *325*
Dussel, R., 86, *99*
Dutt, P., 196, 202, *210*
Dutta, A., *251*
Duxson, M. J., 131, 132, *149*
Dwarki, V. J., 258, *277*
Dwivedi, R. S., 318, *328*
Dyer, R., 354, 355, *364,* 451, *489*
Dymecki, S. M., 109, 110, *123,* 228, *229*
Dyson, E., 341, 352, *368*
Dyson, J., 378, 387, *390*
Dyson, N., *483*
Dziadek, M., 314, *327*
Dzierzak, E., 192, 194, 195, 196, 199, 200, 201, 202, 204, *207, 208, 209,* 220, *231,* 317, *327*

E

Eagleson, G. W., 77, *99*
Eagleson, K. L., 84, *101*
Eales, N. B., 473, *484*
Easley, K., 582, 583, *585*
Easter, S. S., 85, *102*
Easter, S. S., Jr., 85, *102*
Eastman, Q., 579, *585*
Eatock, R. A., 555, 558, 559, *561, 563*
Eaves, C. J., 242, 245, *251, 252*
Ebendal, T., 91, *105*
Ebensperger, C., 140, *144, 145, 147,* 255, 263, *273*
Eberhardt, C., 203, *207,* 223, *228*
Eberspaecher, H., 281, *295,* 473, 476, *498,* 573, *588*
Ebron, M. T., 161, *178*
Eccles, M. R., 398, *419,* 453, *487*
Echelard, Y., 78, *99,* 121, *123,* 141, *145,* 470, *484*
Echenne, B., 94, *99*
Eckert, R. L., 581, *588*
Ecochard, V., 303, *324*
Economides, A. N., 44, *50,* 338, *362*
Economou, A., 377, *390*
Eddison, M., 553, 559, *561*
Eddy, E. M., 27, *33*
Ede, D. A., 128, 137, 139, *146,* 470, *484,* 534, *535,* 577, *585*
Edelberg, J. M., 219, *228, 229*
Edelman, G. M., 134, *148*
Edelmann, W., 43, *50,* 415, *420*
Edgar, B. A., 169, 170, *175*
Edlund, H., 319, 320, 321, *322, 323, 325, 326*
Edlund, T., 78, 79, 80, 81, 85, 87, *99, 104,* 108, 120, *123, 124,* 141, *148,* 319, 320, 321, *323, 325,* 504, 505, 506, *514*

Edmondson, D. G., 260, 262, *273, 274,* 333, 342, *360*
Edom-Vovard, F., 266, *273*
Edwards, J. G., 343, *360*
Edwards, M. C., 468, *485*
Edwards, R. G., 7, *16*
Ee, H. C., 321, *328,* 451, 475, *494*
Egarter, C., 174, *177*
Eggermont, E., 281, *293,* 381, *391*
Eglitis, M. A., 26, *33, 35,* 206, *207*
Eguchi, G., 525, 529, 531, 534, *535, 537*
Ehmke, H., 86, *99*
Ehrlich, J. S., 451, 476, *493*
Ehrlich, M. E., 91, *101,* 505, *517*
Ehrman, L. A., 339, *360*
Eichele, G., 287, *294,* 336, 342, 346, *360, 368, 370,* 469, 470, *486*
Eicher, E. M., 311, *323,* 378, 382, *389, 390, 392*
Eichmann, A., 200, *207, 208*
Eichmuller, S., 574, *588*
Eickholt, B. J., 425, *484*
Eidne, K. A., 512, *514*
Eiholzer, U., 509, *517*
Eimon, P. M., 44, *50,* 338, *362*
Eisen, J. S., 265, 269, *273*
Eisenbach, L., 467, *492*
Eisenberg, L. M., 339, *360*
Eisenman, R. N., 580, *586*
Eisenstat, D. D., 79, 87, 88, 89, 90, 91, 95, *97, 99,* 450, 470, 471, *481, 484*
Eizaguirre, J., 506, *514*
Eizuru, Y., 473, *498*
Ekblom, P., 407, 410, 411, *417, 418, 420*
Ekker, M., 89, *106,* 450, 470, 471, 473, *481, 496, 498,* 546, *561*
Ekker, S. C., 470, *489*
ElAmraoui, A., 504, *514*
El-Amraoui, A., 550, *563*
el Bahi, S., 476, 477, *496*
Elbaum, M., 94, *104*
Elchebly, M., 510, *514*
Eldadah, Z. A., 347, *360,* 453, *495*
El-Deiry, W. S., 341, *370*
Elder, F. F. B., 352, 354, *366*
Elefanty, A. G., 203, *209,* 343, *369*
Elenius, K., 156, *178*
Elger, M., 395, *417*
El Ghouzzi, V., 452, 453, *484*
Elia, A. J., 136, 137, 139, *145,* 341, *370,* 468, *495*
Elias, P. M., 570, 580, 581, 582, 583, *584, 585, 586, 587, 588*
Elinson, R. P., 335, *362*
Elkabes, S., 502, *514*
Elkahloun, A. G., 347, *366*
Elkins, J. A., 357, *358*
Elkins, R., 347, *360*

Elledge, S. J., 287, *294,* 468, *485,* 528, 532, *538*
Ellies, D. L., 450, 471, *496*
Elliot, D. A., 341, 348, 349, 356, *358, 362, 367*
Elliott, C., 173, *179*
Elliott, D. A., 348, 349, 352, 353, 357, *358*
Ellis, K., 15, *15*
Ellisman, M., 453, *486,* 509, *513,* 580, 581, 583, *586*
Elmer, E., 247, *249*
Elsemore, J., 79, *106,* 529, *538*
Elsholtz, H. P., 509, *515*
Elwood, N. J., 203, *209*
Emerson, C. P., Jr., 140, 141, *144,* 258, 260, 263, 264, 265, 267, *272, 274, 276*
Emmen, J. M., 386, *393*
Emmert-Buck, M. R., 173, *176*
Emmons, S., 185, *189*
Emonard, H., 573, *587*
Emoto, H., 311, 312, 313, 314, 315, *323, 325*
Emoto, N., 465, *498*
Emura, M., 315, *324*
Enderich, J., 170, 173, *179,* 238, *252*
Enders, A. C., 156, *175*
Enders, G., 188, *190*
Endo, N., 288, *293*
Enerback, S., 479, *487,* 527, 530, 532, *535*
Engel, J., 408, *418*
Engel, W., 386, 388, *392, 393*
Engelkamp, D., *391,* 528, *537*
Engels, L. J., 193, *207*
Engemann, S., 9, *18*
Engsig, M. T., 291, *292*
Enomoto, H., 289, *294*
Enomoto-Iwamoto, M., 289, *294*
Ensini, M., 109, *123*
Enver, T., 194, 199, 200, 201, *207, 210*
Enwright, J. F. III, 435, 469, 476, *484*
Enyeart-VanHouten, D. L., 381, *392*
Ephrussi, A., 185, *189*
Epifano, O., 14, *16*
Episkopou, V., 2, 4, 91, *102,* 527, 530, 531, *537*
Eppig, J. J., 8, 10, 11, 12, *17, 18*
Epstein, D. J., 78, 79, *99, 102,* 115, 120, 121, *123, 124,* 141, *145,* 428, 470, 475, *484, 491*
Epstein, E. H., 573, 579, 583, *585, 586, 588*
Epstein, E. H., Jr., 287, *293*
Epstein, J. A., 268, *273,* 353, 357, *365,* 528, *535*
Erdelyi, M., 185, *189*
Erdjument-Bromage, H., 405, 407, *416, 417*
Eren, R., 196, *207*
Erickson, C. A., 237, *249,* 423, 424, *484*
Erickson, R. L., 271, *276*
Erickson, R. P., 14, *15*
Ericson, J., 43, *50,* 78, 79, 80, 81, 82, 85, 88, *98, 99,* 120, *123,* 474, *494,* 504, 505, 506, *514*

Eriksson, P. S., 237, 247, 248, *249*
Erkman, L., 556, 558, *562*
Erlander, M. G., 12, *17*
Erlebacher, A., 280, *292*
Erler, T., 167, *174*
Ernfors, P., 559, *562*
Ernsberger, U., 238, *251*
Ernst, H., 350, *370*
Ernst, M. K., 7, *19*, 188, *189*
Ernst, P., 376, *390*
Ernst, R. E., 194, 206, *209*
Erven, A., 546, *565*
Erway, L. C., 550, *563*
Escalante-Alcalde, D., 43, *50*
Escary, J. L., 205, *207*
Esni, F., 322, *324*, 409, *419*
Essner, J. J., 47, *50*
Esteban, L., 171, 173, *178*
Esteve, P., 546, *562*
Esumi, E., 260, *276*
Esumi, H., 310, *327*
Etheridge, A., 351, *364*
Etkin, A., 318, *328*
Etkin, L. D., 8, *17*
Eto, K., 422, 424, 425, 435, 437, 438, 451, 453, 475, 476, *487, 492, 493*, 527, 530, *535*, 572, 588
Eto, Y., 321, *324*
Evans, H. M., 211, *229*
Evans, J. P., 7, *16*
Evans, M. J., 7, *16*, 26, 32, *35*, 158, 160, 166, 167, *178, 179*, 203, *210*, 303, 313, *324, 327*, 453, *482*
Evans, R. M., 172, 173, *174*, 341, 346, 352, 354, 357, *358, 363, 367, 368*, 468, 469, *489, 490*, 511, *513, 516*
Evans, S., 334, 336, 339, 342, *360, 361, 368*
Evans, T., 281, *294*, 316, *326*, 339, 342, 343, 346, 347, *361, 362, 363*
Evrard, Y. A., 136, 137, 138, *145*
Evsikov, S. V., 26, 27, *33*
Exalto, N., 174, *177*
Externbrink, A., 347, *363*
Ezer, S., 441, *488*
Ezine, S., 205, *207*
Ezrin, C., 511, *515*

F

Fabian, B. C., 337, *361*
Fabri, L., 40, 44, *50, 323*, 338, *358*
Faccenda, E., 512, *514*
Factor, S., 353, 357, *365*
Faerman, A., 258, 263, 267, *274*
Faessler, R., 183, 187, 188, *189*
Fagan, A. M., 91, *99*, 559, *562*
Fagerström, C., 91, *105*, 120, *125*

Fahrig, M., 203, *207*, 223, *228*
Faiella, A., 422, 427, 433, *487*
Fainzilber, M., 402, *420*
Fairclough, L., 303, *323, 329*
Fairén, A., 84, *105*
Falchetto, R., 226, *230*
Falck, P., 173, *178*
Fallon, J. F., 471, *493*
Fallon, J. R., 11, *19*, 121, *125*
Faloon, P., 217, *232*
Falzarano, P. R., 557, *563*
Familari, M., 14, *16*
Fan, C. M., 40, 44, *51*, 85, 95, *99, 102*, 140, 141, 142, *145, 147*, 225, 229, 260, 263, 264, 265, 266, *274, 275, 276*, 287, *292*, 506, *516*
Fan, H., 578, 579, *585*
Fan, M. J., 415, *418*
Fan, W., 226, *232*
Fan, X., 415, *418*, 525, *537*
Fancher, K. S., 15, *16*
Fang, G.-H., 218, *230*
Fang, J., 279, *292*, 427, *484*
Fang, M., 402, *417*
Fangman, J., 2, *4*
Fannon, A. M., 408, *419*
Fantes, J., *361*
Fantuzzi, G., 314, *324*
Faq, C.-M., 141, *147*
Faravelli, L., 235, 242, 244, *250*
Farese, R. V., Jr., 161, *175*
Farih-Sips, H., 285, *293*
Farinas, I., 91, 92, *100, 101*, 172, *175*, 239, *249*, 402, *418*, 440, 471, *488, 497*, 549, 559, *562*, 583, *589*
Farkash, Y., 510, *515*
Farlie, P. G., 453, *486*
Farrance, I. K., 262, *274, 277*
Farrell, A. P., 350, *360*
Farrell, M. J., 14, *17*
Farrington, S. M., 200, *206*
Fassler, R., 204, *207*, 284, *293*, 317, *324*, 427, *484*, 574, 583, *584, 585, 589*
Fatkin, D., 343, 350, 351, 352, 357, *358, 359*
Faust, C., 31, *34*
Faust, N., 194, *207*
Faustinella, F., 94, *103*
Fausto, N., 318, *324, 329*
Favier, B., *585*
Favor, J., 94, *104*, 451, 475, *486*, 522, 527, 528, 530, *535, 536*, 549, *562*
Fawcett, E., 442, 456, *484*
Fawcett, J. W., 134, *148*
Fazeli, A., 318, *325*
Fazleabas, A. T., 156, *175*
Featherstone, M., 118, *124*

Federici, R., 314, *330*

Fedtsova, N. G., 504, 505, *515*

Feger, G., 136, *145*

Fegert, P., 316, *323*

Fehon, R. G., 136, *145*

Feijen, A., 47, *50, 92, 106,* 467, *484*

Feingold, K. R., 580, 581, *584, 585, 586, 587*

Feingold, M., 507, *514*

Feinstein, S. C., 91, *98*

Fekany, K., 78, *104*

Fekete, C. N., 375, *389*

Fekete, D. M., 527, *535,* 547, 548, 550, 551, 552, 553, 560, *561, 562, 563*

Feldman, B., 166, *175,* 313, 314, 315, *324*

Feldmeyer, D., 92, *104*

Felice, M. D., 311, *326*

Felix, R., 443, *485,* 511, *514*

Fellous, M., 375, 376, 382, 387, *389, 391, 392*

Feneley, M., 348, 349, 352, 353, 357, *358*

Feng, X. H., 226, *229*

Fenner, M. H., 339, *368*

Ferguson, B., 441, *488, 491, 497*

Ferguson, C., 434, 437, 438, 440, 451, *484, 496*

Ferguson, E. L., 336, *370*

Ferguson, M. W. J., 434, 437, 441, 467, *490, 493, 496,* 568, 582, 583, *586*

Fernandez-Salguero, P., 311, 312, *326,* 474, *488*

Fernandez-Teran, M., 349, *360*

Fernaud-Espinosa, I., 525, *535*

Fernyak, S. E., 269, *273*

Fero, M. L., 558, *563*

Ferrante, E., 378, 387, *389*

Ferrara, N., 203, *207,* 223, *229, 230,* 291, *292*

Ferrari, C., 509, *517*

Ferrari, G., 206, *207,* 246, *249,* 271, *274*

Ferre-D'Amare, A. R., 525, *538*

Ferreira, V., 203, *207,* 223, *228*

Ferreiro, B., 77, *99*

Ferrero, G. B., 357, *361*

Ferretti, E., 443, 449, 478, 479, 482, *484*

Ferretti, P., 422, 426, 449, *487*

Ferri, G. L., 86, *103,* 188, *190*

Ferrier, M. L., 403, *418*

Fessler, L. I., 205, *209*

Fésüs, L., 582, *588*

Fetka, I., 173, *179*

Fetka, L., 428, 475, *497*

Fidler, I. J., 172, 173, *180*

Fielder, T. J., 286, *294*

Fiering, S., 157, *180*

Figdor, M. C., 75, *99*

Filosa, S., 40, 43, *50,* 477, *484*

Filvaroff, E. H., 280, *292*

Fine, J., 512, *513*

Fine, L. G., 404, *420*

Finegold, M. J., 44, *51,* 246, *250,* 315, 318, *326, 327,* 374, 385, *389*

Finger, J. N., 6, *17*

Fink, I. L., 343, *360*

Fiocco, R., 246, *250*

Fiolet, J. W. T., 343, 345, 348, *366*

Fiorenza, M. T., 10, *16*

Fire, A., 14, *16, 17,* 185, *190,* 243, *252,* 341, *360*

Firmin, D. N., 351, *363*

Firulli, A. B., 168, *175,* 342, 343, 344, 347, 349, 352, *361, 365*

Firulli, B. A., 347, 349, *361*

Fischer, A., 353, 356, *361*

Fischer, K.-D., 203, *209,* 217, 224, *231, 232,* 318, *327*

Fischer, L., 280, *292*

Fischer, W. H., 286, *292*

Fischman, D. A., 219, 228, *231*

Fishell, G., 78, 81, 84, 88, 89, *99, 100, 101, 102, 103,* 110, 121, *123, 124,* 241, 244, *249, 251*

Fisher, A., 376, *390*

Fisher, C., 575, *586*

Fisher, D. E., 525, *538*

Fisher, G., 84, *101,* 451, 475, *486,* 527, 528, 530, *536*

Fisher, L. J., 247, *252*

Fisher, M., 424, *495*

Fisher, P., *106*

Fisher, S. A., 341, *361, 369*

Fisher, S. J., 163, 170, 173, 174, *174, 175, 176, 177*

Fishman, D. A., 341, *362*

Fishman, G. I., 334, 348, *367*

Fishman, M. C., 213, *232,* 339, 344, 345, 351, 355, *359, 361, 368,* 395, *417*

Fissore, R. A., 8, *16*

FitzGerald, P. G., 527, 530, 531, *537*

Fjose, A., 546, *563*

Flaherty, L. A., 451, 476, *493*

Flamme, I., 211, 215, 220, 228, *229, 230, 232*

Flanagan, J. N., 381, *392*

Flanders, K. C., 226, *230,* 314, *325*

Flandin, P., 10, *18*

Flannery, M. L., 168, 169, *175, 179,* 349, *360*

Flavell, R. A., 79, 87, *100, 101, 106,* 529, *538*

Flax, J. D., 86, *104*

Fleenor, J., 14, *16*

Fleigel, L., 343, 347, *361*

Fleisch, H. A., 443, *485*

Fleischman, R. A., 204, *207,* 245, *249*

Fleming, J., 556, *565*

Fleming, R. J., 136, *145*

Fleming, S., 399, 416, *419*

Fleming, T. P., 23, *34,* 156, 158, *175*

Fleming, W. H., 244, *252*

Flenniken, A., 119, *125,* 423, 424, *494*

Fletcher, C. F., 85, *99*, 321, *328*

Fletcher, J. M., 407, *417*

Fliedner, T., 195, *208*

Fliering, S., 356, *360*

Fliniaux, I., *585*

Flint, A. F., 193, 206, *207*, 246, *250*, 270, 271, *274*

Flint, O. P., 128, 137, 139, *146*, 470, *484*

Flores-Delgado, G., 311, *329*

Floridia, G., 378, 387, *389*

Florkiewicz, R. Z., 453, *483*

Flower, W. H., 442, *484*

Flugge, G., 247, *250*

Flynn, J., 263, *274*

Flynn, L. M., 161, *178*

Flynn, S. E., 506, 508, 509, *514, 515, 518*

Fode, C., 83, 88, 89, 90, 95, *98, 99*, 435, *484, 490*, 526, *536*

Foged, N. T., 291, *292*

Fogels, H. R., 507, *514*

Fogo, A., 305, *329*, 403, *418*

Fogo, A. B., 453, 467, *496*

Foidart, J. M., 573, *587*

Foitzik, K., 574, *588*

Foley, A. C., 48, *50*, 427, *484*

Foley, K. P., 580, *586*

Folkman, J., 220, *229*

Follette, P. J., 170, *175*

Follettie, M., 40, *49, 323*

Fong, G.-H., 217, 223, 227, *229*

Fong, T. A., 217, 220, *229*

Fontaine-Perus, J., 267, *274*, 335, 337, 340, *358*

Forbes, A., 185, 186, *189, 190*

Ford, L. P., 11, *16*

Forehand, C. J., 469, *482*

Foreman, R. K., 468, *488*

Forlani, S., 9, *16*

Forrest, D., 510, *515*

Forsberg, H., 134, 135, 138, 139, *146*

Forslind, B., 580, *586*

Forster, A., 203, *210*

Forstrom, J. W., 467, *495*

Fortini, M. E., 136, *146*, 553, *561*

Fossett, N., 334, 343, *361*

Foster, D. N., 349, *362*

Foster, F. S., 109, *125*

Foster, J. A., 7, *16*

Foster, J. W., 281, *292*, 377, 380, 381, 386, 388, *390, 391, 393*

Foti, G., 263, *274*

Fougerousse, F., 525, *535*

Foulquier, F., 303, *324*

Fournier-Thibault, C., 264, 266, *273*

Fox, C., 8, *17*

Fox, C. H., 311, 312, *326*, 474, *488*

Fox, G. M., 402, *417*

Fox, H., 322, *326*

Fox, J. W., 94, *100*

Fox, M., 47, *50*, 451, 476, *483*, 508, *514*

Fraboulet, S., 134, 138, 139, *147*

Fraccaro, M., 378, 387, *389*

Frain, M., 118, 119, *124, 125*, 468, *492*

Francesca, T., 114, *122*

Francis, F., 94, *99*

Francis, M., 398, *416*, 550, *561*

Francis, N., 239, *249*

Francis, P. J., 533, *535*

Francis-West, P. H., 140, 141, 142, *147*, 238, *251*, 263, 264, 266, *276*, 422, 467, *481, 484*, 577, 578, 579, *587*

Franco, D., 341, 343, 344, 345, 347, 348, 350, 351, 355, 356, *359, 361, 362*

Franco Del Amo, F., 135, 136, 137, *148*

Francomano, C. A., 286, *292*, 452, 453, *481*

Frank, E., 237, *249*

Franke, W. W., 570, *586*

Frankel, W. N., 136, 137, *147*

Franklin, G., 173, *178*

Frankovsky, M. J., 342, *359*

Franz, T., 83, 84, *99*, 258, 263, 264, 266, *272, 274*, 475, *482*, 522, 527, 532, *535*

Franzoso, G., 442, *484*

Frasch, M., 334, 336, 340, *358, 361, 363, 364, 370*

Fraser, F. C., 443, *495*

Fraser, P., 202, *207, 210*

Fraser, S. E., 40, *52*, 118, 119, *122, 123*, 237, 243, 244, *249, 252*, 424, 425, *488, 495*, 543, *565*

Fraulob, V., 310, *329*, 557, *562*

Frawley, L. S., 511, *514, 515*

Frazer-Abel, A. A., 505, *517*

Frazier-Cierpial, L., 237, *249*

Frederich, M. E., 307, *324*

Frederick, L., 287, *294*

Fredga, K., 388, *390, 391*

Freedman, M., 202, *208*

Freeman, A., 575, 576, 577, 578, *588*

Freeman, G., 313, *324*

Freeman, M., 466, *482*

Freeman, T. B., 247, *251*

Freer, A. M., 121, *124*, 451, 453, 470, 471, *491*

Freidman, T. B., 557, *561*

Freidrich, G. A., 341, 352, *359*

Freie, B., 243, 244, *249*

French, M. C., 398, *419*

Freshney, R. I., 317, *327*

Freund, J. N., 308, 310, *324*

Freund, N., 403, *418*

Frick, H., 442, *485*

Friday, R. V., 166, *176*, 303, *325*

Fridell, R. A., 556, 557, *561, 564*

Friedman, R., 453, 467, *494*, 550, *563*

Friedman, T. B., 475, *481,* 556, *564*
Friedrich, R. W., 557, *564*
Friedrich, U., 281, *293,* 381, *391*
Friend, D. S., 318, *328*
Fries, R., 264, *274*
Friesel, R. E., 453, *492*
Frischauf, A. M., 377, 388, *393,* 415, *417*
Frisen, J., 236, 237, 244, 245, 246, 248, *249, 250,* 318, *324,*
    402, *419*
Frist, N. L., 26, *34*
Fritsch, R., 83, 85, 91, *105*
Fritschy, J. M., 84, *103*
Fritz, D. T., 11, *16*
Fritzsch, B., 84, *101,* 116, 117, *124, 489,* 549, 559, *561,*
    562, 563
Froelick, G. J., 374, 385, *389*
Froesch, E. R., 506, *517*
Frohman, M. A., 79, *98,* 243, *249*
Frojdman, K., 511, *516*
Frolichsthal, P., 235, 242, 244, *250*
Fromental-Remain, C., 310, *329*
Fromm, S. H., 437, *498*
Frost, D., 558, *563*
Frost, H. M., 442, *485*
Frotscher, M., 92, *104*
Fruth, E. E., 318, *324*
Frutiger, S., 321, *326*
Fu, X. Y., 286, *293, 294*
Fu, Y., 334, 348, *361*
Fuchs, E., 247, *250,* 271, *274,* 440, *498,* 570, 571, 572, 574,
    577, 579, 580, 581, 583, *584, 585, 586, 588, 589*
Fuchs, H., 551, 552, 557, *563*
Füchtbauer, A., 10, 12, *16,* 135, *146*
Füchtbauer, E. M., 10, 12, *16*
Fugo, N. W., 22, *33*
Fuhrmann, G., 160, *175,* 183, *190*
Fujii, H., 47, *51,* 89, 95, *99, 102,* 506, *516,* 525, *535*
Fujii, S., 343, 344, 351, *367*
Fujii, T., 374, *390,* 508, *517*
Fujii-Kuriyama, Y., 525, *535*
Fujimori, K. E., 90, *105*
Fujimori, T., 133, 135, 137, *149*
Fujimoto, S., 305, 307, *328*
Fujisawa, H., 223, *230,* 528, *536*
Fujisawa-Schara, A., 258, 271, *272, 274, 275*
Fujita, J., 189, *189*
Fujita, K., 189, *189*
Fujita, S. C., 557, *565*
Fujita, T., 336, *365,* 443, 453, *487*
Fujiwara, M., 312, *328,* 422, 424, 435, 437, 451, 475, 476,
    492, 505, *517,* 527, 530, *535*
Fujiwara, Y., 203, 204, *207, 209, 210,* 219, 224, 228, *232,*
    341, 343, 352, *369*
Fukai, N., 292, *294*

Fukamachi, H., 308, *324*
Fukiishi, Y., 437, *485*
Fukuda-Taira, S., 316, 317, *324*
Fukumoto, S., 347, *359*
Fukumura, D., 218, 223, 225, *229, 230*
Fukushige, T., 165, *175, 177,* 316, *325*
Fukuyama, M., 165, *180,* 316, *330*
Fuller-Pace, F., 84, *102*
Fulton, D., 225, *231*
Fults, D. W., 287, *294*
Fumagalli, F., 511, *513*
Funahashi, J., 112, 114, 115, 116, *123, 124*
Funayama, N., 571, *586*
Fundele, R., 9, *17, 18*
Fung, S., 403, *418*
Funke, B., 353, 357, *365*
Furley, A., 78, 81, *99*
Furlonger, K., 318, *327*
Furness, D. N., 558, *563*
Furness, J. B., 167, *174,* 305, *323*
Furukawa, M., 321, *324,* 506, *516,* 549, *564*
Furukawa, T., 521, 525, 527, *535, 537*
Furumoto, T., 135, *146*
Furuta, Y., 7, *16,* 82, 86, *99,* 287, *292,* 305, 312, 314, 315,
    323, 329, 453, 467, *496,* 504, 505, 507, 508, 510, *518,*
    530, 532, *535*
Furutani-Seiki, M., 78, *104,* 532, 534, *536,* 546, *565*
Fuse, N., 169, *175*
Fushiki, S., 581, 582, 583, *587*

G

Gabel, C. A., 27, *33*
Gabow, P. A., 414, *418*
Gabriel, H. D., 173, *175*
Gad, J. M., 129, 142, *146,* 162, *174*
Gadi, I., 156, *179*
Gaffield, W., 78, *99, 101*
Gage, F. H., 235, 236, 237, 241, 244, 246, 247, 248, *249,*
    250, 251, 252
Gage, P. J., 353, 354, 355, *361,* 433, 478, *485,* 507, 508,
    509, 511, *514, 515*
Gagliardi, P. C., 506, *515*
Gaiano, N., 78, 89, *99, 100,* 110, *123*
Gailani, M. R., 287, *292,* 470, *486,* 579, *586*
Gaillard, D., 271, *277*
Gainer, H., 435, *498*
Gainetdinov, R. R., 511, *513*
Gaio, U., 353, 356, *361*
Gajewski, K., 334, 343, *361, 363*
Galacteros, F., 376, *391*
Galante, P. E., 310, 321, *324, 325, 327*
Galceran, J., 83, 86, 92, *100,* 172, *175,* 440, 471, *488*
Galindo, J. E., 12, *17*

Gall, J. A., 194, *208*
Gallagher, B. C., 542, *562*
Gallahan, D., 226, *230*
Gallardo, M. E., 522, 525, *535, 537*
Galli, R., 235, 242, 244, 246, *250*
Gallicano, G. I., 8, 13, *16, 17,* 27, *33, 34*
Gallione, C. J., 227, *231*
Galliot, B., 475, *485*
Gallo, M., 40, *49, 323*
Galloway, A. C., 224, *232,* 504, *517*
Galvin, B. D., 453, *485*
Galvin, N., 314, *324*
Gambarotta, A., 89, *106,* 450, 471, *498*
Gan, L., 136, 137, 138, *145,* 527, *535, 536,* 556, 558, *565*
Gan, O. I., 199, *208*
Gandarillas, A., 580, *586*
Gandolph, M. A., 415, *420*
Gangadharan, U., 281, *295*
Gannon, F. H., 271, *276*
Gannon, M., *324*
Gans, C., 421, 427, 457, *485*
Gansmuller, A., 40, 43, *50,* 94, *100,* 451, 453, 469, 477, *484, 489*
Gao, J., 219, *233*
Gao, L. Y., 284, *294,* 442, 443, *497*
Gao, M., 11, *16*
Gao, Q. S., 166, *179*
Gao, W. Q., 555, 556, *565, 566*
Gao, X., 136, *149,* 548, 551, 552, *561, 562*
Gao, Y. H., 288, *293,* 442, 443, 453, *488*
Garber, M., 91, *99*
Garbutt, C. L., 26, *34*
García, C., 84, 95, *102, 105*
Garcia, L. F., 547, *562*
Garcia Bellido, A., 136, *145*
Garcia-Cardena, G., 225, *231*
Garcia-Castro, M., 121, *125,* 187, *189,* 355, 356, *361*
Garcia Delgado, C., 453, *486*
Garcia-Fernández, J., 305, *323*
Garcia-Garrido, L., 219, 228, *231*
Garcia-Martinez, V., 317, *328,* 336, 344, *361, 362*
Garcia-Porrero, J. A., 195, 196, 200, *207*
García-Verdugo, J. M., 90, 95, *102, 106*
Garcia-Wijnen, C. C., 194, *208*
Gardahaut, M.-F., 335, 337, 340, *358*
Gardiner, A., 334, 335, 336, 337, 339, *365*
Gardiner, C. S., 509, *518*
Gardner, A., 352, 354, *366*
Gardner, D. G., 343, *364*
Gardner, J., 292, *294*
Gardner, R. L., 25, 26, 27, 28, 29, 30, 31, 32, *33, 34,* 156, 157, 158, 161, 166, *176, 178,* 182, 183, *189,* 202, *210*
Garel, C., 507, *514*
Garel, S., 91, *100*

Garfinkle, J., 292, *294*
Garland, D., 528, 533, *535*
Garnier, J. M., 469, *485, 489*
Garofalo, S., 286, *294*
Garratt, A. N., 353, 356, *361*
Garrels, J. I., 11, *17*
Garretson, D. C., 307, *324*
Garrett, K. L., 260, 269, 271, *274, 275, 277*
Garrick, D., 376, *390*
Garrity, P. A., 399, *419,* 530, 533, *537*
Garrod, D. R., 405, *417*
Gärtner, A., 83, *98*
Garzia-Verdugo, J. M., 236, 237, 244, 245, 248, *249*
Gasca, S., *50,* 121, *123*
Gashier, A., 400, *418*
Gaspar, M. L., 386, *390*
Gassmann, M., 173, 174, *175,* 225, *230,* 424, 425, *485*
Gat, U., 577, 579, *584, 586*
Gato, A., 547, *564*
Gattner, K., 415, *420*
Gattone, V. H. II, 415, *417*
Gaub, M. P., 469, *490, 491*
Gaudenz, K., 452, 453, *481*
Gaunt, S. J., 40, *50,* 376, *390,* 477, *482, 485*
Gautam, M., 411, *419*
Gauthier, Y., 510, *518*
Gautier, R., 198, 199, 200, 201, *207, 208*
Gavalas, A., 117, 118, 119, *123, 125,* 427, 433, 443, 449, 469, 474, 479, *485, 494,* 545, *562*
Gavin, A. C., 7, *16*
Gavis, E. R., 6, *16*
Gavrilina, G., 510, *516*
Gawantka, V., 138, *146*
Gay, N. J., 10, *18*
Gaztambide, S., 506, *514*
Gearhart, J. D., 225, *230*
Gebauer, F., 7, 10, *16*
Gebbia, M., 357, *361, 363*
Gebre-Medhin, S., 313, *323, 326,* 413, *418,* 467, *482, 489*
Gebuhr, T., 357, *364*
Gecz, J., 376, *393*
Gee, H., 266, *274*
Gehring, W. J., 533, *535, 536*
Geiger, B., 408, *418*
Geiger, H., 194, *207*
Geisendorf, S., 532, *537*
Geiser, A., 226, *230*
Gelbart, W., 343, *360*
Gemza, D. L., 526, 527, *535*
Genbacev, O., 173, 174, *176*
Gendron-Maguire, M., 136, *149,* 224, *232,* 427, 432, 433, 443, 449, 473, 474, 477, *485, 494*
Generoso, W. M., 22, *34*

Geng, L., 415, *417, 418*

Geng, X., 579, *589*

Genin, A., 347, *364, 366*

Geoffroy, V., 288, 289, *292*

Georgakopaoulis, D., 343, 350, 351, 352, *359*

George, E. L., 222, *229,* 340, 352, *361*

George, F. W., 373, *390*

George, S., 44, *50,* 227, *231*

Georges, M., 264, *274*

Georges-Labouesse, E., 94, *100,* 183, 187, 188, *189,* 572, 583, *585, 586*

George-Weinstein, M., 263, 264, *274*

Georgiev, O., 317, *325*

Georgopoulos, K., 91, *103,* 204, *210*

Georgopoulou, A., 204, *210*

Geraedts, J. P., 174, *179*

Gerald, W., 400, *418*

Gerber, A. N., 356, *359*

Gerber, H. P., 223, *229,* 291, *292*

Gerety, S. S., 218, 225, *229*

Gerhart, J., 40, 41, 44, *50,* 263, 264, *274*

Gerken, S. C., 412, *417*

Gerlach, L. M., 550, *562*

Germain, L., 317, *324*

German, M. S., 321, *325, 328, 329,* 451, 474, 475, *494,* 509, *515*

German, R., 453, *483*

Germiller, J. A., 550, *562*

Germino, F. J., 414, *419*

Germino, G. G., 414, 415, *419, 420*

Gerrard, B., 287, *292,* 470, *485, 486*

Gershon, M. D., 244, *251*

Gertenstein, M., 168, 169, 172, 173, 174, *174, 178,* 203, *207, 209,* 223, *228,* 349, 355, *367*

Gertner, J. M., 443, *497*

Gertner, R., 286, *294*

Gerton, G. L., 7, *16*

Gertz, B. J., 512, *515*

Gerwe, E. A., 84, *97,* 435, 469, 476, *481*

Gerwins, P., 504, *515*

Geschind, D. H., 247, *251*

Geske, R., 442, 443, *495*

Gespach, C., 271, *277*

Gessler, M., 347, *363,* 415, *417*

Gesteland, K., 38, *52*

Ghahremani, M., 399, *417*

Ghatas, I., 79, 89, *99*

Ghatpande, A., 343, 347, *361*

Ghatpande, S., 343, 347, *361*

Ghattas, I., 79, 89, *102,* 422, 437, 450, 451, 471, 472, 473, 474, *484, 489, 493*

Gherardi, E., 317, *328,* 404, *419, 420*

Ghosh, A., 86, 92, *100, 102, 103*

Ghosh, S., 318, *323*

Ghyselinck, N., 91, *101,* 341, 346, 347, 352, *362,* 469, *484, 485,* 525, *536,* 545, *562*

Giancotti, F. G., 410, *417*

Gibbons, R. J., 376, *390*

Gibson, M., 334, *361*

Giguere, V., 167, *177, 179,* 346, *363*

Gilardi-Hebenstreit, P., 119, *125,* 135, *147,* 546, *564*

Gilbert, D. J., 168, *176,* 411, *418*

Gilbert, T., 403, *418, 420*

Gilbert, W., 525, *537*

Gilbride, K., 314, *328*

Gilgenkrantz, H., 342, *368*

Gill, G. N., 135, *146*

Gillemans, N., 343, *369*

Gillies, S., 287, *292, 470, 486*

Gilmore, E. C., 94, *100*

Gilmour, K. C., 288, *294,* 442, 443, 453, *493*

Giniger, E., 185, *189*

Ginsburg, M., 183, *189*

Ginzberg, R. D., 560, *563*

Giocanazzi, S., 376, *393*

Giraldez, F., 422, 466, *494, 496,* 542, 543, 545, 546, 559, *562, 564, 565*

Girbal-Neuhauser, E., 581, *588*

Girgis-Gabardo, A., 260, 270, 271, 272, *276*

Giros, B., 511, *513*

Giros, R., 341, 352, *362*

Giroux, S., 171, 173, *176,* 318, *324*

Gisselbrecht, S., 340, *361*

Gitelman, I., 135, *146*

Gitelman, S. E., 280, *292*

Gitlin, J. D., 313, *325*

Giton, Y., 84, *100*

Gittenberger-de Groot, A. C, 219, 226, *229, 232*

Gittenberger-de Groot, A. C., 333, 341, *360, 366,* 453, 467, *494*

Gittes, G. K., 319, 321, *324,* 452, *491*

Giuli, G., 381, *390*

Glaser, T., 375, *392,* 415, 416, *417,* 526, 527, 528, *535*

Glasser, S. W., 313, 314, *329*

Gleeson, J. G., 94, *100*

Gleiberman, A. S., 478, *496,* 504, 505, 506, 507, 508, 509, *514, 515, 518*

Glickman, S. E., 388, *393*

Glimcher, L. H., 527, 530, 531, *536*

Glimcher, M. J., 442, *485*

Gliniak, B. C., 173, *179*

Glinka, A., 44, *50,* 325, 338, *361, 362*

Globerson, A., 196, *207,* 245, *251*

Glomski, C. A., 219, 220, *232*

Glotzer, J., 185, *189*

Glowacki, J., 285, *293,* 442, 443, *488*

Glynn, M., 287, *292,* 579, *586*

Gnarra, J. R., 173, *176,* 227, *229*

Gocza, E., 42, *51,* 158, 160, *177, 178*
Goda, H., 511, *515*
Goda, P., 510, *516*
Goddard, A., 402, *420,* 579, *589*
Goddard, J. M., 466, *490,* 545, *561*
Goddell, M. A., 193, *207*
Godfraind, C., 166, *176,* 321, *325*
Godfrey, P., 443, *485,* 508, *515*
Godin, I., 185, 187, 188, *189,* 195, 196, 199, 200, *207, 208, 209,* 220, *229*
Godin, R. E., 171, *175,* 408, *417*
Godsave, S. F., 263, *274*
Goebbels, S., 92, *104*
Goeddel, D. V., 223, *230,* 341, *370*
Goedecke, S., 317, *328,* 404, *419*
Goering, L., 342, *360*
Goetinck, P., *588*
Goff, D. J., 308, *328*
Goff, J. P., 246, *251*
Goff, S. C., 204, *207*
Goff, S. P., 205, *208*
Goffinet, A. M., 94, *100*
Gofflot, F., 132, *146*
Goh, K. L., 173, *176*
Goins, T. L., 237, *249*
Gold, L. I., 467, *493*
Golden, J. A., 79, 82, 86, *100,* 117, *122*
Golden, K., 336, 337, *366, 370*
Goldfarb, M., 166, *175,* 313, 314, 315, 317, *324, 325*
Goldhamer, D. J., 258, 260, 261, 263, 267, *274, 275, 277*
Golding, J. P., 424, 425, *485*
Goldman, D., 132, *146, 148,* 422, 427, *496,* 527, *535*
Goldman, S. A., 95, *100*
Goldowitz, D., 94, *104*
Goldstein, A. M., 287, *292,* 339, *361,* 470, *486*
Goldstein, E. S., 341, *358*
Goldstein, M. M., 347, *362*
Goldstein, R. S., 131, *146*
Goldstein, S. A., 443, *482*
Goldstein, S. L., 314, *326,* 410, *418*
Golosow, N., 319, *325*
Goluboff, L., 511, *515*
Gomez, A. P., 40, 43, *50,* 477, *484*
Gomez, R. A., 412, *420*
Gomez-Pardo, E., 373, 374, 385, *393,* 398, *420,* 475, *496,* 544, *565*
Gomez-Skarmeta, J. L., 550, *561*
Gomperts, M., 186, 187, *189,* 579, *584*
Goncharov, T., 341, *369*
Gong, Q., 84, *100*
Gonzalez, F. J., 311, 312, *326,* 474, *488*
Gonzalez, G. A., 286, *292*
Gonzalez-Ramos, M., 453, *486*
Gonzalez-Sanchez, A., 337, 340, 341, *361, 364, 365*

Good, A., 322, *326*
Good, D. J., 512, *515*
Goodall, H., 26, *34*
Goodard, A., 121, *123*
Gooday, D., 87, 94, *98*
Goodell, M. A., 193, 206, *208,* 270, 271, *275,* 318, *325*
Goodfellow, P. N., 281, 282, *292, 293,* 377, 380, 381, 384, 386, 387, 388, *390, 391, 392, 393*
Goodman, D. S., 468, 469, *495*
Goodman, L., 402, *420*
Goodrich, E. S., 421, 442, 449, 456, 471, *485*
Goodrich, L. V., 78, 80, *100,* 121, *123,* 287, *293,* 470, *485,* 579, *586*
Goodship, J., 352, 354, *366*
Goodyear, P., 379, *391*
Goodyear, R., 555, *562*
Goossens, M., 508, *516*
Gorczyka, M. G., 334, *361*
Gordaze, P. R., 122, *122*
Gordon, D. F., 509, 510, *515*
Gordon, J. I., 309, 310, *325, 329*
Gordon, J. S., 581, *586*
Gordon-Thomson, C., 337, *361*
Goridis, C., 239, *250,* 435, *484*
Goring, D. R., 528, *536*
Gorivodsky, M., 167, 171, *174, 176,* 312, 314, 315, *323*
Gorlin, R. J., 292, *294,* 442, *485*
Goronowitsch, N., 423, *485*
Gorry, P., 443, 451, 453, 469, 474, *485, 489, 490,* 545, *563*
Gortz, J., 578, 583, *586*
Gosden, C., 375, *391,* 399, *419*
Gossett, L. A., 342, *361*
Gossler, A., 132, 134, 135, 136, 137, 138, 139, 140, *144, 145, 146,* 552, 553, 555, 558, 559, *563, 564*
Goszczynski, B., 165, *177,* 316, *325*
Gotay, J., 226, 227, *233*
Gothe, S., 510, *515*
Goto, J.-I., 338, *368*
Goto, K., 531, *536*
Goto, S., 91, *101*
Goto, T., 92, *105*
Gotoh, O., 525, *535*
Gotoh, S., 506, *516,* 549, *564*
Gotsch, U., 226, *228*
Gött, P., 316, *323*
Gotthardt, M., 94, *105*
Gottschalk, M., 311, *329*
Götz, M., 83, 85, 87, 91, 94, *98, 100, 105*
Gould, A., 117, 118, 119, *123, 124,* 544, *563*
Gould, D. B., 453, 479, *488*
Gould, E., 237, 247, 248, *250*
Gould, S. J., 442, *485*
Goulding, M., 260, 263, 264, 266, 268, *274, 275,* 410, *418*
Goumans, M. J., 222, *229,* 467, *484*

Goumnerov, B., 227, *232*

Gourdie, R. G., 219, 228, *231,* 334, 339, *360, 368*

Gourdji, D., 510, *518*

Gove, C., 343, *361*

Gowen, J. W., 453, *488*

Grabel, L. B., 7, *19*

Grabowski, C. T., 161, *178*

Gradwohl, G., 136, *147,* 166, *176,* 310, 321, *325,* 435, *484,* 526, *536*

Graham, A., 422, 425, 428, 432, 433, 434, 435, *481, 484, 485, 490, 497,* 543, *562*

Graham, C. F., 26, *33*

Graham, H., 572, 583, *585*

Graham, J. M., 470, *487*

Grahovac, M. J., 11, 12, *17*

Grail, D., 194, *208,* 453, *486*

Grainger, R. M., 435, 469, 476, *484,* 530, *536,* 542, *562*

Grainger, R. M., Jr., 529, *538*

Grammatopoulos, G. A., 433, *485*

Granadino, B., 525, *535*

Granato, M., 546, 557, *564, 565*

Grange, D. K., 292, *294*

Granholm, A.-C., 402, *419*

Grant, D. B., 506, *514*

Grant, J. H., 468, *485*

Grant, J. W., 356, *366*

Grant, P., 435, *498*

Grant, S. R., 343, *365*

Grapin-Botton, A., 426, 427, 433, *483*

Grass, S., 258, *274*

Grau, E., 40, 42, *51,* 165, *177,* 303, *327*

Graves, J. A., 386, 387, 388, *390, 391, 392, 393*

Graveson, A., 422, *484*

Graw, J., 527, 532, *537*

Gray, C., 402, *420*

Gray, N. K., 10, 11, *15, 16*

Gray, S. W., 312, *328*

Grayson, D. R., 316, *324*

Grayzel, D., 357, *358*

Graziadei, P. P., 77, 79, 84, *100, 105*

Graziano, M. S. A., 237, *250*

Greaves, M., 194, *207*

Greco, T. L., 137, *146,* 511, 512, *518*

Green, E. D., 453, *486*

Green, H., 582, 583, *585*

Green, J. B. A., 167, *179*

Green, M. C., 479, *485*

Greenbaum, L. E., 318, *324*

Greenberg, M. E., 86, *100*

Greenberger, J. S., 246, *251*

Greene, E. C., 442, *485*

Greene, R., 424, 442, 443, *482*

Greenfield, A., 281, *295,* 377, *390, 391*

Greenhaw, G. A., 286, *292*

Greenlund, A. C., 287, *293*

Greenspan, D., 305, *329,* 453, 467, *496*

Greenwood, A. L., 237, *250*

Greer, I. A., 174, *177*

Greer, J. M., 580, *584*

Greewald, J., 354, *367*

Gregg, B. C., 128, *148*

Gregory, M. C., 412, *417*

Gregory, W. K., 421, 473, *485, 493*

Grenier, G., *585*

Grenier, J. K., 471, *493*

Grepin, C., 342, 343, *361*

Gretz, N., 415, *419*

Greve, J. M., 14, *16*

Gridley, T., 135, 136, 137, 138, *146, 148, 149,* 224, 226, 227, *229, 230, 232,* 423, 427, 432, 433, 443, 449, 473, 474, 477, *485, 487, 494,* 546, 553, 555, *563, 565*

Griffin, K. J., 167, *176,* 334, *361*

Griffiths, B. L., 377, 388, *393*

Griffiths, D., 416, *420*

Grigoriadis, A. E., 442, 443, *485, 497*

Grigorieva, E., 260, 267, *275*

Grigoriou, M., 88, 89, *101,* 402, 420, 432, *485*

Grigoriuo, M., 432, *497*

Grim, M., 258, 264, 266, *274*

Grimaldi, P. A., 271, *277*

Grimes, P., 522, 527, *535,* 549, *562*

Grimmond, S., *390*

Grinberg, A., 83, 87, 91, *103, 106,* 173, *176, 177,* 376, *389,* 402, *419,* 432, *498,* 508, *517,* 521, 522, 527, *537*

Grindley, J. C., 84, 85, *100,* 311, 312, 313, 314, 315, *323, 325,* 435, 451, 470, 475, 476, *485,* 509, *513,* 521, 522, 527, 530, *536*

Grinspan, J. B., 82, 86, *100*

Gripp, K. W., 468, *485*

Gritsman, K., 45, 47, *50, 51,* 78, *100,* 338, 353, 354, 356, 357, *361, 370*

Gritti, A., 235, 242, 244, 246, *250*

Grobet, I., 264, *274*

Grobstein, C., 319, *325,* 404, *417*

Groffen, J., 314, *325,* 434, 467, *487*

Grogg, K. M., 227, *231*

Grompe, M., 246, *250,* 318, *326, 327*

Grondona, J. M., 341, 346, 347, 352, *362,* 525, *536*

Gronenborn, A. M., 377, *393*

Groome, A., 528, 533, *535*

Gros, D., 348, *360*

Gros, P., 79, *99,* 266, *277,* 475, *484*

Grose, R., 574, 583, *584*

Groskopf, J. C., 162, *176*

Gross, C. G., 237, *250*

Gross, M. K., 160, *175,* 183, *190,* 260, 266, *274*

Gross, R. E., 239, *250*

Grosschedl, R., 83, 86, 91, 92, *100*, 172, *175*, 440, 471,
        488, *497*, 579, 583, *585*, *587*, *589*
Grosskortenhaus, A., 336, *367*
Grosveld, F., 165, *177*, 195, 196, 199, 200, 202, *207*, *208*,
        210, 343, *369*
Grosveld, G., 201, 204, *209*, 220, *231*
Groudine, M., 169, *175*, *177*
Grounds, M. D., 269, 271, *274*
Grove, E. A., 82, 83, 84, 86, 87, *100*, *101*, *105*
Groves, A. K., 238, 239, 241, *251*, *252*, 543, 546, *562*
Groves, N., 345, 348, *366*
Grow, M. W., 334, 348, *361*
Grubber, J. M., 467, *488*
Grubel, G., 408, *419*
Gruber, P. J., 344, 352, *361*
Grun, F., 525, *536*
Grund, C., 570, *586*, *587*
Grüneberg, H., 132, 137, *146*, 441, 442, 453, 456, 475, 479,
        485
Grunfeld, J. P., 375, *389*
Gruning, W., 415, *418*, 579, *585*
Grunwald, D. J., 214, *230*, 557, *564*
Grunz, H., 44, *52*, 338, *366*
Grusby, M. J., 527, 530, 531, *536*
Gruss, P., 79, 81, 83, 85, 87, 91, 92, 94, 95, *97*, *100*, *102*,
        103, *105*, 113, 115, *125*, 129, 137, 140, *145*, *147*, 160,
        173, *178*, *179*, 183, *190*, 258, 260, 263, 264, 265, 266,
        270, 272, *272*, *273*, *276*, 311, 321, *327*, *329*, 373, 374,
        385, *393*, 398, *417*, *420*, 432, 451, 453, 473, 475, 477,
        481, *482*, *484*, *487*, *490*, *495*, *496*, *498*, 519, 521, 522,
        525, 527, 530, 532, 533, *535*, *536*, *537*, *538*, 544, 547,
        550, *561*, *564*, *565*
Grussendorf, E. I., 570, *584*
Gruters, A., 508, *516*
Grutzner, F., 388, *392*
Gu, D., 318, 322, *325*, *326*
Gu, H., 14, *16*, 319, *325*
Gu, Q., 579, *589*
Gu, S., 382, *389*
Gu, T.-L., 201, 204, *209*, *210*, 220, *231*, *233*
Gu, Y., 87, *101*, 136, *145*, 341, *362*
Gu, Z., 47, *50*
Gualandris, A., 224, *232*, 504, *517*
Gualdi, R., 316, 317, *325*
Gubbay, J., 377, *390*, *391*
Gubler, M. C., 375, *389*
Gubler, P., 10, *18*, *19*
Gudas, L. J., 469, 470, *484*
Guenet, J. L., 386, *390*, 476, 477, *496*
Guerrier, D., 381, 382, *389*, *390*
Gui, Y.-H., 340, *364*
Guichet, A., 185, *189*
Guille, M., 343, *361*
Guillemot, F., 79, 83, 88, 89, 90, 95, *98*, *99*, *100*, *105*, 136,

        146, *147*, 166, 168, 174, *174*, *176*, *178*, 239, *250*, 310,
        321, *325*, 435, *484*, *490*, 526, 527, *536*, *538*, 555, *566*
Guioli, S., 281, *292*, 378, 380, 381, 382, 387, 388, *389*,
        390, *391*, *392*, *393*
Gukovsky, I., 506, 508, 509, *516*, *518*
Gulisano, M., 79, 83, *100*, *104*, 290, *294*, 376, *393*, 422,
        427, 428, 433, 450, 471, 472, 473, 478, *481*, *487*, *495*
Gull, K., 15, *15*
Gullberg, D., 409, *419*, 574, *589*
Gultig, K., 558, *563*
Gumbiner, B. M., 408, *417*, *420*
Gumeringer, C. L., 341, 352, *368*
Gummer, A. W., 558, *563*
Gundersen, K., 269, *274*
Günes, C., 317, *325*
Gunther, T., 130, 132, 140, *148*, 264, 265, *277*, 423, *492*
Guo, H., 12, *17*
Guo, L., 343, 347, *361*
Guo, Q., 122, *122*, 226, 227, 229, 336, 352, *359*
Guo, W., 378, 379, *389*, *393*
Guo, Y.-J., 432, *498*
Gupta, R., 31, *33*, 47, *50*, 451, 476, *483*, 508, *514*
Gupta, S., 318, *324*
Gurdon, J. B., 203, *209*, 263, 264, *274*, *275*, 303, *330*
Guris, D. L., *361*
Gurley, C. M., 268, 269, *274*
Gurney, A., 121, *123*
Gurtner, G. C., 162, 170, *176*
Gussoni, E., 193, 206, *207*, 246, *250*, 270, 271, *274*
Gutch, M. J., 91, *100*
Guthrie, S., 427, *485*, 545, *563*
Gutierrez-Hartmann, A., 505, *517*
Gutierrez-Ramos, J. C., 205, *206*
Guttmacher, A. E., 227, *229*, *231*

H

Haaf, T., 9, *17*, 388, *392*
Haar, J. L., 317, *327*
Haas, O. A., 281, *293*, 381, *391*
Haas, T. L., 223, *229*
Haase, I., *589*
Habal, M., 292, *294*
Habara-Ohkubo, A., 336, *365*
Habener, J. F., 322, *328*
Haber, D. A., 374, 375, *392*, 400, 415, *417*, *418*
Habets, P. E. M., 341, 345, 347, 348, 350, *359*
Habib, R., 416, *419*
Hackenmiller, R., 286, *292*
Hacker, A., 377, 378, 380, 381, 384, 386, 387, *390*, *392*,
        393
Hackett, B. P., 313, *323*, *325*
Hackney, C. M., 558, *563*
Hacohen, N., 466, *485*, *488*, *491*

Hadchouel, M., 166, *178*, 263, 264, *273*

Haddon, C. M., 547, *562*

Hadjantonakis, A.-K., 3, *4*, 23, *35*, 38, *51*, 132, *147*, 166, 174, *174*, *179*, 347, 352, *360*

Hadrys, T., 549, *562*

Hadzic, D. B., 347, 349, *361*

Haendel, M. A., 170, *176*

Haenig, B., 137, *147*

Haenlin, M., 343, *360*, *361*

Hafen, E., 470, *484*

Haffen, K., 308, *325, 326*

Haffner-Kraus, R., 167, 171, *174, 176*, 312, 314, 315, *323*

Haffter, P., 78, *103*, 351, 354, 355, *359*, 546, *565*

Hafner, M., 415, *419*

Haftek, M., 581, *588*

Hagan, B., 530, 532, *535*

Hagan, D. M., 387, *390*

Hage, M., 183, *190*

Hagedorn, L., 236, *250*

Hagenbuchle, O., 321, *324, 326*

Haghighat, A., 11, *16*

Hagman, J., 91, *100*

Hagoort, J., 343, 345, 348, *366*

Hagopian-Donaldson, S., 79, *106*, 529, *538*

Hah, C. S., 347, *362*

Hahn, H., 287, *292*, 470, *485, 486*, 579, *586*

Hahn, S., 468, *495*

Haines, C. F., 84, *97*, 435, 469, 476, *481*

Haines, L., 266, *276*

Haitjema, T., 227, *231*

Hajihosseini, M., 312, *324*, 505, *514*, 546, *562*

Hajra, A., 202, *208*

Hake, L. E., 7, *17*

Halasz, B., 504, *516*

Halata, Z., 258, 264, 266, *274*

Hald, J., 310, 321, *325*

Halder, G., 533, *535, 536*

Hale, L. P., 227, *231*

Halevy, O., 341, *361*

Halford, M. M., 453, *486*

Hall, B. K., 279, *292*, 421, 422, 423, 424, 426, 427, 432, 442, 443, 449, 465, 471, *482, 483, 484, 486, 489, 497*

Hall, D. E., 262, *277*

Hall, F. L., 314, *329*

Hall, J., 470, *486*

Hall, M., 84, *106*, 132, *146*, 469, *497*

Hall, S. E., 387, *393*

Hall, S. H., 452, *487*

Halladay, A. K., 91, *106*

Hallais, M. F., 194, 201, *209*

Hallas, G., 173, *177*

Halliday, A. L., 87, *100*

Hallmann, R., 219, *232*

Hallonet, M., 78, 79, 81, *100*, 113, *123*, 522, 527, *536*

Halmekyto, M., 228, *230*

Halmesmaki, E., 162, *179*

Halpern, M. E., 78, *104*, 339, 341, 349, 351, *364, 370*

Hamada, H., 47, *50, 51*, 89, 95, 99, *102*, 183, *190*, 244, *251*, 338, 352, 354, 356, 357, *365, 367, 368, 370*, 506, *516, 525, 535*

Hamada, Y., 136, *146*, 316, 317, *325*

Hamburger, V., 40, *50*, 568, 572, *586*

Hamersma, H., 292, *294*

Hamilton, D., 15, *18*

Hamilton, H., 568, 572, *586*

Hamilton, R. L., 161, *175, 178*

Hammer, G. D., 379, 381, *390, 392*

Hammer, R. E., 94, *105*, 432, 465, 478, *483*, 498, 511, *516*

Hammerschmidt, M., 287, *294*, 336, *363*, 442, 443, 470, *486, 496*, 546, *565*

Hammitt, D. G., 22, *35*

Hammond, K. L., 530, 533, *536*

Hammond, V. E., 160, 167, *175*

Hamosh, A., 347, *360*

Hampl, A., 8, 10, *18*

Han, J., 228, 229, 246, *249*, 426, *482*

Han, M. J., 3, *4*

Han, Y., 341, *362*

Hanada, K.-I., 244, *251*

Hanafusa, T., 321, *327*

Hanahan, D., 223, *232*, 281, 291, *294*, 320, 321, *323, 324*, 580, *586*

Hanai, A., 557, *564, 565*

Hanakawa, Y., *588*

Hanamure, Y., 473, *498*

Hanaoka, K., 260, *276*, 318, *328*

Hanauer, A., 506, *517*

Hancock, S., 132, *145*

Handyside, A. H., 158, *176, 180*

Hanken, J., 139, *148*, 421, 422, 423, 432, 441, 457, *482, 486*

Hankin, M. H., 527, *535*

Hankinson, O., 172, 173, *177*, 225, *230*

Hanks, M. C., 109, 115, *123*

Hanley, K., 580, 581, *585, 586*

Hanley, N. A., 387, *390*

Hannon, G. J., 341, *361*, *483*

Hannon, K., 258, *274*

Hansen, C. S., 44, *50*

Hansen, J., 571, *586*

Hansen, L. A., 169, *179*

Hansen, S. E., 322, *328*

Hanset, R., 264, *274*

Hanson, I. M., 451, 475, *486*, 527, 528, 530, 533, *536*

Hanson, R. D., 376, *390*

Hantsoo, L., 351, *364*

Happle, R., 443, *486*

Haqq, C. M., 381, *390*

Hara, M., 375, 384, *392*

Hara, T., 376, *392*

Hara, Y., 311, 312, *326*, 474, *488*

Harada, A., 38, *51*

Harada, T., 531, *538*

Harbison, M. D., 443, *497*

Hardcastle, Z., 86, *100,* 121, *123,* 437, 439, 451, *486, 496*

Hardin, J. D., 205, *208*

Harding, G. W., 286, *292,* 558, *561*

Harding, J. D., 321, *328*

Hardman, M. J., 568, 582, 583, *586, 587*

Hardman, P., 404, *420*

Hardy, M. H., 573, 574, 575, 576, *585, 586, 589*

Hardy, R. R., 245, *250*

Hargett, L. K., 85, *100,* 132, *149,* 305, *329,* 453, 467, 479, *496, 498,* 506, *516*

Hargrave, M. R., 281, *294*

Harkes, I. C., 343, *369*

Harland, R. M., 40, 41, 44, 48, *49, 50, 51, 52,* 78, *98,* 141, *147,* 238, *252,* 260, 264, 265, *275,* 304, *323,* 338, *362,* 468, *482, 498,* 550, *561,* 579, *587*

Harlap, S., 22, *34*

Harley, V. R., 282, *293,* 377, 380, 381, 384, 387, *390, 391, 392*

Harlow, E., *483*

Harpal, K., 43, *50, 51,* 133, 135, *145, 149,* 158, 165, *175,* 203, *207,* 223, *228,* 303, 304, 316, 317, *324*

Harper, J. W., 287, *294,* 528, 532, *538*

Harper, S., 570, *587*

Harris, E., 109, 114, 115, *123, 124, 126,* 228, *233,* 428, *491*

Harris, M. J., 423, *486, 487*

Harris, P. C., 415, *420*

Harris, P. E., 500, *518*

Harris, R. C., 169, *179*

Harris, W. A., 77, *99,* 113, *125,* 522, 525, 533, *538*

Harrison, D. E., 245, *250*

Harrison, D. J., 416, *419*

Harrison, K. A., 319, *325*

Harrison, R. W., 475, *481*

Harrison, S. M., 135, *145*

Harrison, W., 352, 354, *366,* 380, *389*

Harry, J. L., 387, *392, 393*

Harsh, G., 512, *516*

Hart, C. E., 467, *495*

Hart, C. P., 443, 474, *490*

Hart, K. C., 453, *485*

Hartenstein, A. Y., 136, *146*

Hartenstein, V., 136, *146,* 205, *209, 367*

Hartigan, D. J., 78, 79, 85, 86, 88, *104*

Hartigan-O'Connor, D. J., 80, 82, 88, *98,* 321, *329*

Hartley, L., 31, *34,* 203, *209,* 317, *325,* 333, 334, 343, 345, 348, 349, 352, 356, *364*

Hartley, R., 575, *586*

Hartmann, D. J., 573, *587*

Harvey, R. B., 31, *34*

Harvey, R. P., 40, 44, *50,* 203, *209,* 317, *325,* 333, 334, 338, 339, 341, 343, 344, 345, 347, 348, 349, 350, 351, 352, 353, 355, 356, 357, *358, 359, 362, 364, 366, 367, 369*

Hasegawa, A., 11, 12, *17*

Hasegawa, G., 195, *208*

Hasel, K. W., 511, *514*

Hashiguchi, H., 338, 354, 356, *367, 368*

Hashimoto, C., 579, *587*

Hashimoto, G., 313, 315, *327*

Hashimoto, K., 188, *189,* 512, *518,* 588

Hashimoto, N., 7, *16,* 376, *391*

Hashimoto, O., 577, *586*

Hashimoto, R., 557, *564*

Hassan, B. A., 555, 558, 559, *561*

Hassan, M., 507, *514*

Hasson, T., 556, *565*

Hastie, N. D., 302, *327,* 341, *365,* 374, 375, 380, *389, 391, 393,* 399, 400, 416, *416, 417, 419, 420,* 451, 475, *486,* 527, 528, 530, *536, 537, 538*

Hasty, P., 110, *123,* 260, *274*

Hata, N., 135, 137, 138, *148*

Hatanaka, Y., 84, *98, 100*

Hatano, O., 375, 376, 380, *390, 392*

Hatch, G., 89, *106,* 450, 471, *498*

Hatcher, C. J., 347, *362*

Hatini, V., 81, 84, 86, *100, 101,* 408, *417,* 479, *486,* 522, 525, 527, *536*

Hatta, K., 528, *536*

Hatta, T., 547, *564*

Hatten, M. E., 94, *100, 106,* 122, *122,* 453, *491*

Hatton, K. S., 10, *16*

Hattori, N., 170, *176*

Hatva, E., 162, *179*

Hatzistavrou, T., 334, *358*

Haud, O., 2, *4*

Haudenschild, C., 220, *229*

Haugen, B. R., 509, 510, *515*

Hauri, H. P., 308, *325*

Hauschka, S. D., 140, *148,* 257, 262, 263, 264, 269, *276, 277*

Hawes, N. L., 527, *535*

Hawkes, R., 94, *101*

Hawkes, S. G., *391*

Hawkins, J. E., Jr., 443, *482*

Hawkins, J. R., 377, 388, *390, 393*

Hawkins, M. G., 165, *175*

Haworth, I. S., 452, 453, 474, *487*

Hay, M. F., 454, *486*

Hayakawa, K., 245, *250*

Hayasaka, M., 260, *276*

Hayashi, D., 336, *365*

Hayashi, H., 14, *18*

Hayashi, S., 169, *175,* 442, *498,* 531, *536*

Hayashi, T., 414, *418*

Hayashi, Y., 311, *329,* 577, *586*

Hayashida, C. Y., 509, *517*

Hayashizaki, Y., 11, 12, *18,* 511, *516*

Haydar, T. F., 87, *100, 101*

Hayes, N. L., 245, *249*

Hayflick, A. J., 512, *516*

Haynes, J. I. II, 532, *535*

Hayward, D., 136, *148,* 262, *276*

Hayward, R., 453, *494*

Hayward, R. D., 453, 466, 474, *498*

He, B., 402, *419*

He, C.-Z., 344, *362*

He, T. C., 86, *100,* 579, *586*

He, X., 89, *100,* 506, *516*

Headon, D. J., 441, *486, 491, 497*

Healy, L., 199, 200, 201, *210*

Heaney, A. P., 504, *515*

Heaney, S., 290, *294,* 398, 399, *420,* 437, 450, 451, 466, 474, *495,* 527, 530, 533, *538,* 550, *565*

Heanue, T. A., 268, *274,* 399, *417*

Heasman, J., 14, *16,* 183, 184, 185, 186, 187, 188, *189, 190,* 303, *323, 330*

Heath, J. K., 167, *174,* 313, *326,* 467, *482*

Heath, M. J., 86, *102*

Heaton, M. B., 435, *491*

Heavey, B., 204, *209*

Hebert, J. L., 313, *323*

Hebert, J. M., 578, 583, *586*

Hebrink, D., 287, *294*

Hebrok, M., 319, 320, *325, 326*

Hecht, F., 506, *517*

Hecht, J. T., 286, *292*

Heckscher, E., 45, 47, *50, 51,* 78, *100,* 338, *361*

Hedstrand, H., 313, *323,* 467, *482*

Heemskerk, J., 78, *104,* 141, *148*

Hefti, F., 91, *98,* 402, *420*

Hehr, A., 433, *492*

Heid, P. J., 165, *180,* 316, *330*

Heikinheimo, M., 165, *176, 179,* 333, 342, *362,* 466, *486*

Heikkila, M., 384, 385, *393*

Heilig, R., 398, *416,* 550, *561*

Heimann, G., 509, *517*

Heimfeld, S., 192, *209,* 244, 248, *252*

Heimrich, B., 92, *104*

Heine, U. I., 314, *325*

Heintz, N., 94, *106*

Heinz, P., 311, *324*

Heisenberg, C. P., 78, *102,* 546, *565*

Heisterkamp, N., 314, *325,* 434, 467, *487*

Heitzler, P., 343, *360, 361*

Helbling, P. M., 225, *229*

Held, M., 281, *293, 294,* 380, 381, *391, 393*

Heldin, C. H., 226, *230,* 468, *486*

Helgren, M., 559, *561*

Heller, N., 546, *562*

Heller, P., 579, *585*

Heller, R. S., 310, 321, *325*

Hellmich, H. L., 402, *417*

Hellstedt, P., 375, *392*

Hellstrom, M., 173, *178,* 313, *323, 326,* 414, *418, 467, 482*

Helman, L. J., 268, *275*

Helmbold, E. A., 227, *231*

Helminen, H., 284, *293*

Helms, J. A., 223, *232,* 281, 291, *294,* 469, 470, *486*

Helwig, U., 263, *272*

Hemati, N., 271, *276*

Hemberger, M., 171, *176*

Hemmati, H. D., 243, 245, *251*

Hemmati-Brivanlou, A., 303, *325,* 533, *535, 585*

Hemming, F., 504, *514*

Henderson, C. E., 402, *420*

Henderson, D. J., 334, 350, *362, 365,* 475, *483*

Henderson, J. A., 469, 470, 475, *486*

Henderson, N., 15, *18*

Henderson, R., 287, *292,* 312, 313, 314, 315, *323*

Hendricks, M., *251*

Hendrickson, W. A., 408, *419*

Henegar, J. R., 237, *249*

Heng, H. H., 121, *124,* 451, 453, 470, 471, *491*

Heninger, G. R., 247, *249*

Henion, P. D., 243, 244, *250*

Henke, F. S., 219, *232*

Henkel, G. W., 168, *179*

Henkmeyer, M., 135, *149*

Henn, W., 288, *293*

Hennings, H., 580, *586, 587*

Henricksen, K., 291, *292*

Henrique, D., 134, 135, 138, 139, *145, 146, 147,* 553, 559, *561,* 577, *585*

Henry, G. L., 166, *176,* 303, *325*

Henry, J. J., 530, 534, *536, 537,* 542, *562*

Henry, M. D., 161, *179*

Hentges, K., 79, 86, *100,* 243, *250,* 379, 384, *391*

Hentges, S., 504, 511, *515*

Hentsch, B., 317, *325*

Hentschel, H., 395, *417*

Her, H., 530, 533, *538*

Herbert, J. M., 225, 226, *229*

Herken, R., 574, 583, *584*

Hermeking, H., 86, *100,* 579, *586*

Hermesz, E., 45, *50,* 508, *515*

Hermiston, M. L., 309, 310, *325*

Hernandez-Verdun, D., 170, *176*

Herregodts, P., 502, *515*

Herrera, D. G., *106*

Herrera, P. L., 321, *328*

Herrick, T. M., 94, *101*

Herring, S., 427, 452, *486*

Herrman, J., 506, *517*

Herrmann, B. G., 45, 47, *50,* 132, 133, 135, 137, 140, *146, 147, 149,* 167, *179, 393,* 468, *483*

Herrup, K., 94, *100,* 112, 115, *123, 124,* 169, *179,* 318, *326,* 428, *491*

Herskowitz, I., 158, *176*

Hertle, M. D., 572, *586, 589*

Hertwig, P., 544, *562*

Hertz, R., 164, *176*

Herva, R., 412, *418*

Herz, J., 94, *100, 105*

Herzenberg, L. A., 157, *180,* 194, *207,* 245, *250*

Herzig, A., 170, *179*

Herzlinger, D., 400, 401, 405, 408, *417, 418, 419,* 571, *588*

Herzog, W., 577, 583, *584*

Hess, H., 160, *175, 178*

Hess, J. L., 376, *390, 393*

Hetherington, C. M., 84, *101,* 451, 475, *486,* 527, 528, 530, *536*

Heuchel, R., 317, *325*

Heutink, P., 227, *231*

Hevner, R., 83, 84, 85, 90, 92, 95, *98, 102, 103, 104*

Hewick, R. M., 467, *498*

Hewson, J., 236, *252*

Hextall, P. J., 377, *390*

Heyberger, S., 506, *517*

Heyer, B. S., 12, *16*

Heyer, J., 43, *50,* 353, 357, *365*

Heyman, H. C., 260, 268, *272,* 468, *482*

Heyman, I., 425, *485*

Heyman, R. A., 511, *513*

Heymann, S., 266, *274*

Heymer, J., 83, *98,* 470, *482*

Heynen, A., 11, *19*

Heyworth, C., 194, *207*

Hiatt, K., 196, 202, *210*

Hibbs, M. L., 453, *486*

Hibi, M., 171, 173, *176*

Hibino, H., 220, *231*

Hicklin, D. J., 227, *233*

Hicks, C., 92, *103,* 136, *149*

Hidaka, K., 262, 268, *274*

Hidaka, M., 217, *230*

Hidalgo-Sanchez, M., 112, *123*

Hidenori, I., *588*

Hiebert, S. W., 201, 204, *209,* 220, *231*

Hieda, Y., 436, *486*

Hiemisch, H., 42, *51*

Hiesberger, T., 94, *100, 105*

Higashinakagawa, T., 376, *391*

Higgins, K. M., 78, 80, *100,* 121, *123,* 579, 583, *588*

Higgs, D. D., 202, *207*

Higgs, D. R., 376, *390*

Hilberg, F., 318, *325*

Hildebrand, J. D., 79, *100*

Hildebrand, M., 449, 471, *486*

Hilding, D. A., 560, *563*

Hill, C. S., 342, *362*

Hill, D. P., *50*

Hill, J., 468, *486*

Hill, R. E., 84, 85, 87, 94, *98, 100, 105,* 302, *327,* 435, 451, 453, 468, 474, 475, 476, *485, 486, 489, 491,* 521, 522, 527, 528, 530, 533, *536*

Hill, R. J., 165, *180,* 316, *330*

Hill, S., 349, 355, *358*

Hillan, K. J., 203, *207,* 223, *229*

Hilton, D., 40, 44, *50,* 323, 338, *358*

Himeno, M., 288, *293*

Himes, D., 559, *563*

Hindmarsh, P. C., 31, *33,* 47, *50,* 376, *389,* 451, 476, *483,* 508, *514*

Hinds, J. W., 237, *250*

Hinterberger, T. J., 258, *274*

Hinuma, S., 258, 260, *276*

Hioki, T., 321, *326*

Hirabayashi, T., 341, 352, *370*

Hirakow, R., 334, *362*

Hirano, S., 91, *101,* 421, 475, *488,* 492, 525, *538*

Hirano, T., 171, 173, *176*

Hirashima, M., 228, *233*

Hirata, K., 84, *98*

Hirata, M., 341, 352, *370*

Hirato, J., 512, *518*

Hiratsuka, S., 217, *230*

Hirchenhain, J., 410, *418*

Hiroi, H., 504, *516*

Hiroi, Y., 336, 343, *365, 368*

Hirokawa, N., 38, *51*

Hirokawa, Y., 381, *392*

Hiromi, Y., 466, *485, 488*

Hirose, J., 350, *362*

Hirose, M., 350, *362*

Hirose, S., 169, *175*

Hirose, Y., 525, *536*

Hirota, A., 451, 475, 476, *492*

Hirota, H., 341, *362*

Hirsch, B., 382, 387, 388, *392*

Hirsch, J. C., 356, *362*

Hirsch, M. R., 239, *250*

Hirschi, K. K., 226, *230*

Hirsinger, E., 136, 140, 141, 142, *146, 147,* 262, 263, 264, 266, *273, 274, 276*

Hirth, K. P., 217, 220, *229*

Hirvonen, H., 528, *538*

His, W., 211, *230*

Hitchcock, P. F., 531, *536*

Hixson, D. C., 318, *324*

Hla, T., 227, *230, 231*

Hnadkiski, B., 574, *588*

Ho, C., 22, *34,* 214, *231,* 285, *293,* 442, 443, *489*

Ho, I. C., 527, 530, 531, *536*

Ho, J., 203, *209,* 217, *232*

Ho, R. K., 14, *17,* 118, *125,* 339, 341, 349, *370*

Ho, S. Y., 352, 354, *367*

Hobson, J. P., 227, *231*

Hobson, K., 236, 240, 241, *250*

Hochedlinger, K., 87, *104*

Hocking, A. M., 405, *418*

Hockley, A. D., 453, 466, 474, *498*

Hodges, M., 174, *174*

Hodgetts, C., 552, 553, 555, 558, 559, *564*

Hodgkin, J., 382, 387, *392*

Hodgkinson, C. A., 525, *536*

Hodgman, R., 7, 14, *19*

Hodgson, G., 194, *208*

Hodivala-Dilke, K. M., 223, *230,* 574, *585*

Hoefen, P. S., 196, 198, *210*

Hoeffler, J. P., 226, *230,* 511, *515*

Hoepker, V., 525, *536*

Hoffarth, R. M., 94, *101*

Hoffer, B. J., 402, *419*

Hoffman, J. I., 333, 356, *362*

Hoffman, S., 134, *148,* 350, *370*

Hoffmann, I., 408, *417*

Hofmann, C., 408, *418,* 451, 467, *490*

Hofmann, U., 574, *588*

Hofstetter, W., 443, *485*

Hogan, B. L., 40, 42, 43, *49, 52,* 82, 84, 85, 86, *99, 100,*
    101, 105, 106, 132, 135, *148, 149,* 161, *176,* 183, *190,*
    203, *210,* 287, *292,* 302, 305, 309, 312, 313, 314, 315,
    319, 320, 321, *326, 327, 329,* 402, 404, *418,* 439, 451,
    453, 467, 470, 475, 479, *485, 486, 488, 490, 494, 495,*
    496, 497, 498, 506, *516,* 527, 528, 530, *536*

Hogan, B. L. M., 188, 189, *190,* 238, 246, *251,* 303, 308,
    311, 312, 313, 314, 315, *323, 325, 326, 328,* 336, 338,
    355, *363, 370,* 403, *418,* 504, 505, 507, 508, 510, *518,*
    570, 573, *589*

Hogan, P., 469, *498*

Hohl, D., 582, 583, *584, 587*

Hokfelt, T., 559, *564*

Holash, J., 224, *230*

Holbrook, K. A., 568, 573, 575, 580, 583, *584, 586, 588*

Holdener, B. C., 31, *34*

Holder, N., 119, *123,* 342, *367,* 468, *486,* 521, 522, *537*

Holland, L. Z., 475, *486*

Holland, N. D., 475, *486*

Holland, P. W., 266, *277, 478, 498*

Holland, P. W. J., 305, *323*

Hollander, G., 195, *210*

Hollemann, T., 78, 79, 81, *100,* 521, 522, 525, 527, 533,
    535, 536

Hollenberg, S., 257, 258, *277*

Hollenberg, S. M., 349, *362*

Holleran, W. M., 582, *584*

Holley, M. C., 553, *563*

Holliday, M., 85, *101*

Holliday, R., 6, *16*

Holloway, J. M., 502, *518*

Hollweg, G., 570, *584*

Holmberg, C., 412, *418, 419*

Holmberg, E., 287, *292,* 470, *486, 497*

Holmes, D. F., 572, 583, *585*

Holmes, G., 375, *391*

Holmes, R. L., 504, *514*

Holmgren, R., 470, *487, 497*

Holmyard, D., 170, 173, *174*

Holst, P. L., 402, *417*

Holstein, A. F., 386, *393*

Holt, C. E., 264, *274*

Holthofer, H., 412, *417*

Holtmann, H., 341, *369*

Holtzer, H., 140, *144,* 263, *274*

Holtzer, S., 263, *274*

Holtzman, D. M., 91, *99, 101*

Holzman, L. B., 412, *417*

Hom, D. B., 343, *369*

Homayouni, G., 94, *99*

Homburger, S. A., 547, *562*

Homgren, R., 121, *123*

Honda, S., 375, 376, 384, *392*

Honjo, T., 136, 137, *147,* 320, 321, *323,* 509, *514*

Hood, L., 347, *364*

Hoodless, P. A., 31, *35,* 164, 165, *175, 179,* 303, 316, *329,*
    468, *497*

Hoogerbrugge, J. W., 385, *389*

Hooghe-Peters, E., 502, *515*

Hooper, J. E., 470, *481*

Hooper, M. L., 416, *419*

Hooshmand, F., 506, 509, *514,* 556, 558, *562*

Hoover, F., 527, *535*

Hopson, A., 442, 443, 449, 456, 470, *481*

Horcher, P. G., 314, *329*

Horigome, N., 421, *488*

Horner, J. W., 532, *536*

Horner, M. A., 316, *325*

Horner, N. A., 165, *176*

Horning, S. E., 376, *393*

Horowitz, M. C., 173, *179*

Horsburgh, G., 84, *101,* 451, 475, *486,* 527, 528, 530, *536*

Hörstadius, S., 434, *486*

Horton, D. B., 6, *17*

Horton, H. F., 318, *328*

Horton, S., 89, *101*

Horton, W. A., 286, *292, 294*

Horvath, E., 512, *515*

Horvitz, H. R., 158, *176*
Horwitz, G. A., 504, *515*
Hosking, B. M., 342, 343, *359*
Hosoda, K., 432, 465, 478, *483*
Hosoda, T., 336, *365*
Hosoda, Y., 258, *274*
Hostikka, S. L., 412, *417*
Hou, H., Jr., 415, *420*
Hou, Z.-H., *251*
Houart, C., 22, *34*, 77, 78, *101*
Houghton, D. C., 416, *419*
Houghton, L., 335, 343, 344, 345, 346, 347, *370*
Houliston, E., 23, *34*
Housman, D., 341, 352, *363*, 374, 375, *391, 392*, 399, 400,
    415, 416, *417, 418, 419*
Houssaint, E., 317, *325*
Houston, D. W., 303, *330*
Houweling, A. C., 341, 345, 347, 348, 350, *359*
Houzelstein, D., 267, *274*
Hovens, C. M., 453, *486*
Howard, K., 186, 187, 188, *189, 190*
Howard, L., 201, 204, *209*, 220, *231*
Howard, M., 349, *362*, 549, 553, *565*
Howard, T. D., 453, *486*
Howell, B. W., 94, *100, 101, 104*
Howells, N., 318, *325*
Howes, G., 203, *207*
Howles, P. N., 434, 453, 467, *483, 493*
Hoyer, J. R., 415, *419*
Hoyer, P. E., 373, *390*
Hoyes, A. D., 570, *586*
Hoyner, A. L., *102*, 471, *486*
Hrabé de Angelis, M., 31, *35*, 132, 136, 137, 139, 140, *145,
    146*, 292, *292*, 320, 321, *323*, 551, 552, 553, 555, 557,
    558, 559, 560, *563, 564*
Hromas, R., 316, *325*
Hrstka, R. F., 161, *179*, 527, 532, *537*
Hseih, C.-M., 347, *359*
Hsiao, J., 580, *585*
Hsieh, J. J., 136, *148*, 262, *276*
Hsieh-Li, H. M., 79, 82, *101*
Hsu, D. R., 44, *50*, 338, *362*
Hsu, L. C., 525, *538*
Hsu, S. C., 579, *587*
Hsu, Y.-P. P., 7, *19*
Hu, B., 399, *419*, 530, 533, *537*
Hu, D., 469, 470, *486*
Hu, G., 267, *272*, 450, 471, 472, 473, *498*
Hu, M., 194, *207*
Hu, N., 318, *326*
Hu, Q. L., 217, *232*
Hu, S., 402, *417*
Hu, T., 347, *366*
Hu, Y., 437, 453, *486, 498*, 580, 581, 583, *586*

Hu, Z., 402, *417, 579*, 583, *588*
Huang, B., 380, *390*
Huang, H. P., 310, 321, 322, *327*
Huang, K., 400, *418*
Huang, L., 225, *230*
Huang, R., 131, *146*, 266, *277*
Huang, S., 376, *389*, 432, *498*
Huang, S.-P., 83, 87, 91, *103, 106*, 130, 131, 132, *145*, 402,
    419, 508, *517*, 521, 527, *537*
Huang, W., 281, 282, 285, *292, 293*
Huang, X., 204, *210*, 220, *233*
Huang, Z., 527, *536*
Huarte, J., 10, 14, *18, 19*, 321, *328*
Huber, G. C., 25, *34*
Huber, M., 582, 583, *587*
Hubner, K., 183, *190*
Hudson, A. J., 271, *275*
Hudson, B. G., 412, *417*
Hudson, C., 166, *176*, 303, *325*
Hudson, J. B., 336, *370*
Huebner, K., 290, *294*, 450, 471, 472, 473, *495*
Huetz, F., 165, *174*, 303, *323*
Hug, B., 214, *230*
Huggins, G. S., 341, 343, 347, 352, *362, 368*
Hughes, G. J., 321, *326*
Hughes, S. M., 265, 268, 269, *272, 274, 276*
Huh, C. G., 94, *103*, 226, *230*
Huh, N., 580, *587*
Huh, N.-H., 580, *588*
Huh, S., 81, 86, *101*, 408, *417*, 522, 527, *536*
Huhtaniemi, I., 511, *516*
Hui, C. C., 79, 85, *101, 103*, 115, 120, 121, *123, 124, 125*,
    310, 312, 318, *326, 327*, 439, 451, 453, 470, 471, *486,
    487, 491*, 505, *517*, 522, *537*
Huisman, T., 506, *517*
Hukriede, N. A., 136, *145*
Hulander, M., 479, *487*
Hume, C. R., 45, *52*
Hummler, E., 314, *324*
Humphrey, J. S., 227, *229*
Humphries, C., 269, *273*
Humphries, R. K., 242, *252*
Hungerford, J. E., 226, *230*
Hunsicker, P., 31, *34*
Hunt, C. V., 166, *176*
Hunt, D. F., 226, *230*
Hunt, J. S., 173, *179*
Hunt, P., 422, 425, 426, 427, 433, 449, 467, 474, *482, 487*,
    571, *586*
Hunter, P. J., 170, *174, 176*
Hunzelman, J., 285, 287, *293, 294*, 442, *488*
Huot, C., 512, *514*
Huot, J., 171, 173, *176*
Hurlin, P. J., 580, *586*

Hurston, E., 318, *324*
Hurt, C. R., 260, 267, *277*
Husslein, P., 174, *177*
Hustad, C. M., 342, *365*
Hustert, E., 281, *294, 380, 393*
Huszar, D., 560, *561*
Hutcheson, A. M., 527, 530, 531, *537*
Hutchins, G. M., 312, *329*
Huth, J. R., 377, *393*
Hutson, M. R., 550, *562*
Huttner, A., 510, *518*
Huxham, I. M., 161, *176*
Huyhn, A., 194, *207*
Huylebroeck, D., 222, 226, 227, *229*, 336, 352, *359*
Huynh, T., 353, 357, *364*
Hwang, J. Y., 580, 581, 583, *587*
Hwang, S. G., 318, *324*
Hwang, S.-Y., 8, 10, 12, 13, *16, 17, 18*
Hyink, D. P., 412, 414, *417, 419*
Hymer, W. C., 511, *517*
Hynes, A.M., 85, *106*
Hynes, M., *101, 121, 123*, 579, *589*
Hynes, M. A., 110, 121, *126*
Hynes, R. O., 170, 173, *176, 180*, 183, 187, 188, *189,* 204, *206*, 219, 222, 223, *228, 229, 230, 233*, 340, 352, *361, 366*, 424, *487*, 571, 572, 574, 581, 583, *585, 589*
Hynes, T. O., 223, *230*
Hyslop, P. S. G., 347, 352, *360*

I

Iangaki, T., 38, *52*
Iannaccone, P., 470, *497*
Iavarone, A., 474, *489*
Ibañez, C. F., 402, *420*
Ibraghimov-Beskrovnaya, O., 161, *179*
Ibrahim, S. N., 512, *515*
Icardo, J. M., 333, 351, 352, 354, 355, 356, *359, 362, 366*
Ichikawa, I., 403, *418*
Igarashi, P., 313, *328*, 413, *419*
Iida, H., 376, *392*
Iida, K., 471, 479, *487*
Iimura, T., 451, 475, 476, *492*
Ikawa, M., 136, *146*, 581, 583, *589*
Ikawa, Y., 7, *16*
Ikeda, H., 502, 511, *515*
Ikeda, K., 506, *516*, 549, *564*
Ikeda, M., 47, *51*, 347, 352, *360*
Ikeda, T., 271, *275*
Ikeda, Y., 375, 376, 379, 380, 381, 384, *390, 391, 392*, 510, *515, 516*
Ikematsu, S., 350, *370*
Ikemura, T., 226, *232*
Ikenaka, K., 94, *103*

Ikeo, K., 533, *536*
Ikeya, M., 260, 264, *275*, 428, *487*
Ikura, T., 465, *492*
Ikuta, K., 192, 194, 196, *207, 209*, 245, *250*
Ikuta, T., 580, *587*
Ilgren, E. B., 166, *176*
Illmensee, K., 570, *586*
Imai, H., 437, 438, *487*
Imai, K., 263, *272*, 451, 475, *493, 498*
Imai, T., 376, 380, *390*
Imaizumi, K., 398, *419*
Imakado, S., 581, *586*
Imaki, J., 531, *538*
Imamoto, A., 157, *180, 361*
Imanaka, T., 308, *329*
Imhof, B. A., 201, *209*
Inada, M., 288, *293*, 442, 443, 453, *488*
Inadera, H., 271, *277*
Inagaki, T., 336, 344, *362*
Incardona, J. P., 78, *99, 101*
Ingenhoes, R., 174, *177*
Ingham, P. W., 78, *101*, 264, 265, 269, *272, 273, 275, 276*, 470, *497*, 534, *536*
Ingraham, H. A., 89, *100*, 376, 379, 380, 381, 384, *390, 391, 392*, 502, 509, *515, 518*
Inoue, H., 321, *325*
Inoue, K., 511, *515*
Inoue, S., 504, *516*
Inoue, T., 77, *101*, 340, 352, 354, *363, 367, 370*, 408, *417, 418*
Ion, A., 376, *391*
Iozzo, R. V., 570, 572, 583, *584, 585*
Ip, H. S., 165, *177*, 303, *327*, 342, *366*
Ip, N. Y., 559, *561*
Irintchev, A., 271, *275*
Irion, U., 336, *367*
Iruela-Arispe, M. L., 219, 226, *230, 233*
Irvine, K. D., 136, *146*, 555, *563*
Irvine, S. M., 471, *493*
Irving, C., 110, 112, 115, *123*, 428, 449, *487*
Isaac, A., 354, 355, 356, *362, 366*
Isaacs, H. V., 167, *176*
Isaacs, I., 240, 244, *252*
Isacson, O., 89, *99*
Iscove, N. N., 193, *207*
Iseki, S., 422, 423, 424, 435, 437, 451, 452, 475, 476, 478, *487, 490, 492*
Isenmann, S., 173, *175*, 260, 266, *272*, 317, *323*, 404, *417*
Ish-Horowicz, D., 134, 135, 138, 139, *146, 147*, 553, 559, *561*, 577, *585*
Ishibashi, M., 136, *146, 147*, 310, 321, *325*, 526, 527, *538*
Ishida, Y., *537*
Ishida-Yamamoto, A., 581, 582, 583, *587*
Ishii, S., 121, *123*

Ishii, Y., 308, *325*
Ishikawa, H., 512, *515*
Ishikawa, T., 172, 173, *176,* 465, *489*
Ishimaru, Y., 47, *50,* 354, *370*
Ishino, K., 580, *588*
Isner, J. M., 220, *228*
Israel, A., 136, *146*
Israel, M. A., 174, *177*
Isselbacher, K. J., 339, *368*
Itami, S., 452, *496,* 580, 581, 583, *589*
Itasaki, N., 117, 118, *123,* 355, *362*
Iten, L. E., 548, 550, 551, *561*
Itie, A., 468, *495,* 510, *514*
Ito, C. Y., 415, *417*
Ito, M., 6, *19,* 376, 378, 379, *389, 393*
Ito, T., 171, *178,* 314, *327*
Ito, Y., 220, 228, *229, 231,* 246, *249,* 288, *293,* 318, *325*
Itoh, H., 228, *233,* 354, *370*
Itoh, M., 11, 12, *18,* 171, 173, *176*
Itoh, N., 311, 312, 313, 314, 315, 321, *323, 324, 325, 327, 328,* 505, *517*
Itskovitz-Eldor, J., 303, *329*
Ittmann, M., 505, *517*
Ivanova, N., 194, 206, *209*
Ivanyi, E., 42, *51,* 158, 160, *177, 178*
Ivins, K. J., 469, *498*
Ivkovic, S., 91, *101*
Iwamoto, M., 289, *294*
Iwana, Y., 504, *517*
Iwasaki, T., 512, *518*
Iwata, T., 286, *293*
Iyer, N. V., 225, *230*
Izac, B., 194, *207*
Izawa, M., 11, 12, *18*
Izpisua-Belmonte, J. C., 39, *51,* 338, 351, 353, 354, 355, 356, 357, *359, 364, 367,* 433, 478, *489, 496,* 507, 508, *518,* 580, 581, 583, *587*
Izumo, S., 219, *232,* 334, 341, 343, 345, 347, 348, 349, 350, 351, 352, *359, 361, 363, 368, 369*

J

Jaber, M., 341, 352, *362,* 511, *513*
Jabs, E. W., 452, 453, 474, *486, 487*
Jacenko, O., 443, *487*
Jäckle, H., 316, *329,* 469, *483*
Jacks, T., 318, *325, 483,* 528, *537*
Jackson, B. W., 570, *586*
Jackson, C. E., 227, *231,* 453, *487*
Jackson, D. I., 377, *390*
Jackson, I. J., 31, *34,* 560, *561, 565*
Jackson, K. A., 193, 206, *208,* 270, 271, *275*
Jackson, M. R., 12, *17*
Jacob, M. P., *585*

Jacobs, H., 170, *179*
Jacobs, H. C., 312, 316, *323, 324*
Jacobs, J., 440, *498,* 577, *589*
Jacobs, M. L., 349, *369*
Jacobs, Y., 443, 449, 476, 479, *482, 487, 495*
Jacobsen, S. E., 11, *17*
Jacobson, A. G., 335, 337, 339, *362, 367,* 422, 423, 435, *487, 492,* 542, *563*
Jacobson, K., 23, *33,* 166, 167, *175, 178*
Jacobson, L., 510, 512, *516*
Jacobson, R. D., 530, *536*
Jacquemin, P., 166, *176,* 321, *325*
Jadjantonakis, A.-K., 357, *368*
Jaenisch, R., 79, *104,* 258, 260, *273, 276,* 314, *326,* 341, 352, *363,* 375, *391,* 400, 410, *418,* 427, 443, 467, 469, *484, 489, 491, 495,* 529, *537,* 559, 560, *561, 562,* 573, 574, 583, *584, 585, 587*
Jaffredo, T., 199, 200, 201, *207, 208,* 1987
Jagla, K., 260, 266, *273, 274, 276*
Jaglarz, M., 186, 187, 188, *189*
Jahoda, C. A., *585*
Jain, M. K., 347, *359*
Jain, R. J., 218, 223, *230*
Jain, R. K., 225, *229*
Jaing, T. X., 577, *589*
Jaiswal, R. K., 206, *209*
Jalanko, H., 412, *419*
Jalic, M., 416, *420*
Jalife, J., 334, 348, *367*
James, B., 185, *189*
James, C. D., 287, *294*
James, R., 160, 167, *174, 175,* 509, 510, *515*
Jameson, J. L., 6, *19,* 376, 378, 379, *389, 393,* 500, *518*
Jami, J., 320, *324*
Jamrich, M., 521, 522, 527, 532, *536, 537*
Jan, L. Y., 136, *145,* 185, *189,* 257, *276,* 334, 349, *358, 365,* 553, *563*
Jan, Y. N., 87, *97,* 136, *145,* 185, *189,* 334, 349, *358, 365,* 553, *561, 563*
Janis, L. S., 91, *101*
Jankovski, A., 95, *101*
Janne, J., 228, *230*
Janney, M. J., 271, *276*
Janse, M. J., 348, *360*
Jantapour, M. J., 174, *177*
Janumpalli, S., 247, *251*
Japon, M. A., 502, *515*
Jarnik, M., 582, 583, *587*
Jarriault, S., 136, *146*
Jarry-Guichard, T., 348, *360*
Jarvis, L. A., 466, *491*
Jaskoll, T., 436, *487*
Jat, P. S., 404, *420*
Jaubert, F., 375, 387, *389, 392*

Jaw, T. J., 91, *103*
Jaworowski, A., 161, *179*
Jaye, M., 453, *487*
Jean, D., 522, *536*
Jeannotte, L., 171, 173, *176*, 312, *323*
Jeffreys, R. V., 505, *513*
Jeffs, P., 264, *277*
Jegou, B., 382, *389*
Jeltsch, M., 218, 223, *230*
Jen, W., 138, *146*
Jenkins, N. A., 78, 79, 81, 85, *99, 100, 103,* 168, 169, 173,
      175, 176, 179, 193, 204, *208,* 321, *328,* 341, 342, 345,
      348, 349, 352, *359, 360, 365, 366,* 411, *418,* 443, 467,
      476, *482, 485, 488,* 508, *515,* 521, 525, 533, *536, 537,*
      538, 556, *565*
Jenkins, R. B., 287, *294*
Jenkinson, E. J., 166, *177*
Jenkinson, J. W., 442, *487*
Jensen, A. M., 86, *101*
Jensen, J., 166, *176,* 310, 321, *325, 327*
Jensen, P. J., 570, *589*
Jensen, U. B., 570, *586*
Jepeal, L., 268, *273,* 528, *535*
Jerabek, L., 244, *252*
Jernvall, J., 441, *487, 493, 497*
Jerome, L. A., 353, 357, *362*
Jeske, Y. W., 377, *391*
Jessee, J., 11, *18*
Jessell, T. M., 3, *4,* 31, *34,* 43, 46, 47, 48, *50, 52,* 78, 79, 80,
      81, 82, 85, 87, 88, *98, 99, 101, 102, 104, 105, 106,*
      108, 109, 120, 121, 122, *122, 123, 124, 125,* 141, *148,*
      165, *179, 250,* 316, 319, 320, 321, *323, 326, 329,* 397,
      420, 423, 474, *489, 494,* 504, 505, 506, *514*
Jetton, T. L., 309, 319, 320, *327,* 506, *516*
Ji, Q., 453, 476, *493*
Ji, X., 531, *536*
Jiang, F., 165, *177,* 303, *327,* 341, 343, 352, *368*
Jiang, R., 423, *487,* 553, 555, *563*
Jiang, T. X., 574, 577, 578, 579, *585, 586, 587*
Jiang, X., 228, 229, *230,* 246, *249, 250,* 334, 348, *362,* 426,
      482
Jiang, Y., 339, 342, 346, 347, *362, 363,* 581, *586*
Jiang, Y.-J., 139, *146,* 546, *565*
Jiang, Z., 90, 95, *106*
Jiangming, L., 346, *363*
Jimenez, M. J., 289, *293*
Jimenez, R., 388, *391*
Jin, D., 416, *420*
Jin, O., 31, *33,* 42, 43, 46, *49, 51,* 116, *122, 323,* 451, 478,
      481, 521, 527, *535*
Jin, P., *105*
Jing, S., 402, *417*
Jing, X., 502, *515*
Jishage, K., 95, *102,* 506, *516*

Job, C., 310, *327*
Jobe, S. M., 343, *363*
Jobert, A. S., 285, *293*
Jochum, W., 87, *104,* 173, *179*
Johansen, F.-E., 342, *369*
Johansen, K. M., 136, *149*
Johansen, T., 546, *563*
Johansen, V. A., 452, *487*
Johansson, C. B., 236, 237, 244, 245, 246, 248, *249, 250,*
      318, *324*
Johansson, S., 574, *589*
John, R., 311, *324*
Johnson, D., 11, 12, *18*
Johnson, D. K., 31, *34*
Johnson, D. R., 453, 471, *487*
Johnson, D. W., 227, *231*
Johnson, G. R., 195, 201, *208,* 314, 317, *325, 328*
Johnson, J., 8, *17*
Johnson, J. A., 168, *177*
Johnson, J. D., 321, *328*
Johnson, J. E., 84, 88, 89, *101, 102,* 168, *176, 177,* 239,
      250, 428, *487*
Johnson, K. R., 10, 12, *16,* 94, *105,* 550, *563*
Johnson, M. C., 356, 357, *358, 366*
Johnson, M. D., 467, *493*
Johnson, M. H., 5, *17,* 23, 26, *34,* 156, 158, 166, *175, 177,*
      178, 180
Johnson, M. L., 442, *487*
Johnson, R., 453, *486,* 541, 542, 546, 552, 555, 558, 559,
      564, 580, 581, 583, *586*
Johnson, R. L., 136, 137, 138, 141, *145, 146,* 262, 264,
      265, *275,* 280, 287, *293,* 305, *328,* 354, 355, 356, *364,*
      451, 470, *485, 487, 489, 490, 494,* 571, 579, *586*
Johnson, R. S., 225, *232,* 442, *487*
Johnston, J. G., 94, *101*
Johnston, M. C., 423, 468, *487, 497*
Jollie, M., 421, *487*
Jollie, W. P., 161, *177*
Joly, J. S., 116, *125*
Jones, B., 453, *494*
Jones, B. M., 453, *493*
Jones, C. J., *251,* 415, *417*
Jones, C. M., 40, 43, *49*
Jones, D. H., 156, *177*
Jones, E. G., 84, *106,* 322, *326*
Jones, E. M., 318, *326*
Jones, J. E., 553, *562*
Jones, J. M., 303, *329*
Jones, K. R., 91, *101,* 549, 559, *562*
Jones, K. T., 8, *16*
Jones, M. C., 338, *362*
Jones, M. J., 2, *4,* 171, 173, *174*
Jones, P. F., 224, *231, 232*
Jones, P. H., 570, *587*

Jones, R. C., 410, *418*
Jones, R. O., 195, *208,* 314, *326*
Jones, S., 451, 453, 471, *498,* 506, *518*
Jones, W., 416, *420*
Jones, Y. Z., 172, 173, *174,* 341, 357, *358*
Jones-Seaton, A., 21, *34*
Jonsson, J., 319, 320, *322, 325*
Jorcano, J. L., 574, 583, *584*
Jordan, C. T., 192, *208*
Jordan, S. A., 31, *34*
Jordan, T., 451, 475, *486,* 527, 528, 530, *536, 538*
Jorgenson, R., 507, *515*
Josephson, R., 89, *101*
Josso, N., 372, 374, 387, *391, 392*
Jost, A., 372, 374, *391,* 507, *515*
Jostes, B., 475, *487*
Jouanneau, J., 223, *230*
Joubin, K., 41, 45, *51*
Joukov, V., 162, *179,* 218, 223, *230*
Joulin, V., 382, *389*
Joutel, A., 227, *230*
Jouve, C., 134, 135, 138, 139, 142, *146,* 262, 263, 264, *274*
Jowett, A. K., 437, 439, *490, 497*
Joy, K. C., 12, *17*
Joyner, A. L., 79, 84, 85, 88, *100, 101, 102, 103, 106,* 109, 110, 111, 112, 113, 114, 115, 116, 120, 121, *123, 124, 125, 126,* 168, *176,* 228, *230, 233,* 239, *250,* 428, 451, 453, 470, 471, *487, 489, 491, 493, 497*
Juhasz, M., 91, *97*
Jullien, D., 9, 14, *15*
Jünemann, G., 443, *486*
Jung, D., 173, *175*
Jung, H. S., 441, *493,* 574, 577, 578, 579, *585, 586, 587, 588*
Jung, J., 316, 317, 320, *324, 325*
Junien, C., 416, *419*
Juppner, H., 285, *293, 294,* 442, 443, *489*
Jurata, L. W., 135, *146*
Jurecic, V., 353, 357, *364*
Jürgens, G., 316, *329,* 469, *483*
Juriloff, D. M., 423, *486, 487*
Juriscova, A., 13, *17*
Jussila, L., 218, 223, *229*
Just, W., 388, *391*
Justice, M. J., 292, *293*
Justus, M., 85, *103*

K

Kaartinen, V., 314, *325,* 434, 467, *487*
Kabl, J. M., 165, *177*
Kablar, B., 258, 260, 261, 262, 267, 271, *275*
Kachar, B., 557, *561*
Kachinsky, A. M., 132, *148,* 263, *276*

Kadesch, T., 257, *275*
Kadler, K. E., 572, 583, *585*
Kadokawa, Y., 136, *146*
Kaestner, K. H., 40, 42, *51,* 135, *146,* 165, *177,* 303, 310, *325, 327,* 479, *487*
Kageyama, R., 136, *146, 147,* 310, 321, *325,* 526, 527, *538,* 555, *566*
Kahane, N., 131, 132, 140, 141, *144, 145, 146,* 255, 256, *273, 275*
Kahn, A., 262, 268, *276,* 342, *368*
Kaibuchi, K., 95, *105*
Kaipainen, A., 162, *179,* 218, 223, *230,* 580, *589*
Kaitila, I., 286, *292,* 453, *481*
Kakeya, T., 511, *515*
Kakinuma, H., 471, 479, *487*
Kakizuka, A., 581, *588*
Kakoi, H., 557, *563*
Kaku, M., 443, 453, *487*
Kalamaras, J., 321, *328, 329,* 451, 470, 475, *494, 497*
Kalantry, S., 2, *4*
Kalatzis, V., 398, *416,* 550, *561, 563*
Kalb, J. M., 316, *325*
Kalb, R. G., 225, *231*
Kalcheim, C., 131, 132, 140, 141, *144, 145, 146,* 255, 256, *273, 275,* 422, 423, 424, *489*
Kalen, M., 414, *418*
Kalka, C., 220, *228*
Kalla, K. A., 89, 95, *104,* 506, 509, 510, *514, 516, 517*
Källen, B., 75, *98*
Kalman, F., 345, 356, *369*
Kalnins, V. I., 527, *535*
Kaltenbach, J. A., 557, *563*
Kalter, H., 442, *487*
Kalvakolanu, D. V., 343, *365*
Kalyani, A. J., 236, 240, 241, *250, 251*
Kamachi, Y., 283, *293,* 381, *391,* 529, 530, 531, 534, *536, 538, 565*
Kamada, N., 432, 450, 465, *489*
Kamada, S., 399, *419*
Kamada, T., 557, *565*
Kamino, K., 343, *362*
Kamisago, M., 346, *358*
Kamiya, A., 318, *325*
Kamps, M. P., 478, 479, *488, 489*
Kan, M., 287, *293*
Kanai, Y., 38, *51,* 380, *391*
Kananen, K., 511, *516*
Kanatani, N., 289, *294*
Kanatsu, M., 203, *208*
Kandel, E. R., *250*
Kane, D. A., 546, *565*
Kanekar, S., 526, 527, *535*
Kaneko, K. J., 5, 9, 10, *17*
Kaneko, S., 525, *535*

Kaneko, Y., 189, *189*, 202, *208*
Kang, I.-S., 354, *365*
Kang, S., 470, *487*
Kania, A., 31, *34*, 43, 46, 47, 48, *52, 104, 397, 420*
Kantor, A. B., 245, *250*
Kanzawa, N., 334, *368*
Kanzler, B., 468, *488*, 571, 572, 574, *585, 587*
Kao, K. R., 335, *362*
Kapiloff, M., 510, *515*
Kaplan, D. H., 287, *293*
Kaplan, D. R., 91, *101*
Kaplan, F. S., 271, *276*
Kaplan, M. S., 237, *250*
Kappel, A., 228, *230*
Kappen, C., 312, *325*
Kapur, R. P., 78, *99, 101*, 465, *498*
Karagogeos, D., 553, *562*
Karaplis, A. C., 285, *293*, 442, 443, *488, 489*
Karashima, T., 581, *585*
Karasuyama, H., 87, *101*
Karavanov, A., *104*, 508, *517*
Karavanova, I. D., 405, *417*, 439, *497*, 577, 578, 583, *589*
Karels, B., 385, *389*
Kargul, G. J., 11, 12, *17*
Karin, M., 87, *104*, 453, *486*, 504, 509, 510, *513, 514, 516*, 580, 581, 583, *586*
Karis, A., 343, *369*
Karl, J., 373, *391*
Karlsson, L., 313, *326*, 414, *418*
Karlsson, S., 203, *207*, 226, 229, *230*, 467, *484*
Karlstrom, H., 246, *249*, 318, *324*
Karlstrom, R. O., 85, *101*, 505, *515*, 532, 534, *536*
Karolyi, J., 14, *15*
Karp, S. J., 287, *293*, 442, *488*
Karpen, S. J., 316, *324*
Karperien, M., 285, *293*, 442, 443, *489*
Karr, D., 286, *292*
Karram, K., 88, *98*
Karsenty, G., 280, 288, 289, *292, 293, 294*, 408, *418*, 451, 467, *490*
Kartasova, T., 581, 582, 583, *587, 588*
Kartori, Y., 506, *516*
Kas, E., 9, 14, *15*
Kasahara, H., 343, 347, *361, 363*
Kasai, N., 511, *516*
Kashiwagi, M., 580, *587, 588*
Kashtan, C. E., 412, 416, *418, 419*
Kaspar, P., 156, *179*
Kass, D. A., 343, 350, 351, 352, *359*
Kastner, P., 91, *101*, 341, 346, 347, 352, *362, 368*, 451, 468, 469, 470, *484, 489, 490*, 525, *536*, 557, *562*
Kastschenko, N., 423, *488*
Kastury, K., 290, *294*, 450, 471, 472, 473, *495*
Katagiri, C., 198, *208*

Katagiri, T., 271, *275*
Katahira, T., 112, 114, 115, *123*
Kataoka, H., 200, *209*
Kataoka, Y., 471, 479, *487*
Katayama, A., 553, *562*
Kathiriya, I. S., 349, *360*
Kato, A., 525, *538*
Kato, K., 263, *274, 275*, 557, *565*
Kato, N., 471, 479, *487*
Kato, S., 505, *517*, 525, *535*
Katoh, M., 22, *34*
Katoh-Fukui, Y., 353, 354, 355, 356, *363*, 376, *391*
Katori, Y., 549, *564*
Katsev, S., 336, *368*
Katsura, Y., 200, *209*
Katz, A., 194, *207*
Katz, B. Z., 408, *418*
Katz, L. C., 94, *102*
Katz, R. W., 509, *514*
Kau, C. L., 198, *208*
Kaufman, M. H., 127, 128, 131, *146*, 195, 196, 199, *208*, 333, *362*, 395, *418*, 451, 475, *488, 493*, 500, 504, *515*, 541, 542, *563*, 568, 571, 574, 579, 581, *587*
Kaufmann, E., 165, *177*, 453, 479, *488*
Kaur, P., 570, *589*
Kaur, S., 79, *105*
Kavanaugh-McHugh, A., 357, *358*
Kawabata, M., 356, *367*
Kawabe, K., 379, *391*
Kawabe, S., 580, *587, 588*
Kawachi, Y., 582, 583, *587*
Kawai, T., 452, *496*, 580, 581, 583, *589*
Kawaichi, M., 136, 137, *147*
Kawakami, A., 43, *50*, 88, *99*
Kawakami, K., 399, *419*
Kawamoto, H., 200, *209*
Kawamura, K., 95, *102*, 471, 479, *487*, 504, 506, *515, 516*
Kawano, H., 95, *102*, 506, *516*
Kawasaki, T., 223, *230*
Kawase, E., 188, *189*, 226, *232*
Kawase, T., 506, *516*, 549, *564*
Kawashima, H., 350, *362*
Kawasoko, S., 443, 453, *487*
Kawata, T., 443, 453, *487*
Kawate, T., 222, *231*
Kawauchi, H., 547, *564*
Kawauchi, S., 525, 527, 530, 531, *536*
Kayden, H. J., 161, *175*
Kazanietz, M. G., *585*
Kazarov, A., 200, *207*
Kaznowski, C. E., 92, *102*
Kazuyuki, O., 183, *190*
Kearne, M., 220, *228*
Kedinger, M., 308, 310, *324, 325, 326*

Keene, D. R., 572, 573, 583, *584, 587*
Keeney, D., 453, 476, *493*
Kehrl, J. H., 319, *325*
Keighren, M., 23, *34*
Keijser, A. G. M., 341, 345, 347, 348, 350, *359*
Keiper-Hrynko, N., 6, *17*
Keithley, E. M., 556, 558, *562*
Kelcher, J., 76, 77, 82, 83, *103*
Kelemen, M., 195, *208*
Keller, G., 195, 200, 203, *207, 209,* 510, *518*
Keller, G. A., 223, *229*
Kelley, C. M., 198, 203, *208, 210*
Kelley, M. W., 553, 555, 557, *562, 563, 564, 565*
Kelly, D., 166, *179*
Kelly, R., 140, 141, *145, 148,* 258, 263, 264, *273, 277,* 341, 343, 344, 345, 356, *361, 362*
Kelly, T., 507, *515*
Kelly, W. A., 534, *535*
Kelsey, A., 375, *391*
Kelsh, R. N., 546, *565*
Kelvin, D. J., 342, *361*
Kemler, R., 15, *16,* 408, 409, 410, *418, 419, 420*
Kemp, C., 31, *33,* 44, 48, *49, 50,* 78, *98,* 304, *323,* 338, *358*
Kemper, O. C., 341, *369*
Kempermann, G., 237, *250*
Kendall, S. K., 511, 512, *515, 518*
Kennedy, M., 195, 200, *207, 209*
Kennedy, T. E., 510, *514*
Kennerdell, J. R., 14, 15, *17*
Kenny, D. A., 135, *146*
Kenny-Mobbs, T., 140, *146*
Kent, J., 380, 387, *391*
Kenyon, K. L., 532, *536*
Keränen, S., 441, *487, 497*
Kere, J., 441, *488, 493*
Kern, M. J., 433, 476, 477, 478, *489*
Kern, S. E., 468, *495*
Kerr, P., 341, *359*
Kerr, W. G., 157, *180*
Kerrebrock, A. W., 136, 137, *147*
Kertesz, N., 40, *49, 323*
Keshet, E., 193, 204, *208, 210,* 225, *229*
Keshvara, L., 94, *99*
Kessel, M., 39, 41, 46, *51,* 78, 81, *103,* 129, *147,* 427, *488,* 546, *564*
Kessler, D. S., 45, *52,* 239, *250,* 303, *325*
Kessler, J. A., 239, *251*
Kestevan, S., 348, 349, 352, 353, 357, *358*
Kesteven, H. L., 442, *488*
Kestila, M., 412, *418, 419*
Ketting, R. F., 14, *18*
Kettlewell, J. R., 382, 387, 388, *391, 392*
Kettunen, P., 441, *487, 493*
Kettunen, P. J., 452, *488, 494*

Keutel, J., 281, *294,* 380, *393*
Keverne, E. B., 435, *488*
Keynes, R., 119, *123,* 225, 229, 422, 425, 428, *490*
Keynes, R. J., 134, *147, 148*
Khan, K. M., 547, *564*
Khan, M. K., 193, 206, *207,* 246, *250,* 270, 271, *274*
Khavari, P. A., 578, 579, 580, 583, *585, 588*
Khoo, P., 3, *4,* 422, 427, *482*
Khoo, W., 341, *370*
Khurana, T. S., 94, *104*
Kidd, S., 136, *147*
Kidder, G. M., 5, *17,* 156, *177*
Kido, M., 38, *51*
Kidwell, J. F., 453, *488*
Kieboom, D., 192, *210*
Kieboom, K., 376, *389,* 453, *481*
Kieckens, L., 203, *207,* 223, *228*
Kiecker, C., 338, *362*
Kiefer, P., 545, *563*
Kier, A. B., 204, *208,* 318, *327,* 453, *483*
Kiernan, A. E., 546, 548, 551, 552, 557, *561, 563, 565*
Kiessling, A. A., 14, *18*
Kikuchi, K., 560, *563*
Kikuchi, T., 506, *516,* 549, *564*
Kikuchi, Y., 303, *326*
Kikuyama, S., 504, *515*
Kil, J., 558, *563*
Kilby, M. D., 174, *179*
Kilkenny, C., 402, *420*
Killary, A. M., 267, *277*
Kilner, P. J., 351, *363*
Kilpatrick, T. J., 86, *101*
Kim, C. H., 469, 470, *486*
Kim, C.-Y., 354, *365*
Kim, E., 375, *390,* 415, *418*
Kim, H. J., 452, *488*
Kim, I. S., 288, *293*
Kim, J.-H., 342, *360*
Kim, J. I., 527, 530, 531, *536*
Kim, J.-S., 173, *177,* 348, *363,* 379, *391*
Kim, M.-S., 347, *362*
Kim, P. C., 310, *326*
Kim, S.-H., 40, 44, *51,* 243, *249,* 338, *359*
Kim, S. K., 319, 320, *325, 326*
Kim, Y., 334, 343, *361, 363,* 474, *488*
Kim, Y. H., 217, 220, *229*
Kimata, K., 284, *294,* 350, *370,* 442, 443, *493, 497*
Kimberling, W. J., 414, *418*
Kimble, J., 10, *19,* 165, *176,* 316, *325, 326,* 340, *358,* 471, *493*
Kimble, R. B., 292, *292*
Kimelman, D., 44, 45, *51,* 167, *176,* 334, *361,* 579, *584*
Kimmel, C. B., 118, *125,* 167, *176,* 465, *491*
Kimmel, R. A., 109, *123,* 228, *230*

Kimmel, S. G., 310, *326*

Kimura, C., 31, *34*, 46, *51*, 114, *124*, 451, 478, *488, 490*, 521, 527, *537*

Kimura, S., 81, 82, 88, 89, 90, 91, *105*, 194, *208*, 311, 312, *326, 327*, 474, *488*, 504, 505, 507, 508, 510, *518*

Kimura, T., 288, *293*, 443, *492*

Kimura, Y., 8, *17*, 408, *418*

Kina, T., 192, 194, *207*, 245, *250*

Kindblom, J. M., 510, *515*

Kinder, S. J., 3, *4*, 38, *50, 51*, 132, *147*, 397, *420*

King, A., 267, *274*

King, C. Y., 381, *390*

King, J. A., 82, *101*, 315, *326*

King, M. L., 6, *17*, 303, *330*

King, T., 353, 356, *363*

Kingbury, B. F., 421, *488*

Kingdom, J. C. P., 174, *177*

Kingsley, D. M., 82, *101*, 315, *326*, 467, *488, 496*

Kinloch, R. A., 15, *18*

Kinnon, C., 201, *208*

Kinoshita, C., 581, 582, 583, *587*

Kinoshita, N., 338, *363*

Kinoshita, T., 318, *325*, 581, 583, *589*

Kintner, C. R., 138, *146*, 226, *232*, 408, 409, 410, *417, 418*

Kinzler, K. W., 86, *100*, 579, *586, 587*

Kioschis, P., *327*

Kioussi, C., 353, 354, 355, *364*, 398, *420*, 433, 478, *489*, *496*, 500, 509, *515*

Kioussis, D., 31, *34*, 311, 316, *327*

Kiraly, K., 284, *293*

Kirby, C., 22, *34*, 157, *177*

Kirby, M. L., 14, *17*, 334, 348, 357, *363*, 451, 473, *495*

Kirchhoff, S., 173, *177*

Kirillova, I., 318, *329*

Kirsch, I. R., 512, *515*

Kirschner, M. W., 338, *363*

Kishida, T., 227, *229*

Kishimoto, T., 204, *209*, 220, *232*, 288, *293*, 318, *325*, 341, 352, *370*, 442, 443, 453, *488*

Kishimoto, Y., 336, *363*

Kispert, A., 45, 47, *50*, 135, 137, 140, 141, *147*, 336, 338, 355, *359, 363*, 384, 385, *393*, 405, 406, 407, *418, 419*, 468, *483*

Kissa, K., 9, 14, *15*

Kisslinger, J. A., 136, *148*

Kist, R., 380, *392*

Kitagawa, M., 286, *293, 294*

Kitaguchi, T., 357, *363*

Kitajewski, J., 140, 141, *147*, 226, *230, 232*, 262, 264, *275*

Kitajima, K., 317, *327*

Kitajima, S., 340, 352, *363, 367*

Kitamoto, Y., 413, *418*

Kitamura, K., 353, 354, 355, 356, *363, 557, 563*

Kitamura, N., 174, *179*, 404, *420*

Kitamura, T., 186, *190*

Kitamura, Y., 288, *293*, 442, 443, 453, *488*, 555, *565*

Kitanowski, F., 451, 453, 474, *489*

Kitchen, J. R., 11, 12, *17*

Kitos, P., 200, 201, *209*

Kitsukawa, T., 223, *230*

Kittappa, R., 200, *209*

Kizumoto, M., 200, *209*

Klagsbrun, M., 156, *178*, 223, *230, 232*

Klaine, M., 200, *208, 209*

Klamt, B., 347, *363*

Klausing, K., 504, 510, *516*

Klausner, R. D., 173, *176*, 227, 229

Klein, G., 411, *418*

Klein, P. S., 405, *418*

Klein, R., 91, *105*, 119, *123*, 171, 173, 174, *174*, 266, *275*, 559, *564*

Klein, R. D., 402, *418*

Klein, W. H., 204, *209*, 260, 261, 262, *274, 276, 277*, 289, *294*, 527, *535, 536*, 556, 558, *565*

Kleinman, H. K., 415, *420*, 424, 442, 443, *482*

Klewe-Nebenius, D., 23, *34*, 160, 166, *178*

Klima, M., 473, *488*

Kline, A. D., 453, *486*

Kline, D., 8, *17*

Klingensmith, J., 40, 42, 43, 45, 48, 49, 50, *51, 52*, 78, *98*, 217, *229*, 304, *323*

Klisak, I., 452, 453, 474, *487*

Kloc, M., 1, *4*, 8, *16, 17*

Kloos, J., 423, *493*

Kloter, U., 533, *535*

Kluk, M., 227, *230*

Klymkowsky, M. W., 423, *482*

Knapp, L. T., 507, 508, 511, *514*

Kneppers, S. L., 285, *293*

Knezevic, V., 40, *51*

Knöchel, W., 165, *177*, 453, 479, *488*

Knoedlseder, M., 94, *104*

Knoepfler, P. S., 478, 479, *488, 489*

Knoetgen, H., 41, 46, *51*, 427, *488*

Knofler, M., 174, *177*, 321, *326*

Knoll, J. H., 288, *293*

Knowles, B. B., 6, 8, 10, 11, 12, 13, 15, *16, 17, 18*

Knowles, H. J., 313, *323*

Knudsen, K. A., 264, *274*, 340, *364, 366*

Knudson, C. M., 196, 198, *210*

Knüsel, B., 91, *101*

Knust, E., 136, *145, 149*

Ko, K., 219, *229, 231*

Ko, M. S. H., 11, 12, *17*

Koban, M. U., 343, *366*

Kobayashi, A., 397, *420*

Kobayashi, S., 186, *190*

Kobelt, A., 281, *293*, 381, *391*

Koblar, S. A., 173, *179*
Kobler, J. B., 353, 357, *365*
Koch, A. W., 408, *418*
Koch, C. J., 225, *229*
Koch, P. J., 582, 583, *587*
Koch, W. J., 341, 352, *362, 363*
Kochanowski, H., 12, *16*
Kodama, T., 432, 450, 465, *489*
Koder, A., 172, 173, *174,* 341, 357, *358*
Kodjabachian, L., 40, *51*
Koehler, G., 205, *206*
Koehler, M., 388, *392*
Koentgen, F., 31, *34,* 345, 348, *366*
Koentges, G., 422, 425, *485*
Koester, S., 264, 268, *275*
Koetgen, F., 348, 349, 352, 353, 357, *358*
Kofron, M., 14, *16*
Koga, Y., 312, *328,* 505, *517*
Kohan, R., 543, *563*
Kohl, L., 15, *15*
Kohn, G., 174, *178*
Kohtz, J. D., 78, 81, *99, 101,* 110, 121, *123, 124*
Koishi, K., 269, *274*
Koiussis, D., 46, 47, 48, *51*
Koivisto, T., 314, *327*
Koji, T., 504, *516*
Kojima, I., 321, *324, 327*
Kojima, M., 317, *327*
Kokaia, M., 247, *249*
Kokaia, Z., 247, *249*
Kola, I., 22, *34*
Kolatsi-Joannou, M., 404, *420*
Kollar, E. J., 421, 438, 439, *481, 488, 491*
Kollet, O., 341, *369*
Kolluri, S. K., 346, *364*
Kolquist, K. A., 400, *418*
Kolski, C., 506, *517*
Komaki, M., 271, *275*
Kominami, R., *537*
Komiya, T., 258, *272*
Komori, T., 204, *209,* 220, *232,* 288, 289, *293, 294,* 442, 443, 453, *488*
Komura, I., 343, *368*
Komuro, I., 334, 336, 343, 347, *361, 363, 365*
Komuves, L., 580, 581, 582, 583, *585, 586, 588*
Kon, Y., 511, *516*
Kondaiah, P., 467, *491*
Kondo, M., 194, *208,* 241, *250*
Kondo, S., 317, *327,* 353, 354, 355, 356, *363*
Kondo, T., 241, 246, *250,* 305, 310, *326*
Kondoh, H., 47, 49, *51, 52,* 120, 121, *125,* 283, *293,* 318, *328,* 357, *365,* 381, *391,* 522, 527, 528, 529, 530, 531, 532, 533, 534, *536, 537, 538, 565*
Konieczny, S. F., 258, *274*

Konno, H., 11, 12, *18*
Kontgen, F., 156, 160, 167, *175, 179,* 203, *209*
Köntges, G., 266, 277, 422, 423, 424, 425, 428, *488*
Kontos, C. D., 225, *230*
Kooh, P. J., 136, *145*
Koopman, P., 281, *293, 295,* 377, 380, 381, 387, *389, 390, 391, 392*
Kopan, R., 136, *146, 147, 148,* 262, 263, *275*
Kopen, G. C., 246, *250*
Kopf, G. S., 7, 8, 13, *16, 17, 19*
Kopf, M., 205, *206*
Kopp, J. B., 408, *420*
Kordower, J. H., 247, *251*
Korematsu, K., 91, *101*
Korfhagen, T. F., 313, 314, *329*
Korhonan, J., 228, *230*
Korhonen, J., 218, *230*
Kornberg, T. B., 470, *481*
Korniszewski, L., 380, *392*
Korsmeyer, S. J., 376, *390, 393*
Korving, J., 183, *190,* 451, *496,* 549, *565*
Korving, J. P. W. F. M., 336, *363*
Korzh, V., 78, *104,* 141, *148,* 546, *563*
Kos, L., 402, *417*
Kosaki, K., 47, *51,* 357, *363*
Kosaki, R., 47, *51,* 357, *363*
Koseki, H., 135, 137, 138, 140, *145, 147, 148,* 398, *417,* 471, 475, 479, *483, 487, 492*
Koshiba, K., 354, *370*
Koshiba-Takeuchi, K., 525, *536*
Koshida, S., 41, *51*
Kostakopoulou, K., 266, *273*
Kostamarova, A. A., 22, *34*
Kostas, S. A., 14, *16,* 341, *360*
Koster, M., 266, *274*
Koster, R., 533, *537*
Kostetskaia, E., 346, 347, *363*
Kostetskii, I., 346, 347, *363*
Kotanides, H., 227, *233*
Kotch, L. E., 225, *230*
Kotecha, S., 40, 44, *50, 323,* 336, 338, 339, 342, 355, *358, 368*
Kothapalli, R., 47, *51*
Kothary, R., 23, *35,* 169, *179,* 258, 264, 266, *274*
Koudrova, M., 266, *274*
Koutsourakis, M., 165, *177*
Kouzarides, T., 342, *368*
Kovacs, C. S., 285, *294*
Kovacs, K., 511, 512, *513, 515, 518*
Kovari, I. A., 412, *417*
Kowalski, J., 223, *229,* 291, *292*
Kowk, G., 281, *292,* 380, *390*
Koyama, E., 423, 424, 451, 475, 476, 478, *490*
Koyama, H., 23, *35*

Koyama, Y., 531, *538*
Kozak, C. A., 284, *294,* 442, 443, *497,* 521, 525, *535, 537*
Kozak, K. R., 172, 173, *177,* 225, *230*
Kozian, D., 264, *277*
Kozu, T., 202, *208*
Krahl, T., 318, 322, *326*
Krakowski, M. L., 318, 322, *326*
Kral, A., 415, *417*
Kramer, S., 466, *485, 488*
Krane, S., 583, *587*
Kraner, S. D., 243, *249*
Krantz, I. D., 347, *364, 366*
Krapp, A., 321, *326*
Krasnow, M. A., 311, *327,* 466, *485, 488, 491*
Krastel, K., 258, 260, 261, 262, 267, *275*
Kratochwil, K., 440, 441, 451, 471, 475, *481, 488, 493*
Krause, D., 194, *207*
Krause, M., 257, *277,* 341, *360*
Krause, W. L., 453, *495*
Krauss, S., 31, *33,* 47, *50,* 78, *101,* 451, 476, *483,* 508, *514,*
    546, *563*
Kraut, N., 169, *175, 177*
Kraybill, A. L., 415, *417*
Krebs, C., 174, *177*
Krebs, L. T., 226, *230*
Kreiborg, S., 292, *294*
Kreidberg, J., 183, 187, 188, *189*
Kreidberg, J. A., 94, *98,* 314, *326,* 341, 352, *363, 365,* 375,
    376, *389, 391, 392,* 400, 410, *418,* 572, 574, *585*
Kremser, T., 334, *363*
Kress, C., 166, *178*
Krezel, W., 91, *101, 104,* 525, *536*
Krieg, P. A., 334, 339, 347, 348, *359, 360, 361, 369*
Krieger, J., 346, *358*
Krieglstein, K., 314, *324*
Kriegstein, A. R., 86, *102*
Krimpenfort, P., 583, *589*
Kritzik, M. R., 318, 322, *326*
Kriz, R. W., 467, *498*
Kromer, L. F., 91, *101*
Kronenberg, H. M., 285, 286, 287, *292, 293, 294,* 442, 443,
    470, *488, 489, 497*
Kronenberg, M. S., 452, *484*
Kros, C. J., 553, *563*
Kroumpouzou, E., 357, *358*
Krude, H., 311, *326,* 508, *516*
Kruger, O., 173, *177*
Krull, C. E., 426, 427, *494*
Krumlauf, R., 45, *52,* 117, 118, 119, *123, 124, 125,* 142,
    147, 422, 424, 425, 426, 427, 428, 433, 449, 453, 468,
    469, 474, 479, *484, 485, 487, 488, 490, 492, 494, 496,*
    544, 545, *562, 563,* 571, *586*
Krushel, L. A., 94, *101*
Krust, A., 469, *489, 491, 494*

Krylova, I., 379, *390*
Kuan, C. Y., 87, *100, 101*
Kuang, W.J., 223, *230*
Kubalak, S. W., 341, 344, 346, 352, *359, 361, 363, 367*
Kubiak, J. Z., 7, *18*
Kubo, H., 14, *17*
Kubo, T., 555, *565*
Kucera, J., 559, *562*
Kuch, C., 264, *277*
Kucherlapati, R., 43, *50,* 353, 357, *358, 365,* 415, *420*
Kudlow, J. E., 511, *516*
Kudoh, S., 336, 343, *365, 368*
Kuehn, M. R., 338, 352, 354, 355, 356, *359, 363, 364,*
    370
Kugler, J. D., 357, *358*
Kuhl, M., 337, *363*
Kuhlman, J., 401, *419,* 571, *588*
Kuhn, H., 421, 442, 457, *488*
Kuhn, H. G., 237, *250*
Kuida, K., 87, *101*
Kulbacki, R., 205, *206*
Kulesa, P. M., 424, *488*
Kulesh, D. A., 570, *589*
Kulikowski, R. R., 355, *364*
Kuliszewski, M., 173, 174, *175,* 313, *329*
Kulkarni, A. B., 94, *100, 103,* 203, *207,* 226, *229, 230,* 467,
    484
Kulkarni, S., 171, *175*
Kumada, M., 432, 465, 478, *483*
Kumanogoh, A., 341, 352, *370*
Kumar, A., 338, *363*
Kumar, S., 318, *328*
Kumar, T., 44, *51,* 441, 467, 468, *490, 491,* 500, 510, *516*
Kume, R., 137, *147*
Kume, T., 402, 404, *418,* 453, 479, *488*
Kun, J., 281, *294*
Kunath, T., 23, *35,* 166, 167, *179,* 357, *368*
Kundu, R. K., 428, 452, *489*
Kunisada, T., 442, *498*
Kunkel, L. M., 193, 206, *207,* 246, *250,* 270, 271, *274*
Kuno, J., 95, *102,* 122, *122,* 173, *179,* 220, 222, *231,* 404,
    420, 506, *516*
Kuno, T., 217, *230*
Kuo, A., 286, *292,* 453, *484*
Kuo, C. J., 164, *177*
Kuo, C. T., 165, *177,* 316, *326,* 342, 343, 352, *363*
Kuo, F. C., 203, *209,* 510, *518*
Kuo, H.-C., 342, 343, 352, *363*
Kuo, T. S., 91, *103*
Kuo, W. L., 347, *364*
Kupper, T. S., 570, *587, 588*
Kupperman, E., 340, 341, *363*
Kupriyanov, S., 168, *179*
Kurant, E., 91, *103*

Kuratani, S., 31, *34,* 46, *51,* 83, *106,* 114, *124,* 305, 307, *328,* 374, *392,* 403, *418,* 421, 422, 424, 426, 435, 437, 442, 451, 457, 475, 476, 478, *488, 490, 492, 496,* 521, 527, *537*
Kurihara, H., 349, *369,* 432, 450, 451, 465, 473, *489, 496*
Kurihara, Y., 349, *369,* 432, 450, 451, 465, 473, *489, 496*
Kurisu, K., 289, *294*
Kurkinen, M., 161, *176*
Kurland, S., 467, *482*
Kuroiwa, A., 41, 47, *51,* 78, *104,* 117, *124,* 427, *485*
Kuroki, T., 580, *587, 588*
Kurosumi, M., 511, *515*
Kuroume, T., 506, *518*
Kurzrock, E. A., 301, *326*
Kusakabe, M., 11, 12, *18,* 557, *565*
Kusumi, K., 136, 137, *147*
Kutsche, L. M., 350, *369*
Kuttenn, F., 375, *389*
Küttner, F., 316, *329,* 469, *483*
Kutunumu, N., 581, *589*
Kuwaki, T., 432, 450, 465, 478, *483, 489*
Kuwana, E., 76, 77, 82, 83, 85, *99, 103*
Kuzmin, I., 227, *229*
Kwan, C. T., 118, 119, *124,* 544, *563*
Kwan, K. M., 31, *34,* 43, 46, 47, 48, *52, 104,* 397, *420*
Kwee, L., 162, 170, *177,* 219, *230*
Kwok, C., 380, 381, *391*
Kwon, Y. T., 94, *98, 101*
Kwong, P. D., 408, *419*

L

Labastie, M. C., 201, *208*
LaBonne, C., 422, 423, *481, 489*
Labosky, P. A., 82, *106,* 203, *210,* 305, 309, 319, 320, *327, 329,* 439, 453, 467, 468, *488, 496, 498,* 506, *516*
Labouesse, M., 14, 15, *16,* 165, *176,* 316, *325*
Labow, M. A., 162, 170, *177,* 219, *230*
Lacerda, D. A., 442, 443, *489*
Lachman, R. S., 286, *294*
Lacktis, J. W., 355, *364*
Lacombe, D., 398, *416,* 550, *561*
Lacroix, B., 308, *325*
Lacy, E., 2, *4*
Ladd, A. N., 334, 335, 336, 337, *363, 370*
Ladher, R., 422, *484*
Lagace, G., 512, *514*
Lagasse, E., 246, *250,* 318, *326*
Lah, M., 40, 44, *50, 323,* 338, *358*
Lahn, B. T., 387, *391*
Lahtinen, I., 228, *230*
Lai, A. F., 9, *18*
Lai, C. C., 318, *326*

Lai, E., 78, 81, 82, 84, 85, 86, *99, 100, 101, 102, 105, 106,* 164, 165, *177, 179,* 241, 243, 244, *251, 252,* 408, *417, 479, 486, 498,* 522, 525, 527, *536*
Lajeunie, E., 452, 453, *484, 489*
Lake, R. J., 136, *144,* 263, *272,* 555, *561*
Lakkaraju, S., 427, 433, 443, 449, 474, 479, *494*
Lakso, M., 218, 223, *230*
Lala, D. S., 375, 376, *390, 391,* 510, *516*
Lallemand, Y., 31, *33,* 46, 48, *49,* 114, *122, 322,* 428, 451, 478, *481,* 521, 527, *535*
Lalli, E., 378, 379, 384, *391, 393*
Lalor, P. A., 194, *207*
Lam, C. W., 579, *589*
Lam, G., 15, *17*
Lam, P., 268, *273*
LaMantia, A. S., 84, *97, 106,* 435, 469, 476, *481, 483, 497*
LaMantia, C., 169, *179*
Lamb, A. N., 380, *390*
Lamb, L. C., 467, *494*
Lambie, E. J., 316, *326*
Lamerdin, J., 412, *418*
Lamers, J. M. J., 343, 345, 348, *366*
Lamers, W. H., 173, *177,* 334, 341, 343, 344, 345, 347, 348, 350, 351, 352, 356, *359, 360, 361, 363, 365, 366, 369*
Lammer, E. J., 468, *497*
Lammert, E., 321, *326*
Lamolet, B., 478, *489,* 507, *516*
Lamond, A. I., 400, *417*
Lamperti, E. D., 94, *100*
Lampron, C., 469, *489, 491*
Lampugnani, M. G., 226, *229*
Lan, Y., 423, *487,* 553, 555, *563*
Lance-Jones, C., 269, *277*
Lanctôt, C., 478, *489,* 507, 508, 510, *514, 516, 518*
Lander, E. S., 136, 137, *147,* 380, *392*
Landis, C. A., 512, *516*
Landis, S., 239, *249*
Landolt, A. M., 505, *513*
Landry, C., 321, *326*
Landry, J., 171, 173, *176*
Lane, A. H., 381, 382, *393*
Lane, J., 416, *417*
Lane, N., 9, *18*
Lane, P. W., 442, *490*
Lane, W. S., *251*
Lanford, P. J., 553, 555, *563*
Lang, H., 547, *563*
Lang, P., 357, *358*
Lang, R. A., 219, *229, 230, 231,* 533, *535*
Langegger, M., 411, *418*
Langeland, J. A., 478, *496*
Langenbach, R., 156, *177*
Langeveld, A., 165, *177*

Langille, B. L., 341, *361*
Langille, R. M., 422, 423, 424, 443, 449, 471, *489*
Langley, E., 342, *368*
Langtimm, C. J., 236, *249*
Lankes, W., 353, 356, *361*
Lansdorp, P. M., 242, 245, *251, 252*
Lanske, B., 285, 287, *293, 294,* 442, 443, 470, *489, 497*
Lapi, P., 311, *326*
Lapic, S., 581, *588*
Lapidot, T., 341, *369*
Lapointe, F., 264, 266, *273, 277*
Lapointe, L., 167, *179*
Lapvetelainen, T., 284, *293*
Lardelli, M., 135, *149*
Larder, R., *390*
Largman, C., 242, *252*
Larkin, S. B., 262, *277,* 469, *491*
Laroia, G., 11, *17*
Larouche, L., 171, 173, *176*
Larson, C., 415, *418*
Larsson, E., 413, *418,* 467, *489*
Larsson, L. I., 321, *327*
Larsson, S. H., *391,* 400, *417*
Larsson-Blomberg, L., 402, *419*
Larue, L., 409, 410, *418, 419*
Lash, J. W., 263, *274,* 333, 336, 340, *364*
Lasko, P., 182, 185, 186, *190*
Lasky, L. A., 193, *210*
Lasorella, A., 474, *489*
Lassar, A. B., 140, 141, 142, *147, 148,* 257, 258, 262, 263, 264, 268, *273, 274, 275, 276, 277,* 282, *293, 328,* 334, 335, 336, 337, 339, 341, *361, 365, 367, 369,* 399, *417*
Latham, K. E., 5, 9, 10, 11, 12, 13, *16, 17, 18, 19,* 168, *178*
Latif, F., 227, *229*
Latinkic, B., 336, 339, 342, *368*
Lattanzi, L., 271, *273*
Lau, A. L., 44, *51,* 440, *484*
Lau, C. K., 84, *102*
Lau, K. K., 165, *177,* 316, *325*
Lau, Y. F., 378, *390*
Laub, F., 581, *588*
Laufer, E., 141, *146,* 262, 264, *275,* 470, *487, 494*
Laughner, E., 225, *230*
Laurent, M., 579, *587*
Laurie, G. W., 415, *420*
Lauster, R., 577, 583, *584*
Lavdas, A. A., 88, 89, *101*
Laverriere, A. C., 316, *326*
Lavker, R. M., 570, 578, *585, 587, 589*
Law, D. J., 357, *364*
Lawitts, J. A., 228, *232*
Lawler, A. M., 129, *147,* 225, *230,* 260, 264, *275,* 353, 356, *367*
Lawlor, P., 553, *563*

Lawrence, H. J., 242, *252*
Lawrence, P. A., 550, *563*
Lawshé, A., 466, *486*
Lawson, K. A., 38, 39, 44, 45, 47, *51, 52,* 158, *177,* 183, *190,* 303, 304, *326,* 333, 336, 338, *363*
Layne, M. D., 347, *359*
Layton, W. M., 352, 354, *363*
Lazarus, J. H., 311, *324*
Lazzarini, R. A., 170, 173, 174, *174, 178*
Lazzaro, D., 85, *103,* 189, *190,* 311, *326,* 478, *481,* 508, *513*
Le, T., 379, *389*
Le, T. C., 415, *420*
Leardkamolkarn, V., 412, *417*
Le Bail, O., 136, *146*
Lebeche, D., 313, 315, *326, 327*
Leberquier, C., 308, 310, *324*
Leblanc-Straceski, J., 357, *358*
Lecerf, F., 271, *277*
Lechleider, R. J., 339, *368*
Lechner, M. S., 399, 402, *417, 420*
Ledan, E., 7, *18*
Ledbetter, D. H., 94, *103*
Leder, P., 40, *52,* 167, 171, 173, *179,* 286, *292,* 338, *368,* 453, *484*
Ledermann, B., 321, *326*
Le Douarin, G. H., *250*
Le Douarin, N. M., 39, 40, *50,* 77, 84, 85, *98,* 113, *123,* 128, 140, 141, 142, *145, 147, 148,* 196, 200, 201, *207,* 236, 237, 241, 244, *249, 250,* 256, 261, 263, 264, 266, 271, *273, 276, 277,* 316, 317, 320, *326,* 422, 423, 424, 426, 427, 433, 435, *483, 489,* 502, 507, *514,* 569, 571, 572, *585, 587*
Leduque, P., 320, *324*
Lee, C. S., 141, *147*
Lee, D., 506, *517,* 532, *537,* 549, *564*
Lee, E., 43, *50,* 78, 80, 83, 85, 86, 87, 91, *98, 103, 106,* 121, *123,* 141, *145,* 260, 265, *273,* 319, *323,* 439, 451, 470, *482,* 508, *517,* 521, 527, *535, 537*
Lee, E. J., 402, *419,* 432, *498,* 508, *517*
Lee, E. Y., 318, *326*
Lee, H.-G., 2, *4*
Lee, H.-H., 336, *363*
Lee, H.-J., 354, *365*
Lee, J., 314, 318, *326, 330,* 579, *585*
Lee, J. C., 161, *179*
Lee, J. E., 92, *102,* 121, *124,* 321, *328*
Lee, J. J., 470, *489*
Lee, J.-S., 12, *17,* 341, *360*
Lee, K., 285, 287, *294,* 442, 443, 465, 470, *489, 491, 497*
Lee, K. F., 94, *97,* 285, *293, 294,* 469, *489,* 559, *562*
Lee, K.-F., 580, 581, 583, *587*
Lee, K. H., 42, *51,* 349, 355, *358*
Lee, K.-H., 336, *363*

Lee, K. J., 45, *52,* 78, *101,* 120, 121, 122, *122, 124,* 343, *366*
Lee, M.-E., 347, *359*
Lee, M. H., *483,* 528, *537*
Lee, M. J., 227, *230,* 314, 315, *329*
Lee, M.-J., 227, *231*
Lee, M. O., 525, *537*
Lee, R. K., 344, *368*
Lee, R. Y., 346, *363*
Lee, S. B., 400, *418*
Lee, S. H., 219, 226, *230, 233,* 580, *587*
Lee, S. J., 129, *147,* 227, *229,* 260, 264, *275,* 353, 356, *367*
Lee, S. L., 510, *516*
Lee, S. M., 82, 84, 86, *101, 104,* 428, *487, 489*
Lee, S. M. K., 116, 117, *124*
Lee, W.-H., 243, 244, *249,* 318, *326*
Lee, Y., 343, *363*
Leeming, G. L., 437, *490*
Lefebvre, V., 281, 282, 283, 284, 289, *292, 293, 294, 295*
Leffell, D. J., 287, *292,* 470, *486,* 579, *586*
Lefkowitz, R. J., 341, 352, *362*
Legeai-Mallet, L., 286, *294,* 453, *494*
Leger, J., 507, *514*
Leger, S., 337, *367*
Legouy, E., 5, 9, 14, *15, 19*
Legrand, J. F., 408, *419*
Lehman, A. L., 6, *17*
Lehmann, M. S., 408, *419*
Lehmann, R., 14, *19,* 185, 186, 188, *189, 190*
Lehner, C. F., 170, *175, 179*
Lehrer, M. S., 570, *589*
Lehtonen, E., 559, *564*
Lei, H., 47, *50,* 376, *393*
Lei, L., 312, 319, *326*
Leibel, R. I., 443, *497*
Leibovici, M., 398, *416,* 550, *561*
Leibowitz, R. T., 247, *251*
Leiden, J. M., 165, *177, 179,* 316, *326,* 341, 342, 343, 347, 352, *362, 363, 366, 368*
Leimeister, C., 347, *363*
Leinwand, L. A., 269, *275*
Leinwand, M., 314, 315, *329*
Leiser, R., 174, *177*
Leiter, A. B., 310, 321, 322, *327*
Leitgeb, S., 570, *587*
Leitges, M., 137, *147*
Le Lievre, C. S., 426, *489*
Lelievre-Pegorier, M., 403, *418*
Lemaigre, F. P., 166, *176,* 321, *326*
Lemaire, L., 39, 40, *51*
Lemaire, P., 263, 264, *274,* 303, *324, 330*
Lemarchandel, V., 200, *208*
Lembo, G., 341, *363*
Le Merrer, M., 285, 286, *293, 294,* 452, 453, *484, 489, 494*

Le Meur, M., 40, 46, 47, *52,* 478, *494,* 583, *586*
LeMeur, M., 114, *125,* 310, 312, 321, *325, 327,* 435, 443, 469, 474, *484, 489, 490, 491,* 544, 545, *563*
Lemieux, M., 312, *323*
Lemischka, I. R., 192, 194, 200, 205, 206, *208, 209*
LeMouellic, H., 571, *587*
Lendahl, U., 135, *149,* 236, 237, 244, 245, 246, 248, *249, 250,* 318, 320, 321, *323, 324*
Lenhard, T., 291, *292*
Lenkkeri, U., 412, *418, 419*
Lentz, S. I., 411, *418*
Leon, Y., 466, *494,* 545, *564*
Leonardi, A., 442, *484*
Leonardo, E., 91, *102*
Leong, L. M., 335, 337, 339, 355, *365, 366*
Leopold, P. L., 579, *588*
Lepori, N. G., 356, *363*
Leptin, M., 336, *367*
Lerman, M. I., 227, *229, 231*
Lerner, C. P., 245, *250*
Lerone, M., 346, *358*
Leroy, J. G., 281, *293,* 381, *391*
Leroy, P., 469, *484, 489,* 557, *562*
Lesot, H., 455, *489, 493*
Lessey, B. A., 156, *175*
Letarte, M., 227, *231*
Lettice, L. A., 451, 453, 474, *489,* 530, 533, *536*
Lettieri, K., *104,* 508, *517*
Leung, D. W., 223, *230*
Leung, K. K., 380, *389*
Leung, S. W., 225, *230*
Leveen, P., 313, *323,* 413, *418,* 467, 482, *489*
Levenberg, S., 408, *418*
Levi, G., 290, *292,* 450, 453, 472, 473, *481,* 510, *518,* 546, 549, *561, 564*
Levi, T., 339, 343, 347, 357, *358, 359*
Levin, L. S., 442, *485,* 507, *515*
Levin, M. E., 94, *106,* 356, *363, 364,* 473, *489*
Levitt, P., 84, *101*
Levorse, J. M., 451, 476, *493*
Levy, D. E., 286, *292, 294*
Lew, D., 504, 510, *516*
Lewandoski, M., 15, *17,* 31, *35,* 84, *106,* 114, 115, *124, 125,* 337, 340, 352, *365, 368,* 453, 466, *489, 496, 497*
Lewin, M., 47, *51,* 357, *363*
Lewis, A. F., 204, *210,* 220, *233*
Lewis, J., 139, *146,* 466, *491,* 545, 546, 547, 549, 552, 553, 555, 558, 559, *561, 562, 563, 564, 565*
Lewis, K. E., 534, *536*
Lewis, P., 441, *483*
Lewis, P. M., 312, *328,* 577, 578, 583, *589*
Lewis, R. S., 8, *16*
Lewis, S., 509, 510, *515*
Lewis, W. H., 26, *34,* 415, *417,* 529, *536*

Leyns, L., 40, 44, *49, 51, 52,* 141, *144, 323,* 338, *366,* 473, 477, *482*

Leys, A., 398, *419*

Leysens, N. J., 478, *495,* 507, *517*

Li, C., 167, 171, 173, *179,* 226, 227, *233,* 286, *292, 293,* 411, *418*

Li, C. W., 542, 550, *563*

Li, D. Y., 226, *230, 232,* 556, 558, *565*

Li, E., 47, *50,* 94, *98,* 353, 356, 357, *366,* 468, 469, *489, 492*

Li, F., 225, *231,* 375, *392*

Li, H., 79, 82, 90, 94, 95, *101, 104, 105, 106,* 162, 170, *176,* 319, 320, 321, *323, 326,* 339, *368,* 502, 507, 508, 509, *516, 517,* 521, 530, *536*

Li, J., 317, *326,* 577, 578, 583, *589*

Li, L., 228, *230,* 347, *364,* 587, *589*

Li, N., 356, *366*

Li, P., 89, 95, *104,* 240, *250,* 283, 284, 289, *293, 294,* 506, *517*

Li, Q., 343, 357, *364,* 574, 580, 581, 583, *587, 588*

Li, R., 203, *209,* 334, 343, 345, 348, 349, 352, 356, *364*

Li, S., 3, *4,* 78, 81, 82, 86, *99,* 260, *276,* 509, *516*

Li, S. C., 81, 86, *101,* 522, 527, *536*

Li, S. W., 284, *293*

Li, T., 220, *228,* 527, 530, 531, *535, 536*

Li, W., 239, *250*

Li, X., 290, *292,* 452, 453, 474, *487*

Li, X. F., 174, *179*

Li, Y., 91, *101,* 172, 173, *177,* 424, 442, 443, *489, 493*

Li, Y.-X., 14, *17*

Li, Z., 266, *277*

Lia, M., 43, *50,* 353, 357, *365*

Liang, I., 235, 244, *251*

Liang, J. O., 78, *104,* 351, *364*

Liang, L.-F., 14, *16*

Liang, M. Y., 527, *535*

Liang, Y., 11, 12, *17,* 556, *564*

Liao, E. C., 214, *230*

Liao, W., 214, 222, *230, 231*

Liao, X., 245, *249,* 312, *328*

Liberatore, C. M., 342, 347, *364, 367*

Licht, J. D., 375, *392*

Licht, P., 388, *393*

Lichter, P., 479, *487*

Lichti, U., 169, *179*

Lichtlen, P., 317, *325*

Lichtler, A. C., 452, *484*

Lidral, A. C., 292, *294*

Lieber, T., 136, *147*

Liebhaber, S. A., 443, *493*

Liegeois, N. J., 532, *536*

Liem, K. D., Jr., 423, *489*

Liem, K. F., *102*

Liem, K. F., Jr., 121, *124*

Lien, C.-L., 342, *364*

Lieschke, G. J., 194, *208*

Liggitt, D., 173, *179,* 322, *328*

Lightfoot, P., 453, *483*

Lightman, S. L., 506, *514*

Ligon, K. L., 135, *145,* 267, *273*

Liguori, G., 338, 352, *370*

Lilly, B., 262, *273,* 341, *364,* 476, *483*

Lim, C., 509, *517*

Lim, D. A., 236, 237, 244, 245, 248, *249*

Lim, D. J., 551, 552, 557, *563*

Lim, H., 156, *177*

Lim, L., 82, *102,* 321, *328*

Lim, M. K., 11, 12, *17*

Limon, J., 83, *100*

Lin, B., 346, *364*

Lin, C. H., 271, *275*

Lin, C. R., 353, 354, 355, *364,* 433, 478, *489,* 508, 509, *515, 516*

Lin, D. M., 435, *489*

Lin, H., 91, *100,* 204, *208,* 341, 343, 352, *368*

Lin, P., 227, *231*

Lin, Q., 165, 168, *175, 177,* 316, *327,* 340, 341, 342, 343, 346, 347, 349, 350, 352, *361, 364, 365, 368,* 451, 473, *495,* 580, 583, *588*

Lin, S., 135, *147*

Lin, S.-C., 508, *516*

Lin, V., 258, *276*

Lin, W., 450, 470, 473, *481,* 546, *561*

Linask, K. K., 333, 336, 340, 356, *364, 369*

Linchan, W. M., 227, *231*

Lindahl, M., 403, *419*

Lindahl, P., 173, *178,* 313, *323, 326,* 414, *418,* 467, *482*

Lindberg, M., 580, *586*

Linder, G., 577, 583, *584*

Linders, K., 376, *393*

Lindout, D., 288, *293*

Lindroth, J. R., 341, *369*

Lindsay, E. A., 353, 357, *364*

Lindsay, R. M., 91, *97, 98,* 434, *489,* 549, *561*

Lindsay, S., 387, *390*

Lindsell, C., 136, 137, *148, 149,* 262, *276,* 553, 555, *563, 565*

Lindvall, O., 247, *249*

Line, S., 284, *294,* 442, 443, *497*

Linehan, W. M., 173, *176,* 227, *229*

Ling, G., 574, *588*

Ling, K. W., 380, *389*

Linney, E., 469, *483*

Linnoila, R., 313, *323*

Lints, T. J., 78, 79, 80, 81, 85, *99,* 120, *123,* 317, *325,* 333, 334, 339, 356, *364, 369*

Linzer, D. I., 162, *176*

Lipkin, S. M., 510, *516*

Lira, S. A., 402, *419*

Lis, W. T., 158, *178*

Lissauer, M. E., 322, *328*

Litingtung, Y., 43, *50,* 78, 80, 85, 86, *98,* 121, *123,* 141, *145,* 260, 265, *273,* 312, 319, *323, 326,* 439, 451, 470, *482,* 521, *535*

Litovsky, S. H., 219, *232,* 341, 343, 352, *360, 369*

Little, C. D., 223, 226, *229, 230*

Little, M., 375, *391*

Little, M. H., 416, *419*

Little, P., 174, *174*

Littlepage, L. E., 7, *17*

Littlewood Evans, A., 557, *563*

Litvin, J., 337, 340, 341, *364, 365*

Liu, A., 84, *102,* 110, 112, 115, 116, *124,* 428, 453, *489*

Liu, C. H., 227, *230, 231*

Liu, F., 353, 354, 355, *364,* 433, 478, *489, 496*

Liu, F. C., 507, 508, *518*

Liu, J., 312, *327*

Liu, J. K., 79, 88, 89, *99, 102, 103,* 121, *124,* 290, 292, 435, 450, 451, 453, 471, 472, 473, 474, 477, 478, *484, 489,* 549, *562*

Liu, M., 92, *102,* 559, *563*

Liu, P., 43, 45, *51,* 202, 204, *208, 210,* 220, 226, 227, *233*

Liu, Q., 531, *536*

Liu, R., 14, *17*

Liu, S., 79, 89, *102,* 473, 474, *489*

Liu, W. H., 31, *34*

Liu, X., 219, *233,* 573, 583, *584, 587*

Liu, Y., 168, *178,* 227, *231,* 289, *294,* 428, 452, *484, 489*

Liu, Z., 116, *126*

Ljungberg, P., 412, *419*

Llacer, R., 341, *360*

Lloyd, R., 504, 511, 512, *515, 516, 518*

Lo, C. W., 157, *177,* 357, *369*

Lo, D. C., 94, *102, 106*

Lo, J., 225, *232*

Lo, L., 236, 239, 242, 243, *250, 251*

Lo, L.-C., 79, 88, 89, *102, 103,* 168, *176, 177,* 239, *250*

Lo, P. C. H., 334, *364*

Lo, S. H., 572, *586*

Lobel, R., 240, 243, *251*

Locascio, A., 355, *359,* 423, *482*

Lock, L. F., 188, *189*

Lockwood, W. L., 341, *358*

Lodish, M. B., 343, 352, *360*

Logan, C., 116, *125*

Logan, M., 337, 339, 343, 347, 354, *359, 364*

Logeat, F., 136, *146*

Loh, Y. P., 502, *514*

Lohikangas, L., 574, *589*

Lohnes, D., 312, *327,* 451, 453, 468, 469, *489, 490*

Lohning, C., 415, *417*

Lohr, J., 349, 355, 356, *358, 364*

Lois, C., 90, 95, *102,* 237, *251*

Lonai, P., 167, 171, *174, 176,* 312, 314, 315, *323,* 341, *369,* 467, *492*

Long, C. R., 8, *16*

Long, F., 286, *293*

Long, J. E., 79, *99,* 290, *292,* 435, 450, 451, 453, 471, 472, 473, 477, 478, *484,* 549, *562*

Long, Q., 89, *106,* 450, 471, *498*

Longaker, M. T., 452, *491*

Longenecker, G., 94, *103*

Longhi, M. P., 227, *231*

Longley, M. A., 582, 583, *587*

Longo, K. A., 271, *276*

Loomis, C. A., 109, 115, *123,* 228, *230*

Loosli, F., 521, 525, 531, 533, *537*

Lopes da Silva, S., 506, *513*

Lopez, J. M., 289, *293*

Lopez, M., 353, 357, *365*

Lopez, S., 268, *276*

Lopez-Fernandez, J., 510, *516*

López-Mascaraque, L., 84, 90, 95, *99, 102, 105*

Lopez-Otin, C., 289, *293*

Lopez-Rios, J., 522, 525, *535, 537*

Lóra, J., 316, 320, *324*

Loring, J. F., 443, *493,* 559, *562*

Loring, J. M., 341, 352, *363,* 375, *391,* 400, *418*

Losken, H. W., 433, *492*

Losos, K., 84, *102,* 110, 115, 116, *124,* 428, 453, *489, 491*

LoTurco, J. J., 86, *102,* 239, *250*

Lou, L. J., 312, *323*

Lough, J., 317, *330,* 335, 336, 337, 342, *360, 364, 368, 370*

Loughan, S., 227, *231*

Loughna, S., 222, *231*

Louis, J.-C., 402, *417*

Loune, E., 511, *516*

Lourenco, D. M., 509, *517*

Louvet, S., 7, *18*

Lovell-Badge, R., 372, 373, 377, 378, 379, 380, 381, 383, 384, 386, 387, 388, *389, 390, 391, 392, 393,* 527, 529, 530, 531, 534, *536, 537*

Lovicu, F. J., 532, *537*

Low, M. J., 502, 510, *513, 515*

Lowe, C., 471, *493*

Lowe, L. A., 338, 352, 354, 355, 356, 357, *359, 364, 370*

Lowe, S. W., 341, *370*

Lowell, S., 570, *586*

Lowenheim, H., 558, *563*

Lowenstein, D. H., 92, *102,* 247, 248, *251*

Lowik, C. W., 285, *293*

Lowry, O. H., 269, *274*

Lowy, D. R., 171, 173, *178*

Lu, B., 269, *275,* 338, *359*

Lu, J., 342, *365, 370*

Lu, J.-R., 342, 343, 353, *364, 365*

Lu, L., 203, *207,* 223, *229,* 506, *517,* 549, *564*

Lu, M. F., 354, 355, *364,* 433, 451, 476, 477, 478, *489*

Lu, M. M., 165, *177,* 263, *274,* 303, 316, *326, 327,* 342, 343, 352, 353, 357, *363, 365, 366*

Lu, N., 468, *491,* 500, *516*

Lu, Q., 478, *489,* 580, 581, 583, *587*

Lu, W., 415, *417, 418*

Lu, Y., 171, 173, *178*

Lucas, P. C., 271, *276*

Lucero, M. T., 240, *250*

Ludgate, M., 311, *324*

Ludlow, J. W., 173, 174, *176*

Lufkin, T., 290, *292,* 312, *327,* 443, 451, 453, 469, 474, *485, 489, 490,* 497, 544, 545, 549, *563, 565*

Lui, H., 284, *293*

Lumsden, A., 116, 119, *123, 125,* 260, 263, 264, 265, 266, *273, 274,* 422, 423, 424, 425, 427, 428, 433, 443, 466, 468, 469, 474, *485, 488, 490, 491, 493,* 544, 545, 546, *562, 563, 564, 565*

Lumsden, A. G., 422, 432, 438, 439, 454, *490*

Lumsden, A. G. S., 438, *490*

Lun, Y., 136, 137, 138, *145*

Lund, L. R., 291, *292*

Lundh-Rozell, B., 470, *497*

Lundrigan, B. L., 377, 387, *393*

Lunsden, A., 119, *122, 123*

Lunsford, L., 186, 188, *190*

Luo, G., 408, *418,* 451, 467, *490*

Luo, J., 167, *177*

Luo, K., 268, *275*

Luo, L., 12, *17,* 556, 558, *562*

Luo, W., 428, 452, 453, 474, *487, 489*

Luo, X., 375, 376, *390, 391,* 510, *515, 516*

Luoh, S. M., 579, *589*

Lupo, G., 525, *535*

Lupu, F., 226, *229*

Luria, A., 8, *18*

Luria, V., 341, *369*

Lusis, A. J., 508, *516*

Luskin, M. B., 95, *100,* 237, *251*

Luteijn, T., 199, 200, 201, 202, 204, *207*

Lutgens, E., 226, *229*

Luton, D., 200, 201, *209,* 220, *232*

Lutz, Y., 343, *361*

LuValle, P., 443, *487*

Luyten, F. P., 408, *420*

Luz, A., 285, *293, 294,* 442, 443, *488, 489*

Lyapunova, E., 388, *391*

Lye, S. J., 173, 174, *175*

Lyght, M., 226, *230*

Lyman, S. D., 188, *189,* 193, 204, *208*

Lymboussaki, A., 162, *179,* 218, 223, *229,* 580, *589*

Lynch, J., 343, 347, *361*

Lynn, A., 479, *482*

Lyon, M. F., 6, *17, 391,* 451, 475, *486,* 527, 528, 530, *536*

Lyon, M. G., 84, *101*

Lyonnet, S., 357, *364*

Lyons, G., 132, *147,* 170, *176,* 258, 262, 267, 269, *272, 273, 275, 276,* 333, 341, 342, 349, *360, 364, 365,* 473, *483*

Lyons, I., 203, *209,* 317, *325,* 333, 334, 343, 345, 348, 349, 352, 356, *364*

Lyons, J., 512, *516*

Lyons, K. M., 45, 47, *50,* 82, *99, 105,* 238, *251,* 308, 321, *326,* 338, 355, *359,* 408, *417,* 451, 467, 468, *483, 484, 490, 495*

Lyons, R. H., 556, *564*

Lyons, W. E., 559, *562*

Lysakowski, A., 555, 558, 559, *561, 563*

M

Ma, L., 290, *294,* 341, *363,* 437, 450, 451, 452, 453, 466, 474, *487, 495*

Ma, N., 453, *486*

Ma, Q., 83, 89, 95, *99, 102,* 435, *490,* 526, *538,* 559, *563*

Ma, X., 201, 204, *207*

Maas, R., 268, *273,* 290, *294,* 398, 399, *420,* 437, 440, 441, 450, 451, 453, 466, 474, 475, *481, 482, 483, 493, 495,* 527, 528, 530, 532, 533, *535, 537, 538,* 550, *565*

Maatta, A., 581, *585*

Mabie, P. C., 239, *250, 251*

Mabry, M., 313, *323*

Macara, L. M., 174, *177*

MacArthur, C., 115, 116, *123*

MacArthur, C. A., 165, *179,* 466, *486*

MacArthur, C. T. P., 116, *124*

MacAuley, A., 163, 169, 170, *177*

Macchia, P. E., 311, *324, 326,* 478, *481,* 508, *513*

Macchia, V., 478, *481,* 508, *513*

Macdonald, R., 78, *102,* 521, 522, *537*

MacDonald, R. J., 321, *328*

MacDougald, O. A., 271, *276*

MacGregor, G., 183, *190*

Macias, D., 219, 228, *231*

Mackarehtschian, K., 84, *102,* 205, *208*

Mackay, A. M., 206, *209*

Mackem, S., 40, 45, *50, 51,* 508, *515*

MacKenzie, A. L., 437, 441, *490*

Mackenzie, S. L., 425, 428, 432, *484, 490*

Mackool, R. J., 452, *491*

MacLellan, P. P., 342, *364*

MacLellan, W. R., 341, *364*

MacNaughton, K. A., 415, *417*

MacNeil, I., 192, 194, *207,* 245, *250*

MacNeill, C., 316, *326*

Maconochie, M., 117, 118, *124,* 449, 468, 479, *484, 492*

MacPhee, D. J., 156, *177*

Madan, A., 225, *231*
Madden, S. L., 400, *418*
Madel, R., 314, *328*
Madison, K. C., 581, *589*
Madri, J. A., 223, *229*
Madsen, O. D., 166, *176*, 310, 321, *325, 327*
Maeda, M., 341, 352, *363*, 375, *391*, 400, *418*
Maeda, T., 92, *102*, 290, *294*, 437, 450, 451, 466, 474, *495*
Maeda, Y., 415, *420*
Maekawa, A., 531, *538*
Maekawa, T., 121, *123*
Maemura, K., 347, *359*, 432, 450, 465, *489*
Maeno, M., 198, *208*
Magdaleno, S. M., 315, *326*
Magenheim, J., 164, *176*
Maggio-Price, L., 468, *485*
Maggs, A. M., 269, *274*
Maghnie, M., 508, 509, *514, 516*
Magin, T. M., 570, 572, 583, *587*
Magli, M. C., 206, *206*, 246, *249*
Magner, M., 220, *228*
Magnuson, M. A., 309, 319, 320, *327*, 453, 476, *493*
Magnuson, T., 31, *34*, 169, *179*
Magram, J., 228, *232*
Magre, S., 372, 387, *392*
Maguire, C. T., 343, 350, 351, 352, *359*
Maguire, M., 226, *230*
Mah, S., 408, 409, 410, *417*
Mah, S. P., 410, *418*
Mahadevaiah, S. K., 377, 386, *389*
Mahan, M. A., 194, 206, *209*
Mahboubi, K., 225, *231*
Maheshwari, H., 511, *513*
Mahlapuu, M., 527, 530, 532, *535*
Mahmood, R., 116, *125*, 545, 546, *563*
Mahon, J. A., 470, *484*
Mahon, K. A., 45, *50*, 94, *104*, 173, *177*, 402, *417*, 450,
    471, *494*, 502, 504, 505, 507, 508, 509, 510, 511, 512,
    514, 515, 517, 518, 521, 522, 527, *537*
Mahony, M., 556, *565*
Maida, J. M., 117, *122*
Mailhos, A., 79, *103*, 521, 533, *537*
Mailutha, K. G., 312, *327*
Maina, F., 266, *275*
Maioli, M., 337, 338, *369*
Maire, P., 262, 268, *276*
Maisonpierre, P. C., 224, 225, 227, *230, 231, 232*
Majesky, M., 342, *359*
Majumdar, A., 395, *417*
Majumder, K., 352, 354, *366*
Majumder, S., 5, 9, *17*
Majzoub, J. A., 510, 512, *516*, 581, *586*
Mak, T. W., 136, 137, 139, *145, 147*, 341, *370*, 468, *495*,
    509, *514*

Makarenkova, H., 219, *229*, 574, 577, *588*
Maki, K., 511, *516*
Maki, R. A., 168, *179*
Maki, Y., 79, *99*
Makino, T., 167, *180*
Malach, S., 405, 407, *416, 417*
Malapert, P., 142, *146*, 262, 263, 264, *274*
Malas, S., 91, *102*
Malcolm, S., 433, 452, 453, 466, *482, 492, 493, 494*
Maleszewski, M., 8, *17*
Maliakal, J., 416, *420*
Malicki, J., 546, *563*
Mallamaci, A., 40, 46, *52*, 79, 92, *102, 104*, 290, *294*, 450,
    471, 472, 473, 478, *495*
Mallo, M., 427, 432, 433, 443, 449, 468, 469, 473, 474,
    477, *485, 488, 490, 494*
Malpel, S., 312, 315, *326*
Maltais, L. J., 311, *323*
Maltepe, E., 172, 173, 174, *174*, 225, *231*
Malter, H. E., 23, 27, *35*
Maltosy, A. G., 568, *588*
MaMahon, A. P., 84, *101*, 246, *249*
Mamuya, W. S., 219, *229*
Manabe, M., 581, *587*
Manaia, A., 200, *208*
Manasek, D. M. D., 352, 354, *363*
Manasek, F. J., 354, 355, *364*
Manasek, M. D., 352, 354, *363*
Mancilla, A., 423, *490*
Mandel, G., 243, *249*
Mandel, J., 283, 284, 289, *294*
Mangalam, H. J., 509, *515*
Mangelsdorf, D. J., 468, *490*
Mangia, F., 10, *16*
Mango, S. E., 165, *176*, 316, *325, 326*
Maniatis, T., 136, *145*, 226, *233*
Manivel, J. C., 416, *419*
Mankoo, B. S., 260, 267, *275*
Manley, N. R., 305, 311, *327*, 545, *561*
Mann, J. R., 443, *482*
Mann, R. S., 91, *103*, 478, *490*
Mann, S. J., 574, *587*
Mannens, M., 174, *174*, 375, *391*
Mannikko, M., 412, *418*
Manning, S., 2, *4*
Mannisto, A., 511, *516*
Manova, K., 2, *4*, 164, *175*, 303, *323*
Manseau, L., 185, *189, 190*
Mansour, S., 281, *292*, 380, 381, *390, 391*, 466, *490*, 545,
    563
Mansouri, A., 83, *103*, 135, 137, *147*, 260, 270, 272, *276*,
    311, 321, *327, 329*, 432, 451, 473, 475, 477, *490, 498*
Mansukhani, A., 286, *294*
Mantero, M. P., 290, *292*

Mantero, S., 450, 453, 472, 473, *481*, 549, *561*
Manzanares, M., 118, 119, *124*, 453, *490*, 544, *563*
Mao, J. I., 453, *487*
Mar, J. H., 262, *274, 277*
Marasso, M. I., 580, 583, *587*
Marazzi, G., 226, *232*
Marceau, N., 317, *324*
Marcelle, C., 141, *147*, 200, *207*, 262, 263, 264, 265, *275*,
    435, 475, *481, 496*
Marchac, D., 453, *489*
Marchionni, M. A., 94, *97*, 240, 244, *252*
Marchuk, D. A., 227, *229, 231, 232*
Marcil, A., 510, *518*
Marco, S., 91, *98*
Marcos, E., 508, *516*
Marcos, M. A., 196, *207*
Marcotti, W., 553, *563*
Marcus, R., 81, 86, *101*
Marcus, R. C., 85, *102*, 522, 527, *536*
Mardon, G., 268, *274*, 399, *417*, 530, 533, *535*
Marekov, L. N., 570, 581, 582, *584, 588*
Margolis, B., 94, *105*
Mariani, M., 525, *535*
Mariano, A., 311, *324*, 478, *481*, 508, *513*
Marigo, V., *105*, 265, *275*, 470, *490*
Marín, F., 91, *100*, 428, *490*
Marin, O., 81, 82, 85, 88, 89, 90, 91, *102, 105*
Marin, X., 435, *484*
Marinkovich, M. P., 572, *587*
Marino, B., 356, 357, *359, 365*
Marino, S., 317, *325*
Marin-Padilla, M., 201, 204, *209, 210*, 220, *231, 232, 233*
Marion, C. D., 44, *50*
Marital, J., 502, *515*
Mark, M., 94, *100*, 117, *123*, 172, 173, *179*, 305, 312, *327*,
    329, 341, 346, 347, 352, 357, *362, 368, 369*, 427, 433,
    443, 449, 451, 453, 468, 469, 474, 479, *484, 485, 489*,
    490, 491, 494, 525, *536*, 544, 545, *562, 563*
Mark, T., 15, *18*
Markakis, E. A., 236, *251*
Markel, D. S., 227, *231*
Marker, P. C., 82, *101*, 315, *326*, 467, *488*
Markert, C. L., 7, *17*
Markham, B. E., 343, *363, 365*
Markkula, M., 42, *51*, 158, *177*, 511, *516*
Marklin, R. J., 354, *365*
Markov, L. N., 581, 582, *587*
Markova, N. G., 580, 583, *587*
Markovich, D., *323*
Markowitz, G., 415, *420*
Marks, S. C., Jr., 442, *490*
Markwald, R. R., 334, 336, 342, 344, 345, 347, 348, 350,
    351, 352, 357, *365, 368, 369, 370*
Markwell, R. R., 357, *369*

Marlton, P., 202, *208*
Maro, B., 7, *18*
Maron, B. J., 357, *367*
Marone, M., 188, *190*
Maroteaux, P., 286, *294*, 453, *494*
Maroto, M., 141, 142, *148*, 262, 264, 268, *275, 276*
Marovitz, W. F., 547, 552, *564*
Marquardt, T., 519, 530, 533, *535*
Marques, S., 31, *33*, 44, *50*, 338, *358*
Mars, W. M., 246, *251*
Marshak, D. R., 206, *209*
Marshall, C. J., 201, *208*
Marshall, D. M., 582, *587*
Marshall, H., 117, 118, 119, *124, 125*, 422, 425, 427, 449,
    468, 479, *484, 485, 487, 490, 492*, 544, *563*
Marshall, V. S., 303, *329*
Marshall Graves, J. A., 376, *392*
Marshall-Graves, J. A., 387, *392*
Marsh-Armstrong, N., 525, *537*
Marsushima, J., 557, *565*
Martens, E., 268, *275*
Mårtensson, I. L., 47, *50*, 451, 476, *483*, 508, *514*
Marth, J. D., 14, *16*
Martí, E., 470, *490*
Martidale, M. Q., 471, *493*
Martin, C., 196, 200, 201, *206, 207, 208*, 219, *229*
Martin, D. I. K., 356, 357, *360, 369*
Martin, D. M., 557, *561*
Martin, G. R., 3, *4*, 15, *17*, 31, *35*, 45, *50*, 79, 84, 85, 86,
    98, 99, 102, 105, 106, 113, 114, 115, 116, *123, 124*,
    125, 132, *145, 146*, 166, *178*, 317, 321, *324*, 328, 337,
    340, 352, *360, 365, 368*, 415, 420, 422, 428, 432, 440,
    449, 450, 451, 453, 466, 475, *483, 489, 490, 491, 492*,
    494, 496, 497, 578, 583, *586*
Martin, G. V., 555, *565*
Martin, J. F., 262, *273*, 333, 342, 354, 355, *360, 364, 365*,
    433, 451, 476, 477, 478, *489, 490, 496*, 549, *565*
Martin, J. S., 203, *207*, 226, *229*, 467, *484*
Martin, L. J., 94, *103*, 264, *274*
Martin, P., 547, *564*
Martineau, J., 373, 383, *391*
Martínez, C., 91, *102*
Martínez, S., 77, 78, 79, 84, 85, 86, 88, 89, 95, *99, 102*,
    103, 104, 106, 113, 114, 116, *123, 124, 125*, 422, 428,
    453, *490, 494, 497*
Martinez-Arias, A., 136, *145*
Martinez Barbera, J. P., 31, *34*
Martinez-Barbera, J.-P., 31, *33*, 46, 47, 48, *50, 51*, 311, 316,
    327, 451, 476, *483, 490*, 508, *514*
Martinez-De Velasco, J., 559, *562*
Martinsen, B. J., *490*
Maruniak, J. A., 237, *249*
Maruthainar, K., 118, *124*
Marvin, M. J., 334, 335, 336, 337, 339, *365*

Marxer, A., 308, *325*
Marziali, G., 189, *190*
Masai, I., 78, *102*
Mascara, T., 443, *495*
Maseki, N., 202, *208*
Mashima, H., 321, *327*
Masiakowski, P., 292, *292*
Mason, A. J., 512, *516*
Mason, A. K., 512, *516*
Mason, C. A., 85, *102*
Mason, I., 84, *102,* 110, 112, 115, 116, *123, 125,* 428, 449, *487,* 545, 546, *563, 565*
Mason, J. O., 87, 94, *105*
Mason, S., 11, 12, *17*
Massabanda, J., 264, *274*
Massaddeq, N., 109, *122*
Massagué, J., 47, *51,* 468, *485*
Massalas, J. S., 87, 91, *103,* 508, *517,* 521, 527, *537*
Masson, N., 512, *514*
Masson, R., 162, *179*
Mastick, G. S., 85, *102*
Masuda, H., 220, *228*
Masui, T., 303, *327*
Masui, Y., 7, *17*
Masuno, M., 398, *419*
Masuya, H., 83, *102*
Mata de Urquiza, A., 91, *105,* 120, *125*
Mathers, P. H., 521, 522, 527, *537*
Mathews, L. S., 511, *513*
Mathieu, M., 506, *517*
Mathis, J. M., 506, *516*
Matise, M. P., 85, *102, 103,* 121, *124, 125,* 505, *517*
Matsubara, Y., 525, *537*
Matsubasa, T., 528, *536*
Matsuda, A., 531, *538*
Matsuda, H., 531, *538*
Matsuda, Y., 223, *230*
Matsui, D., 312, *328,* 505, *517*
Matsui, T., 318, *325*
Matsui, Y., 188, 189, *190,* 246, *251*
Matsuk, M. M., 44, *51,* 440, *484*
Matsuki, M., 581, 582, 583, *587*
Matsumoto, K., 10, *17,* 466, *496,* 525, *536,* 571, *586*
Matsumoto, M., 94, *106*
Matsumoto, N., 581, *588*
Matsunami, H., 340, *366,* 408, *417, 418*
Matsunashi, T., 204, *209,* 220, *232*
Matsuno, K., 553, *561*
Matsuo, H., 466, *496*
Matsuo, I., 31, *34,* 46, 49, *51,* 83, 84, *105, 106,* 114, *124, 125,* 374, *392,* 403, *418,* 421, 451, 457, 478, *488, 490, 496,* 521, 527, *537*
Matsuo, N., 423, 424, 451, 475, 476, 478, *490*
Matsuo, T., 423, 424, 451, 475, 476, 478, *490*

Matsushima-Hibiya, Y., 531, *538*
Matsushita, S., 303, *329*
Matsuzaki, F., 95, *105*
Matsuzaki, T., 557, *565*
Mattei, M. G., 85, 91, *100, 103,* 311, *324,* 372, 387, *392*
Matthews, A. N., 550, *563*
Matthews, K. L., 437, 450, *497*
Matthews, L., 352, 354, *366*
Matthyssen, A., 167, *177*
Mattiacci, M., 263, *274*
Mattot, V., 226, *229*
Mattsson, R., 167, *178*
Matzuk, M. M., 122, *122,* 226, 227, *229,* 336, 352, *359,* 441, 467, 468, *490, 491,* 500, 510, *516*
Mauch, T. J., 397, *418*
Mauer, S. M., 416, *419*
Mauger, A., 572, 573, *587*
Maulik, C., 292, *294*
Maurer, M., 574, *588*
Maurer, R. A., 508, *517*
Mauro, T., 580, 582, 583, *585, 587, 588*
Maury, M., 31, *33,* 46, 48, 49, *49,* 114, *122, 322,* 428, 451, 478, *481,* 521, 527, *535*
Mavilio, F., 206, *207,* 246, *249,* 271, *273, 274,* 443, 449, 478, *482*
Max-Audit, L., 200, *208*
Maxeiner, S., 173, *177*
Maxson, R., 290, *294,* 428, 437, 450, 451, 452, 466, 474, *484, 489, 494, 495*
Maxwell, P., 225, *229*
May, G., 199, 200, 201, *210*
May, L. L., 260, 267, *275*
May, N. R., 95, *102,* 141, *147,* 506, *516*
May, R. T., 248, *251*
May, S. R., 48, *49,* 78, 98, 304, *323*
Mayen, A. E., 193, *207*
Mayer, B., 31, *35*
Mayer, E. L., 203, *209*
Mayer, M., 583, *585*
Mayer, W., 9, *17, 18*
Mayer-Proschel, M., 240, 241, *251*
Mayo, K. E., 443, *485,* 508, 511, *515, 516*
Mayo, M. L., 428, *491*
Mayor, R., 423, *490*
Mayordomo, R., 541, *564*
Mazan, S., 31, *33,* 44, 46, 48, *49, 52,* 114, *122, 322,* 428, 451, 478, *481,* 508, *513,* 521, 527, *535*
Mazzalupo, S., 570, *587*
Mbamalu, G., 171, *175*
Mbiene, J. P., 552, *564*
McAllister, A. K., 94, *102*
McAllister, D., 317, *330*
McAllister, K. A., 227, *231*
McAndrew, J., 511, *516*

McAvey, B. A., 7, *16*
McAvoy, J. W., 532, *537*
McBurney, M. W., 236, 245, *251*
McCabe, E. R., 378, 379, 387, *389, 393*
McCaffery, P., 345, 346, 347, *366*, 469, *482, 492, 498*, 525, *537*
McCaffery, R. J., 346, *367*
McCall, A. E., 122, *122*
McCarthy, J. G., 452, *491*
McCarthy, K. J., 511, *516*
McCarthy, M., *102*, 241, 244, *251*
McCarthy, M. M., 500, *516*
McCarthy, M. T., 427, *484*
McCarty, M., 172, 173, *180*
McCaw, P. S., 257, *275, 276*
McClain, J., 223, 224, 227, *231, 232*, 559, *561*
McClearn, D., *491*
McClive, P. J., 382, 388, *393*
McClure, M. H., 316, *324*
McComb, D. J., 512, *515*
McComb, R. D., 166, *179*
McConnell, S. K., 84, 87, 92, 94, *98, 102, 106*
McCormick, F., 86, *105*, 512, *516*, 579, *589*
McCormick, M. K., 227, *231*
McCormick, R. J., 525, *538*
McCoy, M. J., 162, 170, *176*
McCready, P., 412, *418*
McCurrach, M. E., 341, *370*
McDade, J. R., 347, 349, *361*
McDermott, M., 509, 510, *515*
McDevitt, D. S., 534, *538*
McDonald, D. M., 223, 224, 227, *232*
McDonald, L., 348, 349, 352, 353, 357, *358*
McDonough, B., 357, *358*
McDowall, S. G., 381, *391*
McDowell, T. L., 376, *390*
McElreavey, K., 375, 376, 382, 387, *389, 390, 391, 392*
McEvilly, R. J., 89, 95, *104*, 506, *517*, 556, 558, *562*
McEwen, B. S., 247, 249, 250
McEwen, D. G., 286, *292*, 558, *561*
McFadden, D. G., 168, *175*, 342, 344, 347, 349, 352, *359, 361, 365, 366*
McFadden, K. A., 82, 86, *100*
McGaughey, R. W., 13, *16*, 27, *33, 34*
McGhee, J. D., 165, *175, 177*, 316, *325*
McGrew, M. J., 134, 138, 139, *147*
McHugh, K. P., 223, *230*
McInnes, L., 341, *365*
McInnes, R. R., 527, *535*
McIntosh, I., 286, *292*, 453, *481, 495*
McIntosh, T. J., 582, *587*
McIntyre, J. II, 136, 137, *146*, 553, *563*
McKay, I. J., 466, *491*, 545, 546, *564*
McKay, R. D., 88, 89, *98, 101*, 235, 247, 249, *251*

McKearn, J. P., 192, *208*
McKeehan, W. L., 287, *293*
McKendry, R., 579, *587*
McKenna, H. J., 173, *179*
McKercher, S. R., 168, *179*
McKinnon, W. C., 227, *231*
McKinsey, T. A., 269, *277*, 342, 343, 353, *364, 365, 369, 370*
McLain, K., 204, *208*, 318, *327*
McLaren, A., 26, *34*, 183, 189, *189*, 382, 384, *389, 391*
McLaughlin, J., 10, 12, 13, *18*
McLaughlin, K. A., 339, *367*
McLoon, S. C., 88, *105*
McMahon, A. P., 40, 44, 48, *49, 51*, 78, 79, 82, 85, 86, *98, 99, 101, 102, 104, 105, 106*, 112, 116, 117, 120, 121, *123, 124, 125*, 132, 133, 135, 137, 140, 141, *145, 147, 148, 149*, 166, 167, *175*, 228, 229, *230, 232*, 238, 246, *249, 250*, 260, 262, 264, 265, *275*, 287, 292, *293, 294*, 304, 308, 310, 312, 315, 317, *323, 324, 328, 329*, 334, 336, 348, *362, 363*, 384, 385, 386, *389, 392, 393*, 405, 406, 407, *418, 419*, 423, 426, 428, 439, 440, 441, 442, 443, 451, 453, 468, 470, 471, *482, 483, 484, 486, 487, 488, 489, 490, 491, 496, 498*, 504, 505, 506, *518*, 529, *538*, 545, *565*, 577, 578, 583, *584, 585, 589*
McMahon, G., 217, 220, *229*
McMahon, J. A., 23, *33*, 78, 85, *99, 101*, 132, 137, 141, *145, 147, 148*, 166, 167, *175*, 260, 264, 265, *275*, 304, 317, *323, 329*, 428, 468, 470, *482, 485, 491*, 504, 505, 506, *518*, 577, 578, 583, *584, 589*
McNamara, J. O., 247, *251*
McNiff, J. M., 579, *584*
McNoe, L. A., 398, *419*
McPherron, A. C., 129, *147*, 260, 264, *275*
McQuinn, T., 342, 357, *359, 369*
Mead, P. E., 198, 203, *208, 210*
Mecklenberg, L., 574, *588*
Medberry, S. L., 15, *17*
Medlock, E. S., 317, *327*
Medvinsky, A., 194, 195, 196, 199, 200, *207, 208*, 220, *231*, 317, *327*
Meegdes, B. H., 174, *177*
Meehan, M. A., 452, *492*
Meehan, R. R., 302, *327*
Meeson, A. P., 219, *231*
Megeney, L. A., 258, 260, 271, *275*
Mehler, M. F., 239, *250, 251*
Mehlmann, L., 8, *17*
Mehrara, B. J., 452, *491*
Mei, W. Y., 579, *589*
Mei, Y. L., 402, *417*
Meier, P., 79, *104*, 529, *537*
Meier, S., 424, 428, *491*
Meijers, J. H., 385, *389*
Meijers-Heijboer, H., 528, *538*

Meijlink, F., 451, 476, 477, 478, *491, 496,* 549, *565*

Meindl, A., 378, *392*

Meinecke, P., 468, *485*

Meiners, L. C., 94, *99*

Meinhardt, G., 174, *177*

Meinhardt, H., 138, 139, *147*

Meininger, C. J., 189, *190*

Meins, M., 334, *365*

Meisler, M. H., 428, 475, *497*

Melancon, E., 265, 269, *273*

Melega, W., 239, 241, *251*

Mellitzer, G., 119, *124, 126,* 225, *231, 233*

Mello, C., 14, *16,* 185, *190,* 240, 243, *249, 251, 252*

Mellon, P. L., 504, 509, 510, 512, *513, 516, 517, 518*

Mellon, S. H., 510, *513*

Melmed, S., 510, 512, *513, 518*

Melnick, M., 436, *487*

Meloy, D., 31, *34,* 46, 47, 48, *51,* 311, 316, *327*

Melton, D. A., 3, *4,* 166, *176,* 303, 304, 308, 310, 319, 320, 321, *325, 326, 328, 329,* 405, *418*

Melton, D. W., 570, *587*

Meltzer, P. S., 347, *366*

Mendell, J. R., 271, *273*

Mendelsohn, C., 312, *326, 327,* 403, *418,* 451, 453, 468, 469, *489, 490, 491*

Mendelsohn, M., 122, *124*

Mendez, R., 7, 10, 11, *17, 18*

Mendis, D., 11, *19*

Mendosa, E. S., 543, *564*

Meneses, J. J., 38, 39, 45, 47, *51,* 79, 83, 87, 89, 90, 91, 92, 95, *97, 98, 99, 103,* 290, *292,* 303, 304, 321, *326, 329,* 333, *363,* 410, *418,* 422, 435, 437, 450, 451, 453, 470, 471, 472, 473, 477, 478, *481, 484, 493,* 549, *562,* 570, *589*

Meng, A., 135, *147*

Meng, X., 218, 223, *230,* 403, *419*

Menissier, F., 264, *274*

Mennerich, D., 266, *275*

Meno, C., 47, *51,* 354, 357, *365, 370*

Menon, G. K., 580, 581, 582, *584, 585, 587*

Mentink, M. M., 219, *232,* 333, *360*

Menzel, P., 286, *292*

Meo, T., 386, *390*

Meraz, M. A., 287, *293*

Mercader, N., 91, *102*

Mercer, B., 228, *230,* 269, *277,* 342, *364, 365, 369*

Mercer, E. H., 109, 114, *126,* 228, *233*

Merchant-Larios, H., 382, 384, 387, *391, 392, 393*

Mercier, Y., 9, *15, 18*

Mercola, M., 219, *229,* 334, 335, 336, 337, 338, 339, 340, 356, *363, 366, 367,* 473, *489*

Mercurio, S., 92, *102*

Meredith, A., 89, *101*

Meredith, M. M., 25, *34*

Meredith, M. R., 158, *176*

Merentes-Diaz, E., 160, *178*

Mericksay, M., 266, *277,* 335, 337, 340, *358*

Merker, H. J., 570, *584*

Merlet-Benichou, C., 403, *418, 420*

Merlie, J. P., 411, *419*

Merlino, G., 268, *275*

Merlo, G. R., 290, *292,* 450, 453, 472, 473, *481,* 549, *561*

Mermer, S., 573, *584*

Mershcer, S., 353, 357, *365*

Mertelsmann, R., 288, *293*

Mertin, S., 381, *391*

Mertineit, C., 382, *393*

Meskenaite, V., 453, *486*

Mesnard, D., 27, *33*

Mesquita, S. F., 346, *358*

Messaddeq, N., 94, *100,* 341, 346, 347, 352, *362, 366,* 469, *485,* 583, *586*

Messdeq, N., 341, 352, *368*

Metcalf, D., 192, 193, 194, 195, 196, 198, 203, *208, 209*

Metsaranta, M., 573, *588*

Metsäranta, M., 424, 442, 443, *494*

Mett, I. L., 341, *369*

Mettei, M. G., 40, 47, *51*

Metzger, D., 109, *122*

Metzger, R. J., 311, *327*

Meyer, A. N., 453, *485*

Meyer, B. I., 547, *564*

Meyer, D., 349, *365*

Meyer, J., 281, *293, 294,* 380, 381, *391, 393*

Meyer, M., 204, *207,* 317, *324*

Meyer, T., 225, *230*

Meyerowitz, E. M., 11, *17*

Meyers, E. N., 31, *35,* 114, 115, *124,* 337, 340, 352, *365, 368,* 466, *489, 496*

Meyers, G., 453, *487*

Meza, U., 511, *514*

Mezey, E., 206, *207*

Mi, T., 193, 206, *208,* 270, 271, *275*

Mi, Y., 227, *231*

Miano, J. M., 228, *230,* 342, *365*

Miao, H. Q., 223, *232*

Mic, F. A., 525, *537*

Michael, D., 580, *586*

Michalak, M., 343, 347, *361*

Michaliszyn, E., 510, *514*

Michalopoulos, G. K., 318, *327*

Michaud, J., 95, *97, 102,* 121, *124, 125,* 451, 453, 470, 471, *491, 493,* 506, *513, 516*

Michel, E., 281, *293,* 381, *391*

Michel, R. N., 269, *277,* 342, *369*

Michels, A., 91, *98*

Michelson, A. M., 136, *145,* 165, *178,* 316, *328,* 340, *361*

Michle, C., 263, *274*

Michon, A., 185, *189*
Miettinen, P., 314, *327*, 465, *491*
Mignatti, P., 224, *232*, 504, *517*
Mikawa, T., 219, 228, *231*, 334, 347, *362, 365, 368*
Mikkola, I., 521, 522, *537*
Mikoshiba, K., 94, *103, 104, 106*, 122, *122*, 357, *363*, 423, *492*
Milbrandt, J., 510, *516*
Milenkovic, L., 78, 80, *100*, 121, *123*, 470, *485*
Miles, C., 416, *419*
Milhalek, R. M., *251*
Milla, P. J., 305, *328*
Millan, F. A., 467, *491*
Millauer, B., 227, *231*
Millen, K. J., 115, 122, *124*, 428, 453, *491*
Miller, A. P., 506, 508, 509, *518*
Miller, C. L., 245, *251*, 385, *391*
Miller, C. P., 322, *328*
Miller, C. T., 465, *491*
Miller, D., 475, *485*
Miller, G., 470, *486*, 579, *586*
Miller, I., 188, *189*
Miller, J. B., 132, *148*, 263, 269, *275, 276*
Miller, J. D., 204, *210*, 220, *233*
Miller, J. R., 337, *363*, 405, *418*
Miller, S. A., 355, *365*
Miller, T. A., 204, *208*, 318, *327*
Miller, W., 12, *16*, 438, *491*
Millet, S., 114, 115, *124*, 428, *491*
Millonig, J. H., 122, *124*, 453, *491*
Mills, A. A., 571, 581, 583, *587*
Min, H., 312, *327*
Min, J.-Y., 354, *365*
Mina, M., 439, *491*
Minamino, N., 466, *496*
Minasi, M. G., 246, *250*
Mincheva, A., 479, *487*
Min Deng, J., 31, *34, 328*
Mine, T., 321, *327*
Miner, C., 466, *494*, 545, *564*
Miner, J. H., 260, *276*, 411, 412, *418, 419*
Ming, J. E., 470, *491*
Minichiello, L., 559, *564*
Minkoff, R., 469, 470, *486*
Minnerath, S. R., 94, *100*
Minoo, P., 312, 314, *327*
Minowa, O., 122, *122*, 173, *179*, 217, *230*, 404, *420*, 506, *516*, 549, *564*
Minowada, G., 115, 116, *123*, 466, *491*
Minshull, J., 338, *363*
Mintz, B., 245, *249*
Mione, M., 79, 88, 89, 90, 94, 95, *97, 99*, 450, 471, *484*
Miralles, F., 321, *327*
Miranda, B., 313, 315, *327*

Miranda, M., 9, 10, *17, 19*
Mirtsos, C., 468, *495*
Mishina, Y., 336, 352, *365*, 385, *391*
Miska, E. A., 342, *368*
Missale, C., 511, *513, 516*
Missero, C., 311, *326*
Mitchell, K., 341, *370*
Mitchell, P. J., 79, *104*, 529, *537*
Mitra, J., 7, *17*
Mitsiadis, T., 509, *518*
Mitsialis, S. A., 311, 312, *323*
Mitsock, L. M., 467, *498*
Mitulla, B., 281, *293*, 381, *391*
Miura, H., 353, 354, 355, 356, *363*
Miura, N., 471, 479, *487, 491*
Miwa, H., 528, *538*
Miyagawa, J., 321, *327*
Miyagawa, K., *391*, 416, *419*
Miyagawa-Tomita, S., 340, 352, 353, 354, 355, 356, *363, 367*
Miyake, H., 506, *518*
Miyake, S., 288, *293*
Miyake, T., 442, *486*
Miyama, S., 92, *105*
Miyamoto, K., 466, *496*
Miyamoto, M., 84, *98, 100*
Miyamoto, N., 83, *106*, 374, *392*, 403, *418*
Miyasaka, M., 350, *362*
Miyashita-Lin, E. M., 83, 84, 85, 86, 92, 95, *100, 102, 104*
Miyata, K., 10, *17*
Miyata, T., 92, 94, *102, 103, 106*
Miyazaki, J., 160, *178*, 340, 352, *367*, 443, *492*
Miyazaki, S., 8, *16*
Miyazaki, T., 408, *418*
Miyazaki, Y., 403, *418*
Miyazono, K., 226, *230, 231*, 356, *367*, 468, *486*
Miyoshi, H., 202, *208*
Miyoshi, I., 511, *516*
Mizuno, K., 580, *587*
Mizuno, M., 521, *537*
Mizuno, N., 529, 534, *537*
Mizuno, S., 388, *392*
Mizuno, T., 41, *51*, 303, 308, *324, 329*, 343, *368*
Mizutani, Y., 140, *145, 147*
Mizutani-Koseki, Y., 471, 479, *487*
Mjaatvedt, C. H., 334, 344, 347, 348, 350, 351, 352, 357, *365, 370*
Mo, R., 121, *123, 124*, 310, 312, *326, 327*, 439, 451, 453, 470, 471, *486, 491*
Moak, J. P., 357, *367*
Mobley, W. C., 91, *101*
Mochida, K., 47, *51*, 338, 354, 356, 357, *365, 367, 368*
Mochii, M., 237, *252*, 525, 529, 534, *537*
Mochizuki, T., 352, 354, *365*, 414, *418*

Modolell, J., 550, 555, *561*
Moegelin, A., 452, *498*
Moens, C. B., 118, *125,* 341, 352, *365*
Moens, G., 223, *230*
Moers, A., 508, *516*
Mohandas, N., 443, *493*
Mohanty, N., 14, *17*
Mohiaddin, R. H., 351, *363*
Mohler, J., 78, *99,* 141, *145,* 470, *484*
Möhle-Steinlein, U., 442, 443, *497*
Mohr, S., 341, *369*
Mohri, T., 8, *16*
Mohun, T., 10, *15,* 40, 44, *50,* 337, 338, 339, 345, 348, *358,*
    366
Mohun, T. J., 262, *275,* 334, 335, 336, 339, 342, 355, *359,*
    361, 365, 368
Moisan, M. P., 375, *390*
Molinar-Rode, R., 313, *327*
Molkentin, J. D., 165, *177,* 257, 262, *275,* 316, *327,* 334,
    340, 341, 342, 343, 347, 352, *360, 361, 365, 367*
Moll, I., 570, *587*
Moll, R., 570, *587*
Mollard, R., 314, *327*
Molnár, Z., 92, *102*
Molotkov, A., 525, *537*
Mombaerts, P., 435, *491*
Momma, S., 236, 237, 244, 245, 248, *250*
Momose, T., 525, *536*
Monaco, L., 376, *393*
Monaghan, A. P., 40, 42, 44, *50, 51,* 165, *177,* 303, 314,
    324, 325, 327, 479, *487*
Monaghan, P., 528, *538*
Moncada, S., 315, *324*
Moniot, B., 381, *390*
Monkley, S. J., 172, 173, *177,* 336, 337, *365*
Monod, C., 9, 14, *15*
Monreal, A. W., 441, *491*
Monsoro-Burq, A.-H., 140, 141, *147*
Montesano, R., 223, *230,* 321, *328*
Montezin, M., 581, *588*
Montgomery, C., 442, 443, *495*
Montgomery, M., 341, *364*
Montgomery, M. K., 14, *16, 17*
Montgomery, M. O., 337, 340, 341, *365*
Montgomery, R. A., 347, *360*
Monti-Graziadei, A. G., 77, 79, 84, *100*
Montminy, M. R., 286, *292, 293*
Montonen, O., 441, *488*
Montrocher, C., *585*
Monzen, K., 336, 343, *365, 368*
Moody, S. A., 435, *491, 532, 536*
Mookerjee, B., 225, *231*
Moon, R. T., 44, 45, *51,* 337, *363,* 405, *418,* 579, *584*
Mooney, J. R., 577, *587*

Moons, D., 165, *177,* 316, *325*
Moons, L., 203, *207,* 223, 225, 226, *228, 229*
Moorby, C., 404, *420*
Moore, A. T., 533, *535*
Moore, A. W., 334, 341, 349, *365*
Moore, B., 135, *147*
Moore, C. C., 380, 381, 384, *392*
Moore, G. D., 8, *17*
Moore, K. A., 194, 205, 206, *208, 209*
Moore, K. J., 525, *536*
Moore, K. L., 568, *587*
Moore, L., 185, 186, 188, *190,* 568, 582, 583, *586*
Moore, M. A., 195, 196, 198, 201, *208,* 317, *325*
Moore, M. W., 203, *207,* 223, 229, 402, *418, 420*
Moore, R., 15, *16*
Moore, R. L., 201, *208*
Moore, W. A., 194, *207*
Moore, W. J., 421, 442, 449, 456, 457, 471, *491*
Moorman, A. F. M., 334, 341, 343, 344, 345, 347, 348,
    350, 351, 352, 355, 356, *359, 360, 361, 362, 363, 365,*
    366, 369
Moorman, M. A., 206, *209*
Moos, J., 8, *17*
Moos, M. C., Jr., 14, *16*
Mora, M., 246, *250*
Morais da Silva, S., 380, 381, 384, 387, *392*
Moran-Rivard, L., 260, 266, *274*
Morasso, M. I., 173, *177*
Morassutti, D. J., 235, 236, 242, 244, 245, *250, 251*
Morata, G., 91, *102*
Moreau, A., 478, *489*
Moreau, E., 403, *418, 420*
Morell, R., 475, *481*
Morell, R. J., 556, *564*
Moreno, T. A., 423, *481*
Moreno-Mendoza, N., 382, 387, *391, 392*
Moreno-Rodriguez, R. A., 344, 345, 348, *360, 365*
Morgan, B. A., 305, *328,* 575, 576, 577, 578, 579, *587, 588*
Morgan, D., 352, 354, *366,* 441, *488*
Morgan, E. P., 452, *492*
Morgan, J. I., 313, *327*
Morgan, M. J., 351, 356, *360*
Morgan, R. A., 14, *16*
Morgan, W. C., 137, *147*
Morgenbesser, S. D., 528, *537*
Mori, C., 136, 137, *147,* 173, *179,* 341, 352, *370,* 404, *420*
Mori, K., 237, *252,* 435, *491*
Mori, M., 512, *518,* 528, *536*
Morikawa, Y., 318, *325*
Morimitsu, T., *537*
Morin, P. J., 86, *100,* 579, *586*
Morin, S., 342, 343, *366*
Morishima, M., 353, 357, *364*
Morishima, Y., 581, 582, 583, *587*

Morita, M., 527, 530, 531, *536*
Morita, T., 412, *418*
Moriuchi, T., 11, 12, *18*
Moriwaki, K., 83, *102*
Moriyama, H., 195, *208*
Moriyama, K., 94, *104,* 502, 508, 509, *517,* 557, *564*
Morkin, E., 343, *360*
Morley, G. E., 334, 348, *367*
Moro, J. A., 547, *564*
Morohashi, K., 375, 376, 378, 379, 380, 384, *390, 391, 392*
Morozova, L. M., 26, 27, *33*
Morrell, C., 11, 12, *17*
Morrice, D. R., 534, *536*
Morris, J. F., 400, *418, 419*
Morris, J. H., 260, *274, 276, 277*
Morris, N. P., 442, 443, *489*
Morris, R. J., 570, *587, 589*
Morrisey, E. E., 165, *177,* 303, 316, *326, 327,* 342, 343, 352, *363, 366*
Morrish, T. A., 349, *358*
Morrison, A., 117, 118, 119, *124,* 544, 552, 553, 555, 558, 559, *563, 564*
Morrison, A. D., 469, *491*
Morrison, M., 118, *123*
Morrison, S. J., 235, 236, 238, 239, 240, 241, 243, 244, 245, 248, *249, 251*
Morriss-Kay, G. M., 132, *146,* 199, 200, 201, *210,* 424, 437, 452, 468, 469, *484, 485, 487, 491, 494, 496,* 545, 546, 557, *562, 563, 564*
Morris-Wiman, J., 443, *482*
Morrow, B. E., 353, 357, *365*
Morrow, E. M., 527, *535*
Morrow, K. E., 378, *390*
Morshead, C. M., 235, 236, 244, 245, *251*
Morsli, H., 478, *491,* 541, 542, 550, 551, 552, 555, 558, 559, *564,* 579
Mort-Hopkins, T., 453, *483*
Morton, C. C., 346, *358*
Morton, S., 43, *50*
Mortrud, M. T., 510, *513*
Mosca, J. D., 206, *209*
Mosca, R. S., 356, *362*
Moser, C., 378, *393*
Moses, H. L., 467, *493*
Moshnyakov, M., 559, *564, 565*
Mosinger, B. J., 508, *517*
Moss, J. B., 345, 346, 347, *366*
Moss, M. L., 452, *492*
Mossman, H., 14, *16*
Mostachfi, H., 173, 174, *175*
Motegi, M., 511, *514*
Motegi, Y., 353, 354, 355, 356, *363*
Motoike, T., 222, *231*
Motoyama, J., 121, *123, 124,* 317, 321, *327*

Motro, B., 193, 204, *208*
Motta, P., 7, *19*
Motte, J., 94, *99*
Mouly, V., 344, *362*
Mourton, T., 169, *179*
Moury, J. D., 423, *492*
Mouse Genome Database, *516*
Moussa, S. M., 512, *515*
Moutard, M. L., 94, *99*
Mucenski, M. L., 204, *208,* 318, *327*
Muchamore, I., 422, 425, 427, 468, *485, 487, 490*
Mueller, C., 316, *326*
Muenke, M., 78, *102,* 271, *276,* 357, *359,* 433, 452, 453, 468, 470, *481, 485, 491, 492*
Muglia, L. J., 581, *586*
Muhr, J., 78, 79, 80, 81, 85, *99,* 108, 120, *123, 124*
Mujtaba, T., 240, 241, *250, 251*
Mukai, M., 186, *190*
Mukai, T., 262, 268, *274*
Mukherjee, P., 196, 202, *210*
Muller, A. M., 194, 195, 196, 199, 200, *207, 208,* 220, *231*
Muller, B. K., 525, *535*
Muller, C. W., 346, *358*
Muller, F., 338, *362*
Müller, K.-H., 317, *325*
Müller, M., 134, *147*
Müller, R., 89, *101*
Muller, T. S., 131, *146,* 266, *273,* 353, 356, *361*
Muller, U., 410, *418,* 452, 453, 474, *487,* 557, *563*
Muller, W., 341, *362*
Muller, W. J., 171, *175*
Muller-Rover, S., 574, 578, *588*
Mulligan, G., *483*
Mulligan, R. C., 193, 205, 206, *207, 208,* 246, *250,* 270, 271, *274,* 285, *293,* 442, 443, *488, 489*
Mulliken, J. B., 227, *232,* 288, 292, *293, 294,* 452, 453, 474, *487*
Mullins, M. C., 546, *565*
Mulnard, J. G., 27, *34*
Mumby, M. C., 94, *100*
Mummery, C. L., 222, *229*
Mundlos, C., 288, *293, 294*
Mundlos, S., 281, 288, *293,* 442, 443, 453, *492, 493*
Mundschau, G., 574, 583, *588*
Muneoka, K., 3, *4*
Munnich, A., 286, *294,* 452, 453, *484, 494*
Munoz, E. F., 314, *325*
Munoz, F., 441, *488*
Munoz-Chapuli, R., 219, 228, *231*
Münsterberg, A. E., 140, 141, *147,* 262, 263, 264, 268, *275,* 377, 381, *390, 392*
Muragaki, Y., 427, 443, *484, 492*
Murakami, M., 512, *518*
Murakami, S., 287, *293*

Muramatsu, H., 350, *370*
Muramatsu, M., 11, 12, *18,* 183, *190,* 504, *516*
Muramatsu, T., 350, *370*
Murase, N., 246, *251*
Murata, T., 49, *51*
Murillo-Ferrol, N. L., 337, *359*
Murohara, T., 220, *228*
Murone, M., 579, *589*
Murphy, D., 502, *515*
Murphy, M. W., 382, *392*
Murphy, P., 468, *491*
Murray, D., 163, *175*
Murray, J. C., 354, *370, 478, 495,* 507, *517,* 527, 532, *537*
Murray, M. J., 467, *495*
Murray, P., 200, *208*
Murre, C., 257, *275, 276*
Murrell, J., 227, *231*
Murtaugh, L. C., 282, *293*
Muscat, G. E., 281, *293,* 381, *392*
Muscat, G. E. O., 342, 343, *359*
Muscatelli, F., 378, *392, 393*
Musci, T. S., 117, *125,* 443, 474, *483,* 544, *561*
Muskavitch, M. A., 136, *145*
Mustonen, T., 218, 223, *229, 230,* 441, *493*
Muthukumar, S., 553, *562*
Mutoh, H., 310, 321, 322, *327*
Muzio, L., 92, *102*
MvKinley, M., 31, *34*
Myat, A., 553, 559, *561*
Myer, D., 357, *366*
Myers, R. M., 287, *293,* 579, *586*
Myglia, L. J., 510, 512, *516*
Myokai, F., 423, 424, 451, 475, 476, 478, *490*

N

Na, E., *102,* 241, 244, *251*
Na, S., 87, *101*
Näär, A. M., 510, *516*
Nabeshima, Y., 258, 260, *272, 274, 276*
Nachtigal, M. W., 376, 381, *391, 392*
Nadal-Ginard, B., 262, *277*
Nadeau, J., 263, *272,* 313, 314, 315, *324*
Nagai, K., 378, *390*
Nagai, R., 432, 450, 465, *489*
Nagai, T., 122, *122,* 357, *363,* 423, *492*
Nagakubo, D., 350, *362*
Nagamine, C. M., 378, *390, 392*
Nagano, M., 376, *393*
Nagao, H., 435, *491*
Nagaoka, S., 11, 12, *18*
Nagase, H., 223, *231*
Nagashima, K., 506, *518*
Nagata, A., 95, *102,* 506, *516*
Nagata, E., 94, *106*

Nagayama, Y., 504, *516*
Nagayoshi, M., 7, *16*
Nagorcka, B. N., 577, *587*
Nagy, A., 3, *4,* 23, *35,* 38, 42, *51,* 132, *147,* 158, 160, 164,
    166, 168, 172, 173, 174, *174, 175, 176, 177, 178, 179,*
    203, *207,* 223, *228,* 302, *324,* 357, *368*
Nagy, L., 469, *484*
Nagy, R., 42, *51*
Naito, M., 195, 196, *208, 209,* 228, *233*
Nait-Oumesmar, B., 170, 174, *174, 178*
Nakafuku, M., 95, *105,* 120, 121, *123, 125,* 451, 475, 476,
    492, 522, *537*
Nakagata, N., 10, *17*
Nakagawa, M., 343, 347, *366*
Nakagawa, O., 343, 347, *366*
Nakagawa, S., 408, *418*
Nakagawa, Y., 84, *102*
Nakahara, K., 526, 527, *538*
Nakahara, Y., 353, 354, 355, 356, *363,* 376, *391*
Nakahata, T., 341, 352, *370*
Nakai, S., 95, *102,* 506, *516,* 549, *564*
Nakajima, K., 94, *103, 104, 106*
Nakajima, O., 527, 530, 531, *536*
Nakamura, A., 186, *190,* 355, *364*
Nakamura, H., 112, 114, 115, 116, *122, 123, 124,* 355, *362,*
    525, *536*
Nakamura, K., 343, 347, *361*
Nakamura, M., 577, 583, *584*
Nakamura, S., 77, 95, *101, 105,* 473, *498*
Nakamura, T., 476, *482,* 577, *586*
Nakamura, Y., 479, *482*
Nakanischi, S., 136, *147*
Nakanishi, S., 136, *146,* 237, *252,* 526, 527, *538*
Nakanishi, Y., 436, *486*
Nakano, H., 14, *17*
Nakano, T., 49, *51,* 136, 137, *147,* 509, *514*
Nakao, K., 228, *233*
Nakaoka, T., 336, *365*
Nakase, T., 555, *565*
Nakashima, K., 318, *325*
Nakashima, M., 85, *103,* 121, *125,* 505, *517*
Nakata, K., 289, *294,* 357, *363,* 443, *492*
Nakatake, Y., 315, *324*
Nakatsu, M. N., 260, 266, *274*
Nakatsu, Y., 6, *17*
Nakatsuji, N., 188, *189,* 226, *232,* 352, 354, *365*
Nakauchi, H., 244, *251*
Nakayama, A., 525, *536*
Nakayama, H., 168, 169, 170, *174, 178,* 189, *189*
Namioka, S., 511, *516*
Nanda, I., 388, *392*
Nanobashvili, A., 247, *249*
Napoli, J. L., 469, *497*
Narasimhaswamy, S., 342, *366*
Narcy, F., 286, *294*

Narimatsu, M., 171, 173, *176*

Narita, N., 45, *50*, 160, 164, 165, *175, 178, 179, 323*, 341, 343, 352, *366*

Narumiya, S., 581, *588*

Narvaez, V., 372, 378, 379, 387, *393*

Nascone, N., 334, 335, 336, 337, 340, *366*

Nash, A., 40, 44, *50, 323*, 338, *358*

Naski, M. C., 167, 171, 173, *179*, 286, *293*, 453, *492*

Nathans, J., 527, *535*, 556, 558, *565*

Nations, M., 509, *517*

Nau, P., 580, *585*

Nava, V. E., 227, *231*

Navankasattusas, S., 345, *367*

Navaratnam, V., 333, *362*

Navascues, J., 547, *561*

Nave, K.-A., 92, *104*

Nawata, H., 379, *391*

Naya, F. J., 92, *102, 104*, 269, *277*, 310, 321, 322, *327*, 342, *369*

Nayeem, S. M., 345, 346, 347, *366*

Nayernia, K., 386, *393*

Naylor, S., 440, *494*

Nebreda, A. R., 171, 173, *174*

Neel, V., 136, *145*

Neeman, M., 225, *229*

Nef, S., 386, *392*

Neff, C. D., 415, *419*

Neff, L. A., 285, *294*

Negus, K., 287, *292*, 470, *486*

Nehls, M. C., 160, 167, *178*

Neidhardt, L., 135, 137, *147*

Neihrs, C., 2, *4*

Neilson, E. G., 415, *419*

Neilson, K. M., 453, *492*

Nelkin, B., 313, *323*

Nellen, D., 577, *587*

Nelson, A. R., 509, *515*

Nelson, C. E., 305, *328*

Nelson, D. L., 357, *361*

Nelson, M. C., 172, 173, *174*, 341, 357, *358*

Nemali, M. R., 318, *328*

Nemer, G., 342, 343, *359, 361*

Nemer, M., 342, 343, 352, *359, 360, 361, 363, 366, 367*, 382, *393*

Nemes, Z., 581, 582, *587, 588*

Nemeskery, A., 504, *516*

Nemeth, A., 504, *516*

Nessmann, C., 387, *392*

Nestler, E. J., 247, *249*

Netchine, I., 508, *516*

Neubüser, A., 131, *146*, 263, 272, 321, *328*, 440, 441, 451, 466, 475, *491, 492, 493, 494*

Neufeld, G., 223, *232*

Neuhaus, J., 219, *231*

Neuhauser-Klaus, A., 522, 527, *535*, 549, *562*

Neuhauss, S. C., 395, *417*, 546, *563*

Neumann, H. P., 415, *420*

Neumann, J., 453, *483*

Neumann, P. E., 117, *125*

Neve, R. L., 415, *417*, 469, *498*

Neville, C. M., 335, 343, 344, 345, 346, 347, *370*

New, D., 110, *125*

Newbury-Enob, R. A., 357, *364*

Newgreen, D. F., 453, *486*

Newman, C., 340, *358*

Newton, J. M., 6, *17*

Newton, V. E., 525, *538*, 560, *564*

Neyt, C., *102*, 241, 244, *251*, 265, 266, *276*

Nezanov, N., 570, *589*

Ng, L. J., 281, *293*, 380, 381, *389, 392*, 510, *515*

Ng, M., 341, *370*

Ngai, J., 435, *489*

Nghiem, M., 318, *327*

Ngo-Muller, V., 3, *4*

Nguyen-Luu, D., 550, *562*

Nguyenphuc, Q., 355, *364*

Niaudet, P., 375, *389*

Nichogiannopoulou, A., 204, *210*

Nichols, A., 321, *328*

Nichols, D. H., 424, 425, 437, *492*

Nichols, J., 23, *34*, 158, 160, 166, *178*

Nickel, D. D.7, 235, 242, 244, *250*

Nickolls, G. H., 282, *294*

Nicol, R. L., 342, *364*

Nicolas, J.-F., 9, *16*

Nicolis, S., *390*

Nicoll, C. S., 500, *513*

Nicolson, T., 557, *564*

Nieburgs, A., 502, *514*

Niederlander, C., 2, *4*

Niederreither, K., 341, 346, 347, 352, *366*, 469, *492*, 545, *564*, 573, *588*

Niehrs, C., 44, *50, 51*, 78, *103*, 138, *146*, 325, *327*, 338, *361, 362*

Niessen, C. M., 408, *420*, 583, *589*

Nieto, M. A., 135, *147*, 354, 355, *359, 367*, 423, 428, *482, 490, 492, 495*, 546, *564*

Nieuwkoop, P., 108, *124*, 185, *190*

Nigam, S. K., 415, *417*

Nigam, V., 342, 343, *367*

Nigro, V., 40, 46, *52*, 79, *104*, 478, *495*

Nihei, H., 352, 354, *365*

Nihonmatsu, I., 84, *98*

Nii, A., 525, *538*

Nijhof, W., 245, *249*

Nijjar, S., 343, *361*

Nijweide, P., 285, *293*

Niki, M., 220, *231*

Nikolics, K., 512, *516*

Nikovits, J. W., 343, *369*

Nikovits, W., Jr., 335, 343, 344, 345, 346, 347, *370*
Nilson, J. H., 376, *391,* 510, *515*
Nilsson, A.-S., 402, *420*
Nilsson, E., 246, *249,* 318, *324*
Nilsson, M., 467, *482*
Nilsson, S. K., 194, *208*
Nimpf, J., 94, *105*
Ning, G., 317, *326*
Ning, Y., 380, *390*
Ninomiya, C., 292, *294*
Ninomiya, Y., 376, *393,* 422, 424, 425, 435, 437, 438, 442, 443, 451, 453, 475, 476, *487, 489, 492, 493,* 572, *588*
Nirenberg, M., 334, *363,* 474, *488*
Nishi, M., 95, *102,* 506, *516*
Nishi, S., 531, *538*
Nishida, K., 171, 173, *176*
Nishiguchi, S., 527, 530, 531, *537*
Nishigushi, S., 91, *102*
Nishihara, E., 504, *516*
Nishikawa, S., 172, 173, *176,* 188, *190,* 196, 200, *209,* 228, *233,* 442, *498*
Nishikawa, S.-I., 173, *178,* 188, *190,* 200, 203, *208, 209*
Nishikawa, T., 570, *584*
Nishikori, T., 547, *564*
Nishina, H., 318, *327*
Nishino, Y., 338, 354, 356, *368*
Nishi-Takeshima, M., 506, *516,* 549, *564*
Nishiyama, T., 79, *99*
Nishizaki, Y., 121, *125,* 522, *537*
Nishizawa, K., 290, *294,* 437, 450, 451, 466, 474, *495*
Nishizawa, M., 527, 530, 531, *536*
Nissinen, M., 412, *418*
Niswander, L., 166, *178,* 401, *419,* 571, 577, *585, 588*
Niswender, K. D., 453, 476, *493*
Niveleau, A., 9, *17*
Niwa, H., 23, *34,* 160, 166, *178*
Niwa, O., 376, *392, 393*
Noakes, P. G., 411, *419*
Noben-Trauth, K., 556, *564*
Noble, M., 240, 241, *251,* 404, *420*
Noda, M., 525, *538*
Noda, T., 122, *122,* 173, *179,* 217, 220, *230, 231,* 404, *420,* 506, *516,* 549, *564*
Nodasaka, Y., 557, *565*
Noden, D. M., 254, *276,* 345, *366,* 421, 422, 424, 425, 426, 428, 433, 434, 435, 449, *483, 491, 492, 497,* 558, *562*
Noebels, J. L., 92, *102*
Nofziger, D., 136, *148,* 262, *276*
Nogawa, H., 171, *178,* 314, *327*
Noguchi, K., 376, *391*
Noji, S., 47, *50,* 116, *123,* 354, 357, *365, 370,* 423, 424, 451, 471, 475, 476, 478, 479, *487, 490, 492*
Nolan, P. M., 560, *561*
Nolen, A. A., 452, *492*
Noll, M., 398, *419,* 470, *481*

Nomura, M., 47, *50,* 356, *366,* 375, 376, *390, 392, 393,* 468, *492*
Nomura, S., 288, *293,* 310, *327,* 442, 443, 453, *488,* 555, *565*
Nonaka, I., 260, *276*
Nonaka, K., 282, *294,* 465, *492*
Nonaka, S., 38, *51*
Nonchev, S., 117, 118, 119, *124,* 453, 468, *490, 492,* 544, *563*
Noramly, S., 575, 576, 577, 578, 579, *587, 588*
Nordborg, C., 237, 247, 248, *249*
Nordheim, A., 342, *358*
Nordlund, M. L., 341, 352, *359*
Nordqvist, K., 373, 383, *391*
Norlin, S., 78, *104,* 141, *148,* 504, 505, 506, *514*
Norman, D. J., 506, 508, 509, *518*
Nornes, H. O., 398, *417*
Norris, D. P., 354, *367,* 386, 387, *390*
Norris, W. E., 40, *52,* 265, *276*
North, T., 201, 204, *209,* 220, *231*
Northcutt, R. G., 421, 427, 433, 457, *485, 492*
Norton, C. R., 136, *149,* 226, *230,* 423, *487*
Norton, J. D., 10, *17*
Norwood, V. F., 412, *420*
Noseda, M., 474, *489*
Nothias, F., 84, *103*
Nothias, J.-Y., 5, 9, 10, *17, 19*
Nothwang, H., 388, *392*
Nougues, J., 271, *273*
Novacek, M. J., 442, 457, *492*
Novak, J., 527, *535*
Novikoff, P. M., 318, *324*
Novitch, B. G., 341, *361*
Novoselov, V., 338, *363*
Nowak, S. J., 351, *364*
Nowakowski, R. S., 87, 92, *105,* 245, *249*
Nowling, T., 166, *179*
Nowotschin, S., 354, 355, *359*
Nuckolls, G. H., 465, *492*
Nunes, F. D., 550, *561*
Nusse, R., 45, *51, 52,* 82, *106,* 405, *420,* 577, 579, *588*
Nusslein-Volhard, C., 351, 355, *359,* 546, 557, *564,* 565
Nutt, S. L., 204, *209*
Nuyens, D., 226, *229*
Nye, J. S., 89, *100,* 136, *147,* 262, 263, *275*

O

Oakey, R. J., 6, *17*
Oates, A. C., 14, *17,* 214, *230*
Oba, K., 379, *391*
Obara-Ishihara, T., 401, *419,* 571, *588*
Obinata, T., 258, *274*
O'Brien, M. A., 341, *358*

O'Brien, M. J., 26, *34*
O'Brien, T. X., 343, *366*
Ochi, T., 288, *293*
Ochikubo, F., 557, *563*
O'Connell, M. L., 10, *19*
O'Connell, S. M., 89, 95, *104,* 353, 354, 355, *364,* 433, 478, *489, 496,* 504, 505, 506, 507, 508, 509, *514, 517, 518,* 556, 558, *562*
O'Connor, D. S., 225, *231*
Oda, H., 321, *326,* 432, 450, 465, *489*
Oda, M., 167, *180*
Oda, S., 8, *16*
Oda, T., 347, *366*
Oda, Y., 580, 583, *588*
Odenthal, J., 546, *565*
Odland, G. F., 568, 583, *586*
Ody, C., 201, *209*
Oelgeschlager, M., 342, *358*
Oellig, C., 558, *564*
Oesterle, E. C., 558, *565*
O'Farrell, P. H., 170, *175*
Ofer, L., 166, *178*
Offield, M. F., 309, 319, 320, *327*
Ogasawara, M., 475, *492*
Ogasawara, N., 557, *565*
Ogawa, M., 94, *103, 106,* 196, *209,* 228, *233,* 442, *498*
Ogino, H., 527, 530, 531, *536, 537*
Ogle, R. C., 452, *492*
O'Gorman, S., 443, 449, 479, *495*
O'Guin, W. M., 581, *587*
Ogunrinu, G., 353, 357, *364*
Ogura, K., 525, *536*
Ogura, M., 506, *516,* 549, *564*
Ogura, T., 525, *536,* 571, *586*
Ogura, Y., 173, *178*
Oh, B., 8, 10, 12, 13, *16, 17, 18*
Oh, M. H., 354, *365*
Oh, S. H., 546, *564*
Oh, S. P., 353, 356, 357, *366*
Oh, S. R., 92, *103*
Ohba, M., 580, *587, 588*
Ohfuji, Y., 47, *51*
Ohgane, J., 167, *180*
Ohishi, S., 47, *51,* 356, *367*
Ohkawara, T., 525, *538*
Ohki, M., 202, *208*
Ohlsson, C., 510, *515*
Ohlsson, R., 167, 173, *178*
Ohmori, T., 289, *294*
Ohmoto, T., 247, *249*
Ohnishi, H., 321, *327*
Ohnishi, S., 581, *588*
Ohno, S., 580, *587*
Ohsaki, K., *537*
Ohshima, H., 290, *294,* 437, 450, 451, 466, 474, *495*

Ohshima, T., 94, *100, 103*
Ohsugi, M., 410, *418*
Ohto, H., 399, *419*
Ohtsuka, T., 136, *147*
Ohuchi, H., 47, *50,* 116, *123,* 312, *328,* 353, 354, 355, 356, *363, 370,* 423, 424, 451, 471, 475, 476, 478, 479, *487, 490, 492,* 505, *517*
Ohyama, K., 313, *329*
Oikawa, T., 313, 314, *323*
Oishi, C., 531, *538*
Oiski, S., 357, *365*
Ojeda, J. L., 415, *419*
Oka, C., 136, 137, *147*
Oka, S., 379, *391*
Oka, T., 336, 343, *365, 368*
Okabe, M., 136, *146,* 466, *488,* 581, 583, *589*
Okabe, S., 88, 89, *98, 101*
Okada, H., 220, *231*
Okada, T. S., 529, 531, 534, *536, 537*
Okada, Y., 38, *51*
Okafuji, T., 112, 114, 115, 116, *123, 124*
Okajima, K., 347, *366*
Okamoto, K., 183, *190,* 442, 443, 453, *488*
Okamoto, R., 288, *293*
Okamura, A., 91, *101*
Okamura, H., 442, *498*
Okamura, R. M., 440, *497,* 583, *589*
Okamura, Y., 79, *99*
O'Kane, S., 434, *496*
Okazaki, K., 7, *16*
Okazaki, S., 136, 137, *147*
Okazaki, Y., 11, 12, *18*
Okazawa, H., 183, *190*
O'Keefe, S. J., 14, *18*
Okkema, P. G., 165, *177,* 316, *325*
Okuda, A., 183, *190*
Okuda, T., 49, *52,* 201, 204, *209,* 220, *231*
Olanow, C. W., 247, *251*
Oldenettel, I., 85, *103*
Oldmixon, E. H., 224, 227, *232*
Oldridge, M., 453, 466, 474, *494, 498*
O'Leary, D. D., 84, *102, 103,* 119, *125,* 525, *537*
O'Leary, D. D. M., 92, 94, *98*
Olek, A., 9, *18*
Oliner, J. D., 400, *418*
Oliphant, A. R., 412, *417*
Oliver, G., 79, *103,* 218, 219, *233,* 268, *274,* 317, 321, *329,* 399, *417,* 451, 475, *495,* 521, 527, 530, 532, 533, 534, *537, 538,* 571, *585*
Oliver, J. A., 405, 407, *416, 417*
Olivo, J. C., 85, *103,* 443, 474, *490,* 545, *563*
Olney, A. H., 470, *487*
Olsen, A., 412, *418*
Olsen, B. R., 227, *232,* 280, 281, 288, 292, *293, 294,* 424, 427, 442, 443, 453, *484, 487, 489, 492, 493*

Olson, E. N., 135, 136, 137, 138, *145, 148,* 165, 168, *175, 177,* 228, *230,* 250, 257, 258, 260, 261, 262, 267, 269, *273, 274, 275, 276, 277,* 287, 289, *292, 294,* 316, *327,* 333, 334, 336, 340, 341, 342, 343, 344, 346, 347, 349, 350, 352, 353, 355, *358, 359, 360, 361, 364, 365, 366, 367, 368, 369, 370,* 451, 465, 473, 476, 477, *483, 490, 495, 496*

Olson, L., 509, *518*

Olsson, M., 88, 89, *98, 103,* 110, *125,* 241, 244, *249*

Oltz, E. M., *106*

Olwin, B. B., 271, *273*

O'Malley, B. W., Jr., 556, 558, *565*

Omi, M., 3, *4*

Omori, A., 84, *100*

Omura, T., 375, 376, 380, 384, *390, 392*

O'Neil, J., 525, *537*

Ong, A. C., 415, *420*

Ong, E. S., 172, 173, *174,* 341, 357, *358*

Ong, L., 580, *585*

Ong, S.-H., 171, 173, *178*

Onichtchouk, D., 338, *361*

Onigata, K., 506, *518*

Ono, K., 443, *492*

Ono, T., 525, *537*

Ontell, M., 269, *275*

Onuchic, L. F., 414, *419*

Onuma, Y., 338, *368*

Oosterveen, T. C., 476, 477, 478, *491*

Oostra, B. A., 227, *231*

Oostuyse, B., 226, *229*

Opperman, L. A., 452, *492*

Opthof, T., 348, *360*

Orban, P. C., 14, *16*

Orci, L., 321, *328*

Ordahl, C. P., 128, 130, 131, 132, 140, 141, *145, 147, 149,* 219, *229,* 254, 255, 256, 258, 261, 262, 263, 264, 266, 268, 271, *273, 274, 276, 277,* 334, *360*

Oreal, E., 372, 387, *392*

Orellana, S. A., 415, *419*

Orkin, R. W., 577, 578, 579, *587*

Orkin, S. H., 202, 203, 204, *207, 209, 210,* 219, *232,* 271, *276,* 318, *328,* 341, 343, 352, *360, 369,* 415, *417,* 510, *518*

Orlic, D., 196, 202, *210*

Ormsby, I., 453, 467, *494*

Ornitz, D. M., 116, *126,* 167, 171, 173, *179,* 286, *292, 293,* 313, 314, 315, *324,* 453, *492,* 558, *561*

Ornoy, A., 174, *178*

Oro, A. E., 579, *585*

Oropeza, V., 578, *587*

O'Rourke, D. M., 376, *390*

Orr-Urtreger, A., 467, *492*

Orr-Weaver, T. L., 169, *175*

Ortega, S., 505, *517*

Ortiz de Luna, R. I., 453, *486*

Osawa, K., 525, *535*

Osawa, M., 244, *251*

Osawa, Y., 376, *393*

Osborn, J. W., 437, *492*

Osborne, L., 246, *250,* 318, *326*

Osborne, N., 340, 341, *363*

O'Shea, K. S., 223, *229*

Oshima, K., 403, *418*

Oshima, R. G., 168, *179,* 570, *589*

Oshima, T., 506, *516,* 549, *564*

Oshimura, M., 478, *488*

Oskarsson, M., 7, *18*

Oster, A., 321, *327*

Ostrer, H., 387, *390*

O'Sullivan, M. G., 382, *392*

Osumi, N., 77, 95, *101, 105,* 451, 475, 476, *492,* 525, *538*

Osumi-Yamashita, N., 422, 423, 424, 425, 435, 437, 438, 451, 453, 475, 476, 478, *487, 490, 492, 493,* 527, 530, *535,* 572, *588*

Oswald, J., 9, *18*

Otani, H., 547, 557, *564*

Oto, A. E., 574, 577, 579, 583, *588*

O'Toole, B. A., 453, *483*

Ott, M. O., 132, *147,* 165, *174,* 258, *276,* 303, *323*

Otten, B., 509, *517*

Otto, F., 288, *293, 294,* 442, 443, 453, *493*

Ottolenghi, C., 382, *392*

Ou, C. N., 318, *327*

Oudejans, C., 174, *174*

Oudhof, H. A., 452, *493*

Oulad-Abelghani, M., 40, 47, *51*

Overbeek, P. A., 349, 352, 354, 355, *364, 365, 366, 369,* 380, *389,* 441, *486, 491, 497,* 527, 530, 531, 532, *537*

Overton, G. C., 194, 206, *209*

Overtruf, K., 318, *327*

Ovitt, C., 183, *190,* 311, *324,* 442, 443, *497*

Ow, D. W., 15, *17*

Owen, D. A. J., 26, *33*

Owen, M. J., 288, 289, *293, 294,* 442, 443, 453, *493*

Owens, D. F., 86, *102*

Oxford, J. T., 442, 443, *489*

Ozaki, H., 399, *419*

Ozaki, K., 321, *327*

Ozato, K., 183, *190*

Ozawa, M., 408, *418, 419*

Ozcelik, C., 353, 356, *361*

P

Pabst, O., 310, *327*

Pace, J. M., 424, 442, 443, *493*

Pachnis, V., 88, 89, *101,* 260, 267, *275,* 402, 403, *419, 420,* 432, *485, 497*

Pacifico, F., 338, *360*
Packard, D. S., Jr., 134, *147*
Packer, A. I., 312, *327*
Pagan, S., *364*
Paganessi, L., 337, 354, *364*
Pagan-Westphal, S. M., 337, 354, *364*
Page, D. C., 387, *391*
Pai, C. Y., 91, *103*
Paige, C. J., 193, *207,* 318, *327*
Pajusola, K., 218, 223, *229*
Pakarinen, L., 452, *494*
Pakzaban, P., 89, *99*
Palacios, D., 510, *516*
Paldi, A., 320, *324*
Paleari, L., 290, *292,* 450, 453, 472, 473, *481,* 549, *561*
Palfrey, H. C., 91, *105*
Palgi, J., 559, *564, 565*
Palis, J., 195, *209*
Palko, M. E., 559, *562*
Pallister, P. D., 506, *517*
Pallon, J., 580, *586*
Palmeirim, I., 134, 135, 138, 139, *146, 147*
Palmer, D. A., 348, *358*
Palmer, M., 219, *231*
Palmer, M. S., 377, 388, *393*
Palmer, R., 400, *418*
Palmer, S., 341, 345, 347, 348, 350, *359, 366*
Palmer, S. J., 377, 378, 383, 386, *389, 392*
Palmer, T. D., 236, 237, 241, 244, *251, 252*
Palmes, C., 318, *328*
Palmieri, S. L., 160, *178*
Palmiter, R. D., 318, *328,* 374, 385, *389,* 511, *513*
Palomino, M. A., 344, 348, *360*
Palomino, T., 510, *517*
Pan, D., 136, *147*
Pan, L., 559, *561*
Pan, S. S., 188, *190*
Pandita, R. J., 353, 357, *365*
Panelli, S., 171, 173, *174*
Pang, K., *105*
Panganiban, G., 471, *493*
Panic, S., 307, *329*
Panitz, F., 521, 522, 525, 533, *535*
Pannell, R., 203, *210*
Pannese, M., 290, *294,* 450, 471, 472, 473, *495*
Pant, H. C., 94, *103*
Paolucci, E., 206, *207,* 246, *249,* 271, *274*
Paonessa, P. D., 11, 12, *17*
Papadapoulos, N., 224, *231*
Papadimitriou, J. C., 200, *207*
Papageorge, A. G., 171, 173, *178*
Papaioannou, V. E., 132, 133, 135, 137, *145,* 157, 166, *175, 176, 178,* 353, 357, *362,* 442, *487*
Papalopulu, N., 86, *100*

Papaloucas, A., 570, *584*
Papapetropoulos, A., 225, *231*
Papapoulos, S. E., 285, *293*
Papayannopoulou, T., 173, *179*
Papiernik-Berkhauer, E., 22, *33*
Papkoff, J., 140, 141, *148,* 263, 264, *277*
Paquette, A. J., 243, *249, 252,* 263, 264, 266, *274*
Parada, L. F., 193, 204, *208,* 341, 352, *359, 365,* 386, *392*
Paradies, N. E., 550, *563*
Paradis, G., 193, *207*
Paradis, P., 342, 343, *359*
Parameswaran, M., 38, *52,* 333, *366*
Parati, E. A., 235, 242, 244, *250*
Pardanaud, L., 199, 200, 201, *209*
Pardini, C., 83, 92, *100, 102*
Parent, J. M., 247, 248, *251*
Paria, B. C., 156, *177, 178*
Park, B. K., 89, *106,* 450, 471, *498*
Park, H. L., 85, *102, 103,* 121, *124, 125,* 505, *517*
Park, J. H., 415, *420*
Park, L., 292, *294*
Park, M., 266, *277,* 337, 341, *360, 363, 366*
Park, W. Y., 313, 315, *327*
Parker, J., 465, *491*
Parker, J. D., 510, *513*
Parker, K. L., 375, 376, 379, 380, 381, 384, *389, 390, 391, 392,* 510, *515, 516*
Parker, S. B., 287, *294*
Parks, D. R., 194, *207*
Parks, J., 500, 504, 506, 509, *517*
Parlow, A. F., 506, 509, 512, *513, 515, 516*
Parmacek, M. S., 165, *177,* 303, 316, *326, 327,* 342, 343, 352, *363, 366*
Parmar, M., 91, *105*
Parnavelas, J. G., 88, 89, *101*
Parr, B. A., 385, 386, *392*
Parravicini, E., 405, 407, *416, 417*
Parrington, F. R., 442, *493*
Parrington, J., 9, *18*
Parrott, J. N., 10, *18*
Parslow, T. G., 440, *497,* 583, *589*
Parsons, L. M., 333, 334, 343, 345, 348, 349, 352, 356, *364*
Parsons, S. M., 260, 267, *277*
Partanen, J., 224, *231*
Partington, G., 343, *361*
Parton, L. A., 10, *19*
Parvinen, M., 580, *589*
Pasantes, J., 281, *294,* 380, *393*
Pask, A., 376, 386, 387, *392*
Pasko, D., 530, *536*
Pasquali-Ronchetti, I., *588*
Passarelli, R. W., 452, *492*
Pastor, F., 547, *564*
Pastorcic, M., 416, *420*

Pastore, C., 220, *228*
Pasyk, K. A., 227, *232*
Pászty, C., 443, *493*
Patan, S., 224, *232*
Patapoutian, A., 260, *276*
Patek, C. E., 416, *419*
Patel, B., 12, 13, *18*
Patel, K., 264, 266, *272*, 354, 355, *366*, 574, 577, *588*
Patel, S., 224, *232*, 504, *517*
Patel, Y., 23, *33*, 166, 167, *175*, *178*
Paterson, A. J., 511, *516*
Paterson, B. M., 341, *364*
Patient, R., 165, *177*, 198, 200, *207*, 342, 343, *361*, *367*
Paton, I. R., 388, *392*, 534, *536*
Patrene, D. K., 246, *251*
Patten, B. M., 354, *366*
Patterson, C., 421, *493*
Patterson, K. D., 347, 348, *359*, *360*
Patterson, L. T., 408, 409, 410, *417*
Patton, B. L., 411, *418*
Pauken, C. M., 13, *18*
Paukku, T., 511, *516*
Paul, J., 317, *327*
Paulding, C., 400, *418*
Paulin, D., 266, *277*
Paus, R., 574, 577, 578, 583, *588*, *589*
Pause, A., 227, *229*
Pavlova, A., 415, *417*, *418*
Paw, B. H., 214, *230*
Pawling, J., 203, *207*, 223, *228*
Pawlowski, S. A., 434, 467, *493*
Pawson, A. J., 171, *175*
Pawson, T., 171, 173, *178*
Paydar, S., 474, *494*
Payne, C., 303, *330*
Payne, R. M., 356, *366*
Paynton, B. V., 11, *18*
Payrieras, N., 338, *366*
Paysan, J., 84, *103*
Paznekas, W. A., 453, *486*
Pearce, J. J., 40, 44, *52*, 160, 166, 167, 168, *178*, *179*, 303,
    327, 376, *390*
Pearlman, A. L., 94, *104*, 237, *251*
Pearse, A. G., 320, *327*
Pease, W., 357, *367*
Peault, B., 192, 194, 200, 201, *207*, *208*, *209*, 220, *232*,
    245, *250*
Pedersen, A. C., 291, 292
Pedersen, E. E., 310, 321, *325*
Pedersen, R. A., 26, 32, *35*, 38, 39, 45, 47, *51*, 79, 83, 87,
    89, 90, 91, 92, 95, *97*, *98*, *99*, *103*, 158, *177*, *179*, 290,
    292, 303, 304, 321, *326*, *329*, 333, 336, 338, *363*, 410,
    418, 422, 435, 437, 450, 451, 453, 470, 471, 472, 473,
    477, 478, *481*, *484*, *493*, 549, *562*, 570, *589*

Peeters, L. L., 174, *177*
Peeters, M. C. E., 196, 199, 200, 201, 202, *207*
Pei, Y. F., 519, *537*
Peifer, M., 443, 478, *493*, 575, *588*
Peissel, B., 415, *418*
Pekna, M., 313, *323*, 414, *418*, 467, *482*
Pekny, M., 313, *323*, 413, 414, *418*, 467, *482*, *489*
Pelet, A., 286, *294*, 453, *494*
Pellas, T. C., 188, *190*
Pellegrini, M., 83, *103*
Pelletier, J., 341, 352, *363*, 374, 375, 379, *391*, *392*, 399,
    400, 415, 416, *417*, *418*, *419*
Pelliniemi, L. J., 511, *516*
Peltarri, A., 284, *293*
Pelto-Huikko, M., 527, 530, 532, *535*
Pelton, K. M., 308, *326*
Pelton, R. W., 308, *326*, 467, *493*
Peltonen, L., 412, *418*
Pena, P., 510, *517*
Peng, J., 217, 227, *229*
Penman-Splitt, M., 352, 354, 357, *361*, *366*
Penner, J., 570, 583, *584*
Pennisi, D. J., 172, 173, *177*, 336, 337, *365*
Penny, G., 40, 44, *52*
Pennypacker, J. P., 424, 442, 443, *482*, *493*
Penttila, T. L., 580, *589*
Penzes, P., 469, *497*
Pepicelli, C. V., 312, *328*, 407, *419*
Pepling, M. E., 6, 7, *18*
Pera, E. M., 78, 81, *103*, 546, *564*
Perantoni, A. O., 405, 407, *417*, *419*
Percival, A. C., 320, *328*
Perea-Gomez, A., 41, 42, 43, 44, 47, *52*
Pereira, A., 346, *358*
Pereira, F. A., 92, *106*, 334, 343, 345, 346, 352, *366*, 559,
    563
Perens, E., 222, *231*
Perez, A., 577, 578, 579, *587*
Perez, L., 171, 173, *174*
Perez-Miguelsanz, J., 527, *535*
Perez-Moreno, M. A., 355, *359*, 423, *482*
Pérez-Navarro, E., 91, *98*
Perez-Pomares, J. M., 219, 228, *231*
Perfilieva, E., 237, 247, 248, *249*
Pericak-Vance, M. A., 227, *231*
Perkett, E. A., 467, *493*
Perkins, A., 318, *328*, 341, 352, *359*
Perkins, D., *325*, 470, *485*
Perkins, G., 341, 352, *367*
Perl, A.-K., 311, 322, *324*, *328*
Perlmann, T., 91, *105*, 120, *125*, 509, *518*
Permutt, M. A., 321, *325*
Pernasetti, F., 509, *517*
Perreau, J., 205, *207*

Perrimon, N., 45, *52*

Perrin-Schmitt, F., 452, 453, *482, 484*

Perris, R., 423, 424, *484, 493*

Perron, M., 522, 525, 533, *538*

Perry, M., 262, *276*

Persico, M. G., 162, *179*, 338, 352, *360, 370*

Persing, J. A., 452, *492*

Persson, H., 91, *98*

Pertz, O., 408, *418*

Pesce, M., 86, *103*, 188, 189, *189, 190*

Peschon, J. J., 318, *329*

Pestell, R., 86, *104*, 579, *588*

Peter, W., 160, *178*

Peterka, M., 454, 455, *489, 493*

Peterkova, R., 454, 455, *489, 493*

Peteropoulos, H., 342, *368*

Peters, C., 577, 583, *584*

Peters, D., 305, *329*

Peters, D. J., 414, *418*

Peters, D. M., 453, 467, *496*

Peters, G., 545, *565*

Peters, H., 290, 292, *294*, 398, 399, *420*, 437, 440, 441,
    450, 451, 466, 474, 475, *492, 493, 495, 498*, 527, 530,
    533, *538*, 550, *565*

Peters, K. G., 214, 225, 227, *230, 231*, 312, 314, *328, 330*

Peters, M. A., 525, *537*

Peterson, A., 79, 84, 86, 92, *98, 100*, 236, 243, *250, 252*

Peterson, B. E., 246, *251*

Peterson, C. A., 268, 269, *274*

Peterson, C. L., 166, *179*

Peterson, D. A., 237, 241, 247, 248, *249, 252*

Petit, C., 398, *416*, 550, *561, 563*

Petkovich, M., 469, 470, *484*, 525, *537*

Petrenko, O., 200, *209*

Pettersson, K., 167, *178*

Pettersson, R. F., 558, *564*

Pettit, J., 185, *190*, 243, *252*

Peverali, G., 382, 387, 388, *390*

Pevny, L. H., 204, *209*, 260, 267, *275*, 377, *392*, 531, *537*

Pexieder, T., 341, *366*

Pey, R., 415, *419*

Peyrol, S., *585*

Pfaff, D. W., 435, *495*

Pfaff, M., 574, *589*

Pfaff, S. L., *104*, 319, 320, 321, *323, 325*, 504, 505, 507,
    508, 510, *517, 518*

Pfäffle, R., 509, *517*

Pfeifer, D., 380, *392*

Pfeiffer, R. A., 281, *293*, 381, *391*

Pham, T. D., 244, *251*

Phan, H., 185, *189, 190*

Phelps, A. L., 357, *369*

Phelps, C., 511, *517*

Phelps, D. E., 398, *419*

Philbrick, W. M., 285, *294*

Philips, R. J. S., 441, *493*

Philipsen, S., 343, *369*

Phillips, H. S., 402, *418, 420*, 512, *516*

Phillips, P. E., 467, *495*

Phillips, R. I., 194, 206, *209*

Phillips, R. J., 443, *486*

Phillips, S. J., 311, *330*

Philp, N., 356, *369*

Philpott, A., 522, 525, 533, *538*

Philpott, M., 578, *588*

Phinney, D. G., 246, *250*

Phippard, D., 506, *517*, 549, *564*

Phyllis, R. W., 334, *361*

Piatigorsky, J., 532, *535, 538*

Picard, J. J., 321, *326*

Picard, J. Y., 372, 374, 387, *391, 392*

Piccoli, D. A., 347, *364, 366*

Piccolo, S., 31, *33*, 40, 44, *50, 51, 52*, 338, *358, 366*

Pichel, J. G., 374, *390*, 402, *419*

Pickel, J., 89, *101*

Pictet, R. L., 319, 321, *328*

Pieau, C., 83, *104*, 372, 387, *392*

Piedra, E., 351, 355, 356, *359*

Piedra, M. E., 349, 354, *360, 366*

Piehl, F., 559, *564*

Pieler, T., 81, *100*, 521, 522, 525, 527, 533, *535, 536*

Pierani, A., 80, 82, 88, *98*, 120, 121, *125*

Pierides, A., 414, *418*

Pierpont, M. E. M., 346, 347, *358, 364*

Pietryga, D. W., 204, *208*, 318, *327*

Pignoni, F., 399, *419*, 530, 533, *537*

Pike, B. L., 347, *366*

Pilia, G., 357, *361*

Pinard, J. M., 94, *99*

Pinchera, A., 311, *326*

Pineau, T., 311, 312, *326*, 474, *488*

Pinto, S., *251*

Pintucci, G., 224, *232*, 504, *517*

Pintus, G., 338, *369*

Piotrowska, K., 1, *4*

Piper, D., 240, *250*

Pirottin, D., 264, *274*

Pirro, A., 285, *293, 294*, 442, 443, *489*, 574, 583, *584*

Pirro, M. T., 311, *326*

Pirvola, U., 558, 559, *564, 565*

Pisano, M. M., 528, *537*

Pischetola, M., 311, *330*

Pispa, J., 441, *493*

Piston, D. W., 82, 86, *99*

Pitera, J. E., 305, *328*

Pittenger, M. F., 206, *209*

Pitts, S. L., 322, *328, 512, 516*

Pitts-Meek, S., 121, *123*

Piussan, C., 506, *517*
Pixley, S. K., 82, *105*
Plaas, A. H., 350, *362*
Placzek, M., 78, 79, 80, 81, 85, *99, 104,* 120, *123,* 141, *148,* 531, *537*
Plasterk, R. H. A., 14, *18*
Plate, K. H., 227, *231*
Platt, J. B., 423, *493*
Platt, K. A., 85, *102, 103,* 121, *123, 124, 125,* 470, *487, 493,* 505, *517*
Playford, R. J., 167, *174,* 305, *323*
Pleasure, S. J., 83, 90, 92, *102, 103*
Plehn-Dujowich, D., 79, *106,* 529, *538*
Ploder, L., 527, *535*
Ploemacher, R. E., 192, 193, *207, 210,* 343, *369*
Plowden, J., 357, *358*
Plowman, G. D., 314, *328*
Plum, A., 173, *177*
Plump, A. S., 164, *175,* 303, *323*
Poea, S., 271, *277*
Poelmann, R. E., 219, 226, *229, 232,* 316, *326,* 333, 341, *360, 366*
Pohl, T., 85, *103*
Poirier, F., 183, *190*
Pokrywka, N., 185, *190*
Polani, P., 183, *189*
Polanski, Z., 7, *18*
Polarkis, P., 575, *588*
Poli, V., 318, *324*
Polish, J. A., 510, *516*
Poljak, L., 442, *484*
Polk, C. E., 343, *368*
Pollard, J., 173, *178*
Pollefeyt, S., 203, *207,* 223, *228*
Pollet, N., 138, *146*
Pollock, R. A., 402, *420*
Polonskaia, O., 91, *101*
Pompa, J. L. D. I., 318, *327*
Poncelet, D., 264, *274*
Ponsot, G., 94, *99*
Pontiggia, A., 377, *392*
Pontoglio, M., 166, *178*
Ponzetto, C., 266, 271, *273, 275*
Poole, M. D., 453, 466, 474, *498*
Popp, R. A., 204, *208*
Popperl, H., 45, *52,* 117, 118, *124,* 449, 479, *484*
Porcher, C., 203, *209*
Porras, A., 171, 173, *174*
Porteous, D. J., 399, 416, *419*
Porteous, M. E., 227, *231*
Porter, F., 508, *517*
Porter, F. D., 87, 91, *103,* 173, *176,* 376, *389,* 512, *515,* 521, 527, *537*

Porter, J. A., 78, *98, 101,* 120, *125,* 141, *145,* 287, *292,* 470, *483, 489, 493*
Porter, T. E., 511, *518*
Porteu, A., 262, 268, *276*
Porteus, M. H., 79, 88, 89, *98, 103*
Portillo, F., 355, *359,* 423, *482*
Posadino, A. M., 338, *369*
Post, L. C., 313, *328*
Post, M., 121, *124,* 173, 174, *175,* 312, 313, *327, 328, 329,* 385, *389,* 413, *419*
Postiglione, M. P., 95, *97,* 290, *292,* 450, 453, 472, 473, 478, *481, 491,* 506, *513,* 549, 550, 551, 552, *561, 564*
Postlethwait, J. H., 214, *230, 231*
Postlethwaite, M., 91, *102*
Postmus, J., 174, *174*
Pote, J., 581, *586*
Potten, C. S., 570, *587*
Potter, S. S., 79, 82, 83, 94, 95, *101, 104, 105,* 204, *208,* 318, *327,* 352, 354, 355, *364, 368,* 433, 476, 477, 478, *489,* 502, 507, 508, 509, *516, 517*
Poueymirou, W. T., 166, *175,* 292, *292*
Poulat, F., 381, *390*
Poulin, G., 509, *517*
Poulsen, K. T., 402, *420*
Pournin, S., 546, *565*
Pourquié, O., 134, 135, 136, 138, 139, 140, 141, 142, *146, 147,* 262, 263, 264, 266, *273, 274, 276,* 347, *366*
Poustka, A., *327,* 415, *417*
Powell, B. C., 578, *588*
Powell, T. J., 217, 220, *229*
Powell-Braxton, L., 91, *98,* 203, *207,* 223, *229,* 341, *363*
Pownall, M. E., 167, *176,* 263, *276*
Powrie, J. K., 507, *514*
Prakkal, P. F., 568, *588*
Pramparo, T., 353, 357, *364*
Pratt, H. P., 158, *178*
Pratt, S. J., 214, *230*
Pratt, T., 87, 94, *105*
Prawitt, D., 379, *391*
Prescott, A. R., 527, 530, 531, *537*
Presley, R., 79, *99,* 290, *292,* 422, 432, 435, 437, 442, 450, 451, 453, 456, 471, 472, 473, 477, 478, *484, 493,* 549, *562*
Pressman, C., 287, *292,* 354, 355, *364,* 451, 470, *486, 489,* 579, *586*
Pretsch, W., 522, 527, *535,* 549, *562*
Prezioso, V. R., 164, 165, *175, 177, 179,* 303, 316, *323, 329*
Price, B. M. J., 167, *179,* 203, *207*
Price, D. J., 85, 87, 94, *98, 105*
Price, J., 83, 85, 91, *105,* 237, 239, *252,* 341, 352, *368*
Price, M., 85, *103,* 311, *326,* 450, *484*
Price, S. D., 555, 558, 559, *561, 563*
Price, S. M., 338, 353, 354, 356, 357, *370*

Prideaux, E. M., 158, *177*
Prideaux, V., 23, *35,* 42, *51,* 169, *179*
Priess, J. R., 165, *180,* 185, *190,* 240, 243, *251, 252,* 316, *330*
Prigent, M., 200, *209*
Primig, M., 263, 264, *273*
Prin, F., 572, 574, *585*
Prince, V. E., 118, *125,* 427, *493*
Prinz, M. P., 341, *363*
Pritchard, J., 506, *514*
Pritchard-Jones, K., 374, 375, *389,* 399, *416, 419*
Probst, F. J., 556, 557, *561, 564*
Prockop, D. J., 246, *250,* 284, *293*
Proctor, J., 128, 137, *146*
Proesmans, W., 398, *419*
Proetzel, G., 434, 467, *493*
Proia, R. L., 227, *231*
Prokscha, A., 336, *370*
Prosser, J., 451, 475, *486,* 527, 528, 530, *536*
Prybyla, A. E., 321, *328*
Prywes, R., 342, *360, 369*
Puech, A., 353, 357, *365*
Puelles, L., 75, 76, 77, 78, 79, 82, 83, 84, 85, 86, 88, 89, 92, *98, 99, 103, 104,* 422, 428, 450, 471, *484, 490, 494*
Puissant, F., 27, *34*
Pulleyn, L. J., 433, 453, 466, 474, *492, 494, 498*
Punnett, H. H., 380, 381, *391*
Purcell, K., 186, *190*
Purcell, P., 532, *538*
Purdie, L. A., 451, 453, 474, *489*
Puri, M. C., 224, *231*
Püschel, A. W., 475, *493*
Putaala, H., 412, *418*
Puzis, R., 258, 263, *274*
Pytowski, B., 227, *233*

Q

Qi, H., 136, *144, 148*
Qi, M., 347, *364*
Qi, P., 94, *104*
Qian, F., 414, *419*
Qian, X., 171, 173, *178*
Qiao, J., 405, *417*
Qiao, W., 286, *292*
Qin, M., 15, *17*
Qin, Y., 224, 228, *232,* 380, *389*
Qiu, M., 78, 83, 85, 87, 89, 90, 91, 95, *97, 103,* 422, 437, 450, 451, 470, 471, 472, *481, 493, 496*
Qiu, Y., 310, 321, 322, *327,* 334, 343, 345, 346, 352, *366*
Qu, S., 451, 453, 476, *493*
Quadrelli, R., 346, *358*
Quaegebeur, A., 511, *517*

Quaggin, S. E., 313, *328,* 413, *419*
Quarmby, J., 185, *189*
Quere, P., 201, *209*
Quinlan, E., 11, *19*
Quinlan, G. A., 3, *4,* 30, *35,* 38, 39, 40, 41, 47, *51, 52,* 132, *147*
Quinlan, R. A., 527, 530, 531, *537*
Quinn, A. G., 287, *293,* 579, *586*
Quinn, J. C., 451, 475, *493*
Quintana, D. G., *251*
Quintana-Murci, L., 382, *392*
Quintin, S., 165, *176,* 316, *325*
Quo, R., 440, *497,* 583, *589*
Quondamatteo, F., 574, 583, *584*

R

Raabe, M., 161, *178*
Raats, J., 185, *190*
Rabbitts, T. H., 203, *210*
Rabejac, D., 556, *562*
Rabl, W., 378, *392*
Racey, P. A., 388, *393*
Radcliffe, J., 505, *513*
Radeke, M. J., 91, *98*
Radice, G. L., 340, *366*
Radich, J., 316, *325*
Radovick, S., 509, *514, 517*
Radtke, F., 223, *229*
Radziejewski, C., 224, 227, *231*
Raeburn, A., 357, *364*
Raff, M., 241, 246, *250,* 341, *359*
Raffel, C., 287, *294*
Raffin, M., 337, 339, *366, 367*
Raftery, L. A., 226, *231*
Raghavan, S., 574, 583, *588*
Rago, C., 86, *100*
Ragsdale, C. W., 83, 87, *100, 105*
Rahal, J. O., 443, *485,* 508, *515*
Rajewsky, K., 14, *16,* 341, *362*
Rakic, P., 84, 87, 92, 94, *97, 98, 99, 100, 101, 103,* 136, *148,* 248, *251*
Ralphs, J. R., 468, *497*
Ram, P. T., 12, *19*
Ramachandran, B., 188, *190*
Ramain, P., 343, *360, 361*
Ramakrishna, N., 405, *417*
Ramalho-Santos, M., 308, 310, *328*
Raman, S., 376, *390*
Rambhatla, L., 12, 13, *18*
Ramesar, R., 292, *294*
Ramirez, A., 574, 583, *584*
Ramirez, F., 442, 443, *489,* 573, 581, 583, *584, 588*

Ramirez-Weber, F.-A., 470, *481*
Ramón y Cajál, S., 248, *251*
Ramos, C., 525, *535*
Ramsdell, A., 334, 344, 350, 351, 357, *365*, 473, *493*
Rand, E. B., 136, *149*, 347, *364*
Rand, M. D., 136, *144, 148*, 263, 272, 555, *561*
Ranganayakulu, G., 334, 341, *367*
Ranganayakulu, R., 341, *364*
Rangell, L., 223, *229*
Rangini, Z., 395, *417*, 546, *563*
Rankin, C. T., 353, 356, *367*
Rao, C. V., 510, *518*
Rao, M. S., 236, 239, 240, 241, *250, 251, 252*, 318, *328*
Rao, Y., 90, 95, *106*, 136, *149*, 521, *536*
Rapaport, D. H., 556, 558, *562*
Raphael, Y., 556, 557, *561, 564*
Rapola, J., 22, *34*
Rapolee, D. A., 23, *33*
Raposo do Amaral, C. M., 292, *294*
Rappolee, D. A., 166, 167, *175, 178*
Rasberry, C., 378, 387, *390*
Rashbass, P., 88, *99*, 132, 133, *144, 149*, 528, *537*
Rassner, U., 581, 582, *585, 586*
Rassoulzadegan, M., *391*
Rastan, S., 376, 378, 386, 387, *390, 393*
Ratcliffe, P., 225, *229*, 415, *420*
Rathbun, G., *106*
Ratner, N., 341, 352, *359*
Ratty, A. K., 502, *515*
Rau, W., 388, *391*
Rauch, M., 382, *389*
Rauchman, M., 532, *537, 538*
Raulet, D. H., 205, *208*
Rausa, F., 321, *328*
Rauscher, F. J. III, 400, *418, 419*
Rauskolb, C., 443, 478, *493*
Rauvala, H., 218, 223, *230*
Raven, C. P., 423, *493*
Raven, H. C., 473, *493*
Ravi, R., 225, *231*
Rawls, A., 136, 137, 138, *145, 148*, 250, 260, 262, 267, *273, 276, 277*, 287, *294*
Rawson, E. J., 509, *516*
Ray, D. W., 512, *518*
Ray, J., 241, *252*
Ray, M., 309, 319, 320, *327*
Rayburn, H., 170, *180*, 204, *206*, 219, 223, *228, 229, 230, 233*, 340, *366*, 571, 581, 583, *589*
Raymond, C. S., 382, 387, 388, *391, 392*
Raynaud, A., 139, *148*
Raz, Y., 557, *564*
Read, A. P., 525, *538*, 560, *564*
Readhead, C., 512, *513*
Reardon, W., 433, 453, *492, 493, 494*

Réaume, A. G., 135, 136, 137, *145, 148*
Rebagliati, M. R., 78, *103*
Rebay, I., 136, *145*
Rebel, V. I., 245, *251*
Reber, M., 165, *174*, 303, *323*
Rebrikov, D., 341, *369*
Recan, D., 378, *392*
Reda, D., 169, *179*, 349, *367*
Reddi, A. H., 467, *494*
Reddy, J. C., 375, *392*
Reddy, P. H., 247, *251*
Redmon, J., 452, *492*
Redmond, L., 92, *103*
Reed, K. J., 382, 388, *393*
Reed, R., 263, *274*
Reeders, S. T., 412, 415, *417*
Reeve, W. J., 158, *178*
Reeves, A., 237, *250*
Reeves, R. H., 509, *514*
Refetoff, S., 311, *329*
Regan, C. L., 136, *145*
Reginato, A. M., 280, *294*
Rehorn, K.-P., 165, *178*, 316, *328*
Reichardt, L. F., 91, *101*, 239, 249, 402, 410, *418*, 549, *562*
Reichenberger, E., 292, *294*
Reichmann, V., 336, *367*
Reid, C. B., 94, *105*, 235, 244, *251*
Reid, L. M., 314, *324*
Reid, S. W., 173, *179*, 341, 352, *359*
Reifers, F., 115, *125*, 337, *367*
Reik, W., 9, *18*
Reimold, A. M., 318, *328*
Reiner, C., 271, *273*
Reiner, O., 94, *103, 104*
Reintjes, M., 452, *492*
Reissman, E., 238, *251*
Reiter, J. F., 303, *326*, 342, *367*
Reiter, R., 478, *495*, 507, *517*, 527, 532, *537*
Reith, A., 193, 204, *208*
Reitsma, M., 246, *250*, 318, *326*
Reitz, P., 322, *328*
Relaix, F., 268, *276*
Remak, R., 131, *148*
Ren, C. J., 224, *232*, 504, *517*
Renard, J.-P., 5, 9, 14, *15, 18, 19*
Renaud, D., *250*
Renault, B., 357, *358*
Renfree, M. B., 372, 376, 386, 387, *389, 392, 393*
Renier, D., 452, 453, *484, 489*
Renkawitz-Pohl, R., 334, *363*
Rennels, E. G., 512, *515*
Rennke, H., 314, *326*, 410, *418*
Renshaw, B. R., 173, *179*
Rentschler, S., 334, 348, *367*

Repaske, D. R., 506, *515*
Repérant, J., 83, *104*
Represa, J., 466, *494*, 545, 547, 559, *564*
Resau, J. H., 405, *417*
Reshef, R., 141, 142, *148*, 262, 264, 268, *274, 275, 276*, 399, *417*
Resnick, J. L., 7, *19*, 188, 189, *189, 190*
Retaux, S., 113, *125*
Reutebuch, J., 468, *485*
Reuter, R., 165, *178*, 186, *190*, 316, *328*
Revest, J., 546, *562*
Revest, J.-M., 312, *324*, 505, *514*
Rex, M., 308, *325*
Rey, S., 303, *324*
Reynaud-Deonauth, S., 10, *15*, 342, *368*
Reyne, Y., 271, *273*
Reynolds, B. A., 86, *103*, 206, *206*, 236, 245, 246, 248, *249, 251, 252*
Reynolds, D. M., 414, 415, *418, 420*
Reynolds, K., 89, *101*
Rhee, J., 341, *361*
Rhee, J. T., 344, 345, *370*
Rhee-Morris, L., 292, *294*
Rhim, J. A., 318, *328*
Rhinn, M., 31, *33*, 40, 44, 46, 47, *49, 50, 52*, 114, *125*, 323, 338, *358*, 451, 478, *481, 494*, 521, 527, *535*
Rhodes, S. J., 258, *274*, 509, *517*
Rhodin, J. A., 519, *537*
Ricci, S., 94, *99*
Rice, D. P., 452, *488, 494*
Rice, D. S., 94, *99, 103, 104*, 510, *516*
Richard, S., 117, 118, *124*
Richard-Parpaillon, L., 116, *125*
Richards, A., 188, *190*
Richards, J., 527, 530, 531, *537*
Richards, W. G., 15, *18*, 226, *232*
Richardson, C. D., 222, *231*
Richardson, G., 555, *562*
Richardson, J. A., 94, *105*, 260, 261, 262, 269, *273, 276, 277*, 342, 343, 344, 347, 353, *364, 365, 366*, 432, 465, 478, *483, 498*, 532, *537*
Richardson, M. K., 139, *148*, 236, *249, 252*
Richarson, J. A., 89, *101*
Richieri-Costa, A., 468, *485*
Richter, J. D., 7, 10, 11, 14, *16, 17, 18, 19*
Richtsmeier, J. T., 442, *494*
Rickles, R. J., 10, 14, *18, 19*
Rickmann, M., 134, *148*
Ridall, A. L., 288, 289, *292*
Riddle, R. D., 117, *122*, 141, *146*, 262, 264, *275*, 470, *487, 494*
Ridgeway, A. G., 342, *368*
Ridgway, E. C., 504, 509, 510, *513, 515*
Rieckhof, G. E., 91, *103*

Riedl, A. E., 547, *562*
Rieff, H. I., 94, *104*, 343, *369*
Riethmacher, D., 170, 173, *175, 179*, 260, 266, *272*, 317, *323*, 404, *417*
Riethmacher-Sonnenberg, E., 170, 173, *179*
Rietze, R. L., 206, *206*, 246, *249*
Rifkin, D. B., 224, *232*, 504, *517*
Rigby, P. W. J., 183, *190*, 258, 260, 267, *275, 276*, 475, *496*
Riggs, S., 357, *358*
Rijli, F. M., 117, 118, 119, *123, 125*, 427, 433, 443, 449, 451, 468, 469, 474, 479, *485, 490, 494*, 545, *562*
Riley, B. B., 543, 557, *564*
Riley, J. K., 287, *293*
Riley, P., 168, 169, *178, 179*, 349, 350, 352, 355, *367*
Riley, R. P., 343, 349, *360, 367*
Rimoin, D. L., 286, *294*
Ring, B. D., 312, *327*
Ring, B. Z., 527, 530, 531, *537*
Ring, M., 91, *103*
Rinkwitz-Brandt, S., 85, *103*, 549, 550, *562, 564*
Rintala, M., 424, 442, 443, *494*
Rio, C., 94, *104*
Ripoche, M. A., 320, *324*
Riquet, J., 264, *274*
Risau, W., 203, *207*, 211, 212, 215, 217, 219, 220, 222, 223, 224, 226, 227, 228, *228, 229, 230, 231, 232*
Risling, M., 236, 237, 244, 245, 248, *250*
Ristoratore, F., 116, *125*
Ritchie, H. E., 443, 453, *495*
Rittenhouse, E., 442, 443, *494*
Rivas-Plata, K., 239, *249*
Rivera, A. J., 271, *277*
Rivera-Pérez, J. A., 3, *4*, 40, 43, *50*, 422, 427, 432, 473, 477, *482, 484, 494*
Rivolta, M. N., 553, *563*
Rizzino, A., 166, *179*
Robanus-Maandag, E., 376, *393*
Robb, D., 185, *190*
Robb, L., 31, *34*, 203, *209*, 334, 343, 345, 348, 349, 352, 353, 356, 357, *358, 364*
Robbins, E. S., 224, *232*, 504, *517*
Robbins, J., 343, *365*
Roberson, M. S., 508, *517*
Robert, B., 267, *274*, 412, 414, *419*, 476, 477, *496*
Robert, C., 570, *588*
Roberts, A. B., 226, *230*, 314, *325*, 339, *368*, 467, 468, 469, *494, 495*
Roberts, A. W., 194, *208*
Roberts, D. J., 305, 308, *328, 364*
Roberts, H. M., 183, *189*
Roberts, J. M., 558, *563*
Roberts, R. C., 451, 475, *494*
Roberts, S., 44, *50*
Roberts, V. J., 467, *494*, 509, *513*

Robertson, E. J., 30, 31, *33, 35,* 37, 42, 43, 44, 45, 46, 47,
    49, 50, 51, 78, 82, *98, 99, 105,* 133, 135, 137, *145,
    148,* 160, 171, *175,* 204, *209,* 238, *251,* 303, 321, *326,
    329,* 336, 338, 339, 354, 355, 357, *358, 359, 367, 368,*
    408, *417,* 422, 426, 427, 451, 467, 468, *481, 483, 484,
    490, 495, 497,* 532, *538*
Robertson, K. E., 534, *536*
Robertson, S., 195, *209*
Robey, E., 136, *148*
Robin, N. H., 433, *492*
Robins, P., 579, *585*
Robinson, G., 173, *177,* 450, 471, *494*
Robinson, I. C., 31, *33,* 47, *50,* 451, 476, *483,* 500, 506,
    509, *514*
Robinson, I. C. A. F., 508, *514*
Robinson, N. A., 581, *588*
Robinson, V., 119, *126,* 225, *233,* 423, 424, *494*
Robison, W. G., Jr., 528, 533, *535*
Robitaille, L., 342, 343, *366*
Robl, J. M., 8, *16*
Robson, L. G., 266, 269, *273, 276*
Robson, S., 357, *364,* 387, *390*
Rocancourt, D., 258, 260, 263, 268, *277*
Rocco, M. V., 415, *419*
Rochette-Egly, C., 469, *485, 489*
Rockman, H. A., 341, 352, *362, 363*
Rockwell, K., 203, *209*
Rockwell, P., 227, *233*
Rodaway, A., 342, *367*
Roder, J. C., 42, *51*
Roderick, T. H., 527, *535*
Rodig, S. J., 287, *293*
Rodrigo, I., 355, *359,* 423, *482*
Rodriguez, T., 31, *34,* 46, 47, 48, *51,* 311, 316, *324, 327,*
    339, 348, *360,* 476, *490*
Rodriguez de Cordoba, S., 522, 525, *535, 537*
Rodriguez-Esteban, C., 354, *367,* 478, *496*
Rodriguez-Gallardo, L., 541, *564*
Rodriguez-Mallon, A., 311, *324*
Rodriguez-Rey, J. C., 349, 354, *360, 366*
Rodriquez-Estaban, C., 507, 508, *518*
Roehl, H., 471, *493*
Roelen, B. A., 183, *190,* 222, *229*
Roelen, B. A. J., 336, *363*
Roelink, H., 43, *50,* 78, *99, 101, 102, 104,* 120, 121, *124,
    125,* 141, *148,* 423, *489*
Roeser, T., 39, *51*
Roessler, E., 357, *359,* 468, 470, *485, 491*
Roessner, A., 443, *486*
Rogers, A., 374, 375, *392*
Rogers, G., 578, *588*
Rogers, I., 22, *34*
Rogers, J. M., 161, *178*
Rogers, K. E., 12, *17*

Rogic, D., 416, *420*
Rohdewohld, H., 160, *178*
Rohovsky, S. A., 226, *230*
Rohrer, H., 238, *251, 252,* 434, *489*
Roisen, F., 235, 242, 244, *250*
Rojas, J., 292, *292*
Rolink, A. G., 204, *209*
Roller, M. L., 507, 509, *515*
Rollhäuser-ter Horst, J., 423, *494*
Roman, B. L., 222, *231*
Romand, R., 557, *564*
Romeo, G., 346, *358*
Romeo, P. H., 200, 201, *208*
Romer, A. S., 421, *494*
Rones, M. S., 337, 339, *366, 367*
Rong, P. M., 140, *148*
Rongo, C., 185, *190*
Ronicke, V., 228, *230*
Rönning, O., 424, 442, 443, *494*
Roomans, G. M., 580, *586*
Roop, D. R., 467, 468, *490, 491,* 571, 580, 581, 582, 583,
    584, 587, 588
Ros, M. A., 349, 351, 354, 355, 356, *359, 360, 366*
Rosa, F. M., 303, *328,* 338, *366*
Rose, E. A., 415, *417*
Rosen, E. D., 289, *294*
Rosen, V., 271, *275,* 467, *498*
Rosenberg, P., 409, *419*
Rosenberg, R. D., 219, *228, 229*
Rosenblatt, M., 203, *209,* 387, *392,* 510, *518*
Rosenfeld, M. G., 85, 89, 91, 95, *97, 99, 100, 104, 105,*
    353, 354, 355, *364, 367,* 433, 478, *489, 496,* 502, 504,
    505, 506, 507, 508, 509, 510, 511, *514, 515, 516, 517,
    518,* 556, 558, *562*
Rosenfeldt, H. M., 227, *231*
Rosenquist, G., 344, *367*
Rosenquist, G. C., 303, *328*
Rosenquist, T., 95, *102,* 506, *516,* 578, 583, *586*
Rosenthal, A., 85, *101, 106,* 110, 121, *123, 126,* 402, *418,
    420*
Rosenthal, N. S., 31, *34,* 335, 339, 343, 344, 345, 346, 347,
    350, 351, 352, *359, 366, 367, 370*
Rosewell, I., 288, *294,* 312, *324,* 442, 443, 453, *493,* 505,
    514, 546, *562*
Rosner, M. H., 183, *190*
Ross, A., 85, *100, 391,* 528, *537*
Ross, F. P., 223, *230*
Ross, J., 341, 352, *362, 363*
Ross, M. E., 94, *100*
Ross, R. S., 345, *367,* 467, *495*
Ross, S. A., 346, *367*
Ross, S. E., 271, *276*
Rossant, J., 23, 26, 31, *33, 35,* 40, 42, 43, 44, 46, 48, *49, 50,
    51, 52,* 78, *98,* 110, 112, 115, 116, 121, *122, 123, 125,*

133, 135, 136, 137, 139, *145, 148, 149,* 157, 158, 160, 165, 166, 167, 168, 169, 174, *174, 175, 176, 177, 178, 179,* 182, 183, *189,* 193, 203, 204, *208, 209,* 217, 223, 224, 228, *229, 231, 232,* 303, 304, 313, 316, 317, *323, 324, 328,* 341, 352, 357, *365, 368,* 413, *419,* 451, 468, 478, *481, 483, 495,* 509, *514,* 521, 527, *535*
Rossel, M., 119, *125,* 545, *564*
Rosselli-Austin, L., 237, *252*
Rossier, C., 479, *482*
Roth, W., 577, 583, *584*
Rothbacher, U., 579, *587*
Rothman, A. L., 287, *293,* 579, *586*
Rothman, J. H., 165, *180,* 316, *330*
Rothman, T. P., 244, *251*
Rothnagel, J. A., 580, 581, *584, 586*
Rothstein, J. L., 11, 12, *18, 19*
Rottapel, M., 189, *190*
Rouaud, T., 267, *274,* 335, 337, 340, *358*
Roume, J., 285, *293*
Rousseau, F., 286, *294,* 453, *494*
Rousseau, G. G., 166, *176,* 321, *326*
Rousseau, S., 171, 173, *176*
Roussel, M., 556, 558, *561*
Roux, I. L., 553, 559, *561*
Rowe, D. W., 452, *484*
Rowe, L. B., 11, 12, *17*
Rowitch, D. H., 112, *125,* 228, *229, 230,* 246, *249, 250,* 334, 336, 348, *362, 363,* 407, *418, 419,* 426, *482*
Rowley, J. D., 202, *209*
Rowlitch, D. H., 86, *104*
Rowning, B., 44, *50*
Royer-Pokora, B., 528, *538*
Rozet, J. M., 286, *294,* 453, *494*
Rubel, E. W., 558, *563, 565*
Ruben, R. J., 542, 550, 552, 558, *563, 564*
Rubenstein, J. L., 110, 114, 116, 121, *124, 125, 126,* 290, *292,* 321, *329,* 422, 427, 428, 432, 434, 435, 437, 449, 450, 451, 453, 466, 470, 471, 472, 473, 474, 475, 477, 478, *481, 484, 489, 490, 493, 494, 495, 496, 497, 498,* 549, *562*
Rubenstein, J. L. R., 75, 76, 77, 78, 79, 80, 81, 82, 83, 84, 85, 86, 87, 88, 89, 90, 91, 92, 94, 95, *97, 98, 99, 100, 102, 103, 104, 105, 106,* 443, 449, 472, 479, *484, 495*
Ruberte, E., 443, 469, 470, 474, *484, 490, 491, 494,* 545, 557, *562, 563*
Rubin, E. M., 443, *493*
Rubin, G. M., 136, *147*
Rubinstein, A. L., 78, *104,* 351, *364*
Rubinstein, M., 502, *515*
Rubinstein, S., 8, *18*
Rubock, M. J., 479, *482*
Ruch, J. V., 454, 455, *489, 493*
Ruddle, F. H., 157, *175,* 376, *390,* 450, 471, *496*
Ruderman, J. V., 7, *17*

Rudnick, A., 321, *325*
Rudnicki, M. A., 258, 260, 261, 262, 267, 269, 270, 271, 272, *273, 274, 275, 276*
Rueda, J., 551, 557, *563*
Rueger, D., 238, *251*
Ruffins, S., 423, 426, *494*
Rugendorff, A., 136, *146, 367*
Rugh, R., 319, *328,* 442, *494,* 500, *517*
Ruhin, B., 426, 427, 433, *483*
Ruili, L., 317, *325*
Ruiz, I., 579, *585*
Ruiz i Altaba, A., 78, 83, 84, 86, *99, 103, 104,* 121, *124, 125,* 141, *148,* 165, *179,* 316, *329*
Ruiz-Lozano, P., 172, 173, *174,* 341, 342, 343, 352, 357, *358, 363, 367*
Ruland, S. L., 161, *175*
Rulleyn, L. J., 453, *493*
Rumyantsev, P. P., 341, *367*
Running, M. P., 11, *17*
Ruohala-Baker, H., 185, *189*
Ruoslahti, E., 410, *417*
Ruotsalainen, V., 412, *418, 419*
Rupp, R., 257, *277*
Ruppersberg, J. P., 557, *564*
Ruppert, S., 183, *190*
Rusch, A., 557, *564*
Rusconi, J. C., 136, *148*
Rush, A. J., 247, *249*
Rush, M. G., 287, *292,* 312, 314, 315, *323*
Russ, A. P., 160, 167, *178*
Russell, A., 167, *174*
Russell, E. S., 194, 195, 199, *209*
Russell, L. B., 467, *488,* 525, *538*
Russell, R. G., 353, 357, *365*
Rustin, P., 341, 352, *368*
Rüther, U., 83, 85, *98, 103, 105,* 342, *358,* 442, 443, 470, *482, 497*
Rutishauser, J., 506, *517*
Rutland, P., 433, 453, 466, 474, *492, 493, 494, 498*
Rutledge, J. C., 22, *34*
Rutter, W. J., 168, 169, *175,* 319, 321, *324, 328, 329,* 349, *360,* 509, *515*
Ruvinsky, I., 339, 341, 349, *370*
Ryan, A., 541, 542, 552, 555, 558, 559, *564,* 579, *589*
Ryan, A. F., 556, 558, *562*
Ryan, A. K., 89, 91, 95, *104,* 354, *367,* 478, *496,* 506, 507, 508, 509, *516, 517, 518,* 556, 558, *562*
Ryan, A. M., 223, *229,* 291, *292*
Ryan, G., 204, *210,* 220, *233,* 400, *419*
Ryan, H. E., 225, *232*
Ryan, K., 160, 167, *178*
Ryan, S., 416, *420*
Ryans, A. M., 402, *418*
Ryazanov, A., 15, *18*

Ryder, E. F., 527, *535*
Ryoo, H. D., 91, *103*
Rzhetsky, A., 525, *538*

S

Säämänen, A. M., 424, 442, 443, *494*
Saarialho-Kere, U., 441, *488*
Saarma, M., 402, 403, *419, 420*, 559, *564, 565*
Sabapathy, K., 87, *104*
Sabbagh, W., 354, *367*
Sabin, F., 200, *209*
Sabin, F. R., 211, *232*
Sablitzky, F., 10, *17*
Sabourin, L. A., 260, 270, 271, 272, *276*
Sachs, M., 266, *276*
Sack, R. A., 510, *516*
Sadaghion, B., 425, *494*
Sadeghi-Nedjad, A., 507, *517*
Sadikot, A. F., 86, *104*
Sadl, V., 118, *124*
Sadler, T. W., 110, *125*
Sadovsky, Y., 510, *516*
Safar, F., 236, *251*
Saffitz, J. E., 165, *179*
Saffman, E., 182, 185, *190*
Saga, Y., 135, 137, 138, *146, 148*, 340, 352, *363, 367*, 512, *518*
Sagai, T., 83, *102*
Sagata, N., 7, *16, 18*
Saha, M., 347, *360*
Sahai, E., 342, *369*
Sahenk, Z., 271, *273*
Sahly, I., 398, *416*, 550, *561, 563*
Sahni, M., 286, *294*
Saiardi, A., 511, *517*
Said, J., 510, *518*
Saijoh, Y., 47, *51*, 352, 354, 356, 357, *365, 367*
Sainio, K., 375, *392*, 403, *419*
Saito, M., 341, 352, *370*
Saito, T., 88, *102*, 168, *177*, 239, *250*
Saitou, M., 581, *588*
Sakagami, M., 555, *565*
Sakai, L. Y., 572, *587*
Sakai, M., 183, *190*, 531, *538*
Sakai, N., 451, 475, *493*
Sakai, R., 171, *175*
Sakai, T., 136, 137, *147*, 511, *515*
Sakai, Y., 338, 354, 356, *368*
Sakano, S., 282, *295*
Saksela, E., 405, *419*
Sakuma, R., 338, 354, 356, *367, 368*
Sakuma, S., 350, *370*
Sakuta, H., 525, *538*

Salama, R. H., 350, *370*
Salamon-Arnon, J., 174, *178*
Salas-Cortes, L., 387, *390, 392*
Salazar-Grueso, E., 402, *420*
Saldivar, J. R., 426, 427, *494*
Salentijn, L., 452, *492*
Salinas, P. C., 45, *52*, 268, 269, *274*
Salminen, M., 262, 268, *276*, 547, *564*
Salomon, D. S., 338, *367*
Saltiel, A. R., 162, *176*
Salunga, R. C., 12, *17*
Salvatoni, A., 509, *514*
Salzberg, A., 91, *103*
Samad, T. A., 91, *101, 104*
Samadani, U., 321, *328*
Samanta Roy, D. R., 87, *101*
Samoylina, N. L., 196, 199, 200, *208*, 220, *231*
Sampath, K., 78, *104*, 354, 355, *364*
Sampath, T. K., 78, 81, *99*, 408, 416, *420*, 467, *494*
Samper, E., 509, *514*
Samson, D., 398, *416*, 550, *561*
Samuelson, L. C., 511, 512, *515*
Sanak, M., 286, *294*
Sanbo, M., 223, *230*
Sanchez, A., 388, *391*
Sanchez, M. P., 402, *419*
Sánchez, S., 354, *367*, 423, *495*
Sanchez de Vega, M. J., 333, 352, 354, *362*
Sanchez-Gomez, C., 344, 348, 355, *360*
Sanchez-Pacheco, A., 510, *517*
Sandell, L. J., 282, *295*
Sander, 88, *104*
Sander, M., 321, *328*, 451, 474, 475, *494*
Sandgren, E. P., 318, *328*
Sandilands, A., 527, 530, 531, *537*
Sands, A. T., 287, *294*
Sandulache, R., 522, 527, *535*, 549, *562*
Sanes, J. R., 237, *249, 251*, 411, 412, *418, 419*
Sanford, L. P., 453, 467, *494*, 550, *563*
Sangha, R. K., 174, *179*
Sangiorgi, F., 428, 452, *489*
Sanicola, M., 403, *420*
Sanjo, H., 452, *496*, 580, 581, 583, *589*
San Jose, I., 559, *564*
Sanne, J. L., 469, *483*
Sano, Y., 11, 12, *17*
Sans, A., 552, *564*
Santanna, C., 292, *294*
Santella, L., 23, *34*
Santerre, R. F., 258, *274*
Santoro, S. A., 314, *329*
Santos, E., 171, 173, *178*
Santos, V., 428, *491*
Santschi, L., 239, *250*

Sanvito, F., 321, *328*
Sanyanusin, P., 398, *419*
Sanz, R., 525, *535*
Sapin, V., 557, *564*
Sapir, T., 94, *104*
Saplakoglu, U., 357, *359*
Sarapura, V. D., 509, *515*
Sarasa, M., 337, *359*
Sargent, M. G., 355, 356, *362*, 423, *492*
Sargent, T. D., 173, *177*, 580, 583, *587*
Sariola, H., 341, 352, *363*, 375, *391, 392,* 400, 402, 403, *418, 419, 420,* 453, 467, *494*
Saris, J. J., 414, *418*
Sarkar, D. K., 504, 511, *513, 515*
Sarkar, L., 440, *494*
Sarvetnick, N., 318, 322, *325, 326, 328*
Sasai, Y., 44, *52,* 338, *359*
Sasaki, F., 504, *517*
Sasaki, H., 40, 41, 42, 43, 47, *50, 52,* 120, 121, *123, 125,* 135, *148,* 303, *328,* 338, 355, *370,* 376, *393,* 470, 477, 479, *484, 490, 494,* 522, *537*
Sasaki, K., 204, *209,* 220, *232,* 288, *293,* 442, 443, 453, *488*
Sasaki, N., 11, 12, *18,* 511, *516*
Sasaki, T., 318, *327,* 574, 583, *584*
Sasse, J., 317, *330,* 337, *370*
Sasseville, R., 86, *104*
Sassone-Corsi, P., 379, 384, *391, 393*
Sassoon, D., 226, *232,* 258, 262, 267, 269, *273, 274, 275, 276, 277,* 334, 348, *367,* 385, *391,* 476, *483*
Satake, M., 220, *231*
Sater, A. K., 335, 337, 339, *362, 367*
Satin, J., 343, 344, 351, *367*
Sato, B., 466, *496*
Sato, K., 84, *100,* 311, *330*
Sato, M., 288, *293,* 442, 443, 453, *488*
Sato, N., 579, *588*
Sato, S., 266, *277,* 399, *419*
Sato, T., 112, 114, 115, *123,* 312, *328,* 505, *517,* 525, *535*
Sato, T. N., 222, 223, 224, 226, 227, 228, *231, 232*
Sato, Y., 571, *586*
Satoh, N., 266, *277,* 475, *492*
Satoh, T., 512, *518*
Satokata, I., 290, *294,* 437, 440, 450, 451, 453, 466, 474, *482, 495*
Sauer, B., 15, *18*
Sauer, H., 402, *418*
Saueressig, H., 410, *418*
Saugier, P., 286, *294*
Saulnier, D. M., 225, *229*
Sauls, A. D., 11, 12, *17*
Saumhueter, S., 193, *210*
Saunders, G. F., 451, 475, *486,* 527, 528, 530, *536, 538*
Saunders, J. C., 506, *517,* 549, *564*

Saunders, T. L., 6, *19,* 378, 379, *393,* 443, *482,* 507, 509, 511, 512, *514, 515,* 556, *564*
Sauvageau, G., 242, *252*
Savary, R., 265, *272*
Sawada, Y., *588*
Sawai, S., 318, *328*
Sawchenko, P. E., 89, 95, *104,* 506, 508, 511, *513, 516, 517*
Sawel, L., 579, *586*
Sawyer, D., 314, *324*
Sawyer, H., 214, *230*
Saxen, L., 22, *34,* 375, *392,* 405, *419*
Saxton, T. M., 171, 173, *175, 178*
Sayama, K., *588*
Sayre, E., 94, *103*
Scaal, M., 266, *276*
Scalera, L., 338, *360*
Scambler, P. J., 353, 357, *364, 365, 367*
Scandrett, J. M., 165, *176,* 333, 342, *362*
Scapoli, L., 382, *392*
Scarlett, L. M., 507, 508, 509, 511, *514, 515*
Scarpa, A., 271, *273*
Scarpelli, D. G., 318, *328*
Schach, U., 546, *565*
Schaefer, J. J., 534, *537*
Schafer, K., 260, 266, *275, 276*
Schaffner, W., 317, *325*
Schaible, K., 183, 187, 188, *189*
Schaller, S. A., 3, *4*
Schalling, M., 313, *323,* 374, 375, *392,* 467, *482*
Scharer, E., 582, 583, *587*
Scharfmann, R., 321, *327*
Schartl, M., 388, *392*
Schatteman, G., 220, *228*
Schauer, G., 47, *51,* 357, *363*
Schechter, J., 512, *517*
Schecterson, L. C., 558, 559, *564, 565*
Schedl, A., 88, *99,* 341, *365,* 380, *393,* 528, *537*
Schedl, T., 242, *252*
Scheel, D. W., 321, *328*
Scheele, J., 478, *489*
Scheffer, I., 94, *100*
Schell, U., 433, *492*
Schelling, M. E., 223, *232*
Scherer, G., 281, *293,* 380, 381, *390, 391, 392, 393*
Scherson, T. Y., 423, 425, *495*
Schier, A. F., 45, 47, *50, 51, 52,* 78, *100,* 338, 353, 354, 355, 356, 357, *359, 361, 368, 370,* 395, *417,* 428, 468, *495,* 505, *515,* 534, *536,* 546, *563*
Schieren, G., 415, *419*
Schilling, T. F., 422, 433, 465, *491, 495, 497*
Schimmang, T., 543, 545, 546, 559, *564, 565*
Schimmenti, L. A., 398, *419*
Schindeler, A., 345, 348, *366*
Schinke, M., 219, *232,* 341, 343, 352, *369*

Schinko, I., 560, *565*
Schinzel, A., 281, *293,* 381, *391,* 433, *492*
Schipani, E., 285, 286, 287, *292, 293, 294,* 442, *488*
Schlaeger, T. M., 228, *232*
Schlange, T., 340, *367*
Schleinitz, M., 343, *369*
Schler, A. F., 85, *101*
Schlessinger, D., 357, *361,* 441, *488*
Schlondorff, J., 42, *51,* 479, *487*
Schmahl, J., 382, *392*
Schmahl, W., 94, *104,* 263, *272*
Schmid, M., 388, *392*
Schmid, W., 314, *324*
Schmidt, C., 45, *52,* 130, 131, 132, *145, 146,* 317, *328,* 404, *419*
Schmidt, H., 281, *293,* 381, *391*
Schmidt, J. V., 225, *231*
Schmidt, L., 227, *229*
Schmitt, E. M., 318, *325*
Schmitt, J. P., 350, 353, 357, *359*
Schmitz, M. L., 281, *293,* 381, *391, 393*
Schmoll, M., 264, *277*
Schnabel, C. A., 449, 479, *487*
Schnabel, D., 508, *516*
Schnapp, B. J., 6, *18*
Schnegelsberg, P. N., 258, 260, *276,* 427, *484*
Schneider, C., 238, *252*
Schneider, M. D., 341, 342, *364*
Schneider, P., 441, *497*
Schneider, R. J., 11, *17*
Schneider, V. A., 336, 337, 338, 339, 340, *367*
Schneider-Maunoury, S., 118, *125,* 510, *518,* 546, *564, 565*
Schnurch, H., 226, *228*
Schoderbek, W. E., 508, *517*
Schoeberlein, A., 264, *274*
Schoeller, H., 183, 184, 185, *189*
Schoenherr, C. J., 243, *252*
Schoenwolf, G. C., 38, 40, *50, 52,* 132, *148,* 317, *328,* 333, 334, 335, 336, 344, *361, 362, 368, 370,* 397, *418,* 422, 427, 433, *496, 497*
Scholer, H., 23, *34,* 160, 166, *175, 178,* 183, *190,* 311, *324*
Schonemann, M. D., 89, 95, *104,* 506, *517*
Schopp, J., 441, *497*
Schorderet-Slatkine, S., 7, *16*
Schorle, H., 79, *104,* 353, 357, *365,* 529, *537*
Schorpp-Kistner, M., 173, *179*
Schott, J.-J., 343, 345, 348, 349, 350, 351, 352, 357, *359, 367*
Schowing, J., 432, *495*
Schreck, R., 217, 220, *229*
Schreiber, G., 161, *179*
Schreiber, J., 170, 173, *179*
Schreiber, M., 173, *179*
Schreiber, R. D., 287, *293*

Schreiner, C. M., 204, *208,* 318, *327*
Schroder, R., 570, *587*
Schroeter, E. H., 136, *146, 147, 148*
Schubert, C., 240, 243, *251, 252*
Schubert, F. R., 140, *145,* 263, 264, 265, 266, *273, 277*
Schuchardt, A., 402, *419*
Schuchmann, M., 341, *369*
Schuger, L., 314, *328*
Schughart, K., 130, 132, 140, *148,* 264, 265, *277,* 397, *420,* 423, *492,* 522, 527, *535,* 549, *562*
Schuh, A. C., 203, *209,* 217, 223, *232*
Schuhbaur, B., 341, 346, 347, 352, *366,* 469, *492,* 545, *564*
Schulte, D., 525, *537*
Schulte, T., 547, *564*
Schulte-Merker, S., 336, *363*
Schultheiss, T. M., 263, *274, 328,* 335, 336, 339, *367*
Schultz, G. A., 5, 9, *18, 19,* 310, *325*
Schultz, J. R., 89, *106,* 450, 471, *498*
Schultz, M., 432, 473, 477, *498*
Schultz, R. A., 334, 343, *361*
Schultz, R. M., 5, 7, 8, 9, 11, 12, 13, 14, *15, 16, 17, 18, 19*
Schulz, R. A., 262, *276,* 334, 341, *363, 364, 367*
Schumacher, C. A., 343, 345, 348, *366*
Schuster, G., 352, 354, *366*
Schutz, G., 40, 42, *51,* 165, *177,* 303, 314, *324, 327,* 479, *487*
Schwab, M. H., 92, *104*
Schwanzel-Fukuda, M., 435, *495*
Schwartz, C., 470, *481*
Schwartz, J. H., *250*
Schwartz, K., 343, *366*
Schwartz, L., 43, *50,* 133, *145,* 158, 165, *175,* 203, *209,* 217, 224, *231, 232,* 303, 304, 313, 316, 317, *324, 328,* 413, *419,* 468, *495*
Schwartz, R. J., 336, 342, 343, 344, *359, 360, 364, 366, 367, 369, 370*
Schwarz, J., 342, 350, 352, *364*
Schwarz, M., 115, *125,* 311, *329*
Schwarz, S. M., 27, *34*
Schweickert, A., 351, 353, 354, 355, 356, *359, 361*
Schweppe, R. E., 505, *517*
Schwind, J. L., 433, *495,* 500, *517*
Schwitzgebel, V. M., 321, *328*
Scofield, R. M., 271, *273*
Scott, A., 318, *328,* 453, *487*
Scott, B. W., 247, *252*
Scott, I. C., 169, 170, *178, 179,* 349, *367*
Scott, M. P., 78, 80, 86, *100, 106,* 121, *123,* 287, *293, 294,* 470, *485,* 574, 577, 579, 583, *585, 588*
Scott, W. J., Jr., 204, *208,* 318, *327*
Scotting, P. J., 308, *325*
Scully, K. M., 502, *514*
Scully, S., 226, *232,* 312, *327*
Seale, P., 260, 269, 270, 271, 272, *276*

Searcy, R. D., 342, 346, *367*
Searcy-Schrick, R. D., 347, *364*
Seasholtz, A. F., 502, 511, *513*
Sebbag, M., 581, *588*
Sechrist, J., 423, 425, 426, 435, 475, *481, 494, 495, 496*
Seckl, J. R., 506, *514*
Seeburg, P. H., 512, *516*
Seed, J., 269, *276*
Seegmiller, R. E., 424, 442, 443, *493, 495*
Sefton, M., 354, *367*, 423, *495*
Segal, Y., 415, *417*
Seghezzi, G., 224, *232*, 504, *517*
Segil, N., 556, 558, *561*
Segré, G. V., 285, 287, *293, 294*, 442, 443, 470, *489, 497*
Segré, J. A., 271, *274*, 570, 572, 581, 583, *586, 588*
Seibel, W., 161, *179*
Seidah, N. G., 136, *146*
Seidman, C. E., 339, 343, 345, 346, 347, 348, 349, 350, 351, 352, 353, 357, *358, 359, 367*
Seidman, J. G., 339, 343, 345, 346, 347, 348, 349, 350, 351, 352, 353, 357, *358, 359, 367*
Seifert, E., 316, *329*
Seifert, K. J., 382, 387, *392*
Seifert, R. A., 467, *495*
Seike, M., 94, *103*
Seitanidou, T., 118, *124, 125*, 468, *492*, 546, *564, 565*
Seitz, C. S., 580, 583, *588*
Seker, M., 573, *584*
Seki, T., 582, *584*
Sekimoto, T., 305, 307, *328*
Sekine, K., 312, *328*, 505, *517*
Selby, P. B., 288, *294*, 442, 443, 453, *493, 495*
Seldin, M. F., 172, 173, *176*, 476, *498*
Seleiro, E. A., 45, *52*
Self, T., 556, *565*
Selin, L. K., 415, *418*
Sellars, S., 292, *294*
Selleck, M. A., 39, 40, 41, *52*, 121, *123, 125*, 423, *484, 495*
Seller, M. J., 525, *535*
Sellheyer, K., 468, *491*
Semb, H., 322, *324*, 409, *419*
Semba, I., 282, *294*
Semenova, M. L., 199, *208*
Semenza, G. L., 225, *230, 231, 232*
Semina, E. V., 354, *370*, 478, *495*, 507, *517*, 527, 532, *537*
Senba, E., 318, *325*
Senft, E., 522, 527, *535*, 549, *562*
Sengel, P., 568, 571, 572, 573, 574, *587, 588*
Senior, B., 507, *517*
Senior, R. M., 223, *232*, 281, 291, *294*
Seno, M., 321, *327*
Seo, J.-W., 352, 354, *365, 367*
Sepulveda, J. L., 342, 343, *367*
Sepulveda, V. N., 342, *366*

Serbedzija, G. N., 79, 85, *102, 106*, 237, *252*, 339, *368*, 424, 425, *495*, 529, *538*
Serrano, A., 91, *102*
Serre, G., 581, *588*
Serreri, L., 443, 449, 479, *495*
Serup, P., 80, 82, 88, *98*, 310, 321, *325, 327*
Sesay, A. K., 9, *18*
Sessa, W. C., 225, *231*
Sestan, N., 136, *148*
Setalo, G., 504, *516*
Seth, P. K., 580, *585*
Seth, R., 314, *329*
Seto, M., 376, *390*
Seufert, D. W., 395, *420*
Seung, K. J., 82, *101*
Seung, K. S., 315, *326*
Seuntjens, E., 511, *517*
Seydoux, G., 185, *190*, 243, *249, 252*
Sha, W. C., 318, *323*
Shabanowitz, J., 226, *230*
Shackleford, G. M., 466, *486*
Shafritz, A. B., 271, *276*
Shafritz, D. A., 318, *324*
Shah, N. M., 235, 238, 239, 240, 244, *251, 252*
Shah, S. B., 45, *52*
Shahar, S., 22, *34*
Shahinian, A., 341, *370*
Shalaby, F., 203, *209*, 217, 223, *232*
Sham, M. H., 117, 118, 119, *124*, 380, *389*, 422, 427, 433, 468, *487, 490, 492*, 544, *563*
Shamim, H., 116, *125*, 546, *565*
Shamu, C. E., 382, 387, *392*
Shan, Z., 388, *392*
Shani, M., 258, 263, 267, *274*
Shanley, S., 287, *292*, 470, *486*
Shanmugalingam, S., 78, 86, *104*
Shannon, J. M., 313, *324*
Shapiro, B. M., 27, *33*
Shapiro, D. N., 268, *273*
Shapiro, L., 408, *419*
Shapiro, M. D., 335, 343, 344, 345, 346, 347, *366, 370*
Shapiro, R. L., 224, *232*, 504, *517*
Shapiro, S. D., 223, *232*, 281, 291, *294*
Shapiro, S. S., 303, *329*
Sharara, R., 11, 12, *17*
Sharkey, A., 357, *358*
Sharkis, S., 194, *207*
Sharma, A., 322, *328*
Sharma, K., *104*, 504, 505, 507, 508, 510, *517, 518*
Sharon-Friling, R., 532, *537*
Sharp, P. A., 14, *19*
Sharpe, A., 162, 170, *176*
Sharpe, A. H., 201, 204, *210*, 220, *232, 233*
Sharpe, M., 317, *328*, 404, *419*

Sharpe, P. T., 121, *123*, 422, 432, 434, 437, 438, 439, 440, 441, 450, 451, 467, 471, 472, 478, *484, 485, 486, 490, 493, 494, 495, 496, 497*
Shashikant, C. S., 376, *390*
Shatz, C. J., 92, *97, 102*
Shaw, G., 372, 386, *389, 393*
Shaw, J. P., 22, *34*, 451, 475, *488*
Shawber, C., 136, *148*, 262, *276*
Shawlot, W., 31, *34*, 40, 41, 42, 43, 44, 46, 47, 48, *52, 104*, 114, *125, 328*, 374, *392, 397, 420*, 478, *494*
Shawver, L. K., 217, 220, 227, *229, 231*
Shea, M. J., 31, *34*, 43, 45, *51*, 132, 137, *148*
Sheehan, K. C., 287, *293*
Sheldahl, L. C., 337, *363*
Sheldon, M., 94, *99, 104*
Shelton, J. M., 94, *105*, 269, *273, 277*, 342, *369*
Shen, H. M., 162, 170, *177*, 219, *230*
Shen, L., 78, *99*, 141, *145*, 336, *363*, 384, *389*, 402, 407, *418, 419*, 470, *484*
Shen, M. M., 31, *33*, 40, 45, *50, 52*, 267, *272*, 338, 353, 354, 356, 357, *359, 360, 368, 370*, 382, 387, *392*, 428, 450, 468, 471, 472, 473, *495, 498*
Shen, W. H., 376, 380, 381, 384, *390, 391, 392*
Shen, Y., 94, *103*
Sheng, G., 546, *565*
Sheng, H., 502, 508, 509, *517*
Sheng, H. Z., 83, 94, *104, 106*, 402, *419*, 500, 504, 505, 507, 508, 510, *517, 518*
Shepard, A., 416, *420*
Shepherd, K., 314, *326*, 410, *418*
Sheppard, A. M., 94, *104*
Sher, A. E., 442, 473, *495*, 542, 552, 553, *565*
Sherbon, B., 471, *493*
Sherman, M. I., 156, *174*
Sherr, C. J., 170, *179*
Sheth, A. N., 87, *104*
Shevchenko, V. I., 9, *18*
Shi, H., 15, *18*
Shi, L., 84, 88, 89, 90, 92, *97, 104*, 450, 470, *481*
Shi, X. M., 121, *124*, 451, 453, 470, 471, *491*
Shi, Y., 336, *368*
Shiang, R., 286, 292, *294*, 453, *495*
Shiba, H., 376, *393*
Shibata, Y., 376, *392*
Shibusawa, N., 512, *518*
Shibuya, m., 217, *230*
Shiere, F., 507, *514*
Shigetani, Y., 475, *492*
Shih, J., 40, *52*
Shih, M. S., 416, *420*
Shiino, M., 512, *515*
Shikayama, T., 379, *391*
Shilo, B.-Z., 340, *358, 370*
Shim, E. Y., 243, *249*

Shim, H. H., 398, *419*
Shimamura, K., 76, 77, 78, 79, 81, 82, 83, 84, 85, 86, 88, 89, 92, *98, 102, 103, 104, 106*, 110, 121, *125, 126*, 408, *418*, 422, 427, 428, *494, 495*
Shimasaki, S., 577, *586*
Shimazaki, T., 89, 91, *104*
Shimizu, H., 570, 583, *584*
Shimizu, K., 202, *208*
Shimizu, Y., 288, *293*, 442, 443, 453, *488*
Shimon, I., 510, *518*
Shimono, A., 47, 49, *51, 52*, 318, *328*, 357, *365*
Shimouchi, K., 581, *588*
Shin, J. J., 556, *565*
Shin, T. H., 243, *249*
Shinanura, K., 83, *103*
Shinmei, Y., 531, *538*
Shinoda, K., 376, *393*
Shinomura, T., 350, *370*
Shinpock, S. G., 204, *208*
Shintani, T., 525, *538*
Shinya, M., 41, *51*, 78, *104*
Shinya, T., 427, *484*
Shioda, T., 339, *368*
Shioi, T., 343, *363*
Shiojima, I., 336, 343, *365, 368*
Shiojiri, N., 317, *328*
Shiota, K., 136, 137, *146, 147*, 167, 173, *179, 180*, 341, 352, *370*, 404, *420*
Shipley, J. M., 223, *232*, 281, 291, *294*
Shipley, M. T., 84, *100*
Shirakata, Y., *588*
Shirakawa, H., 8, *16*
Shiratori, H., 338, 354, 356, *368*
Shirayoshi, Y., 226, *232*, 352, 354, *365*
Shiroishi, T., 83, *102*, 376, *391*
Shivdasani, R. A., 203, *209*
Shmitt, D., 581, *588*
Shoji, H., 577, *586*
Shoji, R., 557, *565*
Shope, C., 559, *563*
Shore, E. M., 271, *276*
Shores, E. W., 442, *484*
Shors, T. J., 237, *250*
Short, R. V., 372, *392, 393*
Shou, J., 555, *566*
Shou, W., 44, *51*
Shown, E. P., 378, *390*
Shtein, G. I., 169, *180*
Shtutman, M., 86, *104*, 579, *588*
Shu, H.-B., 341, *370*
Shugar, J. M., 552, *564*
Shuler, C., 314, *325*, 434, 465, 467, *487, 491*
Shultz, L. D., 195, *208*, 442, *498*
Shum, L., 282, *294*, 314, *329*, 428, 465, *491, 492*

Shutter, J. R., 226, *230, 232*
Sibilia, M., 169, *179*
Siciliano, M. J., 202, *208*
Sideras-Haddad, E., 580, *585, 587*
Sidman, C., 194, *207*
Sidman, R. L., 92, 94, *97, 103*
Siebel, K. E., 193, *207*
Siebenlist, U., 442, *484*
Sieber, B.-A., 402, *420*
Sieber-Blum, M., 236, 237, *249, 252*
Siegel-Bartelt, J., 478, *495*, 507, *517*
Sigala, S., 511, *516*
Siggers, P., *390*
Sigrist, K., 165, *177*, 303, 316, *326, 327*, 342, 343, 352, *363*
Silberbach, G. M., 357, *358, 367*
Silberg, D., 310, *325*
Silengo, M., 346, *358*
Sillence, D. O., 443, 453, *495*
Silos-Santiago, I., 91, *99*, 402, *419*, 559, *562*
Silve, C., 285, *293*
Silver, J., 542, 559, *561*
Silver, L. M., 132, *145*, 339, 341, 349, *370*
Silver, M., 220, *228*, 532, *535*
Simak, E., 264, *274*
Simard, A. R., 269, *277*, 342, *369*
Simcha, I., 86, *104*, 579, *588*
Simeone, A., 31, *33*, 40, 46, 48, *49, 52*, 79, 83, 84, 95, *97, 103, 104*, 112, 114, *122, 123, 124*, 266, *275*, 290, *292, 294*, 311, *322, 330*, 338, *360*, 428, 450, 451, 453, 471, 472, 473, 478, *481, 491, 495*, 508, *513*, 521, 527, *535*, 549, 550, 551, 552, *561, 564*
Simmons, D. M., 89, *100*, 502, 506, 509, *514, 515, 516, 518*
Simmons, W. W., 219, *228*
Simon, M., 581, *588*
Simon, M. C., 165, 172, 173, 174, *174, 179*, 204, *209*, 225, *231*, 286, *292*
Simon, R., 10, *18*
Simon-Assmann, P. M., 308, *325, 326*
Simonet, W. S., 312, *327*
Simonetti, D. W., 206, *209*
Simpson, B. B., 47, *50*
Simpson, E. H., 31, *34*, 378, 387, *390*
Simpson, K., 379, *390*
Simpson, P., 136, *148*, 343, *360, 361*
Simpson, T. H., 557, *563*
Sinclair, A. H., 379, 382, 387, 388, *393*
Sinclair, A. M., 192, *209*, 377, 380, 387, 388, *391, 393*
Sinclair, R. A., 194, *208*
Sineone, A., 506, *513*
Singal, A., 341, *369*
Singh, G., 79, *105*
Singley, C. T., 424, *495*

Singson, R., 504, *515*
Sinickas, V., 194, *208*
Sinning, A. R., 342, *368*
Siracusa, G., 86, *103*, 188, *190*
Siracusa, L. D., 157, *178*
Sirard, C., 468, *495*
Sisi, P., 568, 582, *586*
Sissman, N. J., 341, 354, 355, *368*
Siwik, E., 341, *369*
Sjodin, A., 409, *419*
Skaer, H., 171, *179*
Skandalakis, J. E., 312, *328*
Skapek, S. X., 341, *361*
Skeath, J. B., 340, *361*
Skerjanc, I. C., 342, *368*
Skoldenberg, E., 504, *515*
Skolnick, M. H., 412, *417*
Skoultchi, A. I., 353, 357, *365*
Skowronski, J., 11, 12, *18*
Skromne, I., 45, 48, *50, 52*, 427, *484*
Slack, J. M., 27, *34*, 167, *176*, 263, *274*, 320, *328*
Sladek, F. M., 164, *177, 179*
Sladek, N. E., 525, *537*
Sladek, R., 167, *177*
Slaney, S. F., 453, 466, 474, *494, 498*
Slavkin, H. C., 282, *294*, 314, *329*, 428, 465, *491, 492*
Slepecky, N. B., 557, 558, *565*
Slieker, W. A. T., 192, *210*
Sloviter, R. S., 247, *251*
Slusarski, D., 121, *123*, 470, 487, *497*
Small, K., 95, *101*, 507, *516*
Small, K. M., 82, *105*
Small, K. W., 478, *495*, 507, *517*
Smalley, M., 440, *494*
Smalls, O., 357, *358*
Smart, I. H., 87, *104*, 572, *588*
Smets, G., 502, *515*
Smeyne, R., 313, *327*, 559, *562*
Smidt, M. P., 506, *513*
Smiga, S. M., 76, 77, 82, 83, 84, 92, *98, 103*
Smith, A., 23, *34*, 160, 166, *178*, 423, 424, *494*
Smith, A. G., 160, *178*
Smith, A. J., 203, *210*
Smith, B., 341, 352, *362*
Smith, C. A., 379, 382, 387, 388, *393*
Smith, C. K. D., 258, *274*
Smith, C. K. II, 271, *276*
Smith, D., 397, *418*
Smith, D. E., 135, *148*
Smith, D. M., 308, 309, *328, 329*, 337, 354, *364*
Smith, E. A., 286, *292*
Smith, G. H., 226, *230*
Smith, J. C., 40, *49*, 167, *179*, 203, *207, 209*, 303, *323, 329*, 337, 338, *362, 368*

Smith, J. M., 467, *494*
Smith, K., 223, 227, *232*
Smith, L. J., 25, 26, *34, 390,* 432, *496*
Smith, L. T., 573, 583, *584, 588*
Smith, M. J., 377, 388, *393*
Smith, M. M., 433, 436, *495, 497*
Smith, P., *251*
Smith, R., 26, *34,* 84, *102*
Smith, R. A., 219, 220, *232*
Smith, S., 314, 315, *329, 330*
Smith, S. M., 341, 342, 352, *367, 368*
Smith, T. H., 132, *148,* 263, *276*
Smith, T. W., 219, *228*
Smith, V. V., 305, *328*
Smith, W. C., 44, *50, 52,* 238, *252*
Smithers, L., 139, *146*
Smith-Fernández, A., 83, *104*
Smits, P., 283, 284, 289, *294*
Smyth, I., 287, *292,* 470, *486*
Snead, M. L., 428, 452, *489*
Snell, W. J., 8, *18*
Snider, L., 169, *177,* 258, *277*
Snider, W. D., 411, *418*
Snoeck, I., 94, *99*
Snow, M. H. L., 128, *148,* 183, 188, *189, 190*
Snyder, E. Y., 86, *104,* 527, *538*
Snyder-Keller, A., 91, *98*
Soares, M. B., *390*
Soares, V. C., 78, 81, *106,* 243, *252,* 408, *417,* 479, *498*
Sobe, T., 341, *369,* 556, *565*
Sobrier, M. L., 508, *516*
Sockanathan, S., 88, *105,* 377, 386, 387, *390,* 531, *536, 537*
Soffer, D., 341, *369*
Sofroniew, M., 376, *393*
Sohal, G. S., 426, *495*
Sohocki, M., 408, *418,* 451, 467, *490*
Soker, S., 223, *232*
Sokol, S. Y., 415, *418,* 579, *585*
Solloway, M. J., 82, 86, *105,* 135, *148,* 336, *368,* 426, 467, *495*
Solnica-Krezel, L., 78, *104,* 395, *417,* 546, *563*
Solomin, L., 509, *518*
Solomko, A. P., 26, 27, *33*
Solter, D., 6, 8, 9, 10, 11, 12, 13, *16, 17, 18*
Solum, K., 443, *487*
Solursh, M., 424, 473, *495, 498*
Somerset, D. A., 174, *179*
Somlo, S., 414, 415, *418, 419, 420*
Sommer, A., 292, *294*
Sommer, L., *102,* 236, 239, *250, 251, 252,* 320, 321, *323,* 526, *538*
Sommer, P., 316, *323*
Sonenberg, N., 11, *16*
Soneoka, Y., 193, 206, *207,* 246, *250,* 270, 271, *274*

Song, D. L., 113, *125*
Song, H., *588*
Song, K., 267, *277*
Song, Y., 437, *498*
Sonksen, P. H., 507, *514*
Sonnenberg, A., 572, 583, 585, *589*
Sonnenberg-Riethmacher, E., 260, 266, *273*
Sood, S., 453, *495*
Sorensen, L. K., 226, *230, 232*
Soria, J. M., 84, *105*
Soriano, P., 79, 94, *100, 101,* 157, 173, *178, 180,* 183, *190,* 228, *229, 230,* 246, *249, 250,* 334, 341, 348, 352, *359, 362,* 413, 414, *418, 419,* 426, 442, 443, 467, *482, 495*
Sornson, M. W., 506, 508, 509, *518*
Sorrentino, V., 189, *190*
Sosa-Pineda, B., 317, 321, *329,* 451, 475, *495,* 532, *537*
Sotelo, C., 95, *101,* 113, *122*
Souabni, A., 311, *326,* 533, *535*
Soudais, C., 165, *177, 179,* 316, *326,* 342, 343, 352, *363*
Souleyreau-Therville, N., 375, 382, *389, 392*
Soulez, M., 342, *368*
Soults, J., 357, *358*
Southard, J. L., 137, *149,* 311, *323*
Southee, D., 384, *391*
Southwell, B. R., 161, *179*
Souza, P., 313, *329*
Sowden, J. C., 334, *365*
Spada, F., 9, *18*
Spagnuolo, R., 226, *229*
Spana, E. P., 532, *537*
Spangrude, G. J., 192, *209,* 244, 248, *252*
Spano, P., 511, *516*
Sparkes, R., 452, 453, 474, *487*
Sparks, A. B., 86, *100,* 579, *586*
Sparrow, D. B., 335, 336, 337, 339, 342, 345, 348, 355, *365, 366, 368*
Speck, N. A., 196, 199, 200, 201, 202, 204, *207, 209, 210,* 220, *231, 232, 233*
Speicher, S. A., 136, *149*
Spemann, H., 40, 48, *52,* 427, *495, 538*
Spence, A., 42, *52,* 110, *125,* 158, *178*
Spence-Dene, B., 312, *324*
Spencer, J. B., 136, 137, *147*
Spencer-Dene, B., 505, *514,* 546, *562*
Sperbeck, S., 532, *537*
Spicer, D. B., 341, *361*
Spiegel, S., 227, *231*
Spiegelman, B. M., 289, *294,* 442, *487*
Spiegelman, M., 442, 443, *494*
Spindle, A. I., 158, *179*
Spinner, N. B., 347, *364, 366*
Spirina, O. M., 510, *518*
Spiro, A. C., 88, *98*
Spitz, F., 262, 268, *276*

Splitt, M., 357, *359*
Spohr, G., 10, *15*, 342, *368*
Spooner, B. S., 319, *329*
Spörle, R., 130, 132, 140, *148*, 264, 265, *277*, 423, *492*, 522, 527, *535*, 549, *562*
Sporn, M. B., 226, *230*, 314, *325*, 467, 468, 469, *494, 495*
Spotila, J. R., 387, *393*
Spotila, L. D., 387, *393*
Spradling, A. C., 6, 7, *18*
Sprawson, N., 425, *490*
Sproat, G., 118, *123*
Sretavan, D. W., 85, *102*, 506, *514*
Srivasta, A. K., 441, *488*
Srivastava, D., 168, *175*, 341, 342, 343, 344, 346, 347, 349, 350, 352, 355, *358, 360, 361, 365, 366, 368, 369, 370*, 451, 465, 473, *495, 496*
St. Jacques, B., 78, 86, *99, 104*, 141, *145*, 287, *293, 294*, 442, 443, 470, *484, 488, 496*, 577, 578, 583, *589*
St. John, P. L., 412, 414, *417, 419*
St. Jore, B., 353, 357, *365*
St. Onge, L., 321, *329*
St. Vil, D., 307, *329*
Staccini, L., 271, *277*
Stacey, A., 443, *495*
Stacker, S. A., 162, *174*, 453, *486*
Stackhouse, T., 227, *229*
Stacy, T., 201, 204, *209, 210*, 220, *231, 232, 233*
Staeheli, P., 388, *392*
Stahl, J., 511, 512, *518*
Stahle-Backdahl, M., 287, *292*, 470, *497*, 579, *586*
Staines, W. A., 236, 245, *251*
Stainier, D. Y., 115, *125*, 213, 214, 222, *230*, 231, 232, 303, *323, 326*, 340, 341, 342, *363, 367*, 546, *563*
Stainier, D. Y. R., 334, 337, 339, 341, 344, 345, 349, *361, 367, 368, 370*
Stall, A. M., 194, *207*, 245, *250*
Stalsberg, H., 341, 344, 355, *368*
Stamp, G. W., 288, *294*, 442, 443, 453, *493*
Stanford, W. L., 203, *209*, 217, *230, 232*
Stanley, E., 31, *34*, 40, 44, *50*, 194, *208, 323*, 338, 348, 349, 352, 353, 357, *358*
Stanton, B. R., 341, 352, *365*
Stappenbeck, T. S., 310, *329*
Stark, K. L., 132, 137, *148*, 226, *230, 232*, 260, *276*, 317, *329*, 407, *419*
Stark, M. R., 141, *147*, 262, 263, 264, 265, *275*, 435, 475, *481, 496*
Starr, E., 202, *209*
Starzinski-Powitz, A., 264, 271, *275, 277*
Stauber, C., 504, 510, 512, *516, 517*
Staudt, L. M., 183, *190*
Stauffer, T. P., 225, *230*
Stavlijenic, A., 416, *420*
Stavnezer, E., 268, *273*

Stayton, C. L., 376, *393*
Stead, R. H., 258, 260, *276*
Stebbins-Boaz, B., 10, 11, *18*
Steding, G., 386, *393*
Steel, F. L. D., 442, *493*
Steel, K. P., 546, 551, 552, 556, 557, 560, *561, 563, 565*
Steele, K., 44, *50*
Steele-Perkins, V., 400, *419*
Stefana, B., 512, *518*
Stefanadis, J. G., 161, *178*
Stefaneanu, L., 511, 512, *513, 518*
Stein, P., 9, 14, *18*
Stein, R. W., 309, 319, 320, *327*
Steinbeisser, H., *323*, 354, 355, *359*, 477, 482, 483
Steinbeisser, K. W., 40, *50*
Steiner, D. Y., 22, *34*
Steiner, K., *50*
Steiner, K. A., 30, 31, *35*, 39, 40, 41, 46, *52*
Steinert, P. M., 570, 580, 581, 582, *584, 586, 587, 588*
Steingrimsson, E., 168, 169, 173, *175, 179*, 349, *360*, 525, *536, 538*
Steininger, T. L., 91, *105*
Steinmetz, H., 341, *363*
Stemple, D. L., 235, 248, *252*, 395, *417*, 546, *563*
Stennard, F. R. K., 203, *209*
Stensas, L. J., 94, *103*
Stephens, R. J., 313, *324*
Stephenson, E., 185, *190*
Stern, C. D., 2, *4*, 39, 40, 41, 45, 48, *50, 51, 52*, 75, *99*, 134, 138, *147, 148*, 303, *329*, 356, *364*, 387, *393*, 427, *484*, 546, *565*
Stern, H. M., 140, *148*, 262, 263, 264, *277*
Sternberg, E. A., 342, *361*
Sternberg, N., 15, *18*
Sternfeld, D. C., 204, *208*
Sternglanz, R., 349, *362*
Stevanovic, M., 281, *292*, 380, 386, 387, *390*
Steven, A. C., 582, 583, *587*
Stevens, K. A., 165, *174*, 303, *323*
Stevens, L. C., 137, *149*, 532, *538*
Stevens, M. E., 443, *493*
Stevenson, B. J., 321, *324*
Stevenson, L., 31, *33*, 46, *49*, 451, 478, *481*, 521, 527, *535*
Stewart, A. F., 262, *277*
Stewart, C. L., 43, *50*, 156, 162, 170, *175, 177, 179*, 219, *230*
Stewart, M. E., 582, *587*
Stewart, R., 44, *50*
Stewart, T. A., 322, *328*, 512, *516*
Stifani, S., 168, *178*
Stocco, D. M., 379, *393*
Stock, D. W., 450, 471, *496*
Stock, J. L., 79, *105*

Stockdale, F. E., 140, *144*, 254, 263, 268, 269, *273, 275,*
    277, 335, 343, 344, 345, 346, 347, *369, 370*
Stockinger, W., 94, *105*
Stoenaiuolo, A., 478, *495*
Stoetzel, C., 453, *482*
Stoller, J., 334, *361*
Stone, D. M., 121, *123,* 402, *420*
Stone, J. S., 558, *565*
Stoner, C. M., 469, 470, *484*
Storey, K. G., 40, *52*
Storm, E. E., 467, *496*
Stornaiuolo, A., 40, 46, *52,* 79, 83, *104,* 206, *207,* 246, *249,*
    271, *274,* 290, *294,* 450, 471, 472, 473, *495*
Stott, D., 166, *176,* 185, *189,* 303, *325*
Stott, N. S., 577, *589*
Stout, R. P., 84, *105*
Stoykova, A., 83, 85, 87, 91, 94, *100, 105,* 451, 475, *490*
Strachan, T., 357, *364,* 387, *390*
Strachen, T., 352, 354, *366*
Stradler, J., 453, *488*
Strahle, U., 338, *362, 366*
Strain, A. J., 174, *179*
Strand, F. L., 501, *518*
Strates, B. S., 271, *277*
Strauss, A. W., 356, 357, *358, 366*
Strawn, L. M., 227, *231*
Street, S. L., 441, *491*
Strehle, M., 353, 356, *361*
Streit, A., 44, *52*
Streit, W. J., 206, *209*
Streuli, M., 188, *189*
Strickland, C. D., 193, 206, *207,* 246, *250,* 270, 271, *274*
Strickland, K. P., 271, *275*
Strickland, S., 10, 14, 15, *18, 19*
Striegel, J. E., 416, *419*
Strom, T. M., 378, *392, 393*
Strome, S., 185, *190*
Strong, C. F., 136, *146*
Strong, D., 284, *294,* 442, 443, *497*
Strong, L. C., 528, *538*
Stroobant, P. W., 240, 244, *252*
Stroschein, S. L., 268, *275*
Strouboulis, J., 195, 196, 199, 200, *208*
Strubin, M., 321, *324*
Struhl, G., 136, *148,* 550, *563,* 577, *587*
Strunk, K. E., 263, 265, *272, 276*
Strunski, V., 506, *517*
Struwe, M., 130, 132, 140, *148,* 264, 265, *277*
Stuart, E. T., 398, *420,* 432, 473, 477, *498*
Studer, M., 117, 118, 119, *124, 125,* 427, 469, *485,* 545,
    562
Stühmer, T., 89, *106,* 450, 471, *498*
Stutz, A., 7, 10, *16, 18*
Su, B., 172, 173, *180*

Su, G., 312, *327*
Su, H., 346, 353, 357, *364*
Su, M. S., 87, *101*
Su, W. C., 286, *294*
Subbarao, V., 318, *328*
Subbarayan, V., 341, 346, 347, 352, *366, 368,* 469, *492,*
    545, *564*
Subramanian, S. V., 262, *277*
Sucov, H. M., 228, *229, 230,* 246, *249, 250,* 334, 341, 346,
    348, 352, *362, 363, 367, 368, 369,* 426, 469, *482, 489*
Suda, T., 271, *275*
Suda, Y., 83, 84, *105, 106,* 114, *125,* 451, 478, *496*
Sudbeck, P., 281, *293,* 381, *390, 391, 393*
Sudo, T., 442, *498*
Suehiro, A., 353, 354, 355, 356, *363*
Suematsu, S., 341, 352, *370*
Suffolk, R., 31, *34*
Suga, Y., 582, 583, *587*
Sugai, S., 581, *588*
Sugaya, K., 226, *232*
Sugi, Y., 335, 336, 337, *364, 368*
Sugihara, A., 531, *538*
Sugihara, M., 354, *370*
Sugimoto, A., 165, *180,* 316, *330*
Sugimoto, K., 189, *189*
Sugimoto, T., 506, *516,* 549, *564*
Sugimura, K., 473, *498*
Sugino, H., 577, *586*
Sugitani, Y., 506, *516,* 549, *564*
Sugiyama, S., 112, 114, 115, *123*
Sugiyama, T., 471, 479, *487*
Suh, H., 353, 354, 355, *361,* 433, 478, *485*
Suh, Y., 243, *249,* 354, *365*
Suhonen, J. A., 241, *252*
Sukhatme, V. P., 400, *418*
Sulik, K. K., 38, *52,* 341, 352, *362,* 468, *497*
Sullivan, A., 220, *228,* 246, *251*
Sullivan, M. J., 398, *419*
Sumida, H., 355, *362*
Summerton, J., 14, *18*
Sumoy, L., 579, *584*
Sun, B. I., 470, *489*
Sun, E. S., 136, 137, *147*
Sun, L., 204, *210,* 217, 220, 229, *588*
Sun, Q.-Y., 8, *18*
Sun, T., 11, 12, *17*
Sun, T.-T., 570, *585, 587, 589*
Sun, X., 31, *35,* 337, 340, *368,* 466, *491, 496*
Sun, Y. H., 91, *103*
Sunada, Y., 161, *179*
Sundberg, J. P., 226, *230,* 423, *487*
Sundin, O., 346, *368,* 522, 530, 532, *536, 538*
Sunthornthepvarakul, T., 311, *329*
Suoqiang, Z., 558, *564*

Supp, D. M., 352, 354, 355, *364, 368*
Surani, M. A., 158, *176,* 258, 264, 266, *274*
Surani, M. A. H., 26, *35*
Suri, C., 223, 224, 227, *231, 232*
Susasurya, L., 185, *190*
Sussel, L., 80, 81, 82, 83, 88, 89, 90, 91, *98, 103, 105,* 321, *328, 329*
Sussman, D. J., 45, *52*
Sutcliff, K. S., 312, *329*
Sutcliffe, J. G., 511, *514*
Sutcliffe, M. J., 377, 386, *389*
Suter, U., 236, *250*
Sutherland, D. J., 226, *231,* 466, *485*
Sutherland, H. F., 353, 357, *364*
Sutter, C. H., 225, *231*
Suvanto, P., 403, *419*
Suzuki, A., 336, 352, *365*
Suzuki, H., 23, *35,* 432, 450, 465, *489*
Suzuki, M., 478, *488,* 506, *516,* 549, *564*
Suzuki, N., 183, *190,* 305, *329,* 453, 467, *496*
Suzuki, R., 353, 354, 355, 356, *363,* 525, *538*
Suzuki, S. T., 91, *101*
Suzuki, T., 226, *232*
Svensson, E. C., 341, 343, 352, *368*
Svensson, K., 167, *178*
Svensson, M., 574, 583, *584*
Svoboda, M. E., 532, *535*
Svoboda, P., 14, *18*
Swain, A., 372, 378, 379, 380, 384, 387, *390, 391, 392, 393*
Swalla, B. J., 471, *493*
Swann, K., 9, *18*
Swanson, B. J., 170, *176*
Swanson, G. J., 547, 549, 553, *564, 565*
Swanson, L. W., 89, *97, 100,* 502, 506, 509, 511, *514, 515, 516, 518*
Swaroop, A., 528, 531, 532, *536, 537, 538*
Swartz, M., 218, 223, *230*
Swartzendruber, D. C., 581, *589*
Swat, W., 203, *209*
Sweeney, T. M., 452, *492*
Sweeney, W. E., 415, *419*
Sweet, H. O., 94, *105*
Swerdlow, S. H., 204, *208,* 318, *327*
Swiatek, P. J., 136, 137, *148,* 546, 553, *565*
Swiergiel, J. J., 303, *329*
Swirnoff, A. H., 510, *516*
Swolin, B., 413, *418,* 467, *489*
Symbas, P., 312, *328*
Syrop, C. H., 22, *35*
Syu, L. J., 162, *176*
Szabo, J., 452, 453, *481*
Szabo, K., 504, *518*
Szendo, P. I., 336, *370*

Szeto, D. P., 353, 354, 355, *364,* 433, 478, *489, 496,* 504, 505, 506, 507, 508, 509, *514, 518*
Szollosi, D., 1, *4,* 8, *16*
Szonyi, E., 512, *516*
Szucsik, J. C., 79, 82, *101, 105*

T

Tabata, K., 581, 582, 583, *587*
Tabata, M. J., 441, *493*
Tabibzadeh, S., 47, *51*
Tabin, C. J., 3, *4,* 141, *146,* 262, 264, 265, 268, *274, 275,* 280, 285, 287, *293, 294,* 305, 308, 309, *328, 329,* 337, 338, 339, 343, 347, 354, 356, 357, *359, 364,* 399, *417,* 470, *487, 490, 494, 497,* 571, *586*
Tada, M., 303, *323, 329,* 337, *368*
Tadaka, N., 31, *34*
Tadic, T., 452, *484*
Tadolini, B., 338, *369*
Taga, T., 318, *325,* 341, 352, *370,* 405, 407, *416*
Tagher, P. H., 341, *358*
Tagishita, N., 312, *328*
Tagle, D. A., 247, *251*
Tago, K., 399, *419*
Tahta, S. A., 349, *369*
Taigen, T., 342, *365*
Tainsky, M., 571, 572, 580, *584*
Taipale, J., 218, 223, *229*
Taira, M., 83, *106,* 374, *390*
Tajbakhsh, S., 131, 140, 141, *144, 145, 148,* 258, 260, 262, 263, 264, 267, 268, *273, 274, 275, 277*
Tajima, N., 305, 307, *328*
Takabatake, T., 521, *537*
Takada, K., 289, *294*
Takada, R., 470, *490*
Takada, S., 40, 44, *51,* 132, 133, 135, 137, 141, *147, 148, 149,* 260, 264, 265, *275,* 423, 428, *487, 492*
Takaesu, N., 45, 47, *50,* 338, 355, *359,* 468, *483*
Takagi, A., 340, 352, *363, 367*
Takagi, C., 529, *537*
Takagi, S., 528, *536*
Takahara, K., 305, *329,* 453, 467, *496*
Takahashi, A., 10, *17*
Takahashi, E., 511, *516*
Takahashi, I., 282, *294,* 465, *492*
Takahashi, J., 236, 237, 241, 244, *251, 252*
Takahashi, K., 195, 196, *208, 209,* 282, *294,* 465, *492*
Takahashi, M., 241, 244, *252,* 402, *420,* 581, *589*
Takahashi, N., 271, *275,* 525, *536, 537*
Takahashi, S., 338, *368,* 511, *515,* 527, 530, 531, *536*
Takahashi, T., 87, 92, *105,* 220, 228, 521, 529, 534, *537*
Takahashi, Y., 10, *17,* 308, *329,* 571, *586*
Takai, S., 352, 354, *365*
Takakura, N., 173, *178*

Takakusu, A., 375, 376, 380, *390, 392*
Takano, H., 220, *231*
Takano, K., 338, *368*
Takano, T., 308, *329*
Takano, Y., 290, *294,* 437, 450, 451, 466, 474, *495*
Takaoka, K., 555, *565*
Takasaka, T., 506, *516,* 549, *564*
Takasashi, Y., 142, *149*
Takashima, S., 223, *232*
Takauji, R., 90, *105*
Takayama, K., 376, 380, *390*
Takayama, S., 308, *324*
Takebayashi, K., 555, *565*
Takebayashi-Suzuki, K., 334, *368*
Takeda, H., 41, 47, *51,* 78, *104,* 532, 534, *536*
Takeda, J., 581, 583, *589*
Takeda, K., 452, *496,* 531, *538,* 580, 581, 583, *589*
Takeda, M., 466, *496*
Takeda, N., 7, *16,* 46, *51,* 83, *106,* 114, *124,* 451, 478, *488,*
        490, 521, 527, *537*
Takeda, S., 38, *51,* 289, *294*
Takeichi, M., 226, *232,* 340, *366,* 408, 409, *417, 418, 419,*
        528, *536, 538*
Takemura, T., 555, *565*
Takenawa, J., 189, *189*
Takeshima, K., 521, *537*
Taketo, M. M., 135, 137, 138, *148,* 172, 173, *176,* 512, *518*
Taketo, T., 384, *393*
Takeuchi, J. K., 525, *536*
Takeuchi, O., 452, *496,* 580, 581, 583, *589*
Takeuchi, S., 511, *515*
Takeuchi, T., 317, *327*
Takiguchi-Hayashi, K., 84, *98, 100*
Takimoto, E., 336, 343, *365, 368*
Takke, C., 138, *148*
Takuma, N., 504, 505, 507, 508, 510, *518*
Talansky, B. E., 23, 27, *34, 35*
Talarico, D., 162, *179*
Talbot, W. S., 45, 47, *50, 51,* 78, 85, *100, 101,* 338, 353,
        354, 356, 357, *361, 370,* 505, *515,* 534, *536*
Taljedal, I. B., 322, *324*
Talmadge, R. J., 268, 269, *277*
Talreja, D. R., 553, *563*
Tam, P. L., 333, 335, *368*
Tam, P. P., 37, 38, 39, 40, 41, 45, 46, 47, 49, *50, 51, 52,*
        129, 142, *146,* 281, *293,* 333, *366,* 380, 381, *389, 392,*
        397, *420,* 422, 424, 425, 427, 428, *482, 491, 496*
Tam, P. P. L., 2, 3, *4,* 30, 31, *35,* 40, 43, 44, *49, 50,* 128,
        130, 131, 132, 133, 135, 139, *146, 147, 148,* 182, 183,
        190
Tamaddon, H., 334, 348, *367*
Tamai, K. T., 379, 384, *391*
Tamai, S., 506, *518*
Tamai, Y., 172, 173, *176*

Tamamaki, N., 90, *105*
Tamara-Lis, W., 166, *178*
Tamemoto, H., 7, *16*
Tamir, R., 402, *417*
Tampanaru-Sarmesiu, A., 512, *513*
Tamura, K., 354, *367*
Tan, S. S., 47, *52,* 132, 134, *148,* 310, *327,* 424, 425, 428,
        437, *496*
Tan, X., 318, *328*
Tanabe, Y., 43, *52,* 78, *104,* 120, *125,* 141, *148*
Tanaka, A., 466, *496*
Tanaka, H., 116, *123*
Tanaka, K., 479, *491*
Tanaka, M., 168, *178,* 219, *232,* 341, 343, 345, 348, 349,
        350, 351, 352, *359, 368, 369*
Tanaka, S., 23, *35,* 166, 167, *179, 180,* 357, *368*
Tanaka, T., 11, 12, *17,* 341, 352, *370,* 531, *538,* 581, *588*
Tanaka, Y., 38, *51,* 121, *123*
Tanapat, P., 237, 247, 248, *250*
Tanegashima, K., 338, *368*
Taneja, M., 322, *328*
Tang, C., 217, 220, *229*
Tang, H., 135, *147*
Tang, J., 512, *514*
Tang, K., 509, *518*
Tang, M. J., 403, *420*
Tang, T. K., 7, *19*
Tang, T. S., 8, *18*
Tang, Z., 165, *177,* 290, *294,* 303, *327,* 342, *366,* 428, 437,
        450, 451, 452, 466, 474, *489, 494, 495*
Tani, H., 570, *589*
Tani, K., 220, *231*
Tani, Y., 204, *209,* 220, *232*
Taniguchi, M., 471, 479, *487*
Taniguchi, S., 423, 424, 451, 475, 476, 478, *490*
Tannahill, D., 225, *229*
Tanne, K., 443, 453, *487*
Tanswell, A. K., 313, *329*
Tao, H., 47, *50*
Tao, Q. H., 579, *589*
Tao, W., 78, 81, 84, 86, *100, 105, 106,* 165, *177,* 243, *252,*
        479, *498,* 525, *536*
Tapia-Ramirez, J., 243, *249*
Tapscott, S. J., 169, *177,* 257, 258, 260, 261, 262, 264, 265,
        267, 268, 269, *272, 274, 275, 277,* 356, *359*
Tara, D., *361*
Tarcsa, E., 570, *584*
Tarkowski, A. K., 22, *35*
Tarle, S. A., 202, *208*
Tarpley, J. E., 312, *327*
Tarutani, M., 581, 583, *589*
Tashiro, K., 350, *362*
Tassabehji, M., 525, *538*
Tassi, V., 311, *326*

Tata, F., 305, 307, *323*
Tatla, T., 467, *484*
Taub, M., 415, *420*
Taub, R., 271, *276*, 318, *324*
Tavazoie, S. F., 94, *105*
Tavernarakis, N., 15, *18*
Tavian, M., 194, 200, 201, *209*, 220, *232*
Tay, J., 7, 11, 14, *19*
Taya, C., 87, *101*, 352, 354, *365*
Taya, Y., 434, *496*
Taylor, B. A., 311, *330*, 424, 442, 443, *493*, 527, *535*
Taylor, D. G., 226, *230*
Taylor, G., 570, *589*
Taylor, J. M., 268, 269, *274*
Taylor, M. V., 262, *275*
Taylor, N. R., 432, *498*
Tchervenkov, C. I., 349, *369*
Teboul, L., 271, *277*
Teesalu, T., 162, *179*
Tefft, D., 311, *329*
Tefft, J. D., 314, 315, *329*
Teichmann, G., 228, *232*
Teichmann, U., 41, *51*, 427, *488*
Teillet, M. A., 39, 40, *50*, 140, *147, 148, 250*, 263, 264, *276, 277*, 320, *326*
Teitelbaum, S., 223, *230*
Teitelman, G., 320, *323*
Teixeira, J., 385, *393*
Telford, N. A., 5, 9, *19*
Telvi, L., 376, *391*
Temeles, G. L., 12, *19*
Temme, A., 173, *175*
Temple, S., 235, 236, 237, 245, *249, 252*
Tempst, P., 405, 407, *416, 417*
ten Berge, D., 451, 476, 477, *496*, 549, *565*
ten Cate, A. R., 437, 439, *496*
ten Dijke, P., 338, *363*, 468, *486*
ten Dujike, P., 226, *230*
Tenenhaus, C., 243, *252*
Ten Have-Opbroek, A. A. W., 312, *329*
Tennenbaum, T., 169, *179, 589*
Tepass, U., 136, *146*, 205, *209*
Terada, M., 581, *588*
Terada, N., 452, *496*, 531, *538*, 580, 581, 583, *589*
Terasaki, M., 8, *17*
Terhorst, C., 195, *210*
te Riele, H., 376, *393*
Ternet, M., 313, *328*
Terrett, J. A., 357, *364*
Tessarollo, L., 7, *19*, 173, *179*, 559, *562*
Tessier-Lavigne, M., 85, *99, 101, 102*, 140, 141, 142, *145, 147*, 263, 264, 266, *274, 276*, 287, *292*
Testaz, S., 424, *496*

Tetsu, O., 86, *105*, 579, *589*
Tetzlaff, W., 86, *103*
Teuscher, C., 424, 442, 443, *493*
Tevosian, S. G., 219, *232*, 341, 343, 352, *369*
Texido, G., 86, *99*
Tezuka, T., 581, 583, *589*
Thal, L. J., 247, *252*
Thaler, J., 319, *325*
Thaller, C., 469, 470, *486*
Thangada, S., 227, *230*, 318, *328*
Thayer, M. J., 257, 258, 267, *277*
Theele, D. P., 206, *209*
Theil, T., 83, *105*, 119, *125*
Theiler, J., 568, 570, 571, 573, 574, 579, 581, 582, 583, *589*
Theiler, K., 128, 131, 137, 140, *149*, 541, 542, *565*
Theill, L. E., 504, 509, 510, *513, 514, 516*
Thelen, H., 165, *178*, 316, *328*
Thelu, J., *585*
Thepot, D., 165, *175*
Therkauf, W., 185, *190*
Therkidsen, B., 291, *292*
Thesleff, I., 422, 439, 441, 452, 456, *487, 488, 493, 494, 496, 497*, 578, *587*
Theveniau-Ruissy, M., 348, *360*
Thibaud, E., 375, *389*
Thiébaud, C. H., 425, *494*
Thiery, J. P., 201, *207*, 223, *230*, 315, *324*
Thiery, M., 22, *33*
Thise, B., 78, *105*
Thisse, B., 47, *50, 52*
Thisse, C., 47, *50, 52*, 78, *105*, 266, *276*
Thoma, B., 173, *179*
Thomas, B. L., 437, 451, *496*
Thomas, H. D., 378, *390*
Thomas, K. R., 117, *125*, 428, 471, *496*
Thomas, L. J., 442, *496*
Thomas, N., 441, *488*
Thomas, P., 31, *34*, 46, 47, 48, *51*, 311, 316, *327*
Thomas, P. Q., 30, 31, *33, 35*, 45, 46, 47, 48, *50, 52*, 304, 316, *329*, 338, *362*, 451, 476, *483*, 508, *514*
Thomas, T., 95, *97*, 137, *147*, 161, 173, *179*, 341, 346, 349, 350, 352, 355, *368, 369*, 451, 453, 465, 473, *495, 496*
Thomas, U., 136, *149*
Thomazy, V., 469, *484*
Thompson, A., 408, *419*
Thompson, D. A. W., 442, *496*
Thompson, E. M., 5, 9, 14, *15, 18, 19*
Thompson, K., 79, 86, *100*, 243, *250*
Thompson, L. M., 286, *294*, 453, *495*
Thompson, R. P., 341, 342, 357, *368, 369*
Thompson, S. R., 10, *19*
Thomson, J. A., 303, *329*
Thorey, I. S., 570, *589*
Thorn, L., 560, *565*

Thornell, A. P., 288, *294,* 442, 443, 453, *493*

Thorners, J., 45, *50*

Thorogood, P., 140, *146,* 201, *208,* 305, *328,* 421, 422, 423, 424, 426, 432, 435, 441, 442, 449, 457, 467, 468, *481, 482, 486, 487, 496*

Thorsteindottir, S., 131, 132, *149*

Thorsteinsdottir, U., 242, *252*

Thrasher, A. J., 201, *208*

Threadgill, D. W., 169, *179*

Threat, T. A., 11, 12, *17*

Thummel, C. S., 15, *17*

Thurberg, B. L., 343, 352, *360*

Thurston, G., 223, 224, 227, *232*

Ticho, B., 339, 341, 349, 355, *358, 370*

Tickle, C., 280, *294,* 435, 468, *483, 497,* 534, *536,* 577, 578, 579, *587*

Tierney, C., 521, *536*

Tilghman, S. M., 168, *176*

Tilmann, C., 373, 383, *391, 393*

Tilney, L. G., 557, *565*

Tilney, M. S., 557, *565*

Timmons, P. M., 183, *190,* 475, *496*

Timpl, R., 574, 583, *584*

Timpl, T., 411, *418*

Ting, S. A., 571, *585*

Ting-Berreth, S., 577, 578, 579, *585, 587, 589*

Tiveron, M. C., 239, *250*

Tiveron, M.-C., 239, 242, 243, *251*

Tiziani, V., 292, *294*

Tochinai, S., 198, *208*

Todaro, G. J., 465, *484*

Toftgard, R., 287, *292,* 470, *485, 486, 497*

Tohyama, M., 479, *491*

Tokimasa, C., 443, 453, *487*

Tokooya, K., 353, 357, *365*

Toksoz, D., 188, *190*

Tokunaga, H., 413, *418*

Tole, S., 82, 83, 84, 86, 87, *100, 101, 105*

Toledo, S. P., 509, *517*

Toledo-Aral, J. J., 243, *249*

Toles, J. F., 202, *209*

Tolon, R. M., 510, *516*

Tomac, A. C., 432, *498*

Toman, P. D., 573, *588*

Tomarev, S., 268, *274,* 399, *417,* 532, *538*

Tomasiewicz, H., 228, *229*

Tomihoara-Newberger, C., 2, *4*

Tominaga, K., 204, *209,* 220, *232*

Tominaga, S. I., 399, *419*

Tomita, K., 413, *418,* 526, 527, *538*

Tomsa, J. M., 478, *496*

Ton, C. C., 451, 475, *486,* 527, 528, 530, *536, 538*

Tonegawa, A., 142, *149*

Tong, C. X., 115, *123*

Tonini, G., 378, 387, *389*

Tonissen, K. F., 339, *369*

Toole, L., 433, *485*

Topilko, P., 510, *518,* 546, *564*

Topley, G., 580, *585*

Topouzis, S., 342, *359*

Torban, E., 379, *391*

Torchard, D., 382, *392*

Toresson, H., 47, *50,* 82, 83, 91, *105,* 120, *125,* 451, 476, *483,* 508, *514*

Torii, M., 95, *105*

Törnell, J., 467, *482*

Torres, M., 91, *102,* 321, *329,* 373, 374, 385, *393,* 398, *420,* 422, 432, 451, 473, 475, 477, *490, 495, 496, 498,* 542, 543, 544, 546, 550, *561, 565*

Toshikawa, Y., 133, *149*

Tosi, M., 386, *390*

Toso, R. D., 511, *516*

Totorice, C. G., 171, *175*

Touchman, J. W., 556, *564*

Tournier-Lasserve, E., 227, *230*

Towbin, J., 357, *359, 363*

Towers, N., 262, *275,* 336, 339, 342, 355, *368*

Toy, J., 522, *538*

Toyama, J., 335, *358*

Toyama, R., 78, *103,* 374, *390*

Toyoda, Y., 47, *51*

Tozawa, Y., 224, *232*

Traber, P., 310, *325*

Traill, T. A., 357, *358*

Trainor, P. A., 119, *124,* 132, 135, *148,* 424, 425, 426, 427, 428, 453, *485, 490, 496,* 544, *563*

Tran, B., 433, 476, 477, 478, *489*

Tran, C. M., 341, *369*

Tran, T., 442, *484*

Trask, B. J., 347, *364*

Trasler, J. M., 382, *393*

Traub, O., 173, *175, 177*

Trazami, S., 342, *362*

Treacy, M. N., 89, *100*

Treanor, J. J. S., 402, *420*

Treier, M., 85, *105,* 502, 504, 505, 506, 509, *514, 518*

Treisman, R., 342, *362, 369*

Tremaine, R., 432, *486*

Tremblay, G. B., 167, *179,* 478, *489*

Tremblay, J. J., 478, *481,* 508, 510, *513, 518*

Tremblay, K. D., 303, *329*

Tremblay, M., 171, 173, *176,* 312, *323,* 510, *514*

Tremblay, P., 258, 263, 264, 266, *272, 277,* 475, *482*

Trembleau, A., 510, *518*

Tremml, G., *102,* 121, *124,* 423, 489, 508, *517*

Tribioli, C., 451, 453, 474, *497*

Trinh, L. A., 303, *326*

Trommsdorff, M., 94, *100, 105*

Trounson, A., 22, *34*

Truck, T., 344, 345, 348, *365*

Trumpp, A., 86, *105*, 422, 432, 449, 450, 453, 466, 475, *497*

Truong, V. B., 400, *418*

Trupp, M., 402, *420*

Truslove, G. M., 453, *497*

Tryggvason, K., 412, *417, 418, 419*

Trzaskos, J. M., 156, *177*

Tsai, E., 443, *482*

Tsai, F.-Y., 203, *209*, 510, *518*

Tsai, L. H., 94, *98, 101*

Tsai, M. J., 92, *102, 106*, 310, 321, 322, *327*, 334, 343, 345, 346, 352, *366*, 559, *563*

Tsai, M.J., 92, *104*

Tsai, S. F., 204, *209*

Tsai, S. Y., 92, *106*, 334, 343, 345, 346, 352, *366*

Tsakiris, D. A., 223, *230*

Tsang, A. P., 343, *369*

Tsang, S. H., 505, *517*

Tsang, T. E., 3, *4*, 38, 43, *49, 51*, 132, *147*, 397, *420*

Tschudi, C., 15, *18*

Tseng, T.-C., 7, *19*

Tseu, I., 313, *329*

Tsuboi, H., 379, *391*

Tsuchida, T. N., 109, *123*

Tsuchiya, K., 352, 354, *365*

Tsuchiya, R., 376, *391*

Tsuda, T., 356, *369*

Tsui, L. C., 121, *124*, 451, 453, 470, 471, *491*, 528, *536*

Tsujimoto, Y., 136, *146*

Tsujimura, T., 452, *496*, 531, *538*, 580, 581, 583, *589*

Tsukiyama, T., 376, *392*

Tsukui, T., 354, *367*

Tsuruo, Y., 79, *99*

Tu, C. L., 580, 583, *588*

Tuan, R. S., 280, *292*

Tucker, A. S., 432, 433, 434, 437, 438, 439, 440, 441, 450, 451, 467, 478, *484, 485, 496, 497*

Tucker, D. C., 412, *417*

Tucker, P. K., 377, 387, *393*, 526, 527, *535*

Tucker, R. W., *587*

Tucker, S. C., 451, 453, 476, *493*

Tuerk, E. E., 170, 173, *179*

Tufro, A., 412, *420*

Tufro-McReddie, A., 412, *420*

Tufts, R., 341, 343, 352, *368*

Tuil, D., 342, *368*

Tully, T., *251*

Tumas, D. B., 244, *252*

Tuorto, F., 48, *49*, 84, *97*, 114, *122*, 478, *481, 491*, 508, *513*, 549, 550, 551, 552, *564*

Turecková, J., 455, *489*

Turgeon, B., 509, *517*

Turley, H., 415, *420*

Turnball, D., 110, *123*

Turnbull, D. H., 78, *99*, 109, 110, *123, 124, 125*, 228, *230*

Turner, B. M., 9, *19*

Turner, D. C., 134, *147*, 257, *277*, 334, 344, 350, 351, 357, *365*

Turner, D. L., 527, *538*

Turner, E. E., 237, *250*

Turner, P. A., 89, *101*

Turnpenny, L., 352, 354, *366*

Turoto, F., 478, *481*

Turpen, J. B., 196, 198, 203, *208, 210*

Tuschl, T., 14, *19*

Tybulewicz, V. L., 285, *293*, 442, 443, *488*

Tyler, M. S., 432, *497*

Tymowska-Lalanne, Z., *390*

Tyrell, J. B., 512, *515*

Tzahor, E., 334, 335, 337, 339, *369*

Tzartos, S., 271, *277*

Tzelepis, D., 194, *208*

U

Uchida, N., 192, 194, *207*, 241, 243, 244, 245, *250, 251, 252*, 408, *418*

Uchida, T., 527, 530, *535*

Uchida, Y., 582, *584*

Uchikawa, M., 283, *293*, 529, 530, 531, 532, 534, *536, 538, 565*

Uchiyama, M., 290, *294*, 437, 450, 451, 466, 474, *495*

Udy, G. B., 7, *16*

Ueda, E., 581, 582, 583, *587*

Uehara, Y., 173, *179*, 404, *420*

Ueki, Y., 292, *294*

Uemura, M., 473, *498*

Ueno, F., 471, 479, *487*

Ueno, K., 473, *498*

Ueno, N., 336, 352, *365*

Ueno, T., 408, *418*

Ueta, C., 289, *294*

Uilenbroeck, J. T., 385, *389*

Uitto, J., 573, *589*

Ullman-Cullere, M., 223, *230*

Ullrich, A., 217, 220, 227, *229, 231*

Ullu, E., 15, *18*

Umeda, S., 195, *208*

Umemoto, M., 555, *565*

Umesono, K., 525, *536*

Umezu, H., 195, *208*

Unabia, G., 512, *514*

Unden, A. B., 287, *292*, 470, *486, 497*, 579, *586*

Uno, K., 525, *536*

Unsicker, K., 314, *324*

Upadhyaya, C., 171, 173, *178*

Uratani, Y., 84, *98, 100*
Urbanek, P., 115, 118, *125,* 428, 475, *497*
Urist, M. R., 271, *277*
Urness, L. D., 226, *230, 232*
Urrutia, I., 506, *514*
Urushihara, H., 226, *232*
Ushio, Y., 91, *101*
Usman, M. I., 318, *328*
Usuda, H., 195, *208*
Utset, M. F., 174, *177*
Uyttendaele, H., 226, *232*

V

Vaage, J., 428, 475, *497*
Vaahtokari, A., 439, 441, 456, *497*
Vaccarino, F. M., 95, *97,* 506, *513*
Vaessin, H., 257, *276*
Vaglio, A., 346, *358*
Vaidya, D. M., 334, 348, *367*
Vaigot, P., 201, *209*
Vainchenker, W., 194, *207*
Vainio, S., 336, *363, 384,* 385, *393,* 405, 406, 407, *418, 419,* 439, *497*
Valayer, J., 376, *391*
Valdez, M. R., 260, 261, 262, *276, 277*
Vale, W. W., 286, *292*
Valencia, A., 345, 348, *360*
Valenzuela, D. M., 292, *292*
Valerius, M. T., 79, 95, *101, 105,* 507, *516*
Valladares, A., 171, 173, *174*
Vallotton, M. B., 506, *517*
Valverde, F., 84, 90, 95, *99, 102, 105*
Valverde, J., 91, *98*
van Bebber, F., 354, 355, *359*
Van Beneden, E., 31, *35*
Van Blerkom, J., 7, *15, 19, 33, 33*
Van Den Berghe, H., 22, *33*
van den Eijnden-van Raaij, J., 47, *50,* 467, *484*
van den Heuvel, M., 470, *497*
Vandenhoeck, A., 203, *207,* 223, *228*
van den Hoff, M. J. B., 343, 345, 348, *366*
van der Harten, H. J., 285, *293*
van der Kooy, D., 94, *101,* 193, 204, *208,* 235, 236, 244, 245, *251*
Van der Loo, J. C. M., 192, *210*
van der Lugt, N. M., 376, *389, 390, 393,* 453, *481*
van der Meer, R., 453, 476, *493*
van der Neut, R., 572, 583, *585, 589*
VanderPlas-de Vries, I., 333, *360*
van der Schoot, P. J., 385, *389*
van der Valk, M., 376, *393*
van der Veen, C., 574, *588*
van der Zee, R., 220, *228*
van Deursen, J., 201, 204, *209,* 220, *231*

van de Velde, H., 313, *323*
Vande Vijver, V., 511, *517*
Van De Water, T., 474, *497,* 542, 549, 550, 553, 555, 559, *562, 563, 565, 566*
Vande Woude, G. F., 7, *18*
Vandlen, R., 402, *420*
Van Doren, M., 185, 186, 188, *190*
van Eeden, F. J., 546, *565*
van Eeden, J. M., 351, 355, *359*
van Ewijk, W., 195, *210*
Van Exan, R. J., 432, *486,* 573, 576, *589*
van Genderen, C., 440, *497,* 583, *589*
Vanhaelst, L., 502, *515*
Van Hateren, N., *390*
van Helmond, M. J., 385, *389*
van Heyningen, V., 88, *99,* 375, *391,* 399, *419, 420,* 451, 475, *486,* 527, 528, 530, *536, 537, 538*
van Hinsbergh, V. W., 218, *230*
Vankelecom, H., 511, *517*
van Lijnschoten, G., 174, *179*
van Lohuizen, M., 376, *389, 390, 393,* 453, *481*
Van Maldergem, L., 281, *293,* 381, *391*
van Mierop, L. H. S., 343, 350, *369*
van Oostrom, C. G., 425, 435, *497*
Van Regemorter, N., 398, *416,* 550, *561*
van Rooijen, M. A., 222, *229*
van Roon, M., 376, *393*
van Schaick, H., 506, *513*
van Schooten, R., 285, *293*
Vansomphone, D., 581, *586*
Van Swearingen, J., 269, *277*
Van Vliet, G., 512, *514*
Van Voorhis, B. J., 22, *35*
van Wijk, I., 174, *174*
van Wyk, J. J., 532, *535*
VanZant, G., 245, *249*
Varfolomeev, E. E., 341, *369*
Vario, G., 343, *369*
Varlet, I., 43, 44, 45, 46, *50,* 354, 355, 357, *359,* 468, *497*
Varmuza, S., 13, *17,* 22, 23, *34, 35,* 169, *179*
Varnum, D. S., 137, *149,* 532, *538*
Vasa, S. R., 318, *324*
Vasavada, H., 312, *323*
Vasicek, R., 174, *177*
Vasilyev, V. V., 509, *517*
Vass, W. C., 171, 173, *178*
Vassalli, A., 10, 14, *18, 19,* 467, *490*
Vassalli, J. D., 10, 14, *18, 19,* 321, *328*
Vassileva, G., 121, *123,* 132, 137, *148,* 407, *419*
Vassiliauskas, D., 138, *148*
Vassilli, J. D., 7, *16*
Vässin, H., 136, *149*
Vastrik, I., 580, *589*
Vaz, C., 318, *327*
Vazquez, E., 559, *564*

Vazquez, M. E., 91, *105*
Veeranna, 94, *103*
Vega, Q. C., 402, *420*
Veile, R., 321, *325*
Veitch, E., 433, *497*
Veitia, R., 382, *392*
Vekemans, M., 79, *99*, 475, *484*
Velasquez, T., 260, 266, *274*
Veldhuisen, B., 414, *418*
Velinzon, K., *251*
Velkeniers, B., 502, *515*
Vendrell, V., 543, 545, 546, *565*
Veniant, M. M., 161, *178*
Vennstrom, B., 510, *515*
Venters, S. J., 131, 132, *149*
Ventura, C., 337, 338, *369*
Venuti, J. M., 260, *274, 277*
Verano, N. R., 354, *365*
Verbeek, F. J., 345, 356, *369*
Verbout, A. J., 131, 134, *149*
Veress, B., 246, *249*, 318, *324*
Verkerk, A., 343, *369*
Verlhac, M.-H., 7, *18*
Verma, I. M., 258, *277*, 580, 581, 583, *587*
Verma, R., 412, *417*
Vermeulen, J. L. M., 343, 345, 356, *366, 369*
Vermot, J., 341, 346, 347, 352, *366*, 469, *492*, 545, *564*
Verrotti, A. C., 10, *19*
Verschueren, K., 226, 227, *229*, 336, 352, *359*
Verwoerd, C. D., 425, 435, *497*
Vescovi, A. L., 206, *206*, 235, 242, 244, 246, *249, 250*
Vesque, C., 78, 81, 91, *99, 100*, 118, *124*, 468, *492*
Vestlund, K. N., 512, *514*
Vestweber, D., 226, *228*, 410, *420*
Vetter, M. L., 526, 527, *535*
Viallet, J. P., 571, 572, 574, *585, 587*
Vicaner, C., 379, *391*
Victor, J. C., 550, *562*
Vidgen, D., 527, *535*
Viebahn, C., 26, 31, *35*, 41, 46, *51*, 354, 355, *359*, 427, *488*
Vielmetter, E., 355, 356, *361*
Vien, L., 352, 354, *366*
Vigano, M. A., 183, *190*
Viger, R. S., 382, *393*
Vigh, S., 504, *516*
Vigneron, J., 398, *416*, 550, *561*
Vigneron, P., 271, *273*
Vijh, M., 157, *178*
Vikkula, M., 227, *232*
Vilar, J., 403, *418, 420*
Villar, J. M., 337, *359*
Villavicencio, L. G., 345, 348, *360*
Vincent, A., 91, *100*
Vincent, C., 141, *147*, 398, *416*, 550, *561*
Vincent, E. B., 342, *367*

Vincent, M., 9, *15*
Vinós, J., 466, *482*
Vintersten, K., 171, 173, *174*
Viragh, S., 334, 341, 345, 348, 356, *360, 363, 369*
Viragh, S. Z., 348, 351, *369*
Viriot, L., 455, *489*
Visconti, P. E., 8, *17*
Visser, P., 199, 200, 202, *207*
Visvader, J. E., 203, *210*
Vitelli, F., 353, 357, *364*
Vivarelli, E., 140, 141, *145, 148, 149*, 258, 263, 264, *273, 277*
Vivian, J. L., 260, *277*
Vivian, N., 377, 378, 387, *390, 391*
Vize, P. D., 395, *420*
Vlietnck, R., 22, *33*
Vogan, K. J., 266, *277*, 338, 354, 356, 357, *359*
Vogel, H., 468, *491*, 571, 581, 583, *587*
Vogel, K. S., 237, *252*, 341, 352, *359*, 435, *497*
Vogel, W., 171, *175*, 388, *391*
Vogelstein, B., 86, *100*, 579, *586, 587*
Vogt, T. F., 451, 476, *493*
Volk, T., 340, *358*
Volkel, V., 10, *18*
Vollrath, M. A., 555, 558, 559, *561*
von Bartheld, C. S., 558, 559, *565*
Voncken, J. W., 314, *325*, 434, 467, *487*
von Ebner, C., 129, *149*
Vonesch, J. L., 109, *122*, 443, 451, 454, 455, 468, 469, 474, *489, 490, 493*, 545, *563*
von Kessler, D. P., 470, *489*
Von Ohlen, T., 470, *481*
von Schack, D., 141, *144*
Vorbusch, B., 336, *358*
Vorechovsky, I., 287, *292*, 470, *485, 486, 497*
Voronova, A., 257, *275*
Vortkamp, A., 265, *275*, 285, 287, *293, 294*, 442, 443, 470, *489, 490, 497*
Voss, A. K., 95, *97*, 137, *147*, 173, *179*, 453, *496*
Voss, J. W., 502, *518*
Vrablic, T., 38, *52*
Vrancken Peeters, M. P., 219, *232*
Vriend, G., 160, *175*
Vu, T. H., 223, *232*, 281, 290, 291, *292, 294*
Vukicevic, S., 408, 416, *420*
Vuorela, P., 162, *179*
Vuorio, E., 424, 442, 443, *494*, 572, 573, *588, 589*

W

Wada, H., 266, *277*
Wada, R., 227, *231*
Waddington, C. H., 542, *565*
Wagenaar, G. T. M., 343, 345, 348, *366*
Wagner, D. S., 527, *535*

Wagner, E. F., 87, *104*, 169, 173, *179*, 318, *325*, 442, 443, *485*, *497*, 525, *537*
Wagner, G. C., 91, *106*, 439, *497*
Wagner, J., 173, *176*, 510, *514*
Wagner, M. A., 525, *537*, 555, 557, *563*
Wagner, T., 281, *293*, *294*, 380, 381, *391*, *393*
Wagner, T. E., 509, *518*
Waid, D. K., 88, *105*
Wainer, B. H., 91, *105*
Wainwright, B., 172, 173, *177*, 287, 292, 336, 337, *365*, 375, *391*, 470, *485*, *486*
Wai-Sum, O., *393*
Wajnrajch, M. P., 443, *497*
Wakabayashi, K., 321, *327*, 512, *518*
Wakamatsu, Y., 237, *252*, 318, *328*
Wakamiya, M., 3, *4*, 31, *34*, 43, 44, 45, 46, 47, 48, *49*, *51*, *52*, *104*, 328, 422, 427, 477, *482*, *494*
Wakeham, A., 136, 137, *147*, 341, *370*, 468, *495*, 509, *514*
Wakkach, A., 271, *277*
Waknitz, M. A., 303, *329*
Waldrip, W. R., 31, *35*, 468, *497*
Waldron, S., 340, 341, *363*
Wales, J. K., 31, *33*, 47, *50*, 451, 476, *483*, 508, *514*
Walker, A. P., 378, *392*, *393*
Walker, M., 471, *493*
Wall, C., 195, *209*
Wall, N. A., 470, *497*
Wallace, V. A., 86, *101*, *105*
Wallach, D., 341, *369*
Waller, R. R., 357, *369*
Wallin, J., 140, *147*, 475, *496*
Wallingford, J. B., 395, *420*
Walmsley, M., 198, 200, *207*, 343, *361*
Walsh, C., 86, 94, *97*, *100*, *105*, 235, 243, 244, *251*, *252*
Walsh, E. C., 115, *125*, 337, *367*
Walsh, F. S., 425, 428, 432, *484*, *490*
Walsh, J., 2, *4*, 556, *565*
Walsh, P., 375, *391*
Walter, A., 83, *105*
Walter, J., 9, *17*, *18*
Walter, M. A., *390*, 453, 479, *488*
Walterhouse, D., 470, *497*
Walther, B. T., 319, *329*
Walther, C., 83, 85, 91, *105*, 475, *487*, 530, *538*
Walton, D. S., 528, *535*
Walz, G., 415, *418*, 579, *585*
Wan, J. W., 12, *17*
Wanaka, A., 479, *491*
Wandycz, A. M., 243, 245, *251*
Wang, A., 556, *564*
Wang, B., 121, *125*, 195, *210*
Wang, C.-C., 40, 44, *50*, *323*, 338, 345, 348, *358*, *366*
Wang, D., 410, *418*
Wang, D.-Z., 343, 353, *364*

Wang, E. A., *105*, 467, *498*
Wang, F., 90, 92, 95, *98*
Wang, G. F., 335, 343, 344, 345, 346, 347, *369*, *370*
Wang, H., 40, *52*, 218, 225, *229*, *232*, 338, *368*, 450, 471, 472, 473, *498*
Wang, J., 204, *210*, 313, *329*, 509, *515*
Wang, L., 342, *369*
Wang, M. Z., *105*
Wang, Q., 10, *19*, 201, 204, *209*, *210*, 220, *231*, *232*, *233*, 453, *492*
Wang, S. L., 15, *18*, 247, *252*, 260, *276*, 380, *390*
Wang, S. W., 527, *536*
Wang, W., 136, *148*, 268, *275*, 280, 290, *292*, *294*, 341, 352, *370*, 474, *497*, 549, *565*
Wang, X., 11, 12, *17*, 44, *50*, 217, 220, *229*, 246, *250*, 318, *326*, 338, *362*, 381, 382, *393*, 469, *497*
Wang, X.-J., 571, 581, 583, *586*, *587*
Wang, X. Y., 549, *562*
Wang, Y., 262, *273*, 318, *326*, 476, *483*, 500, *516*, *588*
Wang, Z., 504, 510, *515*
Wang, Z. Q., 173, *179*, 442, 443, *485*, *497*
Wanke, E., 235, 242, 244, *250*
Wanninger, F., 570, *587*
Warburton, D., 311, 314, 315, *325*, *327*, *329*, *330*, 434, 467, *487*
Warchol, M. E., 553, 555, 557, *562*, *563*
Ward, C. J., 415, *420*
Ward, J. M., 94, *103*, 171, 173, *176*, *178*, 226, *230*, 311, 312, *326*, 474, *488*, 504, 505, 507, 508, 510, *518*, 525, *538*
Ward, S., 341, 346, 347, 352, *362*, 468, *491*, 545, *564*
Ward, T. A., 398, *419*
Wardle, F. C., 239, *249*
Ware, C. B., 173, *179*, 405, 407, *416*
Ware, M. L., 94, *105*
Waring, M. T., 547, *562*
Waring, P., 167, *174*, 305, *323*
Warkany, J., 346, *369*
Warman, M. L., 227, *232*, 442, 443, 452, 453, 474, *487*, *489*
Warot, X., 109, *122*, 310, *329*
Warren, A. J., 203, *210*, 342, 343, *360*
Warren, K. S., 351, 355, *359*
Warren, N., 85, 87, 94, *105*
Warren, P. D., 156, *175*
Warrior, R., 186, *190*
Warsowe, J., 12, *16*
Wartiovaara, J., 403, 412, *419*
Wartiovaara, K., 403, *419*
Washburn, L. L., 378, 382, *389*, *390*, *392*
Wasmuth, J. J., 286, *294*, 453, *495*
Wassarman, K. M., 8, 15, *17*, *19*, 84, 85, 95, *102*, *106*, 114, *125*, 453, *497*
Wassarman, P. M., 14, 15, *16*, *18*

Wassef, M., 83, 84, *98, 100, 104,* 111, 112, 113, 115, 117, *122, 125*
Wasser, M., 240, *250*
Wassif, C., 87, 91, *103,* 376, *389,* 508, *517,* 521, 527, *537*
Watanabe, H., 284, *294,* 442, 443, 453, *497*
Watanabe, K., 381, 382, *393,* 506, *516,* 549, *564*
Watanabe, M., 338, 341, 347, 354, 356, *359, 367, 368, 369*
Watanabe, N., 7, *16*
Watanabe, S., 292, *294*
Watanabe, T., 422, 424, 435, 437, 475, 476, *492*
Watanabe, Y., 140, 141, 142, *147,* 263, 264, 266, *276, 277*
Watanabe, Y. G., 511, *514*
Waterman, M. R., 375, 376, 380, 384, *390, 392*
Waterman, R. E., 433, *497*
Watkins-Chow, D. E., 85, *106,* 502, 511, 512, *518*
Watnick, T. J., 414, 415, *419, 420*
Watson, A. J., 5, 9, *19,* 156, *177*
Watson, M. S., 357, *358*
Watt, F. M., 242, *252,* 570, 572, 573, 580, 581, *585, 586, 587, 589*
Wattler, S., 160, 167, *178*
Wawersik, S., 532, *538*
Wawrousek, E. F., 528, 533, *535*
Weaver, M., 312, 313, 314, 315, *329*
Weaver, R. G., 398, *419*
Webb, J. F., 422, 434, 435, *497*
Webb, R., 342, *364*
Webb, S., 334, 348, 351, *369*
Webber, E. M., 318, *329*
Weber, H., 415, *420*
Weber, R. J., 26, 32, 35, 158, *179,* 348, 349, 352, 353, 357, *358*
Weber, T. J., 379, 384, *391*
Webster, M. K., 452, 453, 466, *485, 497*
Webster, W. S., 468, *497*
Wechsler-Reya, R. J., 86, *106,* 287, *294*
Wedden, S. E., 468, *497*
Wegner, J., 450, 470, 473, *481,* 546, *561*
Wegner, M., 170, 173, *179,* 238, *252,* 281, *294,* 502, 510, *514, 515*
Wehnert, M., 94, *103*
Wehr, R., 78, 79, 81, *100, 103,* 521, 533, *537*
Wei, L., 342, *369*
Wei, M. H., 227, *229*
Weich, N., 205, *206*
Weichenhan, D., 378, *390*
Weigel, D., 167, *179,* 316, *329*
Weigel, N. L., 379, *390*
Weil, D., 398, *416,* 550, *561*
Weil, M., 266, 272, 528, *538*
Weiler-Guettler, H., 219, *228*
Weimann, J. M., 94, *106*
Weinberg, R. A., 318, *325, 483*
Weiner, R. I., 504, *518*

Weinhold, B., 342, *358*
Weinmaster, G., 92, *103,* 136, 137, *148, 149,* 262, *276,* 553, 555, *563, 565*
Weinstein, B. M., 213, 222, *231, 232*
Weinstein, D. C., 164, 165, *175, 179,* 303, 316, *323, 329*
Weinstein, M., 79, 82, *101, 105,* 167, 171, 173, *179,* 226, 227, *233,* 313, *329*
Weintraub, B. D., 509, *514, 517*
Weintraub, H., 136, *147,* 169, *175,* 257, 258, 262, 263, 272, *273, 274, 275, 276, 277,* 349, *362*
Weir, E. C., 285, *294*
Weir, G. C., 322, *328*
Weiss, A., 170, *179*
Weiss, D. W., 428, 475, *497*
Weiss, J., 500, *518*
Weiss, K. M., 450, 471, *496*
Weiss, L. W., 568, *589*
Weiss, M., 203, *209,* 381, *390,* 510, *518*
Weiss, S., 86, 89, 91, *103, 104,* 236, 245, 248, *251, 252*
Weissenbach, J., 281, 292, 380, *390,* 398, *416,* 550, *561*
Weissman, I. L., 192, 194, 202, *207, 208, 209, 210,* 241, 243, 244, 245, 246, 248, *249, 250, 251, 252,* 318, *326*
Weizäcker, E., 134, *147*
Welker, P., *588*
Wellauer, P. K., 321, *324, 326*
Weller, D., 14, *18*
Weller, P. A., 281, *292,* 380, 381, *390, 391*
Wells, C., *390*
Wells, D., 11, *19*
Wells, J. M., 3, *4,* 304, 319, *329*
Welter, J. F., 581, *588*
Wen, D., 402, *417*
Wen, L., 521, *536*
Wendel, D. P., 226, *230*
Wendling, O., 109, *122,* 172, 173, *179,* 305, *329,* 341, 346, 347, 352, 357, *362, 369,* 545, *562*
Weng, Y., 227, *229*
Wenger, R. H., 225, *230*
Weninger, W. J., 335, 355, *365*
Wenink, A. C., 334, *369*
Wennerberg, K., 574, *589*
Wenstrom, K. D., 22, *35*
Wentworth, J. M., 311, *324*
Werb, Z., 163, 166, 168, 169, 170, *175, 177, 178, 179,* 223, *232, 233,* 281, 290, 291, 292, *294,* 314, *327, 329,* 349, *360,* 465, *491*
Weremowicz, S., 346, *358*
Werner, M. H., 377, *393*
Werner, S., 312, *328,* 574, 583, *584*
Werner-Linde, Y., 580, *586*
Wernig, A., 271, *275*
Wert, S. E., 312, 313, 314, *328, 329, 330*
Wertz, K., *393*
Wertz, P. W., 581, *589*

Wessels, A., 334, 342, 344, 345, 348, 350, 351, 356, 357, *360, 363, 365, 369, 370*
Wessels, N. K., 319, *329*
Wessels, R. B., 574, *589*
West, J. D., 23, *34*, 451, 475, *493*
Westerfield, M., 77, 78, *101*, 265, 269, *273*, 435, 450, 470, 473, 475, *481, 493, 498*, 546, *561*
Westerman, C. J., 227, *231*
Western, P. S., 379, 382, 387, 388, *393*
Westerveld, A., 174, *174*
Westlund, B., 242, *252*
Westoll, T. S., 442, *493*
Weston, A. P., 504, *513*
Weston, J. A., 237, 243, 244, *250, 252*
Westphal, H., 43, *50*, 78, 80, 83, 85, 86, 87, 91, 94, *98, 103, 104, 106*, 121, *123*, 141, *145*, 173, *176*, 260, 265, *273*, 312, 319, *323, 326*, 374, 376, *389, 390*, 402, *419*, 432, 439, 451, 470, *482, 498*, 500, 502, 504, 505, 507, 508, 509, 510, 512, *515, 517, 518*, 521, 527, *535, 537*
Wetsel, W. C., 511, *513*
Wharton, K. A., 136, *149*
Wheatley, M., 236, *252*
Wheatley, S., 281, *293*, 380, 381, 387, *389, 391, 392*
Wheeler, E. F., 558, 559, *565*
Whitcomb, R. W., 500, *518*
White, J. M., 8, *18*, 287, *293*, 424, *481*
White, M. C., 505, *513*
White, P. A., 241, *252*
White, P. M., 235, 236, 238, 239, 240, 241, 243, 244, 245, 248, *251*
White, R. A., 118, *123*
White, R. D., 355, *365*
White, R. I., Jr., 227, *229*
Whitelaw, E., 356, 357, *360, 369*
Whitesides, J., 84, *106*, 469, *497*
Whitfield, L. S., 377, 387, *393*
Whitfield, T. T., 546, *565*
Whiting, J., 422, 425, 427, *487*
Whitington, P. M., 185, *190*
Whitlock, K. E., 435, *498*
Whitman, M., 226, *233*, 338, 354, 356, *367, 368*, 479, *482*
Whitney, J. B. D., 378, *390*
Whitsett, J. A., 311, 312, 313, 314, *328, 329, 330*
Whitten, W. K., 157, *175*
Whitters, M. J., 467, *498*
Whittingham, D. G., 8, *16*
Whittle, M. J., 174, *179*
Whitworth, D. J., 281, 282, *292*, 380, 388, *389, 393*
Whyatt, D. J., 343, *369*
Whyte, D. B., 504, *513*
Wianny, F., 14, *19*, 26, 32, *35*, 158, *179*
Wicherle, H., *106*
Wichmann, W., 506, *517*
Wicht, H., 238, *252*

Wickens, M. P., 10, 11, *15, 16, 19*
Wicking, C., 287, *292*, 470, *485, 486*
Wickramaratne, G. A., 479, *485*
Wickramasinghe, D., 7, *19*
Wicks, I. P., 31, *34*
Wictorin, K., 89, *103*
Widelitz, R. B., 574, 577, 578, 579, *585, 586, 587, 589*
Widmer, D. A., 91, *106*
Widmer, H. R., 91, *98*
Wieand, S. J., 91, *97*
Wieduwilt, M., 465, *483*
Wiegand, S. J., 91, *97*, 224, *230, 231*
Wiekowski, M., 9, *19*
Wierda, A., 165, *174*, 303, *323*
Wieschaus, E., 136, *146*, 443, 478, *493*
Wiese, R. J., 343, *363*
Wiest, W., 570, *587*
Wigle, J. T., 218, 219, *233*, 317, *329*, 527, 530, 532, *538*
Wijgerde, M., 202, *210*
Wilbertz, J., 246, 249, 318, *324*
Wilby, O. K., 128, 137, *146*
Wilcox, J. N., 512, *516*
Wilcox, S. A., 379, 387, *393*
Wilcox, W. R., 286, *294*
Wilde, A. A. M., 348, *360*
Wilder, P. J., 166, *179*
Wiles, M. V., 434, 467, *493*
Wiley, L. M., 26, *33, 35*
Wilkes, A. J., 351, *363*
Wilkes, D., 452, 466, *482*
Wilkie, A. O., 452, 453, 466, 474, *487, 498*
Wilkin, D. J., 286, *294*
Wilkins, A. S., 387, *393*
Wilkinson, D. G., 118, 119, *124, 125, 126*, 133, 135, *147, 149*, 225, *231, 233*, 264, 272, 422, 423, 424, 425, 427, 433, *487, 492, 494, 498*, 525, *537*, 544, 545, 546, *563, 564, 565*, 571, *586*
Wilks, A. F., 162, *174*, 453, *486*
Willecke, K., 173, *175, 177*, 348, *360*
Willetts, K., 313, *323, 326*, 467, *482*
Willhoite, A. R., 236, *251*
Williams, B. A., 141, *149*, 194, *208*, 255, 256, 258, 264, 266, 268, 271, *276, 277*
Williams, B. O., 528, *537*
Williams, B. P., 237, 239, *252*
Williams, C. J., 13, *19*
Williams, C. L., 405, 407, *419*
Williams, D. E., 188, *189, 190*, 193, 204, *208*
Williams, E. A., 47, *52*
Williams, J., 237, *252*
Williams, L., 312, *328*
Williams, M., 247, *251*, 580, *585*
Williams, M. C., 311, 312, *323*
Williams, M. L., 581, *584, 586*

Williams, N. A., 478, *498*
Williams, R., 135, *149*
Williams, R. H., 319, *328*
Williams, R. S., 269, *273, 276, 277,* 342, *369*
Williams, S. C., 432, 465, 478, *483, 498*
Williams, T., 79, *106,* 529, *538*
Williamson, K. A., 375, *391*
Williamson, R. A., 161, *179*
Wilm, B., 451, 475, *493, 498*
Wilming, L. G., 385, *389*
Wilson, C. M., 312, *323*
Wilson, D., 568, *584*
Wilson, D. B., 45, *50,* 160, 164, 165, *175, 176, 178, 179,*
    323, 333, 341, 342, 343, 352, *362, 366,* 466, *486,* 504,
    518
Wilson, D. I., 357, *364,* 387, *390*
Wilson, G. N., 339, *369*
Wilson, J. D., 373, *390*
Wilson, J. G., 346, *369*
Wilson, R., 336, *367*
Wilson, S. W., 22, *34,* 77, 78, 79, *98, 101, 102, 106,* 521,
    522, *537*
Wilson, V., 45, *52,* 132, 133, *144, 149,* 160, 167, *178,* 304,
    329, 407, *417*
Wilson Berry, L., 242, *252*
Wilson-Heiner, M., 570, 583, *584*
Wilson-Rawls, J., 136, *148,* 260, 267, *277*
Wilting, J., 130, 131, 132, 140, *144, 145, 146, 147,* 255,
    263, 266, *273, 277*
Wilton, S., 342, *368*
Wilusz, J., 11, *16*
Windle, J. J., 504, 512, *513, 517, 518*
Windsor, D., 388, *392*
Winer-Muram, H. T., 351, 354, *369*
Winfrey, V., 453, 479, *488*
Wingate, R. J., 119, *122*
Winkler, J., 247, *252*
Winkler, S., 521, 525, 531, 533, *537*
Winkling, H., 378, *390*
Winnier, G., 82, *106,* 203, *210,* 287, *292,* 313, 314, *323,*
    439, 467, 479, *498*
Winnier, G. E., 132, *149,* 506, *516*
Winokur, S. T., 286, *294*
Winter, B., 342, *369*
Winter, J., 173, 174, *175*
Winter, R. M., 453, *493, 494*
Winter, S., 570, *586*
Winterhager, E., 173, *175, 177*
Wirth, J., 281, *294,* 380, *393*
Wirtz, E., 15, *18*
Wisdom, R., 451, 453, 476, *493*
Wise, S. P., 84, *106*
Wisniewski, S., 581, *586*
Wistow, G., 532, *537*

Wit, J., 509, *517*
Wittbrodt, J., 116, *125,* 521, 525, 531, 533, *537*
Witte, D. P., 79, 82, *101, 105,* 352, 354, *368*
Witte, L., 219, 227, *231, 233*
Wittler, L., 41, *51*
Witzenbichler, B., 220, *228*
Wlaczak, V. R., 581, *586*
Wodarz, A., 82, *106,* 405, *420*
Woessner, J. F., Jr., 223, *231*
Wohl, C., 236, *252*
Wohlschlegel, J. A., *251*
Wojnowski, L., 470, *486,* 579, *586*
Wojtowicz, J. M., 247, *252*
Wolburg, H., 219, 224, *231, 232*
Wolburg-Buchholz, K., 224, *232*
Wold, B. J., 239, 241, *251,* 258, 260, 271, *273, 276, 277*
Wolfes, H., 14, *18*
Wolfman, N. M., 338, *363*
Wolfsberg, T. G., 424, *481*
Wolgemuth, D. J., 312, *327*
Woloshin, P., 267, *277*
Wolpert, L., 356, *359,* 577, 578, 579, *587*
Wondisford, F. E., 509, *514, 517*
Wong, A., 271, *277*
Wong, B., 347, *362*
Wong, C., 528, 532, *538*
Wong, D., 165, *174,* 303, *323*
Wong, J. S., 161, *175, 178,* 400, *418*
Wong, K., 90, 95, *106*
Wong, P. C., 137, *149*
Wong, V., 91, *98*
Wong, Y.-M. M., 341, *369*
Woo, I., 290, *294,* 437, 440, 450, 451, 466, 474, *482, 495,*
    530, 533, *538*
Woo, K., 543, *565*
Wood, C. R., 217, *229*
Wood, H., 91, *102,* 199, 200, 201, *210,* 527, 530, 531, *537*
Wood, R. I., 511, 512, *515*
Wood, W., 185, *190*
Wood, W. B., 243, *252*
Wood, W. M., 509, 510, *515*
Woodland, H. R., 166, *176,* 303, *325*
Woods, K. L., 511, *518*
Woolf, A. S., 404, *420*
Woolf, T., *105*
Worby, C. A., 402, *420*
Worley, D., 403, *420*
Worley, K. C., 378, 387, *389*
Worrad, D. M., 9, *15, 19*
Wotton, D., 468, *485*
Wozney, J. M., 271, *275,* 467, *498*
Wrana, J. L., 31, *35,* 468, *497*
Wray, G. A., 471, *493*
Wray, S., 435, *498,* 502, *514*

Wreden, C., 10, *19*
Wright, B. D., 223, *229*
Wright, C. V., 40, 43, 47, *49, 50,* 78, *104, 105,* 133, 137, *145,* 183, *190,* 260, 267, *275,* 309, 319, 320, 322, *326, 327,* 477, *482*
Wright, C. V. E., 336, 354, 355, *363, 364*
Wright, D. D., 478, *489*
Wright, E. S., 26, *34,* 281, *293, 294,* 381, *392*
Wright, G. M., 139, *148*
Wright, M., 397, *418*
Wright, M. E., 475, *498*
Wright, S. J., 8, *19*
Wright, W. E., 257, 258, *275, 276*
Wroblewska, J., 22, *35*
Wu, C., 342, *364*
Wu, D. K., 469, 478, *483, 491,* 541, 542, 546, 550, 551, 552, 555, 557, 558, 559, *561, 564,* 579
Wu, G., 226, *232,* 414, 415, *418, 420*
Wu, H., 219, *233,* 269, *273, 277,* 342, *369,* 583, *587*
Wu, J. E., 314, *329*
Wu, J. Y., 90, 95, *106,* 136, *149,* 521, *536*
Wu, L., 11, *19,* 204, *210,* 428, 452, *489*
Wu, M. T., 227, *230*
Wu, N., 133, *149*
Wu, Q., 226, *233*
Wu, S. Q., 47, *51*
Wu, W., 44, *50,* 90, 95, *106,* 219, *229,* 325, *327,* 338, *361, 362,* 506, 508, 509, *518*
Wu, X., 136, *148,* 336, 337, *366, 370*
Wuenschell, C. W., 88, *102,* 168, *177*
Wunderle, V. M., 380, *393*
Wunsch, A. M., 342, *368*
Wünsch, K., 136, 137, 139, *145*
Wurst, W., 109, 114, 115, *122, 123, 124, 126,* 228, *230,* 428, 479, *487, 491,* 522, 527, *535,* 549, *562*
Wuytack, F., 343, 345, 348, *366*
Wylie, C. C., 14, *16,* 182, 183, 184, 185, 186, 187, 188, *189, 190,* 303, *323, 330*
Wynne, J., 342, *362*
Wynshaw-Boris, A., 31, *33,* 40, 45, *50,* 204, *210,* 220, *233,* 286, *292,* 338, 353, 357, *360, 365,* 453, *484*

X

Xavier, R. J., 353, 357, *365*
Xavier-Neto, J., 335, 339, 343, 344, 345, 346, 347, 350, 351, 352, *359, 366, 367, 370*
Xenophontos, S. L., 414, *418*
Xiang, M., 527, *535,* 556, 558, *565*
Xiao, D., 346, *364*
Xiao, H., 12, *17*
Xiao, J., 399, *419,* 530, 533, *537*
Xie, B., 286, *294*
Xie, J., 287, *293,* 579, *586, 589*

Xie, W. F., 282, *295*
Xing, L., 442, *484*
Xu, C., 338, 352, *370*
Xu, H., 343, 353, *364*
Xu, J., 116, *126,* 453, *492,* 521, 527, *537*
Xu, K., 8, *16*
Xu, P. X., 398, 399, *420,* 527, 530, 533, *538,* 550, *565*
Xu, Q., 119, *124, 126,* 225, *231, 233,* 521, 522, *537*
Xu, S., 14, *16, 17*
Xu, T., 136, *145, 148, 149*
Xu, W., 7, *16*
Xu, X., 167, 171, 173, *179,* 226, 227, *233,* 286, 292, 313, *329,* 336, *370*
Xu, X. M., 555, 557, *563*
Xu, Y., 87, 91, *103, 106,* 508, *517*
Xuan, S., 78, 81, *106,* 243, 252, 479, *498*
Xue, N., 286, *294*
Xue, X. J., 263, *277*
Xue, Y., 136, *149,* 226, *230*
Xue, Z. G., 263, *277*
Xy, H., 353, 357, *365*
Xydas, S., *328,* 335, 336, 339, *367*

Y

Yablonka-Reuveni, Z., 271, *277*
Yacoub, M. H., 341, 351, *359, 363*
Yacoubian, T. A., 94, *106*
Yagi, H., 204, *209,* 220, *232,* 288, *293,* 442, 443, 453, *488,* 506, *518*
Yagi, T., 223, *230*
Yaginuma, H., 122, *122*
Yaginuma, K., 504, 510, *516*
Yagishita, N., 505, *517*
Yagyu, K., 94, *103*
Yahata, T., 339, *368*
Yamaai, T., 260, 266, *273*
Yamada, E., 580, *587*
Yamada, G., 432, 473, 477, *482, 497, 498*
Yamada, H., 321, *327*
Yamada, K., 352, 354, *365,* 581, 582, 583, *587*
Yamada, K. M., 408, *418*
Yamada, M., 94, *106,* 186, *190,* 336, *370,* 512, *518*
Yamada, S., 338, 352, *364*
Yamada, T., 534, *538*
Yamada, Y., 203, *210,* 284, *294,* 318, *329,* 442, 443, 453, *497,* 557, *565*
Yamagami, T., 341, 352, *370*
Yamagata, M., 525, *538*
Yamagishi, H., 349, 355, *369, 370,* 451, 465, 473, *496*
Yamagiwa, H., 288, *293*
Yamaguchi, A., 271, *275,* 288, *293,* 442, 443, 453, *488*
Yamaguchi, T. P., 133, 135, *145, 148, 149,* 203, *209,* 217, 223, *232,* 303, 317, *324,* 451, 453, 471, *498,* 506, *518*

Yamamoto, H., 94, *103,* 168, *179,* 188, *189,* 266, *277*
Yamamoto, I., 262, 268, *274*
Yamamoto, K., 286, *292,* 335, *358*
Yamamoto, M., 527, 530, 531, *536*
Yamamoto, T., 195, *208,* 534, *537*
Yamamura, F., 196, *209*
Yamamura, H., 334, 344, 347, 348, 350, 351, 352, 357, *365, 370*
Yamamura, K., 305, 307, *328,* 443, *492*
Yamanaka, H., 506, *516,* 549, *564*
Yamanishi, K., *588*
Yamasaki, K., *588*
Yamasaki, M., 312, *328,* 505, *517*
Yamasaki, N., 348, 349, 352, *368*
Yamashita, F., 581, 582, 583, *587*
Yamashita, J., 228, *233*
Yamashita, S., 504, *516*
Yamashita, T., 94, *104,* 227, *231,* 502, 508, 509, 511, *516, 517*
Yamazaki, T., 343, *368*
Yan, C., 313, *330*
Yan, J., 167, *180*
Yan, Q., 91, *101,* 226, *232*
Yan, W., 334, *361*
Yan, Y. L., 214, *231*
Yan, Y.-T., 31, *33,* 40, 45, *50,* 338, 353, 354, 356, 357, *360, 370*
Yanagimacht, R., 8, *17*
Yanagisawa, H., 347, *366,* 432, 465, 478, *483, 498*
Yanagisawa, M., 334, *368,* 432, 465, 478, *483, 498*
Yanase, T., 379, *391*
Yanazawa, M., 353, 354, 355, 356, *363*
Yancopoulos, G. D., *106,* 223, 224, 225, 227, *229, 230, 231, 232,* 292, *292,* 549, *561*
Yang, A., 571, 581, 583, *589*
Yang, B. B., 350, *370*
Yang, B. L., 350, *370*
Yang, D. D., 87, *101*
Yang, G., 397, *418*
Yang, G.-Z., 351, *363*
Yang, J., 405, 407, *416,* 579, *589*
Yang, J. M., 530, 532, *536, 538*
Yang, J. T., 170, 172, 173, *176, 180,* 204, *206,* 219, *233,* 571, 581, 583, *589*
Yang, L., 31, *33,* 40, 45, *50,* 338, *360,* 450, 471, 472, 473, *498*
Yang, Q., 269, *273*
Yang, W. P., 225, *230*
Yang, X., 23, *35,* 226, 227, *233,* 240, *250,* 266, *277*
Yaniv, M., 165, 166, *175, 178*
Yano, H., 189, *190*
Yap, A. S., 408, *420*
Yashiro, K., 356, 357, *365, 367*
Yassine, F., 200, *209*

Yasuda, K., 525, 527, 528, 530, 531, 534, *536, 537*
Yasuda, M., 355, *362*
Yasugi, S., 303, 308, *325, 329*
Yasui, K., 473, *498*
Yasui, N., 288, *293*
Yasui, T., 288, *293*
Yasuno, H., 581, 582, 583, *587*
Yasuo, H., 303, *330*
Yatskievych, T. A., 334, 335, 336, 337, *363, 370*
Yazaki, Y., 336, 343, 349, *365, 368, 369,* 451, 465, 473, *489, 496*
Ybot-Gonzalez, P., 475, *486*
Ye, H., 321, *328*
Ye, W., 85, *106,* 110, 121, *126*
Ye, X., 479, *486*
Yee, D., 169, *179*
Yeger, H., 415, *417*
Yeh, W.-C., 341, *370*
Yehia, G., 583, *586*
Yelandi, A., 318, *328*
Yelon, D., 303, *326,* 334, 339, 341, 342, 349, *361, 367, 370*
Yeo, C.-Y., 356, *367*
Yeoh, T., 345, 348, *366*
Yeom, Y., 183, *190*
Yeoman, E., 506, *514*
Yi, C. H., 357, *364*
Yi, W., 388, *393*
Yienger, K., 171, 173, *178*
Yin, M., 434, 467, *493*
Yin, Z., 334, 336, *370*
Ying, C., 258, 260, 261, 262, 267, *275*
Yingling, J. M., 311, 313, 314, 315, *325, 329*
Yip, L., 83, *100*
Ylikoski, J., 558, 559, *564, 565*
Yntema, C. L., 542, 543, *565*
Yoda, H., 532, 534, *536*
Yoder, F., 507, *515*
Yoder, M. C., 196, 202, *210*
Yokata, C., 338, *368*
Yokayama, M., 353, 354, 355, 356, *363*
Yokayama, T., 352, 354, *365*
Yokoi, M., 237, *252*
Yokota, Y., 137, *147*
Yokoyama, M., 7, *16,* 47, *51*
Yokoyama, T., 47, *51,* 354, 355, *364*
Yoneda, K., 570, 583, *584*
Yoneda, Y., 341, 352, *370*
Yonei-Tamura, S., 354, *367*
Yonekawa, H., 352, 354, *365*
Yoneshima, H., 94, *104*
Yoneshita, H., 94, *106*
Yonezawa, S., 557, *564, 565*
Yoon, J. K., 260, *276, 277*
Yoon, S. K., 400, *418*

York, J. D., 225, *230*
Yoshiba, K., 453, *482*
Yoshida, A., 525, *538*
Yoshida, C., 289, *294*
Yoshida, H., 172, 173, *176, 178,* 200, *209,* 442, *498*
Yoshida, K., 318, *325,* 341, 352, *370,* 405, 407, *416,* 531, *538*
Yoshida, M., 83, *106,* 305, 307, *328,* 374, *392,* 403, *418*
Yoshida, N., 341, 352, *370,* 471, 479, *487,* 531, *538*
Yoshida, O., 189, *189*
Yoshida, T., 581, *588*
Yoshida, Y., 171, 173, *176*
Yoshie, O., 350, *362*
Yoshihara, Y., 435, *491*
Yoshii, H., 376, *393*
Yoshikawa, K., 452, *496,* 580, 581, 583, *589*
Yoshikawa, Y., 133, 135, 137, *149,* 557, *563*
Yoshiki, A., 11, 12, *18,* 557, *565*
Yoshiki, S., 288, *293,* 442, 443, 453, *488*
Yoshimoto, T., 502, 511, *515*
Yoshinaga, K., 156, *175*
Yoshinobu, K., 305, 307, *328*
Yoshioka, H., 47, *50,* 354, *370,* 442, 443, 471, 479, *487, 489*
Yoshizaki, N., 557, *564*
Yoshizawa, T., 312, *328,* 505, *517*
Yost, H. J., 47, *50,* 349, 351, 355, 356, *358, 359, 364,* 473, *493*
Young, I. D., 281, *292,* 357, *364,* 380, *390*
Young, J. K., 23, *35*
Young, K. E., 43, *50,* 78, 80, 85, 86, *98,* 121, *123,* 141, *145,* 260, 265, *273,* 319, *323,* 439, 451, 470, *482, 483, 493,* 521, *535*
Young, M. W., 136, *145, 147*
Young, P. E., 193, *210*
Young, R. W., 452, *498*
Young, S. G., 161, *175, 178*
Young, W. S. D., 502, *518*
Young, W. S. III, 512, *516*
Younger-Shepherd, S., 136, *145*
Younossi-Hartenstein, A., *367*
Youssef, E. H., 442, *498*
Yu, A. T., 11, *16*
Yu, A. Y., 225, *230*
Yu, B. D., 376, *390*
Yu, D. B., 376, *393*
Yu, G., 79, 89, *99,* 450, 471, *484, 498*
Yu, Q.-C., 572, 583, *585, 586*
Yu, R. N., 6, *19,* 378, 379, *393*
Yu, T. W., 247, *251*
Yu, Y., 402, *417*
Yuan, B., 135, *147*
Yuan, H., 160, *180*
Yuan, S., 334, 335, 336, 346, 347, *363, 370*

Yuasa, J., 525, *538*
Yuasa, Y., 226, *232*
Yudate, T., 95, *102,* 506, *516*
Yue, Y., 91, *106*
Yun, J. S., 509, *518*
Yun, K., 88, 89, 90, 94, 95, *97*
Yurugi, T., 228, *233*
Yuspa, S. H., 169, *179,* 580, *585, 586, 587, 589*
Yutani, S., 506, *518*
Yutzey, K. E., 342, 344, 345, 346, 347, *364, 367, 370*

Z

Zabel, B. U., 281, 288, *293, 294,* 379, 380, *391, 393,* 478, *495,* 507, *517*
Zabski, S. M., 313, *324*
Zackai, E. H., 452, 453, 468, *481, 485*
Zagouras, P., 136, *144*
Zagzag, D., 224, *230*
Zákány, J., 305, 309, 310, *326, 330*
Zakin, L., 44, *52*
Zalokar, M., 185, *190*
Zalzman, M., 551, 552, 557, *563*
Zambrowicz, B. P., 157, *180,* 183, *190*
Zammit, P. S., 344, 345, 356, *361, 362*
Zamore, P. D., 14, *19*
Zanaria, E., 378, 379, 380, 384, 387, *389, 391, 392, 393*
Zanchi, M., 271, *273*
Zanellato, A., 511, *516*
Zanetti, A., 226, *229*
Zang, Z., 239, *250*
Zangen, D. H., 322, *328*
Zanger, U. M., 375, 384, *392*
Zanin, M. K. B., 350, *370*
Zannini, M., 311, *330*
Zaphiropoulos, P. G., 287, *292,* 470, *485, 486, 497,* 579, *586*
Zappavigna, V., 443, 449, 478, *482*
Zarbl, H., 15, *15*
Zaret, K. S., 165, *174,* 303, 307, 316, 317, 320, *323, 324, 325, 330*
Zarkower, D., 382, 387, 388, *391, 392, 393*
Zasloff, M. A., 271, *276*
Zatloukal, K., 570, *587*
Zavitz, K. H., 399, *419,* 530, 533, *537*
Zawel, L., 86, *100*
Zazopoulos, E., 379, *393*
Zbar, B., 227, *229, 231*
Zbieranowski, I., 163, *175*
Zeeman, E. C., 139, *145*
Zehentner, B. K., 282, *295*
Zehetner, G., 378, *392*
Zeinstra, L., 183, *190,* 336, *363*
Zeitler, P. S., 507, *516*

Zekower, D., 382, 387, *392*
Zelenka, P. S., 532, *535*
Zelent, A., 469, *489, 491, 494*
Zelickson, A. S., 568, *589*
Zeller, R., 86, *99*
Zeller, U., 421, 442, 456, 457, 463, 471, *488, 498*
Zelzer, E., 340, *370*
Zeng, Q., 225, *231*
Zeng, X., 314, *330*
Zenner, H. P., 558, *563*
Zerega, B., 290, *292*, 450, 453, 472, 473, *481*, 549, *561*
Zernicka-Goetz, M., 1, *4*, 14, *19*, 26, 27, 32, *33, 35*, 158, *179*
Zerucha, T., 89, *106*, 450, 471, *498*
Zeschnigk, M., 264, 271, *275, 277*
Zetter, B. R., 189, *190*
Zetterstrom, R. H., 509, *518*
Zevnik, B., 23, *34*, 160, 166, *178*
Zhadanov, A. B., 508, *517*
Zhang, C. L., 342, *365, 370*, 579, *589*
Zhang, D., 239, *251*
Zhang, H., 82, *106*, 336, 339, *370*, 439, 450, 467, 471, 472, 473, *498*
Zhang, J., 45, *50*, 78, 79, *100, 106*, 186, *190*, 303, 318, *328, 330*, 338, *361*, 529, *538*
Zhang, L., 217, 227, *229*
Zhang, M., 347, 348, 350, 352, *370*, 427, 433, 443, 449, 474, *485*
Zhang, N., 135, 136, 137, 138, *149*, 186, *190*, 555, *565*
Zhang, P., 285, 287, *293, 294*, 453, *486*, 528, 532, *538*, 580, 581, 583, *586*
Zhang, Q., 334, 343, *361*
Zhang, R., 288, 289, *292*
Zhang, W., 240, 243, *251*, 260, 262, *276, 277*
Zhang, X., 282, *295*, 414, *419*, 437, *498*
Zhang, X.-K., 346, *364*
Zhang, Y., 350, *370*, 379, *390*, 437, *498*
Zhang, Y. A., 94, *106*, 226, 229, 357, *358*
Zhang, Y.-H., 398, *419*
Zhang, Y.-Y., 243, 244, *249*
Zhang, Z., 12, *16*, 281, 282, 283, 284, 289, *292, 294*, 380, *389*, 437, *498*, 530, 533, *535*
Zhao, B., 341, *364, 367*
Zhao, D., 469, *498*
Zhao, G. Q., 473, 476, *498*
Zhao, J., 314, 315, *327, 329, 330*
Zhao, Q., 79, *106*, 281, *295*, 451, 453, 476, *493, 498*
Zhao, S., 92, *104*, 473, *498*
Zhao, X., 437, *498*
Zhao, Y., 83, *106*, 376, *389*, 432, *498*
Zhao, Z., 450, 471, *496*
Zharhary, D., 196, *207*
Zheng, B., 571, 581, 583, *587*
Zheng, C., 94, *106*

Zheng, H., 347, 352, *360*
Zheng, J. L., 555, 556, *566*
Zheng, M., 316, 317, 320, *324, 325*
Zheng, R. Z., 134, *147*
Zheng, T. S., 87, *101*
Zheng, Y., 243, *249*, 506, *516*, 549, *564*
Zhi, Q., 131, *146*, 266, *277*
Zhong, W., 79, 89, *99*, 164, *179*, 450, 471, *484*
Zhou, C., 92, *106*
Zhou, F., 286, *292*, 453, *484*
Zhou, G., 334, 343, 345, 346, 352, *366*
Zhou, H., 82, *102*, 224, 227, *232*
Zhou, J., 412, 415, *417, 418*
Zhou, L., 90, *106*, 313, *330*, 527, *535*, 556, 558, *565*
Zhou, P., 440, *498*, 577, *589*
Zhou, Q., 268, *275*
Zhou, R., 91, *106*
Zhou, S. X., 30, *35*, 39, 40, 41, *52*, 182, 183, *190*, 268, *275*, 380, *389*
Zhou, W., 342, *369*
Zhou, X., 95, *97*, 338, 355, *370*, 473, *498*, 519, 521, 522, 525, 530, 533, *535*
Zhou, Y., 6, *17*, 173, 174, *176*
Zhou, Z., 582, 583, *587*
Zhu, A. J., 242, *252*, 589
Zhu, D., 428, 452, *489*
Zhu, J., 165, *180*, 316, *330*
Zhu, L., 387, *393*
Zhu, W., 269, *273*
Zhu, X., 317, *330*, 337, *364, 370*
Zhu, Y., 90, *106*
Zhu, Y. Z., 286, *294*
Zhu, Z., 227, *233*
Zhuang, Y., 172, 173, *180*
Zhurinsky, J., 86, *104*, 579, *588*
Zidehsarai, M. P., 509, *513*
Zietler, P. S., 95, *101*
Zile, M., 343, 346, 347, *361, 363*
Zile, M. H., 346, 347, 356, *369, 370*
Ziller, C., 140, *148*, 236, *249*, 569, *587*
Zimmer, A., 89, *101*, 402, *417*, 470, *486*, 579, *586*
Zimmer, J., 281, *294*, 380, *393*
Zimmerman, B., 452, *498*
Zimmerman, L. B., 40, 44, *51*, 141, *147*, 238, *252*, 260, 264, 265, *275*, 468, *498*
Zimmermann, J. W., 12, *19*
Zimmermann, S., 386, *393*
Zindy, F., 556, 558, *561*
Zine, A., 555, *566*
Zinn, C., 558, *563*
Zinyk, D. L., 109, 114, 121, *124, 126*, 228, *233*, 451, 453, 470, 471, *491*
Ziomek, C. A., 23, *34*, 158, *177, 178, 180*
Zipursky, S. L., 399, *419*, 530, 533, *537*

Zirngibl, R., 135, *148*
Ziyadeh, F. N., 415, *419*
Zlot, C. H., 161, *178*
Zock, C., 235, 236, 238, 239, 240, 241, 243, 244, 245, 248, *251*
Zoeller, R. H., 512, *516*
Zoeller, R. T., 502, *518*
Zoerkler, N., 321, *326*
Zoghbi, H. Y., 122, *122,* 555, 558, 559, *561*
Zon, L. I., 198, 199, 203, *208, 210,* 213, 214, *230, 232,* 336, *363*
Zonana, J., 441, *488, 491*
Zorn, A. M., 172, 173, *176,* 303, *330*
Zou, K., 350, *370*

Zou, Y., 342, 343, 352, *363*
Zschiesche, W., 317, *328,* 404, *419*
Zsebo, K., 185, 188, 189, *189, 190,* 246, *251*
Zuber, M. E., 522, 525, 533, *538*
Zuffardi, O., 380, 381, *391*
Zulch, A., 550, *561*
Zuniga, A., *327*
Zuo, L., 506, 508, 509, *518*
Zwartkruis, F., 395, *417,* 546, *563*
Zweigerdt, R., 310, *327*
Zwijsen, A., 222, 226, 227, *229,* 336, 352, *359*
Zybina, E., 23, *35,* 163, 169, *180*
Zybina, T. G., 23, *35,* 163, 169, *180*
Zygar, C. A., 529, *538*

Subject Index

A

ACTH. *see* Adrenocorticotrophic hormone
Activin, 337
Activin receptor-like kinase-1, 225–226
ActRIIa, 68
Acvrl1. *see* Activin receptor-like kinase-1
Adrenocorticotrophic hormone
    characterization, 499
    deficiency, development, 509
    functions, 500
AGM. *see* Aorta gonads mesonephroi
Alport syndrome, 412
Alzheimer's disease, 247
Amh
    Müllerian duct differentiation, 385–386
    production, 385
    regulation, 381–382
AML1
    expression, 201
    liver hematopoiesis, 204
Amphibians, 197
Ang-1, 227
Angiogenesis
    arteries segregation, 225–226
    cell-cell junctions, 226
    clinical implications, 227
    complexity, 227–228
    definition, 212
    pericyte recruitment, 226–227
    smooth muscle cell recruitment, 226–227
    vascular
        channel formation, 222–225
        channel regression, 227
        polarity, 227
    vein segregation, 225–226

Angiopoietins, 224
Animal-vegetal, 27
Anterior patterning, 44–48
Anterior-posterior axis
    CNS
        DV pattern and, 108–109
        mes/met domain
            development, identification of mutants, 115
            fate mapping, 113–114
            gene expression, 112–113
            organization, role of FGF8, 115–16
            patterning models, 116–117
            regulation, 110–112
        Otx2/Gbx2 mutants, 114–115
    diencephalon, 84–85
    epithelial cup, 301
    fetus
        blastocyst, 29
        conserved properties, 40
        duplication, 22
        EGO response, 41–42
        node development, 38–40
        organizer function, 42–44
        polarity specification, 29–32
        relationship, 25–26
    gut tube regionalization, 305
    hindbrain, 117–119
    nephric duct, 401
    prosencephalic neural plate, 78
    somite, 134
    telencephalon, 83–84
Aorta gonads mesonephroi
    CFU-S in, 196
    description, 192
    fate mapping, 202

Aorta gonads mesonephroi (*continued*)
   hematopoietic progenitor source, 199
   hematopoietic stem cell source, 199
   mesodermal precursors, 200–201
AP axis. *see* Anterior-posterior axis
Arginine vasopressin, 95
Aristaless 3, 476
ARNT. *see* Arylhydrocarbon receptor nuclear translator
Artery segregation, 225–226
Arylhydrocarbon receptor nuclear translator
   angiogenesis, 225
   labyrinth branching morphogenesis, 172–173
Asymmetry
   blastocyst, 27–29
   development
     early, 23–27
     overview, 21–23
     preimplantation, 1–2
   enhancer location, 62–63
   fertilized oocytes, 1
   left-right
     diversity among vertebrates, 68–69
     furin, 67–68
     future challenges, 69–70
     GlcNac-TI, 67
     initial determination, 58–59
     models, 55
     morphological, 56–57
     mutations, 58–59
     node flow, 59–61
     Pitx2 on left side, 65–66
     positive/negative feedback loops, 62–64
     retinoic acid, 67
     Sna on right side, 65–66
     TGF-β signals, 59–61
     Zic3, 67
   oocytes, 7–8
AV axis. *see* Animal-vegetal
AVE. *see under* Endoderm
AVP. *see* Arginine vasopressin
Axin, 45

B

BA. *see* Branchial arch
Bapx1, 474
Barx1, 437, 439
Basic helix-loop-helix factors. *see also specific factors*
   angiogenesis, 225
   crainofacial development, 473–474
   MRFs as, 256–258
   oocyte maturation, 10
   retina development, 526–527
   sensory differentiation, 555

   spongiotrophoblast development, 168
   ventricular specification, 348–349
BCC, 579
B cells, 204
BDNF. *see* Brain-derived neurotrophic factor
BF-1
   neural stem cell renewal, 243
   telencephalon, 81
BF-2, 408
bHLH. *see* Basic helix-loop-helix factors
Blast colony forming cell
   assay, 217
   description, 217
Blastocyst
   asymmetry, 23, 25–29
   axes, origins, 26
   lineages
     molecular specification, 159–160
     setting aside, 158
Blastomeres, 151
BL-CFC. *see* Blast colony forming cell
Blood. *see* Hematopoietic system
Blood islands, 211
BMP. *see* Bone morphogenetic protein
Bone
   cartilage replacement by, 279–280
   formation
     Cbfa, 287–290
     Dlx, 290
     gelatinase B, 290–291
     Ihh, 287–288
     Msx, 290
     process, 279–281
     VEGF, 290–291
   ossification
     description, 279
     types, 280
   skull development
     regulation
       calvarium formation, 451–452
       molecular evidence, 442–443
       neurocranium formation, 451
       skeletogenesis, 442
     studies, 441–442
   skull organization
     characterization, 456–457
     chondrocranium, 457
     morphology
       dermatocranium, 464–465
       nasoethmoidal, 457–460
       occipital region, 463
       orbitotemporal, 460, 462
       otic region, 462–463
       splanchnocranium, 463–464

units/divisions, 457
Bone marrow progenitors, 246
Bone morphogenetic protein. *see also* Transforming growth
  factor-β
  BMP2
    cardiac induction, 339
    lung development, 314
    neuronal differentiation, 238–239
    pituitary development, 505, 506
    skull formation, 452
  BMP4
    expression pattern, 183–185
    hematopoietic specification, 203
    inner ear development, 550
    lens development, 532–533
    lung development, 314
    mesoderm formation, 203
    myotome formation, 264
    neuronal differentiation, 238–239
    pituitary development, 505, 506
    skull formation, 450, 452
    tooth development
      bud-to-cap transition, 440–441
      dentition patterning, 437
      initiation, 439
    ureteric bud, 314
  BMP7
    metanephros development, 408
    pancreatic expression, 321
  cardiac induction, 336, 337
  cardiac looping, 355
  crainofacial development, 467–468
  expression in LPM, 64
  follicle formation, 577
  follicle spacing, 557–558
  forebrain morphogenesis, 85–87
  pallial telencephalon patterning, 82–83
  prosencephalic neural plate patterning, 78–79
  signaling, role of chordin, 44
  somite patterning, 141
Brain. *see also specific regions*
  early patterning, 109–110
  endothelial cells, 219
Brain 2/4, 506–509
Brain-derived neurotrophic factor
  inner ear neurogenesis, 559
  neural stem cell survival, 239
Branchial arch
  characterization, 422
  dermatocranium, 464
  fate mapping, 424–426
  intra identity, establishing, 449
Branchio-oto-renal system, 550
Brn3b, 527

C

Cadherins
  cell-cell junctions and, 226
  epithelium mesenchyme conversion, 408–410
  gene, effects on heart, 340
  muscle differentiation, 263–264
  -positive, vascular endothelial cells, 200
Caecum. *see* Intestines
CAKUT syndrome, 403
Calcium, 8
Calcium/calmodulin-dependent protein kinases, 342
CaMKs. *see* Calcium/calmodulin-dependent protein kinases
Campomelic dysplasia
  characterization, 281
  XY reversal and, 380
Cardiac jelly, 350
Cartilage
  ECM function, 280–281
  replacement, 279–280
Cartilage homeoprotein I, 476
β-Catenin
  dermal tumorigenesis, 579
  follicle formation, 555, 557
  neural stem cell renewal, 242
CBFs. *see* Core-binding factors
CD. *see* Campomelic dysplasia
cdc5, 169–170
Cell fate node, 38–40
Cell lineage analysis, 156–158
Cell movement node, 38–40
Cellular retinoid acid binding proteins 1, 91
Central nervous system
  AP patterning
    description, 31
    mes/met domains
      centrally located organizer, 110–112
      development, 115
      fate mapping, 113–114
      gene expression defining, 112–113
      organization, FGF8 role, 115–116
      patterning models, 116–117
    Otx2/Gbx2 mutant role, 114–115
  development, flat-top mutation, 243
  development/patterning, 107–109
  dorsal ventral patterning
    Gli cells role, 121
    SHH regulation, 120–121
    TGF-β regulation, 121–122
  skull support, 457
  stem cells
    adult/fetal, comparison, 245
    cell cycle status, 245
    characterization, 235

Central nervous system (*continued*)
    differentiation
        process, 241
        role of BMP, 238–239
        function in adults, 237
        hematopoietic stem cells, comparison, 243–244
        lineage, 237–238
        neuronal potential, 241–242
        nonneural derivative potential, 245–247
        occurrence, 235–236
        other nerve cells, association, 237
        outstanding issues, 247–248
        response to nervous system injury, 247–248
        survival, role of neutotrophins, 239
        in vivo function, 244–245
Cerberus, 338
CerL, 31
CFC. *see* Cripto-FRL-1-Cryptic factors
CFU. *see* Colony forming factors
CHDs. *see* Congenital heart defects
Cholesterol sulfate, 580
Chondrocytes
    FGF, 286–287
    PTH/PTHrP, 284–286
Chondroitin sulfate proteoglycans, 350
Chordin
    characterization, 44
    crainofacial development, 468
    function, 44
Chorioallantoic branching, 170
Chorioallantoic fusion
    description, 161
    initiating, 170
Chromatin, 9
Chx10, 527
Cited2, 339
Cleft palate, 434
Cloche, 213–215
c-myb, 204
CNC. *see* Cranial neural crest
CNS. *see* Central nervous system
Collagens
    fibrils, dermal development, 572–573
    type II genes, as chrondrocyte markers, 282, 283
    type IV renal development, 412
Colon. *see* Intestines
Colon cancer, 506
Colonization, 201, 205
Colony forming factors
    activity in mammals, 198–200
    hematopoietic system, 193–196
    HSC proliferative potential, 205
    mesodermal precursors, 200–201
Congenital heart defects, 356–357

Core-binding factors
    bone formation, 287–290
    liver hematopoiesis, 204
Cornified envelope precursor proteins, 581
Cortex
    histogenesis, 91–92
    regionalization, 84
Corticotropin-releasing hormone
    in diencephalon, 95
    expression, 512
CRABP. *see* Cellular retinoid acid binding proteins
Cranial neural crest
    cells
        olfactory development, 435
        tooth development, 436–437
    characterization, 423
    migration, 423–424
    pattern formation
        ectoderm, 432–433
        endoderm, 433
        experimental evidence, 426–427
        mesoderm, 428, 432
    studies, 421–422
Cranial suspensory ligament, 386
Craniofacia. *see also* specific regions
    pattern/development
        bHLH, 473–474
        genes encoding
            BMPs, 467–468
            chordin, 468
            EGF, 465
            endothelin, 465
            FGF, 465–466
            follistatin, 468
            hedgehogs, 470
            Lef1, 471
            Noggin, 468
            PDGF, 466–467
            Ph, 466–467
            RA, 468–469
            TGF, 465, 467
            Wnt, 471
        genes regulating
            Alx, 476–477
            Cart1, 476–477
            Dix, 471–472
            goosecoid, 477
            Hand, 473–474
            Hox, 473–474
            Nkx, 447
            Otx, 477
            Pbx, 478–379
            Pitx, 477
            Prx, 476–477

Twist, 473–474
WH, 479
studies, history, 421–422
Cre/*lox* recognition system, 15
CREM. *see* Cyclic AMP responsive element modulator
CRH. *see* Corticotropin-releasing hormone
Cripto-FRL-1-Cryptic factors, 338
Crystallins
   characterization, 528
   lens development, 531
CSL. *see* Cranial suspensory ligament
CSPGs. *see* Chondroitin sulfate proteoglycans
Cyclic AMP responsive element modulator, 286
Cyclin E, 170
Cytochrome P450, 526

D

Dach, 268
Dax1
   conservation, 387
   Leydig cells, 384
   sex determination, 378–380
Delta-like homolog 1
   inner ear neurogenesis, 559
   sensory differentiation, 553
Delta/Notch pathway, 347
Depression, neurogenesis-associated, 247
Dermomyotome axes, 139–140
Desert hedgehog
   characterization, 384
   crainofacial development, 470
Development. *see also specific organs*
   asymmetry
      early, 23–27
      overview, 21–23
   functional analysis
      mutant genes, 13–14
      in vitro approaches, 14
      in vivo approaches, 14–15
   MRF gene expression during, 258
   preimplantation, 10 11
   prepattern, 21–23
Dhh. see Desert hedgehog
Diabetes mellitus
Diencephalon, 84–85
Differentiation
   blastomeres, 151
   cardiac
      GATA factors, 342–343
      MEF2, 341–342
      modulation in cardiac chamber, 343
      SRF, 342

chrondrocyte, 281–284
diencephalon, 95
digestive tract, 307–308
   embryonic, 155–156
endoderm, 301–302
follicle, 578
forebrain
   olfactory bulb, 95
   pallium, 91–95
   subpallial, 89–91
gonadotrope, 509–501
mechanisms, 152
olfactory bulb, 95
ovary, 384
proximal-distal, 313–314
sensory, inner ear
   hair cell, 555–557
   lateral inhibition, model, 553, 555
   organ of corti, 558
   pattern regulation, 552–553
   supporting cell, 557–558
Sertoli cell, 380–382
skeletal muscle
   cell-cell interactions in, 263–264
   embryonic origins, 254–256
   inductive mechanisms, 262–263
   migratory progenitor cells, 265–266
   MRFs in
      characterization, 256–258
      functional groups, 258, 260–261
      gene expression, 258
      gene regulation, 262
   negative/postive factors, 264
   study, history, 253
   terminology, 254
   transcription factors regulating, 266–269
somite, 128–129
stem cells
   alternative models, 240–241
   BMP-induced, 238–239
testis, 380–382
urogenital, 374–375
visceral endoderm, 164–166
yolk sacs, 161
Diphtheria toxin, 510–511
DKK1, 44
DLX transcription factors
   crainofacial development, 471–472
   Dlx1/2
      forebrain differentiation, 89–92
      neuronal subtype specification, 89
      skull formation, 450
      tooth development, 437–438
   Dlx3, dermal development, 580

DLX transcription factors (*continued*)
  Dlx5
    bone formation, 290
    suballial neuron differentiation, 89–91
  Dlx6, bone formation, 290
Dmrt1
  conservation, 387
  testis differentiation, 382
DNA complementary, libraries, 11–13
Dorsal-ventral axis, 108–109
Dorsoventral axis, 139–140
Dreher mutants, 122
Duodenum. *see* Intestines
DV axis. *see* Dorsal-ventral axis
Dynorphin B peptide, 337–338
Dysplasia. *see* Campomelic dysplasia

E

Ear
  inner
    anatomy, 540–541
    cell proliferation, 547
    compartment boundaries, 552
    complexity, 539
    development, 541–452
    morphogenesis
      description, 547
      expression domains, 547, 549–550
      semicircular canal, 547
    neurogenesis, 558–559
    otic placodes
      gene expression, 546
      patterning, 543–546
      specification, 542–543
    regionalized expression, 550–552
    sensory differentiation
      hair cell, 555–557
      lateral inhibition, model, 553, 555
      organ of corti, 558
      pattern regulation, 552–553
      supporting cell, 557–558
    sensory organ specifications, 552
    stria vascularis, 559–560
Early gastrula organizer
  definition, 40
  patterning signals, 41
Ebf1, 91
ECM. *see* Extracellular matrix
Ectoderm
  CNC patterning, 432–433
  dental, 440

dermal, 570–571
myotome formation, 262–263
oral, 440
Edg1, 227
EGF. *see* Epidermal growth factors
EGO. *see* Early gastrula organizer
EGR transcription factors, 510
Embryo
  AP axis
    blastocyst, 29
    conserved properties, 40
    duplication, 22
    EGO response, 41–42
    node development, 38–40
    organizer function, 42–44
    polarity specification, 29–32
    relationship, 25–26
  chondrocranium
    characterization, 457
    morphology
      dermatocranium, 464–465
      nasoethmoidal, 457–460
      occipital region, 463
      orbitotemporal, 460, 462
      otic region, 462–463
      splanchnocranium, 463–464
  gene expression, 11–13
  hematopoiesis, 202–205
  lineages, differentiation, 155–156
  nonneural ectoderm, 570–571
  patterning, 2–3
  periderm, 583–584
  stem cells, in organizer function, 42–43
Emx2 on, 374–375
En2, 115
Endoderm
  CNC patterning, 433
  definition, 301
  development, 156, 164–166
  differentiation, 301–302
  formation
    digestive tract, 307–310
    formation
      analysis, 303–304
      AP patterning, 304
      during gastrulation, 303–304
      gastrulation to midgestation, 304–305
    liver, 316–318
    lung
      basis, 311–312
      differentiation, 313–314
      models, 314
      primary buds, 312
      respiratory system, 312–313

stem cells, 315
    trachea, 312
    vascular, 314
-mesoderm, spatial relation to, 304–305
pancreas
    cell lineages, 320–322
    dorsal/ventral buds, 319–320
    origins, 318–319
    stem cells, 322
thyroid gland, 310–311
visceral, characterization, 161
yolk sac, 302–303
role of cardiac induction, 334–336
visceral
    anterior region
        activity source, 2
        anterior patterning, 45–48
        cardiac induction, 338–339
        CNC patterning, 427
        differentiation, 30–31
        early patterning, 304
        function, model, 47–48
        Lim1 expression, 43
    nascent, AP axis formation, 29–30
    nascent, net movement of cells, 32
Endothelial cells
    arterial, 218
    cadherin-positive, 200
    cell-cell junctions, 226
    generation
        cell number regulation, 215, 217–218
        lineage specification, 213–215
        origin, 212–213
    hemogenic, 219–220
    lymphatic, 218–219
    organ specific, 219
    type specification, 220
    venous, 218
Endothelin-1
    CNC patterning, 432
    crainofacial development, 465
    skull formation, 449–450
Engrailed1/2, 112–117
Eomes, 167
Ephrins
    B1, retina patterning, 525
    B2
        endothelial cell expression, 218
        retina patterning, 525
        vascular development, 225
    B4
        expression, 218
        vascular development, 225
    -eph, rhombomere border formation, 119

Epidermal growth factors
    crainofacial development, 465
    labyrinth branching morphogenesis, 171–173
    skull formation, 452
Epithelium
    lens maintenance, 532
    to mesenchyme conversion
        aggregation, 408–410
        cell adhesion, 408–410
        cell matrix interactions, 410
        epithelial basement membranes, 410–412
        stroma, 407–408
    pigmented, eye, 523, 525
    thickening stage, dental, 454–455
ERRβ, 167
Erβ4, 424
Escargot, 169–170
Estrogen
    dermal development, 581
    pituitary vascularization, 504
Esx1, 172
Ets2, 168
Ewing's sarcomas, 247
Excretory system. *see* Kidney
Extracellular matrix
    cartilage-specific, 280–281
    CNC, 424
    dermal development, 572–574
    epithelial to mesenchyme conversion, 410
    germ cell association with, 187
Extracellular proteinases, 223
Extraembryonic lineages. *see also specific lineages*
    analysis, 156–158
    blastocyst, 159–161
    chorioallantoic placenta development, 161–162
    membranes, development, 162–164
    overview, 155–156
    primitive endoderm, 156
    spongiotrophoblast, 168–169
    trophoblast
        development, 156
        giant cell development, 169–170
        signaling pathways, 166–168
    yolk sacs, 161
Extraembryonic tissues, 2
Eya. *see* Eyes Absent genes
Eye
    basic structure, 519
    development
        conservation, 533
        Pax6 effects, 527–528
    ectopic, 533–534
    endothelial cells, 219
    lens development

Eye (*continued*)
    epithelial cell maintenance, 532
    fiber, 532
    gene regulation, 531
    induction, 528–530
    morphological/biochemical events, 528
    multiple pathways, 534
    placode to vesicle, 530–531
  optic stalk, development, 523, 525
  pigmented epithelium, 523, 525
  retina development
    cell type specifications, 526–527
    dorsoventral/nasotemporal patterning, 525
    genetic determination, 521–522, 525
    RA role, 525–526
Eyes Absent genes, *Eya1*
  conservation/divergence, 533
  inner ear development, 549–550
  lens development, 530–531
  myotome formation, 268
  renal mesoderm, 398–399

F

Fate mapping
  embryo, advantage, 191
  gastrulation studies, 39
  hematopoietic system, 201–202
  Mes/Met domains, 112–113
  testing stem cell potential, 246
Feedback loops, 62–64
Fertilization, 5, 8
Fetus. *see* Embryo
FGF. *see* Fibroblast growth factors
Fibroblast growth factors
  chondrocytes, 286–287
  crainofacial development, 465–466
  dorsal CNS patterning, 121–122
  endoderm formation, 303
  FGF-1
    angiogenic function, 223–224
    lens development, 532–533
  FGF-2
    angiogenic function, 223–224
    endothelial differentiation, 215
    hematopoietic cell differentiation, 215
    lens development, 532–533
    metanephros development, 408
    neural stem cell renewal, 242
    trophoblast development, 167
  FGF-3
    inner ear development, 543, 545–546
    sensory differentiation, 558
  FGF-4, 166–167

FGF-8
  anteroposterior telencephalon patterning, 84
  AP axis polarity, 31
  cardiac induction, 337
  CNC patterning, 432
  forebrain morphogenesis, 86
  mes/met development, 115–116
  mes/met pattern development, 112–113
  pallial telencephalon patterning, 82–83
  pituitary development, 506
  skull development, 449
  tooth development, 438, 440
  genes, effects on lung development, 314
  heart, 340
  hematopoietic specification, 203
  hormonal expression, 511
  labyrinth branching morphogenesis, 171–173
  mesoderm formation, 203
  pituitary development, 505
  pituitary vascularization, 504
  renal tubule induction, 407
  skull formation, 452
Fibronection
  description, 222
  gene, effects on heart, 340
Flat-top mutation, 243
Flk-1
  glomerular development, 412–414
  hematopoietic specification, 203
  mesoderm formation, 203
F molecule model, 55
Fn1. *see* Fibronection
FOG-2
  cardiac myogenesis, 343
  vascular development, 219
Follicle
  cycling, 578–579
  differentiation, 578
  formation, 574–575, 577
  morphogenesis, 578–579
  spacing, 577–578
Follicle-stimulating hormone
  functions, 500
  transcription factors, 510
Follistatin, 468
Forebrain
  anterior neural plate, 79
  cell-type specifications, 87–88
  diencephalon
    cell specification, 95
    differentiation, 95
    induction/patterning, 84–85
  differentiation
    neuronal, 91–95

pallium, 91–95
    subpallial neurons, 89–91
morphogenetic mechanisms, 85–87
neurogenesis, control of, 87–88
neuronal subtype specification, 88–89
organization, 75–77
prosencephalon
    early morphogenesis, 79
    neural plate, patterning subdivisions, 78–79
    organization, 75, 77
telencephalon
    anteroposterior, patterning, 83–84
    organization, 77
    pallial, patterning, 82–83
    subpallial, patterning, 79–82
Forkhead/winged helix genes, 404
Foxc. *see* Forkhead/winged helix genes
FRIZZLED, 82–83
FSH. *see* Follicle-stimulating hormone
Furin, 67–68
Fzd5, 172

G

Gap junctional communication, 60–61
Gastrointestinal tract. *see* specific regions
Gastrulation
    conserved properties, 40
    description, 38
    node development, 38–40
    vertebrate organizers
        embryological studies, 41–42
        function, 42–44
        genetic studies, 41–42
        inhibitory signals, 44
        specification, 44–45
GATA transcription factors
    Amh regulation by, 381–382
    -FOG2, vascular development, 219
    Gata1, hematopoietic specification/differentiation, 204
    Gata2
        hematopoietic specification/differentiation, 204
        liver hematopoiesis, 203–204
        pituitary development, 381–382
    Gata4
        endoderm development, 165
        -FOG, cardiac myogenesis, 342–343
    Gata6, endoderm development, 165
Gbx2
    CNS AP patterning, 114–115
    mes/met domain
        expression, 114
        pattern development, 112–113
        patterning, model, 116–117

GDF
    type 1
        node flow, 60
        pallial telencephalon patterning, 82–83
    type 7, dorsal CNS patterning, 121–122
    type 8 (*see* Myostatin)
GDNF. *see* Glial-derived neurotrophic factor
Germ cells
    characterization, 181
    early appearance, 182–183
    in gonad development, 383–384
    migration
        adverse behavior during, 187–188
        guidance of, 186–187
        process, 185–186
        proliferation/survival during, 188–189
    specification, 183–185
Germinal vesicle. *see* Oocytes
GFP. *see* Green fluorescent protein
GH. *see* Growth hormones
GHRH. *see* Growth hormone-releasing hormone
GJC. *see* Gap junctional communication
Glial cells
    derived neurotrophic factor, 398, 401–403
    missing-1, 170
    neuron generation, 87–88
Glial growth factor. *see* Neuregulin
Gli zinc finger transcription factors
    Gli1
        dermal tumorigenesis, 579
        diencephalon, 84–85
    Gli2
        CNS patterning, 121
        diencephalon, 84–85
        lung development, 312
        tooth development, 439
    Gli3
        CNS patternin, 121
        lens fiber development, 532
        lung development, 312
        optic stalk development, 522
        pallial telencephalon patterning, 83
        tooth development, 439
Glomerular, 412–414
Gonad
    development
        genes involved in, 374–376
        germ cells, 383–384
        Leydig cells, 384
        origins, 372–373
        studies, 382–383
    differentiation, 375–376
    major cell lines, 373
    mesonephros, 373–374

Gonadotrope, 509–510
Gonadotropin-releasing hormone, 510
Goosecoid, 477
Gorlin syndrome, 579
Grb2, 171–172
Green fluorescent protein
   expression, 182
   guidance of, 186–187
   motility, 185
Growth hormone-releasing hormone
   in diencephalon, 95
   expression, 511–512
Growth hormones
   expression, 511–512
   functions, 500
   transcription factors, 510
Gut tube regionalization, 305

H

Hairy genes, 138–139
Hand genes
   cardiac looping, 355
   crainofacial development, 355
   spongiotrophoblast development, 168–169
   ventricular specification, 348–349
HB-EGF, 156
Head
   CNC
      fate mapping, 424–426
      migration, 423–424
      pattern formation
         ectoderm, 432–433
         endoderm, 433
         experimental evidence, 426–427
         information source, 427–428
         mesoderm, 428, 432
      primordial cells, 423
   primordial cells, 422
   skull development
      regulation
         calvarium formation, 451–452
         intra-BA identity, 449–451
         molecular evidence, 442–443
         neurocranium formation, 450
         skeletogenesis, 442
      studies, 441–442
   skull organization
      characterization, 456–457
      chondrocranium, embryonic, 457
      morphology
         dermatocranium, 464–465

         nasoethmoidal, 457–460
         occipital region, 463
         orbitotemporal, 460, 462
         otic region, 462–463
         splanchnocranium, 463–464
      units/divisions, 457
Heart
   apoptosis, 341
   cellular proliferation, 341
   chamber formation, 347–348, 351
   development, 332–334
   disease, congenital, 356–357
   endothelial cells, 219
   induction
      activin, 337
      BMP, 336, 337
      cerberus, 338
      CFC, 338
      conserved pathway, 334
      endoderm, 334–336
      FGF, 337
      timing/stability, 340
      visceral endoderm, 338–339
      Wnts/anti-Wnts, 336–337
   inflow/outflow patterning
      Delta/Notch pathway, 347
      early model, 345
      RA signaling, 346–347
   left-right axis, 351, 354–356
   morphogenesis
      conserved pathway, 334
      field, 339
      looping, 351, 354–356
      segmental model, 344
   myogenesis
      cardiac chamber, 343
      GATA factors, 342–343
      MEF2, 341–342
      SRF, 342
   precursors, migration of, 340–341
   regionality
      development, 343–344
      plasticity of, 344
   structure, 332–334
   ventricular specification
      hand genes, 349–350
      Irx4, 350–351
      MEF2 proteins, 350
      Nkx2-5 homolog, 348–349
      versican, 350
Helix transcription factors
   HNF3, endoderm development, 165–166
   HNF4, endoderm development, 164–166

HNF6, pancreatic expression, 321
HNF3β
    anterior patterning, 47
    endoderm formation, 303
    liver development, 316–317
    organizer function, 42–44
Hemangioblasts, 214
Hematopoiesis
    colonization theory of, 201
    definitive, 203–204
    genes affecting
        cell migration, 204
        fetal liver, 203–204
        lineage specification/differentiation, 203–204
        proliferative potential, 205
    regulation, 202
Hematopoietic system
    characterization, 191
    colonization, 205
    compartments, origins
        amphibians, 196, 197
        colonization theory, 201
        fate mapping, 201–202
        mammals, 198–199
        mesodermal precursors, 200–201
        multiple, 199–200
    evolution, 205
    genetic programing, 205–206
    growth factors, response to, 193–194
    hierarchy, 192
    progenitors
        identification, 192–194
        location, 195–196
    specification, genes involved, 203
    stem cells
        fate of, redirecting, 206
        identification, 192–194
        location, 195–196, 199–201
        proliferative potential, 205
Heparan sulfate proteoglycan, 187
Hepatocyte growth factor
    function, 317
    -SF, role in muscle development, 266
    ureteric bud growth, 404
Hepatocyte growth factors, 171–173
Hereditary hemorrhagic telangietasia, 227
Herpes simplex virus 1, 510
*Hes*1
    pancreatic expression, 321
    paraxial mesoderm expression, 138–139
    sensory differentiation, 555
Hesx1
    anterior neural development, 47

    expression in AVE, 45
    pituitary development, 507–509
Heteroxtaxia, 354
Hex
    characterization, 217
    expression, 30–31
HGF. *see* Hepatocyte growth factor; Hepatocyte growth factors
HGM proteins, 9, 377
Hhex
    AP axis polarity, 31
    expression in *cloche,* 214–215
HHT1. *see* Hereditary hemorrhagic telangietasia
HIF-1α, 225
Hindbrain, 117–119
Histones, 9
Hlxb9, 319–320
Hmx3, 549
HNF. *see* Helix transcription factors
Hoxa1, 544–545
Hox genes
    CNC patterning
        mesoderm, 428, 432
        sources, 427–428
    crainofacial development, 474
    ectodermal development, 433
    gut specification, 305
    hindbrain patterning, 117–118
    neural stem cell renewal, 242
    rhombomere formation, 119
    skull formation, 443, 449
    small intestine, 309
    thyroid growth, 311
HSPG. *see* Heparan sulfate proteoglycan
HSV1. *see* Herpes simplex virus 1
Huntington's disease, 247
Hyaluronic acid, 350
Hyd, 68
Hypothalamic-releasing hormones, 502
Hypothalamus, 504
Hypoxia, 173

I

ICM. *see* Inner cell mass
IKK, 580
Ileum. *see* Intestines
I-mfa], 169
Indian hedgehog
    bone formation, 287–288
    characterization, 285
    crainofacial development, 470
    stomach development, 308, 310

Inner cell mass
  derivatives, 157
  description, 156
  differentiation, 23, 25–26
  isolation, stages, 158
  Oct4, 160
  surface, 28
  trophoblast development, 166
Insl3, 386
Integrins
  dermal development, 573–574
  description, 222–223
  germ cell migration, 188
  hematopoietic cell migration, 204
Integumentary structures. *see specific organs*
Intestines
  development, 309–310
  differentiation, 307–308
  patterning, 307–308
Intrauterine growth restriction, 174
Inversin gene, 354
Irx, 350–351
Isl1
  pancreatic bud formation, 319–320
  pituitary development, 505
Isomerism, 354
Itga-4, 170
IUGR. *see* Intrauterine growth restriction
iv mutation, 58–59

J

Jagged gene, 553, 555
Jak2, 204
Jejunum. *see* Intestines
JNK. *see* c-Jun N-terminal kinase
c-Jun N-terminal kinase, 405

K

Keratin, 579–580
Kidney
  development
    classical models, 404–405
    glomerular, 412–414
    intermediate mesoderm, 397–399
    mesenchymal induction, 405, 407
    PKD-association, 414–415
    vascularization, 412–414
    Wilms' tumor-association, 415–416
    Wnt signaling, 405, 407
  epithelial to mesenchyme conversion
    aggregation, 408–410
    cell adhesion, 408–410
    cell matrix interactions, 410

    epithelial basement membranes, 410–412
    stroma, 407–408
  metanephric mesenchyme, 399–400
  nephric duct initiation, 400–401
  origins, 372, 395–396
  tubules, induction, 407
  ureteric bud outgrowth
    inhibitors/enhancers, 403–404
    RET/GDNF pathway, 401–403
Klf4, 580–581
Kreisler, 544
KROX20, 117–118

L

Labyrinth
  branching morphogenesis, 171–173
  description, 161
  development, 161–162, 170
LacZ, 267
Laminar patterning, 91–95
Laminins, 411
Lateral ganglionic eminence
  description, 77
  development, 81–82
  forebrain differentiation, 89–90
  olfactory bulb differentiation, 95
  specification, 82
Lateral plate mesoderm
  GDF1 expression, 60
  left-side expression in, 62
  node and, line between, 64
  node to, signal transfer, 64
LCR. *see* Locus control region
LDL gene family, 94
Lef1
  crainofacial development, 471
  labyrinth branching morphogenesis, 172
  tooth development, 439–440
LEF/TCF, 86–87
Lefty1
  characterization, 47
  left-right asymmetry, 59–61
  midline, 64–65
  retinoic acid effects, 67
Lefty2
  characterization, 47
  expression, retinoic acid effects, 67
  expression in left-right asymmetry, 59–61
  -nodal, regulatory relationship, 62–64
Legless mutation, 59
Leukemia inhibitory factor
  expression, 512
  implantation process, 156

Leukocyte inhibitory factor, 407
Leydig cells, 373, 384
Lfng
  inner ear neurogenesis, 559
  paraxial mesoderm, 138–139
  sensory differentiation, 555, 558
LGE. *see* Lateral ganglionic eminence
LH. *see* Luteinizing hormone
Lhx
  CNC patterning, 432
  pituitary development, 507–508
  retina development, 521–522
LIF. *see* Leukemia inhibitory factor; Leukocyte inhibitory
    factor
LIM homeodomain, *Lim1*
  anterior neural development, 46–47
  gonad development, 376
  organizer function, 43–44
  renal mesoderm, 397
  urogenital differentiation, 374–375
Lineage specification, 151
Liver
  fetal, hematopoiesis, 203–204
  formation
    cell interactions, 317–318
    endodermal competence, 316–317
    mesodermal signals, 317
    nonendodermal tissues, 318
    regenerative capacity, 318
LMP. *see* Lateral plate mesoderm
Locus control region, 202
Lung
  development
    basis, 311–312
    buds, 312
    mesoderm differentiation, 314
    models, 314
    proximal-distal differentiation, 313–314
    respiratory system, 312–313
    stem cells, 315
    trachea, 312
    vascular, 314
Luteinizing hormone
  expression, 502
  functions, 500
  transcription factors, 510
Luteinizing hormone-releasing hormone, 435

M

MAF, 531
MAP. *see* Mitogen-activated protein
MAP kinases. *see* Mitogen-activated protein kinases
Mash1, 88–89

Mash2, 168–169
Math1, 555, 558
Math5, 526
Matrix metalloproteinases, 223
Matrix remodeling, 223
Maturation promoting factor
  description, 7
  inactivation, 8
Medial ganglionic eminence
  description, 77
  development, 81–82
  forebrain differentiation, 89–90
  specification, 82
Medial-lateral axis, 78
MEF2 proteins
  cardiac myogeneis, 341–342
  ventricular specification, 350
Meiosis, 7–8
Melanocytestimulating hormone, 499
Mental illness, 247
Mes. *see* Mesencephalon
Mesencephalon
  AP patterning, 110–112
  domains, gene expression patterns, 112–113
  fate mapping, 113–114
Mesenchymal cells
  renal development, 405, 407
  renal function, 386
Mesenchyme
  epidermal appendage morphogenesis, 574
  to epithelial conversion
    aggregation, 408–410
    cell adhesion, 408–410
    cell matrix interactions, 410
    epithelial basement membranes, 410–412
    stroma, 407–408
  pituitary development, 504–506
Mesoderm
  cardiac, 335
  cells, 263
  CNC patterning, 428, 432
  differentiation, 314
  -endoderm, spatial relation to, 304–305
  endothelial cell origin and, 212–213
  formation, 203
  hepatic signals, 317
  paraxial
    molecular clock, 138–139
    progenitor cells, 132
  presomitic, cell transition, 132–133
  progenitor cells, 254
  renal, 397–399
Mesp1, 340
Met. *see* Metencephalon

Metanephric mesenchyme, 399–400
Metanephros. *see* Kidney
Metencephalon
   AP patterning, 110–112
   domains, gene expression patterns, 112–113
   fate mapping, 113–114
Mgat, 67
MGE. *see* Medial ganglionic eminence
Microinjection of proteins, 14
Migration
   germ cells
      adverse behavior during, 187–188
      guidance of, 186–187
      process, 185–186
      proliferation/survival during, 188–189
   hematopoietic cell, 204
Mitogen-activated protein, 287
Mitogen-activated protein kinases, 8, 167
Mixer, 166
Morphogenesis
   branching
      chorioallantoic, 171–173
      organ formation, 298–299
   budding, 298–299
   cardiac, 334, 344
   endoderm, 304–305
   epidermal appendage, 574
   follicle, 578–579
   forebrain, 85–87
   inner ear, 547, 549–550
   node, 38–40
   olfactory bulb, 84
   prosencephalon, 78–79
Mouth development, 433–434
Mox2, 267
MPCs. *see* Myogenic precusor cells
MPF. *see* Maturation promoting factor
MRFs. *see* Myogenic regulatory factors
MSH. *see* Melanocytestimulating hormone
Msx transcription factors
   crainofacial development, 474
   Msx1
      formation, 290
      myotome formation, 267
      skull formation, 450
      tooth development
         bud-to-cap, 440–441
         initiation, 439
         patterning, 437–438
   Msx2
      formation, 290
      skull formation, 452
      tooth patterning, 437–438

mTEAD-2 n, 9–10
Mueller cells, 526
Müllerian duct
   characterization, 373–374
   differentiation, 385–386
Muscle. *see* Skeletal muscle
MyoD. *see* Myogenic regulatory factors
Myogenesis. *see under* Differentiation
Myogenic precursor cells, 269
Myogenic regulatory factors
   cell-cell interactions, regulation of, 263–264
   characterization, 151–152, 253
   conservation, 256–258
   ectoderm, 263
   expression, 254
   functional groups, 258, 260–261
   gene expression, 258
   gene regulation, 262
   muscle specification, 268–269
   myotome formation, 265, 266–269
   satellite cells, 269, 271
Myosin, 556
Myostatin, 264
Myotome formation
   ectoderm role, 262–263
   genetic analysis, 264–265
   positive/negative signals, 264
   transcription factors regulating, 266–269
   waves of, 131

N

NCSC. *see* Neural crest stem cells
Nephric duct, 400–401
Nephron, 396
Nerve growth factors
   expression, 511
   neural stem cell survival, 239
Netrin-1
   inner ear development, 547
   pituitary development, 506
Neural axis, 2–3
Neural crest stem cells
   adult/fetal, comparison, 245
   cell cycle status, 245
   characterization, 235
   childhood cancer and, 247
   differentiation
      alternative models, 240–241
      BMP role, 238–239
   hematopoietic stem cells, comparison, 243–244
   lineage, 237–238
   neuregulin, 239–240

neuronal potential, 241–242
nonneural derivative potential, 245–247
occurrence, 236–237
other nerve cells, association, 237
outstanding issues, 248
response to nervous system injury, 247
self-renewal, regulation, 242–243
survival, role of neutotrophins, 239
in vivo function, 244–245
Neural plate
    anterior, 78–79
    prosencephalic, 78–79
Neuregulin, 239–240
NEUROD1, 509
Neurod1, 559
Neuron restrictive silencing factor, 243
Neurons
    development, 247
    differentiation in forebrain, 89–91
    laminar patterning, 94
    subtype, forebrain, 88–89
Neuropilin-1, 223
Neurosphere cell, 246
Neurotrophin 3
    inner ear neurogenesis, 559
    neural stem cell survival, 239
Neurulation, 78–79
Neutotrophins, 239. *see also specific Neutotrophins*
NGF. *see* Nerve growth factors
Ngn2, 526–527
Nkx genes
    Nkx2.1, 474
    Nkx2-3
        L-R asymmetry, 68
        nodal regulation, 66–67
        small intestine, 310
    Nkx2-5
        cardiac looping, 356
        cardiac morphogenesis, 339
        CHD, 357
        heart formation, 348–349
    Nkx5.1, 474
    Nkx6.1, 474
Nodal
    gene, cardiac looping, 355
    left-right asymmetry, 59–61
    Lefty2 regulation, 62–64
    Nkx2-3 regulation, 66–67
Node
    development, 38–40
    flow, 59–61, 64
    genes regulating, 66–67
    and LPM, line between, 64

Noggin
    crainofacial development, 468
    myotome formation, 265
Nose. *see* Olfactory
Notch signaling
    cardiac patterning, 347
    myogenic differentiation, 263
    sensory differentiation, 553
    somite borders and AP polarity, 136–138
    vascular development, 226
NRSF. *see* Neuron restrictive silencing factor
NT-3. *see* Neurotrophin 3

O

Oct4
    germ cell expression, 183
    ICM role, 160
Olfactory bulb
    development, 435
    differentiation, 95
Oligodendrocyte precursor cells, 246
Oocytes
    asymmetry, 7–8
    development
        initial stages, 6–7
        transcription during, 8–10
    fertilized
        asymmetry in, 1
        morphological study, 1
    maturation, 10–11
OPCs. *see* Oligodendrocyte precursor cells
Organogenesis. *see also* specific organs
    functional compartments, 299
    molecular control of, 299–300
    morphogenesis, 298–299
    process, 297–298
    progenitor tissue remodeling, 298
Orthodenticle transcription factor
    Otx1, inner ear development, 549
    Otx2
        anterior neural development, 46–47
        anterior patterning, 47
        CNS AP patterning, 114–115
        expression in Mes/Met domains, 113–114
        function, 44
        inner ear development, 549
        mes/met domain patterning, model, 116–117
        mes/met pattern development, 112–113
        retina development, 521
Ossification
    description, 279, 281
    types, 280

OT. *see* Oxytocin
Ovary differentiation, 384
Oxytocin
   diencephalon, 95
   expression, 502
   functions, 500

P

PA. *see* Posterior anterior interfaces
Paired-box genes
   crainofacial development, 475–476
   Pax2
      inner ear development, 549
      mes/met domain patterning, model, 116–117
      mes/met pattern development, 113
      metanephros development, 408
      optic stalk development, 522
      renal mesoderm, 397–399
      reproductive tract, 385
   Pax3
      muscle development, 266
      myotome formation, 267–268
   Pax5
      mes/met development, 115
      mes/met domain patterning, model, 116–117
   Pax6
      CNC migration, 423–424
      conservation/divergence, eye, 533–534
      dosage effects, eye, 527–528
      ectopic eye development, 533–534
      laminar patterning, 94
      lens development, 530–531
      optic stalk development, 522
      pituitary development, 509
      retina development, 521–522, 526
      tooth development, 440
   Pax7
      muscle development, 266
      muscle satellite cells, 271–272
      myotome formation, 267–268
   Pax8, thyroid growth, 311
   Pax9
      gut tube regionalization, 305
      hematopoietic specification/differentiation, 204
      tooth development, 441
Paired-related genes, 476–477
Pallium, 91–95
Pancreas
   formation
      cell lineages, 320–322
      dorsal/ventral buds, 319–320

      origins, 318–319
      stem cells, 322
   function, 319
Paneth cells, 310
Para-aortic splanchnopleura
   angiogenesis, 225
   CFU-S in, 196
   hematopoietic progenitor source, 199
   hematopoietic stem cell source, 199
   mesodermal precursors, 200–201
ParaHox genes
   Cdx1/2
      characterization, 167
      gut tube regionalization, 305
      role in gut tube regionalization, 305
   Pdx1
      gut tube regionalization, 305
      skull formation, 443, 449
      small intestine formation, 309
Parathyroid hormone, 284–286
Parathyroid hormone related peptides, 284–286
Paraxis, 267
Parkinson's disease, 247
PAS. *see* Para-aortic splanchnopleura
Patch, 466–467
Patterning
   AP
      AVE role, 44–48
      in endoderm, 304
      mes/met, regulation, 110–112
   CNS, 107–109
   diencephalon, 84–85
   digestive tract, 307–308
   dorsal ventral in CNS, 120–122
   DV, in forebrain, 88
   establishment, 1–3
   inflow/outflow, cardiac, 346–347
   laminar, 91–95
   mes/met domain, models, 116–117
   prosencephalic neural plate, 78–79
   retina, 525
   telencephalon
      anteroposterior, 83–84
      pallial, 82–83
      subpallial, 79–82
Pb. *see* Polar body
PBX genes, 478–479
Pdyn. *see* Preprodynorphin gene
Pericytes, 226–227
Periderm, 583–584
Peripheral neuroectodermal tumors, 247
PFP. *see* Prospective floor plate
Pie-1, 243

Pit1, 512
Pituitary gland
  anatomy, 499–500
  development
    cell proliferation, 502, 504
    committed cell type expansion, 510–512
    genetic roles, 506–509
    induction, 504–506
    morphology during ontogeny, 502
    multihormonal cells, 512
    transcription factors
      cell specification synergy, 509–510
      signaling molecule interactions, 506
    vascularization, 502
  hormones
    appearance, 502
    functions, 499–500
PITX gene family
  Pitx1, crainofacial development, 478
  Pitx2
    crainofacial development, 478
    expression on left side, 65–66
    heart looping, 354–355
    mouth development, 433–434
    pituitary development, 506–507
  Pitx3, lens development, 532
  Pitx4, pituitary development, 506–507
PKA. *see* Protein kinase A
PKD. *see* Polycystic kidney disease
Placenta
  chorioallantoic
    development, 161
    function, 161–162
  development, pregnancy complications and, 173–174
  evolution, 163
  species differences, 162–163
Placodes, neurogenic, 434–435
Platelet-derived growth factors
  crainofacial development, 466–467
  glomerular development, 413–414
  lung development, 313
  pericyte recruitment, 226
  smooth muscle cell recruitment, 226
PNET. *see* Peripheral neuroectodermal tumors
Pod1, 413
Podocyte cells, 411
Polar body, 26–27
Polarity
  AP axis
    somitogenesis, 134, 136–138
    specification, 29–31
  vascular, 227
Polyadenylation, 10

Polycystic kidney disease, 414–415
Polycystin-1, 415
POMC. *see* Pro-opiomelanocortin
Posterior anterior interfaces, 138
POU
  inner ear development, 549
  neuronal subtype specification, 89
  pituitary development, 509
  Pou5f1 (see Oct4)
  sensory differentiation, 556, 558
Pregnancy complications, 173–174
Preprodynorphin gene, 338
PRL. *see* Prolactin
Progenitor cells
  allocation to paraxial mesoderm, 132
  migratory muscle, 265–266
Progenitor tissues, 298
Prolactin
  characterization, 500
  expression, 511
Pro-opiomelanocortin
  characterization, 499
  expression, 502
  HSV1, 510
Prop1, 509
Prosencephalon
  description, 75, 77
  early morphogenesis, 79
Prospective floor plate, 64–65
Protein kinase A, 285–286
Protein kinase C, 580
Prox1
  inner ear development, 549
  lens development, 532
Prox2, 549
Proximal-distal differentiation, 313–314
PTF1, 321

R

RA. *see* Retinoic acid
RAR, 403
Rathke's cysts, 512
Rax. *see* Rx
Receptor type tyrosine kinase, 401–403
Rectum development, 310
Remodeling, bone, 298
Reproductive tract development, 385–386
Respiratory system. *see under* Lung
RET. *see* Receptor type tyrosine kinase
Retinaldehyde dehydrogenase 2, 545
Retinoic acid
  crainofacial development, 468–469

Retinoic acid (*continued*)
   Hox expression, 67, 119
   inner ear development, 545
   L-R asymmetry, 67
   lung development, 312
   retina development, 525–526
   sensory differentiation, 557
   signaling in cardiac patterning, 346–347
   trophoblast, 167
Rhombomeres
   border formation, 119
   junctions, 118–119
   segmentation, 119
   transcription factors, 117–118
Rieger syndrome, 507
RNA
   double-stranded, 14
   messenger
      inhibiting translation, 14–15
      oocyte maturation, 10–11
      preimplantation development, 10–111
Rpx, 45
Rx, 521–522

S

Satellite cells
   in adults, 269
   MRFs, 269, 271
   proliferation, 271–272
Sca-1, 192–193
Sclerotome
   axis, generation, 139–140
   cells, function, 128–129
   delineation, 140–141
   formation, 131
   subdivision, 129, 131
Scl/tal-1
   characterization, 216
   expression in *cloche,* 214–215
   hematopoietic specification, 203
   mesoderm formation, 230
Secreted factors of the Frizzled-related protein, 83
Segmentation
   description, 127
   rhombomere regulation, 119
   somite, rate of, 130–131
Sensory neurons, 434–435
Sensu stricto, 77
Serrate-1, 557
Serrate-2, 557
Sertoli cells
   differentiation, 380–382
   genotypes, 382–383

Serum response factor, 342
Sex cord, 188
Sex determination. *see also* Gonad; Ovary
   evolution and, 386–387
   genes involved in
      Dax1, 378–380
      SRY, 376–378
      Tda, 378
Sexual development, 384–386
SF1. *see* Steroidogenic factor 1
SFRP. *see* Secreted factors of the Frizzled-related protein
Shaker2, 556–557
SHH. *see* Sonic hedgehog
Signal transducer and activator of transcription, 286–287
Six-homeodomain transcription factors
   Six1, myotome formation, 268
   Six3
      lens development, 530–531
      retina development, 521–522
Skeletal muscle
   differentiation
      cell-cell interactions in, 263–264
      migratory progenitor cells, 265–266
      MRFs in
         characterization, 256–258
         functional groups, 258, 260–261
         gene expression, 258
         gene regulation, 262
      negative/postive factors, 264
      study, history, 253
      transcription factors regulating, 266–269
   embryonic origins, 254–256
   mesodermal progenitors, 253–254
   myogenesis, inductive mechanisms, 262–263
   regeneration, 269, 271–272
   specification, 268–269
   transcriptional regulation, specificity, 262
Skeleton. *see* Bone
Skin
   development
      factors, 572–574
      tumorigenesis and, 579
   embryonic ectoderm, nonneural, 570–571
   mature
      cells, 570
      characterization, 570
      function, 570
   stratification, 571–572
   terminal differentiation
      early, 579–580
      late, 581–583
      late stage, transit to, 580–581
   tumorigenesis, 579

SMAD2. *see also* Transforming growth factor-β
  anterior patterning, 47
  endoderm formation, 303
  mutants, 43
Small proline-rich region proteins, 581
SMG. *see* Submandibular gland
Smooth muscle cell recruitment, 226–227
Sna, 65–66, 169–170
Snail gene, 354
Snell's waltzer, 556
SOM. *see* Somatostatin
Somatic cells, 373
Somatostatin, 95
Somites
  AP compartments, 143
  AP polarity, role of Notch signaling, 134, 136–138
  borders, role of Notch signaling, 134, 136–138
  compartments, lineage
    dorsal, 141
    generation, 139–140
    lateral, 140–141
    ventral, 140–141
  development, staging, 130
  formation
    kinetics, 138–139
    process, 127–132
  number, variation, 130
  prepatterning, 133–134
  segmentation, rate of, 130–131
  skeletal muscle origins, 254
Sonic hedgehog
  crainofacial development, 470
  dermal tumorigenesis, 579
  diencephalon, 84–85
  follicle development, 578
  forebrain morphogenesis, 86
  L-R asymmetry, 68
  lung development, 312
  mutants, in organizer function, 43
  myotome formation, 264, 266
  pancreatic buds, 319–320
  prosencephalic neural plate patterning, 78
  scletotome delineation, 140–141
  stomach development, 308
  tooth development
    bud-to-cap transition, 440–441
    ectoderm regionalization, 439–440
SOX transcription factors
  chrondrocyte differentiation, 281–284
  ectopic eye development, 533–534
  forebrain differentiation, 91
  lens development
    fiber, 532
    induction, 529

    placode to vesicle, 530–531
    regulation, 531
  sex determination, 91
  testis differentiation, 380–381
Species differences, 162–164
Spongiotrophoblast, 168–169
SPRRs. *see* Small proline-rich region proteins
SRF. *see* Serum response factor
SRY gene
  characterization, 376–378
  conservation, 386–387
Stem cells
  CNS
    adult/fetal, comparison, 245
    cell cycle status, 245
    characterization, 235
    differentiation
      BMP role, 238–239
      process, 241
    function in adults, 237
    hematopoietic stem cells, comparison, 243–244
    lineage, 237–238
    neuronal potential, 241–242
    nonneural derivative potential, 245–247
    occurrence, 235–236
    other nerve cells, association, 237
    outstanding issues, 248
    response to nervous system injury, 247–248
    survival, role of neutotrophins, 239
    in vivo function, 244–245
  embryonic
    characterization, 152–153
    differentiation, 160
    for hemangioblast study, 200
    mouse, 158
    trophoblast development, 166
  epidermal, 570
  hematopoietic
    description, 191
    identification, 192–194
    location, 195–196, 199–201
    proliferative potential, 205
    rate of, redirecting, 206
  hepatic, 318
  ligands, hematopoietic specification, 203
  mesoderm formation, 203
  neural
    adult/fetal, comparison, 245
    cell cycle status, 245
    characterization, 235
    childhood cancer, 247
    differentiation
      alternative models, 239
      BMP role, 238–239

Stem cells (*continued*)
    hematopoietic stem cells, comparison, 243–244
    lineage, 237–238
    neuregulin effects, 239–240
    neuronal potential, 241–242
    nonneural derivative potential, 245–247
    occurrence, 235–236
    other nerve cells, association, 237
    outstanding issues, 248
    response to nervous system injury, 247–248
    self-renewal, regulation, 242–243
    survival, neutotrophin role, 239
    in vivo function, 244–245
  pluripotent muscle-derived, 271–272
  pulmonary, 315
  source, pancreatic ducts as, 322
  trophoblast
    isolating, 166–167
    proliferation, 160
Stereocilia production, 557
Steroidogenic factor 1
  gonad differentiation, 375–376
  hormonal development role, 509–510
  testosterone production, 384
Stomach development, 308–309
Submandibular gland, 436
Subventricular zone
  description, 87
  properties, 88–89
Superovulation, 21
SVZ. *see* Subventricular zone

T

Tabby gene, 441
Tal-1, 216
TATA box binding proteins, 9
T-box transcription factors
  TBr1, in laminar patterning, 92
  Tbx1
    CHD role, 357
    retina patterning, 525
T cells
  CD31, HSC expression, 192–193
  CD34, HSC expression, 192–193
  hematopoietic cell migration, 204
  hematopoietic specification/differentiation, 204
Tcf1, 172
Tcf15. *see Paraxis*
Tda gene, 378
Tdgf1, 31
Telencephalon
  organization, 77
  pallial, 82–83
  subpallial, patterning, 79–82

Testis
  descent, 386
  differentiation, 380–382
Testosterone, 581
Tfm, 386
Thyroid gland development, 310–311
Thyroid hormones, 510
Thyroid-specific enhancer binding protein
  lung development, 312
  pituitary development, 504, 507–508
  thyroid development, 311
Thyroid stimulating hormone
  expression, 502
  functions, 500
  receptor, 311
Thyrotrope, 512
Thyrotropin-releasing hormone
  diencephalon, 95
  expression, 512
Tie1
  receptors, 224
  vascular polarity, 227
  venous malformation disease and, 227
Tinman, 334
Tissue nonspecific alkaline phosphatase, 183
Titf1. see Thyroid-specific enhancer binding protein
TNAP. *see* Tissue nonspecific alkaline phosphatase
Tooth
  characterization, 436
  development
    bud stage, 455
    bud-to-cap transition, 440–441
    cap stage, 455–456
    cells, origins, 436–437
    dentition patterning, 437–438
    ectoderm regionalization, 440
    enamel knot, 441
    epithelial thickening stage, 454–455
    initiation, 438–440
    molar, 456
    morphogenesis, 574
Trachea. *see under* Lung
Transcription factors. *see also specific factors*
  laminar patterning, 92, 94
  during oogenesis, 8–10
Transforming growth factors
  crainofacial development, 465
  hematopoietic specification, 203
  mesoderm formation, 203
  secondary palate development, 434
Transforming growth factor-β
  anterior neural development, 46–47
  crainofacial development, 467
  dorsal CNS patterning, 121–122
  germ cells, 187

myotome formation, 264
node flow, 60
organizer formation, 44–45
pallial telencephalon patterning, 82–83
pancreatic expression, 321
pericyte recruitment, 226
signaling, prosencephalic neural plate, 78–79
signaling, vasculogenesis, 222
smooth muscle cell recruitment, 226
vascular development, 225–226
TRH. *see* Thyrotropin-releasing hormone
Trophoblasts
    branching, 161
    development, 156, 166–268
    giant cell development, 169–170
    proliferation, 160
    stem cells proliferation, 160
TSH. *see* Thyroid stimulating hormone
TSHR. *see* Thyroid stimulating hormone receptor
Tumor angiogenesis, 227
Tumorigenesis, dermal, 579
Tumor necrosis factor, 441
Twist
    characterization, 452
    crainofacial development, 473–474
Tyrosine kinase receptors, 91

U

UDP-*N*-acetylglucosamine:α-3-sDS-mannosideβ-1,2-*N*-acetylglucosaminyl transferase I, 67

V

Vascular channel regression, 227
Vascular endothelial growth factor
    angiogenic function, 223
    bone formation, 291
    description, 216
    endothelial number regulation, 215, 217–218
    expression in hemangioblasts, 214
    function, 200
    glomerular development, 412–413
    pituitary vascularization, 504
    receptors
        expression, 200
        VEGF-R1, 223
        VEGF-R2, 215, 223
        VEGF-R3, 218–220
    VEGF-C, 218–220
Vascular permeability factor, 216
Vascular system. *see also specific elements*
    channel formation, 222–225

development
    associated endothelial cells
        cell number regulation, 215, 217–218
        lineage specification, 213–215
        origin, 212–213
    blood islands, 212
    definition, 211
    process, 220, 222
    pulmonary, 314
    morphogenesis (*see* Angiogenesis)
Vasopressin, 500
Vax1, 522, 525
Vax2, 525
Vcam-1, 170
VCFS/DGS. *see* Velocardiofacial/DiGeorge syndrome
VEGF. *see* Vascular endothelial growth factor
Vein segregation, 225–226
Velocardiofacial/DiGeorge syndrome, 357
Venous malformation disease, 227
Ventral ectodermal ridge, 132
Ventricular zone
    cardiac specification, 349–351
    CNS, 109
    forebrain, 87
VER. *see* Ventral ectodermal ridge
Versican, 350
Vertebrate
    hematopoietic system
        characterization, 191
        hierarchy, 192
    organizers
        gastrulation
            function, 42–44
            inhibitory signals, 44
            specification, 44–45
        in gastrulation
            embryological studies, 41–42
            genetic studies, 41–42
VLDLR in laminar patterning, 94
von Willebrand factor, 219
VPF. *see* Vascular permeability factor
VZ. *see* Ventricular zone

W

Waardenburg syndrome, 268
WAGR, 375
Wilms' tumor
    renal development association, 415–416
    suppressor gene
        Amh regulation by, 381–382
        as metanephric mesenchyme marker, 399–400
        sex determination, 379
        urogenital differentiation and, 374–375

Winged helix/forkhead genes, 479
Wingless factors
   cardiac induction, 336–337
   CNC patterning, 432
   crainofacial development, 471
   dorsal CNS patterning, 121–122
   follicle formation, 555, 557
   forebrain morphogenesis, 86
   labyrinth branching morphogenesis, 172
   mes/met development, 115
   Müllerian duct differentiation, 385–386
   myotome formation, 264–265
   organizer formation, 44–45
   pallial telencephalon patterning, 82–83
   pituitary development, 505–506
   renal development, 405, 407
   somite patterning, 141
WNT. *see* Wingless factors
Wolffian ducts
   characterization, 373
   differentiation, 386
WT1. *see under* Wilms' tumor

X

X chromosome
   characterization, 371–372
   evolution, 386–387

Xgsc, 41
Xotx2, 41
Xsox17, 303–304

Y

Y chromosome
   characterization, 371–372
   evolution, 386–387
   role in maleness, 377
Yolk sacs
   differentiation, 161
   HSC in, 195–196
   visceral endoderm, 302–303

Z

Zic3, 67
Zona pellucida 3 gene, 14–15
Zygote, 26–27